Walford's Guide to Reference Material

WALFORD'S GUIDE TO REFERENCE MATERIAL

SIXTH EDITION • VOLUME 1

Science and Technology

EDITED BY

MARILYN MULLAY MA ALA

Assistant Librarian, Edinburgh School of Agriculture Library,
at the Scottish Agricultural College

and

PRISCILLA SCHLICKE BA ALA

Lecturer, School of Library and Information Studies,
The Robert Gordon University, Aberdeen

LIBRARY ASSOCIATION PUBLISHING
LONDON

Published by
Library Association Publishing Ltd
7 Ridgmount Street
London WC1E 7AE

First published 1959
Supplement 1963
Second edition 1966
Third edition 1973
Fourth edition 1980
Fifth edition 1989
This sixth revised edition 1993

British Library Cataloguing in Publication Data

Walford's Guide to Reference Material. –
Vol.1: Science and Technology. – 6Rev.ed
  I. Walford, A. J. II. Mullay, Marilyn
  III. Schlicke, Priscilla

  ISBN 1-85604-015-1

Computer production software and typesetting by LIBPAC (Computer Services) Ltd, Whittle le Woods, Lancs.
Printed and made in Great Britain by Bookcraft (Bath) Ltd.

Contents

Introduction

From its first edition the purpose of *Walford* has been to identify and evaluate the widest possible range of reference materials. In addition to the expected bibliographies, indexes, dictionaries, encyclopedias, directories, etc., a number of important textbooks and manuals of general practice are included. While the majority of items are reference 'books', *Walford* is a guide to reference 'material'. Thus periodical articles, microforms, online and CD-ROM sources are all represented. In this volume the first index of online and database services is introduced. The objective is for *Walford* to become a 'one-stop' source of information on all types of reference material, regardless of form. Targeted users include librarians developing and revising reference collections, staff on enquiry desks needing advice on further sources when local stock has been checked, research workers in the preliminary stages of projects, and students of library and information studies.

To be of manageable proportions a guide such as this must inevitably be selective. Most major reference tools are included, whenever originally published, provided they remain useful. Geographic scope is international, but with an emphasis on English-language material. A special effort has been made to ensure that the output of small and specialist publishers is not neglected.

Entries in *Walford* follow a subject arrangement based on the Universal Decimal Classification International Medium Edition of 1985 (BS1000M). Subject access for users unfamiliar with UDC can be gained either by checking the contents page to find the relevant section and then browsing through the entries, which are subdivided by form, place etc., or by using the subject index. The classified arrangement of the text has been further clarified in this edition. The first level of subheading within each broad subject class is typographically highlighted by the use of rules (see sample entry below).

Terms in the subject index are generated by the classification numbers given to the entries. Each entry has been allocated a running serial number in the text to provide easy access from both the subject and author/title indexes. Full instruction on the structure and use of the indexes can be found in their introductions.

Individual entries in *Walford* are nearly always based on examination of the actual item and include full bibliographical detail, ISBN and, if in print, the price when it can be ascertained. Brief critical annotations are provided in most cases, giving summary publishing history, outline of contents, comparison with other works, especially notable features and a brief general assessment of overall value, often illustrated by quotations from or references to reviews. The example below shows the general layout of entries.

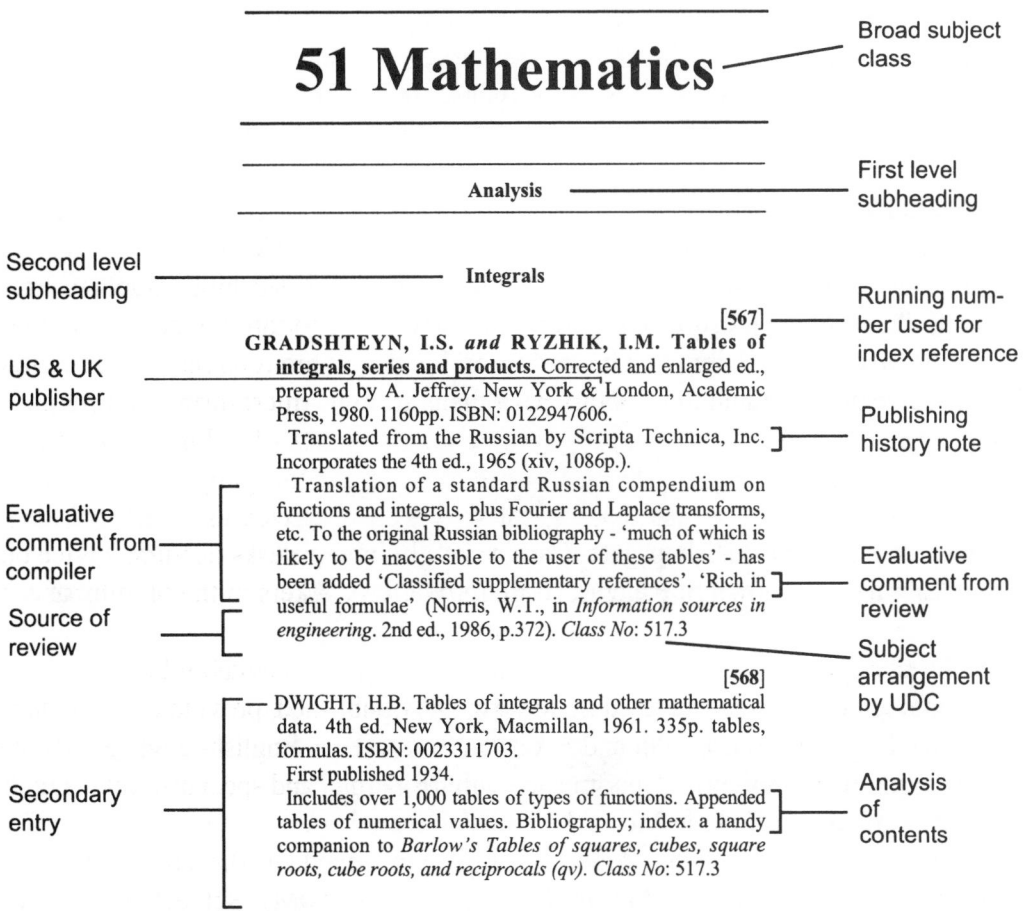

Work on this volume was completed in early 1993. Although no cut-off date for new material was specified, the aim has been to include as many 1992 publications as possible, together with some items published early in 1993. Intimation of planned new and revised editions is also given where possible.

A thorough revision has been undertaken for the 6th edition of volume 1. Many entries have been dropped to make way for new material, while those that have been retained have been revised and updated where necessary. The number of entries has increased by over 1,000 with some sections expanded to reflect increased interest, e.g. wildlife, conservation, solar heating, artificial intelligence, and microcomputers.

This volume is the work of two compilers: Marilyn Mullay was responsible for Classes 5/6, 52/529, 55, 56, 572, 573, 58/59, 61, 63, 64, and 654/656. Priscilla Schlicke was responsible for Classes 51, 53, 54, 6, 608, 62, 629, 66, 67/68 and 69. The whole project was overseen by Barbara Jover and Kathryn Beecroft of Library Association Publishing. Joan Bibby ably constructed a thesaurus based on the UDC classification scheme. Martin Harrison of LIBPAC Computer Services automated the data capture, indexing and typesetting.

The compilers are indebted to John Walford for his continuing interest in this work and his contributions in the form of advice and notes on new material. We should also like to thank the following for their invaluable assistance: Dawson UK Ltd Book Division for information on new titles; A. J. Mullay (Class 58/59, Nature Study; Class 654/656, Communication); Ruth M. Johnson; and the staff of the National Library of Scotland.

Libraries consulted

(a) *General and academic libraries*

British Library, Heriot-Watt University Library, London Library, National Library of Scotland, The Robert Gordon University Library, University of Cambridge Library, University of Glasgow Library, University of Leeds Library, University of Aberdeen Library, University of Edinburgh Library, University of London Library, University of York Library.

(b) *Public libraries*

Aberdeen City Libraries, Bishopsgate, Cambridge, Edinburgh City Libraries, Kensington, Mitchell Library (Glasgow), St Pancras, Holborn, Watford, Westminster, Marylebone.

(c) *Special libraries*

British Geological Survey, Chartered Institute of Surveyors, Darwin Library (University of Edinburgh), Department of Health and Social Security, Department of Energy, Department of Trade (Export Promotion), Departments of the Environment and Transport, Edinburgh School of Agriculture, Erskine Medical Library, Imperial War Museum, Imperial College (Lyon Playfair Library), Institute of Geological Science, Institution of Electrical Engineers, Institution of Civil Engineers, Marine Laboratory (Aberdeen), Ministry of Agriculture, Fisheries and Food, Ministry of Defence, Royal Observatory (Edinburgh), Royal Dick Veterinary College, Royal Botanic Garden (Edinburgh), Royal Institute of Internal Affairs, Royal Institute of British Architects, Royal Anthropological Society, Royal College of Veterinary Surgeons, Royal Astronomical Society, Royal Chemical Society, Science Museum Library, Science Reference and Information Service (Holborn; Aldwych), Scottish National Heritage, Scottish Office Agriculture and Fisheries Department, Scottish Science Library, Shell Oil Library (Aberdeen), Sports Council, United Nations Information Centre, Watford College (on Printing).

Abbreviations

Listed below are the chief bibliographical abbreviations used in the *Guide*. Generally accepted abbreviations such as Co., Corpn. *e.g.*, *i.e.*, Inc., Ltd. and *q.v.* are not included.

AG [German] – Aktiengesellschaft: Co.
ALA – American Library Association
Abt. [German] – Abteiling: part
ampl. [Italian] – ampliata: enlarged
    [Spanish] – ampliado: enlarged
Aufl. [German] – Auflage: edition
augm. [French] – augmenté: enlarged
aum. [Portuguese] – aumentada: enlarged
    [Spanish] – aumentado: enlarged
Ausg. [German] – Ausgabe: printing, edition

BS – British Standard
Bd [German] – Band: volume
BFr – Belgian francs
bearb. [German] – bearbeitet: compiled, edited
Belg. – Belgian

c – copyright date
C. Cd. Cmd. Cmnd. – Command papers
ch. – chapter(s)
chron. – chronology
col. – colour, coloured
cols. – columns
comp. – compiler
corr. – corrected
    [French] – corrigé: corrected
    [Spanish] – corregido: corrected
    [Portuguese] – corrigida: corrected

DFl – Dutch florins
Dan. – Danish
diagrs. – diagrams
distr. – distributed
DM [German] – Deutsche Mark
druk [Dutch] – edition

ea. – each

ed. – edition, editor(s)
    [Italian] – edizione
    [Spanish] – edición
    [French] – édition
    [Portuguese] – ediçáo: edition
enl. – enlarged
erw., erweit. [German] – erweiterte: enlarged
exp. – expanded

FID – Fédération Internationale de Documentation
facsim(s) – facsimile(s)
fasc(s) – fascicule(s)
fig. – figures
fldg. – folding
FFr – French francs
front. – frontispiece

ganz. [German] – gänzlich: complete
glav.red. [Russian] – glavnyi redaktor: editor-in-chief
Gld. [Dutch] – guilders
GmbH [German] – Gesellschaft mit beschränkter Haftung: Ltd
Gosud. [Russian] – Gosudarstvo: State

HMSO – Her Majesty's Stationery Office
Hft [German] – Heft: part number
hrsg. [German] – herausgegeben: edited, published

illus. – illustrations, illustrated
imp. [French] – imprimé, imprimerie: printed, printing firm
izd. [Russian] – izdanie: edition
    [Serbo-Croat] – izdanje: edition
Izdat. [Russian] – Izdatel': publisher

Jahrg. [German] – Jahrgang: annual publication

kiad. [Hungarian] – kiadás: edition
Kr [Danish, Norwegian, Swedish] – kroner

L. [Italian] – lire
l. – leaves
Lfg. [German] – Lieferung: number, part

m. fl. [Danish] – med flere: and others

n.d. – no date
neubearb. [German]: – neubearbeitet: revised
no. – number
Nor. – Norwegian
nouv. [French] – nouvelle: new (edition)
n.p. – no place of publication
Nr [Danish, German] – Nummer: number
n.s. – new series
NV [Dutch] – naamloze vennootschap: limited
   company

o.p. – out of print
omarb. [Swedish] – omarbetad: revised
opl. [Danish] – oplag: edition

p. – page(s)
pl. – plate(s)
port. – portrait(s)
pt(s) – part(s)
pub. – publisher
pubn. – publication

réd. [French] – rédigé: edited, compiled
repr. – reprinted
rev. – revised, revision
   [French] – révisé: revised
   [Portuguese] – revistada: revised

riv. [Italian] – riveduto: revised
R. [South Africa] – rands
Rs – rupees

Sch – Schillings
ser. – series
sér. [French] – série: series
SFr – Swiss francs
supp. – supplementary, supplement(s)
Sw. – Swiss
Swed. – Swedish

t. [French] – tome(s): volume(s)
T. [German] – Teil(e): part(s)

u. [German] – und: and
UDC – Universal Decimal Classification
udg. [Danish] – udgave: edition
uit. [Dutch] – uitgaaf: publication
uitg. [Dutch] – uitgegeven: published
umgearb. [German] – umbearbeitete: revised
Univ. – University
unveränd. [German] – unverändert: unaltered
uppl. [Swedish] – upplaga: edition
utg. [Norwegian] – utgave: edition

v. – volume(s)
v. p. – various pagings
VEB [German] – Volkseigener Betrieb: People's
   Concern
verand. [German] – verandert: revised
verb. [German] – verbesserte: improved
verm [German] – vermehrte: enlarged
vyd. [Czech] – vydání: edition
   [Slovak] – vydanie: edition

wyd. [Polish] – wydanie: edition

5/6 Science & Technology

India

[1]
Modern India and progress in science and technology. New Delhi, Vikas Publishing House, 1979. [vii], 164p. ISBN: 0706907434.
 9 chapters (2. The role of universities - 3. Scientific surveys and research - 4. Learned societies and institutions - 5. Medical education and research - 6. Engineering and industry - 7. Technical education and research - 8. Agricultural education and research - 9. Conclusion). Non-analytical index, p. 161-4. *Class No: 5/6(540)*

Government Policies

Dictionaries

[2]
Wörterbuch der Wirtschaftswissenschaft: russisch, deutsch, englisch. Dobrov, G., *and others*. 2 überarb & erweit. Aufl. Berlin, Verlag der Wirtschaft, 1984. 560p.
 Dictionary of science policy studies. About 8000 Russian-base terms, with German and English equivalents and indexes. *Class No: 5/6.0(038)*

Worldwide

[3]
Longman guide to world science and technology. Harlow, Essex, Longman Group, 1982-.
 Science and technology in the United Kingdom, by R. Nicholson. 1991. xv,312p. £95. ISBN 0582900514.
 Science and technology in Eastern Europe, edited by G. Darvas. 1988. xx,283p. £85. ISBN 0582900549.
 Science and technology in France and Belgium, by E.W. Kellerman. 1988. 140p. £85. ISBN 058290084X.
 Science and technology in the USSR, edited by M.J. Berry. 1988. 424p. £85. ISBN 0582900530.
 Science and technology in Scandinavia, by G. Férne. 1988. 175p. £85. ISBN 0582018927.
 Science and technology in China, by T.B. Tang. 1984. 278p. £85. ISBN 0582900565.
 Science and technology in Japan, by A.M. Anderson. 1991. xviii,382p. £95. ISBN 0582900158.
 Science and technology in the Middle East, by Z. Sardar. 1982. 335p. ISBN 0582900522.
 Science and technology in Africa, by J.W. Forje. 1989. 281p. £85. ISBN 0582000866.
 Science and technology in the USA, by A.H. Teich and J.H. Pace. 1986. 427p. £85. ISBN 0582900611.
 Science and technology in Latin America, by Latin American Newsletters, Ltd. 1983. 373p. £85. ISBN 0582900573.
 Science and technology in Australasia, Antarctica and the Pacific Islands. 1989. 335p. £85. ISBN 0582900603.

....(contd.)
 Science and technology in the Federal Republic of Germany, by F. Krahmer. 1990. 208p. ISBN 0582054397.
 Science and technology in India, Pakistan, Bangladesh and Sri Lanka, by A. Rahman. 1990. xviii,245p. £95. ISBN 0582064694.
 Chapters in each volume concern the organizational sectors, followed by specific areas (*e.g.* agriculture). Appended directory of sources. Establishments and subject index. *Class No: 5/6.0(100)*

[4]
ORGANIZATION FOR ECONOMIC COOPERATION AND DEVELOPMENT. Reviews of national science and technology policy. Paris, OECD, 1965-.
 United Kingdom and Germany. 1967. *Ireland.* 1974. *German Democratic Republic.* 1986. *Austria.* 1988. £8.50. *France.* 1966. *Italy.* 1992. £14. *Spain.* 1971. *Portugal.* 1986. *Finland.* 1987. £9.50. *Norway.* 1985. *Sweden.* 1987. £6. *Iceland.* 1984. *Netherlands.* 1987. £9.50. *Belgium.* 1966. *Switzerland.* 1989. £12. *Greece.* 1984. *Yugoslavia.* 1976. *China.* 1979. *Japan* 1967. *Canada.* 1969. *USA.* 1968. *New Zealand.* 1987. *Australia.* 1986. £7.50.
 Finland (154p.) is OECD's first evaluation of science and technology policy in that country. Sections: Reasons for decision making; Planning; Inputs from university research; Interactions between R & D and industrial policy. *Class No: 5/6.0(100)*

[5]
Science and technology policy: a review of recent developments. London, British Library Science Reference and Information Service, 1991-. 6pa. £90. ISSN: 09529616.
 Contents of June 1991 issue: News and views - World views - Recent publications (160 annotated entries). Subject index for 'Recent publications'. *Class No: 5/6.0(100)*

[6]
Science and technology policy in the 1980s and beyond. Gibbons, M., *and others eds.* London & New York, Longman, 1984. xxvi, 346p. ISBN: 0582902002.
 18 chapters: 1. Science and technology in India: policy, management implications (with 11 notes) - 2. Science policy in the UK; a view from the centre ... 10. The problematic relationship between research and policy ... 17. Some science and technology issues in global disarmament and development - 18. Science and technology in the New International Economic Order (with 32 notes and references). Assembles some of the papers presented at a workshop held in Manchester, 27-29 April 1983. *Class No: 5/6.0(100)*

[7]
UNITED STATES. National Research Council. **Outlook for science and technology:** the next five years. A report of the National Research Council. San Francisco, W.H. Freeman, in collaboration with the National Academy of Sciences, 1982. xx, 788p. illus., diagrs., graphs. ISBN: 0716713454.

Preceded by the NASA report, *Science and technology: a five-year outlook* (Freeman, 1979).

17 chapters, by numerous hands. 2. Some major human diseases ... 5. Ecology and systematics ... 7. Water resources ... 11. Chemical synthesis of new materials - 12. On some recent developments in mathematics ... 14. Research in industry ... 17. Prospects for new technologies (9 contributors). Each chapter has 'Summary and outlook' and references/bibliography/additional reading. Detailed, analytical index, p.761-88. A panorama of scientific and technological achievements and prospects.
Class No: 5/6.0(100)

Europe

[8]
BRAZIER, P.C. *and* SHOULTS, T.A.S. **Guide to European collaboration in science and technology.** London, Science and Energy Policy Studies Unit; Royal Society; Fellowship of Engineering, 1990. 213+4p. insert. £30. ISBN: 0854034293.

First published 1987 as *European collaboration in science and technology.*

3 parts (104 entries): 1. European Community programmes - 2. Non-EC multinational organizations, programmes and facilities - 3. Grant schemes for exchange, fellowships. Appendix: Computer networks, online databases and other sources of information. Detailed index, p.201-13.
Class No: 5/6.0(4)

Europe—Eastern

[9]
Science and technology in Eastern Europe. Darvas, G., *ed.* Harlow, Essex, Longman Group, 1988. xx,283p. £85. (*Longman guide to world science and technology.*) ISBN: 0582900549.

Science and technology policy in each of seven former Eastern bloc countries: Bulgaria, Czechoslavakia, German Democratic Republic, Hungary, Poland, Romania and Yugoslavia. Includes data on relevant associations and bodies. *Class No:* 5/6.0(401)

Great Britain

[10]
Science and technology in the United Kingdom. Nicholson, R., *Sir, and others, eds.* Harlow, Essex, Longman, 1991. xv,312p. £95. (*Longman guide to world science and technology, v.14.*) ISBN: 0582900514.

13 sections: 1. Overview of the UK economy, society and politics and the role of S & T - 2. History, development and organization of UK science and technology up to 1982 - 3. Central government organization and policy making for UK science and technology since 1982 ... 7. Agriculture and food - 8. Environmental and natural resources ... 11. Science and technology in sectors dominated by the private sector ... 13. International collaboration. 'Directory of

....(contd.)
selected research establishments', p.287-312. Short bibliography appended to each chapter.
Class No: 5/6.0(410)

[11]
UK science policy: a critical review of policies for publicly funded research. Goldsmith, M., *ed.* Harlow, Essex, Longman Group, 1984. xxii, 275p. ISBN: 0582902568.

Two parts. 1. Critical, documented reviews (1. Agricultural research policy; 9. Transport research) - 2 . Comparative studies (10. 'France: a logical attitude to science'; 11. 'The Netherlands: towards a national science policy'). Acronyms. Biographical notes. Index, p.267-75.
Class No: 5/6.0(410)

Federal Republic of Germany

[12]
MEYER-KRAHMER, F. **Science and technology in the Federal Republic of Germany.** Harlow, Essex, Longman, 1990. viii,208p. figs., tables. (*Longman guide to world science and technology, v.12.*) ISBN: 0582054397.

Gives an overview of science and technology, showing its importance for the economic development of the country. 9 sections: 1. Introduction - 2. Science and technology in the economy - 3. Structure of science and technology ... 6. Industrial research network and technology transfer - 7. International science and technology - 8. Regional development and science and technology ... 9. Impact of science and technology on economic change. Useful 'Directory of selected establishments', p.180-95. List of abbreviations, p.204-6. Brief subject index.
Class No: 5/6.0(430.1)

France

[13]
Science and technology in France and Belgium. Kellermann, E.W., *ed.* Harlow, Essex, Longman Group, 1988. viii, 131p. £85. (*Longman guide to world science and technology, v.7.*) ISBN: 058290084x.

Two parts: France; Belgium. Separate chapers on each concern science policy, industrial R & D, institutions, etc., followed by specific chapters on individual sectors of science and technology, and a list of useful addresses.
Class No: 5/6.0(44)

USSR

[14]
FORTESCUE, S. **Science policy in the Soviet Union.** London, Routledge, 1990. xi,230p. tables. £42.50. ISBN: 0415023793.

10 chapters (3. The Academy of Sciences ... 6. The research establishment ... 10. Main themes and current developments). Chapter notes, p.187-213. Glossary and abbreviations. References, p.217-25. Detailed, analytical index. Author is a professor of politics. *Class No:* 5/6.0(47)

[15]
Science and technology in the USSR. Berry, M.J., *ed.*
Harlow, Essex, Longman Group, 1988. xvi, 405p. diagrs.,
maps. £85. (*Longman guide to world science and
technology, v.6.*) ISBN: 0582900530.

27 chapters, of which the initial 12 provide historical
background, organizational structure fostering Soviet
scientific research, etc. Chapters 13-27, by specialists,
concern individual branches of science in the USSR,
scientific associations and the like. Index of establishments.
Class No: 5/6.0(47)

Scandinavia
[16]
FERNE, G. Science and technology in Scandinavia. Harlow,
Essex, Longman, 1989. 175p. £85. (*Longman guide to
world science and technology, v.9.*) ISBN: 0582018927.

8 sections: 1. Introduction - 2. Decision-making ... 4.
Research in higher education - 5. Research in the service of
collective needs ... 7. International dimension - 8.
Conclusion. Each section covers Denmark, Finland, Iceland,
Norway and Sweden. Bibliography, p.155-57. Appendix,
'Directory of major research establishments', p.158-70.
Index of establishments. *Class No:* 5/6.0(48)

China
[17]
Guide to China's science and technology policy. 1989.
Beijing, International Academic Publications, 1991.
xii,472p. diagrs., tables. (*White paper on science and
technology, no.4.*) ISBN: 7800031357.

7 parts. 2 appendices (A: State laws and regulations).
Class No: 5/6.0(510)

[18]
Science and technology in China. Tang, T.B., *ed.* Harlow,
Essex, Longman Group, 1984. xii, 269p. £85. (*Longman
guide to world science and technology, v.3.*) ISBN:
0582900565.

The main chapters focus on specific branches of Chinese
science and technology, with a separate chapter on Taiwan.
Statistical appendices are followed by a directory of major
research establishments. Indexes of establishments and
subjects. Presentation is 'clear, concise and in general easily
readable' (*Nature*, 13 December 1986, p.649). It nicely
compliments *Science in contemporary China*, edited by L.A.
Orleans (Stanford Univ. Press, 1980), which goes to the end
of 1974. *Class No:* 5/6.0(510)

Japan
[19]
SIGURDSON, J. *and* **ANDERSON, A.M.** **Science and
technology in Japan.** 2nd ed. Harlow, Essex, Longman,
1991. xviii,382p. figs., tables. £95. (*Longman guide to
world science and technology, v.4.*) ISBN: 0582098637.

First published 1984.
21 chapters: 1. Science and technology in Japan ... 4.
Technology and industrial restructuring ... 9. The
information technologies ... 13. Defence ... 17. Environment
and research ... 21. Technology and society. 4 appendices:
1. General guideline for science and technology policy ... 3.
Universities - 4. Directory of projects within national R & D
programmes. *Class No:* 5/6.0(52)

Indian Subcontinent States
[20]
**Science and technology in India, Pakistan, Bangladesh and
Sri Lanka.** Rahman, A., *and others, eds.* Harlow, Essex,
Longman, 1990. xviii,259p. tables. £95. (*Longman guide to
world science and technology, v.13.*) ISBN: 0582064694.

Gives an overview of science in each country. Part 1.
India (1. Introduction ... 6. New technologies ... 8. Scientific
societies and periodicals ... 10. Epilogue) - Part 2. Pakistan
(11. Introduction ... 14. National science and technology
policy ... 16. Status and progress in important areas - 17.
Conclusions) - Part 3. Bangladesh (18. Introduction ... 23.
International cooperation in science and technology) - Part 4.
Sri Lanka. Short bibliographies throughout. 2 appendices: 1.
Directory of selected establishments - 2. Organizational chart
of science and technology in India. No index.
Class No: 5/6.0(54.)

Islamic World
[21]
Science and technology in the Middle East: a guide to issues,
organizations and institutions. Sardar, Z., *ed.* Harlow,
Essex, Longman Group, 1982. x, 324p. tables, map.
(*Longman guide to world science and technology, v.1.*)
ISBN: 0582900522.

Part 1: Overview - Part 2: Regional organizations (*e.g.*
Arab League Educational, Cultural and Scientific
Organization (ALESCO)), p.83-115 - Part 3: Country
profiles (19; Afghanistan ... People's Democratic Republic
of Yemen), p.119-285. Appendix: 'Directory of major
establishments in the Middle East'. Index of establishments.
Detailed, analytical index, with country subdivisions. Omits
Israel. *Class No:* 5/6.0(5.297)

Africa
[22]
FORJE, J.W. Science and technology in Africa. Harlow,
Essex, Longman, 1989. xiii,281p. tables. £85. (*Longman
guide to world science and technology, v.10.*) ISBN:
0582000866.

16 sections: 1. The continent and its people ... 3. Science
and government - 4. National science and technology policy
... 9. Industrial research and development ... 13.
Cooperation in science and technology - 14. Militarization:
technology and development ... 16. Transport and
communications. 3 appendices (*e.g.* 'Directory of research
establishments'). A bibliography is appended to each
chapter. *Class No:* 5/6.0(6)

USA
[23]
Science and technology in the USA. Teich, A. *and*
Pace, J.H., *eds.* Harlow, Essex, Longman Group, 1986.
xix, 408p. tables. £85. (*Longman guide to world science and
technology, v.5.*) ISBN: 0582900611.

18 chapters: 1. Demographic, economic and political
overview ... 4. Government science and policy ... 6.
Industrial research and development ... 8. Agricultural
science and technology ... 9. Biomedical research and
devlopment ... 14. Associations, societies and academies in
science, engineering and higher education ... 17.
International cooperation in science and technology - 18.

....(contd.)

Scientific and technical information. Appendix 1: Sources of information in US science and technology (p.359-64; grouped); 2. Federal research establishments and index. Non-analytical subject index, p.396-408. The result of a team effort. *Class No:* 5/6.0(73)

America—South

[24]

Science and technology in Latin America. Roper, C. *and* Silva, J., *eds.* Harlow, Essex, Longman Group, 1983. ix, 363p. diagrs., tables, map. £85. *(Longman guide to world science and technology.)* ISBN: 0582900573.

25 country chapters (Argentina ... Venezuela), p.1-248. Each chapter has sections on societies and organizations, information sources in science and technology. Index of establishments, p.314-51. Analytical subject index, p.352-63. *Class No:* 5/6.0(8)

Australasia & Oceania

[25]

RONAYNE, J. *and* **BOAG, C. Science and technology in Australasia, Antarctica and the Pacific Islands.** Harlow, Essex, Longman, 1989. xi,335p. tables, graphs. £85. *(Longman guide to world science and technology, v.11.)* ISBN: 0582900603.

4 sections: 1. Australia (1. Geography, demography, politics and economics ... 8. State and territory governments ... 10. Academic science and research ... 12. Learned and professional societies) - 2. New Zealand (13. Demographic, political and economic background ... 17. Health and medical science ... 20. Other scientific activities) - 3. The South Pacific - 4. Antarctica (22. Geography, resources, climate and politics ... 24. Scientific and research activities). Brief list of references appended to each chapter, 'Directory of selected establishments', p.307-335. *Class No:* 5/6.0(9)

Thesauri

[26]

UNITED NATIONS. Educational, Scientific and Cultural Organization. **SPINES thesaurus:** a controlled and structured vocabulary of science and technology. Rev. ed. Paris, Unesco, 1988. 2v. (xxii,1122p.). £38.

The *Thesaurus* contains *c.*10,500 terms, of which some two-thirds are descriptors linked by a network of more than 74,000 semantic relations. *Class No:* 5/6.0:025.43

Science

Abbreviations & Symbols

[27]

ABBR: abbreviations for scientific and engineering terms. Montreal, Canadian Standards Association, [*c.*1983]. 280p.

Consists of 2 main A-Z sequences: 7,000 terms; 6,800 abbreviations. Appended: a list of 250 symbols and a bibliography. Excludes abbreviations for organizations, agencies, computer-language etc., concentrating on

....(contd.)

'abbreviations for words and word combinations for use in drawings and in engineering and scientific writing'. *Class No:* 5(003)

[28]

BRITISH STANDARDS INSTITUTION. Specification for quantities, units and symbols. London, BSI, 1979-82. 14 pts. (*BS 5775.*)

Replaces BS 1991: *Letter symbols, signs and abbreviations* (1961-).

0. *General principles.* 1982. 16p. £40. 1. *Space and time.* 1979. 20p. £40. 2. *Periodic and related phenomena.* 1979. 12p. £23. 3. *Mechanics.* 1979. 24p. £40. 4. *Heat.* 1979. 20p. £40. 5. *Electricity and magnetism.* 1980. 28p. £40. 6. *Light and related electromagnetic radiations.* 1982. 16p. £40. 7. *Acoustics.* 1979. 20p. £40. 8. *Physical chemistry and molecular physics.* 1982. 20p. £40. 9. *Atomic and nuclear physics.* 1982. 20p. £40. 10. *Nuclear reactions and ionizing radiations.* 1982. 24p. £40. 11. *Mathematical signs and symbols for use in the physical sciences and technology.* 1979. 36p. £40. 12. *Dimensionless parameters.* 1982. 12p. £28. 13. *Solid state physics.* 1982. 24p. £40.

Class No: 5(003)

[29]

—**ROYAL SOCIETY.** Quantities, units and symbols: a report by the Symbols Committee of the Royal Society, representing the Royal Society, the Chemical Society, the Institute of Physics. 2nd ed. London, Royal Society, 1975. 54p. Addendum, 1981. 2p. £1. ISBN: 0854030719.

Has 14 sections: 1. Introduction - 2. Physical quantities and symbols for physical quantities - 3. Units and symbols for units... 6. Chemical elements, nuclides and particles... 7. Quantum states... 8. Nuclear physics... 12. Recommended values of physical constants... 13. Sources - 14. Bibliography (p.50-54). 'A very useful summary' (*RSR*, v.5(2), April/June 1977, p.18). *Class No:* 5(003)

[30]

DE SOLA, R. Abbreviations dictionary. 8th ed. Boca Raton, CRC Press, 1992. xx,1300p. $69.95. ISBN: 0849342473.

Not a specialist science and technology abbreviations dictionary but contains many terms encountered by the scientist. Includes non-English abbreviations. Abbreviations, p.3-943, followed by sections which include: Airlines of the world; Chemical element symbols, atomic numbers, and discovery data; Earthquake data (Richter Scale); Numeration; Winds of the world. *Class No:* 5(003)

[31]

Glossary of Russian abbreviations and acronyms. Washington, Library of Congress, 1967. x, 806p.

First published 1952.

23,600 entries selected from 40,000 abbreviations and acronyms in files of the Lexicography and Terminology Section, Aerospace Technology Division. Restricted to abbreviated words in general use by 20th-century authors, scientists, journalists, teachers, editors, librarians and archivists, and acronyms used in Soviet publications since World War II, emphasising aerospace, scientific and technical literature. *Class No:* 5(003)

[32]
LUINGMAN, C.G. **Dictionary of symbols.** Andover, Hants., Gale Research, 1993. 608p. figs. £47.50. ISBN: 1873477902.

Includes all kinds of graphic symbols. 3 indexes: Graphic search index; Graphic index; Index of terms (lists meanings, names, descriptions, and sign systems). *Class No:* 5(003)

[33]
MURITH, J. *and* BOCABEILLE, J.-M. **Dictionnaire des abréviations et acronymes scientifiques, techniques, médicaux, économiques, juridiques.** Paris, Technique et Documentation., 1992. 949p. FFr950. ISBN: 2852066815.

About 25,000 abbreviations and acronyms, including those of organizations and companies, with their names and addresses. In French and English. *Class No:* 5(003)

[34]
OCRAN, E.B. **Ocran's acronyms:** a dictionary of abbreviations and acronyms used in scientific and technical writing. London, Routledge & Kegan Paul, 1978. xiv, 262p. (reprinted with corrections, 1980).

About 7,000 entries in the main A-Z sequence, p.1-184, followed by 47 subject sections (Aeronautics... Transportation). Excludes most associations and government bodies. Aeronautics and computer services figure prominently. 'Compiled to fulfil a subject approach' (*Introduction*). 'Reference librarians will find this volume very useful' (*Choice*, v.16(1), March 1979, p.58). *Class No:* 5(003)

[35]
Pugh's dictionary of acronyms and abbreviations: abbreviations in management, technology and information science. Pugh, E., *comp.* 5th ed. London, Library Association Publishing, 1987. v, 366p. £60. (£48 to LA members). ISBN: 0853655375.

First published 1968.

More than 34,000 entries, stating country of origin or organization, as considered necessary. Unlike the wide-ranging *Acronyms, initialisms and abbreviations dictionary* (15th ed. Detroit, Mich., Gale, 1991, 3pts.), with its 500,000 odd terms, plus *International acronyms, initialisms & abbreviations dictionary* (2nd ed. Gale, 1987), for terms of foreign origin, Pugh concentrates largely on management, technology and information science. A new edition, edited by P. Auger, is scheduled for 1994. *Class No:* 5(003)

[36]
SHEPHERD, W. **Shepherd's glossary of graphic signs and symbols.** London, Dent, 1971. x, 599p.

About 5,000 signs and symbols, covering engineering, mathematics, electricity, physics, numismatics, meteorology, etc., are interpreted. Bibliography, p.583-90. Index of signs with names, p.591-7. 'Believed to be the first extensive treatment of the subject to be published' (*Preface*). *Class No:* 5(003)

[37]
—ARNELL, A. **Standard graphical symbols:** a comprehensive guide for use in industry, engineering and science. New York, etc., McGraw-Hill, 1963. x, 534p. diagrs.

A compendium of symbols approved as standards by leading US engineering and scientific institutions. Some 9,000 illustration-symbols, with descriptive captions, numbers and titles of official standards. Symbols are grouped in 13 sections, each with a detailed contents-list: Electro-electronic ... Flags and pennants. Appendix 1:

....(contd.)
Symbols available from typographers and printers; 2: Abbreviations to be used on drawings (but only one page on topographical features). Lists of abbreviations and subject index, p.527-34. *Class No:* 5(003)

[38]
WENNRICH, P. **Anglo-American and German abbreviations in science and technology.** [Anglo-amerikanische und deutsche Abkürzungen in Wissenschaft und Technik.] New York, Bowker; Munich, K.G. Saur, 1976-77, 1980. 3v. (xx, 2276p.) and Supplement (vi, 618p.). $125. (*Handbook of international documentation and information, v.14.*)

V.1-3 form a single sequence (A-E, F-O, P-Z) of over 150,000 Anglo-American and German abbreviations. Although entries for journals are categorized as such, plus country of origin, this is an exception to the general rule. The abbreviation 'U' occupies 2 columns. Number of entries is somewhat inflated (*e.g.* 'WHTHS': William Howard Taft High School'!). *Class No:* 5(003)

[39]
ZALUCKI, H. **Dictionary of Russian technical and scientific abbreviations,** in Russian, English and German. Amsterdam, etc., Zalucki, 1968. xii, 387p. $92. ISBN: 0444406573.

About 7,300 entries. Primarily intended for readers of Russian scientific publications. *Class No:* 5(003)

Databases

[40]
Brit-line: directory of British databases. V.4, no.1. London, McGraw-Hill Book Company, 1989. xiii,408p. ISBN: 0077071689.

First published 1986.

4 sections: 1. Databasics - 2. Database directory - 3. Indexes - 4. Prepaid reply card. The database directory contains details (producer, subject, type, database content, host, file details, restrictions, printed version, sample record) of 375 databases. Subjects covered include agriculture, aviation, chemistry, earth sciences, energy, engineering, food science and technology, general science and technology, health and safety, medical science, patents, shipping and veterinary medicine. There are five indexes: producer; host; cross; entry; address - the last two being the most reliable. *Class No:* 5(003.4)

[41]
Gale directory of databases. Marcaccio, K.Y., *ed.* Andover, Hants., Gale Research, 1993. 2v. in 4. (2300p.). £200, set; £145, v.1; £85, v.2. ISBN: 0810357461, set.

Formed by the merger of *Computer readable databases, Directory of online databases,* and *Directory of portable databases.*

V.1. *Online databases.* V.2. *CD-ROM, diskette, magnetic tape, batch access, and handheld database products.*

V.1 describes over 5,300 databases worldwide, while v.2 profiles *c.*3,000 'portable' databases. Each entry includes address; telephone and fax numbers; contact name; subjects covered; languages used; size; format; availability; price; vendors and suppliers. Subject, master, and geographical indexes. Volumes to be published every 6 months to ensure speedy communication of new products, etc. New and amended product entries will be published in *Online & CDROM review. Class No:* 5(003.4)

[42]

HALL, J.L. **Online bibliographic databases.** 4th ed. London, Aslib, 1986. ISBN: 0851422020.

First published 1979.

A 63-page commentary precedes the directory of 250 bibliographic databases that are readily available and mostly in English. Data on each include subject index, printed and online versions, online file details, online service suppliers, indicative access charges, references. Appended bibliography of 1,011 references. Appendices: Data producers; Online service suppliers; Most important word(s) in bibliography; Broad subject headings. Author index; detailed, analytical general index, p.469-508. Prelims. feature sample search templates for 9 leading systems, *eg*, BLAISE, DIALOG, BRS, INFOLINE. A well organised and documented guide.

5th ed. is planned. *Class No:* 5(003.4)

[43]

NTIS. Springfield, National Technical Information Service, US Department of Commerce, 1964-.

Has *c.*1,500,000 records, covering government-sponsored research, development, and reports prepared by state and federal agencies. Multidisciplinary, subject areas include agriculture; chemistry; civil engineering; medicine; biology; transport. Updated every 2 weeks and also available on CD-ROM. *Class No:* 5(003.4)

[44]

SCISEARCH. Philadelphia, Pa., Institute for Scientific Information, 1974-.

Contains *c.*11 million records, taken from *Science citation index (q.v.)* and some from the *Current contents (q.v.)* series which are not included in SCI. Multi-disciplinary, claiming to index 90 percent of the world's 'significant' scientific and technical literature. Provides citation indexing. Updated weekly, with *c.*58,000 items added pa. Available on CD-ROM. *Class No:* 5(003.4)

Bibliographies of Bibliographies

[45]

CONRAD, E. **'LC science tracer bullet: an index'.** In *Reference services review,* v.16(3), 1988, p.49-56. Ann Arbor, Mich., Pierian Press. ISSN: 00907324.

Earlier list in *Reference services review,* v.11, Winter 1983, p.75-80.

The *Tracer bullet* series (Library of Congress, Science and Technology Division) has appeared since 1972. It covers a wide variety of subjects in the natural and physical sciences and technology. In the form of 'library pathfinder' handouts (about 10pa. and each about 4p. in length), it lists texts, handbooks, encyclopedias, dictionaries, bibliographies, government documents, and journal articles, *plus* addresses and telephone numbers of relevant organizations. *Class No:* 5(009)

[46]

Internationale Bibliographie der Fachbibliographien für Technik, Wissenschaft und Wirtschaft. [International bibliography of bibliographies in technology, science and economics.] 9.-2. Ausg. Munich, K.G. Saur, 1969. 2v. *(Handbuch für Dokumentation, 3.)*

A. *Einmalig erscheinende Bibliographien.* 9. Ausg. viii, 887p. B. *Periodisch erscheinende Bibliographien.* 2. Ausg. viii, 817p. Volume A lists *c.*2,200 systematically arranged entries (in 10 classes) for non-periodical bibliographies,

....(contd.)

catalogues and other lists covering science, technology and economics. The indexes, which precede, are a feature: authors and editors, countries (subdivided by subjects), subjects, and publishers. *Class No:* 5(009)

[47]

—BESTERMAN, T. Physical sciences: a bibliography of bibliographies. Totowa, N.J., Rowman & Littlefield, 1971. 2v. (672p.).

A reprint of appropriate sections (*e.g.* Astronomy, Atomic energy, Chemistry, Electricity, Geology, Hydrography, Mathematics, Meteorology, Physics, Science, Statistics) of Besterman's *World bibliography of bibliographies* (4th ed. 1965-66. 4v. & index). No index. *Class No:* 5(009)

[48]

SCIENCE MUSEUM LIBRARY, LONDON. **Science Library bibliographical series.** Permuted key word index to the titles of the Science Museum Library's bibliographical series. London, Science Museum Library, 1980. 25 sheets. (*Science Museum Bibliographical series, no. 808.*)

The index covers bibliographies no.1-807 (1931-1978). An ad hoc mimeographed series of lists, prepared on request in a variety of scientific and technology fields. No.807, *Some references on lipids in the lungs of vertebrates, including man, 1958 to 1976* (1978) ran to 26p. *Class No:* 5(009)

Bibliographies

[49]

AAAS science book list, 1978-86. Wolff, K., *and others, comps. and eds.* Washington, American Association for the Advancement of Science, 1986. 568p. $25. ISBN: 0871683156.

Contains entries for *c.*2,100 science and mathematics books, primarily suitable for libraries or science teachers. Selected from the AAAS's *Science books and films.* Updates the 1972 (3rd ed.) list and its supplement (1978). *Class No:* 5(01)

[50]

—The AAAS science book list: a selected and annotated list of science and mathematics books for secondary school students, college undergraduates and nonspecialists. Deason, H.J., *comp.* 3rd ed. Washington, American Association for the Advancement of Science, 1972. 439p. Supplement, 1978. 457p. $25.

The main volume has critical annotated entries for *c.*1,500 books in Dewey classification order. Ratings are given. Author and subject-title indexes. The *Supplement* covers 2,850 books published since 1969, with increased entries for the behavioural sciences. *Class No:* 5(01)

[51]

A Bibliography of the philosophy of science, 1945-1981. Blackwell, R.J., *comp.* Westport, Conn., Greenwood Press, 1983. 585p. $95. ISBN: 0313231249.

Cites more than 7,000 books, papers, symposia and review articles on 47 topics. Excludes doctoral theses. Index of personal names, but not of subjects. *Class No:* 5(01)

[52]

Bibliotheca chemico-mathematica: catalogue of works in many tongues on exact and pure science, with a subject index. Zeitlinger, H. *and* Sotheran, H.C., *comps*. London, Sotheran, 1921. 2v. Supplements, 1931, 1937, 1952. illus.

53,489 annotated entries in all. Compiled from Sotheran's sale catalogues of scientific and technical works. Details on each item include author, full title, description, date, sale price. Under authors, A-Z, with a detailed subject index. Records many old and rare books important in the history of science. *Class No:* 5(01)

[53]

Books for college libraries: a core collection of 50,000 titles. V.5: Psychology, science, technology, bibliography. 3rd ed. Chicago, American Library Association, 1988. 422p. $90. ISBN: 0838933580.

Lists over 7,500 books, arranged by LC classifications. Omits textbooks unless considered essential.

Class No: 5(01)

[54]

CAMBRIDGE UNIVERSITY. Scientific Periodicals Library. **Union catalogue of scientific libraries in the University of Cambridge: Books published before 1801.** London, Mansell, 1977. 9 fiches.

Author index, with *c.*5,800 entries, plus a chronological index of authors and short titles, covering 1478-1800.

Class No: 5(01)

[55]

CHEN, C.-C. Scientific and technical information sources. 2nd ed. Cambridge, Mass., and London, M.I.T. Press, 1987. xxx, [1], 824p. £76.50. ISBN: 0262031205.

3,650 entries, grouped into 23 form categories (1. Selection tools ... 4. Encyclopedias ... 15. Abstracts and indexes, and current awareness services ... 19. Patents and standards ... 21. Non-print materials ... 22. Professional societies and their publications ... 23. Data bases). Brief annotations; citations to *c.*150 review sources. General index to authors only. Omits medicine. Patchy coverage of certain fields (*e.g.* geology, agriculture). Admits to a heavy U.S. emphasis; very few non-English titles. 'Intended primarily as a reference guide for science and engineering librarians and their assistants and as a textbook for library school students engaged in the study of the structure, properties and output of science literature' (*Preface*). Lack of an analytical subject index is a definite drawback. *Class No:* 5(01)

[56]

DAVIDGE, R. *and* **WOODING, E.R. The Use of scientific information services.** 7th ed. Egham, Surrey, Royal Holloway and Bedford New College, Library and Physics Department, 1988. 87p. charts. £5. ISBN: 0900145765.

Sections, with introductory section and subsection notes, include: Reference material; Introduction to the subject; Bibliography; Searching the literature; Communication with other workers; Foreign literature - 9. Libraries, booksellers, information centres; 10. Computer based information retrieval; Documentation of information; List of abbreviations. Index. Sound introduction and advice addressed to both undergraduate and researchers.

Class No: 5(01)

[57]

Encyclopedia of physical sciences and engineering information sources: a bibliographic guide to approximately 16,000 citations for publications, organizations, and other sources of information on 425 subjects relating to the physical sciences and engineering. Wasserman, S., *and others, eds*. Detroit, Gale, 1989. 736p. $140. ISBN: 0810324989.

Information sources listed include abstracts; yearbooks; bibliographies; monographs; professional associations; handbooks; databases; periodicals. Prices are given but there are no comments on level and extent of coverage. Many cross-references in the index and throughout. Some primary sources are omitted. 'Nevertheless, the directory is a helpful supplement' (*Science and technology libraries*, v.11(1), Fall 1990, p.141). *Class No:* 5(01)

[58]

GROGAN, D. Science and technology: an introduction to the literature. 4th ed. London, Bingley, 1982. 400p. £9.95. ISBN: 0851573150.

First published 1970.

Concerns the kinds of sources. 22 chapters (usually with 'Further reading'): 1. The literature - 2. Guides to the literature - 3. Encyclopedias - 4. Dictionaries - 5. Handbooks - 6. Directories and yearbooks - 7. Books 'in the field' - 8. Bibliographies - 9. Periodicals - 10. Indexing and abstracting services - 11. Computerized databases - 12. Reviews of progress - 13. Conference proceedings - 14. Research reports - 15. Patents and trademarks - 16. Standards - 17. Translations - 18. Trade literature - 19. Theses and research in progress - 20. Non-print media - 21. Microforms - 22. Biographical sources. Analytical index of subjects and forms; titles mentioned in text - merely examples of their kind - are not included. An impressive survey, 'written primarily for students, not practitioners' (*Introduction*). *Class No:* 5(01)

[59]

Handbooks and tables in science and technology. Powell, R.H., *ed*. 2nd ed. London, Library Association, 1983. vi, 297p. ISBN: 0853657262.

First published 1970.

Covers 3,403 handbooks and tables in science, technology, and medicine. Data on each has complete bibliographic citation and many annotations which briefly describe the intellectual content of the handbook and tables. Subject, keyword, author/editor, and title indexes.

Class No: 5(01)

[60]

HURT, C.D. Information sources in science and technology. Englewood, Colo., Libraries Unlimited, 1988. 362p. £22.50. (*Library science text.*) ISBN: 0872875814.

Gives brief citations and annotations for *c.*2,000 English language reference works. Indexes for authors/titles and subject. A useful tool. *Class No:* 5(01)

[61]

International catalogue of scientific literature, 1st - 14th annual. Published for the International Council by the Royal Society of London. London, Harrison, 1902-21. 238v. Reprinted New York, Johnson Reprint Corpn.; London, Johnson Reprint Co., Ltd., 1968. 32v.

Unique coverage for 1901-14 (some sections to 1920). Each annual issue consists of 17v.: A. Mathematics; B. Mechanics; C. Physics; D. Chemistry; E. Astronomy; F.

....(contd.)

Meteorology; G. Mineralogy; H. Geology; J. Geography; K. Palaeontology; L. General biology; M. Botany; N. Zoology; O. Human anatomy; P. Anthropology; Q. Physiology; R. Bacteriology. Each volume has (1) schedules of the classification and subject indexes in English, French, German and Italian; (2) an author catalogue; (3) a classified subject catalogue. Like the Royal Society's *Catalogue of scientific papers, 1800-1900 (qv)*, of which it is a continuation, *International catalogue* also indexes papers in pamphlets, memoirs and books. 4,673 journals were indexed in the 1903 annual volumes, as against 1,535 journals indexed in the Royal Society's *Catalogue*. The only part of the *International catalogue* with a continued existence is Zoology, as covered in *Zoological record*. The *Bibliographie scientifique française* (monthly, 1900-42), included in the *International catalogue*, carried on until 1942 on its own. The *List of journals with abbreviations used in the Catalogue as references* (1903, 312p.; Supplement, 1904, 680p.) includes over 5,500 periodicals from more than 20 countries. *Class No:* 5(01)

[62]
JOHN CRERAR LIBRARY. Catalog. [John Crerar Library]. Boston, Mass., G.K. Hall, 1967. 77v.
Author-title catalog. 35v. $3,800 *Classified subject catalog.* 42v. $4,560. Subject index. 1967. 66p. The John Crerar Library collections (including Illinois Institute of Technology holdings) comprised more than 1,100,000 v. of current and historical research materials in pure and applied sciences. The *Author-title catalog* (563,000 cards, photolithographically reproduced) includes titles of 13,000 current serials. The *Classified subject catalog* (704,000 cards) includes a subject index of 47,000 entries.
Class No: 5(01)

[63]
McGraw-Hill basic bibliography of science and technology.
Recent titles on more than 7000 subjects, compiled and annotated by the editors of the *McGraw-Hill Encyclopedia of science and technology.* New York, etc., McGraw-Hill, 1966. ix, 738p. $25. ISBN: 0070452180.
Aims to direct the purposive reader with an enquiring mind 'to reading beyond the articles in the *McGraw-Hill Encyclopedia of science and technology...*' (*Foreword*). More than 8,000 briefly annotated entries for books; items starred are recommended for young people and school science collections. Annotations state contents and level. Cross-references. 'Topical guide to subject areas' (p.699-738) lacks precision. (*eg*, 'Animal pathology' has over 600 unspecified references). Replaces R.R. Hawkins '*Scientific, medical and technical books*' (2nd ed. Washington, National Research Council, 1958). Valuable, but dated. No author index. *Class No:* 5(01)

[64]
MALCLÈS, L.-N. Les Sources du travail bibliographique.
Geneva, Droz, 1950-58. 3v. in 4. $22.50, v.1; $46.50, v.2; $22.50, v.3.
1-2. 1950-52. 3. *Bibliographies spécialisées (sciences exactes et techniques).* 1958. x, 575, [1]p. V.3 still has importance for the specialist in pure and applied science and medicine. Mlle. Malclès contributed 6 of the 13 chapters. Others deal with chemistry, earth sciences, general and animal biology, medical sciences, zoology, botany and plant physiology, and pharmacy. No chapter on agriculture, and certain aspects of engineering are also wanting (*Journal of*

....(contd.)

documentation, v.15(3), September 1959, p.156-9). The index (p.521-75) has *c.*10,000 entries. Some of the chapters still constitute our best and most detailed conspectus of the scientific and technical literature of the 19th and earlier 20th-century in their fields. *Class No:* 5(01)

[65]
MALINOWSKY, H.R. Best science and technology reference books for young people. Phoenix, Ariz., Oryx, 1991. 216p. $24.95. ISBN: 0897745809.
669 reference books for the age group 5-18 are listed. Arranged, in 12 chapters, by subject, *e.g.* 'Astronomy'; 'Physics'; 'Science'; 'Chemistry', the annotated entries provide full bibliographic information, review citation and recommended age group. 4 indexes: title; author; subject; grade level. The author is bibliographer for science and technology at the University of Illinois-Chicago library. *Class No:* 5(01)

[66]
—WILMS, D.M. Science books for children: selections from *Booklist,* 1976-1983. Chicago, Ill., American Library Association, 1985. viii, 183, [1]p. $15.
About 600 annotated entries. 15 sections: Introduction - Methods - Mathematics - Computers - Astronomy and space science - Physics and chemistry - The earth and how we use it - Prehistoric life - The living world - Plants - Animals - The human body at work - Science at work - Science experiments - Scientists. Author-title index; Subject index. 'The books in this collection range from picture books aimed at preschoolers to works for junior high school students through grade nine. Occasionally an adult title will appear...' (*Introduction*). *Class No:* 5(01)

[67]
MALINOWSKY, H.R. *and* **RICHARDSON, J.M. Science and engineering literature:** a guide to reference sources. 3rd ed. Littleton, Colo., Libraries Unlimited, 1980. 342p.
First published 1967 as *Science and engineering reference sources.*
About 2,500 annotated entries in 14 sections (1. Introduction - 2. Forms of literature - 3. Multidisciplinary sources of information - 4. History of science - 5. Mathematics (Guides to the literature; Abstracts, indexes and bibliographies; Encyclopedias; Dictionaries; Handbooks and tables; Directories)... 13. Biomedical sciences; 14. Bibliography (mostly periodical articles, grouped)). The preference given to form as opposed to further subject division is irritating, but - fortunately - the index (p.287-342) excels. For the library student and librarian, general/special. *Class No:* 5(01)

[68]
The Museum of Science and Industry basic list of children's science books, 1973-1984. Richter, B. *and* Wenzel, D., *comps.* Chicago, Ill., & London, American Library Association, 1985. xi, 154p.
1st supp. 1986. 80p. $6.95; 2nd supp. 1987. 84p. $8.95; 3rd supp. 1988. 80p. $11.95.
The parent list covers 1,400 children's science books up to junior high-school level, in 17 broad subject areas (*e.g.* Earth sciences). Ratings range from recommended to not recommended. 10-30 word annotations indicate reading level and price; other reviews are cited. Appended directory of publishers, list of children's science magazines and review journals, and an annotated bibliography of source books for

....(contd.)

teachers. Author and title indexes. The 1986 supplement adds 500 titles in 16 subject areas. An excellent US 'buying guide and reference tool' (*Reference books bulletin, 1985/1986. p.87*). *Class No:* 5(01)

[69]

PARKER, C.C. *and* **TURLEY, R.V. Information sources in science and technology:** a practical guide to traditional and online use. 2nd ed. London, etc., Butterworths, 1986. [viii], 328p. diagrs., tables, charts. £35. ISBN: 0408014679.

First published 1975.

Main sections: Choosing sources of information and their guides - People - Organizations - The literature (p.37-138) - Information services (p.139-210) - Searching: the literature and computer databases - Obtaining literature in a usable form - Organizing and presenting information - Current awareness. Appendix: Helping the library user. Detailed, analytical index. A feature is the graphic use of charts for comparison and rating purposes (*e.g.* chart 1, Information sources and guides versus coverage). The index (p.302-28) includes authors and titles. Good use of bold type. Admirably meets the authors' objective: 'to produce a text which is suitable for beginners, but which will also serve as a handy reference work for the more experienced' (*Preface*). *Class No:* 5(01)

[70]

—**PRIMACK, A.L.** Finding answers in science and technology. New York, Van Nostrand Reinhold, 1984. 364p. facsims. ISBN: 0442282279.

13 chapters: 1. Finding the answers. Search strategy. Getting the materials - 2. Computer searching - 3. General science... 13. Engineering. Energy. Appendix: Libraries (large science and technology libraries in the U.S.); U.S. government depository libraries; NASA information centers; U.S. publications depository libraries. Detailed subject index, p.339-62. Not a bibliography, but a discussion of literature-searching techniques, followed by running commentary on sources. *Class No:* 5(01)

[71]

Pure and applied science books, 1876-1982: subject index; author index. New York & London, Bowker, [1982]. 6v. (7784p.). $345, set. ISBN: 0835214370.

Lists *c.*220,000 titles. V.1-5: Subject index, using over 56,000 Library of Congress subject headings; v.6: author and title indexes. Despite the dates of coverage cited, over 5,000 titles are pre-1876. Coverage: all areas of physical and biological sciences, omitting introductory level textbooks and government publications. Medicine is allotted a separate listing, - *Health science books, 1876-1982* ([1982]. 4v.) (*qv*). The first comprehensive retrospective bibliography in the area of science publishing. *Class No:* 5(01)

[72]

REUSS, J.D. Repertorium commentationum a societatibus litterariis editarum. Secundum disciplinarum ordinem digessit I.D. Reuss. Göttingen, Dieterich, 1801-21. 16v. Reprinted New York, B. Franklin, 1963.

1. *Historia naturalis, generalis et zoologia.* 2. *Botanica et mineralogia.* 3. *Chemia et res metallicae.* 4. *Physica.* 5. *Astronomia.* 6. *Oeconomia.* 7. *Mathesis; mechanica; hydrostatica; hydraulica; hydrotechnia; aerostatica; pneumatica; technologia; architectura civilis; scientia navalis; scientia militaris.* 8. *Historia.* 9. *Philologia; linguae; scriptores graeci; scriptores latini; litterae*

....(contd.)

elegantiores; poesis; rhetorica; ars antiqua; pictura; musica. 10-16. *Scientia et ars medica et chirurgica.* A valuable classified index to publications of scientific, historical, literary and medical learned societies from their inception in the 17th century, to 1800. The Royal Society's *Catalogue of scientific papers, 1800-1900 (qv)* provides a continuation, for the sciences. *Class No:* 5(01)

[73]

ROYAL SOCIETY. Book catalogue of the library of the Royal Society. Clark, A.J. Frederick, Md., University Publication of America, 1982. 5v. $550. ISBN: 0890935203.

About 62,500 entries for *c.*45,000 books and pamphlets. Basically an author catalogue, with numerous cross-references from names of editors, of scientists as biographers, etc. Pagination is omitted, but locations are given, with a list of 50 named classes and groupings shelves. Entries have been photographically reproduced from catalogue cards. A particularly significant collection for study of the history of science during the Society's first two centuries (1660-). *Class No:* 5(01)

[74]

ROYAL SOCIETY. Catalogue of scientific papers, 1800-1900. London, H.M. Stationery Office, 1867-72; London, Clay, 1877-1902; Cambridge Univ. Press, 1914-25. 19v. Reprinted New York, Johnson Reprint Corpn.; London, Johnson Reprint Co., Ltd., 1968.

V.1-6. 1st series, 1800-63; v.7-8. 2nd series, 1864-73; v.9-11. 3rd series, 1874-83; v.12. Supplementary v., 1800-83; v.13-19. 4th series, 1884-1900.

The standard index to 19th-century scientific papers. An author index to 1,555 periodicals, including transactions of European academies and other learned societies. A separate A-Z sequence runs through each series. Includes medical but excludes purely technical papers. Titles of papers appear in the original language; Russian titles are followed by English, French or German translation. *Class No:* 5(01)

[75]

—**ROYAL SOCIETY.** Catalogue of scientific papers, 1800-1900. Subject index. Cambridge, Cambridge University Press, 1908-14. 3v. in 4.

V.1: *Pure mathematics.* V.2: *Mechanics.* V.3: *Physics.* (2pts.)

Indexes papers for the entire period in each volume. Classified as in the *International catalogue of scientific literature*, although only 3v. of the 17v. planned were published. Details given for each paper are much abbreviated but sufficiently indicative for the subject index to be used independently of the *Catalogue*. Each volume carries a classification schedule, a list of serials (with abbreviations used) and locations in 23 major British libraries. *Class No:* 5(01)

[76]

Sale catalogues of libraries of eminent persons. Volume 11: Scientists. Feisenberger, H.A., *ed.* London, Mansell/Sotheby Parke Bernet, 1975. 302p.

The scientists concerned: Elias Ashmole, Edmund Halley, Robert Hooke, John Ray. This is the final volume in a series of facsimile reprints of catalogues of the libraries and effects of persons prominent in their respective fields. *Class No:* 5(01)

[77]
SCICAT: Science Reference and Information Service Catalogue. London, British Library, Science Reference and Information Service, 1990-. 4 issues pa. Negative microfiche.

Author/title catalogue. Classified catalogue.

Contains additions to SRIS stock from 1974, plus many items published earlier, including all serials held by SRIS. The *Classified catalogue* is arranged according to the SRIS classification scheme and has an index. *Class No:* 5(01)

[78]
Scientific and technical books and serials in print. New York, Bowker, 1972-. Annual. 3v. $249.95.

1. *Books: Subject index.* 2. *Books: Author index.* 3. *Books: Title index. Serials: Subject index. Serials: Title index.*

The 1991 volumes (4850p.) provide bibliographical and ordering information for over 130,000 books and 18,000 periodicals. More than 11,000 new entries. Directory of over 4,000 publishers and distributors. V.1 employs over 13,000 subject categories. Includes serials translated from Russian. Online, as part of BOOKS IN PRINT database, which is also available on CD-ROM. *Class No:* 5(01)

[79]
SUBRAMANYAM, K. **Scientific and technical information resources.** New York, Dekker, 1981. 416p. (*Books on library and information science, 33.*)

Sections on indexes, primary journals, bibliographies, congresses, theses and technical reports; also on secondary journals, commercial catalogues, patents and standards. Bibliography, p.361-81; index, p.387-416. For library and information science students, librarians and documentalists. A development, to some extent, of his lengthy contribution to *Encyclopedia of library and information science* v.26 (1979), p.376-548, on 'Scientific literature', with its 271 references and bibliography, p.522-416. *Class No:* 5(01)

[80]
UNITED STATES. Department of the Interior. **Dictionary catalog of the Department Library.** Boston, Mass., G.K. Hall, 1967. 37v. (29,573p.). Supplements, 1968-. $4,000, set; $910, 4th supp. ISBN: 0816107157, set; 0816100160, 4th supp.

The main catalogue and supplements 1-4 (1968-75) total 992,200 photolithographed catalogue cards. The largest specialised collection in the U.S.A. on mines, minerals, petroleum and coal, plus a good working collection on geology, and unequalled holdings on fish, fisheries and wildlife. *Class No:* 5(01)

Encyclopaedias

[81]
The Encyclopaedia of ignorance: everything you ever wanted to know about the unknown. Duncan, R. *and* Weston-Smith, M., *eds.* Oxford, Pergamon Press, 1977. 2v. in 1. £17. ISBN: 0080224241.

1. *The physical sciences.* 203, xiiip.

2. *Life sciences and earth sciences.* p.208-433, xiiip.

51 scientific theories discussed. V.1. has 23 contributors. The essay, 'Is space curved?', p.85-91, has 7 references. V.2 with 27 contributors, includes 'The limitations of evolutionary theory', p.235-42; 9 references. Each volume has an analytical index. 'The usual encyclopaedia states what

....(contd.)
we know. This one contains papers on what one does not know or matters which lie on the edge of knowledge' (*Editorial preface*). *Class No:* 5(031)

[82]
Encyclopedia of physical sciences and technology. Meyers, R.A.M., *ed.* San Diego, Academic Press, 1987-88. 15v. 11000p. $2,500. ISBN: 0685181383.

Coverage includes chemistry, physics, geology, astronomy, computer science, meteorology and mineralogy. 4,200 glossaries accompany the articles. Includes 6,000 figures and 2,000 tables. Bibliographies consist of 4,000 titles. V.15. *Subject index* contains 45,000 entries. A scholarly encyclopaedia, recommended for academic, rather than school, libraries. 2nd ed. (18v. 13788p. £1,550) due late 1992/93.

Encyclopedia of physical sciences and technology yearbook (1991 ed. £126) updates between editions. *Class No:* 5(031)

[83]
The Encyclopedic dictionary of science. Hunt, C., *ed.* New York, Facts on File, 1988. 256p. illus. $29.95. ISBN: 0816020213.

Explains and defines *c.*7,000 terms. Descriptive entries vary from one to three paragraphs in length. Outstanding feature is 50 pages of full-colour illustrations, Popular dictionary intended for school and public libraries. *Class No:* 5(031)

[84]
McGraw-Hill encyclopedia of science and technology. Parker, S.P., *ed.* 7th ed. New York, etc., McGraw-Hill, 1992. 19v. & index. illus. (some col. pl.), diagrs., tables, maps. $1,900. ISBN: 0079092063, set.

About 3,000 contributors. 7,500 entries arranged A-Z, on all branches of pure and applied science. Areas where most revision for this ed. has taken place include medicine, chemistry, environmental science and engineering, and earth sciences. Signed contributions, normally with short bibliographies appended (*e.g.* 'Acid rain': v.1, p.59-62; 9 references to books and articles; 3 diagrams; 11 cross-references). The most recent bibliographic citations appear to be from 1988. V.20 (636p.) lists contributors and has sections: Scientific notation - Study guides (*e.g.* 'Biology'; 'Physics'). Topical and analytical indexes, p.127-636. A major US encyclopaedia. Annual updates, between editions: *McGraw-Hill yearbook of science and technology (q.v.).* The *Encyclopedia* (6th ed. 1987) has provided several encyclopedia spin-offs, *e.g.* geological sciences. *Class No:* 5(031)

[85]
—McGraw-Hill concise encyclopedia of science and technology. Parker, S.P., *ed.* 2nd ed. New York, McGraw-Hill, 1989. lxxvi, 2222p. illus., figs., diagrs., tables. $110. ISBN: 0070455120.

First published 1984.

400 more articles than previous ed., with 7,700 alphabetically arranged, signed articles derived from the 1987 ed. of the *McGraw-Hill encyclopedia of science and technology (q.v.).* Many illustrations and diagrams. Bibliographies, p.2087-80. Databases, p.2081-5. 49-page appendix includes: Greek alphabet; mathematical notation; elementary particles; biographical listing. 30,000-entry index. Available on CD-ROM. There is some duplication of

....(contd.)
Van Nostrand's scientific encyclopedia (q.v.), which has greater depth, but some entries are unique in both. Libraries on small budgets may well invest in both single volume works rather than the 20v. set ($1,900). *Class No:* 5(031)

[86]
Magill's survey of science: life science series.
Magill, F.N., *ed.* Pasadena, Calif., Salem Press, 1992. 6v. 2763p. $475. ISBN: 0893566128.

163 academic contributors. 370 articles on topics covering plant anatomy, genetics, biology, ecology, physiology, and biochemistry, arranged A-Z. Data in each article include: definitions of key terms; methods of study; context in relation to other aspects of the life sciences; annotated bibliography; cross-references. Cumulated glossary in the final volume, followed by a detailed index. No illustrations. Suitable for those with some basic knowledge. 'A solid contribution' (*Library journal,* 15th September 1992, p.62). *Class No:* 5(031)

[87]
Van Nostrand's scientific encyclopedia.
Considine, D.M., *ed.* 7th ed. New York, Van Nostrand Reinhold, 1988. 2v. xvi,1628+1629-3180p. illus., diagrs., graphs, tables, maps. $195 set. ISBN: 0442217501, set; 0442318146, v.1; 0442318162, v.2.

First published 1968.

The 6,773 alphabetically arranged entries include 800 which are new or revised. Subject coverage includes chemistry, biosciences, plant sciences, physics, mathematics and medicine. Over 3,000 black-and-white illustrations and 500 tables. New to this edition is a 109-page, 45,000-entries index. Brief bibliographies. *Class No:* 5(031)

Handbooks & Manuals

[88]
Composite index for CRC handbooks. 3rd ed. Boca Raton, Fla., CRC Press, Inc., 1991. 3v. [vii], 1111p. $1,295. ISBN: 0849302846.

First published 1971.

Covers *CRC handbooks* published up to 1990. Subject areas include life, physical, engineering, mathematical, and environmental sciences. This set includes a computer laser optical disk, a CD-ROM user's guide, and 2 computer disks. *Class No:* 5(035)

[89]
COTTERILL, R.M.J. The Cambridge guide to the material world. Cambridge Univ. Press, 1985. vii, 352p. illus., diagrs., tables, chemical formulas. ISBN: 0521246407.

18 chapters, on electrons, nuclei and quanta, the chemical bond, states of matter, crystals, water, minerals, ceramics, metals, conductors and insulators, glass, carbon, liquid crystals, polymers, biopolymers, the cell, the plant, the animal. Chapter bibliographies, p.348; detailed index (in tiny type), p.320-2. Intended primarily for the non-specialist. 'A reliable first resource for a student at any level' (*Scientific American,* September 1985, p.22). *Class No:* 5(035)

[90]
HARRÉ, R. The Philosophies of science: an introductory survey. 2nd ed. Oxford, University Press, 1985. [viii], 203p. diagrs. £7.99. (*An OPUS book.*) ISBN: 0192892010.

First published 1972.

7 chapters: 1. The philosophy of science - 2. The forms of reasoning in science - 3. Scientific knowledge - 4. Metaphysical theories - 5. The corpuscularian inheritance - 6. Explanation - 7. Science and society. Further reading, p.196-8 (by chapters). Partly analytical index. Aims 'to bring out the way problems which appear specific to science are actually species of wider philosophical issues' (*Preface to the second edition*). *Class No:* 5(035)

[91]
Science and technology desk reference. Andover, Hants, Gale Research, 1993. 575p. figs., maps. £27. ISBN: 0810388847.

Question and answer entries, divided by topics, *e.g.* 'Mathematics'; 'Plants'. Useful for school and public libraries as a quick reference tool. *Class No:* 5(035)

[92]
The World Book encyclopedia of science. Rev. ed. Chicago, Ill., World Book, Inc., 1991. 8v. (1200p.) col. illus., diagrs., chemical structures. ISBN: 0716632268, set.

Originally published by Verlagsgruppe Bertelsmann International GmbH, Munich 1984. First English ed., 1985.

8v. (ea.140-160p.): *The heavens - Physics today - Chemistry today - The planet earth - The plant world - The animal world - The human body. Men and women of science index. Chemistry today* (159p.) has sections: Elements and molecules; Major groups of elements; Organic chemistry; Biochemistry; Analytical chemistry. Glossary, p.144-51. Detailed analytical index, p.152-58. Illus., are keyed and captioned. *Class No:* 5(035)

Dictionaries

[93]
Catalogue of the Translator's Library of the Department of Trade and Industry: dictionaries, glossaries, encyclopedias, books about Europe. Hamilton, G.E., *ed.* Dobbs Ferry, New York, Oceana, 1975. 3v. $270. ISBN: 0379003759.

1. *Authors.* xxxvii, [1], 319p. 2. *Subjects.* vii,[1], 540p. 3. *Languages.* vii,[1], 555p.

About 5,000 items, covering 68 languages, figure in v.1. V.2 arranges the items under subjects, Abbreviations ... Zoology. V.3 features entry under languages, Afrikaans ... Yiddish. *Class No:* 5(038)

[94]
The Dictionary catalogue. Molho, E., *ed.* New York & Los Angeles, The French & Spanish Book Corpn., 1988. xviii, 174p. $14.95. ISBN: 0828801509.

'Pure and applied sciences' section, p.78-138, lists *c.*1,800 entries, with pagination and prices. Sections: General works - Abbreviations, symbols and units - Agricultural sciences - Astronomy - Biological sciences - Chemistry - Data processing & information systems - Earth sciences - Engineering - Mathematics - Medical sciences - Miscellaneous technologies - Physics - Psychology & psychiatry. *Class No:* 5(038)

[95]

McGraw-Hill dictionary of scientific and technical terms.
Parker, S.P., *ed.* 4th ed. New York, McGraw-Hill, 1989.
xvii,2088, 49p. illus., figs., diagrs. $95. ISBN: 0070452709.
First published 1974.

Contains *c.*100,000 terms (7,600 new to this edition).
Pronunciation given for all terms for the first time. Entries
include synonyms, abbreviations and acronyms. Brief
definitions. *c.*3,000 illustrations in the margins. Appendices
include SI conversion tables, the Greek alphabet, the
periodic table of the elements, schematic electronic symbols
and the classification of living organisms. The last appendix
is a biographical dictionary of over 1,200 eminent historical
and contemporary scientists. Thumb-indexed.
Class No: 5(038)

[96]

UNITED NATIONS. Service Linguistique. Section de
Terminologie et de Documentation Technique. **Catalogue
des ouvrages disponibles au service linguistique.** 5th ed.
Geneva, United Nations, 1987.

Lists *c.*5,000 titles of language dictionaries, etc. Two
parts: General works (abbreviations; general monolingual
dictionaries; general bilingual; atlases; encyclopedias;
science and technology dictionaries: monolingual, bilingual;
linguistics; translation and related subjects; miscellaneous -
Specialized works (aerospace ... transport), p.69-506.
Subject indexes (French; English). One of the few relatively
recent lists of language dictionaries. *Class No:* 5(038)

[97]

World dictionaries in print, 1983: a guide to general and
subject dictionaries in world languages. New York, Bowker,
1983. xii, 579p. $99.50. ISBN: 0835216152.

Lists over 13,600 in-print dictionaries, wordbooks,
glossaries and thesauri in more than 235 languages. Four
indexes: Subject index (using *c.*2,000 Library of Congress
subject headings); Title index (with equally full information);
Language index (grouping titles by language); Author,
editor, compiler index. A key to publishers and distributors
concludes. The fullest list of its kind, but confined to in-print
dictionaries. *Class No:* 5(038)

[98]

**WÜSTER, E. International bibliography of standardized
vocabularies.** 2nd ed., rev. and edited by INFOTERM.
Munich, etc., K.G. Saur, 1979. 540p. ISBN: 3598055021.

First published 1955, as *Bibliography of monolingual
scientific and technical glossaries,* by E. Wüster (Paris,
Unesco), v.1: *National standards.* 219p.

More than 11,000 references to standardized
nomenclature: definitions and symbols. *Class No:* 5(038)

[99]

—International bibliography of specialized dictionaries.
Lengenfelder, H., *ed.* 6th ed. Munich, K.G. Saur, 1979.
xxi, 470p. ISBN: 3598205015.

Not to be confused with E. Wüster's *International
bibliography of standardized vocabularies.* Its 5,719
numbered entries concern monolingual and multilingual
dictionaries that have appeared since 1970. Three of the 9
sections concern science and technology: 1. Science and
medicine - 2. Applied science, technology - 3. Agriculture.
Selected bibliography, p.373-7. Indexes of names and
subjects; directory of publishers. *Class No:* 5(038)

Polyglot

[100]

BUECKEN, F.J. Vocabulário técnico: portugués-inglés-
francês-alemão. 5. ed. São Paulo, Ediçoes Melhoramentos,
1979. 600p.
First published 1946; 4th ed. [1961] (600p.).

Portuguese terms (*c.*15,000 entries), with English, French
and German equivalents, p.15-241. Combined English,
French and German index, giving the Portuguese
equivalents. Appended glossary of motor-vehicle terms,
similarly arranged. *Class No:* 5(038)=00

[101]

**ETTINGEN, S.G. Dictionary of technical terms in five
languages:** Hebrew-English-French-German-Russian. Tel-
Aviv, Yavneh Publishing House, 1961 (reprinted 1983). 2v.
(64, 1188p.). illus.

The dictionary has reverse indexes in English, French,
German and Russian. Introduction and explanatory notes are
in Hebrew. *Class No:* 5(038)=00

[102]

FEUTRY, M. Technological dictionary/ Dictionnaire
technologique/Technologisches Wörterbuch. Paris, La
Maison du Dictionnaire, 1976-78. 3v.
1. *Mechanics, metallurgy, hydraulics and related
industries.* 1976. *Supplemento español.* 1976. 2. *Les
ressorts.* 1978. 3. *Aeronautique.* 1978.

V.1 has 14,051 numbered entries, English-French-
German, in 3 columns, with French and German indexes.
Well spoken of. *Class No:* 5(038)=00

[103]

HOYER, E. von *and* **KREUTER, F. Technologisches
Wörterbuch.** 6. vollkommen neubearb. Aufl., hrsg. von A.
Schlomann. Berlin, Springer, 1932 (reprinted New York,
Ungar, 1944). 3v.
1. *Deutsch-Englisch-Französisch.* 2. *English-German-
French.* 3. *French-German-English.*

About 100,000 terms in each of three languages.
Coverage: technology and allied subjects such as trade and
banking. Overdue for revision, but a thorough and reliable
set of dictionaries. *Class No:* 5(038)=00

[104]

MEDEIROS, M.F. da S. de. Dicionário técnico poliglota:
portugues, espanhol, francés, italiano, inglés, alemão.
Lisbon, Gomez & Rodrigues, 1949-53. 8v.

V.1-3 (1949-51) comprise 68,193 numbered Portuguese
base terms and synonyms, with equivalents in Spanish,
French, Italian, English and German. V.4-8 comprise the
Spanish, French, Italian, English and German equivalents,
plus Latin. An ambitious project, coverage of the Romance
languages being particularly generous.
Class No: 5(038)=00

[105]

—Dictionary of scientific and technical terminology: English,
French, Dutch, Russian. Markov, A.S., *ed.* The Hague,
Nijhoff, 1984. 496p. £35.95; $34.50. ISBN: 0318016613.

Has *c.*9,000 English-base terms, with equivalents in the
other languages. Intended for 'scientists and engineers'. But
too many concepts are introduced for the dictionary 'to be of
more than partial use to anyone but a professional translator,
who would, presumably, prefer a specialized dictionary for
each topic' (*Nature*, v.312, 13 December 1986, p.670).
Class No: 5(038)=00

[106]
QUÉMADA, G. **Dictionnaire de termes nouveaux des sciences et des techniques.** [Dictionary of new scientific and technical terms.] Paris, Conseil International de la Langue Française (CILF), [1983]. xix, 602p. £57.50. ISBN: 2853191117.

Part 1 provides *c.*7,000 French-base terms, with equivalents (where available) in German, English and Spanish. The French terms are then grouped (p.325-87), followed by a single A/Z sequence of English, German and Spanish equivalents. Part 2, 'Répertoire des formants morphosémantiques' (French prefixes and suffixes), p.523-602. *Class No:* 5(038)=00

English

[107]
Academic Press dictionary of science and technology. San Diego, Academic Press, 1991. xxxii,2432p. illus., diagrs., tables. £68. ISBN: 0122004000.

Contains *c.*133,000 fully-defined entries in 122 areas of science. Claims to include more medical terms overall than any other general science dictionary presently available. 122 essays give concise information on each scientific discipline. Over 2,000 illustrations, with 24 colour plates. Pronunciation guides for difficult or phonetically irregular terms. Data in appendix include standard weights and measures periodical table, and chronology of science. Claims to be 'the largest scientific dictionary ever compiled in the English language' (*Preface*). However, criticized in *Nature*, v.361 (no.6411), 4th February 1993 for inconsistencies and poor illustrations. *Class No:* 5(038)=20

[108]
BALLENTYNE, D.W.G. *and* LOVETT, D.R. **A Dictionary of named effects and laws in chemistry, physics and mathematics.** 4th ed. London, Chapman & Hall, 1980. viii,346p. diagrs., tables, chemical structures £17.95. ISBN: 0412223902.

First published 1958.

Several thousand entries, A-Z, for effects and laws best known by the name of the person(s) who discovered or propounded them (*e.g.* Guy Lussac's law, - under 'G'). Formulas and chemical structures; cross-references. Definitions are clearly stated, but omit biographical data and bibliographies.

Ruffner, J.A. *Eponyms dictionaries index* (Detroit, Mich., Gale, 1977. xxxi, [1], 734p. Supplement no.1, 1985. 248p.) covers a much wider field. *Class No:* 5(038)=20

[109]
BRAGONIER, R. *and* FISHER, D. **What's what: a visual glossary of the physical world.** Rev. ed. Maplewood, N.J., Hammond Inc., 1990. viii,581p. illus. $14.95. ISBN: 0843733225.

First published 1981.

Displays thousands of clear illus. of physical objects, duly keyed. Arrangement is by broad categories (*e.g.* Earth; Living things). The 1981 ed. was considered an improvement on the *Oxford Duden Pictorial English dictionary* series (*Choice*, v.19(7), March 1982, p.888). *Class No:* 5(038)=20

[110]
—CRISPIN, F.S. Dictionary of technical terms. 11th ed. New York, Bryce Publishing Co., 1970. vi, 455p. diagrs.

Provides 10,000 brief definitions of technical terms 'in use in all sections of the United States' (*Preface*). Intended for students, draughtsmen, mechanics, builders, electricians, and workmen generally. Syllabification, hyphenation, stress and capitalization shown. Terms are categorized. About 1 diagram per page. *Class No:* 5(038)=20

[111]
BRENNAN, R.P. **Dictionary of scientific literacy.** New York, John Wiley, 1992. xiv,334p. illus., diagrs., tables. £14.95. ISBN: 0471532142.

Lists *c.*700 terms A-Z (Absolute zero ... Zygote). Many clear illustrations and diagrams. Cross-references in bold, upper-case type within the text. Pronunciation is given for some terms. Aimed at 'the intelligent non-scientist' (*Preface*) who wishes to understand scientific terminology which is now part of everyday life. *Class No:* 5(038)=20

[112]
—Harrap's illustrated dictionary of science. London, Harrap, 1988. 256p. figs., diagrs. £19.95. ISBN: 0245547800.

Covers *c.*7,500 terms (Aa ... Zymase). 34 colour illustrations: 1. Nobel Prizes in science, 1901-1987 ... 14. Hormones ... 27. Senses ... 34. Universe-scale and distance. Short biographies, *e.g.* Mendel (¼ column), Darwin (1/3 column). American spelling, with cross-references from that of the English. Written in clear, non-specialist language. 'Reference book for those wanting to understand scientific terms and the meanings behind them' (*British book news*, February 1991, p.103). *Class No:* 5(038)=20

[113]
Chambers science and technology dictionary. Cambridge, Chambers, 1988. xvi,1008p. figs., tables. £17.50; £7.99. ISBN: 0198661673.

First published as *Chamber's technical dictionary* in 1940.

About 45,000 entries, arranged A-Z, followed by 10 appendices, (Paper tables ... Animal kingdom ... Physical constants, standard values and equivalents). Alternative forms, terms derived from the headword and variables in mathematical formulae are in italics while bold type is used for cross-references and vector notation. Trade names are also indicated in the text. General, not intended to replace specialist dictionaries.

A Spanish ed. appeared in 1979 (*q.v.*).
Class No: 5(038)=20

[114]
Code names dictionary: a guide to code names, slang, nicknames, journalese and similar terms,- aviation, rockets and missiles, military, aerospace, meteorology, atomic energy, communications, and others. Ruffner, F.G. *and* Thomas R.C., *eds.* Detroit, Mich., Gale, 1963. 555p. $55. ISBN: 0810306859.

Deciphers more than 8,500 code names (including acronyms), mainly drawn from government publications and originating from World War II, the Korean War, space probes, etc. Chief use is for identifying passwords, operations and projects, not for defining terms.
Class No: 5(038)=20

[115]

Concise science dictionary. Isaacs, A., *and others, eds.* 2nd ed. Oxford, University Press, 1991. vi, 758p. illus., tables. £17.50; £7.99. ISBN: 0198661673.

About 7,500 definitions of terms and concepts in physics, chemistry, biology, the earth sciences, and astronomy. Aims 'to provide school and first-year university students with explanations of unfamiliar words they might come across' (*Preface*). 7 appendices (*e.g.* 4. Geological time scale). Criticised in *Choice* (v.22(9), May 1985, p.1303) for inadequate coverage of agriculture, engineering, medicine and the applied sciences in general. However, it may be assumed that 'science' concerns the physical, earth and biological sciences, in the strict sense of the word. A good general-purpose science dictionary. *Class No:* 5(038)=20

[116]

FLOOD, W.E. Scientific words, their structure and meaning: an explanatory glossary of about 1,150 word elements (roots, prefixes, affixes) which enter into the formation of scientific terms. London, Greenwood Press, 1960. xix, 220p. £32.50. ISBN: 0837175410.

Word elements, A-Z 'Hecter, hetero-': 1 column; 'Pyro, pyr-': 1½ columns. Entries state the meaning of each word and also its origin, usually Latin or Greek. Chemical formulas are given whenever they serve a useful purpose. *Class No:* 5(038)=20

[117]

HARTMANN-PETERSEN, R. *and* **PIGFORD, J.N. Dictionary of science.** London, Edward Arnold, 1984. [iv], 332p. illus. £6.95. ISBN: 0713181680.

Over 6,000 short entries, giving simple, direct explanations. Good coverage of physics, chemistry and biology, plus terms used in agriculture, geology, astronomy and medicine. 200 keyed line-drawings. 'Specially prepared for those studying Biology, Chemistry, General Science, Physical Science and Physics' (*Preface*). Goes beyond 'O' level. Good layout; a moderately priced dictionary with much to recommend it. *Class No:* 5(038)=20

[118]

—The Longman dictionary of scientific usuage. Godman, A. and Payne, E.M.F. Harlow, Essex, Longman, 1979. xxxx, 684p. diagrs., graphs, tables, chemical structures. ISBN: 058252587x.

Definitions of over 10,000 terms in two main sections. The first handles over 1,300 basic terms. The second treats 8,500 'special terms' grouped in thesaurus fashion into *c.*100 subject sets in biology, chemistry, physics etc. 12 appendices. Index, p.615-84. The non-technical definitions (not always accurate) and many illus. give it some appeal to schools and general readers. *Class No:* 5(038)=20

[119]

KNIGHT, D. A Companion to the physical sciences. London, Routledge, 1989. 177p. £7.50. ISBN: 0415009014.

Dictionary of 125 terms, varying in length from a quarter to two pages. Numerous cross-references. 4-page list of individuals discussed in the main sequence is appended. Supplements Bynum, W.F., *The Dictionary of the history of science* (Princeton, 1981). Undergraduate level. *Class No:* 5(038)=20

[120]

A List of scientific terms with Scottish connections. Edinburgh, Scottish National Dictionary Association, 1976.

Supplement to *Scottish national dictionary,* v.10, pt.iv, 1976, p.575-91. *Class No:* 5(038)=20

[121]

UVAROV, E.B. *and* **ISAACS, A. The Penguin dictionary of science.** 6th ed. Harmondsworth, Middx., Penguin Books, 1986. [v], 466p. diagrs., tables, chemical structures. £5.99. ISBN: 0140511563.

5th ed., 1979.

About 5,000 terms are defined, with numerous cross-references. Entries average *c.*50 words in length. 11 appended tables (*e.g.* 6-figure conversion factors). 'In this edition, as in previous editions, the general principle of selecting predominantly scientific terms, as opposed to technical words, has been maintained' (*Foreword to the 1986 edition*). A long-established 'standard work of reference' (*The History of science,* v.24(4), December 1986, p.442). *Class No:* 5(038)=20

[122]

—EMILIANI, E. Dictionary of the physical sciences: terms, formulas, data-physics, chemistry, geology, cosmology. New York, Oxford University Press, 1987. 416p. illus., tables. £27.50; £17.50. ISBN: 0195036514.

Over 5,000 definitions of fundamental terms. 70 tables (p.233-365; 'Isotope chart', p.288-300). *Class No:* 5(038)=20

German

[123]

DE VRIES, L. *and* **HERRMANN, T.M. English-German technical and engineering dictionary.** 2nd ed. New York, London, etc., McGraw-Hill, 1966. [xv], 1154p.

First published 1954.

More than 200,000 entries, including compounds. Lengthy entries are sectionalized. 76 specialist collaborators. New subjects added include: astronautics, automation, climatology, control engineering, data processing, inertial navigation, measurement systems, sail mechanics. 51 categories. Abbreviations figure in main sequence. Badly needs updating. *Class No:* 5(038)=30

[124]

DE VRIES, L. German-English science dictionary, updated and expanded by L. Jacolev and P.L. Bolton. 4th ed. New York, London, etc., McGraw-Hill, 1978. xxxviii, 628p. $37.95. ISBN: 0070166021.

First published 1939.

About 65,000 entries; Addendum, p.559-628: *c.*7,000 entries. Coverage: mathematics, physics, chemistry, biology, botany, zoology, anatomy, agronomy, horticulture. Sub-entries in bold type. 'Keim' (germ) and compounds: 1½ columns. Preliminary essay: 'Suggestions for translators'. Appended lists of geographical names, abbreviations, chemical elements, weights and measures, etc. *Class No:* 5(038)=30

[125]
DE VRIES, L. *and* HERRMANN, T.M. **German-English technical and engineering dictionary.** 2nd ed., completely rev. and enl. New York, London, etc., McGraw-Hill, 1965. ix, 1178p. $95. ISBN: 0070166315.

First published 1950; supp., 1959.

Over 225,000 entries, 100,000 more than in 1st ed. Terms are categorized. Medicine and agriculture are not covered in any depth. Lengthy entries are sectionalized (*e.g.* 'Festigkeit' has 32 compounds, in 3 groups). Includes *c.*25,000 new terms in such fields as nuclear physics, space flight and plastics. Badly needs updating.
Class No: 5(038)=30

[126]
DORIAN, A.F. **Dictionary of science and technology, German-English.** 2nd rev. ed. Amsterdam, etc., Elsevier, 1981. x, 1119p. $202.75. ISBN: 0444419977.

First published 1970.

About 100,000 entries. Terms in the 1st ed. 'regarded as ballast' have been dropped and 20,000 new headwords added. *Class No:* 5(038)=30

[127]
—DORIAN, A.F. Dictionary of science and technology, English-German. 2nd rev. ed. Amsterdam, etc., Elsevier, 1978. xii, 1402p. $195. ISBN: 0444416498.

First published 1967.

Has *c.*100,000 entries, with definitions and explanations of many of the English terms. Well thought of by translators, although the tendency to enter under the qualifying adjective and not the noun qualified should be noted.
Class No: 5(038)=30

[128]
ERNST, R. **Wörterbuch der industriellen Teknik.** [Dictionary of engineering and technology.] 5. Aufl. Wiesbaden, Brandstetter, 1985-89. 2v. £79.95ea.

First published 1948.

1. *Deutsch-Englisch.* 5. Aufl. 1989. 1258p. 2. *Englisch-Deutsch.* 5. Aufl. 1985. 1422p.

V.1 has *c.*157,000 entries. V.2 has an introduction on 'The nature of technical English'. Regarded as the standard work in the field. *Class No:* 5(038)=30

[129]
KUČERA, A. **The Compact dictionary of exact science and technology.** Wiesbaden, Brandstetter, 1982-89. 2v. DM140, v.1; DM95, v.2. ISBN: 0685483045, v.1; 387097088x, v.2.

1. *English-German.* 2nd ed. 1989. xxxi,1460p. 2. *German-English.* 1982. xiv,826p.

V.1 offers *c.*50,000 headword entries; v.2, over 67,000. Gives not merely equivalents but also frequently refers to *Chambers Dictionary of science and technology* (for fuller definitions) and to the German DIN standards. Appendices on German, British and US standards, etc. 'Few dictionaries succeed in putting so much solid, reliable information in so little space. Invaluable not only for anyone working in the field, but also can serve as a model for other bilingual terminological dictionaries' (*ARBA,* 1984, entry 1232).
Class No: 5(038)=30

[130]
WALTHER, R. **Dictionary of technology.** 5th ed. Amsterdam, etc., Elsevier, 1985. 2v. $187.25ea. ISBN: 0444995919, v.1; 0444995900, v.2.

Rev. ed. of *Polytechnisches Wörterbuch.*

1. *English-German.* 1985. 1126p. 2. *German-English.* 1985. viii, 1046p.

Each volume has *c.*100,000 entries for terms in technology and industrial production. Subentries are given separate lines. 40 categories apply to terms. German genders given. Appendices list abbreviations. 'This fundamental work' (*New technical books,* v.70(8), October 1985, item 1051). *Class No:* 5(038)=30

Dutch

[131]
JANSONIUS, H. **Technish Engels woordenboek:** Nederlands-Engels en Engels-Nederlands. Leiden, Nederlandsche Uit., 1976. 555p. $109.25. ISBN: 9061100291.

Dutch-English (p.7-406): *c.*4,000 headwords (many subentries); English-Dutch (p.409-55): *c.*1,500 headwords. *Class No:* 5(038)=393

[132]
STEKHOVEN, G.S. **Kluwer's universal engineering dictionary, Dutch-English.** Deventer, Antwerp, Kluwer Technische Boeken B.V., 1977. xii, 775p.

A continuation of *Ten Bosch technisch woordenboek.*

About 50,000 entry-words, including abbreviations. Many compounds, usually given separate entries. Language areas: British and American English, Flemish, Dutch and Afrikaans. *Class No:* 5(038)=393

Norwegian

[133]
ANSTEINSSON, J. **English-Norwegian technical dictionary.** [Engelsk-norsk teknisk ordbok.] 5th rev. ed. Trondheim, Brun; Kinderhook, N.Y., IBD Ltd., 1991. 461p. $80. ISBN: 8270284920.

First published 1948.

About 20,000 entries. Clear layout; compounds are picked out in bold type. *Class No:* 5(038)=396

[134]
ANSTEINSSON, J. **Norsk-engelsk teknisk ordbok.** 5th ed. Trondheim, Brun; Vero Beach, Fla, Vanous, 1990. 358p. $80. ISBN: 8270284904.

About 25,000 entries for Norwegian terms, with English equivalents.

The two dictionaries are intended for Norwegian technical colleges and do not cater for the foreigner's lack of knowledge of stress, pronunciation or syllabification of Norwegian words. The specialized vocabularies are, however, recommended by translators.
Class No: 5(038)=396

Swedish

[135]
ENGSTRÖM, E. **Engelsk-svensk teknisk ordbok.** 13. uppl. Stockholm, Travaru-Tidning Förlaget; New York, French and European, 1978. 1026, [2]p. $250. ISBN: 0828806756, US ed.

....(contd.)

Over 100,000 entries (including subentries). 'Safety': nearly 2 columns. Includes abbreviations. About 200 categories, etc. are applied to terms. American technical terms receive special attention. *Class No:* 5(038)=397

[136]
ENGSTRÖM, E. Svensk-engelsk teknisk ordbok. [Swedish-English technical dictionary.] 10. uppl. Stockholm, Svensk Trävarutidning Förlaget; New York, French and European, 1977 (1984 printing). 973, [1]p. $250. ISBN: 0828806764, US ed.

About 100,000 entries and subentries. 'Kraft' (power): 3 columns. Appended list of sources. *Class No:* 5(038)=397

[137]
GULLBERG, I.F. Svensk-engelsk fackordbok för näringsliv, fövaltning, undevisning och forskning. [A Swedish-English dictionary of technical terms used in business, industry, administration, education and research.] 2. revid. uppl., med suppl. Stockholm, Norstedt, 1977. xix, [1], 1722p. $225.

First published 1964 (xvi, 1246p.).

The 2nd ed. has an extensive supplement of *c.*40,000 new terms, giving a total of *c.*175,000 categorized terms, A/Z. Rich in idioms. Includes many proper names ('Internationella...': nearly 36 columns). British and US usages not differentiated. Includes abbreviations. List of important dictionaries in 50 languages. Well-produced and a fine example of its kind. *Class No:* 5(038)=397

Danish

[138]
WARRERN, A. Dansk-Engelsk teknisk ordbog. 6. udg. Copenhagen, Clausen Bøger, 1981. 393p. ISBN: 8711040270.

First published 1949; 5th ed., 1976 (385p.).

About 15,000 word-for-word entries. Many compounds. British and US usage differentiated; terms categorized. Entry under qualifying adjective. Abbreviations not included. *Class No:* 5(038)=398

[139]
WARRERN, A. Engelsk-dansk teknisk ordbog. 9. udg. Copenhagen, Clausen Bøger, 1986. 400p. $75. ISBN: 8711041544.

15,000 word-for-word entries, compounds being given separate entries. *Class No:* 5(038)=398

French

[140]
BELLE-ISLE, J.-G. English-French general technical dictionary. [Dictionnaire technique general.] 2nd ed. London, Routledge & Kegan Paul; New York, French and European, 1978. xii, 555p. ISBN: 068656913x.

First published 1965.

About 50,000 English terms, with *c.*126,000 French equivalents. Some 30 categories are applied to terms. Canadian and French phrases are differentiated. Includes expressions; compounds are given separate entries. Bibliography, p.554-5. *Class No:* 5(038)=40

[141]
CUSSET, F. Vocabulaire technique, anglais-français, français-anglais: électricité, mécanique, industries extractives et annexes, métallurgie, sciences. 9. éd. Paris, Berger-Levrault. 1977. 434p.

First published 1946 as *English-French, French-English* dictionary. 8th ed. 1968 (434p.).

About 7,000 entry-words in each half. English-French, p.5-220; French-English, p.221-424. Compiled by a mining engineer. Terms are strictly technical, with separate entries for compounds. Conversion tables, p.427-34. *Class No:* 5(038)=40

[142]
DE VRIES, L. French-English science and technology dictionary. 4th ed., rev. and enl. by S. Hochman. New York, etc. McGraw-Hill, 1976. xiii, 683p. $46.10. ISBN: 0070166293.

First published 1940; 3rd ed; 1962.

About 52,500 entries, including a supplement of *c.*4,500 new terms added to 3rd ed. Emphasises new developments (*e.g.* in automotive technology, astronautics, electronics, electronic data processing). 'All the terms published by the French Ministry of National Education, in January 1973 have been included' (*Library journal*, v.101(13), July 1976, p.1516). *Class No:* 5(038)=40

[143]
Dictionnaire des racines scientifiques. Cailleux, A. *and* Komorn, J., *eds.* 3. éd., rev. et augm. Paris, Société d'Éditions d'Enseignement Supérieur [S.E.D.E.S.], 1981. 263p.

First published 1961 (246p.).

Explains more than one thousand scientific word roots. Could be used in conjunction with W.E. Flood's *Scientific words* (*qv*). *Class No:* 5(038)=40

[144]
DORIAN, A.F. Dictionary of science and technology, French-English. Amsterdam, etc., Elsevier, 1980. x, 1085p.

Claims to have over 150,000 terms relating to more than 150 differing fields. *Class No:* 5(038)=40

[145]
—Dictionary of science and technology, English-French. Dorian, A.F., *comp.* Amsterdam, etc., Elsevier, 1979. [x], 1586p. ISBN: 0444418296.

Covers *c.*150,000 terms, including trivia. Some 100 categories are applied. Entry under qualifying adjective (*e.g.* 'mobile crane' - entered under 'mobile', but not under 'crane'). Criticised for errors of commission and omission. 'It is considered of only marginal use to the professional linguist' (Picken, C. *The Translator's handbook*, p.208, 1983). *Class No:* 5(038)=40

[146]
ERNST, R. Comprehensive dictionary of engineering and technology, with extensive treatment of the modern techniques and processes. Cambridge, University Press; New York, French and European, 1982-1984. 2v. $135ea. ISBN: 052130377x, v.1; 0521303788, v.2; 082880611x, v.1, US ed; 0828806101, v.2, US ed.

1. *Français-Anglais*, viii, 1085p. 2. *English-French*. xxiii, 1399p.

V.1 has 159,142 entries; v.2 195,546 entries. Each term is categorized. 'Deals with all branches of modern industry involving chemistry, electrical engineering, electronics,

....*(contd.)*

commerce, space travel, telecommunications, data processing, & microprocessors. Though expensive, the set belongs in all comprehensive science and technology collections and in the hands of serious translators' *(New technical books,* v.70(8), October 1985, item 1056). *Class No:* 5(038)=40

[147]
Harrap's French and English science dictionary. Hathway, D.E., *consultant ed.* London, Harrap, 1985. ix, 320 + 302p. £19.95. ISBN: 0245540725.

About 30,000 headwords, with *c.*100 categories applied. Part 1: English-French; part 2: French-English. Mostly equivalents. 'Nucleus': 11 sub-entries. 'Selected subject-matter contains terms in the physical and earth sciences, including chemistry, biochemistry, physics, mineralogy, geology and meteorology, as well as the broad sweep of the life sciences ...' *(Preface). Class No:* 5(038)=40

[148]
—Harrap's visual French-English dictionary. London, Harrap, 1987. 924p. £14.95. ISBN: 0245545964.

Has 28 sections (Astronomy ... Symbols), a number of which concern science and technology. Thus, 'offshore platform' is illustrated on p.643, with 18 keyed parts. General, thematic and specialized indexes in both English and French. *Class No:* 5(038)=40

[149]
KETTRIDGE, J.O. French-English and English-French dictionary of technical terms and phrases used in civil, mechanical, electrical and mining engineering and allied sciences and industries, with a method of telegraphic coding. Rev. ed. London, Routledge & Kegan Paul, 1980 (reprinted 1989). 2v. £60, set; £35ea. ISBN: 0415036542, set; 0415043093, v.1; 0415027683, v.2.

First published 1925.

1. *French-English.* 640p. 2. *English-French.* 672p.

Defines about 100,000 terms and phrases, with examples. The *English-French dictionary* is largely an inverted counterpart of the French-English volume. *Class No:* 5(038)=40

Italian

[150]
DENTI, R. Dizionario tecnico, italiano-inglese, inglese-italiano. 11. ed. riv., ampliata ed aggiornata. Milan, S.F. Vanni, 1985. [xvi], 2023p. $100. ISBN: 882031052x.

First published 1946.

About 105,500 entries. Italian-English, p.1-835; English-Italian, p.837-1916. Many sub-entries are given separate entries. 'Leva' (lever), over 1½ columns. Appendices include technical abbreviations in common use, conversion factors, mathematical symbols, British and American measures and metric equivalents. *Class No:* 5(038)=50

[151]
—RAGAZINI, G. *and* RUSSI, D. Nuovissimo dizionario tecnico, inglese-italiano, italiano-inglese. [Milan], Mursia, 1983. 1278p.

Has *c.*50,000 headwords. Italian-English, p.1-641; English-Italian, p.679-1278. About 150 subject categories are applied. Appended list of abbreviations and tables. *Class No:* 5(038)=50

[152]
MAROLLI, G. Dizionario tecnico, inglese-italiano, italiano-inglese. [Technical dictionary, English-Italian, Italian-English.] 11. ed. interamente riv. ed ampl. Florence, Le Monnier; New York, French and European, 1980. 2215p. illus., tables. $275. ISBN: 0828833672, US ed.

10th ed. 1972 (2047p.).

About 135,000 technical terms, including abbreviations. Some 200 categories are applied to terms. Whereas Denti sets out compounds as separate entries, Marolli masses them under the headwords and is less easy to use. Folding diagrams are placed in a loose folder. *Class No:* 5(038)=50

[153]
—Dizionario enciclopédico scientifico e tecnico, inglese-italiano, italiano-inglese. Bologna, Zanichelli, 1980. 1650p.

Based on the *McGraw-Hill dictionary of scientific and technical terms (q.v.).* 102 categories are applied to its 96,000 terms. *Class No:* 5(038)=50

[154]
—GATTO, S. Dizionario tecnico scientifico illustrato italiano-inglese, inglese-italiano. 2. ed., riv. e coretta. Milàn, Ceschina, 1965. xx, [1], [1], 1381p. illus.

Intended for Italians. About 60,000 terms, with separate entry for each application. About 150 categories feature. The small line-drawings are infrequent and hardly merit inclusion. Bibliography, p.vii-viii. *Class No:* 5(038)=50

Rumanian

[155]
Dictionar tehnic: român-englez. Dumitrescu, D., *and others.* Bucharest, Editura Tehnicâ, 1970. xi, 1118p.

About 120,000 Rumanian entry-words, with English equivalents. 51 categories applied. Bibliography of 21 items, p.1118. *Class No:* 5(038)=590

[156]
DUMITRESCU, D., *and others.* **Dictionar tehnic: englez-român.** Bucharest, Editura Tehnicâ; New York, Irvington, 1967. xi, 1301p. $124.50. ISBN: 0829009876, US ed.

About 115,000 Rumanian equivalents of English scientific and technical terms. *Class No:* 5(038)=590

Spanish

[157]
[Castilla's] Diccionario politécnico de las linguas española e inglesa. 3. ed. aumentada y actualizada. Madrid, Ediciones Castilla, S.A., 1965. 2v.

First published 1958; 2nd ed. 1961. Original Spanish as *Diccionario politécnico de las linguas española e inglesa.*

1. *Inglés-español.* xii, 1755p. 2. *Español-inglés.* [v], 1464p.

Although concentrating on engineering technology, the 3rd ed. claims to cover latest developments in industry and commerce, banking and jurisprudence. V.1 has *c.*200,000 entries, with separate lines for sub-entries. Many idioms are compounds. Appendices of conversion tables, signs and symbols. V.2 has under 200,000 entries. Conversion tables appended. *Class No:* 5(038)=60

[158]
Chambers diccionario cientifico y tecnológico i español / inglés / francés / alemán. Collocott, T.C. Barcelona, Omega, 1979. 2v. (xii, 1739p.; 1007p.). ISBN: 8428205310.

V.1 consists of *c.*40,000 Spanish terms and explanations, plus English, French and German equivalents. About 65 categories are applied. V.2 has 3 sequences, - English-Spanish, French-Spanish, German-Spanish.
Class No: 5(038)=60

[159]
COLLAZO, J.L.   English-Spanish,   Spanish-English encyclopedic dictionary of technical terms. 3v. New York, McGraw-Hill, 1980. $195.

1. (2pts). *English-Spanish.* lxcxxv, 1187p. 2. *Spanish-English.* [xv], 430p.

V.1 has *c.*100,000 extended entries with definitions, phrases, synonyms, variant expressions, etc. V.2 has Spanish terms and their English equivalents.
Class No: 5(038)=60

[160]
—GUINLE, R.L. A Modern Spanish-English and English-Spanish technical & engineering dictionary, suitable for Spain and all the Spanish-speaking countries of Central America and South America. Mexico, Continental, 1980. xv, [2], 311p. $27.95.

First published 1938 and frequently reprinted, is by a consulting engineer. *Class No:* 5(038)=60

[161]
Diccionario de términos cientificos y técnicos. Lapedes, D.N., *redactor jefe.* Barcelona, Marcombo; New York, French and European, 1981. 5v. illus. ISBN: 0070791724; 0828806683, US ed.

Spanish translation of *McGraw-Hill Dictionary of scientific and technical terms* (2nd ed., 1978), with its categorized 108,000 entries. V.1-4 provide the translation, adding English term. V.5 provides the English original, with Spanish equivalent. *Class No:* 5(038)=60

[162]
SELL, L.L. Spanish-English technical dictionary of aircraft - automobile - railways - highways - electricity - electronics - radio - television - machine tools - agricultural implements - paints and dyes - petroleum - steel products. New York, etc., McGraw-Hill, 1949. xi, 1706p.

700,000 Spanish and English technical terms, entries being set solid under headwords (*e.g.* 'conducts' (passageway): 5½ columns (somewhat relieved by use of bold type for the Spanish)). About 120 categories are applied. Many idioms.
Class No: 5(038)=60

[163]
—SELL, L.L. English-Spanish comprehensive technical dictionary. New York & London, McGraw-Hill, 1944. xii, 1478p.

Contains 525,000 Spanish and English technical terms. 20 subject-categories are applied. Distinguishes between British and American English. Forbidding layout of type, set solid.
Class No: 5(038)=60

[164]
—SELL, L.L. English-Spanish comprehensive technical dictionary... Section 2. New York, McGraw-Hill, 1959. [3], iv, 1079p.

Adds a further 400,000 entries, including a supplement, p.933-1079. 'Extensive, but rather difficult to use' (Finch, C.A. *An Approach to technical translation* (1970), p.58).
Class No: 5(038)=60

Portuguese

[165]
DE PINA ARAÚJO, A. De Pina's technical dictionary / Dicionário técnico. São Paulo, McGraw-Hill do Brasil, 1978. 2v. (616p. + 495p.).

1. *English-Portuguese.* 616p. 2. *Portuguese-English.* 493p. Reprint of the 1964 ed.

About 75,000 translated terms in all, mostly equivalents. American English used (*e.g.* 'aluminium'). Tendency to enter under qualifying word instead of noun (*e.g.* 'high tide' - under 'high' and not 'tide'). No fewer than *c.*500 categories applied to terms. *Class No:* 5(038)=690

[166]
FÜRSTENAU, E. Dicionario de termos técnicos (inglés-portugués). 11. ed. Porto Alegre, Brazil, Ed. Globo; New York, French and European, 1980. 1157p. illus. $195. ISBN: 0828806527.

First published 1947. *Class No:* 5(038)=690

Greek

[167]
PAPANIKOLAOU, E.N. Tehnikon ellēno-agglikon kai agglo-ellēnikon lexikon meta proforas. Piraeus, Multilēnaiou, 1962. 684p. tables.

'Greek-English and English-Greek technical dictionary'. About 20,000 entries and sub-entries in each half. Coverage: mechanics, electricity, radio, mathematics, physics and chemistry. *Class No:* 5(038)=77

Russian

[168]
Anglo-russkii politekhnicheskii slovar': 87,000 terminov. Chernukhin, A.E., *ed.* Izd. 4. Moscow, Ruskii Iazyk, 1979. 687p. ISBN: 0080219365.

3rd ed. 1976 (647p.).

'English-Russian polytechnical dictionary' has entries for *c.*87,000 terms. Coverage excludes medicine, mathematics, pure and natural sciences. Appendices: abbreviations; conversion tables. *Class No:* 5(038)=82

[169]
CALLAHAM, L.I. Russian-English chemical and polytechnical dictionary. 3rd ed. New York, London, etc., Wiley, 1975. xxvii, 852p. £81. ISBN: 0471129984.

First published 1947; 2nd ed., 1962.

'A technical as well as a chemical dictionary...' (*Preface,* 1st ed.). About 100,000 Russian entries and sub-entries. Terms are categorized, genders given; abbreviations, prefixes and common Russian word-endings included. Good typography for cyrillic characters. Particularly helpful on pure and applied chemistry. *Class No:* 5(038)=82

[170]
—ALFORD, M.H.T. *and* ALFORD, V.L. Russian-English scientific and technical dictionary. Oxford, Pergamon, 1970. 2v. (xxxvi, [1], 764p.; vii, 765-1423p.).

Over 100,000 entries, drawn from 94 different disciplines. Stress shown. Endings of certain cases provided, to illustrate their paradigms. All irregular forms are shown with, where necessary, their own alphabetical entries. Separate entry-lines for compounds, etc. 'Supplementary dictionary of Russian word endings', v.2, p.1809-23. 'Reference sources', v.1, p.xxi-xxiii. Specially designed for people with weak or moderate knowledge of Russian grammar. *Class No:* 5(038)=82

[171]
KUZNETSOV, B.V. **Russian-English dictionary of scientific and technical usage.** [Moscow], Russky Yazyk (distributed by Pergamon), 1986. 655p. $175. ISBN: 0080325513.

About 30,000 entry-words. Not confined to providing English equivalents to Russian terms. Emphasises usage in everyday language as well as giving examples of specialized application. *Class No:* 5(038)=82

[172]
Russko-angliiskii politekhnicheskii slovar'. Chernukhin, A.E., *ed.* Moscow, Ministerstvo Oborony, 1971. 1027p.

Companion volume to the English-Russian polytechnical dictionary and also carrying *c.*80,000 entries. Separate lines for sub-entries. Wide coverage of military terms, reflecting the nature of the publishing body, - Ministry of Defence. *Class No:* 5(038)=82

[173]
UNITED STATES. Library of Congress. Reference Department. Aerospace Information Division. **Soviet Russian scientific and technical terms:** a selective list. Washington, Library of Congress, 1963. [v], 668p.

26,000 entries. 'Generally speaking, those terms are included which cannot be found in standard dictionaries, or which have a special meaning when used in a particular field' (*Introduction*). Incorporates frequently used and important abbreviations, including names of Russian organizations. 31 categories are applied to terms. *Class No:* 5(038)=82

[174]
ZIMMERMAN, M. *and* VEDENEEVA, C. **Russko-angliiskii nauchno-tekhnicheskii slovar perevodchika.** [Russian-English translator's dictionary: a guide to scientific and technical usage.] 3rd ed. Moscow, Nauka Publishers; Chichester, Wiley, 1992. 735p. £65. ISBN: 0471933163.

First published 1967 (294p.).

Not a dictionary of terms, but a collection of *c.*9,000 typical examples from scientific and technical sources, *i.e.,* to give equivalents in full context. Of the 2nd ed., 'This new, updated edition will come as a boon to the younger generation of translators' (*Nature,* v.312, 13 December 1984, p.670). *Class No:* 5(038)=82

Polish

[175]
English-Polish dictionary of science and technology. Czerni, S. *and* Skrzynska, M., *eds.* 9th ed. Warsaw, Wydawnictwa Naukowo-Techniczne, 1990. 1032p. ISBN: 8320409217.

.... (contd.)
First published 1962.

Has *c.*77,000 categorized entry-words; different applications are numbered. English terms that have no Polish equivalents are explained. Appended list of abbreviations of technical terms. organizations, etc. *Class No:* 5(038)=84

[176]
Polish-English dictionary of science and technology. Czerni, S. *and* Skrzynska, M., *eds.* 5th ed. Warsaw, Wydawnictwa Naukowo-Techniczne, 1983. 846p.

About 63,000 entry-words. List of abbreviations appended.

These two dictionaries are of high standing; accurate, up-to-date and with well-balanced coverage of a wide range of subject fields (*Library journal,* v.89(16), 15 September 1964, p.3296, on the 1962 ed.). *Class No:* 5(038)=84

Czech

[177]
Česko-anglický technický slovník. Feigl, J. *and* Klinger, E., *eds.* Prague, Státni Nakladatelství Technické Literatury, 1963. 930p.

A Czech-English technical dictionary. About 60,000 entry-words, covering engineering, chemistry, mineralogy and other related fields. Some 50 specialist contributors. Entry for 'acid' (Kyselina), p.282-90 is followed by names of specific acids. Needs updating.

The approximate counterpart is: *Anglicko-český techniký slovník* (Prague, SNTL, 1969, 1026p.). *Class No:* 5(038)=850

[178]
NOVÁK, J. *and* BINDER, R. **Vreckový anglicko-slovenský a slovensko-anglický technický slovnik.** 2. prep. vyd. Bratislava, Slovenskó Vydavetel'stvo Technickej Literatury, [1964]. 610p.

'Concise English-Slovak and Slovak-English dictionary'. *Class No:* 5(038)=850

[179]
Velký anglicko-cesky technický slovník. Pěkarek, O., *and others.* Prague, Státni Nakladatelství Technické Literatury, 1956-70. 3v. (*c.*3000p.).

'Large English-Czech technical dictionary'. Over 100,000 English entry-words. *Class No:* 5(038)=850

Serbo-Croat

[180]
Rečnik technićkih 13 izrago srpskohrvatso-engleski, francuski-nemaćki; englesko-francusko-nemaćko-srpskohrvatso. Ristić, S., *and others.* 3.izd. Belgrade, Tehnicka Knjica, 1971. 2v. ([iv], 612p. + [ii], 633p.). illus.

'Dictionary of technical terms, Serbo-Croatian/English/French/German, English/French/German/Serbo-Croatian'. V.1 has *c.*25,000 Serbo-Croatian entry-words, with equivalents in the other languages. V.2 *c.*30,000 English entry-words, with equivalents. About 50 categories are applied to terms. V.1 only has small illus. *Class No:* 5(038)=86

Arabic

[181]
A New dictionary of scientific and technical terms, English-Arabic, with illustrations. Al-Khatib, A.Sh., *comp. and ed.* 6th ed. Beirut, Librairie du Liban, 1985. xv, 751p. illus., tables.
First published 1971 (xv, 751p).
About 50,000 entries; 3 columns, with about one illus. per column. 17 appendices (conversion tables, etc.). Bibliography, p.750. A reprint of the 1971 edition in quarto format. *Class No:* 5(038)=927

Turkish

[182]
ÖZBALKAN, N. English-Turkish dictionary. Istanbul, Ozbalkan Yayinlari, 1972. 952p. *Class No:* 5(038)=943.5

[183]
ÖZBALKAN, N. Turkçe-Ingilizce teknik terimler sözlügü. [Turkish-English dictionary of technical terminology.] 2. bask. Istanbul, Ozbalkan Yayinlari, 1984. 1152p. $150. *Class No:* 5(038)=943.5

Hungarian

[184]
Angol-magyar müszaki szótár. Nagy, E. *and* Klar, J., *eds.* 5. vaitozotlan kiad. Budapest, Akadémiai Kiadó; New York, French and European, 1980. viii, 791p. $65. ISBN: 9630523310; 0828806748, US ed.
A reprint of the 2nd ed. (1959).
'English-Hungarian technical dictionary'. About 100,000 entry-words, including meanings in common use. Many compounds. Appendix: English technical abbreviations, p.743-79; weights and measures, p.780-91.
Class No: 5(038)=945

[185]
Magyar-angol müszaki szótár. Nagy, E., *and others, eds.* 5th ed. Budapest, Akadémiai Kiadó, 1983. viii, 752p. ISBN: 9630532735.
Reprint of the 1957 ed.
Companion to the English-Hungarian dictionary. Concerning these two Nagy dictionaries, C.A. Finch (*An approach to technical translation* (1970), p.58) states: 'Of doubtful technical accuracy in some subjects. L. Orzagh's *Angol-magyar szótár* (9th ed. 1979) and *Magyar-angol szótár* (5th ed. 1977. 2v.), both published by Akadémiai Kiadó, can also be consulted with profit for technical terms. *Class No:* 5(038)=945

Finnish

[186]
Suomalais-englantilainin teknikan ja kaupan sanakorjo. Talvitie, J.K., *and others.* 2. uusittu laitos. Espoo, Tietoteos, 1987. 699p. ISBN: 9519035818.
About 50,000 Finnish technical and commercial terms, with equivalents in English. *Class No:* 5(038)=945.41

Chinese

[187]
Han ying kung cheng chi shu tzu hui. [Chinese-English dictionary of engineering technology.] Pei-ching, Hai yang chu pan she, 1986. 892p.
Includes *c.*107,000 terms on technological subjects, including astronautics and telecommunications. Radical index, p.7-13 and index of syllables of Hanyupinyin, p.14-21. *Class No:* 5(038)=951

[188]
Modern Chinese-English technical and general dictionary. New York, etc., McGraw-Hill, 1963. 3v.
1. *Tables* (Wade-Giles - Pinyin conversion table, etc.). Radical stroke index, in alphabetical sequence, - the real key to the dictionary. 2. *Standard telegraphic code, in numerical sequence.* 1900p. 3. *Pinyin romanization in alphabetical sequence* (permuted index). 188p.
About 212,000 entries, over 80% of them common scientific and technical terms. 'The main emphasis is on information not readily available in existing dictionaries' (*Introduction*). Proper names and vocabulary of Chinese Communism are included. Each Chinese character is represented by a 4-digit code (0001-), adapted from the Standard Telegraphic Code. V.2-3 both give English equivalents. Recent linguistic reforms are included. *Class No:* 5(038)=951

[189]
—A Chinese-English dictionary of scientific and technical terms. Harbin, Heilongjiang Renmin Chuban She, 1985. 2v. (1976p.)
Has *c.*100,000 entries for terms in Chinese characters, with English equivalents. *Class No:* 5(038)=951

Japanese

[190]
Inter-Press dictionary of science and engineering. English-Japanese; Japanese-English. [Tokyo], Inter-Press, 1984. 2v. (1927p. + 1888p.). ISBN: 4870870010.
About 120,000 entries in each volume.
Class No: 5(038)=956

Theses

[191]
Comprehensive dissertation index, 1861-1972. Ann Arbor, Mich., Xerox Univ. Microfilms. 1973. 37v. Annual supplements, 1973-. (*1973-82 cumulation. 38v.*)
V.1-16 concern science and technology:
1-4. *Chemistry.*
5. *Mathematics and statistics.*
6-7. *Astronomy and physics.*
8-10. *Engineering.*
11-13. *Biological sciences.*
14. *Health and environmental sciences.*
15. *Agriculture.*
16. *Geography and geology.*
V.11. *Biological sciences: Biology and zoology* (1973. xix, 1210p.) has prelims.: 'Sources consulted', 'How to obtain dissertation copies'. Biology, p.1-218; Zoology, p.219-1210, - under subjects A-Z. About 80,000 entries. Subject index. *Class No:* 5(043)

[192]

Dissertation abstracts international. B: The sciences and engineering. Abstracts of dissertations available on microfilm or as xerographic reproductions. Ann Arbor, Mich., University Microfilms, 1969-. 12pa. $395. ISSN: 01566777.

Originally as *Microfilm abstracts* (v.1-11, 1938-51), then *Dissertation abstracts* (1952-69). As from 1966, the latter appeared in two sections: A. *Humanities and social sciences;* B. *The sciences and engineering*, - both 12pa. In 1977, C. *European abstracts* followed.

The title, *Dissertation abstracts international,* reflects inclusion of disssertations from foreign universities. The abstracts are in subject groups, A-Z (*e.g.* Aerospace studies ... Zoology). Keyword, title and author indexes pa. Available online and on CD-ROM. On microfilm.
Class No: 5(043)

[193]

Masters theses in the pure and allied sciences accepted by colleges and universities of the United States and Canada. New York & London, Plenum Press, 1974-. v.18-. Annual. ISSN: 02442861.

V. 1-17 (1957-73) published by Thermophysical Properties Research Center, Purdue Univ., Lafayette, Ind., covering US colleges and univs. only.

Over 10,000 titles of masters theses pa., covering mainly chemistry, physics and engineering subjects. Arranged under major study disciplines, A-Z, subdivided by universities.
Class No: 5(043)

[194]

Retrospective index to theses of Great Britain and Ireland, 1716-1950. Bilboul, R.R. *and* Kent, F.L., *eds.* Santa Barbara, Calif., ABC-Clio, 1975-77. 5v.

1. *Social sciences and humanities.* 1975. ix, 393p.
2. *Applied sciences and technology.* 1976. xi, 159p.
3. *Life sciences.* 1977. xi, 327p.
4. *Physical sciences.* 1976. xi, 99p.
5. *Chemical sciences.* 1976. xx, 251p.

A checklist, up to 1950 (continued in *Aslib index to theses accepted for higher degrees by the universities of Great Britain and Ireland ...* v.1, 1950-51- (1953-)). 2pa. Subject and author sections for each volume. Available on microfiche and online. *Class No:* 5(043)

Reports Literature

[195]

AUGER, C.P. Information sources in grey literature. 2nd ed. London, Bowker-Saur, 1989. ix,175p. figs. £30. (*Guides to information sources.*) ISBN: 0862918715.

First published 1975 as *Use of reports literature* (London, Butterworths).

12 chapters: 1. The nature and development of grey literature ... 4. Report writing - 5. Theses, translations and meetings papers ... 7. Aerospace - 8. Life sciences ... 11. Energy - 12. Science and technology. 3 appendices: A. Keys to report series codes - B. Trade literature - C. Addresses of organizations mentioned in the text. Short bibliography appended to each chapter. *Class No:* 5(047)

[196]

Dictionary of report series codes. Godfrey, L.E. *and* Redman, H.F., *eds.* 2nd ed. New York, Special Libraries Association, 1973. vi, [1], 645p.

First published 1962.

Relates technical report series codes to US Department of Defense agencies, Atomic Energy Commission, their contractors and other US government agencies. Also lists agencies and the codes they use. Includes a bibliography on the report code literature. *Class No:* 5(047)

[197]

Government reports announcements & index. Springfield, Va., National Technical Information Service (NTIS), 1975-. Semi-monthly. $1,000 (North America), $3,000 (overseas). ISSN: 00979007.

Succeeds *Government research announcements* (1971-75). Orignally as *Bibliography of scientific and industrial reports* (1946-49), then as *Bibliography of technical reports* (1949-54), then as *US government research reports* (1954-64) and then as *US government research and development reports* (1965-71). Title and frequency vary.

About 2,500 abstracts per issue, including patents. 22 subject fields, further divided. Indexes: subject, personal author, corporate author, contract number, accession/report number. Covers government sponsored R & D reports, government analyses, etc. NTIS database is available online and CD-ROM. *Class No:* 5(047)

[198]

Information work with unpublished reports. Holloway, A., *and others.* London, Deutsch, in association with the Institute of Information Scientists, 1976. 302p. illus.

Part 1: Work in large national information centres. Part 2: Work in company-based information units. 15 chapters, each with select bibliography appended. 'Other relevant literature', p.287. Index 'Will surely become a basic text of work in this field for many years to come'.
Class No: 5(047)

Reviews & Abstracts

[199]

Aslib book guide: an international guide to scientific and technical books in English. London, Aslib, 1992-. v.57-. 12pa. £80 (£70 to members) pa. ISSN: 00012521.

Quarterly until March 1948. Formerly *Aslib book list* (1935-91).

V.58(2), February 1993 contains 226 short reviews, by subject specialists, in DC order. Books are graded: 'A', elementary level; general readership; 'B', intermediate level; university textbook; 'C', advanced level; specific readership; 'D', reference books. Broad subject index and author or title index per issue and cumulated annually. A reliable current book-selection tool, normally restricted to books in English. *Class No:* 5(048)

[200]

Bulletin signalétique. Paris, Centre National de la Recherche Scientifique (CNRS), 1956-83. Monthly or quarterly.

Formerly *Bulletin analytique* (v. 1-16. 1940-55).

The following 49 series concerned science and technology: 101. *Information scientifique et technique.* 110. *Information automatique. Recherche opérationnelle. Gestion.* 120. *Astronomie* ... 130, 140. *Physique* I,II. 145. *Electronique.* 150. *Physique. Chimie et technologie nucléaires.* 160. *Physique des solides* ... 161.

....(contd.)

Cristallographie. 165. *Atomes et molécules.* 170. *Chimie.* 220-6. *Bibliographie des sciences de la terre: Minéralogie; Gisements; Roches cristallines; Roches sédimentaires; Stratigraphie; Tectonique; Hydrologie.* 310. *Génie biomédical.* 320. *Biochimie, Biophysique.* 330. *Sciences pharmacologiques.* 340. *Microbiologie. Virologie. Immunologie.* 346. *Ophthalmologie.* 347. *Otorhinolaryngologie.* 348. *Dermatologie.* 349. *Anesthesie. Réanimation. Choc.* 350. *Pathologie.* 351. *Cancer.* 352. *Maladie de l'appareil respiratoire ...* 354. *Maladie de l'appareil digestif ...* 355. *Maladie des reins.* 356. *Maladie du système nerveux.* 357. *Maladie des os.* 359. *Biologie et physiologie animales.* 361. *Endocrinologie.* 362. *Diabéte ...* 363. *Génétique.* 365. *Physiologie des vertebrés.* 370. *Biologie et physiologie végétales.* 380. *Agronomie. Zootechnie.* 390. *Psychologie. Psychopathologie.* 730. *Combustibles. Energie thermique.* 740. *Métaux. Métallurgie.* 745. *Soudage ...* 761. *Microcopie életronique.* 780. *Polymer.* 880. *Génie chimique. Industries chimique et parachimique.* 885. *Nuisances.* 890. *Industries mécaniques. Bâtiment. Travaux publiques. Transports.*

Now published as PASCAL (*qv*). *Class No:* 5(048)

[201]
CHEN, C.C. Biomedical, scientific & technical book reviewing. Metuchen, N.J., Scarecrow Press, 1976. xv, 186p.

Identifies and assesses 500 major biomedical, scientific and technical reviewing journals as quantitative sources of book reviews. One finding was that 9 general science reviewing journals provided reviews of more than 50% of the total output of new science books in the period studied, - 1970 or 1971. *Class No:* 5(048)

[202]
COPLEY, E.J. A Guide to 'Referativnyĭ zhurnal'. 3rd ed. London, British Library, Science Reference Library, 1975. [ii], 26p. facsim. (*Occasional publication.*)

First published 1970.

Lists 65 titles in the *Referativnyĭ zhurnal* series and includes an A-Z subject index. An example is given of a *Referativnyĭ Zhurnal* abstract. Sub-series of *RZh* are not listed. On these, see the more detailed *Rubrikatov referativnykh izdaniĭ SSSR* (Moscow, VINITI), -updated at intervals.

For details of translations from *RZh*, consult announcements in the British Library's *British reports, translations and theses* and US *Government reports, announcements and index* (with NTIS as its database). *Class No:* 5(048)

[203]
Current contents. Philadephia, Pa., Institute for Scientific Information, 1958-. (*6 series.*)

Current contents: Social and behavioral sciences. Weekly. $420.

Current contents: Physical, chemical and earth sciences. Weekly. $420.

Current contents: Life sciences. Weekly. $420.

Current contents: Clinical medicine. Weekly. $420.

Current contents: Engineering, technology and applied sciences. Weekly. $420.

Current contents: Agriculture, biology and environmental sciences. Weekly. $420.

Each of the series provides facsimiles of contents pages of current issues of journals in its field. Accompanying these is

....(contd.)

a list of journal issues covered, title-word index, author index and address directory, publishers address directory, plus the customary request-a-print form. Although offering prompt current awareness, these services do depend for their adequacy on sufficient detail being provided on the various contents pages. Online, CURRENT CONTENTS SEARCH database, containing 950,000 records, with 18,000 records added weekly, and on diskette. *Class No:* 5(048)

[204]
Current contents: Physical, chemical and earth sciences. Philadephia, Pa., Institute for Scientific Information, 1979-. v.19-. Weekly. $420. ISSN: 01632574.

Previously as *Current contents: Physical and chemical sciences* (formed by the merger of *Current contents: physical sciences* and *Current contents: chemical sciences*), 1961-78.

Reproduces contents-pages of more than 1,000 journals in the fields concerned. Each weekly number covers *c.*100 journal issues. Online on CURRENT CONTENTS SEARCH, and available on diskette. *Class No:* 5(048)

[205]
General science index. New York, H.W. Wilson, 1978- 10pa., with annual cumulative index. ISSN: 01621963.

Covers over 100 general English-language science journals, mostly indexed in the more comprehensive *Applied science and technology index* and *Biological and agricultural index.* Arranged under subjects, A-Z, with appended 'Book reviews'. Mainly US journals. Online via Wilsonline, BRS, CD-ROM, May 1984-; updated quarterly. *Class No:* 5(048)

[206]
A Guide to the culture of science, technology and medicine. Durbin, R.T., *ed.* New York, Free Press. 1980 (reprinted 1984, with a bibliographic update). 723p. $75; $24.95.

Nine documented state-of-the-art contributions by specialists. More than 3,200 citations. For graduate students and scholars. 'One of the most important references of its kind for integrating the cross-disciplinary approaches necessary in their endeavours' (*Choice,* v.18, November 1980, p.416). *Class No:* 5(048)

[207]
Index to scientific book contents. Philadelphia, Pa., Institute for Scientific Information, 1985-. 4pa. Annual cumulation. $1,150.

Compiled in 2v.: 1. Contents of books; 2. Author/editor index; Corporate index; Permuterm subject index. The author/editor index is a key to specific book editors and chapter authors. 'Corporate index' is the term used for geographic and organization sections. Online via DIMDI. *Class No:* 5(048)

[208]
LC science tracer bullets. Washington, Library of Congress, Science and Technology Division, Reference Section, 1972-. Irregular.

A series of brief guides to information sources (bibliographies; organizations; persons) on specific topics of current interest. *Geothermal energy* (April 1973. 5 leaves) has 13 unannotated sections: Scope, Introductory text - Subject headings - Basic texts - Additional text - Bibliographies - Government publications - State-of-the-art reviews and conference proceedings - Journal articles - Primary journals - Reports - Selected technical reports - Selected materials - Additional sources of information (names and addresses of two specialists). *LC science tracer*

....(contd.)

bullets issued in 1985 included *Sharks*, *Optical disc technology* and *Acupuncture*.

Compilation volume, containing 173 of these guides published between 1972 and 1989, entitled *Science tracer bullets: a reference guide to scientific, technological, health, and environmental information sources* (H. Henderson, *ed.* Omnigraphics, 1990. 4v. 1687p. $160). *Class No:* 5(048)

[209]

New technical books: a selective list with descriptive annotations. New York Public Library, Science & Technology Research Center, 1915-. 6pa. ISSN: 00286869.

Each issue contains *c.*500 annotated entries for noteworthy books, mostly in English. Entries, in broad Dewey Decimal Classes feature two paragraphs: 'Content' and 'Note', the latter indicating readership, level and standing. Parts of continuing multi-volume works (*e.g.* Kirk-Othmer) and annual review series (*e.g. Advances in ...*) receive attention. The annual author and catchword indexes occupy middle pages of the December issue each year. A valuable selection tool, although somewhat late in appearance.

Class No: 5(048)

[210]

PASCAL [formerly *Bulletin signalétique*]. Vandoeuvre-Les-Nancy, Centre National de la Recherche Scientifique, 1984-.

65 sections, replacing the 49 of the *Bulletin signalétique*, with its monitoring of over 5,000 periodicals worldwide. The PASCAL series has been on line since 1973 (accessible on ESA-IRS and TÉLÉSYSTÉMES-QUESTEL), - a bilingual database having 'more than five million entries with an annual update of over half a million' (Anthony, L.J. *Information sources in engineering.* 2nd ed., 1985, p.148). The series appears to be undergoing revision at present.

PASCAL has three main groups: 'Explore', 'Folio', 'Théma', with sections as follows:

PASCAL Explore, pt. 11. *Physique atomique et moléculaire, Plasmas.* 12. *État condensé.* 13. *Structure des liquides et des solides, cristallographie.* 20. *Électronique et télécommunications.* 27. *Méthodes de formation et traitement des images.* 30. *Microscopie électronique et diffraction électronique.* 32. *Metrologie et appareillage en physique et physiochimie.* 33. *Informatique.* 34. *Robotique, automatique automatisation des processus industriels.* 36. *Pollution de l'eau, de l'air et du sol.* 48. *Environnement cosmique terrestre, astronomie et géologie extraterrestre.* 49. *Météorologie.* 60. *Génétique.* 61. *Microbiologie.* 62. *Immunologie.* 63. *Toxicologie.* 64. *Endocrinologie humaine.* 65. *Psychologie, psychopathologie, psychiatrie.* 71. *Ophthamologie.* 72. *Otorhinolaryngologie, stomatologie, pathologie cerviofaciale.* 73. *Dermatologie, maladies sexuellement transmissibles.* 74. *Pneumologie.* 75. *Cardiologie et appareil circulatoire.* 76. *Gastroentérologie, foie, pancreas, abdomen.* 77. *Néphrologie. Voies urinaires.* 78. *Neurologie.* 79. *Pathologie et physiologie ostéoarticulaires.* 80. *Hématologie.* 81. *Maladies métaboliques.* 82. *Gynécologie, obstetrique, andrologie.* 83. *Anesthésie et réanimation.* 84. *Genie biomédical. Informatique biomédicale.* 99. *Congrès. Rapports. Thèses.*

Folio, pt. 10. *Mécanique et acoustique.* 16. *Chimie analytique, minérale et organique.* 17. *Chimie générale, minérale et organique.* 21. *Electrotechnique.* 23. *Génie chimique, industries chimique et parachimique* 24. *Polymères, peintures, bois* 25. *Transports terrestres et maritimes.* 40. *Mineralogie.* 41. *Gisements métalliques et non-métalliques, économie miniere.* 42. *Roches cristallines.* 43. *Roches*

....(contd.)

sedimentaires. Geologie marine. 44. *Stratigraphie. Géologie régionale. Géologie générale.* 45. *Tectonique. Geophysique interre.* 46. *Hydrologie. Géologie de l'ingénieur. Formations superficialles.* 47. *Paléontologie.* 52. *Biochimie.* 53. *Anatomie et physiologie des vertébrés* 54. *Reproduction des vertébrés, embryologie.* 55. *Biologie végétale.* 56. *Écologie animale et végétale.* 70. *Pharmacologie. Traitements médicamenteux.*

Théma, pt. 195. *Bâtiment, travaux publics.* 205. *Sciences d'information, documentation.* 210 *Industries agrimentaires.* 215. *Biotechnologie.* (French ed.). 216. *Biotechnology.* (English ed.). 230. *Énergie.* 235. *Médicine tropicale.* 240. *Métaux, métallurgie.* 245. *Soudage, brasage et techniques connexe.* 251. *Cancérologie.* 260. *Zoologie fondamentale et appliqué des invertébrés.*

PASCAL available online via DIALOG. *Class No:* 5(048)

[211]

Referativnyĭ zhurnal. Moscow, Akademiya Nauk SSSR, (Institut Nauchnoĭ Informatsii, VINITI), 1953-. Frequency varies. Annual, etc. indexes.

The most comprehensive of abstracting services, producing over 1,000,000 abstracts and references pa., in *c.*70 main series. Coverage: periodical articles, monographic series, patents, standards and specifications. Originals may be in any of some 70 languages. (In fact, one disadvantage for the Western reader may be inability to see originals). Time lag in appearance of abstracts may average 6-7 months. Diagrams and maps are sometimes used to elucidate abstracts. Duplication between the various series is sensibly permitted.

Topics covered in the present series include (in UDC order) Information ... Mathematics ... Energy ... Automation ... Air transport ... Food industry machinery ... Photography and cinematography. Those selected for inclusion in this volume of the *Guide* can be traced by consulting *Referativnyĭ zhurnal* in the Author/Title index. *Class No:* 5(048)

[212]

Science abstracts. London, INSPEC (Institution of Electrical Engineers), 1898-. V.1-.

V.1-5 (1898-1902), reprinted London, Butterworth, 1964, entitled *Science abstracts: Physics and electrical engineering.* From 1903 as Section A: *Physics abstracts* and B: *Electrical engineering abstracts.* Now in three sections: *Physics abstracts.* 6pa. £1,370pa. *Electrical and electronic abstracts.* 12pa. £1,090pa. *Computer and control abstracts,* 1966-, (*q.v.*) 12pa. £720pa.

Available online as INSPEC. *Class No:* 5(048)

[213]

Science and technology annual reference review. Malinowsky, H.R., *ed.* Phoenix, Ariz., Oryx Press. Annual.

1991 issue (368p.) contains over 800 extensive reviews of new publications, written by librarians. Each review is 200-500 words in length. Arranged under general subject headings, *e.g.* medicine, technology, and agriculture, then further subdivided within the subject. 4 indexes: title; author; subject; type of library for which the item is recommended. Useful for those intending to develop science and technology collections. US-slanted. *Class No:* 5(048)

[214]

Science citation index. Philadelphia, Pa., Institute for Scientific Information, 1961-. 6pa., with annual and 5-yearly cumulations for the period 1969-79. $10,175. ISSN: 0036827x.

Citation index. Source index. Permuterm subject index.

The *Citation index* records, A-Z, authors of papers, monographs, etc., followed by a list of later writers who have cited/quoted those originals. The *Source index* has entries for those citing/quoting writers, while the *Permuterm subject index* based on relevant words in titles of papers etc., provides the subject approach. *SCI* covers *c*.3,800 journals, monographs etc. SCISEARCH database.

Available online via DIALOG, Data-Star; CD-ROM (Institute for Scientific Information). *Class No:* 5(048)

Bibliographies

[215]

Abstracting services. Vol.1: Sciences, technology, medicine, agriculture. 2nd ed. The Hague, International Federation for Documentation, 1969. 284p. (*FID pubn. 455.*)

First published 1965. V. 2: *Social sciences and humanities* (1965; 2nd ed., 1969).

About 1,300 abstracting services (journals and cards), plus *c*.100 primary journals with important abstracts sections. Section 1: A/Z list of abstracting services - Section 2: UDC arrangement by subjects. A/Z list of subjects in English, French, Russian and Spanish. Titles, under 39 countries, A/Z. Data on each title: frequency, price, number of abstracts pa., arrangement, indexed, journal and subject coverage. *Class No:* 5(048)(01)

[216]

Index to scientific reviews: an international interdisciplinary index to the review literature of science, medicine, agriculture, technology and the behavioral sciences. Philadephia, Pa., Institute for Scientific Information, 1974-. 2pa. $830. ISSN: 03606661.

The two major parts are: *Source index*, listing review articles (*not* book reviews as such) under authors, and *Permuterm subject index* (running to 4872 columns of tiny print in the 1986 first semiannual). More specialized review material is covered by Research front speciality index. Sources, *c*.2,200 journals and over 200 series, are listed in 'Guide and list of source publications' (on blue pages). Based on review articles included in *Science citation index*. *Class No:* 5(048)(01)

[217]

OWEN, D.B. Abstracts and indexes in science and technology: a descriptive guide. 2nd ed. Metuchen, N.J., Scarecrow Press, 1985. ix, 235p. £17.50; $22.50. ISBN: 0810817128.

First published 1974.

223 abstracting and indexing services. 11 sections, in Dewey order: General - Mathematics, statistics and computer science - Astronomy - Chemistry and physics - Nuclear science and space science - Earth sciences, archaeology, and anthropology - Engineering and technology - Energy and environment - Biological sciences - Agricultural engineering - Health science. Data on each service: arrangement; coverage; scope (*e.g.* international); locating material (entry numbering); abstracts (language,

....(contd.)

etc.); indexes; other material (*e.g.* review articles); periodicals scanned (lists); databases. Title index, p.215-35. *Class No:* 5(048)(01)

[218]

Scientific abstracting and indexing periodicals in the British Library: a guide to SRIS holdings and their use. 4th ed. London, British Library, Science Reference and Information Service, 1991. [iii],64p. £20. ISBN: 071230780x.

3rd ed. 1985 (iii,149p.).

404 entries in all, with emphasis on abstracting services. Introductory matter highlights major abstracting periodicals (nos.1-27). The main sequence (nos.28-389), conforming to the SRS classification scheme (not outline), is followed by a section on research projects and reprints, citing nos.390-404. The necessary title and subject indexes follow. The drop in the number of entries from 1,880 in the 3rd ed. to 404 in this ed. is a cause for concern. Even allowing for cessations, cuts in titles must have been drastic. Roads and road transport are poorly represented, and the user will miss such titles as *Meteorological and geoastrophysical abstracts, Military science index,* and *Photographic abstracts.* Nevertheless, this 4th ed. records much of the basic scientific abstracting material. *Class No:* 5(048)(01)

[219]

STEPHENS, J. Inventory of abstracting and indexing services produced in the UK. 3rd ed. London, British Library, 1986. viii, 238p. £29. (*British Library Information guide, 2.*) ISBN: 0712330801.

Previous eds., 1978, 1983.

430 numbered entries, with titles of services, A-Z. Systematic data on each include subject, scope, sources, language, cost, start, frequency, number of entries, online-access, availability, indexes. Science and technology services are prominent (agriculture: 51 entries; biology: 39; medicine: 36). Broad subject and specific subject indexes; index of authorities (issuing bodies); index of online databases according to processor. Specimen report form questionnaire. Omits ISBN/ISSNs. A valuable research tool. *Class No:* 5(048)(01)

[220]

Technical book review index. Pittsburgh, Pa., JAAD Publishing Co., 1935-. 12pa. $45pa. ISSN: 00400890.

About 300 entries per issue in 5 sections: Pure sciences - Life sciences - Medicine - Agriculture - Technology. A feature is the inclusion of an excerpt from a review in a leading scientific or trade journal, with the reviewer's name and length (in columns or pages) of the original review. Annual author and anonymous title index in the December issue each year. Very largely English - language books are dealt with. A valuable selection tool, particularly as the time-lag in publication of *TBRI* is comparatively speedy. *Class No:* 5(048)(01)

Periodicals

[221]

Nature. Basingstoke, Hants., Macmillan Journals, 1869-. V.1-. Weekly. £118. ISSN: 00280836.

The leading science journal, carrying articles, news and views, short reports of research, correspondence and reviews. V.361, 11 February 1993 issue contains 7 signed

....(contd.)

book reviews. Spring and autumn book supplements provide some 30 signed reviews apiece, prefaced by introductory articles. *Class No:* 5(051)

[222]
Science & technology libraries. Binghamton, N.Y., Haworth Press, 1980-. 4pa. $36 (individuals); $105 (libraries). ISSN: 0194262x.

Each issue has contributions on a particular theme (*e.g.* 'Role of computers in sci-tech libraries', v.6(4); 'Role of translations in sci-tech libraries', v.3(2)). Also featured are special papers, *e.g.* 'Information sources in laser science and technology'. A regular section: 'New reference works in science and technology' (*e.g.* v.7(1), Fall 1986, devotes 40p. to *c*.80 initialled reviews in 5 subject groups). Reviews are descriptive but do indicate readership. *Class No:* 5(051)

[223]
The Scientific journal. Meadows, A.J., *ed.* London, Aslib, 1979. [vii], 300p. graphs, charts, tables. £18; £11. (*Aslib Reader series, v.2.*) ISBN: 0851421180.

50 contributors. 7 parts (24 documented contributions): 1. The development of the scientific journal [1665-1798] - 2. Economics of journal publication (*e.g.* 'Some effects of delay in publication of information in medical journals and implications for the future', with 26 references) ... 4. The refereeing of scientific journals - 5. Characteristics and interrelationships of journals - 6. An alternative viewpoint (*e.g.* 'The deposition of scientific works (the experience of VINITI)') - 7. The future of the scientific journal (*e.g.* 'An on-line scientific journal'). 'Further reading', p.297-8. 'Author index' (*i.e.,* list of contributors). No subject index. An impressive and authoritative survey. *Class No:* 5(051)

Abbreviations & Symbols

[224]
Bibliographic guide for editors & authors. Prepared through the combined efforts of BIOSIS, Chemical Abstracts Service, and Engineering Index, Inc. Washington, American Chemical Society, 1974. 23, 362p.

3 Sections: 1. Guidelines for the use of the coded bibliographic strip - 2. Bibliographic standards - 3 (362p.). Serial titles, abbreviations and codes (complete titles, title abbreviations and ASTM CODEN abbreviations for *c*.27,700 scientific and technical serials). *Class No:* 5(051)(003)

[225]
International CODEN directory. Columbus, Ohio, Chemical Abstracts Service, 1980-. Five yearly (with annual supplements). $900 (plus $240 per supplement). ISSN: 03643670.

First published 1963 as *CODEN for periodical titles* (including non-periodical titles and deleted CODEN).

CODEN - a six-character code, the first five characters being letters (*e.g., Government research announcements*: GVRAA; *Technical book review index*: TBRIA), designed for titles of periodicals for computer retrieval, etc. *Class No:* 5(051)(003)

Bibliographies

[226]
BOLTON, H.C. Catalogue of scientific and technical periodicals, 1665-1895, together with chronological tables and a library check-list. 2nd ed. Washington, Smithsonian Institution, 1897 (reprinted New York, Johnson Reprint Corporation, 1965). vii, 1247p. (*Smithsonian Misc. Coll., v.40, no.1076.*)

First published 1885.

An A-Z title catalogue; pt. 1, a corrected reprint of the 1st ed., listing 4,954 titles; pt. 2, listing over 3,600 new titles. Entry is under earliest title, with later titles listed in chronological order, plus cross-references from later titles, well-known short titles, and names of principal editors. Excludes transactions of learned societies unless later as an independent periodical. Appended chronological table giving publication date of each volume of several hundred journals, 1728-1895. Title index; general subject index; location list for 3,000 titles in *c*.200 US libraries. A valuable checklist for early periodicals not included in the *World list*. *Class No:* 5(051)(01)

[227]
—SCUDDER, S.H. Catalogue of scientific serials of all countries, including the transactions of learned societies in the natural, physical and mathematical sciences, 1633-1876. Cambridge, Mass., Harvard Univ., 1879. xii, 358p.

Reprinted New York, Kraus Reprints, 1965.

Lists over 4,400 titles, arranged by country and town of origin, with indexes of towns and titles, and a brief subject index. Excludes serials devoted entirely to medicine, agriculture or technology. Less complete than, but complementary to, H.C. Bolton's *Catalogue*, which includes applied sciences but not publications of learned societies. *Class No:* 5(051)(01)

[228]
BRITISH LIBRARY. DOCUMENT SUPPLY CENTRE. Journals in translation. 5th ed. Boston Spa, British Library Document Supply Centre, and Delft, International Translations Centre, 1991. viii, 286p. ISBN: 0712320733.

First published 1976; 2nd ed. 1978.

1,352 numbered items, p.1-146 - originals being mostly Russian, plus some Japanese, Chinese, etc. Data on each: start; holdings; brief description; frequency; publisher; former title(s); to be ordered from; held by. 7 appendices, *e.g.* 1. Abstracts of patents and inventions; 8. Current titles, etc. Keyword index; Original titles index; Key to publishers and/or distributors. *Class No:* 5(051)(01)

[229]
CANADA. National Science Library, Ottawa. Union list of scientific serials in Canadian libraries. [Catalogue collectif des publications scientifiques dans les bibliothèques canadiennes.] 8th ed. Ottawa, Canada Institute for Scientific and Technical Information, 1979. Biennial. 2v.

First published 1957.

The 8th ed. lists more than 48,000 scientific serial titles held by 253 Canadian libraries. *Class No:* 5(051)(01)

[230]
Current serials received. April 1991. Boston Spa, British Library Document Supply Centre, [and] London, Science Reference and Information Service. 1986-. Annual. 654p. £45. ISSN: 03090655.

First published 1965 as *Current serials received by the N.L.L.*.

....(contd.)

Lists *c*.72,000 titles, differentiating (by symbols) between those received at the Document Supply Centre (Boston Spa) and those received at SRIS (London). 3 parts: 1. Current serial titles, except for cyrillic, p.7-609; 2. Cyrillic titles; 3. Cover-to-cover translations of cyrillic serials (p.649-54). *Class No:* 5(051)(01)

[231]
INTERNATIONAL COUNCIL OF SCIENTIFIC UNIONS. Abstracting Board. **International serials catalogue.** Paris, International Council of Scientific Unions Abstracting Board, 1978. 2v. ([x], 521p.).

1. *Catalogue.* 2. *Index/concordance.*

A/Z list of titles of over 25,000 scientific and technical journals 'abstracted and indexed by the member services of the International Council of Scientific Unions Abstracting Board'. Entries include CODEN, ISSN and initials of issuing service, Pt. 2 provides ISSN-CODEN and CODEN-ISSN indexes to the titles in pt. 1. *Class No:* 5(051)(01)

[232]
Journals with translations held at the Science Reference Library. Alexander, B.A. London, Science Reference Library, 1985. [iii], 65p. ISBN: 0712307176.

List of 542 titles held, with subject index and index of original titles. Shelf marks indicated. *Class No:* 5(051)(01)

[233]
Keyword index to serial titles (KIST). Boston Spa, British Library Document Supply Centre, 1980-. 4pa. Set of 48x microfiche. £289pa. plus VAT. ISSN: 01439553.

A keyword union list of current and non-current holdings of serials at the BLDSC, Boston Spa, and at the Science Reference and Information Service, plus the Science Museum, London. Other libraries' holdings are to be included. *Class No:* 5(051)(01)

[234]
OXFORD UNIVERSITY. Bodleian Library. **Union list of serials in the science area, Oxford.** Stage II (incorporating Stage I). Oxford, Bodleian Library, 1970. xvii, 598p. (*1st supplement 1971. v, 240p.*)

Stage I, 1968.

Stage II lists *c*.15,000 titles in 50 Oxford libraries. 'This stage of the *Union list* does NOT record all the serials in the Radcliffe Science Library (RSL) but those on the open shelves together with a considerable proportion of those on the stack' (prefatory note). Planned in 3 stages, each to include and amplify its predecessor.

The four richest collections of scientific serials in the UK are cited as the Radcliffe Science Library, BLLD, Science Museum Library and Science Reference Library. *Class No:* 5(051)(01)

[235]
Scientific serials in Australian libraries. Conochie, J.A., *ed.* Melbourne, Commonwealth Scientific and industrial Research Organization [CSIRO], 1976. 3v. Loose-leaf.

Supersedes *Union catalogue of scientific and technical periodicals in the libraries of Australia* (2nd ed., edited by E.R. Pitt. 1951) and *New titles: 1946-52 supplement* (compiled by A.L. Kent. 1954).

Lists *c*.60,000 scientific and technical periodicals and serial titles held in 180 Australian libraries. *Class No:* 5(051)(01)

[236]
Ulrich's international periodicals directory. New York, etc., Bowker. Annual. 3v. $339.95. ISSN: 00000175.

First published 1932. Incorporates *Ulrich's irregular series and annuals.*

The 31st ed., 1992-93 (xlviii,6836p.) is an international directory of *c*.126,000 periodicals in 788 subject groups. 'Abstracting and indexing services'...'Women's interests'. Much detail given in concise form, including presence of book reviews, whether indexed/abstracted or online. Also an appended section, 'Cessations' and a full title index in v.3. Scientific and technical titles figure prominently. Available on CD-ROM, microfiche, and online via DIALOG. A vital reference tool, updated by supplements 3 times a year. *Class No:* 5(051)(01)

[237]
World list of scientific periodicals published in the years 1900-1960. 4th ed. London, Butterworth, 1963-65.

First published in 1925-27.

The 4th ed. 'the last to appear in this format' listed 59,961 numbered titles (A-E, F-P, Q-Z, plus 'Periodic international congresses'). Although thereafter incorporated into *British union-catalogue: New periodical titles;* the *World list* appeared as an annual supplement for a time, with 1960-68 (1970) and 1969-73 (1976).

Entry is under title as published, with adequate cross-references. Holdings of nearly 300 libraries, including BLDSC, Boston Spa. Unlocated items are also listed. A major location list in its day. *Class No:* 5(051)(01)

Histories

[238]
HOUGHTON, B. Scientific periodicals: their historical development, characteristics and control. London, Bingley, 1975. 156p. diagrs.

In 3 parts: 1. History and development of scientific periodicals, 1663 to the present day - 2. Select list of periodicals current, grouped by subject area and international in origin - 3. Evaluation of select list by reference to usage, citations and length of useful life. 'Scientific periodical literature has not hitherto been analysed from the librarian's point of view on a comparable scale' (*British book news*, July 1975, p.480). *Class No:* 5(051)(091)

[239]
KRONICK, D.A. A History of scientific and technical periodicals: the origin and development of the scientific and technical press, 1665-1790. 2nd ed. Metuchen, N.J., Scarecrow Press, 1976. xvi, 356p.

First published 1962.

A study of origins and development in the social and intellectual setting of the time, as well as treating the various types of periodical, - the substantive journal, proceedings of scientific societies, abstracts and review journals, plus the bibliographical apparatus affording access to them. Bibliography. Subject and title indexes. *Class No:* 5(051)(091)

Progress Reports

[240]

UNITED NATIONS. Educational, Scientific and Cultural Organization. **List of annual reviews of progress in science and technology.** [Liste de "mises au point" annuelles sur les progrès de la science et de la technologie.] 3rd ed. Paris, Unesco, 1981. 43p.

First published 1965.

Classed arrangement of several hundred 'annual series containing papers which review fairly narrow topics within the broader subject designated by the serial title' (*Introduction*). Appended list of publishers. No index.

Class No: 5(055)

Yearbooks & Directories

[241]

Consultants and consulting organizations directory. 12th ed. Detroit, Mich., Gale, 1991. 2v. $430, set. ISBN: 0810371227.

First published 1966.

More than 10,000 entries for independent consultants and consulting firms in the US. 14 broad subject classes (Agriculture, forestry, and landscaping ... Social issues and concern). 10 categories of data on each consultancy. Over 900p. of indexes (geographic; subject; industries served; personal name; consulting firms).

New consultants (2pa.) is issued between editions of the main directory (Supp. 13, 1993. $375). *Class No:* 5(058)

[242]

Current contents address directory: Science & technology, 1984-87. Philadelphia, Pa., Institute for Scientific Information, 1985-88. Annual. 4v. ISSN: 08822360.

Partly replaces *Current bibliographic directory of the arts & sciences* (Institute for Scientific Information, 1978-83).

Over 400,000 addresses of scientists and scholars, covering authors (and co-authors) of 4,000 books and articles in 4,400 journals. 3 parts: 1. Directory of authors and source citations to their works (31,272 columns) - 2. Organizations represented by authors - 3. Geographical index to the directory of authors. Tiny print throughout.

Class No: 5(058)

[243]

Directories held by the Science Reference and Information Service. 2nd ed., compiled by Business Information Service. London, British Library, SRIS, 1986. [iii], 101, [1]p. £7. ISBN: 0712307419.

About 2,000 entries, under title A-Z. Call marks are given. *Class No:* 5(058)

[244]

Directory of technical and scientific directories: a world bibliographic guide to medical, agricultural, industrial and natural science directories. 6th ed. London, Longman; Phoenix, Ariz., Oryx Press, 1989. 302p. £99. (*Reference on research.*) ISBN: 0582044294; 0897746376, US ed.

First published as *Directory of scientific directories* in 1969.

In addition to directories, dictionaries, yearbooks, encyclopaedias, handbooks and bibliographies are included. Grouped according to subject headings: general science and technology; agriculture and the environment, chemical and materials technology; computer and electronic sciences; earth and space studies; engineering; industry; medical and biological sciences; physics and energy technology. 3

.... (contd.)

indexes: author, editor and compiler; directory titles; publishers' addresses. The entries have been compiled from information gleaned from questionnaires, publishers' catalogues and direct enquiries. Where contents of directories are listed there is no evaluative comment.

Class No: 5(058)

[245]

INTERNATIONAL COUNCIL OF SCIENTIFIC UNIONS. Year book 1991. Paris, ICSU Secretariat, 1991. 428p. ISBN: 093035723x. ISSN: 00744387.

First published 1954.

Sections include: Standing committees and ad hoc groups - National participation ... Interdisciplinary ICSU bodies - Joint programmes. Addresses of international organizations... Alphabetical address list, p.267-406. Index. *Class No:* 5(058)

[246]

McGraw-Hill yearbook of science and technology: Comprehensive coverage of the important events of the year, as compiled by the staff of the McGraw-Hill Encyclopedia of Science and Technology. New York, etc., McGraw-Hill, 1961-. Annual. illus., diagrs., graphs, tables, maps. $80. ISSN: 00762016.

The 1992 annual (1991. [xii],523p.) has *c.*250 contributors. Entries, A-Z (Acid rain ... Zooplankton). 'Queuing theory' (p.377-83) has 15 references. Detailed, analytical index. *Class No:* 5(058)

[247]

Register of consulting scientists: contract research organizations and others scientific and technical services. Copp, D.J.B., *ed.* 6th ed. Bristol, Hilger, for the Council of Science and Technology Institutes (CSTI), 1984. [x], 72p.

Classified register: independent full-time; independent organizations; university departments/bureaux; part-time consultants; useful addresses. Subject and name indexes precede. Restricted to practices or organizations that have a corporate member of a CSTI body. *Class No:* 5(058)

[248]

Science year: the World Book science annual - a review of science and technology during the ... school year. Chicago, Ill., World Book-Childcraft International, Inc., 1965-. illus. (mostly col.). ISSN: 00807621.

First published 1965.

Science year, 1986 (1985. 400p. $29.95) has 14 'special reports' (*e.g.* 'Cleaning up the River Thames'), followed by 'Science file', - 39 articles, A-Z, plus 6 'close-ups' (*e.g.* 'Compact disk players'). Appendices include 'People in science' and 'World Book supplement' (2 reviewed articles from the 1985 ed. of *World Book encyclopedia*). Detailed index, p.385-396. Fully illustrated and captioned.

Class No: 5(058)

[249]

Yearbook of science and the future. Chicago, Ill., London, etc., Encyclopaedia Britannica, Inc., 1972-. col. illus., diagrs., tables. ISSN: 00963291.

The *1987 Yearbook* (1986. 542p. $24.95) offers 17 feature articles (*e.g.* 'The Centre of the Milky Way', 'Twenty-five years of lasers'), a few with 'For additional reading', and including two articles on institutions (*e.g.* 'The Royal Greenwich Observatory'). Then follows 'Encyclopaedia Britannica science update' (3 pieces, *e.g.* 'Photosynthesis'), 'The science year in review', p.305-494 (A-Z articles,

....*(contd.)*

signed but no references), 'Scientists of the year' (honours and awards; obituary), 'A science classic' (James Watt's steam engine), and 'Science for the future' (2 pieces, *e.g.* 'Future impact of information technologies'). Detailed analytical index, p.521-42. *Class No:* 5(058)

Teaching Materials

[250]

New encyclopedia of science. 6th ed. London, Encyclopaedia Britannica, Inc., 1986. 16v. (viii, 2296p.). col. illus., maps. ISBN: 081725000x.

5th ed., 1982. Earlier as *New junior encyclopaedia of science.*

'Study guides' in v.1 precedes the A-Z sequence of articles (which are unsigned and lack bibliographies). Each page carries illustrations, mostly in colour but deficient in explanatory captions. V.16 includes an index that incorporates a dictionary. Evidently intended for secondary school pupils. 'The *New Encyclopedia of Science* has too many inadequacies of style and content to be recommended except for well-stocked reference collections that would welcome a set useful for browsing, or as a pictorial adjunct to more carefully prepared sources' (*Reference books bulletin*, 1985-1986, p.90). *Class No:* 5(072)

[251]

The New illustrated science and invention encyclopedia. Rev. ed. London, Marshall Cavendish, 1988. 27v. illus. (incl. col.).

Based on *How it works: the illustrated encyclopedia of science and technology* (1978), but simplified and Americanised for children. Formerly *Growing up with science: the illustrated encyclopedia of inventions* (1984. 25v.).

Concerns the 'hows', 'whats' and 'whys' of the scientific world. V.1-23 consist of an A-Z sequence of *c.*40 topics. V.24 contains brief biographical sketches of *c.*100 scientists; v.25 looks at inventors and inventions; v.26 contains a glossary and index to the set; v.27, careers in science. Of the previous ed., 'An up-to-date inviting set of scientific and technical information for elementary and junior high school children, with a good emphasis on graphics' (*Reference books bulletin*, 1984/1985, p.66). *Class No:* 5(072)

[252]

Oxford illustrated encyclopedia. Oxford, University Press, 1985-93. 8v. illus. (mostly col.).

8v. 1. *The physical world;* edited by V. Fuchs. 1985. 384p. £19.50. 2. *The natural world;* edited by M. Coe. 1985. 384p. £19.50. 3. *World history - from earliest times to 1800.* 400p. 1988. £22.50. 4. *World history - from 1800 to the present day.* 400p. 1988. £22.50. 5. *The arts.* 512p. 1990. £25. 6. *Invention and technology.* 392p. 1992. £25. 7. *Peoples and cultures.* 392p. 1992. £25. 8. *The universe.* 300p. 1993. £19.95.

V.1-2 each carry *c.*2,500 entries, A-Z. V.1 concerns mathematics, chemistry, physics, geology and physical geography. V.2, living and extinct animal and plant species, plus specific representative organisms. Biographical entries figure; cross-references. Very well illustrated. No index or bibliographies. Aimed at parents and their children of 15 or over. For school library reference collections, easing the

....*(contd.)*

load of demand for articles in general encyclopedias.

Replaces *Oxford junior encyclopaedia* (2nd ed. OUP, 1964. Corrected reprint, 1974. 13v.). *Class No:* 5(072)

[253]

Science experiments on file experiments, demonstrations and projects for school and home. Bruce, K., *and others.* New York, Facts on File, 1989. 100p. illus., looseleaf. $145. ISBN: 081601888x.

Contains 84 science experiments, demonstrations and projects written by science teachers. 4 sections: earth sciences, biology, physical science/chemistry, and physics. Description of each experiment contains: introduction, time and materials required, safety precautions, step-by-step procedure and analysis. Many line drawings and charts. Appendices index the experiments by grade level, supervision requirements, setting (school or home), number of participants and lists experiments which can be completed in less than an hour. List of sources of materials and equipment. Subject index (no cross-references). For public and school libraries. *Class No:* 5(072)

Bibliographies

[254]

Appraisal: science books for young people. Boston, Mass., Children's Science Book Review Committee, 1967-. V.1-. 4pa. $39.

Critical reviews of 40-50 books per issue, on both pure and applied science. Each book is assessed by a subject specialist as well as a librarian; ratings extend from 'excellent' to 'unsatisfactory'. Age range is also indicated. A special section deals with series. 'Candid authoritative and scrupulous analysis' (*RQ,* v. 7(4), Summer 1968, p.187-8). Addressed to children's librarians, teachers and other providers of books for young people. *Class No:* 5(072)(01)

[255]

School science review: a journal of the Association for Science Education. Hatfield, Herts., The Association, 1920-. v.1-. 4pa. £46. ISSN: 00366811.

V.73(262), September 1991 (176p.) has 50 signed book reviews, p.159-71, which give indication as to level of readership, *e.g.* primary class teacher, students, young readers. 11 signed reviews of visual aids with notes of content, appraisal, and audience. *Class No:* 5(072)(01)

[256]

Science & technology in fact and fiction: a guide to children's books. Kennedy, D.M., *and others.* New Providence, N.J., Bowker, 1990. 319p. $35. ISBN: 0835227081.

*c.*350 fiction and non-fiction books for children are listed. Entries arranged alphabetically by the author's surname. Bibliographic details and an authoritative evaluation are given. Five indexes: author; title; illustrator; subject; readability. Contains fewer entries than *Museum of Science and Technology basic list of children's science books* but includes fiction, items for younger children and has lengthier annotations.

By the same author, *Science & technology in fact and fiction: a guide to children's books and young adult books* (1990. 2v. 331p. $59.95) and *Science & technology in fact and fiction: a guide to young adult books* (1990. 363p. $36). *Class No:* 5(072)(01)

Handbooks & Manuals

[257]

ASE science teachers' handbook. Nellist, J. *and* Nichol, B., *eds.* London, Hutchinson, on behalf of the Association for Science Education, 1986. vi, 298p. diagrs., charts, tables. ISBN: 0091563402.

7 documented chapters by experienced science educators, teachers and others. Chapter 1: Why teach science and why to all? Chapters 2-6 contain suggested activities - pointers to purposeful discussion, investigations or enquiries which can be carried out by individuals or departmental teams. Chapter 7 discusses special problems (*e.g.* Science education and ethnic minorities; Girls in science). Glossary, p.294. Analytical index, p.295-8 'The book has been written for all science teachers but directed particularly to those teaching youngsters in the 7-14 age range' (*Introduction*). *Class No:* 5(072)(035)

Reviews & Abstracts

[258]

Science books & films. Washington, American Association for the Advancement of Science, 1975-. V.11-. 5pa. $35pa. ISSN: 0098342x.

Continues *AAAS science books: a quarterly review* (V.1-10, 1965-75), with assessments and ratings of scientific and technical books and non-print material (films, filmstrips, video cassettes). *Class No:* 5(072)(048)

Yearbooks & Directories

[259]

Sourcebook for out-of-school science and technology education. Paris, UNESCO, 1986. 202p. illus. ISBN: 9231023861.

Translated into 27 languages.

Describes 'various activities and programmes in out-of-school science and technology education, including clubs, societies and congresses for young scientists, science fairs, project work, olympiads and other contests, visits, science camps and excursions' (*Unesco Publications catalogue Supplement* 1986, p.18). *Class No:* 5(072)(058)

Quotations

[260]

Isaac Asimov's book of science and nature quotations. Asimov, I. *and* Shulman, J.A., *eds.* New York, Weidenfeld and Nicolson, 1988. 360p. £12.95. ISBN: 1555841112.

*c.*2,000 quotations in 86 different subject areas (Aeronautics ... Zoology). Quotations arranged chronologically within each subject by the date of the quotation or, if unavailable, by the date of birth of the speaker. Brief description of the speaker is also given. Index of speaker names. *Class No:* 5(082.2)

[261]

MacKAY, A.L. A Dictionary of scientific quotations. Bristol, Adam Hilger, 1991. xii,297p. £9.50. ISBN: 0750301066.

Second and expanded edition of *A Harvest of a quiet eye* (Institute of Physics, 1977).

2,130 quotations, epigrams and aphorisms, under authors, Abelson ... Zuckerman, plus entries for 'Anonymous' and 'Bible'. Authors' dates and sources are given. Quotations in

....(contd.)
foreign languages appear in translation, with or without the original. Subject and first word of title index. *Class No:* 5(082.2)

Tables & Data Books

[262]

BALLENTYNE, D.W.G. *and* **LOVETT, D.R. A Dictionary of named effects and laws in chemistry, physics and mathematics.** 4th ed. London, Chapman & Hall, 1980. viii, 346p. diagrs., tables, chemical structures. £17.95. ISBN: 0412223902.

First published 1958.

Defines and describes precisely 1,500 effects, laws, units and other phenomena. The dictionary, p.1-320 (Abbe number ... Zinkl's Rule). 'Mossbauer effect': ½p. Appendix: A list of named units; Table of organic compounds. In this edition most entries have received some revision. In particular, units are now in SI units, wherever possible. *Class No:* 5(083)

[263]

BRITISH STANDARDS INSTITUTION. Basis of tables. Conversion factors. Part 1. London, BSI, 1974. 1045p. AMD 4153. 1974. Supplement, no. 1. 1967. AMD 342,1255. £52; £52, supp. (*BS 350. Part 1: 1974.*)

Part 1 provides conversion factors for metrology, mechanics, heat, extended basic information units (including SI, metric technical, imperial). Tables of conversion factors showing inter-relationship of units. Letter symbols for avery units. Omits purely electrical units. *Supplement no. 1* adds tables for 51 conversions, plus 36 tables for conversion to conversion to and from SI units. BS 350. Part 2 withdrawn. *Class No:* 5(083)

[264]

CODATA. Inventory of data sources in science and technology: a preliminary survey. Paris, Unesco, 1983. [vii], 229p. $14. ISBN: 923102048x.

An inventory of 652 international sources in 94 countries. 6 sections: 1. Renewable energy sources - 2. Fertilizers - 3. Hydrological sciences and water resources - 4. Nutrition - 5. Pesticides - 6. Soil science. Indexes to institutions, centres, etc.; to centres by country; to personal names; to subjects. Aimed at scientists, engineers and information specialists, particularly those in developing countries.

CODATA bulletin (Oxford, Pergamon Press, 1984. 6pa.) incorporates the *Inventory*.

CODATA international conferences (*e.g.* 10th Conference, on generation, compilation, evaluation and dissemination of data for science and technology. *Proceedings*. North Holland, 1987) are fruitful sources of new data. *Class No:* 5(083)

[265]

Handbooks and tables in science and technology. Powell, R.H., *ed.* 2nd ed. London, Library Association, 1983. vi,297p. ISBN: 0853657262.

First published 1979.

A partly annotated list of 3,403 handbooks and tables in science, technology and medicine, with emphasis on physics, chemistry, engineering, agriculture and biology. Main entries are under titles, with subject, author/editor and publisher indexes. Full bibliographical data. Adequate cross-references. 'A very useful reference guide and acquisitions

....(contd.)

tool for academic, special and large public libraries' (*Reference books bulletin*, 1983/84 (1984), p.100). *Class No:* 5(083)

[266]

International critical tables of numerical data, physics, chemistry and technology; prepared under the auspices of the International Research Council and the National Academy of Sciences. New York, etc., McGraw-Hill, 1926-33. 7v. and index. tables, graphs, charts.

Each volume has explanatory text and contents lists in four languages, English, French, German and Italian. Data are selected from v. 1-7 (1910-26) of *Tables annuelles de constants et données numériques de chimie, de physique et de technologie* (Paris, Gauthier-Villars; Chicago, Univ. of Chicago Press (v. 1-11, pt. 1. 1912-37). Now largely obsolete and incomplete, but important as the work of experts who critically examined the data before including it. The accompanying index covers subject data in detail. *Class No:* 5(083)

[267]

JERRARD, H.G. *and* **McNEILL, D.B. Dictionary of scientific units,** including dimensionless numbers and scales. 6th ed. London, Chapman & Hall, 1992. 255p. tables. £15.95. ISBN: 0412467208.

First published 1963.

Contains *c.*950 units, supported by *c.*650 references. The dictionary, Abampere ... Zhubov scale, p.9-190. 'Beaufort scale': nearly 1p. 5 appendices (*e.g.* 1. Fundamental physical constants; 2. Standardization committees and conferences). References, p.221-39. Detailed analytical index, p.241-55. *Class No:* 5(083)

[268]

LANDOLF, H.H. *and* **BÖRNSTEIN, R. Zahlenwerte and Funktionen aus Naturwissenschaften und Technik.** Neue Serie. [Numerical data and functional relationships in science and technology.] Berlin, Springer Verlag, 1961-. tables, graphs.

Continues Landolf-Börnstein's *Zahlenwerte und Funktionen aus Physik, Chemie, Astronomie, Geophysik und Technik* (6. Aufl. 1950-80. 4v. in 26).

The *Neue Serie* runs to 7 groups, compared with the previous four. Also the text is in English (but sometimes in both German and English): 1. *Nuclear and particle physics.* 2. *Atomic and nucleic physics.* 3. *Crystal and solid state physics.* 4. *Macroscopic and technical properties of matter.* 5. *Geophysics and space research.* 6. *Astronomy, astrophysics and space research.* 7. *Biophysics.*

Tabular and graphic data on the grand scale, the series now running to about 100v., with new volumes appearing as subjects develop or new fields arise. *Class No:* 5(083)

[269]

MOSES, A.J. The Practicing scientist's tables: a guide for physical and terrestrial scientists and engineers. New York, London, etc., Van Nostrand Reinhold, 1978. x, [i], 1292p. tables. ISBN: 0442255845.

12 chapters (1. Introduction - 2. Properties of the elements - 3. Properties of organic compounds ... 10. Properties of superconductors - 11. Properties of the environment - 12. Properties of miscellaneous materials (*e.g.* man-made organic fibres)). Appendix A. List of registered trademarks; B. Table of isotopes. Virtually all the text is in tabular data form. Analytical index. *Class No:* 5(083)

[270]

Tables of physical and chemical constants, and some mathematical functions. Kaye, G.W.C. *and* Laby, T.H., *comps.* 15th ed. London & New York, Longman, 1986. [viii], 477p. graphs, tables. £28.99; $52.95. ISBN: 0582463548.

59 contributors. 4 main sections: 1. General physics - 2. Chemistry - 3. Atomic and nuclear physics - 4. Mathematical functions. Analytical index. Subsection references. All tabulated values appear in SI units. Essential for any reference library. *Class No:* 5(083)

Illustrations

[271]

The Guild handbook of scientific illustration. Hodges, E.R.S., *ed.* New York, Van Nostrand Reinhold, 1989. 575p. illus. $85.95. ISBN: 0442236816.

Book sponsored by the Guild of Natural Science Illustrators. Includes illustration techniques using various media, *e.g.* ink, watercolour, pencil and acrylics. Suggestions on how to handle, measure, preserve and draw various subjects, *e.g.* plants, mammals, birds, fossils, amphibians and reptiles. List of material and equipment suppliers is given. Lengthy bibliography. Well illustrated. *Class No:* 5(084.1)

Chronologies

[272]

ASIMOV, I. Asimov's chronology of science and discovery. London, Grafton Books, 1990. 707p. $29.95. ISBN: 0246136693.

1,450 entries listed in chronological order (Bipedality, 4,000,000 BC ... Shroud of Turin, Distant Galaxies, Greenhouse Effect, 1988). Entries vary from a few lines to two pages - in general, longer than those in *Milestones in science and technology,* by E. Mount and B.A. List (Phoenix, Ariz., Oryx Press, 1987. 141p.) Subject and name indexes. No bibliographies or illustrations. *Class No:* 5(090)

[273]

HELLEMANS, A. *and* **BUNCH, B. The Timetables of science:** a chronology of the most important people and events in the history of science. London, Sidgwick & Jackson, 1989. vii,656p. tables. ISBN: 0283999268.

Records *c.*10,000 separate events, arranged chronologically, under broad subject areas, *e.g.* astronomy, life science, mathematics, physical science, technology. Nine broad overviews: Science before there were scientists: 2,400,000-599 BC ... The Newtonian epoch: 1660-1734 ... Science after World War II: 1946-1988. Name and subject indexes. *Class No:* 5(090)

Histories

[274]

Album of science. New York, Scribner's, 1978-89. 5v. illus., facsims. ISBN: 0684190745, set.

1. *Album of science: Antiquity and the Middle Ages,* by J.E. Murdoch. 1984. xii, 402p.$80.

2. *Album of science: From Leonardo to Lavoisier, 1450-1800,* I.B. Cohen, general editor. 1980. xiii, 333p. $80.

3. *Album of science: The nineteenth century,* by L.P. Williams. 1978. xv, 413p. $80.

Album of science: The biological sciences in the twentieth

....*(contd.)*

century, by M. Borrell. 1989. 306p. $80.

Album of science: The physical sciences in the twentieth century, by O. Gingerich. 1989. 306p. $80.

V. 3 has 6 parts: 1. The environment of science - 2. The world of nineteenth century science - 3. Man - 4. The living world - 5. Atoms and molecules and sources - 6. The marvels of science. Bibliography, p.383. Index (to text & illus.), p.389-413. Essentially a pictorial record, with nearly 600 illus., and facsims. *Class No:* 5(091)

[275]

BRUNO, L.C. The Tradition of science: landmarks of Western Science in the collection of the Library of Congress. Washington, Library of Congress, 1987. xi,351p. illus., facsims. $30. ISBN: 0844405280.

8 chapters: Astronomy - Botany - Zoology - Medicine - Chemistry - Geology - Mathematics - Physics. Epilogue. Chapter bibliographies, p.299-333. Detailed, analytical index, p.385-51. *Class No:* 5(091)

[276]

DARMSTAEDTER, L. Ludwig Darmstaedter's Handbuch zur Geschichte der Naturwissenschaften und der Technik in chronologischer Darstellung. 2. ungearb. und verm. aufl. Berlin, Springer, 1908. x,ii, 1262. 127p. (Reprinted New York, Kraus, 1960).

Brief descriptions, chronologically arranged, of 13,000 scientific discoveries and inventions, from 3500 BC to the year of publication. Name and subject indexes.

Class No: 5(091)

[277]

—WEISS, B. Wie finde ich Literatur zur Geschichte der Naturwissenschaften und Technik. Berlin, Verlag Arno Spitz, 1985. 289p. front., charts. ISBN: 3870612118.

A guide to the history of the literature on the natural sciences and technology. 7 sections deal with types of libraries and organizations, catalogues, bibliographies, directories, histories, leading journals and reference material. 1,653 unannotated entries. Detailed index. Some slant towards German-language countries, but not confined to the historical aspects of scientific and technical literature. *Class No:* 5(091)(01)

[278]

LOSEE, J. A Historical introduction to the philosophy of science. 2nd ed. Oxford Univ. Press, 1980. [v], 245p. £6.99. ISBN: 019219156x.

First published 1972.

14 documented chapters. 1. Aristotle's philosophy of science ... 7. The seventeenth century attack on Aristotelian philosophy (Galileo; Francis Bacon; Descartes) - 8. Newton's axiomatic method - 9. Analysis of the implications of the new science for a theory of scientific method ... 14. Alternatives to orthodoxy (38 references). Selected chapter bibliographies, p.221-41. Index of proper names; analytical index of subjects, p.243-8. The emphasis is on views of the development of scientific method prior to 1940. *Class No:* 5(091)

[279]

RONAN, C.A. The Cambridge illustrated history of the world's science. Cambridge, University Press; London, Newnes, 1983. 543p. illus., facsims, maps. ISBN: 0521258448.

10 chapters: 1. The origins of science - 2. Greek science - 3. Chinese science - 4. Hindu and Indian science - 5. Arabian science - 6. Roman and medieval science - 7. From Renaissance to scientific revolution ... 10. Twentieth-century science. Bibliography (by chapters), p.528-9. Index, p.532-43, includes references to the illus. *Class No:* 5(091)

[280]

SARTON, G. A History of science. Cambridge, Mass., Harvard Univ. Press; London, Oxford Univ. Press, 1953. 2v. illus.

1. *Ancient Greece through the golden age of Greece.* 1953, xxvi. 646p.

2. *Hellenistic science and culture in the last three centuries BC.* 1959. xxvi, 554p.

V. 1 'is not written for classical philologists, but rather for students of science whose knowledge of antiquity is rudimentary, who may never have studied Greek ...'. V. 2, despite its title, 'deals with Roman culture and Latin letters as well as with Greek letters and the culture of Eastern Europe, Egypt and Western Asia' (*Preface*). Both volumes are well illustrated and amply documented.

Class No: 5(091)

[281]

SARTON, G. Introduction to the history of science. Baltimore, Md., Williams and Wilkins, for Carnegie Institution of Washington, 1927-48 (reprinted, Krieger, 1975). 3v. in 5. $425, set. (*Carnegie Institution publication 376.*) ISBN: 0882751727.

1. *Homer to Omar Khayyam.* xi, 839p.

2. (2 pts.). *From Rabbi ben Ezra to Roger Bacon.* xxxv, xvi, 1251p.

3. (2 pts.). *Science and learning in the 14th century.* xxxv, xi, 2155p.

Treatment is by time-periods. Each chapter normally surveys a half-century, summarizing events in science and learning, with notes on lives and works, criticism and bibliographical history of the principal personalities concerned. Each volume has an index, combined in v.3. A monumental bibliographical history, embracing all branches of knowledge. Invaluable source book for students of the early history of science.

Addenda and errata are included in the *ISIS 'Critical bibliography of the history and philosophy of science'.*

Class No: 5(091)

[282]

SINGER, C. A Short history of scientific ideas to 1900. Oxford, Clarendon Press, 1959. xviii, [2], 525p. illus.

'Based on *A short history of science*, first drafted in 1929 and published in 1941' (*Preface*).

A concisely written, straight-forward history, with 188 illus., and many brief biographical notes. Chapters 1-4 (p.6-136) concern scientific ideas in ancient times; chapters 7-9 (p.218 onwards), the 17th-19th centuries. Index of names, p.517-25. The best of the short histories of science. *Class No:* 5(091)

[283]
TATON, R., *ed.* **A General history of the sciences.** Translated by A.J. Pomerans. London, Thames & Hudson, 1963-66. 4v. illus. (incl. pl.), diagrs.

Translated from *Histoire générale des sciences* (Paris, Presses Universitaires de France, 1957-64. 3v. in 4).

1. *Ancient and medieval science, from pre-history to A.D. 1450.* 1963. xvi, 552p.
2. *The beginnings of modern science, from 1450 to 1800.* 1964. xx, 665p.
3. *Science in the nineteenth century.* [1965]. xix, 623p.
4. *Science in the twentieth century.* 1966. xxiv, 638p.

V.1 has 21 notable French contributors. 3 pts.: Ancient science in the East - Science in the Greco-Roman world - The Middle Ages. Chapter bibliographies; name and subject indexes. Covers developments in Phoenicia, pre-Columbian America and medieval India, - all subjects frequently neglected in general histories of science. Readable and well illustrated (v.1: 52 plates; 43 figures). *British book news* (April 1971, p.261) refers to the *General history* as a 'collaborative work which has come under heavy fire'.
Class No: 5(091)

[284]
THORNDIKE, L. History of magic and experimental science. New York, Columbia Univ. Press, 1923-58. 8v. $93ea.

1-2 *First 13 centuries.* 3-4. *14th-15 centuries.* 5-6. *16th century.* 7-8. *17th century.*

'Based on an examination of the original MSS and documents of the early writers, scientists and historians in library colleltions all over the world'. The word 'magic' as used in the title 'is understood in its broadest sense as including all occult arts and sciences, superstitions and folklore' (Hawkins, R.R. *Scientific, medical and technical Books ... 1930-44,* p.4). Well documented. V. 1-2 include a general index, bibliographical index, and index of MSS; v. 3-4 and 5-6, general index, index of MSS and index of incipits; v. 7-8, general index and index of MSS. A monumental survey. *Class No:* 5(091)

Bibliographies

[285]
AMERICAN PHILOSOPHICAL SOCIETY. Library. **Catalog of books in the American Philosophical Society Library.** Westport, Conn., Greenwood Press, [1970]. 28v. ISBN: 0837132665.

Photolithographed reproduction of the Library's card catalogue. Concentrates 'on particular aspects of the history of science and selected fields' (*Introduction*).
Class No: 5(091)(01)

[286]
CORSI, P. *and* **WEINDLING, P. Information sources in the history of science and medicine.** London, Butterworth Scientific, 1983. xvi, 531p. facsims., diagrs. £43. (*Butterworths guides to information sources.*) ISBN: 0408107642.

21 contributors; 24 documented chapters, in 4 parts. Part 1 consists of general chapters on the history of science and relations with associated areas (*e.g.* anthropology, religion). Part 2, on major institutions and research methods. Part 3 (chapters 11-19), historical range of modern sciences and developments. Part 4 reviews status of non-European developments - US, Indian, Islamic and Chinese, African.

....(contd.)
Chapter 8: 'Guide to bibliographical sources', p.137-56; Chapter 19: Medicine since 1500, p.378-407, has bibliography, p.393-407. List of journals, p.501-8. Analytical subject index, including names of authors and institutions, p.509-31. 'Highly recommended' (*Choice,* v. 21 (1), September 1983, p.64). *Class No:* 5(091)(01)

[287]
Horus: a guide to the history of science: a first guide for the study of the history of science, with introductory essays on science and tradition. Sarton, G., *comp.* Waltham, Mass., Chronica Botanica Co., 1952. xviii, 316p.

Greatly expanded from the bibliography in his *Study of the history of science* (Cambridge, Mass., Harvard Univ. Press, 1936), p.53-70, with annotations in Sarton's characteristic style.

Includes sections for the history of science in special countries, in special cultural groups, and for the history of special sciences. Also a list of journals on the history and philosophy of science, and lists of institutes, museums, libraries and international congresses. An important bibliographic guide, not least for the introductory notes to each chapter. Well documented. *Class No:* 5(091)(01)

[288]
ISIS cumulative bibliography: a bibliography of the history of science, formed from *ISIS* critical bibliographies, 1-90, 1913-65. Whitrow, M., *ed.* London, Mansell, in conjunction with the History of Science Society, 1971-84. 6v. £270.

1. *Personalities, A-J.* 1971. £90.
2. *Personalities, K-Z, and institutions, A-Z.* 1971.
3. *Subjects.* 1976. 772p. £90.
4. *Civilizations and periods, Prehistory to Middle Ages.* 1982.
5. *Civilizations and periods, 18th to 19th centuries.* Addenda to v.1-3, 1982. £176 (v.4/5).
6. *Author index.* 1984. 350p. £70.

'The result of weeding, tidying and sorting the *ISIS* Critical Bibliographies 1-90' (Jayawardene, S.A. *Reference books for the historian of science*). The *Personalities* section has nearly 40,000 entries for 10,000 individuals. 'Aristotle': *c.*450 entries; 'Royal Society': 123 entries. 'The whole series will be an essential purchase for any library catering for the needs of historians of events' (*British book news,* August 1977, p.598). *Class No:* 5(091)(01)

[289]
—**ISIS cumulative bibliography, 1966-75:** a bibliography of the history of science formed from ISIS critical bibliographies 91-100, indexing literature published from 1965 through 1974. Neu, J., *ed.* London, Mansell, in conjunction with the History of Science Society, 1980-85. 2v.

1. *Personalities and institutions.* 1980. 514p. £90.
2. *Subjects, periods and civilizations.* 1985. 720p. £90.

Follows the pattern of the earlier set. V.2 includes a bibliographic index to book reviews. 'Its thoroughness, coupled with its layout and indexes, confirm both its scholarly standing and its usefulness for more general enquiries' (*Library review,* v.35(2), Summer 1986, p.149). *Class No:* 5(091)(01)

[290]

—ISIS cumulative bibliography, 1976-85: a bibliography of the history of science formed from ISIS critical bibliographies 101-110, indexing literature published from 1975 through 1984. Neu, J., *ed.* London, Mansell in conjunction with the History of Science Society, 1990. 2v.

V.1. *Persons and institutions.* 800p. £60. V.2. *Subjects, periods, and civilizations.* 80p. £60.

Continues this essential work. *Class No:* 5(091)(01)

[291]

JAYAWARDENE, S.A., comp. Reference books for the historian of science: a handlist. London, Science Museum, 1982. xi, 229p. £2.50. (*Science Museum Library, Occasional publications, 2.*) ISBN: 0901805149.

1,034 numbered entries, some of them annotated, in 3 parts (44 chapters): 1. The history of sciences and its sources - 2. History related subjects - 3. General reference. Author/title index; analytical subject index. Wide-ranging, *e.g.* on catalogues of scientific books, manuscripts and archives, societies. Based on Science Museum Library holdings. 'This handlist provides an invaluable foundation for researchers' (*Natural history book reviews*, v.6 (3/4), 1983, p.145). *Class No:* 5(091)(01)

[292]

—RIDER, K.J. History of science and technology: a select bibliography for students. 2nd ed. London, Library Association, 1970. 75p.

First published 1967.

Includes 353 numbered and annotated entries. 'History of science', items 1-173; 'History of technology', items 174-353; systematic subdivision by periods and subjects. Includes medicine for the first time, but excludes commercial history of industry. Scarcity of material on agricultural and textile history. Index, p.71-75. Largely material in English. *Class No:* 5(091)(01)

[293]

ROLLER, D.H.D. and GOODMAN, M.M. The Catalogue of the history of science collections of the University of Oklahoma Libraries. London, Mansell, 1976. 2v. (1212p).

39,000 main entries; under authors, A-Z; also subject entries and cross-references. 'A very important list for the history of science' (*British book news*, December 1976, p.900). *Class No:* 5(091)(01)

[294]

THORNTON, J.L. and TULLY, R.I.J. Scientific books, libraries and collectors: a study of bibliography and the book trade in relation to science. 3rd rev. ed. London, Library Association Publishing, 1971. iv, 508p. Supplement to third edition, 1969-75, 1978, vi, 172p. ISBN: 0853654247; 0853659206.

First published 1954.

12 chapters, providing 'an introductory history of the production, distribution and storage of scientific literature from the earliest times' (*Preface*, 1954). The rise of scientific societies, growth of periodical literature, scientific publishing and bookselling, and scientific libraries. Extensive author bibliography, p.381-465; detailed index, p.467-508.

The 1969-75 *Supplement* updates, in 12 chapters. Bibliography, p.117-56 (*c.*800 items). Detailed index. 'A marvel of compression and orderly discipline' (*Annals of*

....(contd.)

science, v.36, 1979, p.417), although almost restricted to Western science (especially British and American authors and collections) and medicine. *Class No:* 5(091)(01)

Encyclopaedias

[295]

Dictionary of the history of science. Bynum, W.F., *ed.* Princeton, N.J., Princeton Univ. Press; London, Macmillan, 1981. xxxi, 494p. $75; $17.95. ISBN: 0691023840.

700 articles by 95 scholars, plus 10 subject editors, mostly British. Coverage: pure science, human science and medicine. Basically a dictionary of the leading ideas and concepts of Western science developed over the last five centuries. Entries usually carry brief references. Appended list of 800 scientists, cross-referenced to entries. Bibliography of 130 reference works and monographs. 'This scholarly, high quality dictionary' (*Wilson library bulletin*, March 1982, p.542). *Class No:* 5(091)(031)

[296]

MAYERHÖFER, J. Lexikon der Geschichte der Naturwissenschaften: Biographien, Sachwörter and Bibliographien. Vienna, Hollinek, 1959-76. Lfg.1-8.

Bd.1 (Lfg.1-6), Lfg.7-8. A - Edelstein.

Lfg.1. has a 110p. introduction to the age of scientific research, with appended bibliography. Signed and documented articles follow, on scientists and scientific topics (*e.g.* 'Aristotle' p.259-75, with 7p. of bibliography, in small type; 'Darwin', p.718-26, with 4p. of bibliography). Numerous cross-references. The final article, on 'Edelstein', left unfinished, was completed in the continuation to the *Lexikon. Class No:* 5(091)(031)

[297]

—Archiv der Geschichte der Naturwissenschaften: Biographien, Sachwörter, und Bibliographien. Vienna, Hollinek, 1980-. Heft 1-. Loose-leaf. Irreg.

The continuation of the *Lexikon. Class No:* 5(091)(031)

Reviews & Abstracts

[298]

Francis: Histoire des sciences et des techniques. 522. Vandoeuvre-Les-Nancy, Centre National de la Recherche Scientifique, 1940-. 4pa. ISSN: 00075574.

About 2,000 references pa. Classes: 1 Généralités - 2. Sciences et techniques mathématiques - 3. Sciences et techniques physiques (physics, astronomy, chemistry) - 4. Technologie - 5. Sciences et techniques de la terre - 6. Sciences et techniques de la vie. Analytical subject index and author index per issue and pa. Annual list of periodicals scanned. *Class No:* 5(091)(048)

Periodicals

[299]

Archive for history of exact sciences. Berlin, etc., Springer Verlag, 1960-. 8pa. DM856; $510. ISSN: 00039519.

Focuses mainly on mathematics and natural philosophy, but also includes the physical sciences. V.36(1), 1986 has 5 well footnoted or documented articles, - 4 in English, 1 in German. (Languages used may be English, French, German, Italian, Latin and Spanish.) Claims 'historical research meeting the standards of the mathematical sciences'. The

....*(contd.)*

article 'On square roots and their representations' has 2p. of references; 'Charles Rugeley Bury and his contributions to physical chemistry' carries 35 footnotes.
Class No: 5(091)(051)

[300]
British journal for the history of science. Cambridge, University Press, for the British Society of the History of Sciences, 1962-. 4pa. $125. ISSN: 00070874.
V.25(3), September 1992, carries 3 scholarly articles. Book reviews, p.355-92 (39 signed reviews; affiliations of reviewers are stated); 'Books received', p.393-6. 'Long recognized as a prominent publication in the history of science' (*Magazines for libraries*, 5th ed., 1986, p.883).
Class No: 5(091)(051)

[301]
History of science. Chalfont, St. Giles, Berks., Science History Publications, Ltd., 1962-. 4pa. £61. ISSN: 00732753.
Originally as *History of science: an annual review of literature research and teaching* (Cambridge, Heffer, 1962-71. v.1-10. Annual; v.11-. 4pa.).
V.30(4), December 1992, has 3 well-documented articles (*e.g.* 'The layers of chemical language, II', p.397-437, has 129 references). Annual index of authors-titles. 'Aims to provide analysis, review, and assessment, of specific topics in the history of science, medicine, and technology'.
Class No: 5(091)(051)

Films

[302]
Discovery & invention: a review of films & videos on the history of science and technology. Grant, C. *and* Ballantyne, J., *eds*. London, British Universities Film & Video Council, 1985. £7 (£3.50, to members). ISBN: 0901299391.
53 reviewers, listing 283 items currently available in the UK. 43 subject sections (Agriculture ... Transport - Water); 3 appendices: 1. Late additions; 2. Materials not yet reviewed; 3. Some materials for secondary education. Title index. Distributors' index. *Class No:* 5(091)(084.122)

[303]
EASTWOOD, B. Directory of audio-visual sources: history of science, medicine, and technology. New York, Science History Publications, 1979. v.p. $20. ISBN: 0882021850.
Audio-visual directory of 543 numbered entries, under titles A-Z. List of major academic rental libraries; list of distributors cited. Brief descriptions plus technical data on each title and order number. Topical index. An all-American directory. *Class No:* 5(091)(084.122)

Middle Ages

[304]
KREN, C. Medieval science and technology: a selected, annotated bibliography. New York & London, Garland, 1985. xix, 369p. $53. (*Bibliography of the history of science and technology, v.11. Garland reference library of humanities, v. 492.*) ISBN: 0824089693.
1,470 briefly annotated entries in 13 sections. Included: Physical science and natural philosophy - Philosophy, metaphysics, methodological consideration, logic - Mathematics - Medicine and Veterinary medicine -

....*(contd.)*

Psychology - Natural history - Technology - Education: schools and universities - Quasiscience (*e.g.* alchemy). Appended: Reports on manuscripts. Detailed, non-analytical index, p.341-69. *Class No:* 5(091)"01/14"

[305]
Science in the Middle Ages. Lindberg, D.C., *ed.* Chicago, Ill., & London, Univ. of Chicago Press. 1978. 549p. illus., facsims. $40. ISBN: 0226452324.
16 contributors. 15 chapters: 1. Science, technology and progress in the early Middle Ages - 2. The transmission of Greek and Arabic learning to the West ... 5. Mathematics (with 84 references) ... 9. Astronomy ... 12. Medicine ... 13. Natural history - 14. The nature, scope and classification of the sciences - 15. Science and magic (with 52 references). 'Suggestions for further reading' (by chapters), p.507-15. Notes on contributors. A well documented and illustrated textbook by the author of *A catalogue of medieval and Renaissance optical manuscripts* (1975).
Class No: 5(091)"01/14"

[306]
—GRANT, E. Physical science in the Middle Ages. Cambridge, University Press, 1977. xi, [1], 128p. diagrs. £10.50. ISBN: 0521292948.
First published 1971 (Wiley).
In 6 chapters, plus a 'Bibliographical essay' (p.91-116), - a grouped running commentary, emphasizing the topics covered, but including other aspects of medieval science. Analytical index. Supports his earlier *A Source book in mediaeval science* (Cambridge, Mass., Harvard Univ. Press, 1974. xviii, 864p.), in 'Source books in the history of science' series. *Class No:* 5(091)"01/14"

[307]
STILLWELL, M.B. The Awakening interest in science during the first century of printing: an annotated checklist of first editions viewed from the angle of their subject content: astronomy, mathematics, medicine, natural science, physics, technology. New York, Bibliographical Society of America, 1970. xxix, 390p. $25. ISBN: 0914930001.
Pt.1 has lengthy, valuable annotations on 900 numbered items (plus interpolations). Abbreviations of sources cited, p.xvii-xxvi. Pt.2A: Analytical lists (commentators; editors; places of printing; printers and publishers; translators). Pt. 2B: Chronologies (author, by periods; chronological table of first editions). Detailed index. Well produced.
Class No: 5(091)"01/14"

Modern Times

[308]
Companion to the history of modern science. Olby, R.C., *and others, eds.* London, Routledge, 1990. 1081p. £85; $79.95. ISBN: 0415019885.
61 contributors; 67 documented essays present a guide to Western science from 1500 onwards. Part 1 describes methods and problems of research. Part 2 applies those methods to a wide range of fields, from Newton to relativity, and genetic engineering. Subject and name indexes which are described as 'adequate' in *Choice*, October 1990, p.332. Some typographical errors. *Class No:* 5(091)"15/19"

[309]
KNIGHT, D. Sources for the history of science, 1660-1914.
London, Sources of History, Ltd., 1975. 223p.

A bibliographical essay. Chapters: 1. The history of science - 2. Histories of science - 3. Manuscripts - 4. Journals - 5. Scientific books - 6. Non-scientific books - 7. Surviving physical objects. Many footnotes. Non-analytical index. Focuses on materials available in Britain, and particularly helpful on 19th century.
Class No: 5(091)"15/19"

[310]
PLEDGE, H.T. Science since 1500: a short history of mathematics, physics, chemistry, biology. Ministry of Education, Science Museum. 2nd ed. London, HMSO, 1966. 360p. illus., (pl.), graphs, maps.

First published 1939.

2nd ed. adds a prefatory note on the period 1939-59 and provides further dates. A brief but handy survey of modern science, mainly intended as a background study for the manuals on the history of special science subjects published by the Science Museum. Footnotes and a note on further reading. Valuable name and subject indexes. 16p. of plates.
Class No: 5(091)"15/19"

20th Century

[311]
Harrap's illustrated history of the 20th century - Science: invention and discovery in the 20th century. London, Harrap, 1990. 236p. illus., ports. £18.95. ISBN: 0245600248.

Covers 6 periods: 1900/1919 ... 1973/1989. Appended: Biographies; Glossary; Further reading; Index, p.252-6. About 350 illus. (incl. col.). Quarto format. Succinct; clearly presented for the layperson. *Class No:* 5(091)"19"

[312]
Scientific thought, 1900-1960: a selective survey. Harré, R., *ed.* Oxford, Clarendon Press, 1969. viii, 277p. illus., diagrs. £11.95. ISBN: 0198581262.

13 contributors; 12 surveys: 1. Logic - 2. Relativity and cosmology - 3. Matter and radiation - 4. Geophysics - 5. Chemistry - 6. Biochemistry - 7. Molecular biology - 8. Ecological genetics - 9. Hormones and transmitters - 10. Cell biophysics -11. The viruses - 12 Ethnology. Each survey carries appended references. Author and subject indexes. 8 plates. Aims 'to capture something of the main movements of scientific thought since the beginning of this century' (*Preface*). *Class No:* 5(091)"19"

Libraries

[313]
'Library and archive resources in the history of science'. In *British journal for the history of science*, v.6, 1972-3. p.336-7, 459-61; v.7, 1974. p.100-103, 201-2; v.8, 1975. p.94-99; v.10, 1977. p.89-92; v.11, 1978. p.191-5.

1. 'The Turner Collection of the history of mathematics at the University of Keele'. 2. 'The Crawford Collection of books and manuscripts on the history of astronomy, mathematics, etc., at the Royal Observatory, Edinburgh'. 3. The Mathematical Association Library at the University of Leicester'. 4. 'Private manuscript collections in Scotland'. 5. 'History of science in Durham libraries'. 6. 'Manchester'. 7. 'University College, London'.

....(contd.)
The contribution on Durham libraries (*British journal ...*, v.8, 1975, p.94-99) is by A.J. Burnett and D.M. King. Sections: Introduction - University library - Other local libraries - Conclusion. 16 footnotes.
Class No: 5(091):061:026/027

Biographies

[314]
ASIMOV, I. Asimov's biographical encyclopedia of science and technology: the lives of 1510 great scientists from ancient times to the present, chronologically arranged. 2nd rev. ed. Garden City, N.Y., Doubleday, 1982. 941p. illus. $29.95. ISBN: 0385177712.

First published 1967 (revised 1972).

The 2nd ed. adds 310 biographies (half of them contemporaries) to the 1,200 of the 1972 ed. The chronological sequence begins with Imhotep (fl. 2980-2950 BC). Two-thirds of entries cover 19th-20th century scientists. The index, p.901-41, finds a place for names of individuals other than the 1,510 whose biographies are included. No bibliographies. Very readable; for the layman.
Class No: 5(092)

[315]
The Biographical dictionary of scientists. Abbott, D., *ed.* London, Blond Education, 1984-5. 6v. diagrs.

Mathematicians. 1984. [vii], 175p. *Astronomers.* 1984. 204p. *Physicists.* 1984. 212p. *Chemists.* 1984. 203p. *Biologists.* 1984. 194p. *Engineers and inventors.* (Mullen) 1985. 188p.

Each volume provides *c.*200 biographical sketches of scientists, past and present. No recommended further reading, but major primary works are cited. Glossary and index for each volume. No criteria for inclusion are stated. Very readable, like Asimov's *Biographical encyclopedia of science and technology (q.v.)*. Mainly for school use.
Class No: 5(092)

[316]
A Biographical dictionary of scientists: the lives and work of 1100 scientists and technologists, by several contributors. 3rd. ed. London, A. & C. Black; New York, Wiley, 1982. xiv, 674p. ISBN: 0713622288.

First published 1969.

A-Z sequence of entries for deceased scientists, p.1-575; additional biographies, p.577-616; anniversaries, p.617-49. Appended list of scientists referred to but not given entries. Some references to sources. ('Koch': 2 columns, 4 references; 'Zeiss': 1 column, 3 references.) Subject index. 64 contributors. Designed for the general reader as well as for the serious student. *Class No:* 5(092)

[317]
Chambers biographical encyclopedia of scientists. Daintith, J., *and others, eds.* Edinburgh, Chambers, 1983. viii, 599p. $125.

A large format version of *A Biographical encyclopedia of scientists*; edited by J. Mitchell and others (New York, Facts on File, 1981. 2v.).

15 contributors. About 2,000 entries on important scientists - or on people who have made important contributions to science - from the earliest times to the present day. 'Leibig': nearly 2 columns; 'Lysenko': nearly 1 column; 'Hippocrates': 1 column. Notes works *by*, but not *on* the person concerned. Appended chronology (of their

....*(contd.)*

discoveries and publications). Book list, under titles, A-Z, p.581-8. Index of subjects, followed by names of persons. 'As much about science itself as about scientists' (*Preface*). More comprehensive than *Asimov's biographical encyclopedia of scientists* (*qv*), with its 1,500 entries, but at nearly four times the price. *Class No:* 5(092)

[318]
Chambers concise dictionary of scientists. Millar, D., *and others.* Edinburgh, Chambers; Press Syndicate of the University of Cambridge, 1989. 461p. illus., diagrs. £17.95; £9.95. ISBN: 1852963549.

1,000 entries arranged alphabetically by surname, Abbe ... Zworykin. Concisely written entries, which vary in length from a paragraph to several pages, give dates of birth and death, nationality, field of study, significance and a brief biography. Cross-references in bold and italic type. Name and subject index. No bibliographies. Recommended for school libraries or those libraries which need a general overview. *Class No:* 5(092)

[319]
Concise dictionary of scientific biography. New York, Scribner's, 1981. 773p. $95. ISBN: 068416630x.

Based on the parent *Dictionary of scientific biography* (*qv*). Retains entries for the 5,000 scientists, but with a text reduced to about 10% of the original. 'Highly recommended' (*Choice*, January 1982, p.603). *Class No:* 5(092)

[320]
Dictionary of scientific biography. Gillespie, C.C., *ed.* New York, Scribner's, 1970-80. 16v. (8v. reprint, 1981). $1,080, set; $80ea. ISBN: 0684169622, set.

Edited under the auspices of the American Council of Learned Societies.

1-14. A-Z. 1970-76.

15. *Supplement* 1. *Biographies and topical essays.* 1978. xi, [1], 518p.

Supplement II. 1990. 2v.

16. *Index.* 1980. 510p.

Bibliographical entries for *c.*5,000 deceased scientists, from Thales to Bernal. A panel of 39 consultants for the main volumes; several hundred contributors per volume. Contributions focus on the biographee's place in the history of science rather than on his or her life story. All entries are signed and fully documented. 'Bacon, Francis': v.1, p.372-85, with 51 references (bibliography of 2½ columns: 1. Original works; 2. Secondary); 'Wallace, Alfred Russell': v.14. p.133-40; 1½ columns bibliography). V.15 includes 6 topical essays on Indian, Babylonian, Assyrian, Egyptian, early Japanese and Mayan science. *Supplement II* adds more 20th-century figures; v.2 has an index for both volumes. V.16, name and subject index, includes persons not the subject of biographical articles, mustering 45,000 names in all. Separate lists of contributors and biographies, societies, periodicals, scientists (by field). At least comparable, for the historiographer of science, to the *Dictionary of American biography* and *Dictionary of national biography*. The 'single most valuable reference work in the history of science' (Jayawardene, S.A. *Reference books for the historian of science* (1982), p.19). *Class No:* 5(092)

[321]
Eponyms dictionaries index: a reference guide to persons, both real and imaginary, and the terms derived from their names, providing basic biographical identification, and citing dictionaries, encyclopaedias, word books, journal articles, and other sources for additional information. Ruffner, J.H., *and others, eds.* Detroit, Mich., Gale, 1977. Supplements, 1984-. 760p. $135; $94, 1st supp. ISBN: 0810306883; 0810306891, 1st supp.

About 20,000 eponym entries and 13,000 personal names on which they are based, in one sequence, A-Z. Over 500 biographical sources are cited in all. Fields include agriculture, anthropology, botany, chemistry, engineering and industry, fine arts, law, linguistics, mathematics, mythology, politics, religion, weaponry and zoology. 'The definitive source for eponyms' (*Library journal*, 15 April 1978, p.816). Supplements (1984-.) each carry *c.*9,000 new entries. (1st supp. 1984. 248p.). *Class No:* 5(092)

[322]
FELDMAN, A. *and* **FORD, B. Scientists and inventors.** New York, Facts on File, 1986. 352p. illus. (incl. col.), ports, facsims, diagrs.

More than 150 biographies, each with a two-page spread. Chronological order, from Empedocles (*c.*490-430 BC.) to Christian Barnard (1922-). Index of names; detailed subject index. About 3 illus. per page. An attractive quarto pictorial record, for schools and the general public. *Class No:* 5(092)

[323]
McGraw-Hill modern scientists and engineers. 2nd ed. New York, London, etc., McGraw-Hill, 1980. 3v. (1366p.). ISBN: 0070452660.

Revised and updated ed. of *McGraw-Hill modern men of science* (1966-68. 2v.).

1,140 biographies of scientists, both living and deceased (A-G, H-Q, R-Z and index). 'Mead, Margaret': 3½ colums. Include references to the parent work, *McGraw-Hill encyclopedia of science and technology,* to which it functions as a biographical supplement. V.3 includes a field-of-interest index, p.375-85 (Aeronautical engineering ... Zoology). *Class No:* 5(092)

[324]
POGGENDORFF, J.C. Poggendorff's biographisch-literarisches Handwörterbuch zur Geschichte der exakten Wissenschaften. Leipzig, Barth, 1863-1904; Verlag Chemie, 1923-40; Berlin, Akademie-Verlag, 1955-. (Reprint of v.1-6. Ann Arbor, Mich., Edwards, 1945).

Bd.1-2, to 1857; Bd.3, 1858-83; Bd.4, 1883-1904; Bd.5, 1904-22; Bd.6, 1923-31, Bd.7, 1932-.

Interprets the 'exact' sciences in a broad sense, to cover mathematics, astronomy, physics and geophysics, chemistry, crystallography, mineralogy, and the like. Arrangement under periods is A-Z. Author entries comprise short biography and full bibliography embracing contributions to journals. Living scientists included. Bd.7A carries biobibliographies of German, Austrian and Swiss scientists, technologists and medical men for the period 1932-53. Bd.7B, non-German language scientists, 1932-62 (*e.g.* Needham, Othmer, Bertrand Russell). Bd.7D. Teil 8, Lfg.1-3, 1985-87, has reached Tiselius (p.5482). A most valuable source on the lives and writings of scientists of all countries. *Class No:* 5(092)

[325]
ROYAL SOCIETY. Biographical memoirs of Fellows of the Royal Society. London, Royal Society of London, 1955-. v. 1-. ports. (*A new series, continuing Obituary notices of Fellows of the Royal Society*, v. 1-9, 1932-54. 9v.) ISBN: 0854032584.

Before 1860 the obituary notices were published annually in the Society's *Proceedings*; during 1860-99 they appeared there irregularly. *Proceedings*, v.5, 'contains obituaries of deceased Fellows chiefly for the period 1898-1904', with a general index to obituaries in *Proceedings*, v.10-64, 1880-99. From 1900 they appeared in the *Yearbook*, as well as in *Proceedings*. In 1932 they began separate issue as *Obituary notices*.

Lengthy, authoritative signed articles on deceased Fellows, including foreign members. V. 31, 1985 ([v], 666, [1]p.), includes 23 contributions (*e.g.* on P.L. Kapitza, p.325-74, with bibliography, p. 366-74: scientific papers; patents; references to other authors of scientific papers, books and articles about Kapitza). V.33, 1987, ([v], 735p.) carries 25 biographies, including one of Harold Macmillan. *Class No:* 5(092)

[326]
Scienziate e tecnologi contemporairei. Milan, Mondadori, 1968-74. 3v. illus. (incl. col.), ports.

1. *Adams, Roger to Hodgkin, Dorothy.* 1968. 527p.

2. *Hoffman, Samuel K. to Stanley, Wendell M.* 1974. 527p.

3. *Starr, Chauncey to Zworykin, Vladimir K.* 1974. 738p.

The bulk of v.3 is devoted to 'Annali della scienza e della tecnica contemporanee, 1875-1975' (p.209-705; by decades). 'Teilhard de Chardin,Pierre': 2p., with 2 references and cross-references. V.3 also carries an index to authors and their works in all 3 volumes. A well-illustrated set. *Class No:* 5(092)

[327]
The Who's who of Nobel prize winners. Schlessinger, B.S. *and* Schlessinger, J.H., eds. 2nd ed. Phoenix, Ariz., Oryx, 1991. 256p. $39.50. ISBN: 089774599x.

First published 1986.

First edition covered prizewinners from 1901 to 1985. Coverage is now extended to 1990. Entries have been updated, *e.g.* death dates have been added, publications list extended. Indexes of nationality, name, religion, and educational institution. *Class No:* 5(092)

Bibliographies

[328]
BARR, E.S. An Index to biographical fragments in unspecialized specific journals. Alabama, Univ. of Alabama Press, 1973. vii, [1], 294p. ISBN: 0817396039.

Notes on *c.*7,000 individuals, - 'distinguished persons who, prior to about 1920, were active in the sciences' (*Foreword*). Arranged A-Z (d'Abbadie ... Zarhellen, W). Data: name; specialization; dates; country. About 15,000 citations. 8 coded sources (*e.g. Nature, Philosophical magazine, Popular science monthly, Science*). *Class No:* 5(092)(01)

[329]
GASCOIGNE, R.M. A Historical catalogue of scientists and scientific books. New York, Garland, 1984. 1177p. (*Garland Reference Library of the humanities, v.495.*) ISBN: 0824087526.

More than 13,000 authors are identified. 3 main sections (further subdivided): Classical and medieval periods (p.3-68) - Early modern period, c.1450-1700 (p.71-222) - Eighteenth and nineteenth centuries. Main index: individuals; index of selected titles; index of dictionaries and encylcopedias; index of bibliographies and book catalogues. Titles are based on a survey of holdings in Australian libraries, with an English-language slant. Includes most periodical material. 'A must for any history of science collection' (*Reference books bulletin, 1984-1985*, p.132). *Class No:* 5(092)(01)

[330]
IRELAND, N.O. Index to scientists of the world, from ancient to modern times: biographies and portraits. Boston, Mass., Faxon, 1962. 662p. $13. ISBN: 0873050908.

An index to biographies, portraits and chief scientific contributions of over 7,475 scientists, as contained in 338 biographical dictionaries (*e.g. Chambers's dictionary of scientists, Current biography, Biography index*). Entries, A-Z, state name, dates, description, source and page number. 'Ampère': 26 references; 'Galileo': 60 references; 'Réamur': 12 references; 'Cockroft, Sir J.D.': 4 references. Many entries muster 1 or 2 references only. References are asterisked if sources are especially suitable for younger readers. *Class No:* 5(092)(01)

[331]
JAYAWARDENE, S.A. *and* LAWES, J. 'Biographical notices of historians of science: a checklist'. *Annals of science.* v.36, 1979, p.315-94.

'This is a first attempt at consolidating and extending the lists of biographies of historians of science compiled by George Sarton, Aldo Mieli and Francis Russo.' Lists *c.*3,000 biographical notices (800 of them historians of science), drawn from obituaries, bibliographies, articles in biographical dictionaries, monographs, etc. *Class No:* 5(092)(01)

[332]
—Prominent scientists: an index to collective biographies. Pelletier, P.A., *ed.* 2nd ed. New York, Neal-Schuman, 1985 (3rd ed. 1992. $49.95). 380p. $45. ISBN: 0918212782.

'An index to the biographies of scientists that appear in books' (*Introduction*). It lists *c.*10,500 scientists, representing the contents of *c.*170 books. 'Who's who' type books are excluded. Coded list of collective works in English published primarily between 1960 and 1979, following on from N.O. Ireland's *Index* (*qv*), covering 1930-60. 'Faraday, Michael': 15 references. Indexes of scientists by surnames and by fields (Aeronautics ... Zoology). *Class No:* 5(092)(01)

Women

[333]
HERZENBERG, C.L. Women scientists from antiquity to the present: an index, an international reference listing and biographical dictionary. West Cornwall, Conn., Locust Hill, 1986. 200p. $30. ISBN: 0933951019.

Purports to be a name index of more than 2,500 women who have made their mark as zoologists, chemists,

....(contd.)

mechanical engineers, inventors, cartographers, nurses, midwives, sociologists, etc. Draws on 130 biographical sources. Entries are very brief and, at times, risible. 'Cleopatra, cosmetic chemist' raises doubts. 'Many of the sources are notoriously unreliable' (*Choice*, v.23 (11/12), July/August 1986, p.1656). *Class No:* 5(092)-0055.2

[334]
HØYRUP, E. Women of science, technology and medicine: a bibliography. Roskilde, Denmark, Roskilde Univ. Library, [1987]. viii, 132p. (*Skriftserie fra Roskilde Universitetsbibliotek, 15.*) ISBN: 8773490962.

About 1,000 biographies of women scientists, and to a lesser extent women technologists (inventors and engineers), women physicians, and a few nurses. Entries A-Z; data: dates; claim to fame. Appended groups: women in mathematics, science, psychology and medicine, respectively. General reference works, p.1-4.
Class No: 5(092)-0055.2

[335]
OGILVIE, M.B. Women in science: antiquity through the nineteenth century; a biographical dictionary, with annotated bibliography. Cambridge, Mass., & London, MIT Press, 1986. xi, [1], 254p. $12.50. ISBN: 026265038x.

Biographical accounts of more than 150 women scientists, p.22-187. The annotated bibliography has sections A-G (A. Bibliographies and reference works; abstracts, bibliographies, catalogues, guides and biographical dictionaries and encyclopedias, p.189-96 ... G. Nineteenth and early twentieth centuries, p.225-9). List of subjects of biographical accounts, p.240-5. Detailed index, p.246-54. Supplements the major multi-volume biographical dictionaries. *Class No:* 5(092)-0055.2

Manuscripts & Incunabula

[336]
AMERICAN PHILOSOPHICAL SOCIETY. Library. **Catalog of manuscripts in the American Philosophical Society Library,** including the archival shelflist. Westport, Conn., Greenwood Press, [1970]. 10v. ISBN: 0837149754.

Photographed reproduction of the Library's card catalogue of manuscripts. *Class No:* 5(093)

[337]
JAYAWARDENE, S.A. 'Western scientific manuscripts before 1900: a check list of published catalogues'. In *Annals of science,* v.35, 1978, p.148-72.

386 numbered entries in 10 sections: 1. Science (general) ... 7. Medicine - 8. Miscellaneous (technology; geography; metrology) ... Index of names of persons; index of cities. Uses of P.O. Kristeller's *Latin manuscripts before 1600* (New York, 1965) as a guide.

A supplement to Thorndike and Kibre (*qv*).
Class No: 5(093)

[338]
THORNDIKE, L. and KIBRE, P. A Catalogue of incipits of mediaeval scientific writings in Latin. Rev. and augm. ed. London, The Mediaeval Academy of America, 1963. xxiip., 1938 cols. (*Mediaeval Academy of America publication no. 29.*)

First published 1937.

Entries are arranged A-Z under the initial word. In each case, a brief indication of author and title of the work to

....(contd.)

which the incipit belongs, plus a note of 'one or more manuscripts or editors where it may be found, or some secondary work giving this information, or some other sources for the incipit'. Abbreviations of sources, p.ix-xxii. Detailed, analytical index, cols. 1718-1938.
Class No: 5(093)

[339]
—RONAN, C.A. The Shorter 'Science and civilization in China': an abridgement by C.A. Ronan of J. Needham's original text. Cambridge Univ. Press, 1978-. V.1-. illus., diagrs. £15.95, v.1; £19.95, v.2 £42.50; £17.95, v.3. ISBN: 0521292867, v.1; 0521315360, v.2; 0521315603, v.3.

The work so far has reached v.1-3, covering v.1-2,3 and parts 4-1 and 4-3 of the major series. V.3 (1986. [xi], 298p.) includes a table of Chinese dynasties and a bibliography, p.274-82. This abridgement has been prepared with a general non-scientific readership in view. 'This set cannot be praised too highly' (*New technical books*, March 1987, p.324). *Class No:* 5(510)

Europe

[340]
Scientific Europe: research and technology in 20 countries. Maastricht, Formation Science Europe, 1990. 508p. illus. (incl. col.), diagrs., maps. ISBN: 9073035066.

9 parts: 1. Europe on the move - 2. New partnerships ... 8. Machinery of life - 9. Heirs of Galileo. 17 editors; 80 contributors from 20 countries. Fully illustrated and well captioned. No bibliography. Detailed index. *Class No:* 5(4)

Biographies

[341]
Who's who in science in Europe: a biographical guide in science, technology, agriculture and medicine. 7th ed. Harlow, Essex, Longman, 1992. 4v. (2784p) £525. ISBN: 0582086582.

First published 1967; 3rd ed. (Hodgson), 1978; 6th ed., 1989.

About 21,000 biographical entries for scientists from 30 countries (USSR: only 2 names). Data include titles of major publications and research interests. The main sequence, A-G, H-P, Q-Z is supported by a country and subject index, in which the scientists feature under the country in which they work, subdivided into 8 subject specializations.
Class No: 5(4)(092)

Great Britain

Theses

[342]
British reports, translations and theses: a current awareness bibliography. Boston Spa, British Library Document Supply Centre, 1981-. Monthly. £84pa. ISSN: 01447556.

Supersedes *BLLD announcement bulletin.*

About 1,200 items per issue. Includes Canadian and Australian technical reports. Monthly and quarterly keyterm and author indexes on microform. European SIGLE database (1981-), available online through BLAISE-LINE.
Class No: 5(410)(043)

Reports Literature

[343]

British reports, translations and theses: a current awareness bibliography. Boston Spa, British Library Document Supply Centre 1981-. Monthly. £84pa. ISSN: 01447556.

Supersedes *BLLD announcement bulletin*.

Each issue lists over 1,200 British reports and translations produced by British government organizations, industry, universities and learned bodies, plus doctoral theses (1970-), etc. Includes reports and unpublished translations from Eire. Classes 01 Aeronautics ... 22 Space technology. Key term index per issue. Quarterly cumulating microfiche insert, providing author, report number and key term indexes. Includes Canadian and Australian reports. European SIGLE database (1981-), available through BLAISE.

Class No: 5(410)(047)

Biographies

[344]

MacLEOD, R.M. *and* FRIDAY, J.R. **Archives of British men of science.** London, Mansell, 1972. 65 fiches (COSATI 60 frame format) plus printed index and guide. ISBN: 0720102812.

Each microfiche has biographical data on 60 scientists, A-Z by name. An accompanying booklet indexes the complete work. This venture represents 'the first stage of a survey of the biographical and conceptual history of science carried out at Sussex University' (*Library of Congress information bulletin*, v.31 (11), 17 March 1972, p.121-2).

Class No: 5(410)(092)

[345]

Who's who of British scientists, 1980/81. Boff, S., *ed.* 3rd ed. Dorking, Simon Book Directories 1980. xv,[1], 589p. ISBN: 0882290015.

First published by Longman, 1971.

The 1980/81 ed. has entries for *c.*5,500 British scientists working in British universities, polytechnics, research establishments and industry (Lovell, Sir Bernard: 13½ lines). *Class No:* 5(410)(092)

Boyle

[346]

FULTON, J.F. **A Bibliography of the Honourable Robert Boyle,** Fellow of the Royal Society. 2nd ed. Oxford, Clarendon Press, 1961. xxvi, 217, [ii]p. port., facsims.

First published 1932. Originally in Oxford Bibliographical Society *Proceedings and papers* (v.3 (1), 1932, p.1-172; v.3 (3), 1933, p. 339-65.

Sections: A. Separate works (42); chronological order - B. Contributions to other works - C. Collected works - D. Biography and criticism (p. 155-96). Locations in *c.*100 libraries in USA, Britain, Canada, Sweden, etc. 26 facsims. *Class No:* 5(410)(092)BOY

Dalton

[347]

SMYTH, A.L. **John Dalton, 1760-1844:** a bibliography of works by and about him. Manchester, Univ. Press, 1966. xvi, 114p. facsims.

Part 1: Works by Dalton (separately published works; translations and edited texts; scientific manuscripts ...

.... *(contd.)*

lectures) - Part 2: Works (including periodical articles) about Dalton, p.74-104. 771 numbered items, with bibliographical notes. Locations in 14 libraries. *Class No:* 5(410)(092)DAL

Faraday

[348]

JEFFREYS, A.E. **Michael Faraday:** a list of his lectures and published writings. London, Chapman & Hall, on behalf of the Royal Institution of Great Britain, 1960. xxviii, [1], 86p. front., illus. (pl.).

489 numbered items, in one chronological sequence. Covers (i) all books and separate publications, including later editions and reprints; (ii) all articles, papers, etc., in journals, including letters to *The Times*; (iii) all lectures; (iv) all the manuscript lecture notes in the Faraday Collection at the Royal Institution. Brief bibliographical notes. Biographical references, p.xxv-xxvi. Analytical index. *Class No:* 5(410)(092)FAR

Hooke

[349]

KEYNES, G. **A Bibliography of Dr Robert Hooke.** Oxford, Clarendon Press, 1960. xiv, [1], [1], 115p. illus., facsims.

40 sections, - Individual works; Cutlerian lectures; Philosophical collections; Posthumous works; contributions to books; Hooke's Diaries. 25 libraries were consulted. 4 appendices (3. Biography and criticism, p.87-96). *Class No:* 5(410)(092)HOO

Newton

[350]

GJERTSEN, D. **The Newton handbook.** London & New York, Routledge & Kegan Paul, 1986. xiv, 665p. $59.95. ISBN: 0710202792.

A dictionary of topics, personalities and works, Aberration ... Young, Thomas, p.1-613 (over 500 entries). '*Principia*': p.455-502; 'Priority dispute': p.502-7. Works: (Newton's more important contributions; 119 items). Bibliography, p.626-9. Analytical index, p.648-65. 'A quite thorough research' (*Choice*, v.24(9), May 1987, p.1378). *Class No:* 5(410)(092)NEW

[351]

WALLIS, P. *and* WALLIS, R. **Newton and Newtoniana, 1672-1975:** a bibliography. Folkestone, Kent, Dawson, 1977. xxiv, 362p. £40. ISBN: 0712907696.

About 5,000 entries: *c.*600 library locations. 10 sections (1. Collected works, correspondence, bibliography - 2. The *Principia* ... 8. Report on coinage - 9. Works edited by Newton - 10. Biographies and general works, including Newton's minor scientific papers). Includes material on Newton. Indexes of names; anonyma; printers, publishers and booksellers; places of publication. List of authorities, p.xvii. *Class No:* 5(410)(092)NEW

Priestley

[352]
CROOK, R.E. A Bibliography of Joseph Priestley, 1733-1804. London, Library Association, 1966. xiv, 201p.

Based on a thesis for the LA Fellowship in 1965. 7 sections: theological and religious; political and social; educational and psychological; philosophical and metaphysical; historical; scientific; and 'Works about Priestley' (636 items, mostly biographical). The published work adds a section 8: 'Periodical works of Priestley' (items 637-686). A valuable contribution.
Class No: 5(410)(092)PRI

Manuscripts & Incunabula

[353]
ROYAL COMMISSION ON HISTORICAL MANUSCRIPTS. The Manuscript papers of British scientists, 1600-1940. London, HMSO, 1982. 118p. £3.95. (*Guides to sources for British history.*) ISBN: 0114401225.

Inventory of extant manuscript papers of 635 British scientists, including mathematicians, astronomers, physicists, botanists, medical men and engineers. The papers range from personal correspondence and diaries to laboratory notebooks and other working papers. Index of locations.
Class No: 5(410)(093)

Germany

Biographies

[354]
Kürschners deutschen Gelehrten-Kalender. Bio-bibliographisches Verzeichnis deutschsprachiger Wissenschaftler der Gegenwart. Berlin & New York, de Gruyter, 1925-. About 4-yearly. ISBN: 3110104962.

First published 1925.

The 15th ed. (1987. 3v. (5811p): A-H, I-R, S-Z. DM548) has *c.*40,000 *Who's who*-style entries for contemporary German, Austrian and Swiss scholars, with some prominence given to their writings. V. 3 has appended obituaries, anniversary calendar, subject categories of biographees (p.5527-5751), and list of publishers.
Class No: 5(430)(092)

France

Biographies

[355]
ACADÉMIE DES SCIENCES. Paris. Index biographique des membres et correspondents de l'Académie des Sciences, du 22 décembre 1666 au 1 Octobre 1978. Paris, Gauthier-Villars, 1979. 515p.

A complete list of names of all who have appeared on the rolls of the Academy since its inception. Gives dates of birth and death, election to and offices held in the Academy, details of career, and main scientific interests; indicates biographical notices read or deposited and documents, portraits, etc. held at the Academy. *Class No:* 5(44)(092)

Italy

[356]
Scientific books in Italy: subject guide. Milan, Editrice Bibliografica, 1989. lxv,1004p. ISBN: 8870752267.

2 sections: 1. Contributions - 2. Subject guide. Part 1 contains several badly translated articles on research, libraries, publishing, and science in Italy. Part 2 begins with a 42-page list of Italian publishers' addresses and telephone numbers. A list of books arranged alphabetically by subject follows. Author, title, date of publication, pagination, price and publisher are given for each. *Class No:* 5(450)

USSR

[357]
Guide to Russian reference books. Volume 5: Science, technology and medicine. Maichel, K., *and others.* Stanford, Calif., Hoover Institution Press, Stanford Univ., 1967. 387p. (*Hoover Institution, Bibliographical series, 32.*) ISBN: 0817923217.

Nearly 3,000 items in Russian, English, French and German. Systematically arranged, mainly subjects A-Z. General science - General technology - Agriculture - Astronomy - Mathematics - Physics - Medicine (p.256-302: Bibliographies of bibliographies; bibliographies; bibliographies of periodicals; abstracts and indexes; congresses; encyclopedias; dictionaries; biographies; handbooks; legislation). Index, p.303-84. Excellent annotations; introductory notes to sections. *Class No:* 5(47)

Bibliographies

[358]
Bibliografiia izdanii Akademii Nauk SSSR. Ezhegodnik, 1956-. [Bibliography of publications of the Academy of Sciences of the USSR.] Moscow, Izd-vo Akad. Nauk SSSR; Rockville, Md., Victor Kamkin Inc., 1957-. Annual. ISSN: 05685222.

Annual bibliography of publications of the Soviet Academy of Sciences. 'The most useful Russian bibliography of science available' (*Library trends,* v.15(4), April 1967, p.800). *Class No:* 5(47)(01)

Reviews & Abstracts

[359]
USSR report. Foreign Broadcast Information Service. Springfield., Va., National Technical Information Service, 1975-. 12pa. (*Joint Publications Research Service/JPRS.*)

Formerly *USSR and East European scientific abstracts* service.

USSR report: Chemistry. Cybernetics, computers and automation technology. Earth sciences. Engineering and equipment. Life sciences. Biomedical and behavioral sciences. Materials sciences and metallurgy. Space biology and aerospace medicine.

About 100 abstracts per service, monthly, from Soviet sources. 'JPRS publications contain information primarily from foreign newspapers, periodicals and books, but also from news agency transmissions and broadcasts. Materials from foreign-language sources are translated or reprinted with the original phrasing and other characteristics retained' (*Note*). Some of the series are available in microform.

Current JPRS publications are announced in *Government reports announcements* and index (*qv*), issued semi-monthly

....(contd.)
by National Technical Information Service, and are listed in *Monthly catalog of US Government publications*.
Class No: 5(47)(048)

Periodicals

[360]
Scientific and technical serial publications of the Soviet Union, 1945-1960. Zikeev, N.T. Washington, Library of Congress, 1963. [vii], 347p.
5,091 numbered titles, A-Z. Entries state issuing body, title romanized and translated, date first received by the Library, frequency and Library of Congress call number. Cover-to-cover English translations are included. Keyword index to institutes. The 'Guide to subject content' needs to be far more specific. As it is, under 'Medicine' is a sequence of about 500 numbers, with no titles specified or subject breakdown. *Class No:* 5(47)(051)

Biographies

[361]
Biograficheskiĭ slovar' deyateleĭ estestvoznaniya i tekhnicki. Zvorykin, A.A., *ed.* Moscow, Bol'shaya Sovetskaya Entsiklopediya, 1958-59. 2v. (A-Lm, M-Ya).
'Biographical dictionary of scientists in the natural and technical sciences.' Based on material in *Bol'shaya Sovetskaya Entsiklopediya* (2nd ed. 1949-60), revised and expanded. V.2 (p.427-642) includes *c.*4,500 additional biographies, as well as a review and bibliography of Russian and foreign biographical dictionaries. (Based on Maichel, K. *Guide to Russian reference books*, v.5, entry ST47).
Class No: 5(47)(092)

[362]
Lycedi russkoĭ nauki: ocherki o vydayushchikaya deyatelya'kh estestvoznaniya i tekniki. Kuznetsov, I.V., *ed.* Moscow, Gos. Izd-vo Fiziko-Matematicheskoĭ Literatury, 1961-65. 4v. ports.
1. *Matematika, mekhanika, astronomiya, fizika, kimiya.* 1961. 2. *Geologiya, geografiya.* 1962. 3. *Biologiya, meditsina, sel'skokhozyaistvennie. 1963.* 4. *Tekhnika* 1965.
'Brief biographies of prominent workers in the field of natural science and technology'. Covers only deceased scientists. V.1.(599p.) contains 55 signed biographical sketches, 20 of them chemists. The article on S.V. Lebedev (1874-1934), p.582-94 has a bibliography and portrait; that on A.S. Favorskii, p.516-29, has 1 page of bibliography with material on him as well as by him.
Class No: 5(47)(092)

China

[363]
Chinese science abstracts. Beijing, Science Press, 1982-. 12pa. $310, pt.A; $310, pt.B. ISSN: 02545179.
Part A (issued in odd-numbered months) covers mathematics, mechanics, astronomy, physics and technical sciences. *Part B* (issued in even-numbered months) covers chemistry, life sciences and earth sciences.
Part A, v.9, no.6, November 1990 carried 613 abstracts. More than 100 Chinese scientific periodicals are used in CSA. Keyword and author indexes at the end of each issue of both parts. *Class No:* 5(510)

[364]
NEEDHAM, J. *and* **LING, Wang** *and* **GWEI-DJEN, Lu. Science and civilization in China.** Cambridge Univ. Press, 1954-.
1. *Introductory orientations.* 1954. xxxviii, 318p. £55.
2. *History of scientific thought.* 1956. xxiv. 697p. £85.
3. *Mathematics and the sciences of the heavens and the earth.* 1959. xlvii, 877p. £110.
4. *Physics and physical technology.* Pt.1. *Physics.* 1962. xxxiv, 434p. £70. - Pt.2. *Mechanical engineering.* 1965. iv. 759p. £100 - Pt.3. *Civil engineering and nautics.* 1971. vii. 931p. £110.
5. *Chemistry and chemical technology.* Pt.1. *Paper and printing* by T. Tsuen-Hsuin 1980. 512p. £75. - Pts.2-5. *Spagyrical discovery and invention.* 4v. £335. - Pt.7. *Military technology.* 1987. 736p. £80. - Pt.9. *Textile technology: Spinning and reeling,* by Dieter Kunn. 1988. xxxiv, 520p. £75.
6. *Biology and biological technology.* Pt.1. *Botany.* 1986. xxxii, 718p. £80. - Pt.2. *Agriculture.,* by F. Bray. 1984. 760p. £80.
A comprehensive, well-produced history of Chinese science, scientific thought and technology in relation to the history of civilization, and especially comparative development of Asia and Europe. V.1 concerns geographical and historical background, the Chinese language and cultural contacts with the West. Valuable annotated bibliographies of Chinese and Western books and articles are a feature. Thus, the bibliography in v.4, pt.3 occupies p.700-830, with general index, p.831-905. Profusely footnoted. (An interim list of editions of Chinese texts used for v.1-4 is available gratis.) 'A monumental work of great importance to history of science collections and oriental history collections' (*New technical books*, October 1976, entry 1059).
The Shorter 'Science and civilization in China', an abridgement by C.A. Ronan, has so far reached v.1-3 of the major work. Prepared with a general, non-scientific readership in view. 'Cannot be praised too highly' (*New technical books*, March 1987, p.324). *Class No:* 5(510)

Periodicals

[365]
Periodicals current in mainland China held by the Science Reference Library. Kyang, R.K., *comp.* 2nd ed. London, Science Reference Library, 1984. [ii], 91p. £4.50. ISBN: 0712307117.
First published 1982 as *Periodicals from mainland China.*
A/Z subject and title lists of *c.*450 periodicals held, plus a list of periodicals consisting of abstracts of translations. 'Unless otherwise stated, all journals are in Chinese. English summaries or contents lists are separately noted' (*Introduction*). *Class No:* 5(510)(051)

Japan

[366]
Information from Japan. Papers presented at a seminar on Japanese scientific, technical and commercial information. King, S.V., *ed.* London, British Library, Science Reference Library, 1985. [1], [1], [1], 128p. £12. ISBN: 0712307206.
8 papers (*e.g.* 'The scientific and technical information sciences in Japan: recent trends and development'), plus discussion, conclusions and recommendations. General bibliography, p.115-28 (1. Books; 2. Serials from the

....(contd.)

Western world; 3. Japanese serials; 4. Grey literature; 5. Translations and the language barrier; 6. Dictionaries and guides to the Japanese language). *Class No:* 5(52)

Periodicals

[367]

Directory of Japanese scientific periodicals. Tokyo, National Diet Library, 1984. 1861p. ISBN: 4875820860.

Previous ed. 1974 (1000p.).

Classed arrangement of *c*.12,000 titles of current serials in science and technology published in Japan.

Class No: 5(52)(051)

[368]

GIBSON, R.W. *and* **KUNKEL, B.K. Japanese scientific and technical literature:** a subject guide. Westport, Conn., Greenwood Press, 1981. 560p. $125. ISBN: 0313229295.

Part 1 analyses information activities and bibliographical control of scientific and technical literature in Japan. Part 2 is a subject guide (using modified UDC scheme) to 9,116 journals. Data on each item: title and publisher's imprint (romanized Japanese plus English translation), language, and where abstracted/indexed. *Class No:* 5(52)(051)

[369]

Japanese journals in English: scientific, technical and commercial journals held by the British Library. Science Reference Library and/or the British Library Lending Division. Smith, B. *and* King, S.V., *comps.* Boston Spa, BLLD & London, SRL, 1985. [vii], 138p. £12. ISBN: 0712307214.

Contents: 1. Subject keyword list of English-language journals published in Japan, p.1-118 (*c*.1,000 titles) - 2. Subject keyword list of translated journals and journals with translations - 3. Original titles of Japanese journals appearing in (2), above. Covers titles published 1960-84. The main groups of translated journals are: 1. Cover-to-cover translations; 2. Journals containing selected articles. *Class No:* 5(52)(051)

[370]

SMITH, B. The Scientific and technical serial literature of Japan: a study and bibliography. [Thesis for Library Association diploma], 1975. [xi], 7, 812p., 12l.

Pts. 1-2: 10 preliminary chapters. Pt. 3: main A-Z listing of over 2,000 titles, under actual/Englished title, p.93-485, plus keyword-in-context index. Pt. 4: identification of Japanese language serial titles (including glossary of words used). Pt. 3 includes annuals and ceased journals; it states frequency and notes whether periodicals carry review articles and the like; many cross-references.

Class No: 5(52)(051)

India

[371]

JAGGI, O.P. History of science and technology in India. Delhi, Atma Ram and Sons, 1969-80. illus. (pl.), facsims., tables.

1. *Dawn of Indian technology.* 1969. 160p. 2. *Dawn of Indian science.* 1969. 264p. 3. *Folk medicine.* 1973. 276p. 4. *Indian system of medicine.* 1973. 276p. 5. *Yogic and Tantric medicine.* 1973. 190p. 6. *Indian astronomy and mathematics.* 7. *Science and technology in medieval India.* 8. *Medicine in medieval India.* 1977. 9. *Science in modern*

....(contd.)

India. 10. *Technology in modern India.* 11. *Impact of science and technology in modern India.* 12/15. *Western medicine in India.* 1980. 214p.

A detailed survey. V.15, Social impact (1980, 214p.) has 10 documented chapters. 8 appendices. Bibliography, p.204-7. Analytical index. *Class No:* 5(540)

[372]

Science and technology in medieval India: a bibliography of source materials in Sanskrit, Arabic and Persian. Rahman, A., *and others.* New Delhi, Indian National Science Academy, 1982. xxxi, 719p.

13 chapters: 1. Medicine - 2. Astronomy - 3. Mathematics - 4. Alchemy - 5. Physics - 6. Agriculture - 7. Botany - 8. Zoology - 9. Geography - 10. Gemology - 11. Architecture - 12. Encyclopaedia - 13. Dictionary. Chronological arrangement within chapters. Many works are usually allotted annotations. Numerous cross-references. A detailed survey of manuscripts and books. Unfortunately, no index. *Class No:* 5(540)

Reviews & Abstracts

[373]

Indian science abstracts. New Delhi, Indian National Documentation Centre, 1965-. 24pa. $600. ISSN: 00196339.

Each issue contains *c*.1,000 abstracts and references in UDC order (340.6, Forensic science; 51, Mathematics ... 67/68, Manufactures). Author and keyword index per issue and annually. *Class No:* 5(540)(048)

Islamic World

[374]

An Annotated bibliography of Islamic science. Nasr, S.H., *and others.* Tehran, Imperial Iranian Academy of Philosophy, 1975-78 (paperback ed. 1985). 2v. (432p. + xxiii, 317p.).

V.1 has entries for 2,770 items, in 2 parts: 1. General works - 2. Biographical and bibliographical studies of Muslim men of science. V.2 (parts 3-7): 3. Sciences influential in the formation of the Islamic sciences - 4. Translation of scientific texts into Islamic languages - 5. Classification of the sciences, scientific encyclopedias and bio-bibliographies - 6. Cosmology and cosmography - 7. Logic. Both volumes have an index of writers. *Class No:* 5(5.297)

[375]

NASR, S.H. Islamic science: an illustrated study. London, World of Islam Festival Publishing, 1976. 288p. illus. £15.

A well-documented, finely illustrated coverage of 'the entire spectrum of the subject from mathematics, astronomy, physics and medicine to astrology and alchemy' (*British book news*, August 1976, p.584). An excellent introduction for both students and scientists to medieval Islamic science. *Class No:* 5(5.297)

[376]

NASR, S.H. Science and civilisation in Islam. 2nd ed. Cambridge, Islamic Texts Society, 1987. xvi,388p. figs. £14.95. ISBN: 094662111x.

First published by Harvard University Press, 1968.

13 chapters: 1. The universal figures of Islamic science ...

....*(contd.)*

4. Physics ... 7. Medicine, p.184-229 - 8. The sciences of Man ... 13. The gnostic tradition. Selected bibliography, p.355-66. Index. *Class No:* 5(5.297)

[377]

SEZGIN, F. Geschichte des arabischen Schrifttum. Leiden, Brill, 1967-79. 7v.

1. Qur'an-Wissenschafter. 1967. xiii, [1], 936p.

2. Poesie. 1975. cxii, 807p.

3. Medizin. Pharmazie. Zoologie. Tierheilkunde, bis ca. 430H. 1970. 520p.

4. Alchemie-chemie. Botanik-Agrikultur, bis cas. 430H. 1971. 411p.

5. Mathematik, bis ca. 430H. 531p.

6. Astronomie, bis ca. 430H. 1978. 536p.

7. Astrologie, Meteorologie and Verwandtes, bis ca. 430H. 1979. 502p.

A detailed analysis of Arabic writings, chiefly scientific, up to about AD 1052 (430 Hegira). V.7, on astrology, meteorology and kindred subjects, concerns the works of 54 authors. Bibliography, p.417-22. Indexes to authors and titles. A major survey in its field. *Class No:* 5(5.297)

America — North

[378]

American men and women of science. Editions 1-14. Cumulative index. Press, J.C., *comp.* New York, Bowker, 1984. 847p. ISBN: 0835212386.

Forms an index of over 270,000 scientists, living and dead, who appeared in editions 1-14 (1906-79). Editions 10-13 were in 2 sections: *The physical and biological sciences* and *The social and behavorial sciences*. Of the latter section only psychiatry, public health and statistics were retained in edition 14. *Class No:* 5(71+73)

Biographies

[379]

American men and women of science, 1992-93: a biographical directory of today's leaders in physical, biological and related sciences. 18th ed. New York, Bowker, 1992. 8v. (8228p.). $750. ISBN: 0835230740.

First published 1906, as *American men of science.*

Biographical sketches on over 125,000 active US and Canadian scientists working in over 164 sub-specialities. Almost 4,000 new entries. Separate discipline index (A-Z by specialty). Of the 17th ed., 'A truly indispensable directory for reference collections of all levels' (*Science and technology libraries,* v.10(4) Summer 1990, p.123). Available online via Dialog. *Class No:* 5(71+73)(092)

[380]

Who's who in frontiers of science and technology. 2nd ed. Chicago, Ill., Marquis Who's who, 1985. xvi,606p. $84.50. ISBN: 0837957028.

Approx. 14,000 entries for scientists and technologists 'currently working in North America in the frontier areas of their respective specialities' (*Preface*). Index of 420 subspecialties arranged under 23 major fields (*e.g.* biology, mathematics, engineering). Available on *Marquis Who's who* database. *Class No:* 5(71+73)(092)

USA

Bibliographies

[381]

ALURI, R. *and* **ROBINSON J.S. A Guide to US government scientific and technical resources.** Littleton, Colo., Libraries Unlimited, 1983. 259p. tables. ISBN: 0872873773.

13 documented chapters: 1. Introduction - 2. Grants, awards, fellowships and scholarships - 3. Research in progress - 4. Technical reports - 5. Periodicals - 6. Patents - 7. Scientific translations - 8. Standards and specifications - 9. Audio-visual and non-book resources (*e.g.* maps) - 10. Indexes and abstracts - 11. Data bases (p.165-221) - 12. Information analysis centers - 13. Reference sources. Detailed analytical index, p.247-59. 'Its value lies in its integrated presentation of the government's role in information. Recommended especially for library schools' (*RQ,* Fall 1984, p.106). *Class No:* 5(73)(01)

[382]

BATSCHELET, M.W. Early American scientific and technical literature: an annotated bibliography of books, pamphlets, and broadsides. Metuchen, N.J., Scarecrow Press, 1990. xi,136p. $20. ISBN: 0810823187.

Lists scientific material published in the US between 1665 and 1799. Excludes periodicals, almanacs, some government documents and legislative acts. 3 sections: 1. Medical titles - 2. Technical science titles - 3. Natural and physical science titles. 833 entries. Arranged chronologically within each section. Author and subject indexes. 'A useful research tool, though it falls short of being the authoritative reference work on the subject' (*Choice,* v.28(3), November 1990, p.450). *Class No:* 5(73)(01)

Periodicals

[383]

Encyclopedia of associations: Association periodicals. V.2: Science, medicine, and technology. Detroit, Mich., Gale, 1987. 495p. ISBN: 0810340622.

Information on 2,717 US association periodicals arranged under subject categories, A-Z. Data include description of contents, special issues, where indexed/abstracted, circulation, year established, former title(s), frequency, price, ISSN, online or microform availability, whether advertising is accepted ... Title/keyword Association indexes. *Class No:* 5(73)(051)

Histories

[384]

ROTHENBERG, M. The History of science and technology in the United States: a critical and selective bibliography. New York, Garland, 1982. 262p. $56. (*Bibliography of the history of science and technology, 2. Garland reference library of the humanities, 308.*) ISBN: 0824092783.

832 annotated entries in 6 sections, on bibliographies and general studies; the physical sciences; the biological sciences; the social sciences; technology and agriculture. Most of the references concern items published 1940-80. Author and subject indexes. 'An excellent overview of the role science and technology have played in American history' (*Choice,* v.20(8), April 1983, p.114). *Class No:* 5(73)(091)

Biographies

[385]

ELLIOTT, C.A. Biographical dictionary of American science: the seventeenth through the nineteenth centuries. Westport, Conn., Greenwood Press, 1979. xvii, 360, [1]p. tables. ISBN: 0313204195.

About 900 entries, including nearly 600 scientists never included in *American men of science* (11th ed., 1968), to which it serves as a retrospective companion. Most of the scientists died before 1906. Appendices provide data under 5 heads, A-E: year of birth; place of birth; education; occupation; fields of science. 'This tool fills an important gap for all science collections' (*Choice*, v.16(8), October 1979, p.992). *Class No:* 5(73)(092)

[386]

ELLIOTT, C.A. Biographical index to American science: the seventeenth century to 1920. Westport, Conn., Greenwood Press, 1990. 300p. $55. (*Bibliographies and indexes in American history, no. 16.*) ISBN: 0313265666.

Lists *c.*2,800 American scientists who died prior to 1920. Data include occupation, field of study, birth and death dates, and citations to other sources of biographical information. Supplements *Biographical dictionary of American science: the seventeenth through the nineteenth centuries* (Greenwood Press, 1979) (*q.v.*) which provides actual biographies. Covers some less obvious collections of biographies. Useful for research collections in the history of science. *Class No:* 5(73)(092)

[387]

NATIONAL ACADEMY OF SCIENCES OF THE UNITED STATES OF AMERICA. Biographical memoirs. Washington, National Academy Press (Initially, New York, Columbia Univ. Press), 1877-. ports. ISBN: 0309032873.

Contains memoirs of deceased members of the National Academy, each with a portrait and bibliography of his, or her, published scientific contributions. Clearly written statements on each scientist's achievements. Reminiscent of the Royal Society's *Biographical memoirs*, but less formal. 'The goal of the Academy is to have these memoirs serve as a contribution towards the history of American science' (*Preface*). *Class No:* 5(73)(092)

[388]

SIEGEL, P.J. and FINLEY, K.T. Women in the scientific search: an American bio-bibliography, 1724-1979. Metuchen, N.J., & London, Scarecrow, 1985. xvii, 399p. ISBN: 0810817551.

1,517 bio-bibliographies. 1-49: General biographies of women in science. 50-1,517: Specialists, in 29 classes (Archaeologists, anthropologists, ethnologists, and folklorists ... Zoologists). Chronological sub arrangements. Detailed index, p.379-99. *Class No:* 5(73)(092)

Translations

[389]

British reports, translations and theses: a current awareness bibliography. Boston Spa, British Library Document Supply Centre, 1981-. 12pa. £84pa. ISSN: 01447556.

Supersedes *BLLD announcement bulletin*.

Each issue lists over 1,200 British reports and translations produced by British government organizations, industry, universities and learned bodies, plus docotoral theses (1970-), etc. Includes reports and unpublished translations from Eire. Classes 01, Aeronautics ... 22, Space technology.

.... (contd.)

Key term index per issue. Quarterly cumulating microfiche insert, providing author, report number and key term indexes. Includes Canadian and Australian reports. European SIGLE database (1981-), available through BLAISE. *Class No:* 5=03

[390]

CONGRAT-BUTLAR, S. Translation & translators: an international directory and guide. New York, Bowker, 1979. xi, 244p. ISBN: 0835211584.

8 sections. 1. Recent history & breakthroughs - 2. Associations, centers - 3. Awards, fellowships, grants, prizes - 4. Training and access to the profession - 5. Guidelines: codes of practice, model contracts, copyright legislation - 6. Journals, books (p.75-87, with brief annotations) - 7. Register of translators and interpreters - 8. Translators' & interpreters' market place. Detailed index, p.225-41. Large format. An invaluable guide for the would-be translator. *Class No:* 5=03

[391]

Internationale Bibliographie der Übersetzung. [International bibliography of translations.] Hoof, H. van, *[comp]*. Munich, K.G. Saur, 1973. xv, 591p.

Contents: A. Bibliography of translation (p.3-521), including theory of tanslation; teaching in translation; translator's profession; typology of translation; bibliographies - B. Organizations (transalators' associations; study and research centres; terminology control) - C. Publications (journals; publications on mechanical translation; technological publications; bibliographical publications). Indexes of authors and periodicals. *Class No:* 5=03

[392]

Technical translation bulletin. Oxford, Learned Information Ltd, 1961-. 3pa. $39. ISSN: 04970489.

Originally Aslib Engineering Group's *Electrical translators' bulletin* (1955-60).

Articles concern machine translation, role of translators, language teaching and research, the foreign-language barrier, etc. Bibliography section (bilingual, multilingual and monolingual dictionaries); terminology section; book reviews. *Class No:* 5=03

[393]

Transdex index. Ann Arbor, Mich., University Microfilms International, 1975-. 12pa. Annual cummulation on microfiche. $885. ISSN: 00411116.

Replaces *Transindex: bibliography and index to the United States Joint Publications Research Services (JPRS) translations* (v.9-12. 1970/71-74), which continues *Bibliography-index to current US JPRS translations* (v.1-8. 1962-70).

The key to US JPRS translations. 4 divisions: a series and ad hoc title section; a bibliographic section; a keyword index section; and a personal names section. The JPRS series are mainly translations from Communist Bloc countries' publications. *Class No:* 5=03

[394]
TRANSLATORS' GUILD. Index of members of the Translator's Guild and Register of freelance translators. London, the Guild, 1985. vii, [130]p. Loose-leaf. ISSN: 0267663x.

First published 1962.

List of members (LM 1-73, yellow pages) - List of subjects; main subject index (MSI 1-27, green pages) - Less common languages index (LCL 1-7, blue pages) - List of associate members (AML 11, orange pages) - Geographical list of members (GL 1-7, grey pages). *Class No:* 5=03

[395]
The Translator's handbook. Picken, C., *ed.* 2nd ed. London, Aslib, 1989. vi,382p. £30. ISBN: 0851422357.

First published 1983.

20 chapters, each with further reading: 1. Introductory survey ... 5. Checking, revision and editing ... 9. Relations between languages ... 14. Indexes of translations ... 18. Bibliography of technical translation ... 20. Para-translation activities. Index. Authoritative. *Class No:* 5=03

[396]
World translations index. Delft, International Translations Centre, 1987-. 10pa. DFl1,150. ISSN: 02598264.

Continues *Translations register index* and *World transindex.*

Announces translations of literature relating to all fields of science and technology. Source and author indexes per issue and annually. Translations announced from 1977 can be searched on the *World translations index* database on ESA/IRS. *Class No:* 5=03

Writing & Lecturing

[397]
BEACH, D.P. *and* ALVAGER, T.K. Handbook for scientific and technical research. Englewood Cliffs., N.J., Prentice-Hall, 1992. xi,255p. diagrs., tables, graphs. $40. ISBN: 0134310403.

5 units (15 sections, further divided): 1. What is research? - 2. Research planning and design - 2. Methodologies of research design ... 4. Measurement, data analysis, and models - 5. Presenting the results of research. Appendices: A. Grant proposal guidance - B. A sample research report (with bibliography), p.228-46. Index, p.247-55. 'For students in scientific or technical disciplines' (*Preface*). *Class No:* 5:001.81

[398]
BRITISH STANDARDS INSTITUTION. Specification for the presentation of research and documentation reports. London, BSI, 1972. AMDs 1121, 2694 (1978). 20p. £23. (*BS 4811 1972.*)

Sections: Presentation - Format - Production. *Class No:* 5:001.81

[399]
DAY, R.A. How to write and publish a scientific paper. 3rd ed. Cambridge, University Press, 1989. 211p. £22.50; $25. ISBN: 0521365724.

First published 1979.

30 chapters: 1. What is scientific writing? ... 12. How to cite the references ... 16. Where and how to submit the manuscript ... 22. How to write a book review ... 27. Use and misuse of English ... 30. A personalized summary. 6

....(contd.)
appendices (*e.g.* Words and expressions to avoid). Glossary of technical terms, p.193-97. References, p.199-201 (52 listed). Agreeably written. *Class No:* 5:001.81

[400]
EBEL, H.F. The Art of scientific writing, from student reports to professional publications in chemistry and related fields. Weinheim, VCH, 1987. xix, 493p. diagrs., graphs, tables. ISBN: 3527264698.

Part 1. Scientific writing: aims and forms (reports, theses, papers (journal articles), books) - Part 2. Scientific writing: materials, tools and methods. Appendices A-K (A. Oral presentations, organization and vital materials). References and further reading, p.441-8, 450. Index, p.451-93. *Class No:* 5:001.81

[401]
Handbook of technical writing practices. Jordan, S., *and others, eds.* New York, London, etc., Wiley; Interscience 1971. 2v. (1374 + 25p.). illus., diagrs., tables. $160ea. ISBN: 0317108654, v.1; 0317108662, v.2.

32 chapters (6. Technical reports, p.345-423; bibliography of 16 items). V.2 has a lengthy glossary (p.1035-65) and a selected bibliography of 32 items on technical and scientific writing. Each volume carries an index. A comprehensive manual. *Class No:* 5:001.81

[402]
The ISTC handbook of technical writing and publication techniques: a practical guide for managers, engineers, scientists and technical publishers. Austin, M., *ed.* London, Heinemann, 1985. 213p. ISBN: 043490354x.

8 contributors. Chapter 1: 'Choosing language for effective technical writing'. Other chapters on preparation for writing, editing , illustration, translations and printing. British Standard requirements are considered. Index. 'A work which should be available in all scientific and professional libraries' (*Library review*, v.35(2), Summer 1986. p.149-50). *Class No:* 5:001.81

[403]
—ISTC bibliography on technical communication. Kirkman, J. *and* Peake, J., *eds.* Hatfield, Herts., Institute of Scientific and Technical Communications, 1978. 72p.

Lists 1,300 references in 17 sections, on editing, typography, copyright, style, standards (BSI, ISO, etc). Includes information retrieval. *Class No:* 5:001.81

[404]
KIRKMAN, J. Good style: writing for science and technology. London, E. & F.N. Spon, 1992. viii,221p. £10.95. ISBN: 0419171908.

8 chapters: 1. Which style do technical readers prefer? - 2. On style in general ... 7. Writing for international audiences - 8. On avoiding ambiguity. Appendices A-E. 38 references, p.217-8. Index, p.219-21. *Class No:* 5:001.81

[405]
The McGraw-Hill style manual: a concise guide for writers and editors. Longyear, M., *ed.* New York, McGraw-Hill, 1989. 333p. $17.95. ISBN: 0070886846.

Guidelines for creating publishable manuscripts, with hints on grammatical points such as capitalization, hyphenation, split infinitives. Includes chapters on technical writing in mathematics, electronics, computer science, chemistry, and the life sciences. *Class No:* 5:001.81

[406]
O'CONNOR, M. *and* **WOODFORD, F.P. Writing scientific papers in English:** An ELSE-Ciba Foundation guide for authors. Amsterdam, etc. Elsevier, 1975. vii, 108p. ISBN: 0272795151.

'ELSE': European Association of Editors of Biological Periodicals.

Provides a step-by-step approach. 1. Planning - 2. Preparing - 3. Writing the first draft - 4. Revising - 5. Refining - 6. Typing - 7. Submitting the final version - 8. Responding to the editor - 9. Correcting proofs. 5 appendices (*e.g.* 'Expressions to avoid'). Bibliography, p.99-101. Takes 'both British and American English usage into account as well as certain European printing practices that differ from those in the USA' (*Introduction*). Practical; clearly written. *Class No:* 5:001.81

[407]
—KATZ, M.J. Elements of the scientific paper: a step-by-step guide for students and professionals. New York & London, Yale Univ. Press, 1985. xii, [1], 130p. tables. $25. ISBN: 0300035322.

20 short chapters, with examples (*e.g.* 4. Writing with a computer; 8. Organizing the new material; 9. Results; 16. References; 19 Gestation and rewriting; 20. Editors, referees and revisions). Appendices A-I, include 'Books for further reading'. Brief index. *Class No:* 5:001.81

[408]
WILKINSON, A.M. The Scientist's handbook for writing papers and dissertations. Englewood Cliffs, N.J., Prentice Hall, 1991. 522p. illus., tables. $45. ISBN: 0139694110.

Gives writing examples and techniques for the physical, biological, and social sciences. Describes common errors; includes writing abstracts, and the printing process. 'Thoroughly recommended for inclusion in the science reference collection of all libraries' (*New technical books,* May/June 1991, item 619). *Class No:* 5:001.81

Thesauri

[409]
Cambridge illustrated thesaurus of science & technology. Godman, A. *and* Denney, R. Cambridge, University Press, 1985. 256p. col. illus., tables. £7.95. ISBN: 032126362x.

One of a series of thesauruses, on physics, chemistry, computer science, etc. This thesaurus concentrates on physical sciences and technology, and biological sciences. 'Physical sciences and technology' (p.8-137) has 4 sections and 22 subsections, within each of which are grouped definitions of terms related to the same topic. About 2,000 definitions in all, with cross-references and an index. A feature is the plentiful use of coloured diagrams. Small format. Might prove helpful to 'O' and 'A' level students. *Class No:* 5:025.43

[410]
Thesaurus of scientific, technical, and engineering terms. Philadelphia, Hemisphere, 1988. 2v.(in one) (841+376p.). $125. ISBN: 0891167943.

V.1. *Hierarchical listings.* V.2. *Access vocabulary.*

V.1 contains *c.*20,000 main terms. Includes all terms appearing in the NASA thesaurus and its supplements published through 1986. 3 column text; small type. Aims to deal with ambiguities in technology 'by emphasizing

.... *(contd.)*
uniqueness and developing single concepts and then grouping these into related areas' (*Foreword*). *Class No:* 5:025.43

Information Services

[411]
European sources of scientific and technical information. 9th ed. Harlow, Essex, Longman, 1991. vi,402p. £140. (*Reference on research.*) ISBN: 058207150x.

First published 1957.

A guide to more than 1,300 organizations, including patent and standard offices, national and special bodies and their library facilities. 25 subject sections, subdivided under countries. Subject index; organization and keyword index. Includes Eastern Europe but excludes the former USSR. 10th ed. due 1992/93 (£165). *Class No:* 5:061:025.5

Libraries

Worldwide

[412]
KLESHCHUK, S.E. *and* KORENEVA, S.B. **Directory of libraries and information centers of the Academies of Sciences of Socialist countries.** Moscow, Nauka Publishers, 1986. 120p.

At head of title-page: 'Library of the USSR Academy of Sciences'. Russian ed., 1984.

Directory of 37 libraries: Bulgaria (1); Hungary (1); Vietnam (1); German Democratic Republic (2); Mongolia (1); Poland (6); Romania (1); USSR (20); Czechoslovakia (2); Yugoslavia (2). Data on each library under 15 heads, *e.g.* date of establishment; present status and functions; stock; catalogues; reader services; library publications; activities/programmes. Small format.
Class No: 5:061:026/027(100)

[413]
Libraries, information centres and databases in science and technology: a world guide. Lengenfelder, H., *ed.* 2nd ed. Munich, K.G. Saur, 1988. [xxi], 665p.

First published 1984 as a supplement to *World guide to special libraries* (1983), also edited by H. Lengenfelder.

Contains 13,687 brief entries, arranged A-Z by country and then alphabetically by body. Data include address, subject coverage, holdings, publications, databases. Data bank, organization, and subject indexes. Not comprehensive, but would be useful in a reference library.
Class No: 5:061:026/027(100)

[414]
—World guide to technical information and documentation services. [Guide mondial des centres de documentation et d'information scientifiques.] 2nd ed. Paris, UNESCO, 1975. 514p.

First published 1969. Companion to *World guide to scientific information and documentation services.*

Lists 476 centres (1st ed.: 273) in 93 countries. International centres, p.29-49; national (Algeria ... Zambia), p.53-473. India receives 40 entries, whereas USSR, UK and US muster only 11 among them. Appended: list of international, regional and national directories to technical information and documentation services; alphabetical list of institutions; subject index. *Class No:* 5:061:026/027(100)

Europe

[415]
LEWANSKI, R.C. Subject collections in European libraries. 2nd ed. New York & London, Bowker, 1978. xiii, 495p. ISBN: 0859350118.

First published 1965.

Arranged by subject, based on the 18th ed. of Dewey Decimal Cl. Usual directory information on type of collection, availability, facilities. Subject index, p.483-95. Derived from a mailing of 10,000 libraries, of which c.25% responded, with best results from Great Britain, Scandinavia, Low Countries, and Germany, (1965 ed. had data on c.6,000 libraries). Intended as a companion to L. Ash's *Subject collections* (5th ed. 1978), covering USA and Canada. *Class No:* 5:061:026/027(4)

Great Britain

[416]
The Aslib directory of information sources in the United Kingdom. Codlin, E.M. *and* Reynard, K.W., *eds.* 7th ed. London, Aslib, 1992. 2v. £250. ISBN: 0851422926.

First published 1928.

V.1 (vii,1071p.) has 6,869 entries, including cross-references. Data include: Enquiries to ...; Subject coverage; Special collections; Publications. Featured in this ed. is 'Information sources on the EC Single Market' p.1045-71. V.2 (iii,284p.) contains the subject index (p.1-209) and a 72-page list of organizations' abbreviations and acronyms. *Class No:* 5:061:026/027(410)

[417]
Guide to libraries and information units in government departments and other organisations. Dale, P., *ed.* 30th ed. London, British Library, Science Reference and Information Service, 1992. Annual. £30. ISBN: 0712307877. ISSN: 90525416.

Formerly *Guide to government departments and other libraries* (1948-88).

About 500 entries arranged A-Z. Data for each library include address, telephone/fax number, staff, stock and subject coverage, hours of opening, services, publications. Subject and organization indexes. 2 appendices: 1. Executive agencies - 2. Other sources of information. Formerly updated every 2 years but will now be updated annually. *Class No:* 5:061:026/027(410)

Ireland

[418]
Directory of libraries and information centres in Ireland. 3rd ed. Dublin & Belfast, Library Association of Ireland and Library Association, Northern Ireland Branch, 1990. 277p.

Supersedes *Directory of Northern Ireland libraries* (LA, N. Ireland Branch, 2nd ed., 1977) and Directory of libraries in Ireland (2nd ed., 1988).

c.200 entries. Indexes to libraries and special collections and subjects. Similar format to LA/RSIS *Regional guides to library resources. Class No:* 5:061:026/027(415)

Federal Republic of Germany

[419]
GERMANY. FEDERAL REPUBLIC. ARBEITSGEMEINSCHAFT DER SPEZIALBIBLIOTHEKEN. Verzeichnis der Spezialbibliotheken in der Bundesrepublik Deutschland, einschliesslich West-Berlin. 2. erw. Aufl., neu bearb. von F. Meyen. Braunschweig, Vieweg, 1970. [viii], 207p.

First published 1965.

Directory of c.500 special libraries in the Federal Republic of Germany, and West-Berlin. Dewey Classification grouping. Data: name, address, name of librarian, stock, facilities, co-operation. Many cross-references. Subject, locality and name indexes. Needs to be updated. *Class No:* 5:061:026/027(430.1)

France

[420]
CHAULEUR, A. Bibliothèques et archives: comment se documenter? Guide pratique à l'usage pratique à des étudiants, des professeurs, des documentalistes et archivistes, des chercheurs... 2. éd. Paris, Economica, 1980. 334p.

First published 1978.

Separate parts for libraries and archives, each part with addresses. 'Introduction bibliographique' precedes. Extensive subject index; index of names, book titles and places. Selective listing, with a bias towards the humanities. *Class No:* 5:061:026/027(44)

Italy

[421]
ASSOCIAZIONE ITALIANA BIBLIOTECHE. Guida delle biblioteche scientifiche e tecniche e dei Centri di documentazione italiani. Pavesi, R.P. *and* Salimei, M., *eds.* Rome, Consiglio Nazionale delle Ricerche, 1965. viii, 610p.

Provisional eds., 1955-1961.

Approx. 1,600 returns. *Class No:* 5:061:026/027(450)

Japan

[422]
Directory of information sources in Japan, 1980. Tokyo, Nichigai Association, 1979. 300p.

Guide to 1,468 libraries and services of all kinds, excluding many smaller public libraries, but including non-library information sources. In Japanese and English. Index of libraries in English and Japanese. Subject index. *Class No:* 5:061:026/027(52)

Egypt

[423]
ARAB REPUBLIC OF EGYPT. NATIONAL INFORMATION AND DOCUMENTATION CENTRE. Directory of scientific and technical libraries in the United Arab Republic. Cairo, the Centre, 1970. 242p.

Data on 208 libraries in Egypt.
Class No: 5:061:026/027(620)

America—North & Central

[424]

Directory of special libraries and information centers. DeMaggio, J., *ed.* 16th ed. Detroit, Mich., Gale, 1992-93. 3v. ISSN: 0731633x.

First published 1963.

1. *Directory.*, In 2 pts. $399. 2. *Geographic and personnel indexes.* 1037p. $290. 3. *New special libraries* (published between editions of v.1). $345.

V.1 has over 20.000 entries (A-Z, by name) for special libraries in US, Canada, plus more than 370 special libraries in 80 other countries. Subject index (3,500 subjects). Appendices (*eg*, Networks and consortia). V.2 consists of two separate indexes - geographic and personnel. V.3 acts as an interim supplement between editions. *Class No:* 5:061:026/027(71/73)

[425]

—ASH, L. Subject collections: a guide to special book collections and subject emphases as reported by university, college, public and special librarians and museums in the United States and Canada. 6th ed. New York, Bowker, 1985. 2v. (2144p.). $175. ISBN: 0835219178.

First published 1958; 5th ed. (xi, 1184p.), 1978.

A directory of 20,000 collections, adding *c*.5,000 to entries in the 5th ed. Arranged under Library of Congress subject-headings, A-Z subdivided by US states and Canadian provinces. Less easy to use than *Subject directory of special libraries and information centers*, V.3 & 5, supported as they are by v.2 (Geographic and personnel indexes) of the parent *Directory of special libraries and information centers* (*qv*).

7th ed., revised by W.G. Miller, (Bowker, 1992. 2v. 2500p. $275; £145). *Class No:* 5:061:026/027(71/73)

Canada

[426]

CANADA. National Research Council for Scientific and Technical Information. **Science and technology collections in Canadian government libraries:** a guide. Parkkari, J., *ed.* Ottawa, the Council, 1977. vp.

6 sections: agriculture; defence; energy; environment; health; science and technology. For each collection, data on scope, access and availability of accessions list. In English and French. *Class No:* 5:061:026/027(71)

USA

[427]

A Directory of information resources in the United States: physical sciences, engineering. Washington, U.S. Government Printing Office, 1971. vi, 803p.

1965 ed. as *A directory... : physical sciences, biological sciences, engineering.*

2,891 entries, for a wide range of organizations, - libraries, information centres, professional societies, universities and industrial firms willing to extend their information services beyond their own organization, as well as Federal, state and central government offices. Subject index. *Class No:* 5:061:026/027(73)

[428]

—Sci-tech libraries in museums and aquariums. New York, Columbia Univ., School of Library Service; London, Haworth Press, 1986. xiv,204p. ISBN: 0866564845.

Has also been published as *Science & technology libraries,* v.6(1/2), Fall 1985/Winter 1985/86.

Highlights 'some outstanding sci-tech libraries serving museums and aquariums around the country'. Data on each include history and functions, collections, services to users, catalogue and classification, special features, policies and programmes. *Class No:* 5:061:026/027(73)

Australia

[429]

Directory of special libraries in Australia. Cree, J., *ed.* 7th ed. Ultimo, New South Wales, Library Association of Australia, Special Libraries Section, 1988. 552p. ISBN: 0868040592.

First published 1954.

Lists *c*.1,000 special libraries, arranged by area. Index of names, subjects. *Class No:* 5:061:026/027(94)

[430]

—AUSTRALIA. SCIENTIFIC AND TECHNOLOGICAL INFORMATION SERVICES ENQUIRY COMMITTEE. The STISEC Report to the Council of the National Library of Australia. Canberra, National Library of Australia, 1973-. V.1-.

1. *Scientific and technological information services in Australia.* 1973. 2. *Procedures, evidence examined, findings and appendices.* 1975.

V.2 (376p.) includes summaries of the library survey of 1971. *Class No:* 5:061:026/027(94)

Institutions & Associations

Worldwide

[431]

Scientific and technical organizations and agencies directory: a guide to approximately 15,000 organizations and agencies providing information in the physical sciences, engineering and technology. Young, M.L., *ed.* 2nd ed. Detroit, Mich., Gale, 1987. 2v. (1670p) $195. ISBN: 0810321033.

First published 1985.

19 chapters on US, Canadian, and international scientific and technical organizations and agencies. Each chapter deals with a particular type of organization. Master name and keyword index. *Class No:* 5:061:061.2(100)

[432]

World guide to scientific associations and learned societies. [Internationales Verzeichnis wissenschaftlicher Verbände und Gesellschaften.] Sachs, M., *ed.* 5th ed. Munich, K.G. Saur, 1990. xvi,672p. $275. (*Handbook of international documentation and information, v.13.*) ISBN: 3598205309.

First published 1974, as *World guide to scientific organizations.*

A guide to more than 17,000 associations and societies. Contains information on national and regional associations from 132 countries, in addition to international organizations. Data include name/abbreviation; address; year founded; membership; chief officials. Coverage: science, technology, humanities and social sciences. Subject index

....(contd.)

with 177 headings. An alphabetical index to associations names (including abbreviations) is included for the first time. Other indexes: German-English concordance to areas of specialization; subject index; publications index. Some overlap with *World of learning*. *Class No:* 5:061:061.2(100)

Great Britain

[433]

Centres & bureaux: a directory of concentrations of effort, information and expertise. Sellar, L., *ed.* Beckenham, Kent, CBD Research, Ltd., 1987. ix, 214p. £42. (*CBD Research publication.*) ISBN: 0900246472.

Nearly 1,000 entries for specialist bodies - self-styled 'centre', 'bureau' and otherwise - in the UK and Ireland (*e.g.* Centre for Medical Research, Waste Management Information Bureau, CAB International). A-Z directory and addenda, p.1-172. Index of sponsors and participating oranizations. Subject index (Abortion ... Youth studies), p.195-214. Fills a gap in the CBD series.
Class No: 5:061:061.2(410)

Federal Republic of Germany

[434]

Directory of international cooperation in science and technology: index of institutions in the Federal Republic of Germany cooperating with developing countries. Gottstein, K., *ed.* 2nd enl. and rev. ed. Munich, K.G. Saur, 1989. 256p. £55. ISBN: 3598106262.

595 institutions involved in transfer of information in science, technology or the arts. Partner countries/institutions are listed separately, after the description of activities for each institution. *Class No:* 5:061:061.2(430.1)

China

[435]

Directory of selected scientific institutions in mainland China. Prepared by Surveys and Research Corpn. Stanford, Calif., Hoover Institution Press, for the National Science Foundation, 1976. xxii, 469p.

The directory, p.1-374 (1. Chinese Academy of Sciences ... 490. Yünnan University). 3 appendices include 'A supplementary list of 1,127 additional research and development institutions', p.381-424. 7 indexes, - of institutions (by English translation; by Wade-Giles and P'in Yin romanizations), scientists and other key personnel, fields and sub-fields, major publications, and geographic (province or city). *Class No:* 5:061:061.2(510)

South Africa

[436]

SOUTH AFRICAN COUNCIL FOR SCIENTIFIC AND INDUSTRIAL RESEARCH. Scientific and technical societies in South Africa. Pretoria, the Council, 1967-. Annual.

Includes data on aims, membership requirements, meetings, branches and publications.
Class No: 5:061:061.2(680)

America — North

[437]

Research centers directory: a guide to approximately 12,000 university-related and other non-profit research organizations established on a permanent basis and carrying on continuing research programs. Piccirelli, A., *ed.* 17th ed. Detroit, Mich., Gale, 1992. 2v. $400.

First published 1960.

Entries for over 12,300 research units in the US and Canada are arranged in 17 subject sections under 5 broad headings (Life sciences; Physical sciences and engineering; Private and public policy and affairs; Social and cultural studies; Multidisciplinary and coordinating centres). 5 indexes: detailed subject, alphabetical, acronyms, institutional, and special capabilities (*e.g.* facilities, collections, databases). Available online via DIALOG File 115. *Class No:* 5:061:061.2(71+73)

USA

[438]

BATES, R.S. Scientific societies in the United States. 3rd ed. Cambridge, Mass., M.I.T. Press; Oxford, etc., Pergamon Press, 1965. [xv], 326p.

First published 1945; 2nd ed. 1958.

'The history and influence of the scientific societies of the United States' (*Preface*, 1st ed.). Historical survey in 7 chapters: 1. Scientific societies in eighteenth century America ... 6. The atomic age - 7. Scientific societies in the space age. Chronology; bibliography, p.245-93 (systematically arranged, including 'Special references to American scientific agencies other than scientific societies'). Index, p.295-326. *Class No:* 5:061:061.2(73)

Brazil

[439]

Guia des sociedades e associacões científicas e technológicas do Brasil. Brasília, Instituto Brasileiro de Informacão en Ciencia e Tecnologia, 1984. 293p.

Directory of nearly 230 organizations, plus *c.*40 universities. Subject groups: General - Geology - Biological science - Engineering - Medicine - Agriculture - Applied social sciences - Humanities - Linguistics - Literature & arts. Index of societies and associations.
Class No: 5:061:061.2(81)

Australia

[440]

Australian scientific societies and professional associations. Crump, I.A., *ed.* 2nd ed. East Melbourne, Commonwealth Scientific and Industrial Research Organization (CSIRO), 1978. iv, 226p. ISBN: 0643002820.

First published 1971.

390 numbered entries, plus 400 'other locations' (names only), covering all fields of science and technology, including agriculture and medicine. Data on each note history, purpose, field of interest, activities, publications, library facilities, awards and membership. Indexes for society names, publications, awards, and subjects (Aborigines, Australian ... Zoology).
Class No: 5:061:061.2(94)

Conferences

[441]
Forthcoming international scientific and technical conferences. London, Aslib, 1924-. 4pa. (Main list is in February, updated by supplement in May and cumulative supplements in August and November). £75 (£60, to members). ISSN: 00464686.

Records more than 1,000 conferences pa., arranged in chronological order. Data on each: date; title of conference; location; address for enquiries. Subject, location and organization indexes. *Class No:* 5:061:061.3

[442]
Mind: the meetings index: Series SEMT, science, engineering, medicine, technology. Harrison, N.Y., InterDok, 1984-. 6pa. $425. ISSN: 07395914.

Lists forthcoming SEMT meetings, under keywords. Supporting indexes of sponsors, locations and dates. Details of contacts. *Interdok (qv)* lists the resultant published proceedings. *Class No:* 5:061:061.3

[443]
Scientific meetings; describing future meetings of technical, scientific, medical, and management organizations and universities. San Diego, Calif., Scientific Meetings Publications, 1957-. 4pa. $75. ISSN: 04878965.

Meetings A-Z, with chronological and keyword indexes. Details of contacts. Pronounced US slant. Available on microfilm from University Microfilms International. *Class No:* 5:061:061.3

[444]
TORRENCE, S.R. How to run scientific and technical meetings. New York, Van Nostrand Reinhold, 1991. xix,887p. figs., tables. $39.95. ISBN: 0442236751.

19 documented chapters: 1. Defining and conceptualizing the technical meeting - 2. Organization ... 5. Selecting your meeting site ... 7. Liability, insurance, and the law ... 17. Enhancing your meeting with exhibitions... 19. International meetings. 45 appendices (*e.g.* 14. Sample contract with a hotel; 24. Creating good visuals). Glossary, p.835-54. Index. US-slant, but for anyone involved in organizing a technical meeting. *Class No:* 5:061:061.3

[445]
World meetings: outside United States and Canada. New York, Macmillan Publishing, 1968-. 4pa. $175. ISSN: 00438677.

A two-year-span register of future medical, scientific and technical meetings. Data include information on availability of abstracts or papers. Indexes: by date of meeting, keyword, location, deadline for abstracts or papers and sponsors. *Class No:* 5:061:061.3

[446]
World meetings: United States and Canada. New York, Macmillan Publishing, 1963-. 4pa. $175.

Like its companion register, *World meetings: outside United States and Canada*, provides a two-year-span record of future medical, scientific and technical meetings. Similar data. *Class No:* 5:061:061.3

Bibliographies

[447]
Bibliographic guide to conference publications, 1975-. Boston, Mass., G.K. Hall, 1976-. Annual. $280. ISSN: 03602729.

Annual dictionary catalogue compiled by the Research Libraries of the New York Public Library, plus Library of Congress MARC tapes additional entries. Indexes *c.*26,000 private and government conference publications. *Class No:* 5:061:061.3(01)

[448]
CAMBRIDGE UNIVERSITY LIBRARY. Union catalogue of scientific libraries in the University of Cambridge: Scientific proceedings, 1644-1974. London, Mansell, 1975. 2v. (1232p).

About 25,000 entries for the proceedings of some 6,000 conferences, symposia and lecture meetings. Arranged by name and keyword. *Class No:* 5:061:061.3(01)

[449]
Conference papers index. Bethesda, Md., Cambridge Scientific Abstracts, 1978-. v.6-. 6pa. $975. ISSN: 0162704x.

Formerly *Current programs* (v.1-5, 1973-77).

Lists authors and titles of papers given at scientific conferences. Available online via DIALOG and ESA. *Class No:* 5:061:061.3(01)

[450]
Directory of published proceedings. Harrison, N.Y., InterDok Corpn., 1966-.

Series SEMT - *Science/Engineering/Medicine/Technology*. 1967-. 10pa. $495.

Series SSH - *Social sciences/Humanities*. 1968-. 4pa. $325.

Series PCE - *Pollution Control/Ecology*. 1974-. 2pa. Annual. $150.

The directory records preprints and published proceedings of congresses, conferences, symposia, meetings seminars and summer schools that have been held from 1964 to date.

Series SEMT (September-June) - like the other two series - lists entries chronologically by date of conference and of proceedings, with publishers' and distributors' addresses. Editor, location and subject/sponsor indexes. The annual cumulative volume includes cumulating indexes. *Class No:* 5:061:061.3(01)

[451]
Index to scientific & technical proceedings. Philadelphia, Pa., Institute for Scientific Information, 1978-. 4pa., cumulated 2pa. $1325. ISSN: 01498088.

Indexes over 120,000 conference papers pa. Each issue contains category index (Accoustics ... Zoology); contents of proceedings; author and editor index; sponsor index; meetings location index; permuterm subject index; corporate index (geographic section; organization section). Online via DIALOG. *Class No:* 5:061:061.3(01)

[452]
Proceedings in print. Arlington, Mass., Proceedings in Print, Inc., 1964-. 6pa. $610. ISSN: 00329568.

About 2,500 entries pa. under title of conference, A-Z. Author and subject index. The December issue includes cumulative annual index. *Class No:* 5:061:061.3(01)

[453]
Scientific, engineering and medical societies publications in print. New York, Bowker, 1974-. Two-yearly. $65. ISSN: 02777355.

An in-print service listing societies' available publications. *Class No:* 5:061:061.3(01)

Research Establishments

Worldwide

[454]
International research centers directory. Detroit, Mich., Gale, 1982-. Biennial. ISSN: 02782731.

Subtitle: *A world guide to government, university, independent non-profit and commercial research and development centers, institutes, laboratories, bureaus, test facilities, experiment stations, research parks and data collection and analysis centers, as well as foundations, councils and other organizations which support research.*

5th ed. 1990-91 (1990. Smith, D.L., *ed.* v,1327p.). A guide to *c.*6,500 research facilities in some 145 countries. Entries are arranged by country, with an international section. Name and keyword index; country index; subject index. *Class No:* 5:061:061.62(100)

Europe

[455]
European research centres: a directory of scientific, technological, agricultural and biomedical laboratories. 8th ed. Harlow, Essex, Longman, 1990. 2v. (1751p.). £325. ISBN: 0582061245.

7th ed. 1988.

Directory of *c.*12,000 research laboratories in 31 European countries. Omits former USSR. Data include organization title, address, director's name, graduate research staff, annual expenditure and lists of organizational activities and publications. Major laboratories are highlighted with a star before the title. Subject and establishment indexes. *Class No:* 5:061:061.62(4)

Great Britain

[456]
Industrial research in the United Kingdom: a guide to organizations and programmes. 14th ed. Harlow, Essex, Longman, 1991. 592p. £190. ISBN: 0582082730.

First published 1946 as *Industrial research in Britain* (Guernsey, Hodgson). 13th ed., 1989.

Details of *c.*3,000 laboratories active in agricultural and environmental sciences, chemical and materials sciences, earth and astronomical sciences, electronics and computer sciences, pharmaceutical, biomedical and biological sciences, and engineering and transportation. 6 sections: 1. Industrial firms - 2. Research associations and consultancies - 3. Government departments and their laboratories - 4. Universities and polytechnics - 5. Trade and development associations ... Personal name index. Title of establishment index. Detailed, non-analytical index. *Class No:* 5:061:061.62(410)

USSR

[457]
UNITED STATES. Central Intelligence Agency. **Directory of Soviet research organizations:** a reference aid. Washington, 1978. (Xerox copy was available from Library of Congress). 290p. $81. ISBN: 0835729141.

Part A: A-Z list of 'over 1,700 selected major Soviet educational and scientific research organizations and is based on information received as of 1 January 1978' (*Preface*). Data: name, city, parent body, names of directors and deputy directors, abbreviation/acronym, street address. Part B: Keyword list. Part C: A-Z list of organization directors and deputy directors. *Class No:* 5:061:061.62(47)

Asia—Far East

[458]
Pacific research centres: a directory of scientific, industrial, agricultural, and biomedical laboratories. 3rd ed. Harlow, Essex, Longman, 1990. vi,475p. £198. ISBN: 0582052874.

First published 1986.

Arranged alphabetically by country and including details of *c.*3,000 organizations. 300 new entries since previous edition (1988). Countries covered include Australia, People's Republic of China, Indonesia, Japan, Malaysia, New Zealand, Philippines, Singapore, Taiwan, Thailand, Vietnam. Data note size of organization, key personnel and main research activities. Titles of establishments and subject indexes. *Class No:* 5:061:061.62(51/52+57)

China

[459]
WANNACK-NUNN, S. Directory of scientific research institutes in the People's Republic of China. Washington, National Council for US - China Trade, 1977-78. 3v.

1. Agriculture, fisheries, forestry 2. Chemicals, construction 3. (2pts). Electrical and electronics; light industry; machinery, including metals and mining; transportation.

Entries under subject areas. Does not claim to be definitive and omits universities.
Class No: 5:061:061.62(510)

Israel

[460]
ISRAEL. National Centre of Scientific and Technical Information. **Directory of research institutes and industrial laboratories in Israel.** 3rd ed. Tel Aviv, COSTI, 1980. 230p.

Data on 175 institutions. In Hebrew and English.
Class No: 5:061:061.62(569.4)

USA

[461]
Government research directory: a guide to approximately 3,000 US government research and development centers. Piccirelli, A., *ed.* 6th ed. Detroit, Mich., Gale, 1990. 1100p. $390. ISBN: 0810369222.

First published 1980.

Arranged under sponsoring government agencies. Data on each centre include description of research, special facilities,

....*(contd.)*

publications and information services. Indexes: master name, keyword and agency; geographic; subject. *Class No:* 5:061:061.62(73)

Research Projects

[462]
Current research in Britain. 7th ed. Harlow, Essex, Longman, 1992. Annual. 4v. in 6. £320, set. ISBN: 058220920x, set.

Previously as *Research in British universities, polytechnics and colleges* (RBUPC).

1. *Physical sciences.* 1992. 2v. £125.
2. *Biological sciences.* 1992. 2v. £125.
3. *Social sciences.* 1992. £125.
4. *The humanities.* 5th ed. 1992. Two-yearly. £95.

Register of current research in all British universities and polytechnics, many colleges and other academic institutions, and government departments and other institutions. Pt. 2 consists of name index, study area index, and keyword index. A basic research tool.

Physical sciences covers mathematics, astronomy, physics, chemistry and technology (*e.g.* engineering, accountancy). Details of *c.*65,000 projects.
Class No: 5:061:061.62.005

[463]
SMITHSONIAN SCIENCE INFORMATION EXCHANGE, INC. Information services on research in progress: a world inventory. 2nd ed. Paris, Unesco, 1982. 330p. illus. $20. ISBN: 9231020714.

First published 1978.

Profiles of 230 individual information systems, with international, regional or national coverage. Five indexes, - of subjects, countries, organizations, information systems, and individual names. *Class No:* 5:061:061.62.005

Museums

[464]
DANILOV, V.J. America's science museums. Westport, Conn., Greenwood Press, 1990. 496p. $65. ISBN: 0313258651.

Contains entries for 587 museums, covering areas such as space and aviation, transport and botany. 12 sections, arranged alphabetically by subject, with museums listed alphabetically within each section. Data include address, hours and days of operation, admission fees, history and collection description. Bibliography and index. More detailed than the *Official museum directory* (American Association of Museums, 1980-. Annual).
Class No: 5:061:069

[465]
GOLDBECK, G. Museen in Deutschland (West): Technische Museen. Führer zu der Museen für die Geschichte von Naturwissenschaft, Technik, Gewerbe und Industrie in den Bundesrepublik Deutschland und West Berlin. Stuttgart, J. Fink Verlag, 1975. 71p. illus. ISBN: 3771801974.

West German science/technology museums devoted to the history of natural history, technology, business and industry. Arranged under places, Altera ... Wuppertal. Appended notes on railway historical museums, and list of 8 other museums. Name and subject indexes. *Class No:* 5:061:069

[466]
HUDSON, K. *and* NICHOLLS, A. The Cambridge guide to the museums of Britain and Ireland. Cambridge, University Press, 1989. x,452p. illus. (some col.) maps. ISBN: 0521378923.

First published 1987.

Covers over 2,000 collections, arranged A/Z (Abbots Bromley ... Wollaston). Separate section on museums new to this edition. Data on each collection include hours of opening; facilities for the disabled; refreshments; tours; parking. Index of museum names; index of subjects (*e.g.* agriculture; botany; medicine); index of museums associated with individuals (*e.g.* Charles Darwin). 16 pages of location maps. 400 illustrations. *Class No:* 5:061:069

Scientific Expeditions

[467]
TEREK, F. Scientific expeditions. Jamaica, N.Y., Queens Borough Public Library, 1952. xvii, 176p.

Lists major scientific expeditions, with data on objectives, personnel, equipment, sponsorship, location and the like. Bibliography of source material; index of members and sponsors of expeditions. *Class No:* 5:061:910.4

51 Mathematics

Mathematics

Databases

[468]

MathSci. Providence, R.I., American Mathematical Society.

MATHSCI is a database equivalent of *Mathematical reviews,* and other subfiles, with records from 1959 onwards and available on DIALOG (file 239) and BRS hosts. A rival is MATHEMATICS ABSTRACTS database, corresponding to *Zentralblatt für Mathematik;* available on INKA. Available on CD-ROM (Silver Platter) as *MathSci Disc,* 1982-.
Class No: 510(003.4)

Bibliographies

[469]

DORLING, A.R., *ed.* **Use of mathematical literature.** London, Butterworths, 1977. xii, 260p. (*Butterworths Guides to information sources.*)

16 contributors. 14 Documented chapters: 1. The role of the literature in mathematics - 2. Major organisations and journals - 3. Reference material (encyclopedias; handbooks; abstracting journals) - 4. Mathematical education - 5. History of mathematics - 6. Logic and foundations - 7. Combinations - 8. Rings and algebras - 9. Group theory - 10. Measure and probability - 11. Complex analysis and special functions - 12. Convexity - 13. Topology - 14. Mathematical programming (88 references). Each chapter provides running commentary on items. Author index; subject and title index. 'A valuable guide to the subject for students and scientists' (*New technical books,* March 1978, entry 334).
Class No: 510(01)

[470]

—**GAFFNEY, M.P.** *and* **STERN, L.A.** Annotated bibliography of expository writing in the mathematical sciences. Washington, D.C., Mathematical Association of America, 1976. xi, 282p.

Includes 'over 1,000 articles and a few books of general interest' (*Preface*). Sections: 1. General - 2. Foundations - 3. Algebra - 4. Analysis - 5. Geometry - 6. Statistics and computing - 7. Applications. Author index.
Class No: 510(01)

[471]

—**HØYRUP, L.** Books about mathematics: history, philosophy. education, models, system theory and works of reference, etc. Roskilde, Denmark, Roskilde Univ. Center, 1979. viii, 182p.

A bibliography [unannotated] of 'books ABOUT mathematics - and not "technical" textbooks or research monographs on mathematics ...' (*Preface*). Pt.1: History and philosophy of mathematics and mathematics education, etc. -

....(contd.)
Pt.2: Models and systems theory - Pt.3: Works of reference (1. Bibliographies...Addresses). Author index.
Class No: 510(01)

[472]

DRAKELEY, J., *comp.* **Mathematics for the under-16s.** London, Library Association, Public Libraries Group, 1980. 19p. (*Readers' guide, no. 22.*) ISBN: 065365721.

About 100 briefly annotated entries. Sections: General mathematics (including history, biography) - Primary school level - Secondary school level (general textbooks and revision books; mathematical tables; arithmetics; graphs; statistics; algebra; geometry; trigonometry) - Everyday maths. - Periodicals. Annotations are couched in simple language. *Class No:* 510(01)

[473]

GRINSTEIN, L.S. Mathematical book review index, 1800-1940. New York, Garland, 1992. 448p. $72. (*Garland reference library of social science, 527.*) ISBN: 0824041143.

Lists reviews of books on pure and applied mathematics at all levels. Limited to reviews in English of English-language works published and/or distributed in North America between 1800 and 1940. List of 'Periodicals surveyed' and 'References consulted'. Alphabetical by author; subject index. 'This valuable reference work' (*Choice,* November 1992, p.442). *Class No:* 510(01)

[474]

PARKE, N.G. Guide to the literature of mathematics and physics, including related works on engineering science. 2nd rev. ed. New York, Dover, 1958; London, Constable, 1959. xviii, 436p.

First published 1947.

The 2nd ed. is double the size, although Pt.1 'General considerations' (p.3-74; including principles of making a literature search) remains unrevised. In Pt.2 'The literature', over 5,000 entries appear under *c.*120 subject headings (Acoustics ... Writing-technical), in double columns. One heading is 'Bibliographic aids', p.132-42. Subject headings are defined and notes provided on some titles. Emphasis is on applied mathematics. Mathematics and physics often figure together, but not always (*eg,* 'Tables (mathematical)'; 'History - Mathematics'. Peripheral fields include chemistry, machine and general engineering. Author and subject indexes, p.383-436. Dates of many of the references do not go beyond 1955. *Class No:* 510(01)

[475]

—**DICK, E.M.** Current information sources in mathematics: an annotated guide to books and periodicals, 1960-72. Littleton, Colo., Libraries Unlimited, 1973. x, 281p.

Includes over 1,600 annotated entries with 'emphasis on monographic material' (*Preface*) - books published or reprinted in English or in English translation, 1960 to mid-1972. The 37 chapters cover: 1. General elementary mathematics ... 4. History, biography and reprints of

....(contd.)

classics ... 7. Theory of numbers ... 14. Topology ... 29. Probability and statistics - 30. Bibliographies - 31. Dictionaries ... 34. Periodicals - 35. Directories and guides - 36. Professional organizations and government agencies - 37. A selected list of publishers. Author and subject indexes. 'This work updates nicely N.G. Parke's standard work in this field' (*Wilson library bulletin*, v.47(10), June 1973, p.875, 891). *Class No:* 510(01)

[476]
PEMBERTON, J.E. **How to find out in mathematics:** a guide to sources of information. 2nd ed. Oxford, etc., Pergamon Press, 1969. xiv, 193p. facsims. (*The Commonwealth and international library of science, technology, engineering and liberal studies.*)

First published 1964.

Chapters: 1. Careers for mathematics - 2. The organisation of mathematical information - 3. Mathematical dictionaries, encyclopedias and theses - 4. Mathematical periodicals and abstracts - 5. Mathematical societies - 6. Mathematical education - 7. Computers and mathematical tables - 8. Mathematical history and biography - 9/10. Mathematical books (bibliographies; evaluation and acquisition) - 11. Probability and statistics - 12. Operational research and related techniques. Exercises are appended to chapters. Appendices (*e.g.* Sources of Russian mathematical information). The index, p.181-93, covers authors of works cited. A helpful, comprehensive survey, mainly in the form of a running commentary. *Class No:* 510(01)

[477]
—SCHAEFER, B.K. Using the mathematical literature: a practical guide. New York & Basle, Dekker, 1979. ix, [1], 141p. $22.75.

Offers 'an insight into the vast and varied amount of mathematical literature and to act as a guide to its exploitation' (*Preface*). 15 chapters of running commentary, with appended cross references: 4. The nature of the mathematics literature (forms; subjects; languages; sources) ... 6. The organization of mathematics literature in libraries - 7. Journals - 8. Access to journals: bibliographies, indexes and abstracts - 9. Abstracting services in mathematics - 10. Books - 11. Access to books: bibliographies and reviews - 12. Reference books - 13. Applied mathematics - 14. Statistics - 15. Operations research. Analytical index, p.137-41. 'Not a comprehensive bibliography' (*New technical books*, June 1979, item 693). *Class No:* 510(01)

Encyclopaedias

[478]
Encyclopaedia of mathematics. Hazewinkel, M., *ed.* Dordrecht, Netherlands, Reidel, 1988-. v.1-. Formulas; equations. ISBN: 0155608010, set.

To be completed in 10v. (*e.g.* v.7. Orb-Ray ... v.10. Index).

A translation, with updates and editorial comments, of The Soviet *Mathematical encyclopaedia* ('Soviet Encyclopaedia Publishing House', 1977-85. 5v.).

Three kinds of articles: 1. Surveys of main directions in mathematics - 2. Detailed discussion of concrete problems, results and techniques - 3. Short (reference) definitions. Many include annotations or editorial comment and

....(contd.)

references (to both Soviet and Western literature). Entries classified by AMS numbers from the 1980 scheme. *Class No:* 510(031)

[479]
Encyclopedic dictionary of mathematics. Mathematical Society of Japan. 2nd ed., edited by K. Itó. Cambridge, Mass., & London, MIT Press, 1987. 4v. (2,120p.). diagrs, graphs, tables, formulas. $350. ISBN: 0262090260.

First published 1934. 2nd ed. is English translation of the 3rd Japanese ed. (1985), with references to English-language textbooks in place of the original Japanese textbooks.

V.1-3 (A-E, F-N, O-Z) comprise 450 articles (Abel ... Zeta functions). V.4 includes 23 appendices (1. Algebraic functions ... 23. Statistical estimation and statistical hypothesis testing), as well as numerical and statistical tables, a list of journals, systematic and A-Z lists of the articles. Name index; detailed, analytical subject index (p.1917-2148). 'An essential buy for every library and for many private bookdealers' (*Nature*, v.330, 26 November 1987 p.322). *Class No:* 510(031)

[480]
GELLERT, W., *and others, eds.* **The VNR concise encyclopedia of mathematics.** 2nd ed. New York, etc., Van Nostrand Reinhold, 1989. 776+56p. illus. (partly col.), diagrs., graphs, tables. $30. ISBN: 0442205902.

First published by VEB Bibliographisches Institut (Leipzig, 1973) as *Kleine Enzyklopädie der Mathematik.*

Three parts (42 sections): 1. Elementary mathematics - 2. Steps towards higher mathematics - 3. Brief reports on selected topics (*e.g.* topology, game theory). Closely written, colours being used extensively to help the reader (yellow background for important definitions and groups of formulae; blue, for examples; red, for theorems); red arrows indicate more complicated calculations. The reviewer in *Choice* (October 1990, p.289) found 'almost no change' from the 1977 edition and concluded that the work was 'much more like a reprint than a new edition'. But the *New Scientist* urged readers to 'save up your pennies for a few months and buy' (2 February 1991, p.66). *Class No:* 510(031)

[481]
NAAS, H. *and* SCHMID, H.L. **Mathematisches Wörterbuch,** mit Einbeziehung der theoretischen Physik. Im Auftrag des Instituts für Reine Mathematik an der Deutschen Akademie für Wissenschaft zu Berlin. Berlin, Akademie Verlag GmbH; London & New York, Pergamon Press, 1961. 2v. (A-K. xi, 1043p.; L-Z. vi, 952p).

127 contributors, 96 of them German. Entries concern terms, phrases, methods and trends of research plus brief biographies of 400 deceased mathematicians and the history of mathematics. Some articles carry short bibliographies of texts and papers. Treatment in depth (*e.g.* 'Legendre functions': 18p.). Many formulae. V.2 has appended list of more common symbols and of 82 branches and applications of mathematics. Very favourably reviewed in *Nature*, v.196, no.4855, 17 November 1962. *Class No:* 510(031)

[482]
—Enzyklopädie der Mathematischen Wissenchaften, mit Einschluss ihrer Anwendungen, Hrsg. im Auftrage der Akademien der Wissenschaften zu Berlin: 2 völlig neubarb. Aufl. Leipzig, Teubner, 1960-. v.1-.

First published 1898-1935 (6v. in 23: 1. *Arithmetik und Algebra* (2 pts.) - 2. *Analyse* (2 pts.) - 3. *Geometrie* (6 pts.) - 4. *Mechanik* (4 pts.) - 5. *Physik* (3 pts.) - 6. *Geodäsie und Geophysik. Astronomie*). Noted for its comprehensive treatment and well-documented, scholarly articles, aimed at the specialist. Detailed attention to applications in mechanics, physics, geodesy, geophysics and astronomy. The 2nd ed. is only slowly replacing the 1st ed. *Class No:* 510(031)

[483]
The Prentice-Hall encyclopedia of mathematics. West, B.H., *and others, eds.* Englewood Cliffs, NJ., Prentice-Hall, 1983. 638p. diagrs., graphs. £42.25. ISBN: 0136960138.

An encyclopedia of 'basic definitions, forumlas, examples, applications and occasional projects' (*Preface*). 8 broad topics (*e.g.* algebra; infinity; rational and irrational numbers; probability; zero). For scientists, teachers and non-professionals. *Class No:* 510(031)

[484]
SNEDDON, I.N., *ed.* **Encyclopaedic dictionary of mathematics for engineers and applied scientists.** Oxford etc., Pergamon Press, 1976. viii, 800p. diagrs., tables. ISBN: 0080167675.

More than 2,000 entries, A-Z; over 150 contributors. Longer articles are signed, with short bibliographies (*e.g.* 'Phase-integral methods': 9 columns; 8 references). Numerous cross-references. Appendix: includes addenda, 'Calculators, Electronic'. Editor: Professor of Mathematics, Glasgow University. Full index and good layout, - 'a boon to the librarian as well as the pressured engineer, researcher or student' (*RSR*, v.6(3), July/September 1978, p.19). *Class No:* 510(031)

[485]
The Universal encyclopedia of mathematics. London, Allen & Unwin, 1964; Pan Books, 1980. 715p. diagrs., tables. £3.50. ISBN: 0330243969.

Translated and adapted from Meyer's *Rechenduden* (Mannheim, Bibliographies Institut, 1960).

The clearly set-out dictionary of mathematical topics (p.11-491) is followed by mathematical formulae (p.492-548), and tables of functions with explanations (p.599-715). Some articles in the dictionary run to 20 or more pages (*e.g.*), 'Circles', 'Calculating instruments and machines'). No mention of Boolean algebra, nor of variants. 'Geared to average requirements' (*Foreword*). *Class No:* 510(031)

Handbooks & Manuals
[486]
KORN, G.A. *and* KORN, T.M. **Mathematical handbook for scientists and engineers:** definitions, theorems, and formulas for reference and review. 2nd ed., rev. and enl. New York, London, etc., McGraw-Hill. 1968. xvii, [i], 1130p. tables.

First published 1961.

21 chapters (1.Real and complex numbers. Elementary algebra - 2. Plane analytical geometry ... 20. Numerical calculations and finite differences - 21. Special functions). More detailed discussion and advanced topics appear in

....(contd.)
small type. Appendices A-F (F: Numerical tables, p.989-1089); bibliography. p.1089; Glossary of symbols and notations, p.1090-95. Analytical index, p.1097-1130. *Class No:* 510(035)

[487]
LEDERMANN, W. **Handbook of applicable mathematics.** New York, etc., Wiley, 1980-84. 6v. in 8.

1. *Algebra.* 2. *Probability.* 3. *Numerical methods.* 4. *Analysis,* edited by W. Ledermann and S. Vajda. 1982. 5. (2 pts.): *Combinations and geometry,* edited by W. Ledermann and S. Vajda. 1984-. 6 (2 pts.): *Statistics,* edited by E. Lloyd. 1984.

Each volume has chapter bibliographies and an analytical index. The set is designed specifically for the needs of the professional, although 'it aims to cater for all levels' (*The plan of the handbook*).

V. 7, *Supplement,* appeared in 1990; *Contents and general index* in 1991. *Class No:* 510(035)

[488]
PEARSON, C.E., *ed*. **Handbook of applied mathematics:** selected results and methods. 2nd ed. New York, Van Nostrand Reinhold, 1983. 1307p. ISBN: 0442238665.

First published 1974 (1266p.).

20 contributors. 21 documented chapters (2. Elements of analysis ... 7. Special functions ... 12. Asymptotic methods ... 18. Numerical analysis ... 21. Probability with statistics). Worked examples. The topics dealt with 'involve results and techniques which experience has shown to be of utility in a very broad survey of applications' (*Preface*). Detailed index. *Class No:* 510(035)

[489]
TUMA, J.J. **Engineering mathematics handbook.** 3rd ed. New York, Chichester, etc., McGraw-Hill, 1987. 512p. tables. £44.50. ISBN: 0070650433.

First published 1970.

A compendium, in logical sequence of sections, from algebra, geometry and trigonometry to Laplace transforms and tables of definite integrals. 3rd ed. adds 2 chapters, on plane curves and areas, space curves and surfaces. Bibliography; index. For engineers, architects, teachers and students. *Class No:* 510(035)

[490]
—ASM handbook of engineering mathematics, by Faculty members of the Dept. of Mechanical Engineering, Univ. of Akron, Metals Park, Ohio, American Society for Metals. Ohio, ASM, 1983. 697p. illus. ISBN: 087176157x.

In 2 parts. Pt.1 contains basic operations and theories of algebra, trigonometry, geometry, analytical geometry, calculus, etc. in ascending order of difficulty. Pt.2 gives mathematical equations, focusing on those equations of practical problems in mechanical analysis and design. Index. *Class No:* 510(035)

Dictionaries

Polyglot
[491]
EISENREICH, G. *and* SUBE, R. **Dictionary of mathematics in four languages:** English, German, French, Russian. Amsterdam, etc., Elsevier, in co-edition with VEB Verlag, Berlin, 1982. 2v. (1458p.). $146.25. ISBN: 0444997067.

.... *(contd.)*

1: English-German-French-Russian. 2: German, French and Russian indexes.

About 25,000 base terms. The thesaurus-type approach involves major subject-category groupings: fundamentals of mathematics; logic; algebra; topology; mathematical analysis, calculus, statistics, geometry. Each term is assigned to a subject category (about 60 in all). The emphasis is on 'pure' mathematics; logic, algebra and topography are covered in particularly full detail. *Class No:* 510(038)=00

[492]
—MESCHKOWSKI, H. Mehrsprachen Wörterbuch Matematischer Begriffe. Mannheim, Bibliographisches Institut, 1972. 131, [4]p.

Lists *c.*2,500 German terms, with English-German, French-German, Russian-German and Italian-German equivalents. Genders are stated. *Class No:* 510(038)=00

English

[493]
BOROWSKI, E.J. *and* BORWEIN, J.M. Collins dictionary of mathematics. London, Harper Collins, 1989. xi,659p. diagrs., graphs. £7.99. ISBN: 0004343476.

Over 9,000 terms, briefly defined (*e.g.* 'epicycloid': 7 lines). 400 diagrams. Cross-references. 3 appendices: 1. Symbols and convention - 2. Derivatives and integrals - 3. Hilbert problems. Undergraduate level; good value. *Class No:* 510(038)=20

[494]
CLAPHAM, C. Concise Oxford dictionary of mathematics. Oxford Univ. Press, 1990. 202p. illus., formulas. £15; $29.95. ISBN: 0198661568.

About 1,000 terms, briefly defined (*e.g.* 'dummy variable': 8 lines). Includes brief biographical sketches of leading mathematicians. Cross-references. 'Useful ready reference book suitable for all libraries' (*New technical books,* May/June 1991, item 624). *Class No:* 510(038)=20

[495]
DAINTITH, J. *and* NELSON, R.D. Penguin dictionary of mathematics. Harmondsworth, Penguin, 1986. 304p. £4.99. ISBN: 0140511199.

Covers mathematical terms and theories in both pure and applied mathematics. *Class No:* 510(038)=20

[496]
GLENN, J. *and* LITTLER, G., *ed*. A Dictionary of mathematics. London, etc., Harper & Row, 1984. [viii], 230p. diagrs., graphs. ISBN: 0063182920.

Defines *c.*2,500 terms, half of them being combinations of two or more words, as found in mathematics textbooks of today. 2 appendixes (2. The symbols of mathematics). Aimed at sixth-formers and undergraduates. *Class No:* 510(038)=20

[497]
—HORRILL, P.J.F. Mathematics A-Z. Harlow, Essex, Longman Group, 1986. vi, 166p. diagrs., tables, formulas. ISBN: 0582555854.

Explains more than 1,000 terms and concepts 'encountered in public examinations in mathematics' (*Foreword*). Appended 'Mathematical notations', p.163-6. Intended chiefly for those studying the subject at school or college. *Class No:* 510(038)=20

[498]
JAMES, G. *and* JAMES, R.C. Mathematics dictionary. 5th ed. New York, etc., Van Nostrand Reinhold, 1992. viii,548p. diagrs., equations. £19.95. ISBN: 0442012411.

First published 1942; 4th ed. 1976.

8 contributors. Over 4000 terms, some defined at length (*e.g.* logarithm: 2 cols.). Numerous topics (*e.g.* fractals, fuzzy sets, robust statistics) have been updated. Multilingual index: French, German, Spanish, Russian. Appendices include denominate numbers, mathematical symbols, differential formulas, integral tables, Greek alphabet. *Class No:* 510(038)=20

[499]
McGraw-Hill dictionary of physics and mathematics. Lapedes, D.N., *ed*. New York, etc., McGraw-Hill, 1978. xvi, 1074+45p. illus., tables. ISBN: 0070454809.

Defines more than 20,000 terms, the entries being mostly extracted from the *McGraw-Hill dictionary of scientific and technical terms* (2nd ed. 1978), but also written specially for this work. Definitions average 15-80 words; 55 subject categories applied; 700 marginal illus.; 15 appendices. Well produced. 'Recommended for every high school or college library' (*Choice,* v.164, March 1979, p.56). *Class No:* 510(038)=20

[500]
The Pan dictionary of mathematics. Gibson, C., *ed*. Rev. ed. London, Pan, 1990. [vi],234p. diagrs., graphs. £6.99. ISBN: 0330314556.

First published 1981. This ed. published in the US as *Facts on File Dictionary of mathematics,* 1988.

About 1,500 terms, briefly defined (*e.g.* 'great circle': 5 lines). Cross-references. 12 appendices. Secondary school level. *Class No:* 510(038)=20

[501]
SELKIRK, K. Longman mathematics handbook: the language and concepts of mathematics explained. Harlow, Essex, Longman; Beirut, York Press, 1990. 312p. diagrs. (col.). £6.99. ISBN: 0582012618.

About 3,000 terms, briefly defined (*e.g.* 'integral norm': 9 lines). Arranged by subject (*e.g.* algebra; vectors; probability, etc.), with terms listed in index. Effective use of colour in diagrams. For senior secondary pupils and undergraduates. *Class No:* 510(038)=20

[502]
Webster's new world dictionary of mathematics. Karush, W., *ed*. New York, Prentice-Hall, 1989. 317p. illus., diagrs., graphs, tables. $11.95. ISBN: 0131926675.

Defines over 1400 mathematical terms. Includes arithmetic, algebra, geometry, trigonometry and analytical geometry. Appended: Brief biographies of famous mathematicians, a chart of mathematical symbols and tables of powers, roots, logarithms and trigonometric functions. Much overlap with *Facts on file dictionary of mathematics* (2nd ed., 1988). 'For high school and college students' ('Reference books bulletin', *Book list,* December 15, 1989, p.860). *Class No:* 510(038)=20

German

[503]

HERLAND, L. Dictionary of mathematical sciences. 2nd ed., rev. and enl. London, Hafner, 1965. 2v. (323p.; 349p.). First published 1951-54. 1: *German-English.* 2: *English-German.*

Each volume covers *c.*9,000 terms, with many subentries. Areas include mathematical logic, statistics and commercial arithmetic, plus mathematics as applied to physics and astronomy. Two features: generous use of cross-references, and abundance of model usages. V.1 introduces more than 4,000 revised entries and additions; v.2 has a supplement of new words. *Class No:* 510(038)=30

[504]

—MACINTYRE, S. *and* WITTLE, E. German-English mathematical vocabulary. 2nd ed. Edinburgh & London, Oliver & Boyd, 1966. ix, 95p.

First published 1956.

Includes *c.*1,500 German terms, with English equivalents. Irregular verbs are asterisked. Abbreviations; sketch of grammar. Confined to pure mathematics. 'Designed to help English-speaking mathematicians to read German' (*Preface*). *Class No:* 510(038)=30

French

[505]

BOUVIER, A., *and* *others.* **Dictionnaire des mathématiques.** Paris, Presses Universitaire de France, 1983. xiv, 834p. illus. ISBN: 2130354270.

7,600 terms are defined, covering geometry, analysis, algebra, arithmetic, topology, probability, etc. Supporting line drawings. Intended for a wide range of users (*Bulletin critique du livre français,* no.410, February 1980, entry 10459). *Class No:* 510(038)=40

[506]

—LYLE, W.D. Dictionnaire français et anglais de terminologie mathématique. Ottawa, Didier, 1970. xiv, 137p.

In two parts, French-English and English-French, each with *c.*1,500 entry-words. Appended list of cardinals, ordinals, fractions, decimals and larger numbers. *Class No:* 510(038)=40

Spanish

[507]

GARCÍA-RODRIGUEZ, M. Diccionario matemático, español-inglés/inglés-español. [Mathematics dictionary, Spanish-English, English-Spanish.] New York, Hobbs, Dorman, 1965. [x], 78p.

About 1,500 entry-words in each half. Spanish-English, p.1-40; English-Spanish. p.41-78. *Class No:* 510(038)=60

Russian

[508]

LOHWATER, A.J. A.J. Lohwater's Russian-English dictionary of the mathematical sciences. Boas, R.P., *ed.* 2nd ed. Providence, R.I., American Mathematical Society, 1990. 343p. $50. ISBN: 0821801233.

First published 1961.

More than 10,000 entries, preceded by a short Russian grammar. Idiomatic translations, with stress markings.

....(contd.)

Includes a listing of Russian versions of mathematician's names. Good use of bold cyrillic type. 'Highly recommended as a reference source' (*New technical books,* Sept./Oct. 1991). *Class No:* 510(038)=82

[509]

—ALEXANDROV, P.S., *and others.* Anglo-russkiĭ slovar' matematicheskiĭ terminov. Akademiya Nauk SSR. Matematicheskiĭ Institut. Moscow, Izd-vo Inostr. Library, 1962. 369p.

The English-Russian counterpart of Lohwater and Gould's *Russian-English dictionary of the mathematical sciences.* *Class No:* 510(038)=82

[510]

—MILNE-THOMSON, L.M. Russian-English mathematical dictionary: Words and phrases in pure and applied mathematics, with roots and accents arranged for easy reference. Madison, Univ. of Wisconsin Press, 1962. xiv, 191p.

Includes *c.*10,000 entries (p.3-160), plus outline of Russian grammar and brief list of Russian abbreviations. Good use of bold type; welcome indication of Russian roots of foreign derivation. Not confined to purely mathematical terms. *Class No:* 510(038)=82

Arabic

[511]

LEAGUE OF ARAB STATES. Education, Culture and Sciences Organization. Permanent Bureau of Arabisation. **Lexicon of mathematics:** English-French-Arabic. Rabat, Morocco, Bureau Permanent de Coordination de l'Arabisation, 1971. 192, xxivp.

1,862 English-base terms, with French equivalents and Arabic equivalents/explanations. French and Arabic indexes. *Class No:* 510(038)=927

Chinese

[512]

DE FRANCIS, J. Chinese-English glossary of the mathematical sciences. Providence, R.I., American Mathematical Society, 1964. iv, 275, [1]p.

16,540 terms. 'Characters are arranged on the basis of first, the radical of a character and second, the number of strokes in the character, apart from the radical' (*Preface*). The author is Research Professor of Chinese, Seton Hall Univ., N.J. *Class No:* 510(038)=951

Reviews & Abstracts

[513]

Current contents: CompuMath. Philadelphia, Pa., Institute for Scientific Information. 1981-. Monthly. ISSN: 0276220x.

Contents-list of *c.*250 issues of journals per issue (*c.*120p.). Subjects: science & technology; mathematics; operations; research & management science; statistics & probability. Available online via DIALOG, file 151/ CompuMath. *Class No:* 510(048)

[514]
Current mathematical publications. Providence, R.I., American Mathematical Society, 1969-. v.1(1)-. Bi-weekly. $200 pa. ISSN: 03614794.

Formerly *Contents of contemporary mathematical journals and new publications.*

About 5,000 references per issue. Classified by a modified AMS scheme (00. General ... 08. General mathematical systems ... 20. Group theory and generalizations ... 40. Sequences, series ... 51. Geometry ... 60. Probability theory and stochastic processes ... 70. Mechanics of particles and systems ... 90. Information and communication circuits). Entries for book items are asterisked. Appended list of serials represented per issue; serial additions and changes; etc. Author and keyword indexes per issue, cumulated half-yearly. *Class No:* 510(048)

[515]
Mathematical reviews. Providence, R.I., American Mathematical Society, 1940-. v.1-. Monthly. $2,750pa.

About 45,000 abstracts pa. 94 sections (as in *Zentralblatt für Mathematik,*): 00. General ... 01. History and biography ... 03. Mathematical logic and foundations ... 11. Number theory ... 20. Group theory and generalizations ... 26. Real functions ... 40. Sequences, series ... 51. Geometry ... 60. Probability theory and stochastic processes ... 70. Mechanics of particles and systems ... 80. Chemical thermodynamics, heat transfer ... 90. Economics, operational research, programming games ... 94. Information and communication circuits. Author index; keyword index; serial additions and changes. Annual index: *Index of mathematical papers* (1973-). *Mathematical reviews cumulative index, 1973-1979* covers v.45-58 in 12v., with *c.*300,000 entries. The basic reviewing service in mathematics. Database: MATHSCI; CD-ROM: *MathSci Disc. Class No:* 510(048)

German

[516]
Zentralblatt für Mathematik und ihre Grenzgebiete. [Mathematical abstracts.] Berlin, etc., Springer., 1931-. Irregular. DM 166 per Bd. ISSN: 00444235.

Preceded by *Jahrbuch über die Forschritte der Mathematik* (Berlin, de Gruyter, 1868-1942).

About 2,000 abstracts pa. 94 sections (00-94), as in *Mathematical reviews (qv)*. Abstracts range in length from 1½ lines to 20 lines, signed, and in English, French or German. Author and subject indexes per issue; cumulated. Online database. *Class No:* 510(048)=30

French

[517]
Bulletin signalétique. 110: Analyse numérique. Informatique. Automatique. Statistique et probabilité. Recherche opérationnelle. Gestion. Économie. Paris, Centre National de la Recherche Scientifique, 1940-. v.(1)-. 10pa.

About 18,000 abstracts pa. 10 main classes: A. Analyse numérique - B. Combinatoire. Logique mathématique - C. Informatique - D. Automatique théorique. Systèmes - E. Intelligence artificiel - F. Applications d'informatique et d'automatique - G. Probabilité. Statistique mathématique - H. Recherche opérationelle et modèles formalisés de gestion - I. Sciences économiques et problèmes de gestion - J. Thèses et ouvrages fondamentaux des mathematiques. Subject and

....(contd.)

author indexes per issue, cumulated annually. Annual outline of classifiction scheme and list of sources. *Class No:* 510(048)=40

Russian

[518]
Referativnyĭ zhurnal. Matematika. Akademiya Nauk SSSR. Nauchoĭ Informatsii. Moscow, VINITI, 1953-. Monthly.

About 25,000 abstracts pa. Three main sections: General. Mathematical logic. Theory of numbers. Algebra. Topology. Geometry - Mathematical anlysis - Numerical methods. Probability theory and mathematical statistics. Cybernetics. Monthly author index, and English translation of contents of main sections. Three separate parts are also available. *Class No:* 510(048)=82

Periodicals

[519]
LONDON UNIVERSITY. Library. **Union list of periodicals on mathematics and allied subjects in London libraries.** 2nd ed. London, the University Library, 1968. [viii], 139p.

First published 1958.

926 titles in 45 libraries in London, 15 of the libraries not connected with the University. Entry is under latest title. *World list of periodicals* (3rd ed.) numbers are cited. Appendix: List of abbreviated and alternative titles sometimes found in references. Excludes ephemera, congress proceedings and mongraph series, 'and journals primarily devoted to the teaching of mathematics at the lower level' (*Introduction*). Supplement lists 45 titles held by the Science Reference Library. *Class No:* 510(051)

[520]
—**BRITISH LIBRARY.** Science Reference Library. **Mathematics and statistics periodicals held by Science Reference Library.** London, SRL, 1977. vii, 47p.

In two sections: Mathematics and statistical periodicals (p.1-46; *c.*500 titles) - Abstracting and indexing periodicals. Call numbers are given, with a summary of the classification used. *Class No:* 510(051)

Progress Reports

[521]
Progress in mathematics. Boston, Mass., (originally New York & London), Plenum Press. 1967-. v.1.

Each volume deals with a specific area. Thus, v.60 (1985. 329, [3]p.) is entitled 'Geometry today' and reprints 18 papers presented at the International Conference, Rome, June 4-11, 1984. 29 contributors. One paper concerns 'Algebraic curves and solution equations', with 22 references. Appended list of titles of v.1-59 of *Progress in mathematics. Class No:* 510(055)

Yearbooks & Directories

[522]
World directory of mathematicians. 8th ed. Helsinki, the University, 1986. (Distributed by the American Mathematical Society). xix, 950p.

Published under the auspices of the International Mathematical Union. 7th ed., 1982.

....(contd.)

Lists *c*.30,000 mathematicians from 93 countries. Data on each include posts held, university and its address. List of mathematical organizations precedes the A-Z directory (p.1-789). Geographical list (Algeria ... Zimbabwe).
Class No: 510(058)

Teaching Materials

[523]
Handbook of research on mathematics, teaching, and learning. Grouws, D.A., *ed.* New York, Macmillan, 1992. 771p. $65. ISBN: 0029223814.

29 chapters covering the history of research in mathematics education, teaching methods and the international view. Produced by the National Council of Teachers of Mathematics (US). *Class No:* 510(072)

Quotations

[524]
MORITZ, R.E. Memorabilia Mathematica, or, The philosopher's quotation book. New York, Macmillan Co., 1914. (Reprinted Dover, 1958.) xiii, 410p. diagrs.

Over 2,000 quotations, some of them lengthy, with sources cited. Foreign quotations usually appear only in translation. 21 chapters (*e.g.* 1. Definitions and objects of mathematics; 9/10. Persons and anecdotes; 21. Paradoxes and curiosities). Detailed, analytical index, p.385-410 (*e.g.* 'Liebnitz': ½ column). *Class No:* 510(082.2)

Tables & Data Books

[525]
BARTLETT, B.R. *and* **FYFE, D.J. Handbook of mathematical formulae for engineers and scientists.** London, Denny Publications, Ltd., 1974. [ix], 216p. tables. ISBN: 0950315400.

Over 1,600 mathematical formulas, definitions and methods. Main sections: Mathematics - Statistics - Numerical methods. 'Intended to cover, in standard notation, the definitions and formulae relating to the Mathematical techniques which are required by students in Technical Colleges, Polytechnics and Universities' (*Preface*).
Class No: 510(083)

[526]
COOK, J.L. Conversion factors. Oxford Univ. Press, 1992. 160p. tables. $12.95. ISBN: 0198563523.

More than 3,500 conversion factors, arranged by areas of application. SI units. 'Well formatted ... reasonably priced ... a valuable addition to a reference collection' (*Choice*, September 1992, p.74). *Class No:* 510(083)

Models

[527]
Mathematical modelling: a source book of case studies. Huntley, I.D. *and* James, D.J.G. Oxford Univ. Press, 1990. xiv,462p. £40. ISBN: 0198536577.

28 case studies (some with references) in 4 parts: 1. Simple illustrative examples - 2. Models for class development - 3. Extended problems - 4. Further problem

....(contd.)

statements. Intended for application to undergraduate mathematics teaching, with suggestions made for adaptation, changes to problems. *Class No:* 510(086.48)

Histories

[528]
BOYER, C.B. A History of mathematics. 2nd ed. New York, London, etc. Wiley. 1989. 762p. illus., ports., diagrs., maps. $53.95. ISBN: 0471097632.

First published 1968. This ed. revised by U.C. Merzbach.

27 chapters (1. Primitive origins - 2. Egypt ... 26. The rise of abstract algebra - 27. Aspects of the twentieth century), each with appended references and graded exercises, plus footnote references. Few changes have been made to the material in this new edition, but coverage of the twentieth century has been expanded and the bibliographies and references have been updated. A well-written and admirably documented survey. *Class No:* 510(091)

[529]
—**BALL, W.W.R.** A Short account of the history of mathematics. 4th ed. London, Macmillan, 1908. xxiv, 522p. diagrs.

First published 1888.

The best known of the short histories of mathematics 'illustrated by the lives and discoveries of those to whom the progress of the science is mainly due' (*Preface*). 19 chapters, to the beginning of the 20th century. Footnote references; analytical index, p.499-522. *Class No:* 510(091)

[530]
DUNHAM, W. Journey through genius: the great theorems of mathematics. New York, Wiley, 1990. xiv,300p. illus., diagrs. $19.95; £14.95. ISBN: 0471500305.

Provides an historical, biographical and 'creative' treatment of mathematical proofs. 12 chapters covering Hippocrates, Euclid, Archimedes, Heron, Cardano, Newton, Euler and Cantor. Index, p.295-300. *Class No:* 510(091)

[531]
EVES, H. An Introduction to the history of mathematics. 5th ed. New York, Saunders College Publishing, 1982. xviii, 593p. illus., port., diagrs., formulas. ISBN: 0030620643.

First published 1953.

Pt. 1: Before the seventeenth century (chapters 1-8; 1. Numerical systems - 2. Babylonian and Egyptian mathematics ... 7. Chinese, Hindu and Arabian mathematics - 8. European mathematics, 500-1600). Pt.2 The seventeenth century and after (chapters 9-15; 10. Analytic geometry and other precalculus developments ... 15. Abstraction and the transition into the twentieth century). Each chapter features 'Problem studies', 'Essay topics' and 'Bibliography'. General bibliography, p.512-3; chronological table. Answers and suggestions for the solution of the problem studies. Analytical index, p.553-93. A standard textbook for undergraduates. *Class No:* 510(091)

[532]
HEIJENOORT, J. van. From Frege to Gödel: a sourcebook in mathematical logic, 1879-1931. Cambridge, Mass., Harvard Univ. Press, 1967. *c.*665p. (*Source books in the history of the sciences.*)

51 papers (some of them translated) and letters between mathematicians (*e.g.* Hilbert's lecture, 'The foundations of mathematics' (1927)). References, p.629-55. Detailed, non-

....(contd.)

analytical index. The series aims to provide 'collections of classical papers that have changed the structure of the various sciences'. *Class No:* 510(091)

[533]
—STRUIK, D.J., *ed.* A source book in mathematics, 1200-1800. Cambridge, Mass., Harvard Univ. Press., 1969. xiv, 427p.

Includes 75 excerpts from the writings of Western mathematicians, with references. Criticized in Dorling's *Use of mathematical literature* (*qv*), p.65, for attempting to cover too much ground, leaving a blurred impression, and compared unfavourably to Heijenoort's selection of core papers. *Class No:* 510(091)

[534]
—The World of mathematics: a small library of the literature of mathematics, from A'h-mosé the Scribe to Albert Einstein, presented with commentaries and notes. Newman, J.R., *comp.* New York, Simon & Schuster, 1956; London, Allen & Unwin, 1960. 4v. illus., diagrs.

Covers in popular fashion the whole literature of mathematics, although extracts are inevitably arbitrary in their selection. Footnote references. Detailed, analytical index (v.4, p.2471-2535). *Class No:* 510(091)

[535]
KLINE, M. Mathematical thought: from ancient to modern times. Oxford Univ. Press, 1972. xvii, 1238p. diagrs. (*£45..*)

Two-thirds of this impressive history is devoted to concepts that acquired significance after 1700. Includes bibliographies. Despite its length, 'even his [Kline's] 1250 pages are silent on many things' (Dorling, A.E., ed. *Use of mathematical literature* (1977), p.64). *Class No:* 510(091)

[536]
SMITH, D.E. History of mathematics. New York, Dover; London, Constable, 1958 (reprint). 2v. (xxxiv, 1321p.). illus., tables.

Originally published 1923-5 (Boston, Ginn). 1. *General survey ... 2. Special topics of elementary mathematics.*

A textbook on the history of elementary mathematics 'through the first stages in the calculus' (*Preface*). V.1 contains many brief biographical sketches, chronological tables (to 1850) and a selective bibliography. V.2 has numerous footnote references. 510 illus. in all. *Class No:* 510(091)

[537]
—TEMPLE, S. 100 years in mathematics. London, Duckworth, 1981. xvi, 316p. ISBN: 0715611305.

In 3 parts: 1. Number - 2. Space - 3. Analysis. 17 chapters (10. The concept of functionality; 11. Derivative integrals; 12. Distributions; 13. Ordinary differential equations; 14. Calculus of variations; 15. Potential theory; 16. Mathematical logic; 17. Conclusion). Bibliography, p.283. References, p.285-307. Index of authors; analytical general index. Detailed contents. 'This account is, of necessity, a personal selection of its [mathematics'] major developments' (*Preface*). *Class No:* 510(091)

[538]
STILLWELL, J. Mathematics and its history. New York, etc., Springer-Verlag, 1989. x,371p. illus., diagrs. $49.80. ISBN: 0387969810.

Approaches mathematics 'through its history' (*Preface*). 20 chapters, each with 'Biographical notes' (*e.g.* 9. 'Infinite series' includes notes on James Greogry and Leonhard Euler). References, p.333-62. Index, p.362-71. *Class No:* 510(091)

Bibliographies

[539]
DAUBEN, J.W. The History of mathematics, from antiquity to the present: a selective bibliography. New York & London, Garland, 1985. xxxv, 467p. $80. ISBN: 0824092848.

2,400 entries, mostly annotated, with numerous cross-references. 6 sections: 1. General reference works - 2. Source materials - 3. General histories of mathematics - 4. The history of mathematics: chronological periods (p.29-215) - 5. The history of mathematics: sub-disciplines (algebra ... topology) - 6. The history of mathematics: selected topics (*e.g.,* Mathematics education; Regional studies; Women in mathematics). Author and subject (non-analytical) indexes. 'An excellent addition to existing bibliographical sources in the history of mathematics' (*ISIS*, v.76, December 1985, p.595). The author is editor of *Historia mathematica. Class No:* 510(091)(01)

[540]
LORIA, G. Guida allo studio della storia delle matematiche; generalità didattica, bibliografia. Appendice: Questioni stouche concernenti le scienze esatte. 2.ed., rif & aum. Milan, Hoepli, 1946. xix, 385p.

About 1,600 titles, with comments. 10 chapters (2. survey of histories of mathematics; 3. analysis of the contents of the leading journals for the history of mathematics). 4 appendices (*e.g.* research aids, - manuscripts, biographies, treatise). Index of persons cited. Full contents list, p.355-81. Addressed to students. *Class No:* 510(091)(01)

[541]
MAY, K.O. Bibliography and research material of the history of mathematics. Toronto, Univ. of Toronto Press, 1973. [ix] 818p.

Two parts: 1. Research manual (Information retrieval; Personal information retrieval; Historical analysis and writing) - 2. Bibliography (Biography; Mathematical topics; Epimathematical topics; Historical classifications; Information retrieval). About 30,000 unannotated entries in all. Many cross-references. Appendix 2: 'Coded list of serials' (p.709-818). International in scope. 'This bountiful and authoritative guide' (*Choice*, v.10(10), December 1973, p.1531). Weak on the literature before 1868 and on Oriental studies (*ISIS*, v.65, 1974, p.514-6). *Class No:* 510(091)(01)

[542]
—Historia mathematica. Toronto, Univ, of Toronto Press, 1974-. 4pa.

A publication of the International Commission on the History of Mathematics.

Carries articles, reports on meetings, reviews, and abstracts of current literature on the history of mathematics. Thus, V.13(2), May 1986, has abstracts on p.196-209. Annual index of authors of works reviewed and abstracted. *Class No:* 510(091)(01)

Ancient Greece

[543]

GERICKE, N. **Mathematik in Antike und Orient.** Berlin, etc., Springer, 1984. xii, 292p. diagrs., tables, maps. ISBN: 0387110478.

Four parts: 1. Pre-Greek mathematics (*e.g.* Babylonia; Egypt) - 2. Greek mathematics (p.70-168) - 3. Mathematics in the East (China; India; Islam) - 4. Biography. Biographical notes. Chronology and maps follow. Short catchword index. 140 figures and tables; 4 maps.
Class No: 510(091)(37)

[544]

HEATH, Sir T.L. **A History of Greek mathematics.** Oxford, Clarendon Press, 1921. 2v. diagrs., formulas.

1. *From Thales to Euclid.* xv, 446p. 2. *From Aristarchus to Diophantus.* xi, 586p.

21 chapters, the major mathematicians (*e.g.* Euclid, Archimedes) being given a chapter apiece. The final chapter deals with commentators. Careful exposition is allotted to the main writings; many formulas. V.2 carries an index of Greek words as well as an English index. The standard work in English. *Class No:* 510(091)(37)

Biographies

[545]

ABBOTT, D., *ed.* **Mathematicians.** London, Blond Educational, 1985. [vii], 175p. (*The Biographical dictionary of scientists.*) ISBN: 0584700059.

About 200 entries, Abel ... Zermerlo. Historical introduction precedes. Glossary, p.145-60. Grouped index, with references to name entries, p.161-75. Criticised in *Choice* (July/August 1986, p.1651) for lack of criteria in selection of entries and also for lack of citation of sources. *Class No:* 510(092)

[546]

Biographical dictionary of mathematicians. New York, Scribner's, 1991. 4v. illus. $200. ISBN: 0684192829.

Entries 'borrowed' from the *Dictionary of scientific biography* and its two supplements (Scribner's, 1970-90). Includes illus. Bibliographies not updated. Simple but serviceable index. *Class No:* 510(092)

[547]

GRINSTEIN, L. *and* CAMPBELL, P.J., *eds.* **Women of mathematics:** a bibliographic sourcebook. Westport, Conn., Greenwood Press, 1987. 288p. ISBN: 0313248494.

Bibliographical studies of 43 women mathematicians, from antiquity to the 20th century, in A/Z order. *Library journal* (1 September 1987, p.175) queries selection policy, -'more often attainment of an advanced mathematics degree'. Entries vary in content and quality. Nevertheless, considered (*ibid*) 'a major contribution to the history of mathematics and of women'. *Class No:* 510(092)

[548]

HOWSON, A.G. **A History of mathematics education in England.** Cambridge Univ. Press, 1982. x, 294p. tables. £27.50. ISBN: 0512124061.

9 chapters: 1. Robert Recorde - 2. Samuel Pepys - 3. Philip Doddridge - 4. Charles Hutton - 5. Augustus De Morgan - 6. Thomas Tate - 7. James Wilson - 8. Charles Godfrey - 9. Elizabeth Williams. Appendix: 'A selection of examination papers, syllabus, etc.' Notes p.239-80. Name

....(contd.)

and subject indexes. A biographical account, 'even though it requires frequent scenic changes and "flashbacks"' (*Preface*). *Class No:* 510(092)

[549]

MESCHKOWSKI, H. **Mathematiker - Lexikon.** 3., überarb. & eng. Aufl. Mannheim, Bibliographischer Institut, 1980. 342p. ports. ISBN: 3411015764.

First published 1968.

Entries, 'Abel, Niels Henrik'...'Zu Chong-zhi'; 'Newton': 3½p., 'Laplace': 2p. Bibliography, p.311-42 (A. Works by the mathematicians; B. Literature on individual works; C. General works on the history of mathematics). *Class No:* 510(092)

Bibliographies

[550]

DIODATO, V. *and* TOLAN, M. **'Mathematical biographies: profiles and sources of information on eighteen mathematicians'.** In *Science and technology libraries,* V.11(4), Winter 1991, p.9-42. ISSN: 0194262x.

Brief, non-scholarly and mostly twentieth-century; 6 or 7 references for each mathematician. *Class No:* 510(092)(01)

England

[551]

TAYLOR, E.G.R. **The Mathematical practitioners of Tudor and Stuart England.** Cambridge Univ. Press, for the Institute of Navigation, 1954. xv, 443p. illus.

Three parts: 1. description of the progress and work of practical mathematicians, 1485-1715; 2. short biographical notes on 582 mathematical practitioners; 3. chronological annotated bibliography of 628 mathematical treatises published during the period. An essential reference work on the history of science and mathematics.
Class No: 510(092)(420)

[552]

—TAYLOR, E.G.R. An Index to 'The mathematical practitioners of Hanoverian England, 1714-1800'; compiled by K. Bostok [and others]. Cambridge Univ. Press; London, H. Wynter, 1980. 23p.

Acts as a selective A-Z index of major practitioners, complementing the grouping of names into 14 period sections in the main volume. *Class No:* 510(092)(420)

[553]

—TAYLOR, E.G.R. The Mathematical practitioners of Hanoverian England, 1714-1800: a sequel to *The mathematical practitioners of Tudor and Stuart England.* Cambridge Univ. Press, for the Institute of Navigation, 1966. xv, 503p. illus.

Two parts. 1. Narrative (p.5-106) - 2. Biographies of inventors, makers and users of mathematical instruments, teachers, writers and theorists, by decades (p.107-483). Indexes: 1. General; 2. Secondary practitioners. 'Sources and references', p.xiv-xvi. *The Marine observer* (v.37(46), April 1967, p.88) offers a minor criticism: the grouping of names in 14 separate periods in the biographies, and the two methods of indexing names. *Class No:* 510(092)(420)

Computation Devices

[554]
MADDISON, F. 'Early astronomical and mathematical instruments - a brief survey of sources and modern studies'. In *History of science*, v.2, 1963, p.17-50.

Includes a lengthy bibliography, p.35-50. *Class No:* 510·3

Recreational Mathematics

[555]
EISS, H.E. **Dictionary of mathematical games, puzzles and amusements.** New York, Greenwood Press, 1988. xiv,278p. illus. $49.95. ISBN: 0313247145.

A-Z sequence of mathematical games, both classical and modern, with detailed explanations, line drawings and cross-references. Mathematic theories, underlying the games, feature throughout; chapters include brief bibliographies. Criticized in *Choice* (July/August, 1988, p.1273) for omission of 'significant terms' and poor illustrations. *Class No:* 510·8

[556]
SCHAAF, W.L. A Bibliography of recreational mathematics. Washington, National Council of Teachers of Mathematics, 1970-78. 4v. facsims., diagrs. ISBN: 0873531209.

First published 1955.

V.4 (1978. xii, [ii], 172p.) has 12 chapters: 1. Arithmetical and algebraic recreations - 2. Number theory as recreation - 3. Geometric recreations - 4. Topological recreations - 5. Magic squares and related configurations - 6. Pythagorean recreations - 7. Classical recreations - 8. Combinatorial recreations - 9. Games and puzzles - 10. Miscellaneous mathematical recreations - 11. Mathematics in the arts - 12. Mathematical recreations and education. 3 appendices (A: Contemporary general works on mathematical recreations). Supplementary glossary, p.146-72. Fills a gap. *Class No:* 510·8

[557]
WELLS, D. **The Penguin book of curious and interesting puzzles.** Harmondsworth, Penguin, 1992. viii,382p. illus., diagrs. £6.99. ISBN: 0140148752.

568 puzzles, with solutions, from ancient times, the nineteenth and twentieth centuries. Bibliography, p.373-77; index, p.379-82. *Class No:* 510·8

Logic

[558]
GREENSTEIN, C.H. **Dictionary of logical terms and symbols.** New York, Van Nostrand Reinhold, 1978. xii, [1], 188p. diagrs., tables.

Contains 'a list of technical terms and expressions most likely to be encountered in currently studied works, along with the ways in which they are commonly characterized, used or defined' (*Preface*). Another aim is 'to reduce the difficulty in formalizing key English expressions in alternative ways'. The notational systems are arranged by functions. Glossary of logical terms. Bibliography (authors, A-Z), p.179-88. *Class No:* 510·6

Arithmetic

[559]
SMITH, D.E. **Rara arithmetica:** a catalogue of the arithmetics written before the year MDCI, with a description of those in the library of George Arthur Plumpton of New York. Boston, Mass., & London, Ginn, 1908. xiii, [ii] 507p. facisms.

Addenda. 1939. x 52p. facsims.

The main work lists *c.*750 items published 1478 - *c.* 1600, giving bibliographical descriptions plus brief notes. Index of dates (of printed books); index of names, places and subjects.

Reprinted 1970. *Class No:* 511.1

[560]
—DE MORGAN, A. Arithmetical books, from the invention of printing to the present time, being brief notices of a large number of works drawn up from actual inspection. London, Taylor & Walton, 1847. [xiii], xxviii, 124p.

About 300 titles covering the period 1491-1800. Chronologically arranged and annotated. *Class No:* 511.1

Number Theory

[561]
FLEGG, G. **Numbers, their history and meaning.** London, Deutsch, 1983. 294p. illus., facsims. ISBN: 0023397282.

9 chapters: 2. Counting with numbers - 3. Writing with numbers - 4. Calculating with numbers - 5. Computing with numbers ... 7. Recreational numbers ... 9. Teaching and learning numbers. Bibliographic note, p.291. Indexes (in tiny print): (a) Names quoted; (b) Works quoted; (c) General. Gives interesting examples of the ways in which numerals were represented through the ages. *Class No:* 511.2

[562]
HOPKINS, N.J., *and others*. **The Numbers you need.** Detroit, Mi., Gale, 1992. 354p. illus., tables. $29.95. ISBN: 081038373x.

9 chapters, closely subdivided (*e.g.* health; gambling; sports; weather (temperature scales, pollution, earthquakes, etc.)). 3 appendices: 1. Units of measurement - 2. Money matters - 3. Use of tables, graphs and statistics. 'Further reading' at end of each chapter. Index. 'A welcome addition to general reference collections' ('Reference books bulletin', in *Booklist,* September 1, 1992, p.90). *Class No:* 511.2

[563]
IFRAH, G. **Histoire universelle des chiffres.** Paris, Seghers, 1981. [viii], 568p. illus., diagrs., charts, tables.

Published with the assistance of the Centre National de la Recherche Scientifique.

6 parts (30 chapters), following an introduction: 1. La conscience des nombres - 2. Les comptes concretes - 3. L'invention des chiffres - 4. Les chiffres et les lettres - 5. Les hybrides - 6. Le stade ultime de la notation numérique (including chapter 30: L'origine des chiffres 'arabes'). Appended: Répertoire chronologique; Orientation et notes bibliographiques (by parts and chapters), p.527-65, - 508

.... (contd.)

numbered references. A scholarly study of the history of numerals, ancient and modern, East and West. *Class No:* 511.2

[564]
LEHMER, D.H. Guide to tables in the theory of numbers. Washington, National Research Council, National Academy of Sciences, 1941. xiv, 177p.

Part 1: Descriptive account of existing tables (p.5-83) - Part 2: Bibliography (p.85-125; under authors A/Z, - *c.*500 items, including manuscripts, with locations) - Part 3: Errata. Index, p.173-7. *Class No:* 511.2

Algebra

[565]
HOWSON, A.G., *comp*. **A Handbook of terms used in algebra and analysis.** Cambridge Univ. Press, 1972. ix, 238p. £20; £8.

38 sections (2. Sets and functions ... 21. Topological spaces and continuity ... 37. Measure and Lebesque integration - 38. Fourier series). Appendix 1: Some named theorems and perspectives; 2: Alphabets used in mathematics. Index of symbols. Analytical subject index ('Function(s)': 2/3column), p.225-38. *Class No:* 512

Analysis

Differentials

[566]
COHEN, M.S., *and others*. **Student research projects in calculus.** Washington, D.C., Mathematical Association of America, 1992. 216p. $20. ISBN: 0883855038.

Includes a list of more than 100 project problems, most class tested, with level of difficulty indicated. *Class No:* 517.2

Integrals

[567]
GRADSHTEYN, I.S. *and* **RYZHIK, I.M. Tables of integrals, series and products.** Corrected and enlarged ed., prepared by A. Jeffrey. New York & London, Academic Press, 1980. 1160p. ISBN: 0122947606.

Translated from the Russian by Scripta Technica, Inc. Incorporates the 4th ed., 1965 (xiv, 1086p.).

Translation of a standard Russian compendium on functions and integrals, plus Fourier and Laplace transforms, etc. To the original Russian bibliography - 'much of which is likely to be inaccessible to the user of these tables' - has been added 'Classified supplementary references'. 'Rich in useful formulae' (Norris, W.T., in *Information sources in engineering*. 2nd ed., 1986, p.372). *Class No:* 517.3

[568]
—**DWIGHT, H.B. Tables of integrals and other mathematical data.** 4th ed. New York, Macmillan, 1961. 335p. tables, formulas. ISBN: 0023311703.

First published 1934.

Includes over 1,000 tables of types of functions. Appended tables of numerical values. Bibliography; index. A handy companion to *Barlow's Tables of squares, cubes, square roots, cube roots, and reciprocals (qv). Class No:* 517.3

[569]
TALLARIDA, R.J. Pocket book of integrals and mathematical formulas. 2nd ed. Boca Raton, Fl., CRC Press, 1992. [xiv],225p. graphs, mathematical formulas. £12.40. ISBN: 0849301424.

11 chapters: 1. Elementary algebra and geometry ... 3. Trigonometry ... 7. Integral calculus ... 9. Special functions ... 11. Statistics. Table of derivatives; table of integrals. 5 Appendices: A.2: 'Poisson distribution'. Index, p.219-25. Handy format, 'a companion to the calculator and the computer' (*Preface to the first edition*). *Class No:* 517.3

Functions

[570]
ABRAMOWITZ, M. *and* **STEGUN, I.A. Handbook of mathematical functions,** with formulas, graphs, and mathematical tables. US National Bureau of Standards. Washington, US Government Printing Office; New York, Dover, 1964. xiv, 1045p. graphs, tables, formulas. (*National Bureau of Standards, Applied mathematics series, no. 55.*) ISBN: 0486612724.

29 chapters by 26 specialists. Bibliographies are brief, but helpful; 'a little more annotation would not have come amiss' (*Nature*, v.205, 2 January 1965, p.9). 'Here, for the first time, the numerical tables of mathematical functions normally useful to engineers, physicists, chemists etc., are found in one text that is authoritative' (*The Military engineer*, no. 373, September/October 1964, p.390). *Class No:* 517.5

[571]
CRC handbook of mathematical sciences. Beyer, W.H., *ed.* 6th ed. Boca Raton, Florida, CRC Press, 1987. 860p. diagrs., graphs, tables, formulas. £52.

5th ed. 1978, and formerly as *CRC Handbook of tables for mathematics*.

Compendium on mathematical and statistical functions, logically arranged in 13 sections. 1. Constants and conversion factors - 2. Algebra - 3. Combinatorial analysis - 4. Geometry - 5. Trigonometry - 6. Logarithmic exponential and hyperbolic functions - 7. Analytical geometry - 8. Calculus - 9. Differential equations - 10. Special functions - 11. Numerical methods 12. - Probability and statistics - 13. Aerodynamics. Appendix: Mathematic symbols and abbreviations. Detailed index, p.249-60. *Class No:* 517.5

[572]
HANSEN, E.R. A Table of series and products. Englewood Cliffs, N.J., Prentice-Hall, 1975. 523p. tables. ISBN: 0138819386.

Systematic collection of tables of mathematical series and infinite products (*e.g.* trigonometric series; logarithmic functions). Bibliography for further reading. Index. *Class No:* 517.5

Probabilities

[573]
BURINGTON, R.S *and* MAY, D.C. **Handbook of probability and statistics, with tables.** 2nd ed. New York, London, etc., McGraw-Hill, 1970. xiv, 462p. tables.

First published 1953.

Designed to complement Burington's *Handbook of Mathematical tables and formulas* (*qv*). Pt.1: 18 sections on elementary statistics and probability theory (p.1-346), with section or sub-section references. Pt.2: Tables (Discrete distribution functions ... Natural trigonometric functions, natural and common logarithms). References, p.427-30. Index of symbols; index of Greek symbols; index of numerical tables. Subject index. *Class No:* 519.2

[574]
CRC handbook of tables for probability and statistics. Beyer, W.H., *ed.* 2nd ed. Cleveland, Ohio, Chemical Rubber Co., 1968. xiv, 642p.

First published 1966.

'Many of the tables have been expanded and increased in effectiveness' (*Preface*). 13 parts: 1. Probability and statistics - 2. Normal distribution - 3. Binominal, Poisson, hypergeometric and negative binominal distributions ... 10. Non-parametric statistics - 11. Quality control - 12. Miscellaneous statistical tables - 13. Miscellaneous mathematical tables. Index, p.637-40. *Class No:* 519.2

Statistics

Bibliographies

[575]
KENDALL, M.G. *and* DOIG, A.G. **Bibliography of statistical literature pre-1940,** with supplements to the volumes for 1940-49 and 1950-58. Edinburgh & London, Oliver and Boyd, 1968. vi, [1], 356p.

Four parts: 1. Pre-1900 - 2. 1900-1939 - 3. Supplement to the 1940-49 volume - 4. Supplement to the 1950-58 volume. About 10,000 entries, under authors A-Z. To avoid confusion, the reference numbers of entries in the pre-1900 section are given in prefix A. In the supplementary section, prefix S is used. Part 3 begins at 7,001 and part 4 at 10,000. No subject index. The 3v. in all contain over 27,000 entries, up to 1958. *Class No:* 519.22(01)

[576]
—KENDALL, M.G. *and* DOIG, A.G. Bibliography of statistical literature, 1940-49. Edinburgh & London, Oliver & Boyd, 1965. [vi], 190p.

Lists 6,036 items under authors A-Z, then chronologically. Full bibliographical details. Two defects: lack of a subject index and failure to translate many of the titles in foreign languages. *Class No:* 519.22(01)

[577]
—KENDALL, M.G. *and* DOIG, A.G. Bibliography of statistical literature, 1950-58. Edinburgh & London, Oliver & Boyd, 1962. xii, 297p.

Lists 9,802 items, under authors A-Z. 26 contributors. For 12 journals almost all the articles are indexed; for 42 others, relevant articles only. On its limitations as a finding tool, see *Librarian-statistician relations in the field of economic statistics* (London, Library Association, 1965) p.42-44. The 3v. set was reprinted in 1981 (New York, Arno). Coverage of the literature is continued in *Statistical theory and method abstracts* (*qv*). *Class No:* 519.22(01)

[578]
LANCASTER, H.O. **'Problems in the bibliography of statistics'.** Royal Statistical Society. *Journal*, series A, v.144, 1970. pt.3, p.409-41.

Part 1: Aims and techniques of statistical bibliography. Part 2: Bibliography and information retrieval (74 references appended). A basic bibliographical survey. *Class No:* 519.22(01)

[579]
—LANCASTER, H.O. Bibliography of statistical bibliographies. Edinburgh & London, Oliver & Boyd, for the International Statistical Institute, 1968. ix, 103p.

Comprises: 1. Personal bibliographies - 2. Subject bibliographies (up to 1965, with some for 1966) - 3. National bibliographies. Subject index to 'Personal bibliographies'; author index to subject and national bibliographies. Aims to cover the theory of statistics at university level. *Class No:* 519.22(01)

Encyclopaedias

[580]
KOTZ, S. *and* JOHNSON, N.L., *eds*. **Encyclopaedia of statistical sciences.** New York, Chichester, etc., Wiley, 1982-. v.1-. illus., diagrs., formulas. £850 (set).

V.1. *A - Circular probable error.* 1982. V.9. *Strata chart - Zyskind.* 1988.

V.1 has 100 contributors. Signed articles, including biographies (*e.g.* Bienaymé I.J.: 3½ columns; 10 references; 14 cross-references). The final article in v.3, 'Hypothesis testing', occupies p.712-22, with 17 references, 15 cross-references. Included are entries for leading organizations and core journals. A scholarly encyclopaedia on theoretical and applied statistics. Supplement published 1989. *Class No:* 519.22(031)

Handbooks & Manuals

[581]
TIETJEN, G.L. **A Topical dictionary of statistics.** London, Chapman & Hall, 1986. ix, [i] 171p. £17.50. ISBN: 0412012014.

15 topical chapters, each with short annotated bibliography: Summarizing data - Random variables and probability distribution - Some useful distributions - Estimation and some hypothesis testing - Regression - The design of experiments and the analysis of variants - Reliability and survival analysis - Order statistics - Stochastic processes - Time series - Categorical data - Epidemiology - Quality control and acceptance sampling - Multivariate analysis - Survey sampling. Index of specific terms. Detailed, analytical index. *Class No:* 519.22(035)

[582]
—WALL, J.J. Statistical data analysis handbook. New York, etc., McGraw-Hill, 1986. 136+[2]+[8]p. tables. £49.50. ISBN: 0070679312.

In 5 parts: 1. Simple comparison - 2. Linear relationship - 3. Comparison of relationships - 4. Supporting Topics - 5. Tables, A-T (*e.g.* T: Random digits). 2-page bibliography. 8-page index. *Class No:* 519.22(035)

Dictionaries

Polyglot

[583]
HUNGARY. Központi Statisztikai Hivatel. **Statisztikal szótái.** 1700 stativztikai kifejezés hat nyelven. [Statistical dictionary. 1700 statistical terms in seven languages.] [2 kaid]. Budapest, Statisztikai Kiadá Vállalat, 1961. viii, 171p.

First published 1960.

The basic sequence is of Russian terms, with Bulgarian, Czech, English, German, Hungarian and Polish equivalents and indexes. *Class No:* 519.22(038)=00

[584]
PAENSON, I., *with others*. **Glossaire systématique, anglais-français-espagnol-russe de la terminologie des methodes statistiques.** Oxford, etc., Pergamon Press, 1970. xxxviii, 517p. tables.

About 1,200 statistical terms are given in context over 20 subject chapters (p.2-413). English, French, Spanish and Russian indexes, p.415-7. Bibliography of sources, p.xxiv. 26 tables. A companion to I. Paenson's *Systematic glossary, English/French/Spanish/Russian of selected economic and second terms* (1963). *Class No:* 519.22(038)=00

English

[585]
KENDALL, M.G. *and* BUCKLAND, W.R. **A Dictionary of statistical terms.** 5th ed. by F.H.C. Marriott. London, Longman, for International Statistical Institute, 1990. [viii],223p. tables, formulas. ISBN: 0582019052.

First published 1957 by M.G. Kendall and W.R. Buckland. 4th ed. 1982.

Defines about 3,400 terms, with some 400 new additions. Entries range from 25 to 150 words. Author and date citations are given for more recent terms. Formulae and cross-references figure. *Class No:* 519.22(038)=20

[586]
KURTZ, A.K. *and* EDGERTON, H.A. **Statistical dictionary of terms and symbols.** New York, Wiley, 1939; London, Chapman & Hall, 1940. xiii, 191p. (Reprinted New York, Hafner, 1967).

Concise definitions of *c.*2000 terms. Aimed primarily 'to provide the user with clear and accurate definitions of each of the various meanings of the statistical terms which he encounters in his reading of scientific literature' (*Preface*). 'Multiple ratio correlation', *c.*250 words. *Class No:* 519.22(038)=20

Russian

[587]
KOTZ, S. **Russian-English/English-Russian glossary of statistical terms.** London, Oliver & Boyd, for the International Statistical Institute, 1972. 88p.

Russian equivalents of *c.*2,500 terms included in the 3rd ed. of Kendal & Buckland's *A dictionary of statistical terms,* etc. Some terms are given brief explanations. 'Not a book for the beginner, either in statistics or in Russian' (*Nature*, v.239, 8 September 1972, p.114-5). *Class No:* 519.22(038)=82

Reviews & Abstracts

[588]
Current index to statistics: applications, methods and theory. Washington, American Statistical Association and Institute of Mathematical Statistics, 1976-. v.1-. Annual.

V.10 (1985. [vii], 683p.) features articles from *c.*60 core journals, selected monographs and conference proceedings of 1984. 'This annual series aspires to provide comprehensive index coverage on a timely basis. Here statistics is viewed in a very broad sense and ranges from probability theory to such topics as the details of how to increase the response rate of mail surveys' (*Introduction*). About 1,400 items are listed. Author index, p.1-175; subject index (words-in-context), p.177-655. Index of journal and publisher abbreviations. International coverage. *Class No:* 519.22(048)

[589]
JOINER, P.L., *and others*. **An Author and permuted title index to selected statistical journals.** U.S. National Bureau of Standards. Washington, U.S. Government Printing Office, 1970. iv, 506p.

About 30,000 entries in permuted title index, p.1-354; author index. Appended contents-lists of selected handbooks of mathematical tables. 'This volume must be one of the most comprehensive works of indexing ever undertaken in scientific work' (*Nature*, v.202, 25 April 1964, p.330). *Class No:* 519.22(048)

[590]
Statistical theory and method abstracts. Voorburg, Netherlands, International Statistical Institute, 1959-. v.1(1)-. Quarterly. £46 ($65) pa. ISSN: 00390518.

Continues *International journal of abstracts: Statistical theory and method* (1954-63. 10v. Annual). Incorporates *International journal of abstracts: Application of statistical methods in industry*.

About 3,000 abstracts pa. Classification scheme (00 ... 15), closely subdivided: 00. Mathematical methods - 01. Probability theory ... 08. Analyses of variance - 09. Sampling ... 12. Statistical inference for stochastic processes - 13. Operations research and related topics ... 15. General features. Brief abstracts (20-150 words). Author index precedes in each issue. Annual author index and index to book reviews. Continues Kendall and Doig bibliographies (*qv*). *Class No:* 519.22(048)

Tables & Data Books

[591]
GREENWOOD, J.A. *and* HARTLEY, H.O. **Guide to tables in mathematical statistics.** Princeton, N.J., Princeton Univ. Press; London, Oxford Univ. Press, 1962. lxii, 1014p. diagrs.

Sponsored by the Committee on Statistics of the Division of National Academy of Sciences, National Research Council (USA).

A descriptive catalogue in 16 main sections. Appendices: 1. Supplement to the descriptive catalogue - 2. Contents of books of tables - 3. Mathematical tables listed in this *Guide*, and in Fletcher, etc. Author and subject indexes. Near-printed; a bulky volume. *Class No:* 519.22(083)

[592]
—OWEN, D.B. Handbook of statistical tables. Reading, Mass., Addison-Wesley Publishing Co., Inc.; London, Pergamon Press, 1962. xii, 580p. tables.

Intended for the student statistician, the practicing statistician, quality-control personnel, research worker, etc. 20 sections (1. Normal distribution ... 20. Orthogonal polynomials, random numbers and constants). Bibliography of 251 items (p.541-53). *Class No:* 519.22(083)

[593]
NEAVE, H.R. **Statistical tables for mathematicians, engineers, economists and the behavioural and management sciences.** London, Allen & Unwin, 1978. 87, [1]p. tables.

9 sections: 1. Discrete probability distributions - 2. The normal distribution - 3. Continuous probability distributions - 4. Analysis of variance - 5. Non-parametric tests - 6. Correlation - 7. Random numbers - 8. Quality control - 9. Miscellaneous. References to appropriate books or articles figure for the less familiar tables. *Class No:* 519.22(083)

[594]
PEARSON, E.S. *and* HARTLEY, H.O., *eds*. **Biometrika tables for statisticians.** London, Biometrika Trust, 1976. 2v. tables.

First published 1954-72. Based on *Tables for statisticians and biometricians,* edited by K. Pearson (1924; 3rd ed., 1948).

Tables in applied statistics, v.2 covering more specialised functions, with examples. (Biometrics is concerned with the statistical study of biological problems). *Class No:* 519.22(083)

[595]
POWELL, F.C. **Statistical tables for the social, biological and physical sciences.** Cambridge Univ. Press, 1982. 96p. diagrs., tables. £15, £3.95. ISBN: 0521241413.

Extends the range of the Fisher-Yates tables with 13 line drawings and numerous tables. *Class No:* 519.22(083)

Stochastic Processes

[596]
WOLD, H.O.A. **Bibliography on time series and stochastic processes.** Edinburgh & London, Oliver & Boyd, for the International Statistical Institute, 1965. xv, 516p.

Nearly 6,000 entries, limited - for the period up to 1930 and to a lesser extent 1931-50 - to items with a basic value for further developments. No abstracts, although each entry carries a 6-category code (including 'field of application'). The bibliography, ending with 1959, continues in

....(contd.)
International journal of abstracts: Statistical theory and method (1954-63) and *Statistical theory and method abstracts* (1959-) *(qv)*. *Class No:* 519.246

Computation

[597]
FISHER, D., *ed*. **Rules of thumb for engineers and scientists.** Rev. ed. Houston, Tx., Gulf, 1991. 242p. tables. $28.95. ISBN: 0872017869.

Several hundred published rules, by proponent, A-Z. Indexes of dependent and independent variables. *Class No:* 519.254

[598]
Mathematics of computation: a journal devoted to advances in numerical analysis, the application of computational methods, mathematical tables, high-speed calculators, and other aids to computation. Washington, National Research Council, National Academy of Sciences, 1941-. Quarterly.

Previously (1943-59) as *Mathematical tables and other aids to compilation.*

This journal *(MTAC)* 'serves as a clearing house for all sorts of mathematical work, containing articles, reviews of new books and reviews of new tables'. Lists of errors in old and new tables are published as received from contributors worldwide. The journal's editorial office also maintains a repository of unpublished mathematical tables (UMT). *Class No:* 519.254

Mathematical Tables

---

[599]
BARLOW, P. **Barlow's Tables of squares, cubes, square roots, cube roots, and reciprocals of all integers up to 12,000.** 4th ed., edited by L.J. Comrie. London, Spon, 1941. xii, 258p.

First published 1814. 'In 1840 De Morgan edited a new edition, the stereo plates for which were used for all subsequent editions' *(Preface to the third edition).*

Invaluable to the mathematician and designed to facilitate mathematical processes. Essential for any scientific reference library and decidedly cheaper than the *CRC Handbook of mathematical sciences (qv)*. *Class No:* 519.66

[600]
BRITISH ASSOCIATION FOR THE ADVANCEMENT OF SCIENCE. **Mathematical tables.** Cambridge Univ. Press, 1931-52. v.1-10.

Under the auspices of the Royal Society.

1. *Circular and hyperbolic functions.* 3rd ed. 1951. 2. *Emden functions.* 1932. 3. *Minimum decompositions into fifth powers.* 1933. 4. *Cycles of reduced ideals in quadratic fields.* 1934. 5. *Factor table.* 1935. 6. *Bessel functions.* pt.1. 1937. 7. *The probability integral.* 1939. 8. *Number-division tables.* 1940. 9. *Tables of powers giving integral powers of integers.* 1940. 10. *Bessel functions.* pt.2. 1952. Part Vol.A: *Legendre polynomials.* 1946; B. *The Airy integrals.* 1946. *Auxiliary tables,* nos. 1-2. 1946. Continued in the Royal Society mathematical tables *(qv)*. *Class No:* 519.66

[601]

BURINGTON, R.S. Handbook of mathematical tables and formulas. 5th ed. New York, etc., McGraw-Hill, 1973. x, [2], 500p. diagrs., tables. $24.50.

Part 1 summarizes the more important formulas, definitions and theorems. Part 2: 5-place logarithmic and trigonometrical tables and more specialised tables. References, p.450-1. Glossary of symbols. Index of numerical tables. Analytical subject index, p.459-500. Aims to meet the needs of students and workers in mathematics, engineering, physics, chemistry, science. A well-known handbook. *Class No:* 519.66

[602]

COMRIE, L.J. Chambers six-figure mathematical tables. Edinburgh & London, Chambers, 1948-49. 2v.

V.1: Logarithmic tables; V.2: Natural values (trigonometrical functions of angles in degrees; circular volume; functions; exponential and hyperbolic functions). Each volume has an appendix of physical and other constants, plus a bibliography of more extended tables (p.568-73). *Class No:* 519.66

[603]

CRC standard mathematical tables. Beyer, W.H., *ed.* 29th ed. Boca Raton, Florida, CRC Press, Inc., 1991. *c.*600p. tables., formulas. £29. ISBN: 0849306299.

First published 1964; 28th ed., 1987.

Sections: Constants and conversion factors - Algebra - Combinatorial analysis - Geometry - Trigonometry - Logarithmic, exponential, and hyperbolic functions - Analytical geometry - Calculus - Differential equations - Special functions - Numerical methods - Probability and statistics - Financial tables. Detailed index. Extends beyond the usual mathematical tables, but for fuller coverage see such other CRC publications as *CRC Handbook of mathematical sciences* and *CRC Handbook of tables for probability and statistics. Class No:* 519.66

[604]

FLETCHER, A., *and others, eds.* **An Index of mathematical tables.** 2nd ed. Oxford, Blackwell, for Scientific Computing Service, Ltd., 1962. 2v. (994p.).

First published 1946 (London, Scientific Computing Service, Ltd).

V.1, pt.1: Historical introduction. Index of mathematical tables (with descriptions and comparisons), according to functions (24 series) - V.2, pt.2: Bibliography (p.609-780; over 4,000 references; under authors, A-Z) - pt.3: Errors (in published tables) - pt.4: Index to introduction and pt.1. The 2nd ed., more than twice the size of the 1st, adds the feature, the list of errors. 'This standard work of reference became an indispensable classic from the moment of its publication in 1946' (*Nature,* v.201, 8 February 1964, p.537). *Class No:* 519.66

[605]

LEBEDEV, A.V. *and* **FEDEROVA, R.M. A Guide to mathematical tables.** English ed., prepared from the Russian by D.G. Fry. Oxford, etc., Pergamon Press, 1960. xlvi, 586p. *Supplement* no. 1. 1960. xxxviii, 190p.

Translated from the Russian ed. of 1956.

Tables are arranged by mathematical function or type of operation. International coverage, but its emphasis on Russian tables makes it complementary to Fletcher's *An index of mathematical tables* (*qv*). *Class No:* 519.66

[606]

ROYAL SOCIETY. Mathematical Tables Committee. **Royal Society mathematical tables.** Cambridge Univ. Press, for the Royal Society, 1950-68. 11v.

Continuation of the British Association for the Advancement of Science's *Mathematical tables* (Cambridge Univ. Press, 1931-52. 10v.) (*qv*).

1. *The Forey series of order 1025.* 1950. 2. *Rectangular-polar conversion tables.* 1956. 3. *Tables of binomial coefficients.* 1953. 4. *Tables of partitions.* 1958. 5. *Representation of primes by quadratic forms.* 1960. 6. *Tables of the Riemann zeta function.* 1960. 7. *Bessel functions.* pt.3: *Zeros.* 1960. 8. *Tables of natural and common logarithms to 110 decimals.* 1964. 9. *Tables of indices and primitive roots.* 1968. 10. *Bessel functions.* pt.4: *Kelvin functions.* 1964. 11. *Coulomb wave functions.* 1964. *Auxiliary tables.* 1950. *Class No:* 519.66

Ready Reckoners

[607]

Collins' national decimal reckoner. London, Collins, 1970. 832p.

Follows very much the pattern of *Collins' Instant reference reckoner* (1954), based on sterling. Contents: Main price tables (405p.) - Profit and discount - Profits on sales or mark-up - Interest (170p.) - Ton rates - Cut rates - Wages - Overtime rates - Salary - Multiplication tables - Conversions - Weights and measures. *Class No:* 519.661

Logarithms

[608]

THOMPSON, A.J. Logarithmetica Britannica, being a standard table of logarithms to twenty decimal places of the numbers 10,000 to 100,000. Cambridge, Univ. Press, 1952. 2v.

Issued by the Department of Statistics, University College, London, to commemorate the tercentenary of Henry Briggs' publication of *Arithmetica logarithmica,* 1624.

V.1: Numbers 10,000 to 50,000; v.2: Numbers 50,000 to 100,000. V.1 contains a lengthy general introduction (p.xii-lxxxiii), including a list of errors in Briggs. *Class No:* 519.662

Astronomy

Databases

[609]

INSPEC. Stevenage, Herts., Institution of Electrical Engineers, 1969-. Updates, Updates weekly.

There is at present no online database wholly devoted to astronomy. INSPEC is the best available online because of the close relationship between astronomy and physics. Coverage embraces periodical articles, conference proceedings, books, technical reports, theses, etc.

Other databases that contain astronomical information are: CA Research (for stellar spectra, etc.), PASCAL and *Meteorological and geoastrophysical abstracts,* (*Database.* v.8(4), December 1985, p.15). *Class No:* 520(003.4)

Bibliographies

[610]

HOUZEAU, J.C. *and* **LANCASTER, A. Bibliographie générale de l'astronomie,** ou, Catalogue méthodique des ouvrages, des memoires et des observations astronomiques publiés depuis l'origine de l'imprimerie jusqu'en 1880. Brussels, Hayez. 1882-89. 2v. in 3.

Reprinted London, Holland Press, 1964.

V.1: *Ouvrages imprimés et manuscrits* (1887-89): 15,880 items, systematically arranged; bibliographical notes; locations of MSS are given. The lengthy introduction (to p.310) covers the historical background. V.2: *Mémoires et notices insérés dans les collections académiques et les périodiques* (1882) concerns articles in periodicals, prefixed by a list of periodicals scanned. Systematically arranged; author index, citing short titles of articles. The primary bibliographical science for the history of astronomy. The 1964 reprint includes an author index to v.1.

Class No: 520(01)

[611]

—Astronomischer Jahresbericht. Berlin, de Gruyter, 1900-68. v.1-68.

Continues Houzeau and Lancaster, with a gap for 1880-98. The unpublished material for those years - over 60,000 slips - has been microfilmed by University Microfilms, Ltd. on 18 reels of 35mm. film (1970), the original being at the Observatoire Royale de Belgique, Brussels.

Class No: 520(01)

[612]

—LALANDE, J.J.F. de. Bibliographie astronomique, avec l'histoire de l'astronomie depuis 1781 jusqu'à 1802. Paris, Imprimerie de la République, 1803. viii, 966p.

Lists 5,000 titles and is arranged chronologically, with author and subject indexes (unlike v.1) of Houzeau and Lancaster. Reprinted 1970 (Amsterdam, Gieben).

Class No: 520(01)

[613]

KEMP, D.A. Astronomy and astrophysics: a bibliographical guide. London, Macdonald, 1970. xxiii, 584p. (*Macdonald bibliographical guides.*) ISBN: 0356030113.

Annotated entries for 3,642 numbered items (with interpolations). 75 sections, systematically arranged (1. Reference media - 2. Star catalogues, ephemerides, etc.... 25. Origin of the solar system... 49. Stellar evolution... 73. Cosmology - 74. Abundance and origin of the elements. - 75. Cosmic rays and gamma - and X-ray astronomy). Periodical articles are included. Entries (in column form): running number, title, author, etc; number of items and years covered; annotations (sometimes evaluative). Entries go up to 1969. Author and subject indexes, p.533-84. The author was at one time librarian of the Royal Observatory, Edinburgh. 'An invaluable reference source for astronomers, other scientists working in the fringe areas of astronomy, and librarians' (*The Recorder,* no.271, 15 February 1971, p.2). *Class No:* 520(01)

[614]

LUSIS, A. Astronomy and aeronautics: an enthusiast's guide to books and periodicals. London, Mansell, 1986. xx, 292p. ISBN: 072011795x.

968 annotated and numbered entries. 7 sections: 1. General astronomy - 2. Practical astronomy - 3. History of astronomy - 4. Astronomy plus ... (multi-and interdisciplinary works) - 5. Astrophysics - 6. The solar system - 7. Astronautics (p.198-255). Items asterisked were not available for inspection. Author index; Title index: books; Title index: periodicals; Subject index (analytical). Annotations are often evaluative and may cite reviews in *Sky and telescope,* etc. Attempts to provide a comprehensive annotated list in English 'for anyone ... below professional or undergraduate level' (*Introduction*). Claims to be in some ways a successor to Seal's *Guide* (*qv*). *Class No:* 520(01)

[615]

SEAL, R.A. A Guide to the literature of astronomy. Littleton, Colo., Libraries Unlimited, 1977. 306p.

578 numbered entries, with evaluative comments of a high standard. 4 sections: 1. Reference material in astronomy - 2. General materials (*e.g.* history; textbooks) - 3. Descriptive astronomy - 4. Special topics (*e.g.* space science and aerodynamics). Appended 'Bibliography of basic reference material in astronomy' (p.238-9) - 80 items, including 31 atlases and star catalogues. Items are coded as recommended for A. Astronomy libraries; B. Public libraries; C. College/ university libraries. Author-title and subject indexes. Excludes astronautics. For the non-specialist.

Class No: 520(01)

[616]
—ROYAL ASTRONOMICAL SOCIETY. Catalogue of the Library. [Royal Astronomical Society]. London, the Society, 1886-1924. 3pts.

An author catalogue. Pt.1 (1886) includes entries for much early material. Included are form headings for 'Bibliographies', 'Ephemerides' and sea charts. *Class No:* 520(01)

[617]
—SEAL, R.A. *and* MARTIN, S.S. A Bibliography of astronomy, 1970-1979. Littleton, Colo., Libraries Unlimited, 1982. 407p. ISBN: 0872872807.

Provides a ten-year continuation to Kemp. The 2,119 entries are in 4 parts (18 sections): Reference works - Subject sections - Appendices (International Astronomical Union Colloquia; I.A.U. symposia, 1969-1979; Abbreviations) - Indexes (author; monographic title or conference proceeding index; subject). Annotations average 50 words, conference proceedings being labelled 'C'. Periodical articles are included. Coverage is based on 15 U.S. and Canadian libraries' holdings. Intended as an introductory overview and selection tool for all types of libraries and users, not as a comprehensive reference tool for the decade of the 1970s. *Class No:* 520(01)

[618]
UNITED STATES. Naval Observatory Library, Washington. **Catalog of the Naval Observatory Library.** Boston, Mass., G.K. Hall, 1976. 6v. (4118p.). $655. ISBN: 0816100314.

89,300 photolithographed catalogue-cards, representing a collection of 75,000 items in astronomy and mathematics. The Library was founded in 1842, and the catalogue includes entries for pre-1960 periodical articles. *Class No:* 520(01)

Encyclopaedias

[619]
The Astronomy and astrophysics encyclopaedia. Maran, S.P., *ed.* New York, Van Nostrand Reinhold, 1992. xxix,1002p. illus., figs., tables. £60. ISBN: 0521417449.

403 entries arranged A-Z, written by more than 400 contributors. The articles are very technical and go into great detail. Extensive cross-references. Bibliographies appended to each chapter. Black-and-white illustrations. Index. Attractively presented. 'This volume is now the premier encyclopedia of astronomy' (*Choice*, March 1992, p.1037), although the reviewer in *Reference reviews*, v.6(3) 1992, p.6 thinks more illustrations would have been useful. For the academic library. *Class No:* 520(031)

[620]
The Astronomy encyclopaedia. Moore, P., *ed.*,. London, Mitchell Beazley, 1987. 464p. illus. (incl. col.), ports., diagrs., tables. ISBN: 0855336048.

Published as *The International encyclopedia of astronomy* (Crown, 1987).

About 100 contributors. A-Z sequence of 2,500 entries (50-3,000 words in length), including 7 essays on major topics (*e.g.* 'The big bang'; 'Exploring space'; 'Interstellar matter'; 'Pulsars'). 570 illus. (170 in colour) well captioned. Cross-references. No bibliographies. 'A volume that will remain a popular reference source for many years to come' (*New scientist,* no.1589, 3 December 1987, p.67). For school and public libraries. *Class No:* 520(031)

[621]
Encyclopedia of astronomy and astrophysics. Meyers, R.A., *ed.* San Diego, Academic Press, 1988. xiv,807p. illus., diagrs., tables. $59.95. ISBN: 0122266900.

41 articles, arranged A-Z (Astronomy, Infrared ... X-ray Astronomy), averaging 20 pages in length. Each article has a glossary and bibliography appended. *c.*300 photographs and illustrations. Cross-references and analytical index. *Class No:* 520(031)

[622]
McGraw-Hill encyclopedia of astronomy, by the staff of the McGraw-Hill Encyclopedia of science and technology. New York, McGraw-Hill, 1983. 450p. illus. (incl. col. pl.), tables. £69. ISBN: 0070452512.

230 articles, A-Z and briefly documented, drawn unabridged from the parent *McGraw-Hill Encyclopedia of science and technology* (5th ed., 1982). 103 contributors; extensive cross-references; detailed index. 402 illus., with 13 colour plates added. Well and attractively produced. Claims to be the first technical treatment of astronomy. The reviewer in *Astronomy* (v.11, September 1983, p.39) finds the *Encyclopedia* 'heavily slanted towards theoretical and experimental areas'. *Class No:* 520(031)

[623]
WEIGERT, A. *and* ZIMMERMAN, H. Concise encyclopedia of astronomy. Dickson, J.H., *trans.* 2nd English ed., revised by H. Zimmerman. Bristol, Hilger, 1975. 532p. illus., diagrs., tables.

Based on *ABC of astronomy* (translated from the 2nd ed. of *ABC der Astronomie*).

About 2,000 entries (A ... Zone time), with many cross-references. Includes brief biographies (Kepler: 2 columns; Galileo: 2/3column). 66 plates appended. 'For the advanced amateur and is also an excellent reference book' (*Nature*, v.264, 9 December 1976, p.591-2). *New scientist* (9 December 1976, p.609) finds the photography 'still poorly produced'. *Class No:* 520(031)

[624]
—The Encyclopaedia of astronomy and space. Ridpath, I., *ed.* London, Macmillan, 1976 (rev. ed. 1980).

Over 1,000 short but adequate entries, including brief biographies, as well as more than 100 tables, charts and photographs, many in colour. Decidedly more simple text than Weigert and 'seems aimed at people in their early teens' (*Nature,* v.264, 9 December 1976, p.591-2). *Class No:* 520(031)

Handbooks & Manuals

[625]
ABELL, G.O. **Exploration of the universe.** 5th ed. Philadelphia, Pa., Saunders College Publishing, 1987. x,748p. (incl. col.), ports., diagrs., tables, star charts. ISBN: 0030051436.

First published 1964.

38 chapters, descriptive and historical, with exercises appended. 21 appendices (1. Bibliography; 2. Glossary (p.699-721)... 21. Star maps). Double-column page; analytical index (in tiny print). An attractively produced textbook, well organised and clearly written. Aimed at the average college student, taking a one-year or even a one-semester course. Highly regarded by R.A. Seal (*A Guide to the literature of astronomy*). 6th ed., 1991. 682p. $37.50. *Class No:* 520(035)

[626]

The Cambridge encyclopaedia of astronomy.
Mitton, S., *editor-in-chief*. London, Cape, 1977. 481, [14]p.
illus., sky maps. Page size, 25 x 25cm. £60. ISBN:
0517528061.

16 contributors. 23 chapters (1. A survey of the
universe... 9. The solar system... 15. Our local group of
galaxies... 20. Life in the universe. 21. Major trends in the
history of astronomy... 23. Astronomy in space). Appendix
1: Star atlas of Smithsonian Astrophysical Observatory sky
maps (p.457-71). 'Material has been gathered into cohesive
theories in order to present a more accurate and
understandable guide to the new universe' (*Introduction*).
Analytical index, in tiny type. Over 500 illus. (150 in full
colour), with descriptive captions. Data up to 1975. No
supplementary readings or bibliography. Written primarily
for readers in the northern hemisphere. However, the
volume has a wealth of astronomical data and fine
illustrations, with a detailed index. The editor-in-chief was
then Secretary of the Institute of Univ. of Cambridge and
most of the contributors were associated with the Institute.
Class No: 520(035)

[627]

Encyclopedia of astronomy: a comprehensive survey of our
solar system, galaxy and beyond. Ronan, C., *ed*. London,
New York, etc., Hamlyn, 1979. 240p. illus. (incl. coloured),
star charts, diagrs., graphs.

6 contributors. Chapters : Introduction - The stars - The
solar system - Extragalactic astronomy - Theories of the
universe - Observing the universe - The amateur observer -
Star charts. 13 appendices (6. Star catalogues; 10.
Comparative data on the planets; 12. Names of stars).
Appended glossary, short bibliography and index (in tiny
print). Like the *Cambridge Encyclopedia of astronomy*, fully
illustrated, making good use of a large format. Praised in
New scientist (29 November 1979, p.710) for its in-depth
treatment of galaxies. 'The pictures and diagrams are as
good as the text'. *Class No:* 520(035)

[628]

**HARTUNG, E.J. Astronomical objects for southern
telescopes:** with an addendum for northern observatories.
Cambridge, University Press, 1968 (reprinted 1984). 238p.
illus. tables. £13.95. ISBN: 0521318874.

5 introductory chapters: 1. Radiation - 2. Stars - 3. Star
clusters - 4. Galactic nebulae - 5. Extra-galactic systems - 6.
Amateur observing. Followed by: General list of telescopic
objects for southern observers; Description of constellations
and objects (p.66-227) (over 1,000 objects). Addendum for
northern observatories (80 objects). Bibliography. Index.
16p. of black-and-white plates. For the amateur observer.
Class No: 520(035)

[629]

**LANG, K.R. and GINGERICH, O., *eds*. A Source book in
astronomy and astrophysics, 1900-1975.** Cambridge,
Mass., & London, Harvard Univ. Press, 1979. xx, 922p.
£63.95. ISBN: 0674822005.

Collects 132 seminar papers in astronomy and
astrophysics, - reproduced as originally published, but with
editorial comments. 8 chapters: 1. New windows on the
universe - 2. The solar system - 3. Stellar atmospheres and
stellar spectra - 4. Stellar evolution and nucleosynthesis - 5.
Variable stars and dying stars - 6. The distribution of stars
and the space between them - 7. Normal galaxies, radio
galaxies and quasars - 8. Relativity and cosmology. Author

.... *(contd.)*

and analytical subject indexes. 'Some papers are technical
but well within the purview of the undergraduate science
major' (*Choice*, v.19(1), September 1981, p.34).
Class No: 520(035)

[630]

Larousse astronomy. Morris, M.R., *ed*. Twickenham,
Middx., Hamlyn, 1987. 326p. illus. (incl. col.). facsims.,
diagrs., tables, star charts. ISBN: 0600501086.

A revised translation of *Astronomie, les astres, l'univers*,
originally published by Larousse in 1948 and subsequently,
with supplement, in 1956. Previous ed. *Larousse
encyclopedia of astronomy* (1966).

4 sections (25 chapters): 1. Man and the sky - 2. The
empire of the sun - 3. The realm of the stars - 4. Toward
infinity. Well written and lavishly illustrated, with numerous
plates, diagrams, etc. Bibliography, p.321-2. Index (tiny
print). Essentially for the lay person. *Class No:* 520(035)

[631]

MOORE, P. The Guinness book of astronomy. 3rd ed.
Enfield, Guinness Publishing, 1988. 288p. illus. (incl. col.),
star charts, tables. £12.95. ISBN: 0851123759.

First published 1979. Before 3rd edition, known as *The
Guinness book of astronomy facts and feats*.

Sections: The solar system (p.5-133) - The stars - The star
catalogue (p.163-254) - Telescopes and observatories - The
history of astronomy - Astronomers - Glossary - Index.
Clearly and attractively produced. A good introduction for
the beginner. 4th ed. due to be published 1992/93.
Class No: 520(035)

[632]

MUIRDEN, J. The Amateur astronomer's handbook. 3rd
ed. New York, Harper & Row, 1983. [viii], 472p. illus.,
diagrs., tables. $16.95; £7.95. ISBN: 0061816221.

First published 1968.

5 parts (29 chapters). 1. Equipment - 2. The solar system -
3. The stars and nebulae - 4. Autobiography - 5. Optical
work for amateurs (a major new section). 8 appendices (*e.g.*
Tables of precision for ten years). Glossary; chapter
bibliographies, p.457-62; analytical index. p.463-72. Written
primarily for readers in the northern hemisphere, and for the
novice who wants to make telescope observations on his or
her own. It tries 'to bridge the gap between works that deal
with the spectacular 'wonders of the sky' and the formal
handbook full of facts'. Not to be confused with Sidgwick's
similarly titled book. *Class No:* 520(035)

[633]

—**NEWTON, J. and TEECE, P. The Guide to amateur
astronomy.** Cambridge, University Press, 1988. viii,327p.
illus., diagrs. £17.50. ISBN: 0521340284.

Written by amateur astronomers and intended mainly for
the beginner. 5 sections:- 1. Amateur fundamentals: getting
started - 2. Guide to the night sky - 3. Amateur observing
advanced: projects and techniques - 4. A complete guide to
astrophotography - 5. The build-it-yourself astronomer. 4
appendices, *e.g.* 1. Sky nomenclature ... 4. Star colours and
spectral classes. Good quality illustrations and diagrams.
Class No: 520(035)

[634]
SIDGWICK, J.B. Amateur astronomer's handbook. 4th ed., prepared by J. Muirden. London, Pelham Books, 1979. xxix, 508p. diagrs., tables. ISBN: 0720711649.

1st ed., 1955.

41 sections on the instrumental and theoretical background of practical astronomy (40. Manufacturers and suppliers; 41. Bibliography, p.521-52, under 29 heads). Index, partly analytical, p.555-68. A classic in its field. Although the 4th ed. shows extensive revision, the text retains its original flavour. *Class No:* 520(035)

[635]
—SIDGWICK, J.B. Observational astronomy for amateurs. 4th ed., prepared by J. Muirden. London, Pelham Books, 1982. xix, 348p. ISBN: 0720713781.

Intended as a sequel to Sidgwick's *Amateur astronomer's handbook,* and first published in 1957. The 19 sections (1. Solar observation... 19. Nebulae and clusters) are followed by a lengthy bibliography (p.299-337) under 27 heads, and an analytical index. 'It is a sound reference for established and would-be practising astronomers, even if they have an earlier edition, and also for all amateurs interested in the real basics of solar system astronomy' (*Sky and telescope,* v.65, January 1983, p.40). *Class No:* 520(035)

Dictionaries

[636]
HOPKINS, J., *ed.* **Glossary of astronomy and astrophysics.** 2nd ed. Chicago, Ill., & London, Univ. of Chicago Press, 1980. ix, 196p. £19.95; £10.50. ISBN: 0226351718.

1st ed., 1976. Published under the auspices of *Astronautical journal.*

Briefly defines *c.*2000 of the astronomical and astrophysical terms used in current research. Eponyms and abbreviations also figure. Variant definitions given by different authors are included. A handy compendium. *Class No:* 520(038)

[637]
—The Penguin dictionary of astronomy. Harmondsworth, Middx., Penguin, 1993. 432p. £6.99. ISBN: 0140512268.

About 2,500 entries, covering all aspects of historical and modern astronomy, names of stars, etc., and of observatories. *Class No:* 520(038)

[638]
KLECZEK, J. Astronomical dictionary. New York & London, Academic Press; Prague, Publishing House of the Czechoslovak Academy of Sciences, 1962. 972p. ISBN: 0124119506.

Originally published in Prague, 1961.

An English-Russian-German-French-Italian-Czech dictionary of *c.*10,000 terms in astronomy and related sciences (*e.g.* mathematics, spectroscopy). The main part (p.16-574) consists of English base terms, with equivalents across the double page, in 34 sections. English, Russian, German, French, Italian and Czech indexes, p.577-972. British and American usages are not differentiated, but genders of nouns in the foreign languages are given. *Class No:* 520(038)

[639]
The Macmillan dictionary of astronomy. Illingworth, V., *ed.* 2nd ed., (1st ed., 1979, has U.S. title *The Facts on File dictionary of astronomy*). London, Macmillan, 1985. [v], 437p. diagrs., tables. £30; £12.99. (*Macmillan reference books.*) ISBN: 0333390628.

1st ed., 1979.

19 contributors. Over 2,300 entries, generous in length (*e.g.*'Saturn' : 2 columns; 'Black hole' : 2½ columns). Ample cross-references. 12 appended tables (*e.g.* 'Planetary satellites; 12. Recent successful planetary probes). 'The book walks the fine line between dictionary and encyclopaedia, but it does it very well' (*New scientist,* 5 September 1985, p.57). *Class No:* 520(038)

[640]
MEL'NIKOV, A.O. Anglo-russkiĭ astronomicheskiĭ slovar'. Moscow, Izdat. 'Sovetskaya Entisklopediya'. 1971. 504p.

English-Russian astronomical dictionary. About 20,000 entry-words. *Class No:* 520(038)

[641]
MITTON, J. A Concise dictionary of astronomy. Oxford, University Press, 1991. vii,423p. figs., tables. £12.95. ISBN: 0198539673.

Contains *c.*2,500 entries. Includes acronyms and many cross-references. Bibliography, p.410-12. 7 tables (*e.g.* 'The characteristics of planetary ring systems'). Aimed mainly at the amateur and the general reader. Author is editor of the *Journal of the British Astronomical Association.* *Class No:* 520(038)

[642]
PLANT, M. Dictionary of space. Harlow, Essex, Longman Group, 1986. xi, 270p. illus. £7.95. ISBN: 0582892953.

Over 2,500 entries, including brief biographical sketches, 'Aberration'... 'Zond'. 'Hubble space telescope': 1p., plus cross-references; 'Galileo' : 2/3p. : 'Mars' : ½p. : 'Mission 41B'... 'Mission 61C', p. 143-8. 22 keyed line-drawings. Intended for 'casual readers or keen expert' (*Introduction*). *Class No:* 520(038)

[643]
ROOM, A. Dictionary of astronomical names. London, Routledge, 1988. [vi],282p. illus. £19.99. ISBN: 0415012988.

*c.*600 entries on stars, minor planets, constellations, planets, planetary constellations, and lunar features. 3 sections: 1. Introduction - 2. Astronomical glossary - 3. Dictionary of astronomical names (Asteroid ... Zubeneschamali). 2 appendices: 1. Names of craters on the moon - 2. Names of minor planets (1,000 are listed). Entries in the dictionary are variable in length. 2-page bibliography. Profuse cross-references. A valuable introduction. *Class No:* 520(038)

Reviews & Abstracts

[644]
Astronomy and astrophysics abstracts. Berlin, Springer-Verlag for Astronomisches Rechen-Institut, Heidelberg, 1969-. v.1(1)-. 2pa.

Prepared under the auspices of the International Astronomical Union. Continues *Astronomisches Jahresbericht* (v.1-68, 1900-68).

V.51B: Literature, 1990, pt.1 has 566p. of entries. Parts A and B of v.51 comprise together 11,423 brief abstracts and

....(contd.)

references. The 13 main classes (001-161) comprise: 001-015: Periodicals, proceedings, books, activities - 021-022: Applied mathematics, physics - 031-036: Astronomical instruments and techniques - 041-046: Positional astronomy, celestial mechanics - 051-053: Space research - 061-067: Theoretical astrophysics - 071-080: Sun - 081-085: Earth - 091-107: Planetary system - 111-126: Stars - 131-134: Interstellar matter, nebulae - 041-144: Radio sources, X-ray sources, cosmic rays - 151-161: Stellar systems, galaxy, extragalactic objects, cosmology. Author, subject and object indexes. A major abstracting and indexing service. *Class No:* 520(048)

[645]
PASCAL Explore. E48: Environment cosmique terrestre, astronomie et geologie extraterrestre. Vandoeuvre-Les-Nancy, Centre National de la Recherche Scientifique, 1984-. 10pa. FFr1,180. ISSN: 07612109.

Formerly (v.1-16. 1940-55) *Bulletin analytique,* continuing *Bibliographie mensuelle de l'astronomie. Bulletin signalétique. 120: Astronomie. Physique spaciale. Géophysique* (1940-83).

About 20,000 abstracts and references pa. Classes A-H (A. Physique du globe solide... E. Astronomie fondementale et astronomique. Instruments et techniques. Observations astronomiques et extragalactiques. L'univers). Monthly and annual author and subject indexes. Available online and in microform. *Class No:* 520(048)

[646]
Referativnyǐ zhurnal. 51. Astronomiia. Moscow, Akademiia Nauk SSSR, Institut Nauchnoǐ Informatsii, VINITI, 1953-. 12pa. 96 rubles. ISSN: 04862236.

Previously part of *Referativnyǐ zhurnal. Astronomiia i geodeziia* (1953-62. Monthly).

Each issue contains *c.*1,000 indicative and informative abstracts. Sections: General astronomy - Theoretical astronomy - Astronometry - Astrophysics - Theoretical astrophysics - Stellar astronomy - Radio astronomy - The sun - Solar system - Cosmogony and cosmology - Astronomical optics, telescopes and observations. Exploration of space. Monthly author index. Annual author and subject indexes. *Class No:* 520(048)

[647]
Vistas in astronomy: an international review. London, Pergamon Press, 1955-. v.1-. (Annual; quarterly, 1976-). $145 (1986) pa. ISSN: 00836656.

Coverage: history and philosphy; instrumentation; solar system; nearby stars; stellar spectra; astrophysics; galaxies; galactic clusters; cosmic rays. V.28(4), 1985 carries an editorial and 3 documented contributions (*e.g.* 'The accuracy of Halley's cometary orbit', 18 references) and list of contents for v.22-27. Special issues are a feature (*e.g.* v.17: 'Copernicus, yesterday and today'; v.18 : 'Kepler : four hundred years'; v.28(1/2): 'The longitude zero symposium, 1984'). *Class No:* 520(048)

Periodicals & Progress Reports

[648]
Annual review of astronomy and astrophysics. Palo Alto, Calif., Annual Reviews, Inc., 1963-. v.1-. illus., port. ISSN: 00664146.

V.30, 1992 has 39 contributors. 26 well-documented articles (*e.g.* 'Astronomical masers'; 'Radio emission from

....(contd.)

normal galaxies'; 'Smooth particle hydrodynamics'). Analytical subject index; cumulative index of contributing authors; v.20-30; cumulative index of chapter titles, v.20-30. One virtue of this annual review is the regular inclusion of an introductory autobiographical sketch (*e.g.* 'A half-century of astronomy'. by A.E. Whitford, in v.24, 1986), throwing light on the current state of the art and providing biographical data on a personality. *Class No:* 520(05)

[649]
BRITISH LIBRARY. Science Reference Library. **Periodicals on astronomy held by the SRL.** Jackson, G. *and* Bradley, O., *comps.* London, Science Reference Library, 1976. [vi], 26p. ISBN: 0902914340.

Lists *c.*150 journals, with SRL (Bayswater, now Kean Street) call-numbers. Appended list of 'Abstracts and bibliographies'. *Class No:* 520(05)

Yearbooks & Directories

[650]
The Air & space catalog: the complete sourcebook to everything in the universe. Makower, J., *ed.* New York, Random/Vintage, 1989. 336p. illus. $27.50; $16.95. ISBN: 0394583272.

Lists organizations, equipment, software and activities in the fields of astronomy, aviation meteorology and space flight. Useful for public libraries. *Class No:* 520(058)

[651]
BRITISH ASTRONOMICAL ASSOCIATION. The Handbook of the British Astronomical Association 1992. London, the Association, 1991. 92p. graphs, tables. Annual. £8.25. ISSN: 0068130x.

Contents include: Visibility of planets ... Eclipses ... Time ... Radio signals ... Earth and Sun ... Moon. Lunar occultations - Planets ... Comets - Meteor diary ... Ephemerides of double stars ... Variable stars ... Bright stars ... Elements of the planetary orbits - Satellites - Astronomical and physical constants - Miscellaneous data and telescope data - Conversion formulae. List of articles in earlier editions of the *Handbook,* on back cover. *Class No:* 520(058)

[652]
Yearbook of astronomy. Moore, P., *ed.* London, Sidgwick & Jackson, 1961-. Annual. illus., graphs, star charts. ISBN: 0283060964.

The *1993 Yearbook* (1992. 256p.) has 8 contributors. 3 parts: 1. Monthly charts and astronomical phenomena (*e.g.* 'Some events in 1994')- 2. 8 articles (*e.g.* 'The invention of the reflecting telescope'); 'Some interesting double stars'). Appended: 'Astronomical societies in Great Britain', p.249-560. Small format. *Class No:* 520(058)

Tables & Data Books

[653]
CURTIS, A.R. Space almanac: facts, figures, names, dates, places, lists, charts, tables, maps covering space from Earth to the edge of the universe. Woodsboro, Md., Arcsoft Publishers, 1990. 955p. illus., tables, charts, maps. $19.95. ISBN: 0866680659.

8 major sections: Astronauts - Space stations - Shuttles - Rockets - Satellites - Solar system - Deep space - Calendar of events. 33-page index. Many photographs, drawings and

....(contd.)

illustrations, although these are of variable quality. However, full of useful information, and a handy reference tool. *Class No:* 520(083)

[654]

JASCHEK, C. Data in astronomy. Cambridge, University Press, 1989. x,198p. tables. £35. ISBN: 0521340942.

Aimed at all those who handle astronomical data. 13 sections: 1. Observations ... 4. Archiving of observations ... 6. Designation of astronomical objects ... 9. Data banks and data bases - 20. Data centres - 11. The publication of scientific information ... 13. International data organizations. Particularly useful are the bibliographies appended to each section. (*e.g.* 7. Catalogues (4p. of references)). Index. *Class No:* 520(083)

Films

[655]

A Catalogue of videos and films on astronomy. Zucker, E., *comp.* Dewsbury, West Yorks., Association for Astronomy Education, 1988. 35p.

Lists *c.*300 titles, arranged by distributor, Each entry gives title; format; duration; level; details of whether the item may be hired or bought. Brief subject index. Supplementary lists, updating this catalogue, appear in the Association's newsletter, *Gnomon. Class No:* 520(084.122)

Histories

[656]

Astrophysics and twentieth-century astronomy to 1950. Gingerich, O., *ed.* Cambridge Univ. Press, 1984. 198, lvip. illus., ports., facsims, graphs, tables. £17.50. (*The General history of astronomy, v.4A.*) ISBN: 0521242568.

Two parts (11 sections, each with further reading): 1. The birth of astrophysics and other late nineteenth century trends (*c.*1850-*c.*1950) - 2. Observatories and instrumentation. Appendix: The world's largest telescopes, 1850-1950. Index, p.li-lvi. Well illustrated, and intended for a wide audience.

V.4B is to cover: 3. Modern astrophysics - 4. The structure of the universe (1900-1950) - 5. The sociology of astronomy. V.2. *Planetary astronomy from the Renaissance to the rise of astrophysics: Pt. A. Tycho Brahe to Newton,* edited by R. Taton and C. Wilson. 1989. ix,274p. 'These are going to be books that every astronomer and many non-astronomers will wish to possess for themselves'. (*History of science,* v.23(1), no. 59, March 1985, p.122). *Class No:* 520(091)

[657]

DEVORKIN, D.H., *ed.* **The History of modern astronomy and astrophysics:** a selected, annotated bibliography. New York, Garland. 1982. 434p. illus. $102. (*Bibliographies of the history of science & technology, v.1.*)

Over 1,400 annotated entries, in subject categories, with numerous cross-references. Emphasis is on English-language material, and range of items extends from the popular to the scholarly; also on mid-19th to mid-20th century literature. Index. 'Imperfect, but a useful starting point' (*Journal of the history of astronomy,* v.4, 1983, p.146). *Class No:* 520(091)

[658]

EVANS, D.S. Under Capricorn: a history of southern hemisphere astronomy. Bristol, Philadelphia, Adam Hilger, 1988. x,376p. illus. £35. ISBN: 082564384x.

Covers astronomy in the area from 1600. 11 sections: 1. The expeditionary era (17th and 18th centuries) ... 3. Some international projects ... 6. The new century: Africa ... 8. The post-war renaissance ... 11. Coda. 2 appendices: 1. Some frequently mentioned instruments - 2. Southern hemisphere observatories. References and bibliography, p.329-52. Index. Many black-and-white illustrations. Lines in the text are double-spaced, making the publication bulkier than it needs to be. A good, general introduction. *Class No:* 520(091)

[659]

HERRMAN, D.B. The History of astronomy from Herschel to Hertzsprung. Krisciunas, K., *trans. & rev.* English ed. Cambridge Univ. Press, 1984. x, 220p. illus., ports., diagrs. £22.95. ISBN: 0521257336.

Original German as *Geschichte der Astronomie von Herschel bis Hertzsprung* (1973).

4 sections: 1. Construction and motion of the heavens - Classical astronomy - 2. The origin of astrophysics - 3. Microcosmos - Macrocosmos - 4. Technology and the organization of research. The author is director of the Archenhold Observatory, Berlin Treplow. *Class No:* 520(091)

[660]

History of oriental astronomy: proceedings of International Astronomical Union colloquium no. 91, held in New Delhi, India, 13-16 November 1985. Swarup, G., *and others, eds.* Cambridge, University Press, 1987. 288p. illus., tables. ISBN: 0521346592.

3 sections (Ancient astronomy and its characteristics; Ancient elements and planetary models; Medieval astronomy), comprising 36 papers, *e.g.* 'The characteristics of ancient China's astronomy'; 'Greek astronomers and their neighbours'; 'Indian calendars'; 'An unknown Arabic source for star names'. 4 indexes: subject; manuscript; astronomers; author. Brief bibliographies appended to some of the papers. *Class No:* 520(091)

[661]

MOORE, P. Patrick Moore's history of astronomy. 6th rev. ed. London, Macdonald, 1983. 327p. illus. (mostly black-and-white), facsims. ISBN: 0356086070.

First published 1961.

6 parts (40 chapters): 1. General - 2. Useful tables - 3. The solar system - 4. Stars, nebulas and galaxies - 5. Telescopes and accessories. Appendix : Useful addresses - 6. Star charts (data on backs). Appendix: Some landmarks in the history of astronomy (B.C.585-A.D.1982); List of illustrations; Acknowledgements. Non-analytical index. Well-captioned illus. *Class No:* 520(091)

[662]

World archaeoastronomy: selected papers from the 2nd Oxford International Conference on Archaeoastronomy, held in Merida, Yucatan, Mexico, 13-17 January 1986. Aveni, A.F., *ed.* Cambridge, University Press, 1989. xiii,504p. illus., figs., tables. £80.

4 sections: 1. The boundaries of archaeoastronomy: overviews and syntheses (11 papers, *e.g.* 7. The cosmic temples of old Beijing) - 2. Archaeoastronomy: the textual basis (16 papers, *e.g.* 13. Comets and meteors in the last

....(contd.)

Assyrian Empire) - 3. Archaeoastronomy: an interdiscipline in practice (11 papers, *e.g.* 38. Star walking - the preliminary report) - 4. Additional abstracts submitted. Each of the 38 papers has a short bibliography appended. *Class No:* 520(091)

Biographies

[663]

ABBOTT, D., ed. The Biographical dictionary of scientists: astronomers. New York, Bedrick Books (distrib.: Harper), 1984. 204p. $28. ISBN: 0911745807.

About 200 readable biographies of astronomers, some still living. No recommended readings are given, but major works are cited and briefly described in text. Glossary, p.163-78. Analytical index. The series is considered to be at a higher level than Asimov, I. *Biographical encyclopedia of science and technology* (1982) (*Library journal,* 15 January 1985, p.52-53). *Class No:* 520(092)

[664]

BARANOWSKI, H., comp. Bibliografia Kopernikowska, 1509-1955. Warsaw, Państwowe Wydawnictwo Naukowe. 1958. 449p. illus.

3,750 entries for works by Copernicus and for books and articles about him in many languages, including specific papers on him in more general works. Sections: 1. Works of Copernicus (chronological order, with descriptive notes and locations) - 2. Bibliographies - 3. Monographs - 4. Biographies - 5. On his work and activities - 6. On his significance - 7. Copernicus in literature - 8. Copernicus in the plastic arts - 9. On Copernicus museums and exhibitions - 10. On anniversary celebrations. Indexes of Copernicus' writings and of proper names. List of abbreviations and libraries cited. *Class No:* 520(092)

[665]

Bibliography of astronomers: books and pamphlets in English by and about astronomers. Luther, P., comp. Richmond, Va., Willmann-Bell, 1990-. ISBN: 094339628x.

V.1. *The Spirit of the nineteenth century.* 208p. $34.95.

A projected 3v. set. V.1 covers nearly 30% of all 19th century books with a planned v.2 continuing in that period, adding over 3,000 entries. V.3 will include an index plus an essay on collecting secondhand books, bookseller terminology, and the secondhand book trade. A useful reference source. *Class No:* 520(092)

Institutions & Associations

[666]

Earth and astronomical sciences research centres: a world directory of organizations and programmes. 2nd ed. Harlow, Essex, Longman, 1991. v [vi],623p. £250. (*Reference on research.*) ISBN: 0582082749.

First published 1984.

Guide to over 3,000 technology and research laboratories in more than 100 countries, arranged A-Z (Algeria ... Zimbabwe). Subject coverage includes geochemistry; petrology; mineralogy; planetary and galactic observations; meteorology. Data include title of establishment in original language, acronym and English translation; senior staff; research expenditure; publications; research activities; affiliation. New to this edition is a list of the main national

....(contd.)

and international trade associations and societies. (466 entries). Titles of establishments index; subject index. *Class No:* 520:061:061.2

[667]

Handbook for astronomical societies. Jones, B., *ed.* Plymouth, Federation for Astronomical Societies, 1977-. Annual. diagrs. tables.

Contents of the 12th ed., 1989: 4 articles - FAS list of officers - The Federation of Astronomical Societies - FAS slide list - Other astronomical organizations - List of local astronomical societies - Regional groups - Periodicals (26) - Equipment suppliers - Places to visit - Speakers - Information sources - Visual aid sources - Sources of astronomical software - Additions and amendments - Index to articles in previous handbooks. A very useful quick-reference tool. *Class No:* 520:061:061.2

[668]

History of the Royal Astronomical Society. Dreyer, J.L.E. *and* Turner, H.H., *eds.* Oxford, Blackwell Scientific Publications, 1987. 2v. illus. ISBN: 063202173x, v.1; 0632017910, v.2.

V.1. *1820-1920,* edited by J.L.E. Dreyer and H.H. Turner, vii,258p. V.2. *1920-1980,* edited by R.J. Tayler. vii,262p.

Each decade of the Society's existence is covered by a chapter written by various contributors and this is therefore a strictly chronological account with little to give an overall view. V.2 has 5 appendices: 1. Officers of the Society, 1920-80 ... 3. The Society's medallists 1920-80 ... 5. Statistical information 1920-80. Index. *Class No:* 520:061:061.2

[669]

International directory of astronomical associations and societies together with related items of interest. [Répertoire international d'associations et de sociétés astronomiques ainsi que d'autres entrées d'intérêts général.] Heck, A., *comp.* 8th ed. Strasbourg, Observatoire Astronomique, 1989. 707p. (*Observatoire Astronomique de Strasbourg publication spéciale du CDS, no. 13.*) ISBN: 2908064111. ISSN: 07649614.

7th edition published in 1987.

Intended to complement *International directory of professional astronomical institutions 1990 (IDPAI 1990) (qv).* Lists *c.*3,200 entries from 87 countries, Algeria ... Zimbabwe. Data include date of foundation, membership, activities, publications observatories, coordinates and awards. Index (yellow pages, p.599-707). *Class No:* 520:061:061.2

[670]

International directory of professional astronomical institutions together with related items of interest. [Répertoire international des institutions astronomiques professionnelles ainsi que d'autres entrées d'intérêt général.] Heck, A., *comp.* Strasbourg, Observatoire Astronomique, 1989. 649p. (*Observatoire Astronomique de Strasbourg publication spéciale du CDS, no. 14.*) ISBN: 290806412x. ISSN: 07649614.

First published 1983.

Entries under 87 countries, Algeria ... Zimbabwe (USA, by states, p.319-479). *c.*3,500 entries listed alphabetically.

....(contd.)
Data on each include date of foundation, activities, staff, publications and coordinates. Index (yellow pages, p.545-650). A greatly enlarged revision. *Class No:* 520:061:061.2

Conferences
[671]
INTERNATIONAL ASTRONOMICAL UNION. Transactions of the General Assembly. Pt.A. Reports on astronomy. v.1-. Dordrecht, Kluwer, 1922-. 3-yearly. ISSN: 00801372.
Reports on astronomy form one part of the I.A.U. *Transactions,* the other part being the *Proceedings of the General Assembly. Reports,* covering a wide range, are sectionalized, with valuable bibliographical notes.
Class No: 520:061:061.3

Astrophysics
[672]
ALLEN, C.W. Astrophysical quantities. 3rd ed. London, Athlone Press, 1973. 310p. tables. £50. ISBN: 0485111500.
First published 1955.
15 sections, data being largely numerical and tabular: 2. General constants and units - 3. Atoms - 4. Spectra - 5. Radiation - 6. Earth - 7. Planets and satellites - 8. Interplanetary matter - 9. Sun - 10. Normal stars - 11. Stars with special characteristics - 12. Star population and the solar neighbourhood - 13. Nebulae, sources and interstellar space - 14. Clusters and galaxies - 15. Incidental tables (*e.g.* annual variations). Analytical index, p.303-30. 'The most widely used source of astrophysical data' (Coblans, H., ed. *Use of physics literature* (1975), p.124). Allen : Emeritus Professor of Astronomy, Univ. of London. *Class No:* 520-1

[673]
LANG, K.R. Astrophysical formulae: a compendium for the physicist and astrophysicist. 2nd ed. Berlin & New York, Springer, 1980. 759p. $49. ISBN: 0387099336.
5 major sections: Continuum radiation - Monochrome (Line) radiation - Gas processes - Nuclear astrophysics and high energy particles - Astrometry and cosmology. References are cited for each formula and its derivatives, - 'a total of well over 2,000 references' (Seal, R.A., and Martin, S.S. *A bibliography of astronomy, 1970-1979,* p.49).
Class No: 520-1

Astronomical Instruments
[674]
BENNETT, J.A. The Divided circle: a history of instruments for astronomy, navigation and surveying. Oxford, Phaidon Christie's, 1987. 224p. illus. (some col.), figs. ISBN: 0714880388.
14 chapters: 1. Foundations in astronomy - 2. The beginnings of oceanic navigation ... 6. The heroic age - 7. The growth of observatories ... 11. Astronomical circles ... 14. A practical postscript. Bibliography, p.215-17. Index of makers; index of technical terms; general index.
Class No: 520.0

[675]
—MADDISON, F. 'Early astronomical and mathematical instruments: a brief survey of sources and modern studies.' in *History of science,* v.2, 1963, p.17-50.
Includes a lengthy bibliography, p.34-50. *Class No:* 520.0

Observatories
[676]
Directory of European observatories 1988. Vercoutter, P.A.J., *comp.* 3rd ed. Ieper, Astronomical Contact Group, 1988. 432p. tables. (*Philippe's astronomical series, v.3.*) ISBN: 907247001x. ISSN: 07757085.
729 entries for observatories and institutions (1. Austria ... 27. Yugoslavia). 'Great Britain', 42 entries; 'France', 57 entries. Omits the former USSR. 6 types of observatory are differentiated. Data on each observatory appear under 19 heads (*e.g.* library; regular publications; observatory information; instruments). *Class No:* 520.1

[677]
GREENWICH OBSERVATORY. The Royal Observatory at Greenwich and Herstmonceux, 1675-1975. London, Taylor and Francis, 1975. 3v. £70. ISBN: 0850660939.
V.1: *Origins and early history (1675-1835),* by E.G. Forbes. v.2: *Recent history (1836-1975),* by A.J. Meadows. v.3: *The buildings and instruments,* by D. Howse.
Class No: 520.1

[678]
—SHENTON, D.E. The Royal Greenwich Observatory and the Astronomers Royal, 1675-1976: a bibliography. London, Library Association, 1977. xxv, 723p.
Library Association F.L.A. thesis, accepted 27 September 1977.
Chiefly concerned with bio-bibliographies of the Astronomers Royal. Subject and author indexes.
Class No: 520.1

[679]
KIRBY-SMITH, H.T. US observatories: a directory and travel guide. Princeton, N.J., Van Nostrand Reinhold, 1976. 173p. ISBN: 0442244517.
Data, state by state, on 300 observatories and planetaria: brief description; equipment; public access. 15 of the largest observatories have more detailed treatment. Two brief bibliographies. Index of institutions. *Class No:* 520.1

[680]
KRISCIUNAS, K. Astronomical centers of the world. Cambridge, University Press, 1988. x,320p. illus. £20. ISBN: 0521302781.
A selection of 'the observatories and dynamic individuals that were involved in the collaborative progress of institutionalized astronomy' (*Preface*). 9 sections 1. The Alexandrian Museum - 2. Astronomical capitals of the Moslem world ... 4. Paris and Greenwich - 5. Pulkovo Observatory ... 7. Mt. Wilson and Palomar - 8. The present - 9. The future. Appendix: The electromagnetic spectrum. Notes, p.281-310, is a list of references. Index.
Class No: 520.1

[681]

—MARX, S. *and* PFAU, W. Observatories of the world. Mitton, S., *ed. & rev.* English ed. Poole, Dorset, Blandford Press, 1982. 200p. illus. (incl. col.), diagrs., maps. ISBN: 0713711914.

Originally as *Sternwarten der Welt* (Leipzig, 1979).

Describes 40 selected telescopes from around the world, Abastumani Astrophysical Observatory... Special Astrophysical Observatory, Zelenchukskaya, USSR; plus NASA's 24-metre telescope. Bibliography, p.195-7; non-analytical index; sources of illus. Fully illustrated, with 40 colour plates, 76 black-and-white illus. and 18 line-drawings. *Class No:* 520.1

[682]

RIGAUX, F. Les Observatoires astronomiques et les astronomes. Brussels, Observatoire Royal de Belgique, for the International Astronomical Union, 1959. 452p.

Provisional ed., 1955.

Observatories are listed by country and town. Entries give information on personnel, publications, instruments and type of work carried out by each observatory. No longer current, but would now be of historical interest. *Class No:* 520.1

Telescopes

[683]

BARLOW, B.V. The Astronomical telescope. London, Wykeham Publications (London), 1975. viii, 213p. illus., tables. ISBN: 0851094600.

11 chapters (11: New developments). Chapter references; short glossary. Half-tone illus. and good line-drawings. 'He is very basic in his description of this mainstay of astronomical observation' (*Choice,* v.19(1), September 1981, p.34). 'The astronomy community has needed a book like this for a long time' (*Nature,* v.255(no.5507), 1975, p.432). *Class No:* 520.2

[684]

—LEARNER, R. Astronomy through the telescope. London, Evans, 1982. 224p. illus. (incl. col.), facsims., ports. ISBN: 0217456443.

Devotes 17 chapters to 500 years' progress of astronomy, as seen in the use and development of the telescope. The book (written by a senior lecturer at the Imperial College of Science and Technology, London) is well illustrated, with an analytical index. For the layman; for reading rather than reference (*New scientist,* 28 January 1982, p.251). *Class No:* 520.2(035)

Solar System

Handbooks & Manuals

[685]

BEATTY, J.K. *and* CHAIKIN, A., *eds.* The New solar system. 3rd ed. Cambridge, University Press; Cambridge, Mass., Sky Publishing Corporation, 1990. viii,326p. illus. (incl. col.), maps (incl. col.). £25; $39.95. ISBN: 0521361621.

First published 1982.

Major revision of the previous edition, published in 1982, and incorporating new information collated from *Voyager* and the 1985/86 Halley's Comet visit, among other things. 23 chapters: 1. The golden age of solar-system exploration ... 7. Surfaces of the terrestrial planets ... 12. Planetary

....(contd.)

rings ... 17. Comets ... 20. Small bodies and their origins ... 23. Putting it all together. Also includes planet, satellite and small-body characteristics, glossary, author biographies and suggested further reading, planetary maps and index. Excellent illustrations. Recommended for public and academic libraries. *Class No:* 523(035)

[686]

BODECHTEL, J. *and* GIERLOFF-EMDEN, H.G. The Earth from space. Mayhew, M. *and* Evans, L., *trans.* Newton Abbot, Devon, David & Charles, 1974. 176p. illus. (mostly col.), maps. ISBN: 0715160760.

Sections: Space experiments as milestones in geography, geology, meteorology and cartography - Satellite techniques and technology - Flying spies [satellites] - Pictures of the earth: atmosphere, oceans, continents - Satellites in meteorological research and weather forecasting - The movements of the earth's crust - Landscape types on earth's deserts, mountains, coasts, islands - Our environment in relation to its surroundings - The moon, satellite of the earth. No index, but detailed contents list. A fine production. *Class No:* 523(035)

[687]

CARR, M.H. The Surface of Mars. New Haven, Conn., Yale Univ. Press, 1981. 232p. $29.95. ISBN: 0300032420.

An historical perspective is followed by chapters on Martian craters, atmosphere, volcanoes, sections, surface chemistry, and the two Martian moons (Phobos and Deinos). Aimed at 'the informed scientific reader', and assumes some knowledge of geology, physics and chemistry. 'Science at its very best' (*American scientist,* v.70 March/April 1982, p.274). *Class No:* 523(035)

[688]

The Geology of the terrestrial planets. Carr, M.H., *ed.* Washington, NASA, 1984. 317p. $16. (*Special publication SP 469.*) ISBN: 0160041627.

Takes the terrestrial planets (Venus, Mars, Mercury, Moon, Earth) in solar system order. Data: orbit, magnetic fields (if any), surface characteristics, craters, tectonics, etc., summarizing what has been learned since 1962. The Earth is given a brief chapter. 'Intended for the interested layman' (*Astronomy,* v.13, October 1985, p.88). *Class No:* 523(035)

[689]

MINER, D. Uranus: the planet, rings and satellites. New York, Ellis Horwood, 1990. 334p. illus., figs., tables. (*Ellis Horwood library of space science and space technology.*) ISBN: 0139468803.

13 sections: 1. The discovery of Uranus ... 4. The discovery of the rings of Uranus ... 6. The saga of Voyager 2 ... 10. The magnetosphere of Uranus ... 13. Future studies of Uranus. Incorporates data from the Voyager 2 space mission. Notes, references and bibliographies appended to each chapter. Index. The author worked on the Voyager project. *Class No:* 523(035)

[690]

Planetary exploration. Hughes, D., *ed.* New York, Facts on File, 1990. 6v. $65, set $13.95ea.

V.1: *The distant planets.* v.2: *Jupiter.* v.3: *Mars.* v.4: *The moon.* v.5: *Our sun and the inner planets.* v.6: *Saturn.*

Aimed at children aged 10-14. Many colour illustrations based on recent NASA photographs and scientific knowledge. Index in each volume. Clearly and attractively presented. *Class No:* 523(035)

[691]

The Solar system. Kuiper, G.P. *and* Middlehurst, B.M., *eds.* Chicago, Ill. (and later, London), Univ. of Chicago Press, 1953-63. 4v. illus., diagrs., maps, charts, tables. ISBN: 0685157520, v.2; 0226459276, v.3; 0226459284, v.4.

V.1: *The sun.* 1953. xx, 746p. v.2: *The earth as a planet.* 1955. xviii, 752p. $160. v.3: *Planets and satellites.* 1961. xx, 601p. $50. v.4: *The moon, meteorites and comets.* 1963. xxii, 820p. $60.

V.4 has 22 chapters, each with its own contributors; references appended. Thus, chapter 1: 'The lunar surface: introduction (p.1-56) has references (p.47-50) up to 1962. The analytical 'index of subjects and definitions' (p.799-810) has 76 sub-entries under 'Lunar craters'. A standard work of reference, now somewhat dated. *Class No:* 523(035)

Maps & Atlases

[692]

The Cambridge photographic atlas of the planets. Briggs, G.A. *and* Taylor, F.W. Cambridge University Press, 1982. 255p. ISBN: 0521239761.

6 sections: Mercury - Venus - Earth and moon - Mars - The Jovian system - The Saturnian system. Some 200 photographs (*c.*50% in full colour) obtained from space vehicles over the past two decades. Draws on the NASA collection of photographs, many of them not previously published in books. 'An annotated snapshot album of the planets and their satellites' (*New scientist*, 18 November 1983, p.455). 'The style is highly readable and intended for an educated general public' (*Geology*, v.15, September 1987, p.882). *Class No:* 523(084.3)

Moon

[693]

Lunar sourcebook: a user's guide to the moon. Heiken, G., *and others.* Cambridge, University Press, 1991. xix,736p. illus. (some col. pl.), figs., tables. £50. ISBN: 0521334446.

11 chapters: 1. Introduction - 2. Exploration, samples, and recent concepts of the moon - 3. The lunar environment ... 5. Lunar minerals ... 8. Chemistry ... - 11. Afterword. Bibliography, p.655-715. Sample and subject indexes. 'Summarizes what we know about the Moon as a result of US and USSR lunar missions' (*Preface*). *Class No:* 523.3

[694]

RÜKL, A. Atlas of the moon. Rackham, T.W., *ed.* London, Hamlyn, 1991. 224p. maps. ISBN: 0600571904.

76 maps of sections of the near-side of the moon. Each map is accompanied by summaries giving brief biographical details of the eminent people whose names have been given to lunar formations and a brief description of the formations themselves. Nomenclature is as authorized by the International Astronomical Union. Intended for observers and amateur astronomers. *Class No:* 523.3

[695]

—KOPAL, Z. A New photographic atlas of the moon. London, Hale, 1971. vii, 311p. illus., tables. ISBN: 0709123701.

First published 1965.

3 parts: 1. The moon, an alien world - 2. Atlas of the moon and description of the plates (p.63-293) - 3. Appendices (Tables: 1. List of lunar aircraft (1959-1969) ...

.... (contd.)

3. Kinematic characteristics of the artificial lunar satellites). Bibliographical notes; glossary; index. Magnificent close-up photographs, some taken from spacecraft. *Class No:* 523.3

[696]

—The Times atlas of the moon. Lewis, H.A.G., *ed.* London, Times Newspapers, Ltd., 1969. xxxvii,110p. illus. (incl. col.), maps.

The introductory articles: 'The moon; 'The far side'; 'Mapping the moon'; 'The lunar landscape'; 'Techniques of lunar flight'. 'Index to names' and 'Key to the maps' (*c.*800 entries, with separate lists of promontories, peaks, mountain ranges, bays, marshes, lakes, oceans and seas, rilles (clefts), rimae, valleys) are followed by the maps, p.1-110. The atlas is essentially a reproduction of the US Air Force 1: 1,000,000 lunar charts of pre-Apollo vintage; the most serious weakness is the quality of reproduction of some photographs (*The Listener*, v.83(2134), 19 February 1970, p.256). Page size, 30 x 32.5cm. *Class No:* 523.3

Comets

[697]

Comet Halley: investigations, results, interpretations. Mason, J.W., *ed.* Chichester, Ellis Horwood, 1990. 2v. (*Ellis Horwood library of space science and space technology.*) ISBN: 0131710753, v.1; 0131710834, v.2.

V.1. *Organization, plasma, gas.* xix,295p. V.2. *Dust, nucleus, evolution.* xix,275p.

A selection of review articles dealing with cometary research conducted during the 1985/86 return of Halley's Comet. 39 chapters, 54 contributors. Chapter titles include: 'The dust trail of comet Halley'; 'Models of cometary nuclei'; 'The ageing of comet Halley and other periodic comets'; 'The International Halley Watch'. Each volume contains a list of names, acronyms and abbreviations and a bibliography is appended to each chapter. Many figures, tables and some coloured plates. *Class No:* 523.64

[698]

FREITAG, R.S. Halley's comet: a bibliography. Washington, Library of Congress, 1984. 555p. $26. ISBN: 0844404594.

Annotated bibliography of over 3,200 entries, based on Library of Congress holdings to mid-1974. English- and European-language material. Arrangement is by authors A-Z, with author and subject indexes. Entries cite Library of Congress numbers and library locations. Bibliographical sources are listed. As to the entries themselves, 'daily cartoons jostle thick tomes of dry calculations' (*Scientific American*, v.235(5), November 1985, p.22). *Class No:* 523.64

[699]

—MORTON, B. Halley's comet, 1755-1984: a bibliography. Westport, Conn., Greenwood, 1985. 280p. $36.95. ISBN: 0313240221.

Has 1,301 annotated entries, arranged chronologically, Such an arrangement - complementary to Freitag's - sets 'citations in relation to others as part of the cultural response of its time... *Halley's Comet* is recommended for college, university, and large public libraries' (*RQ*, v.25(2), Winter 1985, p.259). *Class No:* 523.64

[700]
KRONK, G.W. **Comets: a descriptive catalogue.** Hillside, N.J., Enslow, 1984. viii, 331p.

First section: Long-period and non-periodic comets (over 500, listed in order of discovery). Second section: Short period comets (A/Z by common names). Notes on comet watching. References to books and periodical articles (years: -371 to 1800; years 1800 to 1982). Index to common names of comets, p.331. *Class No:* 523.64

[701]
YEOMAN, D.K. **Comets:** a chronological history of observation, science, myth and folklore. New York, Wiley, 1991. x,485p. illus., figs., tables. £24.95. ISBN: 0471610119.

11 sections: 1. The origin of cometary thought - 2. Medieval views ... 6. The return of comet Halley ... 9. The physics of comets ... 11. The births and deaths of comets. Appendix: Naked-eye comets reported through A.D. 1700, p.361-424. Extensive bibliography, p.425-74. Detailed index. *Class No:* 523.64

Sun

[702]
Illustrated glossary for solar and solar-terrestrial physics. Bruzek, A. *and* Durrant, C.J., *eds.* Norwell, Mass., Kluwer, 1977. xviii, 204p. illus., diagrs. $67. ISBN: 9027708258.

14 contributors. 14 sections (1. Solar interior ... 4. Quiet photosphere and chromosphere ... 6. Solar corona ... 11. Solar radio emission ... 14. Solar-terrestrial physics). Sub-section references. Striking illus. Subject index. *Class No:* 523.9

[703]
ZIRIN, H. **Astrophysics of the sun.** Cambridge, University Press, 1988. x,433p. illus., tables, graphs. £45; £15.95. ISBN: 0521302684.

A new edition of *The solar atmosphere* (1966).

12 sections: 1. Looking at the sun - 2. Observing the sun ... 5. Atomic spectra - 6. The interior and the photosphere ... 9. Prominences - 10. Solar activity ... 12. Questions. Abbreviations, p.411. Bibliography, p.412-28. Index. Many black-and-white illustrations. Aimed at the academic and amateur astronomer. *Class No:* 523.9

Stars & Galaxies

Handbooks & Manuals

[704]
HODGE, P.W. **Galaxies.** Cambridge, Md., Harvard Univ. Press, 1986. 174p. illus., maps. $25.50. ISBN: 0674340655.

Contents: Galactic structure - The formation and evolution of galaxies - The missing mass - The nearest spirals - Clusters and superclusters of galaxies - Disturbed galaxies - Quasars. About 100 photographs and maps. Index. A compendium of facts and problems, with highly readable text. *Class No:* 524(035)

[705]
MOTZ, L. *and* NATHANSON, C. **The Constellations:** an enthusiast's guide to the night sky. London, Aurum Press, 1988. xix,411p. illus., diagrs. £19.95. ISBN: 1854100882.

Looks at the myths attached to the constellations, in addition to giving details of galaxies, nebulae, variable stars, quasars, etc. 6 sections, subdivided in 15 chapters: 1. Two bears and a dragon ... 4. The hunter ... 8. Three tragic myths ... 14. Phoenix to Antlia - 15. South circumpolar constellations. Bibliography. Index. *Class No:* 524(035)

[706]
RIDPATH, I. *and* TIRION, W. **Collins' guide to stars and planets.** London, Collins, 1984. 384p. illus., (incl. colour), star charts. £9.99. ISBN: 0002190672.

88 star charts by W. Tirion, depicting the constellations A-Z, form the core of this pocket-sized volume, with descriptive text facing. A further section provides over 35 coloured plates. Analytical index. 'Warmly recommended (in a fairly crowded field) to any would-be astronomer' (*British book news,* April 1985, p.226).

The Night sky, by I. Ridpath and W. Tirion (London, Collins, 1985. 240p.) also makes good use of the 88 constellation charts. *Class No:* 524(035)

[707]
—EICHER, D.J. **The Universe from your backyard:** a guide to deep-sky objects from *Astronomy* magazine. Cambridge, University Press; Milwaukee, Wis., Astromedia, 1988. 188p. illus. (some col.), tables, maps. £15.95. ISBN: 0913135054, US ed; 0521362997, UK ed.

46 articles, each covering a constellation, giving details of 690 deep-sky objects. Arrangement is alphabetical (Andromeda, Triangulum ... Virgo). Many colour illustrations. Bibliography. Index. For school and public libraries. *Class No:* 524(035)

[708]
—MENZEL, D.H. *and* PASACHOFF, J.M. A Field guide to the stars and planets. 2nd ed. Boston, Mass., Houghton Mifflin, 1983. 473p. illus., star charts, maps. $19.95. (*Peterson Field guides no.15.*) ISBN: 039534641x.

Tight-packed, pocket-sized manual - comprises 10 moon maps, 72 monthly sky maps and 52 crowded charts (for the use of observers equipped with a small telescope). 83 full-colour photographs, 10 moon maps, and 11 graphic annual timetables (tracking paths of planets, 1984-89). 'Has much to recommend it' (*Astronomy,* October 1984, p.39).

A revised ed. was due to be published in 1992.
Class No: 524(035)

[709]
—MURDIN, P. *and* ALLEN, D. Catalogue of the universe, with original photographs by D. Malin. Cambridge, University Press. 1979. 288p. illus. (incl. col. pl.). $29.95. ISBN: 052122859x.

Comprises 3 parts: 1. The universe of galaxies - 2. Stars and nebulae - 3. The solar system. Appendix: Messier catalogue. Notes on special photographic techniques; Glossary. Partly analytical index, p.253-6. 53 colour subjects, 200 half-tones and line subjects. Aims 'to illustrate a comprehensive selection of objects in the Universe and to write about the examples chosen, bringing the basic principles out of the objects themselves' (*Preface*).
Class No: 524(035)

[710]
Stars and stellar systems: compendium of astronomy and astrophysics. Kuiper, G.P. *and* Middlehurst, B.M., *eds.* Chicago, Ill., & London, Univ. of Chicago Press, 1961-75. 9v. illus., tables, maps.

V.1: *Telescopes.* 1961 (reprinted 1977). v.2: *Astronomical techniques.* 1962. v.3: *Basic astronomical data.* 1963. v.4: *Clusters and binaries.* (Cancelled). v.5: *Galactic structures.* 1965. v.6: *Stellar atmospheres.* 1960. v.7: *Nebulae and interstellar matters.* 1968. v.8: *Stellar structure.* 1965. v.9: *Galaxies and the universe.* 1975.

Intended as an extension to the 4-v. *Solar system* series *(qv),* to cover astrophysics and stellar astronomy. Separate editors are appointed for each volume. V.2 has 24 main chapters, each by an expert, with extensive references (chapter 17, 'Photographic photometry', has 111 references). Subject index. 'The aim of the series is to present stellar astronomy and astrophysics as basically empirical sciences, co-ordinated and illustrated by the application of theory' *(Preface to the Series).* 'A landmark series' *(New technical books,* July 1978, entry 777). Supported in part by the National Science Foundation. *Class No:* 524(035)

[711]
The Webb Society deep sky observer's handbook. Hillside, N.J., Enslow, 1980-90. 8v. ISBN: 0894901222, v.1; 0894900285, v.2; 089490034x, v.3; 0894900501, v.4; 0894900668, v.5; 0894901338, v.6; 0894901346, v.7; 0894902083, v.8.

1. *Double stars.* 2nd ed. 1986. £14.95; $18.95. 2. *Planetary and gaseous nebulae.* 1980. £13.95; $17.95. 3. *Open and globular clusters,* 1980. £15.95; $19.95. 4. *Galaxies.* 1982. £16.95; $21.95. 5. *Clusters of galaxies.* 1982. £16.95; $21.95. 6. *Anonymous galaxies.* 1987. £13.95; $17.95. 7. *The southern sky.* 1987. £15.97; $21.95. 8. *Variable stars.* 1990. £15.95; $19.95.

The *Handbook's* aim is 'to provide a series of observer's manuals that do justice to the equipment that is available today and to cover fields that have not been adequately covered by other organizations of amateurs. The manuals are designed primarily for the more experienced amateur' *(General preface).* Intended to replace Revd. T.W. Webb's classic *Celestial objects for common telescopes* (first published 1839, reaching 6 editions by 1917, and still in print). The new *Handbook,* with its appendices and bibliographies, 'will justly have a special place in any amateur's bookshelf as the most comprehensive guide' *(British book news,* August 1982, p.466). *Class No:* 524(035)

[712]
—CORLISS, W.R. Mysterious universe: handbook of astronomical anomalies. Glen Arm, Md., The Sourcebook Project, 1979. 710p. $19.95. ISBN: 0915554054.

Assembles over 500 short contributions, 'anomalies' ranging from auroras in comet tails to zodiacal light-rifts and black and white holes to X-ray bursts. 'It is strong meat... but in short bursts the book is a joy... The book is to be highly recommended. It has been well produced, extremely well indexed and is excellent value for money' *(Nature,* v.280 (5717), 5 July 1979, p.92). *Class No:* 524(035)

[713]
ZOMBECK, Martin V. Handbook of space astronomy and astrophysics. 2nd ed. Cambridge, University Press, 1990. [viii],440p. graphs, diagrs., tables. £50; £20. ISBN: 0521314550.

First published 1982.

19 chapters: 1. General data - 2. Astronomy and astrophysics ... 6. X-ray astronomy - 7. Gamma-ray astronomy ... 13. Plasma physics ... 19. Astronomical catalogs. Aims 'to provide a ready reference and working tool for the practicing *[sic]* space astronomer and astrophysicist' *(Preface).* A standard handbook. *Class No:* 524(035)

Tables & Data Books

[714]
DIXON, R.S. and SONNEBORN, G., comps. A Master list of nonstellar optical astronomical objects. [Athens], Ohio State University Press, 1980. 835p. tables. ISBN: 0814202500.

Includes *c.*185,000 listings from 270 catalogues. Data for each object include 1950.0 position, angular diameter, magnitude and description. Bibliography (p.39-47) is under authors A-Z. Appendices A & B (A: 'The Luyten search for faint blue stars'). *Class No:* 524(083)

[715]
—BURNHAM, R., *Jr.* Burnham's celestial handbook: an observer's guide to the universe beyond the solar system. New York, Dover, 1985. 3v. (2138p.) illus., diagrs., graphs. $13.95. ISBN: 048623567x, v.1; 0486235688, v.2; 0486236730, v.3.

Lists over 7,000 objects available to observers with telescopes in the 2-inch to 12-inch range. Objects are arranged A-Z. Appended to v.3 'Index to the Messier objects'; 'The brightest stars'; 'Index to the 25 nearest stars'; 'Bibliography - a partial list of works consulted'. Intended to be a standard catalogue and detailed descriptive handbook. *Class No:* 524(083)

[716]
HIRSHFELD, A. and SINNOTT, R., eds. Sky catalogue 2000.0: stars to visual magnitude 2000.0. Cambridge, University Press, 1982-5. 2v. (604 + 385p.). tables. £45; £25, v.1; £60; £35, v.2. ISBN: 0521417430, v.1; 0521258189, v.2.

1: *Stars to magnitude 8.0.* 2nd ed. 1992. 2: *Double stars, variable stars and nonstellar objects.* 1985.

Companion volumes to Tirion's *Sky atlas 2000.0.* V.1 tabulates 50,071 stars, listed in order of increasing sight ascension. 91% of the stars in the catalogue have their distances given, apart from other data. Both the *Sky atlas* and the *Catalogue* (v.1) are considered 'essential acquisitions for all active astronomical observers' *(Nature,* v.298, 5 August 1982, p.588). V.2 (15 main data sections) covers more than 28,000 celestial objects not in the previous ed. of v.1. Useful for both amateur and professional astronomers. *Class No:* 524(083)

[717]
JONES, K.G. Messier's nebulae and star clusters. 2nd ed. Cambridge, University Press, 1991. xvii,427p. illus., diagrs., tables, maps. $49.50. ISBN: 0521370795.

First published 1968.

Covers the 102 nebulous objects in Messier's catalogue of 1771, plus 7 additional objects. Very well illustrated. 5 sections: 1. Astronomical background to the Messier Objects: the 'missing' and 'additional' objects ... 4. The Messier Objects in detail (p.51-306) - 5. Biographical and historical. 6 appendices, *e.g.* 'Catalogue of stars in the Pleiades'; 'Bibliography'. Indexes of names; objects; subjects. Of the earlier ed., 'An excellent aid for the observer' (Seal, R.A. *A Guide to the literature of astronomy* (1977), p.50). 'The revised edition is tighter than the first' (*Sky and telescope*, v.84(1), July 1992, p.45).
Class No: 524(083)

[718]
NGC 2000.0: the complete new general catalogue and index catalogues of nebulae and star clusters by J.L.E. Dreyer. Sinnott, R.W., *ed.* Cambridge, University Press; Cambridge, Mass., Sky Publishing Corporation, 1988. xxiv,273p. tables. $19.95. ISBN: 0933346514, US ed; 0521378133, UK ed.

An update of J.L.E. Dreyer's *New general catalogue of nebulae and clusters of stars (1888), Index catalogue (1895)* and *Second index catalogue (1908).*

3 sections: 1. NGC 2000.0 - 2. Right ascensions of NGC objects - 3. Right ascensions of IC objects. Information about specific objects not as detailed as Hirshfeld's *Sky catalogue 2000.0* but covers all 13,226 NGC and IC objects, while *Sky catalogue 2000.0* covers fewer than 3,500 of those studied most. 'This volume is certain to be a standard reference for professional and amateur astronomers for years to come' (*Science and technology libraries*, v.11(1), Fall 1990, p.139). *Class No:* 524(083)

[719]
Second reference catalogue of bright galaxies: containing information on 4,364 galaxies, with references to papers published between 1964-1975. De Vaucouleurs, G., *and others.* Austin, Texas, Univ. of Texas Press, 1976. 396p. $101.

Supplement to *Reference catalogue of bright galaxies,* by G. and A. De Vaucouleurs (Univ. of Texas Press, 1964. 268p. $40).

175p. of catalogue data and 147p. of notes, reference and source material for the period January 1, 1964 to July 1, 1975 (covering 4,364 bright galaxies). 66p. of introduction and explanatory matter are added. Recommended for the relatively large and specialized science library. Continued by: *Class No:* 524(083)

[720]
—Third reference catalogue of bright galaxies. Berlin, Springer-Verlag, 1991. 3v. $198, set; $69, v.1; $79, v.2; $79, v.3. ISBN: 0387975497, v.1; 0387955008, v.2; 0387955109, v.3.

V.1 *Introduction, references, notes and appendices.* 714p. V.2 *Data for galaxies between 0h and 12h.* 723p. V.3 *Data for galaxies between 12h and 24h.* 632p.
Class No: 524(083)

[721]
SMITHSONIAN INSTITUTION ASTROPHYSICAL OBSERVATORY. Star catalog. Washington, Smithsonian Institution, 1966 (reprinted 1974). 4v.

Data on the proper position and motion of 258,997 stars of the epoch 1950.0 A comprehensive catalogue that combines earlier star catalogues of that epoch, including the FK4, FK3, AGK2, AGK1 and GC. Available on magnetic tape. 'An indispensable tool for the working astronomer' (Seal, R.A. *A guide to the literature of astronomy,* entry 103).
Class No: 524(083)

Illustrations

[722]
The Cambridge atlas of astronomy. Audouze, J. *and* Israël, G., *eds.* 2nd ed. Cambridge, University Press, 1988. 432p. illus., charts, tables. £40. ISBN: 0521363608.

Translation of *Le grand atlas de l'astronomie* (Encyclopaedia Universalis, 1986).

26 contributors, plus 3 consultant editors. Main sections: Sun (28p.); The solar system (180p.); Stars and the Galaxy (82p.); The extragalactic domain (64p.). 770 photographs (350 in colour) and 350 colour diagrams. There are no great changes from the previous edition, although the introduction is new and there are results of recent space missions, *e.g. Voyager 2* and the *Giotto* mission to Halley's comet. There are only 2p. of maps covering the entire sky, treatment being in the form of 2, 4 or 6-page topics. Of the earlier edition, 'It is an unexcelled reference for the astronomy enthusiast' (*British book news,* September 1985, p.541). Page size: 36.8 x 26.8cm. *Class No:* 524(084.1)

[723]
HENBEST, N. *and* **MARTEN, M. The New astronomy.** Cambridge, University Press, 1983. 240p.

Has 267 full-colour illus., ranking it 'among the most comprehensive of modern astronomical picture albums. Topics and text develop around roughly three-dozen celestial objects' (*Sky and telescope,* v.65, October 1982, p.323). 'Images of astronomical objects taken over the whole band starting with gamma-rays, X-rays and ultra-violet radiation and finishing with the infrared, millimetre waves and radio...' (*New scientist,* 12 January 1984, p.43).
Class No: 524(084.1)

[724]
LAUSTSEN, S. Exploring the southern sky: a pictorial atlas from the European Southern Observatory (ESO). Berlin & New York, Springer, 1987. 274p. illus. $49. ISBN: 0387177353.

Observations taken at ESO, La Silla, Chile. Well over 200 excellent photgraphs, well captioned. 'A valuable reference guide to the southern sky for its organization and completeness. This atlas deserves a wide audience' (*Sky and telescope,* December 1987, p.608). *Class No:* 524(084.1)

Maps & Atlases

[725]
NORTON, A.P. Norton's 2000.0: star atlas and reference handbook (epoch 2000.0). Ridpath, I., *ed.* 18th ed. Harlow, Essex, Longman Scientific & Technical; New York, John Wiley, 1989. xii,179p. maps, diagrs., tables. ISBN: 058203163x.

First published 1910. 17th ed. entitled *Norton's star atlas*

....(contd.)
and reference handbook.

18 maps, plus Northern and Southern hemisphere index map, together indicate the position of *c*.8,700 stars and 600 nebulae. 5 sections: 1. Star charts - 2. Position and time - 3. Practical astronomy - 4. The solar system - 5. Stars, nebulae and galaxies. Appendix includes units and notation, astronomical constants, symbols and abbreviations and useful addresses. Text has been almost entirely rewritten for this edition. The best atlas for amateurs.
Class No: 524(084.3)

[726]
—MOORE, P. The New atlas of the universe. London, Mitchell Beazley, 1984. 272p. illus., (mostly col.), charts, diagrs., maps. ISBN: 0855335378.

Based on his *The atlas of the universe* (1970. 272p.).

Nearly 1,500 maps and photographs, many in colour. Incorporates the new discoveries from the Pioneer, Viking, Voyager and Solar Maximum missions. Accompanying the explanatory text are a beginner's guide to the sky, a catalogue of stellar objects and a glossary of astronomical terms. Addressed to the amateur astronomer.
Class No: 524(084.3)

[727]
—SMITHSONIAN INSTITUTION ASTROPHYSICAL OBSERVATORY. Star atlas of reference stars and nonstellar objects. Cambridge, Mass., MIT Press, 1969. 13p. charts.

Intended for use with the Observatory's *Star catalog*. There are 152 boxed charts, with a transparent location grid.
Class No: 524(084.3)

[728]
—TIRION, W. Sky atlas 2000.0. Cambridge, Cambridge University Press, 1982. 26 star colour charts. £25. ISBN: 0521244676.

Competes with Norton for the quality of its star maps and its recency, including *c*.43,000 stars down to the 8th magnitude for epoch 2000. *Class No:* 524(084.3)

[729]
TIRION, W. Cambridge star atlas 2000.0. Cambridge, University Press, 1991. 74p. tables, maps. £10.95. ISBN: 0521263220.

Contains 12 monthly sky-maps, in addition to a series of 20 detailed star-charts. Designed for use in both hemispheres, and for the amateur or more advanced observer. *Class No:* 524(084.3)

[730]
Uranometria 2000.0. Tirion, W., *and others.* Richmond, Va., Willmann-Bell, 1987. 2v. $39.95ea. ISBN: 094339614x, v.1; 0993396158, v.2.

V.1: *The Northern hemisphere to -6 degrees.* xliii,259p. maps. V.2: *The Southern hemisphere to +6 degrees.*

473 star charts in all, showing stars to 9.5 magnitude. A three-volume companion star catalog is planned.
Class No: 524(084.3)

[731]
WRAY, J.D. The Color atlas of galaxies. Cambridge, University Press, 1988. xii,189p. col.pl. £60. ISBN: 0521322367.

Colour images of more than 600 galaxies, in 94 plates. Table of galaxies illustrated in the atlas, p.189.
Class No: 524(084.3)

Navigation

Handbooks & Manuals

[732]
Admiralty manual of navigation. Rev. ed. London, HMSO, 1954-73. 5v. illus., diagrs., tables, maps. (*B.R.45*.) ISBN: 0117707686; 0117714674.

First published 1914.

V.1 (rev. consolidated ed., incorporating changes 1-4. 1971) is a practical guide for executive officers; v.2 (rev. ed. 1973) provides a textbook of nautical astronomy and off-shore navigation; v.3 (1954) is based on the navigation syllabus for officers qualifying in navigation and direction, and deals with advanced subjects and mathematical proofs not included in v.1-2. Admiralty charts and publications are covered in v.1, chapter 2. V.4 (1962), on certain navigational equipment and techniques used by H.M. ships, is not available to the public. V.5 (6 bound parts) provides exercises on the use of tables (*e.g.* astronomical, Great Circle, tidal and tidal streams, time and chronometer, relative velocity questions).

The 1987 revised ed. of v.1 (xviii, 697p.) is entitled *General navigation, coastal navigation and pilotage,* with 19 chapters, a bibliography (p.675-7) and index (p.677-97).

The Admiralty issues updates and amendments for mariners in its weekly publication, *Admiralty notices to mariners. Class No:* 527(035)

[733]
—Brown's star atlas, showing all the bright stars, with instructions how to find and use them for navigational purposes and Board of Trade examinations. New and rev. ed. Glasgow, Brown, Son and Ferguson, 1977. [v], 49p. charts, tables. ISBN: 0851742718.

2 parts: 1. How to find the stars; 2. Star problems.
Class No: 527(035)

[734]
—COTTER, C.H. 'A Brief historical survey of British navigational manuals', *Journal of navigation,* v.36, May 1983, p.237-48.

47 references appended. *Class No:* 527(035)

[735]
Air navigation plans. 12th-30th ed. Montreal, International Civil Aviation Organization, 1983-85. 5v. loose-leaf in binders, figs., charts, tables.

European region. Provisional 23rd ed. 1985 (Doc. 7457). *Middle East and South East Asia region. Africa-Indian Ocean region.* 24th ed. 1983. (Doc. 7474). *North American, and Pacific regions.* 12th ed. 1984. (Doc. 8755). *Caribbean and South American regions.* 30th ed. 1985. (Doc. 8733/13).

European region has 10 parts: 1. Assumed operating parameters - 2. General planning aspects - 3. Aerodrome operations - 4. ATS routes and associated navigation means - 5. Air traffic management - 6. Air traffic services - 7. Aeronautical information services - 8. Meteorological services - 9. Search and rescue services - 10. Aeronautical communications. *Class No:* 527(035)

Dictionaries

[736]

HARBORD, J.B. Glossary of navigation. 4th ed., rev. and enl. by C.W.T. Layton. Glasgow, Brown, Son & Ferguson, 1938 (reprinted 1977). x, [1], 451p. diagrs. £10. ISBN: 085174277.

First published 1862.

Defines *c*.1,000 terms, some at length. Entries are grouped under the qualified noun (*e.g.* types of 'log' - 'Dutchman's log'. 'ground log', 'current log', 'screw log' - under 'Log'). *Class No:* 527(038)

Almanacs

[737]

The Astronomical almanac for the year 1992: data for astronomy, space sciences, geodesy, surveying, navigation and other applications. Washington, U.S. Government Printing Office; London, H.M. Stationery Office, 1991. Tables. Annual. ix,546p. tables. £14. ISBN: 0118869426.

'Beginning with the edition for 1981, the title *The Astronomical Almanac* replaced both the title *The American Ephemeris and Nautical Almanac* and the title *The Astronomical Ephemeris,* unifying the two series' (*Preface, 1992*).

'The principal ephemerides in this Almanac have been computed from fundamental ephemerides of the planets and the moon prepared at the Jet Propulsion Laboratory, California, in cooperation with the U.S. Naval Observatory' (*Preface 1992*). Sections A-N. (D. Moon, 1992 ... E. Saturn, 1992 ... J. Observatories, 1992 ... N. Index). *Class No:* 527(059)

[738]

—Astronomical phenomena for the year 1992. Washington, US Government Printing Office; London, HMS0, 1991. Annual. 87p. £4; $3.50. ISBN: 0160021162.

Prepared jointly by the Nautical Almanac Office, US Naval Observatory, and HM Nautical Almanac Office, Royal Greenwich Observatory. Printed in advance of the main volume, this contains extracts from *The Astronomical Almanac,* including movements and visibility of planetary and lunar phenomena, etc. 11 sections: Phenomena; International time zones, Sunrise and sunset tables, Moonrise and moonset tables, and Eclipses. *Class No:* 527(059)

[739]

—Planetary and lunar co-ordinates for the years 1984-2000. London, H.M. Stationery Office, 1984. 336p. £10. ISBN: 0118669175.

Issued by HM Nautical Almanac Office in advance of the annual *Astronomical Almanac,* to provide low-precision data for planning purposes. *Class No:* 527(059)

[740]

—YALLOP, B.D. *and* HOHENKERK, C.Y. Compact data for navigation and astronomy for the years 1991-1995. Cambridge, University Press, 1990. xxviii,70p. tables. ISBN: 0521387310.

Prepared by HM Nautical Almanac Office at the Royal Greenwich Observatory and intended primarily for use in the practice of astronomical navigation, hence the more durable cover of this edition. 9 sections: 1. Introduction - 2. Sun and planets ... 6. Refraction and dip - 7. Sight reduction formulae - 8. Times of rising and setting and transit - 9. Hints for programmers. 6 tables: 1. Sun and planets ... 4. Phases of the moon and seasons ... 6. Apparent places of

.... *(contd.)*

stars. Convenient summary of basic astronomical information needed for navigation at sea. *Class No:* 527(059)

Histories

[741]

WATERS, D.W. The Art of navigation in England in Elizabethan and early Stuart times. 2nd ed. Greenwich, Trustees of the National Maritime Museum, 1978. 3v. facsims.

First published 1958.

A specialized study, well supported by illustrations, bibliography and index. Chart reproductions. Bibliography; index of navigational MSS. and works printed before 1640. Detailed index. *Class No:* 527(091)

Biographies

[742]

TAYLOR, E.G.R. The Mathematical practitioners of Hanoverian England, 1714-1800: a sequel to *The mathematical practitioners of Tudor and Stuart England.* Cambridge, University Press, for the Institute of Navigation, 1966. xv,503p. illus.

2 parts: 1. Narrative - 2. Biographies of inventors, makers and users of mathematical instruments, teachers, writers and theorists, by decades. Indexes: 1. General; 2. Secondary practitioners. 'Sources and references', p.xiv-xvi. *Class No:* 527(092)

[743]

TAYLOR, E.G.R. The Mathematical practitioners of Tudor and Stuart England. Cambridge, University Press, for the Institute of Navigation, 1954. xv,443p. illus.

3 parts: 1. Description of the progress and work of practical mathematicians, 1485-1715; 2. Short biographical notes on 582 mathematical practitioners; 3. Chronological annotated bibliography of 628 mathematical treatises published during the period. An essential work. *Class No:* 527(092)

Air & Nautical Almanacs

[744]

The Air almanac. London, H.M. Stationery Office, 1933-. Now annual. (*A.P. 1602.*)

Produced jointly by H.M. Nautical Almanac Office, Royal Greenwich Observatory, and Nautical Almanac Office, U.S. Naval Observatory.

'The object of the *Air Almanac* is to provide in a convenient form the astronomical data for air navigation' (*Explanation*). Tables show daily position of the sun, Aries, planets and moon. A folding navigational star chart is appended. Aimed to meet the general requirement of the military forces in the UK, US, Australia, Canada and New Zealand. *Air almanac 1992* (1991. 168p. £39). *Class No:* 527.05

[745]
The Nautical almanac for the year 1993. London, H.M. Nautical Almanac Office; Washington, Nautical Almanac Office, U.S. Naval Observatory, 1992. 336p. £16.50. ISBN: 0117726184.

The detailed calendar of tables for the sun, moon, planets and stars for the year is followed by 'Explanation', standard times, star charts, etc. 'This Almanac, with minor modifications and changes of language, has been adopted for the Brazilian, Danish, Greek, Indian, Indonesian, Italian, Korean, Mexican and Norwegian almanacs' (*Preface*).
Class No: 527.05

[746]
Reed's nautical almanac, 1993. 62nd year. Sunderland, Reed, 1992.

Editions: *Mediterranean*; *European*; *Baltic*; *North America*.

Sections covering nautical astronomy, coastal navigation, radio aids, tides and tidal streams, navigable distances, sea signalling, flags and flag etiquette, weather forecasting, safety at sea, visual aids and port information, U.K. & continent, harbours and marinas, etc. 1993 ed. contains 400 new harbour chartlets and expanded weather services and radio information. Tides, port entry, lights, anchorages index. General index. Advertiser's index. A mine of information, termed 'the yachtsman's Bible'.

Companion volume, *Reed's nautical companion*, contains information which needs little updating. New data include fast boat navigation; satellite communications. Also published in German as *Reed's nautisches handbuch*.
Class No: 527.05

[747]
—Brown's nautical almanac; daily tide tables for 1991. Brown, T.N., *ed.* Glasgow, Brown, Son and Ferguson, Ltd., 1990. 857 + 39p. tables.

First published 1858.

Has as main contents: Nautical information - Planetary information - Tables of navigable distances - Legal section, meteorology, beacons, buoys, etc. Less extensive than *Reed's Nautical almanac*. *Class No:* 527.05

[748]
—Olsen's fisherman's nautical almanack, 1993, containing tide tables and directory of British fishing vessels. Scarborough, Dennis, 1992. 700p. tables., maps. £17.99. ISBN: 0951181157.

Includes a guide to light houses & light vessels (U.K., Eire, and 5 continental countries), daily tide tables, port facilities and a list of British fishing vessels over 15 tons gross. *Class No:* 527.05

Tide Tables

[749]
Tide tables for the year ... London, Hydrographic Department, 1977-83. 3v. illus., forms, tables, maps. (*NP 201-203*.)

First published 1833.

V.1: *European waters, including Mediterranean Sea.* 1983. 1, 434p. v.2: *Atlantic and Indian Oceans, including tidal stream tables.* 1980. xlviiiip. v.3: *Pacific Ocean and adjacent seas, including tidal stream tables.* 1981. [469]p.

V.1 has 3 parts: 1. Prediction of high and low water for

.... (contd.)
standard ports; 2. Non-harmonic data; 3. Harmonic constants. Geographical index, p.415-20. List of tidal publications. *Class No:* 527.08

Surveying. Geodesy

Bibliographies

[750]
NATIONAL MARITIME MUSEUM. Catalogue of the Library. v.3: *Atlases & cartography.* London, H.M. Stationery Office, 1971. 2v. (catalogue; index).

V.1, p.597-654, 'Cartography & historiography' comprises 188 entries, with appended contents-listing of *Imago mundi: yearbook of old cartography,* v.1-21 (1935-67) and the Map Collectors Circle series. *Class No:* 528(01)

Handbooks & Manuals

[751]
BOMFORD, G. Geodesy. 4th ed. Oxford, Clarendon Press, 1980. vii, 855p. diagrs., tables. ISBN: 019851946x.

First published 1952.

7 chapters: 1. Triangulation, traverse and trilateration (field work) - 2. Computation of triangulation, traverse and trilateration - 3. Heights above sea-level - 4. Geodetic astronomy - 5. Gravity observations - 6. Physical geodesy - 7. Artificial satellites. 10 appendices. Bibliography (620 items), p.800-28. Analytical index. Chapter 7, on Satellite geodesy, because of the great advances during the last ten years, has been much expanded. A standard text.
Class No: 528(035)

[752]
Surveying. Bannister, A., *and others.* 6th ed. Harlow, Essex, Longman Scientific & Technical, 1992. xii,482p. illus., diagrs., tables. £16.99. ISBN: 0582076889.

13 chapters: 1. Introductory - 2. Tape and offset surveying - 3. Levelling - 4. The theodolite and its use ... 8. Analysis and adjustment of measurements ... 10. Setting out ... 13. Photogrammetry. Chapter exercises. Partly analytical index. Many photographs and diagrams. Emphasises modern equipment and techniques. A standard text.
Class No: 528(035)

Dictionaries

[753]
FÉDÉRATION INTERNATIONALE DES GÉOMÈTRES. Commission de Dictionnaire Technique. Dictionnaire multilingue de la Fédération International de Géomètres. Edition trilingue: français-alemand-anglais. Amsterdam, 'Argus', [1963]. xix, 501p.

Prepared with Unesco financial aid and technical aid of the Institut Géographique National, Paris, the Institut für Angewandte Geodäsie, Frankfurt/Main, and the Royal Institute of Chartered Surveyors, London.

About 5,500 terms in French, with definitions and examples of usage, followed by German and English/American equivalents. Genders are stated. French, German and English indexes. Bibliography p.xviii-xix.

....(contd.)

Dutch index: *Dictionnaire multilingue. Index néerlandais* (The Hague, Rijkscommisie voor Geodesie, 1966, 92p.). *Class No:* 528(038)

[754]
Glossaries for surveyors. Minnick, R., *comp.* Rancho Cordova, Calif., Landmark Enterprises, [1985]. [xi], 517p.

Consists of 4 glossaries: *Glossary of mapping, charting and geodetic terms* (U.S. Dept. of Defense, 3rd ed. 1973. p.1-281); *Definitions of terms used in geodetic and other surveys* (U.S. Dept. of Commerce. H.C. Mitchell, 1945); *Tide and current glossary* (U.S. Dept. of Commerce. P. Schureman; revised by S.D. Hicks, 1975, p.283-371); and *Glossary of BLM surveying and mapping terms* (U.S. Dept. of Interior. Bureau of Land Management). Prepared by Cadastral Survey Training Staff, Denver Service Center, 1978).

2nd ed., 1989. 607p. $30. *Class No:* 528(038)

[755]
MARTON, G. Dictionar poliglot de geodezie, fotogrammetrie si cartografie: englezā, românē, germânē, francezā, rusā. Bucharest, Editura Tehnica, 1976. xv, 325, [2]p.

About 4,000 English-base terms, with equivalents and indexes in Rumanian, German, French and Russian. Appended bibliography. *Class No:* 528(038)

[756]
MITCHELL, H.C. Definitions of terms used in geodetic and other surveys. Washington, U.S. Coast and Geodetic Survey, 1945 (reprinted 1969). 121p.

Includes definitions compiled by the Committee on Definitions of the Federal Board of Surveys and Maps. *Class No:* 528(038)

Reviews & Abstracts

[757]
Referativnyĭ zhurnal. 52. Geodeziia i aeros'emka. Akademiia Nauk SSSR, Institut Nauchnoĭ Informatsii. Moscow, VINITI, 1963-. 12pa. 36 rubles. ISSN: 03759717.

Formerly *Referativnyĭ zhurnal. Geodeziia, 1963-69*, previously part of *Referativnyĭ zhurnal. Astronomiia i geodeziia* (1953-).

About 3,000 indicative and informative abstracts pa. on geodesy and aerial surveying, in 15 sections. Monthly author index; annual author and subject indexes. *Class No:* 528(048)

[758]
RICS Library Information Service abstracts and reviews. London, Royal Institution of Chartered Surveyors, 1965-. v.1(1)-. 12pa. £135pa., to non-members. ISSN: 00336939.

Contains *c.*1,500 abstracts pa. Abstracts, 50-400 words in length are arranged under subject A-Z. Items include films and videos. Headings include Arbitration - Auctions - Buildings and construction... Landlord and tenant... Property... Taxation... Water. On back cover: Statistical data (Financial indicators... Agricultural wages). Annual indexes. 'A monthly bulletin of selected references for Chartered Surveyors'. *Class No:* 528(048)

[759]
—RICS Library Information Service Weekly briefing: a digest of news selected from the press. London, Royal Institution of Chartered Surveyors, 1965-. Weekly. £180. ISSN: 00336947.

Uses arrangement and headings as in *RICS Library Information Service abstracts and reviews* (*q.v.*), plus 'Departmental circulars', 'Statutory Instruments', 'Bills in progress' and 'Forthcoming conferences'. *Class No:* 528(048)

Yearbooks & Directories

[760]
ROYAL INSTITUTION OF CHARTERED SURVEYORS. Directory 1993. Basingstoke, Hants., Macmillan, 1992. 2v. £80. ISBN: 0333558685, set.

Formerly RICS *Yearbook.*

V.1 (905p.) contains a list of members, A-Z (p.3-905). V.2 (856p.) is a geographical directory (p.1-856), arranged A-Z by town/city. *Class No:* 528(058)

Great Britain

[761]
GREAT BRITAIN. Ordnance Survey. The History of the retriangulation of Great Britain, 1935-1962... London, HMSO, 1968. 446p. diagrs., graphs. tables. £15.75. ISBN: 0117004936.

Provides a detailed technical account of the retriangulation. 26 illus., 20 diagrs. Bibliography of 51 items. 20 folded diagrs. Well produced. *Class No:* 528(410)

Land

Handbooks & Manuals

[762]
Textbook of topographical surveying. London, HMSO, 1965. 388p. + 16pl. illus., diagrs., forms, maps.

First published 1905.

19 chapters (19. British and foreign maps). Appendices A-J. Detailed analytical index. *Class No:* 528.4(035)

Almanacs

[763]
GREAT BRITAIN. Nautical Almanac Office. The Star almanac for land surveyors... 1993; prepared on behalf of the Science and Engineering Research Council. London, H.M. Stationery Office, 1992. xvi, 80p. tables, star charts. £5.50. ISBN: 0118869450.

Includes ephemeris of the sun, sunrise, sunset, moon's phases; apparent places of 650 stars; radio time signals, auxiliary tables and star charts. Index of star names and numbers. *Class No:* 528.4(059)

Biographies

[764]
Dictionary of land surveyors and local cartographers of Great Britain and Ireland, 1550-1850; compiled from a variety of sources. Eden, P., *ed.* Folkestone, Dawson, 1975/76. 3 pts. (A-F, G-R, S-Z). Supplement. 1979. ISBN:

....(contd.)

0712906576; 0712906819; 0712907386.

Part 1: A-F (108p.) records 'all persons likely to have measured land or made maps of land in areas of less than a complete county in Great Britain or Ireland between the indicated dates'. 2,395 entries (name; dates of birth; death and activity; area; class of maps produced; alternative occupation; partnerships; bibliography; genealogy). Appended to pt.3 is a 'Directory of surveyors by callings', A-Z; topographical index. 'A very significant basic source of reference' (*Geographical journal*, v.142(1), March 1976, p.179). *Class No:* 528.4(092)

[765]

—Dictionary of land surveyors and local cartographers of Great Britain and Ireland, 1550-1850. Supplement. Eden, P., *ed.* Folkestone, Dawson, 1979. p.[vi], 379-528. ISBN: 0712908994.

A project supported by the Social Science Research Council.

Similarly arranged. 3,010 entries, bringing the total for the *Dictionary* to *c.*10,000. Appendix 2: 'Dictionary of surveyors (Irish Plantation surveyors)'. General index and topographical index to the *Supplement*.

Class No: 528.4(092)

Hydrographic Surveying

Handbooks & Manuals

[766]

GREAT BRITAIN. Hydrographic Department. **Admiralty manual of hydrographic surveying. 1965-73.** [New ed.]. London, Hydrographer of the Navy, 1973. 2v.

Previously published 1938 and 1948.

V.1 covers principally the theory and the practical mathematics necessary to enable the surveyor to carry out this work. V.2 deals more specifically with the hydrographic surveyor's practical work and the special methods applicable to marine surveying (*Foreword*). V.1 has 9 chapters (2. Geodesy - 3. The measurement of distance... 9. Heights and levelling; bibliography and analytical index). Of v.2 (1969-) only chapters 1-5 have apparently been published (1969-73). *Class No:* 528.47(035)

Photogrammetry

Handbooks & Manuals

[767]

A Dictionary of photogrammetry. Slama, C.C., *ed.* 4th ed. Falls Church, Va., American Society for Photogrammetry and Remote Sensing, 1980. 1056p. $40. ISBN: 0937294012.

'The definitive work on the use of photogrammetric techniques to extract quantitative information from vertical and oblique aerial photography' (Harris, C.D. *ed. Geographical bibliography for American libraries* (1985), entry 375). *Class No:* 528.7(035)

Dictionaries

[768]

INTERNATIONAL SOCIETY OF PHOTOGRAMMETRY. Multi-lingual dictionary for photogrammetry. Amsterdam, 'Argus', [1969]. 6v.

One volume for each of the 6 languages,- English, French, German, Italian, Polish and Swedish. Cross-references between volumes. *Class No:* 528.7(038)

Remote Sensing

Databases

[769]

RESORS (Remote Sensing On-line Retrieval System). Ottawa, Canada, Centre for Remote Sensing.

Covers 1900 to date. A bibliographic and referral database providing references to the literature of remote sensing, photogrammetry and image analysis and processing, with emphasis on instrumentation techniques and applications. Sources: English- and French-language reports, periodical articles and conference papers. File size: 78,000 documents and 8,000 35mm slides. Updated monthly. 5,000-6,000 additions pa. Available via the Centre. Also available as RESORS Microfiche and on CD-ROM (updated twice pa.). *Class No:* 528.71(003.4)

Bibliographies

[770]

HYATT, E. Keyguide to information sources in remote sensing. London, Mansell, 1988. xiv, 274p. £40. ISBN: 0720118549.

The largest and most up-to-date English-language guide to its subject. 3 parts: 1. Survey of remote sensing and its literature. 2 chapters outline the development of remote sensing, the technology, the systems and the satellites, in general and regionally by continent. 2 further chapters provide an in-depth analysis of the literature of the subject, covering the conventional categories (books, reports, trade literature, etc.) and the secondary sources (bibliographies, abstracts, reviews, databases). Chapter 5 covers language factors and dictionaries; chapter 6 details sources of remotely sensed imagery; chapter 7, 'Audio-visual material', includes slides, videocassettes, maps and atlases). Part 2 (p.127-249) is an annotated bibliography of 819 items. Of the 493 published information sources, over 100 journals and 100 books are listed, and more than 20 entries each for literature guides, abstracts and indexes and databases. Finally, 326 organizations, grouped under 81 countries, with addresses, are listed. This is an immensely impressive guide, well written, clearly set out - the work of an information specialist in a remote sensing research unit. *Class No:* 528.71(01)

[771]

—BRYAN, M.L. Remote sensing of earth resources: a guide to information sources. Detroit, Mich., Gale, 1979. xv,188p. $68. (*Geography and Travel Information guides, v.1.*) ISBN: 0810314134.

378 annotated entries, with extensive cross-references, covering general literature, journals, bibliographies, maps, guides and education and training facilities. Acronyms and abbreviations list. Indexes of author, title, subject. *Class No:* 528.71(01)

Handbooks & Manuals

[772]
LILLESAND, T.M. and KIEFER, R.W. Remote sensing and image interpretation. 2nd ed. New York, John Wiley & Sons, 1987. xiv,721p. illus. (some col.), figs., diagrs., tables. $62.95. ISBN: 0471845175.

First published 1979.

Provides a broad introduction to remote sensing. 10 sections: 1. Concepts and foundations of remote sensing - 2. Elements of photographic systems (p.37-111) ... 5. Photogrammetry ... 7. Thermal and multispectral scanning - 8. Microwave sensing ... 10. Digital image processing. Appendix: Image sources. Index. Many black-and-white illustrations and 24 excellent colour plates. For the student and anyone involved in measuring and managing earth resources. *Class No:* 528.71(035)

[773]
Manual of remote sensing. Colwell, R.N., *ed.* 2nd ed. Falls Church, Va., American Society of Photogrammetry and Remote Sensing, 1983. 2v. (xxxiii, 2440p.) illus., diagrs., (mostly col.). ISBN: 0937294527.

The classic manual. More than 200 contributors. 36 chapters in 2v., with theoretical and systems material in v.1 and practical applications in v.2. Comprehensive coverage of core subject and related fields, with full attention to the technology and equipment. Extensive bibliographies. Glossary, p.1183-98. Combined index, p.1209-32. *Class No:* 528.71(035)

Dictionaries

[774]
RABCHEVSKY, G.A., ed. Multilingual dictionary of remote sensing and photogrammetry: English glossary and dictionary, equivalent terms in French, German, Italian, Portuguese, Spanish, Russian. Falls Church, Va., American Society of Photogrammetry, 1984. xx,[1],343p. $20. ISBN: 0937294462.

1,717 numbered terms defined in English, with equivalents in French, German, Italian, Spanish, Portuguese and Russian. Indexes in other languages refer to English base. Bibliography and acronyms list. *Class No:* 528.71(038)

Yearbooks & Directories

[775]
CARTER, D.J. The Remote sensing source book: a guide to remote sensing products, services, facilities, publications and other materials. London, McCarta/Kogan Page, 1986. 175p. diagrs., maps. ISBN: 1850910340.

Provides detailed guidance on the main products, services and facilities available in the UK and gives details of equipment, imagery collections and personnel in commercial, government and academic institutions. Also covers educational material and maps. Bibliography (p.65-93) is a guide to the current literature.

'Fills a gap in the remote sensing literature that has become conspicuous since the recent explosion... in the volume of remotely sensed data gathered on a global basis... In addition to providing a comprehensive index to the remotely sensed data available within Europe and North America, it provides a reference to textbooks and journals on different applications of remote sensing...' *Journal of physics - E. Scientific instruments,* v.19, October 1986, p.757).

....(contd.)

'It is unfortunately let down by a poor layout that tends to obscure the substantial amount of information contained; this problem is compounded by a scanty index and lack of accurate cross-referencing... [but] the volume is quite informative and fairly comprehensive, especially on British products ... (E. Hyatt, *Keyguide to information sources on remote sensing.* (Mansell, 1988), p.169). It needs to be updated soon if it is to retain its usefulness. *Class No:* 528.71(058)

[776]
Remote sensing yearbook 1990. Cracknell, A. *and* Hayes, L., *eds.* London, Taylor & Francis, 1990. Annual. vii,330p. figs., maps, tables. £50. ISBN: 0850668085. ISSN: 02676133.

First published 1986.

4 sections: 1. Developments in remote sensing - 2. Remote sensing societies, p.139-52 - 3. Satellite data - 4. Directory p.175-330, (12 sections: Establishments receiving and supplying satellite data ... Consultants ... Research establishments ... Societies ... Directory index). The section, Bibliography, which was in previous issues, has been excluded from this edition. *Class No:* 528.71(058)

[777]
United Kingdom remote sensing directory. Farnborough, Hants., National Remote Sensing Centre (Royal Aircraft Establishment), 1986.

An update of the Department of Industry's *Remote sensing of earth resources* (3rd ed., 1976).

Facilities and activities of over 200 organizations in 5 sections: academic, commercial, industrial, consultancy, government and learned bodies. Lists staff and their geographical or subject interests. *Class No:* 528.71(058)

Cartography

Bibliographies

[778]
Bibliographia cartographica ... Hrsg. von der Staatsbibliothek Preussischer Kulturbesitz in Verbindung mit der Deutschen Gesellschaft für Kartographie. [International documentation of cartographical literature.] Munich, etc., K.G. Saur, 1974-. v.1-. Annual. $66. ISSN: 03400409.

Supersedes *Bibliotheca cartographica* (v.1-29/30. 1946/47).

Contains *c.*2,500 numbered items. 12 sections: 1. Bibliographies. Map collection. Documentation - 2. General publications - 3. History of cartography - 4. Institutions and organizations of cartography - 5. Theoretical cartography - 6. Cartographic technology - 7. Topographic and landscape cartography - 8. Thematic maps and cartograms - 9. Atlas cartography - 10. Use and application of maps. General purpose maps - 11. Reliefs and other forms of cartographic representation - 12. Globes. List of periodicals (*c.*300); List of abbreviations; Concordance of the UDC and of the chapter headings of the *Bibliotheca cartographica.* Index of authors. *Class No:* 528.9(01)

[779]
Catalogue of cartographic materials in the British Library 1975-1988. London, Saur, 1989. 3v. ISBN: 0862917654, set; 0862917662, v.1; 0862917670, v.2; 0862917689, v.3.

'Contains bibliographic records for atlases, single-sheet maps, map series, maritime charts, plans, globes and related reference materials acquired by the British Library Map Library since 1974 ... includes a number of pre-1974 and antiquarian items' (Explanatory note by publisher). V.1 and 2 contain names/titles, while v.3 lists geographic names and subjects. Also available on microfiche. *Class No:* 528.9(01)

[780]
HODGKISS, A.G. *and* **TATHAM, A.F. Keyguide to information sources in cartography.** London, Mansell, 1986. x, 253p. £40. (*Keyguides.*) ISBN: 0720117682.

4 parts: 1. Cartography and maps: the subject and its literature - 2. Annotated bibliography of reference sources: history of cartography - 3. Annotated bibliography of reference sources: contemporary cartography - 4. Directory of organizations (nos.797-1397). Detailed analytical index. Items 1-796, in parts 2-3, are usually annotated, frequently indicating readership. A refreshing approach. Aimed at curators of map collections, libraries, cartographers, students and other map-users. *Class No:* 528.9(01)

[781]
Information sources in cartography. Perkins, C.F. *and* Parry, R.B., *eds.* London, Bowker-Saur, 1989. xiii,540p. illus., maps. $96. ISBN: 0408024585.

33 contributors; 6 parts (30 well-documented chapters): 1. General - 2. The history of cartography - 3. Map production - 4. Map librarianship - 5. Types of mapping - 6. Map use and promotion. 3 appendices: 1. Cartographic periodicals - 2. Cartographic societies - 3. Map publishers. Subject and, unlike the other volumes in this series, name indexes. An outstanding volume. *Class No:* 528.9(01)

[782]
PARRY, R.B. *and* **PERKINS, C.R. World mapping today.** London, Butterworths, 1987. [x],583p. diagrs., maps. £145. ISBN: 0408028505.

7 General chapters (*e.g.* 2. The state of world mapping - 3. Map acquisition - 4. Map evaluation - 5, Maps and remote sensing - 6. Digital mapping - 7. Future trends in digital mapping). World mapping, p.61-568: The world; Africa; The Americas; Australia; Europe; The oceans. ('Philippines', p.340-2: Mapping - Further information - Addresses; Catalogue; Atlases and gazetteers; General. 2 map keys, 1:50,000). 'List of geographic indexes' [*i.e.,* of 200 map keys] precedes text. Glossary, p.569-72. Geographical index. Publishers index. Well organized and well produced. For all map libraries as well as for most non-specialist libraries. 2nd ed. planned for 1992.
Class No: 528.9(01)

[783]
UNITED STATES. Library of Congress. Geography and Map Division. **The Bibliography of cartography.** Boston, Mass., G.K. Hall, 1973, 5v.; First supplement, 1980. $615, set; $285, supp. ISBN: 0816102597.

The main work reproduces 91,600 photolithographed catalogue cards. Author, title and subject entries for books and periodical articles on maps, mapmakers, and the history of cartography. Particularly strong on history of cartography and cartobibliography for 1895-1922. Available on microfilm. *Class No:* 528.9(01)

Handbooks & Manuals

[784]
CANTERS, F. *and* **DECLEIR, H. The World in perspective:** a directory of world map projections. Chichester, New York, John Wiley, 1989. x,181p. maps. $74.95; £69. ISBN: 0471921475.

Illustrated overview of 68 different projections for world maps. Part 1. Principles of map projections and their application (The mathematical theory of map projections; The classification of map projections; The evaluation and choice of a map projection) - Part 2. Directory of world map projections, p.39-176 (Introduction; Polyconic projections; Pseudocylindrical projections; Pseudoconical projections; Cylindrical projections). References, p.177-78. Index.
Class No: 528.9(035)

[785]
LOCK, C.B.M. Modern maps and atlases: an outline guide to twentieth century production. London, Bingley, 1969. 619p. ISBN: 0851570720.

A manual in five sections: 1. The techniques of modern cartography - 2. International maps and atlases - 3. National and regional maps and atlases (p.109-403) - 4. Thematic maps and atlases - 5. Map librarianship. The detailed analytical index has *c*.6,000 main entries. A solid contribution on 20th-century mapping (despite absence of map specimens) and an essential reference.

Modern maps and atlases is not fully incorporated into Dr. Lock's later *Geography and cartography* (Bingley, 1976); 'only a few entries (*i.e.,* map librarianship) have been included' (*RQ,* v.16(3), Spring 1977, p.259).
Class No: 528.9(035)

[786]
Manual of map reading and land navigation. 2nd ed. London, HMSO, 1989. various pagings, 5 folding maps. £7.95. (*Army code 70947.*) ISBN: 0117726117.

1st ed. published as *Manuel of map reading,* 1973.

13 chapters: 1. Introduction - 2. Scale - 3. Marginal information - 4. Representation of detail - 5. Representation of relief - 6. Referencing systems - 7. Direction - 8. Compasses and their use - 9. Land navigation - 10. Tactical information printing systems - 11. Map substitutes - 12. Special maps - 13. Map supply. 'Produced primarily to provide instructors in the Armed Services with a comprehensive source of reference on which to base training' (*Publisher's description*). *Class No:* 528.9(035)

[787]
Planetary mapping. Greeley, R. *and* Batson, R.M., *eds.* Cambridge, University Press, 1990. ix,296p. illus., diagrs., tables, maps. (*Cambridge planetary science series.*) ISBN: 0521307740.

Outlines methods and techniques used in making maps of planetary objects. 7 sections: 1. Introduction - 2. History of planetary cartography ... 4. Planetary nomenclature ... 6. Topographic mapping - 7. Geologic mapping. 3 appendices: 1. Map formats and projections used in planetary cartography - 2. Halftone processes for planetary maps - 3. Digital planetary cartography. Bibliography appended to each section. *Class No:* 528.9(035)

[788]
ROBINSON, A.H., *and others*. **Elements of cartography.** 5th ed. New York, Wiley, 1984. 554p. illus. (incl. col.), facsims., diagrs., graphs, tables. $51.95. ISBN: 0471098779.

First published 1953.

17 chapters, systematically covering statistics, photogrammetry, geometry and art, plus technical information ranging from remote sensing and computer-assisted cartography to lettering and construction methods. Appendices A-L; analytical index. The standard work for students in cartography. *Class No:* 528.9(035)

[789]
Surveying and mapping for field scientists. Ritchie, W., *and others*. Harlow, Essex; New York, John Wiley, 1988. ix,180p. illus., diagrs. £14.99. ISBN: 058230086x.

New edition of *Mapping for field scientists* (David & Charles, 1977).

5 chapters: 1. The problem and how to approach a solution - 2. Ground surveying techniques - 3. Aerial surveying techniques - 4. Remote sensing techniques - 5. Cartographic presentation. 2 appendices: 1. Examples of Ordnance Survey information - 2. Sources of aerial photography and earth satellite imagery. Bibliography, p.173-75. Index. 'This book is designed as a comprehensive practical introduction and guide to a range of sources and techniques that can be used by a wide group of people' (*Preface*). *Class No:* 528.9(035)

Dictionaries

[790]
BARGILLIAT, A. **Vocabulaire pratique, anglais-français et français-anglais des termes techniques concernant la cartographie** (géodesie, topographie, dessin, photomécanique, impression). Paris, Institut Géographique National, 1944. 397p.

Part 1: Anglais-français (p.38-217); pt.2: Français-anglais (p.220-397). Conversion tables, p.12-35. *Class No:* 528.9(038)

[791]
BONACKER, W. **Karten-Wörterbuch:** eine Verdeutschung fremdsprachiger Kartersignatur-Bezeichnungen... Bonn, Kirschbaum, 1970. xx, 276p. $59.95. ISBN: 3781207048.

This ed. is a facsimile of 1941 ed.

A polyglot glossary of terms (physical features, etc.) occurring on maps. No fewer than 56 languages are involved, including Breton, Cambodian, Asset and Tatar. Under each language: c.450 words are given in the native tongue, first translated and then in the original script, plus German meaning and notes on pronouncing the original. About 21,000 terms are so covered. *Class No:* 528.9(038)

[792]
Dictionary of mapping, charting and geodetic terms. Madison, Wisc., Map Division IV, 1978. [i], 271p.

About 5,000 technical terms are defined, with numerous cross-references. *Class No:* 528.9(038)

[793]
EDSON, D.T., *ed.* **Glossary of terms in computer assisted cartography.** [Glossaire des termes en cartographie assistée par ordinateur.] 2nd ed. Falls Church, Va., American Congress on Surveying and Mapping, for the International Cartographic Association's Commission on Cartographic Terminology, 1980. 157p.

English/French glossary. *Class No:* 528.9(038)

[794]
GALPERIN, G.L., *comp.* **Anglo-russkiĭ slovar' po kartografii, geodezii, aerofototopografiĭ.** Moscow, Gos. Izd-vo Fiziko-matem. Literatury, 1958. 548p.

An English-Russian dictionary on cartography, geodesy and aerial photography, with an index of Russian terms. *Class No:* 528.9(038)

[795]
Multilingual dictionary of technical terms in cartography. Meynen, E., *comp.* Wiesbaden, Steiner Verlag GmbH.; New York, French and European Publications, 1984. 700p. maps. $135. ISBN: 0828814643, US ed.

First published 1973.

Defines c.1,000 cartographic terms in German, English, Spanish, French and Russian, with equivalents in 9 other languages (Czech, Italian, Japanese, Hungarian, Dutch, Portuguese, Polish, Swedish and Slovak). 2 appendices list names (in the 5 major languages) of individual grouped map projections. 3 sample maps illustrate some of the more important terms. Bibliography. Of the 1973 ed., 'The production of this dictionary has itself done much to standardise cartographic terms' (*Geographical journal,* v.140(2), June 1974, p.339). *Class No:* 528.9(038)

[796]
—BRITISH NATIONAL COMMITTEE FOR GEOGRAPHY. CARTOGRAPHIC SUBCOMMITTEE. Glossary of technical terms in cartography. London, Royal Society of London, 1966. 84p.

Defines 400 terms. 3 appendices, one on named map projections. References, p.79-84. 'Represents the first British contribution to the multilingual dictionary' (*Geographical journal,* v.133(2), June 1967, p.221). *Class No:* 528.9(038)

[797]
—Glossary of cartographic terms and manual of symbols and abbreviations used on latest navigational charts of the various countries... 3rd ed. Monaco, International Hydrographic Bureau, 1951. 186p. Amendments, 1951-.

French and English text. Covers 17 languages. *Class No:* 528.9(038)

[798]
—UNITED STATES. Department of Defense. Army Topographic Command. Glossary of mapping, charting and geodetic terms. 3rd ed. Washington, 1973. 281p.

First published by Defense Intelligence Agency in 1967 as a 'DOD glossary'.

About 5,000 entries. *Class No:* 528.9(038)

Periodicals & Progress Reports

[799]
The Cartographic journal. London, British Cartographic Society, 1964-. v.1(1)-. 2pa. £25pa. ISSN: 00087041.

The December 1992 (v.29(2)) issue carries 8 articles, *e.g.* 'Spatio-temporal maps and cartographic communication'; 'A

....*(contd.)*

user perspective on atlas content and design'. 8 signed reviews, p.187-94; 4 shorter notices; Recent literature, p.196-8. A major journal in its highly specialized field. *Class No:* 528.9(05)

[800]
Progress in contemporary cartography. Taylor, D.R.F., *general ed.* New York, Wiley, 1980-85. v.1-3. illus., diagrs., maps. ISBN: 0471276995; 0471103160; 0471903051.

V.1. *The computer in contemporary cartography.* 1980. xv, 252p. V.2. *Graphic communication and design in contemporary cartography.* 1983. xviii, 314p. £29.50. V.3 *Education and training in contemporary cartography.* 1985. 324p.

V.3 has 19 contributors. 3 sections: 1. Emerging trends and challenges - 2. Specific responses to education for a new cartography - 3. Cartography and the developing nations, (*e.g.* 'Remote sensing, remote development and education in Africa', p.189-225. 71 references). Author and subject indexes. *Class No:* 528.9(05)

[801]
World cartography. New York, United Nations, Dept. of Technical Co-operation and Development, 1951-. Annual. (*ST/CSA/SER.L/17.*) ISSN: 00841471.

Contains statistics, analyses and reports on recent and planned activities in cartography. V.21, 1990 has 2 parts: 1. Local data banks - 2. Organization and functions of a national geographical names standardization programme. Very specialized, but poorly printed in places. *Class No:* 528.9(05)

Yearbooks & Directories

[802]
MAIZLISH, A. *and* **HUNT, W.S. The World map directory, 1989.** Santa Barbara, Map Link, 1988. 278p. maps. $29.95. ISBN: 0929591003.

A catalogue of maps available from the publisher. Lists 46,000 titles, including thematic and topographic maps. International in scope though not all-inclusive - coverage of Eastern Europe, the former USSR, and developing countries is limited. Each entry includes sheet name, Map Link code, price, scale, date and an estimation of the quality of the map. Annotations of variable length are provided for most maps. Contains a list of publisher abbreviations (no addresses) and a geographic index. '*The World Map Directory* is a must for all map collections and map libraries' (*Reference books bulletin*, 1 May 1989, p.1531). *Class No:* 528.9(058)

Maps & Atlases

[803]
Guide to U.S. map resources. Cobb, D.A., *comp.* 2nd ed. Chicago, Ill., American Library Association, 1990. 495p. $65. ISBN: 0838905470.

First published 1986.

Covers 975 map collections in academic, public, private, geoscience, federal and state libraries. Arranged alphabetically by state, city and institution. Data include special collections, hours, holdings and chronological coverage. 128 pages of indexes, *e.g.* Library/Institution Index, Collection Strengths Index. Appendices: 1. Earth science national information centers - 2. State information

....*(contd.)*

resources - 3. State mapping advisory committees. Page layout is better than that of the previous edition and there is more information on each collection than in *Map collections in the United States and Canada (qv)* 'With the improvements in this edition, ALA's *Guide to U.S. map resources* is the preferred directory to U.S. map collections and cartographic resources' (*Booklist*, v.87(4), 15 December 1990, p.881). *Class No:* 528.9(084.3)

[804]
Map collections in the United States and Canada: a directory. Carrington, D.K. *and* Stephenson, R.W., *eds.* 4th ed. New York, Special Libraries Association, 1985. 178p. $35. ISBN: 0871113066.

First published 1954, 3rd ed., 1978.

Details of 804 map collections: USA (states, A-Z); Canada (by provinces). Data include size of collection, annual accessions, special collections, availability, publications. Detailed index. *Class No:* 528.9(084.3)

[805]
WATT, I., *comp.* **A Directory of UK map collections.** 2nd ed. London, McCarta, Ltd., for the British Cartographic Society, [1985]. vi, 248p. £4.50. (*Map Curators Group publication no.3.*) ISBN: 0904482081.

Two sections: The map collections of the copyright libraries (p.1-3) - Other map collections (Allan Ramsay Library... York City Council). Data: number of maps; type and scale; scope; atlases; publications. Indexes: 1. Alphabetical listing; 2. Geographical location; 3. Keywords, cartographers, cartographic collections, areas of coverage, etc. Small format. A guide to *c.*400 collections. *Class No:* 528.9(084.3)

[806]
World directory of map collections. Compiled by the [IFLA] Section of Geography and Map Libraries. Wolter, J.A., *and others, eds.* 2nd ed. Munich & London, K.G. Saur, 1986. xliii, 405p. £35. (*IFLA publication, v.31.*) ISBN: 3598203748.

Lists 670 collections, in 65 countries. Arrangement is by countries. Data on each collection include size and nature, the type and availability of catalogues. Name index of institutions and person in charge. *Class No:* 528.9(084.3)

Histories

[807]
BAGROW, L. History of cartography. Revised and enlarged by R.A. Skelton. Translated from German by D.L. Paisey. London, Watts; Cambridge, Mass., Harvard Univ. Press, 1964. 312p. illus., facsims.

Original as *Die Geschichte der Kartographie* (Berlin, Safari Verlag, 1951.).

Includes an extensive list of cartographers who practised before 1750 (p.227-80). Select bibliography (278 grouped references), p.283-300. 137 plates (21 in colour). 76 line-drawings in text. A handsome volume. 'The mapping of the New World is largely ignored in this work' (Harris, C.D., ed. *Geographical bibliography for American libraries* (1985), entry 325). *Class No:* 528.9(091)

[808]

Cartographical innovations: an international handbook of mapping terms to 1900. Wallis, H.M. *and* Robinson, A.H., *eds.* Tring, Herts., Map Collector Publications (1982), Ltd., in association with the International Cartographic Association, 1987. 353p. facsims., maps. £42. ISBN: 0906430046.

Terms and their history feature in 8 groups (1. Types of maps... 8. Atlases). 'Tidal map', p.134-6, has 1 facsimile and 11 references. General index, p.328-48. Index to bibliographies, p.347-53. Aims to identify on a worldwide scale the main points of advance and change in the science and art of cartography. *Class No:* 528.9(091)

[809]

CRONE, G.R. Maps and their makers: an introduction to the history of cartography. 5th ed. Folkestone, Kent, Dawson, 1978. [v], 152p. facsims., maps. £16. ISBN: 0712907564.

First published 1953.

14 brief chapters (*e.g.* 1. The Classical and medieval heritage - 2. The evolution of the medieval seachart - 3. The cartography of the great discoveries - 8. Hydrographic charts and atlases - 11. The British contribution - 12. National surveys and world maps, 1800-1975 - 13. Atlases and thematic maps - 14. Contemporary cartography). Bibliography, p.144-6 (1. General works on cartography, cartographical periodicals and map catalogues; 2. Reproductions of early maps and charts). Detailed, non-analytical index. Larger format and well illustrated (unlike editions 1-4). 'A good overview of the history of cartography' (*Special libraries,* v.70(7), July 1979, p.309) by the former Librarian and Map Curator of the Royal Geographical Society. *Class No:* 528.9(091)

[810]

The History of cartography. Chicago, Ill., Univ. of Chicago Press, 1987-. v.1-. illus., maps. ISBN: 0226316335, v.1; 0226316351, v.2.

1. *Cartography in prehistoric, ancient and medieval Europe, and the Mediterranean.* 1987. 599p. £79.95. 2. *Cartography in the traditional Islamic and south Asian societies.* 1992. 704p. £99.95.

To be in 6v. V.1 carries the narrative up to *c.*1470. V.2-6 are to extend coverage both chronologically and geographically. 'No large library can afford to be without this' (*Library journal,* 15 September 1987, p.79). *Class No:* 528.9(091)

[811]

Imago mundi: the journal of the International Society for the History of Cartography. Haywards Heath, Sussex, I.M. Ltd. London, 1935-. Annual. ISSN: 03085694.

V.44 (1992) comprises 7 articles (*e.g.* 'Medieval Hebrew manuscript maps'); 1 obituary; 6 reports and notices; chronicle for 1992; a review article; and 12 signed reviews (p.141-49). Appended: '*Imago mundi* bibliography of literature mainly published in 1989-91', including an index, p.153-72. *Class No:* 528.9(091)

[812]

International directory of current research in the history of cartography and in carto-bibliography. Campbell, E.M.J., *and others, comps. and eds.* No.4. Norwich, Geo Books, 1983. vi, 104p. ISBN: 0950401628.

First published 1974; 3rd ed. 1981.

Nearly 300 entries from 34 countries, UK and USA

....(contd.)

predominating. The register of contributors covers only investigations in hand, to yield results 'within three to five years of their first entry in the *Directory*'. Indexes. 'This is the major reference source into current work in the field of carto-bibliography and the history of cartography' (Hodgkiss, A.G., and Tatham, A.F. *Cartography* (1986), p.53). *Class No:* 528.9(091)

[813]

RISTOW, W.W. Guide to the history of cartography: an annotated list of references on the history of maps and mapmaking. Washington, Library of Congress, Geography and Map Division, 1973. iii, 96p.

Expansion of *A guide to historical cartography,* by W.W. Ristow and C.F. LeGear (1954; 2nd ed., 1960).

398 numbered entries, under authors, A-Z. Includes works on individual countries and specialised aspects of cartography. Index to subjects, geographical areas, etc. *Class No:* 528.9(091)

[814]

TOOLEY, R.V. Maps and map-makers. 6th ed. London, Batsford, 1978. xii, 140p. illus., facsims. £29.95. ISBN: 0713413956.

First published 1949.

Treatment is by schools of geography. Chapters: Pre-Christian geography - The Arabs and medieval Europe - Italy - Germany - Holland and Belgium - French cartography - English map-makers. English maritime atlases - The country maps of England and Wales - Scotland and Ireland - Africa - Asia - America - Australia - Scandinavia. Extensive chapter bibliographies; many reproductions of maps. 140 illus. Detailed index. A popular work on the history of cartography. *Class No:* 528.9(091)

Biographies

[815]

TOOLEY, R.V. Tooley's dictionary of mapmakers. Tring, Herts., Map Collector Publications, 1979. xii, 684p. Supplement, 1985. ISBN: 0845117012; 0845117033, supp.

Brief biographical data on over 20,000 mapmakers, from earliest times to 1900: name; date of birth and death (when known), titles of honour (if any), working address and changes of address (which often enables the user to establish the approximate date of publication of a map), main output of maps or atlases, with dates and editions whenever known. Includes a list of 'Works consulted', revealing surprising omissions. The *Supplement* adds *c.*4,000 names to the dictionary. Nevertheless, a standard work and attractively produced. *Class No:* 528.9(092)

[816]

—SKELTON, R.A. Maps: a historical survey of their study and collecting. Chicago, Ill., & London, University of Chicago Press, 1972. 138p. $15. ISBN: 0266761665.

Consists of 4 lectures at the Newberry Library, Chicago. 'Unsurpassed as the best general introduction' (*RQ,* v.12(4), Summer 1973, p.409). Included is a list of R.A. Skelton's published works. *Class No:* 528.9(092)

Worldwide

[817]
Cartography, past, present and future: a Festschrift for F.J. Ormeling. Rhind, D.W. *and* Taylor, D.R.F., *eds*. London, New York, Published on behalf of the International Cartographic Association by Elsevier Applied Science Publishers, 1989. xx,193p. $63. ISBN: 1851663363.

A short biography of F.J. Ormeling is followed by 2 main sections: 1. Contemporary regional issues in cartography - 2. Thematic issues in cartography. Section 1 consists of 7 documented chapters covering cartography in specific countries, *e.g.* 'The status of cartography in Nigeria'; 'China's cartography: the present situation and future perspectives'. The 7 chapters in section 2 cover major issues and developments in cartography, *e.g.* 'Cartography as an art'; 'Education and training and cartography'; 'The revolution in cartography in the 1980s'.
Class No: 528.9(100)

Marine Environment

[818]
ROBINSON, A.H. **Marine cartography in Britain:** a history of the sea chart to 1855. Leicester Univ. Press, 1962. 222p. diagrs., charts.

Sub-title should read: 'a history of British marine cartography of the British Isles down to 1855'. Appendix A: 'Additional biographical notes on some of the sixteenth century surveyors and chart-makers'; B to L: lists of charts and hydrographic surveys (p.152-202). The authoritative text is marred by a disappointing short bibliography and inadequate index. *Class No:* 528.9(26)

Great Britain

[819]
SEYMOUR, W.A., *ed*. **A History of the Ordnance Survey.** Folkestone, Dawson, 1980. xiv, 394p. maps. £35. ISBN: 0712909796.

35 chapters by various hands (1. The origins of the Ordnance Survey... 35. The restoration of the Survey - a new home and new aims. 6 appendices (*e.g.* Recommendations of the Davidson Committee (1938)). Analytical index, p.379-94. A handsome tribute.
Class No: 528.9(410)

Libraries

[820]
LONDON UNIVERSITY. Library Resources Co-ordinating Committee. Geography Sub-committee. **A Guide to geography and map collections.** London, the University, 1985 (2nd ed., 1990). [i], 40p. ISBN: 0718797125.

1. Geography collections (12 libraries) - 2. Map collections (10 libraries). Data: name and address; telephone no.; location; hours of openings; access; readers places; name of map curator; staff availability; catalogue(s); classification; holdings; special collections or strengths; publications; additional facilities).
Class No: 528.9:061:026/027

Map Printing

[821]
WOODWARD, D., *ed*. **Five centuries of map printing.** Chicago, Ill., & London, Univ. of Chicago Press, for the Hermon Dunlop Smith Center for the History of Cartography, the Newberry Library, 1975. xi, 177p. illus., facsims., maps. $7.95. ISBN: 0226907244.

8 contributions, all by American geographers, although developments in Britain and other European countries are well covered. Select bibliography, p.167-70. 'The first book to be devoted to the technical history of the processes used for printing maps'. (*Aslib proceedings,* v.28(6/7), June/July 1976, p.252). Well received in *Cartographic journal,* v.13(2), 1976, p.197-8. *Class No:* 528.9:655.1

Time

[822]
AVENI, A.F. **Empires of time:** calendars, clocks and cultures. London, I.B. Taurus & Co, 1990. ix,371p. £19.95. ISBN: 1850432155.

4 parts: 1. Serving and marking time - 2. Our time: the imposition of order - 3. Their time: following the order of the skies (Maya, Aztec, Inca, Chinese) - 4. A world of time (building on the basic rhythms). Notes, p.343-58. Detailed, analytical index, p.359-71. *Class No:* 529

[823]
The **Book of calendars.** Parise, F., *ed*. New York, Facts on File, 1982. [vii], 387p. tables. $35. ISBN: 0871964678.

Main sections: Ancient calendars (p.1-122) - Africa - Modern Near East - India - Southeast Asia - Far East - Central America - Western calendars (p.291-379, including Calendar of saints; French Revolutionary calendar; Soviet calendar [discontinued in 1940]). Non-analytical index. A handbook that offers basic information on the structure of the calendar and extensive tables for quick conversions. More than 40 ancient peoples' calendars; over 60 calendars 'enabling the reader to translate one calendar date to its appropriate Julian or Gregorian date' (*Introduction*).
Class No: 529

[824]
KRUDY, E.S., *and others, comps. and eds*. **Time: a bibliography.** London, Information Retrieval, 1977. 218p. ISBN: 0904147053.

3,000 references to papers 'mainly published over the past 20 years'. Items are grouped under subjects ranging from philosophy, religion, art, sociology, economics, psychology, biology to the physical sciences. Key-word and subject indexes. Includes some trivia. 'Both the publishers and the compilers who assembled the bibliography are to be congratulated on a fine work' (*British book news,* June 1977, p.438). *Class No:* 529

[825]
SAMUEL, A.E. **Greek and Roman chronology:** calendars and years in Classical antiquity. Munich, Beck, 1972. 325p. (*Handbuch der Altertumswissenschaft. 1. Abt. 7. Teil.*)

A monograph-section of the massive co-operative work on Classical antiquity and part of section 1: *Einleitende und Hilfsdisziplinen*. Well documented. *Class No:* 529

[826]

—FREEMAN-GRENVILLE, G.S.P. The Muslim and Christian calendars; being tables for the conversion of Muslim and Christian dates from the Hijra to the year 2000. London, Oxford University Press, 1963. vii, 87p. tables. (Reprinted London, Collings, 1977). £6.50. ISBN: 0860360598.

Enables the user to ascertain the corresponding days of the week. Tables 6-8 cover the principal Muslim festivals and the principal fixed and movable Christian festivals.
Class No: 529

53 Physics

[827]
BALLENTYNE, D.W.G. *and* LOVETT, D.R. A dictionary of named effects and laws in chemistry, physics and mathematics. 4th ed. London, Chapman & Hall, 1980. viii, 346p. diagrs., tables, chemical structures £17.95. ISBN: 0412223902.

First published 1958.

Records several thousand events, in year order, under broad disciplines. Thus, for 1687: astronomy (3 entries), physics, chemistry, earth sciences, mathematics, meteorology. The main aim is 'to provide, in chronologic order, concise, clear summaries on events or accomplishments contributing to the development of Western science' (*Preface*). *Class No:* 530

Abbreviations & Symbols

[828]
INTERNATIONAL UNION OF PURE AND APPLIED PHYSICS. Commission for Symbols, Units and Nomenclature. **Symbols, units, nomenclature and fundamental constants in physics.** London, The Union, through the Royal Society of London, 1987. 67p. tables. (*Document IUPAP 25.*)

First published 1978 as *Symbols, units and nomenclature in physics.*

As well as thoroughly covering physics, lists embrace mathematical terminology and symbols. 'Strongly recommended for libraries serving physical sciences' (*Science and technology libraries,* V.11(1), Fall 1990, p.740). *Class No:* 530(003)

[829]
POLON, D.D., *ed.* DPMA: **Dictionary of physics and mathematical abbreviations, signs and symbols.** New York, Odyssey Press, 1965. xvii,333p.

One of a series, each with numerous sequences, A-Z plus definitions. Abbreviations for use in text - Letter symbols - Mathematical signs and symbols - Abbreviations for scientific and learned societies - Abbreviations for government annd military departments, agencies and offices. Generous use of white space. Of value to the specialist in one field who must become familiar with publications in another (*The Recorder,* no. 170, 1 December 1966, p.6). *Class No:* 530(003)

Databases

[830]
INSPEC. Stevenage, Herts., Institution of Electrical Engineers, 1969-. Updated weekly.

INSPEC database provides online access to *Physics abstracts, Electrical and electronic abstracts,* and *Computer and control abstracts,* through DIALOG from 1969, and

....(contd.)
through BRS since 1970. Available on CD-ROM. D.T. Hawkins notes both bibliographic (INSPEC, PHYS, SPIN, Atomicindex and Nuclear science) and numeric (Superindex, PHYSCOMP, OECD Nuclear Energy Agency database, and GAPHYOR) databases covering physics (*Database,* v.8(1), February 1985, p.15-16, with a comparison table and references, p.17-18). *Class No:* 530(003.4)

[831]
SPIN (Searchable physics information notices). New York, American Institute of Physics.

Coverage of AIP journals, Russian translations and some additional journals. Author-prepared abstracts. About 485,000 records; updated monthly. Available on DIALOG, file 62. *Class No:* 530(003.4)

Bibliographies

[832]
MELTON, L.R.A. An **Introductory guide to information sources on physics.** Bristol & London, Institute of Physics, 1978. [iv], 44p. £2.50. ISBN: 0854880334.

3 parts: 1. Introduction and the use of libraries - 2. The search technique and the sources - 3. Guide to report writing. 'Characteristics of information sources and their guides' (p.18-38) covers reference material, abstracting and indexing periodicals, computer retrieval services, theses, reports and patents, standards, etc., with helpful commentary. Written primarily with the student or recent graduate in mind, for quick reference. *Class No:* 530(01)

[833]
SHAW, D.F., *ed.* **Information sources in physics.** London, Butterworth, 1985. xii, 456p. graphics, tables. £40.00. (*Butterworths guides to information sources.*) ISBN: 0408014741.

Supersedes *Use of physics literature,* edited by H. Coblans (Butterworth, 1975. xii, 290p.). 21 contributors, 20 documented chapters. 1. Introduction - 2. The scope and control of physics and its literature - 3. Science libraries, reference material and general treatises - 4. Abstracting, indexing and on-line services - 5. Atomic and molecular physics ... 8. Crystallography ... 10. Electronics and computer hardware ... 12. Geophysics, astrophysics and meteorological physics ... 15. Nuclear and particle physics ... 19. Grey literature (*e.g.* reports) - 20. Patent literature. 3 appendices (A: the most important physics journals according to INSPEC; C: Publishers; abbreviations and addresses). Name index; subject and title index. Editor is Keeper of Scientific Books, Radcliffe Science Library, Oxford University. 'The standard work on the literature of physics for both the librarian and the physicist for some considerable time to come' (*Library review,* v.35(2), Summer 1986, p.137). *Class No:* 530(01)

Encyclopaedias & Dictionaries

[834]

THEWLIS, J., *ed*. **Encyclopaedic dictionary of physics:** general, nuclear, solid state, molecular, chemical, metal and vacuum physics, geophysics, biophysics and related subjects. Oxford, Pergamon Press, 1961-64; 1966-75. 9v., + Supplementary v.1-5. illus., diagrs.

V.1-8: A-Z and index; v.9: *Multilingual glossary* (see entry at 530(038)=00).

A major work, covering not only physics but other sciences that have a physical basis. V.1 has over 100 contributors, longer signed articles carrying bibliographies. 'Alpharay, Measurement techniques for' runs to 5½ columns, is in 3 sections and has 5 references. Historical background figures when considered necessary. The subject index in v.8 (p.1-481) has *c*.45,000 entries; author index, p.485-95; errata, p.497-8. References to articles of 500 words or more are given volume number in bold type. 'The articles are, in general, of graduate or near-graduate level' (*Foreword*). An essay-review in *Chemistry and industry* (3 August 1963, p.1273-5) pointed out errors, misprints and insufficiency of diagrams. Compared with Sir. R. Glazebrook's *Dictionary of applied physics* (1922-23. 5v; reprinted Gloucester, Mass., Smith, 1950 - now of historical value only - Thewlis's set lacked adequate planning; the Glazebrook volumes each dealt with a group of subjects and were self-indexing. *Class No:* 530(03)

Encyclopaedias

[835]

BESANÇON, R.M., *ed*. **The Encyclopedia of physics.** 3rd ed. New York, Van Nostrand Reinhold, 1985. [xix], 1378p. diagrs., graphs, formulas. ISBN: 0442257783.

First published 1966; 2nd ed., 1974 (xv, 1067p.).

More than 300 A-Z entries, by *c*.250 contributors. 'Quantum theory': p.985-1000, with 21 references, 7 cross-references. Detailed index, p.1347-78. A substantial revision, aimed at a wide public: physicists, teachers and librarians, students, engineers and scientists. Handy for basic, up-to-date information. *Class No:* 530(031)

[836]

—**LERNER, R.G.** *and* **TRIGG, G.L.**, *eds*. Encyclopedia of physics. 2nd ed. New York, VCH, 1991. 1408p. illus. $150. ISBN: 0895737523.

First published 1981.

Contains 509 signed articles by specialist contributors, including Nobel prize winners. Bibliographies usually indicate readership level. Numerous illus. Complementary to Besançon, according to the reviewer in 'Reference books bulletin' (*Booklist*, August 1991, p.2171). *Class No:* 530(031)

[837]

Encyclopedia of applied physics. Trigg, G.L. New York, VCH, 1991-. V.1-. $255; £160.

V.1. 'Accelerators linear' to 'Analytic methods'. 1991. 666p, ISBN: 156081521.

A major work, with 20v. planned. Articles are subdivided and frequently include glossary, bibliography and 'Additional' readings. Indexes are projected after every third volume. *Class No:* 530(031)

[838]

KNIGHT, D. A Companion to the physical sciences. London, Routledge, 1989. 177p. £7.50. ISBN: 0415009014.

An encyclopedic dictionary (terms, concepts, theories and individuals). Includes some citations to additional in-depth readings. *Class No:* 530(031)

[839]

McGraw-Hill encyclopedia of physics. Parker, S.P., *ed*. 5th ed. New York, etc., McGraw-Hill, 1983. [xix], 1378p. illus., graphs, charts. ISBN: 0070452539.

Selects 760 articles (*c*.400 contributors) from the parent *McGraw-Hill encyclopedia of science and technology* (5th ed., 1982, 15v.) ranging from 'Abbreviation (optics)' to Zeeman effect'. The entry 'Radioactivity' (p.921-38) has 3 tables, 16 diagrams and 14 references. Appendix: 'Scientific notation in the Encyclopedia' (*e.g.* Fundamental constants). Detailed analytical index, p.1317-43, of *c*.7,000 entries. *Nature* (v.307, 26 January 1984, p.393) finds many of the articles excellent. It is aimed at high schools and colleges, 'laid out in an attractive and easily digestible form'. *Class No:* 530(031)

[840]

Magill's survey of science: physical science series. Magill, F.N., *ed*. Pasadena, Ca., Salem Press, 1992. 6v. 2796p. $475. ISBN: 0893566187.

380 entries on classical mechanics, unified field theory, molecular structure, chaos, black holes, expert systems, methodologies of computing and mathematics. Articles are 7 to 8p. and most are written by US academics. Cross-references. Volume indexes; glossary (*c*.500 terms), V. 6. 'Readable ... should be eminently useful' (*Choice*, July/August 1992, p.1659). *Class No:* 530(031)

[841]

MEYERS, R., *ed*. **Encyclopedia of modern physics.** San Diego, Ca., Academic Press, 1990. 773p. $89.50. ISBN: 0122266927.

34 entries, A-Z, surveying the state-of-the-art, traditional and applied physics (*e.g.* holography, quasi-crystals and superconductivity). Each entry has cross-references; brief glossary; index. 'Suitable for advanced undergraduate and graduate students as well as professional scientists and engineers' (*New technical books,* Jan/Feb, 1991, p.1045). *Class No:* 530(031)

Handbooks & Manuals

[842]

American Institute of Physics handbook. Gray, D.E., *ed*. 3rd ed. New York, the Institute, 1972. [2342]p. illus., tables. ISBN: 007001485x.

First published 1957.

About 120 contributors. 9 sections, each with its own editor: 1. Mathematics (with bibliography of 71 annotated entries) - 2. Mechanics - 3. Acoustics - 4. Heat - 5. Electricity and magnetism - 6. Optics - 7. Atomic and nuclear physics - 8. Nuclear physics - 9. Solid-state physics. Many tables (*e.g.* Properties of nuclides, p.8-6 to 8-91). An authoritative compendium. While the *CRC handbook of chemistry and physics* is to be preferred as a first choice, the AIP *Handbook* has a stronger focus on theoretical summaries and references (*Choice*, March 1973, p.55). *Class No:* 530(035)

[843]
FLÜGGE, S., *ed*. **Handbuch der Physik.** [Encyclopedia of physics.] [2.Aufl.]. Berlin, Springer, 1955-.
First published 1926-29 (24v. and index).
Over 70v. have so far been published of the 2nd ed. 1-2: *Mathematical methods.* 1955-56. 3: *Principles of classical mechanics and field theory.* 1956-65. 3 pts. 4: *Principles of thermodynamics and relativity.* 1962. 5: *Principles of quantum theory.* 1958-. pt.1-. 6: *Elasticity and plasticity.* 1958. 6a: *Mechanics of solids.* 1972-74. 4v. 7: *Crystal physics.* 1955-58. 2 pts. 8: *Fluid dynamics.* 1960-63. 2 pts. 9: *Fluid mechanics.* 3v. 1960. 10: *Structure of fluids.* 1960. 11: *Acoustics.* 1961-62. 2 pts. 12: *Thermodynamics of gases.* 1958. 13: *Thermodynamics of liquids and solids.* 1962. 14-15: *Low-temperature physics.* 1956. 16: *Electric fields and waves.* 1958. 17: *Dielectrics.* 1956. 18: *Magnetism. Ferromagnetics.* 1966-68. 2 pts. 19-20: *Electrical conductivity.* 1956-57. 21-22: *Electron emission. Gas discharges.* 1956. 23: *Electrical instruments.* 1967. 25/1: *Crystal optics. Diffraction.* 1961-. pt. 1-. 25/2,26: *Light and matter.* 1958-. 27-28: *Spectroscopy.* 1957-64. 29: *Optical instruments.* 1967. 30: *X-rays.* 1957. 31: *Corpuscles and radiation in matter.* pt.1. 1982. 32: *Structural research.* 1957. 33: *Corpuscular optics.* 1956. 34: *Corpuscles and radiation in matter.* pt.2. 1958. 35-37: *Atoms, Molecules.* 1956-61. 5 pts. 38/1: *External properties of atomic nuclei.* 1958. 38/2: *Neutron and related gamma ray problems.* 1959. 39: *Structure of atomic nuclei.* 1957. 40-42: *Nuclear reactions. Beta decay.* 1957-62. 4 pts. 43: *Mesons* (in preparation). 44-45: *Nuclear instrumentation.* 1958-59. 46: *Cosmic rays.* 1961-67. 2 pts. 47-49: *Geophysics.* 1956-. 50-54: *Astrophysics.* 1958-62. 55: *General index* (authors & subjects) to v.1-54. 1987.
A systematic treatise, the only major current work covering the whole field of physics. Contributions may be in English, German or French, with full references. Thus V.27: *Spectroscopy,* pt.1 (1964. vi, 507p.) has 2 contributions in English, on Line width and Atomic spectra, and one, on Electronic nuclear spectroscopy, in French, with 133 references. The 'numerous beautifully drawn figures' are praised in *Nature* (v.204, 10th October 1964, p.114). Subject indexes act as German and English glossaries. An authoratative series. *Class No:* 530(035)

[844]
LANDAU, L.D. *and* LIFSHITZ, E.M. **Course of theoretical physics.** Oxford, Pergamon Press, 1959-73. 9v. in 10.
Translation of *Teoreticheskaya fizika* (Moscow, Nauka, 1954-67).
1. *Mechanics* 2. *Classical theory of fields.* 3. *Quantum mechanics.* 4. (1-2). *Relativistic quantum theory.* 5. *Statistical physics.* pt.1. 6. *Fluid mechanics.* 7. *Theory of elasticity.* 8. *Electrodynamics of continuous media.* 9. *Statistical physics.* pt.2. 'A magnificent work, written with great clarity and emphasizing the unity of the physical ideas that interweave the whole set. The course can be started at the advanced undergraduate level' (Shaw, D.F. *Information sources in physics* (2nd ed., 1985. p.343)).
Class No: 530(035)

[845]
A Physicist's desk reference. Anderson, H.L., *ed*. New York, American Institute of Physics, 1989. 356p. $60. ISBN: 0883186202.
Updated edition of *Physics vade mecum.* (1981).
22 fields of physics (*e.g.* acoustics, cryogenics, crystallography, elementary particles, fluid dynamics, spectroscopy and structure, nuclear physics, optics, plasma physics, surface physics). Chapters on fundamental constants, conversion factors, basic mathematical and physics formulas. Each chapter by an expert. 'An indispensable handbook for all libraries serving readers who need physics information' (*Science and technology libraries,* v.11(1), Fall 1990, p.139). *Class No:* 530(035)

Dictionaries

Polyglot

[846]
RYDNIK, V.J., *and others, eds*. **Dictionary of physics in four languages:** Russian, English, German and French. Amsterdam, Elsevier, 1989. 392p. £78. ISBN: 0444704906.
About 3,000 English-base terms, with equivalents and indexes in the other languages. *Class No:* 530(038)=00

[847]
SUBE, R. *and* EISENREICH, G. **Wörterbuch Physik:** Englisch/Deutsch/Französisch/Russisch. Berlin, VEB Verlag-Technik, 1973. 3v. (2895p.).
About 75,000 English-base entry-words, with German, French and Russian equivalents. V.3 is the German, French and Russian index volume. Scope includes astrophysics, biophysics, geophysics and mathematical affinities of physics. *Class No:* 530(038)=00

[848]
—THEWLIS, J. **Multilingual glossary:** English, French, German, Spanish, Russian, Japanese. Oxford, Pergamon Press, 1964. xvi, 988p.
Lists 13,675 numbered terms in English, mostly drawn from titles of articles in the *Encyclopaedic dictionary of physics* (*qv.*). The English terms carry number references to French, German, Spanish, Russian and Japanese equivalents, separately listed A-Z, with back-references to the numbered English base-terms. A cumbersome departure from the normal layout of polyglot dictionaries. *Class No:* 530(038)=00

English

[849]
DAINTITH, J. **The Pan dictionary of physics.** London, Pan, 1990. 240p. illus. £6.99. ISBN: 0330314521.
Defines 3000 commonly used physics terms.
Class No: 530(038)=20

[850]
Dictionary of physics. Gray, H.J. *and* Isaacs, A., *eds*. 3rd ed. Harlow, Essex, Longman, 1991. [iv], 635p. diagrs., graphs, tables. £29.50. ISBN: 0582037972.
First published 1958; 2nd ed. 1975.
About 6,000 terms, some defined at length (*e.g.* 'diffraction of light': 5 columns; 7 diagrs.). 14 tables (14: Nobel prizewinners in physics). A standard dictionary. *Class No:* 530(038)=20

[851]

ILLINGWORTH, V., *ed.* **The Penguin dictionary of physics.** Harmondsworth, Penguin, 1991. 544p. illus., diagrs., graphs, tables. £5.99. ISBN: 0140512365.

An abridgement of the *Dictionary of physics* (Harlow, Essex, Longman, 1991).

About 4,500 entries, some defined at length (*e.g.* 'mass': 1 col.). 11 tables appended (7: 'Long-lived elementary particles'). *Class No:* 530(038)=20

[852]

LORD, M.P. Macmillan dictionary of physics. London, Macmillan, 1986. xx, 330,[1]p. diagrs., tables. £29.50; £12.95. ISBN: 0333390660, HB; 0333423771, PB.

Enlarged and updated version of the physics content of *Dictionary of physical science*, edited by J. Daintith (Pan Books, 1978).

Over 4,000 entries, averaging *c.*50 words but occasionally of column length (*e.g.* 'Laser'). 7 tables precede (*e.g.* 'Symbols and SI units for some physical quantities'). 'Further reading suggestions' (undergraduate level: 2 items; pre-postgraduate level: 2). Adequate cross-references; *c.*200 small diagrams. 'Will be of use to Physics students at first year undergraduate and lower levels, and also to amateur enthusiasts and lay persons' (*Preface*).

Class No: 530(038)=20

[853]

McGraw-Hill dictionary of physics and mathematics. Lapedes, D.N., *ed.* New York, McGraw-Hill, 1978. xvi, 1074 + 45p. illus., tables. ISBN: 0070454809.

Defines more than 20,000 terms, the entries being mostly extracted from the *McGraw-Hill dictionary of scientific and technical terms* (2nd ed., 1978), but also written specially for this work. Definitions average 15-80 words; many sub-entries (*e.g.* 'Wave': 4 columns). 55 subject categories are applied. 700 small marginal illus. 15 appendices (*e.g.* SI units; semiconductor symbols and abbreviations). Well produced. 'Recommended for every high school or college library' (*Choice*, v.164, March 1979, p.56).

Class No: 530(038)=20

[854]

—McGraw-Hill dictionary of physics. Parker, S.P., *ed.* New York, etc., McGraw-Hill, 1985. 646p. $13.95. ISBN: 0070454183.

Lists *c.*14,000 physics terms from *McGraw-Hill dictionary of scientific and technical terms* (3rd ed., 1984), (*q.v.*). For students and public libraries. *Class No:* 530(038)=20

[855]

SCHUBERT, J. Dictionary of effects and phenomena in physics: descriptions, applications, tables. Weinheim, New York, VCH, 1987. 140p. $25. ISBN: 0895734877.

Concise descriptions of *c.*400 physical effects, with references. Eponymous terms carry biographical notes on originators. Appendices comprise tables (*e.g.* Cluster courses, effects). More detailed in its field than *Dictionary of named effects and laws in chemistry, physics and mathematics* by D.W.G. Ballentyne and D.R. Lovett (4th ed. 1980). 'Recommended for all physics collections' (*Science and technology libraries*, v.10(3), Spring 1990, p.113).

Class No: 530(038)=20

[856]

THEWLIS, J. Concise dictionary of physics and related subjects. 2nd ed. Oxford, etc., Pergamon Press, 1979. ix, 370p. tables. ISBN: 0080230482.

First published 1973 (viii,361p.).

About 7,200 terms are briefly defined. Entries extend at times to 150 words (*e.g.* 'Carbon fibre'). Numerous cross-references. 3 appendices (3. Values of some fundamental physical constants). The 1973 ed. showed considerable reliance on the *Encyclopaedic dictionary of physics* (*qv*). No illus. or diagrams. *Class No:* 530(038)=20

[857]

—Concise dictionary of physics. Oxford Univ. Press, 1985. [v], 292, [3]p. diagrs., tables. £7.95. ISBN: 0198661428.

Derived from *Concise Oxford dictionary* (O.U.P., 1984), like the companion vols., *Concise dictionary of chemistry* and *Concise dictionary of biology*. Its 2,500-3,000 entries, in small type, include somewhat lengthier entries than the *Macmillan dictionary* (*qv*) (*e.g.* 'Elementary particles': nearly 3 columns), but only 3 appendices (2. Fundamental constants). It carries some entries relating to astronomy and physical chemistry. Small format. *Class No:* 530(038)=20

[858]

—RICHARDS, T. Cambridge illustrated thesaurus of physics. Cambridge Univ. Press, 1985. 256p. diagrs. £4.95.

One of four illustrated thesauri (science and technology; physics; chemistry; computer science) that combine the dictionary definitions of *c.* 2,500 terms with grouping of topics, plus cross-references. A feature is the full use of colour line diagrams, *c.*500 per volume. A useful reference for 'O' and 'A' level students. *Class No:* 530(038)=20

German

[859]

HYMAN, C.J. *and* **IDLIN, R.,** *eds.* **Dictionary of physics and allied sciences.** [Wörterbuch für Physik und verwandter Wissenschafterv.] New York, Ungar, 1978. 2v.

1. *German-English*, by C.J. Hyman. 696p. 2. *English-German*, by R. Idlin. 663p. First published 1958/62.

V.1 has *c.*30,000 entries, employing some 70 categories. Translations are mostly equivalents. V.2, 'by no means an inversion of the first volume' (*Preface*), has *c.*25,000 entries, with sample phrases and sentences. German genders are shown. The 1978 ed. of the 2 volumes simply adds a supplement at the end of each dictionary.

Class No: 530(038)=30

[860]

—DE VRIES, L. *and* CLASON, W.E., *comps.* Dictionary of pure and applied physics. Amsterdam etc., Elsevier, 1963-64. 2v.

V.1: *German-English*. 1963. viii, 367p. V.2: *English-German*. 1964. viii, 341p.

Includes *c.*31,000 entries per volume, including terms from related fields as well as words in common use. Normally gives equivalents only. 'De Vries's habit of omitting any distinction of meanings, although mitigated by the restriction to physics, is still in evidence' (*Nature*, v.207, 28 August 1965, p.905). *Class No:* 530(038)=30

French

[861]

MATHIEU, J.-P., *and others*. Dictionnaire de physique.
Paris, Masson/Eyrolles, 1983. 568p. diagrs., graphs. F250.
ISBN: 2225744483.

Several thousand entries for terms in fundamental physics,
omitting technical aspects, history and biography. Clearly
written and well illustrated. Earmarked in *Bulletin critique
du livre français* (no.460, April 1984, item 126179) as
suitable for specialists and those on graduate and postgraduate courses. *Class No:* 530(038)=40

Russian

[862]

TOLSTOI, D.M., *ed*. Anglo-russkiĭ fizicheskiĭ slovar'.
Moscow, Izdat. 'Sovetskaya Entsiklopediya', 1968; Oxford,
Pergamon, 1978. 848p.

About 60,000 English entry-words, with Russian
equivalents; separate lines for sub-entries. The Dictionary
needs using with care; 'principle of entropy increase'
appears under 'Principle' and not under 'Entropy'.
Abbreviations, p.832-48. Bibliography of sources, p.9-10.
Class No: 530(038)=82

[863]

—**ERNIN, L., *and others*.** Russian-English physics dictionary.
New York, etc., Wiley, 1963. xxx, 554p.

Lists 25,000 entries, including terms in allied fields and in
common usage, abbreviations and also proper names. Word-
for-word entries. Where many compounds are massed under
an entry-word, the entry is sectionized. Layout is clear, with
bold type for sub-entries, as in Callaham's *Russian-English
chemical and polytechnical dictionary* (*qv*). The helpful
reference section (p.xviii-xxix) contains notes on the
alphabet, common Russian technical word endings, verb
formation, and the like. Sources (dictionaries,
encyclopaedias, etc.) p.ix-xii. Ernin's collaborators were the
Consultant Bureau of Physicists - Translators. Well
produced; 'edge' index. *Class No:* 530(038)=82

Arabic

[864]

LEAGUE OF ARAB STATES. Education, Culture and
Sciences Organization. Permanent Bureau of Arabisation.
Lexicon of physics. [Lexique de physique. English-French-
Arabic.] Rabat, Morocco, Bureau Permanent de
Coordination de l'Arabisation, 1971. 190, xlvii p.

2,062 numbered English base-terms, with French and
Arabic equivalents or explanations. French-terms index to
numbered English entries. *Class No:* 530(038)=927

Reviews & Abstracts

[865]

Bulletin signalétique. 40. Physique. Paris, Centre National de
la Recherche Scientifique, 1940-. Monthly.

Sections: 130. *Physique mathématique. Optique.
Acoustique. Mécanique - Chaleur.* 160. *Physique de l'état
condensé.* 161. *Structure de l'état condensé
crystallographique.* 165. *Atomes et molécules. Plasmas.* 171.
Chimie générale et chimie physique.

The five sections together carry *c.*30,000 brief abstracts
and references pa. Each section has subject and author
indexes per issue; annual cumulative author index. PASCAL

.... (contd.)

series, replacing the *Bulletin signalétique* in 1984, and
online, has sections: 11. *Physique atomiqe et moléculaire.
Plasmas.* 12. *État condensé.* 27. *Méthodes de formation et
traitement des images.* 32. *Metrololgie et appareillage en
physique et en pyrchochimie.* Folio 10. *Mécanique et
acoustique.* *Class No:* 530(048)

[866]

Contemporary physics: a review of physics and its
applications. London, New York etc., Taylor & Francis,
1960-. v.1-. 6 pa. ISSN: 00107514.

V.27(5), September/October 1986, features 3 documented
articles (*e.g.* 'The electron-hole liquid in semiconductors'
(81 references); 'Numerical weather prediction' (2
references)). Also, an essay-review ('Polycrystalline
semiconductors'); book reviews; short notices; books
received. No.6 issue of a volume has an appended list of
contents of that volume. *Class No:* 530(048)

[867]

Current contents: Physical, chemical and earth sciences.
Philadelphia, Pa., Institute for Scientific Information, 1979-.
v.19-. Weekly. ISSN: 01632574.

Previously as *Current contents: Physical and chemical
sciences* (formed by the merger of *Current contents:
Physical sciences* and *Current contents: Chemical sciences*),
1961-78.

Reproduces contents pages of more than 1,000 journals in
the fields concerned. Each weekly number covers *c.*100
journal issues. Online on CURRENT CONTENTS
SEARCH, and available on diskette. *Class No:* 530(048)

[868]

Current papers in physics. London, INSPEC (The Institution
of Electrical Engineers), 1966-. Bi-monthly. ISSN:
00113786.

About 75,000 references pa, using the INSPEC
classification scheme. A current awareness service, without
indexes, that aims 'to provide a convenient means for
engineers and scientists to keep abreast of newly published
information'. Companion to *Current papers in electrical and
electronic engineering* (CPE) and *Current papers in
computers and control* (CPC) (*qqv*). *Class No:* 530(048)

[869]

Current physics index. New York, American Institute of
Physics, 1975-. Quarterly. ISSN: 00989819.

Replaces *Current physics advance abstracts* and *Current
physics titles* (1972-).

About 4,000 abstracts per quarter of articles in the
primary journals published by the American Institute of
Physics. Uses the INSPEC classification scheme (outlined in
the *Physics Abstracts* entry). Profuse cross-references.
Author index per issue; annual subject and author indexes.
Class No: 530(048)

[870]

Physics abstracts. Science abstracts. Section A. Stevenage,
Herts, Institution of Electrical Engineers, 1898-. Twice
monthly. £920pa. ISSN: 00368091.

1991: 151,994 abstracts and references. Main sections
(Closely subdivided): 00.00 General - 10.00 The physics of
elementary particles and fields - 20.00 Nuclear physics -
30.00 Atomic and molecular physics - 40.00 Classical areas
of phenomenology (*e.g.* electricity and magnetism) - 50.00
Fluids, plasmas and electric discharge - 60.00 Condensed
matter: structure, thermal and mechanical properties - 70.00

....(contd.)

Condensed matter: electronic structure, electrical, magnetic, and optical properties - 80.00 Cross-disciplinary physics and related areas of science and technology - 90.00 Geophysics, astronomy and astrophysics. Author index: subsidiary indexes: bibliography; book; conference; corporate author. Subject indexes, 2pa. List of journals scanned in half-yearly author index. Four or five-yearly cumulative indexes. The major English-language physics abstracting service. Available online via DIALOG, BRS and Data-Star; CD-ROM. *Class No:* 530(048)

[871]

Physics briefs. [Physikalische Berichte. Edited by Deutscher Physikalische Gesellschaft & Fachinformationszentrum Energie, Physik, Mathematik, in cooperation with American Institute of Physics.] New York, American Institute of Physics, 1979-. Bi-monthly. ISSN: 01707434.

Succeeds *Physikalische Berichte* (v.1-57. 1920-78. Monthly), originally (1845-1918) as *Fortschritte der Physik*.

About 125,000 abstracts and references pa, covering all fields of physics and related topics. Follows the INSPEC classification scheme. Cross-references; subject guide and author index per issue. Half-yearly author and subject indexes. An alerting service for *Physics abstracts.* Online via STN and INKA. *Class No:* 530(048)

[872]

Referativnyĭ zhurnal. Fizika. Akademiya Nauk SSSR, Institut Nauchnoĭ Informatsii. Moscow, VINITI, 1954-. Monthly. ISSN: 00342343.

About 80,000 abstracts pa. in 11 sections (General aspects of Physics ... Acoustics). Covers over 20,000 journals in various languages, as well as theses, reports, patents and some conferences. Author index per issue; annual cumulated author and classified subject index. Parts are available separately. *Class No:* 530(048)

[873]

Reviews of modern physics. New York, American Physical Society, 1929-. v.1-. 4 pa. ISSN: 00346861.

V.57(3), pt.1, July 1985, carries 7 documented articles (*e.g.* 'Surface-enhanced spectroscopy' (6 columns of references); 'Wetting: statics and dynamics' (264 references)). Appended: 'Some review articles appearing in other journals and serial publications'; 'Forthcoming articles'; 'Cumulative list of reprints'. *Class No:* 530(048)

Progress Reports

[874]

Reports on progress in physics. London, Institute of Physics, 1934-. Monthly. £190 pa. ISSN: 00344885.

V.49(6), June 1986, contains 2 documented articles ('Surfaces of rare earth metals' with 4½p. of references; 'Fine structure in ionisation cross sections and applications to surface science', with over 1p. of references). No index. 'Forthcoming articles', listed on back cover. *Class No:* 530(055)

Anthologies

[875]

MAGIE, W.F. A Source book in physics. Cambridge, Mass., Harvard Univ. Press. xiv, 620p. (*Source books in the history of science.*)

More than 100 examples from the writings of physicists in

....(contd.)

the period 1600-1900. Sections: Mechanics - Properties of matter - Sound - Heat - Light - Magnetism and electricity. Biographical notes precede excerpts. *Class No:* 530(082.21)

[876]

The New physics. Davies, P., *ed.* Cambridge Univ. Press, 1989. ix,516p. illus., col. pl., diagrs., graphs. £16.94 (Pbk). ISBN: 0521304202; 0521438314, Pbk.

19 international contributors. 18 chapters survey recent developments in physics. Chaper 1 provides a synthesis of these; other chapters cover general relativity, low temperature physics, astrophysics, chaos, quantum optics, etc. Detailed, analytical index, p.505-16.

Class No: 530(082.21)

[877]

SAMBURSKY, S., *ed.* **Physical thought from the Presocratic to the quantum theory:** an anthology. London, Hutchinson, 1974. xv, 584p. ISBN: 0091028507.

6 sections: 1. Antiquity - 2. The Middle Ages - 3. Copernicus to Pascal - 4. The Royal Society to Laplace - 5. Dalton to Mach - 6. Planck to Pauli. Each section comprises introduction and extracts. List of sources (by sections). Index of authors (bio-bibliographies). Analytical subject index; general index. Aims to depict the development of physical thought 'as a chapter in the history of ideas' (*General introduction*). *Class No:* 530(082.21)

Tables & Data Books

[878]

CONDON, E.U. *and* **ODISHAW, H.,** *eds.* **Handbook of physics.** 2nd ed. New York, etc., McGraw-Hill, 1967. xxix, 1680p. ISBN: 0070124035.

First published 1958.

Nine separately paged parts, each with several chapters by different authors, and references appended. 1. Mathematics - 2. Mechanics of particles and rigid bodies - 3. Mechanics of deformable bodies - 4. Electricity and magnetism - 5. Heat and thermodynamics ... 9. Nuclear physics. Detailed subject index. A standard work, similar in scope to the *American Institute of Physics handbook* (*qv*). *Class No:* 530(083)

[879]

CRC handbook of chemistry and physics. Lide, D.R., *ed.* 71st ed. Boca Raton, Florida, CRC Press, 1990. v.p. diagrs., charts, tables, chemical structures. ISBN: 0849304711.

First published 1914.

This new edition of a classic reference source (the 'Rubber Bible') has been substantially reorganized by a new editor into an arrangement which facilitates browsing. About 20% of the tables are completely new/updated. Includes more tables on health and the environment than earlier versions. An invaluable mine of information which 'remains the essential reference tool for physical data, required in virtually all libraries' (*Choice,* September 1991, p.56).

New editions tend to appear in September.

Class No: 530(083)

[880]
FISCHBECK, H.J. *and* **FISCHBECK, K.H. Formulas, facts and constants** for students and professionals in engineering, chemistry and physics. 2nd rev. and enlgd. ed. Berlin, etc., Springer, 1987. x, 260p. diagrs., graphs, tables. ISBN: 3540176106.

First published 1972.

5 sections: 1. Basic mathematical facts and figures - 2. Units, conversion factors and constants - 3. Spectroscopy and atomic structures - 4. Basic wave mathematics - 5. Facts, figures and data. Useful in the laboratory. Detailed, analytical subject index. *Class No:* 530(083)

[881]
JAMES, A.M. *and* **LORD, M.P. Macmillan's chemical and physical data.** London, Macmillan, 1992. 448p. tables. £35. ISBN: 0333511670.

15 subject areas. About 430 tables, most with references for further information and/or additional data.
Class No: 530(083)

[882]
JERRARD, H.G. *and* **McNEILL, D.B. Dictionary of scientific units,** including dimensionless numbers and scales. 6th ed. London, Chapman & Hall, 1992. 255p. tables £15.95. ISBN: 0412467208.

First published 1963.

Contains *c*.950 units, supported by *c*.650 references. The dictionary, Abampere … Zhubov scale, p.9-190. 'Beaufort scale': nearly 1p. 5 appendices (*e.g.* 1. Fundamental committees and conferences). References, p.221-39. Detailed, analytical index, p.241-55. *Class No:* 530(083)

[883]
KAYE, G.W.C. *and* **LABY, T.H.,** *comps.* **Tables of physical and chemical constants,** and some mathematical functions. 15th ed. London & New York, Longman, 1986. [xiii], 477p. graphs, tables. £28.99; $52.95. ISBN: 0582463548.

14th ed. 1971 (xi,386p).

59 contributors. 4 main sections: 1. General physics (1.Units … 10. Miscellaneous engineering data) - 2. Chemistry (The elements … Miscellaneous data) - 3. Atomic and nuclear physics - 4. Mathematical functions (5. Statistical methods for the treatment of experimental data). Analytical index. Subsection references. All tabulated values appear in SI units. Essential for any reference library. *Class No:* 530(083)

[884]
NORDLING, C. *and* **ÖSTERMAN, J. Physics handbook:** elementary constants and units, tables, forumlae and diagrams, and mathematical formulae. Translated from Swedish. Bromley, Kent, Chartwell-Brett, 1980. 430p. diagrs., graphs, tables, forumlas. ISBN: 0862380006.

Four parts: E. Elementary constants and units - T. Physical tables - F. Physical formulae and diagrams - M. Mathematical formulae. Addenda. Detailed, analytical index, p.401-30. 'It has been used [in Sweden] as a standard source of numerical data and formulae for exercise, problems and … as a companion in problem solving by both students and teachers' (*Foreword*). *Class No:* 530(083)

[885]
TUMA, J.J. Handbook of physical calculations. 2nd ed. New York, McGraw-Hill, 1983. 512p. illus., diagrs., tables, changes. ISBN: 0070654395.

First published 1976 (x,370p).

10 sections of definitions, formulas, charts and tables of elementary and intermediate technological physics, with revised data in appendices. 'It serves as a referencee source for practicing engineers, architects and technologists' (*New technical books*, October 1983, item 915).
Class No: 530(083)

Histories

[886]
CAJORI, F. A History of physics in its elementary branches, including the evolution of physical laboratories. Rev. and enl. ed. New York, Macmillan Co., 1929. xiii 424p. illus.

First published 1899.

Arranged chronologically under periods, each subdivided by the main divisions of physics. Concentrates mainly on period from 17th to 20th century (up to 1925). Includes a bibliography. Intended mainly for students and teachers of physics. *Class No:* 530(091)

[887]
JUNGNICKEL, C. *and* **McCORMMACH, R. Intellectual mastery of nature:** theoretical physics from Ohm to Einstein. Chicago, Univ. of Chicago Press, 1986. 2 vols. (xxviii,350p.; xx,435p.).

V.1: *The torch of mathematics 1800-1870.* ISBN: 0226415813.

V.2 *The now mighty theoretical physics 1870-1925.* ISBN: 0226415848.

A scholarly treatment concentrating on German universities and research laboratories. Extensive notes and bibliography. Analytical index. *Class No:* 530(091)

[888]
NOBEL FOUNDATION. Nobel lectures including speeches and laureates' biographies. Physics. Amsterdam, etc., Elsevier, for the Nobel Foundation, 1942-64. 3v. illus., diagrs., tables.

1. *1901-21.* 1942. xii, 500p. 2. *1922-41.* xii, 458p. 3. *1942-62.* 1964. xiii, 621p.

The Nobel lectures are followed by brief biographies (*e.g.* v.1, p.417-90: Einstein lecture; p.491-2: biography). Each volume has name and subject indexes, plus index of biographies. *Class No:* 530(091)

[889]
PEDERSEN, O. *and* **PIHL, M. Early physics and astronomy:** a historical introduction. New York, American Elsevier, 1974. 413p.

Revised and updated version of the original Danish.

Stresses major aspects of ancient and medieval science history. Numerous sub-sections. Praised in *Nature* (v.252, 8 November 1974, p.130) for its clear and scholarly style; an excellent translation. 'Will generally appeal to science undergraduates reading the books as a part of an introductory survey course in the history of the exact sciences'. *Class No:* 530(091)

Bibliographies

[890]

BRUSH, S.G. *and* BELLONI, E. **The History of modern physics:** an international bibliography. New York, Garland, 1983. xviii, [1], 334p. (*Bibliographies of the history of science and technology, 4. Garland reference library of the humanities, 420.*) ISBN: 0824091175.

About 2,300 entries. Section A covers general and miscellaneous works on the history of modern physics. Section M, research on the history and its use in education, etc., Name index; partly analytical subject index; institutional index. 'Modern physics' is defined here as physics since the discovery of X-rays in 1895. Intended as a companion to both the *ISIS cumulative bibliography* (1913-65) and to Heilbron and Wheaton (*qv.*).

Class No: 530(091)(01)

[891]

—Resources for the history of physics. Brush, S.G., *ed.* Hanover, N.H., The University Press of New England, 1972. 202p.

In 2 sections, - a guide to books and audio visual material; and a guide to original texts of importance and their translations into other languages. Intended primarily for teachers of physics. *Class No:* 530(091)(01)

[892]

HEILBRON, J.L. *and* WHEATON, B.R. **Literature on the history of physics in the twentieth century.** Berkeley, Calif., Office for History of Science and Technology, Univ. of California, 1981. ix, 485p. $20. (*Berkeley Papers on history of science, v.5.*) ISBN: 0918102057.

Contents: General - Biography - Institutions - Mechanics - Electricity and magnetism - Relativity - Radiation - Atomic physics - Nuclear physics - Particles and beams - Quantum physics - Solid states - Geo-, astro-, and biophysics - Physical chemistry - Techniques and apparatus - Physics and society - Philosophy of physics. Useful abbreviations for major sources. Index. For university librarians, doctoral candidates and readers with a serious interest in this subject. *Class No:* 530(091)(01)

[893]

HOME, R.W. **The History of classical physics:** a selected, annotated bibliography. New York & London, Garland, 1984. xix, 324p. (*Bibliographies of the history of science and technology, 8.*)

1,210 briefly annotated entries. 7 chapters: 1. General bibliographical works, biographical dictionaries - 2. Biographies of individual scientists - 3. Collected works, correspondence and bibliographies of individual scientists - 4. General histories - 5. Eighteenth-century physics - 6. Physics in transition, 1790-1820-. 7. Nineteenth-century physics. Non-analytical index, p.293-324.

Class No: 530(091)(01)

Anthologies

[894]

WEAVER, J.H. **The World of physics:** a small library of the literature of physics from antiquity to the present. New York, Simon & Schuster, 1987. 3v. $30 each.

V.1. *The Aristotelian cosmos and the Newtonian system.* 889p. ISBN: 0671499262.

V.2. *The Einstein universe and the Bohr atom.* 941p. ISBN: 0671499309.

V.3. *The evolutionary cosmos and the limits of science.*

....(contd.)

1118p. ISBN: 0671499319.

A collection for the general reader of the major writings in physics. Explanatory introductions help to define the discipline and trace its history. Occasionally heavy going for non-mathematicians, but 'a greater understanding of current physics is to be gained here than could be obtained from works of lesser scope ... a standard' (*Choice,* January 1988, p.793). *Class No:* 530(091)(082.21)

Biographies

[895]

GAMOW, G. **The Great physicists, from Galileo to Einstein.** New York, Dekker, 1988. 338p. illus., diagrs. $7.95. ISBN: 0486257673.

A reprint of the 1961 ed.

Each chapter consists of a biography of a seminal figure in physics history. Source index. Primary aim: 'to inspire young readers to study physics' (*Introduction*).

Class No: 530(092)

[896]

HEATHCOTE, N.H. de V. **Nobel Prize Winners in physics, 1901-1950.** New York, Schuman, 1953. xvi, 473p. illus.

Gives a biographical sketch, a description of the prize-winning work, and a note on its consequences for each Nobel Laureate in Physics (54 in all, up to 1950). Arranged chronologically, it provides a clear picture of the most significant discoveries in physics during 50 important years. Does not include references to other publications. It is based on the official reports in *Les Prix Nobel.*

Class No: 530(092)

[897]

—The Biographical dictionary of scientists: physicists. Abbott, D., *ed.* London, Blond Educational, 1984. [vii], 212p. $18.95. ISBN: 0584700032.

Lists *c.*200 biographies of varying length (*e.g.* 'Maxwell, James Clark': 6 columns; 'Galileo': 3½ columns). Glossary, p.179-193. Analytical index (subjects only), p.195-212.

Class No: 530(092)

[898]

HERMANN, A. **Lexikon Geschichte der Physik, A-Z:** Biographien, Sachwörter, Originalschriften und Sekundärliteratur. Cologne, Deubner, 1972. [vi], 423p. illus., diagrs., facsims.

Signed articles. Particularly valuable for biographical entries. 'Leibniz': 7½ columns (1 column of bibliography; works by; material on); 'Newton': 10 columns (3½ columns of bibliography; works by; material on, *ie*, commentaries, biographies, periodical articles, bibliographies). 'Relativitätstheorie': nearly 8 columns (2/3 column of bibliographic sources, references, bibliographies).

Class No: 530(092)

[899]

The **Nobel prize winners: physics.** Magill, F.N., *ed.* Pasadena, Ca., Salem Press, 1989. 3v., 1346p. ports. $210. ISBN: 0893565571.

Winners are arranged chronologically, with a 6-10p. article for each. Includes: summaries of presentation and acceptance speeches, account of scientific reputation, personal and scientific biography, 'incomplete' (*Choice,* February 1990, p.934-5) list of publications. For undergraduates. *Class No:* 530(092)

Bibliographies

[900]

CROMER, D.E. 'Biographies of physicists: an annotated bibliography'. In *Science and technology libraries*, V.11(4), Winter 1991, p.75-98. ISSN: 0194262x.

16 physicists, mostly twentieth-century; excludes Einstein. *Class No:* 530(092)(01)

20th Century

[901]

GINGERICH, O. The Physical sciences in the twentieth century. New York, Scribner, 1989. 306p. illus. $75. (*Album of science*.) ISBN: 0684154978.

Covers physics, chemistry, astronomy, geology and computers. Includes brief outline of major discoveries, principles and participants within each discipline and numerous captioned illustrations, about 400 in all. Bibliography of science histories; Nobel prize winners in physics and chemistry to 1988. A 'well-conceived and handsomely-executed volume'. (*Choice*, January, 1990, p.826). *Class No:* 530"19"

Writing & Lecturing

[902]

AMERICAN INSTITUTE OF PHYSICS. AIP Style manual. 4th ed. New York, American Institute of Physics, 1990. 64p. illus.

3rd ed., 1978, as *Style manual for guidance in the preparation of papers for journals published by the American Institute of Physics.*

Guide to authors in physics by the AIP Publication Board. *Class No:* 530:001.81

Thesauri

[903]

INSPEC thesaurus, 1991. 9th ed. London, Institution of Electrical Engineers, 1991. v, 547,61p. ISBN: 0852964897.

First published 1973.

The 9th edition contains *c*.13,000 terms (6,000 'preferred terms' and 7,000 cross-references). Terms range over fields of physics, electronics, communications, electrical engineering, information technology, computing and control systems. Alphabetical display of terms (547p.) is followed by a hierarchical display of used terms. *Class No:* 530:025.43

Research Establishments

[904]

CROWTHER, J.G. The Cavendish Laboratory, 1874-1974. London, Macmillan, 1974. xvi, 464p. illus.

33 short chapters on the history of the Cavendish Physics Laboratory, Cambridge. Includes valuable biographical material. References, p.447-53 (Maxwell; Rayleigh; J.J. Thomson; Rutherford; Bragg; Mott; Pollard). *Class No:* 530:061:061.62

[905]

PYATT, E. The National Physical Laboratory. Bristol, Hilger, 1983. x, 270p. illus., ports., graphs, tables.

4 parts: 1. The founding of NPL [1900] - 2. Early development of NPL - 3. The inter-war period - 4. The post-war period. Each part has appended notes. 6 appendices (1. Biographical notes of principal characters; 2. Calendar of events, 1900 ... 4. Highlights in the history of standards). Detailed, non-analytical index, p.264-70. *Class No:* 530:061:061.62

Research Projects

[906]

INSTITUTE OF PHYSICS. Research fields in physics at United Kingdom universities and polytechnics. 8th ed. Bristol & London, the Institute, 1987. 2-3 yearly. xvi, 349p.

First published 1970.

Universities (Aberdeen ... York); Polytechnics (Brighton ... Ulster). Data on each: Research fields, academic personnel; Higher degrees and diplomas; Scholarship and Fellowship sources; Correspondence (contact). Name and subject indexes. Does not record complete work. *Class No:* 530:061:061.62.005

Physical Measurement

[907]

DRAZIL, J.V. Quantities and units of measurement: a dictionary and handbook. London, Mansell; Wiesbaden, Brandstetter, 1983. [v], 313p. £15; £8.95. ISBN: 0720116651.

Supersedes his *Directory of quantities and units* (London, L. Hill, 1971).

3 parts: 1. A dictionary of units of measurements, their symbols and abbreviations - 2. A dictionary of quantities and selected constants - 3. Symbols denoting quantities and constants. 4 appendices (2. UK and US units; 4. Bibliography, p. 287-94). French and German indexes, with references to corresponding English terms in pt. 2. 'The book is comprehensive and all scientists should have access to a copy' (*New Scientist*, 2 April 1984, p.48). *Class No:* 530.08

[908]

Key abstracts: Physical measurements and instrumentation. London, INSPEC, 1976-. Monthly. (*Key abstracts*.)

About 150 abstracts per issue. 9 major sections: 1.0. Measurements and metrology - 2.0 Mechanical measurements and techniques - 3.0 Acoustics and ultrasonics - 4.0 Thermal instruments and techniques - 5.0 Pressure measurement and techniques - 6.0 Optical instruments and techniques - 7. Magnetic and electrical instruments and techniques - 8. Elementary particle and nuclear instruments and techniques - 9.0 Other instrumentation and techniques (*e.g.* X-ray and gamma ray techniques). No index. *Class No:* 530.08

[909]

MELARAGNO, M. Quantification in science: the VNR dictionary of engineering units and measures. New York, Van Nostrand Reinhold, 1991. 333p. illus., tables. $42.95. ISBN: 0442006411.

Covers historical background of measurement and systems

....(contd.)

of units. Non-technical language. Includes chapter on conversion of units. A 'wealth of information' (*Choice, September 1992, p.84). Class No:* 530.08

[910]
ROSSINI, F.D. CRC fundamental measures and controls for science and technology. Cleveland, Ohio, CRC Press, Inc., 1974. [ix], 132p. diagrs., tables. ISBN: 0578190511.

14 Chapters (2. The fundamental units of measurement ... 4. The basic scale of temperature ... 6. The scale of pressure - 7. The scale of atomic weights ... 9. The mechanical equivalent of heat; 'wet' calories ... 11. The fundamental physical constants ... 13. The numerical data for science and technology (with 32 references) - 14. Presentation and analysis of numerical data in the scientific literature). Analytical index, p.123-32. *Class No:* 530.08

[911]
—DIAGRAM GROUP. Comparisons; of distance, size, area, volume, mass, weight, density, energy, temperature, time, speed and number throughout the universe. London, St. Martins Press, 1980. 240p. illus., graphs, tables. ISBN: 0312154844.

A compendium of records and lists, illustrated and compared. The two-page spreads consist of drawings and/or graphs, with succinct text. 'Designed to help lay people comprehend and easily understand differences between objects and quantities' (*Reference sources for small and medium-sized libraries.* 4th ed. A.L.A., 1984, entry 1047). *Class No:* 530.08

[912]
Sensors: a comprehensive survey. Göpel, W., *and others, eds.* New York, VCH, 1989-. V. 1-. illus., diagrs., graphs, tables.

1. *Fundamentals and general aspects.* 1989. ISBN: 3527267670. xiii,641p.

2/3. *Mechanical sensors.*

4. *Thermal sensors.*

5. *Magnetic sensors.*

6. *Optical sensors.*

7/8. *Chemical and biochemical sensors.*

V.1 has 23 contributors, 22 documented chapters in 4 sections: I. Sensor fundamentals - II. Basic sensor technologies - III. Sensor interfaces - IV. Sensor applications. Detailed, analytical index, p.603-29. *Class No:* 530.08

[913]
SYDENHAM, P.H., *and others.* **Handbook of measurement science.** New York, Wiley, 1982-83. 2v. (1413p.).

1. *Theoretical fundamentals.* 1982. xxiv, 654p.

2. *Practical fundamentals.* 1983. xxii, p.655-1413.

3. *Elements of change.* 1992. 584p.

V.1 has 18 contributors; v.2, 17. 32 documented chapters, in all (*e.g.* 7. Pattern recognition (6 contributors), v.1, p.277-330, including 5p. of references; 32. Sources of information measurement, v.2, p.1363-89, with 7 references). V.3 concentrates on factors responsible for the emergence of new measurement systems. Index. 'No measurement or control enquirer should be without it' (*Measurement and control*, v.16, February 1983, p.47-48). *Class No:* 530.08

Relativity & Einstein

[914]
D'INVENO, R.A. Introducing Einstein's relativity. Oxford, Clarendon Press, 1991. 540p. illus. £50. ISBN: 0198596537.

Main topics: black holes, gravitational waves, cosmology. 213 illus. *Class No:* 530.12

Quantum Theory

[915]
KUHN, T.S., *and others, eds.* **Sources for the history of quantum physics:** an inventory and report. Philadelphia, Pa., American Philosophical Society, 1967. Supplement, 1973. [xi], 176p. + 7p. (Supplement, 1973). (*A.P.S. Memoirs, v.68.*)

Funded by the National Science Foundation.

Inventory of materials on the history of quantum physics and related developments in theoretical physics. Sections: 1. Activities and procedures - 2. Author catalogue of principle sources (Abraham, Max ... Zeeman) - 3. Inventory of oral records - 4. Inventory of microfilm records. 3 appendices (1. Sample biographies). Indexes: Source materials by type; Index of names; Index of Institutions; Depositories of relevant sources (p.176). The result of three years intensive work locating unpublished papers, notebooks and letters of more than 250 physicists. *Class No:* 530.145

Mechanics

[916]
POLSKA KOMITET NORMALIZACY. Vocabulary of mechanics in five languages: English, German, French, Polish, Russian. Oxford, etc., Pergamon Press; Warsaw, Wydawnictwa Naukowo - Techniczne, 1962-67. 2v. (viii, 190.; x,452p.) diagrs.

Group 05: *Theoretical mechanics.* 10: *Strength of materials.* 15: *Mechanics of fluids.*

About 3,000 specialized terms in classified order (in line with ISO/TC37 recommendations). The English entry is defined, with equivalents in German, French, Polish and Russian. 5 language indexes. Well produced. *Class No:* 531/534

[917]
Three hundred years of gravitation. Hawking, S.F. *and* Israel, W., *eds.* Cambridge Univ. Press, 1987. xiii,684p. diagrs., tables. £45. ISBN: 0521343127.

16 contributions with extensive references from celebrated physicists to mark the anniversary of the publication of Newton's *Principia.* Most of the articles 'review current developments and future prospects' (*Preface*). Topics covered include 'Dark stars' (W. Israel), 'Astrophysical black holes' (R.D. Blandford), 'Gravitational radiation' (K. Thorne), 'Superstring unification' (J.H. Schwarz). Not for the beginner in physics, but lucid accounts of the forefront of

....(contd.)
research. *No* index, a serious drawback in a work of this nature. 'A guide to current thinking on gravity' (*New Scientist,* 17 September 1987, p.77). *Class No:* 531/534

Solids

Kinematics

[918]
SINGER, L., *comp.* **Russian-English-French-German kinematics dictionary.** London, Scientific Information Consultants, Ltd., 1966. [ii], 128p.

1,169 numbered entry-words in Russian, with English, French and German equivalents and indexes. Genders are not given. *Class No:* 531.1

Barometers

[919]
MIDDLETON, W.E.K. **The History of the barometer.** Baltimore, MD., Johns Hopkins Univ. Press; London, Oxford Univ. Press, 1964. xx, 489p. illus. $127.80. ISBN: 0317084461.

Broadly chronological order of chapters, with footnote references. 'A detailed and wide-ranging account of an instrument whose use now extends far beyond the simple measurement of barometric pressure ... A well-documented and well-written account' (*Geographical journal,* v.131(4), December 1965, p.567-8). *Class No:* 531.78

Fluids

[920]
Annual review of fluid mechanics. Palo Alto, Calif., Annual Review, Inc. 1969-.

V.24, (1992. x,546p.) contains 14 documented articles (*e.g.* 'Contour dynamics methods': 8 columns of references). Subject index. Cumulative indexes: Contributing authors, v. 1-24; chapter titles, v. 1-24. *Class No:* 532

[921]
Encyclopedia of fluid mechanics. Cheremisinoff, N.P. Houston, Texas, Gulf Publishing Co., 1985-. graphs, formulas. ISBN: 0872015130.

1. *Flow phenomena and movement.* 1985. 2. *Dynamics and single-fluid flows and mixing.* 1985. 3. *Gas-liquid flows.* 1986. 4. *Solids and gas-solids flows.* 1986. 5. *Slurry flow technology.* 1987. 6. *Complex flow phenomena and modeling.* 1987. 7. *Rheology and non-Newtonian flows.* 1988. 8. *Aerodynamics and compressible flows.* 1989.

V.8 (xiii,1216p.) has 45 contributors. 2 parts: 1. Turbulence phenomena and modelling - 2. Selected engineering problems. 29 documented chapters (*e.g.* 'Wind wave phenomena', p.589-635, 176 references). Detailed, analytical index, p.1203-16. For the specialist. *Class No:* 532

[922]
NOVECK, S. **Russian-English glossary of physics of fluids and meteorology.** New York, Interlanguage Dictionaries Publishing Corp., 1969. [iii], 93p.

More than 4,000 Russian entry-words, with English equivalents. *Class No:* 532

Rheology

[923]
BRITISH STANDARDS INSTITUTION. **Glossary of rheological terms.** London, BSI, 1975. 18p. £23.90. (*BS5168:1975.*)

About 300 terms on deformation and flow of matter are defined, 'Activation energy' ... 'Young's modules'. *Class No:* 532.135

[924]
Rheology abstracts: a survey of world literature. Oxford, Pergamon Press, for the British Society of Rheology, 1958-. 4pa. ISSN: 0035452x.

About 500 abstracts pa, plus list of new patents. Sections : Theoretical - Instruments and techniques - Polymers and viscoelastic materials - Solutions, pastes and suspensions - Liquids - Flow in porous media - Conferences - Book reviews - Erratum - New patents. Annual author and subject indexes. *Class No:* 532.135

[925]
Solid liquid flow abstracts. Cranfield, Beds., British Hydromechanics Research Association, 1973-. V.1, (1)-. Quarterly. £84pa.

Originally published by Gordon and Breach (v.1-3, 1969-1972), comp. by G.F. Round. About 500 informative abstracts pa. on hydraulic and pneumatic transport of solids, fluid mechanics of flow, equipment and applications and industrial processes. Issue and cumulative annual indexes of personal and corporate authors and of subjects. *Class No:* 532.135

Hydrodynamics

[926]
Fluid flow measurements abstracts. Cranfield, Beds., British Hydromechanics Research Association, 1974-. 6pa. £90.00 pa. ISSN: 00412694.

About 600 informative abstracts pa. of items drawn from 1000 journals, plus reports, conference papers, books and selected British patents. Also includes measurements of physical properties of fluids. Issue and cumulative annual indexes of personal and corporate authors and of subjects. Accessible on-line via FLUIDEX (*qv*). *Class No:* 532.5

[927]
Science research abstracts journal. Part A. Superconductivity, magnetohydrodrynamics, plasmas, theoretical physics and superconductivity research. Bethesda, Md., Cambridge Scientific Abstracts, 1978-. Quarterly.

Originally as *Science research abstracts journal. Part A: Magnetohydrodynamics and plasmas.*

About 6,000 abstracts pa. Main sections: Superconductivity - Magnetohydrodynamics and plasmas - Theoretical physics - Basic and applied physics. Subject, author and sources indexes per issue. *Class No:* 532.5

[928]
Surface wave abstracts. Brentwood, Essex, Multi-Science Publishing Co., Ltd., 1970-. Monthly.

About 350 abstracts pa. 10 sections : Generation and reception - Propagation, guidance and interaction - Measurements - Amplification - Materials and fabrication - Devices - Device applications - Acoustic optics - Non-destructive testing - General surface wave physics. Coverage

....(contd.)
of the world's major periodical literature, conference proceedings and unpublished reports. No index. *Class No:* 532.5

Gases

Humidity

[929]
WEXLER, A., *ed*. **Humidity and moisture:** measurement and control in science and industry. New York, Reinhold; London, Chapman & Hall, 1965. 4v.

Based on papers presented at the 1963 International Symposium on Humidity and Moisture, Washington. 1. *Principles and methods of measuring humidity in gases.* 2. *Applications.* 3. *Fundamentals and standards.* 4. *Measurement and control.*

V.1 comprises 6 sections, 68 subsections (each with literature references). Author and analytical subject indexes. *Class No:* 533.275

Aerodynamics

[930]
GUNSTON, B. **The Guinness book of speed.** Enfield, Middx., Guinness, 1984. 192p. illus. ISBN: 0851122671.

Pt.1: The living world. Pt.2: The technical world. Index to text and illus., p.189-92. *Class No:* 533.6

[931]
Industrial aerodynamics abstracts: world literature of non-aeronautical aerodynamics. Cranfield, Beds., British Hydromechanics Research Association, Fluid Engineering Centre, 1970-.

About 1,200 abstracts pa. Coverage includes fluid mechanics; the atmosphere; wind structure; air pollution; aerodynamics of surface features; instrumentation. Annual reviews. Accessible online via FLUIDEX (*qv*). *Class No:* 533.6

[932]
Recent advances in aerodynamics. Proceedings of an International Symposium held at Stanford University, August 22-26, 1983. Krothpalli, A. *and* Smith, C.A., *eds*. New York, Berlin etc., Springer, [1984?]. xii, 755p. illus., diagrs., graphs, tables.

23 contributors. 21 documented papers. 8 parts (*e.g.* 1. Unsteady dynamics; 2. Jets and shear layers; 3. V/STOL aerodynamics; 6. Experimental techniques; 8. Rotor dynamics and aerodynamics. 'Advances in ejection thrust augmentation', p.375-405; 48 references). *Class No:* 533.6

Plasma Physics

[933]
PRINCETON UNIVERSITY. Plasma Physics Laboratory Library. **Dictionary catalog.** Boston, Mass., G.K. Hall, 1971. 4v.

Photolithographic reproduction of *c.*62,000 cards - 21 cards per page. The catalogue lists journal articles, reports, reprints and books, plus microfiches of all USAEC reports on controlled thermonuclear fusion and dissertations of the Plasma Physics Laboratory doctoral candidates. *Class No:* 533.9

[934]
Reviews in plasma physics. Translated from Russian. New York & London, Consultants Bureau.

V.10, 1986 (original text, 1980), viii, 517p., has 7 contributors. Documented articles include 'Cyclotron instability of the earth radiation belts' (with 113 references); 'Dynamic nonlinear electromagnetic phenomena in plasmas' (with 156 references). No index. *Class No:* 533.9

Sound. Vibrations

Bibliographies

[935]
'References to contemporary papers on acoustics'. Stern, R., *ed*. *The Journal of the Acoustical Society of America,* v.90, No., 2, Pt.2, August 1991, 560p.

Published since v.11 as a supplement.

Cites papers published between 1988 and early 1990. Broad classified sequence; topics include non-linear acoustics, ultrasonics, noise effect and control, physiological and psychological acoustics, speech communication, measurement and instrumentation, and transduction. Entries derived from INSPEC database. *Class No:* 534(01)

Handbooks & Manuals

[936]
Acoustics source book. Parker, S.P., *ed*. New York, etc., McGraw-Hill, 1988. [viii],333p. illus., diagrs., graphs, tables. £35. (*McGraw-Hill science reference series*.) ISBN: 0070455082.

All material previously appeared in *McGraw-Hill encyclopedia of science and technology,* 6th ed. (1987). 9 chapters: 3. Physics of sound ... 5. Sound transducers ... 8. Sound reproduction. A basic treatment, more suitable for schools and beginning students. An unnecessary purchase for libraries holding the full encyclopedia. *Class No:* 534(035)

[937]
HARRIS, C.M., *ed*. **Shock and vibration handbook.** 3rd ed. New York, etc., McGraw-Hill, 1987. v.p. diagrs. £75.

First published 1961.

44 sections, each with references. Section 44: Effects of shock and vibration on man (55p.; 111 references, plus 2p. of additional references). 29p. index. Particularly intended to be used as a working reference by engineers and scientists in the mechanical, civil, acoustical, aeronautical, electricity, air-conditioning, transportation and chemical fields' (*Preface*). *Class No:* 534(035)

Dictionaries

Polyglot

[938]
STEPHENS, R.W.B., *ed*. **Sound.** London, Crosby, Lockwood, Staples, 1975. 854p. (*International dictionaries of science and technology*.)

3,556 English-base terms (with definitions), and

....*(contd.)*

equivalents and indexes in German, French, Italian, Dutch, Portuguese, Russian and Spanish. 'The definitions appear accurate and they are certainly crisp and informative' (*New scientist*, 15 May 1975, p.404). Stresses applied and technical aspects of sound. *Class No:* 534(038)=00

English

[939]

BRITISH STANDARDS INSTITUTION. Glossary of electrotechnical, power, telecommunication, electronics, lighting and colour terms. Part 3. Terms particular to telecommunications and electronics. Group 08. Acoustics and electroacoustics terminology. London, BSI, 1985. 33p. (*BS 4727: Part 3: Group 08: 1985.*)

Supercedes BS 661: 1969.

About 450 terms in a systematic order. Alphabetical index. *Class No:* 534(038)=20

French

[940]

PIRAUX, H. Dictionnaire général d'acoustique et d'electroacoustique. Paris, Eyrolles, 1964. 330p. illus., diagrs.,tables.

Defines nearly 1,000 French terms, arranged A-Z. Not confined to acoustics and electroacoustics. Terms that are loosely or inaccurately applied are also listed, together with the correct words. One of the longest entries is 'Caracteristiques des tubes électroniques' (4p.; 8 diagrs.). English equivalents are given. Appendix tables. *Class No:* 534(038)=40

Reviews & Abstracts

[941]

Acoustics abstracts. Brentwood, Essex, Multi-Science Publishing Co., Ltd., 1967-. pts. A,B. ea. 6pa. ISSN: 00014974.

About 3,000 abstracts pa. Sections: Fundamental acoustics - Solid state acoustics - Liquid state acoustics - Gaseous state acoustics ... Ultrasonic applications - Audio frequencies - Noise - Architectural acoustics - Vibration and shock - Review and miscellaneous. Annual indexes. *Class No:* 534(048)

Histories

[942]

HUNT, F.V. Origins in acoustics: the science of sound from antiquity to the age of Newton. New Haven, Conn., Yale Univ. Press, 1978. 197p.

A major feature of the book is its detailed account of the Arab contribution to the science of acoustics during Western Europe's Dark Ages. An appendix considers the contribution of Newton and his followers. Notes and references. Detailed index. A scholarly text. *Class No:* 534(091)

Optics

Bibliographies

[943]

BRITISH OPTICAL ASSOCIATION. Library catalogue. [British Optical Association]. Mitchell, M., *ed.* London, the Association, 1932-57. 3v.

V.1-2 (1932-35) had the title *Library and Museum Catalogue* and included facsimiles of title-pages. V.3 (1957) is an annotated author catalogue (p.1-226), plus a list of journals, a broad subject index (p.229-55) and 'Books for professional and honours examinations'. The BOA Library contains practically every work on opthalmic optics in English, and many on ocular pathology, anatomy, physiology and related topics, plus foreign works. *Class No:* 535(01)

Handbooks & Manuals

[944]

DRISCOLL, W.G. and VAUGHAN, W. Handbook of optics. New York, London etc., McGraw-Hill, 1978. [xi,1159p.]. diagrs., graphs, tables. £89.95.

Sponsored by the Optical Society of America.

27 contributors. 17 sections, mostly with lengthy lists of references. 1. Radiometry and photometry - 2. Image formation - 3. Optical radiators and sources ... 7. Properties of optical materials (157p.; 3p. of references) ... 11. Spectroscopy (164p.; references, p.154-64) ... 13. Fibre optics ... 15. Optical properties of water - 16. Optical instruments for metrology - 17. Optical modulators. Appendices A-D (C. Lens data). Index of 28p. 964 illus. Standard text. Intended for optics specialists as well as chemists, engineers and medical students. *Class No:* 535(035)

[945]

HAWKES, P.W. and KASPER, E. Principles of electron optics. London, etc., Academic Press, 1989-. diagrs. 3 vols.

Vol. 1: *Basic geometrical optics*. 1989. ISBN: 0123333512.

Vol. 2: *Applied geometrical optics*.

Vol. 3: Forthcoming.

An update of *Grundlagen der Elektronenoptik* (1952) and *Handbuch der Physik*, (1955-, *qv*). Vol. 1 contains 34 chapters in 6 parts: 1. Classical mechanics - 2. Calculation of static fields - 3. The paraxial approximation - 4. Abbreviations - 5. Deflection systems - 6. Computer-aided electron optics. Notes and references, p.577-623. Detailed, analytical index. A scholarly treatment. *Class No:* 535(035)

[946]

PALIK, E.D., ed. Handbook of optical constants of solids. Orlando, Fl., Academic Press, 1985. xviii,804p. diagrs., graphs, tables. (*Academic Press Handbook*.) ISBN: 0125444206.

27 contributors. In two parts: 1. Determination of optical constants (11 chapters); 2. Critiques (sections covering metals, semi-conductors, insulators). Part 1 summarizes methodology; Part 2 presents data, synthesized from published sources, on the optical properties of 37 solids. Well-documented, *e.g.* chapter 11 has 121 references. Extensive tables. No index. *Class No:* 535(035)

Dictionaries

Polyglot

[947]
SCHULZ, E. Wörterbuch der Optik und Feinmechanik.
Wiesbaden, Brandstetter; London, Pitman, 1960-61. 3v.
1. *German-English-French*. 1960. 112p. 2. *English-French-German*. 1961. 124p. 3. *French-German-English*. 1961. 110p.

V.1 covers *c*.3,000 terms in optics and precision mechanics and gives compounds and phrases, with German and French genders. A section of general terms is followed by specialised chapters (each with subdivisions): The eye - Eyeglass optics - Instruments and apparatus of precision mechanics and optics (*e.g.* barometer, compass, microscope) - The workshop. Little on spectroscopy, no general index. Sources, p.vi. 'Clearly directed towards opticians and optical instrument makers' (*Science reference notes*, v.10 (3/4), July/October 1963, item 92). *Class No:* 535(038)=00

English

[948]
Dictionary of visual science Cline, D., *and others*. 3rd ed.
Radnor, Pa., Chilton Book Co., 1980. xviii, 711p. illus., diagrs., charts.
2nd ed., 1968, edited by M. Schapero, and others.
68 contributors. About 20,000 entries, some of them lengthy (*e.g.* 'Lens': p.346-65). Coverage spans illumination and physical optics. Entries include eponyms, organizations and abbreviations.
4th ed., 1989 (xxi,820p. $55). *Class No:* 535(038)=20

German

[949]
BINDMANN, W. Fachwörterbuch Optic und Optischer Gerätebau. Hanau am Main, W. Dausien, 1975. 2v.
1. *English-German*. 408p. 2. *Deutsch-English*. 432p.
Technical dictionary of optics and optical instrument making. 32,000 terms in each volume.
Class No: 535(038)=30

Progress Reports

[950]
Progress in optics. Amsterdam, North-Holland Publishing Co., 1961-. illus., tables. ISBN: 0444867619.
V. 28 (1990. 470p. £84.55) covers digital holography, quantum mechanical limit in optical precision measurement and communication, quantum coherence properties of stimulated Raman a scattering, advanced evaluation techniques in interferometry and quantum jumps.
Class No: 535(055)

Tables & Data Books

[951]
CRC handbook of tables of functions for applied optics.
Levi, L. Cleveland, Ohio, Chemical Rubber Co., 1974. [x], 624p. tables.
31 tables, in sections: A. Elementary functions - B. Error functions. Fresnel integrals - C,D,E. Blackbody radiation tables - F,G. Functions for perfect lens, circular aperature,

.... *(contd.)*
monochromatic radiation - H. Miscellaneous tables. 'A good reference book at the college level' (*Choice*, April 1977, p.198). *Class No:* 535(083)

Infra-Red Rays

[952]
WOLFE, W.L. *and* ZISSIS, G.J., *eds*. The Infrared handbook. Rev. ed. Washington, Office of Naval Research, Dept. of the Navy, 1978. v.p. diagrs., graphs, tables, forumlas.
Prepared by the Infrared Information and Analysis (IRIA) Center, Environmental Research Institute of Michigan.
25 documented chapters (1. Radiation theory ... 25. Ranging, communication and simulation systems). Detailed analytical index of 42p . 'Both a comprehensive review of the field and a detailed reference work' (Shaw D.F., *ed*. *Information sources in physics*. 2nd ed., 1985, p.307). *Class No:* 535-1

Microscopy

[953]
Advances in optical and electron microscopy. Orlando, Florida, & London, Academic Press. 1966-. illus., diagrs., tables. ISBN: 0120299097.
V.9, 1984 (xiii, 370p.) has 7 contributors. The 6 documented articles include: 'Practical problems in high-resolution electron microscopy' (3½p. of references); 'Laser microanalysis' (4½p. of references). Author and subject (analytical) indexes. *Class No:* 535.083.98

[954]
Dictionary of light microscopy. Bradbury, S., *and others, comps*. Oxford Univ. Press; Royal Microscopical Society, 1989. x,139p. diagrs., tables. £6.95 (pbk). (*Microscopy handbooks 15*.) ISBN: 0198564139, pbk.
Covers *c*.1250 terms: A (symbol for absorbance, etc.) ... Zoom. Entries range from 2 to *c*.15 lines. Non-preferred terms are indicated as such. Numerous cross-references. Appendices give equivalent terms in English, French and German. Compiled by the Nomenclature Committee of the Royal Microscopical Society. Well-produced.
Class No: 535.083.98

[955]
Electron microscopy abstracts. London, PRM Science & Technology Agency, 1973-. 4pa. £45. ISSN: 03066869.
About 600 abstracts pa. Sections: Electron optics and instrumentation - Theory of image formation - Non-biological specimen preparation - Biological specimen preparation - Bacteriology - Biological macromolecules - Cells and tissues - Metals and alloys - Semiconductors - Non-metals - Petrology, mineralogy and geochemistry - Miscellaneous. Subject and author indexes per issue, cumulated annually. *Class No:* 535.083.98

[956]
FREUND, H. *and* BERG, A., *eds*. Geschichte der Mikroskopie. Frankfurt am Main, Umschau Verlag, 1963-66. 3v. illus.
1. *Biologie*. 1963. xv, 375p. 2. *Medizin*. 1964. xv, 506p. 3. *Angewandte Naturwissenschaft und Technik*. 1966. xvi, 550p.
History of microscopy, stressing applications. Well documented. *Class No:* 535.083.98

[957]
GRAY, P., *ed*. **The Encyclopedia of microscopy and microtechnique.** New York, etc., Van Nostrand Reinhold, 1973. xi, 638p. illus., diagrs., graphs, tables.

Not a 2nd ed. of *The encyclopedia of microscopy*, edited by G.L. Clark (1961), although one or two contributors are common to both.

Nearly 200 contributors. Entries A-Z. The signed article 'X-ray microscopy' (p.603-11) has sections: 'Image formation; Microanglography; Technique; Historadiography; Conclusion; Commercial sources', with 8 illus., diagrs., graphs, 70 references. Many micrographs (*e.g.* 'Kidney', p.284-8: 6 micrographs, 29 references). Index. *Class No:* 535.083.98

[958]
LOCQUIN, H. *and* LANGERON, M. **Handbook of microscopy.** Translation edited by H. Hillman. [Manuel de miscroscopie.] London, Butterworths, 1983. xii, 322p. illus., diagrs., tables, chemical structures. ISBN: 0408106794.

Originally published Paris, Masson, 1978.

6 parts: 1. Instruments and techniques - 2. Methods of fixation, examination, cutting and mounting - 3. Stains - 4. Staining and impregnation - 5. Topological stains - 6. Non-specific cytological stains. Appendix 1: Physicochemical constants. Glossary of stains and dyes, p.179-87. Bibliography, p.303-8 (covering 1981/82-1948, in reverse chronological order). Detailed analytical index. Adds about 60 new techniques to the original ed., including a lengthy section on organic matter in microfossils. *Class No:* 535.083.98

Light

[959]
DITCHBURN, R.W. **Light.** 3rd ed. London, New York etc., Plenum Press, 1976. xxv, 775, 19p. illus. (incl. col.), diagrs.

Chapters 1-12 provide a course on physical optics for students reading for a first degree in physics; chapters 13-20, - for the postgraduate worker. The chapters include: 2/3. Wave theory; 12. Polarized light; 13. The electromagnetic theory; 20. The limitations of optical instruments. No bibliographies, 19p. index. 6 plates. *Class No:* 535.1

Reflection. Emission.

[960]
CSELT (CENTRO STUDI E LABORATORI TELECOMMUNICAZIONI), TURIN. **Optical fibre communication.** New York etc., McGraw-Hill, 1981. xxxv, 883p. illus., ports. ISBN: 0070148821.

15 documented chapters in 5 parts, by specialists: 1. The transmission medium - 2. Sources and detectors - 3. Cables and connections - 4. Systems - 5. Integrated optics. 'Light propagation theory in optical fibres', p.47-144: 108 references. Detailed, analytical index, p.871-83. *Class No:* 535.3

Holography

[961]
LEHMANN, M. **Holography:** technique and practice. London & New York, Focal Press, 1970. 148p. illus. (pl). diagrs. (*Focal library*.)

9 documented chapters: 1. Introduction and history - 2. Basic principles of optics - 3. Principles of holography - 4. The practice of holography - 5. Holographic systems - 6. Devices and instruments - 7. Special techniques - 8. Applications (*e.g.* 8.3: Data storage and retrieval systems) - 9. The future of holography. 'The emphasis is on practical detail, the theoretical and mathematical content being minimal' (*Aslib book list*, v.36(5), May 1971, entry No. 190). *Class No:* 535.32

Lasers

[962]
Encyclopedia of lasers and optical technology. Meyers, R.A., *ed*. San Diego, Ca., Academic Press, 1991. xii,764p. illus., diagrs., graphs, tables. $69.95. ISBN: 0122266935.

56 contributors. 38 documented entries, A-Z ('Chemical lasers' ... 'Ultrashort laser pulse chemistry and spectroscopy'). Each article begins with brief glossary (*e.g.* 'Micro optics': 6 terms). Cross-references; detailed, analytical index, p.741-64. *Class No:* 535.33

Spectral Data

[963]
BRITISH LIBRARY. Science Reference Library. **A Guide to the published collections of spectral data held by the SRL.** de Hamel, C., *comp*. London, Science Reference Library. 1983. iii, 52p. £3. ISBN: 0712307087.

Supersedes the 1977 ed., by R. Coman.

Grouped, first by spectroscopic technique (17 sections) and then by chemical class (subsections). The sections (Compendia of spectral data ... Mass spectral data) are usually subdivided: indexes and bibliographies; collections of spectral data. Index of subjects. *Class No:* 535.330

[964]
CRC atlas of spectral data and physical constants for organic compounds. Grasselli, J.G. *and* Ritchey, W.M. 2nd ed. Cleveland, Ohio, CRC Press, 1975. 6v. (4688p.). tables, formulas. ISBN: 0878193170.

First published 1973.

1. *Name. Synonym directory. Structures. Spectroscopic aids. Special cross-correlation tables. 2/4. Data tables: Compounds , A-B, C-D, P-Z. 5. Indexes for molecular formulas, molecular weights, physical constants, WLN [Wiswesser Line Notation], Mass spectra 6. Special data indexes.*

Spectral data and physical constraints for *c.*21,000 organic compounds. Bibliographic references, for further reading. *Class No:* 535.330

[965]
ODEGARD, G. *and* HURD, J.M. **'Spectra: a bibliography of sources'.** In *Science and technology libraries*. V.11(4), Winter 1991, p.173-208. ISSN: 0194262x.

About 125 briefly annotated entries. Sections: Books - Carbon-13 NMR - Infrared - Mass - Nuclear Magnetic Resonance - Ultraviolet - Miscellaneous. *Class No:* 535.330

Colours

Encyclopaedias

[966]
HOPE, A. *and* WALCH, M. **The Color compendium.** New York, etc., Van Nostrand Reinhold, 1990. 360p. illus. $49.95.

An encyclopedia of topics on colour, ranging from art and history to the social sciences and technology. Each entry is cross-referenced. Includes bibliographies. Colour illus. and photographs throughout. Wide appeal. *Class No:* 535.6(031)

Handbooks & Manuals

[967]
KORNERUP, A. *and* WANSCHER, J.H. **Methuen handbook of colour.** 3rd ed., introduced and revised by D. Pavey. London, Eyre Methuen, 1978. 252p.

First published 1961 as *Farver i farver* (Copenhagen, Politiken Farlag); English ed., 1963.

A handbook on colour identification, colour systems and colour harmony and contrast, plus a colour dictionary - 30 double-page plates, each with 24 hues, and descriptive glossaries of colour names, with cross-references. A-Z list of names, with notes p.146-92. Appended list of manufacturers and organizations. *Class No:* 535.6(035)

Tables & Data Books

[968]
Colour index. 3rd ed. (2nd revision). Bradford, Society of Dyers and Colourists, 1971-82. Additions and amendments, 4pa. 7v. (7304p). tables.

First published 1924-28.

V.1-5 (1971 ed.) record 7,898 generic names of individual dyes and pigments. V.1-3: Colorants classified according to their usage; v.4: Colorants classified according to their chemical constitution; index of intermediate compounds; index of formulae. V.5: Code letters for colorant manufacturers; index of *Colour index* generic names, with colorants listed under each; index of commercial names of colorants. V.6-7 are supplements to V.1-4, 6. Supplement 1988: Pigments and solvent dyes. *Class No:* 535.6(083)

[969]
SMITHE, F.B. **Naturalist's color guide.** New York, American Museum of Natural History, 1974-81. 3 parts. ISBN: 0913474056.

Part 1 (24p., loose-leaf) comprises 86 named and numbered colour samples/swatches. Part 2 (229p.) acts as a supplement, analyses each colour and correlates it to other colours. A further supplement appeared in 1981.
Class No: 535.6(083)

[970]
WYSZECKI, G. *and* STILES, W.S. **Colour science:** concepts and methods, quantitative data and formulae. 2nd ed. New York, etc., Wiley, 1982. 950p. graphs, charts, tables.

First published 1967 (xiv, [1], 628p.).

Contents: Physical data - The eye - Colorimetry - Photometry - Visual equivalence and visual matching - Uniform colour scales - Visual thresholds - Theories and models of colour vision. Appendix. References. Author and subject indexes. *Class No:* 535.6(083)

Maps & Atlases

[971]
ICI colour atlas. London, Butterworths, 1971. Loose-leaf.

Previously limited availability through ICI Dyestuffs Division.

Illustrates 1,379 colours, permuted into 27,580 by using a provided set of 20 optically graded filters. Page size, 35 x 30cm. *Class No:* 535.6(084.3)

Heat

[972]
Heat bibliography, 1948/52. National Engineering Laboratory. East Kilbride. Heat Division. Edinburgh, H.M. Stationary Office, 1954-. Annual, since 1957.

About 3,000 references pa., with keywords/descriptors added. Sections, subdivided, on applications and plant equipment; energy conversion and thermodynamic cycles; experimental and operational techniques; fluid mechanics; heat and mass transfer processes; physical and thermal properties; nuclear reactors; etc. Subject index.
Class No: 536

Heat Transfer

[973]
Advances in heat transfer. Orlando, Florida, etc., Academic Press. 1969-. illus., diagrs., tables.

V. 22 (1992. 448p.) focusses on biothermal engineering and bioheat transfer. *Class No:* 536.2

[974]
BEGELL, W., *ed*. **Glossary of terms in heat transfer, fluid flow, and related topics.** Washington, Hemisphere, 1983. 176p. $37.95. ISBN: 0891162615.

304 English base terms (plus concise definitions), with equivalents and indexes in Russian, German, French and Japanese. Included are a recommended nomenclature and list of recommended journal title abbreviations. *Class No:* 536.2

[975]
HTFS digest. Heat Transfer and Fluid Flow Service. Harwell, Oxon HTFS, 1968-. 6pa. ISSN: 03091953.

A current awareness service in 23 sections (1. Boilers & evaporation ... 7. Enhanced heat and mass transfer ... 10. Heat exchanges ... 13. Instruments and techniques - 14. Multiphase fluid systems ... 21. Single phase fluid systems ... 23. Miscellaneous. Index per issue. HTFS Library database. *Class No:* 536.2

[976]
ROHSENOW, W.M. *and* HARTNETT, J.P., *eds*. **Handbook of heat transfer applications.** 2nd ed. New York, etc., McGraw-Hill, 1985. 1440p. illus. £85. ISBN: 0070535531.

First published 1975.

Sections : Basic concepts of heat transfer - Mathematical methods - Thermophysical properties - Conduction - Numerical methods in heat transfer - Natural convection - Forced convection - Internal flow in ducts ... Rarsfield gases - Electric and magnetic fields - Condensation - Boiling - Two-phase flow - Radiation. 691 illus. Numerous references. *Class No:* 536.2

Heat Effects on Bodies

[977]

CHANEY, J.F., *and others, eds*. **Thermophysical properties research literature retrieval guide, 1900-1980**. New York, Plenum Press, 1982. 7v. tables. (*First published 1964-70. 2v*.)

    1. *Elements*. 2. *Inorganic compounds*. 3. *Organic compounds and polymeric materials*. 4. *Alloys, intermetallic compounds and cermets*. 5. *Oxide mixtures and minerals*. 6. *Mixtures and solutions*. 7. *Coatings, systems, composites, foods, animal & vegetable products*.

    Covers world literature on 14 thermodynamic and physical properties of 44,300 substances. Over 75,000 references. Each volume has a materials directory (A-Z), search parameters (substances, numerically), a bibliography and author index. An important research tool for the physical chemist. *Class No:* 536.4

[978]

TOULOUKIAN, Y.S., *and others*. **Thermophysical properties of matter**: a comprehensive compilation of data by the Thermophysical Properties Research Center. New York, IFI/Plenum Press, 1970-. 14v. graphs, tables.

    1-3. *Thermal conductivity*. 1970. 4-6. *Specific heat*. 1970. 7-9. *Thermal radiative properties ...* 1970-72. 10. *Thermal diffusivity*. 1974. 11. *Viscosity*. 1972. 12-13. *Thermal expansion*. 1974. 14. *Master index to materials and properties*. 1979.

    Each volume contains introductory text, the body of numerical data (with source references) and a materials index. Primarily intended for designers, researchers, experimentalists and theoreticians. Each volume was to be revised and updated every fifth year. *Class No:* 536.4

Low Temperatures

[979]

AMERICAN INSTITUTE OF PHYSICS. **Glossary of terms frequently used in cryogenics**. New York, the Institute, [1966]. [8], 46, [1]p.

    About 100 terms are defined ('Entropy': 1p.). 13 references to books and *Scientific American* articles. *Class No:* 536.48

[980]

Journal of low temperature physics. New York & London, Plenum Press, 1923-. 12pa. illus., graphs, formulas. $158.75 per v.

    V.64(5/6), September 1986, has 8 documented articles by various hands. 'The structure and modes of a compressible superflow film' carries 14 references; 'Thermal conductivity of normal liquid heat finite temperatures', 14 references. Author and subject indexes for v.64. *Class No:* 536.48

[981]

Progress in low temperature physics. Amsterdam, North Holland Publishing Co, 1955-. diagrs., tables. ISBN: 0444852107, set.

    V. 7A-7B (1978. xxxiii, 746p.) has 15 contributors, 9 papers (*e.g.* Two-dimensional physics, p.371-433, with 202 references). Author and subject indexes. *Class No:* 536.48

Measurements

[982]

PLUMB, H.H., *ed*. **Temperature: its measurement and control in science and industry**. Pittsburgh, Pa., Instrument Society of America, 1972. 4v. illus., graphs, tables.

    V.1 (sections 1-3): Temperature scales and fixed points; Radiation thermometry; Omnibus thermometry - V.2 (sections 4-7): Resistance and electronic thermometry; Magnet and quantum electronic thermometry; Temperature control and calibration; Bridges and potentimetric methods for temperature measurement - V.3: Thermoelectric thermometry; Biology and medicine; Geophysical and astrophysical temperature measurements - V.4, pt.1: Basic methods; Scales and fixed points; Radiation; pt.2: Resistance, electronic and magnetic thermometry; Control and calibration; Bridges. 144 documented chapters in all. General author and subject indexes. *Class No:* 536.5

Thermodynamics

[983]

JAMES, A.M. **A Dictionary of thermodynamics**. London, Macmillan Press, 1976. xi, 262p. diagrs., graphs, tables.

    'Absolute molar entropy' ... 'Zeroth law of thermodynamics'. 'Acids and bases': 5½p. Cross-references. 'Tables of useful data', p.247-59; 'Recommended reference books and textbooks', p.260-1. SI units. 'An interesting book which undergraduate university students will find useful ... an excellent "aide memoire", incorporating all the important terms one usually comes across in a basic course on thermodynamics' (*Aslib book list*, v.42(3), March 1977, item 138). *Class No:* 536.7

[984]

RAZNJEVIC, K. **Handbook of thermodynamic tables and charts**. New York, Hemisphere Publishing Corpn., McGraw-Hill, 1976. viii, 392p. tables, charts. ISBN: 0070812701.

    Sections: Solids - Liquids - Vapors - Gases - Units and measures. More than 100 tables (SI units) on solids, liquids, vapour. 16 folded charts. Includes steam tables up to 1000 degrees and 1000 bar. Comprehensive treatment of basic data for engineering thermodynamics as well as practical and theoretical thermodynamics. *Class No:* 536.7

Electricity

[985]

CONNOLLY, T.F., *ed*. **Electrical properties of solids, surface preparation and methods of measurement**. New York & London, IFI/Plenum, 1972. xxiv, 96p. (*Solid state literature guides, v.4*.)

    A comprehensive work of reference, covering semiconductors, metals, alloys, insulators and dielectrics. International scope, up to 1971, in 5 sections. For the scientific and industrial research worker. *Class No:* 537

[986]

HEILBRON, J.L. Electricity in the 17th and 18th centuries. Berkeley, etc., Univ. of California Press, 1979. xiv, 606p. illus., facsims., diagrs. ISBN: 0520034783.

5 footnoted parts (21 chapters): 1. Early modern physics and its cultivators - 2. Electricity in the seventeenth century - 3. The great discoveries - 4. The age of Franklin - 5. Quantification. Bibliography (A-Z authors), p.501-69. Analytical index, p.571-606. Outstanding scholarly history of the period. *Class No:* 537

[987]

MOTTELAY, P.F., *comp.* **Bibliographical history of electricity & magnetism,** chronologically arranged. Researches into the domain of the early sciences, especially from the period of the revival of scholasticism, with biographical and other accounts of the most distinguished philosophers throughout the Middle Ages. London, Griffin, 1922. xvii, 678p. illus.(pl.), facisms.

Covers the period 2637 B.C.-A.D. 1821. Faraday: p.483-99, with biographical data, including 1p. of references. 6 appendices (4. 'Names of additional electrical and magnetic works, published up to 1800'). Detailed index, p.565-678. A valuable source that can be used to supplement F. Cajori's *History of physics* (*qv*), but the chronological sequence needs to be revised. *Class No:* 537

[988]

WHITTAKER, Sir E.T. A History of the theories of aether and electricity. London, Nelson, 1951-53. 2v.

V.1 (1951) deals with the classical physics theories up to 1900. V.2 (rev. and enl. ed., 1953), with the modern theories, 1900-1926. The standard history on the subject for physicists. *Class No:* 537

Measurements

[989]

Key abstracts: Electrical measurements and instrumentation. London, INSPEC; New York, Institute of Electrical and Electronic Engineers, 1976-. Monthly. (*Key abstracts.*) ISSN: 03077977.

About 150 abstracts per issue. 7 main sections: 1. Measurement science - 2. Measurement equipment and instrumentation systems - 3. Measurement of specific electric and magnetic variables - 4. Measurement of specific nonelectric variables - 5. Biomedical instrumentation and measurement - 6. Aerospace instrumentation - 7. Geophysics instrumentation and measurement. No indexes. A current awareness service. *Class No:* 537.083

Radiation. Discharges

Radiology

[990]

BRITISH STANDARDS INSTITUTION. Glossary of electrotechnical, power, telecommunication, electronics, lighting and colour terms. Part 5: Terms particular to electromedical equipment. Group 01: Radiology and radiological physics terminology. London, BSI, 1985. 58p. (*BS 4727, Part 5. Group 01: 1985.*) ISBN: 0580146146.

Supersedes BS 2597: 1955, withdrawn. Corresponds to IEC Publication 50(881): 1984.

Some 660 numbered terms and definitions. Main sections:

....(contd.)

General terms - Ionizing radiations - Radiological apparatus and techniques - Dosimetry - Radiation protection - Biological effects. Index. *Class No:* 537.53

X-rays

[991]

CLARK, G.L., *ed.* **The Encyclopedia of X-rays and gamma-rays.** New York, Reinhold; London, Chapman & Hall, 1963. xxvii, 1149p. illus., diagrs., tables.

302 contributors; about 350 signed articles, A-Z (entry-word being the keyword in the title, *e.g.* 'Historical milestones for X-rays and gamma rays 1600-1962', - under 'Historical'). References appended to articles are nummbered (*e.g.* 'Microradiography in biology': p.607-18; 78 references; 26 illus. and diagrs.). Occasional uneven coverage (*e.g.* 12p. on 'Portland cement compounds'; under 3p. on 'Biological consequences of exposure to ionizing radiation'). Cross-references; index. *Class No:* 537.531

[992]

X-ray diffraction abstracts. London, PRM Science & Technology Agency, Ltd., 1972-. Quarterly.

About 600 abstracts pa. Sections: Theory and instrumentation - Polymers and glasses - Ceramics - Metals and alloys - Organic compounds - Single crystal studies: organic and organometallic compounds - Lanthanides and actinides. Subject and author indexes per issue. *Class No:* 537.531

Thermionics

[993]

ORGANIZATION FOR ECONOMIC COOPERATION AND DEVELOPMENT. Joint ENEA-IAEA Liaison Group on Thermionic Electrical Power Generation. **Glossary of terms and symbols in thermionic conversion.** Paris, O.E.C.D., 1971. 115p.

Internationally agreed definitions of more than 200 technical terms and expressions. List of preferred symbols. Index of terms in English, French and Russian. *Class No:* 537.58

Magnetism

[994]

RADO, G.T. *and* **SUHL, H.,** *eds.* **Magnetism.** New York & London, Academic Press, 1963-73. 5v. illus., diagrs., tables.

1, *Magnetic ions in insulators, their interactions, resonances and optical properties.* 1963. xv, 688p. 2. *Statistical models, magnetic symmetry, hyperfine interactions and metals.* 1965-66. 2pts. 3. *Spin arrangement and crystal structure, domains and micromagenetics.* 1963. 623p. 4. *Exchange interactions among itinerant electrons* 1966. 407p. 5. *Magnetic properties of metallic alloys.* 1973. [xvi], 400p.

V.5 of this general treatise has 16 chapters by specialists, with chapter references, and author and subject indexes. *Class No:* 537.6

Nuclear Magnetic Resonance

[995]

HOMANS, S.W. **A Dictionary of concepts in NMR.** Oxford, Clarendon Press, 1989. [viii],343p. illus., diagrs., graphs. £40. ISBN: 0198552742.

A-Z (*Absolute-value mode* to *Zero-quantum coherence*), detailed (*c.*1-page) definitions. Emphasis is towards two-dimensional NMR methods in liquids reflecting 'present experimental trends' (*Preface*). Cross references; 'further reading' after many entries. 4 appendices: Properties of cartesian product operators; Trigonometric identities; Matrix algebra; Rotation operators. *Class No: 537.635*

Solid State Physics

[996]

CONNOLLY, T.F., *ed*. **Solid state physics literature guides.** Prepared under the auspices of the Research Materials Information Center, Oak Ridge National Laboratory. New York, IFI/Plenum, 1970-79. 10v. in 11.

1. *Ferro-electric materials and ferro-electricity.* 1970.
2. *Semiconductors - Preparation, crystal growth, and selected properties.* 1972.
3. *Groups iv,v and vi transition metals and compounds - Preparation and properties.* 1972.
4. *Electrical properties of solids - Surface preparation and methods of measurement.* 1972.
5. *Bibliography of magnetic materials and tabulation of magnetic transition temperatures.* 1972.
6. *Ferroelectric literature index.* 1974.
7. *Scattering of thermal neutrons: a bibliography.* 1974.
8. *Bibliography of microwave optical technology.* 1976.
9. *Laser window and mirror materials.* 1977.
10a,b. *Crystal growth bibliography.* 1979.

V.10a: *Bibliography* carries more than 5,000 references on the crystal growth of organic materials covering 1972-77. V.10b provides permuted title index and author index (ix, 430p.). *Class No: 538.9*

[997]

LERNER, R.G. *and* TRIGG, G.L., *eds*. **Concise encyclopedia of solid state physics.** Reading, Mass., Addison-Wesley Publishing Co., Inc., 1983. [xv], 311p. illus., diagrs., graphs, tables. $39.50. ISBN: 020114204x.

About 140 contributors; *c.*120 documented main entries (Acoustoelectric effect ... Work function). The article 'Liquid Crystals' (5½p.) has 14 notes and references and 4 figures. Analytical index, p.305-11. Aims 'to provide in one volume, a convenient and readily accessible set of resource materials in the area of sold state physics' (*Preface*). The articles are drawn from the Lerner and Trigg *Encyclopedia of physics* (1st ed.). *Class No: 538.9*

[998]

Solid state abstracts journal. Bethesda, Md., & Oxford, CSA/Cambridge Science Abstracts, 1960-. 7pa. $628 pa. ISSN: 0438108x.

Incorporates *Science research abstracts journal* (six titles). About 10,000 abstracts pa. 4 classes (closely subdivided): SCM. Solid state chemistry and metallurgy - SMP. Solid state metallurgy and physics - SSP. Solid state physics - SMED. Solid state measurements, effects and devices. Analytical subject and author indexes per issue, cumulated annually. *Class No: 538.9*

[999]

Solid state physics: advances in research and applications. New York, London, etc., Academic Press, 1948-. diagrs., graphs, tables. ISBN: 0126077398.

V.45, (1991. x,295p.) has 3 contributors. The 3 papers include 'Structural ordering in collodial suspensions' (168 footnote references), 'Crystal nucleation in liquids and gases' (344 footnote references) and 'Glass transition and relaxation of disordered structures' (117 footnote references). Author and subject index. Cumulative author index, v.1-45. *Class No: 538.9*

Structure of Matter

[1000]

Advances in atomic and molecular physics. Orlando, Fla., Academic Press, 1965-. Annual. ISBN: 0120038218.

V.28, 1990 (320p.) has 7 contributors. The 4 papers include 'The theory of fast ion-atom collisions' and 'squeezed states of the radiation field'. Author and subject index. Cumulative author index, v.1-28. *Class No: 539*

Nuclear Physics

[1001]

Advances in nuclear physics. New York etc., Plenum Press. 1971-. Annual. ISBN: 0306418649.

V. 20 (1991. 480p.) has two documented articles on single-particle properties of nuclear and relativistic quantum many-body theory. *Class No: 539.1*

[1002]

Annual review of nuclear and particle science. Palo Alto, Calif., Annual Reviews, Inc. 1950-. Annual.

V.41, 1991 (550p.) has 14 articles including 'Decays of B Mesons' and 'Storage rings for nuclear physics'. Cumulative indexes: contributing authors; chapter titles. *Class No: 539.1*

[1003]

BEHRENS, H. *and* RITTBERGER, W. 'Nuclear and particle physics'. In *Information sources in physics*. 2nd ed. Editor, D.F. Shaw. 1985, p.255-78. London, etc., Butterworth, 1985.

Sections: Primary publication in journals and books - Nuclear physics - Particle physics - Abstracts journals - Conferences and proceedings. 10 references. A feature is the profusion of supporting tables (*e.g.* 13: 'Bibliographic databases and abstracts journals for nuclear and particle physics'). *Class No: 539.1*

Particles

[1004]

CINDA 86 (1982-1986): the index to literature and computer files in microscopic neutron data. Vienna, International Atomic Energy Agency, on behalf of the USA National Nuclear Data Center, [etc.,], 1986. 28, 444, 46p. tables.

Supersedes previous issues. CINDA = Computer index of neutron data.

Prelims. include 'Selected literature scanned for the present edition'; 'Neutron data handbooks'. Entries (4), p.1-

....(contd.)

439; a. Elements in isotopes; c. Molecules and mixtures. Annex (A1-A46) of 5 tables (*e.g.* Abbreviations for molecules and mixtures). *Class No:* 539.12

[1005]

Current awareness in particle technology. Particle Science and Technology Information Service, Univ. of Technology, Loughborough, 1976-. Monthly. ISSN: 03764842.

Continues *Particulate information*, 1974-75.

About 9,000 references pa, covering the world literature of particle science and technology: journal articles, conference papers, reports, patents, theses, books. 18 subject sections (Particle properties ... Powder and compact properties ... Aerosols ... Gas filtration ... Crystallisation ... Granulation and pelletisation - Miscellaneous applications). General; Books; Conference index. Appended: 'Forthcoming conference meetings and exhibitions'. Quarterly author and subject indexes, cumulated annually. *Class No:* 539.12

[1006]

McCRANE, W.C. *and* **DELLY, J.C. The Particle atlas:** an encyclopedia of techniques for small particle identification. 2nd ed. Ann Arbor, Mich., Ann Arbor Science Publishers, 1973-80. 6v. (1703p.). micrographs. ISBN: 0250400189.

First published 1967.

1. *Principles and techniques.* 1973. 2. *The light microscopy atlas.* 1973. 3. *The electron microscopy atlas.* 1973. 4. *The particle analyst's handbook.* 1979. 5. *Light microscopy techniques and atlas.* 1979. 6. *Electron-optical techniques and atlas.* 1980.

4,500 micrographs, for the identification of 1,020 substances, *e.g.* fertilizers, insecticides, pigments, industrial chemicals, explosives, minerals. V.3 has 9 illus. of micrographs per page. Each volume carries an index, cumulating. *Class No:* 539.12

[1007]

Neutron activation analysis abstracts. London, PRM Science & Technology Agency, Ltd., 1973-. Quarterly. ISSN: 00479448.

About 1,000 abstracts pa. Sections: Theory - Instrumentation - Analysis of metals and their alloys - Non-metallic elements - Inorganic compounds - Organic materials - Reactor materials - Meteorites - Minerals - Water - Biological materials. Reviews. Subject and author indexes per issue. *Class No:* 539.12

[1008]

Progress in particle and nuclear physics. Oxford, Pergamon Press, 1971-. Annual. ISBN: 0080936671. ISSN: 01466410.

V. 27 (1991. 336p.) covers physics of precision experiments with Z's, structure of nuclear matter and QCD sum rules, colour screening and colour transparency in hard nuclear processes, the Nambu and Jona-Lasinio model. *Class No:* 539.12

Atoms

[1009]

BOORSE, H.A., *and others*. **The Atomic scientists:** a biographical history. New York, Wiley, 1989. viii,472p. ports. £19.95. ISBN: 0471504556.

Brief (*e.g.* Max Planck (1858-1947), p.123-30) narrative biographies of significant contributors to atomic science.

....(contd.)

Chronological. Index, p.469-72 'Highly readable' (*New technical books*, v.23(2), March/April 1990, p.394). *Class No:* 539.18

Energy Levels

[1010]

High energy physics index. Compiled at Deutsches Elektronen-Synchroton DESY, Hamburg. [Hochenergiephysik index.] Karlsruhe, Fachinformationszentrum, 1963-. Bi-monthly. ISSN: 00181447.

About 15,000 references pa. 3 sections: Experimental physics - Instrumental physics - Theoretical physics. Book index. Conference index. Author and affiliation index. Preprint and report number index. Subject index. *Class No:* 539.184

[1011]

SUBE, R. Dictionary of high-energy physics, in English, German, French, Russian. Amsterdam etc., Elsevier, 1987. 164p. £58.56. ISBN: 0444989838.

About 5,000 English-base terms, with equivalents and indexes in German, French and Russian. *Class No:* 539.184

High Pressure

[1012]

Bibliography on high-pressure research. Salt Lake City, Utah, Norten Christensen, Inc., 1970-. 6pa.

Previously as *Bibliography on high pressure research in chemistry and physics*, 1968-70.

About 3,500 references pa. 10 major subject classes, *e.g.* 1. Condensed matter. Physical properties - 2. Thermodynamic properties - 3. Chemical and biological properties ... 5. Geology/geophysics resources ... 7. Superhard materials. Ceramics. Abrasives ... 10. Shock wave investigations. A current awareness bulletin. *Class No:* 539.893

54 Chemistry

Chemistry

Abbreviations & Symbols

[1013]
WOHLAUER, G.E.M. *and* GHOLSTON, H.D., *comps. and eds*. **German chemical abbreviations.** New York, Special Libraries Association, 1965. iv, 63p.

About 3,000 abbreviations, mostly taken from *Chemisches Zentralblatt*, Beilstein, Lenk-Börner, Hunnius (*Pharmazeutisches Wörterbuch*, 1959) and De Vries. Entries state German abbreviations, full form and English translation. Includes abbreviations for technical terms not specifically confined to chemistry. *Class No:* 540(003)

Databases

[1014]
CA SEARCH. Columbus, Ohio, American Chemical Society, Chemical Abstracts Service; 1967-.

CA SEARCH is a major database, derived from the weekly *Chemical abstracts*. It now carries *c*.6 million citations, giving CA Registry Numbers, but omitting the abstracts (except on DIALOG). The latter are covered by CAS ONLINE.

Hawkins, D.T. 'A review of online physical sciences and mathematics databases. Part 2: Chemistry' (*Database*, v.8(2), June 1985, p.31-41) consists of 1. Introduction; 2. Bibliographic databases (2.1 Chemical abstracts; 2.2 Other bibliographic databases); 3. Name directories and substructure searching systems. 4. Numeric databases; 5. Full text databases. Appended chart of relevant databases, with headings: Chemical databases; Major subject(s), Producer, Starting date. 41 references.
Class No: 540(003.4)

Software

[1015]
Directory of chemistry software, 1992. Warr, W., *and others, eds.* Washington, D.C., Cherwell Scientific/ American Chemical Society, 1992. 204p. $34.95. ISBN: 0951823604.

Over 170 software packages, A-Z by program name. Categories: structure drawing, DBMS, expert systems, 3-D molecular modeling, quantum mechanics, utility packages, terminal emulators for on-line searching, communications and off-line query formulation. Data include product description, features and capabilities; company and distributor contact; hardware and software requirements; level of support; price; references to product reviews. 'Recommended' (*Choice*, October 1992. p.270).
Class No: 540(003.42)

Bibliographies

[1016]
ANTONY, A. **Guide to basic information sources in chemistry.** New York, Halstead Press, 1979. [vii], 219p. diagrs. ISBN: 0470265876.

23 chapters (2. Guide to the literature - 3. *Chemical abstracts* - 4. Abstracts and indexes other than *Chemical abstracts* - 5. Bibliographic searching by computer - 6. Periodicals and lists of periodicals - 7. Access to primary publications other than journals - 8. Bibliographies - 9. Nomenclature - 10. Dictionaries and encyclopedias - 11. Language dictionaries ... 13. Specialized data compilations (p.93-138) - 14. Beilstein and Gmelin ... 16. Safety manuals and guides ... 18. Biographies and directories of people ... 21. Monographs, textbooks and treatises - 23. Chemical information search strategy given 11 case examples of types of chemistry questions). Primarily intended for 'the student of chemistry from college freshmen through graduate level' (*Preface*). *Class No:* 540(01)

[1017]
BOLTON, H.C. **A Select bibliography of chemistry, 1492-1902.** Washington, Smithsonian Institution, 1893-1904. 4v. (*Smithsonian Miscellaneous Collections, v.35, 39, 41(3), 44(5)*.)

Reprinted New York, Kraus, 1969.

About 18,000 items. The main two volumes (1893. ix. 1,212p.) have 7 sections (Bibliography - Dictionaries and tables - History of chemistry - Biography - Chemistry, pure & applied - Alchemy - Periodicals). Section 8, an international list of academic dissertations, appeared in the 3rd volume (1st suppt.). Some entries have brief, usually bibliographical, annotations. Subject indexes.
Class No: 540(01)

[1018]
—COLE, W.A. **Chemical literature, 1700-1860: a bibliography** with annotations, detailed descriptions, comparisons and locations. London, Mansell, 1988. *c*.600p. £98. ISBN: 0720119677.

Records *c*.1,450 items, chosen as major contributions to the period of 'chemical revolution'. Bibliographical descriptions, annotations and references. Brief biographies for most authors. Library locations in USA, Canada, UK and France. *Class No:* 540(01)

[1019]
—DUVEEN, D.I. **Bibliotheca alchemica et chemica: an annotated catalogue of printed books.** London, Weil, 1949. vii, 609p. illus.

Facsimile reprint. London, Dawsons, 1965.

Another valuable private collection catalogue.
Class No: 540(01)

[1020]
—FERGUSON, J. Bibliotheca chemica: a catalogue of the alchemical, chemical and pharmaceutical books in the collection of the late James Young of Kelly & Davis. Glasgow, Maclehose, 1906. 2v.

More than a catalogue of 2,500 items. It contains concise and critical biographical accounts of the authors, and appraisals of their works, making it an encyclopaedia of the history, biography and bibliography of early chemistry. *Class No:* 540(01)

[1021]
BOTTLE, R.T. *and* ROWLAND, J.F.B., *eds*. Information sources in chemistry. 4th ed. New York, London, Bowker-Saur, 1992. 320p. $75; £40. (*Guides to information sources.*) ISBN: 1857390164.

First published 1962 as *Use of chemical literature.* 3rd ed. 1979.

An extensive and welcome revision of a reliable work. Wide range of chemical subjects (*e.g.* pharmaceuticals) and source literature. Special attention to and guidance for the complexities of handling chemical structures in electronic searching. *Class No:* 540(01)

[1022]
MAIZELL, R.E. How to find chemical information: a guide for practicing chemists, educators and students. 2nd ed. New York, Chichester, etc., Wiley, 1987. xvii,402p. £55.

First published 1979.

19 chapters: 1. Basic concepts - 2. Information flow and communication patterns in chemistry - 3. Search strategy - 4. Keeping up to date - current awareness programs - 5. How to get access to articles, patents, translations, specifications, and other documents quickly and efficiently - 6. 'Chemical abstracts' service - history and development - 7. Essentials of 'Chemical abstracts' use - 8. Other abstracting and indexing services of interest to chemists - 9. Some US government technical information centers and sources - 10. Online databases, chemical structure searching, and related topics - 11. Reviews - 12. Encyclopedias and other major reference books - 13. Patents - 14. Safety and related topics - 15. Location and using physical property and related data - 16. Chemical marketing and business information sources - 17. Process information - 18. Analytical chemistry - a brief review of some of the literature sources - 19. Summary of representative major trends and development in chemical information. Indexes (name; subject), p.377-402. A practical approach, with 'pros and cons' for many sources described. *Class No:* 540(01)

[1023]
MELLON, M.G. Chemical publications, their nature and use. 5th ed. New York, etc., McGraw-Hill, 1982. xii, 419p. ISBN: 0070415145.

First published 1958.

Part 1 (chapters 1-12): Publications, kinds and nature (Primary sources - periodicals; technical reports; patents; etc.; Secondary sources - periodicals; *Chemical abstracts*; bibliographies; reference works; Tertiary sources - guides and directories). Part 2 (chapters 13-16): Publications, storage and use; Libraries and information centers (US slanted); Searches, manual and computer; Library problems. Author and analytical subject indexes, p.395-419. Well-documented chapters. *Class No:* 540(01)

[1024]
NOWAK, A. Fachliteratur des chemikers. Einfüchrung in ihre Systematik und Benutzung mit einer übersicht über wichtige Werke. 3. Aufl. Darmstadt, Steiner, 1976. 220p. tables. ISBN: 3798504229.

First published 1962 (156p.).

In the 1st ed., 2 of the 6 chapters concern searching the literature, given a particular type of query (*e.g.* identification of substances, production methods, a given chemical reaction). Chapter 5, 'Bibliography' covers systematic treatises, abstracting and indexing services, primary sources, patents, theses, institution publications, laboratory records, etc. Subject index. *Class No:* 540(01)

[1025]
SKOLNIK, H. The Literature matrix of chemistry. New York, etc., Wiley, 1982. xi, [1], 297p. ISBN: 0471705453.

11 chapters: 1. Books -2. Encyclopedias and treatises - 3. Numerical data compilations - 4. Patents - 5. The journal literature - 6. Secondary publication - 7. Chemical Abstracts Service - 8. Other indexing/abstracting services - 9. Computer-based information services - 10. Evolution of the literature, from antiquity to the early twentieth century - 11. Contributions of American chemists to the literature. Name and analytical subject indexes. 'Aimed at students majoring in chemistry and chemical engineering and to scientists and engineers employed by the chemical industry in research and development and in plant operations' (*Preface*). *Class No:* 540(01)

[1026]
—WOLMAN, Y. Chemical information: a practical guide to utilization. 2nd ed. New York, Wiley, 1988. xiv, 291p. $44.95.

First published 1983.

16 sections (2. The scientific journal; 3. The library; 4. Keeping updated - the current awareness programme; 5. How to conduct a literature search; 6. Obtaining numerical data; 7. Chemical reactions and concepts; 8. Selecting a bibliographical database; 9. Synthetic reaction search; 10. Structural and substructural searches; 11. Environmental impact and control; 12. The patent literature; 13. Chemical marketing, processes and engineering information; 14. Gathering information about individuals and organizations; 15. Expert systems; 16. Perspectives. A 'very good publication' according to *Choice* (May 1989, p.1547), which also noted that 'many subjects are treated too briefly'. Glossary of acronyms. Detailed index, partly analytical. *Class No:* 540(01)

[1027]
—WOODBURN, H.M. Using the chemical literature: a practical guide. New York, Dekker, 1974. viii, 362p.

In 18 sections: 3. US government publications ... 6. Collections of physical data ... 10. *Gmelin's Handbuch der organische Chimie* - 11. Abstracting services - 12. Scope and use of indexes on *Chemical abstracts* ... 15. Microform publications - 16. The availability of translations - 17. Patent literature - 18. Computer-readable material. Glossary of acronyms, p.279-81. Index of subjects, forms, titles. 'A practical guide, not a bibliography of sources' (*Preface*). Compared with Mellon, Skolnik, and Bottle (3rd ed.), this guide is considered fullest on major reference books (*Choice, v.21(6),* February 1984, p.808). *Class No:* 540(01)

[1028]

WIGGINS, G. **Chemical information sources.** New York, McGraw-Hill, 1991. 352p. 2 disks (IBM-PC). $42.35. ISBN: 0079099394.

Covers databases including citation, subject and patent searching, with emphasis on online sources. Includes more traditional printed ones (dictionaries, handbooks, etc.) as well. An innovation in this type of work is the floppy disks (3.5 inch) containing over 2100 records with citations for, among other things, the complete reference collection of the Indiana University Chemical Library. 'Recommended very highly' (*Choice*, February 1992, p.920). *Class No:* 540(01)

Teaching Materials

[1029]

JENKINS, E.W. **References for chemistry teachers.** Leeds, Centre for Studies in Science and Mathematics Education, Univ. of Leeds, [1984]. [1], [1], 110p. £1.65.

A sequel to his *Bibliography of resources for chemistry teachers* (1973).

Lists under 36 heads (Analytical chemistry and stoichiometry ... Structure and bonding) 'some of the material that has appeared in three journals': - *Education in chemistry*, *The Journal of chemical education* and *The School science review*. It covers the period ending June 1983. *Class No:* 540(01)(072)

Encyclopaedias

[1030]

CONSIDINE, D.M. *and* CONSIDINE, G.D., *eds.* **Van Nostrand encyclopedia of chemistry.** 4th ed. Princeton, N.J., Van Nostrand Reinhold, 1984. [viii], 1082p. diagrs., tables. $89.50. ISBN: 0442225725.

About 1,300 entries, 'Absolute zero' ... 'Zymolytic reaction'. 112 contributors, although entries are not usually signed. Longer articles are documented (*e.g.* 'Phosphorus': 9½ columns, with 7 references). Analytical index, p.1039-82. Nearly 80% of the text is completely new. 'Designed for ready comprehension by anyone with a general background in chemistry' (*Journal of chemical education*, December 1984). *Class No:* 540(031)

[1031]

—CLARK, G.L., *ed.* The Encyclopedia of chemistry. 2nd ed. New York, Reinhold; London, Chapman & Hall, 1966. xxi, [i], 1144p.

First published 1957, with supplement 1958.

About 2,000 entries, including *c.*100 biographies and some 60 notes on societies, trade association, etc. Some 600 contributors. Special emphasis on theoretical aspects (*e.g.* co-ordination theory and developments in analysis). Systematic treatment (*e.g.* 'Asphalt': a signed article, p.103-6, including major uses; chemical composition; rheology; trends in research). Cross-references, but no appended bibliographies. Analytical index, p.1114-44. For the biologist and engineer as well as the chemist. Well presented. *Class No:* 540(031)

[1032]

—Kingzett's chemical encyclopaedia: a digest of chemistry and its industrial applications. Hey, D.H., *ed.* 9th ed. London, Ballière, Tindall and Cassell, 1966. xi, 692p.

First published, 1919.

A selective encyclopaedic dictionary, with 15 British and US contributors. All references to periodicals in the 8th ed. (1952) seem to have been dropped and date of publication of cited works is omitted. Clark's *Encyclopedia of chemistry* (*qv*) offers about 75% more material, but for quick-reference purposes, Kingzett is considered more useful, although too elementary for the professional chemist. Both encyclopaedias certainly need updating. *Class No:* 540(031)

[1033]

FALBE, J. *and* REGITZ, M. **Römpp's Chemie-Lexikon.** 9, erweiterte und bearbeitete Auf. Stuttgart, Thieme, 1989-. illus., diagrs., tables, chemical structures.

First published 1947; 8th ed., 1979-88.

To be compiled in 9v. Well-documented (*e.g.* 'Recycling', p.3808-10, with 1 column bibliography). Includes brief biographies and some trade names. Cross references. Etymology given. The major German general encyclopedia of chemistry and chemical products. *Class No:* 540(031)

[1034]

McGraw-Hill encyclopedia of chemistry. Parker, S.P., *ed.* New York, etc., McGraw-Hill, 1984. x, 1195p. illus., diagrs., tables, charts. $76.50.

790 articles, extracted intact from the parent work, *McGraw-Hill encyclopedia of science and technology* (5th ed. 1982. 15v.). 'Spectrochemical analysis': 12 columns, 6 diagrs., 5 references. Up-to-dateness and depth of treatment make the *McGraw-Hill encyclopedia of chemistry* superior to Considine and other encyclopedias and dictionaries in the field. Cross-references; index. *Class No:* 540(031)

Dictionaries

Polyglot

[1035]

Dictionary of chemistry and chemical technology. Hulmin, C. *and* Wenchu, B., *eds.* Beijing, Chemical Industry Press; Amsterdam, Elsevier, 1989. 126p. £134.90.

Terms in English, Chinese, Japanese.

Class No: 540(038)=00

[1036]

DORIAN, A.F. **Elsevier's dictionary of chemistry,** including terms from biochemistry, in English, French, Spanish, Italian and German. Amsterdam, etc., Elsevier, 1983. vi, 586p. $109.25; DFl.295. ISBN: 0444442307.

9,013 numbered English-base chemical terms, with equivalents and indexes in French, Spanish, Italian and German. Genders are omitted. The inclusion of biochemical terms provides a link between organic chemistry, physiology, genetics and medicine. *Class No:* 540(038)=00

[1037]

—FOUCHIER, J. *and* BILLET, F. Chemical dictionary in English, French and German. 3rd ed. Amsterdam etc., Elsevier, 1970. xxi, 1464p. D109.25; DFl.295. ISBN: 0444410907.

About 20,000 terms, in three sequences: English-French-German, French-German-English, and German-English-French. *Class No:* 540(038)=00

[1038]

—KRYT, D., *ed*. Dictionary of chemical terminology, in English, German, French, Polish and Russian. Amsterdam, etc., Elsevier, 1980. [xiii], 600p. DFl.195. ISBN: 0444997681.

Lists 3,805 English-base terms and definitions, with equivalents and indexes (on tinted paper) in German, French, Polish and Russian. 15 Contributors. Nearly 100 categories, plus symbols denoting the pure substance, standard state, and activated complex. Omits nomenclature of chemical compounds and terms from chemical terminology and technology. Intended for students, researchers, scientists and engineers.
Class No: 540(038)=00

[1039]

SOBECKA, Z. *and* CHOINSKI, W., *eds*. **Dictionary of chemistry and chemical technology in six languages:** English, German, Spanish, French, Polish, Russian. Oxford, etc., Pergamon Press; Warsaw, Wydawnictwa Naukowo-Techniczne, 1966. vii, 1325p.

First published 1962.

11,987 numbered entries in English, with German, Spanish, French, Polish and Russian equivalents across the double page, and German, etc., indexes, on tinted paper. Categorizes terms, notes differences in meaning and gives genders of nouns. It omits current English terms that have no corresponding expressions in one or other of the foreign languages. No abbreviations. 'An excellent work of reference for any European chemist' (*Chemistry & industry*, 3 December 1966, p.2035). *Class No:* 540(038)=00

English

[1040]

BENNETT, H., *ed*. **Concise chemical and technical dictionary.** 4th ed. London, E. Arnold, 1986. xxxviii, 1271p. £95. ISBN: 0713135840.

First published 1974.

Some 85,000 brief definitions, with many cross-references. Includes abbreviations and chemical structures. Prelims. on nomenclature, pronunciation, names and formulas of radicals occurring in organic compounds. The 4th ed. is enlarged with definitions of new trademark products, chemicals, drugs and terms coined to cover the latest technical developments. Prelims. have a section, 'The pronunciation of chemical words'. Intended for use by both professionals and laymen. *Class No:* 540(038)=20

[1041]

—BEVAN, S.C., *and others*. Concise etymological dictionary of chemistry. London, Applied Science Publishers, Ltd., 1976. ix, 140p.

Defines *c*.2,500 chemical terms. Brief historical notes (*e.g.* where words were coined or proposed, and by whom). Eponyms included. 8 appendices on prefixes, formation of plurals, etc. Helpful introduction, including list of works consulted. Intended for a wide public.
Class No: 540(038)=20

[1042]

—FLOOD, W.E. The Origin of chemical names. London, Oldbourne, 1963. xxxi, 238p. diagrs.

American ed. (New York, Philosophical Library) has title: *The dictionary of chemical names.*

Pt.1: The chemical elements - pt.2: Chemical compounds, minerals and other substances of interest (p.25-233). Pt.2 has more than 1,700 headwords, giving definitions and derivations in 50-250 words. Included are affixes, suffixes and combining units from which many other names have been constructed; also chemical structures. 'Biographical notes', p.234-8. *Class No:* 540(038)=20

[1043]

GODMAN, A. **Longman chemistry handbook.** Harlow, Essex, Longman; Beirut, York Press, 1992. 256p. diagrs., (col.). £5.99. ISBN: 0582088100.

About 1,500 terms, briefly defined (*e.g.* 'equilibrium': 13 lines). Arranged by subject (*e.g.* chemical reactions; colloids; electrolysis; organic reactions, etc.), with terms listed in index. 6 appendices. Effective use of colour in diagrams. For senior secondary pupils and undergraduates. *Class No:* 540(038)=20

[1044]

Grant and Hackh's chemical dictionary. Grant, R.L. *and* Grant, A.C., *eds*. 5th ed. New York, etc., McGraw-Hill, 1987. xii, 641p. diagrs., charts, tables. £74.

First published 1929; 4th ed. 1969.

More than 55,000 entries, including terms in such related fields as physics, medicine, biology, agriculture, pharmacy and mineralogy. Emphasis on chemical nomenclature. A standard chemical dictionary. *Class No:* 540(038)=20

[1045]

HAMPEL, C.A. *and* HAWLEY, G.C., *eds*. **Glossary of chemical terms.** 2nd ed. New York, Van Nostrand Reinhold, 1982. ix [1], 306p. $25.50. ISBN: 0442238711.

First published 1968 (vi, 281p.).

Definitions, by many hands, of *c*.2,000 'essential terms used in chemistry and the process industries' (*Preface to second edition*). 'Oxygen': nearly 1 column; 'Nomenclature': 2/3 column. Includes brief biographical sketches. Cross-references. Intended for students and a wide public. *Class No:* 540(038)=20

[1046]

HAWLEY, G.C., *ed*. **Hawley's condensed chemical dictionary.** 11th ed., revised by N.I. Sax and R.J. Lewis, Sr. New York, etc., Van Nostrand Reinhold, 1987. xv, 1288p. chemical structures. £49.95. ISBN: 0442280971.

First published 1919; 10th ed. 1981.

'A compendium of technical data and descriptive information covering many thousand chemicals' (*Introduction*). 'Uranium': chemical data name, synonym, (CAS Registry number, formula); properties; occurrence; derivatives; forms available; hazards; use. 3 appendices (3. Manufacturers of trademarked products). *Class No:* 540(038)=20

[1047]

HIBBERT, D.B. *and* JAMES, A.M. **Macmillan dictionary of chemistry.** London, Macmillan, 1987. viii, 532p. diagrs., tables. £45; £14.95. ISBN: 0333390814.

5,200 chemical terms and concepts, with emphasis on physical chemistry. Omits extensive lists of organic and inorganic compounds and also physical data of compounds. Entries range in length from 15 words to 1p. Appendices: SI

....(contd.)

units; tables of atomic masses of elements. Intended for students of science, researchers and lay persons.
Class No: 540(038)=20

[1048]
—GODMAN, A. *and* DENNY, R. Cambridge illustrated thesaurus of chemistry. Cambridge Univ. Press, 1985. 256p. illus., diagrs. £5.50.

Groups 2,000 terms and definitions into 31 topics, with cross-references. 550 col. illus. Criticized for omissions. Many definitions are 'either too short to be useful, misleading because they are incomplete, or just wrong' (*British Book News,* April 1986, p.230).
Class No: 540(038)=20

[1049]
—McGraw-Hill dictionary of chemistry. Parker, S.P., *ed.* New York, etc., McGraw-Hill, 1984. [iv], 665p. $32.59. ISBN: 0070454205.

Entries for *c.*10,000 terms in both theoretical and applied chemistry. Terms and definitions were selected from the *McGraw-Hill dictionary of scientific and technical terms* (3rd ed., 1984), for the chemist. *Class No:* 540(038)=20

[1050]
Miall's dictionary of chemistry. Sharp, D.W.A., *ed.* 5th ed. London, Longman, 1981. ix, 501p. diagrs., tables, chemical structures & formulas. £25. ISBN: 0582351529.

First published 1940; 4th ed. edited by L.M. Miall and D.W.A. Sharp, with others.

8 contributors. About 8,000 entries. 'Chromatography': nearly 1 column; 'Molecular distillation': ½ column. Cross-references. Explains terms, 'with brief accounts of important substances, chemical operations and chemists themselves' (*Preface to the fifth edition*). Special attention is paid to industrial processes and the use of particular chemicals. Intended for use in schools, colleges and universities, etc.
Class No: 540(038)=20

Children

[1051]
WERTHEM, J., *and others.* **The Usborne illustrated dictionary of chemistry.** London, Usborne, 1987. 128p. illus., chemical structures. £3.95. ISBN: 0860208214.

Introductory dictionary with useful information arranged thematically so that words are defined in context. 2 indexes: substances, symbols and formulae; terms. Good for schools.
Class No: 540(038)=20-0053

German

[1052]
DE VRIES, L. *and* KOLB, H. **Wörterbuch der Chemie und der chemischen Vefahrenstechnik.** [Dictionary of chemistry and chemical engineering.] 2., überarb. & erw. Aufl. Weinheim/Bergstr., Verlag Chemie; New York & London, Academic Press, 1978-79. 2v. ISBN: 3527256350; 3527257810.

First published 1970-72.

1. *Deutsch-Englisch.* 1978. iv, 849, [1]p.
2. *Englisch-Deutsch.* 1979. ix, 779, [1]p.

V.1 has *c.*80,000 German entry-words, with English equivalents. About 80 categories are applied and German genders given. In v.2 the entry for 'Water' and its

....(contd.)

compounds occupies 3½ columns. Each volume has appendices of conversion tables, abbreviations, etc.
Class No: 540(038)=30

[1053]
—ERNST, R. *and* MORGENSTERN, I.E.V. Dictionary of chemistry, including chemical engineering and fundamental allied science. London, Pitman, 1961-63. 3v.

1: *German-English.* 1961. 727p. 2: *English-German.* 1963. 1056p.

Later reprinted without revision.

V.1 has nearly 45,000 entry-words. 'Ambiguous words are in every case explained by synonyms.' (*Foreword*). About 40 of the many categories apply to chemistry and chemical engineering. British and US usages are differentiated. Many compounds; genders of German nouns are stated. V.2 has 65,000 entry words; abbreviations, p.1047-56. *Class No:* 540(038)=30

[1054]
—FROMHERZ, H. *and* KING, A. Englische und deutsche chemische Fachausdrücke: Ein Leitfaden der Chemie in Englischer und deutscher Sprache. 5 neubarb. Aufl. Weinheim/Bergstrasse, Verlag Chemie GmbH, 1958. xxi, 2-588p.

First published 1934.

Five parts: 1. Elementary chemistry - 2. Inorganic chemistry - 3. Organic chemistry - 4. Physical chemistry - 5. Structure of matter and chemical forces. English text and German translation on facing pages, for easy comparison. Over 10,000 technical terms, italicized in the text, are indexed in each language. English and German abbreviations, p.425-7. *Class No:* 540(038)=30

[1055]
—NEVILLE, H.H., *and others.* A New German-English dictionary for chemists. London, Blackie, 1964. xviii, 330p.

Lists *c.*42,000 terms and equivalents, including a large selection of non-technical words. Differentiates between British and US usage. Abbreviations included. Adequate cross-references. Small but clear type. Considered (*Chemistry & industry,* 7 November 1964, p.1874) to be an advance on Patterson; includes terms in general technology and chemical engineering. *Class No:* 540(38)=30

[1056]
Dictionary of chemistry and chemical technology: English-German. Gross, H., *ed.* 4th ed. Amsterdam, Elsevier; Berlin, VEB Verlag Technik, 1989. 752p. £97.40. ISBN: 0444988637.

English into German only. About 60,000 terms. Coverage is claimed for all branches of chemistry, chemical technology and related fields of science.
Class No: 540(038)=30

[1057]
—Fachwörterbuch chemie und chemische: Deutsch-Englisch. 4th ed. Berlin; Verlag Alexandre Hatier, 1992. 760p. ISBN: 3861170353.

German into English only. About 62,000 terms. The English-German volume [see above] appeared in 1989.
Class No: 540(038)=30

[1058]
PATTERSON, A.M. Patterson's German-English dictonary for chemists. Cox, J. *and* Condoyannis, G., *eds*. 4th ed. New York, etc., Wiley, 1992. liii,890p. £52. ISBN: 0471669911.

First published 1917; 3rd ed. 1950.

About 65,000 German terms with their English equivalents. A welcome new edition (after 40 years) of a standard dictionary, with increased emphasis on recent developments (*e.g.* biochemistry, food processing, the environment). Excellent editorial guidance on use. 'Highly recommended ... Welcome back, old friend' (*Choice*, July/ August 1992, p.1659). *Class No:* 540(038)=30

[1059]
WENSKE, G. Wörterbuch chemie: Englisch/Deutsch. [Dictionary of chemistry: English/German.] Weinheim, VCH, 1992. xiv,1546p. £121. (*Parat.*) ISBN: 3527264280. ISSN: 09306862.

About 50,000 entries in each half. Genders given and obsolete terms and phrases are indicated. Preference is for American rather than British spelling. A major dictionary. *Class No:* 540(038)=30

Afrikaans
[1060]
SUD-AFRIKAANSE AKADEMIE VIR VETENSKAP EN KUNS. Chemiewoordeboek. [Chemical dictionary. Edited by the Vaktaalburo and a Committee for Chemical Terms.] Voortrekkerpers, 1968. 418p.

About 15,000 terms: English-Afrikaans, p.1-196; Afrikaans-English, p.197-395. Equivalents; no genders. Appended list of institutional abbreviations. *Class No:* 540(038)=393.6

French
[1061]
DUVAL, C *and* DUVAL, R. Dictionnaire de la chimie et de ses applications. Paris, Technique et Documentation, 1978. 1087p.

The leading French chemical dictionary. *Class No:* 540(038)=40

[1062]
PATTERSON, A.M. A French-English dictionary for chemists. 2nd ed. New York, Wiley, 1954. 476p.

First published 1921.

'The vocabulary has grown from 35,000 terms to an estimated 42,000, in spite of the fact that many terms having the same spelling and meaning in French and English have been omitted in order to save time and space' (*Preface* to 2nd ed). Some of the technical vocabularies consulted are mentioned in the Preface. Well spoken of. *Class No:* 540(038)=40

[1063]
—CORNUBERT, R. Dictionnaire chimique: anglais-français; mots et locutions fréquemment rencontrés dans les textes anglais et américains. 6th ed. Paris, Dunod, 1977. 215p.

5th ed. 1970.

Includes nearly 6,000 entry-words. Appendices of abbreviations, weights and measures. *Class No:* 540(038)=40

[1064]
—FROMHERZ, H. *and* KING, A. French-English chemical terminology: an introduction to chemistry in French and English. Translated by J. Jousset. Weinheim/Bergstrasse, Verlag-Chemie GmbH; Paris, Gauthier-Villars, 1968. xix, 2-561p.

Similar in layout to their *Englische und deutsche chemische Fachausdrücke* (5th ed. 1958). Index of French terms (*c.*8,000), p.463-508; index of English terms, p.509-61. Intended for students, research scientists, industrial chemists and translators. *Class No:* 540(038)=40

Spanish
[1065]
Diccionario de química y de productos químicos. Español-inglés, inglés-español. 2nd. ed. Barcelona, Omega, 1975. 1017p.

First published 1959.

The 2nd ed. is based on *The Condensed chemical industry* (then in its 8th ed., 1966, revised by G.G. Hawley) and translated by G. Ramos.

The 1959 ed. (1080p. + 130p.) has *c.*30,000 entries. The Spanish-English section (p.29-1080) gives explanations (properties; analysis; uses), with cross-references, whereas the 130-p. English-Spanish section acts as an index of entry-words. *Class No:* 540(038)=60

Russian
[1066]
CALLAHAM, L.I. Russian-English chemical and polytechnical dictionary. 3rd ed. New York, etc., Wiley, 1975. xxviii, 852p. £81. ISBN: 0471129984.

First published 1947; 2nd ed., 1962.

'A technical as well as a chemical dictionary. Inorganic and organic chemistry, chemical technology, and chemical engineering are naturally given the most complete coverage. Mineralogy, metallurgy, mining and geology, general engineering, machinery and mechanics, electrical engineering, pharmacy and botany are also comprehensively covered' (*Preface*, 1st ed.). About 100,000 Russian entries and sub-entries. Terms are categorised, genders given; abbreviations, prefixes and common Russian word-endings included. Good typography for cyrillic. Particularly helpful in the fields of pure and applied chemistry. *Class No:* 540(038)=82

[1067]
HOSEH, M. *and* HOSEH, M.I. Russian-English dictionary of chemistry and chemical technology. New York, Reinhold; London, Chapman & Hall, 1964. xiii, [1], 522p.

About 40,000 entries, many of them compounds. Multiword terms are grouped under the keyword, with the modifier following, A-Z. Abbreviations are included. Categories (and sometimes explanations) appear italicized, in parentheses. Genders of Russian nouns given, but not stress. Well produced on good paper, in clear cyrillic type. Compared with Callaham (also with 40,000 entries, but with compounds subsumed), Hoseh gives more strictly chemical meanings and not the wider annotations that make Callaham such a valuable general technical Russian-English dictionary. *Class No:* 540(038)=82

[1068]
—CARPOVICH, E.A. *and* CARPOVICH, V.V. The Russian-English chemical dictionary. 2nd ed. New York, Technical Dictionaries Co., 1963. 352p.

First published 1961.

About 24,000 Russian entry-words. Coverage extends not only to chemistry and chemical engineering but also to fuels, foods and pharmacology. Includes many formulas and adopts US nomenclature. Good layout but considered less reliable than Callaham (*qv*) by Neiswender (*Guide to Russian reference and language aids*, p.29).

Class No: 540(038)=82

[1069]
—DIVSIC, L.B. *and* MIKHAILOV, V.V., *eds*. Anglo-russkiĭ Khimiko-tekhnologiches Kiĭ. 7th ed. Moscow, 1971. 756p.

English-Russian dictionary of pure and applied chemistry. 30,000 English entry-words, with Russian equivalents.

Class No: 540(038)=82

Arabic

[1070]
LEAGUE OF ARAB STATES. Education, Culture and Sciences Organization. Permanent Bureau of Arabisation. **Lexicon of chemistry.** [Lexique de chimie.] Rabat, Bureau Permanent de Coordination de l'Arabisation, 1971. [1] xv, 141p.

985 Arabic terms, with English and French equivalents, plus indexes to numbered Arabic entries.

Class No: 540(038)=927

Oriental

[1071]
Chinese-English-Japanese glossary of chemical terms. Hong Kong, Joint Publishing Co., 1980. 611p. ISBN: 9620400263.

About 10,000 Chinese-character entry-words, with English and Japanese equivalents and indexes.

Class No: 540(038)=95

Japanese

[1072]
JAPAN: Ministry of Education. Japanese-English scientific terms: chemistry. 4th ed. Tokyo, Chemical Society of Japan, 1987. 689p.

Part 1: Japanese-English (romanised form; Japanese character; English equivalent) - Part 2: English-Japanese (English-Japanese; romanised form). About 8,000 terms.

Class No: 540(038)=956

Reviews & Abstracts

[1073]
British abstracts. London, Bureau of Abstracts, 1945-53. pts.A; B (12pa.); C (4pa.).

Previous titles: *Abstracts of chemical papers.* A: *Pure chemistry* (1924-25); *British chemical abstracts* (1926-37); *British chemical and physiological abstracts* (1938-44). As *British abstracts,* totalled *c.*50,000 abstracts pa. Of value on BIOS and FIAT reports on World War II German chemical industry (see *Guide to reference material* (3rd ed.), v.1, p.58).

.... (contd.)

British abstracts, now superseded as an overall service by *Chemical abstracts,* is being continued as follows: A I-II, - by lists of papers in pure chemistry in *Current chemical papers* (Chemical Society, 1954-69). A III, - by *British abstracts of medical science,* 1954-56; subsequently *International abstracts of biological sciences,* 1956-*c.*82; later *Current awareness in the biological sciences,* 1983- (*qv*). B I-II, - by abstracts in *Journal of applied chemistry,* 1954-71. B III, - by abstracts in *Journal of the science of food and agriculture,* 1954-. C, - by *Analytical abstracts,* 1954- (*qv*). *Class No:* 540(048)

[1074]
Bulletin signalétique. 170. Chemie. Paris, Centre National de la Recherche Scientifique, 1940-. Monthly.

About 35,000 abstracts and references pa. Sections: A. Chimie générale et chimie physique - B. Chimie minérale - C. Chimie analytique - D. Chimie organique: théories - E. Chimie organique: préparations et propriétés - F. Évolution moléculaire; Origine de la vie. Monthly and annual subject and author indexes. *PASCAL Folio;* F17. *Chimie générale* and F16. *Chimie analytique minérale et organique* take over from 1984. Online. *Class No:* 540(048)

[1075]
Chemical abstracts. Columbus, Ohio, American Chemical Society, 1907-. v.1 Weekly. $8,500pa. ISSN: 00092258.

Chemical abstracts provides over 500,000 abstracts pa., taken from over 14,000 journals worldwide, plus patents, conference proceedings, government research reports, books and dissertations. Coverage in 80 sections. Odd-numbered weekly issues feature organic and biochemical sections; even-numbered issues cover applied chemistry, chemical engineering. Weekly keyword, patent and author indexes; half-yearly author, general subject, chemical substance, formula and patent indexes. The earlier 10-yearly cumulative indexes (1907-56) became 5-yearly 1957-. (The 7th Collective index, for 1962-66, occupied 24v., with extremely detailed indexing, - *c.*650 entry-words per 1,000 words in the abstracts).

Chemical abstracts is also available in 5 fortnightly index groupings (1963-): *Applied chemistry and chemical engineering* sections (CAAS), *Biochemical sections* (CABS), *Macramolecular sections* (CAMS), *Organic chemistry* sections (CAOS), and *Physical and analytical chemistry* sections (CAPS). (No volume indexes are included in section subscriptions). Other features available: *Index guide; Ring index handbook;* and a chemical-compound *Registry* (1974), assigning CAS Registry Numbers to unique compounds. *Using chemical abstracts* comprises a cassette and workbook.

CA selects offers a current-awareness service, reproducing 100-200 CA abstracts per bi-weekly issue in 36 areas (*e.g. Chemical hazards, Forensic chemistry, Liquid crystals,* and *New books in chemistry*). *Chemical Abstracts Service source index* (CASSI) lists periodicals abstracted. Other titles include *Chemical titles.*

Chemical abstracts is the most widely used abstracting service in English that covers the whole field of chemistry, pure and applied, and including biochemistry. All science libraries should provide at least some parts of the Chemical Abstracts Service.

CA SEARCH corresponds as a database to *Chemical abstracts. Class No:* 540(048)

[1076]

—Chemical titles. Columbus, Ohio, Chemical Abstracts Service, 1961-. Bi-weekly. $320 pa.

A current awareness service listing c.5,000 research papers from some 750 periodicals worldwide on pure and applied chemistry. A keyword-in-context index precedes a 'Bibliography' of the 'contents tables' of journals A-Z. Appended author index. A list of journals covered appears on the inside cover of each issue, with CODEN and abbreviated titles. *Class No:* 540(048)

[1077]

Chemischer Informationsdienst. Weinheim/Bergstrasse, Verlag Chemie GmbH, 1970-. 2 series, ea. weekly. ISSN: 00012975.

An abstracting service in 2 parts: *Organische Chemie* and *Anorganische und physikalische Chemie,* together covering c.40,000 research papers in organic, inorganic and physical chemistry. Systematically arranged. Abstracts appear about 2 months after the original publication. Replaces *Chemischer Zentralblatt (qv). Class No:* 540(048)

[1078]

—Chemischer Zentralblatt. Berlin, Akademie Verlag GmbH; Weinheim/Bergstrasse, Verlag Chemie, 1856-1969. Weekly.

Initially *Pharmaceutisches Centralblatt* (1830-49); subsequently *Chemisch-pharmaceutisches Centralblatt* (1850-55).

In its heyday it carried c.100,00 indicative abstracts pa., - fuller than *Chemical abstracts,* its rival in world coverage of chemistry and allied subjects, particularly for pre-1937 material. Available on microfilm from Princeton Microfilm Corp., Princeton, N.J. *Class No:* 540(048)

[1079]

Current abstracts of chemistry and index chemicus. Philadelphia, Pa., ISI/Institute for Scientific Information, 1970-. Weekly. chemical structures. ISSN: 0161455x.

Current abstracts in chemistry absorbed *Index chemicus (1960-69)* in 1970.

Concentrates on c.100 key research journals (c.25 titles per issue), chemical structure accompanying each abstract. The synthesis, isolation and identification of new compounds are thus reported. *Index chemicus molecular formula index* (green paper insert) has sections: molecular formula; author; subject; biological activity; corporate; and an alert to labelled compounds, plus corporate index. Back cover lists the journals covered each week. Quarterly and annual cumulated indexes. A 22-year microform cumulation contains the abstracts, 1960-81. The service can thus be used for both current awareness and retrospective search. 'This is one of the most important keys to the recent literature of organic chemistry' (Katz, B. and L.S. *Magazines for libraries* (5th ed. 1986), p.13). *Class No:* 540(048)

[1080]

Current contents: Physical, chemical and earth sciences. Philadelphia, Pa., Institute for Scientific Information, 1979-. v.19-. Weekly.

Previously as *Current contents: Physical and chemical sciences* (formed by the merger of *Current contents: Physical sciences* and *Current contents: Chemical sciences*), 1961-78.

Reproduces contents pages of more than 1,000 journals in the fields concerned. Each weekly number covers c.100 journal issues. Online on CURRENT CONTENTS SEARCH, and available on diskette. *Class No:* 540(048)

[1081]

Referativnyĭ zhurnal. Khimiya. Moscow, Izdat. Akademii Nauk SSSR, 1953-. 24pa. ISSN: 04862325.

About 100,000 abstracts pa. from world literature. Sections (in Russian): General problems of chemistry - Physical chemistry - Inorganic chemistry - Analytical chemistry - Laboratory equipment - Chemistry of earth and space - Organic chemistry - General problems of chemical technology - Corrosion and corrosion protection - Technology of inorganic substances - Chemistry of processing wood - Natural fuels and natural gases - Chemistry and technology of high polymers. Author index per issue; annual author, subject, patent and formula indexes. Also available in separate parts. *Class No:* 540(048)

Periodicals

[1082]

BRITISH LIBRARY. Science Reference Library. **Periodicals on chemistry held by the Science Reference Library.** Compiled by C.L. de Hamel. 2nd ed. London, Science Reference Library, 1983. 73p. £5. ISBN: 0712307001.

About 2,400 titles, in A-Z, word-by-word sequence. Reports, bulletins, etc., are listed under names of issuing bodies. Entries give starting dates of holdings and SRL classification number. *Class No:* 540(051)

[1083]

—ROYAL SOCIETY OF CHEMISTRY. Periodicals in the library of the Royal Society of Chemistry. 4th ed. London, the Society, 1986. [i], [i], 82p. £2.50. ISBN: 0851864694.

Lists c.1,600 title entries, A-Z, with cross-references. Includes amendments up to March 1986. *Class No:* 540(051)

Bibliographies

[1084]

Chemical Abstracts Service [CASSI] source index, 1907-1989 cumulative. Columbus, Ohio, Chemical Abstracts Service, 1990. 2v. (A-L,M-Z); Quarterly supplements, cumulated annually.

Formerly as *Chemical abstracts list of periodicals.*

About 125,000 entries, drawn from 65,000 serials and other sources. Data include former titles, language, history, frequency, price, publisher. CASSI is also available in computer readable form. Online (DIALOG). *Class No:* 540(051)(01)

Progress Reports

[1085]

Annual reports on the progress of chemistry. London, Royal Society of Chemistry, 1905-. 3 sections, each annual. ISSN: 02601818; 00693030; 06701826.

A. *Inorganic chemistry.* £64. B. *Organic chemistry.* £74. C. *Physical chemistry.* £62.

Each annual section volume comprises about a dozen well-documented, authoritative contributions, with author index. Provides for the general reader critical coverage of the significant advances in the major areas of chemistry during the previous year. *Class No:* 540(055)

[1086]
ROYAL SOCIETY OF CHEMISTRY. Specialist periodical reports: reviews of the chemical literature. London, the Society [formerly Chemical Society], 1967-. 45 series, usually annual.

Alicyclic chemistry.
Aliphatic, alicyclic and saturated heterocyclic chemistry.
Aliphatic and related natural product chemistry.
Aliphatic chemistry.
The alkaloids.
Amino-acids, peptides and proteins.
Aromatic and heteroaromatic chemistry.
Biosynthesis.
Carbohydrate chemistry. 2 pts.
Catalysis.
Chemical physics of solids and their surfaces.
Chemical thermodynamics.
Colloid science.
Dielectric and related molecular processes.
Electrochemistry.
Electron spin resonance.
Electronic structure and magnetism of inorganic compounds.
Environmental chemistry.
Fluorocarbon and related chemistry.
Foreign compound metabolism in mammals.
Gas Kinetics and energy transfer.
General and synthetic methods.
Heterocyclic chemistry.
Inorganic biochemistry.
Inorganic chemistry of the main group elements.
Inorganic chemistry of the transition elements.
Inorganic reaction mechanisms.
Macromolecular chemistry.
Mass spectroscopy.
Molecular spectroscopy.
Molecular structure by diffraction methods.
Nuclear magnetic resonance.
Organic compounds of sulphur, selenium and tellurium.
Organometallic chemistry.
Organophosphorous chemistry.
Photochemistry.
Radiochemistry.
Reaction kinetics.
Saturated hetrocyclic chemistry.
Spectroscopic properties of inorganic and organometallic compounds.
Statistical mechanics.
Surface and defect properties of solids.
Terpenoids and steroids.
Theoretical chemistry.

The reports are critical in-depth accounts of the progress in specific areas by acknowledged authorities, normally appearing within 12 months after the period of literature coverage. *Class No:* 540(055)

[1087]
Survey of progress in chemistry. New York, London, Etc., Academic Press, 1974—. Annual. illus., graphs, chemical structures. ISBN: 0126105103.

V.10, 1983 (xiii, [1], 266p.) has 4 documented articles by 5 contributors. 'Catalysis from the point of view of surface chemistry' has 2¼p. of references; 'Metal clusters and metal surfaces', 6p. of references. Analytical subject index, p.259-66. *Class No:* 540(055)

Yearbooks & Directories

[1088]
ROYAL SOCIETY OF CHEMISTRY. Directory of consulting practices in chemistry and related subjects, 1986. London, the Society, [1986?]. 162p.

Part 1: Guide to services (Abrasives ... Yeast and yeast products) - 2. Subject guide - 3. Alphabetical index of consulting practices. Coverage includes related fields such as physics, engineering, computers, biological sciences, medicine, metallurgy. Consulting practices are those 'in which Professional Members of the Royal Society of Chemistry are principals'. *Class No:* 540(058)

Tables & Data Books

[1089]
CRC handbook of chemistry and physics. Lide, D.R., *ed.* 71st ed. Boca Raton, Florida, CRC Press, 1990. v.p. diagrs., charts, tables, chemical structures. ISBN: 0849304711.

First published 1914.

This new edition of a classic reference source (the 'Rubber Bible') has been substantially reorganized by a new editor into an arrangement which facilitates browsing. About 20% of the tables are completely new/updated. Includes more tables on health and the environment than earlier versions. An invaluabe mine of information which 'remains the essential reference tool for physical data, required in virtually all libraries' (*Choice,* September 1991, p.56).

New editions tend to appear in September.

Class No: 540(083)

[1090]
—CRC handbook of chemistry and physics: 1st student edition. Weast, R.C., *ed.* Boca Raton, Florida, CRC Press, 1988. v.p. diagrs., charts, tables, chemical structures. $29.95. ISBN: 0849307406.

The student ed. includes about 2/3 of the full handbook, omitting much of the mathematics, physics, nutrition and earth sciences. 'All libraries and students would be better served by any recently superseded regular edition' (*Choice,* June 1988, p.1532). *Class No:* 540(083)

[1091]
JAMES, A.M. *and* LORD, M.P. Macmillan's chemical and physical data. London, Macmillan, 1992. 448p. tables. £35. ISBN: 0333511670.

15 subject areas. About 430 tables, most with references for further information and/or additional data.

Class No: 540(083)

[1092]
KAYE, G.W.C. *and* LABY, T.H., *comps.* Tables of physical and chemical constants, and some mathematical functions. 15th ed. London & New York, Longman, 1986. [xiii], 477p. graphs, tables. £28.99; $52.95. ISBN: 0582463548.

59 contributors. 4 main sections; section 2. Chemistry (The elements ... Miscellaneous data). Includes statistical methods for the treatment of experimental data. Analytical index. Subsection references. All tabulated values appear in SI units. Essential for any reference library.

Class No: 540(083)

[1093]

LANGE, N.A., *ed*. **Lange's handbook of chemistry.** 13th ed., by J.A. Dean. New York, etc., McGraw-Hill, 1985. v.p. tables, chemical structure. £65. ISBN: 0070161925.

First published 1934.

11 sections, mainly tables: 1. Mathematics - 2. General information and conversion tables - 3. Atomic and nuclear structure - 4. Inorganic chemistry - 5. Analytical chemistry - 6. Electrochemistry - 7. Organic chemistry - 8. Spectroscopy - 9. Thermodynamic properties - 10. Physical properties - 11. Miscellaneous (*e.g.* Glossary). 38p. detailed analytical index, including formulas. Section 7 (153p.) includes description of 7,000 organic compounds. A classic tool for chemists.

The 14th ed. was published in 1992 (ISBN: 0070161941. $79.50). The review in *Choice* (December 1992, p.603) noted numerous changes, including SI units, updated IUPAC nomenclature, new information on laboratory methods, expanded information in many areas (*e.g.* spectroscopy; bond lengths and strengths; potentials of organic and inorganic compounds, etc.) and re-organization of the contents, finding the revised edition overall an 'excellent reference'. *Class No:* 540(083)

[1094]

SHUGAR, G.J., *and others, eds*. **Chemical technicians' ready reference handbook.** 3rd ed. New York, etc., McGraw-Hill, 1990. 864p. illus., diagrs., tables. ISBN: 007037183x.

First published 1973 (463p.). 2nd ed. 1981.

Emphasis on step-by-step procedures. In explaining frequently-used laboratory techniques, adds a new chapter on spectroscopy techniques (*e.g.* NMR, infrared). New sections stress laboratory safety. *Class No:* 540(083)

[1095]

—SHUGAR, G.J. *and* DEAN, J.A. The Chemist's ready reference handbook. New York, McGraw-Hill, 1990. xiv,640p. illus., diagrs., tables. £63.50. ISBN: 0070571783.

Concentrates on the actual questions and problems faced in the chemical laboratory. While deriving relevant material from Shugar's *Chemical technicians' ready reference handbook,* a major portion is new, with sections on chromatography, thermal methods, determination of physical properties and much else, plus practical checklists and trouble-shooting hints. 288 illus. *Class No:* 540(083)

Libraries

[1096]

SWEENEY, J.M. A Guide to sources of physical property data held by the Science Reference and Information Service. London, British Library, 1988. iv,128p. ISBN: 0712307508.

Sources of data, largely chemical. Classified arrangement. Entries include author, title, date of publication and brief annotation. *Class No:* 540(083):061:026/027

Nomenclatures

[1097]

CAHN, R.S. *and* **DERMER, O. Introduction to chemical nomenclature.** 5th ed. London & Boston, Mass., Butterworths, 1979. [vii], 200p. tables, formulas. ISBN: 0408106085.

First published 1959.

....(contd.)

9 documented sections: 1. The development of chemical nomenclature - 2. Inorganic - 3/6. Organic - 7. Stereovisomesism - 8. Natural products - 9. Miscellaneous nomenclature. Appendix: 'Important recent changes from IUPAC Nomenclature by *Chemical Abstracts Indexes'*. Analytical subject index, p.191-200. *Class No:* 540(083.72)

[1098]

INTERNATIONAL UNION OF PURE AND APPLIED CHEMISTRY. Compendium of chemical technology: IUPAC recommendations. Compiled by V. Gold, and others. Oxford, Blackwell Scientific, for IUPAC, 1987. viii, 456p. diagrs., chemical structures. £34.95. ISBN: 0632017678.

Definitions of *c*.3,000 terms ('Absolute zero' ... 'Z value'), with references to sources 'published up to the end of 1985 from the Physical, Inorganic, Organic, Macromolecular and Analytical Chemistry Divisions of IUPAC' (*Preface*). Source documents, p.454-6. Chemical structures are a feature. *Class No:* 540(083.72)

[1099]

—LEES, R. *and* SMITH, A.F. Chemical nomenclature usage. Chichester, E. Horwood for the Laboratory of the Government Chemist, 1983. 172p.

With 22 contributors. 18 chapters (*e.g.* 1. Problems of use of chemical nomenclature in specialized areas; 8. The use of chemical nomenclature in *Chemical abstracts*; 11. Biochemical nomenclature; 18. Nomenclature, the way ahead). Further reading, p.164-8. Index, p.169-72. *Class No:* 540(083.72)

Films

[1100]

ROYAL SOCIETY OF CHEMISTRY. Index of chemistry films, 1984. 7th ed. London, the Society, 1984. 480p. £21. ISBN: 0851864996.

First published 1959.

Comprehensive list of *c*.2,500 films, filmstrips, filmloops, videotapes, slides, soundtapes and overhead transparancies. 7 parts: 1. Physical chemistry - 2. Inorganic chemistry - 3. Organic chemistry - 4. Energy - 5. Biochemical topics - 6. Applied chemistry - 7. Miscellaneous topics. A-Z indexes. Classified index precedes parts. Appended addresses of distributors. *Class No:* 540(084.122)

Patents

[1101]

GRUBB, P.W. Patents in chemistry and biotechnolgy. 2nd ed. Oxford, Clarendon Press, 1986. xii, 335p. £40; £20. ISBN: 019855222x, HB; 0198552211, PB.

First published 1982 as *Patents for chemists*.

A guide to the lay reader of the patent systems of the UK, US and Europe. Contains sections on general principles of patenting, inventions in chemistry, pharmaceuticals and biotechnology, patenting for the chemical inventor and the politics of patents. Also has a glossary of terms and list of cases. *Class No:* 540(088.8)

Chronologies

[1102]
BALLENTYNE, D.W.G. *and* LOVETT, D.R. **A dictionary of named effects and laws in chemistry, physics and mathematics.** 4th ed. London, Chapman & Hall, 1980. viii,346p. diagrs., tables, chemical structures. £17.95. ISBN: 0412223902.

First published 1958.

Records several thousand events, in year order, under broad disciplines. Thus, for 1687: astronomy (3 entries), physics, chemistry, earth sciences, mathematics, meteorology. The main aim is 'to provide, in chronologic order, concise, clear summaries on events or accomplishments contributing to the development of Western science' (*Preface*). *Class No:* 540(090)

[1103]
NEUFELDT, S. **Chronologie Chemie, 1800-1970.** Weinheim & New York, Verlag Chemie, 1977. vi, 359p. ports., tables. DM78. ISBN: 3527256768.

Includes 19 portraits, 4 tables. *Class No:* 540(090)

Histories

[1104]
BROCK, W.H. **The Fontana history of chemistry.** London, Fontana, 1992. xxiii,744p. illus., tables. £8.99. (*Fontana history of science.*) ISBN: 0006861733.

16 chapters, subdivided: 1. On the nature of the universe and the hermetic museum ... 3. Elements of chemistry ... 6. Chemical method ... 12. The chemical news ... 16. At the sign of the hexagon. Extensive 'Bibliographical Essay', chapter-by-chapter, p.674-714, a useful feature. Detailed, analytical index, p.715-44. An early volume in a new series, intended for non-specialists. *Class No:* 540(091)

[1105]
FINDLAY, A. **A Hundred years of chemistry.** 3rd ed., revised by T.I. Williams. London: Duckworth; Atlantic Highlands, N.J., Humanities Press, 1965. 335p.

First published 1937 (reviewing the period 1825-1925).

The 3rd ed. has 13 well-footnoted chapters (2. The development of organic chemistry, 1835-65 ... 8. The rise and development of the fire chemicals industry ... 10. The discovery of new elements. The rare gases ... 13. The development of industrial chemistry). Appendix: Biographical notes, p.300-24. Detailed analytical index. 'Interesting and readable, but in spite of clever revision with the help of Dr. T.I. Williams, the impression remains that the hundred years of chemistry reviewed is 1825-1925, and not 1860-1960' (*Pharmaceutical journal,* v.195(5312), 21 August 1965, p.187). *Class No:* 540(091)

[1106]
HUDSON, J. **The History of chemistry.** London, Macmillan, 1992. 320p. £40. ISBN: 0333535502.

Includes results of recent research into the history of chemistry. *Class No:* 540(091)

[1107]
LEICESTER, H.M. *and* KLICKSTEIN, H.S. **A source book in chemistry, 1400-1900.** New York, etc., Wiley, 1952. xvi, 554p. (*Source books in the history of science.*)

Selections of writings of famous chemists; restricted to material that illustrates the development of chemical theory

....(contd.)
only. Each section, documented, contains explanatory matter on the period concerned, followed by material on the chemists. *Class No:* 540(091)

[1108]
LEICESTER, H.M., *ed.* **A Source book in chemistry, 1900-1950.** Cambridge, Mass., Harvard Univ. Press, 1968. xvii, [1], 408p.

A collection of chemical papers that have shaped the structures of the various sciences (*General editor's preface*). Four parts (subdivided); 1. Techniques - 2. General and physical chemistry - 3. Organic chemistry - 4. Biochemistry. Bibliography of 63 biographies, p.393-402. Index, p.403-8. *Class No:* 540(091)

[1109]
Nobel lectures including presentation speeches and laureates' biographies. Chemistry. Amsterdam, etc., Elsevier, for the Nobel Foundation, 1964-72. 4v. illus., diagrs., tables.

1. *1901-1921.* 1966. xii, 411p.
2. *1922-1941.* 1966. xii, 508p.
3. *1941-62.* 1964. xiii, 712p.
4. *1963-70.* 1972. x, 271p.

The 1963-70 volume contains 8 lectures (*e.g.* 'Recent advances in the chemistry of natural products', by R.B. Woodward, p.100-121, with 8 references), plus biography (p.122-3). 11 biographies in all. Name index; subject index; index of biographies. The Nobel lectures subsequently appear in the annual publication of *Les prix Nobel.* *Class No:* 540(091)

[1110]
PARTINGTON, J.R. **A History of chemistry.** London, Macmillan, 1961-70. v.1 (pt.1), v.2-4. illus., ports.

Enlargement of his *A short history of chemistry* (2nd ed. London, Macmillan, 1948. xiv, 386p.).

V.1 (pt.1. 1970. xiv, 370p.) has 20 chapters (1. Early Greek philosphy ... 20. The Quabalah); name and subject indexes. Final stages of the manuscript were left incomplete when the author died, in 1965. V.2-4 (1961-64) concern the period from 1500 to the 20th century and assess achievements and influence of individual chemists. Practically the whole text is based on primary sources and is profusely footnoted. V.4, on the development of chemistry in the 19th and part of the 20th century, is in 5 parts (electrochemistry; physics, organic and inorganic chemistry; and radioactivity and atomic structure). Each volume has name and subject indexes. Publication of v.1 (pt.2) has been abandoned. *Class No:* 540(091)

[1111]
RUSSELL, C.A., *ed.* **Recent developments in the history of chemistry.** London, Royal Society of Chemistry, 1965. x, 333p. ISBN: 0851869173.

12 chapters: 1. Introduction - 2. Chemical biographies - 3. Chemical education and chemical institutions - 4. Chemistry to 1800 - 5. General and inorganic chemistry - 6. Organic chemistry - 7. Physical chemistry - 8. Analytical chemistry - 9. Biochemistry - 10. Instruments and apparatus - 11. Industrial chemistry - 12. Chemistry by location in Western and Central Europe (with 187 footnotes). Appendices: 1. Periodicals for the history of chemistry: 2. Some useful addresses. Author index; subject index - people; subject index - theories. *Class No:* 540(091)

Bibliographies

[1112]
PENNSYLVANIA UNIVERSITY. Catalog of the Edgar Fahs Smith Memorial Collection in the History of Chemistry. Boston, Mass., G.K. Hall, 1960. 524p. ISBN: 0816105227.

Photolithographic reproduction of the author-title catalogue of 11,000 cards. This collection, began by Edgar Fahs Smith, Professor of Chemistry and Provost of the University of Pennsylvania, comprises more than 13,000 volumes, prints and manuscripts. *Class No:* 540(091)(01)

Biographies

[1113]
Chemistry: Who's who. Agarwal, R.K., *ed*. Sahibabad, India, Chemico Publishing Co, 1987. 336p. ports. $60.

Over 200 entries on chemists in 14 countries: Austria, Brazil, China, Denmark, Egypt, Greece, Hong Kong, Hungary, India, Japan, Saudi Arabia, Spain, UAE, Yugoslavia. Entries include a list of biographers' publications. Index, p.335-6. *Class No:* 540(092)

[1114]
FARBER, E. Great chemists. New York & London, Interscience, 1961. xxvi, 1642p. illus., ports.

Nearly 100 contributors; more than biographies. 59 sections (1. Babylonian chemists - 2. Arabic chemists ... 9. Robert Boyle ... 20. Antoine Laurent Lavoisier ... 34. Leopold Gmelin ... 52. Dmitrii Ivanovitch Mendeleev ... 58. American chemists at the turn of the century - 59. Otto Wallach). Some sections have brief references. Index, p.1619-42. *Class No:* 540(092)

[1115]
—ABBOTT, D., *ed*. The Biographical dictionary of scientists: chemists. London, Blond Education, 1983. [vi], 203p. £8.50. ISBN: 0058470008.

Contains *c*.200 readable biographies (*e.g.* Bunsen, Robert Wilhelm: 1½ columns). The series is designed for school and first-year University students who need quick reference information. Criteria for inclusion are not stated. *Class No:* 540(092)

[1116]
—FARBER, E. Nobel Prize Winners in chemistry, 1901-1961. Rev. ed. London, New York, etc., Abelard Schumann, 1963. ix, 341p. ports., diagrs.

Contains 65 biographical sketches, with small portraits. It quotes from the prize-winning lecture and assesses its consequence in theory and practice. Appended bibliography includes publications on the winners concerned. *Class No:* 540(092)

[1117]
—SMITH, H.M. Torchbearers of chemistry: portraits and brief biographies of scientists who have contributed to the making of modern chemistry. New York, Academic Press, 1949. 270p. ports.

223 biographical sketches (from the Middle Ages to 1900), and a helpful bibliography of biographies by R.E. Oesper. *Class No:* 540(092)

Great Britain

[1118]
CAMPBELL, W.A. *and* GREENWOOD, N.N. Contemporary British scientists. Londona, Taylor & Francis, 1971. [viii], 286p. illus., ports.

Summarizes the life and achievements of *c*.130 distinguished contemporary British chemists. *Class No:* 540(092)(410)

USA

[1119]
MILES, W.B., *ed*. American chemists and chemical engineers. Washington, American Chemical Society, 1976. x, [1], 544p.

517 biographical sketches of deceased chemists and chemical engineers, including editors of journals, writers of influential texts and founders of chemical companies. 162 contributors. Entries, signed, average *c*.700 words; references appended. *Class No:* 540(092)(73)

[1120]
—American men and women of science: chemistry. Press, J. C., *ed*. New York, Bowker, 1977. 1290p.

Brief data on 23,232 scientists in analytical chemistry, biochemistry, chemistry geochemistry, inorganic chemistry, organic chemistry and physical chemistry. *Class No:* 540(092)(73)

20th Century

[1121]
GINGERICH, O. The Physical sciences in the twentieth century. New York, Scribner, 1989. 306p. illus. $75. (*Album of science.*) ISBN: 0684154978.

Covers physics, chemistry, astronomy, geology and computers. Includes brief outline of major discoveries, principles and participants within each discipline and numerous captioned illustrations, about 400 in all. Bibliography of science histories; Nobel prize winners in physics and chemistry to 1988. A 'well-conceived and handsomly-executed volume'. (*Choice,* January, 1990, p.826). *Class No:* 540"19"

Writing & Lecturing

[1122]
DODD, J.S., *ed*. The ACS style guide: a manual for authors and editors. Washington, American Chemical Society, 1986. 264p. diagrs., tables. ISBN: 0841209170.

Contents: The scientific paper - Grammar, style and usage - Illustrations and tables - Copyright and permissions - Manuscripts submissions in machine-readable form - The literal literature: becoming part of it and using it - Making effective oral presentations. 'This is a true *vade mecum,* in the best sense of the word' (*New technical books,* January 1986, entry 831).

The ACS also publishes *Handbook for authors of papers in American Chemical Society publications* (Washington, 1978, vi, 122p). *Class No:* 540:001.81

Information Management

[1123]

ASH, J., *and others*. **Communication, storage and retrieval of chemical information.** Chichester, E. Horwood; New York, Halsted Press: Wiley, 1985. 297p. ISBN: 0853125716.

11 chapters (1. The information need of chemists - 2. Classical methods of communicating non-structural chemical information - 3. Online access to the chemical literature - 4. Databanks - 5. Methods of structure representation and registration ... 8. Reaction indexing - 9. Techniques of structure manipulation - 10. Developments in computing - 11. Trends in the communication of chemical information. Appendices: 1. Glossary of acronyms, trade and product names; 2. Organization addresses. Analytical index, p.278-97. *Class No:* 540:025.4

Institutions & Associations

[1124]

INTERNATIONAL UNION OF PURE AND APPLIED CHEMISTRY. IUPAC handbook 1987-1989. Oxford, Blackwell, 1988. ix,277p. ISBN: 063202383x. ISSN: 02698757.

Extensive data on the structure, activities, officers, etc. of the Union. *Class No:* 540:061:061.2

Research Establishments

[1125]

AMERICAN CHEMICAL SOCIETY. Committee on Professional Training. **Directory of graduate research:** faculties, publications, and doctoral and master's theses in departments or divisions of chemistry, chemical technology, biochemistry, pharmaceutical/medicinal chemistry, clinical chemistry, and polymer science at universitites in the United States and Canada. Washington, the Society, 1953-. Biennial. 1260p. (1986 ed.). £44. ISBN: 0841209359.

Includes names of faculty members, biographical data, their research interests, and titles of their recent publications. Also, statistics of numbers of faculty, postdoctoral appointments, graduate enrolments and masters and Ph.D. degrees granted. *Class No:* 540:061:061.62

[1126]

POLLOCK, G.L., *ed*. **Chemical research faculties:** an international directory. Washington, American Chemical Society, 1984. 407p. $129.95.

Information on institutions in 62 countries other than US and Canada (on which see the ACS *Directory of graduate research*). Aims to provide data to facilitate location of a colleague by chemical specialization, country, academic institution, on name. Faculty and research subject indexes. Covers 727 departments and *c.*8,900 individual faculty members. (*Based on Science & technology libraries*, v,7(6), Fall 1986, p.141). *Class No:* 540:061:061.62

Solids

[1127]

HANNAY, N.B., *ed*. **Treatise on solid state chemistry.** New York & London, Plenum Press, 1976. 6v. graphs, tables. ISBN: 0306350505.

1. *The chemical structure of solids.* xv, 540p. 2. *Defects in solids.* xiii, 527p. 3. *Crystalline and noncrystalline solids.* xvi, 774p. 4. *Reactivity of solids.* xiii, 721p. 5. *Changes of*

.... (contd.)
state. xvi, 600p. 6. A,B. *Surfaces.* xii, 491p.; xiii, 418p.

A well-documented text. V.1, chapter 8: 'Structural characterization of solids', has 115 references. *Class No:* 540-16

Reagents

[1128]

100 modern reagents. Simpkins, N.S. Nottingham, Moss Publishing for the Royal Society of Chemistry, 1989. ix,203p. chemical structures. £19.95. ISBN: 0851868932.

Reagents, A-Z. Data include CAS Registry name and number; molecular formula and weight; reactions; safety and handling; availability; preparation. Examples of reaction schemes and literature references for each reagent. *Class No:* 540-4

[1129]

AMERICAN CHEMICAL SOCIETY. Committee on Analytic Reagents. **Reagent chemicals.** American Chemical Society Specifications: official from January 1, 1987. 7th ed. Washington, the Society, 1986. xiii, 713p. $89.95. (*American Chemical Society Specifications*.) ISBN: 084120991x.

First published 1950; 6th ed. 1981.

Definitions, test procedures, and standards support the actual specifications, of which 32 are new (6th ed.: 333 specifications). For the first time CAS Registry Numbers are included to aid online searching. Index, p.703-13. *Class No:* 540-4

[1130]

FIESER, L.F. *and* **FIESER, M. Reagents for organic synthesis.** New York, etc., Wiley, 1967-. V.1-.

Data on each reagent include structural formula, atomic weight, preparation, uses, sources of supply, critical comments and references. Index of reagents according to use; author and subject indexes per volume. A comprehensive treatise. *Class No:* 540-4

[1131]

MUNDY, B.P. *and* **ELLERD, M.G. Name reactions and reagents in organic synthesis.** New York, Wiley, 1988. 546p. $39.95. ISBN: 0471836365.

Data on over 120 common named reactions and over 145 of the most commonly used reagents in organic synthesis. Journal references for the examples and to Feisers's *Reagents for organic synthesis*. 'This indispensable volume' (*Science and technology libraries*, V.10(3), Winter 1989, p.140). *Class No:* 540-4

Quantum Chemistry

[1132]

Advances in quantum chemistry. New York, London, etc., Academic Press, 1967-. Annual. ISBN: 0120348225.

V.22 (1992. 401p.) covers molecular orbital theory, the Feynman path integral formalism, the coupled cluster based polarization propagator method and bounds to atomic and molecular energy functionals. *Class No:* 540.062

[1133]
ATKINS, P.W. Quanta: a handbook of concepts. 2nd ed. Oxford Univ. Press, 1991. 434p. diagrs., tables. $95. ISBN: 0198555725.

First published 1974.

Brief explanations (most about 1p.) of over 200 quantum concepts, A-Z. Detailed, analytical index; extensive bibliography (with references to 1991 publications). *Class No:* 540.062

Theoretical Chemistry

[1134]
Advances in free radial chemistry. Tanner, D., *ed.* Greenwich, Conn., London, JAI Press, 1990-. graphs, tables. ISBN: 0892328622.

V.1 (1990. xiv,295p.) has 10 contributors. The 6 documented articles include 'Two decades of spin trapping', p.253-95, with 201 references. No index. The new series covers 'a wide variety of topics which are of current interest' (*Preface*, p.xiv). *Class No:* 541

Physical Chemistry

[1135]
Advances in chemical physics. New York, Interscience: Wiley, 1925-. Annual. ISBN: 0471820555.

V.79, 1990 (xi,322p.) has 8 contributors. The 3 papers include 'Dynamics of entangled polymer liquids' (with 202 references) and 'Ultrafast processes and transition-state spectroscopy' (with 116 references). Author and subject indexes. *Class No:* 541.1

[1136]
Annual reports on the progress of chemistry. Section C: Physical chemistry. London, Royal Society of Chemistry (formerly Chemical Society). ISBN: 085186872x.

V.82, 1985 (1986. xii, 324p.) has 8 chapters: 1. Introduction - 2. Nuclear magnetic resonance spectroscopy - 3. Reactions of silicon intermediates - 4. Isobaric heat capacities - 5. Physical aspects of photochemistry - 6. Matrix isolation - 7. Physical chemistry of solids - 8. Electro processes in the films and novel conductors (with 131 footnote references). Author index. *Class No:* 541.1

[1137]
Annual review of physical chemistry. Palo Alto, Calif., Annual Reviews, Inc., 1950-. illus., graphs, tables, chemical structures. ISSN: 0066426x.

V.42, 1991 (x,819p.) has 39 contributors. 24 documented articles, *e.g.* 'Photochemistry and spectroscopy of organic ions and radicals' (p.55-81): 164 references; 'Reactions on transition metal surfaces' (p.251-78): 184 references. Author and subject indexes. Combined index of contributing authors, v.38-42; cumulative index of chapter titles, v.38-42. 'Some related articles in other *Annual reviews*,' p. ix-x. *Class No:* 541.1

[1138]
ATKINS, P.W. Physical chemistry. 4th ed. Oxford Univ. Press, 1990. xii,995p. illus., diagrs., tables. £22. ISBN: 019855284x.

First published 1978. 3rd ed. 1986.

30 chapters, each with 'further reading' and problems, in 3 parts: 1. Equilibrium - 2. Structure - 3. Change. Index, p.981-95. A standard, comprehensive textbook. *Class No:* 541.1

[1139]
BORG, R.J. *and* **DIENES, G.J. The Physical chemistry of solids.** Boston, Mass., Academic Press, 1992. xii,584p. diagrs., graphs, tables. £46.50. ISBN: 012118420x.

14 chapters (closely subdivided), each with 'additional reading'. Appendices, A-H. Author and subject indexes. 'For university chemistry and materials science departments' (*Preface*). *Class No:* 541.1

[1140]
INTERNATIONAL UNION OF PURE AND APPLIED CHEMISTRY. Recommended reference materials for the realization of physicochemical properties. Marsh, K.N., *ed.* Oxford, etc., Blackwell Scientific, 1987. x, 500p. ISBN: 063201718x.

20 contributors. 18 sections (2. Density; 5. Pressure, volume, temperature; 9. Enthalpy; 10. Thermal conductivity; 15. Optical refraction; 18. Relative molecular mass (19 references)). Subentry references and sources. 'Reference materials with known properties that can be used as references for checking the correct operation of or for the calibration of equipment used for physicochemical measurement'. *Class No:* 541.1

[1141]
MILLS, I., *and others.* **Quantities, units and symbols in physical chemistry.** Oxford, Blackwell for the International Union of Pure and Applied Chemistry, 1988. ix,134p. tables. ISBN: 0632017732.

First published 1970 as *Manual of symbols and terminology for physicochemical quantities and units.*

8 sections: 1. Physical quantities and units - 2. Tables of physical quantities - 3. Definitions and symbols for units - 4. Recommended mathematical symbols - 5. Fundamental physical constants - 6. Properties of particles, elements and nuclides - 7. Conversion of units - 8. References. Index of symbols, p.125-34. *Class No:* 541.1

Kinetics

[1142]
BAMFORD, C.H., *and others, eds.* **Comprehensive chemical kinetics.** Amsterdam, etc., Elsevier, 1969-. graphs, tables, formulas.

1. *The practice and theory of kinetics.* 3v.
2. *Homogeneous decompostion and isomerization reactions.* 2v.
3. *Inorganic reactions.* 2v.
4. *Organic reactions.* 6v.
5. *Polymerization reactions.* 2v.
6. *Oxidation and combustion reactions.* 2v.
7. *Selected elementary reactions.*
8. *Heterogeneous reactions.* 4v.
9. *Kinetics and chemical technology.*
10. *Modern methods, theory and data.* 3v. (*i.e.*, v.24-26).

V.29 is entitled *New techniques for the study of electrodes and their reactions.* 1989. 504p. £137. *Class No:* 541.124

[1143]
HOCHSTIM, A.R., *ed*. **Bibliography of chemical kinetics and collision processes:** an annotated bibliography of gas-phase reaction rates and low energy cross sections of atoms, ions and small molecules ... Prepared by M. Berman [and others]. New York, Plenum Press, 1969. ix, 953p. tables.

'About 20,000 entries, each specifying the study of a collision process, drawn from about 7,000 selected articles' (*Preface*). Covers the literature 1900-66. Main index, supplement; reviews. 'List of sources used in the literature search', p.22. Computer-compiled. *Class No:* 541.124

[1144]
JACOBSON, C.A. *and* HAMPEL, C.A. **Encyclopedia of chemical reactions.** New York, Reinhold; London, Chapman & Hall, 1946-59. 8v.

Records more than 15,000 reactions of the elements (excluding carbon and oxygen). Arranged A-Z, first as to the formulas of the reactants, and then as to the reagents. Each volume has lists of abstractors and journals used by them; also indexes to reagents and to substances obtained. Addendum volumes were planned. *Class No:* 541.124

Electrochemistry

[1145]
Advances in electrochemical science and engineering. Gerischer, H. *and* Tobias, C.W., *eds*. Weinheim; New York, VCH, 1990-.

A new series, formerly published by Wiley as *Advances in electrochemistry and electrochemical engineering*.

V. 2 (1992, 269p.) has 4 articles, all extensively documented. 'Progress in cathode activation', p.1-85, carries 594 references. Detailed, analytical index, p.265-69. *Class No:* 541.13

[1146]
CRC handbook series in organic electrochemistry. Meites, L. *and* Zeiman, P. Cleveland, Ohio (later Boca Raton, Fla.), CRC Press, 1977-83. 6v. tables, chemical structures. ISBN: 0849372208.

V.6 ([viii], 538p.) is largely tabular, - 11 main tables (1. Electrotechnical data, p.5-333), including a key to literature citations. Author index. 'Although these six volumes contain the largest collection of organic electrochemical data ever made, ... they include only a fraction of the data that have appeared in the original literature' (*Preface, v.6*). Thus, electrochemical techniques are excluded. *Class No:* 541.13

[1147]
Encyclopedia of electrochemistry of the elements. Bard, A.J., *ed*. New York, Dekker, 1973-. graphs, tables.

V.1-10: Inorganic chemistry; v.11-: Organic chemistry.

V.9, pt.B (1986, xvi, [1], 674p.) has 13 contributors. 5 chapters: Alkali metals - Bismuth - Zirconium and hafnium - Chromium - Uranium. Systematic treatment of each element: 1. Introduction and standard potential; 2. Voltametric characteristics; 3. Kinetic parameters and double-layer properties; 4. Electrochemical studies; 5. Applied electrochemistry. Well-documented (Chromium: 396 references). Detailed, analytical index, p.667-74. Aims 'to provide a critical, systematic, and comprehensive review of the electrochemical behavior of the elements and their compounds' (*Introduction*). *Class No:* 541.13

[1148]
HIBBERT, D.B. *and* JAMES, A.M. **Dictionary of electrochemistry.** 2nd ed. London, Macmillan, 1984. ix, 308p. diagrs., graphs, tables, chemical structures. ISBN: 0333349830.

First published 1976.

Several hundred electrochemical terms and concepts. 'pH': 3½p. Numerous cross-references. 'References and methods', p.298-9; 'Electrochemical journals', p.300; 'Tables of useful data', p.301-8. Further reading, on occasion. Eponyms. 130 diagrams. *Class No:* 541.13

[1149]
KUHN, A.T., *ed*. **Techniques in electrochemistry, corrosion and metal finishing:** a handbook. Chichester, Wiley, 1987. xviii,567p. illus., tables. ISBN: 047191407x.

8 contributors. In 3 parts: A. Introduction to experimental methods - B. Specialized techniques (Section I: Methods based on light; Section II: Spectroscopic methods; Section III: Other methods (*e.g.* physiochemical; radio-chemical; acoustic emission; dyes, etc.) - C. The literature (37p. bibliography of post-1960 monographs, most from the British Library Science Reference and Information Service). 28 chapters, most with extensive references (*e.g.* 'Raman spectroscopy', p.237-49, includes 166). Subject index. Dedicated - gracefully - to the 'Library' and its 'staff'. *Class No:* 541.13

Radiation Chemistry

[1150]
CRC handbook of radiation chemistry. Tabata, V., *and others*. Boca Raton, Fl., CRC Press, 1990. 936p. graphs, tables. £228.50. ISBN: 0849429957.

18 chapters, some with references. Much tabular data. Index, p.921-36. *Class No:* 541.15

Colloids

[1151]
BECKER, P. **Dictionary of colloid and surface science.** New York, M. Dekker, 1990. 202p. $80. ISBN: 0824783263.

Brief definitions (2 lines ... 1/3 page) of terms not found in general chemical dictionaries. Features names of chemists, plus concise biographical data, who have contributed to the fields over the last 200 years. *Class No:* 541.18

Aerosols

[1152]
SANDERS, P.A. **Handbook of aerosol technology.** 2nd ed. New York, etc., Van Nostrand Reinhold, 1979. xi, 528p. illus., diagrs., micrographs, graphs, tables. ISBN: 0442273487.

First published 1970.

3 main parts: 1. Homogeneous systems and their properties - 2. Emulsions, foams and suspensions - 3. Miscellaneous. 25 documented sections (23. Toxicity of fluorocarbon propellants: 189 references, plus 4½p. of supplementary references). Detailed, analytical index, p.513-28. An extensive revision. *Class No:* 541.182.2

Atomic Theory

Radiochemistry

[1153]
The Radiochemical manual. 2nd ed. Amersham, Radiochemical Centre, 1966. 327p. tables.

First published 1962.

Contents: Physical characteristics - How radioisotopes and related products are made - Some characteristic problems and features of radioactive materials - Associated administrative topics. 'Intended as a guide to all who use radioactive substances professionally' (*Foreword*). *Class No:* 541.28

Chemical Compounds

[1154]
Comprehensive coordination chemistry: the synthesis, reactions, properties and applications of coordinate compounds. Wilkinson, Sir G., *ed*. Oxford, etc., Pergamon Press, 1987. 7v. ISBN: 0080262325.

1. *Theory and background.*
2. *Linards.*
3. *Main group & early transition elements.*
4. *Middle transition elements.*
5. *Late transition elements.*
6. *Applications.*
7. *Index.*

V.1 has 18 contributors. Well-documented articles (*e.g.* 'substitution reactions', p.281-329.; 559 references). V.7 (xi, 642p.) comprises 'Index of review articles and specialist texts' (1,707 entries); detailed and analytical cumulative subject index; and formula index. (Co-ordination compounds are compounds in which the atoms or groups are united by co-ordinate bonds). *Class No:* 541.4

[1155]
Handbook of existing and new chemical substances. Chemical Products Safety Division [Japan], *ed*. 5th ed. Tokyo, The Chemical Daily Co., Ltd., 1992. [iv], 833p.

Lists about 20,000 substances registered for import and manufacture in Japan. Includes an English translation of the 'Law, Enforcement Order, Ministerial orders and notifications which we hope will be helpful to business' ('Recommendation', p.(1)). *Class No:* 541.4

Oxides

[1156]
SAMSONOV, G.V., *ed*. **The Oxide handbook.** Translated from Russian by R.K. Johnston. 2nd ed. New York, Plenum Press, 1982. 463p. tables. $75. ISBN: 0306651777.

First published 1973 (xi, 524p.).

Sections: General data, stoichiometry and crystal-chemical properties - Thermodynamics and thermal properties - Molecular properties - Mechanical properties - Electrical and magnetic properties - Optical properties - Nuclear properties and radiation effects - Clinical and catalytic properties - Refractory properties - Phase diagrams and binary element - Oxygen systems. References. Index of elements and their oxides. The oxides in the tables are listed in increasing order

....(contd.)
of atomic number of the element. 644 reference citations, mainly from Russian literature. 'Excellent' (*New technical books,* October 1982, item 984). *Class No:* 541.45

Chemical Structures

Solubility

[1157]
SEIDELL, A. Solubilities: inorganic and metal-organic compounds: a compilation of solubility data from the periodical literature. 4th ed., by W.F. Linke. New York, etc., Van Nostrand, 1958-65. 2v. ISBN: 0841200971; 084120098x.

1. *Argon-Iridium.* 1958, iv, 1487p.
2. *Kalium-Zirconium.* 1965. iii, 1914p.

The elements are arranged A-Z by chemical symbol, and compounds listed A-Z by the chemical symbols of their anions or radicals. The 4th ed. contains for the first time a complete compound index. V.1 has author and subject indexes; v.2, a cumulative inorganic and metal-organic index for v.1-2. V.3 was to have covered organic compounds. *Class No:* 541.8

[1158]
STEPHEN, H. *and* **STEPHEN, T.,** *eds*. **Solubilities of inorganic and organic compounds.** Oxford, etc., Pergamon Press, 1963. 3v. in 7 (7300p.). tables.

1. (2pts.) *Binary systems.*
2. (2pts.) *Ternary systems.*
3. (3pts.) *Multi component systems.*

Translated from Russian. Over 27,000 tables in all, the data - uncritically presented - being based on a survey of periodical literature in all languages, on the solubility of elements, inorganic compounds, metallo-organic and organic compounds. References; index of chemical component formulas and table numbers. *Class No:* 541.8

Laboratory Tools & Techniques

[1159]
COYNE, G.S. The Laboratory handbook of materials, equipment, and technique. Englewood Cliffs, N.J., 1992. xvii,468p. illus., tables. £48.15. ISBN: 0131262289.

8 chapters, closely subdivided: 1. Materials - 2. Measurement ... 5. Compressed gases ... 7. Vacuum systems ... 8. Gas-oxygen torch. 4 appendices (*e.g.* B. 'Polymer resistance'). Detailed, analytical index, p.445-67. Well-produced with good use of bold-face type. Essentially practical. *Class No:* 542

[1160]
JUNGE, H.D., *ed*. **Pocket dictionary of laboratory equipment,** English/German. Weinheim, VCH Verlagsgesellschaft, 1987. viii, 201p.

Two parts: German-English; English-German. Includes combinations of words in basic expressions. American

....(contd.)

spelling. 'A useful work' (*Chemistry and industry*, 2 November 1987). For students, research workers, technicians. *Class No: 542*

[1161]
OCKERMAN, H.W. Illustrated chemistry laboratory terminology. Boca Raton, Fl., CRC Press, 1991. 211p. illus. $22.50. ISBN: 0849301521.

Lists several hundred items, most of them standard pieces of equipment. Table of contents and index in English, Chinese, French, German, Polish, Spanish and Turkish. *Class No: 542*

[1162]
ROYAL SOCIETY OF CHEMISTRY. Working Party. Guidance on laboratory fume cupboards. London, Royal Society of Chemistry, 1990. 35p. £15. ISBN: 0851863299.

Provides practical guidance for users in accordance with Control of Substances Hazardous to Health (COSHH) Regulations (UK). Includes a review of the literature on factors affecting the efficiency of fume cupboard systems. *Class No: 542*

[1163]
YOUNG, J.A., *ed.* **Improving safety in the chemical laboratory :** a practical guide. 2nd ed. New York, Wiley, 1991. 406p. $75. ISBN: 0471530360.

Chapters cover safety organization, labels and safety data sheets, summary of causes of accidents, inspection, chemical hygiene planning. 5 appendices: hazards; safety regulations (US, Canada, UK); personal protection; design; resources. 'A plainly written, thorough guide ... [which] should be in the hands of every manager responsible for laboratory safety' (*Choice*, March 1992, p.1109). *Class No: 542*

Work with High Temperatures

[1164]
KOSOLAPOVA, T.Y., *ed.* **Handbook of high temperature compounds:** properties, production, applications. Translated from Russian by S.N. Gorin and others. New York, Hemisphere, 1990. xxvi,933p. tables. £132. ISBN: 0891168494.

Originally published as *Svoĭstva, poluchemie i primenenie tugoplavkika soedineni,* Moscow, Mettalurgiya, 1986.

Covers borides, carbides, nitrides, silicides, sulfides, selenides and tellurides. 11 chapters, each with brief introduction and numerous tables: 1. Crystal chemical properties - 2. Thermodynamic - 3. Electrical, thermophysical, and magnetic - 4. Optical - 5. Mechanical - 6. Chemical - 7. Preparation of powders - 8. Production of materials - 9. Powder ignition and explosion hazards - 10. Toxicity - 11. Applications. 1138 references, many from Russian literature. A wealth of data. *Class No: 542.46*

Work with Liquids

[1165]
Science and practice of liquid-liquid extraction. Thornton, J.D., *ed.* Oxford, Clarendon Press, 1992. 2v. diagrs., graphs, tables. (*Oxford engineering science.*) ISBN: 0198561784, set.

1. *Phase equilibria; Mass transfer and interfacial phenomena; Extractor Hydrodynamics, selection and design.* xviii,603p. £60. ISBN: 0198562365.

2. *Process chemistry and extraction operations in the*

....(contd.)

hydrometallurgical, nuclear, pharmaceutical, and food industries. xviii,436p. £60. ISBN: 0198562373.

12 contributors, 13 chapters, closely subdivided, with extensive references (*e.g.* v.2, chapter 1 has 10 sections, 9p. of references). Volume indexes. A scholarly treatment. *Class No: 542.6*

[1166]
VISWANATH, D.S. *and* **NATARAJAN, G. Data book on the viscosity of fluids.** New York, Hemisphere, 1989. 990p. $275. ISBN: 0891167781.

Includes organic compounds, inorganic liquids, liquid metals, plasticizers and foods like soybean oil. For each compound one primary journal reference and *Chemical abstracts* CAS Registry number are given. Common name index. 'A major source for viscosity data' (*Science and technology libraries,* v.10(2), Winter 1989, p.138). *Class No: 542.6*

Work with Gases

[1167]
HAAR, L., *and others.* **NBS/NRC steam tables:** thermodynamic and transport properties and computer programs for vapor and liquid states of water in SI units. Washington, New York, McGraw-Hill, Hemisphere Publishing Corpn., 1984. xii, 320p. $34.50 ISBN: 0891163549.

Contents: Thermodynamic property values - Transport and other themophysical property values. Appendices, A-C (C. Equations for transport and other themophysical properties). 70 references, p.317-20. *Class No: 542.7*

Catalysis

[1168]
Advances in catalysis. Orlando, Florida, Academic Press, 1953-. diagrams, graphs, tables. ISBN: 012007835x.

V.35 (1987. 433p.) has 11 contributors. The 6 documented articles include 'Chemical design surfaces for active solid catalysis', p.187-204, with 217 references. Detailed analytical index, p.423-33. *Class No: 542.97*

[1169]
—**Catalysts in chemistry.** Braintree, Essex, R.H. Chandler, 1976-. Monthly. £170pa.

An alerting service, with *c.*4,000 entries pa., monitoring journals, monographs, patents and conference proceedings in several languages. *Class No: 542.97*

[1170]
EMMETT, P.H., *ed.* **Catalysis.** New York, Reinhold; London, Chapman & Hall, 1954-60. 7v. diagrs., tables.

1/2. *Fundamental principles.* 1954-55.

3. *Hydrogenation and dehydrogenation.* 1955.

4. *Hydrocarbon synthesis hydrogenation and cyclisation.* 1956.

5. *Hydrogenation ...* 1957.

6. *Alkylation, isomerization, polymerization, cracking and hydroreforming.* 1958.

7. *Oxidation, hydration, dehydration and cracking catalysts.* 1960.

A comprehensive and authoritative treatise for research chemists, fully documented. Thus v.7 (with 7 chapters by

....(contd.)

specialists) has 183 references for chapter 1, 'Cracking catalysis'. Author index (including references) and brief subject indexes. *Class No:* 542.97

[1171]

TWIGG, M.V., *ed*. **Catalyst handbook.** 2nd ed. London, Wolfe, 1989. 608p. illus., diagrs., graphs, tables. £50. ISBN: 0723408572.

First published 1970.

15 contributors. 10 chapters: 1. Fundamental principles - 2. Process design, rating and performance ... 4. Feedstock purification - 5. Steam reforming ... 8. Ammonia synthesis - 9. Methanol synthesis - 10. Catalytic oxidations. 21 appendices (1. 'Further reading'; 16. 'Specific heats of catalysts'). Extensive references; analytical subject index (18p.). Intended 'primarily for people working in the chemical industry' (*Preface*, p.16). *Class No:* 542.97

Analytical Chemistry

Bibliographies

[1172]

JAMES, S. **Using literature:** analytical chemistry by open learning. Chapman, N.B., *ed*. Chichester, Wiley, 1987. xvi,598p. illus., facsims. ISBN: 0471912204.

An 'open learning' text (part of a complete course developed at Thames Polytechnic) on the chemical literature, including online services and using libraries. Numerous facsimile pages. No index and only a brief contents list. *Class No:* 543(01)

Handbooks & Manuals

[1173]

CRC handbook of basic tables for chemical analysis. Bruno, T.J. *and* Svornos, P.D.N. Boca Raton, Fl., 1989. [viii],517p. tables, chemical structures, graphs. £84.50. ISBN: 0849339359.

11 chapters, closely sub-divided (*e.g.* 'Thin layer chromatography' (p.129-59): 5 sections; 243 references). Index, 28p. *Class No:* 543(035)

[1174]

A Guide to materials characterization and chemical analysis. Sibilia, J.P., *ed*. New York, VCH, 1988. x,318p. illus., diagrs., graphs, tables. ISBN: 0895732696.

Includes over 100 materials methodologies including evaluation, chemical analysis and physical testing techniques. 13 chapters, closely subdivided: 2. Molecular spectroscopy ... 6. X-ray analysis - 7. Microscopy ... 9. Thermal analysis ... 13. Scientific computation. Index, non-analytical, p.307-18. *Class No:* 543(035)

[1175]

KOLTHOFF, I.M., *and others, eds*. **Treatise on analytical chemistry.** New York, London, etc., Interscience: Wiley, 1959-. illus., diagrs., graphs, tables.

1. *Theory and practice.* 1959-76. 11v. & index.

2. *Analytical chemistry of inorganic and organic compounds.* 1959-76. 15v.

3. *Analytical chemistry in industry.* 1967-.

....(contd.)

Comprehensive. Pt.3(4), 1977 (xxxviii, [1], 686p.) has 14 contributors. It contains section D, 'Physical testing methods for the characterization of materials: Surface properties', p.167-227, with 91 references. Each part has cumulative author and subject indexes. Thorough treatment of the elements, one by one. 'Valuable as a source book of ideas as well as facts' (*Chemistry & industry,* no. 47, 23 November 1963, p.1867).

The 2nd rev. ed. (1978-) reflects the extensive development (*e.g.* in computing) and increased specialization of analytical chemists. Pt.1, v.1-14, so far. *Class No:* 543(035)

[1176]

POTTS, L.W. **Quantitative analysis:** theory and practice. New York, etc., Harper & Row, 1987. 710p. diagrs., graphs, tables, formulas. ISBN: 0060452714.

18 chapters (2. Errors in chemical analysis and the standard treatment of data; 8. Solubility equilibria; 9. Precipitation; 13. UV-visible absorption spectrometry and colorimetric methods and anlysis; 15. Potentiometry; 18. Chromographic equipment). 7 appendices, p.645-89 (*e.g.* 7. Some compounds of analytical importance). Answers, by chapters, to selected problems. Detailed, analytical index. Aimed at 'second and third-year undergraduate chemistry, biology, pre-medicine and engineering majors' (*Preface*). *Class No:* 543(035)

[1177]

WILSON, C.L. *and* WILSON, D.W., *eds*. **Wilson and Wilson's Comprehesive analytical chemistry.** Amsterdam, etc. Elsevier, 1959-. illus., diagrs., graphs, tables. ISBN: 0444417354.

V.26: *Radionuclide x-ray fluorescence analysis with environmental applications* (1990. xv,254p.) has 3 authors, 7 closely-subdivided chapters, 589 references. Analytical index. The series 'aims to provide a work which, in many instances, should be a self-sufficient reference work; but where this is not possible, it should at least be a starting point for any analytical investigation' (*Preface*). *Class No:* 543(035)

Dictionaries

[1178]

MALUDZINSKA, G., *ed*. **Dictionary of analytical chemistry:** English, German, French, Polish, Russian. Amsterdam, etc., Elsevier, 1990. x,391p. diagrs., graphs. £86.35. ISBN: 0444987290.

7 contributors. About 2,000 English-base entries, with equivalents and indexes in other languages. Cross-references. About 300 symbols applied. Genders shown. *Class No:* 543(038)

Reviews & Abstracts

[1179]

Analytical abstracts. London, Royal Society of Chemistry, 1954-. 12pa. ISSN: 00032689.

About 12,000 abstracts pa. Classes A-J: A. General analytical chemistry - B. Inorganic chemistry - C. Organic chemistry - D. Biochemistry - E. Pharmaceutical chemistry - F. Food - G. Agriculture - H. Environmental chemistry - J.

....(contd.)

Apparatus and technique. Cross-references. Subject index. Annual subject and author indexes. *Analytical abstracts* is online, January 1982-. *Class No:* 543(048)

[1180]
Index of reviews in analytical chemistry, 1983. London, Royal Society of Chemistry, 1984. [xiv], 68p. £11. (*IRAC 83.*) ISBN: 0851865860.

Lists 1,064 review articles published in English, German or French during 1983. Section 1 covers industrial substances or materials and substance-related topics. Section 2 largely records reviews on analytical techniques or instrumentation. *Class No:* 543(048)

Nomenclatures

[1181]
INTERNATIONAL UNION OF PURE AND APPLIED CHEMISTRY. Analytical Chemistry Division. **Compendium of analytical nomenclature:** definitive rules. Prepared for publication by H.M.N. Irving [and others]. Oxford, etc., Pergamon Press, 1978. viii, 223p. £8. ISBN: 0080223478.

The IUPAC's 'Orange book'. *Class No:* 543(083.72)

Electroanalysis

[1182]
Electroanalytical abstracts: international journal dealing with the documentation of all aspects of fundamental physico-chemical and analytical electrochemistry. Basle, etc., Birkhäuser Verlag AG, 1963-. 6pa. $442pa. ISSN: 00134775.

Abou 4,800 abstracts pa. 6 sections: 1. Electrochemical techniques and instrumental - 2. Electrochemical analyses and syntheses - 3. Chemical equilibrium and reactions (studies by using electrochemistry) - 4. Electrochemical processes and reactions - 5. Electrochemical energy technology - 6. Electrochemical corrosion and surface treatments. Annual author and subject index. *Class No:* 543.25

Gas & Air

[1183]
Encyclopédie des gas. [Gas encyclopaedia.] Amsterdam, Elsevier, for L'Air Liquide, 1976. viii, 1150p. graphs, tables. ISBN: 0444414924.

5 chapters on physical properties and flammability of gases, their safe handling, general metrology, symbols and units. Over 100 monographs (p.61-1150), each with bibliography appended. 'Ammonia' (p.951-72; 26 references) has data under 6 heads: physical properties; flammability; biological properties; precautions in handling and storage; materials of construction; uses. *Class No:* 543.27

[1184]
—KEENAN, J.H., *and others*. Gas tables. 2nd ed. New York, London, etc., Wiley, 1985. 2v. tables. $82. ISBN: 0471842400.

First published 1948.

The main tables include: Air at low pressures (for one pound) ... Standard atmosphere ... Physical constants ...

....(contd.)

Conversion factors. Sources of data and calculation methods. Examples provided to illustrate the use of the tables. Bibliography. *Class No:* 543.27

Optical Methods

Spectroscopy

Abbreviations & Symbols

[1185]
WENDISCH, D.A.W. Acronyms and abbreviations in molecular spectroscopy: an encyclopedic dictionary. Berlin, Springer Verlag, 1990. v,315p. $60; £35. ISBN: 3540513485.

Definitions (¼p. to 2p.). Readable, but more compressed than Honan's *Dictionary of concepts in NMR*. Each entry has at least one reference for further reading (many have 5 or even 10 such references). 'This is the most extensive listing of spectroscopic acronyms' (*Science and technology libraries*, v.11(1), Spring 1991, p.148). *Class No:* 543.42(003)

Encyclopaedias

[1186]
CLARK, G.L., *ed.* **The Encyclopaedia of spectroscopy.** New York, Reinhold; London, Chapman & Hall, 1960. xvi, 787p. illus., diagrs.

Over 100 contributors. 23 main subjects, A-Z, from 'Absorption spectroscopy (spectrophotometry)' to 'X-ray emission spectrometry', these subjects being in turn subdivided, when required, A-Z, to form the final, 185 signed and documented monographs. Adequate cross-references. Some of the monographs have been extracted from or are revised contributions from G.L. Clark's *Encyclopedia of chemistry* (New York, Reinhold, 1957-58). *Class No:* 543.42(031)

Handbooks & Manuals

[1187]
CRC handbook of spectroscopy. Robinson, J.W., *ed.* Cleveland, Ohio, Chemical Rubber Co., 1974-75. 2v. ([xi], 913p. [ix], 578p.). diagrs., charts, graphs, tables.

30 contributors. Aims to cover the major areas of spectroscopy, - 'NMR, IR, Raman, UV (absorption and fluorescence), ESCA, X-ray (absorption, diffraction, fluorescence), mass spectrometry, atomic absorption, flame photometry, emission spectrography, and flame spectroscopy' (*Preface*). V.1 carries 9 tables and detailed index; v.2: 139 tables and detailed index. References are appended to tables. *Class No:* 543.42(035)

Dictionaries

[1188]
DENNEY, R.C. A Dictionary of spectroscopy. 2nd ed. New York, Macmillan, 1982. xiii, 205p. diagrs., graphs, tables, chemical structures. ISBN: 0333138805.

First published 1972.

Defines *c*.1,000 terms; 10 categories. 'Fluorescence' : ½p.; 'Lasers' : over 1p. + diagr. 275 references, p.194-

....*(contd.)*

205. 55 diagrs. Usefully 'deals with some of the newer electron spectroscopies - such as Mossbauer and Auger spectroscopy - that are less well covered in other dictionaries which tend to concentrate exclusively on the more familiar molecular, atomic and electron spectroscopy techniques' (*Nature, v.*300, 3 November 1982, p. 464).

Class No: 543.42(038)

[1189]
MORITZ, H. *and* TÖRÖK, T. **Technical dictionary of spectroscopy and spectral analysis.** Oxford, etc., Pergamon Press; Berlin, VEB Verlag Technik, 1971. 188p. illus. (pl.).

An English-French-German-Russian dictionary of *c.*3,500 specialist terms. Adds a 5-page supplement in Spanish. German, French and Russian indexes.

Class No: 543.42(038)

[1190]
—INTERNATIONAL UNION OF PURE AND APPLIED CHEMISTRY. Commission on Molecular Structure and Spectroscopy. Multilingual dictionary of important terms in molecular spectroscopy. Ottawa, National Research Council of Canada, 1966. 221p.

Languages covered: English, French, German, Japanese, Russian. *Class No:* 543.42(038)

Reviews & Abstracts

[1191]
Atomic absorption and emission spectrometry abstracts. London, PRM Science & Technology Agency, Ltd., 1974-. 6pa. ISSN: 03091813.

1983: 1,001 abstracts. Sections: Theory - Instrumentation - Experimental techniques - Elemental analysis - Physico-chemical studies - Clinical chemistry, pharmacology and toxicology - Forensic sciences - Agriculture and food sciences - Metallurgy - Ceramics, cement and glass technology - Industrial chemistry - Biological sciences - Geological sciences- Elemental analysis index - Environmental sciences - Archaeology - Commercial instrumentation - Books and reviews. Subject and author indexes. *Class No:* 543.42(048)

[1192]
Electron spin resonance spectroscopy abstracts. London, PRM Science & Technology Agency, Ltd., 1973-. 4pa. ISSN: 03017575.

1983: 414 abstracts. V.12(3), 1984 has sections: Theory - Instrumentation and technique - Organic chemistry: Organic chemicals - Organic chemistry: polymers - Organic chemistry: organometallics - Inorganic chemistry (3 sections) - Physical chemistry: inorganic systems - Biological - Applictions - Review. Subject and author indexes per issue. *Class No:* 543.42(048)

[1193]
Laser-Raman & infrared spectroscopy abstracts. London, PRM Science & Technology Agency, Ltd., 1972-. 6pa. ISSN: 03095320.

From 1975 incorporates *Infrared absorption spectra abstracts.*

1983: 745 abstracts. Sections: Theory - Instrumentation and technique - Gases - Matrix-isolation studies - Stimulated, resonance, Beilstein & hyper-Raman effects - Surface-enhanced Raman scattering - Solid state: solid state

....*(contd.)*

phenomena - Inorganic & organometallic chemistry - Organic chemistry - Biochemistry - Polymer science. Subject and author indexes. *Class No:* 543.42(048)

[1194]
Mass spectrometry bulletin. Nottingham University, Royal Society of Chemistry, Mass Spectrometry Data Centre, 1966-. Monthly. ISSN: 00254738.

About 10,000 references pa., 8 sections (*e.g.* 2. Isotopic analysis, precision mass measurement, isotope separation and age determination - 3. Chemical analysis - 4. Organic chemistry - 5. Atomic and molecular processes - 6. Surface phenomena and solid state studies - 7. Thermodynamics and reaction kinetics - 8. Other references). 5 indexes: subject, compound classification, elements, general, author. Cumulated subject and author indexes, 2pa.

Class No: 543.42(048)

[1195]
Mössbauer spectroscopy abstracts. London, PRM Science & Technology Agency, Ltd. 1978-. 4pa. ISSN: 01419066.

About 650 abstracts pa. Sections: Theoretical - Instrumentation and methodology - Studies of chemical compounds - Metals & alloys - Amorphous materials - Geological science - Biological applications - Surface & catalysis studies - Impurity & source studies - Conversion electron Mössbauer spectroscopy - Books and reviews. Isotope index, Author index. *Class No:* 543.42(048)

[1196]
Nuclear magnetic resonance spectrometry abstracts. London, PRM Science & Technology Agency, Ltd, 1970-. 6pa. ISSN: 00481033.

1983: 820 abstracts. Sections: Theoretical studies - Instrumentation and methodology - Shift reagents - Coupling constants - Liquid crystals - Solids - Gas phase and surface absorbed mols - Relaxation and Noe effects - CIDNP effects - Mol stereodynamics and exchange - Mol interactions - Fluorine NMR - Nitrogen NMR - Other nuclei - Inorganic and organometallic chemistry - Organic chemistry - Biochemistry - Polymer science. Subject and author indexes per issue. *Class No:* 543.42(048)

[1197]
X-ray fluorescence spectrometry abstracts. London, PRM Science & Technology Agency, Ltd., 1970-. 4pa. ISSN: 00439851.

1984: 548 abstracts. Sections: Theory - Elemental analysis summary - Metallurgy - Mining and mineral processing - Geological science - Biological science - Environmental sciences - Earth sciences - Industrial chemistry - Miscellaneous services. Subject and author indexes. *Class No:* 543.42(048)

Progress Reports

[1198]
Annual reports on NMR spectroscopy. New York, London, etc., Academic Press, 1968-. graphs, tables, chemical structures. ISBN: 0125053177.

V.1-2 as *Annual review of NMR spectroscopy.*

V.23, 1991 (xi,425p.) has 9 contributors. The 6 documented articles include 'The Cinderella nuclei' (with 215 references) and 'Nuclear spin relaxation in diamagnetic

....*(contd.)*

fields, Part 2' (with 794 references). Analytical subject index. State-of-the art accounts of aspects of nuclear magnetic resonance. *Class No:* 543.42(055)

Chromatography

Handbooks & Manuals

[1199]

Chromatography: fundamentals and applications of chromatography and related differential systems methods. Heftmann, E., *ed.* Amsterdam, Elsevier, 1992. 2v. (*Journal of chromatography library, V.51A,51B.*)

Pt.A. *Fundamentals and techniques.* 547p. ISBN: 0444882367. £76.

Pt.B. *Applications.* 564p. ISBN: 0444882375. £120.

Chapters carry extensive references, most upwards of 200. *Class No:* 543.54(035)

[1200]

POOLE, C.F. and POOLE, S.K. Chromatography today. Amsterdam, Elsevier, 1991. x,1026p. illus., diagrs., graphs, tables. £91 (HB); £46.50 (PB). ISBN: 0444884920; 0444891617, Pbk.

9 chapters, with sub-chapter references. Subject index. Near print. *Class No:* 543.54(035)

Dictionaries

[1201]

ANGELÉ, H.P. Dictionary of chromatography, English, German, French and Russian. Heidelberg, Hutheg, 1984. 132p. $27.

First ed. as *Technical dictionary of chromatography in four languages* (Pergamon, 1970. 119p.).

About 3,000 technical terms in English, with German, French and Russian equivalents and indexes. German and French genders are given. Considered a useful and well-produced dictionary, although there are surprising omissions. *Class No:* 543.54(038)

[1202]

DENNEY, R.C. A Dictionary of chromatography. 2nd ed. London, Macmillan, 1982. [xiv], 229p. diagrs., formulas. £15. ISBN: 0333316673.

First published 1976.

About 1,000 entries ('Absolute detector sensitivity' ... 'Zwitterions'). 'Refractive index detector': 1½p., including 2 keyed diagrams. 55 diagrams in all. Adequate cross-references. 506 references. Aims 'to provide a source of reference for students, technicians and those newly interested in these particular areas of science' (*Introduction*). 'Gives the impression of being more sure-footed than that [his dictionary] on spectroscopy' (*Nature,* v.300, 2 December 1982, p.464). *Class No:* 543.54(038)

[1203]

SERRE, R. Dictionnaire contextuel anglais-français de la chromatographie. Ottawa, the Author, 1981. 106p.

375 English terms used in chromatography are defined, with sources. The French equivalents are shown in correct context, using sentence form, plus references to further readings. 'This dictionary was prepared for translators and

....*(contd.)*

aims to give all the information necessary for a correct translation' (*Journal of chromatography,* v. 236, 5 February 1982, p.262). *Class No:* 543.54(038)

Reviews & Abstracts

[1204]

Chromatography abstracts. Amsterdam, Elsevier, for Chromatographic Society, 1986-. 7pa. £175pa. ISSN: 02686287.

Formerly *Gas and liquid chromatography abstracts* (1958-84).

About 1,250 abstracts pa. Two sections only: 'Gas chromatography abstracts' and 'Liquid chromatography abstracts'. Annual subject index (as issue no. 7 each year) for each section. *Class No:* 543.54(048)

[1205]

—Liquid chromatography abstracts. London, PRM Science & Technology Agency, Ltd., 1973-. 6-weekly. ISSN: 03062104.

About 150 abstracts per issue. Sections: Theory and techniques - Biochemistry - Botanical science - Clinical medicine - Pharmaceutical science - Organic chemistry - Food science - Agriculture - Industrial and inorganic chemistry - Review. Subject and author indexes. *Class No:* 543.54(048)

[1206]

—SIGNEUR, A.V. Guide to gas chromatography literature. New York, Plenum Press, 1964-74. 3v. 1089p.

Usefully enlarges, in depth, the coverage of *Gas & liquid chromatography abstracts* (1958-84). V.1 has 7,577 references under author, with an infuriating subject index (*e.g. c.*800 unspecified entries under 'Lipids'), although this major fault is rectified in v.2. 31,037 references in all. *Class No:* 543.54(048)

[1207]

—Thin-layer chromatography abstracts. London, PRM Science & Technology Agency, Ltd. 1971-. 4pa. ISSN: 00493724.

Carries *c.*600 abstracts pa. Sections: Techniques and apparatus - Botanical science - Pharmaceutical and clinical medicine - Organic chemistry - Food and agricultural chemistry - Industrial and inorganic chemistry. Cross-references. Subject and author indexes per issue and pa. *Class No:* 543.54(048)

Progress Reports

[1208]

Advances in chromatography. New York, Dekker, 1962-. Annual. ISBN: 0824775465.

V.25, 1986 (xvii, [1], 391p.) has 14 contributors. The 8 articles include 'Liquid chromatography of carbohydrates' (p.279-307, with 258 references). Contents of other volumes [v.1-14]. Author and subject indexes. *Class No:* 543.54(055)

[1209]
Recent advances in thin-layer chromatography. Dallas, F.A.A., *and others, eds.* New York, Plenum, 1988. xiii,247p. illus., graphs, tables. ISBN: 0306429349.

26 chapters in 4 sections: Instrumentation - Radio-thin-layer chromatography: instruments and applications - Sorbents and modifiers - Applications. Compound and subject indexes. *Class No:* 543.54(055)

Tables & Data Books

[1210]
CRC handbook of chromatography. Zwieg, G. *and* Sherma, J., *eds.* Cleveland, Ohio, CRC Press, 1972. 2v. tables. ISBN: 0849305615; 0878195629.

Chromatographic data for *c.*12,000 chemical components in all. Tables cover gas, liquid, paper and thin-layer chromatography, plus compound index, in v.1. V.2 has 2 sections: Principles and techniques; Practical applications. Appended: International chromatography book directory. Detailed, analytical subject index. *Class No:* 543.54(083)

Biological Reactions

[1211]
Chemoreception abstracts: chemical senses & applied techniques. Bethesda, Md., Cambridge Scientific Abstracts, 1973-. 4pa. $209pa.

About 1,500 abstracts pa., drawing on *c.*5,000 primary journals, etc. Sections: Peripheral & central sensory mechanisms ... Animal behavior & chemical communication, including phenomenes ... Special products (*e.g.* perfumery) - Odor control - Chemosensory aspects of food ... Miscellaneous - Book notices - Notification of proceedings. Author and analytical subject indexes, cumulated annually. *Class No:* 543.9

Inorganic Chemistry

Encyclopaedias

[1212]
SITTIG, M. Inorganic chemical and metallurgical process encyclopedia. Park Ridge, N.J., Noyes Development Corpn., 1968. 883p. illus., diagrs.

Nearly 2,000 entries for chemicals A-Z (Alumunia, alpha, from aluminium trihydrate ... ziranium, tetrachloride, purified). Footnote references to sources. US patents. Diagrams of apparatus. *Class No:* 546(031)

Handbooks & Manuals

[1213]
Comprehensive inorganic chemistry. Bailou, J.C., *Jr., and others, eds.* Oxford, etc., Pergamon, 1973. 5v. diagrs., tables. ISBN: 008017275x.

1. H. Noble gases. Group IA, IIB, IIIB, C and Si.
2. Ge, Sn, Pb. Group VB, VIB, VIIB.
3. Groups IB, IIB, IIIA, IVA, VA, VIA, VIII.
4. Lenthanides. Transition metal compounds.
5. Actinides. Master index..

Data on physical, chemical and biological properties of

....(contd.)

elements. V.2 (xvii, 1613p.) has 13 contributors. Section 23: Sulphur (p.798-933) carries 579 footnote references. Analytical index, p.1594-1613. *Class No:* 546(035)

[1214]
GMELIN-INSTITUT. Gmelin's Handbuch der anorganischen chemie. 8. vollig neu bearb. Aufl. Leipzig, Berlin, Weinheim/Bergstrasse,Verlag chemie GmbH, 1922-.

First published as Gmelin's *Handbuch der theoretischen Chemie* (1817-1819). 8th ed. under auspices of the German Chemical Society until 1945; then by the Gmelin-Institut.

More than 70v. have appeared, plus a new supplement series. Thus, the main volume 'Water desalting' (1974) had Supplement 1 in 1979. The classification system is based on the Gmelin system of the numerical sequence of elements (system number). Each element has a systematic monograph (history, occurrence, formation, preparation, with a summary of latest information, plus a full bibliography of source material). 'Biel'(Lead): C4. System-Nummer 47 (xix, p.1213-1851) covers literature up to 1969. Index. Formula index AC-NS(11.), 1975-79.

Noted: Lippert, W. 'Adapting the Gmelin Handbook to modern information requirements' (*Journal of chemical information and computer science,* v.19(4), November 1979, p.201-5). *Class No:* 546(035)

[1215]
MELLOR, J.W. A Comprehensive treatise on inorganic and theoretical chemistry. London, Longmans, Green, 1922-37. 16v. Supplements 1-7, 1956-80. 16v. illus., diagrs., tables.

Surveys the elements in periodic-table order. V.1 is mainly theoretical; v. 16 consists largely of the general index. The scope of articles on each element is similar to that in Gmelin. The Supplements update, with copious literature references (*e.g.* v.8, Suppt. 2, final section, 'Determination of nitrogen in natural products', occupies p.655-73, with 683 references). Indispensable in all chemistry libraries. *Class No:* 546(035)

Progress Reports

[1216]
Advances in inorganic chemistry and radiochemistry. New York, London, etc. Academic Press, 1957-. v.1-. Annual. chemical structures. ISBN: 012023629x.

V.38, 1992 (xi,487p.) has 16 contributors. The 10 papers include 'EPR spectroscopy of iron-sulfer proteins' (184 references) and 'Structural and functional diversity of ferredoxins and related proteins' (374 references). Subject index. *Class No:* 546(055)

[1217]
Annual reports on the progress of chemistry. Section A: Inorganic chemistry. London, Royal Society of Chemistry, 1905-. v.1-. Annual. ISBN: 0851861806.

V.83, 1986 (1987. xiii, 441p.) has 14 contributors. Chapters 2-12 concern the elements, in turn (p.3-379); chapter 13, 'Radiochemistry', has sections on isotope production, labelled compounds, chemistry of nuclear transformations, the environment, and miscellaneous. Chemical structures. Well documented throughout. Author index. *Class No:* 546(055)

[1218]
Progress in inorganic chemistry. New York, etc., Wiley, 1959-. v.1-. Annual. graphs, tables, chemical structures. ISBN: 0471803340.

V.39, 1992 (vi,536p.) has 12 contributors. The 7 papers include 'Molybdenum oxygen chemistry' (with 391 references) and 'Applications of scanning tunnelling microscopy to inorganic chemistry' (375 references). Analytical subject index; cumulative author index, v.1-39.
Class No: 546(055)

Tables & Data Books
[1219]
CRC handbook series in inorganic electrochemistry. Meites, L., *and others, eds.* Boca Raton, Florida, CRC Press, 1980-81. 2v. ISBN: 0849303613.

13 tables of data on the electrochemical behaviour of ions in solvents, etc. Key to citations. Companion set: *CRC Handbook series in organic electrochemistry* (*qv*).
Class No: 546(083)

[1220]
PUDDEPHATT, R.J., and **MONAGHAN, P.K. The Periodic table of elements.** 2nd ed. Oxford, Clarendon Press, 1986. x, 102p. diagrs., graphs, tables. £15; £7.50. (*Oxford chemistry series.*) ISBN: 0198555156.

First published 1972.

8 Chapters, on atomic structure; building the periodic table; periodicity of main group elements (atomic, bonding, physical and structural properties); transition elements; lanthanides and actinides. Most chapters have problems appended, with answers on p.97. Bibliograpy of 5 items; subject index. Presents 'the periodic behaviour of the elements in a non-theoretical way' (*Preface*). Suitable for senior secondary pupils and first-year undergraduates.
Class No: 546(083)

[1221]
—WILLIAMS, R.A. Handbook of the atomic elements. London, Vision Press, 1971. 124, [1]p. tables.

First published New York, Philosophical Library, 1970.

Provides more than 30 data on 100 elements (Actinium ... Zirconium), p.13-115. Details include principal quantum number, atomic number, x-ray notation, density, melting point, ground state, electron configuration, radioactive isotopes. Such information is based on Carbon-12 and values released by IUPAC. 5 appended tables (*e.g.* 1. Percentage composition of the elements in the earth's surface).
Class No: 546(083)

Nomenclatures
[1222]
Inorganic chemical nomenclature: principles and practice. Block, B.P., *and others.* Washington, DC, American Chemical Society, 1990. 210p. tables. $59.95; $39.95. ISBN: 0841216975; 0841216983, Pbk.

16 chapters, from single-atom species to inorganic polymers. 13 tables. 'Sufficient for nearly all uses' (*Choice, February 1991, p.958*). *Class No:* 546(083.72)

[1223]
Nomenclature of inorganic chemistry: recommendations 1990. Leigh, G.J., *ed.* Oxford, Blackwell for the International Union of Pure and Applied Chemistry, 1990. xxi,289p. ISBN: 0632023198; 0632024941, Pbk.

First published 1959; 2nd ed. 1972 known as IUPAC's 'Red book'.

States the 1990 recommendations on nomenclature, including formulas and names of compounds in general, names for ions and radicals. Tables note prefixes or affixes used in the inorganic nomenclature index. 'More comprehensive than the ACS volume on the subjects covered' (*Choice, February 1991, p.958*).
Class No: 546(083.72)

Non-metals
[1224]
REID, R.C., *and others.* **The Properties of gases and liquids.** 4th ed. New York, etc., McGraw-Hill, 1987. x. 742p. illus. ISBN: 0070517991.

First published 1958.

12 sections: 1. The estimation of physical properties -2. Pure-component constants -3. Pressure-volume-temperature relations of pure gases and liquids -4. Volumetric properties and mixtures -5. Thermodynamic properties -6. Thermodynamic properties of ideal gases -7 Vapour pressures and enthalpies of vaporization of pure fluids 8. Fluid-phase equilibria in multicomponent systems -9 Viscosity -10. Thermal conductivity -11. Diffusion coefficients -12. Surface tension. 2. appendices (A. Property data bank: property data values for over 600 pure chemicals). Detailed, analytical index. *Class No:* 546.1/.2

Fluorine
[1225]
SIMONS, J.H., *ed.* **Fluorine chemistry.** New York & London, Academic Press, 1950-64. 5v. tables.

V.4 (1964. xv, 505p.) has 4 contributors (all US). 4 chapters: 1. General chemistry of fluorine-containing compounds - 2. Physical chemistry of fluorocarbons - 3. Radiochemistry and radiation chemistry of fluorine - 4. Industrial and utilitarian aspects of fluorine chemistry (190 references). Analytical subject index. *Class No:* 546.16

Oxygen
[1226]
Advances in oxygenated processes. Baumstark, A.L., *ed.* Greenwich, Conn., London, JAI Press, 1989-. diagrs., tables. ISBN: 0892329505.

V.2 (1990. xii,236p.) has 8 contributors. The 6 documented articles include 'Generation and reactivity of dioxirane intermediates', p.1-59, with 114 references.
Class No: 546.21

Water
[1227]
AMMON, F. von. Wörterbuch der Wasserchemie. [Dictionary of water chemistry. Dictionnaire de la chimie de l'eau.] Deerfield Beach, Florida, VCH Verlagsgesellschaft, 1985. 203p. $58. ISBN: 0895734966.

A German-English-French dictionary, promoted by the

....*(contd.)*

German Chemical Society's Water Chemistry Section. Main coverage: Water composition and quality, water purification, and all aspects of water chemistry. US terms are cross-referenced to British. *Class No:* 546.212

[1228]
FRANKS, F., *ed.* **Water: a comprehensive treatise.** New York, etc., Plenum Press, 1972-. diagrs., tables.

1. *The physics and physical chemistry of water.* 1972. xx, 596p.

2. *Water in crystalline hydrates, aqueous solutions of simple non-electrolytes.* 1973. xix, 683p.

3. *Aqueous solutions of simple electrolytes.* 1973. vii, 472p.

4. *Aqueous solutions of amphiphiles and macromolecules.* 1975. xxi, 839p.

5. *Water in disperse systems.* 1975. xvi, 366p.

6. *Recent advances.* 1979. 466p.

7. *Water and aqueous solutions at subzero temperature.* 1982. 400p.

Each volume carries its own bibliography of references (*e.g.* v.1, p.545-78) and index. V.6 concerns recently-developed methods of testing chemistry and physics of aqueous systems, solutions and hydrates. 'This treatise should be available to all serious research workers with chemical or biological problems in a primarily aqueous world' (*Nature,* v.261, 24 June 1976, p.729, on v.4-5). *Class No:* 546.212

[1229]
TEKNISKA NOMENKLATURCENTRALEN. **Vattenordlista.** [Glossary of water.] Stockholm, the Centre, 1968-74. 3v. (*TNC 41, 45, 54.*) ISBN: 9171960414.

Swedish-base terms and definitions, with French, German and English equivalents and indexes for this water glossary, covering hydrological terms. *Class No:* 546.212

Carbons

[1230]
KOSOLAPOVA, T. **Carbides:** properties, production and applications. New York & London, Plenum Press, 1971. xiv, 298p. diagrs., tables.

States physico-chemical properties and analyses the carbide of each metal by reference to its structure and equilibrium diagram. Separate chapters on methods of production and application. Over 900 references. 'An outstanding book of reference, covering both fundamental and technical aspects' (*Aslib book list,* v.36 (9), September 1971, entry 38). *Class No:* 546.26

[1231]
RODD, E.H., *ed.* **Rodd's chemistry of carbon compounds:** a modern comprehensive treatise. 2nd ed., edited by S. Coffey. Amsterdam, etc., Elsevier, 1964-77. Supplements, 1973-. 4v. and suppts. illus., tables.

First published 1951-62 (10v.).

1. *General introduction. Aliphatic compounds.* 11v.

2. *Aliphatic compounds.* 6v.

3. *Aromatic compounds.* 11v.

4. *Heterocyclic compounds.* v.1-10.

5. *Miscellaneous. General index.*

Full references to original sources. Each volume has an index of compounds, named reactions, reagents and methods. Part E, *Six-membered monohetrocyclic compounds*

....*(contd.)*

... was published as a supplement to Vol. 4 in 1990.

The 1st ed. bridged the gap between the encyclopaedic surveys like Beilstein (*qv*) and textbooks. *Class No:* 546.26

Metals

[1232]
ROBB, C. **Metals databook.** London, The Institute of Metals, 1987. [xvi],500p. diagrs., tables. £90. ISBN: 0904357694.

Data, largely in tabular form, on the chemical, physical and mechanical properties of commercially available metallic materials. Much of the data is specific to British Standards. Covers welding, heat treatment, test house and laboratory data. 4 appendices: 1. S.I. units - 2. Hardness testing - 3. Material specifications - 4. Equilibrium diagrams. No index, but detailed contents list. Small, handy format, but very expensive. *Class No:* 546.3/.9

Rare Earths

[1233]
GSCHNEIDNER, K.A., *Jr and* EYRING, Le R., *eds..* **Handbook of the physics and chemistry of rare earths.** Amsterdam, etc., North-Holland Co., 1978-. graphs, tables.

1. *Metals.* 1978, xxv, 894p.

2. *Alloys and intermetallics.* 1979. xiii, 628p. £110.

3./4. *Non-metallic compounds.* 1979. xiii, 604p; xiii, 602p. £115; £105.

5. *Rare earth alloys and compounds.* 1982. x, 701p. £132.

10. *High energy spectroscopy.* 1987. 611p. £113.

11. *Two hundred year impact of rare earth on science.* 1988. xiv, 594p. £113.

13. *Rare earth elements.* 1990. xii,473p.

Each volume carries documented chapters by specialists. V.5 is an expansion of K.A. Gschneidner's earlier *Rare earth alloys: a critical review of the alloy systems of the earth, scandium and yttrium metals* (1961). Chapter 71, in v.10: 'X-ray absorption and emission spectra,' p.453-549, has 6 pages of references. 'Indispensible for academic libraries and chemistry collections' (*New technical books,* June 1980, item 273). *Class No:* 546.65

[1234]
—McCARTHY, G.J., *and others, eds.* The Rare earths in modern science and technology. New York, Plenum Press, 1978-82. 3v. illus., graphs, tables.

Assembles papers read at the 13th, 14th and 15th Rare Earth Conferences, 1977, 1979 and 1981. Each volume has author and subject indexes. V.1 (1978. xv, 629p.) contains 88 documented papers in 10 subject areas (*e.g.* 'Optical properties of rare earth metals', with 71 references). *Class No:* 546.65

[1235]
—Rare earth bulletin: an interdisciplinary abstracts journal. London, Multi-Science Publishing Co., 1972-. 6 pa. £395pa. ISSN: 03078531.

Includes *c.*1,000 abstracts pa. Sections: Extraction; Chemical properties; Miscellaneous; Phase equilibrium; Crystallography; Nuclear properties; Solid state; Mechanical and accoustic properties; Applications. No indexes. *Class No:* 546.65

[1236]
JEZOWSKA-TRZEBIATOWSKA, B., *and others*. The
Rare elements: occurrence and technology. Amsterdam,
Elsevier; Warszawa, PWN-Polish Scientific Publishers,
1990. x,536p. diagrs., graphs, tables. £131.25. (*Topics in
inorganic and general chemistry, 23.*) ISBN: 0444988777.
Rev. and enlarged translation of the 1976 Polish original.
7 chapters, each with appended bibliography (6: The
technology of rare elements, p.238-496 (copper ... uranium;
each element, with references)). Indexes (subject; name),
p.523-36. *Class No:* 546.65

Organic Chemistry

Databases

[1237]
Beilstein online. Frankfurt, Beilstein Institute.
The online version of *Handbuch der organischen chemie.*
Coverage: 1779-. (*c.*3,500,000 records). Available via
DIALOG and STN; updated 2-4 times per year. The
reviewer in *Choice* (July/August 1991, p.1768) found 'some
improvements over the printed version', but noted that
'almost all the 5th supplement data are merely cited by
literature reference'. DIALOG was found less expensive for
complete records; otherwise the two hosts were competitive
in price. *Class No:* 547(003.4)

[1238]
—'Beilstein online.' *Science & technology libraries,* v.10(4)
Summer 1990, p.125-7.
A discussion of *Beilstein online. Class No:* 547(003.4)

[1239]
—HELLER, S.R. The Beilstein online database:
implementation, content, and retrieval. Washington, DC,
American Chemical Society, 1990. vii,168p. graphs, tables,
chemical structures. $34.95. (*ACS symposium series, 436.*)
ISBN: 0841218625.
Invaluable guide to 'Beilstein online'. Provides an
introduction and overview of the database and hints for
search techniques and strategies. Detailed, analytical index,
p.157-68. *Class No:* 547(003.4)

[1240]
Chapman & Hall chemical database. London, Chapman &
Hall. *c.*100,000 records.
Formerly *HEILBRON.*
Data on over 200,000 substances, drawn from the
publisher's sources, *e.g. Dictionary of organic compounds,
Dictionary of organometallic compounds, carbohydrates,
amino acids and peptides,* Updated semi-annually. DIALOG,
File 303. *Class No:* 547(003.4)

CD-ROM

[1241]
**The Chapman & Hall chemical database: the natural
products on CD-ROM.** London, Chapman & Hall, 1992.
£2950, full subscription; £1750, renewal.
Over 80,000 substances in 33,000 entries. Data: name;
synonyms; structure display; hazards; toxicity; CAS Registry
number; colour; melting/boiling points; refractive index;

.... *(contd.)*
pharmacological and other uses, etc. Each entry carries a
bibliography. IBM PC-AT or compatible micro.
Class No: 547(003.40)

Encyclopaedias

[1242]
Dictionary of organic compounds: the constitution of
physical, chemical, and other properties of the principal
carbon compounds and their derivatives, together with the
relative literature. Buckingham, J., *ed.* 5th ed. London,
Chapman & Hall, 1982. 7v. (7848p.). Annual supplements,
1983- . £1,175. ISBN: 0412170000.
First published 1934-37; 3rd ed., edited by Sir I. Heilbron
and H.M. Bunbury (1953. 4v.).
The main work has data on 150,000 of the most common
and important organic compounds for chemists, biochemists,
pharmacologists and biologists. Documented. Five indexes:
Name index; Molecular formula index; Heteroatom index;
Chemical Service Registry number; Cumulative structure.
Known as 'DOC 5'. 'An essential adjunct wherever organic
compounds are used' (*Nature,* v.300, 25 November 1982,
p.336). Supplements have cumulative indexes.
Available online in the *Chapman and Hall Chemical
database* (formerly *HEILBRON*); DIALOG, file 303.
Class No: 547(031)

Handbooks & Manuals

[1243]
Beilstein's Handbuch der organischen chemie. 4. Aufl. hrsg.
und bearb. von F. Richter (from 1948; previously by
Deutsch Chemische Gesellschaft). Berlin, etc., Springer-
verlag, 1918- .
The 4th ed. of the definitive work (*Hauptwerk*) covered
the literature on *c.*200,000 compounds up to the end of
1909. Supplements (*Ergänzungswerke*): 1-4 cover 1910-19,
1920-29, 1930-49 and 1950-59 respectively. A 5th
supplement, for 1960-69 began publication in English in
1964. Each supplement consists of 27v. Subject and formula
indexes to both main work and supplements are proceeding.
Data (where applicable): name(s), formula, structure(s),
history, occurrence, formation, preparation, properties
(physical, chemical and physiological, technology, analysis,
etc. Each volume contains indexes, list of abbreviations for
sources, and classifed table of contents. Beilstein is the most
extensive reference work to the literature of organic
chemistry, including patents. *Class No:* 547(035)

[1244]
—BRITISH LIBRARY. Science Reference Library. An
Introduction to Beilstein's Handbuch der organischen
Chemie. Rev. ed. London, SRL, 1985. 9p. *Gratis.* ISSN:
03064301.
In 6 sections: 1. Introduction - 2. Information given in
Beilstein - 3. Structure - 4. Indexes - 5. Search procedures:
based on the indexes - 6. Future developments. Appendix 1:
List of the more common abbreviations used in Beilstein and
their meaning in English - 2. A bibliography of guidebooks
to Beilstein's Handbook. *Class No:* 547(035)

[1245]

—How to use Beilstein: Beilstein's Handbook of organic chemistry. Berlin etc., Springer, [1984]. 34p. chemical structures. *gratis*.

In 4 sections: 1. Questions about Beilstein; 2. How to search for and find compounds in Beilstein; 3. The history of Beilstein; 4. Literature on Beilstein, p.34.

Class No: 547(035)

[1246]

HOUBEN, J. *and* WEYL, T. **Methoden der organischen Chemie.** 4. vollig neue gestaltete Aufl., hrsg. von E. Müller. Stuttgart, Thieme, 1952. illus., tables.

1, pt.1: *Allgemeine Laboratorium spraxis*. 1958.

2: *Analytische Methoden*. 1953.

3, pt.1: *Physikalische Forchungsmethoden*. 1955.

4, pt.1: a-b: *Oxidation*. 1981; c-d: *Reduktion*. 1980.

4, pt.2: *Allgemeine chemische Methoden*. 1955.

4, pt.3: *Carbocyclische Driering-Verbindungen*. 1971.

4, pt4: *Isocyclische Driering-Verbindungen*. 1971.

5, pt.1: a-d *Kohlenwasserstoffe*. 1970-72.

5, pt.2a: *Alkine d-und Polyine Allene, Kumulene*. 1977.

5, pt.2b: *Arene und Arine*. 1951.

5, pt.3-4: *Halogen-Verbindungen*. 1962, 1960.

6, pt.1, a-b: *Alkohole*. 1979-84.

6, pt.1, c: *Phenole*. 1976.

6, pt.1, d: *Enole*. 1978.

6, pt.2-4; 7, pt.1,4; 8: *Sauerstoff-Verbindungen*. 1963-66, 1954, 1968, 1952.

7, pt.2, a-c: *Ketone*. 1973-77.

7, pt.3, a-c: *Chinone*. 1977-79.

9: *Schwefel-, Selen-, Tellur-Verbindungen*. 1955.

10, pt.1-4; 11, pt. 1-2: *Stickstoff-Verbindungen*. 1965-71.

12, pt.1-2; E, pt. 1-2: *Organische Phosphor-Verbindungen*. 1963-64, 1982.

13, pt. 1-2a, pt. 4-8, pt. 9b: *Metallorganische-Verbindungen*. 1970-84.

13, pt. 3 a-c: *Organobor-Verbindungen*. 1982-83.

14, pt. 1-2: *Makromolekulare Stoffe*. 1961-63.

E 3: *Aldehyde*. 1983.

E 4: *Kohlinsaure Derivate*. 1983.

E 5, pt. 1-2: *Carbonsauron and Carbonsäure Derivative*. 1965.

E 11, pt.1-2: *Organische Schwefel-Verbindungen*. 1985.

Subject index for v. E4: author index for v. E4, E5, E11, (1987).

The 4th ed. is to be in 16v. (v.16 index), comprising 57 parts by 1987, with supplementary volumes (Erweiterungs- und Folgebände), *i.e.*, v.E3,4,5,11, to cover developments of the past two decades. *Class No:* 547(035)

[1247]

MARCH, J. **Advanced organic chemistry:** reactions, mechanisms, and structure. 4th ed. New York, Wiley, 1992. xv,1495p. diagrs., tables, chemical formulas. £45. ISBN: 0471601802.

19 chapters; 2 appendices: A: 'The Literature of organic chemistry' (p.1239-68); B: 'Classification of reactions by type of compound synthesized' (p.1269-1300). Indexes: Author (p.1301-1431); Subject (p.1433-95). Primarily a textbook, but valuable for the more than 10,000 references included. *Class No:* 547(035)

Dictionaries

[1248]

GOLD, V., *ed.* **'Glossary of terms used in physical organic chemistry'.** Provisonal ed., 1979. In *Pure and applied chemistry*, v.51, 1979, p.1725-1801. diagrs., chemical structures.

About 130 contributors. A dagger against the title of an entry 'implies that we recommend discontinuation of use'. 2½p. of references appended. *Class No:* 547(038)

[1249]

PATAL, S. **Glossary of organic chemistry** including physical organic chemistry. New York & London, Interscience: Wiley, 1962. xiv, 227p. chemical structures.

Concise definitions; references frequent. Many structural formulas; also, reactions. Deliberately omits most analytical methods. 'Writen mainly for ... those who are not specialists (or not yet specialists) in Organic and Physical Organic Chemistry, but who have need and occasion to use the relevant literature' *(Preface)*. *Class No:* 547(038)

Reviews & Abstracts

[1250]

Index of reviews in organic chemistry. London, Royal Society of Chemistry, 1968 -. Annual.

Originally as *Index of reviews in organic chemistry*, compiled by D.A. Lewis (Welwyn Garden City, Herts., ICI Ltd) with cumulate issue (1971) and annual supplements by London, Chemical Society - now Royal Society of Chemistry.

IROC 85 (1986. ix, 55p. £22) records *c*.1,000 review articles in English, German or French during 1985. Two sections: 1, covering individual compounds and classes of compounds; 2, covering chemical processes and phenomena (*e.g.* reaction types; spectra). List of *specialist periodical reports*, books cited, and conference papers. *Class No:* 547(048)

Progress Reports

[1251]

Advances in physical organic chemistry. New York, London etc., Academic Press, 1965-, v.1-. Annual. ISBN: 0120335212.

V.27, 1992 (viii,312p.) has 5 contributors. The 4 papers include: 'Effective change and transition-state structure in solution' (with 5p. of references) and 'The principle of non-perfect synchronization' (also with 5p. of references). Author index, to references. Cumulative indexes of authors and titles, v.1-27. *Class No:* 547(055)

[1252]

Progress in physical organic chemistry. New York, Wiley: Interscience, 1971-. v.1-. Annual. ISBN: 0471814741.

V.15, 1985 (ix, [1], 362p.) has 8 contributors. The 4 articles include 'Structural principles of unsaturated organic compounds: evidence by quantum chemical calculations' (with 132 references). Author and subject indexes; cumulative index (titles of articles), v.1-15. *Class No:* 547(055)

Tables & Data Books

[1253]
CRC handbook series in organic electrochemistry.
Meites, L. *and* Zeiman, P. Cleveland, Ohio (later Boca Raton, Fla.), CRC Press, 1977-83. 6v. tables, chemical structures. ISBN: 0849372208.

V.6 ([viii], 538p.) is largely tabular, - 11 main tables (1. Electrotechnical data, p.5-333), including a key to literature citations. Author index. 'Although these six volumes contain the largest collection of organic electrochemical data ever made, ... they include only a fraction of the data that have appeared in the original literature' (*Preface*, v.6). Thus, electrochemical techniques are excluded.
Class No: 547(083)

Nomenclatures

[1254]
INTERNATIONAL UNION OF PURE AND APPLIED CHEMISTRY. Nomenclature of organic chemistry. 4th ed. Oxford, etc., Pergamon Press, 1979. xix, 359p. tables, chemical structures. ISBN: 0080223699.

Sections, A,B,C,D,E,F and H. (Sections A and B were first published in 1958, section C, in 1965, and sections D,E,F,H, in 1979).

Sections: A Hydrocarbons - B. Fundamental heterocyclic systems ... D. Conductivity ... H. Isotopically modified compounds. *Class No:* 547(083.72)

[1255]
NICKON, A. *and* SILVERSMITH, E.F. Organic chemistry: the name game: modern coined terms and their origins. New York, etc., Pergamon, 1987. xii,347p. illus., chemical structures. ISBN: 008034481x; 0080351573, Pbk.

Covers several hundred distinctively named reactions and structures. 20 chapters, each on a common theme; *e.g.* 9. 'Join organic and see the world' discusses, among others, the plant alkaloid 'istanbulamine' (collected in western Anatolia, Turkey), and 'manxane' (synthesized by researchers in Northern Ireland and Scotland, who were struck by its structural similarity to the coat of arms of the Isle of Man). Extensive chapter references. 6 appendices (A. Etymology of some traditional chemical names ... F. Nobel prizes in the sciences). Name index, 9p. An unusual, humorous treatment. Highly recommended in *Choice*, May 1988, p.1429. *Class No:* 547(083.72)

Reactions

[1256]
GOWAN, J.E. *and* WHEELER, T.S. Name index of organic reactions. 2nd ed. London, Longmans, 1960. v, 293p. diagrs.

First published 1950, by The Society of Chemical Industry.

739 named and numbered reactions, A-Z, showing chemical structures. Colour reactions are not shown (see *Merck index* for these). For each reaction, reference to the original literature, *Organic reactions, Organic syntheses,* etc. Type of reaction index; general (analytical index). No biographical data (for which see A.R. Surrey's *Name reactions in organic chemistry* (2nd ed. New York, Academic Press, 1961). Considered singularly free from mistakes (*Chemistry & Industry*, 24 June 1961, p. 937-8). *Class No:* 547.03

[1257]
Organic reactions. New York, Wiley, 1942-. Chemical formulas.

V.42 (1992. xix,669p. £71. ISBN: 0471544108) has 2 reviews ('The Mitsunobu reaction' p.335-656, carries 887 references). 'Cumulative chapter titles by volume' p.vii-xviii. Indexes to vols. 1-42: author; chapter and topic.
Class No: 547.03

Synthesis

[1258]
Comprehensive organic synthesis: selectivity, strategy and efficiency in modern oroganic chemistry. Trost, B.M., *ed.* Oxford, Pergamon Press, 1991. 9v. tables, chemical formulas. $3000. ISBN: 0080359229, set.

V. 1-2. *Additions to C-X -bonds.* xxi,989p.; xxiii,1233p.
V. 3. *Carbon-carbon -Bond formation.* xxi,1186p.
V. 4. *Additions to and substitutions at C-C -bonds.* xxi,1299p.
V. 5. *Combining C-C -bonds.* xxv,1333p.
V. 6. *Heteroatom manipulation.* xxi,1194p.
V. 7. *Oxidation.* xxv,1012p.
V. 8. *Reduction.* xxv,1139p.
V. 9. *Cumulative indexes.* xv,811p.

Over 250 international contributors. V. 1 has 22 closely subdivided chapters, each with extensive references. A major work. *Class No:* 547.057

[1259]
Organic syntheses; an annual publication of satisfactory methods for the preparation of organic chemicals. New York, Wiley; London, Chapman & Hall, 1971-.

The series' main object is to state the most convenient laboratory methods for preparing various substances in ½ to 5lb. lots, but as far as possible also adaptable to large-scale development. V.69 (1990) records '30 checked procedures of value to the modern practicing chemist'. Important in all areas of analytical chemistry as reference sources.

Collective volumes, with cumulative indexes, act as revised editions of v.1-9, 10-19, etc. *Class No:* 547.057

[1260]
The Total synthesis of natural products. Ap Simion, J., *ed.* New York, Wiley, 1973-.

9 vols. have appeared so far. Each contains one or more articles on an aspect of organic synthesis. V.9 (1992. ISBN: 0471551899. £98) contains 'The synthesis of insect pheromones, 1979-1989', by K. Mori of 534p. and 1,229 references. *Class No:* 547.057

Ring Index

[1261]
PATTERSON, A.M., *and others, eds*. The Ring index: a list of ring systems used in organic chemistry. 2nd ed. Washington, American Chemical Society, 1960. Supplements 1-3, 1963-65.

First published 1940.

The main volume (1452p.) has 7,272 numbered entries, arranged from one-ring to twenty-two ring systems. Structural formulas are provided. Index of names (p.1307-1410); numerical index. *Supplements* 1-3 record systems 7,728-14,265.

'Completeness of coverage of the literature is the result of the co-operative efforts of the indexing staff of *Chemical*

....(contd.)

abstracts who record all new ring systems as they are encountered in their indexing' (*Preface*, Supplement 1). The Chemical Abstracts Service's *Parent compound handbook* (Columbus, Ohio 1976, plus supplements, 6pa.) provides current coverage. *Class No:* 547.12

Compounds

[1262]

BARTON, Sir D. *and* **OLLIS, W.D. Comprehensive organic chemistry;** the synthesis and reactions of organic compounds. Oxford, etc., Pergamon Press, 1979. 6v. (*c.*8,000p).

1. Stereochemistry, hydrocarbons, halo compounds, oxygen compounds. 2. Nitrogen compounds, carboxylic acids, phosphorous compounds. 3. Sulphur, silenium, silicon, boron, organometallic compounds. 4. Heterocyclic compounds. 5. Biological compounds. 6. Author, formula, subject, reagent, reaction indexes.

Emphasises throughout the properties and reactions of all the important classes of organic compounds, including those compounds prepared by synthesis as well as natural products created by biosynthesis. Over 100 contributors; over 20,000 literature references. Actual coverage of v.6 (xii, 1,628 p.): formula, subject, author, reaction and reagent. 'A must for all chemistry collections of importance' (*New technical books*, v.64(10), December 1979, entry 1493). *Class No:* 547.2

[1263]

CRC handbook of data on organic compounds. Weast, R.C., *ed.* 2nd ed. Boca Raton, Florida, Chemical Rubber Co., 1989. 9v. tables. $2,400. ISBN: 0849304202.

First published 1985.

Greatly expanded 2nd ed. Physical properties and spectral data for 30,000 of the most common organic compounds. Data: molecular weight, boiling point, melting point, density, refractive index, specific rotation, colour, solubility. Spectral data include infra-red, Raman, ultraviolet, nuclear magnetic resonance, mass spectroscopy, plus CAS Registry Number, CAS index no., synonyms, chemical structure, line and molecular entries. Indexes for melting points, etc. Annual supplements planned. *Class No:* 547.2

[1264]

CRC handbook of tables for organic compound identification. Rappoport, Z., *comp.* 3rd. ed. Cleveland, Ohio, CRC Press, 1967. ix, 564p. graphs, tables, chemical structures.

First published 1960 (Jerusalem, Science Press) as *Tables for identification of organic compounds.*

Identifies more than 8,150 parent compounds in all, A-Z (p.1-427). Appended tables of constants, etc. Index of organic compounds, p.479-560. A further index lists names of tables and major subjects. *Class No:* 547.2

[1265]

DREISBACH, R.R., *comp.* **Physical properties of chemical compounds.** Washington, American Chemical Society, 1957-61. 3v. tables.

Treats 1,421 substances. V.1 has sub-title: 'A systematic tabular representation of accurate data on the physical properties of 511 cyclic compounds'. V.2 continues the survey, dealing with 476 organic straight-chain compounds. V.3 presents data on the physical properties of 414 alipathic

....(contd.)

compounds, 22 miscellaneous compounds and elements of these. Each volume has an index, cumulating in v.3. *Class No:* 547.2

[1266]

TIMMERMANS, J. Physico-chemical constants of pure organic compounds. v.1. London, Cleaver Hume; v.2. Amsterdam, etc., Elsevier, 1965. 2v.

'The results of determination up to the end of 1964 for a large number of classified compounds' (*Aslib book list*, v.30(10), October 1965, entry 513). *Class No:* 547.2

[1267]

UTERMARK, W. *and* **SCHICKE, W. Melting point tables of organic compounds.** 2nd new and supplemented ed. New York, Interscience, 1963. xxxii, 715p.

Tabulates names, formulas, molecular weights, specific gravity and boiling points of 3,213 compounds, arranged in order of melting points. *Class No:* 547.2

Hydrocarbons

[1268]

GALLANT, R.W. Physical properties of hydrocarbons. Houston, Texas, Gulf Publishing, 1968-70. 2v. diagrs., graphs, tables.

V.1 (Chapters 1-22) states in graphic form the temperature variations of vapour pressure, heat of vaporization, heat capacity, density, viscosity, thermal conductivity and surface tensions of hydrocarbons, with references to late sources. V2 (Chapters 23-44) had a 2nd ed. in 1984 (viii, 214p.): Aldehydes ... Water. Appended conversion tables; 304 references. Index by formula; index by compound names. *Class No:* 547.21

Organometallic Compounds

[1269]

Advances in organometallic chemistry. New York, Academic Press, 1963-. diagrs., chemical structure, tables. ISBN: 012031133x.

V.33 (1992. 401p.) covers boron atoms in transition metal clusters and the analogy between structure and bonding in parent hydrides and multiple bonded silicon and germanium compounds. *Class No:* 547.25

[1270]

Dictionary of organometallic compounds. Buckingham, J., *ed.* London, etc., Chapman & Hall, 1984. Supplements 1985-. 3v. (*Supplements, by J.E. Macintyre.*) ISBN: 0412247100.

1. *Ag-Mn.* xviii, 1278p.

2. *Mo-Zr.* viiip, p.1279-2468.

3. *Name index. Molecular formula index. CAS registry Number index.* vii, 764p.

Supplements 1-3. 1985-7.

V.1-2 cover, in element sections, carefully selected useful and important organometallic compounds and their derivatives, - 16,000 entries, describing *c.*40,000 compounds. *First supplement* lists *c.*2,300 entries, each *Supplement* carrying name, etc., indexes. Data on entries: chemical structure, chemical and physical properties, and literature references. The work is designed as a companion to the *Dictionary of organic compounds* (*qv*), earlier edited

....(contd.)

by I.M. Heilbron. Annual suppts. 'It may be regarded as the Red Bible' (*Nature*, v.314, 21 March 1985, p.296). Heilbron online, via DIALOG. *Class No: 547.25*

[1271]
Organometallic compounds: abstracts of literature and patents relating to compounds which contain at least metal, carbon and hydrogen atoms. Braintree, Essex, R.H. Chandler, Ltd., 1961-. Biweekly. tables, chemical structures. ISSN: 00305136.

About 8,500 abstracts (mostly brief) pa., under compounds A-Z. Author index per issue. Six-monthly subject, author and patent name indexes. *Class No: 547.25*

Carbohydrates

[1272]
Carbohydrates. Collins, P.M. London, Chapman & Hall, 1987. xi,720p. chemical structures. £95. (*Chapman & Hall chemistry sourcebooks.*) ISBN: 0412269600.

Data largely derived from the *Dictionary of organic compounds* (5th ed. 1982) and supplements, but with revision and update. Indexes: Name; Molecular formula; CAS Registry number; Type of compound. *Class No: 547.454*

Amino Acids

Organophosphorous

[1273]
EDMUNDSON, R.S., *ed. and comp*. **Dictionary of organophosphorus compounds.** London, Chapman & Hall, 1988. xxi, 1347p. tables, chemical structures.

Entries (p.1-880) in alphanumeric order. Introduction includes 'Sources of further information', p.xiii-xxii. Name index, p.881-987; Molecular formula index, p.989-1118; CAS Registry Number index, p.1119-120. Type of compound index, p.1203-1347. Many compounds are included as derivatives of main-entry compounds. Heilbron (file 303) available online via DIALOG. *Class No: 547.5*

Heterocyclic Compounds

[1274]
Advances in heterocyclic chemistry. New York, London, etc., Academic Press. 1949- Annual. ISBN: 0120206382.

V.53, 1992 (vii,429p.) has 7 contributors. The 3 documented papers include 'N-A minrazoles' (p.86-231) with about 700 references. Cumulative indexes (author, title, subject) for v.46-53. *Class No: 547.7*

[1275]
Comprehensive heterocyclic chemistry: the structure, reactions, synthesis and uses of hetrocyclic compounds. Katritzky, A.R. *and* Rees, C.W., eds. Oxford, etc., Pergamon, 1984-85. 9 vols. diagrs., tables. ISBN: 0080262008, set.

Vol. 1. *Introduction, Nomenclature, review literature, biological aspects, industrial uses, less-common heteroatoms.* O. Meth-Cohn, ed. xvi,731p.

Vol. 2. Part 2A. *Six-membered rings with one nitrogen atom.* A.J. Boulton, A. McKillop, eds. xiv,689p.

Vol.3. Part 2B. *Six-membered rings with oxygen, sulfur or*

....(contd.)

two or more nitrogen atoms. A.J. Boulton, A. McKillop, eds. xiv,1210p.

Vol. 4. Part 3. *Five-membered rings with one oxygen, sulfur or nitrogen atom.* C.W. Bird, G.W. Cheeseman, eds. xiv,1195p.

Vol. 5. Part 4A. *Five-membered rings with two or more nitrogen atoms.* K.T. Potts, ed. xiv,994p.

Vol. 6. Part 4B. *Five-membered rings with two or more oxygen, sulfur or nitrogen atoms.* K.T. Potts, ed. xiv,1171p.

Vol. 7. Part 5. *Small and large rings.* W. Lwowski, ed. xiv,867p.

Vol. 8. Part 6. *Indexes.* C.J. Drayton, ed. p.xvi,1111p.

Vol. 9. *Handbook of heterocyclic chemistry.* 1985. xxii,542p.

The aim of the work is two-fold: to give an up-to-date overview of the subject as a whole and to provide enough detail to answer specific questions and to direct attention to detailed reviews. Numerous contributors. Contents include general reviews of the literature and monograph studies of specialist topics. Indexes: physical data, authors, subjects, rings. V. 9, *Handbook*, is a digest of the general chapters in V. 1-8, and therefore suitable for libraries not requiring the full text. *Class No: 547.7*

Natural Substances

[1276]
DEVON, T.K. *and* **SCOTT, A.I., eds**. **Handbook of naturally occurring compounds.** New York, London, etc., Academic Press, 1972-75. 2v. diagrs., chemical structures. ISBN: 0122130012; 0122136020.

1. *Acetogenins, shikimates, and carbohydrates.* 1975, xi, 644p.

2. *Terpenes.* 1972. xi. 576p.

Each volume consists of structural chemical formulas, with A-Z, molecular weight, and molecular formula indexes. A proposed v.3 and annual supplements never appeared. *Class No: 547.9*

Fats

[1277]
GUNSTONE, F.D., and *others*. **The Lipid handbook.** London, Chapman & Hall, 1986. 900p. £175. ISBN: 0412244802.

A monograph on fatty acids and lipids. Occurrence and characteristics of oils and fats precede a reference section on the physical and chemical properties of *c*.2,000 lipids (A-Z) and their derivatives. Data derive from the *Dictionary of organic compounds* database. Name index. Molecular formula index. CAS Registry Number index. *Class No: 547.915*

Alkaloids

[1278]
GLASBY, J.S. Encyclopedia of the alkaloids. New York & London, Plenum Press, 1975-83. 4v. ISBN: 0306308452.

V.1-2 (A-H, I-Z. 1423p.) provide data on the occurrence, structure and physical constants of more than 3,000 alkaloids, their salts and derivatives, chemical structure; references. V.3 (1977) and v.4 (1983), both A-Z, record compounds reported in the literature (to the end of October

.... *(contd.)*

1976 and for 1977-81 respectively). V.3 includes a formula index for v. 1-3 and v. 4 also carries a formula index. *Class No:* 547.94

Proteins

[1279]

Advances in protein chemistry. New York, London, etc., Academic Press, 1949-. Annual. ISBN: 012034243x.

V.43 (vi,392p. 1992) has 7 contributors. The four papers include 'Structure and stability of bovine casein miscelles' (p.63-152) with 9p. of references. *Class No:* 547.96

Crystallography

Bibliographies

[1280]

MEGAW, H.D., *with others*. **Crystallographic book list.** Cambridge, H.D. Megaw, c/o Cavendish Laboratory, 1965. Supplement, 1966. xii, 83p.

5 lists: 1. Books, in English, French, German, Spanish and Russian, - but not Japanese - (authors, A-Z; data include reviews; subject reference; teaching level) -2. Conferences (in chronological order) -3. Serial publications (titles, A-Z) - 4. Subject classification of books -5. Appendix (additional items).

Second supplement in *Journal of applied crystallography* (v.5, 1972, p. 148-62). Latest supplement, - in the same journal (v.15, 1982, p.640-76) as 'Current crystallographic books, 1970 through 1981', edited by J.H. Robertson. *Class No:* 548(01)

[1281]

—**MACKAY, A.L.** 'Crystallography'. In Shaw, D.F., *Information sources in physics.* London, Butterworth, 1985. p. 133-45.

Provides a concise and recent survey of the literature under 16 heads: Historical - Crystallographic book lists - Periodicals - Databooks - Services and databases - Serials - Geometry, symmetry and mathematics - Computing - Diffraction theory - Structure analysis and general crystallography - Crystal chemistry - Crystal physics - Imperfections, morphology and growth - Electron microscopy and diffraction - Neutron diffraction and synchroton radiation - Biological structure - References (7). *Class No:* 548(01)

Handbooks & Manuals

[1282]

HAMMOND, C. Introduction to crystallography. Oxford Univ. Press, 1990. 104p. £9.95. ISBN: 0198364236.

A step-by-step, user-friendly approach to the basics of crystallography. Includes biographical notes, a further reading list and exercises. 'For microscopists and science students'(*New technical books,* March/April 1991, entry 393). *Class No:* 548(035)

[1283]

VAINSTEIN, B.K., *and others, eds.* **Modern crystallography.** Berlin, etc., Springer, 1979-84. 4v.

First published Moscow, Nauka, 1979.

1. *Symmetry of crystals, methods of structural crystallography*, by B.K. Vainstein.

2. *Structure of crystals*, by B.K. Vainstein, and others.

3. *Crystal growth*, by A.A. Chemov, and others.

4. *Physical properties of crystals*, by L.A. Shumov, and others.

By 3 members of the Institute of Crystallography of the USSR Academy of Sciences. Describes all branches of crystallography in their interrelation, presenting 'a unified science' (*Foreword*). Each volume carries bibliography, references and subject index. Chapters vary in length (*e.g.* 'Crystalline state', v.1, p. 1-26; 'Structure analysis of crystals', v.1, p. 223-376). Intended for research workers, scientists, engineers and students. *Class No:* 548(035)

Dictionaries

[1284]

BACKHAUS, K.O. Technik-Wörterbuch Kristallographie: Englisch-Deutsch-Französisch-Russisch. Berlin, VEB Verlag Technik, 1972. 132p.

About 3,600 English-base terms, with German, French and Russian equivalent and indexes. Coverage: geometric crystallography, physico-chemical crystallography, crystal chemistry and physics. *Class No:* 548(038)

Reviews & Abstracts

[1285]

Bulletin signalétique. 161. Structure de l'état condensé. Cristallographie. Paris, Centre National de la Recherche Scientifique, 1940-83. Monthly.

About 9,000 abstracts and references pa. Main sections: A. Structure des liquides et des solides. Cristallographie - B. Surfaces et interfaces, couches minces et trichites (structure et propriétés non electriques) - C. Techniques et croissance cristalline et de depôt de couches minces. Monthly and annual author and subject indexes.

As *Pascal Explore* 12. État condensé; 13. *Structure des liquides et des solides. Cristallographie*, since 1984. Online. *Class No:* 548(048)

[1286]

Zentralblatt für Mineralogie. Stuttgart, E. Schweizerbart'sche Verlagsbuchhandlung, 1950-. Monthly. ISSN: 05147123.

Preceded by *Jahrbuch für Mineralogie* (1830-32), *Neues Jahrbuch für Mineralogie* (1833-1942) and *Zentralblatt für Mineralogie, Geologie und Paläontologie* (1943-49).

Teil 1: *Kristalographie und Mineralogie.*

2: *Petrographie, techische Mineralogie, Geochemie und Lagerstättenkunde.*

Teil 1 (7pa., pt.7: annual author & subject indexes) has *c.*3,000 abstracts and references pa, in sections. Teil 2 (13pa., pt.13: annual author & subject indexes) has *c.*5,000 abstracts and references pa., in sections A-N. *Class No:* 548(048)

Yearbooks & Directories

[1287]

World directory of crystallographers, and of other scientists employing cyrstallographic methods. 7th ed. Dordrecht, Reidel, for the International Union of Crystallography, 1986. 270p. ISBN: 9027720940.

First published 1957; 6th ed., 1981.

A directory of 8,968 crystallographers in 69 countries, A-Z (Algeria ... Zimbabwe, plus German Democratic Republic). Data include name, address, year of birth, status, and major fields of scientific interest. Name index.

Class No: 548(058)

Tables & Data Books

[1288]

International tables for crystallography. 2nd rev. ed. Dordrecht, Reidel, for the International Union of Crystallography, 1987-. (*3rd series of International tables*; 1st series, 1935.) ISBN: 9027722809.

A. *Space group symmetry.* 1987. xvi, 878p. Brief teaching edition (120p.) was published in 1988.

Volume A (pt.1: Tables for plane groups and space groups has 15 documented sections, plus subsections) treats the symmetries of one-, two- and three-dimensional space groups and point groups in direct space. Detailed features of each space group appear under 13 heads.

Volumes B (*Reciprocal space*) is in preparation.

C. *Mathematical and physical tables.* 1992. 500p.

Class No: 548(083)

[1289]

International tables for X-ray crystallography. Dordrecht, Reidel, 1952-83. 4v. tables. ISBN: 902271956x, v.2.

1. *Symmetry groups*; edited by N.F.M. Hendry and K. Lonsdale. 1952.

2. *Mathematical tables*; edited by J.B. Kasper and K. Lonsdale. 1982.

3. *Physical and chemical tables*; edited by C.H. MacGillavry and other 1983.

4. *Revised and supplementary tables*; edited by J.A. Ivers and W.C. Hamiton 1982.

Each volume carries a dictionary of crystallographic terms/phrases in English with equivalents in French, German, Russian and Spanish. V.3 has a subject index; v.4 applies to tables in v. 2-3. 'The crystallographer's Bible' (Coblans, H., ed., *Use of physics literature* (1975), p.237).

Will gradually be replaced by the 2nd rev. ed.

Class No: 548(083)

[1290]

PORTER, M.W. *and* **SPILLER, R.C. The Barker index of crystals:** a method for the identification of crystalline substances. Cambridge, Heffer, for the Barker Index Committee, 1951-65. 3v. (7 pts.).

1. (2 pts.). *Crystals of the tetragonal, hexagonal, trigonal and orthorhombic systems.* 1951.

2. (3 pts.). *Crystals of monoclinic systems.* 1956.

3. (2 pts.). *Crystals of the anorthic systems.* 1965.

An index based on geometric form rather than on crystal structure as described in T.V. Barker's *Systematic crystallography* (London, Murby, 1930). V.1 describes 2,991 compounds; v.2, *c.*3,500. *Class No:* 548(083)

Maps & Atlases

[1291]

LIMA-DE-FARIA, J., *ed.* **Historical atlas of crystallography.** Boston, Kluwer Academic, 1990. 158p. $36. ISBN: 079230649x.

Includes bibliography on the history of crystallography. Indexes. 'An important contribution ... particularly useful for the teaching of crystallography' (*Choice,* April 1991, p.1339). *Class No:* 548(084.3)

Liquid Crystals

[1292]

Advances in liquid crystals. New York, London, etc., Academic Press, 1978- Annual. diagrs. tables. ISBN: 0120250055.

V.5, 1982 (xv, 240p.) has 8 contributors. The six documented papers include 'Investigation of liquid crystals by Mossbauer spectroscopy' (with 59 references) and 'Neutron incoherent scattering studies of liquid crystals' (with 30 references). Contents of previous volumes (v. 1-4) precedes. Analytical index. *Class No:* 548.56

Crystal Structure

[1293]

Structure reports: a critical guide to papers on crystallography of metal alloys and compounds. Utrecht, Oostboek, for the International Union of Crystallography, 1940/41-.v.8-. Annual. ISSN: 01667033.

Continues *Strukturbericht,* 1913-39 (Leipzig, Akademie Verlagsgesellschaft, 1931-43. 7v.).

Each annual now (1965v.-) has 2 parts: A. Metals and inorganic sections; B. Organic compounds (including organometallic compounds). The 2-part *Report* v.48, for 1981 (1852p.) appeared in 1986. Data normally cover name, formula, papers reported, unit cell and space group, analysis, atomic positions, plus detailed description and discussion of the structure. Subject, formula and author indexes. Cumulative subject/formula indexes (1913-1980); ten-year author index (1971-1980). 60-year structure index (1913-1973). *Class No:* 548.7

[1294]

—**WYCKOFF, R.W.G.** Crystal Structures. 2nd ed. New York, London, etc., Interscience; Wiley. 6v. diagrs., tables.

First published 1948 (4v.).

V.2-3 (1964-66) cover inorganic compounds. The volumes contain data on each structure, plus a bibliography and name and formula indexes. Now somewhat dated.

Class No: 548.7

Mineralogy

Bibliographies

[1295]

O'DONOGHUE, M. The Literature of mineralogy. London, British Library, Science Reference and Information Service, 1986. 85p. diagrs. £12. ISBN: 0712307370.

10 sections: Major reference monographs - Abstracts - Databases - Current awareness publications and review

.... *(contd.)*

literature - General comprehensive surveys - Crystals - Mineral formation and deposits - Mineral identification - Journals: general; for the collector - Regional mineralogy - Clay minerals - Mineral names - Glossaries and dictionaries - Directories - Maps - Addresses. Detailed index, p.78-85. 9p. of line-drawings (142 figures). A valuable, up-to-date reference tool, with concise annotations. *Class No:* 549(01)

Encyclopaedias & Dictionaries

[1296]

CARR, D.D. *and* **HERZ, N.,** *eds*. **Concise encyclopedia of mineral resources.** Cambridge, Mass., MIT Press, 1989, Oxford, Pergamon, 1989. 440p. illus. $145; £80. (*Advances in materials science and engineering, v.8*.) ISBN: 0262031558, US; 0080347347, UK.

Articles, A-Z. Subject index. Derived from the *Encylopedia of materials science and engineering* (*qv*), with new chapters and revisions; cross-references. *Class No:* 549(03)

Encyclopaedias

[1297]

DEESON, A.F.L., *ed*. **The Collector's encyclopedia of rocks and minerals.** Newton Abbot, Devon, David & Charles, 1975. 288p. col. illus., tables.

Introductory tables list rocks (*c*.400) by type, and minerals (*c*.350), with basic data. The minerals section (A-Z) notes chemical formula, composition, crystal system, habit, colour, lustre, streak, fracture, cleavage, hardness test, specific gravity, special features, methods of identification, occurrence, environment, varieties and synonyms. 'Designed to assist the amateur collector'. *Class No:* 549(031)

[1298]

The Encyclopedia of gemstones and minerals. Holden, U. *and* Mathez, E.A., *eds*. New York, Facts on File, 1992. 320p. illus. £24.95. ISBN: 0876021775.

Properties/values of over 200 rocks, stones, jewels and other mineral substances. 400 illus. Data on characteristic form, specific gravity, symmetry, system, refraction, location, use and commercial value. 'Highly recommended for all types of libraries' (*Library Journal,* 15 February 1992, p.160). *Class No:* 549(031)

[1299]

FRYE, K., *ed*. **The Encyclopedia of mineralogy.** Stroudsburg, Pa., Hutchinson Ross, 1982. xxi, 794p. $95. (*Encyclopedia of earth sciences, v.4B.*)

Comprises 148 articles by over 100 specialists, with an A-Z mineral glossary of *c*.3,000 entries. Readership is likely to be mainly professional scientists and technologists; 'too complex for the non-mineralogist' (*Nature,* v.297, 24 June 1982, p.710). *Class No:* 549(031)

[1300]

ROBERTS, W.L., *and others*. **Encyclopedia of minerals.** New York, Van Nostrand Reinhold, 1990. xxiii,979p. illus. (some col. pl.). £77; $99.95. ISBN: 0442276818.

First published 1974.

Item-by-item description of more than 2,500 minerals, Abelsonite ... Zykaite. Data include crystal system, space group, lattice constants, hardness, density, cleavage, habit,

.... *(contd.)*

colour-lustre, mode of occurrence, strongest diffraction lines and selected references. Glossary, p.xix-xxiii. 237 colour photos. *Class No:* 549(031)

Handbooks & Manuals

[1301]

DANA, E.S. Dana's manual of mineralogy. 20th ed., edited by C.S. Hurlburt, Jr. and C. Klein. New York, London, etc., Wiley. 1989. 596p. illus., diagrs., tables. $52. ISBN: 0471805807.

First published 1848.

Sections cover crystallography (structure, chemistry, etc.) and minerals (physical and optical properties). Systematic mineralogy and determinative tables for some 200 common minerals. Mineral and general indexes; extensively illustrated. A classic text. *Class No:* 549(035)

[1302]

DANA, J.D. *and* **DANA, E.S. The System of mineralogy.** 7th ed., entirely rewritten and greatly enl. by C. Palache [and others]. New York & London, Wiley, 1944-62. 3v.

First published 1837; 6th ed. 1892.

1. *Elements; sulfides; sulfosalts; oxides.*

2. *Halides; nitrates; borates; carbonates; sulphates; phosphates; arsenates; tungstates; molybdates, etc.*

3. *Silica minerals.*

Systematic and detailed descriptions of minerals, with references to the literature in each case. Minerals are arranged in 50 classes. Each volume carries a subject index. Authoritative classification of minerals. *Class No:* 549(035)

[1303]

DEER, W.A. Rock-forming minerals. 2nd ed. Harlow, Essex, Longman Group, 1979-. illus. (incl col.), diagr., graphs, tables.

1A. *Orthosilicates.* 2nd ed. 1982. 936p. £85.

1B. *Disilicates and ring silicates.* 2nd ed. 1986. £91.

2A. *Chain silicates.* 1978. 668p. £34.

3. *Sheet-silicates.*

4. *Framework silicates.*

5. *Non-silicates.*

V.2A has two parts: Pyroxenes (p.3-543; including 19p. of references) - Non-pyroxenes (Wollastonite ... Serendibite). The monograph on ilmenite in v.5 (p.28-33) covers properties, structure, chemistry, optical and physical properties, distinguishing features, paragenesis; 31 references. Non-analytical index. *Class No:* 549(035)

[1304]

—**Elsevier's mineral and rock table.** Amsterdam, Elsevier, 1982. 70 x 133 cm. coloured wall chart.

Illustrates 74 rock-forming minerals and 40 ore minerals. 'A wealth of information concerning the microscopic properties of igneous, metamorphic, sedimentary and ore-containing rocks' (*Nature,* v.302, 10 March 1983, p.183). *Class No:* 549(035)

[1305]

GERMAIN, G. *and* **SCHUBNEL, H.J. Larousse des minéraux.** Paris, Larousse, 1981. 384p. F235. ISBN: 2035182018.

Concise description of 2,900 species of minerals, A/Z. Data include notes on stratigraphy and nomenclature. Numerous quality photographs. The introduction states the

....(contd.)

basics of mineralogy and its history. Favourably reviewed in *Bulletin critique des livres français*, no. 432, December 1981, entry 117183. *Class No:* 549(035)

[1306]
Industrial minerals and rocks (nonmetallics other than fuels). Lefond, S.J., *ed*. 5th ed. New York, American Institute of Mining, Metallurgical and Petroleum Engineers, 1983. 2v. (1,446p.). illus.

First published 1937.

4 parts (with extensive chapter references by specialists): 1. Introduction - 2. Industrial minerals grouped by uses - 3. Sources of information for industrial minerals - 4. Commodities. Concerns the origin, occurrence, nature, properties, winning and uses of commercially important minerals. Statistics mostly relate to USA. Each volume carries an index. *Class No:* 549(035)

[1307]
NICKEL, E.H. *and* **NICHOLS, M.C. Mineral reference manual.** New York, Van Nostrand Reinhold, 1991. 250p. $14.95. ISBN: 0442003447.

A-Z listing of valid minerals. Data: name, formula, current status, crystal system, appearance, hardness, measured and calculated density, type, locality, mineral classification, name origin, indication of related species and selected literature references for each species; also, crystal structure determination, when available. 'An excellent reference addition to mineralogical collections in all libraries' (*New technical books,* May/June 1991, 722). *Choice* (June 1991, p.1620) was less enthusiastic: 'merely a listing allowing someone in mineralogy to put a few pieces of information onto a name'. *Class No:* 549(035)

[1308]
O'DONOGHUE, M., *ed*. **The Encyclopedia of minerals and gemstones.** London, Orbis; New York, Putnam, 1976. 304p. illus. (incl. col.), ports, facsims, tables. ISBN: 0856232098.

8 chapters: The chemistry of minerals - The crystalline state - Geology for the collector - Minerals valuable to man - Identifying minerals - The fashioning of stones - Conserving and displaying minerals - The mineral kingdom (p.148-287; data on over 1,000 different minerals). Appendix: Some organic materials (*e.g.* amber). Glossary, p.288. Mineral identification tables. Conversion tables. Bibliography, p.296-7 (grouped). Detailed, non-analytical index. Superb illus., a number in full-page colour. 'The general public and many professionals will welcome this technically precise and aesthetically pleasing introduction to the world of minerals' (*Library journal,* 15 November 1976, p.2359). *Class No:* 549(035)

[1309]
O'DONOGHUE, M. An Illustrated guide to rocks and minerals. Limpsfield, Surrey, Dragon's World, 1991. 192p. col. illus., diagrs. £6.95. ISBN: 1850251114.

Part 1: Rocks. Part 2: Minerals (including 'Mineral descriptions', p.71-169). Appended: Deposits; Museums. Bibliography. Glossary. Detailed index. About 100 minerals are covered, with superb coloured photographs. 'This book is intended to introduce some of the major rock and mineral species' (*Preface*). *Class No:* 549(035)

[1310]
PASCAL, P. Nouveau traité de chimie minérale. Paris, Masson, 1956-64. 33v. illus. (pl.), diagrs., tables.

Previous ed. as *Traité de chimie minérale* (1931-34. 13v.).

A massive, comprehensive treatise, fully documented, and running to well over 20,000 pages of text. Fasc. 2 of v.17, 'Cobalt, nickel' (1963. xi, 896p.) carries 8,422 references to the literature.

Compléments au Nouveau traité de chimie minérale, by P. Pascal, began publication in 1974. *Class No:* 549(035)

[1311]
PELLANT, C. Rocks, minerals and fossils of the world. Boston, Mass., Little, Brown, 1990. 175p. illus. $17.95. ISBN: 0316697966.

Contents: Glossary - Geological time scale - Igneous rocks - Metamorphic rocks - Sedimentary rocks - Fossils. Further reading. Index. Over 300 colour illus. 'A basic reference work for college and research libraries' (*New Technical Books,* Nov./Dec. 1990, 1734). *Class No:* 549(035)

Children
[1312]
COOK, D. *and* **KIRK, W. Field guide to the rocks and minerals of the world.** London, Kingfisher Books, 1991. 192p. col. illus., tables. £7.95. ISBN: 088272699x.

A general overview of rocks and minerals, their formation and the qualities for identification. Glossary; further reading (p.192); index. Suitable for school libraries. *Class No:* 549(035)-0053

Dictionaries
[1313]
NAKHIMIZHAN, O.E., *comp.* **Pyatiyazychnyĭ slovar' mineralogicheskikh nazvanii.** Moscow, Glav. Redakt. Inostr. Naucho-Tekh. Slovarei Fizmatgiz, 1962. 347p.

Polyglot dictionary of mineralogical species and varieties, in five languages. English, French, German and Spanish terms are given in one sequence, with Russian equivalents. No Russian language index. *Class No:* 549(038)

[1314]
THRUSH, P.W., *ed.* **A Dictionary of mining, minerals, and related terms.** Washington, US Dept. of the Interior, 1968. vii, [1], 1269p.

About 55,000 terms (including abbreviations), with *c.*150,000 definitions. Quotes sources (listed on p.1259-68). Cross-references. Compared with A.H. Fay's *A glossary of the mining and mineral industry* (Washington, USGPO, 1920. 745p.), which Thrush revises and enlarges. Fay (*c.*20,000 terms) should be retained for its several thousand Spanish American mining terms. *Class No:* 549(038)

Reviews & Abstracts
[1315]
Bulletin signalétique. 220. Minéralogie. Géochimie. Géologie extraterrestre. Paris, Centre National de la Recherche Scientifique, 1940-83. Monthly.

About 5,500 abstracts and references pa. Main sections: Minéralogie - Géochimie - Géologie isotopique. Géochronologie - Cosmochimie. Géologie extraterrestre. Author, subject and geographical indexes; English-language

....(contd.)

subject index - per issue and annually.

PASCAL Folio 40: *Mineralogie. Géochimie. Géologie extraterrestre,* as from 1984. Online. *Class No:* 549(048)

[1316]

Mineralogical abstracts. London, Mineralogical Society of Great Britain, and Mineralogical Society of America, 1920-. Quarterly. £95.

Earlier (1920-58) published in *The mineralogical magazine.*

About 5,500 abstracts and references pa. 18 sections, A-V: Age determination - Apparatus and techniques - Book notices ... Economic minerals and mineral deposits ... Experimental mineralogy ... Geochemistry ... Meteorites and tektites ... Mineral data ... Petrology ... Various topics. Index of authors per issue; annual author and subject indexes. *Class No:* 549(048)

[1317]

Zentralblatt für Mineralogie. Stuttgart, E. Schweizerbart'sche Verlagsbuchhandlung, 1950-. Monthly. ISSN: 05147123.

Preceded by *Jahrbuch für Mineralogie* (1830-32), *Neues Jahrbuch für Mineralogie* (1833-1942) and *Zentralblatt für Mineralogie, Geologie und Paläontologie* (1943-49).

Teil 1: *Kristalographie und Mineralogie.*

2: *Petrographie, techische Mineralogie, Geochemie und Lagerstättenkunde.*

Teil 1 (7pa., pt.7: annual author & subject indexes) has *c.*3,000 abstracts and references pa, in sections. Teil 2 (13pa., pt.13: annual author & subject indexes) has *c.*5,000 abstracts and references pa., in sections A-N. *Class No:* 549(048)

Yearbooks & Directories

[1318]

World directory of mineralogists. Cesbon, F., *ed.* Orléans, Bureau de Recherches Géologiques et Minières; Marburg, West Germany, International Mineralogical Association, 1985. 361p. illus. ISBN: 2531509001.

First published 1962.

Compiled with the help of the representatives of the national mineralogical societies. *Class No:* 549(058)

Nomenclatures

[1319]

EMBERY, P.G. *and* **FULLER, J.P.,** *eds.* **A Manual of new mineral names, 1892-1978.** London, British Museum (Natural History) and Oxford Univ. Press, 1980. x, 467p. £32. ISBN: 0198585012.

A compilation from 30 lists published in the *Mineralogical magazines,* 1897-1976. About 5,500 'new' mineral names, A-Z, followed by an author list. Complements J.D. Dana's *System of mineralogy (qv). Class No:* 549(083.72)

[1320]

FLEISCHER, M. Glossary of mineral species. [4th] ed. Tuscon, Arizona, Mineralogical Record, 1983. [vi], 202p. $8.

First published 1971; previous ed. 1980.

About 2,000 one or two line entries. An A-Z list of names, symmetry and chemical compositions of mineral species. Includes only the most commonly used synonyms and obsolete names. Cross-references. Lists all known minerals,

....(contd.)

with their chemical composition, crystal system, colour, grouping by mineral family, plus references. *Class No:* 549(083.72)

[1321]

—**MITCHELL, R.S.** *and* **HENRY, J.R.** Mineral names: what do they mean?. New York, Van Nostrand Reinhold, 1979. xv, [1], 229p.

In two parts: 1. Mineral names: a discussion (*e.g.* names from persons, places) - 2. Mineral names: an alphabetical list. 3 appendices (1. Glossary). General bibliography, p.235-7 (authors, A-Z). Aims 'to bring together the derivations of all minerals' (*Preface*). *Class No:* 549(083.72)

Worldwide

[1322]

DUDA, R. *and* **REIJL, L. Minerals of the World.** London, Hamlyn, 1986. 520p. col. illus. ISBN: 0600333762.

602 minerals. Descriptive text and diagram, facing col. illus. (plates, p.31-479). Identification tables (p.482-508). List of reference books (p.511-13). Non-analytical index (p.514-20). Over 500 col. photographs and 300 line drawings. *Class No:* 549(100)

Marine Environment

[1323]

MERO, J.L. The Mineral resources of the sea. Amsterdam, etc., Elsevier, 1965. xiii, 312p. illus., maps.

8 chapters. 1. Introduction - 2. Marine beaches - 3. Minerals from sea water - 4. The continental shelves - 5. Strata underlying the soft sea-floor sediments - 6. The deep sea-floor - 7. Ocean mining methods - 8. Some economic and legal aspects of ocean mining. References, p.296-304. Non-analytic index. *Class No:* 549(26)

Europe

[1324]

BOISSONNAS, J. *and* **OMENETTO, P. Mineral deposits within the European Community.** Berlin, Springer-Verlag, 1988. xxiii,558p. illus., maps. $144. ISBN: 0387182012.

3 parts: 1. Tungsten (tin, lithium, etc.) - 2. Platinum-group elements - 3. Base metals. *Class No:* 549(4)

Great Britain

[1325]

GREG, R.P. *and* **LETTSOM, W.G. Manual of mineralogy of Great Britain and Ireland.** Broadstairs, Kent, Lapidary Publications, 1977. xvi, 483, lxvii p. illus. £11.

Facsimile reprint, with 4 supplementary lists of British minerals. First published 1858.

Systematic arrangement, with detailed mineralogy and localities of occurrence. G.P. Embrey's foreword carries biographical notes on collectors and others mentioned in the text. It remains the only comprehensive mineralogy of Great Britain and Ireland. *Class No:* 549(410)

France

[1326]
Inventaire minéralogique de la France. Paris, Bureau de Recherches Géologiques et Minières, 1977-. v.1. maps.
Arranged by départements (*e.g.* 2. Hautes Alpes; 3. Finistére 4. Alpes Maritimes; 5. Côtes du Nord ...). Details of mineralogical excursions, with maps, access points, types of deposit, and mineralogy. Each guide has general and mineral indexes, plus a bibliography. *Class No:* 549(44)

USA

[1327]
CHESTERMAN, C.W. *and* **LOWE, K.E. The Audubon Society field guide to North American rocks and minerals.** New York, Knopf, 1978. 850p. illus. (col.). ISBN: 0394502698.
For each of *c.*232 mineral species, this pocket guide provides data (mineralogical name ... occurrence) and a colour plate; also relevant data on 40 types of rocks. Appendices include a glossary, bibliography (p.799-801) and guide to mineral environments. Name and locality indexes. *Class No:* 549(73)

Museums

[1328]
HEY, M.H. An Index to the collection of minerals in the British Museum (Natural History); with an alphabetical index of accepted mineral names and synonyms. 2nd ed. London, British Museum (Natural History), 1971. 3v. ISBN: 0565000977; 0565005782; 0565007254.
First published 1950.
Based on a card index used in the Department of Mineralogy. Gives literature references.
Appendix to the second edition ..., by M.H. Hey (1963. xi, 135p. £5.25); *A second appendix to the second edition* ..., by M.H. Hey and P.G. Embrey (1974. xii, 160p. £8.50). *Class No:* 549:061:069

[1329]
World directory of mineral collections. International Mineralogical Association. 2nd ed. Copenhagen, Geological Museum, 1977. n.p.
Lists collections in 32 countries. Data: name of collection (in original and in English), address, name of person in charge, total number of specimens (usually divided into minerals, rocks, ores, gems, meteorites/tektites), uses of the collection, speciality, loan facilities, exchange arrangements, catalogue, times of admission. *Class No:* 549:061:069

Gemmology

Bibliographies

[1330]
GILL, J.O. Gill's Index to journals, articles and books relating to gems and jewelry: works in the English language, arranged chronologically and by type. Santa Monica, Calif., Gemological Society of America, 1978. vii, 420p. tables. ISBN: 0873110099.
7 sections: 1. *Minerals yearbook* (chapters on gemstones) - 2. *The Gemmologist* (August 1931 - December 1962) - 3. *Gems & Geology* - 4. *The Journal and Gemmology* - 5. *Lapidary journal* - 6. *The Australian Gemmologist* - 7. *Gem*

....(contd.)
Library bibliography (1652-1978). Table of subjects (Gems & gem materials; Gem locations; Gemology; Jewelry). *Class No:* 549.091(01)

[1331]
O'DONOGHUE, M. The Literature of gemstones. London, British Library, Science Reference and Information Service, 1986. [ii], 75, [2]p. £12. ISBN: 0712307389.
16 Sections: Encyclopedic works - General comprehensive surveys - Gem testing - Gem testing instruments - Quick reference publications and tables - Textbooks - Synthetic stones - Gemstones - Fashioning - Works on individual species - Jewelry - Gemstones of particular localities - Bibliographies - Abstracts - Journals - Gemstone prices - Guide to collections - Regalia. A list 'based largely on the collection of Science Reference and Information Service of the British Library' (*Preface*). *Class No:* 549.091(01)

Encyclopaedias

[1332]
AREM, J.E. Color encyclopedia of gemstones. 2nd ed. New York, Van Nostrand Reinhold, 1987. 288p. illus. $53.95. ISBN: 0442208332.
Data on all mineral species (organic, occurrence, size) which are cut and polished for ornamental purposes. Over 360 col. illus. Index. *Class No:* 549.091(031)

[1333]
BARIAND, P. *and* **POIROT, J.P. Larousse des pierres précieuses, fines, ornamentales, organiques.** Paris, Larousse, 1985. 264p. illus. (col.), diagrs.
The bulk of this attractive record of precious stones, A-Z, is preceded by chapters on symbolism of gemstones, their origin and historical use, their properties, fashioning and synthesis. 350 col. illus. (some of them previously published). 'The book is worth buying for its illustrations alone' (*Journal of gemmology,* v.20, January 1986, p.57).
An English translation was published by Van Nostrand Reinhold, 1992 ($45; ISBN: 0442302894). 'Highly recommended' (*Choice,* September 1992, p.72). *Class No:* 549.091(031)

Handbooks & Manuals

[1334]
CAVENARGO-BIGNAMI MONETA, S. Gemmologia. 4th ed. 3v. illus., tables.
First published 1958.
Extensive coverage of all types of precious stones, including artificial types,. V.3 concerns gem testing, identification tables and a bibliography. Fully illustrated and well presented, though not made for easy handling. 'As with the Webster book the term gemstones is widely interpreted' (O'Donoghue, M. *The literature of gemstones,* p.10). *Class No:* 549.091(035)

[1335]
CIPRIANI, C. The Macdonald encyclopedia of precious stones. 3rd ed. London, Macdonald, 1986. 384p. illus. (col.), diagrs., table. £7.95.
English translation of the Italian (Milan, Mondadori, 1984).
Three parts: Natural stones (introduction, p.10-72; gem descriptions) - Organic gems - Synthetic and artificial

....(contd.)

products. Synoptic table. Glossary. Bibliography, p.381. Analytical index. A handy, small-format guide, with over 300 well-coloured illus. *Class No:* 549.091(035)

[1336]
HURLBUT, C.S. *and* **KAMMERLING, R.C. Gemology.** 2nd ed. Chichester, Wiley, 1991. 336p. illus., diagrs., tables. $58. ISBN: 0471526673.

First published 1979.

Covers the occurrence, mineralogy, identification and fashioning of gemstones. Gem determinative tables. Chapter references and detailed, analytical index.
Class No: 549.091(035)

[1337]
SMITH, G.F.H. Gemstones. Revised by F.C. Phillips. 14th ed. London, Chapman & Hall, 1972. xii, 580p. illus. (incl. col.), diagrs., tables. ISBN: 0412108909.

First published 1912.

Four parts (41 footnoted chapters): 1. Physical character - 2. Technology and history - 3. Description (A. Principle stones; B. Other gem materials; C. Ornamental stones; D. Organic products, *e.g.* pearl, ivory, resin) - 4. Classification tables. Appendices include bibliography (12 groups), p.544-56. Index, p.559-80. Still the most complete scholarly work in English. *Class No:* 549.091(035)

[1338]
WEBSTER, R. Gems: their sources, descriptions and identification. 4th ed., revised by B.W. Anderson. London, etc., Butterworths, 1983. xxii, 1,006p. illus. (incl. col. pl.), diagrs., tables. ISBN: 0408011485.

First published 1962.

Two parts (38 chapters), Pt. 1 (1-24): Description of gemstones (diamond ... ivory and tortoise shell) - Pt. 2 (25-38): (How gemstones are identified, *e.g.* crystals: specific gravity, colour, measurement of refraction index, luminescence. Identification, p.895-929. Appendices (*e.g.* 'Famous diamonds and those large diamonds to which names have been applied'. 'Suggested list of permissive names for gemstones'). Bibliography, p.965-70. Name and subject (partly analytical) indexes. 16 colour plates.
Class No: 549.091(035)

Dictionaries

[1339]
READ, P.G. Dictionary of gemmology. 2nd ed. Oxford, Butterworth, 1988. 256p. illus., diagrs., tables. £25. ISBN: 0408029520.

First published 1982.

About 3,000 entries, giving concise descriptions of the principal gem materials, plus definitions of associated scientific terms, and notes on gemmological instruments. Includes trade names of natural and man-made materials. Cross-references from disused/misleading terms. Aimed at both professional gemmologist and the student.
Class No: 549.091(038)

[1340]
—**SHIPLEY, R.M.,** *and others.* Dictionary of gems and gemology, including ornamental, decorative and curio stones. 6th ed. Los Angeles, Calif., Gemological Institute of America, 1974. 230p. ISBN: 0873110072.

First published 1945.

Subtitled: 'A glossary of over 4,000 English and foreign

....(contd.)

words, terms and abbreviations which may be encountered in English literature or in the gems, jewelry or art trades'. Brief entries give location, historical background and pronunciation. *Class No:* 549.091(038)

Tables & Data Books

[1341]
SCHUMANN, W. Stones and minerals of the world. Translated by F.H. Steger. New York, Sterling Publishing Co., 1977. 224p. col. illus., tables. ISBN: 0806955260.

German original published by BLV Verlagsgesellschaft mbH. Munich.

Descriptive text faces colour plates in 4 sections: Minerals (function and structure; properties; classification; rock-forming minerals. 11 colour plates) - Precious stones and gemstones (characteristics; descriptions; trade names and mineralogical names; cutting of stones. 8 colour plates) - Rocks (igneous; sedimentary; metamorphic rocks; meteorites. 38 colour plates) - Ore deposits (12 colour plates). Fossils. Hints to collectors. Mineral identification tables. Aids to identification. A concise compendium in small format. *Class No:* 549.091(083)

[1342]
SINKANKAS, J. Gemstones and mineral data book: a compilation of data, recipes, formulas, and the instructions for the mineralogist, gemologist, lapidary, jeweller, craftsman and collector. New York, Winchester Press, 1972. vi, 346p. tables. ISBN: 0876910673.

Data on *c.*1,000 gems, stones and ornamental minerals, their characteristics and chemical properties. 12 sections. Much use of tables. 'An excellent reference book for the "rock hound" and lapidary, it will be very useful in a mineralogy laboratory or museum' (*Choice*, v.9(11), January 1973, p.1435). *Class No:* 549.091(083)

Elements

[1343]
EMSLEY, J. The Elements. Oxford, Clarendon Press, 1989. vii,256p. tables. £28; £9.95. ISBN: 0198552386; 0198552378, Pbk.

Key facts (discovery, names, formulae, properties, etc.) about the chemical elements (A-Z). Appended list of 16 properties in element name and numerical order. A handy book. *Class No:* 549.2/.9

[1344]
FERGUSON, J.E. The Heavy elements: chemistry, environmental impact and health effects. Oxford, Pergamon Press, 1990. vii,614p. diagrs., graphs, tables. ISBN: 0080348602; 0080402755, Pbk.

4 parts (16 documented chapters): 1. Introduction and history - 2. Chemistry - 3. Heavy elements in the environment - 4. Heavy elements in human beings. 10 heavy elements: antimony, arsenic, bismuth, cadmium, indicum, lead, mercury, selenium, tellurium, thallium. About 1400 references; detailed, analytical index, p.599-614.
Class No: 549.2/.9

55 Earth Sciences

Earth Sciences

Databases

[1345]

BURK, C.F., Jr. 'A Worldwide list of sources and reference databases in the geosciences'. In *Database*, v.5(2), June 1982, p.11-21.

Tabular databases under 6 heads: General geoscience databases (18); Mineral databases (24); Fuel databases (22); Geology databases (21); Geochemistry databases (14); Geophysics databases (12). Headings; Name / Organization / Country / Type / Public online access / Comments / References. *Class No:* 550(003.4)

[1346]

GeoArchive. Didcot, Oxon., Geosystems, 1974-.

Contains *c*.800,000 records, adding more than 100,000 records pa. Indexes more than 5,000 serials, books, conferences, technical reports, and doctoral dissertations. Subject coverage includes geophysics; geology; palaeontology; geochemistry; while also providing coverage of mineral and petroleum production and resources, names of new taxa, new minerals, and new stratigraphic names. Also available on CD-ROM. *Class No:* 550(003.4)

[1347]

GEOBASE. Norwich, Geo Abstracts Ltd., 1980-.

Over 380,000 references on geography, geology, ecology and related disciplines, with 40,000 additional records each year. Includes abstracts as well as bibliographic citations. Covers over 2,000 science and technology journals and books; reports; monographs; theses. Includes *International development abstracts; Geographical abstracts; Mineralogical abstracts; Ecological abstracts.* Available via DIALOG and ORBIT, and on CD-ROM. *Class No:* 550(003.4)

[1348]

GeoRef. Alexandria, Va., American Geological Institute; GeoRef Information System, 1785-.

Contains *c*.1.8 million records, adding 80,000 annually, and scans *c*.4,500 serials and other publications, such as theses and dissertations; state, federal and non-US government documents; technical reports. Corresponds to the print publications *Bibliography and index of North American geology, Bibliography of theses in geology, Geophysical abstracts, Bibliography and index of geology exclusive of North America,* and *Bibliography and index of geology (q.v.).* Up-dated monthly. Available on CD-ROM. *Class No:* 550(003.4)

Bibliographies

[1349]

Bibliography and index of geology. Alexandria, Va., American Geological Institute, 1969-. 12pa. $1,345. ISSN: 00982784.

Succeeds *Bibliography of North American geology* (U.S. Geological Survey, 1931-72, which included literature 1785-1970), and *Bibliography and index of geology exclusive of North America* (Geological Society of America), 1933-68. 32v.

References photocomposed from citations in GeoRef. *(qv),* the AGI database. 29 fields of interest (01. Mineralogy and crystallography - 02. Geochemistry - 03. Geochronology - 04. Extraterrestrial geology - 05/06. Petrology - 07. Marine geology and oceanography - 08/11. Paleontology - 12. Stratigraphy, historical geology and paleoecology... 16. Structural geology - 17/20. Geophysics - 21. Hydrogeology and hydrology - 22. Engineering and environmental geology - 23/25. Surficial geology - 26/29. Economic geology). Profuse cross-references. Omits geography and meteorology, both well covered elsewhere. List of serials precedes. Analytical subject and author indexes; annual cumulative index in 4 pts. Available on microfiche, online and CD-ROM. The major indexing service for geology. *Class No:* 550(01)

[1350]

—**MULVIHILL, J.,** *ed.* User guide to the *Bibliography and index of geology.* Falls Church, Va., American Geological Institute, 1982. 160p. $8.95.

Has 6 sections: 1. Scanning and searching - 2. Documents covered - 3. Fields of interest - 4. Subject indexing - 5. Alphabetical term list ... (p.24-57) - 6. Special lists (*e.g.* Commodities; Fossils; Soils; Geographic terms). Analytical index, p.139-60. *Class No:* 550(01)

[1351]

Books in the earth sciences and related topics: a bibliography and index of non-periodical literature (base-line January 1976). Worthing, Sussex, Bibliographic Press, 1977-. no.1-. 4pa. £46pa. ISSN: 01407805.

About 700 author entries, with full bibliographical details. Coverage extends to economics of minerals and energy resources. Subject index, index to conference symposia volumes, and title-keyword index per issue. Annual author index. A helpful check-list that cites items in press and indicates readership (*e.g.* professional/research/reference).

A complementary serials index from the same publishers is *British geological literature (qv). Class No:* 550(01)

[1352]

Geologic reference sources: a subject and regional bibliography of publications and maps in the geological sciences. Ward, D.C., *and others.* 2nd ed. Metuchen, N.J., Scarecrow Press, 1981. $49.50.

First published 1972.

Contains 4,324 entries, - mostly references. Three main sections; General (in which some of the principal information sources are discussed); Subject (p.75-318): earth sciences;

....(contd.)

physical geography; environmental geology; Regional (World; North America ... Antarctic). Computer databases are identified in the General section. Non-analytical subject index; geographic index. Intended as a ready-reference guide, although the subject index falls short here. *Class No:* 550(01)

[1353]
Information sources in the earth sciences. Wood, D.N., *and others, eds.* 2nd ed. London, Bowker Saur, 1989. 518p. £45. (*Guides to information sources.*)

First published as *Use of earth sciences literature* (Butterworth, 1973).

20 chapters: 1. Introduction - 2. Earth science libraries and their use ... 6. Computerized information services and geological databases/databanks ... 9. Palaeontology ... 12. Petrology ... 16. Engineering and environment geology ... 18. Meteorology and climatology ... 20. History of geology. The chapter on computerized bibliographic and data retrieval is new to this edition. Aimed at students and practising geologists. *Class No:* 550(01)

[1354]
—GUELPA, P.M. 'Sciences de la terre'. In Malclès, L.-N. *Les sources du travail bibliographique*, v.3. Geneva, Droz; Paris, Librairie Minard, 1958. p.236-96.

Systematically surveys the chief reference material, bibliographies and periodicals in the earth sciences. Contents lists and annotations are given for the most important items. Regional geology covers p.272-87. Still of value. *Class No:* 550(01)

[1355]
UNITED STATES. Department of the Interior. **Dictionary catalog of the Department [of the Interior].** Boston, Mass., G.K. Hall, 1967-. 37v.; Supplements 1-4. 1968-75. 18v. $4,000, set; $910, supp. 4 (8v.). ISBN: 0816107157, set; 0816100160, supp. 4.

724,000 + 268,200 catalogue cards, reproduced photolithographically on 29,573 + 12,818p. The Department of the Interior has the largest specialized collection in the US on mines, minerals, petroleum and coal, as well as a good working collection on geology. The collection covers both published and unpublished material. *Class No:* 550(01)

[1356]
UNITED STATES. Geological Survey. Library. **Catalog of the United States Geological Survey Library.** Boston, Mass., G.K. Hall, 1970. 25v. Supplements 1-3. 1972-76. 21v. $2,720, set; $950, supp.1 (11v.); $475, supp.2 (4v.); $690, supp.3 (6v.). ISBN: 0816107122, set; 0816108765, supp.1; 081611031x, supp.2; 0816100519, supp.3.

416,000 + 350,000 catalogue cards reproduced photolithographically. The largest geological library in the world, intended to be as complete as possible in geology, palaeontology, petrology, mineralogy, ground and surface water, cartography and mineral resources. In general only books and monographs are individually catalogued, under authors, titles (including periodical titles) and subject. Included are a number of old or rare items. *Class No:* 550(01)

Encyclopaedias

[1357]
Encyclopedia of earth sciences. Fairbridge, R.W., *ed.* New York, Van Nostrand Reinhold.

1. *The Encyclopedia of oceanography.* 1966. 2. *The Encyclopedia of atmospheric sciences and astrogeology.* 1967. 3. *The Encyclopedia of geomorphology.* 1968. 4A. *The Encyclopedia of geochemistry and environmental sciences.* 1972. 4B. *The Encyclopedia of mineralogy*, edited by K. Frye. 1981. 6. *The Encyclopedia of sedimentology*, edited by R.W. Fairbridge and J. Bourgeois. 1978. 7. *The Encyclopedia of paleontology*, edited by R.W. Fairbridge and D. Jablanski. 1979. 8(1). *The Encyclopedia of world regional geology: Western hemisphere.* 1975. 11. *The Encyclopedia of climatology*, edited by R.W. Fairbridge and J.E. Oliver. 1987. 12(1). *The Encyclopedia of soil science: Physics, chemistry, biology, fertility and technology*, edited by R.W. Fairbridge and C.W. Finkl, Jr. 1979. 14. *The Encyclopedia of field and general geology*, edited by C.W. Finkl, Jr. 1988. 15. *The Encyclopedia of beaches and coastal environments*, edited by M.L. Schwartz. 1982. 16. *The Encyclopedia of solid earth geophysics*, edited by D.E. James. 1989. *The Encyclopedia of igneous and metamorphic petrology*, edited by D.R. Bowes. 1989. (Unnumbered volume).

Planned in 24v. An ambitious, comprehensive series of encyclopaedias, alphabetically-arranged. Lengthy, documented articles preponderate, each beginning with a simple explanation, expanding into detailed treatment. Inter-volume cross-references. Detailed indexes. *Class No:* 550(031)

[1358]
Encyclopedia of earth system science. Nierenberg, W.A., *ed.* San Diego, Academic Press, 1992. 4v. (2500p.). illus., diagrs., graphs, tables. $950, set. ISBN: 0122267222, set; 0122267206, v.1-2; 0122267214, v.3-4.

257 international contributors. 229 entries, A-Z. Each article includes a bibliography containing books and journal articles, and a glossary. Useful cross-references. The excellent illustrations consist of colour plates and several thousand black-and-white photographs, maps, graphs, charts, and tables. V.4 contains a 170-page. double column subject index and a 12-p. (15,000 entries) relational index linking closely related articles. This is a basic overview, aimed at students, the public, and professionals. While *The Encyclopedia of earth sciences (q.v.)* has greater depth, this has longer articles than *McGraw-Hill encyclopedia of the geological sciences (q.v.).* For academic libraries, although the reviewer in *Library journal,* 15 February 1992, p.156 comments, 'small undergraduate and public libraries can skip this and make do nicely with a more general encyclopedia such as the *McGraw-Hill encyclopedia of science and technology*'. *Class No:* 550(031)

[1359]
McGraw-Hill encyclopedia of the geological sciences. Parker, S.P., *ed.* 2nd ed. New York, McGraw-Hill, 1988. 722p. illus. $85. ISBN: 0070455007.

First published 1978.

Derived from *McGraw-Hill encyclopedia of science and technology* (6th ed. 1987). 520 signed articles, with 235 revised since the first edition. Subject coverage includes mineralogy, geochemistry, geology, geophysics and petrology. A-Z arrangement, with cross-references. Length

.... *(contd.)*

of entries varies from one paragraph to several pages. *c.*600 illustrations. Includes list of contributors. Subject index. For large public, and university, libraries. *Class No:* 550(031)

[1360]

The Marshall Cavendish illustrated encyclopedia of plants and earth sciences. Moore, D.M., *ed.* London, Marshall Cavendish, 1988. 10v. ISBN: 0863079016.

V.1-2: *Dictionary of plants.* V.3-5: *Flowering plant families.* V.6-7: *Plant ecology.* V.8-9: *Earth sciences.* V.10 includes indexes, glossary and bibliography. International in scope though the majority of the 120 contributors are British or European. Numerous colour photographs and diagrams throughout. Many cross-references. For school and public libraries. *Class No:* 550(031)

Handbooks & Manuals

[1361]

The Cambridge encyclopedia of earth sciences. Smith, D.C., *editor-in-chief.* Cambridge Univ. Press, 1982. 496p. illus., (mostly col.), diagrs., graphs, tables, maps. ISBN: 0521239001.

32 contributors. 6 parts (27 chapters): 1. The earth sciences in perspective - 2. Physics and chemistry of the earth - 3. Crustal processes and evolution - 4. Surface processes and environments - 5. Evaluation of earth resources and hazards - 6. Extraterrestrial geology (chapter 27: 'The geology of the solar system'). Glossary, p.468-78; Further reading, by chapters. Analytical index, p.481-94. Concentrates on those aspects in which the most significant advances have recently been made. 'The aim is to provide ready access to the wealth of geological information which has been gleaned over the years and to show how it fits into the framework of plate tectonics' *(Foreword)*. *Class No:* 550(035)

[1362]

—**The New Larousse encyclopedia of the earth.** Bertin, L., *ed.* Rev. ed. London, Hamlyn, 1972. 424p. illus. (incl. col.), maps.

First published 1961.

Has 3 parts: 1. The present; 2. Earth in the service of man - 3. The past (geological history and former life on earth; fossil man). Over 500 black-and-white illus.; 30 col. plates. Despite some misspellings and errors, 'the book is in all senses good value' *(Geographical magazine,* February 1973, p.396-7). For the general reader. *Class No:* 550(035)

[1363]

Magill's survey of science: earth science series. Magill, F.N., *ed.* Englewood Cliffs, N.J., Salem Press, 1990. 5v. $400. ISBN: 0893566063.

377 articles, written by professional geologists. Each chapter contains cross-references and a short bibliography. Subject index and glossary appear in the final volume. Not so detailed as *Encyclopedia of earth sciences* (Van Nostrand Reinhold. 1966-) *(qv).* Aimed at the layperson. *Class No:* 550(035)

Dictionaries

English

[1364]

Chambers earth sciences dictionary. Edinburgh, Chambers, 1991. vi,250p. figs., tables. £20; £9.99. ISBN: 0550132449.

Over 6,000 definitions based on entries in *Chambers science and technology dictionary (q.v.).* Subject coverage includes geology, mineralogy, crystallography, chemistry, physics and biology. Major feature is special articles which give fuller treatments of important topics, *e.g.* 'Bivalves'; 'Iron'; 'Quaternary'. Alternative forms and cross-references are indicated. *Class No:* 550(038)=20

[1365]

The Concise Oxford dictionary of earth sciences. Allaby, A. *and* Allaby, M., *eds.* Oxford, University Press, 1990. xxi,410p. £6.99. ISBN: 0198661460.

One-third of the *c.*8,000 entries are taken from *The Oxford dictionary of natural history,* (1985). Entries, AA ... Zosterophyllophytina, include abbreviations, *e.g.* FFT, and short biographies, *e.g.* Holmes, Buffon, Thomson. Many cross-references. 11-page bibliography. 'A mine of information' *(Nature,* v.344, 29 March 1990, p.391). *Class No:* 550(038)=20

[1366]

McGraw-Hill dictionary of earth sciences. Parker, S.P., *editor-in-chief.* 3rd ed. New York, London, etc., McGraw-Hill, 1984. [viii], 837p. $46.95; £38. ISBN: 0070452520.

Brief definitions of *c.*15,000 terms, most of the entries having been previously published in the *McGraw-Hill Dictionary of scientific and technical terms* (1983). Apart from earth sciences and related fields of science and engineering, coverage extends to crystallography, mapping, mineralogy, mining engineering, palaeontology and petroleum engineering. Terms are categorised in 18 fields. No illus. 'The many cross-references tend to distract' *(Geographical journal,* March 1986, p.122). *Class No:* 550(038)=20

[1367]

McNEIL, M. Earth sciences reference. Carlsbad, Calif., Flamingo Press, 1991. 709p. $52. ISBN: 0938905015.

Concise, informative definitions covering geology, geophysics, ecology, geography, meteorology, and oceanography. Biographies included. Subject and geographical indexes. *Class No:* 550(038)=20

German

[1368]

Dictionary of geosciences. English/German, German/English. [Wörterbuch Geowissenschaft.] Watznauer, A., *ed.* 3rd ed. Amsterdam, etc., Elsevier, with Berlin, VEB Verlag Technik; New York, French and European, 1987-88. 2v. (388p. + 372p.). tables, charts, formulas. $95ea.; £52.50ea. ISBN: 3871441392, v.1; 3871441406, v.2.

First published 1973.

Each volume carries *c.*38,000 terms. Omits hydrogeology and soil mechanics, - the subject of two other Elsevier dictionaries. *Class No:* 550(038)=30

[1369]

MURAWSKI, H. Geologisches Wörterbuch. Erklärung der geologischen Fachausdrücke der deutschen literatur für Geologen, Paläontologen, Mineralogen, Geographen, Geophysiker, Bodenkundler, Bau-und Bodeningenieure, Studierende und alle Freunde der Geologie. 7. erg. und. erw. Aufl. Stuttgart, Enke; New York, French and European, 1977. 280p. diagrs., tables. $29.95. ISBN: 3432841078, US ed.

First published 1937.

The standard German-German dictionary of terms in geology and allied fields. About 2,000 entries, some of them definitions, others, fuller explanations. Complicated terms are dealt with in simple language, plus diagrs. Many cross-references. *Class No:* 550(038)=30

French

[1370]

Dictionary of earth sciences: English-French, French-English. [Dictionnaire des sciences de la terre: anglais-français, français-anglais.] 2nd ed. Chichester, John Wiley; Paris, Masson, 1992. xiii,299p. £24.95. ISBN: 0471935352, US ed; 2225823952.

First published 1980.

Lists the most common terms in the geological sciences and related disciplines. French section contains *c.*25,500 terms, while the English-French part lists *c.*16,000. Ignoring the badly translated preface, this is a useful list of equivalents. *Class No:* 550(038)=40

Russian

[1371]

Anglo-russkii geologicheskii slovar': okolo 52,000 terminov [English-Russian dictionary of geology.] Timofeev, P.P., *and others*. Moscow, Russkiiiazyk, 1988. 540p. ISBN: 5200002842.

52,000 terms, covering geology, geophysics, hydrogeology, mineralogy, stratigraphy, tectonics, palaeontology, geochemistry, etc. Short list of abbreviations and a list of the Russian spelling of the names of foreign geologists that frequently occur in the literature. Appended tables for geological time periods, etc. *Class No:* 550(038)=82

[1372]

—SOFIANO, T.A., *comp.* Russko-angliiskii geologicheskii slovar'. Pod red. A.P. Ledeva i V.E. Khaina. Moscow, Glav. Red. Inostr. Nauchno-Tekhn. Slovarei Fizmatgiza, 1960. 559p.

American supplement. New York, Telberg, 1964. 148p.

35,000 Russian-English entries, with similar coverage to the foregoing. *Class No:* 550(038)=82

[1373]

BHATNAGAR, K.P. Elsevier's dictionary of geosciences: Russian-English. Amsterdam, Elsevier, 1991. xii,1023p. £125. ISBN: 0444884254.

Contains *c.*56,000 terms used in geology, meteorology, palaeontology, sedimentology, geochemistry, etc. Brief list of abbreviations. *Class No:* 550(038)=82

[1374]

USSR. VSESOYUZNYI NAUCHNO-ISSLEDOVATEL'SKII GEOLOGICHESKII INSTITUT (VSE-GEI) MINISTERSTVA GEOLOGII I OKHRANY NEDR SSSR. Geologicheskii slovar'. 2nd ed. Moscow, Gosud. Nauchno-Tekhnicheskoe Izdat. Lit. po Geologii i Okhrane Nedr, 1960. 2v. (402p. + 445p.). tables.

First published 1955.

Entries for over 12,000 Russian geological terms, stating origins. Coverage includes volcanology, geomorphology, geochemistry, hydrogeology, crystallography, mineralogy, palaeontology, petrology, as well as economic geology. Entries are included for individual minerals. *Class No:* 550(038)=82

Reviews & Abstracts

[1375]

Current contents: Physical, chemical and earth sciences. Philadelphia, Pa., Institute for Scientific Information, 1961-. v.1(1)-. Weekly. $420. ISSN: 01632574.

Each issue contains the title pages of *c.*150 journals. Also has current book contents; title word index; author index and address directory; publishers address directory. Online, CURRENT CONTENTS SEARCH and available on diskette. *Class No:* 550(048)

[1376]

Geoscience documentation: journal for geoscience information. Didcot, Oxon., Geosystems Publications, 1969-. 6pa. £50. ISSN: 00168483.

Each issue contains *c.*250 references. Classes 100000-900000: Regional geology; Applied geology; Mineral deposits; General geology (including mineralogy, palaeontology); Geochemistry; Physical geology (including geomorphology, history of geology); Methodology (including equipment and instrumentation); Geoscience information. Subject, stratigraphical, geophysical and author indexes. *Class No:* 550(048)

[1377]

—GeoSciTech citation index: an international interdisciplinary index to the literature of geology, earth science, mineralogy atmospheric sciences and related disciplines. Philadelphia, Pa., Institute for Scientific Information, 1976-. Annual, with 2 interim issues pa.; 5 year cumulation, 1976-80. ISSN: 07320558.

Consists of 3 parts: 1. Citation index (authors; anonymous; patents) - 2. Corporate index; geographic sections; source index - 3. Permuterms. Subject index. *Class No:* 550(048)

[1378]

Geotitles: geoscience bibliography. London, Geosystems Publications, 1960-. 12pa. ISSN: 00168564.

Each issue contains *c.*3,000 references in a closely classified sequence, 100000-900000 (General publications - Regional geology... Equipment and instrumentation - Geoscience information). Stratigraphic, geophysical and author indexes. Online via DIALOG. *Class No:* 550(048)

[1379]

PASCAL folio 40/47. Vandoeuvre-Les-Nancy, Centre National de la Recherche Scientifique, 1984-.

   40. *Minéralogie. Géochimie. Géologie extraterrestre.* Monthly.

   41. *Gisements métalliques et non-métalliques.* Monthly.

   42. *Roches cristallines.* 6pa.

   43. *Roches sedimentaires. Géologie extraterrestre.* Monthly.

   44. *Stratigraphie. Géologie régional et général.* Monthly.

   45. *Tectonique.* 6pa.

   46. *Hydrologie. Géologie de l'ingénieur. Formations superficielles.* Monthly.

   47. *Paléontologie.* Monthly.

   Part 40: *Minéralogie,* etc., has *c.*4,000 brief abstracts and references pa. Main sections: A. Minéralogie; B. Géochimie; C. Géochemie isotopique. Géochronologie; D. Cosmologie. Géologie extraterrestre. Coverage included theses, reports, conference proceedings, monographs. Author, permuted keyword, geographical and mineral name indexes per issue and annually. Available online.

Class No: 550(048)

[1380]

Referativnyĭ zhurnal. 08. Geologiia. Moscow, VINITI; Rockville, Md. Victor Kamkin Inc., 1982-. 12pa. ISSN: 04862309.

   Formerly part of *Referativnyĭ zhurnal. Geologiya, geografiya,* 1954-55. *Referativnyi zhurnal. Geologiia/ Akademiia Nauk SSSR, Institut Nauchnoi Informatsii,* 1956-81.

   About 35,000 abstracts pa. 10 sections: General geology - Stratigraphy. Palaeontology - Anthropogene period and geomorphology of land and sea bed - Geochemistry. Mineralogy. Petrography - Ore deposits - Nonmetalliferous deposits - Geological and geochemical methods of research for minerals - Exploration methods and valuation of mineral deposits - Exploration geophysics and logging - Techniques of geological prospecting - Hydrogeology. Engineering geology. Frost action research. Annual author and subject indexes. *Class No:* 550(048)

Progress Reports

[1381]

Annual review of earth and planetary sciences. illus., diagrs., charts, maps. Palo Alto, Calif., Annual Reviews, Ltd., 1973-. Annual. ISSN: 00846597.

   V.20, 1992 has some 28 contributors. The 20 documented papers include 'Atmospheric circulation and deep-sea sediments'; 'The first radiation of land vertebrates'; 'Planetary rings'. Subject index. Cumulative indexes of contributing authors, v.8-20, and chapter titles, v.8-20. Prelims. list 'some selected articles in other *Annual reviews*'.

Class No: 550(055)

Tables & Data Books

[1382]

CLARK, S.P., Jr. ed. Handbook of physical constants. Rev. ed. New York, Geological Society of America, 1966. 587p. tables. (*Geological Society of America. Memoir 97.*) ISBN: 0813710979.

   First published 1942.

   Data, in 27 sections, on the physical constants of all subjects and substances relating to earth sciences. 30

....(contd.)

contributors; over 200 tables. 'A "must" for all geological and chemical libraries and reference shelves' (*Economic geology,* v.61(6), September/October, 1966, p.1164).

Class No: 550(083)

[1383]

CORLISS, W.R., comp. Handbook of unusual natural phenomena. Glen Aron, Md., The Sourcebook Project, 1977 (reprinted 1988, New York, Outlet). [vi], 542p. illus.

   8 chapters: 1. Luminous phenomena - 2. Optical and radio anomalies in the atmosphere - 3. Unusual weather phenomena - 4. Mysterious natural sounds - 5. The strange phenomena of earthquakes - 6. Phenomena of the hydrosphere - 7. Falling material (*e.g.* dust, thunderstorms, and sundry inorganic matter) - 8. Magnetic disturbances. Cites journal article in each case. Subject and source indexes. Aims to provide 'a comprehensive collection of reliable eye accounts of unusual natural phenomena' (*Preface*). *Class No:* 550(083)

Photographs

[1384]

The Home planet. Kelley, K.W., comp. London, Queen Anne Press; Reading, Mass., Addison-Wesley, 1988. 149p. illus. (col.). $22.95. ISBN: 0356159481.

   A stunning collection of colour photographs of Earth taken from space vehicles and satellites. *Class No:* 550(084.12)

Histories

[1385]

PORTER, R. The Earth sciences: an annotated bibliography. New York & London, Garland, 1983. xxxiv, 192p. illus. $37. (*Bibliography of the history of science and technology, v.3.*) ISBN: 0824092678.

   808 briefly annotated entries in 10 sections: 1. Bibliography and reference works - 2. General histories - 3. Specialist histories (C. Geomorphology; D. Palaeontology; I. Mineralogy) - 4. Cognate sciences (*e.g.* Natural history) - 5. Studies by area - 6. Biographical studies - 7. Institutional histories - 8. The social dimension - 9. Geology, culture and the arts. Entries average *c.*50 words. Author and subject index. 'A valuable bibliographical source book... About 1/3 of titles are in a major Western language other than English' (*Natural history book reviews,* v.7(3), 1984, p.171).

Class No: 550(091)

Information Services

[1386]

Keeping current with geoscience information. Proceedings of the 15th meeting of the Geoscience Information Society, November 16-20, 1980 (*Proceedings,* v.11, 1981). Pruett, N.J., ed. Alexandria, Va., Geoscience Information Society, c/o American Geological Institute, 1981. v, 214p. $20. ISBN: 0934485089.

   Part 1. Symposium: Keeping current with geoscience information (7 papers, *e.g.* 'GeoRef coverage and improvements in the *Bibliography and Index of Geology*', by J.G. Mulvihill, p.55-64) - Part 2. Contributed papers (*e.g.* 'A directory of information resources in the United States: geosciences and oceanography', by J.F. Price, p.81-88). *Class No:* 550:061:025.5

Libraries

[1387]

CROSSFIELD, N.L., *and others, eds.* **Directory of geoscience libraries, United States and Canada.** 3rd ed. Alexandria, Va., Geoscience Information Society, 1986. 99p. $20.

2nd ed., 1974.

Data on 407 libraries (2nd ed.: 327). Entries are arranged A-Z by US states, then by Canadian provinces. Asterisked entries are for libraries making no returns. Organization index. 'This is a timely comprehensive and relatively inexpensive directory, especially useful for geoscience collections in industry, academic, government and the larger public libraries' (*Science & technology libraries,* v.7(4), Summer 1987, p.98). *Class No:* 550:061:026/027

Institutions & Associations

[1388]

Earth and astronomical sciences research centres: a world directory of organizations and programmes. 2nd ed. Harlow, Essex, Longman, 1991. v [vi],623p. £250. (*Reference on research.*) ISBN: 0582082749.

First published 1984.

Guide to over 3,000 technology and research laboratories in more than 100 countries, arranged A-Z (Algeria ... Zimbabwe). Subject coverage includes geochemistry; petrology; mineralogy; planetary and galactic observations; meteorology. Data include title of establishment in original language, acronym and English translation; senior staff; research expenditure; publications; research activities; affiliation. New to this edition is a list of the main national and international trade associations and societies. (466 entries). Titles of establishments index; subject index. *Class No:* 550:061:061.2

[1389]

UNITED STATES. National Referral Center for Science and Technology. **Geosciences and oceanography:** a directory of information resources in the United States. Washington, Library of Congress, 1981. xx, 375p.

Contents; Organization locator - Directory, A/Z (p.1-275). Data on each body include area of interest, holdings, publications and information services. Extensive analytical subject index, p.277-375. 'For the purpose of this directory, "geoscience" and "oceanography" have been defined quite broadly' (*Preface*). Thus, palaeontology and astronomy both figure in the index. *Class No:* 550:061:061.2

Geophysics

Encyclopaedias

[1390]

The Encyclopedia of solid earth geophysics. James, D.E., *ed.* New York, Van Nostrand Reinhold, 1989. xvi,1328p. figs. $124.95. (*Encyclopedia of earth sciences, 16.*) ISBN: 0442243669.

160 articles, arranged A-Z (Absolute age determination: radiometric ... Viscous remanent magnetization (VRM) and viscous remagnetization) covering seismology, gravity, geodesy tectonophysics, geomagnetism and related subjects.

.... (contd.)

Short bibliographies and cross-referenced list of related entries. Subject and author citation indexes. Aimed at both the academic and layperson. *Class No:* 550.3(031)

[1391]

Encyclopedic dictionary of exploration geophysics. Sheriff, R.E., *comp.* 3rd ed. Tulsa, Oklahoma, Society of Exploration Geophysicists, 1991. 376p. illus., diagrs., graphs, tables, maps, formulas. $59. ISBN: 1560800186.

First published 1972.

About 5,000 brief unsigned entries, 'ABC method' ... 'Z transform'. 'Satellite navigation': ½ column. Cross-references. Appendices A-N (*e.g.* C. Symbols used in geophysical exploration; N. Publishing houses). References, p.375-6. 'Compiled for "practical" geophysicists rather than for researchers or other specialists' (*Excerpts from the preface to the first edition*). *Class No:* 550.3(031)

[1392]

International dictionary of geophysics, seismology, geomagnetism, aeronomy, oceanography, geodesy, gravity, marine geophysics, meteorology, the earth as a planet, and its evolution. Runcorn, S.K., *and others, eds.* Oxford, etc. Pergamon Press, 1967. 2v. (1728p.). illus., diagrs., charts, maps. $740, set; £315, set. ISBN: 0080118348, set.

220 contributors: over 700 signed articles, A-Z. An encyclopaedic dictionary that omits biographies but includes eponyms (*e.g.* Kepler's laws). Entries vary in length - and quality - from 20-word definitions to lengthy articles (*e.g.* 'Magnetic properties of minerals', p.861-75, with 21 figures, 3 tables and 3 columns of references). Many cross-references. Index in v.2 (p.1689-1728). 'The dictionary provides an authoritative source of reference for students and for scientists and engineers working in unfamiliar areas of geophysics' (*Science journal,* v.4(6), June 1968, p.94). *Class No:* 550.3(031)

Handbooks & Manuals

[1393]

FOWLER, C.M.R. **The Solid earth:** an introduction to global geophysics. Cambridge, University Press, 1990. xviii,472p. tables, diagrs. £40. ISBN: 0521370256.

9 chapters, each with appended 'problems and bibliography'. Glossary, p.454-62. 7 appendices. A textbook intended for students and geologists. 'Well-written, accurate, comprehensive and up-to-date ... a book to be used in conjunction with a course, not given to the average student to read for himself' (*Geophysical journal international,* v.105(2), May 1991, p.557). *Class No:* 550.3(035)

Dictionaries

[1394]

GUSEV, B.V., *and others.* **English-Russian dictionary of applied geophysics.** Moscow, Russkiĭ Vazyk', Pergamon Press; New York, State Mutual, 1984. ix, 488p. £43; $50. ISBN: 0080281680, UK ed; 0317594192, US ed.

About 25,000 English terms with Russian equivalents. Abbreviations included. Appendices 1-2: Trade names for main types of geophysical equipment. Index of Russian terms, p.479-88. *Class No:* 550.3(038)

Reviews & Abstracts

[1395]
Meteorological and geoastrophysical abstracts. Boston, Mass., American Meteorological Society, 1960-. v.1(1)-. 12pa. $750. ISSN: 00261130.

Formerly *Meteorological abstracts and bibliography*, 1950-59.

Indexes over 200 sources, including technical journals, monographs, proceedings, reviews, and annual publications. Each issue has *c.*600 abstracts and references. 6 sections: A. Environmental sciences - B. Meteorology - C. Astrophysics - D. Hydrosphere. Hydrology - E. Glaciology - F. Physical oceanography. Author, subject and geographic indexes. Available online, 1972-, with database containing *c.*150,000 records. *Class No: 550.3(048)*

Progress Reports

[1396]
Advances in geophysics. New York, etc., Academic Press, 1952-. v.1-. ISSN: 00652687.

V.33, 1991 ($79.50) contains 4 papers, (*e.g.* 'Source mechanism of earthquakes', p.81-140), each followed by a considerable number of references. Detailed, analytical index, p.307-21. *Class No: 550.3(055)*

Yearbooks & Directories

[1397]
The Geophysical directory. 4th ed. Houston, Texas, The Geophysical Directory, Inc., 1990. 526p. $35.

Lists *c.*3,600 companies providing geophysical equipment, supplies or services, plus mining and petroleum companies using geophysical techniques. Also, a list of geophysicists and geologists directly involved with geophysical operations - plus a list of companies supplying geophysical software. Separate section on the activities of national and international geophysical societies. Extensive company index. *Class No: 550.3(058)*

Published Series

[1398]
Annals of the IQSY. International Years of the Quiet Sun. Cambridge, Mass., & London, MIT Press, 1968-70. 7v.

1. *Geophysical measurements: techniques, observation schedules, and treatment of data.* 1968.
2. *Solar and geophysical events, 1960-65 (Calendar record).* 1968.
3. *The Proton flare project (The July 1966 event).* 1969.
4. *Solar-terrestrial physics: solar aspects.* 1969.
5. *Solar-terrestrial physics: terrestrial aspects.* 1969.
6. *Survey of IQSY observations and bibliography.* 1970. 608p.
7. *Sources and availability of IQSY data.* 1970. 368p.

Annals of an international programme of scientific research and study aimed at investigating how the sun's 11-yearly quiet cycle influences the solar system.

Annals of the International Geophysical Year (Oxford, Pergamon Press, 1957-70. 48v. illus., diagrs., tables) record studies of the sun at maximum activity during the 1957-78 IGY of the International Council of Scientific Unions. *Class No: 550.3(082.1)*

Tables & Data Books

[1399]
UNITED STATES. Air Force. Air Force Geophysics Laboratory, Air Force Systems Command. **Handbook of geophysics and the space environment.**

Jursa, A.S., *scientific editor*. 4th ed. Hanscom Air Force Base, Air Force Geophysics Laboratory, 1983. vp. illus., diagrs., tables, maps.

88 contributing editors. 25 documented chapters (1. The sun... 25. Infrared astronomy). Chapter 21: 'Atmospheric composition': 67p., includes 10p. of references. Appendix: units, constants and conversion factors. Aims 'to provide Air Force designers, engineers and systems operators with facts and data about the environment in which the Air Force operates' (*Preface*). *Class No: 550.3(083)*

Histories

[1400]
BRUSH, S.G. *and* **LANDSBERG, H.E. The History of geophysics and meteorology:** an annotated bibliography. New York & London, Garland, 1985. 450p. ISBN: 0824091167.

'Geophysics' is here interpreted to include the origin and development of the solar system and the formation of the earth's surface features. About 2,000 annotated entries in sections A-W (A. General histories; B. Biographies ... G. Geodesy, terrestrial gravitation, size and shape of the earth - H. Formation of the earth's surface features ... M. Meteorology: general histories ... R. Atmospheric dynamics and thermodynamics ... V. Upper atmosphere, ionosphere, magnetosphere, aurora). Author and subject indexes. *Class No: 550.3(091)*

Marine Environment

[1401]
CRC handbook of geophysical exploration at sea. Geyer, R.A., *ed.* Boca Raton, Florida, CRC Press, 1983. [vi], 445p. illus., diagrs., graphs, tables, maps. $139.95. (*CRC series in marine sciences.*) ISBN: 0849302226.

24 contributors. 3 parts: General considerations (8 documented papers); Instrumentation systems (3); Case histories (3, *e.g.* 'Tectonic origin, Gulf of Suez', as deduced from gravity data, p.417-32. 42 references).

1992 ed., *Geophysical exploration at sea* (424p. £109), is planned. *Class No: 550.3(26)*

Research Projects

[1402]
O'REILLY, W., *comp.* **International directory to geophysical research.** Amsterdam, etc., Elsevier, 1986. [v], 122p. £32.81; $48.75. ISBN: 0444425918.

About 2,700 entries, A-Z (name, address, specialization). Subject index in 8 sections (1. Solid earth and core - 2. The oceans - 3. Terrestrial atmosphere and magnetosphere - 4. Extra-terrestrial - 5. Engineering, resources and exploration - 6. Physical properties of materials - 7. Methods, techniques and instrumentation - 8. Mathematical/computational techniques and theoretical geophysics). Aims 'to provide a means of reaching geoscientists who have certain general or specific scientific interests' (*Preface*). *Class No: 550.3:061:061.62.005*

Earthquakes (Seismology)

Bibliographies

[1403]

HOLLIS, E.P. Bibliography of earthquake engineering. 3rd ed. [Oakland, Calif.], Earthquake Engineering Research Institute, 1971. x, 247p. $5. ISBN: 0318163179.

First published 1954 as *Bibliography of engineering seismology.*

4 parts: Miscellaneous sources of data (bibliographies, periodicals, maps, directories, etc.) - Seismology, seismic geology and seismometry (by country) - Dynamics of soils, rocks and structures - Design and construction in seismic regions. Author index. *Class No:* 550.34(01)

[1404]

TYCKSON, D.A. Earthquake prediction. Phoenix, Arizona, Oryx, 1986. 67p. $18.75. ISBN: 0897742281.

246 annotated entries for material published between 1976-1983. English-language items only. Author index. Aimed at the general public. *Class No:* 550.34(01)

Handbooks & Manuals

[1405]

BRYANT, E. Natural hazards. Cambridge, University Press, 1991. xviii,294p. illus., figs., tables, maps. £40; £14.95. ISBN: 052137295x.

3 documented sections: Section 1 examines storms; wind; the oceans; drought and flood; precipitation; fires: Section 2, earthquakes; volcanoes; tsunami; land instability: Section 3 looks at responses to hazards. Many small illustrations. Glossary, p.275-86. Index. *Class No:* 550.34(035)

[1406]

WHITTOW, J. Disasters: the anatomy of environmental hazards. Harmondsworth, Middlesex, Penguin Books, 1980. 411p. illus., diagrs. (*Pelican Books.*)

5 parts: 1. Introduction (environmental hazards) - 2. Subterranean stress (2. Earthquakes; 3. Volcanoes, Sinking coastlines) - 3. Surface instability (5. Landslides and avalanches; 6. Ground surface collapse) - 4. The restless atmosphere (7. High winds; 8. Floods; 9. Drought; 10. Snow and fog) - 5. Will we ever learn? (11. Hazard city: a case study of Los Angeles; 12. Disaster research). Chapter references, p.399-402. Further reading, p.403-5, Analytical index, p.406-11. Praised, especially for its illus. and index, in *Geographical journal,* v.152(2). July 1986, p.258. *Class No:* 550.34(035)

[1407]

—CORNELL, J. The Great international disaster book. 3rd ed. New York, Scribner's, 1982. 472p. illus. $5. ISBN: 068417345x.

First published 1976.

Claims to cover all types of disasters, both natural and man-made (but not war or holocausts) throughout history. Each chapter concerns a particular kind of catastrophe. 'One of the most comprehensive sources of its kind', the author being 'press liaison for the Smithsonian Institution Center for Short-lived Phenomena' (*Reference Services Review,* v.12(2), Summer 1984, p.83-84). *Class No:* 550.34(035)

[1408]

—HARRIS, S.L. Agents of chaos: earthquakes, volcanoes, and other natural disasters. Missoula, Mont., Mountain Press Publishing Company, 1990. 260p. illus., maps. $12.95. ISBN: 0878422439.

3 sections: 1. Earthquake hazards in the US - 2. Volcanic hazards in the West - 3. Ice and fire, myth and reality. Each of the 30 chapters has a bibliography appended. Glossary. Good illustrations. *Class No:* 550.34(035)

Dictionaries

[1409]

JAPAN. MINISTRY OF EDUCATION. Japanese scientific terms: Seismology. [Tokyo], Japanese Society for the Promotion of Science, 1974. 182, [1]p.

About 5,000 entries. Appendices: Magnetic scale; Seismic intensity scale. *Class No:* 550.34(038)

Reviews & Abstracts

[1410]

Abstract journal in earthquake engineering. Berkeley, Calif., Earthquake Engineering Research Center, Univ. of California, 1971-. 2pa. $100. ISSN: 03635732.

About 800 abstracts per issue. 9 sections: 1. General topics of conference proceedings - 2. Selected topics in seismology - 3. Engineering seismology - 4. Strong motion seisometry - 5. Dynamics of soils, rocks and foundations - 6. Dynamics of structures - 7. Earthquake-resistant construction and hazard reduction - 8. Earthquake effects - 9. Earthquakes and natural disasters. Title, author and analytical subject indexes. Abstracts cover technical papers, research reports, books, codes and conference proceedings. *Class No:* 550.34(048)

[1411]

Bibliography of seismology / Bibliographie de séismologie / Bibliographia seismologii. Newbury, Berks., International Seismological Centre, 1965-. 2pa. £90.

Published with Unesco financial assistance.

About 2,000 references per issue. 3 parts: Subject index; author index; list of citations to important events. Appended: 'A cumulative list of source abbreviations' (*c.*500 sources). *Class No:* 550.34(048)

Geochemistry

[1412]

CHESTER, R. Marine geochemistry. London, Unwin Hyman, 1990. xviii,698p. diagrs., tables, maps. £92; £34.95. ISBN: 004551108x.

17 chapters: 1. Introduction - 2. The input of material to the ocean reservoir ... 8. Dissolved gases in sea water ... 12. Down-column fluxes and the benthic boundry layer ... 17. Marine geochemistry: an overview. Bibliography appended for each chapter. Aimed at intermediate and advanced students of chemical oceanography, marine geochemistry and the earth sciences. Criticized in *Choice,* January 1991 for badly reproduced tables due to poor paper quality, but 'provides the best descriptive survey of ocean geochemistry since the 1970s'. The author is Professor of Oceanography at the University of Liverpool. *Class No:* 550.4

[1413]

The Encyclopedia of geochemistry and environmental sciences. Fairbridge, R.W., *ed.* New York, etc., Van Nostrand Reinhold, 1972. xxi, 1321p. diagrs., graphs, tables, maps. (*Encyclopedia of earth sciences, v.4A.*)

378 articles, A-Z, by 237 specialists. Broad scope, embracing all elements in nature, all the more important cycles and processes, mineral groups, and all economic ore deposits. Each contribution has references and cross-references (*e.g.* 'Groundwater motion in draining basins', p.478-87, has 3 columns of references, plus cross-references to other volumes in the series. 'Hydrocarbons', p.495-503; 2p. of references). A major reference source. *Class No:* 550.4

[1414]

Geochemical reference material compositions: rocks, minerals, sediments, soils, carbonates, refractions & ores used in research & industry. Potts, P.J., *and others.* Latheronwheel, Caithness, Whittles Publishing; Boca Raton, Fla., CRC Press, 1992. xiv,313p. tables. £60. ISBN: 1870325400.

Sections: 1. Reference materials indexed by sample name - 2. Reference materials indexed by sample description - 3. Compilation tables (p.22-189) - 4. Concentration ladders - 5. References to source data. For the analyst, geochemist, and those in the mining and metallurgical industries involved in the exploitation of raw materials. *Class No:* 550.4

[1415]

INSTITUTE OF GEOLOGICAL SCIENCES. Regional geochemical atlases. Oxford, Blackwell, 1978-84.

Shetland. 1978. *Orkney.* 1979. *South Orkney and Caithness.* 1980. *Sutherland.* 1982. *Hebrides.* 1984. £40.

Each area has 20 or 30 folding maps and 12p. of text, and each atlas of maps covers *c.*16 elements (barium ... zirconium). Designed to provide systematic data on trace element levels in stream sediments as a basis for mineral exploration. *Class No:* 550.4

[1416]

WEDEPOHL, K.M., *ed.* **Handbook of geochemistry.** Berlin, New York, Springer-Verlag, 1969-74. 2v. in 6. diagrs., graphs, tables, Loose-leaf.

About 70 contributors. V.1 'contains fundamental facts of geochemistry, geophysics and cosmochemistry, together with definitions, dimensions, methods of evaluation, etc.' (*Preface*). V.2 (5v., *c.*3300p.) has data on 'each element, with the exception of the noble gases, the lanthanides and the platinum elements'. *Class No:* 550.4

[1417]

The Wolfson geochemical atlas of England and Wales. Compiled by the Applied Geochemistry Research Group, Imperial College of Science and Technology, London. Webb, J.S., *and others.* Oxford, University Press, 1978. 70p. maps. £90. ISBN: 0198911130.

55 maps (48 in colour) plus 3 maps in back cover. Scale: 1:2,000,000. The computer-produced maps depict data on 21 elements from the analysis of 50,000 stream samples. Introductory text provides examples of potential uses of the atlas in agriculture, ecology and public health, as well as mineral exploration. *Class No:* 550.4

Prospecting & Exploration

[1418]

Applied geology for engineers. Great Britain. Ministry of Defence. London, HMSO, 1976. xxxv, 378p. illus., diagrs., graphs, tables, maps. £9. (*Military engineering, v.15.*) ISBN: 0117715972.

Sponsored jointly by the Ministry of Defence and the Institution of Civil Engineers.

11 chapters: 1. The scope of geology - 2. General geological theory - 3. The formation of rocks - 4. Properties and identification of rock types - 5. Landforms and rocks - 6. Geophysical methods - 7. Geological maps and other information sources (p.147-70)- 8. Terrain evaluation - 9. Site selection and evaluation - 10. Engineering in rocks and soil - 10. Construction materials - 11. Groundwater and its exploitation. Annex A: Unified classification system for soils. Detailed index, p.364-78. *Class No:* 550.8

[1419]

Engineering geology abstracts. Alexandria, Va., American Geological Institute, 1984-. 4pa. $35. ISSN: 07423101.

Produced in co-operation with the Association of Engineering Geologists, Canadian Geotechnical Society, International Association of Engineering Geology and Geological Society of America: Engineering Geology Division.

Contains *c.*4,000 references and abstracts pa. to engineering and geology literature published worldwide. 22 subject headings (01. General engineering geology - 02. Rock mechanics - 10. Foundation excavation, treatment, grouting - 14. Land use, planning: environmental - 21. Geophysical methods, instrumentation - 22. Miscellaneous). Abstracts up to 200 words in length. Numerous cross-references. No indexes. *Class No:* 550.8

[1420]

McLEAN, A.C. *and* **GRIBBLE, C.D. Geology for civil engineers.** London, Allen & Unwin, 1985. xviii, 314p. illus., diagrs., tables, maps. £45. ISBN: 0046240055.

8 chapters: 1. Introduction - 2. Minerals and rocks - 3. Superficial deposits - 4. Distribution of rocks at and below the surface - 5. Subsurface (ground) water - 6. Geological exploitation of an engineering site - 7. Rocks and civil engineering - 8. Principal geological factors affecting certain geological projects. 'References and selected reading' appended to each chapter. Appendices A-H (A. Descriptions of some important soil groups; F. Quality of aggregates). Analytical index, p.305-14. Well tailored to meet the needs of qualified engineers. *Class No:* 550.8

[1421]

NELSON, A. *and* **NELSON, K.D. Dictionary of applied geology, mining and civil engineering.** London, Newnes, 1967. vii, 421p. diagrs.

Concise definitions, with cross-references. Detailed list of minerals, gemstones and rock types. Other topics include building materials, rock structures in relation to civil engineering projects, water resources and supplies, soil mechanics and site investigation, geophysics, exploratory drilling and sampling. Appendix 1: 'Recommended symbols for rock and soils'. 'Intended primarily for school and college students, and engineers in the mining and civil engineering professions' (*Preface*). *Class No:* 550.8

Remote Sensing

[1422]
HYATT, E. **Keyguide to information sources in remote sensing.** London, Mansell, 1988. xiv,274p. £40. ISBN: 0720118549.

3 parts: 1. Survey of remote sensing and its literature - 2. is an annotated bibliography of 819 items which includes books, journals, literature guides, abstracts and indexes, and databases - 3. is a list of 326 organizations, grouped under 81 countries, with addresses. Index. A well-presented, impressive guide. *Class No:* 550.81

[1423]
Manual of remote sensing. Colwell, R.N., *ed.* 2nd ed. Falls Church, Va., American Society of Photogrammetry, 1983. 2v. (xvii, xvii, 2440p.). illus. (incl. col.). ISBN: 0937294527.

First published 1975.

V.1: Theory, instruments and techniques; v.2: Interpretation and applications. 36 documented chapters; 279 col. plates. 'The most comprehensive work on remote sensing to date' (Harris, C.D. *Geographical bibliography for American libraries* (1985, item 356). *Class No:* 550.81

[1424]
RABCHEVSKY, G.A., *ed.* **Multilingual dictionary of remote sensing and photogrammetry:** English glossary and dictionary, equivalent terms in French, German, Italian, Portuguese, Spanish, Russian. Falls Church, Va., American Society of Photogrammetry, 1984. xx, [1], 343p. $20. ISBN: 0937294462.

7 parts, Part 1: English glossary and dictionary (1,716 numbered terms, with equivalents and indexes in the other 6 languages). Appended selected bibliography, p.341-3. *Class No:* 550.81

[1425]
Remote sensing of earth resources: UK groups and individuals engaged in remote sensing, with a brief account of their activities and facilities. Lindsay, E.J., *comp. and ed.* 5th ed. London, Department of Industry, 1981. [v], 386p.

First published 1972.

'Remote sensing activity' enumerates 10 applications (*e.g.* 6. Meteorology; 7. Environmental pollution), with data: name and address of organization; present activity; persons involved; future activites; facilities and equipment available; publications. Appendices include 'Facilities for education and training in Remote Sensing ...' Index of organizations. Grouped subject entries. Addenda. *Class No:* 550.81

[1426]
—BRYAN, M.L. Remote sensing of earth resources: a guide to information sources. Detroit, Mich., Gale, 1979. xv, 188p. $68. (*Geography and Travel Information guides, v.1.*) ISBN: 0810314134.

Carries 378 annotated entries in 8 chapters: 1. General literature; 2. Proceedings; 3. Manuals and guides; 4. Catalogs; 5. Maps; 6. Bibliographies; 7. Journals; 8. Workshops, university and training courses; [U.S.]. Each chapter has an introduction. Author, title and subject index; NTIS (National Technical Information Service, Springfield, Va.); Series index. *Class No:* 550.81

Geothermal Exploration

[1427]
ARMSTEAD, H.C. **Geothermal energy:** its past, present and future contributions to the energy needs of man. 2nd ed. London & New York, Spon, 1983. xxxviii, 404p. illus., diagrs., tables. £49; $59.95. ISBN: 0419122206.

First published 1978.

19 chapters (1. Historical note... 4. The heat energy of the earth - 5. The nature and occurrence of geothermal fields - 6. Exploration - 7. Drilling - 8. Bore characteristics and their measurement - 9. Fluid collection and transmission - 10. Electric power generation from geothermal energy - 11. Direct applications of earth heat... 14. Some economic considerations... 18. The future - 19. Epilogue: the frame of the geothermal picture). 257 references, p.377-88. Detailed, partly analytical index, p.389-404. A comprehensive treatment; the standard textbook. *Class No:* 550.836

[1428]
BOWEN, R. **Geothermal resources.** 2nd ed. London, New York, Elsevier, 1989. xii,485p. illus., figs., maps, tables. £74; $52. ISBN: 1851662871.

First published 1979.

7 well-documented chapters: 12. The origins of earth heat - 2. Geothermal systems and models - 3. Geothermal exploration ... 6. Environmental impact - 7. Uses of geothermal energy. Brief glossary. 4 appendices: 1. Geothermal miscellanea - 2. Companies of geothermal interest - 3. Organizations of geothermal interest - 4. World geothermal localities. Author and subject indexes. *Class No:* 550.836

[1429]
Geothermal energy: a current awareness bulletin. Springfield, Va., Office of Scientific and Technical Information, US Dept. of Energy, 1983-. Semi-monthly. $135pa. (domestic), $270 (outside the N. American continent). ISSN: 08966257.

Formerly *Geothermal energy update,* 1977-82.

International in scope, each issue carries *c.*20 abstracts. General coverage: Resource status and assessment - Geology and hydrology of geothermal systems - Geothermal exploration - Legal and institutional aspects - Environmental aspects and waste disposal - By-products - Geothermal power plants - Geothermal engineering - Direct energy utilization - Geothermal data and theory. Energy database. *Class No:* 550.836

[1430]
Geothermal energy - the potential in the United Kingdom. Downing, R.A. *and* Gray, D.A., *eds.* London, HMSO, 1986. xiii, 187p. diagrs., charts, graphs, tables, maps. ISBN: 0118843663.

Programme sponsored by the Department of Energy and the Commission of the European Communities.

11 contributors. Assessment of results of an investigation of Britain's geothermal potential by the British Geological Survey, 1977-84. 10 chapters (*e.g.* 6. 'Geochemistry of geothermal craters in the UK'; 9. 'Engineering and economic aspects of low anthalopy development'). References, p.162-70. 4 appendices (3. Glossary of technical terms). Partly analytical index. Geological map of the UK, in back pocket. The first comprehensive review of its kind for the UK. *Class No:* 550.836

[1431]

Handbook of geothermal energy. Edwards, L.M., *and others, eds.* Houston, Texas, Gulf Publishing Co., 1982. ix, 613p. illus., diagrs., graphs, tables. £67; $89. ISBN: 0872013227.

15 contributors. 10 chapters: 1. Introduction - 2. Worldwide geothermal resources - 3. Geology of geothermal systems - 4. Exploration for geothermal energy - 5. Geothermal well drilling and completion - 6. Casing and tubular design concepts - 7. Geothermal cementing - 8. Formation evaluation - 9. Reservoir engineering concepts - 10. Energy conversion and economic issues for geothermal energy. Statistics, to 1982. Detailed, analytical index, p.599-610. Aims at 'a comprehensive review of significant developments in the location and production of geothermal energy' (*Foreword*). Complementary to Armstead in stressing practical aspects. *Class No:* 550.836

[1432]

SUMMERS, W.K. Annotated and indexed bibliography of geothermal phenomena. Socorro, New Mexico, New Mexico State Bureau of Mines and Mineral Resources, 1971. [vi, 665p.].

14,177 references, some with abstracts, to all the natural, physical and chemical aspects of the earth's heat. Claimed to be 95% complete to December 1969, with selected references from 1970. Computer-produced bibliography, in rundown order. Manually produced subject index; also geographic and author indexes. *Class No:* 550.836

Geology

Bibliographies

[1433]

'Geological literature'. In *Encyclopedia of library and information science* v.9, 1973, p.309-44. Lea, G, *and others.*

A survey of geology documentation. Sections on the primary literature, nomenclature and terminology, secondary sources (p.324-52). Detailed accounts of international, national and regional, and specialized services, guides to literature and reference works (75 items, unannotated); select list of 180 abbreviations. 74 references. 'Geological libraries and collections' (*Encyclopedia*, v.9, p.283-309) describes the development of geological libraries and collections, with references and bibliography (21 items). Of historical interest only. For more current information, the reader should consult the relevant chapters in D.N. Wood's *Information sources in the earth sciences (q.v.)* (London, Bowker Saur, 1989). *Class No:* 551(01)

[1434]

—MACKAY, J.W. Sources of information for the literature of geology: an introductory guide. 2nd ed. Edinburgh, Scottish Academic Press, 1974. v, 59p.

Originally (1971) prepared for internal use in the University College, London, Department of Geology; now 'slightly revised and edited for a more general audience' (*Preface*). 9 chapters (2. General bibliographical services ... 4. Special subject services (*e.g., Mineralogical abstracts*)- 5. Regional and minor subject bibliographies in geology ... 8. A miscellany of useful sources of information (*e.g.* library

.... (contd.)

catalogues; theses; maps) - 9. Using information services. 13 references. Valuable commentary on abstracting and indexing services. Admits to a bias towards metropolitan sources. *Class No:* 551(01)

[1435]

—MARGERIE, E. de. Catalogue des bibliographies géologiques. Paris, Gauthier-Villars, 1896. (Reprinted Amsterdam, Meridian, 1968. 2v.). xx, 733p.

1,667 numbered entries for bibliographies, in general and regional sections. Author, subject and geographical indexes. A major source for bibliographies "hidden" in books, as appendices, or in periodicals. *Class No:* 551(01)

[1436]

—MATTHEWS, E.B. Catalogue of published bibliographies in geology, 1898-1920. Washington, National Research Council, 1923. 228p. Bulletin of the National Research Council, v.6(5), October 1923.

Continues Margerie in part, omitting regional bibliographies. Subject arrangement; author index. *Class No:* 551(01)

Encyclopaedias

[1437]

The Encyclopedia of applied geology. Finkl, C.W., *ed.* Princeton, N.J., Van Nostrand Reinhold, 1984. xxviii, 832p. illus., diagrs., graphs, tables, maps. $122.95. (*Encyclopedia of earth sciences, v.13.*) ISBN: 0442225377.

82 contributors. 87 main entries, well subheaded and documented (*e.g.* 'Geochronology': 8p.; subsections, 3 tables, 3 graphs; 55 references. 'Geomorphology, applied': p.243-70; nearly 5p. of references). Valuable preface, with sections on abstracting and indexing services, computerised databanks,. periodicals, texts and reference works. Adequate cross-references. Samples a wide range of interrelated topics in geology in the service of man. *Class No:* 551(031)

[1438]

Encyclopedia of field and general geology. Finkl, C.W., *ed.* New York, Van Nostrand Reinhold, 1988. 911p. $89.95. (*Encyclopedia of earth sciences, v.14.*) ISBN: 0442224900.

Intended as a companion volume to *Encyclopedia of applied geology* (*q.v.*), this volume includes biogeochemistry, geoanthropology, and geobotany. As with the other volumes in the series, there are many internal cross-references, and to v.13. Good index. 'A must in any geology collection at all levels' (*New technical books,* Nov/Dec, 1988, item 1579). *Class No:* 551(031)

Handbooks & Manuals

[1439]

LAMBERT, D. *and* **THE DIAGRAM GROUP. The Field guide to geology.** New York, Facts on File, 1988. 256p. diagrs., maps. $22.95. ISBN: 0816016976.

Contains many illustrations with over 500 maps, diagrams, and charts. Appendices list achievements of famous geologists, and geological displays around the world. Brief bibliography. Subject index. For public libraries. *Class No:* 551(035)

[1440]

PUTNAM, W.C. Putnam's geology. Birkeland, P.W. *and* Larson, E.E. 5th ed. New York, Oxford University Press, 1989. 646p. illus. (incl. col.), diagrs., maps. $49.95; $28.95; £22.50. ISBN: 0195056302.

First published 1964.

19 chapters: 1. The planetary system ... 8. Volcanism ... 11. Mass movements and related geological hazards ... 15. Shore processes and landforms ... 19. Resources and energy. References appended to each chapter. Conversion table and glossary. Analytical index, p.635-46. A detailed textbook, with numerous black-and-white illustrations. *Class No:* 551(035)

[1441]

WILHELMS, D.E. The Geologic history of the moon. Washington, US Government Printing Office, 1987. 302p. illus. (some col. pl.).

A review of lunar science from the viewpoint of historical geology. 14 chapters: 1. General features ... 3. Crater materials ... 6. Structure - 7. Relative ages ... 9. Nectarian system ... 12. Eratostherian system ... 14. Summary. References cited, p.283-94. Index. 12 colour plates and many black-and-white illustrations. For geoscientists, planetologists and geologists. *Class No:* 551(035)

Dictionaries

Polyglot

[1442]

Geostatistical glossary and multilingual dictionary. Olea, R.A., *ed.* New York, Oxford University Press, 1991. 177p. $39.95. ISBN: 0195066898.

Part 1: The glossary (Absolute dispersion ... Z-score). Part 2: The multilingual dictionary (English, Chinese, French, German, Greek, Portuguese, Russian, Spanish). Selected bibliography, p.85-88. *Class No:* 551(038)=00

[1443]

ZYLKA, R. Geological dictionary: English, Polish, Russian, French, German. Warsaw, Wydawnictwa Geologiczne, 1970. 1439p.

About 25,000 numbered English-base entries, with Polish, Russian, French and German equivalents (and synonyms) and indexes. 'One of those rare multilingual dictionaries of technical terms that will please a geologist, a translator, and a graduate student faced with any important piece of geological literature from Europe' (*Choice,* v.9, November 1972, p.1116). *Class No:* 551(038)=00

[1444]

—Multilingual thesaurus of geosciences. Rassam, G.M., *and others, eds.* Oxford, etc., Pergamon Press, 1987. 500p. £115; $120. ISBN: 0080364314.

A six-language geological dictionary, with *c.*5,000 terms, mostly taken from the American Geological Institute's *Glossary of geology (q.v.),* in English, French, German, Italian, Spanish and Russian. Has a field index listing terms by subfields within geology. *Class No:* 551(038)=00

English

[1445]

BATES, R.L. *and* JACKSON, J.A. Glossary of geology. 3rd ed. Amsterdam, Elsevier, 1987. 754p. £60. ISBN: 0913312894.

First published 1957. Published in the US and Canada by the American Geological Institute.

Contains *c.*37,000 terms, covering subject areas which include geology; geophysics, sedimentology; hydrogeology; mineralogy; plate tectonics. For the first time, pronunciation is given. Includes abbreviations and dates when terms were first used. There are *c.*2,000 bibliographic references. The definitive geology glossary. *Class No:* 551(038)=20

[1446]

CHALLINOR, J. Challinor's dictionary of geology. 6th ed., edited by A. Wyatt. Cardiff, Univ. of Wales Press, 1986. xviii, 374p. £25; £9.95. ISBN: 0708309135.

First published 1961.

Defines over 2,000 terms, many at length. Numerous quotations; valuable commentary. 'Landslide': nearly 1 columm. 63 categories are applied. Classified index, p.345-74. Probably the most comprehensive source of geologic terminology for the general public. Sources of quotations are chiefly British in origin. *Class No:* 551(038)=20

[1447]

—WATT, A. Longman illustrated dictionary of geology: the principles of geology explained and illustrated. Harlow, Essex, Longman Group, 1982. 192p. col. illus., diagrs., tables, maps. £5.99. ISBN: 0582555493.

Over 1,500 entries, supported by more than 350 illus. Adequate cross-references. Appendix: Common abbreviations in geology; International system of units; Understanding scientific words. Detailed analytical index, p.171-82. 'It is an ideal book for first-year undergraduate students' (*Bulletin of the American Association of Petroleum Geologists,* v.67, June 1983, p.1045). *Class No:* 551(038)=20

[1448]

LAPIDUS, D.F. Collins dictionary of geology. London, Harper Collins, 1990. 565p. figs. £5.95. ISBN: 0004341481.

First published as *Facts on File dictionary of geology and geophysics* (Facts on File Publications, 1987).

Contains over 4,500 entries, Aa ... Zweikanter. Length of entries varies from a few words, *e.g.* 'Dorsum' to several pages, *e.g.* 'Carboniferous'. Many cross-references. 5 appendices: 1. SI units ... 4. Some techniques used in the analysis of rocks and minerals - 5. Suggested further reading. Useful for school, public and academic libraries, and complements *Challinor's dictionary of geology (qv).* *Class No:* 551(038)=20

[1449]

WHITTEN, D.G.A. *and* BROOKS, J.R.V. The Penguin dictionary of geology. London, Allen Lane, 1978. [516]p. illus., diagrs., tables. £5.99. ISBN: 0140510494.

Paperback ed., 1972.

About 3,500 terms are defined, with commentary (*e.g.* 'Basalt' : 1p.; 'Karst' : 1½p.). Brief biographies are included. Appended table of minerals, grouped bibliography, p.515-6. Intended for the amateur geologist and for scientists in other fields who need quick reference to the vocabulary of geology. *Class No:* 551(038)=20

Czech

[1450]
ZEMAN, O.S. Anglicko-Český geologický slavník s rejstříken Českých Nžviů. 2nd ed. Prague, Academia Praha, 1985. 367p.

English-Czech dictionary of geological terms, with Czech reverse index. 18,241 numbered English entries. Czech index, p.307-479. List of sources, p.9-11.

Class No: 551(038)=850

Arabic

[1451]
ARAB LEAGUE EDUCATIONAL, CULTURAL AND SCIENTIFIC ORGANIZATION. Lexicon of geology: English-French-Arabic. Rabat, Morocco, Bureau Permanent de Co-ordination de l'Arabisation, 1971. 238, xlvp. *Class No:* 551(038)=927

Theses

[1452]
Comprehensive dissertation index, 1861-1972. Volume 15: Geography and geology. Ann Arbor, Mich., Xerox: Univ. Microfilms, 1973. xxi, 728p. Cumulation for 1973-77, with annual suppts. ISBN: 0835700941.

Arranged by subject keyword. Data: title, author, degree, university, number of pages, date of degree, *Dissertation abstracts international* citation, order number. No indexes. Difficult to use. *Class No:* 551(043)

Reviews & Abstracts

[1453]
Geological abstracts. Norwich, Elsevier, Geo Abstracts, 1989-. 12pa. $612; £360. ISSN: 09540512.

Previously *Geophysics and tectonic abstracts* (1982-85), originally *Geophysical abstracts* (1977-81).

1986-1988, split into 4 series: *Economic geology*; *Sedimentary geology*; *Palaeontology & stratigraphy*; *Geophysics & tectonics*.

Divided into 16 subject areas: Mineralogy - Geochemistry - Geochronology - Igneous and metamorphic geology - Sedimentary geology - Stratigraphy - Quaternary geology - Palaeontology - Regional geology - Geophysics - Structural geology and tectonics - Seismology - Engineering and environmental geology - Economic geology - Extra-terrestrial geology - General texts. Monthly, annual and regional indexes. Available online on the GEOBASE database. Available on microfilm. *Class No:* 551(048)

[1454]
Zentralblatt für Geologie und Paläontologie. Teil 1 : Allgemeine, angewandte, regionale und historische Geologie. Stuttgart, E. Schweizerbart'sche Verlagsbuchhandlung, 1950-. 13pa. $75 per issue. ISSN: 03405109.

Each issue carries 1,400 references. 6 sections: 1. General geology; 2. Applied geology; 3. Regional geology; 4. Historical geology; 5. Miscellaneous; 6. Geophysics. Entries include descriptors. Keyword index; geographical index; index of topographical maps. Annual catchword and geographical indexes; topographic index for maps. Teil 2: *Paläontologie (qv).* *Class No:* 551(048)

Periodicals

[1455]
BRITISH LIBRARY. Science Reference Library. Periodicals on geology held by the Science Reference Library. Bradley, O. *and* Bird, S., *comps.* London, the Library, 1978. [2], iv, 144p.

Nearly 2,500 periodicals, under titles A-Z, with SRL call numbers. Alternative, ceased and continuation titles find a place. Appended list of 116 abstracting and indexing periodicals. *Class No:* 551(051)

[1456]
GEOLOGICAL SOCIETY, London. List of serial publications held in the library of the Geological Society, London. London, the Society, 1978. [v], 172p. £10. ISBN: 0903317230.

Lists 2,700 titles of serial publications currently received up to August 1977, and discontinued titles. Sections: Non-Russian periodicals A-Z - Russian titles - Loan sets of geological maps. *Class No:* 551(051)

Yearbooks & Directories

[1457]
Directory of geo-analytical facilities in British universities, colleges, polytechnics and research institutions. Potts, P.J. *and* Tindle, A.G., *comps.* Latheronwheel, Caithness, Whittles Publishing, 1990. 153p. tables. £40. ISBN: 1870325451.

Directory tables, p.2-124. Index of specialists; index of techniques. Addresses of contributing departments. References, p.145-53. *Class No:* 551(058)

[1458]
The Geologist's directory: a guide to geological services, equipment and sources of geological information. Reeves, G.M. 5th ed. London, Institution of Geologists, 1992. 224p. £20. ISBN: 0903317737.

4th ed., 1988.

9 sections: 1. The Institution of Geologists - 2. Geology in government (*e.g.* 'Local government in England', 'County Councils in Ireland') - 3. Geology in education - 4. Geology in industry - 5. Consultants (*e.g.* 'Consulting geochemists', 'Consulting mining engineers') - 6. Specialist services - 7. Buyers guide - 8. Geological information sources - 9. Companies working overseas. *Class No:* 551(058)

[1459]
—Geological directory of the British Isles: a guide to information sources, 1978. Diment, J.A., *ed.* London, Geological Society of London, Geological Information Group, 1978. 109p. £3. ISBN: 0903317206.

Previously published 1970.

3 sections: 1. Geographical index - 2. Directory of organizations (p.18-98) - 3. Name index. Data on organizations (Aberdeen ... Zennor): name and address; telephone; date founded; library (including main subjects; special collections); publications. *Class No:* 551(058)

Teaching Materials

[1460]
THOMPSON, D.B. *and* SCRAGG, H. 'Resources for geology teachers', in *Geology teaching,* v.6(1), March 1981, p.5-19.

Previous lists of resources in *Geology teaching:* v.1.(3), 1976, p.64-66, 94; v.2(1), 1977, p.16-19; v.3(1), 1978,

....(contd.)
p.13-17; v.5(2), June 1980, p.59-67.

March 1981 listing has 22 sections, entries being
unannotated but coded as to level (I = infants; S =
secondary; F = university and research; G = general adult
reference). Films are covered in 3 groups.
Class No: 551(072)

Tables & Data Books

[1461]
The Geology of the terrestrial planets. Carr, M.H., *ed.*
Washington, NASA, 1984. 317p. $16. (*Special publication
SP 469.*) ISBN: 0160041627.

Takes the terrestrial planets (Venus, Mars, Mercury,
Moon, Earth) in solar system order. Data: orbit, magnetic
fields (if any), surface characteristics, craters, tectonics, etc.,
summarizing what has been learned since 1962. The Earth is
given a brief chapter. 'Intended for the interested layman'
(*Astronomy,* v.13, October 1985, p.88). *Class No:* 551(083)

Nomenclatures

[1462]
Geological nomenclature: English-Dutch-French-German.
Schieferdecker, A.A.G., *ed.* Gorinchem, J. Noorduijn, for
the Royal Geological and Mining Society of the Netherlands,
1959. xvi, 523p.

First published 1929.

5,409 numbered English-base terms, with explanations in
English and equivalents in Dutch, French and German.
Classified arrangement: 10 sections, with subdivisions.
Combined index in all four languages. A valuable polyglot
dictionary. *Class No:* 551(083.72)

[1463]
—Geological nomenclature. Visser, W.A., *ed.* Utrecht, Bohn,
Scheltema & Holkema; Norwell, Md., Kluwer, 1980. xxvi,
[1], 540p. tables. $164.50. ISBN: 9024724031.

A later edition of Schieferdecker's *Geological
nomenclature,* includes Spanish in the languages covered.
4,876 numbered English-base terms, with equivalents in
Dutch, French, German and Spanish. Thematic
arrangement. 5 language indexes, 14 indexes.
Class No: 551(083.72)

Films

[1464]
**AMERICAN ASSOCIATION OF PETROLEUM
GEOLOGISTS. Index to films related to geology and
energy exploration.** Tulsa, Oklahoma, the Association,
1976. 74p. *Class No:* 551(084.122)

Maps & Atlases

[1465]
DERRY, D.R. World atlas of geology and mineral deposits.
London, Mining Journal Books, 1980. [iv], 110p. diagrs.,
tables, maps. ISBN: 0470269960.

3 parts: the text (Introduction; Landscape and geology;
Structure and history of the earth; Distribution of
earthquakes and volcanoes; Distribution of mineral deposits)
- the world atlas (p.25-88; mostly in colour: The Americas
... Arctic) - Tables: world mineral resources production and

....(contd.)
reserves. Suggested reading, p.102. Sources of additional
information. Glossary, p.108-10. 'Written for the layman'
(*Geological magazine,* v.118, July 1981, p.436).
Class No: 551(084.3)

[1466]
—Geological world atlas. Compiled for Unesco and the
Commission for the Geological Map of the World by the
International Geological Mapping Office, 1984. 22 sheets.

On the scale of 1:10,000,000 (1:16,000,000 in the case of
ocean sheets). Sheet 1: General; Sheets 2-15, 17; Continents;
Sheets 16, 18-22; Oceans. Each sheet or group of sheets has
brief explanatory note in English and French.
Class No: 551(084.3)

[1467]
—Index to maps in earth science publications, 1963-1983. Van
Balen, J., *comp.* Westport, Conn., Greenwood Press, 1985.
400p. $55; £58.50. ISBN: 0313249636.

Aims at listing *c.*4,900 geological maps in serials and
monographs published during the period. Maps are recorded
A-Z, with author index. Periodicals were searched for the
years 1951-82, but only 60 monographs selected for 1963-
83. The lack of selection criteria, the absence of scales in the
source index, and the emphasis on larger-area rather than
smaller-area maps are criticised in *Choice,* v.23(4),
December 1985, p.587. *Class No:* 551(084.3)

[1468]
MALTMAN, A. Geological maps: an introduction. Milton
Keynes, Open University Press, 1990. viii,184p. illus. (col.
pl.), figs., maps. £14.50. ISBN: 0335152228.

Covers fundamental principles, with data on sources of
geological maps, a short history, new techniques and forms
of geology maps. 15 documented chapters: 1. Some
fundamentals of geological maps ... 5. Geological cross-
sections - 6. Visual assessment of outcrop patterns - 7.
Unconformities ... 12. Geological history from maps ... 15.
Current trends in geological maps. For first-year geology
undergraduates but could also be useful for the more
advanced student. A knowledge of rocks and geological
processes is assumed. *Class No:* 551(084.3)

Chronologies

[1469]
**THOMPSON, S. A Chronology of geological thinking from
antiquity to 1899.** Metuchen, N.J., Scarecrow Press, 1988.
vii,320p. $29.50. ISBN: 0810821214.

Chronology, p.1-270. List of sources cited and their
abbreviations. Bibliography of sources cited, followed by an
author index. A subject index would have been useful.
Class No: 551(090)

Histories

[1470]
**ADAMS, F.D. The Birth and development of the geological
sciences.** Baltimore, Md., Williams & Wilkins, 1938;
reprinted, New York, Dover, 1954. v. 506p., facsims.,
ports. $10.95. ISBN: 048626372x.

A history in 14 footnoted chapters 'to the earlier part of
the nineteenth century'. Chapter 1. Introduction,
bibliography and sources; 2. Geological science in Classical
times ... 3/5. Middle Ages (p.51-169); 10. The origin of
mountains (p.329-98, with 75 references); 13. Quaint stories

.... *(contd.)*

and beliefs. Profuse quotations. Analytical index, p.495-506. 'In preparing the present history the writer has in all cases gone back for his material to the original authorities' *(Introduction)*. By no means superseded.
Class No: 551(091)

[1471]
—LAUDAN, R. From mineralogy to geology: the foundations of a science, 1650-1860. Chicago, Ill., Univ. of Chicago Press, 1987. 278p. illus., facsims. $27.50; £21.95. ISBN: 0226469506.
10 chapters (*e.g.* 2. Mineralogy and cosmogony in the late seventeenth century; 4. The botanical model rejected; 7. Historical geology; 9. Lyell's geological logic). Chapter notes, p.229-35. Extensive bibliography, p.236-68. Detailed, analytical index, p.269-78. Stresses concepts. '...it deserves to be widely read' (*TLS*, no.4442, 20 May 1988, p.562).
Class No: 551(091)

[1472]
ZITTEL, K.A. von. **Geschichte der Geologie und Paläontologie bis Ende des XIX Jahrhunderts.** [History of geology and palaeontology.] Munich, Oldenbourg, 1899. vii, 868p. $35.
English translation by M.M. Ogilvie-Gordon (London, Walter Scott, 1905; reprinted New York, Hafner, 1962).
A masterly survey of ancient, medieval and early modern geology in four parts. 'Most complete history of geology' (Smith, H.W. 'Guide to geological literature', in *Journal of geological education,* v.18(1), January 1970, p.25).
Class No: 551(091)

Biographies
[1473]
LA ROCQUE, A. **Contributions to the history of geology.** Columbus, Ohio, State Univ., Dept. of Geology, 1964. 3v. Mimeographed.
Publication began in parts. 1. *Biographies of geologists.* 2. *Bibliography of the history of geology.* 3. *Biographic index.*
V.3 [iii], 217p.) has *c.*20,000 very brief entries, giving name; dates; description; nationality; specialization; post held. List of sources. *Class No:* 551(092)

[1474]
SARJEANT, W.A.S. **Geologists and the history of geology:** an international bibliography from the origins to 1978. London, Macmillan, 1980. 7v. in 6 (4526p.) + Supplement, 1979-84. 2v. (1691p.) and additions (Malabar, Fla., Krieger, 1987). $450; $175, supp. ISBN: 0333293638, set.
V.1 (5 sections) provides an introduction and records histories of geology and related sciences; v.2-3 (sections 6-7): The individual geologists, A-Z; v.4 (sections 8-9): Index of geologists by nationality and country; index of geologists by speciality; v.5 (section 10): Index of authors, editors and translators, p.3535-4515. 'This bibliography attempts to bring together details of all those works in languages using the Latin alphabet which deal with the history of geology' (section 1, *General introduction*). Bibliographies of about 10,000 geologists - a mammoth undertaking. 'It merits unqualified recommendation to those libraries serving users interested in the history of science' (*Choice,* v.18(4), December 1980, p.510). New supplements are planned at 5-yearly intervals. *Class No:* 551(092)

[1475]
—GUNTAU, M. Biographien bedeutender Geowissenschaftler der Sowjetunion. Berlin, Akademie Verlag, 1979. 199p. illus., facsims., ports., diagrs. (*Schriftenreihe für geologische Wissenschaften, Heft 14.*)
Provides 19 biographical sketches of Russian and Soviet geologists of the 18th-20th centuries, with summaries in Russian and English (*e.g.* M.V. Lomonosov (1711-1765): p.7-18. 1 port., 1 facsim. 1p. of bibliography - material by and on). 36 illus. in all. Supplements Sarjeant regarding Soviet geoscientists. *Class No:* 551(092)

Worldwide
[1476]
Beiträge zur regionalen Geologie der Erde. Berlin & Stuttgart, Borntraeger; New York, Lubrecht & Cramer, 1961-. illus., maps.
1. *Die Geologie Mittelamerikas.* R. Weyl. 1961. xvi, 226p.
2. *Die Geologie von Paraguay.* H. Poutzer. 1962. xii, 184p.
3. *Geologie von Chile.* W. Zeil. 1964. xi. 233p.
4. *Geologie der Antillen.* R. Weyl. 1966. viii, 410p.
5. *Rocky Mountains.* D.H. Roeder. 1966. xiv, 316p. $99.50.
6. *Geologie von Syrien und dem Libanon.* R. Wohlfart. 1967. xii, 326p.
7. *Geologie von Jordanien.* F. Bender, 1968. xi, 230p. Supplementary ed. in English, with revisions. 1974, xi, 196p. $110.95.
8. *Geologie von Ungarn.* L. Trunko. 1969. x, 258p.
9. *Geologie von Brasilien.* K. Beurlen. 1970. viii, 444p.
10. *Geology of the South Atlantic Islands,* R.C. Mitchell-Thomé. 1970. x, 368p. $105.
11. *Geologie von Ecuador.* W. Sauer. 1971. ix, 316p.
12. *Geology of the Middle Atlantic Islands.* R.C. Mitchell-Thomé, 1976. ix, 382p. $138.60.
Volumes 13-19 deal with the Andes (1979), Afghanistan (1980), Central America (2nd ed. 1980. $105), Burma (1983), the USSR (1985), N.W. Germany (1986) and Greece (1986), respectively. An excellent series of regional geologies. *Class No:* 551(100)

[1477]
The Encyclopedia of world regional geology. Part 1: Western Hemisphere (including Antarctica and Australia). Fairbridge, R.W., *ed.* Stroudsburg, Pa., Dowden, Hutchinson & Ross, 1975. xv,704p. illus., tables, maps. $129. (*Encyclopedia of earth sciences, v.8.*) ISBN: 012786461x.
88 contributors; *c.*150 signed entries, with references and cross-references. Entries are under countries A-Z. 'United States of America', p.502-641 ('Alaksa', p.513-22: climatology and biogeography; physiography and geomorphology; tectonic framework; stratigraphy; geological history; mineral resources. 52 references, 2 illus., 2 maps). 'West Indies', p.658-66; 20 cross-references. Author and subject indexes. Well illustrated. A major work of reference of considerable value to all levels of enquirer.
Class No: 551(100)

[1478]

Overseas geology and mineral resources. London, HMSO, 1950-.

New series in print: nos.41-63 (1973-86). Some of the original series (v.1 1950-71) were entitled *Colonial geology and mineral resources*. The range is now worldwide (*e.g. no.49: Outline of the geology and mineral deposits of the Democratic Republic of the Sudan and adjacent areas*, by J.R. Vail. 1978. no.58: *An annotated bibliography of Ecuadorian geology*, by C.R. Bristow. 1981.).

The *Overseas memoirs* series (no.1-. 1976-) supplements the above as more specialised monographs (*e.g. The geology of Guadalcanal, Solomon Islands*, by B.D. Hackman. 1980.). *Class No:* 551(100)

Marine Environment

[1479]

BROOKS, J.R.V. *and* **THOMAS, P. 'A Selected bibliography of North Sea geology (up to 1974)'.** In *Proceedings, Geologists Association*, v.85, 1974. p.103-11.

141 items, under 12 subject headings (*e.g.* Forties Field). Updated by their 'A classified bibliography of northwest European continental shelf geology, 1974 and 1975', *Scottish journal of geology*, v.13(1), 1977, p.1-10. *Class No:* 551(26)

[1480]

KENNETT, J.P. Marine geology. Englewood Cliffs, N.J., Prentice-Hall, 1982. xv, 813p. illus., diagrs., graphs, tables, maps. ISBN: 0135569362.

Introduction: History of marine geology - 1. The structural and oceanographic setting - 2. The ocean margins - 3. Oceanic sediments and microfossils - 4. Ocean history. References, p.752-87. Author index; analytical subject index (p.795-813). The text, 'intended for a first course in maritime geology at the advanced undergraduate or graduate level, assumes little background in oceanography' (*Preface*). *Class No:* 551(26)

[1481]

The Sea ... v.3: The earth beneath the sea. History. Hill, M.N., *ed.* New York & London, Interscience: Wiley, 1963 (Reprinted 1980. Melbourne, Fla., Krieger. $88). xvi, 963p. ISBN: 0898740991. *Class No:* 551(26)

[1482]

—**SHEPARD, F.P.** Geological oceanography. New York, Crane Russak & Co., 1977. 214p. illus. (some col.), diagrs.

'Written for students beginning to study geology or anyone interested in important developments in oceanography, - this text provides 'a clear and simple account, by an authoritative author' (*Marine geology*, v.28(3/4), 1978, p.293). *Class No:* 551(26)

[1483]

SHEPARD, F.P. Submarine geology ... with chapters by D.L. Inman and E.D. Goldberg. 2nd ed. New York, Evanston, 1963, & London, Harper and Row, 1964. (3rd ed. 1974). xviii, 537p. illus., diagrs., fldg. chart, maps. (*Harper's Geoscience series*.)

First published 1948.

Successive chapters deal with waves; sedimentation; beaches; continental shelves; coral reefs; deep ocean floor topography; deposits and stratigraphy; mineralogy; etc. 222 illus., diagrs. and maps. References (authors, A-Z), p.489-534. Appendix A: units and symbols. Detailed index.

....(contd.)

'General, but by no means exclusive, emphasis on shore and nearshore aspects of submarine geology' (*Geological magazine*, v.101(5), September/October 1964, p.477). *Class No:* 551(26)

Europe

[1484]

AGER, D.V. The Geology of Europe. New York, London, etc. McGraw-Hill, 1980. xix, 535p. diagrs., table, maps. $27.95. ISBN: 0070841152.

4 parts (17 chapters): 1. Eo-Europe - 2. Palaeo-Europe - 3. Meso-Europe - 4. Neo-Europe. Prelims. include 'Keys to the geological divisions of Europe; Stratigraphical table; General references'. Each chapter has selected references (*e.g.* 16. Southern Alpides: 2¼p. of grouped references). Non-analytical index. 'It is Europe as seen by a stratigraphic palaeontologist' (*Preface*). Author, at University College, Swansea. *Class No:* 551(4)

[1485]

GEOLOGICAL SOCIETY, London. **Sources of information on the geology of the continental EEC countries.** London, the Society, 1980. [1], [1], 150p. (*Miscellaneous paper no. 12*.)

European Economic Community countries, Belgium ... Netherlands (*e.g.* 'Geological mapping and geological surveys of France', p.48-63). Appended: 'Selected bibliography of references, available in London geological libraries on the geology and mineral resources of the continental EEC countries', p.129-36. *Class No:* 551(4)

[1486]

RUTTEN, M.G. The Geology of Western Europe. Amsterdam, etc., Elsevier, 1969. xviii, 520p. illus., diagrs., tables, maps.

18 documented chapters, closely subdivided: 1. The main divisions; 2/3. Northern Europe; 4/7. Hercynian Europe; 8. The Massif Central; 9/16. Alpine Europe; 17, Lowlands and low plateaus; 18. Younger volcanoes. Addendum: Sources for geological information on Western Europe, p.475-6. Reference [names] index; detailed, non-analytical subject index, p.485-520. Helpful introduction, with 15 references. A well-produced survey. *Class No:* 551(4)

[1487]

ZIEGLER, P.A. Geological atlas of western and central Europe. 2nd ed. The Hague, Shell Internationale Petroleum Maatschappij BV, 1990. 239p. figs., maps. 56 enclosures. £50. (*International Lithosphere Program publication, no.148*.) ISBN: 9066441259.

First published 1982.

9 chapters: 1. Introduction ... 3. Caledonian suturing of Laurussia ... 5. Permo-Triassic development of Pangaea ... 9. Plate tectonics and basin evolution: geodynamics processes. References, p.195-233. Geographic index, p.234-37. Geodynamic processes, p.238-39. A major revision, with 56 enclosures in this edition, compared to 40 in the previous. The number of text figures has increased from 29 to 100. A standard work. *Class No:* 551(4)

Great Britain

[1488]
BRITISH GEOLOGICAL SURVEY. British regional geology. 2nd-4th ed. London, HMSO, 1935-.
Scotland:
Orkney and Shetland, by W. Mykura. 1976.
The Northern Highlands of Scotland, by G.S. Johnstone and W. Mykura. 4th ed. 1990. £6.
Grampian Highlands. 3rd ed., by G.S. Johnstone, 1966.
The Tertiary volcanic districts. 3rd ed., by J.E. Richey. 1961.
The Midland Valley of Scotland. 3rd ed., by I.B. Cameron and D. Stephenson. 1985-. £5.
The South of Scotland. 3rd ed., by D.C. Greig and others. 1971.
England:
Northern England. 4th ed., rev. by B.J. Taylor, and others. 1971.
The Pennines and adjacent areas. 3rd ed. by W. Edwards and F.M. Trotter. 1954.
Eastern England from the Tees to the Wash, 2nd ed., by P. Kent, 1980.
Central England. 3rd ed., by B. Hains and A. Horton. 1969.
East Anglia and adjoining areas. 4th ed., by C.P. Chatwin, 1961.
Bristol and Gloucester district. 2nd ed., by G.A. Kellaway and F.B.A. Welsh. 1948.
London and Thames valley. 3rd ed., by R.L. Sherlock, 1930.
The Wealden district. 4th ed., by R.W. Gallois. 1965.
The Hampshire basin and adjoining areas, by R.V. Melville and E.C. Freshney. 1982.
South-west England. 4th ed., by E.A. Edmonds and others. 1975.
Wales:
North Wales. 3rd ed., by T.N. George. 1961.
South Wales. 3rd ed., by T.N. George. 1970.
The Welsh Borderland. 3rd ed., by J.B. Earp and B.A. Hains. 1971.
A series of pamphlets and handbooks that include maps based on the 1:250,000 scale maps of the Geological Survey. *Class No:* 551(410)

Bibliographies

[1489]
British geological literature (new series): a bibliography and index of geology (and related topics) of the British Isles and adjacent sea areas. Worthing, Sussex, Bibliographic Press, 1972-. 4pa. £48; $90 pa. ISSN: 01407813.
In original form, last appeared in 1968.
About 500 abstracts pa. (each *c.*60 words in length). The 26 sections (History of geology ... Engineering geology) include geomorphology and palaeontology. Quarterly author index; annual author, subject and locality indexes.
Class No: 551(410)(01)

Handbooks & Manuals

[1490]
ANDERSON, J.G.C. Field geology in the British Isles: a guide to regional excursions. Oxford, etc., Pergamon, 1983. xi, 324p. maps. £16. ISBN: 0080220541.
10 chapters (*e.g.* 4. Precambrian terrains; 5/6 Caledonian

....(contd.)
terrains; 7. Hercynian terrains; 8. Alpine terrains). 3 appendices (1. Stratigraphical grouping of excursions). Detailed index, p.320-4. Describes 194 geological itineraries, complementing J.G.C. Anderson's and T.R. Owen's *The structure of the British Isles* (2nd ed. Pergamon, 1980) by providing local data. *Class No:* 551(410)(035)

Theses

[1491]
HODGSON, A.V. *and* LAMING, D.J.C., *comps.* Titles of research theses, 1960-1975: Geology of the British Isles and offshore areas, with classified subdiscipline/regional list. 2nd ed. Worthing, Sussex, Bibliographic Press, 1976. [iii], 128p.
Compiled in co-operation with Computer Centre, Sunderland Polytechnic.
About 1,350 titles. Chronological author list; classified subdiscipline list (arranged by year of award, then by authors A-Z); author index. 2-letter code for geological systems; 4-letter code for geophysical subdisciplines. *Class No:* 551(410)(043)

Government Publications

[1492]
Government publications. Sectional list 45, revised November 1986. Natural Environment Research Council. British Geological Survey. London, HMSO, 1986. 47p. *gratis.*
24 sections, including 'Bulletins of the Geological Survey of Great Britain', 'Reports of the Institute of Geological Sciences and British Geological Survey', 'Mineral assessment reports' and 'Memoirs (England and Wales)'. *Class No:* 551(410)(061.1)

Maps & Atlases

[1493]
BRITISH GEOLOGICAL SURVEY. 1:50,000 maps [England & Wales; Scotland; Northern Ireland]. Southampton, Ordnance Survey, for the British Geological Survey.
Replaces the 1:63,360 (one inch to one mile) map series.
The 1:50,000 series of maps appears in two forms: 'solid' (showing the solid rocks at, or immediately beneath, the earth's surface), and 'drift' (showing detritus or drift materials such as broken rock, sand, gravel, alluvium and glacial clays). Some sheets combine the 'solid' and 'drift'. National Grid is given.
Other scales include the 1:250,000, covering the UK and its continental shelf.
A series of 1:50,000 sheet memoirs accompanies the geological maps. Those in print are listed in HMSO's sectional list 45: *Government Publications (qv).*
Class No: 551(410)(084.3)

Histories

[1494]
CHALLINOR, J. The History of British geology: a bibliographical study. Newton Abbot, Devon, David & Charles, 1971. 224p.
Part 1: The primary literature: a chronological listing (to

....(contd.)

1969), 659 items - 2. The major themes (86; *e.g.* 16. The Western counties of England; 74. Pleistocene geology; 86. Some recent collaborative volumes). Appendices: A. Secondary and associated literature; B. Index of authors, with some biographical detail; C. Index of names: places, counties, stratigraphical divisions, fossils. Well-documented and providing 'lively and most interesting reading' (*TLS,* no.3645, 7 January 1972, p.21). *Class No:* 551(410)(091)

Institutions & Associations

[1495]

WILSON, H.E. Down to earth: one hundred and fifty years of the British Geological Survey [1835-1985]. Edinburgh & London, Scottish Academic Press, 1985. [iv], 190p. illus., ports., facsims. ISBN: 0701304733.

18 chapters, on the origins, development and expansion of the British Geological Survey, at home and overseas. Chapter 18: 'What next: the future of the Geological Survey'. References, p.159; Index, p.190. A personal, not an official account. *Class No:* 551(410):061:061.2

Scotland

[1496]

Geology of Scotland. Craig, G.Y., *ed.* 3rd ed. London, Geological Society, 1991. xii,612p. illus., diagrs., charts, tables, maps. £65; £29. ISBN: 090331763x.

First published 1965.

19 contributors. 16 chapters: 1. The growth and structure of Scotland - 2. The Lewisian complex ... 4. Moine ... 12. Permian and Triassic (p.421-38; nearly 5p. of references) ... 16. Economic geology. Partly analytical index, p.597-612. Well produced, with over 200 illustrations. For all those with an interest in geology. *Class No:* 551(411)

Ireland

[1497]

CHARLESWORTH, J.K. Historical geology of Ireland. Edinburgh & London, Oliver & Boyd, 1963. vii, 565p. diagrs., maps.

18 chapters; 140 maps and diagrams. Lengthy bibliography (authors, A-Z), p.500-30. Non-analytical index 'A most important and carefully compiled work, and will hold the field as a standard reference for many a day' (*Nature,* v.199, 7 September 1963, p.941). 'The Irish equivalent of the "Handbooks on the regional geology of Great Britain". (*Geographical journal,* v.129(4), December 1963, p.514-5). Criticised in *Geography* (v.49(4), November 1964, p.447) for lack of a large coloured geological map of Ireland and indifferent illus. Designed for the advanced student of Irish geology or the professional geologist. *Class No:* 551(415)

[1498]

A Geology of Ireland. Holland, C.H., *ed.* Edinburgh, Scottish Academic Press, 1981. x, 335p. illus., diagrs., graphs, maps. £15. ISBN: 0707302692.

9 contributing authors. 19 documented chapters (*e.g.* 3. The Orthotectonic Caledonides; 12. Permiam and Mesozoic; 16. The Quaternary - until 10,000 BP; 18. Economic geology; 19. The history of Irish geology). Analytical index, p.317-35. Deals primarily with historical geology, but also

....(contd.)

offers a short survey of geophysical evidence, of Ireland's now very important economic geology and 'the fascination of Irish geology itself' (*Preface*). *Class No:* 551(415)

England & Wales

[1499]

Geology of England and Wales. Duff, P. McL. D. *and* Smith, A.J., *eds.* London, The Geological Society, 1992. xix,651p. illus., figs., tables. £75; £34. ISBN: 0903317702.

22 contributors. Comprehensive account including recent advances in offshore drilling, plate tectonics, age dating, structure geology and sedimentary environments. 19 documented sections: 1. The growth and structure of England and Wales - 2. Precambrian ... 8. Silesian - 9. Permian ... 12. Cretaceous - 13. Tertiary ... 15. Offshore geology - 16. Igneous rocks ... 18. Deep geology - 19. Economic geology. Many black-and-white illustrations and figures. Index, p.639-51, is in very small print. *Class No:* 551(42)

Wales

[1500]

BASSETT, D.A. Bibliography and index of geology and allied sciences for Wales and the Welsh borders, 1536-1896. Cardiff, National Museum of Wales, 1963. [x], 246p. £4.

7 sections: 1. Introduction; 2. Publications of the Geological Survey and Museum; 3. Palaeontological Society monographs; 4. List of books, papers, etc. (p.12-126; chronological order of publication); 5. List of books and papers written in Welsh; 6. Author index; 7. Analytical subject index. *Class No:* 551(429)

[1501]

—BASSETT, D.A. Bibliography and index of geology and allied sciences for Wales and the Welsh borders, 1897-1958. Cardiff, National Museum of Wales, 1961. 376p. £4.

8 sections, including 2: Reference books, bibliographies, indexes, etc., plus papers and books containing comprehensive lists of references; and 6: Theses accepted for higher degrees. Author and detailed, analytical subject indexes.

Updated in issues of *Geological journal* and *Welsh geological quarterly. Class No:* 551(429)

China

[1502]

The Geology of China. Zunyi, Y., *and others.* Oxford Univ. Press, 1986. 303p. tables, maps. £55. (*Oxford Geological Science series.*) ISBN: 019854460x.

Sections: Background - Stratigraphy - Magnetic and metamorphic rocks of China since 1939 - Geotectonic development of China. Includes regional stratigraphical tables. Selected references. Stratigraphic index. Subject index. *Class No:* 551(510)

Asia—Middle & Near East

[1503]
AVNIMELECH, M.A. Bibliography of Levant geology.
Jerusalem, Israel Program for Scientific Translations, 1965-69. 2v.

Covers the 'Fertile Crescent', including Cyprus, Hatay, Israel, Jordan, Lebanon, Sinai and Syria. V.1: c.4,500, up to 1963; v.2 adds c.900 pre-1963 references and 1,100 for 1963-68. A-Z list, with chronological and subject indexes. List of quoted periodicals and serials.

Current bibliography of Middle East geology (Jerusalem, Geological Survey of Israel, 1976-) supplements Avnimelech. *Class No:* 551(53+56)

Asia—South East

[1504]
HUTCHISON, C.S. Geological evolution of south-east Asia.
Oxford, Clarendon Press, 1989. xv,368p. illus., diagrs., maps. £77.50. (*Oxford monographs on geology and geophysics, 13.*) ISBN: 0198544391.

9 sections: 1. Introduction - 2. Late Mesozoic and Cainozoic tectonic features ... 5. Terrains of Cathaysian affinity (p.133-200) ... 7. Ophiolites and sutures ... 9. Granite and associated plutonic rocks. Many maps and figures. References, p.318-49. Index. *Class No:* 551(59)

Africa

[1505]
FURON, R. The Geology of Africa. Hallam, A. *and* Stevens, L.A., *trans.* Edinburgh & London, Oliver & Boyd, 1963. xii, 377p. illus., maps.

French original first published 1950; 2nd ed. (Paris, Payot, 1960), - used for this translation.

Part 1: General stratigraphy; part 2: Regional ecology (25 chapters on individual regions, exclusive of North Africa). Numerous footnotes reference. Combined author and subject index (inadequate on subjects). 'Highly useful for scholars and students. The volume, though dated, remains the best source available' (Harris, C.D. *ed., Geographical bibliography for American libraries* (1985), entry 2613). *Class No:* 551(6)

[1506]
—Bulletin d'information et de liaison. [Information and liaison bulletin.] Orleans, Association of African Geological Surveys, 1976-. no.1-. 4pa. FFr100. ISSN: 03968863.

Has a major feature, - 'Bibliographie analytique africaine' (*e.g.* issue 1, p.11-127: literature, 1973-1975). Extensive abstracts, arranged by country, then topic. Issue 1 also contains reports on progress in geology in African countries, international project meetings. No index. *Class No:* 551(6)

Africa—West

[1507]
Geology and mineral resources of West Africa.
Wright, J.B., *and others.* London, Allen & Unwin, 1985. xiii, 187p. diagrs., maps. £54. ISBN: 0045560013.

4 sections (18 chapters): 1. The Precambrian of West Africa. 2. Sedimentary basins in West Africa - 3. Mesozoic to Cenozoic igneous activity in West Africa - 4. The Quaternary of West Africa. Glossary, p.163-71. Index of

....(contd.)
place names. Subject index and bibliographies. 'The principal aim is to provide a broad view of West African geology as a whole' (*Preface*). *Class No:* 551(66)

America — North

[1508]
Archaeological geology of North America. Lasca, N.P. *and* Donahue, J., *eds.* Boulder, Colo., Geological Society of America, 1990. 633p. illus., diagrs., tables, maps. $62.50. (*Centennial special volume, 4.*) ISBN: 081375304x.

Published as part of the celebration of the centenary of the Society. Parts: Eastern Canada - Western Canada - Eastern United States - Central United States - Costa Rica and Mexico - North America general - Techniques. Documented chapters range from site specifics to regional studies. Some look at laboratory and field methodology, applications of geophysics to archaeological investigations, and isotope chemistry. Detailed, analytical index, p.619-33. For research and academic libraries. *Class No:* 551(71+73)

[1509]
GEOSCIENCE INFORMATION SOCIETY. Guidebooks Committee. **Union list of geologic field trip guidebooks of North America.** 5th ed. Washington, American Geological Institute, 1989. xxix, 201p. $60. ISBN: 0913312975.

Union list, p.1-172 (Abilene Geological Society (Texas) ... Wyoming Geological Association. Miscellaneous guide books). Participating libraries, p.xiii-xvi. Geographic index. *Class No:* 551(71+73)

Australasia & Oceania

[1510]
The Geological evolution of Australia & New Zealand. Brown, D.A., *and others.* Oxford, etc., Pergamon Press, 1968. x, 409p. diagrs., charts, map. ISBN: 008203186x.

12 sections (*e.g.* The Precambrian system ... The Quaternary system). References, p.365-86; unpublished references, p.387-8. Map references, p.389-92. Detailed, non-analytical index. Aims to be an 'update teaching aid in the stratigraphy of Australia and New Zealand' (*Preface*). *Class No:* 551(9)

Greenland

[1511]
Geology of Greenland. Escher, A.S. *and* Watt, W.S., *eds.* Copenhagen, Grønlands Geologiske undesøgelse [Geological Survey of Greenland], 1976. 603p. illus., diagrs., tables, maps. ISBN: 8798040405.

29 contributors. 21 documented sections (*e.g.* Economic minerals; Petroleum geology; Coal geology; Fossil flora of Greenland). Detailed, analytical index. Fully illustrated and mapped. Aerial photographs are a feature. An admirable example of a country geological survey. *Class No:* 551(988)

Antarctic

[1512]
Geological evolution of Antarctica. Thomson, M.R.A., *and others, eds.* Cambridge, University Press, 1991. xi,722p. illus., diagrs., graphs, tables, maps. £55. ISBN: 0521372666.

.... *(contd.)*

The text of 138 papers presented at the 5th International Symposium on Antarctic Earth Sciences, in Cambridge, UK, 23-28 August 1987.

6 topics: 1. Crustal development: the craton - 2. Crustal development: the Transantarctic Mountains - 3. Crustal development: Weddell Sea-Ross Sea region - 4. Crustal development: the Pacific margin - 5. Crustal development: Gondwana break-up - 6. Evolution of Cenozoic palaeo-environments. Papers carry references. Analytical index, p.707-22. Many illustrations with clearly written captions. A valuable compendium. *Class No:* 551(99)

Information Management

[1513]

Geoscience information: a state-of-the-art review. Proceedings of the 1st International Conference on Geological Information, London, 10-12 April, 1978. Harvey, A.P. *and* Diment, J.A., *eds.* Heathfield, Sussex, The Broad Oak Press, 1979. iii, 287p. £53.95. ISBN: 0906716004.

31 papers, mostly with references appended. 5 parts: Review of geological documentation - Aspects of geological documentation - Application of information handling to applied geology - Documentation in special areas (*e.g.* bibliographical control of geological maps) - User viewpoints. No index. *Class No:* 551:025.4

Thesauri

[1514]

GeoRef thesaurus and guide to indexing. Shimomura, R., *ed.* 5th ed. Alexandria, Va., American Geological Institute, 1989. 752p. Also available in microfiche. $75; $25, microfiche. ISBN: 091331983x.

First published 1977.

Thesaurus of 20,242 terms, A-Z, 4,284 of the new to this ed. A 'guide to indexing', provides a three-level hierarchical index. Categories: A. First order terms; F. Fossils; H. Igneous rocks; M. Soils; N. Sediments; O. Areas. Appended lists of authorities and abbreviations. *Class No:* 551:025.43

Libraries

[1515]

ROBINSON, E., *comp.* **A Guide to geology libraries in London.** 3rd ed. London, Geology Subject Sub-committee of the Library Resources Committee, Univ. of London, 1982. 24p.

First published 1974 as *Libraries for the geologist in and around London* (London, Imperial College, Lyon Playfair Library, 1974).

A student's guide to 22 libraries, including 3 public libraries. Entries are in narrative from, outlining facilities available (*e.g.* reference/loan, seating, photocopying, postal address, hours of opening, name of librarian, etc.). No index. A companion guide covers geography and map collections. *Class No:* 551:061:026/027

Museums

[1516]

Geology in museums: a bibliography and index. Sharpe, T., *comp.* Cardiff, National Museum of Wales, 1983. 128p. (*Geological series, no. 6.*) ISBN: 0720002818.

'Over one thousand references relating to geological curation' (*Introduction*). Under authors, A/Z. Keywords appended to entries. Keyword index ('Palaeontology': 8½ columns). *Class No:* 551:061:069

Earth Structure

[1517]

HOLMES, A. Holmes' principles of physical geology. Duff, P. McL. D., *ed.* 4th ed. London, Chapman and Hall, 1992. xvi,791p. illus. (some col.), diagrs., maps. ISBN: 0412438305.

First published 1944.

31 sections, each with selected references (*e.g.* 5. Igneous rocks: volcanic and plutonic ... 12. Igneous intrusions (p.176-206), 46 illus. and diagrs.; 3 references ... 21. Ice Ages and climatic change ... 31. Orogenic belts). Analytical index. Very well illustrated. Excellent value; standard text. *Class No:* 551.1

[1518]

—Hutchinson encyclopedia of the earth. Smith, P.J., *ed.* London, etc., Hutchinson, 1986. 248 + 8p. illus. (incl. col.), diagrs., maps. £14.95. ISBN: 0091658608.

Has 6 parts (23 chapters): The earth today - The solid earth - The fluid earth - Shaping the earth (*e.g.* climate and climate change) - Surface environment - Geology and man (*e.g.* engineering geology). Further reading and credits, p.24. Glossary of geological terms. Index. Fully illustrated in colour. 'In this book we see the Earth in the exciting light of the plate-tectonic revolution' (*Introduction*). A popular treatment. *Class No:* 551.1

[1519]

JEFFREYS, Sir H. The Earth: its origin, history and physical constitution. 6th ed. Cambridge Univ. Press, 1976. x, [2], 574p. illus. (pl.), diagrs., tables, maps. £70. ISBN: 0521206480.

First published 1924.

12 chapters (1. The mechanical properties of rocks ... 3. Observational seismology - 4. The theory of the figures of the earth and moon ... 9. The age of the earth ... 11. The origin of the earth's surface features). Appendices A-H (G. Statistical methods). Bibliography and author index, p.535-66. Undergraduate level. *Class No:* 551.1

[1520]

LAMBERT, D. *and* **THE DIAGRAM GROUP. The Cambridge guide to the earth.** Cambridge, University Press, 1988. 256p. figs., diagrs., maps. £27.95; £4.49. ISBN: 0521333652.

12 chapters: 1. Sizing up the earth - 2. The restless coast - 3. Fiery rocks ... 7. How rivers shape the land - 8. The work of the sea ... 11. The last 600 million years - 12. Rocks and man. 1-page bibliography. Index. Large type face and many diagrams, figs., etc. For those with no previous knowledge of the subject. *Class No:* 551.1

[1521]

Planet earth: an encyclopaedia of geology. Hallam, A., *ed.* Oxford, Elsevier-Phaidon, 1977. 320p. illus. (incl. col.), diagrs., tables, maps.

9 sections: The earth and its neighbours - Processes that shape the earth - Landscapes - The ocean floor - Economic geology - The rocks of the earth - Geological history - History of life - Making of geology. Glossary. Index. Most chapters consist of a narrative account, but there is A-Z order for minerals, rocks, fossils and biographies. Authoritative text; readable, with excellent illus. and art work. 'A long and impressive list of contributors... A good general introduction to earth sciences for the layman, and is well worth reading by the professional' (*Journal of natural history,* v.19(1), January/February 1979, p.128). *Class No:* 551.1

[1522]

SKINNER, B.J. *and* **PORTER, S.C. Physical geology.** New York, John Wiley, 1988. xii,750p. illus. (many col.), figs., diagrs. $51.95. ISBN: 0471056685.

5 sections: 1. Planet earth and its materials - 2. Time and the changing landscape - 3. The dynamic earth - 4. The earth's resources - 5. Beyond planet earth. Brief list of references appended to each chapter. 5 appendices: A. Organization of matter - B. Identification of common minerals - C. Identification of common rocks - D. Maps, cross sections, field measurements, interpretations - E. Units and their conversions. Glossary, p.723-38. Index. Many colour illustrations. *Class No:* 551.1

Geodynamics

Volcanoes

[1523]

INTERNATIONAL ASSOCIATION OF VOLCANOLOGY. Catalogue of the active volcanoes of the world, including Solfatara fields. Rome, the Association, 1951-. pt.1-. 22v. illus., tables, maps.

1. *Indonesia.* 1951.
2. *Philippine Islands and Cochin China.* 1953.
3. *Hawaiian Islands.* 1955.
4. *Africa and the Red Sea.* 1957.
5. *Melanesia.* 1957.
6. *Central America.* 1958.
7. *Kurile Islands.* 1958.
8. *Kamchatka and continental areas of Asia.* 1959.
9. *United States of America.* 1960.
10. *Antarctica.* 1960.
11. *Japan, Taiwan and Marianas.* 1962.
12. *Greece.* 1962.
13. *Kermadec, Tonga and Samoa.* 1962.
14. *Archipelago de Colón, Isla San Felix and Islas Juan Fernández.* 1962.
15. *The Chilean continent.* 1963.
16. *Arabia and the Indian Ocean.* 1963.
17. *Caucasus. Turkey.* 1963.
18. *Italy.* 1965.
19. *Colombia, Ecuador and Peru.* 1966.
20. *West Indies.* 1966.
21. *Atlantic Ocean.* 1967.
22. *New Zealand.* 1975.

Partly financed by Unesco. Pt.16 (xvii, 64p.) includes for each volcano: 1. Name and location - 2. Form and structure

....(contd.)

- 3. Volcanic activity - 4. Petrography - 5. Bibliography (chronologically arranged). The entry on Piton de la Fournaise (p.39-52) includes a list of outbursts (with symbols indicating lava flows, normal explosions, etc.), 2 maps and 1½p. bibliography. Pt. 16 also has a general bibliography for the area (p.xv-xvii).

Updates appear in *Bulletin of volcanic eruptions* (Tokyo, Volcanological Society of Japan, 1961-. no. 1-), which provides an 'Annual report of the world volcanic eruptions'. *Class No:* 551.21

[1524]

MACDONALD, G.A. Volcanoes. Englewood Cliffs, N.J., Prentice Hall, 1972. xiii, 510p. illus., maps. ISBN: 0139422196.

16 chapters (*e.g.* Lava flows, p.66-107; Kinds of volcanic eruption, p.198-254; Fumaroles, hot springs, geysers and thermal power, p.323-43; Volcanoes and man, p.400-28). Appendices list active volcanoes and describe the igneous rocks. Bibliography, p.468-92; index, p.493-510. Prepared primarily for students of geology, but readily understandable to the layperson. *Class No:* 551.21

[1525]

SIMKIN, T., *and others.* **Volcanoes of the world:** a regional directory gazetteer and chronology of volcanism during the last 10,000 years. Stroudsburg, Pa., Hutchinson Ross, and the Smithsonian Institution, 1981. vii, 233p. ISBN: 0879334088.

3 parts: a directory of volcanoes (1,343 active; 5,564 dead); chronology of eruptions; gazetteer. Based on the computerized database at the Smithsonian Institution. The 30p. introduction outlines the catalogues' background, sources and significance of the various categories of information utilised. 709 references. 'This is a truly monumental work which, it is hoped, will be updated from time to time' (*Natural history book reviews,* v.7(4), 1985, p.258). *Class No:* 551.21

[1526]

Volcanoes of North America: United States and Canada. Wood, C.A. *and* Kienle, J., *eds.* Cambridge, University Press, 1990. 354p. illus., maps. $70. ISBN: 0521364698.

81 contributors give short, but detailed entries on 262 volcanoes in the region. Data include volcano type; elevation; location; eruptive history; composition; selected references; maps and photographs. 'Loaded with pertinent, very helpful information for the scientist as well as for the intelligent general reader' (*Choice,* v.28(7), March 1991, p.1110). *Class No:* 551.21

Tectonics

[1527]

DENNIS, J.G., *ed.* **International tectonic dictionary: English terminology.** Tulsa, Okla., American Association of Petroleum Geologists, Committee on Structural Nomenclature, 1967. xi, 196p. illus. (*A.A.P.G. memoir 7.*)

At head of title: International Geological Congress, Committee for the Geological Map of the World.

About 400 terms are dealt with (*e.g.* Nappe, p.111-2: derivation; definitions; history and usage; recommendation; synonyms and partial synonyms). Bibliography, p.164-84 (A-Z authors). *Class No:* 551.24

[1528]
The Encyclopedia of structural geology and plate tectonics. Seyfert, C.K., *ed.* New York, Van Nostrand Reinhold, 1987. 912p. illus., diagrs., graphs, maps. $114.95. (*Encyclopedia of earth sciences, v.10.*) ISBN: 0442281250.

About 90 contributors. 119 main entries, A-Z, Many cross-references 'Gondwanaland', p.309-14; 6 references. 'Rift valleys', p.671-88; 3 columns of references. Author citation index. Detailed, non-analytical subject index. A well-organized treatment, typical of this continuing series. 'Highly recommended' (*Choice* v.23(7), March 1988, p.1062). *Class No:* 551.24

[1529]
—DE SITTER, L.U. Structural geology. 2nd ed. New York, etc., McGraw-Hill, 1964. xii, 511p. diagrs.

3 parts (36 chapters): 1. Theoretical structural geology - 2. Comparative structural geology - 3. Geotechnics. References. 'It stands apart from most other books dealing with this subject in presenting facts and hypotheses with distinctive clarity' (*Geological journal*, v.123(2), June 1957, on the 1st ed.). It makes great use of comparative aspects, whereas other books have a distinctive regional slant. *Class No:* 551.24

[1530]
Geological abstracts. Geophysics & tectonics. Norwich, Geo Abstracts Ltd., 1982-1988. 6pa.

Continues *Geophysics abstracts* (1977-82).

About 500 abstracts per issue. 24 sections (General geophysics - Earth's rotation, shape and structure ... Age determination ... Geothermal systems ... Earthquake theory, measurement and prediction ... Tectonics ... Structure ... Volcanology ... Geophysics and resources ... Well logging [bore holes]). Abstracts, signed, average 100 words apiece. Annual author and subject indexes. Continued in *Geological abstracts, (qv)*, 1989-. *Class No:* 551.24

[1531]
KASBEER, T. Bibliography of continental drift and plate tectonics. Boulder, Colo., Geological Society of America, 1972-75. xi, 96p. + vi, 151p. (*GSA Special papers no.142, 164.*) ISBN: 0515711628.

2,784 references, in all. V.1 has 2 sections: 1. Origin and development (chronological order; selected cross-references; author index), citing general articles as well as material on sea floor spreading and mid-ocean ridges, paleomagnetism and rock magnetism. Section 2: Convection currents, other lines of evidence (glacial, evaporites, palaeontological) expansion theory, monographs and symposia. Author index. V.2 provides a similar sequence, citing older references as well as new items. *Class No:* 551.24

Sedimentation

[1532]
The Encyclopedia of sedimentology. Fairbridge, R.W. *and* Bourgeois, J., *eds.* Stroudsburg, Pa., Dowden, Hutchinson & Ross, Inc., 1978. xvi, [1], 901p. illus., graphs, tables, maps. (*Encyclopedia of earth sciences, v.6.*) ISBN: 0879331526.

193 contributors from 23 countries. Entries. 'Abrasion pH' ... 'Zebra dolomite' - each signed, with references and cross-references. 'Estuarine sedimentation': p.288-93, with 11/3 cols. of bibliography and 6 drawings. Analytical index, p.875-90. 'An essential reference book for geologists,

.... (contd.)
oceanographers, and scientists, hydrologists and others in related fields' (*Choice*, v.16(2), April 1979, p.202). *Class No:* 551.3

[1533]
PETTIJOHN, F.J. *and* **POTTER, P.E. Atlas and glossary of primary sedimentary structures.** New York, Springer-Verlag, 1964. ISBN: 0387963502.

198 excellent photographs. Facing each are title, location, geological age and explanatory statement in English, German, French and Spanish. 70-page glossary of 300 terms, with definitions in English, and equivalents and indexes in German, French and Spanish. *Class No:* 551.3

Floods

[1534]
WARD, R. Floods: a geographical perspective. London, Macmillan, 1978. 244p. illus., diagrs., graphs, maps. (*Focal problems in geography series.*) ISBN: 0333148924.

12 chapters (*e.g.* 2/3. Floods caused by precipitation; 6. Flood problems; 8/10. Human response to the flood hazard; 11. Economic response to the flood hazard; 12. Prospect). Bibliography of works cited. Detailed, analytical index, p.234-44. *Class No:* 551.311

Glaciology

Bibliographies

[1535]
AMERICAN GEOGRAPHICAL SOCIETY. Department of Exploration and Field Research. **Author/title, subject and geographic catalogs of the Glaciology Collection.** Boston, Mass., G.K. Hall, 1971. 3v. $330. ISBN: 0816109222.

About 48,900 author, subject and regional catalogue cards, photolithographically reproduced. Classified order, based on the Scott Polar Research Institute's *Universal decimal classification for use in Polar libraries.* *Class No:* 551.32(01)

[1536]
SCOTT POLAR RESEARCH INSTITUTE, Cambridge. **The Library catalogue of the Scott Polar Research Institute, Cambridge.** Boston, Mass., G.K. Hall, 1976. Supplement. 1981. 19v.; 5v. ISBN: 0816112169.

The main catalogue has 300,000 photolithographed catalogue cards. Three separate catalogues; author, subject (UDC order, with special regional classification) and regional. The largest collection of its kind. *Class No:* 551.32(01)

Handbooks & Manuals

[1537]
EMBLETON, C. *and* **KING, C.A.M. Glacial geomorphology.** 2nd ed. London, Edward Arnold, 1975. [x], 573p. illus. (incl. plates), tables, maps. ISBN: 0470238941.

Fully revised version of parts 1-3 of *Glacial and periglacial geomorphology.*

3 parts: 1. Basic conceptions of glaciation and glacial behaviour - 2. Glacial and fluvioglacial erosion - 3. Glacial

....(contd.)
and fluvioglacial depositions. 19 documented chapters (*e.g.*
7. 'Corques': 4p. of references). Index. Well illustrated.
Class No: 551.32(035)

[1538]
—EMBLETON, C. *and* KING, C.A.M. Periglacial
geomorphology. 2nd ed. London, Edward Arnold, 1975. x,
203p. illus. (incl. pl.), graphs, tables, maps.
A fully revised ed. of part 4 of *Glacial and periglacial
geomorphology* (1968).
7 documented chapters (2. Frozen ground phenomena,
with 5½p. of references; 5. The action of snow; 7.
Periglacial wind action). Index, p.199-203.
Class No: 551.32(035)

[1539]
LLIBOUTRY, L. **Traité de glaciologie.** Paris, Masson, 1964-
65. 2v. (iv, 1040p.). illus. (incl. pl.), graphs, maps.
1. *Glace, neige, hydrologie, nivale.* 1964. 2. *Glaciers,
variations du climat, sols gelés.* 1965.
23 chapters in all, with appended bibliographies (Ch. 20:
Variations séculaires et millénaires du climat, p.824-72;
references, p.870-2). 'A complete and critical review of
glaciology, these volumes will be equally useful to the
geomorphologist and the engineer, to the ski instructor and
the meteorologist' (*Nature and resources,* v.2(3), September
1966, p.19). *Class No:* 551.32(035)

Dictionaries

Polyglot
[1540]
Elsevier's dictionary of glaciology in four languages:
English (with definitions), Russian (with definitions), French
and German. Kotlyakov, V.M. *and*
Smolyarova, N.A., *comps.* Amsterdam, Elsevier, 1990.
336p. £62; $114.50.
*c.*1,200 terms, with French, German, and Russian
equivalents and indexes. *Class No:* 551.32(038)=00

[1541]
Illustrated glossary of snow and ice. Armstrong, T., *and
others.* 2nd ed. Cambridge, Scott Polar Research Institute,
1973. v, 60p. illus. £2.50. ISBN: 0901021016.
First published 1966. Issued with UNESCO support.
Classified summary of terms, p.7-8. Definitions and
linguistic equivalents (Ablation ... Young ice) in English,
Danish, Finnish, French, German, Icelandic, Norwegian,
Russian and Spanish. Language indexes. 78 illus.
Class No: 551.32(038)=00

Progress Reports
[1542]
Advances in periglacial geomorphology. Edited by M.J.
Clark, for the International Geophysical Union Commission
on the Significance of Periglacial Phenomena. Chichester,
Wiley, 1988. xxiv, 481p. illus., diagrs., graphs, maps. £90.
ISBN: 0471909815.
18 contributors. 4 parts (17 documented chapters): 1.
Weathering, erosion and related sedimentary features - 2.
Frozen ground and active layer processes - 3. Process and
form: the example of frost mounds - 4. Perspectives on the
periglacial system. Chapter 6.: Ground ice and permafrost;

....(contd.)
p.113-49, has 6½p. of references. Abstracts in English,
French and German precede each chapter text. Detailed,
analytical index, p.475-81. Well organized; authoritative.
Class No: 551.32(055)

[1543]
Periglacial geomorphology; proceedings of the 22nd Annual
Binghamton Symposium in Geomorphology. Dixon, J.C.
and Abrahams, A.D., *eds.* Chichester, John Wiley, 1992.
xii,354p. illus., graphs, tables. ISBN: 0471933422.
23 contributors. 14 documented papers: 1. Periglacial
geomorphology: what, where, when? ... 5. Mechanical
weathering in the Antarctic: a maritime perspective (with 3p.
of references) ... 14. Recent ground warming inferred from
the temperature in permafrost near Mayo, Yukon Territory.
Brief index. *Class No:* 551.32(055)

Nomenclatures
[1544]
**WORLD METEOROLOGICAL ORGANIZATION. WMO
sea-ice nomenclature:** terminology, codes and illustrated
glossary. Geneva, WMO, 1970. 147p. illus.
Terms are arranged first by subject, then in A-Z order.
162 supporting illus. Quality of paper, illus. and binding are
criticized in *The Polar record,* v.16(101), 1972, p.279-80.
Class No: 551.32(083.72)

Polar Regions
[1545]
DARTMOUTH COLLEGE LIBRARY. Hanover, N.H.
**Dictionary catalogue of the Stefansson Collection on the
Polar regions.** Boston, Mass., G.K. Hall, 1967. 8v. $790.
ISBN: 0816106762.
About 115,000 photolithographed catalogue cards,
representing *c.*20,000 volumes and *c.*20,000 pamphlets and
many manuscript items. The collection's main emphasis is
now historical, with primary concern for the history of Polar
exploration. *Class No:* 551.32(98/99)

[1546]
KOSACK, H.P. **Die Polarforschung.** Ein Datenbuch über die
Natur-Kultur-Wirtschaftsverhältnisse und die
Erforschungsgeschichte der Polarregionen. Braunschweig,
Vieweg, 1967. xvi, 472p. tables, maps.
Tables list Polar organizations and scientific stations;
chronology of expeditions. Cartographic display of basic
data; short bibliography; comprehensive index. 'A
remarkable condensation into a single volume of all the
general data relating to both polar regions' (*Geographical
journal,* v.134(2), June 1968, p.290).
Class No: 551.32(98/99)

[1547]
Polar and glaciological abstracts. Compiled by the Scott
Polar Research Institute Library and Information Service,
Cambridge. Cambridge, University Press, 1990-. 4pa.
$95pa. ISSN: 09575073.
Continues *Recent Polar and glaciological literature* (1981-
89).
Each issue contains *c.*1,300 abstracts, each 30-100 words
in length. Class A-Z (A. Geophysical sciences (General) ...
E. Glaciology ... T. Social anthropology and ethnography ...

....(contd.)

Z. Miscellaneous). Contains subject-geographic and author indexes. Annual cumulative index. For the research library. *Class No:* 551.32(98/99)

Arctic

[1548]
Arctic bibliography. Prepared by the Arctic Institute of North America with the support of government agencies in the United States and Canada. Montreal & London, McGill-Queen's Univ. Press, v.1-16, 1953-75. 16v.

V.1-16 carried 108,723 signed abstracts of books, periodical articles, government reports, etc., covering all aspects of the Arctic-administration and governments. Eskimos, archaeology, economic and social conditions, physics, geology, life sciences, mapping, population, communications, psychology, education, colonization, transportation and ethnography. Of the 8,011 entries in v.13, 43% are Russian, 42% English, and 5% Scandinavian in origin. Arrangement is by authors A-Z, with a very detailed subject and locality index (*c.*35,000 entries in v.12 index). 'It would be difficult to imagine polar research today without this scholarly resource' (*Polar record,* v.14(88), January 1968, p.71). *Class No:* 551.32(98)

[1549]
ARCTIC INSTITUTE OF NORTH AMERICA Library. **Catalogue. [Arctic Institute of North America Library].** Boston, Mass., G.K. Hall, 1969. Supplements, 1971-. 4v. $410, set; $130, supp. 1; $265, supp. 2; $455, supp. 3. ISBN: 0816108234, set; 0816108307, supp. 1; 0816110301, supp. 2; 0816111626, supp. 3.

The main catalogue has *c.*70,000 author and subject catalogue cards, photolithographically reproduced. Entries cover books, pamphlets and reprints, as well as microfilms, audio-tapes and recordings, but not current Arctic maps. Analytical entries for *c.*4,500 articles in non-polar serials. Concentrates on the Polar regions, especially the Arctic or sub-Arctic, plus cold weather research, snow and ice studies. Supplements 1-3 (1971-80) add *c.*70,000 cards. *Class No:* 551.32(98)

[1550]
ASTIS current awareness bulletin. Calgary, Alberta, ASTIS, Arctic Institute of North America; Univ. of Calgary, 1979-. 6pa. $75Can. ISSN: 07058434.

About 2,500 abstracts (each 25-300 words). 22 classes (A. Geography, geomorphology, and cartography - B. Geology, mineralogy, geochemistry, and palaeontology - C. Soils and permafrost... G. Ice (except glacier and ground ice)... Q. Petroleum, natural gas and pipelines... T. Native people (except archaeology)- U. Archaeology - V. History - X. General). Broad geographic index. Author index.

Annual cumulation, *ASTIS bibliography,* on microfiche. *Class No:* 551.32(98)

Antarctic

[1551]
AMERICAN GEOPHYSICAL UNION. Antarctic research book series. Washington, American Geophysical Union, 1964-. v.2-. irreg.

Formerly *Antarctic research series.*

....(contd.)

Provides a major series of research reports (*e.g.* no.55, 1992: Stilwell, J.D., ed. *Molluscan systematics and biostratigraphy* (202p.)). *Class No:* 551.32(99)

[1552]
Antarctic bibliography. Prepared by the Library of Congress. Washington, Library of Congress, 1965-. v.1-. Annual (irregular). $41. ISSN: 00664626.

Each issue contains *c.*2,000 abstracts. 13 subject areas: A. General - B. Biological sciences - C. Cartography - D. Expeditions - E. Geological sciences - F. Ice and snow - G. Logistics, equipment and supplies - H. Medical scientists - I. Meteorology - J. Oceanography - K. Atmospheric physics - L. Terrestrial physics - M. Political geography. Author-title, analytical subject, geographic and grantee indexes.

The *Antarctic bibliography, 1951-1961 (1970)* extends cover retrospectively.

Antarctic bibliography (1962-) is the printed version of COLD database (Cold Regions Research and Engineering Laboratory). *Class No:* 551.32(99)

[1553]
Antarctic earth science. Oliver, R.L., *and others.* Cambridge, University Press, 1983. 719p. illus., diagrs., graphs, tables, maps. £55. ISBN: 0521258367.

The text (full or abstracted) of 174 papers presented at the 4th International Symposium on Antarctic Earth Sciences in Adelaide, South Australia, 16-20 August 1982.

15 Symposium topics: Precambrian East Antarctic craton; East Antarctica; West Antarctica boundary and the Ross Orogen, including Northern Victoria Land; ... Maritime geology; Antarctic resources; ... Crustal structure of Antarctica; ... Subantarctic islands; Canozoic igneous activity. Papers carry references. Analytical index, p.681-97. Illus. are well captioned. A valuable compendium. *Class No:* 551.32(99)

[1554]
—**Antarctic science.** Walton, D.W.H., *ed.* Cambridge, University Press, 1987. viii, 280p. illus, (incl. col.), graphs, tables, maps. £7.99. ISBN: 052126233x.

6 contributors. 5 parts (18 chapters): 1. Geography, politics and science - 2. Life in a cold environment - 3. Antarctic ice and rocks - 4. The Antarctic atmosphere - 5. Cooperation or confrontation. Select bibliography, p.272-3. Detailed, analytical index, p.274-8. Striking illus. Reflects contributors' experience of British Antarctic findings. *Class No:* 551.32(99)

[1555]
Current Antarctic literature. Prepared by Cold Regions Bibliography Project, Science and Technology Division, Library of Congress, for the Division of Polar Programs, National Science Foundation. Washington, Library of Congress, 1961-. 12pa. *Gratis.* ISSN: 0096879x.

'Prepared for later formal publication in the *Antarctic bibliography*'

Over 3,000 abstracts (50-250 words) pa. 13 classes: A. General - B. Biological sciences - C. Cartography - D. Expeditions - E. Geological sciences - F. Ice & snow - G Logistics, equipment and supplies - H. Medical sciences - I. Meteorology - J. Oceanography - K. Atmospheric physics - L. Terrestrial physics - M. Political geography. Available on CD-ROM. *Class No:* 551.32(99)

[1556]
FOGG, G.E. **A History of Antarctic science.** Cambridge, University Press, 1992. xxi,483p. illus., figs., maps. £55. (*Studies in polar research.*) ISBN: 0521361133.

Coverage spans 3 centuries, beginning with Halley. 13 sections: 1. Introduction - 2. The science of the early explorations - 3. The national expeditions of 1828 to 1843 ... 7. The sciences of the Antarctic seas - 8. The earth sciences ... 11. Man and the Antarctic environment ... 13. Postscript. Extensive bibliography, p.425-63. Index, p.464-83.
Class No: 551.32(99)

[1557]
HEADLAND, R.K. **Chronological list of Antarctic expeditions and related historical events.** Cambridge, University Press, 1989. 730p. illus., graphs, maps. £75. (*Studies in Polar research.*) ISBN: 0521309034.

A revision of Roberts, B. 'Chronological list of Antarctic expeditions'. In *Polar record,* v.9(60), September 1958.

Gives information on 3,342 expeditions. 8 sections: 1. Introduction - 2. Structure - 3. Observations - 4. Statistical examination - 5. Information - 6. Chronological list (p.51-603) - 7. Bibliography - 8. Index (27,000 entries). For all libraries concerned with Antarctic science.
Class No: 551.32(99)

[1558]
Report on United Kingdom Antarctic research, 1991 report. London, Royal Society, 1991. viii,62p. ISBN: 0854034528.

Covers the period April 1990 - September 1991: Record of activities; October 1991 - September 1992: Planned activities. Sections: Science highlights - List of permits for entry into SPAs and SSSIs - University research in Antarctica - Prospectus of field activities planned for 1991/92 - Future activities, planned and funded - Principal investigators and addresses - Bibliography, p.43-62.
Class No: 551.32(99)

Geomorphology (Earth's physical forms)

Bibliographies

[1559]
AMERICAN GEOGRAPHICAL SOCIETY. **Research catalogue of the American Geographical Society.** Boston, Mass., G.K. Hall, 1962. 15v. (10436p.) and map supplement. Supplements 1-2, 1972-78. 6v. $1,595, 15v. set. ISBN: 0816106282, set.

1-2. *General.*
3. *Regional. North America.*
4-5. *United States.*
6. *Mexico, Central America, Bermuda, West Indies, South America.*
7. *South America.*
8-10. *Europe.*
11. *Africa.*
12-13. *Asia.*
14. *Australasia.*
15. *Polar regions, Oceania, Tropics.*

219,000 entries (21 card entries per page). A photolithographic reproduction of the largest geographical collection in the Western hemisphere, and the *Research catalogue* is particularly strong in periodical article references. The 'Map supplement' ([i], 24p.) is a map key to the classification key used.

.... *(contd.)*
First supplement (1972-74. 4v.) and *Second supplement* (1978. 2v.) each divide into regional and topical catalogues.
Class No: 551.4(01)

[1560]
—Current geographical publications: additions to the 'Research catalogue of the American Geographical Society'. New York, the Society, 1938-. v.1-. 10pa., omitting July and August. $58pa. ($93pa. overseas). ISSN: 00113514.

5 sections: 1. Topical (aids to geographical research: biogeography; climatology; economic geography...politcal geography) - 2. Regional - 3. Selected maps - 4. Selected books and monographs - 5. Selected reviews. About 750 references per issue. A major current-awareness source in its field. Available on microfilm. *Class No:* 551.4(01)

[1561]
—VAN BALEN, J. Geography and earth science publications: an author, title and subject guide to books reviewed, and an index to the reviews. Ann Arbor, Mich., Pierian Press, 1978. 2v. (313p. + 232p.). $62.50, set; $39.50ea. ISBN: 0876500904, v.1; 0876500912, v.2; 0685040429, set.

Devotes v.1 to years 1968-72 (3,400 author entries, 4,900 citations to book reviews from 21 geographical journals), v.2: 1973-75 (2,500 author entires, 3,700 citations to book reviews). *Class No:* 551.4(01)

[1562]
A **Geographical bibliography for American libraries.** Harris, C.D., *and others, eds.* Washington, Association of American Geographers, 1985. xxiii, 437p. $20. ISBN: 089291193x.

A joint project of the Association of American Geographers and the National Geographic Society.

71 contributors. 7 parts: 1. General aids and sources - 2. History, philosophy and methodology, - 3. Systematic fields of physical geography - 4. Systematic fields of human geography - 5. Applied geography (*e.g.* Planning; Military geography) - 6. Regional geography (worldwide) - 7. Publications suitable for school libraries. 2,903 concise, well-annotated entries. Analytical index, p.384-437. An important aid, 'to assist libraries in the United States, Canada, and other countries to identify, select and secure publications of value in geography that are appropriate to the purposes and resources of each collection' (*Introduction*).
Class No: 551.4(01)

[1563]
—GODDARD, S. A Guide to information sources in the geographical sciences. London, Croom Helm, 1983. xiii, 273p. diagrs., tables, maps. ISBN: 038920403x.

13 contributors. Chapters 1-9 concern 'The communication of man's spatial experience', 'The systematic approach' (*e.g.* geomorphology) and 'Sources of regional information'. Chapters 10-13, on 'Tools for the geographer' cover cartographic items, aerial photographs and satellite information, statistics, and archives. Running commentary. No index. *Class No:* 551.4(01)

[1564]
—WRIGHT, J.K. *and* PLATT, E.T. Aids to geographical research: bibliographies, periodicals, atlases, gazetteers and other reference books. 2nd ed., completely revised. New York, Columbia Univ. Press, for the American Geographical Society, 1947. xii, 331p. $38.50. (*A.G.S. Research series, no.22.*) ISBN: 083713384x.

Reprinted Westport, Conn., Greenwood Press, [1971] and

....(contd.)
was first published 1921.

1,174 numbered entries, with detailed, evaluative annotations. Sections: Introduction - General aids - Topical aids - Regional aids (by areas and countries) and general geographical periodicals. Detailed index of systems cited. A valuable tool for items published up to 1945, but now rather dated. *Class No:* 551.4(01)

[1565]
HARRIS, C.D. Bibliography of geography. Chicago, Ill., Univ. of Chicago, 1976-. ISBN: 0890650861, v.1; 0890651124, v.2.

Pt.1. *Introduction to general aids.* 1976. ix, 276p. $12. 2. *Regional.* v.1: *The United States of America.* 1984. viii, 178p. $12.(v.2-4 are to deal with Soviet Union, Europe; and Africa, Asia, Australia and the Pacific).

Pt.1 has 585 annotated entries in 16 sections (1. Bibliographies of bibliographies... 5. Books - 6. Serials... 10. Maps and atlases... 13. Dictionaries - 14. Encyclopedias - 15. Statistics - 16. Methodology in geography). Appendix 2: 'A small geographical reference collection' (entries 530-580). Non-analytical index. V.2 (974 entries, some briefly annotated) opens with a general section: 'World-wide bibliographies and guides'. 'The Americas as a whole. North America as a whole. The United States of America' has 4 sections: 1. General aids - 2. Physical geography, related earth sciences, the environment and resources - 3. Human geography and related social sciences - 4. Regions of the United States. Index of subjects, authors and short titles. *Class No:* 551.4(01)

Encyclopaedias
[1566]
The Encyclopedia of geomorphology. Fairbridge, R.W., *ed.* New York, London, etc., Van Nostrand Reinhold, 1968 (reprinted 1982). xvi, 1298p. illus. diagrs., tables, maps. $126.95. (*Encyclopedia of earth sciences, v.3.*)

About 410 entries, A-Z; 150 contributors from 20 countries. Each article has an appended list of references. 'Geography: concept, growth and status', p.384-8 with 6 references. 'Submarine geomorphology', p.1079-97; with 10 references. 12½ lines of cross-references. Index of *c.*6,000 terms. A standard source, now in need of revision. *Class No:* 551.4(031)

Handbooks & Manuals
[1567]
CHORLEY, R.J. *and* **SUGDEN, D.E. Geomorphology.** London, Chapman & Hall, 1984. 632p. illus., diagrs., charts, tables, maps. £18.99. (*Geographies for advanced studies.*) ISBN: 0416325904.

4 parts (20 sections each with references): 1. Introduction - 2. Geological geomorphology - 3. Geomorphic processes and landforms - 4. Climatic geomorphology. Appendix: Applied geomorphology. Index. A textbook, well illustrated and documented, for sixth-formers and undergraduates. R.J. Chorley is Professor of Geography, Univ. of Cambridge. *Class No:* 551.4(035)

[1568]
—**BRIDGES, E.M.** World geomorphology. Cambridge, University Press, 1990. x,260p. illus., diagrs., maps. £37.50; £14.95. ISBN: 0521383439.

10 chapters (*e.g.* 2. Continental drift and plate tectonics ... 9. The Pacific Ocean basin - 10. Geographical implications of major geomorphological features). Further reading, p.253-5. Index. Textbook, aimed at undergraduates. *Class No:* 551.4(035)

[1569]
SPARKS, B.W. Geomorphology. 3rd ed. London, Longman, 1986. xxi, 561p. illus., diagrs, tables. £16.99. ISBN: 0582306711.

First published 1960.

17 chapters (*e.g.* 3. Weathering - 7. The effects of rocks as relief - 12. Landforms in the humid tropics - 17. The estimation of denudation). References (by chapters), p.521-43. Index, p.547-61. About 250 illus. 'Written as a university textbook' (Brewer, J.G. *The literature of geography* 2nd ed. 1978, p.169). *Class No:* 551.4(035)

[1570]
The Student's companion to geography. Rogers, A., *and others, eds.* Oxford, Blackwell, 1992. 386p. illus., tables. £50; £14.95. ISBN: 063117088x.

50 contributors. 6 sections: 1. Introductory essays - 2. What is geography? Past, present and future - 3. How to study geography (*e.g.* 'Remote sensing') - 4. What's what and who's who in geography (*e.g.* 'The literature of physical geography'; 'Biographical dictionary') 5. A geographical directory (*e.g.* 'World libraries and museums'; 'On-line data sources') - 6. What next? (*e.g* 'Postgraduate studies in Australia'; 'Overseas work and independent travel'). Index. A good overview of the subject. Suitable for those beginning, or contemplating, a career in geography. *Class No:* 551.4(035)

Dictionaries
[1571]
Bibliography of mono- and multilingual dictionaries and glossaries of technical terms used in geography, as well as in related natural and social sciences. Meynen, E., *comp.* Wiesbaden, Steiner, for International Geographical Union, Commission on International Geographical Terminology, 1974. xx, 246p. $58.50. ISBN: 3515018468.

Lists 3,211 dictionaries, arranged by fields of geography and related disciplines. *Class No:* 551.4(038)

[1572]
Dictionary of concepts in physical geography. Huber, T.P., *and others.* New York, Greenwood Press, 1988. 301p. $49.95. (*Reference sources for the social sciences and humanities no.5.*) ISBN: 0313253692.

Contains 88 entries. Definitions are in the style of bibliographic essays, averaging 3 pages in length. Bibliography and list of sources (many annotated) follow the entries. Includes 'Outline of concepts', listing 185 terms. Index. Numerous cross-references. *Class No:* 551.4(038)

Polyglot

[1573]
Chertyrekyh' yazychuyy entsiklopedicheskiy slovar' torminov po fiziacheskoy geografii: Russko-Anglo-Nemetsko-Frantsuzskiy. [Four language encyclopaedic dictionary of terms in physical geography: Russian-English-German-French.] Shehukin, I.S. Moskva, Sovetskaya Entsiklopediya, 1980. 703p.

'Compared with an English-language counterpart, *The Encyclopaedic dictionary of physical geography (q.v.)* edited by Andrew S. Goudie, it includes nearly three times as many terms (5,700 compared with 2,000) and thus substantially more definitions, but has less full treatment of some of the current research, concepts, or methods in physical geography' (C.D. Harris, *'Recent Soviet geographical dictionaries'*, in *Soviet geography*, v.29, no.8 October 1988, p.775-782). *Class No:* 551.4(038)=00

[1574]
International geographical glossary. [Internationale geographische Terminologie.] Meynen, E., *ed.* Deutsche Ausgabe. Wiesbaden, F. Steiner; New York, French and European, 1985. 1479p. $350. ISBN: 0828809577, US ed.

About 2,400 German-base geographical terms, with equivalents in English, French, Italian, Spanish, Russian and Japanese. Terms are also grouped by categories for English, German and French. 'The product of international collaboration' (*A Geographical bibliography for American libraries;* edited by C.D. Harris, 1985, entry 102). *Class No:* 551.4(038)=00

English

[1575]
CLARK, A.N. Longman dictionary of geography, human and physical. Harlow, Essex, Longman, 1985. ix, 724p. £33; £12.50.

About 8,000 shortish entries for terms 'commonly used in geographical writing over the past 100 years' (*Preface*), with claims to be the first dictionary to deal with the major aspects of geography in one volume. Entries for terms covered by Stamp and Clark are cited as 'G' and those in Chisholm's *Handbook of commercial geography* (20th ed. Longman, 1980) as 'C'. Includes terms for commodities; also, brief biographies. 'The book is a valuable addition to geographical literature and will be an asset to many people, by no means only geographers' (*Geographical journal,* v.152(2), July 1986, p.257).

There is an abridged and revised 1990 ed., published by Penguin. *Class No:* 551.4(038)=20

[1576]
The Encyclopaedic dictionary of physical geography. Goudie, A., *ed.* Oxford, Blackwell Reference, 1985. xvi, 528p. illus., diagrs., tables. £50; £15.95. ISBN: 0631132929.

55 contributors. About 2,500 entries, ranging from 10-word definitions to articles of 3 columns. All but very brief entries are documented (*e.g.* Hydrology: 11 'Reading and references'), but sources of terms are not included - unlike Stamp's *Glossary (qv)*. Adequate cross-references and a fully analytical index, p.483-528. 'Abbreviations in physical geography', p.xi-xvi. 'We have designed this dictionary for professional geographers and for earth, environmental and life scientists who work on the boundaries of our discipline. It is also intended for use by tertiary-level students and

....(contd.)
secondary/school teachers' (*Editor's Introduction*). Companion to *The dictionary of human geography,* edited by R.J. Johnston, and others (Blackwell, 1981). *Class No:* 551.4(038)=20

[1577]
—SMALL, J. *and* WITHERICK, M. A Modern dictionary of geography. 2nd ed. London, Edward Arnold, 1989. 232p. diagrs., graphs, charts. £30; £10.99. ISBN: 0340431978.

Has well over 1,800 entries, - definitions plus commentary - covering both physical and human geography. It thus supplements Goudie (*qv*) and also *The dictionary of human geography,* edited by R.J. Johnston and others (1981). Adequate cross-references, but no index. For 'A'-level students and first-year undergraduates. *Class No:* 551.4(038)=20

[1578]
A Glossary of geographical terms, based on a list prepared by a Committee of the British Association for the Advancement of Science. Stamp, Sir D. *and* Clark, A.N., *eds.* 3rd ed. London, Longman, 1979. xxix, [1], 571p.

First published 1961.

57 correspondents and collaborators. Coverage: physical, human, economic and political geography. Terms are given various definitions, drawn freely from leading works, with acknowledgements, and 'Comment' added, as necessary. Thus the entry 'Savannah, Savanna, Savana' cites 6 sources plus comments on derivation, spelling, meaning etc. Appendix 2 lists words in foreign languages that have been absorbed into English literature; Appendix 3: 'Some stratigraphical terms'. 'List of standard works', p.xvii-xxv. 'A valuable work of reference, with its most useful features preserved' (*Geographical magazine,* January 1980, p.316). *Class No:* 551.4(038)=20

[1579]
—The Penguin dictionary of physical geography. Whittow, J.B., *ed.* London, Lane, 1984. 591p. diagrs., tables, maps. £7.99. ISBN: 014051094x.

Claims to have over 5,000 entries, differentiating between British and US usage wherever they occur. 'Dry Valley': nearly 1 column; 'Tornado': ¾ column. 'See' and 'see also' reference to other Penguin science dictionaries. No bibliography. Aims to include all the terms currently used by physical geographers working within the 16-19-age school groups. A few photographs would have helped. *Class No:* 551.4(038)=20

[1580]
LOCK, C.B.M. Geography and cartography: a reference handbook. 3rd ed. London, Bingley, 1976. 720p.

An integration of *Geography: a reference handbook* (1968; 2nd ed. 1972) and *Modern maps and atlases* (1969).

1,400 entries, A/Z (titles of works, forms of literature, biographies, topics), with a strong bibliographical and cartobibliographical slant. Extended entries on 'Cartography', (p.159-95), 'Audio-visual aids', 'Bibliographies, national', 'Classification', 'Education in geography and cartography', 'Globes', 'Map librarianship', 'Maps (historical)', 'Abstracts'. Detailed index, p.635-762. The work does not fully integrate *Modern maps and atlases,* which retain for its coverage of national, regional and thematic maps and atlases (*RQ,* v.15(3), Spring 1977, p.258-9). A/Z order is apt to scatter related material and choice of

....*(contd.)*

entry-words can be capricious (*e.g.* 'Man and wildlife'). Illus. and samples would have been appreciated. A bulky but rewarding volume. *Class No:* 551.4(038)=20

French

[1581]
Dictionnaire de la géographie. George, P., *ed.* 4e éd. Paris, Presses Universitaires de France, 1990. 520p. diagrs. FFr320. (*Grands dictionnaires.*) ISBN: 2130429815.

1970 ed.: 448p.

More than 3,000 clear definitions of terms used in French geography. About two-thirds of the terms concern physical geography. References are appended to lengthier entries. *Class No:* 551.4(038)=40

Russian

[1582]
KALESNIK, S.V., *ed.* **Entsiklopedicheskiĭ slovar' geograficheskikh terminov.** Moscow, Sovetskaya Entsiklopediia, 1968. 435p.

'Encyclopaedia dictionary of geographical terms'. A Russian-Russian dictionary defining *c.*4,200 terms, with special emphasis on physical geography. *Class No:* 551.4(038)=82

Reviews & Abstracts

[1583]
Bibliographie géographique internationale. Paris, Centre National de la Recherche Scientifique, Laboratoire d'Information et de Documentation en Géographie (INTERGEO), 1891-. v.1-. 4pa. FFr560.

About 6,500 abstracts pa. 11 sections: 01. General - 02. Historical - 03. Physical geography - 04. Human geography - 05. World. Multiregional geography - 06. Europe - 07. Asia - 08. Africa - 09. America - 10. Oceania - 11. Polar regions. Appended: Index of periodicals. List of principal English terms. Subject index. Place index. Author index. Online via Questel. 'It is unexcelled in its coverage of the field of geography since the rise of this field as an academic and research discipline' (Harris, C.D. *Bibliography of geography,* Pt.1. Univ. of Chicago, Dept. of Geography, 1976, p.29). *Class No:* 551.4(048)

[1584]
Geographical abstracts. Physical geography. Norwich, Elsevier; Geo Abstracts, Ltd., 1989-. 12pa. £420pa. ISSN: 09540504.

Formerly *Geo abstracts.* A. *Landforms and the Quaternary,* 1966-85 and *Geographical abstracts. A. Landforms and the Quaternary,* 1986-88. Merged with other titles in the series - B. *Climatology and hydrology*; E. *Sedimentology*; G. *Remote sensing, photogrammetry and cartography.* Continues *Geological abstracts: palaeontology and stratigraphy.*

Covers 'over 1,000 leading geographical journals, plus books, conference proceedings, reports and theses' (*Editorial introduction*). Subject areas covered include sedimentology; landforms and the Quaternary; hydrology; meteorology and climatology; remote sensing, photogrammetry and cartography. Each issue contains a regional index, and there is a separately published index

....*(contd.)*

issue arranged by subject, author, and geographical area. Included in the GEOBASE (*q.v.*) database. *Class No:* 551.4(048)

[1585]
Referativnyĭ zhurnal. Geografiia. Moscow, VINITI, 1956-. 12pa. 320 rubles. ISSN: 00342378.

About 50,000 abstracts pa. 10 sections, each with an editor: Theoretical - Cartography - Anthropology. Geomorphology - Oceanography. Hydrology - Meteorology - Biogeography - Geography: USSR - Geography: Europe - Geography: Asia and Africa - Geography: America, Australia, Oceania and Antarctica. Annual author, subject and geographical indexes. The definitive Soviet bibliographic source of geographical material. International coverage, with an estimated 25% of abstracted items being Eastern European in origin. *Class No:* 551.4(048)

Periodicals & Progress Reports

[1586]
International list of geographical serials. Harris, C.D. *and* Fellmann, J.D., *comps.* 3rd ed. Chicago, Ill., University of Chicago, Dept. of Geography, 1980. [iv], 457p. $12. (*Research paper no.193.*) ISBN: 0890651000.

First published 1960.

'A comprehensive retrospective inventory of 3,445 geographical serials from 107 countries [International. Afghanistan ... Zimbabwe] in 55 languages with locations in union lists' (sub-title). A list of principal sources precedes and 'Lists of titles in non-Roman scripts' follows the definitive inventory. Index and cross-reference, p.399-457. *Class No:* 551.4(05)

[1587]
—HARRIS, C.D. Annotated world list of selected current geographical serials. 4th ed. Chicago, Ill., Univ. of Chicago, Dept. of Geography, 1980. [iv], 165p. $12. (*Research paper no.194.*) ISBN: 0890651019.

Records '443 current geographical serials from 72 countries with a study of serials most cited in geographical bibliographies' (sub-title). Criteria for selection included 'frequency, regularity, and longevity of publication, availability in major libraries, linguistic accessibility and frequency of citation in major geographic bibliographies'. Index, p.143-85. Location maps on inside covers. A valuable check-list. *Class No:* 551.4(05)

[1588]
Progress in physical geography: an international review of geographical work in the natural and environmental sciences. London, Edward Arnold, 1969-. v.1-. diagrs., tables, maps, 4pa. ISSN: 03091333.

Initially annual (v.1-9, 1969-76).

Most recent volume, V.10, March 1986, contains 3 documented articles (*e.g.* 'The application of weather radar to transport', with 33 references). 5 progress reports (*e.g.* 'Physical hydrology', with 4½p. of references; Sea levels, with 3p. of references). 'Books reviewed', p.147-56 (18 signed reviews, some with references). *Class No:* 551.4(05)

Teaching Materials

[1589]

Handbook for geography teachers. Long, M., *ed.* 6th ed., rewritten and enlgd. London, Methuen Educational, 1974. [xii], 744p. £25. ISBN: 0423888307.

First published 1932. At head of title-page: 'University of London, Institute of London.

24 contributors, 7 book reviewers. 9 sections: 1. The teaching of geography - 2. Outdoor geography - 3. Indoor geography - 4. Atlases, globes and maps - 5. Visual aids - 6. Geographical societies and other organizations - 7. Books for the primary stage - 8. Books for the secondary stage (annotated; 22 sections, p.401-592) - 9. Books for teachers and sixth form (26 sections). No index. A valuable source for teachers and school librarians that needs to be updated. *Class No:* 551.4(072)

Maps & Atlases

[1590]

BLUME, H. Das Relief der Erde: ein Bildatlas. [Colour atlas of the surface forms of the earth.] Stuttgart, Ferdinand Enke Verlag, 1992. 140p. illus. (mostly col.), figs., maps. DM98. ISBN: 1852932066, English ed; 3432992416.

English translation published by Belhaven Press, London.

Describes and illustrates the most important processes and forms in geomorphology. 11 sections: 1. General classification - 2. Processes: weathering and erosion ... 4. Volcanic landforms ... 8. Glacial landforms ... 11. Anthropogenic landforms. Bibliography and sources of figures and tables, p.138-39. Index. Many colour illustrations. For the layman and geoscience student. *Class No:* 551.4(084.3)

[1591]

Fiziko-geograficheskiĭ atlas mira. Moscow, Akademiya Nauk SSSR, 1964. 298p. ISBN: 0471800708.

A physico-geographical atlas of the world, with 240p. of maps and 45p. of text. Pt.1 devotes more than 70 maps to geology, geomorphology, and soils of the world. Pt.2, on the continents; pt.3 concentrates on the USSR (over 60 maps). Page size, 50 x 35cm. Unsurpassed as a physical world atlas. *Class No:* 551.4(084.3)

Histories

[1592]

History of geomorphology: from Hutton to Hack. Tinkler, K.J., *ed.* Boston, Unwin Hyman, 1989. xi,344p. figs., maps. $75. (*Binghamton symposia in geomorphology: international series, no. 19.*) ISBN: 0045511381.

15 well-documented essays: 1. On the nature of geo-history, with reflections on the historiography of geomorphology ... 3. Worlds apart: eighteenth century writings on rivers, lakes and the terraqueous globe ... 12. Different aspects of Polish geomorphology: palaeogeographic, dynamic and applied ... 15. Afterword. Analytical index, p.333-44. *Class No:* 551.4(091)

[1593]

—TINKLER, K.J. A Short history of geomorphology. London & Sydney, Croom Helm, 1985. xviii, 317p. ISBN: 0709924410.

Four parts: 1. Introduction. The frames of reference - 2. Birthpangs of a discipline - 3. The principles: a century [19th] of debate - 4. A system and its feedback - 5. One

.... *(contd.)*

more century: making the system work. Bibliography (authors, A-Z), p.241-79. Analytical index ('Weathering': nearly 1 column), p.281-317. Written for students and earth science professionals. *Class No:* 551.4(091)

[1594]

The History of the study of landforms or, The development of geomorphology. Chorley, R.J., *and others.* London, Methuen; Routledge, 1964-91. 3v. illus., ports., facsims. ISBN: 0416268900, v.2; 0415056268, v.3.

1. *Geomorphology before Davis.* 1964. xvi, 678p. 2. *The life and work of William Morris Davis.* 1973. xxii, 874p. £65. 3. *Historical and regional geomorphology,* 1890-1950. 1991. xxiii,496p. £65.

V.1 has 4 pts. (2. The age of Lyell: 1820-1845 - 3. Marine versus subaerial erosionists: 1846-1875 - 4. The western explorations). Many questions; 26p. of references; 133 illus. and maps, and an excellent 'Informative index', p.649-78. V.2 has a bibliography of Davis (p.793-825), general references, p.828-39, and a bibliography. Subject and place index, plus an index of persons. *Class No:* 551.4(091)

Worldwide

[1595]

The Evolution of geomorphology: a nation by nation summary of development. Walker, H.T. *and* Graham, W.E., *eds.* New York, Wiley, 1993. n.p. illus. £85. ISBN: 0471938580.

Covers 53 countries and describes each country's national policy on landscapes. *Class No:* 551.4(100)

Tropics

[1596]

THOMAS, M.F. Tropical geomorphology: a study of weathering and landform development in warm climates. New York, Wiley; London, Macmillan, 1974. xii, 332p. illus. ISBN: 0470858970.

Most of the examples of characteristic landforms of the tropics are taken from West Africa, Papua & New Guinea, and Australia. Bibliography, p.291-311. Well illustrated. *Class No:* 551.4(213)

Arid Zones

[1597]

Arid zone geomorphology. Thomas, D.S.G. London, Belhaven Press; New York, Halsted Press, 1989. vi,372p. illus., figs., tables. £50; £22.50; $64.95. ISBN: 1852930209.

16 contributors. 4 sections, made up of 16 chapters: 1. The nature of arid environments ... 4. Slope and pediment systems ... 9. Playas, pans and salt lakes ... 13. Wind erosion forms ... 16. Perspectives on arid zone geomorphology. References are appended to each chapter. Index. *Class No:* 551.4(213.52)

[1598]
HOPKINS, S.T. *and* JONES, T.E. **Research guide to the arid lands of the world.** Phoenix, Arizona, Oryx, 1983. 391p. maps. ISBN: 0897746661.

3,199 entries, arranged geographically and by subject, (further subdivided). Descriptive rather than evaluative annotations. Author and subject index.
Class No: 551.4(213.52)

Europe

[1599]
Geomorphology of Europe. Embleton, C., *ed.* New York, Wiley; London, Macmillan. 1984. [x], 465p. illus., maps. £25. ISBN: 0471800708, US ed; 0333379632, UK ed.

20 chapters by members of the Commission on Geomorphological Survey and Mapping of the International Geographical Union. 1. Structural and tectonic framework of the continent of Europe - 2. Principal structural and tectonic features of ocean floors around Europe - 3. Exogenic landforms of Europe - 4. Iceland ... 16. Balkan Peninsula ... 19. Ural mountains - 20. Submarine morphology around Europe (8 maps; 1 cross-section). References (A-Z authors), p.431-47. Detailed, analytical index, p.449-65. 'The first comprehensive survey of the geomorphology of Europe' (*New technical books,* February 1986, item 285).
Class No: 551.4(4)

Great Britain

[1600]
The Geomorphology of the British Isles. Brown, E.H. *and* Clayton, K., *general eds.* London, Methuen, 1976-81. 5v. illus., tables, maps. ISBN: 0416839908.

Scotland, by J.B. Sissons. 1976. *Ireland,* by G.L.H. Davies and N. Stephens 1978. *Southeast and Southern England,* by D.K.C. Jones. 1981. *Eastern and Central England,* by A. Straw and K. Clayton. 1979. *North England,* by C.A.M. King. 1976.

The *Ireland* volume (xii, 250p.) has 7 chapters (7. Coastline), has references on p.222-42 and a non-analytical index. Volumes on the *Midlands* and *Wales and Southwest England* have not so far appeared. *Class No:* 551.4(410)

Information Management

[1601]
Geographic information systems. Antenucci, J.C., *and others.* New York, Van Nostrand Reinhold, 1991. xiii,301p. illus. (some col. pl.), figs., diagrs., tables. $59.95. ISBN: 0442007566.

3 sections: 1. Technology for the information age (Introduction ... Applications; Benefits and costs) - 2. System components (Data base concepts; Data types ... Geographic information system software; System configurations and data communications) - 3. Managing change (Implementation; Legal issues; Horizons). Glossary, p.276-92. Index. References appended to each chapter.
Class No: 551.4:025.4

Libraries

[1602]
LONDON UNIVERSITY. Library Resources Co-ordinating Committee. Geography Sub-committee. **A Guide to geography and map collections.** London, the University, 1985 (2nd ed., 1990). [i], 40p. ISBN: 0718707125.

1. Geography collections (12 libraries) - 2. Map collections (10 libraries. Data: name and address; telephone no.; location; hours of opening; access; readers' places; name of map curator; staff availability; catalogue(s); classification; holdings; special collections or strengths; publications; additional facilities). An important guide.
Class No: 551.4:061:026/027

Islands

[1603]
Oceans and islands. Talbot, F.H. *and* Stevenson, R.E., *eds.* London, Merehurst, 1991. 240p. illus. (col.), figs., maps. ISBN: 1853911569.

40 contributors. 3 sections: 1. The miracle of the sea - 2. Islands: worlds apart - 3. The future of oceans and islands. Facts about oceans and islands, p.228-29. Glossary. Further reading. Index. Many colour illustrations. For school and public libraries. *Class No:* 551.42

[1604]
Standard encyclopedia of the world's oceans and islands. Huxley, A., *ed.* London, Weidenfeld & Nicolson, 1962. 383p. illus. (incl. col.), maps.

A-Z sequence of more than 350 articles (Aden ... Zuyder Zee), with geographical and background data on the world's oceans, channels, straits, gulfs, currents, islands, capes, etc. (Greenland is ranked as an island, but Australia not). 'Galapagos Islands': location, area, map reference, plus 1½ columns of description. Gazetteer of *c.*2,000 entries, p.325-64. 10 location maps; 16 col. plates. No bibliographies. Evidently for popular consumption. *Class No:* 551.42

Mountains

[1605]
Geology of the Himalayas. Beijing, Geological Publishing House, 1991. 2v. illus., diagrs., tables, graphs, maps. ISBN: 7116007385, v.1; 7116007482, v.2.

V.1. *Papers on geophysics.* v,298p. V.2. *Papers on geology.* v,544p.

Based on an International Symposium of the Geology of the Himalayas. V.1 has 21 documented chapters. It contains 'achievements of earth sciences research on the Himalayas and its adjacent areas since 1980' (*Foreword*).
Class No: 551.432

[1606]
Standard encyclopedia of the world's mountains. Huxley, A., *ed.* London, Weidenfeld & Nicolson, 1962. 383p. illus. (incl. col.), maps.

Introductory essays, a brief glossary and location maps precede an A-Z sequence of unsigned articles on 'The world's mountains' (p.55-320). Gazetteer of over 1,500 entries (p.323-57) plus acknowledgements and index. 'Aconcagua', 'Eiger', - each 2½ columns. Entries comprise a brief statement, location, height, map reference, description, history. Stresses human rather than physical

....*(contd.)*

aspects. Article on the Andes (3p.) is less extensive and informative than that in the *Columbia-Lippincott gazetteer.* No bibliography. *Class No:* 551.432

Coastlines

[1607]
CARTER, R.W.G. **Coastal environments:** an introduction to the physical, ecological and cultural systems of coastlines. London, San Diego, Academic Press, 1988. xv,617p. illus., figs., tables, maps. £50; $95. ISBN: 0121618552.

An overview of the subject. Sections include: Waves and wave dominated coasts - Shoreline morphodynamics ... Sea-level changes ... Structures and organization ... Coastal hazards. References, p.561-609. For academic libraries. *Class No:* 551.435.36

[1608]
Dunes of the European coasts: geomorphology-hydrology-soils. Cremlingen-Destedt, CATENA Verlag, 1990. vi,223p. illus., facsims., diagrs., graphs, tables, maps. DM139. (*CATENA supp., 18.*) ISBN: 3923381239.

20 contributors, each with references. 6 sections ('Geohydrology', p.109-83). Profuse maps. No index. *Class No:* 551.435.36

[1609]
The Geology of continental margins. Burk, C.A. *and* Drake, C.L., *eds.* New York, Springer, 1974. 1009p. $69.30. ISBN: 038706866x.

Over 100 contributors. 'An indispensable reference for those specializing in the geology of continental margins'. *Class No:* 551.435.36

[1610]
KING, C.A.M. **Beaches and coasts.** 2nd ed. London, Edward Arnold, 1972. 570p. £25. ISBN: 0713156090.

First published 1959.

A comprehensive and systematic survey of coastal geomorphology. 'Incorporates quantitative approaches in coastal engineering with exploratory descriptions of coastal landforms and beach features. Provides the best coverage of international literature' (Harris, C.D. ed., *A geographical bibliography for American libraries* (1985), item 687). *Class No:* 551.435.36

[1611]
SCHWARTZ, H.L. **The Encyclopedia of beaches and coastal environments.** Stroudsburg, Pa., Van Nostrand Reinhold, 1982. 940p. illus., maps. $149.95; £109. (*Encyclopedia of earth sciences, v.15.*) ISBN: 0879332131.

184 contributors. Articles A-Z, usually documented. Cross-references. Coverage: geomorphology; ecology; coastal engineering; continental, regional and specific types of coast. Well illustrated. 'The most thorough glossary of its kind' (*A geographical bibliography for American libraries,* edited by C.D. Harris (1985), p.96). *Class No:* 551.435.36

[1612]
STEERS, J.A. **The Coastline of England and Wales.** 2nd ed. Cambridge, University Press, 1964. xxviii, 762p. diagrs., tables, maps.

First published 1946.

8 of the initial 19 chapters give detailed geomorphological data on the coast, working anticlockwise from 'The Solway to the Dee', and ending with 'The North-East Coast'. Chapters 20-27 add data on the same regional pattern, noting

....*(contd.)*

the great damage done to the east coast on 31 January and 1 February 1933. 163 diagrs., tables and maps. Numerous footnotes, including references to the plates in Steers' *The coast of England and Wales in pictures* (2nd ed., 1960). Index, p.721-50. 'He has given us what is, perhaps, the most important physiographic monograph since geography attained full status in the British universities' (*Geographical journal*, v.109(1/3), July 1947, p.109, on the 1946 ed.).

The Coastline of Scotland, by J.A. Steers (C.U.P., 1973. xvi, 325p. illus., maps) is the companion volume. *Class No:* 551.435.36

Deserts

[1613]
Desertification: a world bibliography. Prepared for the 22nd International Congress, Moscow, 1976. Paylore, P., *comp. and ed.* Tucson, Arizona, Office of Arid Lands Studies, Univ. of Arizona, 1976. 644p.

1,409 references (often annotated) in English, French and German, plus 217 Russian references (translated). Geographical arrangement, by continent then country.

Followed by *Desertification world bibliography update, 1976-1980* (with imprint as above), adding 400 annotated entries. (Based on C.D. Harris's *A geographical bibliography for American libraries* (1985), entries 2762-3). *Class No:* 551.435.77

[1614]
MIDDLETON, N.J. *and* THOMAS, D.S.G. **World atlas of desertification.** London, Edward Arnold, 1992. 80p. 297 x 420mm. £89.50. ISBN: 0340655122.

Large-format atlas which shows the extent and severity of desertification throughtout the world's arid regions. Compiled by the UN Environmental Program from recent data collected and interpreted by more than 25 regional experts. For academic and research libraries covering earth sciences, climatology, geography, and environmental sciences. *Class No:* 551.435.77

[1615]
PETROV, M.P. **Deserts of the world.** Translated from the Russian by the IPST staff. Jerusalem, Israel Program for Scientific Translations; New York, etc., Wiley, 1976. viii, 647p. illus., diagrs., tables, maps.

Russian edition, 1973.

A global survey of arid regions. Three parts: 1. Physical features of world deserts - 2. Specific environmental features of deserts; 3. Natural resources of deserts and prospects for their investigation and exploitation. Each part carries a bibliography. General index. Illus. are helpful but lack precision. 'The book should be easily available, because it contains so much of value to university and sixth-form students, as well as to the general reader' (*Geographical journal*, v.144, March 1978, p.137). *Class No:* 551.435.77

Coral Reefs

[1616]
Coral reefs of the world. Nairobi, United Nations Environment Programme; Gland, International Union for Conservation of Nature and Natural Resources, 1988. 3v. $100, set; $55ea. ISBN: 2880329434, v.1; 2880329442, v.2; 2880329450, v.3.

V.1. *Atlantic and eastern Pacific.* xlvii,373p. maps. V.2.

.... (contd.)

Indian Ocean, Red Sea and Gulf. l389p. maps. V.3. *Central and western Pacific.* xlix,329p. maps.

Arrangement is by country within each volume. Data for each reef include geographical location; area, depth, altitude; physical features; reef structures and corals; scientific importance and research. Lists of references are given for each section. An excellent, well presented work which, although biased towards environmental aspects of coral reefs, also gives much useful information about their physical features. *Class No:* 551.438

[1617]
GUILCHER, A. **Coral reef geomorphology.** Chichester, John Wiley, 1988. vi,228p. illus., figs., maps. $89.95. ISBN: 0471917559.

5 sections: 1. Distribution and ecology - 2. Surface features of coral reefs - 3. The origin of coral reefs - 4. Types of reefs - 5. Reef and man. Bibliography, p.203-22. Location index. The many black-and-white illustrations could be of better quality. *Class No:* 551.438

Caves

[1618]
Atlas of the great caves of the world. Courbon, P., *and others.* St. Louis, Mo., Cave Books, 1989. 369p. maps. $30. ISBN: 0939748215.

Translation and revision of *Atlas des grands gouffres du monde* (1986).

Gives statistics and maps of all the major caves throughout the world. Information up to date as of 1988. Maps can be of variable quality. However, packed with useful information. The definitive work. *Class No:* 551.44

[1619]
Current titles in speleology: bibliographical details of papers published throughout the world on caving topics. London, British Cave Research Association, 1972-. Annual.

Continues *Speleological abstracts* (1964-70).

No. 23: The literature of 1990 (1991. [x],112p. £11) has *c*.3,000 entries in 6 main sections: Generalia - Africa (Algeria ... Zaire) - America (Anguilla ... Venezuela) - Asia (Burma ... Vietnam) - Australasia and Pacifica (Australia ... Pitcairn Island) - Europe (Austria ... Yugoslavia). A list of periodicals consulted and addresses precedes the sections. '1990 sees the smallest number of titles in *CTS* since 1982' (*Editorial*). *Class No:* 551.44

[1620]
FORD, T.D. *and* CULLINGFORD, C.H.D. **The Science of speleology.** New York & London, Academic Press, 1976. xiv, 593p. illus., diagrs., graphs, tables. $99.95. ISBN: 0122625501.

22 contributors. 14 chapters (*e.g.* 2. The geology of caves; 8. Cave minerals and speleotherms; 10. Cave faunas; 11. Cave flora; 12. Bats in caves; 14. The computer in speleology). Cave and fissure index. Subject index. The standard work. 'Reference of inestimable value to anyone engaged in speleological research ... a primer of caving' (*Palaeogeography, palaeoclimatology, palaeoecology,* v.24(4), 1978, p.568-9). *Class No:* 551.44

[1621]
MIDDLETON, J. *and* WALTHAM, T. **The Underground atlas:** a gazetteer of the world's cave region. London, Hale, 1986. 239p. illus., tables, maps. $16.95. ISBN: 0709027982.

Main sections: The continents - The countries (Afghanistan ... Zimbabwe, p.25-228). Appendices: 1. Longest, deepest and largest caves - 2. National caving organizations - 3. Further reading (running commentary, p.234-5) - 4. Cave map credits. Glossary, p.238-9. A popular account, by the Senior Lecturer in Geology at Trent Polytechnic. *Class No:* 551.44

Oceanography

CD-ROM

[1622]
Oceanographic and marine resources. Baltimore, Md., National Information Services Corporation, 1992. 2v. $970, v.1; $445, v.2.

V.1 of this 2v. CD-ROM set contains *Oceanographic literature review,* (1976 to present; 110,000 abstracts and citations); Institute of Oceanographic Sciences Deacon Laboratory; Proudman Oceanographic Laboratory; Plymouth Marine Laboratory (1985 to present; 240,000 records from this consortium of 3 major oceanographic institutes). V.2 has *Sea Grant abstracts* (1968 to present; over 30,000 bibliographic records) and *c*.135,000 records from the National Oceanic and Atmospheric Administration's collection covering hydrographic surveying, oceanography, meteorology, hydrology, living marine resources, and meteorological satellite applications. *Class No:* 551.46(003.40)

Bibliographies

[1623]
FREEMAN, C. **A Guide to the literature of oceanography** and an enquiry into the density, distribution, and use of the periodical literature of the discipline. A thesis presented for the degree of Master of Science in Information Studies, University of Sheffield, Postgraduate School of Librarianship and Information Science. University of Sheffield, 1971. 306p.

14 chapters (3. Information flow in the marine sciences - 4. Guides to the literature (13 items, briefly annotated) - 5. Directories - 6. Guide to meetings - 7. Abstracting and indexing services (p.64-132) - 8. Bibliographies - 9. Concentration of information (*e.g.* Encyclopaedias; Tables and handbooks) - 10. Information analysis entries - 11. Translations ... 14. Literature use). 4 appendices (*e.g.* Institutions involved in the marine sciences and contacted for information - 28). *Class No:* 551.46(01)

[1624]
Oceanographic index. Woods Hole Oceanographic Institution, Mass. Sears, M., *comp.* Boston, Mass., G.K. Hall, 1971-72.

Author cumulation, 1946-1970. 3v. 58,600 cards. $330.
Author cumulation, 1972-1974. 1v. 18,100 cards. $130.
Regional cumulation, 1946-1970. 1v. 28,900 cards. $110.
Regional cumulation, 1971-1974. 1v. 11,500 cards. $120.
Subject cumulation, 1946-1971. 4v. 95,500 cards. $420.
Subject cumulation, 1971-1974. 2v. 40,000 cards. $265.

....(contd.)

Organismal cumulation, 1946-1973. 3v. 48,400 cards. $240.

These cumulative indexes are working guides to journal articles, monographs, etc. on the marine sciences held at the Library of the Marine Biological Laboratory and at the Woods Hole Oceanographic Institution. They also serve as cumulative indexes to the bibliographic abstracts section of *Deep sea research (qv)*. *Class No:* 551.46(01)

[1625]
SCRIPPS INSTITUTION OF OCEANOGRAPHY, Library. **Catalogs of the Scripps Institution of Oceanography Library. University of California at San Diego.** Boston, Mass., G.K. Hall, 1970. 12v. Supplement 1-. 1973-. $270.

Author/title catalog. 7v. (5212p.). $730.
Subject catalog. 2v. (1759p.). $220.
Shelf list. 2v. (1299p.). $205.
Shelf list of documents, reports and translations collection. 592p. $105.

The 12v. comprise 175,500 photolithographically reproduced card catalogue entries for 80,000 bound volumes and 13,000 reports. *Class No:* 551.46(01)

[1626]
WOODS HOLE OCEANOGRAPHIC INSTITUTION. **Catalog of the Library of the Marine Biological Laboratory and the Woods Hole Oceanographic Institution.** Boston, Mass., G.K. Hall, 1972. 12v. $1,215. ISBN: 0816109370.

354,000 author-catalogue cards, photolithographically reproduced. v.13: *Journals catalog* (over 4,000 titles). The main catalogue includes nearly 300,000 periodical articles. *Class No:* 551.46(01)

Encyclopaedias

[1627]
The Encyclopedia of oceanography. Fairbridge, R.W., *ed.* New York, Van Nostrand Reinhold, 1966. 1056p. illus. (incl.pl.), diagrs., tables, maps. (*Encyclopedia of earth sciences, v.1.*)

245 signed articles, A-Z, by 135 specialists, on all aspects of oceanography and submarine geology. 'Atlantic Ocean': 30p.; 'Sea ice': 5p.; 'Mineral potential of the oceans': 8p.; 'Gulf Stream': 5p.). Short selected bibliograhies follow most articles; well illustrated. The first of a series, each volume being autonomous but cross-indexed. The general editor is Professor of Geology, Columbia Univ. Aimed at high-school students, teachers and specialists. *Class No:* 551.46(031)

[1628]
GROVES, D.G. *and* HUNT, L.M. **Ocean world encyclopedia.** New York, etc., McGraw-Hill, 1980. xv, 443p. illus., facsims, maps. $54.50. ISBN: 0070250103.

Over 400 entries. Coverage: physical, geological, chemical, biological, meteorological and oceanographic aspects. Entries: 'Abalone (molluscs)' ... 'Yellow Sea'. The sectionalized article 'Instrumentation, oceanographic', p.165-78, has 16 illus. Striking, well-captioned photographs, but no bibliographies. Detailed, analytical index, p.419-43. Aimed at being 'both useful and interesting to high school, and college students as well as to interested nonspecialists from many walks of life' (*Preface*). Both authors are on the professional staff of the National Academy of Sciences. *Class No:* 551.46(031)

[1629]
McGraw-Hill encyclopedia of ocean and atmospheric sciences. Parker, S.P., *editor-in-chief.* New York etc., McGraw-Hill, 1980, [ix], 580p. £65; $79.95. ISBN: 0070452679.

230 articles; much of the material in them was previously published in the *McGraw-Hill Encyclopedia of science and technology* (4th ed. 1979). Over 200 contributors. 'Marine sediments': p.254-68; 9 references; 17 illus., diagrs., graphs, maps. About 500 illus. in all. Analytical index. Some US slant. *Class No:* 551.46(031)

[1630]
Standard encyclopedia of the world's oceans and islands. Huxley, A., *ed.* London, Weidenfeld & Nicolson, 1962. 383p. illus. (incl. col.), maps.

A-Z sequence of more than 350 articles (Aden ... Zuyder Zee), with geographical and background data on the world's oceans, channels, straits, gulfs, currents, islands, capes, etc. (Greenland is ranked as an island, but Australia not). 'Galapagos Islands': location, area, map reference, plus 1½ columns of description. Gazetteer of *c.*2,000 entries, p.325-64. 10 location maps; 16 col. plates. No bibliographies. Evidently for popular consumption. *Class No:* 551.46(031)

Handbooks & Manuals

[1631]
Chemical oceanography. Riley, J.P. *and* Skirrow, G., *eds.* 2nd ed. New York, London, etc., Academic Press, 1975-. graphs, tables, map. ISBN: 0125886012.

V.10, 1989 (xviii,404p.) is devoted entirely to SEAREX (the Sea-Air Exchange Program). 12 documented chapters, *e.g.* 'Atmospheric and oceanic cycling of mercury' (with 5p. of references); 'Mineral aerosol transport to the Pacific Ocean' (with 4p. of references). Index, p.393-400. Contents of v.1-9. *Class No:* 551.46(035)

[1632]
GROSS, M.G. **Oceanography:** a view of the earth. 4th ed. Englewood Cliffs, N.J., Prentice-Hall, 1987. xi, 406p. illus. (incl. col.), diagrs., maps. ISBN: 0136298920.

First published 1972.

14 chapters: 1. History of oceanography - 2. Ocean basins - 3. Plate tectonics - 4. Sediments - 5. Seawater - 6. Atmosphere and climate - 7. Ocean structure - 8. Ocean circulation - 9. Waves - 10. Tides and tidal currents - 14. Bottom-dwelling organisms. Chapter questions and references. 7 appendices (*e.g.* 2. Useful data about the earth and ocean). Glossary, p.383-96. Index, p.397-406. Boxed information. Textbooks for sixth-formers and undergraduates. *Class No:* 551.46(035)

[1633]
The Ocean basins and margins. Nairn, A.E.M. *and* Stehli, F.G., *eds.* New York, Plenum Press, 1973-85. 7v. illus., diagrs., graphs, maps. $110ea.

1. *The South Atlantic.* 1973. xv, 583p.
2. *The North Atlantic.* 1974. xiv, 598p.
3. *The Gulf of Mexico and the Caribbean.* 1975. xvi, 503p.
4a. *The Eastern Mediterranean.* 1977. xv, 503p.
4b. *The Western Mediterranean.* 1978. xiv, 447p.
5. *The Arctic Ocean.* 1981. xiv, 447p.
6. *The Indian Ocean.* 1982. xvii, 796p.
7. *The Pacific Ocean.* 1985. xiv, 733p.

V.6, *The Indian Ocean,* has 21 contributors. The 15

....(contd.)

documented chapters include: 1. Sedimentation and sedimentary deposits; 5. The Red Sea Region; 10. South East Asia; 15. The Antarctic margin (with 1½p. of references). Partly analytical index. Each volume includes authoritative reviews by specialists. Detailed geological and geophysical coverage. *Class No:* 551.46(035)

[1634]

The Sea: ideas and observations in progress in the study of the sea. Hill, M.N., *ed.* New York, Interscience: Wiley, 1962-83. 6v. in 10. illus., diagrs., tables, maps.

1. *Physical oceanography.* 1962. xv, 564p.

2. *The composition of sea water. Comparative and descriptive oceanography.* 1963. xv, 554p.

3. *The earth beneath the sea. History* 1963. xvi, 963p.

4.(2pts.) *New concepts of sea floor evolution* edited by A.E. Maxwell. 1971.

5. *Marine chemistry,* edited by E.D. Goldberg. 1974. xiv, 895p.

6. *Marine modeling;* edited by E.D. Goldberg. 1977. xxv, 1045p.

7.(2pts.) *The oceanic lithosphere,* edited by C. Emiliane. xii, [i], 1728p.

8. *Deep-sea biology;* edited by G.T. Rowe. 1983. ix[i], 560p. $95. *(qv).*

9. *Ocean engineering science;* edited by B. Le Méhauté and D.M. Hanes. 1990. 2v. xii,1317p.

V.1 has 33 contributors; v.2:29. V.3 (39 contributors) is the first comprhensive work of its kind. 3 sections: 1. Geophysical exploration - 2. Topography and structure - 3. Sedimentation. Chapter 34, 'The Pleistocene period' has 6½p. of references. Well illustrated throughout. Each volume has author and subject indexes. An authoritative, critical and detailed survey that attempts to be 'a balanced account of how oceanography and the thoughts of oceanographers were moving' *(Preface).*

Class No: 551.46(035)

[1635]

SHEPARD, F.P. Geological oceanography. New York, Crane Russak & Co., 1977. 214p. illus., (some col.), diagrs.

'Written for students beginning to study geology or anyone interested in important developments' in oceanography, - this text provides 'a clear and simple account, by an authoritative author' *(Marine geology,* v.28(3/4), 1978, p.293).

Class No: 551.46(035)

Dictionaries

[1636]

AGENCE DE COOPÉRATION CULTURELLE ET TECHNIQUE and CONSEIL INTERNATIONAL DE LA LANGUE FRANÇAISE. Vocabulaire de l'océanologie. Paris, the Agence; C.I.L.F.; New York, French and European, 1976. 431p. $49.95.

French-base terms, with definitions and eqivalents and indexes in English, German and Russian. Genders are shown.. Appended are classification tables.

Class No: 551.46(038)

[1637]

—**PERMANENT INTERNATIONAL ASSOCIATION OF NAVIGATION CONGRESSES.** Dictionnaire technique illustré en six langues. [Illustrated technical dictionary in six languages.] Brussels, the Association [1957]-. illus., diagrs.

La mer/The sea... by A. Rouville. [1957]. 272p.

A systematically arranged dictionary of 2,284 French terms, with equivalents and reverse indexes in English, German, Spanish and Italian. 6 sections: 1. Sea-water - 2. Conditions of sea-waves - 3. Tides. Currents - 4. Winds - 5. Meteorology - 6. Marine charts. Soundings. Appendix, on charts, etc., p.181-98. *Class No:* 551.46(038)

[1638]

FISCHER, K. Fachwörterbuch der Meereskunde-Meerestechnik. Deutsch-Englisch, Englisch-Deutsch. 2. erw. Aufl. Rostock-Warnemünde, Institut für Meereskunst der AdW der DDR, 1979. 329p.

An East German dictionary (German and English) of oceanography and ocean technology. *Class No:* 551.46(038)

[1639]

Glossary of oceanographic terms. Baker, B.B. *Jr.,, and others, eds.* 2nd ed. Washington, US Naval Oceanographic Office, 1966. vi, 204p. diagrs. *(US Naval Oceanographic Office, Special publication SR35.)*

First published 1960 (US Hydrographic Office).

Definitions of *c.*4,500 terms (*e.g.* 'ship motion') - List of 74 sources. Appendix A: Abbreviations and acronyms; B: List of oceanographic institutions, agencies, activities and groups (mostly US). *Class No:* 551.46(038)

Reviews & Abstracts

[1640]

Aquatic sciences and fisheries abstracts. Bethesda, Md., Cambridge Scientific Abstracts, 1978-. 12pa. 3pts. $1,410, set; $885, pt.1; $635, pt. 2, $295, pt. 3 (6pa.).

Formerly part of *Aquatic sciences and fisheries abstracts,* 1971-77.

Pt.1: *Biological sciences and living resources.*

Pt.2: *Ocean technology, policy and non-living resources.*

Pt.3: *Aquatic pollution and environmental quality.*

Pt.2 carries *c.*8,800 abstracts pa. Coverage: oceanography, limnology, geochemistry, underwater optics and acoustics, geology and geophysics, marine technology and engineering resources (including oil, minerals, desalination and energy, pollution policy and sea law, documentation). Monthly and annual subject and author indexes. Magnetic tapes available. ASFA database available on CD-ROM. Available online via BRS(CSAL), CISTI, DIMDI, DIALOG and ESA. *Class No:* 551.46(048)

[1641]

Deep-sea research. Pt.B: Oceanographic literature review. Oxford. etc., Pergamon Press, 1979-. 12pa. £435pa. ISSN: 01980254.

Formerly *Oceanographic abstracts and bibliography,* 1977-78.

About 7,000 abstracts and references pa. Sections A-G: A. Physical oceanography - B. Marine meteorology - C. Chemical oceanography - D. Submarine geology and geophysics - E. Biological oceanography - F. General - G. New patents. Sections are closely subdivided. Annual author and subject indexes. An authoritative source, scanning *c.*3,500 journals. Available online and on CD-ROM.

Class No: 551.46(048)

[1642]

Marine science contents tables. [Actualités des sciences de la mer.] Rome, Food and Agriculture Organization, 1966-. v.1-. 12pa. *Gratis.* ISSN: 00253205.

Supported by UNESCO Division of Marine Sciences and the Aquatic Sciences and Fisheries Information Systems.

Reproduces contents-pages of 129 journals in the area. Coverage: biology, fisheries, biochemistry and physiology, deep-sea research, hydrography, coastal research, marine ecology, marine geology, marine geophysics, marine pollution. Included in each issue is a list of periodical parts covered, a directory of publishers and a list of forthcoming marine science conferences. *Class No:* 551.46(048)

[1643]

Oceanic abstracts. Bethesda, Md., Cambridge Science Abstracts, 1972-. 6pa. $995. ISSN: 07481489.

*c.*12,000 abstracts pa. 8 sections: Marine biology, oceanography, ecology - Physical and chemical oceanography and meteorology - Marine geology, geophysics and geochemistry - Marine pollution, environmental protection - Marine resources, living - Marine resources, non-living - Ships and shipping - Books and conferences. Subject, organism, geographic and author indexes. An essential bibliographical tool in its field. Online and available on magnetic tape.

McCusig, H. 'Navigating your way through *Oceanic abstracts* database' (*Database,* v.1(1), December 1978, p.26-41) describes the system as 'the foremost database in the oceanographic field'. *Class No:* 551.46(048)

[1644]

Underwater information bulletin. Guildford, Surrey, IPC Science and Technology Press, Ltd., 1969-83. 6pa.

Formerly *Underwater journal information bulletin,* 1971-73.

Carried *c.*1200 citations pa., on marine geology and biology, oceanography and offshire engineering. For current awareness, in its day. No cumulative indexes. Now part of *Marine and petroleum geology* (February 1984-. Quarterly). *Class No:* 551.46(048)

Periodicals & Progress Reports

[1645]

INSTITUTE OF OCEANOGRAPHIC SCIENCES. Serial holdings of UK marine and freshwater sciences libraries. Godalming, Surrey, the Institute, [1977]. 185p.

Lists 6,000 serials by titles, A-Z. Data: library location (9 UK libraries) and years of holdings. *Class No:* 551.46(05)

[1646]

Oceanography and marine biology: an annual review. Aberdeen Univ. Press, 1963-. Annual. ISSN: 00783218.

V.29, 1991 (581p.) has 8 documented articles. They include 'The biology of hydrothermal vents: ecology and evolution' (with 18p. of references) and 'The role of fluid mechanics in the ecology of marine birds' (with 12p. of references). Author, systematic and analytical subject indexes. *Class No:* 551.46(05)

Almanacs

[1647]

MANGONE, G.J. Mangone's concise marine almanac. 2nd rev. and expanded ed. New York, Taylor & Francis, 1991. viii,199p. tables. ISBN: 0844816744.

7 sections: 1. Measurements of marine forms - 2. Physical-political marine features (*e.g.* 'Warships of the minor powers') - 4. Merchant marines and ports (*e.g.* 'Major ports of the world by tonnage') - 5. Fisheries - 6. Marine and seabed minerals - 7. Pollution of the marine environment (*e.g.* 'Biologic agents'). Index. Many statistical tables. Useful as a quick reference tool. *Class No:* 551.46(059)

Festschriften

[1648]

The Ocean floor: a Bruce Heezen commemorative volume. Scrutton, R.A. *and* Talwani, M., *eds.* New York, etc., Wiley, 1982. 318p. illus., diagrs., maps. $250. ISBN: 0471100919.

A Festschrift by 38 authors, the 18 papers covering aspects of most oceans, from global to microscopic scale. A tribute to Heezen and a bibliography of his writings (over 300) precede. Index. 'Largely a reference book and will be of great value in many libraries' (*Geographical journal,* v.149(2), July 1982, p.231). *Class No:* 551.46(082.20)

Tables & Data Books

[1649]

CRC handbook of geophysical exploration at sea. Geyer, R.A. Boca Raton, Florida, CRC Press, 1983. [vi],445p. illus., diagrs., graphs, tables, maps. $139.95. ISBN: 0849302226.

24 contributors. 3 main parts (13 documented sections): General considerations (*e.g.* Satellite contributions to geophysical exploration at sea) - Instrumentation systems (*e.g.* Hydrophone cables for seismic exploration) - Case histories (*e.g.* Tectonic origin, Gulf of Suez, as deduced from gravity data. 42 references). Analytical subject index, p.435-45.

1992 ed., *Geophysical exploration at sea* (424p. £109), is planned. *Class No:* 551.46(083)

[1650]

CRC handbook of marine science. Cleveland, Ohio, CRC Press, 1974. 2v. ([viii], 627p. + [viii], 390p.). graphs, tables, maps. $149, v.1; $75, v.2. ISBN: 0849302110, v.1; 0878193901, v.2.

V.1 (mainly tabular data), edited by F.G.W. Smith, has 6 sections, on chemical and physical oceanography, atmospheric science, geology, ocean engineering, tables of conversion and constants. Analytical index. V.2, edited by F.G.W. Smith and F.A. Kalber, has 4 sections: 1. Primary productivity; 2. Fishery statistics; 3. Zooplankton populations; 4. Phytoplankton populations. 4 folding maps. Analytical index.

A v.3 was to deal with other facets of marine productivity. *Class No:* 551.46(083)

[1651]

—International oceanographic tables. Paris, UNESCO, 1968-73. 2v. ISBN: 9230009067, v.1; 9230010448, v.2.

Prepared by a Joint Panel on Oceanographic Tables and Standards, and published, with Unesco, by the National Institute of Oceanography, Wormley, Godalming, Surrey. V.1 appeared in loose-leaf form, with English, French, Spanish and Russian text, in 1968. V.2 (in binder, xvi, 141p.), 1973. *Class No:* 551.46(083)

Gazetteers

[1652]

UNITED STATES. Board on Geographic Names. **Gazetteer of undersea features.** 4th ed. Washington, Defense Mapping Agency, 1990. xi,146,152p.

2nd ed., 1971.

Official standard names approved by the Board on Geographic Names. Includes index.

Standardization of undersea features, produced by GEBCO (General Bathymetric Chart of the Oceans) [*c*.1983], is in English/French, English/Russian and English/Spanish versions. *Class No:* 551.46(083.86)

Maps & Atlases

[1653]

Atlas of the seas round the British Isles. Lee, A. *and* Ramster, J., *eds.* Lowestoft, MAFF, Directorate of Fisheries Research, 1981. Loose-leaf. £10. ISBN: 0907545001.

75 large coloured charts, including explanatory text, sources used and further sources available. Coverage: water movements, seabed features, dissolved chemicals, plankton, fisheries, oil and gas deposits, ferry routes, telephone cables, dumping areas, marine safety systems. 'Brings together for the first time the nature of the seas themselves' - resources, pollution, dangers and safety measures (*Library review,* v.31, Summer 1982, p.158-9). *Class No:* 551.46(084.3)

[1654]

INTERNATIONAL HYDROGRAPHIC ORGANIZATION. **The General bathymetric chart of the oceans.** (GEBCO). 5th ed. [Monaco], International Hydrographic Organization, the Intergovernmental Oceanographic Commission, 1983.

Co-sponsored by the International Hydrographic Organization, and the Intergovernmental Oceanographic Commission.

GEBCO (5th ed.) comprises 16 sheets, Mercator projection, on a scale of 1:10 million at the equator, plus 2 Polar sheets on a scale of 1:6 million at 75° latitude. The 18 map sheets are available boxed, with a reduced world map (1:35 million) and a supporting manual, (Based on review in *Geographical journal,* November 1986, p.86, p.430-1). *Class No:* 551.46(084.3)

[1655]

The Mitchell Beazley atlas of the oceans. London, Mitchell Beazley, 1977. 208p. illus., diagrs., maps.

Sections: The ocean realm (geomorphology) - Man's ocean quest (history) - Life in the oceans - The great resource (fish, oil, minerals, energy) - The face of the deep (60p. of coloured maps, mineral resources; living resources) - Encyclopedia of marine life (p.172-86). Tables. Glossary. Index, *c*.1,000 entries. 'The weakness of this book lies not

.... *(contd.)*

with the material presented but in the presentation' (*Library journal,* 1 February 1978, p.355). A lavishly illustrated account, but not an atlas. *Class No:* 551.46(084.3)

[1656]

STOMMEL, H. *and* FIEUX, M. **Oceanographic atlases:** a guide to their geographic coverage and contents. Woods Hole, Mass., Woods Hole Press, 1978. vi, [114p.]. $15; $7.50. ISBN: 091517622x.

Examines 97 atlases (usually 1 per page) as to scope, scale, measurements, data plotted and contours, sources of data; dates. 5 areas: Antarctic Ocean, Atlantic Ocean and adjacent seas; Indian Ocean and adjacent seas; Pacific Ocean and adjacent seas; Whole ocean studies. Small key maps show area of ocean covered. For oceanographers, hydrographers and geographers. *Class No:* 551.46(084.3)

[1657]

The Times atlas and encyclopaedia of the sea. Couper, A.D. 2nd ed. London, Times Books, 1989. 272p. illus. (col.), diagrs., tables, maps. ISBN: 0723003181.

First published as *The Times atlas of the oceans* (Times Books, 1983).

Comprehensive but unconventional (neither a detailed atlas nor an alphabetically-arranged encyclopaedia) treatment of the world's oceans, written and cartographed by 28 contributors. 5 sections: The ocean environment (incl. the ocean basins, p.26-43) - Resources of the ocean (incl. fisheries, p.90-101) - Ocean trade (incl. shipping routes, p.146-59) - The world ocean (incl. strategic use of the oceans, p.178-91). 11 appendices, glossary, index, p.258-72. With *c*.400 illustrations and maps, all in colour, it should serve a wide readership. *Class No:* 551.46(084.3)

[1658]

World ocean atlas. Translated from Russian by D.A. Brown. Gorshkov, S.G., *ed.* Oxford, etc., Pergamon Press, 1975-83. 3v. illus., maps. £90, v.1; £310; $620; v.2; £380; $760, v.3. ISBN: 0080199992, v.1; 0080219535, v.2; 0080287352, v.3.

1 *Pacific Ocean.* 1975. 350p. 2. *Atlantic and Indian Oceans.* 1978. 350p. 3. *Arctic Ocean.* 1983. 184pl.

Presents oceanological and meteorological data. Contents of v.2: History of ocean exploration; Ocean bed; Climate; Hydrology; Hydrochemistry; Biogeography; Reference and navigation geographical charts. V.3 has 184 plates and a 4-page index. Introduction and index in English; map plates in Russian. Highly priced, but important.

Class No: 551.46(084.3)

Biographies

[1659]

FOOD AND AGRICULTURE ORGANIZATION. **International directory of marine scientists.** Rome, F.A.O., 1977. [viii], 33p. $25. ISBN: 9250003676.

Computer-prepared guide. Names are arranged under countries (A-Z), plus statement on specialization. Name index. 3rd ed., 1983. *Class No:* 551.46(092)

[1660]
Who's who in ocean and freshwater science. Varley, A., *ed.*
F. Hodgson; Longman Group, 1978. 336p. $84. ISBN:
0582900506.
About 4,000 entries, A-Z. Data include present post,
affiliations, specializations and full address. No subject
category or geographical indexes. *Class No:* 551.46(092)

Great Britain

[1661]
**Estuaries and coastal waters of the British Isles: a
bibliography of recent scientific papers.** Roberts, E.K., *ed.*
Plymouth, Marine Biological Association of the United
Kingdom, Library and Information Services, 1979-. Annual
(irregular). maps. ISSN: 03093964.
Formerly *Estuaries of the British Isles: a bibliography of
recent scientific papers,* 1977-78.
48 sea areas, plus British Isles (general), England
(general), Ireland (general), Scotland (general) and Wales
(general). No.16, 1992, carries 1,223 references. Subjects
include biology, geochemistry, sedimentation, fisheries.
Author index. *Class No:* 551.46(410)

Polar Regions

[1662]
Polar oceanography. Smith, W.O., *Jr., ed.*   San Diego,
Academic Press, 1990. 2v. $69.50, Pt.A; $65, Pt.B. ISBN:
0126530319, pt.A; 0126530327, pt.B.
Pt.A. *Physical science.* xviii,406p. Pt.B. *Chemistry,
biology, and geology.* xvi,354p.
13 chapters in the set: 1. Meteorology - 2. Sea ice in the
Polar regions ... 6. Small-scale processes ... 10. Polar
zooplankton ... 13. Particle fluxes and modern sedimentation
in the Polar oceans. A bibliography is appended to each
chapter. *Class No:* 551.46(98/99)

Equipment & Instruments

[1663]
McCONNELL, A. Historical instruments in oceanography:
background to the Oceanography Collection at the Science
Museum. London, HMSO, 1981. [iv], 52p. illus., facsim.,
map. £3.95. ISBN: 011290324x.
Sections: 1. General introduction (seas and oceans) - 2.
Historical investigations (p.4-46; *e.g.* 'Sounding voyages for
laying telegraph cables. 1850-1900'; 'International fisheries
research') - 3. Modern oceanography. 'Further reading',
p.51. End-paper maps. *Class No:* 551.46:002.50

Information Services

[1664]
Guide to information services in marine technology.
Myers, A., *comp.* 3rd ed. Riccarton, Edinburgh, Institute of
Offshore Engineering, Heriot-Watt University, 1979. 136p.
First published 1973.
Contents: Summary of organizations listed, grouped
(Government depts. and libraries ... United States of
America - International organizations) - Organizations, p.12-
126 (1 per p.) - Principal marine technology journals -
Abstracting and indexing journals and computerized
databases (annotated). Data on services include: person to be
contacted; background; subject interests; library facilities;

....(contd.)
information services; publications. Bibliography, p.132.
Analytical subject index, p.133-6.
Class No: 551.46:061:025.5

[1665]
—**Directory of library and information facilities.**
Moulder, D.S., *comp.* 2nd ed. Plymouth, Marine Biological
Association Library and Information Services, 1986. 37p.
Scope: oceanography, marine environment, marine
biology, freshwater biology, hydrography, fisheries. Typical
data include hours of opening, affiliation, availability, loan
service, information services, catalogue, stock, publications
(institution; library), special collections, facilities.
Class No: 551.46:061:025.5

Institutions & Associations

[1666]
**Annotated acronyms and abbreviations of marine science
related international organizations.** Ashby, C.M. *and*
Flesh, A.R., *eds.*   3rd ed. Washington, US Dept. of
Commerce, National Oceanic & Atmospheric
Administration, Environmental Data and Information
Service; National Oceanographic Data Center, 1981. 349p.
First published 1969. *Class No:* 551.46:061:061.2

[1667]
UNITED STATES. National Referral Center for Science and
Technology. **Geosciences and oceanography:** a directory of
information resources in the United States. Washington,
Library of Congress, 1981. xx, 375p.
Contents; Organization locator - Directory, A/Z (p.1-275).
Data on each body include area of interest, holdings,
publications and information services. Extensive analytical
subject index, p.277-375. 'For the purpose of this directory,
"geoscience" and "oceanography" have been defined quite
broadly' (*Preface*). Thus, palaeontology and astronomy both
figure in the index. *Class No:* 551.46:061:061.2

Research Establishments

[1668]
Ocean research index: a guide to ocean and freshwater
research, including fisheries research. Varley, A., *ed.* 2nd
ed. Guernsey, F. Hodgson, 1976. 637p.
Mainly a directory of research establishments, p.9-501
(international; 125 countries, A-Z; UK, p.356-401; USA,
p.402-87; USSR, p.342-55). Data: name and address;
direction of research; affiliations; scope of interests.
'Oceanographic research literature: a bibliography of
selected guides and periodicals', p.503-48, - an author list,
compiled by A.P. Harvey. Index of original-language titles;
index of English-language titles. Needs to be updated.
Class No: 551.46:061:061.62

Meteorology

Encyclopaedias

[1669]

The Encyclopedia of atmospheric sciences and astrogeology. Fairbridge, R.W., *ed.* New York, London etc., Reinhold, 1967. xv, 1200p. illus., diagrs., tables, maps. (*Encyclopedia of earth sciences, v.2*). ISBN: 012786458x.

About 150 contributors. Signed articles, A-Z, with references appended. Lengthier articles preponderate, beginning with a simple explanation and proceeding to detailed technical data (*e.g.* 'Radar astronomy', p.786-91; 3 figures, 1 table, ½ col. of references, 25 cross-references). Well illustrated. Intended for all scientists, for those still in high school to the emeritus professor (*Preface*). A comprehensive reference work on astronomy, climatology, meteorological and related sciences. *Class No:* 551.5(031)

Handbooks & Manuals

[1670]

Handbook of applied meteorology. Houghton, D.D., *ed.* New York, etc., Wiley, 1985. xv, 1461p. illus., diagrs., graphs, charts, tables, maps. $105; £102. ISBN: 0471084042.

54 contributors. 5 parts: 1. Fundamentals - 2. Measurements - 3. Applications - 4. Societal impacts - 5. Resources (data; books and journals; education; research centers and libraries; directory sources). 46 documented chapters in all (*e.g.* ch. 4: 'Weather forecasting', p.205-79, including 4p. of references and bibliography). Appendices: A. Glossary and units; B. Climatic data (monthly; 147 locations, worldwide), p.1369-1431. A massive compendium; North American slant. 'Designed for professionals and technicians outside the meteorological profession.' (*New technical books,* January 1986, p.101). *Class No:* 551.5(035)

[1671]

Handbook of aviation meteorology. 2nd ed. London, HMSO, 1971. xvii, 404p. £18. (*Met. O. 818 A.P. 3340.*)

Includes 40 references, 32 plates. *Class No:* 551.5(035)

[1672]

HEASTIE, H. A Course in elementary meteorology. 2nd ed. London, HMSO, 1978. xii, 208p. illus. (pl.) diagrs. £7.50. (*Met. O. 911.*) ISBN: 0114003122.

First published 1962.

Part 1: Physical meteorology; part 2: Synoptic meteorology. 12 documented chapters. Appendices: A. Radar and meteorology; B. Weather satellites. 'Books for further reading', p.402. Analytical index. An introductory textbook of theoretical meteorology, 'intended for the reader whose knowledge of physics is roughly equivalent to those of upper science forms in schools, although parts ... are of a rather higher standard' (*Preface to the first edition*). *Class No:* 551.5(035)

[1673]

Meteorology source book. Parker, S.P., *ed.* New York, McGraw-Hill, 1988. 304p. figs., maps. $45. (*McGraw-Hill Science reference series.*) ISBN: 0070455112.

Derived from the *McGraw-Hill encyclopedia of science and technology (q.v.).* Introduction, followed by 8 sections: 1. The atmosphere - 2. Atmospheric electrical phenomena ... 7. Instrumentation, observation and forecasts - 8. Weather modification. Brief chapter bibliographies. Numerous illustrations and charts. Subject index. For public and academic libraries. *Class No:* 551.5(035)

[1674]

Observer's handbook. 4th ed. London, HMSO, 1982. vi, 220p. illus., diagrs. £13.95. (*Met. O. 933.*) ISBN: 0114003297.

First published 1952.

11 chapters of instructions in the exposure of meteorological instruments and in the making of weather observations, both instrumental and non-instrumental, at all types of stations. 1. Observational routine - 2. Clouds - 3. Visibility - 4. Weather - 5. Wind - 6. State of ground and concrete slab - 7. Atmospheric pressure - 8. Temperature and humidity - 9. Precipitation - 10. Sunshine - 11. Special phenomena (*e.g.* electrometeors). 3 appendices (*e.g.* Recording of observations ...). Bibliography, p.213. Analytical index, p.214-20, 31 pl., 23d. Well organized; for both amateur and professional meteorologists. *Class No:* 551.5(035)

[1675]

—**The Marine observer's handbook.** London, HMSO, 1977. 157p. illus. (pl.), diagrs., tables. £10.50. (*Met. O. 887.*) ISBN: 0114002975.

Has 4 parts (14 chapters): 1. Instrumental observations - 2. Non-instrumental observations - 3. Phenomena - 4. Summary of meteorological work at sea. 'Ice terms, arranged by subject', p.65-74, plus 12 pl. 24 cloud pl., 23 tables. Analytical index, p.153-7. *Class No:* 551.5(035)

[1676]

—**Meteorology for mariners:** a text-book of elementary theoretical meteorology for Merchant Navy officers which also presents the practical application of meteorology to safe and economic ship operations. 3rd ed. London, HMSO, 1978. x, 215p. diagrs., graphs, maps. £14.70. ISBN: 0114003114.

First published 1957 (19 chapters).

6 parts: 1. The meteorological elements - 2. Climatology - 3. Weather systems - 4. Weather forecasting - 5. Ocean surface currents - 6. Ice and exchange of energy between sea and atmosphere. Appended bibliography of 10 items. *Class No:* 551.5(035)

Dictionaries

[1677]

Anglo-russkiĭ meteorologicheskiĭ slovar'. Ainbinder, M.I., *and others.* Moscow, Fizmetgiz, 1959. 244p.

An English-Russian meteorological dictionary of *c.*7,000 terms, with a Russian index. Covers 'general, synoptic and (in part) dynamical meteorology, and also climatology' (Maichel, K. *Guide to Russian reference books,* v.2, edited by J.S.G. Simmons (1964), entry F353). *Class No:* 551.5(038)

[1678]
Dictionnaire encyclopédique d'agrométéorologie: française-anglais-espagnol. [Encyclopaedic dictionary of agrometeorology: French-English-Spanish.] Versailles, INRA Editions, 1990. 323p.

Gives definitions of *c.*500 currently used agrometeorological terms. Related expressions are given. English-French and Spanish-French indexes of main terms and a French index of all terms. *Class No:* 551.5(038)

[1679]
Glossaire de météorologie et de climatologie. Villeneuve, G.-O., *and others*. 2nd ed. Quebec, Presses de l'Université Laval, 1980. xix,626p. tables. ISBN: 2763768962.

About 6,000 French terms defined in 10-40 words, with English equivalents and categorized. English-French index. List of works consulted, p.xv-xix. *Class No:* 551.5(038)

[1680]
Glossary of meteorology. Hüschke, R.E., *ed.* Boston, Mass. American Meteorological Society, 1959. viii, 638p. $25. ISBN: 0933876351.

Sponsored by the US Weather Bureau, and others.

Attempts to define *c.*7,000 terms, - 'Every important meteorological term likely to be found in the literature today' - in words 'that are understandable to the generalist and yet palatable to the specialist' (*Preface*). Cites short phrases to illustrate usage. Includes mathematical approach. *Class No:* 551.5(038)

[1681]
Meteorological glossary. Lewis, R.P.W., *comp.* 6th ed. London, HMSO, 1991. 335p. illus. (incl. col. pl.). £20. ISBN: 0114003637.

About 2,300 technical terms used in meteorology are concisely defined and discussed, with profuse cross-references. The publisher claims that 'many new terms and definitions have been introduced as a result of research and developments in meteorology, climatology, hydrology and from increasing research into the greenhouse effect (given 110 words) and global warming', (although there is no entry for latter term). Attractively produced, although the use of lower case typeface slightly irritates.

The 3rd ed., (1938-39) carried tables of equivalents in English, Danish, Dutch, French, German, Italian, Norwegian, Portuguese, Spanish, and Swedish. *Class No:* 551.5(038)

[1682]
ROGOYSKI, D.A., *comp.* **Glossary of Polish-English meteorological terms.** Washington, Library of Congress, 1968. vi, 301p.

Prepared by the Aerospace Technology Division, Library of Congress.

About 10,000 terms and phrases. Patterned after the American Meteorological Society's *Glossary of meteorology*, edited by R.H. Hübschke (1959) (*qv*). *Class No:* 551.5(038)

[1683]
WORLD METEOROLOGICAL ORGANIZATION. International meteorological vocabulary. Geneva, Secretariat of the WMO, 1966. xvi, 276p. (*WMO/OMM/BMO, no.182.TP91.*)

Multilingual meteorological nomenclature, p.1-56; international meteorological definitions, p.57-194. About 2,000 terms are defined. English, French, Spanish and Russian indexes. Appendix A: Abridged international nomenclature. *Class No:* 551.5(038)

Reviews & Abstracts

[1684]
Meteorological and geoastrophysical abstracts. Boston, Mass., American Meteorological Society, 1960-. v.1(1)-. 12pa. $750. ISSN: 00261130.

Formerly *Meteorological abstracts and bibliography,* 1950-59.

Each issue contains *c.*600 abstracts and references. 6 sections: A. Environmental sciences - B. Meteorology - C. Astrophysics - D. Hydrosphere. Hydrology - E. Glaciology - F. Physical oceanography. Author subject and geographic indexes, cumulated annually and at longer intervals, *e.g.* for v.21-26. Online on DIALOG.

Cumulated bibliography and index to 'Meteorological and geoastrophysical abstracts', 1950-1969: classified subject and author arrangements (Boston, Mass., G.K. Hall, 1972. 8v.) is made up of a 4-v. subject sequence of 138,000 entries, classified by UDC and a corresponding 4-v. author sequence. *Class No:* 551.5(048)

[1685]
—**GREAT BRITAIN.** HM Stationery Office. Government publications. Sectional list 37: Meteorological Office. Revised 1 September 1985. London, HMSO, [1986]. 9p. *gratis.*

Has sections: Handbooks, text-books and tables - Journals - Meteorological data: current periodic issues - Climatological normals or averages - Charts of marine meteorology, sea-surface currents and ice - Weather over the oceans and coastal regions - Researches and applied meteorology (*e.g. Geophysical memoirs*) - Mineralogical reports - Registers, diagrams, forms and charts. Index, p.9. *Class No:* 551.5(048)

Tables & Data Books

[1686]
Smithsonian meteorological tables. List, R.J., *ed.* 6th rev. ed. Washington, Smithsonian Institution, 1966 (reprinted 1984). 540p. $30. (*Smithsonian Miscellaneous collections, 114.*) ISBN: 0874741157.

The most comprehensive available set of tables for meteorological calculations. Supplementary conversion tables, etc. *Class No:* 551.5(083)

Maps & Atlases

[1687]
Atlas of meteorology: a series of over four hundred maps. Under the patronage of the Royal Geographical Society. Buchan, A., *ed.* Edinburgh, Bartholomew, 1889. 40, xivp. with 34pl. of maps.

Formed v.3 Bartholomew's *Physical atlas*. Text: 1. General introduction - 2. Description of the maps in the atlas - 3. Appendices (meteorological services and societies, their publications; bibliography; glossary; meteorological tables). The 400 maps comprise world maps, with insets for Britain, France, Europe, etc. Index to letterpress and maps. Page size: 45 x 30cm. Of historical value. *Class No:* 551.5(084.3)

Histories

[1688]
BRUSH, S.G. *and* **LANDSBERG, H.E. The History of geophysics and meteorology:** an annotated bibliography. New York & London, Garland, 1985. 450p. ISBN: 0824091167.

'Geophysics' is here interpreted to include the origin and development of the solar system and the formation of the earth's surface features. About 2,000 annotated entries in sections A-W (A. General histories; B. Biographies ... G. Geodesy, terrestrial gravitation, size and shape of the earth - H. Formation of the earth's surface features ... M. Meteorology: general histories ... R. Atmospheric dynamics and thermodynamics ... V. Upper atmosphere, ionosphere, magnetosphere, aurora). Author and subject indexes. *Class No:* 551.5(091)

Practical Work

Instruments

[1689]
Handbook of meteorological instruments. 2nd ed. London, HMSO, 1980-82. 7v. illus., diagrs., graphs, tables. (*Met. O. 919.*)

1. *Measurement of atmospheric pressure.* 1980.
2. *Measurement of temperature.* 1980, vii. 70p.
3. *Measurement of humidity.* 1981.
4. *Measurement of surface wind.* 1981.
5. *Measurement of precipitation and evaporation.* 1981.
6. *Measurement of sunshine and solar and terrestrial radiation.* 1982.
7. *Measurement of velocity and cloud height.* 1982.

Booklets intended primarily for Met. Office personnel about the instruments used at official stations. 'Particulars of some other types are included to illustrate different principles' (*Introduction*). Each volume has bibliography, glossary and index. *Class No:* 551.508

[1690]
MIDDLETON, W.E.K. The History of the barometer. Baltimore, Md., Johns Hopkins Press, 1964. xx, 489p. illus. $127.80. ISBN: 0317084461.

Broadly chronological order of chapters, with footnote references. 'A well documented and well written account' (*Geographical journal,* v.131(4), December 1965, p.568). *Class No:* 551.508

Vapours

Clouds

[1691]
International cloud atlas. Geneva, WMO, 1956-87. 2v. illus. (pl.). $78, v.2.

Replaces *International atlas on clouds and types of sky* (1934).

V.1 provides a comprehensive text, with detailed descriptive study of clouds, and the methods and techniques of observing them. V.2 is a collection of 224 excellent plates, 103 in colour, supporting the v.1 text.

International cloud atlas - abridged atlas (Geneva, WMO, 1956. 210p.) selects 72 plates from v.2, plus 'brief

....(contd.)
description and explanatory text to meet day-by-day needs of meteorological observers at surface stations' (*Weatherwise,* v.16(4), August 1963, p.182). *Class No:* 551.576

[1692]
—Cloud types for observers. Rev. ed. London, HMSO, 1982. 38p. col. illus. £11. ISBN: 0114003343.

Explains the classification, based on 10 main groups of clouds (divided into 3 levels, - low, medium and high). Each level is accorded a section. *Class No:* 551.576

[1693]
SCORER, R. Clouds of the world: a complete colour encyclopedia. Melbourne, Victoria, Lothian Publishing Co. (Pty.), Ltd.; Newton Abbot, Devon, David & Charles, 1972. 176p. illus. (incl. col. pl.).

14 sections (1. Cumulus ... 4. Cirrus ... 7. Altocirrus (p.100-5; 4 black-and-white illus.; 14 col. illus.) ... 8. Warm sector cloud ... 11. Condensation trails ... 13. Optical phenomena - 14. Rotation. Appendix: Photogrammetry and stereoscopic photography. Bibliography, p.172. Detailed, non-analytical index. 14 col. plates, each with 5-8 photos., lengthy descriptive captions facing. 'The text is admirably clear ... The book can be heartily recommended to all those whose business involves clouds, particularly meteorologists and aviators' (*British book news,* January 1973, p.35). *Class No:* 551.576

Droughts

[1694]
Drought in Africa. 2. [Sécheresse en Afrique. 2.] Dalby, D. *and* Church, R.J.H., *eds.* 2nd rev. and expanded ed. London, International African Institute, in association with the Environment Training Programme, UNEP-IDEP-SIDA, 1977. viii, 200p. diagrs., tables, maps. $54.10. (*African Environment Special report 6.*) ISBN: 0853020566.

First published 1973.

22 contributors. 3 parts (20 chapters, with appended notes, references and French résumés): Introduction - General studies - Special studies. Chapter 14: 'The extent and intensity of the 1969-73 drought in Nigeria: a provisional analysis', p.114-26 (notes; 27 references: résumé). No index. A *Drought in Africa 3* would certainly now be timely. *Class No:* 551.577.3

Avalanches

[1695]
Avalanche atlas: illustrated international avalanche classification. International Commission on Snow and Ice of the International Association of Hydrological Sciences. Paris, Unesco, 1981. 265, [1]p. £24; $16.95. (*Natural hazard series, 2.*) ISBN: 9230016969.

Presents a systematic scheme of classification for avalanches and snow data, supported by a 'Photographic guide' (descriptive data, in English, French, Spanish, Russian and German, facing photographs). *Class No:* 551.578.48

Climate

Bibliographies

[1696]

RIGBY, M. Bibliography of climate textbooks and studies. Geneva, World Meteorological Organization (World Climate Programme), 1984. 170 + 6p.

Records 1,171 references to climate textbooks, atlases, reviews and handbooks, mostly post-1950 publications. Excludes periodical articles and the like, but includes some WMO *Technical notes* and conference reports. Numerous indexes: personal author, corporate author and title entry, subject and geographical name, as well as index of atlases, maps and charts, and index of handbooks, textbooks and training material. *Class No:* 551.58(01)

Encyclopaedias

[1697]

The Encyclopedia of climatology. Oliver, J.E. *and* Fairbridge, R.W., *eds.* New York, Van Nostrand, 1986. 986p. $109.95; £80. (*Encyclopedia of earth sciences, 11.*) ISBN: 0879330090.

Over 200 documented articles, A-Z, mainly on developments in climatology during the past years. Many cross-references. Author, subjects and geographic indexes. Includes entry on recent issues such as acid rain. The coverage of palaeontology has been significantly extended, but an important topic such as ozone is given short shrift (*New technical books,* January/February 1988, item 286). Supplemental to R.W. Fairbridge's *Encyclopedia of atmospheric science and astrogeology (qv).* *Class No:* 551.58(031)

Handbooks & Manuals

[1698]

The Guinness book of weather facts and feats. 2nd ed., by I. Holford. Enfield, Middlesex, Guinness Superlatives, 1984. 253p. illus. (incl. col.), ports., diagrs., tables, maps. £8.95. ISBN: 0851122434.

First published 1977 (240p.).

20 sections (*e.g.* 1. The radiating sun; 2. The invention of thermometers; 4. The pressure and character of the atmosphere; 5. The nature of wind; 6. Weather charts; 13. Rain, flood and drought; 14. Snow and thaw flood; 18. Optical phenomena; 19. Microclimate; 20. Forecasting). Bibliography (by chapters), p.243-5; detailed index, p.247-55. The 2nd ed. updates records, with 'further examples of weather which overstep average conditions' (*Introduction*). *Class No:* 551.58(035)

[1699]

—ROTH, G.D. Collins' guide to the weather. London, Collins, 1981. 256p. illus. (incl. col.), maps. ISBN: 0002190109.

A basic introduction, translated from the German, *Wetterkunde für Alles* (1979). Chapters include 'All about weather maps', 'Typical European weather and conditions', 'More advanced meteorology', 'Weather observation' and 'Climatic statistics'. Glossary, p.245-9; detailed index, p.251-6. Small format. *Class No:* 551.58(035)

[1700]

SCHWARZBACH, M. Climates of the past: an introduction to paleoclimatology. Translated and edited by R.O. Muir. Princeton, N.J., etc., Van Nostrand, [1963] (reprinted 1977). xii, 328p. illus. diagrs., charts, tables, maps. $24. ISBN: 0404162150.

Translated from the 2nd completely revised and enlarged German ed.

The standard work on the subject. Supported by an extensive bibliography, p.265-300. *Class No:* 551.58(035)

Maps & Atlases

[1701]

WALTER, H. and LIETH, H. Klimadiagram Weltatlas. Jena, Fischer, 1960-67. 3 Lfg. Loose-leaf.

33 main and 22 subsidiary maps, displaying statistics of 8,000 climate stations. Lfg.1 (1960) comprises 12 maps and 65 supplementary sheets, with 2,836 diagrams of climate stations of the world. Page size, 60 x 45cm. Aimed to solve in graphic form climatic analogues 'from the perhaps limited and rather specialised point of view of plant ecology' (*Geographical journal*, v.129, December 1963, p.562-3). *Class No:* 551.58(084.3)

[1702]

—Weltkarten zur Klimakunde. [World maps of climatology.] Rodenwald, E. *and* Jusatz, H.J., *eds.* 2. Aufl. Berlin & London, Springer-Verlag, 1965. 28p. diagrs., maps.

Edited under the sponsorship of the Heidelberger Akademie der Wissenschaften. As a collection of 5 world and climatic maps, it continues the maps in v.1-3 of the *World atlas of epidemic diseases (qv)*, on the same projection and scale, *i.e.*, 1:45,000,000. Coverage: mean January and July sunshine (hours), annual total hours of sunshine, general isolines of global radiation, and seasonal climates of the earth. *Class No:* 551.58(084.3)

Histories

[1703]

LAMB, H.H. Climate: present, past and future. London, Methuen, 1972-77. 2v. illus., graphs, tables, maps. ISBN: 0416115306, v.1; 0416115403, v.2.

1. *Fundamentals and climate now.* 1972. xxxi, 613p. 2. *Climatic history and the future.* 1977. xxx, 835p.

Highlights physical and dynamic climatology. A feature is the provision of extensive bibliographies (chapter references: v.1, p.555-96; v.2, p.715-802). Each volume also carries geographical and analytical subject indexes. A major review of climate change. *Class No:* 551.58(091)

Worldwide

[1704]

Monthly climatic data for the world. Washington, US National Climatic Data Center, 1948-. 12pa. $36. ISSN: 00270296.

Sponsored by the World Meteorological Organization, in co-operation with [The US] Weather Bureau.

Data: mean values (at the surface) of pressure, temperature, relative humidity and precipitation (with departures from normal of these three elements) are issued with comparatively little delay for *c.*1,500 stations. Mean values of upper-air temperature, dew point and wind are

....*(contd.)*

included for *c*.400 stations. A thoroughly comprehensive monthly data service. The subscription includes an annual summary. Available on microfiche. *Class No:* 551.58(100)

[1705]
PEARCE, E.A. *and* **SMITH, C.G. The World weather guide.** 2nd ed. London, Hutchinson, 1990. 480p. graphs, tables, maps. £12.95. ISBN: 0091745357.

First published 1990.

Continent and country data schedules: Africa - North America - Central and South America - Asia - Australia - Caribbean islands - Europe - Oceanic islands. (Australia, p.298-308, is divided into 4 climatic regions, with 19 weather stations listed. Data: temperature (highest recorded; average daily; lowest recorded); relative humidity; precipitation. Analytical index, p.473-80). Impressive, useful for both the actual and armchair traveller.
Class No: 551.58(100)

[1706]
Tables of temperature, relative humidity, precipitation and sunshine for the world. London, HMSO, 1958-.

1. *North America and Greenland,* (including Hawaii and Bermuda). 1981. (Met. O. 856a). £23.

3. *Europe and the Azores.* (Met.O. 856c).

4. *Africa; the Atlantic Ocean south of 35 degrees north and the Indian Ocean.* 1983. (Met. O. 856d). £23.

V.2, 5 & 6 bear the older series title, - *Tables of temperature, relative humidity and precipitation of the world.*

2. *Central and South America, the West Indies and Bermuda.* 1959. (Met. O. 617b). £1.50.

5. *Asia.* 2nd ed. 1966. (Met. O. 617c).

6. *Australasia and the south Pacific Ocean, the corresponding sectors of Antarctica.* 1958.

Monthly data (average and extremes for about 1,000 carefully selected and well distributed stations throughout the world). *Class No:* 551.58(100)

[1707]
—World weather records, 1971-1980. Washington, National Climate Data Center, 1987-89. 2v.

V.1. *North America.* 1989; v.2. *Europe.* 1987.

Tabulates average readings of sea-level pressure, monthly and annual mean temperatures and precipitation. 'Lake and river levels and dates of freezing and thawing are noted for some locations' (*Reference services review,* v.13(3), Fall 1985, p.92).

Earlier records appear in the Smithsonian Institution's *Miscellaneous collection* publication no.79 (1929, covering pre-1921 data), no.90 (1934, covering 1921-30) and no.105 (1947, covering 1931-40). The US Weather Bureau issued *World weather records, 1941-50* (USGPO, 1959) and *World weather records, 1951-60* (1965-68), while Environmental Data and Information published *World weather records, 1961-70,* (v. 1-3, 6), 1979-. *Class No:* 551.58(100)

[1708]
The Weather handbook: a summary of climatic conditions and weather phenomena for selected cities in the United States and around the world. Conway, M. *and* Liston, L. L., *eds.* Rev. ed. Norcross, Ga., Conway Data, 1990. 548p. illus., maps. $39.95. ISBN: 0910436290.

Gives basic climatic information on 250 US cities and over 600 other cities throughout the world.
Class No: 551.58(100)

[1709]
World survey of climatology. Landsberg, H.E., *ed.-in-chief.* New York, Elsevier, 1969-. ISBN: 0444422056, v.1A; 0444417761, v.3; 0444413367, v.6.

1A. *General climatology: heat balance climatology,* edited by H.E. Landsberg. 1985. $84.75.

2-3. *General climatology,* edited by H. Flohn, H.E. Landsberg. 1969-81. $164, v. 3.

4. *Climate of the free atmosphere,* edited by D.F. Rex. 1969.

5. *Northern and Western Europe,* edited by C.C. Wallén. 1970.

6. *Central and Southern Europe,* edited by C.C. Wallén, 1977. $146.25.

7. *Soviet Union,* edited by P.E. Lydolph. 1977.

8. *Northern and Eastern Asia,* edited by H. Arakawa. 1969.

9. *Southern and Western Asia,* edited by H. Arakawa and K. Takahashi. 1981.

10. *Africa,* edited by J.F. Griffiths. 1972.

11. *North America,* by R.A. Bryson and F.K. Hare. 1974.

12. *Central and South America,* by Schwerdtfegar. 1976.

13. *Australia and New Zealand,* by J. Gentilli. 1971.

14. *Polar regions,* edited by S. Orvig. 1970.

15. *Oceania,* by H. Van Loon. 1984.

Each volume consists of contributions by specialists, with appended bibliography of references, plus author and subject indexes. *Class No:* 551.58(100)

Europe

[1710]
Climatic atlas of Europe. [Atlas climatique de l'Europe.] Paris, UNESCO; Geneva, World Meteorological Organization, [1970]. Loose-leaf, maps. $164.

Two sets of 13 maps (scale: 1:5,000,000 and 1:10,000,000). Each set shows distribution, monthly and annual, of the mean atmospheric temperature and precipitation, plus a map representing annual temperature ranges. Based on data collected 1931-60 from several thousand stations. 'The first atlas to include annual as well as monthly temperature and precipitation maps for the entire European region' (*Stechert-Hafner book news,* v.26(1), September 1971, p.3). Page size: 60 x 42cm.

Companion atlases: *Climatic atlas of Asia* (1981. 28 maps); *Climatic atlas of North and Central America* (1979), and *Climate atlas of South America,* (v.1. 1975).
Class No: 551.58(4)

Great Britain

[1711]
BRAZELL, J.H. London weather. London, HMSO, 1968. 249 + 21p. illus. (incl. pl.), diagrs., maps. (*Met. O. 783.*)

Revised and updated version of W.A.L. Marshall's *A century of London weather* (1952), adding statistics of weather from earliest times in A.D.4 up to 1964. A systematic study of London's weather by the month, season, public holiday and type of weather, - fog, warm and cold spells, flood and drought. Bibliography, p.242-3. 23 plates.
Class No: 551.58(410)

[1712]
The Climate of the British Isles. Chandler, T.J. *and* Gregory, S., *eds.* London, Longman, 1976. xvii, 396p. graphs, tables, maps.

15 recognised contributors; 15 chapters (2. Synoptic climatology; 4. Radiation and sunshine; 10. Recent climatic change; 12. Upland climates; 15. Regional climates). References, p.343-69 (authors, A-Z). General index (p.377-96); station index. Attractively produced. 'Provides authoritative and comprehensive coverage of weather information for the British Isles' (*New technical books,* November 1977, item 1545). *Class No:* 551.58(410)

[1713]
PAGE, J. *and* **LEBEN, S.R. Climate in the United Kingdom:** a handbook of solar radiation, temperature and other data in the thirteen principal cities and towns. London, HMSO, 1986. x, 391p. tables, maps. £19.95. ISBN: 0114123012.

Produced under contract to the Energy Technology Support Unit of the UK Department of Energy.

5 sections: 1. Solar geometry and solar energy - 2. Other meteorological data - 3. Physical properties of materials - 4. Algorithms - 5. Appendices (including Glossary, p.373-9; 41 references). *Class No:* 551.58(410)

[1714]
—ATKINSON, D.W. Weather. Oxford, Pergamon Press, for the Royal Statistical Society and Social Research Council, 1985. p.1-105, diagrs., graphs, tables.

No.29 in the *Reviews of United Kingdom statistical sources* series, sharing the volume (xiii, 224p.) with no.30 (*Water,* by E.C. Penning-Russell and D.J. Parker). Surface data, upper air data and organizational data are supported by a bibliography of 163 references. *Class No:* 551.58(410)

[1715]
STIRLING, R. The Weather of Britain. London, Faber & Faber, 1982. 270p. illus., diagrs., tables, maps. ISBN: 0571116957.

A highly factual account of the vagaries of British weather, but much more readable than the somewhat academic *Climate of the British Isles,* by T.J. Chandler and S. Gregory (1976) (*qv*). Excellent illus. 'All in all an enjoyable book' (*Geographical journal,* v.150, March 1984, p.111). *Class No:* 551.58(410)

[1716]
WHITE, E.J. *and* **SMITH, R.I. Climatological maps of Great Britain.** Penicuik, Midlothian, Institute of Terrestrial Ecology, 1982. 14p. + 23p. of maps. £2. ISBN: 0904282694.

Bibliography, p.6. 28 illus. (23 in colour).
Class No: 551.58(410)

USA

[1717]
RUFFNER, J.A. *and* **BAIR, F.E. Weather almanac.** 5th ed. Detroit, Mich., Gale, 1987 (6th ed. 1991. 800p.). viii, 811p. tables, charts. $110.

4th ed., 1984.

About 50% of this compendium gives detailed weather records in narrative and tabular form for 109 key US cities, with newly released 1951/80 climatic normals, revised climatic extremes up to and including 1985. Special feature:

....(contd.)
the section on the problem of upper-atmosphere ozone. The remainder of the book concerns worldwide weather and climate, records, etc., plus a glossary. *Class No:* 551.58(73)

Research Projects

[1718]
BAKER, D.J. Planet earth: the view from space. Cambridge, Mass., Harvard University Press, 1990. 194p. illus. $26. ISBN: 0674670702.

A survey of international research projects to predict global climate changes. Focuses mainly on *Mission to Planet Earth,* the US long-range study of the planet as a whole. Glossary. References. 'Strongly recommended for all libraries whose collections encompass environmental issues' (*New technical books,* May/June 1991, item, 656). *Class No:* 551.58:061:061.62.005

Stratigraphy (Historical Geology)

[1719]
DOTT, R.H., *Jr and* **BATTEN, R.L. Evolution of the earth.** 3rd ed. New York, etc., McGraw-Hill, 1981. vii, 573p. + [13]p. illus., facsims., diagrs., tables, maps. $39.56. ISBN: 0070176256.

First published 1971.

9 chapters, each with 'Readings' (*e.g.* 4. The numerical dating of the earth; 11. Earliest Paleozoic history; 18. Pleistocene glaciation and the rise of man). Glossary. Analytical index (13p.). 4th ed., 1988. 704p. $40.44. *Class No:* 551.7

[1720]
International stratigraphic guide: a guide to stratigraphic classification, terminology and procedure. Hedberg, H.D., *ed.* New York, etc., Wiley, 1976. xvii, 200p. illus. $32.50. ISBN: 0471367435.

Prepared by the International Subcommission on Stratigraphic Classification of IUGS Commission on Stratigraphy.

8 chapters, - introduction; principles, definitions and procedures; stratotypes; lithostratigraphic units; biostratigraphic units; chronostratigraphic units; relation between different units. Bibliography. Standard work on stratigraphic nomenclature. *Class No:* 551.7

[1721]
Lexique stratigraphique international. Publié sur l'initiative de la Commission de Stratigraphie du Congrès Geologique International. Paris, Centre National de la Recherche Scientifique, 1956-77. 8v., each in fascicules (*c.*17,000p.).

1. *Europe.*
2. *URSS.*
3. *Asie.*
4. *Afrique.*
5. *Amérique latine.*
6. *Océanie.*
7. *Amérique du nord.*
8. *Termes stratigraphiques majeurs.* 1963-77.

Contributions by *c.*100 specialists. Each country fascicule (*e.g.* v.4. fasc. 96: *Iran.* 1972. 376p.) has a detailed description of stratigraphic formations (A-Z), followed by a

....(contd.)
list of references, a stratigraphic index and folding maps. The dictionary of stratigraphic terms (v.8) is particularly helpful. A standard source of reference. *Class No:* 551.7

[1722]
Geologic time scale 1989. Harland, W.B., *and others.* Cambridge, Cambridge University Press, 1990. xvi,263p. figs., tables. £25. ISBN: 0521383617.

Revised edition of *A geologic time scale 1982.*

7 sections: 1. Introduction - 2. The chronometric (numerical) scale - 3. The chronostratic scale - 4. Isotopic methods, dates, precision and database - 5. Chronometric calibration of stage boundaries - 6. The magnetostratigraphic time scale - 7. Geologic events and the time scale. 6 appendices, *e.g.* 2. Recommended three-character abbreviations for chronostratic names with alternative symbols. References and selected bibliography, p.223-46. General index, p.243-48. Stratigraphic index, p.249-63. *Class No:* 551.7

Quaternary

[1723]
CATT, J.A. Soils and Quaternary geology: a handbook for field scientists. Oxford, Clarendon Press, 1986. x, 267p. illus., diagrs., tables. £55. (*Monographs on soil and resource surveys, no.11.*) ISBN: 0198545681.

7 well-documented and illustrated sections (*e.g.* Climatic change during the Quaternary; Processes of erosion and disturbance during the Quaternary; Stratigraphic nomenclature and classification; Soil formation during the Quaternary; Effects of the Quaternary on soils & present land surfaces). Bibliography, p.233-61. Assumes some prior knowledge of soils, although is not an introduction to soil science but to Quaternary geology. *Class No:* 551.79

Petrology

[1724]
Dictionary of petrology. Tomkeieff, S.I., *and others.* New York, etc., Wiley, 1983. x, 680p. tables. $250. ISBN: 0471101591.

Definitions (Aa lava ... Zwiller rock) are followed by reference[s] to first or early use, as well as any change in meaning or usage; actual derivation; reference to synoptic table. Synoptic tables regroup the terms into different subject areas of petrology. Index to synoptic tables. *Class No:* 552

[1725]
The Encyclopedia of igneous and metamorphic petrology. Bowes, D.R., *ed.* New York, Van Nostrand Reinhold, 1989. xviii,666p. illus., figs., graphs. $125.95. (*Encyclopedia of earth sciences.*) ISBN: 0442206232.

Important alphabetically-arranged reference work aimed at petrologists working with igneous and metamorphic rocks. Rock nomenclature listed (e.g. granite ... granulite, etc), with extensive bibliographical and cross-references. Over 100 contributors from 18 countries. Many illustrations, graphs, and charts. Author and subject indexes. *Class No:* 552

[1726]
Zentralblatt für Mineralogie. Teil 2. Petrographie, Technische Mineralogie, Geochemie und Lagerstättenkunde. Stuttgart, E. Schweizerbart'sche Verlagsbuchhandlung, 1950-. 13pa. ISSN: 05147123.

About 1,200 abstracts per issue, with profuse cross-references. Coverage: rocks, technical aspects of mineralogy, geochemical and mineral deposits. Closely classified entries. Annual author and subject-geographical index. *Class No:* 552

Rocks

Handbooks & Manuals

[1727]
DEER, W.A. Rock-forming minerals. 2nd ed. Harlow, Essex, Longman Group, 1978-. illus. (incl. col.), diagrs., graphs, tables.

First published 1963.

1A. *Orthosilicates.* 2nd ed. 1982. 936p. £85.

1B. *Disilicates and ring silicates.* 2nd ed. 1986. £91.

2A. *Chain silicates.* 1978. 668p. £34.

3. *Sheet silicates.*

4. *Framework silicates.*

5. *Non-silicates.*

V.2A: *Chain silicates* (2nd ed. 1978. [viii], 668p.) is well sectionalized and documented (*e.g.* Non-pyroxenes: 9 sections, Wollastonite ... Serondibite). Non-analytical index. Primary aim of the volumes: 'to provide a work of reference useful to advanced students and research workers in the geological sciences'. (*Preface to first edition*). *Class No:* 552.1/.4(035)

[1728]
The Macdonald encyclopedia of rocks and minerals. London & Sydney, Macdonald, 1983. 608p. illus. (incl. col.). £10.99. ISBN: 0356091473.

First published by Mondadori (Milan), 1977.

Part 1: Minerals (text facing col. illus. on right), nos.7-276 (*e.g.* silicates, nos.157-274) - Part 2: Rocks, nos.277-377 (*e.g.* Igneous rocks, nos.277-323). A lengthy introduction (p.7-83; p.43-39) precedes each part. Symbols indicate grade of rarity (very rare; rare; common; very common). Over 1,000 col. illus. Index of entries, p.594-604. Small format. A remarkably cheap compendium. *Class No:* 552.1/.4(035)

[1729]
POUGH, F.H. A Field guide to rocks and minerals. 4th ed. Boston, Mass., Houghton Mifflin, 1976. 317p. illus. (incl. col.), diagrs. $19.45; $13.45. ISBN: 0395240476.

First published 1953.

Part 1: An introduction to the study of rocks and minerals (collection; locations; physical properties of minerals; crystal classifications; chemical classification of minerals; texts; techniques and tips) - Part 2: Mineral descriptions. Uranium ores. Glossary, p.333-8. Annotated bibliography. Detailed index. A practical book, designed for the amateur. Photographs are mainly of specimens in the American Museum of Natural History's collection. *Class No:* 552.1/.4(035)

Dictionaries

[1730]

MITCHELL, R.S. **Dictionary of rocks.** New York, Van Nostrand Reinhold, 1985. xi, 228p. illus. (incl. col.). $43.95. ISBN: 0442263287.

Has c.5,000 definitions, categorized. Appended glossary, p.219-28. Recent igneous-rock definitions published by the International Union of Geological Sciences are included. Claims to be 'the first dictionary in the English language devoted exclusively to the names of rocks' (*Preface*). A fitting companion to the earlier book, *Mineral names: what do they mean?*, by R.S. Mitchell, with S.R. Henley (Van Nostrand Reinhold, 1979). *Class No:* 552.1/.4(038)

[1731]

Rocks and minerals. Deeson, A.F.L., *ed.* Newton Abbot, David and Charles, 1973. 288p. illus., charts. ISBN: 0715363301.

Well over 2,000 different types (AA or Aphrolith ... Zussmanite). Data include composition, physical properties, distinguishing characteristics, methods of identification, environment, variety. 1,000 small col. illus. 'Designed primarily to assist the amateur collector, whether he is a collector already or somewhere along the line' (*Introduction*). *Class No:* 552.1/.4(038)

Reviews & Abstracts

[1732]

'**Geomechanics abstracts**'. In *International journal of rock mechanics and mining sciences & geomechanics abstracts.* Oxford, etc., Pergamon Press, 1964-. 6pa. £525. ISSN: 01489062.

Over 2,000 abstracts pa. Sections: Geology - Hydrogeology - Properties of rocks and soils - *In situ* stress - Site investigation and field observation - Analysis techniques and design methods - Rock breakage and excavation - Rock and soil reinforcement and support - Surface structures - Underground excavations - Subjects peripheral to geomechanics (*e.g.* Snow and ice mechanics). Online since 1977 via ORBIT. *Class No:* 552.1/.4(048)

Tables & Data Books

[1733]

CRC **handbook of physical properties of rocks.** Carmichael, R.S., *ed.* Boca Raton, Florida, CRC Press, 1982-84. 3v. (1089p.). tables, graphs. $129.95, v. 1; $119, v. 2; $110, v. 3. ISBN: 0849302269, v.1; 0849302277, v.2; 0849302285, v.3.

16 contributors. V.1: Mineral composition of rocks. Electrical and spectroscopic properties of rocks and minerals - V.2: Seismic velocities. Magnetic and engineering properties of rocks and minerals - V.3 Density of rocks and minerals. Elastic constants of minerals. Inelastic properties of rocks and minerals: strength and sheology. Radioactivity properties of minerals and rocks. Seismic attenuation. All chapters have appended bibliography. Each volume carries an index. 'The intent is to bridge the gap between individual reports with only specific limited data, and massive assemblies of data which are uncritically presented' (*Preface*).

An updated and abridged edition of this work appeared in 1989, *CRC practical handbook of physical properties of rocks and minerals* (741p. CRC Press). *Class No:* 552.1/ .4(083)

[1734]

Physical properties of rocks and minerals. Touloukian, S., *and others.* New York, etc., McGraw-Hill, 1981 (reprinted New York, Hemisphere, 1989). xx, 548p. diagrs., graphs, tables. $112. (*CINDAS data series on material properties, Group 2, v.2/2.*) ISBN: 0891168834.

31 contributors. 13 chapters (most of them with bibliographies appended). Chapters 3-12, on the physical, mechanical, electrical, magnetic and thermophysical properties of rocks. Chapter 13: Heat flow in the earth's crust. Detailed table of contents rather than an index. Emphasis is on data useful to geothermal applications. *Class No:* 552.1/.4(083)

Great Britain

[1735]

Igneous rocks of the British Isles. Sutherland, D.S., *ed.* New York, etc., Wiley, 1982. 645p. £78.60; $245; $160. ISBN: 0471278106.

37 contributors. 7 sections, reviewing the major occurrences of igneous rocks in Great Britain and Ireland. 'An obligatory purchase for any geological institution library, and will find itself constantly thumbed through by students, teachers and research workers' (*Geological magazine,* v.120, July 1983, p.407). *Class No:* 552.1/ .4(410)

America — North

[1736]

The Audubon Society field guide to North American rocks and minerals. Chesterman, C.W. New York, Knopf, 1979. 850p. illus. (col. pl.), diagrs., tables. $17.95. ISBN: 0394502698.

A pocket-sized guide to 232 mineral species and 40 types of rocks, in 4 parts. Pt.1, Minerals and pt.2, Rocks, both include a guide to identification, a visual key, and colour plates 1-702 and 703-94 respectively. Pt.3, Mineral collecting; pt.4, Appendices (Glossary, p.785-97; Bibliography, p.799-801; Rock forming minerals; localities; etc.). Index, p.831-50. *Class No:* 552.1/.4(71+73)

Meteorites

[1737]

BRITISH MUSEUM (NATURAL HISTORY). **The Catalogue of meteorites,** with special reference to those represented in the collection of the British Museum (Natural History). 4th ed., edited by A.L. Graham, and others. London, British Museum (Natural History), with the Univ. of Arizona Press, Tucson, 1985. 464p. £20. ISBN: 0565009419.

3rd ed., 1966; Appendix, 1977. M.H. Hey was the compiler of the 3rd ed.

Data on each meteorite name; place where found (or place of impact), plus co-ordinates; chemical content (with citations to the original literature); weight; present owner; specimen number (if appropriate). The British Museum (Natural History)'s collection runs to c.2,784 meteorites (*Astronomy,* v.14 , March 1986, p.72). A standard work of reference, the definitive list of all well-documented meteorites known worldwide.

Next edition of the *Catalogue* is not until 1992. Meanwhile an updating set of floppy discs is available. *Class No:* 552.6

[1738]

—HUGHES, D.W. 'The History of meteors and meteor showers.' (In *Vistas in astronomy*, v.2, 1982, p.325-45).

Effectively starts with the work of Edward Halley (1686-). 14 chronology-sections extend to 1947. 81 references. *Class No:* 552.6

Economic Geology

[1739]

Bibliography of economic geology. Didcot, Geosystems, 1982-. 6pa. £85; $170. ISSN: 00167053.

Formerly *Geocom bulletin*, 1968-81.

Over 5,000 references pa. Classes 100000-900000, 'arranged by subject according to the economic geology subset of *Geosaurus: Geosystems thesaurus of geoscience,*' in 32 sections (General publications ... Geoscience information). Each issue has subject, locational, stratigraphical, geographical and author indexes. Available online via DIALOG, and on CD-ROM. *Class No:* 553

[1740]

World resources 1990-91: a report. New York, Oxford, Oxford University Press, 1990. (1992-93 ed. Oxford, University Press, 1992. 384p. $32.50; $19.95). xiv,383p. tables, maps. $17.95. ISBN: 0195062299.

4 sections: 1. A global perspective - 2. Special focus chapters - 3. Conditions and trends (population and health; human settlements; ... oceans and coasts; atmosphere; ... policies and institutions: natural resources accounting) - 4. World resources data tables, p.241-364. World map. Index. *Class No:* 553

Minerals & Ores

Encyclopaedias

[1741]

CARR, D.D. *and* **HERZ, N.,** *eds.* **Concise encyclopedia of mineral resources.** Oxford, Pergamon Press, 1989. xxiii,426p. diagrs., tables. £80; $145. (*Advances in materials science and engineering, v.8.*) ISBN: 0080347347, UK; 0262031558, US.

Mainly a compilation of articles from the *Encyclopedia of materials science and engineering* (Pergamon Press, 1986), arranged A-Z (Abrasives ... Zirconium and hafnium resources). Larger topics are subdivided, *e.g.* coal: geology; mining; world resources. Short bibliographies are appended to the articles. Subject index. *Class No:* 553.2/.4(031)

Handbooks & Manuals

[1742]

JENSEN, M.L. *and* **BATEMAN, A.M. Economic mineral deposits.** 3rd ed. New York, etc., Wiley, 1979. viii, 593p. illus., diagrs., graphs, maps. ISBN: 0855331739.

First published 1942.

3 parts (29 documented sections): 1. Principles - 2. Processes of formation of mineral deposits - 3. Metallic mineral deposits - 4. Nonmetallic mineral deposits. General references on economic geology, p.579-80. Analytical index,

....(contd.)

p.381-93. 'Instead of following a classification of mineral deposits, processes of mineral formation and concentration are emphasized' (*Preface*). *Class No:* 553.2/.4(035)

Tables & Data Books

[1743]

FINLAY, C.J. *and* **CRIDDLE, A.J. The Quantitative data file for ore minerals** of the Commission on Ore Microscopy of the International Mineralogical Association, 1986. 2nd ed. London, British Museum (Natural History), 1986. xlix, 420p. (oblong). ISBN: 0565009974.

Contents: Introduction; Historical background ... keys for identification; bibliography (p.xvi-xvii); COM wavelength key. Data file, p.1-420 (Acanthite ... Zinchenite), providing chemical formula, chemical composition, X-ray data, etc. *Class No:* 553.2/.4(083)

[1744]

World mineral statistics. Lofty, G.J., *and others*. Keyworth, Nottingham, British Geological Survey, 1990. 373p. tables. £40. ISBN: 0852721889.

Data include figures for production, export and import, with additional information being given for the more important minerals, *e.g.* commodity prices. A supplement giving preliminary data for 1985-1989 is sent free to subscribers. *Class No:* 553.2/.4(083)

Maps & Atlases

[1745]

DIXON, C.J. Atlas of economic mineral deposits. London, Chapman & Hall, 1979. 142p. diagrs., tables, maps. £89. ISBN: 0412143801.

5 sections: 1. Deposits in geological environments at the earth's surface - 2. Mineral deposits in sedimentary rocks - 3. Mineral deposits associated with felsic magmatic environment - 4. Mineral deposits in basic and ultrabasic magmatic rocks - 5. The world distribution of mineral deposits. Glossary of mineral names; Units of measurement; Key to stratigraphic names. Index, p.141-2. A valuable aid for the teaching of economic geology. The author, senior lecturer in Mining Geology, Imperial College, London. *Class No:* 553.2/.4(084.3)

Worldwide

[1746]

Overseas geology and mineral resources. London, HMSO, 1950-.

New series in print: nos.41-63 (1973-86). Some of the original series (v.1 1950-71) were entitled *Colonial geology and mineral resources*. The range is now worldwide (*e.g.* no.49: *Outline of the geology and mineral deposits of the Democratic Republic of the Sudan and adjacent areas,* by J.R. Vail. 1978; no.58: *An annotated bibliography of Ecuadorian geology,* by C.R. Bristow. 1981).

The *Overseas memoirs* series (no.1-. 1976-) supplements the above as more specialised monographs (*e.g. The geology of Guadalcanal, Solomon Islands,* by B.D. Hackman. 1980). *Class No:* 553.2/.4(100)

[1747]
RENSBURG, W.C.J., van. **Strategic minerals.** Englewood Cliffs, N.J., Prentice-Hall, 1986. 2v. (xvi, [1], 552p. + xiii, 362p.). diagrs., graphs, tables. $149.80ea; £160.30ea. ISBN: 0138513872, v.1; 0138514119, v.2.

V.1: *Major mineral exporting regions of the world: issues and strategies* has 6 parts: Canada; Southern Africa; Australia; Latin America; China; The oil-crop economies. Bibliography (6 parts), p.527-42. Detailed index. V.2: *Major mineral consuming regions of the world* covers 4 parts: United States; Japan; Western Europe; USSR. Bibliography (4 parts), p.331-47. Detailed index. p.349-62. A major aim was to analyze 'the effects of government policies, laws and regulations on the availability, reliability, and cost of producing minerals' (*Preface*). *Class No:* 553.2/.4(100)

[1748]
RIDGE, J.D. **Annotated bibliographies of mineral deposits in Africa, Asia (exclusive of the USSR) and Australasia.** Oxford etc., Pergamon Press, 1976. viii, 546p. maps. $200. ISBN: 0080204597.

Similar treatment to the *Europe* volume. Main parts: Africa (Morocco ... Zambia), Asia (Burma ... Turkey), Australasia (Australia, New Zealand, Papua, New Guinea, New Caledonia, Fiji). Appendix 1: Classification of ore deposits. Index of authors; A-Z list of deposits; Index of deposits (according to metals and minerals produced; age of mineralization; modified Lindgren classification). 'I have tried to include as many references as possible to languages other than English' (*Abstract*). *Class No:* 553.2/.4(100)

[1749]
WOLFE, J.A. **Mineral resources:** a world review. London, Chapman and Hall, 1984. xv, 293p. ISBN: 0412251906.

Part 1 consists of short essays on various aspects of the mineral industries (*e.g.* exploration). Part 2 includes reports on 21 metals and 18 non-metals. Glossary; briefly annotated bibliography; index. 'This reference tool is designed for those interested in mineral resources and economy but with training in another field' (*New technical books*, November 1985, item 1185). *Class No:* 553.2/.4(100)

Marine Environment

[1750]
EARNEY, F.C.F. **Marine mineral resources.** London and New York, Routledge, 1990. xxiv,387p. illus., graphs, tables, maps. £65. (*Ocean management and policy series.*) ISBN: 041502255x.

2 parts: 1. Implications of Law of the Sea negotiations - 2. Resources of continental shelves (petroleum, construction aggregates, minerals, etc.). Extensive chapter references; 30p. bibliography. *Class No:* 553.2/.4(26)

Europe

[1751]
Mineral deposits of Europe. Dunning, F.W., *and others, eds.* London, the Institution of Mining and Metallurgy; the Mineralogical Society, 1978-1989. 5v. in 4. £30, v.1; £42, v.2; £65, v.3; £120, v.4/5. ISBN: 0900488441, v.1; 0900488638, v.2; 0900488905, v.3; 1870706072, v.4/5.

V.1. *Northeast Europe.* 1978. 362p. V.2. *Southeast Europe.* 1982. 304p. V.3. *Central Europe.* 1986. 355p. V.4/5. *Southwest and eastern Europe, with Iceland.* 1989. 454p.

....(contd.)
V.4/5 arranged by country, *e.g.* Spain, Italy. Bibliographies appended to each chapter. Name and subject indexes. 'Some national chapters are landmarks, being the first comprehensive descriptions in English (or any other language) of their ore deposits', (*Foreword*). *Class No:* 553.2/.4(4)

[1752]
RIDGE, J.D. **Annotated bibliographies of mineral deposits in Europe.** Oxford, etc., Pergamon Press, 1984-1989. ISBN: 0080302424, v.1; 0080302432, v.2; 0080240224, set.

Pt.1. *Northern Europe including examples from the USSR in both Europe and Asia.* 1984. viii,778p. £125. Pt.2. *Western and south central Europe.* 1989. viii,473p. £76.

Pt.1 covers Ireland; Great Britain; Norway; Sweden; Finland; Poland; USSR. Pt.2 covers Portugal; Spain; France; Belgium-Germany-Netherlands; Switzerland; Italy; Iran. References in Pt.1 are mostly in English, while Pt.2 references are made up of the major European languages. Pt.2 has only an author index, thus greatly reducing its usefulness, while Pt.1 has an index to authors; A/Z index of deposits; index of deposits (according to metals and minerals produced; according to age of mineralization; according to the modified Lindgren classification). *Class No:* 553.2/.4(4)

America

[1753]
RIDGE, J.D. **Annotated bibliographies of mineral deposits in the Western Hemisphere.** Boulder, Colo., Geological Society of America, 1972. xiv, 681p. maps. Supplement, 1974. 8p. $160. (*Geological Society of America. Memoir no.131,* - revision and expansion of *Memoir no.75, 1958.*) ISBN: 083575491x. ISSN: 00107053.

Covers North and South America; arrangement, under country/US state/Canadian province, is by deposit. For each deposit: minerals mined; age and position in the Lindgren classification, plus bibliography and notes. Indexes of authors and deposits (A-Z, by age, by metals and minerals produced, by Lindgren classification). Appendices provide a classification of ore deposits, and topics for consideration. Aimed, like the other two volumes, at the economic geologist in his study of specific deposits. *Class No:* 553.2/.4(7)

Inorganic Minerals

Salt

[1754]
LEFOND, S.J. **Handbook of world salt resources.** New York, Plenum Press, 1969. xxiii, 384p. tables, charts, maps.

Concise country data under continents (North America, p.1-136 - Central America - South America - Europe, p.173-260 - Africa - Asia - Oceania) on rock salt deposits and solar salt production. Location of salt mines; output statistics. 'Sources of information' and 'Bibliography' appended to each country entry. Subject index. 76 maps, 162 charts, tables. *Class No:* 553.631

Hydrocarbons

Coal

[1755]
Coal geology and coal technology. Ward, C.R., *ed.* Oxford, Blackwell Scientific, 1984. [ix], 345p. illus., diagrs., graphs, tables. ISBN: 0867930969.

7 contributors. 9 chapters (1. Chemical analysis and classification of coal - 3. Coal petrology and petrographic analysis - 4. Coal utilization - 5. Geology of coal - 6. Coalfield exploration - 7. Coal-mining geology - 8. Coal properties and marketing - 9. Coal and the environment). Appendix: List of national and international standards for coal and coke analysis and testing. References (authors, A/Z), p.319-34. Analytical index, p.335-45. *Class No:* 553.94

[1756]
TODD, A.H.J. Lexicon of terms relating to the assessment and classification of coal resources. London, Graham & Trotman, 1982. 136p. tables. £42.95. ISBN: 0860104036.

A dictionary of terms used in 15 countries and in international organizations. Terms, where necessary, are translated into English. Appended list of sources. *Class No:* 553.94

Petroleum

[1757]
BEYDOUN, Z.R. The Middle East regional geology and petroleum resources. Beaconsfield, Bucks., Scientific Press, 1988. 292p. diagrs., maps. ISBN: 090136021x.

'Geological evolution of the region, overview of main producing regions and fields, national surveys of producing fields, unproducing discoveries, and future reserves' (*Books in the earth sciences,* no.43, April 1985, p.862). 7 folding maps, cross-sections. *Class No:* 553.982

[1758]
EARNEY, F.C.F. Petroleum and hard minerals from the sea. V.H. Winston & Sons, and L.E. Arnold, 1980. 291p. illus., diagrs., graphs, tables, maps. ISBN: 0470270098.

Part 1: The continental margin - Part 2. The deep seabed. Chapter 3: Offshore petroleum, p.33-61, with 6 maps, 3 graphs, 3 tables, 2 diagrs., 592 references, p.261-81. Analytical index, 283-91. Worldwide coverage. *Class No:* 553.982

[1759]
Elsevier's oil and gas field dictionary in English/American, French, Spanish, Italian, Dutch, German and Arabic. Chaballe, L.Y., *and others.* Amsterdam, etc., Elsevier, 1980. xii, 672p. $200; £128.12. ISBN: 0444418334.

4,843 English/American base-terms, their synonyms and variants, with equivalents and indexes in French, Spanish, Italian, Dutch and German. A supplementary section deals with Arabic terms. *Class No:* 553.982

[1760]
—STOHAROV, D.E., *ed.* Russian-English oil-field dictionary. Oxford, etc., Pergamon Press, 1983. 431p.

30,000 entries, terms being drawn from scientific literature published in the 1970s. Usefully meets a deficiency in the language coverage of the Elsevier dictionary. *Class No:* 553.982

[1761]
HOBSON, G.D. *and* **TIRATSOO, E.N. Introduction to petroleum geology.** 2nd ed. Beaconsfield, Scientific Press, 1981. 352p. illus., diagrs., charts, tables, maps. £18. ISBN: 0901360120.

10 documented chapters: 1. The nature of petroleum - 2. The origin of petroleum - 3. The migration of petroleum - 4. The nomenclature of petroleum - 5. Surface exploration for petroleum - 6. Subsurface ... 7. Formation evaluation - 11. Delineation of the reservoir - 9. Petroleum in space and time (p.269-307; 120 references) - 10. Petroleum production and units. Statistics, to 1979. 50 tables; nearly 100 figures. Analytical index, p.343-52. *Class No:* 553.982

Hydrology

Bibliographies

[1762]
WATER RESOURCES CENTER ARCHIVES, University of California. **Dictionary catalog. [Water Resources Center Archives].** Boston, Mass., G.K. Hall, 1970. 5v. (3373p.). Supplements, 1-. 1971-. $545, set; $285, supp. 6. ISBN: 0816108846, set; 0816102449, supp. (6th. 1978).

The main catalogue has 97,000 photolithographically reproduced catalogue-card entries. Coverage embraces water quality, pollution and reclamation, waste disposal, water as a natural source and its uses, flood control, sediment transport and coastal engineering. *Class No:* 556(01)

Handbooks & Manuals

[1763]
UNITED NATIONS. Educational, Scientific and Cultural Organization. **Ground-water studies: an international guide for research and practice.** Brown, R.H., *and others, eds.* Paris, UNESCO, 1972-. (*Studies and reports in hydrology,* 7.)

The original 8 chapters (now o.p.) have been added to by 4 supplements (to 1984; only supp. 3 (1979. $12) in print), including further chapters (9,10,14,15). *Class No:* 556(035)

Dictionaries

Polyglot

[1764]
Elsevier's dictionary of environmental hydrogeology. Pfannkuch, H.O., *comp.* Amsterdam, Elsevier, 1990. 332p. £85.93. ISBN: 0444872698.

Previously published, 1969, as *Elsevier's dictionary of hydrogeology in three languages.*

Defines 5,422 terms in English, with German and French equivalents and indexes. Coverage extends beyond ground water and includes interdisciplinary terms from hydro-meteorology, water resources planning, water quality control, etc. *Class No:* 556(038)=00

[1765]
Hydrographic dictionary. 3rd ed. Monaco, International Hydrographic Bureau, 1974. 370p.
2nd ed. 1951(259p.).
About 1,000 terms in each of 12 languages. - English, French, Danish, Dutch, Portuguese, Spanish, Argentine, Uruguayan, Italian, Norwegian, Swedish, German. The left-hand page gives numbered keywords in English and French, with definitions; the right-hand page gives equivalents in the other 10 languages. 11 language indexes.
Class No: 556(038)=00

[1766]
TEKNISKA NOMENKLATURCENTRALEN.
Vattenordlista. [Glossary of water.] Stockholm, Tekniska Nomenklaturcentralen, 1968-74. ISBN: 9171960414.
1. *Hybrobiologiska termer.* 1968. 84p. 2. *Geologiska, hydrologiska, meteorologiska termer.* 1970. 80p. 3. *Vattenordlistregister.* 1974. 132p.
1,410 Swedish terms and definitions, in all, with equivalent terms in English, French, German and Finnish. Notes often contain 'cross-references to cognate and/or contrasting terms listed elsewhere in the glossary' (*The Incorporated linguist,* v.10(3), July 1971, p.91, on v.1).
Class No: 556(038)=00

[1767]
WORLD METEOROLOGICAL ORGANIZATION.
International glossary of hydrology. Paris, UNESCO, 1974. xxiv, 393p. tables. £25. (*WMO/OMN/BMO no.385.*) ISBN: 9263003858.
1,277 English-base terms, with Russian, French and Spanish equivalents (identified by serial numbers), plus UDC numbers. Subject scope: hydrology, surface water, hydrometeorology, soil moisture, hydraulics, water resources management. List of abbreviations of sources precedes. Appendix B: symbols and units.
Class No: 556(038)=00

English

[1768]
VEATCH, J.O. *and* **HUMPHREYS, C.R. Water and water use terminology.** Kaukauna, Wisconsin, Thomas Printing and Publishing, 1966. xvi, 375, [v]p. illus.
About 1,500 definitions in terms relating to lakes, surface water and wetlands, with references appended. Numerous cross-references. Good and plentiful illus. General bibliography of 7 items. *Class No:* 556(038)=20

Scandinavian

[1769]
Nordic glossary of hydrology: English - Danish - Finnish - Icelandic - Norwegian - Swedish, with definitions in English. Johansson, I., *ed.* Stockholm, Almqvist, 1984. 224p. £22.50. ISBN: 9122006923. *Class No:* 556(038)=395

Russian

[1770]
DEEV, G.N. *and* **LOSEV, K.S. Anglo-russkii gidrologicheskii slovar'.** Moscow, Izdat. 'Sovetskaya Entsiklopediya', 1966. 299p. illus., diagrs.

.... (contd.)
14,000 English hydrological terms, with Russian equivalents, Russian word index, p.267-99.
Class No: 556(038)=82

Reviews & Abstracts

[1771]
Hydro-abstracts. Minneapolis, Minnesota, Environmental Hydrology Corpn., 1980-. 12pa. $120.
Formerly *Water resources abstracts.*
About 2,000 abstracts pa., coverage including reports, proceedings, bulletins and other monographs. 40 sections (*e.g.* 1. Properties of water; 8. Groundwater; 15. Saline water conversion; 25. Identification of water pollutants; 31. Water quality control; 35. Water demand; 40. Data acquisition). Abstracts state key terms. Three 5" x 3" entries per page, for filing. *Class No:* 556(048)

[1772]
Selected water resources abstracts (SWRA). Reston, Va., Water Resources Scientific Information Center, U.S. Geological Survey, 1968-. 12pa. $145. ISSN: 0037136x.
About 2,000 abstracts per issue. Subject fields and groups: 01. Nature of water. 02. Water cycle; 03. Water supply augmentation and conservation; 04. Water quantity management and control; 06. Water resources planning; 07. Resources data; 08. Engineering works; 09. Manpower, grants and facilities; 10. Scientific and technical information. Subject and author indexes; organizational index; accession number index. Available on CD-ROM and also online, via DIALOG. *Class No:* 556(048)

Progress Reports

[1773]
Advances in hydroscience. Orlando, Florida, Academic Press, 1964-.
V.14, 1986 (x, 344p.). has 4 contributors. Three well-documented articles: 'Hydraulics of sewers' (7½p. of references) - 'Passive microwave remote sensing of soil moisture' - 'Modeling of unsteady open-channel flow'. Index, p.335-44. *Class No:* 556(055)

Tables & Data Books

[1774]
The Water encyclopedia. Leeden, F. van der, *and others, eds.* Chelsea, Mich., Lewis Publishers, 1990. 808p. illus., tables, maps. $149.95. (*Geraghty and Miller groundwater series.*) ISBN: 0873711203.
First published 1970.
Substantially revised and expanded from previous ed. More than 700 tables, diagrams, and charts. 11 chapters (*e.g.* 'Climate'; 'Surface and ground water'; 'Water management'). Subject index. US-slanted.
Class No: 556(083)

Histories

[1775]
BISWAS, A.K. History of hydrology. Amsterdam, etc., North-Holland Publishing Co., 1970. xxi, 336p.
Documented chapters (Hydrology prior to 600 BC ... Rain gauges in the 17th, 18th and 19th centuries). Subject index.

....*(contd.)*

The first comprehensive account up to the end of the 19th century, although more space could have been devoted to the period during the 19th century when the first countrywide instrument networks were being installed (*Nature*, v.228, 12 September 1970, p.1119). *Class No:* 556(091)

Europe

[1776]

International hydrogeological map of Europe. Paris, UNESCO, with Bundesanstalt für Geowissenschaften und Rohstoffe, Hanover, 1970-. Maps.

49 sheets, in up to 55 colours, plus 1 general sheet. Legend in English, French and German, but notes in English only. Sheet size: 63 x 63cm. *Class No:* 556(4)

Great Britain

[1777]

Hydrological data United Kingdom: 1989 yearbook: an account of rainfall, river flows, groundwater levels and river quality, January to December 1989. Wallingford, Oxon., Institute of Hydrology, 1990. 200p. tables. £18. ISBN: 0948540257.

First published 1985. Contains data previously published in the discontinued *Surface water: United Kingdom* and *Groundwater: United Kingdom.*

8 sections: 1. Scope and sources of information - 2. Hydrological review - 3. The 1988/89 drought - 4. River flow data, p.45-142 - 5. The surface water data retrieval service - 6. Groundwater level data - 7. The groundwater data retrieval service - 8. Surface water quality data. Directory of measuring authorities, p.196-8.

Every five years, a catalogue of river flow gauging stations and groundwater level recording sites together with statistical summaries, is published. *Class No:* 556(410)

Rivers & Lakes

Worldwide

[1778]

CZAYA, E. Rivers of the world. Translated from the German by S. Furness. Cambridge, University Press, 1983. 248p. illus. (incl. col.), diagrs., tables, maps. $22.95. ISBN: 0521258359.

8 chapters: 1. Arteries of the continents - 2. Rivers shape their valley - 3. Cataclysmic events in river history - 4. Waterfalls and rapids - 5. Pluvial lakes - 6. Inland drainage - 7. Lowland rivers and river mouths - 8. Rivers in harness. Bibliography, p.233-5 (authors, A-Z). Detailed, non-analytical index, p.236-48. 192 illus. A mass of factual information, clearly presented. *Class No:* 556.5(100)

[1779]

Discharge of selected rivers of the world. Paris, UNESCO, 1969-86. 3v. in 6. £3.75, v.3, pt.3; £8.25, v.3, pt.4. (*Studies and reports in hydrology, 5.*)

1. *General and regime characteristics of stations selected.* 1969. 70p. 2. *Monthly and annual discharges recorded at various selected stations [up to 1964].* 3. *Mean monthly and extreme discharges.* 4 pts. [covering 1965-79].

Title and text in English, French, Spanish and Russian. *Class No:* 556.5(100)

[1780]

GRESSWELL, R.K. *and* **HUXLEY, A. Standard encyclopedia of the world's rivers and lakes.** London, Weidenfeld & Nicolson, 1962. 384p. illus. (pl.), maps.

40 contributors. Main section consists of nearly 500 entries, each of *c.*200 words. Supported by 11 location maps, and extensive gazetteer (with data on rivers, etc., length, course, etc.) and an index. Well illustrated; 16 colour plates. 'The maps let the book down' (*Geographical journal*, v.12(3), September 1966, p.425): distance scales and projections are not given. *Class No:* 556.5(100)

[1781]

Rand McNally encyclopedia of world rivers. Chicago, Ill., Rand McNally, 1980. 352p. illus. (incl. col.), maps. ISBN: 0528810680.

Data on 1,750 rivers, continent by continent. Profiles state source, length, tributaries, natural features, dams, hydroelectric power stations, industrial activity, agriculture, fauna and flora, history, etc. Major rivers are treated in depth. No index or bibliographies. *Class No:* 556.5(100)

Great Britain

[1782]

British rivers. Lewin, J., *ed.* London, Allen & Unwin, 1981. [x], 216p. illus., diagrs., graphs, tables, maps. ISBN: 0045510474.

7 contributors. 6 chapters: 1. River systems and river regimes - 2. Contemporary erosion and sedimentation - 3. Mountain streams - 4. Channel forms and channel changes - 5. Water quality - 6. River management. References, p.196-211. Analytical index. 'This book is intended for a wide readership' (*Preface*). *Class No:* 556.5(410)

[1783]

—**MUIR, R.** *and* **MUIR, N. Rivers of Britain.** Exeter, Devon, Webb & Dower, in association with Michael Joseph, London. 1986. 225p. illus., (incl. col.). ISBN: 0863501109.

11 sections (3. Rivers of the Wetlands ... 5. Rivers in service - 6. To get to the other side: bridges; fords and ferries ... 10. Rivers in danger - 11. River management - which way forward). Includes notes on flora and fauna, canals, boxed information on nature reserves, ports, villages, buildings, etc. Partly analytical index. No bibliography. *Class No:* 556.5(410)

America — North

[1784]

Rolling rivers: an encyclopedia of America's rivers. Bartlett, R.A., *ed.* New York, etc., McGraw-Hill, 1984. 398p. illus. $36.50. ISBN: 0070039100.

Signed, documented essays on 117 major North American rivers, with their tributaries and other rivers. Data head each essay: origin, length, geologic history, discovery and settlement, events in its history, dams, reservoirs, agricultural products and industries, etc. Essays fall into 6 geographical sections. The collection lacks maps and the index fails to include tributaries of most rivers, events, and the like 'There is no other work that covers American rivers at a comparable level of detail' (*RQ*, v.24(1), Fall 1984, p.109-110). *Class No:* 556.5(71+73)

56 Palaeontology (Fossils)

Palaeontology

Bibliographies

[1785]
LUM, A. 'Palaeontology'. in *Information sources in the earth sciences,* edited by D.N. Wood *and others.* 2nd ed. London, Bowker-Saur, 1989. p.236-273.

Readable and authoritative review of major works in the subject. Headings, after general historical survey, are: Taxonomy - Palaeontological texts (including techniques, and popular up to advanced textbooks) - Palaeoecology & palaeoclimatology - Evolution & the fossil record - Micropalaeontology - Palynology - Palaeobotany - Invertebrate palaeontology - Vertebrate palaeontology (including brief review of popular books on dinosaurs) - Abstracting & indexing services. Appendix lists 132 currently published palaeontological periodicals. The author is on the staff of the Palaeontology library at the British Museum (Natural History). *Class No:* 560(01)

Encyclopaedias

[1786]
FAIRBRIDGE, R.W. *and* JABLONSKI, D., *eds.* **The Encyclopedia of palaeontology.** Stroudsburg, Pa., Dowden, Hutchinson and Ross, 1979. xii, [1], 886p. illus., tables, maps. (*Encyclopedia of earth sciences, v.7.*)

122 contributors. Most of the signed articles carry bibliographies. 'Vertebrate palaeontology' (p.843-51) has sections: 'Preservation of vertebrate fossils'; 'Fossil vertebrates and organic evolution'; 33 references; 19 cross-references. 'The entries are written at several conceptual levels, from the most basic and inclusive articles such as *Invertebrate Paleontology* and *Evolution,* to specialized, technical treatments such as *Morphology, Constructional* and *Taxonomy, Numerical'* (*Preface*). The index, partly analytical, has *c.*5,000 entries. According to *Bulletin des bibliothèques de France* (July 1981, entry 909), the *Encyclopedia* almost entirely ignores the work done in any language other than English. Thus, Lamarck is omitted in the 'History of palaeontology before Darwin' article. *Class No:* 560(031)

[1787]
—The Macdonald encyclopedia of fossils. Arduini, P. *and* Teruzzi, G. London, Macdonald, 1986. 317, [3]p. col. illus., diagrs. £9.95. ISBN: 035612367x.

Has text and illus. on 260 fossils; Plants, entries 1-16; invertebrate fossils, 17-186; vertebrate fossils, 187-260. Text, including symbols for geological eras and habitats, faces text. Glossary, p.307-17. Index. A neat, small-format guide. *Class No:* 560(031)

[1788]
—STEEL, R. *and* HARVEY, A.P. The Encyclopedia of prehistoric life. London, Mitchell Beazley, 1979. 218p. illus., charts. ISBN: 0070609209.

Has 23 contributors (13 of them at the British Museum (Natural History)). The initialled articles - 'Academia Sinica' to 'Zone fossils' - include biographies but lack bibliographies, although a grouped bibliography appears on p.213. Glossary, p.214-5; index (in tiny print), p.216-8. Careful, captioned line-drawings (nearly 300). 14p. of coloured evolutionary charts (Invertebrates ... Primates). An attractively-produced, popular treatment. *Class No:* 560(031)

Handbooks & Manuals

[1789]
FORTEY, R. Fossils: the key to the past. 2nd ed. London, British Museum (Natural History), 1991. 187p. illus. (incl. col.). £12.95. ISBN: 0565011073.

First published 1982.

Provides a lively yet authoritative, fully-illustrated survey for the amateur palaeontologist. The 10 chapters include 'How to recognise fossils', 'Making a collection' (with notes on collecting, preserving and identifying), and a new chapter recounting the identification of a new dinosaur from recently-discovered fossil remains. 'In this book I have tried to show how fossils, far from being mere dry bones, can be used to reconstruct the history of the Earth' (*Preface*). *Class No:* 560(035)

[1790]
KUMMEL, B. *and* RAUP, D. Handbook of palaeontological techniques. Prepared under the auspices of the Palaeontological Society. San Francisco & London, W.H. Freeman, 1965. xiii, 852p.

80 contributors. 5 parts: 1. General procedures and techniques applicable to major fossil groups - 2. Description of specific techniques - 3. Techniques in palynology - 4. Bibliography on palaeontological techniques (collecting; labeling and storage; preservation ...) - 5. Compilation of bibliographies (p.768-832; 739 unannotated entries). *Class No:* 560(035)

[1791]
LAMBERT, D. *and* THE DIAGRAM GROUP. The Cambridge field guide to prehistoric life. Cambridge University Press, 1985. 256p. illus., diagrs. £27.95.

'A concise key to prehistoric animals, plants and other organisms' (*Foreword*). 10 chapters: 1. Fossil clues to prehistoric life - 2. Fossil plants - 3. Fossil invertebrates - 4. Fossil fishes - 5. Fossil amphibians - 6. Fossil reptiles - 7. Fossil birds - 8. Fossil mammals - 9. Records in the rocks - 10. Fossil hunting. 'Further reading', by chapters, p.248; index, p.249-56. Marginal 'field guide' illus.; a neat production. Uses both popular and scientific terms and a balanced account aimed at the enquiring eleven-year old as well as the budding scientist. *Class No:* 560(035)

[1792]

—CROUCHER, R. *and* WOOLLEY, A.R. Fossils, minerals and rocks: collection and preservation. London, British Museum (Natural History), with Cambridge Univ. Press, 1982. 60p. illus., diagrs. £4.25. ISBN: 0521247365.

Devotes p.27-52 to fossil collection and preservation. The other short sections concern field equipment, information sources (guides and memoirs; journals: books; maps), minerals and rocks. Appended: 'Further reading and sources of reference'; 'Materials, manufacturers and suppliers'; 'Addresses'. Analytical index. *Class No:* 560(035)

[1793]

—Fossils of the world: a comprehensive, practical guide to collecting and studying fossils. Turek, V., *and others.* London, Hamlyn, 1988. 495p. illus. (col.). ISBN: 0600550141.

Consists mainly (p.36-483) of colour illustrations of fossils, arranged into plant, protista, and animal, kingdoms. The last is sub-divided into porifera, archaeocyatha, coelenterata, bryozoa, brachiopoda, mollusca, annelida, arthropoda, echinodermata, hermichordata, and chordata. Nearly all the specimens are of central European origin, and the picture quality is excellent. The book's section on collecting fossils is perhaps less valuable (p.484-6) and there is a 1-page bibliography. *Class No:* 560(035)

[1794]

—NIELD, E.W. *and* TUCKER, V.C.T. Palaeontology: an introduction. Oxford, Pergamon Press, 1985. xx, 178p. illus., diagrs. £28. ISBN: 0080238548.

12 chapters surveying the field (*e.g.* 2. The origin of life and the earliest fossils ... 11. Vascular plants - 12. The meaning of fossils) with 'Suggested further reading', plus brief evaluative notes. Author, systematic and non-analytical subject indexes. Fine line-drawings, keyed/captioned, are a feature. Quarto-format. Designed for students embarking upon a study of palaeontology. *Class No:* 560(035)

[1795]

MACFALL, R.D. *and* WOLLIN, J.C. Fossils for amateurs: a handbook for collectors. Rev. ed. New York, Van Nostrand Reinhold, 1983. 374p. illus., maps. $24.95; $19.95. ISBN: 0442263481.

First published 1972.

Expansion and updating of the 1972 ed. with a lengthy appended list of books and magazines, and an adequate index. Relates fossils to other aspects of geology. 'Should be in every fossil collector's library' (*Technical book review index*, v.50(6), June 1984. p.207). *Class No:* 560(035)

[1796]

NEAVERSON, E. Stratigraphical palaeontology: a study of ancient life-provinces. 2nd ed., rev. and enl. Oxford, Clarendon Press, 1955. xii, 806p. illus. (incl. pl.).

First published 1928.

The introduction notes some of the more important works, and the 18 chapters each have an evaluated reading list appended. 90 illus., apart from 18 plates. Palaeontological, stratigraphical and geographical indexes, p.741-806. A standard work. *Class No:* 560(035)

[1797]

PIVETEAU, J. Traité de paléontologie; publié sous la direction de J. Piveteau. Paris, Masson, 1952-69. 7v. in 10, illus., diagrs., tables.

1. *Les stades inférieurs du règne animal. Généralités. Protistes. Spongiaires. Coelentérés. Bryozoaires.* 1952. 782p. FFr653.

2. *Problèmes d'adaptation et de phylogenèse. Brachiopodes. Chétognathes. Annélides. Mollusques.* 1952. 790p. FFr653.

3. *Les formes ultimes d'invertébrés: morphologie et évolution. Onychophores. Anthropodes. Échinodermes. Stomochordés.* 1953. 500p. FFr848.

4. *Agnathes. Placodermes, etc.* 1964-69. 3pts. (pts.1-2. FFr1,032).

5. *La sortie des eaux. Naissance de la tétrapodie, L'exubérance de la végétative. La conquête de l'air. Amphibiens. Reptiles. Oiseaux.* 1955. 114p. FFr830.

6. *L'origine des mammifères et les aspects fondamentaux de leur évolution. Mammifères (évolution).* 1958-61. 2v. FFr1,553.

7. *Vers la forme humaine. Le problème biologique de l'homme. Les époques de l'intelligence. Primates. Paléontologie humaine.* 1957.

Each chapter, by a specialist, covers systematic and biological aspects of fossil groups, their ecology and the evolution of species. Vertebrates, especially mammals and humans, are given particular attention. An important and comprehensive work of reference. *Class No:* 560(035)

[1798]

STEARN, C.W. *and* CARROLL, R.L. Paleontology: the record of life. New York, Wiley, 1989. 453p. illus. £52. ISBN: 0471845280.

Arranged into 3 parts: 1. Introduction - 2. The fossil record - 3. Lessons from the record. Well illustrated with photographs and line drawings, the book has a North American bias (the authors are Canadian) and 'the book is designed as a textbook for a first course in paleontology ... for a typical student who has a general science education up to college level but little knowledge of geology'. Suggested reading after each chapter, and an index, p.433-53. *Class No:* 560(035)

Reviews & Abstracts

[1799]

Geographical abstracts. Physical geography. Norwich, Elsevier; Geo Abstracts, 1989-. £420pa. ISSN: 09540504.

Formerly *Geo abstracts. A. Landforms and the Quaternary*, 1966-85 and *Geographical abstracts. A. Landforms and the Quaternary*, 1986-88. Merged with other titles in the series - B. *Climatology and hydrology;* E. *Sedimentology;* G. *Remote sensing, photogrammetry and cartography.* Continues *Geological abstracts: palaeontology and stratigraphy.*

This journal, which abstracts 'over 1,000 leading geographical journals, plus books, conference proceedings, reports and theses' (*Editorial introduction*) includes in its subject coverage such relevant headings as Palaeogeography; Sedimentation and tectonics; Systematic palaeoecology; Quaternary. Mainly author abstracts listed, up to 300 words each. 12 issues pa., each with regional index, plus separately published index issue arranged by subject, authors, and geographical areas. Online GEOBASE, available on DIALOG. *Class No:* 560(048)

[1800]
PASCAL folio. F47: Paléontologie. Vandoeuvre-Les-Nancy, Centre National de la Recherche Scientifique, 1985-. 10pa. FFr815.

Previously as part of *Bulletin signalétique*. 11: *Sciences de la terre*. 2. *Physique du globe. Géologie. Paléontologie* and *Bulletin signalétique. Sciences de la terre*. 227: *Paléontologie* (1972-83).

About 4,000 abstracts and references pa. Main sections: 01. Paléontologie: généralités - 02. Paléobotanique - 03. Paléontologie des invertébrés - 04. Paléontologie des vertébrés. Appended list of new serials, dissertations, new books. Author, subject, geographical and palaeontological indexes, plus English-language subject index, per issue and annually. Available online. *Class No:* 560(048)

[1801]
Zentralblatt für Geologie und Paläontologie. Teil 2: Paläontologie. Stuttgart, E. Schweizerbart'sche Verlagshandlung, 1950-. 7pa. $46ea. ISSN: 00444189.

Formerly *Zentralblatt für Geologie und Paläontologie. Teil 2: Historische Geologie und Paläontologie*.

Contains *c*.900 abstracts pa. 3 main sections: Allgemeine Paläontologie - Paläozoologie (Evertebrata; Vertebrata) - Paläobotanik. Each bimonthly issue carries one or more review articles and a subject index. Annual author index. *Class No:* 560(048)

[1802]
—The Zoological record. York, BIOSIS, UK, and London, Zoological Society, 1865 (covering 1864)-. v.1-. 20 annual parts. ISSN: 01443607.

Includes palaeontology in its coverage, each of the annual parts carrying a palaeontological index.

Described as 'the most important life science database for palaeontology' (A. Lum, in D.N. Wood's *Information sources in the earth sciences* (2nd ed. 1989), p.269). Online via DIALOG. *Class No:* 560(048)

Tables & Data Books

[1803]
Fossilium catalogus. Berlin, W. Junk, 1913/14-(with reprints). 2 series.

1. *Animalia*. Edited by F. Frech, and others. 1913-. (pt.127. 1983).

2. *Plantae*. Edited by W.J. Jongmans, and others. 1914-. (pt.91. 1985).

Each of the two series comprises catalogues by specialists on individual fossil orders or genera (*e.g. Bryozoa (genera)*, by R.S. Bassler, *Fossilium catalogus*, series 1, pt.67. 1934), with index and bibliography. Series 1. pt.72, 1938, *Catalogus bio-bibliographies* is a worldwide list of palaeontologists, past and present. 'By far the most comprehensive and most important index to published generic and specific names' (Lum, A. 'Palaeontology'; In Wood, D.N., and others, eds. *Information sources in the earth sciences*, 2nd ed. London, Bowker-Saur, 1989, p.236-73). *Class No:* 560(083)

Maps & Atlases

[1804]
HALLAM, A., *ed*. Atlas of palaeobiogeography. Amsterdam, Elsevier, 1973. xii, 531p. illus., maps. $154. ISBN: 0444409750.

47 contributions, by specialists, on distribution of a wide variety of fossil groups during the last 600 million years (*e.g.* Ondovician trilobites; articulate brachiopods; graptolites; corals; conodonts, p.13-58. Jurassic plants, p.329-38; Jurassic and cretaceous dinosaurs, p.339-52). Chapter bibliographies. Index of genera and references. Authoritative; assumes some geological and palaeontological knowledge. *Class No:* 560(084.3)

Histories

[1805]
EDWARDS, W.N. The Early history of palaeontology. London, British Museum (Natural History), 1967. vii, 58p. illus., facsims. £1.50. ISBN: 0565006584.

Based on *Guide to an Exhibition illustrating the early history of palaeontology* (British Museum (Natural History). 1931. Special guide no.8).

Succinct account up to the early 19th century, plus chapters on geological terminology, the rise of museums and the Department of Palaeontology, British Museum (Natural History). Illus. include 4 plates, mainly from early books. *Class No:* 560(091)

Biographies

[1806]
Dictionary of palaeontologists of the world. Doescher, R.A., *ed*. 5th ed. Washington, International Paleontological Association, 1990. 447p. $20.

4th ed., 1984, *Directory of paleontologists*.

Arranged A-Z. Data: name; address; telephone number; subject specialties. Taxonomical index; geographic index. Includes China and Russia. *Class No:* 560(092)

[1807]
Palaeontologi. Catalogus bio-bibliographicus. In *Fossilium catalogus*. 1. Animalia. Pars 72. 1938. Lambrecht, K., *and others*. xxii,495p. $40.

Reprinted 1978 as *Paleontologi: a biographical and bibliographical register of paleontologists* (Salem, N.H. Ayer).

'Contains sentence-length biographical notes and references to biographies and memorials of over three thousand palaeontologists and geologists of all countries; it is an important reference work and the point of beginning for biographical searching for non-Americans and even for some Americans' (*Stechert-Hafner book news*, v.18(3), November 1963, p.30). A major source of information, not limited to palaeontologists; includes some geologists. *Class No:* 560(092)

Great Britain

[1808]
BRITISH MUSEUM (NATURAL HISTORY). British Caenozoic fossils (Tertiary and Quaternary). 5th ed. London, British Museum (Natural History), 1975. vi, 132p. illus. (pl.), map. £4.95. ISBN: 0565055402. *Class No:* 560(410)

[1809]
—BRITISH MUSEUM (NATURAL HISTORY). British Mesozoic fossils. 6th ed. London, British Museum (Natural History), 1983. xi, 207p. illus. (pl.), map. £4.95. ISBN: 0565008722. *Class No:* 560(410)

[1810]
—BRITISH MUSEUM (NATURAL HISTORY). British Palaeozoic fossils. 4th ed. London, British Museum (Natural History), 1975. vi, 203p. illus. (pl.), map. £4.95. ISBN: 0565056247.

Each volume includes explanatory notes on the plates, a coloured distribution-map, bibliography and index. Simply written, attractively produced and inexpensive booklets, 'to enable the young, or those without experience, to know about fossils they may expect to find and to identify for themselves those they have collected' (Preface to *British Mesozoic fossils*). *Class No:* 560(410)

[1811]
PALAEONTOGRAPHICAL SOCIETY. Directory of British fossiliferous localities. London, the Society, 1954. xiv, 268p.

Three sections: England; Wales; Scotland. Arrangement is A-Z by counties, then towns, in each section. A guide to some 2,000 sites. Each entry states exact location, 1-inch Geological Survey and Ordnance Survey (New popular ed.) map-references, and a brief summary of geological and palaeontological data. Appended is a list of section references. Now much dated for South, South East and Midland England, but still of value for remote areas. *Class No:* 560(410)

America—North & Central

[1812]
THOMPSON, I. The Audubon Society field guide to North American fossils. New York, Knopf, 1982. 846p. illus. (incl. col.), diagrs., geol. maps. $17.95. ISBN: 0394524128.

Fossils found in the US have priority. Over 500 excellent colour plates. 'The geologic maps are less successful' and text descriptions are difficult for the casual collector to follow (*Library journal*, 1 March 1983, p.490). Appendices; index. Addressed to the serious amateur palaeontologist. *Class No:* 560(71/73)

Museums

[1813]
CLEEVELY, R.J. World palaeontological collections. London, British Museum (Natural History) and Mansell, 1983. 365p. £50. ISBN: 0565008501.

Main contents: Bibliography (p.17-23: the history of palaeontology and fossil collecting; Collections and collecting; Biographical and other reference sources) - Bibliography of published catalogues of cited material - Alphabetical list of collections (persons, museums, etc., p.38-328) - Index of museums and collection holdings (British Isles; Europe; N. America; Latin America; Australia and New Zealand; Africa; Asia (including Middle East)). A revision of C.D. Sherborn's *Where is the ... collection?* (1940). The Cleevely directory is, however, limited to fossil (and mineral) collections and excludes 'rock' collections (for which see *World directory of mineral collections* (2nd ed., 1977)). *Class No:* 560:061:069

[1814]
—BASSETT, M.G. 'Bibliography and index of catalogues of type, figured and cited fossils in museums in Britain' *Palaeontology*, v.18(4), 1975, p.753-73).

Lists 109 catalogues, with taxonomic, stratigraphic and museum indexes. Supplementary details of reference aids for the curator/researcher who is trying to locate old collections and specimens. Prepared as an aid to tracing type collections and individual specimens. *Class No:* 560:061:069

Palaeoclimatology

[1815]
CROWLEY, T.J. *and* NORTH, G.R. Paleoclimatology. Oxford, University Press, 1991. 339p. diagrs. £45. ISBN: 0195039637.

Aimed at graduate-level readers, this book summaries observational and modelling studies as aids to understanding climatic change. Includes detailed list of scientific references (p.277-322), but the authors work on the possibly controversial assumption that the Cretaceous period ended with the impact of an extra terrestrial body, such as a comet or meteor. 'Those who wish to anticipate future climatic change will be well served if they read *Paleoclimatology*' (*Science,* 16 August, 1991). *Class No:* 560:55

[1816]
SCHWARZBACH, M. Climates of the past: an introduction to paleoclimatology. Translated and edited by R.O. Muir. Princeton, N.J., etc., Van Nostrand, [1963] (reprinted 1977). xii, 328p. illus. diagrs., charts, tables, maps. $24. ISBN: 0404162150.

Translated from the 2nd completely revised and enlarged German ed.

The standard work on the subject. Supported by an extensive bibliography, p.265-300. *Class No:* 560:55

Palaeobiology

[1817]
BRIGGS, D.E.F. *and* CROWTHER, P.R. Palaeobiology: a synthesis. Oxford, Blackwell, 1990. 583p. illus. £89.50. ISBN: 0632025255.

Comprises 120 chapters by a large number of international experts on all aspects of the subject, arranged into following subdivisions - 1. Major events in the history of life - 2. Evolutionary process and the fossil record - 3. Taphonomy - 4. Palaeoecology - 5. Taxonomy, phylogeny, and biostratigraphy - 6. Infrastructure of palaeobiology. 'Each chapter stands on its own as a well-crafted analysis of a portion of science ... highly recommended for any undergraduate or university collection' (*Choice*, April, 1991, p.1336). *Class No:* 560:57

Palaeoecology

[1818]
LADD, H.S., *ed.* **Treatise on marine ecology and paleoecology. Volume 2: Paleoecology.** New York, Geological Society of America, 1957. x, 1077p. illus., maps. (*Memoir of the Geological Society of America.*)

Important for its 'Annotated bibliography of marine palaeoecology' (p.691-1032), with 54 divisions (*e.g.* 'Diatoms': introduction; 27 well-annotated references. 'Nautiloids': introduction, p.829-46; bibliography of 44 items). Index, p.1033-77. *Class No:* 560:574

[1819]
—The Ecology of fossils: an illustrative guide. McKerrow, W.S., *ed.* London, Duckworth, 1978. 384p. illus., diagrs., maps. ISBN: 0715609440.

7 contributors. Following introduction and classification of organisms, chapters deal with geological periods (Precambrian ... Cretaceous - Cenozoic - present day). Glossary, p.365-9; references, p.370-6. Non-analytical index in tiny print. 125 illus. and diagrs. *Class No:* 560:574

Micropalaeontology

[1820]
Bibliography and index of micropaleontology. New York, Micropaleontology Press, American Museum of Natural History, 1972-. v.1-. Monthly. $74; $550 (corporate). ISSN: 03007227.

About 2,500 references pa., in 12 classes. (01. Algae ... 06. Foraminifera ... 09. Palynomorphs ... 11. Radiolarea - 12. Miscellaneous (including papers that mention more than one microfossil group)). Entries record illus., sheet maps, tables and number of references. Author index per issue and annually. Highly relevant to petroleum exploration. Online, GEOREF. *Class No:* 560.086

[1821]
BRASIER, M.D. **Microfossils.** London, Allen & Unwin, 1980. xii, 193p. diagrs. £16.45. ISBN: 0045620024.

10 documented chapters on general aspects of micropalaeontology. Includes naming of parts; naming and classification of specimens; general history of groups; geological applications; preparatory methods. Bibliography, p.169-175. Appendix on reconnaissance methods, p.162-78. Systematic and general indexes. 'Invaluable and long-awaited textbook' (Nield, E.W. and Tucker, V.C.T. *Palaeontology: an introduction* (Pergamon Press, 1985, p.164)). *Class No:* 560.086

[1822]
POKORNÝ, V. **Principles of zoological micropalaeontology.** Allen, K.A., *trans and* Neale, J.W., *ed..* Oxford, Pergamon Press, 1963-65. 2v.

Translated from the German, *Grundzüge der zoologischen Mikropaläontologie* (Berlin, VEB Deutsche Verlag der Wissenschaften, 1958. 2v.).

18 chapters on micropalaeontology, concentrating on a systematic approach (*e.g.* Foraminifera, v.1 p.91-471; Ostracoda. v.2, p.69-348). Chapter bibliographies at end of each volume, plus English - French - German - Russian glossary of terms. V.1 has fossil, subject and author indexes;

.... (contd.)
v.2 has author and subject indexes only. 'Intended as a textbook and as a manual for practising micropalaeontologists' (*Foreword*). *Class No:* 560.086

Palaeobotany

[1823]
ANDREWS, H.N., *Jr.* **Index of generic names of fossil plants, 1820-1965.** Washington, US Geological Survey, 1970. 354p. (*US Geological Survey bulletin 1300.*)

Revises and updates *Andrews' Index ... 1820-1950.* 1955 (bulletin 1013).

Based on the US Geological Survey's working Compendium index of Paleobotany, now comprising over 200,000 cards. Gives full coverage of all known genera and species of fossil plants, but omits diations, spores and pollen. Other groups of microfossils are dealt with, but not completely. Full references for all citations. Not a critical study, but 'intended rather as an information source concerning the origin of the respective generic contents' (*Geographical abstracts*, 1971/4, abstract 71B (1978)).

Supplement, A.M. Blazer, *Index of generic names of fossil plants, 1966-1973* (US Government Printing Office, 1975. 54p.). *Class No:* 561

[1824]
BOUREAU, E. **Traité de paléobotanique.** Paris, Masson, 1964-. v.1-. illus., tables.

A comprehensive treatise, planned in 9v.:

1. *Introduction. Généralités Thallophyta.*
2. *Bryophyta. Psilophyta. Lycophyta.* 1967. 846p. FFr990.
3. *Sphénophyta, Noeggerathiophyta.* 1964. 544p. FFr700.
4(1). *Filicophyta.* 1970. 520p. FFr646.
4(2). *Pteridoyphylla.* 1975. 772p. FFr1,510.
5. *Périodospermophyta.*
6. *Glossopteriodophyta. Cycadophyta. Cordaitophyta. Ginkgophyta. Coniferophyta.*
7. *Chiamydospermophyta, Angiospermophyta.*
8. *Sporologie.*
9. *Paléophytogéographie. Conclusions générales du traité.*

V.4(1), ([iv], 519p.) has 378 figures, section bibliographies and a partly analytical index. A standard work. *Class No:* 561

[1825]
—BOUREAU, E. Rapport sur la paléobotanique dans le monde, 1950/1974. [World report on palaeobotany, 1950/1974.] Utrecht, Oosthoek's Uitg. B.V., 1956-73. 9v. (*Legnum vegetabile, 7, etc.*)

Edited by the International Organization of Palaeobotany and published by the International Bureau for Plant Taxonomy and Nomenclature. V.9 (1973) carries 1,647 references in 5 sections (1. Généralités, biographies, méthodes, techniques, travaux généraux... 5. Tertiaire (et Quaternaire)). No index, but does include a directory of *c.*1,148 palaeobotanists and their addresses. *Class No:* 561

[1826]

ERDTMAN, G. **Handbook of palynology:** morphology - taxonomy - ecology; with 'Appendix on groups other than spores and pollen', by W.A.S. Sarjeant. Riverside, N.J., Hafner, 1968. 486p. illus. $36.95. ISBN: 0028442504.

Bibliography, p.245-6. A comprehensive textbook and a corrective to those handbooks of palynology that 'very often prove to be concerned exclusively with pollen and spores' (Sarjeant, W.A.S. and Harvey, A.P. in Wood, D.N. *Use of earth sciences literature* (1973), p.269). *Class No:* 561

[1827]

HUNTLEY, B. *and* BIRKS, H.J.B. **An Atlas of past and present pollen maps for Europe 0-13000 years ago.** Cambridge Univ. Press, 1983. 667p. + 34 overlaps as a separate volume, maps.

Maps show relative percentage for 40-50 species of pollen at various intervals of time over the past 13,000 years. Included are 'maps of vegetational reconstruction and of refugial areas' (*Choice*, v.21(10), June 1984, p.1446). *Class No:* 561

[1828]

MEYEN, S.V. **Fundamentals of palaeobotany.** London, Chapman & Hall, 1987. xxii, 432p. diagrs. ISBN: 0412271109.

Translated from Russian under author's supervision.

8 chapters: Preservation types and techniques of study of fossil plants - Principles of typology and of nomenclature of fossil plants - Fossil plants systematics - Palaeopalynology - Epidermal-circular studies - Plant palaeoecology - Palaeofloristics - Relationship between palaeobotany and other fields of natural history. Good reference bibliography, p.383-414. Index covers illustrations as well as text. 'A useful basic text in any university college or public library' (*New technical books,* January/February 1988, item 2987). *Class No:* 561

[1829]

TRALAU, H. **Bibliography and index to palaeobotany and palynology, 1950-70.** Stockholm, the Author, Stockholm Museum of Natural History, 1974. 2v.

About 23,000 references under authors in v.1, v.2 being a KWIC index. A continuation covering 1971-75 (2v.) appeared in 1983. *Class No:* 561

Fossil Invertebrates

[1830]

CLARKSON, E.N.K. **Invertebrate palaeontology and evolution.** 3rd ed. London, Allen & Unwin, 1993. ix,434p. illus., diagrs., maps. £19.95. ISBN: 0412479907.

First published 1979.

Documented chapters; 2 parts. Part 1: General palaeontological concepts (chapters 1-3) - Part 2: Invertebrate phyla (chapters 4-12). Chapter 11: 'Arthropods', p.339-95. 27 figures; 5p. of bibliography. Detailed analytical index, p.425-34. Numerous carefully drawn examples. Systematic index, p.417-24. 'A course book for second and third year students reading Geology or Earth Science' (*Preface*). *Class No:* 562

[1831]

MOORE, R.C., *ed*. **Treatise on invertebrate paleontology.** Prepared under the guidance of the Joint Committee on Invertebrate Paleontology. New York, Geological Society of America; Laurence, Kansas, Univ. of Kansas Press, 1953-.

Sponsored by the Geological Society of America, the Paleontographical Society, the Paleontological Society and the Society of Economic Paleontologists and Mineralogists.

1. A. *Introduction.* B-D *Protista* (C. pt.2, 1964; D. pt.3, 1954). E. *Archaeocyatha, porifera.* 1955. Rev. ed. 1972. F. *Coelenterata.* 1956. G. *Bryozoa.* 1953. H. *Brachiopoda.* 1965. 2v. I-N. *Mollusca* (I. 1960; K. 1964; L. 1957; N. 1969-71. 3v.). O-R. *Arthropoda.* 1955-69. 5v. S-U. *Echinodermata* (S. 1967. 2v; U. 1966. 2v.). V. *Graptolithina.* 1955. 2nd ed. 1970. W. *Miscellanea.* 1962. Suppt. 1: *Trace fossils and problematica.* 2nd ed., 1975. Suppt. 2. 1981. X. *Addenda. Index.*

Aims 'to present the most comprehensive statement of knowledge concerning invertebrate fossil groups' up to 1950 (*Editorial preface*). Each volume contains detailed descriptions; section references; detailed index. A feature is the wealth of line-drawings and diagrams. Part F. *Coelenterata* (1956. xx, 498p.) has 358 diagrs. According to P.-M. Guelpa (Malclès, L.-M. *Les sources du travail bibliographique,* v.3, p.250), Moore is stronger on systematics than is Piveteau (*Traité de paléontologie) (qv).* The standard work on invertebrate palaeontology, written by the world's leading authorities (about 150 contributors).

The revised and enlarged editions of certain of the volumes (Boulder, Colo., Geological Society of America) began in 1970. *Class No:* 562

Foraminifera

[1832]

ELLIS, B.F. *and* MESSINA, A.R. **Catalogue of Foraminifera.** New York, American Museum of Natural History, 1940. 30v. (+ loose-leaf supplements, 1945-), illus.

Special publication (official report no.65-1-97-21 WP16) conducted under the auspices of the Work Progress Administration.

Systematic descriptions: genera, species and varieties arranged A-Z by genera, etc. Index to synonyms. V.30: 'Index to taxonomic changes and Bibliography'. A monumental work. 'Since foraminifera are the most economically important group of fossils, the literature is vast and bibliographic aids are especially needful' (W.A.S. Sarjeant and A.P. Harvey, in D.N. Wood's *Use of earth sciences literature* (1973, p.266)). *Class No:* 563.12

[1833]

—ELLIS, B.F. *and* MESSINA, A.R. Catalogue of index Foraminifera. New York, American Museum of Natural History, 1965-67. 3v. illus. *Class No:* 563.12

[1834]

—ELLIS, B.F., *and others.* Catalogue of index Smaller Foraminifera. New York, American Museum of Natural History, 1968-69. 3v. illus. *Class No:* 563.12

[1835]

HAYNES, J.R. **Foraminifera.** London, Macmillan, 1981. x, 433p. illus., ports., maps. £50. ISBN: 0333286812.

About two-thirds of the book is devoted to descriptive morphology and classification. Extensive bibliography, p.340-89. Comprehensive, well illustrated, 'Despite its price, this book has no real competition, and it must be recommended very highly' (*New scientist,* 15 July 1982, p.166). *Class No:* 563.12

[1836]

Stratigraphical atlas of fossil foraminifera. Jenkins, D.G. and Murray, J.W., *eds.* 2nd ed. Chichester, Ellis Horwood Ltd; Halsted Press, 1989. 593p. illus., tables, maps. £69.95. (*British Micropalaeontological Society series.*) ISBN: 0745801536, Ellis Horwood; 0470212268, Halsted Press.

First published 1984.

25 contributors. Arrangement is by litho-stratigraphical systems in 13 documented chapters. In this edition, range charts have been added and there is new information for the Cambrian to Devonian, Triassic, Jurassic and for the Cretaceous and Cenozoic periods of the North Sea. Important foraminiferal taxa are illustrated with taxonomic and morphological data. General index and index of general and species. British in scope. *Class No:* 563.12

Fossil Vertebrates

[1837]

BENTON, M.J. **Vertebrate palaeontology:** biology and evolution. London, Unwin Hyman, 1990. xii, 377p. figs. ISBN: 0045660018.

10 sections: 1. Vertebrate origins - 2. Early fishes - 3. The amphibians … 7. The age of dinosaurs - 8. The birds … 10. Human evolution. 2 appendices (Geological time; Classification of the vertebrates). Glossary, p.335-39. Bibliography, p.340-62, and systematic index. Many illustrations. Aimed at both the layperson and the specialist. *Class No:* 566

[1838]

CAMP, C.L., *and others.* **Bibliography of fossil vertebrates, 1928/1933-.** New York, Geological Society of America, 1940-73. 9v. 5 yearly.

Preceded by O.P. Hay's *Bibliography and catalogue of the fossil vertebrates of North America* (Washington, Geological Survey, 1902. *Bulletin* no.179) and his *Second bibliography* (Washington, Carnegie Institution, 1929-30. Publication no.390).

Camp's *Bibliography of fossil vertebrates, 1969-72* (1973, xlvi, [i], 733p. Geological Society of America. *Memoirs*) is, like Romer's (*qv*), an author catalogue subarranged chronologically, with *c.*7,500 entries. Subject index (p.388-575) has detailed subdivision under countries and includes obituaries. Systematic index, p.577-733. Supplementary list of serials, p.xi-xxxix.

Continued in *Bibliography of fossil vertebrates, 1973-1977,* by M. Green and others (South Dakota School of Mines and Technology, 1979), and now annually by *Bibliography of fossil vertebrates,* 1978-, (1985 ed. Berkeley, Calif., Society of Vertebrate Paleontology. 624p.). *Class No:* 566

[1839]

LEAKEY, L.S.B., *ed.* **Fossil vertebrates of Africa.** Orlando, Florida, Academic Press, 1969-76. 4v. ISBN: 0124404014, v.1; 0124404022, v.2; 0124404030, v.3; 0124404049, v.4.

V.4 (1976. 402p. $81), edited by R.J.G. Savage and S. Coryndon, has 6 specialized chapters,- on aspects of elephant shrews (E. and S. Africa), rodents, primates from East Africa, pigs from Miocene African sites, etc. 'Although too detailed for the general reader, the volume will become an established reference work for those working in the field of African prehistory, vertebrate palaeontology and evolution' (*British book news,* October 1976, p.754). *Class No:* 566

[1840]

ROMER, A.S., *and others.* **Bibliography of fossil vertebrates, exclusive of North America, 1509-1927.** New York, Geological Society of America, 1962. $160. (*Geological Society of America, Memoir 87.*) ISBN: 0835771830.

A comprehensive bibliography of published work, particularly periodical articles. Nearly 40,000 entries, under authors A-Z, then chronologically. No subject or locality index, - a severe handicap to non-specialists, at least. About 2,000 periodicals are cited (p.xiii, lxxxix). *Class No:* 566

[1841]

ROMER, A.S. **Vertebrate palaeontology.** 3rd ed. Chicago, Ill., Univ. of Chicago Press, 1966. viii, 468p. illus. $45. ISBN: 0226724883.

First published 1933.

Includes an outline classification (p.347-96) and an extensive bibliography (p.397-418). The standard one-volume work.

A companion volume is his *Notes and comments on vertebrate palaeontology* (Univ. of Chicago Press, 1968. viii, 304p.). *Class No:* 566

Fish

[1842]

MOY-THOMAS, J.A. **Palaeozoic fishes.** 2nd ed., rev. by R.S. Miles. London, Chapman and Hall, 1971. xi, 259p. illus.

First published 1939.

Extensively revised and enlarged. Sections on the probable life-history patterns of the major groups of fishes dealt with are added. 'This work is of major importance and contains excellent bibliographies' (Sarjeant, W.A.S. and Harvey, A.P., in Wood, D.N., ed. *Use of earth sciences literature* (1973), p.279). *Class No:* 567

Reptiles & Amphibians

[1843]

SWINTON, W.E. **Fossil amphibians and reptiles.** 5th ed. London, British Museum (Natural History), 1973. ix, 133p. illus. (incl. pl.). £1.40. ISBN: 056500543x.

First published 1954.

A compact work, with 17 plates and 68 keyed line-drawings. Appended are a geological time scale, a classification, and a glossary (p.127-9). Detailed, non-analytical index. 'Asterisks after the names of fossils in the text indicate that specimens are on exhibition' (*Preface*). *Class No:* 568.1

Dinosaurs

[1844]
The Macmillan illustrated encyclopedia of dinosaurs and prehistoric animals: a visual who's who of prehistoric life. Cox, B., *and others*. London, Macmillan, 1988. 312p. illus. (col.). ISBN: 0333486994.

7 sections, covering over 600 species: Fishes - Amphibians - Reptiles - Ruling reptiles - Birds - Mammal-like reptiles - Mammals. Data in each entry give name, period of existence, locality, size and general description. 2-page glossary. Classification of vertebrates, p.300-3. Brief bibliography, p.304. List of international museums, p.305. Well illustrated, with an extensive index. *Class No:* 568.19

[1845]
THULBORN, T. Dinosaur tracks. London,. Chapman and Hall, 1990. 410p. illus. £37. ISBN: 0412328909.

A comprehensive summary of, and introduction to, dinosaur trace fossils, including chapters: 4. Dinosaur tracks in the field and laboratory - 5. Identify the track-maker ... 7. Problematical and anomalous tracks - 8. Estimating the size of the track-maker. 'Will appeal to a variety of palaeontologists, sedimentologists, and biologists. It should even prove attractive to amateur dinosaur fanatics ... a remarkably original book' (*Geological magazine,* v.128(1), 1991, p.84). *Class No:* 568.19

Encyclopaedias & Dictionaries

[1846]
GLUT, D.F. The New dinosaur dictionary. Secaucus, N.J., Citadel Press, 1982. 288p. $19.95. ISBN: 0806507829.

First published 1972 as *The dinosaur dictionary*.

An up-to-date catalogue of all dinosaur genera. Abbreviations indicate suborder, family, etc. and the Mesozoic period in which the dinosaurs lived. Notes on physical characteristics and identification of sites. Well illustrated. Although discoverers' names and dates of their first reports are cited, a bibliography of published reports is lacking. *Class No:* 568.19(03)

[1847]
LAMBERT, D. *and* **THE DIAGRAM GROUP. Dinosaur data book:** the definitive, fully illustrated encyclopedia of dinosaurs and other prehistoric reptiles. New York, Facts on File with the British Museum (Natural History), 1990. 320p. figs., maps. £7.95. ISBN: 0816024316.

7 chapters: 1. The age of dinosaurs - 2. The A to Z of dinosaurs - 3. Dinosaurs classified - 4. Dinosaur life - 5. Dinosaurs worldwide - 6. Dinosaurologists - 7. Dinosaurs revived. Bibliography, p.306-10. Authoritative and lively text, with illustrations and maps on virtually every page. *Class No:* 568.19(03)

[1848]
NORMAN, D. The Illustrated encyclopedia of dinosaurs. London, Salamander Books, 1985. 208p. illus. (mostly col.), ports., facsims., diagrs., maps. £14.95. ISBN: 0861012259.

35 sections, including 'Controversies', 'Geography of finds', 'Museums and displays'. Glossary, p.202-3. Index (over 2,000 entries, in tiny print), p.204-8. *Author's foreword* has notes on 6 'serious books of recent years'. A quarto volume, copiously illustrated, with 25 double-page spreads. Claims to provide 'one of the most detailed and comprehensive layman's guides to the dinosaurs yet published'. Author is a lecturer in zoology at Brasenose College, Oxford. *Class No:* 568.19(03)

Handbooks & Manuals

[1849]
LAMBERT, D. *and* **THE DIAGRAM GROUP. Collins' guide to dinosaurs.** London, Collins, 1985. 256p. illus. (two-colour), ports., diagrs., charts. £6.95. ISBN: 0001953877.

First published 1983.

6 chapters: 1. What were the dinosaurs? - 2. Dinosaurs in the making - 3. Dinosaurs identified - 4. Their changing world - 5. Discovering dinosaurs - 6. Dinosaurs displayed (listing over 70 museums with exhibits from the Age of Dinosaurs; Argentina ... USSR). Brief entries for over 300 kinds of dinosaurs, - 'all currently known dinosaur genera, one in five of which have been named since 1970' (*Foreword*). Further reading, by chapters, p.248; detailed, analytical index (dinosaur names no longer used are followed by current names). Illus. include 'field guide' silhouettes and family trees. 'This book can be recommended to the seriously interested beginner or amateur' (*Natural history book reviews,* v.7(3), 1984, p.149). *Class No:* 568.19(035)

[1850]
—**Dinosaurs and their living relatives.** 2nd ed. London, British Museum (Natural History), with Cambridge Univ. Press, 1985. 72p. col. illus. £5.95. ISBN: 056506426x.

Explains dinosaurs and how they evolved into the crocodiles and birds alive today. Copiously illustrated. 'The work will probably appeal to those in their early teens' (*Children's books (British book news),* June 1986, p.32). *Class No:* 568.19(035)

[1851]
—**HALSTEAD, L.B.** *and* **HALSTEAD, J. Dinosaurs.** Poole, Dorset, Blandford Press, 1981. 170p. illus. (incl. col.). ISBN: 0713710179.

10 short chapters, describing *c.*125 species.'The colour section [p.61-108] illustrates all the major types of dinosaurs, including many that have never previously figured in books' (*Preface*). Includes a chapter on the classification of 12 families. 'An entirely new approach, along the lines of a field identification manual' (*Natural history book reviews,* v.6(3/4), 1983, p.149-50). *Class No:* 568.19(035)

Birds

[1852]
SWINTON, W.E. Fossil birds. 3rd rev. ed. London, British Museum (Natural History), 1975. v, 81p. illus. (incl. col. front.), chart. £1.50. ISBN: 0565053973.

First published 1958.

Companion volume to his *Fossil amphibians and reptiles (qv)*. Contents: Introduction - The history of flight - How birds fly - Jurassic birds - Cretaceous birds - Tertiary birds - Origin of ratites. Geological chart; Classification; Glossary (p.71-74). Analytical index. 27 line-drawings; col. front. Good value. *Class No:* 568.2

Mammals

[1853]
HOPWOOD, A.T. *and* **HOLLYFIELD, J.P. An Annotated bibliography of the fossil mammals of Africa (1742-1950).** London, British Museum (Natural History), 1954. 194p. (*Fossil mammals of Africa, no.8.*)

About 1,000 entries, with annotations, especially taxonomic and stratigraphic. Geographic, systematic and

....(contd.)

nominal indexes. 'Designed for zoologists interested in the comparative anatomy, distribution and history of the mammalian faunas of Africa, rather than for geologists and archaeologists' (Introduction). Class No: 569

[1854]
KURTÉN, B. **Pleistocene mammals of Europe.** London, Weidenfeld & Nicolson 1968. viii, 367p. illus., maps. (*World naturalist.*)

'Has become a classic; it breaks new ground and remains a frequently quoted text' (*Nature,* v.290, 12 March 1981, p.168). Class No: 569

[1855]
KURTÉN, B. *and* ANDERSON, E. **Pleistocene mammals of North America.** New York, Columbia Univ. Press, 1980. 442p. tables. $82. ISBN: 0231037333.

Part 1 lists *c.*250 faunal sites, with notes on geology, geography, ecology and faunal composition. Part 2 is a critical and annotated list for the Blancan and Pleistocene mammal species in Canada, USA and Northern Mexico. 'It will be an invaluable reference source for those working not only on Pleistocene faunas, but on problems of evolution, extinction and biogeography' (*Nature,* v.290, 12 March 1981, p.168). Class No: 569

[1856]
LILLEGRAVEN, J.C., *and others, eds.* **Mesozoic mammals:** the first two-thirds of mammalian history. Berkeley, Calif., & London, University of California Press, 1979. 320p. illus. $65. ISBN: 0520055828.

Covers the period *c.*240 to 64 million years before the present. 'As such it will be of great value to those actively involved in research in this group, as well as of those generally interested in a technical account of early mammalian evolution' (*Library journal,* 1 November 1979, p.2360). Class No: 569

[1857]
SAVAGE, D.E. *and* RUSSELL, D.E. **Mammalian paleofaunas of the world.** Reading, Mass., & London, Addison-Wesley, 1983. 432p. ISBN: 0201064944.

Lists more than 3,000 genera and at least twice that number of species over a 200-million year time-span. 'A highly reliable work, an indispensible reference source and a monumental data base from which to test our theories, be they evolutionary, faunal or geographical' (*Nature,* v.304, 4 August 1983, p.471). Class No: 569

Man

[1858]
DAY, M.H. **Guide to fossil man.** 4th ed. London, Cassell; Chicago, University Press, 1986. xv, [i], 432p. illus., diagrs., tables, maps. ISBN: 0304312886; 0226138895, US ed.

First published 1965.

Three parts. 1: The anatomy of fossil man - 2. The fossil hominids (Europe, plus country division; Near East; Northwest Africa; East Africa; Southern Africa; The Far East; Oceania), p.16-401 - 3. Essays on fossil man (with references). Records new evidence from 34 sites featured in previous editions, plus details of 15 new sites. Appended 'Geologic and palaeomagnetic time scales'; Glossary. Partly

....(contd.)

analytical index. Author is Professor of Anatomy, United Medical and Dental Schools of Guy's and St. Thomas's Hospital, Univ. of London. Class No: 569.9

[1859]
JOLLY, C.J., *ed.* **Early hominids of Africa.** London, Duckworth, 1978. x, 598p. illus., diagrs., graphs, tables. ISBN: 0715609289.

4 parts: 1. The sites - 2. Geological, fauna and archeological evidence - 3. Anatomical evidence - 4. The interpretation of hominid diversity. 24 contributions, - papers read at a New York conference in 1974. Coverage is Africa south of the Sahara, over a specific span of time: the Plio-Pleistocene. General and author indexes. Class No: 569.9

[1860]
KENNEDY, G.E. **Paleoanthropology.** New York, McGraw-Hill, 1980. xiii, [1], 479p. illus., tables. $39.95. ISBN: 0070340463.

10 chapters: 1. Evolutionary biology, taxonomy and paleoanthropology - 2. The order primates - 3. Early primates: Cretaceous to Oligocene - 4. Miocene hominoid radiation - 5. The family hominidae - 6. Hominids of the Plio-Pleistocene - 7. Plio-Pleistocene hominid localities - 8. Hominids of the Middle Pleistocene - 9. Hominids of the Early Pleistocene - 10. Summary. Each chapter has 'Summary' and 'Suggestions for further reading', Appendix: Dating techniques. Glossary, p.426-30; bibliography (authors, A-Z; p.431-67). Analytical index, p.469-79. 'Intended for students who have completed an introductory course in physical anthropology or biology' (*Preface*). The emphasis is on the fossil record itself. The author, - Assistant Professor of Anthropology, Univ. of California, Los Angeles. Class No: 569.9

[1861]
—LAMBERT, D. Cambridge guide to prehistoric man. Cambridge Univ. Press, 1987. 256p. illus., (incl. col.), tables. £9.50. ISBN: 0521336449.

A companion to the popular *Cambridge Field guide to prehistoric life (qv).* Its brief non-technical text on prehistoric man's biological and cultural evolution is followed by copious line-drawings and tables. 'Further reading', p.248. Index. Class No: 569.9

[1862]
OAKLEY, K.P., *and others, eds.* **Catalogue of fossil hominids.** London, British Museum (Natural History), 1971-77. 3v. illus., (incl. pl.), maps. £70. ISBN: 0565006614; 0565007114; 056500767x.

1. *Africa.* 1977. xiv, 210p. £20. 2. *Europe.* 1971. xiv, 370p. £30. 3. *America, Asia & Australasia.* xiv, 228p. £20.

Each part is arranged by country/location. Data at place-name designate the fossil, describe site and geographical location, name of discoverer, geological deposit, burial (or not), stratigraphic age, archaeological context, palaeontological context, relative dating of bone, antler and dentine, absolute data, relationship of bones, first published report, anatomical description, most recent revision of the morphological evaluation, reference to best illustrated account, any other important publications, postal address of repository of fossil, and of moulds for casts. Index and location maps per volume. Class No: 569.9

[1863]

SMITH, F.H. *and* **SPENCER, F. The Origins of modern humans:** a world survey of the fossil evidence. New York, Wiley, 1985. $93.95. ISBN: 047183419x.

11 contributors, of which 9 concern skeletal remains found in Europe, Western Asia, East Asia and Australasia, China, and the Americas. While conceding that the text is refreshingly free from the Eurocentric bias, the review in *Nature* (v.314, 18 April 1985, p.649) quarrels with the maps and quality of photographs and drawings. Nevertheless, 'An invaluable source of reference for those already familiar with the fossil record, anatomy and population genetics'.

Class No: 569.9

572 Anthropology

Anthropology

Databases

[1864]
BRADY, F. *and* LAMBERT, M. 'Alternative databases for anthropology searching'. In *Database*, v.7(1), February 1984, p.28-35. ISSN: 01624105.

The *Database* article cites PSYCINFO, SOCIAL RESEARCH, LLBA, BIOSIS and MEDLINE as relevant bibliographical databases. Droessler, J.B. and Wilke C. ('Physical anthropology literature: online access', *Reference Services Review,* v.12(2), Summer 1984, p.22-26) cite BIOSIS REVIEWS, LIFE SCIENCES COLLECTIONS, MEDLINE and EMBASE, there being no database specifically devoted to anthropological literature.
Class No: 572.0(003.4)

Bibliographies of Bibliographies

[1865]
LIBRARY-ANTHROPOLOGY RESOURCE GROUP.
Anthropological bibliographies: a selected guide.
Smith, M.L. *and* Damien, Y.M., *eds.* 2nd ed. South Salem, New York, Redgrave, 1981. 307p.

About 3,200 entries for material mainly published 1957-77. 7 chapters; 1-6: Geographical (Africa, America, Asia, Europe, Oceania, USSR); 7: Topical coverage includes filmographies, discographies, books, article-length bibliographies and 'hidden' bibliographies appended to articles and books. Some entries are annotated. Detailed index, insufficiently analytical. 'The work is definitely not as useful as it first appears' (*Reference Services Review,* v.11(3), Fall 1983, p.67), but 'Of first importance' (Webb, W.H., and associates. *Sources of information in the social sciences,* item F457). *Class No:* 572.0(009)

[1866]
—CURRIER, M. 'Problems in anthropological bibliography'. In *Annual review of anthropology*, no.5, 1976, p.15-34.

Surveys 100 years of trends in anthropological indexing and abstracting services, including the relevance of early general indexes. Sections embrace computer indexing and abstracting, and compilation and publishing of bibliographies. Limited attention to monographic bibliographies. Bibliography of 30 items. No index.
Class No: 572.0(009)

Bibliographies

[1867]
Author and subject catalogues of the Tozzer Library. 2nd ed. Boston, G.K. Hall, 1988. Microfiche. ISBN: 0816117314.

Successor to the *Catalogue of the Library of the Peabody*

.... *(contd.)*
Museum of Archaeology and Ethnology. Consists of 1,122 negative microfiches, accompanied by a printed fiche guide. Supplements appear annually. *Class No:* 572.0(01)

[1868]
—HARVARD UNIVERSITY. PEABODY MUSEUM OF ARCHAEOLOGY AND ETHNOLOGY. Library. Catalogue. [Peabody Museum of Archaeology and Ethnology, Library]. Authors, 1963. 27v. Supplements 1-4. 1970-79. 14v. Subjects. 1963. 27v. Supplements 1-4. 1970-79. 16v. Boston, Mass., G.K. Hall, 1963-. 54v. (29302p.) Supplements-. ISBN: 0816106479.

Over one million photolithographed catalogue cards in all. International in scope, the Peabody Museum Library has one of the most comprehensive collections of anthropological material. The *Catalogue* indexes periodicals, books and collections of essays, symposia, congresses and festschrifts. 'The best retrospective index to the anthropological literature' (*Reference Services Review,* v.12(4), Winter 1984, p.37). *Class No:* 572.0(01)

[1869]
—HARVARD UNIVERSITY. PEABODY MUSEUM OF ARCHAEOLOGY AND ETHNOLOGY. Library. Index to subject headings. [Peabody Museum of Archaeology and Ethnology, Library]. 2nd rev. ed. Boston, Mass., G.K. Hall, 1981. 177p. ISBN: 0816104080.

First published 1963 (117p.).
Comprises *c.*18,000 headings and cross-references.
Class No: 572.0(01)

[1870]
Bibliographic guide to anthropology and archaeology 1987. Boston, G.K. Hall, 1988. 382p. ISBN: 081617069x.

The first annual supplement to the *Author and subject catalogues of the Tozzer Library,* on microfiche (G.K. Hall, 1988).

Lists material catalogued between June 1986 and August 1987 by Harvard University's Tozzer Library. Subject areas include linguistics, cultural anthropology, physical anthropology and archaeology. *Class No:* 572.0(01)

[1871]
Ecce homo: an annotated bibliographic history of physical anthropology. Spencer, F., *comp.* New York & London, Greenwood Press, 1986. xiii, [1], 459p. illus., facsims. £45; $95. (*Bibliographies and indexes in anthropology, no.2.*) ISBN: 0313240566.

2,340 annotated entries. 4 parts: 1. Ancient, medieval, Renaissance and early modern literature - 2. Eighteenth century - 3. Nineteenth century (945 entries) - 4. Twentieth century (1,168 entries: Human population biology; Primatology; Paleoprimatology; Paleo-anthropology - Old World; New World). Name and subject indexes.
Class No: 572.0(01)

[1872]

GRAVEL, P.B. *and* RIDINGER, R.B.M. **Anthropological fieldwork:** an annotated bibliography. New York, Garland, 1988. 241p. $33. (*Garland reference library of social science, 419.*) ISBN: 0824066421.

700 entries to journal articles, monographs, book chapters, and collected essays in English, French and German. Arranged A-Z by author, with annotations of *c*.100 words. Geographical index. Very brief subject index (2p.). Not comprehensive but covers the most important aspects of anthropological fieldwork. *Class No:* 572.0(01)

[1873]

International bibliography of social and cultural anthropology. [Bibliographie internationale d'anthropologie sociale et culturelle.] London, Tavistock Publications, 1955-. v.1-. Annual. (*International bibliography of the social sciences.*)

Prepared by the International Committee for Social Science Information and Documentation.

Each issue contains *c*.1,700 entries. 10 sections: A. Anthropology: general studies; B. Materials and methods of anthropology; C. Morphological foundations; D. Ethnographic studies of people and communities; E. Social organisation and relationships; F. Religion, magic, witchcraft; G. Problems of knowledge, arts and science, folk traditions; H. Studies of culture and personality. 'National character'; I. Problems of acculturation and social change; contact situations; J. Applied anthropology. Author and analytical subject indexes. *Class No:* 572.0(01)

[1874]

MEXICO. BIBLIOTECA NATIONAL DE ANTROPOLOGÍA E HISTORIA, MEXICO CITY. **Catalogo.** [Biblioteca National de Antropologia e Historia, Mexico City. Boston, Mass., G.K. Hall, 1977. 10v. (7081p.). $960.

143,000 photolithographed catalogue cards, in one author, title, subject and added entry sequence, A-Z. *Class No:* 572.0(01)

[1875]

WESTERMAN, R. 'Anthropology'. In *Sources of information in the social sciences: a guide to the literature* (3rd ed, edited by W.H. Webb and associates. Chicago & London, American Library Association, 1986), p.332-402. ISBN: 083890405x.

Two main sections: Survey of the field (F1-F149) - Survey of the reference works (F420-F1072). The first section has 15 subsections each with an introduction and unannotated lists of items. The second section, also closely subdivided, has short introductions highlighting major material, plus annotated entries or running commentary. A revision and updating of chapter 6, Anthropology, in *Sources of information in the social sciences* (2nd ed., edited by C.M. White and associates. 1973). A basic bibliographical tool in this field. *Class No:* 572.0(01)

[1876]

—BROWN, S., *comp.* Finding the source in sociology and anthropology: a thesaurus - index to the reference collection. Newport, Conn., Greenwood Press, 1987. 269p. $49.95. (*Finding the source series.*) ISBN: 0313252637.

A subject bibliography of 578 titles - selected from Sheehy's *Guide to reference books* and C. White's *Sources of information in the social sciences* (2nd ed. 1973), with title and author indexes, - plus a 'thesaurus-index' of the

....(contd.)

relevant descriptors, synonyms, related and narrower terms. '*Finding the source* is useful for someone who lacks the intellectual context of reference genres and who is looking for specific information'. (*Wilson library bulletin*, v.61(10), June 1987, p.83). *Class No:* 572.0(01)

Encyclopaedias & Dictionaries

[1877]

HUNTER, D.E. *and* WHITTEN, P., *eds.* **Encyclopedia of anthropology.** New York, Harper & Row, 1976. xi, 411p. illus., tables, maps. ISBN: 0060470941.

About 1,400 entries, by 87 contributors. 'Monotheistic religions': 2 columns; 4 references. Includes brief biographies (*e.g.* Levi-Strauss: nearly 1 column). Well illustrated, with *c*.200 photographs and line drawings. Cross-references, and subject index, less adequate. Some US slant. Aims to be a 'compact, comprehensive, accessible reference source' (*Preface*) and largely succeeds.

Fuller in coverage and more comprehensive than *Dictionary of anthropology,* edited by C. Winwick (New York, Philosophical Library, 1956. vii, 579p). *Class No:* 572.0(03)

[1878]

SEYMOUR-SMITH, C. **Macmillan dictionary of anthropology.** London, Macmillan Press, 1986. [vi], 305p. ISBN: 0333365909.

About 2,000 entries (*e.g.* 'Kinship terminology': nearly 2 columns). Includes short biographies (*e.g.* 'Mead, Margaret: nearly 2 columns). Ample cross-references. 'Selective bibliography and further reading' (authors, A/Z), p.293-305. Stresses social and cultural aspects, and conceptual issues. 'There are few fixed definitions in anthropology. Instead ours is a discipline which advances by constantly revising and interrogating current concepts and their use' (*Foreword*). Lacks the supporting illus. in Hunter-Whitten's *Encyclopedia* (*qv*). Praised in *RQ* (v.27(1), Fall 1987, p.139) for the quality of its definitions. *Class No:* 572.0(03)

[1879]

—DAVIES, D. A Dictionary of anthropology. London, F. Muller, 1972. 197p. illus. (incl. 37 plates). ISBN: 0584100647.

Gives several hundred definitions, plus comment. Includes very brief biographies. Bibliography (authors, A-Z), p.195-7. Written in simple language; amateurish line-drawings, in addition to the photographs. Stresses ethnological aspects. For the lay person. *Class No:* 572.0(03)

[1880]

STEVENSON, J.C. **Dictionary of concepts in physical anthropology.** Westport, Conn., Greenwood Press, 1991. 448p. $85. ISBN: 0313247560.

74 concepts are examined with coverage extending to current usage, development, bibliography, and additional sources of information. Name and subject indexes. Criticized in *Choice,* March 1992 for being an expensive undergraduate book and for not providing clearer definitions. *Class No:* 572.0(03)

Handbooks & Manuals

[1881]
HONIGMANN, J.J., *ed.* **Handbook of social and cultural anthropology.** Boston, Houghton Mifflin, 1973. viii, 1295p. tables. $67. ISBN: 0395306310.

33 contributors. 28 review articles on cognitive anthropology, symbolism, structuralism and semiotics (*e.g.* 1. History of cultural anthropology; 5. Ecological anthropology and anthropological ecology; 15. The structural anthropology of Claude Lévi-Strauss (references, p.704-16); 21. Cultural psychiatry). Chapter 'credits and acknowledgements', p.1251-5. Name index; analytical subject index. *Class No:* 572.0(035)

[1882]
MARTIN, R. Lehrbuch der Anthropologie in systematischer Darstellung, mit besonderer Berücksichtigung der anthropologischen Methoden. 3. Aufl. hrsg., von K. Saller. Stuttgart, Fischer, 1957-62. 3v. (2416p.). illus., tables.

2nd ed., 1928 (1816p.).

A standard treatise, systematically arranged, with numerous section bibliographies. Profusely illustrated with *c.*1,000 illus. in all. V.3 included a 600-page bibliography, plus subject and author indexes. *Class No:* 572.0(035)

Theses

[1883]
SOCIAL SCIENCE RESEARCH COUNCIL. Social Anthropology Committee. **Research in social anthropology, 1975 - 1980:** register of theses in social anthropology accepted for higher degrees at British universities. Webber, J., *comp.* London, Royal Anthropological Institute, 1983. xxxix, 404p. £14. ISBN: 090063233x.

425 entries, with abstracts and note of availability of theses at British Library Lending Division. Prelims. include: 'List of authors in this Register'; 'List of Univ. depts., presented ...'; university and county indexes; subject and culture, and subject indexes. *Class No:* 572.0(043)

Reviews & Abstracts

[1884]
Abstracts in anthropology: archaeology/cultural & physical anthropology/linguistics. Farmingdale, N.Y., Baywood Publishing Co., 1970-. 4pa. $239. ISSN: 00013455.

V.25(2), 1992 (189p.) has 750 abstracts. Two sections: Archaeology - Physical anthropology (General; Fossil record and evolution; Living humans: population, genetics; calcified tissues; growth and development; body composition and adaptation, living non-human primates). Abstracts average 100 words in length; *c.*300 periodicals scanned. The quarterly subject-author index fails to cumulate annually, - a drawback in view of wide subject coverage. *Class No:* 572.0(048)

[1885]
Abstracts in German anthropology. Göttingen, Edition Re, 1980-. 2pa. ISSN: 01732986.

About 250 abstracts in German per issue. Covers *c.*50 German periodicals on aspects of anthropology worldwide. Sections: General/theoretical/historical/studies - Regional studies (Africa, The Americas, Asia, Australia, Oceania,

....(contd.)
Europe). Includes theses in German. Author and subject indexes and list of periodicals seen, per issue. *Class No:* 572.0(048)

[1886]
Anthropological index to current periodicals in the Museum of Mankind Library (incorporating the former Royal Anthropological Institute Library). London, the Museum, 1963-. 4pa. £65; $103. ISSN: 00035467.

As *Index to current periodicals received in the Library of the Royal Anthropological Society,* 1963-67.

About 10,000 numbered references pa. Sections: 1. General - 2. Africa - 3. Americas - 4. Asia - 5. Australasia. Pacific - 6. Europe. Subdivisions (where appropriate): General; Physical anthropology; Archaeology; Cultural anthropology; Ethnography; Linguistics. The first issue of each volume lists additions and amendments to current periodicals, while a revised list of current periodicals, with abbreviations, is published separately. Annual author index, but no subject index. Excellent for current awareness. *Class No:* 572.0(048)

[1887]
Anthropological literature: an index to periodical articles and essays; compiled by Tozzer Library. Cambridge, Mass., Tozzer Library, 1979-. $125. ISSN: 01903373.

Each volume contains a minimum of 8,000 references; v.13, 1991, contains 8,575 entries taken from 988 journals and 92 edited works. 5 sections: Archaeology - Biological/ physical - Cultural/social - Linguistics - Research in related fields and topics of general interest. 4 indexes per issue (author; subject; list of journals indexed; list of edited works), all of which cumulate in the fourth issue of each volume. Some overlap with *Anthropological index* (*qv*), 'but each has material not in the other' (University of London Institute of Archaeology. *Bulletin* no.17, 1980, p.194) and this publication uses Library of Congress subject headings rather than geographic and broad topic headings. *Class No:* 572.0(048)

[1888]
Excerpta medica. 1. Anatomy, anthropology, embryology and histology. Amsterdam, Excerpta Medica, 1947-. 16pa. (2v.). DFl1,358. ISSN: 00144053.

Section 3, 'Anthropology', has nearly 150 abstracts pa. on physical aspects. Subsections: 1. Techniques and apparatus; 2. Paleoanthropology; 3. Evolution; 4. Variability; 5. Heredity of normal features; 6. Body build and body composition; 7. Ecologic factors and adaptations. Online, EMBASE and also available on microfilm from University Microfilms International. Available on CD-ROM. *Class No:* 572.0(048)

Periodicals

[1889]
American anthropologist: journal of the American Anthropological Association. Washington, the Association, 1888-. 4pa. $90pa. ISSN: 00027294.

Sections: Applied anthropology - General/theoretical anthropology - Social/cultural anthropology - Biological anthropology - Linguistic anthropology - Archaeology - Film review. Annual title and author indexes. A major reviewing journal in its field. *Class No:* 572.0(051)

[1890]

LIBRARY-ANTHROPOLOGY RESOURCE GROUP.
Serial publications in anthropology. Grollig, F.V. *and*
Tax, S., *eds*. 2nd ed. South Salem, N.Y., Redgrave, 1982.
177p.

First published 1973 (Univ. of Chicago Press).
4,387 titles, A-Z, of periodicals and other serials that
carry anthropological material. Cross-references. Subject
index includes names of issuing bodies.
Class No: 572.0(051)

[1891]

—WILLIAMS, J.T. Anthropology journals and serials: an
analytical guide. Westport, Conn., Greenwood Press, 1986.
152p.

Has fewer than 4,387 titles listed in Grollig and Tax, but
its arrangement is complementary. 4 of the 5 chapters
concern archaeology, cultural anthropology, linguistics,
physical anthropology. Chapter 5 covers appropriate indexes
and abstracts. Title, subject and geographic indexes.
Class No: 572.0(051)

[1892]

Man: the journal of the Royal Anthropological Institute.
London, the Institute, 1966-. 4pa. £71; $115pa. ISSN:
00251496.

V.27(3), September 1992 contains 8 documented papers,
e.g. 'Cognitive evolution in primates: evidence from tactical
deception', p.609-28. Summaries of each article in French.
This issue has 51 signed book reviews and a 'Books
Received' section. A leading academic journal on social
anthropology. Also available in microform.
Class No: 572.0(051)

Progress Reports

[1893]

Annual review of anthropology. Palo Alto, Calif., Annual
Reviews, Inc., 1972-. Annual. ISSN: 00846570.

V.21, 1992, has 28 contributors. 21 documented articles:
Overview - Archaeology of the Iroquois ...
Palaeoepidemiology ... The social anthropology of
technology ... The anthropology of African landholding ...
Human rights and anthropology. Author and subject indexes.
Cumulative index of contributing authors, v.14-21;
cumulative index of chapter titles, v.14-21. For the
specialist-researcher. *Class No:* 572.0(055)

[1894]

Research in economic anthropology: a research annual.
Greenwich, Conn., Jai Press, 1978-. v.1-. Annual. ISSN:
01901281.

V.6, 1984, has 5 parts (10 papers): 1. Hunter-gatherers -
2. Slavery, trade and production - 3. Property of the dead -
4. East Africa: Turkana and Gezera - 5. Marxism analysis.
Bibliography appended to each paper. 'A state-of-the-art
yearbook that should reflect what anthropologists specifically
have to say about the economy' (*Preface*). Most recent
volume in print is v.9, 1987 ($49.50). *Class No:* 572.0(055)

Teaching Materials

[1895]

ROYAL ANTHROPOLOGICAL INSTITUTE. Teachers'
resource guide. 4th ed. London, Royal Anthropological
Institute, 1990. 58p. ISBN: 0900632380.

First published 1973.

....(contd.)

3 sections: A. Anthropology in education (1. School
examinations in anthropology ... 6. Some libraries with
anthropological content) - B. Anthropological resources (*e.g.*
visual aids, museums) - C. Bibliography (1. General
introductions ... 6. Family, marriage and gender ... 15.
Teachers' guides and sourcebooks). No index. A helpful
tool. *Class No:* 572.0(072)

Films

[1896]

HEIDER, K.G. Films for anthropological teaching. 7th ed.
Washington, American Anthropological Association, 1983.
[iii], 312p. $15. (*Special publication of the AAA, no.16.*)
ISBN: 0913161002.

A-Z title list of 1,575 films, p.26-280, with italicized
annotations summarizing content. References to reviews.
Grouping of films by geographical areas and by topics
precedes. People index, p.285-96; film title index, p.297-
312. List of major distributions. *Class No:* 572.0(084.122)

[1897]

WOODBURN, J., *ed*. The Royal Anthropological Institute
Film Library catalogue. London, the Institute, 1982. viii,
86p. £4.

A/Z title list of films, p.1-79. Data include type of film (in
colour/black-and-white), length, time/name of director, etc.,
annotation, and references to reviews. Analytical index,
p.81-88. Insert: 'Hire charges'. *Class No:* 572.0(084.122)

Histories

[1898]

EVANS-PRITCHARD, E.E. A History of anthropological
thought. London, Faber & Faber, 1981. xxxvi, 218p.
£16.50. ISBN: 0571117620.

15 chapters: 1. Montesquieu (1689-1775) ... 15. Hertz
(1882-1915), p.3-183. Appendix; Notes and comment on
Max Miller (1823-1900) ... L.A. White (1900-1975).
Bibliography, p.205-9. Detailed, analytical index, p.211-8.
Class No: 572.0(091)

[1899]

KEMPER, R.V. *and* PHINNEY, J.F.S., *eds*. The History of
anthropology: a research bibliography. New York, Garland,
1977. xvi, 212p. (*Garland reference library of social*
science, v.3.)

2,439 references. 4 sections: 1. General sources on the
history of anthropology - 2. Background - 3. Modern
anthropology (sections A-J, p.47-177) - 4. Related social
sciences - 5. Bibliographical sources. Author index. Focuses
on secondary and primary sources. A starting point for
students of the subject. *Class No:* 572.0(091)

Biographies

[1900]

GACS, U., *and others*. Women anthropologists: a
biographical dictionary. Westport, Conn., Greenwood
Press, 1988. 448p. $59.95. ISBN: 0313244146.

Biobibliographies of *c*.58 women anthropologists. The
period covered encompasses approximately a hundred years.
Entries are 3-7 pages in length, with a list of references by,
or about, the individual. *Class No:* 572.0(092)

[1901]
International dictionary of anthropologists. New York,
Garland, 1991. xl,823p. $75. (*Garland reference library of
the social sciences, v.638.*) ISBN: 0824050940.
   First published as *Biographical directory of
anthropologists born before 1920* (Garland, 1988).
   Includes only individuals born before 1920, thus excluding
many important figures. *c.*725 contributions from over 500
countries. Brief glossary and good name/subject index.
Class No: 572.0(092)

[1902]
Fifth International directory of anthropologists.
Spurr, H., *managing ed.* 5th ed. Chicago, Ill., Univ. of
Chicago Press, 1976. x, 496p. $29.95. ISBN: 0226790770.
   First published 1938.
   Very brief biographical data on 4,373 anthropologists
(name; date of birth; position and affiliation; sex; research
interests; publications; languages). Geographical,
chronological and subject/methodological indexes. The
subject index has four sections: cultural anthropology;
archaeology; physical anthropology; linguistics.
Class No: 572.0(092)

[1903]
KENNA, M.E. *and* **KENNA, J.C.** 'Published portraits of
anthropologists and workers in allied fields'. In *Current
anthropology*, v.14, 1973, p.83-101; v.16, 1975, p.271-81.
   Lists a total of *c.*1,250 individuals (names; dates of birth
and death; references to published portraits in journal
articles and monographs). *Class No:* 572.0(092)

Somatology (Anatomy of Man)

[1904]
BROTHWELL, D.R. Digging up bones: the excavation,
treatment and study of human skeletal remains. 3rd ed.
London, British Museum (Natural History); Oxford
University Press, 1981. 208p. illus. ISBN: 0198585047.
   1. Notes for guidance in excavating and reporting on
human remains - 2. Description and study of human bone -
3. Demographic aspects of selected biology - 4.
Measurement and morphological analysis of human bones -
5. Injuries and marks on bone - 6. Ancient disease - 7.
Concluding remarks. For the layperson as well as the
specialist. Bibliography, p.179-99. Index, p.201-8. 17 plates.
Class No: 572.5

[1905]
**FRISANCHO, A.R. Anthropometric standards for the
assessment of growth and nutritional status.** Ann Arbor,
University of Michigan Press, 1990. 189p. graphs, tables.
$75. ISBN: 0472101463.
   5 sections: 1. Necessity for new anthropometric standards
- 2. Methods and materials - 3. Anthropometric classification
- 4. Anthropometric standards (p.37-118) - 5. The standard
in practice: examples. Literature cited, p.133-42.
Appendices: A. Anthropometric tables for blacks - B.
Anthropometric tables for whites. Subject index.
Class No: 572.5

Ethnology (Races)

Bibliographies

[1906]
**Demos: internationale ethnographische und folkhistorische
Informationen.** Zurich, Kunst & Wiessen Erich Bieber,
1960-. 4pa. DM60. ISSN: 0011832x.
   Annotated bibliography of books, periodical articles and
pamphlets, cumulated 10-yearly. Texts in English, French
and German. *Class No:* 572.9(01)

Encyclopaedias

[1907]
Encyclopedia of world cultures. V.1: North America.
O'Leary, T.J. *and* Levinson, D., *eds.* Riverside, N.J., G.K.
Hall, 1991. 424p. maps. $100. ISBN: 0816118086.
   V.1. *North America.* 424p. maps.
   First volume of a projected 10-volume set due to be
completed in 1993. 223 articles arranged alphabetically on
cultures in the US, Canada and Greenland, varying in length
from a few lines, *e.g.* 'Koyukon', to six pages, *e.g.*
'Apache'. Data in each entry include ethnonyms, history and
cultural relations, linguistic and geographic orientation,
economy, kinship, marriage and family, and religion. Short
bibliographies of six or seven items are appended to each
entry. Appendix lists extinct native American tribes.
Glossary and ethnonym index. Filmography lists *c.*300 films
and videos. 'When complete, the
Encyclopedia of world cultures will be unique in its
provision of concise summaries for almost all cultural
groups' (*Booklist,* v.87(21), July 1991 p.2062).
Class No: 572.9(031)

[1908]
The Illustrated encyclopedia of mankind. Carlisle, R., *and
others, eds.* 3rd ed. London, Marshall Cavendish, 1990.
22v. illus., maps. ISBN: 1854350323.
   First published 1978, 2nd ed. 1984.
   V.1, 2-15 contain A-Z treatment of racial, ethnic and
cultural groups and nationalities. V.16-20 cover general
themes (unchanged from the 1984 ed.), *e.g.* 'Costume';
'Social organization'; 'Beliefs of the world'. There is a new
Origins of mankind volume with its own index. V.22
contains general, thematic and geographic indexes;
population charts; bibliography; glossary. Over 3,000 colour
photographs. Recommended for public and school libraries
which do not already possess the 1984 ed. Otherwise the
1984 ed. would suffice. *Class No:* 572.9(031)

[1909]
—**Family of man: people of the world, how and where they
live.** Widdows, R., *ed.* London, Marshall Cavendish, 1974-
76. 7v. (2748p.) illus. (incl. col.), maps.
   Appeared in 98 weekly parts, arranged A-Z, with
interposed articles (*e.g.* 'The future of mankind'). Part 98
consists of index and 5p. of 'Further reading'. Well
illustrated. *Class No:* 572.9(100)

ANTHROPOLOGY

Handbooks & Manuals

[1910]

COLE, S. Races of man. 2nd ed. London, British Museum (Natural History), 1965. 131p. front., illus., diagrs., maps.

First published 1961.

Chapters: 1. Definitions of race - 2. Physical characters of race - 3. Genetics and evolution - 4. Blood groups and other biochemical characters - 5. Racial origins - 6. Caucasoids - 7. Australoids and Pacific islanders - 8. Mongoloids (including American Indians) - 9. Negroids. Chapter references, p.125-7; well illustrated with half-tones and line-drawings. A sound introduction for beginners.
Class No: 572.9(035)

Dictionaries

[1911]

Dictionnaire de l'ethnologie et de l'anthropologie. Bonte, P. and Izard, M., *eds*. Paris, Presses Universitaires de France, 1991. xii,755p. FFr496. ISBN: 2130433839.

Encyclopaedic format with entries ranging from *c.*200 words to several pages in length. Bibliography appended to each entry. 83 biographies. Cross-references could be improved and an effective index is absent. However, a useful reference tool. *Class No:* 572.9(038)

[1912]

WINTHROP, R. Dictionary of concepts in cultural anthropology. Westport, Conn., Greenwood, 1991. 347p. $65. (*Reference sources for the social sciences and humanities, 11.*) ISBN: 0313242801.

80 concepts are discussed, each beginning with a definition followed by a detailed examination of the use and history of the concept. Each entry has a list of further reading appended. Many cross-references. Name and subject indexes. *Class No:* 572.9(038)

Reviews & Abstracts

[1913]

Francis: 529: Ethnologie. Nancy, Institute de l'Information Scientifique et Technique, 1947-. 4pa. FFr393.09. ISSN: 07851473.

Formerly *Bulletin signalétique. 529: Ethnologie.*

About 2,500 brief annotations and references pa. 8 classes: 01. General studies - 02. Sources and methods - 03. Morphological source materials - 04. Monographs - 05. Social structure and relations - 06. Religion. Magic. Witchcraft - 07. Cognitive problems. Arts and science - 08. Acclimatization. Social change. List of journals monitored during the quarter. Index of concepts, geographic index, ethnic index, author index per issue. Available online.
Class No: 572.9(048)

Tables & Data Books

[1914]

MURDOCK, G.P. Atlas of world cultures. Pittsburgh, Univ. of Pittsburgh Press, 1981. 151p. $18.95. ISBN: 0822934329.

Revision of his *Ethnographic atlas* (1967).

Not an atlas but coded data on 563 cultures 'most fully described in the ethnographic literature', applying 76 factors. (The 1967 *Atlas* coded 862 cultures). 'Although the coded

....(contd.)

format requires skill to utilize, the mass of information presented makes the effort worthwhile' (*Library journal,* 1 June 1981, p.1210). *Class No:* 572.9(083)

[1915]

—**MURDOCK, G.P.,** *and others*. Outline of cultural materials. 5th ed. New Haven, Conn., Human Relations Area Files Press, 1982. 247p. $25. ISBN: 0875366546.

Delineates the basic HRAF classification scheme placing data into 79 major and 619 major subject divisions, plus cross-references. *Class No:* 572.9(083)

[1916]

—**MURDOCK, G.P.,** *and others*. Outline of world cultures. 6th ed. New Haven, Conn., Human Relations Area Files Press, 1983. xii, 359p. $25. ISBN: 0875366643.

The companion volume, takes a geographical/historical stance. 8 main regions: World; Asia; Europe; Africa; Middle East; North America; Oceania. Detailed index, p.195-259. Aims to provide 'an outline organization and classification of the known cultures of the world'. *Class No:* 572.9(083)

Maps & Atlases

[1917]

Atlas narodov mira. Moscow, Akademiya Nauk SSSR, 1964. 184p. maps, tables.

'Atlas of peoples of the world'. 112p. of coloured maps, (average scale: 1,10M.), with 72p. of text. Data on world population, races, languages, religion, ethnic groups, etc., as of 1961. Index of names of groups, in cyrillic and romanized. 'The finest collection of ethnographic maps of all areas of the globe' (Soviet Information Service. *Bulletin* no.2188, 10 August 1964, p.1). Page size: 34 x 24cm.
Class No: 572.9(084.3)

[1918]

—**SPENCER, R.F.** *and* JOHNSON, E. Atlas for anthropology. 2nd ed. Dubuque, Iowa, W.C. Brown, 1968. [61]p. maps.

First published 1960.

Provides 15 distribution maps, - 1 general, 5 for N. America, S. America, Africa, Eurasia and Oceania respectively; 5 tribal and linguistic maps for those areas, and 4 for Old World and New World prehistory.
Class No: 572.9(084.3)

[1919]

PRICE, D.H. Atlas of world cultures: a geographical guide to ethnographic literature. Newbury Park, Calif., Sage Publications, 1989. 155p. maps. $48; $17.95. ISBN: 0803932405.

Published in cooperation with the Human Relations Area Files.

'A geographical guide to ethnographic books, articles, reports, archival materials, maps and atlases, and other materials of use to anthropologists' (*Introduction*). Bibliographical and geographical information on over 3,500 cultural groups. Contains 41 maps. 1,237 references cited in the bibliography. Culture index, p.127-55, contains cross-references to listings of cultures in *Human Relations Area Files* and Murdock's *Atlas of world cultures (q.v.).*
Class No: 572.9(084.3)

Worldwide

[1920]
Ethnologie régionale. Poirier, J., *ed.* Paris, Gallimard; New York, French & European, 1972-8. 2v. (xv, [1], 1608p.). illus., diagrs., tables, maps. $59.95, v.1; $92.60, v.2. (*Encyclopédie de la Pléiade, 33, 42.*) ISBN: 0686564413, v.1; 0686564421, v.2.
    1. *Afrique-Océanie.* 1972. 2. *Asie-Amérique.* 1978.
    Documented essays (*e.g.* Mozambique, p.922-65; 4p. of bibliography). Particularly rich in indexes: names of people; geographical; authors; subjects (p.1509-58); detailed contents list; list of maps and illus. An outstanding contribution to the Encyclopédie series.
Class No: 572.9(100)

[1921]
Peoples of the world. Stacey, T., *editorial director.* New York, Dansbury Press: Grolier, 1972. 20v. (each 148p.). col. illus., maps. ISBN: 0854682759.
    1. *Australia and Melanesia* (incl. New Guinea). 2. *Africa, from the Sahara to the Zambesi.* 3. *Europe.* 4. *Mexico and Central America.* 5. *Islands of the Atlantic, and the Caribbean.* 6. *Amazonia and pampas.* 7. *Andes.* 8. *The Pacific - Polynesia and Micronesia.* 9. *Southern Africa and Madagascar.* 10. *Indonesia, Philippines and Malaysia.* 11. *South-east Asia.* 12. *The Indian subcontinent* (incl. Ceylon). 13. *China* (incl. Tibet), *Japan and Korea.* 14. *USSR east of the Urals.* 15. *Western and Central Asia.* 16. *The Arctic.* 17. *The Arab world.* 18. *North America.* 19. *Man the craftsman.* 20. *The future of mankind. General index.*
    20 well-illustrated, slim volumes. Numerous contributors.
Class No: 572.9(100)

[1922]
Sixty cultures: a guide to the HRAF probability sample files. Lagacé, R.O., *ed.* New Haven, Conn., Human Relations Area Files. 1977-.
    Part A (1977, 506p. Loose-leaf) comprises profiles of the 60 cultures (Amhara-Yangama) on which HRAF has files. Appendix ('Sample unit concordance') displays 4 types of listing: 1, by cultural units; 2, by G.P. Murdock's *Ethnographic atlas* codes; 3, by 'Outline of world cultures' codes; 4, by 'Standard cross-culture numbers'. 'Major subject index'. Part B promises to cover 60 more cultures (Webb, W.H. *Sources of information in the social sciences,* 3rd ed., 1986, p.391). *Class No:* 572.9(100)

[1923]
SWEET, L.E. *and* O'LEARY, T.J., *eds.* **Circum-Mediterranean peasantry: introductory bibliographies.** New Haven, Conn., Human Relations Area Files, 1969. xxv[i], 106p. map. $15. (*Behavior science bibliographies.*) ISBN: 0875362346.
    Short sections on the countries of North Africa, the Middle East and southern Europe that border on the Mediterranean (*e.g.* Albania, p.29-35, including 3p. of references). No index. An introduction 'for the student who is approaching this area' (*Preface*). *Class No:* 572.9(100)

Developing Countries

[1924]
KURIAN, G.T. **Glossary of the Third World:** words for understanding Third World peoples and cultures. New York, Facts on File, 1989. 300p. $35. ISBN: 0816018421.
    About 10,000 words useful for cultural or anthropological

.... (contd.)
research. Over 70 languages from 120 countries. Appended 'Language families of the world'. No bibliography.
Class No: 572.9(4/9-77)

Europe

[1925]
East European peasantries: social relations. An annotated index of periodical articles. Sanders, I.T., *and others, comps.* Boston, Mass., G.K. Hall, 1976. vi, 179p. $18. ISBN: 0816178607.
    Lists over 800 periodicals on East European peasantries during the 20th century. While the bibliography concentrates on social relations, the articles also concern the demographic, economic and cultural aspects of rural life in the area before and after World War II. Arrangement is by countries: Bulgaria, Czechoslovakia, Greece, Hungary, Poland, Rumania and Yugoslavia, - but not Albania.
Class No: 572.9(4)

[1926]
KUTER, L. **The Anthropology of Western Europe:** a selected bibliography. Bloomington, West European Studies, Indiana Univ., 1978. 132p.
    Comprises 1,112 numbered, unannotated entries in 2 sections. Section 1 lists apposite journals; section 2 groups by geographical area, with A-Z author subdivision. General author index. 'Strong in French and English sources' (Webb, W.H., and associates. *Sources of information in the social sciences,* item 864). *Class No:* 572.9(4)

[1927]
THEODORATUS, R.J. **Europe: a selected ethnographic bibliography.** New Haven, Conn., Human Relations Area Files Press, 1969. xi, 544p. $25. (*Behavior science bibliography.*) ISBN: 0875362397.
    About 8,000 entries, arranged under major European ethnic groups. Cyprus, Iceland, the Azores, the Canaries, Cape Verde Islands and Tristan da Cunha appear because of their European cultural connections. Main emphasis is on 19th and 20th centuries. *Class No:* 572.9(4)

USSR

[1928]
AKINER, S. **Islamic peoples of the Soviet Union:** an historical and statistical handbook. 2nd ed. London, KPI, 1986. 462p. maps. £65. ISBN: 071030188x.
    First published 1983.
    5 sections: 1. Introduction - 2. General information - 3. European USSR and Siberia - 4. Transcaucasia and northern Caucasus - 5. Central Asia and Kazakhstan. Appendix on non-Muslim Turkic peoples of the Soviet Union. Gives details of history and origin, occupations, education and religious diversions. Chronological table. Bibliography, p.446-52. Of the previous ed., 'A helpful supplement to Zev Katz's *History of the major Soviet nationalities'* (*Choice,* v.21(7), March 1984, p.945). *Class No:* 572.9(47)

[1929]
Handbook of major Soviet inhabitants. Katz, Z., *ed.* New York, Free Press, 1975. 481p. *Class No:* 572.9(47)

[1930]
HORAK, S.M. Guide to the study of the Soviet nationalities: non-Russian peoples of the USSR. Littleton, Colo., Libraries Unlimited, 1982. 365p. $35. ISBN: 087287270x.

11 contributors. 1,345 numbered, annotated entries, averaging *c.*100 words apiece. Sections (with introductions): Balts; Belorussians; Ukrainians; Moldavians; Jews; Peoples of the Caucasus (non-Islamic); Islamic peoples (p.174-229); Germans; People of Siberia. Index of authors, compilers and editors, p.255-65. *Class No:* 572.9(47)

[1931]
TITOVA, Z.A. Etnografiya. Bibliografiya russkikh bibliografii po etnografii narodov SSSR (1831-1969). [Ethnography. Bibliography of Russian bibliographies on the ethnography of the peoples of the USSR.] Moscow, Kniga, 1970. 143p.

Annotated list of 734 titles, in historico-ethnographical sections. Index of ethnic appellations, names and titles, and geographical names. *Class No:* 572.9(47)

[1932]
—BENNIGSEN, A. *and* WIMBUSH, S.E. Muslims of the Soviet empire: a guide. Bloomington, Indiana Univ. Press, 1986. 294p. tables, maps. $29.95. ISBN: 0253339588.

3 parts. Part 1 traces the historical roots of Islam. Part 2 examines Muslim groups by regions. Part 3 is a selected bibliography of sources. 88 statistical tables; 3 maps; extensive index. For students and general readers. Very favourable review in *Choice,* v.24(8), April 1987, p.1194. *Class No:* 572.9(47)

Asia—Middle & Near East
[1933]
The Central Middle East: a handbook of anthropology and published research on the Nile Valley, the Arab Levant, southern Mesopotamia, the Arabian Peninsula, and Israel. Sweet, L.E., *ed.* New Haven, Conn., HRAF Press, 1971. 323p. $9. ISBN: 0875361072.

First published 1968.

5 chapters, one on each area concerned. (*e.g.* The Arabian Peninsula, p.199-266). Each chapter has two parts: a narrative, and a selected, annotated bibliography (*e.g.* Israel; text - anthropology, religion and culture, p.267-303; bibliography, 46 items). Non-analytical index, p.313-23. *Class No:* 572.9(53+56)

[1934]
FIELD, H. Bibliography on Southwestern Asia. Coral Gables, Florida, Univ. of Miami Press, 1953-62. 7v. *Supplement* (8v. in 7), 1968-72.

31,254 of the 47,165 entries concern anthropology, drawing on material in about 40 languages. 'Southwestern Asia', in this context, ranges from Egypt to West Pakistan, Turkey and the Caucasus to the Arabian peninsula. Systematic arrangement (of the other entries, 11,505 deal with zoology, and 4,406 with botany). A monumental work.

The *Supplement* (Coconut Grove, Florida, Field Research Projects) is by H. Field and E.M. Laird. *Class No:* 572.9(53+56)

Asia—South & South East
[1935]
Ethnic groups of insular Southeast Asia. LeBar, F.M., *ed. and comp.* New Haven, Conn., Human Relations Area Files Press, 1964. 2v. $25, v.1; $25, v.2. ISBN: 0875364039, v.1; 0875364055, v.2.

1. *Indonesia, Andaman Islands, and Madagascar.* 226p. 2. *Philippines and Formosa.* 167p.

A series of descriptive ethnographic summaries. Thus, v.1, with 19 contributors, has 3 main regions. Short appended bibliographies under tribal groups. Bibliography (authors, A-Z), p.196-219. Index of ethnic names, p.221-6. 10 maps. A companion volume to *Ethnic groups of mainland Southeast Asia (qv).* *Class No:* 572.9(54+59)

[1936]
Ethnic groups of mainland Southeast Asia. LeBar, F.M., *and others.* New Haven, Conn., Human Relations Area Files, 1964. [xiii], 288p. maps. $75.

9 contributors, 4 parts: 1. Sino-Tibetan - 2. Austroasiatic - 3. Tai-Kandai - 4., Malayo-Polynesian. 151 tribal groups, each with appended references. Bibliography (general), p.267-79. Index of names. Country-name concordance. *Class No:* 572.9(54+59)

[1937]
FÜRER-HAIMENDORF, E., von. An Anthropological bibliography of South Asia: together with a directory of recent anthropological field work. Paris & The Hague, Mouton, 1958-70. 3v. $112, v.1; $78.75, v.2; $102, v.3. ISBN: 0686225317, v.1; 9027962065, v.2; 9027963029, v.3.

12,603 numbered references. V.1 (1958) covers the period up to 1954; v.2 (1964), 1955-59; v.3 (1970), 1960-64. 'South Asia' implies the Indian sub-continent. V.1 has 20 sections, 19 of them on particular areas (each section is subdivided: A. Select bibliography of works issued prior 1940; B. Bibliography of works issued 1940-45; C. Field research, 1940-1954. A & B include periodical articles and unpublished theses). Author index. Continued in: *Class No:* 572.9(54+59)

[1938]
—An Anthropological bibliography of South Asia, ... together with a directory of anthropological field research. Kanitkar, H.A., *comp.* Paris & The Hague, Mouton, 1976-.

Continues Fürer-Haimendorf, but drops physical anthropology and prehistoric archaeology. V.1 (1976) covers a 5-year period, 1965-9. *Class No:* 572.9(54+59)

India
[1939]
NATIONAL LIBRARY, Calcutta. Bibliography of Indology. v.1: Indian anthropology. Kanitkar, J.M., *ed.* Rev. and enl. ed. by D.L. Banerjee and A.K. Obdedar. Calcutta, 1960. 290p.

2,067 entries, mostly annotated, in 10 geographical regions. Covers books and selected periodical articles. Author and subject indexes. *Bibliography of anthropology of India (including index to current literature)* (Calcutta, Anthropological Survey of India, 1976-) provides a more detailed continuation in its v.1: 1960-1964. *Class No:* 572.9(540)

Islamic World

[1940]

Muslim peoples: a world ethnographic survey. Weekes, R.V., *ed.* 2nd ed. Westport, Conn., Greenwood Press, 1984. 2v. maps. $95, v.1; $95, v.2; $135, set. ISBN: 0313246394, v.1; 0313246408, v.2.

First published 1978.

Extensively revised and expanded. 139 contributors. 110 documented groups, Acehnese ... Yoruk, (*e.g.* 'Bengalis', p.137-43, including 2p. of bibliography - books, articles, unpublished manuscripts). 3 appendices: 1. Muslim nationalities of the world; 2. Muslims and their ethnic groups (1983); 3. Major Muslim ethnic groups (1983). Index, p.931-53. 8 maps showing locations of Muslim groups. Of the 1978 ed., 'A timely reference work for both public and academic libraries' (*Library journal,* 15 April 1979, p.587). A scholarly reference tool. *Class No:* 572.9(5.297)

Africa

[1941]

COMITÉ INTERNATIONAL DU FILM ETHNOGRAPHIQUE ET SOCIOLOGIQUE. Premier catalogue sélectif international de films ethnographiques sur l'Afrique noire. Paris, UNESCO, 1967. 408p.

Catalogue of 467 films, with annotations and, occasionally, brief critiques. A country-by-country survey, following Africa in general. Indexes of films, subjects, ethnic groups, and film directors. 'This outstanding catalogue' (*Guide to research and reference works on sub-Saharan Africa* [1971], edited by P. Duignan, item 1110). *Class No:* 572.9(6)

[1942]

GIBSON, G.D. 'A Bibliography of anthropological bibliographies: Africa'. In *Current anthropology,* v.10(5), December 1969, p.527-66. map.

872 numbered and closely classified items, with annotations and cross-references. Section 2: General encyclopaedic - 2.04. European dependencies (former and present) - 2.1. North Africa, general ... 2.8. Malagasy Republic. Further subdivided by subjects. 43 library locations 'A research aid of exceptional importance' (Webb, W.H., and others. *Guide to sources of information in the social sciences.* 3rd ed., 1986, item F459). Updated, to 1975, in *African journal* (New York, Africana Publishing), v.8, 1977, p.232-42; v.9, 1978, p.293-206. *Class No:* 572.9(6)

[1943]

JONES, R., *comp.* **African bibliographies:** ethnography, sociology, linguistics and related subjects. London, International African Institute, 1958-61.

West Africa. 1958. v, 116p. *North-east Africa.* 1959. ii, 51p. *East Africa.* 1960. [1], iii, [3], 62p. *South-east Central Africa and Madagascar.* 1961. 53p. *West Central Africa.* 1963. 60p.

Sub-title of each volume: 'general, ethnography/sociology, linguistics'. Three regional bibliographies are based on the Institute's library card-index. The main source is the 'Bibliography of current publications' in each issue of *Africa* (quarterly), 1929-70, - now issued separately as *International African bibliography* (quarterly. 1971-). Aims to list all significant works rather than to produce an exhaustive bibliography. Stresses the cultures, languages and institutions of tropical Africa. *Class No:* 572.9(6)

[1944]

WIESCHHOFF, H.A. Anthropological bibliography of negro Africa. New Haven, Conn., American Oriental Society, 1948. Reprinted, Millwood, N.Y., Kraus Reproduction. xi, 461p. $48. (*American Oriental series, v.23.*) ISBN: 0527026972.

A 'Tribal index' of *c.*18,000 entries, arranged under tribes, A-Z, further subdivided to accommodate travellers' accounts, history, geography, etc. Numerous cross-references for tribal names. Periodical article sources, drawn from over 200 scholarly journals, as well as books and pamphlets. *Class No:* 572.9(6)

Nigeria

[1945]

ITA, N.O. Bibliography of Nigeria: a survey of anthropological and linguistic writings, from the earliest times to 1966. London, Cass, 1971. xxxv, 273p. ISBN: 0714624586.

5,411 annotated entries. Part 1: Nigeria - general (4 sections) - Part 2: Ethnic divisions (A-Z), p.79-245, subdivided into 'General and ethnographic studies' and 'Linguistic studies' and further into Edo, Fulani, Hausa ... Author and ethnic indexes; index of Islamic studies. *Class No:* 572.9(669)

Africa—Southern

[1946]

The Bantu-speaking peoples of Southern Africa. Hammond-Tooke, W.D., *ed.* 2nd ed. London, Routledge & Kegan Paul, 1974. xxii, 525p. illus., maps.

First published 1937; edited by I. Schapera.

Contributions by specialists. Extensive bibliography, p.473-510. 22p. of plates. Index. *Class No:* 572.9(68)

[1947]

Bibliographie de l'Afrique sud-saharienne, sciences humaines et sociales. Tervuren (Belgium), Musée Royal de l'Afrique Centrale, 1978-. Annual. BFr12.

Earlier (1925-59) as *Bibliographie ethnographique du Congo Belge* and *Bibliographie ethnographique de l'Afrique sud-saharienne* (1960-77).

About 1,500 annotated entries for books and articles pa. Authors A-Z, with subjects and tribes indexes. Locations in 4 Belgian libraries. *Class No:* 572.9(68)

[1948]

The Peoples of Southern Africa. Hammond-Tooke, W.D. and West, M., *eds.* Cape Town, Philip, 1978-.

A 14-volume series that began with *The peopling of Southern Africa,* by R. Inskeep in 1978. A second volume, *Peoples of South West Africa,* by W. Pendleton has been followed by others on Khoisan peoples; Xhosa-speaking peoples; Southern Sotho and Tewara peoples; Northern Sotho peoples; Venda, Tsonga and Ndebele peoples; Afrikaans-speaking white people; English-speaking white people; Coloured peoples; Indian peoples; Rural life and urban life. (Source: Musker, R. *Guide to South African reference books. Supplement 1970-1976* (1977), entry no.72). *Class No:* 572.9(68)

South Africa

[1949]

SCHAPERA, I., *comp*. **Select bibliography of South African native life and problems.** Compiled for the Inter-University Committee for African Studies. London, Oxford Univ. Press. 1941. xii, 249p. Supplements. Cape Town, 1956-74.

About 2,500 briefly but critically annotated items in 6 sections: Physical anthropology - Archaeology - Ethnography (p.36-109; 714 items) - Modern status and conditions (p.110-204) - Linguistics, includes periodical articles and government reports. Author index only. The 4 supplements cover 1939-70. *Class No:* 572.9(680)

America

[1950]

GIBSON, G.D. **'A Bibliography of anthropological bibliographies: the Americas'.** In *Current anthropology*, v.1, January 1960, p.61-75.

Annotated bibliography of 200 items, with locations in 13 libraries, Geographical sections: 1. America, general - 1.1 North America - 1.2. Latin America ... 1.22. Antilles ... 1.23. Western South America ... 1.24. Eastern South America. Index of authors, subjects, journal titles. *Class No:* 572.9(7)

[1951]

PAN AMERICAN INSTITUTE OF GEOGRAPHY AND HISTORY. **Boletín bibliográfico de antropologia americana.** Tocubaya, Mexico. Instituto Panamericano de Geografia e Historia, 1937-72. Annual (irregular).

About 2,000 items pa. Apart from reviews of books and periodicals, covers physical, social and applied anthropology, archaeology, art and prehistory, ethnography, folklore, linguistics. Index volume covers 1937-67.

Continued in 'Revista de libros y revistas' (*Bibliografia anthropológica*, v.36(45)-. 1973-). *Class No:* 572.9(7)

America — North

[1952]

Harvard encyclopedia of American ethnic groups. Thernstrom, S., *and others, eds.* Cambridge, Mass., Belknap Press of Harvard Univ. Press, 1980. xxv, 1076p. tables, maps. $95. ISBN: 0674375122.

120 contributors; 106 signed articles (Acadians ... Zoroastrians); 29 thematic essays. The 'Acadians' article has sections: Acadia [East Canada] and diaspora; The Acadians in Louisiana; Social and family life; Group maintenance; Bibliography of 21½ lines. Cross-references. 2 appendices (1. 'Methods of estimating the size of groups'). 87 maps. Lacks an index. Nevertheless, 'a compact, authoritative encyclopedia' (*Choice*, v.15(5), January 1981, p.636). *Class No:* 572.9(71+73)

[1953]

MURDOCK, G.P. *and* O'LEARY, T.J. **Ethnographic bibliography of North America.** 4th ed. New Haven, Conn., Human Relations Area Files, 1975. 5v. $35, v.1; $74, v.2; $73.80, v.3; $35, v.4; $111.30, v.5. ISBN: 0875362052, v.1; 0835728021, v.2; 0317106155, v.3; 0875362117, v.4; 0317106163, v.5.

First published 1941; 3rd ed., 1960.

Supplement, 1973-87, by M.M. Martin and T.J. O'Leary, published 1990. 3v. (v.1 *Indexes;* v.2-3 *Citations*).

....(contd.)

1. *General North America.* 2. *Arctic and Subarctic.* 3. *Far West and Pacific coast.* 4. *Eastern United States.* 5. *Plains and Southwest.*

About 40,000 entries, arranged by geographical areas, with ethnic group subdivisions. No index of authors or subject listings. 'Its value and long-term use are well worth its cost' (*Choice*, v.13(10), December 1976, p.1276). *Class No:* 572.9(71+73)

America—North & Central

[1954]

BERNAL, I. **Bibliografia de arqueologia y etnografia: Mesoamerica y norte de México, 1514-1960.** Mexico, Instituto Nacional de Antropologia e Historia, 1962. 634p. maps. (*Memoria, 7.*)

13,134 items (books and periodical articles), arranged by area. Described in *Handbook of Latin American studies* (v.25, 1963, p.6-7) as 'the most complete bibliography ... ever assembled on the pre-Hispanic culture of Mesoamerica and northern Mexico'. Includes references to reviews. Rather uneven; weak on southern Mesoamerican ethnography. *Class No:* 572.9(71/73)

Canada

[1955]

The Native peoples of Canada: an annotated bibliography of population biology, health and illness. Meiklejohn, C. *and* Rokala, D.A., *eds.* Ottawa, Canadian Museum of Civilization, 1988. 564p. $29.95. (*Archaeological Survey of Canada, no. 134.*) ISBN: 0660107740.

2,100 citations to books, articles and theses. Length of annotations ranges from one line to half a page. Scope is mostly English-language publications, with a few French and German. Entries arranged alphabetically by author. Subject coverage includes culture, medical history, demography, genetics, and environment. Author and subject indexes. *Class No:* 572.9(71)

[1956]

PROULX, J.-R., *comp*. **Review of ethnohistorical research on the native peoples of Quebec.** Quebec, Consulting Services on Social Sciences, Development and Cultural Change, Ministère des Affaires Culturelles du Québec.

1. *Assessment of research.* vp. 2. *Bibliography of ethnohistorical works, 1960-1983.* xvii, 163p. 3. *Bibliography of published works.* xiii, 172p.

V.2 covers Amerindians of Quebec (12 sections) and Inuit of Quebec. V.3 concerns primary and secondary sources on 12 tribes (Amerindians ... Inuit). *Class No:* 572.9(71)

USA

[1957]

ALLEN, J.P. *and* TURNER, E.J. **We the people: an atlas of America's ethnic diversity.** New York, Macmillan, 1988. 315p. tables, maps. (some col.). $105. ISBN: 0029014204.

Distribution of 67 ethnic groups is shown using 115 maps (111 in colour). 13 chapters, the first 3 of which discuss the methods and techniques used in compiling the data, while the remaining 10 look at geographical ancestry origin. Summary table for each group includes data on multiple ancestry; number of persons claiming single ancestry; the 5 US counties with the largest population of that group. Almost

....(contd.)

one third of the book is made up of appendices of county ethnic census data. Brief ethnic populations and places indexes. 'Outstanding reference source' (*Booklist,* 15 April 1988, p.1399). *Class No:* 572.9(73)

[1958]
—'Ethnics in American society'. In *Booklist,* July 1990, p.2109-12.

49 briefly annotated entries, covering references sources on ethnic minorities. Sections: General; African Americans; Asian Americans; Hispanic Americans. *Class No:* 572.9(73)

[1959]
JOSEY, E.J. *and* DELOACH, M.L. **Ethnic collections in libraries.** New York, Neal Schuman, [c.1983]. xvi, 361p. $39.50. ISBN: 0918212634.

18 original essays on non-European ethnic collections (i.e. Hispanic American, Afro-American, Asian-American and native American) in US academic and public libraries, with data on accessibility, special items, publications, etc. 'An excellent starting point' (*Reference books bulletin, 1983-1984,* p.87). *Class No:* 572.9(73)

[1960]
WASSERMAN, P. *and* MORGAN, J. **Ethnic information sources of the United States.** 2nd ed. Detroit, Mich., Gale, 1983. 2v. (1380p.). $175. ISBN: 0810303671.

First published 1976.

Covers more than 90 ethnic groups, A-Z, with data on each under 26 heads. Omits such major groups as blacks, American Indians and Eskimos, - adequately treated elsewhere. Indexes. A 3rd ed. is planned.
Class No: 572.9(73)

Black Races

[1961]
Black Americans information directory: a guide to approximately 4,500 organizations, agencies, institutions, programs, and publications concerned with black American life and culture. Smith, D.L., *ed.* Detroit, Gale Research, 1990. 424p. ISBN: 0810374439. ISSN: 10458050.

1990-91 ed.: 17 sections: 1. National associations ... 4. Library collections (143 entries) - 5. Museums and other cultural organizations ... 14. Publications (newspapers, periodicals, newsletters and directories) ... 15. Videos. Master name and keyword index.
Class No: 572.9(73)(=96)

[1962]
HOWARD UNIVERSITY LIBRARY. Washington, D.C. **Dictionary catalog of the Jessie E. Moorland Collection of negro life and history.** Boston, Mass., G.K. Hall, 1970. 9v. Supplement (3v.), 1976. $980. $400, supp. ISBN: 0816108714; 0816109443, supp.

150,000 photolithographed catalogue cards. Subject, title and author entries. Appended index to African and American periodicals. *Class No:* 572.9(73)(=96)

[1963]
Index to black periodials. Boston, Mass., G.K. Hall, 1984-. Annual. $101.72. ISSN: 08996253.

Originally *Index to selected periodicals received in the Hallie Q. Brown Library. Decennial cumulation, 1950-1959* (1962); then as *Index to periodical articles by and about negroes. Cumulation, 1960-1970* (1971); then *Index to periodical articles by and about blacks* (1973-83).

....(contd.)

Indexes *c.*3,000 articles pa. in 37 journals. Late in appearance, states *Magazines for libraries* (6th ed. 1989, p.23). *Class No:* 572.9(73)(=96)

[1964]
NEW YORK PUBLIC LIBRARY. **Dictionary catalog of the Schomburg Collection of negro literature and history.** Boston, Mass., G.K. Hall, 1962. 9v. (8474p.). Supplements, 1967-.

The Schomburg Collection, states the prospectus, is international in scope 'covering every phase of Negro activity wherever Negroes have lived in significant numbers'. *Class No:* 572.9(73)(=96)

[1965]
—Afro-American reference. Davis, N., *ed.* Westport, Conn., Greenwood Press, 1985. 288p. $55. (*Bibliographies and indexes in Afro-American studies, no.9.*) ISBN: 031324930x.

An annotated guide to 105 general reference works on black studies, plus topical chapters on 537 books covering slavery, sociology, family, etc. 'A boon to public and academic libraries' (*Wilson library bulletin,* April 1986, p.62). *Class No:* 572.9(73)(=96)

[1966]
—A Bibliography of the negro in Africa and America. Work, M.N., *comp.* New York, H.W. Wilson, 1928. xxi, [ii], 698p. Reprinted in New York, Octagon, 1965.

Lists more than 17,000 entries for books, pamphlets and articles in different languages issued before 1928. Pt.1: 'The negro in Africa'; (p.1-247); pt.2: 'The negro in America'. Systematic subdivision. In the author index, names of negro authors are asterisked.

Bibliographies in the quarterly *Journal of negro history* (Washington, Association for the Study of Negro Life, 1916-) update. *Class No:* 572.9(73)(=96)

[1967]
—STEVENSON, R.M. Index to Afro-American reference resources. Westport, Conn., Greenwood Press, 1988. 315p. $39.95. (*Bibliographies and indexes in Afro-American and African Studies, no. 20.*) ISBN: 0313245800.

Covers 181 sources, mainly published in the US, but also including South America, Canada, and the Caribbean. Author and title indexes. *Class No:* 572.9(73)(=96)

America—South

[1968]
O'LEARY, T.J. **Ethnographic bibliography of South America.** New Haven, Conn., Human Relations Area Files, 1963. xxiv, 387p. $27.50. (*Behavior science bibliographies.*)

24,000 entries, arranged into cultural areas, subdivided by tribal groups, A-Z. Significant works are asterisked.
Class No: 572.9(8)

[1969]
OLIEN, M.D. **Latin Americans: contemporary peoples and their cultural traditions.** New York, Holt, Rinehart & Winston, 1973. vii, 408p. illus., tables, maps. ISBN: 0030862525.

Features an extensive bibliography, p.321-96, helpfully subdivided. 'Designed to give the student and general reader an understanding of the temporal, spiritual and cultural setting of the region' (*Handbook of Latin American Studies: Social sciences.* v.37, item no.518). *Class No:* 572.9(8)

[1970]

—Handbook of Latin American studies: Social sciences. Austin, Texas, Univ. of Texas Press, v.27-, 1965-. Two-yearly.

Carries a regular section on anthropology. Thus, v.45 (1983) carries p.15-233, 'Anthropology', -1,649 annotated entries in 5 sections : General - Archaeology - Ethnology - Anthrolinguistics - Human biology. The volume has subject and author indexes, plus a list of journals covered.
Class No: 572.9(8)

Indonesia

[1971]

KENNEDY, R. **Bibliography of Indonesian peoples and cultures.** Rev. ed., edited by T.W. Maretzki and H. Th. Fischer. New Haven, Conn., Southeast Asia Studies, Yale Univ., by arrangement with Human Relations Area Files, 1955. 2v. (xxviii, 663p.). maps. (1962 ed. in 1v.). $57.30.
First published 1945.

About 12,000 unannotated entries. Area coverage: Indonesia, Sumatra, Bornea, Celebes, Java (and Madura), Lesser Sunda Islands, Moluccas. No indexes.

Repertorium op de literatuur betreffende de Nederlandiche Kolonien is cited (v.1, p.1-3) as listing all publications issued between 1595 and 1932.

The 1962 ed. of Kennedy's *Bibliography* adds no new titles since 1951. Important items are asterisked.
Class No: 572.9(910)

Melanesia

[1972]

ELKIN, A.P. **Social anthropology in Melanesia:** a review of research. London, Oxford Univ. Press, 1953 (reprinted Greenwood Press, 1976). xiii, 166p. maps.
Issued under the auspices of the South Pacific Commission.

3 parts: 1. Types of ethnographical record and research, up to 1950 (chapters 1-3) - 2. Survey of anthropological knowledge of the region, with suggestions for research projects (chapters 4-10; many references) - 3. Principles of a plan of anthropological research in Melanesia, related to native welfare and development. *Class No:* 572.9(932)

Papua—New Guinea

[1973]

AUSTRALIAN NATIONAL UNIVERSITY, Canberra. Department of Anthropology and Sociology. **An Ethnographic bibliography of New Guinea.** Canberra, Australian National Univ. Press, 1968. 3v.

V.1: Author index; v.2: District index; v.3: Proper names index. About 10,000 entries, ranging from mid-19th century publications to those of 1964. *Class No:* 572.9(954)

Polynesia

[1974]

BERNICE P. BISHOP MUSEUM, Honolulu. **Dictionary catalog of the Library of Bernice P. Bishop Museum.** Boston, Mass., G.K. Hall. 9v. (6443p.). First and second supplements, 1967-69. 676p. +239p. ISBN: 0816106797.

The main catalogue comprises 135,000 cards, photolithographically reproduced. Concentrates on Polynesia

....(contd.)

and the adjacent areas of Micronesia and Melanesia. 'Emphasis is on materials relating to the network of relationships in culture and nature throughout the island area'. *Class No:* 572.9(96)

[1975]

TAYLOR, C.R.H., *comp.* **A Pacific bibliography:** printed matter relating to the native peoples of Polynesia, Melanesia and Micronesia. 2nd ed. London, Oxford Univ. Press, 1965. xxx, 692p. map.
First published 1931.

More than 16,000 references (to periodical articles and books), classified into island groups and by subjects. Covers every island group of Polynesia, Melanesia and Micronesia, plus New Zealand. Over 13% of entries concern the Maori, over 5%, Hawaii, and 4%, the Fiji Islands (Ad Orientem. *Catalogue nine,* 1967, p.147). *Class No:* 572.9(96)

Amerindians, North

Bibliographies

[1976]

HOXIE, F.E. *and* MARKOWITZ, H. **Native Americans: an annotated bibliography.** Englewood Cliffs, N.J., Salem Press, 1991. 325p. $40. ISBN: 0893566705.

Over 1,000 annotated entries. 4 sections: 1. General studies and references - 2. History - 3. Culture areas - 4. Contemporary life. Annotations are non-evaluative. Updates and complements Hirschfelder's *Guide to research on North American Indians (qv)* (Chicago, ALA, 1983). Author index but no subject index. *Class No:* 572.9(=97)(01)

[1977]

Index to literature on the American Indian, 1970-. San Francisco, Indian Historian Press, 1972-. Annual.
Published for the American Indian Historical Society.

An author and subject index of monographs and periodical articles, initially from US and Canadian journals.
Class No: 572.9(=97)(01)

[1978]

Indians of the United States and Canada: a bibliography. Smith, D.L., *ed.* Santa Monica, Calif., 1974-83. 2v. (*Clio bibliography series 3, 9.*) ISBN: 0874361494.

4,905 annotated entries in all, drawn from *America: history and life abstracting service,* 1954-78, on native North American history from pre-Columbian times to date. 4 period sections, divided by regions, then tribes. Subject and author indexes. *Class No:* 572.9(=97)(01)

[1979]

NEWBERRY LIBRARY CENTER FOR THE HISTORY OF THE AMERICAN INDIAN. **Bibliographical series.** [Newberry Library Center for the History of the American Indian]. Bloomington, Indiana Univ. Press, 1976-. v.1-. maps.

A series of critical bibliographies (more than 30 titles, to date). *The Yakimas: a critical bibliography,* by H.H. Schuster (1983. 158p.) includes a bibliographical essay, a basic reading list, suggested titles for a core library collection, and a discussion of Yakima tribal publications. ('An excellent introduction to the ethnicity of the Pacific Northwest population' *Choice,* v.20(2), October 1982, p.247). *Class No:* 572.9(=97)(01)

[1980]

SMITHSONIAN INSTITUTION. National Museum of Natural History. National Anthropological Archives. **Catalog to manuscripts at the National Anthropological Archives.** Boston, Mass., G.K. Hall, 1975. 4v. (2855p.). $435, set. ISBN: 0816111944.

22,800 photolithographed catalogue cards, on the languages, technology and history of North American Indians. About 5,000 main entries. Includes transcripts of oral history and music, drawings, manuscript maps, and maps with manuscript annotations.
Class No: 572.9(=97)(01)

[1981]

WOLF, C.E. *and* CHIANG, N.S. **Indians of North and South America:** a bibliography based on the collection at the Willard E. Yager Library-Museum, Hartwick College, Oneonta, N.Y. Supplement. Metuchen, N.J., Scarecrow Press, 1988. 654p. $59.50. ISBN: 0810821273.

Supplements the original bibliography (Scarecrow Press, 1978).

Contains 3,542 entries. Listed items may be borrowed from the library. Indexes to titles, series and subjects. Criticized in *Choice,* v.26(4), December 1988 for typographical errors, 'otherwise, it is an excellent volume of great value to research libraries.' *Class No:* 572.9(=97)(01)

Encyclopaedias

[1982]

WALDMAN, C. **Encyclopedia of native American tribes.** New York; Oxford; Facts on File, 1988. xiii,293p. illus. (col.). ISBN: 0816014213.

Alphabetical list of tribes and peoples (Abnaki ... Zuni), p.1-263. Glossary, p.265-73. 4-page list of further reading. Index. Suitable for school and public libraries.
Class No: 572.9(=97)(031)

Handbooks & Manuals

[1983]

Handbook of Middle American Indians. Wauchope, R., *ed.* Austin, Univ. of Texas Press, 1964-76. 16v. Supplements, 1981-.

1. *Natural environment and early cultures.* 1964. 2-3. *Archaeology of Southern Mesoamerica.* 1965. 4. *Archaeological frontiers and external connections.* 1966. 5. *Linguistics.* 1967. 6. *Social anthropology.* 1967. 7-8. *Ethnology.* 1969. 9. *Physical anthropology.* 1970. 10-11. *Archaeology of Northern Mesoamerica.* 1971. 12-15. *Guide to ethnohistorical science.* 1973-75. 16. *Sources cited and artifacts illustrated.* 1976.

Each volume consists of specialist contributions on the life, customs, culture, milieu, etc., plus an extensive bibliography and detailed index. Lacks any preface or formal introduction to the work as a whole. Covers the area South of F.W. Hodge's *Handbook of American Indians (qv),* and updates J.H. Steward's *Handbook of South American Indians (qv).* 'It would be hard to exaggerate the importance of this series' *(American anthropologist,* v.67, 1965, p.1333).

Supplements are published irregularly.
Class No: 572.9(=97)(035)

[1984]

Handbook of North American Indians. Sturtevant, W.C., *ed.* Washington, Smithsonian Institution, 1978-. illus., maps.

1. *Introduction.* 2. *Indians and Eskimos in contemporary society.* 3. *Environment, origin and population.* 4. *History of Indian-white relations.* 1991. 5. *Arctic;* edited by D. Damas, 1984. 6. *Subarctic;* edited by J. Helm. 1981. 7. *The Northwest coast;* edited by W.C. Sturtevant and W. Suttles. 1990. 8. *California;* edited by R.F. Heizer. 1978. 9/10. *Southwest;* edited by A. Ortiz, 1979-83. 11. *The Great Basin;* edited by W.L. D'Azevedo. 1989. 12. *The Plateau.* 13. *The Plains.* 14. *The Southeast.* 15. *Northeast;* edited by B.G. Tripper. 1978. 16. *Technology and the visual arts.* 17. *Native languages.* 18/19, *Biographical dictionary.* 20. *Consolidated index.*

Planned in 20v. V.7, *The Northwest coast,* (1990), comprises 58 chapters, made up of introductory chapters on human biology, language, history of research, early prehistory, followed by 24 chapters on particular peoples. Profusely, and well illustrated; well documented (v.7 has a 100-page bibliography) and indexed. Authoritative. 'The unrelenting excellence of these volumes marks an editorial achievement of the first rank' (Webb, W.H., and associates. *Sources of information in the social sciences* (3rd ed., 1986. item F989)). *Class No:* 572.9(=97)(035)

[1985]

JENNESS, D. **The Indians of Canada.** 7th ed. Toronto, Univ. of Toronto, with National Museum of Canada, 1977. 432p. illus. (incl. col. pl.), ports., maps. $118.10. *(National Museum bulletin, no.65. Anthropological series, no.15.)* ISBN: 0835781801; 0802022863.

Part 1 concerns social and cultural aspects; part 2, individual tribes. Well documented; general index.
Class No: 572.9(=97)(035)

Tables & Data Books

[1986]

STUART, P. **Nations within a nation: historical statistics of American Indians.** Westport, Conn., Greenwood Press, 1987. x,251p. tables. $50. ISBN: 0313238138.

9 sections: 1. Introduction, scope and purpose ... 3. Population ... 8. Employment, earnings, and income - 9. Indian resources and economic development. Bibliography, p.233-41. A quick reference tool for historical statistics.
Class No: 572.9(=97)(083)

Maps & Atlases

[1987]

WALDMAN, C. **Atlas of North American Indians.** New York, Facts on File, 1985. 276p. illus. (incl. col.), maps. $29.95; $16.95. ISBN: 0871968509.

Over 100 two-colour maps. Appendices include a chronology of North American Indian history, a list of US and Canadian tribes, with locations, reservations, a directory of Indian museums and archaeological sites, and a list of major Indian place-names in the US and Canada.
Class No: 572.9(=97)(084.3)

Research Methods

[1988]

HAAS, M.L. **Indians of North America:** methods and sources for library research. Hamden, Conn., Shoestring Press, 1983. xii, 160p. $29.50. ISBN: 0208019804.

3 parts: 1. Library methodology and research library reference tools - 2. Annotated bibliography of books by topic - 3. A list of books on individual tribes. Adequate index. Well organized. 'A first-rate general guide to researching the Indians of North America' (*RQ*, v.24(1), Fall 1984, p.108). *Class No:* 572.9(=97):001.891

Research Projects

[1989]

Guide to research on North American Indians. Hirschfelder, A.B., *and others*. Chicago, Ill., American Library Association, 1983. 330p. $32.50. ISBN: 0838903533.

27 topical chapters, in 4 major areas: general studies; history; economic and social aspects; religion, arts and literature. About 1,100 annotated entries for books, periodical and government documents. Author-title and subject indexes. 'A good guide to basic tools published before 1980' (*Wilson library bulletin,* November 1983, p.225-6). *Class No:* 572.9(=97):061:061.62.005

Amerindians, South

[1990]

SALZANO, F.M. *and* CALLEGARI-JACQUES, S.M. **South American Indians:** a case study in evolution. Oxford, Clarendon Press, 1988. xii,259p. tables, maps. ISBN: 0198576358.

Describes what is known about the origin and differentiation of the South American Indians. 10 chapters: 1. Origins and early differentiation ... 3. Population structure - 4. Ecology, nutrition and physiological adaptation ... 6. Morphology ... 9. Discontinuous genetic variability: multivariate analysis - 10. Synthesis. 3 indexes: author; population; subject. Bibliographies appended to each chapter in addition to an appendix listing *c.*400 items and 14 statistical tables. *Class No:* 572.9(=98)

[1991]

STEWARD, J.H., *ed*. **Handbook of South American Indians.** Washington, Smithsonian Institution, 1946-59. 7v. illus., maps.

1. *The marginal tribes.* 1946. 2. *The Andean civilizations.* 1946. 3. *The tropical forest tribes.* 1948. 4. *The circum-Caribbean tribes.* 1948. 5. *The comparative ethnology of South American Indians.* 1949. 6. *Physical anthropology, linguistics and cultural geography of South American Indians.* 1950. 7. *Index.* 1959.

A truly comprehensive study, with extensive chapter bibliographies. V.7, the general index, has *c.*20,000 entries. Some of the volumes have glossaries. *Class No:* 572.9(=98)

Maoris

[1992]

BARLOW, C. **Tikanga whakaaro: key concepts in Maori culture.** Oxford, University Press, 1991. xvii,187p. illus. £13.50. ISBN: 0195582128.

Explanations in English and Maori of 70 terms which are

....*(contd.)*

important in Maori culture. Entries, which are arranged A-Z, vary in length from ½-page to several pages. Brief bibliography, p.187. *Class No:* 572.9(=992.32)

[1993]

METGE, J. **The Maoris of New Zealand: Rautatu.** Rev. ed. London, etc., Routledge & Kegan Paul, 1976. xx, 382p. illus., diagrs., tables. ISBN: 0710083521.

First published 1967.

18 chapters: 1/2. The Maoris before 1800 ... 6. The bases of daily living ... 14. Leadership and social control ... 17. Literature and art - 18. Maori and Pekeha. Appendix: Maori spelling and pronunciation. Glossary. Bibliography (grouped), p.351-67. Non-analytical index. 16 plates; 11 tables. *Class No:* 572.9(=992.32)

[1994]

TAYLOR, C.R.H. **A Bibliography of publications on the New Zealand Maori** and the Moriori of the Chatham Islands. Oxford, Clarendon Press, 1972. 152p. ISBN: 0198181566.

A revision of the New Zealand and Maori section of the author's *Pacific bibliography (qv).* 42 sections. Includes books and periodical articles. Index. *Class No:* 572.9(=992.32)

[1995]

—COPPELL, W.G. World catalogue of theses and dissertations about the Australian aborigines and Torres Strait islanders. Sydney Univ, Press, 1977. x, 113p. ISBN: 0424000393.

An author list of *c.*700 entries. Subject index. *Class No:* 572.9(94)

Aborigines

[1996]

GREENWAY, J. **Bibliography of the Australian aborigines and the native peoples of Torres Strait, to 1959.** Sydney, London, etc., Angus E. Robertson, 1963. xv, 420p. map.

10,283 numbered entries under authors A-Z. A list of abbreviated titles of *c.*1,000 periodicals precedes. The subject index (p.371-420) includes a list of tribes and a key map locating 554 tribles, but as a subject index it is poorly constructed. Under 'Language : Linguistic studies; General' there are *c.*700 undifferentiated references to entry numbers. The heading 'General works' features on p.398-420, and there are insufficient specific headings throughout. *Class No:* 572.9(=995)

[1997]

—TINDALE, N.B. Aboriginal tribes of Australia: their terrain, environmental controls, distribution, limits and proper names. Berkeley, Calif., University of California Press, 1974. 2v. (Text; Maps). $150. ISBN: 0520020057.

Text (xii, 404p) has 4 parts: 1. The people and the land - 2. Catalogue of Australian aboriginal tribes (by states, then by tribes, A-Z, with references, p.161-262) - 3. Alternatives, variant spellings and invalid terms. Appendix: 'Tasmanian tribes'. Bibliography, p.389-401. Non-analytical index, p.389-401. Non-analytical index, p.389-401; Tasmanian tribe index, p.402-4. V.2 consists of 4 folding location maps. *Class No:* 572.9(=995)

[1998]
THAWLEY, J. *and* **GAUCI, S. Bibliographies on the Australian aborigine:** an annotated listing. 2nd ed. Bundoora, Vic., Borchardt Library, La Trobe University, 1987. 44p. (*La Trobe University Library publication no. 32.*) ISBN: 0858166623.

First published 1979.

163 entries, arranged alphabetically by author. Scope covers English language monographs, bibliographical articles in scholarly Australian journals, and some unpublished bibliographies available in Australian libraries. Appendix: 'List of useful Australian periodicals'. Subject and geographic index. *Class No:* 572.9(=995)

573 Biology

Biology

Abbreviations & Symbols

[1999]

DUPAYRAT, J. Dictionary of biomedical acronyms and abbreviations. 2nd ed. Chichester, John Wiley & Sons, 1990. 162p. £19.50. ISBN: 0471926493.

First published 1984.

About 7,000 definitions for the 4,000 most common biomedical acronyms. Increase in the number of basic acronyms is 30% up on the previous edition. Intended primarily for researchers, physicians and translators. *Class No:* 573.0(003)

[2000]

Units, symbols, and abbreviations: a guide for biological and medical editors and authors. Baron, D.N., *ed.* 4th ed. London, Royal Society of Medicine Services, 1988. 64p. £5. ISBN: 0905958780.

First published 1971.

Sections: 1. Metrication and SI units - 2. Symbols and nomenclature (p.21-51; 23 references) - 3. Layout of references - 4. Proof correction. 'Many journals quote this booklet as a guide in their Instructions to Authors', (*Introduction*). *Class No:* 573.0(003)

Databases

[2001]

BIOSIS PREVIEWS. Philadelphia, Pa., BIOSIS, 1969-.

Contains almost 8,000,000 citations from *Biological abstracts, Biological abstracts/RRM,* and *Bio-research index.* The *Biological abstracts* records, 1976- have abstracts while most *BA/RRM* records do not contain abstracts; *BioResearch index* records have abstracts. Updated weekly, and available on CD-ROM from Silverplatter. *Class No:* 573.0(003.4)

[2002]

SNOW, B. 'Data selection in the life sciences'. In *Database* v.8(3), August 1985, p.15-44.

5 sections: 1. Databases with life sciences information (over 100 entries) - 2. Comparisons in coverage (BIOSIS PREVIEWS; CA SEARCH; EMBASE (*Excerpta Medica*) International Pharmaceutical Abstracts (IPA); MEDLINE; SCISEARCH) - 3. Ease of access-indexing of secondary concepts in search modifiers - 4. Chemical substance indexing in life science databases: comparison of nomenclature standards - 5. Data availability (databank; years of coverage; update frequency).

Serial sources for the BIOSIS data base (Bio Sciences Information Service, 1990. 415p.) lists more than 29,000 titles.

....(contd.)

CABS (*Current awareness in biological sciences*) database contains abstracts and references in 12 different printed services. *Class No:* 573.0(003.4)

Bibliographies

[2003]

Biologie-Dokumentation: Bibliographie der deutschen biologischen Zeitschriftenliteratur, 1796-1965. Scheele, M. *and* Natalis, G. Munich, K.G. Saur, 1981-82. 24v. $2,200. ISBN: 3598302959.

274,752 author entries, A-Z, occupy v.1-19. Catchword index, v.20-22. Systematic index, v.23-24. Bibliographical data in entries include illus., tables, and language of original. *Class No:* 573.0(01)

[2004]

DAVIS, E.B. Using the biological literature. New York & Basle, Dekker, 1981. xx, 286p. $69.75. ISBN: 0824772091.

11 subject chapters, subdivided by form: 1. Introduction to biological literature - 2. Subject headings - 3. General biology - 4. Biochemistry, biophysics and molecular biology - 5. Botany, including fungi - 6. Biology - 7. Entomology - 8. Genetics - 9. Microbiology and immunology - 10. Physiology - 11. Zoology. Annotations, where given, are usually brief. Chapter-end references. Analytical index, p.249-86. Praised at the time for up-to-dateness and comprehensiveness (*Choice,* v.19(5), January 1982, p.603), but now overtaken by Wyatt (*qv*). *Class No:* 573.0(01)

[2005]

Information sources in the life sciences. Wyatt, H.V., *ed.* 3rd ed. London, etc., Butterworths, 1987. xiv, 191p. tables. (*Butterworths Guides to information sources.*) ISBN: 040811472x.

First published 1966, as *Use of biological literature;* edited by R.T. Bottle and H.V. Wyatt; 2nd ed. 1971 (392p.).

9 contributors. Introduction; 13 chapters (6 by H.V. Wyatt): 1. Reading for profit: current awareness - 2. Literature searching by computer - 3. Secondary sources: abstracts, indexes and bibliographies - 4. Databanks; collections [of specimens] - 5. Guides to the literature - 6. Biochemical sciences - 7. Microbiology - 8. Biotechnology - 9. Genetics - 10. Zoology - 11. Ecology - 12. Botany - 13. History of biology. Index, p.187-91.

This 3rd ed. is only half the size of the 2nd, and users are referred to the 2nd ed. on foreign serials, patents, botanical taxonomy, history and biography of biology. Aimed at working scientists and information specialists; 'a guide rather than a compendium'. Good on databanks. Up-to-date, but marred by an inadequate index, and exorbitantly priced. *Class No:* 573.0(01)

[2006]

—KIRK, T.G., *Jr*. Library research guide to biology: illustrated research strategy and sources. Ann Arbor, Mich., Pierian Press, 1978. 83p. facsims. $25. (*Library research guide series, no.2.*) ISBN: 087650098x.

Stresses methodology, with numerous examples. Appended lists of recent serials, guides to the literature, and 'Basic reference sources for biology courses', plus 'Using *Chemical abstracts*' and '*Zoological record*'. Brief index of titles. Addressed to students, teachers and librarians. *Class No:* 573.0(01)

[2007]

—KRONICK, D.A. The Literature of the life sciences. Philadelphia, Pa., ISI Press, 1985. 219p. $29.95. ISBN: 0894950452.

Provides an overview of the literature, primary and secondary, rather than a detailed examination of titles (484 are listed) and subject areas. Methods of searching the literature, via bibliographies and other retrieval services (including online) are discussed. Index. Addressed mainly to librarians, students and scientists. *Class No:* 573.0(01)

[2008]

SMITH, R.C., *and others*. **Smith's guide to the literature of the life sciences.** 9th ed. Minneapolis, Minnesota, Burgess Publishing Co., 1980. xi, 223p. ISBN: 0808755762.

First published 1943, as *Guide to the literature of the zoological sciences;* 8th ed. 1972.

About 2,000 annotated entries in 11 sections: 1. The scientist and research librarian - 2. The mechanics of the library and book classification - 3. Secondary literature: textbooks and reviews - 4. Bibliographic form; bibliographies and indexes of the life sciences - 5. Abstracts and abstracting journals - 6. Ready reference works - 7. Serials and primary research journals - 8. Literature of taxonomy - 9. Searching the literature (including databases) - 10. Preparations for writing theses and dissertations - 11. Preparation for scientific writing. Designed to help young scientist-aspirants to become familiar with the searching, reviewing, reporting, publishing and indexing systems in the life sciences. Primarily a library guide. *Class No:* 573.0(01)

[2009]

WARREN, K.S., *ed*. **Coping with the biomedical literature:** a primer for the scientist and the clinician. New York, Praeger, 1981. xi, 233p. illus., diagrs., facsims., tables. $45. ISBN: 0275913554.

12 contributors. 4 parts (12 contributors): 1. The structure of the information system - 2. Producing biomedical information - 3. Utilizing biomedical information - 4. Sources of biomedical information (*e.g.* 'The National Library of Medicine'). Each contribution has appended references (*e.g.* 'Ecology of the biomedical literature', 21 references). Detailed analytical index. Stresses research techniques. *Class No:* 573.0(01)

Encyclopaedias

[2010]

Encyclopedia of human biology. Dulbecco, R., *ed*. San Diego, Academic Press, 1991. 8v. (6722p.), illus. (incl. pl.), tables. $1,950. ISBN: 0122267516, v.1; 0122267524, v.2; 0122267532, v.3; 0122267540, v.4; 0122267559, v.5; 0122267567, v.6; 0122267575, v.7; 0122267583, v.8.

Includes coverage of anthropology, biophysics, ecology, genetics, physiology and toxicology. More than 700

.... (contd.)

contributors, 2,000 illus., 3,600 bibliographic entries. Glossary contains 4,800 entries. Separate subject index volume. 900 tables. Each entry consists of a glossary, text and bibliography and is many pages in length (*e.g.* 'Mood disorders'; 8p.: 'Artificial kidney'; 16p.). Review in *Choice*, October 1991 states that this will be the definitive work for many years and continues 'with an improved index, and perhaps some greater concern for rational article titling, this would be a masterpiece'. *Class No:* 573.0(031)

[2011]

—The Encyclopedia of the biological sciences. Gray, P., *ed*. 2nd ed. Melbourne, Fla., Krieger, 1981 (reprint of 1970 ed.). xxv, [1], 1027p. illus., diagrs., tables, maps. $74.50. ISBN: 0898743265.

First published 1961.

800 signed articles by nearly 500 contributors. Articles, except when they are only brief definitions, are signed, sectionalized and documented (*e.g.* 'Microscope', p.555-8; 4 references, 8 figures). About 100 biographies are included. Scope: the developmental, ecological, functional, genetic, structural and taxonomic aspects of the biological sciences. Particularly useful to reference librarians are the entries under 'Name, group, gender and juvenile' (for group terms) and 'Zoological gardens' (citing leading publications). Many diagrams, carefully drawn. The analytical index (p.1002-27) features bold-face type for main articles or passages giving particular detail. Cross-references are inadequate and headings inconspicuous. Now outdated on such rapidly changing subjects as molecular biology and recent discoveries in primate and human evolution. *Class No:* 573.0(031)

Handbooks & Manuals

[2012]

The Cambridge encyclopedia of life sciences. Friday, A. *and* Ingram, D.S., *general eds*. Cambridge Univ. Press, 1985. 432p. illus. (incl. col.), diagrs., tables, maps. £25. ISBN: 0521256968.

Three major parts (15 chapters), each with further reading. 1. Processes and organization (1. The cell ... 5. The geology) - 2. Environments (6. Marine environments ... 10. Living organisms and environments) - 3. Evolution and the fossil record (11. The evolutionary process ... 15. Recent history of the fauna and flora). 31 contributors. Species index; subject index (in tiny print). The book's central theme is 'the adaptations of plants and animals to their environments and their relations with other organisms within those environments' (*Foreword*). Well integrated and well illustrated. *Choice*, (v.23(5), January 1986, p.724) finds coverage uneven, although the book - as much a textbook as an encyclopaedia - is 'highly recommended for undergraduate and public libraries'. *Class No:* 573.0(035)

[2013]

KEETON, W.T. *and* GOULD, J.L. **Biological science.** 4th ed. New York, W.W. Norton, 1986. xxiv, 1267p. illus., (incl. col.), diagrs. ISBN: 0393953858.

First published 1967.

5 sections (43 chapters): 1. The chemical and cellular basis of life - 2. The biology of organisms - 3. The perpetuation of life - 4. The biology of populations and communications - 5. The genesis and diversity of organisms. More coloured illus. than in 3rd ed. Appendix: A

....(contd.)

classification of living things. Suggested reading, p.A7-A19. Glossary, p.A27-A45. Index, p.A46-A91.

A 5th ed. was due to be published in 1992.
Class No: 573.0(035)

Dictionaries

Polyglot

[2014]
HAENSCH, G. *and* **HABERKAMP DE ANTON, G. Dictionary of biology** in English, German, French, Spanish. 2nd ed. Amsterdam, Elsevier, 1981. xii, 680p. $169.25; DM128. ISBN: 0444419683.

First published 1976.

12,809 numbered English-base terms, with equivalents and indexes in German, French and Spanish. The aim is 'to obtain as proportionate a selection of terms as possible from all areas of biology' (*Preface*), although topics such as environment, ethology and ecology figure more prominently than in the 1976 ed. Genders are stated.
Class No: 573.0(038)=00

English

[2015]
BLINDERMAN, C. Biolexicon: a guide to the language of biology. Springfield, Ill., C.C. Thomas, 1990. 363p. $39.75. ISBN: 0398056714.

Lists several hundred Greek and Latin elements which form the base of terms in biology, medicine and science. Many illustrations. Index. *Class No:* 573.0(038)=20

[2016]
Cambridge illustrated thesaurus of biology. Gutteridge, A.C. Cambridge Univ. Press, 1983. 245, [12]p. col. illus., graphs, tables.

Over 2,5000 definitions of terms, including other words related to the same topic. 38 broad headings/sections (*e.g.* cells; bacteria; plants (p.45-79); animals, nervous system), combining to some extent functions of dictionary and thesaurus. A/Z index of words. Much use of colour; keyed illus. Small format. *Class No:* 573.0(038)=20

[2017]
—**HALE, W.G.** *and* **MARGHAM, J.P.** Collins' dictionary of biology. London, Collins, 1988. 576p. diagrs. £6.99. ISBN: 0004343514.

Has more than 5,600 entries and includes brief biographies. 285 diagrams. The authors are both at Liverpool University, Dept. of Biology.
Class No: 573.0(038)=20

[2018]
Chambers biology dictionary. Walker, P.M.B., *ed.* Cambridge, W.R. Chambers; University Press, 1989. xii,324p. diagrs. £17.50. ISBN: 185296152x.

Derived from *Chambers science and technology dictionary* (*q.v.*). Contains *c.*10,000 definitions, including *c.*3,000 in zoology, *c.*2,500 in botany, and *c.*1,200 in biochemistry, molecular biology and genetics. Definitions are concise, with some terms given fuller treatment in a series of over 100 special articles, *e.g.* genetic manipulation, poison. Synonyms and cross-references are indicated in italic and bold type. Largely a duplication of the same subject areas in the parent

....(contd.)

dictionary but has a good, clear layout and would be useful for the more specialized life sciences library.
Class No: 573.0(038)=20

[2019]
GRAY, P. The Dictionary of the biological sciences. Melbourne, Fla., Krieger, 1982 (reprint of 1967 ed.). xx, 602p. $52.50. ISBN: 0898744414.

An outgrowth of *The encyclopedia of the biological sciences (qv).* More than 40,000 very concise definitions in the field of organisms, their anatomy, function and mutual interactions. A drawback is the massing of words, derived from the same root (*e.g.* 'Metabolic' is under '-bolic'). 'Acquire, even if Henderson (which it complements) is available' (*Library journal,* v.93(8), 15 April 1968, p.1585).
Class No: 573.0(038)=20

[2020]
HENDERSON, I.F. *and* **HENDERSON, W.D. Henderson's dictionary of biological terms.** 10th ed., by E. Lawrence. Harlow, Longman Scientific and Technical, 1989. xi,637p. £5.99. ISBN: 0582463629.

First published 1920; as *A directory of scientific terms.* (1st-8th eds.).

Over 18,000 headwords. Updates existing definitions and incorporates many new terms *e.g.* AIDS, gene therapy, biotechnology, ribozyme. Revised and simplified outline classification of the Plant and Animal Kingdoms, the Fungi and Monera (prokaryotes). New appendix of *c.*60 structural formulae of important biotechnical compounds.
Class No: 573.0(038)=20

[2021]
International dictionary of medicine and biology. Landau, S.I., *ed.-in-chief.* New York, Wiley, 1986. 3v. (li, 3200p.). $395. ISBN: 047101849x.

Over 100 contributors; more than 150,000 entries, thanks to the inclusion of biology. Among the subjects most extensively covered are ecology, environmental health, infectious diseases, occupational medicine, toxicology. A concise guide to usage precedes. Highly recommended as 'indispensable for medical and science libraries' (*Library journal,* 15 June 1986, p.61). *Class No:* 573.0(038)=20

[2022]
JAEGER, E.C. A Source-book of biological names and terms. 3rd ed. Springfield, Ill., Thomas; Oxford, Blackwell, 1955 (reprinted 1978). xxxv, 317p. illus. $48.50. ISBN: 0398009163.

First published 1944; 2nd ed. 1950.

Lists nearly 15,000 elements from which scientific biological names and terms are made. 32p. supplement adds *c.*1,000 terms to those in the 2nd ed. Final section of 300 short biographies of 'persons commemorated in botanical and zoological generic names' (*Preface*).
Class No: 573.0(038)=20

[2023]
McGraw-Hill dictionary of the life sciences. Lapedes, D.N., *editor-in-chief.* New York, etc. McGraw-Hill, 1976. 907, 38p. ISBN: 0070452628.

Over 20,000 concise entries, either drawn from *McGraw-Hill Dictionary of scientific and technical terms* (1974) or else specially prepared. Embraces agriculture, analytical chemisty, veterinary medicine, virology, zoology and other related fields of biology. Appendix (38p.) of measurement systems, antibiotics, clinical pathology, animal taxonomy,

.... *(contd.)*

etc. 1,200 small marginal illus. 'Highly recommended' (*Library journal,* v.102(5), 1 March 1977, p.546). *Class No:* 573.0(038)=20

[2024]
MARTIN, E. **Dictionary of life sciences.** 2nd ed. London, Macmillan Press, 1983. [iv], 396p. illus., diagrs., chemical structures. ISBN: 0333348672.

First published 1976.

14 contributors. Over 3,000 entries ('Radioactive dating': 1 column; 'Photosynthesis': 2½ columns). Entries are defined under the most commonly used term, with synonyms shown in brackets. Companion volume to the Macmillan *Dictionary of physical sciences* and *Dictionary of earth sciences.* 'A valuable aid for the A-level biologist and undergraduate' (*New scientist,* 12 April 1984, p.48). *Class No:* 573.0(038)=20

[2025]
—The Facts-on-File dictionary of biology. Tootill, E., *ed.* Rev. ed. New York, Facts on File, 1988. 326p. illus. $24.95. ISBN: 0816018650.

Defines *c.*3,500 terms used in the life sciences. Keyed line-drawings. For students of medicine/biology. *Class No:* 573.0(038)=20

[2026]
The New Penguin dictionary of biology. Abercrombie, M., *and others.* 8th ed. London, Penguin Books, 1990. 599p. diagrs. £5.99. ISBN: 0140511776.

First published 1951 as *The Penguin dictionary of biology.*

Contains *c.*6,000 of the more common biological terms. Length of entries varies from a few words to more than a page, *e.g.* 'Species'. Abbreviations for categories indicate application to botany and/or zoology. Numerous cross-references. *Class No:* 573.0(038)=20

French

[2027]
VAILLANCOURT, J. **Lexique anglais-français:** termes techniques à l'usage des biologistes. Ottawa, Editions de l'Université d'Ottawa, 1978. xii, 427p.

Over 10,400 numbered English terms are given French equivalents. The index of French terms is keyed by number to the main section. English terms are categorized and French terms are given genders. Invaluable for the student and professional alike. *Class No:* 573.0(038)=40

Russian

[2028]
CARPOVICH, E.A. **Russian-English biological and medical dictionary.** 2nd ed. New York, Technical Dictionaries Co., 1960. 400p. $25. ISBN: 0911484019.

First published 1958.

32,650 entries, giving direct adjective-noun approach, - very acceptable to the inexperienced translator and occasional reader of Russian scientific papers. Particularly valuable on biological terms that frequently appear in the medical literature, - terms not found in other dictionaries. Applications of terms appear in parentheses. *Class No:* 573.0(038)=82

[2029]
English-Russian biological dictionary. Chibisova, O.I. *and* Koziar, L.A., *eds.* 3rd ed. Oxford, etc., Pergamon Press, 1978. 732p. £120. ISBN: 0082316329.

About 60,000 entries, including abbreviations and Latin names. 'Nerve': 1½ columns. The dictionary is to be used with caution, since terms consisting of attribute and a key word will be found under the key word. For example, the term 'Chromosome complex' will be found under the key term 'complex'. *Class No:* 573.0(038)=82

Theses

[2030]
BILBOUL, R.R. *and* KENT, F.L., *eds.* **Retrospective index to theses of Great Britain and Ireland, 1716-1950. Volume 3: Life sciences.** Oxford, European Bibliographic Center-Clio Press, 1977. [vii], xi, 327p. £66. ISBN: 0903450054.

About 10,000 entries on the biological and medical sciences. Subject index (Abdomen ... Zygorhynchus). Author index (Name - Title of thesis - Degree - Date - University). *Class No:* 573.0(043)

Reviews & Abstracts

[2031]
Biocontrol news and information: a quarterly journal of news items, review articles and abstracts from the CAB abstracts database, prepared by the Commonwealth Institute of Biological Control. Wallingford, Oxon., CAB International, 1980-. 4pa. £97; $174pa. ISSN: 01431404.

Contains *c.*2,800 abstracts pa, covering *c.*2,500 journal articles, in addition to reports, books, and conferences. Subject coverage includes crop pests; weeds; medical and veterinary pests; techniques; biology; taxonomy and catalogues; ecology. Mini-reviews of literature on topical subjects included in each issue which also contains author and subject indexes. Available online as part of the *CAB Abstracts (qv)* database, on the CD-ROM equivalents and, from 1991, on floppy disk. *Class No:* 573.0(048)

[2032]
Biological abstracts. Philadelphia, Pa., BioScience Information Service (BIOSIS), 1926-. Semi-monthly. $4,425 (USA), pa. ISSN: 00063169.

The world's leading abstracting service in biology and biomedicine. Over 235,000 abstracts pa. from *c.*9,500 journals. Arranged in 84 sections, A-Z, including Aerospace and underwater biological effects; Cardiovascular system; Ecology (Environmental biology); Genetics and Cytogenetics; Immunology; Neoplasms and neoplastic agents; Nervous system; Pharmacology; Public health; Toxicology; Virology. Author index; Biosystematics index; Generic index; Subject index - per issue. Cumulative indexes, 2pa., and 5-yearly (on microfiche, *e.g.* 1975-79).

Reprints from *Biological abstracts* include *Abstracts of mycology* and *Abstracts of entomology.*

Biological abstracts/RRM (qv) is complementary to *Biological abstracts* in its coverage. Both services feed into *BIOSIS PREVIEWS* database (1969-). *Class No:* 573.0(048)

[2033]
Biological abstracts/RRM: reports, reviews, meetings. Philadelphia, Pa., BioScience Information Services (BIOSIS), 1980-. Fortnightly. ISSN: 01926985.

As *Bioresearch index,* v.1-17 (1965-79).

About 235,000 references pa. (No abstracts, despite the title, but keywords are appended to entries). Coverage: symposia, meetings, reviews, monographs, book chapters, etc., - thus complementary to the coverage of *Biological abstracts.* Author, biosystematic, generic and subject indexes. *BIOSIS PREVIEWS* database (1969-).
Class No: 573.0(048)

[2034]
Biological and agricultural index: a cumulative subject index to periodicals in the fields of biology, agriculture and related sciences. New York, H.W. Wilson, 1964/65-. 11pa. ISSN: 00063177.

Continues *Agricultural index,* 1919-64.

Covers 226 key English-language journals, selected by subscriber vote. Includes experimental station publications and book reviews. Entry under topic. US slanted, as compared with *Bibliography of agriculture* and *Agrindex.* About 40,000 entries pa. Available online (WILSONLINE), CD-ROM (WILSONDISC) in addition to magnetic tape and diskette format. *Class No:* 573.0(048)

[2035]
Biology digest. Medford, N.J., Plexus Publishing, Inc., 1974-. 9pa. $125. ISSN: 00952958.

About 4,000 abstracts pa. from *c.*200 periodicals (including *Nature* and *Scientific American*). Sections: Viruses, microflora and plants; Animal kingdom; The human organism; Infectious diseases; Population and health; Cell biology and biogenesis; Environmental quality; General topics. Occasionally carries a feature article; some book reviews. Author and keyword indexes per issue and annually. *Class No:* 573.0(048)

[2036]
Current awareness in biological sciences (CABS). Oxford, etc., Pergamon, 1983-. 12pa. £2,620. ISSN: 07334443.

Formerly *International abstracts of biological sciences,* 1954-82.

About 15,000 references pa., with authors' addresses. Sections (closely subdivided): Biochemistry - Cell and developmental biology - Ecological sciences - Genetics and molecular biology - Microbiology - Plant science - Pharmacology and toxicology - Immunology - Physiology. Journal list, p.viii-xvi. Author index, but no subject index. Available online CABS database via BRS. Also available in microform. *Class No:* 573.0(048)

[2037]
Current contents: Agriculture, biology and environmental sciences. Philadelphia, Pa., Institute for Scientific Information, 1970-. Weekly. $560. ISSN: 00900508.

Previously (1970-72) subtitled 'Agriculture, food and veterinary sciences'.

Issue for 20 May, 1991 contains title-pages of 98 journals. Sections include: Biology ... Environment/Ecology ... Plant sciences ... Entomology/Pest control ... Veterinary medicine/Animal health. Title word index; author index and address directory; publishers' address directory. Available on diskette and online on the CURRENT CONTENTS SEARCH database. *Class No:* 573.0(048)

[2038]
Current contents: Life sciences. Philadelphia, Pa., Institute of Scientific Information, 1958-. Weekly. $950. ISSN: 00113409.

Provides contents lists of *c.*250 journal titles per issue, 1,000 journals altogether. Subject coverage includes: multidisciplinary; chemistry ... neurosciences & behavior; animal & plant science. Author, title and address directory; Publishers address directory. Request-a-print service available. Available online, on CURRENT CONTENTS SEARCH, and on diskette. *Class No:* 573.0(048)

Periodicals & Progress Reports

[2039]
Advances in applied biology. London, New York, etc., Academic Press, 1975-. ISSN: 03091791.

V.10, 1984 (ix, [1], 300p.) has 10 contributors. The 5 papers include 'Plant somatic hybridization' (p.1-69; nearly 12p. of references), and 'Seed quality in grain legumes' (p.217-85; 14p. of references). Detailed, non-analytical subject index. Cumulative list of chapter titles, v.1-10. 'The main purpose of the series is to draw together subject matter from currently important fields of applied biology to produce a synthesis for students, teachers and specialists from other fields' (*Preface*). No further volumes have been published since 1984. *Class No:* 573.0(05)

[2040]
Serial sources for the BIOSIS database. Philadelphia, Pa., BioSciences Information Service, 1971-. Annual.

The 1990 list (vii,415p.) records *c.*29,700 serials in 43 languages covered in *Biological abstracts* (*q.v.*) and *Biological abstracts/RRM* (*q.v.*). Prelims. note new serial titles and ceased/changed serial titles. Data: title, CODEN, frequency, publisher. Appended key to frequency codes and list of publishers. *Class No:* 573.0(05)

Yearbooks & Directories

[2041]
DIXON, A., *comp. and ed.* **Useful addresses for biologists.** 2nd ed. Hatfield, Herts., Association for Science Education, 1984. [i], 38p. £0.30.

First published 1977.

About 400 briefly annotated entries, in 6 sections: 1. Audio-visual aids - 2. General laboratory materials - 3. Living material - 4. Microscopes and accessories - 5. Publishers, scientific societies, etc. - 6. Organisations and commercial bodies offering educational material. Confined to the UK. *Class No:* 573.0(058)

Tables & Data Books

[2042]
ALTMAN, P.L. *and* **DITTMER, D.S. Biology data book.** 2nd ed. Bethesda, Md., Federation of American Societies for Experimental Biology, 1972-74. 3v. (2123p.) diagrs., tables.

First published 1964.

V.1 has tables on genetics, cytology, reproduction, development and growth, properties of biological substances; materials and growths. 5,941 references. - V.2, on biological regulations and toxins, environment and survival, parasitism, sensory and neuro-biology. 5,970 references. - V.3, on

....*(contd.)*

nutrition, digestion and excretion, metabolism, respiration and circulation, bood and other body fluids. Each volume has its own index. *Class No:* 573.0(083)

[2043]
FISHER, Sir R.A. *and* **YATES, F. Statistical tables for biological, agricultural and medical research.** 6th ed. Edinburgh, Oliver & Boyd, 1963. x, 146p. tables.

First published 1938; 5th ed. 1956.

50 tables (numbered 1-34, with interpolations). Main additions to the 6th ed. are: v, 'Fiducial limits for a variance component', and xvii, 'Balanced incomplete blocks-combinatorial solutions'. The valuable introduction give examples of the ways in which the tables can be used and includes 82 references to books, monographs and papers on statistic methods up to 1963. No index. A standard source. *Class No:* 573.0(083)

[2044]
PARKER, S.P., *ed.* **Synopsis and classification of living organisms.** New York, etc., McGraw-Hill, 1982. 2v. (1282p.) illus., diagrs. $295. ISBN: 0070790310.

8,200 articles by nearly 200 contributors. V.2 has a taxon and common name index. 'The compilation that covers both the characteristic and taxonomic position of all taxa down to and including families of viruses, bacteria, plants and animals in a single source' (*RQ,* v.21(4), Summer 1982, p.421-2). *Class No:* 573.0(083)

Nomenclatures

[2045]
JEFFREY, C. Biological nomenclature. 3rd ed. London, Edward Arnold, 1989. ix,86p. ISBN: 0713129832.

First published 1973.

8 sections (The systematic background - Names and codes - Scientific names ... Name changes and synonymy - Authorities and their citation - Special cases). Notes to the text. Bibliography, glossary/index. Of the 2nd ed. (1977), 'The author [Principal Scientific Officer. Royal Botanic Gardens, Kew] has made an excellent job of reducing them [the rules] to the level where they should be understandable by the teacher...' *The School science review,* v.10(208), March 1978, p.570). For all schools with a sixth-form library. *Class No:* 573.0(083.72)

[2046]
WOODS, R.S. The Naturalist's lexicon: a list of Classical Greek and Latin words used, or suitable for use, in biological nomenclature; with abridged English-Classical supplement. Pasadena, Calif., Abbey Press, 1944. xviii, 282p. Addendum, 1947. 47p.

A Greek and Latin-English lexicon, covering more than 15,000 terms. The English-Classical supplement (p.259-82) is classified, dealing in turn with nouns, adjectives, verbs and prefixes. *Class No:* 573.0(083.72)

Patents

[2047]
CRESPI, R.S. Patenting in the biological sciences. Chichester, Wiley, 1982. 211p. £50. ISBN: 0471101516.

An introduction to patenting procedures for research workers in biotechnology and the pharmaceutical and

....*(contd.)*

agrochemical industries. The laws of the UK, USA and other countries, as well as the European Patent Convention and other treaties, are discussed. *Class No:* 573.0(088.8)

Histories

[2048]
ALLEN, G.E. Life science in the twentieth century. Cambridge Univ. Press, 1979. xxv, 258p. illus. £20; £5.95. ISBN: 0471023361.

7 chapters (1. The influence of Darwinian thought on late nineteenth-century biology ... 7. The origin and development of molecular biology). Bibliography, with running commentary, by chapters, p.229-49. Index. 'The history of a few selected areas whose growth and development is characteristic of the vast field that has become general biology'. *Class No:* 573.0(091)

[2049]
DAWES, B. A Hundred years of biology. London, Duckworth, 1952. 429p. diagrs. (*Hundred years series.*)

19 chapters (including 15. Marine biology - 16. Parasites and parasitic diseases - 17. Antibiotics - 18. Agricultural history - 19. Some research institutes [very largely British] and their work). Chapter bibliographies, p.385-418. Index, p.419-29, lists *c.*2,000 names of biologists. *Class No:* 573.0(091)

[2050]
MAGNER, L.N. A History of the life sciences. New York & Basle, Dekker, 1979. xi, 489p. $49.75. ISBN: 082478071x.

14 chapters: 1. In the beginning - 2. The Greeks: natural philosophers and scientists ... 14. Genetics in the twentieth century: molecular biology, or all the world's a phage. Notes and references appended to chapters. Detailed, non-analytical index, p.457-89. Biographical sketches are a feature. *Class No:* 573.0(091)

[2051]
NORDENSKIÖLD, E. The History of biology. Translated from the Swedish by L.B. Eyre. London, Kegan Paul, 1929; reprinted New York, Tudor, 1960. xii, 629, xvp. illus.

Based on a series of lectures, originally published in 3v. (Stockholm, 1920-29). Bibliography, p.617-29. 'Still the best one-volume reference history of biology' (*Stechner-Hafner book news,* v.20(3), November 1965, p.34). *Class No:* 573.0(091)

[2052]
SMIT, P. History of the life sciences: an annotated bibliography. London, A & C Black, 1974.; Forest Burgh, N.Y., Lubrecht & Cramer. xiv,1071p. $96.

About 5-6,000 annotated entries. Chapter 1: General reference tools - 2. Historiography of the life and medical sciences (A. Ancient and medieval period; B. Renaissance and later periods). Appended: 'Selected list of biographies, bibliographies, etc. of famous biologists, medical men, etc., including some modern reviews of their publications'. Index of personal names, p.1037-71. 'Originated as a plan to provide an extension of those parts of Sarton's *Guide to the history of science* that dealt with the life sciences' (*Introduction*). *Class No:* 573.0(091)

[2053]

—OVERMIER, J.A. The History of biology: a selected, annotated bibliography. New York, Garland Publishing, 1989. xviii,157p. $21. (*Bibliographies of the history of science and technology, v.15.*) ISBN: 0824091183.

619 annotated entries covering biology from the 18th to the mid-20th century. Cited items published prior to 1985 and mostly English language. Aimed at students and those new in the field. Does not claim to be comprehensive. *Class No:* 573.0(091)

Biographies

[2054]

ABBOTT, D., *ed*. **The Biographical dictionary of scientists: biologists.** London, Blond Educational, 1983. [vi], 182p. ISBN: 0584700016.

About 200 highly selected, very readable, entries ('Lysenko': 12/3p.; 'Alfred Russel Wallace': 12/3p.). Historical introduction, p.1-6; glossary, p.141-55; index, p.157-82. Useful at school and layperson level. Major general encyclopaedias provide more background. *Class No:* 573.0(092)

[2055]

American men and women of science, 1992-93: a biographical directory of today's leaders in physical, biological and related sciences. 18th ed. New York, Bowker, 1992. 8v. (8228p.). $750. ISBN: 0835230740.

First published 1906, as *American men of science*.

Biographical sketches on over 125,000 active US and Canadian scientists working in over 164 sub-specialities. Almost 4,000 new entries. Separate discipline index (A-Z by specialty). Of the 17th ed., 'A truly indispensable directory for reference collections of all levels' (*Science and technology libraries,* v.10(4) Summer 1990, p.123). Available online via Dialog. *Class No:* 573.0(092)

Writing & Lecturing

[2056]

COUNCIL OF BIOLOGY EDITORS. **Council of Biology Editors style manual: a guide for authors, editors and publishers in the biological sciences.** 4th ed. Arlington, Va., Council of Biology Editors, 1978. xvii, 265p. tables. ISBN: 0914340026.

First published 1960.

12 chapters (2. Writing the article: the first draft and revisions - 3. The final draft - 4. Editorial review of manuscripts ... 6. Proof correction - 7. Indexing - 8. General style conventions - 9. Style in special fields - 12. Annotated bibliography, p.231-40). Analytical index, p.241-65. *Class No:* 573.0:001.81

Libraries

[2057]

Life sciences libraries in and around London. Yeadon, J., *and others, eds*. 6th ed. London, Imperial College: King's College, 1986. 60p. ISBN: 0852871597.

5th ed., 1982.

4 sections: A. Library provision in the life sciences at Imperial College - B. Library provision in the life sciences at King's College (KQC) - C. Libraries in London of interest to life scientists - D. Libraries within reach of London of interest to life scientists. Data on each library include

.... (contd.)

opening hours; admission and access; sizes and nature of stock; other services offered, *e.g.* photocopying. 'Library searching' (p.5-8) notes on computer based searches, current-awareness journals, abstracting and indexing journals, etc. Index lists 92 main libraries. *Class No:* 573.0:061:026/027

[2058]

World directory of biological and medical sciences libraries. Poland, U.H., *ed*. Munich, New York, etc., K.G. Saur, 1988. 203p. $30. ISBN: 3598217722.

Lists libraries from over 100 countries. Appendices cite union lists; computer service centres; national and international library directories; addresses for biological and medical library associations. *Class No:* 573.0:061:026/027

Institutions & Associations

[2059]

DARNAY, B.T. *and* YOUNG, M.L., *eds*. **Life sciences organizations and agencies directory:** a guide to approximately 8,000 organizations and agencies providing information in the agricultural and biological sciences. Detroit, Mich., Gale. 1988. xxi, 864p. $175. ISBN: 0810318261.

7,662 numbered entries. 18 chapters (*e.g.* 5. Computer information services: international; 6. Consulting firms; 10. Research centers: international; 16. US federal government research centers). Basic data on each organization include date of foundation and name of contact, research activities and fields. Master name and keyword index. US emphasis. *Class No:* 573.0:061:061.2

[2060]

WYATT, H.V. **A Directory of information resources in biology in the UK.** Boston Spa, British Library, 1981. v, 83p. £8. (*British Library R & D report 5606.*) ISBN: 090598465x.

Data on 77 organizations (government departments, museums, research bodies and institutions, private societies and universities) in biology, botany, zoology, medicine, veterinary medicine, agriculture, food, etc. Data on each include availability, scope, stock, special collections, catalogues, publications, affiliations. Subject index. *Class No:* 573.0:061:061.2

[2061]

—BIOLOGICAL COUNCIL. Handbook of UK biological societies. London, the Council, 1991. 48p. £1.50. ISSN: 09513884.

Has sections: Officers of the Council. Reports of Sub-committees, Panels and Representatives. Publications of the Council. Addresses of societies affiliated to the Biological Council (p.9-20). Provisional dates of meetings of biological societies, 1991-1993. International meetings to be held in the UK during 1991-1994. Universities in England, Wales, Scotland and Northern Ireland. Polytechnics in England and Wales. *Class No:* 573.0:061:061.2

Research Projects

[2062]

Current research in Britain. Biological sciences. Harlow, Essex, Longman. Annual. ISSN: 02671956.

Replaces *Research in British universities, polytechnics and colleges* (RBUPC). First published 1985.

.... *(contd.)*

One of a set covering 4 subject areas, registering research in every British university and polytechnic, many colleges and other institutions, with details of *c.*60,000 projects in all. *Biological sciences* (7th ed. 1992. 2v. £125) lists *c.*20,000 projects. 'Research in progress' (p.1-500) lists universities, polytechnics, colleges (other than univ. colleges), and other institutions (including government depts.) numerically. Data number, name, and dept., address, head of dept., researchers and title of project, dates, bequests, etc. Name, study area and keyword indexes. The standard British research tool in biological sciences.
Class No: 573.0:061:061.62.005

Practical Work & Techniques

Collections

[2063]
World directory of collections of cultures of micro-organisms. McGowan, V.F. *and* Skerman, V.B.D., *eds.* 2nd ed. [Brisbane], World Data Centre, Queensland University, 1982.

First published 1972 (New York, Wiley-Interscience), compiled by S.M. Martin and others.

Lists 356 culture collections in 56 countries (1972 ed.: 349 collections).

Mycological Institute's *Directory of collections of micro-organisms* (CAB, 1979) is confined to locations in the UK.
Class No: 573.082

Microscopy

[2064]
Dictionary of light microscopy. Compiled by the Nomenclature Committee of the RMS. Bradbury, S., *and others.* Oxford, University Press; Royal Microscopical Society, 1989. ix,139p. figs., tables. £6.95. (*Microscopy handbooks 15.*) ISBN: 0198564139.

Gives definitions for over 1,250 terms (A ... Zoom). Many cross-references. 4 appendices, p.63-139: 1. Figures and tables - 2.-4. English-French-German equivalent terms.
Class No: 573.086

[2065]
Electron microscopy abstracts. London, PRM Science & Technology Agency, Ltd., 1973-. 4pa. £45. ISSN: 03066869.

*c.*600 abstracts pa. Sections: Electron optics and instrumentation - Theory of image formation - Non-biological specimen preparation - Biological specimen preparation - Bacteriology - Biological micromolecules - Cells and tissues - Metals and alloys - Semiconductors - Non-metals - Petrology, mineralogy and geochemistry - Miscellaneous. Subject and author indexes per issue.
Class No: 573.086

[2066]
GRAY, P., *ed.* **The Encyclopedia of microscopy and microtechniques.** New York, etc., Van Nostrand Reinhold, 1973 (reprinted 1982). xi, 638p. illus., diagrs., graphs, tables. $49.50. ISBN: 0898743354.

Nearly 200 contributors. Entries A-Z. The signed article 'X-ray microscopy' (p.603-11) has sections: Image formation - Microanglography - Technique - Historadiography - Conclusion - Commercial sources (8

.... *(contd.)*

illus., diagrs., graphs. 70 references). Many micrographs (*e.g.* 'Kidney', p.284-8; 6 micrographs. 29 references).
Class No: 573.086

[2067]
LOCQUIN, H. *and* **LANGERON, M. Handbook of microscopy.** Translation edited by H. Hillman. [Manuel de microscopie.] London, etc., Butterworths, 1983. xii, 322p. illus., graphs, diagrs., tables, formulas. ISBN: 0408106794.

Original French as *Manuel de microscopie* (1978).

6 chapters (subdivided): 1. Instruments and techniques - 2. Methods of fixation, examination, cutting and mounting - 3. Stains - 4. Staining and impregnation - 5. Topological stains - 6. Non-specific cytological stains. Appendix 1: Physical constants. Bibliography, p.303-8. Detailed, analytical index, p.309-22. *Class No:* 573.086

Staining

[2068]
CLARK, G. *and* **KASTEN, F.H. History of staining.** 3rd ed. Baltimore, Williams & Wilkinson, 1983. x,304p. illus. ISBN: 0683017055.

2nd ed. published 1948.

24 chapters: 1. Ralph Dougall Lillie ... 5. Cochineal dyes ... 9. Aniline dyes in histology ... 15. The staining of blood and parasitic protozoa ... 22. Immunological staining ... 24. A history of protein and nucleic acid histochemistry. Bibliography, p.253-90. Index. 'Provides the most useful historical summary now available in English in the development of stain technology' (*Science*, v.223, 17 February 1984, p.694). *Class No:* 573.086.16

[2069]
H.J. Conn's biological stains: a handbook on the nature and uses of the dyes employed in the biological laboratory. Conn, H.J., *and others.* 9th ed., edited by R.D. Lillie, with others. Baltimore, Md., Williams & Wilkins, 1977. xi, 692p. diagrs., tables, chemical structures.

First published 1925; 8th ed., 1969 (xii, 498p.).

20 chapters (1. Use and standardization of dyes as biological stains - 2. General nature and classification of dyes ... 19. Tables 1-9 - 20. Methods for testing biological stains). Bibliography, with brief notes, p.613-56. Detailed index, p.657-92. *Class No:* 573.086.16

Tissue & Cell Cultures

[2070]
DONNELLY, D.J. *and* **VIDAVER, W.E. Glossary of plant tissue culture.** London, Belhaven Press, 1988. 141p. £40. ISBN: 1852930616.

Short definitions of *c.*1,500 terms (ABA ... Zygotic embryo). Many cross-references. 3-page list of sources. A few entries list individuals important in plant tissue culture science. Includes eponyms but excludes plant species names.
Class No: 573.086.8

[2071]
GIBCO search: an information service to our customers in tissue culture laboratories. University of Sheffield Biomedical Information Service. Paisley, GIBCO, Ltd., 1981-. 6pa.

Replaces *Tissue culture abstracts,* 1966-79.

About 200 references per issue. 10 sections (1. Books,

.... *(contd.)*

reviews, symposia ... 3. Growth factors ... 6. Epithelial cells - 7. Carcinoma cells - 8. Nerve cells - 9. Muscle cells). A current awareness service. *Class No:* 573.086.8

Exobiology

[2072]

SABLE, M.H. Exobiology: a research guide. Brighton, Mich., Green Oak Press, 1978. xi, 324p.

The bibliography section has 3,633 references in 16 subsections (Astronomy and origin of life and of planets ... Life in individual planets ... Unidentified flying objects, including flying saucers). Directory section: Organizations; periodicals. Author and subject indexes; index to directory section. *Class No:* 573.5

[2073]

—ANGELO, J.A. The Extraterrestrial encyclopedia: our search for life in outer space. 2nd ed. New York, Facts on File, 1991. 272p. illus. (some col.), tables. $40. ISBN: 0816022763.

First published 1985.

Explains major space technologies and developments in the search for extraterrestrial life. Articles, arranged A-Z, vary in length from a paragraph to several pages. Many tables and cross-references. Black-and-white drawings and photographs, some of which are in colour. An improvement on the previous edition is the inclusion of an index. For public, academic and special libraries. *Class No:* 573.5

Ecology

Abbreviations & Symbols

[2074]

WENNRICH, P., *comp.* **Anglo-American and German abbreviations in environmental protection.** Munich, K.G. Saur, 1980. vi, 618p. $35. ISBN: 3598100604.

About 30,000 German and Anglo-American abbreviations plus full forms, in one sequence. Coverage: environmental protection and areas with related problems: biology, chemistry, agriculture, medicine and physics. *Class No:* 574(003)

[2075]

—DAWSON, E. Directory of environmental abbreviations. London, Environment Council, 1988. 69p. ISBN: 0903158345.

Contains *c.*1,000 abbreviations. Each entry includes an expansion of the abbreviations and a line or two on its meaning or function. Short bibliography. *Class No:* 574(003)

Databases

[2076]

COX, J. Environment databases 1991. London, Aslib, 1991. 65p. (*Aslib online guide.*) ISBN: 0851422616.

58 databases listed, Acidoc ... Westlaw Multistate Environmental Law Database. Data include producer; telephone, fax and telex numbers; host; content and coverage; number of references; frequency of updating; type of database; language; user aids. 4-page index of hosts. Covers air, water, land, minerals, climatology, acid rain, ozone depletion, etc. *Class No:* 574(003.4)

[2077]

ECO directory of environmental databases in the United Kingdom 1992. Barlow, M., *eds.* Bristol, ECO Environmental Education Trust, 1992. 290p. £35. ISBN: 1874666008.

Lists *c.*300 databases. Data include main speciality, subject range, area and timespan, methods of access, and charges. 3 main sections: 1. Databases in non-profit-making organizations - 2. Databases from commercial organizations - 3. Services, projects, and useful information. Abbreviations, p.281-2. Bibliography. Indexes: database name; organization name; subject; area; organization categories; media; users. *Class No:* 574(003.4)

Bibliographies

[2078]

Environment impact assessment: a bibliography, with abstracts. Clark, B.D., *and others.* London, Mansell; New York, Bowker, 1980. vi, 516p. ISBN: 0720108993.

About 1,000 entries, some with lengthy annotations. 5 sections: 1. Aids to impact assess - 2. Critiques and reviews of environmental assessments - 3. Environmental income assessment and other aspects of planning - 4. Environmental impact assessment in selected countries (USA, Canada, Australia, Continental Europe, other countries), p.249/483 - 5. Information sources, p.487-94. Selected periodicals. Author and subject indexes. As this is a topic which is now generating more interest, this publication needs to be updated soon if it is to retain its value. *Class No:* 574(01)

[2079]

HAMMOND, K.A., *and others, eds.* **The Sourcebook of the environment:** a guide to the literature. Chicago, Ill., Chicago Univ. Press, for American Association of Geographers, 1978. 613p. $27.50. ISBN: 0226315223.

26 contributors on aspects of the environment literature. About 3,800 references. Critical introductions to the basic material, 'with directions for examination of more advanced and more specialized works' (*Preface*). 'Basic guide' (*A geographical bibliography for American libraries,* edited by C.D. Harris (1985), p.113). *Class No:* 574(01)

[2080]

SARGENT, F., II., *comp.* **Human ecology: a guide to information sources.** Detroit, Mich., Gale, 1983. xi, 293p. $68. (*Health affairs information guide studies, v.10.*) ISBN: 0810315041.

An annotated guide to books and periodical articles. Each section (subdivided) has an introductory essay. Subject coverage, - 'nature and scope of human ecology, the setting (abiotic and biotic), human-environment manipulations, human-environment interactions, environmental quality,

....(contd.)

community health, and disease control' (*Reference books bulletin,* 1983-1984, p.104). Author, title and subject indexes. *Class No:* 574(01)

[2081]
—JONES, O.J. *and* JONES, E.A. Index of human ecology. London, Europa, 1974. ix, 169p. £5.50. ISBN: 0900362669.

Has 3 parts. 1. Principal secondary journals (28 subjects, Agriculture ... Theology) - Part 2. Secondary journals; ancillary services, (*i.e.,* abstracting journals; reviews) - Part 3. Consolidated index (*c.*7,000 terms, coded to pt. 1). *Class No:* 574(01)

Encyclopaedias

[2082]
McGraw-Hill encyclopedia of environmental science. Parker, S.P., *editor-in-chief.* 2nd ed. New York, etc., McGraw-Hill, 1980. x, 858p. illus., diagrs., graphs, tables, maps. $72.50. ISBN: 0070452644.

First published 1974.

Over 250 contributors. 'Surveying the environment' (5 essays) precede the A/Z sequence of signed articles on environmental science and engineering. 'Sea water': p.634-57 (with 15 figures, 7 references); 'Radiation biochemistry': p.588-92; 6 references). Many of the articles were written specially for this volume; some were taken from the *McGraw-Hill Encyclopedia of science and technology* (4th ed. 1977). Analytical index, p.839-58. 650 illus. in all. 'A very valuable basic reference for libraries without the multi-volume encyclopedia' (*Library journal,* 1 April 1975, p.653, on the 1974 ed.). *Class No:* 574(031)

Handbooks & Manuals

[2083]
Ecological engineering: an introduction to ecotechnology. Mitsch, W.J. *and* Jorgensen, S.E., *eds.* New York, Wiley, 1989. xiv,472p. figs. $59.95. ISBN: 0471625590.

2 sections: 1. Basic principles (1. Introduction to ecological engineering ... 4. Principles of ecological modeling ... 7. Agriculture and ecotechnology) - 2. Case studies of ecological engineering (12 chapters, *e.g.* 16. Ecotechnological approaches to the restoration of lakes). Chapter references. Index. For the academic library. *Class No:* 574(035)

[2084]
GRZIMEK, B., *and others, eds.* **Grzimek's encyclopedia of ecology.** English ed. New York, London, etc., Van Nostrand Reinhold, 1976. 705p. illus. ISBN: 0442229488.

Originally published Zurich, Kindler Verlag AG, 1971. A supplement to *Grzimek's animal life encyclopedia (qv).*

42 contributors. Part 1: The environment of animals (1. Adaptations to the abiotic environment; 3. Habitats and their fauna; 4. Man as a factor in the environment of animals) - Part 2. The environment of man. Many coloured plates; excellent coloured maps. 'A worthy purchase' (*New technical books,* November 1977, entry 1564). *Class No:* 574(035)

[2085]
ODUM, E.P. Fundamentals of ecology. 3rd ed. Philadelphia, Pa., Saunders, 1971. 574p. $37. ISBN: 0721669417.

First published 1953.

Includes sections on applied and human ecology. A 40-page bibliography; detailed index. 'Perhaps the best all-round treatment of ecology' (Brewer, J.G. *The literature of geography.* 2nd ed., 1973, p.183). *Class No:* 574(035)

Dictionaries

Polyglot

[2086]
EUROPEAN PARLIAMENT. Ecological terminology. Luxembourg, European Parliament, 1982. 259p. $24.95. ISBN: 0828809437.

Based in part on *Terminology of the Environment* (European Parliament, 1974).

Entries in French; equivalents in English, German, Danish, Dutch and Italian. Indexes in the six languages, plus a bibliography. *Class No:* 574(038)=00

[2087]
PAENSON, I. Environment in key words: a multilingual handbook of the environment, English-French-German-Russian. Oxford, Pergamon Press, 1990. 2v. xxxiv,662p.,268p. figs., tables. £150. ISBN: 0080245242.

First published in 1972 as *Multilingual systematic glossary of environmental terms.*

3 chapters, each in 4 languages: 1. The ecological balance - 2. Disruption of the ecological balance - 3. Some measures for the re-establishment of the ecological balance. Words and phrases in bold type in the text appear in the index (v.2). The chapters are written in half-page column style so that the reader can see all 4 language equivalents spread over 2 pages and in context. 'The author considers each term in the context of a given discipline or even group of disciplines' (*Preface*). *Class No:* 574(038)=00

English

[2088]
ALLABY, M. Macmillan dictionary of the environment. 3rd ed. London, Macmillan, 1988. [v], 423p. £35. ISBN: 0333455614.

First published 1977 ([v], 532p.).

About 5,000 interdisciplinary terms, straying into many fields, - botany, zoology, chemistry, physics, mathematics, economics (*e.g.* ½p. each on Ice Age, nuclear reactor, mangrove). Includes a few biographical entries, for Darwin, Malthus, Lamarck, etc. 'Soil classification': 1p. Profuse cross-references. *Class No:* 574(038)=20

[2089]
Dictionary of ecology and environment. Collin, P.H., *and others.* Teddington, Middx., Peter Collin Publishing, 1988. 220p. £6.95. ISBN: 0948549076.

Covers *c.*5,000 English terms and their German equivalents. Many examples of usage are given. German-English glossary, p.231-94. Appendix consists of outline classification of living organisms; geological time scale and radiation of some flora and fauna; estimated human population growth since 8000 B.C.; endangered species; recent major natural and man-made disasters. *Class No:* 574(038)=20

[2090]

A Dictionary of ecology, evolution and systematics.
Lincoln, R.J., *and others.* Cambridge Univ. Press, 1982.
[viii], 293p. charts, tables, maps. £15.95. ISBN:
0521269024.

About 10,000 rather short working-definitions, with
particular emphasis on evolution and taxonomy and special
attention to principles, processes and classification (*Preface*).
Produced by British Museum (Natural History) staff. 21
appendices (1. Geological time scale; 18. Latin
abbreviations; 21. The Beaufort wind scale). 'Highly
recommended' (*Library journal,* 1 April 1983, p.731).
Class No: 574(038)=20

[2091]

The Green book. Pope, S., *and others.* London, Hodder &
Stoughton, 1991. viii,337p. £12.95. ISBN: 034053298x.

Arranged A-Z (Acid rain ... Zoos). Length of entries
ranges from a few lines to several pages (*e.g.* Chernobyl;
4p.). Many cross-references. Appendix of useful addresses,
p.333-37. No bibliographies. Useful for school and public
libraries. *Class No:* 574(038)=20

[2092]

—CRUMP, A. Dictionary of environment and development:
people, places, ideas and organizations. London, Earthscan
Publications, 1991. 272p. £15. ISBN: 185383078x.

Over 800 definitions, arranged A-Z (Aborigines ... Zinc).
Definitions are more concise than *The Green book (qv)* (*e.g.*
Chernobyl; ½-page) but the book covers a wider range of
issues. Many cross-references. No bibliographies. Suitable
for academic and public libraries. *Class No:* 574(038)=20

[2093]

MONKHOUSE, F.A. *and* **SMALL, J. A Dictionary of the
natural environment.** London, Edward Arnold, 1978. [vi],
320p. illus., chart., maps.

About 4,000 terms defined. Based on Monkhouse's *A
dictionary of geography* (2nd ed. 1970). 'For students of
geography and the environmental sciences, and for those
non-professionals working within this field ...
Comprehensive reference work' (*New technical books,*
October 1978, entry no.1260). *Class No:* 574(038)=20

Reviews & Abstracts

[2094]

Current advances in ecological and environmental sciences.
Oxford, etc., Pergamon Press, 1989-. 12pa. £480pa. ISSN:
09556648.

One of a series of 12 Pergamon *Current advances.*
Formerly *Current advances in ecological sciences* (1975-
88).

About 18,000 references pa. 58 sections (*e.g.* 4.
Distribution; 6. Community ecology - 7. Population
dynamics - 8. Animal behaviour - 9. Productivity and growth
rates (plant, animal, microbiological systems); 17.
Interactions with the physical environment; 23.
Reproduction; 27. Nutrition; 29. Autoecology; 43.
Pollution). Author index per issue. 'Designed by working
scientists for the working scientist'. Also available on
microfiche. *Class No:* 574(048)

[2095]

**Current contents: Agriculture, biology and environmental
sciences.** Philadelphia, Pa., Institute for Scientific
Information, 1970-. Weekly. $560. ISSN: 00900508.

Previously (1970-72) subtitled 'Agriculture, food and
veterinary sciences'.

Issue for 20 May, 1991 contains title-pages of 98 journals.
Sections include: Biology ... Environment/Ecology ... Plant
sciences ... Entomology/Pest control ... Veterinary
medicine/Animal health. Title word index; author index and
address directory; publishers' address directory. Available
on diskette and online on the CURRENT CONTENTS
SEARCH database. *Class No:* 574(048)

[2096]

Ecological abstracts. Norwich, Geo Abstracts, Ltd., 1974-.
12pa. £478. ISSN: 0305196x.

1,500 abstracts per issue. Sections: Global and general
ecology - Marine ecology - Tidal and estuarine ecology -
Freshwater ecology - Terrestrial ecology (plants & animals) -
Applied ecology - Evolution, palaeoecology and historical
ecology - General theory, methods and techniques. Subject
index per issue. Annual subject, author and regional indexes.
List of journals. Online GEOBASE (1987), available on
DIALOG (Search file 292). Also available on microfilm.
Class No: 574(048)

[2097]

Ecology abstracts. Bethesda, Md., Cambridge Scientific
Abstracts, 1980-. 12pa. $770pa. ISSN: 01433296.

Formerly *Applied ecology abstracts,* 1975-79.

25 sections: Contains *c.*11,000 abstracts pa. Sections
include: Ecosystem studies - General; Arid zones;
Microorganisms; Paleoecology; Human settlements;
Recreation/Landscaping. Author and subject indexes per
issue, plus cumulative annual subject and author indexes.
Available online (LIFE SCIENCES COLLECTION), on
magnetic tape and CD-ROM. *Class No:* 574(048)

[2098]

PASCAL folio 56. Écologie animale et végétale.
Vandoeuvre-Les-Nancy, Centre National de la Recherche
Scientifique, 1985-. 10pa. FFr1,080. ISSN: 02461153.

Replaces in part *Bulletin signalétique. 360 Biologie
animale. Physiologie et pathologie des invertébrés* (1940-
83).

About 6,000 abstracts pa. Subject, thematic, systematic,
and author indexes per issue and annually. Online.
Class No: 574(048)

Periodicals

[2099]

Environmental periodicals bibliography. Baltimore, Md.,
National Information Services Corporation, 1990. $546.

Compiled by the Environmental Studies Institute of the
International Academy of Santa Barbara. International in
scope, covering scientific, technical and popular journals.
Contains over 400,000 citations collected since 1972.
Subject areas include human ecology, water and land
resources, nutrition, air and energy. Available on CD-ROM.
Class No: 574(051)

Progress Reports

[2100]

Advances in ecological research. London, etc., Academic Press, 1972-. v.1-. Annual. diagrs., graphs. ISSN: 00652504.

V.23, 1992 (xiii,355p.) has 8 contributors. The 5 papers include 'Mechanisms of microarthropod-microbial interactions in soil' (with 10p. of references); 'Positive-feedback switches in plant communities' (with 16p. of references). Analytical subject index. Cumulative list of titles, v.1-23. *Class No:* 574(055)

Yearbooks & Directories

[2101]

Directory for the environment: organisations, campaigns and initiatives in the British Isles. Frisch, M., *ed.* 3rd ed. London, Green Print, 1990. 320p. £13.99. ISBN: 1854250361.

Previously published as *Directory for the environment,* 2nd ed. 1986.

Lists local and national organizations, campaigns and government bodies concerned with the human and natural environment. Coverage includes consultancies, charities, research institutes, campaigning groups, learned societies, and commercial enterprises throughout Britain and Ireland. *Class No:* 574(058)

[2102]

Directory of environmental consultants 1990. 2nd ed. London, Environmental Data Services, 1990. xxviii,256p. £40. ISBN: 090767304x.

First published 1988.

Details of 225 organizations in the UK (AB Consultancy Ltd ... Yard Ltd). Data include address; telephone/telex/fax numbers; contact name; number of UK projects; qualified staff; skills; facilities; client groups; areas of interest. 3 appendices: 1. Areas of expertise - 2. Regional guide - 3. Late entrants. *Class No:* 574(058)

[2103]

ENDOC directory: environmental information and documentation centres of the European Communities. Hitchin, Herts., Peter Peregrinus, for the Commission of the European Communities, 1981. vi, 600p. ISBN: 0906048400.

519 numbered entries, under countries, - Belgium, Denmark, France, German Federal Republic, Ireland, Netherlands, UK. Data on each centre includes date of foundation, aims and geographical coverage, environmental subject areas, information activities. Organization index; index of named services; subject index (in six languages), p.421-600. ECHO database. Now somewhat dated. *Class No:* 574(058)

[2104]

—DEZIRON, M. *and* BAILEY, L. A Directory of European environmental organizations. Oxford, Blackwell, 1991; Cambridge, Mass., Three Cambridge Center, 1992. 177p. £40. ISBN: 0631183868.

2 sections: 1. Governmental organizations - 2. Non-governmental organizations. Data for each entry cover name; address of headquarters; telephone/fax; director and principal staff; contact person; foundation date; purpose; organization; activities; publications; funding; members. Alphabetical and country indexes. Well laid out and easy to read. *Class No:* 574(058)

[2105]

The Environmental yearbook. Northampton, Taylor Marketing Services, 1989. 390p. £58. ISBN: 0946749019.

UK in scope. 2 sections: 1. Organizations involved with the environment - 2. Local government environmental personnel. Section 1 is arranged alphabetically by subject, *e.g.* 'Bird organizations'; 'Horticulture'. Data consists only of address and telephone number. Section 2 is arranged by council. Useful for its wide-ranging coverage but it should be regularly updated, and typographical errors corrected. *Class No:* 574(058)

[2106]

FITCH, J.M., *ed.* Environmental and international sciences research centres: a world directory of organizations and programs. Harlow, Essex, Longman, 1984. 742p.

Data on *c.*3,500 industrial, governmental and academic organizations, plus libraries and observations in *c.*130 countries. Index of establishments and an extensive subject index. *Class No:* 574(058)

[2107]

FORRESTER, S. Environmental grants: a guide to grants for the environment from government, companies and charitable trusts. London, Directory of Social Change, 1989. 309p. £12.50. (*Directory of Social Change publication.*) ISBN: 0907164471.

2 sections: 1. Grant sources (1. Local authority support - 2. Central government departments and statutory agencies - 3. Grant-making trusts - 4. Grants from independent organizations - 5. Company charitable donations - 6. Awards and competitions) - 2. Sources of information and help (1. Partnerships, brokerage, environmental entrepreneurism - 2. Technical and financial services - 3. Information and advice). British in scope. Index [of organizations], p.301-309. *Class No:* 574(058)

[2108]

TRZYNA, T.C. World directory of environmental organizations. 3rd ed. Claremont, Calif., California Institute of Public Affairs, 1989. 176p. ISBN: 091201287x.

2nd ed., 1973.

Arranged in 7 parts, the largest (p.91-164) listing organizations alphabetically under country, and including government departments, learned societies, pressure groups, etc. Preceded by subject and geographical indexes, and section on 'The United Nations system' (p.29-46), lists of intergovernmental and international non-governmental organizations, and an index of organizations and major programmes. A useful international directory which is now more regularly updated - a 4th ed. is due to be published in 1993. *Class No:* 574(058)

Tables & Data Books

[2109]

UNITED NATIONS. Environment Programme. Environmental data report. 3rd ed. Oxford, Blackwell, 1991. viii,408p. figs., tables, maps. £50. ISBN: 0631180834. ISSN: 09569324.

First published 1987.

International in scope and contains 'a wealth of up-to-date information at regional, national and local levels' (*Introduction*). 10 parts: 1. Environmental pollution ... 4. Population/settlements ... 6. Energy ... 10. International co-

....(contd.)

operation. 3 appendices, covering contributors, country names, and abbreviations. References are given throughout. *Class No:* 574(083)

Maps & Atlases

[2110]
LEAN, G. *and* **HINRICHSEN, D. Atlas of the environment.** London, Helicon, 1992. 195p. diagrs., tables, maps. £12.99. ISBN: 0091774349.

42 sections: Major biomes, climatic regions and land use ... Deserts and desertification ... Tropical forest destruction ... The greenhouse effect ... Biological diversity and genetic resources ... Major conservation effects ... Migration ... Antarctica. Only includes data from the 1980s and 1990s. Over 200 full colour maps and diagrams. Select bibliography, p.187-92. No index. A general overview; for school and public libraries. *Class No:* 574(084.3)

[2111]
MASON, R.J. *and* **MATTSON, M.T. Atlas of United States environmental issues.** New York, Macmillan, 1991. 252p. illus., diagrs., tables, maps. $90. ISBN: 0028972619.

Includes maps and information on agricultural land, forests, natural hazards, solid wastes and pollution. More than 130 maps, tables and diagrams. Glossary, bibliography, and index. Although sometimes lacking in detail, the information is well written and precise. Recommended for school, public and academic libraries. *Class No:* 574(084.3)

[2112]
MIDDLETON, N. Atlas of environmental issues. New York, Facts on File, 1989. 63p. illus., maps. $16.95. (*World contemporary issues.*) ISBN: 081602023x.

Consists of double-page spreads devoted to environmental themes, *e.g.* noise, soil erosion, irrigation. In addition to colourful charts and maps, a few paragraphs discuss the problem in general. Brief index. Aimed at young adults and suitable for school and public libraries. *Class No:* 574(084.3)

Worldwide

[2113]
Ecosystems of the world. Goodall, D.W., *ed.* Amsterdam, etc., Elsevier, 1977-. illus., diagrs., tables, maps.

1. *Wet coastal ecosystems*; edited by V.J. Chapman, 1977. 4. *Mires: swamp, bog, fen and moor*; edited by A.J.P. Gore. 4A. *General studies*, 1983. 4B. *Regional studies*, 1983. 5. *Temperate deserts and semi-deserts*; edited by N.E. West. 1983. 9. *Heathlands and related shrublands*; edited by R.L. Specht. 2v. 1979-81. 10. *Temperate broad-leaved evergreen forests*; edited by J.D. Ovington. 1983. 11. *Mediterranean-type shrublands*; edited by F. Di Castri, etc. 1981. 12. *Hot deserts and arid shrublands*; edited by M. Evenari. 1985-6. 2v. 13. *Tropical savannas*; edited by F. Bourlière. 1983. 14A. *Tropical rain forest ecosystems*; edited by F.B. Golley. 1983. 14B. *Tropical rain forest ecosystems;* edited by H. Lieth and M.J.A. Werger. 1989. 15. *Forested wetlands*; edited by A. Lugo. 1989. 17A. *Managed grasslands: regional studies*; edited by A. Breymeyer. 1989. 17B. *Managed grasslands*; edited by R.W. Snaydon. 1987. 18. *Field-crop ecosystems;* edited by C.J. Pearson. 1991. 21. *Bioindustrial ecosystems*; edited by D.J.A. Cole and G.C. Brander. 1986. 23. *Lakes and reservoirs*; edited by F.B.

....(contd.)

Taub. 1984. 26. *Estuaries and enclosed seas*; edited by B.H. Ketchum. 1983. 27. *Continental shelves*; edited by H. Postma and J.J. Zijlstra. 1987. 29. *Managed aquatic ecosystems;* edited by R.G. Michael. 1987.

Each volume is the work of a team of specialists. Extensive bibliographies.

Monumental compendium of information about processes and characteristics of ... ecosystems of the world ... 'The most thorough treatment of its kind to date' (*Geographical bibliography for American libraries;* edited by C.D. Harris (1985), p.76-77). *Class No:* 574(100)

Marine Environment

[2114]
DAVIES, B.R. *and* **WALKER, K.F. The Ecology of river systems.** The Hague, W. Junk; Norwell, Mass., Kluwer, 1986. 793p. $232.

Concentrates on the ecology of 13 major rivers outside Europe: Africa (6), Americas (5), Australia (1). (*Nature* v.325, 26 February 1987, p.769) queries omission of the Indus, Ganges, Irrawaddy, Yangtze, Amur, Lene, Yenirei, Ob, plus the Mississipi and Orinoco). Attention is rightly directed towards draining basins and dams. *Class No:* 574(26)

[2115]
Directory of wetlands of international importance: sites designated for the list of Wetlands of International Importance. 2nd ed. Gland, Switzerland, Cambridge, IUCN for the Ramsar Convention Bureau; Washington, D.C., Island, 1990. 782p. $50. ISBN: 2831700140.

First published 1987.

Covers 52 countries, Algeria ... Yugoslavia. The entry for the Volga delta, p.631-2, has data on location; area; degrees of protection; site description; international and national importance; changes in ecological character; management practices; scientific research and facilities; principal reference material. Introductory notes on each country precede data on individual wetlands. Italy, Austria and Denmark have the most entries. Lack of funds means that this edition is not a thorough revision, and the publishers accept that some inconsistencies will occur and information may be out of date. *Class No:* 574(26)

[2116]
KINNE, O., *ed.* **Marine ecology:** a comprehensive integrated treatise on life in oceans and coastal waters. New York, etc., Wiley, 1970-84. 5v.

1. *Environmental factors.* 1970. 3pts. 2. *Physiological factors.* 1975. 2pts. 3. *Cultivation.* 1976. 3pts. 4. *Dynamics.* 1978. 5. *Ocean management.* 1984. 4pts.

Many specialist contributors. Emphasis on laboratory methods. Bibliographies. Author, taxonomic and subject indexes. 'Highly recommended' (*Nature*, v.272, 23 March 1978, p.380-1). *Class No:* 574(26)

[2117]
The Shore environment. Price, J.H., *and others, eds.* London, New York, Academic Press, for the Systematics Association, 1980. 2v. £76.50, v.1; £109, v.2. (*Systematics Association special volumes.*) ISBN: 0125647018, v.1; 0125647026, v.2.

1. *Methods.* xx, 321, xlip. 2. *Ecosystems.* xx, 622, cp. Papers from an international meeting. 44 contributors, The 27 papers include 'Field teaching methods in shore ecology',

.... *(contd.)*

and 'Strategies of data collection and analysis of subtidal vegetation'. Both volumes carry a taxonomic index. Substantial subject index. Appended list of Systematics Association publications. *Class No:* 574(26)

[2118]

—BARNES, R.S.K. *and* MANN, K.H. Fundamentals of aquatic ecology. 2nd ed. Oxford, Blackwell, 1991. vii,270p. illus., diagrs., graphs, tables, maps. £39.50. ISBN: 0632029811.

2nd ed. of *Fundamentals of aquatic ecosystems* (1980).

9 contributors; 5 parts, with each of the 13 chapters followed by 'Further reading'. Pt.1. Introduction - 2. Aquatic ecosystems - 3. Aquatic individuals and communities - 4. Habitat types peculiar to aquatic systems - 5. Human effects. Detailed, analytical index.
Class No: 574(26)

[2119]

—BAXTER, J.M. The Mitchell Beazley pocket guide to the seashores of Britain and Northern Europe. London, Mitchell Beazley, 1990. 144p. illus. (col.). £2.99. ISBN: 0855337699.

Intended as a companion and field guide for identification. Includes animals and plants of the shore, the more common coastal flowering plants, the birds of the shore and coastal waters, and some mammals. Habitat symbols are given for each entry. 400 illustrations. Brief glossary, index and bibliography. For public libraries. *Class No:* 574(26)

Europe

[2120]

European environmental yearbook: nature conservation, protection of the environment, town and country planning in Belgium, Denmark, the Federal Republic of Germany, France, Greece, Ireland, Italy, Luxembourg, the Netherlands, Portugal, Spain and the United Kingdom, with special surveys on Australia, Japan, USA and USSR. 4th ed. London, Docter International, 1990. xxii,897p. charts, tables, maps. $150. ISBN: 0871796910.

At head of title page: 'DOCTER. Institute for Environmental Studies, Milan'. Prepared with the co-operation and financial assistance from the Commission of the European Communities, Brussels.

188 contributors. Part 1 deals with 22 topics in nature conservancy and town and country planning in the then 12 member countries of the European Economic Community. Bibliographical references. Part 2 comprises special surveys (*e.g.* 'Australia'). Part 3: Environmental institutes and associations. Legislation. Part 4, Documentation.
Class No: 574(4)

India

[2121]

MANI, M.S., *ed*. Ecology and biogeography in India. The Hague, Junk, 1974. xix, 773p. illus., diagrs., graphs, tables, maps. (*Monographiae biologicae, v.23.*) ISBN: 9061930758.

24 documented chapters (1. Introduction - 2. Physical features - 3. Geology - 4. Weather and climate patterns ... 19. Biography of the Peninsula ... 23. Biogeography of the Indo-Gangetic plains - 24. Biogeographical evolution in India). Chapter 11, 'The tribal man in India: a study in the

.... *(contd.)*

ecology of the primitive communities', p.281-321, has 56 references. Detailed index, p.725-73. Numerous tables and maps. *Class No:* 574(540)

Africa

[2122]

ROGERS, D.J., *comp*. A Bibliography of African ecology: a geographically and topically classified list of books and articles. Westport, Conn., Greenwood Press, 1979. 499p. $85. (*African Bibliographic Center. Special bibliographic series, n.s,.6.*) ISBN: 0313205523.

About 8,000 references, in 5 geographical regions, each with 10 subject divisions (plants; animals; aquatic ecology; abiotic environment; palaecology; anthropology; human health; history and geography; conservation; general and miscellaneous). Two lists of relevant serials: those published in Africa; those published outside Africa. No author index. A well organized bibliography. *Class No:* 574(6)

Antarctic

[2123]

Antarctic ecology. Laws, R.M., *ed*. Orlando, Florida, & London, Academic Press, 1984. 2v. (368p. + 544p.). $112, v.1; $118, v.2. ISBN: 0124395015, v.1; 0124395023, v.2.

Updates M.W. Holdgate's *Antarctic ecology* (1930).

1. *Terrestrial environment and ecology.* 2. *Marine environment and ecology.* 13 contributors. 15 chapters in all (1. The terrestrial environment ... 4. Introduced mammals ... 15. Conservation and the Antarctic). Each chapter has an extensive bibliography, and each volume a composite 25-p. subject index. An authoritative survey. *Class No:* 574(99)

[2124]

MAY, J. The Greenpeace book of Antarctica: a new view of the seventh continent. London, Dorling Kindersley, 1988. 192p. illus. (col.), figs., maps. £16.99. ISBN: 0863182836.

4 sections: 1. Terra incognita (14 chapters: 1. Antarctica evolving ... 7. Beneath the ice ... 12. Meteorites ... 14. Map of Antarctica) - 2. Life at the end of the world (9 chapters: 1. Land flora and fauna ... 4. Krill ... 9. Whales) - 3. The human presence (12 chapters: 1. Exploration ... 5. Protected sites ... 12. A scientist's view) - 4. Greenpeace perspectives. 2 appendices (*e.g.* 'Antarctic bases & refuges'). Bibliography. Brief list of abbreviations. Gazetteer; index. Many colour illustrations. For public libraries.
Class No: 574(99)

Hydrobiology (Marine Biology)

Bibliographies

[2125]

FRESHWATER BIOLOGICAL ASSOCIATION, Cumbria. Catalogues of the Library of the Freshwater Biological Association. Boston, Mass., G.K. Hall, 1979. 6v. $725, set. ISBN: 0816102899.

About 126,000 author-catalogue cards, photolithographically reproduced. Coverage: biology, physics and chemistry of inland waters worldwide, with

....*(contd.)*

emphasis on freshwater algae, planktonic and benthic invertebrates, fishes and lake sediments. Represents one of the world's finest collections of limnological literature. *Class No:* 574.5(01)

[2126]
MARINE BIOLOGICAL ASSOCIATION OF THE UNITED KINGDOM, Plymouth. **Catalogues of the Library. [Marine Biological Association of the United Kingdom].** Boston, Mass., G.K. Hall, 1977. 16v. (12246p.). ISBN: 0816100764.

The Main catalogue is an A/Z author/name listing of *c.*13,000 books, 40,000v. of periodicals, 1,450 current periodicals and serials, and over 50,000 reprints and pamphlets. Analytical entries for relevant periodical articles, technical reports, conference proceedings, etc. are included since 1968. The Subject catalogue primarily lists books and monographs but also covers review-type papers. Marine and estuarine pollution material has been collected since 1970. *Class No:* 574.5(01)

[2127]
WOODS HOLE OCEANOGRAPHIC INSTITUTION. **Catalog of the Library of the Marine Biological Laboratory and the Woods Hole Oceanographic Institution.** Boston, Mass., G.K. Hall, 1971. 12v. (9339p.). ISBN: 0816109370.

About 339,000 author-catalogue cards for books, reports of expeditions, and articles, photolithographically reproduced. V.12, *Journal catalog* (418p.), lists over 4,000 titles. Coverage includes biology, zoology, botany, physiology, microbiology, medicine, physics, chemistry, mathematics, geology, meteorology, fisheries and oceanography. *Class No:* 574.5(01)

[2128]
—Oceanographic index. Cumulation 1946-1973: Marine organisms, chiefly planktonic. Sears, M., *comp.* Boston, Nass., G.K. Hall, [1974?]. 3v, (2307p.). $240, set. ISBN: 0816109338.

An outgrowth of the Woods Hole Oceanographic Institution's *Catalog. Subject cumulation. Class No:* 574.5(01)

Handbooks & Manuals

[2129]
BAKER, J.M. *and* WOLFF, W.H.J., *eds.* **Biological surveys of estuaries and coasts.** Cambridge Univ. Press, 1987. xx, 449p. diagrs., graphs, tables, maps. ISBN: 0521324076.

18 contributors. 15 documented chapters (1. Planning biological surveys ... 4. Flora and microfauna of intertidal sediments ... 10. Bacteria and fungi ... 13. Birds ... 15. Safety). Appendices throughout the work (*e.g.* Appendix 11.2: Estimation of a flushing time-scale for an estuary). Mainly European material. *Class No:* 574.5(035)

[2130]
—MAITLAND, P.S. Biology of fresh waters. 2nd ed. Glasgow, Blackie; New York, Chapman and Hall, 1990. xi,276p. illus. £37. ISBN: 0216929881.

First published 1978.

9 sections: 1. The aquatic environment - 2. Plants and animals of fresh waters ... 5. Field studies: sampling in fresh

....*(contd.)*

waters ... 7. Communities and energy flow - 8. Fresh water and humans - 9. A global view. Bibliography, p.257-68. Index. Suitable for academic libraries. *Class No:* 574.5(035)

[2131]
HORSMAN, P.V. **The Seafarer's guide to marine life.** London, etc., Croom Helm, 1985. xiv, 256p. illus. (incl. col.), diagrs., tables, maps. ISBN: 0709937156.

13 chapters: 1. The ecology of the sea ... 4. The coelenterates ... 6. The crustacea ... 8. The fish - 9. The reptiles - 10. The mammals - 11. The Sargassum community - 12. The seashore and marine fouling - 13. The future of the marine world. Bibliography, p.246-7. Glossary, p.248-52. Detailed, analytical index, p.253-6. 'Written mainly at sea as a result of my observations and the data accumulated in ships' logbooks' (*Introduction*). 160 small drawings; 80p. of distribution maps. Well produced. *Class No:* 574.5(035)

[2132]
HUTCHINSON, G.E. **A Treatise on limnology.** New York, Wiley; London, Chapman & Hall, 1957-75. 3v. illus., diagrs., graphs, tables.

1. *Geography, physics and chemistry.* 1957. xiv, 1015p. 2. *Introduction to lake biology and the limnoplankton.* 1967. xi, 115p. 3. *Limnological botany.* 1975. x, 660p.

32 chapters in all. Each volume carries a bibliography and index of authors; index of lakes (giving co-ordinates); index of genera and species of organisms; general index. *Class No:* 574.5(035)

Reviews & Abstracts

[2133]
Aquatic sciences and fisheries abstracts. Bethesda, Md., Cambridge Scientific Abstracts, 1978-. 12pa. 3pts. $1,410, set; $885, pt.1; $635, pt.2; $295, pt.3 (6pa). ISSN: 01405373.

Replaces *Aquatic sciences and fisheries abstracts,* 1971-77 (an amalgamation of *Aquatic biology abstracts* and *Current bibliography for aquatic sciences and fisheries*).

Pt.1. *Biological sciences and living resources.* Pt.2. *Ocean technology, policy and non-living resources, (qv).* Pt. 3. *Aquatic pollution and environmental quality.*

Pt. 1 contains *c.*24,000 abstracts pa. Main parts: 1. General aspects; 6-7. Biology; 74-194. Ecology, ecosystems and pollution; 195-245. Fisheries. Author and subject indexes per issue and annually. CD-ROM and online BRS(CSAL), CISTI, DIMDI, DIALOG, ESA. Magnetic tape also available. *Class No:* 574.5(048)

[2134]
Marine science contents tables. Compiled by Fishery Information, Data and Statistics Service, Fisheries Dept., Food and Agriculture Organization. Rome, FAO, 1966-. 12pa. Gratis. ISSN: 00253308.

Reproduction of title pages of *c.*100 journal issues on earth and life sciences, as well as on marine titles. Appended list of journals covered, alongwith the publishers and addresses. Also includes a listing of forthcoming marine science conferences. Key to periodicals listed. Headings in English, French, Spanish and Russian. *Class No:* 574.5(048)

Progress Reports

[2135]
Oceanography and marine biology: an annual review. Aberdeen Univ. Press, 1963-. ISSN: 00783218.

V.29, 1991 (581p.) has 11 contributors. The 8 papers include 'The biology of hydrothermal vents: ecology and evolution' (with 18p. of references) and 'The role of fluid mechanics in the ecology of marine birds' (with 12p. of references). Author, systematic and subject (analytical) indexes, p.523-81. *Class No:* 574.5(055)

Tables & Data Books

[2136]
CRC practical handbook of marine science. Kennish, M.J., *ed.* Boca Raton, Fla., CRC Press, 1989. vii,710p. graphs, tables. $54.95. ISBN: 0849337003.

9 sections: 1. Air-sea interaction - 2. Chemical oceanography - 3. Physical oceanography ... 6. Phytoplankton - 7. Primary productivity - 8. Zooplankton - 9. Compounds from marine organisms. Numerous references. Subject index, p.693-710. 'Represents a systematic collection of selective physical, chemical and biological reference data on the ocean' (*Preface*). For marine scientists, administrators and other professionals. *Class No:* 574.5(083)

Underwater

[2137]
The Encyclopedia of underwater life. Banister, K. *and* Campbell, A. C., *eds.* London, Allen & Unwin, 1985. xxxii, 287p. illus. (mostly col.), diagrs., maps. $25. (*Unwin Animal library*.) ISBN: 0045000352.

Contributions by 17 specialists, apart from the two editors. Equal attention to fishes and aquatic invertebrates, but omitting aquatic insects. Well-illustrated, medium-length articles. Short glossary and bibliography. Detailed index, in tiny print. For the general reader. *Class No:* 574.5(204)

[2138]
ROWE, G.T., *ed.* **Deep-sea biology.** New York, etc., Wiley-Interscience, 1983. [x], 560p. illus., diagrs., graphs, tables, maps. (*The sea*, v.8.) ISBN: 0471044024.

19 contributors. 12 chapters, well documented: 1. Problems of deep-sea biology: an historical perspective (with 10½p. of references) ... 3. Recent advances in instrumentation in deep-sea biological research (3½p. of references) ... 7. Biochemical and physiological adaptations of deep-sea animals (12p. of references) ... 12. Parasitism in the deep-sea (over 11p. of references). Analytical index, p.553-60. Aims 'to reflect the progress in broad general subjects made in the last few years' (*Preface*). Authoritative. *Class No:* 574.5(204)

Biogeography (Geographical Scatter of Organisms)

[2139]
BROWN, J.H. *and* **GIBSON, A.C. Biogeography.** St. Louis, Missouri, C.V. Mosby, 1983. 653p. $39.95. ISBN: 0801608244.

Documented chapters on both historical and ecological biogeography (*e.g.* the distribution of plant and animal taxa).

.... (contd.)
Glossary. Over 1,800 bibliographical references in all. More than 600 illustrations and maps. Detailed index. University student level. 'Praised in *Science* (v.222, 14 October 1983, p.157) for 'the wealth of information presented and its accumulation into a coherent framework'. *Class No:* 574.9

[2140]
PEARS, N. Basic biogeography. 2nd ed. London, Longman, 1985. x, 358p. illus., diagrs., charts, tables. £13.99. ISBN: 0582301203.

First published 1977.

Two parts (10 documented chapters): 1. Basic considerations (*e.g.* Plant dynamics and the nature of vegetation. Ecosystems) - 2. Selected examples (*e.g.* The vegetation. The soils. The impact of man). General index, p.343-53; Species index. *Class No:* 574.9

[2141]
VINCENT, P. The Biogeography of the British Isles: an introduction. London, Routledge, 1990. xv,315p. figs., tables, maps. £40. ISBN: 0415034701.

10 chapters: 1. The history and scope of British biogeography - 2. The physical environment ... 6. Limitations of the environment - 7. Geographical relationships - 8. Environmental change ... 10. The consequences of urban and industrial growth. Glossary, p.271-77. Bibliography, p.278-91. Aimed at the layperson and academic. *Class No:* 574.9

Genetics

Dictionaries

[2142]
BIASS-DUCROUX, F. *and* **NAPP-ZINN, K.,** *comps.* **Glossary of genetics in English, French, Spanish, Italian, German, Russian.** Amsterdam, etc., Elsevier, 1970. xvi, 436p. $91. (*Glossaria interpretum, 16.*) ISBN: 044440712x.

2,938 entries in English, with French, Spanish, Italian and German equivalents. Russian numerical index; French, Spanish, Italian, German and Russian indexes. Avoids mere transposition of words from one language into the others. *Class No:* 575(038)

[2143]
Glossary of genetics: classical and molecular. Rieger, R.O., *and others.* 5th rev. ed. Berlin, etc., Springer Verlag, 1991. 553p. diagrs., tables. $39. ISBN: 0387520546.

First published 1954 as *Genetisches und cytogenetisches Wörterbuch.*

About 4,000 entries, stating first use of term (author; date). Bibliography. 100 diagrams, etc. Of the 4th ed. (1976), 'Remains the outstanding reference glossary of genetics' (*Heredity*, v.41(1), 1978, p.115-6). The *Glossary* has also been translated into Russian and Polish. *Class No:* 575(038)

[2144]

—OLIVER, S. *and* WARD, J.M. A Dictionary of genetic engineering. Cambridge Univ. Press, 1985. [v], 153p. diagrs., tables. £19.95. ISBN: 0521260809.

Defines several hundred technical terms and jargon, with 43 diagrs. 6 short appendices, on genetic maps, genotype nomenclature, etc. The terms are carefully selected, 'to avoid those found in ordinary dictionaries of genetics'.
Class No: 575(038)

[2145]

KING, R.C. *and* STANSFIELD, W.D. **A Dictionary of genetics.** 4th ed. New York, Oxford University Press, 1990. 406p. illus., diagrs., chemical structures. £32. ISBN: 0195063708.

First published 1968.

7,100 concise definitions, 20 per cent of which are new or updated. 250 illustrations. 4 appendices: A. Classification - B. Domesticated species - C. Chronology; Index of scientists (1,010 entries); Bibliography - D. Periodical list; Multijournal publishers; Foreign words in scientific titles. A German edition (Verlagsgesellschaft, 1990) contains a German/English index. *Class No:* 575(038)

[2146]

MACLEAN, N. **Macmillan dictionary of genetics & cell biology.** London, Macmillan Press, 1987. xxi, 422p. diagrs., tables, chemical structures. £12.99. ISBN: 0333394631.

About 4,000 entries (*e.g.* 'Extraembryonic membranes': over 2 columns; & diagrs.). Many cross-references. 5 appendices precede (*e.g.* 2. Chromosome numbers in various species; 5. Classification of vicious organisms). For students and those working in the related fields of microbiology and biotechnology. *Class No:* 575(038)

Reviews & Abstracts

[2147]

Excerpta medica. 22: Human genetics (including developmental defects). Amsterdam, Excerpta Medica Foundation, 1963-. 24pa. DFl1,1709. ISSN: 00144266.

About 200 abstracts (150-300 words) per issue. 39 sections (1. General sections … 13. Heredity … 17. Blood and hemopoiectic system … 19. Metabolism: inborn errors … 30. Demography). Subject and author indexes per issue. Online, EMBASE and available on microfilm and CD-ROM. *Class No:* 575(048)

[2148]

Genetics abstracts. Bethesda, Md., Cambridge Scientific Abstracts, 1968-. 12pa. $860. ISSN: 0016674x.

Each issue carries *c.*1,200 abstracts, monitoring some 5,000 primary and other journals. 21 sections (Molecular genetics … Invertebrate genetics, Human genetics - Medical genetics). Book notices. Author and subject indexes. Online, LIFE SCIENCES COLLECTION. Also available on CD-ROM and magnetic tape. *Class No:* 575(048)

Progress Reports

[2149]

Advances in genetics. New York & London, Academic Press, 1947-. Annual. illus., graphs, tables. ISSN: 00652660.

V.29, 1991 (ix,367p.) has 6 contributors. The 5 documented articles include 'The structure and biogenesis of

….(contd.)

yeast ribosomes' (16p. of references); 'Evolutionary genetics of fish' (with 27p. of references). Analytical subject index. *Class No:* 575(055)

[2150]

Annual review of genetics. Palo Alto, Calif., Annual Reviews Inc., 1967-. v.1-. ISSN: 00664197.

V.25, 1991 (viii,682p.) has 41 contributors. The 24 documented papers include 'Import of proteins into mitochondria' (with 147 references) and 'Mechanisms and biological effects of mismatch repair' (with 143 references). Analytical subject index, p.661-75. Cumulative indexes: Contributing authors, v.21-25; chapter titles, v.21-25. 'Some related articles in other *Annual reviews*', p.vii-viii. *Class No:* 575(055)

Yearbooks & Directories

[2151]

Genetic engineering and biotechnology yearbook. Amsterdam, etc., Elsevier, 1986. 1233p. $800. ISBN: 0444427147.

Part 1: USA (643p.): US genetic engineering corporations; Non-genetic engineering/biotech. corporations with activities in the field. Part 2: Genetic engineering activities outside USA (p.645-1091p.) 2 appendices (*e.g.* Fields of focus, - product index). Index, p.1127-1233. Directory data on organizations include products, research and financial information. *Class No:* 575(058)

Evolution

[2152]

CHAPMAN, R.G. *and* DUVAL, C.T. **Charles Darwin, 1809-1882:** a centennial commemorative. Wellington, Nova Pacifica, 1982. vii, 376p. illus. (incl. col). ISBN: 090860310x.

Limited ed. of 750.

11 contributors. 4 sections, each subdivided: 1. Darwin biography - 2. Darwin and the 19th century - 3. Darwin and the sciences - 4. Darwin and the 20th century. Many illus., including 27 colour plates. Bibliography, p.355-67. *Class No:* 575.8

[2153]

—Biological journal of the Linnean Society, v.17(1), February 1982. Linnean Society of London. 135p.

A special issue devoted to Charles Darwin on the occasion of the centenary of his death. 9 documented contributions (*e.g.* 'Turning points in Darwin's life' (with 37 references); 'Darwin and the historian' (with 83 footnotes); 'Dogma and doubt' (27 references); 'Epilogue' (18 references)). Analytical index, p.127-35. *Class No:* 575.8

[2154]

Encyclopedia of human evolution and prehistory. Tattersall, I., *and others, eds.* New York, Garland, 1988. 900p. illus., diagrs., tables. $95. ISBN: 1558621172.

*c.*2,000 entries, arranged A-Z. Extensive coverage of all aspects of evolutionary theory, primatology, evolutionary and prehistoric hominid research, genetics, primate palaeontology and palaeontology. Many cross-references and references for further reading. Numerous clear illustrations.

....(contd.)
Useful for lay persons or undergraduates. 'An excellent reference tool, and the only one of its type' (*Science and technology libraries,* v.10(1), 1989, p.125).
Class No: 575.8

[2155]
MILNER, R. **The Encyclopedia of evolution:** humanity's search for its origins. New York, Facts on File, 1991. 481p. illus., figs., maps. $45. ISBN: 0816014728.
Contains more than 600 articles (500-1,000 words in length) covering topics which include adaptation, natural selection, theorists and animal models of evolution. Length of entries varies from a single paragraph to several pages. More than 200 drawings, photographs and charts. Maps are not of high quality. Good cross-references, brief bibliographies and detailed index. Aimed at a general audience but would also be suitable for academic libraries.
Class No: 575.8

[2156]
Oxford surveys in evolutionary biology. Oxford, University Press, 1984-. v.1-. Annual. diagrs., charts, graphs, tables. ISBN: 0198542305.
V.8: 1992 (vii,392p.) has 15 contributors. The 10 documented papers include 'Coevolution among competitors', p.63-109 (with 8p. of references); 'A causal analysis of stages in allopatric speciation', p.219-57 (12p. of references). Index, followed by contents lists of previous volumes. The series aims to stimulate discussion and review progress in evolutionary studies. *Class No:* 575.8

[2157]
SAVAGE, R.J.G. *and* LONG, M.R. **Mammal evolution:** an illustrated guide. London, British Museum (Natural History), 1986. vi, 259p. illus. (incl. col.), diagrs. £19.50. ISBN: 0565009427.
13 sections (*e.g.* 1. Bones into stones ... 5. Insectivores ... 5. Early roosters and browsers ... 13. Of men and monkeys). Systematic tables on endpapers. Bibliography, p.251-4. Colourful illus. Index, p.255-9. Takes a popular approach and 'contains a glossary to help reading for the novice'. *Class No:* 575.8

[2158]
—PATTERSON, C. Evolution. London, British Museum (Natural History), with Cornell Univ. Press, and others, 1978. vii, 197p. illus., ports.
1970 ed. as *A handbook on evolution.*
15 chapters (1. Propositions; 2. Species; 3. Variations within species; 4. Heredity; 5. Genetics and variations; 6. Mutation; 7. Natural selection theory; 8. Selection in action; 9. The origin of species; 10. Speciation in the Galapagos Islands; 11. Evolution beyond the species level; 12. Proof and disproof; 13. The origin and early evolution of life; 14. Evolution and man; 15. Who's who (17 short biographies)). Further reading (21 titles); Glossary (88 terms). Aims to be 'simple enough to be comprehensible to those with little or no technical knowledge of biology' (*Foreword*), - and succeeds. An excellent introduction. *Class No:* 575.8

[2159]
SMITH, F.H. *and* SPENCER, F. **The Origins of modern humans:** a world survey of the fossil evidence. New York, Wiley, 1985. $93.95. ISBN: 047183419x.
11 contributors, of which 9 concern skeletal remains found in Europe, Western Asia, East Asia and Australasia, China, and the Americas. While conceding that the text is refreshingly free from the Eurocentric bias, the review in *Nature* (v.314, 18 April 1985, p.649) quarrels with the maps and quality of photographs and drawings. Nevertheless, 'An invaluable source of reference for those already familiar with the fossil record, anatomy and population genetics'.
Class No: 575.8

Cells (Cytology)

[2160]
Annual review of cell biology. Palo Alto, Calif., Annual Reviews Inc., 1985-. v.1-. illus. ISSN: 07434634.
V.8 (1992) has 43 contributors. 19 documented reviews (*e.g.* 'Regulation of translation in eukaryotic systems'; 'The interaction of bacteria with mammalian cells)'. Subject index. Cumulative indexes of contributing authors and chapter titles, v.4-8. Commenting on v.1, *Nature* (v.320, 13 March 1986, p.117) noted pleasing use of illus. and readability of many of the articles. *Class No:* 576

[2161]
DE ROBERTIS, E.D.P. *and* DE ROBERTIS, E.M.F. **Cellular and molecular biology.** 8th ed. Philadelphia, Pa., Lea & Febiger, 1987. xxix,734p. illus., diagrs., graphs, tables. ISBN: 0812110129.
24 sections, each with references and summary: 1. The cell-structural organization ... 4. Cell membrane and permeability ... 13. The interphase nucleus, chromatin, and the chromosome ... 19. The genetic code and genetic engineering ... 24. Cellular and molecular neurobiology. 'Literary sources in cell and molecular biology', p.25-26. Detailed analytical index, p.721-34. Well illustrated with micrographs, etc. A popular college textbook that has been translated into 7 languages. *Class No:* 576

[2162]
The Dictionary of cell biology. Lackie, J.M. *and* Dow, J.A.T., *eds.* London, Academic Press, 1989. 262p. tables. £20.50. ISBN: 0124325602.
Brief entries (A23187 ... Zymosan) supplemented by cross-references and 27 tables. Criticized for errors in *Nature* (v.343, 18 January 1990, p.225) but 'the authors and editors have made a good start in their praiseworthy attempt to provide easy access to words in common usage'.
Class No: 576

[2163]
International review of cytology: a survey of cell biology. Orlando, Calif., Academic Press, 1965-. v.1-. ISSN: 00747696.
V.133: 1992 (ix,319p.) has 11 contributors. The 6 documented articles include 'Annulate lamellae: a last frontier in cellular organelles', p.43-120 (9p. of references); 'Small GTP-binding proteins', p.187-230 (9p. of references). Detailed, analytical index. *Class No:* 576

Parasites (Parasitology)

[2164]

Advances in parasitology. New York & London, Academic Press, 1963-. v.1-. ISSN: 0065308x.

V.30, 1991 (x,261p.) has 9 contributors. The 5 documented articles include 'Cultivation of helminths in chick embryos' (with 10p. of references); 'Influence of pollution on parasites of aquatic animals' (with 10p. of references). Analytical index, p.239-61. *Class No: 576.8*

[2165]

FOSTER, W.D. A History of parasitology. Edinburgh & London, Livingstone, 1965. vii, 202p. illus., ports.

Chapter 1 traces the general development of parasitology from ancient times to about 1850. Chapters 2-13 deal with certain parasites 'chosen because investigations related to them opened up new fields in parasitology' (*Preface*), e.g. 2. The Cestodes (tapeworm) ... 6. Hookworm. Chapter 14 'Parasitology established', covers the period 1860 to 1932. 284 references. Index to personal names. 15 plates, mostly portraits. Aims 'at a level comprehensive to the non-professional parasitologist' (*Preface*). *Class No: 576.8*

[2166]

Index-catalogue of medical and veterinary zoology. US Bureau of Animal Industry. Zoological Division. Authors A-Z. Washington, US Government Printing Office, 1932-52. 12v. Supplement 1-. 1953-.

Index to world literature 'on parasites and parasitisms of man, of domestic animals, and of wild animals whose parasites may be transmitted to man and domestic animals' (*Bulletin of the Medical Library Association, v.41, no.2, April 1953, p.110*). In the Author catalogue, each author's papers are arranged chronologically.

Supplements 1-6 (1953-56) covered backlog. Since then, supplements (authors, A-Z) are annual. Beginning with suppt.15, *Parasite subject catalogues* (containing indexes to the author references in the *Author catalogues* and including a catalogue of hosts) have been issued in addition. *Class No: 576.8*

Biochemistry

Encyclopaedias & Dictionaries

[2167]

HOOD, W. A-Z of clinical chemistry. New York, Halsted: Wiley, 1980. vii, 386p. illus. ISBN: 0470270292.

Concise definitions of *c.*1,300 terms on techniques, tests, diagnostic methods, etc. Cross-references; additional references where considered appropriate, plus a short list of general textbooks. 'This volume is a welcome source for undergraduates and trainees starting any kind of course in clinical chemistry' (*Choice, v.18(9), May 1981, p.1236*). *Class No: 577.1(03)*

[2168]

SCOTT, T. *and* EAGLESON, M. Concise encyclopedia of biochemistry. 2nd ed. Berlin, Walter de Gruyter, 1988. 649p. diagrs., tables, charts, formulas. $89.90. ISBN: 0899254578.

First ed. (1983) based on 2nd ed. of *Brockhaus ABC Biochimie* (1981).

Extensive revision brings in new material covering enzymology, molecular biology, metabolism, metabolic regulation and natural products. Entries, A-Z, vary in length from one paragraph to several pages ('Evolution': 4 columns, 2 charts, 1 table; 'Recombinant DNA technology': 8p., 12 diagrs.). Each enzyme is given its Enzyme Commission number. Adequate cross-references. An excellent aide-mémoire. *Class No: 577.1(03)*

[2169]

STENESH, J. Dictionary of biochemistry and molecular biology. 2nd ed,. New York, Wiley, 1989. vii,525p. $69. ISBN: 0471840890.

Originally *Dictionary of biochemistry,* published in 1975.

Approximately 16,000 entries 'representing an increase of about 30% over the first edition ... drawn from over 500 books and 1,000 articles' (*Preface*). Includes abbreviations, cross-references, names of specific compounds, and additional information inserted into definitions (average 40 words) where appropriate. Designed for scientists and students in the life sciences. 'Will prove a valuable addition to most research and academic collections' (*New technical books,* November/December 1990, item 1747). *Class No: 577.1(03)*

Encyclopaedias

[2170]

GLICK, D.M. Glossary of biochemistry and molecular biology. New York, Raven Press, 1990. 194p.

Brief definitions of over 1,500 terms. Includes most new biotechnology terms but omits specific enzyme and gene names. Handy, small, format. *Class No: 577.1(031)*

Handbooks & Manuals

[2171]

FLORKIN, M. *and* STOTZ, E.H. Comprehensive biochemistry. Amsterdam, etc. Elsevier, 1962-. illus., tables, formulas.

Sections: 1. (v.1-4): *Physico-chemical and organic aspects of biochemistry.* 2. (v.5-11): *Chemistry of biological compounds.* 3. (v.12-16): *Biochemical reaction mechanisms.* 4. (v.17-21): *Metabolism.* 5. (v.22-29): *Chemical biology.* 6. (v.30-32): *A history of biochemistry.*

V.35 (1983. xviii, 404p.): Selected topics in the history of biochemistry: Personal recollections. Pt. I, has 12 well-documented chapters (*e.g.* 5. Sir Frederick Gowland Hopkins, p.103-128). 24 references. Name index, p.390-404. *Class No: 577.1(035)*

[2172]

ISI atlas of science. Biochemistry and molecular biology, 1978/80, including mini-reviews of 102 research fund specialties. Philadelphia, Pa., Institute for Scientific Information, 1981. xvi, 540p, charts.

11 contributors. Cluster maps/charts are used to trace the historical development of an area of research and identify the papers that made the most important contributions to the

....*(contd.)*

subject. Each map is accompanied by a short essay or 'mini-review'. Journal, author, subject and specialty title word indexes. *Class No:* 577.1(035)

[2173]

Varley's practical clinical biochemistry. Gowenlock, A.H., *and others, eds.* 6th ed. London, Heinemann Medical Books, 1988. xv,1050p. illus. tables. £75. ISBN: 043338067.

First published 1954.

36 chapters: 1. Hazards in the clinical biochemistry laboratory ... 12. Quality control ... 14. Collection of specimens ... 22. Enzymes ... 30. Thyroid function tests ... 36. Drugs and poisons. Chapter references. Appendices, p.1017-24. Analytical index. For use as a 'bench book' by practising clinical biochemists. A standard text. *Class No:* 577.1(035)

Reviews & Abstracts

[2174]

Biochemistry abstracts. Part 1: Biological membranes. Bethesda, Md., Cambridge Scientific Abstracts, 1972-. 12pa. $635pa. ISSN: 87567504.

Formerly *Biological membrane abstracts* (1972-79).

Contains *c.*8,800 abstracts pa, 5,000 primary and other journals. Sections: Reviews - Membrane components - Receptors - Physical properties ... Transport and related phenomena (single cells & tissues) ... Cell interactions ... Membrane bound enzymes - Bioenergetics - Miscellaneous - Book notices - Notification - Proceedings. Author index and analytical subject index per issue and annually. Online, LIFE SCIENCES COLLECTION. Also available on magnetic tape and CD-ROM. *Class No:* 577.1(048)

Periodicals & Progress Reports

[2175]

Annual review of biochemistry. Palo Alto, Calif., Annual Reviews, Inc., 1932-. v.1-. Annual. ISSN: 00664154.

V.61, 1992 (viii,1359p.) contains 36 papers, including 'Inositol phosphate biochemistry'; 'Animal cell cycles and their control', p.441-70 (with 261 refs.); 'Small catalytic RNAS'. Author, subject indexes. Cumulative list of contributing authors, v.57-61. Cumulative index of chapter titles, v.57-61. Some related articles in other *Annual reviews,* p.vii-viii. *Class No:* 577.1(05)

[2176]

JACKSON, G., *and others.* **Periodicals on biochemistry held by the SRL.** London, Science Reference Library, 1977. v, 23p.

Lists more than 400 titles of periodicals and annual reviews. Appended list of 28 abstracting and indexing periodicals. *Class No:* 577.1(05)

[2177]

Scientific serials review: biomedicine. New York, Wiley/ASFRA, 1986-. 6pa. $148. ISSN: 08848319.

V.1, 1986 includes 'Trends in genetics'; 'Health libraries review'; as well as articles on vaccine, clinical nutrition, alternative medicine, etc. The aim is to evaluate each biomedical journal every 3 years. Entries are based on a matrix of 60 questions/data: previous titles; languages; inclusion of book reviews, abstracts; online availability; changes of subject coverage. Annual list of journals covered.

....*(contd.)*

'Developed to provide libraries and scientists with time, updated reviews of both new and existing periodicals' (*Preface*). *Class No:* 577.1(05)

Tables & Data Books

[2178]

CRC handbook of biochemistry and molecular biology. Fasman, G.B. 3rd ed. Cleveland, Ohio, CRC Press, [1976-77]. 9v. tables.

First published 1968.

A. Proteins - amino acids, peptides, polypeptides, proteins. 3v. B. Nucleic acids - purines, pyrimidines, nucleotides, oligonucleotides ... 2v. C. Lipids, carbohydrates, steroids. D. Physical and chemical data, miscellaneous - ion exchange, chromatography, buffers, miscellaneous. 2v. Cumulative series index.

The cumulative index (1977. [x], 295p.) has subject entries, chemical substances entries, errata and addenda. A comprehensive source of data. *Class No:* 577.1(083)

[2179]

DAWSON, R.M.C., *and others.* **Data for biochemical research.** 3rd ed. Oxford, Clarendon Press, 1986. 580p. tables, chemical structures. £37.50. ISBN: 0198553587.

Features 25 tables, arranging compounds in functional groups, *e.g.* amino acids, amines, amides, peptides and their derivatives, vitamins and co-enzymes. Within each table, compounds are arranged A-Z. Data cover synonyms, molecular formula and weight, melting point, solubility, plus reference to the relevant literature on methodology. 'Directed towards the scientist's laboratory, but should be a very handy reference source for chemistry & biology library collections' (*Science and technology libraries,* v.7(4), Summer 1987, p.118). *Class No:* 577.1(083)

Histories

[2180]

TEICH, M. *and* **NEEDHAM, D.M. A Documentary history of biochemistry 1770-1940.** Leicester, University Press, 1992. xxxvii,579p. figs., tables. £90. ISBN: 071851341x.

Contains a selected collection of reprints and covers the period beginning with the work of Priestley and Lavoisier and ending with the split from molecular biology in the 1940s. 7 sections: 1. Enzymes ... 3. Respiration - 4. Carbohydrates ... 6. Lipids - 7. Conceptual and disciplinary issues. Many footnotes. Select bibliography, p.571-79. *Class No:* 577.1(091)

Biographies

[2181]

FRUTON, J.S. A Bio-bibliography for the history of the biochemical sciences since 1800. Philadelphia, Pa., American Philosophical Society, 1982. xii, 885p. $20. ISBN: 0871699834.

Nearly 10,000 entries (name; dates; references), *e.g.* 'Mendel, Gregor': 4 book references, 10 periodical references; 'Pasteur, Louis': 10 book references, 7 periodical references. Abbreviations for sources: reference works (*e.g.* Poggendorff), p.848-58; serial publications, p.859-85. 'This reference work is most valuable, primarily for the less well known biochemists' (*Choice,* v.20(11/12), July/August 1983, p.1574). *Class No:* 577.1(092)

Proteins

[2182]

Biochemistry abstracts. Part 3: Amino-acids, peptides and proteins. Bethesda, Md., Cambridge Scientific Abstracts, 1980-. 12pa. $695. ISSN: 87567520.

Continues *Amino-acids, peptides and protein abstracts* (1972/79).

About 9,900 abstracts pa. Numbered sections (Amino acids ... Conformation & solution studies of peptides & proteins ... Peptide synthesis .,. Evolutionary aspects - Miscellaneous - Book notices - Notification of proceedings). Author and subject indexes per issue and annually. Sources include *c.* 5,000 primary journals. Online, LIFE SCIENCES COLLECTION. Also available on CD-ROM and magnetic tape. *Class No:* 577.112

Nucleic Acids

[2183]

Biochemistry abstracts. Part 2: Nucleic acids. Bethesda, Md., Cambridge Scientific Abstracts, 1970-. 12pa. $695. ISSN: 87567512.

Continues *Nucleic acid abstracts* (1970-79).

Contains *c.*11,000 abstracts pa. Sections: Reviews - Purines, pyrimidines & analogs - Nucleosides & analogs ... Transfer RNA - Protein biosynthesis - RNA - DNA - Enzymes - Immunological aspects - Protein-nucleic acid association - Book notices, author and subject indexes per issue and annually. Online, LIFE SCIENCES COLLECTION. Also available on magnetic tape and CD-ROM. *Class No:* 577.113

Lipids

[2184]

The Lipid handbook. Gunstone, F.D., *and others, eds.* London, Chapman & Hall, 1986. 900p. £175. ISBN: 0412244802.

A major compendium on the occurrence, isolation, chemical identification and technical applications of fatty acids. A reference section covers the physical properties and literature references for *c.*2,000 lipids and their derivatives (extracted from the database of Heilbron's *Dictionary of organic compounds,* 5th ed. (*qv*)). Name, molecular formula and CAS Registry Number indexes. For more advanced students in chemistry, biochemistry, food science, etc. *Class No:* 577.115

Enzymes

[2185]

Enzyme handbook. Schomburg, D. *and* Salzmann, M., *eds.* New York, Springer-Verlag, 1990. 2v. $150ea. ISBN: 0387525793, v.1; 0387525807, v.2.

First published 1969.

V.1: *Class 4: Lyases.* V.2: *Class 5: Isomerases, Class 6: Ligases.*

About 2,000 enzymes arranged according to the 1984 Enzyme Commission list of enzymes and subsequent supplements. Data for each enzyme is divided into 7 sections: 1. Nomenclature - 2. Reaction and specificity - 3. Enzyme structure - 4. Isolation and preparation - 5. Stability - 6. Cross-references to structure databanks - 7.

....(contd.)

Bibliography. Looseleaf format. Data sheets for about 200 enzymes will be published quarterly. A projected 10-volume series. *Class No:* 577.15

[2186]

The Enzymes. Boyer, P.D., *ed.* 3rd ed. New York & London, Academic Press, 1970-.

First published as *The enzymes: chemistry and mechanism of action,* edited by J.B. Sumner and K. Myrback (1950-52. 2v. in 42).

V.19 (1991), *Mechanisms of catalysis* (459p.), has 7 chapters: 1. Binding energy and catalysis ... 3. Steady-state kinetics ... 7. Stereochemistry of enzyme-catalyzed reactions at carbon. Chapter references. Author and subject indexes. An authoritative treatise. *Class No:* 577.15

[2187]

NOMENCLATURE COMMITTEE OF THE INTERNATIONAL UNION OF BIOCHEMISTRY. Enzyme nomenclature, 1978. New York & London, Academic Press, 1979. vi, 606p. ISBN: 0122271602.

Recommendations of the NCIUB on the nomenclature and classification of enzymes. Sections: 1. Historical introduction - 2. The classification and nomenclature of enzymes - 3. Enzyme list - 4. References (3,859) to the enzyme list. Appendix : Nomenclature of electro-transport proteins. *Class No:* 577.15

Vitamins

[2188]

Handbook of vitamins: nutritional, biochemical and clinical aspects. Machlin, L.J., *ed.* New York, Dekker, 1990. 616p. $125. (*Food science and technology series.*) ISBN: 0824783514.

Detailed data on all known vitamins, including chemistry, food content, metabolism, functions and deficiency symptoms. New to this ed. are state-of-the-art reviews of each vitamin. An authoritative and comprehensive source of information. Of the first ed., 'For nutritionists, biochemists and even the interested lay person' (*Science & technology libraries,* v.5(4), Summer 1985, p.98). *Class No:* 577.16

[2189]

The Vitamins: chemistry, physiology pathology, methods. Sebrell, W.H., *Jr and* Harris, R.S., *eds..* 2nd ed. New York & London, Academic Press, 1967-72. 7v. tables.

1-5, edited by W.H. Sebrell, Jr., and R.S. Harris. 6-7, edited by P. Gyorgy and W.H. Pearson.

V.1 (xiii, 570p.) musters 17 contributions; author and subject indexes. Chapter 2 of v.1, 'Ascorbic acid requirements of animals', has 70 references. v.3 (xv, 601p.) has 30 contributors. The standard treatise on vitamins. *Class No:* 577.16

Antibiotics

[2190]

Dictionary of antibiotics and related substances. Bycroft, B. W., *ed.* London, Chapman & Hall, 1988. 944p. figs. £475. ISBN: 0412254506.

Partially derived from the 5th ed. of *Dictionary of organic compounds* (Chapman & Hall, 1982) and its annual supplements, although each entry has been reviewed and updated. 7 sections: 1. Introduction - 2. Descriptions of main antibiotic types (*e.g.* 'Nucleoside antibiotics') - 3. Entries in

....(contd.)

alphabetical order, p.1-748 - 4. Name index - 5. Molecular formula index - 6. CAS Registry Number index - 7. Type of compound index. Short bibliographies are appended to each entry. *Class No:* 577.18

[2191]

GLASBY, J.S. Encyclopedia of antibiotics. 2nd ed. New York, etc., Wiley, 1979. [v], 467p. chemical formulas. £126. ISBN: 0471097226.

First published 1976.

About 1,800 entries for antibiotics, with data on formula, melting point, elaborating organism, method of preparation and purification, those organisms against which it is effective, and toxicity. 'Penicillic acid': 1 column, 9 references, formula, chemical structure. Also, references to major background papers in scientific and patent literature. Adds over 400 new antiboiotics to the 1976 ed.

A 3rd ed. was due to be published in 1992.

Class No: 577.18

Molecules

[2192]

EVANS, A. Glossary of molecular biology. London, Butterworths, 1974. [iv], 55p. ISBN: 0408706406.

About 400 terms, A-Z, plus cross-references. The appended bibliography, p.510-55, consists of 139 references. *Class No:* 577.2

Biophysics

[2193]

Annual review of biophysics and biomolecular structure. Palo Alto, Calif., Annual Reviews, Inc., 1992-. ISSN: 10568700.

Formerly *Annual review of biophysics and bioengineering* (1972-85) and *Annual review of biophysics and biophysical chemistry* (1986-91).

V.21, 1992 (508p.) contains 18 papers, including: 'Structure and function of actin', p.49-76 (133 references); 'Protein involvement in transmembrane lipid asymmetry', p.417-41 (144 references). Analytical subject index; cumulative index of contributing authors, v.17-21; cumulative index of chapter titles, v.17-21. 'Some related articles appearing in other *Annual reviews*' (p.viii-ix). *Class No:* 577.3

[2194]

McAINSH, T.F., *ed.* **Physics in medicine & biology encyclopedia:** medical physics, bioengineering and biophysics. New York, Pergamon Press, 1986. 2v. $420. ISBN: 0080264972.

More than 200 documented articles (1 to 7 pages in length), by experts. A bonus is a classified list of the articles into 25 broad topics. Adequate cross-references. A glossary of terms associated with anatomy, physiology and pathology is appended. The intended readership is described as 'hospital physicians, medical technicians and clinicians' (*Science & technology libraries,* v.7(4), Summer 1987, p.109). *Class No:* 577.3

[2195]

RASCH, D. Biometrisches Wörterbuch: mit Stichworten in neun Sprachen. 3. völlig neu gefasste u. erw. Ausg. Frankfurt, Deutsch, Harri, 1988. 2v. diagrs. ISBN: 3817100529.

1. *Erläuterungen.* 2. *Register.*

A dictionary of biometrics in 9 languages, with *c.*3,000 German entry-words. *Class No:* 577.3

Viruses (Virology)

[2196]

A Dictionary of virology. Rowson, K.E.K., *and others.* Oxford, Blackwell Scientific, 1981. [v], 230, [2]p. tables, formulas. ISBN: 0632006978.

Several hundred entries for viruses, using names approved by the International Committee for the Taxonomy of Viruses. 'Simian viruses': 2p., 3 references. Title of the book should be 'A dictionary of vertebrate viruses'. A pioneer effort. 'This is certainly a good buy for the veterinary and medical virologist' (*British book news,* December 1981, p.738). *Class No:* 578

[2197]

—Virology: directory and dictionary of animal, bacterial and plant viruses. Hull, R., *and others.* London, Macmillan, 1989. x,325p. figs. £39.50. ISBN: 0333390636.

Has data on all the viral species classified by the International Committee on the Taxonomy of Viruses. Includes equations, formulae, definitions of units, and references to recent reviews of particular topics. 6 appendices give lists of insects infected by certain viruses. Extensively cross-referenced. *Class No:* 578

[2198]

NICHOLAS, R. *and* **NICHOLAS, D. Virology: an information profile.** London, Mansell, 1983. viii, 236p. £30. ISBN: 0720116732.

3 parts. 1. Overview of virology and its literature (history and scope; organizations; conferences; the literature; searching the literature), p.1-168 - 2. Bibliography (392 unannotated entries), p.169-95 - 3. Directory of organizations, culture collections and libraries, p.197-223. Analytical index, p.227-36. Excludes biochemistry and genetics. Despite some omissions, it is a 'very good analytical critical work that fills a gap in the literature available' (*British book news,* March 1984, p.162). *Class No:* 578

[2199]

Virology and AIDS abstracts. Cambridge, University Press, 1988-. 12pa. $820. (*Cambridge scientific abstracts.*) ISSN: 08965910.

Originally as *Virology abstracts* (1967-87).

1991: *c.*8,800 abstracts drawn from *c.*5,000 primary sources. Sections include: Acquired Immunodeficiency Syndrome - Virus taxonomy and classification ... Immunology - Antiviral agents ... Viral infections of man ... Viral infections of higher plants. Author and subject indexes per issue and annually. Online: DIALOG, ORBIT, etc. *Class No:* 578

Microbiology

Handbooks & Manuals

[2200]
PELCZAR, J., Jr and **CHEN, E.C.S. Microbiology.** 5th ed.
New York, etc. McGraw-Hill, 1986. x, 918p. illus., diagrs.,
tables. £48.36. ISBN: 0070492344.

8 sections, each with appended bibliography (1.
Introduction - 2. Microorganisms - 3. Microbial physiology
and genetics - 4. The world of bacteria - 5. Microorganisms;
fungi - 6. Control of microorganisms - 7. Environmental and
industrial microbiology - 8. Microorganisms and disease).
Glossary. Name, organism and subject indexes. A student
textbook. *Class No:* 579(035)

[2201]
WISTREICH, G.A. and **LECHTMAN, M.D. Microbiology.**
5th ed. New York, Macmillan; London, Collier Macmillan,
1988. xx,989p. illus., diagrs., tables. ISBN: 0024289507.

8 chapters, each with short appended bibliography. 1.
Introduction to microbiology ... 7. Practices of immunology
... 8. Microorganisms and infectious diseases. 6 appendices.
Glossary. Guide to organism pronunciation. Index.
Class No: 579(035)

Dictionaries

[2202]
COWAN, S.T. A Dictionary of microbial taxonomy.
Hill, L,.R., *ed.* Cambridge Univ. Press, 1978. xxii, 285p.
ISBN: 052121890x.

First published 1968.

4 preliminary chapters (*e.g.* 2. 'Source material for
taxonomy') are followed by the dictionary, p.33-281
(*c.*2,500 entries). Some entries are short essays (*e.g.*
'Patronymic prefixes': over 5½p.). Includes brief
biographies and organizations. Frequent references to
International Code of nomenclature of bacteria and
International Code of botanical nomenclature. References,
p.269-85. 'An accurate guide to taxonomic practice'
(*Choice*, v.16(1), June 1979, p.506). *Class No:* 579(038)

[2203]
SINGLETON, P. and **SAINSBURY, D. Dictionary of
microbiology and molecular biology.** 2nd ed. Chichester,
New York, Wiley, 1987. xii, 1019p. illus., diagrs., charts,
tables, chemical structures. £87.50; $160. ISBN:
0471911143.

First published 1978 as *Dictionary of microbiology* (viii,
481p.).

About 10,000 terms; many cross-references. 'AIDS': 3
columns; 'Salmonella': over 2 columns; 'Electron
microscope': over 4p. plus keyed diagrams. 5 appendices
(*e.g.* Biosynthesis). Key to journal-title abbreviations. Key to
book references (202 in all), p.1011-19. For undergraduates
and postgraduates in this field, primarily.
Class No: 579(038)

Reviews & Abstracts

[2204]
Microbiology abstracts. Bethesda, Md., Cambridge
Biological Abstracts, 1972-. 12pa. $1,955, set; $780, sect.A;
$880, sect.B; $740, sect.C.

A. *Industrial and applied microbiology.* B. *Bacteriology.*
C. *Algology, mycology and protozoology.*

Section A. contains *c.*7,700 abstracts pa. Sections include
'Products of microorganisms'; 'Mineral microbiology';
'Quality control'; 'Methodology'. Author and subject
indexes per issue and annually.

Section B. *Bacteriology* carries *c.*13,200 abstracts pa.
Sections are numbered (1-3. Methodology, media and
culture apparatus ... 191-203. Ecology and destruction -
205. Miscellaneous topics). Books. Proceedings. Author and
subject indexes per issue and annually.

Section C. *Algology* carries *c.*9,900 abstracts pa. in 17
sections (Taxonomy, structure and function ... Spoilage &
biogeneration - Pollution - Miscellaneous topics). Book
notices. Notification of proofs. Author and subject indexes
per issue and annually.

All 3 sections are available online (LIFE SCIENCES
COLLECTION), on magnetic tape, and on CD-ROM.
Class No: 579(048)

Progress Reports

[2205]
Advances in applied microbiology. New York, etc. Academic
Press, 1956-. Annual. ISSN: 00652164.

V.37, 1992 (vii,385p. £49.50) has 13 contributors. The 7
documented papers include 'Microbial degradation of
nitroaromatic compounds' (17p. of refs.); 'An evaluation of
bacterial standards and disinfection practices used for the
assessment and treatment of stormwater' (5p. of refs.); 'The
sensitivity of biocatalysts to hydrodynamic shear stress'
p.166-226 (18p. of refs.). Index, p.363-81. Contents of
previous volumes, v.27-36. *Class No:* 579(055)

[2206]
Annual review of microbiology. Palo Alto, Calif., Annual
Reviews, Inc., 1947-. Annual. illus., diagrs., tables. ISSN:
00664227.

V.46, 1992 (ix,757p.) contains 23 papers which include
'Science and politics'; 'Molecular biology of methanogens',
p.165-92, (with 188 references); 'Signaling and host range
variation in nodulation'; 'Metabolism and functions of
trypanothione in the kinetoplastida'. Subject index.
Cumulative indexes to contributing authors and chapter
titles, v.42-46. *Class No:* 579(055)

[2207]
Methods in microbiology. New York, etc., Academic Press,
1969-. Annual.

V.24 (1992. x,450p. $99) concerns 'methodologies used in
mycorrhizal research' (*Preface*). 42 contributors. 22 papers
(*e.g.* 'Pathogenic and endomycorrhizal associations', with
3p. of references). Detailed, analytical index, p.433-50.
Class No: 579(055)

Tables & Data Books

[2208]
CRC handbook of microbiology. Laskin, A.I. *and* Lechevalier, H.A., *eds.* 2nd ed. Cleveland, Ohio, CRC Press, 1977-. tables.

First published 1973.

1. *Bacteria.* 1977. 2. *Fungi, algae, protozoa and viruses.* 1978. 3/4. *Microbial composition.* 1981-2. 5. *Microbial products.* 1984. 6. *Growth and metabolism.* 1984. 7. *Microbial transformation.* 1984. 8. *Genetics and immunology.* 1987. 9. Pt.A. *Antibiotics*; Pt.B. *Antimicrobial inhibitors.* 1988.

V.1 ([xi], 787p.) has 30 contributors. Bacteria are considered in 33 sections, each with tables and bibliography. Appended: 'General information' (including immunological and immunochemical classification of micro-organisms); Glossary, p.519-49; Important culture collections; Literature guide for microbiology, p.597-647. Taxonomic and topical indexes. *Class No:* 579(083)

Marine Environment

[2209]
RHEINHEIMER, G. Mikrobiologie der Gewässer. [Aquatic microbiology.] 4th ed. Jena, Gustav Fischer Verlag, 1991; Chichester, Wiley, 1992. viii,363p. illus. $79.50. ISBN: 0471926957.

First published 1971.

15 chapters: 1. Introduction ... 3. The microorganisms of waters ... 7. Pathogens in aquatic plants and animals and their control ... 11. Geomicrobiological processes in waters ... 14. The economic significance of aquatic microorganisms - 15. Outlook. References, p.313-50. Index of names of genera. Subject index. Aimed at undergraduates and professionals. *Class No:* 579(26)

Bacteria

[2210]
Bergey's manual of systematic bacteriology. 2nd ed. Baltimore, Md., Williams & Wilkins 1984-89. 4v. illus., tables. ISBN: 0683041088, v.1; 0683078933, v.2; 0683079085, v.3; 0683090615, v.4.

Replaces *Bergey's Manual of determinative bacteriology* (8th ed. 1974).

V.1 by N.R. King. 1984, xxvii, 964p. V.2 by H.A. Sneath, 1984. xviii, 634p. V.3 by J.T. Staley, 1989. xxviii, 714p. V.4 by S.T. Williams, 1989. xxii,420p.

205 contributors and 17 main sections in all. Each volume contains an extensive bibliography (*e.g.* v.1: p.837-942) and an index of scientific names of bacteria. V.1 covers the Gram-negative bacteria; v.2, the Gram-positive bacteria; v.3, the archaebacteria, cyanobacteria and remaining Gram-negatives; v.4, the actinomycetes. *Class No:* 579.8

[2211]
—Index Bergeyana: an annotated alphabetical listing of names of the taxa of the bacteria. 7th ed,. Edinburgh & London, Livingstone, 1966. xiv, 1472p.

A companion volume to Bergey's *Manual of determinative bacteriology* 'It represents an earnest effort to assess the significance of more than twenty thousand names of bacterial taxa that have been found in the literature' (*Preface*). Bibliography (A-Z authors), p.1211-1472. *Class No:* 579.8

[2212]
International code of nomenclature of bacteria. Sneath, P.H., *and others.* Rev. ed. Washington, American Society for Microbiology, 1992. 232p. $47. ISBN: 155581039x. *Class No:* 579.8

[2213]
Microbiology abstracts. Section B: Bacteriology. Bethesda, Md., Cambridge Scientific Abstracts, 1972-. 12pa. $740. ISSN: 03008398.

Formerly as Section B: *General microbiology and bacteriology,* 1966-71.

About 13,200 abstracts pa. 16 sections: Methodology, media and culture - Identification, taxonomy and typing - Cell structure and function - Genetics and evolution - Phage-host interactions - Antibiotics - Antibacterial agents - Immunology - Human, Animal, Invertebrate bacteriology - Plant diseases - Microbial symbioses, antibioses and predation - Ecology and distribution - Miscellaneous topics. Books. Proceedings, Author and subject indexes per issue and annually. Available online (LIFE SCIENCES COLLECTION), on magnetic tape, and CD-ROM. *Class No:* 579.8

[2214]
SAMSON, P. Glossary of bacteriological terms. London, Butterworths, 1975. 165p. diagrs.

Not strictly confined to bacteria; rather, a glossary of microbiological terms. 'A very useful book, not only for those first entering a medical microbiological field but as a quick reference for many workers in laboratories' (*British book news,* August 1975, p.570). *Class No:* 579.8

58 Natural History

Nature Study

Bibliographies

[2215]

AMERICAN MUSEUM OF NATURAL HISTORY. Research catalog of the Library. [American Museum of Natural History]. Boston, Mass., G.K. Hall, 1977-78. 13v. + 12v. $1320. ISBN: 0816100640; 0816102384.

Authors. 1977. 13v. (9180p.). *Classed catalog*. 1978. 12v. (8859p.).

The library houses one of the world's leading natural history collections. Its 325,000 volumes comprehensively cover a wide variety of subject areas, 'including anthropology, general natural history, entomology, geology, herpetology, mineralogy and peripheral biological sciences'. The *Authors* catalogue has entries for *c*.17,000 serial titles as well as for rare books, manuscripts, pamphlets, visuals, letters. *Class No:* 58/59:502(01)

[2216]

—ACADEMY OF NATURAL SCIENCES OF PHILADELPHIA. Catalog of the Library. [Academy of Natural Sciences of Philadelphia]. Boston, Mass., G.K. Hall, 1972. 16v. $1755. ISBN: 081610946x.

First published 1980.

Has about 258,000 dictionary-catalogue entries, photolithographically reproduced, in the main *Catalog*. The collection is strongest in systematics, palaeontology and stratigraphic geology, plus limnology, marine biology, ecology and animal behaviour. Many analytical entries. *Class No:* 58/59:502(01)

[2217]

—AMERICAN MUSEUM OF NATURAL HISTORY. The New catalogue of the Library. [American Museum of Natural History]. Boston, Mass., G.K. Hall, 1983. 6v. ISBN: 0816102740.

Catalogues material acquired by the Library since the 1960s. More than 69,000 entries, typeset from computer type, it continues the 1977/78 volumes. *Class No:* 58/59:502(01)

[2218]

—Recent publications in natural history. New York, American Museum of Natural History, Dept. of Library Services, 1983-. 4pa. $17pa. ISSN: 07380925.

Has 26 subject sections and cites monographs in the natural sciences as a whole. *Class No:* 58/59:502(01)

[2219]

Archives of natural history. London, the Society for the Bibliography of Natural History, 1982-. 3pa. £85. ISSN: 02609541.

Formerly *Journal of the Society for the Bibliography of Natural History* (1936-81).

Carries original papers, reviews and bibliographical surveys (*e.g.* 'A check-list of natural history bibliographies and bibliographical scholarship', v.5(6), p.428-67, - a valuable unannotated list of *c*.900 items and supplemented in later issues). V.19(2), 1992 contains 42 signed book reviews, in addition to 12 documented papers. *Class No:* 58/59:502(01)

[2220]

BRITISH MUSEUM (NATURAL HISTORY). Catalogue of the books, manuscripts, maps and drawings in the British Museum (Natural History). London, the Trustees, 1903-15. 5v. (1903-15); Supplement, v.6-8 (1922-40).

Reprinted New York, Stechert-Hafner, 1964.

An author catalogue of the General Library and Departmental libraries (Zoology, Geology, Mineralogy, and Botany). V.1-5: 90,000v.; v.6-8: 30,000v. Four form divisions are used: atlases; dictionaries; encyclopaedias; gazetteers. Within each heading, order is chronological. Periodicals are entered under title or location of issuing society. Maps initially appear under name of country charted, Many cross-references from names to places, etc. V.6 (1922) has 48p. of addenda and corrigenda.

Lists of accessions to the Museum Library (1955-74. 6pa.) has 3 parts: (A) Books, pamphlets and special publications; (B) New periodicals; (C) Selected papers and monographs from periodicals received. *Class No:* 58/59:502(01)

[2221]

FREEMAN, R.B. British natural history books, 1495-1900: a handlist. Folkestone, Kent, Dawson, 1980. 437p. £5. ISBN: 0712909710.

Two parts: 1. Alphabetical list (4,206 numbered entries); 2. List of titles in date order from 1495-1800. Subject index, partly analytical, p.403-37. Numbers in italics indicate 'those works which, in the opinion of the author, are the most important contributions to the subject of the entry'. 'Something more than a short title list and something more than a bibliography' (*Introduction*). It reveals 'an unassailable completeness, scrupulous care over minor details, and a seemingly total freedom from printers' slips' (*British journal for the history of science*, v.14(2), no.46, March 1981, p.86-87). *Class No:* 58/59:502(01)

[2222]

—SEAMONS, G.R., *comp*. Natural history. London, Library Association, Public Libraries Group, 1982. 61p. (*Readers' guide, no.43*.) ISBN: 085365705x.

Supersedes the LA County Libraries Group's *Natural history* (1963). 283 numbered, briefly annotated entries in 10 sections, plus 'Periodicals' (51 entries), 'Series' (9),

....*(contd.)*

'National societies & organizations (62) and 'Statutory bodies' (4). Offers 'a selection of better books on British natural history' (*Introduction*). *Class No:* 58/59:502(01)

Encyclopaedias & Dictionaries

[2223]

ALLABY, M., *ed*. **The Oxford dictionary of natural history.** Oxford University Press, 1985. xiv, 688p. £22.50. ISBN: 0192177206.

56 countributors and consultants. More than 12,000 entries, A-Z, listing taxa of animals and plants down to family level, including 'any lower taxa of particular interest' (*Preface*). Numerous cross-references, especially from common to scientific name. 'Felidae (cats)': 2/3 column; 'Dance language': ¾ column. A few brief biographies (*e.g.* Darwin: ½ column). Failure to give priority to common-name entry is criticized in *New scientist* (10 April, 1986, p.68). However, 'It's an essential book to have about the classroom, laboratory, library or study' (*British book news,* March 1986, p.165). *Class No:* 58/59:502(03)

[2224]

—FITTER, R. *and* FITTER, M. **The Penguin dictionary of British natural history.** London, Allen Lane, 1978. 348p. illus., charts, tables. ISBN: 0713911352.

A reprint, with revisions, of the 1967 ed. About 4,000 entries and cross-references, concerning species, terms, organizations, etc. Common-name entry preferred to scientific name. 'Natural history is interpreted in the broadest sense, including all living things and natural phenomena of the earth and its atmosphere' (*The scope of this book*). Intended for naturalists, country lovers, young people and students working in the field.
Class No: 58/59:502(03)

[2225]

BURTON, M. *and* BURTON, R., *eds*. **The International wildlife encyclopedia.** London, Marshall Cavendish Ltd., 1980. 24v. (2861p.). illus. (mostly col.), maps. £291.45.

Sequence, Aardvark ... Zorro, concluding in v.23. The latter has appended 'Special interests' (*e.g.* Adaptation and longevity; Ecology). V.24 embraces 'A selected list of endangered species'; 'Animals in the wild'; 'International zoos', 'Periodicals bibliography' (8 titles), 'Adult bibliography'; 'Young readers bibliography'; Animal index (p.2783-2840); Subject index; Systematic index, p.2651-61. Fully illustrated in colour (sometimes whole page). A popular natural-history encyclopaedia.
Class No: 58/59:502(03)

[2226]

The Cambridge illustrated dictionary of natural history. Lincoln, R.J. *and* Boxall, G.A. Cambridge, University Press, 1987. 413p. illus. £24.95. ISBN: 0521305519.

Provides *c.*10,000 concise definitions of terms, - 'adequate to explain the meaning or essential concept of a term or to briefly outline the characteristics of a group of organisms' (*Preface*). Entry is under scientific name, with cross-reference from common name. Ecological terms are included. 700 small but clear line-drawings. Both authors are at the British Museum (Natural History).
Class No: 58/59:502(03)

Handbooks & Manuals

[2227]

PALMER, E.L. *and* FOWLER, H.S. **Fieldbook of natural history.** 2nd ed. New York, etc., McGraw-Hill, 1975. xviii, 779p. illus. $42.95. ISBN: 0070481962.

First published 1949 (x, [1], 664p.).

Five parts (astronomy; the earth; rocks and minerals; plant kingdom; animal kingdom), but concentrating on botany and zoology, - using a new classification. Fine line-drawings are a feature. A valuable popular compendium. 'It may be useful as a resource book for elementary-school teachers and others interested in a single volume covering natural history' (*Choice*, v.12(5/6). July/August 1175, p.664,.
Class No: 58/59:502(035)

Reviews & Abstracts

[2228]

Natural history book reviews: an international bibliography. Berkhamstead, Herts., AB Academic Press, 1976-. v.1-. 4pa. (originally 3pa). illus., facsims. £49; $79. ISSN: 1308180x.

V.10(1), 1990 contains a bibliographic essay. 'First readings in charology', p.1-4. Reviews arranged in 14 categories: General (10 items) ... Ecology (3) ... Mammals (5) ... The Young Naturalist (1). Full bibliographical details precede each review. Apparent lack of index for recent volumes reduces the value of this work which is an important current source in its field.
Class No: 58/59:502(048)

[2229]

Wildlife review: an indexing and abstracting publication of wildlife literature. [Washington], US Dept. of the Interior, Fish and Wildlife Service, 1935-. no.1-. 6pa. $26pa. ISSN: 00435511.

About 12,000 references pa., covering over 1,300 periodicals. Sections: General conservation - Plants - Wildlife - Mammals - Birds - Amphibians and reptiles. Entries include names of organizations, etc., for contact purposes. Appended 'Book reviews and new journals'. Author, geographic and subject indexes per issue; annual cumulative index adds a taxonomic index. 'Serves to alert wildlife biologists of current worldwide literature on wildlife management and conservation.' Available on microfilm from UMI, and on CD-ROM from National Information Services Corporation. *Class No:* 58/59:502(048)

Periodicals

[2230]

BRITISH MUSEUM (NATURAL HISTORY). **Serial publications in the British Museum (Natural History) on microfiche.** London, the Museum, 1985-. ISSN: 02673347.

Supersedes *List of serial publications in the British Museum (Natural History) Library* (1975. 3v. (1275p.)). Microfiches are updated at six-monthly intervals.

24,000 entries for serial publications relating to the life and earth sciences. One of the most comprehensive collections in its field. *Class No:* 58/59:502(051)

Yearbooks & Directories

[2231]

MEENAN, A., *comp. and ed.* **Directory of natural history and related societies in Britain and Ireland.** London, British Museum (Natural History), 1983. vii, 407p. £16.50. ISBN: 0565008595.

Nearly 1,000 entries, Aslib Biological and Agricultural Sciences Group ... Zoological Society of Northern Ireland. Data on each society include aims, membership, meetings, publications, additional information (*e.g.* library). Arranged under English and Welsh counties, Scotland, Ulster, and Eire. Appended geographical and subject indexes. A much-needed comprehensive and up-to-date directory, superseding A. Lysaght's *Directory of natural history and other field study societies in Great Britain* (1959). *Class No:* 58/59:502(058)

[2232]

The Naturalists' directory and almanac (international). Gainesville, Fla., Sandhill Crane Press, 1877-. 2-yrly. ISSN: 0277609x.

Title and publisher vary.

45th ed., 1990. 310p. $24.95. Two main sections: a directory of natural scientists worldwide, A-Z, (stating location and special interests), under continents and countries (largely USA); and a listing of natural history museums, societies, associations, etc. *Class No:* 58/59:502(058)

Illustrations

[2233]

CLEWIS, B. **Index to illustrations of animals and plants.** New York, Neal-Schumann, 1991. 217p. $49.95. ISBN: 1555700721.

3 sections: 1. How to use this book - 2. Key to abbreviations of sources - 3. Common-name index (161p.). Data for each entry include code citation, type of illustration and page number. Scientific-name index and 9-page bibliography. 'An outstanding reference for science collections and picture researchers' (*Booklist*, 1 May 1991, p.1740). *Class No:* 58/59:502(084.1)

[2234]

DESMOND, R. **Wonders of creation: natural history drawings in the British Library.** London, British Library, 1986. 248p. col. illus. (pl.), facsims. £25. ISBN: 0712300716.

10 chapters (2. The foundations of natural-history drawing - 3. Flowers in religious art - 4. Herbals - 5. The Renaissance - 6. The seventeenth century - 7. The golden age - 8. Voyages and explorations - 9. The Islamic world and Asia - 10. From Victorian times to the present). 40 colour plates, with captions facing, p.143-201. Select bibliography, p.240. Index of manuscripts, p.242. Detailed index. 'Drawings have been chosen to represent national and regional development, different aspects of natural history studies - plants, mammals, birds, fishes, insects, etc. - and examples of the work of some of the best artists' (*Preface*). A finely produced quarto. *Class No:* 58/59:502(084.1)

[2235]

MUNZ, L.T. *and* SLAUSON, N.G. **Index to illustrations of living things outside North America:** where to find pictures of flora and fauna. Hamden, Conn., Archon/Shoe String Press, 1981. 441p. $55. ISBN: 0208018573.

About 9,000 entries for plant and animal illus. Entry under common name, with reference to source of illus., - 206 books, mostly published since 1963. Index of scientific names. Complementary to J.W. Thompson's *Index to illustrations of the natural world: where to find pictures of the living things of North America* (*qv*). For school libraries. *Class No:* 58/59:502(084.1)

[2236]

THOMPSON, J.W., *comp.* **Index to illustrations of the natural world:** where to find pictures of the living things of North America. Slauson, N., *ed.* Syracuse, N.Y., Gaylord, 1977. (Reprinted Hamden, Conn., Shoe String Press, 1983). 265p. $47.50. ISBN: 0208020381.

About 6,000 entries for plant and animal illus. Entry states common name and letter-code reference to source of illus. Bibliography of sources (178 books published since 1960) under titles, A-Z. Index of scientific names. Follows on from *Natural history index-guide: an index to 3,365 books and periodicals in libraries*, compiled by B. Altsheler (2nd ed., rev. and enl. New York, H.W. Wilson, 1940. 583p.). For school libraries. *Class No:* 58/59:502(084.1)

Films

[2237]

AMERICAN MUSEUM OF NATURAL HISTORY. Department of Library Services. **Catalog of the American Museum of Natural History film archives.** Root, N.J., *ed.* New York, Garland, 1987. 410p. $71. (*Garland Reference library of the humanities, 723*.) ISBN: 0824084594.

Annotated list of 291 films in the archives, produced 1908-1981, on mainly ethnographic or zoological themes. Entries provide descriptive summary and technical data. Detailed indexes by date, genre, subject, personal or corporate name, geographic location, expedition or film title. (Based on annotation in *Choice*, v.25(1), September 1987, p.85). *Class No:* 58/59:502(084.122)

Maps & Atlases

[2238]

The Atlas of the living world. Whitfield, P., *and others.* London, Weidenfeld and Nicolson, 1989. 220p. illus. (col.), figs., maps. ISBN: 0297796429.

6 sections with a 2-page spread on each topic within the section: 1. Global patterns - 2. Habitat patterns - 3. Niche patterns - 4. Changing patterns - 5. The human impact - 6. A catalogue of life (*e.g.* 'Flowering plants'. 'Successful families: birds'; 'Endangered species: invertebrates'). Index. Charts. 1-page bibliography. For school and public libraries. *Class No:* 58/59:502(084.3)

[2239]

—HUXLEY, Sir J. *and* VEVERS, G., *eds.* The Atlas of world wildlife. London, Mitchell Beazley, 1973. 208p. illus. (incl. col.), maps.

Claims over 700 colourful maps, showing location and distribution of animal life worldwide. Brings out relationship

....(contd.)

between animals, vegetation, terrain and climate. Fully illustrated. Page size: 36 x 27cm.
Class No: 58/59:502(084.3)

Histories

[2240]
Annual bibliography of the history of natural history. London, British Museum (Natural History), v.1: Publications of 1982-. 1985-. Irregular.

V.5: 1986 (published 1991) has 2,794 author-title entries, A-Z. Sources: holdings of the British Museum (Natural History) libraries, Science Museum, Victoria & Albert Museum, and Wellcome Institute for the History of Medicine collections. Non-analytical subject index, p.84-100. Index of biographies and persons as subjects; index of institutions. Late in appearance. *Class No:* 58/59:502(091)

Biographies

[2241]
BRITISH MUSEUM. Department of Printed Books. A Catalogue of the works of Linnaeus. 2nd ed. London, Trustees of the British Museum, 1933. xi, 246, 68p. illus., facsims.

First published 1907.

Catalogues of the British Museum (Bloomsbury) and the British Museum (Natural History) (South Kensington). 20 sections (19: Bibliography). Appendices 1-8 (1: Societies, periodicals, gardens, islands, towns, squares, etc. named after Linnaeus). Index to his works, p.67-68. Followed by *An index to the authors (other than Linnaeus) mentioned in the catalogue ...* (2nd ed. London, Trustees of the British Museum, 1953. 59p.). *Class No:* 58/59:502(092)

[2242]
MIALL, L.C. The Early naturalists: their lives and work (1530-1789). London, Macmillan, 1912. xi, 396p.

An account of the naturalists who worked and wrote in the period between the Reformation and the French Revolution. Some 40 are dealt with in detail. Discusses not only their work but their characters and the intellectual climate in which they worked. Works are listed. Footnote references; analytical index. *Class No:* 58/59:502(092)

[2243]
PEATTIE, D.C. Green laurels: the lives and achievements of the great naturalists. London, Harrap, 1937. 384p. illus. (pl.), ports.

15 chapters, providing an historical survey from the Middle Ages to Jean-Henri Fabre. 'Sources and reference material' (by chapters), p.369-75. Most of the 21 plates are portraits. *Class No:* 58/59:502(092)

[2244]
PRAEGER, R.L. Some Irish naturalists: a biographical note-book. Dundalk, Tempest, 1941. 208p. ports.

Brief biographical notes on over 300 professional and amateur geologists, zoologists and botanists, from 12th to 20th centuries (p.38-182); also a section detailing societies, institutions and team-work surveys (p.183-200). *Class No:* 58/59:502(092)

[2245]
RAVEN, C.E. English naturalists from Neckam to Ray: a study of the making of the modern world. Cambridge Univ. Press, 1947. x, 379p.

A survey to about 1660. Brief notes on the naturalists and their writings. Many footnote references. Indexes of subjects, flora, fauna and persons (p.359-79). The Preface states that this work was a preliminary for 'A History of Natural History in Britain'. *Class No:* 58/59:502(092)

[2246]
—**MARTIN, E.A. A Bibliography of Gilbert White,** the naturalist and antiquarian of Selborne, with a bibliography and a descriptive account of the village of Selborne. 2nd ed., reprinted with additions. Folkestone, Kent, Dawson, 1970. viii, 193p. [7]p. illus.

Devotes chapters 1-5 to biography and chapters 6-14 to bibliography. Bibliographical descriptions of various editions of *The natural history of Selborne,* etc., are accompanied by 7 plates and 10 text illus. 'A full-scale revision, including biographical and critical works, not included by the original biographer', is called for (*British book news,* April 1971, p.269). *Class No:* 58/59:502(092)

Manuscripts & Incunabula

[2247]
BRIDSON, G.D.R., *and others, comps.* **Natural history manuscript resources in the British Isles.** London, Mansell, 1980. 812p. £105. ISBN: 0720115590.

A directory of *c.*800 manuscript collections of scientific material held in *c.*400 repositories in the British Isles and Eire. Main subject fields: botany, zoology, biology, ecology, conservation, geology, mineralogy, palaeontology, palaeoanthropology and oceanography. Pre-1600 manuscripts are excluded. Index to persons, places and species. Altogether the book is a remarkable 'achievement of British scholarship' (*British book news,* February 1981, p.78). *Class No:* 58/59:502(093)

Europe

[2248]
CHINERY, M., *ed.* **Field guide to the wildlife of Britain & Europe.** London, Kingfisher Books, 1987. 288p. col. illus., maps. £7.95. ISBN: 0862722098.

4 contributors. 5 chapters: 1. Mammals - 2. Birds - 3. Reptiles and amphibians - 4. Fishes - 5. Invertebrates. Index (small type), p.280-7, Useful addresses. Small location maps. Essentially concerned with Western Europe. *Class No:* 58/59:502(4)

[2249]
HARRIS, T. The Natural history of the Mediterranean. London, Pelham Books, 1982. 224p. ISBN: 0720713919.

Part 1 (50p.) surveys the diverse habitats of the region. Part 2 is a guide to representative flora and fauna (*e.g.* seaweeds, flowering plants, invertebrates, fish, amphibians, reptiles, birds and mammals): Bibliography, p.212-3. Index. Lavishly illustrated in colour. Not a field guide, but highly recommended for the novice or short-stay visitor (*British book news,* September 1982, p.552). *Class No:* 58/59:502(4)

[2250]
The Natural history of Britain and Northern Europe.
London, Hodder and Stoughton; 1978-9. 5v., ea.224p. illus, (mostly col.), maps. ISBN: 0340226145; 0340231548; 0340226153; 0340231556; 034023153x.

1. *Mountains and moorlands,* by A. Darlington. 1978. 2. *Towns and gardens,* by D. Owen, 1978. 3. *Fields and lowlands,* by D. Boatman. 1979. 4. *Coasts and estuaries,* by R. Barnes. 1979. 5. *Rivers, lakes and marshes,* by B. Whitton. 1979.

The 5 volumes cover the whole of the northern half of Europe west of Russia and the Baltic states, and include Iceland. V.3 has 2 parts: 1. Ecological survey, with map of lowland vegetation of Northern Europe; 2. A field guide to a representative selection of plants and animals. Plates (p.74-213) of plants, invertebrates, amphibians, reptiles, birds, mammals. Detailed index. Entry is under the vernacular names of plants and animals. Aimed at the traveller-naturalist. *Class No:* 58/59:502(4)

Great Britain

[2251]
AUTOMOBILE ASSOCIATION. Book of the British countryside. London, Reader's Digest; A.A. Distribution, 1973 (reprinted 1988). 535, [1]p. illus., maps. ISBN: 0276420160.

35 contributors. Mainly an A-Z encyclopaedia, with 1,800 entries on the geomorphology, archaeology, flora and fauna of Britain. Extended articles on 49 topics (*e.g.* Beechwoods; Estuaries). Appendices include notes on Country Code; national parks; organizations. Glossary of natural history terms. Over 3,000 illus. in colour, - a leading feature. No index. Useful to both amateur and professional naturalists, although not always helpful for identifying animals and plants. 'To my mind, still the best illustrated natural history encyclopaedia relating to this country' (*British book news,* October 1980, p.585). Companion volume: *A.A. Wildlife in Britain, (qv). Class No:* 58/59:502(410)

[2252]
—ANGEL, H., *and others.* The Natural history of Britain and Ireland. London, Michael Joseph, 1981. 256p. col. illus., maps. ISBN: 0718119894.

Essentially a pictorial record, taking six different habitats, - coastlands and islands; freshwater wetlands; lowland grasslands and heaths; upland; woodlands and hedgerows; towns and suburbs. 6 contributors to the text. 'Further reading', p.246. Detailed, analytical index (in tiny print). 200 col. illus. and maps. Quarto format.
Class No: 58/59:502(410)

[2253]
—The 'Country Life' book of the natural history of the British Isles. Morris, P., *consultant ed.* London, Country Life Books, 1979. 320p. col. illus., maps. ISBN: 0600315401.

7 sections: The sea - The coast - Woodlands - Lowland grasslands and heaths - The uplands - Wetlands and freshwater - Man-made habitats. Analytical index (tiny print), with *c.*3.500 entries. Over 1,600 well-captioned coloured illus. No bibliography. 'With its superb colour photography, masterly artwork and competent text, this is a book that any naturalist ... might treasure and use with profit' (*British book news,* March 1980, p.158).
Class No: 58/59:502(410)

[2254]
—CRAWFORD, P. The Living isles-Safari UK: a natural history of Britain and Ireland. London, BBC, 1985. 320p. col. illus., maps. £14.95. ISBN: 0563203692.

Written as a companion to the BBC television series. Earlier chapters deal with the various habitats. A gazetteer of 450 nature reserves in 20 areas, open to the public occupies p.227-309. Bibliography, p.310. Detailed index, in tiny print, p.312-20. *Class No:* 58/59:502(410)

[2255]
The New naturalist library. London, Collins.

The series extended to over 70v. by 1992, dealing in detail with flora and fauna, conservation of the land and nature, localities. *The Natural history of Orkney,* by R.J. Berry (1985. 304p. illus, (incl. col). ISBN 0002190621) has 12 chapters (2. Biological history; 4. Habitats and vegetation; 6. Otters, seals and whales; 9. Birds; 10. Ordinary men; 11. Orkney naturalists; 12. Conservation development and the future). 'References and further reading', p.223; bibliography, p.226. Appendix (list of species). Non-analytical index. 120 illus. (20 in colour). Author, Professor of Genetics at University College, London.

Recently published in the series is *The Soil,* by B.N.K. Davis, and others, (London, Harper Collins, 1992. 192p.).
Class No: 58/59:502(410)

[2256]
RACKHAM, O. The History of the countryside. London, Melbourne, Dent, 1986. xvi, 445p. illus., facsims, tables, maps. £9.95. ISBN: 0460125524.

17 chapters: 1. Regions - 2. Historical methods and use of evidence - 3. Conservation - 4. Animals and plants: extinctions and new arrivals - 5. Woodland ... 8. Fields ... 10. Trees of hedgerow and farmland ... 14. Moorland ... 17. Marshes, fens, rivers and the sea. Bibliography, p.396-7. 629 References, p.398-414. Glossary; detailed analytical index. Scholarly; well-documented. Author, lecturer in the Botany Dept. at Cambridge. *Class No:* 58/59:502(410)

[2257]
SEAWARD, M.R.D., ed. A Handbook for naturalists. London, Constable, 1981. 202p. illus., maps. ISBN: 0094623902.

First published 1968.

13 contributors. Contents: The Country Code, plus 11 contributions (1. Natural history in Britain ... 6. Natural history museums; 7. Zoos and wildlife parks; 8. Nature trails; 9. Wildlife and the law; 10. Natural history and wildlife conservation organizations; 11. Books and periodicals (p.167-90; grouped). Appendices: A. Vice-counties; B. Grid references. Partly analytical index. Offers 'practical help for those taking up natural history fieldwork for the first time' (*Preface*). Not aimed solely at the young naturalist. Does not entirely replace the 1968 ed.
Class No: 58/59:502(410)

Maps & Atlases

[2258]
DARLINGTON, A. Natural history atlas. London, Warne, 1970. 112p. illus. (incl. col. plates), maps.

'The maps show various habitats in Great Britain (*e.g.* sand dunes, heathland, moorland). For each type of habitat the colour plates illustrate the animals and plants. General ecology of each habitat is discussed in the text and notes on

....(contd.)

field study methods are given' (*Geographical abstracts,* 1971(4), abstract no. 71B/1112). Prepared for young pupils and amateur naturalists. *Class No:* 58/59:502(410)(084.3)

Wales

[2259]

CONDRY, W. The Natural history of Wales. London, Collins, 1981. 287p. illus., maps. (*The New naturalist.*) ISBN: 0002395682.

9 chapters: 1. Corries, crags and summits - 2. Moorlands, mires and conifers - 3. Rivers, lakes and marshes - 4. The native woodlands - 5. The flora of the limestones - 6. Farmlands, villages and estates - 7. The industrial scene - 8/9. The coast. Appendix 1: Reserves and places of interest; 2. Glossary. Bibliography, p.272-6 (Background notes; Flora and fauna; Conservation). Index, detailed, partly analytical and in tiny print, p.277-87. 65 black-and-white photographs. A comprehensive survey that keeps to the popular themes. *Class No:* 58/59:502(429)

[2260]

—**DAVIES, R.B.** A Guide to the literature of Welsh natural history, including material relating to Wales or written in the Welsh language. London, Library Association, 1981. 254l.

Thesis submitted for Fellowship of the Library Association.

500 numbered entries. 'Historical survey of natural history writers in Wales' is followed by annotated list of the literature, closely subdivided: Natural history (items 1-308) - Botany (items 309-80) - Zoology (items 385-500). Index to authors and works [including periodicals] identified by title; index to locations; index by region and county with systematic subdivisions. A thorough piece of research. *Class No:* 58/59:502(429)

USSR

[2261]

KNYSTAUTAS, A. The Natural history of the USSR. London, etc., Century, 1987. 224p. illus (incl. col.), maps. ISBN: 071261401x.

9 chapters: 1. The shaping of a subcontinent - 2. Vegetation and animal distributions - 3. Conservation - 4. The icy north - 5. The largest forest in the world - 6. Mixed forests - 7. Mountains - 8. Steppes and deserts - 9. Wetlands. Appended tables (*e.g.* 'Typical birds of the wetlands of the USSR', p.218-9). Bibliography, p.220. 272 full-colour photographs; location maps. Index (in tiny print), p.221-4. Quarto format. A popular approach. *Class No:* 58/59:502(47)

China

[2262]

The Natural history of China. Zhao, Ji, *and others.* London, Collins, 1990. 224p. illus. (col.), maps. ISBN: 0002190435.

8 sections: 1. Introduction - 2. The geography of China - 3. Forests ... 6. Grasslands - 7. Deserts - 8. Conservation and nature protection. Pronunciation guide. Appendices (*e.g.* Typical mammals of the forests; Typical reptiles of the steppes). 220 colour plates and maps. No bibliography. 'A fine compendium for nature lovers' (*Booklist,* v.86(19), 1 June 1990, p.1859). *Class No:* 58/59:502(510)

India

[2263]

HAWKINS, R.E., *general ed.* **Encyclopedia of Indian natural history.** Delhi, Oxford Univ. Press, on behalf of the Bombay Natural History Society, 1986. [xiii], 620p. illus. (incl. col.) diagrs., maps. $65. ISBN: 0195616235.

Over 550 signed articles, 'Accentors' ... 'Z.S.I., The Zoological Survey of India'. 'Palms': 7½ columns, 9 illus. Cross-references. Covers natural history of India, Pakistan, Bangladesh, Sri Lanka, Burma, Bhutan, Nepal. Detailed index and glossary, p.609-20. Authoritative; 96 distinguished contributors. *Class No:* 58/59:502(540)

America — North

[2264]

COLLINS, H.H. Harper & Row's complete field guide to North American wildlife. Eastern edition. New York, Harper & Row, 1981. 714p. illus. (incl. col.). ISBN: 0690019696.

Earlier ed. as *Complete field guide to American wildlife, central and north* (1959).

'Covering 1500 species of birds, mammals, reptiles, amphibians, food and game birds of both fresh and salt waters, mollusks, and the principal marine invertebrates occurring in North America east of the 100th meridian from the 55th parallel to Florida north of the Keys' (*Subtitle*). Data include appearance, behaviour, habitat. Index. *Class No:* 58/59:502(71+73)

[2265]

RANSOM, J.E. Harper & Row's complete field guide to North American wildlife. Western edition. New York, Harper & Row, 1981. 809p. illus. (incl. col.). ISBN: 0690019696.

'Covering 1800 species of birds, mammals, reptiles, amphibians, food and game fishes of both salt and fresh waters, mollusks, and the principal marine invertebrates occurring in North America West of the 100th meridian from the 55th parallel to the border of Mexico' (*Subtitle*). Data include appearance, behaviour, habitat. Index. *Class No:* 58/59:502(71+73)

[2266]

RICCIUTI, E. The Natural history of North America: in association with the American Museum of Natural History. New York, Facts on File, 1990. 224p. illus. (col.), maps. ISBN: 0831763140.

10 sections: 1. The evolution of a continent - 2. The high Arctic and tundra ... 5. Mountains - 6. Grasslands - 7. Deserts ... 9. Subtropical wetlands - 10. Conservation. 5 appendices, *e.g.* Canyons ... Bird migration ... Bibliography of field guides and field books. Index. Lavishly illustrated. Suitable for public libraries. *Class No:* 58/59:502(71+73)

USA

[2267]

MEISEL, M. A Bibliography of American natural history, the pioneer century, 1769-1865. Brooklyn, Premier Publishing Co., 1924-29. (Reprinted New York, Hafner, 1967). 3v.

V.1: 'Annotated bibliography of the publications relating to the history, biography and bibliography of American natural history and its institutions during colonial times and the pioneer century, which have been published up to 1924,

....(contd.)

with a classified subject and geographical index and a bibliography of biographies'. V.2-3 cover institutions. 1760-1844 and 1845-1865. *Class No:* 58/59:502(73)

Australia

[2268]
BREEDEN, S. *and* BREEDEN, K. A Natural history of Australia. London, Collins, 1970-77. 3v. illus. (incl. col), maps.

1. *Tropical Queensland.* 1970. 262p. 2. *Australia's Southeast.* 1973. 256p. 3. *Australia's North.* 1977. 308p.

V.1 has 11 chapters: 1. North of Capricorn - 2/6. Tropical rain forest - 7. Streams - 8. Mangroves and paperbarks - 9. Lagoons - 10. Open woodland - 11. Conservation. Bibliography, p.255-6 (A-Z authors). Detailed, partly analytical index. A well-produced quarto, the fully captioned col. illus. being a feature. V.2 includes Tasmania. *Class No:* 58/59:502(94)

[2269]
—KEAST, A. Australia and the Pacific Islands. London, Hamish Hamilton; New York, Random House, 1966. [v], 298p. illus. (incl. col.), maps.

Covers the diverse natural history of Australia, New Zealand, New Guinea and Indonesia. 14 chapters (1. Gum trees, bowerbirds, and koalas ... 14. Fiords, glaciers and Antarctic beeches). Supplementary reading, p.293. Analytical index. Quarto format; strikingly large illus., with descriptive captions. *Class No:* 58/59:502(94)

Research Projects

[2270]
Keyword-index of wildlife research. Zurich, Swiss Wildlife Information Service, 1974-. v.1-. Annual (irregular). SFr70.

Text in English or German. Indexes international publications in wildlife biology and management. *Class No:* 58/59:502:061:061.62.005

Museums

[2271]
DAVIS, P. *and* BREWER, R.C. A Catalogue of natural science collections in North-East England, with biographical notes on the collections. [Durham], North of England Museum Service, 1986. [iv], 333p. ISBN: 0951094507.

Main catalogue (*c.*1,500 entries, under authors, A-Z), p.7-165. Biographical notes on the collections, p.166-266. Taxonomic and geographical indexes. A quarto volume. *Class No:* 58/59:502:061:069

[2272]
STEARN, W.T. The Natural History Museum at South Kensington: a history of the British Museum (Natural History), 1735-1980. London, Heinemann, in association with the British Museum (Natural History), 1981. xxiii, 414p. illus., ports., graph. ISBN: 0434736007.

5 parts (22 chapters): 1. The Bloomsbury years, 1753-1880 - 2. The Museum at South Kensington - 3. The Natural History Museum, 1883-1949 - 4. The Museum Departments (Zoology; Entomology; Palaeontology (formerly Geology); Mineralogy; Botany; Library Services (p.317-30)) - 5. An era of change and independence, 1950 - 1980. Appendices:

....(contd.)

A. Principal Librarians and senior Natural History officials; B. Directors and senior officials; C. Trustees; D. Number of visitors to the Museum [graph]. Detailed analytical index. 88 plates. The definitive history, to 1980. *Class No:* 58/59:502:061:069

Nature Reserves

[2273]
DUFFEY, E. National parks and reserves of Western Europe. London, Macdonald, 1982. 288p. col. illus., col. maps. ISBN: 0356085864.

18 contributors, for 18 countries; Norway and Iceland; Sweden; Finland; Denmark; West Germany; Austria; Switzerland; Netherlands; Belgium and Luxembourg; United Kingdom and Eire; France; Spain; Portugal; Italy; Yugoslavia; Greece. Glossary and map key. Index to plants and animals, p.273-83; General index, p.284-5. Addresses of conservation organizations. Bibliography, p.287 (general; by country chapters). Helpful marginal data. Excellent colour photographs and maps. A worthy quarto for any library. *Class No:* 58/59:502.4

[2274]
International handbook of national parks and nature reserves. Allin, C.W., *ed.* Westport, Conn., Greenwood Press, 1990. 560p. tables, maps. $75. ISBN: 0313249024.

Describes and evaluates national parks and nature reserves in 25 nations and one regional cluster. Data include history of park preservation and administrative structures concerned with park protection. Bibliography appended to each chapter. Suitable for academic libraries. *Class No:* 58/59:502.4

[2275]
INTERNATIONAL UNION FOR CONSERVATION OF NATURE AND NATURAL RESOURCES. IUCN directory of neotropical protected areas. International Union for Conservation of Nature and Natural Resources. Dublin, Tycooly International Publishing, Ltd., for the International Union for the Conservation of Nature and Natural Resources, 1982. xx, 436p. maps.

Region surveyed: Central and South America, Caribbean, and Florida. Protected areas, under countries, Antigua ... Virgin Islands (US). Ecuador has 8 such areas, including Galápagos (21 headings, *e.g.* physical location, physical features, vegetation, noteworthy fauna, principal reference material, staff, budget). Outline maps. No index, but detailed contents. The first of ICUN's new series of protected-area directories, a contribution to the Global Environmental Monitoring System. *Class No:* 58/59:502.4

[2276]
The Macmillan guide to British nature reserves. Hywell-Davies, J., *and others.* 2nd ed. London, Macmillan, 1989. xii,654p. col. illus., maps. ISBN: 0333467906.

First published 1984.

Arranged under English and Welsh counties, and Scottish regions. About 3,000 entries (some regrettably brief) for nature reserves and wildlife sites throughout the British Isles (excluding Eire). Data on each: location, type of reserve; features; visiting seasons; description. List of contributors.

....*(contd.)*

No bibliography. Well illustrated in colour. The most comprehensive of British nature reserve gazetteers, although bulky as a guide. *Class No:* 58/59:502.4

[2277]

A Nature conservation review: the selection of biological sites of national importance to nature conservation in Britain. Ratcliffe, D., *ed.* Cambridge, University Press, 1977. 2v. (xvi, 401p; viii, 320p.); illus., tables, maps. £110 + £85. ISBN: 052121159x, v.1; 0521214033, v.2.

V.1 has 13 sections: 1. The ecological background to the site selection ... 4-10. Types of ecosystem sites (*e.g.* coastland, open waters, upland grassland and heaths). Bibliography, p.389-91. Site index; detailed subject index. V.2 (*Site accounts*) describes the 735 locations, both public and private. Data on location, size, geology, topography, soil, past and present land-use and notable species of plants and animals. Sites are arranged by region under type of ecosystem, as in v.1. Detailed index. Editor: Chief Scientist, Nature Conservancy Council 'Invaluable for general reference' (*Nature,* v.269(5625), 1977, p.271). *Class No:* 58/59:502.4

[2278]

—**RSPB nature reserves.** Hammond, N., *ed.* 2nd ed. Sandy, Beds., Royal Society for the Protection of Birds, 1984. 189p. illus. (incl. col.). ISBN: 0903138115.

Has 12 contributors; 14 sections (*e.g.* Seabird cliffs; Reed swamps; Lowland heathland; Upland oakwoods; Moorland in Scotland and Wales). Data on locations: OS map reference; extent; management; type of bird, mammal, etc. Gazetteer of RSPB reserves (well over 100,000 acres, mostly acquired during the last ten years). Appendix 1: Habitats for birds in the United Kingdom; Reserves acquired during 1983. Detailed, partly analytical, index, p.179-89. *Class No:* 58/59:502.4

[2279]

Protected areas of the world: a review of national systems. Gland, Switzerland; Cambridge, UK, International Union for Conservation of Nature and Natural Resources, 1992. 4v. maps. $150. ISBN: 2831700906, v.1; 2831700914, v.2; 2831700922, v.3; 2831700930, v.4.

1. *Indomalya, Oceania, Australia and Antarctic.* xx,351p. 2. *Palaearctic.* xxviii,556p. 3. *Afrotropical.* xxii,360p. 4. *Nearctic and neotropical.* 400p.

Entry for each country includes area; population; economic indicators; policy and legislation; international activities; administration and management; systems reviews; addresses; references; summary of protected areas. *Class No:* 58/59:502.4

[2280]

Protected landscapes: experience around the world. Gland, Switzerland; Cambridge, UK, International Union for Conservation of Nature and Natural Resources, 1987. x,404p.

Covers 26 countries and discusses over 140 individual sites. Data for each entry include geographical location; date and history of establishment; area; land tenure; altitude; climate; vegetation; fauna; local human population; scientific research and facilities; conservation management; budget; local administration; references. *Class No:* 58/59:502.4

Nature Conservation

Bibliographies

[2281]

DENVER PUBLIC LIBRARY. Catalog of the Conservation Library of the Denver Public Library. Boston, Mass., G.K. Hall, 1974. 6v. (3436p.). First supplement. 1978. 2v. (1154p.). $630. ISBN: 0816111138.

The Conservation Library in Denver, Colo. is considered 'one of the most comprehensive of its kind in the world'. (*RQ*, v.14(3), Spring 1975, p.255). The collection is particularly strong on fish and wildlife management, biology, pollution control, and economics, and international in scope. *Class No:* 58/59:502.6(01)

[2282]

World plant conservation bibliography. [London], Royal Botanic Gardens, Kew; World Conservation Monitoring Centre, 1990. xv,645p. £15. ISBN: 0947643249.

Designed to complement *Plants in danger; what do we know?* (World Conservation Monitoring Centre, 1986) (*qv*). Contains over 10,000 citations to published literature, ranging from specific to general, local to international. Includes data from the late 1970s onwards. Includes references on conservation of rare and threatened plants; erosion and conservation of plant genetic resources; extent and loss of habitats and vegetation; a selection of forestry papers; role and work of botanical institutions in plant conservation; conservation thinking and policy; protected areas. 3 indexes: 1. Plant names - 2. Plant families - 3. Geographical. Supplements and revisions are planned. *Class No:* 58/59:502.6(01)

Dictionaries

[2283]

ALLABY, M. Macmillan dictionary of the environment. 3rd ed. London, Macmillan, 1988. [v], 423p. £35. ISBN: 0333455614.

First published 1977 ([v], 532p.).

About 5,000 interdisciplinary terms, straying into many fields, - botany, zoology, chemistry, physics, mathematics, economics (*e.g.* ½p. each on Ice Age, nuclear reactor, mangrove). Includes a few biographical entries, for Darwin, Malthus, Lamarck, etc. 'Soil classification': 1p. Profuse cross-references. *Class No:* 58/59:502.6(038)

Yearbooks & Directories

[2284]

Directory for the environment: organisations, campaigns and initiatives in the British Isles. Frisch, M., *ed.* 3rd ed. London, Green Print, 1990. 320p. £13.99. ISBN: 1854250361.

Previously published as *Directory for the environment,* 2nd ed. 1986.

Lists local and national organizations, campaigns and government bodies concerned with the human and natural environment. Coverage includes consultancies, charities, research institutes, campaigning groups, learned societies, and commercial enterprises throughout Britain and Ireland. *Class No:* 58/59:502.6(058)

Films

[2285]
JOY, S., *ed.* **Environmental film directory.** 4th ed. London, CoEnCo, 1984. 63p.

Completely revised ed. of *A Directory of natural history and conservation films* (3rd ed. Council for Nature, 1978). *Class No:* 58/59:502.6(084.122)

Maps & Atlases

[2286]
The Atlas of endangered species. Burton, J.A., *ed.* London, Macmillan, 1991. 256p. illus., maps. £16.99. ISBN: 0028370810.

20 contributors. 7 biogeographic regions, each subdivided into sections: Habitats and plants - Wildlife - Species in focus. Many colour photographs and maps. 7 world maps show climatic and biogeographic regions, pollution, population, protected areas, natural disasters, and countries which have signed international environmental conventions. List of endangered animals. List of organizations. Brief bibliography. *Class No:* 58/59:502.6(084.3)

[2287]
The Last rain forests: a world conservation atlas. Collins, M., *ed.* London, Mitchell Beazley, in association with IUCN; the World Conservation Union, 1990. 200p. illus. (col.), maps. £17.99. ISBN: 0855337893.

7 sections: 1. What are rain forests? - 2. Why we need rain forests ... 4. How rain forests work ... 6. Atlas of the rain forests - 7. The challenge of conservation. Excellent photographs and maps, with a well-organized text. Glossary. Index. For all libraries covering this subject. *Class No:* 58/59:502.6(084.3)

Worldwide

[2288]
INTERNATIONAL UNION FOR CONSERVATION OF NATURE AND NATURAL RESOURCES. The IUCN plant red data book. International Union for the Conservation of Nature and Natural Resources; comprising Red Data sheets on 230 selected plants threatened on a world scale. Compiled by G. Lucas and H. Synge for the Threatened Plants Committee of the Survival Service Commission of IUCN. IUCN, Morges, Switzerland, 1978. 540p.

Published with the financial assistance of World Wildlife Fund (WWF), United Nations Environment Programme (UNEP).

The distribution of taxa in the *Red data book* is arranged by regions. The 250 chosen species were selected from an estimated 20-25,000 vascular plants in the IUCN categories of Extinct, Endangered, Vulnerable or Rare. Status data in each case: distribution; habitat and ecology; conservation measures taken; biology and potential value; cultivation; description; references. *Class No:* 58/59:502.6(100)

[2289]
Plants in danger: what do we know? Davis, S.D., *and others.* Gland, Switzerland; Cambridge, UK, International Union for Conservation of Nature and Natural Resources, 1986. xlv,461p. $30. ISBN: 2880327075.

Aim is to show how to find information on threatened plants. Entries arranged alphabetically by country (Afghanistan ... Zimbabwe). Data include area; population;

....(contd.)
floristics; vegetation; checklists and floras; information on threatened plants; laws protecting plants; voluntary organizations; botanic gardens; useful addresses; additional references. 3 appendices: 1. General and regional references - 2. Index to bibliography - 3. The implementation of conservation conventions relevant to plants. Geographical index. *Class No:* 58/59:502.6(100)

Great Britain

[2290]
British red data books. Peterborough, English Nature, 1977-.

1. *Vascular plants.* 1977. 2nd ed., 1984. 128p. £5.50. 2. *Insects.* 1987. 448p. 3. *Invertebrates other than insects.* 253p. 1991. £8.

Vascular plants (1977 ed. [3], xxvi, 98p. maps) covers 321 species, - about 18% of the native flora, and it adopts IUCN categories (endangered; vulnerable; rare). *Class No:* 58/59:502.6(410)

[2291]
ENGLISH NATURE. Bibliography series. Peterborough, English Nature, 1979. ISSN: 01431722.

1. *Nature conservation and agriculture.* 1979. [12]p. £1. 2. *Wildlife in the city.* 1979. [16]p. £1. 3. *The New Forest.* 1979. 34p. £2. 4. *Natural environment of the Severn Estuary and British Channel areas.* 1981. 47p. £3.
Class No: 58/59:502.6(410)

[2292]
EVANS, D. A History of nature conservation in Britain. London, New York, Routledge, 1992. xxv,274p. illus., tables. £45; £14.99. ISBN: 0415066522.

9 chapters: 1. The why and the wherefore ... 3. The 1940s: national parks and nature reserves ... 7. The 1970s: going public and getting places ... 9. The future: unity and universality. Species and general indexes. Bibliography. For public and academic libraries. *Class No:* 58/59:502.6(410)

Asia

[2293]
The Conservation atlas of tropical forests: Asia and the Pacific. Collins, N.M., *ed.* London, Macmillan, 1991. 256p. illus. (col.), tables, maps. £65. ISBN: 0333539923.

2 parts: 1. The issues (*e.g.* 'Shifting cultivation'; 'Tropical timber trade') - 2. Country studies, p.84-249 (*e.g.* 'India', p.126-40). References appended to each chapter. Acronyms and glossary, p.250. Index of species. General index, p.253-56. Illustrations and maps are of excellent quality. More detailed, and perhaps, more technical than the same author's *The Last rain forests (q.v.).* Recommended for academic and special libraries supporting research in this area.

The second volume in the series, *Conservation atlas of tropical forests: Africa* (256p. $85), published 1992. *Class No:* 58/59:502.6(5)

America — North

[2294]
Conservation directory: a list of organizations, agencies, and officials concerned with natural resource use and management. Gordon, R.E., *ed.* Washington, National Wildlife Federation, 1956-. Annual. ISSN: 0069911x.

First published 1950.

....(contd.)

38th ed. 1993 (xvii,430p. $18) contains entries for 2,000 organizations. Main sections concern US government, state and provincial organizations and agencies (*e.g.* Maine; 23 entries), as well as Canadian and international bodies. Data include date founded, descriptions of functions, office-bearers, publications. Subject, name, and publications, indexes. *Class No:* 58/59:502.6(71+73)

Wildlife Conservation

Bibliographies

[2295]

BAKER, S. Endangered vertebrates: selected, annotated bibliography, 1981-1988. New York, Garland, 1990. xv,197p. $30. (*Garland reference library of social science, no.480.*) ISBN: 0824047966.

International in scope and includes a wider range of publications than *Official World Wildlife Fund guide to endangered species of North America, (q.v.).* Coverage includes books, symposia, journal articles, and dissertations. Appendix gives names and serial publications of appropriate organizations. Indexes allow access by common name, scientific name, author or geographic area. Has both academic and general appeal. *Class No:* 58/59:502.7(01)

Handbooks & Manuals

[2296]

SALVADORI, F.B. and FLORIO, P.L. Wildlife in peril. Newton Abbot, Devon, Westbridge Books: David & Charles, 1978. 208p. col. illus. ISBN: 0715376969.

Translated from the Italian (1977), has data on 84 mammals (p.14-129), 31 birds and 8 reptiles endangered: Latin name; class; order; family; French, Italian and German terms; description; geographic distribution; habitat; population. Bibliography, p.199-200; index (in small type), p.201-8. The fine coloured illus. are the main feature. *Class No:* 58/59:502.7(035)

[2297]

Save the birds. Diamond, A.W., *and others.* Cambridge, University Press, 1989. 384p. col. illus., maps. ISBN: 0521343674.

3 main sections: 1. Earth's great ecosystems (*e.g.* 'Coasts and estuaries'; 'Arid lands') - 2. Save the birds - we need them - 3. Saving Britain's wildlife. Lavishly illustrated. Suitable for school and public libraries. *Class No:* 58/59:502.7(035)

Tables & Data Books

[2298]

INTERNATIONAL UNION FOR CONSERVATION OF NATURE AND NATURAL RESOURCES. Red data book. 2nd ed. Morges, Switzerland, IUCN, 1978-. v.1-. Loose-leaf.

Prepared and published with the assistance of UNESCO, the UN Environment Programme (UNEP) and World Wildlife Fund International (WWF).

1. *Mammalia.* 2nd rev. ed., by J. Thornback. 1978. 2.

....(contd.)

Aves. 2nd rev. ed., by W.B. King, 1979. 3. *Amphibians and reptiles.* Rev. ed. by R.E. Honecker. 1979. 4. *Pisces.* 5. *Angiosperms.*

Data: Status and summary. Distribution. Population. Habitat and ecology. Threats to survival. Conservation measures taken. Conservation measures proposed. Remarks. References.

The series is continued by publications on specific topics, *e.g. Lemurs of Madagascar and the Comoros: the IUCN red data book.* 1990. 240p. illus. *Class No:* 58/59:502.7(083)

[2299]

—**GROOMBRIDGE, B.** World checklist of threatened amphibians and reptiles. 4th ed. Peterborough, English Nature, 1988. viii,138p. £16. ISBN: 0861394658.

7 sections: 1. Amphibia: Anura ... 4. Reptilia: Crocodylia ... 7. Reptilia: Serpentes. 37-page bibliography (549 items). Indexes of generic and common names, and specific names, with synonyms. *Class No:* 58/59:502.7(083)

[2300]

—World checklist of threatened mammals. Inskipp, T. *and* Barzdo, J., *comps.* Peterborough, English Nature, 1987. 125p. £6.50. ISBN: 0861393368. *Class No:* 58/59:502.7(093)

[2301]

1990 IUCN red list of threatened animals. Gland, Switzerland; Cambridge, UK, International Union for Conservation of Nature and Natural Resources, 1990. xxiv,192p. $15. ISBN: 2831700310.

Covers mammals; birds; reptiles; amphibians; fishes; invertebrates; insects. Lists over 5,000 taxa, with category indicating the degree of threat. *Class No:* 58/59:502.7(083)

Worldwide

[2302]

BALOUET, J.-C. and ALIBERT, E. Extinct species of the world: lessons for our future. Translated by K.J. Hollyman. London, Charles Letts in association with David Bateman, 1990. 192p. col. illus., maps. $19.95. ISBN: 1852381000.

Descriptions of *c.*300 extinct animals. A broad historical survey is followed by 11 chapters: The Indian Ocean - The North Pacific - The South Pacific - The Atlantic Ocean - The Mediterranean - Europe - Africa - Asia - Australia - North America - South America. Appended is a section on plants and animals that humans have made extinct. Detailed, analytical index. Lavishly illustrated. *Class No:* 58/59:502.7(100)

[2303]

MOUNTFORT, G. Rare birds of the world: a Collins/ICBP handbook. London, Collins, 1988. 256p. illus. (some col.), figs. £12.95. ISBN: 0002198355.

Based on ICBP/IUCN *Red data books.* 8 sections: 1. Birds and man ... 3. The Nearctic region ... 7. The Australasian region - 8. The Pacific islands. Concise entries for each species. 32 colour plates showing habitat and 50 black-and-white drawings. 2 appendices, in tabular form: 1. The world's threatened species - 2. Birds presumed to have become extinct since 1600. Small bibliography. Indexes of vernacular and scientific names. Author one of the founders of the WWF. *Class No:* 58/59:502.7(100)

Europe

[2304]

GRIMMETT, R.F.A. *and* **JONES, T.A.,** *comps*. **Important bird areas in Europe.** Cambridge, International Council for Bird Preservation, 1989. x,888p. tables, maps. £21.50. (*ICBP technical publication no.9.*) ISBN: 0946888175.

Inventories arranged A-Z by country, Albania ... Yugoslavia. Describes 2,444 sites and provides an indication of the extent to which they are unprotected or threatened. The chapter on each country is subdivided into sections on general information, ornithological importance, conservation infrastructure and protected-areas system, international resources relevant to the conservation of sites, overview of the inventory, glossary, inventory and main references. Maps showing the sites are also included. 6 appendices: 1. Numerical criteria used to select sites for migratory species ... 6. English names of species mentioned in the text. 'It is intended for international agencies, governmental and non-governmental nature conservationists, land-use planners and professional and amateur ornithologists' (*Introduction*).
Class No: 58/59:502.7(4)

[2305]

SMIT, C.J. *and* **WIJNGAARDEN, J.N. Threatened mammals in Europe.** Wiesbaden, Akademische Verlagsgesellschaft, 1981. 259p. maps. ISBN: 3400004383.

Supplementary volume of *Handbuch der Säugetiere Europas* and written in English to attract a wider readership.

The Introduction states the case for conserving animals, summarising the threat to their survival. Detailed 'Data sheets on the threatened species' follow, accompanied by distribution maps. Favourably reviewed in *Natural history book reviews,* v.7(1/2), 1984, p.41.
Class No: 58/59:502.7(4)

Great Britain

[2306]

Important bird areas in the United Kingdom including the Channel Islands and the Isle of Man. Pritchard, D.E., *and others, eds*. Sandy, Beds., RSPB, 1992. 540p. maps. £35. ISBN: 0903138468.

Lists 256 sites of international importance which qualify for protection under the EC's Directive on the conservation of wild birds. Site accounts (p.73-475) are arranged geographically, *e.g.* 'Scotland'. Maps show site location. 6 appendices (1. Bird species named in text). Index to sites. Companion volume to *Important bird areas in Europe (q.v.),* 1989. *Class No:* 58/59:502.7(410)

[2307]

SITWELL, N. The Shell guide to Britain's threatened wildlife. London, Collins. 1984. 208p. col. illus. £8.95. ISBN: 0002190503.

Pictorial guide to 172 plants, birds, mammals, reptiles, amphibians, fish and invertebrates considered to be in danger of extinction. 'No similar collection [of natural history photographs] has ever been produced before' (*British book news,* October 1984, p.612).
Class No: 58/59:502.7(410)

America — North

[2308]

Bird conservation 3. Jackson, J.A., *ed*. Madison, Wis., published for the International Council for Bird Preservation, United States Section by Wisconsin, 1989. 177p. $17.50. ISBN: 0299111202.

V.3 devoted to the single theme of North American forest ecosystems and their birds. Covers seven forest types. No subject index but contains index of bird species. No illustrations. *Class No:* 58/59:502.7(71+73)

[2309]

The Official World Wildlife Fund guide to endangered species of North America. Moseley, C. Washington, D.C., Beacham Publishing, 1990-92. 3v. illus. $280.

Describes more than 500 animals and plants. V.1 deals with mammals and plants; v.2 with birds, fishes, insects and other species; v.3 adds species federally listed from August 1989 to December 1991. Entries are 2 pages in length and include a map, black-and-white photograph and a table giving information on habitat, food, reproduction and threats to the species. Each entry concludes with a general essay plus a bibliography. Each volume has a glossary, an index of occurrence by state and an index of common and scientific names. No cumulative index.
Class No: 58/59:502.7(71+73)

USA

[2310]

GARDNER, R.L., *and others*. **'Official state lists of endangered, threatened, or rare species'.** In *Reference services review,* v.19(1), Spring 1991, p.23-38, 48.

Bibliography of lists produced by state government agencies, arranged alphabetically, Alabama ... Wyoming. Each entry gives a brief description of the publication and its arrangements. A 3-page list of responding agencies and their addresses is given. *Class No:* 58/59:502.7(73)

58 Botany

Bibliographies

[2311]
DAVIS, E.B. Guide to information sources in the botanical sciences. Littleton, Colo., Libraries Unlimited, 1987. 175p. $35. ISBN: 0872874397.

3 parts: 1. Bibliographical control - 2. Ready-reference sources (sections 3-7) - 3. Additional sources of information (8-10: Historical materials; textbooks; key publishers, services and important series). 720 annotated entries. Detailed, grouped index, p.153-75. The main focus is on printed materials and computerized databases. A useful survey for students, librarians, avocational and professional botanists. American slant (of 40 botanical societies listed, only 8 are non-US, no index entry for the 2 main British botanical gardens, - Kew and Edinburgh).
Class No: 580(01)

[2312]
HENREY, B. British botanical and horticultural literature before 1800: computing a history and bibliography of botanical and horticultural books printed in England, Scotland and Ireland from the earliest times until 1800. London, Oxford Univ. Press, 1975. 3v. facsims., port.

1. *The sixteenth and seventeenth centuries: history and bibliography.* xxvi, 290p. 2. *The eighteenth century: history.* xvi, 748p. 3. *The eighteenth century; bibliography.* xvii, 142p.

V.3 includes items 377-1524, with addenda of 15. Location in 65 libraries, including USA(5), Canada (1) and Paris (2). Also, a list of private libraries, booksellers, etc. A finely produced work. *Class No:* 580(01)

[2313]
Huntia: a journal of botanical history. Pittsburgh, Pa., Hunt Botanical Library, Carnegie Mellon University, 1964-. Irregular. $50. ISSN: 00734071.

Devoted, states the editor (G.H.M. Lawrence), to 'studies of the literature on systematic botany and horticulture, botanical voyages and exploration, early agriculture, medical botany, and related subjects of botanical biography, iconography and bibliography'. *Class No:* 580(01)

[2314]
KENT, D.H., *comp.* **Index to botanical monographs.** A guide to monographs and taxonomic papers relating to phanerogams and vascular cryptogams found growing wild in the British Isles. London, Academic Press, for the Botanical Society of the British Isles, 1967. xii, 163p.

Nearly 1,900 references to monographs, taxonomic and cytotaxonomic papers published since 1800. Abbreviations of titles of periodicals, etc. (c.350), p.1-24. Index to botanical monographs, systematically arranged, p.25-153; index to families and genera, etc. 'A very useful book for

....(contd.)
amateurs and students, especially those without easy access to a large botanical library' (*Aslib book list,* v.35(4), April 1968, entry no.185). *Class No:* 580(01)

[2315]
PRITZEL, G.A. Thesaurus litteratura botanicae omnium gentium. Ed. nov. reform. Leipzig, Brockhaus, 1872-[1877]; reprinted Milan, Görich, 1950, and Königstein, Koeltz, 1972. viii, 547p. $89. ISBN: 3874290352.

An author list of 10,871 books, followed by anonyma and periodica. Includes a valuable subject index of early floras, monographs, etc. The most important reference work for early botanical literature up to 1870; confined to separately published material. *Class No:* 580(01)

[2316]
—JACKSON, B.D. Guide to the literature of botany: being a classified selection of botanical works, including nearly 6,000 titles not given in Pritzel's *Thesaurus*. London, Longman, for the Index Society, 1881 (reprinted, Koeltz, 1974). xl, 626p. $39. ISBN: 3874290697.

Comprises *c.*9,000 short-title entries, with author index. Supplements Pritzel. *Class No:* 580(01)

[2317]
ROYAL BOTANIC GARDENS, Kew. Library. **Author and classified catalogues. [Royal Botanic Gardens, Kew, Library].** Boston, Mass., G.K. Hall, 1973. 9v.

Author catalogue. 5v. (3347p. $545). *Classified catalogue.* 4v. (2372p. $455).

177,000 photolithographed catalogue cards in all. Particularly rich in early botanical books and in works on plant taxonomy and distribution. *Class No:* 580(01)

[2318]
—ROYAL BOTANIC GARDENS, Kew. Library. Current awareness list. 1975-. Monthly. ISSN: 02634740.

Records about 400 entries per issue. Sections: Systematics; Floristics (divided geographically); Chromosome surveys; Palynology; Bibliography; Botanical institutions. Appended (as required) 'Additions to periodicals publications currently received by the Library'.
Class No: 580(01)

[2319]
STAFLEU, F.A. *and* **COWAN, R.S. Taxonomic literature:** a selective guide to botanical publications and collections, with data, commentaries and types. 2nd ed. Utrecht, Bohn, Scheltema and Holkema, 1976-88. 7v. 6250p. $1360. (*Regnum vegetabile, 94, 98, 105, 110, 112, 115, 116.*) ISBN: 9031302244.

Arranged under authors, A-Z, with brief biographical notes. 16,614 items from 6,186 authors are contained in v.1-6. Also, information on composite works, herbaria with collections, sources of bibliographical and biographical notes, annotated list of authors, most important works, etc. Name and title indexes per volume. *Class No:* 580(01)

[2320]

—Index to European taxonomic literature. Utrecht, International Bureau for Plant Taxonomy.

Published annually in the *Regnum vegetabile* series. *Class No:* 580(01)

[2321]

—LAWRENCE, G.H.M. Taxonomy of vascular plants. New York, Macmillan, 1951. xiii, 823p. illus. $46.95. ISBN: 002368190x.

Includes a lengthy, well annotated and invaluable 'Literature of taxonomic botany', p.284-331. Entries are grouped: General taxonomic indexes - World floras and manuals (by areas) - Monographs and revisions - Bibliographies, catalogues and review serials - Periodicals - Glossaries and dictionaries - Cultivated and economic plants - References to miscellaneous topics (including maps and cartography, biographical references, and outstanding botanical libraries). The volume has an extensive index, p.777-823. *Class No:* 582.35

[2322]

SWIFT, L.H. **Botanical bibliographies:** a guide to bibliographic materials applicable to botany. Minneapolis, Minn., Burgess, 1970. xxxix, 804p.

5 main sections: general bibliography; background literature (*e.g.* biological), botanical literature (general and reference works; taxonomic, ecological and physiological botanical works); literature of applied botany (plant cultivation, etc.); literature of auxiliary studies (style manuals; botanical illustration, etc.). 'Highly recommended for all reference libraries' (*Library journal,* v.96(11), 1 June 1971, p.1963-4). *Class No:* 580(01)

[2323]

UNITED STATES. Department of Agriculture. **Botany subject index.** Boston, Mass., G.K. Hall, 1958. 15v. (15050p.). ISBN: 0816105065.

315,000 photolithographically reproduced catalogue cards. Entries are arranged under subjects (subdivided, as necessary) and scientific names. Includes textbooks (arranged chronologically), voyages and travels, biographies and geographical botany. Worldwide in scope. *Class No:* 580(01)

Encyclopaedias & Dictionaries

[2324]

ARAB LEAGUE EDUCATIONAL, CULTURAL AND SCIENTIFIC ORGANIZATION. **Lexicon of botany: English, French, Arabic.** Rabat, 1971. 172, xxviip. *Class No:* 580(03)

[2325]

BAILLON, H. **Dictionnaire de botanique.** Paris, Hachette, 1876-92. 4v. (xii, 2660p.) illus.

Reputed to be 'the most complete dictionary of the plant kingdom' (Lawrence, G.H.M. *Taxonomy of vascular plants,* p.319). It is fully illustrated and particularly useful for its inclusion of biographies of the great botanists. *Class No:* 580(03)

[2326]

Elsevier's dictionary of botany. Amsterdam, etc., Elsevier, 1979-82. 2v. ISBN: 0444417877, v.1; 0444419772, v.2.

1. *Plant names in English, French, German, Latin and Russian.* 1978. viii, 580p. $169.25. 2. *General terms in English, French, German and Russian.* 1982. vi, 744p.

V.1 has 6,131 English-base terms, with French, German and Russian equivalents and indexes. The Russian (cyrillic) index is separate. V.2 (9,967 terms) includes terms from closely - related fields, *e.g.* biochemistry, agriculture, forestry, grassland research, medical properties of plants. *Class No:* 580(03)

[2327]

JACKSON, B.D. **A Glossary of botanic terms,** with their derivation and accent. 4th ed., rev. and enl. London, Duckworth, 1928. x, [ii], 481p.

First published 1900.

Latin equivalents accompany most of the 10,000 terms defined. Sub-entries are grouped under the generic term (*e.g.* the entry for 'Cryptanthous' occupies one column and has 23 sub-entries). Appendix D: Bibliography of botanical dictionaries (40 annotated items), p.478-81. The standard English-language botanical glossary. *Class No:* 580(03)

[2328]

The Marshall Cavendish illustrated encyclopedia of plants and earth sciences. Moore, D.M., *ed.* London, Marshall Cavendish, 1988. 10v. ISBN: 0863079016.

V.1-2: *Dictionary of plants.* V.3-5: *Flowering plant families.* V.6-7: *Plant ecology.* V.8-9: *Earth sciences.* V.10 includes indexes, glossary and bibliography. International in scope though the majority of the 120 contributors are British or European. Numerous colour photographs and diagrams throughout. Many cross-references. For school and public libraries. *Class No:* 580(03)

[2329]

STEARN, W.T. **Botanical Latin:** history, grammar, syntax, terminology and vocabulary. 3rd ed., rev. Newton Abbot, Devon, David & Charles, 1983. xiv, 566p. illus. £18. ISBN: 0715385488.

First published 1966 (London, Nelson, xiv, 566p.).

About 80p. of concise grammar, extensive vocabulary of *c.*6,000 terms and a general bibliography and list of references. Intended primarily to help botanists to read the extensive literature of their subject, and to describe new species in language that is correct, lucid and unambiguous. A 4th ed. was due to be published late 1992. *Class No:* 580(03)

[2330]

TOOTILL, E., *ed*. **The Facts-on-File dictionary of botany.** New York, Facts-on-File, 1984. 390p. illus. $24.95. ISBN: 0871968614.

Over, 3,000 entries on pure and applied botany, from taxonomy to ecology. Concise and readable definitions, often supported by illus. Extensive cross-references. Intended for students, naturalists, geographers and anyone else interested in plants. *Class No:* 580(03)

[2331]
USHER, G. **A Dictionary of botany,** including terms in biochemistry, soil sciences and statistics. London, Constable, 1966. v, 404p.

About 10,000 terms (all the phyla, classes, orders, orders and families) are briefly described in 1-100 terms; cross-references. 'Compiled to meet the needs of university students and grammar school pupils' (*Author's preface*).
Class No: 580(03)

Handbooks & Manuals
[2332]
Plant science: growth, development and utilization of cultivated plants. Hartmann, H.T., *and others.* 2nd ed. Englewood Cliffs, N.J., Prentice-Hall International, 1988. xii,674p. illus., diagrs., tables, maps. $51. ISBN: 0136810659.

First published 1981.

3 sections (31 documented chapters): 1. Plants: structure, classification, growth, reproduction and utilization (1-12) - 2. An overview of the fruit crops and ornamental plants (13-22) - 3. Major agronomic, vegetable and fruit crops (23-31). Appendix includes 'Nutritive values of the edible part of plants'. Glossary, p.649-66. Index. Intended as a textbook for university and college students. *Class No:* 580(035)

[2333]
STRASBURGER, E. **Strasburger's textbook of botany.**
Denffer, D. von, *and others.* New English ed., translated from the 30th German ed., by P. Bell and D. Coombe. London, Longman, 1976. xvi, 877p. illus., diagrs., tables, maps.

First published 1894.

4 parts: 1. Morphology - 2. Physiology - 3. Systematics and evolution - 4. Plant geography. Chronology, p.xv-xvi. References (by parts and sections), p.798-816. Excellent analytical index, p.817-77. Claimed to be the most comprehensive and up-to-date textbook of plant science in a single volume. *Class No:* 580(035)

[2334]
—The Guinness book of plant facts and feats.. Duncalf, W.G. Enfield, Middlesex, Guinness Superlatives, Ltd., 1976. 204p. illus. (incl. col.), facsims., tables. ISBN: 0900424257.

Has 15 sections: 1. Evolution of plant life - 2. Plants in relation to man ... 5. Glossary ... 7. Trees (p.41-88) ... 9. Flowering plants ... 12. Vegetables and fruits - 13. Plants of death ... 15. Radiocarbon dating and dendrochronology. Bibliography, p.199-200. Index (in tiny print, p.201-4).

Also, *The Guinness guide to plants of the world* (1992. 256p. illus. £19.95. *Class No:* 580(035)

Dictionaries
[2335]
MACURA, P., *ed.* **Russian-English botanical dictionary.**
Columbus, Ohio, Slavica Publishers, Inc.; Oxford, Holden Books, 1982. 678p. $49.95. ISBN: 0893570923.
Class No: 580(038)

Reviews & Abstracts
[2336]
Current advances in plant science (CAPS). Oxford, etc., Pergamon Press, 1972-. 12pa. £490. ISSN: 03064484.

About 25,000 references pa. 48 sections: 1. Photosynthesis ... 3. Cell organization ... 6. Plant molecular biology ... 12. Enzymes ... 17. Reproductive development of seed plants ... 21. Growth regulators ... 26. Taxonomy, systems and evolution ... 30. Environmental physiology - 31. Agronomy and horticulture ... 33. Crop protection - 34. Plant pathology - 35. Ecology... 45. Plants and medicine. Author index per issue. Contributes to CABS database on PERGAMON ONLINE. *Class No:* 580(048)

[2337]
—Botanisches Zentralblatt. Referierendes Organ für das Gesamtgebiet der Botanik In- und Auslandes. Kassel (later Jena, Dresden), Fischer, 1880-1945. v.1-179.

Was the international reporting service that preceded *Excerpta botanica. Class No:* 580(048)

[2338]
Excerpta botanica. Sections A & B. Stuttgart, Fischer, 1959-. DM744, pt.A; DM220, pt.B.

A. *Taxonomica et chorologica.* irreg. B. *Sociologica.* irreg.

Section *A.* has *c.*5,000 abstracts pa. on systematic botany (A. Taxonomica. Phylogenia - B. Chrologica - C. Palaeobotanica - D. Varia (*e.g.* herbaria, botanical gardens, museums, congresses, biographies, bibliographies)- E. Recersiones (reviews). Periodical author and subject indexes.

Section *B.* carries bibliographies on specific topics and selected countries. About 2,000 references pa., systematically arranged, with author index. Edited in connection with the International Association for Plant Taxonomy, Utrecht. *Class No:* 580(048)

[2339]
PASCAL folio. **F55: Biologie végétale.** Nancy, Centre National de la Recherche Scientifique, 1985-. 10pa. FFr1,105.

Formerly *Bulletin signalétique. 370. Biologie et physiologie végétales* (1940-83).

About 12,000 abstracts and references pa. Sections 01-09. 01. Généralités - 02. Cytologie - 03. Morphologie et biologie - 04. Physiologie et biologie - 05. Reproduction, multiplication, sexualité, hybridisation, biologie florale - 06. Embryologie - 07. Parasitisme et symbiose - 08. Origine des espèces et évolution - 09. Ecologie et géographie botanique. Subject, systematic and geographical indexes; author index. Available online and on microform. *Class No:* 580(048)

Periodicals & Progress Reports
[2340]
Advances in botanical research. London, Academic Press, 1963-. ISSN: 00652296.

V.18 (1991) has 8 contributors. 4 chapters: 1. Photosynthesis and stomatal responses to air, and the use of physiological and biochemical responses for early detection and diagnostic tools - 2. Transport and metabolism of carbon and nitrogen in legume nodules - 3. Plants and wind - 4. Fibre optic microprobes and measurement of the light microenvironment within plant tissues. Each chapter has a

....*(contd.)*

lengthy bibliography appended. Author index to bibliographical citations. Analytical subject index. *Class No:* 580(05)

[2341]
Botanical review, interpreting botanical progress. New York, Botanical Garden, 1935-. v.1-. 4pa. illus. $55pa. (US); $61 (non-US). ISSN: 00068101.

1-3 contributions per issue with extensive bibliographies. Annual list of articles. Available in microform from UMI. 'The primary review journal in the field' (Katz, B., & L.S. *Magazines for libraries.* 6th ed., 1989, p.175). *Class No:* 580(05)

[2342]
LAWRENCE, G.H.M., *and others, eds.* **B-P-H: botanico-periodicum-Huntianum.** Pittsburgh, Pa., Hunt Botanical Library, 1968. 1063p. $20. ISBN: 091319610x.

Cites *c.*25,000 titles of botanical periodicals. List of selected reference works, p.15; selected references used in *B-P-H,* p.21-23. 'A compendium of information on all periodical (serial) publications that regularly contain (or, in some period of their history, included) articles dealing with the plant sciences and botanical literature, and with the persons who have contributed to botany and to literature' (*Preface*). *Class No:* 580(05)

[2343]
ROYAL BOTANIC GARDENS, Kew. **List of periodical publications in the Library.** Kew, Royal Botanic Gardens, 1978. 94p. ISBN: 0112411495. *Class No:* 580(05)

[2344]
—BRITISH LIBRARY. Science Reference Library. Periodicals on botany held by the SRL. London, Science Reference Library, 1977. v, 49p.

Has *c.*700 periodical titles, A-Z, with holdings and call numbers. Appended: 'Abstracting & indexing periodicals' (38 titles, A/Z). An outline of the SRL classification scheme for botany precedes. *Class No:* 580(05)

Nomenclatures

[2345]
GLEDHILL, D. The Names of plants. 2nd ed. Cambridge, University Press, 1989. vi,202p. figs., tables. £22.50. ISBN: 0521366682.

First published 1985.

7 sections: The nature of the problem; The size of the problem; Towards a solution to the problem; The rules of botanical nomenclature; The International Code of Nomenclature for Cultivated Plants; Botanical terminology; The glossary. Does not deal with specific names of plants but with individual stems or words which combine to make up the complete name of a plant. Bibliography, p.199-202. *Class No:* 580(083.72)

[2346]
HOLMES, S. Outline of plant classification. London & New York, Longman, 1983. vii, 181p. ISBN: 0582446481.

10 chapters (1. The classification of plants - 2. The organization of the higher taxa - 3. Schizophyta - 4. Algae ... 9/10. Spermophyta). Glossary, p.157-69. References and further reading, p.171-2. Detailed non-analytical index, p.173-81. Aims 'to provide how and why names have been changed and to describe the characteristics of each taxonomic group' (*Preface*). 'A reference work that will be

....*(contd.)*

useful to many botanists and naturalists generally' (*Natural history book reviews,* v.7(3), 1984, p.124-5). *Class No:* 580(083.72)

[2347]
HUNT, T. Plant names of medieval England. Cambridge, D.S. Brewer, 1989. lvi,334p. £29.50. ISBN: 0859912736.

4 sections: 1. Introduction (The arrangement of the materials; The manuscripts; List of incipits; The principal sources of medieval botany; Synonymies; The plant names in the *synonyma herbarum;* Bibliographies) - 2. Text - 3. Index of vernacular names - 4. Index of botanical names. Includes over 1,800 vernacular names, covering over 600 plant species. Arranged A-Z (Aaron ... Zizannia), using Latin headwords. Many entries have references to works which include discussions or indicate relevant material. *Class No:* 580(083.72)

[2348]
International Code of Botanical Nomenclature, adopted by the 14th Botanical Congress, Berlin, July-August 1987. Greuter, W., *and others, eds.* Königstein, Koeltz, 1988. xiv,328p. (*Regnum vegetabile, v.118.*) ISBN: 3874292789.

Consists of authoritative rules for the codifying of plants, the main part of the volume, p.80-289, delineating plant families in detail. English language only. The quarterly *Taxon* reports changes in the *Code. Class No:* 580(083.72)

Illustrations

[2349]
CLEWIS, B. Index to illustrations of animals and plants. New York, Neal-Schumann, 1991. 217p. $49.95. ISBN: 1555700721.

3 sections: 1. How to use this book - 2. Key to abbreviations of sources - 3. Common-name index (161p.). Data for each entry include code citation, type of illustration and page number. Scientific-name index and 9-page bibliography. 'An outstanding reference for science collections and picture researchers' (*Booklist,* 1 May 1991, p.1740). *Class No:* 580(084.1)

[2350]
NISSEN, C. Die Botanische Buchillustration: ihre Geschichte und Bibliographie. Stuttgart, Hierbemann, 1951. 2v. (vii, 264p. + v, 324p). Supplement, 1966, viii, 97p.

Arranged by periods and countries (*e.g.* Germany, p.161-222). V.1: *Geschichte.* V.2: *Bibliographie* (2,207 numbered items, under authors, A-Z, p.1-204), with brief biographical notes; appended list of periodicals, anonymous works and titles in series. Indexes of artists, plants, countries and authors. *Class No:* 580(084.1)

[2351]
—Index Londiniensis to illustrations of flowering plants, ferns and fern allies, being an amended and enlarged edition, continued up to the end of the year 1920, of Pritzel's 'Alphabetical register of representation of flowering plants and ferns' ... prepared under the auspices of the Royal Horticultural Society of London at the Royal Botanic Gardens, Kew. Stapf, O. Oxford, Clarendon Press, 1929-31. 6v. Supplement for 1921-35. 1941. 3v.

An A/Z index, by genus and species. The illustration is listed under the name used in the particular book - a possible

....*(contd.)*

cause of error.

Updated in the asterisked entries (denoting illustrated) in *Index Kewensis* supplements (*qv*). *Class No:* 580(094.1)

Histories

[2352]
CURLE, R.H.P. The Ray Society: a bibliographical history.
London, Quaritch, for the Ray Society, 1954. vi, 101p. ports. £5. ISBN: 0903874075.

The object of the Ray Society (founded 1844) was to publish monographs and memoirs on zoology and fauna, with special reference to British fauna and flora. This history, produced on the occasion of the Society's centenary, includes a list of the Ray Society's publications, 1844-1951 (p.95-101), plus a subject bibliography, with notes on each item, p.32-95. *Class No:* 580(091)

[2353]
GREENE, E.L. Landmarks of botanical history.
Egerton, F.N., *ed.* Stanford, Calif., Stanford Univ. Press, 1983. 2v. (xi, 1139p.). $125. ISBN: 0804710759.

Part 1: Introduction to the philosophy of botanical history (The Rhizotomi ... Valerius Cordatus, 1515-1544). Appendices on pre-Grecian and medieval knowledge of plants - Part 2: Introduction to the Italian forefathers of the fifteenth century. Appendix: Biology in the seventeenth century. Each volume has *c.*50p. of notes appended. Bibliography, p.1039-1109. Combined analytical index, p.1116-39. *Class No:* 580(091)

[2354]
MOORE, D.M., *ed.* Green planet: the story of plant life on earth. Cambridge Univ. Press, 1982. 288p. illus. (mostly col.), ports., diagrs., maps. ISBN: 0521246105.

30 contributors. 7 chapters (1. Charting the green planet; 2. Tools of the trade (*e.g.* nomenclature); 3. Evolution of the green planet; 4. Environmental factors; 5. Vegetation today; 6. The realms of plants; 7. Man and the green planet). Biographies, p.274-7; Glossary, p.278-81. Analytical subject index. Index of plant names, p.286-8. Bibliography of 30 items, p.288. 445 illus. (300 col.). *Class No:* 580(091)

[2355]
MORTON, A.G. History of botanical science: an account of the development of botany from ancient times to the present time. London, New York, etc., Academic Press, 1981. xii, 474p. illus., ports. $39.95. ISBN: 0125083823.

11 chapters (1. The beginnings of the knowledge of plants ... 11. Botany in the twentieth century), plus 4p. of chapter notes, including biographical data. The central theme: the evolution of botanical theory. Addressed to university students, the author being Emeritus Professor of Botany, London University. *Class No:* 580(091)

[2356]
ZOHARY, D. *and* HOPF, M. Domestication of plants in the Old World. Oxford, University Press, 1988. [vii], 249p. illus., charts, maps. £40. ISBN: 0198541988.

10 chapters: 1. Sources of evidence for the origin and spread of cultivated plants - 2. Cereals - 3. Pulses - 4. Oil and fibre crops- 5. Fruit trees and nuts - 6. Vegetables and tubers - 7. Condiments and dyes - 8. Fruit collection from the wild - 9. Plant remains in representative archaeological

....*(contd.)*

sites - 10. Conclusions. References, p.223-42. Chronological chart. 25 site orientation maps. Analytical index. *Class No:* 580(091)

Biographies

[2357]
DESMOND, R. Dictionary of British and Irish botanists and horticulturists, including plant collectors and botanical artists. 3rd ed. London, Taylor & Francis, 1977. xxvi, 747p. ISBN: 0850660890.

The *Dictionary* is the latest revision of *A biographical index of British and Irish botanists,* by J. Britten and G.E.S. Boulger (1893), revised by A.B. Rundle (1931).

About 10,000 entries, A-Z, for deceased people, mainly those born in the British Isles. Data: name, date and place of birth and death; education; qualifications; honours officeship in relevant societies; brief details of career; serial publications; editorship of periodical(s); biographical references in books and periodicals; location of plant collections, herbaria, manuscripts, drawings, portraits, state of collections, any plant commemorating the individual, and that person's name. Subject index classifies many of the entries under profession, plants on the country where the flora were collected and studied. 'This new Dictionary is by any standards a remarkable piece of work, reflecting an immense labour on its compiler's part, and shedding light into many very obscure corners' (*American scientist*, v.35(1) 1978, p.84). *Class No:* 580(092)

[2358]
—**NEW YORK BOTANICAL GARDEN. Library.** Catalog of the manuscript and archival collections and index to the correspondence of John Torrey. Boston, Mass., G.K. Hall, 1973. 482p. ISBN: 0816110182.

Lists *c.*180,000 items from a major US botanical horticultural research library. Scope: events and developments in botany, horticulture and natural history from late 19th century to the present. *Class No:* 580(092)

[2359]
HUNT BOTANICAL LIBRARY. Biographical dictionary of botanists, represented in the Hunt Institute Portrait Collection, Hunt Botanical Library, Carnegie Mellon Univ., Pittsburgh, Pa. Boston, Mass., G.K. Hall, 1972. viii, 451p.

About 10,000 entries, stating name, dates, place of birth and death, and category (*e.g.* plant taxonomist). *Class No:* 580(092)

[2360]
NEW YORK BOTANICAL GARDEN. Library. Biographical notes upon botanists. Boston, Mass., G.K. Hall, 1974. 3v. (1658p.). $370. ISBN: 0816106959.

Notes on *c.*44,000 botanists, arranged A/Z. Details: dates, education, honours, professional positions and memberships, outstanding contributions and writings, as well as printed works referring to the botanists (*e.g.* in Pritzel and national biographical dictionaries). *Class No:* 580(092)

Libraries

[2361]
MATHEW, M.V. The History of the Royal Botanic Garden Library, Edinburgh. Edinburgh, HMSO, 1987. 134p. ISBN: 0114933421.

The history of what is now the United Kingdom's most

.... (contd.)

comprehensive library (dating from 1670) on taxonomic botany, horticulture and related subjects outside London. *Class No:* 580:061:026/027

Botanical Gardens

[2362]

International directory of botanical gardens V. Heywood, C.A., *and others*. 5th ed. Königstein, Koeltz, 1990. 1019p. $150. ISBN: 387429319x.

First published 1963.

1,400 entries, under countries A-Z. Data: status (*e.g.* governmental); date of foundation; co-ordinates; rainfall; taxa (number of species and varieties grown); herbarium; publications; greenhouses; hours open to the public; name of director and staff). 'USSR': p.849-938. The new computerized format makes the book needlessly bulky and there are no page headings for guidance.

Class No: 580:061:061.6

[2363]

—HYAMS, E. Great botanical gardens of the world. London, Nelson, 1969. 288p. illus.

A sumptuous volume, with nearly 100 colour plates. Arranged by continents and countries. 41 gardens, their history, research activities, specialists, etc. (Kew, p.104-21). Index map of the world's botanical gardens (525 locations); index. *Class No:* 580:061:061.6

Practical Work

[2364]

BRITISH MUSEUM (NATURAL HISTORY). Instructions for collectors. No. 10. Plants. 6th ed. (reprinted, with minor amendments). London, Trustees of the British Museum (Natural History), 1965. 72p. illus.

5th ed., 1921.

Includes 'Notes on collecting special groups' (*e.g.* Algae), p.35-60. *Class No:* 580.08

Collections

[2365]

Index herbariorum: a guide to the location and contents of the world's public herbaria. 8th ed. New York, New York Botanical Gardens for the International Association for Plant Taxonomy, 1990. 693p. $70. (*Regnum vegetabile, v.120.*) ISBN: 0893273589. ISSN: 00800694.

First published 1952.

Pt.1: *The Herbaria of the world*. 8th ed., by P.K. Holmgren and others. 1990.

Covers *c*.1,800 herbaria worldwide. Arranged A-Z by country (10 appearing for the first time), with sites arranged alphabetically within each country. Data: addresses, important collections, sponsoring organizations, geographical arrangment, herbarium abbreviations, name of director and staff. Indexes include important collections; staff; cities, p.615-93. Indispensable and well-organized.

Class No: 580.082

[2366]

KENT, D.H. *and* **ALLEN, D.E.,** *comps.* **British and Irish herbaria:** an index to the location of herbaria of British and Irish vascular plants. 2nd ed. London, Botanical Society of the British Isles, 1984. *c*.350p. £8. ISBN: 0901158089.

First published 1958.

8 sections, especially 2: 'Universities, museums and other institutions possessing herbaria': principal collections and contributions (over 630) institutions, under locations, A-Z). Section 3: Collectors. Privately owned herbaria are listed in a further section. Extensive bibliography, p.317-31. Classified index to locations of collections.

Class No: 580.082

[2367]

—HANCOCK, E.G. *and* MORGAN, P.J. A Survey of zoological and botanical material in museums and other institutions of Great Britain. Secretary, Biology Curators Group, c/o City Museum, Sheffield, 1980. 32p. tables. ISBN: 0950709204.

2 sections: 1. Zoological holdings; 2. Botanical holdings. Mainly tables (*e.g.* Section 1, table 2: Zoological holding of national and university museums). Zoological appendix: numbers of zoological speciments of 32 phyla, class or order present in museums - Botanical appendix: numbers of botanical specimens present in museums and universities. Needs updating. *Class No:* 580.082

[2368]

The National Plant Collections directory 1992. Woking, Surrey, National Council for the Conservation of Plants & Gardens, 1992. 119p. illus. (col. pl.). £2.25. ISBN: 0951862308.

Lists over 550 National Plant Collections in the UK. Data include information on size, location, directions, opening times, special open days, publications. *Class No:* 580.082

Plants (Flora)

Bibliographies

[2369]

Flora & fauna of localities: a bibliography of books and articles dealing with plants and animal life of specific countries and regions of the world. Huckin, D., *ed.* Caldicote, Biggleswade, Beds., Clover Publications, 1981. 2v. £12.50, set. ISBN: 0907755003, set.

1. *Great Britain and Europe.* [ix], 118p. 2. *Africa, America, Asia, Australia & USSR.* viii, 158p.

The bibliography excludes school texts, conference papers and symposium proceedings, as well as erudite material only likely to be of use to a specialist. *Class No:* 581.1(01)

[2370]

Key works to the fauna and flora of the British Isles and Northwestern Europe. Kerrich, G.J., *and others, eds.* London, Systematics Association, 1978. xii, 179p.

First published 1942.

About 2,500 entries, with occasional annotations, covering books and periodical articles. Systematic arrangement. Supersedes the Systematics Association's *Bibliography of key works for the identification of the British fauna and flora*, by J. Smart and G. Taylor (1953). No index. For more

....(contd.)
specialized scientific libraries. 'Graduate and professional levels' (*Choice*, v.16(1), March 1979, p.54,56).
Class No: 581.1(01)

Illustrations

[2371]
MUNZ, L.T. *and* SLAUSON, N.G. **Index to illustrations of living things outside North America:** where to find pictures of flora and fauna. Hamden, Conn., Archon/Shoe String Press, 1981. 441p. $55. ISBN: 0208018573.

Companion to *Index to illustrations of the natural world: where to find pictures of the living things of North America; (q.v.)* compiled by J.W. Thompson (Syracuse, N.Y., Gaylord, 1977, 265p.).

Entries for over 9,000 species, drawn from illustrations in 206 books. The main index has entry under common name, plus scientific name and source, indicating type of illus. (coloured or black-and-white). Scientific name index follows. Bibliography, under titles. Wide-ranging, but highly selective (*Choice*, v.19(8), April 1982, p.1046).
Class No: 581.1(084.1)

[2372]
NOVAK, F.A. **The Pictorial encyclopedia of plants and flowers.** Barton, J.G., *ed.* London, Hamlyn; New York, Crown, 1968. 589p. illus. (pl.).

Originally published by Artia (Czechoslovakia), 1965.

Consists of 1,120 photogravure illus. (48 in colour), with descriptive captions, p.9-559. The excellent illus. usually show plants in flower or fruit. Index of botanical and common names. Supplements the line-drawing approach to illustration of flora. *Class No:* 581.1(084.1)

[2373]
Plant, animal and anatomical illustration in art & sciences: a bibliographical guide from the 16th century to the present day. Bridson, G.D.R. *and* White, J.J., *comps.* Winchester, Herts., St. Paul's Bibliographies; Detroit, Omnigraphics, 1990. ix,450p. facsims. £75. ISBN: 0906795818.

7,670 entries arranged chronologically in each section and including articles as well as books. Sections: A. Bibliographies - B. Nature in general - C. Plants - D. Animals - E. The human body - F. Artist biographies - G. Periodicals. Appendices: Books on colour; Organizations. Title, subject, and name indexes. For research libraries.
Class No: 581.1(084.1)

Histories

[2374]
COATS, A.M. **The Quest for plants:** a history of the horticultural explorers. London, Studio Vista, 1969. 400p. ISBN: 0289239852.

10 chapters, geographically arranged (The Mediterranean and Near East ... South America). Bibliography, p.379-85. Detailed, non-analytical index, p.386-400. 26 illus. 'This book is chiefly, although not exclusively, concerned with the exploits of professional gardeners who travelled abroad expressly for the purpose of collecting hardy horticultural plants'. (*Foreword*). *Class No:* 581.1(091)

Worldwide

[2375]
BLAKE, S.F. **Geographical guide to floras of the world:** an annotated list, with special reference to useful plants and common plant names. Washington, US Government Printing Office, 1942-61. (reprinted, Koeltz, 1974). (*US Dept. of Agriculture. Miscellaneous publication, no.401, 797.*) ISBN: 3874290689, v.1; 3874290603, v.2.

1. *Africa, Australia, North America, South America, and the islands of the Atlantic, Pacific and Indian Oceans.* 1942. $40. 2. *Western Europe: Finland, Sweden, Norway, Denmark, Iceland, Great Britain with Ireland, Netherlands, Belgium, Luxembourg, France, Spain, Portugal, Andorra, Monaco, Italy. San Marino and Switzerland.* 1961. $65.

Annotated bibliography based on US Dept. of Agriculture Library's subject card-catalogue of botany. Pt.1: *c.*2,597 main entries, 428 subsidiary; admits periodical articles but omits works of a popular nature and terms now only of historical value. Pt.2 (*c.*4,000 entries).
Class No: 581.1(100)

[2376]
FRODIN, D.G. **Guide to standard floras of the world:** an annotated, geographically arranged systematic bibliography of the principal floras, enumerations, checklists and chronological atlases of different areas. Cambridge, University Press, 1984. xx, 619p. ISBN: 0521236886.

Combined a thorough analysis of floras and floristics with a geographically arranged guide to the most useful floras and checklists of vascular plants of the world (*Nature,* v.315, 2 May 1985, p.2). Systematic bibliography (999 entries) has 10 divisions (0. World floras; 1. North America; 6. Europe; 9. Melanesia and Oceania). Appendix A: Major general bibliographies; B. Abbreviations of serials cited. Geographical and author index. *Class No:* 581.1(100)

[2377]
HEYWOOD, V.H., *ed.* **Flowering plants of the world.** London, Oxford University Press, 1978 (reprinted Croom Helm, 1985). 336p. illus. (incl. col.), maps. £19.95. ISBN: 0709937784.

A taxonomic review of the 305 families of flowering plants; also data on orders. Each family is illustrated with a distribution map, diagnostic features, classification details, below family level, affinities to other families, and economic uses. 218 specially commissioned illus. 'Bibliography', p.336. Pleasing layout. 'Certainly a useful reference work for any library' (*New scientist,* v.80(1126), 28 October 1978, p.204). 'There is a good deal of information packed into this volume' (*Kew magazine,* v.3, November 1986, p.196). *Class No:* 581.1(100)

[2378]
—TAKHTAJAN, A. Floristic regions of the world. Translated from the Russian. Berkeley, Calif., Univ. of California, 1986. 522p. $75. ISBN: 0520040279.

Derived from the Russian text of 1978, uses such chorinomic categories as kingdom, regions, provinces, districts and intermediate areas. 'A remarkable compendium ... The book contains many individual insights into the origin and history of the regional floras' (*Nature,* v.328, 23 July 1987, p.304). *Class No:* 581.1(100)

[2379]
KÜCHLER, A.W., *ed*. International bibliography of vegetation maps. Lawrence, Univ. of Kansas Libraries, 1966-. 4v.

   1. *Vegetation maps of North America*. 2nd ed. 1980. xi,324p. 2. *Vegetation maps of Europe*. 1966, vii, 584p. 3. *Vegetation maps of Union of Soviet Socialist Republics, Asia and Australia*. 1968. 359p. 4. *Africa, South America, and World maps*. 1970.

   Each volume has *c*.1,500 entries for published vegetation maps. Data: date, title of map, colour, scale, legend; author of map, when and where published. *Class No:* 581.1(100)

[2380]
Vegetationsmonographien der einzelnen Grossräume. Walter, H., *ed*. Stuttgart, Gustav-Fischer Verlag, 1965-. maps.

   1. *Vegetation von Nord und Mittelamerika;* edited by R. Knapp. 1965. 373p. 2. *Die Wälder Südamerikas;* edited K. Hueck. 1966. 122p. 2a. *Vegetationskarte von Südamerika;* edited by K. Hueck and P. Siebert. 1972. 71p. DM48. 3. *Die Vegetation von Afrika*. edited by R. Knapp. 1973. 626p. 4. *The Vegetation of Australia;* edited by N.C.W. Beadle, 1980. 690p. 7. *Die Vegetation Osteuropas, Nord- und Zentralasiens*. by H. Walter. 1974. 452p. 10. *Klimadiagramm-Karten der einzelnen Kontinente und die ölogische Klimagliederung der Erde;* edited by H. Walter, and others. 1975. 36p. DM98.

   Detailed survey of regional plant embryology and vegetation. V.4 is in English. 'The most thorough, cohesive treatment of regional vegetation available' (Harris, C.D., *ed*. *A geographical bibliography for American libraries*, 1985. p.78). *Class No:* 581.1(100)

Tropics

[2381]
LÖTSCHERT, W. *and* BEESE, G. Collins' guide to tropical plants: a descriptive guide to 323 ornamental and economic plants, with 274 colour photographs. King, C., *trans*. London, Collins, 1983. 256p. col. illus. £12.95. ISBN: 0002191121.

   Sections: Introduction - Ornamental plants - Colour photographs (p.81-176: 3 or more per page) - Economic plants. Glossary, p.249-50. Index of scientific names; index of English common names. Small format.
Class No: 581.1(213)

Europe

[2382]
BLAMEY, M. *and* GREY-WILSON, C. The Illustrated flora of Britain and Northern Europe. London, Hodder & Stoughton, 1989. 544p. illus. (incl. 446 col. pls.). £25. ISBN: 0340401702.

   A copiously illustrated handbook of *c*.2,500 flowering wild plants from Britain and Northern Europe, comprising specially-commissioned colour paintings to aid identification. (With up to 10 species shown per page, a larger page-size might have been advisable, as the book is already too bulky as a field guide). Detailed family descriptions, preceded by helpful introduction (p.9-30)and glossary (p.31-38). Some tree families included (p.41-59). Key to plant families (p.489-518) and English and Latin indexes. An attractive flora aimed at the serious amateur botanist.
Class No: 581.1(4)

[2383]
FITTER, A. New generation guide to the wild flowers of Britain and Northern Europe. Austin, Texas, Univ. of Texas Press, 1987. 320p. col. illus., tables. $16.95. ISBN: 029275535x.

   New ed. of *The wild flowers of Britain and Northern Europe* (1974).

   3 sections: 1. The evolution of flowering plants - 2. The directory of species - 3. The natural history of wild flowers. Section 1 gives an analysis of structure and anatomy. The directory has detailed descriptions and over 1,000 coloured illus. of all but the rarest species in the area, - over 1,400 wild flowers. Also details of status, conditions, locations, habitat, times and methods of germination, range, flowering, pollination and breeding. Final section, - on each stage in the life of the wild flower. Appendix of scarce species. Index, p.308-20. *Class No:* 581.1(4)

[2384]
Flora europaea. Tutin, T., *and others*. Cambridge, University Press, 1964-. 6v. illus., maps.

   1. *Psilotaceae to Platanaceae*. 1992. 2nd ed. 530p. £100. 2. *Rosaceae to Umbelliferae*. 1968. xxvii, 454p. 3. *Dispensiceae to Myoporaceae*. 1972. xxix, 370p. 4. *Plantaginaceae to Compositae (and Rubiacead)*. 1976. xxix, 505p. 5. *Alismataceae to Orchidaceae*. 1980, xxxviii, 452p., *Checklist and chromosome index*. 1982. 440p. *Consolidated index*. 1982. 212p.

   The first complete flora of Europe. The project is sponsored by the Linnean Society of London and is based on studies in herbaria and in the field. Systematic descriptions of all flowering plants and ferns in Europe and intended primarily for professional botanists. Handsomely produced. The 2nd ed. of v.1 revises the descriptions and updates the first 79 families, incorporating many taxa new to Europe. *Class No:* 581.1(4)

[2385]
—POLUNIN, O. Flowers of Europe: a field guide. London, etc. Oxford, University Press, 1969. xii, 662p. illus. (incl. col. pl.).

   Describes 2,600 wild flowers growing in Europe. Description of species, p.38-585, is followed by 192p. of coloured illus (1,926 in all). Glossary, p.8-22, with marginal line-drawings, and keys to families precede. Bibliography (p.586-91) of regional and country floras. English and Latin indexes. Should appeal at several levels, although bulky for the rucksack or pocket. *Class No:* 581.1(4)

[2386]
—POLUNIN, O. *and* WALTERS, M. A Guide to the vegetation of Britain and Europe. Oxford, University Press, 1985. [ix], 238p. illus. (incl. col.), maps. £25. ISBN: 0192177133.

   Has 3 parts: 1. General introduction; 2. Plant communities of Europe; 3. The National Parks and nature reserves of Europe. Selected bibliography, p.215-8, glossary of terms, p.219-23. Indexes of Latin and English names of illustrated types; index to plant communities of Europe. 170 col. plates of over 1,000 plants; 280 line-drawings. Claims 'to be the first attempt to bring together all the information on the subject at a layman's level' (*Preface*). The authors 'have struck the right balance between the superficial and the overly learned' (*Nature,* v.318, 14 November 1985, p.123). *Class No:* 581.1(4)

[2387]
POLUNIN, O. *and* SMYTHIES, B.E. **Flowers of South-west Europe:** a field guide. Oxford, University Press, 1973. xv, 480p. £12.95. ISBN: 0192881787.

3 chapters: 1. Landform, climate and vegetation - 2. The plant-hunting region (23, Algarve ... Camargue). National parks, nature reserves and national walks - 3. The identification of species, p.167-413. Bibliography, p.476-80 (7 groups). Index of popular names. Index of place names. 64p. of colour plates. Index of plants. 21 terrain colour plates; 12 maps. Indebtedness to *Flora Europaea* is acknowledged. *Class No:* 581.1(4)

Great Britain

[2388]
Atlas of the British flora. Perring, F.H. *and* Walters, S.M., eds. 3rd ed. [S.l.], Botanical Society of the British Isles, 1982 (reprinted 1990). xxiii,444p. £23.50. ISBN: 0901158194.

First published 1962.

Distribution maps (4 per page): Pteridophyta; Gymnospermae; Angiospermae (dicotyledones; monocotyledones), p.1-406. 3 appendices 1. List of aggregate and provisional maps; 2. Vice-county records omitted from the maps. The introduction includes a history of mapping of plant distribution. Bibliography, p.425. Index (tiny print), p.435-42. *Class No:* 581.1(410)

[2389]
—PERRING, F.H. *and* SELL, P.D., eds. Critical supplement to the Atlas of the British flora. EP Publishing, Ltd., 1976. [viii], 159p.

Has distribution maps of 500 flowering plants and ferns, plus explanatory text. Appendix: Key to herbarium abbreviations. Bibliography, p.154-5. Index (in tiny print), p.156-9. *Class No:* 581.1(410)

[2390]
BENTHAM, G. **Handbook of the British flora:** a description of the flowering plant and ferns indigenous to or naturalized in the British Isles. Hooker, Sir J.D., *rev.* 7th ed., revised by A.B. Rendle. Ashford, Kent, Reeve, 1924. 606p.

A descriptive flora, with keys and distributional ranges. The introduction includes explanations of growth, morphology of structures, notes on classification, herbaria techniques, a glossary and key to the families. Bentham's *Handbook,* replaced by Clapham, Tutin and Warburg's *Flora of the British Isles (qv),* will nevertheless remain for many years the familiar work for the amateur botanist. *Class No:* 581.1(410)

[2391]
British plant communities. Rodwell, J.S., *and others, eds.* Cambridge, University Press, 1991-. ISBN: 0521235588, v.1.

V.1. *Woodlands and scrub.* 1991. x,395p. £70. V.2. *Mires and heaths.* 1991. £95. V.3. *Grasslands and montane vegetation.* 1992. £105. V.4. *Aquatic communities, swamps and tall-herb fens.* V.5. *Maritime and weed communities.*

A projected set of 5 volumes, which aims to describe over 250 plant communities in Great Britain (excluding Northern Ireland). V.1 contains an introduction, a key, and 25 community descriptions. Each community description contains data on synonymy, constant species, physiognomy, sub-communities, habitat, zonation and succession, distribution, affinities, floristic tables, and distribution maps.

.... (contd.)
Index of synonyms to woodlands and scrubs, p.369-76. Index of species in woodlands and scrub, p.377-84. Bibliography, p.385-95. Culmination of 15 years' detailed survey and analysis of British vegetation by those involved in the National Vegetation Classification.
Class No: 581.1(410)

[2392]
—TANSLEY, A.G. The British islands and their vegetation. Cambridge, University Press, 1939 (reprinted, with corrections, 1949). 2v. illus. (pl.), diagrs.

Parts: 1. The British Isles as environment of vegetation - 2. History and existing distribution of vegetation - 3. The nature of classification of vegetation - 4. The woodlands - 5. The grasslands - 6. Hydroseres - 7. Heath and moor - 8. Mountain vegetation - 9. Maritime and submarine vegetation. Each volume has 163 plates, 179 figures. Chapter-end references. Index in v.2. *Class No:* 581.1(410)

[2393]
CLAPHAM, A.R., *and others.* **Flora of the British Isles.** 3rd ed. Cambridge, University Press, 1987. ISBN: 0521309859.

First published 1952; 2nd ed., 1962.

Main sections: Pteridophyta - Gymnospermae - Angiospermae (Dicotyledons. Monocotyledons). Data on habit; whether native/introduced; ecology; geographical distribution; evolutionary history; agricultural significance, etc. Bibliography, p.633. Notes on families and life forms. 'Likely to be the standard British flora well into the next century' (*Nature,* v.328, 27 August, p.772).
Class No: 581.1(410)

[2394]
—CLAPHAM, A.R., *and others.* Flora of the British Isles: illustrations. Drawings by S.I. Roles. Cambridge Univ. Press, 1957-65. 4v. illus. ea.£11.95.

Follows the sequence and nomenclature of the parent work. 1,910 black-and-white drawings.
Class No: 581.1(410)

[2395]
GODWIN, Sir H. **The History of the British flora:** a factual basis for phytogeography. 2nd ed. Cambridge, University Press, 1975. x, 541p. illus., graphs, tables, maps.

First published 1956.

7 sections: 1. Introduction - 2. Collection and identification of plant remains - 3. The background scale of Pleistocene events - 4. Recorded sites - 5. The plant record - 6. Pattern of change in the British flora - 7. Conclusion. Bibliography, p.503-21 (A-Z authors). Analytical index.
Class No: 581.1(410)

[2396]
KERRICH, G.J., *and others, eds.* **Bibliography of key works for the identification of the British fauna and flora.** 3rd ed. London, Systematics Association, c/o British Museum (Natural History), 1967. vii, 186p.
Class No: 581.1(410)

[2397]
—KENT, D.H. 'Floras great and small'. In *Natural history book reviews,* v.5(1/2), p.3-15.

Concerns regional and local floras in the British Isles (and Ireland), 1964-80. A valuable running commentary that pays tribute to *Flora europaea.* *Class No:* 581.1(410)

[2398]

PERRING, F. *and* **WALTERS, M. The Macmillan field guide to British wildflowers.** London, Macmillan, 1989. xxx,226p. col. illus. £9.95. ISBN: 0333445228.

3 sections: 1. General identification keys - 2. Special identification keys - 3. Flower species, p.1-221. 750 coloured illustrations accompanied by textual information. Index. Although not claiming to be comprehensive - it omits trees, shrubs and the more rare wildflowers - this is a useful guide. A flexible hard-wearing cover would have added to its value. *Class No:* 581.1(410)

[2399]

—McCLINTOCK, D. *and* FITTER, R.S.R. The Pocket guide to wild flowers. London, Collins, 1956. xii, 340p. illus.

Main purpose is to enable anyone to name any wild flower, grass, sedge, tree or shrub that he or she is likely to see in the British Isles. 112 plates (64 in colour and arranged in colour sequence, blue ... yellow). Glossary. Indexes to English and Latin names 'One of the most successful attempts made in recent years to put down our Flora inside a single manageable volume'. (*Journal of the Royal Horticultural Society,* v.85(8), August 1960, p.375). *Class No:* 581.1(410)

[2400]

—McCLINTOCK, D. Supplement to 'The Pocket guide to wild flowers'. Platt, Kent, D. McClintock, [1959]. ix, 89p.

Describes 450 further species. The *Supplement* also adds a long list of hybrids not recorded in the parent work; a valuable list of albinos, a bibliography and a list of county floras and checklists. *Class No:* 581.1(410)

[2401]

ROSS-CRAIG, S. Drawings of British plants; being illustrations of species of flowering plants growing naturally in the British Isles. London, Bell, 1964-74. 31pts. and index (pts.1-9 in 3v.; pts.10-23).

Systematic arrangement (*e.g.* pts.8-9: Rosaceae; 15-18: Compositae), giving brief description of the plant and its common name. Partly replaces Fitch and Smith's *Illustrations of the British flora* and R.W. Butcher's *Further illustrations of British plants* (1930). The drawings (*c.*30 plates per part) of inflorescence, the plant's constituent parts and its general habit are of a high standard, and the reproductions are uniformly excellent. *Class No:* 581.1(410)

[2402]

SIMPSON, N.D. A Bibliographical index of the British flora, including floras, herbals, periodicals, societies and references relating to the identification, distribution and occurrences of phanerogams, vascular cryptogams and charophytes in the British Isles. Bournemouth, privately printed for the Author, 1960. xix, 429p.

More than 65,000 entries for books, articles and manuscripts relating to the British Isles (including Ireland), from the 15th century to date. Data are included on plant lore, local names, poisonous plants and weeds. A much needed bibliographical tool, even if its coverage is admittedly incomplete, its arrangement dispersive, and full pagination of periodical articles is not given. *Class No:* 581.1(410)

[2403]

STACE, C.A. New flora of the British Isles. Cambridge, University Press, 1991. 1226p. illus. £24.95. ISBN: 0521427932.

Contains 2,990 keyed and number species, nearly half as many again as Clapham's *Flora of the British Isles (q.v.).* Designed to identify plants found in the wild in the British Isles. Preceding the main section are chapters designed to ensure maximum use of the book, *e.g.* 'Geographical scope'; 'Identification keys'. Many black-and-white photographs and line-drawings illustrate over half the taxa. Will be a standard work. 'This eagerly awaited and reasonably priced book fulfils almost all expectations and is essential for anyone with a serious interest in the British and Irish floras' (*Watsonia,* v.19(2), August 1992, p.161). *Class No:* 581.1(410)

Scotland

[2404]

BURNETT, J.H., *ed*. **The Vegetation of Scotland.** Edinburgh & London, Oliver & Boyd, 1964. xiii, 614p. illus., diagrs., tables, maps.

9 contributors. 3 pts: 1. The physical background - 2. The vegetation - 3. History and pattern of Scottish vegetation. Chapter bibliographies (*e.g.* 6. Grasslands of the forest and sub-alpine zones, p.168-215; references, p.209-15). Indexes of authors, place names and plant names; general index. Very readable, but no glossary for non-technical readers. Handsomely produced. *Class No:* 581.1(411)

Ireland

[2405]

WEBB, D.A. An Irish flora. 6th rev. ed. Dundalk, Dundalgan Press; Chester Springs, Penn., 1977. xxi, 277p. illus. $27. ISBN: 085221023x.

First published 1943.

Description of families, genera and species (p.1-233) follows the taxonomic scheme of *Flora europaea.* Prelims, including a bibliography (p.xvii-xviii) and general key to families. Glossary of technical terms. Index to Irish names; index to Latin and English names. Pocket sized; small type. *Class No:* 581.1(415)

Wales

[2406]

ELLIS, R.G. Flowering plants of Wales. Cardiff, National Museum of Wales, 1983. ix, 338p. ISBN: 0720002710.

Catalogue of Welsh flowering plants, p.41-193. Data: Latin name; Welsh name; English name; Status; Habitat; Months of flowering; Vice-county distribution; Frequency. Valuable introductory survey. Distribution maps, p.194-280. 3 appendices: 1. Key to the abbreviations of recorder's initials; 2. List of references (p.285-93); 3. World distribution of Welsh plants. Detailed and analytical index, p.296-338. An impressive production, on stout paper. Definitive. *Class No:* 581.1(429)

France

[2407]
BONNIER, G. **Flore complète illustrée en couleurs de France, Suisse et Belgique.** Paris, Librairie Général de l'Enseignement, n.d. 120 facsicules + Table générale. ea. fasc., FFr39; FFr62, table générale.

First published 1927.

Revised continuously since 1927. Each fascicule consists of 8 plates, with appropriate commentary. Indispensable reference tool for French libraries (Beaudiquez, M., and others. *Ouvrages de référence pour les bibliothèques publiques.* 3rd ed. 1986, item 566). *Class No:* 581.1(44)

USSR

[2408]
Flora SSSR. Moscow & Leningrad, Akademiya Nauk SSSR, 1934-64. 30v. and index. illus. (pl.).

A monumental achievement: 22,000p. of text; 1,250p. of illus. Originally edited by V.L. Komarov and completed under Prof. E.G. Bobrov (*Nature,* v.205 (4,976), 13 March 1965, p.1046-49). 92 contributors.

The English translation by the Israel Program for Scientific Translations, under the sponsorship of the National Science Foundation and the Smithsonian Institution - *Flora of the USSR* - has 24v. 'A valuable source of reference for studies involving the flowering plants of the Northern Hemisphere, and especially of the Eurasian land mass' (*Aslib book list,* v.37(4), April 1972, entry 161).

Class No: 581.1(47)

Asia

[2409]
FIELD, H. **Bibliography on Southwestern Asia.** Coral Gables, Florida, Univ. of Miami Press, 1953-62. Supplement, 1968-72. 8v. in 7.

V.1 ('A first compilation') - 7. ('A seventh compilation'), 1953-62, form a series of bibliographies: 47,165 titles (anthropology: 31,254; zoology: 11,505; botany: 4,406) in about 40 languages, covering some 2,500 journals as well as books. Author index. A monumental work.

Class No: 581.1(5)

[2410]
MERRILL, E.D. *and* WALKER, E.H. **A Bibliography of Eastern Asiatic botany.** Jamaica Plain, Mass., Arnold Arboretum of Harvard Univ., 1938. xiii, 719p. Supplement 1, 1960. (Forestburgh, N.Y., Lubrecht & Cramer. $18.50). ISBN: 0934454116, supp.1.

Sponsored by the Smithsonian Institution. Arnold Arboretum, New York Botanical Garden and Harvard-Yenching Institute.

Annotated bibliography of over 21,000 items, covering the botanical literature of Japan, China, Formosa, Korea, Manchuria, Mongolia, Tibet, E. Siberia and the Far East (but not the East Indies or Pacific area). Author A-Z sequence, with subject and geographical indexes. Appended lists of Chinese and Japanese botanical works of historical interest and serials. The supplement, covering 1936-58, has 11,000 entries. *Class No:* 581.1(5)

Africa—Southern

[2411]
CODD, L.E.W., *and others, eds.* **Flora of Southern Africa:** the Republic of South Africa, Basutoland, Swaziland and South West Africa. Pretoria, Botanical Research Institute, 1963-.

To be completed in 50v. V.1,9,13,22 and 26 (1963-78) have at least been published (Musiker, R. *South Africa* (1979), entry 43). *Class No:* 581.1(68)

America — North

[2412]
The Audubon Society field guide to North American wildflowers. New York, Knopf, 1979.

Eastern region, by W.A. Niering and M.C. Olmstead. 863p. *Western region,* by R. Spellenberg. 1979. 862p.

Together cover North America. Both devote the first part to colour plates, arranged by colour; a detailed descriptive text forms the second part. Generic and English name indexes. *Class No:* 581.1(71+73)

[2413]
SHETLER, S.G. *and* SKOG, L.E. **A Provisional checklist of species for Flora North America.** St. Louis, Missouri, Missouri Botanical Garden, for Man and Biosphere, Flora North America, 1978. xix, 199p. (*Monographs in systematic botany from the Missouri Botanical Garden, v.1.*)

Revision *Flora North America Report 64.*

Revised checklist of 16,274 species, 2,350 genera and 233 families, and a further step towards a Flora North America. Geographical coverage: USA, Canada and Greenland. In taxonomic order. Data include plant characteristics, geographical region of sources of species entry.

Class No: 581.1(71+73)

USA

[2414]
RICKETT, H.W. **Wild flowers of the United States.** New York, McGraw-Hill, 1966-75. 6v. and index. illus. (pl.), maps. ISBN: 0893272744, v.1; 0893272779, v.2; 0893272809, v.4; 0893272841, v.5; 0893272876, v.6.

Publication of the New York Botanical Garden.

1. *The Northeastern states.* 1966. $65. 2pts. 2. *The Southeastern states.* 1967. $85. 2pts. 3. *Texas.* 1969. 2pts. 4. *The Southwestern states.* 1970. 2pts. $92. 5. *The Northwestern states.* 1972. $85. 6. *The Central mountains and plains.* 1973. $92. 3pts. *Complete index.* 1975.

V.1 (2pts. 559p.) has 180 plates, with concise systematic description, plus distribution, facing. End-paper maps. Each volume carries an index. *Class No:* 581.1(73)

[2415]
TORREY BOTANICAL CLUB, New York, *comp.* **Index to American botanical literature.** Boston, Mass., G.K. Hall, 1968. 4v. Supplements, 1977-.

The main *index* has entries for *c.*106,000 author catalogue cards, photolithographically reproduced. It covers the period 1886-1966. The first *supplement* (1977) covers 1967-76. Further supplements are planned to be ten-yearly.

Class No: 581.1(73)

America—South

[2416]
'Bibliografia botanica para América Latina'. In *Boletín de la Sociedad Argentina Botanica*, v.17, 1976, p.368-83.
Class No: 581.1(8)

New Zealand

[2417]
ALLAN, H.H.B. Flora of New Zealand. Wellington, Government Printer, 1961-80. 3v. illus., maps.

V.1 (liv, 1085p.) gives systematic descriptions of indigenous vascular plants (ferns and their allies, conifers and the dicotyledonous flowering plants), except monocotyledons. Glossary, p.977-97. Maori names of plants, p.998-1007. Bibliography, p.xiii-xxxiv. Maps and end-papers only. 'An indispensable tool' (*Nature*, v.192, 23 December 1961, p.1108). *Class No:* 581.1(931)

[2418]
WARDLE, D. Vegetation of New Zealand. Cambridge, University Press, 1991. 672p. illus., diagrs., graphs, tables, maps. £105. ISBN: 0521258731.

16 chapters: 1. The physical and biological environment ... 3. Plant form in relation to habitat ... 5. Description, nomenclature and classification of vegetation, environment and ecological processes ... 9. Grassland and heathfield ... 14. Biomass: growth, nutrition and tolerance ... 16. Disturbances, regeneration and trends in native forest. 2 appendices (2. Common names and Latin equivalents). References, p.615-36. Index of plant genera and species. General index. *Class No:* 581.1(931)

Australia

[2419]
BURBIDGE, N.T. Dictionary of Australian plant names: gymnosperms and angiosperms. Sydney, Angus & Robertson, 1963. xviii, [2], 345p. maps.

About 45,000 entries. Data under each accepted name: citation (author; place of publication), synonymy, family or families; phytogeographic details; bibliographical references. Many cross-references; italicized names are synonyms and homonyms. Genera, arranged under families, p.315-45. Genera asterisked in both gymnosperms and angiosperms are naturalized aliens. *Class No:* 581.1(94)

[2420]
HARRIS, T.Y. Wild flowers of Australia. 8th ed. London, Sydney, etc., Angus & Robertson, 1979. xiii, 207p. illus. (*Australian natural science library*.) ISBN: 0207136440.

First published 1938.

Part 1: Popular descriptions, with plates. Part 2: Key to the families of plants. Technical descriptions of plants, p.76-184. Glossary of botanical terms, p.185-94. Books of reference (24 titles). Names of authors and abbreviations. Analytical index, p.198-207. Descriptions face illus. A standard guide. *Class No:* 581.1(94)

Micronesia

[2421]
SACHET, M.H. *and* **FOSBERG, F.R. Island bibliographies: Micronesian botany.** Land environment and ecology of coral atolls. Vegetation of tropical Pacific islands. Compiled under the auspices of the Pacific Science

.... (contd.)
Board. Washington, National Academy of Sciences, National Research Council, 1955. v, 577p. Supplement. 1971. 427p. *Class No:* 581.1(965)

Physiology

[2422]
Annual review of plant physiology and plant molecular biology. Palo Alto, Calif., Annual Reviews, Inc., 1950-. V.1-. Annual. ISSN: 10402519.

Published as *Annual review of plant physiology*, 1950-1989.

V.42, 1991 (x,762p.) has 42 contributors. 25 documented chapters (*e.g.* 'Fructan metabolism in grasses and cereals', p.77-101 (126 references); 'The role of heat shock proteins in plants', p.579-620 (185 references)). Author and subject indexes. Cumulative index of contributing authors, v.34-42; Cumulative index of chapter titles, v.34-42. 'Related articles of interest to readers', p.ix. *Class No:* 581.10

[2423]
Encyclopedia of plant physiology. Person, A. *and* Zimmermann, M., *eds*. New series. Cambridge, Heffer, 1975-86.

Continues old series, (*Handbuch der Pflanzenphysiologie* (in 3 pts: 1. General foundations (2v.) - 2. Metabolism (v.3-13) - 3. Growth, development, movements (v.14-18)).

The new monograph series is in English. 1-3. *Transport in plants*. 1975-76. 4. *Physiological plant pathology*. 1975. 5-6. *Photosynthesis*. [I-II]. 1977. 7. *Physiology of movements*. 1979. 8. *Secondary plant products*. 1980. 9/11. *Hormonal regulation of development*. 1980-85. 12. A-D. *Physiological plant ecology*. 1981-83. 13. A,B. *Plant carbohydrates*. 1981-82. 14. A,B *Nucleic acids*. 1982. 15.A. *Inorganic plant nutrition*. 1983. 16. A,B. *Photomorphogenesis*. 1983. 17. *Cellular interactions*. 1984. 18. *Higher plant cell respiration*. 1985. 19. *Photosynthesis*. III. 1986. *Class No:* 581.10

Carnivorous Plants

[2424]
LECOUFLE, M. Comment choisir et cultiver vos plantes carnivores. [Carnivorous plants: care and cultivation.] London, Blandford, 1990. 144p. illus. (col.). £18.99; £8.95. ISBN: 0713721855.

3 sections: 1. General principles - 2. Traps and flowers - 3. Principal species in cultivation, p.50-142. Data for each species include etymology; habitat; description; reproduction; pests and diseases; related species; prey. Glossary. Brief bibliography. List of carnivorous plant societies. Many colour illustrations. *Class No:* 581.13

Diseases

[2425]
Annual review of phytopathology. Palo Alto, Calif., Annual Reviews, Inc., 1963-. v.1-. Annual. ISSN: 00664286.

V.29, 1991 (x,516p.) contains 23 documented papers (*e.g.* 'Delignification of wood-decay fungi', p.381-442 (with 68 references); 'Molecular and genetic analysis of toxin production by pathovars of *Pseudomonas syringae*' (p.247-78) (with 128 references)). Subject index. Contributing authors, v.20-29; chapter titles, v.20-29. 'Some related articles in other *Annual reviews*, p.x. *Class No:* 581.2

[2426]
BUCZACKI, S.T. *and* HARRIS, K.M. Collins' guide to the pests, diseases and disorders of garden plants. London, Collins, 1981. 512p. illus., (incl. col.). ISBN: 0002191032.

General introduction - A/Z key to symptoms - 24 col. plates (captions facing) - Pests (eelworms ... mammals), p.132-264 - Diseases (rusts ... viruses), p.267-471. - Disorders (mineral nutrient; non-nutritional disorders (*e.g.* drought; pollution)). Data: symptoms; biology; treatment. Grouped bibliography, p.493-5. Glossary. Index, p.500-12. 'Highly practical, authoritative and free from waffle' (*Natural history book reviews*, v.6 (1/2), 1982, p.78).

Shorter guide to pests, diseases and disorders of garden plants (Collins, 1983. 320p. illus (incl. col.)) is by K.M. Harris, and others. *Class No:* 581.2

[2427]
HOLLIDAY, P. A Dictionary of plant pathology. Cambridge, University Press, 1989. xvi,369p. £32.50. ISBN: 052133117x.

*c.*8,000 entries. Lists authoritative names of pathogens with brief descriptions and bibliographic references. Entries also include references and notes on plant pathologists, names of diseases, crops and their pathology, terminology, names of disorders, taxonomic groups, fungicide names, vectors of viruses and toxins. 'An excellent dictionary with a wealth of bibliographic information' (*Science and technology libraries*, v.10(4) Summer 1990, p.120). *Class No:* 581.2

[2428]
PIRONE, P.P. Diseases and pests of ornamental plants. 5th ed. New York, etc., Wiley; Interscience, 1978. 566p. illus., diagrs. $55. ISBN: 0471072494.

First published 1943.

2 parts: 1. Diseases and pests in general, p.3-102, including selected bibliography. 2. Diseases and pests of particular hosts, arranged A-Z by botanical names of host plants. Involves nearly 500 genera of ornamental and commonly grown plants (*e.g.* Prunus persica: 2 columns on symptoms of disease and control by fungicides, pesticides, etc.). *Class No:* 581.2

Embryology

[2429]
DARLINGTON, C.D. *and* WYLIE, A.D. Chromosome atlas of flowering plants. 2nd ed., rev. and expanded. London, Allen & Unwin, 1955. xix, 519p.

First published 1945.

The 1st ed. dealt with cultivated plants only. The 2nd ed. includes other flowering plants and covers more than 15,000 species. For each species: the somatic number of the chromosomes; the authority who made the count; use made of the plant; and its distribution. Appended exhaustive bibliography, p.459-500. *Class No:* 581.3

[2430]
HORDER, T.J., *and others*. A History of embryology. Cambridge, University Press, 1986. 502p. illus., diagrs., tables. (*British Society for Developmental Biology Symposium*, v.8.) ISBN: 0521259533.

Chapters by specialists on the history since 1800. Appended list of references, sources of information. Chronological table of basic dates precedes. 'Gives a "feel" of the different concepts that followed each other' (*British book news*, July 1986, p.409). *Class No:* 581.3

Spores

[2431]
KREMP, G.O.W. Morphologic encyclopedia of palynology and illustrations of spore and pollen. Tucson, Univ. of Arizona Press, 1965. xiii, 186p. illus. (pl.).

1,280 terms relating to spores and pollen are explained; definitions are quoted from original sources and over 800 are illustrated. Very thorough, although it does not attempt to weed out incorrect definitions (*Nature,* v.209, 12 February 1966, p.651). *Class No:* 581.33

Anatomy

[2432]
Handbuch der Pflanzenanatomie. Begründet von K. Linsbauer. Fortgeführt von G. Tischler und A. Paschler. [Encyclopedia of plant anatomy.] 2. völlig neubearb. Aufl., hrsg von W. Zimmermann, P.G. Ozenda. Berlin, Borntraeger, 1934-. (v.14(1): *Leaf structure of a Venezuelan cloud forest in relation to the microclimate.* 1990. xi,244p. DM148). illus.

First published 1922- (not completed by 1943).

An extensively documented treatise on plant anatomy (*e.g.* v.2, *Allgemeine Pflanzenkaryologie. Erganzunsband* (1953-63. ix, 1227p.) has a massive list of references, p.776-1121). V.8(5), *Die primäre Bau der Angiospermenwurzel* (1968. viii, 472p.) has 372 text illus., bibliography (A-Z authors, p.410-39), and author subject indexes. Contributions are in English or German. *Class No:* 581.4

Seeds

[2433]
MARTIN, A.C. *and* BARKLEY, W.D. Seed identification manual. Berkeley & Los Angeles, Univ. of California Press, 1961. (reprinted 1973). [vii], 221p. illus.

824 seed photographs (farmlands; wetlands; woodlands), p.7-122, followed by 'Identification clues' (systematically arranged; 287 figures). Selected bibliography (largely US), p.207-9. Index, p.211-2. *Class No:* 581.48

[2434]
—BARTON, L.V. Bibliography of seeds. New York & London, Columbia Univ. Press, 1967. [v], 858p. $121.50. ISBN: 0231029373.

Comprises 20,163 references to published literature, under authors A/Z. Plant index, p.706-816; subject index, p.817-58 (usually analytical, but the entry at 'Viability' has *c.*700 unspecified references). Based on the file at the Boyce Thompson Institute for Plant Research, Yonkers, N.Y. *Class No:* 581.48

[2435]
Seed abstracts. Wallingford, Oxon., CAB International, 1978-. 12pa. £136; $251. ISSN: 01410180.

About 4,200 abstracts pa. 14 sections: Seed morphology and anatomy ... Seed chemistry ... Germination ... Seed production ... Pests and diseases - Breeding and selection ... Seed assessment and testing ... Miscellaneous. Reports; Conferences; Books. Author and subject indexes per issue and annually. Cross-reference to other abstract journals published by CAB International. Online, CAB ABSTRACTS and available on CAB ABSTRACTS CD-ROM. *Class No:* 581.48

Ecology

[2436]

Comparative plant ecology: a functional approach to common British species. Grime, J.P., *and others*. London, Unwin Hyman, 1988. ix,742p. diagrs., graphs. £85. ISBN: 0045810281.

Provides standardized Antecological Accounts of the biology and ecology of common vascular plants of the British flora. Gives data sources; contents; interpretation and use. Antecological Accounts, p.53-615. Tables of attributes. Summary and conclusions. Appendix: Patterns of specialization (*e.g.* 'Statistical rationale'; 'Clustering policy'). References, p.670-98. Analytical subject, author and species (generic, binomial and vernacular) indexes. *Class No:* 581.5

Economic Aspects

[2437]

Applied botany abstracts. Lucknow, India, Economic Botany Information Service, National Botanical Research Institute, 1981-. 4pa. $20. ISSN: 09702377.

Formerly *Current literature in plant science.*

About 800 abstracts per issue. 8 sections: 1. Non-traditional/under-utilized plants - 2. Biomass/energy plants (*e.g.* fuel crops) - 3. Ornamental plants - 4. Plants and environmental pollution - 5. Traditional herbal drugs - 6. Ethnobotany - 7. Rare/endangered/threatened plants - 8. Other economic plants. New publications (books and periodicals). Key word index; author index. List of journals consulted (*c.*300). *Class No:* 581.6

[2438]

LEWINGTON, A. Plants for people. London, Natural History Museum, 1990. vii,232p. col. illus. £19.95. ISBN: 0565010948.

Gives an insight into the ways in which plants are used. 7 sections: 1. Starting the day ... 3. From first foods to fast foods ... 6. Getting around - plants that transport us - 7. Recreation - plants that entertain us. Index. No bibliography. A popular treatment, attractively presented, with many colour illustrations. For school and public libraries. *Class No:* 581.6

[2439]

MERINO-RODRIGUEZ, M. Plants and plant products of economic importance: vocabulary - Latinus, English, Français, Español. Deutsch. Rome, Food and Agriculture Organization, 1974. (*FAO Terminology bulletin 25.*) *Class No:* 581.6

[2440]

TANAKA, T. Tanaka's cyclopedia of edible plants of the world. Nakao, S., *ed.* Tokyo, Keigaku Publishing Co., 1976. [ix], 924p.

Lists more than 10,000 edible plants, giving bibliographical source and date (*e.g.* 'Uphof 1968'). Bibliography (authors, A-Z, p.791-80). Detailed index, p.803-924. 'The real aim of the book is to save the peoples of the world' (*Preface*). *Class No:* 581.6

[2441]

UPHOF, J.C.T. The Dictionary of economic plants. 2nd ed. Lehre, Cramer, and New York, Stechert-Hafner, 1968. [viii], 591p. $49. ISBN: 3768200019.

First published 1959.

The 2nd. ed. adds over 3,000 species, bringing total different species of economic plants briefly described to 9,500. Entry is under Latin name, with reference from common name. Data on geographical distribution, products and principal uses. Bibliography, systematically arranged, p.453-91. Helpful group references (*e.g.* from 'Confectionery, Plants used in'). A valued reference tool. *Class No:* 581.6

[2442]

—HOWES, F.N. A Dictionary of useful and everyday plants and their common names. Cambridge Univ. Press, 1974. [x], 290p. ISBN: 0521085209.

Contains *c.*8,000 entries. 'Some useful reference works', p.289-86. No index. Based on material contained in J.C. Willis's *A dictionary of the flowering plants and ferns* (6th ed. 1931) (*qv*). *Class No:* 581.6

[2443]

—The Oxford book of food plants. Harrison, S.G., *and others*. London, Oxford Univ. Press, 1969. viii, 206p. illus. (pl.).

Has accurate and attractive col. illus. (drawn from live specimens, in the main and to scale), with facing descriptive text (*e.g.* banana, p.108-9: plates showing young plant, inflorescence, leaf detail and fruit). Favourably reviewed in *Journal of the Royal Horticultural Society,* v.95(3) 1970, p.130). *Class No:* 581.6

[2444]

—Popular encyclopedia of plants. Heywood, V.H. and Chant, S.R., *eds*. Cambridge, University Press, 1982. 368p. col. illus. ISBN: 0521246113.

Has a misleading title. It concerns plants and plant products useful to human beings. 2,200 entries, A-Z under Latin and common names, provide data on natural distribution, habit, usefulness, etc. Glossary of technical terms. 773 colour illus., adjacent to the pertinent text. *Class No:* 581.6

[2445]

—USHER, G. A Dictionary of plants used by man. London, Constable, 1974. 619p. ISBN: 0094579202.

Comparable in scope to Uphof. Main entries are under genera. While Usher is fuller than Uphof in providing cross-references from common names, Uphof is richer in local names and plants and their uses (*e.g.* by North American Indians). *Class No:* 581.6

Systematic Botany (Categories of Plants)

Herbal & Medicinal Plants

[2446]

LAUNERT, E. The Hamlyn guide to edible and medicinal plants of Britain and Northern Europe. London, Hamlyn, 1981. 288p. col. illus. ISBN: 0600372162.

Published in paperback (Hamlyn, 1989; £6.99) as *Guide to edible and medicinal plants of Britain and Northern Europe.*

Colour illus. accompanied by brief descriptions and

....*(contd.)*

indication of habitat and distribution. Entries also state active ingredients, effect, parts used and application. Indicates differences between species that look alike. Short list of suppliers of medicinal herbs and plant-based preparations. Glossaries of botanical and medical terms, p.278-81. 4 appendices: 1. Notes on the most important ingredients of medicinal plants - 2. Table of edible plants, with their uses - 3. Recipes for edible plants - 4. List of common ailments and plant species recommended for their treatment. Bibliography, p.277. Index p.282-8. *Class No:* 582:615.32

Flowerless Plants (Cryptogams)

[2447]

BOEDIJN, K.B. Plants of the world: the lower plants. London, Thames & Hudson, 1965. 312p. illus. (incl. pl.).

Published in the Netherlands under the title *Lagere Planten.* Translated by A.J. Pomerans.

Covers ferns, fungi, mosses and lichens. Botanical descriptions, copious illus. (157 in colour, 383 in black-and-white) No bibliographies. Analytical index, p.297-312. Complementary to H.C.D. De Wit's *Plants of the world, the higher plants (qv). Class No:* 582.2

[2448]

FARLOW REFERENCE LIBRARY OF CRYPTOGAMIC BOTANY, Harvard, University. **Catalog. [Farlow Reference Library of Cryptogamic Botany].** Boston, Mass., G.K. Hall, 1979. 6v. $760. ISBN: 0816102791.

Lists *c.*60,000 books, periodical titles and reports in one author-title sequence. The library is particularly rich in pre-1850 cryptogamic material. *Class No:* 582.2

[2449]

The Oxford book of flowerless plants: ferns, fungi, mosses and liverworts, lichens and seaweeds. Brightman, F.H. *and* Nicholson, B.E., *illus.* London, Oxford Univ. Press, 1966. viii, 208p. illus. (col.).

Grouped ecologically: Seashore - Grasslands - Uplands - Wet places - Woodlands. 688 plants (including 250 fungi and 78 species of seaweed), drawn from actual specimens. Not to scale. 'Suggestions for further reading', p.201. Index, p.202-8. Text faces well-coloured illus. Primary aim: 'to help the beginner with identification'. *Class No:* 582.2

Algae, Seaweeds, etc.

[2450]

BRITISH MUSEUM (NATURAL HISTORY). Seaweeds of the British Isles. London, British Museum (Natural History), 1977-. diagrs., tables.

1. *Rhotophyta.* 1977-. (pt.1 1977. £13.75; pt.2A. 1983. £13). 2. *Chlorophyta.* (1991-. £27). 3. *Fucophyceae* (Phaeophyceae) (pt.1. 1987. £30). 4. *Tribophyceae* (Xanthophyceae) (1987. £7.50).

The result of many years' research carried out by the British Museum (Natural History) and the British Phycological Society. Bibliographical references. The series is to cover all the British and the majority of northern Atlantic seaweeds. Description of each species incorporates notes on ecology and distribution, plus 1 or more illus. *Class No:* 582.26

[2451]

CHAPMAN, V.J. *and* **CHAPMAN, D.J. The Algae.** 2nd ed. London, Macmillan, 1973. xiv, 497p. illus., tables, maps. ISBN: 0333142705.

First published 1962.

22 documented chapters (1. Classification; 2. Cyanophyta; 3/5. Chlorphyta ... 7/8. Chrytophyta ... 13. Fossil algae (p.309-27, including 1p. of references) ... 18. Fresh water ecology (p.388-403, with 8 tables and ½p. of references) ... 22. Algal physiology). A neat production. 'Intended as a brief introductory text only' (*Preface* to the 2nd ed.). *Class No:* 582.26

[2452]

LÜNING, K. Seaweeds: their environment, biogeography, and ecophysiology. New York, Wiley, 1990. xiii,527p., illus., diagrs., graphs, tables. $89.95. ISBN: 0471624349.

First published 1985 as *Meeresbotanik* (Stuttgart, Thieme).

Pt.1: Distribution and structure of seaweed vegetation (chapters 1-5) - Pt.2: Ecophysiology of seaweeds (chapters 6-8). Bibliography, p.371-483 (A-Z author). Appended: 'Taxonomic overview of genera'. Detailed, analytical index, p.489-527. The German first edition 'was an immediate success, prompting the production of this translation,' for which the text has been updated and expanded by about twenty per cent' (*Journal of the Marine Biological Association*, v.71(1), February 1991, p.246). *Class No:* 582.26

Fungi, Moulds

Encyclopaedias & Dictionaries

[2453]

AINSWORTH, G.C. Ainsworth & Bisby's dictionary of the fungi. 7th ed., by D.L. Hawksworth. Farnham Royal, Commonwealth Agricultural Bureaux, 1983. 412p. illus., diagrs., tables. ISBN: 0851985157.

First published 1943; 6th ed., 1971.

*c.*16,500 entries. Covers more than the taxonomy of the fungi; morphology, structure, spore types and various aspects of physiology, metabolism and ecology are also treated. 'The *Dictionary* is the place to begin any enquiry concerning the fungi' (*Natural history book reviews,* v.7(3), 1984, p.169). *Class No:* 582.28(03)

[2454]

Index of fungi. Wallingford, Oxon., CAB International, 1948-. 2pa. £40pa. ISSN: 00193895.

Formerly *Review of applied mycology: supplement* 1940-47.

Comprises a list of names (A-Z) of new genera, species, infraspecific taxa, new combinations and new names of fungi. About 1,500 entries per issue. Online, CAB ABSTRACTS. *Class No:* 582.28(03)

[2455]

Mykologisches Wörterbuch: 3,200 Begriffe in 8 Sprachen - Deutsch, Englisch, Französisch, Spanisch, Latein, Tschechisch, Polnisch, Russisch. Berger, K., *ed.* Stuttgart & New York, Fischer, 1980. 432p. illus., tables. ISBN: 0685059022.

10 collaborators. 3,190 numbered German terms with English, French, Spanish, Latin, Czech, Polish and Russian equivalents and indexes. Genders are shown. Bibliography, p.393-4. *Class No:* 582.28(03)

[2456]

SNELL, W.H. *and* **DICK, E.A. A Glossary of mycology.**
2nd rev. ed. Cambridge, Mass., Harvard Univ. Press;
London, Oxford Univ. Press, 1971. $17. ISBN:
0674354516.

Expansion of W.H. Snell's *Three thousand mycological
terms* (1936).

Over 7,000 terms on fungi research and its applications.
Includes terms that cannot strictly be called mycological
from their presence in mycological literature or generally of
interest to mycologists. *Class No:* 582.28(03)

Handbooks & Manuals

[2457]

AINSWORTH, G.C. *and* **SUSSMAN, A.S.,** *eds*. **The Fungi:**
an advanced treatise. New York & London, Academic
Press, 1965-73. 4v. in 5.

1. *The fungal cell.* 1965. 748p. 2. *The fungal organism.*
1966. 805p. 3. *The fungal population. Ecology.* 1966. 738p.
4A. *A taxonomic review, with keys. Ascomycetes and fungi
imperfecti.* 1973. 621p. 4B. *A taxonomic review with keys.
Lower fungi and busiodimycetes.* 1973. 536p.

A well-documented treatise, the work of numerous
specialist contributors. V.1, chapter 12. 'Carbohydrate
metabolism' (p.302-47) has references, p.338-47. V.1 has
author and subject indexes, index to fungi and
actinomycetes. *Class No:* 582.28(035)

[2458]

BRESINSKY, A. *and* **BESL, H. A Colour atlas of poisonous
fungi:** a handbook for pharmacists, doctors and biologists.
London, Wolfe Publishing, 1990. xi,198p. illus. (col.),
diagrs. £70. ISBN: 0723415765.

3 sections: 1. Poisoning by fungi - 2. The poisoning
syndromes and the poisonous fungi - 3. Index to the
identification of fungi (including identification keys).
Glossary. Bibliography, p.269-88. Index.
Class No: 582.28(035)

Reviews & Abstracts

[2459]

Abstracts of mycology, reporting worldwide research in
mycology. Philadelphia, Pa., Biosciences Information
Service (BIOSIS), 1967 - 12pa. $200pa. ISSN: 00013617.

About 5,000 abstracts and references pa. Each monthly
issue contains entries from two consecutive issues of
Biological abstracts and from two consecutive issues of
Biological abstracts/RRM (Reports, Reviews, Meetings).
Covers: Fungi in biochemistry; cytology; genetics;
microbiology - medical and industrial; pathology, plant and
animal; systematics. Indexes: author, biosynthetic, generic,
subject. *Class No:* 582.28(048)

[2460]

Bibliography of systematic mycology. Wallingford, Oxon.,
CAB International, 1947-. 2pa. £26. ISSN: 00061573.

About 1,600 references pa. Lists papers and books on all
aspects of the taxonomy of fungi. Systematically arranged,
with author index. *Class No:* 582.28(048)

[2461]

Review of medical and veterinary mycology. Wallingford,
Oxon., CAB International, 1943-. 4pa. £113. ISSN:
00346624.

As *Annotated bibliography of medical mycology,* 1943-50.

About 3,500 records added pa. Covers mycoses of man
and domestic, farm and wild animals, allergic disorders
associated with fungi, and poisoning by fungi or mould-
contaminated foods. Quarterly, annual author and subject
indexes. Online, CAB ABSTRACTS and on CD-ROM.
Class No: 582.28(048)

Yearbooks & Directories

[2462]

International mycological directory. Hall, G.S. *and*
Hawksworth, D.L. 2nd ed. Wallingford, Oxon, CAB
International, 1990. 163p. £15. ISBN: 0851986935.

First published 1971.

Based on responses from 122 organizations. Individual
entries arranged A-Z (All-Union Collection of Microorganisms ... Victoria University Mycology Group). Data
include status; organization type; scope; interests; portrait;
culture collection; herbarium; publications; database; visual
materials; awards; courses. Indexes to awards and honours;
courses; databases; libraries; herbaria; interests; meetings;
visual materials; publications. Sources of further
information, p.154-59 (*e.g.* 'Patents or patent depositing
authorities'). IMA Draft Statutes, p.160-63.
Class No: 582.28(058)

Histories

[2463]

**AINSWORTH, G.C. Introduction to the history of
mycology.** Cambridge Univ. Press, 1976. xi, 359p,. illus.,
facsims., ports. ISBN: 0521210135.

11 chapters and epilogue. A documented outline, taking 'a
number of the important themes and to trace each one from
early times to the present day' (*Preface*). Notes on the text
p.291-307. Chronology and bibliography. Name and subject
indexes. *Class No:* 582.28(091)

[2464]

—AINSWORTH, G.C. Introduction to the history of medical
and veterinary mycology. Cambridge, University Press,
1986. xi,228p. illus. £40. ISBN: 0521307155.

10 sections: 1. Introduction ... 3. Names: problems of
nomenclature ... 5. Epidemiological problems - 6.
Therapeutic problems ... 10. Regional developments. Notes
on the text. Bibliography and chronology, p.181-214. Name
and subject indexes. *Class No:* 582.28(091)

Yeasts

[2465]

ROSE, A.H. *and* **HARRISON, J.S.,** *eds*. **The Yeasts.** 2nd
ed. London & New York, Academic Press, 1987-91. 4v.
£48.50, v.1; £47.50, v.2; £59, v.3; £72.50, v.4. ISBN:
0125964110, v.1; 0125964129, v.2; 0125964137, v.3;
0125964145, v.4.

Previously published in 3v., 1969-71.

1. *Biology of yeasts.* 1987. xii,423p. 2. *Yeasts and the
environment.* 1987. xvi,309p. 3. *Metabolism and physiology
of yeasts.* 1989. xxiv,635p. 4. *Yeast organelles.* 1991.
xxx,765p.

....(contd.)

17 contributors in v.4. Chapter 5: 'Cell walls', p.199-277 includes references (264-277). Author and analytical subject indexes. A detailed, authoritative, treatise. *Class No:* 582.282

[2466]
Yeasts: characteristics and identification. Barnett, J.A., *and others*. 2nd ed. Cambridge, University Press, 1990. ix,1002p. illus. £135. ISBN: 0521350565.

First published 1983.

Describes 597 yeast species recognised by the Centraalbureau voor Schimmelcultures at the beginning of September 1991. 11 sections: 1. Introduction ... 4. Laboratory methods for identifying yeasts ... 6. Summary of specific characteristics ... 8. The keys - 9. Tables for identifying individual species - 10. Register of yeast names - 11. Register of specific epithets. Glossary. References, p.955-98. Subject index. 850 photomicrographs. A yeast identification program for microcomputers, for use with this book, is also available. *Class No:* 582.282

Mushrooms & Toadstools

[2467]
BELS-KONING, H.C. *and* **KUIJK, W.M.,** van. **Mushroom terms:** ... polyglot on research and cultivation of edible fungi. English-German-Dutch-Danish-French, Italian, Spanish & Latin. Wageningen, Centre for Agricultural Publishing and Documentation [PUDOC], 1980. xxii, [1], 312p. £23. ISBN: 9022006735.

Supersedes *Mushroom terms in five languages* (PUDOC, 1966).

1,169 English-base terms (plus synonyms), with equivalents and indexes in the other 7 languages. Genders are given. 'Literature', p.311-2. *Class No:* 582.287

[2468]
BUCZACKI, S. Collins New Generation guide to the fungi of Britain and Europe. London, Collins, 1989. 320p. illus. (col.). £12.95; £8.95. (*Collins New Generation Guide.*)

3 sections: 1. What is a fungus? - 2. The directory of fungus species - 3. The natural history of fungi. Contains over 1,300 colour illustrations. Each entry includes information on appearance, size, colours and features necessary for identification, habitats and substrates. Glossary, p.307-11. Index, p.312-20, restricted to names of species. *Class No:* 582.287

[2469]
Elsevier's dictionary of edible mushrooms: botanical and common names in various languages of the world. Chandra, A., *comp.* Amsterdam, Elsevier, 1989. xxxix,259p. £54.68. ISBN: 0444883886.

Latin names of 628 species of fungi listed in alphabetical order. Each entry gives as many common names as possible from Bulgarian; Chinese; Czech; Danish; Dutch; English; Filipino; Finnish; French; German; Hungarian; Indian languages; Italian; Japanese; Norwegian; Pakistani (Urdu); Polish; Romanian; Russian; Spanish; Swedish. Indexes to common names in each language. References given for each species. *Class No:* 582.287

[2470]
GRUNERT, H. *and* **GRUNERT, R.** **Field guide to mushrooms of Britain and Europe.** Swindon, Wilts., Crowood Press, 1991. 288p. illus. (mostly col.). £8.99. ISBN: 1852235926.

First published in German (Munich, 1984).

A pocket guide, consisting of 324 coloured photographs, with descriptions on facing pages. About 30 line-drawings. 'Collecting fungi', p.280-3. Bibliography, p.284. Index. Covers all the most important species. *Class No:* 582.287

[2471]
PEGLER, D. Field guide to the mushrooms and toadstools of Britain and Europe. London, Kingfisher Books, 1990. 192p. illus. (col.). £7.95. ISBN: 0862725658.

Covers 450 species, illustrating 386 in colour and with accurate detail. Chapters are devoted to different groups and each species has a concise text dealing with appearance, habitat, frequency and season. 'Similar species' distinguishes fungi that are liable to be confused. Pages colour coded for ease of reference. Glossary, index, list of useful addresses + bibliography. 'A cheap, well illustrated and easily used guide' (*Reference reviews*, v.5(1) 1991, p.18). *Class No:* 582.287

[2472]
PHILLIPS, R. Mushrooms and other fungi of Great Britain and Europe. London, Ward Lock & Pan Books, 1981. 288p. illus. (col.). ISBN: 0330264419.

Contents: Introduction; Visual index; Glossary. Agarics - Gomphediaceae - Chanterelles - Bolates - Polypores - Ascomycetes. Bibliography, p.282. Full colour illus., with descriptive captions, identifying over 900 specimens, showing the various stages of growth. A handsome quarto that deserves a place in every public library at least. *Class No:* 582.287

[2473]
RINALDI, A. *and* **TYNDALO, V. The Complete book of mushrooms:** over 1,000 species and varieties of American, European, and Asiatic mushrooms, with 460 illustrations in black and white and in color. Mancinelli, I. *and* Mancinelli, A., *tr.* New York, Crown, 1974. 331p. illus. (incl. col.). ISBN: 0517514931.

Arranged by genera and species. Includes sections on nutritional properties, toxicity, etc. Glossary. 300 colour plates. *Class No:* 582.287

[2474]
—DICKINSON, C. *and* LUCAS, J., *eds*. The Encyclopedia of mushrooms. London, Orbis Publishing, 1979. 280p. illus. (mostly col.) diagrs. ISBN: 0856134155.

Contents: The mushroom in history; The biology of fungi; The lifestyles of fungi; The mushroom habitat; Mushrooms and man; Mushrooms as food; The naming of mushrooms. Reference section, p.118-265 (with index: botanical names; common English names). Glossary. Bibliography, p.269. General index. An identification manual, covering 500 species worldwide, for the layman. A fully illustrated quarto and much more than a coffee-table book. *Class No:* 582.287

Lichens

[2475]

HAWKSWORTH, D.L. *and* **SEAWARD, M.R.D. Lichenology in the British Isles, 1568-1978:** an historical and bibliographical survey. Richmond, Surrey, Richmond Publishing Co., 1977. vii, 231p. £9.95. ISBN: 0855462000.

About 2,695 references, - a valuable, comprehensive bibliography, indexed under vice-county headings, plus a well-illustrated text. In many ways this is a volume for the specialist, but beginners should not be discouraged from using it (*British book news,* June 1978, p.451). *Class No:* 582.29

Mosses

[2476]

CROSBY, M.R. *and* **MAGILL, R.E. A Directory of mosses:** an alphabetical listing of genera, ... together with a systematic arrangement of the families of mosses and a catalogue of family names. St. Louis, Missouri, Missouri Botanical Garden, 1977. vii, 43p.

The first detailed guide to the familial composition of the genera of mosses published for more than 50 years (*Preface*). *Class No:* 582.34

[2477]

SMITH, A.J.E. The Moss flora of Britain and Ireland. Cambridge, University Press, 1978. 706p. illus. £30. ISBN: 052129973x.

A comprehensive, treatise on nearly 700 species of moss. Well illustrated, with identification keys for all genera and species described. *Class No:* 582.34

[2478]

WATSON, E.V. British mosses and liverworts: an introductory work, with full descriptions and figures of over 200 species and keys for the identification of all except the very rare species. 3rd ed. Cambridge, University Press, 1981. xviii, 519p. illus. (incl. pl.). ISBN: 0521285364.

First published 1955.

Main sequence: Musci (p.124-401); Hepaticae (p.401-498). Introduction includes notes on works of reference. Appended habit lists, p.1-24. Detailed index, p.511-9. Well illustrated, with descriptive captions. A standard work. *Class No:* 582.34

Vascular Plants

[2479]

BSBI abstracts: abstracts from literature relating to the vascular plants of the British Isles. London, Botanical Society of the British Isles, c/o Dept. of Botany, British Museum (Natural History), 1971-. Annual. £4.50. ISSN: 03072657.

Replaces 'Abstracts from literature', in *Proceedings of the Botanical Society of the British Isles,* 1954-69.

About 800 abstracts and references pa. Sections: Introduction - History - Herbaria - Biography - Floras, catalogues, etc. (Europe ... North America) - Chromosome surveys, etc. - Pteridophyta - Gymnospermae - Angiospermae - Miscellaneous. *Class No:* 582.35

[2480]

Gray Herbarium index. Harvard University. Boston, Mass., G.K. Hall, 1968. 10v. (8121p.). $945. ISBN: 0816107548.

An index of vascular plants of the Western hemisphere. 259,000 cards are reproduced. These cards note newly proposed names, stating genus, species and subordinate categories. Duplicates *Index Kewensis* (*qv*) for generic names and binomials, 'but it is invaluable to students of New World plants since 1873, irrespective of the category and irrespective of whether published in a New World or Old World publication' (Lawrence, G.H.M. *Taxonomy of vascular plants,* p.287). *Class No:* 582.35

[2481]

KENT, D.H. List of vascular plants of the British Isles. London, Botanical Society of the British Isles, 1992. xvi,384p. £11.50. ISBN: 0901158216.

Updates and revises J.E. Dandy's *List of British vascular plants* (1958).

Includes those plants which are, at the present time, native or have been introduced and established in the UK, Eire, the Isle of Man, and the Channel Islands. Also lists extinct native species. Entries contain information on nomenclature, synonyms, hybrids, species, and subspecies. 2-page bibliography. Will be the standard work and is essential for any library with a collection in this subject. *Class No:* 582.35

[2482]

Kew record of taxonomic literature relating to vascular plants. London, HMSO, 1971-. 4pa. £60. ISBN: 0112412297.

1991(2) issue, for material scanned during May-July 1991, contains 1,279 references. Sections include: General; Systematics; Floristics; Palynology; Personalia; Botanical institutions. Genus and author indexes. Periodical abbreviations. *Class No:* 582.35

[2483]

PERRING, E.M. *and* **FARRELL, Ll.,** *comps.* **British red data books. 1. Vascular plants.** 2nd rev. ed. Nettleham, Lincoln, Royal Society for Nature Conservation, 1984. 128p. £5.50. ISBN: 0902484044. *Class No:* 582.35

Ferns

[2484]

Index filicum supplementum quintum pro annis 1961-1975. Jarrett, F.M., *and others.* Oxford, University Press, 1985. 245p. £45. ISBN: 0198545797.

Continuation of C. Christensen's *Index filicum* (1906).

An index to the family, generic, infrageneric and specific names of ferns and their allies, published 1961-75. This 5th supplement differs from its predecessors by including fern allies. New names are printed in bold and are also listed with full bibliographic references. Catalogus literaturae, p.219-45. *Class No:* 582.350

[2485]

JAHNS, H.M. *and* **MASSFLINK, A.K. Collins' guide to the ferns, mosses and lichens of Britain and Northern and Central Europe.** Launert, E., *and others, tr. and rev.* London, Collins, 1983. 272p. illus. (mostly col. pl.), diagrs., maps. £8.95. ISBN: 0002192543.

Original as *Ferne, Moose, Flichten ...* (Munich, 1980).

Sections: Ferns - Mosses and liverworts - Lichens - Keys. 679 numbered entries for plants (over 750 species

....*(contd.)*

described). Glossary. Further reading and societies. 650 col. photographs. Index of genera; index of common names. 'The result is a very useful handbook for the English-speaking naturalist on British and European criptogens' (*Natural history book reviews*, v.7(1/2), 1984, p.106). *Class No:* 582.350

[2486]
JERMY, C. *and* **CAMUS, J. The Illustrated field guide to ferns and allied plants of the British Isles.** London, Natural History Museum Publications, 1991. xiv,194p. figs. £7.95. ISBN: 0565011723.

Aimed at the amateur botanist. Silhouettes or drawings are supplemented by data on leaves; leaf-base; habitat; habit; distribution; conservation status; similar plants. Indexes to Latin and English names. Both authors are members of the Botany Department of the Natural History Museum, London. *Class No:* 582.350

[2487]
JONES, D.L. Encyclopaedia of ferns: an index to ferns, their structure, biology, economic importance, cultivation and propagation. London, British Museum (Natural History), 1987. xvii, 433p. illus. (incl. col.), diagrs. ISBN: 0565010190.

7 parts (1. Introduction; structure and botany of ferns - 2. The cultural requirements of ferns - 3. Pests, diseases and other ailments of ferns - 4. Propagation and hybridisation of ferns ... 6. Ferns to grow (p.202-385). 7 Appendices (*e.g.* Ferns for wet soils). Glossary. Bibliography, p.417-20. Fern societies and study groups. Detailed index, p.422-33. 400 illus. (250 in colour). Well produced; by a professional Australian horticulturalist. *Class No:* 582.350

[2488]
PAGE, C.N. The Ferns of Britain and Ireland. Cambridge Univ. Press, 1982. xii, 447p. illus., maps. ISBN: 0521232139.

Main text, p.55-412. Prelims. include a glossary, chart key to the main pteridophyta groups, and botanical subdivisions of Britain and Ireland. Appendices: 'Cultivation of ferns from spores'; 'Conservation of native pteridophyta'; 'Further studies most needed' (p.421-8). Bibliography, p.429-37. Index, p.439-47. *Class No:* 582.350

Flowering Plants (Seed Plants & Phanerogams)

[2489]
BELL, A.D. Plant form: an illustrated guide to flowering plant morphology. Oxford, New York, Oxford University Press, 1991. 341p. illus. (col.), diagrs. £25. ISBN: 0198542194.

2 parts: 1. Morphological description (*e.g.* 'Leaf morphology'; 'Vegetative multiplication') - 2. Constructional organization (*e.g.* 'Meristem position'; 'Plant branch construction'). Many coloured illustrations supplemented by clearly drawn diagrams. Bibliography, p.317-20. Index. Many cross-references. Aimed at students and amateur horticulturalists. *Class No:* 582.4

[2490]
DE WIT, H.C.D. Plants of the world: the higher plants. London, Thames & Hudson, 1963-65. 2v. (335p. + 340p.). illus. (incl. pl.).

Translated from the Danish, *Hogere planten* (1963-65).

V.1 concerns monocotyledons and part of the dicotyledons; v.2 completes the dicotyledons. A number of families are discussed in each order, with descriptions of the flower, reference to distribution and habitat, plus such added information as garden species, herbal use and evolution of common name. Magnificent and profuse illus. (in all, 362 in colour, 573 black and white). Each volume has an analytical index. Only 5 items 'For further reading' (v.2. p.317). *Class No:* 582.4

[2491]
Flowering plant index of illustration and information. Isaacson, R.T., *comp.* Boston, Mass., G.K. Hall, 1979. 2v. $230, set; $260, supp. ISBN: 0816103011, set; 0816104034, supp.

50,000 catalogue cards, photolithographically reproduced, for coloured illustrations of flowering plants. The sources represent over 100 books, mostly published since 1930. Indexed by both common and botanical names.

A 2v. supplement was published in 1982. *Class No:* 582.4

[2492]
HUTCHINSON, J. The Families of flowering plants, arranged according to a new system based on their phylogeny. 3rd ed. Oxford, Clarendon Press, 1973. xviii, [1], 908p. illus., diagrs., tables, maps.

First published 1926.

Part 1: Dicotyledones (descriptions of orders and families of dicotyledones, with keys to genera of smaller families), p.153-634 - Part 2: Monocotyledones. Glossaries, p.126-8, p.906-7. 430 figures. Hutchinson (who died in 1972) rearranged the families of plants - under dicotyledones and monocotyledones - in ascending order. *Class No:* 582.4

[2493]
Index Kewensis plantarum phanerogamarum nomina et synonyma omnium generum et specierum a Linnaeo usque a annum MDCCCLXXXV complectens. Oxford, Clarendon Press, 1893-95. 2v. Supplementum, 1901-. (1986-90. 1991. 354p. £95).

Added title-page in English: 'Index Kewensis: an enumeration of the genera and species of flowering plants from the time of Linnaeus to the year 1885 inclusive ...' Includes authors' name, works in which first cited, native countries and synonyms. The present volumes list over 500,000 names, A/Z. '*Index Kewensis* is the reference employed to determine the source of the original publication of a generic name or binomial of a seed plant. It does not account for names of terms nor of any plants in divisions subordinate to the Spermatophytes, nor for names of plants as placed in categories below that of species' (Lawrence, G.H.M. *Taxonomy of vascular plants*, p.286). *Class No:* 582.4

[2494]
MABBERLEY, D.J. The Plant-book: a portable dictionary of the higher plants. Cambridge University Press, 1987. xii, 706p. £25. ISBN: 0521340608.

Dictionary, A-Z. ('Crocus': 15 lines; 'Rosa' : 2/3p.) - *c.*2,500 entries. Sources: 1. Floras and handbooks; 2. Periodicals. Abbreviations: 1. General; 2. Authors' names (full names & dates). Very small type. *Class No:* 582.4

[2495]

WILLIS, J.C. A Dictionary of flowering plants and ferns.
8th ed., student ed., rev. by H.K.A. Shaw. Cambridge Univ.
Press, 1985. xxii, 1245, lxvip. ISBN: 0321313953.

First published 1897.

Over 50,000 entries, giving for every genus an outline of
its distribution, family and often some additional facts. 'It is
good value and all serious students of botany should have a
copy' (*Natural history book reviews,* v.8(2), 1986, p.115-6).
Class No: 582.4

Monocotyledons

[2496]

METCALFE, C.R. Anatomy of the monocotyledons.
Oxford, University Press, 1960-.

1. *Gramineae,* by C.R. Metcalfe. 1960. 2. *Palmae,* by
P.B. Tomlinson. 1961. 3. *Commelinales - Zingiberales,* by
P.B. Tomlinson, 1969. £32. 4. *Juncales,* by D.F. Cutler.
1969. £32. 5. *Cyperaceae,* by C.R. Metcalfe. 1971. £40. 6.
Dioscoreales, by E.S. Ayensu. 1972. £40. 7. *Helobiae
(Alismatidae),* by P.B. Tomlinson. 1982. £70.

A companion work to his *Anatomy of the dicotyledons
(qv).* V.1 includes a valuable introduction, including the
general morphology of the grass plant. Descripton of genera
and species, in A/Z order. is in 2 groups: 1. Genera not in
the Bambuseae (p.1-540); 2. Genera in the Bambuseae
(p.540-89). Bibliography, p.630-50. Index, p.715-31.
Class No: 582.52

Grasses

[2497]

**CLAYTON, W.D. and RENVOIZE, S.A. Genera
graminum: grasses of the world.** London, HMSO, 1986.
389p. figs. £25. (*Kew bulletin additional series XIII.*) ISBN:
0112500064.

2 parts: 1. The grass plant - 2. Enumeration of genera
(651 entries). Data for each entry include bibliographic
references; description; anatomy; habitat; chromosomes;
geographical distribution. Index to genera and names of
higher rank, p.379-89. *Class No:* 582.542

[2498]

**FITTER, R. and FITTER, A. Collins' guide to the grasses,
sedges, rushes and ferns of Britain and Northern Europe.**
London, Collins, 1984. 256p. illus. (col.), maps. £10.95;
£9.99. ISBN: 0002191288.

Data on 26 families of grasses, sedges, rushes, and ferns
(20 species), preceded by keys. Colour illus. of over 420
plants. Glossary, p.220-3. 500 tiny distribution maps.
Further reading, p.250. Indexes of English and scientific
names. Apart from native grasses, etc., the text includes 'a
number of widely established aliens' (*Introduction*).
Class No: 582.542

[2499]

**Grasses, ferns, mosses and lichens of Great Britain and
Ireland.** Phillips, R. *and* Grant, S., *eds.* London, Pan
Books, 1980. 191p. £12.99. ISBN: 0330259598.

Contents: Grasses, p.12-79; Ferns, p.76-117; Mosses
(including liverworts), p.118-63; Lichens, p.164-81. A
glossary precedes. Analytical indexes of botanical and
common names. Full-colour illus. (1-6 per page), with
descriptive captions. 'Nearly 160 species out of the total

.... *(contd.)*

British *bryophyta* flora, around 1,000, have been described'
(*Introduction*). No bibliography. A good popular
introduction. Quarto. *Class No:* 582.542

[2500]

HUBBARD, C.E. Grasses: a guide to their structure,
identification, uses and distribution in the British Isles. 3rd
ed., revised by J.C.E. Hubbard. Harmondsworth, Penguin
Books, 1984. 476p. illus. ISBN: 0140222790.

2nd ed. 1968 (462p.).

Description and illus. of individual grass, p.59-372, with
text facing illus. page. Introduction includes 'Key for naming
wild and agricultural grasses'. Appendices include 'The uses
of grass' and a classification. Bibliography, p.47-52. Indexes
to common and botanical names. A standard work.
Class No: 582.542

[2501]

**ROSE, F. Colour identification guide to the grasses, sedges,
rushes and ferns of the British Isles and north-western
Europe.** London, Viking, 1989. 240p. illus. £35. ISBN:
0670806889.

420 species described in the text, with more tha 350
illustrated by 62 colour plates. Descriptions of each are
followed by distribution and ecological data. Useful feature
is the use of keys to several plant groups. Many clear and
accurate line drawings. Well presented. Useful for both the
beginner and the professional botanist. *Class No:* 582.542

[2502]

SMITHSONIAN INSTITUTION. Index to grass species.
Chase, A. *and* Niles, C.D. Boston, Mass., G.K. Hall,
1970. 3v. (1746p.). $330. ISBN: 081610445x.

62,400 cards reproduced. Entries give name, authority and
complete bibliographic citation for each taxa. Lists the
scientific names of all species, sub-species and varieties of
grasses described since the time of Linnaeus (fl. 1737),
claims the prospectus'. *Class No:* 582.542

Orchids

[2503]

BUTLER, K.P. Steinbachs Naturführer: Orchideen. [Field
guide to orchids of Britain and Europe: the species and
subspecies growing wild in Europe, the Near East and North
Africa.] Rev. English ed. Munich, Mosaik Verlag GmbH,
1986; Swindon, Wilts., The Crowood Press, 1991. 288p.
illus. (col.), diagrs., maps. £9.99. ISBN: 1852235918.

Describes and illustrates, with 562 excellent colour
photographs, every European species and subspecies of
orchid. Gives descriptions of characteristics; habitat;
synonyms; flowering period; distribution; key; variation.
Bibliography, p.278-79. Indexes to English and scientific
names. *Class No:* 582.594

[2504]

**DAVIES, P., and others. Wild orchids of Britain and
Europe.** London, Chatto & Windus: Hogarth Press, 1983.
256p. col. pl. ISBN: 0701126436.

6 sections: 1. Orchid biology - 2. Orchid ecology - 3.
Classification - 4. Species descriptions (p.49-184, with cross-
references to col. plates) - 5. In search of orchids (Great
Britain ... Israel) - 6. Photographing wild orchids.
Bibliography, p.245-7 (books; journals). Analytical index,

....(contd.)
p.249-56. A neat production. Orchids are the second largest family of plants in the vegetable kingdom.
Class No: 582.594

Dicotyledons

[2505]
METCALFE, C.R. *and* **CHALK, L.,** *and others*. **Anatomy of the dicotyledons.** 2nd ed. Oxford, Clarendon Press, 1987-. illus. (pl.), diagrs., tables. ISBN: 0198542534, v.1; 0198545940, v.2; 0198545932, v.3.

First published 1950 (2v.).

1. *Systematic anatomy of the leaf and stem.* 1988. 286p. £27.50 2. *Wood structure and conclusion of the general introduction.* 1989. 297p. £30. 3. *Magnoliales, illiciales and laurales.* 1987. 234p. £60.

V.1-2 of this proposed multi-volume work, provide a general introduction to principles and data. V.1 (12 chapters) includes a brief history of systematic anatomy. Each chapter provides 'Suggestions for further reading'. Bibliography (A/Z authors). Name and analytical subject indexes.
Class No: 582.6/9

Water Plants

[2506]
MÜHLBERG, H. The Complete guide to water plants: a reference book. Translated from the German. East Ardsley, Wakefield, W. Yorkshire, EP Publishing, Ltd., 1982. 392p. illus. (incl. col.). ISBN: 0715807897.

German ed., 1980.

Two parts: 1. The biology of water plants - 2. The families, genera and species of aquatic plants (Liverworts ... Cattails, p.92-376). Data on species: named by ...; distribution; cultivation and propagation. Appended: Explanation of specific epithets; List of authorities (p.381-5). Bibliography, p.386. Index of scientific names of the species, p.387-91. 221 black-and-white and col. illus. Sources of illus. *Class No:* 582.671

[2507]
Water plants of the world: a manual for the identification of the genera of freshwater macrophytes. Cook, C.D.K., *and others*. The Hague, W. Junk; Kinderhook, N.Y., IBD, 1974. viii, 561p. $175.

Aims at exhaustive coverage of water plants (aquatic macrophyta) down to genus level. The initial key is based largely on vegetative character. Glossary of technical terms, for the non-botanist. An authoritative guide.
Class No: 582.671

59 Zoology

Zoology

Databases

[2508]
ZOOLOGICAL RECORD ONLINE. Philadelphia, Pa.,
BIOSIS, 1978-.
  Online version of *Zoological record (q.v.)*. Contains
c.1,500,000 records giving worldwide coverage of
zoological literature, scanning *c*.6,000 journals. Updated
monthly. Available on CD-ROM. *Class No:* 590(003.4)

Bibliographies

[2509]
AMERICAN MUSEUM OF NATURAL HISTORY.
Research catalog of the Library. [American Museum of
Natural History]. Boston, Mass., G.K. Hall, 1977. 13v.
$1320. ISBN: 0816100640. *Class No:* 590(01)

[2510]
BRITISH MUSEUM (NATURAL HISTORY). Animal
identification: a reference guide. London, British Museum
(Natural History); Chichester, Wiley, 1980. 3v. £79.50, the
set. ISBN: 0471277657; 0471277665; 0471277673.
  1. *Marine and brackish water animals;* edited by R.W.
Sims, 1980. vi, 111p. £14.55. 2. *Land and freshwater
animals,* (not insects); edited by R.W. Sims. 1980, x, 120p.
£14.55 3. *Insects;* edited by D. Hollis. 1980. viii, 160p.
£20.95.
  Systematically arranged. V.3 (Insects, general) (Thysanura
... Hymenoptera), with regional grouping as necessary;
c.4,500 references. 'This work should be of immense value
to applied biologists, biologists working a poorly worked
region of the world, expedition zoologists or amateur
naturalist collecting on an overseas holiday' (*Natural history
book reviews*, v.5(3/4), 1982, p.128-9). *Class No:* 590(01)

[2511]
ENGELMANN, W. Bibliotheca historico-naturalis.
Verzeichnis der Bücher über Naturgeschichte, welche in
Deutschland, Scandinavien, Holland, England, Frankreich,
Italien und Spanien in den Jahren 1700-1846 erschienen
sind. Leipzig, Engelmann, 1846. ix, 786p.
  About 10,000 references to books and pamphlets on
natural history, published 1700-1846 in Northern and
Western European countries. Sections cover bibliographies,
general works, comparative anatomy and physiology,
zoology and palaeontologie. Author and subject indexes.
Stresses local material. Continued in: *Class No:* 590(01)

[2512]
—Bibliotheca zoologica. Verzeichnis der Schriften über
Zoologie welche in den periodischen Werken enthalten und
vom Jahre 1846-1860 selbständig erschienen sind. Leipzig,
Engelmann, 1861. 2v. (x, xxiv, 2144p.).
  Similarly arranged. 40,000 entries, including periodical
articles since 1700. *Class No:* 590(01)

[2513]
—Bibliotheca zoologica II. Verzeichnis... 1861-1880. Leipzig,
Engelmann, 1886-1923. 8v. (6620p.).
  Has *c*.160,000 entries. *Class No:* 590(01)

[2514]
HARVARD UNIVERSITY. Catalogue of the Library of the
Museum of Comparative Zoology. Boston, Mass., G.K.
Hall, 1970. 8v. First supplement 1976. 770p. $835; $140,
supp. ISBN: 081610767x; 0816108110, supp.
  169,100 catalogue cards in all, photolithographically
reproduced. Main entries under author, or title, plus added
entries for joint authors, editors, series, etc. All important
monograph series are analysed. No subject entries, except
for biographies, works about institutions, scientific
institutions, and ships used in such expeditions.
Class No: 590(01)

[2515]
McGILL UNIVERSITY, Montreal. A Dictionary catalogue
of the Blacker-Wood Library of Zoology and
Ornithology. Boston, Mass., G.K. Hall, 1970. 9v. $930.
ISBN: 081610719x.
  140,000 dictionary catalogue cards, photolithographically
reproduced. The 60,000 volumes include *c*.2,000 periodical
sets, of which only 500 are current. *Class No:* 590(01)

Encyclopaedias

[2516]
BURTON, M. *and* BURTON, R. Marshall Cavendish
international wildlife encyclopedia. Rev. ed. London,
Marshall Cavendish, 1990. 25v. illus. £291.45; $449.95.
ISBN: 086307734x.
  Covers fish, reptiles and amphibians, birds, and mammals,
with insects, crustaceans and invertebrates given more
selective coverage. Articles arranged A-Z. Animals listed by
common name. V.25 includes a 3-part index - animal;
subject; list of animals classified by phylum. 13-page
bibliography. Many illustrations. *Class No:* 590(031)

[2517]
International encyclopedia of wildlife. Lakeville, Conn.,
Grey Castle Press, 1991. 15v. 3053p. $495. ISBN:
1559050527.
  V.1-5. *Mammals.* 6-8. *Birds.* 9. *Reptiles and amphibians.*
10. *Fish.* 11-15. *Invertebrates.*
  Arrangement is by class, sub-divided by families, each of
which has a chapter devoted to it. The 1,400 colour
drawings and 4,000 colour plates are of high quality and are
this encyclopaedia's outstanding feature. Easier to see

.... *(contd.)*

relationships among animals in a class in this set than in the Marshall Cavendish set, which is arranged alphabetically. V.15 has an index of scientific and common names. An excellent reference tool for school and public libraries. *Class No:* 590(031)

Handbooks & Manuals

[2518]
GRASSÉ, P.-P. Traité de zoologie: anatomie, systématique, biologie. Paris, Masson, 1948-. (v.16; fasc.7, 1982). 17v. (various fascicules).

v.1: *Phylogenie. Protozoâires ...* 8-10. *Insects.* 11: *Echinodermes. Stomocordés. Procordés.* 12: *Vertébrés.* 12. *Agnathes et poissons.* 14: *Reptiles.* 15: *Oiseaux.* 16-17: *Mammifères.*

Each volume/fasc. is fully illustrated with section bibliographies and partly analytical subject index. Shows a remarkable unity and uniformity of a high standard, although bibliographies are uneven. The only major treatise on zoology worthy of the name, according to M.-G. Madier (Malclès, L.-N. *Les sources du travail bibliographique,* v.3, p.327). *Class No:* 590(035)

[2519]
Handbuch der Zoologie. Eine Naturgeschichte der Stämme der Tierreiches. Kukenthal, W. 2. Aufl., hrsg. von J.G. Helmake [u.a.]. Berlin, de Gruyter, 1968-.

First published 1925.

An important, standard treatise on the families of the animal kingdom, comprehensive, with emphasis on systematic zoology. Latest part of 2nd ed: Bd.6(31), 1991. *Class No:* 590(035)

[2520]
—The Cambridge natural history. Harmer, S.F. *and* Shipley, A., *eds.* London, Macmillan, 1895-1909. Reprinted Weinheim, Engelmann, 1958. 10v. illus., maps.

Was, in its day, the standard British treatise on zoology. Systematically arranged (v.1: *Protozoa, Porifera, Coclenterata and Ctenophora, Echinodermata ...* v.10: *Mammalia.*). Well illustrated, with numerous footnote references. Each volume has an index. V.1 (21 chapters, by 4 specialists, contains 291 line-drawings; index, p.625-71). *Class No:* 590(035)

Dictionaries

[2521]
ARAB LEAGUE EDUCATIONAL, CULTURAL AND SCIENTIFIC ORGANIZATION. Lexique de zoologie. Rabat, Morocco, Bureau Permanent de Coordination de l'Arabisation, 1971. 124, xxiip.

English - French - Arabic. *Class No:* 590(038)

[2522]
BILLY, C. Glossaire de zoologie. Paris, Doin, 1985. 240p. *Class No:* 590(038)

[2523]
The Concise Oxford dictionary of zoology. Allaby, M., *ed.* Oxford, University Press, 1991. viii,508p. ISBN: 0198661622.

Intended for the zoologist or zoology student, this reference work consists of *c.*8,000 entries, including cross-references. Definitions vary from 10-400 words, covering taxa (down to familial level for vertebrates, but with less complete coverage of other phyla); earth history; ecology; evolutionary concepts; genetics; physiology; taxonomic principles; zoogeography; as well as some biographical material on individuals distinguished in the science. Taxonomic names are given in Latin with common names cross-referenced. Comprehensive and workmanlike subject dictionary. *Class No:* 590(038)

[2524]
HENDERSON, I.F. and HENDERSON, W.D. Henderson's dictionary of biological terms. 10th ed., by E. Lawrence. Harlow, Essex, Longman, 1989. xi, 637p. £5.99. ISBN: 0582463629.

First published 1920; as *A directory of scientific terms.* (1st-8th eds.).

Over 18,000 headwords. Updates existing definitions and incorporates many new terms *e.g.* AIDS, gene therapy, biotechnology, ribozyme. Revised and simplified outline classification of the Plant and Animal Kingdoms, the Fungi and Monera (prokaryotes). New appendix of *c.*60 structural formulae of important biotechnical compounds. *Class No:* 590(038)

[2525]
KLEMM, M. Zoologisches Wörterbuch. Paläarktische Tiere. Deutsch/Lateinisch/Russisch, Russisch/Lateinische/ Deutsch, mit Registern der wissenschaftlichen Namen. Berlin & Hamburg, Parey, 1973. xiv, 854p. $350. ISBN: 068656474x.

2 parts: 1. Vertebrates; 2. Invertebrates. Each part has *c.*30,000 entries. Index of scientific names follows the German/Latin/Russian and Russian/Latin/German sequences. Bibliography, p.398-833. *Class No:* 590(038)

[2526]
LEFTWICH, A.W. A Dictionary of zoology. 3rd ed. London, Constable; New York [etc.], Van Nostrand, 1973. ix, 478p.

First published as *A student's dictionary of zoology* (1963).

About 6,700 concise definitions (4-100 words). Cross-references. Part 2: 'English names'; Part 3: 'Principles of classification and nomenclature'. Excludes 'such medical and anatomical terms which are to be found in any medical dictionary and such elementary biological terms as are to be found in any school textbook' (*Preface*). Appended simplified classification of the animal kingdom, p.468-74. Bibliography, p.475-8. *Class No:* 590(038)

[2527]
PENNAK, R.W. Collegiate dictionary of zoology. Melbourne, Fla., Krieger, 1987. 594p. $26.50. ISBN: 0898749212.

About 20,000 definitions of terms in zoology and allied fields, - unlike Leftwich. Many cross-references; references from common names to Latin names. Appended: condensed taxonomic outline of the animal kingdom. *Class No:* 590(038)

Reviews & Abstracts

[2528]

The Zoological record, being the record of zoological literature relating to the year... York, BIOSIS, UK, and London, Zoological Society, 1865 (covering 1864)-. v.1-. Annual parts. £1200pa. (UK subscribers). ISSN: 01443607.

V.1-6 as *Record of zoological literature;* v.43-52 (1906-15) formed Section N of the *International catalogue of scientific literature.* Since v.115 (covering 1978) published jointly by BIOSIS UK and the Society.

1. *Comprehensive zoology.* 2. *Protozoa* (1. Recent; 2. Fossil). 3. *Porifera.* 4. *Coelenterata.* 5. *Echinodermata.* 6. *Platyhelminthes, annelida, conodonta* (A,B,C). 7. *Brachiopoda.* 8. *Bryozoa* (Polyzoa). 9. *Mollusca.* 10. *Crustacea* (1. Recent; 2. Fossil). 11. *Trilobita.* 12. *Arachnida.* 13. *Insecta* (A,B,C,D,E,F). 14. *Protochordata.* 15. *Pisces* (1. Recent; 2. Fossil). 16. *Amphibia.* 17. *Reptilia.* 18. *Aves.* 19. *Mammalia.* 20. *List of new genera and subgenera.*

About 80,000 references pa. (27 parts). Each part consists of author list of papers and books, with subject, geographic, palaeontological and systematic indexes. The most important record for retrospective searching in systematic zoology (including palaeontology). Online database, 1978-, via DIALOG, BRS. *Class No:* 590(048)

Nomenclatures

[2529]

GOZMANY, L. **Vocabularium nominum animalium Europae system linguis redactum.** Budapest, Akadémiai Kiadó, 1979. 2v. (lxiii, 1171p. + 1015p.).

1979 ed. published in the US and UK in 1990 as *Seven-language thesaurus of European animals* (Chapman & Hall. £135).

V.1 contains 12,000 numbered entries for terms in Latin, German, English, French, Hungarian, Spanish and Russian. V.2 is an index of vernacular names, linked by the numbers to the scientific names of European animals (including insects) in v.1. *Class No:* 590(083.72)

[2530]

International code of zoological nomenclature. 3rd ed. London, International Trust for Zoological Nomenclature, in association with British Museum (Natural History), etc., 1985. xx, 338p. £15. ISBN: 085301003x.

Adopted by the 20th General Assembly of the International Union of Biological Sciences. Previous ed., 1964.

The Code (in French and English facing) occupies 18 sections, p.3-179. Appendices A-F include A. Code of ethics; D. Recommendations on the formation of names. Glossary, p.251-97. Index of scientific names; English index. The new Code came into force on 1 January 1983. An important key. *Class No:* 590(083.72)

[2531]

NEAVE, S.A., *ed.* **Nomenclator zoologicus:** a list of names of genera and subgenera in zoology, from the tenth edition of Linnaeus, 1758, to the end of 1945. London, Zoological Society of London, 1939-75. v.1-7.

V.4 includes a short supplement (addenda and corrigenda). V.5: 1936-45 (1950) includes further addenda and corrigenda to v.1-4. V.6 (1960) covers 1945-55; v.7 (1975), 1956-65.

.... *(contd.)*

Entries consist of name; namer; data; source (journal reference, etc.); genus. About 270,000 entries, so far. *Class No:* 590(083.72)

[2532]

SHERBORN, C.D. **Index animalium:** sive, Index nominum quae ab A.D. MDCCLVII generibus et speciebus animalium imposita sunt ... Sectio prima a kalendis Ianuariis MDCCLVIII usque ad finem Decembris MDCCC. Cambridge Univ. Press, 1902. liv, 1195p. ISBN: 0565008030. *Class No:* 590(083.72)

[2533]

—SHERBORN, C.D. Sectio secunda a kalendis Ianuariis MDCCCCI usque ad finem Decembris MDCCCL. London, the Trustees of the British Museum (Natural History, 1922-33). 33pts. i.

This set, one of the basic reference works of taxonomic zoology, covering the period 1758-1850, contains 440,000 references, recorded over 43 years. It has been described as one of the most outstanding examples of patient, painstaking and self-sacrificing industry of all times.

Class No: 590(083.72)

Illustrations

[2534]

MUNZ, L.T. *and* SLAUSON, N.G. **Index to illustrations of living things outside North America:** where to find pictures of flora and fauna. Hamden, Conn., Archon/Shoe String Press, 1981. 441p. $55. ISBN: 0208018573.

Companion to *Index to illustrations of the natural world: where to find pictures of the living things of North America; (q.v.)* compiled by J.W. Thompson (Syracuse, N.Y., Gaylord, 1977, 265p.).

Entries for over 9,000 species, drawn from illustrations in 206 books. The main index has entry under common name, plus scientific name and source, indicating type of illus. (coloured or black-and-white). Scientific name index follows. Bibliography, under titles. Wide-ranging, but highly selective (*Choice,* v.19(8), April 1982, p.1046). *Class No:* 590(084.1)

[2535]

NISSEN, C. **Die Zoologische Buchillustration.** Stuttgart, Hiersemann, 1969-78. 2v.

V.1 (*Bibliographie*) records 4,826 numbered items, with bibliographical notes. Detailed indexes, under titles of books, countries, subjects and authors. List of sources cited, p.5-7. V.2 (*Geschichte*) concerns the history of zoological illustration. *Class No:* 590(084.1)

[2536]

—KNIGHT, D. Zoological illustration: an essay towards a history of printed zoological pictures. Folkestone, Kent, Dawson, 1977. 224p. illus. £14. ISBN: 0712907866.

Has 103 illus. of average quality. Each chapter carries a fairly lengthy bibliography. Thorough text; inadequate index (*Journal of natural history,* v.12(5), September/October 1978, p.591-2). *Class No:* 590(084.1)

ZOOLOGY

Histories

[2537]
Turning points in zoological sciences: a Royal Society Discussion organized ... as part of the 150th anniversary celebrations of the Zoological Society of London, 30th September 1976. London, Royal Society, 1977. vi, 107p. ISBN: 085403093x.

First published in *Proceedings of the Royal Society of London,* series B, v.199, no.1138, p.335-443.

8 papers: 1. Concepts in morphongesis - 2. The physiology of wild animals - 3. Chemical communication - 4. Chromosomal changes in vertebrate evolution - 5. Genes and the structure of organisms - 6. Mechanics, fields, and statistical mechanics in development biology - 7. On integration and teleonomy - 8. Hormones and behaviour. Each paper has its own bibliography and was 'chosen to illustrate some of the advances that have taken place over a wide field during the past 150 years' (*Preface*).
Class No: 590(091)

Libraries

[2538]
'Sci-tech libraries serving zoological gardens'. In *Science and technology libraries,* v.8(4), Summer 1988, p.1-60.

The introduction is followed by 7 documented contributions. Confined to US zoological libraries. Final contribution: 'American zoological park libraries: historical considerations and their archival status'.
Class No: 590:061:026/027

Institutions & Associations

[2539]
CURLE, R.H.P. The Ray Society: a bibliographical history. London, Quaritch, for the Ray Society, 1954. vi, 101p. ports. £5. ISBN: 0903874075.

The object of the Ray Society (founded 1844) is to publish monographs and memoirs on zoology and botany, with special reference to British fauna and flora. This history, produced on the occasion of the Society's centenary, includes a list of the Ray Society's publications, 1844-1951, as well as a bibliography, arranged by subjects, with bibliographical notes on each item (p.32-95).
Class No: 590:061:061.2

[2540]
MAKEY, D.G., *and others*. **ISIS world geographic and zoological institution directory:** a listing of the world's zoos and aquariums, natural history museums of the United States and Canada, and world geopolitical divisions. St. Paul, Minn., ISIS (International Species Inventory System), 1974. x, 342, xp. Loose-leaf.

Geographical arrangement (Continental Europe ... Oceania). Institution bibliography, p.339.
Class No: 590:061:061.2

[2541]
ZUCKERMAN, S. The Zoological Society of London, 1826-1976 and beyond. Symposium of the Zoological Society of London. London, etc., Academic Press, for the Zoological Society, 1976. xviii, 353p. illus., graphs, plans, tables, maps. ISBN: 0126133409.

20 documented contributions (*e.g.* 'The Royal Society of London: evolution of a constitution'; 'The Library and scientific publications of the Zoological Society of London'

.... (*contd.*)
... 'The introduction of new species of animals for the purpose of domestication'). General index, p.337-53.
Class No: 590:061:061.2

Zoos

[2542]
International zoo yearbook. London, Zoological Society of London, 1961-. Annual. ISSN: 00749664.

V.29, 1990 (vii,412p.) has 3 sections by various hands: 1. Horticulture in zoos (p.1-52) - 2. New developments in the zoo world (p.53-243: nutrition; exhibits; breeding and reproduction; housing; animal welfare) - 3. Reference section (*e.g.* Species of fishes bred in captivity during 1987 and multiple generation births; Census of rare animals in captivity 1988). Appendix 1: Taxonomic authorities consulted in the *Yearbook.* Author index, v.27-29; Subject index, v.27-29. Articles are documented.
Class No: 590:061.3

[2543]
KIRCHSHOFER, R., *ed*. **The World of zoos:** a survey and gazetteer. 1st English language ed. Translated by H. Morris. London, Batsford, 1968. 327p. illus. (incl. col.).

Originally as *Zoologische Gärten der Welt* (Frankfurt-am-Main, Umschau Verlag). 1966.

Two main sections - apart from introduction and articles - are the illus. (p.33-145), plates, usually in colour, plus action photographs of animals etc., and a gazetteer. The latter (p.219-321) arranges the world's zoological gardens by countries A-Z. Abbreviations denote availability of conducted tours, demonstrations, facilities for filming and photography, research work, size, staff, opening hours, etc.
Class No: 590:061.3

Practical Work (Techniques)

[2544]
MAHONEY, R. Laboratory techniques in zoology. 2nd ed. London, Butterworths, 1973. 516p. illus., diagrs., illus. ISBN: 0408704675.

First published 1966.

12 chapters (each with small bibliography): 1. Phylum technology - 2. Aquarium and vivarium management - 3. Histological technique - 4. Electron microscope technique (new) - 5. Embryological techniques - 6. Fluid preservation - 7. Injection and corrosion techniques - 8. Museum techniques - 9. Techniques for the preparation of vertebrate skeletons - 10. Physiological techniques - 11. Palaeontological techniques - 12. Useful data. The sections on electrophoresis and chromatography are deliberately kept short because of adequate coverage in other textbooks. Index, p.499-518. *Class No:* 590.08

Collections

[2545]
HANCOCK, E.G. *and* **MORGAN, P.J.** A Survey of zoological and botanical material in museums and other institutions of Great Britain. Secretary, Biology Curators Group, c/o City Museum, Sheffield, 1980. 32p. tables. ISBN: 0905709204.

2 sections: 1. Zoological holdings; 2. Botanical holdings. Mainly tables (*e.g.* Section 1, table 2: Zoological holdings of national and university museums). Zoological appendix:

....*(contd.)*

numbers of zoological specimens of 32 phyla, class or order present in museums - Botanical appendix: numbers of botanical specimens present in museums and universities. Needs updating. *Class No:* 590.082

Physical Zoology

[2546]
DAVIS, S.J.M. The Archaeology of animals. London, Batsford, 1987. 224p. £17.99. ISBN: 0713445726.

2 sections, 8 chapters in all: 1. Methods and problems in zoo-archaeology ... 4. In what season was a site occupied ... 6. From hunter to leader ... 8. Britain: a zoo-archaeological use study. Bibliography, p.197-208. Index, p.215-224. *Class No:* 591.04

Animals (Fauna)

Bibliographies

[2547]
QUINN, M.E. 'Animals: a selected list of reference tools'. In *Booklist*, v.88(20), 15 June 1992, p.1869-74.

21 annotated entries covering general works; birds; fish; insects; mammals; reptiles and amphibians. *Class No:* 591.1(01)

Encyclopaedias

[2548]
TREMAIN, R. The Animals who's who: celebrated animals in history, popular culture, literature and lore. London, Routledge & Kegan Paul, 1982. xv, 335p. illus., facsims. ISBN: 0710094493.

Features over 1,100 animals, 'Aardy' [Aardvak] ... 'Zsa Zsa' [a rabbit]. Inclusions: Balaam's ass; Mickey Mouse; The Cheshire cat; Winnie the Pooh; Black Beauty. Entries include information on the significance or derivation of the name or phrase. Selected biography, p.289-93; Picture credits, p.295-304. Detailed index of over 2,000 terms, listing titles, authors, individuals, names of nonbook sources, etc. *Class No:* 591.1(031)

Handbooks & Manuals

[2549]
BALOUET, J.-C. and ALIBERT, E. Extinct species of the world: lessons for our future. Translated by K.J. Hollyman. London, Charles Letts, 1990. 192p. illus. (col.), maps. $19.95. ISBN: 1852381000.

After an introductory section, 'The extinct species', there are 11 sections, arranged geographically, *e.g.* 'The Indian Ocean'; 'North America'. 'Plants and animals that humans have made extinct', p.182-88. Many colour drawings. Index. For school and public libraries. *Class No:* 591.1(035)

[2550]
GRZIMEK, B., *ed*. Grzimek's animal life encyclopedia. New York, London, etc., Van Nostrand Reinhold, 1972-74. 13v. illus., maps. ISBN: 0317388533.

Originally Kindler Verlag AG, Zurich, 1966-70.

About 150 contributors, mostly German. Coverage: v.1. Lower animals - v.2. Insects - 3. Mollusks - 4/5. Fishes. Amphibians - 6. Reptiles - 7/9. Birds - 10/13. Mammals. Each volume has an index and appendices (including a four-language dictionary in English, German, French and Russian; conversion tables; supplementary readings). Well illustrated. 'Emphasis is on what animals do, their habitats, behaviour, feeding and breeding biology ... Anatomy, physiology and ethnology take second place' (*TLS*, no.3669, 23 June 1972, p.729). 'The single most complete source on the animal kingdom' (*RQ*, v.25 (1), Fall 1985, p.146). *Class No:* 591.1(035)

[2551]
Longman illustrated animal encyclopedia. Whitefield, P., *consultant ed*. Harlow, Essex, Longman, 1984. 600p. ISBN: 0582556910.

US edition as *Macmillan illustrated animal encyclopedia.*

Covers *c.*2,000 species of living vertebrate animals: mammals (p.10-193); birds (p.194-401); reptiles; amphibians; fishes. Data: name, range, habitat, size, commentary. Two-page spreads, illus., facing text. Classification, p.584-7. Detailed analytical index. *Class No:* 591.1(035)

[2552]
—The New Larousse encyclopedia of animal life. Burton, M., *ed*. Rev. ed. London, Hamlyn, 1980. 640p. col. illus. ISBN: 0883321327.

1967 ed (also 640p.) as *Larousse encyclopedia of animal life*.

Chief feature: the generous-sized full-colour illustrations. Systematic order of text, from protozoa to mammalia. Glossary. Further reading list (revised and updated). Index. Good value. *Class No:* 591.1(035)

[2553]
Remarkable animals: a unique encyclopaedia of wildlife wonders. Sundén, U., *ed*. Enfield, Middx., Guinness Superlatives Ltd., 1987. 239p. illus. (col.). £12.95. ISBN: 085112867x.

Contents: Mammals (p.7-81) - Birds (p.82-129) - Fishes (p.130-65) - Amphibians and reptiles (p.166-205) - Insects and arachnids (p.206-35). Popular, almost sensationalistic, treatment of zoology, but well-produced and highly readable. *Class No:* 591.1(035)

Illustrations

[2554]
CLEWIS, B. Index to illustrations of animals and plants. New York, Neal-Schumann, 1991. 217p. $49.95. ISBN: 1555700721.

3 sections: 1. How to use this book - 2. Key to abbreviations of sources - 3. Common-name index (161p.). Data for each entry include code citation, type of illustration and page number. Scientific-name index and 9-page bibliography. 'An outstanding reference for science collections and picture researchers' (*Booklist*, 1 May 1991, p.1740). *Class No:* 591.1(084.1)

[2555]
Plant, animal and anatomical illustration in art & sciences: a bibliographical guide from the 16th century to the present day. Bridson, G.D.R. *and* White, J.J., *comps*. Winchester, Herts., St. Paul's Bibliographies; Detroit, Omnigraphics, 1990. ix,450p. facsims. £75. ISBN: 0906795818.

7,670 entries arranged chronologically in each section and including articles as well as books. Sections: A. Bibliographies - B. Nature in general - C. Plants - D. Animals - E. The human body - F. Artist biographies - G. Periodicals. Appendices: Books on colour; Organizations. Title, subject, and name indexes. For research libraries. *Class No:* 591.1(084.1)

Marine Environment

[2556]
BANISTER, K. *and* CAMPBELL, A. **The Encyclopedia of aquatic life.** New York, Facts on File, 1985. 367p. illus (mostly col.). $45. ISBN: 0816012571.

3 sections: Fishes - Invertebrates - Marine mammals. Data concentrate on classification, body structure, habits. A feature: 400 colour photographs and fine line-drawings. Glossary. Index (in tiny print). 'One of the best encyclopedias on this topic' (*Library journal,* 1 March 1986, p.87). *Class No:* 591.1(26)

[2557]
COLEMAN, N. **Encyclopedia of marine animals.** London, Blandford, 1991. 324p. illus. (col.), maps. £12.50. ISBN: 0713722894.

Sections: Sponges - Cnidarians - Flatworms - Segmented worms - Sea mosses - Crustaceans (p.79-109) - Molluscs - Echinoderms - Sea squirts - Fishes (p.189-285) - Reptiles - Mammals. Bibliography, p.306. Over 1,000 colour photographs (mostly small). Index of common names; index of scientific names; index of family names. The photographs are taken from the files of the Australasian Marine Photographic Index (AMPI). *Class No:* 591.1(26)

[2558]
ERWIN, D. *and* PICTON, B. **Guide to inshore marine life.** London, Immel Publishing, 1987. 120p. col. illus. £9.95. ISBN: 0907151345.

Covers 200 of the most common species found in European waters. Includes fish; cnidaria; mollusca; worms; algae. Brief bibliography and index. Colour illustration of each species. Aimed at the enthusiast and suitable for public libraries. *Class No:* 591.1(26)

[2559]
HALSTEAD, B.W. **Poisonous and venomous marine animals of the world.** 2nd ed. Princeton, N.J., Darwin Press, 1988. 1168+288p. illus. (incl. col.). $250. ISBN: 0878500502.

First published 1965-70.

Taxonomic order of well over 500 species; documented chapters. 288 pages of plates. Personal name index. Subject index. Numerous illus. *Class No:* 591.1(26)

[2560]
The Marine fauna of the British Isles and north-west Europe. Hayward, P.J. *and* Ryland, J.S., *eds.* Oxford, Clarendon Press, 1990. 2v. (996p.) figs. £95, v.1; £85, v.2. ISBN: 0198573561, v.1; 0198575157, v.2.

V.1. *Introduction and protozoans for arthropods.* V.2. *Molluscs to chordates.*

....(contd.)

Provides dichotomous keys, illustrations, brief descriptions, and distribution notes for over 2,000 species. Cumulated index of technical terms and taxonomic index appear in each volume. *Class No:* 591.1(26)

Physiology

Reproduction

[2561]
Bibliography of reproduction: a classified monthly list of references compiled from the research literature. Cambridge, Reproduction Research Information Service, 1963-. 12pa. £195. ISSN: 00061565.

Offical publication of the Society of Fertility, the Society for the Study of Reproduction, the Austrian Society for Reproductive Biology, and the Blair Bell Research Society.

About 20,000 references pa. on the vertebrates, including man. 47 sections (*e.g.* 1. Environment, radiations and nutrition ... 16. Ovary ... 27/28. Fetus ... 33/34. Parturition ... 37/38. Reproductive cycles ... 45. Immunology - 46. Genetics - 47. General & miscellaneous. Author and animal index; symbols & abbreviations; abbreviations for authors' addresses. Insert: 'List of future meetings'. Available in microform. *Class No:* 591.16

Diseases

[2562]
Zoo and wild animal medicine. Fowler, M.E., *ed*. 2nd ed. Philadelphia, Pa., etc., W.B. Saunders Co., 1986. xxiv, 1127p. illus., diagrs., tables. $105; £73. ISBN: 0721610137.

Sponsored by Morris Animal Federation, Denver, Colorado.

About 70 contributors. 5 parts: 1. General information - 2. Special medicine: amphibians and reptiles - 3. Special medicine: birds - 4. Special medicine: mammals (p.515-1039) - 5. Invertebrates. 13 appendices (1-6: Common and scientific names of amphibians, etc.). Detailed, analytical index, p.1089-1127. Has been translated into Japanese. *Class No:* 591.2

Embryology

[2563]
A History of embryology. The eighth symposium of the British Society for Developmental Biology, Nottingham, 1983. Horder, T.J., *and others.* Cambridge, University Press, 1986. xxiv, 477p. diagrs. ISBN: 0521259533.

15 documented papers, including 'Embryology and classical zoology in Great Britain'; 'Hans Spemann and the organiser'; 'Primary embryonic induction in retrospect'; 'Regeneration and holism in biology'. Chronological outline precedes, p.xviii-xxiv. Very useful 'References to resources relating to the history of embryology', p.435-62. Index, p.463-77. *Class No:* 591.3

Behaviour

Encyclopaedias & Dictionaries

[2564]

HEYMER, A. Ethological dictionary, German - English - French. New York & London, Garland, 1977. 238p. illus. ISBN: 0824070054.

Several hundred German base-terms, with English and French equivalents and indexes - 'Aggression': 4½ columns in the three languages. Bibliography, p.206-20. 138 illus. *Class No:* 591.5(03)

[2565]

IMMELMANN, K. *and* BEER, C. A Dictionary of ethology. Cambridge, Mass., Harvard University Press, 1989. xiii,336p. $35. ISBN: 0674205065.

Covers *c.*900 terms. Abnormal behavior ... Zoosemiotics. Entries vary in length - 'Companion': 1½ lines, 'Habituation': 1 page. Many cross-references. 3-page bibliography. 'An excellent companion to *The Oxford companion to animal behaviour* (1981) and to *Grzimek's encyclopedia of ethology*' (*Science and technology libraries,* v.11(1), Fall 1990, p.137). *Class No:* 591.5(03)

[2566]

The Oxford companion to animal behaviour. McFarland, B., *ed.* Oxford Univ. Press, 1981. (Re-issued, with new index, 1987). [xii], 657p. illus., diagrs., charts, maps. £25; £12.95.

69 contributors. Over 200 longish articles (*e.g.* 'Facial expression': 11p.; 'History of the study of animal behaviour' : 17p.; 'Migration': p.380-7, with 3 maps). Bibliography, p.609-10. Index of English names of animals; index of scientific names of animals. Carefully drawn illus. The first *Oxford companion* on a scientific subject, and 'an impressive addition to a distinguished series' (*Reference book bulletin, 1983-1984,* p.136). Aimed at the layperson. 'Cramped layout and tiny print' (*New scientist,* 20 August 1981, p.483). *Class No:* 591.5(03)

Handbooks & Manuals

[2567]

The Audubon Society encyclopedia of animal life. Buettmer, G., *ed.* New York, Clarkson Potter (distributed by Crown), 1982. 606p. illus. (incl. col.).

6 major sections (Animals; Birds; Reptiles; Amphibians; Fishes; Invertebrates), subdivided mostly by families. Data on physical characteristics, social habits, range, habitat and reproductive behaviour. Over 1,000 illus., well captioned and cross-referenced from text. 'Even libraries that own the 13v. Grzimek (1972) will want this harmony of words and pictures' (*Wilson library bulletin,* January 1983, p.428). *Class No:* 591.5(035)

[2568]

GRZIMEK, B., *ed*. Grzimek's encyclopedia of ethology. English ed., edited by K. Immelmann. New York, London, etc., Van Nostrand Reinhold, 1977. xx, 708p. illus.

Originally published by Kindler Verlag AG, Zurich, 1973. 43 contributors. 43 chapters on animal behaviour (1. History of ethology ... 8. Behavioural functions of hearing ... 18. Predictable patterns in animal signals ... 26. Forms and the study of animals in captivity ... 43. Phytogenetic adaptations in human behaviour). Supplementary readings,

....(contd.) p.681-4. Dictionary of ethological terms. Very good coloured illus; descriptive captions. Supplements *Grzimek's Animal life encyclopedia. Class No:* 591.5(035)

Reviews & Abstracts

[2569]

Animal behavior abstracts. Bethesda, Md., Cambridge Scientific Abstracts, 1973-. 4pa. $470pa. ISSN: 03018695.

Previously published by Information Retrieval, Ltd., London; initially as *Behavioural biology abstracts. Section A, Animal behaviour.*

About 1,200 abstracts per issue, drawn from *c.*5,000 primary and other journals. Subjects of papers range from neurophysiology to ecology, from genetics to social anthropology, and the biology of particular taxonomic groups. Appended: Book notices; Notices of proceedings. Author and analytical subject indexes per issue and annually. Online (LIFE SCIENCES COLLECTION) and also available on magnetic tape and CD-ROM. *Class No:* 591.5(048)

Maps & Atlases

[2570]

JARMAN, C. Atlas of animal migration. London, Heinemann, 1972. 124p. illus., diagrs., maps. ISBN: 0434371459.

6 chapters: 1. Why animals migrate - 2. Birds - 3. Mammals - 4. Fishes - 5. Reptiles and amphibians - 6. Insects. Also a list of migratory bird refuges in the US and bird migration watchpoints in Europe, - each accompanied by a map. World maps show routes used by various migratory bird species and by migratory land and sea animals. Index to text, illus. (or their captions), and map keys. Colourfully illustrated, with numerous maps and easy-to-read text. *Class No:* 591.5(084.3)

Sound Recordings & Tapes

[2571]

BRITISH BROADCASTING CORPORATION. Catalogue of natural history recordings. 4th ed. London, BBC Recorded Programme Library, 1969. 2v. [vi], 488p.

Previous issue, 1961.

V.1 covers species, with index of scientific names; v.2, habitat (atmosphere).

Wildlife Sound Recording Society, c/o BLOWS (British Library of Wildlife Sounds) has a National Sound Archive, London, that includes a collection of recordings of more than 5,000 species of bird calls and song (Wyatt, H.V., ed. *Information sources in the life sciences.* 3rd ed., 1987, p.158). *Class No:* 591.5(086.7)

Europe

[2572]

The 'Country Life' guide to animals of Britain and Europe: their tracks, trails and signs. London, Country Life Books, 1984. 320p. illus. (mostly col.). $64. ISBN: 0600388042.

Covers tracks and trails of most European mammals and many types of birds and invertebrates. Also a guide to

....(contd.)

feeding remains, pellets, droppings, etc. Glossary, p.311-2; selected bibliography, p.313. Analytical index, p.314-20. Illus. have descriptive captions. *Class No:* 591.5(4)

Histology

[2573]
KRSTIĆ, R.V. **Illustrated encyclopedia of human histology.** Berlin, Springer-Verlag, 1984. 450p. illus. ISBN: 3540131426.

Etymology and definitions of terms are supported by more than 1,500 illustrations (photographs, micrographs, line drawings), with adequate suggestions for further readings and cross-references. 'This excellent volume should be available for all students of biology, lower-division onwards' (*Choice*, v.22(9), May 1985, p.1306). *Class No:* 591.8

Geographic Distribution of Animals

[2574]
BARTHOLOMEW, J.G. **Atlas of zoogeography:** a series of maps illustrating the distribution of over seven hundred families, genera and species of existing animals. Under the patronage of the Royal Geographical Society. Edinburgh, Bartholomew, 1911. vi, 67, xip. + 36 double pl. of maps.

Formed v.5 of Bartholomew's *Physical atlas.*

About 220 (chiefly world) maps, illustrating distribution of 'the higher animals [mammals, birds, reptiles and amphibians, with selected fishes, molluscs and insects] over the surface of the earth' (*Preface*). Sections of text: 1. General principles - 2. Historical and geographical - 3. Zoological explanation of plates - 4. Bibliography, p.57-67 (*c.*1,000 items, arranged by regions and then by animals). General index. Size: 45 x 30cm. *Class No:* 591.9

[2575]
Flora & fauna of localities: a bibliography of books and articles dealing with plants and animal life of specific countries and regions of the world. Huckin, D., *ed.* Caldicote, Biggleswade, Beds., Clover Publications, 1981. 2v. £12.50, set. ISBN: 0907755003, set.

1. *Great Britain and Europe.* [ix], 118p. 2. *Africa, America, Asia, Australia & USSR.* viii, 158p.

The bibliography excludes school texts, conference papers and symposium proceedings, as well as erudite material only likely to be of use to a specialist. *Class No:* 591.9

[2576]
Key works to the fauna and flora of the British Isles and Northwestern Europe. Kerrich, G.J., *and others, eds.* London, Systematics Association, 1978. xii, 179p.

First published 1942.

About 2,500 entries, with occasional annotations, covering books and periodical articles. Systematic arrangement. Supersedes the Systematics Association's *Bibliography of key works for the identification of the British fauna and flora,* by J. Smart and G. Taylor (1953). No index. For more specialized scientific libraries. 'Graduate and professional levels' (*Choice*, v.16(1), March 1979, p.54,56). *Class No:* 591.9

[2577]
LEVER, C. **The Naturalized animals of the British Isles.** London, Hutchinson, 1977. 600p. illus., tables, maps. ISBN: 0091277906.

4 parts (59 sections): 1. Mammals - 2. Birds - 3. Amphibians and reptiles - 4. Fish. Section 29: 'Mandarin ducks', p.282-98 (with 2 illus., 1 location map). Chronological table of introduction of animals. Chapter bibliography, p.521-56; general bibliography, p.557-8. Indexes of people, species and places. End-paper maps. 'The first comprehensive account of the exotic vertebrates of the British Isles; (*Foreword*). 'A work of formidable scholarship' (*British book news*, January 1986, p.37). *Class No:* 591.9

Systematic Zoology (Categories of Animals)

Invertebrates

[2578]
BARNES, R.D. **Invertebrate zoology.** 5th ed. Philadelphia, Pa., W.B. Saunders, 1987. ix, 893p. illus., diagrs. $40. ISBN: 003008914x.

First published 1963.

A textbook on major, minor and even some fossil invertebrate phyla. Reduction of *c.*170p. on the 4th ed., although new material is incorporated, plus an expanded glossary, index and 'boxed essays' on topics. Each of the 20 chapters has an appended bibliography, sometimes briefly annotated. Authoritative, readable and finely illustrated by photographs and drawings. 'So Barnes is better than ever, though it's still the case that only 3 per cent of the space is given to the insect' (*Nature*, v.332, 10 March 1988, p.184). *Class No:* 592.0

[2579]
HYMAN, L.H. **The Invertebrates.** New York, London, etc., McGraw-Hill, 1940-1967. 6v.

1. *Protozoa thru ctenophora.* 1940. 2. *Platyhelminthes and rhynochocoeia.* 1951. 3. *Acanthocephala: aschelminthes and entroproeta.* 1951. 4. *Echinodermata.* 1955. 5. *Smaller coelomate groups.* 1950. 6. *Mollusca,* 1. 1967.

A comprehensive treatise on the morphology, embryology, physiology, etc. of the invertebrates, concentrating on the group rather than on the species. Each volume-section has extensive references to the literature. Fully illustrated. *Class No:* 592.0

[2580]
NICHOLS, D. *and* COOKE, J.A.C. **The Oxford book of invertebrates:** protozoa, sponges, coelemerates, worms, molluscs, echinoderms and arthropods (other than insects). London, Oxford Univ. Press, 1971. viii, 218p. illus. (col.).

Coloured illus. on recto (*c.*5 species per illus.), facing descriptions; arranged phylum by phylum. 'A classification of invertebrates', p.192-207; glossary, p.208-9. 'Sources of further information' includes societies and associations, p.210. Illus. are all drawn to the same scale, 'except where great differences in size have made this impossible' (*Introduction*). Good colour. *T.S.L.* (24 November 1971, p.1611) notes lack of simple, illustrated keys. Intended for the layman. *Class No:* 592.0

[2581]
PEARSE, V., *and others*. **Living invertebrates**. Palo Alto, Calif., Blackwell Scientific Publications, 1987. xi, 848p. illus. (incl. col.), diagrs. ISBN: 0865423121.

31 chapters include Sponges; Nemerteans; Mollusc body plan; Crustaceans; Chelicerates; Myriapods; Insects; Echinoderms; Invertebrate chordates. 'For each taxon we give a proper name, common name if any, brief nontechnical characterization, notes on habits and habitats, and examples of common or representative genera' (*Preface*). Names chosen are those thought best to express relationships between the animals. Small-type notes supplement the text. Bibliography, p.821-9. Index, p.831-48. *Class No:* 592.0

Marine Environment

[2582]
GEORGE, J.D. *and* GEORGE, J.J. **Marine life:** an illustrated encyclopedia of invertebrates in the sea. London, Harrap, 1979. 288p. illus. (col. pl.).

Combines text and 128p. of colour plates (10 illus. per plate). The text includes biological, behavioural and ecological notes. Valuable bibliography (p.264-8), including identification guides. Glossary, p.269-70. Detailed index, p.273-88 (*c.*3,000 entries). *Class No:* 592.0(26)

Protozoa

[2583]
Protozoological abstracts. Wallingford, Oxon., CAB International, 1977-. 12pa. £216. ISSN: 03091287.

About 4,000 abstracts and references pa. Sections: Hosts; General reviews; Associations; Taxonomy and nomenclature; Morphology and ultrastructure; Evolution and genetics; Immunity, serology and resistance; Pathology and pathogenicity; Biochemistry, physiology and behaviour; Epidemiology and epizootiology; Control; Technique; Life-history; Free-living forms; Unclassified. Reports; Conferences; Books. Author and subject indexes per issue and pa. Online, CAB ABSTRACTS. *Class No:* 593.1

Marine Plankton

[2584]
BOLTOVSKOY, E. **Diccionario de la terminologia de plancton marine** en cinco idiomas: inglés, español, alemán, francés y ruso. Buenos Aires, República Argentina, Secretaria de Marina, Servicio de Hidrografia Naval, 1963. xix, 107p.

3,773 numbered items in English, with equivalents in Spanish, German, French and Russian. Groups of organisms (*e.g.* foraminifera) are categorized. Spanish, German, French and Russian indexes. *Class No:* 593.14

[2585]
—WIMPENNY, R.S. The Plankton of the sea. London, Faber, 1966. 3-426p. illus. (incl. pl.), diagrs., tables.

Gives a detailed account of the distribution and behaviour of planktonic animals. Extensive bibliography, p.308-28. *Class No:* 593.14

Molluscs. Shellfish & Shells

[2586]
ABBOTT, R.T. *and* DANCE, S.P. **Compendium of seashells:** a colour guide to more than 4,200 of the world's marine shells. New York, Dutton, 1982 (reprinted 1986, Melbourne, Fl., American Malacologists; London, Letts, 1991). ix, 410p. col. illus. $49.95. ISBN: 0525932690.

Univalves (p.18-283) - Dentaliums and tusk shells - Bivalves - Nautilus and squids. Strange forms of shells and non-mullusks - Taxonomic classification of mollusca, with major bibliographic references, p.379-90. Includes data on 180 species of cowries, 334 cones, 132 volutes, 140 scallops, etc. Captioned col. illus. Index to popular names; index to scientific names. 'Offers a quick guide to those maritime species the amateur conchologist is most likely to encounter when doing fieldwork, exchanging or purchasing from commercial sources' (*Preface*). *Class No:* 594

[2587]
KERNEY, M.P. *and* CAMERON, R.A.D. **A Field guide to the land snails of Britain and north-west Europe**. London, Collins, 1979. 288p. col. illus., tables, maps.

Coverage: Great Britain and Ireland, Scandinavia, East and West Germany, France, Belgium and Holland. Text sections: The biology of slugs and snails; Finding, collecting and preserving slugs and snails; Mapping schemes. Glossary of shell terms. Systematic check-list of the land molluscs of N.W. Europe (26 families), p.39-212. Greenhouse aliens. Bibliography and societies, p.213-4. Index. 24 colour plates; 276 + 116 distribution maps. A compact compendium. *Class No:* 594

[2588]
The Macdonald encyclopedia of shells. London, Macdonald, 1982. 512p. col. illus., tables, maps. ISBN: 0356085759.

5 chapters: Soft surface mollusks (entries 1-133) - Firm surface mollusks (134-221) - Coral dwellers (222-302) - Other marine mollusks (303-35) - Land and fresh water mollusks (336-57). Classified table of the species mentioned in text. Glossary. Detailed index of entries, p.503-12. Data on appearance, size, geographical occurrence and ecological environment; symbols roughly indicate dimensions. Small format. 1,230 col. illus.; distribution maps. *Class No:* 594

[2589]
—ROGERS, J.E. The Shell book: a popular guide to a knowledge of the families of living mollusks, and an aid to the identification of shells, native and foreign. Boston, Mass., C.T. Branford, 1908. (Reprinted 1936). xxi, 503p. illus. (incl. col.).

Rev. ed. 1951 adds preface and appended 'List of modern names'.

5 parts: 1. How to know shells - 2. The univalves and chitons - 3. The tooth shells - 4. The bivalves - 5. The cephalopods. Analytical index, p.467-80. Illustration index, p.481. 104 illus. (8 in colour). *Class No:* 594

[2590]
SMITH, S.M. *and* HEPPELL, D. **Checklist of British marine mollusca**. Edinburgh, National Museums of Scotland, 1991. 114p. (*National Museums of Scotland information series, no.11*.) ISSN: 09527737.

Aims to present an up-to-date taxonomy of the species. Includes deep-water and pelagic species. Systematic list of higher taxa, p.4-8. Notes on the taxonomy and nomenclature, p.80-9. References, p.89-93. Index to genera and species, p.94-114. *Class No:* 594

[2591]
TEBBLE, N. **British bivalve seashells:** a handbook for identification. 2nd ed. London, H.M. Stationery Office, 1976, for the Royal Scottish Museum. v, 212p. illus. (incl. col.). ISBN: 011491401x.

First published 1966 (Trustees of the British Museum (Natural History, v, 212p.).

Descriptions of seashells (Nut-shells: Nuculacea ... Poromyas and cuspidarias: Potomyacea), p.23-206. Index to superfamily, family and specific names, p.207-12. Literature, p.6-7; glossary, p.7-11. 110 illus. *Class No:* 594

[2592]
Wagner and Abbott's standard catalog of shells. Wagner, R.J.L. *and* Abbott, R.T., *ed.* 3rd ed. Greenville, Delaware, American Malacologists, Inc., 1978-. vp. maps. Loose-leaf.

First published 1964 as *Van Nostrand's Standard catalog of shells.*

A systematic account of the molluscs, with biological data on families and genera. Other sections note world size-records and values, shell collection catalogues, maps, glossary, bibliographies, collecting methods, expeditions, history, biography, newsletter, etc. Indexes. *Class No:* 594

[2593]
WYE, K.R. **The Illustrated encyclopaedia of shells.** London, Headline; New York, Facts on File, 1991. 288p. illus. (col.). £19.95. ISBN: 0747204683; 0816027021, US ed.

Published in the US as *The Encyclopedia of shells.*

Data on over 1,000 shells include common name; species name; locality; habitat depth; average size; general description; stylized shape. A colour illustration accompanies each entry. Glossary, indexes to Latin and common names, and brief bibliography. For serious shell-collectors although its coverage is not as extensive as Abbott and Dance's *Compendium of seashells (q.v.). Class No:* 594

Worms

[2594]
Helminthological abstracts. CAB Institute of Parasitology, St. Albans. Wallingford, Oxon., CAB International, 1932-. 12pa. £238; $433.

Continues *Helminthological abstracts. Series A. Animal and human helminthology.*

Contains *c.*5,000 abstracts pa. in 15 sections (Animal hosts - Associations - Morphology, taxonomy and nomenclature ... Cytology, genetics and evolution - Hirudinea - Unclassified). Reports. Conferences. Books. Author and subject indexes per issue and annually. Available online CAB ABSTRACTS, and on CD-ROM.
Class No: 595.1

[2595]
POZNIAK, G.I., *comp.* **Russian-English dictionary of helminthology and plant nematology.** Farnham Royal, Slough, Commonwealth Agricultural Bureaux, 1979. 108,xp. £17.85. (*Commonwealth Institute of Helminthology, TC no.49.*) ISBN: 0851984479.

Defines and explains more than 6,000 terms from the fields of helminthology (animal and human) and nematology (plant and parasites and free-living forms). Intended for scientific translators and abstractors, professional linguists and scientists. *Class No:* 595.1

Spiders

[2596]
ARNETT, R.H. *and* SAMUELSON, G.A. **The Insect and spider collections of the world.** New York, Flora & Fauna, 1986. 220p. $19.95. ISBN: 091684630x.

Data for each entry: number of specimens; curators; region covered; means of storage; loan policy. Taxonomic index. Name index to private owners or curators.
Class No: 595.4

[2597]
HAMMEN, L. van der, *ed.* **Glossaire de la terminologie acarologique.** [Glossary of acarological terminology.] The Hague, W. Junk, 1976-. v.1-.

1. *General.* 2. *Opilioacarida.* 3. *Holothyrida.* 4. *Gamasida.* 5. *Ixodida.* 6. *Actinedida.* 7. *Oribatida.* 8. *Tarsonemida.* 9. *Acaridida.*

V.1 has *c.*2,000 entries for terms in French, English and German. *Class No:* 595.4

[2598]
LOCKET, G.H. *and* MILLIDGE, A.F. **British spiders.** London, Ray Society, 1951-74. 3v. illus. £25, v.1-2; £15, v.3. ISBN: 0903874156, v.1-2; 0903874024, v.3.

A systematic description of the species. Each volume has bibliography and index. V.2 includes addenda to v.1, new species, and a check-list of British spiders. V.3 updates the information in v.1-2, providing descriptions of the additional species discovered in Britain, revising nomenclature and giving much more information on distribution.
Class No: 595.4

[2599]
PRESTON-MAFHAM, R. *and* PRESTON-MAFHAM, K. **Spiders of the world.** Poole, Dorset, Blandford Press; New York, Facts on File, 1984. 191p. illus. (incl. col.). £14.99. ISBN: 0713713771.

8 chapters (2. Structure of spiders - 3. Classification of spiders - 4. Courtships and mating - 5. Life history of spiders - 6. Prey capture - 7. Spider defence mechanisms - 8. Spiders and man). Appendix: Spider families. Glossary. Guide to further reading. Index, p.187-91. 130 illus. (62 in colour). 'It should be very useful to the biologists and a reference for students at all levels' (*Choice,* v.22(3), November 1984, p.449). *Class No:* 595.4

[2600]
ROBERTS, M.J. **The Spiders of Great Britain and Ireland.** Colchester, Essex, Harley Books, 1985-87. 3v. (2v. of text; v.3: Colour plates). £150, set. ISBN: 0946589054, v.1; 0946589062, v.2; 0946589070, v.3; 0946589186, set.

1. *Atypidae - Therdiosomatidae.* 1985. 250p. £50. 2. *Linyphiidae and check list.* 1987. 204p. £50. 3. *Colour plates.* 1985. 250p. £60.

V.1 includes a general introduction to arachnology, including morphology and behaviour, classification and nomenclature. V.3 has 237 colour plates illustrating 307 species (in some cases, both sexes). An authoritative textbook for identification, supplementing both J. Blackwall's pioneer work, *A history of the spiders of Great Britain and Ireland* (1861-64). and *British spiders,* by G.H. Locket and A.F. Millidge (*qv). Class No:* 595.4

[2601]

—JONES, D. A Guide to spiders of Britain and northern Europe. Rev. ed. London, Hamlyn, 1989. 320p. col. illus. £10.95. ISBN: 0600567109.

First published 1983 as *Country Life guide to spiders of Britain and northern Europe*.

Covers over 350 spiders and harvestmen. Each entry contains a description of male and female; habitat; size; season; distribution. Many coloured illustrations. Brief list of further reading and equipment suppliers and societies. Index in very small type. *Class No:* 595.4

Insects

Bibliographies

[2602]

Entomological literature 1800-1864. Gilbert, P., *ed.* Cambridge, Chadwyck-Healey, 1989-91. 49 double-frame, 105mm x 148mm positive monochrome and colour microfiche. £17,950 (complete collection with printed guide and cataloguing on tape/cards).

Approximately 1,800 monographs, in original published language, published between 1800 and 1864, based on Hager's *Bibliotheca entomologica* (Leipzig, 1862). Source of the documents is the British Museum (Natural History) and the editor is that organization's Entomology Librarian. Printed guide provides bibliographic access by author. *Class No:* 595.7(01)

[2603]

GILBERT, P. *and* **HAMILTON, C.J. Entomology: a guide to information sources.** 2nd ed. London, Mansell, 1990. [viii],259p. £40. ISBN: 0720120527.

First published 1983.

8 chapters: 1. Introduction - 2. Naming and identification of insects - 3. Specimens and collections - 4. The literature of entomology (p.72-147: primary journals; review journals; monograph series) - 5. Searching and locating the literature - 6. Keeping up with current events (newsletters, *c.*80 entries) - 7. Entomologists and their organizations - 8. Miscellaneous services (translation services and guides; apicultural information). 1,822 entries for books, periodical articles, journals and organizations. Detailed analytical index, p.243-59. Of the 1st ed., 'A most useful reference work, well ordered, judiciously annotated ...' (*Natural history book reviews*, v.7(3), 1984, p.156). *Class No:* 595.7(01)

[2604]

ROYAL ENTOMOLOGICAL SOCIETY OF LONDON. Catalogue of the Library. [Royal Entomological Society of London]. Boston, Mass., G.K. Hall, 1979. 5v. $565. ISBN: 0816103151.

An author card-catalogue, photolithographically reproduced. About 78,500 entries for 9,000 monographic works, 50,000 pamphlets and 600 journals, as well as manuscripts and drawings. No subject index. Important for earlier material.

Royal Entomological Society of London has also produced *Serial publications in the Library* (1983). *Class No:* 595.7(01)

Handbooks & Manuals

[2605]

BRIAN, M.V. Social insects: ecology and behavioural biology. London, Chapman & Hall; Forestburgh, N.Y., Lubrecht & Cramer, 1983. [x], 377p. illus., diagrs. £45; $24.95. ISBN: 041222920x; 0934454450, US ed.

17 chapters (2. Food - 3. Foraging by individuals ... 6. Nests ... 8. Defence - 9. Food processing ... 12. Reproduction - 13. Evolution of insect societies ... 16. Communities - 17. Two themes (*e.g.* 2. Social organization). References, p.322-61. Author and subject indexes. Well produced. *Class No:* 595.7(035)

[2606]

DAVIES, R.G. Outlines of entomology. 7th ed. London, Chapman and Hall, 1988. 408p. diagrs. £44. ISBN: 0412266709.

9 sections: 1. Introduction - 2. Insect structure and function - 3. Development and metamorphosis - 4. Classification and biology ... 7. The biology of insect populations - 8. Biology and control of injurious insects - 9. Select classified bibliography (p.343-83). Index. More fully illustrated than earlier editions. Aimed at all serious students of entomology. *Class No:* 595.7(035)

[2607]

IMMS, A.D. Imms' general textbook of entomology. 10th ed, by O.W. Richards, and R.G. Davies. London, Chapman & Hall, 1977. 2v. (1354p.). £17.95, v.1; £38, v.2. ISBN: 041215210x, v.1; 0412152304, v.2.

9th ed. 1964 (London, Methuen).

1. *Structure, physiology and development.* 2. *Classification and biology.*

A comprehensive treatise, well documented (*e.g.* v.2. Order 26; Diptera, p.951-1071, has bibliography, p.1046-71). Detailed index to v.2 (p.1281-1304), including authors of articles. 'A workmanlike updating by two of the most experienced general entomologists of the day' (*Journal of natural history*, v.12(2), March/April 1978, p.233). *Class No:* 595.7(035)

[2608]

—COGAN, B.H. *and* SMITH, K.G.V. Insects: instructions for collectors. No.4A. 5th ed. London, British Museum (Natural History), 1974. vi, 169p. ISBN: 056500705x.

Has 37 illus. One of a popular series. *Class No:* 595.7(035)

[2609]

ROYAL ENTOMOLOGICAL SOCIETY OF LONDON. Handbooks for the identification of British insects. London, the Society, 1949-. 11v. illus.

1. 16pts. (General introduction. Thysanura ... Siphonaptera). 1949-57. 2. *Hemiptera.* 8pts. 1965-84. 3. *Lepidoptera.* 4/5. *Coleoptera.* 15pts. 1952-77. 6/8. *Hymenoptera.* 5pts. 1951-78 9/10. *Diptera.* 10pts. 1949-88. 11. *Check list of British insects.* 2nd ed. 5pts. 1964-78.

A detailed, invaluable series, with clear illus., systematic descriptions, and bibliographies. Indispensable for entomologists. *Class No:* 595.7(035)

[2610]
WOOTTON, A. **Insects of the world.** Poole, Dorset, Blandford Press, 1984. 224p. illus. (incl. col.). £14.99. ISBN: 0713713631.

16 chapters (1. What is an insect? ... 4. Anatomy and movement - 5. Senses and communication - 6. Feeding, digestion and excretion ... 8. Life histories - 9. Courtship and reproduction ... 13. Defences - 14. Migration - 15. Insects and man - 16. Nomenclature and classification). Glossary. 'Guide to further reading', p.241-5. Detailed index, p.216-24. *Class No:* 595.7(035)

Dictionaries

[2611]
KÉLER, S., von. **Entomologisches Wörterbuch** mit besonderer Berücktigung der morphologischen terminologie. 3. durchgesehene und erw. Aufl. Berlin, Akademie-Verlag, 1963. xvi, 788p. illus. (incl. pl.), tables.

First published 1956.

About 10,000 terms are defined and discussed, sometimes at length and well-documented (*e.g.* Hymenoptera: 7 columns, including 2 columns of bibliography). Different applications of terms are explained with references to sources. Profuse cross-references. A valuable chronological list of glossaries and handbooks precedes (p.1-24). 67p. of plates (drawings, with commentary facing). *Class No:* 595.7(038)

[2612]
SÉGUY, E. **Dictionnaire des termes techniques d'entomologie elementaire.** Paris, Lechevalier; New York, French & European, 1967. 465p. $175. (*Encyclopédie entomologique, no.41.*) ISBN: 2720504661.

Standard dictionary for French entomological terms. *Class No:* 595.7(038)

[2613]
TORRE-BUENO, J.R. de la. **The Torre-Bueno glossary of entomology.** Nichol, S.W., *comp.* Rev. ed. *Supplement A,* by G.S. Tulloch. New York, New York Entomological Society, in co-operation with the American Museum of Natural History, 1989. xvii,840p. $24.95. ISBN: 0934454450.

First published 1937 (ix,336p.), 2nd ed. 1962.

Greatly expanded from previous eds. with definitions of *c.*16,000 terms. Extensive cross-referencing. Includes a bibliography and classification by orders. Illustrations now omitted. A standard source. *Class No:* 595.7(038)

Reviews & Abstracts

[2614]
Abstracts of entomology, incorporating worldwide research in entomology, pure and applied studies of insecta and arachnida, including chemical, physical, biological controls, economic entomology, physiology, pathology, morphology, systematics, medical entomology, ecology. Philadelphia, Pa., BioScience Information Service, 1970-. 12pa. $190. ISSN: 00013579.

Reprints entries from *Biological abstracts* and *Biological abstracts/RRM.* Author index. Biosystematic index. General index. Subject list. Indexes are in very small type. *Class No:* 595.7(048)

[2615]
—LEFTWICH, A.W. A Dictionary of entomology. London, Constable, 1976. [v], 360p.

Provides concise, non-technical definitions of over 4,000 terms, describing nearly 3,000 species of insect. Cross-references. Classification of insects, p.352-800. Bibliography, p.359-60. Aimed at amateur entomologists, naturalists and students of zoology with a special leaning towards entomology. See *Entomology: a guide to information sources,* by P Gilbert and C.J. Hamilton for many errors in the definitions given (*Entomology,* entry 220). *Class No:* 595.7(038)

[2616]
Entomology abstracts. Bethesda, Md., Cambridge Scientific Abstracts, 1969-. Monthly. $625pa. ISSN: 00138924.

About 950 abstracts per issue. Sections: General. Bibliography - Systematics - Techniques - Morphology - Physiology, anatomy, histology & biochemistry - Reproduction & development - Behaviour, biology & ecology - Genetics - Evolution - Geography & present-day faunas - Fossil forms & faunas. Notification of proceedings. Book notices. Detailed author and subject indexes per issue. Reprinted from *Biological abstracts,* with its BIOSIS PREVIEWS database. Available on CD-ROM. *Class No:* 595.7(048)

[2617]
PASCAL thema. 260: **Zoologie fondamentale et appliquée des invertébrés.** Vandoeuvre-Les-Nancy, Centre National de la Recherche Scientifique, 1985-. 10pa. FFr1,450. ISSN: 07611714.

Replaces in part *Bulletin signalétique. 360. Biologie animale. Physiologie et pathologie des invertébrés* (1940-83).

About 7,000 abstracts and references pa. Subject, thematic, systematic, and author, indexes per issue and annually. Available online. *Class No:* 595.7(048)

[2618]
Review of agricultural entomology. Wallingford, Oxon., CAB International, 1989-. 12pa. £275; $517. ISSN: 09576762.

Formerly *Review of applied entomology. Series A: Agriculture,* (1913-1989).

About 1,000 abstracts per issue. 14 sections (Taxonomy - Anatomy, morphology - Reproduction and development - Physiology and biochemistry - Genetics and sterility - Ecology and behaviour - Geographical distribution, faunas ... Disorders - Protection against arthropods and use of arthropods for biological control - Chemicals, including toxicity tests and environmental effects). Books, reports, conferences. Author and subject indexes per issue and annually. Online, CAB ABSTRACTS. Available on CD-ROM. *Class No:* 595.7(048)

Progress Reports

[2619]
Advances in insect physiology. London, New York, etc., Academic Press, 1963-. v.1-. Annual. illus., diagrs., tables. ISBN: 0120242184.

V.23, 1991, has 6 contributors. 4 documented papers (*e.g.* 'Aerodynamics and the origin of insect flight' (with 3p. of references); 'The genetics of division of labour in honey bee colonies' (with 5p. of references)). Index, p.211-20. *Class No:* 595.7(055)

[2620]
Annual review of entomology. Palo Alto, Calif., Annual Reviews, Inc., 1956-. v.1-. Annual. ISSN: 00664170.

V.35, 1990 (viii,641p. £26.50) has 25 documented papers (*e.g.* 'Sir Boris Uvarov (1889-1970): the father of Acridology'; 'Ecology and management and arthropod pests of poultry', p.101-26 (with 174 references); 'Evolution of the digestive system of insects'; 'Structure and function of insect glia', p.597-621 (94 references)). Subject index. Cumulative index of contributing authors, v.26-35. Cumulative index of chapter titles, v.26-35. An important source on current entomological research.
Class No: 595.7(055)

Biographies

[2621]
GILBERT, P. A Compendium of the biographical literature on deceased entomologists. London, British Museum (Natural History), 1977. vii, 455p. illus. (pl.). ISBN: 0565007866. *Class No:* 595.7(092)

[2622]
—OSBORN, H. A Brief history of entomology, including time of Demosthenes and Aristotle to modern times, with over five hundred portraits. Columbus, Ohio, Spahr and Glenn, 1952. iii, 303p. ports.

A popular outline history, of particular value for its *c.*1,000 biographical sketches and the portraits.
Class No: 595.7(092)

Europe

[2623]
CHINERY, M. A Field guide to the insects of Britain and Northern Europe. 2nd ed. London, Collins, 1976. 352p. illus. (mostly col.). ISBN: 0002192160.

First published 1962.

Preliminary text concerns the biology of insects, collecting and preserving, classification, and a key to the orders of European insects. Two main orders - Apterygote and Pterygote - are then considered in detail. Glossary, p.323-36; bibliography (grouped, p.337-41). Entomological suppliers. Detailed index (in tiny print). Over 1,000 illus. (788 in colour). Small format. 3rd ed. (389p. £14.99) is to be published in 1993. *Class No:* 595.7(4)

Museums

[2624]
ARNETT, R.H. *and* **SAMUELSON, G.A. The Insect and spider collections of the world.** New York, Flora & Fauna, 1986. 220p. $19.95. ISBN: 091684630x.

Data for each entry: number of specimens; curators; region covered; means of storage; loan policy. Taxonomic index. Name index to curators or private owners. 'Includes privately owned collections if registered with a museum or public held collection' (*Choice,* v.24(9), May 1987, p.1371).
Class No: 595.7:061:069

Butterflies & Moths

Worldwide

[2625]
D'ABRERA, B. Butterflies of the world. Faringdon, Oxon., & Melbourne, Lansdowne Editions, 1977-. 5v. illus., maps.
1. *Butterflies of the Australian region.* 2nd ed. 1977. 415p. 2. *Butterflies of the Afro-tropical region.* 1980. 593p. 3. *Butterflies of the Neo-tropical region.* 1981. 4pts. 4. *Butterflies of the Oriental region.* 1982. 5. *Butterflies of the Holarctic region.*

V.1 includes New Zealand, Papua, New Guinea, Timor, Solomon Islands, and surrounding areas. Arranged in taxonomic sequence. Lavishly illustrated. Bibliography, p.397-4001. Index, p.403-15. No geographical index. Only 3 maps. Large format. An identification guide, produced without access to collections held by the Commonwealth Scientific and Industrial Research Organisation, Canberra.
Class No: 595.78(100)

Europe

[2626]
CARTER, D.J. *and* **HARGREAVES, B. A Field guide to caterpillars of butterflies and moths in Britain and Europe.** London, Collins, 1986. 296p. col. illus. tables. £10.95. ISBN: 000219080x.

A systematic guide to more than 500 species of caterpillars and their food. Bibliography, p.270-1. Foodplant lists, p.272-86. General index, p.286-92. Foodplant index, p.293-6. Over 300 coloured illus. The aim is 'to provide an identification guide to caterpillars that are likely to be encountered in gardens, parks and the countryside, and a useful reference work for students and ecologists interested in insect-plant relationships' (*Preface*). *Class No:* 595.78(4)

[2627]
CHINERY, M. Butterflies and day-flying moths of Britain and Europe. London, Collins, 1989. 320p. illus. (col.). £12.95; £8.95. ISBN: 0002197855.

Contains information on all 360 European butterflies and 260 day-flying moths. 3 sections: 1. The evolution of butterflies and moths - 2. The directory of species (p.26-199) - 3. From egg to adult (*e.g.* 'Feeding behaviour'; 'Finding a mate'). Over 1,500 colour paintings of every identified stage in the life-cycle, and distribution maps for every species. Glossary. Index. *Class No:* 595.78(4)

Great Britain

[2628]
EMMET, A.M. The Scientific names of the British lepidoptera: their history and meaning. Colchester, Essex, Harley Books, 1991. 288p. illus. £24.95. ISBN: 0946589283.

Gives the meanings of 2,494 specific names together with the names of genera, families and suborders to which they belong. 4 appendices: 1. People commemorated in the names of lepidoptera - 2. Geographical names - 3. Unresolved names - 4. Apparent errors in R.D. Macleod's *Key to the names of the British butterflies and moths.* Bibliography, p.245-49. 3 indexes: 1. Scientific names - 2. Special index to the Phycitinae - 3. List of monochrome plates. The author is one of Britain's leading lepidopterists.
Class No: 595.78(410)

[2629]
The Moths and butterflies of Great Britain and Ireland.
Heath, J. and others. Colchester, Essex, Harley Books,
1976-. illus. (incl. col.), maps.
V.1. *Micropterigidae to heliozelidae.* 1976. 343p. V.2.
Cossidae to heliodinidae. 1985. 460p. V.7. Pt.1 *Hesperiidae
to nymphalidae.* 1987. 370p. V.7. Pt.2. *Lasiocampidae to
thyatiridae.* 1991. 400p. V.9. *Sphingidae to noctuidae.*
1983. 288p. V.10. *Noctuidae (Cuculliinae to hypeninae) and
agaristidae.* 1983. 459p.
Projected in 11v. V.10: *Noctuidae (part 2) and
Agaristidae* (1983) features 225 distribution maps (although
lacking a situation map covering the whole area) and 13
superb colour plates, with 23 text figures. A check list of
species is not included. 'Despite these criticisms, this volume
is a definitive work and is far superior to anything currently
available on the group' (*Natural history book reviews,*
v.7(3), 1984, p.157-9). V.1,2,7,9 and 10 have so far
appeared. Aims 'to provide a complete, fully illustrated, and
up-to-date work of identification' (*Introduction*). A standard
work. *Class No:* 595.78(410)

[2630]
—THOMAS, J. *and* LEWINGTON, R. The Butterflies of
Britain and Ireland. London, Dorling Kindersley, 1991.
224p. illus. (incl. col. pl.), maps. £16.99. ISBN:
0863185916.
Covers every species of butterfly regularly seen in Britain,
plus some migrants. The fine illustrations display every stage
of the life cycle, including local variations. Area
bibliography and good index. Well written.
Class No: 595.78(410)

[2631]
**SKINNER, B. Colour identification guide to moths of the
British Isles.** Harmondsworth, Middx., Penguin Books;
Viking, 1984. x, 267p. illus (incl. col.), diagrs. £35. ISBN:
0670803545.
Contains 42 colour plates of all species resident in the
British Isles during the last 100 years. Includes recent
changes in distribution, new resident species, and changes in
nomenclature. Omits descriptions of larvae. Cross-
references. Short bibliography, p.246. Scientific names of
plants, p.247-8. Indexes of scientific and English names,
p.249-67. General book, for the nonspecialist.
Class No: 595.78(410)

Vertebrates

[2632]
**BLACKWELDER, R.E. Guide to the taxonomic literature
of vertebrates.** Ames, Iowa, Iowa State Univ. Press, 1972.
xviii, [1], 259p. $10.95. ISBN: 0813816300.
About 12,000 references. Sections: Animalia in general -
Cephalochordata - Pisces - Amphibia - Reptiles - Aves -
Mammalia - Addenda. Subdivisions for general works,
bibliographies, persons and institutions, exploration and
localities, taxonomic methods, nomenclature studies,
regional studies (faunas; checklists; keys); special subject
lists, etc.. No indexes. For beginners in taxonomy.
Class No: 596

[2633]
COLBERT, E.H. Evolution of the vertebrates: a history of
the backboned animals through time. 4th ed. New York,
Wiley-Liss, 1991. xvii,478p. illus. $49.95. ISBN:
0471850748.
31 chapters (1. Introduction ... 6. Early vertebrate faunas
... 8. The advent of the reptiles ... 16. Years of the
dinosaurs ... 19. Marsupials ... 31. The age of mammals),
followed by appendix, 'A classification of the chordates'.
Well illustrated. Bibliography, p.439-46. Index.
Class No: 596

[2634]
PASCAL folio. F53: Anatomie et physiologie des vertébrés.
Vandoeuvre-Les-Nancy, Centre National de la Recherche
Scientifique, 1984-. 10pa. FFr1,870. ISSN: 07611900.
Formerly *Bulletin signalétique. 360. Biologie animale.
Physiologie et pathologie des invertébrés. Ecologie* (1940-
83), and *Bulletin signalétique. 365. Physiologie des
vertébrés* (1940-83).
About 7,000 abstracts and references pa. Three main
sections, as in title. Subject, thematic, systematic and author
indexes per issue and annually. Available online, and in
microform. *Class No:* 596

[2635]
**WOOD, C.A. An Introduction to the literature of
vertebrate zoology,** based chiefly on the titles in the Blacker
Library of Zoology, the Emma Shearer Wood Library of
Ornithology, the Bibliotheca Osleriana, and other libraries of
McGill University, Montreal. London, Oxford Univ. Press,
1931 (reprinted 1974, Lubrecht & Cramer). xiv, 643p. $95.
ISBN: 3487053578.
Three main sections: (1) 19 chapters on the literature of
vertebrate zoology; (2) a geographical index to items in
section (1); and (3), 'a partially annotated catalogue of the
titles on vertebrate zoology in the libraries of McGill
University' (p.173-643). A valuable bibliographical tool.
Class No: 596

[2636]
YOUNG, J.Z. The Life of vertebrates. 3rd ed. Oxford,
University Press, 1981. xx, 645p. illus., diagrs., graphs.
$25. ISBN: 0198571739.
First published 1950.
33 chapters. Includes the embryology, anatomy,
physiology, biochemistry, palaeontology and ecology of all
vertebrates. 'The whole book is organized around the theme
that mechanisms of homeostasis have become increasingly
more complex during vertebrate evolution, allowing life to
continue under conditions not possible before' (*Preface to
the third edition*). Chapter references, p.586-608. Author
index; analytical subject index (p.613-43). *Class No:* 596

[2637]
—The Oxford book of vertebrates: cyclostomes, fish,
amphibians, reptiles and mammals. Nixon, M. *and*
Whitely, D. London, Oxford Univ. Press, 1972. 224p.
illus., maps.
Describes and illustrates *c.*340 species, excluding birds.
'An unusual and useful book (especially for higher school
level, with detailed biological information)' (*British book
news,* February 1973, p.98). *Class No:* 596

Fishes

Bibliographies

[2638]

DEAN, B. A Bibliography of fishes. Eastman, C.R., *rev. and ed.* New York, American Museum of Natural History, 1916-23. 3v. Reprinted New York, Lubrecht & Cramer, 1973.

1. *Author lists, A-K.* 1916. 2. *Author lists, L-Z;* anonymous titles, nos.1-650. 3. *Anonymous titles,* nos.651-712; addend. to v.1-2, etc.

Includes *c.*50,000 titles. V.3 also lists pre-Linnean publications, general bibliographies that include references to fishes, voyages and expeditions, and periodicals relating to fish and fish culture. 3-section subject index. One of the most complete works of its kind. *Class No:* 597(01)

[2639]

—ATZ, J.W. Dean bibliography of fishes, 1968. New York, American Museum of Natural History, 1971. 512p.

Computer-produced, lists items in 5 parts: a systematic index, taxonomically arranged; a subject index; a geographic index; an author index; and the bibliography (in accession number order). The 1969 v., also by J.W. Atz (1973. [xiii], 853p.) brings total entries to 89,101. *Class No:* 597(01)

Encyclopaedias & Dictionaries

[2640]

Piatiiazychnyi slovar nazvanii zhivotnykh. ryby: latiniskii, russkii, angliiskii, nemetskii, frantuzskii. [Dictionary of animal names in five languages: fishes: Latin, Russian, English, German, French.] Sokolov, V.E. Moscow, Russky Yazyk Publishers, 1989. 733p. ISBN: 5200002370.

Lists *c.*11,700 names, arranged in 5 sections: Leptocardii (lancelets) - Cephalapspidomorphi (lampreys) - Myxini (hagfishes) - Chrondrichthyes (cartilaginous fishes) - Osteichthyes (bony fishes). *Class No:* 597(03)

[2641]

WHEELER, A. The World encyclopedia of fishes. London, Macdonald, 1985. xiv, 368p. illus. (incl. col.). ISBN: 0356107159.

Previously published as *Fishes of the world* (New York, Macmillan, 1975).

Describes nearly 1,500 species individually (A-Z), with entry under scientific names. Data include size, distribution and colour. Glossary and plates (96p.) precede. These colour plates - 50 in all - are arranged in systematic order of families. Some 700 fine line-drawings in text. The author, Principal Scientific Officer at the British Museum (Natural History). *Class No:* 597(03)

Handbooks & Manuals

[2642]

Fishes of the north-eastern Atlantic and the Mediterranean. [Poissons de l'Atlantique du nord-est et de la Méditerranée.] Whitehead, P.J.P., *and others, eds.* Paris, Unesco, 1984-86. 3v. figs., maps. $180. ISBN: 9230022152, v.1; 9230023086, v.2; 9230023094, v.3.

A continuation of *Check-list of the fishes of the north-eastern Atlantic and of the Mediterranean* (Unesco, 1973). A fauna of saltwater fish of the palaearctic region. Data on 220 families and 1,250 species. Entries include line-drawings, diagnoses, keys, biological data, synonyms, distribution

....(contd.)

maps and literature. V.3 contains an additional bibliography to *Check-list of the fishes of the north-eastern Atlantic and the Mediterranean,* p.1415-26, as well as an alphabetical index of common names, p.1429-42, and an alphabetical index of scientific names, p.1443-73. *Class No:* 597(035)

Reviews & Abstracts

[2643]

Aquatic sciences and fisheries abstracts. Part 1: Biological sciences and living resources. Compiled by United Nations Dept. of International Economic and Social Affairs, FAO, and the Intergovernmental Oceanographic Commission, with the collaboration of the ASFI Input Partners. Bethesda, Md., Cambridge Scientific Abstracts, 1978-. 12pa. $885pa. ISSN: 01405373.

Sub-title: 'An international journal for the science, technology and management of marine and freshwater environments'.

Contains *c.*24,000 abstracts pa. Main parts: General aspects - Biology - Ecology - Fisheries. 27 sections ('Ichthyology' is subdivided: General; Geographical distribution; Taxonomy and morphology; Reproduction and development; Genetics and evolution; Physiology, biochemistry, biophysics). Author, subject and taxonomic indexes. Available on magnetic tape, online, and CD-ROM. *Class No:* 597(048)

Maps & Atlases

[2644]

BERRA, T.M. An Atlas of distribution of the freshwater fish families of the world. Lincoln, Nebraska, Univ. of Nebraska Press, 1981. xxix, 197p. tables, maps. ISBN: 0803214111.

Includes a bibliography (p.181-91) and index. *Class No:* 597(084.3)

Worldwide

[2645]

NELSON, J. Fishes of the world. 2nd ed. New York, Wiley, 1984. 523p. illus., maps. $59.95. ISBN: 0471864757.

First published 1976.

Systematic treatment of fishes, - orders, sub-orders and families, and listing genera. Anatomical data; ecological status, 45 distribution maps of classes/families worldwide. The clear drawings are well linked to the text. Readership: museum systematists (for the classification) and 'fishery biologists whose trade is population dynamics' (*Nature,* v.312, 13 December 1984, p.679). *Class No:* 597(100)

[2646]

STERBA, G. Freshwater fishes of the world. Rev. ed. Translated [from the German] and revised by D.W. Tucker. Reigate, Surrey, T.F.H. (Great Britain), Ltd., 1973. 2v. (879p. + 282p.) illus. (incl. pl.), tables.

Previous ed. 1962. as *Süsswasserfische aus aller Welt* (Leipzig, 1959).

Basic data, systematically arranged on identification and aquarium biology of *c.*1,300 species of freshwater fishes kept in aquaria in Central Europe. Annotated bibliography, v.1, p.850-9. The 1973 ed. adds 90p. of plates to the 192 in

....(contd.)

the 1962 ed. Author was at the time Director of the Zoological Institute, Univ. of Leipzig. A major reference source in its field. *Class No:* 597(100)

Europe

[2647]

WHEELER, A. The Fishes of the British Isles and Northwest Europe. London, Macmillan, 1969. xvii, 613p. illus. maps.

Has 3 main sections: Marsipobranchii (lampreys and hagfish) - Selachii (sharks, rays and rabbit fishes) - Pisces (Osteichthyes) (bony fishes). Data: common and Latin names; French, Dutch, German, Danish and Swedish names; identification; biology; line-drawings. Small distribution maps. Bibliography (grouped), p.589-601. Index. *Class No:* 597(4)

Africa—Southern

[2648]

Smith's sea fishes. Smith, M.M. *and* Heemstra, P.C., *eds.* 6th ed. Berlin, Springer-Verlag, 1986. xx, 1047p. illus., diagrs. $162. ISBN: 0387168516.

Originally published by Trustees of the Sea Fishes of Southern Africa Book Fund, 1949 as J.L.B. Smith's *The sea fishes of Southern Africa.*. 5th ed., 1977.

Contents: Anatomy of fishes - Biology of fishes - Dangers of the sea - Scientific nomenclature and classification - Common names - Oceanography of the Southern African region - Ichthyology in South Africa - Key to classes - Class Agnatha - Class Chondrichthyes - Class Osteichthyes - Literature list - Numerical key for identification - Glossary - Special numbers list; old and new numbers. Index. Fully illustrated (*c.*1,800 figures). Authoritative for the region. *Class No:* 597(68)

Reptiles & Amphibians

[2649]

ARNOLD, E.N. *and* BURTON, J.A. A Field guide to the reptiles and amphibians of Britain and Europe. London, Collins, 1978. 272p. illus. (incl. col. pl.). £14.99. ISBN: 0002193183.

Coverage is Europe west of Moscow. 5 sections: Salamanders and newts; Frogs and toads; Tortoises, terrapins and sea turtles; Lizards and amphisbaenians (7 areas); Snakes. Glossary, p.243-7; Bibliography (grouped), p.248-52. 126 small distribution maps. 35 illus. (257 in colour). Concise, synoptic text, but remarkably readable. 'This book is a very real contribution to herpetology and will be of great value not only to the travelling naturalist but to all students of the subject' (*Journal of natural history,* v.19(1), January/February 1979, p.126). *Class No:* 598.1

[2650]

—FRAZER, D. Reptiles and amphibians in Britain. London, Collins, 1983. 256p. illus. £11. ISBN: 0002197068.

Rewrites and updates M. Smith's classic *British amphibians and reptiles* (1973). Appended is an annotated 'Taxonomy of British herpetology'. 'An essential book' (*Natural history book reviews,* v.7(3), 1984, p.151), although the reviewer detects 'a certain sloppiness'. *Class No:* 598.1

[2651]

DITMARS, R.L. The Reptiles of North America: a review of the crocodilians, lizards, snakes, turtles and tortoises inhabiting the United States and Northern Mexico. New York, Doubleday, 1936. 476p. illus, (incl. col.). *Class No:* 598.1

[2652]

FROST, D.R., *ed.* Amphibian species of the world: a taxonomic and geographical reference. Lawrence, Kansas, Allen Press, and the Association of Systematics Collections, 1985. 732p. ISBN: 0942924118.

Data on 4,014 named species, with taxonomic index. 'Compiled for the Parties to the Convention on International Trade in Endangered Species of Wild Fauna and Flora to serve as a standard reference to amphibian nomenclature ...' *Class No:* 598.1

[2653]

—HALLIDAY, T.R. *and* ADLER, K., *eds.* Encyclopedia of reptiles and amphibians. London, Allen & Unwin, New York, Facts on File, 1986. 149p. illus. (incl. col.), maps. $24.95. ISBN: 0816013594.

Has 19 contributors. Entries are supported by 150 illus and 70 maps, plus a glossary, bibliography and index (in tiny print). For public and secondary school libraries, primarily. *Class No:* 598.1

[2654]

PETERS, J.A. Dictionary of herpetology: a brief and meaningful definition of words and terms used in herpetology. New York, Hafner, 1964. 392p. illus. (pl.). ISBN: 0685020371.

About 3,000 terms defined. Different meanings stated; citation of sources, including references to articles in periodicals. Author: Associate Curator of Herpetology, US National Museum, Washington. *Class No:* 598.1

[2655]

WORRELL, E. Reptiles of Australia: crocodiles, turtles, tortoises, lizards, snakes; describing all Australian species, their appearance, their haunts, their habits ... Sydney, Angus & Robertson, 1963. xv, 207p.

Over 330 illus., many in full colour. The check-list of Australian reptiles (p.161-98) is claimed to be the most complete ever compiled and published in Australia. Glossary, p.157-60. Index. *Class No:* 598.1

Snakes

[2656]

FITZSIMONS, V.F.M. Snakes of southern Africa. London, Macdonald, 1962. 424p. illus., maps.

Taxonomic aspects are precisely and lucidly defined. Detailed record of localities, with keys and location maps. Valuable field notes. Bibliography of over 750 items, p.383-406. An excellent example of a regional study. 'Undoubtedly one of the great books on snakes' (*Tropical diseases bulletin,* v.60(7), July 1963. p.677-8). *Class No:* 598.12

[2657]

HARDING, K.A. *and* WELCH, K.R.G. Venomous snakes of the world: a checklist. Oxford, Pergamon Press, 1980. x, 188p. ISBN: 0080254950.

Published as Supplement no.1 (1980) in the journal, *Toxicon.*

2 sections: 1. Taxonomy (arranged by families, subfamilies and tribes, genera, and species, listed A-Z). 2.

....*(contd.)*
Geographical distribution, arranged by continent. Bibliography of 4p. Taxonomic references (7p.). Author and subject indexes. *Class No:* 598.12

[2658]
MATTISON, C. Snakes of the world. Poole, Dorset, Blandford Press, 1986. 190p. illus. (mostly col.), chart, maps. £9.99. ISBN: 0713723408.

9 chapters (2. Size, shape and function - 3. Colour and markings - 4. Reproduction - 5. Food and feeding ... 7. Ecology and behaviour ... 9. Snake families (p.124-79)). Bibliography, p.180-3 (5 sections, each with short introduction). Detailed analytical index to text and illus. Well balanced, with good quality illustrations. *Class No:* 598.12

[2659]
PARKER, H.W. Snakes - a natural history. 2nd ed., rev. and enlarged, by A.G.C. Grandison. London, British Museum (Natural History), with Ithaca, N.Y., Cornell Univ. Press, 1977. iv, 108p. illus. (incl. col. pl.). £6.95. ISBN: 0565010107.

Replaces his *Natural history of snakes* (1965).

Includes glossary and bibliography. 16p. of colour plates; 29 figures. 'The illustrations are quite outstanding' (*Journal of natural history,* v.11(8), November/December 1977, p.719). *Class No:* 598.12

[2660]
PHELPS, T. Poisonous snakes. 2nd ed. Poole, Blandford, 1989. viii, 237p. illus. (some col.). £10.95. ISBN: 0713721146.

First published 1981.

10 chapters 1. Introduction - 2. Classification and distribution ... 4. The elapids ... 7. Venom and snakebite - 8. Snakes and Man - 9. Poisonous snakes in captivity - 10. In the field. 2 appendices. Index, p.226-37. *Class No:* 598.12

[2661]
Poisonous snakes of the world. United States. Department of the Navy. Bureau of Medicine and Surgery. Mineola, N.Y., Dover Publications, 1991. xiii,203p. illus. (some col. pl.), figs., maps. $14.95. ISBN: 048626629x.

Reprint of 1966 ed., published by the US Government Printing Office.

9 chapters: 1. General information - 2. Precautions to avoid snakebite ... 4. First aid ... 7. Distribution and identification of poisonous land snakes - 8. The sea snakes - 9. Antivenin sources; glossary; general references; index. Over 170 photographs and illustrations, with 54 in colour. Clearly presented. For the specialist and layman. *Class No:* 598.12

[2662]
WILLIAMS, K.L. and WALLACH, V. Snakes of the world. Melbourne, Fla., Kriefer, 1989-. $31.50, v.1. ISBN: 0894642154.

1. *Synopsis of snake generic names.* 1989.

Based on A.S. Romer's *Oestology of the reptiles,* although expanded and with cross-references added. Aims to provide a list of all snakes, both living and extinct. Arranged alphabetically with brief bibliographic record of the first article describing the genera. Includes invalid as well as valid names, with cross-referencing between each. *Class No:* 598.12

Birds

Bibliographies

[2663]
MILLER, M.A. Birds: a guide to the literature. New York & London, Garland Publishing, Inc., 1986. 912p. $86. ISBN: 0824087100.

1,942 entries, with notes on coverage, description, illus., additional features. 8 sections: 1. Guide to general works - 2. Guide to subjects - 3. Species and group studies - 4. Area studies - 5. Field studies - 6. Children's literature - 7. Biographies - 9. Fiction. Cross-references. Appendix (list of journals, periodicals and organization publications). Index of names (including anonymous works), p.845-87. *Class No:* 598.2(01)

[2664]
—**STRONG, R.M.** A Bibliography of birds, with special reference to anatomy, behavior, biochemistry, embryology, pathology, physiology, genetics, ecology, aviculture, economic ornithology, poultry culture, evolution and related subjects. Chicago, Ill., Natural History Museum, 1939-59. 4pts.

Consists of pts.1-2: author catalog of *c.*25,000 entries in the *Zoological record* section 18, up to *c.*1938; pt.3: subject index; pt.4: finding index (supporting the author index). *Class No:* 598.2(01)

Encyclopaedias & Dictionaries

[2665]
CAMPBELL, B. and LACK, E., eds. A Dictionary of birds. Calton, England, T. & A.D. Poyser, for the British Ornithologists' Union, 1985. 670p. illus. £49.50. ISBN: 0856610399.

Contributed by *c.*280 specialists. Entries, A-Z, are under English names of birds chosen. Data: definition; general characteristics; mode of life, distribution; familial taxonomy, 'Cuckoo': p.123-6, with 6 references; 'Poetry, Birds in': p.475-8, with 15 references. Many cross-references. Over 500 photographs, drawings, etc. (*Nature,* v.317, 3 October 1985, p.392, finds the drawings disappointing, failing to complement the text.) The *Dictionary* is in line of succession from A. Newton and H. Gadow's *Dictionary of birds.* Intended as a greatly expanded updating of H.F. Witherby's *Handbook of British birds* (London, Witherby, 1938-41. 5v. illus., maps). The illus. are a definite improvement on Witherby's tiny quarter-page illus. 'Highly recommended for college, university and public libraries' (*Choice,* v.23(5), January 1986, p.724-5). *Class No:* 598.2(03)

[2666]
PERRINS, C.M. and MIDDLETON, A.L.A. The Encyclopedia of birds. London, Allen & Unwin; New York, Facts on File, 1985. £30; $45. (*Unwin Animal library.*) ISBN: 0045000328.

90 contributors. A series of articles in systematic order (3 parts: 1. Ostriches to Bulton quails - 2. Plovers to woodpeckers - 3. The passerines). Data on each family/ group of families include physical adaptations, distribution, evolutionary history, classification, breeding, diet and feeding habits, spatial organization. 'Woodpecker', p.296-303: 12 col. illus., 5 line-drawings, distribution map. Bibliography, p.xvi-xvii. Glossary. Analytical index (tiny print), p.xx-xxi. Superb illus., mostly in colour - the book's strength. Like Campbell & Lack, Perrins & Middleton has

....(contd.)

much to offer. 'But, if forced to choose, I would regard the Encyclopaedia as desirable and the Dictionary as necessary' (*Nature*, v.317, 3 October 1985, p.392).
Class No: 598.2(03)

[2667]
—THOMSON, Sir A.L., *ed.* A New dictionary of birds. London, Nelson, 1964. 928p. illus.

A centenary publication of the British Ornithologists' Union.

Entries, A-Z, describe bird habits, plumage, distribution and migration, and the relations between birds and man. Many articles are of encyclopaedic length (*e.g.* 'Heron': p.365-8; 10 items of bibliography). Shorter definitions for ornithological terms and British bird-names. Includes American and overseas usage. 48p. of colour and black-and-white plates. Index of generic names. Succeeds *Dictionary of British birds,* by A. Newton and H. Gadow (1896), which must still be retained for its excellent introductory history of ornithology, missing from Thomson. The latter, however, is admirably produced and remains an invaluable encyclopaedic dictionary. *Class No:* 598.2(03)

[2668]
PERRINS, C.M. The Illustrated encyclopaedia of birds: the definitive guide to birds of the world. London, Headline, 1990. 420p. illus. ISBN: 074720277x.

Lists 9,300 species, with 1,200 of them illustrated in colour and keyed to the colour plate. World checklist of species, p.366-411. Indexes of scientific and common names. No bibliography. 'A very good compromise between academic lists and the more popular treatments' (*Reference reviews,* v.5(1) 1991, p.19). *Class No:* 598.2(03)

[2669]
WEAVER, P. The Birdwatcher's dictionary. Calton, England, T. & A.D. Poyser, Ltd., 1981. 155p. illus., maps. £8.50. ISBN: 0856610283.

About 1,000 entries, Abdomen ... Zygodactyl. Includes names of organizations, but omits highly technical terms (*e.g.* anatomical). Cross-references. 4 appendices: A. Abbreviations; B. North American birds; C. Birdwatcher's Code of conduct; D. The British and Irish list of species in families and orders. British English equivalents are given for North American 'English names' of birds.
Class No: 598.2(03)

Handbooks & Manuals

[2670]
The Cambridge encyclopedia of ornithology. Brooke, M. *and* Birkhead, T., *eds.* Cambridge, University Press, 1991. 362p. illus. (incl. col.), diagrs., graphs, tables. £24.95. ISBN: 0521362059.

39 contributors. 11 sections: 1. Introduction - 2. Anatomy and physiology - 3. Movement ... 5. The daily activities of birds ... 8. Bird populations ... 10. Behaviour - 11. People and birds. Ornithological organizations, p.327-30. Further reading, p.334-5. Index of scientific names; index of common names; index of subjects. Authoritative, well presented and clearly written. Its coverage of ecological and topical issues complements that of Perrins' *The Illustrated encyclopaedia of birds (qv).* 'Very good value ... thoroughly recommended' (*Reference reviews,* v.6(1), 1992, p.24).
Class No: 598.2(035)

[2671]
FULLER, E. Extinct birds. London, Rainbird, 1987. 256p. illus. (some col.). ISBN: 0670817872.

Features those birds which have vanished since 1600. 18 sections: 1. Ratites ... 7. Diurnal raptors ... 11. Pigeons and doves ... 14. Owls ... 18. Hypothetical species and mystery birds. 59 colour, and 81 black-and-white, plates. Bibliography, p.245-51. Index, p.252-56. Does not claim to be comprehensive, but attractively presented.
Class No: 598.2(035)

Reviews & Abstracts

[2672]
Recent ornithological literature: supplement to *The Emu, The Auk, Ibis.* [Washington & London], American Ornithological Union, British Ornithologists' Union, and Royal Australasian Ornithologists Union, 1983-. Quarterly.

About 750 references per issue. Sections: General biology (general; African; Australian; Nemotic; Neotropical; Oriental; Palearctic) - Taxonomy. 4th supp. each year includes list of journals received (over 900 journals). Over 150 volunteers co-operate in this project.
Class No: 598.2(048)

Periodicals

[2673]
BRITISH LIBRARY. Science Reference Library. **Periodicals on ornithology held by the Science Reference Library.** Jackson, G. London, Science Reference Library, 1976. 13p.

Lists *c.*200 titles; also 10 abstracting services and bibliographies. Call numbers are given. Classification scheme for ornithology used in the Library is outlined.
Class No: 598.2(051)

Yearbooks & Directories

[2674]
The Birdwatcher's yearbook and diary. Maids Moreton, Bucks., Buckingham Press, Buckinghamshire, 1981-.

The 1992 *Yearbook* (1991. 320p. £10.50) has 9 parts: 1. County directory - 2. National directory - 3. International directory - 4. Quick reference - 5. Diary 1992 - 6. Log charts - 7. Features - 8. Books and articles - 9. Reserves and observatories. Index of reserves and observatories. This ed. also contains 'Contents 1981-1991', listing feature articles and irregular reference sections from that period.
Class No: 598.2(058)

Tables & Data Books

[2675]
HOWARD, R. *and* MOORE, A. A Complete checklist of the birds of the world. 2nd ed. London, Academic Press, 1991. xxxiv,622p. £22.50. ISBN: 0123569109.

Arranged by orders and families, Struthioniformes ... Passeriformes, p.1-542. References, by families, p.8-47. Indexes of Latin and English names, p.543-622. As a checklist, goes down to subspecies. Preceded by J.L. Peters' *Checklist of birds of the world* (1931-87. 16v.) which it, in part, updates, but lacks Peters' taxonomic notes, references and distribution notes. *Class No:* 598.2(083)

Nomenclatures

[2676]

BAUMEL, J.J., *and others, eds.* **Nomina anatomica avium:** an annotated anatomical dictionary of birds. Prepared by the International Committee on Avian Anatomical Nomenclature, a committee of the World Association of Veterinary Anatomists. London, Academic Press, 1979. xxv, 637p. illus. £56.50. ISBN: 0120831503.

Detailed annotations, giving homology, synonymy, structure and species variation. Chapters include Osteologia - Myologia - Systema respiratorium - Organa senoria. Taxonomic list gives scientific and common names. Bibliography, p.535-72. General index, p.573-637, embracing items in the list of terms, annotations, illus. or their legends. *Class No:* 598.2(083.72)

[2677]

LODGE, W. Birds: alternative names: a world checklist. London, Blandford, 1991. 208p. £10.95. ISBN: 0713722673.

List of birds which have one or more names in common usage. Excludes local or dialect names. Bibliography, p.195-97. Indexes of English and generic names. Not comprehensive but good value. *Class No:* 598.2(083.72)

Illustrations

[2678]

LYSAGHT, A.W. The Book of birds: five centuries of bird illustration. London, Phaidon Press, 1975 (reprinted 1984). 208p. illus. (incl. col.).

40 of the 142 illus. are in colour. *Class No:* 598.2(084.1)

Worldwide

[2679]

Handbook of the birds of Europe, the Middle East and North Africa: the birds of the Western Palearctic. Cramp, S. *and* Simmons, K.E.L., *and others, eds.* Oxford Univ. Press, 1977-. col. illus., diagrs., tables, maps.

1. *Ostrich to Ducks.* [ix], 722p. 1977. £85. 2. *Hawks to Bustards.* 696p. 1980. £85. 3. *Waders to Gulls.* 1000p. 1983. £85. 4. *Terns to Woodpeckers.* 970p. 1985. £85. 5. *Tyrant flycatchers to Thrushes.* 800p. 1987. £85. 6. *Warblers.* 736p. 1992. £75.

The Western Palearctic includes all Europe, plus the Middle East to Turkey, Iraq and Jordan, and North Africa down to 19°N. The 7v. will describe all 795 species of birds, of which *c.*600 are breeding species. V.1 has 56 colour plates and 350 text figures. The text consists mainly of accounts of species, under orders. 'Mute swan' (*cygnus olor*) occupies, p.372-9, with 1 col. pl., 8 other illus., 2 distribution maps. Field characteristics: Habitat, distribution, population, movement, social pattern and behaviour, voice, breeding, plumages, etc. Plates include illus. of eggs and nest down. References, p.701-14; glossary; index. *Class No:* 598.2(100)

[2680]

LEVER, C. Naturalized birds of the world. Harlow, Essex, Longman, 1987. xx,615p. illus., tables, maps. ISBN: 0582460557.

Describes 133 species 'imported from its natural range to a new country or region either deliberately or accidentally by human agency' (*Preface*) and the effects on the native biota. Maps compare natural distribution with naturalized

....(contd.)

distribution. Bibliography, p.537-89. Geographical index. Index of vertebrate species. The author is the authority on naturalized animals. *Class No:* 598.2(100)

[2681]

SIBLEY, C.G. *and* **MONROE, B.L.,** *Jr.* **Distribution and taxonomy of birds of the world.** New Haven; London, Yale University Press, 1990. xxiv,1111p. maps. $125. ISBN: 0300049692.

Taxonomic listing covering 9,672 living species of birds. Data in each species entry include scientific name; author of the name and year of description; standardized English name; coded world number; a brief description of preferred habitats; detailed description of species distribution; alternative common names. Also contains a world numbers cross-listing, 24 maps, gazetteer, p.875-906, bibliography, p.907-39 and an index which includes common and scientific names. Authoritative. *Class No:* 598.2(100)

Marine Environment

[2682]

HARRISON, P. Seabirds: an identification guide. Rev. ed. Beckenham, Kent, Croom Helm, 1985. 448p. illus. £24.99. ISBN: 0709937873.

A selection of 88 col. plates, with facing captions, is followed by a systematic section (Penguins ... Seaducks), p.200-409. Data on over 800 species include range, plumage, flight, habit, distribution and movement. Distribution maps, p.410-42; bibliography (cross-referenced to main text). p.443-5. 'This book will be a standard work for many years to come' (*Natural history book reviews,* v.8(3), 1986, p.219-20). *Class No:* 598.2(26)

Europe

[2683]

PETERSON, R., *and others.* **A Field guide to the birds of Britain and Europe.** 4th ed. London, Collins, 1983. 384p. £14.99. ISBN: 0002190737.

First published 1954.

77 colour plates, with brief data facing, depict more than 1100 birds of 505 species. Main sequence: Divers ... Buntings. p.36-210. Prelims. 'How to identify books', 'European check-list', 'The ornithological societies'. 362 small distribution maps. Index, p.227-39. Small format. Excludes CIS. 'Basic stock item for all libraries and schools' (*Library world,* March 1966, p.279, on the 1965 ed.). *Class No:* 598.2(4)

[2684]

—Glossarium Europae avium. Jorgensen, H.J. *and* Blackburne, C.I. Copenhagen, Munksgaard, 1941. 192p.

A list of European birds arranged primarily by Latin names. Equivalents in 17 other languages, - Czech, Danish, German, English, Spanish, French, Icelandic, Italian, Hungarian, Dutch, Norwegian, Polish, Portuguese, Russian, Finnish, Swedish and Turkish. The number of species (bar one) tallies with that in R. Peterson's *A Field guide to the birds of Britain and Europe,* which also gives Dutch, French, German and Swedish names. *Class No:* 598.2(4)

[2685]
PFORR, M. *and* LIMBRUNNER, A. The Breeding birds of Europe: a photographic handbook. London, Croom Helm, 1981-82. 2v. col. illus. £27.50, set. ISBN: 0709920482.

1. *Divers to Auks.* 1981. 327, [6]p. 2. *Sandgrouse to crows*, 1982. 394p.

Over 700 high-quality colour photographs by German, Austrian and Danish wildlife photographers, depicting more than 250 species of birds. Data on habitat, food, breeding, biology and migration. Does not deal with identification. Index of photographers. Indexes of English and Latin names of birds. 'Very highly recommended' (*Natural history book reviews*, v.6(3/4), 1983, p.146-7). *Class No:* 598.2(4)

[2686]
—VOOUS, K.H. Nelson's atlas of European birds. London, Nelson, 1960. 284p. illus., maps.

Translated from the Dutch, *Atlas van de europese vogels* (1960).

Distribution maps show the world breeding-range of 419 species of birds breeding in Europe west of the Urals. Basic data on habitat, food and nesting requirements of each species, with notes on movements. The 335 excellent half-tone supporting plates are included not for identification purposes but to show birds in their characteristic habitat. *Class No:* 598.2(4)

[2687]
Tracks & signs of the birds of Britain and Europe: an identification guide. Brown, R., *and others*. London, Christopher Helm, 1992. 232p. illus. (some col.), figs. £14.99. ISBN: 0713635231.

Covers resident, visiting and vagrant birds of Europe. Includes their feeding and behavioural signs, droppings, pellets, feathers and skulls. Over 1,000 illustrations, including 47 colour plates. Brief bibliography and a list of European bird protection and study organizations. Useful for both the amateur and scientist. *Class No:* 598.2(4)

Great Britain

[2688]
BANNERMAN, D.A. The Birds of the British Isles. Edinburgh & London, Oliver & Boyd, 1952-63. 12v. illus. (pl.), maps.

Systematically arranged. Entry for each bird species includes notes on identification, occurrence in Britain, distribution, habits and migration, breeding and nesting habits. The accurately coloured plates are a feature, each volume containing about 30 such plates. Footnote references. V.12 includes a short comparative list of Celtic bird-names of the British Isles (p.405-23), and an index of scientific names in the 12v. A classic work. *Class No:* 598.2(410)

[2689]
—WITHERBY, H.F., *ed*. The Handbook of British birds. London, Witherby, 1938-41. 5v. illus., maps.

Arranged by families. For each species details concern habitat, characteristics, distribution (in Britain and abroad), and descriptions, plus a col. illus. 520 numbered entries. V.5 includes general indexes of English and scientific names in the 5v., and each volume has an index of English names and a glossary of terms. 'A standard work of reference for British and, indeed, European ornithologists' (*Nature*, v.196, 17 November 1962, p.610). *Class No:* 598.2(410)

[2690]
Checklist of the birds of Great Britain and Ireland. Knox, A.G., *comp*. 6th ed. Tring, Herts., British Ornithologists' Union, 1992. 50p. ISBN: 0907446159.

The standard list, referred to in other bird books; *i.e.*, birds are cited by their check-list number on this work. Systematic list, p.4-24. 2 appendices, *e.g.* 2. List of species. Indexes to scientific and English names. *Class No:* 598.2(410)

[2691]
MULLENS, W.H. *and* SWANN, H.K. A Bibliography of British ornithology, from the earliest times to the end of 1912, including biographical accounts of the principal writers and bibliographies of their published works. London, Macmillan, 1916-17. xv, 691p. Supplement, by H.K. Swann, 1923. xvi, 42p.

Under authors A-Z, only those authors being included who have published separate works on ornithology. The biographies carry notes on published works. Also, lists of works cited, - bibliographies, biographical works, and periodicals. A valuable and informative bibliography. *Class No:* 598.2(410)

[2692]
SHARROCK, J.T.R., *comp*. The Atlas of breeding birds in Britain and Ireland. Berkhamsted, Herts., Poyser, 1976. 477p. maps. £26. ISBN: 0856610186.

Location maps, with text facing, plus references, p.28-452. Past distribution maps, p.454-66. List of former breeding species; additional species. Names of plants and animals mentioned in text. Index of bird names. Fieldwork was done by 10-15,000 observers, for Biological Records Centre of the Natural Environment Research Council, the British Council for Ornithology and the Irish Wild Birds Conservancy. *Class No:* 598.2(410)

[2693]
SWANN, H.K. A Dictionary of English and folk-names of British birds, with their history, meaning and first usage; and their folk-lore legends, etc. relating to the more familiar species. London, Witherby, 1913. (Reprinted London, EP Publishing, Ltd., 1977). xii, 266p.

About 5,000 names of birds. Attempt to combine 'the English "book"-names from past authors, giving the history and first usage of the "accepted" names of the species, and also the provincial, local or dialect names in use now or formerly in the British Isles, indicating the locality and meaning where possible' (*Preface*). 'Raven': nearly 2p. Bibliography, p.ix-xii. *Class No:* 598.2(410)

[2694]
—LOCKWOOD, W.B. The Oxford book of British bird names. Oxford, University Press, 1984. 184p. ISBN: 0192141554.

Contains about 1,500 entries, A/Z, relating to 257 species. ('Throstle': 1 column). Cross-references. Introduction includes sections on linguistic evolution, sources. Bibliography, p.17-20. *Class No:* 598.2(410)

USSR

[2695]
FLINT, V.E., *and others*. A Field guide to birds of the USSR, including Eastern Europe and Central Asia. Translated from the Russian by N. Bourso-Leland. Princeton, N.J., Princeton Univ. Press, 1984. 353p. illus.

....(contd.)

(incl. col.), tables, maps. $85; $29.95. ISBN: 0691082448.

Originally published in the USSR, 1968; updated and revised.

48 colour plates, plus 71 line-drawings, illustrate 654 of the USSR 728 species. The sections for each species state field marks, habits, range and distribution. The maps are hard to read; only breeding ranges are shown. 'A landmark publication, opening the door to the vast avifauna heretofore unavailable to the Western World' (*Choice*, v.22(1), September 1984, p.62). *Class No:* 598.2(47)

Indian Subcontinent States

[2696]
Handbook of the birds of India and Pakistan. Ali, Salim *and* Ripley, S.D., *eds.* 2nd ed. Delhi, Oxford Univ. Press, 1978-. illus. (incl. col.), maps.

1. *Divers to hawks.* 2nd ed. 1979. 440p. £42.50 2. *Megapodes to crab plover.* 2nd ed. 1980. 362p. £42.50. 3. *Stone curlews to owls.* 2nd ed. 1981. 344p. £42.50. 4. *Frogmouths to pittas.* 2nd ed. 1983. 284p., £42.50. 5. *Larks to the grey hypocolius.* 2nd ed. 1987. 278p. £37.50. 6. *Cuckoo-shrikes.* 7. *Laughing thrushes to the mangrove whistler.* 8. *Warblers to redstarts.* 9. *Robins to wagtails.* 10. *Flowerpeckers to Buntings.*

Geographical coverage extends to Bangladesh, Nepal, Bhutan and Sri Lanka. 2nd editions of v.6-10 are in preparation.

The one volume *Compact handbook* (2nd ed. 1987. $98) comprises v.1-4 of the 2nd ed. and v.5-10 of the 1st ed. Also, *Concise handbook of the birds of India and Pakistan* (Oxford, University Press, 1989. $39.95).
Class No: 598.2(54.)

Africa

[2697]
BROWN, L.H. The Birds of Africa. Urban, E.K., *and others, eds.* London, New York, etc., Academic Press, 1982-. illus. (incl. col.), diagrs., maps. ISBN: 0121873016.

V.1 (xii, [1], 521p.) covers 10 orders of birds. Strutheoniformes ... Falconiformes, - the first of 4v. Data: range and status, description, field characteristics, voice, general behaviour, food, breeding, biology. Bibliography, p.479-507: 1. General and regional references; 2. References for each family; 3. Sources of sound recordings. Indexes of scientific names, English names, French names. Quarto format.

V.2 (1986. 552p.) covers orders Galliformes ... Columbiformes. *Class No:* 598.2(6)

Africa—Southern

[2698]
PROZESKY, O.P.M. A Field guide to the birds of Southern Africa. London, Collins, 1971. 350p. illus., maps. £9.95. ISBN: 0002120267.

Emphasis on bird song and bird calls as aids to identification; no information on breeding habits. Clear, concise and well-printed text. According to *TSL* (no.3600, 26 February 1971, p.254), the illustrations fall short of expectations, and the colours are not always true to the written description. *Class No:* 598.2(68)

America — North

[2699]
The Audubon Society encyclopedia of North American birds. Terres, J.K. New York, Knopf, 1980. 1109p. illus. (incl. col.). $75. ISBN: 0394466519.

A compendium on the 847 species of birds recorded in the USA and Canada. About 6,000 entries, A-Z, on the life histories of the bird species, biographical notes on American ornithologists, articles on avian biology (*e.g.* courtship, migration, song), definitions of terms, organizations. Full cross-references. Bibliography. 'By far the most ambitious and extensive book of its kind' (*Reference books for small and medium-sized libraries* 4th ed. ALA, 1984, item 1064). *Class No:* 598.2(71+73)

[2700]
PALMER, R.S. Handbook of North American birds. New Haven & London, Yale Univ. Press, 1962-88. 5v. illus. ISBN: 0300008147, v.1; 0300020783, v.2/3; 0300040598, v.4, pt.1; 0300040601, v.5, pt.2.

Sponsored by the American Ornithologists' Union.

1. *Loons through flamingoes.* 1962. 2/3. *Waterfowl.* 1976. (xi, 521p.; vii, 560p.). 4/5. *Condors & raptors.* 1988. (448p. $45; 400p. $45).

V.2-3 have 32 contributors. 'Atlantic Canadian goose' (v.2. p.190-2) has a distribution map and data on measurements, weight, breeds, nesting, migration, captive geese, introduction abroad, etc. V.3 contains index to v.2-3. Excellent colour plates and line-drawings.
Class No: 598.2(71+73)

[2701]
—GRUSON, E.S. Words for birds: a lexicon of North American birds, with biographical notes. New York, Quadrangle Books, 1972. 305p.

An ornithological, etymological, biographical and historical guide to 800 American birds.
Class No: 598.2(71/73)

America—South

[2702]
The Birds of South America. Volume 1: The oscine passerines. Ridgely, R.S., *and others.* Oxford, University Press, 1989. xvi, 520p. illus (col.pl.), maps. £50. ISBN: 0198572174.

Covers 750 birds in this group. General chapters on habitat, biogeography, migration and conservation are followed by 31 colour plates which group similar genera together. Text opposite the plates refer the reader to entries in the second part of the book which give details of identification, similar species, habitat and behaviour, range, and maps showing distribution. Bibliography, p.496-502. Indexes to English and scientific names. First volume in a projected four. *Class No:* 598.2(8)

[2703]
BLAKE, E.R. Manual of neo-tropical birds. Chicago, Ill., Univ. of Chicago Press, 1977-. v.1-. illus., maps. ISBN: 0226056414.

Planned in 4v. 1. *Spheniscidae (Penguins) to Laridae (Gulls and allies).* 640p. $100.

Aims to provide descriptions of all Central and South American subspecies of birds. Literature references. Favourably reviewed in *Library journal* (v.102(10), 1 June 1977, p.1265). *Class No:* 598.2(8)

Australasia & Oceania

[2704]
Handbook of Australian, New Zealand and Antarctic birds.
Melbourne, Oxford University Press, 1990-. illus. (some col.), tables, maps. ISBN: 0195530683, v.1.

V.1: *Ratites to ducks* (2 pts.: A. *Ratites to petrels;* B. *Australian pelican to ducks*). 1400p.

First volume in a projected series of five to be published under the auspices of the Royal Australasian Ornithologists Union. Species accounts give data on field identification; habitat; distribution; food; movements; population; social organization; social behaviour; breeding; voice; plumage; breeding. A bibliography is appended to each species account. Maps show the distribution of 162 breeding birds. 5 appendices (*e.g.* 'Aboriginal names'). Indexes of scientific and English names. Excellent colour plates.
Class No: 598.2(9)

[2705]
—The Birds of Australia: a book of identification; 758 birds in colour. Simpson, K., *ed.* South Yarrow, Victoria, Lloyd O'Neil Pty.; Wakefield, NH, Hollowbrook, 1984. 352p. illus. (incl. col.). $45. ISBN: 0855504927; 088072059x, US ed.

Contributors: 44, plus Australian Wader Bird Group. Parts: Key to families - Field information (p.16-271) - The handbook. All 758 Australian species are illustrated on 128 colour plates, with range-maps facing. Numerous fine line-drawings. Appendices contain valuable general, ornithological information, *e.g.* breeding seasons. 'Highly recommended', although too large and heavy for a field guide (*Choice,* v.22(8), April 1985, p.1136).
Class No: 598.2(94)

Wildfowl

[2706]
DELACOUR, J. The Waterfowl of the world. London, Country Life, 1954-64. 4v. illus. (incl. col. pl.), tables.

1. *The magpie goose, whistling ducks, swans and geese, sheldgeese and shelducks.* 1954. 2. *The dabbling ducks.* 1956. 3. *Elders, pochards, perching ducks, scotens, golden-eye and mergansers.* 1967. 4. *General habitats ...* 1964.

V.4 (364p.) has 5 contributors, rounding off v.1-3 with chapters of general interest (1. General habits; 2. The reproduction cycle (with 8p. of bibliography); 3. Ecology; 4. Distribution and special relationships; 5. Fowling; 6. Conservation and management; 7. Aviculture; 8. Domestic waterfowl; 9. The anatomy of the waterfowl; 10. Fossil anseriformes - 11. Corrections and additions). Index to chapters 1-9, p.355-64. A definitive work, with well-documented chapters. *Class No:* 598.412

[2707]
MADGE, S. and BURN, H. Wildfowl: an identification guide to the ducks, geese and swans of the world. London, Christopher Helm, 1988. 298p. col. illus., maps. £22.99. ISBN: 0747022011.

47 colour drawings, depicting over 700 individual birds, with range maps accompanying. Text referring to the bird is in a separate section. Data include field identification; voice; description; measurements; habitat; population. 'This is the first field guide to all 155 of the world's waterfowl in virtually every plumage' (*Library journal,* 15 April 1988, p.40). *Class No:* 598.412

[2708]
—SOOTHILL, E. *and* WHITEBREAD, P. Wildfowl of the world. Poole, Dorset, Blandford Press, 1978. iii, viii, 297p. (mainly col.), maps. £14.99. ISBN: 0713721103.

Descriptive data and distribution maps for 128 species of wildfowl. Bibliography, p.292-3; index of Latin and English names of wildfowl. A bonus is the directory of wildfowl collections worldwide. *Class No:* 598.412

Pigeons & Doves

[2709]
GOODWIN, D. Pigeons and doves of the world. 3rd ed. London, British Museum (Natural History), with Ithaca, New York, Cornell Univ. Press, 1983. 363p. illus. (incl. col. pl.), maps. £25. ISBN: 0565008471.

First published 1967 (vi, 446p).

10 chapters (1. Nomenclature, genera, species, sub-species, varieties and breeds ... 3. Coloration and plumage ... 5. Maintenance behaviour ... 7. Display and social behaviour). 24 sections, each with readings. Index of English names. Index of scientific names. *Class No:* 598.65

Eagles & Birds of Prey

[2710]
Birds of prey. Newton, I., *ed.* London, Merehurst Press, 1990. 240p. illus. (col.). £25. ISBN: 1853911313.

3 sections: Raptors of the world - Raptor biology - Relations with man. Checklist of living diurnal raptors, p.226-29. Lavishly illustrated. *Class No:* 598.91

[2711]
BROWN, L. *and* AMADON, D. Eagles, hawks and falcons of the world. London, Country Life Books, 1965. 2v. illus. (incl. col.), maps.

V.1 has 18 chapters on taxonomy, morphology, flight, hunting methods, migration, breeding, etc. V.2 has chapters 19-21, on the genera and species, nomenclature and taxonomy, with grouped supplementary bibliography and 94 maps. Index to v.1-2, p.933-45. 414 plates in all.
Class No: 598.91

[2712]
BURTON, P. Birds of prey of the world. Limpsfield, Surrey, Dragon's World, 1989. 208p. illus. (col.). £18.95. ISBN: 1850280851.

Arranged in 5 sections: 1. New World vultures - 2. The osprey - 3. Kites, vultures, eagles and hawks - 4. The secretary bird - 5. Caracaras and falcons. Distribution maps, and systematic list and measurements, are apart from the text and illustration. Brief bibliography. Index. 256 excellent illustrations. *Class No:* 598.91

Mammals

CD-ROM

[2713]
Mammals of the world: a multimedia encyclopedia. Washington, D.C., National Geographic, 1990. $99.

Combines text, sound illustrations and video. Data on more than 200 animals. Each article contains essays and fact

.... (contd.)

boxes equivalent to *c*.600 pages of text. More than 700 colour photographs, 150 maps and 155 animal sounds. Easy to use and reasonably priced. *Class No:* 599(003.40)

Encyclopaedias

[2714]
Grzimek's encyclopedia of mammals. Grzimek, B., *ed.* New York, McGraw-Hill Publishing Company, 1990. 5v. illus. maps. $500. ISBN: 0079095089, set.

Based on *Grzimek's animal life encyclopedia (q.v.).*

More than 200 contributors. Entries in each volume are arranged by order and suborder. Each entry has an introductory section, with a map showing geographical distribution, followed by chapters discussing specific families or species. Standardized summary charts include data on scientific name; number of species and sub-species; size; life span; food; habitat; enemies. Outstanding feature is the 3,500 colour photographs which have been laser-scanned. Bibliography and index at the end of each volume, although there is no index for the complete set. Recommended for school, public and academic libraries. *Class No:* 599(031)

Handbooks & Manuals

[2715]
The Encyclopedia of mammals. Macdonald, D., *ed.* London, Allen & Unwin; New York, Facts on File, 1984. 2v. (479;480p.). col. illus.; col. diagrs. £30ea. ISBN: 004500028x, v.1; 0045000298, v.2.

About 200 contributors in all. 1: The carnivores (7 families: the cat family ... the hyena family), sea mammals, primates, etc. V.2 covers herbivores, insectivores, bats and marsupials. Each quarto volume carries a grouped bibliography, glossary and analytical index. The entry 'Squirrel': p.612-27; 25 col. illus.; 2 col. diagrs. 1,200 col. illus., superbly presented. 'Its main market is likely to be naturalists, yet the technical detail should appeal to students and biologists' (*British book news,* February 1985, p.100, on v.2). *Class No:* 599(035)

[2716]
YOUNG, J.Z. The Life of mammals: their anatomy and physiology. 2nd ed., with assistance of M.J. Hobbs. London, Oxford Univ. Press, 1975. xv, 528p. £25.50. ISBN: 0198571569.

First published 1957.

53 chapters (6. The framework of the body - 7. Muscles ... 11. The head of mammals ... 15. Respiration ... 21. The heart and circulation ... 30. The automatic nervous system ... 38. Receptor organs ... 48. The pituitary gland ... 53. The development of mammals). Bibliography (by chapters), p.494-507. Author and subject index. *Class No:* 599(035)

Tables & Data Books

[2717]
CORBET, G.B. *and* **HILL, J.E. A World list of mammalian species.** 3rd ed. Oxford, University Press; Natural History Museum Publications, 1991. viii,243p. figs. £30. ISBN: 0198540175.

First published 1980.

Systematic list of 4,327 species. Data: Latin name;

.... (contd.)

common name; distribution. Bibliography (general works; geographical sources; taxonomic sources), p.213-31. Detailed index, p.232-43. *Class No:* 599(083)

[2718]
HONACKI, J.H., *and others, comps. & eds.* **Mammal species of the world:** a taxonomic and geographic reference. With the assistance and cooperation of the American Society of Mammalogists. Lawrence, Kansas, Association of Systematics Collections, in conjunction with Allen Press, 1981. 694p.

Compiled for the Parties to the Convention on International Trade in Endangered Species of Wild Fauna and Flora.

Lists *c*.4,170 specis, in taxonomic order. Data: scientific name and authority; type locality; geographical distribution; legal status of endangered species and controlled species. Bibliography of cited sources. Taxonomic index. Aims to provide a standard reference to mammalian nomenclature. *Class No:* 599(083)

[2719]
—BURTON, M. Systematic dictionary of mammals of the world. 2nd ed. London, Museum Press, Ltd., 1965. 307p. illus.

First published 1962.

Provides data: general characteristics; habits; habitats; food; breeding; range; longevity; special comments; other statistics. Detailed analytical index, p.275-307. Careful drawings. *Class No:* 599(083)

Nomenclatures

[2720]
GOTCH, A.P. Mammals - their Latin names explained: a guide to animal classification. Poole, Dorset, Blandford Press, 1979. 271p. illus. charts. ISBN: 0713709391.

Part 1: The animal kingdom; part 2: The mammal orders. Bibliography, p.231. General index; index of English names; index of Latin ones. Selects over 1,000 of the *c*.4,300 known species of mammals. *Class No:* 599(083.72)

Worldwide

[2721]
BURTON, J.A. *and* **BURTON, V.G. Collins' guide to the rare mammals of the world.** London, Collins, 1987. 240p. col. illus. ISBN: 0002191725.

Sections: Monotremes - Marsupials - Bats - Primates ... Rodents - Carnivora - Seals and sea lions - Elephants - Perissodactyls - Sea cows - Artiodactyls - Cetaceans. Data: description; characteristics (habitat, feeding, breeding); distribution and population; reasons for scarcity; location and captive populations. Bibliography, p.232-3. Index, p.233-40. 98 col. plates illustrate more than 440 species. *Class No:* 599(100)

[2722]
LEVER, C. Naturalized mammals of the world. London, Longman, 1985. xvii, 487p. illus., tables, maps. ISBN: 0582460565.

Describes 66 species 'imported from its natural range to a new country either deliberately or accidentally by human agency' (*Preface*). 9 sections: 1. Marsupiala - 2. Edentata - 3. Insectivora - 4. Primates - 5. Carnivora - 6. Perissodactyla - 7. Artiodactyla - 8. Rodentia - 9.

.... (contd.)

Lagomorpha. Bibliography, p.441-74. Index of vertebrata species, p.475-87. 26 appended country tables, p.413-40. 'Very thorough' (*New scientist,* 2 January 1986, p.44-45). *Class No:* 599(100)

[2723]
NOWAK, R.M. Walker's mammals of the world. 5th ed. Baltimore, Md., Johns Hopkins Press, 1991. 2v. (1629,lxiii), illus., charts, tables. $89.95. ISBN: 080183970x.

First published 1964.

V.1: Monotremata ... Rodentia; v.2: Rodentia ... Artiodactyla. V.2 includes a bibliography of literature cited, p.1501-1614 (A-Z authors), - a major contribution, and greatly expanded from the 4th ed. V.1 has an analytical index of orders, families and genera, with bold type for scientific names. Numerous black-and-white illus. (*c.*1 per page), substantial revision of a classic. Of the 4th ed., 'This monumental work is indispensable' (*Library journal,* 15 April 1984, p.784). *Class No:* 599(100)

Tropics

[2724]
EMMONS, L.H. Neotropical rainforest mammals: a field guide. Chicago and London, University of Chicago Press, 1990. xiv,281p. illus. (some col.), figs., maps. £35.95; £15.95. ISBN: 0226207161.

Includes the lowland rainforest mammals found in Central and South America at elevations below 1,000 metres - *c.*500 species. These consist of opossums, anteaters, sloths and armadillos, bats, monkeys, carnivores, dolphins, tapirs, peccaries and deer, manatees, rodents and rabbits. Data for each species include means of identification, variation, similar species, sounds, natural history, geographic range, status, map showing distribution, and references. 6 appendices: A. Glossary ... D. Tracks of large mammals ... F. Checklist and index of scientific names. 29 colour plates. Index of genera and common names, p.275-81. *Class No:* 599(213)

[2725]
Mammals of the neotropics. Eisenberg, J.F. *and* Redford, K.A. Chicago and London, University of Chicago Press, 1989-92. 2v. illus., tables, maps. £67.95, v.1; £75.95, v.2. ISBN: 0226195406, v.1; 0226706818, v.2.

V.1. *The northern neotropics: Panama, Colombia, Venezuela, Guyana, Suriname, French Guiana.* x,449p. V.2. *The southern cone: Chile, Argentina, Uruguay, Paraguay.* 430p.

V.1 has 16 well-documented sections, v.2, 14. Data include physical description; summary of range; habitat. Colour and black-and-white plates. Each volume has an index of scientific and common names. A third volume covering Brazil, Bolivia, Peru and Ecuador is planned. *Class No:* 599(213)

Marine Environment

[2726]
FOOD AND AGRICULTURE ORGANIZATION. Mammals in the seas. Rome, FAO, 1978-82. 4v.

1. *Report of the FAO ACMRR Working Party on Marine Animals.* 1978. xxv, 264p. 2. *Principal species summaries and report on sirenians.* 1979. xiii, 151p. 3. *General papers*

.... (contd.)

and large cetaceans. 1981. x, 504p. 4. *Small cetaceans, seals, sirenians and otters.* 1982. x, 931p. $85.

V.4 includes a paper 'The status of seals in the United Kingdom' (p.253-265), summaries in French and Spanish preceding. (Sections: Identification; distribution and movements; status of stocks; vital parameters; food habits and relations to fisheries; economic value; conservation legislation; conclusions. 33 references). A detailed study. *Class No:* 599(26)

[2727]
RIDGWAY, S.H. *and* **HARRISON, R.J. Handbook of marine mammals.** New York & London, Academic Press, 1981-89. 4v. illus., tables. ISBN: 0125885016, v.1; 0125885024, v.2; 0125885032, v.3; 0125885040, v.4.

1. *The walrus, sea lions, fur seals and sea otter.* 1981. xiv, 235p. 2. *Seals.* 1981. xv, 359p. 3. *The sirenians and baleen whales.* 1985. xviii, 362p. 4. *River dolphins and the larger toothed whales.* 1989. 442p. $105.

Documented contributions by various hands, highlighting life history, population dynamics, reproduction and diseases. 'The *Handbook* should appeal to all those seeking an introduction to these mammals and will also serve as a useful reference work for marine mammalogists' (*British book news,* February 1982, p.102). Well written. *Class No:* 599(26)

Europe

[2728]
CORBET, G. *and* **OVENDEN, D. The Mammals of Britain and Europe.** London, Collins, 1980. 253p. illus. ISBN: 000219774x.

'Includes all species of mammals that occur in a wild state in Europe West of the USSR, including Spitzbergen and Iceland, and in the northeastern Atlantic north of 35 degrees north' (*Preface*). Excludes extinct species of the region. Colour plates and drawings of 190 species are followed by 15 sections: 1. Tracks and signs - 2. Skulls and teeth - 3. Antlers and horns - 4. Marsupials - 5. Insectivores - 6. Lagomorpha - 7. Rodents - 8. Primates - 9. Carnivores - 10. Pinnipedes - 11. Odd-toed ungulates - 12. Even-toed ungulates - 13. Cetaceans - 14. Checklist of European mammals - 15. Mammals in Britain and Ireland (plus distribution maps of British and Irish species). *Class No:* 599(4)

Great Britain

[2729]
CORBET, G.B. *and* **HARRIS, S. The Handbook of British mammals.** 3rd ed. Oxford, Blackwell Scientific Publications for the Mammal Society, 1991. xiv,588p. illus., maps. £49.50. ISBN: 0632016914.

First published 1964 (edited by H.N. Southern).

An extensive revision of the previous edition. A systematic account of the 98 species known to be feral in the British Isles and surrounding seas. Data include: recognition; sign; description; relationships; measurements; variation; distribution; habitat; feeding; population. 3 appendices: 1. Extinct species - 2. Ephemeral introductions and escapes - 3. Sound recordings. Items in the bibliographies appended to each chapter total 3,186. Glossary, list of abbreviations and

....(contd.)

index. 'It will definitely be the standard work for the next two decades' (*Reference reviews,* v.5(2), 1991, p.24). *Class No:* 599(410)

[2730]
MATTHEWS, L.H. Mammals in the British Isles. London, Collins, 1982. 208p. (*New naturalist library, no.68.*) ISBN: 0002197383.

Devotes the first third of the text on geological, climate and human influence, shaping the range of British mammal habitat. Ecology and behaviour are then considered. 'It is an excellent introduction to the natural history of British mammals for anyone, amateur or professional' (*Nature,* v.300, 9-15 December 1982, p.556). *Class No:* 599(410)

Africa
[2731]
HALTENORTH, T. *and* **DILLER, H. A Field guide to the mammals of Africa, including Madagascar.**
Hayman, R.W., *trans.* London, Collins, 1980. 400p. col. illus. maps. £14.99. ISBN: 0002197782.

Translated from *Säugetiere Afrikas and Madagaskars*.

Describes *c*.324 species, including the largest and middle-sized, but omitting small-sized species (*e.g.* bats). Introduction outlines geological and climatological history of Africa, plus the history of mammalian fauna of Africa and Madagascar. Appended section on introduced fauna. Bibliography, p.1-2.

Superior illus. 'Should become a basic reference for any professional or amateur interested in the wildlife of Africa' (*Natural history book reviews,* v.5(3/4), 1982, p.142-3). *Class No:* 599(6)

Africa—Southern
[2732]
SMITHERS, R.H.N. The Mammals of the Southern African subregion. 2nd ed. University of Pretoria, 1990. xxxii,769p. illus., tables, maps. ISBN: 0869798022.

First published 1983.

340 species of mammals, including sea mammals (whales and seals) are described. Data: colloquial name; taxonomic notes, description, distribution, habitat, habits, food, reproduction, skull. Coverage: South Africa, Namibia, Botswana, Zimbabwe south of the river Zambezi, and adjacent oceans. Of the previous ed., 'A mammoth undertaking' (*Nature,* v.312, 13 December 1985, p.677). *Class No:* 599(68)

America—North & Central
[2733]
HALL, E.R. The Mammals of North America. 2nd ed. New York, Wiley, 1981. 2v. (xv, 690p. + 679p.). illus., diagrs., maps. ISBN: 0471054437; 0471054445.

First published 1959.

Details of 3,607 subspecies and monotypic species of native mammals of the post-Columbian period (1492-1977), in Greenland, Canada, USA, central America (down to and including Panama), the Greater and Lesser Antilles south to, and including, Grenada. Data on each: measurements, description, comparison with other members of the species,

....(contd.)

citations to original descriptions, distribution, etc. Each volume contains the same 80p. index to technical names and 10p. index to vernacular names. *Class No:* 599(71/73)

Australia
[2734]
STRAHAN, R., *ed.* **The Complete book of Australian mammals.** London, Angus & Robertson, 1984. 529p. col. illus. £37.50. ISBN: 0207144540.

107 specialist contributors. A feature: colour photographs of 251 of the 273 species included with illus. of almost all the 129 placental mammals. Bibliographical references for each species. 'Sets a standard, although it does not cover critical identification requiring internal characteristics' (*Nature,* v.312, 13 December 1984, p.677). *Class No:* 599(94)

Whales, etc.
[2735]
MINASIAN, S.M., *and others.* **The World's whales:** the complete illustrated guide. Washington, Smithsonian Books, 1984. 224p. col. illus. ISBN: 0895990148.

Covers all 70 recognized cetacean species, - whales, dolphins and porpoises. Arranged by suborder, then family, genus and species. Data include name and date of first description and year of discovery. Introductory chapter offers basic facts about cetacean biology. Glossary. List of selected readings. A feature: 'nearly 160 colour photographs of most of the species swimming in the open seas' (*Science & technology libraries,* 1985/6, v.8(1/2), p.177-8). *Class No:* 599.5

[2736]
—Whales and dolphins. Martin, A.R., *and others.* London, Salamander Books, 1990. 192p. illus. (col.), maps. £6.95. ISBN: 086101488x.

Covers 76 species of whales, dophins and porpoises. The first section gives a general overview while the second looks at each species. Information includes indication of size; distribution maps; classification; description; habitat; food and feeding; behaviour; life history; world population; man's influence. At least one colour illustration of each species. Suitable for school and public libraries. *Class No:* 599.5

[2737]
WATSON, L. Whales of the world. 2nd ed. London, etc., Hutchinson, 1985. 302p. illus. (col.), charts, maps. £12.99. ISBN: 0091597110.

First ed. published as *Sea guide to whales of the world,* 1982.

7 chapters: 1. Cetology and cetologists - 2. The origin and evolution of whales - 3. Description - 4. Stranded cetaceans - 5. Natural history - 6. Status [population] - 7. Distribution. 3 appendices (3. Biography, p.285-6: 100 selected 2-4 line notes). Bibliography, 287-97 (A/Z; including many periodical articles). 76 species are listed, p.10-11; 96 species illus., 167 col. diagrs., 82 distribution maps (for dolphins and purpoises, as well as whales). Non-analytical index, p.298-302. 'An outstanding work' (*Library journal,* 1 February 1982, p.249). *Class No:* 599.5

[2738]

—FRASER, F.C. British whales, dolphins and porpoises: a guide to the identification and reporting of standard whales, dolphins and porpoises on the British coasts. 5th ed. London, British Museum (Natural History), 1976. xi, 34p. illus. £1.50. ISBN: 0565055496.

First published 1950.

Has notes on 6 whalebone whales and 18 toothed whales, one per page. Data on each: length, colour, form, teeth, general remarks, plus illus. 31 illus. in all. *Class No:* 599.5

Ungulates

[2739]

MOCHI, U. *and* CARTER, T.D. **Hoofed mammals of the world.** New York, Scribner, 1971. xxi, 268p. illus., maps. ISBN: 0684123827.

First published 1953.

Describes 300 species and subspecies of ungulates. Bibliography, p.257-8 (A-Z authors). Index of animals illustrated. Line drawings are mostly silhouettes. A few small distribution maps. *Class No:* 599.6

Horses

[2740]

CLABBY, J. **The Natural history of the horse.** London, Weidenfeld and Nicolson, 1976. 116p. illus. (*The World naturalist.*) ISBN: 0297770497.

15 chapters (1. Perspectives - 2. The ancestors - 3. The making of *Equus* - 4. The horses of the Ice Age - 5. Wild horses - 6. Asiatic wild horses - 7. African wild asses and donkeys - 8. Zebras - 9. Hybrids - 10. The origin of the domestic horse - 11. Some breeds of Europe and America - 12. Some breeds of South America - 13. Social behaviour - 14. Some aspects of horse breeding - 15. The horse crafts. Bibliography (67 items), p.105-7. Analytical index, p.109-16. Author, - Director of the Army Veterinary and Remount Services. *Class No:* 599.723

Deer

[2741]

WHITEHEAD, G. K. **Deer of the world.** London, Constable, 1972. xii, 194p. illus. (incl. pl.), maps.

Concentrates on taxonomy and the worldwide distribution of the family. 32 plates. *Class No:* 599.735

[2742]

—CHAPLIN, R.E. Deer. Poole, Dorset, Blandford Press, 1977. 218p. illus.

Supplementary to Whitehead. 'An excellent introduction to the general biology of ruminants' (*Journal of natural history,* v.12(2), March/April 1978, p.235). *Class No:* 599.735

Carnivores

[2743]

EWER, R.F. **The Carnivores.** London, Weidenfeld & Nicolson, 1973. xv, 494p. illus., tables. (*World naturalist series.*)

10 chapters (2. The skeleton; 3. Anatomy of the soft parts; 4. The special senses; 5. Food and food finding; 6. Signals and social organization; 7. Social organization and living space; 8. Reproduction; 9. Fossil relatives; 10. Classification

.... *(contd.)*

and distribution of the living spaces). References (A-Z authors), p.416-53. Author index; species and analytical subject index. *Class No:* 599.74

Seals

[2744]

KING, J.E. **Seals of the world.** 2nd ed. Oxford Univ. Press, and London, British Museum (Natural History), 1983. 240p. illus. (incl. col.), diagrs., maps. £9.95. ISBN: 0565008684.

First published 1964.

Introduction; Section 1: The diversity of pinnipeds (including fossils); Section 2: Pinniped biology. Section 3: Appendices (Origin of scientific names. Geographical index. References, p.217-35; further reading, p.236). Detailed analytical index, p.237-40. 38 colour plates illustrate nearly every species. 'A completely new, much enlarged and greatly improved book, and like its predecessor it is destined to become a successful standard text for all marine mammalogists' (*Natural history book reviews,* v.7(3), 1984, p.141-2). *Class No:* 599.745

Primates

[2745]

Current primate references. Seattle, Univ. of Washington, Primate Information Center, 1973-. 12pa. $75. ISSN: 05904102.

Contains *c.*150 references per issue. 37 sections: Bacterial infections ... Viral infections. References include names and addresses of authors, for contacting. Book reviews. Primate index. Author index. *Class No:* 599.8

[2746]

NAPIER, J.R. *and* NAPIER, P.H. **A Handbook of living primates.** London & New York, Academic Press, 1967. xiv, 456p. illus. (pl.), tables.

Three parts: 1. Functional morphology of primates - 2. Profiles of primate genera - 3. Supplementary and comparative data (*e.g.* taxonomy and nomenclature; habitats of primates; limbs and locomotion). References (A/Z authors; *c.*900 items), p.417-46. 113 plates. Index to illus. and plates, p.447-56. 'A work of great scholarship' (*Science journal,* v.4(3), March 1969, p.92). *Class No:* 599.8

[2747]

—NAPIER, J.R. *and* NAPIER, P.H. The Natural history of the primates. London, British Museum (Natural History), 1985. 200p. illus., (incl. col.), charts, maps. £15. ISBN: 0565008706.

By comparison with the *Handbook,* lavishly illustrated in a larger format and clearly written. 6 chapters (1. What are primates? ... 5. Profiles of primates - 6. Human evolution). Glossary, p.187-9; Further reading, p.190; References, p.191-5 (A-Z authors). Analytical index, p.196-200. But some definitions are open to questions and citation of sources could be more numerous. 'A book which will better serve the keen amateur rather than the student or professional' (*Nature,* v.316, 18 July 1985, p.323). *Class No:* 599.8

[2748]
WOLFHEIM, J.W. Primates of the world: distribution, abundance and conservation. Seattle, Washington, Univ. of Washington Press, 1983. 832p. illus. $57.50. ISBN: 0295958995.

Assesses the status in the wild of about 150 primate species and discusses the future in terms of their habitats, distributions and, ultimately their population. Over 1,000 references to the literature. 'A well-designed framework ... An excellent job' (*Nature,* v.312, 13 December 1984, p.676). *Class No:* 599.8

Apes & Monkeys

[2749]
KAVANAGH, M. A Complete guide to monkeys, apes and other primates. New York, Viking Press, 1984. 224p. col. illus., maps. ISBN: 0670435430.

7 sections: 1. Primates in perspective; 2. The prosimians; 3. The "almost monkeys"; 3. The odd ones and the dry-nosed prosimians; 4. The monkeys of the New World; 5. The monkeys of the Old World; 6. The tailless primates: apes and man; 7. Prospects for the primates. Outstanding colour illus. Distribution maps. Appended classification of the order primates; glossary; short bibliography; subject index. Claims to have a set of photographs covering every living genus of primates. *Class No:* 599.82

[2750]
TUTTLE, R.H. Apes of the world: their social behavior, communication, mentality and ecology. Park Ridge, N.J., Noyes, 1987. 421p. illus, tables. $55. ISBN: 0815511043.

Data provide readable, succinct accounts of the taxonomy, locomotion, feeding, nesting, tool use, cognition and communication, sociobiology. Extensive bibliography. 'Tuttle has written a scholarly work that could serve as a standard reference for many years'. (*Nature,* v.328, August 1987, p.678). *Class No:* 599.82

6 Technology

Biotechnology

Databases

[2751]
Chemical engineering and biotechnology abstracts. Cambridge, Royal Society of Chemistry.

Covers industrial practice and theoretical chemical engineering. Includes processes, substances, operations, transport, equipment safety. About 125,000 records, updated monthly. Available on DIALOG (file 315) and Data-Star.
Class No: 6:573(003.4)

Bibliographies

[2752]
CRAFTS-LIGHTY, A. Information sources in biotechnology. 2nd ed. London, Macmillan; New York, Stockton Press, 1986. [xii], 403p. tables. ISBN: 0333392806.

First published 1983.

13 sections: 1. What is technology? The science and the business - 2. Information sciences in biotechnology (p.24-32) - 3. Monographs, book series and textbooks (p.33-93) ... 7. Abstracting and secondary sources - 8. Computer databases - 9. Patents and patenting - 10. Market surveys. 16 appendices (*e.g.* 1. Bibliographies in technology). Author is Managing director, Biocommerce Data, Ltd. *Class No:* 6:573(01)

[2753]
EUESDEN, M., *ed.* **Introduction to biotechnology information.** London, British Library, Science Reference and Information Service, 1991. 90p. £19.75. ISBN: 0712307621.

5 sections: 1. Scientific information - 2. Biotechnology - patents - 3. Online databases - 4. British official publications - 5. Business information in biotechnology. 9 appendices, A-I (I: Further reading). Detailed, analytical index, p.83-88.
Class No: 6:573(01)

[2754]
NEWAY, J.M. 'Biotechnology information sources: an update'. *Science & technology libraries,* v.8(2), Winter 1987/88, p.97-117.

Introduction - General aspects - Marketing aspects - Scientific aspects - Regulatory aspects - Patents aspects - Appendix: Dictionaries; Guides - Books: general aspects - Books: business aspects - Books: patents aspects - Books: regulatory aspects - Books: science aspects. The text highlights major sources. 3 references. Appendix source list of *c.*100 items. *Class No:* 6:573(01)

Handbooks & Manuals

[2755]
ATKINSON, B. *and* **MAVITUNA, F. Biochemical engineering and biotechnology handbook.** 2nd ed. Basingstoke, Macmillan, 1991. 1271p. diagrs., graphs, tables. £125. ISBN: 0333424041.

First published 1983.

20 chapters, closely subdivided, with reference and 'further reading'. Covers properties of micro-organisms, production processes, economics, measurement and instrumentation, nomenclature. Numerous tables. Detailed, analytical index. '... wealth of detail and breadth of coverage' (*Reference reviews,* v.6(6), 1992, p.25).
Class No: 6:573(035)

[2756]
BAJAY, Y.P.S., *and others.* **Biotechnology in agriculture and forestry.** Berlin, etc., Springer Verlag, 1986-. illus., micrographs, tables. ISBN: 0387518096.

An important, comprehensive work. So far (1986-91) 16v. have appeared. Trees (1986-91. 3v.); Crops (1986-88. 2v.); Potato (1987); Medical and cosmetic plants (1988-91. 3v.); Plant, protophasts and genetic engineering (1989. 3v.); Legumes and oilseed crops (1990-); Somaclonal variation in crop improvement (1990-); Wheat (1990); Rice (1990). Full sub-section references. Each volume carries a subject index.
Class No: 6:573(035)

[2757]
Biotechnology: a comprehensive treatise in 8 volumes. Rehm, H.-J. *and* Reed, G., *eds.* Weinheim, New York, VCH, 1981-89. 8 vols. + Cumulative indexes. illus., graphs, tables.

V. 1. *Microbial fundamentals.*
V. 2. *Fundamentals of biochemical engineering.*
V. 3. *Biomass, microorganisms for special applications, microbial products I, Energy from renewable resources.*
V. 4. *Microbial Products II.*
V. 5. *Food and feed production with microorganisms.*
V. 6a. *Biotransformations.*
V. 6b. *Special microbial processes.*
V. 7a. *Enzyme technology.*
V. 7b. *Gene technology.*
V. 8. *Microbial degradations.*

V. 7b. (xv,587p. 1989) has 34 contributors, 10 closely subdivided, documented chapters in 4 parts: I. Methods in industrial gene technology - II. Products of gene technology - III. Altered organisms as products of gene technology - IV. Safety. Volume index, p.571-87. *Class No:* 6:573(035)

[2758]
BRITISH LIBRARY. Science Reference Library. Health and safety in biotechnology. London, SRL [1986]. [1], [1], 80, [1]p. *Gratis.*

7 sections: 1. Biotechnology, general (national and international surveys and proposals) - 2. Recombinant DNA - 3. Laboratory safety - 4. Industrial applications - 5. Environmental considerations - 6. General information

....(contd.)

sources (journals and newspapers; abstract journals; online databases) - 7. Organizations. Keyword index (p.47-80). *Class No:* 6:573(035)

[2759]
Comprehensive biotechnology: the principles, applications and regulation of biotechnology in industry, agriculture and medicine. Moo-Young, M., *ed.* Oxford, etc., Pergamon Press, 1985. 4v. illus., diagrs., graphs, tables. £695.

1. *The principles of biotechnology: scientific fundamentals.* A.T. Bull and H. Dalton, volume editors. 1985. xxv, 688p.

2. *The principles of biotechnology: engineering considerations.* C.L. Clooney and A.E. Humphrey, volume editors, 1985. xxv, 633p.

3. *The principles of biotechnology: current commodity products.* H.W. Blanch, and others, volume editors. 1985. xxv, 1136p.

4. *The practice of biotechnology: speciality products and service activities.* C.W. Robinson and J.A. Howell, volume editors. 1985. xxix, 308p.

V.4 has 3 sections (67 subsections): 1. Specialized activities and potential applications - 2. Governmental regulations and concerns - 3. Waste: management and pollution control. Appended: Glossary of terms; Nomenclature. Cumulated subject index to the 4v., p.1165-1308. Nearly 100 contributors. Subsection 18: 'Assay of industrial waste enzymes', p.329-60; 4½p. of references. 'A notable success ... *Comprehensive biotechnology* will be an essential purchase for all departments and institutions, academic or industrial, that claim an interest in any aspect of ... biotechnology' (*Nature*, v.321, 5 June 1986, p.571-2).

A supplementary volume, *Annual biotechnology,* ed. by L.A. Babiuk appeared in 1989. *Class No:* 6:573(035)

[2760]
HIGGINS, I.J., *and others, eds.* **Biotechnology: principles and applications.** Oxford, Blackwell Scientific Press, 1985. [xi], 422p. illus., diagrs., graphs, tables, chemical structures. ISBN: 0632610290.

17 contributors. 10 chapters: 1. What is biotechnology? - 2. Energy and biotechnology - 3. Food, drink and biotechnology - 4. Chemistry and biotechnology - 5. Materials and biotechnology - 6. The environment and biotechnology - 7. Genetics and biotechnology - 8. Medicine and biotechnology - 9. Agriculture and biotechnology - 10. Chemical engineering. Each chapter has appended 'Further reading'. Aims 'to present a broad, informed view of contemporary biotechnology' (*Preface and acknowledgements*). *Class No:* 6:573(035)

[2761]
The UK biotechnology handbook '90. Crafts-Lighty, A., *ed.* 2nd ed. London, BioCommerce Data Ltd., 1990. xvi[ii],728p. ISBN: 1871393019.

First published 1988.

Part 1 includes a discussion of various issues, *e.g.* finance, career opportunities, emerging equipment markets, etc. Part 2 is a directory of companies and organizations in 5 sections: 1. Primary producers - 2. Resource base - 3. Financial sector - 4. Science base - 5. Government network. Indexes: organizations; subjects (p.649-722). *Class No:* 6:573(035)

Dictionaries

[2762]
COOMBS, J. Macmillan dictionary of biotechnology. London, Macmillan, 1986. 330p. £29.95.

Over 3,000 definitions, plus numerous cross-references. Defines biotechnology as 'the application of organisms, biological systems or biological processes to manufacturing and service industries'. Contains terms not found elsewhere, and 'fills a void in this rapidly developing field' (*Science & technology libraries,* v.8(1), Fall 1987, p.170). *Class No:* 6:573(038)

[2763]
Dictionary of biotechnology: English, German. Babel, W., *and others, eds.* Amsterdam, Elsevier, 1989. 113p. ISBN: 0444989005.

About 7300 English-base terms with German or 'Germanized' (*Preface*) equivalents. Genders given. *Class No:* 6:573(038)

[2764]
WALKER, J.M. *and* **COX, M. The Language of biotechnology:** a dictionary of terms. Washington, D.C., American Chemical Society, 1988. 255p. $49.95; $39.95. (*ACS Professional reference books.*) ISBN: 0841214891, (HB); 0841214905, (PB).

Emphasis is on technology rather than biology. 'A necessity' (*Choice*, March 1989, p.1124). *Class No:* 6:573(038)

Glossaries

[2765]
Biotechnology glossary. EEC Translation Service, *ed.* Amsterdam, Elsevier, 1990. vi,1074p. £85. ISBN: 1851665692.

About 700 terms with equivalents in each of the nine EC languages. Intended for translators dealing with technical works. *Class No:* 6:573(038.1)

Reviews & Abstracts

[2766]
Abstracts in biocommerce. Slough, Berks., Biocommerce Data, Ltd., 1982-. 26pa. £135pa.

200 abstracts per issue. Coverage: 'Commercial news in biotechnology, including genetic engineering, monoclonal antibodies, enzymes' (*Inventory of abstracting and indexing services produced in the UK*. Compiled by J. Stephens. 2nd ed., 1986, item 3). Online database. *Class No:* 6:573(048)

[2767]
Biodeterioration abstracts: a world abstract journal covering the fields of materials, biodeterioration and biodegradation of wastes. Prepared from the CAB ABSTRACTS database by the CAB International Mycological Institute. Wallingford, Oxon., CAB International, 1987-. Quarterly. ISSN: 09510621.

1987: 2,038 abstracts. 2 main groups: Biodeterioration (37 sections), and Biodegradation (20 sections). Appended: Reports. Conferences. Books. V.2(3), 1988, includes a review article. Author and subject indexes per issue. *Class No:* 6:573(048)

[2768]
Biotechnology research abstracts. Bethesda, Md., Cambridge Scientific Abstracts, 1984-. Monthly. ISSN: 07335709.

About 120,000 abstracts pa., monitoring c.5,000 primary journals and other sources. Sections: General topics and reviews ... Patents ... Genetic engineering - Immobilization - Cell culture - Products of biotechnology - Applications - Fermentation and process engineering - Proceedings - Books. Author and subject indexes per issue. Online as LIFE SCIENCES COLLECTION. *Class No:* 6:573(048)

[2769]
Current biotechnology abstracts. Nottingham, Royal Society of Chemistry, 1983-. Monthly. £248pa. ISSN: 02643391.

About 5000 abstracts pa. 4 parts (15 sections): Technocommunications (*e.g.* 2. Legal issues and safety) - Techniques (*e.g.* Fermentation technology) - Industrial areas (8. Pharmaceutics; 9. Energy production; 10. Agriculture; 11. Chemical industry; 12. Food) - Information (14. Forthcoming events; 15. Books, reviews, etc.). Subject, substance, and company indexes per issue and annually. No author index. Online access: DIALOG (file 358), DATA-STAR. Worldwide coverage of relevant patents and journals. *Class No:* 6:573(048)

[2770]
Derwent biotechnology abstracts. London, Derwent Pubrs., Ltd., 1982-. 24pa. ISSN: 02625318.

About 400 abstracts per issue. Classes A-M: A. Microbiology; D. Pharmaceuticals; E. Agriculture; H. Other chemicals; M. Waste disposal. Indexes (per issue and cumulated pa.): Author; Corporate Affiliate; Patentee; Subject. 'Covers all aspects of biotechnology from manipulation through biomedical engineering and fermentation to downstream processing' (*Notes*).

Available online via DIALOG, file 357. Over 120,000 records; updated monthly. *Class No:* 6:573(048)

Periodicals

[2771]
DOWNS, L.J. The Biotechnology marketing sourcebook. London, British Library, Science Reference and Information Service, Biotechnology Information Service, 1990. [iv],137p. £30.

Introduction - Record structure - Alphabetical list of journals (p.1-120; 251 titles) - Publishers' addresses - Keyword index. The first of a new series. Worldwide listing of 251 English-language periodicals, newsletters and abstracts for workers in 'biotechnology, the life sciences, health care, biochemistry and related fields' (*Introduction*). Details include advertising contact/scope/readership. For marketing managers, science, technology and medical publishers, conference organizers and information officers. *Class No:* 6:573(051)

Progress Reports

[2772]
Advances in biotechnological processes. New York, Alan R. Liss, Inc., 1983-.

V.6 (1988. xvii, 296p.) has 12 contributors. The 7 papers include: 'New advances in animal and human virus vaccines', p.254-85 (78 references); 'Production,

.... (contd.)
characterization and application of cyclodextrins', p.32-71 (212 references). 'Contents of previous volumes' precedes. Subject index. *Class No:* 6:573(055)

[2773]
Biotechnology and genetic engineering reviews. Russell, G.E., *ed.* Newcastle upon Tyne, Intercept, 1984-. Annual. ea. £90.

Each annual volume encompasses the entire field, from pest control and wastewater treatment to energy production and the genetics of human disease. 'Aims to avoid the two traps into which annual reviews often fall - either surveying particular fields frequently and exhaustively, even when there is nothing useful to say, or pulling together batches of *ad hoc* reviews with no organic coherence' (*British book news*, July 1987, p.392). *Class No:* 6:573(055)

Yearbooks & Directories

[2774]
COOMBS, J. The International biotechnology directory, 1993. Alston, Y.R. 8th ed. Basingstoke, Macmillan, 1992. 632p. £125. ISBN: 0333565533.

Data on more than 8,500 organizations. Covers Western Europe, North America, Brazil, Australasia and Japan. 3 parts: International organizations and information services - Government organizations and societies - Companies and organization; products and areas of research (by country, A-Z). Indexes: Companies; buyers' guide; products. *Class No:* 6:573(058)

[2775]
Directory of British biotechnology, 1989/90. Harlow, Essex, Longman Group, 1988, Biennial. 176p. £65. ISBN: 0582036054.

Profiles of the main UK companies, laboratories and academic groups involved in biotechnology manufacturing, research and consultancy. Data on each include ownership; name of contact; biotechnological activities, budgets, number of graduates employed in biotechnology; links abroad. Subject index groups organizations under headings: agriculture; bioelectronics; commodity chemicals (*e.g.* biomass); environment; pharmaceuticals; speciality chemicals; manufacture of biotechnology equipment. *Class No:* 6:573(058)

Worldwide

[2776]
Industrial biotechnology international, 1988/89: technology advances and corporate developments. Harlow, Essex, Longman Group, 1988. Annual. xii, 317p. £85. ISBN: 0582026350.

First published 1981.

2 main sections: Technological advances and market analyses (including therapeutics; diagnostics; agriculture and botany; waste treatment) - Biotechnology and the corporations (*i.e.*, biotechnology and pharmaceutical industries). Indexes of personnel, organizations and subjects. *Class No:* 6:573(100)

Great Britain

[2777]

DUNNILL, P. *and* **RUDD, M. British biotechnology and British industry.** Swindon, Wilts., Biotechnology Directorate of the Science and Engineering Research Council. 89p. ISBN: 0901660574.

A report, in 12 sections (e.g., The food processing study; Energy supply or saving; Inter-industry linkage). 22 references. *Class No:* 6:573(410)

USSR

[2778]

RIMMINGTON, A. *and* **GREENSHIELDS, R. Technology and transition:** a directory of biotechnology in the post-coup USSR. London, Pinter, 1992. 224p. £75. ISBN: 1855670380.

Includes books and likely developments.
Class No: 6:573(47)

Japan

[2779]

Biotechnology guide Japan, 1990-91: company directory and comprehensive analysis. Nikkei Biotechnology, *eds.* New York, Stockton Press; Basingstoke, Macmillan, 1990. xv,591p. £95. ISBN: 0935859667, US; 0333518055, UK.

Data on about 800 Japanese companies: operations, finance, R & D projects, major patents. Subject index, p.568-91. *Class No:* 6:573(52)

USA

[2780]

DIBNER, M.D. Biotechnology guide U.S.A.: companies, data, analysis. 2nd ed. New York, Stockton Press; London, Macmillan, 1991. xiv,652p. £95. ISBN: 1561590150, US; 0333562801, UK.

First published 1988.

Data on over 800 companies working with the technologies of genetic engineering. 7. Sections: 1. Introduction - 2. Directory of companies ... 5. Company areas of interest (*e.g.* bioseparations, vaccines, fungi, etc.) - 6. Partnerships - 7. Analysis of the industry. 5 appendices (c. 'Dead companies'). Index of companies, p.637-52. *Class No:* 6:573(73)

[2781]

UNITED STATES. Congress. Office of Technology. **Commercial biotechnology:** an international analysis. Washington, US Government Printing Office, 1984. [vi], 612p. illus., diagr., tables, chemical structures. $20. (*OTA-BA-218.*)

Parts 1-3 (chapters 3-10): 1. The technologies - 2. Firms commercializing biotechnology 3. Applications of biotechnology in specific industries (*e.g.* pharmaceuticals; agriculture; specialty chemicals and food additives; environmental applications; commodity chemicals and energy production; bioelectronics). Chapters 11-21: Analysis of US competitiveness in biotechnology. Glossary. Appendices A-K. Detailed, analytical index, p.605-12. *Class No:* 6:573(73)

Information Services

[2782]

BRITISH LIBRARY. Science Reference Library. **Guide to research information sources in biotechnology.** London, SRL, [1986]. [iii], 10, [1]p. *Gratis.*

3 parts: 1. International directories - 2. European directories (A. Europe and the EEC; B. National directories of the EEC countries: Belgium, Britain, Denmark, Eire, France, Italy, the Netherlands, West Germany) - 3. Databases. *Class No:* 6:573:061:025.5

Research Projects

[2783]

Directory of research in biotechnology. 7th ed. Swindon, Biotechnology Directorate of the Science and Engineering Research Council, 1989. xviii,162p. ISSN: 02689421.

8 priority areas, *e.g.* 1. Biochemical engineering; 2. Bioconversions; 4. Whole plant biotechnology; 7. Protein engineering. Data: project title; academic supervisor(s); individual supervisor(s); student. Index of research programmes and studentships, p.vi-xviii. *Class No:* 6:573:061:061.62.005

Technology

Abbreviations & Symbols

[2784]

Abkürzungen in der technik. [Abbreviations and acronyms in technology.] Leipzig, Fachbuchverlag Leipzig, 1991. 372p. ISBN: 3343007315.

About 10,000 entries, with English definitions for about half of them. Bibliography, p.372. *Class No:* 60(003)

[2785]

Pugh's dictionary of acronyms and abbreviations: abbreviations in management, technology and information science. Pugh, E., *comp.* 5th ed. London, Library Association Publishing, 1987. £60 (£48 to L.A. members). ISBN: 0853655375.

First published 1968.

More than 34,000 entries, stating country of origin or organization, as considered necessary. Unlike the wide-ranging *Acronyms, initialisms and abbreviations dictionary* with its 500,000 odd terms, plus *International acronyms, initialisms and abbreviations dictionary* (2nd ed. Gale, 1987), for terms of foreign origin, Pugh concentrates largely on management, technology and information science.

A new edition, edited by Peter Auger, is scheduled for publication in 1994. *Class No:* 60(003)

Databases

[2786]

NTIS [National Technical Information Service]. Springfield, Va., US Department of Commerce.

A wide range of disciplines: management, agriculture, building, business, chemistry, engineering, energy, materials science, medicine, transportation, etc. Lists unclassified US reports and these of other countries. A wealth of information. Over 1,500,000 records; updated monthly.

....(contd.)
Available on DIALOG, file 6.
CD-ROM: DIALOG OnDisc; Silver Platter.
Class No: 60(003.4)

Bibliographies

[2787]
Bibliographic guide to technology. Boston, Mass., G.K. Hall, 1976-. Annual. ISSN: 03602761.

Records not only the relevant publications catalogue by the New York Public Library's Research Libraries but also additions to the Library of Congress MARC tapes. The 1991 v. includes material catalogued between September 1, 1990 and August 31, 1991. *Class No:* 60(01)

[2788]
BIRMINGHAM PUBLIC LIBRARIES. Reference Library. **Technical bibliographies, 1975-83.** Birmingham, the Library, 1975-83.

1. *Vinyl chloride and health.* 2nd ed. 1976. 2. *Electric road vehicles.* 1975; 1st & 2nd suppts. 1976-81. 3. *Lead pollution.* 1975. 4. *Solar energy.* 1975; 4a. 1st supplement, 1977. 5. *Wind power.* 1976. 6. *Hydroponics.* 1976. 7. *Rabies.* 1976. 8. *Geothermal energy.* 1976. 9. *Heat pumps.* 1976. 10. *Tidal power.* 1977. 11. *Health and safety at work.* 1978. 1st, 2nd & 3rd suppts. 1978-80. 12. *Stirling engine.* 1980. 13. *Wave power.* 1970-1980. 14. *Fuel from waste.* 1983.

Solar energy and its supplement together provided 990 references. The series is now discontinued.
Class No: 60(01)

[2789]
Führer durch die technische Literatur: Katalog technischer Werke für Studium und Praxis. Hannover, F. Weidemanns Buchhandlung. Annual. ISBN: 3878910118.

First published 1906/07.

The 68th ed., 1981 (818p.), has over 13,000 entries in 8 classes: 1. Basic sciences (mathematics, astronomy, physics); 2. Architecture, building; 3. Mechanical engineering, mining, energy; 4. Electricity & electronics ... 6. Chemisty & chemical engineering; 7. Management, dictionaries, patents, technical literature as a whole, standards; 8. Technical periodicals. Subject and author indexes. List of publishers cited. *Class No:* 60(01)

[2790]
RINK, E. Technical Americana: a checklist of technical publications printed before 1831. Millwood, N.Y., Kraus International, 1981. xxviii, 776p. ISBN: 0527754471.

Sponsored by the Eleutherian Mills Memorial Library.

6,065 numbered entries for separately published works. 12 sections (each with chronological arrangement): 1. General works - 2. Technology - 3. Agriculture - 4. Crafts and trades - 5. Medical technology - 6. Military technology - 7. Civil engineering - 8. Mechanical engineering - 9. Manufacturing - 10. Mining and mineral production - 11. Sea transportation - 12. Inland transportation. Bibliographic references, p.xvi-xvii. Key to locations (c.170). Detailed, analytical index. 'This is a valuable bibliography, carefully compiled and very thorough in its coverage' (*College & research libraries,* v.43, July 1982, p.339). *Class No:* 60(01)

[2791]
SUBRAMANYAM, K. 'Technical literature'. In *Encyclopedia of library and information science* (New York, Dekker), v.30 (1980), p.144-209. facsim., diagrs., table.

Introduction - Technical reports - Standards and specifications - Trade catalogs. 79 references. Bibliography (p.200-9): Information needs of engineers and technologists; Technical reports; Standards and specifications - Trade catalogs - House journals (13 titles). *Class No:* 60(01)

[2792]
TECHNISCHE INFORMATIONSBIBLIOTHEK, Hannover. **Katalog der Technischen Informationsbibliothek (TIB),** Hannover. Microfiche ed. Mainz-Kastel, N. Gärtner, 1983. (Distributor, K.G. Saur). £1,500. ISBN: 3598304455.

1,254,960 entries on 664 fiches. Includes title entries for monographs, periodicals, all German dissertations and, in particular, published and unpublished research reports. Records the holdings of a leading scientific and technical library. *Class No:* 60(01)

[2793]
UNITED NATIONS. Industrial Development Organization. **Guides to information sources.** New York, United Nations, 1972-. ea. $4. (*UNIDO/LIB/SE.D/1-.*)

1. *Information sources on the meat-processing industry.* Rev. ed. 1976. xii, 88p. 2. *Information sources on the cement and concrete industry.* Rev. ed. 1977. xii, 70p. 3. *Information sources on the leather and leather goods industry.* 1979. xiii, 85p. 4. *Information sources on the furniture and joinery industry.* 1977. xii, 89p. Rev ed. 5. *Information sources on the foundry industry.* 2nd ed. 1977. xii, 87p. 6. *Information sources on industrial quality control.* Rev. ed. 1980. xii, 71p. 7. *Information sources on the vegetable and processing industry.* 1977. xii, 109p. 8. *Information sources on agricultural implements and machinery industry. 1973. xii, 108p.* 9. *Information sources on building boards ...* 1974. 94p. 10. *Information sources on the pesticides industry.* Rev. ed. 1982. xii, 77p. $3. 11. *Information sources on the pulp and paper industry.* 1974. xii, 92p. 12. *Information sources on the clothing industry.* 1974. xii, 116p. 13. *Information sources on the animal feed industry.* 1975. xii, 62p. 14. *Information sources on the printing and graphics industry.* 1975. xii, 65p. 15. *Information sources on the non-alcoholic beverage industry.* 1975. xii, 72p. 16. *Information sources on the glass industry.* 1975. xii, 64p. 17. *Information sources on the ceramic industry.* 1976. xii, 68p. 18. *Information sources on the paint and varnish industry.* 1976. xii, 68p. 19. *Information sources on the canning industry.* 1976. xii, 83p. 20. *Information sources on the pharmaceutical industry.* 1976. xii, 144p. 21. *Information sources on the fertilizer industry.* 1976. xii, 96p. 22. *Information sources on the machine-tool industry.* 1976. vii, 71p. 23. *Information sources on the dairy product manufacturing industry.* 1977. xii, 88p. 24. *Information sources on the soap and detergent industry.* 1977. xii, 67p. 25. *Information sources on the beer and wine industry.* 1977, xii, 81p. 26. *Information sources on the iron and steel industry.* 1977, xii, 103p. 27. *Information sources on the packaging industry.* 1977. xii, 110p. 28. *Information sources on the coffee, coca, tea and spices industry.* 1977. xii, 74p. 29. *Information sources on the petrochemical industry.* 1978. xii, 141p. 30. *Information sources on non-conventional sources of energy.* 1978. xii, 110p. 31. *Information sources on woodworking machinery.* 1978. xii, 90p. 32. *Information sources on the electronics*

....(contd.)

industry. 33. *Information sources on bioconversion of agricultural wastes*. 1979. xii, 94p. 34. *Information sources on the natural and synthetic rubber industry*. 1979. xii, 108p. 35. *Information sources on the utilization of agricultural resources for the production of panels, pulp and paper*. 1977. xii, 98p. 36. *Information sources on industrial maintenance and repair*. 1979. xii, 87p. 37. *Information sources on industrial training*. 38. *Information sources on essential oils*. 1981. xii, 89p. 39. *Information sources on flour milling and the baking products industries*. 1981. xii, 100p. 40. *Information sources on grain storage*. 1982. xii, 97p.

Each guide has *c*.12 sections: 1. Professional, trade and research organizations, learned societies and special information services - 2. Directories - 3. Sources of statistics, marketing and other economic data - 4. Basic handbooks, textbooks and manuals - 5. Monograph series - 6. Current periodicals - 7. Current abstracting and indexing periodicals - 8. Proceedings, papers and reports - 9. Specialized dictionaries and encyclopaedias - 10. Bibliographies - 11. Films and film catalogues - 12. Other potential sources of information (*e.g.* conferences; patents and licences; translation sources). A few of the sections are annotated. Appended (in most cases): 'Bibliographical sources used in compiling the present guide'. A helpful series, if somewhat dated, for researchers. *Class No:* 60(01)

Encyclopaedias

[2794]
LUEGER, O. Lexikon der Technik. Hrsg. von A. Ehrhardt und Hermann Franke. 4. vollst. neubearb. und erw. Aufl. Stuttgart, Deutsche Verlags-Anstalt, 1960-72. 17v. illus., diagrs., tables.

First published 1894.

1. *Grundlagen des Maschinenbaues*. 1960.
2. *Lexikon der Elektrotechnik und Kerntechnik*. 1960.
3. *Lexikon der Werkstoffe und Werkstoffprüfung*. 1961.
4. *Lexikon der Bergbaues*. 1962.
5. *Lexikon der Huttentechnik*. 1963.
6-7. *Energietechnik und Kraftmaschinen*. 1965.
8-9. *Fertigungstechnik und Arbeitsmaschinen*. 1967-68.
10-11. *Bautechnik*. 1966.
12. *Fahrzeugtechnik*. 1967.
13-14. *Feinwerktechnik*. 1969.
15. *Fabrikorganisation, Fördertechnik*. 1970.
16. *Verfahrenstechnik Nahrungsmitteltechnik, Haushaltechnik*. 1971.
17. *Registerband*. 1972.

A series of specialized technological dictionaries, each in A-Z entry order. Chemical engineering is excluded (on which see *Ullmann's Encyclopädie*, at 66.0(031)). V.5 (xii, 824p.; 3,000 entries; 232 illus. and diagrs.; 334 tables) has 22 contributors. All articles are signed, and many carry references. Technical terms mentioned in articles are italicized and appear as entries. Clear, keyed diagrams. Numerous cross-references. V.17, the general index, has *c*.60,000 entries. *Class No:* 60(031)

Dictionaries

[2795]
Elsevier's dictionary of technology: English-Spanish. Thomann, A.E. Amsterdam, Elsevier, 1990. 2v. (1742p.). £159.

Pt. 1: A-Ingot; Pt. 2: Ingot-Z.

About 80,000 entries and sub-entries. Near-print — hardly legible. A Spanish-English companion is in preparation. *Class No:* 60(038)

[2796]
FREEMAN, H.G. Wörterbuch technischer Begriffe mit 4,300 Definitonen nach DIN, Deutsch und Englischer. Berlin, Beuth-Verlag, 1983. 702p. ISBN: 3416108041.

German definitions are given of the 4,300 German headwords, plus English equivalent of the headword and English-German index. *Class No:* 60(038)

[2797]
JOLY, H, *ed*. **Dictionnaire des industries:** 36,000 définitions. Index anglais-français. Paris, Conseil International de la Langue Française [CILF], 1986. xvi, 1082p.

French terms are explained; English equivalents of terms. English-French index, p.633-1082. Quarto. *Class No:* 60(038)

[2798]
SCHLOMANN, A., *ed*. **Illustrated technical dictionaries in six languages.** London, Constable, 1906-32. (Also published Munich, Oldenbourg; New York, McGraw-Hill). 17v. illus., diagrs. (*Schlomann - Oldenbourg series*.)

1. *Elements of machinery and tools*. (1,632 diagrs.). 2. *Electrical engineering, including telegraphy and telephony*. 3. *Steam boilers and steam engines*. 4. *Internal combustion engines*. 5. *Railway construction and operation*. 6. *Railway rolling stock*. 7. *Hoisting and conveying machinery*. 8. *Reinforced concrete*. 9. *Machine tools*. 10. *Motor vehicles*. 11. *Metallurgy of iron*. 12. *Pneumatics*. 13. *Structural engineering*. 14. *Raw materials of the textile industry*. 15. *Spinning*. 16. *Weaving and woven fabrics*. 17. *Aeronautics*. (omits Spanish and Russian terms).

A monumental, much dated series. Compilation has been most thorough, and even the most recondite terms (*e.g.* in railway engineering; weaving) appear. The pattern of classified sections, keyed parts of diagrams, and equivalents in the six languages (English, German, French, Russian, Italian, Spanish supported by language indexes) has proved acceptable for later technical dictionaries. V.2, 14, 17 have been reprinted (Munich, Oldenbourg). *Class No:* 60(038)

Theses

[2799]
BILBOUL, R.R., *ed*. **Retrospective index to theses of Great Britain and Ireland, 1716-1950. Volume 2: Applied sciences and technology.** Oxford, ABC-Clio, 1976. [8], xi, 159p. facsim.

About 1,500 entries under subjects (Acetaldehyde ... Zirconium), p.1-90. Subject cross-references. 'Metal alloys': nearly 4 columns. Author index, p.91-159. *Class No:* 60(043)

Reports Literature

[2800]

Corporate author authority list. Karl, A.V., *ed.* 2nd ed. Detroit, Mich., Gale., in cooperation with National Technical Information Service. 2v. (xiv, 2,103p.). $180, the set. ISSN: 07413270.

A 40,000 entry authority list of main entries for US government-sponsored research, development and engineering reports - as well as foreign technical reports and other analyses prepared by national and local government agencies, their contractors/grantees. *Class No:* 60(047)

[2801]

Report series codes dictionary: a guide to more than 20,000 alphanumeric codes used to identify technical reports. Aranson, E.J., *ed.* 3rd ed. Detroit, Mich., Gale, 1986. x, 647p. $175. ISBN: 0810321475.

Rev. ed. of *Dictionary of report series codes* (2nd ed. Special Libraries Association, 1973; first published 1962).

Includes over 20,000 codes found in a database produced jointly by the US Departments of Commerce and Energy, NASA and the Department of Defense. 2 major sections: report series codes, A-Z; corporate author or issuing agencies, A-Z. *Class No:* 60(047)

Reviews & Abstracts

[2802]

Applied science and technology index. New York, H.W. Wilson, 1913-. 11pa.; quarterly, annual cumulations. ISSN: 00036986.

As *Industrial arts index*, 1913-57.

Subject index to *c*.350 English-language journals, mostly US. Closely subdivided (*e.g.* 'Electronics industry' (December 1991: 15 subject and country subheadings, plus cross-references). Appended 'Book reviews'. Selection of periodicals for indexing ... is accomplished by subscriber vote. Time-lag about 4 months. Also available online in WILSONLINE, covering the H.W. Wilson index series and via BRS; CD-ROM. *Class No:* 60(048)

[2803]

Bulletin signalétique. Paris, Centre National de la Recherche Scientifique (CNRS), 1956-83. Monthly or quarterly.

Formerly *Bulletin analytique* (v. 1-16. 1940-55).

The following 49 series concerned science and technology: 101. *Information scientifique et technique.* 110. *Information automatique. Recherche opérationnelle. Gestion.* 120. *Astronomie* ... 130, 140. *Physique I,II.* 145. *Electronique.* 150. *Physique. Chimie et technologie nucléaires.* 160. *Physique des solides* ... 161. *Cristallographie.* 165. *Atomes et molécules.* 170. *Chimie.* 220-6. *Bibliographie des sciences de la terre: Minéralogie; Gisements; Roches cristallines; Roches sédimentaires; Stratigraphie; Tectonique; Hydrologie.* 310. *Génie biomédical.* 320. *Biochimie, Biophysique.* 330. *Sciences pharmacologiques.* 340. *Microbiologie. Virologie. Immunologie.* 346. *Ophthalmologie.* 347. *Otorhinolaryngologie.* 348. *Dermatologie.* 349. *Anesthesie. Réanimation. Choc.* 350. *Pathologie.* 351. *Cancer.* 352. *Maladie de l'appareil respiratoire* ... 354. *Maladie de l'appareil digestif* ... 355. *Maladie des reins.* 356. *Maladie du système nerveux.* 357. *Maladie des os.* 359. *Biologie et physiologie animales.* 361. *Endocrinologie.* 362. *Diabéte* ... 363. *Génétique.* 365. *Physiologie des vertebrés.* 370. *Biologie et physiologie végétales.* 380. *Agronomie.*

....(contd.)

Zootechnie. 390. *Psychologie. Psychopathologie.* 730. *Combustibles. Energie thermique.* 740. *Métaux. Métallurgie.* 745. *Soudage* ... 761. *Microcopie életronique.* 780. *Polymer.* 880. *Génie chimique. Industries chimique et parachimique.* 885. *Nuisances.* 890. *Industries mécaniques. Bâtiment. Travaux publiques. Transports.*

Now published as Pascal. *Class No:* 60(048)

[2804]

Clover index. Caldecote, Biggleswade, Beds., Clover Publications, 1975-. 4pa. £37 pa.

Title varies.

About *c*.3,000 entries in each issue, A-Z. Draws upon 90 popular science magazines such as *Amateur gardening, Flight, New Scientist, Wireless World* and *Your model railway.* Titles of articles are occasionally given brief explanatory notes. Cheaply produced from typescript. To some extent supplements *Current technology index* for public libraries. Layout is critically reviewed in *Library review* (v.28, Summer 1979, p.120-1). Annual cumulation.

Class No: 60(048)

[2805]

Current contents: Engineering, technology & applied sciences. Philadelphia, Pa., Institute for Scientific Information, 1975-. Weekly. $315 pa. ISSN: 00957917.

Preceded by *Current contents: Engineering & technology.* (1970-74).

Reproduces contents pages of *c*.100 journal issues each week, covering some 700 journals during the year. 13 disciplines (*e.g.* management, aerospace, chemistry, metallurgy, nuclear power, optics and acoustics). Prelim. pages note journal coverage changes, book announcements, etc. Title index; author index, plus address dictionary; publishers address directory. *Class No:* 60(048)

[2806]

Current technology index. London, Bowker-Saur, 1981-. 6pa., with annual cumulation. ISSN: 02606593.

Formerly *British technology index*, 1962-80. Published by the Library Association until 1990.

Indexes *c*.20,000 items pa. Arranged under subjects/topics A-Z. A feature is the use of chain subject headings (*e.g.* 'Petroleum, Pipelines, Repair, Freezing, Carbon dioxide' - virtually a highly compressed descriptive annotation). Coverage: general technology, applied sciences, engineering, chemical technology, manufactures and technical services. 316 journals are currently scanned. Time lag: 5-7 weeks. *CTI* is available on CD-ROM. *Class No:* 60(048)

[2807]

Japanese technical abstracts. Ann Arbor, Mich., Japanese Technical Information Service, University Microfilms International, 1986-. Quarterly. ISSN: 08825246.

Draws on *c*.700 Japanese journals on technology, applied sciences, engineering and business management. 80% in Japanese, according to *RQ*, v.26(1), Fall 1986, p.114,116. *Class No:* 60(048)

[2808]
New technical books: a selective list with descriptive annotations. New York, New York Public Libraries, The Research Libraries, 1915-. 10pa.

1985: 1,500 annotated entries for noteworthy books, mostly in English. Entries, in broad Dewey Decimal classes, feature two paragraphs: 'Contents' and 'Note', the latter indicating readership, level and standing. About 75% of items concern technology. Parts of continuing multi-volume works (*e.g.* Kirk-Othmer) and annual review series (*e.g. Advances in ...*) receive attention. The annual author and catchword indexes occupy middle pages of the December issue each year. Larger format and 6pa. as from 1988. A valuable selection tool, although somewhat late in appearance. *Class No:* 60(048)

Periodicals

[2809]
BALACHANDRAN, S., *ed*. Directory of publishing societies: the researcher's guide to journals in engineering and technology. New York, etc., Wiley, 1982. ISBN: 0471092002.

Lists 224 journals, A-Z. Bibliographical data: aims and scope; manuscript submission procedures; style guide; refereeing procedures; acceptance rate and time between submission and publication. Page-charge information. 'For science library schools' (*New technical books,* v.68(1), January 1983, item 1). *Class No:* 60(051)

Progress Reports

[2810]
New technologies and development. Johnston, A. *and* Sasson, A., *eds*. Paris, Unesco, 1986. 281p. diagrs., tables. (*Notebooks on world problems, 1.*) ISBN: 923102454x.

6 sections. 1. Introduction; society, technology and industry - 2. Information technologies and telecommunication - 3. Biotechnologies - 4. Health and technologies - 5. New energy technologies - 6. Technological advances and environmental change. Selective, grouped bibliography, p.274-81. *Class No:* 60(055)

Yearbooks & Directories

[2811]
Directories held by the Science Reference and Information Service. 2nd ed., compiled by the Business Information Service. London, SRIS, 1986. [iii], 101p. £7. ISBN: 0712307419.

Lists more than 2,000 directories held by SRIS. Arranged under titles, A-Z. Class-marks and availability of earlier issues shown. *Trade directories subject index* (7th ed., 10p., on blue tinted paper) may also be consulted, as part of the SRIS's Business Information Service. *Class No:* 60(058)

[2812]
—Leading consultants in technology. 2nd ed. Detroit, Mich., Gale Research Co., 1989. 2v. (xxii, 1,996p.). ISBN: 089235089x.

A directory of *c.*18,000 scientific and technical consultants active in the USA. V.1 provides who's who data on the specialities in 6 broad categories: electronics, mechanical technology, chemistry, civil engineering/energy technology/

....(contd.)
earth sciences, physics and biotechnology. V.2: 'Expertise index' (1,500 subject headings) and index of names. Selected from *Who's who in technology. Class No:* 60(058)

Standards

[2813]
BRITISH STANDARDS INSTITUTION. BSI Standards Catalogue, 1992. London, BSI, 1992. Annual. xv,747p. £34. ISBN: 0580203069. ISSN: 09530338.

First published 1984; continues *British Standards yearbook,* 1937-83.

British Standards and related publications, p.1-457: General series - BS CECC Publications - BS QC Publications - Codes of Practice - Automobile series - Marine series - Aerospace series - Handbooks - Published documents - Drafts for development - Education publications - British Standards Society publications - European standards - ESII publications - THE publications - Miscellaneous items. Detailed, analytical index, p.536-747. *Class No:* 60(083.74)

[2814]
DIN [DEUTSCHE INDUSTRIE NORMEN]. Katalog für technische regeln. Berlin, Beuth Verlag, Annual. 2v.

Gives details of standards in both German and English. Over 4000 DIN standards have been translated in full into English. Updated monthly (new standards and information on drafts and revisions in *DIN Mittelungen Electronom* (Berlin, Beuth Verlag)). *Class No:* 60(083.74)

[2815]
Standards & specifications. Bethesda, Md., National Standards Association.

US government and industry standards and specifications. Data: issuing organization, Federal Supply Classification code, adoption, ANSI designation. Suppliers of products conforming to standards included. About 125,000 records; updated monthly. Available on DIALOG, file 113. *Class No:* 60(083.74)

Films

[2816]
STEADMAN, H., *ed*. The Industrial film guide. London, Kogan Page, 1974. 219p.

4 parts: 1. Index of film libraries - 2. List of films (*c.*2,000) classified by subject area (39 areas, Accountancy ... Work Study) - 3. A short guide to hiring and purchasing equipment - 4. Film producers. A/Z list of titles. Subject index. Data on each film include producer, date, length (time), hire charge, purchase price, colour, distribution, and brief description. *Class No:* 60(084.122)

Chronologies

[2817]
HARDUP, P., *ed*. The Timetable of technology. New York, Harvest Books, 1982. 240p. $29.50. ISBN: 0878512098.

Compares scientific events for 1900-82 and beyond, in a series of two-page spreads per year. 6 broad topics: Communications and information; Transport; Warfare and space exploration; Energy and industry; Medicine and food production; Fringe benefits (inventions). British emphasis. *Class No:* 60(090)

Histories

[2818]
DAUMAS, A., *ed*. **A History of technology and invention:** progress through the ages. Translated from the French. London, J. Murray, 1980. 3v. illus., diagrs.

1. *The origins of technological civilization, to 1450* vi, 596p. 2. *The first stage of mechanization, 1450-1725.* [x], 794p. 3. *The expansion of mechanization, 1725-1870.* ix, 768p.

Original French ed. as *Histoire générale des techniques* (Paris, Presses Universitaires de France, 1962-68), - the work of 20 contributors, with collaboration of Centre de Documentation Historique des Techniques. M. Daumas: Conservateur du Musée du Conservatoire National des Arts et Métiers. 'It would be a good thing if this work were to find its way into the libraries of our schools' (*The Times,* 16 May 1980, p.9). *Class No:* 60(091)

[2819]
GILLE, B. Histoire des techniques: techniques et civilisations; techniques et sciences. Tours, Gallimard, 1978. xx, 165p. illus., table. (*Encyclopédie de la Pléiade, v.41.*)

5 specialist contributors. Broad subject grouping: Prolégomènes à une histoire - Techniques et civilisations (Ancient world; Greeks, Romans; Middle Ages; Industrial Revolution; Modern) - Techniques et sciences (economics; geography; language; science, politics; etc). Subsections have bibliographies appended. Chronological table. Name index (3,000 persons, with brief designations). Title and subject indexes. *Class No:* 60(091)

[2820]
History of technology. London & New York, Mansell, 1976-. illus. diagrs. ISBN: 0720121140.

V.13 (1991, 272p.) carries a collection of papers on the history of electrical technology, marking the bicentenary of Michael Faraday. Publication schedule is erratic.
Class No: 60(091)

[2821]
McNEILL, I., *ed*. **An Encyclopedia of the history of technology.** London, Routlege, 1990. 1085p. illus., diagrs., charts. £80. ISBN: 0415013062.

22 contributions, each with further reading, by an international team of specialists. Theme: important inventions and their development, from Stone Age to Space Age. Subject and name indexes. 150 black-and-white illus. Editor was formerly Executive Secretary of the Newcomen Society for the Study of the History of Science and Technology. *Class No:* 60(091)

[2822]
SINGER, C., *and others,* eds. **A History of technology.** Oxford, Clarendon Press, 1954-84. 8v. illus. (incl. pl.), tables, maps. v.1-7, ea. £50; v.8, £27.50. ISBN: 019858105x.

1. *From early times to the fall of the ancient empires.* 1954, iv, 627p.

2. *The Mediterranean civilizations and the Middle Ages, c.700 BC to c.1500.* 1956. ix, 802p.

3. *From the Renaissance to the Industrial Revolution, c.1600-c.1750.* 1957. 804p.

4. *The Industrial Revolution, c.1750-c.1850.* 1958. xxxiii, 728p.

5. *The late nineteenth century, c.1850-c.1900.* 1958. xxxviii, 888p.

6-7. *The twentieth century (c.1900-c.1950),* edited by T.I.

.... *(contd.)*
Williams. 1978. 2v.

8. *Consolidated indexes.* 1984. 232p.

Intended as a comprehensive history of the 'methods and skills by which man has attained a gradual easing of his earthly lot through mastery of his natural environment' (*Preface*). Technology as an aspect of history, especially social history, and as an influence on the development of civilization. Each chapter is by a specialist, and each volume is well equipped with illus. (incl. plates), tables, maps and bibliographies. V.8 comprises 4 parts: 1. List of contents for v.1-7; 2. Index of names; 3. Index of place-names; 4. Index of subjects (p.127-232). 'An interesting authoritative account of inestimable value' (*Aslib book list,* May 1979, item 199). *Class No:* 60(091)

[2823]
—**DERRY, T.K.** *and* **WILLIAMS, T.I.** A Short history of technology, from the earliest times to AD 1900. London, Oxford Univ. Press, 1960. xviii, 782p. illus., chron., tables. ISBN: 0198581231.

Owes a good deal to Singer, although it is no abridgement. Whereas Singer is for the specialist student, Derry and Williams is for the general reader. Geographically it is largely confined to the ancient Near East, Western Europe and North America. 353 line-drawings, mainly taken from Singer. Select bibliography (p.750-8), plus chapter bibliographies. Index of subjects; index of persons and place names. *Class No:* 60(091)

[2824]
—**WILLIAMS, T.I.** A Short history of twentieth-century technology. London, Oxford Univ. Press, 1982. xix, 414p. £15. ISBN: 0198581159.

Designed for the general reader, providing a broad conspectus of technological development and 'a glimpse of the social, economic and political factors that influenced the development' (*Nature,* v.302, 3 March 1983, p.90). 30 chapters. Well-written, with many well chosen illus., and excellent bibliographies. The author is one of the general editors of *A history of technology. Class No:* 60(091)

Bibliographies

[2825]
FERGUSON, E.S. Bibliography of the history of technology. Published jointly by the Society for the History of Technology and the M.I.T. Press. Cambridge, Mass. & London, Massachusetts Inst. of Technology Press, 1968. xx, 347p.

About 1,500 annotated entries, in 14 main groups (1. General works - 2. General bibliographies and library lists - 3. Directories of technical, academic and business organizations - 4. Early source books and monographs - 5. Encyclopaedias, compendia and dictionaries of technology - 6. Biography - 7. Government publications and records - 8. Manuscripts - 9. Illustrations - 10. Travel and descriptions - 11. Periodicals and serial publications - 12. Technical societies, education and exhibitions - 13. Technology and culture - 14. Subject fields. Detailed analytical index, p.307-47. Aims 'to provide a reasonably comprehensive introduction to primary and secondary sources' (*Preface*). A valuable bibliography in its field. Updated by annual 'Current bibliography in the history of technology' in *Technology and culture,* 1964. *Class No:* 60(091)(01)

[2826]
JOHN CRERAR LIBRARY. A List of books on the history of industry and industrial arts. Josephson, A.G.S., *comp.* Chicago, Ill., John Crerar Library, 1915 (reprinted 1976). [ix], 486p. (*Bibliographical publications, no.11.*)

Companion volume to *A list of books on the history of science.*

Lists the Library's resources, - *c.*3,300 titles - on the subjects concerned, in classified order. Includes history of commerce and banking, but not of economics. Gives bibliographical descriptions, contents of symposia and similar works, annotating as necessary. Includes bibliographies and periodicals with some titles in French, German and Italian. *Class No:* 60(091)(01)

Reviews & Abstracts

[2827]
Bulletin signalétique. 522. Histoire des sciences et des techniques. Centre National de la Recherche Scientifique, Direction de l'information Scientifique et Technique. Paris, Centre de Documentation Sciences Humaines, 1940-. 4pa. ISSN: 00075574.

About 2500 references pa. Classes: 1 Généralités - 2. Sciences et techniques mathématiques - 3. Sciences et techniques physiques (physics, astronomy, chemistry) - 4. Technologie - 5. Sciences et techniques de la terre - 6. Sciences et techniques de la vie. Analytical subject index and author index per issue and pa. Annual list of periodicals scanned. *Class No:* 60(091)(048)

Periodicals

[2828]
Technology and culture: an international quarterly of the Society for the History of Technology. Chicago, Ill., Univ. of Chicago Press, 1960-. v.1(1).

The April issue each year includes an annual 'Current bibliography in the history of technology', with *c.*1,000 briefly annotated entries. Author and subject indexes. *Class No:* 60(091)(051)

[2829]
—History of technology index, 1991: journal articles in the Science Museum Library. London, Science Museum, 1992. *c.*100p. £16.95. ISBN: 0901805505.

Complements the 'Current bibliography' published annually in *Technology and culture. Class No:* 60(091)(051)

20th Century

[2830]
HERRING, S.D. From the Titanic to the Challenger: an annotated bibliography on technological failures of the twentieth century. New York, Garland, 1989. 486p. $54. ISBN: 0824030435.

Evaluative annotations; indicates inclusions of photographs/drawings. 'Has no comparable equivalent' (*RQ,* v.29(3), Spring 1990, p.440). *Class No:* 60(091)"19"

Biographies

[2831]
Who's who in industry: 1991. Margetts, J., *ed.* London, Fulcrum, 1991. vii,964p. ISBN: 1870452038. ISSN: 09604642.

About 10,000 entries, A-Z. Usual 'Who's Who' biographical information. Indexes: Company; Sector; Regional. A brief check reveals surprisingly little overlap with *Who's Who.* Publishers plan annual volumes. *Class No:* 60(092)

[2832]
Who's who in technology. Unterburger, A., *ed.* 6th ed. Detroit, MI, Gale, 1989. 2v. (2742p.). $380 (set). ISBN: 0810349507 (set).

First published 1979 as *Who's who in technology today* by Technology Recognition Corp.

V.1. *Biographies.* 1830p.

V.2. *Indexes.* 912p.

About 38,000 biographies, giving data on employment, education, special interests, expertise, publications, addresses. 4 indexes: Expertise; Geographic; Employer. Emphasis on US. *Class No:* 60(092)

Worldwide

[2833]
Longman concise world technology policies. Georghin, L. *and* Cunningham, P. Harlow, Essex, Longman, 1992. 480p. tables. £125. (*Longman guide to world science and technology.*) ISBN: 0582057302.

Largely a digest of data from Longman's national *Guide[s] to science and technology,* with some updating. Country-by-country analysis includes details of major organizations and provides possibility of comparison. *Class No:* 60(100)

Ancient Greece & Rome

[2834]
OLESON, J.P. Bronze age, Greek and Roman technology: a select, annotated bibliography. New York, Garland, 1986. xvi,515p. (*Bibliographies of the history of science and technology, v.7. Garland reference library of the humanities, v.646.*) ISBN: 0824086775.

2030 items in 11 chapters (subdivided by topic): 1. Sources (ancient authors; bibliographies, etc.) - 2. General surveys ... 6. Construction and civil engineering - 7. Manufacturing and trade (handtools; metalworking; ceramics, etc.) - 8. Transportation ... 10. Military technology. Each section begins with a brief introduction indicating coverage. Includes record keeping and cultural attitudes to technology as well as traditional subject matter. Entries include a descriptive, occasionally evaluative annotation, and a number of citations. Index of modern authors (p.485-515); no subject index. *Class No:* 60(37/38)

[2835]
WHITE, K.D. Greek and Roman technology. London, Thames & Hudson, 1984. 272p. illus., maps. £20. (*Aspects of Greek and Roman life.*) ISBN: 050040044x.

Part 1 concerns methods of investigation, transmission of technical processes and technical resources of the Classical world. Part 2 analyses technological developments in building, civil engineering, mining and metallurgy, food and agriculture, land and sea transport, and hydraulic

....(contd.)

engineering. 177 illus. closely complement the text. Extensive appendices. Author is Professor Emeritus in Ancient History, Univ. of Reading. *Class No:* 60(37/38)

Developing Countries

[2836]

Applied technology index: an indexing service to the literature of appropriate technology and related fields (ATI). Brighton, Noyce, 1980-. (Temporary mailing address, 1986-: J. Noyce, Burwood, Sydney). Quarterly. £60pa.

About 1,200 references pa. Classified order of entries AA. General; AB. Appropriate technology; AT. Transfer of technology; BA. Agriculture, general ... EA. Biofuels ... MA. Health, general ... XF. Water pumps; Z. Miscellaneous. A current-awareness service. 4th issue pa is an author index. *Class No:* 60(4/9-77)

[2837]

Appropriate technology organizations: a worldwide directory. Compiled by the Center for Business Information, New York. Jefferson, N.C., McFarland, 1984. 149p. $29.95. ISBN: 0899500986.

Lists about 2,000 international organizations that, by their own definition, are involved in appropriate technology. The index is an A-Z key to the directory's geographical arrangement. *Choice* (v.21(11/12), July/August 1984, p.1581) finds that the editor's criteria are too inclusive. *Class No:* 60(4/9-77)

[2838]

GHOSH, P.K. *and* **MORRISON, D.E.,** *eds*. **Appropriate technology in Third World development.** Westport, Conn., Greenwood Press, 1984. xxiv, 494p. (*International development resource books, 14.*)

Part 1: Current issues, trends, analytical methods, strategies and policies, country studies - Part 2: Statistical information and sources - Part 3: Resources bibliography (annotated) of items published after 1965 - Part 4: Directory of information sources. Aims to investigate trends in 'appropriate technology' that might contribute to Third World economic development, and to provide information and data related to appropriate technology policies. Emphasises policy-making for development rather than advocacy or technical feasibility. *Class No:* 60(4/9-77)

[2839]

HEAWOOD, R. *and* **LARKE, C. The Directory of appropriate technology.** London, Routledge, 1989. xi,[i],317p. ISBN: 0415627016.

Originated in the Work of the Alternate Technology Information Group (1976-).

Sections include: General/multi-disciplinary organizations; Ecology/environment; Education. Appended lists cite reference material, serial publications, audiovisual aids, etc. Intended for people - young and old - who want to explore options. *Class No:* 60(4/9-77)

[2840]

PACEY, A. Technology in world civilization. a thousand-year history. Cambridge, Mass., MIT Press, 1990. 238p. $19.95. ISBN: 0262161186.

Includes the transfer of technology to less-common areas — China, Islamic India, and the hunters and farmers of

....(contd.)

South America, Africa and Asia. The thousand-year period runs from 700AD to 1970. Pacey is a leader in the appropriate technology movement. *Class No:* 60(4/9-77)

Europe

[2841]

FREEMAN, C., *and others*. **Technology and the future of Europe:** global competition and the environment. London, Pinter, 1991. 450p. £45. ISBN: 0861870751.

5 parts: 1. The background of European industry and technology - 2. Information and communication in technology - 3. Other industries and technologies (*e.g.* chemical industry) - 4. Energy, environment and regulation - 5. The international context and policy agenda. References (authors A-Z), p.397-414. Detailed, analytical index. *Class No:* 60(4)

Japan

[2842]

TALBOT, D.E. Japan's high technology: an annotated guide to English-language information sources. Phoenix, Az., Oryx, 1991. 171p. $45; £32. ISBN: 0897745280.

About 500 entries for English-language materials. 13 chapters covering guides, conference reports, abstracting and indexing services, bibliographies, subject dictionaries, directories and handbooks, yearbooks, newsletters and magazines, industry reports, newspapers, online databases, patents and translations. Descriptive annotations, with prices and directory data included in citations. 4 appendices: Booksellers - Libraries - Other sources of information (noncommercial organizations) - Online vendors. 'Highly recommended' (*Choice*, May 1991, p.1468). *Class No:* 60(52)

Islamic World

[2843]

AL-HASSAN, A.Y. *and* **HILL, D.R. Islamic technology:** an illustrated history. Cambridge Univ. Press, 1986. xiv, 304p. illus., facsims. £25.

11 chapters: 1. Introduction - 2. Mechanical engineering - 3. Civil engineering - 4. Military technology - 5. Ships and navigation - 6. Chemical technology - 7. Textiles, paper and leather - 8. Agriculture and food technology - 9. Mining and metallurgy - 10. Engineers and artisans - 11. Epilogue. Selected bibliography, p.287-95. Index. Fully illustrated (163 half-tones and line-drawings). A readable introduction to a hitherto largely neglected field. *Class No:* 60(5.297)

Thesauri

[2844]

BRITISH STANDARDS INSTITUTION. ROOT thesaurus. 3rd ed. London, BSI, 1988. 2v. ISBN: 058016991x.

1. *Subject display.* xxxi,588p. 2. *Alphabetical list.* iii,632p.

V.1 comprises an introduction (including symbol notation for descriptions), a subject display contents list, and the subject display schedules. V.2 gives the supporting alphabetical list (*c.*1,600 entries and cross-references) plus a chemical formula index. Schedules mainly concern technology and science. 'ROOT's presentation in the form of

....(contd.)

a subject display with a complementing alphabetical list ensures that experienced and new users alike are able to achieve a new standard of consistency' (*Foreword,* v.1). *Class No:* 60:025.43

[2845]

International technical thesaurus: Noriane/Isonet/databases on standards. English version. Paris, Afnor, 1981. 471p. ISBN: 2121602119.

An 'enriched' edition of the International Organization for Standardization (ISO) thesaurus of 1976. The English version provides 1,500 preferred terms and synonyms. Part 1: a structured A/Z list of terms, with French equivalents; Part 2, a permuted list; part 3, a structured subject list. *Class No:* 60:025.43

[2846]

Thesaurus of scientific, technical, and engineering terms. Science Information Resource Center. Cambridge, Mass., etc., Hemisphere Publishing, 1988. xix,841,376p. $125. ISBN: 0891167943.

Based primarily on the NASA *Thesaurus* and the *DOD Thesaurus of scientific and engineering terms.* 2 v. (in one). V. 1. *Hierarchical listing* contains subject terms and USE cross references; V. 2. *Access vocabulary* lists permuted terms, embedded terms, postable and nonpostable terms and other word entries. *Class No:* 60:025.43

Information Services

[2847]

European technical consultancies. Harlow, Essex, Longman, 1988. 342p. £95. ISBN: 0582931877.

A directory of more than 700 technical consultancies throughout Europe. Data: names of senior executives, type of consultancy work undertaken and categories of industry concerned. *Class No:* 60:061:025.5

[2848]

Technical services in the United Kingdom. Harlow, Essex, Longman, 1989. 964p. £295. ISBN: 0582056640.

A directory of nearly 2,000 commercially available services offering technical support and information in the UK. Subject, industrial application and name of establishment indexes. *Class No:* 60:061:025.5

[2849]

UNITED NATIONS. Educational, Scientific and Cultural Organization. **World guide to technical information and documentation.** 2nd ed., rev. and enl. Paris, Unesco, 1975. 515p.

First published 1969.

Listed 476 centres (1969 ed: 273) in 93 countries. International, p.29-49; national (Algeria ... Zambia). India: 40 entries, whereas USSR, UK and USA muster only 11 between them. Appended: list of international, regional and national directories to technical information and documentation services; A-Z list of institutions; subject index. Largely superseded by the Longman 'Guide to world science and technology' series. *Class No:* 60:061:025.5

Government Organizations

[2850]

Government research directory: a guide to approximately 3,000 US government research and development centers. Piccirelli, A., *ed.* 6th ed. Detroit, Mich., Gale, 1990. 1100p. $390. ISBN: 0810369222.

First published 1980.

Arranged under sponsoring government agencies. Data on each centre include description of research, special facilities, publications and information services. Indexes: master name, keyword and agency; geographic; subject. *Class No:* 60:061:061.1

Research Establishments

Great Britain

[2851]

Industrial research in the United Kingdom: a guide to organizations and programmes. 14th ed. Harlow, Essex, Longman Group, 1991. 592p. £190. ISBN: 058208273x.

First published 1946 as *Industrial research in Britain.* 13th ed. 1989.

Details of *c.*3,000 laboratories active in agricultural and environmental sciences, chemical and materials sciences, earth and astronomical sciences, electronics and computer sciences, pharmaceutical, biomedical and biological sciences, and engineering and transportation. 6 Sections: 1. Industrial firms - 2. Research associations and consultancies - 3. Government departments, and their laboratories - 4. Universities and polytechnics - 5. Trade and development associations - 6. Personal name index. Title of establishment index. Detailed, non-analytical subject index. *Class No:* 60:061:061.62(410)

USA

[2852]

Directory of American research and technology: 1991. Press, J.C., *ed.* 25th ed. New Yorker, Bowker, Annual. ISSN: 08860076.

Formerly published as *Industrial research laboratories of the United States.*

Lists 11,322 organizations, of which 6,544 are parent companies. Geographical, personnel and R & D indexes. *Class No:* 60:061:061.62(73)

Technical Drawing

[2853]

FRENCH, T.E., *and others.* **Engineering drawing and graphic technology.** 13th ed. New York, etc, McGraw-Hill, 1986. 752p. illus. ISBN: 0070221638.

4 parts: 1. Basic graphs - 2. Elements of space geometry - 3. Applied graphics and design - 4. Special topics (*e.g.* maps and technology). Appendices. 1,210 illus. *Class No:* 60:744.42

[2854]
GIESECKE, F.E., *and others*. **Technical drawing.** 8th ed. New York, Macmillan Co., Inc.; London, Collier-Macmillan, 1991. *c*.900p. illus., diagrs., tables. £37.50. ISBN: 0023426505.

Numerous chapters (*e.g.* Instrumental drawing; geometric construction; sectional views; perspective; graphs). Covers computer applications. Analytical index. Intended as a class text and reference book in technical drawing. *Class No:* 60:744.42

Patents & Inventions

Handbooks & Manuals

[2855]
Butterworths intellectual property law handbook. Phillips, J., *ed.* London, Butterworth, 1990. vi,839,12p. £24.95. ISBN: 0406504717.

Covers primarily UK legislation on patents, copyrights, registered/unregistered designs, trade marks, plant varieties, intellectual property, professional status and practice. Detailed, analytical index, 12p. *Class No:* 608(035)

[2856]
CRAWFORD, N.K. *and* MORGAN, E.J. **The Innovators handbook: 1988/89:** a guide for firms seeking external help with innovation, technology transfer, and successful product development. Harlow, Essex, Longman, 1988. v.p. ISBN: 0582025648.

4 parts, 17 chapters: 1. Developing technology - 2. Financing technological development - 3. Purchasing and selling technology - 4. Protecting technological development. Includes brief case studies. Directory of 425 organizations (contact, phone, service, etc.). Index of organizations. *Class No:* 608(035)

[2857]
FOLTZ, R.D. *and* PENN, T.A. **Protecting scientific ideas and inventions.** 2nd ed. Boca Raton, Fl., CRC Press; Cleveland, Oh., Penn Institute, 1990. vi,377p. ISBN: 0944606067.

First published 1987.

Covers patents, trademarks, product warranties, copyright. Emphasis is on US legislation. An essentially practical treatment, written for the layperson. The first edition was 'recommended' by *Choice* (April 1988, p.1222). *Class No:* 608(035)

[2858]
McCARTHY, J.T. **McCarthy's desk encyclopedia of intellectual property.** Washington, D.C., BNA Books, 1991. 385p. $57. ISBN: 0871796821.

About 600 terms, clearly defined, most with references. 5 appendices (*e.g.* 'Registers of copyright, 1897-1985'). 'A mine of information ... highly recommended' (*Choice,* March 1992, p.1052). *Class No:* 608(035)

Trade Names, Trade Marks, Brand Names

Databases

[2859]
Trade names database. Detroit, Mi., Gale Research.

About 250,000 consumer brand names and their owners. Semi annual reloads. Available on DIALOG, file 116. *Class No:* 608(088.7)(003.4)

Bibliographies

[2860]
KASE, F.J. **Trade marks: a guide to official trademark literature.** Dobbs Ferry, N.Y., Oceana; Leiden, Sijthoff, 1974. viii, 420p. ISBN: 0379002299; 9028602143.

Entries A-Z by country (65 countries). Each entry gives: name and address of office in charge of trademark registrations; official trademark journal and any other official publications relating to trademarks; duration of protection. Description of contents of journal, with explanation of heading data, etc. Facsimiles of title pages and representative page provided. International trademark treaties in appendix. Bibliography. Index. *Class No:* 608(088.7)(01)

[2861]
NEWTON, D.C. **Trade names:** an introductory guide and bibliography. London, British Library, 1991. [ii],202p. £30. ISBN: 0712307532.

Substantially based on *Trade marks - a guide to the literature and directory of lists of trade names* by B.M. Rimmer (British Library, 1976).

Part 1 (p.7-59): 8 chapters on legal aspects and international agreements; 6 appendices (6: Periodicals); Index. Part 2: Introduction; Directory of lists of trade marks; Index. Includes some guidance on databases. Sources are based on the Science Reference and Information Service (SRIS) collection of the British Library. *Class No:* 608(088.7)(01)

Dictionaries

[2862]
ROOM, A. **Dictionary of trade name origins.** London, Routledge & Kegan Paul, 1982. 217p. £8.95. ISBN: 0710201745.

Reprinted, with corrections, 1982, 1983.

Famous brand names are listed A-Z, with a description giving the identity of the product, the name of the manufacturer, and the history of the trade name. A new edition was due in 1992. *Class No:* 608(088.7)(038)

Yearbooks & Directories

[2863]
A-Z of European brands. London, Euromonitor Pubs., 1991. 400p. £95. ISBN: 0863384004.

Over 5000 brands, considered by country, product sector and owner. Leading brands sold in the European market, both US and European. *Class No:* 608(088.7)(058)

Laws

[2864]
BLANCO WHITE, T.A. *and* **JACOB, R. Kerly's law of trade marks and trade names.** 12th ed. London, Sweet and Maxwell, 1986. lxxv, 873p. £65. ISBN: 0421350601.

An authoritative text on UK law now including commentary on service marks. Appendices give texts of laws, rules and conventions. *Class No:* 608(088.7)(094.1)

[2865]
—British trade mark law and practice: report of the Committee to Examine British Trade Mark Law and Practice. Chairman, H.R. Mathys. London, H.M. Stationery Office, 1974. vii, 95p. (*Cmnd. 5601.*)

Considers national and international aspects of trade mark registration and gives recommendations for change. *Class No:* 608(088.7)(094.1)

[2866]
—REID, B.C. A Practical introduction to trademarks. London, Waterlow, 1984. viii, 184p. £14.95.

Includes chapters on procedure in the UK Trade Marks Registry, following a historical introduction. Legal remedies are discussed, and there are two chapters on international aspects. *Class No:* 608(088.7)(094.1)

[2867]
—TRADE MARK REGISTRY. Work manual. London, Patent Office, 1984-. Loose-leaf.

Issued in chapters. States the working procedures used by the British Trade Mark Registry in the processing of applications. *Class No:* 608(088.7)(094.1)

[2868]
MEINHARDT, P. *and* **HAVELOCK, K.R. Concise trade mark law and practice.** Aldershot, Gower, 1983. xv, 206p. £21. ISBN: 0566023695.

A survey of law and practice in the UK, of EEC rules and of continental practice relevant to British exporters. *Class No:* 608(088.7)(094.1)

Worldwide

[2869]
JACOBS, A.J. Trade marks throughout the world. 3rd ed. New York, Trade Activities, 1982-. Loose-leaf. ISBN: 0876321260.

Updated three times pa.

Brief details are provided of the trade mark procedures in about 200 countries. Appendices list the international and national classification schemes, international conventions and give at-a-glance tables by country of various features. *Class No:* 608(088.7)(100)

[2870]
Marques internationales. Geneva, World Intellectual Property Organization, 1893-. ea. SF30. ISSN: 00253936.

Contains registrations, renewals etc. of trade marks on the international register maintained by WIPO under the Madrid Agreement. *Class No:* 608(088.7)(100)

[2871]
—Répertoire alphabétique et phonétique des marques internationales. Mortsel, Compu-Mark, 1949/69. 1956/76. 1977-.

Indexes registrations and renewals published in *Marques internationales (qv)*. 42 classes, in 5 groups, kept up-to-date by quarterly supplements that supersede the previous supplement and cumulate annually. *Class No:* 608(088.7)(100)

Great Britain

[2872]
The Trade mark journal. London, Patent Office, 1976-. Weekly. ea. £4.50.

Contains official notices, applications for trade marks advertised before registration, marks registered, renewals and other proceedings. Trade mark applications are arranged in classified order. *Class No:* 608(088.7)(410)

[2873]
UK trade names including imported items 1990-1991. 11th ed. East Grinstead, Kompass Publishers, 1990. 343,201p. ISBN: 0862682061.

A list of 60,000 trade names in alphabetical order; a list of 14,000 companies using those names (on pink pages); a list of foreign companies with UK agents. Lapsed trade names and those for food, drink, tobacco and pharmaceuticals are excluded. Imported items are marked with an asterisk. *Class No:* 608(088.7)(410)

USA

[2874]
DIAMOND, S.A. Trade mark problems and how to avoid them. 2nd ed. Chicago, Crain Books, 1981. xii, 276p. £19. ISBN: 087251059x.

A practical book on US trade mark practice for the non-technical reader. The text includes choosing, clearing and protecting trade marks, slogans, acronyms, characters, nicknames, use in advertising, and licensing and is liberally illustrated with case studies. *Class No:* 608(088.7)(73)

[2875]
HILL, C.R. Trade marks and brand management: selected annotations. New York, United States Trademark Association, 1976. 188p.

An annotated bibliogoraphy, with sections on journals and periodicals and on books subdivided by subject. An appendix lists other sources of information. *Class No:* 608(088.7)(73)

[2876]
Official gazette of the United States Patent and Trademark Office: Trademarks. Washington, US Patent and Trademark Office, 1872-. Weekly.

Since February 1971 published separately from Patents section. Weekly issues include offical notices; trademarks published for opposition; trademark registrations issued; details of renewals and cancellations; index of registrants.

Annual index of registrants published separately from *Annual index of patents* since 1927. *Class No:* 608(088.7)(73)

[2877]
WOOD, D., *ed*. **Trade names dictionary.** Detroit, Mich., Gale, 1974-. 2v. ISSN: 02728818.

7th ed. 1989. 1,952p. $320.

A-Z list of about 206,000 US consumer-oriented trade names, brand names, etc., with names and addresses of 44,000 manufacturers and suppliers. Companion vol. *Trade names dictionary: company index* (7th ed. 1989. $320). *New trade names* is a supplement to both.
Class No: 608(088.7)(73)

[2878]
—Trade marks register of the United States. Washington, Trademarks Registry, 1961-. Annual. ISSN: 00825786.

Lists active trademarks, arranged in classified order. Information now available online from the US Patent and Trademark Office. *Class No:* 608(088.7)(73)

[2879]
—WOOD, D., *ed*. Trade names dictionary: company index. 7th ed. Detroit, Mich., Gale, 1989. 2v. $320.

First published 1979.

An A-Z list of US companies, with addresses and their trade names. *Class No:* 608(088.7)(73)

Worldwide

[2880]
International treaties on intellectual property. Leaffer, M.A., *ed*. Washington, DC, Bureau of National Affairs, 1990. 630p. tables. $58. ISBN: 0871796597.

Lists treaties in force. Text of treaties, preceded by summary. 7 parts: 1. Intellectual property - 2. Patents - 3. Trademarks - 4. Copyright and neighbouring rights - 5. Industrial design - 6. Classification - 7. Organizations. Detailed, analytical index, p.623-36. *Class No:* 608(100)

Inventions

Encyclopaedias

[2881]
The Inventions that changed the world. London, Reader's Digest Association, 1982. 368p. illus., diagrs.

Details of significant inventions arranged alphabetically according to subject-matter. There are shorter lists of great inventors, a chronology of inventions and a comprehensive index. *Class No:* 608.1(031)

[2882]
ROBERTSON, P. The Shell book of firsts. London, Ebury Press & M. Joseph, 1974. 254p. illus., facsims. ISBN: 0718112792.

The well-authenticated entries are arranged A-Z. By no means all the 'firsts' are concerned with technical achievements. Some, like the zip fastener, are famous, but others like the first production of tea bags, would be difficult to find elsewhere. 'The first radio broadcast', p.145-7, includes quotations. Both British and foreign firsts are given, where the information was available. Chronology of the first occurring in each year, 1767-1974. Index of *c*.3,000 entries. Sources of illus. *Class No:* 608.1(031)

Handbooks & Manuals

[2883]
BAKER, R. New and improved: inventors and inventions that have changed the modern world. London, British Museum Publications, 1976. 168p. illus., facsims. £4.50.

The introduction gives a short history of the patents system and the main body of the work provides references to patents, with their numbers, of well-known inventions (mostly British), under 363 headings. Historical notes on later developments, and in all 800 to 900 patents are cited. Drawings from patent specifications. Chronological, inventor and subject indexes. Suggestions for further reading, p.27-29. *Class No:* 608.1(035)

[2884]
EISENSCHITZ, T.S. *and* **PHILLIPS, J. The Inventors' information guide.** 2nd ed. London, Queen Mary College, University of London, 1985. 100p. £6.50. ISBN: 0951066404.

Aimed at the individual inventor with few resources, it contains details of information sources, acquiring a patent, developing an invention, professional services, finance, marketing, design, addresses, publications etc.
Class No: 608.1(035)

[2885]
GISCARD D'ESTAING, V-A., *ed*. **The Book of inventions and discoveries.** London, Queen Anne Press, 1990. 291,[i]p. col. illus., facsims., tables. £9.95. ISBN: 0356188353.

Based on *Le livre mondial des inventions*.

Transport - Warfare - Science - Space - Medicine - Information technology - Ecology and the environment - Power and industry - Everyday life - Games, toys, sport - The bizarre - Media and Communications - The arts. Index of inventors and of inventions. Dates are given throughout; boxed information. Excellent for quick reference, teenagers and browsers. *Class No:* 608.1(035)

[2886]
JEWKES, J., *and others*. **The Sources of invention.** London, Macmillan, 1969. 372p.

Part 1, which occupies half of the volume, concerns invention and inventors, research in organizations and a review of past inventive effort. Part 2 comprises 46 selected case histories from automatic transmission to the zip fastener. Part 3 summarizes 10 new case histories. Most entries carry a list of references. *Class No:* 608.1(035)

Yearbooks & Directories

[2887]
Directory of associations of inventors. 3rd ed. Geneva, World Intellectual Property Organization, 1985. vii, 54p. (*WIPO publication no. (622 EF).*)

A loose-leaf publication listing the inventors' organizations in each country. *Class No:* 608.1(058)

Chronologies

[2888]
The Harwin chronology of inventions, innovations, discoveries from pre-history to the present day. London, Constable, 1987. vp. £10.95. ISBN: 0094661502.

A chronological list of technological achievements from flint-knapping of three million years ago to the variable-geometry vertical-axis wind turbine of 1986. Entries

....*(contd.)*

generally give the year of invention, its title, inventors and their place of origin and a brief note. Contains an extensive 35p. index to subject matter and a short bibliography. *Class No:* 608.1(090)

Histories

[2889]
PETROSKI, H. The Evolution of useful things. New York, Knopf, 1992. 256p. $25. ISBN: 0679412263.

Brief accounts of the origin and evolution of a range of everyday items, *e.g.* safety pins, paper clips, sand paper 'post-it' notes, etc. Index. 'Penetrating' and 'anecdotal' (*Booklist,* December 1, 1992, p.609). *Class No:* 608.1(091)

[2890]
WILLIAMS, T.I. Science: invention and discovery in the twentieth century. London, Harrap, 1990. 256p. illus. (mostly col.), facsims., ports., diagrs., graphs. £15.95. ISBN: 0245600248.

4 consultant editors. 6 periods: 1900/1914 ... 1973/1989. Appended: Biographies, p.228-49 (*c.*200 major Scientists and technologists); Glossary; Further reading, p.250. Detailed, analytical index. Text includes 12 'special features' (*e.g.* 'The structure of the nucleus', p.120-37). Covers physical sciences, biological and medical sciences and technology. About 350 illus., many in colour. For the lay person. *Class No:* 608.1(091)

Patents

Databases

[2891]
Claims/US Patent abstracts. Alexandria, Va., IFI/Plenum Data Corporation.

The online version of the patents listed in the general, chemical, electrical and mechanical sections of the *Official gazette* of the US patent office, as well as US design patents. Over 2,000,000 records; updated monthly. Available on DIALOG, files 23 (1950-1970), 24 (1971-1981), 25 (1982-present), 340 (1950-present). *Class No:* 608.3(003.4)

[2892]
Derwent world patents index. London, Derwent Publications, Ltd.

Covers 30 patent issuing authorities. Full abstracts for patents since 1981. The online version of *World patents abstracts journal.* Over 9,000,000 records; updated weekly or monthly. Available on DIALOG, files 350 (1963-1986), 351 (1987-), 352 (1963-). *Class No:* 608.3(003.4)

[2893]
Inpadoc/Family and legal status. Vienna, European Patent Office.

Lists patents in all areas of technology issued in 56 countries. Corresponds to the COM microfiche publication *INPADOC Patent gazette.* Over 10 million patent families, 17 million patent documents, 19 million legal status documents. Updated weekly. Available on DIALOG, file 345. *Class No:* 608.3(003.4)

[2894]
MARCHANT, P. Online patents and trademarks databases: 1989. 3rd ed. London, Aslib, 1988. [iv],52p. ISBN: 0851422489.

First published 1983 as *Patents databases.* 2nd ed. 1987.

Describes 55 patents and trademark databases, including updating, producer, host, costs, conditions, search aids, language. Also 8 'planned' databases. Directory of hosts and producers. Useful but inevitably dated.
Class No: 608.3(003.4)

[2895]
SIMMONS, E.S. 'Patents'. *Manual of online search strategies.* Editors, C.J. Armstrong and J.A. Large, pp.51-127. 2nd ed. Aldershot, Ashgate, 1992. ISBN: 1857420071.

Useful introduction to the range of patent databases available, their coverage and host systems. Includes information on formulating searchers for effective retrieval. *Class No:* 608.3(003.4)

Bibliographies

[2896]
Information sources in patents. Auger, P., *ed.* New Providence, NJ, Bowker Saur, 1992. *c.*300p. $70. ISBN: 0862919061.

Covers all forms of patent literature, including online and CD-ROM sources. Includes life sciences and the engineering, chemical and pharmaceutical industries. International in scope. *Class No:* 608.3(01)

[2897]
International guide to official industrial property publications. Van Dulken, S., *ed.* 3rd ed. London, British Library, Science, Technology and Industry, 1992. v.p. facsims. £45. (*Key resource series.*) ISBN: 0712307925.

First published 1985, ed. by B. Rimmer; 2nd ed. 1988, updating supplement 1990.

15 sections: 2-13. Countries (in geographical groupings) - 14. International conventions - 15. Computer-based searches. Detailed, analytical index. An 'indispensable volume' (*Reference reviews,* V.7(1), 1993, p.26). *Class No:* 608.3(01)

[2898]
Introduction to patents information. Van Dulken, S., *ed.* 2nd ed. London, British Library Science Reference and Information Service, 1992. 126p. facsims. £25. (*Information in focus.*) ISBN: 0712307907.

Chapters on value of patents, layout, granting procedure, publications, classification and electronic retrieval. 4 countries (UK, US, Germany, Japan) are covered closely. Glossary, country code listing, 'indispensable' (*Reference reviews,* V.7(1), 1993, p.26) bibliography, detailed, analytical index. *Class No:* 608.3(01)

[2899]
NAGANAGOUDAR, I.R., *and others.* **World literature on patents (1900-1981).** [New Delhi]: Council of Scientific and Industrial Research, 1981. 204p.

Prepared for the National Seminar on Patent Systems held at Hyderabad in 1981.

A bibliography of 1,828 books, conference papers and journal articles covering invention, history and development of the patent system, patenting procedure, patent law, patent

....(contd.)

information and patent surveys. Entries are arranged according to date of publication in ten-year periods, and indexed by author and subject. *Class No:* 608.3(01)

[2900]
Patent information and documentation handbook. Geneva, World Intellectual Property Organization, 1981-.

An English/French, multivolume, loose-leaf work updated periodically with replacement sheets. Sections are devoted to standards for patent documentation, the international patent classification, Patent Cooperation Treaty, kinds of patent publications, microforms, and access to patent documentation. *Class No:* 608.3(01)

[2901]
—HILL, M. Patent documentation. London, Sweet and Maxwell, 1979. vii, 196p.

Derived from a German edition by A. Wittman and R. Schiffel, updated and adapted to UK requirements. It provides information on the practical use of the world's patent literature. *Class No:* 608.3(01)

Handbooks & Manuals

[2902]
LEVY, R.C. Inventing and patenting sourcebook: how to sell and protect your ideas. Huffman, R.J., *ed,*. New York, Gale, 1990. 922p. $75. ISBN: 0810348713.

Supplies a wealth of directory-type information from a US angle. *Class No:* 608.3(035)

Dictionaries

Polyglot

[2903]
KASE, F.J. Dictionary of industrial property. Legal and related terms. English, Spanish, French and German. Alphen aan den Rijn, Sijthoff and Noordhoff, 1980. xii, 216p. £29.25. ISBN: 902860619x.

2,416 terms in English, A-Z, with Spanish, French and German indexes. *Class No:* 608.3(038)=00

[2904]
SZENDY, G.L. **Wörterbuch des Patentwesens in fünf Sprachen.** 2nd ed. Düsseldorft, VDI-Verlag, 1985. xxxvii, 1088p. ISBN: 3184005917.

A multilingual dictionary of German, English, French, Spanish and Russian terms covering patent terminology as well as industrial property in the broader sense. German-base terms, with equivalents and indexes in the other four languages. *Class No:* 608.3(038)=00

[2905]
UEKI, E. Six-languages dictionary of industrial properties. Tokyo, Patent Data Center, 1979. Y58,000.

4,000 industrial property and copyright terms are listed in each of the six sections: Japanese, English, French, German, Russian, Spanish. Against each term are the translations in each of the other five languages. *Class No:* 608.3(038)=00

[2906]
—Industrial property glossary. Geneva, World Intellelectual Property Organization, 1980-81. 4v. (WIPO publications 817-8 824-5).

Gives translations of about 300 terms, in Russian, English and French; Portuguese, French and English; Chinese, English and French; German, English and French, respectively. *Class No:* 608.3(038)=00

[2907]
—MIATELLO, A. Trilingual glossary on industrial property. Turin, A. Meynier, 1985. 93p.

Covers 500 terms and phrases. Three parts: French - Italian, English - Italian, and Italian - French - English. Most of the terms are followed by references to provisions of relevant texts in which the terms are used. *Class No:* 608.3(038)=00

German

[2908]
UEXKÜLL, J.D. F. von *and* REICH, H.J. **Dictionary of patent practice.** 3rd ed. Koln, Carl Heymanns Verlag, 1983. xxv, 276p. ISBN: 3452193225.

A German-English, English-German dictionary of patent terms and abbreviations. *Class No:* 608.3(038)=30

[2909]
WITTMANN, A. *and* KLOS, J. **Dictionary of patent terms:** English/German, German/English. 5th ed. Munich, Wila Verlag, 1986. 520p. ISBN: 3870191507.

Rev. and enlgd. ed. of the work established by B. Klaftan (1971).

Appended to each part is a supplement of special terms used in patent drawings and in text processing. *Class No:* 608.3(038)=30

Japanese

[2910]
RUSSELL, R.W. Patents and trademarks in Japan. 3rd ed. Tokyo, Russell, 1974. [5], 401p.

Over 700 Japanese terms are transliterated and arranged in alphabetical order of English meaning, with a full explanation of their significance in Japanese law. *Class No:* 608.3(038)=956

Reviews & Abstracts

[2911]
Abstracts and abridgements of patent specifications. London, Patent Office, 1883-. Weekly. Annual volume. illus.

Abstracts of British patent specifications published under the Patents Act, 1977, plus abridgements published (decreasingly) under the Patents Act, 1979. About 400 items per issue. Subject and name indexes weekly, cumulated. *Class No:* 608.3(048)

[2912]
Auszüge aus den Europäischen Patentanmeldungen. Munich, Wila Verlag, 1980-. Weekly.

Abstracts in English, French or German of applications for European Patents. Published in thee parts: Chemistry; Electricity and physics; and General, with abstracts arranged

....*(contd.)*

according to the International Patent Classification. Earlier abstracts were published by the European Patent Office in a different format. *Class No:* 608.3(048)

[2913]

—European patent bulletin. Munich, European Patent Office, 1978-. Weekly. ea. DM 20.

Contains bibliographical details of patent applications published and patents granted. Details are given of other stages reached by each European patent application. Annual name indexes to granted patents and to published applications are published separately. *Class No:* 608.3(048)

[2914]

Gewerblicher Rechtsschutz und Urheberrecht (GRUR). Internationaler Teil. Weinheim, VCH Verlagsgessellschaft, 1967-. Monthly. ISSN: 04358600.

Abstracts in German but international in scope. Classified coverage of periodical articles, textbooks and case law on various aspects of protection of industrial property and copyright. Sections on decisions in industrial property cases. Includes book reviews. Some items, but not the abstracts, appear in English translation in the quarterly *International Review of Industrial Property and Copyright Law* (IIC) by the same publisher (ISSN 00189855). *Class No:* 608.3(048)

[2915]

World patents abstracts journal. London, Derwent Publications, Ltd., 1975-. Weekly.

Currently divided into four subject-based sections giving abstracts of general and mechanical patents.

Individual country coverage by Derwent abstract journals: *German patents gazette* (1968-), *German patents abstracts* (1953-), *Soviet inventions illustrated* (1961-), *United States patents abstracts* (1971-), *Belgian patents abstracts* (1955-), *British patents abstracts* (1951-), *French patents abstracts* (1961-), *Japanese patents report* (1962-), *Japanese patents gazette* (1975-), *Netherlands patents report* (1964-), *European Patents Report* (1978-), *PCT Patents Report* (1978-). *Chemical Patents Index* (1951-) is a weekly abstracting service dealing with chemical patents and *Electrical Patents Index* (1980-), is the corresponding service covering patents on electrical technology. *Class No:* 608.3(048)

[2916]

World patents index gazette. London, Derwent Publications Ltd., 1974-. Weekly.

Four sections: Chemical, General, Mechanical, Electrical. Currently each section has four indexes: Patentee, IPC, Accession number, Patent number. Separate priority concordance index covering all four sections. Cumulations for all indexes on microfiche. Covers about 30 patenting authorities. *Class No:* 608.3(048)

Periodicals

Bibliographies

[2917]

BARTON, H.M. Industrial property literature: a directory of journals. London, British Library, Science Reference Library, 1981. 60p. £5. ISBN: 0902914588.

Gives details of the contents of the official patent, trade

....*(contd.)*

mark and design gazettes and the unofficial periodicals from 80 countries as well as series with multinational coverage. *Class No:* 608.3(051)(01)

Worldwide

[2918]

Industrial property. Geneva, World Intellectual Property Organization, 1961-. Monthly. ISSN: 00198625.

Source of information on developments to international treaties, at the international bureau of WIPO and within member countries. Contains calendar of events, book reviews, removable sections on new industrial property laws and annual statistics. *Class No:* 608.3(051)(100)

[2919]

PCT [Patent Cooperation Treaty] gazette. Geneva, World Intellectual Property Organization, 1978-. fortnightly. ea. SF18. ISSN: 02507757.

In four sections: 1. Published international applications, giving bibliographic details and an abstract - 2. Notices relating to Applications - 3. Indexes - 4. General information. Annual name and subject-matter indexes. *Class No:* 608.3(051)(100)

[2920]

World patent information. Oxford, Pergamon, Published for the Commission of the European Communities and the World Intellectual Property Organization, 1979-. Quarterly. DM290 pa. ISSN: 01722190.

Covers patent documentation, classification and statistics and contains articles, short communications, details of meetings, and literature reviews. *Class No:* 608.3(051)(100)

Europe

[2921]

Official journal of the European Patent Office. Munich, European Patent Office, 1978-. Monthly. ea. £5.50. ISSN: 01709291.

Contains reports of the EPO Boards of Appeal, general information from the EPO and from the contracting states, lists of professional representatives and a calendar of events. *Class No:* 608.3(051)(4)

Great Britain

[2922]

Official journal (patents). London, Patent Office, 1854-. Weekly. ea. £4.30.

Contents: official notes; details of applications for patents; applications published; granted patents (including European patents designating the UK); details of designs registered and renewed; progress of patents and legal proceedings; receipts of foreign patents in the Science Reference and Information Service. Appended list of additions to the Science Reference Library (Holborn). Annual cumulated name index, annual cumulated subject-matter indexes in each volume of abstracts of patent applications. *Class No:* 608.3(051)(410)

USA

[2923]
Official gazette of the United States Patent and Trademark Office: Patents. Washington, US Patent and Trademark Office, 1872-. Weekly.

Includes offical notices, abstracts of rexamination, reissue, utility, plant and design patents; abstracts of statutory invention registrations.

Annual index published separately. Part I: cumulated lists of applicants for patents, reexaminations, etc. Part II: index to subjects of invention - separate classified listings of patents, design patents, etc., and summary of publication statistics.

Available online as *Claims/US Patent abstracts,* DIALOG, files 23, 24, 25, 340. *Class No:* 608.3(051)(73)

Tables & Data Books

[2924]
100 years of industrial property statistics. Geneva, World Intellectual Property Organization, 1983. 245p. (*WIPO publication no. 876.*) ISBN: 9280501046.

A compilation of statistics from the periodical *Industrial property* (*qv*) and other WIPO publications. Contains synoptic tables on patents, trademarks, designs, utility models and plant varieties. *Class No:* 608.3(083)

[2925]
BOSWORTH, D.L. Intellectual property rights. Oxford, Pergamon Press, 1986. xiii, 288p. (£33. Reviews of United Kingdom statistical sources. V.19. Review 32.). ISBN: 0080339026.

A review of publicly available sources of statistical information on patents, designs, plant protection and royalties. Contains factual information about statistical series in the form of a Quick Reference List, extensive appendices and a subject index. *Class No:* 608.3(083)

Europe

CD-ROM

[2926]
ESPACE-EP. Woodbridge, CT., Reading, Berks., Research Publications International. $3195pa.

Complete collection of full European patent applications. 13 searchable fields, including IPC codes, applicant names, title words in English, German, French. About 60 CD-ROMs pa. *Class No:* 608.3(4)(003.40)

USA

CD-ROM

[2927]
Patent View: US patent information on CD-ROM. Woodbridge, CT., Reading, Berks., Research Publications International. $4995pa.

Full-image US Utility Patents on CD-ROM. Searchable on 22 bibliographic fields. *Class No:* 608.3(73)(003.40)

Information Services

[2928]
EISENSCHITZ, T.S. Patents, trade marks and designs in information work. London, Croom Helm, 1987. 236p. £40. ISBN: 0709909586.

Introductory section on the concepts behind industrial property followed by sections on patents, industrial copyright and designs, trade marks and EEC competition law. The book is intended as a source book for information workers, with details of the documentation particularly of the UK and US. *Class No:* 608.3:061:025.5

[2929]
EUROPEAN COMMUNITIES COMMISSION. Patent information and documentation in Western Europe. 3rd ed. Munich, K.G. Saur, 1988. 219p.

Lists patent literature in 18 Western European countries, with additional sections on Japan, Soviet Union, USA, the World Intellectual Property Organization and the Patent Cooperation Treaty, - plus international standards for patent literature, abstracting services, and databases. *Class No:* 608.3:061:025.5

[2930]
World directory of sources of patent information. Geneva, World Intellectual Property Organization, 1985. Loose-leaf. (*WIPO publication no. 209 (E).*) ISBN: 9280501097.

Lists detailed holdings of Patent Office libraries and other major patent libraries. Listing is by information centre with indexes to classified collections, numerical collections, official gazettes and abstracts by office of publication. *Class No:* 608.3:061:025.5

Government Organizations

[2931]
Directory of national and regional industrial property offices. Geneva, World Intellectual Property Organization, 1986. (*WIPO publication no. 601.*) ISBN: 9280500988.

A loose-leaf publication updated at least once a year containing the addresses, telephone numbers etc., of patent offices. *Class No:* 608.3:061:061.1

Patent Law

Worldwide

[2932]
BAXTER, J.W. World patent law and practice. 2nd ed. London, Sweet and Maxwell; New York, Matthew Bender, 1973. £195.

Vol.1, xiv, 455p, with 1st suppt. (1974), second suppt. (1976). Vol.2, loose-leaf, periodic supplements comprising substitute pages.

A multinational reference work for patent practitioners arranged in subject order so that the position in any topic can be easily obtained for a large number of countries. Additional chapters contain details of conventions and new legislation. *Class No:* 608.3:347.77(100)

[2933]
FYSH, M. *and* THOMAS, R.W. **Industrial Property Citator.** London, European Law Centre, 1982. xiv, 321p. £65. ISBN: 0907451047.

Updated by a supplement published in 1985.

Comprises a list of cases in alphabetical order of the name of the plaintiff with subject indexes for UK and Ireland and for overseas cases. Coverage is reported court cases on patents, trade marks, designs, plant varieties, copyright and neighbouring rights. *Class No:* 608.3:347.77(100)

[2934]
JACOBS, A.J., *ed.* **Patents throughout the world.** 3rd ed. New York, Trade Activities, 1985-. loose-leaf. ISBN: 0876321252.

Updated three times a year.

Brief details are provided of the patent procedures of over 150 countries. Separate tables enable features such as taxes to be compared at a glance for all countries and there is an extensive section on international conventions. *Class No:* 608.3:347.77(100)

[2935]
LADAS, S.P. **Patents, trademarks and related rights.** Cambridge, Mass., Harvard Univ. Press, 1975. 3v. xv, ix, ix, 2115p. ISBN: 0674657756.

Three volumes divided into 10 parts. This work presents a study of the development of national and international aspects of industrial property rights (patents, industrial designs, utility models and trademarks) and as such is not intended as a quick reference book on procedures or legal aspects. Index to texts and to cases cited. *Class No:* 608.3:347.77(100)

[2936]
Manual for the handling of applications for patents, designs and trade marks throughout the world. 2nd ed. Amsterdam, Manual Industrial Property B.V., 1936-. loose-leaf, in 4 v. ISBN: 9071888010.

Updated by supplements about once a year.

For each of over 180 countries details are given of the patent, design and trade mark laws and procedures. There are also sections on international conventions and agreements. This title is commonly referred to as the 'Dutch Manual'. *Class No:* 608.3:347.77(100)

[2937]
MELVILLE, L.W. **Forms and agreements on intellectual property and international licensing.** 3rd ed. London, Sweet and Maxwell; New York, Clark Boardman, 1984. 2v. £160. ISBN: 0421053002.

Loose-leaf, updated anually.

Details of the UK, US and European Community Law on intellectual property: patents, know-how, trade secrets, copyright, and plant variety rights. Model clauses or 'precedents' for use in licensing agreements, tables of statutes, regulations, and agreements are given. *Class No:* 608.3:347.77(100)

[2938]
REVERDIN, A. *and* SCHLAEPFER, F., *eds.* **Katzarov's manual on industrial property.** 9th ed. Geneva, Katzarov S.A., 1981. Loose-leaf, in two binders.

Kept up-to-date with annual or more frequent supplements.

After extensive sections on international agreements there are separate sections for each country giving basic national information and details of patent, trade mark and other

.... *(contd.)*
industrial property law and procedures. For each country there are lists of publications and industrial property offices, societies and agents. *Class No:* 608.3:347.77(100)

Europe

[2939]
BLANCO WHITE, T.A., *and others, eds.* **Encyclopedia of United Kingdom and European patent law.** London, Sweet and Maxwell, 1977. Loose-leaf; supplements. £124. ISBN: 0421233508.

Extensive commentary and annotated text on the UK 1949 and 1977 Patents Acts and Rules; the European Patent Convention, Community Patent Convention and Patent Cooperation Treaty with Rules; the Paris Convention for the protection of Industrial Property; and the Strasbourg Conventions on the unification of patent law and classification. *Class No:* 608.3:347.77(4)

[2940]
CHARTERED INSTITUTE OF PATENT AGENTS. **European patents handbook.** 2nd ed. London, Longman Professional, 1988-. Loose-leaf.

First published 1978.

For patent practitioners contains introductory material, commentary, source material, decisions, legal advice and a noter-up as part of the updating service. It is concerned with the European Patent Convention and the Community Patent Convention. *Class No:* 608.3:347.77(4)

[2941]
—Guidelines for examination in the European Patent Office. Munich, European Patent Office, 1979-. Loose-leaf, updated periodically by replacement pages. £20.

Written for the staff of the EPO, but useful as a guide on practice and procedures for patent practitioners. *Class No:* 608.3:347.77(4)

[2942]
HEARN, P. **The Business of industrial licensing:** a practical guide to patents, trade marks and industrial design. 2nd ed. Aldershot, Gower, 1986. xii, 252p. £25. ISBN: 0566025876.

Written for business executives, the book describes licensing law and practice in the UK and European Community. The final section sets out sample heads of agreement and is referenced to corresponding portions of text. *Class No:* 608.3:347.77(4)

[2943]
—CAWTHRA, B. Patent licensing in Europe. 2nd ed. London, Butterworths, 1986. xxi, 249p. £40. ISBN: 0406148414.

Traces developments in the patent licensing laws in the European Community, citing significant decisions of the European Court. A commentary on clauses in licensing agreements is given, and recent developments in Community patent licensing law are discussed. *Class No:* 608.3:347.77(4)

[2944]
PATERSON, G. **The European patent system.** London, Sweet & Maxwell, 1992. 600p. £70. ISBN: 0421430508.

Commentary on case law developed by the Board of Appeal following introduction of the European Patent Convention. *Class No:* 608.3:347.77(4)

[2945]

WALTON, A.M., *and others*. **Patent law of Europe and the United Kingdom.** London, Butterworths, 1981-82. £120. ISBN: 040641503x.

Loose-leaf, updated to 1982 by substitute pages. 2v.

A commentary on the UK 1949 and 1977 Patents Acts, the European Patent Convention, Community Patent Convention and Patent Cooperation Treaty. Various statutes are reproduced and other sections contain information on forms, practice and precedents. *Class No:* 608.3:347.77(4)

Great Britain

[2946]

ALDOUS, W., *and others*. **Terrell on the law of patents.** 13th ed. London, Sweet and Maxwell, 1982. lxi, 1157p. £71. ISBN: 0421249005.

First published 1884.

After a short history and discussion of the nature of patentable inventions, the 1977 and the surviving parts of the 1949 UK Patents Acts are analysed in detail. The effect of the European Patent Convention (EPC) is considered and appendices give the texts of the EPC, Patent Cooperation Treaty, rules of the Supreme Court, articles from the Treaty of Rome, as well as the 1977 and 1949 UK Patents Acts. *Class No:* 608.3:347.77(410)

[2947]

BLANCO WHITE, T.A. **Patents for invention.** 5th ed. London, Stevens, 1983. xxxv, 304p. £41. ISBN: 0420463100.

A text for practitioners on the previous UK patent law, the 1949 Patents Act. *Class No:* 608.3:347.77(410)

[2948]

CHARTERED INSTITUTE OF PATENT AGENTS. **CIPA Guide to the Patents Act, 1977.** 2nd ed. London, Sweet and Maxwell, 1984. liii, 772p. £98. ISBN: 0421305304.

Updated by supplements.

Reproduces the UK Act section by section, with commentary. There is also commentary on the Patent Rules 1982 and appendices provide text and commentary on the 1949 Patents Act and relevant rules and other statutes. *Class No:* 608.3:347.77(410)

[2949]

CORNISH, W.R. **Intellectual property:** patents, copyright, trade marks and allied rights. London, Sweet and Maxwell, 1989. xlvi,528p. £30. ISBN: 0421379707, ; 0421379804, Pbk.

A detailed text on British law in the context of the EEC written for the relative novice to intellectual property. Two chapters are given on the common ground including enforcement of rights and the remaining fifteen chapters are divided into sections on patents, confidence, copyright and trade marks. There are appendices on control of monopolies & restrictive practices, 'old' patents, joint interests in mark and protection of plant varieties. *Class No:* 608.3:347.77(410)

[2950]

—Intellectual property rights and innovation. London, H.M. Stationery Office, 1986. iv, 78p. £6.70. (*Cmnd. 9712.*)

The UK Government's White Paper containing proposals for reform of patents, designs and copyright, following the Green Papers *Intellectual property rights and innovation* (Cmnd. 9117), *Copyright and design law* (Cmnd. 6732),

.... *(contd.)*

Reform of the law relating to copyright, designs and performer's protection (Cmnd. 8302), and *The recording and rental of audio and video copyright material* (Cmnd. 9445). *Class No:* 608.3:347.77(410)

[2951]

GRALLAFENT, R.J., *and others*. **Intellectual property law and taxation.** 3rd ed. London, Longman, 1989. xxvi, 338p. ISBN: 0851213227.

2nd ed. 1984.

In three parts: intellectual property law by type; the taxation of intellectual property rights; and practical examples in the form of four case studies. Appendices list model agreements, statutes, and relevant tax publications. *Class No:* 608.3:347.77(410)

[2952]

MERKIN, R. **Richards Butler on copyright, designs and patents: the new law.** London, Longman, 1989. xl,693p. £55. ISBN: 0851215653.

A commentary on the current British patent law followed by the full text of the 1988 Act. Detailed, analytical index, p.677-93. *Class No:* 608.3:347.77(410)

[2953]

REID, B.C. **A Practical guide to patent law.** Oxford, ESC Publishing, 1984. xxii, 438p. £19.95; £13.95. ISBN: 0906214165; 0906214167.

Written for the student, lawyer or business reader this book provides a guide to UK patent practice with reference to international conventions. Lists of statutes, appendices and a glossary are given.

A new ed. appeared in 1992 (London, Sweet & Maxwell. ISBN: 042145220x). *Class No:* 608.3:347.77(410)

Korea

[2954]

DRAZIL, J.V. **Guide to Japanese and Korean patents** and utility models. London, British Library, Science Reference Library, 1976. 135p. £5. ISBN: 0902914219.

A historical survey of the protection of industrial property in the two countries with extensive details of the gazettes and specifications published. *Class No:* 608.3:347.77(519)

Japan

[2955]

Japanese laws relating to industrial property. Revised and reprinted 1975. Tokyo, Japanese Group of AIPPI, 1966-. 2 vol. loose-leaf.

Translated by the Japanese Patent Office.

English language translations of the Japanese law and regulations on patents, utility models, designs and trade marks. *Class No:* 608.3:347.77(52)

USA

[2956]

LIPSCOMB, E.B. **Walker on patents.** 3rd ed. Rochester, N.Y. Lawyers Cooperative Publishing; San Francisco, Bancroft-Whitney, 1984-. 11 vol.

The third edition of 'Walker' is being issued in volumes as

....(contd.)

they are prepared. Aimed at the US bench and bar, it provides a detailed commentary on US patent law and procedure. *Class No:* 608.3:347.77(73)

[2957]

ROSENBERG, P.D. Patent law fundamentals. 2nd ed. New York, Clark Boardman, 1986-. Loose-leaf. $155. ISBN: 0876320981.

Two vols. updated by supplements.

The work is an extensive text on US patents, trade secrets, trade marks and copyright with a final part devoted to international patent conventions and comparative foreign patent law. Appendices give samples of forms and documentation, related laws and rules.
Class No: 608.3:347.77(73)

Industrial Designs & Models

[2958]

FRENCH, M.J. Invention and evolution: design in nature and engineering. Cambridge, Univ. Press, 1988. xvi,324p. illus., diagrs. £35. ISBN: 0521307597; 0521314925, Pbk.

Covers numerous elements of design for function. 10 chapters, closely subdivided (*e.g.* 2. Energy ... 3. Materials ... 5. Structures (beams; feathers, etc.) ... 8. Economy, form and beauty ... 10. Designing and inventing). Detailed, analytical index, p.318-24. *Class No:* 608.4

[2959]

FYSH, M. Russell-Clark on copyright in industrial design. 5th ed. London, Sweet and Maxwell, 1974. xxiii, 314p. £40. ISBN: 0421176008.

Though this book does not include recent court decisions on the subject, it remains a definitive commentary on British law for the practitioner. This edition also contains a chapter on EEC legislation as well as tables of cases and statutes. *Class No:* 608.4

[2960]

GREENE, A.M., *ed.* **Designs and utility models throughout the world.** Updated by A.J. Jacobs. New York, Trade Activities, 1983-. Loose-leaf. ISBN: 0876323786.

Updated periodically.

Brief details are provided of the design and petty patent or utility model procedures of over 200 countries. Appendices provide details of international conventions, national and international classification schemes and at-a-glance comparative tables of available protection, type of examination, renewal terms, etc. *Class No:* 608.4

[2961]

KASE, F.J. Designs: a guide to official literature on design protection. Dobbs Ferry, N.Y., Oceana; Leyden, Sijthoff, 1975. 411p. ISBN: 0379003686; 9028603956.

Introduction to official literature on the protection of industrial designs and utility models. Entries are arranged by country (89). Each gives name and address of industrial property office and then information on: kinds of protection and definitions; Official journal and other publications; other official publications relating to designs; duration of protection. Includes facsimile reproductions of official journals. Appendix; international classification for industrial designs. Bibliography. Index. *Class No:* 608.4

[2962]

—**MYRANTS, G.** Protection of industrial designs. London, McGraw-Hill, 1977. xx, 211p. illus. £6.95. ISBN: 007084495x.

A practical reference tool on the procedures involved and advice in selling protection of industrial designs in the UK and abroad. Artistic copyright and design copyright are covered in one chapter. Index. *Class No:* 608.4

61 Medicine

Medicine

Abbreviations & Symbols

[2963]
DUPAYRAT, J. Dictionary of biomedical acronyms and abbreviations. 2nd ed. Chichester, John Wiley and Sons, 1990. 162p. £19.50. ISBN: 0471926493.

First published 1984.

About 7,000 definitions for the 4,000 most common biomedical acronyms. Increase in the number of basic acronyms is 30% up on the previous edition. Intended primarily for researchers, physicians and translators. *Class No:* 610(003)

[2964]
HABER, K. Common abbreviations in clinical medicine. New York, Raven Press, 1988. xii,227p. $17.50. ISBN: 0881673390.

Compilation of abbreviations found in patients' records and on medical charts. 2 parts: 1. Abbreviations, their meanings, and specialties most likely to use these abbreviations - 2. Abbreviations according to the usage of various medical subspecialties (Anesthesiology ... Urology). *Class No:* 610(003)

[2965]
HEISTAR, R. Dictionary of abbreviations in medical sciences: with a list of the most important medical and scientific journals and their traditional abbreviations. London, Springer-Verlag, 1989. viii,287p. $39. ISBN: 0387504877.

Contains *c.*15,000 abbreviations, including some in Latin, French, German, Italian and Spanish. 13-page appendix lists 460 journal title abbreviations. *Class No:* 610(003)

[2966]
LOGAN, C.M. *and* **RICE, M.K. Logan's medical and scientific abbreviations.** Philadelphia, Pa., J.B. Lippincott Co., 1987. xix, 673p. $29.95. ISBN: 0397545894.

Compiled from over 60 sources. 6 sections: 1. Abbreviations - 2. Symbols - 3. Chemotherapy regimes - 4. Latin terms: charting and prescriptions - 5. Cancer staging abbreviations - 6. Elements. Extensive coverage, *e.g.* the single letter A has 53 terms. Useful bibliography, p.671-3. *Class No:* 610(003)

[2967]
MASA: medical acronyms, symbols and abbreviations. Hamilton, B. *and* Guidos, B., *eds.* 2nd ed. New York, Neal-Schuman, 1988. 277p. $55. ISBN: 1555700128.

First published 1984.

Over 32,000 entries, 12,000 added since the previous edition. Terms found in older literature and medical records are included in addition to current abbreviations. Numerous cross-references. *Class No:* 610(003)

[2968]
Medical abbreviations and eponyms. Sloane, S.B., *ed.* Philadelphia, Pa., & London, W.B. Saunders Co., 1985. 432p. $19.45; £19. ISBN: 0721615228.

A small-format compendium in two parts. 'Medical abbreviations', p.1-93, has *c.*6,000 entries ('LA': 10 applications) and includes chemical structures. 'Medical eponyms', p.194-402, carries *c.*7,000 entries. 5 appendices (*e.g.* Symbols; Positions of fetus). *Class No:* 610(003)

[2969]
Stedman's abbreviations, acronyms & symbols. Baltimore, Md., Williams & Wilkins, 1992. ix,664p. $24. ISBN: 0683079263.

About 20,000 entries arranged clearly in 2 columns per page. Derived from the literature studied to enhance the coverage in *Stedman's medical dictionary (q.v.)*. 2 sections: 1. A-Z abbreviations & acronyms, p.1-655 - 2. Symbols (*e.g.* 'Arrows'; 'Numbers'; 'Primes, checks, dots and miscellany'). A useful quick-reference tool. *Class No:* 610(003)

[2970]
STEEN, E.B. Baillière's abbreviations in medicine. 5th ed. London, Baillière Tindall, 1984. 256p. $10. ISBN: 0702010367.

First published 1960.

Lists more than 7,500 common medical and related abbreviations ('CD': 16 different meanings), p.1-248. Appended abbreviations of titles of the principal medical journals. 'Some useful sources', p.254-5; Symbols. Still covers most needs. *Class No:* 610(003)

[2971]
—**JABLONSKI, S. Dictionary of medical acronyms and abbreviations.** Philadelphia, Pa., Hanley & Belfus; St. Louis, etc., C.V. Mosby, Co., 1987. vii, 205p. $14.95. ISBN: 0932883028.

Has *c.*18,000 entries. Small format. *Class No:* 610(003)

[2972]
Units, symbols, and abbreviations: a guide for biological and medical editors and authors. Baron, D.N., *ed.* 4th ed. London, Royal Society of Medicine Services, 1988. 64p. £5. ISBN: 0905958780.

First published 1971.

Sections: 1. Metrication and SI units - 2. Symbols and nomenclature (p.21-51; 23 references) - 3. Layout of references - 4. Proof correction. 'Many journals quote this booklet as a guide in their Instructions to Authors', (*Introduction*). *Class No:* 610(003)

Databases

[2973]
EMBASE. Amsterdam, Excerpta Medica/EMBASE Publishing Group, 1974-.

International in coverage and contains more than 4 million citations, many with abstracts, from more than 3,500

....*(contd.)*

journals. All medical subject areas included. Updated weekly. Corresponds mainly to the *Excerpta medica* series but *c*.25 percent of the citations added to the database each year do not appear in the corresponding hard copy journals. Available on CD-ROM, diskette, magnetic tape.

A *List of journals abstracted* is published annually.

Class No: 610(003.4)

[2974]

LYON, E. Online medical databases 1991. 5th ed. London, Aslib, 1991. 96p. *(Aslib online guide.)* ISBN: 0851422748.

First published 1983.

A list of 135 bibliography or full-text databases, all commercially available online. Data include content; producer; text; costs. The addresses of hosts, producers and suppliers are listed. Bibliography and index.

Class No: 610(003.4)

[2975]

MEDLINE [Medical Analysis and Retrieval Service On-Line]. Washington, National Library of Medicine, 1971-.

Database, with full coverage of *Index medicus* by 1974, with monthly update of *c*.25,000 entries. Search aids: MESH [Medical Subject Headings], an annual thesaurus of headings used in *Index medicus*, and Permuted Mesh headings. The UK contribution to MEDLINE is provided by the British Library, Boston Spa. Available on CD-ROM.

Class No: 610(003.4)

[2976]

Online databases in the medical and life sciences. New York, Cuadra/Elsevier, 1987. xxi, 170p. $29.95. ISBN: 0444012729.

Descriptions of 795 online databases, p.1-128. Data on each: type; subject; producer; online service; content; language; coverage; countries; time span; updating. Addresses of online services and gateways, p.129-36. Subject index; online service/gateway index; master index.

Class No: 610(003.4)

[2977]

World databases in medicine. London, Bowker-Saur, 1993. 350p. £60. ISBN: 0862916135.

A detailed, evaluative guide to databases currently available in medicine and the biosciences.

Class No: 610(003.4)

CD-ROM

[2978]

FRYER, R.K. 'Beyond MEDLINE: a review of ten non-MEDLINE CD-ROM databases for the health sciences'. In *Laserdisk professional,* v.2(3), May 1989, p.27-32.

Evaluates Compact Cambridge Life Sciences; CCINFOdisc; *Science citation index* Compact Disc Edition; ONCODISC; Compact Library; AIDS; Computerized Clinical Information Systems; CANCER-CD; CHEM-BANK; OSH-ROM and PsycLIT. *Class No:* 610(003.40)

Bibliographies of Bibliographies

[2979]

BESTERMAN, T. Medicine: a bibliography of bibliographies. Totowa, N.J., Rowman and Littlefield, 1971. 409p. *(The Besterman World bibliographies.)*

Entries are extracted from Besterman's *A world*

....*(contd.)*

bibliography of bibliographies (4th ed. 1965-66. 4v.). Arrangement is under subjects A-Z (*e.g., Anatomy*) and also specific topics. Extensive language coverage.

Class No: 610(009)

Bibliographies

[2980]

BRANDON, A.N. *and* HILL, D.R. 'Selected list of books and journals for the small medical library'. In *Bulletin of the Medical Library Association,* v.79(2), April 1991, p.195-222.

A revision of the list in *Bulletin of the Medical Library Association,* v.77(2), April 1989, p.139-69.

Book list contains 607 unannotated entries, AIDS ... Urology. 'Dictionaries, directories and encyclopedias', nos. 109-119. Author/editor index. Intended as a selection guide for the small or medium-sized library in a hospital or other medical facility or as a core collection for a group of small hospital libraries. US slanted. *Class No:* 610(01)

[2981]

COLE LIBRARY OF EARLY MEDICINE AND ZOOLOGY. Catalogue of books and pamphlets. [Cole Library of Early Medicine and Zoology]. Eales, N.B. [Oxford], Aldine Press, for the Library, Univ. of Reading, 1969-75. 2pts. ISBN: 070490148x, Pt. 2.

1. *1477 to 1800.* xiv, 425p. 2. *Catalogue of books, monographs and principal papers, 1800 to the present day,* (£7) and *Supplement to part 1.* xvii, 428p.

Part 1 records *c*.2,000 items, arranged chronologically.

Class No: 610(01)

[2982]

Consumer health information source book. Rees, A.M. *and* Hoffman, C., *eds.* 3rd ed. London, Oryx Press, 1990. 224p. £39.50. ISBN: 089774408x.

First published 1981.

Annotated lists of 750 books, 79 popular magazines, more than 700 pamphlets and 49 professional publications, 175 organizations. Topics arranged A-Z with reviews 100-300 words in length. Author, title and subject indexes. 90 percent of the sources listed in the bibliographies are new. For public libraries as well as health and medical collections.

Class No: 610(01)

[2983]

—Encyclopedia of health information sources. Wasserman, P., *ed.* Detroit, Mich., Gale, 1986. 483p. $155. ISBN: 0810321351.

Offers health information (literature and organizations) on over 450 medical subjects, with ample cross-references. Promised interim supplement has not yet been published.

Class No: 610(01)

[2984]

Coping with the biomedical literature: a primer for the scientist and the clinician. Warren, K.S., *ed.* New York, Praeger, 1981. xi, 233p. facsims., graphs, tables. $45. ISBN: 0275913554.

12 contributors. 4 parts (12 chapters, mostly documented): 1. The structure of the information system (*e.g.* 'Selective aspects of the biomedical literature') - 2. Producing biomedical information (*e.g.* 'Reviewing reviews') - 3. Utilizing biomedical information - 4. Sources of biomedical information ('The National Library of Medicine'; 'The

.... *(contd.)*

Institute for Scientific Information'; 'Libraries and how to use them'). Partly analytical index. Addressed, authoritatively, to a specialized readership. Thus, chapter 9, 'Evaluation: requirements for clinical application', 'has been summarized for those clinicians who are behind in their clinical reading'. *Class No:* 610(01)

[2985]
Core collection of medical books and journals. Hague, H. and Jackson, M., *comps.* Manchester, Medical Information Working Party, Haigh and Hochland, 1992. 64p. £2.50. ISBN: 1869888065.

List of over 800 books, arranged under 47 subject headings, and over 200 journals. Unannotated entries but useful for inexperienced medical librarians and aimed at postgraduate medical libraries. *Class No:* 610(01)

[2986]
Current bibliographies in medicine. Washington, D.C., US Government Printing Office, 1991-. $3ea; $52 (20-issue series).

Replaces *Literature search.*

Compiled by the National Library of Medicine. Each issue varies in length from 14 to 50 pages and contains lists of books, periodicals and audio-visual materials on topics of current popular interest. Recent titles include: 'Medical waste disposal'; 'Laboratory animal welfare'; 'Nutrition and AIDS'. *Class No:* 610(01)

[2987]
FRANCIS A. COUNTWAY LIBRARY OF MEDICINE (Harvard Medical Library). **Author-title catalog ... for imprints through 1959. [Francis A. Countway Library of Medicine].** Boston, Mass., G.K. Hall, 1973. 10v. (8104p.). $1,755. ISBN: 0816110247.

Author-title catalogue of *c.*263,000 photolithographed cards - the merged catalogue of Harvard and Boston medical libraries. The collection is rich in medical Hebraica and Judaica, manuscripts, histories of medicine, and clinical medicine. *Class No:* 610(01)

[2988]
GANN, R. The Health information handbook: resources for self-care. Aldershot, Hants, Gower, 1986. xii, 251p. £35. ISBN: 0566035499.

5 chapters, each with references: 1. Responsibility for health - 2. The informed patient - 3. Health information in the UK (p.114-26) - 4. Health information in other countries (p.169-179) - 5. Getting started: collecting and organising health information (books, booklets/leaflets, periodicals. Subject range: 'The aged' ... 'Women's health', p.204-25. Thesaurus of subject headings, p.229-42). Detailed, non-analytical index. A practical guide for librarians and other providers of basic medical information. Some UK slant. *Class No:* 610(01)

[2989]
Health science books, 1876-1982. New York, Bowker, 1982. 4v. (4601p.). $225. ISBN: 0835214478.

Subject index. 3v. (A-Z). *Author index. Title index.*

A comprehensive bibliography of over 132,000 books published in the United States and catalogued by Library of Congress during the last 107 years. 'Subjects covered include all major health disciplines as well as biomedical and related sciences' *(Preface)*. V.4 (p.3989-4601) includes 'Guide to MeSH/LC equivalents in health science books' and 'Guide to LC/MeSH equivalents'. *Class No:* 610(01)

[2990]
Information sources in the medical sciences. Morton, L.T. and Godbolt, S., *eds.* 4th ed. London, Bowker-Saur, 1992. xvi,608p. £50. *(Guides to information sources.)* ISBN: 0862915961.

First published 1974 as *Use of medical literature.*

27 contributors. 24 chapters: 1. Medical libraries and their use - 2. Primary sources of information ... 5. Online and CD-ROM sources - 6. Audio-visual materials ... 9. Public health ... 12. Pathology, clinical and experimental ... 16. Diagnostic radiology ... 21. Dentistry ... 24. Historical, biographical and bibliographical sources. Index to subjects and some titles but not to authors. *Class No:* 610(01)

[2991]
Introduction to reference sources in the health services. Roper, R.W. *and* Boorkman, J.A., *eds.* 2nd ed. Chicago, Ill., Medical Library Association, 1984. 302p. $27.20.

First published 1980.

A well-organized tool, aimed primarily at library school students, librarians and library users in the medical field. Chapters on such areas as indexing and abstracting sources; audiovisual material; conferences, reviews and translations. Computerized databases, and history sources, are backed by further readings. Some US emphasis. *Class No:* 610(01)

[2992]
—CHEN, C.-C. Health sciences information sources. Cambridge, Mass., and London, MIT Press, 1981. xxxviii, [1], 767p. $80; £71.95. ISBN: 0262030748.

Arranges *c.*3,500 annotated entries by type of material (1. Selection tools - 2. Guides to the literature - 3. Bibliographies - 4. Encyclopedias - 5. Dictionaries ... 22. Nonprint materials - 23. Professional societies and their publications - 24. Databases (chart), p.1-632). 'Reference literature' (periodical article references). Title and author indexes; author index to 'Reference literature'. Each class is divided by subject fields, A-Z. Absence of a subject index detracts considerably from the usefulness of this tool, - intended for library schools, with priority for the structure of the literature rather than its contents. *Class No:* 610(01)

[2993]
—Finding the source of medical information: a thesaurus index to the reference collections. Shearer, B.S. *and* Bush, G.L., *comps.* Westport, Conn., Greenwood Press, 1985. 275p. $42.95; £36.95. ISBN: 0313240949.

Lists 447 reference books and textbooks, by titles, as medical sources. Alternate title and author indexes follow. The main part of the work is the thesaurus subject index, p.43-275, A-Z, with cross-references from specific to more inclusive subjects. Useful as a quick-reference source-key. *Class No:* 610(01)

[2994]
—STRICKLAND-HODGE, B. *and* ALLAN, B. Medical information: a profile. London, Mansell, 1986. ix, 145p.

Also intended mainly for library-school students. Six chapters stress printed reference sources, on-line searching, the role of organizations and people as contacts, and the search strategy, with six case-studies, plus an index. 'The book would have been improved by involving someone experienced in the practice of answering medical enquiries in a clinical setting' *(British book news,* July 1986, p.410-11). *Class No:* 610(01)

[2995]
Keyguide to information sources in paramedical sciences.
Hewlett, J., *ed*. London, New York, Mansell, 1990.
xii,270p. £45. ISBN: 0720120578.

3 sections: 1. Overview of paramedical sciences and their literature (1. Information and the paramedical professions ... 9. Radiography, diagnostic and therapeutic) - 2. Bibliographical listing of sources of information, p.105-206 (1,012 items) - 3. List of selected organizations (1. International associations and societies ... 8. Audio-visual and non-book material producers). Index of subjects, titles and authors, p.239-70. Written by, and for, practitioners in the UK. 'It is, however, an exemplary model for anyone who perceives the need for producing a similar book directed toward an American readership' (*Choice*, v.28(9), May 1991, p.1460). *Class No:* 610(01)

[2996]
KURIAN, G. T. **Global guide to medical information.** New York, Elsevier, 1988. 808p. $101.75. ISBN: 0444013008.

International in scope, listing over 13,000 publications and organizations. Largest section lists periodicals. Medical publishers listed according to country. Names and addresses for national, regional, and international associations are given for each medical subject area. Not comprehensive but a good starting point for those looking for international information. *Class No:* 610(01)

[2997]
LONDON SCHOOL OF HYGIENE AND TROPICAL MEDICINE. University of London. **Dictionary catalogue. [London School of Hygiene and Tropical Medicine].** Boston, Mass., G.K. Hall, 1965. 6v. Serials catalogue, 286p. First supplement. 1971. 441p.

A photo-offset reproduction of the 90,000 cards, - author and subject entries for *c*.40,000 monographs, pamphlets and non-serial reports in the library, with additional sequences for periodical titles and serial reports.

The *Serials catalogue* comprises 6,000 cards. The *First supplement, c*.20,300 cards. *Class No:* 610(01)

[2998]
Medical and health care books and serials in print: an index to literature in print. New York, etc, Bowker, 1985-. Annual. 2v. $175. ISBN: 0835230325.

Originally as *Bowker's Medical books in print* (1972-77); editions 1-6, then as *Medical books and serials in print* (1978-84).

1. *Subjects. Authors.* 2. *Titles. Serials. Publishers.*

The 1991 annual lists more than 62,000 books and 12,000 serials. Subject (5,800 headings), author and title indexes. Also available online through DIALOG Information Services. *Class No:* 610(01)

[2999]
Medical reference works, 1679-1966: a selected bibliography. Blake, J.B. *and* Roos, C., *eds.* Chicago, Ill., Medical Library Association, 1967. viii, 343p. Supplements 1-3, 1970-75. $4.35, supp. 2; $3, supp. 3. ISBN: 081082440x, supp.2; 0810824418, supp.3.

2,703 numbered entries, concisely annotated, in the main volume. Nos.1-413: Medicine, general (indexes and abstracts; reviews; bibliographies; translations; theses; congresses; dictionaries; periodicals; directories; encyclopedias. - Nos.414-832: History of medicine - Nos.833-2073: Special subjects, A-Z (Aerospace and submarine medicine ... Zoology)). Excludes bibliographies

....(contd.)
of individuals, etc. Basic reference works for the smaller medical library are asterisked. A primary selection tool in its day.

Supplements 1-3 carry 1,059 items. Further supplements are due from National Library of Medicine computer bases. *Class No:* 610(01)

[3000]
MORTON, L.T. *and* WRIGHT, D.J. **How to use a medical library.** 7th ed. London, Library Association Publishing, 1990. v,88p. £14.95. ISBN: 0851574661.

First published 1934.

10 sections: 1. Introduction - 2. Finding your way around the library - 3. Periodicals - 4. Indexes and abstracts - 5. Searching the literature - 6. Basic reference works - 7. Aids for writers and speakers - 8. Historical and biographical sources - 9. Audio-visual aids - 10. Biomedical library facilities in Britain. References, p.78-82. Analytical index, p.83-88. Written as a guide for medical practitioners, research workers and those who are training for a career in medical librarianship. *Class No:* 610(01)

[3001]
NEW YORK ACADEMY OF MEDICINE. Library. **Author catalog. [New York Academy of Medicine Library].** Boston, Mass., G.K. Hall, 1969. 43v. (38339p.). 1st supplement. 1974. 4v. (3474p.). $4,290 + $545. ISBN: 0816108293; 081610851x.

The catalogue and its supplement comprise 877,900 photolithographically reproduced cards. At the end of 1967 the Library included 365,000 bound volumes and 165,000 pamphlets. *Class No:* 610(01)

[3002]
—NEW YORK ACADEMY OF MEDICINE. Library. **Subject catalog.** [New York Academy of Medicine Library]. Boston, Mass., G.K. Hall, 1969. 34v. 1st Supplement. 1974. 4v. $3,550, set; $545, supp. 1. ISBN: 0816101841, supp. 1; 0816108269, set. *Class No:* 610(01)

[3003]
UNITED STATES. National Library of Medicine. **A Catalog of sixteenth century printed books in the National Library of Medicine.** Dorling, R.J., *comp.* Bethesda, Md., US Dept. of Health, Education and Welfare, Public Health Service, National Library of Medicine, 1967. xii, 698p.

4,808 numbered items, p.1-618. Lengthy titles, concise bibliographical notes. Bibliographical references, p.vii-xii. Indexes of printers and publishers (geographical index; name index). Concordance of *Short-title catalogue* items and serial numbers in this catalogue. Index of vernacular imprints. *Class No:* 610(01)

[3004]
UNITED STATES. National Library of Medicine. **National Library of Medicine current catalog.** Bethesda, Md., US Dept. of Health, Education and Welfare, 1966-. 4pa. (previously 24pa., then 12pa), with annual and 5-yearly cumulation.

Preceded by the *Armed Forces Medical Library catalog ..., 1950-1954* (1955. 6v.) and *National Library of Medicine catalog, 1955-1959* (1960. 6v.) and 1960-1965 (1960. 6v.).

Annual cumulation 1989 (2v. 3211p. $58) has 4 sections: Monographs: name and title section - Serials - Serials, subject section - Medical reference works. *Class No:* 610(01)

[3005]
UNITED STATES. War Department. Surgeon-General's Office. **Index-catalog of the Library of the Surgeon General's Office, United States Army.** Washington, Government Printing Office, 1880-1955. 58v.

Series 1, 1880-95. 16v.; series 2, 1896-1916. 21v.; series 3, 1918-1932, 10v.; series 4, v.1-11 (A-MN), 1936-55.

A dictionary catalogue of 485,334 entries, listing contents of one of the world's largest medical libraries, covering book and periodical articles. Includes many biographical entries. Invaluable as a bibliographical tool in this field.

Concluded in US Dept. of Health, Education and Welfare, Public Health Service's *Index-catalog of the Library of the Surgeon General's Office, National Library of Medicine.* 5th series (Washington, G.P.O., 1959-61. 3v.). 'Entries are for items which are monographic in nature and they are for 1950 or earlier imprints' (*Preface*). V.1 covers Authors; v.2-3, Subjects. *Class No:* 610(01)

[3006]
WALTERS, B. **'Medical reference tools for the layperson'.** In *Reference books bulletin,* 1988/89, p.1-5.

38 annotated entries. Includes dictionaries; directories; drug handbooks; manuals and handbooks; statistics; medical education; physician and medical facility evaluations; indexes. 'This list selectively identifies key reference materials on diseases, drugs, physicians, health facilities and other personal health concerns' (*Introduction*).

Class No: 610(01)

[3007]
WELCH, J. *and* KING, T.A. **Searching the medical literature:** a guide to printed and online sources. With a chapter on drug information services, by D.E. Hands. London & New York, Chapman & Hall, 1985. x, 154p. facsims., diagrs., tables, map. ISBN: 041225610x.

8 chapters: 1. Introduction: the new information environment - 2/3. Searching the periodical literature (print resources; online (*e.g.* MEDLARS)) - 4. Drug information services - 5. Sources of statistical information - 6. The reference collection - 7. Background reading (primary and secondary sources; bibliographies and review articles; book reviews; library networks) - 8. Source of advice for medical communications. Each chapter carries further reading or references. Detailed index, p.147-54. A well organized guide. *Class No:* 610(01)

[3008]
WELLCOME HISTORICAL MEDICAL LIBRARY. **A Catalogue of printed books in the Wellcome Historical Medical Library.** London, the Library, 1962-.

1. *Books printed before 1641.* 1962. xv, 407p. 2. *Books printed from 1641 to 1850. A-E.* 1966. 3. *Books printed from 1641 to 1850. F-L.* 1976. 565p.

Author catalogues. V.1, compiled by F.N.L. Poynter, records 7,000 items in 'perhaps the greatest collection of medico-historical works in the world' (*British book news,* no.270, February 1963, p.91-92). Brief references to items listed in *Catalogue of incunabula* (1951). (*qv*). Indexes follow, to places of publication, and to printer and publishers; concordance with STC numbers for English books.

V.2 lists *c.*18,000 items, some with biographical notes. If an item is not listed in Wing's STC, this is stated. Cross-references. Well-produced. *Class No:* 610(01)

[3009]
WORLD HEALTH ORGANIZATION. **Publications catalogue, 1948-1989.** [World Health Organization]. Geneva, WHO, 1990. *Gratis.*

5 sections: 1. Publications grouped by subject - 2. Nonserial publications - 3. Series - 4. Periodicals - 5. Official publications - 6. Regional publications - 7. Publications of the International Agency for Research on Cancer - 8. Publications of the Council for International Organizations of Medical Sciences.

WHO have also published *Publications catalogue: new books, 1986-90* and *Publications catalogue: new books, 1991. Class No:* 610(01)

USSR

[3010]
PERKINS, L. **A Bibliography of Soviet sources of medicine and public health in the USSR:** a publication of Geographic Health Studies by the John E. Fogarty International Center for Advanced Study in the Health Service. [Bethesda, Md.], US Dept. of Health, Education and Welfare, Public Health Service, National Institutes of Health, 1975. xiv, 235p.

About 10,000 entries. Sections 1-23 (1. Allergy ... 22. Urology and kidney disease - 23. Miscellaneous). Sections 24-31 (24. Biomedical research ... 31. Medicine and society). Entries are in tiny print. *Class No:* 610(01)(47)

Encyclopaedias

[3011]
Black's medical dictionary. 36th ed. London, A. & C. Black, 1990. 738p. ISBN: 0713632089.

First published 1906.

For the general public, with notes on causes, symptoms and treatment, and written in simple English. The 36th ed. claims 30 new entries (*e.g.* for 'Cluster headaches'; 'Cyclical oedema of women') while some sections have been completely rewritten (*e.g.* 'Dentistry'; 'Ophthalmology'). 37th ed., 1992. 744p. £19.99. *Class No:* 610(031)

[3012]
Blakiston's Gould medical dictionary: a modern comprehensive dictionary of the terms used in all branches of medicine and allied sciences. 4th ed. New York, etc., McGraw-Hill, 1979. xxi, 1632p. illus. (pl.), tables. $36. ISBN: 0070057001.

First published 1949 as *Blakiston's New Gould medical dictionary.*

A defining dictionary, with over 80,000 entries, giving syllabification and etymolgy. 34 contributors. 26 keyed anatomical plates (between letters N and O). Appendix tables (p.1499-1632) include 'Chemical constituents of blood', 'Thermoelectric equivalents', 'Weights and measures'.

Also, *Blakiston's Gould pocket medical dictionary* (4th ed. 1979. 992p. $28.50). *Class No:* 610(031)

[3013]
Butterworth's medical dictionary. 2nd ed. Editor-in-chief, M. Critchley. London, Butterworths, 1978. xxxii, 1942p. £29.50. ISBN: 040700193x.

First published as *The British medical dictionary,* edited by Sir A.S. McNally (1961; rev. ed., 1965).

About 58,000 entries, revised by a team of 50 consultants. 'Paralysis': 5½ columns; 'Spasm': 2 columns. Very brief biographical data under eponyms. Pronunciation of entry-

.... *(contd.)*

words; chemical structures. Appendix: 'Anatomical nomenclature', tabulated. Regarded as a landmark in British medical lexicography. '99.9% of the time this lexicon does exactly what one would expect of it' (*The Lancet,* no.8063, 11 March 1978, p.536). *Class No:* 610(031)

[3014]
Dictionnaire français de médecine et de biologie. Manuila, A., *and others.* Paris, Masson; New York, French and European, 1970-73. 4v. $175, v.1; $175, v.2; $195, v.3; $195, v.4. ISBN: 0686570324, v.1; 0686570332, v.2; 0686570340, v.3; 082885856x, v.4.

About 150,000 entries, including lengthy entries as well as definitions. Bibliographical references are provided for descriptions of illnesses and relatively rare syndromes. V.4 is intended as a guide to users of the *Dictionnaire.* Definitions are excellent and clear. A basic work for French and for French special libraries. It also provided the *International dictionary of medicine and biology* (*qv*) with a valuable starting point to build an 'exhaustive list of English terms' (*Acknowledgements,* v.1). *Class No:* 610(031)

[3015]
—Reallexikon der Medizin und ihrer Grenzgebiete. Munich, Urban & Schwartenberg, 1966-77. 6v.

The German equivalent of the *Dictionnaire.* *Class No:* 610(031)

[3016]
Dorland's illustrated medical dictionary. 27th ed. Philadelphia, Pa., W.B. Saunders, 1988. xxxii, 1888p. illus., tables. $31.95; £26.50. ISBN: 0721631541.

1st-23rd eds. as *The American illustrated medical dictionary.*

17 consultants. Over 80,000 entries, including abbreviations. 'Cell' (with subentries for types of cell in bold): 15 columns. Claims to establish 'certain standards of etymological propriety' by including 'Fundamentals of medical etymology', p.xvii-xxxii. Occasional small line-drawings in addition to the more detailed 53 plates. Generally considered to be reliable. *Class No:* 610(031)

[3017]
—Dorland's medical dictionary: shorter edition. Philadelphia, Pa., W.B. Saunders, 1980. 768p. illus. (col.). ISBN: 0721631428.

Has c.45,000 entries on medicine and health care, and is addressed to the general public. Includes anatomical illus. in colour and various tables. For the small library unable to afford the full dictionary. *Class No:* 610(031)

[3018]
—Dorland's pocket dictionary. 24th ed. Philadelphia, Pa., W.B. Saunders, 1989. 688p. illus (col. pl.), charts, tables. $21. ISBN: 072162202x.

First published 1898.

Also gives pronunciation. Preliminary note on 'Combining forms in medical etymology', p.xiii-xxi. *Class No:* 610(031)

[3019]
Melloni's illustrated medical dictionary. Dox, I., *and others, eds.* 2nd ed. Baltimore, Md., Williams & Wilkins, 1985. li, 533p. illus. (col.), tables. $27.95; £21. ISBN: 0683026410.

First published 1979.

Defines 26,000 terms in non-technical language. Includes lengthy contributions (*e.g.* 'Nerves', p.313-32, with much tabular information). A distinguishing feature, - the

.... *(contd.)*

supporting 2,500 illus., usually occupying the top half of each page. Aims to be useful as a reference both for the general public as well as designed for students of the health services. Well produced. 'For non-specialists' (*Scientific American,* v.256, August 1986, p.27). *Class No:* 610(031)

[3020]
—RIGAL, W.A. The Inverted medical dictionary: a method of finding medical terms quickly. Westport, Conn., Technomic Pub. Co., Inc., 1976. v, 261p.

Groups all aspects of a part of the body, etc, for rapid identification (*e.g.* 'face' - enlargement, fissure, pains, spasms. etc.), followed by the correct medical term, plus a brief definition. 'To be used as a supplement' (*Choice,* v.4(9), November 1977, p.1194).

Also, Stanaszek, M.J., *The Inverted medical dictionary* (Lancaster, Pa., Technomic, 1991. $75).
Class No: 610(031)

[3021]
—SLOANE, S.B. The Medical word book: a spelling and vocabulary guide to medical transcription. 3rd ed. Philadelphia, Pa., W.B. Saunders, 1991. xi,1097p. illus., tables. $27.95. ISBN: 0721632432.

First published 1973.

Has 3 parts: 1. General terms; 2. Systems and specialities (15 heads); 3. Guides to terminology (*e.g.* abbreviations). Small format. *Class No:* 610(031)

[3022]
MILLER, B.F. *and* **KEANE, C.B. Encyclopedia and dictionary of medicine, nursing and allied health.** 4th ed. Philadelphia, Pa., W.B. Saunders, 1987. xxii, 1437p. illus. (incl. col.), diagrs., tables. $26.95. ISBN: 0721618154.

First published 1972, as *Encyclopedia and dictionary of medicine* (1089p.).

11 consultants. Over 30,000 entries ('Vocabulary', with anatomical tables and col. plates), p.1-1356. Preceded by 'Combining forms in medical terminology'. Table of muscles, p.790-808. 181 illus., 54 tables. Many cross-references. 13 appendices (2. Desirable weight for men and women; 15. Summary of normal development in the first three years). Thumb-indexed. Draws extensively on work now in progress for the 27th ed. of Dorland's *Illustrated medical dictionary.* Written by a physician and a nurse. A 5th ed. is due in 1992, edited by M. O'Toole (xxx,1780p. £16.50). *Class No:* 610(031)

[3023]
Mosby's medical, nursing, and allied health dictionary. 3rd ed. St. Louis, Mo., C.V. Mosby Company, 1990. xx,1608p. illus. (incl. col.), diagrs., graphs, tables. $25.95. ISBN: 0801632277.

First published 1983.

64 consulting contributors. About 26,000 entries, 3,000 new to this edition. Length of entries varies from a few words to several columns, *e.g.* 'Burn therapy': method; nursing orders; outcome criteria; 2 columns. 44-page colour atlas of human anatomy. 21 appendices, p.1267-1608 (*e.g.* Pharmacology; Directory of nursing organizations and health care workers; Leading health problems and communicable diseases). Pronunciation and stress of terms is given. 10,000 cross-references. Well produced; careful line-drawings; thumb ed. *Class No:* 610(031)

[3024]

The Oxford companion to medicine. Walton, J., *and others, eds.* Oxford, University Press, 1986. 2v. (xxi, [1], [v], 1524p.). illus. ISBN: 0192611917.

About 130 contributors. Documented entries (A-M, N-Z) on a wide range of topics (*e.g.* 'Law and medicine in the UK', p.651-62, with 2 columns of references'; 'Sport and medicine'. p.1326-34, with 11 references; 'Communication between doctors and patients', p.1456-70, with 1 column of references). Also brief biographies and notes on medical institutions, as well as curiosa. Appendix 1: 'Major medical and related qualifications'; 2: 'Medical abbreviations'. Praised in *New scientist* (10 April 1986, p.87) for separating British and US approaches on such topics as dentistry, hospitals and nursing. For *Nature* (v.324, 13 Nov. 1986, p.174), 'a curious, rather unpredictable almost quirky encyclopaedia'. For both reference and browsing. *Class No:* 610(031)

[3025]

—The Faber medical dictionary. Wakeley, Sir C., *ed.* 2nd ed. by J.G. Bate. London, Faber & Faber, 1975. [iv], 483p.

First published 1925, with *Supplement, 1965.*

Carries *c.*30,000 concise entries, including eponyms. Etymology and pronunciation are given. Supplements Dorland in providing anglicized spelling of medical terms. *Class No:* 610(031)

[3026]

Stedman's medical dictionary. 25th ed. Baltimore, Md., Williams & Wilkins, 1990. xxxviii,1784p. illus. (incl. col. pl.). $44.95. ISBN: 0683079166.

First published 1911 as *A Practical medical dictionary.*

About 100,000 entries. Syllabification and stress are shown. Short entries for well-known medical persons, plus cross-reference to eponym. Lengthy prelims. include 'Medical etymology' (re-written for this ed.). 5 appendices: Comparative temperature scales; Temperature equivalents; Weights and measures; Laboratory reference range values; Blood groups. Like *Blakiston's Gould,* a defining medical dictionary omitting notes on treatment. Vocabulary is more technical than in *Dorland* or *Blakiston's Gould.* Thumb ed. *Class No:* 610(031)

[3027]

—STEDMAN, T.L. Illustrated Stedman's medical dictionary. 5th unabridged Lawyer's ed. Cincinnati, Ohio, Anderson, 1982. vp. illus.

Includes a 67-p. "Lawyers' section". *Class No:* 610(031)

[3028]

Taber's cyclopedic medical dictionary. Thomas, C.L., *ed.* 15th ed. Philadelphia, Pa., F.A. Davis Co., 1985 (16th ed. 1989. 2401p. $28.50). xxx, 2170p. illus. (incl. col.), tables. ISBN: 0803683081.

First published 1940.

17 consultants. About 70,000 definitions, with etymology and pronunciation shown. 'Malaria': 2 columns (etiology; incubation; symptoms; treatment; prophylaxis; types of malaria). 19 appendices (*e.g.* 7. Latin and Greek nomenclature; 13. Medical emergencies; 14. Directories (US and Canada)). Written primarily for the North American market. 'Somewhat less coverage than Blakiston, Dorland and Stedman, but the definitions are usually less technical' (*Library journal,* 1 November 1965, p.4718, on the 9th ed., 1962). *Class No:* 610(031)

[3029]

WINGATE, P. *and* **WINGATE, R. The Penguin medical encyclopedia.** 3rd ed. London, Penguin, 1988. viii,519p. diagrs. £5.95. ISBN: 0140512209.

First published 1972.

Entries range from a few words to several pages (*e.g.* 'Immunity': 7 pages, covering general principles, natural and adaptive immunity, immunopathology, hypersensitivity, antoimmunity, immunity and therapeutic). Short biographical non-technical language. Many cross-references. Aimed at patients and those caring for them. *Class No:* 610(031)

Handbooks & Manuals

[3030]

The American Medical Association encyclopedia of medicine. Clayman, C.B., *ed.* New York, Random House, 1989. 1184p. illus., figs. $44.45. ISBN: 0394565282.

A-Z guide to over 5,000 terms, including symptoms, treatments and drugs. Contains more than 2,200 drawings and pictures showing medical techniques, anatomy and disease processes. Entries vary from one paragraph to over a page in length. Informative. 'Reasonably priced and highly recommended for all libraries' (*Choice,* v.27, no.2, 1989 p.278). *Class No:* 610(035)

[3031]

BEVAN, J. Your family doctor: the home medical encyclopaedia. London, Mitchell Beazley, 1980. 2v. (viii, 1030p.). illus. (incl. col.), diagrs., graphs, charts. ISBN: 0855332956.

1. *Reference volume: guide to current medical practice.* 502p. 2. *Practical volume.* p.503-1030p.

V.1 operates on the question-and-answer principle, with numerous cross-references. V.2 covers first aid; treating minor injuries and disorders; What could be wrong?; Family safety; What you should have on hand; Care of the sick at home; Professional care for the sick; Pregnancy and childbirth; The healthy child; the growing child; Age-by-age chart (p.808-933); Taking care of your own body; Physical fitness and exercises; Mental fitness. Detailed analytical index, p.1010-30.

US ed. as *Home medical encyclopedia* (Chicago, Ill., World Book-Childcraft International, 1980. 4v. (1029p.)). *Class No:* 610(035)

[3032]

—Medical update, 1982-: the 'World Book' family medical annual. Chicago, Ill., World Book, [1981]. illus., (incl. col.). ISSN: 07320183.

Supplements Bevan. The 1983 annual ([4], 268p.) has 33 contributors. Apart from 13 signed articles, it carries a 'Medical file' section, reporting current research on aging, alcoholism, birth control, weight control, etc. For schools, public libraries and home use. *Class No:* 610(035)

[3033]

Oxford textbook of medicine. 2nd ed. Oxford, University Press, 1987. 2v. illus. (incl. col.), diagrs., tables. £125. ISBN: 0192615513; 0192618555, CD-ROM.

First published 1983.

1: Sections 1-12 and index. 2: Sections 13-28; appendix and index.

About 450 contributors. 28 documented sections (5: Infections. 639p.). 'Diseases of the eye in general medicine': 21p., 34 col. illus., over 2 columns of references, grouped. Final sections: 26. Sports medicine - 27. Medicine in old age

....(contd.)

- 28. Terminal care. Appendix: 'Normal or reference values for biochemical data'. Detailed analytical index in v.2: 114p. 'A major objective of this *Textbook* is to provide a more global picture of disease than is usually given in books of this type' (*Preface to the second edition*). The leading British postgraduate textbook.

Also available as *Oxford textbook of medicine on CD-ROM*, (OUP, 1989. £300). *Class No:* 610(035)

[3034]
ROYAL SOCIETY OF MEDICINE. Family medical guide.
Harlow, Essex, Longman, 1980 (reprinted 1988). 720p. £16.95.

30 chapters, concentrating on parts of the body (*e.g.* 18: Ear, nose and throat, p.418-41, with outstandingly good drawings, carefully keyed. Encyclopedia of medical terms, p.616-86; glossary, p.687-9.). Analytical index, p.690-719. A well-balanced, well-produced quarto. *Class No:* 610(035)

[3035]
WORLD HEALTH ORGANIZATION. International classification of procedures in medicine. Geneva, WHO, 1978. 2v. *Class No:* 610(035)

Dictionaries

Polyglot

[3036]
Eight-language dictionary of medical technology, comprising approximately 8000 technical terms, English, German, French, Russian, Spanish, Polish, Hungarian, Slovak. Albert, R. *and* Hahnewald, H., eds. Oxford, etc., Pergamon Press; Berlin VEB Verlag Technik, 1978. [vii], 594p. ISBN: 0080237630.

English-base terms, with equivalents (across the double page) and indexes in the other 7 languages. Strictly medical terminology. 'In addition to the quoted sources, catalogues, company brochures and other publications of the medical, technical industry, articles in medical and medical-technical journals have formed the basis of this dictionary' (*Preface*). *Class No:* 610(038)=00

[3037]
—Lexicon medicum: Anglicum, Russicum, Gallicum, Germanicum, Polonicum. Zlotnickiego, B., *pod red.* Warsaw, Panstwowy Zaklad Wydawnictw Lekatskich, 1971. xii, 1603p.

About 15,000 English-based entry-words, with Russian, French, German, Latin and Polish equivalents and indexes. Horizontal arrangement of entries. *Class No:* 610(038)=00

[3038]
Elsevier's encyclopaedic dictionary of medicine. Dorian, A.F., *ed.* Amsterdam, etc., Elsevier, 1987-90. ISBN: 0444872930, Pt.C; 0444428267, Pt.D; 0444428232, Pt.A; 0444428240, Pt.B.

A. *General medicine.* 1987. ix, 1175p. £171.87 B. *Anatomy.* 1987. ix, 601p. £121.56 C. *Biology, genetics and biochemistry.* 1989. xii,708p. £123.43 D. *Therapeutic substances.* 1990. x,492p. £93.75.

In 5 languages: English, French, German, Italian and Spanish. About 17,000 terms in v.A-B. Concise English definitions, with equivalents in the other 4 languages. 'Based

....(contd.)

on the English equivalents of the Nomina Anatomica as set forth by the sixth International Congress of Anatomists in Paris in 1955' (*Preface*). *Class No:* 610(038)=00

[3039]
—Radcliffe European medical dictionary. Oxford, Radcliffe Medical Press, 1991. 1022p. £49.95. ISBN: 1870905652.

Almost 4,000 key medical words and phrases are translated into 5 languages; English; French; German; Italian; Spanish. There is a list of abbreviations at the end of each language's section. Many words relating to new and emerging medical disciplines are included. Pocket reference version also available (208p. £8.50).
Class No: 610(038)=00

[3040]
Europäisches medizinisches Wörterbuch: Deutsch, English, Spanisch, Französisch, Italienisch. Stuttgart, Schattauer, 1991, [vi],1022p. illus.

Nearly 10,000 German-based entries (including abbreviations), with English, Spanish, French, and Italian equivalents and indexes. *Class No:* 610(038)=00

[3041]
NOBEL, A. Dictionary of medical objects. / Medizinisches Sachwörterbuch/Dictionnaire d'objets médicaux. 7th ed. Berlin, Springer-Verlag, 1983. x, 1344p. DM348. ISBN: 0387121382.

A trilingual dictionary of medical terms, with a Latin supplement. 19,114 English-base terms, with German and French equivalents (or explanations) and indexes. Usage of English creaks somewhat and entry-words must be sought with caution. (Why, for example, entry 'Factors, diabetic' and none for 'Diabetic factors'?). *Class No:* 610(038)=00

[3042]
Questionarium medicum. Meyboom, F., *comp.* Amsterdam, Elsevier, 1961. viii, 213p. (*Glossaria interpretum.*)

About 400 terms in 17 languages, - English, French, Italian, Spanish, Portuguese, German, Dutch, Norwegian, Swedish, Finnish, Polish, Russian, Greek, Chinese, Japanese, Malay and Esperanto. Reverse language keys. Intended primarily for physicians who have to deal with foreign patients. Now out of print. *Class No:* 610(038)=00

[3043]
SLIOSBERG, A. Elsevier's medical dictionary in five languages: English/American, French, Italian, Spanish and German. 2nd rev. ed. Amsterdam, etc., Elsevier, 1975. [xi], 1452p. £175.62. ISBN: 0444411038.

First published 1964.

Over 20,000 English/American base entries (numbered 1-18,341, with interpolations, - the same total as in the 1st ed.). French, Italian, Spanish and German equivalents and indexes. Index of English synonyms. Bibliography, p.[x]-[xi]. Omission of Russian continues to be a drawback. *Class No:* 610(038)=00

[3044]
VEILLON, E. Medical dictionary / Medizinisches Wörterbuch/Dictionnaire médical. 5. vollstandig neubearb. und erw. Aufl. von A. Nobel. Berne, etc., Huber, 1969 (6th ed., 1977. 1329p.). xxii, 1320p. Erganzungsband. 1974. iv, 464p.

First published 1950.

A bilingual medical dictionary in 3 sections. English-French-German (40,944 numbered English entries); French-

.... *(contd.)*

German-English, German-English-French. German and French genders given. The supplement adds *c*.10,000 entries. Popular for quick-reference and favourably reviewed in *Lebende Sprachen,* v.20(2), March/April 1974, p.60-61. *Class No:* 610(038)=00

[3045]
—GARRIDO JUAN, A. Diccionario médico-suplemento español. Berlin & New York, Springer, 1970. [xi], 536p.

Adds a fourth language to the Veillon trilingual dictionary (1968), - Spanish. *Class No:* 610(038)=00

English

[3046]
Churchill's illustrated medical dictionary. New York, Edinburgh, Churchill Livingstone, 1989. xxx,2128p. illus. $33.95. ISBN: 0443086915.

Contains *c*.100,000 entries covering 65 subject areas in clinical medicine and related sciences. 32,000 pronunciations are given and specially commissioned line-drawings are included. Good coverage of international usage. Thumb-indexed. *Class No:* 610(038)=20

[3047]
Concise medical dictionary. 3rd ed. Oxford, University Press, 1990. 759p. illus. £15.95; £6.95. ISBN: 0192619314.
First published 1980.

About 10,000 entries. 33 contributors and advisers. 'Each entry contains a basic definition, followed - where appropriate - by a more detailed explanation or description' *(Preface)*. For this edition, many new terms relating to prenatal diagnosis, molecular genetics and infertility treatment have been included. Adequate cross-references. Entries range in length from a few words to 2/3 column (*e.g.* 'Leprosy'; 'Sickness benefit'). Good value.
Class No: 610(038)=20

[3048]
DAVIES, P.M. Medical terminology: a guide to current usage. 4th ed. London, Heinemann Medical Books, 1985. 365p. ISBN: 0433071842.

First published 1969 as *Medical terminology in hospital practice.*

5 parts: 1. Introduction - 2. Medical terms referring to certain general pathological processes - 3. Medical terms referring to certain infectious diseases (5 sections) - 4. Medical terms referring to diseases of the various systems of the body and obstetric terms (17 sections, p.63-305) - 5. Certain other types of diseases (*e.g.* tropical diseases). Appendix: Standard works and journals consulted by the author, p.321-5. Glossary. Some important abbreviations. Detailed index, p.335-65. Good use of bold type.
Class No: 610(038)=20

[3049]
The Dictionary of modern medicine. Segen, J.C., *comp. and ed.* Carnforth, Lancs.; Park Ridge, N.J., Parthenon Publishing Group, 1992. 800p. illus. $75. ISBN: 1850703213.

Lists more than 10,000 terms, many of recent vintage, used in everyday medical life. Does not include terms from classicial anatomy, eponyms and pronunciation guidelines but aims to complement the traditional medical dictionary.
Class No: 610(038)=20

[3050]
International dictionary of medicine and biology. Landau, S.I., *ed.-in-chief.* New York, Wiley, 1986. 3v. (li, 3200p.). $395. ISBN: 047101849x.

Over 100 contributors; more than 150,000 entries, thanks to the inclusion of biology. Among the subjects most extensively covered are ecology, environmental health, infectious diseases, occupational medicine, toxicology. A concise guide to usage precedes. Highly recommended as 'indispensable for medical and science libraries' (*Library journal,* 15 June 1986, p.61). *Class No:* 610(038)=20

[3051]
Medical phrase index: a one-step reference to the terminology of medicine. Lorenzini, J.A., *comp. and ed.* 2nd ed. Oradell, N.J., Medical Economics Books. 1989. xi,948p. $45. ISBN: 0874895391.

Over 100,000 entries, A-Z. 'Bone(s)': 6 columns of compound words of which 'bone' is the qualifying adjective (*e.g.* bone-marrow); from 'b., absorption' to 'b., zygomatic'. *Class No:* 610(038)=20

[3052]
Medical speciality terminology. Young, C.G., *and others.* St. Louis, C.N. Mosby Co., 1971-1972. 2v. illus., tables.
1. *Pathology, clinical cytology and clinical pathology.* 1971. vii, [1], 251p. 2. *X-ray and nuclear medicine.* 1972. ix, 124p.

V.1 has 3 main sections, 20 subsections, each of which carries references and a subject index.
Class No: 610(038)=20

[3053]
ROBERTS, F. Medical terms, their origin and construction. 6th ed., revised by B. Lennox. London, Heinemann Medical, 1980. x, 132p. ISBN: 0433191511.
First published 1954.

5 parts: 1. Our classical roots - 2. Word construction - 3. Exemplary word lists, p.45-88 (49 subheadings, including 'Words not to be confused') - 4. Non-Classical origins (including acronyms and eponyms), p.89-102 - 5. Miscellany (notes). Detailed index, p.119-32. 'One of the best guides available to the meaning and derivation of medical terms for that vast majority of medical students who know 'small Latin and less Greek' (*Times literary supplement,* no.2,990, 19 June 1959, p.374, on the 3rd ed). *Class No:* 610(038)=20

[3054]
SKINNER, H.A. The Origin of medical terms. 2nd ed. Baltimore, Md., Williams & Wilkins, 1961. x, 438p. illus. (Reprinted London, Collier-Macmillan, 1970).
First published 1949.

Lists *c*.4,000 commonly used terms of chemical, anatomical and clinical practice, plus detailed explanation of origin and significance. 'On anatomical nomenclature and the basic sciences there is little to fault, but in clinical and historical matters the work is less satisfactory' (*The Lancet,* 17 March 1962, p.57). *Class No:* 610(038)=20

[3055]
—HAUBRICH, W.S. Medical meanings: a glossary of word origins. San Diego, Harcourt Brace Jovanovich, 1984. 285p. ISBN: 0151585199.

Covers *c*.1,500 medical terms, giving etymolgies and concise definitions, plus historical background. Aims to update Skinner's *The origin of medical terms,* adding new meanings and new words. Groups, phobias, colours, etc.
Class No: 610(038)=20

[3056]
SLOANE, S.B. **The Medical word book:** a spelling and vocabulary guide to medical transcription. 3rd ed. Philadelphia, Pa., W.B. Saunders Company, 1991. xi,1097p. illus. $27.95. ISBN: 0721632432.

First published 1973.

18,000 more entries than previous ed. and a new section on immunology. 3 parts: 1. General terms - 2. Systems and specialties (*e.g.* 'Gastroenterology') - 3. Guide to terminology (*e.g.* 'Abbreviations and symbols'). *Class No:* 610(038)=20

[3057]
Webster's medical desk dictionary. Springfield, Mass., Merriam-Webster, Inc., 1986. 26a, 790p. $21.95. ISBN: 0412298805.

Has *c.*40,000 entries, noting syllabification and pronunciation. Includes brief biographies and eponyms. The prelims. carry own article. 'The history and etymology of medical English' and a list of prefixes, suffixes and combining forms. Thumb-index. *Class No:* 610(038)=20

[3058]
—AMERICAN MEDICAL ASSOCIATION. CMIT / Current medical information and terminology. 5th ed. Chicago, Ill., the Association, 1981. 801p.

First published 1963 as *Current medical terminology.*

3,262 'preferred terms and definitions' ... 'for the naming and description of diseases and conditions in practice and in areas related to medicine' (*cover sub-title*). Includes a section of cross-references from French, German and Spanish names of diseases. *Class No:* 610(038)=20

German

[3059]
BUNJES, W.E. **Wörterbuch der Medizin und Pharmazeutik, Deutsch-Englisch.** 3. neubearb. Aufl. Stuttgart, Thieme, 1981. xxvii, 574p. DM152. ISBN: 3133705032.

First published 1951 (author: F. Lejeune): 2nd ed. (1968-69).

About 50,000 entries. 'Lungen' (pulmonary): 3 columns of subentries. Prelims. include a bibliography of medical reference books (p.x), abbreviations and conversion tables. About 80 categories are applied to terms. Stress of German words is indicated. *Class No:* 610(038)=30

[3060]
—BUNJES, W.E. Medical and pharmaceutical dictionary: English-German. 4th ed. Stuttgart, etc., Thieme, 1981. xxviii, 556 + 140p. tables. DM152.

Has *c.*85,000 entries, of which the 140-p. supplement includes 17,000 new to the 3rd ed of 1974. Subentries appear in bold type. About 80 categories are applied. Prelims. include 'Bibliography of medical reference books' (p.xiii), conversion tables, list of Latin and Greek prefixes. *Class No:* 610(038)=30

[3061]
Medical dictionary, English-German, German-English. [Medizinisches Wörterbuch, Englisch-Deutsch, Deutsch-Englisch.] Eichhorst, C., *ed.* 2nd ed. Berlin, Eschenbach Verlag, 1989. 703p. DM68. ISBN: 3980144801.

This ed. contains *c.*80,000 terms, more than double the previous ed. Covers all branches of medicine. *Class No:* 610(038)=30

[3062]
NÖHRING, J. **Dictionary of medicine, German-English.** Amsterdam, etc., Elsevier, 1987. 846p. £129.68. ISBN: 044498982x.

About 60,000 German terms, covering all branches of medicine, with English equivalents. *Class No:* 610(038)=30

[3063]
—German-English medical dictionary. Schoenewald, T.S., *comp.* London, H.K. Lewis, 1952. viii, 241p. £8.

About 15,000 main entries. Still valued for its citing of words in context. 'A phrase dictionary illustrating the various shades of meaning of a word or expression by giving suitable examples of it' (*Introduction*). *Class No:* 610(038)=30

[3064]
—NÖHRING, J. Dictionary of medicine, English-German. Amsterdam, etc., Elsevier, 1984. 708p. £106.56. ISBN: 0444996419.

Has *c.*55,000 English, terms covering all branches of medicine, with German equivalents. *Class No:* 610(038)=30

[3065]
UNSELD, D.W. **Medizinisches Wörterbuch der deutschen und englischen Sprache.** [Medical dictionary of the English and German languages.] 9th ed., rev. and enl. Stuttgart, Wissenschaftliche Verlagsgesellschaft GmbH, 1988. 695p. tables. DM68. ISBN: 3804709923.

First published 1946 (authors: M. Goertz and D.W. Unseld): 8th ed., 1982.

Englisch-Deutsch, p.3-349; Deutsch-Englisch, p.353-691. About 25,000 entries in each half. American and British English spelling are differentiated. German genders given. Eponyms are included. Appendix of weights, measures and temperatures. Small format. A handy dictionary for translators. *Class No:* 610(038)=30

Dutch

[3066]
MOSTERT, F.J.A. **Medisch woordenboek,** engels/nederlands, nederlands/engels. 3rd ed. Utrecht, Bohn, Scheltema & Holkema, 1990. 206p. ISBN: 9031305669.

First published 1971.

Lists *c.*5,000 terms. Bibliography, p.201-6. *Class No:* 610(038)=393

French

[3067]
DELAMARE, J. *and* DELAMARE-RICHE, T. **Dictionnaire français-anglais des termes de médecine.** [English-French dictionary of medical terms.] 2. éd., rev. et augm. Paris, Maloine, 1986. 510 + 542p. FFr432. ISBN: 2224011202.

First published 1970.

About 35,000 entries in each half, obsolete terms being indicated. Numerous eponyms and cross-references. Grouped headings (*e.g.* Fever: 4½p. of types). Better balanced halves than in Lépine (*qv*). *Class No:* 610(038)=40

[3068]

—LÉPINE, P. Dictionnaire français-anglais, anglais-français des termes médicaux et biologiques. Paris, Flammarion Médicines-Sciences, 1984. 878p. FFr350.

First published 1974.

Has 25,000 French terms, and is primarily intended for French users. *Class No:* 610(038)=40

[3069]

Dictionnaire de médecine Flammarion. 3. éd. Paris, Flammarion Médecine-Sciences, 1989. 948p.

First published 1975.

A French-French medical dictionary, with *c*.19,000 entries. Included is an English-French vocabulary, a list of English and French abbreviations, a list of leading journals, table of biological constants, etc. (4th ed. 1991. 1024p. FFR249). *Class No:* 610(038)=40

[3070]

—GARNIER, M. *and* DELAMARE, J. Dictionnaire des termes techniques de médecine. 21. éd. Paris, Maloine, 1984. 875p. FFr170.

Has *c*.25,000 headwords, with numerous subentries. Includes abbreviations and symbols. Standard French-French medical dictionary. *Class No:* 610(038)=40

[3071]

GLADSTONE, W.J. *and* **ROCHE, P. Dictionnaire anglais-français des sciences médicales et paramédicales.** [English-French dictionary of medical and paramedical sciences.] 3. éd. Saint-Hyacinthe, Québec, Edisem; Paris, Maloine, 1990. 1099p. FFr550. ISBN: 2891301277.

First published 1978.

Contains *c*.85,000 entries in over 70 specialized areas. This edition includes terms from biotechnology; genetics; immunology; molecular biology; AIDS. A definitive and comprehensive reference work of medical terms and expressions (many hitherto unpublished even in monolingual dictionaries), plus, as necessary, contextual information, explanatory notes and categorization of terms. Of the 1978 ed., a model of 'how translators would like a dictionary prepared' (*The incorporated linguist*, v.17(3), p.58, 68). *Class No:* 610(038)=40

Italian

[3072]

LUCCHESI, M. Dizionario medico, inglese/italiano, italiano/inglese. Milan, Rafaello Cortina Editore, 1987. 1456p. ISBN: 8870780376.

English-Italian section, p.1-1179. Italian-English section which follows, p.1185-1456, is a list of Italian words with English equivalents. Abbreviations are included, and homeopathic terms are new to this edition.

Class No: 610(038)=50

Spanish

[3073]

McELROY, O.H. *and* **GRABB, L.L. Spanish-English, English-Spanish medical dictionary.** [Diccionario médico, español-inglés, inglés-español.] Boston, Little, Brown and Company, 1992. xliii,506p. ISBN: 0316555614.

Contains more than 20,000 entries. List of medical abbreviations; sections on English and Spanish grammar;

....*(contd.)*

tips on pronunciation. 3 appendices, *e.g.* 'Signs and symptoms in most common disorders and diseases'.

Class No: 610(038)=60

[3074]

Ruiz Torres diccionario de términos médicos. 6th ed. Madrid, Editorial Alhambra, 1989. 880p.

First published 1957; 3rd ed. 1965 (714p.).

About 35,000 entries in the English-Spanish half (p.1-581), including abbreviations. Many compounds (*e.g.* 'Ligament': over 2 columns). Spanish-English: p.585-880.

Class No: 610(038)=60

[3075]

—GARRIDO JUAN, A. Diccionario inglés-español para medicos y estudiantes de medicina. 3rd ed. Barcelona, Editorial Pediátrica, 1987. 772p. ISBN: 8471930862.

First published 1972.

Simple list of English medical terms with Spanish equivalents. Small, chunky format makes it useful as a vade mecum. *Class No:* 610(038)=60

Classical Languages

[3076]

ANDERSON, D. *and* **BUXTON, R. A Pocket etymology of medical terms:** an introduction to the Greek and Latin roots of medical terminology. Bristol, Bristol Classical Press, 1981. iii, 47p. ISBN: 0862920159.

450 terms. Introduction includes Greek alphabet and its transliteration, common changes in word-forms, stems, common prefixes and suffixes. Will help readers to work out the meaning of other words which, for brevity's sake, we have omitted' (*Preface*). Aimed primarily at medical students. *Class No:* 610(038)=7

Russian

[3077]

Anglo-russkiĭ meditsinskiĭ slovar'. Akzhigtov, G.N., *ed.* Moscow, Russky Yazyk, 1988. 604p.

70,000 English terms, with Russian equivalents. Appendix of abbreviations and symbols. Companion volume:

Class No: 610(038)=82

[3078]

—Russko-angliĭskiĭ meditsinskiĭ slovar'. Moscow, Izdat. Ruskiĭ Yazyk, 1975. 647p.

Has *c*.50,000 entries and sub-entries. The Russian terminology is based on the 2nd ed. of *Bol'shaya meditsinskaya entsiklopediya* (1965) and 2nd ed. *Malaya meditsinskaya entsiklopediya* (1970), among other sources. English equivalents have American spelling.

Class No: 610(038)=82

[3079]

CARPOVICH, E.A. Russian-English biological and medical dictionary. 2nd ed. New York, Technical Dictionaries Co., 1960. 400p. $25. ISBN: 0911484019.

First published 1958 (398p.).

32,650 entries, giving direct adjective-noun approach, - very acceptable to the inexperienced translator and occasional reader of Russian scientific papers. (See *Science reference notes*, v.6(1), January 1959, p.19-21, for comparison with S. Jablonski's dictionary). Particularly valuable on biological terms that frequently appear in the

....(contd.)

medical literature, - terms not found in other dictionaries. Applications of terms appear in parentheses.

Class No: 610(038)=82

[3080]

JABLONSKI, S. Russian-English medical dictionary.
Levine, B.S., *ed.* New York & London, Academic Press, 1958. xi, 423p.

About 29,000 entries, giving genders and parts of speech for Russian words. Jablonski barely gives common usage of terms; Carpovich does. Jablonski, however, lists frequently used Russian medical abbreviations, including discipline concerned. Carpovich does in general categorize terms. Jablonski is particularly good on drug terminology. The two dictionaries complement each other well.

Class No: 610(038)=82

[3081]

PETROV, V.I. *and* **CHUPIATOVA, S.I. Russko-anglĭiskiĭ meditsinkiĭ slovar'-razgovornik.** [Russian-English medical dictionary phrasebook.] Izd. 2-e, ispr. i dop. Moscow, Rus. iaz, 1987. 595p. illus. (some col.).

First published 1983. *Class No:* 610(038)=82

Polish

[3082]

JEDRASZKO, S. Slownik lekarski, angielsko-polski i polski-anglielski. Wyd. 3. Warsaw, Panstwowe Zaklad Wydawnictw Lekarskich, 1969. 683, [1]p.

English-Polish, p.19-520 (*c.*40,000 entries, with pronunciation of English words, but no genders of Polish nouns. 'Zapalenie' (Inflammation): 31/3p. of sub-entries. Abbreviations. Latin words and conversion tables, p.521-2.

Class No: 610(038)=84

Chinese

[3083]

Han Ying i hsueh ta tzu tien. [The Chinese-English medical dictionary.] Peiching Shih, Jen min wei sheng chu pan she; Hsin hua shu tien Peiching fa hsing so fa hsing, 1987. xxviii,1837p. ISBN: 7117004746.

2 sections, each arranged A-Z: Section 1 covers anatomy, physiology, diseases, chemicals, medical equipment - Section 2 includes acupuncture, natural remedies, symptoms, homeopathy. *Class No:* 610(038)=951

Negro-African

[3084]

A Short English-Swahili medical dictionary. White, T.H. *and* Sorsbie, C.E., *eds.* Expanded and edited by C.E. Sorsbie, with others. Edinburgh, etc., Churchill Livingstone, 1978. 105p. £2.50. ISBN: 0443016720.

About 2,500 entries for terms and expressions. 'A short bibliography' (A-Z authors), p.103-5.

Class No: 610(038)=96

Reviews & Abstracts

[3085]

Bibliography of medical reviews. Washington, National Library of Medicine, 1955-67. Annual. 12v.

Lists review articles - as opposed to single book reviews - has subsequently appeared in monthly *Index medicus* and annual *Cumulated index medicus*. *Class No:* 610(048)

[3086]

Excerpta medica. Amsterdam. Excerpta Medica-Embase Publishing, 1947-. 41 sections, 6-32pa. DFl41,161, complete series (includes a semi-annual cumulation on CD-ROM); price varies for individual series.

The most comprehensive medical abstracting service. The 41 sections contain 250,000 abstracts pa. in English from *c.*3,500 medical journals. Monthly and annual author and subject indexes. Computerized indexing since 1969.

1. *Anatomy, anthropology, embryology and histology.* 1947-. 24. *Anesthesiology.* 1966-. 31. *Arthritis and rheumatism.* 1965-. 27. *Biophysics, bio-engineering and medical instrumentation.* 1967-. 16. *Cancer.* 1953-. 18. *Cardiovascular diseases and cardiovascular surgery.* 1957-. 15. *Chest diseases, thoracic surgery and tuberculosis.* 1948-. 29. *Clinical and experimental biochemistry.* 1948-. 13. *Dermatology and venereology,* 1947-. 21. *Developmental biology and teratology.* 1961-. 40. *Drug dependence, alcohol abuse, and alcoholism.* 1973-. 3. *Endocrinology.* 1947-. 46. *Environmental health and pollution control.* 1971-. 50. *Epilepsy abstracts.* 1968-. 49. *Forensic science abstracts.* 1975-. 48. *Gastroenterology.* 1971-. 5. *General pathology and pathological anatomy.* 1948-. 20. *Gerontology and geriatrics.* 1958-. 36. *Health policy, economics and management.* 1971-. 25. *Hematology.* 1967-. 22. *Human genetics.* 1963-. 26. *Immunology, serology and transplantation.* 1967-. 6. *Internal medicine.* 1947-. 4. *Microbiology, bacteriology, mycology, parasitology and virology.* 1948-. 8. *Neurology and neurosurgery.* 1948-. 23. *Nuclear medicine.* 1964-. 10. *Obstetrics and gynecology.* 1948-. 35. *Occupational health and industrial medicine.* 1971-. 12. *Ophthalmology.* 1947-. 33. *Orthopedic surgery.* 1956-. 11. *Otorhinolaryngology.* 1948-. 7. *Pediatrics and pediatric surgery.* 1947-. 30. *Clinical and experimental pharmacology.* 1948-. 2. *Physiology.* 1948-. 32. *Psychiatry.* 1948-. 17. *Public health, social medicine and epidemiology.* 1955-. 14. *Radiology.* 1947-. 19. *Rehabilitation and physical medicine.* 1958-. 9. *Surgery.* 1947-. 52. *Toxicology.* 1983-. 28. *Urology and nephrology.* 1967-.

Excerpta Medica Foundation also issues monthly current bibliographies, and 2 indexing services: *Adverse reaction titles* and *Drug literature index.* Online, EMBASE and available on CD-ROM. *Class No:* 610(048)

[3087]

Index medicus. Washington, National Library of Medicine, 1960-. 12pa. $319. ISSN: 00193879.

Covers *c.*3,000 journals. Each issue has subject and author sections, plus - since 1967 - 'Bibliography of medical reviews' (articles are all well documented surveys of the recent biomedical literature'). A comprehensive index to the world's medical literature. Annual cumulation as *Cumulated index medicus.* The computerized information retrieval service, MEDLINE (*qv*), is associated with *Index medicus.* Available on CD-ROM. Indispensable in its field.

Class No: 610(048)

[3088]

—Cumulated index medicus. Bethesda, Md., National Library of Medicine, 1961-. Annual. ISSN: 00901423.

Compiled in 16v. *Medical subject headings (MeSH),* - the primary listing of subject descriptions that appear in *Index medicus* (January issues) and here reprinted (1963-). V.2: *Journals indexed.* Bibliography of medical reviews; Author index, A-Bs, V.3-7: Author index, Bu-Z. V.8-16: Subject index. Available on microfiche and microfilm.
Class No: 610(048)

[3089]

—STRICKLAND-HODGE, B. How to use 'Index Medicus' and 'Excerpta Medica'. Aldershot, Gower, 1986,. 60p. £25. ISBN: 0566035324.

Mainly intended for medical library staff.
Class No: 610(048)

[3090]

PASCAL (medical series). Vandoeuvre-Les-Nancy, Centre National de la Recherche Scientifique, 1984-.

Formerly *Bulletin signalétique* medical series (nos. 233, 251, 310, 320, 330, 340, 346/349, 352, 354/357, 359, 361/364) now (1984-) appear as parts, respectively, of the PASCAL THEMA, FOLIO and EXPLORE series, online:

T235 *Médecine tropicale.* T251 *Cancérologie.* E84 *Génie biomédical, informatique biomédicale.* F52 *Biochemie, biophysique moléculaire, biologie moléculaire et cellulaire.* E63 *Toxicologie.* F70 *Pharmacologie.* E61 *Microbiologie.* E62 *Immunologie.* E71 *Opthalmologie.* E72 *Otorhinolaryngologie, stomatologie, pathologie cervicofaciale.* E73 *Dermatologie, maladies sexuellement transmissibles.* E83 *Anesthésie et réanimation.* E74 *Pneumologie.* E75 *Cardiologie et appareil circulatoire.* E76 *Gastroentérologie, foie, pancreas, abdomen.* E77 *Nephrologie, voies urinaires.* E78 *Neurologie.* E79 *Pathologie et physiologie ostéarticulaires.* E80 *Hématologie.* F54 *Reproduction des vertébrés, embryologie des vertébrés et des invertébrés.* E64 *Endocrinologie humaine et expérimentale, endocrinopathies.* E82 *Gynécologie, obstetrique, andrologie.* E81 *Maladies métaboliques.* E60 *Génétique. Class No:* 610(048)

[3091]

Popular medical index. Letchworth, Herts., Mede Publishing, 1978/9-. Annual. ISSN: 01421107.

The 1987/1988 annual has *c.*1,200 entries for periodical articles and books under subjects, Abdomen ... Zinc. A list of 58 journals monitored, - 'the most frequently quoted periodicals in the Index' precedes. These include not only *British medical journal* and *Nursing times,* but also *New scientist* and *Women's realm.* Price and ISBN are given for books. *Class No:* 610(048)

Periodicals & Progress Reports

[3092]

Annual review of medicine: selected topics in the clinical sciences. Palo Alto, Calif., Annual Reviews, Inc., 1950-. ISSN: 00664219.

The well-documented papers of v.43, 1992 include: 'The ophthalmopathy of Graves' Disease'; 'Transplantation of pancreas'. Analytical subject index; cumulative index of contributing authors, v.39-43; cumulative index of chapter titles, v.39-43. 'Some selected articles in other *Annual reviews*' precedes. *Class No:* 610(05)

[3093]

'List of journals indexed', in the January issue of *Index medicus* each year (1963-),$bis reprinted in annual *Cumulated index medicus,* and also available separately. About 2,500 titles, listed (1) by abbreviations, (2), by full titles. *Class No:* 610(05)

[3094]

—Vital notes on medical periodicals. Chicago, Ill., Medical Library Association, 1952-82. v.1-30. 3pa.

Listed new, change-of-title, deceased, etc., medical periodicals. 'Vital' refers to birth and death dates. Place of publication and frequency are given. Latterly (v.7-30) *Vital notes* had a section on congresses and list of directories. Since contributed by a variety of libraries, this provided a union list of sorts. *Class No:* 610(05)

[3095]

—World medical periodicals. 3rd ed. New York, World Medical Association, 1961. xli, 407p. *Supplement.* 1968. 68p.

First published 1953 (Paris, WHO and UNESCO); 2nd ed. 1957.

Adds nearly 1,000 titles to the 4,841 of the 2nd ed. Ceased items are asterisked. International, with cross-references and including abstracting journals, plus veterinary medicine. Prefaces and subject and country indexes in English, French, Spanish and German. *Class No:* 610(05)

Yearbooks & Directories

[3096]

ABMS compendium of certified medical specialists. 2nd ed. Evanston, American Board of Medical Specialties, 1988. 7v. (10000p.). $245. ISBN: 0934277125.

Data on *c.*340,000 physicians. 24 sections, one for each speciality, each with entries A-Z, plus geographic cross-index by state and city. V.7 contains a master A-Z index of all specialists.

While *Directory of medical specialists (qv)* is a comparative publication, its entries are arranged geographically within each speciality section, with one master A-Z index for all specialities. *Class No:* 610(058)

[3097]

American medical directory. Chicago, Ill., American Medical Association, 1906-. (32nd ed. 1990. $495.). 4v. ISBN: 0899700225.

1. *Alphabetical index of physicians.*

2-4. *Geographical register of physicians.*

'Directory of physicians in the United States, Puerto Rico, Virgin Islands, certain Pacific Islands and the US physicians temporarily located in foreign countries' *(subtitle).* Brief who's who data on physicians. Also lists leading medical societies, hospitals and medical journals.
Class No: 610(058)

[3098]

—Directory of medical specialists: 1989-90. 24th ed. Chicago, Ill., Marquis Who's Who, for the American Board of Medical Specialties, 1989. 3v. $295. ISBN: 0837905249.

First published 1940.

Has more than 365,000 entries in all, listed by speciality and then geographically. Brief curriculum vitae and official address in each case, - more detailed than in *American medical directory (qv).* Outlines requirements for

....(contd.)

certification in the case of each speciality. Index of all biographees appears at the end of v.3. 25th ed., 1991-92, 6500p. $295. *Class No:* 610(058)

[3099]
COUNCIL FOR PROFESSIONS SUPPLEMENTARY TO MEDICINE. Professions Supplementary to Medicine Act, 1960, section 2 (d). London, the Council. Annual. 7v.

The chiropodists register. The dieticians register. The medical laboratory scientific officers register. The occupational therapists register. The orthoptists register. The physiotherapists register. The radiographers register.

The 25th ed. of the *Medical laboratory scientific officers register,* 1987/88, states qualifications required for inclusion, lists names of the Council members, and records names and addresses of those on the Register (p.19-411).

Class No: 610(058)

[3100]
The Medical annual: the year book of general practice. Bristol, Clinical Press, 1884-.

The 1992 v. (xiv,243p. £35) has 42 well-documented essays. Papers include 'Recent advances in diabetes and its drug therapy'; 'Migraine'; 'Medico-political matters'. Index, p.237-43. *Class No:* 610(058)

[3101]
The Medical directory, 1991. 147th ed. Harlow, Longman, 1991. 2v. (xix,3966+287p.). £85, set (1992 ed. £139). ISBN: 058207665x.

First published 1845.

The main sequence (v.1: A-L, v.2: M-Z) covers medical practitioners registered to practise in the UK. Data include qualifications, appointments (past and present) and publications. V.2 has appended: local lists; hospitals; university and medical schools; Royal Colleges in England and Scotland; postgraduate institutions; postgraduate medical centres; research institutions; coroners (England & Wales); government departments and statutory bodies; royal appointments; medical MPs, and titled members of the profession. Index; index to advertisements. For most purposes, the *Directory* is preferred to the *Medical register,* because of its additional listings. *Class No:* 610(058)

[3102]
The Medical register, 1992 ... comprising the principal list of the Register of Medical Practitioners on 1st January 1992, with other information concerning medical registration. London, General Medical Council, in pursuance of the Medical Act, 1983, 1992. Annual. 3v. (xxii,4632p.). tables.

First published 1859.

The official list of medical men and women, - *c.*100,000 brief entries (name, date of birth, address, medical qialifications, type of registration). Prelims. in v.1 state registrable qualifications, including recognised overseas qualifications. *Class No:* 610(058)

[3103]
Sell's Health Service buyers guide, 1991: a comprehensive buyers guide to a wide range of English and continental suppliers to the National Health Service. 15th annual ed. Epsom, Sell's Publications, Ltd., 1990. 356p. £40. ISBN: 0854996982.

4 sections: 1. Products and services - 2. Alphabetical list of companies - 3. Trade names and brands - 4. Advertisers

....(contd.)

index. Also, index to classified products and services, and a list of the main exhibitions. Refers the reader to advertisements, where applicable. *Class No:* 610(058)

Government Publications

[3104]
Government publications. Sectional list 11, revised February 1986. Department of Health and Social Security. London, HMSO, 1986. 24p. *Gratis.*

Two parts: Health and social services - Social security. The first part (p.3-17) covers 50 topics (Back pains ... Children ... Drugs and pharmaceuticals ... Health and social subjects: reports ... Hospitals, administration and organisation ... Mental health ... National health service - Nurses ... Research ... Whooping cough). This Sectional List is, at time of writing, no longer available. Details of recent and forthcoming books can be found in *Medicine and health catalogue,* issued by HMSO. *Class No:* 610(061.1)

Teaching Materials

[3105]
Health education index: incorporating guide to voluntary and other support organizations and self-help groups concerned with healthcare: 1987/88. Webb, D., *comp.* 9th ed. London, Edsall, 1987. ISBN: 0902623508.

Lists over 2,000 aids to health education, - booklets and pamphlets, books, films, film strips and slides, audiovisual and nonprint material; organizations - voluntary and professional. Subject index; sources of supply index; initials and abbreviations.

Health education index (1980-), revised 2-3 yearly, is aimed at schoolteachers. Definitive reference of health education resources available in the UK.

Class No: 610(072)

Quotations

[3106]
Medical quotes: a thematic dictionary. Daintith, J. *and* Isaacs, A., *eds.* Oxford, New York, Facts on File, 1989. 260p. £13.95. ISBN: 0816020949.

Published in paperback as *Dictionary of medical quotations* (Collins. 1989).

About 1,600 quotations, arranged A-Z by theme (Abortion ... Youth), ranging from old favourites through to film, television and radio broadcasts. Cross-references given to related topics. 2 indexes: keyword-key phrase and author. 'Every waiting room should have one' (*Library Association record,* January 1990, p.68). *Class No:* 610(082.2)

[3107]
STRAUSS, M.B. Familiar medical quotations. London, Churchill, 1968 (reprinted, Little, Brown, 1985). xix,968p. £42. ISBN: 0316819158.

About 7,500 quotations, under subjects A-Z. Index of authors, p.679-705 (entries for authors with more than 50 quotations are asterisked and page numbers omitted). Index of keywords, p.709-968. Fills a gap. *Class No:* 610(082.2)

Tables & Data Books

[3108]
FISHER, Sir R.A. *and* YATES, F. **Statistical tables for biological, agricultural and medical research.** 6th rev. ed. Edinburgh, Oliver & Boyd, 1963. x, 146p. tables.

First published 1938; 5th ed. 1956.

50 tables. The valuable introduction gives examples of the ways in which the tables can be used and includes 82 references, covering books, monographs and papers on statistical methods, up to 1963. No index.
Class No: 610(083)

[3109]
Geigy scientific tables. 8th rev. and enl. ed., edited by C, Lentner. West Caldwell, N.J., Ciba-Geigy Corpn., 1981-. ISBN: 0914168509, v.1; 0914168517, v.2; 0914168525, v.3; 0914168533, v.4; 0914168541, v.5.

7th ed., edited by K. Diem and C. Lentner.

1. *Units of measurement; body fluids; composition of the body; nutrition.* 1981. 295p. 2. *Introduction to statistics; statistical tables; mathematical formulae.* 1982 240p. 3. *Physical chemistry; composition of blood; hematology; somatometric data.* 1984. 359p. 4. *Biochemistry; metabolism of xenobiotics; inborn errors of metabolism; pharmacogenetics and ecogenetics.* 1986. 330p. 5. *Heart and circulation.* 1990. 280p.

Further volumes are in preparation. Ready reference source of basic data in biomedicine. *Class No:* 610(083)

[3110]
World health statistics annual. Geneva, World Health Organization, 1951-. Annual. tables, maps.

The 1991 v. (xxii,349p. £40) has 4 sections: A. Global overview - B. Special topic - C. Vital statistics and life tables - D. Causes of death (by sex and age: Africa, America, Asia; Europe; Oceania; Age-standardized death rates for selected countries, by sex; Causes ...). 12 sets of tables. Two annexes (2. List of member states of WHO, by region). *Class No:* 610(083)

[3111]
—'Medical statistics' In *Aslib proceedings,* v.38 (5), May 1986, p.131-75. diagrs., tables.

Consists of 4 documented papers presented at the Aslib Biological and Agricultural Group seminar, London, 29 November 1985 (*e.g.* 'Handling statistical enquiries in the medical library: some practical examples'). *Class No:* 610(083)

Illustrations

[3112]
THORNTON, J.L. *and* REEVES, C. **Medical book illustration:** a short history. Cambridge & New York, Oleander Press, 1983. 142p. illus. £15; $35. ISBN: 0906672074.

8 Chapters. 1. Materials, methods and general sources - 2. Ancient medical illustration - 3. Mediaeval manuscripts - 4. The invention of printing and the sixteenth century - 5. The seventeenth century - 6. The eighteenth century - 7. The nineteenth - 8. The twentieth century. Bibliography, p.128-35 (authors, A-Z). Detailed index, p.136-42. Helps to fill a gap in the history of medicine. *Class No:* 610(084.1)

Portraits

[3113]
NEW YORK ACADEMY OF MEDICINE. **Portrait catalog of the New York Academy of Medicine.** [New York Academy of Medicine]. Boston, Mass, G.K. Hall. 1971. 5v. $360. ISBN: 0816102333, set.

Includes 10,784 separate portraits, plus 151,792 entries of portraits appearing in books and journals. If biographical data or an obituary accompanies a portrait, the entry is asterisked or obelisked. 3 supps., ($120ea.), 1965-76. The set covers holdings 1959-75. *Class No:* 610(084.10)

[3114]
—NEW YORK ACADEMY OF MEDICINE. **Illustration catalog of the Library.** 3rd ed. Boston, Mass., G.K. Hall, 1976. 264p. $40. ISBN: 0816100381.

Contains *c.*5.500 card references to some 20,000 illus. in histories and journals, and also in early volumes in the Rare Book Dept. A library tool, and as such neither complete nor perfected. *Class No:* 610(084.10)

[3115]
ROYAL COLLEGE OF PHYSICIANS OF LONDON. **Portraits.** Piper, B. *and* Wolstenholme, G., *eds.* London, Churchill, 1964. 468p. illus.

Each portrait is illustrated and accompanied by biographical notes; references to other portraits; notes on provenance and sources of information. Details of over 200 portraits (in oils/chalk/pencil, busts, miniatures, etc.). Coloured frontispiece of College Arms.
Class No: 610(084.10)

[3116]
—RIEDMAN, S.R. *and* GUSTAFSON, E.T. **Portraits of Nobel Laureates in medicine and physiology.** London, Abelard-Schuman, 1964. 343p. *Class No:* 610(084.10)

[3117]
—ROYAL COLLEGE OF PHYSICIANS OF LONDON. **Catalogue of engraved portraits of the Library.** [Royal College of Physicians of London]. Driver, A.H. London, the College, 1952.

Records over 4,500 portraits, principally of British physicians. But surgeons are also represented by eminent medical men of other nations. 'The index of portraits in books', maintained in the College, is available on microfilm. *Class No:* 610(084.10)

[3118]
—WELLCOME INSTITUTE OF THE HISTORY OF MEDICINE. **Portraits of doctors and scientists in the Wellcome Institute:** a catalogue. Burgess, R., *ed.* London, the Institute, 1973. xxiv, 459p. illus. £19. *Class No:* 610(084.10)

Films

[3119]
Catalogue of films and videos in the British Medical Association library. London, Library Association Publishing, 1993. 330p. £35. ISBN: 1856040828.

Detailed annotated list of films and videos in the BMA library as of 31 December 1992. Entries include filmographic details, synopses, target audience, reviewers' comments and distributors' information. Alphabetical listing, followed by classified sequence using MESH. Index. *Class No:* 610(084.122)

Audio-Visual Materials

[3120]

Health media review index: a guide to reviews and descriptions of commercially-available nonprint material for the medical, mental, allied health, human service and related counseling professions. Provan, J.E. *and* Hunter, J.W., *eds.* Metuchen, N.J., Scarecrow Press, 1985. 862p. $59.50. ISBN: 081081739x.

'Indexes more than 4,000 health-related audiovisual items, as reviewed and described in 139 journals, 1980-83, Uses MeSH (National Library of Medicine Subject Headings). Sources of items comprise over 1,000 organizations and individuals. 'Its value is in bringing together nonprint reviews for the broad field of health'. (*Library journal*, August 1985, p.83).

Ed. covering 1984-86 (D.J. McAlpin and others, *eds.* xx,751p. $52.50) was published 1988. *Class No:* 610(086)

[3121]

JONES, M.C. International guide to locating audio-visual materials in the health services. Aldershot, Hants., Gower, 1986. ix. 54p. £25. ISBN: 0566035529.

4 parts 4: 1. Aids to locating health sciences audio-visual materials within the UK (A. Printed research aids; B. Computers and manual search services; C. Current awareness. 64 references) - 2. Aids to locating health sciences audio-visual materials in the USA and Canada. - 3. Aids to locating health sciences audio-visual materials in countries other than the UK, USA and Canada. - 4. Aids to the selection of health services audio-visual materials (includes D: Published reviews). 4 appendices: 3. Guides to assessing audio-visual materials; 4. Glossary of acronyms. Name, country and general indexes. Covers films, videocasettes, slides, overhead projection transparencies, flipcharts and educational computer software. Excludes microfilms. Particularly helpful on selecting materials. *Class No:* 610(086)

[3122]

—STEWART, M.C. 'Audio-visual materials': In *Information sources in the medical sciences*, edited by L.T. Morton and S. Godbolt (4th ed., 1992), p.109-34.

A survey, backed by over 5p. of references. 3 appendices: 1. Addresses of organizations; 2. Directories listing producers/distributors of audio-visual aids; 3. A select list of journals concerned with audio-visual aids in medicine (those asterisked carry reviews of new materials). *Class No:* 610(086)

Histories

[3123]

ACKERKNECHT, C.H. A Short history of medicine. Bethesda, Md., Johns Hopkins University, 1982. 263p. $11.95.

Original German as *Geschichte der Medizin* (4th ed. Stuttgart, Enke, 1979. viii, 236p.).

A well-packed survey, from prehistoric medicine to present times. 'Every chapter has at least one statement that causes one to admire the men and women who preceded us. An extensive bibliography and index included. All in all, a thought-provoking 240 pages!' (*Delaware medical journal*, v.54, December 1982, p.705). *Class No:* 610(091)

[3124]

—CASTIGLIONI, A. A History of medicine. Translated from the Italian and edited by E. B. Krumbhaar. 2nd ed. New York, Knopf, 1947. 1192, xlip. illus. New York, J. Aronson, 1978. 1216p.

Original Italian, 1936.

Of special value for the early history of medicine, particularly Greek and Roman, supplementing Garrison (*qv*). Extensive bibliographies. *Class No:* 610(091)

[3125]

Bibliography of the history of medicine, US Dept. of Health, Education and Welfare, Public Health Service, National Institute of Health, National Library of Medicine. Washington, Superintendent of Documents, US Govt. Printing Office, 1965-. Annual, with 5-yearly cumulations (No. 25, 1985-89. $25). ISSN: 00677280.

Each annual v. has 3 main sections: list of biographies, subject index; author index. Covers *c.*140 of the journals from *Index medicus* on HISTLINE since 1970. Considerable overlap with *Current work in the history of medicine* (4pa) and tardy in appearance. *Class No:* 610(091)

[3126]

—Bibliotheca Osleriana: a catalogue of books illustrating the history of medicine and science, collected, arranged and annotated by Sir William Osler, Bt. and bequeathed to McGill University. Oxford, Clarendon Press, 1929. xxxvi, 786p.

Reprinted Montreal, McGill-Queen's Univ. Press, 1969.

Lists *c.*7,600 bound volumes, representing major contributions to medicine and allied sciences. Entry arrangement renders reference a little difficult, but this is facilitated by the index. Annotations are of special interest, being typically Oslerian. Contains a wealth of bibliographical knowledge and remains readable.

The 1969 reprint (xliii, 792p.) includes a new prologue, addenda and corrigenda. 'A vast resource of accurate bibliographic and historical information' (*TLS*, no.3566, 2 July 1970, p.732). *Class No:* 610(091)

[3127]

CORSI, P. *and* **WEINDLING, P. Information sources in the history of science and medicine.** London, Butterworth Scientific, 1983. xvi, 531p. facsims., diagrs. £43. (*Butterworths guides to information sources.*) ISBN: 0408107642.

Relevant chapters: 2. The historiography of medicine; 19. Medicine since 1500; 20-23. Science and medicine in America, Islam, India and China, respectively. *Class No:* 610(091)

[3128]

Current work in the history of medicine: an international bibliography of references received ... London, Wellcome Institute for the History of Medicine, 1954 - 4pa. £20pa (individuals), £30pa (institutions). ISSN: 00113999.

Issue no.155, October 1992, is a subject index to 977 entries. Appended: Address of authors; Subject index; Author index; New books (p.349-52); Since 1974 compiled partly from MEDLARS (Boston Spa) print-outs. *Class No:* 610(091)

[3129]
McGREW, R.E. *and* McGREW, M.P. **Encyclopedia of medical history.** New York, etc., McGraw-Hill, 1985. xiv, [i], 400p. $41.25. ISBN: 0333288025.

103 lengthy entries, 'Abortion' ... 'Yellow Jack', each with 'Additional readings' (*e.g.* 'Quackery/Charlatanism': 8 columns; 21 lines of readings 'Drug abuse': 10½ columns; 2/3p. of readings, 1 cross-reference.) Includes entries for Chinese, Indian, etc. medicine. Analytical index - a bonus - carries over 4,000 entries, p.365-400. No biographies. A highly readable text, written for lay persons rather than practitioners. *Class No:* 610(091)

[3130]
MORTON, L.T. **Morton's medical bibliography:** an annotated checklist of texts illustrating the history of medicine (Garrison and Morton). 5th ed., by J.M. Norman. Aldershot, Scolar Press, 1991. 1267p. £85. ISBN: 0859678970.

First published 1943.

8927 main entries (1051 of them new), arranged by subject, *e.g.* biology, epidemiology, leprosy, obstetrics, then chronologically. Includes translations and reprints. Sections on medical biography, medical bibliography and medical lexicography. Indexes of personal names and subjects. Unrivalled. *Class No:* 610(091)

[3131]
SIGERIST, H.E. **A History of medicine.** New York, Oxford Univ. Press, 1951-61 (reprinted 1987). 2v. illus. (pl.). $39.95ea. (*Department of the History of Medicine, Yale Univ. Publication nos.27, 38.*) ISBN: 0195001028, v.1; 0195001036, v.2.

1. *Primitive and archaic medicine.* xxi, 564p. 2. *Early Greek, Hindu and Persian medicine.* xvi, 352p.

Each volume has chapter references. In v.1 appendices cover bibliographies of histories of medicine, museums of medical history, and literature on palaeogeography since 1930. Originally conceived as a definitive history in 8v. The author's death in March 1957 brought this design to a premature end. V.2 was drafted but not completed, and has been edited by L. Edelstein. *Class No:* 610(091)

[3132]
—RHODES, P. An Outline history of medicine. London, Butterworths, 1985. 236p. illus., diagrs., maps. £17.50; £11.50. ISBN: 0407004106.

Updates Singer by giving generous coverage to 20th-century changes and stressing such innovations as the use of the stethoscope and the emergence of bacteriology. General social attitudes are also traced. Appended chronology, bibliography and index. 'A lively up-to-date survey of medicine from earliest times to today' (*British book news,* August 1985, p.481). *Class No:* 610(091)

[3133]
SINGER, C. *and* UNDERWOOD, E.A. **A Short history of medicine.** 2nd ed. Oxford, Clarendon Press, 1962. xvi, 854p. illus. pl.

First published 1928.

Over 70% of the 2nd ed. is on the 19th and 20th centuries, and this part has been rewritten and expanded, whereas the earlier part shows only modest changes. Substantial bibliography, p.761-95. The standard general history of medicine, now somewhat dated. *Class No:* 610(091)

[3134]
Thornton's medical books, libraries and collectors: a study of bibliography and the book trade in relation to the medical sciences. Besson, A., *ed.* 3rd ed. Aldershot, Hants., Gower, 1990. xxi,417p. facsims. £55. ISBN: 0566054817.

First published 1949.

10 contributors, 10 chapters: 1. Medical books before the invention of printing - 2. Medical incunabula ... 7. The growth of medical periodical literature ... 9. Private medical libraries - 10. Medical libraries of today. Extensive bibliography, p.342-57. Detailed, analytical index, p.358-417. A survey in depth. *Class No:* 610(091)

[3135]
WELLCOME INSTITUTE FOR THE HISTORY OF MEDICINE. **Subject catalogue of the history of medicine and related sciences.** Munich, Kraus International Pubns., 1980. 18v.

1-9. *Subject section.* 10-13. *Topographical section.* 14-18. *Biographical section.*

Coverage is from 1954 to 1977; acts as an authority list for *Current work in the history of medicine (qv)* for the period concerned. 'Not an index to the entire collection, but is probably one of the most comprehensive guides in existence to the modern secondary literature of the history of medicine and allied sciences' (*Preface*). *Class No:* 610(091)

[3136]
—The Truman G. Blocker, Jr., History of medicine collections: books and manuscripts. Wygant, L.J., *ed.* Galveston, Univ. of Texas Medical Branch, 1986. 432p. facsims. $50. ISBN: 0292780931.

Records 13,000 books and manuscripts in the Univ. of Texas' Moody Medical Library on the subject. Included is a brief but helpful bibliography. 'The collection is the largest resource in the Southwest' (*New technical books,* November 1986, item (259)). Well-produced. *Class No:* 610(091)

Biographies

[3137]
BENDINER, J. *and* BENDINER, E. **Biographical dictionary of medicine.** New York, Facts on File, 1990. 284p. $40. ISBN: 0816018642.

International in scope, with entries varying in length from one paragraph to two pages. Written for the lay person. Cross-references and a brief bibliography. Quick-reference guide which may be useful for school and public libraries. *Class No:* 610(092)

[3138]
Dictionary of American medical biography. Kaufman, M., *and others.* Westport, Conn., Greenwood Press, 1984. 2v. (1027p.). $175. ISBN: 031321378x.

Continues H.A. Kelly and W.L Bussage's *Dictionary of American medical biography* (1928).

Records more than 1,000 medical personalities of the 17th-20th centuries who died before 31 December 1976. Includes references to other sources (*e.g., Dictionary of American biography*). Nearly 100p of appendices, listing biographees by date and place of birth, by state where prominent, by speciality/occupation, by medical school attended. Extensive index. *Class No:* 610(092)

[3139]
ELLIS, H. Bailey and Bishop's notable names in medicine and surgery. 4th ed. London, H.K. Lewis & Co., 1983. xv,272p. illus. ports. £18.95. ISBN: 0718604660.

3rd ed., 1959, as *Notable names in medicine and surgery: short biographies of some whose discoveries (not necessarily the greatest medical discoveries) have become eponymous in the medical and allied professions.*

69 biographical sketches of eponymous medical men, arranged A-Z (Addison ... Winslow). List of biographies for further reading, p.256-62. Index. *Class No:* 610(092)

[3140]
FIRKIN, B.G. *and* WHITWORTH, J.A. Dictionary of medical eponyms. Carnforth, Lancs., Parthenon Publishing Gp., 1987. vii, 591, [1]p. ports. £28. ISBN: 1850701385.

About 1,500 entries. Length varies from single line to 2 pages (*e.g.* 'Down syndrome': ¾p; 'Listeria monocytogenes': 2p.). 'Reference Books that we particularly relied upon ...', p.vii. 'This book is an endeavour to list eponyms used in the practice of internal medicine in Australia and probably in most of the English-speaking countries in the world' (*Introduction*). Well produced. *Class No:* 610(092)

[3141]
—LOURIE, J. Medical eponyms: who was Coudé? London, Pitman Medical, 1982. 220p. £4.95. ISBN: 0272226433.

Has *c.*900 eponymous entries, with brief biographical details, thus supplementing Bailey and Bishop. *Class No:* 610(092)

[3142]
HIRSCH, A. Biographisches Lexikon der hervorragenden Ärzte aller Zeiten und Völker. 2. Aufl. durchgesehen und ergänzt von F. Hübotter und H. Vierordt. Berlin, Urban, 1929-35. 5v. and Supplement. ports. (Reprinted 1962).

First published 1884-88.

A unique universal biographical dictionary of physicians who attained prominence in their profession before 1880. Entries include lists of their writings and sources of additional information. The supplement includes corrigenda and addenda. *Class No:* 610(092)

[3143]
Medical sciences international who's who. 4th ed. Harlow, Essex, Longman Group, 1990. vi,1341p. £325. (*Reference on research.*) ISBN: 0582041937.

First published 1980. 2nd ed. as *International medical who's who.*

Biographical details of *c.*8,000 senior medical and biomedical scientists from over 90 countries. 2 parts: 1. Biographical profiles (A to Z, p.1-1306) - 2. Country and subject index. 5th ed. due to be published 1992/93. *Class No:* 610(092)

[3144]
—The International who's who in medicine. Soham, Ely, Cambs., Melrose Press, 1986. 814p. £95. ISBN: 0900332867.

Includes 5,000 biographies of professionals in medical and related fields. *Class No:* 610(092)

[3145]
MORTON, L.T. *and* MOORE, R.J. A Bibliography of medical and biomedical biography. Aldershot, Hants., Scolar Press, 1989. ix,208p. £39.50. ISBN: 0859677974.

Originally intended as 3rd ed. of J.L. Thornton's *Select bibliography of medical biography,* (2nd ed. Library Association, 1970).

1,600 individuals listed (Abano ... Zuckerman). Biographies restricted to English language and translations. Locations of archival material are also listed. Collective biographies, p.173-77. Short list of books on the history of medicine and related subjects, p.178-88. Discipline index of biographees, p.189-208. *Class No:* 610(092)

[3146]
—GARRISON, F.H. An Introduction to the history of medicine, with medical chronology, suggestions for study, and bibliographic data. 4th ed., rev. and enl. Philadelphia, Pa., Saunders, 1929. 996p. ports. $32.75. ISBN: 0721640303.

Reprinted 1960 as *History of medicine.*

Chiefly of bibliographical interest. Numerous portraits. Details of bibliographies of medicine. Appendices include a chronology of medicine and public hygiene (p.809-72) and bibliographic notes for collateral reading (p.884-922). Recognised as a standard history of medicine. *Class No:* 610(091)

[3147]
The Roll of the Royal College of Physicians of London; comprising biographical sketches of all the eminent physicians whose names are recorded in the annals ... 2nd ed., rev. and enl. London, the Royal College, 1878-. (v.8. 1989). £22ea, v.1-7; £55, v.8.

V. 1-3 (1878) cover the period 1518-1825 and constitute the most authentic source of biographical information for many physicians. V.4 (1955): 1826-1925, records Fellows elected up to 1925 who died before 1 January 1954. V.5 (1968) covers 1954-65; v.6-8 continue the record up to 1988. *Class No:* 610(092)

[3148]
ROYAL COLLEGE OF PHYSICIANS OF LONDON. William Harvey, 1578-1657: an exhibition of books and pamphlets illustrating his life and work. London, The Royal College, 1957. 25p. port.

39 annotated entries, plus notes on the portraits and genealogy of William Harvey, 'In this tercentenary exhibition the chief emphasis has been placed on the life of Wm. Harvey and in particular on his association with the College' (*Foreword*). *Class No:* 610(092)

[3149]
SOURKES, T. Nobel Prize Winners in medicine and physiology, 1901-65. New and rev. ed. New York, London, etc., Abelard-Schuman, 1966. xii, 464p. ports.

First published 1953 (covering 1901-50), by L.G. Stevenson.

About 100 biographical sketches (chronologically arranged) each with small portrait, description of the prize-winning work and its consequences in theory and practice, plus references. Classification of the awards, by topic. Index, p.451-60. *Class No:* 610(092)

[3150]
—Nobel Laureates in medicine or physiology. Fox, D.M., *and others, eds.* New York, Garland, 1990. 595p. $95. (*Garland reference library of the humanities, v.552.*) ISBN: 0824078926.

Articles arranged A-Z by name and covers 1901-89. Index of prizewinners and scientists mentioned in text or bibliographies. No subject index. *Class No:* 610(092)

[3151]
TALBOT, C.H. *and* HAMMOND, E.A. The Medical practitioners in medieval England: a biographical register. London, Wellcome Historical Medical Library, 1965. x, 503p.

Arranged A-Z, each with references. 'List of books consulted and abbreviations' p.427-49. Analytical general index, p.451-503. Aims 'to bring together in convenient form all the discoverable biographical information on the medical profession in medieval Britain' (*Introduction*). *Class No:* 610(092)

[3152]
Women in medicine: a biliography of the literature on women physicians. Chaff, S.L., *and others.* Metuchen, N.J., Scarecrow Press, 1977. 1136p.

Annotated bibliography of 4,000 items written by and about women, 1750 onwards. Sectionalized, - biography, history, psychosocial aspects of female doctors, their wartime activities, and even citations of references in fiction. Includes foreign-language material, 'An excellent reference book for large medical collections and for public and high school libraries which are heavily used for research' (*Library journal*, v.103(4), 15 February 1978, p.451). *Class No:* 610(092)

Manuscripts & Incunabula

[3153]
ISKANDAR, A.Z. A Catalogue of Arabic manuscripts on medicine and science in the Wellcome Historical Medical Library. London, the Library, 1967. xvi, 255, [1]p. front., facsims. £22.

Valuable introductory comments on some of the most important MSS, p.1-72; catalogue, p.73-228 (A-Z titles). Bibliographical references; p.xv-xvi. Index to introduction; indexes of manuscripts, authors, scribes, former owners, and places. Facsims. of *c.*150-200 items. *Class No:* 610(093)

[3154]
KLEBS, A.C. Incunabula scientifica et medica: short-title list. Bruges, St. Catherine Press, 1938 (reprinted, Hildesheim, George Olms, 1963). 359p.

First published as v.4(1) of *Osiris.*

Registers more than 850 editions of medical books published in the 15th century. Under authors A-Z, with editions in chronological order. *Class No:* 610(093)

[3155]
—OSLER, Sir W. Incunabula medica: a study of the earliest printed medical books, 1467-1480. Oxford Univ. Press, for the Bibliographical Society, 1921. 140p.

Contains 217 entries. *Class No:* 610(093)

[3156]
—POYNTER,, F.N. A Catalogue of incunabula in the Wellcome Historical Medical Library. London, Oxford Univ. Press, 1954. 160p. illus (pl.).

Contains 610 main short-title entries, many of which represent unique items. An invaluable illustrated guide to the finest collection of medical incunabula in Britain. *Class No:* 610(093)

[3157]
WELLCOME HISTORICAL MEDICAL LIBRARY. Catalogue of Western manuscripts on medicine and science in the Wellcome Historical Medical Library. Moorat, S.A.J. London, the Library, 1962-73. 2v. in 3.

1. *MSS written before 1650 A.D.* 1962. 2. *MSS written after 1650 A.D.* 1973. 2pts. (768 + 720p.).

V.1 contains detailed bibliographical descriptions of *c.*1,500 individual works, with 11 indexes, - chronological, linguistic, scribes, bindings, owners, book-plates, subjects, etc. *Class No:* 610(093)

Laws

[3158]
WORLD HEALTH ORGANIZATION. International digest of health legislation. 1948-. Geneva, WHO, 1950-. v.1(1)-. 4pa.

Summarizes health laws and regulations, citing appropriate clauses, under countries, Australia ... Zaire. Appended bibliographic section. Microfiche ed., v.1-25, covering 1948-1974. *Class No:* 610(094.1)

Underwater

[3159]
SHILLING, C.W. *and* WERTS, M.F. Underwater medicine and related sciences; a guide to the literature; an annotated bibliography, keyword index and microthesaurus. New York, IFI/Plenum, 1973-75. 2v.

Updates *An annotated bibliography on diving and submarine medicine* (New York, Gordon and Breach, 1971).

Sponsored by the Bureau of Medicine and Surgery, and the Office of Naval Research.

Each volume carries *c.*1,800 annotated entries. Permuted subject index, author index, and microthesaurus. *Class No:* 610(204)

[3160]
—The Underwater handbook: a guide to physiology and performance for the engineer. Shilling, C.W., *and others, eds.* New York, etc., Wiley, 1977. 912p.

Practical and wide-ranging - 'as shown when it covers underwater blast, electric shock, safety precautions in pressure chambers, the design and use of underwater breathing equipment and tools, and the selection and training of divers' (*British medical journal*, 10th September 1976, p.696). *Class No:* 610(204)

Tropics

[3161]
Tropical and geographical medicine. Warren, K.S. *and* Mahmoud, A.A.F., *eds.* 2nd ed. New York, etc., McGraw-Hill, 1990. 1159p. $125. ISBN: 007068328x.

7 parts, by 154 specialists. 1. Clinical, biological and epidemiological considerations in the approach to tropical

.... *(contd.)*

medicine - 2. Protozoan diseases - 3. Metazoan diseases (including anthropods and venomous animals) - 4. Viral and chlamydial diseases - 5. Bacterial, spirocheral and rickettsial diseases - 6. Fungal diseases - 7. Nutritional diseases. 4 appendices and index, p.1115-59. Of the first ed., 'This is the most comprehensive reference on tropical diseases which embodies an ecological perspective' (Harris, C.D., ed. *Geographical bibliography for American libraries* (1985), item 2760). *Class No:* 610(213)

Great Britain

[3162]

NHS data book. Fry, J., *and others.* Lancaster, etc., MTP Press, Ltd., 1984. 271p. diagrs., graphs, tables. ISBN: 0852007353.

18 chapters (*e.g.* 1. Population; 2. Socio-economic factors; 4. Social pathologies; 8. Personnel in NHS [National Health Service]; 11. Prescribing; 12. Psychiatry; 13. Maternity services; 18. Costs). Tables cite sources. Brief index, p.269-71. 'It includes social and demographic data, NHS facts and figures ...' (*Preface*). *Class No:* 610(410)

[3163]

NHS handbook. Connah, B. *and* Pearson, R., *eds.* 7th ed. London, Macmillan, 1991. xxiii,267p. illus., tables. ISBN: 033353736x.

First published 1980.

9 sections: 1. The NHS today - 2. Management - 3. Funding - 4. Trends in health care ... 8. Care in the community - 9. Partnerships with the NHS. Reference section, *e.g.* 'Reference map'; 'Abbreviations and acronyms'. Index. Provides an up-to-date overview. *Class No:* 610(410)

Scotland

[3164]

COMRIE, J.D. History of Scottish medicine. 2nd ed. London, Baillière, Tindall & Cox, for Wellcome Historical Medical Museum, 1932. 2v. illus., maps.

First published 1927.

A marked improvement on the 1-volume ed. of 1927. Noteworthy for its illustrations and biographical information. *Class No:* 610(411)

India

[3165]

JAGGI, O.P. History of science and technology in India. Delhi, Atma Ram & Sons, 1969-80. 15v. illus.

8 of the 15v. relate to Indian medicine. 3. *Folk medicine.* 1973. 3. *Indian system of medicine.* 1973. 5. *Yogic and Tantric medicine.* 1973. 8. *Medicine in medieval India.* 1977. 12-15. *Western medicine in India.* 1979-80.

V.3 (xxxii, 228p.) has 2 parts: 1. Tribal medicine (section 3: 'Causation of diseases', p.37-66, with 42 references) - 2. Village medicine. Each part has appended select bibliography. General index (non-analytical), p.221-8. *Class No:* 610(540)

Islamic World

[3166]

AGHA, Z.M. Bibliography of Islamic medicine and pharmacy. [Bibliographie der Islamischen Medizin und Pharmacie.] London, E.J. Brill, 1983. 108p.

698 entries, often briefly annotated, under authors; A-Z. British Library locations are shown. *Class No:* 610(5.297)

[3167]

—EBIED, R.Y. Bibliography of mediaeval Arabic and Jewish medicine and allied sciences. London, Wellcome Institute of the History of Medicine, 1971. 150p. £12.50. (*Occasional series, 2.*) ISBN: 0822432251.

Is, unlike Agha's *Bibliography,* restricted to the Middle Ages. It also includes reference material not strictly relevant. *Class No:* 610(5.297)

[3168]

ULLMANN, M. Islamic medicine. Edinburgh Univ. Press, 1978. xiv, 138p. illus., facsims. (*Islamic surveys, 11.*)

'An account of a medical system which was introduced into Arab countries in the 9th century A.D. and was practised throughout the Middle Ages and right up to modern times' (*Introduction*). Survey of previous books on the subject, p.xii-xiii. Source, p.115-6; notes, p.117-30. Excludes such topics as surgery, hospital institutions, medical teaching, the doctor's social standing and relations with patients. *Class No:* 610(5.297)

America — North

[3169]

Bibliography of the history of medicine of the United States and Canada, 1939-1960. Miller, G., *ed.* 2nd ed. Baltimore, Md., Johns Hopkins Press, 1965. xv, 425p. $34.50.

Arranged by broad topics, with author index. A re-issue, in consolidated form, of the annual bibliography published in the *Bulletin of the history of medicine* since 1940.

Continued by 4v. set covering the years 1964-1984 (v.1. 1964-1969, v.2. 1970-1974, v.3. 1975-1979, v.4. 1980-1984; Bethesda, Md., DHEW, 1972-85). *Class No:* 610(71+73)

Writing & Lecturing

[3170]

'BRITISH MEDICAL JOURNAL'. How to do it. Volume 1. 2nd ed. London, British Medical Association 1985. xiii, 266p. illus., tables. ISBN: 0727901869.

First published 1979.

Reprints 47 articles from *British medical journal* on a wide variety of practical challenges, - organizing an international medical meeting, writing the MD thesis, becoming a medical journalist/editor, editing a specialist journal, reviewing a book. Some of the contributions carry references. *Class No:* 610:001.81

[3171]

DUDLEY, H. The Presentation of original work in medicine and biology. Edinburgh, Churchill Livingstone, 1977. vii, 99p. diagrs., graphs. ISBN: 044301583x.

5 chapters, each with references. 1. Deciding to communicate - 2. Structure of scientific thought in relation to spoken and written communication - 3. Matters of detail (*e.g.* referees) - 4. English, particularly but not exclusively for written communication - 5. Aspects of speaking at

.... (contd.)

scientific meetings, 15. 'textual acknowledgements'. Appendices: 1. General references and scientific writing (17 items, with brief-evaluative annotations); 2. Reference conventions; 3. Abstracts for the Surgical Research Society; 4. Statistic sources useful to the medical biologist. Author, then Professor of Surgery, Univ. of London.
Class No: 610:001.81

[3172]
LANE, N.D. *and* **KAMMERER, K.L. Writer's guide to medical journals.** Cambridge, Mass., Ballinger; Wiley, 1975. xi, 327p.

Contents include style instructions; classified subject guide; and journal publication data and instructions on over 300 general and research journals. Style manuals and sources of standards, p.323-7. A guide to the best medium for publishing medical articles. *Class No:* 610:001.81

Research Methods

[3173]
BENNINGTON, J.L. Saunders dictionary & encyclopedia of laboratory medicine and technology. Philadelphia, Pa., W.B. Saunders, 1984. xx, 1674p. illus., diagrs., tables. $79.95. ISBN: 072161714x.

46 contributors. Vocabulary (p.1-1621) of *c.*30,000 entries, a mixture of definitions and explanations. 'Enzymes': nearly 3p. Adequate cross references. Eponyms included. 3 appendices: A. Cancer chemotherapy drugs; B. Bacteriologic specimen collection; C. Reference ranges and laboratory values of clinical importance. For 'laboratorians, clinicians, nurses, allied health and public health personnel, basic scientists, and students preparing for courses in health care fields' (*Preface*). 'A rich source for the physician whose primary interest is patient care and teaching' (*Journal of the American Geriatrics Society*, v.33, August 1985, p.572). *Class No:* 610:001.891

[3174]
Dictionary of medical laboratory sciences. Farr, A.D., *ed.* Oxford, Blackwell Scientific Pubns., 1988. viii, 318p. £17.99. ISBN: 0632017627.

About 3,500 entries, Ab ... Zymosan. Includes acronyms (*e.g.* ACB: Association of Clinical Biochemists); trade names (*e.g.* OKT serum) and a few important names in medical history (*e.g.* Edward Jenner). Bold type for entries and cross-references. *Class No:* 610:001.891

[3175]
Introduction to medical laboratory technology. Baker, F.J., *and others.* 6th ed. London, Butterworths, 1984. x, 408p. illus., diagrs., tables. £24.50. ISBN: 0407732527.

First published 1954.

6 sections (44 subsections): 1. General (*e.g.* Health and safety in the laboratory) - 2. Clinical chemistry - 3. Cellular pathology - 4. Microbiology - 5. Haematology - 6. Blood transfusion technique. Systematic treatment, stressing methods and techniques. 'Recommended for further reading' (5 headings), p.383. Appendix 1: Manufacturers' names and addresses; Useful information (*e.g.* Saturated solutions). Analytical index, p.393-408. Standard textbook. *Class No:* 610:001.891

[3176]
—**WORLD HEALTH ORGANIZATION. Manual of basic techniques for a health laboratory.** Geneva, WHO, 1980. [v], 487p.

A revised version of *Basic techniques for a medical laboratory*, by E. Lévy-Lambeth (1974). 3 main parts: 1. General laboratory procedures - 2. A. Parasitology; B. Bacteriology; C. Serology; D. Mycology - 3: A. Examination of urine; B. Examination of cerebrospinal fluid; C. Haematology: D. Blood chemistry; E. Blood transfusion. Reagents and their preparation, p.479-87. Detailed analytical index, p.479-87. *Class No:* 610:001.891

Information Management

[3177]
Handbook of medical library practice. Darling, L., *ed.* 4th ed. Chicago, Ill., Medical Library Association, 1982-1988. 3v. $70, set; $19.40, v.1; $23.80, v.2; $30.80, v.3. ISBN: 0912176113, v.1; 0912176121, v.2; 0912176210, v.3.

First published 1943; 2nd ed. 1970.

1. *Public services in health science libraries.* 1982. 2. *Technical services in health science libraries.* 1983. 3. *Health science librarianship and administration.* 1988.

Notably updates the 3rd ed. by including in v.1, chapters on databases and computerized methods. For 'beginning and experienced medical librarians as well as teachers and students in library schools' (*Library journal*, 15 January 1983, p.114). *Class No:* 610:025.4

[3178]
LIVESEY, B. *and* **STRICKLAND-HODGE, B. How to search the medical sources.** Aldershot, Hants., Gower, 1989. viii,119p. figs. £25. ISBN: 0566035332.

6 sections: 1. Searching the medical sources - 2. Recording information - 3. Sources: background information - 4. Periodicals and periodical articles - 5. Other sources - 6. Organizing and presenting information. 91 figures reproduce extracts from the publications under discussion. List of references, p.113-16. For the library student and medical library staff. *Class No:* 610:025.4

[3179]
Medical librarianship. Carmel, M., *ed.* London, Library Association Publishing, 1981. xi, 359p. £26.50; £14.95. (*Handbooks on library practice.*) ISBN: 0853655022.

23 contributors. 21 chapters (1/3: The professions - 4/5: Information: serials, books, AV ... 11/14: Enquiry service ... 18: Some recent developments in health science libraries in North America - 19. Third World medical libraries - 20. Library and information research in health care - 21. Keeping in touch. Chapter references. Analytical index, p.351-9. Aimed at 'practising medical and health care librarians and seriously interested students' (*Library Association record*, v.84(3), March 1982, p.113). *Class No:* 610:025.4

Libraries

[3180]
Directory of medical and health care libraries in the United Kingdom and Republic of Ireland. Wright, D., *ed.* 8th ed. London, Library Association Publishing, 1992. 300p. £25. ISBN: 1856040666.

First published 1957. Editions 1-4 as *Directory of medical libraries in the British Isles.* 6th ed., 1986.

....(contd.)

Lists *c*.700 libraries concerned with medicine, nursing, occupational therapy, and pharmacology. Data include availability; stock policy, special collections, classification, hours of opening, special services. 3 appendices: List of non-responding libraries; List of members of NHS Regional Libraries Group. Indexes: personal names, establishments, countries, special collections. *Class No:* 610:061:026/027

[3181]

JAMIESON, A.H. Directory of health library resources in Scotland. Edinburgh, Association of Scottish Health Sciences Librarians, 1986. £2.50. ISBN: 187022900.

A comprehensive list of 116 entries. Data on each: location; telephone; accessibility; subject specializations; hardware/software used; classification systems; staff. Appendix: 1. Health Education depts.; 2. Drug information centres; 3. ASMSL Committee and subcommittees; 4. ASHSL members not mentioned in the Directory; 6. List of members. Now needs updating. *Class No:* 610:061:026/027

[3182]

World directory of biological and medical sciences libraries. Poland, U.H., *ed.* Munich, New York, K.G. Saur, 1988. 203p. $30. ISBN: 3598217722.

Libraries from over 100 countries are listed. Data include addresses; subject areas; holdings; availability of materials. Appendices list national and international library directories; union lists; cooperative service centres; addresses of biological and medical library associations. *Class No:* 610:061:026/027

Institutions & Associations

[3183]

Directory of European societies and associations: medicine 1992. London, AG Publishing, 1992. 130p. ISBN: 1874305005.

Lists 200 medical societies and associations in 27 countries, A-Z by name. Data include full address, telephone/fax numbers, contact name, publications, conferences. Indexes: subject; mailing list; 1992/93 conference; country; society. *Class No:* 610:061:061.2

[3184]

The Directory of health centres. Harlow, Essex, Longman, 1981-. Annual.

The 1986 annual (486p.) records more than 1,300 health centres in the UK. Arranged A-Z by towns, then services; practitioners, dentists, etc. *Class No:* 610:061:061.2

[3185]

Directory of international and national medical and related societies. Zeitak, G. *and* Berman, F., *eds.* 2nd ed. Rehovot, Israel, PBZ Informatics, and Oxford, Pergamon Press, 1990. 340p. ISBN: 0080374956.

4 parts: 1. Introduction - 2. Listing of societies (international; national, Afghanistan ... Zimbabwe, p.51-262) - 3. Indexes: Countries list with reference numbers; Alphabetical list of society names; Thesaurus of subject headings; Key to subject codes; Societies listed by subject - 4. Questionnaire used. About 4,500 well-indexed entries. *Class No:* 610:061:061.2

[3186]

Encyclopedia of medical organizations and agencies. 3rd ed. Detroit, Mich., Gale, 1990. 1079p. ISBN: 0810348497.

First published 1983.

69 subject chapters, covering the programs of more than 11,250 of the leading US organizations in medicine and health. 8 functional groups (National and international associations ... Research centres and institutes). Detailed subject cross index; master name and keyword index.

4th ed., 1991. 1000p. $205. *Class No:* 610:061:061.2

[3187]

Medical and health information directory. Backus, K., *ed.* 5th ed. Detroit, Mich., Gale, 1990-91. 3v. $485, set. ISBN: 0810349264, v.1; 0810349272, v.2; 0810349280, v.3; 0810349256, set.

First published 1977.

1. *Organizations, agencies and institutions.* 1990. 1182p. 2. *Publications, libraries, and other information services.* 721p. 1990. 3. *Health services including clinics, treatment centers, care programs and counselling/diagnostic services.* xvii,865p. 1991.

For US medical and health professionals, government officials, librarians, consumers, and others. Covers *c*.48,000 organizations. Each volume is self-contained. *Class No:* 610:061:061.2

Conferences

[3188]

Calendar of congresses of medical sciences. [Calendrier des congrès des sciences médicales.] Geneva, Council for International Organizations of Medical Sciences. Annual.

No. 51, 1990 lists congresses planned for 1990-95. Data for each entry includes date; place; title of congress; contact name and address. Indexes of subjects and countries. *Class No:* 610:061:061.3

[3189]

World meetings: Medicine. Chestnut Hill, Mass., World Meetings Information Center, 1978-. 4pa., cumulated.

Previously part of *World meetings: United States and Canada* (1963-. 4pa.) and *World meetings outside USA and Canada* (1968-. 4pa.), covering scientific, engineering and medical meetings.

Details of more than 1,000 medical meetings pa. *Class No:* 610:061:061.3

[3190]

—BISHOP, W.J. Bibliography of international congresses of medical sciences. Prepared ... under the auspices of the Council for International Organizations of Medical Sciences [CIOMS], with the financial assistance of UNESCO. Oxford, Blackwell, 1958. xxii, 238p.

Lists 363 congresses chronologically under subjects A-Z, and provides bibliographical data on published proceedings. *Class No:* 610:061:061.3

Research Establishments

[3191]

Medical research centres: a world directory of organizations and programmes. 9th ed. Harlow, Essex, Longman, 1990. vi,877p. £290. (*Reference on research.*) ISBN: 0582061237.

Succeeds *Medical research index* (5th ed. Hodgson, 1979); *Medical research centres* (8th ed. Longman, 1988).

Arranged A-Z by country, Angola ... Zimbabwe. A guide

....*(contd.)*

to *c*.6,000 organizations and laboratories which conduct or finance medical or biological research and development. Data: name and address; telephone, telex and fax numbers; names of key personnel; finance; activities; publications; clients. Indexes of establishments and subject. 10th ed. due to be published in 1993. *Class No:* 610:061:061.62

[3192]
MEDICAL RESEARCH COUNCIL. Handbook 1986. [Medical Research Council]. London, the Council, 1986. iv, 319p.

Sections include: Research of MRC establishments and staff; Research supported by grants; Health services research; Research and development contracts; Outside support; International collaboration; Clinical research professorships and senior fellowships; Fellowships and scholarships; Publications. Personal names index; analytical subject index, p.299-319. *Class No:* 610:061:061.62

Research Projects

[3193]
GREAT BRITAIN. Department of Health. DH yearbook of research and development 1990. London, HMSO, 1990. xvi,199p. £14. ISBN: 0113213204.

Formerly *Research and development report and handbook.*

8 sections: 1. Valediction - 2. AIDS theme research ... 4. Research units ... 6. NHS information technology ... 8. Publications (A. Previous research ... D. Procurement Directorate research). Appendix 1: International research: UK representatives. Indexes by name; research agency; subject. *Class No:* 610:061:061.62.005

[3194]
The Medical research directory. New York, Chichester, etc., Wiley, 1983. 730p. ISBN: 0471103357.

The main text consists of research-project abstracts, in 45 subject groups (1. Anatomy ... 45. Occupational medicine). Name index; analytical subject index, p.643-730. *Class No:* 610:061:061.62.005

Medical Schools

[3195]
The Directory of schools of medicine and nursing: British qualifications and training in medicine, dentistry, nursing and related professions. 2nd ed. London, Kogan Page, on behalf of International Hospital Group, 1984. x, 752p. ISBN: 0850387884.

First published 1983.

9 parts: 1. Introduction - 2. Medicine - 3. Dentistry - 4. Nursing, midwifery and health visiting - 5. Professions supplementary to medicine - 6. Professions relating to medicine (*e.g.* clinical psychology; pharmacy - 7. Alternative medicine - 8. Health-care technology - 9. Further information (list of useful addresses; further reading). Index of qualifications and abbreviations. Detailed general index (in small type, p.742-54). Part 2-5 each have subsection: 'Directory of qualifications and training by institutions'. A comprehensive guide. *Class No:* 610:061:378

[3196]
—Directory of British postgraduate medical qualifications. Higson, N. London, Chapman & Hall Medical, 1987. xix, 299p. maps. £35. ISBN: 0412291002.

Arranged under organizing bodies. 1 entry per page. Two indexes precede: Index by type of qualification and subject (certificates; diplomas; master degrees; other degrees and awards) - Index by organizing body. Outline location maps. *Class No:* 610:061:378

[3197]
—Directory of postgraduate medical centres, 1986. [London], Council for Postgraduate Medical Education in England and Wales, and National Association of Clinicals, [1986]. 87p.

Has data on 381 postgraduate medical centres, including those in psychiatric hospitals and in teaching hospitals: England (14 regions; p.4-70), Wales, Northern Ireland, Scotland. Appendices. Index. *Class No:* 610:061:378

[3198]
World directory of medical schools. 6th ed. Geneva, WHO, 1988. 311p. £14.

First published 1953.

Lists institutions of undergraduate medical education in 127 countries and areas, Afghanistan ... Zimbabwe. Data: administration; conditions of admission; curriculum; examinations; number of hospital beds; statistics for academic year; licence to practise; type of degree. Annex: number of medical schools in 1955, 1960, 1970, 1975, and 1985, by country or area.

Regularly updated in *WHO chronicle* (monthly). *Class No:* 610:061:378

Ethics

[3199]
Bibliography of bioethics. New York, The Free Press; London, Collier-Macmillan, 1975-. Annual.

V.15, 1989 (601p.) carries 2,400 subject entries. Topics include abortion, AIDS, euthanasia, genome mapping, organ transplantation, and surrogate motherhood. Descriptions are appended to entries, some of which contain annotations. Bioethics thesaurus, p.241-78. Title and author indexes. One of the ongoing research projects of the Kennedy Institute of Ethics. Also available on a National Library of Medicine database, BIOETHICSLINE. *Class No:* 610.17

[3200]
—GOLDSTEIN, D.M. Bioethics: a guide to information sources. Detroit, Mich., Gale, 1982. 336p. (*Health affairs information guide series, v.8.*)

An annotated bibliography of books, articles and US government documents published 1973-80. Subject arrangement of *c*.1,000 items; author, title and subject indexes. 'A reasonable, convenient and reliable alternative to the annual *Bibliography of bioethics,* 1975-.' (*Wilson library bulletin,* v.56(10), June 1982, p.781). *Class No:* 610.17

[3201]
Dictionary of medical ethics. Duncan, A.S., *and others, eds.* 2nd ed. London, Darton, Longman and Todd, 1981. xxxi, 459p. diagrs., tables. ISBN: 0232514925.

First published 1977.

About 150 contributors. All articles are signed, with references (*e.g.* 'Mental illness, Certification of': 2½ columns, 12 references; 'Psychiatry, Misuse of': 4p. 17 references). 'The work is of high standard and includes

....(contd.)

questions of social as well as purely medical importance ... Can be unreservedly recommended' (*British book news,* March 1978, p.196-7). 'This much needed and valuable book' (*British medical journal,* 5 November 1977, p.1218-9), on the 1st ed. *Class No:* 610.17

[3202]

Encyclopedia of bioethics. Reich, W.T., *ed.* New York, Free Press, 1978. 4v. (xxxix, 1933p.). 2v. ed., 1982. $250, 1978 ed.; $180, 1982 ed. ISBN: 0029260604, 1978 ed; 002925910x, 1982 ed.

285 contributing authors from 15 countries. 315 signed and well-documented articles, with ample cross-references. The lengthy contribution on abortion covers medical, religious and legal aspects. V.4 includes a detailed index as well as a topic-grouping of articles and a statement of various codes of ethics from the medical professions. 'This is primarily an encyclopedia of issues and ideas rather than of events, and therefore ... is not out of date' (*Reference & subscription books reviews, 1982-1983,* p.27). *Class No:* 610.17

[3203]

THOMPSON, W.A.R. A Dictionary of medical ethics and practice. Bristol, J. Wright, 1977. [viii], 244p. ISBN: 0723604541.

About 100 documented entries (*e.g.* 'Transplantation': 10 columns, 15 references; 'Euthanasia': 4 columns, 5 references; 'Jewish medical ethics': 5 columns, 4 references; 'Suicide': 3½ columns, 7 references; 'Abortion': 5½p., 7 references). No index or cross-references. Reflects British professional and lay thinking. Primarily for the general practitioner. *Class No:* 610.17

[3204]

—BRITISH MEDICAL ASSOCIATION. The Handbook of medical ethics. 2nd rev. ed. London, the Association, 1984. 111p. £5. ISBN: 0727901117.

First published 1981.

6 sections: The history of medical ethics - Relationships between doctors and individuals - Relationships between doctors and groups - Etiquette, professional discipline and the law - Ethical dilemmas - Ethical codes and statements (p.69-87) - Resolutions of BMA Annual Representative Meetings. Bibliography (8 sections), p.95-98. Analytical index, p.99-111. A well-received concise survey. *Class No:* 610.17

Anatomy

Handbooks & Manuals

[3205]

Gray's anatomy. Williams, P.L. *and* Warwick, R., *eds.* 37th ed. Edinburgh, Churchill Livingstone, 1989. xvi, 1598p. illus. (incl. col.), tables. £85. ISBN: 0443025886.

First published 1858.

8 subject sections: Introduction - Cytology - Embryology - Osteology - Arthrology - Myology - Angiology - Neurology - Splanchnology. Bibliography (authors, A/Z), p.1477-1548. Analytical index, p.1549-98. Excellent keyed line-drawings,

....(contd.)

well captioned. The classic textbook on anatomy. The two editors are former professors of Anatomy, Guy's Hospital Medical School, Univ. of London. *Class No:* 611(035)

[3206]

—DIAGRAM GROUP. The Human body on file. Swan, R., *ed.* New York, Facts on File, 1983. v.p. illus. Loose-leaf. $145. ISBN: 0871967065.

Consists of black-and-white outline drawings of the various systems/parts of the human body, from nervous system to endocrine system. Each section has a one-page introductory text. Three major uses are claimed: 'as an atlas of anatomy, as a reference source of images for photocopying, and as a basis for examination paper illustration' (*Foreword*). Although a good source for anyone studying or teaching anatomy at a basic level, it lacks adequate text matter. *Class No:* 611(035)

Dictionaries

[3207]

DONÁTH, T. Anatomical dictionary, with nomenclatures and explanatory notes. English ed., edited by G.N.C. Crawford. Oxford, etc., Pergamon Press, 1969. 634p.

Original ed. in Hungarian (1959).

A comparison of the nomenclature of Basle, Jena and Paris (10 sections), with additional terms, p.15-315. Explanatory dictionary, p.317-517. About 5,000 definitions, with a selection of biographical data on 500 prominent anatomists, p.519-47. Index, p.537-634. Of academic rather than general interest (*RQ,* v.10(1), Fall 1970, p.70). *Class No:* 611(038)

[3208]

—FIELD, E.J. *and* HARRISON, R.J. Anatomical terms: their origin and derivation. 3rd ed., rev. and enl. Cambridge, Heffer, 1968. xvii, 312p.

2nd ed. published in 1957.

Nearly 2,000 entries, including biographical notes. *Class No:* 611(038)

[3209]

Stedman's anatomy & physiology words. Baltimore, Md., Williams & Wilkins, 1992. v,450p. $28. ISBN: 0683079417.

Basically a listing of *c.*44,000 words used in anatomy and physiology. Presented in context as one would hear or write them, with preceding words or adjectives. Does not give definitions. A quick-reference tool for 'anyone who transcribes, records, copy edits, or reads medical records, reports, and other material generated by health care professionals' (*Explanatory notes*). *Class No:* 611(038)

Reviews & Abstracts

[3210]

Excerpta medica. 1. Anatomy, anthropology, embryology and histology. Amsterdam, Excerpta Medica, 1947-. 16pa. (2v.) DFl1,358. ISSN: 00144053.

About 4,000 abstracts pa. 9 sections: 1. General aspects - 2. Anatomy (systematic; functional and experimental; typographic)- 3. Anthropology; 4. Histology (methods; all, tissue and organ culture; general cytology; general histology); 5. Microscopic structure of vertebrates; 6. Microscopic structure of invertebrates; 7. Embryology; 8. Transplantation and regeneration; 9. Breeding of

.... *(contd.)*

experimental animals. Annual subject and author indexes. Online, EMBASE and available on CD-ROM. Available on microfilm. *Class No:* 611(048)

Tables & Data Books

[3211]
INTERNATIONAL COMMISSION ON RADIOLOGICAL PROTECTION. Report of the Task Group on Reference Man. Oxford, etc., Pergamon Press, 1975. xix, 480p. *(ICRP publication no.23.)* ISBN: 0080170242.

3 main sections: (1). Anatomical values for Reference Man - (2). Gross and elemental content of Reference Man - (3). Physiological data for Reference Man. Supporting each section: addendum, summary, and references. 'Produced for the purpose of simplifying radiological damages from radionuclides used in medicine ... A particularly useful compilation of quantitative data on the human body' (Welch, J., and King T.A. *Searching the medical literature* (1985), p.125). *Class No:* 611(083)

Nomenclatures

[3212]
Nomina anatomica. 6th ed. Edinburgh, Churchill Livingstone, 1989. 211p. ISBN: 0443040850.

Prepared by the International Anatomical Nomenclature Committee and is systematically arranged, setting the standard for anatomical nomenclature. *Class No:* 611(083.72)

Illustrations

[3213]
Atlas of human cross-sectional anatomy: with CT and MR images. Cahill, D.R., *and others.* 2nd ed. New York, Wiley-Liss, 1990. 251p. illus. $149.95. ISBN: 0471509884.

High quality photographic images. Highly recommended. *Class No:* 611(084.1)

[3214]
McMINN, R.M.H. *and* HUTCHINGS, R.T. A Colour atlas of human anatomy. 2nd ed. London, Wolfe Medical Publications, Ltd., 1988. 351p. col. illus. £32; £17.95. ISBN: 0723415269.

First published 1977 (532p.).

Has 750 coloured and keyed natural-sized photographs of bones and dissections of the human body. Appended lists of vessels and nerves. 'Designed for the undergraduate student, medical student, or practicing physician' (*RQ,* v.17(4), Summer 1978, p.364, on the 1977 ed.). *Class No:* 611(084.1)

[3215]
Plant, animal and anatomical illustration in art & sciences: a bibliographical guide from the 16th century to the present day. Bridson, G.D.R. *and* White, J.J., *comps.* Winchester, Herts., St. Paul's Bibliographies; Detroit, Omnigraphics, 1990. ix,450p. facsims. £75. ISBN: 0906795818.

7,670 entries arranged chronologically in each section and including articles as well as books. Sections: A. Bibliographies - B. Nature in general - C. Plants - D. Animals - E. The human body - F. Artist biographies - G.

.... *(contd.)*

Periodicals. Appendices: Books on colour; Organizations. Title, subject, and name indexes. For research libraries. *Class No:* 611(084.1)

Histories

[3216]
COLE, F.J. History of comparative anatomy: from Aristotle to the eighteenth century. London, Constable; New York, 1975. viii, 524p. $13.25. *Class No:* 611(091)

Physiology

[3217]
Handbook of physiology: a critical comprehensive presentation of physiological knowledge and concepts. Rev. ed. Baltimore, Md., Williams & Wilkins, for the American Physiological Society, 1977-. illus. $50-$395 per v.

First published 1959-83 (10 sections). Rev.

1. *The nervous system.* 5v. (8pts.). 1977-87. 2. *The cardiovascular system.* 4v. (6pts.). 1979-84. 3. *The respiratory system.* 4v. 1985-88. 4. *Adaptation to the environment.* 1964. 6. *The gastrointestinal system.* 4v. 1989-90. 7. *Endocrinology.* 7v. 1972-76. 8. *Renal physiology.* 2v. 1991. 9. *Reactions to environmental agents.* 1977. 10. *Skeletal muscle.* 1983.

The revised ed. was started in 1977 before completion of the earlier ed. The latter's intention was 'to cover the physiological sciences in their entirety and in about 10 years, and to repeat the process periodically thereafter' (*Foreword,* v.1, Section 1). Well documented multivolume treatise. *Class No:* 612

[3218]
O'CONNOR, W.J. Founders of British physiology: a biographical dictionary, 1820-1885. Manchester Univ. Press, 1988. ix, 278p. £45. ISBN: 0719025370.

Arranged chronologically. 3 parts: 1. Physiologists, 1820-35 - 2. Physiologists, 1835-70 - 3. Experimental physiologists, 1870-85. About 100 short biographies, based primarily on obituaries. Contain brief descriptions of institutions involved with physiology during that period. Touches on botanists and zoologists (*e.g.* Balfour; Vines). Short bibliographies appended to each chapter. *Class No:* 612

[3219]
—**O'CONNOR, W.J. British physiologists 1885-1914:** a biographical dictionary. Manchester, University Press, 1991. xvii,582p. £55. ISBN: 0719032822.

Continues *Founders of British physiology: a biographical dictionary, 1820-1885,* by W.J. O'Connor (Manchester, University Press, 1988) (*q.v.*).

13 sections: 1. Physiologists and physiology 1885-1914 - 2. Physiologists in Cambridge 1805-1914 ... 5. London teaching hospitals ... 7. Physiology in Scotland 1885-1914. In addition to a general index, there is an index of physiological topics. *Class No:* 612

[3220]
Textbook of physiology (BDS). Emslie-Smith, D., *and others.* 11th ed. Edinburgh, Churchill Livingstone, 1988. xii,548p. illus., diagrs., tables. ISBN: 0443034125.

First published 1950 as *Textbook of physiology and biochemistry.*

Has 21 contributors. 41 chapters (The blood ... Energy balance and exercise ... Thermoregulation). Numerous half-tones, line-drawings, and tables. Standard text, with many chapters completely rewritten, and with the medical student mainly in mind. Some references follow each chapter. Index p.535-548. 'BDS' is the book's 'affectionate nickname' (*Preface* to 11th ed.) and comprises the initials of the first editors' surnames. *Class No:* 612

Bioengineering

[3221]
Bioengineering and biotechnology abstracts: Formerly *Bioengineering abstracts.* New York, Engineering Information, Inc., 1974-. Monthly. $460 (in North America), $510 (elsewhere). ISSN: 07366213.

Drawn from *Engineering index monthly.* Formerly *Bioengineering abstracts.*

About 2,500 abstracts pa. on engineering disciplines applied to medicines. Follows the A-Z subject arrangement used in *Engineering index monthly* (*qv*), with full cross-references. Each issue covers 25 core journals. Author and subject indexes per issue, cumulated annually. List of acronyms, initials and abbreviations of organizations precedes. Online COMPENDEX (the machine-readable version of *Engineering index*). *Class No:* 612.08

[3222]
Dictionary of medical equipment. Brown, M., *and others.* London, Chapman & Hall, 1986. viii, 288p. £28. ISBN: 0412282909.

Briefly describes equipment (*e.g.* artificial kidney) as used in the Institute of Medical and Dental Bioengineering, Royal Liverpool Hospital. Entries A-Scanner ... X-Y recorder. 3 appendices: 1. List of abbreviations; 2. Grouped minor entries (Anaesthesia ... Medical ultrasound); 3. Reference list, p.288. No illus. *Class No:* 612.08

[3223]
Encyclopedia of medical devices and instrumentation. Webster, J.G., *ed.* New York, Wiley, 1988. 4v. 3022p. illus., tables, graphs. $525. ISBN: 0471829366, set; 0471629685, v.1; 0471629693, v.2; 0471629707, v.3; 0471611352, v.4.

Over 250 documented articles describing every aspect of medical devices and instrumentation. Articles are several pages in length and are well-illustrated, *e.g.* 'Blood rheology': 7p.; 10 graphs; 60 refs. Over 2,000 illustrations and 200 tables in the set. V.4 has a detailed, analytical index, p.2945-3022. For any person with a scientific background and an interest in technology. *Class No:* 612.08

[3224]
Medical equipment: a guide to selected literature and sources of information. Anthony, K., *comp.* London, Science Reference and Information Service, 1986. [ii], 40p. illus. (facsims.). £5. ISBN: 0712307397.

9 sections: Books; Journals; Directories; Market research reports; Statistics; Company information; Abstracting and

....(contd.)
indexing journals (with SRL locations); Patents; Further sources of information. Briefly annotated entries. Index, p.38-40. *Class No:* 612.08

[3225]
SKALAK, R. *and* **CHIEN, S. Handbook of bioengineering.** New York, etc., McGraw-Hill, 1987. 932p. illus. $107.50. ISBN: 0070577835.

Numerous sections (Mechanics of soft tissue - Properties of bone ... Theory and design of implantable pacemakers - Circulatory assist devices - The artificial kidney - Bioengineering of total joint replacement - Biomechanics of the human spine and trunk). 393 illus. *Class No:* 612.08

Digestive System

Nutrition

Dictionaries

[3226]
Dictionary of food and nutrition. Adrian, J., *and others.* Chichester, Ellis Horwood; Weinheim, VCH Verlagsgesellschaft mbH, 1988. v,233p. illus., diagrs., tables. ISBN: 0895734044.

Translation of *Dictionnaire de biochimie alimentaire et de nutrition* (Paris, Tecnique et Documentation, 1981) and revised and updated.

Brief explanations of terms relating to human food science, nutrition, and mineral feedstuffs. Includes tables, chemical formulae, and illustrations. Appendices give composition details of selected foodstuffs and animal feeds. Well presented and easy to use. *Class No:* 612.39(038)

Progress Reports

[3227]
Annual review of nutrition. Palo Alto, Calif., Annual Reviews, Inc., 1981-. v.1-. Annual. ISSN: 01999885.

V.11, 1991 (x,537p.) has 36 contributors. 22 documented papers (*e.g.* 'Current issues in fructose metabolism', p.21-39 (with 110 references); 'Fluorides and osteoporosis, p.309-24 (94 references)). Subject index. Cumulative index of contributing authors, v.7-11; Cumulative index of chapter titles, v.7-11. 'Some related articles in other *Annual reviews'*, p.ix. *Class No:* 612.39(055)

Vitamins

[3228]
Handbook of vitamins: nutritional, biochemical and clinical aspects. Machlin, L.J., *ed.* 2nd ed. New York, Dekker, 1990. 616p. $125; £91.45. (*Food science and technology, v.40.*) ISBN: 0824783514.

Detailed data on all known vitamins, including chemistry, food content, metabolism, functions and deficiency symptoms. New to this ed. are state-of-the-art reviews of each vitamin. An authoritative and comprehensive source of information. Of the first ed., 'For nutritionists, biochemists and even the interested lay person' (*Science & technology libraries,* v.5(4), Summer 1985, p.98). *Class No:* 612.392.5

[3229]

MERVYN, L. The Dictionary of vitamins: the complete guide for vitamins and vitamin therapy. Wellingborough, Northants., Thorsons, and West Byfleet, Surrey, Turner Pubns., 1984. 208p. tables. ISBN: 0722509065.

About 1,000 entries, A ... Yogurt. 'Losses in food processing', p.125-34 (15 tables). Includes very brief biographical notes. Cross-references. Bibliography (books; periodicals), p.207-8. Clearly written text; for lay persons. *Class No:* 612.392.5

[3230]

MERVYN, L. Thorson's complete guide to vitamins and minerals. 2nd ed. Wellingborough, Northants., Thorson's, 1989. 384p. graphs. £5.99. ISBN: 0722521472.

First published 1986.

Arranged A-Z (A ... Zirconium). Gives general information on each mineral and vitamin, in addition to details of supplementation regimes and the use of vitamins and minerals in conjunction with drugs prescribed by the doctor. Author is Technical Director of Booker Nutritional Products. *Class No:* 612.392.5

Death

[3231]

Encyclopedia of death. Kastenbaum, R. *and* Kastenbaum, B., *eds.* Phoenix, Ariz., Oryx, 1989. 320p. $74.50; £59.95. ISBN: 089774263x.

60 contributors, and 131 articles arranged alphabetically by topic. Coverage includes historical, medical, literary, practical, and psychological aspects related to death. A thorough examination of the issues raised. Highly recommended. *Class No:* 612.67

[3232]

SIMPSON, M.A. Dying, death and grief: a critically annotated bibliography and source book of thanatology and terminal care. 2nd ed. Pittsburgh, University of Pittsburgh Press, 1987. xv,259p. $29.95. ISBN: 0822935619.

Annotated entries for material published 1979-87. Coverage extends to key journals, films, audio-visual material available in Great Britain teaching material, kits, etc., and to European literature on the topics concerned. Subject and author indexes. *Class No:* 612.67

[3233]

—The Oxford book of death. Enright, D.J., *ed.* Oxford Univ. Press, 1983. xiii, 351p. £6.99. ISBN: 0192141295.

A wide-ranging book of quotations, with index. *Class No:* 612.67

[3234]

Sourcebook on death and dying. Fruehling, J.A., *ed.* Chicago, Ill., Marquis Who's who, 1982. 788p. illus. ISBN: 0837958016.

Part 1, 'Current issues', reprints in their entirety periodical articles on euthanasia, wills, bereavement, counselling and hospices. Part 2 provides vital statistics, survivor benefits and legal tax information for the US and Canada. A directory of institutions precedes a substantial bibliography (include audiovisual items). Scant coverage of such topics as children, and autopsies. *Class No:* 612.67

Health & Hygiene

Encyclopaedias

[3235]

AMMER, C. The New A-Z of women's health: a concise encyclopedia. New York, Facts on File, 1989. 544p. illus., tables. $29.95. ISBN: 0816020736.

*c.*1,000 entries arranged alphabetically. Many cross-references. Entries vary fron one paragraph to several pages. 2 appendices (Subject index and a list of resources). Easy to use. No bibliographies.

Rev. and exp. ed., 1991 (Alameda, Calif., Hunter House. xiii,472p.). *Class No:* 613(031)

[3236]

Complete family health encyclopedia. Smith, T., *ed.* London, Dorling Kindersley, 1990. 1184p. illus. (inc. col.). £25. ISBN: 0863184383.

Previous ed. (1980) as *Family health encyclopedia* (720p.).

*c.*5,000 entries and 2,000 illus. 2 sections: 1. The A-Z of health and medicines, p.49-1094 - 2. Drugs glossary, p.1095-1114. Detailed, analytical index, p.1115-54. Boxed information and well-captioned illustrations. Aims to be a sound, up-to-date medical reference source for use in the home. A handsome quarto, well produced. *Class No:* 613(031)

[3237]

STOPPARD, M. Everywoman's medical handbook. London, Dorling Kindersley, 1988. 384p. illus., diagrs., tables. £14.95; $19.95. ISBN: 0863182755.

Covers over 2,000 diseases, disorders and conditions. 3 main parts: The healthy female body, (p.9-69) - the A-Z of conditions, diseases, procedures and symptoms (p.71-345; 1-2p. per topic) - Women and medicine (*e.g.* Drugs index; Glossary, p.366-75; Health directory (grouped); Index (p.378-83)). Boxed information on symptoms throughout. *Class No:* 613(031)

Handbooks & Manuals

[3238]

JOLLY, H. Book of child care: complete guide for today's parents. 4th ed. London, Allen & Unwin, 1985. [xv], 569p. illus. £14.95; £8.99. ISBN: 0046490345.

First published 1978.

Part 1: The healthy child (1. Pregnancy and birth ... 7. Looking after your new baby ... 9. Effects of the new baby on family life ... 12. How children grow and develop ... 17. Discipline and punishment ... 18. Play ... 23. The one-parent family - 24. Travel, holidays and living in the tropics). Part 2: The sick child (26. The sick child at home ... 31. Coughs and colds ... 41. Measles and other infectious diseases ... 47. The mentally handicapped child ... 49. Accidents - 50. When a child dies). Useful addresses, p.559-62. Detailed analytical index, p.563-9. Well produced; fully illustrated. Translated into several foreign languages as a basic text. *Class No:* 613(035)

[3239]

The Macmillan guide to child health. Hull, D., *ed.* London, Macmillan, 1985. 382p. illus., diagrs., charts. ISBN: 0333392507.

46 contributors. 4 parts: 1. Care of the child - 2. Care of sick child - 3. Problems, diseases and disorders (p.120-310) - 4. Appendices: resources (support groups and publications, etc.). Emergency actions, p.330-41. Index of syndromes, p.342-3. Analytical index, p.344-52. Over 500 two-colour illus. A guide that spans the child's life from conception to adolescence. For public library and home use. Analytical index. *Class No:* 613(035)

[3240]

The New 'Our bodies, ourselves': a book by and for women, by the Boston Women's Health Book Collective. Rev. ed. New York, Simon & Schuster, 1992. 751p. illus. $17.95. ISBN: 0671791761.

Formerly *Our bodies, ourselves* (1976. 352p.).

A compendium on women's physical and health issues (*e.g.* birth control, menopause, estrogen replacement therapy and surgery, growing old). One section details 40 symptoms, with step-by-step diagrams for each symptom and its treatment. Bibliography and index. Of the previous ed., 'Essential for consultation' (*Choice*, v.22(10), June 1985, p.1478). *Class No:* 613(035)

[3241]

SHEPHERD, B.D. *and* **SHEPHERD, C.A. The Complete guide to women's health.** 2nd ed. New York, Penguin Books, 1990. xxiv,502p. illus., diagrs., tables. $15.95. ISBN: 0452264391.

Ten sections on health strategies, choosing a doctor, birth control, pregnancy planning, pregnancy and childbirth, sexual issues, menopause, and nutrition and physical fitness for a healthy life. A final section, 'What your symptoms mean', has step-by-step diagrams. Appended glossary, further reading, and how to carry out a breast self-examination. Bibliography, p.477-79. Index. 'Comprehensive reference work on diagnoses and treatments' (*RQ*, v.22(4), Summer 1983, p.419). *Class No:* 613(035)

[3242]

SMITH, T. The New Macmillan guide to family health. 2nd ed. London, Macmillan, 1987. 848p. illus. (incl. col.), charts, tables. £24.95. ISBN: 0333455258.

29 contributors. 4 parts: 1. The healthy body - 2. Symptoms and self-diagnosis - 3. Diseases, disorders and other problems - 4. Caring for the sick. Boxed information, a feature. Appended: Helpful organizations, p.762-3; drug index, p.792-801. Glossary. 'Accidents and emergencies', p.815-32. Detailed, analytical index, p.833-46. Designed for use in both sickness and health. *Class No:* 613(035)

[3243]

Travellers' health: how to stay healthy abroad. Dawood, R., *ed.* 3rd ed. Oxford, University Press, 1992. xxiv,472p. £7.99. ISBN: 0192622471.

About 40 contributors. 14 sections. 1. Staying healthy abroad - 2. Diseases spread mainly by food, drink and poor hygiene ... 5. Diseases spread by insects - 6. Animal bites; rabies; venomous bites and stings - 7. Air and sea travel - 8. Environmental and recreational hazards (*e.g.* skiing) - 9. Some common troubles (*e.g.* skin and blisters) - 10. Sex and contraception abroad (*e.g.* AIDS) - 11. Travellers with special needs (*e.g.* the disabled traveller) - 12. Becoming an

....(contd.)

expatriate - 13. Preparation for travel - 14. A special achievement. 9 appendices (6: Medical kit check-list). Further reading, p.446-50. Glossary. Index, p.455-72. 'An anthology of invited, *specialist* opinion on a wide range of problems of concern to travellers' (*Preface*). Eminently practical. *Class No:* 613(035)

Reviews & Abstracts

[3244]

Abstracts on hygiene and communicable diseases. London, Bureau of Hygiene and Tropical Diseases, 1926-. 12pa. £150; $290. ISSN: 02605511.

Produced in co-operation with CAB International. Incorporates *Bulletin of hygiene*.

1986: 4,809 signed abstracts. 3 main sections. Environmental health - Community health - Diseases and their control. Book reviews. Author and subject indexes. Forthcoming meetings. Available online and in microform. *Class No:* 613(048)

[3245]

FAMLI / Family medicine literature index. London, Ontario, World Organization of National Colleges, Academies and Academic Associations of General Practitioners (WONCA), in cooperation with the National Library of Medicine, 1980-. Quarterly. ISSN: 02272393.

About 6,000 references pa. V.6(4), the annual cumulation for 1985, contains MEDLARS subject section (subjects A-Z) p.1-97, with author index. Supplementary subject and author sections list references to family medical journals not in *Index medicus* and therefore not on the MEDLARS database. Appended 'Family medical books published 1980-1985' and 'Publications of member organizations of WONCA'. Prelims. include list of journals indexed, and keywords in FAMLI (a thesaurus). *Class No:* 613(048)

[3246]

Popular medical index. Letchworth, Herts., Mede Publishing, 1978/9-. Annual. ISSN: 01421107.

The 1987/1988 annual has *c*.1,200 entries for periodical articles and books under subjects, Abdomen ... Zinc. A list of 58 journals monitored, - 'the most frequently quoted periodicals in the Index' precedes. These include not only *British medical journal* and *Nursing times,* but also *New scientist* and *Women's realm*. Price and ISBN are given for books. *Class No:* 613(048)

Yearbooks & Directories

[3247]

The Health directory. Macdonald, F., *comp.* 2nd ed. London, Bedford Square Press, 1990. 121p. £6.95. ISBN: 0719912563.

First published 1979.

Lists nearly 1,000 national bodies, advice services, and informal self-help groups in the UK. Subject index. Aimed at the non-specialist but also a useful reference tool for the professional. *Class No:* 613(058)

Tables & Data Books

[3248]

World health statistics annual. Geneva, World Health Organization, 1951-. Annual. tables, maps.

The 1991 v. (xxii,349p. £40) has 4 sections: A. Global overview - B. Special topic - C. Vital statistics and life tables - D. Causes of death (by sex and age: Africa, America, Asia; Europe; Oceania; Age-standardized death rates for selected countries, by sex; Causes ...). 12 sets of tables. Two annexes (2. List of member states of WHO, by region). *Class No:* 613(083)

Food

Bibliographies

[3249]

PRYTHERCH, R. *and* **STANLEY, S. Food, cookery and diet:** an information guide. Aldershot, Gower, 1989. ix,109p. £15. ISBN: 0566035685.

Lists the standard, basic sources. 6 sections: 1. Basic reference sources - 2. The scientific background - 3. Professional catering - 4. Training, education and careers - 5. Home cooking - 6. Healthy eating and special diets. Index. *Class No:* 613.2(01)

Encyclopaedias

[3250]

TVER, D.F. *and* **RUSSELL, P. The Nutrition and health encyclopedia.** 2nd ed. Princeton, N.J., Van Nostrand Reinhold, 1989. vii,639p. illus., diagrs., charts, tables. $44.95. ISBN: 0442233973.

First published 1981.

A-Z definitions and longer articles (150 revised definitions; 86 new terms), the latter on carbohydrates, toxins, drugs, bodily functions, etc. Tabular data on food, caloric and vitamin values have been expanded and updated for this ed. 'An important encyclopedic source of nutritional information' (*New technical books,* v.75(2), March/April, 1990, no.572). 13 appendices, *e.g.* 3. Table of international atomic weights. *Class No:* 613.2(031)

Handbooks & Manuals

[3251]

ROGERS, J. The Encyclopedia of food and nutrition. London, Merehurst, 1990. 480p. illus. (col.), tables. £25. ISBN: 1853911690.

29 sections, *e.g.* 'Fruits'; 'Pulses'; 'Fast foods'. Most entries include a coloured illustration and data cover major nutrients; nutritional information; description; origin and history; buying and storage; preparation and use; processing; varieties. Tables, p.468-473, give figures for recommended daily intake of minerals, vitamins, etc. Index but no bibliography. Atractively presented. Suitable for the layperson, public libraries, and quick reference. *Class No:* 613.2(035)

Dictionaries

[3252]

BENDER, A.E. Dictionary of nutrition and food technology. 6th ed. London, Butterworths, 1990. [vi], 336p. tables. £29.50. ISBN: 0408037539.

First published 1960.

About 4,000 entries and cross-references, p.1-309. Entries for abbreviations and trade names. Includes 300 new terms; 300 entries revised. Bibliography (now separated from entries), p.310-24 (Additives and ingredients ... Toxicology). Tables, p.327-36. Aims 'to assist the specialist from one field to understand the technical terms used by the variety of specialists in the food fields' (*Preface* to 5th ed.). Author was Professor of Nutrition, Queen Elizabeth College, Univ. of London. *Class No:* 613.2(038)

[3253]

—**The A-Z of nutritional health:** a guide to the relation between diet and health. Mayes, A., *comp.* Wellingborough, Northants., Thorsons, 1991. 303p. tables. £7.99. ISBN: 0722524765.

Has *c.*2,000 entries, usually concise, A-Z 'Fibre, dietary': 3p.; 'Proteins': 3p. Final table: 'Dietary restrictions practised by religious and ethnic groups', p.295. Bibliography, p.301-3. Clearly written; good layout. *Class No:* 613.2(038)

[3254]

—**The Penguin encyclopedia of nutrition.** Yudkin, J. Harmondsworth, Middlesex & New York, Penguin; Viking, 1985. 431p. ISBN: 0713916621.

Has entries 'Absorption ... Zinc'. 'Food additives': 2½p. Includes biographies. Numerous cross-references. Detailed, analytical index, p.393-431. Author was Professor of Nutrition and Dietetics at University College, London, 1954-71. *Class No:* 613.2(038)

Reviews & Abstracts

[3255]

Nutrition abstracts and reviews. Series A. Human and experimental. Wallingford, Oxon., CAB International, 1931-. 12pa. £389; $702. ISSN: 03091295.

About 8,5000 abstracts pa. Main sections: Techniques - Foods - Physiology and biochemistry - Human health and nutrition - Disease and therapeutic nutrition. Cross-references. Abstracts not signed. Review article precedes abstracts. Author and subject (analytical) indexes per issue and annually. Online, CAB ABSTRACTS. Available on CAB ABSTRACTS CD-ROM. *Class No:* 613.2(048)

Progress Reports

[3256]

World review of nutrition and dietetics. Basle & London, Karger; Wiley, 1959-. v.1-. Irregular.

Each volume has documented contributions on a particular theme (*e.g.* v.49 (1986): 'Nutrition and the quality of life'), with an adequate index. Intended to provide a forum for critical reviews such as cannot be published in the ordinary journals, nor in textbooks. 'Reviews are necessarily coloured by the views of the authors, but most of them, though not all, succeed in providing a critical survey of the field. They also serve as valuable platforms to promulgate ideas' (*Chemistry & industry,* 6 August 1977, p.659). Highly priced (v.55(1988). 266p.: £96.25). *Class No:* 613.2(055)

Tables & Data Books

[3257]

BOWES, A.D. *and* **CHURCH, C.F.,** *comps.* **Bowes and Church's food values of portions commonly used.** Philadelphia, Lippincott, 1989. 328p. tables. $18.50; £14.50. ISBN: 0397547277.

Rev by J.A.T. Pennington.

Lists both generic and brand names. Includes values for proteins, cholesterol, fats, and vitamins. Tables for biotins, amino acids, caffeine, salicylates, etc. Bibliography of additional sources for food composition data. Good index. 'Possibly the most complete book of this kind' (*Science & technology libraries,* v.10(2), Winter 1989, p.134).
Class No: 613.2(083)

[3258]

Food composition and nutrition tables, 1986/87/ ... Compiled by H. Scherz, and others, on behalf of the Bundesministerium für Ernährung Landwirtschaft und Forsten, Bonn. 3rd ed. Stuttgart, Wissenschaftsliche Verlagsgesellschaft GMbH, 1986. xxvi, 1032p. tables. ISBN: 3804708331.

Coverage: milk and milk products; fats and oil/margarine; meat; fish and molluscs; cereals; vegetables, fruits. Glossary of the food constituents. Data columns: protein, fat, carbohydrates, constituents. Subject and author subjects p.1017-32. *Class No:* 613.2(083)

[3259]

McCANCE, R.A. *and* **WIDDOWSON, E.M.** **The Composition of foods.** 5th rev. and extended ed. London, Royal Society of Chemistry; Ministry of Agriculture, Fisheries and Food, 1991. xiii,462p. tables. £29.95. ISBN: 0851863914.

3 sections: 1. Introduction - 2. Tables, p.21-388 (*e.g.* Milk and milk products; Herbs and spices; Nuts; Beverages) - 3. Appendices (*e.g.* Organic aids; Recipes; References; Index). Supplements are to be published in 1992. A standard work.
1st supp., *Fruit and nuts.* (1992. 144p. £24.50); 2nd supp., *Vegetable dishes.* (1992. 200p. £24.50).
Class No: 613.2(083)

[3260]

—**BRIGGS, D.** *and* **WAHLQVIST, M.** Food facts: the complete no-fads plain facts guide to healthy eating. Harmondsworth, Middlesex, Penguin Books, 1984. [vii], 256p. illus., diagrs., tables, charts. ISBN: 0140465421.

First published by Penguin Books Australia.

Features 50 detailed food charts, p.45-222. Lists of further reading and organizations. Detailed analytical index, p.254-6. A bargain. *Class No:* 613.2(083)

[3261]

—**BRITISH MEDICAL ASSOCIATION.** Diet, nutrition and health: report of the Board of Science and Education, March 1986. London, the Association, [1986]. 69p.

Has 6 sections: 1. General introduction - 2. Food trends - 3. Factors affecting food consumption - 4. Food and health - 5. Dietary objectives for the nation - 6. Implementing dietary objectives. 72 references, p.65-69. No index. Aims at providing guidance on what constitutes a healthy diet.
Class No: 613.2(083)

[3262]

Manual of nutrition. 9th ed. London, HM Stationery Office, 1985. [xi], 131p. £3.25. ISBN: 0112427391.

First published 1945.

2 parts (13 sections). 1. Nutrients and their utilization (*e.g.* 2. Carbohydrates, fats, energy needs and food consumption, digestion and absorption of major nutrients, minerals, vitamins; 9. Recommended intake of nutrients) - 2. Nutritional value of food and diets. 7. Appendices (5. Food additives; 7. Books for further reading (18 items)).
Class No: 613.2(083)

Histories

[3263]

DRUMMOND, J.C. *and* **WILBRAHAM, A.** **The Englishman's food:** a history of five centuries of English diet. New and rev. ed. London, Cape, 1957 (reprinted 1991. London, Pimlico). 482p. illus., tables. £10.

First published 1939. 1991 reprint as *The Englishman's food: five centuries of English diet.*

5 parts (24 chapters): 1. Medieval and Tudor England - 2. The seventeenth century - 3. The eighteenth century - 4. The nineteenth century - 5. The twentieth century. 3 appendices, on diets and nutritional value tables. Many footnote references. Analytical index, p.471-82. A mine of information, dealing with such topics as scurvy, dental caries, prison diets, and English cooking of cabbages.
Class No: 613.2(091)

Vegetarianism

[3264]

DYER, J.C. Vegetarianism: an annotated bibliography. Metuchen, N.J., Scarecrow Press, 1982. xi, 280p. $22.50. ISBN: 081081532x.

1,412 entries. Part 1: Early works (pre-20th century, through 1899; 1900-1959); Part 2: Recent works, 1960-1980 (in 33 subject categories, *e.g.* 'Medical aspects of vegetarian diets'). Brief helpful annotations. Appendix lists 200 vegetarian cookbooks. Author and detailed subject index. 'Because of the wide range of material covered, this will be a useful source in both public and academic libraries' (*Reference books bulletin, 1983-1984,* p.102).
Class No: 613.261

[3265]

The Vegetarian handbook. 17th ed. Altrincham, Cheshire, Vegetarian Society, 1989. 128p. £2.99. ISBN: 0900774304.

Replaces, in part, *The International vegetarian handbook.*

6 sections: Product directory, p.11-62 - Shops and suppliers - Did you know? - Common food additives and contaminants - Taking it further - Facilities for vegetarians. Companion volume is *The International vegetarian travel guide.* *Class No:* 613.261

Industrial Health

Aerospace Medicine

[3266]

Aerospace medicine and biology: a continuing bibliography. Washington, National Aeronautics and Space Administration, Scientific and Technical Information Branch, 1952-. 12pa. $19.50.

....(contd.)

Formerly *Aerospace references in medicine and biology*. About 110 abstracts per issue. 5 categories: 51. Life sciences (general) - 52. Aerospace medicine - 53. Behavioral sciences - 54. Man/system technology and life support - 55. Planetary biology. 7 indexes: subject; personal author; corporate source; foreign technology; contract number; report number; accession number. Annual cumulative index. A selection of annotated references to unclassified reports and journal articles announced in the previous month's *Scientific and technical aerospace reports* (STAR) and *International aerospace abstracts* (IAA). *Class No:* 613.69

Physical Welfare

[3267]

WISEMAN, J. The SAS survival handbook. London, Collins Harvill, 1986. 288p. illus. (incl. col.). £15; £11.99. ISBN: 0002171856.

Introduction-Essentials (*e.g.* equipment) - Strategy - Climate and terrain - Food (*e.g.* temperate edible plants) - Camp craft - Reading the signs (*e.g.* weather prediction) - On the move (jungle travel; etc.) - Health (*e.g.* warm climate diseases; first aid, p.198-222) - Survival at sea (*e.g.* dangerous fish) - Rescue (*e.g.* signals and codes) - Disaster (*e.g.* nuclear explosion and aftermath). Fully illustrated (*e.g.* 'Arctic shelters', p.130-2, has 14 small line-drawings). Boxed information. No index. The author served 26 years in the Special Air Service (SAS). This is the SAS's complete course, - how to survive outdoors, on land or sea, in any weather, in any part of the world. A bargain.
Class No: 613.7

Sports Medicine

[3268]

Encyclopaedia of sports medicine. Dirix, A., *and others, eds.* Oxford, Blackwell Scientific Publications, 1988-. £39.50. ISBN: 0632019638, v.1.

V.1. *The Olympic book of sports medicine.* xii,692p.

Produced by the International Olympic Committee's Medical Commission and the International Federation of Sports Medicine. V.1 has over 50 contributing authors and 13 sections: 1. Introduction ... 3. Assessment of physical and functional capacity - 4. Environmental conditions ... 8. Female athletes - 9. Sport and physical activities in older people ... 13. Doping and doping control. A bibliography is appended to each chapter. Abundant figures and graphs, with easy to read text. *Class No:* 613.70

[3269]

Encyclopedia of sport sciences and medicine. Under the sponsorship of the American College of Sports Medicine, Univ. of Wisconsin, and in cooperation with the American Association of Health, Physical Education and Recreation ... New York, Macmillan Co.; London, Collier-Macmillan, 1971. xlvii, 1707p. illus.

9 areas: 1. Physical activity, general - 2. Physical activity, sports, games and exercises - 3. Environment - 4. Emotions and intellect - 5. Growth, development and ageing - 6. Drugs - 7. Prevention of disease and injury - 8. Special application of physical activity to the handicapped individual - 9. Rehabilitation - 10. Safety and protection. History of the

....(contd.)

development of sports and medicine, p.xxxiii-xlviii, with 80 references. Detailed contents. Subject and author indexes. US-slanted. *Class No:* 613.70

[3270]

PETERSON, L. *and* RENSTRÖM, P. Sports injuries: their prevention and treatment. London, Martin Dunitz, 1983. 488p. illus. (mostly col.). £19.95. ISBN: 0906348919.

Originally published in Sweden, 1983.

9 sections (6. Sports injuries by specific area, p.174-389; 'Tennis elbow'; 42/3p.; 3 col. illus., 1 black-and-white). Final sections: 8. Risks to children and adolescents; 9. Training of different parts of the body. Bibliography, p.463-5 (authors, A-Z). Glossary, p.466-77. Detailed analytical index, p.478-88. Attractively produced. *Class No:* 613.70

[3271]

Textbook of science and medicine in sport. Bloomfield, J.E., *and others, eds.* Melbourne, Blackwell, with support of the Australian Sports Commission, 1992. xv,591p. illus. (col.), diagrs., graphs, tables. £34.50. ISBN: 0867937612.

5 parts (527 sections): 1. The anatomy and biomechanics of sports performance ... 4. Sports medicine, p.139-419) - 5. Special considerations in sports medicine. Detailed, analytical index. *Class No:* 613.70

[3272]

TVER, D.F. *and* HUNT, H.F. Encyclopedic dictionary of sports and medicine. London, Chapman & Hall, 1986. 232p. illus. $32.50. ISBN: 0412013614.

Over 600 entries, A/Z, with a separate glossary (that should have been incorporated). 'Although it cannot compare in scope to the classic *Encyclopedia of sports sciences and medicine* (1971), it fills a need for a quick-reference tool defining illness and injuries related to sports and physical activities' (*Choice*, v.24(4), December 1986, p.611). For example, it includes entries on athlete's foot, altitude sickness, etc. *Class No:* 613.70

[3273]

The Year book of sports medicine. Chicago, Ill., Mosby Year Book Inc. 1979-. Annual. ISSN: 01620908.

A literature review from 97 leading journals, *Acta endocrinologica ... Thorax.* The 1986 Year book (466p. £41) has 6 sections: 1. Exercise physiology and medicine - 2. Biomechanics ... 5. Women in sports - 6. Athletic training. Articles would be of interest to medical practitioners, teachers, sports scientists and physical educationalists. *Class No:* 613.70

Physical Education

[3274]

DIAGRAM GROUP. The Complete encyclopedia of exercises. New York & London, Paddington Press, 1979. 335p. diagrs.

More than 350 exercises of all types. Includes official national fitness programmes of the UK, USA and People's Republic of China. Crudely-drawn diagrams. Brief bibliography and index. *Class No:* 613.72

[3275]

—Trends and developments in physical education. Proceedings of the VII Commonwealth and International Conference on Sport, Physical Education, Dance, Recreation and Health, Glasgow, 18-23 July 1986. London & New York, Spon, 1986. xiii, 493p. diagrs., tables. ISBN: 0419139109.

Comprises 66 papers in 3 sections: 1. Keynote and supporting papers - 2. Theme related presentations (*e.g.* 'Sport within the Chinese educational system'. 18 references) - 3. Allied contributions. Brief index. *Class No:* 613.72

[3276]

Sports documentation monthly bulletin. Birmingham, Sports Documentation Centre, Univ. of Birmingham, 1971-. 12pa. £32pa.(individuals), £40 (institutions). ISSN: 01421794.

About 3,500 abstracts pa. Two main parts. *General* (25 headings: Administration... Women, including Physical education and Sports injuries). *Individual sports and activities* (8 grouped headings: Association football, Australian Rules football ... Triathlon, Volleyball, Weightlifting, Wrestling). Author and subject indexes per issue, cumulated annually. *Class No:* 613.72

[3277]

Sportsearch: the contents of current journals/Titres d'articles courants. Ottawa, Sports Information Resource Centre, 1985-. 12pa. ISSN: 0882550x.

Monitors nearly 250 periodicals in sport and physical education over the world, most of them being published in English. 5 sections: 1. Specific sports - 2. Sport medicine and science - 3. Professional - 4. Abstracts - 5. Forthcoming conferences. Appended list of journal publishers and addresses. Available online via BRS, Data-Star and DIALOG; CD-ROM. *Class No:* 613.72

[3278]

—Bibliographical index on physical and health education, sport and allied subjects. London, Physical Education Association of Great Britain and Northern Ireland, 1975-83. Quarterly.

Two parts: 1. Bibliography of books (*e.g.* on dance; physical education; sports, games and outdoor activities; swimming and water sports; medical and scientific aspects of sport), - about 400 annotated entries pa. Cross-references. Prices given. - 2. Bibliography of articles (*c.*2,000 pa, with similar coverage to pt.1, plus occasional explanatory notes). *Class No:* 613.72

[3279]

VAN DALEN, D.B. *and* **BENNETT, B.L. A World history of physical education,** cultural, philosophical, comparative. 2nd ed. Englewood Cliffs, N.J., Prentice Hall, 1971. x, 694p. ISBN: 0139679197.

First published 1953.

5 parts (28 chapters): 1. Physical education in ancient societies - 2. Physical education in the Middle Ages and early modern times - 3. Physical education in modern Europe (GDR, Swedish, Danish, French, English, Soviet, etc. nationalism) - 4. Physical education in the US (p.365-550). Final chapter 28: Physical education in education for internationalism. Well-footnoted. Bibliography, p.649-78 (by chapter). Name and analytical subject indexes. 'Will continue to be standard work in the field' (*Choice,* v.8(5/6), September 1971, p.702). *Class No:* 613.72

[3280]

—Landmarks in the history of physical education. McIntosh, P.C., *and others.* 3rd ed. London, Routledge & Kegan Paul, 1981. x, 262p. illus. £9.99. ISBN: 0710007965.

Has 9 documented chapters (*e.g.* 'Landmarks in the history of physical education since World War II' with 54 references, plus 9 book items). Detailed, analytical index, p.251-62. Based on lectures given in the University of Birmingham. *Class No:* 613.72

Sleep

[3281]

Sleep research. Los Angeles, Univ. of California, Brain Information Service, Brain Research Institute, 1968-. Annual. ISSN: 00930407.

2 main sections: 'Ongoing research' (including abstracts; 290p.); 'A record of published research', - a 'sleep bibliography'. 20 subject categories are applied in the 'Sleep bibliography', *e.g.* 1. Books, reviews and theoretical discussions; 2. Neurophysiology - general; 16a. Pathology and disorders associated with sleep; 20. Biological rhythms and altered environments. KWIC index; author index. *Class No:* 613.79

[3282]

THORPY, M.J. *and* **YAGER, J. The Encyclopedia of sleep and sleep disorders.** New York, Facts on File, 1990. 356p. $45. ISBN: 0816018707.

880 terms, arranged A ... Z, varying in length from one sentence to several pages. Entries include cross-references and short bibliographies. Some of the information on the diagnostic classification and history of sleep disorders would be difficult to find elsewhere. Appendices include organizations providing information on sleep disorders and a list of sleep laboratories and centres. Intended for the layperson and the professional. *Class No:* 613.79

Alcoholism

[3283]

Alcohol and the public health: a study by a working party of the Faculty of Public Health Medicine and the Royal Colleges of Physicians on the prevention of harm related to the use of alcohol and other drugs. London, Macmillan in association with the Faculty of Public Health Medicine, Royal Colleges of Physicians, 1991. viii,218p. graphs. £30; £8.99. ISBN: 0333547780.

12 documented sections: 1. How much do we drink? - 2. Medical and psychiatric problems related to acohol use - 3. Alcohol and harm to the community ... 8. Provision of adequate and early help ... 11. Educating doctors - 12. Recommendations for actions. Useful addresses, p.204-6. Index. *Class No:* 613.81

[3284]

MADDEN, J.S. A Guide to alcohol and drug dependence. 2nd ed. Bristol, Wright, 1984. xii, 288p. ISBN: 0723607559.

First published 1979, tables.

9 documented chapters (1. Terminology, prevalence, causation ... 3. Disabilities and social features in drug dependence, p.62-114, with 7½p. of references ... 8. Treatment of drug misuse and dependence - 9. Prevention).

....*(contd.)*

Analytical index, p.279-88. Author: Consultant-in-charge, Mersey Regional and Drug Dependence Unit. *Class No:* 613.81

[3285]
O'BRIEN, R. *and* **CHAFETZ, M. The Encyclopedia of alcoholism.** 2nd ed. New York, Facts on File, 1991. 346p. tables. $45. ISBN: 081601955x.

First published 1982.

Over 500 entries, A-Z, covering a wide variety of aspects of alcohol and alcoholism: production, consumption by ethnic groups, regulations and customs worldwide, medical aspects, psychological theories, treatment of alcoholism, organizations, and slang. Many statistical tables; 24p. of bibliographies. Subject index. Of the first ed., 'Clear, concise style; useful to both the layperson and the professional' (*Library journal*, 1 April 1983, p.733). US-slanted. *Class No:* 613.81

Drugs (Narcotics)

[3286]
ABEL, E.L. A Dictionary of drug abuse terms and terminology. Westport, Conn., Greenwood Press, 1984. xi, 187p. $49.95. ISBN: 0313240957.

Over 1,000 words and expressions on the use of drugs likely to be abused. Includes slang terms as well as common and technical words. Bibliography, p.185-7. Extensive, 'but by no means exhaustive' (*Science & technology libraries*, v.6(3), Spring 1986, p.132).

Updates Abel's *Marihuana dictionary: words, terms, events and persons relating to cannabis* (Greenwood Press, 1982. 136p. $39.95). *Class No:* 613.83

[3287]
ANDREWS, T. A Bibliography of drug abuse, including alcohol and tobacco. Littleton, Colo., Libraries Unlimited, 1977. Supplement, 1977-1980. 306p. + 312p. ISBN: 0872872521.

The main volume has 752 annotated entries. Items, mostly in English, are post-1960 publications. Part 1: general reference sources, including journal titles. Part 2: classed list of subject areas (*e.g.* psychology, education, the law, medical aspects, hallucinogens, marihuana, stimulants). Alcohol and tobacco products: 30p. Author/title and subject indexes. The *Supplement, 1977-1981,* ($27.50) similarly arranged, covers 341 items. *Class No:* 613.83

[3288]
The Encyclopedia of drug abuse. Evans, G., *and others.* 2nd ed. New York, Facts on File, 1991. 370p. tables. $45. ISBN: 0816019568.

First published 1984.

Over 500 entries, ranging in length from several lines to several paragraphs. Covers political and legal aspects of drug abuse as well as medical and psychological issues. 4 appendices, including lists of drug slang, statistical tables, and a directory of relevant agencies. Extensive bibliography; index. Designed as a companion to *The encyclopedia of alcoholism*, by R. O'Brien and M. Chafetz (1991) (*qv*). Of the first ed., 'No library should be without this comprehensive and timely resource on drug abuse' (*Choice*, v.22(3), November 1984, p.405-6). *Class No:* 613.83

[3289]
Encyclopedia of psychoactive drugs. Leader, M., *ed.* London, Burke, 1985-.

Amphetamines, by S.E. Lukas. 1988. 101p. *LSD*, by M.E. Trulson. 1988. 121p. *Marijuana*, by M. Cohen. 1985. 109p. *Nicotine*, by J.E. Henningfield. 1985. *Alcohol: customs and rituals*, by T. Babor. 1988. 121p. *Alcohol and alcoholism*, by R. Fishman. 1988. 106p. *Inhalants*, by G.R. Glowa. 1988. 98p. *Barbiturates*, by J.E. Henningfield. 1988. 82p. *Cocaine*, by C.-E. Johanson. 1988. 97p. *Alcohol: teenage drinking*, by A.R. Lang. 1988. 133p. *Escape from anxiety and stress*, by T. McLellan. 1988. 82p. *Over-the-counter drugs*, by P. Sanberg and R.M.T,. Krema. 1988. 118p. *Tranquillizers*, by G. Winger. 1988. 113p. *Heroin*, by F. Zackon. 1988. 129p. *Drugs and civilization*, by S. Freeman. 1988. 111p.

A series of monographs, to be completed in 25v., - 3 on alcohol, 5 on hallucinogens, 3 on narcotics, 1 on non-prescription drugs, 4 on sedative hypnotics (including inhalants), 4 on stimulants, and 5 on understanding drugs (*British book news*, January 1986, p.11). *Class No:* 613.83

[3290]
Excerpta medica. 40. Drug dependence, alcohol abuse and alcoholism. Amsterdam, Excerpta Medica Foundation, 1972-. 6pa. DFl843pa. ISSN: 03044041.

About 1,800 abstracts pa. 12 sections: 1. General aspects - 2. Drugs - 3. Individual response - 4. Diagnosis - 5. Medical treatment - 6. Rehabilitation - 7. Epidemiology - 8. Social aspects - 9. Prevention - 10. Human behavior - 11. Brain metabolism - 12. Drug related diseases. Subject and author indexes per issue and annually. Online, EMBASE and available on CD-ROM. *Class No:* 613.83

[3291]
Multilingual dictionary of narcotic drugs and psychotropic substances. London, HMSO, 1984. 876p. £37. ISBN: 0119080907.

Part 1 consists of monographs on the substances under international control. Data: structure, molecular weight, molecular formula, structural formula. Part 2 is a comprehensive A-Z cross-index of all the names cited in the monographs, in English, French and Russian, with Arabic and Chinese translations of the substances' names. Also *Addendum, 1* (x,65p. £11.50). *Class No:* 613.83

Smoking

[3292]
Fourth report of the Independent Scientific Commmittee on Smoking and Health. Froggatt, P., *Sir, chairman.* London, HMSO, 1988. 68p. £5.70. ISBN: 0113211317.

5 sections: 1. The Product Modification programme - 2. Smoking related diseases and the reduction in sales weighted average tar yields ... 5. Additives to tobacco and cigarette papers. 7 appendices, *e.g.* 'Summary of research programmes supported through the Tobacco Products Research Trust'. The bibliography cites 97 papers. *Class No:* 613.84

[3293]
ROYAL COLLEGE OF PHYSICIANS OF LONDON.
Smoking or health: the third report from the Royal College
of Physicians of London. London, Pitman Medical, 1977.
128p. £5. ISBN: 0272794126.
First published 1962.
10 sections: 1. Tobacco consumption, promotion and
control in Britain ... 3. Pharmacology and toxicology ... 4.
Smoking and cancer ... 6. Smoking, bronchitis and other
conditions ... 9. The smoking habit - 10. Prevention of
disease due to smoking. Appended: 'Summary of
recommendations for action'. 3p. of references, p.125-7.
Class No: 613.84

[3294]
Smoking and health bulletin. Rushville, Md., US Dept. of
Health and Human Services, Public Health Service, Office
on Smoking and Health, 1973-. Semi-monthly. *Gratis.* ISSN:
00810363.
About 2,000 unsigned abstracts pa. 19 sections, *e.g.*
Pharmacology and toxicology ... Smoking prevention and
intervention ... Tobacco economics - Legislation - General.
Author and organizational index; analytical subject index.
Available online via DIALOG. *Class No:* 613.84

[3295]
—Bibliography on smoking and health. US Dept. of Health and
Human Services, Public Health Service, Office on Smoking
and Health; WHO collaborative Centre for Reference on
Smoking and Health, 1910-. Annual. *Gratis.*
Has 17 of the 19 sections in *Smoking and health bulletin,*
with similar indexes. *Class No:* 613.84

Sex

[3296]
NEWMAN, J. *and* **TILKE, A. Sex education:** a guide to
evaluation of materials. 2nd ed. Birmingham, Youth
Libraries Group of the Library Association, 1989. 74p.
£7.50. ISBN: 0946581096.
First published 1978.
7 sections: 1. Introduction - 2. Books for younger children
... 5. AIDS ... 7. Materials for parents, carers and teachers.
Appendix: List of addresses, p.69-70. Author and subject
indexes. *Class No:* 613.88

[3297]
Sex research: bibliographies from the Institute of Sex
Research [Indiana University]. Scherer-Brewer, J. *and*
Wright, R.W., *comps.* Tucson, Arizona, Oryx Press, 1979.
x, 212p. $71.50. ISBN: 0720108330.
A list of 4,267 references on all aspects of human
sexuality, stressing current rather than historical material. 11
major categories (*e.g.* Sex behaviour, sex variations, sex and
society). Citations cover books, book chapters, periodical
articles, conference papers and non-print items. Author and
subject indexes. Well-organized. *Class No:* 613.88

Public Health

[3298]
Elsevier's dictionary of public health in six languages:
English, French, Spanish, Italian, Dutch and German.
Deblock, N.J.L., *comp.* Amsterdam, etc., Elsevier, 1976.
[xii], 196p. £54.06; $82. ISBN: 0444413952.
2,363 English-base terms, (13,000 entries), with French,
Spanish, Italian, Dutch and German equivalents and indexes.
Class No: 614

[3299]
**Excerpta medica. 17. Public health, social medicine and
hygiene.** Amsterdam, Excerpta Medica, 1955-. 24pa.
DFl1,887pa. ISSN: 00144215.
About 6,000 abstracts pa. 15 sections: 1. General aspects -
2. Statistics - 3. Communicable diseases - 4. Social hygiene -
5. Industrial medicine and work - 6. Dental hygiene - 7.
Sanitation - 8. Military medicine - 9. Atomic warfare - 10.
Nutrition - 11. Veterinary medicine - 12. Medical zoology -
13. Developing countries - 14. Ethnography - 15.
Alternative medicine methods. Subject and author indexes
per issue and annually. Online, EMBASE. Available on CD-
ROM. *Class No:* 614

[3300]
Glossary of health care terminology. Hogarth, J., *comp.*
Copenhagen, WHO Regional Office for Europe, 1975. 476p.
An English glossary of *c.*350 headwords with index
entries for numerous other terms. Coverage: health care and
services; resources; finance; manpower; building; drugs;
management; planning; statistics. *Class No:* 614

[3301]
On the state of the public health for the year 1991: the
annual report of the Chief Medical Officer of the Department
of Health ... London, HMSO, 1992. Annual. vi,184p.
£14.50. ISBN: 0113215347.
8 sections: 1. Vital statistics - 2. The nation's health - 3.
Health of black and ethnic minorities - 4. Needs,
effectiveness and outcomes - 5. Communicable diseases - 6.
Other events of interest in 1991 - 7. Control of medicines,
medical equipment, environmental health and toxicology - 8.
International health. *Class No:* 614

[3302]
Oxford textbook of public health. Holland, W.W., *and
others, eds.* 2nd ed. Oxford, University Press, 1991. 3v.
diagrs., graphs, tables. £200, set; £70, ea. ISBN:
0192617060, v.1; 0192617079, v.2; 0192617087, v.3;
0192619268, set.
First published 1984-85 as a revision of W. Hobson's *The
theory and practice of public health* (1961).
1. *Influences of public health.* xix,561p. 2. *Methods of
public health.* xxii,563p. 3. *Applications in public health.*
xix,657p.
Over 100 contributors. Extensive coverage, relating to
developed countries, and including more examples from
non-English speaking countries than in the previous edition.
Includes sections on public health policies and strategies,
provision of public health services, epidemiological
approaches, health service planning and evaluation and
needs of special client groups. Bibliographies are appended
to each chapter and each volume has an index covering the
set. *Class No:* 614

Pollution Control

Air Pollution

[3303]
Air pollution titles: guide to current air pollution literature. University Park, Pa., Center for Air Environmental Studies, Pennsylvania State University, 1966-. 6pa. $130. ISSN: 00022497.

About 600 references per issue, cumulated in the November/December number. A keyword-in-context index precedes the author-title list (on pink paper). Also available in microform and microfiche. *Class No: 614.71*

[3304]
GREAT BRITAIN. Department of the Environment. Air Pollution Monitoring Management Group. **Glossary of air pollution terms.** London, HMSO, 1979. ii, 47p. £2. *(Pollution report no.5.)* ISBN: 0117514322.

Defines about 500 terms in 1-7 lines. Appendix: 'List of specialist and more definitive glossaries, p.47.
Class No: 614.71

[3305]
Handbook of air pollution analysis. Harrison, R.M. *and* Perry, R., *eds.* 2nd ed. London, Chapman & Hall, 1986. xxiii, 634p. *Class No: 614.71*

[3306]
TEKNISKA NOMENKLATURCENTRALEN. Luftbehandlingsordlista. [Glossary of air treatment.] Stockholm, the Centre, 1978. 204p. *(TNC 69.)*

742 numbered Swedish base terms, plus explanations, with eaquivalents and indexes in English, French, German and Finnish. *Class No: 614.71*

[3307]
WORLD HEALTH ORGANIZATION. Regional Office for Europe. **Glossary of air pollution.** Copenhagen, the Office, 1980. 114p. *(WHO Regional publications, European series, no.9.)* ISBN: 9290201096.

Defines *c.*1,000 terms, with cross-references. 'Spectroscopy', 'Hydrocarbons', - each ½p.
Class No: 614.71

Radioactivity

[3308]
CRC handbook of management of radiation protection programs. Miller, K.L. *and* Weidner, W.A., *eds.* Boca Raton, Florida, CRC Press, 1988. 498p. tables. $139.95.

Includes substantial appendices on transport and packaging of radioactive materials. 2nd ed. due to be published 1992/93. *Class No: 614.73*

[3309]
Sources and effects of ionizing radiation. United Nations Scientific Committee on the effects of atomic radiation: 1977 report to the General Assembly, with annexes. New York, United Nations, 1977. [v], 725p. graphs, tables. £33. ISBN: 0119051885.

Sections: 1. Introduction - 2. Effects of radiation - 3. Radiation sources and exposures to radiation (A. General aspects; B. Sources of human radiation exposures). Appended 'List of reports received by the Committee', p.21-725 (with 373 references). *Class No: 614.73*

[3310]
SUBE, R. Dictionary of radiation protection, radiobiology and nuclear medicine. Amsterdam, etc., Elsevier, 1986. 474p. $195; £95.31. ISBN: 0444995498.

An enlargement of R. Sube's *Dictionary of nuclear engineering in English, German, French, Russian* (1965).

About 12,000 English-base terms, with equivalents and indexes in German, French and Russian. *Class No: 614.73*

Public Safety

[3311]
COLEMAN, R.J. *and* **WILLIAMS, K.H. Hazardous materials dictionary.** Lancaster, Pa., Technomic Publishing Co., Inc., 1988. vi, 176p. tables. $39. ISBN: 0877625395.

About 2,500 brief entries. Includes acronyms and abbreviations. 'Hazardous material categories', p.69-70. Aims to 'help document and standardize the nomenclature of hazardous material' *(Introduction)*. *Class No: 614.8*

[3312]
The Health and safety directory, 1991/92. Kingston-upon-Thames, Kluwer, 1991. 950p. ISBN: 1870080432.

6 parts (25 sections): A. Regulatory authorities and national bodies - B. Regional and local public services - C. Organizations world-wide - D. Health & safety in industry - E. Professional bodies, safety organizations and training - F. Miscellaneous (section 19: Research; 21. Bibliography and sources of information; 22: Acronyms, synonyms and abbreviations)). Index of individuals; index of committees; detailed general index, p.863-910; classified section.
Class No: 614.8

[3313]
Health and safety science abstracts. Bethesda, Md., Cambridge Scientific Abstracts, in conjunction with the Institute of Safety and Systems Management, Univ. of Southern California, 1987-. 4pa. $595pa, including annual index. ISSN: 08929351.

Previously (1973-86) as *Safety science abstracts journal.*

Contains *c.*6,400 abstracts pa. 6 main sections: SGO: Monographs, conferences & bibliographies - SI: Industrial and occupational safety - ST: Transportation safety - SA: Aviation and aerospace safety - SM: Medical safety. Analytical subject and author indexes per issue and annually. Available on magnetic tape and online. *Class No: 614.8*

Industrial & Occupational Safety

Databases

[3314]
HSELINE. Sheffield, Health and Safety Executive, 1981-.

A database and major source of bibliographical references to published documents on health and safety at work. About 148,000 references by December 1992, adding *c.*1,000 entries monthly. Coverage emphasizes mining, nuclear technology, explosions, explosives, hazardous chemicals, occupational hygiene, industrial pollution, occupational safety and agricultural safety. Online access, ESA-IRS, Pergamon INFOLINE. Available on CD-ROM.
Class No: 614.8-027(003.4)

Bibliographies

[3315]

Health and safety at work. Winship, I., *comp.* Newcastle-upon-Tyne, Information North, 1989. 18p. £5. ISBN: 0906433053.

2 sections: Books (*e.g.* 'Legislation'; 'Construction'; 'Industrial diseases') - Other sources of information (*e.g.* 'Specialist information'; 'Computerized sources'; 'Organizations'). Most of the references are from the 1980s. *Class No:* 614.8-027(01)

Encyclopaedias

[3316]

INTERNATIONAL LABOUR OFFICE. Encyclopaedia of occupational health and safety. Permeggiani, L., *technical ed.* 3rd (revised) ed. Geneva, ILO, 1983. 2v. (xxiv, xxiii, 2538p.). illus. (incl. col. pl.), diagrs., plans, tables. ISBN: 9221032892.

First published 1930.

V.1: A-K; v.2: L-Z. About 1,000 contributors, on a wide range of topics, from accident causation, ergonomics and occupational cancer to lighting, psychology, and statistics. 'Mines, ventilation of' occupies nearly 3p. in v.1, with 5 references, 5 figures. 'Shift work': 3½p., 7 references. Appendices: Basic data; International documentation. Some operational guidelines. Analytical index, p.2447-2533. Despite uneven treatment of subjects and restricted documentation, a major contribution. 'A practical guide ... particularly [for] those in developing countries and those without access to adequate library facilities' (*Science & technology libraries*, v.5(2), Winter 1984, p.75). *Class No:* 614.8-027(031)

Handbooks & Manuals

[3317]

Environmental and occupational medicine. Rom, W.N., *ed.* 2nd ed. Boston, Little, Brown & Co., 1992. xxv,1493p. illus., diagrs., graphs, tables. $106. ISBN: 0316755605.

Over 150 contributors. 125 documented sections. 3 parts: Diseases (1-53) - Exposures (56-100) - Control of environmental and occupational diseases and exposures (111-125). Detailed, analytical index, p.1449-93. An important contribution. *Class No:* 614.8-027(035)

[3318]

HARRINGTON, J.M. *and* **GILL, F.S. Occupational health.** 3rd ed. Oxford, Blackwell Scientific Publications, 1992. v,350p. £16.95. ISBN: 0632031891.

First published 1983.

13 sections: 1. Introduction - 2. Health services ... 5. Chemical agents - 6. Physical agents ... 9. Control of airborne contaminants ... 11. Some legal background to occupational health - 12. Education - 13. Sources of information. Index. This ed. incorporates the latest European legislative changes and has improved coverage of epidemiology and the evaluation of workplace exposure. *Class No:* 614.8-027(035)

[3319]

HUNTER, D. The Diseases of occupations. 6th ed. London, Hodder & Stoughton, 1978. xxii, 1257p., illus., graphs.

First published 1955.

15 chapters: 1. Man and his work - 2. The Industrial Revolution, 1760-1830 - 3. Social reforms in the 19th century - 4. Health of the worker in the twentieth century - 5. The ancient metals - 6. The other metals - 7. The newer metals - 8. The aromatic carbon compounds - 9. The aliphatic carbon compounds ... 12. Occupational diseases of the skin and occupational cancer - 13. Diseases due to physical agents - 14. The pneumoconioses - 15. Accidents. Detailed analytical index, p.1171-1257. *Class No:* 614.8-027(035)

[3320]

Safety at work. With 25 specialist contributors. In association with the Institution of Occupational Safety and Health. Ridley, J.R., *ed.* 3rd ed. London, Butterworth-Heinemann, 1990. 752p. illus., facsims., diagrs. £70. ISBN: 0750610182.

First published 1983.

4 parts: 1. Law (8 chapters, *e.g.* 'Consumer protection') - 2. The management of risk (7 chapters, *e.g.* 'Records and statistics') - 3. Occupational health and safety (8 chapters, *e.g.* 'Safety on construction sites'). 6 appendices (3. List of abbreviations). Bibliography appended to each chapter. Analytical index, p.740-52. *Class No:* 614.8-027(035)

Reviews & Abstracts

[3321]

Excerpta medica. 35. Occupational health and industrial medicine. Amsterdam, Excerpta Medica, 1953-. 16pa. DFl1,547. ISSN: 00144398.

About 2,5000 abstracts pa. 22 sections: 1. General aspects - 2. Legislation. Social security - 3. Medical examination - 4. Labor, sex and age - 5. Working time, leisure activities - 6. Ergonomics - 7. Physical and mental activity: workload - 8. Micro environmental - 9. Physical environment - 10. Chemical environment ... 20. Accidents - Disablement. Rehabilitation - 22. Medical treatment. Subject and author indexes per issue and annually. Online, EMBASE and available on CD-ROM. *Class No:* 614.8-027(048)

Institutions & Associations

[3322]

INTERNATIONAL SOCIAL SECURITY ASSOCIATION. International Section for Research on Prevention of Occupational Risks, Geneva. **Répertoire international des recherches appliquées à la protection de l'homme au travail.** [International directory of applied research for the protection of men at work.] 2nd ed. Paris, Institut National de Recherche et de Securité (INRS), 1977. 433p.

First published 1971.

Part 1: Research projects (28 topics. 1. Toxic and dangerous substances ... 28. General studies on specific sectors of industry or occupations) - Part 2: The organizations (37 countries, Algarve ... Yugoslavia). General index of organizations. Name index. Analytical index, p.337-418. *Class No:* 614.8-027:061:061.2

Fires & Fire Prevention

[3323]
BRITISH STANDARDS INSTITUTION. Glossary of terms associated with fire. London, the Institution, 1975-90. 9pts. (*BS 4422.*)

1. *General terms and phenomena of fire.* 1987. 8p. £11. 2. *Structural fire protection.* 1990. 4p. £6.50. 3. *Fire detection and alarm.* 1990. 12p. £23. 4. *Fire protection equipment.* 1975. 16p. £28. 5. *Smoke control.* 1989. 4p. £6.50. 6. *Evacuation and means of escape.* 1988. 4p. £6.50. 7. *Explosion detection and suppression means.* 1988. 4p. £6.50. 9. *Marine terms.* 1990. 4p. £6.50. *Class No:* 614.84

[3324]
BURKLIN, R.W. and FURINGTON, R.Q. Fire terms: a guide to their meaning and use. Boston, Mass., National Fire Prevention Association, 1980. vii, 212p. illus. $18.50. ISBN: 087765185x.

Has 'over 4,000 words and phrases in daily use in the United States, plus a selection of terms in current British use' (*Introduction*). 100 small illus. of equipment and procedures. Appendices list government agencies and private organizations. *Class No:* 614.84

[3325]
Dictionary of fire technology. 3rd ed. Leicester, Institution of Fire Engineers, 1979. 170p. £4. ISBN: 090334503x.

Defines *c.*3,500 terms, mostly technical, with cross-references. Entries average 2 lines. Refers, where relevant, to British Standards. 'It is believed that this book contains virtually every word and term a Fire Service Officer or Fire Engineer is likely to encounter' (*Preface*). *Class No:* 614.84

[3326]
—**Fire sciences dictionary.** Kuvshinoff, B.W., *and others.* New York, etc., Wiley; Interscience, 1977. viii, 439p. diagrs. ISBN: 0471511137.

Definitions, some with commentary, of *c.*10,000 terms in firemanship (*e.g.* 'Smoke': nearly 1 column). Draws on British and Canadian as well as US sources. Includes abbreviations. Some 150 diagrams. *Class No:* 614.84

[3327]
Fire protection directory, 1986. Epsom, Surrey, A.E. Morgan Publications, [1986]. 141p.

9 sections: 1. Buyers' guide (p.9-73) - 2. National and governmental authorities - 3. Public fire brigades UK. Fire prevention officers. Fire service stations. Abbreviations - 4. International and private fire brigades UK. - 5. Fire liaison panels. Associations, institutions, societies (UK) - 6. Fire insurance. Fire insurance companies - 7. Fire engineering publications - 8. Public fire organizations/brigades (overseas) - 9. Associates, institutes, etc. (overseas). Index to advertisers. *Class No:* 614.84

[3328]
Manual of firemanship. New series. London, HMSO, 1974-. Books 1-12, illus. (incl. pl.).

1. *Elements of combustion and extinction.* 1974. 128p. £4.95. 2. *Fire brigade equipment.* 1974. 224p. £5.95. 3. *Hand pumps, extinguishers and foam equipment.* 2nd ed. 1988. xiii,102p. £5.50. 4. *Incidents involving aircraft, shipping and railways.* 1985. xxvii, 266p. £7.50 5. *Ladders and appliances.* 1984. 204p. £6.50. 6. *Breathing apparatus and resuscitation.* 2nd ed. 1989. xiii,123p. £5.95. 7. *Hydraulics, pumps and pump operation.* 2nd ed. 1986. xvi,208p. £6.50. 8. *Building construction and structural fire protection.* 1976. 200p. £5.50. 9. *Fire protection of*

....(contd.)
buildings. 2nd ed. 1990. xii,208p. £6.50. 10. *Fire brigade communications and mobilizing.* 2nd ed. 1991. xvi,149p. £5.95. 11. *Practical firemanship. 1.* 1981. 192p. £5.50. 12. *Practical firemanship. 2.* 1983. 276p. £7.95.

Covers both theory and practice. Book 9 has 3 parts: Fire extinguishing systems; Fire alarm systems; Fire-venting systems. 21 chapters. Glossary and terminology. Further reading, p.181-3. 46 plates. Includes concise practical instructions. *Class No:* 614.84

Hazardous Chemicals

Bibliographies

[3329]
LEES, N. Hazardous materials: sources of information on their transportation. London, the British Library, Science Reference and Information Service, 1990. [v],70p. (*British Library bibliography.*) ISBN: 0712307737.

7 sections: 1. Introduction - 4. Landbased transport - 5. Water: sea and inland waterways - 7. Transfrontier movement of hazardous waste. 2 appendices: A. Useful addresses - B. Further information (*e.g.* 'List of abbreviations'). Author and subject indexes. Entries have annotations and state availability of items. Covers the period 1979 to January 1990. *Class No:* 614.878(01)

Handbooks & Manuals

[3330]
Handbook of reactive chemical hazards. 3rd ed. London, Butterworths, 1985. xxvi, 22, 1852p. chemical structures. ISBN: 0406013885.

First published 1975.

1. 'Special chemical section': about 4,900 elements and compounds in formula order, p.1-1414. 2. 'Class, group and topic section' follows, A-Z. 5 appendices: 1. Source title abbreviations used in references; 3. Glossary of abbreviations and technical terms used in the text; 4. Index of chemical names used in section 1; 5. Index of class, group and topic titles used in section 2. 'For research students, practising chemists, safety officers and others concerned with the safe handling of reactive chemicals' (*Introduction*). *Class No:* 614.878(035)

[3331]
—**CRC handbook of laboratory safety.** Furr, A.K., *ed.* 3rd ed. Boca Raton, CRC Press, 1990. 704p. diagrs., tables. $110. ISBN: 0849303532.

First published 1967.

18 contributors. 6 chapters, with documented subsections: 1. Assignment of responsibilities - 2. Emergency programs - 3. Laboratory facilities - design and equipment - 4. Laboratory operations - 5. Nonchemical laboratories - 6. Personal protective equipment. Appendix: Laboratory checklist. Index. *Class No:* 614.878(035)

[3332]
Hazards in the chemical laboratory. Bretherick, L., *ed.* 4th ed. London, Royal Society of Chemistry, 1986. xiii, 604p. graphs, tables. £29.50. ISBN: 0851864899.

First published 1971; edited by G.D. Muir.

10 documented chapters. Introduction - Health and Safety at Work Act, 1974 - Safety planning and management - Fire protection - Reactive chemical hazards - Chemical hazards

.... (contd.)

and toxicology - Health care and first aid - Hazardous chemicals - Precautions against radiation - An American view (legislation and safety measures in the US). Authoritative and practical. 5th ed. was due to be published late 1992/early 1993 (700p. £45). *Class No:* 614.878(035)

[3333]
MARSHALL, V.C. **Major chemical hazards.** Chichester, E. Horwood, Ltd; New York, etc., Wiley, 1987. [xiii], 587p. illus., diagrs., tables, plans, maps. £82. ISBN: 085312969x.

26 chapters (*e.g.* 4. The nature of hazards and risks; 9. The handling and storage of liquids and gases; 15. Case histories of toxic releases; 17. Methods of control of major chemical hazards; 24. Research and consultancy; 26. Summary and conclusions). 4 appendices: 1. Glossary of terms - 2. Abbreviations - 3. Access to databases on chemical accidents - 4. Notes on the TNT equivalence of organic products. References (authors, A/Z), p.540-75. Non-analytical index. p.576-87. *Class No:* 614.878(035)

[3334]
Safe practices in chemical laboratories. London, Royal Society of Chemistry, 1989. [iv],50p. £10. ISBN: 0851863094.

Sections include: 'Organization for safety'; 'Design'; 'Operation'; 'Legislation and bibliography'. Appendix 1: The reporting of injurious diseases and dangerous occurrences regulations 1985. *Class No:* 614.878(035)

[3335]
SITTIG, M. **Handbook of toxic and hazardous chemicals and carcinogens.** 3rd ed. Park Ridge, N.J., Noyes Publications, 1991. 2v. xi,1685p. illus. ISBN: 0815512864.

Each chemical is classified as a carcinogen, hazardous substance, hazardous potential and/or priority toxic pollutants. An excellent source of economic information.

Previously as *Handbook of toxic and hazardous chemicals.* (2nd ed. 1985). *Class No:* 614.878(035)

Reviews & Abstracts

[3336]
Chemical hazards in industry. London, Royal Society of Chemistry, 1984-. Monthly. £215pa.

200 abstracts per issue. A current-awareness monthly on health and safety, chemical and biological hazards, plant safety, legislation, protective equipment and storage related to the chemical and allied industries. Online to DATA-STAR and Pergamon INFOLINE. *Class No:* 614.878(048)

Tables & Data Books

[3337]
Hazardous chemical data book. Weiss, G., *ed.* 2nd ed. Park Ridge, N.J., Noyes Data, 1986. 1069p. $98. ISBN: 0815510721.

Information on 1,015 hazardous chemicals (1 per page): their physical and chemical properties, chemical reactivity, hazard classification, labelling, shipping data, fire, exposure and water pollution, health hazards and toxicity.
Class No: 614.878(083)

[3338]
LEWIS, R.J., *Sr.* **Hazardous chemicals desk reference.** 2nd ed. New York, Van Nostrand Reinhold, 1991. 1100p. $84.95. ISBN: 0442004974.

First published 1987.

Entries for *c.*5,000 hazardous chemicals present poisonous, irritant, corrosive, explosive and carcinogeric properties, hazard ratings, synonyms, CAS, NIOSH & DOT number identifiers, standards for explosive limits & general physical properties. *Class No:* 614.878(083)

[3339]
LEWIS, R.J., *Sr.* **Sax's dangerous properties of industrial materials.** 8th ed. New York, Van Nostrand Reinhold, 1992. 3v. $399.95. ISBN: 0442011326, set; 0442012764, v.1; 0442012772, v.2; 0442012780, v.3.

7th ed. (1989), *Dangerous properties of industrial materials.*

A major revision and updating of the seventh edition. V.1 contains 3 sections: 1. CAS Registry cross-index - 2. Synonym cross-index (contains French, German, Italian, Dutch, Japanese and Polish synonyms, in addition to English) - 3. References (p.693-743). V.2 and v.3 contain chemical data and hazard potential for *c.*20,000 substances (for this ed., 14,000 entries have been revised and 1,500 new entries have been added). Significant increase in entries containing skin and eye irritation data.
Class No: 614.878(083)

[3340]
NORBACK, C.T. *and* NORBACK, J.C., *eds.* **Hazardous chemicals on file.** New York, Facts on File, 1988. Annual update. 3v. loose-leaf. $250, set. ISBN: 0816013535.

327 main entries for common toxic chemical substances. Data include exposure limit; leak and spill procedures; health hazards. More limited in coverage than Sax and Lewis but a quick reference tool for the general reader. US-slant. *Class No:* 614.878(083)

[3341]
PLUNKETT, E.R. **Handbook of industrial toxicology.** 3rd rev. ed. London, Edward Arnold, 1987. 605p. £92. ISBN: 071313626x.

2nd ed., 1976 [iii], 552p.

Intended as a quick reference guide; substances A-Z. Concise data on each: synonyms; description; occupational exposure threshold limit value; toxicity; preventive measures; references (if any). Appendices. Bibliography. Index. *Class No:* 614.878(083)

First Aid

[3342]
First aid manual: the authorised manual of St. John Ambulance, St. Andrew's Ambulance Association, the British Red Cross Society. 5th ed. London, Dorling Kindersley, 1987. 224p. illus. (col.). £4.95. ISBN: 0863182321.

4th ed., 1982.

Sections: The principles and practice of first aid - Major first aid techniques - Action at an emergency - Asphyxia - Wounds and bleeding - Circulatory disorders - Unconsciousness - Fractures (p.106-23) - Back injuries - Muscle and joint injuries - Burns and scalds - Effects of extremes of temperature - Poisoning - Foreign bodies - Aches - Procedure at major incidents - Dressings and bandages - Handling and transport - Emergency childbirth.

....*(contd.)*

Index (analytical), p.218-23. 2 appendices: Voluntary Aid Society observation chart; Manual artificial ventilation. Step-by-step instructions, well illustrated. This joint *First aid manual* 'contains all the information which is necessary for standard First Aid courses'. 6th ed., due 1992/93 (256p. £9.99). *Class No:* 614.88

[3343]
KIRBY, N.G. *and* **MATHER, S.J. Baillière's handbook of first aid.** 7th ed. London, Baillière Tindall, 1985. vii, 360p. illus. £7.95. ISBN: 0702010979.

First published 1941.

3 parts (36 sections): 1. Assessment of the casualty and casualty report - 2. Structure and function - 3. Clinical practice (15. Asphyxia and respiratory disorders ... 36. Transport by stretcher (p.331-52. 27 line-drawings)). Detailed, analytical index, p.353-60. Improved line-drawings. 'This book is intended for instructors and for those who will practise advanced first aid' (*Preface*).
Class No: 614.88

[3344]
ROYAL LIFE SAVING SOCIETY. National Technical Committee. **[Handbook, Royal Life Saving Society].** 4th ed. 8 booklets, in binder. illus. (incl. col.), charts.

First published 1963.

1. *Life saving and water safety.* iii, 31p. 2. *Water rescue skills.* 54p. 2nd ed., 1982. 40p. 3. *Teaching water safety.* 58p. £1.80. 4. *Life saving teachers guide.* 23p. £1.50. 5. *Resuscitation and first aid.* [viii], 38p. £2.50. 6. *The award scheme.* [v], 47p. 7. *Lifeguard manual.* 84p. £3.75. 8. *Examiners' guide and the award scheme.*

Sections 1, 2, 6, and 8 are available in one vol. (1990. £6.50). *Class No:* 614.88

Pharmacology

Databases

[3345]
SNOW, B. 'Online database coverage for pharmaceutical journals.' In *Database,* v.7(1), February 1984, p.12-26. ISSN: 01624105.

8 databases were selected for study and search made in over 650 relevant journals. Titles most frequently indexed are double asterisked, - for core collections. CA SEARCH scored highest for coverage and timeliness. A list of journals, p.16-23. 8 references; 2 tables.
Class No: 615(003.4)

Bibliographies

[3346]
ANDREWS, T. Guide to the literature of pharmacy and the pharmaceutical sciences. Littleton, Colo., Libraries Unlimited, 1986. viii, 383p. tables. ISBN: 0872574206.

958 numbered entries. 10 sections: 1. Guides to the literature and bibliographies - 2. Abstracting and indexing services - 3. Reviews, surveys and yearbooks - 4. Pharmacopoeias and standards - 5. Comprehensive works - 6. Pharmacy - 7. Historical works - 8. Dictionaries and encyclopedias - 9. Databases and catalogs - 10. Handbooks,

....*(contd.)*

manuals and guides. 'The primary intended audience ... is librarians, especially those in pharmaceutical, medical, health science and related libraries'. (*Preface*).
Class No: 615(01)

[3347]
Information sources in pharmaceuticals. Pickering, W.R., *ed.* London, Bowker-Saur, 1990. xvii,566p. tables. £55. ISBN: 0408025182.

International in scope. 28 contributors. 3 parts (20 documented chapters). General appendix to Part 3 contains sections on literature sources; online databases; centres of excellence; health statistics. Index, p.554-66. Editor is in charge of the Wellcome Foundation's central R & D information function. *Class No:* 615(01)

[3348]
Side effects of drugs annual: a worldwide yearly survey of new data and trends. Amsterdam, Elsevier, 1977-. v.1-. ISSN: 03786080.

Succeeds *Side effects of drugs: adverse reactions as reported in the medical literature of the world,* 1955/56-76.

Annual 15 (1991. ix,586p.) has 61 contributors, 51 documented papers. 26 special reviews include, 'Amiodarone and the lungs'; 'Safety and risks of enzyme inhibition'. Indexes of drugs, and side effects.
Class No: 615(01)

[3349]
STRICKLAND-HODGE, B. *and* **JEPSON, M.H. Keyguide to information sources in pharmacy.** London, Mansell, 1989. x,178p. £30. ISBN: 0720119669.

Part 1: Survey of pharmacy and its information sources (4 sections, *e.g.* 3. Finding out about pharmacy; 4. Keeping up to date with the subject) - Part 2: Bibliography (615 annotated entries, under type of literature: directories ... online database directories) - Part 3: Directory of organizations (selected list of national pharmaceutical associations, arranged alphabetically by country). 716 numbered entries in all. Index, combining authors, titles and type of publication, p.163-78. Aimed at the information worker and researcher. *Class No:* 615(01)

Encyclopaedias

[3350]
Encyclopedia of pharmaceutical technology. Swarbrick, J. *and* Boylan, J.C., *eds.* New York, Marcel Dekker, 1988-.

V.1. *Absorption of drugs to bioavailability of drugs and bioequivalence.* V.2. *Biodegradable polyester polymers as drug carriers to clinical pharmacokinetics and pharmacodynamics.* V.3. *Clinical supplies to dermal diffusion and delivery principles.* V.4. *Design of drugs to drying and driers.* V.5 *Economic characteristics of the R & D-intensive pharmaceutical industry to fermentation processes.*

First 5 volumes of a projected 12v. set. Includes illustrations (some coloured), bibliographies, and indexes.
Class No: 615(031)

Dictionaries

[3351]

BOWMAN, W.C. *and* BOWMAN, A. **Dictionary of pharmacology.** Oxford, Blackwell Scientific Publications, 1986. vii, 234p. diagrs. £12.50. ISBN: 0632011319.

About 1,000 terms, including cross-references (*e.g.* 'Abstinence syndrome' *See* 'Withdrawal syndrome'). Many entries contain references to relevant literature. Drug names are generally excluded, although drug classes and prototype drugs have entries. Terms are in large, bold, easily readable type. 'More an encyclopedia than a dictionary in the depth of information given for most entries' (*British book news,* September 1986, p.536). *Class No:* 615(038)

[3352]

—Thesaurus of subject terms and cross-references to 'International pharmaceutical abstracts'. 4th ed. Bethesda, Md., American Society of Hospital Pharmacists, 1987. 195p. ISBN: 0930530772.

Also known as *IPA thesaurus and frequency list.*
Class No: 615(038)

[3353]

Pharmacological and chemical synonyms: a collection of names of drugs, pesticides and other compounds drawn from the medical literature of the world. Marler, E.E.J., *comp.* 9th ed. Amsterdam, Elsevier, 1990. viii,562p. ISBN: 0444904875.

First published 1956.

Entry is under international accepted name, with cross-references from alternate, proprietary or chemical names. 'Intended as a guide to the identification of substances which appear under a bewildering variety of names' (*Preface to 3rd ed.*). *Class No:* 615(038)

[3354]

SLIOSBERG, A. **Dictionary of pharmaceutical science and techniques,** in five languages, English-French-Italian-Spanish-German. Amsterdam, etc., Elsevier, 1968-80. 2v. (xii, 686p. + 552p.). $255.50, set. ISBN: 0444405445, v.1; 0444416641, v.2; 0686859251, set.

1. *Pharmaceutical technology.* 1968 (2nd reprint, 1980). 2. *Materia medica.*

V.1 has 7,507 numbered English terms, with French, Italian, Spanish and German equivalents and indexes. Lacks chemical formulas; inadequate cross-references. V.2 has 4,688 entries for English, French, Italian Spanish, German and Latin equivalents and indexes. *Class No:* 615(038)

Reviews & Abstracts

[3355]

Excerpta medica. 30. Clinical and experimental pharmacology. Amsterdam, etc., Elsevier, 1992-. 32pa. DFl2,468pa. ISSN: 00144347.

Formerly *Excerpta medica. 30. Pharmacology* , 1948-91.

About 8,000 abstracts pa. 37 sections: 1. General aspects - 2. Automotive nervous system - 3. Motor system ... 6. Psychotropic and neurotropic agents ... 11. Cardiovascular system ... 16. Respiratory systems ... 18. Endocrine system ... 20. Anti-infective agents ... 24. Antineoplastic agents and carcinogens ... 27. Immunologic agents ... 31. Vehicles and additives - 32. Toxic substances and antagonists - 33. Drug addiction ... 37. Drugs used in opthalmology. Online, EMBASE and available on CD-ROM. *Class No:* 615(048)

[3356]

International pharmaceutical abstracts: key to the world's literature of pharmacy. Bethesda, Md., American Society of Hospital Pharmacists, 1964-. 24pa. $425. ISSN: 00208264.

About 15,000 abstracts pa. 26 sections (Pharmaceutical technology ... Adverse drug reactions ... Pharmaceutics ... Pharmaceutical chemistry ... Environmental toxicology ... Sociology, economics and ethics - Pharmacy practise). Subject index per issue; semi-annual author and subject indexes. Available online and CD-ROM.
Class No: 615(048)

Yearbooks & Directories

[3357]

Annual register of pharmaceutical chemists. London, Pharmaceutical Society of Great Britain. Annual.

The 1992 *Register* (1024p. £65.) comprises names and addresses of all registered pharmacists, bodies operating retail pharmacy businesses and names of their superintendents, and business titles/addresses of all the registered retail pharmacies in country, county, and town order. *Class No:* 615(058)

[3358]

SITTIG, M. *and* KOWALSKI, J.S. **Drug companies and products world guide.** Kingston, N.J., Sittig and Noyes, 1988. 596p. $60. ISBN: 0800242394.

Entries, arranged A-Z, are drawn from *Pharmaceutical manufacturing encyclopedia* (Sittig and Noyes). Coverage extends to over 2,000 US and 2,000 non-US companies. *Class No:* 615(058)

[3359]

—Pharmacology and pharmacologists: an international directory. Oxford, University Press, 1981. xxii, 387p. ISBN: 0192001019.

National, regional and international pharmaceutical societies, p.xix-xxii. Biographical entries, p.1-334. Research activities index, p.335-87. Covers 12,500 pharmacologists worldwide. The preface includes useful historical notes. Now somewhat dated. *Class No:* 615(058)

Histories

[3360]

COWER, D.L. *and* HELFAND, W.H. **Pharmacy: an illustrated history.** New York, Abrams, 1990. 272p. illus. $75. ISBN: 0810914980.

Traces the evolution of pharmacology from ancient civilizations to modern genetic engineering. Contains 308 high-quality illustrations, including 151 colour plates of sculptures, paintings, documents, manuscripts and institutions. *Class No:* 615(091)

[3361]

KREMERS, E. *and* URDANG, G. **Kremers and Urdang's history of pharmacy.** 4th ed., revised by G. Sonnedecker. Philadelphia, Pa., Lippincott, 1976. xv,571p. illus., ports.

First published 1940.

In 4 sections (18 chapters): 1. Pharmacy's early antecedents - 2. The rise of professional pharmacy in representative countries of Europe - 3. Pharmacy in the United States - 4. Discoveries and other contributions to society by pharmacists. 7 appendices, *e.g.* 1. Representative drugs of the American Indians. Notes and references, p.495-556. Index. *Class No:* 615(091)

Homeopathy

[3362]
SMITH, T. **An Encyclopaedia of homoeopathy:** a comprehensive reference book and survey of the subject from its beginnings to the present day. Worthing, Sussex, Insight Editions, 1983. [viii], 283p. £9.95. ISBN: 0946670013.

About 1,000 entries, A-Z. Highlights history, philosophies, major personalities, remedies and illnesses. Cites major works by homeopathic practitioners. *Class No:* 615.015.32

[3363]
—CASTRO, M. The Complete homeopathy handbook: a guide to everyday health. London, Macmillan, 1990. 257p. figs. ISBN: 0333428889.

4 sections: 1. History, principles and prescribing - 2. The Materia Medicas and Repertories - 3. Prescribing guidelines and follow-through (Stress as a cause of disease ... Repertorising chart - Sample cases) - 4. The appendices (Glossary - First-aid kits, pharmacies and courses - Homeopaths and homeopathic organizations - Useful reading and bookshops). *Class No:* 615.015.32

Pharmacopoeias

Yearbooks & Directories
[3364]
Chemist & druggist directory, and tablet and capsule identification guide. London, Benn, 1972-. Annual. illus., tables. ISSN: 02625881.

1991 ed. (436p. £72.) has 6 sections, all further subdivided: A. Tablet and capsule identification guide - B. Manufacturers and suppliers index - C. Buyers guide - D. Pharmaceutical organisations - E. Hospital pharmacists - F. Wholesale/retail outlets. Section E covers both private and public sectors. *Class No:* 615.11(058)

Worldwide
[3365]
The International pharmacopeia. 3rd ed. Geneva, World Health Organization, 1979-. (v.1-3, 1979-88). $27.50, v.1; $41.95, v.2; $65, v.3. ISBN: 9241541512.

To be in 5v. V.1 describes 42 general methods of analysis. V.2-3 consist of monographs: 'Quality specifications', - to be continued in v.4-5. V.2 (1981, 342p.) has specifications for 126 individual drugs widely used in health care. Although published by WHO, the series, produced by an international group of experts, does not have legal force in itself. *Class No:* 615.11(100)

Great Britain
[3366]
British pharmacopoeia, 1988. Published on the recommendation of the Medicines Commission pursuant to the Medicines Act, 1968. 2nd ed. London, HMSO, 1988. 2v. (740 + 400p.). graphs. ISBN: 0113208375, set.

Contains 2,100 monographs for items and substances used in medicine. V.1 features medicinal and auxiliary substances, plus infra-red spectra. V.2 contains the formulary, blood products, immunological products, radiopharmaceutical preparations, and surgical materials.

.... (contd.)
Formulary extensively revised. 24 appendices, *e.g.* 6 Qualitative sections and tests; 18. Methods of sterilisation. V.2 has index to both volumes. Updated by an annual addendum. (1992 addendum; £29). *Class No:* 615.11(410)

[3367]
DPF/BNF: Dental practitioners' formulary together with British national formulary, for use in the National Health Service. London, British Medical Association & The Pharmaceutical Press. 1990. Irregular. ISBN: 0853692408, DPF; 0853692009, BNF. ISSN: 02636419, DPF; 0260535x, BNF.

Previously published separately up to 1990 as *British national formulary* and *Dental practitioners' formulary*.

No. 20, 1990-92 comprises two formularies in a single volume. DPF (x,36p.) is divided into 4 sections (guidance on prescribing); classified notes on drugs and preparations; appendices (5, interactions ... breast-feeding); and list of dental preparations. BNF (xiv,562p.) consists of 'Guidance on prescribing, emergency treatment of poisoning', then 15 sections of 'Classified notes on drugs and preparations' (12: ear, nose, and oropharynx, p.349-354). 8 appendices and index (p.523-562). Poor cover design over-emphasises dental content.

No. 1, 1992 (580p. £9.95); No. 2, 1992 (620p. £9.95). *Class No:* 615.11(410)

[3368]
MARTINDALE, W. **The Extra pharmacopoeia.** Reynolds, J.E.F. 29th ed. London, The Pharmaceutical Press, 1989. xxx,1896p.

First published 1883.

An extensive revision of the previous edition. 3 sections: 1. Monographs on drugs and ancillary substances (72 chapters, *c.*4,000 substances) - 2. Supplementary drugs and other substances (*c.*800 drugs are covered) - 3. Formulas of British proprietary medicines (*c.*670 medicines are included). Directory of manufacturers, p.1645-1680, has 4,600 entries. 3 indexes: 1. Index to clinical uses - 2. Index to Martindale identity numbers - 3. General index. 'Aims to provide unbiased concise reports on the actions and uses of most of the world's drugs and medicines to aid the practising pharmacist and physician' (*Preface*). Available online and on CD-ROM. *Class No:* 615.11(410)

[3369]
The Pharmaceutical codex, incorporating the 'British pharmaceutical codex'. Prepared in the Department of Pharmaceutical Sciences of the Pharmaceutical Society of Great Britain and published by direction of the Society's Council. 11th ed. London, Pharmaceutical Press, 1979. xxiii, 1101p. graphs, chemical structures & formulas. £29. ISBN: 0853691290.

Formerly as the *British pharmaceutical codex,* - designed to supplement the information in *The British Pharmacopoeia* by providing information on actions, uses, undesirable effects of pharmaceutical substances and preparations and by providing in addition formulas and standards for a range of materials that were not included in the *Pharmacopoeia* (*Preface*).

(Thus, the entry for Glyceryl trinitrate, p.394-5, has data under some 14 heads). Appendix 1: Reagent solutions; 2: Infra-red spectra. Analytical index, p.1049-1101. *Class No:* 615.11(410)

USA

[3370]

Physicians' desk reference / PDR. Oradell, N.J., Medical Economics. Annual.

First published 1947.

The 1989 ed. (2418p. illus.) contains a 'Production information section' (*c.*1700p.). It concerns over 2,500 pharmaceuticals, A-Z. Data: indicates usage, dosage, description, clinical pharmacology, supply, warnings, contraindicators, adverse reactions, overdose and precautions. Appendices: Conversion tables ... Guide to management of drug overdose. Comprehensive. Prepared by drug manufacturers and their consultants. For the American practitioner.

46th ed., 1991; supps. A & B, 1992.

Class No: 615.11(73)

[3371]

—Physicians' desk reference for nonprescription drugs. Oradell, N.J., Medical Economics, 1980-. Annual.

A companion volume to *Physicians' desk reference,* for nonprescriptive drugs. Product identification and data, plus list of useful addresses. 12th ed., 1991. $33.95.

Class No: 615.11(73)

[3372]

—Physicians' generix 1992. New York, Data Pharmaceutica, 1992. 2500p. tables. $68. ISBN: 1880891026.

Covers generic drugs, arranged A-Z by generic name, with numerous cross-references. Data include chemical description; precautions; adverse reactions; usage; brand names. Keyword index. For the physician, pharmacist, and medical, public, and university libraries.

Class No: 615.11(73)

[3373]

—United States pharmacopeia: drug information for the consumer. Rockville, Md., US Pharmacopeial Convention, 1980-. Annual.

Has index-reference for 5,000 generic and brand names. Less comprehensive, less detailed than *Physicians' desk reference,* according to the review in *Choice,* (v.25(1), September 1987, p.96). *Class No:* 615.11(73)

[3374]

The United States pharmacopeia (22nd revision, official from January 1, 1990) [and] *The National formulary* **(17th ed., official from January 1, 1990).** Rockville, Md., US Pharmacopeial Convention, 1984, 5-yearly. lvii, 1683p. tables.

Combines and updates *The Pharmacopoeia of the United States of America* (1820-1975) and *The National formulary* (1888-1975).

The United States pharmacopeia part covers official monographs of USPXXII (drug substances and dosages, A-Z), Reagents. Tables. *The National formulary* part covers official monographs of NFXVII (pharmaceutical ingredients). Combined analytical index.

Class No: 615.11(73)

Medications & Drugs

[3375]

The British Medical Association guide to medicines and drugs. London, Dorling Kindersley, 1988. 432p. illus., diagrs. ISBN: 0863182984.

Structured information (general; quick reference;

.... *(contd.)*

information for users; special precautions; possible adverse effects) on 213 of the most widely used drugs. 5 parts: 1. Understanding and using drugs - 2. Drug finder index - 3. Major drug groups - 4. A-Z of drugs - 5. Glossary and index. Appended drug poisoning emergency guide.

Class No: 615.2

[3376]

Dictionary of drugs; chemical data, structures and bibliographies. Elks, J. *and* Ganellin, C.R., *eds.* London, Chapman and Hall, 1990. 2v. £815. ISBN: 0412273004.

Spin-off of the *Dictionary of organic compounds* (1982) and its annual supplements, plus additional data and references. Entries in v.1 include CAS registry numbers and literature references. V.2 contains name index; molecular formulae; CAS registry number index; type of compound index; structure index. Serves 'as a definitive source of concise data about significant drugs' (*Science & technology libraries,* v.11(2), 1991, p.219). Available online via HEILBRON database. *Class No:* 615.2

[3377]

Drugs handbook, 1992-93. Turner, P., *and others.* 12th ed. London, Macmillan, 1992. xviii,178p. £12.95. ISBN: 0333550935.

First published 1978. Now annual.

Definitions of drug group names and medical terms. Part 1: Approved names (*c.*1,100, A-Z), p.3-100. - 2. Trade names (with cross-references to appropriate approved names), p.103-75. Data: mechanism of action; therapeutic indications; chief unwanted effects. Appendix: Common slang names for misused drugs. P. Turner is Professor of Clinical Pharmacology, St. Bartholomew's Hospital, London. *Class No:* 615.2

[3378]

MIMS: monthly index of medical specialities. London, Haymarket Publishing, Ltd.

Lists prescriptive products. The June, 1991 issue has 21 sections (1. Gastrointestinal tract ... 21. Dressings and appliances). For use only by registered medical practitioners in the UK, to whom it is distributed free. *Class No:* 615.2

[3379]

PARISH, P. Medical treatments: the benefits and risks. London, Penguin, 1991. xii,1310p. tables. £9.99. ISBN: 0140098992.

89 sections: 1. The benefits and risks of drug treatments ... 20. Angina ... 29. Migraine ... 49. Penicillins ... 75. Skin diseases ... 89. Risks of one drug reacting with another drug. 2 appendices (1. Generic prescription drugs and their brand names - 2. Update). Index. Packed with information and written for both the layperson and professional. Good value. *Class No:* 615.2

[3380]

PARISH, P. Medicines: a guide for everybody. 7th ed. Harmondsworth, Middx., Penguin Books, 1992. xvi,615p. £7.99. ISBN: 0140100956.

First published 1976.

Part 1: Basic principles. The use of drugs. Dependence on drugs. Part 2: sections 1-53, p.41-330 (1. Sleeping drugs ... 3. Vaccines). Part 3: A popular pharmacopoeia, p.333-602. Analytical subject index, p.607-15. Of the 6th ed., 'The standard popular pharmacopoeia' (*Reference reviews,* v.3(1), March 1989, p.25). *Class No:* 615.2

[3381]
PEARCE, M.E. **Medicines and poisons guide.** 4th ed. London, Pharmaceutical Press, 1984. 220p. diagrs., tables. ISBN: 0853691738.

First published 1978.

Prepared in Law Dept. of the Pharmaceutical Society of Great Britain.

4 sections: 1. Medicines for human use - 2. Veterinary drugs, p.119-76 - 3. Non-medicinal poisons - 4. Prescribed dangerous substances, p.196-205. The guide is intended primarily as a practical guide to the legal classification of medicinal products and non-medical poisons for the purpose of retail sale or supply. *Class No:* 615.2

Origins of Medicines & Therapy

Plant Medicines

[3382]
A **Bibliography on herbs,** herbal medicine, 'natural' foods, and unconventional medical treatment. Andrews, T., *and others.* Littleton, Colo., Libraries Unlimited, 1982. 339p.

749 entries, with comparative annotations, for both scientific and popular works. 21 chapters, in 2 parts: 4 chapters on reference items; 17 chapters, on subject areas (*e.g.* herb cookery; folk medicine). Appended directory of organizations, associations and groups. Author/title and subject indexes. 'This exhaustive bibliography' (*Choice,* v.20(3), November 1982, p.405). *Class No:* 615.32

[3383]
British herbal pharmacopoeia, 1983. Consolidated ed. Cowling, Nr. Keighley, W. Yorkshire, British Herbal Medicine Association. 255p. £3. ISBN: 0903032074.

First published 1971.

Monographs, A-Z. Data: Synonyms: definition; description; therapeutics (action; indication; specific indication; combinations used; preparations and dosage). Appended index to therapeutic indications. Alphabetical index of botanical, common names and synonyms, p.248-55. *Class No:* 615.32

[3384]
CRC handbook of medicinal herbs. Duke, J.A. Boca Raton, Florida, CRC Press, 1985. [xi], 677p. illus, tables. $275. ISBN: 0849336309.

Catalogue of herbs, p. 1-516, with details of 365 medicinal species (uses, folk medicinal applications, chemistry and toxicity). 5 tables, p.517-98 (*e.g.* 1. Medicinal herbs, toxicity ranking and price list). References, p.585-594. *Class No:* 615.32

[3385]
GRIEVE, M. **A Modern herbal:** the medicinal, culinary, cosmetic and economic properties, cultivation, folklore of herbs, grasses, fungi, shrubs and trees, with all their modern scientific uses... [2nd ed.] edited by Mrs. C.F. Leyel. London, Cape, 1974. 1032p. illus.

First published 1931, in 2v.

Herbs are arranged A-Z under common names, followed by Latin names and orders; synonyms. Index of country names of plants covered. Many line-drawings. The work, an exhaustive compilation, is marred by its lack of reference to the sources of information. *Class No:* 615.32

[3386]
Herbs: an indexed bibliography, 1971-1980: the scientific literature on selected herbs, and aromatic and medicinal plants of the temperate zone. Simon, J.T., *and others.* Hamden, CT, Archon Books, 1984. xviii, 770p.

7,900 citations, covering 64 commercially significant herbs. Ten broad subject categories. Data on each herb: botanical, cultural and chemical uses. A short section includes books on herbs, reports, conferences and symposia, and general references. *Class No:* 615.32

[3387]
HOFFMAN, D. **New holistic herbal.** 3rd ed. Shaftesbury, Element Books, 1990. 284p. £9.99. ISBN: 1852301937.

Describes the most effective herbs for treating the different body systems (digestive, respiratory, etc.). A very good herbal.

Thorson's guide to medical herbalism: a comprehensive and practical introduction by the same author (3rd ed. Thorson's 1991. £7.99) is more medical and technical. It 'would be excellent for any doctor wishing to use herbal remedies' (*British book news,* February 1992, p.78). *Class No:* 615.32

[3388]
LAUNERT, E. **The Hamlyn guide to edible and medicinal plants of Britain and Northern Europe.** London, etc., Hamlyn, 1981. 288p. illus. (col. pl.). £6.99. ISBN: 0600563952.

Colour plates on rectos, with data (descriptions; habitat) facing. 4 appendices (1. Notes on the most important ingredients of medicinal plants; 3. Recipes for edible plants). Selected list of suppliers of medicinal herbs and plant-based preparations, Bibliography (authors, A-Z), p.277. Glossary of botanical terms; glossary of medical terms. Index, p.282-8. *Class No:* 615.32

[3389]
MILLS, S.Y. **The A-Z of modern herbalism:** a comprehensive guide to practical herbal therapy. Wellingborough, Thorson's, 1989. 222p. £6.99. ISBN: 072251882x.

Previously published as *Dictionary of modern herbalism* (1985).

Arrangement is A-Z (Abortifacient ... Zingiber officinale). Data for herbs include synonyms; habitat; constituents; actions; applications; dosage. Brief list of useful addresses. Author is Joint Director of the Centre for Complementary Health Studies at the University of Exeter. *Class No:* 615.32

Therapeutic Treatments

[3390]
ALLAN, B. **Guide to information sources in alternative therapy.** Aldershot, Gower, 1988. viii,216p. £27.50. ISBN: 0566056119.

3 parts: 1. Information sources in alternative therapy (1. Introduction ... 4. Searching for information) - 2. The information sources: a bibliography (5. General information sources ... 8. Specialized sources: rebirthing to yoga) - 3. People, organizations and activities. Index (includes authors, subjects and titles) p.193-216. Not comprehensive but aims to be a starting point for finding required information. *Class No:* 615.8

[3391]
Alternative medicine in Britain. Saks, M., *ed.* Oxford, Clarendon Press, 1992. x,271p. £35; £9.95. ISBN: 0198272782.

18 contributors. 4 parts (18 documented chapters): 1. Health care in Britain before the mid-nineteenth century - 2. The rise of medical orthodoxy and the development of alternative medicine - 3. Alternative medicine in modern Britain - 4. Alternative medicine: the future. Detailed, analytical index. *Class No:* 615.8

[3392]
Reader's Digest family guide to alternative medicine. London, Reader's Digest Association, 1991. 399p. figs., illus. £25.95. ISBN: 0276420101.

Arranged A-Z (About healing ... Zone therapy) and lists alternative methods of treating more than 200 ailments. Entries vary in length, *e.g.* 'Dysentery', 1 column; 'Immune system', 4 pages. Also gives orthodox views of alternative treatments. Bibliography and index. *Class No:* 615.8

[3393]
STANWAY, A. Alternative medicine: a guide to natural therapies. Rev. ed. Harmondsworth, Penguin Books, 1986. 319p. diagrs. ISBN: 0140085610.

Information on 32 therapies, arranged A-Z (*e.g.* Acupuncture; Biofeedback; Naturopathy, Yoga). Each entry carries a definition, background, how it's done, how it works, why does it work, what's it used for, does it work etc. Useful bibliography, p.50-63. Index. *Class No:* 615.8

[3394]
A Visual encyclopedia of unconventional medicine. Hill, A., *ed.* London, New English Library, 1979. 240p. illus. (incl. col.), ports.

49 contributors. Main sections: Comprehensive systems - Diagnostic methods - Physical therapies - Hydrotherapy - Plant-based therapies - Nutrition - Wave, radiation and vibration - Mind and spirit therapies - Self-exercise therapies. 'Acupuncture': p.56-60 (6 illus., 3 col.). Reading list (general and subjects, A-Z). Resources (associations). Analytical index, p.237-40, Cover sub-title: 'A health manual for the whole person'. *Class No:* 615.8

[3395]
WEST, R.W. *and* **TREVELYAN, J.E. Alternative medicine: a bibliography of books in English.** London, Mansell, 1985. xiii, 210p. £35. ISBN: 0720117216.

2,189 entries with one-line annotations. 8 sections: Homoepathy - Herbal medicine - Naturopathy - Clinical nutrition - Osteopathy - Chiropractice - Chinese medicine - Acupuncture. Non-analytical index. Aims to 'provide a good balanced 'first' that is of use to all who have an interest in alternative medicine' (*Introduction*). *Class No:* 615.8

Acupuncture

[3396]
CAMPBELL, A. Acupuncture: the modern scientific approach. London, etc., Faber and Faber, 1987. xi, [2], 151p. $4.95. ISBN: 057114652x.

10 documented chapters: 1. Traditional Chinese medicine -2. Scientific acupuncture - 3. Learning acupuncture - 4. Acupuncture tools and their use - 5. General principles - 6. The treatment of musculoskeletal disorders - 7. Other disorders - 8. Electro-acupuncture and other specialised techniques - 9. Transcutaneous electrical nerve stimulation -

.... (contd.)
10. Clinical research in acupuncture. Annotated bibliography, p.143-5. Detailed, non-analytical index. Author is consultant physician, Royal London Homoeopathic Hospital. *Class No:* 615.814

Hydrotherapy & Spas

[3397]
DENBIGH, K. A Hundred British spas. London, Spa Publications, 1981. xxi, 287p. illus, maps, plans. ISBN: 0150757403.

8 sections: 1. Early spas of noble patronage - 2. London's old spas - 3. A ring of spas - 4. Spas patronised by George III - 5. Regency and early Victorian spas - 6. Early hydropathic spas - 7. Rural and provincial spas - 8. Spas of the north (*e.g.* Harrogate). Pharmaceutical notes (saline, chalybrate and sulphur waters). Bibliography, p.281-2. Detailed analytical index. 87 illus., 42 location maps and plans. *Class No:* 615.838

[3398]
JOSEPH, J. Spa-finders guide to spa vacations at home and abroad. New York, John Wiley, 1990. xi,286p. $12.95. ISBN: 0471515558.

Provides complete descriptions for 120 spas throughout the world, although most are in the US. Covers seven major types of spas: Fitness and beauty spas; Luxury spas; Weight-loss spas; New Age retreats; Mineral spas; Spas abroad. Spa almanac, p.273-9, groups spas according to their strengths, *e.g.* 'Spas for women only'; 'Vegetarian spas'. *Class No:* 615.838

Radiology

[3399]
ETTER, L.E. Glossary of words and phrases used in radiology, nuclear medicine, and ultrasound. 2nd ed. Springfield, Ill., C.C. Thomas, 1970. xxv, 355p. illus., diagrs., tables. $45. ISBN: 0398005265.

First published 1960 as *Glossary of words and phrases used in radiology and nuclear medicine.*
Definitions of *c.*7,000 words and phrases. Footnote references. Prelims. include list of eponyms, and index of radiological and ultrasonic reports (p.52-96). Appendix of keyed figures, p.349-55. 'Prepared from various sources for medical secretaries, x-ray technicians, medical students and residents in radiology ... ' (*Title-page*). *Class No:* 615.849

[3400]
Excerpta medica. 14. Radiology. Amsterdam, Excerpta Medica, 1947- 24pa. DFl1,996.

About 8,000 abstracts pa. 6 sections: 1. General aspects - 2. Physics and techniques - 3. Hazards and protection - 4. Radiology - 5. Radiodiagnosis - 6. Radiotherapy. Analytical subject and author indexes per issue and annually. Online, EMBASE; CD-ROM. *Class No:* 615.849

[3401]
Excerpta medica. 23. Nuclear medicine. Amsterdam, Excerpta Medica, 1964 -. 20pa. DFl1,373pa. ISSN: 00144274.

About 5,000 abstracts pa. 8 sections: 1. General aspects - 2. Nuclear physics in biology and medicine - 3. Radiochemistry - 4. Radiation hygiene - 5. Radiology - 6.

....(contd.)

Tracer techniques - 7. Diagnostics - 8. Therapy. Subject and author indexes per issue and annually. Online, EMBASE; CD-ROM. *Class No:* 615.849

[3402]
ITURRALDIE, M.P. Dictionary and handbook of nuclear medicine and clinical imaging. Boston, Mass., CRC, 1990. 520p. $125. ISBN: 0849332338.

Contains over 500 terms, some of them unique, in Part 1. Part 2 comprises tables, *e.g.* Chemical properties of substances used in imaging; measurements and calculations. *Class No:* 615.849

[3403]
MYERS, P.A. *and* **MARTIN, T.A. Glossary for radiologic technologists.** New York, Praeger, 1981. ix, [i], 194p. illus., diagrs., graphs. $35. ISBN: 0275913511.

About 1,500 concise, 1-4 line entries. Appended (p.165-92) terminology of procedures; electrical symbols; x-ray circuit; weights and measures. References, p.193-4. *Class No:* 615.849

[3404]
—BRITISH STANDARDS INSTITUTION. Radiology and radiological physics terminology. London, BSI, 1985. 58p. £62. (*BS 4727: Pt.5: Group 01:1985.*)

Supersedes BS 2597:1955.

About 400 terms. Sections include: General terms; Ionizing radiations; Radiological appratus; Dosimetry; Radiation protection. Appendix: A. Values of fundamental physical constants. Index. *Class No:* 615.849

Psychotherapy

[3405]
WALROND-SKINNER, S. A Dictionary of psychotherapy. London, Routledge & Kegan Paul/Methuen, 1986. [xi], 379p. £40. ISBN: 0710099789.

About 850 documented entries (*e.g.* 'Cartesian': 1½ columns; 6 references, 2 cross-references; 'Leadership': 1½ columns, 6 references, 3 cross-references) Brief biographies (*e.g.* Jung). Embraces classical concepts as well as new psychotherapies. The author is currently Associate Director of Ordinands and Tutor in Pastoral Studies for the Diocese of Bristol. 'This new dictionary definitely fills a gap in the field' (*Choice*, v.24(2), October 1986, p.288). *Class No:* 615.851

Toxicology (Poisons)

Bibliographies

[3406]
HARGREAVES, J.M. Literature sources in toxicology: a survey. London, British Library, 1980. [v],70p. £8. ISBN: 0905984595.

British Library. Research & Development report, no.5542.

Includes such information sources as databases; databanks; manual sources; centres and research organizations in UK, USA, etc. Also, related programmes in progress of national and international organizations. Appendices: Addresses of

....(contd.)

on-line service suppliers; Sources of information used in the compiling of this survey; Selected references, p.69-70. 'Covers only those aspects of toxicology which relate to man and his environment' (*Introduction*). *Class No:* 615.9(01)

[3407]
WEXLER, P. Information resources in toxicology. 2nd ed. New York, Elsevier, 1988. xxiv,510p. £67.81. ISBN: 0444012141.

First published 1982.

2 sections: 1. United States resources (16 chapters, *e.g.* 'History'; 'Journals'; 'Audiovisuals'; 'Poison control centres') - 2. International organizations (20 chapters, *e.g.* 'International Union of Toxicology'; 'China'; 'Italy'; 'Sweden'). Entries on books, journals, etc. are annotated. Concentrates on health effects. Topics not stressed include management of hazardous wastes, aspects of pollution control, drug, alcohol and tobacco abuse. Heavy US slant. *Class No:* 615.9(01)

Encyclopaedias & Dictionaries

[3408]
COLEMAN, R.J. *and* **WILLIAMS, K.H. Hazardous materials dictionary.** Lancaster, Pa., Technomic Publishing Co., Inc., 1988. vi, 176p. tables.

About 2,500 brief entries. Includes acronyms and abbreviations. 'Hazardous material categories', p.69-70. Aims 'to help document and standardize the nomenclature of hazardous material' (*Introduction*). *Class No:* 615.9(03)

[3409]
Dictionary of toxicology. Hodgson, E., *and others.* New York, Van Nostrand Reinhold, 1988. 395p. $82.95. ISBN: 0442318421.

Brief descriptions rather than detailed definitions. Subject areas include chemicals, drugs, physiology. For the non-specialist scientist who is likely to encounter toxicological terms in the course of a study. *Class No:* 615.9(03)

[3410]
PEARCE, M.E. Medicines and poisons guide. 4th ed. London, Pharmaceutical Press, 1984. 220p. diagrs., tables. ISBN: 0853691738.

First published 1978.

Prepared in Law Dept. of the Pharmaceutical Society of Great Britain.

4 sections: 1. Medicines for human use - 2. Veterinary drugs, p.119-76 - 3. Non-medicinal poisons - 4. Prescribed dangerous substances, p.196-205. The guide is intended primarily as a practical guide to the legal classification of medicinal products and non-medical poisons for the purpose of retail sale or supply. *Class No:* 615.9(03)

[3411]
PLUNKETT, E.R. Handbook of industrial toxicology. 3rd rev. ed. London, Edward Arnold, 1987. 605p. £92. ISBN: 071313626x.

2nd ed., 1976 [iii], 552p.

Intended as a quick reference guide; substances A-Z. Concise data on each: synonyms; description; occupational exposure threshold limit value; toxicity; preventive measures; references (if any). Appendices. Bibliography. Index. *Class No:* 615.9(03)

Handbooks & Manuals

[3412]

Caserett and Doull's toxicology: the basic science of poisons. Amdur, M.O., *and others, eds.* 4th ed. New York, Pergamon Press, 1991. xiii,1033p. figs. ISBN: 0080402917.

First published 1975, *Toxicology: the basic science of poisons,* L.J. Casarett and J. Doull, *eds.*

45 contributors. 4 sections (31 documented chapters): 1. General principles of toxicology - 2. Systemic toxicology - 3. Toxic agents, p.565-818 - 4. Applications of toxicology. Index, p.999-1033. Designed primarily to be used as a textbook for courses in toxicology although research scientists in toxicology will find some sections useful, as will those concerned with community health, food technology, agriculture, pharmacy, and veterinary medicine. *Class No:* 615.9(035)

[3413]

SITTIG, M. Handbook of toxic and hazardous chemicals and carcinogens. 3rd ed. Park Ridge, N.J., Noyes Publications, 1991. 2v. xi,1685p. illus. ISBN: 0815512864.

Each chemical is classified as a carcinogen, hazardous substance, hazardous potential and/or priority toxic pollutants. An excellent source of economic information. Previously as *Handbook of toxic and hazardous chemicals.* (2nd ed. 1985). *Class No:* 615.9(035)

Yearbooks & Directories

[3414]

REGULATORY ASSISTANCE CORPORATION. Directory of toxicological and related testing laboratories. New York, Hemisphere Publishing Corporation, 1991. xvi,102p. $99.95. ISBN: 0891169040.

Directory of US toxicology, ecotoxicology, environmental, analytical, and support service laboratories. 2 parts: 1. Listing of toxicology laboratories (data include: name and address; services offered; tests performed; special expertise; equipment available; key personnel; FDA approval) - 2. Appendices (A. Summary list of laboratory contacts - B. List of studies conducted and/or services provided - C. Types of chemicals tested). *Class No:* 615.9(058)

Tables & Data Books

[3415]

LEWIS, R.J., *Sr.* **Hazardous chemicals desk reference.** 2nd ed. New York, Van Nostrand Reinhold, 1991. 1100p. $84.95. ISBN: 0442004974.

First published 1987.

Entries for *c.*5,000 hazardous chemicals present poisonous, irritant, corrosive, explosive and carcinogeric properties, hazard ratings, synonyms, CAS, NIOSH & DOT number identifiers, standards for explosive limits & general physical properties. *Class No:* 615.9(083)

[3416]

LEWIS, R.J., *Sr.* **Sax's dangerous properties of industrial materials.** 8th ed. New York, Van Nostrand Reinhold, 1992. 3v. $399.95. ISBN: 0442011326, set; 0442012764, v.1; 0442012772, v.2; 0442012780, v.3.

7th ed. (1989), *Dangerous properties of industrial materials.*

A major revision and updating of the seventh edition. V.1 contains 3 sections: 1. CAS Registry cross-index - 2. Synonym cross-index (contains French, German, Italian,

....(contd.)

Dutch, Japanese and Polish synonyms, in addition to English) - 3. References (p.693-743). V.2 and v.3 contain chemical data and hazard potential for *c.*20,000 substances (for this ed., 14,000 entries have been revised and 1,500 new entries have been added). Significant increase in entries containing skin and eye irritation data. *Class No:* 615.9(083)

Europe

[3417]

FROHNE, D. *and* **PLANDER, H.J. A Colour atlas of poisonous plants: a handbook for pharmacists, doctors, toxicologists, and biologists.** London, Wolfe Publishing, 1984. 291p. illus (incl. col.), diagrs., tables, chemical structures. £55. ISBN: 0723408394.

Translated from the 2nd German ed., by N.G. Bisset.

5 sections, of which 3, 'The most important plants with alleged or actual toxic properties (Amaryllidacea ... Verbenaceae)' occupies p.31-231. 'Taxus baccata (yew)', p.223-5, has 1 col. illus., 2 micrographs, 1 diagr., 1 table of boxed data. Bibliography; extensive index. Limited to plants indigenous to/naturalised in Central and Western Europe, plus commonly met decorative plants and house plants. *Class No:* 615.9(4)

Great Britain

[3418]

COOPER, M.R. *and* **JOHNSON, A.W. Poisonous plants in Britain and their effects on animals and man.** London, HMSO, 1984. xiv, 305p illus. (incl. col.). £12.95. *(Reference book 161.)* ISBN: 0112425291.

Complete revision of MAFF Bulletin 161, *British poisonous plants.* (last reprinted 1979).

Over half of the book concerns the families of poisonous plants, ranging form Alliaceae to Umbelliferae. Details involve, in each case, poisonous principles, poisoning in animals, human poisoning and treatment, etc. Complements Frohne and Plander's *A Colour atlas of poisonous plants* (1984) *(qv),* being well-documented but deficient in illus. *Class No:* 615.9(410)

Clinical Medicine

[3419]

Current contents: Clinical medicine. Philadelphia, Pa., Institute for Scientific Information, 1974-. Weekly. $360.

Formerly *Current contents. Clinical practice.*

Each issue reproduces contents-pages of *c.*200 issues of journals as 'Discipline guide', (General medicine ... Dermatology). Title-word index author index; and address directory; publishers address directory. Available online, CURRENT CONTENTS SEARCH, and on diskette. *Class No:* 616

Types of Disorders

Inflammations

[3420]

Excerpta medica. 15. Chest diseases. Thoracic surgery and tuberculosis. Amsterdam, Excerpta Medica, 1948-. 20pa. DFl1,373. ISSN: 00144193.

About 5,000 abstracts pa. 38 sections, *e.g.* 1. General aspects; 2. Tuberculosis; 3. Pneumonia and other respiratory infections; 7. Allergy and bronchial asthma; 8. Neoplastic diseases of the lung; 19. Cardiovascular diseases related. Detailed analytical subject index and author index per issue and annually. Online, EMBASE; CD-ROM.

Class No: 616.002

Tumours. Cancer

[3421]

The Cancer reference book: direct and clear answers to everyone's questions. Levitt, P.M., *and others.* Rev. ed. New York, Facts on File, 1983. 287p. $24.95. ISBN: 0871966840.

First published 1979.

Topical chapters in question and answer form (*e.g.* 'What evidence is there that cigarettes cause cancer?') on the nature of cancer, proven and unproven methods of treatment, terminal care, current research and theories on prevention. A lengthy chapter discusses how mortality rates differ among various forms of cancer. Appended brief glossary, a list of cancer centres worldwide, a list of British and American titles, and a detailed index. 'Suitable for public libraries' (*Reference books bulletin, 1983-84,* p.60).

Class No: 616.006

[3422]

Excerpta medica. 16. Cancer. Amsterdam, Excerpta medica, 1953-. 32pa. DFl2,420. ISSN: 00144207.

About 10,000 abstracts pa. 3 sections, closely subdivided: 1. General aspects - 2. Experimental cancer research - 3. Clinical aspects. Detailed analytical subject index and author index per issue and annually. Online, EMBASE; CD-ROM.

Class No: 616.006

[3423]

International directory of specialized cancer research and treatment establishments. 4th ed. Geneva, UICC, 1986. 356p. ISBN: 2882360002.

Published under the auspices of the Committee on International Collaborative Activities (CICA).

87 countries, Argentina ... Zimbabwe, and their international cancer establishments. Data: director: major depts.; cancer-related personnel; projected expenditure, 1985, in US dollars; annual cancer-patient statistics; fields of activity; annual report; cancer registry ... Index of establishments; index of establishment directors and departmental heads; member organizations of the UICC.

Class No: 616.006

[3424]

WORLD HEALTH ORGANIZATION. International Agency for Research on Cancer (IARC). **Directory of on-going research in cancer epidemiology, 1992.**

Coleman, M.P., *and others, eds.* Lyons, etc, IARC Publications, 1992. xxii,773p. £42. ISBN: 9283221176.

List of projects, Algeria ... Zimbabwe, 7 indexes: investigators; topics; sites; types of study; chemicals;

.... (contd.)

occupations; countries. List of cancer registries. List of biological materials banks. Indexes on diskette.

Class No: 616.006

Deformities

[3425]

Excerpta medica. 21. Developmental biology and teratology. Amsterdam, Excerpta Medica, 1921-. 16pa. DFl1,638pa. ISSN: 00144258.

About 5,000 abstracts pa. 9 sections: 1. Books, reviews, congresses - 2. Demography, epidemiology and geographic pathology - 3. General biology of development - 4. Gonads, gametogenesis and development - 5. Maternal-foetal relations - 6. Pathology and teratology - 7. Experimental developmental biology - 8. Cell, tissue and organ cultures - 9. Techniques. Subject and author indexes per issue and annually. Online, EMBASE. Available on CD-ROM.

Class No: 616.007

[3426]

WYNBRANDT, J. *and* **LUDMAN, M.D. The Encyclopedia of genetic disorders and birth defects.** New York, Facts on File, 1991. 426p. tables. $45. ISBN: 0816019266.

Arranged A-Z (Aarskog syndrome ... Zygodactyly). Length of entries varies from a few paragraphs to more than three pages. Data include prenatal diagnosis, discovery and name, prognosis and incidence in the population. Names and addresses of relevant organizations are given. Cross-references and brief bibliography. Index enables user to locate entries by subject and by racial or ethnic group affected. No illustrations. US/Canada-slanted.

Class No: 616.007

Diseases (Pathology)

Handbooks & Manuals

[3427]

Human health and disease. Altman, P.L. *and* Katz, D.D., *comps. and eds.* Bethesda, Md., Federation of American Societies for Experimental Biology, 1977. xxi, [i], 435p. diagrs., graphs, tables. (*Biological handbooks, 2.*) ISBN: 0913822116.

More than 300 contributors. 7 parts (186 sections): 1. Infectious diseases - 2. Immunological factors - 3. Metabolic disorders - 4. Organ system diseases -5. Neurologic diseases - 6. Endocrinology and endocrinopathies - 7. Radiation. Numerous tables and references ('Adrenal hormones', p. 328-32; 40 references). Detailed, analytical index (p.375-435) includes entries for diagrams, footnote and headnote items). A mine of information; for specialists.

Class No: 616.01(035)

[3428]

—**Disease data book.** Fry, J., *and others.* Lancaster etc., MTP Press, 1986. viii, 405p illus., diagr., graphs, tables. £33.75. ISBN: 0852009224.

Takes 22 important medical conditions, using an economic synoptic style. It follows throughout a sequence of questions and answers for each condition, - 'What is it? Who gets it when? What happens? What to do?' The sections range from

....*(contd.)*

1: High blood pressure to 22: Cancers. Detailed analytical index, p.397-405. For rapid reference as an aide-mémoire on procedure for the medical trainee and others in the profession. *Class No:* 616.01(035)

Reviews & Abstracts

[3429]

Excerpta medica. 5. General pathology. Amsterdam, Excerpta Medica, 1948-. 24pa. DFl2,046pa. ISSN: 00144095.

About 7,5000 abstracts pa. 4 main sections, closely subdivided: 1. General aspects of pathology - 2. General pathology - 3. Organ pathology - 4. Techniques and laboratory methods. Subject and author indexes per issue and annually. Online, EMBASE and available on CD-ROM. *Class No:* 616.01(048)

Nomenclatures

[3430]

International nomenclature of diseases. Geneva, Council for International Organizations of Medical Sciences; World Health Organization, 1979-. 4v.

V.2. *Infectious diseases.* 208p. 1987. V.3. *Diseases of the lower respiratory tract.* 1979. V.4. *Diseases of the digestive system.* 269p. 1991. V.5. *Cardiac and vascular diseases.* 113p. 1991.

'The principal objective of IND is to provide, for every morbid entity, a single recommended name' (*Introduction*). *Class No:* 616.01(083.72)

[3431]

ROYAL COLLEGE OF PHYSICIANS OF LONDON. Joint Committee. **The Nomenclature of disease.** 8th ed. (subject of decennial revision). London, HMSO, 1960. xi, 398p.

15 sections: 1. Etiological classification - 2. Deseases of the body as a whole ... 14. Disease of the ear - 15. List of eponyms (p.229-48). Detailed analytical index, p.249-398. *Class No:* 616.01(083.72)

Maps & Atlases

[3432]

CLIFF, A.D. *and* **HAGGETT, P. Atlas of disease distributions:** analytic approaches to epidemiological data. Oxford, Blackwell, 1988. 300p. £100; £25. ISBN: 0631131493.

International in scope. Primarily describes charting, statistical and mapping techniques but will be of interest to practitioners and students of epidemiology. Bibliographies appended to each chapter as well as a lengthy list of data sources, references and atlases at the end of the volume. 'Recommended for both the mapmaker and the user of epidemiological data' (*Choice*, v.27, no.2, 1989, p.282). *Class No:* 616.01(084.3)

[3433]

—Atlas of diseases. May, J.M., *ed.* New York, American Geographical Society, 1950-5. 17pl.

Coloured maps on 17 sheets, 64 x 97 cm. Each plate forms a unit, complete with text and list of sources on the verso. Aimed to map world distribution of some 25 diseases. 'An extension of the idea behind the Seucher Atlas', - a wartime German production showing epidemic diseases in

....*(contd.)*

the areas for which military operations were planned (*Geographical magazine*, April 1960. p.558). Now somewhat dated. *Class No:* 616.01(084.3)

[3434]

—The Welt-Seuchen Atlas. [World atlas of epidemic disease.] Hamburg, Falk-Verlag, 1952-61. 3v.

Hrsg. in Auftrage der Heidelberger Akademie der Wissenschaften von E. Rodenwaldt unter Mitarbeit von H.J. Jusatz.

An enlargement of the World War II *Seuchen Atlas*. It contains *c.*120 maps. Text in German and English; European slant. An outstanding, detailed atlas. *Class No:* 616.01(084.3)

[3435]

HOWE, G.M. National atlas of disease mortality in the United Kingdom. Rev. and enl. ed., on behalf of the Royal Geographical Society. London, Nelson, 1970. [viii], 197p. tables, maps.

First published 1963.

Part 1 includes 56 maps (27 x 22 cm.) of the 1st ed., covering 1954-58; part 2 has 29 new maps, and covers 1959-63. Maps, all on a 1:3 million scale, show geographical distribution of the Standard Mortality Ratio, 1954-58, in England, Wales, Scotland and Ireland for each sex separately for 14 groups of causes of death and for all causes of death. Maps and text, p.17-159. Transparencies can be superimposed, to reveal place names and administrative units. Helpful historical introduction. 'Intended for geographers and epidemiologists, and workers in public health and social services as well' (*British medical journal*, no.5714, 11 July 1970, p.94). Somewhat critically reviewed in *The Lancet*, 26 March 1970, p.658. *Class No:* 616.01(084.3)

Worldwide

[3436]

A World geography of human diseases. Howe, G.M., *ed.* London, New York, etc., Academic Press, 1977. xxviii, [2], 621p., diagrs., tables, maps. £102.50. ISBN: 0123571502.

22 contributors, nearly all British. States 'what is known about a selected range of important diseases in terms of pattern, distribution and trend' (*Foreword*). 20 sections (1. Schistosomiasis ... 20. Deficiency diseases), with appended references. Some of the maps are adapted from American Geographical Society's *Atlas of diseases*, 'It is a book which will become not only a standard work of reference, but also provoke debate on the major issues of world health and the provision of health services in the future' (*International affairs*, v.54(3), July 1978, p.475-6). 'Should be available to all students of medicine' (*British medical journal*, no.6114, 18 March 1978, p.708). *Class No:* 616.01(100)

Classifications

[3437]

Manual of the international statistical classification of diseases, injuries and causes of death. Based on the Recommendations of the Ninth Revision Conference, 1975, and adopted by the 25th World Health Assembly, 1977-78. Geneva, WHO, 1977-78. 2v. (xxxvii, 773p.; 670p.). tables.

8th revision, 1967-69. 2v.

V.1 tabulates 17 categories of diseases (1. Infectious and

....(contd.)

parasitic diseases ... 17. Injury and poisoning. Supplementary classification of external causes of injury and poisoning and of factors affecting health status and contact with health services. Also, medical participation and rules for classification (p.697-741). V.2, *Alphabetical index* to v.1. *Class No:* 616.01:001.33

[3438]

—International classification of impairments, disabilities and handicaps: a manual of classification relating to the consequences of disease. Geneva, WHO, 1980. 207p. tables. £7.25. ISBN: 9241541261.

Supplements the International classification of diseases. 4 sections, all subdivided: 1. The consequences of disease - 2. Classification of impairments - 3. Classification of disabilities - 4. Classification of handicaps. Only section 1 has any bibliographic references, and only section 2 has an index. *Class No:* 616.01:001.33

[3439]

—ROYAL COLLEGE OF GENERAL PRACTITIONERS. Classification of diseases, problems and disorders. London, the College, 1984. 70p. £4.75. ISBN: 0850840945.

Provides a summary version of the International classification. *Class No:* 616.01:001.33

Tropical Diseases

[3440]

Hunter's tropical medicine. Strickland, G.T. 7th ed. Philadelphia, Pa., W.B. Saunders, 1991. xxix,1153p. illus., micrographs, diagrs., graphs, tables, maps. $150. ISBN: 0721629709.

First published as *A Manual of tropical medicine* (1945). 6th ed., 1984.

116 contributors. 10 parts (1. Clinical practice in the tropics ... 10. Tropical diseases in a temperate climate). Many subdivisions. Extensive sections on parasitic diseases, nutritional deficiencies, and tuberculoses. Sectional and subsectional bibliographies. Detailed, analytical index, p.1097-1153). The most detailed and comprehensive clinical tropical medicine textbook. *Class No:* 616.01-036.21

[3441]

Manson's tropical diseases. Manson-Bahrs, P.E.C. *and* Bell, D.R., *eds.* 19th ed. London, Baillière Tindall, 1987. xxii, 1557p. illus., diagrs., tables, maps. £58.

First published 1898. 18th ed. 1980 (xiv, 843p.).

17 sections, *e.g.* 1. Diseases commonly presenting as fever; 2. Diseases commonly presenting as diarrhoea; 15. Clinical problems in the tropics; 16. Protecting the traveller; 17. Drugs. 4 appendices (3. Medical entomology, p.1381-1488, with 123 references). Detailed analytical index. A well-established classic. *Class No:* 616.01-036.21

[3442]

Tropical diseases bulletin. London, Bureau of Hygiene and Tropical Diseases, 1912-. 12pa. £60 (individuals), £115 (libraries). ISSN: 00413240.

1992: 3,075 abstracts abstracts. Sections: Environmental health - Community health - Diseases and their control - Medical entomolgy. Adequate cross-references. Author and subject indexes per issue and annually. Earlier subtitle: 'selective and critical abstracts and reviews of world

....(contd.)

literature in all aspects of health and disease in tropical and sub-tropical regions'. Available online via DIMDI. *Class No:* 616.01-036.21

Epidemics

[3443]

A Dictionary of epidemiology. Last, J.H., *ed.* 2nd ed. New York, Oxford Univ. Press; International Epidemiology Association, 1988. ix,141p. £25; £11.95. ISBN: 0195054806.

Several hundred terms, drawn from such related areas as biostatics, demography and microbiology. Includes brief sketches of prominent epidemiologists and some references to the primary literature. Appended list of sources. For undergraduates and clinicians. *Class No:* 616.01-036.22

Patient types

Paediatrics

[3444]

Excerpta medica. 7. Pediatrics and pediatric surgery. Amsterdam, Elsevier, 1947-. 24pa. DFl2,067pa. ISSN: 00144118.

Formerly *Pediatrics,* 1947-72.

About 8,000 abstracts pa. 41 sections: 1. General aspects ... 5. Newborn, infant and fetus ... 10. Handicapped child ... 12. Preventive and social pediatrics ... 19. Cardiovascular system ... 21. Respiratory system ... 2. Nervous system ... 25. Metabolism ... 31. Locomotor system ... 35. Accidents ... 40. Children in hospital - 41. Cancer. Subject and author indexes per issue and annually. Online, EMBASE and on CD-ROM. *Class No:* 616.01-053.2

Geriatrics

Databases

[3445]

AGELINE. Washington, DC, American Association of Retired Persons, 1978-.

Database containing over 28,000 references with abstracts of material (English language only) published since 1978 (selective coverage of pre-1978 items). Based on the collection of AARP's National Gerontology Research Center and covers about 500 journals in addition to books, conference papers, government documents etc. Subject coverage includes psychological, political, social and economic aspects of aging. US-slanted. Available through BRS and DIALOG. *Class No:* 616.01-053.9(003.4)

Bibliographies

[3446]

ETHEL PERCY ANDRUS GERONTOLOGY CENTER. Library. **Catalogs of the Ethel Percy Andrus Gerontology Center,** University of Southern California. Boston, Mass., G.K. Hall, 1976. 2v. (1193p.).

1. *Author-title catalog.* 2. *Subject catalog.*

....(contd.)

22,500 photolithographed catalogue cards. The collection comprises *c.*5,500 book titles, conference proceedings, federal documents, hearings and legislation, city and state reports, journals, annual reviews. The library is the most comprehensive in the US on life-span development and aging. *Class No:* 616.01-053.9(01)

[3447]

Health care of the elderly: an information sourcebook. Petersen, M.D. *and* White, D.L., *eds.* Newbury Park, Calif., Sage, 1989. 576p. £49.95. ISBN: 0803933355.

Critcally reviews the research literature on selected aspects of health care for the elderly. Topics include: nursing homes; special support services; case management. Most of the references cited are from the 1980s. *Class No:* 616.01-053.9(01)

Encyclopaedias

[3448]

The Encyclopedia of aging. Maddox, G.L., *editor-in-chief.* New York, Springer Publishing Co., 1987. xix,928p. tables. $96. ISBN: 0826148409.

Over 200 contributors; *c.*450 entries, 'Abilities' ... 'World Health Organization' - US slant (*e.g.* 'American Association of Retired Persons'; 'Medicaid'; 'Native American Aged'), but many of the entries have general application (*e.g.* 'Adult education': 3 columns; 'Filial responsibility': 2 columns; 'Sleep disorders': 5 columns). Extensive bibliography (*c.*5,000 references), p.705-833. Index, p.835-90. *Class No:* 616.01-053.9(031)

Handbooks & Manuals

[3449]

CRC handbook of physiology in aging. Masoro, E.J., *ed.* Boca Raton, Florida, CRC, 1981. 502p. tables. $139.95. ISBN: 0849331439.

A compendium of lists, tables, etc. summarizing data, with full references. 'It deserves a place on the desks of all those who practice geriatric medicine as well as those who teach or study the physiologic processing of aging' (*Journal of the American Geriatric Society,* v.30, March 1982, p.225). *Class No:* 616.01-053.9(035)

[3450]

Oxford textbook of geriatric medicine. Evans, J.G. *and* Williams, T.F., *eds.* Oxford, University Press, 1992. xvii,756p. illus., graphs, tables. £85. ISBN: 0192615904.

26 documented sections: 1. The ageing of populations and communities ... 3. Infections ... 8. Gastroenterology ... 13. Joints and connective tissue ... 17. Skin disease ... 22. Symptom management and palliative care ... 26. Reference values for biological data in older persons. Index, p.729-56. Designed as companion to *Oxford textbook of medicine* and intended for specialists, trainees and other health professionals involved with elderly people. *Class No:* 616.01-053.9(035)

Reviews & Abstracts

[3451]

Excerpta medica. 20. Gerontology and geriatrics. Amsterdam, Excerpta Medica, 1958-. 8pa. DFl858.

About 2,000 abstracts pa. 25 sections: 1. General aspects - 2. General medical treatment - 3. Surgery - 4. Infectious diseases - 5. Cardiovascular system - 6. Respiratory organs - 7. Alimentary tract - 8. Urinary system ... 15. Endocrine system - 16. Nervous system ... 24. Hospitalization, nursing - 25. Social welfare (housing). Subject and author indexes per issue and annually. Online, EMBASE; CD-ROM. *Class No:* 616.01-053.9(048)

Yearbooks & Directories

[3452]

CPA world directory of old age. Crosby, G., *and others, ed.* Detroit, St. James, 1989. 208p. $85. ISBN: 1558620230.

Arranged alphabetically by country. Gives basic statistical data on aging for almost every country, *e.g.* number and percentage of the over-65 population, life expectancy, average individual income. Organizations listed separately, arranged by country. Index of organizations' name. *Class No:* 616.01-053.9(058)

Research Projects

[3453]

Old age: a register of social research 1985-1990. Crosby, G., *ed.* London, Centre for Policy on Ageing, 1991. iv,188p. £18. ISBN: 0904139786.

Lists UK research projects, both current and those completed between 1985 and 1990, being undertaken in universities, central and local government, research institutes, and voluntary bodies. Data include address, contact, abstract, aims, methodology, research methods, funding, public availability of research, publications. Indexes to subjects, funding bodies, authors, and institutions. *Class No:* 616.01-053.9:061:061.62.005

Allergies

[3454]

WINTER, R. The People's handbook of allergies and allergens. Chicago, Contemporary Books, 1984. 168p. ISBN: 0809253917.

Dictionary of concise but thorough definitions. Appendices include questions and guidelines to assist identification of allergens, the causes of allergic contact dermatitis, pollen alerts by region and month, food allergies by family, and sources for further information. 'A useful and needed source for data on allergens and terms related to allergies' (*Library journal,* 1 January 1985, p.88). *Class No:* 616.01-056.4

Syndromes

[3455]
Dictionary of medical syndromes. Magalini, S.I., *and others.*
3rd ed. Philadelphia, Pa., J.B. Lippincott, 1990. xiii,1042p.
ISBN: 0397508824.
First published 1971.
Entries for *c.*3,000 syndromes. 'Riley-Day': synonyms; symptoms; signs; etiology; pathology; diagnostic procedures; therapy; prognosis; bibliography (3 references). Detailed analytical index, p.957-1042. *Class No:* 616.01-06

[3456]
JABLONSKI, S. Jablonski's dictionary of syndromes and eponymic diseases. 2nd ed. Malabar, Fla., Krieger, 1991. ix,665p. illus. $99.50.
1st ed. as *Illustrated dictionary of eponymic syndromes and diseases, and their eponyms* (1969).
Greatly expanded from previous ed. About 3,000 entries, with a wealth of cross-references. 'Alagille syndrome': ½ column, 2 references; 'Aortic arch syndrome': nearly 1 column, 3 references, 14 synonyms; 'Balingall disease': ½ column, 1 reference. Aims 'to gather together in one volume the profusion of eponymous and noneponymous syndromes and eponymous diseases' (*Introduction*). *Class No:* 616.01-06

Symptoms & Diagnosis

[3457]
French's index of differential diagnosis. Hart, F.D., *ed.*
12th ed. Bristol, J. Wright, 1985. xii, 1032p. illus. (incl. col.) tables. ISBN: 0723607850.
First published 1912.
18 contributors. A classic encyclopaedic dictionary of diagnosis. 'Abdomen, rigidity on' ... 'Yawning'. The article 'Joint, affection of' has 14 sections, p.453-73, with 18 illus. 334 of the 739 illus. are in colour. The detailed, analytical index, on blue-tinted paper, p.897-1032, is a quick-reference bonus. *Class No:* 616.01-07

[3458]
GRIFFITH, H.W. Complete guide to symptoms, illness and surgery. Tucson, Ariz., Body Press; HP Books, 1985. 896p. charts. $14.95. ISBN: 039951709x.
3 sections: (1) symptoms, possible problems and recommended action; (2) disorders, A-Z; what to expect and when to see G.P.; (3) surgery, recommendations re. postoperative care. Information is 'An excellent reference book for today's patient' (*American reference book annual,* v.17 1986, p.640).
By the same author, *Complete guide to symptoms, illness and surgery for people over 50* (Body Press, 1992. 896p. $29.95). *Class No:* 616.01-07

[3459]
The Merck manual of diagnosis and therapy. 15th ed. Rahway, N.J., Merck & Co., 1987. 2v. illus., tables. $21.50. ISBN: 0911910069.
First published 1899.
1. *General medicine.* 2. *Gynecology, obstetrics, pediatrics, genetics. Class No:* 616.01-07

[3460]
SHREEVE, C.M. The Alternative dictionary of symptoms and cures: a comprehensive guide to disorders and their orthodox and alternative remedies. London, etc., Century, 1986. viii, 504p. £12.99. ISBN: 0712612920.
Three parts. 1. A-Z of symptoms and conditions - 2. Disorders - 3. Therapies of the body (*e.g.* herbal medicine), of the mind (*e.g.* psychotherapy), of the soul (*e.g.* spiritual healing), p.427-98. Many cross-references. Bibliographical references to parts 1-2, 3. Keyed illus. *Class No:* 616.01-07

Treatment

[3461]
STANWAY, A. A Dictionary of operations. London, Paladin; Grafton Books, 1981. 436p.
3 parts: 1. Having an operation - 2. An A-Z of operations - 3. Procedures and investigations. Amniocentesis ... X-ray computed tomographs. 'Some self-help organizations of interest to surgical patients', p.427-31. Analytical subject index. Aims to make a stay in hospital more understandable and less stressful. *Class No:* 616.01-08

[3462]
TREVELYAN, J. and DOWSON, D. Thorsons guide to medical tests: the complete handbook of diagnostic and preventive tests. Wellingborough, Thorsons, 1989. 256p. illus., diagrs. £7.99. ISBN: 0722515030.
A consumer guide which gives the patient information about medical tests. 9 sections: 1. Introduction - 2. Diagnostic tests - 3. The role of preventative and occupational tests - 4. Major diseases, infections and conditions ... 8. Mental health - 9. Conclusions. For each area information is provided on: What tests are available; What can be asked for and where; What each test entails; How safe are the tests: Are the tests necessary. Index, p.250-56. Some black-and-white illustrations of medical equipment. Useful for public libraries. *Class No:* 616.01-08

Nursing

Bibliographies
[3463]
Core collection in nursing and the allied health sciences: books, journals, media. Peretz, A.A., *and others, eds.*
Longman, Oryx Press, 1990. 256p. ISBN: 0897744640.
Lists *c.*1,000 books, periodicals and nonprint materials. Divided into broad subject headings, *e.g.* AIDS; Computers. Contains media source list and a publishers', distributors', and producers', directory. Author, title and subject indexes. Aimed at both the consumer and the professional.
Class No: 616.01-083(01)

[3464]
ROYAL COLLEGE OF NURSING. A Bibliography of nursing literature. London, Library Association for the Royal College of Nursing, 1968-. ISBN: 0853654700; 085365316x; 0853656231; 0853657467.
1. *1859-1960, with an historical introduction.* Edited and compiled by A.M.C. Thompson. 1968. xx, 132p. £3. 2. *1961-1970.* Edited and compiled by A.M.C. Thompson. 1974. xiii, 223p. £36.50. 3. *1971-1975.* Edited and compiled

....(contd.)

by F. Walsh. 1985. xix, 256p. £68. 4. *1976-1980*. Edited and compiled by F. Walsh. 1986. xxii, [i], 405p. £68.

About 45,000 unannotated entries in all. Subject arrangement. The 1976-1980v. has *c*.20,000 entries in 14 classes (Nursing as a profession ... Nurse and patient - Education and research - Health services and hospitals - Child health ... Medicine and therapeutics - Surgery, accidents and emergencies. Occupational health - Psychiatry - Diseases and treatment). Author index, p.355-405. *Class No:* 616.01-083(01)

[3465]
STRAUCH, K.P. *and* BRUNDAGE, D.-J. **Guide to library resources for nursing.** New York, Appleton-Century-Crofts, 1980. 309p. illus. ISBN: 0838535283.

Section 1, - an annotated bibliography of the most important books, periodicals, etc. Arranged under specializations, *e.g.* women's health, nursing education, nursing research, psychology. Section 2 is an excellent aid to the use of a library and its bibliographical research tools. Author, title and subject indexes. 'The most important volume on a nurse's bookshelf'. (*Choice*, v.18(5), January 1980, p.641).

A 2nd ed. is due to be published by Oryx Press in 1992 (224p. $45). *Class No:* 616.01-083(01)

Encyclopaedias & Dictionaries

[3466]
Baillière's encyclopaedic dictionary of nursing and health care. Weller, B.F., *ed.* London, Baillère Tindall, 1989. xxvi,1042p. figs. £12.95. ISBN: 0702011967.

Aimed at the UK-trained and practising nurse and compiled by 45 specialists. Over 32,000 main and sub-entries (A ... Zymosis). Definitions vary in length, *e.g.* Balm, 1 line; Intracranial, 3 columns, and include pronunciation. 3 appendices: 1. Degrees, organizations and abbreviations - 2. Units of measurement and table of chemical elements - 3. Useful addresses. *Class No:* 616.01-083(03)

[3467]
McGraw-Hill nursing dictionary. New York, etc. McGraw-Hill, 1979. xxvii, [1], 1008 + 30p. graphs, tables. $23.95. ISBN: 0070450196.

39 contributors. About 50,000 entries, rarely more than 100 words in length, but showing pronunciation, stress and syllabification. Anatomical, conversion and dietary appendix tables. A spin-off from the database of *Blakiston's Gould medical dictionary* (4th ed., 1979), plus terms of particular importance in nursing education and practice, plus a 'Language interpreter for nurses'. For students, practitioners and teachers of nursing. *Class No:* 616.01-083(03)

[3468]
MILLER, B.F. *and* KEANE, C.B. **Encyclopedia and dictionary of medicine, nursing and allied health.** 4th ed. Philadelphia, Pa., W.B. Saunders, 1987. xxii, 1437p. illus. (incl. col.), tables. $26.95. ISBN: 0721618154.

First published 1972, as *Encyclopedia and dictionary of medicine*.

11 consultants. Over 30,000 entries ('Vocabulary', with anatomical tables and col. plates), p.1-1356. Table of muscles, p.790-808. 181 illus., 54 tables. Many cross-references. 13 appendicies. Thumb-indexed. Draws extensively on work in progress for the 27th ed. of

....(contd.)

Dorland's *Illustrated medical dictionary* (*qv*). Written by a physician and a nurse. A 5th ed. is due in 1992, edited by M. O'Toole (xxx,1780p. £16.50). *Class No:* 616.01-083(03)

[3469]
Mosby's medical, nursing, and allied health dictionary. 3rd ed. St. Louis, Mo., C.V. Mosby Company, 1990. xx,1608p. illus. (incl. col.), diagrs., graphs, tables. $25.95. ISBN: 0801632277.

First published 1983.

64 consulting contributors. About 26,000 entries, 3,000 new to this edition. Length of entries varies from a few words to several columns, *e.g.* 'Burn therapy': method; nursing orders; outcome criteria; 2 columns. 44-page colour atlas of human anatomy. 21 appendices, p.1267-1608 (*e.g.* Pharmacology; Directory of nursing organizations and health care workers; Leading health problems and communicable diseases). Pronunciation and stress of terms is given. 10,000 cross-references. Well produced; careful line-drawings; thumb ed. *Class No:* 616.01-083(03)

[3470]
Pearce's medical and nursing dictionary and encyclopaedia. 15th ed. London, Faber & Faber, 1983. 413p. £7.99. ISBN: 0571180809.

First published 1933, as *A short encyclopaedia for nurses*.

About 3,000 entries, both 1-2-line definitions and 1-column description (*e.g.* 'Poliomyelites, acute (infantile paralysis: symptoms; nursing and treatment; prophylaxis)). Historical appendix (including 'Old entries'), p.423-38. Select bibliography, p.441-440. Select list of organizations, p.438, 442-3. Cross-references. 'Where treatment is mentioned, it is only to place it in its context in the ward or clinic, it is *not* to be used as providing sufficient information for *undertaking* treatment' (*Preface*). *Class No:* 616.01-083(03)

Handbooks & Manuals

[3471]
The Handbook of nursing. Howe, J., *and others, eds.* New York, Wiley, 1984. 1756p. diagrs., charts, tables. ISBN: 0471895245.

Contribution with short annotated bibliographies, by nurses for practising nurses, focusing on the management of care. Parts of the text were previously published in *McGraw-Hill Handbook of clinical nursing* (1979). Chief themes: maternity, pediatrics, medical-surgical, aging, psychiatric and behavioural problems, and emergency care. 7 appendices of tables and charts. 17-p. of subject index. *Class No:* 616.01-083(035)

Dictionaries

[3472]
Baillière's nurses' dictionary. 21st ed. London, Baillère Tindall, 1990. 594p. illus. £2.50. ISBN: 0702014567.

First published 1912.

Entries vary from a few words to more than a column. Pronunciation guide is given. Many cross-references and illustrations. The 15 appendices are a useful feature (*e.g.* 'Innovative therapies'; 'AIDS'). Pocket-sized format. Good value. *Class No:* 616.01-083(038)

Theses

[3473]

Steinberg collection of nursing research: catalogue. London, Royal College of Nursing, 1989. 57p.

Lists 513 dissertations and theses by nurses, or about nursing, submitted for higher degrees in the UK. *Class No:* 616.01-083(043)

Reviews & Abstracts

[3474]

International nursing index. New York, American Journal of Nursing Co., 1966-. 4pa.; annual cumulation. $250. ISSN: 00208124.

Published in co-operation with National Library of Medicine.

Indexes over 270 nursing journals published worldwide, as well as nursing articles in the 2,600 biomedical titles indexed in *Index medicus.* A list of current books published by and for nurses is appended to each issue and the cumulation. Online access is available through MEDLINE. 'This is the master index to nursing literature' (*Reference service review,* v.9(4), October/December 1981, p.22). Available on CD-ROM. *Class No:* 616.01-083(048)

[3475]

—Cumulative index to nursing and allied health literature. Glendale, Calif., Glendale Adventist Medical Center, 1984-. v.22-. 6pa., with annual cumulation. $220.

Continues annual *Cumulative index to nursing,* 1961-76. v.1-21, with wider scope and more acceptable frequency. Covers the major English language and popular nursing journals, as well as selective indexing of medical periodicals. Appended to each issue is a list of audiovisual items and of book reviews. 'CINAHL would be my first choice for any hospital or school of nursing' (*Magazines for libraries* (5th ed., 1986), p.22). Available online and on CD-ROM. *Class No:* 616.01-083(048)

[3476]

Nursing bibliography: a monthly list of current publications on nursing and allied subjects. London, Royal College of Nursing, 1972-. Monthly. £40pa. ISSN: 03009947.

8-9,000 references pa. under subjects, Abortion ... Wounds. Selectively indexes more than 200 English-language periodicals. Coverage: books, reports, articles, pamphlets and theses. Each issue carries an author index. The separate list of subject headings and of journals indexed is revised annually. *Class No:* 616.01-083(048)

[3477]

Nursing research abstracts. London, Dept. of Health and Social Security, Index of Nursing Research, 1978-. 4pa. £14pa. ISSN: 01413899.

1990: 471 abstracts. 21 sections (Research, methodology and finding ... Social problems and welfare - Information services). Concentrates on British nursing research, coded 'ongoing research' or 'recently completed research'. Author and subject indexes per issue, cumulated annually. Online DHSS-DATA. *Class No:* 616.01-083(048)

Histories

[3478]

Nursing: a historical bibliography. Bullough, B., *and others.* New York, Garland, 1981. xxiv, 408p. (*Garland reference library of social sciences, v.66.*) ISBN: 0824095111.

5,203 numbered, unannotated entries for books and articles. 20 sections, *e.g.* 1. Indexes, bibliographies and abstracts; 2. History of nursing; 7. Biographies and autobiographies; 9. Nursing specialties; 10. Education; 17. Sociology; 20. Nurses and nursing in literature. Author index. Updated through Wellcome Medical Historical Museum and National Library of Medicine bibliographies. *Class No:* 616.01-083(091)

Biographies

[3479]

A Bio-bibliography of Florence Nightingale; completed by S. Goldie. Bishop, W.J., *comp.* London, Dawsons of Pall Mall, for the International Council of Nurses, with which is associated the Florence Nightingale International Foundation, 1962. 160p. port.

10 chapters. 1 Nursing - 2. The army - 3. Indian and colonial welfare - 4. Hospital - 5. Statistics - 6. Sociology - 7. Memoirs and tributes - 8. Religion and philosophy - 9. Miscellaneous work - 10. Selected writings about Florence Nightingale and her times. Appendix: Chronological list of Florence Nightingale's writing. 150 numbered and annotated items. Index. *Class No:* 616.01-083(092)

Education Courses

[3480]

WELL, J. The Directory of continuing education for nurses, midwives and health visitors. London, Newpoint Publishing Company, 1990. 239p. £4.95. ISBN: 0862631297.

14 sections: 1. Introduction - 2. Statutory control of nursing education ... 8. Courses for midwives ... 22. Higher education courses - 12. Teacher preparation courses - 13. Management courses - 14. Open and distance learning. 5 appendices (*e.g.* 'Alphabetical list of courses'; 'Useful addresses'). *Class No:* 616.01-083:377.3

[3481]

—Directory of schools of nursing, 1980. 4th ed. London, HMSO, 1980. xi, 270p. £10. ISBN: 0113207182.

First published 1972. 3rd ed., 1977 (vii, 364p.).

3 parts: 1. Pre-basic and basic training schemes (England, Wales, Isle of Man, Channel Islands, Scotland, Northern Ireland) - 2. Further training and education for qualified nurses - 3. Midwifery. Addresses for further information. General index. Specialized index. *Class No:* 616.01-083:377.3

Pathological Anatomy

Histopathology

[3482]
LAW, J.W. *and* OLIVER, H.J. Glossary of histopathological terms. London, Butterworth, 1973. [v], 138p. diagrs., chemical structures.

Concise definitions of *c*.700 terms likely to be encountered in normal laboratory practice. Includes short biographies.
Class No: 616.01-091.8

Immunology

Bibliographies

[3483]
NICHOLAS, R. *and* NICHOLAS, D. Immunology: an information profile. London, Mansell, 1985. vii, 216p. ISBN: 0720117240.

3 parts. 1: Overview of immunology and its literature (5 sections, including 3. Organisations in immunology; 4. Immunology: its conferences; 5. Immunology: its literature (journals, reviews and monographic series, books, dissertations, patents, reference works, audiovisual aids, library classification schemes as applied to immunology)) - 2. Bibliography, p.133-61 - 3. Directory of organizations and database hosts. Index, p.204-16. 854 numbered entries. Aims 'to provide for the everyday information needs of immunologists'. Mainly restricted to English language sources. *Class No:* 616.01-097(01)

Encyclopaedias & Dictionaries

[3484]
Dictionary of immunology. Rosen, F.S., *and others*. London, Macmillan Reference Books, 1989. 223p. £25; £15.95. ISBN: 0333347420.

c.1,200 definitions, some encyclopedic in length and detail, *e.g.* 'Interleukin': 2p. Many cross-references. Covers terms from molecular biology, cell biology and genetics as well as from immunology. Does not contain as many abbreviations and assumes more clinical knowledge than *Dictionary of immunology* (3rd ed. Oxford, Blackwell Scientific Publications, 1985) *(q.v.)*. 'This is the best immunology dictionary that is currently available', (*Science & technology libraries*, v.10(4), 1990, p.119).
Class No: 616.01-097(03)

[3485]
Dictionary of immunology. Herbert, W.J., *and others, eds*. 3rd ed. Oxford, etc., Blackwell Scientific Publications, 1985. xi,240p. ISBN: 0632009844.

First published 1971.

c.1,700 terms and eponyms defined, with 300 entries added since previous ed. 'Immunoglobulin': over 1p. Extensive cross-references; chemical structures. Explains how terms have been used in the literature and originates 'from the need for a glossary for use in undergraduate teaching' (*Preface*). *Class No:* 616.01-097(03)

[3486]
Encyclopedia of immunology. Roitt, I.M., *ed*. London, Academic Press, 1992. 3v. (1578p.) illus., tables. £300. ISBN: 0122267605, set; 0122267613, v.1; 0122267621, v.2; 012226763x, v.3.

About 700 contributors. Articles normally about 2-3 pages in length, with brief bibliographies appended. Cross-references. Complete list of entries given at the beginning of each volume and each volume contains the index for the set. *Class No:* 616.01-097(03)

Handbooks & Manuals

[3487]
WEIR, D.M. *and* STEWART, J. Immunology. 7th ed. Edinburgh, Churchill Livingstone, 1993. 372p. figs., tables. £11.95. ISBN: 0443046603.

First published 1970.

2 sections (10 documented chapters): 1. Basic immunology - 2. Immunology in action. Theory and application of immunological principles for under- and postgraduate students. *Class No:* 616.01-097(035)

Reviews & Abstracts

[3488]
Excerpta medica. 26. Immunology, serology and transplantation. Amsterdam, Excerpta Medica, 1967-. 32pa. DFl2682. ISSN: 00144304.

About 9,000 abstracts pa. 15 sections: 1. General aspects - 2. Laboratory methods and techniques - 3. Treatment ... 7. Antibodies - 8. Cell-mediated immune responses - 9. Modulation of immune responses ... 15. Transplantation immunology. Subject and author indexes per issue and annually. Online, EMBASE. Available on CD-ROM.
Class No: 616.01-097(048)

Histories

[3489]
SILVERSTEIN, A.M. A History of immunology. San Diego, Academic Press, 1989. xxi,422p. $44.95. ISBN: 012643770x.

12 chapters, with appended bibliographies: 1. Theories of acquired immunity - 2. The royal experiment on immunity, 1721-1722 ... 5. The concept of immunological specificity ... 10. Anti-antibodies and anti-idiotype immunoregulation, 1899-1904 ... 12. Magic bullets and poisoned arrows: the uses of antibody. 3 appendices: A. The calendar of immunological progress (includes 'Important books in immunology, 1892-1968'), p.327-40 - B. Nobel Prize highlights in immunology, p.341-50 - C. Biographical dictionary (Akroyd ... Zinsser). Glossary, name and subject indexes. *Class No:* 616.01-097(091)

Biographies

[3490]
LEFANU, W. A Bibliography of Edward Jenner. 2nd ed. Winchester, St. Paul's Bibliographies, 1985. xv, 160p. illus. £30. ISBN: 0906745192.

First published 1951 as *A bio-bibliography of Edward Jenner*.

8 chapters (1. Experiments and observations ... 5. Posthumous works ... 8. Biographies, dedications and

.... *(contd.)*
portraits). 150 entries, with very detailed notes on each, including collation, contents, and copy used. Notes (p.145-50) include bibliographic references. Index, p.151-60. *Class No:* 616.01-097(092)

Special Pathology

Cardiovascular Diseases

[3491]
Excerpta medica. 18. Cardiovascular diseases and cardiovascular surgery. Amsterdam, Excerpta Medica, 1966-. 24pa. DFl1,796pa. ISSN: 00144223.
Formerly *Cardiovascular diseases,* 1957-65.
About 7,000 abstracts pa. 18 sections: 1. General aspects ... 3. Diagnostic procedures ... 6. Coronary heart disease ... 11. Heart rate and rhythm ... 13. Hypertension - 14. Thrombosis and embolism ... 17. Treatment - 18. Prevention and rehabilitation. Subject and author indexes per issue and annually. Online, EMBASE; CD-ROM. *Class No:* 616.1

Haematology

[3492]
BRACONI, L.R. Hematology: a glossary of terms in English/American, French, Spanish, Italian, German, Russian. Amsterdam, etc., Elsevier, 1964. [vii], 306p. (*Glossaria interpretum, G9.*)
2,210 English/American base terms, with French, Spanish, Italian, German and Russian equivalents and indexes. *Class No:* 616.15

[3493]
Excerpta medica. 25. Hematology. Amsterdam, Excerpta Medica, 1967-. 24pa. DFl.1,196pa. ISSN: 00144290.
About 8,000 abstracts pa. 20 sections: 1. General aspects ... 4. Anemia ... 6. Leukocytes and related cells ... 10. Blood clotting, fibrosolysis ... 15. Blood groups ... 17. Transfusion ... 24. Laboratory techniques. Subject and author indexes per issue and annually. Online, EMBASE. Available on CD-ROM and microfilm. *Class No:* 616.15

Ear, Nose & Throat

[3494]
Excerpta medica. 11. Oto-, rhino-, laryngology. Amsterdam, Elsevier, 1948-. 16pa. DFl.1,293pa. ISSN: 00144150.
Contains *c.*4,000 abstracts pa. 35 sections: 1. Skull - 2. Face and lips ... 5. Salivary glands ... 9. Larynx and esophagus ... 14. Nasal cavity ... 19. Middle ear ... 26. Hearing - 27. Speech disorders ... 35. General aspects. Author and subject indexes per issue and annually. Online, EMBASE. Available on CD-ROM and microfilm.
Class No: 616.21

[3495]
LARRAURI, A. Dictionnaire d'oto-rhino-laryngologie, en cinq langues: français-anglais-espagnol-allemand-italien. Paris, Librairie Maloine; New York, French & European, 1971. 5v. $95. ISBN: 0686569989.
About 3,000 ear, nose and larynx terms in 5 sequences: French - English - Spanish - German - Italian ... Italian - French - English - Spanish - German. *Class No:* 616.21

Hearing (Otology)

[3496]
GALLAUDET COLLEGE, Washington, D.C. **Dictionary catalog on deafness and the deaf.** Boston, Mass., G.K. Hall, 1971. 2v. (1683p.). ISBN: 0816108773.
39,300 catalogue cards, photolithographically reproduced. Entries include dissertations, reports and microfilms. Author, title, series and subject entries. *Class No:* 616.28

[3497]
Gallaudet encyclopedia of deaf people and deafness. Van Cleve, J.V. New York, etc., McGraw-Hill, 1987. 3v. (1400p.). $330. ISBN: 0070792291, set.
Sponsored by Gallaudet College, Washington, D.C.
271 articles, most of them documented, on sign language, audiology, auditory disorders organizations, deaf community publications, rehabilitation, demographics, biographies of distinguished deaf individuals, educational programmes, conditions and status of the deaf community in most major countries, etc. 250 illus. 8,000-entry index. 'Likely to become the standard work in its field' (*Choice,* v.24(8), April 1987, p.1198). *Class No:* 616.28

[3498]
Terminology of communication disorders: speech - language - hearing. Nicolosi, L., *and others.* 3rd ed. Baltimore, Md., Williams and Wilkins, 1989. 354p. diagrs. ISBN: 0683065009.
First published 1978.
Contains *c.*6,000 terms, with 129 entries significantly revised and 389 completely new for this ed. Main section, A-Z. 12 appendices (*e.g.* Mean length of utterance; Language tests and procedures; Diagnostic differences). No bibliography. *Class No:* 616.28

[3499]
TURKINGTON, C. *and* **SUSSMAN, A. Encyclopedia of deafness and hearing disorders.** New York, Facts on File, 1992. 320p. illus. $45. ISBN: 0816022674.
Over 500 entries, by specialists, A-Z. Includes clinical terms, parts of the ear, noted hearing specialists, and deaf personalities, *e.g.* Beethoven. Suggested reading. 13 appendices, including lists of organizations and support agencies. *Class No:* 616.28

Dentistry

Bibliographies

[3500]
CLENNETT, M.A. Keyguide to information sources in dentistry. London, Mansell, 1985. 287p. facsim. £40. (*Keyguide series.*) ISBN: 072011747x.
Part 1: Survey of dentistry and its libraries (8 sections, *e.g.* 5. Finding out about the literature; 7. Sources of special information: patents, audiovisual material, etc., 8. Language

.... *(contd.)*

problem) - Part 2: Bibliography (666 annotated entries, under subjects: anaesthesia ... radiography; miscellaneous subjects) - Part 3: Directory of organizations (selected list of dental libraries; national dental associations; dental associations and societies; schools and university depts.; publications). 1,445 numbered entries in all. Detailed and analytical index, p.273-87. Well researched. Intended for the information worker or non-dental researcher.

Class No: 616.314(01)

Dictionaries

Polyglot

[3501]
FÉDÉRATION DENTAIRE INTERNATIONALE. A Lexikon of English dental terms, with their equivalents in español, Deutsch, français, italiano. The Hague, Sijthoff, 1966 (reprinted 1974). 424p.

Over 7,000 English-base terms, with Spanish, German, French and Italian equivalents and indexes. The multilingual vocabulary is based on WHO terminology and its 'Application of the Internal classification of diseases to dentistry and stomatology', etc.

Class No: 616.314(038)=00

English

[3502]
FAIRPO, J.E.H. *and* FAIRPO, C.G. Heinemann dental dictionary. 3rd ed. London, Butterworth-Heinemann, 1987. xiv, 311p. tables. £12.95. ISBN: 0750600985.

First published 1962 as *Heinemann Modern dictionary for dental students.*

About 5,000 terms ('Cavity': 1 column). Includes eponyms. Lists of abbreviations and signs precede. 4 appended tables on arteries, muscles, nerves and veins of the head and neck. 'Dental periodicals', p.287-311.

Class No: 616.314(038)=20

[3503]
HARTY, F.J. *and* OGSTON, R. Concise illustrated dental dictionary. Bristol, J. Wright, 1987. ix, 291p. illus. £12.50. ISBN: 0723607885.

About 3,000 entries ('Crown': 2 columns, with 2 illus.). 16 appendices (*e.g.* radiographic technique and anatomy; Addresses of selected dental schools; Addresses of national dental organizations). *Class No:* 616.314(038)=20

[3504]
JABLONSKI, S. Illustrated dictionary of dentistry. Philadelphia, Pa., W.B. Saunders, 1982 (reprinted Krieger, 1990). xviii, 919p. illus., tables. $79.50. ISBN: 0894644777.

Attempts to standardize North American usage of dental terminology. 1.3 million words. Entries provide background to definitions as well as indicating pronunciation, synonyms, etymology, etc. 'An authoritative reference' (*British dental journal*, v.153, 16 November 1982, p.356).

Class No: 616.314(038)=20

Reviews & Abstracts

[3505]
Dental abstracts. Chicago, Ill., American Dental Association, 1956-. 12pa. $90pa. ISSN: 00118486.

About 750 longish abstracts pa. More than 100 dental and medical journals are scanned. 5 illustrated sections: Hands-on (diagnosis ... temporomandibular joint) - The front office (staff relations; practice management; academics (undergraduate study)) - Inquiry (anatomy ... research methodology) - The big picture (health; history; perspective). Annual author and subject indexes. *Class No:* 616.314(048)

[3506]
Index to dental literature, 1939-. Chicago, Ill., American Dental Association, 1943-. 4pa. $150. ISSN: 00193992.

Initially (1939-61) as *Index to dental literature in the English language.*

About 10,000 references pa. Contents: Dental books published; Dissertations and theses; Serials indexed; Dental descriptors; Cross-references; Qualifiers; Bibliography of dental reviews (subject and name sections); Index to dental literature (subject and name sections). Annual cumulation. Available online on MEDLINE database and on CD-ROM.

Class No: 616.314(048)

Yearbooks & Directories

[3507]
American dental directory. Chicago, Ill., American Dental Association, 1947-. Annual.

Primarily a local list, with entries under state and city, with an A-Z name index. Data include specialization and date of dental graduation. *Class No:* 616.314(058)

[3508]
The Dentists register, 1992, comprising the names and addresses of dental practitioners registered at 31 January 1992, together with the local list of names so registered, and the list of bodies corporate carrying on the business of dentistry. London, General Dental Council, 1992. 56p. + 526p. tables.

First published 1879.

The register is preceded by text of the Dentists Act 1984, plus table of registrable qualifications. Monthly amendments are issued. *Class No:* 616.314(058)

Histories

[3509]
RING, M.E. Dentistry: an illustrated history. New York, Abrams, 1985. 320p. illus. $75. ISBN: 0810911000.

A well-written and finely illustrated account that 'will become the most up-to-date resource on the history of dentistry in the English language' (*Journal of American Dental Association,* v.112 January 1986, p.84). *Class No:* 616.314(091)

Stomach (Gastric Disorders)

[3510]
Excerpta medica. 48. Gastroenterology. Amsterdam, Excerpta Medica, 1971-. 20pa. Dfl.1393. ISSN: 00313580.

About 5,000 abstracts pa. 9 sections: 1. General aspects - 2, Morphology - 3. Physiology - 4. Diagnostic procedures - 5. Gastrointestinal diseases - 6. Gastrointestinal infections - 7. Liver and biliary system - 8. Pancreas - 9. Peritoneum,

....(contd.)

mesentery and omentum. Analytical subject and author index per issue and annually. Online, EMBASE and available on CD-ROM. *Class No:* 616.33

Endocrinology

[3511]
Excerpta medica. 3. Endocrinology. Amsterdam, Excerpta Medica, 1947-. 24pa. DFl.472pa. ISSN: 0014407x.

About 7,000 abstracts pa. 9 sections: 1. General aspects - 2. Diagnosis, diagnostic procedures and techniques - 3. Treatment - 4. Growth, development and aging ... 6. Endocrine glands ... 8. Reproductive system - 9. Hormone-related metabolism. Subject and author indexes per issue and annually. Online, EMBASE. Available on CD-ROM and microfilm. *Class No:* 616.4

Skin (Dermatology)

[3512]
FRY, L. Illustrated encyclopedia of dermatology. 2nd ed. Lancaster, MTP Press, 1985. viii, 575p. illus., tables. ISBN: 0852008678.

58 sections (1. Acanthosis nigricans ... 11. Eczema ... 25. Lupus erythematosus ... 58. Viral infections and AIDS). Emphasis is on the clinical aspects and skin diseases. The treatment sections of all the chapters have been revised. No bibliography. Inadequate index. *Class No:* 616.5

[3513]
LEIBER, M. *and* **ROSENBLUM, M.A. A Dictionary of dermatological words, terms and phrases.** New York, etc., McGraw-Hill, 1968. xviii, 440p.

Entries for about 1,500 terms and expressions. Some contributions are lengthy, *e.g.* 'Eczema': 2/3p., with a further 2p. on types of eczema. Pronunciation is indicated. Readable. No illus. *Class No:* 616.5

Arthritis

[3514]
Excerpta medica. 31. Arthritis and rheumatism. Amsterdam, Excerpta Medica, 1965-. 8pa. DFl911. ISSN: 00144355.

About 3,000 abstracts pa. 22 sections, *e.g.* 1. General surveys and diagnostic procedures; 2. Anatomy and morbid anatomy; 3. Biochemistry, pharmacology and physiology; 6. Rheumatoid arthritis; 17. Diseases of the spinal and peripheral joint. Subject and author indexes per issue and annually. Online, EMBASE. *Class No:* 616.72

Neurology

[3515]
ADELMAN, G., *ed.* **Encyclopedia of neuroscience.** Berlin, Springer Verlag; Cambridge, Mass., Birkhauser, 1987. 2v. (1565p.) illus., diagrs., tables. $165. ISBN: 3764333359; 0817633359, US ed.

Contributions by 600 specialists. More than 700 original articles, each with brief bibliography and cross-references. Biographies also feature. Includes neurosurgery, psychiatry, artificial intelligence. Anatomical illus. of the brain and nervous system. Authoritative. *Class No:* 616.8

[3516]
Current advances in neuroscience. Oxford, etc., Pergamon, 1984-. 12pa. £400. ISSN: 07411677.

About 1,600 abstracts per issue. 28 sections: 1. Neurochemistry; 8. Physiology of the peripheral nervous system; 11. Central nervous system; 12. Pharmacology of the central nervous system; 17. Neurotransmitters, their receptors, agonists and antagonists; 19. Sensory systems; 25. Invertebrate neurophysiology; 29. General concepts, reviews and symposia. Author index and journals list per issue. Participates on CABS databases, on BRS. *Class No:* 616.8

[3517]
Desk reference for neuroscience. 2nd ed. Berlin, Springer-Verlag, 1992. 306p. figs. $69. ISBN: 0387976299.

A revision of Lockard's *Desk reference for neuroanatomy* (1977).

Brief definitions of *c.*4,500 terms related to neuropathology, neurophysiology and neuropharmacology. 53 line drawings. Bibliography of *c.*100 items. For academic libraries, as a quick-reference tool. *Class No:* 616.8

[3518]
Excerpta medica. 8. Neurology and neurosurgery. Amsterdam, Elsevier, 1948-. 32pa. DFl.2,420pa. ISSN: 00144134.

About 10,000 abstracts pa. v. 46 sections: 1. General aspects. History - 2. Neuroanatomy - 3. Neurophysiology - 4. Neuropathology ... 36. Neurological aspects of mental disease ... 44. Developmental and genetic disorders - 45. Diagnostic processes - 46. Treatment. Author and subject indexes per issue and annually. Online, EMBASE and available on CD-ROM and microfilm. *Class No:* 616.8

[3519]
GARRISON, F.H. Garrison's history of neurology; rev. and enl., with a bibliography of classical, original and standard work in neurology. McHenry, L.C. Springfield, Ill., Thomas, 1969. xiii, [1], 552p. $77.50. ISBN: 039801261x.

Chapters 1-8. From ancient times to the 19th century - 9. Clinical neurology - 10. The neurological examination - 11. Neurological diseases. 411 references; bibliography, p.471-527 (authors, A-Z). *Class No:* 616.8

Epilepsy

[3520]
Dictionary of epilepsy. Part 1: Definitions. Gastaut, H., *and others.* Geneva, World Health Organization, 1973. 76p. $8. ISBN: 9241540273.

More than 1,500 entries, chiefly on clinical aspects. Includes eponyms and syndromes. 'If a term is followed by a definition, this indicates that it is the recommended term for that concept ...' All terms not followed by definitions are 'obsolete, inaccurate or not recommended for some other reason'. *Class No:* 616.853

[3521]
Excerpta medica. 50. Epilepsy abstracts. Amsterdam, Excerpta Medica, 1969-. 6pa. DFl.718pa. ISSN: 03038459.

Covers the literature 1947/67 onwards.

About 1,500 abstracts pa. 9 sections: 1. General aspects - 2. Basic sciences - 3. Seizures - 4. Etiology - 5. Genetics - 6. Systematic changes related to seizures - 7. Diagnostic aids - 8. Psychology, sociology and epidemiology - 9. Treatment.

....(contd.)

Subject and author indexes per issue and annually. Online, EMBASE and available on CD-ROM, and microfilm. *Class No:* 616.853

Psychiatry

Bibliographies

[3522]
GREENBERG, B. **How to find out in psychiatry:** a guide to sources of mental health information. Oxford, etc., Pergamon Press, 1978. xii, [1], 112, [1]p. facsims., tables.

12 chapters (2. Guides to libraries & the psychiatric literature - 3. Primary sources of information: periodicals and books - 4. Secondary sources of information: bibliographies, indexes, abstracts & reviews - 5. Dictionaries, glossaries, encyclopedias and handbooks - 6. Directories - 7. Nomenclature and classification - 8. Education - 9. Mental health statistics - 10. Drugs & drug therapy - 11. Tests and measurements - 12. Nonprint material). References, p.83. Appendix A: Classics in psychiatric literature, p.82-97. Index of titles, authors, 'Highly recommended' (*Choice,* v.16(5/6), July/August 1979, p.648). *Class No:* 616.89(01)

[3523]
GUHA, M. 'Psychiatry'. In *Information sources in the medical sciences,* edited by L.T. Morton and S. Godbolt, Chapter 17 (4th ed. 1992), p.387-412.

Sections on periodicals; abstracts, indexes, bibliographies; dictionaries; reviews; conferences, monographs, books; reference works, and biography and history of psychiatry. 1½p. of bibliography. *Class No:* 616.89(01)

[3524]
MENNINGER FOUNDATION, Topeka, Kansas. **Catalog of the Menninger Clinic Library.** Boston, Mass., G.K. Hall, 1972. 4v. (1704p.). First supplement, 1978. 2v. (914p.). $435, set. ISBN: 0816109613.

56,700 cards reproduced in the main catalogue; 19,100 cards in the *First supplement* Author-title and subject catalogue parts. Includes serial holdings under titles of journals. The Menninger Foundation is the centre for treatment, research and professional education in psychiatry. *Class No:* 616.89(01)

Encyclopaedias

[3525]
BRUNO, F.J. **The Family mental health encyclopedia.** New York, Wiley, 1989. 422p. ISBN: 0471635731.

*c.*500 clearly-written definitions of theories, disorders, people, drugs, and therapies. Designed for the layperson in dealing with mental health professionals. *Class No:* 616.89(031)

[3526]
CAMPBELL, R.J. **Psychiatric dictionary.** 6th ed. New York, Oxford University Press, 1989. 811p. £40. ISBN: 0195052935.

First published 1940.

Encyclopaedic dictionary of *c.*10,000 terms and cross-references, covering relevant terms in both technical language (using DSM-III-R nomenclature - thus, 7p. on different types of 'schizophrenia') and popular usage ('gay',

....(contd.)

'moron'). Very brief biographical entries. Preceded by 'Key to pronunciation', book references (1p.) and abbreviations (9p.). The standard dictionary in its field. *Class No:* 616.89(031)

[3527]
International encyclopedia of psychiatry, psychology, psychoanalysis & neurology. Wolman, B.B., *ed.* New York & London, Van Nostrand Reinhold, for Aesculapius Publishers, Inc., 1977. 12v. (5627p.). diagrs. $675, set. ISBN: 0918228018, set.

1,500 authors, some 300 editors and consultants. Nearly 2,000 survey articles, each documented. 305 biographies (*e.g.* 'Adler': v.1, p.235-40; ½ column of bibliography). 1,605 articles on concepts and issues (*e.g.* 'Aggression in children', v.1, p.348-53; 1½ columns of bibliography). V.12 (410p.) comprises author and subject indexes, although the subject index is non-analytical (*c.*200 unspecififed entries under 'Transference'). 'This eminently readable and impressive encyclopedia' (*Library journal,* 15 April 1978, p.817).

Progress volume 1 (1983. xxxiv, 509p. $89. ISBN 091822828x) continues the tradition of the 12-v. set. 136 signed articles, by 125 specialists. *Class No:* 616.89(031)

Handbooks & Manuals

[3528]
American handbook of psychiatry. Arieti, S., *editor-in-chief.* 2nd ed. New York, Basic Books, Inc., 1974-86. 8v.

First published in 3v., 1959-60.

1. *The foundation of psychiatry.* 2. *Child and adolescent psychiatry.* 3. *Adult psychiatry.* 4. *Organic disorders and psychosomatic medicine.* 5. *Treatment.* 6. *New psychiatric frontiers.* 7. *Advances and new directions.* 8. *Biological psychiatry.*

A much expanded, updated 2nd ed. V.6 has 5 parts, 44 documented chapters (*e.g.* chapter 26. 'Human sexuality research and treatment frontiers, p.665-91, has 137 references; chapter 36. 'Computer applications in psychiatry', p.811-40, 109 references). Each volume carries name and detailed, analytical subject indexes. *Class No:* 616.89(035)

[3529]
Companion to psychiatric studies. Kendall, R.E. *and* Zealley, A.K., *eds.* 4th ed. Edinburgh, Churchill Livingstone, 1988. xii, 828p. diagrs., graphs, tables. £59.50. ISBN: 0443037368.

First published 1973; 3rd ed. 1983.

38 contributors. 41 documented sections (*e.g.* 9. Genetic aspect of mental disorders, p.161-90, with 4p. of references; 27. Psychiatry of old age, p.564-8, with 3p. of references). Detailed analytical index, p.802-28. 'Intended as a general textbook for all postgraduate students of psychiatry' (*Preface*). *Class No:* 616.89(035)

[3530]
Handbook of psychiatry. Shepherd, M., *ed.* Cambridge Univ. Press, 1982-85. 5v.

1. *General psychopathology.* 1983. xi, 307p. 2. *Mental disorders and somatic illness.* 1983. 337p. 3. *Psychoses of uncertain etiology.* 1982. xiv, 313p. 4. *The neuroses and personality disorders.* 1983. xviii, 500p. 5. *The scientific foundations of psychiatry.* 1985. xiv, 359p.

Each volume carries an extensive bibliography (*e.g.* v.3,

....*(contd.)*

*c.*1,300 references). Chapters include such topics as 'The historical background'; 'Neurological disorders'; 'Affective psychoses'; 'Philosophy and psychiatry'. Cross-references between the volumes. Author and subject indexes.
Class No: 616.89(035)

[3531]
Oxford textbook of psychiatry. Gelder, M., *and others.* 2nd ed. Oxford, University Press, 1989. xiv,1079p. tables. £25. ISBN: 0192616307.
First published 1983.
22 chapters, each with 'Further reading' appended: 1. Signs and symptoms of mental disorder ... 4. Aetiology ... 11. Organic psychiatry ... 22. Forensic psychiatry. References, p.911-1028. Appendix: 'The law in England and Wales'. Author and detailed, analytical indexes. An extensive revision of 'an introductory textbook for trainee psychiatrists, and also ... an advanced textbook for clinical medical students' (*Preface* to the first edition).
Class No: 616.89(035)

Dictionaries

Polyglot

[3532]
PETERS, U.H. Wörterbuch der Psychiatrie und medizinischer Psychologie. 3rd ed. Munich, Baltimore, Urban & Schwarzenberg, 1984. 716p. $65. ISBN: 3541049634.
Several thousand German base-terms, with definitions, plus English-German and French-German indexes. Bibliography, p.715-6. *Class No:* 616.89(038)=00

English

[3533]
AMERICAN PSYCHIATRIC ASSOCIATION. Joint Commission on Public Affairs. **A Psychiatric glossary.** 6th ed. Washington, D.C., American Psychiatric Association, 1989. ix, 142p. tables. $39.60. ISBN: 0880480270.
First published 1957.
Defines *c.*1,500 terms, plus brief notes on 'prominent names' and schools of psychiatry. Nearly all the terms derive from 'dynamic psychiatry'. 7 extensive tables on drugs, psychological tests, research terms, etc. Full cross-references. Aimed at personnel in the mental health field, whereas *Chambers' dictionary of psychiatry,* by J.A. Bressel and G. La Fond Cantzlaar (1967. xiv, 234p.), with *c.*2,000 terms, is intended primarily for the general public.
Class No: 616.89(038)=20

[3534]
LANCASTER HEALTH AUTHORITY. Glossary of terms, tests and drugs used in psychiatric practice. 4th ed. Lancaster, the Authority, 1984. [i], 91p.
First published 1962 as *A guide to psychiatric and psychological terminology.*
5 sections: 1. Terminology - 2. List of common phobias - 3. List of psychological tests - 4. Drugs in use for the treatment of psychiatric disorders - 5. The Mental Health Acts. 'Produced to satisfy a demand for this type of literature not only in Psychiatric Hospitals but many other areas concerned with Mental Health Services' (*Preface*).
Class No: 616.89(038)=20

[3535]
—Dictionary of psychiatry. Walton, H., *ed.* Oxford, Blackwell Scientific Pubns., 1985. 170p. £16.50. ISBN: 0632009721.
Has 13 contributors. Entries include brief biographies (*e.g.* Nietzsche: ½p. 'Migraine': 1/3p.; 'Psychoanalytical theory'; 2/3p.). Cross-references, but no bibliography. 'Embodies a response to the expressed need for an authoritative, accessible and compact set of definitions and terms used in psychiatry and cognate subjects' (*Foreword*). Editor is Professor of Psychiatry, University of Edinburgh.
Class No: 616.89(038)=20

[3536]
—THAKURDAS, H. *and* THAKURDAS, L. Dictionary of psychiatry. Revised by B. Thakurdas. Lancaster, MTP, 1979. vii, 111p. diagrs. £9.75; £5.50. ISBN: 0852002351.
Defines *c.*900 terms. 8 useful appendices: A. International Standard Classification of Mental Disorders - B. List of abbreviations - C. A guide to prefixes and suffixes - D. Common phobias - H. Essentials of the Mental Health Act, 1969. Designed as a ready reference for doctors, medical students, psychiatric nurses, social workers 'or anybody involved with psychiatric patients' (*Introductory note*).
Class No: 616.89(038)=20

[3537]
WORLD HEALTH ORGANIZATION. Mental disorders: a glossary and guide to their classification in accordance with the Ninth Revision of the International Classification of Diseases. Geneva, WHO, 1978. 95p. £6. ISBN: 924154377.
Offprint of chapter 5 'Mental disorders' of the *International Classification of Diseases* (9th ed. 1977-78), and successor to WHO *Glossary of mental disorders and guide to their classification* (1974).
Quick reference lists 1-3, p.58-60. 3 annexes. Detailed, analytical index, p.83-95. *Class No:* 616.89(038)=20

[3538]
—Lexicon of psychiatric and mental health terms: Volume 1. Geneva, World Health Organization, 1989. vii,76p. SFr16. ISBN: 924154242x.
Designed for use with Chapter 5 of the Ninth Revision of the *International Classification of Diseases, (ICD-9).*
Contains definitions of over 300 terms in Chapter 5 of ICD-9. Complements *Mental disorders: a glossary and guide to their classification in accordance with the Ninth Revision of the International Classification of Diseases, (q.v.). Class No:* 616.89(038)=20

German

[3539]
HAAS, R. Dictionary of psychology and psychiatry: English/German. Toronto, C.J. Hogrefe, 1980-88. 2v. $78, v.1; $98, v.2. ISBN: 0889370001, v.1; 0889370133, v.2.
Published under the auspices of the Centre for Psychological Information and Documentation at the University of Trier, West Germany.
V.1. *English/German* 1980. 453p. V.2. *German/English.* 1988. 440p.
V.1 contains more than 30,000 English terms, with German equivalents, covering such subjects as neurophysiology, cybernetics, sexology.
Class No: 616.89(038)=30

Reviews & Abstracts

[3540]

Excerpta medica. 32. Psychiatry. Amsterdam, Excerpta Medica, 1948-. 20pa. DFl1,497pa. ISSN: 00144363.

About 6,000 abstracts pa. 36 sections: 1. History and general aspects - 2. Medical psychology - 3. Mental tests - 4. Psychophysiology ... 10. Mental deficiency ... 12. Psychosis - 15. Sexology ... 21. Group therapy ... 23. Psychopharmacology ... 28. Social psychiatry - 29. Child psychiatry ... 32. Suicide - 33. Forensic psychiatry - 34. Military psychiatry - 35. Preventive mental hygiene - 36. Education and training. Annual subject and author indexes. Online, EMBASE; CD-ROM. *Class No:* 616.89(048)

Progress Reports

[3541]

Psychiatry: the state of the art. Proceedings of the 7th World Congress of Psychiatry, July 11-16, 1983. Pichot, P., *and others, eds.* New York & London, Plenum Press, 1985. 8v. graphs, charts, tables. ISBN: 0306416018, v.1; 0306416026, v.2; 0306416034, v.3; 0306416042, v.4; 0306416050, v.5; 0306416069, v.6; 0306416077, v.7; 0306416085, v.8.

1. *Clinical psychopathology, nomenclature and classification.* 2. *Biological psychiatry, higher nervous activity.* 3. *Pharmacopsychiatry.* 4. *Psychiatry and psychosomatic medicine.* 5. *Child and adolescent psychiatry, mental retardation and geriatric psychiatry.* 6. *Drug dependence and alcoholism, forensic psychiatry, military psychiatry.* 7. *Epidemiology and community psychiatry.* 8. *History of psychiatry, national schools, education and transcultural psychiatry.*

More than 100 contributors. V.7 (xxi, 693p.) contains 103 documented papers in 15 sections, plus addendum. The paper 'Psychiatric epidemiology since three years after retirement', p.541-9, carries 22 references. Another, 'On the history of American psychiatry', p.179-81, has 40 references. International in scope. *Class No:* 616.89(055)

[3542]

Yearbook of psychiatry and applied mental health. Chicago, Ill., Mosby Year Book, 1979-. Annual (1990 ed. $54.95). illus., diagrs., graphs, tables.

Continues *Yearbook of neurology, psychiatry and neurosurgery.*

The *1987 Yearbook* (499p. £36.50) has 17 sections, covering the literature received up to June 1986 (*e.g.* 1. Biologic psychiatry; 5. Child and recent psychiatry; 17. Community psychiatry). The topic 'Inpatient care and its alternatives', p.416-24, carries reviews of 8 items. Aims 'to highlight the emerging trends and directions'. Subject and author indexes. *Class No:* 616.89(055)

Histories

[3543]

ALEXANDER, F.G. *and* **SELESNICK, S.T. The History of psychiatry:** an evaluation of psychiatric thought and practice, from prehistoric times to the present. London, Allen and Unwin, 1967. xvi, 471, [1]p. illus.

4 parts (25 chapters): 1. The age of psychiatry - 2. From the ancients through the modern era - 3. The Freudian age - 4. Recent developments. Appendices A-C (C. The organization of psychoanalytic and psychiatric teaching,

....(contd.)

practice and research). Chapter notes, p.415-30. Bibliography (general; 25 chapters), p.431-51. Partly analytical index. *Class No:* 616.89(091)

[3544]

HUNTER, R. *and* **MACALPINE, I. Three hundred years of psychiatry, 1535-1860:** a history present in selected English texts. Hartsdale, N.Y., Carlisle Publishing, Inc., 1982. xxvi, [1], 1107p. facsims. $95. ISBN: 0910177007.

This is a reprint of 1963 ed.

About 300 extracts from original sources, each with a short introduction. Chronological arrangement: Barthomaeus Anglicus, 1535 ... Thomas Laycock, 1860. Lengthy general introduction precedes. Index of names; index of subjects (partly analytical). *Class No:* 616.89(091)

Biographies

[3545]

AMERICAN PSYCHIATRIC ASSOCIATION. Biographical directory of fellows & members of the American Psychiatric Association. New York & London, Bowker; Jaques Cattell Press, 1989. 1956p. $137.50; $93.50. ISBN: 0890421862.

Lists *c.*30,000 American and some foreign psychiatrists, with data on career, specialization and publications. Geographical index. *Class No:* 616.89(092)

Research Methods

[3546]

Research methods in psychiatry: a beginner's guide. Freeman, C. *and* Tyrer, P., *ed.* London, Royal College of Psychiatrists, 1992. 320p. £15. ISBN: 0902241486.

Aimed at all researchers in social and medical science. *Class No:* 616.89:001.891

Communicable Infections & Fevers

[3547]

Abstracts on hygiene and communicable diseases. London, Bureau of Hygiene and Tropical Diseases, 1926-. 12pa. £150; $290.

About 4,500 signed abstracts pa. 3 main sections: Environmental health - Community health - Diseases and their control. Book reviews. Author and subject indexes. Forthcoming meetings. Available online on DIMDI. *Class No:* 616.9

[3548]

A Dictionary of virology. Rowson, K.E.K., *and others.* Oxford, Blackwell Scientific Publications, 1981. v, 230, [2]p. chemical structures. ISBN: 0632006978.

About 2,000 entries ('arbovirus': over 5p.; 'leukaemia': 15 lines, 1 reference). Cross-references. Appended: Taxonomy of vertebrate viruses. Follows the pronouncements of the International Committee for the Taxonomy of Viruses, and intended to provide every vertebrate virus with an approved name. *Class No:* 616.9

[3549]
Infectious diseases in Europe: a fresh look. Velimirovic, B., *and others*. Copenhagen, WHO Regional Office for Europe, 1984. x, 330p. tables, maps. £4. ISBN: 9289010150.

8 review-chapters: 1. Does Europe have a communicable disease problem? ... 3. Figures and trends for selected infectious diseases in Europe, 1950-1980 ... 8. Changes in outlook: communicable as opposed to infectious disease. 10 annexes (*e.g.* on the increase of AIDS in Europe). Many statistical tables and graphs. Bibliographies appended to chapters. No index. *Class No:* 616.9

[3550]
Principles and practice of infectious diseases. Mandell, G.L., *and others eds*. 3rd ed. New York, etc., Churchill Livingstone, 1990. 2340p. illus., diagrs., graphs, tables, chemical structures, maps. $189.95. ISBN: 0443086869.

First published 1979.

Over 240 contributors. 4 parts (297 documented chapters): 1. Basic principles in the diagnosis and management of infectious diseases - 2. Major clinical syndromes - 3. Infectious diseases and their etiologic agents - 4. Special problems. Chapter 149: 'Poliovirus', p.1359-67, with 94 references. Detailed 106-page index. 'This expanded and extensively rewritten 3rd ed. ... all of the chapters have been revised, rewritten and updated' (*Preface*). *Class No:* 616.9

AIDS

[3551]
AIDS information: world literature on human retroviruses and acquired immuno-deficiency systems. Folkestone, Bailey Brothers and Swinfen, 1985-. 12pa. £70. ISSN: 09531580.

V.4(7), July 1988 issue has 154 abstracts and references. 11 sections (*e.g.* Transmission; Treatment; Immunology; Assay methods and virology; 7-cell leukaemia viruses; Public policy, ethical and legal statistics; [worldwide]; Future conferences). Keyword index. Culled from 1,300 international biomedical journals. A current-awareness service; online. *Class No:* 616.98

[3552]
AIDS information sourcebook 1989-1990. Malinowsky, H.R. and Perry, G.J., *eds*. 2nd ed. London, Oryx Press, 1990. 224p. ISBN: 0897445442.

First published 1988.

600 new or updated lists of testing, treatment and counselling centres. Chronology of significant AIDS-related events. Annotated bibliography contains AIDS-related books, pamphlets, films, videotapes and online databases. 3rd ed. 1991-92 (312p. $39.95). *Class No:* 616.98

[3553]
AIDSLINE. Bethesda, Md., National Library of Medicine, 1980-.

Online equivalent of *AIDS bibliography* (1988-. 12pa.). Contains more than 48,000 citations, with abstracts, to international, primarily English-language, literature. Also available on CD-ROM. *Class No:* 616.98

[3554]
Virology and AIDS abstracts. Cambridge, University Press, 1988-. 12pa. $820. (*Cambridge scientific abstracts*.) ISSN: 08965910.

Originally as *Virology abstracts* (1967-87).

1991: *c.*8,800 abstracts drawn from *c.*5,000 primary sources. Sections include: Acquired Immunodeficiency Syndrome - Virus taxonomy and classification ... Immunology - Antiviral agents ... Viral infections of man ... Viral infections of higher plants. Author and subject indexes per issue and annually. Online: DIALOG, ORBIT, etc. *Class No:* 616.98

Surgery

Reviews & Abstracts

[3555]
Excerpta medica. 9. Surgery. Amsterdam, Excerpta Medica, 1947-. 24pa. DFl2,037pa. ISSN: 00144134.

About 7,000 abstracts pa. 26 sections: 1. General aspects ... 5. Nervous system - 6. Head ... 13. Alimentary tract ... 19. Kidney... 21. Circulatory system: vascular disorders ... 24. Instrumentation - 25. Experimental surgery - 26. Surgical development. Subject and author indexes per issue and annually. Online, EMBASE and available on CD-ROM. *Class No:* 617(048)

Progress Reports

[3556]
Recent advances in surgery. Edinburgh, etc., Churchill Livngstone, 1992. illus., diagrs., graphs, tables. ISBN: 0443045690. ISSN: 01438395.

First published 1965.

The 1992 volume (250p.) has 25 contributors. The 14 documented articles include: 'Surgical management of ulcerative colitis' (p.23-38); 'Blood transfusion: indications and hazards' (p.119-36); 'Fluid balance and electrolytic disturbance' (p.209-24). Analytical subject index. *Class No:* 617(055)

Portraits

[3557]
ROYAL COLLEGE OF SURGEONS OF ENGLAND. A Catalogue of the portraits and other paintings and sculpture in the Royal College of Surgeons of England. LeFanu, W. Edinburgh & London, Livingstone, 1960. xii, 119p. illus. (incl. col.).

245 painted or carved portraits of surgeons are recorded, plus a descriptive phrase to indicate their association, if any, with the College. Provenance and sources of information are indicated. Nearly half the entries are illustrated. *Class No:* 617(084.10)

Biographies

[3558]
PLARR, V.G. Plarr's lives of the Fellows of the Royal College of Surgeons of England. Power, Sir D'A., *and others*. Bristol, Royal College; London, Simpkin, Marshall, 1930. 2v.

Records lives of the Fellows from 1843 to those who died before 1930, with citations of their significant publications. Continuation volumes bring the roll of Fellows up to the end of 1973: for 1930-1951, by Sir D'Arcy Power and W.R. LeFanu (1953); for 1952-1964, by R.H.O.B. Robinson and W.R. LeFanu (1970); and for 1965-1973, by J.P. Ross and W.R. LeFanu (1981). *Class No:* 617(092)

[3559]
POWER, Sir D'A. Lives of the Fellows of the Royal College of Surgeons of England, 1930-1951. LeFanu, W.R. London, Royal College of Surgeons of England, 1953. xii, 889p.

Continues Plarr. 'An account of all the Fellows whose death was reported between the beginning of 1930 and the end of 1951'. Entries are detailed and authoritative; information was checked with relatives and the source stated. The biographee's publications are listed and reference made to other published obituaries. Includes lives of some Fellows who died before 1930, omitted from Plarr. *Class No:* 617(092)

[3560]
ROBINSON, R.H.O.B. *and* LEFANU, W.R. Lives of the Fellows of the Royal College of Surgeons of England, 1952-1964. Edinburgh & London, Churchill Livingstone, 1970. [vii], 470p. £3.50. ISBN: 0043006814.

Continues Plarr and Power.

About 800 entries, - brief, accurate and informative, biographical sketches. 'Victor Bonney': *c*.2p., with 12 lines of bibliographical notes. 'Contributions do not always include a succinct account of each individual's contribution to medicine' (*British medical journal*, no.5698, 21 March 1970, p.742). *Class No:* 617(092)

[3561]
ROSS, J.P., *Sir and* LEFANU, W.R. Lives of the Fellows of the Royal College of Surgeons of England, 1965-1973. London, Pitman, 1981. 405p. £10. ISBN: 0272796107.

Continues Plarr, Power and Robinson.

'Contains biographical details of surgeons who have died between 1965 and 1973'. Perhaps of more topical interest than the previous volumes as it includes details of senior colleagues and teachers of surgeons currently in practice. *Class No:* 617(092)

Techniques

Anaesthesia

[3562]
Excerpta medica. 24. Anesthesiology. Amsterdam, Excerpta Medica, 1966-. 10pa. DFl836pa. ISSN: 00144282.

About 2,500 abstracts pa. 6 sections: 1. General aspects (1.21: Resuscitation and intensive care) - 2 General anesthesia 3. Local anesthesia (3.2: Spinal anesthesia) - 4. Special anesthesia - 5. Physiology - 6. Pharmacology. Subject and author indexes per issue and annually. Online, EMBASE and available on CD-ROM. *Class No:* 617-089.5

Orthopaedics

[3563]
Excerpta medica. 33. Orthopedic surgery. Amsterdam, etc., Elsevier, 1956-. 10pa. DFl898pa. ISSN: 00144371.

About 7,000 abstracts pa. 10 sections: 1. General aspects - 2. Experimental orthopedics - 3. Diagnoses, diagnostic procedures and techniques - 4. Treatment - 5. Growth and development - 6. Injury - 7. Infections - 8. Neoplasms - 9. General orthopedics - 10. Orthopedic aspects of systematic disorders. Subject and author indexes per issue and annually. Online, EMBASE; CD-ROM. *Class No:* 617.3

Ophthalmology

Bibliographies

[3564]
BRITISH OPTICAL ASSOCIATION. Library catalogue. London, the Association, 1932, 1935, 1957. 3v.

Consists of author catalogues, with annotated entries. Subject index provides broad groupings. V.3 has *c*.3,000 entries. *Class No:* 617.7(01)

Encyclopaedias & Dictionaries

[3565]
Dictionary of visual science. Cline, D., *and others*. 3rd ed. Radnor, Pa., Chilton Book Co., 1980. xviii, 711p. illus., diagrs., charts.

2nd ed., 1968, edited by M. Schapero, and others.

68 contributors. About 20,000 entries, some of them lengthy (*e.g.* 'Lens': p.346-65). Coverage spans ocular anatomy, ocular physiology, ocular pathology, ocular embryology, illumination and physical optics. Entries include eponyms, organizations and abbreviations.

4th ed., 1989. (xxi,820p. $55). *Class No:* 617.7(03)

[3566]
MILLODOT, M. Dictionary of optometry. 2nd ed. London, Butterworth, 1990. 221p. £16.95. ISBN: 0407022112.

About 3,000 terms used in optometric practice, with definitions, synonyms, abbreviations, examples, and cross-references. Intended for students and practitioners. 3rd ed. to be published 1993 (£19.95). *Class No:* 617.7(03)

[3567]
SARDEGNA, J. *and* PAUL, T.O. The Encyclopedia of blindness and vision impairment. New York, Facts on File, 1991. vii, 329p. $45. ISBN: 0816021538.

Over 500 entries, arranged A-Z. Entries, many of which have bibliographies, vary from a few lines to articles of 1 or 2 pages. Appended list of agencies, institutions, schools, national organizations, rehabilitation centres, disability databases, etc. Extensive bibliography. Name and subject index. For the layperson. *Class No:* 617.7(03)

[3568]

THORNTON, S.P. **Ophthalmic eponyms:** an encyclopedia of named signs, syndromes and diseases in ophthalmology. Birmingham, Alabama, Aesculapius Publishing Co., 1967. 324p.

Entries, A-Z in each of 2 sections: 1. Signs, symptoms and diseases in medical, pediatric, and neuro-ophthalmology - 2. Eponyms in ophthalmic surgery. Sources are often cited. *Class No:* 617.7(03)

Reviews & Abstracts

[3569]

Ophthalmic literature. London, Institute of Ophthalmolgy, 1947-. 4pa. £70. ISSN: 00303720.

More than 4,000 abstracts specially-commissioned pa. V.44(5), 1991 contains *c.*500 abstracts, divided into 2 sections: 1. The eye (18 sections: 1. Lids ... 18. Ocular motility) - 2. General sections (*e.g.* 3. Pathology; 10. Veterinary ophthalmology). Author and subject indexes. New format planned for 1992 onwards and a change of title to *New ophthalmic literature. Class No:* 617.7(048)

Yearbooks & Directories

[3570]

The Opticians register, 1992; comprising the following registers and lists kept by the General Optical Council: The Register of ophthalmic opticians who both test sight and fit and supply optical appliances: The Register of ophthalmic opticians who test sight but do *not* fit and supply appliances, and The Register of dispensing opticians; together with the list of bodies corporate carrying on business as ophthalmic opticians, and the list of bodies corporate carrying on business as dispensing opticians, containing registrations and enrolments up to November 1991. London, General Optical Council. lii,395p.

The lists of opticians are followed by the 'Local list' (p.279-395), in 7 areas, - London postal region, Greater London, England, Wales, Scotland, N. Ireland and outside the UK. *Class No:* 617.7(058)

[3571]

—**International optical year book and diary.** Sutton, Surrey, Consumer Industries Press. Annual. ISBN: 061700451x.

Has its 83rd ed., 1986. (358p. + diary). Contents include 'Professional organisations addresses and activities' p.9-51; 'Company information', p.87-152, and 'Product information', p.167-223. Appended: Trade names; Who's who. Index to articles in *The Optician,* p.335-54. *Class No:* 617.7(058)

Biographies

[3572]

OVIO, G. **Storia dell'oculista.** Genoa, Tipografia Ghibaudo, 1950-52. 2v. (131p.; 1020p.). ports.

1. *Delle origine al 1850.* 2. *Dal 1850 al 1900.*

Biographical data on oculists. Each volume carries an extensive bibliography (v.1: 2,031 references; v.2: 7,795 references) and subject index. *Class No:* 617.7(092)

Gynaecology

Obstetrics

[3573]

ANDERSON, M. **An A-Z of gynaecology,** with comments on aspects of management and nursing. London, Faber & Faber, 1986. 125p. illus. £3.50. ISBN: 0571139663.

Five sections: 1. Anatomy and physiology - 2. Benign disorders - 3. Malignant diseases - 4. Surgical procedures - 5. Hormones and related drugs. Some entries have brief notes on nursing and general management. Further readings, p.117-8: Useful organisations, p.119. Detailed analytical index, p.121-5. *Class No:* 618.2

[3574]

How to find out: information sources in midwifery. Bristol, Midwives Information and Resource Service, 1991. 64p. £2.95. (*MIDIRS fact pack, no. 2.*) ISBN: 0951424106.

14 chapters: 1. Planning your search ... 5. Libraries and information services ... 9. Indexes, abstracting and current awareness services - 10. Computer databases ... 13. Voluntary and self-help organizations - 14. Further reading. *Class No:* 618.2

[3575]

Obstetric-gynecologic terminology, with section on neonatology. Hughes, E.C., *ed.* Philadelphia, Pa., F.A. Davis Co., 1972. xxi, 731p. ISBN: 0803647255.

9 sections, each with an A-Z sequence of entries. 1. Anatomy - 2. Anatomy of generative organs - 3. Diseases and conditions of generative organs - 4. Benign and malignant neoplasmas - 5. Selected gynecologic topics (*e.g.* menopause) - 6. Physiology of reproduction - 7. Obstetrics - 8. Neonatology - 9. Glossary of congenital anomalies. Sources, p.645-6. Abbreviation index. Detailed, analytical index, p.651-731. Well organized. *Class No:* 618.2

[3576]

Obstétrique et gynécologie: glossaire des termes obstétricaux et gynécologiques en français, latin, anglais, russe, allemand, espagnol, italien, grec. Louros, N.C., *and others.* Amsterdam, etc., Elsevier, 1964. [vii], 444p. (*Glossaria interpretum, G8.*)

2,576 French-base terms, with Latin, English, Russian, German, Spanish, Italian and Greek equivalents and indexes. Genders not shown. *Class No:* 618.2

[3577]

PEEL, J. **The Lives of the Fellows of the Royal College of Obstetricians and Gynaecologists, 1929-1969.** London, Heinemann Medical, 1976. xviii, 390p.

An account of the most important activitites of 318 Fellows, as far as it has been possible to collect the necessary details. A list of honorary Fellows precedes. *Class No:* 618.2

Abortion

[3578]

Abortion bibliography for 1986. Goode, P.T., *comp.* Troy, N.Y., Whitston Publishing Co., 1988. 400p. $35. ISBN: 0878753699.

First published 1972 (covering 1970).

2 sequences: Books, government publications and

....(contd.)

monographs; Periodical literature (title index - *c.*2,000 entries; subject index (*e.g.* contraceptives; family planning (both subdivided by countries); sterilization; vasovasectomy). Author index. *Class No:* 618.39

62 Engineering

Engineering

Abbreviations & Symbols

[3579]
Dictionary of engineering acronyms and abbreviations.
Keller, H. *and* Erb, U.   London, Adamintine, 1989.
[v],312p. £55. ISBN: 074490014x.

About 30,000 entries. Aims to cover the major fields of
engineering. Some North American bias (*e.g.* DTI is
identifed as 'Department of Trade and Industry [US]', but
not for the UK). Well-produced with bold type for entry
terms. *Class No:* 620(003)

[3580]
POLON, D.D., *ed.*   **EESS/Encyclopaedia of engineering
signs and symbols.** New York, Odyssey Press, 1965. xviii,
412p. diagrs., tables.

5,000 graphic symbols in standard use by engineers,
draughtsmen and other technicians. Sections: Electrical-
electronics graphic symbols - Material graphic symbols -
Aircraft and spacecraft mechanical graphic symbols -
Architectural and structural graphic symbols ... Engineering
drawings - Color codes - Designations. Index to graphic
symbols, p.403-12. *Class No:* 620(003)

Databases

[3581]
Compendex. New York, Engineering Information Inc., 1970-.

COMPENDEX is the machine-readable database for
Engineering index (now *Engineering index monthly, qv*).
Among its hosts are BRS and DIALOG. Connect charges
are fairly high. V.N. Anderson's 'Searching the engineering
databases' (*Database*, v. 10(2), April 1987) gives a brief
comparative survey. CD-ROM (Dialog).

SHE subject headings in engineering supplies the terms
used in *Engineering index* and its associated COMPENDEX
database. *Class No:* 620(003.4)

[3582]
MOON, M.   'Engineering'. *Manual of online search
strategies.* Editors, C.J. Armstrong and J. A. Large, pp.288-
308. 2nd ed. Aldershot, Ashgate, 1992. ISBN: 1857420071.

Lists 52 engineering databases, with their focus of interest
and host systems. *Class No:* 620(003.4)

Bibliographies

[3583]
ANTHONY, L.J., *ed.* **Information sources in engineering.**
2nd ed. London, Butterworths, 1985. [x], 579p. £45.
(*Butterwoths Guide to information sources.*)

First published as *The use of engineering literature*, K.W.
Mildren (1976. [viii], 621p.).

26 contributors; 28 chapters. 4 main sections: Introduction

.... *(contd.)*
- Primary infomation sources (Reports ... Trade catalogues) -
Secondary information sources (Abstracting and indexing
services; bibliographies and reviews ... standard reference
sources) - Specialized subject fields (chapters 11-28; 11.
Stress analysis ... 28. Nuclear power engineering, with 5p.
of references). 'Index to subjects, information services and
organizations', detailed and analytical, p. 563-79. An
outstanding contribution to the series.'No other source
provides such comprehensive coverage of the literature of
engineering' (*Special libraries*, v. 78(2), Spring 1987,
p.152. *Class No:* 620(01)

[3584]
ENGINEERING SOCIETIES LIBRARY, New York.
Classed subject catalog, Engineering Societies Library.
Boston, Mass., G.K. Hall, 1963. 12 v.; Index (1963. 956p.).
Supplements, 1964-74. 10 v. ISBN: 0816146533.

The main subject catalogue (215,000 cards) is classified by
UDC (modified). It covers all the separate books, pamphlets,
reports and bulletins in the Library. Continued in
Bibliographic guide to technology (*qv*). *Class No:* 620(01)

[3585]
MACKERLE, J.   Finite element methods: a guide to
information sources. Amsterdam, Elsevier, 1991. xii,371p.

Over 800 entries, with extensive annotations. 7 chapters:
1. Books - 2. Conference proceedings ... 5. Journals ... 7.
Databases. 7 appendices (*e.g.* A. 'List of publishers with
addresses', p.295-99). Indexes: author/editor; subject.
Class No: 620(01)

[3586]
TECHNISCHE   INFORMATIONSBIBLIOTHEK,
Hannover.   **Katalog   der   Technischen
Informationsbibliothek (TIB),** Hannover. Microfiche ed.
Mainz-Kastel, N. Gärtner, 1983. (Distribtuion, K.G. Saur).
£1,500. ISBN: 3598304455.

1,254,960 entries on 664 fiches. Includes title entries for
monographs, periodicals, all German dissertations and, in
particular, published and unpublished research reports.
Records the holdings of a leading scientific and technical
library. *Class No:* 620(01)

Encyclopaedias

[3587]
McGraw-Hill encyclopedia of engineering. Parker, S.P., *ed.*
New York, etc., McGraw-Hill, 1983. viii, 1264p. illus.
ISBN: 0070454868.

690 documented articles of specialists, taken from the
McGraw-Hill Encyclopedia of science and technology (5th
ed., 1982) (*qv*), without abridgement. 10 major branches of
engineering are covered: design, electrical, industrial,
mechanical, metallurgical, mining, nuclear, petroleum, and
production. List of contributors; detailed, analytical index.
1,670 good illus. Bibliographies need up-dating.

Electronics and computers are dealt with in a separate
volume (1984). *Class No:* 620(031)

Handbooks & Manuals

[3588]
Eshbach's handbook of engineering fundamentals.
Tapley, B.D., *ed.* 4th ed. New York, Wiley, 1990. [2368p.]
diagrs., charts, tables. $74.95. ISBN: 0471890847.
3rd ed. 1975.

Some 78 contributors. 16 documented sections (*e.g.* 12.
Electronics: 174p. 50 references). Many formulas. Detailed,
analytical index. Covers mechanical, electrical, chemical and
civil engineering. Significantly revised, with an added
chapter on computer science, and a much expanded chapter
on automatic control. Equations, definitions and explanations
are a feature. Bibliographies, although updated, rarely list
post-1980 publications. *Class No:* 620(035)

[3589]
Handbook of reliability engineering and management.
Ireson, W.G. *and* Coombs, C.F., Jr., *eds.* New York,
McGraw-Hill, 1988. v.p. diagrs., graphs, tables. $59.50.
ISBN: 007032039x.

17 contributors. 19 chapters in 4 parts: 1. Introduction to
reliability - 2. Management of reliability - 3. Engineering for
reliability - 4. Mathematics of reliability. Appendices of
tables and charts. Detailed, analytical index, 10p.
Class No: 620(035)

[3590]
Maynard's industrial engineering handbook.
Hodson, W.K., *ed.* 4th ed. New York, McGraw-Hill, 1992.
1872p. illus., diagrs., graphs, tables. £86.95. ISBN:
0070410860.

Data on methods of improvement, work measurement,
planning and control of manufacturing systems, quality
control, facilities and material flow. 800 illus. Detailed,
analytical index. *Class No:* 620(035)

[3591]
SALVENDY, G. Handbook of industrial engineering. 2nd
ed. New York, Wiley, 1992. xxvii,2780p. illus., diagrs.,
graphs, tables. £115. ISBN: 0471502766.

Published in association with the Institute of Industrial
Engineers. First ed. 1982.

Over 150 contributors. 5 parts (108 subsections, with
references): 1. Industrial engineering function - 2.
Technology - 3. Human dimensions - 4. Planning, design
and control - 5. Quantitative methods for decision making.
Detailed, analytical index, p.2705-80. *Class No:* 620(035)

Adolescents

[3592]
MACAULEY, D. *and* ARDLEY, N. The Way things work.
London, Guild Publishing, 1988. 384p. illus (col.). £15.99.
ISBN: 0863183239.

4 parts: 1. The mechanics of movement - 2. Harnessing
the elements - 3. Working with waves - 4. Electricity and
automation (including 'Robots': 2p. spread). Appended:
'Eureka! The invention of machines', p.358-73. Technical
terms, p.374-9; index p.380-4. Suitable for school libraries.
Class No: 620(035)-0053.7

Dictionaries

English

[3593]
Industrial engineering terminology: a revision of ANSI
Z94.0-1982: an American national standard, approved July
10, 1989. Norcross, Ga., Industrial Engineering and
Management Press, Institute of Industrial Engineers, 1990.
535p. $72.95. ISBN: 0898061059.

About 15,000 technical terms, acronyms and
abbreviations. 17 sections, each with bibliography:
Biomechanics ... Manufacturing systems. Detailed,
analytical index, p.489-535. 'Highly recommended' (*Choice*,
April 1991, p.1290).

Also published by McGraw-Hill (1992; £70; 0070317305).
Class No: 620(038)=20

[3594]
McGraw-Hill dictionary of engineering. Parker, S.P., *ed.*
New York, etc., McGraw-Hill, 1985. [xi], 659p. ISBN:
0070454124.

16,300 entries, drawn from the *McGraw-Hill Dictionary of
scientific and technical terms* (3rd ed., 1984). Excludes
chemical, electrical and food engineering terms. 13
categories or 'fields' are applied to terms. Different
applications are numbered. No illus. Has a tight binding, so
that the book does not open flat. *Class No:* 620(038)=20

[3595]
SCHARF, B. Engineering and its language. London, Muller
1971. viii, [1], 398p. illus., tables.

21 subject chapters (1. Engineering materials - 2. Iron and
steel ... 6. Plastics ... 10. Machine tools and cutters ... 15.
Engines ... 20. Electricity and magnetism - 21. Process
instrumentation). Bold type for terms; uses and types
described (*e.g.* 'Switches': p. 278-80). 343 line-drawings.
Appended conversion tables; revision questions. Index
(*c.*4,000 entries). The vocabulary of engineering (shown in
context) for translators and authors. *Class No:* 620(038)=20

German

[3596]
Wörterbuch der industriellen Teknik. [Dictionary of
engineering and terminology.] 5. Aufl. Wiesbaden,
Brandstetter, 1985-89. 2v. £79.95ea.
First published 1948.

1. *Deutsch-Englisch.* 5. Aufl. 1989. 1258p. 2. *Englisch-
Deutsch.* 5. Aufl. 1985. 1422p.

V.1 has *c.*157,000 entries. V.2 has an introduction on
'The nature of technical English'. Regarded as the standard
work in the field. *Class No:* 620(038)=30

French

[3597]
**FORBES, J.R. Dictionnaire des techniques et technologies
modernes:** anglais-francais. [Dictionary of engineering and
modern technology: English-French.] Paris, Technique et
Documentation, 1990. xiv,521p. £53. ISBN: 2852063908.

Over 30,000 entries. *Class No:* 620(038)=40

[3598]
KETTRIDGE, J.O. **French-English and English-French dictionary of technical terms and phrases** used in civil, mechanical, electrical and mining engineering, and allied sciences and industries; with a method of telegraphic coding. Rev. ed., with suppt. by Y.R. Arden. London, Routledge & Kegan Paul, 1980. (reprinted 1989). 2v. £60, set; £35, ea. ISBN: 0415043093, v.1; 0415027683, v.2.

First published 1959.

1. *French-English.* 640p. 2. *English-French.* 672p.

About 100,000 entries for terms and phrases, covering both volumes. Each volume has a supplement and an appendix of abbreviations, symbols and conversion tables. *Class No:* 620(038)=40

Reviews & Abstracts

[3599]
The **Engineering index monthly:** the index to the world's engineering developments. New York, Engineering Index, Inc., (now Engineering Information), 1884-. 12 pa. with annual cumulation. $1,560 pa. ISSN: 01683036.

Until September 1962 as *Engineering index.*

Now produces *c.*22,000 abstracts pa. Coverage includes some conference proceedings. About 75% of articles are English-language material. Author and subject indexes per issue. *The Engineering index annual* (with 4-year cumulations) includes author, author affiliation and cumulative subject indexes, as well as the abstracts themselves. The *Index* is available online as COMPENDEX [COMPuterized ENgineering InDEX], 1970-. A major abstracting service. On CD-ROM (Dialog).
Class No: 620(048)

Periodicals

[3600]
'The **Engineer' index, 1856-1959:** names and subjects. Proctor, C.E., *comp.* London, Morgan, 1964. [vi], 212p.

Index of illustrated articles, plus articles over a ½ column long, excluding editorials, letters, book reviews and other features. Two sequences: 'Names' (p. 1-159) and 'Subjects'. 'Names': names of authors, personal and place names written about, pseudonyms, firms, names of ships, etc. The 'Subjects' index (p. 161-213) is inadequate, - generic rather than specific, and non-analytical. References are to volume and page numbers only. But a representative world-list of 77 libraries holding complete or near-complete sets of *The engineer* is given. Of value to historical researchers in this field. *Class No:* 620(051)

[3601]
The **Innovative engineer:** 125 years of 'The Engineer'. London, Morgan-Grampian, 1981. 232p. illus., diagrs. £8.50.

13 articles to celebrate 125 years of the journal, and of engineering progress (and occasionally disaster). Topics include the London underground, telephones, electric light, the Tay Bridge, Tower Bridge, motor cars and the submarine. Numerous original illustrations are reproduced. *Class No:* 620(051)

Yearbooks & Directories

[3602]
Association of **Consulting Engineers Who's who & year book,** 1990. London, Municipal Journal Ltd., for Association of Consulting Engineers, 1990. 458p. . £34. ISBN: 0900562514.

First published 1946.

Authorized list of members. 12 sections (6. Particulars of individual members, p.27-212; 9. Particulars of firms, p.223-414; 10. Firms with offices overseas; 11. Classification tables; 12. Specialist services and suppliers). *Class No:* 620(058)

[3603]
—International directory of consulting engineers. [Fédération Internationale des Ingenieurs - Conseils.] London, Rhys Jones Marketing Consultants, 1986. 263p. tables.

Includes entries for *c.*200 firm-consultants (1 per page) in 25 countries, Australia ... Yugoslavia. Data on each include international experience, personnel, services offered, fields of specialization, and typical projects. Index of fields of specialization; index of international offices; index of firms (in English and Arabic). *Class No:* 620(058)

[3604]
'The **Engineer' buyers guide,** 1992. 95th ed. Tonbridge, Kent, Morgan-Grampian, 1992. viii,812p. £55. ISBN: 0863821553. ISSN: 01431455.

6 sections: 1. Buyers' guide: general section - 2. Buyers' guide: Machine-tool section - 3. Brand and trade names - 4. Alphabetical index to companies - 5. Foreign companies and their UK agents - 6. Associations, institutions and societies. Indexes: advertisers; products and services. *Class No:* 620(058)

Tables & Data Books

[3605]
COOK, J.L. **Conversion factors.** Oxford Univ. Press, 1992. 160p. tables. $12.95. ISBN: 0198563523.

More than 3,500 conversion factors, arranged by areas of application. SI units. 'Well formatted ... reasonably priced ... a valuable addition to a reference collection' (*Choice,* September 1992, p.74). *Class No:* 620(083)

[3606]
CRC **handbook of tables for applied engineering science.** Bolz, R.E. *and* Tuve, G.L., *eds.* 2nd ed. Cleveland, Ohio, CRC Press 1973. xiii, 1166p. tables.

35 contributors. 11 sections: 1. Engineering materials and their properties - 2. Electrical science and radiation - 3. Chemistry and applications - 4. Nuclides and nuclear engineering - 5. Energy engineering and transport - 6. Mechanics, structures and machines - 7. Environmental and bioengineering - 8. Environmental protection and human safety -9. Communication and computation - 10. Measurement and instrumentation - 11. Processes and control. Appendix: Engineering organizations and publishers (mostly US). Analytical index, p.1119-66. Mostly tables; section references. Omits most of the common mathematical tables. Uses both SI and conventional units. *Class No:* 620(083)

[3607]
GIECK, K. **Engineering formulas.** 5th ed. New York, McGraw-Hill, 1986. v.p. diagrs., tables.

English ed. first published 1967, as *A collection of technical formulae.* 4th. ed. 1983. Original German work as *Technische Formelsammlung* (26. Aufl. 1979).

A classic collection of *c.*2,000 technical and mathematical formulas. 'The material is presented in an extremely terse manner ... [It] may become my most frequently used engineering reference' (*Applied optics*, v. 22, 15 July 1983, p.2123) of the 4th ed. *Class No:* 620(083)

[3608]
—BARNETT, S. *and* CRONIN, T.M. Mathematical formulas for engineering and science students. 4th ed. London, Butterworth, 1986. 77p. $9.95. ISBN: 0582447585.

Covers most of the commonly used mathematical formulas at University and college level. Omits logarithmic tables and other functions, here replaced by some frequently used statistical tables. 'Best for personal purchase' (*New technical books*, January 1987, p.12). *Class No:* 620(083)

[3609]
HICKS, T.G. *and* HICKS, S.D., *eds.* **Standard handbook of engineering calculations.** 2nd ed. New York, etc., McGraw-Hill, 1985. 2v. (xviii, 1468p). illus. ISBN: 007028735x.

First published 1972.

Provides over 5,000 direct and related calculation procedures. Type of engineering dealt with: civil, architectural, mechanical, electrical, elctronics, chemical control, aeronautical and astronautical, marine, nuclear, and sanitary. Also, engineering economics. 793 illus. *Class No:* 620(083)

[3610]
Kempe's engineer's year-book, 1991. Sharpe, C., *ed.* 96th ed. London, Morgan Grampian, [1991]. 2v. [xxii, 3500p.] illus., diagrs., graphs, tables.

About 100 contributors. V.1 has 46 sections, A1-F5 (B6, 'Numerical engineering', has 36p. with ½p. of bibliography). V.2 has 45 sections, F6-L10 (H4, 'Line communication', has 27p. with a brief bibliography). This ed. includes new chapters on 'Brazing', 'Intellectual property law for the engineer' and 'Radio communication underground'. Chapters on 'Microprocessors', 'Radio Communication', 'Airports and air transport' have been dropped. 'Chapter guide to information on environmental issues', p.ix. A major engineering reference tool. *Class No:* 620(083)

[3611]
McGraw-Hill handbook of essential engineering information and data. New York, McGraw-Hill, 1991. 1072p. illus. $89.50. ISBN: 0070227640.

A collection of basic engineering information, data and principles for day-to-day problem solving. Drawn from various McGraw-Hill engineering handbooks, supplemented by explanatory comments. About 50 sections. 760 illus. 'Highly recommended' (*Choice*, March 1991, p.1102). *Class No:* 620(083)

[3612]
TUMA, J.J. **Engineering mathematics handbook.** 3rd ed. New York, etc., McGraw-Hill, 1987. 512p. illus., tables. ISBN: 0070654433.

First published 1970; 2nd ed. 1979.

Main parts: Algebra; geometry; trigonometry; plane analytical geometry; space analytical geometry; elementary functions. Differential calculus; sequences of series; integral calculus; vector analysis; functions of a complex variable; Fourier series; ordinary differential equations; partial differential equations; Laplace transforms. Numerical methods; probability and statistics. Tables of indefinite integrals; tables of definite integrals; plane curves and areas; space curves and surfaces. Aims to serve as a desk-top reference book for engineers, scientists and architects. *Class No:* 620(083)

Standards

[3613]
KVERNELAND, K.O. **World metric standards for engineering.** New York, Industrial Press, 1978. xi, [735p.] diagrs., tables. ISBN: 0831111135.

Aimed to assist engineers, manufacturers, designers, and draughtsmen in North American and other countries where transition to the metric system is taking place; this work provides comparisons of the standards in 8 of the largest industrial countries of the world and is thus of value even in areas where SI units are already in use. 16 of its 20 sections cover components, systems, materials and practices of general engineering interest, but 6 in particular concern mechanical engineering: 8. Seven threads - 9. Fasteners ... 12. Bearings - 13. Mechanical power transmission systems - 14. Fluid power systems and components ... 17. Metal cutting tools. Text, tables and illustrations are adequate for their intended purpose and impressive in coverage. *Class No:* 620(083.74)

Models

[3614]
JACKSON, A. *and* DAY, D. The Model maker's handbook. Driver, E., *ed.* London, Pelham Books, 1981. 352p. illus. (incl. col.), charts, tables.

Cover subtitle: an illustrated manual of over 1,000 techniques for making all types of models, miniatures and diagrams. Sections concern basic techniques (tool kit, materials, etc.); painting and finishing; figures; dioramas and landscapes; railways; motors and engines; radio control; aeroplanes; boats; cars. Useful addresses; glossary. Detailed analytical index, p.344-50. Step-by-step instructions. Hundreds of illus., with descriptive captions. *Class No:* 620(086.48)

Histories

[3615]
ARMYTAGE, W.H.G. A Social history of engineering. [4th ed]. London, Faber & Faber, 1976. 381p. illus. (pl.), ports., tables.

First published 1961.

27 readable chapters, from prehistory onwards, with an amount of biographical data. Appendix A: Chapter bibliographies (p. 337-55); B: List of professional institutions. Index of subjects and plates; index of persons.

.... *(contd.)*

Three aims: 'to chart technological developments with special reference to Britain, to indicate how they affected and seem affected by social life at certain stages and to offer some clues as to the origin of innovations and institutions' (*Preface*). *Class No:* 620(091)

[3616]

CHANNEL, D.F. **The History of engineering science:** an annotated bibliography. New York, Garland, 1989. c.300p. $42. (*Reference library of the humanities, 1150.*) ISBN: 0824066367.

Includes books, articles and dissertations in English, French and German. 'Well-researched ... recommended' (*Library Journal*, January 1990, p.100). *Class No:* 620(091)

[3617]

GARRISON, E. **A History of engineering and technology:** artful methods. Boca Raton, Fl., CRC Press, 1991. 176p. illus., diagrs. $39.95. ISBN: 0849388368.

Engineering advances and their industrial and social significance from antiquity to the present day. Well-illustrated with a useful bibliography. *Class No:* 620(091)

[3618]

HILL, D. **A History of engineering in classical and medieval times.** London, Croom Helm, 1984. [xiv], 263p. illus. (incl. pl.). ISBN: 0709912099.

Describes the important engineering achievements in Europe and western Asia, 600BC-AD1450. 3 parts (12 chapters): 1. Civil engineering - 2. Mechanical engineering - 3. Fine technology (instruments; automata; clocks). Chapter notes (*e.g.* 5. Roads: p.76-82. 29 references). Bibliography, p.248-55. Detailed analytical index. Pays considerable attention to descriptions of techniques and machines. *Class No:* 620(091)

[3619]

NEWCOMEN SOCIETY FOR THE STUDY OF THE HISTORY OF ENGINEERING. **Transactions, 1920/21-.** [Newcomen Society]. London, Newcomen Society, 1922-. v. 1-. illus. (incl. pl.), diagrs., tables.

In addition to historical papers, most volumes carry 'Analytical bibliography of the history of engineering and applied science', arranged under subjects. V.38, for 1965-6 (1968), includes 'Sir Robert Smirke; a pioneer of concrete construction' (p.5-22) and 'Origin of the canning industry' (p.145-52; 28 footnote references). *General index* to v.1-32, 1920-60 (Cambridge, Heffer, 1963) is a name and subject index; items are asterisked if they contain illus. The longest running British journal for the history of engineering and technology. *Class No:* 620(091)

Biographies

[3620]

TURNER, R. *and* GOULDON, S.L., *eds.* **Great engineers and pioneers in technology.** New York, St. Martin's Press, 1982-. v.1-. illus.

1. *From antiquity through the Industrial Revolution.* 1981. 488p.

6 contributors. 373 profiles, in 5 sections: 1. Engineering in the ancient world (prior to 600AD) - 2. Medieval engineering: Islam and Europe - 3. Engineering in the Far East - 4. Engineering in the Renaissance - 5. The Industrial Revolution. Each section comprises an introductory essay, biography and 'further reading'. Chronology, to 1850.

.... *(contd.)*

Bibliographical essay, p.461-5. List of illus. Index of names and subjects, p.475-88. Well produced. 'No other comparable work' (*RQ*, Winter 1982, p.201). *Class No:* 620(092)

[3621]

—The **Biographical dictionary of scientists: engineers and inventors.** Abbott, D., *ed.* London: Blond Educational, 1985. [vi], 189p. diagrs. £12.95. ISBN: 0584700040.

Provides basic facts on about 200 engineers and inventors, p.7-163 (*e.g.* Edison; Brunel - each 2½ columns). Glossary and index follow. Diagram, but no portraits. Aims to fill the gap, at an economic price, between too long and costly or too short and insufficiently informative works of this sort. Criteria for selection of names is queried in *Choice* (v. 24(2), September 1986, p.78). *Class No:* 620(092)

[3622]

—CARVILL, J. **Famous names in engineering.** London, Butterworth, 1981. [x], 93p. illus., ports., table. £8.95. ISBN: 0408005394.

Aims to provide 'brief and ... interesting and informative biographies of 83 famous people whose laws, theories, and inventions form the basis of any advanced course in Engineering' (*Preface*). Names, A-Z. Each biographical note occupies about one page, plus portrait drawing and relevant formula or illus. Index of subjects groups the names. Chronological table, Pythagoras ... Ernst Nussett (died 1957). Bibliography gives further reading on each person. *Class No:* 620(092)

[3623]

—International **who's who in engineering.** Kay, E., *ed.* Cambridge, International Biographic Centre, 1984. [x], 589p. ISBN: 0900332719.

Brief entries, with mention of date of birth, family, education, appointments, memberships, publications, hobbies and addresses. Selected list of engineering institutions, p.583-9, arranged geographically. *Class No:* 620(092)

[3624]

Who's who in science and engineering. New York, Reed Publishing, 1992. 1084p. $199. (*Marquis Who's who.*) ISBN: 0837957516.

22,000 living scientists from 70 countries (80% US). Data: current address, birth date and place, family members, education, career history. *Class No:* 620(092)

Great Britain

[3625]

WALKER, D. *and* PAPADAKIS, A., *eds.* **Great engineers:** a survey of British engineers, 1837-1987. London; Academic Editions/St. Martin's Press, in association with the Royal College of Art, 1987. 288p., illus. (incl. col.). £29.95. ISBN: 0856709174.

22 essays by leading engineers, historians and industrialists. Covers the period from the Industrial Revolution to the present day, including both major 19th century engineers such as Brunel and the Stephensons, as well as 20th-century exponents (*e.g.* Ove Arup and Partners, Felix Samuely and Büro Happold). Added section of biographies of over 60 engineers, a list of great British

....*(contd.)*

inventions, an extensive bibliography, and index. More than 400 illus. (200 in colour). Quarto format. *Class No:* 620(092)(410)

[3626]

—BELL, S.P., *comp.* A Biographical index of British engineers in the 19th century. New York & London, Garland Publishing, 1975. x, 246p. $22.

A key to biographical information (*e.g.* obituary notices) on *c.*3,000 engineers, appearing in 24 periodicals, 1823-1900. 1-2 references each, on average, per entry (Sir Henry Bessemer: 10). *Class No:* 620(092)(410)

[3627]

Who's who of British engineers, 1982-83. 6th ed. London, Simon Books, 1982. 322p. ISBN: 086229004x.

First published 1966.

About 8,000 short entries, with biographical data on engineers in industry, colleges, universities, research and private practice, noting special interests, qualifications, publications and addresses, - supplied by the individuals themselves. Appended list of professional institutions and category index. The 6th ed. gives 'a greater emphasis ... to chemical engineering' (*Preface*). *Class No:* 620(092)(410)

USA

[3628]

Who's who in engineering. Davis, G., *ed.* 6th ed. New York, American Assocation of Engineering Societies, 1985. xv, 846p.

First published 1970 as *Engineers of distinction, including scientists in related fields* (New York, Engineers Joint Council).

About 20,000 entries, A-Z, for North American (mostly US) engineers of distinction. Data include employment, education, achievements, addresses. Engineering societies, p.1-66. Appended indexes by state/province and specialization. *Class No:* 620(092)(73)

[3629]

—ROYSDEN, C. *and* KHARTI, L.A. American engineers of the nineteenth century: a biographical index. New York & London, Garland Publishing, 1978. xx, 247p.

Concentrates - like its companion volume, S.P. Bell's *A biographical index of British engineers in the 19th century* (*qv*) - on obituary notices in journals, transactions and proceedings of professional societies, plus references to such general sources as *Dictionary of American biography*. A quick-reference source. 'Also the obituaries quoted will give some idea of the importance of the person and provide the basic leads for a more detailed investigation' (*Library review*, v. 28, Autumn 1979, p. 189). *Class No:* 620(092)(73)

[3630]

—Who's who in engineering: a biographical dictionary of the engineering profession, 1922/1923-64. New York, Lewis Historical Publishing Co., 1922-64.

The 9th and last ed. (2198p.) appeared in 1964, with 20,000 entries for US engineers and foreign engineers who had undertaken important work in the USA. Preceded by list of US organizations; abbreviations and addenda. Geographical cross-index of persons. *Class No:* 620(092)(73)

Europe

[3631]

Annual review of engineering industries and automation: 1990. Economic Commission for Europe. London, HMSO, 1992. 2v. £35. ISBN: 9211165288.

V.1: General economic developments; analysis of engineering industries (production, investments, manpower, price indices, research and development); national performance; international trade. V. 2: Statistical annex. Covers engineering industries in Japan where appropriate. *Class No:* 620(4)

Writing & Lecturing

[3632]

HASLAM, J.M. Writing engineering specifications. London, Spon, 1988. xvi,172p. diagrs., tables. ISBN: 0419137505; 0419137602, Pbk.

7 chapters, closely subdivided: 1. Context - 2. Presenting - 3. Effect - 4. Writing - 5. Grammar - 6. Models and layout - 7. Claims. Appendix: 'Word processing management'. Index (8p.). Emphasis is practical and commercial, rather than academic. A basic manual. *Class No:* 620:001.81

[3633]

MICHAELSON, H.B. How to write and publish engineering papers and reports. 3rd ed. Phoenix, Az., Oryx Press, 1990. xvi,221p. illus., diagrs., tables. £17.95. ISBN: 0897746503.

Covers writing techniques, parts of the paper (abstracts; tables; citations, etc.), submission to a publisher and technical developments (DTP; networking, etc.). References, p.205-11. Detailed, analytical indedx, p.213-21. *Class No:* 620:001.81

[3634]

VAN EMDEN, J. A Handbook of writing for engineers. London, Macmillan, 1990. 96p. £4.99. ISBN: 0333469429.

Practical guide to writing style for a range of documents in engineering: letters, reports, technical notes, instructions. *Class No:* 620:001.81

Thesauri

[3635]

SHE: subject headings for engineering. New York, Engineering Information Inc., 1987. xviii,218p. ISBN: 0873940190.

First published 1972; supplement, 1977. Previous ed. 1983.

Thesaurus of terms for use with various Engineering Information services, from *Engineering index monthly* to *Energy abstracts* to *Compendex*. Arranged by main headings, followed by subheadings. Numerous 'see' and 'see also' references to preferred indexing terms. *Class No:* 620:025.43

Research Establishments

[3636]

Engineering research centres: a world directory. Archbold, T., *and others eds.* 2nd ed. Harlow, Essex. Longman Group, 1988. viii,599p. £220. ISBN: 0582017785.

Data on more than 3,500 official laboratories, industrial research centres, and educational establishments with research and development programmes. Subject areas extend

....(contd.)

to standards, metrology, computer theory and software, and transport engineering. Arranged by countries, then bodies A-Z. Data include names of senior staff, number of graduate research staff, annual expenditure, activities, publications, products. Index of establishments; subject index.
Class No: 620:061:061.62

Education Institutions

[3637]

UNITED NATIONS. Educational, Scientific and Cultural Organization. **Directory of engineering education institutions:** Africa, Arab States, Asia, Latin America and the Caribbean. 3rd ed. Paris, Unesco, 1986. xiv, 303p. ISBN: 9230023930.

First published 1962.

Information on 781 degree-awarding institutions in 61 countries. *Class No:* 620:061:37

Engineering Design

[3638]

BRITISH LIBRARY. Science Reference Library. **Useful books of reference for designers** (1926-1983) held by the Science Reference Library. Palyza, M.M., *ed.* London, SRL, 1984. 3v. ea.£5.

1. *Metrication; physical units; databooks for various materials; use of computers in engineering; Hand- and databooks for designers in civil engineering.* [iv], 163p.

2. *Hand- and databooks for designers in electrical and electronic engineering.* [iii], 71p.

3. *Hand- and databooks for designers in mechanical engineering.* [iii], 121p.

Each of the 3 parts carries *c.*1,000 references. Subject index precedes. *Class No:* 620.001.6

[3639]

CULLUM, R.D. **Handbook of engineering design.** London, Butterworths, 1988. [vii], 303p. illus., diagrs. graphs. £59.50. ISBN: 0408005580.

19 contributors. Part 1 (chapters 1-11, each with bibliography: 1. Stages in design; 2. Engineering materials; 3. Stress analysis; 4. Bearings; 5. Fastenings; 7. Design aspects of production processes) - Part 2 (chapters 12-16; *e.g.* 13. Structure and organization in design offices). Directory; addresses. Detailed analytical index, p.301-3. Well illustrated. *Class No:* 620.001.6

[3640]

—PERRY, R.H., *ed.* **Engineering manual:** a practical reference of design methods and data in building systems, chemical, civil, electrical, mechanical and environmental engineering and energy conversion. 3rd ed. New York, etc., McGraw-Hill, 1976. x, 965p. graphs, charts, tables.

First published 1959.

19 contributors. 10 sections (4. Building systems, engineering has 111p., 86 tables). Some footnote references. Analytical subject index. Aims 'to summarise practical, easily used design methods and requisite data across the spectrum of engineering'. Stresses practice; lacks references on computers and their application. *Class No:* 620.001.6

[3641]

FRENCH, M.J. **Invention and evolution:** design in nature and engineering. Cambridge, Univ. Press, 1988. xvi,324p. illus., diagrs. £35. ISBN: 0521307597; 0521314925, Pbk.

Covers numerous elements of design for function. 10 chapters, closely subdivided (*e.g.* 2. Energy ... 3. Materials ... 5. Structures (beams; feathers, etc.) ... 8. Economy, form and beauty ... 10. Designing and inventing). Detailed, analytical index, p.318-24. *Class No:* 620.001.6

[3642]

INSTITUTION OF ENGINEERING DESIGNERS. **Official reference book and buyers' guide.** London, Sterling Publications, 1991. 248p. illus. £30.

19 brief articles (*e.g.* 'Designing for quality' (p.46-52)). Classified buyers' guide: Engineering design tools and computers - Materials ... Coatings and finishes - Power transmission ... Production and manufacturing technology - Test measurement and instrumentation - Professional services. *Class No:* 620.001.6

[3643]

MUCCI, P. **Handbook for engineering design using standard materials and components.** 3rd ed. Southampton, P.E.R. Mucci, 1990. vi,240p. illus., tables. ISBN: 0951155520.

First published 1986; 2nd ed. 1987.

10 chapters (*e.g.* 2. Materials ... 5. Bearings ... 9. Mechanical components (seales; springs), most including names and address of manufacturers and suppliers. 7 appendices (D. Fits and tolerances for holes, p.220-21). Index, p.236-40. Much tabular information.
Class No: 620.001.6

Materials Science

Databases

[3644]

Engineered materials abstracts. Metals Park, Oh., ASM International; London, Institute of Metals.

Corresponds to the print publication of the same name. Covers the science of polymers, ceramics and composite materials and the practices of materials science and engineering. Nearly 100,000 records; updated monthly. Available on DIALOG, file 293, and on CD-ROM (DIALOG OnDisc, *METADEX*). *Class No:* 620.02(003.4)

Encyclopaedias

[3645]

BEYER, M.B., *ed.* **Encyclopedia of materials science and engineering.** Oxford, etc. Pergamon Press, 1986. 8v. (6105p.). illus., diagrs.

More than 1,400 contributors. 1,580 signed, documented articles on 45 broad subject areas (*e.g.* metals, ceramics, polymers, magnetic materials, welding). Some articles, designated 'An overview', introduce and survey a subject, furnish systematic and comprehensive bibliographic references to the entire subject, and are extensively cross-referenced (*e.g.* 'Advanced ceramics: an overview', v.1, p.98-105: 4 sections, 3 diagrs., 9 illus., 16 references). V.8, *Index*, includes systematic outline of the *Encyclopedia*, list of

....*(contd.)*

contributors, citation index, subject index (*c.*50,000 entries), materials information sources (p.527-42, supplementing article references), and list of acronyms. Aims to produce a comprehensive work to 'serve the needs of education and industry' (*Introduction*). Supplementary v.2 (1990) carried 130 articles (*e.g.* 'Germanium', p.931-35, with a bibliography of 18 items). *Class No:* 620.02(031)

[3646]
Encyclopedia of materials characterization. Brundle, C., *and others, eds.* Oxford, Butterworth Heinemann, 1992. v.p. illus., diagrs., tables. £75. ISBN: 0750691689.

Covers techniques (description, data obtained, use, future trends) used in modern materials analysis. *Class No:* 620.02(031)

[3647]
GRAYSON, M., *ed.* **Encyclopedia of composite materials and components.** New York, etc., Wiley, 1983. xxviii, 1161p. illus., diagrs., graphs, tables.

Reprints of more than 50 articles from Kirk and Othmer's *Encyclopedia of chemical technology*. Ablative materials ... Sealants. Many data tables, standards, specifications and literature references. Detailed analytical index, p.1109-61. Designed as a definite and comprehensive guide to the properties, manufacturing methods and components. *Class No:* 620.02(031)

[3648]
International encyclopedia of composites. Lee, S.M., *ed.* New York, etc., VCH, 1990. 6v. £156 per v. ISBN: 0895732904, set.

Includes more than 300 contributors. Lengthy, documented articles ('Aramid fiber composites', p.37-56, 120 references), with cross-references. Each volume includes a list of 'main entries'. Topics covered fall into 4 groups: Special types, *e.g.* engineering materials; Behaviour under varying conditions; Those with special properties; Design aspects. Of value to practising scientists and engineers and advanced-level students. A major work. *Class No:* 620.02(031)

Handbooks & Manuals

[3649]
ASM Engineered materials reference book. Metals Park, Oh., Metals Information, 1989. [x],517p. diagrs., graphs, tables. ISBN: 0871703505.

4 sections: 1. Composites - 2. Ceramics - 3. Plastics - 4. Electronic materials. 17 appendices (*e.g.* 'Automated literature searching', p.485-88). Numerous tables. *Class No:* 620.02(035)

[3650]
BOLTON, W. Newnes engineering materials pocket book. Oxford, Heinemann Newnes, 1989. diagrs., tables. xiii,194p. £10.95. (*Pocket Book.*) ISBN: 043490113x.

Covers materials - ferrous; aluminium; copper; magnesium; nickel; titanium; polymeric; ceramic; composite - in separate chapters. Each includes discussion of material itself, details of coding systems and compositions, heat treatment information, properties, typical uses. Appendix: conversion tables. Index, 6p. Small format. Intended for students with project work and practising engineers. *Class No:* 620.02(035)

[3651]
BRADY, G.S. *and* **CLAUSER, H.R. Materials handbook:** an encyclopedia for managers, technicians, supervisors and foremen. 13th ed. New York, etc., McGraw-Hill, 1991. 1056p. [1]p. charts, tables. $74.50; £60. ISBN: 0070070741.

First published 1929, 12th ed. 1986.

More than 15,000 entries, A-Z. Materials covered include butter, cotton, linoleum, plastics, rubbers, tea, water softeners and zirconium. Data: full description; definition; use; relationships to other materials. Index, 81p. 'Highly recommended' ('Reference books bulletin', *Book list,* 1 December 1991, p.722). *Class No:* 620.02(035)

[3652]
—CLAUSER, H.R., *ed.* Encyclopedia/Handbook of materials, parts and finishes. [New ed]. Westport, Conn., Technonic Publishing Co., 1978. iii, 564p. graphs, tables.

First published 1963.

Over 150 contibutors. Covers ceramics, metals, plastics and wood, concentrating on basic characteristics and applications. List of contributors and their articles. Detailed, analytical index, p.553-64. *Class No:* 620.02(035)

[3653]
BROUTMAN, L.J. *and* **KROCK, R.H.,** *eds.* **Composite materials.** New York & London, Academic Press, 1974. 8v. illus.

1: *Interfaces in matrix composites.*
2: *Mechanics of composite materials.*
3: *Engineering application of composites.*
4: *Metallic matrix composites.*
5: *Fracture and fatique.*
6: *Interfaces on polymer matrix composites.*
7-8: *Structural design and analysis.*

V.6 (xiii, 421p.) has 10 sections, 11 contributors. Section 7: 'Effect of interface on fracture' (p.245-85) has 2p. of references. Each volume has its own author and subject indexes. *Class No:* 620.02(035)

[3654]
Elsevier materials selector. Waterman, N.A. *and* Ashby, M.F., *eds.* London, Elsevier Applied Science, 1991. 3v. (2228p.), diagrs., graphs, tables. ISBN: 1851666052, set, UK; 0849377900, set, US.

Includes information on (1) Selection and specification of materials and manufacturing routes for a new product and (2) Evaluation of alternative materials & manufacturing routes for an existing product. Covers a variety of materials (*e.g.* steels, aluminium alloys, ceramics, coatings, nylons, PVC) and testing procedures. *Class No:* 620.02(035)

[3655]
Engineered materials handbook: prepared under direction of the ASM International Handbook Committee. Metals Park, Oh., ASM International, 1987-. v.1-. illus. ISBN: 0871702797, v.1.

V. 1. *Composites.*
V. 2. *Engineering plastics.*
V. 3. *Adhesives and solvents.*

Comprehensive, with charts and graphs. Each volume has its own subject index and glossary. *Class No:* 620.02(035)

[3656]
LUBIN, C., *ed.* **Handbook of composites.** New York, etc., Van Nostrand Reinhold, 1982. x, [1], 786p. illus., graphs, tables, chemical formulas. ISBN: 0442248976.

Sponsored by the Society of Plastic Engineers.

4 sections (28 documented chapters): 1. Raw materials - 2. Processing methods - 3. Design - 4. Applications. Chapter 24, 'Testing of reinforced concretes', has 17 references and 1p. of bibliography. Glossary, p.753-78. 3 Appendices (A. Typical properties of reinforced composites). Non-analytical index, p.779-86. *Class No:* 620.02(035)

[3657]
MALLICK, P.K. *and* NEWMAN, S., *eds.* **Composite materials technology:** process and properties. Oxford Univ. Press, 1990. 480p. illus., diagrs., tables, graphs. $98. ISBN: 0195208471.

13 contributors. 10 documented chapters covering sheet, compression, reaction injection molding, resin transfer molding, filament winding pultrusion, injection molding and quality systems. 'A good reference' (*Choice*, September 1991). *Class No:* 620.02(035)

[3658]
Materials and technology: a systematic encyclopaedia of the technology of materials used in industry and commerce, including foodstuffs and fuels. London, Longman; Amsterdam, de Bussy, 1968-76. 8v. illus. (incl. pl.), diagrs., tables.

Based on J.F. Van Oss's *Warenkennis en Technologie* (5th ed. 1948-50). Updated and rewritten.

US title as *Chemical technology: an encyclopedic treatment* (New York, Barnes & Noble, 1968-76, 8v.).

1. *Air, water, inorganic chemicals and nucleaonics.* 1968.

2. *Non-metallic ores, silicate industries and solid mineral fuels.* 1971.

3. *Metals and ores.* 1970.

4. *Petroleum, organic chemicals and plastics.* 1972.

5. *Natural organic materials and related mineral fuels.* 1972.

6. *Wood, paper, textiles, plastics and photographic materials.* 1973.

7. *Vegetable food products and luxuries.* 1975.

8. *Edible oils and fats, and animal food products.* 1975.

V.2 (xxviii, 828p.) has 12 chapters (1. Rock-forming minerals and rocks ... 6. Glass ... 12. Energy (p.704-96): 7 sections with 39 figures; 15 tables; 10 graphs; references). Index, p.797-828. Comprehensive, but now calls for updating. *Class No:* 620.02(035)

[3659]
SCHWARTZ, M.M. **Composite materials handbook.** 2nd ed. New York, etc., McGraw-Hill, 1992. v.p. $79.50. ISBN: 0070558191.

First published 1983.

Covers component materials, composite systems, fabrication methods, joining and quality control. *Choice* (November 1992, p.500) identifies a lack of 'state-of-the-art techniques for the mechanical analysis of composites' as a shortcoming, but recommends the work as an 'excellent addition'. *Class No:* 620.02(035)

[3660]
SMITH, C.O. **The Science of engineering materials.** 3rd ed. Englewood Cliffs, N.J., Prentice-Hall, 1986. xii, 579p. illus., diagrs., graphs, tables. £49.20. ISBN: 0137948840.

First published 1969. 2nd ed. 1977.

10 chapters on design and selection, bonding, behaviour, applications development and failure, analysis of materials. Most chapters end with a number of 'questions and problems' for solution and discussion. 5 appendices of data tables. Bibliography. Subject index. Covers the basic materials science needs in a first course for engineering students. *Class No:* 620.02(035)

[3661]
Treatise on materials science and technology. Herman, H., *ed.* Orlando, Florida, London, etc., Academic Press, 1972-. (v.1-27. 1972-88). illus., graphs, charts, tables. ISBN: 0123418011.

V. 10, pt. B: *Properties of solid polymeric materials* contains a paper 'Electronic properties of polymers', p.637-76, with 17 charts and graphs, 3 tables and 3p. of references. 4v. (v.12, 17, 22, 26) have so far been devoted to glass. V.27 (xi, 493p.): *Analytical techniques for thin films* (1988). *Class No:* 620.02(035)

Dictionaries

[3662]
HOLMSTROM, J., *and others.* **Trilingual dictionary for materials and structures.** Oxford, etc., Pergamon Press, 1971. xxxvi, 947p.

Published with the assistance of Unesco. At head of title-page: 'RILEM. International Union of Testing and Research Laboratories for Materials and Structures'.

8,063 numbered terms (English terms, equivalents and indexes in French and German, and definitions in all three languages). Grouped: A. General aspects - B. Metals - C. Cement and concrete - D. Siliceous, calcareous, agrilaceous materials - E. Soils - F. Wood - G. Other organic materials. - H. Water. Bibliography of sources, p.821-4. *Class No:* 620.02(038)

[3663]
Macmillan dictionary of materials and manufacturing. John, V., *ed.* London, Macmillan, 1990. vi,431p. diagrs., formulas. £35. ISBN: 0333455584.

6 contributors; *c.*3,500 terms, A-Z. Definitions range from 2 lines to 1 column. Cross-references. Covers metallic, ceramic, polymeric and composite materials; includes processes and manufacturing systems, where appropriate. *Class No:* 620.02(038)

Reviews & Abstracts

[3664]
Engineered materials abstracts. London, Institute of Metals and Metals Park, Ohio, ASM International, 1986-. 12pa. $1075; £750, full price (discount for members). ISSN: 09519998.

Covers polymers, ceramics and composite materials for use in the design, construction and operation of structures, equipment and systems. 7 sections: 1. Fundamental characteristics - 2. Laboratory techniques and quality control - 3. Properties - 4. Raw materials - 5. Fabricating and

.... *(contd.)*

finishing - 6. Engineering design and application - 7. Other coverage. 5 indexes per issue: subject; trade names; materials; author; corporate author. *Class No:* 620.02(048)

[3665]
Key abstracts: Advanced materials. London, Institution of Electrical Engineers, 1987-. Monthly. ISSN: 09504753.

About 3,500 abstracts pa. Covers methods of preparation, surface treatments, and materials testing in 3 major sections: 1.0 Composite materials - 2.0 Ceramics, refractories and glasses - 3.0 Polymers and plastics. Subject index per issue. *Class No:* 620.02(048)

Progress Reports

[3666]
Annual review of materials science. Palo Alto, Calif., Annual Reviews Inc., 1970-. Annual. illus., graphs, tables.

V. 21 (1991. 450p.) includes papers on 'NMR methods for solid polymers', 'Molecular films' and 'Quasicrystals', among others. Cumulative indexes for contributing authors and chapter titles. *Class No:* 620.02(055)

Tables & Data Books

[3667]
CRC handbook of materials science. Lynch, C.T., *ed.* Boca Raton, Florida, CRC Press, 1974-80. 4v. illus., tables.

1. *General properties.* 1974. 782p.

2. *Metals, composites, and refractory materials.* 1975 [vi], 440p.

3. *Nonmetallic materials and applications.* 1975. [xi], 642p.

4. *Wood.* 1980. [xi], 459p.

Mostly tabular data. Thus, v.4 has tables on p.72-402, standard definitions of wood terms, and forest product associations (USA). Each volume has an analytical index, with a general index and an index to specific names. *Class No:* 620.02(083)

[3668]
—Practical handbook of materials science. Lynch, C.T., *ed.* Boca Raton, Florida, CRC Press, 1989. [viii],636p. tables. £33.50. ISBN: 084933702x.

Data largely taken from *CRC Handbook of materials science,* and updated where necessary. Sources identified. 13 sections: 1. The elements - 2. Elemental properties - 3. Physical properties of compounds - 4. Conversion tables - 5. Miscellaneous materials properties - 6. Ceramics - 7. Composites - 8. Electronic - 9. Graphite - 10. Metallic - 11. Nuclear - 12. Polymeric - 13. Material information. Detailed, analytical index, p.619-36. *Class No:* 620.02(083)

[3669]
HOULDCROFT, P.T., *ed.* **Materials data sources.** Compiled for the Materials Group, the Institution of Mechanical Engineers, and sponsored by the Institute of Metals. London, Mechanical Engineering Pubns., for the Institution of Mechanical Engineers, 1987. vi, 111p. tables. £10. ISBN: 085298636x.

6 sections: 1. Metals and alloys - 2. Refractories, ceramics, glasses and hardmetals - 3. Polymers and composites - 4 Timber - 5. Databases and materials selectors - 6. Educational establishments. Appendices: A. Addresses

.... *(contd.)*

of standards organizations; B. Joining and adhesive bonding consumables. Small format. For quick-reference. *Class No:* 620.02(083)

[3670]
MORRELL, R. Handbook of properties of technical and engineering ceramics. National Physical Laboratory. London, H.M. Stationery Office, 1985-. pt.1-. illus. ISBN: 0114800529; 0114800537.

Part 1 (1985. 360p. £33) initiates a multi-part handbook by providing a broad comparison of the different types of ceramic. Part 2, *Data reviews,* section 1: *High alumina ceramics* (1987. 270p. £29.50) deals with the properties of a group of widely used types of materials. *Class No:* 620.02(083)

India

[3671]
The Wealth of India: a dictionary of Indian raw materials and industrial products. New Delhi, Council of Scientific & Industrial Research, 1948-76. 20v. illus.

Raw materials. Industrial products.

Industrial products, v.7: Plastics - shuttles (1971. xvi, [1] 5, 367, 12p.) has 29 lengthy articles and an index. Bibliography, p.xi-xiv. Well illustrated. *Class No:* 620.02(540)

Information Services

[3672]
REYNARD, K.W. UK materials information sources 1989. London, The Design Council, 1989. viii,183p. £18. ISBN: 0850722357.

About 650 entries, A-Z, covering government departments and research laboratories, trade and professional organizations, commercial companies, publishers, higher education and consultants. Includes name, address, telephone, fax, contact and in some, but certainly not all, cases brief note of areas of interest and publications. Index of subjects, acronyms, trade and individual names, published titles. No criteria given for selection of entries; *e.g.* most UK universities are listed, but not Bristol, York, Stirling, Sussex. Handy, but much of what is here is available in the *Aslib directory (qv). Class No:* 620.02:061:025.5

Conferences

[3673]
LOVEDAY, M.S., *and others, eds.* **Measurement of high temperature mechnical properties of materials:** based on the edited proceedings of a symposium held from 3-5 June, 1981, at the National Physical Laboratory, Teddington, England. London, H.M. Stationery Office, 1982. xiii, 351p., illus., diagrs., graphs, tables. ISBN: 0114800499.

25 contributors. 20 sections (1. High temperature and time-dependent mechanical testing: an historical introduction … 20. Specimen manufacture). Documented (*e.g.* 9. Uniaxial testing apparatus and test pieces, p. 128-57: nearly 2p. of references). Author and subject indexes. *Class No:* 620.02:061:061.3

Research Establishments

[3674]

Materials research centres: a world directory of organizations and programs in materials science. 4th ed. Harlow, Essex, Longman Group, 1991. 798p. £250. ISBN: 0582081246.

First published 1983. 3rd ed. 1989.

Data on over 5,000 research units in more than 80 countries. Coverage: industrial chemistry and chemical process engineering, hydrocarbon processing, refined technology, metallurgy, synthetic materials and fibres, composite materials, wood technology and fine chemicals. Subject and title of establishments indexes.

Class No: 620.02:061:061.62

Research Projects

[3675]

BUNSELL, A.R. *and* **KELLY, A.,** *eds.* **Composite materials:** a directory of European research. Sevenoaks, Kent, Butterworth Scientific, 1985. x, 160p. ISBN: 0408221658.

Lists individuals, research establishments and research topics. 9 countries: Great Britain, France, Italy, Belgium, Netherlands, Denmark, Sweden, Germany, Norway. Scope: research on fibres, matrix composite types, testing, analysis and design, fabrication and processing, application, non-destructive testing. For students and recent graduates looking for research oportunities or interesting employment in the composite field. *Class No:* 620.02:061:061.62.005

Materials Testing

[3676]

AMERICAN SOCIETY FOR TESTING AND MATERIALS. Annual book of ASTM standards. Philadelphia, Pa., the Society, 1939-. Annual. 66v. diagrs.

16 sections (*1986 Annual book*): 1. *Iron and steel products.* 6v. 2. *Nonferrous metal products.* 6v. 3. *Metals test methods and analytical procedures.* 6v. 4. *Construction.* 9v. 5. *Petroleum products, lubricants and fossil fuels.* 5v. 6. *Paints, related coatings, and aromatics.* 3v. 7. *Textiles.* 2v. 8. *Plastics.* 4v. 9. *Rubber.* 2v. 10. *Electrical insulation and electronics.* 5v. 11. *Water and environmental technology.* 4v. 12. *Nuclear, solar and geothermal energy.* 2v. 13. *Medical devices.* 1v. 14. *General methods and instrumentation.* 2v. 15. *General products, chemical specialities and end use products.* 9v. 16. *Index.* 1v.

Early 1993 sections include 1. *Iron and steel products.* 7v. (v.01.07: 'Shipbuilding').

Annual series that includes 'all current formally approved ASTM standards and tentative test methods, definitions, recommended practices, classifications, and specifications, and other related material, such as proposed methods' (*Introduction*). Each part has its own index, v.16 being the general index. Internationally recognized for quality and up-to-dateness of its standards (some lengthy). Supplemented by monthly *ASTM standardization news. Class No:* 620.1

[3677]

AMERICAN SOCIETY FOR TESTING AND MATERIALS. Compilation of ASTM standard definitions. Sponsored by ASTM Committee E-8 on Nomenclature and Definitions. 3rd ed. Philadelphia, Pa., the Society, 1976. vii, 731p. diagrs.

Formerly as *Glossary of ASTM definitions.*

A compilation of all terms listed under the heading

.... *(contd.)*

'Definitions' in the 1973 ed. of the *Annual book of ASTM standards.* Numerous cross-references. Excludes symbols and abbreviations. Entries consist of concise definition plus designation of the ASTM standard containing them.

Class No: 620.1

[3678]

AMERICAN SOCIETY FOR TESTING AND MATERIALS. Directory of testing laboratories, 1990-91. Philadelphia, Pa., ASTM, 1989. 274p. $50. ISBN: 0803112211.

Covers US, Canada, other countries. A-Z index and subject indexes. *Class No:* 620.1

[3679]

NATLAS year book. National Testing Laboratories Accreditation Scheme year book, 84/85. 3rd ed. Teddington, Middx., NATLAS Executive, National Physical Laboratory, 1984. 206p.

11 sections: 1. The UK National Testing Laboratory Accreditation Scheme ... 4. Other national schemes ... 11. List of NATLAS Accreditation laboratories. Data: name; address; telephone; telex; registration number. Advertisers index. *Class No:* 620.1

[3680]

NDE Handbook: non-destructive examination methods for condition monitoring. Böving, K.G., *ed.* London, Butterworths, 1989. 417p. illus., diagrs., graphs. £70. ISBN: 040804392x.

Covers 36 tests, each with NDE principle, development method, areas of applications, practical examples, training desired or required and references. Detailed, analytical index, p.413-17. A practical approach. *Class No:* 620.1

Strength

[3681]

TEKNISKA NOMENKLATURCENTRALEN. Hallfasthetsoralista: svenska-engelsk-fransk-tysk-dansk. Stockholm, the Centre, 1977. 230p. ISBN: 9171960686.

850 Swedish terms concerning strength of materials, with equivalents and indexes in English, French, German and Danish. *Class No:* 620.17

Stress, Vibration etc.

[3682]

BRITISH STANDARDS INSTITUTION. Glossary of terms relating to mechanical vibration and shock. London, BSI, 1991. 59p. (*BS 3015: 1991.*)

Defines terms used in studying and testing the effect of mechanical vibration and shock on human beings and on equipment. *Class No:* 620.178.3

[3683]

ROARK, R.J. Roark's formulas for stress and strain. Young, W.C., *ed.* 6th ed. New York etc., McGraw-Hill, 1989. 763p. illus., tables. $36.25. ISBN: 0070725411.

First published 1938. 5th ed. 1976.

3 sections: 1. Definitions - 2. Facts, principles and methods - 3. Formulas and examples. Extensive chapter bibliographies. Appendix lists factors of stress. A classic in the field. *Class No:* 620.178.3

[3684]
The Shock and vibration digest. Willowbrook, Il., Vibration Institute, 1969-. 12pa.

Published by Shock and Vibration Information Center, Naval Research Laboratory, 1969-1986.

About 2500 abstracts pa. 12 sections: Mechanical systems - Statistical systems - Mechanical components - Statistical components ... Analysis and design - General topics. Prelims. include feature article, review article, book reviews, notes of short courses, reviews of meetings. Appended list of periodicals scanned. Annual indexes of feature articles, review aricles, book reviews, author and subject indexes. *Class No:* 620.178.3

Non-Destructive

[3685]
BRITISH STANDARDS INSTITUTION. Glossary of terms used in non-destructive testing. London, BSI, 1984/85. 5 pts. (*BS 3683: 1984/85.*)

1. *Penetrant flaw detection.* 1985. 8p. £10.30.
2. *Magnetic particle flaw.* 1985. 16p. £17.90.
3. *Radiological flaw detection.* 1984. 20p. £23.90.
4. *Ultrasonic flaw detection.* 1985. 20p. £23.90.
5. *Eddy current flaw detection.* 1985 (1983). 12p. £10.30.

Makes particular reference to the terms likely to be mentioned in technical reports. *Class No:* 620.179

[3686]
The NDT year book: the official year book of the British Institute of Non-Destructive Testing, 1987/8. Northampton, The Institute, 1987. 189p. tables. ISSN: 09522395.

10 sections: 1. The British Institute of Non-Destructive Testing - 2. NTD methods and capabilities - 3. Buyers guide - 4. Information (periodicals; books; lectures; conferences and exhibitions; other information sources) - 5. Research - 6. Standards - 7. Education - 8. Certification - 9. The National NDT Centre - 10. Other bodies concerned with NDT. Index of advertisers. *Class No:* 620.179

Metallography

[3687]
SMITH, C.S. A History of metallography: the development of ideas on the structures of metals before 1890. Chicago, Ill., Univ. of Chicago Press, 1960. xxi, 291p. illus.

5 sections: 1. The background of metallography in works of art - 2. The philosophic background - 3. Observations of structure before Sorby - 4. The work of Henry Clifton Sorby (1863-87) - 5. The period of active observation. 110 good illus. ('Kris': p. 35-39, with 8 illus.). Chapter references, p.261-80. 'Highly recommended' (*Nature*, v.192 (4805), 2 December 1961, p.543-4). *Class No:* 620.18

Defects & Deterioration

Corrosion

[3688]
Corrosion abstracts. Houston, Texas, National Association of Corrosion Engineers, 1951-. 6pa. $170 - $250 pa. ISSN: 00109339.

8 sections: 1. General - 2. Testing - 3. Characteristic corrosion phenomena - 4. Corrosion environments - 5. Preventive measures - 6. Materials of construction - 7.

....(contd.)
Equipment - 8. Industries. About 700 abstracts per issue 'of the world's corrosion control literature'. Annual Subject and author indexes per issue. *Class No:* 620.193

[3689]
Corrosion control abstracts. London, Scientific Information Consultants, 1966-. Monthly.

English translation of *Referativnyĭ zhurnal. Korroziya* ...

About 800 abstracts and references per issue. 12 sections (*e.g.* General section; Corrosion theory; Gas corrosion; The corrosion resistance of metals and alloys; Protective metallic coatings and chemical substance treatment; Paint, polymer and other coatings; Corrosion research test methods). *Class No:* 620.193

[3690]
SCHWEITZER, P.A. Corrosion resistance tables: metals, plastics, nonmetallics and rubbers. 2nd ed. New York & Basle, Dekker, 1985. xvi, 1231p. tables. ISBN: 0824775414.

Substances: Acetaldihyde ... Zinc sulfate. Data are tabulated under 12 heads. Temperature range: 60 deg. F. to 5,460 deg. F. Online as *Corrosion database*, available on SDC orbit. *Class No:* 620.193

[3691]
SHREIR, L.L., *ed.* **Corrosion.** 2nd ed. London, Newnes-Butterworths, 1976. 2v. (2300p). diagrs., charts, tables.

First published 1963. 1. *Corrosion of metals and alloys.* 2. *Corrosion control.*

101 contributors. 22 sections (21. Tables; 22. Glossary of terms, symbols and abbreviations). Bibliographies appended to sections of chapters (*e.g.* 20.11, 'Corrosion testing of organic coatings', has 269 references). V.2 carries an index to both volumes. Good layout and typography. *Class No:* 620.193

[3692]
TEKNISKA NOMENKLATURCENTRALEN. Korrosions-ordlista. Norkiska termer... [Glossary of corrosion terms in Scandinavian languages, with equivalents in English, French, German.] Stockholm, the Centre, 1968. 130p. (*TNC 40.*)

Glossary of 326 Swedish terms, with equivalents in Danish, Finnish, Norwegian, German, English and French, and definitions in Scandinavian languages and English. Indexes are in Danish, Finnish, Norwegian, German, English and French. 'Appears to be rather deficient in terms relating to relevant chemical compounds as well as in methods and materials used for the protection of metals and alloys against corrosion' (*The Incorporated linguist*, v.9 (2), April 1970, p.51). *Class No:* 620.193

Bibliographies

[3693]
GABE, D.R. 'Corrosion, corrosion-resistance and protection'. *Information sources in metallic materials.* Editor, M.N. Pattern, p.253-63. London, Bowker-Saur, 1989. (*Guides to information sources.*) ISBN: 0408014911.

Sections: Introduction - Learned societies - Dictionaries and glossaries - Standards organizations - Data sources (handbooks; trade associations, etc.) - Books (63 items) - Current research (abstracts; journals; reviews; conferences). *Class No:* 620.193(01)

Handbooks & Manuals

[3694]

Corrosion handbook: corrosive agents and their interaction with materials. Behrens, D., *ed*. Weinheim, VCH for DECHEMA, 1987-. V.1-.

9 volumes have appeared so far. V.9 (1991. ix,375p.) covers methanol and sulfur dioxide and investigates metallic, non-metallic inorganic, and organic materials as well as those with special properties. Detailed, analytical subject index, p.349-75). *Class No:* 620.193(035)

[3695]

Handbook of corrosion data. Craig, B.D. Metals Park, Oh., ASM International, 1989. xiii,683p. graphs, tables. £117. ISBN: 0871703610.

Data was obtained from the corrosion literature published between 1985 and 1987. In 2 parts: 1. Corrosion of metals and alloys (p.1-101) - 2. Corrosion media (p.105-683). No index, but a list of 'expanded contents'.
Class No: 620.193(035)

[3696]

KUHN, A.T., *ed*. **Techniques in electrochemistry, corrosion and metal finishing:** a handbook. Chichester, Wiley, 1987. xviii,567p. illus., tables. ISBN: 047191407x.

8 contributors. In 3 parts: A. Introduction to experimental methods - B. Specialized techniques (Section I: Methods based on light; Section II: Spectroscopic methods; Section III: Other methods (*e.g.* physiochemical; radio-chemical; acoustic emission; dyes, etc.) - C. The literature (37p. bibliography of post-1960 monographs, most from the British Library Science Reference and Information Service). 28 chapters, most with extensive references (*e.g.* 'Raman spectroscopy', p.237-49, includes 166). Subject index. Dedicated - gracefully - to the 'Library' and its 'staff'.
Class No: 620.193(035)

Protection

[3697]

GREAT BRITAIN. Department of Trade and Industry. **Wear resistant surfaces in engineering:** a guide to their production, properties and selection. London, H.M. Stationery Office, 1986. vi. 171p. illus., diagrs. graphs, tables. £18.95. ISBN: 0115138269.

11 documented sections: 1. Types of wear - 2. Comparison of surface treatment processes - 3. Applications - 4. Thermal hardening - 5. Thermochemical treatment - 6. Electrochemical treatment - 7. Chemical treatment - 8. Chemical vapour deposition - 9. Physical vapour treatment - 10. Spraying processes - 11. Weld hardfacing. 3 appendices (1. Wear test methods). Detailed analytical index, p. 168-71.
Class No: 620.197

Energy (Sources)

Databases

[3698]

Energy line. New York, Bowker.

The online version of *Energy information abstracts*. About 80,000 records; updated monthly. Available on DIALOG, file 69. *Class No:* 620.9(003.4)

[3699]

Energy science and technology. Washington, D.C., US Department of Energy.

Energy topics included: nuclear, wind, fossil, geothermal, tidal, solar. Also covers the environment, energy policy and conservation. Over 2,500,000 records; updated biweekly. Available on DIALOG, file 103. *Class No:* 620.9(003.4)

[3700]

EUROPEAN COMMUNITIES COMMISSION. Directory of energy databases. Luxembourg, the Commission, 1985. (*Eur 9097.*)

Lists more than 100 databases available through European hosts. *Class No:* 620.9(003.4)

[3701]

LUEDTKE, J.R. 'Energy and the environment' *Manual of online search strategies.* Editors, C.J. Armstrong and J.A. Large, pp.236-87. 2nd ed. Aldershot, Ashgate, 1992. ISBN: 1857420071.

Extensive list of databases, both online and CD-ROM, which cover energy and the environment. Hosts and disc suppliers identifed. *Class No:* 620.9(003.4)

CD-ROM

[3702]

World energy CD-ROM. Harlow, Essex, Longman, 1992. £1750.

Covers oil and gas, mining and chemical companies. 30,000 biographical entries, accounts, financial data, major activites, etc. IBM PC. *Class No:* 620.9(003.40)

Bibliographies

[3703]

ANTHONY, L.J., *ed.* **Information sources in energy technology.** London, etc., Butterworths, 1988. xii, 324p. £35. (*Butterworths Guides to information sources.*) ISBN: 040803050x.

11 contributors. 3 parts (16 documented chapters): 1. Energy in general - 2. Fuel technology - 3. Specific energy sources (solid, liquid and gaseous fuels; nuclear energy; solar and geothermal energy; alternative energy sources). Detailed, analytical 'Index to subjects, information services and organizations'. Well up to the Butterworths' series standards. Some overlap with the companion *Information sources in engineering*. *Class No:* 620.9(01)

[3704]

—Current energy information. London, Dept. of Energy, 1986-. Weekly. £25 pa. ISSN: 09512691.

Devotes its first section to reproduced contents pages of *c*.16 selected energy-related periodicals received by the Dept. of Energy Library. Second section, 'New titles', on coloured paper and normally included fortnightly, provides a selected list of new books, technical reports, pamphlets, government reports, plus brief articles. 13 headings: Coal ... Renewable energy resources. Miscellaneous. Appended: Dept. of Energy new releases; forthcoming meetings, conferences and exhibitions; periodical titles currently included. *Class No:* 620.9(01)

[3705]

—GREAT BRITAIN. HM Stationery Office. Energy catalogue, '86. London, H.M. Stationery Office, 1986. [i], 34, [1]p. *gratis.*

Brief annotated entries, with prices, in 7 sections: Statutes in force - Policy and perspective - Energy sources (oil; gas; electricity; petrol; coal) - Nuclear energy - Research and alternative technology - Conservation - Statistics and reference. Index, p.26-29. *Class No:* 620.9(01)

[3706]

—LAMBERT, C.M., *comp.* Energy. London, Library Association, Public Libraries Group, 1981. 32p. £1.80. (*Readers' guide, no. 32.*) ISBN: 0853656746.

Includes 225 mostly briefly annotated entries in 10 sections. A. Energy: general works - B. Energy conservation - C. Conventional energy sources - D. Renewable forms of energy - E. Energy and the environment - F. Energy in other countries - G. Books for children and for schools - H. Reference works (general; bibliographies; research; statistics) - I. Journals - J. Series. A distinctly helpful checklist. *Class No:* 620.9(01)

[3707]

Energy bibliography & index. Houston, Texas, Gulf Publishing Co., 1978-83. 6v. in 7.

Lists material on energy - including synthetic fuel - consisting of documents captured from the Germans in World War II, and available in translation, in the collection of Texas A & M University. V.6 (2pts.) comprises the index. *Class No:* 620.9(01)

[3708]

GREAT BRITAIN. Department of Energy. **Publications in print 1985.** Insert, Additions ... New entries covering the period January-August 1986. Rev. ed. London, Dept. of Energy, 1986. [v], 116p.; [i], 11, 8p. ISBN: 0904552268.

About 800 entries in all. Sections: Alternative energy sources - Coal - Combined heat and power/District heating - Electricity - Energy efficiency and consumption (domestic; general; industrial) - Energy policy - Gas - Nuclear - Offshore activity - Oil - Research and development - Statistics. Appendices: A. Energy papers; B. Offshore technology; C. R & D reports sponsored by ETRU [Energy Technology Research Unit]; D. Severn tidal power reports. Index, p.77-116. *Class No:* 620.9(01)

[3709]

WEBER, R.D. Energy information guide. Santa Barbara, Calif., & Oxford, ABC-Clio, 1982-84. 3v. (xvi, xix, xxi, 1355 [1]p). ea. $39.95.

1. *General and alternative energy sources.* 2. *Nuclear and nuclear power.* 3. *Fossil fuels..*

3,082 annotated entries in all for English language reference works on production, distribution, storage and/or consumption of energy. The chapters in v.1 are subdivided by form. Online bases are cited. Four indexes: author, title, subject and government document index. V.3 carries a cumulative index. US slanted. 'Fills a serious gap in the literature' (*RQ*, Winter 1982, p.200-1). *Class No:* 620.9(01)

[3710]

—WEBER, R.D. Energy update: a guide to current reference literature. San Carlos, Calif., Energy Information Press, 1991. 455p. $42.50. ISBN: 096285185x.

Supplements the author's *Energy Information Guide.* Concentrates on readily available reference literature. 6 categories: 1. Dictionaries and encyclopedias - 2. Handbooks and manuals - 3. Directories - 4. Statistical sources - 5. Indexes, abstracts and bibliographies - 6. Online databases (75). 4 indexes: author; title; subject; document number. Updates are planned every four years. 'Well-researched and carefully planned' (*Choice,* November 1991, p.424). *Class No:* 620.9(01)

Encyclopaedias

[3711]

McGraw-Hill encyclopedia of energy, by the staff of the 'McGraw-Hill Encyclopedia of science and technology'. Parker, S.P., *ed.* 2nd ed. New York, etc., McGraw-Hill, 1980. [ix], 838p. ISBN: 0070452687.

First published 1976.

About 300 contributors. 'Energy perspectives', consisting of 6 articles (*e.g.* 'Outlook for full reserves'), precedes 'Energy technology', p.65-791, an A-Z sequence of signed and documented articles (*e.g.* 'Internal Combustion engine', p.336-49, with 8 references, 21 figures). 3 appendices (3: Energy-related publications). Analytical index, p.823-38. Many of the articles appear in the 1977 edition of *McGraw-Hill Encyclopedia of science and technology.*
Class No: 620.9(031)

[3712]

—GLASSTONE, S. Energy desk book. New York, etc., Van Nostrand Reinhold, 1983. xii, 453p. illus. diagrs.

400 entries A/Z, a combined glossary, encyclopedia. Covers all energy sources, with emphasis on general principles rather than technology. Length of entries ranges from 15 lines ('Benfield process')— to 40p. (on solar energy, including its collection, application, conversion). Cross-references. Intended for scientists, engineers, administrators, government officials. *Class No:* 620.9(031)

Handbooks & Manuals

[3713]

CASSEDY, E. *and* **GROSSMAN, P.Z. Introduction to energy:** resources, technology and society. Cambridge Univ. Press, 1990. 338p. illus., diagrs., tables. £30; $66. ISBN: 0521350913.

Fossil fuels (coal, oil, natural gas) and nuclear fission; economic, health and safety criteria, and decision-making. Diagrs., tables and photos. Chapter notes and references. *Class No:* 620.9(035)

[3714]

DRYDEN, I.G.C. The Efficient use of energy. 2nd ed. Sevenoaks, Kent, Butterworth Scientific, in collaboration with the Institute of Energy, on behalf of the UK Dept. of Energy, 1982. xiii, 602p. illus., diagrs. graphs, tables. ISBN: 0408012501.

First published 1973 (xiii, 602p.), - an update of the Ministry of Power's *Efficient use of energy* (HMSO, 1958).

More than 60 contributors. 6 sections (20 chapters): Production of heat - Use of heat - Production of mechanical and electrical energy - Total energy systems and heat

....*(contd.)*

salvage - Materials and control - Environmental aspects. Chapter references. Appendices A-F (A. Natural fuels and their chief marketable products; E. World energy resources and demands). Analytical index, p.593-602. A major textbook that reflects the British viewpoint, with stress on conservation. *Class No:* 620.9(035)

[3715]
HUNT, V.D. **Handbook of energy technology;** trends and perspectives. New York, etc., Van Nostrand Reinhold, 1982. [viii], 1018p. illus., diagrs., tables, maps. ISBN: 0442225555.

7 parts: 1. Energy technology overview - 2. Fossil energy - 3. Solar, geothermal, electric and storage systems - 4. Fusion energy - 5. Nuclear waste management - 6. Magnetic fusion energy - 7. Reference information, p.965-92 (*e.g.* glossary; abbreviations and acronyms; references - a bibliography, in 9 sections). Detailed, analytical index, p.993-1018. US slanted. *Class No:* 620.9(035)

[3716]
STOUT, B.A. **Handbook of energy for world agriculture.** London and New York, Elsevier, 1990. xiii,504p.

9 chapters (each with summary): 1. Overview - 2. Energy use - 3. Energy flow - 4. Energy arrangement - 5. Energy from biomass (p.197-307) - 6. Solar energy - 7. Wind energy - 8. Hydropower - 9. Alternatives. Appendix 1: Sources of further information (organizations); 2: Conversion of units. Detailed, analytical index. Bibliography (A-Z author), p.466-94. *Class No:* 620.9(035)

[3717]
WILBUR, L.S., *ed.* **Handbook of energy systems engineering:** production and utilization. New York, etc., Wiley, 1985. xiii, 1775p. illus., diagrs., graphs, tables. ISBN: 0471866334.

98 contributors. 16 chapters (most of them with references and bibliographies) cover a wide spectrum of energy applications. These include demographics, coal, nuclear, petroleum and gas technologies, hydroelectric and solar power, electricity, advanced energy systems and engineering mathematics. A massive volume. Intended for professionals as a 'mini primer' for decision-making (*Preface*). *Class No:* 620.9(035)

Dictionaries

Polyglot

[3718]
WORLD ENERGY CONFERENCE, London. **Energy terminology.** [Terminologie de l'energie / Energie terminologie / Terminologia de la energia.] 2nd ed. Oxford, etc., Pergamon Press, 1986. xxiii, 539p. ISBN: 0080340717.

First published 1983.

About 1,500 defined terms and concepts in English, French, German, Spanish, across the double page. Arranged in 19 broad subject areas (general; forecasting methodology; solid fuels; solar energy; conservation; etc.). Alphabetical indexes in the four languages. *Class No:* 620.9(038)=00

[3719]
—TEKNISKA NOMENKLATURCENTRALEN. Energiordlista. [Glossary of energy.] Stockholm, the Centre, 1984. 112p. (*TNC 81*.)

Includes *c.*700 Swedish-base terms (with definitions), plus equivalents and indexes in English, French, German, Danish and Norwegian. *Class No:* 620.9(038)=00

English

[3720]
CRABBE, D. *and* McBRIDE, R., *eds.* **The World energy book:** an A-Z, atlas and statistical source book. London, Kogan Page, 1978. 259p. diagrs., graphs, tables, maps. ISBN: 0850380812.

6 contributors. A-Z dictionary of *c.*1,000 terms and definitions, p.9-194 ('Fusion reactor': 3 columns, 1 diagr., 1 graph). Cross references. 24 energy resources maps. 4 Statistical appendices (1: Figures, to 1976; 4.2: Natural and manufactured hydrocarbons derived from petroleum). 'Intended as a comprehensive reference guide to energy sources, energy related technology, economics and all factors related to the search for, extraction of, and utilization of the major alternative sources of energy' (*Introduction*). *Class No:* 620.9(038)=20

[3721]
KUT, R. **Dictionary of applied energy conservation.** London, Kogan Page; New York, Nichols, 1982. 214p. illus. ISBN: 0850385776.

About 2,000 entries, with numerous cross-references. 'Solar chimney': 1¼ columns; 'Induction heating': 1½ columns. 120 fine line-drawings are a feature. 3 appendices of conversion factors, units and symbols. The author is a consultant engineer. *Class No:* 620.9(038)=20

[3722]
SLESSER, M., *ed.* **Dictionary of energy.** 2nd ed. London, Macmillan, 1988. 318p. diagrs., tables. £29.95; £10.95. ISBN: 0333454618; 0333454626, Pbk.

First published 1982.

About 2000 terms, some defined at length. Includes organization, laboratories and acronyms. Numerous cross-references. Good coverage of economic terms as they apply to energy. *Class No:* 620.9(038)=20

[3723]
—BRACKLEY, P. Energy and environmental terms: a glossary. Aldershot, Hants., Gower, 1988. 200p. diagrs., tables. £18.95. ISBN: 056605759x.

Provides a layman's guide to hundreds of words, acronyms, agreements, organizations, chemical names, etc. *Class No:* 620.9(038)=20

[3724]
—COUNIHAN, M. A Dictionary of energy. London, Routledge & Kegan Paul, 1981. x, 157p. diagrs., graphs, tables. ISBN: 0710008473.

Nearly 1,000 entries, 'Absolute zero' ... 'Zincaloy'. 'Geothermal energy': 3p., including 2 graphs. Appendix: 'International energy statistics', p.150-5. Bibliography (A-Z authors), p.156-7. For the student, general reader and professional. *Class No:* 620.9(038)=20

[3725]

—GILPIN, A. *and* WILLIAMS, A. Dictionary of energy technology. Guildford, Surrey, Butterworth Scientific, 1982. [viii], 392p. diagrs., tables, maps.

First published as *Dictionary of fuel technology*, 1970.

Its 2,500 entries depict energy in its contemporary setting. Includes formulas and adequate cross-references. 2 appendices of conversion factors and a bibliography, p.387-92. For professionals. *Class No:* 620.9(038)=20

[3726]

—HUNT, V.D. Energy dictionary. New York, etc., Van Nostrand Reinhold, 1979. vii, [1], 518p. illus., diagrs., graphs, charts, tables. ISBN: 0442273959.

Contains more than 4,000 concise entries, 'Abatement' ... 'Zone refining'. Has the advantage of a bibliography (9 sections), p.509-18 and is well furnished with diagrams, etc. Appended conversion factors and glossary of acronyms. US slanted. *Class No:* 620.9(038)=20

Reviews & Abstracts

[3727]

Energy information abstracts. New Providence, N.J., Bowker, 1981. 12pa., with annual cumulations. $990 pa. ISSN: 01476521.

About 14,000 abstracts pa. in broad subject order. Subject, author, source indexes per monthly issue. Available on CD-ROM (Bowker) as *Enviro/Energyline abstracts PLUS*. *Class No:* 620.9(048)

[3728]

Energy research abstracts. Oak Ridge, Conn., U.S. Department of Energy, Technical Information Centrer, 1977-. 24pa. $146(domestic); $182.50(foreign). ISSN: 01063604.

Began as *ERDA reports abstracts* (1975), changed to *ERDA research abstracts* (1975-76), then *ERDA energy research abstracts* (1976-77).

About 60,000 abstracts pa., covering 'all scientific and technical reports, journal articles, conference papers and proceedings, books, patents, theses and monographs originated by the US Dept. of Energy, its laboratories, energy centers and contractors'. Sections: Coal and coal products - Petroleum - Natural gas - Oil shales and tar sands - Nuclear fuels ... Solar energy - Geothermal energy - Wind energy ... Energy management and policy ... Materials ... Biomedical sciences, applied studies ... Physics research. Corporate, author, subject, contract number, report number indexes per issue; semi-annual index. *Class No:* 620.9(048)

[3729]

The Engineering index. Energy abstracts: a monthly review of energy technology. New York, Engineering Information, Inc., 1974-. 2pa. ISSN: 00938404.

About 12,000 abstracts pa., drawn from *Engineering index monthly (qv)*. Main sections: 01. Coal and coal products - 02. Petroleum - 03. Natural gas ... 14. Solar energy ... 20. Electric power engineering - 21. Nuclear power plants ... 39. Energy, conservation, consumption, and utilization ... 53. Environmental-social aspects of energy technologies ... 70. Controlled thermonuclear research. Author, author affiliation and subject indexes per issue. Covers both conventional and alternative energy sources. *Class No:* 620.9(048)

[3730]

Fuel and energy abstracts: a bimonthly summary of world literature on all technical, scientific, commercial and environmental aspects of fuel and energy. Guildford, Butterworth Scientific, on behalf of the Institute of Energy, 1978-. 6pa. £170 pa. ISSN: 01406701.

Formerly *Fuel abstracts and current titles/* FACTS (1960-), succeeding monthly *Fuel abstracts* (1945-58).

About 8,000 abstracts pa. Classified sections: 01. Solid fuels - 02. Liquid fuels - 03. Gaseous fuels ... 06. Electric power operation and utilization ... 15. Environment - 16. Fuel science and technology ... 18. Energy conversion and recycling. Subject and author indexes per issue and pa. Coverage: 'World literature on scientific, technical, commercial and environmental aspects of fuel and energy'. Highly regarded. *Class No:* 620.9(048)

Periodicals

[3731]

VIOLA, J., *and others*. Energy research guide: journals, indexes and abstracts. Cambridge, Mass., Ballinger Publishing Co., 1983. xi, 284, [1]p. ISBN: 0884100979.

Master list of over 500 English-language periodicals, indexes and abstracting journals, followed by 7 subject lists (energy use and conservation; renewable and alternative energy services; oil and gas; coal; selective power, including nuclear; environment; world energy). Descriptions of periodicals, indexes and abstracts, A-Z, p.57-284 (data: frequency, scope, price, address and telephone number). *Class No:* 620.9(051)

[3732]

—GREAT BRITAIN. Department of Energy Library. Periodicals list. London, the Dept. Library, 1986. [45]p.

Lists about 800 titles currently held, noting frequency and holdings. 3 sections: Dept. of Energy Library, Thames House South, London; Offshore Suppliers Office Library, Glasgow; Gas and Oil Measurement Branch, Leicester. Entries for abstracting/indexing services are asterisked. Cross-references. *Class No:* 620.9(051)

Progress Reports

[3733]

Annual review of energy. Palo Alto Calif., Annual Reviews, Inc. 1976-. ISBN: 0824323165.

V.16 (1991. 578p. £46) has articles by a range of contributors, *e.g.* 'Clean coal technology and advanced coal-based power plants'; 'Safety and environmental aspects of fusion energy'; 'Soviet energy dilemma and prospects', etc. Subject index. Cumulative indexes of contributing authors; chapter titles. *Class No:* 620.9(055)

Yearbooks & Directories

[3734]

Energy world reference book and buyer's guide. London, Manor House Press, Ltd., for the Institute of Energy. ISSN: 02674939.

The 1985 *Guide* has sections: General information - Useful addresses - Energy conversions - Energy production and supply - Buyer's guide to energy equipment and services (p.109-278) - Manufacturers and suppliers. Index. *Class No:* 620.9(058)

Tables & Data Books

[3735]

BALACHANDRAN, S., *ed*. **Energy statistics:** a guide to information sources. Detroit, Mich., Gale, 1980. xii, 272p. $28. (*Natural world information guide series, v.1*.) ISBN: 0810314193.

A-Z subject/keyword list of *c*.40 'most used national and international energy serials', followed by a section giving full bibliographical data on the serials so analysed. A third section is an annotated guide to additional material on specific sources of energy. Appended directory of publishers and subject index to the third section. US slanted.
Class No: 620.9(083)

[3736]

LOFTNESS, R.L. **Energy handbook.** 2nd ed. New York, etc., Van Nostrand Reinhold, 1984. 703p. illus., diagrs., graphs, tables. ISBN: 0442259921.

First published 1978.

A sourcebook of tabular data, supported by 16 chapters of text on energy resources, consumption patterns, fossil fuel, energy conversion, efficiency and transport, environment, energy costs, futures. No bibliography as such, but footnote references, as required. The author is Director of the Washington Office of the Electric Power Research Institute.
Class No: 620.9(083)

[3737]

—DORF, R.C. The Energy factbook. New York, etc., McGraw-Hill, 1980. vii, [1], 226p. illus., diagrs., graphs, tables. ISBN: 007017623x.

21 short chapters, replete with captioned illus., graphs and statistics (up to 1978). Chapters include: 6. Petroleum; 12. Nuclear fission power and fusion research; 16. Energy storage systems and alternate technologies; 17. Conservation of energy; 21. Energy policy. 'The book is intended for anyone who needs to make decisions that in part require information about energy supplies and uses' (*Preface*).
Class No: 620.9(083)

[3738]

—OSBORN, P.D. Handbook of energy data and calculations, including directory of products and services. Sevenoaks, Kent, Butterworths, 1985. 287p. graphs, charts, tables. ISBN: 0408013273.

Includes data charts and tables, as well as calculation and analysis procedures. Appended bibliography and sources.
Class No: 620.9(083)

Worldwide

[3739]

INTERNATIONAL ENERGY AGENCY. **Energy statistics** and main series from 1960. [Statitstiques de l'energie, 1970-1985.] Paris, Organization for Economic Co-operation and Development, 1987. 2v. (xl, 513p. + xv, 624p.).

V.1. OECD; IEA; North America; Pacific; OECD Europe; EEC; Australia ... Iceland. V.2, Ireland ... Yugoslavia. The tables include all 'commercial' sources of energy, both primary and secondary (coal products, manufactured gases, petroleum products and electricity). 'Other solid fuels' covers peat, wood and waste (*General notes*). *Class No:* 620.9(083)(100)

[3740]

—BP Statistical review of world energy. London, BP, 1987. [i], 38p. illus., diagrs., tables. *gratis*.

Covers, in turn, oil, natural gas, nuclear energy, hydro-electricity and primary energy. Appended conversion factors and definitions. A handy pictorial quick-reference source.
Class No: 620.9(083)(100)

[3741]

UNITED NATIONS. **Energy statistics yearbook.** diagrs., graphs, tables. ISBN: 9210611020.

Formerly *World energy supplies, 1952-1978*; *Yearbook of world energy statistics, 1979-1981*; *Energy statistics yearbook*, 1982-.

The 1984 Yearbook (1986. liv, 441p.) has sections: Commercial energy - Production, trade and consumption, 1981-1984 - Solid fuels - Liquid fuels - Gaseous fuels - Electrical energy - Nuclear fuels, energy resources and prices. 39 tables for 1981-84. Bibliography (countries, A-Z), p.xxx-liv. *Class No:* 620.9(083)(100)

Europe

[3742]

Energy statistical yearbook. Luxembourg, Eurostat. ISBN: 9282568733.

The 1985 *Yearbook* (1986. lxxxvii, 200p. £13.20) provides statistics on coal, oil, gas and electricity for 1980/1985, covering 12 European countries (Germany, France, Italy, Netherlands, Belgium, Luxembourg, UK, Ireland, Denmark, Greece, Spain, Portugal). Main parts: Energy economics: characteristic features - Energy indicators - Balance sheet: energy supplies - Tables by energy sources. Updated in *Energy monthly statistics*.
Class No: 620.9(083)(4)

Great Britain

[3743]

GREAT BRITAIN. Department of Energy. **Digest of United Kingdom energy statistics, 1990.** London, H.M. Stationery Office, 1990. 134p. diagrs., charts, tables, maps. £15.75. ISBN: 0114134235.

Commenced with Ministry of Power *Statistical digest for the years 1948 and 1949* (1950).

8 sections: Energy - Coal and other solid fuels - Oil and gas resources - Petroleum - Gas - Electricity - Prices and values - Foreign trade. Explanatory notes on tables, plus footnotes. Annex: Long term trends in energy.
Class No: 620.9(083)(410)

Biographies

[3744]

Energy and nuclear sciences international who's who. 3rd ed. Harlow, Essex, Longman Group, 1990. 340p. £150. (*Longman Reference on research series*.) ISBN: 0582039681.

Previously as *International who's who in energy and nuclear sciences* (1983).

Part 1 contains professional biographical profiles of about 3,000 scientists and engineers from over 75 countries involved in the generation, storage and efficient use of energy. Part 2 is a country index, subdivided by subject specialists. *Class No:* 620.9(092)

Information Services

[3745]

World directory of energy information. Compiled by Cambridge Information and Research Services, Ltd. Farnborough, Hants., Gower, 1981-84. 3v. ISBN: 0566021956; 0566023741; 0566023873.

    1. *Western Europe.* 1981. [x], 323, [2]p.

    2. *Middle East, Africa and Asia/Pacific.* 1982. xii, 418p.

    3. *The Americas, including the Caribbean.* 1984. x, 360p.

    A guide to organizations and publications worldwide concerning energy development and management. Each volume has 4 parts, - the energy framework; country reviews; energy organizations; energy publications (with indexes for titles, subjects and countries, and publishers). Statistics need to be updated, as available in the United Nations, *Energy statistics yearbook (qv).*

Class No: 620.9:061:025.5

[3746]

—LAMBERT, C.M. *and* LARKE, C., *comps.* Energy: sources of information on energy, with particular reference to the environment. London, Dept. of the Environment & Transport Library, 1979. 117p. (*DOE/DTp information series, no.3.*)

    In 8 sections: 1. Central government depts. - 2. Parliamentary bodies - 3. The nationalized fuel and oil industries - 4. Official and advisory bodies - 5. Other organizations - 6. Research and development - 7. European, international and US energy organizations - 9. Other sources of information (books; abstracting journals; periodicals; statistical sources; directories; reference books and yearbooks). Detailed index. A comprehensive survey, now somewhat dated, for UK responsibilities.

Class No: 620.9:061:025.5

Institutions & Associations

[3747]

World energy and nuclear directory. 8th ed. Harlow, Essex, Longman, 1990. 574p. £250. ISBN: 0582079330.

    Combines former *World energy directory* (2nd ed. 1985) and *World nuclear directory* (7th ed. 1985).

    Records 1,500 research and technology laboratories in the nuclear sciences, plus 2,000 industrial, official, academic and independent organizations, laboratories and consultancies carrying out or promoting development work in non-atomic energy generation and distribution. Establishment and subject indexes.

Class No: 620.9:061:061.2

Conferences

[3748]

Energy and the environment in the 21st century. Tester, J.W., *and others.* Cambridge, Mass., MIT Press, 1991. xix,1006p. $50. ISBN: 0262200783.

    Proceedings of the conference held at the Massachusetts Institute of Technology. The 2 plenary sessions comprise sections A-F: A. Transport systems - B. Industrial processes - C. Building systems - D. Electric power for the developing countries - E. Economics and policy - F. Advanced energy supply technologies. Numerous tables and charts. Detailed, analytical index. 'A valuable contribution to all academic library collections' (*Choice,* March 1992, p.1099).

Class No: 620.9:061:061.3

Energy Conservation & Exploitation

Bibliographies

[3749]

SPURGEON, K. Renewable energy and its environmental impact: an information pack. London, Institution of Electrical Engineers, Technical Information Unit, 1991. *c.*90p. £33. ISBN: 0852964935.

    Abstracts of technical papers from INSPEC database. Sections: Geothermal - Tidal - Ocean thermal energy conversion (OTEC) - Hydroelectric - Solar - Biomass - Wind. Further chapters on books, equipment suppliers, professional and trade association. *Class No:* 620.92(01)

Handbooks & Manuals

[3750]

TWIDELL, J.W. *and* **WEIR, A.D.** Renewable energy sources. London, Spon, 1986. xx, 439p. illus., diagrs., graphs, tables. £14.95. ISBN: 0419120009.

    16 documented chapters: 1. Principles of renewable energy ... 3. Heat transfer ... 5. Solar water heating ... 8. Hydro-power - 9. Power from the wind ... 11. Biofuels ... 13. Tidal power - 14. Ocean thermal energy conservation - 15. Geothermal energy - 16. Energy storage and distribution. Appendices of units, data and formulas. Detailed analytical index, p.425-39. 'Primarily intended to support courses for undergraduates in physical science and engineering beyond first year level' (*Preface*).

Class No: 620.92(035)

Dictionaries

[3751]

EUROPEAN PARLIAMENT. Terminology Office. **Terminology of new and renewable sources of energy.** Luxembourg, the Office. 1990. 99p.

    Previous ed. 1982.

    English-base terms, with French, Italian, German, Dutch, Danish, and now Greek indexes. A list of organizations in the energy field, units of measurement, plus abbreviations, precede. Appended bibliography. *Class No:* 620.92(038)

Reviews & Abstracts

[3752]

Renewable energy bulletin. Brentwood, Essex, Multi-Science Publishing Co., 1980-. Pts. A & B. ISSN: 0306364x.

    Part A (1986: 568 abstracts) deals with solar energy and photoelectric/photochemical devices. Part B (1986: 535 abstracts) concerns other forms of alternative energy, - goethermal, hydropower, ocean, hydrogen and wind power. Both parts cover periodical literautre, conference proceedings, unpublished reports, press reports and book notices. Annual subject and author indexes for each part.

Class No: 620.92(048)

Tables & Data Books

[3753]
INTERNATIONAL ENERGY AGENCY. **Renewable sources of energy.** Paris, Organization for Economic Development and Cooperation (OECD), March 1987. 334p. illus., diagrs., tables.

55 statistical tables - a feature - with figures up to December 1985. 6 preliminary chapters. Annex 1: Nature, status and outlook of renewable energy sources and conversion technologies (p.79-293): solar, wind, biomass, geothermal and ocean energies. Annex 2: Economics of renewable energy technologies. No index.
Class No: 620.92(083)

Maps & Atlases

[3754]
MUSTOE, J.E.H. **An Atlas of renewable energy resources in the United Kingdom and North America.** New York, Wiley, 1984. [v], 202p. diagrs., graphs, tables, maps. ISBN: 0471102938.

10 chapters (2. Solar energy - 3. Wind energy - 4. Wave energy - 5. Ocean thermal energy - 6. River energy - 7. Biofuels - 8. Geothermal energy - 9. Tidal energy - 10. Conclusion). 8 appendices (2. Thermal value of fuels). References, p.189-94; bibliography, p.195. Analytical index, p.197-202. 17 tables; 16 maps. *Class No:* 620.92(084.3)

Developing Countries

[3755]
KRISTOFERSON, L.A. *and* BOKALDERS, V. **Renewable energy technologies:** their applications in developing countries: a study of the Beiger Institute, the Royal Swedish Academy of Sciences. Oxford, etc., Pergamon, 1986. xviii, 319p. illus. $60.

An investigation of the potential of renewable energy sources (*e.g.* solar, tidal, wind) for the economic development of Third World countries.
Class No: 620.92(4/9-77)

Government Organizations

[3756]
GREAT BRITAIN. Department of Energy. **New and renewable sources of energy:** a directory of UK expertise. 32p.

Compiled for the United Nations Conference on New and Renewable Sources of Energy, Kenya, 1981, by the Information Office Energy Technology Support Unit, Atomic Energy Research Establishment, Harwell, Oxon., Feb. 1981.

A directory of 75 bodies, boards and institutes. Sections: A. General - B. Fuels from biomass - C. Fuelwood. Charcoal - D. Solar technology - E. Hydropower - F. Wind - G. Geothermal - H. Ocean. A-Z index to organizations.
Class No: 620.92:061:061.1

Research Establishments

[3757]
International **directory of new and renewable energy information sources and research centres.** 2nd ed. Paris, Unesco, 1986. xvii, 661p. ISBN: 9231022989.

....(contd.)

Information on resources, organizations, institutions and publications (journals, directories and databases) covering new and renewable energies. *Class No:* 620.92:061:061.62

Solar Energy

[3758]
DICKINSON, W. *and* CHEREMISINOFF, P.N., *eds.* **Solar energy technology handbook.** New York, Dekker; London, Butterworths, 1980. 2v. illus., diagrs., graphs, tables. (*Energy, power and environment, 6.*) ISBN: 0824768728; 0824769279.

A: *Engineering fundamentals.* xiii, 882p.

B: *Applications, systems design, and economics.* xii, 805p.

Pt. A has 58 contributors. 6 units (26 documented chapters): 1. The solar resource - 2. Solar thermal collectors - 3. Photovoltaics - 4. Bioconversion - 5. Wind energy - 6. Solar energy storage system. Chapter 22, 'Characteristics and uses of wind machines' (p.665-717) has 53 references and a bibliography of 6 items. Pt. B, sponsored by the American Section of International Solar Energy, concerns unit 7, 'Applications of solar technology' and unit 8, 'Nontechnical issues' (*e.g.* environmental, health and safety issues). *Class No:* 620.92:523.9

[3759]
EGGERS-LURA, A. **Solar energy in developing countries:** an overview and buyers' guide for solar scientists and engineers. Oxford, etc., Pergamon, 1979. vii, 205p. (*The Pergamon European heliostudies, 1.*) ISBN: 0080232531.

7 chapters (2. General information on solar energy activities of interest for developing countries - 3. State of the art - 4. Solar R & D activities in developing countries - 5. Sources of literature and information … 7. Bibliography (annotated), p.89-97). Detailed, analytical index, p.199-205. *Class No:* 620.92:523.9

[3760]
LEWIS, O. **The European passive solar handbook.** London, Batsford, 1992. 304p. illus. £40. ISBN: 0713469188.

Design information on passive solar architecture. Review of the EC solar programme. *Class No:* 620.92:523.9

[3761]
PROPERTY SERVICES AGENCY. **Solar energy:** an annotated bibliography. Pressey, N., *comp.* London, PSA Library Service, 1976. [viii], 115p.

655 numbered entries on aspects of solar energy of particular interest to the building and construction industry, mostly published in 1974-76. 26 sections; 1-5: bibliographies, conference papers, historical studies, directories, periodicals. Includes periodical articles. Author and subject indexes. 'A well-constructed bibliography' (*Library review,* 1977(1), p.60). *Class No:* 620.92:523.9

[3762]
Solar energy update: a current awareness journal. Oak Ridge, Tenn. US Dept. of Energy, Office of Scientific and Technical Information, 1982-86. Monthly.

17 sections (*e.g.* 14. Solar energy … 16. Tidal and wave power - 17. Wind energy). 5 indexes: Corporate author, Personal author, Subject, Contract number, Report number. File number correlation. Provided abstract and indexing coverage of science and technology reports, journal articles, conference papers and proceedings, books, patents, theses and monographs from all sources on solar energy.

....*(contd.)*

'Information on subjects covered will continue to be included in the Energy Abstracts Press'.
Class No: 620.92:523.9

Hydrogen Energy

[3763]
Hydrogen energy quarterly literature review. Albuquerque, New Mexico, Global Resources and Associates, 1983-. Quarterly.

Previously as *Hydrogen energy : a bibliography with abstracts* (Albuquerque, Technical Application Centre, Univ. of New Mexico).

About 1,000 abstracts pa. 6 sections on hydrogen energy: general references; production; utilization; storage, transmission and distribution; safety; materials. Annual author, title and keyword indexes. *Class No:* 620.93

Biomass Energy

[3764]
Biomass abstracts. Dublin, Biomass Conversion Technical Information Service, 1978-. 6pa. ISSN: 03324079.

About 3,500 abstracts pa. Sections: Biomass systems - Biomass resources (forestry, agriculture, municipal wastes, algae) - Harvesting, collection - Materials handling - Photogenesis - Microbial conversion - Thermochemical conversion - Combustion - Pulping - Purification, extracting - Densification, concentration - Properties - Environmental aspects, health and safety - Products - Synthetic fuels - General. Corporate, author and subject indexes. Biomass calendar insert. A comprehensive, worldwide source of information on biofuels. *Class No:* 620.95

[3765]
—Biomass bulletin. London, Multi-Science Publishing Co., 1981-. Quarterly. £70pa.

Carries 50 abstracts per issue on 'current and proposed activity, including laboratory research and field trials, both on fuel and the recycling aspects of biomass'. Author and subject indexes per issue. *Class No:* 620.95

[3766]
—COOMBS, J. Biomass: an international directory of companies, products, processes and equipment. New York, Macmillan, 1986. 243p.

'Will certainly be useful to many members of the biomass community' (*Energy & fuels,* September/October 1987, p.448). *Class No:* 620.95

[3767]
Biomass handbook. Kilani, P. *and* Wexall, C., *eds.* New York, Gordon & Breach, 1989. 963p. $349. ISBN: 0881242693.

Sections: Biomass production - Biomass conversion - Biomass utilization - Biotechnology for biomass - Biomass statistics and properties. Index. A comprehensive source with numerous literature references. *Class No:* 620.95

[3768]
LOWENSTEIN, M.Z., *ed.* **Energy applications of biomass.** New York, Elsevier, 1985. ix, 325p. illus.

Papers from the National Meeting on Biomass R & D for Energy Applications, 1-3 October 1984 at Arlington, Virginia. 3 sections: 1. Issues of importance to biomass energy research - 2. Research interests of biomass sponsors - 3. Biomass energy research projects. *Class No:* 620.95

[3769]
MILNE, T.A., *and others.* **Source book of methods of analysis for biomass and biomass conversion processes.** Amsterdam, Elsevier, 1990. 438p. tables. £56; $104. ISBN: 1851665277.

Identifies international references to standard testing methods. *Class No:* 620.95

Alternative Sources

[3770]
BRITISH LIBRARY. Science Reference and Information Service. **Sources of information on alternative energy technology held at the Science Reference and Information Service.** Dunning, P.H., *comp.* London, British Library, Science Reference and Information Service, 1986. 52p. £7. ISBN: 071230732x.

A briefly annotated guide to the holdings of the Science Reference Library during the last ten years. Coverage: biomass energy, energy recycling, geothermal energy, solar energy, wave and wind energy. Arrangement is by form of material (*e.g.* books, directories, business reports). Shelf-marks are given. Index to list of organizations with an interest in alternative energy technologies. 'A most useful guide in its field' (*Aslib information,* v.14(10), October 1986, p.235). *Class No:* 620.97

[3771]
CLARE, R., *ed.* **Tidal power:** trends and developments. London, Thames Telford, 1992. 334p. £35. ISBN: 072771905x.

Covers both national and international works and studies on tidal power. *Class No:* 620.97

[3772]
GRAYSON, L., *ed.* **Recycling: energy from community waste:** a guide to sources. 2nd ed. London, British Library, Science Reference and Information Service, 1991. iv,143p. £25. ISBN: 0712307618.

International survey in 4 sections: 1. Legislative background - 2. Energy recovery options - 3. Bio-conversion - 4. Thermo-chemical conversion (p.68-126). Indexes of personal authors, corporate bodies, subjects. 576 numbered references under sub-sections. *Class No:* 620.97

[3773]
UNITED NATIONS. Industrial Development Organization. **Information sources on non-conventional sources of energy.** New York, United Nations, 1978. xii, 110p. (*Guides to information sources, no.30.*)

669 numbered entries. 11 sections: 1. Professional, trade and research organizations, learned societies and special information services - 2. Directories - 3. Sources of statistics, marketing and other economic data - 4. Basic handbooks, textbooks and manuals - 5. Monograph series - 6. Current periodicals - 7. Current abstracting and indexing periodicals - 8. Proceedings, papers and reports - 9. Other potential sources of information (*e.g.* A: Consulting and

....*(contd.)*

engineering services ... G. Packaging - H. Patents and licences). 'Bibliographical sources used in compiling the present guide'. Sections 2-3, 6-7, 9-11 have annotated entries. *Class No:* 620.97

[3774]
VEZIROGLU, T.N., *ed.* **Alternative energy sources. VII.** Washington, Hemisphere, 1987. 6v., illus.
  1. *Solar energy.*
  2. *Solar Applications.*
  3. *Indirect solar/nuclear.*
  4. *Bioconversion/hydrogen.*
  5. *Hydrocarbons/energy transfer.*
  6. *Energy economics and planning.*
Proceedings of the 7th Miami International Conference on Alternative Energy Sources, 9-11 December 1985. Includes bibliography and index. *Class No:* 620.97

Mechanical Engineering

Databases

[3775]
ISMEC (Information Service in Mechanical Engineering). Bethesda, Md., Cambridge Scientific Abstracts.
Covers 1973 to date. A bibliographic database providing references, with abstracts from 1982, to the literature of mechanical and production engineering and related fields, including engineering management. File size: 195,000 citations. Updated monthly. 10,000 additions p.a. Hosts: DIALOG; ESA-IRS. Cost per connect hour: $80. Printed version: *ISMEC bulletin. Class No:* 621(003.4)

Bibliographies

[3776]
AMERICAN SOCIETY OF MECHANICAL ENGINEERS. Seventy-seven year index: technical papers, 1880-1956. New York, the Society, 1957. [viii], 382p. $20.
Continuation author and subject indexes have been published by Nichigai Associates Inc., Tokyo, covering 1957-70 and 1971-75 (1976. ix, 542p., XXIIIp.). The latter also includes a paper number index. *Class No:* 621(01)

[3777]
AUGER, C.P. Engineering eponyms: an annotated bibliography of some named elements, principles and machines in mechanical engineering. 2nd rev. ed. London, Library Association, 1975. viii, 122p. £7 (£5.60 to members). ISBN: 0853654379.
First published 1965. Originally a thesis approved for L.A. Fellowship.
Entries for *c.*650 eponyms (Abbott Profiolometer ... Zoelly Turbine). Gives name and dates of inventor; type of invention, machine, etc. (patents sometimes cited); *c.*2 references per entry. Cross-references. List of principal works consulted, p.108-11. Subject index groups entries by types. *Class No:* 621(01)

[3778]
BRITISH LIBRARY. Science Reference Library. **Useful books of reference for designers (1926-1983) held by the Science Reference Library. Part 3: Mechanical engineering.** Dalyza, M.M., *sel.* London, the Library, 1984. i, 121p. £5. ISBN: 0712307141.
A bibliography of about 700 works arranged under *c.*400 headings by the Library's unique classification system. There is a Contents list to indicate the broad groupings and an index of subjects which leads easily to specific groups of relevant publications. The scope is far wider than the conventional interpretation of mechanical engineering, extending to aerodynamics, aircraft, automobile, traffic, marine and sanitary engineering. Clearly set out in reproduced typescript, the entries include an occasional annotation - more would greatly enhance the value of this compilation. Most of the standard works are listed but there are few recent publications and one or two over 50 years old. *Class No:* 621(01)

[3779]
INSTITUTION OF MECHANICAL ENGINEERS. Brief subject and author index of papers published in the *Proceedings,* **1847-1950.** London, the Institution, 1951. 235p.
  1937-1960. 252p.
  1. 1951-64. 335p. £12. ISBN: 0852982062.
  2. 1960-69. 1971. 491p. £15. ISBN: 0852982070.
  3. 1968-72. 1974. 412p. £20. ISBN: 0852982488.
  4. 1973-76. 1979. 154p. £30. ISBN: 0852984235.
Continuation volumes of the index appear at about five-yearly intervals and overlap, considerably in the earlier years. They are now numbered in a series and published by Mechanical Engineering Publications, the Institution's publishing arm. From v.4 the subject headings are based on the EJC *Thesaurus of engineering and scientific terms (TEST)* (Engineers Joint Council, 1969). *Class No:* 621(01)

Encyclopaedias

[3780]
CLARKE, D., *ed.* **Encyclopaedia of how it works:** from abacus to zoom lens. London, Marshall Cavendish, 1977. 247p. illus. (mostly col.). £4.95. ISBN: 0856863046.
225 devices, machines, instruments and tools are described and illustrated, in A-Z order, ranging from simple (*e.g.* screws) to complex (*e.g.* television sets) and from ancient times to the present. An index of alternative names, cross-references and inventors enhances access.
Class No: 621(031)

[3781]
—**CLARKE, D.,** *ed.* Encyclopaedia of how it's made. London, Marshall Cavendish, 1978. 200p. illus., diagrs. £4.95. ISBN: 085685428x.
A companion to *The encyclopaedia of how it works (qv),* from which some of the content is reproduced. 35 entries, from 'Acid manufacture' to 'Xerogoraphy', covering industrial processes concerned with food and drink, medicine, clothing, engineering, entertainment, sports and crafts. Well-illustrated and clearly described. Concise index. *Class No:* 621(031)

Handbooks & Manuals

[3782]

FLÜGGE, W. **Handbook of engineering mechanics.** New York, etc., McGraw-Hill, 1962. xxvi, [1632p.]. diagrs., graphs, tables. £79.95. ISBN: 0070213925.

88 contributors. 7 parts: 1. Mathematics - 2. Mechanics of rigid bodies - 3. Theory of structures - 4. Elasticity - 5. Plasticity and viscoelasticity - 6. Vibrations - 7. Fluid mechanics. Chapter references. Over 1,000 diagrams, charts, tables and graphs. 'This handbook has not been written for readers, but for users, and its place is not on the bookshelf, but on the desk' (*Preface*). Librarians be warned! It is a tribute to the enduring stature of this work that its publishers keep it in print after a quarter of a century. *Class No:* 621(035)

[3783]

KUTZ, M., *ed.* **Mechanical engineer's handbook.** New York, etc., Wiley, 1986. xix, 2,316p. illus., diagrs., graphs, tables. £121. ISBN: 047108817x.

Intended, as an independent work, to complement the last (12th) edition of Kent's *Mechanical engineers' handbook* (Wiley, 1950). 82 contributors. 78 chapters in 6 parts, with emphasis on Materials and mechancial design (630p.) and Energy and power (800p.). Other parts: Digital computers - Manufacturing engineeering - Systems, controls and instrumentation - Management and research. An extremely wide-ranging work, covering theory, practice, engineering, management in all their aspects, some inevitably thinly. Despite the Editor's rightful criticism of wholesale reproduction of tables of data ('Why destroy forests to make paper for printing tables ...?') the work contains many hundreds, of necessity, supported by numerous graphs. Chapter references and bibliographies and a 60-page index complement well-organized and pleasingly displayed text and data. *Class No:* 621(035)

[3784]

TIMINGS, R. *and* MAY, T. **Newnes Mechanical engineers' pocket book.** Oxford, Butterworth-Heinemann, 1990. xviii,596p. diagrs., graphs, tables. £14.95. ISBN: 0750609192.

5 sections, closely subdivided: 1. Engineering mathematics and science - 2. Engineering design data - 3. Engineering materials - 4. Computer aided engineering - 5. Cutting tools. 3 appendices (3. Contributing companies). Index, p.592-96. 'A compilation of useful data' (*Preface, p.xvii*). *Class No:* 621(035)

Dictionaries

Polyglot

[3785]

SCHVARTS, V.V. **Illustrated dictionary of mechanical engineering:** English - German - French - Dutch - Russian. The Hague, Lancaster, Nijhoff, 1984. 416, [6]p. diagrs., tables. $49.50. ISBN: 9020116681.

Based on author's *Kratkii illystrirovannyiĭ russko-angliiskii slovar' po mashinostroeniyu (qv).*

Over 3,600 terms, uniquely numbered and arranged systematically under 14 conventional subheadings of mechanical engineering (1. Engineering mechanics - 2. Strength of materials ... 4. Machine elements ... 9. Welding ... 11. Machining ...). Key language English. Most entries clearly illustrated. Indexes in all six languages. '... designed

....(*contd.*)

for people who have just started studying mechanical engineering terms in a foreign language, particularly for those who have little or no knowledge of either the terms or their meaning ...' (*Foreword*). *Class No:* 621(038)=00

English

[3786]

NAYLER, G.H.F. **Dictionary of mechanical engineering.** 3rd ed. London, etc., Butterworths, 1985. 394p. diagrs. £26. ISBN: 0408015055.

First published 1967. Previous ed. 1975.

'Previous editions have concentrated on terms relating to moving parts, machines, and the production of power and its adoption in transport and mechanisms. This edition has added terms relating to both mechanical engineering design and manufacture, typified by the field of robotics' (*Author's Preface*). *Class No:* 621(038)=20

Russian

[3787]

SHVARTS, V.V. **Kratkiĭ illystrirovannyiĭ russko-angliiskiĭ slovar' po mashinostroeniya:** 3795 terminov. [The Concise illustrated Russian-English dictionary of mechanical engineering: 3795 terms.] Moscow, Russian Language Publishers, 1980. 224p. illus.

'This dictionary ... supplements almost all terms with ... internationally accepted symbols, formulae, diagrams, engineering drawings, sketches, etc.' [*Preface*]. Intended, therefore, to be especially helpful to users with limited technological experience. Russian terms are arranged in 15 groups (*e.g.* Machine elements, Welding), subdivided into sections (*e.g.* Arc welding, Gas welding), numbered sequentially, with English equivalents. Index of Russian terms leads to numbered items. Used as foundation for same author's *Illustrated dictionary of mechanical engineering (qv)*. *Class No:* 621(038)=82

Reviews & Abstracts

[3788]

Applied mechanics reviews. New York, American Society of Mechanical Engineers, 1948-. v.1-. Monthly. $396. ISSN: 00036900.

Each issue contains one or more review articles; book reviews; and a review of the journal literature, containing upwards of 1,000 author or indicative abstracts. (Before 1984 many informative abstracts supplemented the indicative abstracts - all signed - and author summaries were not used.) The abstracts are in one of 10 groups (*e.g.* Mechanics of solids, Thermal sciences, Automatic control), each of which is subdivided (*e.g.* Robotics). Annual author and subject indexes - latter index matches the classification scheme used to group the abstracts and is merely the latest in a series of different methods practised. Specific subject retrieval has not been easy for many years. About 350 core journals are scanned. *Class No:* 621(048)

[3789]
ISMEC: mechanical engineering abstracts. Bethesda, Md., Cambridge Scientific Abstracts, 1973-. v.1(1)-. Bi-monthly. ISSN: 03060039.

Originally published by the Institute of Mechanical Engineering as an index to current literature. Published monthly. 2 vols. per year before 1979. Subsequently published by Data Courier Inc., the present publishers took over in 1981 and, in 1982, v.15, began to publish abstracts. Groups contents into 9 sections, (*e.g.* 1000. Management and production; 3000. Mechanics, materials and devices; 4000. Production processes, tools and equipment) with sub-sections (*e.g.* 3600. Tribology) also sub-divided (*e.g.* 3610. Bearings). Each issue (bi-monthly from 1987) has author and subject indexes. Cumulative annual indexes. Scope: 250 journals, theoretical and applied, plus many conferences and books. Each entry has unique identifier and full citation. Abstracts of variable quality and efficiency - often verbose. Annual content has been diminishing from 15,000 in 1982 to 9,600 in 1991. Also available as a computerized database, ISMEC online via established hosts *e.g.* DIALOG, ESA-IRS, or through leased tapes. *Class No:* 621(048)

[3790]
Referativnyĭ zhurnal. 14. Tekhnologiya mashinostroeniya. Moscow, VINITI, 1963-. Monthly. diagrs. R.224.40. ISSN: 00342599.

About 25,000 signed indicative and informative abstracts pa. 4 sections covering: Metal cutting, machine tools and tools - Technology and equipment for machine tools - Technology and equipment for forging - Technology and equipment for foundry work. Abstracts illustrated with diagrams. English contents list for Part A.
Class No: 621(048)

Yearbooks & Directories

[3791]
Engineering components and materials index. Bracknell, Berks., Technical Indexes, Ltd., 2pa. £40pa. ISBN: 0860413349. ISSN: 03088383.

This data book is the index to the associated microform file of the catalogues of manufacturers and suppliers of a wide variety of mechanical, electrical, hydraulic and pneumatic engineering components, materials and products. Covers iron and steel and non-ferrous metals and alloys; plastics, rubber, ceramics and glass processing and treatment. Includes transducers, transformers, switches, relays, electrical connectors and cables, and clutches, brakes, bearings, gears, belts, plus pumps, compressors, valves, seals. 7 sections. Issue 34, October 1983 - May 1984 (221p.) contains products and trade names (the index to the selector charts) - Selector charts (p.15-154) - Addresses (p.165-209) - Stockholders of steel, non-ferrous metals and plastics. *Class No:* 621(058)

[3792]
Machinery's buyers guide. Horton Kirby, Kent, Findlay Publications, 1987. xxi, 1616p. illus., tables. £20. ISSN: 01420658.

First published 1926. Publisher varies.

'Annual directory of engineering products and services' (*Cover sub-title*). Alphabetical by product, tool, service, etc., followed by addresses section (A-Z list of *c.*6,000 firms).

....(contd.)
Includes trade names and agencies list and miscellaneous tabulated reference material, Many advertisements.
Class No: 621(058)

[3793]
Manufacturing and materials handling index. Bracknell, Berks., Technical Indexes, Ltd., 2pa. £30 pa. ISSN: 02688093.

Previously entitled *Materials handling index.*

This data book is the index to the associated microform file of the catalogues of manufacturers and suppliers of engineering products, tools, machines and equipment for materials handling, packaging and maintenance work, available on separate subscription. 10 sections (A: Buildings and associated fixtures ... E: Production machines and tools ... G: Materials handling equipment ... Z: Services). Issue 27, August 1987 - March 1988 (176p.) contains Products and trade names (the index to the Selector charts) - Selector charts (p.23-115) - Professional and trade organizations - Addresses (p.126-74). *Class No:* 621(058)

Tables & Data Books

[3794]
AMERICAN SOCIETY OF MECHANICAL ENGINEERS. Metal Engineering Board. **ASME handbook.** New York, etc., McGraw-Hill, 1956-1964. 4v. (*Handbook series.*)

Engineering tables. 2nd ed. 1956. 695p., $89. ISBN 0070015163.

Metals engineering: design. 2nd ed. 1965. 619p. $79.50. ISBN 00701518x.

Metals engineering: processes. 1958. 928p. ISBN 0070015147.

Metals properties. 1954. 440p. ISBN 0070015139.

Engineering tables covers, in 15 sections, bearings, gears, keys, bolts, screws, nuts, springs, tubes, gaskets, etc. Has bibliography of sources. *Metals engineering: design* is much enlarged, has over 100 contributors, and arranges its 59 sections in 5 parts, with references, many tables and detailed index. *Metals engineering: processes* has 49 contributors, 40 sections grouped under 9 heads (*e.g.* 1: Heat treatment of steel; 7: Machining; 9: Electro-forming). Many illus., diagrams, tables. Concise index. *Metal properties* gives data on individual metals, such as chemical composition, characteristics, technological, physical and mechanical properties, thermal conductivity, coefficient of thermal expansion. *Class No:* 621(083)

[3795]
CARVILL, J. Mechanical engineer's data book. Oxford, Butterworth-Heinemann, 1991. 352p., illus., tables. £49.50. ISBN: 075061014x.

Extensive use of illus. and tables. Includes information on manufacturing processes. *Class No:* 621(083)

[3796]
Machinery's handbook: a reference book for the mechanical engineer, draftsman, toolmaker and machinist. Oberg, E., *and others.* 24th ed. New York, Industrial Press, 1992. xiv,2543p. diagrs., graphs, tables. ISBN: 0831124245.

First published 1914. Previous edition 1988.

A comprehensive data book. Main sections: Mathematics - Mechanics - Strength of materials - Properties, treatment and testing of materials - Plastics - Dimensions, gaging and measuring - Tooling and toolmaking - Machining operations

....(contd.)

- Manufacturing processes - Fasteners - Threads and threading - Gears, splines and cams - Bearings - Measuring units. Covers American and British practice. Detailed, analytical index, p.2485-2543.

The 24th ed. includes a new section on plastics and revisions of material on wood and machining operations. Remains indispensable. *Class No:* 621(083)

[3797]

—Machinery's handbook guide to the use of tables and formulas. Amiss, J.M., *and others.* 24th ed. companion. New York, Industrial Press, 1992. xii,239p. diagrs. ISBN: 0831124997.

Previously entitled *The use of 'Handbook' tables and formulas.* First published 1931.

Designed for use in conjunction with the 24th ed. of *Machinery's handbook,* with hundreds of examples, solutions, test questions and typical applications to enable *Handbook* users to get to know it thoroughly and to use it effectively and speedily. *Class No:* 621(083)

[3798]

Marks' Standard handbook for mechanical engineers. Avallone, E.A. *and* Baumeister, T., *eds.* 9th ed. New York, etc., McGraw-Hill, 1987. xxii, [1867]p. diagrs., graphs, tables. £89. ISBN: 007004127x.

First published 1916. Previous ed. 1978.

150 contributors. Framework of 19 main sections unchanged, covering conventional physical science and engineering topics, plus transportation, electrical and electronic engineering and building construction and equipment (*e.g.* 3. Mechanics of solids and fluids - 4. Heat ... 6. Materials of engineering (206p.) ... 8. Machine elements (226p.) ... 14. Fans, pumps and compressors ... 17. Industrial engineering). Dual system of units used. Extensive index (12,000 entries). More than 2,000 illus. and over 1,000 tables. Coverage of some topics (*e.g.* machine elements, power generation, fuels) greater than in Parrish (*qv*), which this work usefully complements. '... this new edition is a major revision, with over 80% of its material brought up to date ... The Handbook's proven format of subject classification makes it very easy to use, with detailed referencing ... There is strong emphasis, as in previous editions, on tabular and graphical presentation of data for ease of use' (*Chartered mechanical engineer*, December 1987, p.63). *Class No:* 621(083)

[3799]

Spon's Mechanical and electrical services price book. 19th ed. London, Spon, 1988. Annual. xi, 550p. graphs, tables. £32. ISBN: 041914420x. ISSN: 03054543.

· First published 1968.

Parts 1 and 2 give current wage rates and prices of materials and prices of measured work in mechanical and electrical installations in buildings. Parts 3 and 6 contain cost data for approximate estimating, daywork, fees and contract details for large industrial projects. *Class No:* 621(083)

Standards

[3800]

Mechanical standards: basic standards; complete to 31 August 1983. 1st English ed. Berlin, DIN (Deutsches Institut für Normunge.V.)/Beuth Verlag, 1984. 408p. diagrs., tables. DM128. (*DIN Handbook 1.*) ISBN: 3410116451. ISSN: 07227337.

....(contd.)

English-language translations of 50 basic standards in mehcanical engineering, including ISO threads, metrology, dimensioning, drawings, tolerances, surfaces, fasteners and gearing. Arranged in DIN number order, with alphabetical subject index and a separate classified contents list. Based on the 20th ed. of the *DIN Handbook 1* (*TAB 1*) and closely related to and overlapping with *DIN Handbook 3* (*i.e., TAB 3*): 'Mechanical engineering standards for study and practice'. This overlap will be removed in future editions of *TAB 3.* New editions of the Handbooks normally appear every 3 years. *Class No:* 621(083.74)

Histories

[3801]

BURSTALL, A.F. A History of mechanical engineering. London, Faber & Faber, 1963. 456p. illus. £3; £1.40. (*Technology today and tomorrow.*)

9 chapters, each with bibliographies and references, cover successive periods of history from 4,500 BC to the 1960s. In each chapter significant advances are helpfully pinpointed under 5 standard headings. Author has leaned heavily on Singer and Holmyard's *History of technology* and the *Proceedings* of the Institution of Mechanical Engineers, with obvious benefit. *Class No:* 621(091)

[3802]

STRANDH, S. Machines: an illustrated history, translated by A. Henning. London, Artists House/Mitchell Beazley, 1979. 240p. illus., (incl. col.) diagrs. £14.95. ISBN: 0861340124.

Originally published by A.B. Nordbok (Gothenburg, 1979).

Adopts a wide definition of 'machines' to include all forms of tool, instrument and utensil and covers their origin and use in 9 chapters with different themes, *e.g.* 2. Machine elements and elementary machines - 3. From hand tool to industrial robot ... 8. Computers. Chapters 1 and 9 cover earliest times and the present day respectively. Chapters 4-6 discuss the classes of prime mover. Bibliography, p.234. Good index. Liberal provision of excellent drawings, many coloured. Author is Director, Museum of Science and Technology, Stockholm. *Class No:* 621(091)

[3803]

—VINCE, J. Power before steam. London, Murray, 1985. 160p. illus. £8.95. ISBN: 0719541751.

A brief account of 'the manner in which man has used his own muscle power, and power derived from animals, water, wind and gravity' (*Introduction*). Much emphasis on water and wind mills. Text in author's own calligraphy adorned by his line-drawings. List of places to visit, p.154-5. Bibliography and index. *Class No:* 621(091)

Biographies

[3804]

AMERICAN SOCIETY OF MECHANICAL ENGINEERS. Mechanical engineers in America, born prior to 1861: a biographical dictionary. New York, American Society of Mechanical Engineers, 1980. xix, 330p. ports. $20.

Comprises, first, a list, with dates, of 1,688 people - the great majority male - born before 1861 (a) in the US and active there or abroad, or, (b) born elsewhere and active in America, in areas nowadays recognized as mechanical engineering, from the late 18th to the early 20th century.

.... *(contd.)*

Biographical sketches are provided in the second part of the work for 500 of these regarded as prominent in their own time or since. There are portraits of 50. Sources of information are listed. *Class No:* 621(092)

Turbines

[3805]

Referativnyĭ zhurnal. 49. Turbostroenie. Akademiya Nauk SSSR. Moscow, VINITI, 1963-. Monthly. R.27.40. ISSN: 00342629.

About 2,400 signed indicative and informative abstracts pa. on turbine engineering. 3 main sections: Steam and gas turbines; Hydraulic turbines; Wind turbines. UDC numbers. Annual subject index. English contents summary. *Class No:* 621-135

Precision Engineering

[3806]

DAVIDSON, A., *ed*. Handbook of precision engineering. London, Macmillan, 1970-74. 12v. illus.

Original Dutch ed. (Eindhoven, Philips), 1966-70. 1. *Fundamentals. 2. Materials. 3. Fabrication of non-metals. 4. Physical and chemical fabrication techniques. 5. Joining techniques. 6. Mechanical design applications. 7. Electrical design applications. 8. Surface treatment. 9. Machining processes. 10. Forming processes. 11. Production engineering. 12. Precision measurement.*

This set aims to cover the design and manufacture of fine mechanisms and to relate these explicitly to precision engineered products and parts. V. 3-5 cover production techniques; v. 6, 7, 11, 12, with practical design and the equipment and compounds needed. All chapters are by experts. A review of v.10 in *Production engineering* (v.53(12), Dec. 1974, p.522) commends it as clearly written, amply illustrated, and with an extensive index. *Class No:* 621-187

[3807]

FARAGO, F.T. Handbook of dimensional measurement. 2nd ed. New York, Industrial Press, 1982. xv, 524p. illus., diagrs., tables. £49.50. ISBN: 0831111364.

First published 1968.

Principles and practices of measurements, including instruments such as gauges, microscopes and other optical devices. Separate chapters cover measurement of angles, straightness, flatness, roundness, surfaces, etc. New chapters in this edition deal with screw threads, gears, process control and automated methods. 80 tables with illustrations and synopses are particuarly useful in facilitating speedy and efficient consultation. Many instruments are illustrated in half-tones.

'It must be emphasized that this book is American and aimed primarily at the American reader. The concept of the book and its format ... is (!) very good. Unfortunately, the author does not cover the British or ISO standard ... In general, this is a thoroughly practical and useful book' (*International journal of machine tool design & research,* v.23 (2/3), 1983, p.169). *Class No:* 621-187

Springs

[3808]

CARLSON, H. Spring designer's handbook. New York, Dekker, 1978. xv, 350p. illus., diagrs., graphs, tables. £16.95. (*Mechanical engineering series, 1*.) ISBN: 0824766237.

A compact reference work containing sufficient descriptive, factual information and data to serve as an adequate working guide for engineers concerned with the practical problems of spring design. The 20 chapters cover materials (1-10); design (11-14); manufacture (15-18); plus a glossary and miscellaneous tables. There are 219 illustrations, diagrams, curves and data tabulations. Imperial and SI units are both used. *Class No:* 621-272

[3809]

—CARLSON, H. Spring manufacturing handbook. New York, Dekker, 1982. 362p. illus., tables. $69.00. (*Mechanical engineering series, 15.*) ISBN: 0824716787.

Companion volume to the same author's *Spring designer's handbook (qv)*. Covers all aspects of manufacturing springs of all types and of many materials, supported by numerous illustrations and tables of data and specifications. A third book by Carlson entitled *Springs: troubleshooting and failure analysis* deals with problems of design, materials and manufacturing processes. *Class No:* 621-272

Maintenance Engineering

[3810]

BRITISH STANDARDS INSTITUTION. Glossary of maintenance management terms in terotechnology. 2nd ed. London, BSI, 1984. 20p. £23.90. (*BS3811:1984.*)

First published 1974.

195 technical terms in installation and maintenance of equipment. 9 sections (*e.g.* 4: Reliability). Sources often indicated. Index. *Class No:* 621-7

[3811]

Maintenance engineering handbook. Higgins, L.R. *and* Morrow, L.C., *eds*. 4th ed. New York, etc., McGraw-Hill, 1987. v.p. illus., diagrs., charts, tables. $79.50.

1st and 2nd eds. (1957, 1965), edited by L.C. Morrow. 3rd ed. 1987.

The 4th ed. claims to contain 40% new material and 30% revised material. Special emphasis on computers in maintenance of machinery. Profusion of illustrations and tables; glossary; bibliography. Detailed, analytical index. *Class No:* 621-7

[3812]

Standard handbook of plant engineering. Rosaler, R.C. *and* Rice, J.O., *eds*. New York etc., McGraw-Hill, 1983. xv, [1,824]p. diagrs., graphs, tables. £79.95. ISBN: 0070521603.

Supersedes *Facilities and plant engineering handbook* (1973), edited by B.T. Lewis and J.P. Marron.

678 contributors. 4 parts, 20 sections. A. The basic plant facility: construction, equipment and maintenance - B. Plant operation equipment: selection and maintenance - C. The maintenance function: basic equipment and supplies - D. Supplementary technical data. Appendices contain brief guide to mainly non-published US sources of information and metric conversion tables. Brief glossary and good, 43-page index. Almost 900 diagrs., graphs, tables, etc.

'Presents a wealth of information on planning, construction, operation and maintenance of manufacturing

....(contd.)

and service facilities' (*Welding design & fabrication,* v.57, Mar. 1984, p.232). (Another favourable review in *Civil engineering - ASCE,* v.52, Dec. 1982, p.22). *Class No:* 621-7

[3813]

—ELONKA, S.M. Standard plant operator's manual. 2nd ed. New York, etc., McGraw-Hill, 1975. vii, 386p. illus., diagrs., tables, graphs. ISBN: 0070192960.

First published 1956. Previous ed. (as *Plant operator's manual*) 1965.

A collection of practical information and data drawn very largely from the author's numerous periodical articles covering a wide range of topics and presented in an informal style. 20 chapters (1. Boilers, steam equipment ... 5. Steam engines ... 10. Diesel engines ... 13. Bearings, oil grooving ... 18. Rigging - 19. Fire, accident prevention ...). Good index (essential for this work). *Class No:* 621-7

Protection of Machinery

Fluid Sealing

[3814]

Fluid sealing abstracts. Cranfield, Beds., British Hydromechanics Research Association, 1966-. v.1-. Quarterly. £84pa. ISSN: 00154660.

About 600 informative abstracts pa. of items drawn from over 1,000 journals, plus reports, books, standards and conference proceedings. Selected British patents are included. Main sections: Seals in general - Static seals - Dynamic seals - Reciprocating seals - Rotary seals - Seal materials. Issue and cumulative annual indexes of personal and corporate authors and of subjects. *Class No:* 621-762

[3815]

Seal users handbook. Austin, R.M. *and* Nau, B.S. 2nd ed. Cranfield, B.H.R.A. Fluid Engineering, 1979. 220p, illus., diagrs., graphs, tables. $73. ISBN: 0900983728.

Previous edition 1974.

Technical section (Chapters 1-6, covering selection, economics, materials, standards - British, German, Japanese, Swedish, American - servicing, failures) followed by trade directory - list of manufacturers and product index with specifications chart. List of tradenames, a glossary and a bibliography complete this compact handbook-cum-directory. *Class No:* 621-762

[3816]

Seals and sealing handbook. 2nd ed. Morden, Surrey, Trade and Technical Press, 1986. xviii, 535p. illus., diagrs., graphs, tables. £55. ISBN: 0854611010.

First published 1981, 'Compiled by the editors of the Trade & Technical Press Ltd'.

Aims to provide comprehensive information (basics, principles, selection criteria, applications, data, suppliers). 11 sections (1. Fundamentals and principles - 2. Static seals - 3. Dynamic seals - 4. Oil seals ... 6. Special seal types ... 8. Seals for special applications - 9. Materials - 10. Data (and glossary) - 11. Buyer's guide and indexes of subjects and advertisers). *Class No:* 621-762

Nuclear & Atomic Engineering

Databases

[3817]

EABS. Luxembourg, Commission of the European Communities.

Covers Euratom records 1966 to date. A bibliographic database, providing references, with abstracts (since 1984), to research reports and EEC, ECSC and Euratom conference papers. Covers nuclear science and technology, coal and steel, new energy sources, the environment, radiation protection, etc. File size: 40,000 citations (mid-1986). Updated monthly. Approx. 1,500 additions pa. Host: ECHO Service. Cost per connect hour: *c.*$25. Printed versions: *Euro-abstracts.* (Section I covers Euratom research). In English, French and German. *Class No:* 621.039(003.4)

[3818]

ENSDF (Evaluated Nuclear Structure Data File). Upton, N.Y., Brookhaven National Laboratory National Nuclear Data Center.

Covers 1930 to date. A numeric data bank providing data on nuclear structure and radioactive decay for all known isotopes. Critcally evaluated data is obtained from 11 cooperating countries, including US, UK, USSR, FDR. Filesize: 9000 records. Updated twice yearly. About 500 sets of data added pa. Available via FIZ Karlsruhe (formerly INKA). Printed version: *Nuclear data sheets.* *Class No:* 621.039(003.4)

[3819]

INIS (International Nuclear Information System). Vienna, International Atomic Energy Agency.

Covers 1975 to date. A bibliographic database, providing references, with abstracts, to the literature of nuclear physics and nuclear engineering and related topics (*e.g.* isotopes, radiation protection, safety, the legal social, environmental, economic, medical aspects). File size: 1,100,000 references (September 1986). Updated fortnightly or monthly (varies with host). 90,000 additions pa. Hosts include IAEA, DIALOG, ESA-IRS, FIZ Karlsruhe (formerly INKA). Costs per connect hour vary with host: $50 - 78. Available on magnetic tape and CD-ROM (BRS Europe). Publications: *Thesaurus.* Printed versions: *INIS atomindex.* In English. *Class No:* 621.039(003.4)

Bibliographies

[3820]

CHESTER, K., *comp.* **Nuclear energy and the nuclear industry:** a guide to selected literature and sources of information. 2nd ed. London, Science Reference and Information Service, 1986. 44p. £9. ISBN: 0712307427.

Previous ed. 1982.

A valuable guide, with annotations, to over 100 publications in the subject field: 30 abstracting/indexing services; almost 60 periodicals; 12 directories and yearbooks; 4 publications with data on nuclear reactors; 4 bibliographies series. Also contains list of 35 organizations and 19 suppliers, companies and consultants, plus 12

....(contd.)
publications giving market research and survey information. Notes about books, reports, conference and patents are included, but no entries. No index. *Class No:* 621.039(01)

[3821]
—BEHRENS, H. *and* RITTBERGER, W. 'Nuclear and particle physics'. In *Information sources in physics*. 2nd ed. Editor, D.F. Shaw. 1985, p.255-78. London, Butterworth, 1985.
Headings: Primary publication in journals and books - Data compilations - Abstracts journals - Conferences and proceedings - References. *Class No:* 621.039(01)

[3822]
—GRAHAM, F.A. *and* KING, P.A. 'Nuclear power engineering'. In *Information sources in engineering*, editor, L.J. Anthony. 2nd ed. London, Butterworths, 1985. p.548-62.
Headings: History - Organizational sources - Primary sources of information - Secondary sources of information - References (31 items). *Class No:* 621.039(01)

[3823]
INTERNATIONAL ATOMIC ENERGY AGENCY. **Catalogue 1985: International Atomic Energy publications.** Vienna, IAEA, 1985. [viii], 285p.
Previous ed. 1981/82.
A cumulation listing all sales publications of the IAEA issued from 1958 to October, 1984, which are still available, plus some which are now out of print. Irregular supplements, *e.g.* First supplement, January 1986, covers publications issued November 1984 to December 1985. Part 1 lists all publications by subject, under 7 major headings (1. Life sciences - 2. Nuclear safety and environmental protection - 3. Physics - 4. Chemistry, geology and raw materials - 5. Reactors and nuclear power - 6. Industrial applications - 7. Miscellaneous) with subdivisions for each. Bibliographical details, price and description for each publication. Part 2 comprises indexes of the various series and an alphabetical index of titles and keywords.
Class No: 621.039(01)

[3824]
—INTERNATIONAL ATOMIC ENERGY AGENCY. Bibliographical series. Vienna, IAEA, 1960-80. nos. 1-43.
Each bibliography covered a particular aspect of the peaceful utilization of atomic energy. No. 43, *Peaceful uses of nuclear explosions* (1980, 443p.), was the second of this title. The first (no. 38, 1970, 466p.) recorded 1,759 references. The second listed 1,335 references to the literature published 1969-79, with abstracts in English. Arranged by subject, with author and specific subject indexes. *Class No:* 621.039(01)

[3825]
JEDRUCH, J. **Nuclear engineering databases, standards and numerical analysis.** New York, Van Nostrand Reinhold, 1985. xix, 295p. illus., diagrs., graphs, tables. £64.35. ISBN: 0442245742.
A guide to sources of information, data projects and collections, computerized databases, regulations, standards, computer-aided design methods, property and performance data centres and bases, government bodies and engineering societies and their publications, concerned with nuclear reactor engineering. Heavily orientated to the US. Many useful lists and tabulations. Classified bibliography (p.267-81). Extensive references. 'Nuclear engineers, public

....(contd.)
utilities personnel and government regulatory officials will find this guide to the software of nuclear engineering an indispensable on-the-job tool' (*Mechanical engineering*, v.107, August 1985, p.72). *Class No:* 621.039(01)

[3826]
SCHLACHTER, G.A., *ed.,*. **Nuclear America:** a historical bibliography. Santa Barbara, Calif., ABC-Clio Information Services; Oxford, Clio Press, 1984. xiii, 184p. £21.85. (*ABC-Clio Research guides.*) ISBN: 0874363608.
More than 800 abstracts of publications selected from 2,000 periodicals published in 90 countries, 1973 - 1982, concerned with peaceful and (predominantly) military uses of nuclear energy and restricted to items directly concerned with US. Subject and author indexes. '... anyone involved explicitly in the nuclear debate may find it useful in exploring the American Scene' (*Nuclear energy*, v.23(4), August 1984, p.209). *Class No:* 621.039(01)

[3827]
UNITED KINGDOM. Atomic Energy Authority. **UKAEA list of publications available to the public.** Harwell, Atomic Energy Research Establishment, 1955-. Monthly. Annual cumulation. *Gratis.* ISSN: 00417289.
Periodical articles, books, pamphlets, reports and patents by UKAEA staff, grouped under these categories. Annual cumulation is classified into 21 categories with subdivision into articles, books and pamphlets, and reports. Patents and non-UKAEA reports are separately listed. Cumulation no. 31: January - December 1986 (1987. 90p. £7.25. ISBN 0705811263) lists 1,147 numbered items under headings such as: Nuclear physics; Fuel processing and reprocessing; Health and Safety; Accelerators; Transport and handling of materials; Waste management. Author and report number indexes. *Class No:* 621.039(01)

[3828]
UNITED STATES. Nuclear Regulatory Commission. **Title list of documents made publicly available.** Washington, the Commission, 1979-. v.1-. Monthly. (*1. NUREG-0540.*)
Successor to *Power reactor docket information,* issued by the U.S. Department of Energy until 1979.
Covers documents received or published by the Nuclear Regulatory Commission, categorized as docketed (relating to civilian nuclear power plant and other uses of radioactive material) and non-docketed (material received by or issued by the Commission as a regulatory agency). Documents include all kinds of reports, correspondence, licences, etc. Items are arranged in categories, under docket number where given one, with one-line annotations. Indexes of personal authors and corporate sources and of report numbers. V.9(9), September 1987, 492p.
Class No: 621.039(01)

Handbooks & Manuals
[3829]
Nuclear power: policy and prospects. Jones, P.M.S., *ed.* New York, Wiley, 1987. 416p. $154. ISBN: 0471907324.
16 contributors. 4 sections: 1. Introduction - 2. Areas of controversy - 3. National experience of nuclear power (US, France, Canada, UK, FRG, Japan, India; the former USSR and other planned economies are largely omitted) - 4. Policy and prospects. Chapter references; detailed, analytical index. 'An excellent resource' (*Choice,* May 1988, p.1424).
Class No: 621.039(035)

[3830]
RAHN, F.J., *and others*. **A Guide to nuclear power technology:** a resource for decision making. New York, etc., Wiley, 1984. xi, 985p. illus., diagrs., graphs, tables. £77.50. ISBN: 0471889148.

Aims at comprehensive coverage of the technology associated with nuclear power, 'tuned to the person who has some technical background but is not a specialist in the field' (*Preface*). In addition to reactors and nuclear power plant operation there are chapters on fuel, materials, radioactivitiy and waste disposal, health, safety, regulations, economics and proliferation factors. 10 pages of statistics. 3 glossaries of acronyms. Very useful chapter bibliographies. Index (29 pages). *Class No:* 621.039(035)

Dictionaries

Polyglot

[3831]
ALONSO SANTOS, A., *and others*. **Léxico de términos nucleares.** Madrid, J.E.N., 1973. xl, 730p. (*Publicaciones científicas de la Junta de Energia Nuclear.*) ISBN: 8450062950.

Part 1 contains *c.*2700 terms in Spanish relating to nuclear science and engineering, with English and French equivalents and explanatory definitions in Spanish. Part 2 contains English - Spanish and French - Spanish indexes. Part 3 consists of a glossary of extended discussions of the meaning and use of 125 imported terms in the subject field, with literature references. *Class No:* 621.039(038)=00

[3832]
CLASON, W.E., *comp*. **Elsevier's dictionary of nuclear science and technology in six languages:** English, American, French, Spanish, Italian, Dutch and German. 2nd rev. ed. Amsterdam, etc., Elsevier, 1970. [ix], 787p. $121.50. ISBN: 044440810x.

First published 1958. Russian supplement 1961.

7,806 English/American terms and phrases, with equivalents in the other 5 languages. Each entry consecutively numbered. Indexes in each other language.

Reactor technology (Pt. B, v.1(4), February 1961, p.238) questions the validity of a 'neat row of five foreign equivalents' and hints that some of the equivalents may have been concocted. *Class No:* 621.039(038)=00

[3833]
SUBE, R., *ed*. **Wörterbuch Kerntechnik:** Englisch - Deutsch - Französisch - Russisch. Essen, Girardet; Berlin, VEB Verlag Tehnik; Oxford, Elsevier, 1985. 1,199p. £112.60. ISBN: 3773651252; 0444995935.

18,500 terms in English, with unique alphanumeric codes for each and equivalents in German, French and Russian, in columns. Indexes in German, French and Russian, utilizing codes for referral. Although atomic and nuclear physics are included, emphasis is on nuclear engineering (installations, materials, power, generation and reactors), radiation and weapons. Very thorough coverage. Numerous cross-references. Larger and much-improved version of author's 1961 *Dictionary of nuclear physics and technology*. *Class No:* 621.039(038)=00

English

[3834]
BRITISH STANDARDS INSTITUTION. Glossary of terms used in nuclear science and technology. 2nd ed. London, BSI, 1973. 52p. £36. (*BS3455:1973.*) ISBN: 0580078558.

First published as *Glossary of terms used in nuclear science* (1962).

Scope and size have been reduced to concentrate on the technology and engineering terminology. About 750 terms and cross-references. *Class No:* 621.039(038)=20

[3835]
—HANCHETT, J.G. and HASSELBERG, F.W., *comps*. Glossary of terms: nuclear power and radiation. Singh, M.H., *ed*. Washington, U.S. Nuclear Regulatory Commission, 1981. 52p. $3.75. (*NUREG-0770.*)

300 terms defined to be understandable to the layman. *Class No:* 621.039(038)=20

[3836]
Chambers nuclear energy and radiation dictionary. Walker, P.M.B., *ed*. Edinburgh, Chambers, 1992. viii,260p. diagrs., tables. £20. ISBN: 0550132465.

About 2,500 terms, briefly defined (*e.g.* 'alpha particle': 5 lines) Cross-references. Preceded by 11 chapters on historical development and themes of nuclear energy (*e.g.* 3. Energy from nuclear fission ... 5. Nuclear safety ... 8. Radiation detection ... 11. Biological effects). *Class No:* 621.039(038)=20

French

[3837]
FRANCE. Commisariat a L'Énergie Atomique. **Dictionnaire des sciences et techniques nucléaires.** 3rd ed. Paris, Eyrolles, 1975. 488p. graphs, tables. F175.

First published 1964. Previous ed. 1966.

Present ed. has weeded obsolete terms and added new ones to give a total of 3,400, each defined in French in Part 1, with English equivalents. Part 2. lists the latter and gives French equivalents. Part 3 (Annexes) contains 7 tables (Fundamental particles; SI units; radioactive elements; Periodic table, etc). *Class No:* 621.039(038)=40

Reports Literature

[3838]
UNITED STATES. Nuclear Regulatory Commission. **Regulatory and technical reports:** abstracts index journal. Washington, the Commission, 1976-. v.1-. Quarterly. (*NUREG-0304.*)

Bibliographic data and informative abstracts for the regulatory and technical reports issued by the US Nuclear Regulatory Commission (successor to the Atomic Energy Commission) and its contractors, in NUREG report number order. Issue and cumulated annual indexes of personal authors, subjects, originating organizations, contractors and licencsed facilities. April-June 1987, contains 21p. of abstracts. *Class No:* 621.039(047)

Reviews & Abstracts

[3839]
INIS Atomindex. Vienna, International Nuclear Information System, International Atomic Energy Agency, 1920-. v.1(1)-. 24pa. Sch., 5,100pa. ISSN: 00047139.

Successor to *Nuclear science abstracts* (ceased 30 June 1976). V.1-20 as *List of bibliographies on nuclear energy*.

About 90,000 indicative and informative abstracts pa. Classified order: A00. Physical sciences - B00. Chemistry, materials and earth sciences - C00. Life sciences - D00. Isotopes, isotope and radiation applications - E00. Engineering and technology - F00. Other aspects of nuclear energy. Each section subdivided in detail, *e.g.* E10.00 Engineering - 16.00 Accelerators - E16.30. Components and auxiliaries - E16.31. Ion sources. Each abstract lists subject headings used in indexes. Each issue has personal author, corporate entry, subject, conference date, conference place and report, standard and patent number indexes. (*N.B.*, No index of conference *names*.) Semi-annual indexes cumulative annually. English abstracts and translations of foreign-language titles. Covers reports, conference papers, patents, journal articles and books (in that order for each section). Also available on magnetic tape (from which the printed version is prepared), and online via DIALOG, ESA-IRS, INKA, BELINDIS and STN, and on CD-ROM (BRS Europe). First issue in each volume (1st January) contains list of over 4,300 journals scanned regularly.
Class No: 621.039(048)

[3840]
Nuclear science abstracts. A semi-monthly publication of the United States Atomic Energy Commission (1975-1976: Energy Research and Development Association). Washington, U.S. Government Printing Office, 1948- June 1976. V.1-33, no.12.

First issued 1947 as *Abstracts of declassified documents*.

Covered journal articles, books, patents and reports issued by the US Atomic Energy Commission and other US government agencies and by universities and industrial and research bodies worldwide. 2 vols. pa. 1973-1976. V. 31-32, 1975, contained 66,208 indicative and informative abstracts in 21 alphabetically arranged sections (Chemistry ... Engineering ... Particle accelerators ... Physics (Nuclear) ... Reactor technology and radiation - General and miscellaneous) with subdivisions. Personal and corporate author indexes, subject index and report number index, plus five-year cumulations. With 940,000 abstracts, this remains a unique and invaluable guide to the unclassified literature of the first 30 years of the nuclear age. Succeeded by *INIS Automindex* (qv). *Class No:* 621.039(048)

[3841]
Physics abstracts. Science abstracts. Section A. Stevenage, Herts, Institution of Electrical Engineers, 1898-. v.1-. Twice monthly. £920 pa. ISSN: 00368091.

This section of *Science abstracts* includes nuclear physics, nuclear engineering, radioactivity, nuclear power, atomic and molecular physics, spectra and properties, and thus constitutes a major souce of abstracts derived from the INSPEC database of more than 3,700 journals, plus books, reports, conferences, etc. A significant proportion of the 140,000 abstracts p.a. relate to these subjects. Half-yearly and 4-yearly cumulative indexes of authors, subjects, etc., enhance the value. Available online from a number of hosts. CD-ROM. *Class No:* 621.039(048)

[3842]
—Euro abstracts. Section 1: Euratom and EEC R & D and demonstration projects. Scientific and technical publications and patents. Luxembourg, Commission of the European Communities; Oxford, Learned Information, 1963-. V.1(1)-. Monthly. £54.40pa. ISSN: 00142352.

About 1,200-1,400 informative abstracts pa. 10 sections, subdivided (General and miscellaneous ... Nuclear technology - Physics). Indexes of authors and inventors (of patents cited). In English. Available online *gratis* through ECHO (European Commission Host Organization).
Class No: 621.039(048)

Progress Reports

[3843]
Advances in atomic and molecular physics. New York, Academic Press, 1965-. v.1-. ISSN: 00652199.

V.28 1990 (320p.) has 7 contributors. The 4 papers include 'The theory of fast ion-atom collisions' and 'Squeezed States of the radiation field'. Author and subject index. Cumulative author index, v.1-28.
Class No: 621.039(055)

[3844]
Advances in nuclear physics. New York, Plenum Press, 1968-. v.1-. ISSN: 00652970.

Longer, more penetrating reviews - *c*.3 per vol. V.15, 1985: Analytical insights in intermediate energy hadron-nucles scattering; Recent developments in quasi-free nucleon-nucleon scattering; Energetic particle emission in nuclear reactions.

V.17, 1987 (xv, 369p. $69.50): p-Matrix methods in hadronic scattering; Dibaryon resonances; Skyrmions in nuclear physics; Microscopic description of nucleus-nucleus collisions. *Class No:* 621.039(055)

[3845]
Advances in nuclear science and technology. New York, Plenum Press, 1962-. v. 1-. ISSN: 00652989.

Reviews of selected fields with greater interest for nuclear engineers. Some vols. have special theme (*e.g.* v.17, 1986: Simulators for nuclear power; V.18 contained 5 papers on aspects of nuclear power plant - 3 concerned light water reactors). Occasional Festschrift vols. (*e.g.* v.19 1987. 490p. $85) in honour of Eugene Wigner). V.21 (1990. 300p. £59.20) contains 5 papers, including 'Expert systems and their use in nuclear power plants' and 'Health effects of low level radiation'. *Class No:* 621.039(055)

[3846]
Annual review of nuclear and particle science. Palo Alto, Calif., Annual Reviews Inc., 1978-. v. 28-. ISSN: 01638998.

Formerly (1952-77) as *Annual review of nuclear science*.

V.36 (1986) contains 19 reviews, the work of 29 contributors. It includes a Cumulative Index to the contents of v.27-36, listing chapter titles under 18 headings. There are entries for the reviews in v. 36 under 11 of these headings, *e.g.* Accelerators, Astrophysics, Instrumentation, Nuclear applications, Nuclear structure, Particle theory.
Class No: 621.039(055)

[3847]

Progress in nuclear energy. Oxford, etc., Pergamon Press, 1956-. ISSN: 01491970.

Two series of this title have been published. One presents conventional reviews in different parts of the subject field, sometimes more than once yearly (*e.g.* vols. 12-14 were published in 1985). The other (the 'New series' published a number of reviews in 12 specific series fields (*e.g.* II. Reactors; IX. Analytical chemistry). Additionally a 12v. set (desribed as 'Old series') was published in 1976 to disseminate the proceedings of a European Conference with the title *Nuclear energy maturity*. It is also published as a more conventional 'international review journal covering all aspects of nuclear science and engineering' (*Aims and scope*) intended, from 1987, to give rapid publication by direct reproduction from authors' typescripts.
Class No: 621.039(055)

[3848]

Progress in particle and nuclear physics. Oxford, Pergamon Press, 1971-. Annual. ISBN: 0080936671. ISSN: 01466410.

V. 27 (1991. 336p.) covers physics of precision experiments with Z's, structure of nuclear matter and QCD sum rules, colour screening and colour transparency in hard nuclear processes, the Nambu and Jona-Lasinio model.
Class No: 621.039(055)

Yearbooks & Directories

[3849]

World nuclear industry handbook, 1988. 2nd ed. Sutton, Surrey, Nuclear Engineering International/Reed Business Publishing Ltd., 1987. 304p. graphs, tables, map. ISSN: 00295507.

Published as supplement to November 1987 issue of *Nuclear engineering international*.

Includes statistical tables on nuclear capacity and performance worldwide. Reactor directory surveys progress in each country, with statistics, and lists reactors in one A-Z name index and under each country, with brief information on status. Sections follow giving general information and detailed technical data for each reactor (planned; under construction; operating; or shut down). Buyers' guide lists companies A-Z, and by countries A-Z, with addresses and products/services; also products and services A-Z, with their companies. *Class No:* 621.039(058)

Almanacs

[3850]

GREENHALGH, G. *and* **JEFFS, E.,** *eds.* **Nuclear industry almanac.** V.1. Western Europe. Great Missenden, Bucks., Nuclear Energy Intelligence, 1982. tables. £35. ISBN: 0907643000.

Energy profiles of 18 West European countries (UK, p. 153-68), covering nuclear plant construction and operation, licensing procedures, radiation protection, research activities, uranium supplies, enrichment, reprocessing, spent fuel, waste management). Surveys of nuclear industries of 11 countries, with details of firms. Directory of almost 400 companies and organizations. *Class No:* 621.039(059)

Tables & Data Books

[3851]

BASHKIN, S. *and* **STONER, J.O.,** *Jr.* **Atomic energy level and Grotrian diagrams.** Amsterdam, North-Holland, 1976-82 (v. 1-4).

1. *Hydrogen I - Phosphorus XV.* 1976. xx, 616p. $137.75. ISBN 072043227. Addendum v. 1978. viii, 176p. ISBN 0444852360.

2. *Sulphur I - Titanium XXII.* 1978. xviii, 650p. $137.75. ISBN 0444851496.

3. *Vanadium I - Chromium XXIV.* 1981. xvi, 550p. $137.75. ISBN 0444860061.

4. *Manganese I - Manganese XXV.* 1982. xiv, 354p. $97.75. ISBN 0444864636.

Pictorial representation of energy levels and electron transitions. Special and structural data for atoms and molecules. Extensive bibliographies. 'These two volumes [v. 1-2] are extremely useful to atomic spectroscopists and are highly recommended' (*Applied spectroscopy,* v.33(4), 1979, p. 422-3). *Class No:* 621.039(083)

[3852]

CINDA. Vienna, International Atomic Energy Agency, 1979-85.

A worldwide bibliography of the literature on microscopic neutron data, published on behalf of the US National Nuclear Data Centre, the USSR Nuclear Data Centre, the NEA Data Bank and the IAEA Nuclear Data Section. Contains well over 200,000 entries, with data on microscopic cross sections, angular distributions, energy spectra and resonance parameters, arranged by element and isotope. 2 archival volumes plus supplements: CINDA - A. 2 vols. 1979. 1,929p. Sch.1,050 (Entries for literature 1935-1976). CINDA - B. 1984. 771p. Sch.700 (Entries for literature 1977-81).

Annual cumulated updates began with CINDA 84. CINDA 85. 1985. 394p. Supplement Dec. 1985. Sch.560 (Entries for literature 1982-). Also available online.
Class No: 621.039(083)

[3853]

ERDTMANN, G. Neutron activation tables. Verlag Chemie International, 1976. 146p. $45.90. (*Topical presentations in nuclear chemistry, v.6.*) ISBN: 3527256938.

Includes cross section values of thermal, epithermal and fast reactor neutrons, decay rates of the activation products and data tables on radionuclides and natural isotopes.

'The presentation of the data is clear and practical ... A valuable and useful data tabulation for all those who use the activation with reactor and 14-MeV generator neutrons as an analytical tool or for the production of isotopes' (*Nuclear science & engineering*, v. 63, August 1977, p. 514). *Class No:* 621.039(083)

[3854]

Neutron cross sections. New York, London, etc., Academic Press, 1981-1984. 2v. graphs, tables. ISBN: 0125097018.

V.1 *Neutron resonance parameters and thermal cross sections* by S.F. Mughabghab, M. Divadeenam and N.E. Holden. 2 Parts. 1981-1984. $65.00 Part A. Z = 1-60. Part B. Z = 61-100.

The 4th ed. of a work previously known as BNL-325, *Neutron cross sections, v.1, Resonance Parameters* (3rd ed. 1973). Originally by D.J. Hughes and J.A. Harvey. Also referred to as the 'Barn Book'.

Presents recommended thermal cross sections, resonance properties and resonance parameters for masses Z = 1 to

....*(contd.)*

100. V.1 contains a survey of the physics of thermal and resonance neutrons. V.2 gives data for neutron cross section curves. *Class No:* 621.039(083)

[3855]

Nuclear data sheets. Duluth, Minn., Academic Press, 1969-. Monthly. (3v. pa). tables. . $343.

Produced and edited by the National Nuclear Data Center for the International Network for Nuclear Structure Data Evaluation, assembled from ENSDF (Evaluated Nuclear Structure Data File), the computerised data bank, maintained by the Center, which is based at the Brookhaven National Laboratory, Upton, New York. Provides data on nuclear structure and radioactive decay for all known isotopes, critically evaluated by scientists supplying the data from 11 countries. There are indexes of references every four months - one for each vol. - with annual cumulations since 1978 and cumulations for 1969-74 and 1975-77.
Class No: 621.039(083)

[3856]

Nuclear fact book. Platt, A.M., *and others, eds.* 2nd ed. Chur, Switzerland; London, etc., Harwood, 1985. xv, 176p., graphs, tables. ISSN: 87553376.

Sponsored by the US Department of Energy.

Intended as 'an annual desk-reference ... on energy production, consumptions and costs'. Miscellaneous data, facts, statements, summaries of legislation drawn from many sources (cited). Covers development and use of nuclear power, waste, reprocessing, disposal, proliferation, radiation, environmental and sociopolitical facts, in relation to the US. Glossaries of acronyms and radiation terms.
Class No: 621.039(083)

[3857]

UNITED STATES. National Bureau of Standards. **Atomic data and nuclear data tables.** Duluth, Minn., Academic Press, 1973-, v.12-. 6pa. (3 v. of 2 issues each). diagrs., tables. $302pa. ISSN: 0092640x.

Formerly *Atomic data* (vs.1-5, 1969-73) and *Nuclear data, Section A* (1965-70) and *Nuclear data tables* (vs.1-11, 1965-73). Replaces *NBS Circular 499: Nuclear data.*

Experimental and theoretical data in various areas of atomic and nuclear-structure physics. About 120 categories, from absorption coefficients to x-rays. Some issues contain a single paper, others five or six. Cumulative indexes of authors and subjects covering 1965-87, all vols. to v.37, appeared in v.37(ii), November 1987.
Class No: 621.039(083)

Standards

[3858]

EUROPEAN COMMUNITIES COMMISSION. Directorate - General Research, Science and Education. Nuclear Plant Safety Division. **Nuclear standards:** catalogue and classification. Fichtner, N., *and others, comps.* 3rd ed. Berlin, Cologne, Beuth Verlag, 1981. 230p. (*EUR 7135.*) ISBN: 3410580026.

Compiled under study contract granted to DIN (Deutsches Institut für Normung e.V.). First published 1975 as *Catalogue and classification of technical safety rules for light-water reactors and reprocessing plants* by M. Bloser, N. Fichtner and R. Neider (EUR 5362). Previous ed. 1977 with slightly modified title and compiled by present authors (EUR 5849). Title, issuing body, standard number, date and

....*(contd.)*

status (Amendment, Draft, Published, etc.) for 2,500 documents issued by 35 countries and international bodies. 4,000 entries in 19 main categories (*e.g.* Radiation protection; Containment; Material standards) and subdivisions (*e.g.* Radiological shielding; leakage). Section 1.1 provides a useful bibliography of glossaries of terminology and symbols (*c.*60 items).
Class No: 621.039(083.74)

[3859]

Nuclear IEEE Standards; and American National Standards on nuclear instrumentation published by IEEE. New York, IEEE; Wiley, 1978. 2v. diagrs., graphs, tables. $65. ISBN: 047105321x, v.1; 0471054301, v.2.

Contains the texts of 45 standards issued 1965-78 by ANSI and IEEE. Topics include tests, measuring equipment, instrumentation, control, protection and safety systems, reliability, standby power supplies, storage batteries.
Class No: 621.039(083.74)

[3860]

—BRITISH STANDARDS INSTITUTION. Nuclear energy. Milton Keynes, BSI, 1985. 2p. £1. (*SL 36.*)

One of the Sectional List series, giving the number, title and price group of 62 British Standards - specifications, glossaries, etc. - covering nuclear energy and its applications and other relevant standards. No index. (Full details are to be found in the annual catalogue (*qv*)).
Class No: 621.039(083.74)

Histories

[3861]

LONGSTAFF, M. Unlocking the atom: a hundred years of nuclear energy. London, Muller, 1980. 175p. illus., diagrs., tables. £7.95. ISBN: 058410457x.

A brief account, by an enthusiast from within the industry. Chapters on early experimental work and wartime developments precede the main body of the work which outlines nuclear power's growth, with emphasis on the UK. All kinds of reactor are described and there is discussion of fuels, wastes and safety factors. Bibliography p. 171.
Class No: 621.039(091)

Biographies

[3862]

Energy and nuclear sciences international who's who. 3rd ed. Harlow, Essex, Longman Group, 1990. 340p. £150. (*Longman Reference on research series.*) ISBN: 0582039681.

Previously as *International who's who in energy and nuclear sciences* (1983).

Part 1 contains professional biographical profiles of about 3,000 scientists and engineers from over 75 countries involved in the generation, storage and efficient use of energy. Part 2 is a country index, subdivided by subject specialists. *Class No:* 621.039(092)

Thesauri

[3863]

INIS: Thesaurus. Vienna, International Atomic Energy Agency, 1987. (*INIS reference series.*) ISBN: 9201781873.

First published 1970 and based on the 1969 ed. of *EURATOM Thesaurus.*

Contains 17,243 accepted terms (descriptors) and 5,803 forbidden terms (non-descriptors) from the fields of nuclear physics and reactor technology and related topics. Identifies deleted and added terms (since previous ed.) in separate lists, and arrows in main alphabetical sequence also identify added terms. Scope notes for many terms. Appendix gives full listings of narrower terms in certain cases.
Class No: 621.039:025.43

Conferences

[3864]

Meetings on atomic energy. Vienna, International Atomic Energy Agency, 1969-. Quarterly. Sch.480 pa. ISSN: 00476641.

Details of conferences, exhibitions and training courses. Conferences section has KWIC index of titles and place/date index. International training courses are arranged by date and national courses alphabetically by country. Final section gives detailed information on IAEA meetings, past and planned. *Class No:* 621.039:061:061.3

Research Establishments

[3865]

International directory of nuclear utilities. 4th ed. Lakewood, Ohio, Lotte, 1987. Annual. iv, 165p. $195. ISBN: 0961205431.

'Nuclear utilities' appears to comprise official administrative bodies (*e.g.* UKAEA) and power generating bodies (*e.g.* CEGB). 124 are listed for 30 countries. For USSR and other 'Eastern Bloc' countries only the state authority is named. About 450 nuclear power stations are listed, by country, with an A-Z index, which excludes non-power generating reactors (*e.g.* Dounreay). Apart from address and senior personnel, only type of reactor and power rating are given. Index of *c.*1,600 personnel. 40 per cent of the information in this directory relates to USA. Of limited value and extraordinarily expensive.
Class No: 621.039:061:061.62

Reactors

[3866]

ELECNUC. Paris, Commissariat à l'Energie Atomique (CEA).

Coverage: mid-1950s to date. A textual/numeric databank of information on more than 800 nuclear power stations and nuclear reactors worldwide. Detailed information for each on planning, design, dates, component and plant manufacturers, technical and operational data (power input and output, cooling, reactor type, etc). Updated monthly. Available, with CEA permission, via CISI. *Class No:* 621.039.4

[3867]

GLASSTONE, S. *and* **SESONSKE, A. Nuclear reactor engineering.** 3rd ed. New York, Van Nostrand Reinhold, 1981. xv, 805p. diagrs., graphs, tables. $39.50. ISBN: 0442200579.

First published 1963.

Covers 'the fundamental scientific and engineering principles of nuclear reactor systems, especially those used for the generation of electric power' (*Preface*). Emphasis is on nuclear fission and reactor design. Has chapters on reactor analysis, kinetics and control, materials, fuels, radiation protection, shielding and safety and types of power reactor. Chapter references.

'The authors have achieved a reasonable balance in the treatment of the various topics and have produced a textbook for a large audience' (*American scientist*, v.69, Nov/Dec., 1981, p.685). *Class No:* 621.039.4

[3868]

—NERD, A.V. Guidebook to nuclear reactors. Berkeley, Los Angeles, Univ. of California Press, 1979. xiii, 289p., illus., diagrs., tables, maps. $44.50; $15.95. ISBN: 0520034821.

Describes and illustrates reactors as they exist and indicates possible developments in nuclear systems. Includes background information on many controversial questions, *e.g.* reactor safety; radioactive waste disposal; alternative fuel cycles (uranium - plutonium *versus* thorium - uranium). 4 parts: 1. General introduction - 2. Commercial nuclear reactors - 3. Uranium resources, advanced fuel cycles and nuclear materials - 4. Advanced reactor systems. 6 Appendices of reference material, plus glossary (p. 275-80) and index. *N.B.* has a US perspective and pre-dates the Three Mile Island accident in 1979. *Class No:* 621.039.4

[3869]

INTERNATIONAL ATOMIC ENERGY AGENCY. Nuclear research reactors in the world. Vienna, IAEA, 1986. 105p. graphs, tables. Sch.100. (*Reference data series, no. 3.*) ISBN: 9201592868.

First ed. of a publication which replaces the Agency's previous two guides: *Power and research reactors in member states* and *Research reactors in member states.* Intended to be issued periodically.

17 tables showing research reactors in 55 countries with details of category (test, training, etc.), operating experience, current status (operating, under construction, planned, etc.), types (heavy water, graphite, etc), thermal power, general information and technical data for each. Includes shut-down facilities and covers 1957-85. *Class No:* 621.039.4

[3870]

MARSHALL, W., *ed.* **Nuclear power technology.** Oxford, Clarendon Press, 1983. 3v. illus., diagrs. £150. (*Oxford science publications.*)

1. *Reactor technology.* xv, 503p. £50. ISBN 0198519486.
2. *Fuel cycle.* xv, 456p. £50. ISBN 0198519583.
3. *Nuclear radiation.* xv, 363p. £50. ISBN 0198519591.

32 contributors in 24 chapters cover all aspects of nuclear power from 'How reactors work', through various types of fission reactor (with one chapter on fusion), fuel (design, reprocessing, etc.), waste, with a separate vol. on radiation, its risks and its applications in medicine and dating. Chapter references, suggestions for further reading and glossaries are valuable adjuncts. Separate volume indexes.

Various reviews from within the industry pay tribute to content, style, range, clarity, with isolated criticisms about the book's treatment of waste and costings, but an

....*(contd.)*

independent authority should be cited: 'A noble effort, which has produced a fine work of science' (*Nature*, v. 309, June 7, 1984, p. 565). *Class No:* 621.039.4

[3871]
POCOCK, R.F. Nuclear power: its development in the United Kingdom. London, Institution of Nuclear Engineers/ Unwin, 1977. x, 272p. illus., diagrs., graphs. £12.80; £5.10. ISBN: 0905418158, hardback; 0905418166, paperback.

An account by an engineer employed within the industry relying on public domain sources - government papers, reports, periodicals, newspapers and books.

[The author] 'is to be congratulated on having achieved an extremely readable volume of little more than 250 pages ... It is quite clear that the author recognized the difficulty of covering such a wide canvas in a small book and if one accepts this limitation, the book provides the reader with a good picture of how people have looked on nuclear power, its difficulties and its organizational problems year by year' (*Nuclear energy*, v.17(2), April 1978, p.75-6). *Class No:* 621.039.4

[3872]
Referativnyĭ zhurnal. 50. Yadernye reaktory. Moscow, VINITI, 1964-. Monthly. diagrs. R25. ISSN: 00342653.

About 3,000 signed abstracts pa. on nuclear reactors. 2 main sections: nuclear reactors and thermonuclear reactors, subdivided into 18 and 4 subsections respectively, *e.g.* design, calculation, fuel elements, control, materials. Illustrated with diagrams. UDC numbers. Annual subject index. English contents summary. *Class No:* 621.039.4

Radioactive Elements

[3873]
BADASH, L. Radioactivity in America: growth and decay of a science. Baltimore, Md., Johns Hopkins Univ. Press 1979. xviii, 327p. illus., diagrs. $18.95. ISBN: 0801821878.

Outlines the development of work on the chemical and physical aspects of the science connected with radioactivity, emphasizing the work done in America beween 1900 and 1920, and with particular attention to chemists Bertram Boltwood and Herbert McCoy. Full notes and index. Apart from quibbles about 'America' and 'decay' in the title and criticism of the 'tedious' presentation of the 40-odd naturally occuring radioactive isotopes, this book is warmly welcomed and labelled 'excellent' by a distinguished reviewer in *Physics today* (v.33 (7), July 1980, p.45). *Class No:* 621.039.5

[3874]
MILLER, E.W. *and* **MILLER, R.M. Environmental hazards: radioactive materials and wastes:** a reference handbook. Santa Barbara, Calif., ABC-Clio, 1990. 298p. $39. ISBN: 0874362342.

Various chapters with a range of information (*e.g.* nature and characteristics of radioactive materials; directory of organizations that deal with waste; annotated bibliography of books; listing of journal articles and government documents; select list of journal titles; bibliography of a-v materials). Glossary; index. 'An excellent addition to any reference collection' (*Choice*, April 1991, p.1292). *Class No:* 621.039.5

Isotopes

[3875]
BROWNE, E. *and* **FIRESTONE, R.B. Table of radioactive isotopes.** New York, etc., Wiley, 1986. v.p. illus., tables. ISBN: 047184909x.

The isotopes are tabled in progressive atomic number order. Included: some 262 drawings of the mass-chain decay scheme. 'This massive and highly detailed work should satisfy the demand for *adopted* properties of *all* radiations emitted by nuclei' (*Science & Technology libraries*, v.8 (1), Fall 1987, p.183). *Class No:* 621.039.8

[3876]
—Radiochemistry and the decay of isotopes. Romer, A., *ed.* New York, Dover, 1970. xv, 260p. ports., tables. ISBN: 0486625679.

Contains 26 original papers, with commentary. 'Historical essay', by A. Romer precedes, p.1-60. Appendix: 'Transformation series of the natural radioactive elements: Uranium, Thorium and actium'. List of abbreviations. Index of names; analytical index of topics. *Class No:* 621.039.8

[3877]
ROYAL SOCIETY OF CHEMISTRY. Isotopes: essential chemistry and applications. London, the Society, 1980. 412p. £15. (*Special publications, no.35.*) ISBN: 0851868304.

Covers, in detail, the chemistry of isotopes and their industrial applications. *Class No:* 621.039.8

Steam Power

Handbooks & Manuals

[3878]
GOODALL, P.M., *ed.* **The Efficient use of steam.** Guildford, Surrey, IPC Science and Technology Press, 1980. 469p. illus., diagrs., graphs, tables. £22. ISBN: 0861030184.

Previously published 1947, written by Sir O. Lyle.

23 contributors. Aims to assist industrial steam engineers or designers in safe and cost-effective use of steam. Covers theory, practice and equipment, *e.g.* 1: Properties of water and steam; 4: Shell boilers; 11: Turbines and engines; 12: Heat transfer; 13: Steam heating; 19: Steam costing. Lists relevant British standards. Appendix contains shortened steam tables (0.01 to 374.15 deg. C; 0.01 to 221 bar). References and index. *Class No:* 621.1(035)

Tables & Data Books

[3879]
HAAR, L., *and* **others. NBS/NRC steam tables:** thermodynamic and transport properties and computer programs for vapor and liquid states of water in SI units. Washington, Hemisphere Publishing Corp., New York, McGraw-Hill, 1984. xii, 320p. tables, $34.50. ISBN: 0891163549.

Thermodynamic property values for the specific volume, density, specific internal energy, specific enthalpy, and specific entropy of water and steam for the range 0 - 2000 deg. C and 0 - 30,000 bar. Tables for transport and other thermophysical property values are also given (*e.g.*

....*(contd.)*

viscosity, Prondtl number, dielectric constant). Appendices provide details of the Helmholz function used for derivation of values and of the computer programs used, and the relevant equations. 'These tables have been reviewed and approved by the Office of Standard Reference Data of the [US] National Bureau of Standards ...' (*Foreword*). 70 references. *Class No:* 621.1(083)

[3880]
—RIVKON, S.L., *and others*. Thermodynamic derivatives for water and steam. Translated by J. Kestin. New York, etc., Wiley, 1978. v, 264p. graphs, tables. $25. ISBN: 0470263636.

These tables of derivative values are based on the IFC 1968 Formulation for Scientific and General Use, and cover specific volume, enthalpy, entropy and isobaric specific heat as functions of pressure and temperature, - deemed to be more useful in engineering applications. Entries cover pressure range 0.1 to 50 MPa, and temperature range 0 to 600 deg. C. *Class No:* 621.1(083)

[3881]
IRVINE, T.F. *and* LILEY, P.E. **Steam and gas tables,** with computer equations. Orlando, Florida, Academic Press, 1984. xiv, 185p. graphs, tables. $29.50. ISBN: 0123740800.

Tables cover saturation and superheat properties of steam and properties of air and 12 other gases. Relevant equations are given, which were used in microcomputer programs published by the authors. Appendices contain tabulated conversion factors and list of references. SI Units used. *Class No:* 621.1(083)

[3882]
—IRVINE, T.F. *and* HARTNETT, J.P. Steam and air tables in SI units; including data for the substances and a separate Mollier chart for steam. Washington, Hemisphere Publishing Corp., 1976. [vii], 127 p. tables. $23. ISBN: 0891160043.

Tables 1 to 6 (12 of which give values) cover properties of steam and thermal conductivity of water. Ranges: 0 - 700 deg. C; 0.01 to 970 bar. Coverage: specific volume, density, specific enthalpy, specific entropy.

'... A compilation of what should be useful data, all in SI units ... [which is] ... surely one day to become standard in the United States ... The conversion process will not be easy and less-than-perfect data sets are bound to appear. This book, admirable in its extent, is an example' (*Chem. Eng. Science,* v.31, 1976, p. 857). No justification is offered. *Class No:* 621(083)

[3883]
KEENAN, J.H., *and others.* **Steam tables:** thermodynamic properties of water, including vapor, liquid and solid phases. 2nd ed. New York, Wiley - Interscience, 1978. xxiii, 156p. diagrs., charts, tables. $20.95. ISBN: 0471042102.

Previous ed. 1969.

The Keenan and Keyes Tables were first published in 1936. The present tables appeared in 1969 and extended their range to 1300 deg. C and 100 MPa in pressure, with independent correlation of all new data. This 2nd ed. now provides full conformity with SI units. Values for specific volume, internal energy, enthalpy and entropy are tabulated for pressure and temperature parameters. Mollier and temperature-entropy charts are included. *Introduction to the steam tables* is repeated in French, German, Italian, Japanese, Russian and Spanish. *Class No:* 621.1(083)

[3884]
SCHMIDT, E. **Properties of water and steam, in SI units.** [Thermodynamische Eigenschaften von Wasser und Wasserdampt: 0 - 800 deg. C, 0 - 1,000 bar.] 2nd rev. and updated printing. New York, Berlin, etc., Springer-Verlag, 1979. 190p. tables. $40.20. ISBN: 0387096019.

Previously published 1969.

Parallel texts in English and German. Main tables (on equilibrium properties of water and steam), based on 1967 IFC Formulation for Industrial Use, are repeated, and there are revised tables for transport properties - dynamic viscosity, thermal conductivity, Prandtl number and surface tension. New Skeleton Tables and relevant equations are provided and there is a bibliography of 265 items. *Class No:* 621.1(083)

[3885]
—POTTER, J.H. Steam charts. New York, American Society of Mechanical Engineers, 1976. xiv, 128p. $25. (*ASME SI-10.*)

'Thermodynamic properties of steam, h-v, in graphical form for the superheated, vapor and liquid conditions in both SI (Metric) and US customary units' (*Sub-title*). Enthalpy - volume charts are provided first in SI, then in US Customary Units. *Class No:* 621.1(083)

Histories

[3886]
BRIGGS, A. **The Power of steam:** an illustrated history of the world's steam age. London, Joseph, 1982. 208p. illus. (some col.). £10.50. ISBN: 0718120760.

A non-technical account of the uses of steam power from the earliest days to the present time when nostalgia has nurtured a strong revival of interest, manifesting itself in a steam preservation movement. All applications are described and illustrated (more than 200 photographs and drawings). A list of places (mostly in the UK) where steam engines may be seen is included.

'An illuminating and infuriating book ... It's a coffee-table book with references - although in most instances the illustrations add to, rather than detract from the text' (*New Scientist,* v.93, 11 Mar. 1982, p. 664). *Class No:* 621.1(091)

[3887]
HILLS, R.L. **Power from steam: a history of the stationary steam engine.** Cambridge, Univ. Press, 1989. xv,338p. illus., diagrs., tables. £45. ISBN: 0521343569.

Traces 'the technical development of the steam engine as applied in industry and centred on the cotton textile industry in the Lancashire region' (*Preface*). 15 chapters: 1. The noblest machine ... 10. The internal operation of the machine ... 15 The most economical mode of power. Extensive chapter notes and bibliography, p.292-326. Detailed, analytical index, p.327-38. A scholarly treatment. *Class No:* 621.1(091)

Electrical Engineering

Abbreviations & Symbols

[3888]
BRITISH STANDARDS INSTITUTION. Guide for graphical symbols for electrical power, telecommunications and electronics diagrams. London, BSI, 1985-86. 13 pts. (*BS 3939:1985-1986.*)
Supersedes BS108 and BS530.
Pt. 1: General information, general index. 1986. 35p.
Pt. 2: Symbol elements, qualifying symbols and other symbols having general applications. 1985.
Pt. 3: Conductors and connecting devices. 1985. 8p.
Pt. 4: Passive components. 1985. 8p.
Pt. 5: Semiconductors and electron tubes. 1985. 24p.
Pt. 6: Production and conversion of electrical energy. 1985. 16p.
Pt. 7: Switchgear, controlgear and protective devices. 1985. 28p.
Pt. 8: Measuring instruments, lamps and signalling devices. 1985. 12p.
Pt. 9: Telecommunications: switching and peripheral equipment. 1985. 16p.
Pt. 10. Telecommunications: transmission. 1985. 32p.
Pt. 11: Architectural and topographical installation plans and diagrams. 1985. 12p.
Pt. 12: Binary logic elements. 1985. 88p.
Pt. 13: Analogue elements. 1985. 12p.
Over 2,000 symbols for use in draughting throughout the field. *Class No:* 621.3(003)

[3889]
Index of acronyms and abbreviations in electrical and electronic engineering. Weinheim, etc., VCH, 1989. [vii],538p. £75. ISBN: 0895738120.
About 45,000 entries. More comprehensive than Keller (*qv*), but little to choose between them.
Class No: 621.3(003)

[3890]
INSTITUTE OF ELECTRICAL AND ELECTRONICS ENGINEERS. Electrical and electronics graphic symbols and reference designations. 2nd ed. New York, IEEE, 1987. v.p. diagrs. ISBN: 0471634565.
Includes all IEEE and ANSI standards. Covers logic functions, parts and equipment, diagrams and electrical wiring. *Class No:* 621.3(003)

[3891]
INTERNATIONAL ELECTROTECHNICAL COMMISSION. Letter symbols, including conventions and signs for electrical technology: a handbook for everyday use. Geneva, I.E.C., 1983. vi, 114p. illus. S Fr.45. ISBN: 2287300053.
'This handbook contains ... the principles governing conventions for electric and magnetic circuits ... and the standardization of letter symbols, together with a list of letter symbols in general use ... The terms used are those of the latest edition of the IEV (International Electrotechnical Vocabulary) ...' (*Introduction*). Letter symbols and graphic symbols for quantities are linked to SI units. All text and entries in English and French. Index in each language.
Class No: 621.3(003)

Databases

[3892]
INSPEC. Stevenage, Herts., Institution of Electrical Engineers, 1969-. Updated weekly.
INSPEC database provides online access to *Physics abstracts, Electrical and electronic abstracts*, and *Computer and control abstracts*, through DIALOG from 1969, and through BRS since 1970. Available on CD-ROM. D.T. Hawkins notes both bibliographic (INSPEC, PHYS, SPIN, Atomicindex and Nuclear science) and numeric (Superindex, PHYSCOMP, OECD Nuclear Energy Agency database, and GAPHYOR) databases covering physics (*Database*, v.8(1), February 1985, p.15-16, with a comparison table and references, p.17-18). *Class No:* 621.3(003.4)

[3893]
Merlin-Tech. Grenoble, Merlin-Gerin Documentation.
Covers 1973 to date. A bibliographic database providing references, with abstracts, to world literature on electrical engineering, electronics and telecommunications in periodical articles and books. File size: 33,000 citations. Updated monthly, *c*.2,500 additions pa. Host: ESA-IRS. In French. *Class No:* 621.3(003.4)

[3894]
ZDE (Zentralstelle Dokumentation Elektrotechnik). Frankfurt-am-Main, FDR, FIZ Technik.
Covers 1968 to date. A bibliographic data base, providing references, with abstracts, to world literature of electrical engineering, electronics, communications, computers and control engineering, and including semiconductor physics, plasma physics, superconductivity and other aspects of electronic and magnetic materials. File size: over 750,000 citations. Updated monthly. 60,000 additions pa. Available via FIZ Technik. In English and German.
Class No: 621.3(003.4)

Bibliographies

[3895]
ARDIS, S.B. A Guide to the literature of electrical and electronics engineering. Poland, J.M., *ed*. Littleton, Colorado, Libraries Unlimited Inc., 1987. xx, 190p. £29.95. (*Reference sources in science and technology series.*) ISBN: 0872874745.
The first English-language, wide-ranging and substantial guide to the field since Burkett and Plumb's *How to find out in electrical engineering* (1967), and therefore very welcome. Its subject range extends to communications and computing engineering and includes microwaves, radar, materials science and satellite technology. Excluded are robotics, electrical wiring and construction and applications of electronic devices in music, biomedicine and avionics. Coverage is restricted to English-language reference material published post-1978. Arrangement is by type of material (bibliographies, abstracts, data bases, patents, standards, etc.). 8 parts, 25 chapters, with annotated entries for almost 700 publications and information sources.
British users will note omission, over-brief treatments, mis-spelling of some British sources of information, and will be disadvantaged by the American publisher and price data given, but will still find this a valuable guide overall, especially its strong coverage of the many handbooks in this subject area. *Class No:* 621.3(01)

[3896]
BRITISH LIBRARY. Science Reference Library. **Useful books of reference for designers (1926-1983) held by the Science Reference Library. Part.2: Electrical and electronic engineering.** Palyza, M.M., *sel*. London, the Library, 1984. i, 71p. £5. ISBN: 0712307133.

A bibliography of almost 500 works arranged under *c*.200 headings by the Library's unique classification system. There is a Contents list to indicate the 12 broad groupings and an index of subjects which leads easily to specific groups of relevant publications (despite many surprising omissions, *e.g.* no entries for microelectronics or microprocessors). Strong on circuits and electronic data sources. Clearly set out in reproduced typescript, the entries include an occasional annotation - more would greatly enhance the value of this compilation. *Class No:* 621.3(01)

[3897]
Index to IEEE publications. New York, IEEE Press, 1951-. Annual.

The 1990 ed. (V.1: Author; V. 2: Subject) lists all articles in the Institute's journals, transactions and magazines; published papers presented at IEEE-sponsored conferences; summaries of standards, books, conference records, special issues, etc. *Class No:* 621.3(01)

Handbooks & Manuals

[3898]
CHANG, S.S.L., *ed*. **Fundamentals handbook of electrical and computer engineering.** New York, Wiley, 1982-3. 3v. diagrs., graphs, charts, circuits, tables. £50. ISBN: 0471862150, v.1; 0471862134, v.2; 0471862142, v.3.

1. *Circuits, fields and electronics*. xi, 707p.

2. *Communication, control, devices, and systems*. xi, 737p.

3. *Computer hardware, software, and applications*. xi, 507p.

Aims to provide details of current research and development across a wide range of topics. Claims that the set will enable non-specialists to enter and carry out design work in an unfamiliar research area seem exaggerated. The work is strong, however, on basic principles and theory in all the subjects it treats. Chapter references, bibliographies and glossaries are provided. (Citations later than 1979 are hard to find.) Review in *IEEE Spectrum* complains that 'treatment of subjects is often too brief - and sometimes erroneous or out-of-date' (v.20, August 1983, p.14). *Class No:* 621.3(035)

Dictionaries

Polyglot

[3899]
COMMISSION ÉLECTROTECHNIQUE INTERNATIONALE. **Dictionnaire CEI multilingue de l'électricité.** [IEC multilingual dictionary of electricity.] Geneva, CEI, 1983. 2v. (xvi, 892p.; [vi], 908p. ISBN: 2827300036, v.1; 2827300044, v.2.

'This Dictionary is not intended to replace the IEV [*International electrotechnical vocabulary*, (qv.)] in which terms remain classified by specialized fields, but supplements it by giving in alphabetical order all the terms arranged under key words ...' (*Foreword*). The dictionary

....(contd.)
covers more than 7,500 terms drawn from the 37 currently revised chapters of the IEV. (Terms from subjects undergoing revision - *e.g.* telecommunications - are excluded.) Users must beware this limitation and consult the list of fields covered which appear on yellow pages in the middle of v.1.

V.1 contains entries and definitions of all terms in French, followed by entries and definitions in English, with equivalents in Russian, German, Spanish, Italian, Dutch, Polish and Swedish given for each term.

V.2 contains alphabetical indexes in each language other than English and French, with references to the English and French definitions. *Class No:* 621.3(038)=00

[3900]
IEC multilingual dictionary of electricity. New York, Institute of Electrical and Electronics Engineers/Wiley, 1983. 461p. $49.95. ISBN: 0471807842.

Comprises the part of v. 1 of the *IEC Multilingual Dictionary* which lists terms in English and gives their definitions in English with equivalent terms in French, German, Italian, Dutch, Polish, Spanish, Swedish and Russian. Upwards of 18,000 entries, including 10,000 cross-references. Excludes telecommunications and certain electronics and microelectronics terminology. *Class No:* 621.3(038)=00

[3901]
INTERNATIONAL ELECTROTECHNICAL COMMISSION. **International electrotechnical vocabulary.** 3rd ed. Geneva, the Commission, 1973-. In progress.

First published 1938. Previous ed. 1954-70.

A standard multilingual (8 or 9 languages) dictionary previously issued in 22 subject parts (*e.g.* Machines and transformers; Electric traction; Lighting), with terms arranged in classified lists and including definitions in English, supplemented by indexes in all languages. *Class No:* 621.3(038)=00

[3902]
LUGINSKY, Y.N., *and others*. **Dictionary of electrical engineering;** English-German-French-Dutch-Russian. Deventer, Netherlands, Kluwer; Moscow, Russky Yazyk Publishers, 1985. 480p. ISBN: 9020119109.

About 8,000 terms, uniquely identified by initial letter and index number, are given in English, with equivalents in other four languages (with genders and - where necessary - plurals). Indexes in other four languages enable translations between any combinations of languages. Emphasis is on basic electrical engineering subjects: circuit theory; measurements; power generation, transmission and distribution; electric machines, etc. *Class No:* 621.3(038)=00

English

[3903]
BRITISH STANDARDS INSTITUTION. **Glossary of electrotechnical, power, telecommunication, electronics, lighting and colour terms.** London, BSI, 1971-92. 5 parts. (*BS 4727: 1971-92*.)

Supersedes BS 204, BS 205, BS 233 and BS 1611. Pt. 1: *Terms common to power, telecommunications and electronics*. Groups 01-13. 1971-1986. Pt. 2: *Terms particular to power engineering*. Groups 01-18. 1971-1992.

....(contd.)

Pt. 3: *Terms particular to telecommunications and electronics*, Groups 01-12. 1971-92. Pt. 5: *Terms particular to electromedical equipment*, Group 01. 1985.

'Titles to any further groups will be added in due course' (*Guide to the glossary*, p.4). *Class No:* 621.3(038)=20

[3904]
INSTITUTE OF ELECTRICAL AND ELECTRONICS ENGINEERS. IEEE standard dictionary of electrical and electronic terms. Jay, F., *ed.* 4th ed. New York, IEEE/Wiley, 1988. 1270p. illus.

First published 1972.

About 24,000 entries, including almost 9,000 new or revised terms, with separate list of *c.*15,000 acronyms, abbreviations, etc. All definitions are derived from IEEE or other standards (about 450 sources) and are thus authoritative. *Class No:* 621.3(038)=20

[3905]
JACKSON, K.G. *and* **FEINBERG, R. Dictionary of electrical engineering.** 2nd ed. London, Butterworths, 1981. [8], 350p. illus. £15. (*Butterworths technical dictionaries.*) ISBN: 0408004509.

First published 1965, with reprints.

Upwards of 3,000 terms 'in general use among engineers' (*Preface*), including electronics terms where applications are relevant in electrical engineering areas. 2nd ed. contains 200 new entries and substantial revision of many others. Appendices include SI units, decimal prefixes and common abbreviations. Many clear line drawings.

Class No: 621.3(038)=20

German

[3906]
BUDIG, P.K., *ed.* **Dictionary of electrical engineering and electronics.** Amsterdam, Elsevier, 1985. 2v. (723p.; 690p.). ISBN: 0444995951, v.1; 0444995943, v.2.

V.1: *English-German;* v.2: *German-English.* Each volume contains around 60,000 terms from all fields of electrical engineering, electronics, telecommunications, with special attention to modern developments in, for example, lasers, superconductivity, microelectronics, opto-electronics, computing. Some significant and surprising omissions noted in English-German volume.

'The main aim of the editor and his colleagues was to produce a very up-to-date and comprehensive listing, and, to a large extent, they have succeeded' (*Physical science, measurement and instrumentation, management and education, reviews* (*IEEE Proceedings A*), v.134(4), April 1987, p.342). *Class No:* 621.3(038)=30

[3907]
—**HABERLE, G.D.** *and* **HARBERLE, H.D.** Kurzlexikon der Elektrotechnik: elektrotechnische Begriffe und Ihre Erklärung. Frankfurt-a-m, Frankfurter Fachverlag, 1979. 104p. DM15. ISBN: 3872340549.

1,300 entries, including cross-references (few) and abbreviations, briefly defined for a wide range of topics within electrical engineering and electronics, concentrating thereby on more frequently encountered terms. In German. *Class No:* 621.3(038)=30

[3908]
Fachwörterbuch Energie - und Automatisierungs-technik. [Dictionary of power engineering and automation.] Munich, Siemens, 1985-86. 2v. (546p.; 482p.). £58.30. ISBN: 0471909262 v.1; 0471909394 v.2.

V.1: *German-English.* V.2: *English-German.*

60,000 items in each vol., covering power supply systems, electrical machines, power electronics, electronic and computer control systems and components and relevant terminology in chemical and mechanical engineering. The dictionaries have been compiled in Germany by staff of the Translating Department of Siemens AG and there is emphasis on German abbreviations and standards. Compound names, especially German phrases, cause difficulties in tracing the 'Key' entry where cross references are not comprehensively provided.

'These dictionaries can be confidently recommended' (*Electric power applications* (*IEEE Proceedings B*), v.134(3), May 1987, p.133).

'... a thoroughly scholarly work and well worth the effort to master its intricacies' (*Physical science, measurement and instrumentation, management and education, reviews* (*IEEE Proceedings A*), v.134(4), April 1987, p.342). *Class No:* 621.3(038)=30

[3909]
SCHWENKHAGEN, H.F. *and* **MEINHOLD, H. Wörterbuch elektrotechnik und elektronik:** Deutsch-Englisch/Englisch-Deutsch. [Dictionary of electrical and electronic engineering, German-English, English-German.] 5th ed. Düsseldorf, Cornelson Girardet, 1989. 868p. ISBN: 3464494004.

First published 1959 as *Fachwörterbuch Elektrotechnik*, by H.F. Schwenkhagen.

About 40,000 entries in each half. Entry terms clearly set out in bold face with equivalents in roman. Noun/adjectival compounds and phrases (*e.g. to put a transmitter on the air*) indicated below entry word (*put*). Covers electrical, electronic and control engineering with terminology of related fields. Context of many terms is indicated by UDC numbers. Genders and verb functions not given. A welcome new edition of a standard work. *Class No:* 621.3(038)=30

[3910]
WENNRICH, P. Encyclopedic Dictionary of electronics, electrical engineering and information processing: English-German, German-English. [Enzyklopädisches Wörterbuch der elektrotechnik, elektronik und informations verarbeitung.] Munich, K.G. Saur, 1990-. V.1-. £110 each. ISBN: 3598106815, (V.1); 3598106807, (8 volume set).

V.1: A-C (English-German). xv,431p.

To be completed in 8 v. (English-German, v.1-4; German-English, v.5-8). 200,000 English entries; 600,000 German equivalents. Includes alternative uses, further explanations and differentiates American and British English. A definitive work. *Class No:* 621.3(038)=30

[3911]
—**WENNRICH, P.** Encyclopedic Dictionary of electronics, electrical engineering and information processing. Munich, K.G. Saur, 1990. 2v., each with 400p. £65. each. ISBN: 3595108807, set.

V.1: Englisch-Deutsch.

V.2: Deutsch-Englisch.

A shortened version of his 8-v. set, in preparation. 100

....(contd.)
specialised categories are applied. Entries (*c*.50,000 per volume) mostly consist of headwords and equivalents.
Class No: 621.38(038)=30

French

[3912]
PIRAUX, H. Dictionnaire français-anglais des termes relatifs à l'électronique, l'électrotechnique, l'informatique et aux applications connexes. New ed., rev. and corrected by J. Milsant. Paris, Eyrolles, 1981. 208p.
Over 10,000 terms with *c*.20,000 equivalent English terms, covering electrical engineering, electronics, telecommunications and related subjects in physics. A companion vol. to the author's *Dictionnaire anglais-français relatifs à l'électronique, et électrotechnique* (13th ed. 1980).
Noted: 10th ed. 1984 of the French-English dictionary and 14th ed. 1983 of the English-French dictionary.
Class No: 621.3(038)=40

Spanish

[3913]
IBEAS, F.F. Diccionario technologico inglés-español de electricdad, electronica, telecomunicacion y materias afines con la fisica, la optica y la quimica. Madrid, Alhambra, 1974. xi, 441p. diagrs., graphs, tables. E950. ISBN: 8420504920.
Spanish equivalents for about 38,000 terms and abbreviations in electricity, electronics and telecommunictions, supported by several thousand terms from physics, chemistry, optics and nuclear energy. Separate section for about 1,700 English abbreviations and initialisms. 36 tables of units, conversions and miscellaneous data. 2,000 illus. *Class No:* 621.3(038)=60

Russian

[3914]
MACURA, P. Russian-English dictionary of electrotechnology and allied sciences. With supplement. Malabar, Florida, Krieger, 1986. x, 707p. + 234p. (Supplement). ISBN: 0898748692.
First published 1971.
This reprint, with supplementary material, contains about 60,000 entries, ranging beyond its title scope into related fields, *e.g.* computing, physics, acoustics, control, astronomy. It also includes many non-technical words and proper nouns and names. Grammatical forms and noun genders are given. The bound-in Supplement adds a further 10,000 entries to photoset typescript.
Class No: 621.3(038)=82

Chinese

[3915]
English-Chinese glossary of electronic and electrical engineering. New York, French and European, 1980. 636p. $29.95. ISBN: 068697364x.
Few English-Chinese dictionaries are currently available. This example is reputed to be 'especially strong in electrical and power teminology' (Ardis, S.B. *Guide to the literature of electrical and electronic engineering.* 1987).
Class No: 621.3(038)=951

Reports Literature

[3916]
ERA reports, 1969-1986. 16th ed. Leatherhead, Surrey, ERA Technology Ltd., 1987. v, 104p. (*ERA Report 87-0000.*)
15th ed., 1986.
Details of almost 300 reports, with title, report number, price, availability and informative abstract. Arranged in descending numeric order, *i.e.*, reverse chronological order. Subject index. Coverage: whole fields of electrical and electronic engineering, including materials, devices, and components. *Class No:* 621.3(047)

Reviews & Abstracts

[3917]
CEGB abstracts. London, Central Electricity Generating Board, 1985-. v.1-. Monthly. £50. ISSN: 02670372.
First published 1949, as *CEGB digest*, 1958-84.
About 1,000 indicative abstracts pa. chiefly on the generation and transmission of electricity. 17 categories, some subdivided, *e.g.* Fuels; Nuclear power generation; Transmission and distribution. Other headings include: Flue gases, Chimneys, Plumes, Hydroelectric power, Pumped storage, Reliability, Quality Control. Includes announcements of relevant meetings and conferences. Occasional reading lists (*e.g.* Sep/Oct. 1987 issue has book list on energy strategy). No indexes. *Class No:* 621.3(048)

[3918]
Current papers in electrical and electronics engineering. London, Institution of Electrical Engineers. 1964-. no. 1-. Monthly. £95. ISSN: 00113778.
Originally as *Current papers for the professional electrical and electronics engineer*, then as *Current papers in electrotechnology*. Present title 1969-. An INSPEC alerting service produced jointly by the IEE and the Institute of Electrical and Electronics Engineers Inc., in association with the Institution of Electronic and Radio Engineers.
About 75,000 items pa. giving bibliographical citations only, covering everything published in *Electrical and electronics abstracts* (*qv*), arranged by the same classification, and including periodical articles, reports, books, theses and conference papers. No indexes. Nowadays published virtually simultaneously with the abstracts journal and therefore loses its 'current awareness' value.
Class No: 621.3(048)

[3919]
Electrical and electronics abstracts. Science abstracts. Section B. London, INSPEC/Institution of Electrical Engineers, in association with the Institute of Electrical and Electronics Engineers, Inc., New York, 1898-. v.1-. Monthly. £750 pa.
Currently more than 70,000 abstracts pa. and still growing! Classifed into 8 major categories (*e.g.* 10.00: Circuit theory and circuits - 20.00: Components, electronic devices and materials; ... 80.00: Power systems and applications), subdivided (*e.g.* 12.00: Electronic circuits; 12.70: Filters; 12.70F: Digital filters). Each issue has 6 indexes (subjects, authors, bibliographies, books, conferences, corporate authors). Half-yearly indexes, plus cumulative 4-5 year indexes. The major abstacting service in its field. Available on-line as part of the INSPEC data base. Also INSPEC magnetic tapes are available twice monthly. CD-ROM. *Class No:* 621.3(048)

[3920]
Key abstracts. Hitchin, Herts., INSPEC/Institution of Electrical Engineers, 1975-. 14pts. Monthly. £56, per part; £690, all sections.

Advanced materials. Preparation, structure, properties, testing.

Antennaes and propagation. Includes radiowave propagation, radar, radio navigation.

Artificial intelligence. Includes expert systems and knowledge engineering.

Computer communications and storage. Multiprocessor systems, discs, interfaces, networks.

Computing in electronics and power. Computer applications in electronics, communications, etc.

Electronic circuits. Power electronics, amplifiers, pulse circuits, filters, etc.

Electronic instrumentation. Includes measurement systems, transducers, display devices.

Measurements in physics In mechanics, heat, optics, fluid dynamics, radiation.

Optoelectronics. Fibre optics, lasers, holography, etc.

Power systems and applications. Networks, apparatus, electric machines, industrial applications.

Robotics and control. Includes applications in industry, production, transport.

Semiconductor devices Includes integrated circuits.

Software engineering Includes programming languages, operating systems, data base management.

Telecommunications. Includes information theory, switching, optical communications.

Selections from the INSPEC data base of about 250 key papers which deliberately restrict the subject coverage to match more aptly the interests of researchers and practical engineers. Originally in 8 parts, extended to 11 and now (1988) expanded to 14, with 4 new subject areas and revisions of title and scope of others. *Class No:* 621.3(048)

[3921]
PASCAL folio. Part F21. Électrotechnique. Paris, Centre National de la Recherche Scientifique, 1985-. 10 pa. F570.

First published (1940-) as *Bulletin signalétique. 4. Physique. II. Électricité.*

As *Bulletin signalétique,* Part 140, (1961-84) contained about 5,500 entries, many with informative or indicative abstracts, systematically arranged in 11 main groups, with sections and sub-sections for each, *e.g.* Matériaux - 09. Électrotechnique - 10. Électroénergetique - 11. Applications ... Items drawn from 13,000 journals, conference papers, reports and theses. Subject and author indexes per issue and annually. Available on paper or in microform. Also available on-line as part of PASCAL file, through ESA-IRS or Questel. *Class No:* 621.3(048)

[3922]
Referativnyĭ zhurnal. 21. Elektrotekhnika. Moscow VINITI, 1982-. Monthly. R.206.

Formerly *Referativnyĭ zhurnal Elektrotekhnika i elektroenergetika* (1980-81). Originally *Referativnyĭ zhurnal Elektrotekhnika i energetika* (1961-1979).

Since the hiving off in 1982 of the *Energetika* (power) portion, now contains about 27,000 indicative and informative signed abstracts pa. with UDC numbers, some illustrations and graphs. 11 sections: General problems - Electrotechnical materials, capacitors, wires and cables - Electrical machines and transformers - Electrical switchgear - Power conversion techniques - Electrical drives and industrial automation - Electrotechnology - Electrical

.... *(contd.)*
equipment in transport - Electrification and automation in agriculture - Domestic electrification - Lighting and infra-red techniques. Annual author and subject indexes.
Class No: 621.3(048)

Periodicals

[3923]
List of journals and other serial sources, INSPEC. London, INSPEC/Institution of Electrical Engineers, 1983-. Annual. ISSN: 02647508.

Continues *List of journals (INSPEC).* The 1990 ed. lists about 4,000 titles of serial publications (including monograph series, annual reviews, report series) which are currently scanned for the INSPEC data base. Additionally, publications, which have ceased or which are no longer scanned and cross references (for changed titles) are included. Bibliographical details for each include ISSN, CODEN and British Library Document Supply Centre shelfmark. There are indexes of abbreviated titles, CODENs, ISSNs, and journals abstracted completely, plus an arrangement of titles under country of publication.
Class No: 621.3(051)

Yearbooks & Directories

[3924]
Electrical and electronics trades directory: 1987. 105th ed. Stevenage, Herts., Peregrinus, 1987. Annual. xvi, [615]p. £45. ISBN: 086341091x.

First published 1883.

The 'Blue Book' offers 'a comprenhensive guide to companies in the electrical and electronics industries, showing their main sources of manufacture and supply' *[Foreword].* Lists *c.*4,000 organizations. Covers manufacturers, suppliers, servicing companies, representatives, wholesale distributors. A classified list of products and materials comprises half the content. Indexes of firms and persons and of trade names. Emphasis is British.
Class No: 621.3(058)

[3925]
—**Where to buy electrical equipment and services.** Gale, L., *ed.* 1987 ed. Redhill, Surrey, Where to Buy Ltd., Annual. 170p. £22. ISSN: 02686333.

Previously published as: *Where to buy: electrical plant, suppliers and services.*

A buyers' guide arranged alphabetically under product, component or service (Access control-Yacht light fittings) with name, address, telephone and telex data.
Class No: 621.3(058)

Tables & Data Books

[3926]
Electrical engineering handbook. Berlin, Siemens / London, Heyden, 1976. Reprinted 1981. xii, 750p. diagrs., graphs, tables. £28.35. ISBN: 0855012315.

English translation of *Handbuch der Elektrotechnik,* first published 1969.

63 contributors. 16 chapters. 4 main parts: 1. Electrical engineering fundamentals - 2. Electrical power engineering - 3. Power engineering applications - 4. Control engineering and automation. An abundance of data and factual

....*(contd.)*
information. 361 tables.

1982 ed. published by Wiley Heyden (ISBN: 0471260207). *Class No:* 621.3(083)

[3927]
FINK, D.G. *and* BEATY, H.W., *eds.* **Standard handbook for electrical engineers.** 12th ed. New York, etc., McGraw-Hill, 1987. xvi, [2210]p. illus., diagrs., graphs, circuits, tables. $85.50. ISBN: 0070209758.

First published 1907; 11th ed., 1978.

A companion to Fink and Christiansen's *Electronic engineers' handbook.* 115 contributors. 28 sections. Scope: generation, transmission, distribution, control, conservation and application of electrical power with relevant supporting material. Major sections: 2. Electric and magnetic circuits - 3. Measurements and instruments - 4. Properties of materials - 5. Steam generation ... 10. Power system components - 11. Alternative power sources ... 14. Transmission systems ... 18. Power distribution ... 20. Motors ... 22. Electronics - 23. Electricity in transportation. Considerable revision and much new material. Complete chapter on standards, with full list of US standards in alphabetical subject order. Almost all sections have reference lists and bibliographies, but many citations are elderly or ancient (pre-World War II). The contents of each section are better displayed but identification of topics by paragraph number makes location more difficult. Worse: the index practice is inconsistent in referring - conventionally - to pages. *Class No:* 621.3(083)

[3928]
LAUGHTON, M.A. *and* SAY, M.G., *eds.* **Electrical engineer's reference book.** 14th ed. London, etc., Butterworths, 1985. [xii, 980]p. illus., diagrs., graphs, tables. ISBN: 0408004320.

First published 1945. M.G. Say's editorship has been continuous since then. Previous ed. 1973.

36 chapters by 72 specialist contributors grouped in: Fundamentals - Energy supply - Power plant - Application. Separate chapters on education and training and on standards. Many new chapters - *e.g.* Control systems and analysis; Power electronics; HUDC - and expansion into self-contained chapters of some topics - *e.g.* turbines; nuclear reactor plant; road, railway and marine transportation. Greater depth of coverage of most topics, with updated illustrative matter.

'Impressively produced in a clear, double column format, the book was a pleasure to handle ... close attention to indexing and contents listing ... prevent the reader becoming overwhelmed by the sheer scale. A minor criticism is the uneven distribution of suggested further reading ... This is a book for quick answers and also for more relaxed browsing. It is highly recommended' (*Electronics and power*, v.32(9), Sept. 1986, p.681). *Class No:* 621.3(083)

[3929]
—REEVES, E.A., *ed.* Newnes electrical pocket book. 21st ed. Oxford, Heinemann Newnes, 1992. 526p. illus., diagrs. £12.95.

First published 1937.

Contains chapters of theoretical background, followed by practical applications in: power supply and distribution; wiring; lighting; motors; switch-gear; heating and refrigeration; instrumentation. Succinct information, sometimes superficial. Adequate index; numerous line drawings and half-tones. *Class No:* 621.3(083)

[3930]
SMITH, C.B., *ed.* **Efficient electricity use:** a reference book on energy management for engineers, architects, planners and managers. 2nd ed. Oxford, etc., Pergamon, 1978. xx, 778p., illus., diagrs., graphs, charts, tables. $44. ISBN: 0080232272.

First published 1976.

Based on a research programme sponsored by the (US) Electric Power Research Institute, this handbook aims to summarize 'current technology for improving electrical energy use efficiency'. (*Preface*). Includes many tables of relevant data, more than 500 literature references and index. *Class No:* 621.3(083)

[3931]
SUMMERS, W.I., *ed.* **American electricians' handbook.** 11th ed. New York, etc., McGraw-Hill, 1987. [1664]p. illus., diagrs., tables. $ 64.50. ISBN: 0070139326.

First published 1913. Previous ed. 1981.

Long-established compendium previously edited by T. Croft. Aims to give complete guidance on selection, installation, maintenance and operation of all types of electrical equipment and wiring. Extensively revised to incorporate developments in electronics and solid-state circuits and computing. 11 Divisions, *e.g.* 1. Fundamentals ... 4. General electrical equipment and batteries ... 7. Generators and motors ... 9. Interior wiring (315p.). - 10. Electric lighting (205p.). In addition to the 100's of tables throughout the work, Division 11 collects 116 tables of general application. There are over 1,500 illus., clear and apposite. *Class No:* 621.3(083)

Standards
[3932]
INTERNATIONAL ELECTROTECHNICAL COMMISSION. Catalogue of publications, complete to 1st January 1986: world standards for electrical and electronic engineering. Geneva, IEC, 1960-. Annual.

1986 ed. (333p. SFr.12) lists all IEC standards issued as of 1st January 1986, in order by standard number, giving title, date, pagination, edition, price and identifying the Technical Committee responsible, with annotation varying from a few words to a detailed summary. Standards issued during previous year are highlighted by printing on blue background. A 20-page subject index aids consultation.

A bi-monthly *IEC Bulletin* includes lists of newly issued standards in its regular contents. *Class No:* 621.3(083.74)

[3933]
—BRITISH STANDARDS INSTITUTION. Electrical engineering: power, electronics, telecommunications, acoustics, illuminations, domestic appliances. Milton Keynes, BSI, 1985. 45p. £2. (*SL 26.*)

Gives the number, title and price group of about 1,300 BSI publications - standards, codes of practice, published documents (PD) and drafts for developments (DD) - covering the whole field of electrical and electronic engineering, arranged alphabetically under 50 subject headings. Full details and summaries are to be found in the annual *Catalogue* (*qv*). No index. *Class No:* 621.3(083.74)

[3934]

—INTERNATIONAL ELECTROTECHNICAL COMMISSION. Yearbook. [International Electrotechnical Commission]. Geneva, etc., IEC, 1986. *c.*600p. SFr.48.

Describes work and progress of all Technical Committees, supplemented by complete list of all final drafts of standards about to be published and those recently published.
Class No: 621.3(083.74)

[3935]

—Quick reference to IEEE standards. New York, IEEE, 1980. [vi], 568p.

A complete index of all terms in IEEE standards and ANSI standards published by IEEE. The index section (p.418) lists the terms which appear in the detailed Document Contents section and identifies the standard numbers containing the selected term. The full tables of contents are given for all standards in the second section. Computer-generated, the index throws up many terms which are unlikely to be sought (*e.g.* Even, Event, Evidence, Example).

Also noted: a series of 8 IEEE catalogues known by their colours and covering different areas of interest, *e.g. IEEE Brown book: Recommended practice for power system analysis.* 1980. *Class No:* 621.3(083.74)

[3936]

WILLIAMS, D., *ed.* **Electrical standards in world trade.** Morden, Surrey, IPC Middle East Publishing Co. / IEC, 1982. 109p. illus., diagrs. Sw.Frs.70. ISBN: 0946325006.

Details of main standards and approval marks of 43 member countries of the IEC, with reviews of the standards scene in the countries and regions of the world which constitute the major markets for electrical and electronic equipment. Based on a series of articles published in *Middle East electricity.* Chapters 2-6 provide valuable succinct explanations of the various international standards bodies and their activities, including CMEA (the Council for Mutual Economic Assistance) covering the Eastern Bloc countries, the EEC and PASC (the Pacific Area Standards Congress). 9 chapters give details for the main exporting countries.
Class No: 621.3(083.74)

[3937]

Worlds standards mutual speedy finder. V.2: Electrical and electronic: comparative tables of 5 countries' and international standards for mutual speedy foreign standards search: USA, United Kingdom, West Germany, France, Japan, International. Tokyo, International Technical Information Institute, 1976. [1,150]p. tables. $95. (*World standards mutual speedy finder, v.2.*)

Aimed to assist the identification of equivalent standards of the countries named and of the International Electrotechnical Commission (IEC). More than 18,000 electrical and electronics standards are grouped into 12 categories (*e.g.* 1. Generality; Common rules ... 3. Power; Electric apparatus ... 7. Cables and wires ... 12. Electronic parts), each extensively subdivided (*e.g.* 12.12: Thermistors). Relevant standards of each issuing body are set out for each country at each subheading in numerical (specification number) order in adjacent columns. Equivalents (if any) may be fairly easily identified. There are indexes of standards numbers, products and parts, and keywords. *Class No:* 621.3(083.74)

Patents

[3938]

Electrical patents index: (EPI). London, Derwent Publications, 1980-. Weekly.

One of the 4 major services of the Derwent patents organization. Based on two weekly bulletins which each contain *c.*4,300 abstracts spread over 6 sections of the Derwent classification (S. Measuring; Testing - T. Computing and control - U. Semi-conductors and electronic circuitry - V. Electronic components - W. Communications - X. Electric power engineering). *EPI Bulletin (Country order)* groups the entries under country, arranged by patent number, with abstracts and diagrams. *EPI Bulletin (Classified)* contains essentially the same information, but arranged by EPI class within the 6 sections. Cumulative indexes are also available (on microfiche). *Monthly Profile booklets* cover 50 separate subject classes, enabling scanning of narrower areas (*e.g.* U13. Integrated circuits). The whole system is available online as an integral part of the *World Patents Index* database. *Class No:* 621.3(088.8)

Chronologies

[3939]

DAVIS, H.B.O. **Electrical and electronic technologies: a chronology of events and inventors, to 1900.** Metuchen, N.J., Scarecrow Press, 1981. viii, 213p. £17.50. ISBN: 0810814641.

Potted history from *c.*640 BC to 1900 AD in six chapters. Chapters 1 and 2 cover the centuries before 1500 and the remaining centuries to 1900 have one chapter each. Arrangement in each is strictly by date and each entry is devoted to a particular activity, discovery, publication or event in the life of individual scientists; thus each can receive several entries. Biographical material is given at the birth-year entry for many of them. Bibliography of sources: pp 176-81. Index of persons, subjects and places (31p).
Class No: 621.3(090)

[3940]

—DAVIS, H.B.O. Electrical and electronic technologies: a chronology of events and inventors from 1900 to 1940. Metuchen, N.J., Scarecrow Press, 1983. xi, 208p. £17.50. ISBN: 0810815907.

Continues the 1981 work. It covers the first four decades of the 20th century in 4 chapters, with longer introductions and many entries under subjects (*e.g.* TV standards) and devices (*e.g.* Radio range beacons). Two appendices index vacuum tubes and radio stations. 32p. index of persons, subjects, organizations. Similar format to the 1981 vol., but biographical entries are omitted. *Class No:* 621.3(090)

[3941]

—DAVIS, H.B.O. Electrical and electronic technologies: a chronology of events and inventors, 1940-1980. Metuchen, N.J., Scarecrow Press, 1985. 313p. £25. ISBN: 0810817268.

Continues the foregoing, bringing the story up to the previous decade. *Class No:* 621.3(090)

Histories

[3942]
ATHERTON, W.A. **From compass to computer: a history of electrical and electronics engineering.** London, Macmillan; San Franciso, San Francisco Press, 1983. xiv, 337p. illus., diagrs., graphs, tables. £26; £11.95. ISBN: 0333352661, Macmillan Hardback; 0333352688, Paperback.

The development of electrical engineering and the other technologies it has spawned, from the early investigations of electricity and magnetism to microelectronics and computers. Describes the work of the pioneers - workers and companies - in chapters covering discrete topics, *e.g.* electric lighting, telecommunications, radio. Chapter references. Name and subject indexes. *Class No:* 621.3(091)

[3943]
HOLLISTER-SHORT, G. *and* JAMES, F., *eds.* **History of technology.** London, Mansel, 1991. 272p. illus. £40. ISBN: 0720121140.

Papers on the history of electrical technology, marking the bicentenary of the birth of Michael Faraday. *Class No:* 621.3(091)

[3944]
RYDER, J.D. *and* FINK, D.G. **Engineers and electrons:** a century of electrical progress. New York, IEEE Press, 1984. xix, 251p. illus., diagrs. $29.95. ISBN: 087942172x.

A popularized history of the developments in electrical engineering over two centuries culminating in microelectronics and computing, with emphasis on American contributions and the role of the Institute of Electrical and Electronics Engineers and its predecessors. Chapter bibliographies.

'Delightful reading and sometimes thought-provoking ... Highly recommended'. (*IEEE Proceedings*, v.73, Jan. 1985, p. 169). *Class No:* 621.3(091)

Bibliographies

[3945]
FINN, B. **The History of electrical technology:** an annotated bibliography. New York, Garland, 1991. 342p. $48. (*Garland reference library of science and technology, 18.*) ISBN: 0824091205.

About 1500 entries to secondary sources, largely in English, published through 1986. 4 main sections: 1. Historical works - 2. Communications - 3. Power - 4. Miscellaneous (*e.g.* computers, lasers, music). Author and subject indexes. Author is curator of the Electricity and Modern Physics Division, National Museum of American History. *Class No:* 621.3(091)(01)

Biographies

[3946]
IEEE **membership directory.** New York, Institute of Electrical and Electronics Engineers, 1987. Annual. 1903p.

In addition to standard information about members gives short biographies of IEEE Fellows (p. 17-253). *Class No:* 621.3(092)

[3947]
INSTITUTION OF ELECTRICAL ENGINEERS. **Yearbook and list of members:** 1987-88. London, IEE. Annual. *c.*800p. ISBN: 0852963556.

Provides standard biographical information about members. *Class No:* 621.3(092)

Thesauri

[3948]
INSPEC **thesaurus,** 1991. 9th ed. London, Institution of Electrical Engineers, 1991. v, 547,61p. ISBN: 0852964897. First published 1973.

The 9th edition contains *c.*13,000 terms (6,000 'preferred terms' and 7,000 cross-references). Terms range over fields of physics, electronics, communications, electrical engineering, information technology, computing and control systems. Alphabetical display of terms (547p.) is followed by a hierarchical display of used terms. *Class No:* 621.3:025.43

Institutions & Associations

[3949]
READER, W.J. **A History of the Institution of Electrical Engineers, 1871-1971.** London, Peregrinus, for Institution of Electrical Engineers 1987. x, 327p, illus., (1 col), tables, map. £15. ISBN: 0863411037.

Extends R. Appleyard's *History of the Institution of Electrical Engineers* (1939), which covered the period 1871-1931, to a centenary history, which also contains a chapter giving a brief glance at events since 1971. 17 chapters grouped into 3 parts illustrate the 3 phases of development: telegraph engineers; power engineers; electronics engineers. Appendices contain valuable reference material: Brief chronology; text of Royal Charter and Bylaws; complete list of Presidents and Secretaries to 1986, with portraits and biographical notes for 1939-1986; membership and financial statistics; list of sources. Index of names and subjects. *Class No:* 621.3:061:061.2

Circuits

Encyclopaedias

[3950]
GIBILISCO, S. **International encyclopedia of integrated circuits.** 2nd ed. Blue Ridge Summit, Pa., TAB, 1992. 1086p. illus., diagrs., graphs, tables. $84.95. ISBN: 0830630260.

First published 1989.

Lists over 800 ICs from 10 manufacturers in the US and Canada, arranged by application and manufacturer. Data include electrical and thermal ratings, static and dynamic characteristics, electrical characteristics, applications, recommended operating conditions. Cross-references. Detailed, analytical index. *Class No:* 621.3.04(031)

Handbooks & Manuals

[3951]
BUCHSBAUM, W.H. **Encyclopedia of integrated circuits:** a practical handbook of essential reference data. Englewood Cliffs, N.J., Prentice-Hall, 1981. xxii, 420p. diagrs. $24.95. ISBN: 013275875x.

Covers over 200 IC'S in four categories: Analog, Consumer, Interface, Digital, giving for each a brief description, functional block or logic diagram, key parameters and applications. Six appendices contain dimensional outlines, brief list of US manufacturers, operating and circuit characteristics of selected IC's and a brief glossary. Index. *Class No:* 621.3.04(035)

[3952]

SEIDMAN, A.H., *ed*. **Integrated circuits applications handbook.** New York, etc., Wiley 1983. xxv, 673p. illus., diagrs., graphs, circuits, tables. (*Wiley electrical and elctronics technology handbook series*.) ISBN: 0471077658.

Comprenhensive coverage of all aspects and applications of IC's, arranged in 3 parts, 20 chapters, each contributed by an expert. Part 1 (chapters 1-8) covers digital IC's (TTL, MOS and CMOS logic, semi-conductor memories, microprocessors and microcomputers, etc); Part 2 (chapters 9-17) deals with linear IC's (*e.g.* operational amplifiers, active filters, data conversion, IC's in communications, etc); Part 3 (chapters 18-20), is concerned with fabrication (thick and thin film technologies, etc). 200 examples, numerous tables, circuits and flowcharts make this a practice-oriented handbook, well supported by chapter references and index. *Class No:* 621.3.04(035)

[3953]

—PARR, E.A. Logic designer's handbook. London, Granada, 1984. xvi, 456p. illus., diagrs. £18. ISBN: 0246118881.

14 chapters of theoretical and practical information for designers of digital circuits. Coverage: theory, logic families and application circuits (*e.g.* timers, counters, shift registers). Emphasis is on TTL and CMOS. While the book has been praised as a practical, up-to-date and highly readable survey of current logic engineering, the review in *IEE Proceedings* (v.132(3), May 1985, p.185) points to lack of references, occasionally misleading simplicity and inaccuracies. *Class No:* 621.3.04(035)

[3954]

WILLIAMS, A.B., *ed*. **Designer's handbook of integrated circuits.** New York, etc., McGraw-Hill, 1984. xxvi, 805p. diagrs., graphs, charts, tables. £75. ISBN: 007070435x.

22 contributors provide a large collection of popular and frequently used IC's. Aim is to assist selection of devices and circuits, by providing comparison charts, discussions and design examples. Each chapter concentrates on a family of IC's, *e.g,* operational amplifiers, active filters, telecommunications, timing circuits, analogue (digital and digital/analogue) conversion, logic circuits, microprocessors, optoelectronics. Chapter references and glossaries provide valuable support.

'A very useful compendium of electronic techniques and circuits' (*Electronics & Power*, v.31, Jan. 1985, p.85). *Class No:* 621.3.04(035)

Tables & Data Books

[3955]

IC Master, 1987. New York, Hearst Business Communications, 1987. 2v. [vp]. £85.

Directory and source of data and application notes for integrated circuits, microcomputer boards, microprocessors, linear and memory devices and interfaces available from US manufacturers. Various indexes permit different approaches and facilitate product selection - companies; functions; part numbers. There are also sections covering different product categories - military parts; digital devices; microprocessors; microcomputer boards - an equivalents index and a directory of manufacturers and distributors. *Class No:* 621.3.04(083)

[3956]

—SESSIONS, K.W. IC schematic source master. New York, etc., Wiley, 1978. [vi], 557p. diagrs., graphs, circuits. ISBN: 0471026239.

Displays 1,500 schematic diagrams for electronic circuits using integrated circuits. 16 sections (*e.g.* Television and video circuits; Timers; Optoelectronic circuits; Analogues and digital converters). Brief descriptions. Component values normally given. *Class No:* 621.3.04(083)

Power Supply

Databases

[3957]

Electric power database. Palo Alto, Ca., Electric Power Research Institute.

Corresponds to the print work *Digest of research in the electric utility industry*. Covers US and Canadian research in hydroelectric power, fossil fuels, nuclear power, transmission, economics, advanced power systems and environmental assessment. Over 36,000 records; updated monthly. Available on DIALOG, file 241. *Class No:* 621.311(003.4)

[3958]

Electrical Power Industry Abstracts: (EPIA). Washington, Edison Electric Institute.

Covers 1975 to date. A bibliographic data base, providing references, with abstracts to the literature of electric power plants, emphasising the environmental aspects. Includes siting, transmission, fuel transportation, land use, waste disposal, safety and risk factors. Covers technical reports by utilities and US Federal and State agencies. File size: 31,000 entries (mid-1986), updated bi-monthly. 3,000 additions pa. Hosts include ORBIT. Cost per connect hour: $86. Printed version: *Inforum*. *Class No:* 621.311(003.4)

Bibliographies

[3959]

METZ, K.S. **Information sources in power engineering:** a guide to energy science and technology. Westport, Conn., Greenwood Press, 1975. xiii, 114p. £8.50. ISBN: 0837185386.

Annotated entries in 9 sections: Journals, newsletters, translations - Conferences and published proceedings - Abstracting and indexing services - Organizations - US government organizations - Information centers, services and libraries - Textbooks - Bibliographies - Reference sources (*e.g.* handbooks; directories; thesauri and dictionaries) - Keeping informed (journals). Annotated bibliography, p. 107-9. *Class No:* 621.311(01)

[3960]

—BARBOUR, M., *comp*. Uninterruptible power supplies: a select bibliography. London, Institution of Electrical Engineers, 1986. 77p. ISBN: 0852963343.

193 English-language items from the INSPEC database, 1979-1985, with indicative and informative abstracts, grouped under 9 headings (*e.g,* Systems - Rectifiers - Switches - Batteries). Author and subject indexes. *Class No:* 621.311(01)

Handbooks & Manuals

[3961]

HICKS, T.G. **Power generation:** calculations reference guide. New York, McGraw-Hill, 1987. 366p. illus. $36.50. ISBN: 0070288003.

Calculations for power generation compiled from *Standard book of engineering calculations*. *Class No:* 621.311(035)

[3962]

Modern power station practice. British Electricity International, London. 3rd ed., incorporating *Modern power system practice*. Oxford, etc., Pergamon Press, 1991. 12v. illus., diagrs. £80 each. ISBN: 0080164366.

First published 1963-64. 5v. 2nd ed. 1971. 8v.

A. *Station planning and design.*
B. *Boilers and ancillary plant.*
C. *Turbines, generators and associated plant.*
D. *Electrical systems and equipment.*
E. *Chemistry and metallurgy.*
F. *Control and instrumentation.*
G. *Station operation and maintenace.*
H. *Station commissioning.*
J. *Nuclear power generation.*
K. *EHV transmission.*
L. *System operation.*
M. *Index.*

A revised and expanded series of textbooks (in the broadest sense of the word) with exercises, covering all aspects of power station practice. The impact of privatisation appears as a running theme. Volume indexes, as well as a consolidated index (V.M.) enhance the value of this authoritative work. *Class No:* 621.311(035)

Histories

[3963]

BOWERS, B. **A History of electric light and power.** London, Peregrinus, 1982. vii, 278p., illus., diagrs. (*IEE history of technology series, v. 3.*) ISBN: 0906048710.

The first section deals with early developments in the understanding and use of electricity. The remaining three quarters considers applications, with emphasis on lighting, describing the methods of generation and creating illumination. There are chapters on public supply and power stations; legislation; local vs. national supply (all with particular reference to Britain). Modern developments and practices in lighting, electric heating and traction conclude the history. Generous use of photographs and quotations from contemporary catalogues and reports. *Class No:* 621.311(091)

Great Britain

[3964]

ELECTRICITY COUNCIL. **Electricity supply in Great Briatin:** a chronology - from the beginnings of the industry to 31 December 1976. 2nd ed. London, Electricity Council, 1977. [iii], 99p. ISBN: 0851880533.

First published 1973.

Covers 1831-1976. Includes brief descriptions of Acts, Bills, White Papers. Overseas developments are indicated by asterisk. References, p.84-85. Detailed analytical index. *Class No:* 621.311(410)

[3965]

ELECTRICITY COUNCIL. **Handbook of electricity supply statistics.** London, Electricity Council. Annual. tables, graphs. ISSN: 04401905.

The 1989 ed. ([vi],140p.), is predominantly for England and Wales, with separate sections for Scotland and Northern Ireland, for UK as a whole and for member countries of the EEC. Covers generation, transmission, sales, consumption, finance, prices and tariffs. *Class No:* 621.311(410)

[3966]

—CENTRAL ELECTRICITY GENERATING BOARD. Statistical yearbook: 1984/85. London, the Board, 1985. Annual. 13p. tables, maps. ISBN: 0902543806.

Supplements the information in the CEGB's *Annual report and accounts*. Statistical tables cover finance, manpower, operations and plant, fuel consumption, electricity supplied, transmission, with more detail of power stations commissioned and decommissioned, and output of individual plants. *Class No:* 621.311(410)

[3967]

Electricity supply handbook. 44th ed. Sutton, Surrey, Reed Business Publishing, 1948-. Annual. 309p. illus., maps. £25. ISBN: 061701227x.

Sections: Regional electricity companies - The National Grid Company - Generating Organizations - Bodies related to the supply industry - Other electricity undertakings - Government departments - Other organizations. Appendices: Power stations - Tariffs - Privatization - Gazetteer. *Class No:* 621.311(410)

[3968]

HANNAH, L. **Engineers, managers and politicians:** the first fifteen years of nationalized electricity supply in Britain. London, Macmillan, 1982. xiii, 336p. diagrs., tables. £33. ISBN: 0333220870.

Covers the period 1947-62. Commissioned by the Electricity Council but independent in views expressed. Based on free access to documents and interviews with politicians, civil servants and staff within the industry at all levels. Document citations, liberal notes and references, statistical appendices and a good index help to make this a valuable source. *Class No:* 621.311(410)

[3969]

—HANNAH, L. Electricity before nationalization: a study of the development of the electricity supply industry in Britain to 1948. London, Macmillan, 1979. xiii, 467p. £33. ISBN: 0333220862.

The same author was originally commissioned to write an authoritative history of the industry to cover the period from 1880. Benefitting from access to official documents, he emphasizes the period from 1918 and concentrates on national rather than local developments. Full notes and references. *Class No:* 621.311(410)

Hydroelectric Power

[3970]

BROWN, J.G., *ed.* **Hydro-electric engineering practices.** 2nd ed. London, Blackie, 1958-70. 3v. illus., diagrs., tables.

1. *Civil engineering.* 2nd ed. 1964. xxviii, 1233p. £11.50.
2. *Mechanical and electrical engineering.* 2nd ed. 1970. xxvi, 861p. £20.
3. *Economics, operation and maintenance.* 1958. xxv, 473p. £19.95.

....(contd.)

25 specialist contributors. Chapter bibliographies. Volume indexes. Still the only encyclopaedic coverage of its subject. *Class No:* 621.311.21

[3971]
GULLIVER, J.S. *and* **ARNDT, R.E.A.,** *eds.* **Hydropower engineering handbook.** New York, McGraw-Hill, 1991. illus., diagrs., graphs, tables. v.p. $85. ISBN: 0070251932.

Emphasises design aspects and cost estimates, with some attention to the environmental effects of hydropower plant. Case studies and examples feature. Chapters carry references. Index. *Class No:* 621.311.21

Wind Power

[3972]
BURKE, B.L. *and* **MERONEY, R.N. Energy from the wind:** annotated bibliography. Fort Collins, Colorado, Colorado State University Solar Energy Applications Laboratory, 1975. viii, [172p].

770 references, with abstracts or annotations, from 2 papers by John Smeaton, written in 1759, to 17 publications of 1975. Presented in reverse date order - latest works first - in yearly batches. 50 per cent of the items have appeared since 1960. Includes details of 39 books, 36 conferences, 9 bibliographies and 9 patents. Publications cited range from popular reviews to technical aerodynamics studies. Each item is numbered within its publication year and categorised by subject or type. In addition to an author index of nearly 400 writers there is an index of broad subjects (*e.g.* Small scale generation) and type of material (*e.g.* Historical; Books). Supplements or revisions were planned. *Class No:* 621.311.24

[3973]
FARMER, P. Wind energy, 1975-1985: a bibliography. New York, Springer-Verlag, 1986. 167p. ISBN: 0387161031.

Sections: Research and development - Resource assessment and site selection - Design of wind energy conversion systems - Institutional incentives and controls in wind energy conversion systems - Applications of wind energy conversion systems. Bibliography. Author index. A comprehensive international annotated bibliography. *Class No:* 621.311.24

[3974]
—**LAPEYSEN, E.H.,** *comp.* Wind power systems: a select bibliography. Brussels, Centre National de Documentation Scientifique et Technique, Bibliothèque Royale Albert 1er, 1977. 55p.

Lists 331 items in 5 groups, each arranged by author: Periodical articles (290); bibliographies (7); conference proceedings (11); reports (21); reviews (2). Coverage: wind energy policy; wind conversion; wind technology; wind transfer. Subject index. *Class No:* 621.311.24

[3975]
—Wind energy utilization: a bibliography. Albuquerque, Univ. of New Mexico, 1975. 496p./microfiche (6 fiches). (*Technical Application Centre report no. TAC-W-75-700.*)

Completely covers the period 1944-74, with abstracts largely drawn from relevant large-scale abstracting services. 8 sections: 1. General - 2. Utitlization - 3. Wind power plants ... 5. Wind machines - 6. Wind data.... Indexes of authors, corporate sources, permuted titles, permuted keywords. *Class No:* 621.311.24

[3976]
GOLDING, E.W. The Generation of electricity by wind power: with an additional chapter by R.I. Harris. London, Spon; New York, Halstead Press (Wiley), 1955; reprinted 1976. xviii, 332p. illus., diagrs., graphs, tables. £6.

Widely regarded as a classic text in its field, this work was reprinted in the wake of the 'energy crisis' of the mid-1970s, with an Appendix to update the account of research and technical developments. 19 chapters cover 3 broad areas: wind behaviour and its determination; wind-driven machines; economic use of wind power under different conditions. The first section, especially chapters 3-10, contains a large amount of data, presented in tabulated and graphical form. In addition to chapter bibliographies, there is a selected bibliography of surface wind data, classified geographically, and a short French-English-German glossary. The Subject index relates to the main text and not to the content of the updating Appendix. 'Includes work done on wind surveying in the 1950's ... still thought to be the definitive statement on wind characteristics in the UK' (*Engineer*, v. 243, Sept. 30 1976, p. 243). *Class No:* 621.311.24

[3977]
HUNT, V.D. Windpower: a handbook on wind energy conversion systems. New York, Van Nostrand Reinhold, 1981. xvii, 610p. illus., diagrs., graphs, tables. $39.50. ISBN: 0442273894.

14 chapters (1. Introduction - 2. Historical development - 3. Wind characteristics and their impact ... 6. Towers and systems installation ... 8. Wind energy conversion systems - 9. Applications ... 11. Commercialization (small-scale and large-scale conversion systems) - 12. Environmental, institutional and legal barriers ... 14. The future of wind power). Chapter references. Bibliography (p. 513-20). Glossary (p. 521-7). Points of contact (people and organizations). Reference information (*i.e.,* 50 pages of tabulated information and useful data, *e.g.* units, symbols, conversion factors, average windpower data for North American locations). Index. *Class No:* 621.311.24

[3978]
—**ELDRIDGE, F.R.** Wind machines. 2nd ed. New York, Van Nostrand Reinhold, 1980. xiii, 214p. illus., diagrs., graphs, tables, maps. (*MITRE energy resources and environment series.*) ISBN: 0442261349.

First published 1975 and then sponsored by the US National Science Foundation.

This rev. ed. gives great emphasis on small and wind machines, although large-scale systems remain dominant. 9 chapters: 1. Viability ... 3. Taxonomy ... 5. Applications - 6. Siting - 7. Performance - 8. System design - 9. Future utilization. Appendix contains short glossary; list of US suppliers; references and selected, classified bibliography (p. 199-209). Liberally illustrated and attractively produced. 'An excellent source for the professional engineer' (Park, J. *The wind power book* (1980), p.237). *Class No:* 621.311.24

[3979]
—**PARK, J.** The Wind power book. Palo Alto, Calif., Cheshire Books, 1981. 240p. illus., diagrs., graphs, tables, maps. $21.95; $14.95. ISBN: 091735203x.

An expanded, rewritten version of Park's *Simplified wind power systems for experimenters* (2nd ed. Sylmar, Calif., Helion, Inc., 1975). This book remains an essentially practical manual for non-professional designers and users of wind machines. 7 chapters: 1. Introduction - 2. Wind power systems - 3. Wind energy resources - 4. Wind machine

.... *(contd.)*

fundamentals - 5. Wind power design - 6. Building a wind power system ... Appendix (p. 156-236) of numerical, climatic and design data (including wind-power tables and maps, and degree-day tables, - all for the North American continent - water consumption and electrical load tables, for domestic requirements, etc). Bibliography, p. 237-40. *Class No:* 621.311.24

[3980]
Wind energy abstracts: the industrial wind power abstracts journal. St. Johnsbury, Vt., Windbooks, 1983-. Monthly.

About 7,500 abstracts pa. 21 sections (*e.g.* 01. General; 02. Technology development; 05. Structural dynamics; 09. Wind resource assessments; 14. Performance testing; 18. Construction/operation and maintenance; 21. National/regional programs/policies). *Class No:* 621.311.24

[3981]
Wind engineering abstracts. Brentwood, Multi-Science Publishing Co., 1982-, v.1-. Quarterly. £53. ISSN: 02630915.

Upwards of 250 informative abstracts pa. for relevant articles in the major journals in physics, engineering, the environment, energy and architecture, with coverage also of conference proceedings, reports, occasional books and patents. 9 sections: Meteorology, geography, topography - Aerodynamics, rotors, and blades ... Installations and applications ... Power generation and transmission; Energy conversion and storage ... Economic, environmental and legal aspects - Reviews and miscellaneous. Annual author and subject indexes. *Class No:* 621.311.24

Handbooks & Manuals

[3982]
Wind energy and the environment. Swift-Hook, D.T., *ed.* London, Peregrinus, 1989. xiii,170p. diagrs., graphs, tables. $68. ISBN: 0863411762.

Covers environmental and legal aspects of wind-generated energy from a European point-of-view. 31 contributors. 17 chapters, subdivided, most with references. Index, p.168-70. 'A valuable resource for public officials and environmental groups' (*Choice,* February 1990, p.969). *Class No:* 621.311.24(035)

Machines

[3983]
BEZNER, H. Electrical machines dictionary. [Elektromaschinen Wörterbuch: German-English, English-German.] Wiesbaden, Brandstetter, 1978. [x], 275, 257p. DM80. ISBN: 3870970871.

About 18,000 entries covering all kinds, parts, construction, operation and testing of electrical machines. Standard and most-frequently used equivalents are given first. Clear layout. Grammatical forms and genders indicated. *Class No:* 621.313

[3984]
Electrical motor handbook. Chalmers, B.J., *ed.* London, Butterworths, 1988. [x],546p. illus., diagrs., graphs, tables. £65. ISBN: 0408007079.

Covers rotating machines above 10kw output. 15 contributors. 11 chapters, closely subdivided, with references: 1. Characteristics - 2. Environment - 3. Selection - 4. Variable-speed drives and motor control - 5. Materials

.... *(contd.)*

and motor components - 6. Insulation - 7. Ancillary equipment - 8. Works and acceptance testing - 9. Installation, site testing and commissioning - 10. Maintenance and failures - 11. Units, dimensions and conversion factors. Detailed, analytical index, p.529-46. Intended 'to stand alongside the *J & P Transformer book* and the *J & P Switchgear book.' (Preface). Class No:* 621.313

[3985]
—BISCOE, G.I., *and others.* A Bibliography of published works on stepping motors. 3rd ed. Leatherhead, Surrey, Electrical Research Association, 1979. [189]p. (*ERA 77-1023.*)

Lists more than 1,000 references in chronological order, 1919-1977. Includes patents. Separate chronological listings of patents, selected abstracts and standards (7 only). *Class No:* 621.313

[3986]
—ELECTRO-CRAFT CORPORATION. DC motors, speed controls, servo systems: an engineering handbook. 3rd ed. Oxford, Pergamon Press, 1977. xxv, [504]p. illus., diagrs., graphs, tables. $33. ISBN: 008021715x.

First published 1972. Previous ed. 1973.

A compendium of articles on all aspects of DC motors by about 20 staff of the corporation. 6 chapters of text and data plus final chapter of specifications and data sheets for about 50 products marketed by the company. 1. Terminology - 2. DC motors and generators ... 4. Speed controls and servo systems - 5. Applications ... Appendices of units, conversions, and miscellaneous tables. Index. *Class No:* 621.313

[3987]
LAITHWAITE, E.R. A History of linear electric motors. Basingstoke, Hants., Macmillan, 1987. ix, 389p. illus., diagrs., tables. £33. ISBN: 0333399285.

A highly personalized account of the development of linear induction motors and their applications by an engineer whose work over many years will ensure that his name remains associated with them. Chapters 1-3 deal largely with the 19th century. The remainder of the book is concerned with electromagnetic levitation and applications in high-speed transport systems. Chapter 10 gives a potted history of the decade after 1975. A bibliography of 1,000 items (p. 234-84), classified in 5 groups and arranged chronologically in each, covers 1902-75.

'A conducted tour of the author's personal memory bank, plus some very interesting historical material ... It is all very untidy, but much more interesting than it would have been had the editor tried to order the contents' (*New scientist,* no. 1554, 2 April 1987, p. 53). *Class No:* 621.313

[3988]
SMEATON, R.W., *ed.* Motor application and maintenance handbook. 2nd ed. New York, etc., McGraw-Hill, 1986. xi [676]p. illus., diagrs., graphs, tables. £59.95. ISBN: 0070584486.

First published 1969.

33 contributors, 20 sections. Sections 1-3 cover basic information - load requirements, power supply, motor types - to aid selection for particular applications. Types of motor are dealt with in considerable detail in Sections 4-10, treating motors for special applications in Sections 9 and 10. Sections 11-20 cover components, installation, maintenance, repair and specific problems - *e.g,* insulation, bearings, lubrication,

....(contd.)

noise. Liberal illustration, graphical and tabulated presentation of data relating to design, characteristics, performance, supplemented by extensive reference to American standards make this a very useful reference work. Supported by chapter references and an effective index. Units for dimensions and ratings are 'US Customary', *i.e., not* metric. *Class No:* 621.313

Transformers

[3989]
FRANKLIN, A.C. *and* **FRANKLIN, D.P. J & P transformer book.** 11th ed. London, Butterworths, 1983. 815p. illus., diagrs. £80. ISBN: 0408004940.

First published 1925, by Johnson & Phillips, Ltd. Previous ed. 1973.

Oustanding reference book for design, installation and maintenance of power transformers. Aimed at users, suppliers, students, rather than at specialists, and presents infomation clearly. Takes account of changes in design, materials and standards.

'This 11th ed. lives up to its high reputation ... while the book contains much that is useful it is sometimes scattered over a number of chapters. Some sections are rather wordy and these two factors make searching for details a longer operation than it would have been if a more thorough revision had been undertaken' (*Electronics and power*, v.30, August 1984, p. 641). *Class No:* 621.314

[3990]
SCOLES, G.J. Handbook of rectifier circuits. Chichester, E. Horwood, 1980. 238p. diagrs., circuits. £22. (*Ellis Horwood series in electrical and electronic engineering.*) ISBN: 0853121265.

Presents about 230 circuits with diagrams and brief descriptions, indicating combinations for special applications. 'It should be emphasized ... that the very comprehensive listing of such a huge variety of circuits has necessitated a rather superficial treatment of any one particular type ... A stimulating source of ideas for anyone who is responsible for the design of DC power supplies' (*Electronic circuits and systems (IEE Proceedings G)*, v.129(1), February 1982, p.28). Index. *Class No:* 621.314

Transmission

Cables

[3991]
Electric cables handbook. Bungay, E.W.G. *and* McAllister, D., *eds.* 2nd ed. Oxford, BSP Professional Books, 1990. xiv,977p. illus., diagrs., graphs, tables. ISBN: 063202299x.

First published 1982.

28 contributors, almost all from BICC. 5 parts: 1. Theory, materials, design - 2. Wiring cables - 3. Supply distribution cables - 4. Transmission cables - 5. Submarine cables. Appendices (p. 707-941) contain tables of data on all commonly used cable types. Classified bibliography (25p.), index and chapter references.

'... gives a comprehensive and authoritative account of the state of the art ... [and] ... has the merit of presenting much

....(contd.)

good explanatory material rather than merely giving factual information...' (*Electrical review* of the first ed.). *Class No:* 621.315.2

[3992]
KURTZ, E.B. *and* **SHOEMAKER, T.M. Lineman's and cableman's handbook.** 8th ed. New York etc., McGraw-Hill, 1992. xi, v.p. illus., diagrs., circuits, tables. $59. ISBN: 0070356955.

First published 1928; previous ed. 1986.

A complete, self-contained guide to theory and practice, intended for home study but usable 'on the job' as a reference source. Of 50 sections, 11 cover electricity and its applications; 31 are devoted to construction and maintenance of overhead and underground distribution and transmission lines; 7 deal with specific safety issues. Section 50 has 174 self-test questions. Index, 15p. relates to American practice, National Electrical Code and standards. Illus, on almost every page.

'The definitive work on the elements of an electrical transmission system' (*Public Works*, v. 113, April 1982, p.20, on the 6th ed.). *Class No:* 621.315.2

[3993]
RICHLING, C. *and* **DREWITZ, I. Wörterbuch der kabeltechnik,** deutsch-englisch-französisch. Wiesbaden, Brandstetter, 1976. 616p. DM60. ISBN: 3870970723.

About 5,500 main entries, with equivalents. Three sequences: German-English-French; English-German-French; French-German-English. Parts of speech are stated. *Class No:* 621.315.2

Wires

[3994]
BLACK, R.M. The History of electric wires and cables. London, P. Peregrinus, 1983. [xiii], 290p. illus. £24. (*IEE history of technology series, v. 4.*) ISBN: 0863410014.

Outlines the development of the principal types of electric wires and cables from the 18th century to the present day use of optical fibre telecommuniation links. Electric lighting, submarine telephone and power cables and the specialized needs of collieries, ships and aircraft are covered in separate chapters, and recent advances are dealt with in a final chapter. Chapter references and bibliographies.

'The serious historian will find it less than totally satisfactory because it does not go sufficiently deeply into many details and often does not suggest a line of research to fill the gaps ... If one asks whether this book adds to knowledge, whether it makes available previously inaccessible information, then it clearly succeeds'. (*Physical science, measurement and instrumentation, management and education, and Reviews (IEE Proceedings A)*, v. 132, no. 1, Jan 1985, p. 75).

'Well illustrated, and the contents are well set out to facilitate locating any particular aspect ...' (*Electronics and power*, v. 30, Feb. 1984, p. 158). *Class No:* 621.315.3

[3995]
RICHTER, H.P. *and* **SCHWAN, W.C. Practical electrical wiring:** residential, farm and industrial. 15th ed. New York, McGraw-Hill, 1990. 645p. illus., diagrs. $33. ISBN: 0070523032.

Updated to meet requirements of the 1990 National Electrical Code (NEC). Explores new topics, *e.g.* fluorescent lamps, incendiary circuits for hazardous

....(contd.)

locations. Less technical and detailed than *American electricians' handbook* (McGraw-Hill, 1987). A substantial index; short bibliography; glossary. For novices and professionals; a handy DIY home reference. *Class No:* 621.315.3

Insulation

[3996]

CONFERENCE ON ELECTRICAL INSULATION AND DIELECTRIC PHENOMENA. Committee on Digest of Literature, and National Research Council. Committee on Dielectrics. **Digest of literature on dielectrics.** Washington, National Academy of Sciences. 1937-79, v.1-42. tables. ISBN: 0309029341.

A valuable review series with extensive surveys in the different areas of its subject field, covering the publications of the previous calendar year. Occasionally (as in v. 42, 1979) reproduced tables of data reported in the literature covered. V. 42, 1979, contains 14 reviews (1. Tables of permittivity, dipole moment, and dielectric relaxation time ... 3. Dielectric breakdown of solids ... 5. Electrical breakdown in gases ... 10. Piezoelectricity .... 13. Cryogenics - 14. Applications). The bibliographies for these reviews list an astonishing 3,687 references.

Continued in *IEE Transactions on electric insulations.* 1984-. *Class No:* 621.315.6

[3997]

SHUGG, W.T. Handbook of electrical and electronic insulating materials. New York, Van Nostrand Reinhold, 1986. xv, 598p. diagrs., tables. £53. ISBN: 0442281226.

A tightly organized reference work which groups dielectric materials into product groups (*e.g.* Thermoplastic moulding compounds; thermosetting moulding compounds; magnetic wire enamels; dielectric films; tapes and coated fabrics; mica products; ceramic and glass insulations) and for each insulating material sets out the chemistry, grades, processing methods, properties and end uses. Properties and ranking data are clearly tabulated. Detailed references to American (notably ASTM and Military) standards. Excellent index structured to lead directly to all infomation about each material. *Class No:* 621.315.6

Distribution & Control

[3998]

INSTITUTE OF ELECTRICAL AND ELECTRONICS ENGINEERS. National electrical safety code. 1981. New York IEEE, 1981. 366p. diagrs., tables. (*American National Standards.*)

'Basic provisions for safeguarding of persons from hazards arising from the installation, operation, or maintenance of conductors ... equipment in electric-supply stations and ... electric-supply and communication lines ... [and] ... work rules for ... construction, maintenance and operation ...' An authorized American National Standard, containing definitions, grounding [*i.e.,* earthing] rules and Parts 1 to 4 of the 1981 *Code,* covering: 1. Supply stations - 2. Overhead lines - 3. Underground lines - 4. Work rules. Index.

1984 ed. (Wiley) $20 ISBN: 0471806064. *Class No:* 621.316

[3999]

—CLAPP, A.L. NESC handbook: development and application of the American National Standard Electrical Safety Code: grounding rules, general rules and Parts 1, 2 and 3. New York, Institute of Electrical and Electronic Engineers, 1984. 430p. $22. ISBN: 0471807834.

Aimed to aid users of the 1984 ed. of the US National Electrical Safety Code (American National Standard C2-1984) and intended for use in conjunction with the text of that code. *Class No:* 621.316

[4000]

—McGraw-Hill's National Electrical Code handbook. McPartland, J.F., *ed.* 17th ed. New York, etc., McGraw-Hill, 1981. xvi, 1133, 28p. $24.50. ISBN: 0070456933.

First published 1932.

Based on the 1981 *Code* (ANSI/NFPA 70-1981). Explains and amplifies 'the most commonly encountered rules ...' (*Preface*). Does not include text of *Code* or its tabulated data, but claims 'greater analysis and interpretation' than other handbooks. *Class No:* 621.316

[4001]

—ROSS, J.A., *ed.* National Electrical Code handbook. 2nd ed. Quincy, Mass., National Fire Protection Association, 1980. xxi, 1007p. illus., diagrs., tables. $19.50. ISBN: 0877651868.

Based on the 1981 *Code* (ANSI/NFPA 70-1981). Contains complete text of *Code* with commentary and elucidations where necessary. Supported by illustrations. Covers recommended practices and standards for wiring design, protection, methods and materials, specific equipment (*e.g.* transformers, generators, batteries, lighting, lifts, machine tools), processes (*e.g.* electroplating), environments (*e.g.* garages), and conditions (*e.g.* high voltage). Definitions, index and many tables provide invaluable support. *Class No:* 621.316

[4002]

INSTITUTION OF ELECTRICAL ENGINEERS. Regulations for electrical installations: (IEE wiring regulations). 15th ed. Hitchin, Herts., IEE, 1981. (Reprinted incorporating Amendments, 1983), plus Amendments annually. viii, 220p., diagrs., graphs, tables. £16. ISBN: 0852962355.

First published 1882. 14th ed. 1966, with 7 reprints and 4 amendments issues, 1968-76.

Arrangements accords with IEC Publication 364 '*Electrical installations of buildings*'. 6 parts. (24 chapters): 1. Scope, object and fundamental requirements for safety - 2. Definitions - 3. Assessment of general characteristics - 4. Protection for safety - 5. Selection and erection of equipment - 6. Inspection and testing. Each is subdivided (*e.g,* 53. Switchgear - 531. Devices for protection against electric shock - 531-1 and 531-2. Overcurrent protective devices). Safety is the objective of every regulation. 16. Appendices (1. Cited British Standards. 2. Statutory regulations ... 4. Current carrying capacities and voltage drops for various cables. Plus much other data). *Class No:* 621.316

[4003]

—JENKINS, B.D. Commentary on the 15th edition of the IEE wiring regulations. Rev. 2nd ed., covering all amendments to and including January 1985. Stevenage, Herts., Peregrinus, 1985. xviii, 268p. diagrs., graphs, tables. £14. ISBN: 0863410405.

Gives explanations, further data and, in many cases,

....(contd.)

design data for those involved in installation work at all levels. Particular emphasis on protection against electric shock and against overcurrent. The final chapter relates the *Regulations* to smaller installations, - the great majority. 6 appendices (one reproduces the text of the *Regulations*, 1st ed. (1882)). Subject Index and an index that relates regulation number to text. *Class No:* 621.316

[4004]
—WHITFIELD, J.F. A Guide to the 15th edition of the IEE wiring regulations. Rev. 2nd ed. London, Peregrinus/IEE, 1985. [xiv], 172p. diagrs., graphs, tables.

A Guide and explanation of most of the *c.*550 regulations aimed to help electricians and technicians understand and apply them in practical situations. Clear text and numerous simple diagrams assist. Refers user to B.D. Jenkins' *Commentary* (*qv*) for more technical and detailed information. Subject index and cross-reference index (which relates regulation number to text). *Class No:* 621.316

[4005]
REEVES, E.A., *ed*. **Handbook of electrical installation practice.** 2nd ed. London, Granada, 1983. 2v. (364p.; 380p). illus., diagrs., graphs, tables. £40. ISBN: 0246119497, set.

28 contributors. 20 chapters (*e.g.* 1. Power supplies in the UK ... 4. Distribution in buildings - 5. Electricity on construction sites ... 7. Earthing, cathodic protection and lightening protection ... 10. Standards, specifications and codes of practice ...). Claims to cover all aspects of industrial, commercial and industrial installations, but ignores some (*e.g.* intruder alarms) and gives insufficient treatment to others (*e.g.* heating and ventilating systems; emergency lighting). Cites relevant British Standards and 15th ed. of IEE Wiring Regulations.

Review in *Electronics and power* (Sept. 1983, p.660) criticises: 'considerable bulk and cost'; failure to cover all aspects of subject, as claimed; certain misleading statements; inadequate coverage within chapters; but concludes: 'Nevertheless the handbook will be found to be a useful reference book ...'. *Class No:* 621.316

[4006]
SEIP, G.G., *ed*. **Electrical installations handbook;** translated by K.G. King. 2nd rev. and enl. ed. Berlin, etc., Siemens AG, Chichester, Wiley, 1987. 3v. (1522p). illus., diagrs., graphs, tables. £100. ISBN: 047191343x, Wiley; 3800914670, Siemens.

First German ed. 1971.

59 contributors. 6 major sections, 29 sub-sections, with many more subdivisions: 1. Electrical installation engineering for power supply and distribution 2. Electrical installations engineering for space heating, air conditioning and ventilation - 3. Lift installations - 4. Electrical installations engineering in large buildings and outdoors - 5. Electrical installations engineering for special installations and systems - 6. Installation specifications and safety measures. Section 1. Electrical installation ... for power supply and distribution (p.17-569). Subdivisions: 1-1. Definition of terms (p. 17-101); ... 1-4. System protection - 1.4.5. Protective devices for high-voltage systems - 1.4.5.5. Protection of ring feeders ... 1-12 Earthing systems. Clearly displayed text, illus. and tables, with helpful use of colour and marginal bold face subheadings. Subsection 28 contains references to VDE standards and relationship to CENELEC. *Class No:* 621.316

Switchgear

[4007]
SMEATON, R.W., *ed*. **Switchgear and control handbook.** 2nd rev. ed. New York, etc., McGraw-Hill, 1987. 1056p. illus., diagrs., graphs, tables. £59.95. ISBN: 0070584494.

First published 1976.

Over 30 contributors. Covers design, application and maintenance of switchgear and control equipment, including installation and service requirements, load and safety factors, seismic and other forms of protection, the control and protection of motors and, in this edition, introduces the use of computers and solid state devices.

A review of the 1976 ed. noted 'It will be of use to the engineer or technician needing practical guidance over a very wide field of electric power applications' and drew attention to the emphasis on US practices, standards and literature references, and to 'some limitations in coverage' but concluded: 'I foresee this book finding a place amongst a collection of reference works, or replacing a number of them' (*Electronics & Power*, v. 23(8), August 1977, p.654). *Class No:* 621.316.3

[4008]
—LYTHALL, R.T. J & P switchgear book: an outline of modern switchgear practice, for the non-specialist user. 7th ed. London, Newnes-Butterworth, 1972. xii, 804p. illus., diagrs., circuits, tables. £16.80. ISBN: 0408000694.

First published 1927, by Johnson & Phillips, Ltd. Previous ed. 1963.

Extensive treatment of all types of switchgear and protective systems. 23 chapters, each with references and bibliography (2. Circuit breaking ... 11. Protective gear ... 18. High voltage load-breaking switches). Additional bibliography. Index (p. 801-4). *Class No:* 621.316.3

Protection

[4009]
BRITISH STANDARDS INSTITUTION. Electrical protection relays. Section 1.1. Glossary of protection relay terms. London, BSI, 1982. 12p. diagrs. (*BS 142: Section 1.1:1982 (1991)*.) ISBN: 0580129047.

Over 100 terms defined in systematic order with alphabetical index. *Class No:* 621.316.9

Measurement

[4010]
JUNGE, H.D. Wörterbuch messtechnik und einheiten: Deutsch/Englisch. [Dictionary of measurement engineering and units: English/German.] Weinheim, VCH, 1991. (*Parat.*) ISBN: 3527264221.

About 10,000 entries in each part of the dictionary. Tinted pages list all essential old and new English and German units, together with their conversion factors. *Class No:* 621.317

[4011]
NOLTINGK, B.E., *and others*. **Instrumentation reference book.** London, Butterworth, 1988. v.p. illus., diagrs. £95. ISBN: 0408015624.

6 parts: 1. Mechanical measurement - 2. Measurement of temperature and chemical composition - 3. Electrical and radiation measurement - 4. Instrumentation systems - 5. Further scientific and technical information - 6. Directories and commercial information (1-10. 10: Master address list).

.... *(contd.)*

Index. About 30 main contributors with chapters by various hands. 'Mainly applications oriented' (*Introd.*). Well produced. *Class No:* 621.317

[4012]
O'DELL, T.H. Circuits for electronic instrumentation. Cambridge Univ. Press, 1991. 218p. illus., diagrs. $69.50; £35. ISBN: 0521404282.

Covers a wide range of circuits (*e.g.* comparators, wideband direct couple feedback amplifiers, phased locked loops, voltage controlled oscillators, etc.) at an advanced level. Extensive chapter bibliographies. *Class No:* 621.317

Magnetic Materials

[4013]
EVETTA, J.E. Concise encyclopedia of magnetic and superconducting materials. Oxford, Pergamon Press, 1992. xviii,703p. illus., diagrs. £140. (*Advances in materials science and engineering, 7.*) ISBN: 0080347223.

Incorporates relevant material from the *Encyclopedia of materials science and engineering* (1986) and newly commissioned articles. *Class No:* 621.318

[4014]
MOSKOWITZ, L.R. Permanent magnet design and application handbook. Boston, Mass., Cahners Books International, 1976. vii, 385p., illus., diagrs., graphs, tables. $49.95. ISBN: 0843618000.

Wide-ranging treatment which covers history, terminology, classifications of permanent magnets, manufacturing processes, design, circuit effects, environmental effects, measurement and testing, standards and applications. Substantial appendices contain tables of magnetic and physical properties, an international index of permanent magnet materials and US equivalents, a bibliography (p. 359-71), glossary and index. *Class No:* 621.318

[4015]
WOHLFARTH, E.P., *ed*. Ferromagnetic materials: a handbook on the properties of magnetically ordered substances. Amsterdam, etc., North-Holland, 1980-. V.1-illus., diagrs., graphs, tables. $382. ISBN: 0444853138, set.

The first 3 v. of a planned 5 v. set intended to replace the monumental *Ferromagnetism* by R.M. Bozorth (Princeton, Van Nostrand, 1951). V. 1 contains 7 and v.2 contains 8 lengthy reviews on different materials (*e.g.* v. 1: 1. Iron, cobalt, nickel ... 4. Rare earth compounds ... 6. Amorphous ferromagnets ...) each dealing with structure, the magnetic, electronic, thermal, optical and other properties. Many tables and references to the literature, plus indexes of cited authors, subjects and materials (chemical formulae). V.4 (xvii,653p.) appeared in 1988. *Class No:* 621.318

[4016]
—**DEMAW, M.F. Ferromagnetic-core design and application handbook.** Englewood Cliffs, N.J., Prentice-Hall, 1981. x, 256p., illus., diagrs., graphs, tables. £16.45. ISBN: 0133140881.

Applications chapters cover : rods, bars and slugs; toroids; beads, sleeves and pot cores. Separate chapter on permanent magnets. Practical circuit examples for narrowband and broadband transformers and inductors; ferrite loop antennas, RF chokes, etc. Appendices of design data. Glossaries, Bibliography. Index. *Class No:* 621.318

Batteries

[4017]
APPLEBY, A.J. *and* FOULKES, F.R. Fuel cell handbook. New York, Van Nostrand Reinhold, 1989. xxvi,762p. illus., diagrs., graphs, tables. £67.50. ISBN: 0442319266.

Covers both technology and economics. 5 parts; 19 chapters, closely sub-divided: 1. General aspects of fuel cells - 2. Fuel cell programs, concepts, and areas of application - 3. Design considerations for practical fuel cell systems - 4. State of the art - 5. References (2802 items). Detailed, analytical index, p.749-62. A specialised treatment for the expert; well-produced. *Class No:* 621.355

[4018]
CROMPTON, T.R. Battery reference book. London, etc., Butterworth, 1990. xii,v.p. illus., diagrs., graphs, tables. £120. ISBN: 0408007907.

2 introductory chapters (technology; selection), then chapters 3-63 in 6 parts: 1. Characteristics - 2. Theory and design - 3. Performance evaluation - 4. Applications - 5. Charging - 6. Suppliers (types of batteries). Appendices: suppliers (manufacturers; international; A-Z by country); glossary (*c.*225 terms; 'Activated stand life' - 'yield'); Standards (British; IEC); bibliography (3p.). Detailed, analytical index. 'Of interest to battery manufacturers and users and the manufacturers of equipment using batteries' (*Preface*). 'Fills a gap' (*Choice*, November 1990, p.513). *Class No:* 621.355

[4019]
LINDEN, D., *ed*. Handbook of batteries and fuel cells. New York, etc., McGraw-Hill, 1984. xix, [1,055]p. illus., diagrs., graphs, tables. £61.50. ISBN: 0070378746.

54 contributors. The first comprehensive handbook for all types of battery and fuel cell. 43 sections arranged in 6 parts. 1. Principles of operation -2. Primary batteries (11 types) - 3. Secondary batteries (12 types, with extensive treatment of lead-acid). 4. Advanced secondary batteries (6 types) - 5. Reserve batteries (7 types of delayed activation battery) - 6. Fuel cells. For each type gives general characteristics, chemistry, construction, charging and performance characteristics, applications. 7 appendices contain definitions, properties of materials, symbols, conversion factors, bibliography, manufacturers. Excellent index, many clear diagrams and tabulations. 'Anyone wanting to know which types of batteries are suitable for particular applications will find this book a useful reference volume' (*Electronics and power*, v.30. Nov/Dec. 1984, p.885). *Class No:* 621.355

Electroplating

[4020]
DURNEY, L.J., *ed*. Electroplating engineering handbook. 4th ed. New York, etc., Van Nostrand Reinhold, 1984. x, 790p. illus., graphs, tables. $69.50. ISBN: 0442220022.

First published 1954. Previous ed. 1971 by A.K. Graham.

51 contributors. Coverage includes surface preparation, pre-and post-treatments, engineering data to aid materials selection and choice of equipment. New techniques such as pulse plating and high-speed plating are included.

Part 1. General processing data - Part 2. Engineering fundamentals and practice. Most parts of this work have been completely or substantially re-written. Aim is 'a useable handbook', with 'frequent references to other sources ... The present goal is proper guidance and the provision of the most

....(contd.)

frequently required facts, not everything that is available' (*Preface*). 'The new handbook is a worthwhile addition to the finisher's library...' (*Metal finishing*, v.83, Jan. 1985, p.110). *Class No:* 621.357

[4021]
Tables and operating data for electroplaters. Redhill, Surrey, Portcullis Press, 1975. xiv, 414p. graphs, tables. £17.65. ISBN: 0901994707.

Based on a translation of *Tabellen und Betriebsdaten für die Galvanotechnic*, 4th ed. 1971 (Leuze Verlag, Saulgau, Germany), revised by C.R. Draper. The 4 German eds. were by E. Luther, H. Benninghoff and O.P. Krämer.

Contains data on constants, acids, bases, surface area, electro-deposition, properties, standards, tests, commonly used chemicals, abrasives and troubleshooting. *Class No:* 621.357

Electric Heating

[4022]
HOBSON, L. Electroheat references. London, British National Committee for Electroheat, 1986. iv, 74p.

Supersedes *Electroheat teaching references* (BNCE, 1980).

More than 300 references, with abstracts, derived from a computer-based literature search. Arranged in 13 categories (1. General electroheat references ... 3. Dielectric heating ... 7. Induction heating ... 10. Microwave heating ... 13. Heat pumps), each subdivided into: Design and construction; operation; theory and calculations; power supplies; applications. *Class No:* 621.36

[4023]
—BRITISH STANDARDS INSTITUTION. Glossary of electrotechnical, power, telecommunication, electronics, lighting and colour terms. Part 2. Terms particular to power engineering. Group 10. Industrial electroheating terminology. London, BSI, 1985. 32p. £11. (*(BS 4727: Part 2: Group 10: 1985.)*

Supercedes BS 2759:1956.

About 500 terms defined in systematic order. Alphabetical index. *Class No:* 621.36

Waves & Oscillations

Microwaves

[4024]
Handbook of microwave and optical components. Chang, K., *ed.* New York, Wiley, 1989-91. 4v. illus., diagrs., graphs, tables. £45 each v.

1. *Microwave passive and antenna components.* xiv,907p. ISBN: 0471613665.

2. *Microwave solid-state components.* xiv,635p. ISBN: 0471843652.

3. *Optical components.* xiii,616p. ISBN: 0471613673.

4. *Fiber and electro-optical components.* xi,484p. ISBN: 0471613657.

V.3 has 14 contributors, 9 chapters ('gas lasers', p.451-596, carries 298 references). Volume indexes. *Class No:* 621.37.029

[4025]
HOFFMANN, R.K. Handbook of microwave integrated circuits; translated by G.A. Edis and N.J. Keen. Norwood, Mass., Artech House, 1987. illus., diagrs., graphs, tables. (*Artech House microwave library*.) ISBN: 0890061637.

Originally published in German as *Integrierte Mikrowellenschaltungen* (Springer, 1983).

'This book provides the fundamental information for the design of MICs, with emphasis on the electrical properties of the various forms of strip transmission lines ... [and] ... an introduction to the technology of these circuits and a treatment of the spurious electrical effects ... such as field distortions ... (*Preface*). A comprehensive text with a noteworthy bibliography of 1,256 references. *Class No:* 621.37.029

[4026]
LAVERGHETTA, T.S. Handbook of microwave testing. Dedham, Mass., Artech House, 1981. [xviii], 518p. illus., diagrs., tables. £54. ISBN: 0890060703.

Companion volume to same author's *Microwave measurements and techniques* (1976). Covers the major areas of test equipment for power levels, noise measurements, frequency, etc. 12 appendices contain much useful information. Concentrates on US equipment and excludes all European and Japanese instruments. Bibliography p.511-5. Brief index.

'... A practical microwave engineer working for a firm that is well equipped with modern US microwave test equipment will find this book very valuable' (*Microwaves, optics and antennas (IEEE H)* v.129 (4), August 1982, p.160). But 'omission of such matters as testing mixers, microwave transistors and ... microwave semiconductor devices generally, and of problems involved in testing microcircuits is regrettable. Nevertheless, technicians new to the microwave techniques and wishing to familiarise themselves with equipment and methods still in use will find the book useful if they can come to terms with the garrulous, direct approach. The presentation of text is very good' (*Electronics and power*, v.28, June 1982, p.465). *Class No:* 621.37.029

[4027]
Microwave handbook. Dixon, M.W., *ed.* Potters Bar, Radio Society of Great Britain, 1989-. V.1-. illus., diagrs., graphs, tables.

V.1. *Components and operating techniques.* 1989. v.p. (0990612886).

V. 2. *Construction and testing. 1991. c.200p.*

To be in 3v. V.1 has 5 contributors; 6 chapters. Apart from much referenced information, it is a timely collection of practical diagrams, hints and tips. 2-page index. V. 2: Chapters 7-13. *Class No:* 621.37.029

[4028]
POZAR, D.M. Microwave engineering. Reading, Mass., Addison-Wesley, 1990. xvii[i],726p. illus., forumlas, graphs. £56.95. (*Addison-Wesley series in computer engineering*.) ISBN: 0201504189.

12 documented chapters: Electromagnetic theory ... Lines and waveguides - Network analysis ... Resonators ... Filters ... Circuits - Systems. 10 appendices (C. 'Bessel functions'). Detailed, analytical index, p.719-26. A textbook for advanced undergraduates; problems. *Class No:* 621.37.029

[4029]

SUCHER, M. *and* FOX, J., *eds.* **Handbook of microwave measurements.** 3rd rev. and enl. ed. New York, Halstead Press, 1975. 3v. (1145p.) diagrs., graphs, charts, tables. $59.50. ISBN: 0470835389.

First published 1954, edited by M. Wind and H. Rappaport.

Presents guidance in 15 chapters by 25 contributors on the fundamentals, techniques and devices used in the measurement of microwave parameters and component characteristics, 1. Frequency and wavelength ... 3. Measurement of power ... 7. Attenuation ... 9. Dielectric constant ... 13. Open nonconventional waveguides. Sans serif type face distinguishes procedures from discussion of theory and experiment. *Class No:* 621.37.029

[4030]

THUÉRY, J. **Microwaves:** industrial, scientific, and medical applications. Grant, E.H., *ed.* Boston, Ma., London, Artech House, 1992. xviii,670p. $88; £66. ISBN: 0890064482.

Translated from the French.

4 sections: 1. Microwaves - 2. Industrial applications (drying, treatment of elastomers, etc.) - 3. Applications in the food industry - 4. Biological effects and medical applications. More than 4000 references, most from English-language sources. List of industries, organizations and universities active in the field. 'Highly recommended' (*Choice,* October 1992, p.334). *Class No:* 621.37.029

Quasars

[4031]

ELION, G.R. *and* ELION, H.A. **Electro-optics handbook.** New York, Dekker, 1979. ix, 359p. diagrs., graphs, tables. $39.75. (*Electro-optics series, v.3.*) ISBN: 0824768795.

A compilation of 'fundamental material underlying all of electro-optics' (*Preface*) bringing together from the widely-scattered literature 'the basic data required for evaluation, selection and proper application of electro-optic tests ...' (*Preface*). 14 chapters (1. Fundamental definitions (upwards of 300 terms) - 2. Symbols, designations and conversions - 3. Photometric and radiometric quantities, standards and units ... 5. Lasers - 6. Detectors ... 8. Optical formulas, materials and components ... 14. Computer design of electro-optic components and systems). 76 tables. Numerous diagrams and graphs. Appendix contains properties of the elements. Index. '... Represents one of the first attempts to create a concise, up-to-date source of fundamental information on this specialized field' (*Physics today,* v.33(2), Feb. 1980, p.59). *Class No:* 621.371

Filters

[4032]

DICKENSON, C. **Filters and filtration handbook.** Amsterdam, Elsevier, 1991. xi,778p. illus., graphs, tables. £95. ISBN: 1851670780.

8 sections, subdivided (*e.g.* 8. 'Selective data'). Appendices: Buyers' guide; Trademark index; Classified index; Names and addresses. *Class No:* 621.372

Waveguides

[4033]

MARCUVITZ, N. **Waveguide handbook.** London, P. Peregriuus, 1986. xviii, 428p. diagrs., graphs, tables. (*IEE Electromagnetic waves series, no.21.*) ISBN: 0863410588.

First published 1951.

This edition reprints the original with correction of known typographical and other errors. Chapters 1-3 summarize field and network theory. Chapters 4-8 provide equivalent circuit parameters for a variety of microwave structures, supported by numerous graphs. 'The book is for those requiring a fairly fundamental approach to their waveguide problems. It does not explicitly deal with the more practicable problems and as such the title "Handbook" is a little misleading ... For all that, the work does form a useful reference book for the shelf' (*Electronics and power,* v.33(2), Feb. 1987, p.142). *Class No:* 621.372.8

Pulses

[4034]

BRITISH STANDARDS INSTITUTION. **Guide to pulse techniques and apparatus.** London, BSI, 1989. graphs. £23.90. (*BS 5698: 1989.*)

1. *Pulse terms and definitions.* 25p.

2. *Pulse measurement and anlysis, general considerations.* 15p.

About 150 terms defined in systematic order, with formulae as appropriate. Alphabetical index. *Class No:* 621.374

Amplifiers

Lasers

Bibliographies

[4035]

GOMERSALL, A. **Lasers in materials processing:** a bibliography of a developing technology. Kempston, Beds., IFS (Publications), 1986. vii, 167p. illus. £35. ISBN: 0948507241, UK edition.

Aims to cover all the major applications in addition to publications on the physics and metallurgy of laser systems. Emphasis is (intentionally) on laser processes (welding, cutting, drilling, machining), surface modification (coating, cladding, heat treatment, etching) and alloying. Includes US, European and Japanese sources. Almost 900 items arranged under 5 main heads with 30-odd subdivisions. Author index but no subject index. Includes conference papers. Informative abstracts. *Class No:* 621.375.826(01)

[4036]

High-power lasers: improvements and applications: a literature study. Sigmond, J. *and* Terpstra, M., *eds.* London, Elsevier Applied Sciences, 1988. ix,226p. illus. £50. ISBN: 1851662340.

A review US, Japanese, French, West German, UK, European (Munich) and international (PCT, Geneva) patents, patent applications, technical and scientific literature published between 1980 and 1985. Over 200 references. 'List of patentees', p.209-24. *Class No:* 621.375.826(01)

Handbooks & Manuals

[4037]

ANDREWS, D.L. Lasers in chemistry. 2nd ed. Berlin, Springer-Verlag, 1990. xii,188p. diagrs., graphs, tables. ISBN: 0387517774.

First published 1986.

5 main chapters: 1. Principles of laser operations - 2. Laser sources - 3. Laser instrumentation in chemistry - 4. Chemical spectroscopy with lasers - 5. Liquid-inducted chemistry. Each chapter has applied questions, with references. 113 figures. 4 appendices (3. Selected bibliography). *Class No:* 621.375.826(035)

[4038]

HECHT, J. The Laser guidebook. New York, etc., McGraw-Hill, 1986. x,381p. illus., diagrs., tables. ISBN: 0070277338.

Concentrates on the 'functional characteristics of lasers' (*Preface*). 24 chapters: 1-6 (introductory explanation); 7-24 (types of lasers, *e.g.* 7. Helium-ion ... 10 Carbon dioxide ... 17. Dye ... 21. Ruby). Appendix A: glossary (aberration - YLF); Appendix B: types of lasers by wavelength. Detailed, analytical index, p.369-80. Earlier, shorter versions of much of the material first appeared in the journal *Lasers & applications. Class No:* 621.375.826(035)

[4039]

Laser handbook. Amsterdam, etc., North-Holland, 1972-90. 6v. diagrs., graphs, tables.

1-2., edited by F.T. Arecchi and E.O. Schulz-Dubois. 1972. xxvii, 1947p. £181.25. ISBN 0720402131.

3., edited by M.L. Stitch. 1979. v, 878p. £129.69. ISBN 0444852719.

4., edited by M. Bass and M.L. Stitch. 1985. x, 594p. £90.63. IBSN 0444869271.

5., edited by M. Bass and M.L. Stitch. 1985. ix, 692p. £102.13. ISBN 0444869344.

6., edited by W.B. Colson and C. Pellegrini. 1990, 580p. £107.90. ISBN 0444869530.

A series of 'expository monographs' (*Preface*) on laser technology and applications, no longer (since v.3) limited by length of article. V.3 has 11 articles; v.5 has 5. 'With the increasing use of lasers in such a wide variety of scientific applications, this book [v.3] will be a valuable reference for advanced students and practising scientists in physics, chemistry and engineering' (*Philosophical magazine,* part B, v.41(6), 1980, p.706). A review in *Physics today* (Dec. 1980, p.59-60) draws attention to inevitable omissions of recent information in v.3. but commends the series, and ends: 'Every research group in quantum mechanics should have access to this set of references'. *Class No:* 621.375.826(035)

[4040]

MALLOW, A. and CHABOT, L. Laser safety handbook. New York, etc., Van Nostrand Reinhold, 1978. xiii, 353p. illus., diagrs., graphs. $24.50. ISBN: 0442250924.

Brings together diverse information otherwise scattered in the literature. 15 chapters. Chapters 1-5 introduce lasers (theory, properties, etc.) and the hazards associated with them. Chapter 6 covers protective standards and deals with factors affecting maximum permissible exposure (MPE). Chapter 7 and 8 are concerned with laser beam evaluation and classification, control of beam and other laser hazards. US Federal legislation is extensively described in Chapter 10. Safety programmes and medical surveillance, protective eyewear and atmospheric effects in outdoor use are covered

.... (contd.)

in the remaining chapters. The 3 appendices contain a glossary (p.277-301) and American standard guidance codes. *Class No:* 621.375.826(035)

[4041]

MILONNI, P.W. and EBERLY, J.H. Lasers. Chichester, Wiley, 1988. 731p. $50. ISBN: 0471627313.

18 chapters. 1-9 cover basic principles and 10-18 provide details of laser operations and applications. End-of-chapter problems. 'Outstanding ... highly recommended' (*Choice,* May 1988). *Class No:* 621.375.826(035)

Reviews & Abstracts

[4042]

Journal of current laser abstracts. Littleton, Mass., PennWell Publishing Co/Laser Focus Publications, 1967-. v.4(1)-. Monthly. $360 pa. ISSN: 00220264.

First published as *Laser/maser international* (v.1-3. 1964-66).

V.23, 1986-1987, contained 4,697 indicative abstracts. Entries are in 3 main groups, each subdivided: General laser publications (including textbooks, conference proceedings and bibliographies) - Theoretical aspects of lasers - Experimental laser research. Issue indexes of personal and corporate authors and patents. Annual pesonal and corporate author indexes. *Class No:* 621.375.826(048)

Yearbooks & Directories

[4043]

Industrial laser annual handbook, 1986. Belforte, D. *and* Levitt, M., *eds.* Tulsa, Oklahoma, PennWell Books, 1986. 250p. $110.

4 sections: Laser materials processing data and guidelines - Review of laser processing - Company and product directory - Related products and services. 'In their preface the editors state that it is intended that future editions ... will build upon the present one. It is hoped that they do; this could become a very valuable database' (*Optics and laser technology,* v.18, Dec. 1986, p.324). *Noted*: 1987 ed. 350p. $125. ISBN 0878143203. *Class No:* 621.375.826(058)

Tables & Data Books

[4044]

BENNETT, W.R. Atomic gas laser transition data: a critical evaluation. New York, Plenum Press, 1979. 293p. diagrs., tables. $75. (*IFI database library.*) ISBN: 0306651874.

159p. of tables and 120p. of annotated references (432 citations). The tables contain only critically evaluated data (oscillation criteria, laser wavelength measurements, ion species and transition identification). Other useful data, not evaluated, are contained in the descriptive summaries in the bibliography. No index of authors.

'... a carefully compiled set of tables arranged in alphabetical order for the 51 elements which have been used to create atomic gas lasers ... This is a useful volume ... Which will, one hopes, sound the death-knell on other less carefully prepared sets of tables' (*Microwaves, optics and antennas* (*IEE Proceedings H*), v.127(2), April 1980, p.106). *Class No:* 621.375.826(083)

[4045]

—DONALDSON,, R.W. *and* EDWARDS, J.G. Concise databook of optical materials: a selection of optical and physical data for the laser and optics laboratory. Teddington, Middx., National Physical Laboratory, 1983. [1], 65p. graphs, tables. ISBN: 0946754004.

Data for 20 optical materials useful for laser research and the laser industry. Materials include gallium arsenide, lithium niobate, neodymium YAG, yttrium aluminium garnet (YAG) and potassium chloride. Data for up to 22 properties, including Young's modulus, melting point, specific heat, temperature coefficients of refractive index and optical path, laser damage threshold, electrostriction. Refractive index and transmission data are presented in graphs. 179 literature references listed. Conversion tables facilitate interpretation in units other than SI. *Class No:* 621.375.826(083)

[4046]

CRC handbook of laser science and technology. Weber, M.J., *ed.* Boca Raton, Florida, CRC Press, 1982-87. 5v. diagrs., graphs, tables. (*CRC Handbook series of laser science and technology.*)

Update and expansion of *CRC Handbook of lasers,* with *selected data on optical technology* (1971).

1. *Lasers and masers.* 1982 552p. $133. ISBN 0849335019.

2. *Gas lasers.* 1982. 584p. $133. ISBN 0849335027.

3. *Optical materials.* Part 1. *Nonlinear optical properties/ radiation damage.* 1986. 480p. $125. ISBN 0849335035.

4. *Optical materials.* Part 2. *Properties. Fundamental properties.* 1986. 496p. $135. ISBN 0849335045.

5. *Optical materials.* Part 3 *Applications, coatings and fabrication* . 1987. 544p. $126.50. ISBN 0849335051.

Suppt.1. *Lasers.* 1991. 560p. £151. ISBN 084933506x.

Extensive tables of experimental data on all types of primary lasers with bibliographical citations to original sources and work are contained in v.1-2. Later volumes cover physical properties of laser optical materials and fabrication techniques. Supplement 1 appeared in 1991. *Class No:* 621.375.826(083)

Electronics

Abbreviations & Symbols

[4047]

BROWN, P. Electronics and computer acronyms. 2nd ed. London, etc., Butterworths, 1988. [vii], 272p. diagrs. £24. ISBN: 0408023988.

First published 1985 as *Dictionary of electrical electronics and computer abbreviations.*

Nearly 400 new entries, bringing the total to almost 2,500, shift the emphasis heavily to the title subjects, although telecommunications, microprocessors, lasers, fibre optics, radio, television, audio and general electrical engineering terms are also included. Definitions and explanations range from one line to fifteen, are often illustrated and are clear and concise. Unexpected entries include whole words, *e.g.* jack, tweeter, X-ray, whilst the Appendix (p.270-2) oddly concentrates entirely on battery choice and use. *Class No:* 621.38(003)

[4048]

—Electronic abbreviations, symbols and terms. 2nd ed. Basingstoke, Hants., Macmillan Educational Ltd., 1985. vi, 79p. £1.95. ISBN: 0333391675.

Previous ed. 1984.

Identifies *c.*200 common abbreviations, *c.*120 circuit and component symbols and *c.*70 logic symbols (showing British and US variants). Defines *c.*70 terms. Highly selective and brief, but simple and satisfactory for limited amateur use. *Class No:* 621.38(003)

[4049]

Index of acronyms and abbreviations in electrical and electronic engineering. Weinheim, etc., VCH, 1989. [vii],538p. £75. ISBN: 0895738120.

About 45,000 entries. More comprehensive than Keller (*qv*), but little to choose between them. *Class No:* 621.38(003)

[4050]

INSTITUTE OF ELECTRICAL AND ELECTRONICS ENGINEERS. Electrical and electronics graphic symbols and reference designations. 2nd ed. New York, IEEE, 1987. v.p. diagrs. ISBN: 0471634565.

Includes all IEEE and ANSI standards. Covers logic functions, parts and equipment, diagrams and electrical wiring. *Class No:* 621.38(003)

Databases

[4051]

ELCOM. Bethesda, Md., Cambrige Scientific Abstracts.

Covers 1980 to date. A bibliographic database providing references, with abstracts, to world literature on electronics, computing and communications, including research, commercial and marketing information. Sources include journal articles, government reports, conference papers, theses, patents. File size: 100,000 citations. Updated bi-monthly; *c.*10,000 additions pa. Hosts: ESA-IRS; BRS. Printed versions: *Electronics and communications abstracts* and *Computer and informative systems abstracts journal.* 'ELCOM ... might be the best choice for the undergraduate student or computer hobbyist who is not interested in exotic foreign language materials ... Elcom covers newspapers and trade literature that may not be considered scholarly enough [for INSPEC and COMPENDEX]' (*Database,* v.4(2), June 1981, p.13-29). *Class No:* 621.38(003.4)

[4052]

—VIDEOLOG. Santa Clara, Calif., Videolog Communications, 1985-.

Textual and numeric databank. Contains 10 files relating to the electronics industry. Product data, covering the contents of the Harris Directory: *Who's who in electronics, D.A.T.A. Books,* and product specifications from many manufacturers. Covers 650,000 semiconductors, integrated circuits and microcomputer systems. Trade directory (14,000 companies); 18,000 product categories. Semiconductor prices. Specifications of military devices. US-based. Daily updates. Available through TELENET and TYMNET. *Class No:* 621.38(003.4)

Bibliographies

[4053]

MOORE, C.K. *and* **SPENCER, K.J. Electronics: a bibliographical guide.** London, Macdonald, 1961. xvii, 411p.

In its time, an outstanding bibliography of bibliographies in its field. About 3,000 items in 68 sections, with subdivisions. Section 1: Abstracting, bibliographical and translating services. For each entry bibliographical details, number of references cited and brief annotations are given. Includes a large number of journal articles. Author and title indexes (p.373-411). Invaluable guide to the literature (1945-1959) of a fast growing field despite 'minor blemishes' noted in *The computer bulletin*, v.5(3), Dec. 1961, p.125. *Class No:* 621.38(01)

[4054]

—**CORBETT, L.** Electronics: a bibliographical guide 3. London, Macdonald, 1973. xxiii, 799p. £20.

Continuation of Moore and Spencer's volumes (*qv*), to cover 1965-68. 4,364 numbered items. 330 periodicals cited. Cumulated author/name index (p.631-704) and cumulated subject index (p.705-99) cover all 3v. It is sad that publishing economics have militated against the continuation of this series into the era of the integrated circuit and the microprocessor, despite the massive bibliographical effort required. *Class No:* 621.38(01)

[4055]

—**MOORE, C.K.** *and* **SPENCER, K.J.** Electronics: a bibliographical guide 2. London, Macdonald, 1965. xvi, 369p.

A worthy update of the previous item, with nearly 3,000 items covering July 1959 to December 1964. 70 sections (1. Reference media, nos.1-303). Very favourably reviewed - with justification - in *Post Office electrical engineers' journal* (v.59(4), January 1967, p.298). *Class No:* 621.38(01)

[4056]

UNITED NATIONS. Industrial Development Organization. **Information sources on the electronics industry.** New York, United Nations, 1979. xii, 103p. (*UNIDO Guides to information sources, no.32.*)

822 numbered entries. Standard UNIDO Guide format of 11 sections: 1. Professional, trade and research organizations, learned societies and special information services (regional listing) - 2. Directories (regional and subject listing) - 3. Sources of statistics, marketing and other economic data (regional listing) - 4. Basic handbooks, textbooks and manuals - 5. Monograph series - 6. Current periodicals - 7. Current abstracting and indexing periodicals - 8. Proceedings, papers and reports - 9. Specialized dictionaries and encyclopaedias - 10. Bibliographies - 11. Other potential sources of information (A. Consulting and engineering services - H. Standards and specifications - J. Translation services). Bibliography of sources used. Strengths: Organizations - 268 entries; Periodicals - 227 entries. *Class No:* 621.38(01)

Encyclopaedias

[4057]

DOUGLAS-YOUNG, J. Portable electronics data book. Englewood Cliffs, N.J., Prentice Hall, 1989. vii,360p. diagrs., graphs, tables. £36. ISBN: 0136858279.

Entries, A-Z ('Alternating current' - 'Waveguides and cavities'). Cross-references. Numerous tables. Several

.... (contd.)

entries contain computer programmes for calculations (*e.g.* 'Energy stored in a capacitor'). Handy format. *Class No:* 621.38(031)

[4058]

GIBILISCO, S. *and* **SLATER, N.,** *eds.* **Encyclopedia of electronics.** 2nd ed. Blue Ridge Summit, Pa., TAB Books, 1990. 960p. illus., diagrs., graphs, circuits, tables. $69.50. ISBN: 0830633898.

First published 1985.

A collection of over 3,000 short articles, A-Z, presenting clear definitions and explanations, with appropriate illustrations (nearly 2 per page), from the whole field of electronics and related subjects. The entries are also listed under 17 categories which alphabetically display the scope (Antennas and feed lines - Audio electronics - Broadcasting and communications ... Computers and digital electronics - Electricity and magnetism ... Switching and control ... Wiring and construction). A specific subject index of 5,500 entries facilitates reference to relevant articles for terms not found in the main encyclopaedia section. 'A good book for public, high school and college libraries' (*Choice*, v.28(6), December 1990. p.607). *Class No:* 621.38(031)

[4059]

McGraw-Hill encyclopedia of electronics and computers. Parker, S.P., *ed.* 2nd ed. New York, London, etc. McGraw-Hill, 1988. [x], 964p. illus., diagrs., graphs, tables. £59.50. ISBN: 007045499x.

First published 1984.

'Much of the material ... has been published previously in the *McGraw-Hill Encyclopedia of science and technology*, 6th edition [1987]' (t.p. verso). Contains 520 articles, 120 of them 'completely revised' (*Preface*), by 328 contributors, alphabetically arranged by topic. Conveniently collects articles from three main areas: principles of electronics; science and technology of electronic devices; applications of devices, with emphasis on computing. Extensive analytical index facilitates retrieval. Useful article bibliographies. 1250 illus. *Class No:* 621.38(031)

[4060]

TRAISTER, J.E. *and* **TRAISTER, R.J. Encyclopedic dictionary of electronic terms.** Englewood Cliffs, N.J., Prentice-Hall, 1984. [iv], 604p. illus., diagrs., circuits, tables. £45.65. ISBN: 0132767780.

Fewer than 1,500 descriptive entries, frequently lengthy, well illustrated with excellent line drawings, cast in simple language understandable by the novice. Avoids the temptation to bulk the work by the inclusion of computer terminology. *Class No:* 621.38(031)

Handbooks & Manuals

[4061]

BELOVE, C., *ed.* **Handbook of modern electronics and electrical engineering.** New York, etc., Wiley, 1986. xxviii, 2401p. illus., diagrs., graphs, circuits, tables. £112. ISBN: 0471097543.

105 US and 1 Canadian contributors. 69 chapters grouped in 13 parts (1-3 cover mathematics, materials and components; 4-7 deal with circuits, systems engineering and automobile control; 8-13 contain chapters on medical applications, sound and video recording, communications, navigation and radar, computers and energy engineering). Emphasis is heavy on electronics. Each chapter starts with

....(contd.)

basics, provides overview of its subject, design information and bibliography.

'A handy source of practical, state-of-the-art information on the design, development and manufacture of electronic circuits and systems ... The Handbook's coverage of telecommunications is unprecedented' (*IEEE Power Engineering review,* PER-6, Dec. 1986, p.27). *Class No:* 621.38(035)

[4062]

COLLINS, T.H. Analog electronics handbook. New York, Prentice-Hall, 1989. xxiv,460p. diagrs., graphs, tables. ISBN: 0130331198.

28 chapters, each in numbered sections. Chapters 1-6 cover background theory and mathematics; 7-28 cover applications (*e.g.* 7. Passive components; 15. Active filters; 23. Modulation; 27. Microwaves and radar principles). Worked examples and over 500 diagrs., etc. Includes glossary of symbols (letter and graphic); Tables (*e.g.* Laplace transforms); analytical index. No bibliography or 'further reading'. 'A useful and well-written book' (*Choice,* May 1989, p.1548). *Class No:* 621.38(035)

[4063]

FINK, D.G. *and* **CHRISTIANSEN, D.,** *eds.* **Electronic engineers' handbook.** 3rd ed. New York, etc., McGraw-Hill, 1989. xvi, [2528]p. illus., diagrs., graphs, circuits, tables. (*McGraw-Hill handbooks.*) ISBN: 0070209812.

First published 1975. 2nd ed. 1982.

'New edition now covers the entire range of electronics' (*Preface*). About 170 contributors; 29 sections (including revised, expanded or new sections on power electronics, pulsed circuits, CAD programs and languages, and standards). Detailed, analytical index, 81p. Contains over 2,000 illustrations and over 3,000 references. *Class No:* 621.38(035)

[4064]

Handbook for electronics engineering technicians. Kaufman, M. *and* Seidman, A.S., *eds.* 2nd ed. New York, etc., McGraw-Hill, 1984. xxxiv, [715]p. tables. £32.95. ISBN: 0070334080.

24 chapters, by 8 contributors, cover resistors, capacitors, coils, transformers, meters and measurements, semiconductor devices, integrated circuits, operation amplifiers, power supplies, batteries and (new to this edition): microprocessors, microwaves, fibre optics, active filters and oscilloscopes. Each chapter contains definitions, types and characteristics of components, problems and solutions, data tables and nomographs. 13 appendix tables provide miscellaneous data (some of doubtful value in the age of pocket calculators). Good index.

Condensed soft-cover version as *Electronics sourcebook for technicians and engineers* (New York, etc., McGraw-Hill, 1988. 592p. £19.95. ISBN 0070335591). *Class No:* 621.38(035)

[4065]

Handbook of electronic package design. Pecht, M., *ed.* New York, Marcel Dekker, 1991. xiii,871p. illus., diagrs., graphs, tables. £115. (*Mechanical engineering, 76.*) ISBN: 0824779215.

12 contributors. 12 documented chapters (2. Electronic components ... 4. Electronic assemblies ... 6. Layout ... 9. Design for vibration and shock ... 12. Electronic materials and properties). Chapter 7, 'Thermal design analysis': 91p.,

....(contd.)

9 sections, 3 exercises, 207 references. Appendices, A-E (A: 'Acronyms used in electronics'). Detailed, analytical index, p.861-71. Well-produced. *Class No:* 621.38(035)

[4066]

Handbook of electronics calculations: for engineers and technicians. Kaufman, M. *and* Seidman, A.H. 2nd ed. New York, McGraw-Hill, 1988. v.p. illus., diagrs., graphs, tables. ISBN: 0070335281.

First published 1979.

Several hundred worked-out problems covering a range of topics from analog and digital circuits. Practical rather than theoretical. 25 chapters, closely subdivided, each with brief bibliography. Appendices, A-R (*e.g.* K: 'Exponentials'). Detailed, analytical index. *Class No:* 621.38(035)

[4067]

HARPER, C.A., *ed.* **Handbook of electronic systems design.** New York, etc., McGraw-Hill, 1980. 849p. illus., diagrs., graphs, circuits, tables. £59.95. ISBN: 0070266832.

29 contributors. Covers computers, communications networks and systems, radar measurement and digital systems, with relevant factual information and technical data in appropriate form (graphs, tables, etc.) Glossaries and references, supplemented by detailed index, assist consultation and understanding. *Class No:* 621.38(035)

[4068]

MANDL, M. Electronics handbook. Reston, Va., Reston Publishing Co., 1983. xiv, 360p. diagrs., graphs, circuits, tables. $34.95. ISBN: 0835916030.

'... contains a broad coverage of electronic topics ranging from basic concepts to radio, television, communications and digital circuitry. It is intended as a reference text ...' (*Preface*). Provides description, definition, explanation, circuits, data, symbols and other information for many aspects of electronics and communications, supported by glossary and index. *Class No:* 621.38(035)

[4069]

MATISOFF, B.S. Handbook of electronics packaging design and engineering. New York, etc., Van Nostrand Reinhold, 1982. viii, 471p. illus., diagrs., tables. £27.65. ISBN: 0442201710.

A determinedly practical guide covering all factors in the design and fabrication of electronic components, including project planning, ergonomics, materials selection, thermal control, environmental protection and safety. Contains chapter of reference data and a glossary of terminology. Supported by numerous diagrams and tables and a succinct index. *Class No:* 621.38(035)

[4070]

MAZDA, F., *ed.* **Electronic engineer's reference book.** 6th ed. London, etc., Butterworth, 1989. v.p. illus., diagrs., graphs, circuits, tables. ISBN: 0408054301.

First published 1958 by Haywood Books. 5th ed. 1983.

85 contributors. Follows the five-part arrangement of the 5th ed.: 1. Mathematical and electrical techniques - 2. Physical phenomena - 3. Materials and components - 4. Electronic circuit design and instrumentation - 5. Applications (including communication satellites, networks, computers, videotape). 6 added chapters on newer topics, *e.g.* application specific integrated circuits, computer aided design techniques, digital system analysis, software engineering, local area networks, integrated services digital

.... *(contd.)*
network (ISDN). Clear text; attractively laid out; simple diagrams; excellent analytical index; useful reference and 'further reading' lists. *Class No:* 621.38(035)

[4071]
MAZDA, F. Power electronics handbook: components, circuits and applications. London, Butterworth, 1990. 417p. illus. $79.95. ISBN: 0408030046.
14 chapters in 3 parts: 1. Components - 2. Circuits - 3. Applications. Appended: symbols; glossary; bibliography. Detailed, analytical index. Recommended 'for the wealth and diversity of useful material summarized in a single volume' (*Choice,* November 1990, p.518). *Class No:* 621.38(035)

[4072]
SCROGGIE, M.G. *and* **JOHNSTONE, G.G. Radio and electronic laboratory handbook.** 9th ed. London, etc., Newnes-Butterworths, 1980. xiv, 592p. illus., diagrs., graphs, circuits, tables. £17.95. ISBN: 0408003731.
First published 1938. Previous ed. 1971.
A highly regarded 'classic' in its field. Aimed, as always, to provide guidance for experiments 'without unlimited funds and space ... or substantial qualifications and experience' (*Preface*). The guidance is essentially practical and contained in 13 chapters: 1. The ends and the means - 2. Premises and layout - 3. Fundamental principles of measurement - 4. Sources of power and signals ... 8. Choice and care of equipment - 9. Measurement of circuit parameters - 10. Signal measurements ... 12. Dealing with results - 13. For reference (units, symbols, laws, concepts, data). A commendatory review in *Wireless world* (v.86,(3), March 1980, p.70) referred to 'lavish provision of references in the text'. *Class No:* 621.38(035)

Dictionaries

Polyglot

[4073]
MIRIMANOV, R.G., *and others.* **Dictionary of electronics:** English - German - French - Dutch - Russian. Deventer, Netherlands, Kluwer, 1985. 544p. ISBN: 9020117882.
9,000 entries in English, with foreign-language equivalents, cross-referenced from other language indexes by alphanumeric code for each term.
Class No: 621.38(038)=00

English

[4074]
AMOS, S.W., *ed.* **Dictionary of electronics.** 2nd ed. London, etc., Butterworths, 1987. [vii], 324p. diagrs., graphs, circuits. £35. ISBN: 0408027509.
First published 1981.
2nd ed. includes new definitions, and revisions or expansions of some others. Aims to keep abreast of developments in semiconductor devices, digital techniques, computers and microprocessors. Target users: non-experts and workers in other fields. Deliberate overlap of coverage with *Dictionary of audio, radio and video,* by R.S. Roberts and *Dictionary of telecommunications,* by S.J. Aries. Approximately 2,300 entries (including cross-references) with terms defined informatively, many with lengthy descriptions and illustrations. *Class No:* 621.38(038)=20

[4075]
GRAF, R.F. Modern dictionary of electronics. 6th ed. Indianapolis, Indiana, Sams, 1984. 1152p. diagrs., graphs, circuits. $39.95. ISBN: 0672220415.
First published 1962. Previous ed. 1977.
Definitions for about 25,000 terms (5,000 more than in 5th ed.). Liberal illus. with identifying captions. Clear definitions, as simple as possible. Metric equivalents for cutomary US measurement units. Good presentation, arrangement and cross-reference.
Class No: 621.38(038)=20

[4076]
McGraw-Hill dictionary of electronics and computer technology. Parker, S.P., *ed.* New York, etc. McGraw-Hill, 1984. [x], 582p. £28.50. ISBN: 0070454167.
10,000 terms selected from the *McGraw-Hill dictionary of scientific and technical terms,* 3rd ed. 1984, aimed 'to explain the basic scientific and technical principles of all relevant disciplines - electricity, electronics, solid-state physics, and computers' (*Preface*). Intentional emphasis on computing technology. Includes acronyms, abbreviations and cross-references. Concise but adequate definitions, with few extending beyond *c.*40 words.
'The definitions are concise and don't contain verbiage ... A precise and diversified dictionary that should be utilized by electronic and computer professionals as well as students, libraries, technical writers and general-interest readers ...' (*Sci-Tech news,* v.39, August 1985, p.82).
Class No: 621.38(038)=20

[4077]
SINCLAIR, I.R. Collins dictionary of electronics. London, Collins, 1988. vi,378p. diagrs., graphs, tables. £4.99. ISBN: 000434345x.
Over 2,000 terms, briefly defined (*e.g.* 'logic gate': 11 lines). Covers theoretical and practical applications. 4 appendices: D. graphical symbols. References, p.vi.
Class No: 621.38(038)=20

[4078]
TURNER, R.P. *and* **GIBILISCO, S. Illustrated dictionary of electronics.** 5th ed. Blue Ridge Summit, Pa., TAB Books, 1991. 720p. diagrs., tables. (*TAB Professional and reference book series.*) ISBN: 0830673458.
4th ed. 1988.
More than 27,000 terms and abbreviations defined, briefly (*c.*3 to 4 lines). Numerous cross references, but few illustrations. Appendices: schematic symbols; tables of data (*e.g.* resistor color codes; mathematical functions; electronics abbreviations). 'Good' ('Reference books bulletin', *Booklist,* May 1 1991, p.1738).
Class No: 621.38(038)=20

[4079]
—YOUNG, C. The Penguin dictionary of electronics. 2nd ed. Harmondsworth, Penguin, 1988. [viii], 671p. diagrs., circuits. £7.99. ISBN: 0140511873.
First published 1979 as *New Penguin dictionary of electronics.*
About 3,500 terms defined, some at length (*e.g.* 'loudspeaker': 2p.) Cross-references. 15 tables appended (2. colour Codes). *Class No:* 621.38(038)=20

[4080]
WILSON, F.A. A Reference guide to practical electronic terms. London, Bernard Babani, 1992. [vi],433p. diagrs. £5.95. ISBN: 0859342328.

About 1200 terms, some defined at length (*e.g.* 'Impatt diode': *c.*1p., with diagram). Cross references.
Class No: 621.38(038)=20

German

[4081]
OPPERMANN, A. Wörterbuch der Elektronik: Englisch-Deutsch. [Dictionary of electronics: English-German.] Munich, etc., Saur; Detroit, Mich., Gale, 1980. 691p. $120. ISBN: 068698305x.

German equivalents of *c.*100,000 terms covering electronics and related areas such as computing, radar, nuclear physics.

Later ed., in 2 vols.

1. *Englisch-deutsch.* 1987. *c.*800p. DM180. ISBN: 3980122735.

2. *Deutsch-englisch.* 1984. *c.*794p. DM160. ISBN 3980122700. *Class No:* 621.38(038)=30

French

[4082]
FLEURTY, M. Dictionnaire encyclopédique d'électronique: English-French. [Encyclopedic dictionary of electronics: English-French.] Paris, La Maison du Dictionnaire, 1991. 1072p. FF1200; £127.90. ISBN: 285608043x.

40,000 entries; 20,000 definitions. Ample cross-references. '… a must for any serious technical translator working either into or from French' (*The Linguist,* v.31(5), 1992, p.164). *Class No:* 621.38(038)=40

[4083]
KING, G.G. Elsevier's dictionary of electronics: English/French. Amsterdam, etc., Elsevier, 1987. *c.*600p. £105. ISBN: 0444426426.

30,000 terms, English-based with French equivalents. French/English vol. not yet published.
Class No: 621.38(038)=40

Spanish

[4084]
MATAIX, M. Diccionario de electrónica y energía nuclear, inglés-español. Barcelona, Danae, 1969. 772p. 620pstas.

10,000 English-language terms, ranging over electronics, computing and nuclear energy, with Spanish equivalent and brief Spanish definitions. Index of Spanish terms. Severely limited by age and subject spread.
Class No: 621.38(038)=60

Russian

[4085]
KALUGIN, I.K., *and others.* Modern English-Russian dictionary of radioelectronics. [Anglo-russkiĭ slovar' po sovremennoĭ radioelektronike.] 2nd ed. Moscow, 'Soviet Encyclopaedia' Publishing House, 1972. 448p. diagrs.

Russian equivalents of *c.*20,000 terms drawn from the electronics field, not radio, as its title would seem to suggest. (Although radio, televison and telecommunications are

.... *(contd.)*
included, there are many surprising omissions). Includes lists of abbreviations and symbols. Indexes Russian terms by page and line number references.
Class No: 621.38(038)=82

Reviews & Abstracts

[4086]
Electronics and communications abstracts. London, Multiscience Publishing, 1961-. v.1(1)-. Monthly. £85pa. ISSN: 00135119.

6,000-7,000 indicative abstracts. Covers periodical articles, conference proceedings, reports and books. 6 sections: Electromagnetic wave techniques - Materials, devices and phenomena - Circuits and networks - Communications - Control - Computers - Measuring, recording and miscellaneous applications. Twice-yearly subject index and annual author index. No other indexes.
Class No: 621.38(048)

[4087]
Electronics and communications abstracts journal. Bethesda, Md., Cambridge Scientific Abstracts, 1972/3 - v.4-. 6pa. + Index issue. $462pa. ISSN: 03613313.

Formerly *Electronics abstracts journal,* (v.1-3, 1966-1971/2).

About 10,000 indicative and informative abstracts, unsigned, pa. 5 main sections, subdivided: EP. Electronic physics - ES. Electronic systems - EC. Electronic circuits - ED. Electronic devices (including microelectronics) - ECO. Communications. Sources: periodical articles, reports, conference papers, books, theses, patents. Issue indexes of authors and subject. Annual subject and author indexes. Available on magnetic tape and online as ELCOM (*qv*), accessible via DIALOG, BRS, IRS, ORBIT, Infoline.
Class No: 621.38(048)

[4088]
PASCAL Explore. Part F20. Électronique et Télécommunications. Paris, Centre National de la Recherche Scientifique. 1985-. 10pa. F850.pa.

First published (1940-) as *Bulletin signalétique.* 4. *Physique.* II. *Electricité.*

As *Bulletin signalétqiue,* Part 145 (1961-84) contained about 7,500 indicative and informative abstracts pa. (v.42, 1981, had 11,019 abstracts) systematically arranged in 10 main groups, with sections and sub-sections for each, *e.g.* 03. Dispositifs à l'étatsolide … 05. Circuits … 07. Télécommunications. Items drawn from 13,000 journals, conference papers, theses and reports. Subject and author indexes per issue and annually. Available on paper or microform. Also available online as part of PASCAL file, through ESA-IRS or QUESTEL. *Class No:* 621.38(048)

[4089]
Referativnyĭ zhurnal. 23 Elektronika. Academiya Nauk, SSSR. Moscow, VINITI, 1980-. Monthly. R148. ISSN: 04862287.

Formerly: *Referativnyĭ zhurnal elektronika i ee primenenie* (1963-1979).

More than 20,000 indicative and informative, signed abstracts pa. 4 sections: Electronic tubes and devices - Semiconductor devices - Optoelectronic devices - Electronic materials. Half of the content is in the semiconductors

.... *(contd.)*

section. Issue author and patent indexes and scanned journals list. Annual author and subject indexes. English contents summary. UDC nos. *Class No:* 621.38(048)

Progress Reports

[4090]

Advances in electronics and electron physics. London, Academic Press, 1954-. v.7-. 2 or 3 pa., with supplements. illus., diagrs., graphs, tables. ISSN: 00652539.

Previously *Advances in electronics,* vs. 1-5, 1948-1953.

Several major reviews in each vol. V.84 (1992. 360p. ISBN: 0120147262. £67) deals with image mathematics and image processing, including vector quantization, lattice transforms, invariant pattern representations and ochographic imaging. *Class No:* 621.38(055)

Yearbooks & Directories

[4091]

DATA digests. San Diego, Calif., D.A.T.A., Business Publishing, 1978-. (UK distributor: H.T.I. Ltd., Romford, Essex). diagrs., tables. £85-£125 pa.

The largest collection of device data. Currently published in 19 books (annually with quarterly or semi-annual updates). Separately available. 4 groups: Integrated circuits (5 books, *e.g. Digital ICs,* plus *IC Alternate Source digest*); Discretes (5 books, *e.g. Diodes,* plus *Discrete Alternate source digest*); Special applications (3 books, *i.e., High reliability electronic components; International semiconductor directory; Application notes*); Materials (4 books, *e.g. Adhesives*). The publishing pattern undergoes continuing revision and a major change is taking place in 1988 with the reduction of titles from over 30 to 19, the change of name to *Digest* (from *Book*) and the abandonment of a separate set of books listing discontinued devices. The series aims to provide manufacturer information and technical data and the identification of replacements and alternatives. Magnetic tape versions are also available. *Class No:* 621.38(058)

[4092]

The Directory of electronics, instruments and computers. 21st ed. London, Morgan-Grampian, 1987. [vi], 525p. £28. ISBN: 0862130816. ISSN: 02671441.

Also cited as the *EIC directory,* and formerly published as *IEA directory.*

Section 1 lists 4,100 UK manufacturers with addresses, etc. Section 2 gives UK agents for over 3,000 overseas firms. There are separate classified buyers' guides for electronics and instruments (240p.) and for computers (34p.). Section 6 provides location guide to distributors; overseas firms', distributors'; and distributors' addresses. Includes a trade names index of over 8000 names. *Class No:* 621.38(058)

[4093]

Electronics applications sourcebook. Helms, H.L., *ed.* New York, etc., McGraw-Hill, 1986. 2v. (xxxv, [3,338]p. illus., diagrs., graphs, circuits, tables. £265. ISBN: 0070792631.

A convenient collection of applications literature (*c.*400 items) supplied by 21 electronics and semiconductor US companies. Includes applications notes or technical briefs for particular devices or technologies; data sheets; extracts from handbooks; and reprints from periodicals. Arranged

.... *(contd.)*

alphabetically by company name, items under alphanumeric coding used by company or under title. Indexes by component number and by subject. Additional volumes planned to extend range of companies and to add new material from that already included. *Class No:* 621.38(058)

[4094]

European electronics directory 1991: Components and sub-assemblies: the guide to manufacturers, distributors and agents. Wedgewood, C.G., *ed.* Oxford, Elsevier, 1991. viii,673p.

Previously published as *European electronics component distributors directory.*

6 Sections: 1 Manufacturers, country by country - 2. Agents and representatives, country by country - 3. Index of all companies - 4. Classified list of products and suppliers - 5. Index of products - 6. Glossary of German and French product names. *Class No:* 621.38(058)

[4095]

—International electronics directory '90. 3rd ed. Oxford, Elsevier, 1990. 2v. (1408p.). ISBN: 0444700056.

4 sections: 1. Manufacturers, country by country - 2. Representatives, country by country - 3. Alphabetical list of companies - 4. Classified list of products and services. *Class No:* 621.38(058)

[4096]

INTEL CORPORATION. Handbooks. Santa Clara, Calif., the Corporation, 1985. 10v. diagrs., tables. $120 (set).

Comprising: *Quality/reliability handbook. CHMOS handbook. Memory components handbook. Telecommunication products handbook. Microcontroller handbook. Microsystem components handbook. Development systems handbook. OEM Systems handbook. Software handbook. Military handbook. Class No:* 621.38(058)

[4097]

MULLARD, LTD. Mullard technical handbooks. London, Mullard, Ltd. Annual. diagrs., graphs, circuits.

Comprises 4 sets of books, each consisting of several parts.

1. *Semiconductor devices.* 7 parts, plus sub-parts (*e.g.* Part 4a. Power diodes. Part 4b. Thyristors and triacs).

2. *Electronic tubes.* 4 parts, plus sub-parts (*e.g.* Part 16. Cathode ray tubes).

3. *Components, materials and assemblies.* 7 parts, plus sub-parts (*e.g.* Part 1c. Fixed resistors. Part 6. Loudspeakers).

4. *Integrated circuits.* 10 parts, plus sub-parts (*e.g.* Part 3. ICs for telephony).

Information includes general and functional description, electrical and other data, performance characteristics and ratings. Also: *Quick reference guide* (Annual). Lists alphanumerically and briefly describes all Mullard products.

Many of the large manufacturers of electronic components and devices publish detailed series of handbooks covering all their products, with diagrams, specifications, description, applications notes and tabulated data of characteristics, performance, etc., with aim of assisting designers in the selection and use of their products; equivalents and alternatives are also often listed. Two typical examples are here shown. Similar handbooks are available from other companies, *e.g.* Mostek Corporation, Motorola Semiconductor Products, RCA, Ltd., Signetics Corporation, Texas Instruments, Ltd. *Class No:* 621.38(058)

[4098]
Who's who in electronics, 1987. Graeser, K., *ed.* 39th ed. Twinsburg, Oklahoma, Harris Publishing Co, 1987. Annual. 1096 + 208p. $93. ISBN: 0916512886. ISSN: 97348576.

A directory covering about 11,000 US manufacturers of electronic equipment, devices and components, over 4,000 distributors and about 4,200 representatives, each category arranged first A-Z, then geographically by state and town. Includes a classified products directory (490p.). Often referred to as the *Harris Directory*. *Class No:* 621.38(058)

Tables & Data Books
[4099]
BRAND, J.R. Handbook of electronic formulas, symbols and definitions. 2nd ed. New York, Van Nostrand Reinhold, 1992. 400p. diagrs., tables. £22. ISBN: 0442003021.

First published 1979.

A pocket book intended to provide the majority of formulas and definitions needed for calculations in passive and active circuit technology. About 100p. of new material, with an additional section on operational amplifiers. *Class No:* 621.38(083)

[4100]
BUCHSBAUM, W.H. Complete handbook of practical electronic reference data. 2nd ed. Englewood Cliffs, N.J., Prentice-Hall, 1978. xxxi, 645p. diagrs., graphs, circuits, tables. $19.95. ISBN: 0130846244.

First published 1973.

A handbook aimed to provide 'all the essential information that anyone interested in electronics needs to have ...'. The reader requires 'only the most fundamental knowledge of electronics ...' Contains formulae, reference data, charts and graphs. Covers all major fields of electronics, including opto-electronics, radar, computers. *Class No:* 621.38(083)

[4101]
CLIFFORD, M. Master handbook of electronic tables and formulas. 5th ed. Blue Ridge Summit, Pa., TAB Books, 1990. xx,553p. diagrs., tables. $22.95 (Pbk). ISBN: 083062192x; 0830621911, pbk.

4th ed. 1984.

Aims to provide solutions to problems in electronics and communications by copious provision of tables giving data for the many hundreds of relevant formulas, obviating the need for calculations. 25 chapters cover basic concepts (resistance, inductance, power, etc.), applications (recording, video, antennas, etc.), computers, constants, conversions numbers. One chapter lists all formulas. Extensive tables. Detailed, analytical index, p.541-53. *Class No:* 621.38(083)

[4102]
Electronic databook. Graf, R.F., *ed.* 3rd ed. Blue Ridge Summit, Pa., TAB Books, 1983. viii, 407p. diagrs., graphs, tables, nomograms. $24.95; $19.95. ISBN: 0830601384; 0830615385, pbk.

First published 1971 as *Electronic design databook*. Previous ed. 1974.

A very wide-ranging and useful compilation of information and calculation aids, clearly presented. Organized in 6 sections (1: Frequency data - 2: Communication - 3: Passive components and circuits - 4: Active components and circuits - 5: Mathematical data, formulas, symbols - 6: Physical data). Includes a glossary (p.192-205) of integrated circuits

....(contd.)
and microelectronics terms. Marred only by incorrect index page number references to section 6.

4th ed. 1988. 528p. £19.75. ISBN: 0830629580. *Class No:* 621.38(083)

[4103]
—BRINDLEY, K. Newnes radio and electronic engineer's pocket book. 18th ed. Oxford, Heinemann Newnes, 1992. 341p. illus., diagrs., tables. £9.95. ISBN: 0434901873.

First published 1940. Previous ed. 1987.

This new edition remains a handy source of a wide range of fundamental facts and data, supported by an effective index, but occasionally needing a magnifying glass to extract the required information from the tiny page. *Class No:* 621.38(083)

[4104]
Electronic engineering index. Bracknell, Berks., Technical Indexes Ltd., 1978-. 3pa. illus. £60pa. ISSN: 03088375.

This data book is the index to the associated microform file of the catalogues of manufacturers and suppliers of electronic components, devices and equipment, which is available on separate subscription. 8 sections (A. Passive components - B. Active components ... H. Power sources, miscellaneous equipment and services). Issue 54, November 1987 to April 1988 (304p.) contains Products and tradenames (the index to the Selector charts) - Distributors - Selector charts (p.38-217) - Professional and trade organizations - Addresses (p.232-302). *Class No:* 621.38(083)

[4105]
Handbook of electronic materials. New York, etc., IFI/Plenum, 1971-72. 8v. graphs, tables. $66per v.

1. *Optical materials properties*, by A.J. Moses, 1971, 104p. ISBN: 0306671018.

2. *III-IV semiconducting compounds*, by M. Neuberger. 1971. 115p. ISBN: 0306671026.

3. *Silicon nitride for microelectronic applications*, by J.T. Milek. Part 1. Preparation and properties. 1971. 118p. ISBN: 0306671034.

4. *Niobium alloys and compounds*, by M. Neuberger, D.L. Grigsby and W.H. Veazie Jr. 1972. 70p. 0306671042.

5. *Group IV semiconducting materials*, by M. Neuberger. 1971. 67p. ISBN 0306671050.

6. *Silicon nitride for microelectronic applications*, by J.T. Milek. Part 2. Applications and devices. 1972. 164p. ISBN: 0306671069.

7. *III-V ternary semiconducting compounds, datatables*, by M. Neuberger. 1972. 56p. ISBN: 0306671077.

8. *Linear electrooptic modular materials*, by J.T. Milek and M. Neuberger. 1972. 265p. ISBN: 0306671085.

A set of 8v. compiled by the Electronic Properties Information Center, a US Department of Defense - designated unit based at the Hughes Aircraft Company in California. The vols. present data on the electronic, optical and magnetic properties of all types of materials used in electronics: insulators, semiconductors, metals, superconductors, etc. *Class No:* 621.38(083)

[4106]
HUGHES, F.W. Illustrated guidebook to electronic devices and circuits. Englewood Cliffs, N.J., Prentice-Hall, 1983. xvi, 431p. diagrs., graphs, circuits, tables. ISBN: 0134513282.

The title well reflects the scope of this book, which identifies the basic theory, components and function of each circuit and system and gives advice on testing and troubleshooting methods. 14 chapters cover: components; solid-state devices; optoelectronic devices; vacuum tubes; power supplies; amplifiers; oscillators; audio circuits; radio circuits; television circuits; integrated circuits; operational amplifiers; digital circuits; microcomputers. Excellent text and illustrations but access is impeded by lack of index.
Class No: 621.38(083)

[4107]
MANDL, M. Electronic data reference manual. Reston, Va., Reston, 1979. ix, 323p. diagrs., circuits, tables. $29.95. ISBN: 0835916413.

An updated, extended version of the author's *Handbook of modern electronic data* (Reston, 1973) which makes no reference to its predecessor nor to its reprinting of considerable portions of it without change. Its emphasis has shifted to solid-state devices and integrated and digital circuits, but it continues to offer a wide range of information and data relating to basic electronic components, devices and circuits - power supplies, oscillators, amplifiers, transmission lines, antennas, etc. Brief but serviceable index.
Class No: 621.38(083)

[4108]
PASAHOW, E. Electronic ready reference manual. New York, etc., McGraw-Hill, 1984. xv, 569p. diagrs., tables. $28.95. ISBN: 0070487235.

A pocket book (14 x 10cm) which aims 'to provide a single source of current electronics information' for users at all levels. Presents formulas, equations, tables, diagrams, data and circuits for basic concepts, passive and active components, linear circuits, filters, power supplies, measurement, communications, digital circuits, microprocessors and computers in 10 sections. Final 6 sections cover mathematics, symbols, conversion formulas, properties of materials and safety. Contents page shows range of each section (except micro-processors) but index is too brief. Absence of many page numbers hampers access to required information. *Class No:* 621.38(083)

[4109]
—TURNER, R.P. Electronic conversions, symbols and formulas. Blue Ridge Summit, Pa., TAB Books, 1975. 224p. $8.95. ISBN: 0830657509.

A pocket book of selected information, concentrating on formulas grouped by concept (*e.g.* 1. Mathematics - 2. Resistance - 3. Capitance ... 7. Magnetics) and circuits and components (*e.g.* 9. Semiconductors), With chapters for letter symbols and conversion factors. Over-brief index. 2nd ed., by R.F. Turner and S. Gibilisco. 1988. *c.*£12. ISBN 0830628657. *Class No:* 621.38(038)

Standards

[4110]
Annual book of ASTM standards. Section 10. Electrical insulation and electronics. Vols. 10.04 and 10.05 . Electronics. Philadelphia, Pa., American Society for Testing and Materials, 1992. 2v. £64 each. ISSN: 01922998.

.... *(contd.)*
These 2v. of section 10 of the *Annual book of ASTM standards* contain the texts of over 260 standards in alphanumeric sequence of their designation codes, grouped under 13 headings (*e.g.* Electro-optics; Microelectronic packaging; Silicon, electrical properties; Silicon, mechanical properties; Photolithography). Subject indexes.
Class No: 621.38(083.74)

Histories

[4111]
DUMMER, G.W.A. Electronic inventions and discoveries: electronics from its earliest beginnings to the present day. 3rd rev. and expanded ed. Oxford, etc., Pergamon, 1983. ix, 233p. illus., diagrs., charts. £22; £9.95. (*Pergamon international library*.) ISBN: 0080293549, Hard cover; 0080293530, Flexicover.

First published in 1977 as *Electronic inventions 1745-1946.* Previous ed. 1978.

500 events are described in chronolgical order from 1642-1982 (the event being the first announcement of a discovery or invention wherein a new technique or device is introduced into electronics). The descriptions are taken from the text of the source in the literature. Indexes of subjects, inventors and bibliographies are appended. 9 brief chapters with concise histories of branches of electronics (e.g., radar, television) with illustrative charts. Chapter 10 lists inventions by subject, chronologically for each, under 13 heads. Bibliography.

Review of 1977 ed. noted: '... despite a few slips, hard to avoid in such a work, it was surely well worth compiling and should prove to be a valuable reference book for multitudes' (*Electronics and power,* v.23 (3), March 1977, p.243).
Class No: 621.38(091)

[4112]
—SHIERS, G. Bibliography of the history of electronics. Metuchen, N.J., Scarecrow Press, 1972. xiii, 423p. $8.50. ISBN: 0810804999.

Annotated bibliography of 1,820 periodical articles, books and reports on the history of electronics and communications from 1850. Grouped by subjects. *Class No:* 621.38(091)

Worldwide

[4113]
TODD, D. The World electronics industry. London, Routledge, 1990. [xiv],335p. figs., graphs, maps, tables. £35. ISBN: 0415024978.

8 chapters, closely subdivided: 1. Introduction - 2. Pivotal global players - 3. Structural factors - 4. Defense electronics - 5. Innovation and enterprise - 6. Japanese powerhouse - 7. NIC [Newly-industrialising countries] challenge - 8. Conclusions. Extensive notes and references (*e.g.* chapter 4 has 79). Glossary ('AAM' - 'Wafer'), p.307-14; non-analytical index, p.327-35. A scholarly treatment, marred by poor quality camera-ready copy. *Class No:* 621.38(100)

Research Establishments

[4114]
Electronics research centres: a world directory of organizations and programmes. 2nd ed. London, Longman, 1989. 523p. $325; £180. (*Reference on research series.*) ISBN: 0582036046.

....*(contd.)*

First published 1986.

Profiles of organizations in over 75 countries - private companies, academic institutions, government laboratories, etc. 'Electronics' includes semiconductors, communications, electronic circuits, information science. By country A-Z, then by name of company. Indexes of organizations and subjects. *Class No:* 621.38:061:061.62

Components

[4115]

BRINDLEY, K. Newnes Electronics assembly pocket book. Oxford, Newnes, 1991. vi,298p. illus., diagrs., graphs. £10.95. ISBN: 0750602228.

Covers a wide range of areas in assembly and wiring (*e.g.* safe working practice; specifications; component identification; cables; thin film modules, etc.). Nearly 1000 illus. Small, handy format. *Class No:* 621.38.032

[4116]

CODUS. Sheffield, Codus, Ltd.

A data bank (textual-numeric) for quality assured electronic components, providing comprehensive technical information on all components approved to CECC BS 9000 and (UK) IECQ Specifications. Covers 1967 onwards. File size: 71,000 entries (July 1986). Updated daily. Host. Codus Ltd: Costs: £200 pa., plus £55 per connect hour. Available to subscribers only. *Class No:* 621.38.032

[4117]

ECDB: (Electronic Components Data Bank). Frascati, Italy, European Space Agency.

A data bank containing full descriptions and physical and operational characteristics of electronic components currently available in Europe. Includes passive (*e.g.* resistors) and active (*e.g.* thyristors) components and integrated circuits. Data include manufacturers and distributors. File size: 130,000 items (1981). Updated monthly. 30,000 additions pa. Host: ESA.

Printed equivalent: *Passive components reference book.* (*NB. Active* components are excluded). Not available commercially. *Class No:* 621.38.032

[4118]

Electronics reliability data: a guide to selected components. London, INSPEC/Institution of Electrical Engineers, 1981. tables.

Presents reliability data sheets to show failure rates for over 100 different components (resistors, capacitors, diodes, rectifiers, transistors, thyristors, microcircuits), derived from data in the Systems Reliability Service (SRS) of the National Centre for Systems Reliability. In addition provides a collection of about 720 relevant abstracts, reprinted from the INSPEC data base, classified by type of component, as above. Bibliography, p.151. *Class No:* 621.38.032

[4119]

GINSBERG, G.L. A User's guide to selecting electronic components. New York, Wiley, 1981. xii, 249p., illus., diagrs., tables. £21.80. ISBN: 0471083089.

Intended as an aid for design engineers in the selection of suitable components. Provides information about types, advantages and disadvantages, behaviour, construction, effect on circuits, failure and applications. 6 Parts: Resistors - Capacitors - Electromagnetic components (transformers, etc) - Power sources (including batteries) - Special-function

....*(contd.)*

components (oscillators, filters, etc) - Solid-state devices (*e.g.* diodes, rectifiers, transistors, optoelectronic devices). 130 references. Index. *Class No:* 621.38.032

[4120]

—**JONES, T.H.** Electronic components handbook. Reston, Va., Reston Publishing Co., 1978. xviii, 391p. illus., diagrs., graphs, tables. $24.95.

Aims to provide users with 'a better understanding of component performance characteristics' (*Preface*) as an aid to effective selection. Gives indication of performance, reliability and comparative costs. Emphasizes industrial rather than hobby varieties. Ten chapters cover capacitors, resistors, transformers, relays, switches, connectors and semiconductors, with definitions, types, manufacturers (US) and further reading for each. *Class No:* 621.38.032

Circuits

Encyclopaedias

[4121]

GRAF, R.F. The Encyclopedia of electronic circuits. Blue Ridge Summit, Pa., TAB Books, 1985-1991. 3v. circuits, tables. $29.95 (each v.).

V.1 (1985. 760p. ISBN: 0830609385) presents *c.*1,300 circuits in an A-Z sequence, with a subject index.

V.2 (1988. 838p. ISBN: 083063183) includes alarm and security circuits, smoke, moisture and metal detectors, computer, fibre optic and laser circuits, among much else.

V.3 (1991. 838p. ISBN: 083063348) organizes circuits into 108 categories.

Almost 1,300 circuits presented in 98 alphabetically arranged sections (Alarms ... Metal detectors ... Zero crossing detectors). Brief descriptive notes for most circuits and sources (books, periodicals, trade literature) given for each. Specific index leads directly to required circuits. All 3v. are fully illustrated. *Class No:* 621.38.04(031)

Handbooks & Manuals

[4122]

DI GIACOMO, J., *ed.* **VLSI handbook:** silicon, gallium arsenide, and superconductor circuits. New York, etc., McGraw-Hill, 1989. xiv,v.p. illus., diagrs., graphs. ISBN: 0070169039.

32 contributors, most from US industry. 'A practical manual ... to describe and explain the VLSI design methodology process from the customer's point of view' (*Preface*). 11 parts (32 chaps.): 1. Design - 2. Test - 3. Fabrication ... 7. Packaging - 8. Economics ... 10. Analog circuits - 11. Special topics. Appendices (7) cover definitions, acronyms, constants, scaling factors, etc. Detailed analytical index. *Class No:* 621.38.04(035)

[4123]

LENK, J.D. Handbook of practical electronic circuits. Englewood Cliffs, N.J., Prentice-Hall, 1982. xii, 334p. diagrs., graphs, circuits. £16.45. ISBN: 013380741x.

270 circuits, commonly used in various branches of electronics, *e.g.* audio-frequency, radio-frequency, oscillator, filter, switching, control and operational amplifier circuits, giving the theory, operation and applications of each. Aimed at students, experimenters and technicians.

'It is a refreshing surprise to find that, unlike many

....(contd.)

similarly titled books, it sets the circuits it describes in a framework of background and explanation which makes it unusually instructive and informative ...' (*Electronics and power*, v.28, Sept., 1982, p. 615).
Class No: 621.38.04(035)

[4124]
MARSTON, R.M. Newnes electronic circuits pocket book. Oxford, Butterworth Heinemann, 1991. [viii],336p. diagrs., tables. ISBN: 0750601329.

625 diagrams; 22 chapters (*e.g.*6. 'Norton op-amp circuits', p.76-87). A practical approach in a handy format (for use if not for library shelves).
Class No: 621.38.04(035)

[4125]
O'DELL, T.H. Circuits for electronic instrumentation. Cambridge Univ. Press, 1991. 218p. illus., diagrs. $69.50; £35. ISBN: 0521404282.

Covers a wide range of circuits (*e.g.* comparators, wideband direct couple feedback amplifiers, phased locked loops, voltage controlled oscillators, etc.) at an advanced level. Extensive chapter bibliographies.
Class No: 621.38.04(035)

[4126]
VLSI Electronics: microstructure science. Einspruch, N.G., *ed.* New York, etc., Academic Press, 1981-. v.1-. illus., diagrs., graphs, tables.

A multi-volume treatise on Very Large Scale Integration, each volume independent, which aims to provide 'a comprehensive exposition that describes the state of this science and technology and that assesses trends for the future of VLSI electronics and the scientific base that supports its development ... Includes subjects ... from microscopic aspects of materials behaviour and device performance ... through ... fabrication of VLSI circuits ... to ... VLSI in systems applications'. (*Editor's Preface*).

'The books ... contain a very impressive and well-presented multi-disciplinary cross-section of today's state-of-the-art in microelectronics and should certainly be part of the library of anyone who is seriously interested in the subject' (Review of v.1-6, *Microelectronics J.* v.15(1), Jan. 1984, p.60). *Class No:* 621.38.04(035)

[4127]
VLSI Handbook. Einspruch, N.G., *ed.* Orlando, Florida, Academic Press, 1985. xxvi, 902p. diagrs., graphs, tables. £93. (*Handbooks in science and technology*.) ISBN: 0122341007.

74 contributors. Aims to be 'a comprehensive compilation to provide data, performance application information, and guidelines for the entire range of VLSI [Very Large Scale Integration] technology' - circuits, fabrication and applications (*Preface*). 52 chapters cover circuit design, manufacturing processes, applications in microprocessors, RAM, ROM, computers, electronic warfare, radar, medicine, communications, image and speech processing, automobiles. Chapter references, bibliographies and glossaries vary in provision. Certain chapters contain large numbers of tables and figures. Excellent index.
Class No: 621.38.04(035)

[4128]
WILLIAMS, T. The Circuit designer's companion. Oxford, Butterworth Heinemann, 1991. vi,302p. illus., diagrs., graphs, tables. $75. ISBN: 0750611421.

9 chapters, with emphasis on reliability in analog circuit design: 1. Grounding and wiring - 2. Printed circuits ... 5. Laser integrated circuits ... 9. General product design. Appended: British & related IEC standards. Bibliography; index, p.295-305. 'Highly recommended' (*Choice*, January 1992, p.772). *Class No:* 621.38.04(035)

Dictionaries

Polyglot

[4129]
NAGY, P. *and* **TARJÁN, G. Elsevier's dictionary of microelectronics:** in English, German, French, Spanish and Japanese. Amsterdam, etc., Elsevier, 1988. ISBN: 0444426590.

8,521 English-base terms, with translations into other languages; Romanized Japanese. Indexes: English (in place of cross references in the main list; 12,293 items); German (12,866); French (11,491); Spanish (10,956); Japanese (11,697). *Class No:* 621.38.04(038)=00

English

[4130]
DOUGLAS-YOUNG, J. Illustrated encyclopedic dictionary of electronic circuits. Englewood Cliffs, N.J., Prentice-Hall, 1983. 444p. diagrs., circuits, tables. $27.95. ISBN: 0134507347.

Aims to give guidance on choice, design building, testing and fault identification of several hundred basic electronic circuits. Alphabetical arrangement, with descriptions, schematic diagrams, parts lists and cross references. Appendices of useful data. Index of limited utility.
Class No: 621.38.04(038)=20

[4131]
HOLLAND, R.C. Illustrated dictionary of microelectronics and microcomputers. Oxford, etc., Pergamon Press, 1985. v, 162p. diagrs., tables. £28; £11.75. ISBN: 0080316344, Hardback; 0080316352, Flexicover.

About 850 terms, including cross-references, with definitions and explanations ranging from a few words to several hundred, supported by diagrams. Clear and simple. Includes a number of manufacturing terms from the trade literature. 'For those bewildered by the new language of microelectronics it is a very useful dictionary [which explains] often-used words, phrases and abbreviations' (*Aslib book list*, v.50(9), Sept. 1985, item 441).
Class No: 621.38.04(038)=20

[4132]
—PLANT, M. Microelectronics, A to Z. Harlow, Essex, Longman, 1985. xiv, 209p. £2.95. ISBN: 0582892856.

Upwards of 2,000 entries, including cross-references, with terms defined and, in many cases, explained in simple language. *Class No:* 621.38.04(038)=20

German

[4133]

BINDMANN, W. Dictionary of microelectronics: English-German, German-English. Amsterdam, etc., Elsevier, 1984. 634p. £85. ISBN: 0444996192.

Equivalents of *c.*22,000 terms in the fields of microelectronics, solid state physics, microlithography, integrated circuits, microcomputers, software and peripherals. Appears to lack applications and fabrication technology. *Class No:* 621.38.04(038)=30

[4134]

—ATTIYATE, Y.H. *and* SHAH, R.R. Wörterbuch der Mikroelektronik und Mikroechnertechnick. [Dictionary of microelectronics and microcomputer technology: Deutsch-Englisch, Englisch-Deutsch.] Düsseldorf, VDI Verlag, 1984. vi, 469p.

Gives equivalents for *c.*7,500, with definitions and explanations for *c.*2,000 key terms, - intended to be 'easily understood and yet precise' (*Preface*). This objective has been met. Wide scope, including data processing and application in automation and production. American spellings (*e.g.* fiber optics). *Class No:* 621.38.04(038)=30

[4135]

—Wörterbuch der Mikroelektronik. [Dictionary of microelectronics: English-German, German-English.] Vaterstetten, IWT Verlag; Düsseldorf, VDI Verlag, 1980. 218p. $39.95. ISBN: 3883220019, IWT; 3184190692, VDI.

Records *c.*7,000 terms in current use, with emphasis on industrial needs. Includes entries under abbreviations. A supplementary volume for new technology is planned. *Class No:* 621.38.04(038)=30

Tables & Data Books

[4136]

MARKUS, J., *comp*. Digital circuits ready-reference. New York, etc., McGraw-Hill, 1982. xx, 162p. £21.95. (*Ready-reference series*.) ISBN: 0070404577.

A collection of circuits for clock signals, converters, data transmission, memories, microprocessors, etc., drawn from journals, books and trade literature. Component values, brief description and bibliographical citation for source are given for each. Author and subject indexes. Others in series: *Electronics projects ready-reference* (1982. 181p.) *e.g.* burglar alarms, battery chargers, remote control; *Popular circuits ready-reference* (1982. 216p.) *e.g.* amplifiers, converters, oscillators; *Special circuits ready-reference* (1982. 234p.) *e.g.* audio-control, fibre optics, optoelectronics; *Communications circuits ready-reference* (1982. xvii, 216p.)

Author and McGraw-Hill have published 4 earlier, large collections: *Sourcebook of electronic circuits* (1968. viii, 888p), 3,000 circuits; *Electronic circuits manual* (1971. ix, 988p), 3,180 circuits; *Guidebook of electronic circuits* (1974. 1067p), 3,600 circuits; *Modern electronic circuits reference manual* (1980. 1238p), 2,700 circuits. These 9v. together contain over 14,000 circuits and, it is claimed, virtually no duplication. *Class No:* 621.38.04(083)

[4137]

—MARKUS, J. *and* WESTON, C. Essential circuit reference guide. New York, McGraw-Hill, 1988. 531p. diagrs. $59.50. ISBN: 0070402623.

Selection of 'classic' circuits culled from 3 of Markus's earlier works. Diagrams and schematics of available, still-in-use circuits for industrial and hobby applications. 62 sections, arranged by type of circuit. Author and subject indexes. List of sources. *Class No:* 621.38.04(083)

Printed Circuits

[4138]

BÖHSS, G., *comp*. Dictionary of printed circuit technology; English-German/German-English. Oxford, etc., Elsevier, 1985. 185p. £45.30. ISBN: 0444995552.

About 7,000 terms, including phrases, with equivalents. Genders given. Clear presentation, sans serif typeface used (bold for entry terms). *Class No:* 621.38.049.75

[4139]

COOMBS, C.F., *Jr.*, *ed*. Printed circuits handbook. 3rd ed. New York, etc., McGraw-Hill, 1988. xxxvi, [884]p. illus., diagrs., graphs, tables. $59.50. ISBN: 0070126097.

First published 1967.

Previous ed. 1979. 27 contributors. Aims to provide in one volume the information needed to design, manufacture, test and repair printed wiring boards and assemblies (*Preface*). 8 parts, 35 chapters (1. Introduction to printed wiring - 2. Engineering - 3. Fabrication (the largest part - 8 chapters, 241p.) - 4. Assembly and test - 5. Soldering - 6. Quality and reliability - 7 Multilayer circuits - 8 Flexible circuits). The newest techniques - *e.g.* surface mount technology (SMT) - are generously covered. Numerous tables, chapter references and a glossary are valuable adjuncts to a clear, logically-presented, essentially practical text. Good index. Emphasis is heavy on US sources. *Class No:* 621.38.049.75

[4140]

HINCH, S.W. Handbook of surface mount technology. New York, Wiley; Harlow, Longman, 1988. xvi,481p. illus., diagrs., graphs, tables. ISBN: 0582005175, UK; 047021094x, US.

15 chapters in 4 parts: I. Overview - II. Components - III. Design - IV. Manufacturing technology. 2 appendices (Acceptance criteria for visual inspection; glossary) index (9p.). Surface mount technology (SMT) has made 'dramatic changes' in the manufacture of electronic assemblies; Hinch (of Hewlett-Packard) has attempted a 'thorough, balanced treatment ... from a single source' of the skimpy and fragmented literature (*Preface,* p.v.). For practising engineers. One of the first works in a new field, this is 'highly recommended for technical libraries' (*Choice*, April 1989, p.1360). *Class No:* 621.38.049.75

[4141]

SCARLETT, J.A., *ed*. The Multilayer printed circuit board handbook. Ayr, Electrochemical Publications, 1985. xxiii, 590p. illus., diagrs., tables. £59. ISBN: 0901150150.

24 chapters by 17 UK expert contributors cover the entire field of multilayer PCBs. The editor is responsible for over one third of the text, including a wide-ranging Introduction (60p.). Boards made by printing, etching, bonding and plating processes are covered and there are chapters on quality and inspection, design, repair, presses and machining, with a final chapter on the Multiwire process.

....*(contd.)*

Emphasis throughout is on manufacture. Effective subject index but no glossary.

'... is an essential reference source for designers and provides comprehensive, authoritative guidance on the design, manufacture and use of expensive and complex types of PCBs' (*Microelectronics journal*, v.16(14), July 1985, p.35).

'This is undoubtedly the most authoritative and comprehensive book yet written on multilayer circuit boards and should be on the shelf of every multilayer circuit board manufacturer and user' (*Microelectronics and reliability*, v.25(6), 1985, p.1157). *Class No:* 621.38.049.75

Connectors

[4142]

CLAYTON, P.A., *ed*. **Handbook of electronic connectors.** Ayr, Scotland, Electrochemical Publications, 1982. xix, 297p. illus., diagrs., graphs, tables. £47. ISBN: 0901150118.

14 contributors. 6 chapters on the main groups of connectors (RF interconnection, multiway connectors, printed circuit board connectors, hermetic and underwater connectors, flat cable systems); 2 chapters on fibre optics connectors; 1 each on plating finishes and quality control. *Class No:* 621.38.06

Semi-conductors

Abbreviations & Symbols

[4143]

BRITISH STANDARDS INSTITUTION. **Specification for letter symbols for semiconductor devices and integrated microcircuits.** London, BSI, 1980. 32p. £75.40. (*BS 3363:1980.*)

Supplement 1: 1981. 18p. Supplement 2: 1981. 12p. *Class No:* 621.382(003)

Databases

[4144]

EMIS: (Electronic Materials Information Service). London, INSPEC/Institution of Electrical Engineers.

Covers Dec. 1981 to date. A data bank giving detailed properties of 6 major semiconducting materials (silicon, amorphous silicon, gallium arsenide, indium phosphide, lithium niobate and quartz) and less extensive data for 90 other materials. Textual information includes data reviews, sample preparation, measurement techniques. Properties data are derived from individual scientists and companies, from reviews, handbooks, and other published literature. File size: 10,000 records. Updated monthly. Available on BRS; ESA-IRS. Cost per connect hour: £60. Also contains supplier index (with addresses, etc). Data base available on magnetic tape, updated monthly. *Class No:* 621.382(003.4)

[4145]

SCPR: (Semiconductor Parameter Retrieval). Englewood, Colorado, Information Handling Services.

An alpha-numeric data bank containing technical descriptive data for over 160,000 semiconductor devices available from 150 manufacturers and distributors, with references to location of the maker's data sheet in IHS's Visual Search Microfilm Service (VSMF) and company information. Updated 3 times pa. Available via: BRS. *Class No:* 621.382(003.4)

Encyclopaedias

[4146]

GRAYSON, M., *ed*. **Encyclopedia of semiconductor technology.** New York, etc., Wiley, 1984. xxviii, 940p. illus., diagrs. £94.95. (*Encyclopedia reprint series*.) ISBN: 0471881023.

34 lengthy articles, alphabetically arranged under subject, all reprinted from the 3rd ed. of the Kirk-Othmer *Encyclopedia of chemical technology*. They cover methods of manufacture, properties and uses of semiconductors and elements from which they are derived. A comprehensive index with heavy emphasis on chemical compounds facilitates reference. It must be remembered that many of the articles (from the early part of the alphabet) appeared in Kirk-Othmer as far back as 1980. *Class No:* 621.382(031)

[4147]

MAHAJAN, S. *and* KIMERLING, L.C. **Concise encyclopedia of semiconducting materials and related technologies.** Oxford, Pergamon Press, 1992. xxiii,582p. illus., diagrs. £130. (*Advances in materials science and engineering, 8*.) ISBN: 008034724x.

Incorporates relevant material from the *Encyclopedia of materials science and engineering* (1986) and newly commissioned articles. Topics include growth of bulk crystals, metallization, passivation. *Class No:* 621.382(031)

Handbooks & Manuals

[4148]

Digital systems reference book. Holdsworth, B. *and* Martin, G.R. Oxford, Butterworth Heinemann, 1991. v.p. illus., diagrs. $250; £120. ISBN: 0750610085.

Covers mathematical techniques, integrated circuit technologies and software design to applications. Criticized in *Choice* for the inconsistent use of British, American and international standards, but nevertheless 'an important reference tool' (*Choice*, April 1992, p.1205). *Class No:* 621.382(035)

[4149]

MOSS, T.S., *ed*. **Handbook on semiconductors.** Amsterdam, etc., North-Holland, 1980-82, reprinted 1982-86. 5v. diagrs., graphs, tables. c.£520. ISBN: 0444852980 set.

1. *Band theory and transport properties*. 1982. xviii, 879p.

2. *Optical properties of solids*. 1988. xiv, 634p.

3. *Materials, properties and preparation*. 1980. xiv, 926p.

4. *Device physics*. 1981. xvi, 970p.

5. In preparation.

V.1-4 contain 58 chapters by 82 expert contributors. Extensive references; volume author and subject indexes. *Class No:* 621.382(035)

Dictionaries

[4150]

WRIGHT, H.C. **Semiconductor reference book: from acceptor impurities to zeher breakdown.** Bromley, Kent, Chartwell-Bratt, 1991. 78p. diagrs. £8.95. ISBN: 9144349017.

Brief entries (1 to 3 per page) with definitions, mathematical formulae, diagrams, 'further reading'. School level and upwards. 'A very useful book' (*Reference reviews,* V.6(2) 1992, p.27), particularly at the price.
Class No: 621.382(038)

Tables & Data Books

[4151]

BEADLE, W.J., *and others, ed.* **Quick reference manual for silicon integrated circuit technology.** New York, etc., Wiley, 1985. 736p. diagrs., graphs, tables. £75. ISBN: 0471815888.

15 sections covering fundamental constants, processing data, evaluation of processes, device properties. Presented in many hundreds of tables, curves, graphs, measurements, and drawn from numerous cited sources. Supported by 1,200-item index.

'It is particularly useful to have a concise description of measurement techniques and theory alongside data ... The data which are presented has [sic] been carefully assessed and sifted ... [The volume's] generally high quality and ease of use [justify] the publisher's price' (*Solid-state and electron devices* (*IEEE Proceedings I*), v. 133(1), Feb. 1986, p.28).
Class No: 621.382(083)

[4152]

—Semiconductor general-purpose replacements. Indianapolis, Indiana, H.W. Sams, 1978. 1116p. $14.95. ISBN: 0672215764.

Lists 180,000 replacement devices by reference number, covering 10 US manufacturers and ranging over transistors, diodes, rectifiers and integrated circuits.
Class No: 621.382(083)

Electron Photocells

[4153]

NENTWIG, K., *comp.* **Elsevier's dictionary of opto-electronics and electro-optics:** in four languages: English/German/French/Spanish. Amsterdam, etc., Elsevier, 1986. [vii], 295p. £72. ISBN: 0444426175.

3,500 English-base terms, with German, French and Spanish equivalents, followed by indexes in each language referring to numbered entry in Basic table. No definitions and no genders given. *Class No:* 621.383

[4154]

TEXAS INSTRUMENTS. **Optoelectronics data book:** infrared, imaging and visible products. Berrill, D.J., *ed.* 6th ed. Bedford, Texas Instruments, 1984. [443]p. diagrs., tables. £5. ISBN: 0904047369.

First published 1974.

A collection of data sheets, application reports, etc., for more than 250 standard devices arranged in product groups (*e.g.* Infrared-emitting devices, Photodetectors, Light-emitting diodes), with sections on quality and reliability and applications. An interchangeability index links over 600 devices produced by other manufacturers with TI equivalents. *Class No:* 621.383

Telecommunications

Databases

[4155]

Telecommunications. New York, EIC/Intelligence, Inc.

Formerly TELECOM. Covers Oct. 1984 to date. A bibliographic data base providing references, with abstracts, to world literature on telecommunications, including commercial and economic aspects. Derived from journal articles, conference papers, reports, patents, etc. File size: over 4,000 items. Updated monthly. Hosts: DIALOG, ESA-IRS. Printed version: *Telecommunications abstracts.*
Class No: 621.39(003.4)

[4156]

TELEDOC. Issy les Moulineaux, Centre National d'Études des Télécommunications.

Covers 1972 to date. A bibliographic data base providing references, with abstracts, to world literature on telecommunications, electronics and related subjects, including acoustics, optics, computing. File size: 105,000 items. Updated monthly. About 8,000 additions pa. Available via QUESTEL. Printed version: *Bulletin signalétique des télécommunications.* In English and French.
Class No: 621.39(003.4)

Handbooks & Manuals

[4157]

Electronic communications handbook. Inglis, A.F., *ed.* New York, McGraw-Hill, 1988. v.p. illus., diagrs., graphs, tables. £59.50. ISBN: 0070317119.

22 chapters, closely subdivided, in 2 parts: 1. Transmission and switching technologies - 2. Electronic communications systems. 29 contributors, most from industry. 12p. detailed, analytical index. 'A compendium of reference data' (*Preface*). *Class No:* 621.39(035)

[4158]

PETERSEN, D. **Audio, video and data telecommunications.** New York, McGraw-Hill, 1992. 344p. illus. £17.95. ISBN: 0077074270.

The basic theories of telecommunications, with modern applications. Practical examples and case studies.
Class No: 621.39(035)

Dictionaries

Polyglot

[4159]

CLASON, W.E., *comp.* **Elsevier's telecommunication dictionary in six languages:** English/American, French, Spanish, Italian, Dutch and German. 2nd rev. ed. Amsterdam, etc., Elsevier, 1976. [ix], 604p. £22.12.

First ed. (compiled by A. Visser) 1955, with Danish as the basic language. 7,095 numbered terms, A-Z, in English/American main section, with equivalents in other languages. Alphabetical indexes in other languages refer to numbered entries in main section. Genders are given. Subject area indicated for 35 categories. *Class No:* 621.39(038)=00

English

[4160]
ARIES, S.J. **Dictionary of telecommunications.** London, Butterworths, 1981. [iv], 329p. diagrs., tables. £15. ISBN: 0408003286.

About 3,200 entries, including cross-references. Definitions, many supplemented by extensive explanatory material and some illustrated by diagrams, cover the entire field of telecommunications (defined here as 'the transmission of information from one point to another by wire, radio, optical or other electromagnetic systems').

Deliberate and coordinated overlap with S.W. Amos' *Dictionary of electronics* and R.S. Roberts' *Dictionary of audio, radio and video (qv)*. Aims to include North American terminology where this differs. Appendix (p.317-324) identifies almost 500 abbreviations and acronyms. *Class No:* 621.39(038)=20

[4161]
BRITISH KINEMATOGRAPH SOUND AND TELEVISION SOCIETY. **BKSTS Dictionary of audio-visual terms.** London, Focal Press, 1983. [vi], 138p. diagrs. £10.95. ISBN: 0240512014.

Approx. 2,400 technical terms covering 'the preparation and presentation of pictures and sound by film and video as well as by tape-slide, film-strip and multivision' (*Preface*). Includes computer graphics. Brief entries. Acronyms and abbreviations included. *Class No:* 621.39(038)=20

[4162]
GRAHAM, J. *and* LOWE, S. **The Facts on File dictionary of telecommunications.** 2nd ed. New York, Facts on File, 1991. 240p. diagrs., tables. $24.95. ISBN: 0816020299.

First published 1983.

Covers acronyms, abbreviations, computer-related terms and a few commercial products. Major concepts have extended definitions. 'Highly recommended for academic and technical libraries' (*Library Journal*, 1 July 1991, p.86,88). *Class No:* 621.39(038)=20

[4163]
LANGLEY, G. **Telephony's dictionary.** 2nd ed. Chicago, Ill., Telephony Publishing Corp./London, Pitman, 1986. xi, 402p. . £24.95. ISBN: 0273026690.

Cover title (as published in UK): 'International dictionary of telecommunications'. First published 1982.

16,000 terms covering the entire field of telecommunications, with many terms from computing and data processing, defined or explained, briefly or at considerable length (from a few words to 200). Appendices: acronyms and abbreviations; SI units and the electromagnetic spectrum terminology. *Class No:* 621.39(038)=20

[4164]
ROBERTS, R.S. **Dictionary of audio, radio and video.** London, Butterworths, 1981. 256p. diagrs., graphs, circuits. £15. ISBN: 0408001356.

Definition and explanation of *c.*1700 terms aimed at general readers. Well cross-referenced. Deliberate and coordinated overlap with S.W. Amos' *Dictionary of electronics* and S.V. Aries' *Dictionary of telecommunications (qv)*. Aims to include North American terminology where this differs. *Class No:* 621.39(038)=20

[4165]
WEIK, M.H. **Communications standard dictionary.** New York, Van Nostrand, Reinhold, 1983. ix, 1045p. illus., diagrs. £346.75. ISBN: 0442219334.

Intended as 'a comprehensive compilation of terms and definitions used in the field of communications' - which is widely defined. Over 9,000 entries, abbreviations and cross-references, with definitions based wherever possible on standards. Observed emphasis on fibre optics communications (including illustrations) is due to apparently complete inclusion of entries from same author's *Fiber optics and lightware communications standard dictionary* (Van Nostrand Reinhold, 1981) which is thereby utterly superseded. *Class No:* 621.39(038)=20

French

[4166]
LUCCA, J. de. **Dictionnaire des télécommunications:** anglais-français. [Dictionary of telecommunications: English-French.] Paris, Masson, 1988. 401p. £27. ISBN: 222581003x.

Over 34,000 entries. 'As comprehensive and up-to-date as anyone could possibly hope for' (*Languages international*, v.1 (Jan-Feb. 1989) p.35). *Class No:* 621.39(038)=40

Spanish

[4167]
FREEMAN, R.L. **English-Spanish, Spanish-English dictionary of communications and electronic terms.** Cambridge Univ. Press, 1972. xii, 206p. £7.50. ISBN: 0521080800.

Contains, 8,000 English terms with Spanish equivalents and 9,000 Spanish terms with English equivalents. Coverage extends to data processing, reliability, video and space communications. Additional features: lists of abbreviations and initialisms with equivalents in each language; words and abbreviations deriving from 'decibel'. *Class No:* 621.39(038)=60

Reviews & Abstracts

[4168]
Bulletin signalétique des télécommunications. Issy le Moulineaux, Centre National d'Études Télécommunications, 1946-. Monthly.

Over 11,000 abstracts pa., arranged in 4 sections: books; periodicals; conference papers; CNET publications, with the first two sections grouped systematically, *e.g.* 2. Télécommunications par fil (Télégraphie; Téléphonie, etc.) - 3. Télécommunications sans fil (Télévision, Radiolocalisation, etc.) - 4. Électroniques et structures électroniques ... 6. Physique appliquée et électrotechnique ... 8. Technologie industrielle. *Class No:* 621.39(048)

[4169]
Referativnyĭ zhurnal. 64. Elektrosviaz. Moscow, VINITI, 1968-. Monthly. R64. ISSN: 01347772.

Previously as *Referativnyĭ zhurnal Radiotechnika i electosviaz* (1963-66).

About 9,000 indicative and informative signed abstracts pa. Covers all aspects of telecommunications in 12 sections (*e.g.* Communication theory - Telecommunication equipment ... Telegraphy - Data transmission - Telephony - Facsimile

....(contd.)

transmission ... Power supply). Issue indexes of authors, periodicals abstracted and patents. Annual author and subject indexes. *Class No:* 621.39(048)

Yearbooks & Directories

[4170]
Communications: the most comprehensive guide to the UK telecommunications industry. Egan, M., *ed.* London, Macmillan, 1986. vii, 290p. £65. ISBN: 0333386906.

Directory of 2,000 companies, A-Z, with usual information plus brief descriptions of their activities and products, complemented by classified section listing these firms under one more of 80 category headings, *e.g.* Answering machines, Message switching systems, Viewdata. Also brief listings of trade associations, consultancies, etc., and reviews of major organizations, particularly British Telecom. '... an essential reference source for telecommunications managers, telecommunications and computer companies ... (*Telecommunications journal*, v.54(8), 1987, p.541). *Class No:* 621.39(058)

[4171]
—The Communication users' yearbook. Simpson, A., *ed.* London, National Centre for Information Technology, 1986. Annual-. 496p. £45. ISBN: 0850125588. ISSN: 02691914.

One-quarter of this work consists of reviews of the present state-of-the-art in technology and standards and a collection of miscellaneous information, with emphasis on standards and on regulatory, governmental, professional and trade bodies. A useful glossary (350 terms and abbreviations) is included. A buyer's guide to equipment and services occupies the remainder of the volume and comprises 9 categories (*e.g.* computers, cable and satellite) with names, addresses, etc., of firms listed in alphabetically-arranged sub-sections (*e.g.* Encryption services). Some information on specific models and items is given in tabulated form, including prices. As a reference work this suffers from lack of single alphabetical indexes of either products or firms. *Class No:* 621.39(058)

[4172]
—Telecomm's products and services directory: a directory and buyers' guide to manufacturers and suppliers of products and services in the telecommunications industry. 1987-88 ed. London, Telecommunications Press, 1987. x, 289p. £46. ISBN: 0907401082.

Previous ed. 1985 (1984).

Alphabetical directory of *c.*600 UK firms, ranging from one-man retailers to British Telecom (10,000 employees) and multinationals (*e.g.* IBM), giving company information and activities. Classified products, services and equipment sections, with helpful notes and prices. Index of 400 trade names and glossary of 650 technical terms, both being usefully annotated. Occasional errors noted. *Class No:* 621.39(058)

Tables & Data Books

[4173]
FREEMAN, R.L. Reference manual for telecommunications engineering. New York, etc., Wiley 1985. xv, 1504p., illus., diagrs., graphs, charts, tables, nomograms. $112. ISBN: 0471867535.

Aims 'to provide a central source of basic information' for

....(contd.)

designers and users of all kinds of telecommunications systems. Content is assembled in 26 subject areas (*e.g.* Signaling, Fiber optics transmission, Radio transmission, Electromagnetic interference), each type with introductory material, essential information, references and, often, a bibliography. Most of the material is drawn from authoritative sources, which are identified. Index of more than 3,000 entries. *Class No:* 621.39(083)

[4174]
MARKUS, J. Communications circuits ready-reference. New York, etc., McGraw-Hill, 1982. xvii, 216p. diagrs. (*Ready reference series.*) ISBN: 0070404607.

A collection of about 400 circuits grouped in 15 alphabetically arranged chapters (*e.g.* Antennas, Frequency modulation, Receivers, Telephones, Teleprinters, Televisions, Transmitters). Circuit diagrams, component values, brief descriptions and bibliographical citation for sources are given for each. Author and subject indexes. For details of other four vols. in the 'Ready reference series' see entry at 621.38.04(083). *Class No:* 621.39(083)

Standards

[4175]
MACPHERSON, A. International telecommunication standards organizations. Boston, Mass., London, Artech House, 1990. xiii,317p. diagrs., tables. ISBN: 0890063656.

13 chapters. Focuses on international organizations and the main North American, Asian-Pacific and European standards bodies. Glossary of acronyms, p.309-17. *Class No:* 621.39(083.74)

Institutions & Associations

[4176]
International directory of telecommunications: market trends, companies, statistics and personnel. Robert, S. *and* Hay, T., *eds.* 4th ed. Harlow, Essex, Longman Group, 1990. xiv,386p. tables. £140. ISBN: 0582647382.

Profiles of *c.*800 telecommunications services companies, equipment manufacturers and broadcasting concerns. Data: company structure; address; phone and fax number; activities; turnover; product areas; operating territory. Arranged geographically. Indexes of organizations, products and key personnel. *Class No:* 621.39:061:061.2

Wave Signals

[4177]
ADAMS, M.S. *and* HENNINGS, I.D. Optical fibres and sources for communications. New York, Plenum, 1990. 180p. £35. (*Updates in applied physics and electrical technology.*) ISBN: 0306437112.

Includes abstracts of about 300 papers. Contents: Introduction - Propagation in multimode fibres - Propagation in monomodel fibres - Semiconductor luminescence - Light-emitting diodes - Semiconductor lasers - Selected abstracts. *Class No:* 621.391.6

[4178]
BODSON, D. *and* BOTEZ, D., *eds*. **Electro-optical communications dictionary.** Rochelle Park, N.J., Hayden, 1983. [viii], 168p. £22.50. ISBN: 0810409615.

Contains about 2,500 entries and cross-references covering fibre optic and lightwave communication systems 'consistent with international, federal, industry, and technical society standards' (*Preface*). Definitions are therefore full and are often annotated. An appendix contains almost 200 acronyms and abbreviations. *Class No:* 621.391.6

[4179]
CLARRICOATS, P.J.B., *ed*. **Progress in optical communications.** London, P. Peregrinus, 1980-82. 2v. (*IEE Reprint series, v.3-4.*)
1. 1980. 247p. £14.30. ISBN: 090604832x.
2. 1982. 328p. £25. ISBN: 0906048842.

Collections of letters (brief communications) to *Electronics letters*, 1978/79 and 1980/81 respectively. Each has groupings covering: fibre theory and measurement; fibre properties, design and manufacture; components and splices; systems and instruments. V.2 additionally has a section on sources and detectors. Because *Electronics letters* has played a major role in disseminating research results in this field and practises strict refereeing of submitted letters, these volumes constitute a valuable source of information. Subject and author indexes. Reviews in *Electronics and power* (v.26(12), 1982, p.905; v.30(4), 1984, p.325) commend them for beginners in the subject. *Class No:* 621.391.6

[4180]
GEISLER, J., *and others*. **Optical fibres.** Oxford, etc., Pergamon, 1986. xiv, 635p. illus., diagrs. (*EPO Applied technology series, v. 5.*) ISBN: 0080305776.

Aims to explain the technology 'from a practical point of view by making use of the latest information described in patent and other literature from all over the world' (*Foreword*) based on the files of the European Patent Office by three of its examinees. They present a well-structured literature survey of around 1,100 national and international patents (including about 300 US, 300 Japanese, 100 British, 100 French and 75 German) in simple language, supported by ample illustration. 3 parts: Making optical fibres; Connectors, terminals, branches; Optical fibre transmission. Indexes of patent numbers, patentees and specific subjects. *Class No:* 621.391.6

[4181]
WEIK, M.H. **Fiber optics standard dictionary.** 2nd ed. New York, Van Nostrand Reinhold, 1989. 352p. $32. ISBN: 0444223876.

Updates *Fiber optics and lightwave communications standard dictionary.*

Several thousand terms are defined and explained, with liberal use of examples, illus. and cross-references. 12 pages of bibliography. An essential dictionary in its specific field. *Class No:* 621.391.6

[4182]
WHITE, D.R.J. **Handbook series on electromagnetic interference and compatibility.** Gainesville, Virginia, Don White Consultants, Inc., 1971. 6v. illus., graphs, tables. $339 (set). (*Handbook series on EMI/EMC.*)
1. *Electrical noise and EMI specifications.* 1971, reprinted 1981. [580p.].
2. *EMI test methods and procedures.* 2nd ed. 1980. [450p.].

....(contd.)
3. *EMI control methods and techniques.* 3rd ed. 1981. [800p.].
4. *EMI test instrumentation and systems.* 2nd ed. 1980. [400p.].
5. *EMI prediction and analysis techniques* 1972. [700p.].
6. *EMI specifications, standards and regulations* 2nd ed. 1981. [1035p.]. *Class No:* 621.391.6

[4183]
YEH, C. **Handbook of fiber optics:** theory and applications. San Diego, Calif., Academic Press, 1990. 384p. $89.95. ISBN: 0127104558.

Practical or theoretical basis for the study of optical fibres and how they contact light waves along their length. Index. *Class No:* 621.391.6

Radio & Television Apparatus

Handbooks & Manuals

[4184]
Radio and television servicing, 1986-87 models. Wainwright, R.N., *ed*. London, Macdonald, 1987. Annual. 828p. diagrs., circuits. £29.50. ISBN: 0356144399.

First published 1952.

A guide to circuits and essential servicing information for a wide selection of currently available television sets, computer monitors, radio receivers and cassette tape recorders (including a compact disc player for the first time). Emphasis is on televison receivers. A cumulative index covers all entries, 1952-86, with cross references to similar models and relevant volumes in series. *Class No:* 621.396/.397(035)

Dictionaries

English

[4185]
DIAMANT, L., *ed*. **The Broadcast communications dictionary.** 2nd ed., rev. and enl. New York, Hastings House, 1978. 201p. $9.95. (*Communication arts books*.) ISBN: 0803807880.

Contains about 3,200 terms and abbreviations from the whole field of radio, television, sound and vision recording. 'Designed for every day assistance' (*Introduction*), this work has many technical terms and abbreviations, but also a large number of non-technical entries in use on both sides of the Atlantic. (*e.g.* BBC, rehearse) and colloquialisms (*e.g.* soap, tranny). Clearly intended as a brief and inexpensive source for lay recipients of the broadcasting ouput. *Class No:* 621.396/.397(038)=20

German

[4186]
ARD [Arbeitsgemeinschaft der öffentlich rechlichen Rundfunkanstalten der Bundesrepublik Deutschland], *comp*. **Dictionary of radio and television terms, English-German, German-English;** compiled and edited by ARD ... in collaboration with the BBC ... and the ZDF. 2nd ed. London, New York, etc. Heyden, 1975. xxi, 199p. ISBN: 085501251x.

....(contd.)

First published 1972 by Pitman.

'Comprises about 12,000 items in the German-English part and about 17,000 items in the English-German part' (*Foreword*). Based on a German-French dictionary published in 1970. Includes a brief joint list of abbreviations (p.xiii-xxi). Genders given for German nouns. *Class No:* 621.396/.397(038)=30

Tables & Data Books

[4187]

World radio TV handbook. London, Billboard, 1992. 590p. illus., tables, maps.

V.46, 1992, includes services listed by country, station information, languages, operating times, frequencies, maps of principal transmitter sites worldwide, reception conditions, annual test reports on receivers.

A guide to radio and television broadcasting stations throughout the world, with data on frequencies, radiated power, exact location, operating hours, frequency schedules, languages broadcast, news bulletins, etc. Separate sections for radio and television, each arranged by continent and then alphabetically by country. Separate list of long and medium wave stations and short wave stations. Information about international organizations. Miscellaneous data: world time; solar activity; reception conditions; receivers owned in each country; suitable megahertz bands, etc. User's guide in 4 languages. Equipment test reports. Advertisements. *Class No:* 621.396/.397(083)

[4188]

—AMOS, S.W., *ed*. Radio, TV and audio technical reference book. 5th ed. London, etc., Butterworths, 1977. [xv, 1172]p. illus., diagrs., tables, circuits. £25. ISBN: 040800259x.

First published 1954. Previous ed. 1963.

31 contributors. Aimed at technicians, servicemen and amateur enthusiasts to provide 'an essentially practical account of modern developments ...' (*Preface*). 35 chapters. 9-page index. Fundamentals may change little, but a decade of developments in equipment and the rise of video recording lessen the value of this work. *Class No:* 621.396/.397(083)

Radio

[4189]

The ARRL handbook for the radio amateur. Hutchinson, C.L., *ed*. 62nd ed. Newington, Conn., American Radio Relay League, 1984-. Annual. illus., diagrs., graphs, circuits, tables. US $22.50; elsewhere $24. (*Radio amateur's library, publications no. 6.*) ISBN: 087259162x cloth; 0872590623 paper.

First published as *Radio amateur's handbook* and with this title until 61st ed. 1983.

60th ed. [640p]. had 23 chapters (1. Amateur radio - 2. Electrical laws and circuits ... 4. Solid-state fundamentals ... 6. HF transmitting - 7. VHF and UHF transmitting - 8 Receiving systems ... 14. Specialized communication systems ... 16. Test equipment and measurements ... 23. Vacuum tubes and semiconductors). Glossary (chapter 3, p.17-20); data tables (drill sizes, wire gauges, etc., chapter 17); base and case diagrams for vacuum tubes and semiconductors and operating data for transmitting valves (chapter 23). Index (11p.).

....(contd.)

'Any amateur radio enthusiast would be ill-advised to pursue his hobby without it' (Review of 1980 ed. in *Wireless world*, v.8-6(4), April 1980, p.82). *Class No:* 621.396

[4190]

ORR, W.I. Radio handbook. 22nd ed. Indianopolis, In., Sams, 1982. [1173]p. illus., diagrs., graphs, circuits, tables. $39.95. ISBN: 0672218747.

First published 1967. Previous ed. 1978.

A standard handbook. 35 chapters cover all theoretical aspects of radio, transmitting and receiving, including components and devices, design, interference, power supplies, mobile equipment, propagation (5 chapters on antennas) and construction practices. Chapter 35 contains miscellaneous data. Index (8p.) lacks some expected terms. Little attention to audio and output stages. *Class No:* 621.396

[4191]

RADIO SOCIETY OF GREAT BRITAIN. Radio communication handbook. 5th ed. London, the Society, 1976-77. v.p. [698]p. illus., diagrs, graphs, circuits, tables. £9.88. ISBN: 0900612584.

First published 1938. Previous ed. 1968. 5th ed. originally published in 2v.

Aims to cover everything the active radio amateur needs to know, from first principles and simple components to satellite and image communication in 23 chapters.

'The most comprehensive textbook on the theory and practice of amateur radio ...' (*Telecommunication journal*, v.44, July 1977, p.350). *Class No:* 621.396

[4192]

Referativnyĭ zhurnal. 24. Radiotekhnika. Akademiya Nauk SSSR. Moscow, VINITI, 1967-. Monthly. illus. R280. ISSN: 0034267x.

Formerly: *Referativnyĭ zhurnal radiotekhnika i elektrosviaz* (1963-1966).

35,000 indicative and informative, signed, occasionally illustrated abstracts pa., on radio engineering, widely interpreted to include television, radar and quantum electronics. 7 sections: General and theoretical radio engineering - Aerials - Broadcasting - Radiolocation, radionavigation, radio control, television - Radio transmitters and receivers - Quantum electronics - Manufacture of radio equipment and parts. UDC numbers. English contents summary. List of journals scanned in issue no. 1 of each vol. Annual author and subject indexes. *Class No:* 621.396

[4193]

Reference data for radio engineers. 6th ed. Indianapolis, In., etc., H.W. Sams, for ETT, 1975. [1026]p. diagrs., graphs, charts, tables, circuits. £21. ISBN: 0672206751.

First published in UK 1942. Previous ed. 1968.

Covers all aspects of radio communication in 42 chapters by 52 contributors. New chapters include active filter design, optoelectronics and optical communications. Bibliographies, glossaries and 46-page index. *Class No:* 621.396

[4194]

—GILES, T.G. *and* JESSOP, G.R. Radio data reference book. 4th ed. London, Radio Society of Great Britain, 1977. vii, 192p. diagrs., tables, graphs, charts. £3. ISBN: 0900612320.

First published 1962.

Minimum text - emphasis on tables, curves, charts. 9 sections (1. Units and symbols ... 3. Resonant circuits and

....(contd.)

filters ... 5. Aerials and transmission lines - 6. Radio and TV services - 7. Maps and meteorological data - 8. Materials and engineering data ...). Brief index (2p).
Class No: 621.396

[4195]
SAVESKIE, P.N. Radio propagation handbook. Blue Ridge Summit, Pa., TAB Books, 1980. 499p. diagrs., graphs, tables, maps. $17.95. ISBN: 083069949x.

Presents much information and guidance in a popular style. Covers ground wave, ionospheric, microwave, VHF, UHF and millimetre wave propagation in 8 chapters with hundreds of tables, graphs and maps. The second half of the book contains 33 appendices, including noise maps for different locations and seasons, refractivity and K-factor data for 61 worldwide locations, rain attentuation data for 29 locations. Brief index. *Class No:* 621.396

VHF

[4196]
JESSOP, G.R., *ed*. VHF/UHF manual. 4th ed. London, Radio Society of Great Britain, 1983. [viii], [528]p. illus., diagrs., graphs, circuits, tables. £8.50. ISBN: 0900612630.

First published 1969. Previous ed. 1976.

Covers the spectrum above 30 MHz. 11 chapters, *e.g.* 2. Propagation ... 4. Receivers - 5. Transmitters ... 8. Antennas - 9. Microwaves - 10. Space communications. Extensively illustrated.

'Previous editions ... gained worldwide acceptance as the standard handbook for amateur radio on v.h.f. and u.h.f. and microwaves. This fully revised and greatly expanded fourth edition provides a wealth of design and constructional information for a wide variety of equipments ...' (*Radio and electronic engineer*, v.54(5), May 1984, p.214).
Class No: 621.396.029

Apparatus & Circuits

[4197]
HALL, G.L. The ARRL antenna book. 14th ed. Newington, Conn., American Radio Relay League, 1982. [vi, 321]p. illus., diagrs., graphs, tables, maps. $8. (*Radio amateur's library, publication no. 15.*) ISBN: 0872594149.

First published 1939. Previous ed. 1974.

Aimed at the amateur to enable understanding of the fundamentals of transmission and propagation of radio signals and the design and operation of aerial (antenna) systems. Chapters 1-7 cover principles and performance characteristics; Chapters 8-14 deal with specific designs for amateur frequency bands (including those suitable for space communications). Chapters 15 and 16 concern measurements, test equipment and guidance on determining orientation, with longitude and latitude data for 474 locations worldwide. Appendix contains brief glossary. Adequate index. *Class No:* 621.396.6

[4198]
JOHNSON, R.C. and JASIK, H. Antenna engineering handbook. 2nd ed. New York, etc., McGraw-Hill, 1984. [1408]p. illus., diagrs., graphs, tables. $95. ISBN: 0070322910.

First published 1961.

57 contributors. 46 chapters, grouped into 4 sections (Introduction and fundamentals; Types and design methods;

....(contd.)

Applications; Associated topics). 15 new chapters (*e.g.* Satellite antennas; Microwave propagation). Bibliographies and 11-page index.

'... should be the definitive work on the science of antennas for many years to come' (*Science and technology libraries*, v.5, (3), Spring 1985, p.65).

A new ed. is due in 1993. *Class No:* 621.396.6

[4199]
PHILLIPS, V.J. Early radio wave detectors. London, P. Peregrinus, 1980. xv, 222p., illus., diagrs. £16. (*IEE History of technology series, v. 2.*) ISBN: 0906048249.

Covers the period from the experiments of Heinrich Hertz in the late 1860s to the coming of the crystal detector and thermionic valve in the early 1900s with the work of J.A. Fleming and Lee De Forest. Chapter references.
Class No: 621.396.6

Stations

[4200]
DRAKE, C.F. and ATTFIELD, J.B., *eds*. Amateur radio call book. 1984 ed. Potters Bar, Herts., Radio Society of Great Britain, 1984. 297p. ISBN: 0900612657.

More than 50,000 amateur radio call signs, with holders' names and addresses, compiled from Department of Trade and Industry records and covering England, Isle of Man, Northern Ireland, Jersey, Scotland, Guernsey, Wales and Eire in separate sequences. *Class No:* 621.396.7

[4201]
GREAT BRITAIN. Department of Trade and Industry. United Kingdom table of radio frequency allocations. London, H.M. Stationery Office, 1985. vi, 308p. tables, map. £12. ISBN: 0115138196.

Sets out the internationally agreed frequency allocations and their uses and gives the UK allocations in parallel, with footnote annotations to explain subdivisions or limitations. Annexes provide greater detail for the UK of the subdivisions of certain bands. Glossary, p.5-20.
Class No: 621.396.7

[4202]
Guide to broadcasting stations. Darrington, P., *ed.* 19th ed. London, Butterworth. c.250p. ISBN: 0434903035.

Originally known as *'Wireless world' guide to broadcasting stations* (Butterworth/Newnes. 18th ed. 1980).

World coverage of long, medium, short wave and VHF sound broadcasting stations. *Class No:* 621.396.7

Space Radiocommunication

Satellites

[4203]
The 1987 satellite directory. 9th ed. Potomac, Md., Phillips Publishing, 1987. Annual. xxx, 866p. illus., diagrs., maps. $200. ISBN: 093496033x.

US directory, with 8 chapters presenting information on satellite operators, transponder brokers, transmission services, manufacturers and distributors of components and systems, technical services providers and earth stations facilities. Two chapters cover non-US satellite systems and suppliers of products and services. Company index. Glossary. *Class No:* 621.396.94.04

[4204]

—BROWN, M.P., *ed*. Compendium of communication and broadcast satellites, 1958 to 1980. New York, IEEE Press, 1981. xii, 375p. $40.50. ISBN: 0879421533.

Published for the Communication and Broadcast Satellite Committee of the IEEE Aerospace and Electronic Systems Society. 25 contributors provide pictures, block diagrams of payload, broadcast frequencies, technical data and launch/lifetime dates for all known satellites of the period. *Class No:* 621.396.94.04

[4205]

Cable and satellite yearbook (Europe), 1987. London, 21st Century Publishing, 1986. Annual. 256p. £47.45. ISBN: 0951014722.

Survey of developments and plans in Western European countries, covering communications satellites, direct broadcast satellites, programmes, television receive only (TVRO) systems; followed by directory of organizations, regulating bodies, manufacturers, etc., for each country. *Class No:* 621.396.94.04

[4206]

LONG, M. World satellite almanac. 2nd ed. Indianapolis, In., Sams, 1987. xlix,650p. illus., diagrs., graphs, tables. £35. ISBN: 067222559x.

First published 1985.

11 chapters in 2 parts: 1. The global satellite network - 2. Satellite coverage of International Telecommunication Union (ITU) Regions 1, 2, and 3. Appendices, A-L: D. GMT conversion chart; H. Directory of satellite manufacturers. Glossary, p.619-35; detailed, analytical index, p.637-50. *Class No:* 621.396.94.04

[4207]

MORGAN, W.L. *and* **GORDON, G.D. Communications satellite handbook.** New York, Wiley, 1989. 900p. illus. $69.95. ISBN: 0471316032.

5 main parts covering: teletraffic; communications; satellite systems; multiple access techniques; space craft technology; satellite orbits. Extensive chapter references. Lists of symbols, constants and acronyms. *Class No:* 621.396.94.04

[4208]

UNGER, J.H.W., *ed*. **Literature survey of communication satellite systems and technology.** New York, IEEE Press, 1976. 409p. $14.95. ISBN: 0879420677.

Companion volume to *Communication satellite systems: an overview of the technology*, edited by Y.F. Lum and R.G. Gould, and serves as the bibliography for it. Over 3,600 references from 1914-1975. (50 per cent cover 1969-1975). 3 parts: List of references (chronological); Permuted title index; Author index. Abbreviations index. *Class No:* 621.396.94.04

[4209]

WILLIAMSON, M. The Communications satellite. New York, A. Hilger, 1990. 420p. $70. ISBN: 0852741928.

Includes a case study of direct broadcasting by satellite (DBS). Excludes electronic control systems design and transmission and coding methods. *Class No:* 621.396.94.04

Radar

[4210]

BARTON, D.K., *ed*. **Radars.** Dedham, Mass., Artech House Inc., 1974-1983. 6v. illus., diagrs., graphs. £21 per v. (*Artech Radar Library*.)

Collections of papers, largely American, reprinted from scattered journals and reports, on aspects of radar. 1. *Monopulse radar.* 1974. ISBN 0890060304. 2. *The radar equation.* 1974. ISBN 0890060312. 3. *Pulse compression.* 1975. ISBN 0890060320. 4. *Radar resolution and multipath effects.* 1975. ISBN 0890060339. 5. *Radar clutter.* 1975, rep. 1977. ISBN 0890060347. 6. *Frequency agility and diversity.* 1977. ISBN 0890060673. Annotated bibliographies. *Class No:* 621.396.96

[4211]

SWORDS, S.S. Technical history of the beginnings of radar. London, P. Peregrinus, 1986. xiv, 325p. illus., diagrs. £30. (*IEE History of technology series, v.6.*) ISBN: 086341043x.

Traces developments from the beginning of the 20th century to the Second World War, with separate accounts for the UK, Germany, USA, France, Italy, Japan, Russia, Holland and Hungary, with valuable annotated bibliographies for the first seven. A more comprehensive treatment of work in the UK is given in chapter 5 (p.174-257). Appendices reproduce several historical memoranda, provide maps and mathematical analysis.

'I found little to criticise in this book ... Probably the best single volume on the subject ever published' (*IEEE Spectrum*, v.23, Sept. 1986, p.17). *Class No:* 621.396.96

Television

[4212]

BENSON, K.B. *and* **WHITAKER, J.G.,** *eds*. **Television and audio handbook for technicians.** New York, etc., McGraw-Hill, 1990. v.p. illus., diagrs. $90. ISBN: 0070047871.

Much of the text and many illus. derived from *Television engineering handbook* and *Audio engineering handbook*. Limited chapter references; subject index. Suitable for public libraries and technical colleges. *Class No:* 621.397

[4213]

BENSON, K.B. Television engineering handbook. New York, etc., McGraw-Hill, 1985. 1478p. illus., diagrs. $89.50. ISBN: 0070047790.

Data for all aspects of design and operation, including signal generation, processing, transmission and reproduction.

'A reference source for engineers involved in the design, development, maintenance and operation of television equipment and systems ... Enhanced by 1,091 illustrations' (*SMPTE journal*, v.95, March 1986, p.340). *Class No:* 621.397

[4214]

BURNS, R.W. British television: the formative years. London, P. Peregrinus/Science Museum, 1986. xv, 488p. illus., diagrs., tables. £48. (*History of technology series, No. 7.*) ISBN: 0863410790.

An account of the beginnings and experimental work of Baird and others; early developments, tests and public transmissions; the emergence of competition from EMI Ltd; the establishment, work and findings of the 1934 joint BBC/GPO Television Committee; and the eventual launch of a

....(contd.)
public high-definition television service by the BBC in 1936. Ends with the closure of the service on the outbreak of war in September 1939. Based very soundly on primary source material and supported by 4 appendices and numerous contemporary illustrations. *Class No:* 621.397

[4215]
BURRELL, M., *ed*. Television engineers' pocket book. 7th ed. London, etc., Newnes Technical Books, 1982. [vi], 314p. illus., diagrs., graphs, circuits, tables. £7.95. ISBN: 0408004444.

First published 1954. Previous ed. 1973.

'This is not an instruction manual but a guide' (*Preface*). The author and 7 'specialist contributors' provide a guide to the principles, the circuitry, the components and the varieties of television reception, the test equipment, fault finding, servicing and adjustment techniques in 10 chapters (1. Standards, waveforms and principles ... 4. Television integrated circuits ... 9. Aerials and interference ... 10. Cathode ray tubes). Chapters 11 and 12 cover teletext, viewdata and video cassette recorders and video cameras. Chapter 13 contains colour codes, data and organizations addresses. Overall aim is to act as an aid to servicing technicians. *Class No:* 621.397

[4216]
CLASON, W.E., *comp*. Elsevier's dictionary of television and video-recording in six languages: English/American-French-Spanish-Italian-Dutch-German. Amsterdam, etc., Elsevier, 1975. [viii], 608p. £102.10. ISBN: 0444412247.

Based on *Elsevier's Dictionary of television, radar and antennas* (1955), this polyglot version excludes terms on radar and antennas (other than for television); these will be found in *Elsevier's Telecommunications dictionary* (*qv*).

5,300 terms and phases are listed, numbered and defined in English, with equivalents in French, Spanish, Italian, Dutch and German. For each foreign language there is an A-Z index referring to the English section via the reference numbers. *Class No:* 621.397

[4217]
JACKSON, K.G. *and* TOWNSEND, B. Television and video engineer's reference book. London, Butterworth Heinemann, 1991. 772p. illus., diagrs. £125. ISBN: 0750610212.

Comprehensive, including television transmitters, transformers, sound editing, audio recording, mixers and special effects, studio lighting, lighting control, receiving antennae, semiconductors and microelectronics. *Class No:* 621.397

[4218]
TRUNDLE, E. Newnes Television and video engineer's pocket book. 2nd ed. Oxford, Butterworth Heinemann, 1992. vi,352p. illus., diagrs., circuits, tables. £12.95. ISBN: 0750606770.

First published 1987.

Concentrates on 'entertainment' electronic equipment. 21 chapters: 4. Satellite television ... 9. Nicam stereo sound ... 12. Video on magnetic tape ... 19. Control systems ... 21. Reference data (resistors; connector pinning; equipment marking codes, etc.). Index, p.348-52. *Class No:* 621.397

Apparatus

[4219]
STEPHENSON, D.J. Newnes guide to satellite TV: installation, reception and repair. Oxford, Heinemann Newnes, 1990. xi,210p. diagrs., graphs, tables. £14.95. ISBN: 0434918180.

10 chapters: 2. Antennae ... 4. Satellite receivers ... 6. Installation: surveying the receiving site ... 9. Installation: interior work - 10. Repair of satellite equipment. 4 appendices: 2. Suppliers of satellite equipment and spares. A practical treatment. *Class No:* 621.397.6

Heat Engineering

[4220]
ELLIOTT, T.C., *ed*. Standard handbook of powerplant engineering. New York, McGraw-Hill, 1989. xiv,v.p. illus., diagrs., graphs, tables. £66.95. ISBN: 0070191069.

46 chapters in 6 sections: 1. Steam generation - 2. Turbines and diesels - 3. Fuels and fuel handling - 4. Pollution control - 5. Plant electric systems - 6. Instrumentation and control. Analytical subject index. 40 contributors, from industry. A practical approach. *Class No:* 621.4

[4221]
GUYER, E.C., *ed*. Handbook of applied thermal design. New York, etc., McGraw-Hill, 1989. xv, v.p. illus., diagrs., graphs, tables. £56.95. ISBN: 0070253536.

67 contributors. Intended to provide 'a general perpsective' (*Preface*) on heat transfer analysis, materials performance, heating and cooling technology and instrumentation and controls. 13 parts, each with a number of chapters: 1. Fundamentals of thermal design and analysis ... 4. Structural materials in thermal design - 5. Heat transfer fluids ... 9. Refrigeration ... 12. Thermal sensors - 13. Automatic temperature controls. Appendix: 'Classified unit conversions'. Detailed subject index (12p.). 'Recommended for libraries serving a technical clientele' (*Choice*, Feb. 1989, p.968). *Class No:* 621.4

[4222]
HEWITT, G.F., *ed*. Hemisphere handbook of heat exchanger design. New York, etc., Hemisphere, 1990. xxxviii, v.p., illus., diagrs., graphs, tables. £203. ISBN: 0891168664.

Incorporates the 5 volume *Heat exchanger design handbook* and all supplements, 1983.

73 contributors, most of them practising engineers. 5 parts: 1. Heat exchanger theory - 2. Fluid mechanics and heat transfer - 3. Thermal and hydraulic design - 4. Mechanical design - 5. Physical properties. Analytical subject index (48p.). A wealth of data but a rather unwieldy volume. *Class No:* 621.4

Internal Combustion Engines

[4223]
BRITISH STANDARDS INSTITUTION. Vocabulary for reciprocating internal combusion engines. London, BSI, 1979. Addendum, 1983. 24p. £34.90. (*BS 5676: 1979*.)

130 terms defined in English, French and Russian, in systematic order with indexes in each language. Addendum 1. Symbols. 1983. 6p. *Class No:* 621.43

[4224]
FIRTH, C., *ed*. **World engine digest.** 5th ed. London, PRS Business Publications, 1985. [xiv], 264p. graphs, charts, tables, maps. £164.05. ISBN: 0906237351.

Provides production and market data for petrol and diesel engines; profile of 31 leading manufacturers, A-Z; selective directory of manufacturers worldwide, arranged under project/country/name. Based on a computerized engine data bank developed by the publishers. *Class No:* 621.43

[4225]
Referativnyĭ zhurnal. 39. Dvigateli vnutrennogo sgoraniya. Akademiya Nauk SSSR. Moscow, VINITI, 1963-. Monthly. R48.60. ISSN: 04862279.

About 5,500 indicative and informative, signed abstracts pa., on all aspects and types of internal combustion engines, classified in sections (*e.g.* Construction and calculation ... Diesel engines ...). Annual subject index. English contents summary. UDC numbers. *Class No:* 621.43

Diesel Engines

[4226]
Diesel engines of the world, 1984. Cannon, S., *ed*. London, John Martin, 1984. ii, 195p. tables. [£75?]. (*PRS yearbooks*.) ISBN: 0906237335.

'... The first comprehensive directory of the world's diesel engines' (*Introduction*). Lists *c.*8,000 engines with specifications (including no. of cylinders, configuration, RPM, HP, capacity, piston data and dimensions) drawn from a computerized data base held by the publishers. Covers *c.*200 engine builders and provides alphabetical and geographical listings. Production statistics for 1982, definitions and standards information complete the work. *Class No:* 621.436.1

[4227]
LILLY, L.C.R., *ed*. **Diesel engine reference book.** London, Butterworths, 1984. [707]p. illus., diagrs., graphs. £95. ISBN: 0408004436.

Successor to *Diesel engine principles and practice*, by C.C. Pounder, first published 1955. Previous ed. (2nd) 1962.

32 contributors and chapters, grouped into: Theory; Engine design practice; Lubrication; Environmental pollution; Crankcase explosions; Engine types; Engine testing; Maintenance. Chapter references. Index (*c.*1,700 entries). Lists of engine types and manufacturers, institutions, and fuel injection equipment manufacturers.

'A comprehensive work on the design, operation and maintenance of all types of diesel engines ... Ranges through subjects which will be of long-term use to engine designers, installation engineers and ... enormous number of users ...' (*Automotive-engineering*, v.92, Dec. 1984, p.102). *Class No:* 621.436.1

[4228]
MAHON, L.L.J. **Diesel generator handbook.** Oxford, Butterworth Heinemann, 1992. viii,646p. illus., diagrs., graphs, tables. £75. ISBN: 0750611472.

15 chapters, closely subdivided: 1. Reciprocating internal combustion engines ... 3. A.C. generators - general ... 6. Engine governing ... 9. Switchgear and control gear ... 12. Fuels and lubricating oils ... 15. Operation and maintenance. 4 appendices (*e.g.* B: Batteries). Detailed, analytical index. *Class No:* 621.436.1

Gas Turbines

[4229]
BOYCE, M.P. **Gas turbine engineering handbook.** Houston, Texas, Gulf Publishing Co., 1982. ix, 603p. illus., diagrs., graphs, tables. $79. ISBN: 0872018784.

Information on design, components (compressors and turbine types), materials, fuels, lubrication, operation and maintenance of gas turbines, supported by chapter references and tabulated data. *Class No:* 621.438

Pneumatic Energy Machines

[4230]
BARBER, A. **Pneumatic handbook.** 7th ed. Morden, Surrey, Trade & Technical Pr., 1989. 640p., xv. illus., diagrs., tables. £64. ISBN: 0854611374.

First published 1965. Previous ed. 1982.

Standard reference work for pneumatic and compressed air engineering. 12 sections cover principles, compressors, air lines, valves, actuators, circuits, tools, techniques, devices and machines, miscellaneous data, lists of standards and symbols and buyers' guide, glossary and index. Copious illustrations and many tables. Interspersed advertisements. *Class No:* 621.5

[4231]
Directory of pump, valve and compressor manufacturers in Western Europe. London, European Directories, 1981. 2v. £25. ISBN: 0906685044.

1. *Europe*. xxvii, 479p. 2. *The United Kingdom*. xxvii, 173p.

Published by a division of Inter Company Comparisons Ltd. First edition of directory of manufacturers listing 5,500 companies with informtion on address, etc., executive personnel, sales, turnover, number of employees and details of products. Separate section for each country. Supplemented by indexes for pumps, valves, compressors, industrial applications, each subdivided first by subject (*e.g.* gas compressors), then by country (in v.2). Does not claim to be a 'buyers' guide'. *Class No:* 621.5

[4232]
NEUBERT, G., *ed*. **Technical dictionary of hydraulics and pneumatics:** English/French/German/Russian; with a supplement in Spanish. Oxford, etc., Pergamon, 1973. [vii], 257p. £31. ISBN: 0080169589.

About 3,500 terms are listed in the main, English-language section, each uniquely numbered, with equivalents in French, German and Russian in adjacent columns.

Supplemented by indexes in these three languages and in Spanish, linked by the term number. *Class No:* 621.5

Vacuum Technology

[4233]
BRITISH STANDARDS INSTITUTION. **Vacuum technology - graphical symbols.** London, BSI, 1978. 16p. diagrs. £23.90. (*BS 5543: 1978*.)

50 symbols defined and illustrated. *Class No:* 621.52

[4234]
HOLLAND, L., *and others, eds*. **Vacuum manual.** London, Spon, 1974. x, 427p. diagrs., graphs, tables. £10.75. (*High vacuum series*.) ISBN: 0419107401.

Three parts: 1. Basic data - 2. Vacuum equipment (pumps, instrumentation, plant and systems, with tabulated specifications and data for manufacturers' products) - 3. Recent developments in vacuum science and technology. Indexes of manufacturers, equipment, subjects and advertisers. *Class No:* 621.52

[4235]
WEBER, F., *comp*. **Elsevier's dictionary of high vacuum science and technology in six languages:** German, English, French, Spanish, Italian, Russian. Amsterdam, etc., Elsevier, 1968. [xiv], 539p. $57.

3,692 numbered entries, German-based, with English, French, Spanish, Italian, Russian equivalents and occasional explanatory notes. English, French, Spanish and Italian indexes. Russian numerical and alphabetical indexes. *Class No:* 621.52

[4236]
WEISSLER, G.L. *and* CARLSON, R.W., *eds*. **Vacuum physics and technology.** New York, Academic Press, 1979. 616p. diagrs., graphs, tables. $55. (*Methods of experimental physics, v.14.*) ISBN: 0124759147.

15 contributors to 16 chapters which cover theory, production and measurement of vacua, design of systems protective devices, operation and maintenance of vacuum systems and properties of materials used in vacuum technology. This work 'presents fundamental vacuum-design criteria and practice in a clear manner. On balance it will be a valuable addition to modernize a vacuum laboratory library'. (*Physics Today*, v.33(5), May 1980, p.58.) *Class No:* 621.52

Compressed Air Equipment
[4237]
GREENE, R.W., *ed*. **'Chemical engineering' guide to compressors.** New York, etc., McGraw-Hill, 1984. vii, 264p. $37.50. (*Chemical engineering books*.) ISBN: 0070243123.

29 articles reprinted from *Chemical Engineering*, 1973-1983, aimed to present information on all aspects of compressors. 8 sections (1. Selection - 2. Calculations and evaluations - 3. Surge control - 4. Operation and maintenance - 5. Seals and packings - 6. Prime movers - 7. Adjustable speed drives - 8. Fans and blowers). Index. *Class No:* 621.54

Wind Power & Windmills
[4238]
REYNOLDS, J. **Windmills and watermills.** London, Evelyn, 1970. 196p. illus. (incl. col.), diagrs. £4.20. (*Excursions into architecture*.) ISBN: 0238789438.

An authoritative survey of the origins, development and applications of the various kinds of windmill and watermill used in many parts of the world. Chapters on the machinery, its principles, components and materials. Illustrated on every page by photographs and the author's isometric drawings, and supplemented by extensive glossary (p.183-191) and selected bibliography. Index of names. *Class No:* 621.548

[4239]
SHILLINGFORD, A.E.P. **England's vanishing windmills.** London, Godfrey Cave Associates, 1979. 157p. illus. (mostly col.), diagrs. £7.95. ISBN: 0906223156.

This copiously illustrated book contains a brief history of windmills in England, and sections describing the several types and their structure and machinery. A gazetteer section lists and describes windmills of London and 21 counties. Photographs of 'lost and decayed' windmills. Appendix, listing windmills open to visitors. Bibliography, p.150. Glossary of 130 terms, with index. *Class No:* 621.548

[4240]
—WAILES, R. **The English windmill.** London, Routledge & Kegan Paul, 1954. xxii, 264p. illus., diagrs., maps.

3 parts (22 chapters): 1. The mills - 2. The machinery - 3. The men. 4 appendices. Glossary. Bibliography. Index. 32 plates, 64 diagrs. Mentions over 300 mills in the text (37 in Lincolnshire, 46 in Norfolk, and 47 in Suffolk). *Class No:* 621.54

Refrigeration

Handbooks & Manuals
[4241]
AIR-CONDITIONING AND REFRIGERATION INSTITUTE. **Refrigeration and air-conditioning.** Englewood Cliffs, N.J., Prentice-Hall, 1979. xii, 863p. illus., diagrs., graphs, tables. ISBN: 0137701640.

A comprehensive text presented informally at technician/d.i.y. level, which has some value as a reference work because of the large number of tables, graphs, charts containing relevant data. *NB.* Equipment information and environmental factors relate to North America. Includes glossary, problems/answers and good index.

2nd rev. ed. 1987. (779p. £34.25. ISBN 0137701810). *Class No:* 621.56(035)

[4242]
BOAST, M. **Newnes Refrigeration pocket book.** Oxford, Butterworth Heinemann, 1991. viii,450p. illus., diagrs., graphs, tables. £14.95. ISBN: 0750600799.

14 chapters (2. Evaporators and coolers ... 4. Refrigerants ... 9. Cold room capacities ... 12. Glossary (p.379-93)). Index, p.445-50. A wealth of information in a handy format. *Class No:* 621.56(035)

Dictionaries
[4243]
BOOTH, K.M., *comp*. **Dictionary of refrigeration and air conditioning.** Amsterdam, etc., Elsevier, 1970. viii, 315p. . £21. ISBN: 044420069x.

More than 2,600 terms from both sides of the Atlantic are defined in entries ranging from a few words to more than 200. Copious cross-references. Author's Preface insists 'The book is a dictionary and not an encyclopedia'. Because of the impending changeover to use of SI units at the time of publication, metric equivalents to Imperial units are frequently given. Inevitably lacks terms relating to important recent developments (*e.g.* new refrigerants, use of computer and microprocessor control methods) but sound, descriptive

.... *(contd.)*

definitions of fundamental material make this a brief dictionary of continuing importance in its field. *Class No:* 621.56(038)

[4244]
BRITISH STANDARDS INSTITUTION. Glossary of refrigeration, heating, ventilating and air-conditioning terms. London, BSI, 1984. 84p. £38. (*BS 5643: 1984.*)
Supersedes BS 1584 and BS 5643: 1979.

Contains around 2,300 terms, of which more than one third are cross-references, many to noun/adjective forms from the alternative (*e.g.* dark smoke *see* smoke, dark), apparently assuming that the reader will overlook the clear explanation of policy in the Foreword! Concise definitions. *Class No:* 621.56(038)

[4245]
INTERNATIONAL INSTITUTE OF REFRIGERATION. New international dictionary of refrigeration. Oxford, etc., Pergamon, 1975. xxxvii, 560p. £35.

About 3,500 terms in English with definitions in English and French, followed by equivalents in Russian, German, Spanish, Italian and Norwegian. Classified arrangement. 14 sections (A. Basic data and instrumentation ... Q. International applications - Miscellaneous). A-Z index for each language. *Class No:* 621.56(038)

[4246]
ROZENBERG, M.B. English-Russian dictionary of refrigeration and low-temperature technology. 2nd ed., rev. and enl. Oxford, etc., Pergamon, 1979. 467p. tables. £25. ISBN: 0080247377.

Identical publication with variant title: *English-Russian dictionary of refrigerating and cryogenic engineering/Anglo-russkiĭ slovar' po kholodil'noĭ tekhnike* (Moscow, Russian Language Publishers, 1978).

About 20,000 terms with Russian equivalents and indication of acceptable synonyms and with explanations in Russian of certain translations. Covers refrigeration equipment, manufacture and use, heat transfer equipment, insulating materials, air conditioning and cryogenics. Includes abbreviations list and 18 conversion tables. *Class No:* 621.56(038)

Progress Reports

[4247]
Advances in cryogenic engineering. New York, Plenum Press, 1960-. v.1-. illus., diagrs.

A review series generated by joint biennial conferences. Thus v.31 (1986. xxii, 1346p. $125. ISBN: 0306422913) contained most of the papers of the 1985 Cryogenic engineering conference and covered, *inter alia:* Applications of superconductivity; heat exchangers; heat and mass transfer; refrigeration and liquefaction; cryogenic applications. V.32 (1986. xix, 1120p. $110 ISBN 0306422921 contained papers presented at the 6th International cryogenic materials conference, 1985, on: Austenitic steel developments; ... composites and polymers; refrigeration materials; mechanical properties; ... materials and devices for superconductivity electronics; ... new superconductivity materials and methods. *Class No:* 621.56(055)

Yearbooks & Directories

[4248]
The Refrigeration and air conditioning year book. Croydon, Maclaren, 1987. Annual. 366p. illus. ISSN: 02685361.

Published since 1975 as part of subscription to *Refrigeration and air conditioning and heat recovery;* also available separately. Previously as *Refrigeration and air conditioning year book.*

8 sections: 1. Company information - 2. Classified guide to manufacturers and suppliers - 3. Contractual and specialist services - 4. Geographical guide - 5. Trade names - 6. Foreign brand names and UK agencies - 7. UK agencies for foreign firms - 8. Profession and other organizations. Full-page advertisements by categories: (industrial, commercial refrigeration; air conditioning; instrumentation and control; insulation). *Class No:* 621.56(058)

Tables & Data Books

[4249]
SCHWARTZBERG, F. Cryogenic materials data handbook. Wright-Patterson Air Force Base, Ohio, Air Force Materials Laboratory 1970. 2v. (vi, 750p.). graphs. (*Report AFML-TOR-64-280.*)
1. Sections A-C. 2. Sections D-I, plus References, Testing bibliography and Materials guide.

Mechanical and physical property data and information on 88 metallic and non-metallic materials, arranged in 11 sections, with materials, properties and cumulative indexes and complete list of references. *Class No:* 621.56(083)

Histories

[4250]
THÉVENOT, R. A History of refrigeration throughout the world; translated by J.C. Fidler. Paris, International Institute of Refrigeration, 1979. 476p.

The author modestly claims only 'to present a preliminary sketch', but he ranges comprehensively over the whole field of the production and use of artificial cold from 1755 onwards, identifying the contributions of around 1,000 scientists and engineers to the development of the technologies associated with refrigeration. The descriptive survey in Part 1 is divided into 4 epochs (1755-1875; 1875-1914; 1919-1939; 1945-). Part 2 presents chronologies of inventions, advances, applications under 10 headings (*e.g.* Refrigerated transport, Air conditioning, Cryology). Part 3 contains brief biographical sketches of 65 important men in the history of the subject. *Class No:* 621.56(091)

Research Projects

[4251]
Directory of low temperature research in Europe. McDonald, P., *ed.* London, Adam Hilger, 1992. 408p. illus. £47.50. ISBN: 0750301767.

A survey of research and development. Data: institution, research team members, description of project, specialist equipment. *Class No:* 621.56:061:061.62.005

Heat Pumps

[4252]
HOLLAND, F.A., *and others*. **Thermodynamic design data for heat pump systems.** Oxford, etc., Pergamon, 1982. vii, 347p. graphs, tables. £75. ISBN: 0080287271.

Intended as 'A comprehensive database and design manual' (*sub-title*), the book presents design data (graphical and tabular) for 21 working fluids for heat pump systems, each designated by its R code number based on the International Numbering System detailed in BS 4580: 1970. Two introductory chapters cover theory and derivation of the data and its use.

'This handbook can be a reference and a guide to the process design engineer involved in the conservation, recovery and re-cycling of heat energy in a wide range of industries' (*Heating/Piping/Air conditioning*, v.54, August 1982, p.172). *Class No:* 621.577

[4253]
LOYD, S. **The Heat pump:** an annotated bibliography with a survey of suppliers. 2nd ed. Bracknell, Berks., Building Services Research and Information Association, 1981. iv, 106p. £9. (*Bibliography LB 103/81.*) ISBN: 0860220990.

First published 1975.

505 numbered items, with citations and indicative or informative abstracts, drawn largely from English-language and German sources - periodicals (mainly), books, reports and standards. 21 sections (General Surveys ... Air source - Water source - Earth source ... Testing ... Heat recovery ... District heating). Appendix lists suppliers and their products, trade names and importers. *Class No:* 621.577

Fluids Handling Engineering

[4254]
FLUIDEX. BHRA **Fluid engineering abstracts.** Cranfield, Beds., British Hydromechanics Research Association.

Covers 1973 onwards. References, with abstracts, to literature of fluids, including aerodynamics, fluid flow, fluid engineering, offshore technology and rheology. About 250,000 records updated monthly. Available on DIALOG, file 96 and Data-Star. Cost per connect hour: $69 and £52.65. Available on magnetic tape. Publications: *BHRA Fluid Engineering thesaurus*; *Guide to the Fluidex database*, *Fluidex Notes*, *Fluidex Newsline*. Printed versions: several abstracts journals covering the various fields of interest, *e.g. Tribos*, *Fluid-sealing abstracts*, *Pumps and other fluids machinery abstracts*, *Pipelines abstracts*. *Class No:* 621.6

[4255]
List of periodicals scanned and abstracted for inclusion in the FLUIDEX DATABASE. Cranfield, Beds., BHRA Fluid Engineering, 1981. 18p.

Lists *c.*700 periodicals covering a wide field. Gives abbreviated titles and Codens. *Class No:* 621.6

[4256]
NATIONAL FLUID POWER ASSOCIATION. **Fluid power standards.** 7th ed. Milwaukee, Wisconsin, the Association, 1986. 10v. $380.

Covers pumps, motors, power units and reservoirs, filtration and contamination, cylinders, accumulators, fluids, lubricants and sealing devices, hose, tubes, and fittings, valves and testing procedures. Contains bibliographies,

....(contd.)
glosssary and graphical symbols list. 'The most complete reference work available on design and performance standards for hydraulic and pneumatic fluid power products' (*Mechanical engineering*, v.109, March 1987, p.80). *Class No:* 621.6

[4257]
Thesaurus for fluid engineering. Guy, N.G., *ed.* Cranfield, Beds., BHRA, 1981. xiii, 128p. £15. ISBN: 0906085578.

Two aims: (a) to give guidance to the vocabulary used in the fluid engineering literature; (b) to guide searchers using the BHRA FLUIDEX computerized database. Derived from actual use in that database, the scope is wide and includes some general terms (*e.g.* Introduction; Study). Direct entry is used, with each term (word or phrase) followed by a conventional set of broader, narrower, related, preferred terms. Upwards of 9,000 terms. *Class No:* 621.6

[4258]
TROSKOLANSKI, A.T. **Dictionary of hydraulic machinery:** in English, German, Spanish, French, Italian and Russian. Amsterdam, etc., Elsevier, 1985. xxiii, 737p. illus., diagrs., graphs, tables. £114.06. ISBN: 0444997288.

Revised and enlarged translation from Polish original *Magzynyi urzgdzenia hydrauliczne - Projecia podstawave* (1974).

Contains about 4,300 entries classified in 14 sections (1-4. Concepts - 5-14. Hydraulic machinery types, problems and applications). Base language is English. Concise definitions for most terms, with foreign-language equivalents. Indexes in each language lead to main entries by means of number codes. *Class No:* 621.6

Fans

[4259]
DALY, B.B. **Woods practical guide to fan engineering.** 3rd rev. ed. Colchester, Woods of Colchester, Ltd., 1978. xxxi, 376p. diagrs., graphs, charts, tables. £9. ISBN: 0950294128.

First published 1952. Previous ed. 1960.

Aimed at users and engineers responsible for selection and installation, not at specialists or designers of fans. Nevertheless it is an invaluable assembly of information and data, covering its subject widely in 14 chapters (1. The air and human well-being - 2. General ventilation - 3. Air distribution and circulation ... 7. Fans and fan performance - 8. Fan drives and fan control ... 10. Noise, vibration and fatigue - 11. Fan testing and performance prediction ... 13. Ventilation of tunnels and mines - 14. Useful data (p.312-365)). Over 80 tables and charts, identified easily by alphabetical list (e.g., Beaufort wind scale; Relative humidities). Over 160 diagrams and graphs are similarly identified (*e.g.* Psychrometric chart). Data section includes conversion factors and tables, properties of air, water, etc., climatic data, weights and measures). Bibliography, p.366. *Class No:* 621.63

[4260]
JORGENSEN, R., *ed.* **Fan engineering:** an engineer's handbook on fans and their applications. 8th ed. Buffalo N.Y., Buffalo Forge Co., 1983. xvi, [1030]p. diagrs., graphs, charts, tables. $40.

First published. 1914. Previous ed. 1970.

Pocket-size format. 4 parts. Part 1, Fundamentals, contains 8 chapters. Part 2, Fans, has 11 chapters, covering

....(contd.)

terminology, fan types, laws, testing, control, noise, motors and drives, and selection. Part 3 deals with applications in 10 chapters (*e.g.* ventilation, drying, air cleaning). Part 4 contains 8 appendices (*e.g.* units, nomenclature, miscellaneous data, selected bibliography). Author and subject indexes. 'Will be a very useful reference tool for any engineer involved with the design and operation of fan systems' (*Chemical engineering,* v.91, Dec. 10, 1984, p.197). *Class No:* 621.63

Pipes & Pipelines

[4261]

BRITISH STANDARDS INSTITUTION. **Specification for graphical symbols for general engineering. Part 1. Piping systems and plant.** London, BSI, 1977. 32p. diagrs. £33.60. (*BS 1553: Part 1: 1977.*)

Supersedes BS 974 and BS 1553, Part 4, 1956.

Specifies symbols for use in flow and piping diagrams for process plant and heating and ventilating installations. *Class No:* 621.643

[4262]

—BRITISH STANDARDS INSTITUTION. Glossary of terms and symbols for measurement of fluid flow in closed conduits. London, BSI, 1991. 51p. (*BS 5875: 1991; ISO 4006: 1991.*)

Covers general and special areas, *e.g.* differential pressure devices, critical flow measurement, velocity area methods, tracer methods, flow meters, etc. Over 200 terms defined in systematic order with alphabetical index. *Class No:* 621.643

[4263]

GUY, N.G., *ed.* **Pipe protection bibliography.** London, Elsevier/BHRA Fluid Engineering Centre, 1987. viii, 285p. unpriced. ISBN: 1851660771.

A classified bibliography, with abstracts, of 620 items published 1975-1985 ('a period of considerable activity' (*Introduction*) in developments in pipe protection) derived from the FLUIDEX computerized database. Periodical articles and conference papers in almost equal proportions. Covers pipeline transportation of oil, water, gas, sewage, solids and power and process plant pipework. Contents spread over 38 sections in six main subject categories which cover corrosion, coatings, cathodic protection and applications. Author and corporate source indexes, but no subject index. 18 other bibliographies cited. *Class No:* 621.643

[4264]

HOLMES, E., *ed.* **Handbook of industrial pipework engineering.** New York, etc., McGraw-Hill, 1973. xvii, 570p. illus., diagrs., graphs, tables. $47.50. ISBN: 0070844275.

14 contributors. 21 chapters (1. Plant start-up ... 4. Pipes and pipefittings ... 6. Valves and allied fittings ... 12. Stress analysis ... 14. Piping design: organizations and practice ... 16. Piping design and drafting by computer ... 19. Estimating, measurement, costing and cost control ... 21. Pipeline maintenance). 13 appendices (conversions; mathematical tables; wire gauges; British Standards, etc.). A comprehensive, authoritative reference work for designers and users of pipework in chemical, oil and other industrial plants. *Class No:* 621.643

[4265]

KENTISH, D.N.W. **Pipework design data.** New York, etc., McGraw-Hill, 1982. viii, 239p. diagrs., tables. £44.95. ISBN: 0070845581.

Companion volume to the author's *Industrial pipework*, which was concerned with design and selection of tubes, fittings, valves and other pipework equipment and with basic pipework techniques. This work has 6 chapters on matters related to pipework in plant: plant layout; platforms, stairs and personnel access; pumping design and layout; structural steelwork; estimating; specification writing. Chapter 7 (p.75-215) consists of designer's tables (A. Abbreviations (over 500 are listed) - B. British Standards (about 350 standards, Codes of practice and Handbooks are included) - C. Conversions - D. Dimensions ... F. Flange tables ... N. Nomograms ... P. Properties of tube and pipe ... V. Velocities ... Y. Young's Modulus values - Z. Z values). Glossary (p.217-234 - over 500 terms). Clearly and attractively presented with illustrations reflecting superb draughtsmanship. *Class No:* 621.643

[4266]

Pipe joints: a state of the art review. London, Institution of Mechanical Engineers, 1984-86. 3v. illus., diagrs., graphs, tables.

1. *Gaskets.* 1984. x,47p.

2. *Non-metallic pipe joints.* 1985. xv,93p.

3. *Metallic pipe joints.* 1986. xiii,98p.

Part 1 covers types of gaskets, selection of materials, bolting leads and effect of surface finishes. Part 2, in 9 sections, covers materials (plastics, rubbers, glass), design, joints and linings; Part 3, in 8 sections, covers materials, preparation, welding, joints and reliability. Both 2 and 3 list relevant BSI Standard numbers and corrosion data. No index, but detailed contents pages help overcome this. Intended to provide 'a practical summary' of 'current knowledge and experience' (*Foreword,* p.ii). For practising engineers. Small, handy format. *Class No:* 621.643

[4267]

Pipelines abstracts. Cranfield, Beds., British Hydromechanics Research Association, 1984-. v.1-. 6 pa. £90. ISSN: 02653990.

About 800 abstracts pa. of items drawn from over 1,000 journals, plus reports, books, standards and conference proceedings. Main sections: Hydraulics of pipes - Components and materials - Planning and design - Construction - Operations - Maintenance.

Issue and cumulative annual indexes of personal and corporate authors and of subjects. Accessible online via FLUIDEX (*qv*). *Class No:* 621.643

[4268]

Referativnyĭ zhurnal. 45. Truboprovodnyĭ transport. Moscow, VINITI, 1963-. Monthly. illus. R30. ISSN: 00342610.

About 2,500 signed abstracts on pipelines and pipeline transportation. 12 sections (General problems ... Construction of pipelines ... Operation of gas pipelines ... Oil storage - Gas storage ... Corrosion and anticorrosive protection. Annual subject index. UDC numbers. English contents summary. Some illustrations. *Class No:* 621.643

Valves & Taps

[4269]

BARBER, M.J., *and others, eds.* **Handbook of power cylinders, valves and controls.** Morden, Surrey, Trade and Technical Press, 1986. ix, 364p. illus., diagrs., graphs, charts, tables. £46. ISBN: 0854611002.

Aims to provide 'essential information on the principles and control of hydraulics and pneumatics, together with detailed coverage of all types of hydraulic and pneumatic cylinders, valves and their electronic controls'. 8 sections of text with information on theory, construction, operation, performance, selection, including relvant tables and graphs, followed by data section listing standards and symbols. Concise, but adequate index (8p.). *Class No:* 621.646

[4270]

INSTRUMENT SOCIETY OF AMERICA. ISA handbook of control valves. Hutchison, J.W., *ed.* 2nd ed. Pittsburgh, Pa., Instrument Society of America, 1976. 533p. illus., diagrs., graphs, tables. $35. ISBN: 0876642342.

A collection of design data tables and charts and other information to assist in the solution of control valve problems. Includes guidance on flow characteristics, cavitation, installation, noise. Indexes of subjects and tabulated data.

'Without question, the most up-to-date, comprehensive and authoritative work on control valves yet published ... replete with clear, useful illustrations, graphs, charts and tables, containing virtually every piece of information required for proper valve selection' (*Chemical engineering*, v.83, Sept. 27, 1976. p.11). *Class No:* 621.646

[4271]

LYONS, J.L. and **ASKLAND, C.L. Lyons' Encyclopedia of valves.** New York, etc., Van Nostrand Reinhold, 1975. ix, 290p. illus., diagrs., graphs, tables. £32.45. ISBN: 0442249616.

Part 1 provides brief descriptions of valves used in fluid control under type of closure (ball, globe, poppet, etc.) and a glossary of terms (p.12-84) with numerous illustrations. Part 2 contains data of use to the designer: size determination; spring design; properties of valve materials; fluid power symbols and standards; test methods; noise calculation. Relates to American standards and English (!) units with conversion factors for SI units. A handbook, *not* an 'Encyclopedia'. *Class No:* 621.646

[4272]

LYONS, J.L. Valve designer's handbook. New York, Van Nostrand Reinhold, 1982. xiii, 882p. illus., diagrs., graphs, tables. £70.95. ISBN: 0442244632.

A compilation of 100 articles, papers, data tables and charts by many contributors, some reprinted, others original material, designed to 'alleviate the burden of searching for information that is generally widespread throughout the industry' (*Preface*). 21 sections (*e.g.* Valve engineering and design data - Cavitation - Screw threads and fasteners - Materials - Valve test methods). The many data tables are particularly valuable. Contains a bibliography of books, articles, standards, etc., and an index of variable quality. *Class No:* 621.646

[4273]

—BRITISH VALVE MANUFACTURERS ASSOCIATION. Valve users manual: a technical reference book on industrial valves for the control of fluids. Kempley, J., *ed.* London, Mechanical Engineering Publications, 1980. 103p. illus., diagrs., charts, tables. £8. ISBN: 0852984286.

Supersedes 3 eds. of *Technical reference book on valves for the control of fluids.* Authoritative definition, description and illustration of all commonly used valve types and variants. 18 sections, from the simple ball and butterfly valves to pressure control, automatic action and special purpose valves. Also covers construction materials, corrosion resistance (tabulated), selection, installation, maintenance (with tables of flow data), steam, SI units. Index. *Class No:* 621.646

Pumps

[4274]

Europump terminology. 2nd ed. Morden, Surrey, Trade & Technical Press, 1978-82. 2v. illus.

First published 1968.

1. *Pump names: désignations de pompes/denominazione delle pompe/dénominación de las pompes/ Pumpenbenennungen.* 1982. [319p.]. ISBN 0854610898.

2. *Pumpenanwendung/pump applications/utilisation des pompes/imprègo delle pompe/utilización de las bombas.* 1978. xviii, 427p. £57.30. ISBN 0854610715.

The work of the European Committee of Pump Manufacturers, these dictionaries list terms, with definitions, numbers and brief explanations in German, with equivalents and explanations in English, French, Spanish and Italian. Indexes in other languages lead to entries in main section. *Class No:* 621.65

[4275]

KARASSIK, I.J., *and others eds.* **Pump handbook.** 2nd ed. New York, etc., McGraw-Hill, 1986. xxii, [1351]p. illus., diagrs., graphs, tables. £95. ISBN: 0070333025.

First published 1976.

72 contributors. Contents assembled in 13 sections. Major sections: 2. Centrifugal pumps (330p.) - 3. Displacement pumps - all types (127p.) - 6. Pump drivers (178p.) - 8. Pumping systems (118p.) - 9. Pump services, *i.e.,* applications and areas of use (317p). Other sections cover: materials; controls; intakes; selection and purchase; installation, operation and maintenance; testing. Extensive appendix of general technical data supplements hundreds of specialized tables in the text. Excellent index provides good access to well-organized and clearly presented contents. SI and US Customary units are used and bibliographial references are provided. 'Readers may be overwhelmed by the magnitude of the material. It is truly a matter of having access to more information on pumps than most people care to know or are able to absorb. But that is really the mark of a good reference text ... The editors have done an excellent job of providing an authoritative guide ...' (*Chemical engineering*, v.93, Nov. 10, 1986, p.140). *Class No:* 621.65

[4276]

Pumping manual. Dickenson, C. 8th ed. Morden, Surrey, Trade & Technical Press, 1988. [xi],746p. illus., diagrs., graphs, tables. £75. ISBN: 0854611231.

First published 1964. Previous ed. 1984.

Aims to provide a comprehensive compendium of practical information and data on the selection, installation, operation and maintenance of industrial pumps. 4 sections. Section 2 has 5 sub-sections (Pump types and characteristics - Duty pumps - Pump construction - Pumping practice - Ancillaries). Section 3 covers 21 industrial applications, from water transport to nuclear power. Index. *Class No:* 621.65

[4277]

—McNAUGHTON, K., *ed.* 'Chemical engineering' guide to pumps. New York, etc., McGraw-Hill, 1984. vii, 325p. illus., diagrs., tables, graphs. £39.95. (*Chemical engineering trade books.*) ISBN: 007024314x.

43 articles reprinted from *Chemical Engineering* 1973-1983, aimed to provide an overview for money-conscious designers and users. 5 sections (1. Selection, design and costing - 2. Centrifugal pumps - 3. Positive displacement pumps - 4. Special applications - 5. Drives, seals, packing and piping). Index. *Class No:* 621.65

[4278]

Pumps and other fluids machinery abstracts. Cranfield, Beds., British Hydromechanics Research Association, 1971-. v.1-. 6 pa. £90pa. ISSN: 03022870.

Over 1,100 abstracts pa., of items drawn from 1,000 journals, plus reports, books, conference papers and selected British patents. Main sections: Fluids machinery - Fans, blowers, compressors - Jet ejectors and pumps (all kinds) - Marine craft propulsion - Hydraulic turbines - Instrumentation and experimental techniques. Issue and cumulative annual indexes of personal and corporate authors and of subjects. Accessible on-line via FLUIDEX (*qv*). *Class No:* 621.65

[4279]

Referativnyĭ zhurnal. 61. Nasosotroenie i Kompressorostroenie Kholodil'noe mashinostroenie. Akademiya Nauk SSSR, Moscow, VINITI, 1970-. Monthly. diagrs. R36.

Formerly *Referativnyĭ zhurnal nasosostroenie i Kompressorostroenie* (1964-69).

About 4,000 signed abstracts pa. on pumps, compressors, refrigeration and air conditioning. Annual subject index. *Class No:* 621.65

[4280]

WARRING, R.H., *ed.* **Pumps: selection systems and applications.** 2nd ed. Morden, Surrey, Trade & Technical Press, 1984. [iv], 271p. illus., diagrs., graphs, tables. £17. ISBN: 0854610960.

15 chapters covering theory and practice of pumping fluids (1. Types of fluids ... 3. Fundamentals of pump performance), kinds of pumps (4. Centrifugal pumps ... 7. Hand pumps), products handled, materials, pipework calculations, pumping systems design, testing and selection of appropriate pumps, followed by a data section (p.254-263) and efficient index.

'This compact, well-devised and very practical work is completed with a comprehensive index, and should prove of

....(contd.)

particular value to those involved with pumping problems in mining/mineral processing' (*Mining magazine*, v.151, August 1984, p.145). *Class No:* 621.65

Production Engineering

Databases

[4281]

CAD/CAM online. New York, EIC/Intelligence Inc.

Covers 1983 to date. A bibliographic database providing references, with abstracts, to world literature on computer-integrated manufacturing, covering computer-aided design and and manufacture (CAD/CAM) and commercial and economic aspects. Sources include periodical articles, conference papers, research reports, government reports, patents and books. File size: more than 20,000 items. Updated monthly. 1,500 additions pa. Hosts: DIALOG; ESA-IRS. Printed version: *CAD/CAM abstracts*. *Class No:* 621.7/.9(003.4)

Bibliographies

[4282]

COX, J., *and others.* **Keyguide to information sources in CAD/CAM.** London, Mansell; Lawrence, KS, Ergosyst Associates, 1988. xxi,257p. ISBN: 072011974x, UK; 0916313158, US.

3 parts. I. Survey of CAD/CAM and its information sources (7 documented chapters, p.3-110). II. Bibliography (514 items). III. Directory of organizations (217 research centres, online hosts; international in scope). Detailed, analytical index, p.245-57. Glossary (AGV-Workstation), p.xv-xxi. An authoritative survey. *Class No:* 621.7/.9(01)

[4283]

HORNICK-LOCKARD, B. 'Computer-aided design: a selected bibliography for community college libraries'. *Choice,* V.30(1), September 1992, p.57-62.

About 50 annotated entries in 10 sections, with running commentary. Material suitable for community college and further education libraries. *Class No:* 621.7/.9(01)

[4284]

NATIONAL ENGINEERING LABORATORY, East Kilbride. **Automation of small-batch production:** bibliography. (Cover title; 'Automated small-batch production'). East Kilbride, Glasgow, NEL, 1978. 138p.

The second of three documents prepared as part of a Department of Industry commissioned study. Its bibliography was compiled largely by searches on the DIALOG and RECON computerized data base systems. Almost 900 items, many with brief annotations, are presented under 3 headings: Manufacturing systems; Work on related topics; Robotics. *Class No:* 621.7/.9(01)

[4285]

—INSTITUTION OF MECHANICAL ENGINEERS. Library and Information Services. Flexible manufacturing systems: an annotated bibliography. London, IME, 1984. 17p.

176 references drawn from British, American, Australian, New Zealand and (a few) German periodical and conference sources, covering the period 1973-1983. Three-quarters of the items are dated 1979-82. Author and subject/geographic/company name indexes. *Class No:* 621.7/.9(01)

Laws

[4286]

DEVINE, J. *and* WATTS, A., *comps*. **Information pack on product liability.** London, Information and Library Service, Institution of Mechanical Engineers, 1987. vii,80p. £14.50. (*Information pack, 6.*) ISBN: 0852986505.

3 parts: 1. Product liability information (abstracts and indexes - text books), p.1-41; 2. Recent liability coverage (26 items), p.42-48; Product liability references (122 items; aeronautical engineering - welding), p.49-73. Author and subject indexes. A useful series. *Class No:* 621.7/.9(01)(094.1)

Handbooks & Manuals

[4287]

BENEDICT, G.F. **Nontraditional manufacturing processes.** New York, Dekker, 1987. xiii, 381p. illus., diagrs., graphs, tables. $95. (*Manufacturing engineering and materials processing, no. 19.*) ISBN: 0824773527.

21 chapters cover the use of energy in new ways or the application of forms of energy preivously not used in manufacturing processes concerned with machining (removal processes), joining and forming. Each chapter deals with a different process in the same format: principles and equipment; capabilities and parameters; application; summary (succinct pros and cons); references. There are 8 chapters each for mechanical and thermal processes. 2 Appendices list US equipment manufacturers and acronyms. Each chapter has a list of references. Excellent illustrations.

'For those seeking a single volume of information, defining the principles and capabilities, together with advantages to be gained and pitfalls to be avoided, this book is to be recommended' (*Production engineer* v.66(7), July/August 1987, p.11). *Class No:* 621.7/.9(035)

[4288]

BRALLA, J.G., *ed*. **Handbook of product design for manufacturing:** a practical guide to low-cost production. New York, etc., McGraw-Hill, 1986. xiii, [1120]p. illus., diagrs., tables. £95. ISBN: 0070071306.

The aim of this handbook is easy and economical production. Emphasis is on products, not processes or functional design, which are fully covered in related handbooks. 55 contributors present information and recommendations in 74 chapters grouped into 8 sections (Introduction - Raw materials - Formed metal components - Machined components - Castings - Nonmetallic parts - Assemblies - Finishes). Each chapter is tightly organized: processes are summarized; applications outlined; economical production quantities, suitable materials, design recommendations, dimensional factors and tolerances are indicated. Adequately supported by illustrations and relevant data and 43-page index. Vast majority of material has appeared elsewhere in other forms, but this is a highly

.... (*contd.*)

convenient compilation of it, enhanced by important new material. There is a very useful quick-reference chapter. *Class No:* 621.7/.9(035)

[4289]

—BOLZ, R.W., *ed*. Production processes: the productivity handbook. 5th ed. New York, Industrial Press, 1977. xiii, [1089]p. illus., diagrs., tables. $45. ISBN: 0831110880.

First published 1949. Previous ed. 1974.

Includes contributions from 8 US experts for new or updated chapters. Intended as 'a broad study of the many manufacturing processes utilized in the production of any and all types of products and components' (*Preface* to 1st ed.), but contains few references to the literature. 69 chapters, 12 sections (1. Producibility, automation and design principles - 2. Metal removing methods - 3. Metal forming methods ... 8. Fabricating methods ... 12. Materials (2 sections of data and conversion factors)). 18-page index. *Class No:* 621.7/.9(035)

[4290]

DROZDA, T.J., *ed*. **Tool and manufacturing engineers handbook.** 4th ed. Dearborn, Mich., Society of Manufacturing Engineers, 1983-86. 5v. illus., diagrs., graphs, tables. $99 per v.

First published as *Tool engineers handbook* in 1949. Previous ed. 1976.

1. *Machining*. 1983. [1484]p. ISBN 0872630854.

2. *Forming*. 1984. [900]p. ISBN 0872631354.

3. *Materials and finishing*. 1985. [864]p. ISBN 0872631761.

4. *Quality control and assembly*. 1986. [750]p. ISBN 087263177x.

5. *Manufacturing management*. 1988. [551]p. ISBN 0872633063. 6 sections, subdivided: 1. Operations and strategic planning - 2. Managerial leadership and its foundations - 3. Human resources - 4. Manufacturing/engineering interface - 5. Resource utilization - 6. Quality. Detailed, analytical index, 9p. *Class No:* 621.7/.9(035)

[4291]

—LE GRAND, R., *ed*. Manufacturing engineers' manual: 'American Machinist' reference book sheets. New York, etc., McGraw-Hill, 1971. ISBN: 0070370664.

A collection of reference information originally published over many years in *American machinist*. 24 chapters: 10 cover metal cutting techniques; 3 on metal forming; 2 on welding. Others cover metal finishing, heat treatment, plating, etc. High proportion of tabulated data. *Class No:* 621.7/.9(035)

[4292]

Flexible manufacturing systems handbook; prepared by the staff of the Automation Management Systems Division of the Charles Stark Draper Laboratory, Cambridge, Mass. Park Ridge, N.J., Noyes Publications, 1984. xviii, 392p. diagrs., charts, tables. $ 48. ISBN: 0815509839.

Based on a 5-volume set of publications produced under contract to the US Army. (v.5 of the original set, containing user's manuals for software packages, not included.)

Defines, describes and illustrates FMS technology. Provides practical design and operational guidance. Glossary of terms. Bibliography. No index. *Class No:* 621.7/.9(035)

Dictionaries

[4293]

Dictionary of CAD. Clarke, J. *and* Mayer, T., *eds.* Oxford, Butterworth Heinemann, 1991. 484p. illus. £40. ISBN: 0750612398.

About 2,500 terms, most defined briefly.
Class No: 621.7/.9(038)

[4294]

HUNT, V.D. Dictionary of advanced manufacturing technology. Amsterdam, etc., Elsevier, 1987. xi, 431p. illus., diagrs. £34.53. ISBN: 0444012087.

Over 5,800 terms, acronyms and abbreviations defined and illustrated, from the fields associated with the application to manufacturing of computers, robots, artificial intelligence and automatic control. Preceded by a chapter giving an overview of advanced manufacturing technology. Brief directory of organizations and companies, a list of abbreviations and acronyms and a short bibliography complete the work. *Class No:* 621.7/.9(038)

[4295]

INTERNATIONAL INSTITUTE FOR PRODUCTION ENGINEERING RESEARCH (CIRP). Wörterbuch der Fertigungstechnik. [Dictionary of production engineering.] Pahlitzsch, G., *ed.* Essen, Girardet, 1962-1984. 9v. illus. DM350 per set. ISBN: 3773656901.

1. *Forging and deep forging.* 1962. 108p. 2. *Grinding - surface roughness.* 1963. 139p. 3. *Sheet metal forming.* 1965. 136p. 4. *Fundamental terms of cutting.* 1969. 123p. 5. *Cold extrusion and upsetting.* 1969. 157p. 6. *Planing, slotting, broaching, turning.* 1972. 194p. 7. *Drilling, countersinking, counterboring, reaming, thread production.* 1977. 220p. 8. *Milling, sawing, gear manufacturing.* 1979. 359p. 9. *Electrochemical, electrodischarge, electron beam, photonen beam (laser) and chemical machining.* 1984. 170p.

A German-English-French dictionary edited by expert committees of CIRP (Collège International Recherche Production). V.1 has entries for *c.*400 German-base terms with English and French indexes and nearly 100 illus. V.1 has 2 other versions: V.1N (covering Danish, Norwegian, Swedish, Finnish) and V.1R (German, Spanish, Italian, Portugese). *Class No:* 621.7/.9(038)

[4296]

TVER, D.F. *and* BOLZ, R.W. Encyclopedic dictionary of industrial technology: materials, processes and equipment. New York, London, Chapman and Hall, 1984. [vi] 353p. diagrs., tables. £39. ISBN: 0412005018.

Wide-ranging coverage of materials, machine tools, computers, robots, manufacturing processes and treatments, and equipment, handling, inspection and testing. Definition and description of *c.*1,800 terms vary from one line to fifty. Cross-references. Select bibliography p.349-53.
Class No: 621.7/.9(038)

Yearbooks & Directories

[4297]

Automated manufacturing directory. Weston, R., *ed.* 2nd ed. London, Morgan-Grampian, 1986. x, 357p. £25. ISBN: 0862130700.

Covers robotics, computer-integrated manufacturing, CAD, numerical control, machine and production control systems and process control. Index of companies, associations, societies. *Class No:* 621.7/.9(058)

Workshop Practice

[4298]

AVITZUR, B. Handbook of metal-forming processes. New York, etc., Wiley, 1983. xxxi, 1020p. illus., diagrs., graphs, tables. £100. ISBN: 0471034746.

A descriptive but analytical approach to the wide variety of metal-forming processes now available. For each process discussed the relevant principles and concepts are introduced to enable the engineer to gain complete understanding. Problems are appended to assist in this. 18 chapters (*e.g.* 1. Basic concepts - 2. Forging ... 6. Impact extrusion - 7. Hydrostatic extrusion ... 10. Deep drawing ... 13. Rolling ... 17. Soft tooling - 18. Replacing brute force). Copious references and illustrations. *Class No:* 621.7

[4299]

—**AMERICAN SOCIETY FOR METALS.** Periodical Publication Department. Source book on cold forming: a discriminating selection of outstanding articles from the periodical literature. Metals Park, Ohio, American Society for Metals, 1975. viii, 367p. illus., diagrs., tables, graphs. $93.80. ISBN: 0317111515.

53 articles reprinted from 6 US, UK and Japanese journals and from the 9th ed. of the ASM *Metals handbook*, mainly published 1965-75, plus tables of units, conversion factors and equivalence data for the several hardness numbers. 6 sections: 1. State of the technology - 2. Part design and applications - 3. Cold forming materials (*NB* Cold forming and extrusion of steel - a two-part review article (p. 155-206) with 267 references) - 4. Inspection and testing - 5. Cold forming processes and equipment - 6. Useful tables for cold formers. Includes cold forging data sheets issued by the International Cold Forging Group. Good index.
Class No: 621.7

[4300]

ROSS, R.B. Handbook of metal treatments and testing. 2nd ed. London, Chapman and Hall, 1988. xix,568p. Tables. £35. ISBN: 0412313901.

A-Z listing of processes, treatments and testing methods with brief definition and, either, a cross-reference to one of 8 major sections for fuller treatment, or extended explanation with sub-sections. The major sections are: blasting; electroplating; hardness testing; heat treatment; mechanical testing; non-destructive testing; painting; welding. 'List of processes covered', p.xi-xix. 9 appendices of miscellaneous data. *Class No:* 621.7

Dies & Patterns

[4301]

STREET, A., *ed.* The Diecasting book. 2nd ed. Redhill, Surrey, Portcullis Press, 1986. [xi],774p. illus., diagr., tables. ISBN: 0861082354.

First published 1977.

A comprehensive handbook on all aspects of the technology of diecastings covering processes, the various alloys and ferrous metals used, management and related topics. *Class No:* 621.7.07

Forge Work

[4302]

BYRER, T.G., *ed*. **Forging handbook.** Cleveland, OH, Forging Industry Association, and Metals Park, OH, ASM, 1985. ix,296p. illus., diagrs., graphs, tables. ISBN: 0871701944.

Numerous contributors. 4 sections, each with a number of parts: 1. Introduction and applications - 2. Forging design - 3. Material characteristics - 4. Manufacture of forgings. Extensively illustrated. Glossary, p.275-87; analytical subject index, p.289-96. A very practical treatment for the 'expert and newcomer' (*Preface*, p.v). *Class No:* 621.73

[4303]

OPEN DIE FORGING INSTITUTE. Metallurgical and Research Committee. **Open die forging manual.** 3rd ed. Cleveland, Ohio, Forging Industry Association, 1982. viii, 200p. illus., diagrs., tables.

A 'reference book of basic technical information on open die forging metallurgy, equipment, testing, quality assurance, design and purchasing considerations' (*Preface*). Includes chapters on manufacturing methods, materials, service life factors. Glossary of about 300 terms, 3 appendices and index supplement the text. There are charts of heat and temper colours. *Class No:* 621.73

Foundry Work

Bibliographies

[4304]

UNITED NATIONS. Industrial Development Organization. **Information sources on the foundry industry.** New York, United Nations, 1973. xix, 95p. (*UNIDO Guides to information sources, no. 5.*)

565 numbered items. 13 sections. 1. Professional, trade and research organizations, learned societies and special information services - 2. Directories - 3. Sources of statistics and economic data - 4. Basic handbooks, textbooks and manuals - 5. Mono-encyclopedias - 6. Monograph series - 7. Specialized dictionaries and encyclopaedias - 8. Bibliographies - 9. Current periodicals - 10. Current abstracting and indexing periodicals - 11. Proceedings, papers and reports - 12. Films on foundry practice - 13. Other potential sources of information (*e.g.* patents, standards). Worldwide coverage of organizations and publications. Bibliography of sources used.
Class No: 621.74(01)

Dictionaries

[4305]

BRUNHUBER, E. **Giesserei fachwörterbuch.** [Foundry dictionary.] 4th ed. Berlin, Schiele and Schön, 1988. 14,1157p.

4 parts (German, English, French, Italian). Each part has about 8,500 headwords with equivalents in the other 3 languages. Genders given; US usage indicated.
Class No: 621.74(038)

[4306]

SIMONS, E.N. **Dictionary of foundry work.** London, Crosby Lockwood, 1972. viii, 227p. illus., tables. £2.95. ISBN: 0258968516.

About 900 entries and cross-references. Definitions are explanatory, clear and understandable, ranging from one line to two pages ('sands'); most occupy a few lines. 35 clear line-drawings. *Class No:* 621.74(038)

[4307]

STÖLZEL, K., *ed*. **Dictionary of metallurgy and foundry technology.** Amsterdam, etc., Elsevier, 1984-87. 2v.

1. *English-German.* 1984. 418p. £79.58. ISBN 0444996125.

2. *German-English.* 1987. 464p. £75.08. ISBN 0444995145.

Wide-ranging scope. Brief entries, with equivalents and occasional definition or explanation.

Review of v.1: 'It will be of particular value to linguists, documentation experts and scientists as well as teachers and students ...' (*Iron and steelmaker*, v.14, August 1987, p.65).
Class No: 621.74(038)

Reviews & Abstracts

[4308]

BCIRA abstracts of international literature on metal castings production. Birmingham, British Cast Iron Research Association, 1986-. v.34-. 6pa. £56pa. ISSN: 02683393.

First published 1923 as *BCIRA bulletin and foundry abstracts*. Subsequently part of *BCIRA journal* (to 1968). Then *BCIRA abstracts of foundry literature* (1969-77) and latterly *BCIRA abstracts of international foundry literature* (1978-85).

About 1300 indicative and informative abstracts pa., covering books, patents and articles from *c*.50 journals published in 30-odd countries. Classified arrangement: 11 sections (Raw materials and energy - Melting and casting - Moulding and coremaking - Quality control ... Management and organization ... Environmental pollution and protection - General). Annual author and subject indexes, but none in issues. *Class No:* 621.74(048)

Yearbooks & Directories

[4309]

Foundry directory and register of forges: UK and Western Europe. 14th ed. Worcester Park, Surrey, Metal Bulletin Books, 1987. xxv, 297p. £30. ISBN: 0947671099. ISSN: 00718130.

First published 1959. Present publisher since 1983.

Predominantly UK coverage but Western Europe section introduced in previous ed. now expanded to *c*.90p. Contains: Alphabetical index to all foundries and forges included; alphabetical list of all UK foundries with company information and details of casting methods, dimensions, products, followed by UK buyers' guide of products (castings and forgings); similar treatment of foundries and forges in 17 European countries, and buyers' guide. Buyers' guide to suppliers of equipment, supplies and services.
Class No: 621.74(058)

[4310]
Foundry yearbook and casting buyer's directory, 1988.
Redhill, Surrey, Foundry Trade Journal, 1988. Annual.
374p. £35. ISBN: 0861082826. ISSN: 02645319.

2-part directory. Part 1: Trade directory, including details
of UK foundry industry, organizations in UK and overseas,
guide to UK and Eire foundries and forges. Part 2: Buyers'
guide (alphabetical list of suppliers; classified list of plant,
equipment, services; trade marks; UK agents, etc.). Index to
advertisers. *Class No:* 621.74(058)

Tables & Data Books

[4311]
**FOSECO (F.S.) LTD. Foseco Foundryman's handbooks,
facts, figures and formulae.** Burns, T.A., *rev. and ed.* 9th
ed. Oxford, etc., Pergamon, 1986. x, 435p. diagrs., graphs,
tables. £26. ISBN: 0080325491 Hard cover.
Previous ed. 1975.

Practical guidance and recommendations on making
castings by all common casting methods. 10 sections: 1.
Tables and general data (p.1-67) - 2. Sands and sand
bonding systems ... 4. Light casting alloys ... 6. Iron
castings ... 8. Steel casting specifications ... 10. Principal
FOSECO products (including publications). SI Units. Many
tables. Good index. *Class No:* 621.74(083)

Smelting

[4312]
World survey of non ferrous smelters. Taylor, J.C. *and*
Traulsen, H.R., *eds.* Warrendale, Pa., Metallurgical
Society, 1987. ix,398p. diagrs., tables. ISBN: 0873390261.

Covers copper, nickel, lead and zinc. Gives company
names, addresses and operating data. *Class No:* 621.745

Casting

[4313]
WALTON, C.F., *ed*. Iron castings handbook; covering data
on gray, malleable, ductile, white, alloy and compacted
graphite irons. 3rd ed. Rocky River, Ohio, Iron Castings
Society, 1981. [xxv], 831p. illus., diagrs., graphs, charts,
tables. $27.50.
First published 1958. Previous ed. 1971.

Introductory chapters on the casting process,
specifications, testing, metallurgy, design. Chapters on
mechanical properties of irons listed in title with tables of
relevant data. Chapters on corrosion, wear, heat treatment,
welding, machining. Chapter references. Glossary of terms
(p.801-18). 17 plates. Index. *Class No:* 621.746

[4314]
—HERRMANN, E. *and* HOFFMAN, D. Handbook of
continuous casting. Düsseldorf, Aluminium-Verlag, 1980.
ix, 742p. diagrs. DM1250. ISBN: 3870171340.

Part 1: a 61-page review of the state-of-the-art worldwide
from the reports of 32 firms. Part 2 - the main part - contains
brief details of 3,711 patents, in classified arrangement, each
numbered, from 17 countries, taken out 1958-74, with illus.
from the specification. Part 3 is a tabulation of these to
facilitate identification of equivalents in other countries. Part
4 lists patents applied for in each country in numerical
(patent no.) order, and includes many from the period 1974-
79. There is good cross-referencing between sections.
Class No: 621.746

[4315]
WIESER, P.F., *ed*. Steel castings handbook. 5th ed. Rocky
River, Ohio, Steel Founders' Society of America, 1980.
[525]p. illus., diagrs., graphs, tables. $35. ISBN:
0960467408.
First published 1941, edited by C.W. Briggs. Revised
every decade.

Intended primarily for designers, materials engineers and
purchasers of steel castings. 6 parts: (1. General information
- 2. Designing castings - 3. Manufacture - 4. Material
selections (A. Mechanical and chemical properties. B.
Processing properties. C. Physical properties). - 5.
Specifying - 6. Buying castings). Emphasis is on mechanical
and chemical properties, especially of structural carbon and
low alloy steels. Abundant tables, graphs and illustrations,
supplemented by extensive glossary (c.1,350 terms).
Class No: 621.746

[4316]
—Steel castings abstracts. Sheffield, Steel Castings Research
and Trade Association. 1968-. v.17-. 6pa. £55pa. ISSN:
00390909.
Formerly *BSCRA abstracts* (1952-1968).

1,500 abstracts pa. from 250 worldwide periodicals,
books, reports, patents. 18 sections (General ... Patterns and
pattern-making ... Casting processes ... Heat treatment ...
Health and safety, dust and fumes). Annual author and
subject indexes. *Class No:* 621.746

Powder Metallurgy

Bibliographies

[4317]
COLEMAN, D.S. 'Powder metallurgy'. *Information sources
in metallic materials.* Editor, M.N. Patten, p.217-31.
London, Bowker-Saur, 1989. (*Guides to information
sources.*) ISBN: 0408014911.

Sections: Introduction (process; advantages; production
statistics; products - 16 references) - Information sources
(*e.g.* reference journals (20 references); books (31
references); conferences; associations; standards (40).
Class No: 621.762(01)

Handbooks & Manuals

[4318]
**FAYED, M.E. *and* OTTEN, L., *eds*. Handbook of powder
science and technology.** New York, etc., Van Nostrand
Reinhold, 1984. xiv, 850p. illus., diagrs., graphs, tables.
$79.50. ISBN: 0442226107.

26 contributors from 7 countries. 19 chapters cover
particle size analysis and measurement, fundamental
properties of powders and their behaviour, size enlargement
and reduction, mixing, storage, fluidization technology,
pipeline transport, sedimentation, filtration, cyclones, wet
scrubbers, etc. Chapter references.

'I definitely can recommend this handbook to all chemical
engineers who are involved with the storage, handling and
processing of particulate solids and powders' (*Chemical
engineering*, v.92, Jan. 7, 1985, p.113).
Class No: 621.762(035)

[4319]

HAUSNER, H.H. *and* **MAL, M.K. Handbook of powder metallurgy.** 2nd ed. Chemical Publishing Co., 1982. 608p. $85. ISBN: 0820603015.

21 chapters. Principles of powder metallurgy and characteristics of powders and processes, Chapters 1-11. Remainder cover currently-produced materials. An appendix covers statistics, applications, physical and chemical data, and terminology.

'... the authors have compiled a concentrated handbook consisting of graphs and tables, practically without text, useful not only for practicing (!) powder metallurgists but all materials engineers and scientists. ... this handbook should be in the hands of everyone interested in metal powders and powder metallurgy' (*International journal of powder metallurgy*, v.19(3), July 1983, p.233-4).
Class No: 621.762(035)

[4320]

Metals handbook; v.7. Powder metallurgy: prepared under the direction of the ASM Handbook Committee. Klar, E., *ed.* 9th ed. Metals Park, Ohio, American Society for Metals, 1984. xviii, 897p. illus., diagrs., graphs, tables. $92. ISBN: 0871700131.

The first completely new volume in the 9th ed., rather than a revision of an 8th ed. volume. About 300 contributors/reviewers. Four major sections: Production of metal powders - Characterization and testing of metal powders - Consolidation of metal powders - Powder systems and applications. Detail much greater than elsewhere in the literature 'such that the book serves as an introduction ... for the technically trained individual and as a reference book for the experienced engineer' (*Preface*). Includes glossary (p.1-13) and brief history. Analytical index (p.852-97).

'It is an excellent and stimulating book of current technology and a standard work for many years' (H.H. Hausner. *International Journal of powder metallurgy and powder technology*, v.20(4), Oct. 1984, p.362).
Class No: 621.762(035)

[4321]

—Metals handbook; v.9. Metallography and microstructures. Metals Park, Ohio, American Society for Metals, 1985. $104. ISBN: 0871700158.

Contains much information of importance to the powder metallurgist, including metallographic preparation techniques and many high quality micrographs relating to powder metallurgy microstructures. '... this is a high-quality volume which will be of genuine assistance to anyone involved in P/M. Together, vols. 7 and 9 [of *Metals Handbook*] constitute a complete treatise on modern powder metallurgy technology' (*International journal of powder metallurgy and powder technology*, v.22(2), April 1986, p.128).
Class No: 621.762(035)

Dictionaries

[4322]

BRITISH STANDARDS INSTITUTION. Glossary of terms relating to powders. London, BSI, 1958. Amendment PD 5673. 1965. 16p. diagrs., tables. £11.70. (*BS 2955: 1958.*)

6 sections: 1. General - 2. Types of powder - 3. Particle size - 4. Particle shape - 5. Powder properties - 6. Powder processing. 2 appendices. Index. *Class No:* 621.762(038)

[4323]

—BRITISH STANDARDS INSTITUTION. Powder metallurgical materials and products. Part 1, Section 1.2. Glossary of terms. London, BSI, 1981. 14p. £17.90. (*BS 5600: 1981.*)

About 150 terms defined in systematic order with A-Z index. *Class No:* 621.762(038)

Reviews & Abstracts

[4324]

Metal powder report: the monthly journal for the PM industry. Shrewsbury, Salop., MPR Publishing Services, 1946-. v. 1-. £48. ISSN: 00260657.

Issues contain short 'Literature review' with citations and informative abstracts of recent periodical and conference papers, arranged under subjects. August 1987 issue covers 60 papers from the May 1987 Annual Powder Metallurgy Conference and 4 papers from 3 journals.
Class No: 621.762(048)

[4325]

Powder metallurgy science and technology: abstracts of current journals, report and patent literature. Swarthmore, Pa., Peters Technology Transfer, 1968-. v. 1-. Monthly. ISSN: 00485020.

Coverage includes conference papers and book reviews. About 1,400 informative abstracts pa. Numbered consecutively from no.1, 1968. Classified arrangement: General information - Powder production - Powder properties - Compaction - Sintering - Forging, rolling and extrusion - Physical properties of sintered compacts - Ferrous metals and alloys - Non ferrous metals and alloys ... Fibre metallurgy - Refractories - Porous and friction materials -Nuclear applications ... Cermets and cemented carbides. *Class No:* 621.762(048)

Progress Reports

[4326]

Modern developments in powder technology. Princeton, N.J., Metal Powder Industries Federation/American Powder Metallurgy Institute 1973-. v.6-. (v.1-5, 1966-71 published by Plenum Press). illus., diagrs., graphs. ISSN: 00972223.

The proceedings of International Powder Metallurgy Conferences have been published under this title since the second conference, 1965, whose papers appeared in v.1-3 of the series. Proceedings of the latest conference (Toronto, 1984) appear in v.15-17 (1985. 592p. $240 per set. ISBN: 0918404673). *NB.* Proceedings of these International Conferences also comprise certain vols. in the *Progress in powder metallurgy* series (*qv*). - *e.g.* v.36 comprises proceedings of the 6th International Conference, Washington, 1980. *Class No:* 621.762(055)

[4327]

Progress in powder metallurgy. Princeton, N.J., Metal Powder Industries Federation, 1947-. v.1-. illus., diagrs., graphs. ISSN: 00796719.

The Proceedings of the Annual Powder Metallurgy Conference are published under this title - *e.g.* v.39 comprises the proceedings of the 1983 Annual Conference held at New Orleans (1984. 688p. ISBN: 0918404614; ISSN: 00796719). *NB.* Proceedings of the International Powder Metallurgy Conferences held every few years are

....(contd.)

also published as vols. in this series - *e.g.* v.36 comprises the proceedings of the 6th International Conference, Washington, 1980. *Class No:* 621.762(055)

[4328]

—International journal of powder metallurgy and powder technology. V.16(3), July 1980, p.291-330.

These pages provide a chronological list of articles, supplemented by author and subject indexes. Covering 385 articles, this cumulated index constitutes a useful bibliography in the field. *Class No:* 621.762(055)

Maps & Atlases

[4329]

HUPPMANN, W.J. *and* **DALAL, K. Metallographic atlas of powder metallurgy.** Freiburg, West Germany, Verlag Schmid GmbH, 1986. 190p. illus., diagrs. $100.

A short, comprehensive and illustrated overview of processes and methods is followed by over 460 micrographs illustrating powder particles, microstructure of selected powder metallurgy materials (*e.g.* various steels, non-ferrous metals and alloys) and microstructural features of components (*e.g.* surface roughness) and defects in materials and their effects. Classified arrangement. No index.

'The quality of the micrographs ... is excellent. The atlas is highly recommended to those involved in production, testing, quality control, research and development, in the field of powder metallurgy (*International journal of powder metallurgy and powder technology*, v.22(4), Oct. 1986, p.282-3). *Class No:* 621.762(084.3)

Boiler Making

[4330]

BEDNAR, H.H. Pressure vessel design handbook. 2nd ed. New York, Van Nostrand Reinhold, 1986. xi, 431p. diagrs., graphs, charts, tables. £46.50. ISBN: 0442213859.

First published 1981.

Aimed as a practical aid to design in conformity with section 8 of the ASME Boiler and Pressure Vessel Code, there are chapters on loads and stresses; design for tall and short vertical vessels and supports for horizontal vessels; welds; materials; numerical methods for stress analysis. 10 Appendices include 172 bibliographical references, a list of relevant US standards and a glossary. Imperial units used.

'Should be a valuable resource not only to junior mechanical and vessel design engineers but also to practicing professionals who wish to expand their knowledge of the subject. Bednar has done a fine job in organizing and condensing much widely scattered information ...' (*Chemical engineering*, v.93, Oct. 13 1986, p.95). *Class No:* 621.772

[4331]

CHUSE, R. *and* **EBER, S.M. Pressure vessels:** the ASME Code simplified. 6th ed. New York, etc., McGraw-Hill, 1984. x, 267p. illus., diagrs., tables. £29.95. ISBN: 0070108749.

First published 1954. Previous ed. 1977.

Concentrates on the requirements of Section 8 of the 1980 ed. of the ASME Boiler and Pressure Vessel Code, which deals with pressure vessels, and aims to assist the designer, fabricator or inspector who may use the Code infrequently. Provides various charts and tables to enable such users to apply and interpret the Code's requirements. 6 appendices of

....(contd.)

fundamental data relating to wall thicknesses, dimensions, volumes, corrosion. Two new chapters cover power boilers and nuclear vessels.

'Brings many frequently overlooked items to the reader's attention and suggests methods for better design, fabrication and inspection control that will lead to more efficient and economical operations and better quality control systems' (*Mechanical engineering*, v.106, Nov. 1984, p.106). *Class No:* 621.772

[4332]

MEGYESY, E.F. Pressure vessel handbook. 7th ed. Pressure Vessel Handbook Publishing, 1986. 480p. illus., diagrs., tables, maps. $38. ISBN: 0914458124.

First published 1972. Previous ed. 1983.

An invaluable assembly of formulas, data, guidance on design and construction for engineers concerned with pressure vessels. Relates to the provisions of section 8 of the ASME Boiler and Pressure Vessel Code, 1983 ed. In parts: 1. Design and construction - 2. Geometry and layout - 3. Measures and weights - 4. Design of steel structures - 5. Miscellaneous (*i.e.,* Abbreviations; Codes and standards; Laws; Literature; Definitions; Index). Hundreds of tables.

'Serves as an important and needed reference for those involved in the practical aspects of pressure vessels' (*National safety news*, v.129, Feb. 1984, p.62, - Review of previous ed.). *Class No:* 621.772

[4333]

Referativnyĭ zhurnal. 42. Kotlostroenie. Moscow, VINITI, 1963-. Monthly. diagrs. R18pa. ISSN: 00342424.

About 1,200 signed indicative and informative abstracts pa. on all aspects of boilers. 8 sections, covering general questions, boiler theory, design and calculations, control, plant and special materials. Illustrated with diagrams. UDC numbers. Annual subject index. English contents summary. *Class No:* 621.772

Tube Making

[4334]

Tube industry yearbook: 1992. Leamington Spa, Warwickshire, International Tube Association, 1992. 492p. illus. ISBN: 095114880x.

Product category list; Product and equipment directory. Indexes in English, German, French, Spanish, Italian. *Class No:* 621.774

Wire Drawing & Working

[4335]

CLAYTON, P.A., *ed*. **The Encyclopaedia of wire.** Oxted, Surrey, Magnum Publications, 1979. 168p. illus., diagrs. £7.25.

A *Wire industry* publication. More than 1,500 terms relating to the manufacture and uses of all types of wire defined, sometimes briefly, but frequently at considerable length with descriptions of processes, etc. *Class No:* 621.778

[4336]
Wire industry yearbook: 1992. 32nd ed. Oxted, Surrey, Magnum Publications, 1992. 464p. £44. ISSN: 00840424.

World coverage. 900 product groups, each listed in 5 (English, German, French, Spanish, Italian) languages, under 7 headings (Ferrous wire - Non-ferrous wire - Machinery and plant for wire manufacture - Machinery and plant for making wire products - Dies, tools, etc. - Lubricants, raw materials & services - Wire products). A-Z list, with addresses, phone and Telex nos., of about 3,500 firms. Index of brand names. *Class No:* 621.778

Welding

Databases

[4337]
WELDASEARCH. Abington, Cambridge, Welding Institute.

Covers 1967 to date. A bibliographical data base providing references, with abstracts, to the literature of all aspects of welding: design, metallurgy, fabrication, performance of welds, soldering, brazing, thermal cutting, joining of metals and plastics. Derived from journal articles, research reports, books, standards, patents, theses. File size: 87,000 reference (1986). Updated monthly; *c.*5,000-6,000 additions pa. Host: DIALOG; cost per connect hour: $84. Printed version: *Literature abstracts.* Available on magnetic tape. *Class No:* 621.791(003.4)

Bibliographies

[4338]
DUMPER, L. *and* **LOADER, J.** 'Welding, brazing and soldering'. *Information sources in metallic materials.* Editor, M.N. Patten, p.278-315. London, Bowker-Saur, 1989. (*Guides to information sources.*) ISBN: 0408014911.

Sections: Information sources - Types of literature (*e.g.* bibliographic services; journals (53 titles); standards; reports) - Subject literature (*e.g.* processes; metallurgy; markets; consumables; health and safety; nondestructive testing; design) - Education and training. *Class No:* 621.791(01)

[4339]
FISCHER, K. 'Welding: a core list for community college libraries'. *Choice,* V.28(10), June 1991, p.1599-1608.

Running commentary in 10 sections. Annotated entries; for community college and further education libraries. *Class No:* 621.791(01)

Handbooks & Manuals

[4340]
AMERICAN WELDING SOCIETY. Welding handbook.
Kearns, W.H., *ed.* 7th ed. Miami, Florida, American Welding Society/London, Macmillan, 1975-85. 5v. illus., diagrs., graphs, tables.

1. *Fundamentals of welding,* ed. by C. Weisman. 1975. 373p. £34. ISBN 0333198743.

2. *Welding processes - arc and gas welding and cutting, brazing, and soldering.* 1978. 592p. £17.50. ISBN 0333256662.

3. *Resistance and solid-state welding and other joining processes.* 1980. 459p. £29. ISBN 0333273958.

4. *Metals and their weldability.* 1983. 582p. £39. ISBN

....(*contd.*)
0333293428.

5. *Engineering, costs, quality and safety.* 1985. 456p. £68. ISBN 0333293436.

The major work of reference, with contributions by specialists from many countries. Revised volume by volume. Whole set currently in 7th ed. Extensively provided with bibliographical references. Generously illustrated. Separate indexes.

Publication of the 8th ed. began in 1987.

1. *Welding technology.* 1987. xiii,638p. £66.

2. *Welding processes.* 1988. *c.*600p.

Class No: 621.791(035)

[4341]
Metals handbook; v.6. Welding, brazing and soldering: prepared under the direction of the ASM Handbook Committee. Nippes, E.F., *ed.* 9th ed. Metals Park, Ohio, American Society for Metals, 1983. xvii, 1152p. $92. ISBN: 0871700077.

Recent developments, a widening of scope and a high proportion of new sections have led to a 50 per cent increase in size of this volume since the previous (8th) ed. of 1971. The work of *c.*200 contributors is presented under 13 main headings, of which 2 are introductory, 8 relate to welding, 1 to brazing and 1 to soldering. Welding accounts for over 50 of the articles, brazing has 14 and soldering 1. A glossary (p.1-20), index (p.1113-52), list of abbreviations and symbols and many references effectively enhance this work. *Class No:* 621.791(035)

[4342]
—**STEWART, J.P.** The Welder's handbook. Reston, Va., Reston Publishing Co., 1981. ix, 534p. $21.95. ISBN: 0835986055.

A practical guide to processes and essential information and data for welders, designers, engineers and managers. 20 chapters (*e.g.* 1. Basic welding consumables ... 6. TIG welding ... 11. Welding special alloys - 12. Brazing guide ... 19. Comparative welding costs - 20. Pipe welding). Appendix ('Useful information'), p.491-527. Index (inconsistent in construction). Numerous tables and illus.

'Gives operating data for many welding processes, helpful to welders; solutions to shop problems, which supervisors often need; cost data ... useful for estimators and designers; and technical data ... required by metallurgists and welding engineers' (*Welding design and fabrication* v.54, Oct. 1981, p.174). *Class No:* 621.791(035)

Dictionaries

[4343]
BRITISH STANDARDS INSTITUTION. Welding terms and symbols. London, BSI, 1980-91. diagrs., charts. £72.20. (*BS 499: 1980-1991.*)

Pt.1: *Glossary for welding, brazing and thermal cutting.* 1991. 113p.

Pt.2: *Specification for symbols for welding.* 1980. 20p. AMD 7439, Nov. 1982.

Pt.2c: *Welding symbols.* (Chart). 1980. 1p.

The *Glossary* defines over 850 terms, arranged in systematic order and supported by A-Z index. *Class No:* 621.791(038)

Reviews & Abstracts

[4344]

Referativnyĭ zhurnal. 63. Svarka. Moscow, VINITI, 1968-. Monthly. R72.60. ISSN: 01313525.

About 9,000 signed indicative and informative abstracts pa. on welding. 14 sections (General problems ... Arc welding and cutting ... Beam welding ... Gas welding and cutting ... Soldering; brazing ... Welded metal constructions). Annual author and subject indexes. UDC numbers. English contents summary.
Class No: 621.791(048)

[4345]

Welding abstracts. Oxford, Pergamon Press, for the Welding Institute, 1988-. Monthly. Dm825pa. ISSN: 09520287.

About 500 abstracts per issue. Sections: Processes - Production - Equipment and materials - Properties and testing - Manufacturing and construction - General. Author and subject indexes per issue. Includes patents and standards. Online. *Class No:* 621.791(048)

Standards

[4346]

BRITISH STANDARDS INSTITUTION. Welding. Milton, Keynes, BSI, 1986. 3p. £1. (*SL 7.*)

One of the Sectional List series, giving the number, title and price group of 63 British Standards specifications, published documents (PD) and drafts for development (DD), covering all types of welding and brazing and including relevant standards from other fields. No index. (Full details are to be found in the annual *Catalogue* (*qv*)).
Class No: 621.791(083.74)

[4347]

International standards index: welding and related processes. Abington, Woodhead, 1991. 646p. £300. ISBN: 1855730219.

Loose-leaf. Bi-annual supplements planned.

Lists nearly 5000 standards on welding and allied techniques from 16 countries. Includes European and international standards. *Class No:* 621.791(083.74)

Thesauri

[4348]

INTERNATIONAL INSTITUTE OF WELDING. The International welding thesaurus. 3rd ed. Abington, Cambridge, Welding Institute, 1989. 195p. £45.

First published 1974, based on 2nd ed. of *A thesaurus for welding and allied processes* (Welding Institute, 1971). Prepared by a joint British-French-German study group. Contains about 1,500 descriptors and synonyms, broader, narrower and related terms, in English, with trilingual lists of descriptors in each language. Tables show the 'families' of descriptors (*e.g.* Properties; Phenomena). Preface and introduction in English and French.
Class No: 621.791:025.43

Metal Finishing

Bibliographies

[4349]

RUDZKI, G.J. Surface finishing systems: metal and non-metal finishing handbook guide. Metals Park, Ohio, American Society for Metals / Teddington, Middx., Finishing Publications, 1983. 421p. charts, tables. £35. ISBN: 090447707x.

The third part of this book - half of it - comprises a cumulative index to *Metal finishing abstracts*, 1960-82, in which entries are arranged alphabetically under chemical element, Actinium - Zirconium, with subdivision for processes, *e.g.* Spray coatings.

'This section will prove invaluable to researchers and information scientists undertaking literature searches. There are, however, two limitations. The first that ... the references are limited to new techniques or conditions which are not generally known. The second ... is that ... it is not complete. The references were ... carefully chosen.'
Class No: 621.793(01)

[4350]

Surface treatment plant and processes; incorporating *Metal finishing plant and processes.* Teddington, Middx., Finishing Publications, Ltd., May/June 1986-. v. 22(3)-. 6 pa. £67. ISSN: 09505202.

As *Metal finishing plant and processes*, (v.1-22, 1965-86). 900 brief descriptions of new products and processes pa., in 21 sections (Cleaning and degreasing ... Electroplating ... Vacuum and gas coatings ... Waste treatment and recovery ... General and miscellaneous). Index to company names with annual cumulative indexes of company names, trade names and subjects. *Class No:* 621.793(01)

Handbooks & Manuals

[4351]

Advanced surface coatings: a handbook of surface engineering. Rickerby, D.S. *and* Matthews, A., *eds.* Glasgow, Blackie, 1991. xiii,368p. illus., diagrs., graphs. £75. ISBN: 0216928990.

16 contributors. 13 chapters, closely subdivided: 2. Use of plastics in deposition technologies ... 4. Evaporation - 5. Spatter disposition (129 references) ... 9. Thermal spraying ... 12. Evaluation of coatings - 13. Market perspective and future trends. Detailed, analytical index, p.365-68.
Class No: 621.793(035)

[4352]

The Canning handbook: surface finishing technology. 23rd ed. Birmingham, W. Canning / London, Spon, 1982. xiv, 1094p. illus., diagr., tables. ISBN: 0419129006.

First published 1889 as *Practical instructions for gold, silver, nickel, brass and copper, plating and polishing.* Previously published (1st ed. 1901; 22nd ed. 1978) as *Handbook on electroplating.*

35 chapters: 1-5. Polishing; 6-32. Electroplating; 33-34. Lacquering and bronzing; 35. Safety precautions. Index: p.1068-94. A standard handbook produced by a company with a 200-year history in the field whose recent diversification is matched by the widening of scope of this edition. *Class No:* 621.793(035)

[4353]

KUHN, A.T., *ed*. **Techniques in electrochemistry, corrosion and metal finishing:** a handbook. Chichester, Wiley, 1987. xviii,567p. illus., tables. ISBN: 047191407x.

8 contributors. In 3 parts: A. Introduction to experimental methods - B. Specialized techniques (Section I: Methods based on light; Section II: Spectroscopic methods; Section III: Other methods (*e.g.* physiochemical; radio-chemical; acoustic emission; dyes, etc.) - C. The literature (37p. bibliography of post-1960 monographs, most from the British Library Science Reference and Information Service). 28 chapters, most with extensive references (*e.g.* 'Raman spectroscopy', p.237-49, includes 166). Subject index. Dedicated - gracefully - to the 'Library' and its 'staff'.
Class No: 621.793(035)

Reviews & Abstracts
[4354]

Surface treatment technology abstracts; incorporating *Metal finishing abstracts*; including *Printed circuits and elctronics coatings abstracts*. Teddington, Middx., Finishing Publications, Ltd., 1986-. v. 28(1)-. 6 pa. £290pa. ISSN: 09505199.

As *Metal finishing abstracts*, v. 1-27, 1959-85. Previously appeared in *Electroplating and metal spraying* (1947-1959).

Over 7,000 indicative and informative abstracts and citations pa. covering surface treatment, in 20 classes (Cleaning and degreasing ... Electroplating ... Metal spraying and hard facing ... Corrosion and electrochemistry ... General and miscellaneous). Covers journal articles, books, standards, reports and patents - full coverage for UK, US, GDR and USSR patents.

Separate section for *Printed circuits and electronics coatings abstracts*. 4 sections (Printed circuit boards; semi conductors and integrated circuits; Electrical connectors; Miscellaneous). Over 2,000 abstracts pa. Separate annual indexes for each part, of authors, subjects and patent numbers. Abstracts appear 3-12 weeks after publication.
Class No: 621.793(048)

Yearbooks & Directories
[4355]

Finishing handbook and directory. Bean, J.E., *ed*. London, Sawell Publications. *c*.500 illus., diagrs., tables. ISSN: 00715182.

Includes technical chapters on electroplating and painting in addition to an encyclopedia covering all finishing subjects. Directory of manufacturers and services, with lists of trade names and associations. Reference tables and bibliography of books on finishing. (p. 236-8). *Class No:* 621.793(058)

[4356]

—The Anti-Corrosion handbook and directory. Bean, J.E., *ed*. 2nd ed. London, Sawell Publications, [1989]. 195p. illus., tables.

12 chapters on corrosion in various industries and of particular materials. 'Encyclopaedia', p.6-20; 'Reference tables', p.115-21. Directory of associations and buyers' guide. Handy format. *Class No:* 621.793(058)

[4357]

'Metal finishing' guide book - 1987 directory. 55th ed. Hackensack, N.J., Metals and Plastics Publications, 1987. Annual. 1038p. illus., diagrs., graphs and tables. $75 pa., including the monthly journal. ISSN: 00260576.

Published each January as the 13th issue of the volume. This ed. is v.85(7A), mid-January 1987. Comprises a comprehensive handbook of finishing, plating and treatments processes and a brief directory of the industry. Technical section (p.1-898) has 9 sections: (Mechanical surface preparation ... Electroplating solutions ... Surface treatments - Organic finishing ... Finishing plant engineering - Tables and data sheets). Directory contains classified buyers' guide for equipment, material and processes, alphabetical list of manufacturers, regional list of US distributors, and trade names index. Subject index. Many tables.
Class No: 621.793(058)

Tables & Data Books
[4358]

BENNINGHOFF, H. **Index of chemicals used for the treatment of metal surfaces.** [Chemikalien-Index für die Oberflächentechnik der Metalle.] Amsterdam, etc., Elsevier, 1974. xxiv, 1260 columns. diagrs., tables. ISBN: 0444410759.

1,034 chemicals and chemical products, named in accordance with IUPAC recommendations in German, English and French, with empirical and structural formulae, characteristics, solubility in water and acids, reactions and hazard classification and data. Indexes of chemicals in each language, including popular names, and of formulae.
Class No: 621.793(083)

Packaging

Databases
[4359]

Packaging Science and Technology Abstracts. Frankfurt-a-M., FRG, International Food Information Service.

Covers 1982 to date. A bibliographic database providing references, with abstracts, to the literature of all aspects of packaging, with emphasis on foodstuffs. Includes aerosols, coatings, filling, handling, storage, transport, labelling, law, materials, marketing, waste. Sources: periodical articles, books, conference papers, standards, reports, trade literature. About 28,000 records; updated 5 times a year. Available on DIALOG, file 252. Available on magnetic tape and CD-ROM. *Class No:* 621.798(003.4)

[4360]

PIRA Abstracts. Leatherhead, Surrey, PIRA, the Research Association for the Paper and Board, Printing and Packaging Industries.

Covers 1975 to date. A bibliographic database providing references, with abstracts, to the literature of paper and pulp manufacture, printing and packaging. Packaging materials, operations and equipment, products, markets and companies are covered. Sources: over 700 periodicals, books, reports, standards and conference papers. File size: 110,000 items (1986). Updated fortnightly. 10,000 additons pa. Available via INFOLINE. Cost per connect hour: $70. Printed version

....(contd.)

split between: *International packaging abstracts*; *Printing abstracts; Paper and board abstracts.*
Class No: 621.798(003.4)

Bibliographies

[4361]
UNITED NATIONS. Industrial Development Organization. **Information sources on the packaging industry.** New York, United Nations, 1977. xii, 110p. $4. (*UNIDO guides to information sources, no. 27.*)

858 entries. 12 sections: 1. Professional, trade and research organizations, learned societies and special information services - 2. Directories (88 listed) - 3. Sources of statistics, marketing and other economic data - 4. Basic handbooks, textbooks and manuals - 5. Monograph series - 6. Current periodicals (119 listed) - 7. Current abstracting and indexing periodicals - 8. Proceedings, papers and reports - 9. Specialized dictionaries and encyclopedias - 10. Bibliographies - 11. Films and film catalogues - 12. Other potential sources of information (*e.g.* G. Meetings and conferences; I. Standards and specifications). Bibliography of sources used. *Class No:* 621.798(01)

Handbooks & Manuals

[4362]
HANLON, J.F. Handbook of packaging engineering. 2nd ed. New York, etc., McGraw-Hill, 1984. viii, [536]p. illus., diagrs., graphs, tables. $74. ISBN: 0070259941.

First published 1971.

Aims 'to contain in a single volume all the pertinent data on packaging that is likely to be needed' (*Preface*). New packaging materials - *e.g.* stretch and metallized films - and developments in established practices - *e.g.* aerosol use, labelling and marking systems, etc. - have been included. 20 chapters (1. Elements of packaging - 2. Paper and paperboard - 3. Films and foils ... 8. Plastics ... 13. Closures, applicators, fasteners, and adhesives ... 19. Quality control - 20. Laws and regulations). Detailed coverage of films and plastics - Chapters 3 and 8 account for over a quarter of the book. Numerous illustrations and tables of data. 20-page index is thorough, but access if often indirect - *i.e.,* 'propellants' are traced via 'Aerosols'; 'polyethyene' is found via 'films' or 'plastics'.
Class No: 621.798(035)

Dictionaries

[4363]
BRITISH STANDARDS INSTITUTION. Glossary of packaging terms. London, BSI, 1973-. 8 pts. diagrs. (*BS 3130: 1973-1986.*)

pt. 1: *General.* 1990. 16p.
pt. 2: *Paper and board packaging.* 1990. 14p.
pt. 3: *Glass containers and closures.* 1974. 18p.
pt. 4: *Metal containers and aerosols.* 1976. 24p.
pt. 5: *Plastics and flexible packaging* (excluding paper). 1976. 12p.
pt. 6: *Wooden packaging.* 1981. 8p.
pt. 7: *Paper sacks* 1986. 16p.
pt. 8: *Plastic sacks.* 1987. 16p.

Over 1,000 terms are defined and illustrated in systematic order with alphabetical indexes. *Class No:* 621.798(038)

Reviews & Abstracts

[4364]
International packaging abstracts. Oxford, etc., Pergamon, 1981-. v.1-. Monthly. DM1,465. ISSN: 02607409.

Compiled by the Information Section of Pira (the Research Association for the Paper & Board, Printing & Packaging Industries). More than 4,000 informative abstracts pa., derived from over 250 periodicals, and from books, reports, standards, conference proceedings and newspapers. 8 sections (1. Packaging in general - 2. Materials - 3. Components - 4. Retail unit packages - 5. Transit packs - 6. Packaging for specific products - 7. Packaging operations and machinery - 8. Distribution), each subdivided (*e.g.* 3.1 Labels; 6.13 Beverages). Author and subject indexes in each issue. Mean delay time for abstracts less than 3 months. Also available online through INFOLINE.
Class No: 621.798(048)

[4365]
Packaging Science and Technology Abstracts. [Referatedienst Verpackung.] Frankfurt-am-Main, International Food Information Service, in cooperation with Fraunhofer-Institut für Lebensmittel technologie und Verpackung. 1982-. v.1-. 6pa. DM 500 pa. ISSN: 07223218.

About 3,200 informative abstracts pa., in English, for non-German originals, and in English and German for German originals. Covers scientific, technical, marketing and management aspects of packaging. Derived from worldwide literature: 400 periodicals, books, pamphlets, standards, specifications, patents, conference papers, research reports and trade literature. Each abstract liberally indexed with descriptors which are 'rotated' to provide multi-access subject indexes in each issue, in English and in German. Author index in each issue. Cumulated annual subject and author indexes. Classified in 25 sections covering 10 subject groups (Packaging economy - Packaging material ... Packaging machines ... Transport, storage). Available on magnetic tape. Online version: PSTA (*qv*).
Class No: 621.798(048)

Yearbooks & Directories

[4366]
Packaging industry directory, 1992. 16th ed. Tonbridge, Kent, Benn, 1992. Annual. vi,400p. £62. ISBN: 086382126x. ISSN: 02699834.

Previously entitled '*Packaging review' buyers' guide.*

Lists almost 2,300 manufacturers and suppliers. Classified buyers' guide in 5 parts (p.131-232); brand names index (p.233-53). Includes UK agents, consultants, trade organizations and institutes, packaging terms.
Class No: 621.798(058)

Containerization

[4367]
'Containerisation international' year book 1986. London, National Magazine Co., 1986. 684p. illus., tables. £87. ISBN: 0852233795. ISSN: 03057402.

A world directory in a fast-growing service area. Lists with details, 340 container ports, by continent; services (combined transport operators, road hauliers, etc.); equipment and trade names; repair companies; 4,000 container carrying vessels. General index and ships index.
Class No: 621.798.12

[4368]
Jane's containerisation directory, 1992-93. 25th ed. Coulsdon, Surrey, 1992. 520p. illus., diagrs., tables. £135. ISBN: 0710609930.

Grouped under headings: Ports and inland terminals - Container operators - Handling systems - Container Services - Container industry overview. *Class No:* 621.798.12

Machine Elements

[4369]
BÖGELSACK, G., *and others.* 'Terminology for the theory of machines and mechanisms'. In *Mechanism and machine theory,* v.18(16), 1983. p.379-408.

A draft produced by Commission A of the International Federation for the Theory of Machines and Mechanisms as the first step to the publication of a standard terminology. Over 800 terms are defined and presented in a classified arrangement - 4 major sections with 17 sub-sections: Structure of machines and mechanisms; Kinematics; Dynamics; Machine control and measurements. A-Z index facilitates quick access to any definition. *Class No:* 621.8

[4370]
Handbook of mechanical power drives. 3rd ed. Morden, Surrey, Trade & Technical Press, 1980. vi, 680p. illus., diagrs., graphs, charts, tables. £52. ISBN: 0854610855.

Previous ed. 1977. Five sections: 1. Systems and fundamentals - 2. Power transmission equipment (*e.g.* belts, bearings, drives, clutches, brakes, gears) - 3. Related mechanisms and systems (*e.g.* pulleys, springs, speed reducers, seals) - 4. List of relevant British Standards - 5. Buyers' guide (25p.). Index. Well illustrated and supported by data tables. *Class No:* 621.8

[4371]
Handbook of rotordynamics. Ehrich, F.F. New York, McGraw-Hill, 1992. v.p. illus., diagrs., graphs, tables. $59.95. ISBN: 0070193304.

4 chapters, closely subdivided, with references: 1. Vibration - 2. Analytical prediction of response - 3. Rotor balancing - 4. Performance verification. Useful practical examples. Well up to the standard of other McGraw-Hill *Handbooks. Class No:* 621.8

[4372]
PARMLEY, R.O., *ed.* **Mechanical components handbook.** New York, etc., McGraw-Hill, 1985. xii, [757]p. illus., diagrs., tables. £59.95. ISBN: 0070485143.

Many contributors. 14 chapters covering all types of component, *e.g.* gears, chains, belts, bearings, clutches, springs, fasteners and retaining and locking devices. Supported by hundreds of diagrams and tables of data. *Class No:* 621.8

[4373]
ROTHBART, H.A., *ed.* **Mechanical design and systems handbook.** 2nd ed. New York, etc., McGraw-Hill, 1985. xii, [1812]p. illus., diagrs., graphs, charts, tables. £96.50. ISBN: 0070540209.

Previous ed. 1964.

52 contributors. This comprehensive handbook for machine design and dynamic analysis of mechanical systems groups its 43 sections into 5 main groups: Mechanical engineering fundamentals - Systems analysis and synthesis - Mechanical design fundamentals - Mechanical fastener components - Power-control components and subsystems.

....(contd.)

The influence of the computer is pervasive and has necessitated the re-writing of almost half the sections. Pertinent data for fabrication, function, stability, reliability, performance, machine life, are in profusion. Extensive reference lists. Detailed index (62p).

'Contains a wealth of information for mechanical, electrical, civil and chemical engineers ... An extensive and comprehensive reference for any engineering office' (*Public works,* v.117, August 1986, p.20). *Class No:* 621.8

[4374]
SHIGLEY, J.E. *and* **MISCHKE, C.R.,** *eds.* **Standard handbook of machine design.** New York, etc., McGraw-Hill, 1986. xii, [1632]p. diagrs., graphs, charts, tables. £96. ISBN: 0070568928.

42 contributors cover the entire field of machine design. 34 sections deal with specific machine elements (springs, clutches, cams, etc.) and design techniques (*e.g.* load cycle analysis) whilst the remaining 13 cover factors such as wear, stress, etc., or measurement, numerical methods, etc. Almost 1,000 illustrations of various kinds, numerous tables, formulae and data. 25-page index of specific terms.

'New materials, components, processes and analytical tools are discussed, along with computer-aided design methods and other machine computational techniques' (*Mechanical engineering,* v. 108, August 1986, p.69). *Class No:* 621.8

Bearings & Shafts

[4375]
BRITISH STANDARDS INSTITUTION. Glossary of terms for rolling bearings. London, BSI, 1984. 44p. illus., diagrs. (*BS 6560: 1984.*)

About 350 terms and definitions in systematic order with illustrative diagrams linked by index numbers, also used in the A-Z index. *Class No:* 621.833.2

[4376]
—BRITISH STANDARDS INSTITUTION. Metrical spherical plain bearings. Part 6. Glossary of terms. London, BSI, 1983. 16p. £17.90. (*BS 5983: 1983.*)

About 60 terms defined and illustrated with A-Z index. *Class No:* 621.833.2

Gears & Clutches

[4377]
BRITISH STANDARDS INSTITUTION. Glossary for gears. London, BSI, 1976. 2 parts. diagrs. £43.90. (*BS 2519: 1976.*)

pt. 1: *Geometrical definitions.* 1976. 36p. pt. 2: *Notation.* 1976. 4p.

Over 300 terms defined and illustrated in systematic order. 85 symbols listed and identified. *Class No:* 621.833.3

[4378]
DRAGO, R.J. Fundamentals of gear design. Boston, Mass., etc., Butterworth, 1988. xiii,560p. illus., diagrs., graphs, tables. ISBN: 040990127x.

Intended as a 'general design reference guide' for the 'beginning or occasional' gear designer (*Preface,* p.xi). 11 chapters (with references and sometimes bibliography) in 4 parts: I. Introduction (history; types of gears; theory) - II. Fabrication and inspection - III. Failure and load capacity evaluation - IV. Lubrication. 3 appendices (gear geometry;

.... *(contd.)*

nomenclature; quality designations); index (15p.). Author is a practising and consulting engineer. Well-produced; extensively illustrated. 'An excellent general reference to the basics of gear design' (*Choice*, Feb. 1989, p.968).
Class No: 621.833.3

[4379]

HOUGHTON, P.S. **Gears:** spur, helical, bevel, internal, epicyclic and worm. 3rd ed. London, Technical Press, 1970. [v, 612]p. illus., diagrs., graphs, tables. ISBN: 0291394019.
First published 1952. Previous ed. 1961.

Now emphasizes practical problems of gear design and use. Includes chapters on lubrication, bearing loads on gear teeth, tooth modification and gear failure. 26 chapters (1. Types of gears; notation (and definitions) - 2. Spur gears ... 12. Gear materials ... 17. Converting to SI Metric units ... 21. Machining times ... 23. Inspection ... 26. Tables; bibliography; index). Lengthier chapters cover horse power ratings of various gear types. More than 200 tables.
Class No: 621.833.3

[4380]

—DUDLEY, D.W., *ed.* Gear handbook: the design, manufacture and application of gears. New York, etc., McGraw-Hill, 1962. [937]p. illus., diagrs., charts, tables. £34.45.

46 contributors. A practical handbook on gear elements and formulae, gear teeth design, gear calculations, loading and rating, gear cutting and forming, finishing and grinding, gear efficiency and lubrication. 24 chapters. Data tables. Chapter references to US standards. *Class No:* 621.833.3

[4381]

ORTHWEIN, W.C. **Clutches and brakes:** design and selection. New York Dekker, 1986. xii, 358p. illus., diagrs., graphs, tables. $55. (*Mechanical engineering series, no. 50.*) ISBN: 0824773934.

13 chapters cover various types of brake and clutch, friction materials, vibration and heat dissipation factors. Chapter 11 lists SAE standards, reproducing the text of 15. Aim is to derive relevant formulas, display examples of use and offer access to appropriate flow charts and computer programs.

'This reference work contains the formulas for the design and selection of a variety of brakes and clutches ... simplifies the design selection process by deriving lengthy formulas and accelerating iteration procedures' (*Mechanical engineering*, v.109, Oct. 1986, p.81). *Class No:* 621.833.3

[4382]

STOKES, A. **Gear data handbook:** design and calculations. Oxford, Butterworth-Heinemann, 1992. 290p. illus., diagrs., tables. £40; $95. ISBN: 0750611499.

Covers spur gears, helical gears, straight, spiral, zerol and hypoid gears and worm gears in a practical approach. 'A valuable reference book' (*Choice*, December 1992, p.651).
Class No: 621.833.3

Materials Handling

[4383]

BRITISH STANDARDS INSTITUTION. **Glossary of terms used in materials handling.** London, BSI, 1964-75. 9v. diagrs. £165. (*BS 3810: 1964-1975.*)

pt. 1: *Terms used in connection with pallets, stillages, hand and powered trucks.* 1964. 36p.

.... *(contd.)*

pt. 2: *Terms used in connection with conveyors and elevators (excluding pneumatic and hydraulic handling).* 1965. 28p.

pt. 3: *Terms used in connection with pneumatic and hydraulic handling.* 1967. 28p.

pt. 4: *Terms used in connection with cranes.* 1968. 28p.

pt. 5: *Terms used in connection with lifting tackle.* 1971. 36p.

pt. 6: *Terms used in connection with pulley blocks.* 1973. 20p.

pt. 7: *Terms used in connection with aerial ropeways and cableways.* 1973. 20p.

pt. 8: *Terms used in connection with lifts, lifting platforms and inclined haulage.* 1975. 12p.

pt. 9: *Terms used in connection with rail transport.* 1975. 12p.

600 items defined and illustrated. *Class No:* 621.86

[4384]

Materials handling handbook. Kulwiec, R.A., *ed.* 2nd ed. New York, etc., Wiley, 1984. xviii, 1,458p. illus., diagrs., graphs, tables. £81.25. ISBN: 0471097829.
First published 1958.

83 contributors. Sponsored by the American Society of Mechanical Engineers and the International Material Management Society. The development of computer-controlled automatic plant and warehouses has made this new edition imperative. The objective of the Joint Committee of the Societies was 'a useful ... working tool for users and designers of materials handling systems and equipment' (*Foreword*) with emphasis 'on presenting data that the working engineer can use to select, size or apply equipment and systems' (*Preface*). 5 Parts, 36 Chapters and 2 Appendices: 1. Introduction - 2. Unit materials handling - 3. Bulk materials handling - 4. Transportation interface - 5. Safety, environment and human factors. Appendix B lists sources of information. Index (p.1,437-58).

'Offers a good deal of problem-solving guidance ... It will prove to be a very useful reference source for engineers involved in unit-handling and/or bulk-handling operations' (*Chemical engineering*, v.93, March 3, 1986, p.134).
Class No: 621.86

Fasteners

[4385]

Industrial fasteners handbook. 3rd ed. Morden, Surrey, Trade & Technical Press, 1985. vi, 650p. illus., diagrs., tables. £57.50. ISBN: 0854610979.

Complete revision and updating of earlier editions to take account of new manufacturing techniques and new types of fasteners and materials. Attempts to provide detailed description and data for all known types, arranged logically in sections. 1. Review of mechanical fasteners (including design, performance, materials, manufacture) - 2. Bolts and machine screws - 3. Special duty fasteners (*e.g.* vibration-resistant, fasteners for plastics, furniture, buildings) - 4. Special purpose fasteners (*e.g.* rivets, plugs, pins, clips, staples) - 5. Tools used for fastening - 6. Data, standards, trade associations - 7. Buyers' guide (including trade names, classified index, manufacturers' addresses). Well illustrated, copious data tabulations and effective index. Advertisements.
Class No: 621.88

[4386]

—PARMLEY, R.O., *ed*. Standard handbook of fastening and joining. 2nd ed. New York, etc., McGraw-Hill, 1989. [672]p. illus., tables. £55. ISBN: 0070485224.

First published 1977.

15 sections: Threaded fasteners - description and standards - Standard pins ... Rope splicing and tying - Metrics and general data. The 2nd ed. contains over 50% new or revised material. For engineers, machinists and technicians in all fields. 880 illus. *Class No:* 621.88

[4387]

SIDDERS, P.A., *ed*. **Guide to world screw threads**. 1st American ed. New York, Industrial Press, 1969. 318p. diagrs., tables. $19.50. ISBN: 0831110929.

Previously published in England under the title: *Machinery's screw thread book* (20th ed. 1969. Machinery Publishing).

American ed. contains additional data, largely of US concern. 12 sections covering all types of screw threads - British Whitworth and non-Whitworth; Unified and American; American pipe threads; Continental series; Horological screws; Inspection; Nut, bolt and screw dimensions; Tapping sizes. Details of ISO metric screw threads where available. (Work on metrication was proceeding when book was published.) 235 tables of relevant data and dimensions. Brief glossary (English-French-German). *Class No:* 621.88

[4388]

—BRITISH STANDARDS INSTITUTION. Glossary of terms for cylindrical screw threads. London, BSI, 1984. 14p. diagrs. £17.90. (*BS 6528: 1984*.)

34 fundamental terms and definitions in English and French, with appendices listing Russian, German and Italian equivalents. *Class No:* 621.88

Lubrication

Databases

[4389]

TRIBO (Tribology Index). Berlin, BAM (*i.e.,* Bundestanstalt für Materialprüfung).

Covers 1972 to date. A bibliographic database, providing references to the literature of all aspects of tribology - friction, wear and lubrication of moving parts. Includes stressing conditions, materials, machine parts, biomechanics. Derived from journal articles, conference papers, standards, books, reports, theses. Citations in English (some titles in vernacular). File size: Over 60,000 items. Updated three times pa. *c.*7,000 additions pa. Hosts include FIZ Karlsruhe (formerly INKA), STN. Cost per connect hour: *c.*$60. Printed version: *Documentationtribology: Wear, friction and lubrication. Class No:* 621.89(003.4)

[4390]

—Documentationtribology: wear, friction and lubrication. Berlin, BAM (*i.e.,* Bundestantalt für Materialprüfung). $150.

Previously *Dokumentation Versch Leiss, Reibung und Schmierung*.

Printed version, TRIBO database. Classified arrangement - Sections, sub-sections, alphabetical by author. Indexes of personal, corporate authors, report numbers, standards numbers, subjects and conference titles.

Class No: 621.89(003.4)

Bibliographies

[4391]

FOX, M.F., *and others, comps*. **Bibliography on engine lubricating oil**. Aldershot, Hants., Gower Technical Press, 1987. xv, 223p. £45. ISBN: 0291397069.

More than 3,000 citations, without abstracts, to the literature of 20 countries, drawing on over 200 periodicals, on conference papers, research reports, books and standards, in 16 subject sections (A. Composition and consumption of lubricating oil ... D. Degradation of lubricating oils ... G. Lubricating oils for diesel engines ... P. Alternative bearings and dry lubricants) plus Q. Future trends and S. Standards (almost 300 from 9 countries are listed). Items arranged chronologically by year and A-Z by author, uniquely numbered. Language is indicated. Scope notes introduce each section. Cross-references are given. Journals list, with abbreviations (but not always consistent with those used in entries). Author index. Sadly, no subject index. *Class No:* 621.89(01)

Encyclopaedias

[4392]

KAJDAS, C., *and others, eds*. **Encyclopedia of tribology**. Amsterdam, Elsevier, 1990. viii,478p. £100. (*Tribology, v.15*.) ISBN: 0444884017.

About 1500 terms (Abrasion - Zirconium ...) are briefly described (*e.g.* 'detergent', *c.*1p.). Most entries carry references, 1,750 in all. *Class No:* 621.89(031)

Handbooks & Manuals

[4393]

CRC handbook of lubrication; (Theory and practice of tribology). Booser, E.R., *ed*. Boca Raton, Florida, CRC Press, 1984. 2v. illus., diagrs., graphs. $253. ISBN: 0849339014, v.1; 0849339022, v.2.

1. *Application and maintenance*. [xiv], 600p. 2. *Theory and design*. [xiv], 689p.

Sponsored by the American Society of Lubrication Engineers as successor to McGraw-Hill *Standard handbook of lubrication engineering* (1968).

80 contributors. V.1. gives approximately equal treatment to: applications in automobiles, aircraft and machine tools; other industries; maintenance. V.2. is similarly divided between: friction, wear and lubrication theory; lubricants and their properties and uses; design principles. Chapter bibliographies. *Class No:* 621.89(035)

[4394]

CROOK, S., *and* MORTIER, R.M., *eds*. **Chemistry and technology of lubricants**. Glasgow, Blackie Academic, 1991. 320p. illus. ISBN: 0216929210.

Designed for chemists and chemical engineers working with lubricants and as an information source for research and academic institutions. *Class No:* 621.89(035)

[4395]

KRAGELSKY, I.V. *and* ALISIN, V.V., *eds*. **Friction wear, lubrication, tribology handbook;** translated by F. Palkin and V. Palkin. Moscow, Mir Publishers; Oxford, Pergamon, 1981. 3v. diagrs., graphs, tables. ISBN: 0080275915.

Published originally as *Trenie, iznashivanie i smazka* (Moscow, Mashinostroenie Publishers, 1978).

This, the first tribology handbook to be published in the USSR, is a compilation of contributions by more than 50

....(contd.)

experts, 'conceived as an aid to design engineers in choosing materials and lubricants', and 'summarizes the results of ... research on the friction and wear of machine parts and gives data on lubricants and lube additives' (*Preface*). 'Excellent translation. ... a good appreciation of the present-day status of the subject within the USSR' (Barwell, F.T. 'A semi-centennial perspective of lubrication'. Paper C251/87 in *Tribology - friction, lubrication and wear: fifty years on*. London, Inst. Mech. E., 1987. v.1, p.13).
Class No: 621.89(035)

[4396]
NEALE, M.J., *ed.* **Tribology handbook.** London, Butterworths, 1973 (Rev. impression, 1975). [500]p. illus., diagrs., graphs, tables. £19.50. ISBN: 0408000821.

Sponsored by Ministry of Technology Committee on Tribology. 128 contributors. 6 sections (Component selection, design and performance - Lubrication - Properties of materials for tribological components and surfaces - Environmental factors - Failures and repair - Basic information). Helpful presentation ('... since the information is all in graphical or tabular form the reader can see instantly whether what he needs is included ...' *Editor's Preface*).

'... its treatment of some important topics is rather superficial' (F.T. Barwell. 'A semi-centennial perspective of lubrication'. Paper C251/87 in *Tribology - friction, lubrication and wear: fifty years on*. London, Inst. Mech. E. 1987. v.1, p.13). *Class No:* 621.89(035)

[4397]
PETERSON, M.B. *and* **WINER, W.O.,** *eds.* **Wear control handbook.** New York, American Society of Mechanical Engineers, 1980. 1358p. illus., diagrs., graphs, tables. $75.

34 contributors. This handbook, with its pioneering attempt to collect all wear data for machinery design and maintenance, was sponsored by the ASME's Research Committee on Lubrication. The content is presented in 36 chapters under 6 headings: 1. Introduction - 2. Wear fundamentals - 3. Materials - 4. Designing for wear - 5. Component wear - 6. Appendices. Section 5, Component wear, has 16 sections (*e.g.* Bushings; Rolling bearings; Gear wear; Cutting tool wear). Emphasis throughout is on wear, rather than lubrication and lubricants. Appendices contain useful 8-language glossary prepared for the OECD and covering 650 terms. Index (p.1323-58).
Class No: 621.89(035)

Reviews & Abstracts

[4398]
Tribos (Tribology abstracts). Cranfield, Beds., British Hydromechanics Research Association, 1968-. v.1-. £90pa. ISSN: 00412694.

Continues *Tribology* (quarterly).

About 2,000 informative abstracts pa. of items drawn from over 1,000 journals, plus reports, standards and conference proceedings. Main sections: Testing and instrumentation - Materials - Friction - Wear - Lubrication - Lubricants - Machine components - Machinery and other applications.

Issue and cumulative annual indexes of personal and corporate authors and of subjects. Available online, as part of FLUIDEX database. *Class No:* 621.89(048)

Histories

[4399]
DOWSON, D. History of tribology. London, etc., Longman, 1979. xvii, 677p. illus., diagrs., graphs, tables. £70. ISBN: 0582447664.

A comprehensive history of studies, concepts and devices associated with friction, lubrication, wear and bearings from prehistoric times to the present when the subject has acquired a specific term to isolate it: tribology. The treatment is chronological; chapters 1-8 cover the period before the Industrial Revolution; chapters 9-11 deal with the period since. Each chapter is subdivided by topics (*e.g.* Bearings), facilitating concentration on particular aspects of the subject. An appendix provides biographical sketches of 23 men important for their contributions, from Leonardo da Vinci onwards. Bibliography and references, p.617-660. Subject and name indexes. *Class No:* 621.89(091)

Machine Tools

Abbreviations & Symbols

[4400]
BRITISH STANDARDS INSTITUTION. Symbols for machine tools. London, BSI, 1971-83. 3 pts. £115. (*BS 3641: 1971-83*.)

pt. 1: *General symbols*. 1971 (1983). 36p.

pt. 2: *Specification for numerical control symbols*. 1980. 20p.

pt. 3: *Additional general symbols*. 1983. 30p.

Supplement: *Guide to application of additional symbols (and index) for machine tools*. 1980. 16p.
Class No: 621.9(003)

Bibliographies

[4401]
UNITED NATIONS. Industrial Development Organization. Information sources on the machine tool industry. New York, United Nations, 1976. xii, 71p. £2.60. (*UNIDO Guides to information sources, no. 22*.)

486 items. 11 sections (some annotated): 1: Professional, trade and research organizations, learned societies and special information services - 2: Directories - 3: Sources of statistics, marketing and other economic data - 4: Basic handbooks, textbooks and manuals - 5: Current periodicals - 6: Current abstracting and indexing periodicals - 7: Proceedings, papers and reports - 8: Specialized dictionaries and encyclopaedias - 9: Bibliographies - 10: Films, transparencies and film catalogues - 11: Other potential sources of information (*e.g.* meetings and conferences, patents, standards, translation services). Worldwide coverage of organizations and publications. Bibliography of sources used. *Class No:* 621.9(01)

Handbooks & Manuals

[4402]
JACKSON, A. *and* **DAY, D. The Complete book of tools.** London, M. Joseph, 1978. 352p. illus. ISBN: 0718117579.

Describes and illustrates over 600 different hand and power tools. Carefully drawn (often two-colour) illus.; well-captioned. 'Saws': p. 74-117; 'planes': p. 134-49. Glossary, p. 343-4. Detailed analytical index, p. 345-52.
Class No: 621.9(035)

[4403]

WECK, H. Handbook of machine tools; translated by H. Bibring. New York, etc., Wiley, 1984. 4v. illus., diagrs., graphs. £97.50. ISBN: 0471262269 set.

1. *Types of machines, forms of construction and applications.* xx, 285p.

2. *Construction and mathematical analysis.* xx, 296p.

3. *Automation and controls.* xx, 451p.

4. *Metrological analysis and performance tests.* xvi, 145p.

Originally published in German; derived from lectures at the Technical University in Aachen. Comprehensive treatment of all types of machine tools and processes, including developments in automation and control technology.

Lists of references in each volume, all predominantly German. *Class No:* 621.9(035)

Dictionaries

[4404]

CLASON, W.E. Elsevier's dictionary of tools and ironware; in English/American, French, Spanish, Italian, Dutch and German. Oxford, Elsevier, 1982. viii, 298p. £61.56. ISBN: 0444420851.

Over 2,500 terms in English, with equivalents and indexes in other languages. *Class No:* 621.9(038)

[4405]

GUTIÉRREZ, M.F., *comp.* Elsevier's dictionary of machine tools and elements: in three languages, English, German and Spanish. Amsterdam, Elsevier, 1990. [viii],298p. £95. ISBN: 0444886974.

About 5,500 English-base terms with equivalents and indexes in the other languages. Genders given.
Class No: 621.9(038)

[4406]

HEILER, T. Technisches Bildwörterbuch für spanende Werkzeuge zur 'Metallbearbeitung': Französisch, englisch, deutsch, italienisch, spanisch und einem Anhang in russisch, von Paul Hüter. 2. überarb. Aufl. Munich, Hanser, 1970. [xi], 580p. illus., diagrs.

First published 1965.

1,800 terms in French, with English, German, Italian, Spanish equivalents and brief descriptions, p. 1-409 (and Russian supplement, p. 475-580). Systematically arranged (A. Chisels - A1. General terms - A2. Types of chisels ... M. Grinding wheels). About 1,500 illus., and very helpful diagrs. of tools. Also cites the appropriate standard, *e.g.* BS or DIN. *Class No:* 621.9(038)

[4407]

JUNGE, H. D. Machine tools and mechanical engineering: Englisch/Deutsch, Deutsch/Englische. Weinheim, VCH, 1991. *c.*500p.

About 10,000 entries in each half. German genders given.
Class No: 621.9(038)

Tables & Data Books

[4408]

MACHINABILITY DATA CENTER. Machining data handbook. 3rd ed. Cincinnati, Ohio, Metcut Research Associates, 1980. 2v. $150. ISBN: 0936974001 Set.

First published 1966.

Aims to provide starting recommendations for the important material-removing operations for each of 61 work

.... (contd.)

materials, including certain non-metals, in both SI and 'English' (!) Units. Sections 1-8 cover conventional operations (*e.g.* turning, milling, drilling, tapping, planing, sawing); Sections 9-13 deal with non-traditional processes (mechanical, electrical, thermal, chemical). Other sections (in v.2) cover tool materials, tool geometry, cutting fluids, power requirements, etc.

Data for each specific task are easily located step-by-step to determine feed, speed, tool material and geometry, cutting fluid and power required for each process for each material for specified conditions. *Class No:* 621.9(083)

Standards

[4409]

BRITISH STANDARDS INSTITUTION. Machine tools: engineers' tools and cutting tools. Milton Keynes, BSI, 1985. 6p. £1. (*SL 20.*)

One of the Sectional List series, giving number, title and price group of 135 British standards, specifications, glossaries and published documents (PD) covering all types of tools and related topics, in 17 sections. (For full details and summaries, see the BSI annual *Catalogue* (*qv*)).
Class No: 621.9(083.74)

Histories

[4410]

ROLT, L.T.C. Tools for the job: a history of machine tools to 1950. Rev. ed. London, H.M. Stationery Office, 1986. 274p. illus., facsims., ports. £9.95. ISBN: 0112964335.

First published 1985. At head of title-page: 'Science Museum'.

11 sections: 1. Machines in the craftman's workshop - 2. Industrial machine-tools of the eighteenth century ... 5. Clement, Fox, Roberts, Nasmyth and Whitworth ... 8. American machine tools and their makers ... 10. Metal-cutting becomes a science - 11. Into the twentieth century. Bibliography, p.256-9. Detailed, analytical index, p.260-74.
Class No: 621.9(091)

Conferences

[4411]

Industrial Machine Tool Design and Research Conference. Proceedings. London, Macmillan, 1961 -. Annual. illus., diagrs., graphs.

The proceedings of an important series of highly-regarded annual conferences, originally held alternately in Birmingham and Manchester and published under the title *Advances in machine tool design and research* and edited by S.A. Tobias and F. Koenisberger. The Proceedings of the 26th International ... Conference, ... Manchester, 17th-18th September 1986, ed. by B.J. Davies (ISBN: 0333395344. xvii, 488p. £80.) contains 52 papers grouped under 10 headings (*e.g.* CADCAM integration; Computer aided techniques in design, manufacture and error measurement; Machine tool accuracy and control; Robotics; Flexible manufacturing systems; Machine tool design and performance ...). An excellent source for international developments in design and research with less emphasis on industrial practice. *Class No:* 621.9:061:061.3

Numerical Control

[4412]

AMKREUTZ, J.J. Wörterbuch numerische Steuerung und Werk-zeugmaschinen. [Dictionary of numeric control and machine tools. [Dictionnaire de commande numérique et machines-outils].] Cologne, Datakontext-Verlag, 1985. 254p; 251p; 251p. DM120. ISBN: 3921899397.

Three separately-paged vocabularies bound together: German-English-French; English-German-French; French-German-English. Each contains *c.*10,000 words with equivalents in other languages, without definitions. Produced by word-processing programs. *Class No:* 621.9-52

Cutting Tools

[4413]

BROOKES, J.K.A. World directory and handbook of hard metals. 3rd ed. London, Engineers' Digest, 1982. 372p. illus., diagrs., graphs, tables. (*Engineers' Digest publication.*) ISBN: 0950493120.

First published 1975. Previous ed. 1979.

Part 1: *Handbook* (p.9-118), comprising 12 chapters and an index (1. Historical; 2. Manufacture of sintered carbides; 4. Properties of sintered carbides; Design of carbide components; 9. Carbides in the machine shop; 10. Non-machinery applications; Hard metal specifications). A good lavishly illustrated guide to the technology. - Part 2: *Directory* (p.119-371), providing company and product information and technical data for a claimed 95% of the world's suppliers of carbide tools and other hard metal products. Chapter 15 has unique tabulations of applications data for the various grades manufactured. Indexes of companies, trade names and hard metal grades.

4th ed. 1987 (500p. £95. ISBN 0950899518), - 'now completely revised and greatly expanded, this invaluable work of reference to the hard metal industry should be of immense use to all who specify, purchase or use cutting tools or wear-resistant materials' (*Chartered mechanical engineer,* Feb. 1988, p.53). *Class No:* 621.9.02

[4414]

HIGGINBOTHAM, R., *ed.* **Jet cutting and cleaning bibliography.** London, etc., Elsevier, 1988. ix,258p. £58. ISBN: 185166226x.

About 600 items, all from the FLUIDEX (*qv*) database, covering the literature of water jetting from 1980-87. In 11 sections: 1. Fluid mechanics of jets ... 3. Equipment ... 7. Mining and tunnelling ... 10. Safety - 11. Bibliography. Most references are to periodical articles and conference papers. Personal and corporate author indexes. Effective editing of downloaded records with good use of bold type, but at this price an expensive alternative to online searching. *Class No:* 621.9.02

[4415]

STEKHOVEN, G.S. *and* **VALK, W.B.,** *comps.* **Elsevier's dictionary of metal-cutting tools** in seven languages: English/American - German - Dutch - French - Spanish - Italian - Russian. Amsterdam, etc., Elsevier, 1970. [viii], 458p. diagrs. Dfl.275. ISBN: 0444408568.

More than 1,900 terms listed, numbered and defined in English (some definitions extensive), with equivalents in other languages. Separate indexes in these languages keyed by numbers to English section. Brief bibliography and 12p. of figures to illustrate selected terms. *Chartered mechanical*

....(contd.)

engineer (v.18(6), June 1971, p.232) queries selection and notes inaccuracies and criticizes use of 'outdated British and metric units'. *Class No:* 621.9.02

Industrial Diamonds

[4416]

Industrial diamond review. Ascot, Berks., De Beers Industrial Diamond Division, 1965-. v.25-. 6 pa. £49 pa. ISSN: 00198145.

Formerly *Bibliography of diamond tools and related subjects* (1944-57); *Industrial diamond abstracts* (1958-1964). Part of trade journal.

About 1,100 indicative and informative abstracts and references to journal articles, plus book notices, patents lists, lists of other publications, films and videos. 19-20 sections (A. Properties of diamonds - C. Properties of ceramics - H. Truing and dressing - Q. Gem working - V. Non-traditional machining processes ...). *Class No:* 621.9.025.7

[4417]

SMITH, N.R. Users' guide to industrial diamonds. London, Hutchinson, Benham, 1974. 352p. illus., diagrs., graphs, tables. £4.50. ISBN: 0091170508.

11 chapters covering the nature of diamonds (varieties, structure and characteristics), methods of working, range of products and tool making. 8 chapters cover applications in the transport, electrical and electronics industries, glass, stone, building and civil engineering fields, mineral development, jewellery, and watchmaking. Citations of appropriate standards and patents, with useful chapter bibliographies. Data tabulations and index (p. 334-52).

'... Contains a wealth of up-to-date and explicit information in a compact form ... It is well balanced, clear, very readable and ... something of a model for a book of its type' (*Production engineer* v.54(12), Dec. 1975, p.648). *Class No:* 621.9.025.7

Drills

[4418]

MOUREAU, M. *and* **BRACE, G. Dictionnaire du forage et des puits:** anglais/français; français/anglais. [Dictionary of drilling and boreholes: English/French; French/English.] Paris, Éditions Technip, 1990. [xii],403p. ISBN: 2710805825.

About 4,000 entries, including abbreviations, in each half. *Class No:* 621.95

Sheet Metal Working

[4419]

Sheet metal industries year book. 14th ed. Redhill, Surrey, Sheet Metal Industries, 1986. Annual. 276p. £28. ISBN: 0861082125. ISSN: 03057798.

8 chapters (1. General information - addresses and memberships of relevant trade associations - 2. Agencies - 3. Tables of technical data for metals, alloys and processes - 4. Glossaries - derived from BS 1968 (now withdrawn by BSI) - 5. Conversion factors and tables - 6. Trade marks - trade names - 7. Buyers' guide: alphabetical and classified sections - 8. Stockholders of steel and other materials). Indexes of tables and advertisers. *Class No:* 621.98

Mining

Databases

[4420]

MinSys. Didcot, Oxon, Geosystems, 1980-. Updated weekly.

Online database covering both practical and economic aspects of mining. About 10,000 items added pa.

Stark, M.M. 'Databases', in his *Mining and mineral industries*, has no less than 48 annotated entries for online services, p.28-45. Of these 6 or 7 concern British systems - Coal database; GEOARCHIVE; Geomechanics Abstracts; HSELINE; IMMAGE and Minsys. *Class No:* 622(003.4)

Bibliographies

[4421]

STARK, M.M. Mining and mineral industries: an information sourcebook. Phoenix, Arizona, Oryx Press, 1988. [ix], 124p. $35.50. (*Oryx sourcebook series in business and management*.) ISBN: 0097742958.

558 numbered entries, in 12 form sections: Core library collection (general; journals) - Dictionaries and encyclopedias - Handbooks and manuals- Directories - Bibliographies and literature reviews - Indexes and abstracts - Databases (items 221-68) - Industry surveys - Conferences and symposia - Texts and treatises - Journals - Maps and Atlases. 'Associations and research centers'. Sources. Section title and subject indexes. 'The literature selected varies from introductory to extremely technical and scholarly' *(Introduction)*. *Class No:* 622(01)

[4422]

—KAPLAN, S.R., *ed.* A Guide to information sources in mining, minerals and geosciences. New York, etc., Interscience: Wiley, 1965. xiv, 599p. (*Guides to information sources in science and technology, v.2.*)

In two parts: 1. *Organisations* (1. International - 2. North America ... 5. Europe ... 9. Oceania), p.1-445 - 2. *Literature* (Geography; geology; geophysics and geochemistry ... Mineralogy - Mining and metallography ... Coal; gas, iron and steel; petroleum). Over 500 publications in pt.2, covering abstracts, dictionaries, directories, journals, etc. Index of geographical areas; index of literature; index of organizations. *Class No:* 622(01)

Handbooks & Manuals

[4423]

SME mining engineering handbook. Given, I.A., *ed.* New York, Society of Mining Engineers, 1973. 2v. [v.p.]. illus.,diagrs., graphs.

Supercedes R. Peele and J.A. Church's *Mining engineers' handbook* (3rd ed. New York, Wiley, 1941, reprinted 1970), not only updating on traditional underground and surface-mining practices, but introducing such subjects as surface mining, solution mining, Arctic mining, nuclear applications, as well as rock mechanics, fragmentation techniques, systems engineering, and the use of computers (*Foreword*).

V.1 has 16 sections, each with extensive references (12. 'Underground mining systems and equipment': 262p., 237 line drawings, over 9p. of references and bibliography). V.2 (sections 17-35), ranges from 17, 'Open pit and strip mining systems and equipment' to 27, 'Mineral processing', ending

....(contd.)

with 34, 'Engineering tables', 35, 'Sources of information' (34p.) 51-p. index to both volumes. Definitions index. About 250 contributors. A model of its kind. *Class No:* 622(035)

Dictionaries

Polyglot

[4424]

AMSTUTZ, G.C., *with others*. Glossary of mining geology: in English, Spanish, French and German. Stuttgart, Enke, 1971. xii, 196. illus., tables.

About 900 English-base terms and definitions, with Spanish, French and German translation of terms and definitions. Keyed illus. Cites authorities. 4 language indexes. 10 appendices (*e.g.* 4. Major part of a (coal) mine; 5. Geologic and mine map symbols). 50 figures. *Class No:* 622(038)=00

[4425]

ETNER, F.F., *and others*. Russo-anglo-nemetsko-frantsazki gorny slovar'. [Russian-English-German-French mining dictionary.] Moscow, Ruskiĭ Iazyk, 1987. 420p.

About 3000 Russian-based terms, with eqivalents and indexes in English, German and French. *Class No:* 622(038)=00

[4426]

WYLLIE, R.J.M., *and* **ARGALL, G.O., *eds*. World mining: glossary of mining, processing and geological terms.** English/Svenska/Deutsch/français/español. Translated by R. Lambert. Rev. ed. San Francisco, Calif., Miller Freeman Pubns., 1975. 432p. ISBN: 0879300310.

First published 1968 (G.O. Argall, ed.).

11,406 numbered entries. Glossary of English terms, p.9-270. Swedish, German, French and Spanish equivalents and indexes. No genders given. Index to advertisers. Covers 'terms mostly commonly used for geology, hydrography, exploration, underground and open pit mining, minerals benification, hydrometallurgy, smelting and refining' (*subtitle*). *Class No:* 622(038)=00

English

[4427]

BRITISH STANDARDS INSTITUTION. Glossary of mining terms. London, BSI, 1969-74. 11 sections. (*BS 3618.*)

1. *Planning and surveying*. Rev. ed. 1969. 12p. £10.20.

2. *Ventilation*.Rev. ed. 1971. 16p £11.

3. *Boring and ventilation*. 1971. 16p. £11.

4. *Drainage*. 1971. 12p. £10.20.

5. *Geology*. 1971. 20p. £11.

6. *Drilling and blasting*. 1972. 24p. £15.20.

7. *Electrical engineering and lighting*. 1973. 12p £15.20.

8. *Winning and working*. 1974. 32p. £23.90.

9. *Shifts and associated equipment*. 1974. 20p. £23.90.

10. *Transport*. 1967. 20p. £11.

11. *Strata control*. 1967. 16p. £11.

Primarily concerned with coal mining, although account has been taken of other forms of mining and quarrying in Britain. *Class No:* 622(038)=20

[4428]
NELSON, A. **Dictionary of mining.** London, Newnes, 1964. [vii], 523p. illus.

About 5,000 entries, some lengthy (*e.g.* 'Rockhurst': 9½p.). Includes eponyms; cross-references. Covers terms in applied geophysics, mechanics, and automation. 'Prepared for the convenience of mining men - the student, the official and the miner' (*Preface*). *Class No:* 622(038)=20

[4429]
THRUSH, P.W., *ed.* **A Dictionary of mining, minerals, and related terms.** Washington, US Dept. of the Interior, 1968. vii, [1], 1269p.

About 55,000 terms (including abbreviations), with *c.*150,000 definitions. Quotes sources (listed on p.1259-68). Cross-references. Compared with A.H. Fay's *A glossary of the mining and mineral industry* (Washington, USGPO, 1920. 745p.), which Thrush revises and enlarges. Fay (*c.*20,000 terms) should be retained for its several thousand Spanish American mining terms. *Class No:* 622(038)=20

German

[4430]
LENK, G. *and* BÖRNER, H. **Technisches Fachwörterbuch der Grundstoff - Industrien.** 2. Aufl. Göttingen, Vandenhoeck & Ruprecht, 1958-62. 2v. tables.

First published 1949-54. 1. *Englisch-Deutsch.* 1958. 638p. 2. *Deutsch-Englisch.* 1962. 732p.

Sub-title: 'Mining, non-metallic mineral industry; dressing; metallurgy; metal-working industry; building materials in industry'. Each volume has *c.*30,000 main entries, with compounds subsumed. V.2 carries a lengthy supplement, p.579-682, also available separately. Abbreviations are appended. US terminology is differentiated from British. Symbols, a feature. A strictly technical dictionary, giving the context of each term instead of - as in De Vries - a string of equivalents. *Class No:* 622(038)=30

French

[4431]
AUGER, P. *and* ROUSSEKU, L.J. **Lexique anglais-français de l'industrie minière.** Quebec, Office de la Langue Française, Government de Quebec, 1981. 89p. + 81, [1]p.

Two parts: 1. L'Exploration; 2. La minéralogie. Part 1 features 7,355 terms, English-French, with index of French terms. Part 2 has 1,066 terms, English French, with index of French terms. Each part has an appended 2-p. bibliography. *Class No:* 622(038)=40

Russian

[4432]
KOSMINSKII, B.M., *and others, comps.* **Anglo-russkii govnotekhnicheskii slovar'.** Moscow, Ugletekhizdat, 1958. 478p.

An English-Russian mining-technology dictionary with *c.*30,000 terms. Abbreviations, p.462-78; bibliography, p.6-7. More strictly technical than *Anglo-russkii gornii slovar'*, by Baron, L.I. and Erishov, M.N. (1958). *Class No:* 622(038)=82

[4433]
IMM **abstracts:** a survey of world literature on economic geology, mining, mineral dressing, extraction metallurgy [and] allied subjects. London, Institution of Mining and Metallurgy, 1950-. 6pa. £75pa. ISSN: 00190020.

About 3,000 abstracts per issue. 13 sections: Generalities - Mineral industry - Mathematical methods and computing - Physical and chemical studies - Analysis and instrumentation - Economic geology - Mining - Mineral processing - Metallurgy - Health and safety - Environment - General engineering - Civil engineering. No indexes. Microfiche and microfilm available. Online through IMM. *Class No:* 622(048)

[4434]
UNITED STATES. Bureau of Mines. **New publications of the Bureau of Mines.** Monthly list. Washington, U.S. Government Printing Office, 1910-. Monthly cumulated annually.

Four types of annotated entry: 1. Sales publications (information circulars) - 2. Free publications (Bulletin; Reports of investigators; Information circulars; Mineral commodity summaries; Reprints from *Mineral yearbook*; Mineral industry surveys) - 3. Open file reports; NTIS (Mineral land assessment reports etc.) - 4. Outside publications. Annual *List of Bureau of Mines publications and articles* started in 1961. *Class No:* 622(048)

[4435]
'Financial Times' **Mining international year book, 1992.** Harlow, Essex Longman Group, 1991. xxiii,572p. tables. ISBN: 0582067138.

Previously as *Walter Skinner's Mining international year book* (founded 1887 as *Mining Year book*).

Companies, A-Z (entries 1-728; data: business; properties and operations; latest ore reserves; financial summary)- Associations (729-819). Geographical, product indexes. Tables p.542-45. International professional sources. Suppliers' directory and business guide. Advertisers' index. Directory of over 700 companies worldwide.

World mines register: the worldwide directory of mining and mineral processing operations (San Francisco, Calif., Miller Freeman) has not been recorded since v.33: 1981/82. *Class No:* 622(058)

[4436]
Mining annual review, 1985. London, Mining Journal, 1985. Annual (1987. £32). 556p. illus., tables, maps. ISBN: 0900117397.

Main sections: Metals and minerals (Precious metals ... Gemstones, p.39-149) - Technical progress reports (Extractive metallurgy; Mineral exploration; Mineral processing; Surface mining; Underground mining) - Countries (*c.*150; by regions: North America ... Eastern Europe). Appended: Professional directory; Buyers guide; Index to mines and companies; Index to advertisers. *Class No:* 622(058)

[4437]

The Mining directory, mines and mining equipment companies worldwide. 5th ed. London, Don Nelson Pubns., 1990. 722p. £49. ISBN: 0946004021.

Main portion: A-Z listing of mines and mining equipment companies. Appended buyers' guide lists companies by subject. Geographical and company indexes. 5th ed. includes more companies and company specific data than did previous editions. *Class No:* 622(058)

Histories

[4438]

GREGORY, C.E. **A Concise history of mining.** Oxford, etc., Pergamon Press, 1980. xviii, 259p. illus., maps. ISBN: 0080238823.

5 parts: 1. What is mining? - 2. The eight ages of man (Paleolithic ... The uranium age, from 1950) - 3. Chronological development of particular aspects of mining (Mine drainage ... Working conditions) - 4. Traditions, customs and folklore of mines - 5. Epilogue (*e.g.* The impact of mineral production on national economies). Appendix: Significant dates in mining history. Glossary of mining terms. Bibliography (A-Z authors), p.245-50. Non-analytical index, p.251-8. A popular account, with 43 illus. *Class No:* 622(091)

Developing Countries

[4439]

BOSSON, R. *and* VARON, B. **The Mining industry and the developing countries.** London, Oxford Univ. Press, 1977. xii, 292p. illus., tables. £20; £9.95. ISBN: 0199200963.

Published for the World Bank.

Excludes fuel and major construction materials. Bibliography. 'A valuable examination of the world's mining industry, its structure, operations, objectives and the various outside influences that are likely to affect its development' (*Trade and industry,* 13 January 1978, p.52). *Class No:* 622(4/9-77)

Great Britain

[4440]

Directory of mines and quarries. London, British Geological Survey.

Records all known operational mineral workings in the UK. Data include name, location, map reference, commodity, company operating and basic statistics. *Class No:* 622(410)

Africa

[4441]

African mining, '91. London, New York, Elsevier, 1991. 346p. ISBN: 1851666540.

Many contributors; 11 sections, including mining, metallurgy, geology. Near-print. *Class No:* 622(6)

Thesauri

[4442]

INSTITUTION OF MINING AND METALLURGY. **IMMAGE Thesaurus.** London, The Institution, 1984. [178]p.

Terms for use with *IMM abstracts,* both print and online. *Class No:* 622:025.43

Conferences

[4443]

CHINA INSTITUTE OF MINING AND TECHNOLOGY. **Mining science and technology.** Proceedings of the International Symposium on Mining Technology and Science, September 1985. Beijing, People's Republic of China, China Coal Industry Publishing House, and Clausthal-Zellerfeld, FDR, Trans Tech. Pubns., 1987. [xviii], 1413p. ISBN: 0878490590.

Symposium papers in 7 sections: Underground mining engineering - 2 Surface mining engineering - 3. Mine ventilation and safety - 4. Mine construction - 5. Coal geoscience and mine surveying - 6. Mining machinery - 7. Mining electro-technology. No index, but detailed contents. A comprehensive survey. *Class No:* 622:061:061.3

Methods

Oil extraction

[4444]

HYNE, N.J. **Dictionary of petroleum exploration, drilling and production.** Tulsa, Oklahoma, Penn Well Books, 1990. [v],623p. illus. ISBN: 0878143521.

About 20,000 terms. Definitions range from a single word to one column (*e.g.* 'Sucker-rod pump'). Well-produced. *Class No:* 622.276

[4445]

Petroleum engineering handbook. Bradley, H.B., *and others, eds.* Richardson, Tx., Society of Petroleum Engineers, 1987, second printing 1989. xxiv,v.p. illus., diagrs., graphs, tables. ISBN: 1555630103.

First published 1962 as *Petroleum production handbook.*

59 chapters, closely subdivided, in 3 sections: 1. Mathematics (basic tables; calculation procedures - 1 chapter) - 2. Production engineering (materials; methods; tools - 18 chapters) - 3. Reservoir engineering (properties; recovery data and methods; reserves; formation evaluation; etc. - 40 chapters). Author index (14p.); detailed, analytical index (59p.). Numerous tables and illustrative material; extensive chapter references. Comprehensive and invaluable. *Class No:* 622.276

Offshore

[4446]

CHRYSSOSTOMIDIS, M. **Offshore petroleum engineering:** a bibliographic guide to publications and information sources. New York, Nichols Publishing Co.; London, Kogan Page, 1978. 367p. maps. ISBN: 089397045x.

About 2,500 references. Class A-X (*e.g.* E. Offshore/ Ocean engineering; F. Oceanography and ocean environment description; G. Offshore structure engineering; J. Offshore petroleum engineering and development; Q.

....(contd.)

Diving, underwater and deep ocean operations; T. Oil spills; W. Maps). Appendices include 'Supplementary sources & suggestions for keeping up to date'; 'Sources consulted'. Directory of publishers and other organizations. Permuted topic, author and titles indexes. *Class No:* 622.276.04

[4447]
—MACLACHLAN, M. An Introduction to marine drilling. Ledbury, Herefordshire, Oilfield Publications, 1987. 346p.

Includes an insert 'The North Sea: oil and gas activity and concession map'. *Class No:* 622.276.04

[4448]
—TAYLOR, G.E. 'Offshore engineering'. *Information sources in engineering,* (2nd ed. Editor L.J. Anthony. 1985), p.474-90. 2nd ed. London, Butterworths, 1985.

Includes 'Primary sources of documentary information' (p.480-6) and 'Secondary sources of information' (p.486-90: abstracting and indexing services; directories; books). *Class No:* 622.276.04

[4449]
Offshore abstracts. Hove, Sussex, Offshore information Literature, 1974-. 6pa. £110pa. ISSN: 03050513.

1987: 641 abstracts. V.15(1) January/February 1988: 17 headings, A-W: Arctic environment - Buoys and mooring systems - Cables - Corrosion - Drilling - Economic surveys - Environmental hygiene - Oceanology - Pipes and pipelines - Production - Structural engineering - Transport and plant hire - Underwater equipment - Vibration - Wave energy - Welding. Coverage includes conference papers, research reports, standards and patents. *Class No:* 622.276.04

[4450]
Offshore engineering abstracts. Cranfield, Beds., BHRA, The Fluid Engineering Centre, 1986-. 6pa.

125 abstracts per issue. Coverage: design, construction, operation and maintenance of offshore structures and associated equipment. Sources: articles, monographs, reports, conference proceedings, patents. Author and subject indexes per issue and annually. FLUIDEX database, available on DIALOG and ESA-IRS. *Class No:* 622.276.04

[4451]
Register of offshore units, submersibles & diving systems; 1987-88. London, Lloyd's Registry of Shipping, (1987). Annual. 659p. tables. ISSN: 01414143.

Sections: Mobile drilling rigs - Submersibles - Diving systems - List of support ships, barges, rigs - Work units (*e.g.* pipe handling) - Owners of offshore equipment - Owners and managers - Offshore support ships. Closely tabulated data. No index. *Class No:* 622.276.04

[4452]
WHITEHEAD, H. An A-Z of offshore oil & gas: an illustrated glossary and reference guide to the offshore oil & gas industries and their technology. 2nd ed. London, Kogan Page, 1983. 438p. illus., diagrs., tables, chemical structures. £27.50. ISBN: 0850386276.

First published 1976.

About 4,000 terms and abbreviations defined, plus cross-references. 30 appendices, p.315-426 (*e.g.* 1. Offshore oil and gas regions of the world; 29. Bibliography, p.433-4; 30. Abbreviations commonly used in relation to offshore oil and gas. Well illustrated (200 line-drawings, 20 full-page maps); keyed diagrams. 'An indispensable reference for engineers,

....(contd.)

economists and geoscientists concerned with petroleum resources' (*Choice,* v.21, October 1983, p.258). *Class No:* 622.276.04

Types of Mine

Oilfields

[4453]
Elsevier's oil and gas field dictionary, in English/American, French, Spanish, Italian, Dutch, German and Arabic. Chaballe, L.Y., *and others.* Amsterdam, etc., Elsevier, 1980. xii, 672p. $200; £128.12. ISBN: 0444418334.

4,843 English/American base-terms, with French, Spanish, Italian, Dutch and German equivalents and indexes. Arabic supplement. *Class No:* 622.323

[4454]
RÅDET FOR TEKNISK TERMINOLOGI. Norwegian-English-French dictionary of petroleum production: boring and production, geophysics and geology. Oslo, R.T.T., 1976. 170p.

Classified arrangement of *c.*1,550 terms in all, with Norwegian, English and French indexes. *Class No:* 622.323

[4455]
SOCIETY OF PROFESSIONAL WELL LOG ANALYSTS. Glossary of terms and expressions used in well logging. 2nd ed. Houston, Texas, the Society, 1984. 116p. illus., diagrs., graphs.

About 500 terms, p.7-104. 'Introduction log': 1 column. Sources, p.5-6. Appended: Nomenclature for well-logging services used by service organizations. *Class No:* 622.323

[4456]
STOHAROV, D.E., *ed.* **Russian-English oil-field dictionary.** New York, etc, Pergamon Press, 1982. 431p. $27. ISBN: 0080281699.

About 30,000 entries. Appendix of conversion tables. *Class No:* 622.323

[4457]
TIRATSOO, E.N. Oilfields of the world. 3rd ed., rev. Beaconsfield, Berks., Scientific Press, 1986. xvi, [416]p. illus., maps, tables. ISBN: 090136018x.

First published 1973; 3rd ed., 1983 (xvi, 392p.).

Extensive data on the geology, history and economic geography of *c.*2,500 oilfields in some 90 countries. The 3rd ed., rev., includes a 24p. *Supplement* which incorporates new developments and updated statistical information. Appendices cover properties of crude oil and world oil reserves. Chapter references. 183 tables, 83 line-drawings, 4 plates and 8p. of maps. *Class No:* 622.323

Coal Mines

[4458]
BRITISH STANDARDS INSTITUTION. Glossary of coal preparation terms. London, BSI, 1982. [ii], 28p. £40. (*BS 3552: 1982 (1989)*.) ISBN: 0580130878.

Previously as *Glossary of terms used in coal preparation.* (BS 3552: 1962).

About 400 terms defined in 11 sections (*e.g.* 4. sizing; 5. cleaning; 11. Automatic control). Appended: 'Publications containing additional relevant definitions'. *Class No:* 622.33

[4459]
Concise guide to world coalfields. Compiled by World Coal Resources and Reserves Data Bank Service, IEA Coal Research. London, IEA Coal Research, 1983 -. Loose-leaf. chart, graphs, maps.

4 Sections: 1. Contents and introduction - 2. Major coal-bearing countries (Afghanistan ... Zimbabwe) - 3. Indexes, classifications, and remaining coal-bearing country descriptions - 4. Glossary of terms. Appendix. For each major coal-bearing country an outline map is provided. *Class No:* 622.33

[4460]
Guide to the coalfields, 1989. Redhill, Surrey, Colliery Guardian, 1988. Annual. 528p. ISBN: 086108312x.

2 parts: 1. Government departments; Inspectors; Deep and surface coal mines; Coal preparation; Associations and societies, universities and technical colleges; Museums, etc., - 2. Buyer's guide; Suppliers, A-Z; Trade marks; Classified guide to 'Who supplies what'. Small format. *Class No:* 622.33

Gold & Silver Mines

[4461]
Silver: exploration, mining and treatment. Institution of Mining and Metallurgy. London, Institute of Mining and Metallurgy, 1988. x,344p. illus., maps, tables. £46.80. ISBN: 1870706048.

31 papers presented at the international conference on 'Silver-exploration, mining and treatment' organized by the IMM, in association with the Cámara Minera de México and The Silver Institute, and held in Mexico City, 21-24 November 1988. Sections: Geology and mineralogy - Case histories - Silver in complex sulphide minerals - Flotation - Extraction and refining. *Class No:* 622.342

Quarries

[4462]
INSTITUTION OF MINING AND METALLURGY. Surface mining and quarrying. London, the Institution, 1983. [vii], 449p. illus., diagrs., graphs, tables. ISBN: 0900438662.

38 documented papers presented at the 2nd International Surface Mining and Quarrying symposium organized by IMM in association with the Institute of Quarrying, held in Bristol, 4-6 October 1983. 'Developments in remote sensing for mineral exploration' has 17 references. *Class No:* 622.35

Drainage

[4463]
Mine drainage bibliography, 1910-1976. Gleason, V.E. *and* Russell, H.H., *comps.* Washington, United States Environmental Protection Agency; Pennsylvania Dept. of Environmental Resources, 1976. vii, 288p.

About 1,200 indicative and informative abstracts, arranged chronologically. Author and analytical subject index, p.238-88. *Class No:* 622.5

Mineral Dressing

[4464]
TAGGART, A.F. Handbook of mineral dressing: ores and industrial minerals. New York, Wiley; London, Chapman and Hall, 1945. [1915]p. tables. ISBN: 0471843482.

2nd ed. by A.G. Taggart. 1964.

With 13 contributors. 22 sections, on crushing, grinding, screen sizing, clarification with water, washing and scrubbing, concentration, dewatering, filtration, drying, storage and mill storage, sampling and testing, design and construction of ore-treatment plants, mathematics. Tables. 47-p. index. Some sections carry bibliographies.

V.2, to deal with preparation of fuels and the methods whereby metalliferous and non-metallic concentrates are rendered into primary consumer products, was not published. *Class No:* 622.7

Safety

[4465]
RLSD bibliography. Sheffield, Health and Safety Executive, Library and Information Services, 1986-. Annual.

Continues *SMRE* [Safety in Mines Research Establishment] *bibliography* (3rd ed. 1968), *SMRE/HSL* bibliography, 1969-78, and HSL bibliographies, 1979, 1980, 1981/82, 1983, 1984, 1985.

1986 annual contains *c.*70 abstracts, headed 'List of papers arranged alphabetically by name of author'. Cross-references from joint authors. No subject index.'RLSD': Research and Laboratory Services Division of the Health and Safety Executive.

HSELINE online database (1981-) indexes *RLSD bibliography* and all other Health and Safety Executive publications, updating monthly with 1000 items. Hosts: ESA-IRS, PERGAMON INFOLINE. *Class No:* 622.8

Military & Naval Engineering

Abbreviations & Symbols

[4466]
PRETZ, B. Dictionary of military and technological abbreviations and acronyms. London, Routledge & Kegan Paul, 1983. [iv], 496p. £60. ISBN: 0710092741.

Records *c.*50,000 acronyms and abbreviations used by the British, American, German and Soviet military. Bibliography, p.495-6. *Class No:* 623(003)

Bibliographies

[4467]
Jane's index of indices, 1991-92. 2nd ed. Coulsdon, Surrey, 1992. *c.*300p. £135. ISBN: 0710610343.

Covers 21 1991-92 Jane's yearbooks. Entries A-Z, equipment and manufacturers. Indexes: country; *Jane's Defence Weekly; Jane's Intelligence Review.* An index to many of the Jane's publications. *Class No:* 623(01)

[4468]

UNITED STATES. Military Academy. West Point, N.Y. Library. **Subject catalog of the military art and science collection,** with selected author and added entries, including a preliminary guide to the manuscript collection. Westport, Conn., Greenwood Press, 1970. 4v.

A dictionary catalogue with *c.*66,000 photolithographed cards. 'Catalog of one of the most comprehensive collections on military engineering' (Chen, C-C. *Scientific and technical information sources.* 1977, p.8). *Class No:* 623(01)

Handbooks & Manuals

[4469]

CHANT, C. A Compendium of armaments and military hardware. London, Routledge & Kegan Paul, 1987. 568p. £45. ISBN: 0710207204.

4 Sections: 1. Land weapons - 2. Warships - 3. Aircraft - 4. Missiles. Data: type, dimensions, weight, propulsion, performance. No illus. *Class No:* 623(035)

Dictionaries

Polyglot

[4470]

Brassey's Multilingual military dictionary. London, etc., Brassey's Defence Publishers, 1987. xvii, 815p. £48. ISBN: 0080270328.

6,960 entries in American English (with British English index), plus equivalents and indexes in French, Spanish, German, Russian and Arabic. 6 appendices: A. Ranks; B. Units/formations; C. Numerals/notation; D. Colors; E. Points of the compass; F. Tools, etc. Cross references. Fills a need. *Class No:* 623(038)=00

English

[4471]

UNITED STATES. Joint Chiefs of Staff. **Department of Defense Dictionary of military and associated terms,** incorporating the NATO and IADB dictionaries. Washington, Government Printing Office, 1984. 403p.

Initially as US Department of Defense. *A dictionary of military terms* (1963).

Defines *c*5,000 standardized military terms, including abbreviations. *Class No:* 623(038)=20

German

[4472]

KROLLMAN, F. Langenscheidt's Fachwörterbuch Wehrwesen: Englisch-Deutsch; Deutsch-Englisch. Berlin-Schönberg, Langenscheidt; London, Methuen, 1957. 769p.

More than 46,000 military terms and equivalents, with skilful use of symbols and abbreviations. Extensive section of 1,500 British and US abbreviations. Terms on structure and equipment of modern armed forces are well represented. Appendices include table of interrelation of units of measurement. The editor worked in the German Ministry of Defence's Language Section. More useful for purely military (as opposed to colloquial) terms than Eitzen, K.H. *German-English, English-German military dictionary* (1957. 549p.). *Class No:* 623(038)=30

French

[4473]

CHABALLE, J.H. *and* **DAVIAUT, P. Military dictionary, English-French, French-English.** Ottawa, Cloutier, 1943. xix, 1016p.

Prepared under the direction of the Chief of General Staff, Department of National Defence, Canada.

Includes many aeronautical, radar and other terms. Appendices: Abbreviations (p.951-68) - Army trades - Grammar - Conversion tables, etc. Bibliography, p.xvii. *Class No:* 623(038)=40

[4474]

DOBENIK, R.H. *and* **HARTLINE, G.W. Dictionnaire technique de la marine:** Anglais/Français et Français/Anglais. [Naval technical dictionary: English/French and French/English.] Paris, La Maison du Dictionnaire, 1989. [xxii],646,[19]p. illus. ISBN: 2856080316.

About 35,000 entries in each half. Includes nautical, military, technical terms. Brief bibliography (28 items). *Class No:* 623(038)=40

Spanish

[4475]

WELLS, R.A. English-Spanish, Spanish-English modern dictionary of military technology. Fairfax, Va., Lexicon Press, 1977. viii, 323p.

Entries emphasize technology. *Class No:* 623(038)=60

Russian

[4476]

JOINT TECHNICAL LANGUAGE SERVICE, CHELTENHAM. Russian-English, English-Russian military dictionary. London H.M. Stationery Office, 1983. x, [1], 688p. ISBN: 0112300189.

About 30,000 entries in each half. About 200 categories are applied to terms. Many compounds. Appended list of abbreviations. The dictionary contains vocabulary encountered in a wide variety of modern Russian texts covering military, technical and general subjects. *Class No:* 623(038)=82

[4477]

SUDZILOVSKII, G.A., *and others.* **Anglo-russkii voennyi slovar'.** [English-Russian military dictionary.] Izd.2.,perer. i dop. Moscow, Voennoe Izdat, Ministerstva Oborony SSSR, 1968. 1060, [3]p.

First published 1960.

About 59,000 entry-words ('Gun': 4 columns; 'Launching': 1¼ columns). Abbreviations (mostly American), p.917 - 1004. Terms are categorized. *Class No:* 623(038)=82

[4478]

—**NADYSEV, M.A.** Russko-angliiskii voehnotekhicheskii slovar' [Russian-English military and technical dictionary.] Moscow, Voenore Izdat, 1975. 622p.

Over 30,000 terms, abbreviations and symbols. *Class No:* 623(038)=82

Arabic

[4479]

KAY, E. **Arabic military dictionary: English-Arabic, Arabic-English.** London, Routledge & Kegan Paul, 1986. 320p. £30. ISBN: 0710204582.

About 10,000 terms, abbreviations and symbols.
Class No: 623(038)=927

Japanese

[4480]

CRESWELL, H.T., *and others*. **A Dictionary of military terms, English-Japanese, Japanese-English.** Tokyo, Kaitakusha, [1973]. iv, 1226p.

American edition: Univ. of Chicago Press, [1973].

Japanese-English part (p.513-1226) has *c.*17,500 entries and sub-entries. Headwords are romanized, followed by Japanese characters. Many compounds and idioms.
Class No: 623(038)=956

Reviews & Abstracts

[4481]

Air University Library index to military periodicals. Alabama, Maxwell Air Force Base, Air University Library, 1949-. Quarterly. With annual cumulative issues.

Originally as *Air University periodical index.*

V.37(4), cumulative issue, January/December 1986 (vii,644p.) has *c.*15,000 entries under subjects A-Z, with cross-references. It forms 'a subject index to significant articles, news items and editorials appearing in 78 English language military and aeronautical periodicals' (*Preface*).
Class No: 623(048)

[4482]

Current military literature. Oxford, The Military Press, 1983-. 6pa. £60pa. ISSN: 02041674.

Section J, *Technology* (323 abstracts and references in the v.3(6), 1985 issue) has headings: Missiles & rockets; Armoured vehicles; Artillery; Infantry; Engineer; Electronics, computers; Logistic; Ammunition, mines & explosives; Surveillance, intelligence, optronics; Nuclear, biological & chemical (NBC) warfare; Power, fuels & engines; Naval; Aviation; Navigation; Internal Security; Air defence; Directed enemy weapons; Space; Environment.
Class No: 623(048)

[4483]

Military science index. Shrivenham, Wilts., Royal Military College of Science, 1962-. 12pa., with annual cumulation.

1985: *c.*2,700 references to periodical articles, with brief notes on article titles where considered necessary. Arranged under topics, alphabetically, from 'Aerodynamics' to 'Weather'. Periodicals are chiefly British and American.

Defence news, a daily newspaper current-awareness service, reproducing cuttings, is circulated by the library as back-up. *Class No:* 623(048)

[4484]

—GREAT BRITAIN. Ministry of Defence. HQ Library Services. **MOD Library accessions: new books and articles.** London, Ministry of Defence Whitehall Library. Weekly. Limited circulation.

A current awareness service; Issue 24 of 1987 carries 176 entries under *c.*80 subject headings, 'Aeronautics and aeronautical engineeing' ... 'Weapon systems and equipment'. Concise annotations are given, where

....(contd.)

considered necessary. Ministry of Defence Library locations and UDC classmarks appended. Differs from *Military science index* in including books as well as articles.
Class No: 623(048)

Periodicals

[4485]

IMPERIAL WAR MUSEUM. Department of Printed Books. **List of current journals, corrected to August 1980.** London, the Museum, 1981. [ii], 49p. £1.

About 400 titles, A-Z, noting subtitles, publishers, addresses and frequency. Appended 8 subject groups: 1. General and military history, politics and international affairs - 2. Army and joint service publications - 3. Naval - 4. Aviation - 5. Ex-servioce - 6. Heraldry and collecting - 7. Weaponry and modelling - 8. Museum and professional.
Class No: 623(051)

Museums

[4486]

UNITED STATES. Department of the Army. Center of Military History. **Guide to US army museums and historic sites.** Cary, N.M., *Jr.* Washington, US Government Printing Office, 1975. vii, 116p. illus., maps.

7 parts: 1. US army and army national guard museums - 2. Other Dept. of Defense museums - 3. Federal museums and historic areas with military connections (*e.g.* Smithsonian Institution) - 4. Private, state and municipal military museums - 5. Historic sites on US army property - 6. Sources of information on military history - 7. Bibliography. 2 appendices (2. US army museum locations). Local maps. 'This volume is concerned with artifacts and landmarks' (*Foreword*). *Class No:* 623:061:069

[4487]

WISE, T. **A Guide to military museums.** 6th ed. Town Moor, Doncaster, S.Yorks., Athena Books, 1988. [80]p. £1.95, illus.

First published 1969.

Over 200 entries, under place-names, Abingdon ... York, Aberdeen ... Stirling, Brecon ... Tenby, Armagh ... Enniskellen, Gorey ... St. Peter's (Jersey), plus 'Other places of interest'. Data: description; admission; hours; facilities; approaches. Index of regiments & their museums; index of special museums (name, subject). Small illus. of regimental insignia. *Class No:* 623:061:069

Forts

[4488]

HOGG, I.V. **Fortress: a history of military defence.** London, Macdonald and Jane's, 1975. 160p. illus., facsims., plans, maps. ISBN: 0356081223.

7 chapters: 1. Ramparts and palisades - 2. The great leveller - 3. The entry of the masters (*e.g.* Vauban) - 4. The drawing board approach - 5. The years of plenty (17th-18th centuries) - 6. The testing time (19th century) - 7. The ferro-concrete revival (*e.g.* Maginot Line). Glossary, p.155-8. Index, p.159-60. No bibliography. Very fully illustrated quarto. *Class No:* 623.1

[4489]
ROBERTS, R.B. **Encyclopedia of historic forts;** the military, pioneer and trading posts of the United States. London, Macmillan, 1987. 894p. $95. ISBN: 002926880x.

Entries A-Z by state and then by post name within each state. Limited to military posts, omitting Navy and Air Force establishments (with some exceptions). Data include dates and construction details, history and fate. Prelims. include a glossary of fortification terms. 'This is the first comprehensive effort to treat this subject for the entire country' (*Choice*, v.25(7), March 1988, p.1070).
Class No: 623.1

Weapons & Arms

Encyclopaedias

[4490]
CHANT, C. **World encyclopedia of modern air weapons.** Wellingborough, P. Stephens, 1989. 312p. illus. £19.95; $29.95. ISBN: 1852600284.

A companion to his *World encyclopedia of modern aircraft armament* (1988). Includes a section on 'variants and applications'. Covers air-to-air surface, bombs, guns, etc., with technical data. Good captions and good black-and-white illus. Inferior to Gunston's *Illustrated encyclopedia of aircraft armament* in this respect, but superior in its background information. The full index gives official and common names, weapons by manufacturer.
Class No: 623.4(031)

[4491]
HOGG, I.V. **The Encyclopedia of weaponry.** London, Quarto, 1992. 224p. illus. (col.). £19.95. ISBN: 0851125212.

Also known as *The Guinness Encyclopedia of weaponry.*
8 chapters, closely subdivided: 1. The first weapons - 2. Medieval developments - 3. The age of gunpower - 4. Mobile warfare - 5. The age of invention - 6. The First World War - 7. The Second World War - 8. Today and tomorrow. Detailed, analytical index, p.216-22. Lavishly illustrated; a popular treatment. *Class No:* 623.4(031)

[4492]
TARASSUK, L. *and* BALAIR, C., *eds*. **The Complete encyclopedia of arms and weapons.** 1st UK ed., translated from the Italian by S. Mulcahy and others. London, Batsford, 1982. 544p. illus. (incl. col.). £29.95. ISBN: 0713415959.

Originally Mondadori (Milan), 1979, as *Enciclopedia ragionata delle armi.*
9 contributions, from 6 countries. 'The guiding principle behind this encyclopedia has been to provide extensive information on offensive and defensive weapons and armour over a wide period of history and a vast range of countries' (*Preface*). Entries A-Z. That on bayonets (p.78-79) describes 19 types of bayonet, with keyed line-drawings. Cross-references. 'Select bibliography' (grouped) p.534-44. The 1,250 illus. include 90 colour plates. 'Incomparable' (*Wilson library bulletin*, March 1983, p.606). *Class No:* 623.4(031)

Handbooks & Manuals

[4493]
FRIEDMAN, R.S. **An Illustrated examination of advanced technology warfare:** detailed study of the latest weapons and techniques for warfare today and into the 21st century. London, Salamander Books, Ltd., 1985. 208p. col. illus., maps. ISBN: 0861012550.

6 British and American contributors. 7 sections: 1. Electronic warfare - 2. Straegic warfare -3. Space warfare - 4. Air warfare - 5. Land warfare - 6. Naval warfare - 7. Conventional warfare (*e.g.* anti-terrorism weapons). Index (in tiny print), p.206-8. Quarto sized. *Class No:* 623.4(035)

[4494]
HARTCUP, G. **The Silent revolution:** development of conventional weapons, 1945-1985. London, Brassey's Publications Ltd., 1992. 192p. £26. ISBN: 008036702x.

Surveys weapons against the background of rational strategic priorities. Includes personal accounts and views of officers, scientists and officials. *Class No:* 623.4(035)

[4495]
Jane's land-based air defence: 1992-93. 5th ed. Coulsdon, Surrey, Jane's Information Group, 1992. *c.*300p. illus. £135. ISBN: 0710609795.

Covers SAMs and anti-aircraft guns, as well as installation. Data: development history, description, variants, specification, manufacturer details. Entries illustrated. Available on CD-ROM. *Class No:* 623.4(035)

[4496]
Jane's weapon systems, 1988-89. Blake, B. 19th ed. London, Jane's Publishing Co., 1988. Annual. [51], 1008p. illus., diagrms., tables. ISBN: 0710608551.

First published 1969.
3 main sections (p.34-904), subdivided by country: Strategic weapons and systems (land based) - Shipborne weapons and systems - Airborne weapons and systems. Glossary precedes. Analysis (tables, p.906-57): missiles; radar; sonar equipment; torpedoes; electronic warfare equipment; NATO designations of Soviet systems and equipment. Numerical list of entries. Index, p.995-1008. Online via DIALOG. *Class No:* 623.4(035)

[4497]
RHEINMETALL. **Handbook on weaponry.** 1st English ed. Dusseldorf, Rheinmetall GmbH, 1982. xxxv, 752p. illus., diagrs., graphs, tables.

Translation of *Waffentechnisches Taschenbuch* (5th ed.) 1980).
15 sections by various hands. 1. Explosives - 2/4. Ballistics - 5. The application of probability theory - 6. Sighting and aiming - 7. Automatic weapons - 8. Guns (p.215-412) - 9. Gun mechanics - 10. Gun and gun turret test rigs - 11. Ammunition - 12. Rockets - 13. Fuzes and propelling charge igniters - 14. Ballistics and weapon testing methods - 15. Tables (24). Detailed, analytical subject index, p.731-52. Small format and illus. *Class No:* 623.4(035)

Dictionaries

[4498]
QUICK, J. **Dictionary of weapons and military terms.** New York, etc., McGraw-Hill, 1973. xii, 515p.

About 15,000 terms are briefly defined in 10-100 words, with many cross-references. 'Cover': 6 meanings. Includes

....*(contd.)*

abbreviations, acronyms, code words and slang. Tables list weapon types. Over 1,000 illus. US slanted. *Class No:* 623.4(038)

Histories

[4499]

DIAGRAM GROUP. Weapons: an international encyclopedia, from 5000 BC to 2000 AD. Harding, D., *ed*. Rev. ed. London, Macmillan, 1990. 336p. illus., facsims., diagrs. £12.95. ISBN: 0333540069.

First published 1980.

Progressive survey in 7 sections. 1. Arming the hand (clubs, daggers, swords, bayonets, etc.) - 2. Hand-thrown missiles - 3. Hand held missile throwers - 4. Mounted missile throwers - 5. Positioned weapons - 6. Bombs and self-propelled missiles - 7. Chemical, nuclear and biological weapons. New developments in weaponry. Regional and historical indexes. Famous names; grouped bibliography; analytical index. Over 2,500 scaled drawings, well captioned. *Class No:* 623.4(091)

Underwater

[4500]

Jane's underwater warfare systems. Watts, A.J., *ed*. Coulsdon, Surrey, Jane's Information Group, 1992. 320p. tables. £135. ISBN: 0710609841.

Data on submarine and marine warfare, weapons and their launch systems, etc. *Class No:* 623.4(204)

Scotland

[4501]

CALDWELL, D.H. Scottish weapons and fortifications. Edinburgh, D. Donald Publishers, Ltd., 1981. x, 452p. illus., facsims., ports. ISBN: 0859760472.

14 contributors. 17 chapters (3. 'Late medieval defences in Scotland', p.21-54; 7. 'The early basket-hilt in Britain', p.153-252, with 167 references and notes; 9. 'The birth of the Scottish pistol', p.315-38; 11. The word *claymore*, p.378-87, with 27 references and notes. Detailed analytical index, p.442-52. 236 illus. in all. Not 'confined to matters primarily relating to Scotland' (*Preface*). *Class No:* 623.4(411)

USSR

[4502]

BONDS, R., *ed*. **The Illustrated directory of modern Soviet weapons.** London, Salamander Books, Ltd., 1986. 480p. illus. (incl. col.), maps. ISBN: 0861012526.

3 sections: 1. Land weapons and equipment (p.6-159) - 2. Naval weapons and equipment (p.161-319) - 3. Air weapons and equipment (p.321-480). No bibliography or index, but contents list is detailed. *Class No:* 623.4(47)

Equipment & Instruments

[4503]

British Army Equipment Exhibition, 22-27 June 1986, Aldershot. London, Defence Export Services Organisation, [1986]. cx, 272p. illus.

Catalogue of exhibitors' entries (1 per page), 'Aardvark

....*(contd.)*

Clear Mines, Ltd. ...' 'Zengrange, Ltd.'. Exhibitors index; Product index (in English, French, Spanish, Chinese Arabic, p.ix-cx). Well illustrated.

British Defence Equipment catalouge, 1987, 1988 (17th ed. 1987). comprises 3v. V.1/2 contain descriptions of the products and services offered by all participants, in 59 sections. V.3, Products and series index consists of company directory, trade names directory and abbreviations index. *Class No:* 623.4:002.50

Artillery & Rockets

[4504]

HOGG, I.V. The Illustrated encyclopedia of artillery: an A-Z guide to artillery techniques and equipment throughout the world. London, Stanley Paul, 1987. 256p. illus. (incl. col.), tables. ISBN: 0091726549.

'A short history of artillery from 1326 to the present' precedes an A-Z guide to artillery, with tabulated or boxed information. 14 nationalities: American/US, Austrian, Belgian, British, Chinese, Czech, French, German, Israeli, Italian, Japanese, Russian, South African, Swiss. *Class No:* 623.41

[4505]

Jane's armour and artillery, 1992-93. Foss, C.F. 8th ed. London, Jane's Publishing Co., 1987. Annual. *c*.800p. illus., diagrs., tables. £135. ISBN: 0710609973.

Technical data on: MBTS - Reconnaissance vehicles - Armoured personnel carriers - Armoured fighting vehicle families. Tank destroyers - Self propelled guns and howitzers - Self-propelled anti-aircraft guns and surface-to-surface missiles - Towed anti-tank guns - Towed guns - Multiple rocket launchers - Coastal artillery guns and missiles. Addenda. Armour and artillery in service. Index; 1-3 illus. per page. *Class No:* 623.41

Armoured Vehicles

[4506]

Jane's military vehicles & logistics, 1992-93. 13th ed. Coulsdon, Surrey, Jane's Information Group, 1992. *c*.700p. illus. £135. ISBN: 0710609817.

Formally *Jane's Military vehicles and ground support equipment* (8th ed., 1987).

Sections: Armoured engineer vehicles - Recovery vehicles and equipment - Bridging systems - Mine warfare equipment - Transport equipment - Miscellaneous equipment (*e.g.* Rapid runway repair equipment/portable runways). Fully illustrated; technical data, a feature. Glossary precedes. Index. *Class No:* 623.42

[4507]

—**VANDERVEEN, B.** World directory of modern military vehicles: unarmoured vehicles from 1970. London, Warne, 1983. 256p. illus. ISBN: 0723231656.

Covers such types as motorcyles and three-wheelers, cars and light transports, personnel carriers and buses, trucks, tractors and wreckers, amphibians and tracked vehicles, handling and construction equipment. Over 1,000 plates, many of them small. Non-analytical index, p.254-6. *Class No:* 623.42

[4508]

SENGER und Etterlin, F.M., von. **Taschenbuch der Panzer.** [Tanks of the World, 1983.] 6th ed. Munich, Bernard & Graefe Verlag, 1983. 827p. illus., diagrs., tables.

First published 1960.

Part 1: Type data sheets (19 countries, p.25-715); Part 2: Tables. Appendices A-M (A. Soviet exports; H. Distribution of armoured carriers). Data in Part 1 include development and production; distinguishing features; role; assessment. 'There is no comparable source in English' (Luttwak, E. *A dictionary of modern war* (1971). p.11). Fully illustrated. Small format. *Class No:* 623.42

[4509]

—The Guinness book of tank facts and feats: a record of armoured fighting vehicle achievements. Macksey, K., *comp. and ed.* 3rd ed. Enfield, Middx., Guinness Superlatives, Ltd, 1980. 256p.

First published 1972.

In 7 period-sections (1. From Chariot to armoured cars ... 6. Armour in limited war, 1946 to the present). 8 appendices, including 'Tank nomenclature', 'Tank strengths of the nations'. Sources and bibliography, p.249. Index, p.251-6. *Class No:* 623.42

Small Arms

[4510]

EZELL, E.C. *and* PEGG, T.M. **Small arms of the world:** a basic manual of small arms. 12th ed. London, Arms and Armour Press, 1983. 894p. illus., diagrs, tables. £25.

First published 1943 as *Basic manual of military small arms.* A completely new and revised version of the classic work by W.H.B. Smith.

Part 1. Small arms development since 1945 (Rifles and carbine development; Machine gun development; Submarine gun development; Handgun developments; Special purpose weapons development (*e.g.* Grenade launches). Part 2: Description of small arms by nations (44) and a basic manual of recent weapons: Argentina ... Yugoslavia, p.194-843. Characteristic of each given. Appended: Small arms for outdoor sports. Small arms ammunition. Selected bibliography, p.883-5. Detailed, analytical index, p.889-94. 2,000 illus. *Class No:* 623.44

[4511]

GREAT BRITAIN. Department of Trade and Industry. Translation Service. **Small arms glossary, English-French-German.** London, the Department, 1986. [*c.*130]p. £6.

Revised version of the 'Small arms glossary' produced for the 1984 Small Arms Conference.

3 sequences, each with *c.*500 entries: English/German/French (on white paper) - French/English/German (on green paper) - German/French/English (on yellow paper). Terms on which decision was not definitive are asterisked. Oblong format. *Class No:* 623.44

[4512]

HOGG, I.V. *and* WEEKS, J. **Military small arms of the 20th century:** a comprehensive illustrated encyclopedia of the world's small calibre firearms. 5th ed. London, Arms & Armour Press, 1985. 303p. illus., tables. ISBN: 0853687080.

First published 1973.

Coverage: handguns; submachine guns; bolt-action rifle;

.... (contd.)

automatic rifles; machineguns; anti-tank rifles; ammunition. 40 countries involved. Technical descriptions. About 700 illus. *Class No:* 623.44

[4513]

Jane's infantry weapons, 1992-93. Hogg, I.V., *ed.* Coulsdon, Surrey, Jane's Information Group, 1992. *c.*800p. £135. ISBN: 0710609868.

Over 1700 weapons and accessories from 252 manufacturers in 69 countries. Technical data on: personal weapons; crew-served weapons; ammunition; ancillary equipment. Online vial DIALOG, available on CD-ROM. *Class No:* 623.44

[4514]

—Brassey's infantry weapons of the world. Bowen, J.H., *ed.* 2nd ed. London, Brassey's Publications, Ltd., 1979. [viii], 480p. illus. ISBN: 0904609103.

First published 1975.

In 12 well-organized sections: Pistols and revolvers - Submachine guns - Rifles and carbines - Grenade launchers - Missiles and rockets - Mortars - Grenades - Grenade Launchers - Flame throwers - Mines - Combat aids (*e.g.* bayonets; radar) - Infantry support vehicles. Addenda. Glossary and abbreviations. Table of countries and their infantry weapons. List of manufacturers and their addresses. Index, p.475-80. Numerous black-and-white illus. *Class No:* 623.44

Pistols

[4515]

HOGG, I. V. *and* WEEKS, J. **Pistols of the world:** the definitive illustrated guide to the world's pistols and revolvers. London, Arms and Armour Press, 1992. 352p. illus. ISBN: 1854090364.

Covers about 2,000 pistols. 3 parts: 1. Introduction (including glossary and bibliography) - 2. Directory (pistols by manufacturer, A-Z (*e.g.* Colt, about 96 models)) - 3. Appendices (Databank, pistol names A-Z; Ammunition). Well-designed with excellent illustrations. *Class No:* 623.443

[4516]

SMITH, W.H.B. **Book of pistols and revolvers;** completely updated by J.E. Smith. Secaucus, N.J., Castle Books, 1979. xvi, 816p. illus., diagrs.

First published 1946. 1968 ed. (Harrisburg, Pa., Stackpoole Books) reprinted 1979.

Part 1.: Handguns, to include World War II (13 chapters, p.1-590; 3 appendices: 3. Glossary of pistol and revolver terms) - Part 2: Pistol and revolver trends, 1945-1968 (countries, A-Z). Index. Covers handguns worldwide, their identification, calibre and operation, illustration, description and history, construction, ammunition, dismounting and assembling. The standard reference work on handguns of the world. *Class No:* 623.443

Ammunition

[4517]

HOGG, I.V. **The Illustrated encyclopedia of ammunition.** London, Apple Press, 1985. 256p. illus. (incl. col.), facsims. ISBN: 1850760438.

Sections: A history of ammunition, p.8-64; A/Z of ammunition; Small arms; Mortars and grenades; Artillery;

....(contd.)
New developments (*e.g.* Laser guidance), General index, p.[254]; Index to headwords, p.255. Fully illustrated. *Class No:* 623.45

[4518]
Jane's ammunition handbook, 1992-93. Coulsdon, Surrey, Jane's Information Group, 1992. *c.*400p. illus., diagrs. £135. ISBN: 0710610009.
Covers: Small arms; Cannon; Air defence; Tank and artillery; mortar; grenades; pyrotechnics. Data include: description; history; specifications; equivalent projectiles by country; list of equipment. *Class No:* 623.45

Atomic Weapons

[4519]
COCHRAN, T.B., *and others.* **Nuclear weapons databook.** National Resources Defense Council, Inc. Cambridge, Mass., Ballinger, 1984-. V.1-. illus., maps. ISBN: 088410172x, v.1.
1. *U.S. nuclear forces and capabilities.* 1984. xix, 340p.
2. *U.S. nuclear warhead production.* 1987. xv, 223p.
3. *U.S. nuclear warhead facility profiles.* 1987. xv, 132p.
4. *Soviet nuclear weapons.* Harper & Row, 1989. 433p.
The first of a series of 8v. V. 1-3 concentrate on US weapons; V.4 on Soviet weapons; v.5-8, to be on, the environment, command and control of nuclear weapons and strategy, arms control in general , and a history of nuclear weapons. V.1 has 9 chapters (1. The nuclear weapons system: an overview ... 9. Army nuclear weapons. 75p. of statistics. Glossary. Index, p.329-40. 150 photographs); V.4 has 10 chapters. *Class No:* 623.454.8

Missiles

[4520]
GUNSTON, B. The Illustrated encyclopedia of the world's rockets and missiles: a comprehensive technical directory and history of the military guided-missile systems of the 20th century. London, Salamander Books, Ltd., 1979. 264p. col. illus.
Sections: General introduction - Surface to surface missiles (land tactical; land strategic; sea tactical; sea strategic) - Air to surface missiles (tactical; strategic) - Surface to air missiles (land; sea) - Air to air missiles - Anti-tank missiles - Anti submarine missiles. Data on each type: description; dimensions; launch weight; range. Descriptive captions. *Class No:* 623.46

[4521]
'World missile directory'. Richardson, D., *comp.* In *Flight international,* 10 October 1988, p.33-71. illus. (incl. col.), tables.
Sections: Strategic missiles - Cruise missiles - Surface-to-surface - Air-to-surface - Anti-ship - Surface-to air - Air-to-air - Anti-tank. Country subdivision in each section. Comparative data tables appended. *Class No:* 623.46

Field Engineering

Communications

[4522]
Jane's military communications, 1992-93. Williamson, J., *ed.* 13th ed. Coulsdon, Surrey, Jane's Information Group, 1992. *c.*800p. £135. ISBN: 0710609809.
First published 1979 (for 1979/80).
Contents: Radio communications - Line communications. Overview of each country's systems. Abundant technical data. Available on CD-ROM. *Class No:* 623.61

Radar Countermeasures

[4523]
HALL, P.S., *and others.* **Radar.** London, Brassey's, 1991. xiii,170p. £22.50. (*New battlefield weapons systems and technology, 9.*) ISBN: 0080377106.
Specific function and operation of radar equipment in a military setting. Includes electronic warfare. *Class No:* 623.62

[4524]
The International countermeasures handbook. Wiseman, C.H., *ed.* 11th ed. Palo Alto, Calif., E.W. Communications, Inc., 1986. Annual. 444p. ISBN: 0918994128.
First published 1975.
5 main parts: Electronic warfare budgets - [Electronic] systems of the West - Guided missiles of the world - Soviet weapons and electronics, p.197-282 - Analysis and technology (17 sections, including 'Electric warfare bibliography', p.393-412). Indexes and references, p.437-43. *Class No:* 623.62

[4525]
Jane's radar and electronic warfare systems. Blake, B., *ed.* Coulsdon, Surrey, Jane's Information Group, 1992. 600p. illus. £135. ISBN: 0710609892.
Covers systems in development, production and service. *Class No:* 623.62

[4526]
WALDMAN, H. The Dictionary of SDI [Strategic Defense Initiative]. Wilmington, Delaware, SR Books, 1988. 173p. illus., diagrs., tables. ISBN: 0842022813.
Over 1,000 entries on Strategic Defense Initiative, including persons, laboratories. 'Antisatellite' (ASAT) weapon: over 1p., illus. Appended: 'Text of Treaty between the United States of America and the Union of Soviet Socialist Republics on the limitation of anti-ballistic missile systems, 1972'. Aims 'to classify and simplify SDI's complex terrain, lay out the technology involved, and identify those areas of debate'. *Class No:* 623.62

Aircraft

[4527]
CHANT, C. Encyclopaedia of modern aircraft armaments. Wellingborough, P. Stephens, 1988. 304p. £19.95. ISBN: 0850598621.
Includes helicopter armament. *Class No:* 623.7

Encyclopaedias

[4528]
ANGELUCCI, E., *ed*. **The Rand McNally encyclopedia of military aircraft, 1914-1980.** Translated by S.M. Harris. Chicago, Ill., Rand McNally, 1981. 546p. col. illus., diagrs., charts. ISBN: 0528815474.

Original as *Atlantec encilopedico degli aerei militari del mondo* (Milan, Mondadori, 1980).

A splendidly illustrated record of 800 types of military craft flown between 1914 and 1980. 4 chapters: 'The First World War'; 'Military aviation between the two World Wars'; 'The Second World War'; and 'Military aviation from 1945 to the present day'. Subdivision by types. Country index; general index. Colour illus. of over 1,700 individual aircraft; 772 line-drawings (usually three-view); charts of national markings. Appended 3p. bibliography. Omits helicopters. Supplements *Jane's all the world's aircraft* by giving historical treatment. 'Indispensable for aviation history and military science collections' (*Wilson library bulletin*, March 1982, p.545-6).
Class No: 623.7(031)

[4529]
The encyclopedia of world air power. Gunston, B., *ed*. 2nd ed. Twickenham, Middx., Temple Press, 1986. 319p. illus. (incl. col.). £17.50. ISBN: 0600351653.

First published 1980.

The world's air forces (Western Europe ... South America), p.7-66. The world's military aircraft, A-Z, p.67-301. Data on each type: description; specification; illus. - cutaway drawings and silhouettes. The world's air-launched missiles, p.302-15. Detailed index, p.316-9.
Class No: 623.7(031)

Handbooks & Manuals

[4530]
GUNSTON, B. **An Illustrated guide to modern fighters.** London, Salamander, 1984. 159p. col. illus., diagrs. ISBN: 0861010858.

Details of 57 fighter aircraft, Aeritalia G81 ... Yakovlev. Data: origin; type; engines; dimensions; weight; performance; armament; history; users; development. Col. illus.; 3-way view line-drawings. *Class No:* 623.7(035)

[4531]
Jane's avionics, 1992-93. Brinkman, D. 11th ed. Coulsdon, Surrey, Jane's Information Group, 1992. *c.*600p. illus., diagrs. £135. ISBN: 0710609906.

Over 2,000 avionics products and systems. Sections: Sensors - Data processing and displays - Radio communications. Directory of manufacturers. 1200 illus. Available on CD-ROM. *Class No:* 623.7(035)

Worldwide

[4532]
AP REES, E. **World military helicopters.** London, Jane's Publishing Co., 1986. 192p. £12.50. ISBN: 0710603746.

Survey of the world's military helicopters, in alphabetical order of manufacturers, Aerospatiale ... Westland. 70 basic types, either those in current service, or flying as prototypes, or under construction. Each type is given 1 black-and-white illus. per page. Appendix: 'World military helicopter fleet', p.187-92. *Class No:* 623.7(100)

[4533]
'Military aircraft of the world'. Hatch, P.E., *comp*. In *Flight International*, 29 July - 4 August 1992, p.31-57. illus. (incl. col.), tables.

Reviews the past 12 months' sales and developments of the world's (Afghanistan - Zimbabwe) current and future military aircraft. Comparative data tables appended.
Class No: 623.7(100)

Great Britain

[4534]
EVERETT-HEATH, J. **British military helicopters.** London, Arms and Armour Press, 1986. 224p. illus. £14.95.

A survey of military helicopter types, from Cierva helicopters and Bristol helicopters to Westland 30 and EH-101. 4 appendices (*e.g.* helicopter projects and designs). List of abbreviations. Glossary. Bibliography, p.220. Over 200 photographs. Detailed, non-analytical index.
Class No: 623.7(410)

USSR

[4535]
GUNSTON, B. **Aircraft of the Soviet Union:** the encyclopedia of Soviet aircraft since 1917. Baldry, D., *ed*. London, Osprey, 1983. 414, [2]p. illus., ports, diagrs. ISBN: 085045445x.

Includes biographies. Entries: 'Alexandrov-Kalinin ... Zlokazov'. Prelims include 'Soviet aircraft designations', 'Air weapons', 'Materials', 'Engines', 'Soviet organizations', 'Miscellaneous data'. Appended list of grouped light aircraft, p.408-14. Cross-sections, a feature.
Class No: 623.7(47)

USA

[4536]
WATERS, A.W. **All the US Air Force airplanes, 1907-1983.** New York, Hippocrene Books, [*c.*1983]. [vii], 413p. ISBN: 0882548825.

'Historical index section' and 'Historical aircraft section' precede the main part, on US aircraft, 1944 to the present, p.59-311, plus 'Space vehicles, 1974 to the present'. 7 appendices include Bibliography (p.316-7); Air Force bases. Detailed index, p.399-413. *Class No:* 623.7(73)

Navies

Abbreviations & Symbols

[4537]
WEDERTZ, B. **Dictionary of naval abbreviations.** 3rd ed. Annapolis, Md., Naval Institute Press, 1984. 330p.

First published 1970; 2nd ed. 1977 (360p).

Explains *c.*45,000 abbreviations 'commonly used by the naval establishment and ... unavailable in standard dictionaries' (*Preface*). Pruned of entries in the 2nd ed. not related to strictly naval matters. *Class No:* 623.8(003)

Bibliographies

[4538]
SMITH, M.J. **Battleships and battle-cruisers, 1884-1984:** a bibliography and chronology. New York, Garland Publishing Co., 1985. 691p. (*Garland reference library of social science, 203.*) ISBN: 0814089022.

5,500 references to books, periodical articles, scholarly papers, government publications, etc. Chapter and section introductions. The chronology has appended index of ship names, major battles, conferences and treaties. 'This outstanding bibliography, the most extensive compilation to date' (Choice, v.23(5), January 1986, p.730).
Class No: 623.8(01)

Handbooks & Manuals

[4539]
FRIEDMAN, N. **The Naval Institute guide to world naval weapons systems.** Anapolis, MD, Naval Institute Press, 1989. 511p. photos, diagrs. $89.95. ISBN: 0870212953.

Arranged by type of system. Sections: Surveillance and control - Strategic strike systems - Strikes - Surface warfare - Anti-aircraft warfare - Anti-submarine warfare - Mine warfare. Index. *Class No:* 623.8(035)

Dictionaries

[4540]
PALMER, J., *comp.* **Jane's dictionary of naval terms.** London, Macdonald & Jane's. 1975. [iv], 342p.

Over 6,000 entries for naval military terms as well as those in general seamanship, plus some purely mercantile terms. Different meanings are numbered. Many cross-references. Sources (p.2) include Royal Navy, US Navy and NATO manuals. *Class No:* 623.8(038)

[4541]
—NOEL, J.V. Naval terms dictionary. 5th ed. Annapolis, Md., Naval Institute Press, 1986. 352p.

First published 1952; 4th ed. 1978.

About 6,000 entry-words, - US Navy colloquialisms, slang and technical terminology, including abbreviations. 5 appendices (*e.g.* ship, aircraft and missile designation systems; electronic nomenclature). *Class No:* 623.8(038)

[4542]
UNITED STATES. Naval Institute. **Naval phraseology:** English-French-Spanish-Italian-German-Portuguese. 2nd ed. Annapolis, Md., the Institute, 1953. [ix], 326p.

First published 1944 (v, 326p.).

4 parts: 1. Naval terms (*c.*1,800) in sections A-H (Administration ... War), with indexes of English terms - 2. Colloquies (*i.e.,* dialogues) in English, with translations in French, Spanish, Italian, German and Portuguese - 3. Correspondence - 4. Tables of ranks, with sleeve markings. *Class No:* 623.8(038)

Reviews & Abstracts

[4543]
Naval abstracts. Arlington, Va., Center for Naval Analysis, 1977-. Quarterly.

About 1,000 concise abstracts pa. from c.300 periodicals. International in scope, although mostly US and British journal sources. Each issue includes a list of periodicals consulted and an author index. Annual broad subject index.

....(*contd.*)
RSR (v.9(1), January/March 1981, p.97-99) notes some overlap with Air University Library index to military periodicals. (*qv*). Online database was contemplated. *Class No:* 623.8(048)

Anthologies

[4544]
Warships, 1992. Gardiner, R., *ed.* London, Conway Maritime, 1992. 256p. illus. £24. ISBN: 0851776025.

Reprints of original articles; 275 illus.
Class No: 623.8(082.21)

Histories

[4545]
Conway's all the world's fighting ships. London, Conway Maritime Press, 1979-83. 4v. in 5. illus., diagrs. ISBN: 0870219235.

Conway's All the World's fighting ships, 1860-1905. 1979. [vii], 440p.

Conway's Fighting Ships, 1906-1921. 1983. [viii], 439p.

Conway's All the World's fighting ships, 1922-1946. 1980. [viii], 456p.

Conway's All the World's fighting ships, 1947-1982. pt.1: *The Western Powers;* pt.2: *The Warsaw Pact and non-aligned nations.* 1983. 2v.

The two 1947-1982 volumes aim to provide a coherent overview of the post-World War II naval revolution, using newly released information where possible (Foreword). Provides an historical survey (beginning with the introduction of ironclad ships), complementing *Jane's Fighting ships,* as well as correcting data in the latter's volumes. Each of the Conway volumes provides technical data on types of ships, photographs and line drawings, plus a detailed index. 'Conway's is an invaluable addition to naval reference collections of any size' (*Reference books bulletin, 1984-1985* (1986). p.42). *Class No:* 623.8(091)

[4546]
—COWIN, H.W. Conway's directory of modern naval power, 1986. London, Conway Maritime, 1985. 288p. illus. £25. ISBN: 0851773621.

In 4 sections: 1. Navies - 2. Warships - 3. Naval aircraft - 4. Naval missiles and guns. 600 illus. Indexes, p.285-88. Helps to supplement *Conway's All the World's fighting ships* series. *Class No:* 623.8(091)

[4547]
GOODWIN, P. **The Construction and fitting of the English sailing men of war, 1650 - 1850.** London, Conway Maritime Press, 1987. xi, 276p. illus., diagrs., tables. ISBN: 0851773265.

10 chapters (1. The construction of a ship of War in frame ... 10. Hull protection). 8 appendices (2. The quantities of timber used in ship construction; 4e. The 1745 Establishment list). Bibliography, p.267. Detailed, analytical index, p.269-76. Fully illustrated; meticulously drawn diagrams, to scale. *Class No:* 623.8(091)

[4548]

—LAVERY, B. The Arming and fitting of English ships of war 1600-1815. London, Conway Maritime Press, 1987. 319p. illus., diagrs., tables. ISBN: 0851774512.

11 parts (47 chapters): 1. Steering ... 5. Guns ... 9. Internal fittings. 11 appendices, each linked to a part (2. Anchors, capstans). Extensive chapter references to contemporary documents. Intended 'in the tradition' of Peter Goodwin (*Preface*) (*qv*). *Class No:* 623.8(091)

[4549]

LAVERY, B. The Ship of the line. London, Conway Maritime Press, 1983-84. 2v. illus., diagrs., tables. ea. £20. ISBN: 0851772528.

1. *The development of the battleship, 1650-1850.* 1983.

2. *Design, construction and fittings, 1650-1850.* 1984.

V.1 (224p.) covers the general historical background, and includes extensive tables listing all ships of 50 guns and above, divided by period, rate, class and design, with full technical data. 130 plates, 65 line-drawings. V.2 concerns technical developments in hull design, construction, rigging, armament and fittings. *Class No:* 623.8(091)

[4550]

—LAMBERT, A. Battleships in transition: the creation of the steam battlefleet, 1815-1860. London, Conway Maritime Press, 1984. 168p. illus. £11.95. ISBN: 085117315x.

Bridges the gap between B. Lavery's *The ship of the line, 1650-1850* and Conway's *All the World's fighting ships, 1860-1982*. 95 illus. *Class No:* 623.8(091)

Worldwide

[4551]

BISHOP, C., *ed*. The encyclopedia of world sea power. Lonsdon, Temple Press, 1988. 319p. illus. (incl. col.). £17.50. ISBN: 0600332748.

Sections: Missile submarines - Nuclear attack sumbarines - Diesel submarines - Aircraft carriers - Cruisers - Destroyers - Western frigates - Frigates - Patrol craft - Amphibious assault ships - Military hovercraft - Mine warfare vessels - Anti-ship missiles - Naval artillery - Naval surface-to-surface missiles - Naval anti-aircraft weapons - Underwater weapon systems - Underwater weapons. World navies, (p.291-319), Albania ... Zaire. Detailed captions. Fully illustrated. *Class No:* 623.8(100)

[4552]

Jane's fighting ships, 1992-93. Sharpe, R., *ed*. Coulsdon, Surrey, Jane's Information Group, 1992. Annual. *c*.800p. illus., diagrs., tables. £135. ISBN: 0710609833.

Founded in 1879 by F.T. Jane.

Covers over 8,000 warships from 150 nations. Data include: main machinery, speed, missiles, countermeasures, combat data systems, radars, modernisation, sales. 3,000 illus. The leading authority on current navies worldwide. Online via DIALOG; available on CD-ROM. *Class No:* 623.8(100)

[4553]

—COUHAT, J.L. *and* FREZELIN, B., *eds*. Combat fleets of the world, 1988/89: their ships, aircraft and armament. Annapolis, Md., Naval Institute Press, 1988. 896p. ISBN: 0870211943.

First published 1897 as the French biennial *Les flottes de combat*.

.... (contd.)

As equally comprehensive in scope as *Jane's Fighting ships*. Although less expensive than 'Jane', it provides less detail and less illustrative matter. *Class No:* 623.8(100)

[4554]

—Weyers Flottentaschenbuch. Munich, Lehman, Biennial.

First published 1900 as *Taschenbuch der deutschen Kriegsflotte*.

The German counterpart of 'Jane'. *Class No:* 623.8(100)

[4555]

Warships of the world: an illustrated encyclopedia. Gallipini, G. New York, Times Books, 1986. 320p. col.illus. $70. ISBN: 0812911296.

Each section covers one class of ships; includes submarines. Technical data. Indexes: Country; name of ship. *Class No:* 623.8(100)

USSR

[4556]

POLMAR, N. Guide to the Soviet navy. 4th ed. London, Arms & Armour Press, 1986. xiv, 536p. £35. ISBN: 0853685214.

3rd ed. 1983 ([xv], 465p.).

Detailed analysis of the Soviet navy of the late 1980s, the ships, their weaponry and electronic systems. 425 illus., including many photographs previously unpublished. Detailed drawings of new ships. Index. *Class No:* 623.8(47)

[4557]

—MOORE, J. Warships of the Soviet Navy. London, etc., Jane's Publishing Co., 1981. 192p. illus. ISBN: 0710601034.

Describes 33 types of vessel (nuclear-powered, ballistic missile submarines ... Tugs/floating docks). Data include: name; displacement; aircraft; missiles, guns; propulsion; range; complement; radar. Glossary of abbreviations used. Class index. General (detailed) index. *Class No:* 623.8(47)

Asia

[4558]

BLECHMAN, B.M. *and* **BERMAN, R.P. Guide to Far Eastern navies.** Annapolis, Md., Naval Institute Press, 1978. xiv, 586p. illus., tables. ISBN: 0870217976.

Part 1: The navies in Far Eastern waters (7 chapters; references). Part 2: Ships and aircraft of Far Eastern navies (China, Japan, Taiwan, South Korea, North Korea, Philippines). Data: Characteristics; Armaments; Electronics; Individual ships; number; builders; in service; etc. Detailed name index, p.581-6. Oblong format. *Class No:* 623.8(5)

USA

[4559]

Dictionary of American naval fighting ships. Mooney, J.L., *ed*. Washington, Government Printing Office, for the U.S. Naval History Division. 8v. $142, the set.

Descriptions and histories of *c*.10,000 US fighting ships since 1775; entries under ships names, A-Z, through the 8 volumes. Appendices scattered throughout, list the ships by

.... *(contd.)*

type and class. No general index. Well illustrated. Revision of the historical sketches of ships in the earlier volumes is in hand. *Class No:* 623.8(73)

Submarines

[4560]
MILLER, D. Submarines of the world. London, Salamander, 1991. 189p. illus. £18.95. ISBN: 0861015622.

A technical directory of the major submarines from 1888 to the present. *Class No:* 623.827

[4561]
'The Submarine'. In *Navy International*, May 1988, p.224-75.

Includes data on international submarine forces; design; combat information; sonar; periscopes; electronic warfare; machinery; weapons. *Class No:* 623.827

Naval Defence

[4562]
Jane's naval weapon systems. Coulsdon, Surrey, Jane's Information Group, 1992-. 300p. Looseleaf binder + updates (3pa.). £250. ISBN: 0710608934.

Data on 300 weapons and systems, 110 manufacturers. Contents: handling systems; equipment (missiles, guns, launchers, etc.); display systems. Supplements published April, August, December. Available on CD-ROM. *Class No:* 623.9

[4563]
KIELY, D.G. Naval electronic warfare. London, Brassey's, 1988. xiii, 122p. £9.95. (*Brassey's Sea power: naval weapons, weapon systems and technology series, no. 5.*)

A survey of naval electronic warfare systems. *Class No:* 623.9

Civil Engineering

Databases

[4564]
DIBLEY, S. 'Online databases in civil engineering (Construction; building)'. In *UK Online User Group Newsletter*, no. 30, March 1984, p.5-6.

Observes that 'information on civil engineering is widely regarded as a gap in databases'. ACOMPLINE (Greater London Council. Research library), BRIX (Building Research Establishment) and IBSEDEX (Building Services Research and Information Association), plus PICA (Property Services Agency Information on Construction and Architecture) provide part-answer. *Class No:* 624(003.4)

Bibliographies

[4565]
INSTITUTION OF CIVIL ENGINEERS. Index to publications. London, the Institution.

Title varies.

A series of five-year indexes to I.C.E. publications (*e.g.* January 1965- December 1969, 1971. 114p.). This was

.... *(contd.)*

continued by J.E. Holmstrom as *Analytical index to the publications of the Institution of Civil Engineers. January 1970- December 1974* (1976. x, 149p.), which covered both *Proceedings* and *Géotechnique*. Author, subject, place and project name index. The latest index covers 1975-79.

Earlier indexes include *Subject and author index to the Institution's publications, November 1948 to December 1959 inclusive* (1960. 138p.). *Class No:* 624(01)

[4566]
SKEMPTON, A.W. British civil engineering literature, 1640-1840: a bibliography of contemporary printed reports, plans and books. London, Mansell, 1987. xvii, 302p. illus., facsims, maps. £45. ISBN: 0720112461.

1,914 numbered entries. Topics covered: fen drainage, river navigation, canals, docks and harbours, bridges, roads, railways and water supply, 'with a selection of related pamphlets, and works on mechanism, strength of materials and hydraulics' (*Preface*). 'Guide to the literature' (running commentary), p.ix-xiv. Bibliography, p.1-275; under authors, A-Z, with annotations and locations; addenda. Subject index: Aqueduct ... Water supply, p.287-302. *Class No:* 624(01)

Handbooks & Manuals

[4567]
BELL, F.G. Ground engineer's reference book. London, Butterworth, 1987. 1264p. illus., tables. £99.50. ISBN: 0408011734.

About 70 contributors. 5 parts (59 chapters): 1. Properties and behaviour of the ground - 2. Investigation in ground engineering - 3. Treatment of the ground - 4. Construction in ground engineering - 5. Numerical methods and modelling in ground engineering. Chapter 11: 'Stability of soil slopes' (16p. 19 references; 2 items of further reading; 10 figures). 1,028 illus. in all. For practising civil engineers. *Class No:* 624(035)

[4568]
—**The Guinness book of structures:** bridges, towers, tunnels, dams. Enfield, Middx., Guinness Superlatives, Ltd., 1976. 288p. illus., ports, tables, maps.

Provides historic data and facts on the largest, greatest, earliest and oldest structures. 4 chapters: 1. Above ground - 2. Below ground - 3. Hydraulic works - 4. Engineers and construction (great civil engineers; great schemes ... nicknames). 54 tables. Index. p.277-88. *Class No:* 624(035)

[4569]
BLAKE, L.S., *ed.* **Civil engineer's reference book.** 4th ed. London, Butterworth, 1990. v.p. illus., diagrs., graphs, tables. £90. ISBN: 0408012080.

First published 1951 as *Civil engineering reference book;* 3rd ed. 1975.

52 contributors. 45 documented sections (2. Strength of materials - 3. Theory of structures ... 11. Site investigation ... 18. Dams ... 24. Airports ... 32. Tunnelling ... 44. Offshore construction). Section 5, 'Hydraulics': 28p., 62 line drawings, 15 tables, 44 references, brief bibliography. Detailed, analytical index, 19p. *Class No:* 624(035)

[4570]
Construction materials reference book. Doran, D.K., *ed.*
Oxford, Butterworth-Heinemann, 1992. v.p. illus., diagrs.,
tables. £95; $205. ISBN: 0750610042.
2 parts: 1. Metals - 2. Non-metals. Topics include cement,
concrete, ceramics, timber, stone, aluminium, steel, etc.
Chapters identify relevant British standards. Several hundred
literature references. 'Recommended' (*Choice*, November
1992, p.500). *Class No:* 624(035)

[4571]
MERRITT, F.S., *ed.* **Standard handbook for civil
engineers.** 3rd ed. New York, etc., McGraw-Hill, 1983.
[1632]p. illus., diagrs., graphs, tables, formulas. ISBN:
0070415153.
First published 1968.
26 contributors, 23 sections (Structural design ...
Specification ... Construction materials ... Structural steel
design and construction ... Wood design and construction ...
Earthwork ... Building engineering - Highway engineering -
Bridge engineering - Airport engineering - Tunnel
engineering - Water engineering - Environmental
engineering - Harbour engineering). Section references.
Detailed, analytical index of 58p. 'Important for all
technology and engineering reference collections' (*New
technical books,* February 1984, entry no. 253).
Class No: 624(035)

[4572]
—MERRITT, F.S. *and* KURTZ, M., *eds.* Civil engineering
reference guide. New York, etc., McGraw-Hill, 1986.
[608]p. illus. $49.50. ISBN: 0070415226.
A condensation of F.S. Merritt's *Standard handbook for
civil engineers.* 11 contributors. 9 sections: 1. Structural
theory - 2. Structural steel design - 3. Concrete design and
construction - 4. Wood design and construction - 5. Bridge
engineering - 6. Geotechnical engineering - 7. Water
engineering - 8. Environmental engineering - 9. Surveying.
Analytical index (24p.). *Class No:* 624(035)

[4573]
PARMLEY, R.O. Field engineer's manual. New York,
McGraw-Hill, 1981. xxvii, 611, [1]p. diagrs., tables. ISBN:
0070485135.
20 sections (1. Pocket electronic calculator ... 5.
Construction: materials, data and safety ... 8. Mechanical
formulas and data ... 10. Structural ... 13. Drainage ... 16.
Water supply, water storage and fire protection ... 18.
Energy and fuels ... 20. First aid). Partly analytical index,
p.597-611. Small format. *Class No:* 624(035)

[4574]
Structural engineering handbook. Gaylord, E.H., Jr. *and*
Gaylord, C.N., *eds.* 2nd ed. New York, McGraw-Hill,
1990. v.p. diagrs., graphs, tables.
46 contributors. 30 sections. 763 illus. An authoritative
reference work, for engineers, architects and students in
these fields. Analytical index (over 2,000 entries).
Class No: 624(035)

Dictionaries

Polyglot

[4575]
Elsevier's dictionary of civil engineering in four languages:
English, German, Spanish and French. Guitterez, H.F.
Amsterdam, Elsevier, 1991. [x],392p. £85. ISBN:
0444889876.
Over 6,000 terms. Bibliography. *Class No:* 624(038)=00

[4576]
GROS, E. *and* **SINGER, L.,** *comps.* **Russian-English -
French-German constructional engineering dictionary.**
London, Scientific Information Consultants, Ltd., 1965. ii,
194p.
About 1,500 terms 'from structural mechanics and related
fields most frequently used in building, bridge construction
and civil engineering, with particular emphasis on steel
structures' (back of title-page). Russian entry-words, with
English, French and German equivalents, but no indexes.
Genders not stated. *Class No:* 624(038)=00

[4577]
**TEKNISKA NOMENKLATURCENTRALEN.
Kommunaltekniska ordlista.** [Glossary of municipal
engineering.] Stockholm, the Centre, 1976. 214p. (*TNC 61.*)
ISBN: 9171960627.
472 Nordic language terms, with definitions in Swedish
and equivalents in English, French and German. Indexes in
English, Danish, Finnish, French, German, Norwegian and
Swedish. *Class No:* 624(038)=00

English

[4578]
SCOTT, J.S. Dictionary of civil engineering. 4th ed.
Harmondsworth, Penguin, 1991. 608p., illus. £8.99. ISBN:
0140512462.
3rd ed. 1981.
About 5,000 terms are concisely defined. Includes a wide
range of terms in soil mechanics, heavy construction and
mining. 'Retaining wall': 1p. with illus. US usage is
covered. For the layman and non-specialist engineer.
Class No: 624(038)=20

German

[4579]
**BUCKSCH, H. Wörterbuch für Bautechnik und
Baumaschinen.** [Dictionary of civil engineering and
construction machinery and equipment.] 8. Aufl. Wiesbaden,
Bauverlag GmbH, 1982-87. 2v.
First published 1955.
V. 1: German-English; V. 2: English-German. About
65,000 entry-words in each volume. British and US usage
differentiated. Entries are grouped *e.g.* 'Tunnel and
Stollenbau' (construction of tunnels and galleries), v.1, with
separate lines for subentries. Many compounds. German
genders shown. Appended conversion tables.
Class No: 624(038)=30

[4580]
SOUSA FERREIRA, K. de. **Terminologie des Hoch- und Brückenbaues.** Englisch-deutsch, deutsch-englisch. [Glossary of building and bridge construction.] Berlin, Ernst, 1969. viii, 216p. ISBN: 3433001235.

About 3,000 terms, each shown in context. 10 chapters: 1. Strength of materials - 2. Structures - 3. Building materials - 4. Timber construction - 5. Stone, brick and concrete construction - 6. Steel construction - 7. Composite construction in steel and concrete - 8. Structural elements of building construction - 9. Bridges - 10. Foundations, mechanics of soil. Bibliography. German and English indexes. *Class No:* 624(038)=30

Swedish

[4581]
TEKNISKA NOMENKLATURCENTRALEN. **Byggordsamling.** [Construction vocabulary: Swedish-English, English-Swedish.] Stockholm, the Centre. 1979. 351p. (*TNC 72.*)

About 9,000 terms, Swedish and English. 'Contains many terms from other TNC glossaries on house-painting, concrete, industrial engineering, planning and building, geotechnics, timber construction and municipal engineering' (*Incorporated linguist,* Summer 1977, p.75). *Class No:* 624(038)=397

French

[4582]
BUCKSCH, H. **Dictionnaire pour les travaux publics.** [Dictionary of civil engineering and construction machinery and equipment.] 9th ed. Paris, Eyrolles, 1982.

French-English. 547p. *English-French.* 420p.

Each volume has *c.*30,000 entry-words. There are German-French and German-Spanish volumes, apart from the German-English volumes. *Class No:* 624(038)=40

Russian

[4583]
BHATNAGAR, K.P. **Elsevier's dictionary of civil engineering:** Russian-English. Bhattacharya, S.K., *ed.* Amsterdam, etc., Elsevier, 1988. x, 694p. £98.43. ISBN: 0444429611.

About 36,000 Russian entry-words, covering civil and structural engineering. Includes eponyms and trade names. 'Reference works', p.vii-x. *Class No:* 624(038)=82

Arabic

[4584]
KAY, E. **Arabic dictionary of civil engineering:** English-Arabic, Arabic-English. London, Routledge & Kegan Paul, 1986. 272p. £30. ISBN: 0710204299.

About 1,500 terms, covering civil engineering and the construction industry, from theory and planning through on-site requirements to office practices. *Class No:* 624(038)=927

Reviews & Abstracts

[4585]
ASCE publications information. New York, American Society of Civil Engineers, 1966-. 6pa. ISSN: 07341962.

January/February 1988 number displays contents of 21 issues of 16 ASCE journals. 'Special publications and manuals & reports' follows, plus *c.*200 abstracts of these items. Subject and author indexes. Annual combined index. *Class No:* 624(048)

[4586]
International civil engineering abstracts. Dublin & London, CITIS, Ltd., 1982-. 10pa. ISSN: 03324085.

Formerly *ICE abstracts,* 1974-81.

About 3,000 abstracts pa. 10 sections: Structural engineering - Bridges - Buildings; materials; construction - Tunnels; underground structures; mechanics - Miscellaneous - Hydraulic power stations; dams - Coastal engineering; ports; waterways - Public health engineering; water supply; irrigation; hydrology - Highway and traffic engineering; transportation; airports. *Class No:* 624(048)

Yearbooks & Directories

[4587]
Ground engineering yearbook, 1991. Sanders, D.M. London, Thomas Telford, 1991. 712p. illus. ISBN: 0727716131.

Annual.

Usual directory type information on seven areas: 1. General information on ground engineering - 2. Standards and codes of practice - 3. Institutions and associations - 4. Materials and products - 5. Plant and equipment - 6. Contractors and specialist contracting services - 7. Consulting engineers. Indexes: Trade names; Company and personnel; Subject. *Class No:* 624(058)

Tables & Data Books

[4588]
Construction and civil engineering index. Bracknell, Berks., Technical Indexes, Ltd.

Issue 26, June to October 1988 (248, [1]p.) comprises a data book, - the index to the associated microfile of manufacturers' catalogue. Sections: Products and trade names - Selector charts - Foreign Trade offices - What's new - Addresses (p.156-243) - Professional and trade organizations. *Class No:* 624(083)

[4589]
HICKS, T.G. **Civil engineering calculations reference guide.** New York, etc., McGraw-Hill, 1987. xii, 291, [1]p. diagrs., graphs, tables. £35. ISBN: 0070287988.

A condensation of Hick's *Standard handbook of engineering calculations* (2nd ed. 1985).

18 sections, closely subdivided (*e.g.* Principles of statics; Geometric properties of areas; Analysis of stress and strain; Steel beams and girders; Timber engineering; Reinforced concrete, Prestressed concrete; Fluid mechanics; Soil mechanics). Coverage includes surveying and photogrammetry. References, p.286. Index, p.287-91. Calculations are in dual units. *Class No:* 624(083)

[4590]

Spon's civil engineering and highway works price book, 1992. 6th ed. London, Spon, 1991. xxii,779p. £57.50. ISBN: 0419173803. ISSN: 0957171x.

First published 1984.

13 parts: 1. General - 2. Preliminaries and general terms - 3. Resources - 4. Unit costs (civil engineering works) - 5. Unit costs (highway works) - 6. Oncosts and profits - 7. Costs and tender prices indices - 8. Daywork - 9. Professional fees - 10. Approximate estimates - 11. Outputs - 12. Tables and memoranda - 13. Updates. Detailed, analytical index, p.765-79. *Class No:* 624(083)

Histories

[4591]

NORRIE, C.M. Bridging the years: a short history of British civil engineering. London, E. Arnold, 1956. [xii], 212p. illus.

A general survey from 1760 to 1914 (160p.) is followed by description of some notable tunnels, harbours and bridges, river works, canals and dams. Included: a list (up to 1950) of professional engineers who were Fellows of the Royal Society, and a list of founder members of the Federation of Civil Engineering Contractors. *Class No:* 624(091)

[4592]

STAPLETON, D.H. The History of civil engineering since 1600: an annotated bibliography. New York, Garland, 1984. xxxiii, 232p. (Bibliographies of the history of science and technology, 14.). ISBN: 0824089480.

1,283 entries, most with brief, primarily descriptive annotations. Arranged in chronological sections, subdivided by topics (*e.g.* government agencies ... Dams and river control - Waterways and canals ... City planning - Highways - Bridges, etc.). Emphasis is on American and European scholarly material; literature for the 'amateur or buff' is avoided (*Introduction*, p.xiv). Bibliographic essay, p.xix-xxxiii. Author-name index; no subject index. *Class No:* 624(091)

[4593]

STRAUB, H. A History of civil engineering an outline from ancient to modern times. English translation by E. Rockwell. London, Hill, 1952. Reprinted Cambridge, Mass. & London, MIT Press, 1964. xviii, 256p. illus.

German original: *Die Geschichte der Bauingerieurkunst* (Basle, Verlage Birkhauser, 1949).

The English translation introduces some amendments, and corrections. 78 illus. A chronological table (p.242-7) precedes a selected bibliography, p.248-50. Biographical details are given of the principal engineers whose theories are described. *Class No:* 624(091)

England

[4594]

BARBEY, H.F. Civil engineering heritage: Northern England. London, Telford, 1981. xx, 178p. illus., maps. £5.75. ISBN: 072370326x.

Organized by the Panel for Historical Engineering Works of the Institution of Civil Engineers.

Covers the region of Northern England from South Yorkshire up to the border beween England and Scotland. Includes Isle of Man. 8 chapters (1. The Border ... 8.

....(contd.)

Merseyside and Greater Manchester). Chapter bibliographies, p.167-76. Detailed index. Index of additional sites. Intended for civil engineers and general readers. *Class No:* 624(420)

Wales

[4595]

SIVEWRIGHT, W.J. Civil engineering heritage: Wales & Western England. London, Telford, 1986. 232p. £6.95. ISBN: 072270326x.

Organized by the Panel for Historical Engineering Work of the Institution of Civil Engineers.

Covers the civil engineering heritage in Wales and Western England to just south of Bristol. Chapter bibliographies. Detailed index. Companion volume to H.F. Barbey's book on *Northern England* (*qv*). *Class No:* 624(429)

USA

[4596]

SCHODEK, D.L. Landmarks in American civil engineering. Cambridge, Mass., MIT, 1987. 383p. illus. $50. ISBN: 026219256x.

A survey of nearly 100 projects. Sections: canals, roads, railroads, bridges, tunnels, water supply and control, environmental engineering, dams, buildings, urban planning, power systems, surveying and mapping, coastal facilities, airports. 250 illus. Index. 'Recommended' (*Choice*, January 1988, p.792). *Class No:* 624(73)

Information Services

[4597]

CONSTRUCTION INDUSTRY RESEARCH AND INFORMATION ASSOCIATION. The CIRIA UK construction information guide. Richardson, B. London, E. & F.N. Spon, 1989. xi,361p. £37.50. ISBN: 0419148906.

First published 1979.

Part 1: Alphabetical list of UK local authorities (but not identifying the relevant specific departments); part 2: Alphabetical list of UK construction information sources (p.39-283). No references to HMSO publications (including The Building Regulations); no mention of BRIX, etc. databases. Subject and acronym keys. *Class No:* 624:061:025.5

Measurements

[4598]

INSTITUTION OF CIVIL ENGINEERS. Civil engineering standard method of measurement. 2nd ed. London, Telford, 1985. 95p. tables. ISBN: 0727702181.

First published 1976.

8 sections: 1. References - 2. General principles - 3. Application of the work classification - 4. Coding and numbering of items - 5. Preparation of the Bill of Quantities - 6. Completion, pricing and use of the Bill of Quantities - 7. Method-related charges - 8. Work classification. Classes A-Y. *Class No:* 624-18

Safety Devices

[4599]

KING, R.W. Construction hazard: safety handbook. London, Butterworth, 1983.

Focuses on on-site safety in building, civil engineering, chemical and process plant construction and offshore engineering. Sufficiently detailed to cover hazards of individual trades. Based on British examples and experience. *Class No:* 624-7

Structures

[4600]

HOLMSTROM, J., *and others*. Trilingual dictionary for materials and structures. Oxford, etc., Pergamon Press, 1971. xxxvi, 947p.

Published with the assistance of Unesco. At head of title-page: 'RILEM. International Union for Testing and Research Laboratories for Materials and Structures'.

8,063 numbered entries, English-French-German. Classified arrangement; English, French and German indexes. Classes A-H: A. General aspects - B. Metals - C. Cement and concrete - D. Siliceous, Calcareous and argillaceous material - E. Soils - F. Wood - G. Other organic materials - H. Water. Each class has set subdivisions: 1. Materials; 2. Manufacture; 3. Properties and phenomena; 4. Measurements; 5. Constructional use. Bibliography of sources, p.821-4. *Class No:* 624.01

[4601]

JACKSON, N, *ed*. Civil engineering materials. 3rd ed. New York, Scholium, 1984. 429p. illus., diagrs., graphs, tables. $39.50. ISBN: 0333347919.

7 contributors. 5 parts (26 chapters): 1. Metals - 2. Timber - 3. Concrete - 4. Bituminous materials - 5. Soils (with 23 references). Some chapters (*e.g.* 11. Types of timber) carry further reading as well as references. Detailed, analytical index. Addressed to students of civil engineering, emphasizing the factors that affect the engineering decisions. *Class No:* 624.01

[4602]

LIU, H. Wind engineering: a handbook for structural engineers. Englewood Cliffs, N.J., Prentice-Hall, 1991. 209p. illus., diagrs., graphs, tables. $45.20. ISBN: 0139602798.

Covers general characteristics of wind, wind pressures, scale mode/testing, current building codes and standards (US). *Class No:* 624.01

Concrete

[4603]

Concrete structures reference guide. Gaylord, E.H., Jr. *and* Gaylord, C.N., *eds*. New York, McGraw-Hill, 1988. v.p. illus., diagrs., graphs, tables. £25.95. ISBN: 0070230676.

Material taken from the editors' *Structural engineering handbook* (1979).

8 contributors; 6 sections, closely subdivided: 1. Design of reinforced concrete structural members - 2. Design of prestressed concrete structural members - 3. Concrete construction methods - 4. Masonry construction - 5. Thin-shell concrete structures - 6. Reinforced concrete bunkers and silos. Detailed, analytical index. *Class No:* 624.012.3

[4604]

FINTEL, M.J., *ed*. Handbook of concrete engineering. 2nd ed. New York, Van Nostrand Reinhold, 1985. xiv [1], 892p. illus., diagrs., graphs, tables. £99.50. ISBN: 0442226233.

First published 1974.

25 contributors. 27 mostly documented sections (6. Properties of materials for reinforced concrete; 8. Prestressed concrete (58 references); 11. Tabular structures for tall buildings; 16. Silos and bunkers (36 references); 19. Concrete pipes; 26. Computer applications; 27. Structural plain concrete). Detailed analytical index, p.575-92. *Class No:* 624.012.3

[4605]

KONG, F.K., *and others*, *eds*. Handbook of structural concrete. London, Pitman, 1983. v.p. illus., diagrs., graphs, tables, formulas. ISBN: 0273085557.

6 parts (41 documented chapters): 1. Introduction - 2. Material - 3. Design and analysis - 4. Construction - 5. Structures - 6. Practical construction. Chapter 15, 'Earthquake: resistant structures' has 78p. and includes 84 references and a glossary; chapter 41, 'Structural design: practical guidance and information sources', 44p., 85 references. *Class No:* 624.012.3

[4606]

REYNOLDS, C.E. *and* STEEDMAN, J.C. Reinforced concrete designer's handbook. 10th ed. London, Spon, 1988. xii,436p. diagrs., graphs, tables. ISBN: 0419145303; 0419145400, Pbk.

First published 1932; 9th ed. 1981.

In 2 parts. Part I includes general and descriptive material (*e.g.* structural analysis; materials and stresses (p.36-48); electronic computational aids). Part II gives tables, data, worked examples for specific structures (*e.g.* continuous beams (p.150-72); columns (p.340-76)). 3 appendices: A. Mathematical formulae and data; B: Metric/imperial length conversions; C: Metric/imperial equivalents for common units. Analytical index, p.433-6. *Class No:* 624.012.3

[4607]

WANG, Chu-Kia *and* SALMON, C.G. Reinforced concrete design. 4th ed. New York, Harper & Row, 1985. x, 947p. illus., diagrs., graphs, tables, formulas. ISBN: 0066468963.

22 sections (with 'Selected references'), *e.g.* 1. Introduction, materials and properties; 5. Shear strength and shear reinforcement; 6. Development of reinforcement; 16. Members in compression and bending; 15. Length effects on columns; 19. Torsion; 21 Introduction to prestressed concrete; 22. Composite construction.

Reflects changes in design rules arising from publication of the 1983 American Concrete Institute *Building code and commentary*. *Class No:* 624.012.3

Metal

[4608]

HAYWARD, A. *and* WEARE, F. Steel detailers' manual. Oxford, etc., BSP Professional Books, 1989. xii,203p. diagrs., illus., graphs, tables. ISBN: 0632018453.

6 sections (1. Use of structural steel - 2. Detailing practice - 3. Design guidance - 4. Detailing data - 5. Typical connection details - 6. Examples of structures). 'Further reading', 'Bibliography', Index. Frequent reference to British Standards and Codes of Practice. Of value to practising draughtsmen, engineers, students. Highly detailed. *Class No:* 624.014

[4609]

The Sections book. Redcar, Cleveland, British Steel Corporation, 1982. 10pts. (Loose-leaf). diagrs., graphs, tables.

Published jointly by BSC Sections and Commercial Steels and the British Constructional Steelwork Association (BCSA).

The 10 parts comprise sections A-V (A: Explanatory notes; H: Safe load tables for grade 42 steels (78p.); U: Cranes; V: Other steel products). Detailed, analytical index. Collections of relevant standards and technical data for the structural steelwork designer. *Class No:* 624.014

[4610]

Steel designers' manual. 5th ed. Oxford, Blackwell Scientific, 1992. 1000p. illus. £69.50. ISBN: 0632024887.

First published 1955 (Crosby, Lockwood).

Based on limit state design and UK code of practice, BS 5950. Covers single and multi-storey buildings, industrial steelwork and bridges. *Class No:* 624.014

Plant

[4611]

Plant engineer's reference book. Snow, D.A., *ed.* Oxford, Butterworth-Heinemann, 1991. v.p. illus., diagrs., graphs, tables. £100. ISBN: 0750610158.

38 contributors, most from UK industry. 38 chapters, ranging from site location, building design and specification, energy conservation, insurance cover, health and safety and education and training. Detailed, analytical index of 15p. *Class No:* 624.05

Rock Engineering

[4612]

TEKNISKA NOMENKLATURCENTRALEN. Bergteknisk ordlista. [Glossary of rock engineering.] Stockholm, the Centre, 1979. 306p. (*TNC 73.*) ISBN: 9171960732.

791 Swedish base-terms, with explanations, followed by equivalents and indexes in 7 other languages: English, French, Spanish, German Danish, Norwegian and Finnish. *Class No:* 624.12

Soil Engineering

Encyclopaedias

[4613]

VOLLMER, E. Encyclopaedia of hydraulics, soil and foundation engineering. Amsterdam, etc., Elsevier, 1967 (Reprinted 1983). [viii], 398p. illus., tables. Dfl.225. ISBN: 0444406158.

About 4,500 brief entries, definitions ranging from 2 to *c.*100 words. Subjects: hydraulic engineering; hydromechanics; river training weirs and dams; water resources; soil mechanics; soil extraction; soil deposition; slope protection; foundation engineering; soil mechanics; foundation excavation in dry soil; foundation materials behaviour in soil. 4 appendices (symbols, abbreviations and conversion factors). Bibliography, p.[viii]. 48 line drawings. 16 tables. *Class No:* 624.13(031)

Handbooks & Manuals

[4614]

BELL, F.G. Fundamentals of engineering geology. London, Butterworths, 1983. [viii], 648p. illus., graphs tables, maps. ISBN: 0408011696.

14 documented chapters (*e.g.* 'Mechanical properties of soils', p.488-527, with 2½p. of references). Detailed non-analytical index, p.342-8. Written as a companion to the author's *Engineering geology and geotechnics* (1980). Primarily written for engineering geologists, civil engineers and mining engineers, plus students of engineering geology (*Preface*). *Class No:* 624.13(035)

[4615]

CARTER, M. Geotechnical engineering handbook. London, Pentech Press, 1983. [xx], 226p. illus., diagrs., tables. ISBN: 0727307029; 0412003414, US.

14 Chapters on site investigation procedure and geotechnical design practice (1. Soil description and classification ... 3. Laboratory testing - 4. Field testing ... 11. Retaining Walls ... 14. Flexible pavements). Includes a number of design charts in the form of data tables, extensively illustrated, each covering a specific topic (*e.g.* Data sheet 11-4: Construction for active wedges). Subject index. Intended for practising engineers, postgraduates and final year undergraduate students. *Class No:* 624.13(035)

[4616]

LEGGET, R.F. *and* KARROW, P.F. Handbook of geology in civil engineering. New York, etc., McGraw-Hill, 1982. [1184]p. illus., diagrs. ISBN: 0070370613.

5 parts (50 chapters): 1. Geological background - 2. Preliminary studies - 3. Civil engineering works (chapters 18-36) - 4. Special problems - 5. Geology and the environment. 'Dam foundations' (chapter 25) has 3p. of references, plus 'Suggestions for further reading'. Appendices: A. Glossary of geological terms; B. Geological Surveys of the world; C. Geological societies of the world; D. Some useful journals; E. Illustration credits. Subject index; name index.

Geology and engineering, by R. Legget and A.W. Hatheway (1988. 832p.), plus *Solutions manual*, is based on the *Handbook*. *Class No:* 624.13(035)

Dictionaries

[4617]

SOMERVILLE, S.M. *and* PAUL, M.A. Dictionary of geotechnics. London Butterworths, 1983. [ix], 283p. illus., diagrs., graphs, tables. £32.50. ISBN: 0408004371.

About 3,000 terms defined. Accommodates abbreviations, eponyms and formulas. Sources are cited. 20 appended tables (*e.g.* 10. 'Restivities of rocks and soils'). 100 illus. Scores over J.A. Barker's *Dictionary of soil mechanics* because of its fuller definitions and inclusion of illus. 'The definitions are geared for students and practitioners and may well be beyond the general reader's grasp' (*Choice*, v.21(1). September 1983, p.68). *Class No:* 624.13(038)

[4618]

—**JENKINS, J.D. *and* SMITH, A.M., *comps.* Thesaurus of rock and soil mechanics terms.** Oxford, etc. Pergamon Press, 1984. 62p. £9.25. ISBN: 0018031038.

Provides a structured list of 'controlled terms' for keywords used to index material in *Geomechanics abstracts*

....(contd.)

database (at Imperial College of Science and Technology, London). Terms are arranged A-Z, with preferred terms in bold type. *Class No: 624.13(038)*

[4619]
TEKNISKA NOMENKLATURCENTRALEN. Geoteknisk ordlista. [Glossary of geotechnics.] Stockholm, the Centre, 1975. 176p. (*TNC 59.*)

503 Swedish base-terms, with explanations, plus equivalents and indexes in Finnish, German, French, English and Russian. Classified list of terms precedes the glossary. *Class No: 624.13(038)*

Reviews & Abstracts
[4620]
Geotechnical abstracts. International Society for Soil Mechanics and Foundation Engineering. Cologne, German Society for Soil Mechanics and Foundation Engineering. 1970-. Monthly. DM 430pa.

1,530 abstracts pa., covering more than 500 international journals, series and conference proceedings. Sections: A. General - B. Geological and environmental aspects - C. Site investigations - D. Soil properties - E. Analysis of soil-engineering problems - F. Rock properties - G. Analysis of rock-engineering problems - H. Design, construction and behaviour of engineering works - K. Construction methods and equipment - S. Snow and ice mechanics and engineering. An index-card service, incorporated into GEODEX SYSTEM/G. *Class No: 624.13(048)*

Soil Mechanics
[4621]
ALVIN, R.G. *and* **SMOOTS, V.A. Construction guide for soils and foundations.** New York, Wiley, 1988. xxi,276p. illus., diagrs., tables. $50. ISBN: 047180486x.

30 chapters in 2 parts: 1. Concepts (soil behaviour, etc.) - 2. Applications (subsurface exploration; excavations; foundations; linings and membranes, etc.). Chapters 6 and 7 cover information sources. Detailed, analytical index, p.267-76. A practical treatment. *Class No: 624.131*

[4622]
BARKER, J.A. Dictionary of soil mechanics and foundation engineering. London, Construction Press, 1981. [v], 210p. ISBN: 086095885x.

Defines over 2,500 terms, with commentary. US terms and acronyms are included. Numerous cross-references, but lacks diagrams. Aimed at a wide audience. Has clearer, if more concise definitions than Somerville and Paul's *Dictionary of geotechnics* (qv). *Class No: 624.131*

[4623]
Developments in soil mechanics: the second Ten Rankine lectures. London, Institution of Civil Engineers, 1983. [vii], 566p. illus., ports., diagrs., charts, graphs, tables. ISBN: 0727701800.

10 documented lectures (*e.g.* 'Friction of rocks and stability of rock slopes', with 3p. of references; 'Considerations in the earthquake-resistant design of earth and rockfill dams', with 2p. of references'). No index.

The first Ten Rankine lectures, entitled *Milestones in soil mechanics* (ICE, 1970) were published annually in June issues of *Géotechnique. Class No: 624.131*

[4624]
Geomechanics abstracts. Compiled by Rock Mechanics Information Service, Imperial College of Science and Technology, London. In *International journal of rock mechanics.* Oxford, etc., Pergamon Press, 1970-. 6pa.

Formerley *Rock mechanics abstracts.*

3-400 abstracts per bi-monthly issue of *International journal of soil mechanics.* Coverage: rock and soil mechanics, engineering geology, mining, tunnelling, foundation engineering, site investigation, analysis techniques, and design methods. Online database: Pergamon INFOLINE (1977-). *Class No: 624.131*

Excavation
[4625]
CHURCH, H.K. Excavation handbook. New York, etc. McGraw-Hill, 1981. xvi, [962]p. illus., diagrs., tables, maps. ISBN: 0070108404.

18 chapters, with each section summarized, covering the 'methods, machinery, and costs for excavating the earth's mantle' (*Preface*, p.xv). Excludes underground excavation, *e.g.* mining and tunnelling, except where supplementary to surface excavation. Glossary; bibliography; index. Intended for students, practising engineers and equipment suppliers. *Class No: 624.133*

Foundations
[4626]
PILOT, G., *ed.* **Foundation engineering.** Paris, Presses Ponts et Chaussées, 1985-, v.1-. illus. ISBN: 2859780394.

1. *Soil properties. Foundation design and construction.*

V.1 (358p.) covers soil properties and site investigation, shallow and deep foundations. V.2 is to deal with retaining structures, improvement of soil properties, techniques and earthworks. Emphasizes French practice. *Class No: 624.15*

[4627]
WINTERKORN, H.F. *and* **FANG, H.-Y.,** *eds.* **Foundation engineering handbook.** New York, etc., Van Nostrand Reinhold, 1975. xvi, 751p. illus., charts, tables. £59.95.

27 contributors. 25 sections (1. Subsurface explorations and sampling ... 25. Earthquake effects on soil-formation systems). 'Foundation vibrations': p.673-99, including 3½ columns of references. Analytical index. *Class No: 624.15*

Earthquake Engineering
[4628]
Abstract journal in earthquake engineering. Berkeley, Calif., National Information Service for Earthquake Engineering, Earthquake Engineering Research Center, Univ. of California, 1971-. 2pa. $80pa. ISSN: 03635732.

About 1,000 abstracts per issue. 9 sections: 1. General topics and conference proceeedings - 2. Selected topics in seismology - 3. Engineering seismology - 4. Strong motion seismology - 5. Dynamics of soils, rocks and foundations - 6. Dynamics of structures - 7. Earthquake-resistant design and construction, and hazard reduction - 8. Earthquake effects - 9. Earthquakes as natural disasters. Title, author and subject indexes. *Class No: 624.159.1*

[4629]

Soil dynamics and earthquake engineering: proceedings of the 2nd International Conference, on board the liner, the Queen Elizabeth 2, New York to Southampton, June/July 1985. Brebbin, C.A., *and others*, eds. Berlin, etc., Springer-Verlag, 1985. [xiv, 633]p. illus., diagrs. (Computational mechanics publication). ISBN: 0905451341.

Includes 54 papers on the 'State of the Art' and the sessions covered at the conference: 1. Seismicity, ground motion and site response - 2. Soil behaviour under dynamic loads: analytical and experimental procedures - 3. Liquefacation of soils - 4. Soil structure interaction - 5. Earth dams and stability of slopes under dynamic loads - 6. Dynamic earth pressures and earth retaining structure.

Computational Mechanics Publications (Ashurst Lodge, Ashurst, Southampton SO4 2AA) aslo produce a quarterly journal, *Soil dynamics and earthquake engineering*. *Class No:* 624.159.1

Tunnels

[4630]

BICKEL, J.O. *and* **KUESEL, T.R. Tunnel engineering handbook.** New York, Van Nostrand Reinhold, 1982. viii, 690p. illus., diagrs., graphs. £63.25. ISBN: 0442281277.

25 sections (7. Rock tunnels, p.123-207: 53 references; 10. Tunnel-boring machines; 15. Subway construction; 19. Tunnel ventilation; 23. Fibre protection; 25. Tunnel operation and maintenance). Detailed, analytical index, by chapters, p.657-90. *Class No:* 624.19

[4631]

—**HOWIE, W.** *and* **CHRINES, M.**, *eds.* **Thames Tunnel to Channel Tunnel: 150 years of civil engineers.** Selected pages from the *Journal of the Institution of Civil Engineers*, published to celebrate its 150th anniversary. London, Telford, 1987. 294p. illus., diagrs., tables, plans. £15. ISBN: 072770396x.

Reprints 13 papers, some of them facsimiles (*e.g.* Sydney Harbour Bridge, p.180-239). Chronology, 1916-79, plus 'Retrospects'. *Class No:* 624.19

[4632]

SANDSTRÖM, G.E. The History of tunnelling: underground workings through the ages. London, Barrie & Jackson, 1963. xii, 427p. illus., diagrs., maps.

19 chapters, with separate chapters on the St. Gothard, Lotschberg and Simplon tunnel operations, and ending with the Mont Blanc tunnel. Glossary, p.411-4. Bibliography, p.415-7. Analytical index, p.419-27; end-papers maps. Lavishly illustrated (151 illus.). *Class No:* 624.19

[4633]

Tunnelling 91. Institution of Mining and Metallurgy. Oxford, etc., Elsevier, 1991. 523p. illus. £85. ISBN: 1851668249.

Proceedings of the 6th International Symposium, organized by the IMM, London, 14-16 April 1991. Rail and sewer tunnel work. International aspects. 465 illus. *Class No:* 624.19

Bridges

[4634]

JERVOISE, E. The Ancient bridges [of England and Wales]. London, Architectural Press, 1932-36. 4v, illus.

The ancient bridges of the south of England. 1930. xvi, 138p.

.... (contd.)

The ancient bridges of the north of England. 1931. xii, 146p.

The ancient bridges of mid and eastern England. 1932. xi, 164p.

The ancient bridges of Wales and western England. 1936. xii, 180p.

Written on behalf of the Society for the Preservation of Ancient Buildings. Many brief descriptions, with dates. Each volume has 78-90 illus. No bibliographies. Detailed indexes. *Class No:* 624.2

[4635]

LEONHARDT, F. Brücken: Ästhetik and Gestallung. [Bridges: aesthetics and design.] London, Architectural Press, 1982; Stuttgart, Deutsche Verlags-Anstall GmbH. 308p. illus. (incl. col.), diagrs. ISBN: 0851397646.

German and English text.

13 chapters (4. Guidelines for the aesthetic design of bridges; 5. Influence of alignment; 6. Influence of building materials; 7. Old stone bridges; 8. Pedestrian bridges; 10. Elevated bridges; 11. Large beam bridges; 12. Large arch and frame bridges; 13. Suspension bridges). Bibliography (952 items). Detailed, analytical index, p.299-308. Excellent illus. *Class No:* 624.2

[4636]

RICHARDS, J.M. The National Trust book of bridges. London, Cape, 1984. x, 214p. illus. (incl. col.). £12.50. ISBN: 0224021060.

10 chapters: 1. The Middle Ages - 2. The improved highway: sevententh to nineteenth centuries - 3. Private and ornamental bridges; 4. Early iron bridges - 5. Canal bridges and aqueducts - 6. Railway bridges and viaducts - 7. Suspension bridges - 8. Footbridges - 9. New styles, techniques and purposes: the last hundred years - 10. Bridges never built. Glossary. Bibliography, p.203-4. Analytical index, p.207-14. For the general reader. Well produced. *Class No:* 624.2

[4637]

WITTFOHT, H. Building bridges: history, technology, construction. Düsseldorf, Beton-Verlag 1984. 327p. illus. $99. ISBN: 3764001703.

Translated and revised ed. of *Triumph der Spannweiter*.

Sections: Use of timber - Reinforced concrete - Launching the entire bridge - Prestressed concrete's limits - Lightweight concrete - Bridges of tomorrow. About 500 photographs. Portrays 'a level of professionalism and scholarship which makes it ideal not only for the most technical collections but also for a large readership who are non-specialists but just plain bridge lovers' (*New technical books*, v.71(3), March 1986, entry 464). *Class No:* 624.2

Railway Engineering

Encyclopaedias

[4638]

NOCK, O.S., *ed.* **Encyclopedia of railways.** London, Octopus Books. 480p. illus., ports., maps.

By various hands. Sections: History (p.8-99) - Passenger travel (p.100-47) - Locomotive development (p.148-207) - Civil engineering - Great trains - Railway operation ...

....(contd.)

Railways at war (p.258-397) - Biography (p.398-427; 39 portraits) - Records - Disasters - Chronology - Facts and figures - Glossary. Index, p.466-80. 1,300 illus. 600 in colour). No bibliography. Editor: Past President, Institution of Railway Signal Engineers. *Class No:* 625.1(031)

Handbooks & Manuals

[4639]
BAKER, R.F., *ed.* **Handbook of railway engineering.** Malabar, Florida, R.F. Krieger, 1982. x, 894p. illus., diagrs., tables. ISBN: 0818744822.

4 parts (22 sections): Policies, administration & planning - Traffic control, design & construction - Bridges, tunnels, lighting - Cost, maintenance. Detailed, analytical index. 51 contributors. US slanted. Quarto. *Class No:* 625.1(035)

[4640]
HAY, W.W. Railroad engineering. 2nd ed. New York, Chichester, etc., Wiley, 1982. xvi, 758p. illus., diagrs., tables. ISBN: 0471364002.

First published 1953.

The 2nd ed. again covers 'fundamentals' of 'location, construction and maintenance of a modern railroad' from the 1st ed. (*Preface to the first edition*, p.xi), as well as up-to-date developments including track-train dynamics and economic problems. Name and subject indexes. Intended for students and the railroad industry. *Class No:* 625.1(035)

[4641]
WESTWOOD, J.N. The Railway data book. Cambridge, P. Stephens, 1985. 224p. diagrs., tables, maps. £8.95. ISBN: 0850596297.

23 chapters (2. Railway ownership and control; 4. Track and its terminology; 8. Signals; 11. Steam locomotive trends and trend-setters; 12. Electrification; 14. The diesel loco; 26. The passenger train; 23. A note on stations). 3 appendices (1. Preserved steam railways - 2. Major railway museums - 3. Railway enthusiast societies (British, French, German, US and Canadian)). Further reading (magazines, books, timetables). Non-analytical index. Concentrates on Western Europe (especially Britain) and North America. For the railway enthusiast. *Class No:* 625.1(035)

Dictionaries

[4642]
JACKSON, A.A. The Railway dictionary: an A-Z of railway terminology. Stroud, Glouces., Alan Sutton, 1992. xiii,340p. ISBN: 0750900385.

About 6,000 terms and abbreviations (including acronyms and companies). 'Track circuits': 1 column. British slant. Sources, p.xi-xiii. *Class No:* 625.1(038)

[4643]
UNION INTERNATIONALE DES CHEMINS DE FER. Lexique général des termes ferroviaires. 3. éd. Paris, The Union, 1975. 1602p.

First published 1957.

About 13,000 French-based railway terms, with German, English, Italian, Spanish and Dutch equivalents and indexes. The standard dictionary in its field. *Class No:* 625.1(038)

[4644]
—**Dictionnaire technique des chemins de fer.** Berlin, VEB Verlag Technik, 1984. 399p.

Covers English, French, German and Russian terms. *Class No:* 625.1(038)

[4645]
—**HERBERT, J.,** *comp.* **Glossary of railway terms,** English-French-German-Italian-Spanish-Swedish. Amsterdam, etc., Elsevier, 1960. [v], 413p. (Glossaria interpretum).

Covers 1,995 terms, including American English vocabulary. Main sequence is French-based, with equivalents and indexes in the other languages. Whereas the *Lexique des termes ferroviaires* was intended primarily for use in railway offices, this handy *Glossary* - now dated - is most suited for translation of railway terms at international meetings. *Class No:* 625.1(038)

Reviews & Abstracts

[4646]
British Rail Research Monthly review of technical literature. Derby, British Rail Research Division, Railway Technical Centre, 1936-. Monthly.

About 1,000 abstracts pa. 112 journals were monitored for the July 1987 issue. 2 parts: 1. Articles (with short abstracts); 2. Reports and papers. International in scope. Annual subject index. 'Up to date and well presented' (Styles, B.R., In *Information sources in engineering* (2nd ed., 1985) p.416). *Class No:* 625.1(048)

[4647]
Information Eisenbahn. Dokumentation des Fachschrifttums. Mainz, Deutscher Bundesbahn, Dokumentationsdienst, 1980-. Monthly. ISSN: 01702947.

Formerly *Kurzauszuge aus dem Schrifttum für des Eisenbahnwesen,* 1956-76.

About 3,500 abstracts. 9 sections, with supplement of patents, translations, etc. *Class No:* 625.1(048)

[4648]
Railway research & engineering news. Section B: Abstracts. Limburg, Netherlands, Railway Research Index Division, Railroad Engineering Index Institute, 1951-. Quarterly.

Supplement to Section A: *Reports and conferences.* Marked 'For internal circulation only'.

About 30 abstracts per issue, in numerous sections, 'Axles and bearings' ... 'Workshop and workshop equipment'. Intended for current awareness. *Class No:* 625.1(048)

[4649]
Selection of international railway documentation. Paris, International Union of Railways; Brussels, International Railway Congress, 1964-. 10pa.

English, French and German editions.

About 1,500 abstracts pa. Sections: Management. Administration - Research. Environment - Fixed installations. Selection of films. Index of key words per issue precedes. *Class No:* 625.1(048)

Yearbooks & Directories

[4650]
Railway directory & year book, 1988. 93rd year. Sutton, Surrey, Reed Business Publishing , Ltd., 1987. Annual. 770p. £32. ISBN: 0617005053.

Railways and officials (International; Great Britain; Africa

....(contd.)

... Middle East) - Metro and light railway systems - Trade directory (Locomotives & rolling stock ... Wagon hiring/leasing). General index; Manufacturers and suppliers index; Personal index; Advertisers' index. *Class No:* 625.1(058)

Maps & Atlases

[4651]
HOLLINGSWORTH, B. Atlas of the world's railways. New York, Everest House, 1980. 350p. illus. ISBN: 0896960489.

A country-by-country account, detailing brief history of railway system, important companies involved, current status, route mileage, etc. Sources are not stated for statistics, and supporting maps are often small scale. However, the illustrations are well chosen. No less than three indexes: general; locomotives and trains; railroads. A useful source of information by a professional railwayman (*Wilson library bulletin,* 55, December 1980, p.301). *Class No:* 625.1(084.3)

[4652]
—NOCK, O.S. World atlas of railways. London, Mitchell Beazley, 1978. 224p. illus. (col.), ports., maps.

With *c.*25 contributors. Country-by-country treatment, the main aim being to show the influence of geography on railways. Appendices include chronology, who's who, glossary, and bibliography (p.218-9). Detailed index. Critically evaluated in *Library review* (v.29, Summer 1980, p.150) for omissions in the who's who section, misleading captions, and journalistic approach to information, although the illustrations are found excellent. For sixth-form college libraries. *Class No:* 625.1(084.3)

Biographies

[4653]
MARSHALL, J. A Biographical dictionary of railway engineers. Newton Abbot, Devon, David & Charles, 1978. 247, [2]p. ISBN: 0715374895.

Concise data on more than 600 railway engineers of the past (Stephenson, George: 2 columns; 8 references. Brunel: 1½p.; 5 references). Important North American and European railway engineers are included. 2-p. non-analytical index. Intended as a work of reference for the railway historian. *Class No:* 625.1(092)

Worldwide

[4654]
FONTGALLAND, B. The World railway system. Cambridge Univ. Press, 1984. xiv, 209p. diagrs., maps. ISBN: 0521245419.

Originally published as *Le Système ferroviaire dans le monde* (Paris, CELSE, 1980).

A theoretical discussion of railways in terms of systems analysis and cybernetics. 7 chapters: 1. Principles - 2. Technology - 3. Production [of a transport system] - 4. Sales - 5. Management - 6. Macrosystems - 7. International Organizations. The author was Honorary Secretary General of the International Union of Railways. Brief bibliography; no index. *Class No:* 625.1(100)

[4655]
Jane's world railways 1992-93. 34th ed. Coulsdon, Surrey, Jane's Information Group, 1992. *c.*850p. illus., diagrs., tables. £135. ISBN: 071060999x.

Manufacturers - Private freight car leasing companies - International rail services - International associations - Consultancy services - Railway systems. *Class No:* 625.1(100)

Europe

[4656]
'Europe's railways: an asset to the environment' *Rail International,* October 1987. p.1-70. illus., diagrs., tables.

Reproduces papers given at the International Environment symposium, 22 October 1974, in Mannheim. The Group of Twelve European Railways participated.

Centres on three themes: the railway as an environmentally friendly mode of service to society and the economy; state-of-the-art technical developments of railways; the health aspects of rail services. Articles include an analysis of long-distance traffic in Austria; railways as a safe means for carriage of dangerous goods; a review of railway noise; energy-saving trains in West Germany. *Class No:* 625.1(4)

Great Britain

[4657]
AWDRY, W. *and* COOK, C. A Guide to the steam railways of Great Britain. London, Pelham Books, 1978. 240p. illus., maps.

1. The South-west - 2. London and the Home Counties - 3. East Anglia - 4. The Midlands - 5. Wales - 6. Lancashire and the North-west - 7. Yorkshire and the North-east - 8. Scotland. Gazetteer (p.221-38) and list of minor railway sites in Britain and Ireland. Description and history. Index, p.239-40. Well illustrated, popular treatment. *Class No:* 625.1(410)

[4658]
—NOCK, O.S. O.S. Nock's Pocket encyclopaedia of British steam railways and locomotives. Poole, Dorset, Blandford Press, 1983. 195 + 192p. col. illus., maps. ISBN: 0713713127.

Combines the two volumes, *The pocket encyclopaedia of steam railways of Britain in colour* (1967) and *The pocket book of British steam locomotives in colour* (1964).

Each part has historical introduction, descriptions, coloured illus. and index. Good use of colour. A popular account. *Class No:* 625.1(410)

[4659]
BAKER, S.K., *comp.* Rail atlas of Britain and Ireland. 4th ed. Oxford Publishing Co., 1984. iii, 115p. maps. ISBN: 0860932818.

First published 1977.

86 pages of maps, preceded by key map. Colours, symbols, etc. indicate surface, tunnels, depots, etc. Gazetteer (nearly 2,500 entries), p.89-115. *Class No:* 625.1(410)

[4660]

COOPER, B.K. **British Rail handbook.** 2nd ed. London, Ian Allan, 1984. 172p. illus., diagrs., maps. ISBN: 0711013640.

First published 1981.

12 sections of data: 1. BR a national railway system - 2. Numbering and wheel arrangement - 3. Permanent way - 4. Diesel traction - 5. Generators and motors - 6. Diesel locomotives - 7. Electric traction and motive power - 8. The air brake - 9. 125 mile/h and above - 10. Signalling - 11. Passenger rolling stock - 12. Freight services and rolling stock. No index. Factual; accurate. *Class No:* 625.1(410)

[4661]

JOHNSON, J. *and* LONG, R.A. **British railways engineering, 1948-80.** Bond, R.C., *ed*. London, Mechanical Engineering Publications, Ltd., 1981. 636p. ISBN: 0852984464.

15 chapters (1. End of an era, 1923 to 1947; 2. Nationalization; 5. The final years of steam traction; 6. Diesel traction; 6. Civil engineering; 11. The contribution of technological research and development; 15. Past, present and future). Appendix: Headquarters and Regional Chief Engineers, 1948-80. Bibliography, p.604-6. Index of engineers. Detailed, analytical index, p.623-36. Records the achievements of the engineering and research departments of British Railways during the first thirty years of nationalization since January 1, 1948. *Class No:* 625.1(410)

[4662]

—JAMES, L. A Chronology of the construction of Britain's railways, 1778-1855. London, Ian Allan, 1983. 120p. illus., maps. £10.95. ISBN: 0711012776.

Begins with the earliest tramway, the Froghall Railway in Staffordshire, and ends with the mid-19th century railway boom. A series of maps plots the growth of track in England, Wales and Scotland. The author was Principal of the British Railways School of Transport, Derby, 1968-1976. Index. *Class No:* 625.1(410)

[4663]

JONES, E. **The Penguin guide to the railways of Britain.** London, Allen Lane, 1981. x, 377p. illus., maps. ISBN: 0713911379.

12 sections: 1. The coming of railways - 2. The path of technical development - 3. The railways of London - 4. The Great Western ... 7. The Midland ... 11. Scotland - 12. Conclusions: the new age of railways. 3 appendices: 1. Steam, Diesel and electric wheel arrangement; 2. Railway company liveries; 3. Railway museums. Glossary of architectural terms. Guide to further reading (running commentary), p.363-5. Detailed index, p.361-77. 41 plates; 24 maps. A comprehensive historical survey of the 9 regional mainland networks. *Class No:* 625.1(410)

[4664]

OTTLEY, G. **A Bibliography of British railway history.** 2nd ed. London, H.M. Stationery Office, 1983. 683p. ISBN: 0112903347.

First published 1965 (683p.). 2nd ed., - a reprint. At head of title-page 'Science Museum/National Railway Museum'.

7,950 entries, many annotated briefly, with locations in 16 contributing libraries (notably the British Museum) and private collections. Excludes children's books (usually) and highly technical specialized books, but includes spotter's books and others not specifically railway books yet containing useful railway material. Detailed classification. Main classes A-T (B. Railway transport at particular periods;

.... (contd.)

C. Railway transport in particular areas; R. Research and study of railways and railway history; T. General directories, gazetteers, atlases, guide books, lists of stations, distance tables, time-tables). Entries state number of illus. Appended lists and 'genealogies' of railway companies. Extensive author-title and subject index (p.477-683). A model of its kind.

Supplement published 1988. *Class No:* 625.1(410)

[4665]

WIGNALL, C.J. **Complete British railway maps and gazetteers, 1825-1985.** Fully rev. ed. Poole, Dorset, Oxford Publishing Co., 1985. [iv], 71 + 79p. maps. ISBN: 086093294x.

First published 1983.

Detailed maps (p.2-71), indicating by colour distinction 6 types of lines (*e.g.* lines open/closed to passenger; lines underground at this point). 79p. gazetteer of stations. Coverage includes Isle of Man but not Northern Ireland. A well-constructed atlas. *Class No:* 625.1(410)

USA

[4666]

FREY, R.L. **Railroads in the nineteenth century.** New York, Facts on File, 1988. 491p. illus., maps. $75. (*Encyclopedia of American business history and biography,* 2.) ISBN: 0816020124.

48 contributors; 126 articles, most fairly lengthy (3,000 to 5,000 words). Emphasis is biographical, but also covers brakes, coupling, locomotives, etc. Bibliographies follow most articles and locate archival material. Extensively illustrated. 'A must' for scholars and railway enthusiasts (*Choice,* April 1989, p.1312). *Class No:* 625.1(73)

[4667]

HUBBARD, F. **Encyclopedia of North American railroading:** 150 years of railroading - the United States and Canada. New York, etc., McGraw-Hill, 1981. vi, 376, [1]p. illus., ports., maps. ISBN: 0070308284.

A-Z arrangement of entries (mostly documented: Air conditioning ... Zulu cars. Traces US and Canadian railway history from the 1830s onwards. Includes biographies and railway companies. Cross-references. Detailed, analytical index. Dramatizes 'the interesting instead of sticking too closely to the nuts and bolts' (*What this book is all about*). *Class No:* 625.1(73)

Model Railways

[4668]

SMEED, V., *ed*. **The Complete railway modeller.** London, Ebury Press, 1982. 192p. illus. ISBN: 0852231962.

8 contributors. The 12 fully-illustrated chapters concern: Scales, space & selection; track design; track construction; twin track wiring; locomotives; rolling stock; new developments; tools; etc. Glossary, p.182-5. Detailed index, p.187-91. *Class No:* 625.1-181.4

Track & Permanent Way

[4669]

WHITEHOUSE, P.B. *and* **SNELL, J.B. Narrow gauge railways of the British Isles.** Newton Abbot, Devon, David & Charles, 1984. 160p. illus., facsims., maps. £10.95. ISBN: 0715385232.

Describes narrow-gauge railways with gauges of 15in.-3 feet, now mainly used as tourist routes. Despite the title, this volume includes Ireland, where 3ft. was the norm for this type of railway. Bibliography, p.158. Index. Quarto format. *Class No: 625.14*

Underground Railways

[4670]

CARRINGTON, B., *and others*. **Subways of the world.** In *Mass transit.* V.12(11), November 1985; v.12(12), December 1985. illus., diagrs.

Mass transit. v.12(11), November 1985, p.10-58, provides an A-Z guide to the world's operating metro systems. Details give population served, network length, system status, fare system, equipment used, etc. *Mass transit,* v.12(12), December 1985, p.44-54, lists A-Z world metro systems that are under construction and proposed systems. *Class No: 625.42*

[4671]

DAY, J.R. A Source book of underground railways. London, Ward Lock, 1980. 128p. illus. ISBN: 0706358945.

A brief survey covering the beginning, electrification, growth, automation and future of the underground, as well as the design of trains and stations. Emphasis is on Europe and North America. Profusely illustrated. *Class No: 625.42*

[4672]

—**HOWSON, N.F.** London's Underground. London, Ian Allan, 1981. 160p. illus., maps. ISBN: 0711010439.

With 9 chapters (1. Two historic railways: Metropolitan; Inner circle - 2. The first tube railways ... 9. The Victoria and Jubilee lines, and the future). Appendices: 1. Chronology of principal events in London's underground railways (1863-1969); 2. List of underground lines and routes. Illustrations, p.74-97. *Class No: 625.42*

Tramways

[4673]

DUCKLEY, R.J. History of tramways from horse to rapid transit. Newton Abbot, Devon, David & Charles, 1987. 184p. ISBN: 0713306416.

8 chapters, international in scope. 1. Horse tramways - 2. Mechanical traction - 3. The early electrical tram - 4. The standard era (American, British and European) - 5. Beyond the city boundary - 6. Modernization - 7. The rapid tramway. Bibliography, p.178-9. Detailed index (countries; subjects). *Class No: 625.46*

Cable Railways & Ropeways

[4674]

Internationale Seilbahnterminologie. [International ropeway terminology.] Vienna & Heidelburg, Bohmann, 1965. 8pts. illus.

German-based terms, with English, French and Italian equivalents. 8 parts: 1. Ropeway systems - 2. General data -

....(contd.)
3. Ropes - 4. Stations - 5. Line equipment - 6. Rolling stock - 7. Operating equipment - 8. Operations. No index. *Class No: 625.5*

Roads

Databases

[4675]

IRRD/International Road Research Documentation. Paris, OECD.

A database produced by the Organization for Economic Cooperation and Development/OECD, Paris. Available through hosts ESA-IRS and DATA-STAR.

COMPENDEX (1969-), the database for *Engineering index* (now *Engineering index monthly*) also covers road engineering and has a number of hosts, including both ESA-IRS and DATA-STAR. *Class No: 625.7(003.4)*

Bibliographies

[4676]

STYLES, B.R. 'Highway, transport and constructional engineering'. Information sources in engineering (2nd ed. Editor, L.J. Anthony. 1985), p.419-28. London, Butterworth, 1985.

Organizational sources of information - Highway and transport engineering - Documentary sources of information (abstracting and indexing services; journals, conferences, monographs and textbooks). *Class No: 625.7(01)*

Handbooks & Manuals

[4677]

BAKER, R.F., *ed*. **Handbook of highway engineering.** Malabar, Florida, Robert E. Kreiger Publishing Co., 1975. 894p. illus., diagrs., tables, maps. ISBN: 0898744822.

51 contributors. 4 parts; 26 sections (*e.g.* 4. Urban transportation planning (with 27 references); 10. Traffic control; 16. Pavement design, construction and reconstruction (with 79 references); 24. Maintenance procedures; 26. Snow and ice control). Detailed, analytical index, p.883-94. US slanted. *Class No: 625.7(035)*

[4678]

MONTGOMERIE, G., *ed*. **The Transport engineer's handbook.** 2nd ed. London, Kogan Page, 1984. 231p. illus. ISBN: 0850387930.

Covers road transport only. Topics include UK legislation and its effects on commercial road haulage, vehicle design, construction and road testing., technical developments, commercial driving techniques. Directory of manufacturers, distributors and trade associations (p.195-227). Index. *Class No: 625.7(035)*

[4679]

OGLESBY, C.H. *and* HICKS, R.G. **Highway engineering.**
4th ed. New York, London, etc., Wiley, 1982. xiii, [1],
844p. illus., charts, tables. £74; £21.95. ISBN: 047102936x,
HB; 0471871451, PB.

First published 1954; 3rd ed. 1975.

The new edition gives 'added attention' (*Preface,* p.vi) to
rehabilitation and maintenance, efficient management, public
transport, energy shortages and costs, safety, new materials
and problems of developing nations. 21 chapters, each with
problems: 3. Highway and urban transporation planning ...
8. Driver, vehicle, traffic and road characteristics - 9.
Highway design ... 15. Constructing the road bed ... 19.
Bituminous pavements ... 21. Highway maintenance and
rehabilitation. Index, p.809-44. For practising engineers and
administrators, for junior, senior and fifth-year college
students, and as a starting point for advanced courses and
individual study. A neat production. *Class No:* 625.7(035)

[4680]

SALTER, R.J. **Highway design and construction.** 2nd ed.
London, Macmillan Education, 1984. x, 288p. illus.,
diagrs., graphs, tables. ISBN: 0333459289.

First published 1979.

8 documented chapters. 1. Highway pavement materials -
2. The production and testing of bituminous mnaterials - 3.
Flexible pavement - 4. Concrete pavement - 5. Dressings -
6. Earthworks - 7. Permanent construction - 8. Pavement
maintenance (p.220-83. 31 references). Detailed, analytical
index. The author is at the University of Bradford.
Class No: 625.7(035)

Dictionaries

[4681]

BRITISH STANDARDS INSTITUTION. **Glossary of
building and civil engineering terms. Part 2. Civil
engineering. Subsection 2.4.1 Highway engineering.**
London, BSI, 1986. 22p. (*BS 6100: Subsection 2.4.1:1986.*)

Supercedes BS 892: 1967.

Terms defined in systematic order. Alphabetical index.
Class No: 625.7(038)

[4682]

INTERNATIONAL ROAD RESEARCH
DOCUMENTATION. **Thesaurus 1975: English
alphabetical list: English-French-German.** Paris, IRRD:
OECD, [1975?]. [v], 105, 7p.

About 3,00 English-base terms, with selected equivalents
in French and German. Coverage: Highway engineering and
administration - Design of roads and related structures -
Materials - Soils and rocks - Construction and supervision of
construction - Maintenance - Traffic - Accident studies -
Vehicles. *Class No:* 625.7(038)

[4683]

—Ordbok for veg- og traffikk teknikk. Norsk-dansk-engelsk-
fransk-svensk-tysk. Oslo, Universitetsforlaget, 1973. 137p.
illus. (*Radet for teknisk terminololgi RTT31.*)

Records *c.*1,000 Norwegian-based terms on road and
traffic engineering, with Danish, English, French, Swedish
and German equivalents and indexes. *Class No:* 625.7(038)

Reviews & Abstracts

[4684]

Documentation strasse. Cologne, Farschungsgesellschaft für
Strassen und Verkehrswesen, 1961-. Monthly.

About 100 abstracts per issue, covering administration,
legal aspects, planning, materials, trial methods, road
building in developing countries, road bridges and tunnels.
Indexes per issue. *Class No:* 625.7(048)

[4685]

HRIS abstracts. Washington, Highway Research Information
Service, National Research Council, 1968-. Quarterly. ISSN:
00176222.

About 1,800 abstracts per issue, covering research
reports, technical papers in conference proceedings, and
journal articles. Classified order (*e.g.* 11. Administration;
17. Energy and environmental; 32. Cement and concrete;
51. Safety; 54. Operations of traffic control; 62. Foundations
(soil); 90. Highway research, general). Source, author and
retrieval term indexes.

TRIS (Transportation Research Information Service) is the
online service, hosted by DIALOG, and available to US and
Canadian users. *Class No:* 625.7(048)

[4686]

Road abstracts: a review of world literature on roads and
traffic. Crowthrone, Berks., Technical Information and
Library Group, Road Research Laboratory, 1934-68. v.1-
35(4). Monthly.

Over 1,000 abstracts pa. 24 sections: 1-11. Construction
and executive - 12. Bridges, tunnels and other structures -
13. Surface water drainage and culverts - 14. Traffic
engineering (general) - 15. Safety and accidents - 16.
Planning and economics - 17. Road lay-out - 18. Direction
and control of traffic - 19. Traffic studies and parking - 20.
Methods and apparatus for traffic studies - 21. Road user
characteristics - 22. Surface characteristics - 23. Street
lighting - 24. Vehicles and vehicle lighting. A feature is the
grouping of references within a single entry. Covers *c.*170
journals, plus specifications and books. Annual subject and
author indexes.

Superseded by International Road Documentation (IRRD)
service (Paris, OECD; English language version.
Crowthorne, Berks., Transport and Road Research
Laboratory, 1965-), producing *c.*8,500 abstracts pa. from
magnetic tape. *Class No:* 625.7(048)

[4687]

Road documentation for developing countries. Crowthorne,
Berks., Overseas Unit, Transport and Road Research
Laboratory, 1974-. Annual.

500 abstracts pa. Coverage: Design of roads and related
structures; road materials; soils and rocks; road construction
and maintenance; traffic and transport; road accident studies;
vehicles; economics and administration.

Contributes to OECD's INTERNATIONAL ROAD
RESEARCH DOCUMENTATION database through
DATA-STAR and ESA-IRS. (Based on entry 351 in H.J.
Stephens. *Inventory of abstracting and indexing services
produced in the UK.* 2nd ed., 1983.). *Class No:* 625.7(048)

Histories

[4688]

SCHREIBER, H. **The History of roads, from amber route to motorway.** Translated by S. Thomson. London, Barrie & Rockliff, 1961. vii, 311p. illus., map.

Chronological treatment, in 17 chapters. Chapters 1-9 deal with ancient and Roman roads. Bibliography, p.293-5. Analytical index. *Class No:* 625.7(091)

Great Britain

[4689]

BALLAN, D. **Bibliography of road making and roads in the United Kingdom.** London, King, 1914. xviii, 281p.

Lists 4,000 items, covering the period 1535 to the beginning of the 20th century, - an enlargement of a bibliography prepared in 1906 by Beatrice and Sidney Webb. Sections: General works - Great Britain - England and Wales - London - Wales - Scotland - Ireland - Construction and repair of roads - Traffic. Author and subject indexes. Full bibliographical details, plus British Museum shelf numbers and locations in the British Library of Political and Economic Science, or, failing that, elsewhere. *Class No:* 625.7(410)

Research

[4690]

INTERNATIONAL ROAD FEDERATION. **IRF research and development.** Washington, International Road Federation (IRF). Annual.

1980 ed. as *World survey of current research and development on roads and road transport.*

Sponsored by the US Federal Highway Administration, OECD, and the Transportation Research Board.

An inventory of ongoing road research in 77 countries. Criticized by B.R. Styles (in *Information sources in engineering.* 2nd ed, 1985, p.417) for loose editing, an inadequate index and 'an awkward-to use subject arrangement', despite its invaluable coverage. *Class No:* 625.7.001.5

[4691]

—GREAT BRITAIN. Department of the Environment *and* DEPARTMENT OF TRANSPORT. **Register of research, 1977. Pt.3: Roads and transport.** London, DOE, HQ Library, 1977. vi, 362p. gratis.

The last of its kind to be published. The 1977 ed. recorded 1,477 numbered entries for road and transport projects in the UK, based on information collected from a survey of *c.*2,300 organizations (government depts., local authorities, research associations, universities, nationalized and private industry; and consultants). Appended: Research organizations; list of addresses; indexes of research sponsors, research workers and research organizations. Subject index. *Class No:* 625.7.001.5

Hydraulic Engineering

Databases

[4692]

FLUIDEX. **BHRA Fluid engineering abstracts.** Cranfield, Beds., British Hydromechanics Research Association.

Covers 1973 onwards. References, with abstracts, to literature of fluids, including aerodynamics, fluid flow, fluid engineering, offshore technology and rheology. About 250,000 recordsp updated monthly. Available on DIALOG, file 96 and Data-Star. Cost per connect hour: $69 and £52.65. Available on magnetic tape. Publications: *BHRA Fluid Engineering thesaurus*; *Guide to the Fluidex database*, *Fluidex Notes*, *Fluidex Newsline*. Printed versions: several abstracts journals covering the various fields of interest, *e.g. Tribos*, *Fluid-sealing abstracts*, *Pumps and other fluids machinery abstracts*, *Pipelines abstracts*. *Class No:* 626(003.4)

Bibliographies

[4693]

MYERS, A. **Guide to information services in marine technology.** 3rd ed. Edinburgh, Institute of Offshore Engineering, Heriot-Watt University, 1979. 136p. £3. ISBN: 0904046060.

2nd ed. 1976.

Directory of 115 organizations (Aberdeen City Libraries... Y-ARD, Ltd.), p.12-126. Data include: initial enquiries to ... ; background; subject interests; information services; publications. Appended lists: Printed marine technology journals; Abstracting and indexing journals and computerised databases (annotated). Bibliography, p.132. Non-analytical subject index. 'Written with the needs of UK users primarily in mind' (*Introduction*). *Class No:* 626(01)

Encyclopaedias

[4694]

VOLLMER, E. **Encyclopaedia of hydraulics, soil and foundation engineering.** Amsterdam, etc., Elsevier, 1967 (Reprinted 1983). [viii], 398p. illus., tables. Dfl.225. ISBN: 0444406158.

About 4,500 brief entries, definitions ranging from 2 to *c.*100 words. Subjects: hydraulic engineering; hydromechanics; river training weirs and dams; water resources; soil mechanics; soil extraction; soil deposition; slope protection; foundation engineering; soil mechanics; foundation excavation in dry soil; foundation materials behaviour in soil. 4 appendices (symbols, abbreviations and conversion factors). Bibliography, p.[viii]. 48 line drawings. 16 tables. *Class No:* 626(031)

Handbooks & Manuals

[4695]

MYERS, J.J. **Handbook of ocean and underwater engineering.** New York, etc., McGraw-Hill, 1969. [1100p.] illus., diagrs., graphs, tables.

12 sections: Basic oceanography - Basic hydrodynamics - Underwater fields and instrumentation - Tools, rigging and machinery - Cable technology - Underwater power sources -

....*(contd.)*

Materials and equipment testing - Fixed structures - Vessels and floating platforms - Diving - Ocean operations - Wind and wave loads. Includes bibliographies. Refs. 673 illus. 'A vital engineeing reference book' (*Library journal,* v.94(13), July 1969, p.2592). *Class No:* 626(035)

[4696]
PINCHES, M.J. *and* ASHBY, J.G. **Power hydraulics.** Englewood Cliffs, N.J., Prentice Hall, 1989. 400p. diagrs. $54.67. ISBN: 0136874436.

Covers pumps, hydraulic valves, actuators, fluids, hydraulic system design, maintenance and control systems. Detailed, analytical index. Appendix includes exercises and answers. 'An excellent practical reference' (*Choice,* May 1989, p.1548). *Class No:* 626(035)

[4697]
WARRING, R.H. **Hydraulic handbook.** 8th ed. Morden, Surrey, Trade & Technical Press 1983. xviii, 506p. illus., diagrs., tables. ISBN: 0854610944.

First published 1958.

Sections: 1. Principles - 2A/E. Components - 3. Systems - 4. Machines - 5. Instrumentation and testing - 6. Applications - 7. Surveys (*e.g.* of hydraulic pumps and motors) - 8. Buyers guide. Subject index. Advertisers' index. The standard handbook, reflecting current developments in oil hydraulic engineering. *Class No:* 626(035)

[4698]
ZIPPARRO, V. *and* HASEN, H. **Handbook of applied hydraulics.** 4th ed. New York, etc., McGraw-Hill, 1992. 992p. illus. £77.95. ISBN: 0070730024.

Coverage: Hydrology and basic hydraulics - Water conveyance - Dams - Water use, control and disposal. 41 documented sections. Over 900 illus. US slant. *Class No:* 626(035)

Dictionaries

Polyglot

[4699]
ACADEMY OF THE HEBREW LANGUAGE. **Dictionary of hydraulics:** Hebrew - English - French - German. Jerusalem & Haifa, Academy of the Hebrew Language & Technion - Israel Institute of Technology, 1961. 175, 15, 14, 15p.

About 450 terms. English, French, German and Hebrew sequences, plus number key to classified list (Hebrew, with English, French and German equivalents). *Class No:* 626(038)=00

[4700]
INTERNATIONAL COMMISSION ON IRRIGATION AND DRAINAGE. **Multilingual technical dictionary on irrigation and drainage.** English, French, German. Stuttgart, Franckh'sche Verlagshandlung, 1971. 948p., illus.

12,153 terms in one sequence of 13 chapters. English, French and German indexes. 315 line drawings. Supplementary volumes in French, Spanish, Arabic, Indian, Japanese, Persian and Portugese. *Class No:* 626(038)=00

[4701]
NEUBERT, G., *ed.* **Technical dictionary of hydraulics and pneumatics:** English/French/German/Russian; with a supplement in Spanish. Oxford, etc., Pergamon, 1973. [vii], 257p. £31. ISBN: 0080169589.

About 3,500 terms are listed in the main, English-language section, each uniquely numbered, with equivalents in French, German and Russian in adjacent columns.

Supplemented by indexes in these three languages and in Spanish, linked by the term number. *Class No:* 626(038)=00

[4702]
TROSKOLANSKI, A.T. **Dictionary of hydraulic machinery:** in English, German, Spanish, French, Italian and Russian. Amsterdam, etc., Elsevier, 1985. xxiii, 737p. illus., diagrs., graphs, tables. £114.06. ISBN: 0444997288.

Revised and enlarged translation from Polish original *Magzynyi urzgdzenia hydrauliczne - Projecia podstawave* (1974).

Contains about 4,300 entries classified in 14 sections (1-4. Concepts - 5-14. Hydraulic machinery types, problems and applications). Base language is English. Concise definitions for most terms, with foreign-language equivalents. Indexes in each language lead to main entries by means of number codes. *Class No:* 626(038)=00

English

[4703]
TAYLOR, P.A. **Dictionary of marine technology.** Oxford, Butterworth, 1989. 244p. $105.

Defines terms in marine and offshore engineering, naval architecture, shipbuilding, shipping and ship operations. Intended to displace the *Dictionary of marine engineering and nautical terms* (Newnes, 1965), emphasizing marine technology terms. 'See' references; some illus. *Class No:* 626(038)=20

German

[4704]
VOLLMER, E. **Wasserwesen, Erd- und Grundbau;** Deutsch-Englisch. 2.Aufl. Stuttgart, Fischer, 1973. 415p. illus., tables. ISBN: 3437201182.

First published 1967 (415p.).

German-English, dictionary of hydraulics, soil and foundation engineering. About 5,000 German terms are given concise definitions. English index, p.332-413. Bibliography, p.414-5.; conversion tables. 52 illus. *Class No:* 626(038)=30

French

[4705]
DOBENIK, R.H. *and* HARTLINE, G.W. **Dictionnaire technique de la marine:** Anglais/Français et Français/Anglais. [Naval technical dictionary: English/French and French/English.] Paris, La Maison du Dictionnaire, 1989. [xxii],646,[19]p. illus. ISBN: 2856080316.

About 35,000 entries in each half. Includes nautical, military, technical terms. Brief bibliography (28 items). *Class No:* 626(038)=40

Reviews & Abstracts

[4706]

Civil engineering hydraulic abstracts. Cranfield, Beds., British Hydraulics Research Association - BHRA Fluid Engineering, 1968-. Monthly. £98 pa.

Prepared in collaboration with Hydraulics Research Station, Wallingford. Formerly *Channel*, 1968-73.

About 200 abstracts per issue. 4 main sections: Fluid mechanics - Hydraulics - Instrumentation and experimental techniques - Operations and utilities. Author and subject indexes per issue. Annual author, corporate author and subject indexes. *Class No:* 626(048)

Periodicals

[4707]

BARNETT, J. B. Marine science journals and serials: an analytical guide. London, etc., Greenwood Press, 1986. 171p. (*Annotated bibliography of serials: a subject approach, 7.*) ISBN: 030324717x.

Lists 327 items under titles, A-Z. International in scope (*c.*30 countries). Excludes titles concerning maritime commerce and transport, seafood preservation, and fish marketing. Geographical index; index of publishers; subject index; index of variant titles. *Class No:* 626(051)

Progress Reports

[4708]

Developments in hydraulic engineering. Novak, P., *ed.* Amsterdam, etc., Elsevier, 1983 - Annual. illus.

V.3 (1986. 318p.) concentrates on river and coastal engineering. V.4 (1987. 350p. £44) takes other aspects. Its documented contributions deal with lake hydraulics, tidal power generation, groundwater flow, and other aspects of groundwater engineering. *Class No:* 626(055)

Yearbooks & Directories

[4709]

Lloyd's international marine equipment guide. London, Lloyd's, 1992. 372p. £75. ISBN: 1850444889.

Section 1: Product categories with suppliers, A-Z by area. Section 2: *c.*7,000 suppliers. *Class No:* 626(058)

Tables & Data Books

[4710]

MORGAN, N., *ed.* **Marine technology reference book.** Boston, Mass., Butterworth, 1990. v.p. tables. $195. ISBN: 0408027143.

13 sections, with 'references and further reading': 1. Ocean environments - 2. Offshore structures ... 12. Electronic navigation and radar - 13. Maritime law. Well-produced. 17p. index. 109 photos; 750 line-drawings. 19 contributors. 'Designed to serve as a first point of reference' (*Preface*). *Class No:* 626(083)

Histories

[4711]

GARBRECHT, G., *ed.* **Hydraulics and hydraulic research:** a historical review. Rotterdam & Boston, Mass., Balkema, 1987. ix, 362p. illus., ports., facsims., graphs, tables. ISBN: 9061316216.

....(contd.)

At head of title-page: 'International Association for Hydraulic Research, 1935-1985'. Jubilee volume.

35 contributors. Papers include 'Birth of hydraulics during the Renaissance period', p.55-79 (16 references); 'Great names in the development of hydraulic machinery', p.251-60 (28 references). No index. *Class No:* 626(091)

Libraries

[4712]

Directory of library and information facilities. Moulder, D.S., *comp.* Plymouth, Marine Biological and Information Services, 1986. 37p.

Provides information on libraries. Subheadings include hours: of opening; affiliation; availability; loan services; catalogue; stock; Institute publications; Library publications. *Class No:* 626:061:026/027

Underwater Work

[4713]

GREAT BRITAIN. Ministry of Defence. Director of Naval Warfare. **Diving manual.** London, the Ministry, 1972 (reprinted 1976, incorporating changes). v.p. illus., diagrs., tables. Loose-leaf. (BR.2806).

Supersedes BR.155, *Diving manual* (1964) and *Diving regulations* (1965).

7 chapters: 1. Theory of diving - 2. Regulations - 3. Conduct of diving operations - 4. Breathing apparatus: drill and operation - 5. Decompressor - 6. Divers' illness and injuries - 7. Civilian and expedition diving. Bibliography (8 items). 5-p. index. *Class No:* 626.02

[4714]

HAUX, G.F.K. Subsea manned engineering. London, Bailliere, Tindall, 1982. x, 538p. illus., diagrs. ISBN: 0702007498.

12 sections (1. Hyperbaric diving simulators and test facilities; 3. Deep diving systems; 7. Submersibles; 8. Submarines; 11. Underwater welding, repair and construction habitats; 12. Animal experiment and equipment test chambers). 4 appendices: Research centres, institutes, government agencies and training schools; Selected diving contractors; Equipment manufacturers; Scientific and technical journals, newsletters and trade magazines. Bibliography, p.527-8. Analytical index, p.531-8. Well illustrated. An authoratitive survey by the director, Haux-life-support GmbH, Waldbronn, West Germany. *Class No:* 626.02

[4715]

LARN, R. *and* **WHEELER, R. Commercial diving manual.** Newton Abbot, Devon, David & Charles, 1984. vp. illus., diagrs., tables. £20. ISBN: 0715381447.

2 parts (24 chapters): 1. Basic diving practice - 2. Surface and underwater skills (chapter 24: Conversion factors and tables). Index of 7p. *Class No:* 626.02

[4716]

MILNE, P. H. Underwater engineering surveys. London, Spon, 1980. ix, 366p. illus., diagrs., graphs, tables. ISBN: 041911310x.

10 documented chapters (*e.g.* 3. Underwater navigation systems; 4. Hydraulic surveying (with 3p. of references); 7. Underwater photography and television; 8. Underwater work systems; 10. Underwater inspection and maintenance). Partly

.... (contd.)
analytical index, p.332-6. Quotes examples, with good illus. 'May be recommended as a pioneer work for both students and practitioners' (*British Book news,* June 1981, p.353). *Class No:* 626.02

[4717]
SOCIETY FOR UNDERWATER TECHNOLOGY. Advances in underwater technology, ocean science and offshore engineering. Dordrecht, etc., Kluwer, 1985-. illus., diagrs., tables.
25v. to date.
Volumes consider the entire field of offshore technology. V.25, *Safety in offshore drilling: the role of shallow gas surveys* (ed. D.A. Ardus and C.D. Green. 1990. viii,293p. £54.50. ISBN: 0792308991) has 13 chapters in 5 parts (Introduction - Acquisition - Processing, analysis and interpretation - Application of geophysical data - Procedures, regulations and guidelines). *Class No:* 626.02

Canals

[4718]
CALVERT, R. Inland waterways of Europe. London, Allen & Unwin, 1963. 259p. illus, maps.
13 chapters (1. The Rhine; 3. The Danube; 4/6. France; 6. Benelux; 7. Germany; 8. Russia; 9. Scandinavia; 10. Finland and the Baltic states; 11. Central Europe; 12. Southern Europe; 13. Conclusion). Appendices: 1. Maps and level profiles; 2. Passenger services. Bibliography, p.248-50. Author index. 48 illus. *Class No:* 626.1

[4719]
McKNIGHT, H. The Shell book of inland waterways. 2nd ed. Newton Abbot, Devon, David Charles, 1981. 493p. illus. (incl. col.), maps. £12.95. ISBN: 071538233x.
First published 1975.
18 chapters (*e.g.* 3. Water supply; 4. Locks and lifts; 5. Bridges; 6. Aqueducts; 7. Towns; 8. Railways and design; 11. Operation and maintenance; 12. Working boats; 13. Commercial carriers; 14. Pleasure cruising; 15. Boatyards and marinas; 17. Waterway museums; 18. Associations and boat clubs). Selected bibliography, p.208. Gazetteer (Note on derelict waterways; England and Wales; Scotland; Ireland), p.212-484. Detailed, non-analytical index. Well illustrated. *Class No:* 626.1

Docks & Harbours

[4720]
BRUUN, P. Port engineering. 4th ed. Houston, Texas, Gulf Publishing Co., 1988. 2v. illus., diagrs., charts, tables. $350, set.
First published 1973. 3rd ed. 1981.
V. 1. *Harbour planning, breakwaters, and marine terminals.*
V. 2. *Harbour transportation, fishing ports, sediment transport, geomorphology, inlets, and dredging..*
A comprehensive reference work. V. 2 carries an index. *Class No:* 627.2/.3

[4721]
World ports and harbours abstracts. Cranfield, Beds., British Hydromechanics Research Association, 1983-. 6pa. £75pa. ISSN: 02640775.
Incorporates *International dredging abstracts* (1976-82).
1986: 1,011 abstracts. Sections: Hydraulics - Development/maintenance - Environmental aspects - Port operations - Vessels (general; rigs; workboats; dredgers; hoppers). Author, corporate author and subject indexes per issue and annually. *Class No:* 627.2/.3

Cargo Handling

[4722]
Cargo handling abstracts. London, International Cargo Handling Coordination Association. (ICHCA), 1978-. Quarterly. £20pa.
About 30 abstracts per issue, covering journals, technology reports, and conference proceedings. *Class No:* 627.35

[4723]
INTERNATIONAL CARGO HANDLING COORDINATION ASSOCIATION. A Multilingual glossary of cargo handling terms. 3rd ed. London, the Association, 1987. 205p. illus. ISBN: 0906297974.
First published 1982; 2nd ed. 1984.
462 terms in 6 sections: 1. Transport vehicles - 2. Handling and securing equipment - 3. Goods and unit loads - 4. Infrastructure and terminals - 5. Cargo Administration - 6. Abbreviations. English terms with Dutch, Finnish, French, German, Italian, Portuguese, Spanish and Swedish translations. Each entry has a brief definition in English. Language indexes. *Class No:* 627.35

[4724]
International distribution and handling review. Cranford, Beds., Cranfield Institute of Technology, 1971-. 6pa. £50 pa.
120 abstracts per issue. Coverage: distribution systems, freight transport, storage and warehousing, packaging safety, and training of workers. *Class No:* 627.35

Floods & Flood Control

[4725]
HANDMER, J., ed. Flood hazard management: British and continental perspectives. Norwich, Geo Books, 1987. viii, 297p. ISBN: 0860942082.
6 sections (with subsection references): 1. Introduction: the British urban flood hazard - 2. Flood related institutions and policy in Britain - 3. Implementation of land use policy: local and international experience - 4. Hazard response - 5. Project appraisal and risk assessment - 6. Conclusion. The product of research and policy development activity at the Middlesex Polytechnic Flood Hazard Research Centre. *Class No:* 627.51

[4726]
NATURAL ENVIRONMENT RESEARCH COUNCIL. Flood studies report. London, the Council, 1975. 5v. charts, graphs, tables, maps.
1. *Hydrological studies.* 2. *Meteorological studies.* 3. *Flood routeing studies.* 4. *Hydrological data.* 5. *Maps.*
V.5 ([viii], 541p.) describes the collection, appraisal and processing of hydrological data used in the investigation and presents the basic data. Apart from flood record statistics, it

....*(contd.)*

provides a master list of gauging stations, catchment characteristics and flood statistics. The 24 maps in v.5 illustrate v.1-2. The work of hydrologists at the Institute of Hydrology. Confined to Great Britain and Ireland. *Class No:* 627.51

[4727]

River and flood control abstracts. Cranfield, Beds., British Hydromechanics Research Association, 1985-. 6 pa. £84. ISSN: 02669870.

1986: 603 abstracts. Sections: General - Control structures - Flow routeing structures - Isolated structures. Author, corporate author and subject indexes per issue and annually. FLUIDEX database. *Class No:* 627.51

[4728]

UNITED NATIONS. Educational, Scientific and Cultural Organization. **World catalogue of very large floods.** Paris, Unesco, 1976. 424p. (*Studies and reports in hydrology.*)

This overview of flood situations throughout the world includes information enabling more elaborate methods to be used for treatment of data. English, French, Spanish and Russian versions available. *Class No:* 627.51

Navigation Aids

Lighthouses

[4729]

HAGUE, D.B. *and* **CHRISTIE, R. Lighthouses:** their architecture, history and archaeology. London, Gower, 1975. xiv, [i], 307p. illus., maps.

The setting is mainly Great Britain and Ireland. 8 chapters: 1. History - 2. British lighthouse administration - 3. General design and construction - 4. Rock towers - 5. Illumination and fog signals - 6. Light vessels and floating aids - 7. Unit beams, daymarks or seamarks - 8. Lighthouse builders. Glossary. Appendix (*e.g.* Light dues; building estimates). Bibliography (grouped, including Acts of Parliament, manuscript sources). Analytical index. 29 plates; 47 line-drawings. *Class No:* 627.715

[4730]

—STEVENSON, D.A. The World's lighthouses before 1820. London, Oxford Univ. Press, 1959. xxiv, 310p. illus., diagrs., tables, maps.

Records the history of development of seamarks 'from wood fires of antiquity to the reflector lights of 1819'. *Class No:* 627.715

Dredging

[4731]

International dredging abstracts. Cranfield, Beds., British Hydromechanics Research Association, BHRA Fluid Engineering, and International Association of Dredging Companies, 1976-82. Quarterly.

About 500-600 abstracts pa. World literature on soil mechanics, site hydraulics, dredging craft and components, plus spoil disposal operations, dredging policy, legislation, environmental aspects and economics.

Incorporated in *World ports and harbours abstracts,* 1983-. *Class No:* 627.74

Dams

[4732]

GOLZE, A.R., *ed.* **Handbook of dam engineering.** New York, etc., Van Nostrand Reinhold, 1977. [xiii], 793p. illus., diagrs., charts.

14 sections, by various hands, with references or bibliographies. 1. Planning and environmental studies - 2. Hydrological studies - 3. Materials suitable for construction ... 7. Design of rockfill dams ... 14. Public safety controls for dams and reservoirs. Reflects current American practice. Recommended 'unreservedly for students, graduates and practising engineers' (*New Scientist,* 27 April 1978, p. xviii). *Class No:* 627.8

[4733]

INTERNATIONAL COMMISSION ON LARGE DAMS. Technical dictionary on dams. 2nd ed. Moscow, the Commission, 1962. 380p.

Prepared for publication by the Committee for the USSR Particiapation in International Power Conferences.

About 4,500 terms in Russian, French, English, German, Spanish, Italian, Portugese, Rumanian, Czech, French. 4 Subject sections: 1. General; 2. Dams; 3. Appurtenant works; 4. Construction works. Languages index.

Norwegian edition - *Onbok for dambygging* (2nd ed. Oslo, RTT, 1968. [ix], 270p. illus.), similarly arranged, has *c.*4,500 Norwegian terms, with French, English, German, Spanish, Italian, Portugese equivalents. 7 language indexes. *Class No:* 627.8

[4734]

World register of dams. [Registre mondial des barrages.] Paris, International Commission on Large Dams, 1984. 752p. illus., tables.

Earlier ed. 1964; supplement 1966.

List of dams (member countries, Corée ... Zimbabwe) p.147-728. List of dams (non-member countries). p.729-51. Data on each include location: high and crest length of dam; volume; gross capacity of reservoir; owner; engineering by; construction by. Bibliography, p.652. Classified and statistical tables precede lists. 25 illus. A 'large dam' is defined as one of at least 15 metres (50 ft.) high, from lowest general foundations to crest. *Class No:* 627.8

Public Health Engineering

Abbreviations & Symbols

[4735]

WENNRICH, P. Anglo-American and German abbreviations in environmental protection. Munich, Verlag Documentation, 1979. 550p. £56. ISBN: 3598100604.

A supplement to Wennrich's one A-Z sequence *Anglo-American and German abbreviations in science and techology* (1976-77. 3v.). *Class No:* 628(003)

Databases

[4736]

ENVIROLINE. New York, Environment Information Center, 1971-.

ENVIROLINE is the database corresponding to *Environment abstracts*. Available through hosts DIALOG, ESA-IRS, ORBIT and SDC.

European Communities Commission's ENGUIDE (Munich, Franklin, 1980. EUR 6842) is a guide to bibliographical databases for users of environmental information. *Class No:* 628(003.4)

Encyclopaedias

[4737]

ASHWORTH, W. The Encyclopedia of environmental studies. New York, Facts on File, 1991. 424p. £35. ISBN: 0816015317.

Describes related scientific terminology, regulatory agencies and environmental laws, individual environmentalists and events that have profoundly affected the environment. *Class No:* 628(031)

[4738]

Encyclopedia of environmental control technology. Cheremisinoff, P.N., *ed.* Houston, Texas, Gulf Publishing Co., 1989-. V.1-. diagrs., charts, tables.

V.1. *Thermal treatment of hazardous wastes.* 1989. 828p. $155.

V.2. *Air pollution control.* 1989. 1066p. $155.

V.3. *Wastewater treatment technology.* 1989. 684p. $155.

V.4. *Hazardous waste containment and treatment.* 1990. 776p. $155.

Each volume has chapter references and an index. V.1 covers current treatment methods in the field and projects future trends in research and technologies. Vols. 5-10 are forthcoming. *Class No:* 628(031)

[4739]

Encyclopedia of environmental health: law and practice. Hawke, N., *ed.* London, Sweet & Maxwell, 1968-. 4v. loose-leaf. (*The Local government library*.)

A comprehensive catalogue of statute law, instruments and government (UK) circulars. Regular updates issued. *Class No:* 628(031)

[4740]

McGraw-Hill encyclopedia of environmental science. 2nd ed. New York, etc., McGraw-Hill, 1980. 858p. illus., diagrs., graphs, tables, maps. $72.50. ISBN: 0070452644.

First published 1974.

Articles mainly drawn from the *McGraw-Hill Encyclopedia of science and technology*. Added section of feature articles on broad topics of special importance. *Class No:* 628(031)

[4741]

—TUDGE, C., *ed.* The Encyclopedia of the environment. London, Croom Helm, 1988. 248, [8]p. illus. (col.), ports, diagrs., maps.

Has 6 contributors. 9 sections: 1. Observation and explanation - 2. Animal abilities - 3. Animal behaviour - 4. Ecology and ecosystems - 5. The rise of humanity - 6. Support from the law - 7. Support from the water (*e.g.* fish-farming) - 8. Destruction and pollution - 9. Conservation. Further reading and credits. Glossary of terms. Detailed

....(contd.)

analytical index of 5p. The main feature: the many and striking col. illus. Main theme: conservation of the natural world. *Class No:* 628(031)

Handbooks & Manuals

[4742]

CLAY, H.M. Clay's handbook of environmental health. 15th ed., edited by W.H. Basset and F.G. Davies. London, H.K. Lewis, 1981. viii, [2], 851p., illus., tables. ISBN: 0718604512.

First published 1933; 14th ed. 1977.

36 chapters in 6 parts: 1. Administration - 2. Construction technology - 3. Housing - 4. Health and safety - 5. Pollution control - 6. Food safety and hygiene. Most chapters have 'Further reading'. Changes in the 15th ed. include a new chapter on food hygiene and a rewritten chapter on occupational health and safety. Index, p. 821-51. Cites Acts in full. A basic handbook.

The 16th ed. was published in 1991. *Class No:* 628(035)

[4743]

JACKSON, M.H., *and* *others.* **Environmental health reference book.** London, Butterworths, 1988. *c.*600p. illus. £70. ISBN: 0408026006.

15 sections: Introduction - Food poisoning and hygiene - Food control - Pathology of food animals - Housing administration - Building science and technology - Health and safety - Atmospheric pollution - Water quality and water supply - Sewerage and waste water disposal - Solid waste management - Noise control - Radiation - Pestology - Communicable diseases. Well illustrated. Emphasis is on the technical aspects rather than on legislation. *Class No:* 628(035)

[4744]

Standard handbook of environmental engineering. Corbitt, R.A., *ed.* New York, McGraw-Hill, 1990. v.p. illus., diagrs., graphs, tables. $89.50. ISBN: 0070131589.

Comprehensive, covering technical aspects of air and water quality control standards and treatment, wastewater and solid waste disposal, and hazardous waste management. Separate legislation/regulations section. Over 900 tables, charts and diagrams. Extensive chapter references; lengthy subject index. US slanted. *Class No:* 628(035)

Dictionaries

Polyglot

[4745]

Dictionary of environmental technology: English/French/German/Russian. Amsterdam, Elsevier, 1988. 527p.

4-column page; compressed layout. 'This appears to be an excellent dictionary' (*Language monthly,* May 1988, p.22). *Class No:* 628(038)=00

[4746]

TEKNISKA **NOMENKLATURCENTRALEN. Miljöordlista.** svensk-engelsk-fransk-tysk. [Glossary of environmental terms. Swedish-English-French-German.] Stockholm, the centre, 1972. vi, 86p.

772 terms relating to human environments. In Swedish, with English, French, and German equivalents and indexes. *Class No:* 628(038)=00

English

[4747]

CRUMP, A. Dictionary of environment and development: people, places, ideas and organizations. London, Earthscan, 1991. 272p. £15. ISBN: 185383078x.

Entries range from brief definitions to short articles. Includes current and retrospective statistics. 'Recommended' (*Choice*, March 1992, p.1042). *Class No:* 628(038)=20

[4748]

GILPIN, A. Dictionary of environmental terms. London & Henley, Routledge & Kegan Paul, 1978. 199p. illus., maps. First published 1976.

Clear definitions of over 900 terms, with ample cross-references. 'Radioactivity': 2p., including map of Windscale area; 'Los Angeles smog': nearly 2p. Appendix: 'The United Nations Conference on the Human Environment, 1972'. *Class No:* 628(038)=20

[4749]

PORTEOUS, A. Dictionary of environmental science and technology. Milton Keynes, Open Univ. Press, 1991. xii,403p. diagrs., graphs, tables. ISBN: 0335092314; 0335092306, Pbk.

Entries, A-Z (Abatement ... Zoo plankton), defined, sometimes at length (*e.g.* 'Pesticides': *c.*1p.). Cross references. 4 appendices (1. 'Pollution and the environment: organizations'). Author is professor of environmental engineering at the Open University. *Class No:* 628(038)=20

[4750]

STEVENSON, H. *and* WYMAN, B. The Facts on File dictionary of environmental science. New York, Facts on File, 1991. 294p. $25; £18.50. ISBN: 0816023174.

A popular treatment, with about 3,000 current terms and some acronyms (*e.g.* 'Not-in-my-backyard syndrome: NIMBY'). 'Useful in undergraduate libraries' (*Choice,* March 1992, p.1056), but not detailed enough for the specialist. *Class No:* 628(038)=20

Spanish

[4751]

VILLATE, J.T. Dictionary of environmental engineering and related science. [Diccionario de ingenieria ambiental y ciencias afines. English-Spanish, Spanish-English.] Miami, Florida, Ediciones Universal, 1979. xvi, 445p. tables. ISBN: 089729209x.

Each half has *c.*1,500 headwords. English-Spanish, p. 1-216 (plus tables of measures; prefixes). Spanish-English, p. 217-445. Includes phrases eponyms. 'References consulted', p. 445. *Class No:* 628(038)=60

Russian

[4752]

MILOVANOV, E.V. *and* WEIZMAN, E.A. English-Russian dictionary of environmental control. Moscow, Russian language Publishers, 1980. 367p.

About 14,000 English terms, A-Z, with Russian equivalents. Appended: List of international organizations (p. 326-56); Regulations and legislation; International treaties and conventions. *Class No:* 628(038)=82

Reviews & Abstracts

[4753]

Environment abstracts. New York, Environment Information Center, 1971-. 10 pa. ISSN: 00933287.

About 5,000 abstracts pa. Sections (01. Air pollution ... 03. Energy ... 05. Environmental design & urban ecology... 09. Land use and misuse... 14. Radiological contamination ... 15/16. Renewal resources (terrestrial; water) ... 19. Water pollution ... 21. Wildlife. Conferences. Subject, industry and author indexes, cumulated in annual *Environment Index.* The leading abstracting service in the environment field. ENVIROLINE database. *Class No:* 628(048)

[4754]

Excerpta medica. 46: Environmental health and pollution control. Amsterdam, etc., Elsevier, 1971-. 10 pa. ISSN: 03005194.

About 3,000 abstracts pa. 16 sections: 1. General aspects - 2. Types and sources of pollutants - 3. Effects of pollution - 4. Measurement of pollution - 5. Meteorological aspects of pollution - 6. Dangerous goods - 7. Legal and administrative aspects - 8. Criteria, and standards of pollution - 9. Treatments, prevention and control of pollution - 10. Socioeconomic aspects of pollution - 11. Noise and vibration - 12. Pesticides and herbicides - 13. Radiation - 14. Thermal pollution - 15. Biometeorological research - 16. Disaster control. Subject and author indexes per issue, and annually. *Class No:* 628(048)

[4755]

Green engineering: a current awareness bulletin. London, Information and Library Service, Institution of Mechanical Engineers, 1991-. 12pa. £99pa.; $193 pa. ISSN: 09608796.

Briefly annotated entries arranged by type of publication (*e.g.* 'articles', 'news', etc.), then alphabetically by subject (*e.g.* 'aerosols' - 'waste management'). About 70 entries per issue. Annual author index. Likely to have a broad appeal. *Class No:* 628(048)

[4756]

Library bulletin: abstracts of current literature on social and environmental planning, roads, traffic and transport, countryside and recreation, housing and local government, water supply and waste disposal, pollution and conservation. Department of the Environment and Department of Transport. London, DOE/Dtp. Library Services, 1972-. Fortnightly.

About 150 brief abstracts and references per issue. Issue no. 12, 1 July 1988, has 6 main sections: Public administration - Transport - Science and technology - Planning, land and construction - Environmental pollution and public health (Air pollution ... Water) - Social sciences and management. DOE/DTp shelf-marks given. Annual name and subject indexes. Insert: Department of the Environment and Department of Transport *Publications monthly list* (of Acts, regulations, reports, etc., arranged by subjects). *Class No:* 628(048)

Yearbooks & Directories

[4757]

Who's who in the environment: England. Cowell, S., *and others, eds.* 2nd ed. London, The Environment Council, 1992. viii,392p. ISBN: 0903158388.

First published 1990.

Covers organizations at regional and national level, A-Z.

....(contd.)

Information includes name, address, telephone, contact, scope, number of staff and volunteers, membership, local branches. List of organizations which provide speakers and photographic libraries. Subject index, p.341-71. *Class No:* 628(058)

Quotations

[4758]

RODES, B.K. *and* ODELL, R. **A Dictionary of environmental quotations.** New York, Simon & Schuster, 1992. 335p. $30. ISBN: 0132105764.

About 3,700 identified quotations in English or English translation. Alphabetical subject arrangement; chronological within categories. Indexes: author; subject. 'Highly recommended' (*Choice*, December 1992. p.600). *Class No:* 628(082.2)

Tables & Data Books

[4759]

CRC handbook of environmental control. Bond, R.G. *and* Straub, C.P., *eds.* Cleveland, Ohio, CRC Press, 1972-78. 5v. diagrs., graphs, tables. ISBN: 0878192700.
  1. *Air pollution.* 1972. xii, 876p.
  2. *Solid waste.* 1973. ix, 880p.
  3. *Water supply and disposal.* 1973. [x], 835p.
  4. *Waste water treatment* 1974. [xi], 905p.
  5. *Hospital and health care facilities.* 1975. 440p.
  Cumulative series index for volumes 1-5. 1978. xii, 101p.

V.1 has 4 sections: 1. The atmosphere and air pollutants - 2. Effects of air pollution - 3. Emission sources - 4. Air pollution and control measures. Volumes consist largely of tables, with cross-references. Each volume has an analytical subject index. The *Cumulative series index* numbering system is, however, faulty; entry numbers do not necessarily correspond to volume numbers. *Class No:* 628(083)

[4760]

—Practical handbook of environmental control. Straub, C.P., *ed.* Boca Raton, Fl., CRC Press, 1989. 537p. tables. $45. ISBN: 0879337070.

Based on data in the *CRC Handbook of environmental control* (5v.). Many tables; numerous references. Substantial subject index. For professionals and students in the field of environmental control. *Class No:* 628(083)

[4761]

GREAT BRITAIN. Department of the Environment. **Digest of environmental protection and water statistics.** London, H.M. Stationery Office, 1979-. Annual. ISBN: 0117519324.

9th ed., 1987 (ix, 71p. £8) has 11 sections: 1. Trends (graphical presentation) - 2. Air quality - 3. Water quality - 4. Radio-activity - 5. Noise - 6. Blood lead concentrations - 7. Solid waste - 8. Water supply and use - 9. Landscape and nature conservation - 10. Public attitude to environment - 11. Environmental monitoring in the United Kingdom. Appended: Statistical bulletins; Discontinued tables; DOE pollution papers and reports. *Class No:* 628(083)

[4762]

UNITED NATIONS. Department of International Economic and Social Affairs. **Directory of environment statistics.** v, 305p. £30. (*ST/ESA/STAT/Ser M/75.*)

Part 1: National publications (p. 6-283): Albania ... Zaire. Part 2: Subjects. Includes environmental statistical publications published by other international organizations. Data collected by 150 member states. 2 annexes (1. Environmental areas and parameters). *Class No:* 628(083)

Maps & Atlases

[4763]

LEAN, G., *and others.* **Atlas of the environment.** London, Hutchinson, 1990. 192p. illus., tables, maps. £17.99. ISBN: 0091747104.

Introduction and 42 articles (Major biomass - Climatic regions and land - Human numbers ... The Arctic-Antarctic). Compiled in collaboration with WWF/World Wide Fund for Nature. Only data from the 1980s and 1990 are included. Appended unit and conversion tables. Bibliography, p.187-92 (authors A-Z). Detailed contents but no index. Well illustrated in colour throughout. *Class No:* 628(084.3)

America — North

[4764]

World environmental directory: North America. Gough, B.E., *ed.* 5th ed. Silver Spring, Md., Business Publishers, Inc., 1989. v.p. ISBN: 0916742059.

15 sections covering a range of areas: products; professional services; government and independent agencies; environmental databases; education institutions; grants. Separate section covers Canada. *Class No:* 628(71 + 73)

Information Services

[4765]

ENDOC directory: environmental information and documentation centres of the European Communities. Hitchin, Herts., P. Peregrinus, for the Commission of the European Communities, 1981. vi,600p.
  Pilot ed. 1978.

519 centres, in Belgium, Denmark, Federal Republic of Germany, France, Ireland, Netherlands, UK. Data on each: name and address, contact, 'phone-number;' nature (*e.g.* governmental); aims; environmental subject areas; information activities. Subject indexes (in six languages). Online on ECHO. *Class No:* 628:061:025.5

Institutions & Associations

[4766]

Directory of the environment: organisations in Britain and Ireland, 1986-7. Barker, M.C., *ed.* 2nd ed. London, Routledge & Kegan Paul, 1986. xvii, 277p.
  First published 1984.

Directory of nearly 1,400 organizations concerned with the physical and human environment. Data on international, national, regional and local bodies: contact, aims, activities, status, publications. Quangos and organizations that no longer exist are indicated. Appended list of journals, p. 231-48. Bibliography, 1983-85 (Agriculture ... Wildlife). Subject index. Well organized, with good use of bold type, for quick-reference. *Class No:* 628:061:061.2

Conferences

[4767]

Environmental engineers: Proceedings of the 1985 Speciality Conference, sponsored by the Environmental Engineering Division of the American Society of Civil Engineers [and others], Boston, Mass., July 1-5, 1985. O'Shasughnessy, J.C., *ed.* New York, ASCE, 1985. xvi, 1137p. ISBN: 0872624684.

About 200 documented contributions at 29 technical sessions (1. Land application of sludge ... 3. Water supply ... 6. Wastewater collection systems ... 9. Low level radioactive waste control ... 12. Water treatment ... 24. Air quality problems ... 27. Ground water contamination ... 29. Small systems). Subject and author indexes, p. 1,131-37. *Class No:* 628:061:061.3

Research Projects

[4768]

FARRELL, S. World environmental research directory. Leatherhead, Surrey, Pira International, 1992. viii,299p. ISBN: 0902799959.

Organizations, by country A-Z, in broad subject groupings: Manufacture and process control - Energy - Environmental management - Air pollution - Water and effluent - Hazardous waste management - Solid waste management and recycling - Contaminated land - Noise pollution - Agriculture - Nature conservation - Aquatic environment - Flora conservation - Wildlife conservation. Data: name; research project titles; research personnel; length of project; funding; published papers. Indexes: project leaders; subject; organizations. *Class No:* 628:061:061.62.005

Water Supply

Databases

[4769]

Water resources abstracts. Reston, Va., US Department of the Interior, Geological Survey.

Covers water planning (demand, economics, cost allocations), water cycle (precipitation, snow, groundwater, lakes, erosion) and water quality (pollution, waste treatment). About 250,000 records; updated monthly. Available on DIALOG, file 117. *Class No:* 628.1(003.4)

[4770]

—Waternet. Denver, Co., American Water Works Association.

Covers drinking water, treatment, water quality, groundwater, surface water, development, reuse, regulations, pollution, wastewater disposal and treatment. About 23,000 records; updated bimonthly. Available on DIALOG, file 245. *Class No:* 628.1(003.4)

Bibliographies

[4771]

WATER RESOURCES CENTER ARCHIVES, University of California. **Dictionary catalog. [Water Resources Center Archives].** Boston, Mass., G.K. Hall, 1970-. 5v. (3373p.). Supplements, 1-. 1971-. $545, set; $285, supp. 6. ISBN: 0816108846, set; 0816102449, supp. (6th, 1978).

The main volumes comprise 97,000 photolithographed catalogue cards. The Water Resources Center Archives is a

....(contd.)

research library on engineering, economic, social and legal aspects of water. It largely concentrates on US report material, government publications (federal, state, regional, municipal), publications of water-related bodies, conferences and symposia; photographs and maps. *Class No:* 628.1(01)

[4772]

—PICKFORD, J., *ed.* Indexed bibliography of publications on water and water engineering for developing countries. Loughborough, Univ. of Technology, Dept. of Civil Engineering, WEDC Group 1977. 52p. ISBN: 0906055016.

About 200 references. Author list of publications, with cross-references to other authors. *Class No:* 628.1(01)

Handbooks & Manuals

[4773]

DE ZUANE, J. Handbook of drinking water quality: standards and control. New York, Van Nostrand Reinhold, 1990. 523p. illus., diagrs., tables. $69.95. ISBN: 0442239092.

Chapters cover basic physical and chemical parameters of water, potential contaminants, standards for water analysis and treatment. Includes US federal and state acronyms and table of conversions. 'An especially handy book' (*Choice*, May 1991, p.1508). US slanted. *Class No:* 628.1(035)

[4774]

FRESENIUS, W., *and others, eds.* **Water analysis:** a practical guide to physico-chemical, chemical and microbiological water examination and quality assurance. Berlin, etc., Springer-Verlag, 1988. xxv, 804p. illus., tables. ISBN: 354017723x; 038717723x US.

Highly detailed survey, including the physics and chemistry of water, methods for analysis in laboratories with simple equipment or in the field (*e.g.* for third world countries), theoretical investigation and evaluation of analysis techniques. Sections: 1. Sampling, local testing - 2. Methods of water analysis - 3. Inorganic parameters - 4. Organic parameters - 5. Biological analysis - 6. Evaluation of data analysis. Detailed index; brief bibliography. *Class No:* 628.1(035)

[4775]

LINSLEY, R.K. and FRANZINI, J.B. Water resources engineering. 3rd ed. New York, etc., McGraw-Hill, 1979. 720p. tables. ISBN: 0070379653.

First published 1964; 2nd ed., 1972.

Chapters: Descriptive hydrology ... Groundwater - Reservoirs - Dams, spillways, gates, and outlet works ... Hydraulic machinery ... Irrigation - Water supply systems - Hydroelectric power - River navigation - Drainage - Sewerage and wastewater treatment - Flood damage mitigation - Planning for water resources development. Appendix: 'useful tables'. Name and subject indexes. Comprehensive and wide-ranging. 4th ed. due. *Class No:* 628.1(035)

Dictionaries

Polyglot

[4776]

MEINCK, F. *and* MÖHLE, H. **Dictionary of water and sewage engineering;** in German, English, French and Italian. 2nd rev. and enlarged ed. Amsterdam, etc., Elsevier, 1977. 738p. Dfl.370. ISBN: 0444427015.

8,844 German-base terms, with English, French and Italian equivalents and indexes. Genders are stated. International conversion tables and international measurement units. *Class No:* 628.1(038)=00

[4777]

RÅDET FOR TEKNISK TERMINOLOGI. **Andbok for vann og avlop.** [Dictionary of water and waste water.] [Oslo], Universitetsforlaget, 1973. 274p. (*RTT 38.*)

About 2,500 Norwegian-base terms and definitions, with English, French and German equivalents and indexes. *Class No:* 628.1(038)=00

[4778]

—TEKNISKA NOMENKLATURCENTRALEN. Va-teknisk ord lista. [Glossary of water supply and sewerage.] Stockholm, the Centre, 1977. 236p. diagrs. (*TNC 65.*)

A revision of TNC 51 (1973).

Lists 791 Swedish-base terms, with equivalents in Danish, English, Finnish and French, plus indexes in the five languages. *Class No:* 628.1(038)=00

[4779]

VAN DER TUIN, J.D. **Elsevier's dictionary of water and hydraulic engineering, in five languages:** English, French, Spanish, Dutch, German. Amsterdam, etc. Elsevier, 1987. xiv, [2], 449p. ISBN: 0444427686.

5,117 English-base terms, with French, Spanish, Dutch and German equivalents and indexes. Coverage: Water, hydraulic engineering, water management, experimental aspects. *Class No:* 628.1(038)=00

[4780]

—AMMON, F. von. Wörterbuch der Wasserchemie. [Dictionary of water chemistry. Dictionnaire de la chimie de l'eau.] Deerfield Beach, Florida, VCH Verlagsgesellerschaft, 1985. 203p. $58. ISBN: 0895734966.

A German-English-French dictionary, promoted by the German Chemical Society's Water Chemistry Section. Main coverage: Water composition and quality, water purification, and all aspects of water chemistry. US terms are cross-referenced to British. *Class No:* 628.1(038)=00

English

[4781]

NELSON, A. *and* NELSON, K.D. **Dictionary of water and water engineering.** London, Butterworths, 1973. vi, 271p. illus. £17.50. ISBN: 0408000902.

About 3,000 brief definitions, with many cross-references. 'Levee': *c.*100 words. Includes terms in hydraulics, hydrology, hydrogeology and meteorology. Appendix: Organizations, p.267-71. Both SI and imperial units are used. *Class No:* 628.1(038)=20

[4782]

SCOTT, J.S. *and* SMITH, P.G. **Dictionary of waste and water treatment.** London, etc., Butterworths, 1981. vi, 359p. illus. ISBN: 0408004959.

About 2,500 terms, with definitions ranging from 1 line to *c.*½p. Appendix lists conversion factors and maximum contaminants in drinking water by various authorities. *Class No:* 628.1(038)=20

French

[4783]

Dictionnaire de l'eau. Quebec, Assocation québécoise des techniques de l'eau, Office de la langue francaise, 1981. xxiv, 544p. tables. (*Cahiers de l'office de la langue francaise.*) ISBN: 2551042291.

3,694 numbered French terms, with English equivalents and French explanations. Index of English terms, p. 471-534. Appendix of weights and measures. *Class No:* 628.1(038)=40

[4784]

JOVULOT, W. **Lexique trilingue de l'eau.** Paris, Johenet, 1989. 230p.

Terms in French, English, German. *Class No:* 628.1(038)=40

Reviews & Abstracts

[4785]

Aqualine abstracts. Oxford, etc., Pergamon Press, on behalf of Water Research Centre, Medmenham, 1927-. Bi-weekly. DM595 (two-yearly). ISSN: 02635534.

Formerly *WRC information* (1974-84), incorporating *Water pollution abstracts*, 1949-73.

1991: 4,815 abstracts. 8 sections: 1. Water resources and supplies - 2. Water quality - 3. Monitoring and analysis of water and wastes - 4. Water treatment - 5. Underground services and water use - 6. Sewage - 7. Industrial effluents - 8. Effect of pollution (sewage; heated discharges; metals; pesticides; fossil fuels; radio-activity). Annual author, subject and KWOC indexes.

AQUALINE database, available on Pergamon INFOLINE 1960-. 'The world's technical literature at your fingertips' by P. Russel and B. Wilkinson (*Water*, no. 37, March 1981, p. 26-28) discusses online information services on the water industry, particularly AQUALINE database. *Class No:* 628.1(048)

[4786]

Selected water resources abstracts: (SWRA) Reston, Va., Water Resources Scientific Information Center, US Geological Survey, 1968-. $145. ISSN: 0037136x.

About 1,500 abstracts and references per issue. Subject fields and groups: 01. Nature of water - 02. Water supply - 03. Water supply augmentation and conservation - 04. Water quality management and control ... 07. Resources data - 08. Engineering works ... 10. Scientific and technical information. Entries include descriptors. Subject, author, organizational and accession number indexes per issue. Available on CD-ROM and also online via DIALOG. *Class No:* 628.1(048)

Progress Reports

[4787]

Advances in water engineering. Tebbult, T.H.Y., *ed.*
London, New York, Elsevier Applied Science Publishing,
1985. ix, 361p. graphs, tables. ISBN: 0853343748.

41 documented papers in 5 parts: Surface water hydrology
- Ground water hydrology - Water quality, supply and
treatment - Wastewater treatment and pollution control -
River engineering. ('Advances in water quality, supply and
treatment', p.129-38. 34 references.) List of contributors.
Detailed, analytical subject index. *Class No:* 628.1(055)

Yearbooks & Directories

[4788]

Who's who in the water industry, 1987. Rickmansworth,
Herts., Turret-Wheatland, Ltd., for Water Authorities
Association. 224p. maps.

Contents: Water industry map - Water undertakings: water
authorities - Water authorities: organizations - Water
Research Centre - Government departments - Water industry
in Scotland - Water industry in Northern Ireland - Water
industry organizations (UK; international) - Water people,
102/16 - Water publications, 117/26. Suppliers guide. A-Z
directory, plus trade marks, p. 129-214. Subject index, A-Z.
Advertiser's index. *Class No:* 628.1(058)

Tables & Data Books

[4789]

PARKER, D.J. Water. (Review of the United Kingdom
statistical sources. Review no. 30). Oxford, Pergamon Press,
for the Royal Statistical Society and Science Research
Council. p. 107-226.

7 sections: 1. Introduction - 2. Data sources - 4. Water
supply statistics - 4. Water pollution and related statistics - 5.
Water recreation and amenity statistics - 6. Flood allevation -
7. Evaluation. Quick-reference list, p.183-201; bibliography
(49 references), p.203-4. Appendices. Subject index (non-
analytical), p.223-6. *Class No:* 628.1(083)

Histories

[4790]

ROBINS, F.W. The Story of water supply. London, Oxford
Univ. Press, 1946. 207p. illus. (incl pl.).

A survey, in non-technical language, of the means
employed by peoples in many parts of the world, to obtain
supplies of water for drinking, irrigation and sanitation. 45
illus. (including plates). Index to subjects and plates;
topographical index. Footnotes and text references.
Class No: 628.1(091)

Thesauri

[4791]

Aqualine thesaurus 2. Smith, J.G. *and*
Jennings, P.G., *comps.* 2nd ed. Chichester, Ellis Horwood
Ltd., for the Water Research Centre, 1987. 994p. . £75.
ISBN: 085312938x.

First published 1980.

43,000 index terms (13,000 preferred; 30,000 non-
preferred) for searching both the WRC AQUALINE
database and *Aqualine abstracts.* In three parts: general
subject terms; trade names; taxonomic names. Intended for

....(contd.)
librarians/information officers and subject specialist end
users; probably essential for efficient searching of the
database. *Class No:* 628.1:025.43

Conferences

[4792]

**INTERNATIONAL WATER SUPPLY ASSOCIATION.
Uniting the world of water:** proceedings of the 16th World
Congress of the International Water Supply Association, held
in Rome, Italy, 3-7 November 1986. In *Water supply*, v.
5(3/4), 1987. v.p. diagrs., tables.

Documented papers on progress over the past decade;
state-of-the-art of microtechnoogy in water services;
consumer service and public relations; water tariffs; acid
rain and its influence on water quality; sludge disposal and
the recovery of water; drinking water quality and its
significance for health; databanks and mapping for
distribution networks. *Class No:* 628.1:061:061.3

[4793]

World water '86: water technology for the developing world.
Proceedings, International Conference organized by the
Institution of Civil Engineers, London, 14-16 July 1986.
London, T. Telford, 1987. 260p. illus., diagrs., graphs,
tables. ISBN: 0727703609.

30 papers (9. 'Water resources for irrigation', p.131-6, 18
references; 7. 'Recent developments in wastewater
treatment', p.105-10, 20 references). Appended: Discussion
on papers - General discussion - Summary of conference;
Closing address. Quarto. *Class No:* 628.1:061:061.3

Treatment

[4794]

LORCH, W., *ed.* **Handbook of water purification.** 2nd ed.
Chichester, E. Horwood, 1987. 777p. illus., diagrs., tables.
ISBN: 0853129916.

First published 1981.

19 editors and contributors. 3 parts (22 documented
chapters): 1. Background to water purification - 2.
Purification processes (membrane; distillation; disinfection) -
3. Purified water practice. Chapter 18: 'Sterile and
apyrogenic water', p.640-92; 124 references. Detailed
index, p.773-7. *Class No:* 628.16

[4795]

Water treatment handbook. 5th English ed., translated from
French, by Language Consultants. New York, Halsted:
Wiley, 1979. xxx, 1,186p. ISBN: 0470267106.

Previously published in Paris (Degrémont), 1973. First
French ed., 1950.

Sections: 1. General aspects of water and water treatment -
2. Treatment plant and processes - 3. Treatment methods
according to the nature and final use of the water - 4.
General information - 5. Formulae - Legislation (including
'Useful addresses': France; other countries). Bibliographic
data: p. 1,151-58. Index. Well designed. *Class No:* 628.16

Desalination

[4796]
Desalination abstracts. Tel Aviv, National Center of Scientific and Technological Information. Ministry of Energy and Infrastructure, 1966-. Quarterly. ISSN: 00119172.

Over 1,000 abstracts pa. 13 sections: 1. Desalination, general - 2. Distillation - 3. Osmotic processes - 4. Freezing - 5., Electrodialysis - 6. Ion exchange - 7. Other processes - 8. Corrosion and sealing - 9. Heat transfer - 10. Chemical and physical properties - 11. Water supply and requirements - 12. Energy sources - 13. Miscellaneous. Subject, author and patent indexes. *Class No:* 628.165

Pollution

[4797]
INSTITUTE OF WATER POLLUTION CONTROL. Glossary of terms used in water pollution control. Maidstone, Kent, the Institute, 1975. 153p.

Nearly 2,000 terms defined, p.9-150. Appended list of acronyms, p.151-2. Publications consulted, p.6. One of a series of manuals on British practice in water pollution control published by the Institute. *Class No:* 628.19

[4798]
[Water pollution]: literature review. *In Journal of Water Pollution Control Federation*, v.56(6), June 1984, p.491-806.

Literature review of water pollution in the US. Sections (each with several papers, plus references and tables, highlighting various aspects): administration; chemical analysis of water and wastewater; wastewater treatment; industrial wastes; fate and effects of pollutants. *Class No:* 628.19

Sewerage

[4799]
METCALF & EDDY, Inc. **Wastewater engineering:** treatment, disposal, refuse. New Delhi, Tata; McGraw-Hill 1979. xviii, 929p. illus., diagrs., tables.

14 sections (*e.g.* 1. Wastewater engineering: an overview; 5. Fundamentals of process analysis; 10. Design of facilities for the biological treatment of wastewater; 13. Land-treatment systems; 14. Effluent disposal and reuse). Discussion topics and problems. References (21). Appended conversion tables, etc. Name and analytical indexes. *Class No:* 628.2/.3

[4800]
STANBRIDGE, H.H. History of sewage treatment in Britain. Maidstone, Kent, Institute of Water Pollution Control, 1976-77. 3v. (16 pts). illus., diagrs.

pts. 1-5: *Introduction of the water-carriage system.* 1976.

pt. 6: *Biological filtration.* 1976.

pts. 7-16: *Activated sludge process.*1977.

A well-documented, mimeographed survey. Part 12, 'Utilization and disposal of sewage sludge' (86p.) has 186 references, 11 illus. Part indexes. *Class No:* 628.2/.3

Marine Pollution

[4801]
CHAMP, M.A. *and* **PARK, P.K. Global marine pollution bibliography:** ocean dumping of municipal and industrial wastes. New York, etc. IFI/Plenum, 1982. xiv, 399p. $69.50. ISBN: 0306652056.

Lists 1,742 entries, some with abstracts, most dating from the 1970s. Arranged A-Z by author, with a broad subject index (*e.g.* c.270 undifferentiated items under 'municipal wastes - sewage sludge') and an author index. Topics covered include municipal and industrial wastes, legislation, international conventions, ocean dumping criteria, dump sites, waste management strategies. A new edition 'in three years' indicated in the *Preface* appears not to have been published. *Class No:* 628.394

[4802]
Marine pollution research titles. Plymouth, Marine Pollution Information Centre, Marine Biological Association of the United Kingdom, 1974-. Monthly. £47 pa. ISSN: 02048059.

1986: 2,642 references on marine and estuarine pollution. 15 sections: 1. General - 2. Detection - 3. Analysis - 4. Oil - 5. Oil removal - 6. Metals - 7. Pesticides - 8. Radioactivity - 9. Polychlorinated biphenyls - 10. Other chemicals - 11. Domestic sewage - 12. Pulp and paper - 13. Food processing - 14. Heat - 15. Solids. No indexes. A current awareness service. *Class No:* 628.394

[4803]
Wastes in the ocean. New York Wiley, 1983. 6v.

1. *Industrial and sewage wastes in the ocean;* edited by I.W. Duedall and others. 1983. xviii, 431p.

2. *Dredged material disposal in the ocean;* edited by K.P. Parr and others. 1983. xxii, 522p.

3. *Radioactive wastes in the ocean;* edited by K.P. Parr. 1983. xxii, 522p.

4. *Energy wastes in the ocean;* edited by I.W. Duedall. 1985. 818p.

5. *Deep-sea waste disposal,* edited by by D.H. Kester. 1985. 346p.

6. *Near-shore waste disposal,* edited by B.H. Ketchum and others. 1985. 534p. *Class No:* 628.394

Waste Disposal

[4804]
Biotechnology of waste treatment and exploitation. Sidwick, J.M. *and* Holdon, R.S., *eds.* Chichester, Ellis Horwood, New York, Halsted Press, 1987. 332p. diagrs., graphs, tables. £55. ISBN: 0853129177, UK; 0470210311, US.

18 contributors. 13 chapters, closely subdivided and extensively documented (*e.g.* chapter 3, 'Immobilized whole-cell biocatalysts', p.32-80, includes 248 references). Detailed, analytical index, p.326-32. *Class No:* 628.4

[4805]
CRC handbook of incineration of hazardous wastes. Rickman, W.S., *ed.* Boca Raton, Fl., 1991. [viii],593p. illus., diagrs., tables. £176. ISBN: 0849305578.

Covers incinerator technology, regulatory requirements, siting issues and includes case histories. 19 contributors; 9 chapters, some extensively documented (*e.g.* 'Regulatory requirements and the permitting process': 116p.; 344 references). Index, 16p. *Class No:* 628.4

[4806]

Croner's hazardous waste disposal guide. Kingston upon Thames, Surrey, Croner Publications, 1988. [viii],368p. £30. ISBN: 0900319623. ISSN: 09542922.

Information for the United Kingdon in 3 parts: 1. Commentary (legal aspects; waste management; disposal contractors, etc.) - 2. Disposal information (charts; index) - 3. Directory. *Class No:* 628.4

[4807]

ISAAC, P.C.G. A Glossary of sanitation engineering terms. Newcastle-upon-Tyne, University of Durham, 1955. iv. 72p. tables. (*Univ. of Durham, King's College, Dept. of Civil Engineering. Bulletin no. 4.*)

Concise definiton of *c*.3,000 terms. Appendix on hardness factors; also conversion tables and abbreviations. *Class No:* 628.4

[4808]

—BRITISH STANDARDS INSTITUTION. Glossary of building and civil engineering terms. Part 2. Civil engineering. Section 2.7 Public health. Environmental engineering. London, BSI, 1992. 26p. (*BS 6100: Section 2.7:1992.*)

Supercedes BS 6100: Section 2.7:1988.

Defines 289 terms in systematic order, Solid waste ... Disposal ... Processing ... Pollution and odour. Alphabetical index. *Class No:* 628.4

[4809]

Waste and environment today: bibliographic journal. Harwell, Oxon., A E A Technology, 1992-. 12pa. £216pa. ISSN: 09654496.

Formerly *Waste Management today,* 1988-91.

Covers disposal, recycling/recovery, environmental hazards, legislative and policy aspects, business opportunities, etc. About 200 entries per issue in 11 broad areas (*e.g.* land and sea disposal, surveys and models, etc.). Non-analytical subject index per issue. *Class No:* 628.4

[4810]

WERTZ, C.A. Hazardous waste management. New York, McGraw-Hill, 1989. xiii,461p. illus., diagrs., maps, tables. $46.95; £16.95. ISBN: 0070692912.

14 chapters, each with 'questions' and brief bibliography: 2. Risk assessment - 3. Environmental legislation (US) ... 6. Waste minimization and resource recovery - 7. Chemical, physical and biological treatment ... 10. Ground water contamination ... 14. Site remediation. 4 appendices (*e.g.* A. Priority pollutants); detailed, analytical index, p.451-61. For students and practitioners. *Class No:* 628.4

Radioactive Waste

[4811]

CROMER, D.E. *and* **THOMAS, D.R. 'Radioactive waste management and disposal information sources'.** *Science and technology libraries,* v.11(3), Spring 1991, p.119-38.

15 sections, including 'Statistical information', 'Encyclopedias', 'Conference proceedings', 'Journals', 'Technical reports', 'Indexes (online and printed)', 'Social and political perspectives'. 6 references. *Class No:* 628.4.047

[4812]

LAMBERT, C.M., *comp.* **Radioactive waste:** a select list of material. London, Departments of the Environment and Transport, 1982. 61p. £1.90. (*Bibliography series, no.190E.*) ISBN: 0718402057.

230 items, published since 1978, frequently with indicative or informative abstracts, arranged in 7 sections: 1. Official publications - 2. Administrative and environmental - 3. Technological and scientific - 4. General and background material ... Plus sections for series; annual reports and surveys; further sources of information (bibliographies, abstracting and indexing journals, periodicals). Excludes press and Parliamentary coverage and DOE-sponsored research reports. *Class No:* 628.4.047

[4813]

MILLER, E.W. *and* **MILLER, R.M. Environmental hazards: radioactive materials and wastes:** a reference handbook. Santa Barbara, Calif., ABC-Clio, 1990. 298p. $39. ISBN: 0874362342.

Various chapters with a range of information (*e.g.* nature and characteristics of radioactive materials; directory of organizations that deal with waste; annotated bibliography of books; listing of journal articles and government documents; select list of journal titles; bibliography of a-v materials). Glossary; index. 'An excellent addition to any reference collection' (*Choice,* April 1991, p.1292). *Class No:* 628.4.047

[4814]

Waste management research abstracts: information on radioactive waste programmes in progress. Vienna, International Atomic Energy Agency, 1970-. Annual. *Gratis,* on request.

No. 16, 1985 (408p.) consists of research data sheets (ending August 1985) sent by governments of 33 member states. 6 sections: 1. Studies related to the treatment of radioactive wastes and effluent - 2. Studies concerning the storage and disposal of radioactive wastes ... 4. Safety and environment studies ... 6. Miscellaneous. Alphabetical descriptor index; author index. *Class No:* 628.4.047

Environmental Pollution Measures

[4815]

Control of pollution encyclopedia. Garner, J.F., *ed.* Sevenoaks, Kent, Butterworths, 1977. 2 loose-leaf binders. tables. ISBN: 0406205108.

9 sections. V.1: 1. General introduction - 2. Waste on land - 3. Pollution on land - 4. Pollution of the sea - 5. Atmospheric pollution - 6. Pollution by noise. V.2: A. Pollution by radioactive and other hazardous substances - 7. Machinery provisions - 8. Forms and precedents - 9. EEC legislation and other documents. Appended tables and index. Supplements include 'current topics'. J.F. Garner is Emeritus Professor of Public Law, Univ. of Nottingham. For risk managers, safety advisers and other specialists in pollution control. *Class No:* 628.5

Databases

[4816]

LUEDTKE, J.R. 'Energy and the environment'. *Manual of online search strategies.* Editors, C.J. Armstrong and J.A. Large, pp.236-87. 2nd ed. Aldershot, Ashgate, 1992. ISBN: 1857420071.

....(contd.)
Extensive list of databases, both online and CD-ROM, which cover energy and the environment. Hosts and disc suppliers identifed. *Class No:* 628.5(003.4)

Encyclopaedias
[4817]
FRANCK, I. *and* **BROWNSTONE, D. The Green encyclopedia.** Englewood Cliffs, NJ, Prentice Hall, 1992. 512p. illus. $35; $20. ISBN: 0133656853; 0133656772, Pbk.
Over 1000 brief entries on people, specific animals, organizations, government agencies, pesticides, parks and wildlife preserves. International in scope. Cross-references. Bibliography; glossary. *Class No:* 628.5(031)

Handbooks & Manuals
[4818]
BENARDE, M.A. Global warning ... global warming. New York, Wiley, 1992. [xiii],317p. illus., graphs, maps. $29.95. ISBN: 0471519237.
5 parts (1. Reasons for seasons ... 5. Approaches to energy information), well-documented. Helpful introduction, p.1-13, with 11 references. Glossary; appendix; index. *Class No:* 628.5(035)

[4819]
NATIONAL SOCIETY FOR CLEAN AIR. NSCA reference book. Dunmore, J., *ed.* Brighton, Sussex, the Society, 1985. viii, 294p. ISBN: 090347428x.
First published 1978.
4 sections: 1. Air pollution - 2. Noise - 3. Water pollution - 4. Wastes. Each section features an outline of the law; British Standards are appended. Annex 1: Community environmental measures. References, p.27-84. Detailed analytical index (in small type), p.285-90. *Class No:* 628.5(035)

Dictionaries
[4820]
JOHNSON, C. Green dictionary. London, Optima, 1991. 320p. £9.99. ISBN: 0356195688.
Simple explanations of over 400 terms, issues and ideas connected with the environmental debate. *Class No:* 628.5(038)

[4821]
NATIONAL SOCIETY FOR CLEAN AIR. NSCA environmental glossary. Dunmore, J., *ed.* Brighton, the Society, 1985. 89p. ISBN: 0903474298.
About 1,000 terms, concisely defined. Sections: Air pollution terms; Noise terms; Waste terms; Radiation terms. 'Additional references', p.89. Covers vocabulary of technical articles and lectures. *Class No:* 628.5(038)

English
[4822]
JONES, G., *and others.* **Collins dictionary of environmental science.** Glasgow, Harper Collins, 1990. vi,473p. illus., graphs, maps, tables. £6.99. ISBN: 0004343484.
Published in the US as *The Harper Collins dictionary of environmental science* (New York, Harper Perennial, 1992.

....(contd.)
453p. ISBN: 006271533x ($25); ISBN: 0064610403 ($13)).
Over 1800 terms, briefly defined (*e.g.* 'epeirogenesis': 9 lines). Covers topics from the biological, physical and social sciences. Cross-references. *Class No:* 628.5(038)=20

Reviews & Abstracts
[4823]
Acid rain abstracts annual. New York, Bowker A & I Publishing, 1989-. $149.
1988,222p. (ISBN: 0835226409).
Bi-monthly issues. The 1988 annual reprints the 733 abstracts for the year. Focuses on the sources, causes and effects of acid deposition, and the economic, political, health and natural resource issues. Geographic index, geographic key term list and expanded subject keyterm list (plus author, subject and source indexes - as in the monthly issues). The annual appends a list of 'Conferences and events' and 2 articles. Available on CD-ROM. *Class No:* 628.5(048)

[4824]
Pollution abstracts; with indexes. Bethesda, Md., Cambridge Scientific Abstracts, 1970-. 6 pa.
1,500 abstracts per issue. Sections: Air pollution - Marine pollution - Freshwater pollution - Sewage and wastewater treatment - Waste management - Lead pollution - Toxicology and health - Noise - Radiation - Environmental action. Analytical subject and author indexes per issue, cumulated annually. Available online via DIALOG, file 41. About 175,000 records, updated bimonthly. *Class No:* 628.5(048)

Progress Reports
[4825]
Global warming: the Greenpeace report. Leggatt, J., *ed.* Oxford Univ. Press, 1990. xi,554p.
19 Sections: 1-4: Science - 5-7: Impacts - 8-19: Policy responses. 20 contributors. Notes and chapter references, p.481-544. Detailed, non-analytical index. *Class No:* 628.5(055)

Tables & Data Books
[4826]
Statistical record of the environment. Darney, A.J., *ed.* Detroit, Mi., Gale, 1992. 855p. graphs, tables, maps. $89.50. ISBN: 0810383748.
851 statistical tables in 10 chapters, closely subdivided. Includes data gathered between 1987 and 1991, largely from US government sources. Appendices include list of sources, abbreviations and acronyms. Keyword index. *Class No:* 628.5(083)

Worldwide
[4827]
World guide to environmental issues and organizations. Harlow, Essex, Longman, 1990. xxviii,386p. ISBN: 0582002705.
21 contributors. 4 parts: 1. Issues (*e.g.* nuclear power and the environment; vehicle emissions) - 2. Politics - 3. Conventions, reports, directives and agencies - 4. Organizations. Further reading and references, p.357-65. Detailed, analytical index, p.366-86. *Class No:* 628.5(100)

Research Establishments

[4828]

Pollution research index: a guide to world research in environmental pollution. Sors, A.I. *and* Coleman, D., *eds.* 2nd ed. Guernsey, F. Hodgson, 1979. 555p. ISBN: 0582900069.

First published 1975.

Directory of more than 2,000 pollution research centres in over 100 countries. Includes government departments, universities, research institutes and manufacturing industries. Entries indicate scope of activities. Arranged by countries A-Z. Index. *Class No:* 628.5:061:061.62

Industrial Pollution

[4829]

Acid deposition in the United Kingdom, 1981-1985: a second report of the United Kingdom Review Group on Acid Rain. Prepared at the request of the Department of the Environment. Stevenage, Herts., Warren Spring Laboratory, 1987. vi, 101p. diagrs., tables, maps. ISBN: 0856244570.

14 sections (*e.g.* 2. Methods of measurement and data sources; 3. Precipitation composition and wet deposition; 4. Gases and aerosols; 7. Emissions (sulphur dioxide; oxides of nitrogen; ammonia). 14 references, p.75. Appendixes A-F (F. Maps reproduced from the First Report of the Group covering the period 1978-80, p.97-101). 48 diagrs. 16 tables. *Class No:* 628.51

[4830]

Acid rain and the environment 1988-1991: a select bibliography. Grayson, L., *Comp.* London, British Library Science Reference and Information Service; Letchworth, Technical Communications, 1991. iv,217p. ISBN: 0946655421.

Acid rain and the environment 1980-84: a select bibliography. P. Farmer, ed.

Acid rain on the environment 1984-88: a select bibliography. L. Grayson, ed.

Over 900 entries in 6 sections: 1. The issue - 2. The causes - 3. Research - 4. Surveys - 5. The effects - 6. Mitigation strategies. Descriptive annotations. Indexes: authors and editors; corporate bodies. *Class No:* 628.51

[4831]

Air pollution. Stern, A.C. 3rd ed. New York, Academic Press, 1976-77. 5v. (*Environmental sciences.*)

V. 1. *Air pollutants, their transformation and transport.*

V. 2. *The effects of air pollution.*

V. 3. *Measuring, monitoring, and surveillance of air pollution.*

V. 4. *Air quality management.*

V. 4 (xxii,946p.) is the work of 29 contributors. 21 chapters, with references, in 3 parts: A. Control concepts - B. Control devices - C. Process emissions and their control. Detailed, analytical index, p.931-46.

3 supplementary volumes were published, 1986-88. *Class No:* 628.51

[4832]

STOPP, G.H., Jr. Acid rain: a bibliography of research, annotated for easy access. Metuchen, N.J., Scarecrow, 1985. xiv, 174p. ISBN: 0810818221.

886 annotated entries, under authors A-Z. Items - books, periodical articles, conference proceedings and reports - are all in English. Annotations indicate technical level of item.

....(contd.)

US and Canadian references predominate. The partly analytical index is 'seriously flawed', states *Reference books bulletin 1985-1986*, p. 22. *Class No:* 628.51

Noise Abatement

[4833]

BRAGDON, C.R. Noise pollution: a guide to information sources. Detroit, Mich., Gale, 1979. xvi, 524p. ISBN: 0810313456.

Annotated entries for more than 4,000 periodicals, reports, books, etc. 9 chapters: 1. Physiological effects - 2. Behavioral effects - 3/4. Abatement - 5. Community noise - 6. Environmental impact - 7. Acoustics - 8. Noise legislation - 9. References (primary periodicals; secondary periodicals; Indexes). Addendum p.443-66. Author, title, and subject indexes. US slanted. *Class No:* 628.517

[4834]

Handbook of noise and vibration control. 5th ed. Morden, Surrey, Trade & Technical Press, 1983. ix, 498p. illus., charts, tables.

4th ed., 1979.

7 sections: 1. Basic accoustics - 2a. Hearing conservation, protection, health and safety - 2b. Vibration: effect on people - 3a. Measuring techniques - 3b. Vibration measuring and testing - 4. Types of noise (fan, factory, aircraft, etc). - 5a. Sound insulation and absorption - 5b. Vibration control - 6. Legislation (UK and European) - 7. Buyers' guide (classified). Trade names index. Advertisers index. Subject index. *Class No:* 628.517

[4835]

NELSON, P.M., *ed.* **Transportation noise reference book.** London, Butterworths, 1987. [492]p. illus., diagrs., graphs, tables. £90. ISBN: 0408014466.

27 contributions. Main parts (24 documented sections): Introduction and physical assessment - The effects of transportation noise on man - Train noise - Aircraft noise - Decision making methods for transport noise control. 4 appendices (3. standards relating to traffic noise and its control; 4. list of addresses). Index. 405 illus., including 20 half-tones. 'Highly recommended' (*Choice*, March 1988, p.1126). *Class No:* 628.517

[4836]

NVB (Noise & vibration bulletin). Brentwood, Essex, Multi-Science Publishing, 1970-. Monthly. ISSN: 00290047.

Issues include about 20 'abridgements', plus 'News and comment'. Coverage: industrial; aircraft; road and other transport; general; meetings; products and information. Each issue carries a review article. *Class No:* 628.517

[4837]

NVI/Noise and vibration in industry: an international quarterly digest. Brentwood, Essex, Multi-Science Publishing, 1986-. 4pa. £78. ISSN: 09508163.

About 50 fairly lengthy abstracts per issue. 8 sections (Measuring and recording instrumentation and analysis ... Audiology and audio - Standards, legislation and regulations - Contested cases). No index. Current awareness bulletin for factory managers, occupational health and safety officers, etc. *Class No:* 628.517

[4838]
SÁENZ, A.L. *and* STEPHENS, R.W.B., *eds.* **Noise pollution:** effects and controls. New York, etc., Wiley, published on behalf of the Scientific Committee on Problems of the Environment (SCOPE) of the International Council of Scientific Unions (ICSU), 1986. xix, 446p. illus., graphs, tables, formulas. (*SCOPE 24.*) ISBN: 0471903256.

3 parts (18 documented chapters): 1. Fundamentals of noise and hearing - 2. Effects of noise on man - 3. Sources of noise and control. (Chapter 10: 'Noise pollution during the night', p. 265-82, including 3½ p. of references.) Author index. Detailed, analytical subject index. *Class No:* 628.517

[4839]
VULKAN, G. *and* GOMERSALL, A. **Traffic noise:** a review and bibliography on surface transportation noise. Kempston, Beds., I.F.S. Publications Ltd., 1979. [216]p. ISBN: 0903608103.

1,561 annotated entries for the period 1964-78. Sections A-L (*e.g.* A. General information on surface transportation noise - B. Traffic noise and its abatement - C. Railway noise and its abatement - D. Vehicle design for noise reduction G. Social and economic impact of traffic noise ... J. Legislation regulations and general standards for surface transportation noise control - K. Literature reviews and bibliographies on surface transportation noise - L. Reports). Bibliographic sources. Author index. *Class No:* 628.517

Industrial Effluents & Wastes
[4840]
Directory of waste disposal and recovery. Pratt, A., *ed.* London, Godwin, 1978. 232p. diagrs. ISBN: 0711434123.

For the manager dealing with 'everyday problems of recovery or disposal of used materials or wastes' (*Editor's introduction*, p.5). Part 1: 9 articles on various aspects (legal, pollution control, incentives, etc.) of waste management; part 2: annotated directory (Local authority waste disposal executives ... Courses ... Recoverers ... Consultants ... Laboratories, analysts and research organizations ... Publications and trade journals). Index. *Class No:* 628.54

[4841]
KENWORTHY, L. *and* SCHAEFFER, E. **A Citizens guide to promoting toxic waste reduction.** New York, INFORM, 1990. 122p. $15. ISBN: 0918780543.

Aimed at environmental groups, not individuals, seeking to develop strategies to reduce waste from industrial plants. Based heavily on *Toxics Release Inventory* produced by the US Environmental Protection Agency. Appendices include state government contacts, health effects, right-to-know entitlements, glossary. An obvious US slant, should be of value to similar groups in other countries. *Class No:* 628.54

[4842]
PATRICK, P.K., *comp.* **Glossary of solid waste.** Copenhagen, World Health Organization, Regional Office for Europe, 1980. v, 92p. £4.80. ISBN: 9290201991.

About 700 definitions ('Incinerator: 2/3p). Cross-references. Interdisciplinary. For use professionally or otherwise. *Class No:* 628.54

[4843]
ROBINSON, W.D., *ed.* **The Solid waste handbook:** a practical guide. New York, Chichester, etc., Wiley, 1986. xxvii, 811p., illus., diagrs., maps, tables. £89. ISBN: 0471877115.

Concerned with solid waste management rather than engineering design. The emphasis is on public issues, legislation, regulation, finance, economics, waste recovery, and on the situation in the US. 21 chapters in 3 parts: 1. Public issues - 2. Implementation issues: systems, hardware, operations - 3. Hazardous solid wastes. 42 contributors. Index. Intended for use at all levels of government, national to local. *Class No:* 628.54

Lighting Engineering

Bibliographies
[4844]
BARNETT, D.A. **The Literature of lighting:** an annotated bibliography of the literature of illuminating engineering up to 1965. University Microfilms Ltd., 1968. 173p.

Thesis accepted for Fellowship of the Library Association.

A selective listing, mainly for 1945-1965. Covers periodical titles, but not periodical articles. Historical introduction, followed by 10 sections: 1. Abstracts and indexing services ... 3. Handbooks and sources of data - 4. Books (68p.; *c.*150 titles) ... 6. Conferences, preprints of papers ... 8. Standards, codes of practice, government publications, and patents - 9. Education and research in illuminating engineering; theses - 10. Historical supplement: books published before 1945. Index of authors and titles. *Class No:* 628.9(01)

Handbooks & Manuals
[4845]
CAYLESS, M.A. *and* MARSDEN, A.M., *ed.* **Lamps and lighting:** a manual of lamps and lighting prepared by members of staff of Thorn EMI lighting, Ltd. 3rd ed. London, E. Arnold, 1983. 522p. ISBN: 0713134879.

First published 1966.

5 parts (29 sections, each with further reading): 1. Fundamentals - 2. Lamps - 3. Luminaries and circuits - 4. Interior lighting - 5. Exterior lighting (*e.g.* roads; navigation). 2 appendices: 1. Lamp data; 2. Glossary. References (A-Z authors), p.489-99. Author and subject indexes. 'Written as a reference source for building service engineers, architects and specialists in the lighting industry, as well as others ...' (*Preface*). 'A delightfully practical book' (*Lighting research and technology* v.17(4), 1985, p.194). *Class No:* 628.9(035)

[4846]
MURDOCH, J.B. **Illumination engineering** - from Edison's lamp to the laser. New York, Macmillan Publishing Co.; London, Collier Macmillan, 1985. xiv, [i], 541p. diagrs., graphs, tables. ISBN: 0029485800.

11 documented chapers, *e.g.* 2. Lighting calculations and measurements; 5. Vision and color; 6. Lamps; 7. Interior lighting design: average illuminance; 9. Daylighting; 10. Optics and control of light; 11. Exterior lighting (11 references). Problems, p.491-526, with answers to selected

....(contd.)

problems, p.527-9. Detailed, analytical index, p.533-41. 'A scientific and technical treatment of lighting, rather than an architectural and design treatment' *(Preface)*.
Class No: 628.9(035)

Dictionaries

[4847]

ZIMMERMANN, R., *comp.* **Dictionary of lighting engineering in four languages:** English, German, French, Russian. [Wörterbuch lichtechnik.] Amsterdam, etc., Elsevier, 1990. 426p. £86; $141. ISBN: 0444988505.

Intended to replace 8-language dictionary by the same author published in 1977. Approximately 11,000 entries with equivalents. German, French and Russian indexes. Cross references. Genders given. *Class No:* 628.9(038)

Reviews & Abstracts

[4848]

Lighting research and technology. London, Illuminating Engineering Society of Great Britain, 1969-. Quarterly. ISSN: 00243426.

Formerly *Illuminating Engineering Society Transactions.*

Includes abstracts and occasional reviews. V.20(3), 1988, carries 30 abstracts (sections: Photometry, radiometry, colour - Vision and visual conditions). *Class No:* 628.9(048)

Histories

[4849]

BOWERS, B. A History of electric light and power. London, Peregrinus, 1982. vii, 278p., illus., diagrs. (*IEE history of technology series, v. 3.*) ISBN: 0906048710.

The first section deals with early developments in the understanding and use of electricity. The remaining three quarters considers applications, with emphasis on lighting, describing the methods of generation and creating illumination. There are chapters on public supply and power stations; legislation; local vs. national supply (all with particular reference to Britain). Modern developments and practices in lighting, electric heating and traction conclude the history. Generous use of photographs and quotations from contemporary catalogues and reports.
Class No: 628.9(091)

629 Transport Vehicles

Transport Vehicles

Models

[4850]
WILLIAMS, G.R. **The World of model[s].** London, A. Deutsch, in association with Rainbow Reference Books, 1970-76. 4v. illus., (incl. col). ea. £9.95.

The world of model aircraft. 1973. 224p.
The world of model cars. 1976. 256p. (See entry at 629.114.6(086.48)).
The world of model ships and boats. 1971. 256p.
The world of model trains. 1970. 256p.

Each volume has 32 pages of colour plates and over 200 black-and-white illus. *Class No:* 629.0(086.48)

Carriages & Carts

[4851]
BERKEBILE, D.H. **Carriage terminology:** an historical dictionary. Washington, Smithsonian Institution Press, and Liberty Corp. Books, 1978. 487p. illus.

Carriage types, p.13-305; Carriage nomenclature, p.307-409; Harness nomenclature, p.411-66; William Fulton's 1796 glossary, p.467-82. Selected bibliography, p.483-7. Profusely illustrated. Well produced. *Class No:* 629.11

Motor Vehicle Engineering

Bibliographies

[4852]
Cumulative index of SAE Technical papers (1906-1991). Warrendale, Pa., Society of Automotive Engineers, 1992. 4v. 3722p. £271.

A set comprising the 11th ed. (1965-1991) and the historical ed. (1906-1964).

About 40,000 SAE technical papers in 1800 subject areas. Papers listed by subject and author. *Class No:* 629.113(01)

[4853]
LUTON CENTRAL LIBRARY. **Automobile engineering:** additions to the special collection of the Luton Central Library. Luton, Beds., Public Library, 1969. [vii], 129p. Supplements, [1977] & [1984].

First published 1962.

Luton Public Library [now Luton Central Library] is the specializing library in this subject in the South-Eastern region. The main catalogue lists *c.*6,000 itmes (unannotated). Sections: Lists of periodicals - Classified sequence (motor vehicle engineering; driving; cars; commercial vehicles; motor cycles). Omits bicycles, caravans and sporting cars. Supplements cover July 1971-December 1976 and January 1977-December 1983 respectively. *Class No:* 629.113(01)

[4854]
SHIELD, M.J. *and* ROBINSON, M.S. 'Automotive engineering'. *Information sources in engineering.* Editor, L.J. Anthony (2nd ed., 1985), p.270-93. London, Butterworth, 1985.

Sections: Legislation - Standards - Organizational sources of information (learned societies and institutions; research institutes; cooperative organizations) - Documentary sources of information (abstracting and indexing services; directories and buyers' guides; commercial information; journals). *Class No:* 629.113(01)

Handbooks & Manuals

[4855]
BOSCH, R. **Automotive handbook.** 2nd ed. Düsseldorf, Robert Bosch, GmbH, 1986. 760p. diagrs., graphs, tables.

By a team of contributors. A compendium of technical data ('Internal combustion engine': p.282-337), from 'Bases and fundamentals' to 'Road traffic legislation'. Index of headings (*c.*2,000 entries), p.686-706. Cites DIN standards. *Class No:* 629.113(035)

[4856]
NEWTON, K. *and* STEEDS, W. **The Motor vehicle;** for students, draughtsmen and owner-drivers. 11th ed. Oxford, Butterworth-Heinemann, 1991. 888p. illus., diagrs., graphs. £35. ISBN: 0750604077.

First published 1929. 10th ed. 1983.

Sections: 1. Fundamentals - 2. The engine - 3. Transmission - 4. The carriage units. Covers both petrol and diesel engines. Carefully drawn illus. *Class No:* 629.113(035)

[4857]
WASHINO, S. **Automobile electronics.** New York, Gordon and Breach, 1989. xvi,127p. diagrs., graphs, tables. $49. (*Japanese technology review.*)

Valuable for the English-language coverage of Japanese research (*New technical books,* v.75(2), March/April 1990, p.501). *Class No:* 629.113(035)

Dictionaries

Polyglot

[4858]
BLOK, C. *and* JEZEWSKI, W. **Illustrated automobile dictionary in six languages:** Nederlands/Ruskiĭ/English/Français/Deutsch/Italiano. Deventer, Holland, Kluwer Technische Bocken, and Warsaw, Wydawnictwa Komunikacj i Lacznosci, 1978. 502p. illus. ISBN: 9020110705.

About 3,000 terms in each language across the page, with line-drawings facing. 8 sections: Vehicles; Drive line (US),

....*(contd.)*

Transmission (GB); Chassis; Body; Electrical equipment; Appendix; Service terms. Alphabetical index. *Class No:* 629.113(038)=00

[4859]

—BLOK, C. *and* JEZEWSKI, W. Six-language illustrated dictionary of the motor car: Polish/Russian/English/French/ German/Italian. Warsaw, 1976. 867p. illus.

Includes *c.*7,000 Polish terms, with equivalents and indexes in other languages. *Class No:* 629.113(038)=00

[4860]

DE COSTER, J. Dictionary for automotive engineering: English-French-German. 2nd enl. ed. Munich, K.G. Saur, 1986. 619p. ISBN: 3598105916.

1,972 English-base terms, with French and German definitions. Cross-references. Index of French and German terms. Bibliography, p.619. *Class No:* 629.113(038)=00

[4861]

KONDO, K., *comp*. Elsevier's dictionary of automotive engineering in five languages: English, German, French, Italian and Spanish. Amsterdam, etc., Elsevier, 1977. [vii], 639p. illus. Dfl.360. ISBN: 0444415904.

Previous ed. (1960) as *Elsevier's Automobile dictionary in eight languages (qv).* The 8 languages included Russian, Japanese and Portugese.

6,425 English-base terms, with German, French, Italian and Spanish equivalents and indexes. Illustrations, p.540-639, nos.1-91 (keyed parts, *e.g.* 26: Diesel engine, injection system: 25 parts). 'An unusual, excellent and one-of-a-kind reference book' (*Choice*, v.15(1), March 1978, p.44). *Class No:* 629.113(038)=00

[4862]

—STEKHOVEN, G.S. Elsevier's automobile dictionary; in English/American (with some definitions), French, Italian, Spanish, Portuguese, German, Russian, Japanese. Amsterdam, etc., Elsevier, 1960 (2nd reprint 1983). viii, 946p. Dfl.375. ISBN: 0444405178.

Covers 5,225 terms, with equivalents and indexes. Differs from the Kondo dictionary in concentrating on the motor car, its parts and upkeep, as well as its extended language cover. *Class No:* 629.113(038)=00

English

[4863]

GOODSELL, D. Dictionary of automotive engineering. Warrendale, Pa. Society of Automotive engineers; Oxford, Butterworth-Heinemann, 1989. 182p. illus. $48. ISBN: 0408007834.

Some 2,500 technical terms, including some slang. Distinguishes US and UK terminology and usage. Definitions are 'brief but accurate' ('Reference books bulletin', *Booklist*, July 1990, p.2114) and the illustrations helpful. A paper ed. appeared in 1991 (Butterworth-Heinemann. £12.95. ISBN: 0750602872). *Class No:* 629.113(038)=20

[4864]

SAE glossary of automotive terms. 2nd ed. Warrendale, Pa., Society of Automotive Enginers, 1992. 460p. £61. ISBN: 1560911980.

Nearly 800 terms, extracted from new and revised technical reports between 1986 and 1992. Emphasis is on materials, components and systems used in land and sea vehicles. *Class No:* 629.113(038)=20

German

[4865]

BOSCH, R. Technical dictionary for automotive engineering. Brookfield, Vermont, Renouf USA, 1976-77. 2v.

1. *German-English.* 1976. 354p. 2. *English-German* 1977. 369p.

Each volume has *c.*15,000 strictly technical entries. Cross-references for variant spelling. *Class No:* 629.113(038)=30

[4866]

JUNGE, H.D. *and* LUKHAUP, D. Wörterbuch kraftfahrzeugtechnik: Englisch/Deutsch; Deutsch/Englisch. [Dictionary of automotive engineering: English/German; German/English.] Weinheim, VCH, 1991. [viii],388p. £42. (*Parat.*) ISBN: 3527281711. ISSN: 09306862.

About 10,000 entries in each half. Genders given. American spelling preferred. *Class No:* 629.113(038)=30

Reviews & Abstracts

[4867]

Automobile abstracts: a monthly survey of worldwide technical literature. Nuneaton, Warwickshire, Motor Industry Research Association, 1955-. Monthly. £135pa. ISSN: 09090817.

Originally as *Monthly summary of automobile engineering literature*, 1955-67; as *MIRA abstracts* 1972-74.

About 2,700 abstracts pa., in 6 main sections: Vehicles (motor industry; operation; design; performance) - Components - Fluids - Materials (metals, plastics, glass, etc.) - Production - Research (R & D principles and organizations). Appended: Other SAE papers; Supplementary information; Forthcoming conferences, courses and motor shows. 'Periodicals held in MIRA Library' (*c.*350); occasional book reviews and surveys, and list of MIRA publications. *Class No:* 629.113(048)

[4868]

—Business news index: a bimonthly list of worldwide commercial information. Nuneaton, Warwickshire, Motor Industry Research Association, 1978-. 24 pa. ISSN: 0206194x.

'Contains references [*c.*140 per issue] of interest to the commercial side of the motor and component industries, taken from a variety of economic and technical journals in the MIRA Library'. Arranged A-Z by companies' names; also, where necessary, by names of country, sections of industry, or broad headings (*e.g.* 'Energy', 'Research and development'). A current-awareness source. *Class No:* 629.113(048)

[4869]

Bulletin mensuel de documentation UTAC. Paris, Union Technique de l'Automobile, du Motocycle et du Cycle, 1947-. Monthly. ea. FFr.35. ISSN: 0041705x.

Parts: 1. 'Documentation analytiqué (8 sections; *c.*2,700 abstracts pa.) - 2. Patents (US, German, French, British) - 3. Translations - 4. Bibliography - 5. Standards (Bureau des Normes de l'Automobile, BNA) - 6. News items.
Class No: 629.113(048)

[4870]

SOCIETY OF AUTOMOTIVE ENGINEERS. Technical literature abstracts. Warrendale, Pa., the Society, 1975-. Quarterly.

The abstracts are of papers presented at SAE conferences. Subjects cover aircraft, diesel engines, fuels and lubricants, transportation systems, trucks and buses.
Class No: 629.113(048)

Yearbooks & Directories

[4871]

Automotive consultants directory, 1992. Warrendale, Pa., Society of Automotive Engineers, 1992. 141p. £19. ISBN: 1560912693.

Listings for 198 consulting engineers, support personnel and test facilities. International. *Class No:* 629.113(058)

Tables & Data Books

[4872]

Automotive technical data book: tune-up data for petrol-engined cars and light commercial vehicles from 1983 to 1992. McGeoch, J., *comp.* Sparkford Nr. Yeovil, Somerset, Haynes, 1992. v.p. diagrs., tables. £30. ISBN: 1850108005.

About 60 makes of vehicle. Data include engine capacity, cooling system, fuel system, ignition system, brakes, tyres, suspension, torque wrench settings, capacities.
Class No: 629.113(083)

[4873]

Gale's auto sourcebook, 1991: a guide to information on 1987-91 cars and light trucks. Hill, K. *and* Palmisano, J.M., *eds.* Detroit, Mich., Gale, 1991. 469p. $89.50. ISBN: 0810383128.

Data include model descriptions, features, dimensions, general specifications, evaluations, recalls, safety and repair reports, repair manuals. *Class No:* 629.113(083)

[4874]

SAE handbook. Warrendale, Pa., Society of Automotive Engineers, 1924-. Annual.

1. *Materials.* 2. *Parts & components.* 3. *Engines, fuels, lubricants, emissions, and noise.* 4. *On-highway vehicles & off-highway machinery. Index.*

Originally in 1v., now runs to 4v. Includes over 2,000 specifications and more than 1,000 standards, plus recommended practices and information reports on ferrous and nonferrous metals, nonmetallic metals, threads, etc.
Class No: 629.113(083)

Histories

[4875]

DETROIT PUBLIC LIBRARY. The Automotive history collection of Detroit Public Library: a simplified guide to its holdings. Boston, Mass., G.K. Hall, 1966. 2v.

22,900 photolithographed catalogue cards reproduced in page form. Pt. 1: Dictionary catalog of books - pt. 2: Periodical shelf-list - pt. 3: Check-list of automobile catalogs - pt. 4: Descriptions of special collections. This collection is the largest of its kind in the US. *Class No:* 629.113(091)

Great Britain

[4876]

SOCIETY OF MOTOR MANUFACTURERS AND TRADERS. Motor industry of Great Britain 1986: world automotive statistics. London, Statistical Dept., SMMT, 1986. Annual. 256p., tables. £45 (non-members).

4 chapters: 1. Production - 2. Registrations: new registrations; vehicles in use - 3. Overseas trade - 4. Miscellaneous (*e.g.* International registration letters). The SMMT constitution and organization. Short list of SMMT publications. Organizations connected with the motor industry: 1. UK; 2. Overseas (p.247-86). 105 tables of statistics, 1976-85. An essential source in its field.

SMMT Buyers guide to the motor industry is issued annually. *Class No:* 629.113(410)

Thesauri

[4877]

ASLIB. Motor Industry Information Group. **Automotive engineering terms: a thesaurus.** [London, Aslib, the Group], 1974. [ii], 38p.

For indexing managerial, scientific and technical information applied to the motor industry as a whole. Indicates what term to use, and not to use.
Class No: 629.113:025.43

Museums

[4878]

NICHOLSON, T.R. The World's motor museums. London, Dent, 1970. 143p. illus. ISBN: 0460039296.

Arranged under 26 countries, Argentina ... USA. Great Britain, p.76-94, with 21 illus.; USA, p.114-43. Includes brief description of the collection, with a list of outstanding specimens, and dates and times of opening. Good half-tone illus. (8 in colour). *Class No:* 629.113:061:069

Lorries & Vans

[4879]

GEORGANO, N. World truck handbook. 2nd ed. London, Jane's Publishing Co., 1986. [vi], 328p. £7.95. ISBN: 0710602154.

6 sections (p.11-331): Two-axle trucks - Three-axle trucks - Four-axle trucks - Articulated trucks - Dump trucks - Road tractors. Glossary. Bibliography. Index. 1 illus., per page. Oblong format. *Class No:* 629.114.4

[4880]

MILLER, D., *with others*. **The illustrated encyclopedia of trucks and buses.** London, Hamlyn. ISBN: 0600388204.

15 historical chapters, tracing developments, 1769-. A-Z sequence of trucks and buses follows, p.65-308, describing over 4,000 individual marques. Glossary. More than 1,000 illus., usually small. *Class No:* 629.114.4

Buses & Coaches

[4881]

JENKINSON, K.A. **Preserved buses.** 2nd rev. ed. London, Ian Allan, 1978. 304p. ISBN: 0711008329.

Bodywork - Abbreviations of ownership - Types of buses (p.16-301), listing over 1,530 preserved passenger vehicles of *c.*270 different models. Data: original and current owners; year of manufacture; make and type of bodywork; etc. 1 photograph per page. Small format. *Class No:* 629.114.53

[4882]

—HIBBS, J. 'The History of the motor bus industry: a bibliographical survey'. *Journal of transport history*, v.2, February 1973, p. 41-55. 270 references.

Supplements his *The history of British bus services* (Newton Abbot, Devon, David & Charles, 1968. 280p. illus). *Class No:* 629.114.53

Private Cars

Encyclopaedias

[4883]

GEORGANO, N. **The New encyclopedia of motorcars,** 1885 to the present. New York, Dutton, 1983. 688p., illus. $45.

First published 1968 as *The complete encyclopedia of motorcars 1885 to the present*. 2nd ed. 1973 (751p).

Includes profiles of about 4,300 makes of cars built over the last century worldwide. One illustration per make, as far as possible. Data on each make: nationality; date(s); principal place of manufacture; history; technical information (selective). 'An excellent contribution to existing automotive historical materials' (*Antique Automobile* v.47, November/ December 1982, p.17). *Class No:* 629.114.6(031)

Handbooks & Manuals

[4884]

AUTOMOBILE ASSOCIATION. **Know your car.** Basingstoke, Hants. 319p. illus., diagrs.

Explains the modern motor car, how to service it, and how to diagnose faults and put them right. Some 16 sections: Engine - Transmission - Cooling - Electrics - Suspension - Steering - Brakes - Tyres - Bodywork - Heating and ventilation - The MOT test - Second-hand buying - Accessories - Music on the move - Curing faults - Weekend workshop. Index, p.316-9. Keyed and captioned illus. *Class No:* 629.114.6(035)

[4885]

BALDWIN, N., *and others*. **The World guide to automobile manufacturers.** New York, Facts on File, 1987. 584p. illus. (incl. col). $50. ISBN: 0816018448.

Restricted to 1,000 entries. About one-third of the *c.* 950 illus., are in colour, captions state make, model and year. More selective than Georgano's *Encyclopedia of motorcars*,

....(contd.)

concentrating on the manufacturers. 'A reference book for a broad audience' (*Wilson library bulletin*, April 1988, p.104). *Class No:* 629.114.6(035)

[4886]

CURTIS, A. **The Penguin book of the car.** Harmondsworth, Middx., Penguin Books, 1985. 336p. illus., graphs. ISBN: 0140463119.

14 chapters: 1. The basic elements - 2. The conventional petrol engine - 3. More about the petrol engine ... 7. The electrical system ... 10. Tyres, handling and suspension - 11. The brakes - 12. Styling, structure and safety - 13. The practical side - 14. Legal and insurance. Index, p.325-36. Keyed illus. *Class No:* 629.114.6(035)

Dictionaries

[4887]

SAE **glossary of automotive terms.** 2nd ed. Warrendale, Pa., Society of Automotive Enginers, 1992. 460p. £61. ISBN: 1560911980.

Nearly 800 terms, extracted from new and revised technical reports between 1986 and 1992. Emphasis is on materials, components and systems used in land and sea vehicles. *Class No:* 629.114.6(038)

Tables & Data Books

[4888]

The Guinness world car record. Berg, I. London, Guinness Publishing, 1992. 768p. illus. £14.99. ISBN: 0851125298.

Performance and technical data of cars offered for sale since 1885. Entries under various headings; cross-references to makers and models. *Class No:* 629.114.6(083)

[4889]

World motor vehicle data. Detroit, Mich., Motor Vehicle Manufacturers Association of the United States, 1964-. Annual. tables.

Tabulated data; international coverage. Details concern models, fuel types, weight, cylinder, and age of vehicle in use. Retrospective statistics on production, sales, registrations, imports and exports. *Class No:* 629.114.6(083)

Models

[4890]

WILLIAMS, G.R. **The World of model cars.** London, A. Deutsch, in association with Rainbow Reference Books 1976. 256p. illus. (incl col). £9.95. ISBN: 0233962875.

11 chapters (*e.g.* 1. The world of model cars and other road vehicles; 5. Die-cast model cars; 7. The age of plastics; 11. Radio-controlled model cars). Glossary and abbreviations. 'For further reading', p.199-200. Index. 32 col. pl.; over 200 black-and-white illus. *Class No:* 629.114.6(086.48)

Histories

[4891]

FLOWER, R. *and* JONES, M.W. **One hundred years of motoring:** an RAC social history of the car. Croydon, Royal Automobile Club, in association with McGraw-Hill, 1981. [iv], 224p. illus., (incl. col). ISBN: 0862110181.

....*(contd.)*

Sections: The first spark, 1885-1905 - The challenge of the years, 1900-1914 - Middle class motoring, 1914-1935 - Town and countryside, 1922-1945 - The triumph of technology, 1945-1960 - Countries fit for cars, 1960-1980. Traces the impact of motoring on the British way of life since 1885. Fully illustrated. Partly analytical index, p.221-4. *Class No:* 629.114.6(091)

[4892]
RUIZ, M. One hundred years of the motor car, 1882-1986. London, Collins: Willow, 1987. 280p. illus. (incl. col.). £30. ISBN: 0002181940.

Translated from the Italian (Mondadori) by M. Piòtrowska.

2 main parts: One hundred years of progress; One hundred years of different marques (Alfa Romero ... Volvo, plus other countries, A-Z. Porsche: 2p., 8 illus). Includes history of rallying and motor racing. A handsome quarto. *Class No:* 629.114.6(091)

Worldwide
[4893]
World cars, 1985. Lösch, A., *ed.* New York, Herald Books: Pelham, [1985]. Annual. 439p. illus. (incl. col). tables. ISBN: 0910714177.

Published and edited annually by l'Editrice dell' Automobile LEA, publishing company of the Automobile Club of Italy.

Motor racing - Special bodies - World automobile production (p.42-409: Europe ... Australia). Data: engine; transmission; performance; steering; electrical equipment; dimensions and weight; optionals. Indexes: Name of car; Maximum speed; Makes, models and prices. *Class No:* 629.114.6(100)

Great Britain
[4894]
CULSHAW, D. *and* **HORROBIN, P. The Complete catalogue of British cars.** New York, W. Morrow, 1974. 511p. illus., diagrs., tables.

Lists all production models by nearly 700 manufacturers, 1895-1974. Excludes racing cars, commercial vehicles and special cars. Tabulated technical data. 9 appendices on electric cars, names and addresses of manufacturers, motoring clubs and organizations, etc. Detailed, non-analytical index. Profusely illustrated (c.1,100 illus). *Class No:* 629.114.6(410)

[4895]
NICHOLSON, T.R. The Birth of the British motor car. London, Macmillan, 1982. 3v. (506p.). illus. (pl). ISBN: 0333327179.

1. *A new machine, 1769-1842.* 2. *Revival and defeat, 1842-93.* 3. *The last battle, 1894-97.*

Each volume has notes and an index. V. 3 also carries a select bibliography p. 499-506, in 3 sections. Well researched. A work of 'tremendous scope and unquestionable thoroughness' (*Antique automobile,* v.47, May/June 1983, p.3). *Class No:* 629.114.6(410)

Japan
[4896]
CUSUMANO, M.A. The Japanese automobile industry: technology and management at Nissan and Toyota. Cambridge, Mass., Council on East Asian Studies, Harvard Univ., 1985. xxi, 487p. graphs, charts, tables. ISBN: 0674472551.

6 chapters (1. Company origins and truck technology transfer ... 6. Quality control: manufacturing and design). Appendices, p.385-401. Chapter notes, p.403-33. Bibliography, p.434-52. Detailed, analytical index, p.453-87. *Class No:* 629.114.6(52)

USA
[4897]
GUNNEL, J.A., *ed.* **Standard catalog of American cars, 1946-1975.** 2nd ed. Motor Book International, 1987. 800p. illus. $24.95. ISBN: 0873410459.

First published 1982.

A companion volume, edited by J. Flammang, covering the period 1976-86, appeared in 1988. *Class No:* 629.114.6(73)

Sports Cars
[4898]
COTTON, M. Directory of world sportscars. Bourne End, Bucks., Aston Pubns., 1988. 216p. illus. £11.95. ISBN: 094662738x.

Over 70 distinct car makes and models appearing since 1982 in world championships. 165 illus. *Class No:* 629.114.62

[4899]
HOUGH, R. A History of the world's sports cars. London, Allen & Unwin, 1961. 223p. illus. (incl. col).

Part 1: The world's sports cars: an outline history. p.13-96. Appended: Specifications of the world's sports cars, p.97-140. Part 2: Sports car racing (trophies; races), p.141-218. Index includes references to illus. *Class No:* 629.114.62

[4900]
—**BATCHELOR, D.** *and* **ROBSON, G. The Great book of sports cars.** Yeovil, Somerset, Foulis, 1988. 420p. illus. £29.95. ISBN: 0854296948.

Mainly concerned with post-World War II period, dovetailing into Hough's *History.* Includes more than 200 profiles of personalities. *Class No:* 629.114.62

Motor Cycles
[4901]
AYTON, C., *ed.* **World motorcycles.** Yeovil, Somerset, Foulis, 1983. 238, [6]p., illus.

Covers 16 countries, Austria ... USSR (Italy, p.52-104). Technical data: manufacturer; engine; transmission; suspension; brakes; dimensions, etc. Appended 'Model finding index' (6p.): c.500 models. 1-2 illus. per page. *Class No:* 629.118

[4902]
CHRISTENSEN, R.D. Motorcycles in magazines, 1895-1983. Metuchen, N.J., Scarecrow Press, 1985. [viii], 342p. illus. ISBN: 081081756x.
2,503 numbered entries, usually with brief annotations. 3 parts: 1. General interest and popular technical magazines, 1895-1983 - 2. Automotive magazines, 1950-1983 - 3. *Cycle* magazines, 1978-83. Indexes: A. Motorcycle tests, impressions, descriptions and model announcements; B. Competition reports; C. General, p.333-42.
Class No: 629.118

[4903]
FRANCIS, G. *and* PROST, P. The Penguin book of the motorcycle: a guide to maintenance and problem solving. Harmondsworth, Middx., Penguin Books, 1985. 302p. ISBN: 0140465898.
7 sections: 1. Engines: the basic difference - 2. Maintenance and servicing - 3. Solving engine problems - 4. Engine failure problems - 5. Engine performance problems - 6. Electrical problems - 7. Tips and tricks. Glossary of motorcycling terms, p.273-85. Useful addresses. Your bike's specifications. Analytical index, p.299-302. Designed for all motorcycle riders, regardless of mechanical knowledge or motorcycling experience. *Class No:* 629.118

[4904]
—COUNTER, C.F. Motorcycles: a historical survey, as illustrated by the collection of motorcycles in the Science Museum. 2nd ed. London, H.M. Stationery Office, 1970. xiii, 117p. illus.
First published 1956.
10 chapters (1. The pioneers, 1869-1894 ... 10. Accessories and ancillaries, from 1956). Bibliography, p.109. *Class No:* 629.118

[4905]
SAE motorcycle safety and environmental terminology. Warrendale, Pa., Society of Automotive Engineers, 1977. v, 179p.
About 1,500 terms defined (*e.g.* 'Ignition system': ½p). Entries make references to SAE standards. The latter are listed in Appendix 1, p.169-77. Appendix 2: 'Use and example of permuted-term index'. US vocabulary. *Class No:* 629.118

[4906]
TRAGATSCH, E., *ed*. The Illustrated encyclopedia of motorcycles. London, etc., Hamlyn, 1977. 320p. illus. (incl. col.). tables.
'The A-to-Z of the world's motorcycles', with 3 or 4 illus. per page. A history survey precedes (p.30). Also 'The great designers' and 'The classic bikes through the decades', with good colour photographs. Appended: 'The speed records', 'The leading models (from the 1920s onwards). Glossary. No index. *Class No:* 629.118

Bicycles

[4907]
COLES, C.W. *and* GLENN, H.T. Glenn's complete bicycle manual: selection, maintenance, repair. New York, Crown, 1973. 339p. illus.
Each chapter deals with a part of the bicycle (*e.g.* sprocket cluster; brake). 'The owner or servicer of American, European or Japanese bicycles will find this a useful manual to determine bicycle repairs needed and how to do them.

....(contd.)
Fully illustrated, with step by step instructions' (*Library journal*, 15 April 1974, p.1100). No index.
Class No: 629.118.3

[4908]
WATSON, R. *and* GRAY, M. The Penguin book of the bicycle. London, Allen Lane, 1978. 333p. illus. ISBN: 0713910992.
7 chapters: 1. The bicycle in fashion again - 2. The bicycle today; naming the parts - 3. The aesthetics of it all - 4. Some more history, technical and social - 5. A visit to the industry - 6. Cycle sport - 7. The bicycle in fashion again: another look. Books consulted and further reading, p.313-8. Useful addresses. Well illustrated. Detailed, analytical index, p.325-33. *Class No:* 629.118.3

Shipbuilding

Bibliographies

[4909]
KAHLER, R.C. 'Marine technology'. *Information sources in engineering.* Editor, L.J. Anthony. (2nd ed. 1985) p.325-41. London, Butterworth, 1985.
Sections: Classification societies - Learned societies - Research institutes - Periodicals - Trade directories and yearbooks - Recent monographs and textbooks. *Class No:* 629.12(01)

Encyclopaedias

[4910]
BLACKBURN, G. The Illustrated encyclopedia of ships, boats, vessels and other water-borne craft: comprising an alphabetical directory of all types of craft past and present, containing much discussion of the development of hulls and rigging, together with mention of some of the most outstanding warships and commercial vessels. Supplemented by a glossary, bibliography and index. London, Murray, 1978. 447p. illus.
Over 700 types of craft, A-Z, p.19-392. Glossary p.395-441. Bibliography of relevant books (13 items, with brief evaluative annotations). Index of names of vessels in text, p.445-7. Includes hovercraft and hydrofoils. An essentially personal approach, being reproduced from longhand script, with over 600 elegant line-drawings by the author. Although 'ideal for college and public libraries in general ... [it] does not fill the need for an authoritative nautical dictionary' (*Choice*, v.16(4), June 1979, p.505). *Class No:* 629.12(031)

[4911]
DUDSZUS, A. *and* HENRIOT, E. Dictionary of ship types: ships, boats and rafts, under oar and sail. Translated from the German ed., Das Schiffstypen Lexikon (1983), by K. Thomas. London, Conway Maritime Press, 1986. 252p. illus. (incl. col.), facsims, tables. ISBN: 0851773605.
Dictionary of ship types, A-Z, p.38-239, describing 1,300 vessels. Chronology (to 1964), p.8-14. Bibliography (A-Z, authors), p.241-5. 6 appendices (*e.g.* Wind strength; 4. Sail area and displacement of significant sailing ship types).

....(contd.)
Splendidly illustrated, with 800 photographs and line drawings. The *Dictionary* retains strong emphasis on German and Dutch craft. *Class No:* 629.12(031)

[4912]
The Oxford companion to ships and the sea. London, Oxford Univ. Press, 1976. (reissued 1988). £14.95 (paperback). ISBN: 0192820842.

3,700 articles on nautical terms, famous ships, seamen, navigators, ports, explorers. etc. 'Diving' (unassisted, assisted and free diving), p.250-5; 5 illus., references to Bathyscope, Bathysphere. Biographical sketches (*e.g.* 'Tasman', 'Von Spee', 'Conrad'). Also, Maritime museums, naval instuments. Generally omits material on oceanography and small-boat pleasure sailing. 5 appendices (1. Equivalent ranks; 3. Rules of the road). 226 half tones; 187 line-drawings. 'Highly recommended' (*Library journal*, 15 January 1977, p.188). *The Economist* (23 October 1976, p.142) finds that 'Modern technology has largely been sacrificed for the sake of history'. *Class No:* 629.12(031)

Handbooks & Manuals

[4913]
The Guinness book of ships and shipping facts and feats. Hartman, T. Enfield, Middx., Guinness Superlatives, 1983. 265p. illus. (incl. col.), facsims., ports., tables, maps. ISBN: 0851122698.

Sections: Experiment and exploration - Warships and warfare - Trade and transport - Distress and disaster - Odd facts. Appendices: Types of vessel. Glossary of nautical terms, p.242-54. Bibliography, p.255. Index of ships' names; Index of proper names. *Class No:* 629.12(035)

[4914]
MUCKLE, W. Naval architecture for marine engineers. London, Newnes-Butterworths, 1975. [v], 407. ISBN: 0408001690.

Sections: The function of the ship - Ship types - Definitions, principal dimensions, form coefficient - Classification societies and governmental authorities - Ship calculations - Buoyancy, stability and trim - The sea and ship matters - Structural strengths - Resistance - Propulsion - Powering - Vibration - Ship design. Extensive chapter references. Detailed index. 'A welcome addition to the limited number of up to date books in the field' (*Marine engineers review*, October 1975, p.51).
Class No: 629.12(035)

[4915]
TAGGART, R., ed. Ship design and construction: written by a group of authorities. New Society, Society of Naval Architects and Marine Engineers, 1980. xiv, 734p. illus., diagrs., tables.

Evolved from the *Design and construction of steel merchant ships* (1955), and a 1969 revision.
18 sectionalised chapters (1. Mission analysis and basic design ... 4. Load-line assignments ... 8. Hull materials and welding ... 16. Ship construction - 17. Launching - 18. Trials and preparation for delivery). Glossary, p.717-28. Detailed analytical index. 'Concerned with the practical aspects of ship design as they relate to the requirements of the owner and operator, and as they relate to the characteristic of the mission that the ship is to perform' (*Introduction*). *Class No:* 629.12(035)

Dictionaries

Polyglot

[4916]
PAASCH, H. From keel to truck: dictionary of naval terms; English-French-German-Spanish-Italian. Based upon the original dictionary by Captain Paasch, by L. Bataille and M. Burnet. 5th ed., rev. and enl. London, Philip, 1937. [*c.*1000]p. pl.

First published 1885.
About 17,500 English entry-words and definitions, arranged by subjects, with translations and equivalents in the other 4 languages; 5 language indexes. Plates are on verso pages; that on the schooner shows 50 keyed parts. Many compounds (*e.g.* 'Ballast-donkey-valve-rod stuffing-box-gland', p.820). A classic dictionary, unrivalled for its definitions in all 5 languages covered and for its many plates (103 in all) showing plans and detailed parts of ships.
Class No: 629.12(038)=00

[4917]
Schiffahrts-Wörterbuch: deutsch-englisch-französisch-spanisch-italienisch. Hamburg, Horst Kamm-Verlags-und Handelsgesellschaft GmbH, 1987. 837p.

About 8,000 German base-terms, with equivalents in English, French, Spanish and Italian, across the double page. Appended section: 'Ships papers', in German, Portugese, Danish, Norwegian, Swedish and Dutch. Abbreviations, p.791-837. *Class No:* 629.12(038)=00

[4918]
—BERSCH, E., *ed.* Shipbuilding, seafaring technical dictionary; in Russian, English and German. East Berlin, Verlag Technik, 1980. 880p.

Includes several indexes. *Class No:* 629.12(038)=00

[4919]
VANDENBERGHE, J.P. *and* **CHABALLE, L.V., comps.** **Elsevier's nautical dictionary in six languages:** English/American, French, Italian, Dutch and German. 2nd completely rev. ed. Amsterdam, etc., Elsevier, 1978. x, 949p. Dfl.465. ISBN: 0444416943.

First published 1965-66 (3v.) by P.E. Segditsas.
18,782 entries, representing *c.*24,000 English terms or expressions - 50% more than in the 1st ed. French, Italian, Dutch and German equivalents and indexes. Coverage: 'shipbuilding, types of sea-going vessels, equipment aboard, seamanship, navigational aids and their use on board, sea-going staff and port-personnel, active or passive beaconing, new shipping techniques, goods handling, insurance and trading, commercial and financial documents, international law of the sea, yachting and racing, port terminology' (*Preface*). Thumb-indexed. *Class No:* 629.12(038)=00

English

[4920]
INSTITUTE OF MARINE ENGINEERS. Glossary of marine technology terms. London, Heinemann, in association with the Institute of Marine Engineers. 434p. ISBN: 0434908401.

Defines *c.*1,500 terms ('Ignition': 2/3p.; 'Propellor': 1p). Involves seamanship, ship construction, marine engineering, electricity and electronics. Amusing explanation of why a ship is called 'she' (quoting *Annual report*, RINA, 1976, v.118, p.liv). *Class No:* 629.12(038)=20

[4921]
LAYTON, C.W.T. Dictionary of nautical words and terms:
8,000 definitions in navigation, seamanship, rigging, meterology, astronomy, naval architecture, average, ship economics, hydrography, cargo, stowage, marine engineering, ice terminology, buoyage, yachting, etc. Rev. 3rd ed. Glasgow, Brown, Son, & Ferguson, 1987. 397p. £22. ISBN: 0851745369.

First published 1955.

8,000 definitions in navigation, seamanship, rigging, metrology, astronomy, naval architecture, ship economics, hydrography, cargo stowage, marine engineering, ice terminology, buoyage, yachting, etc. (*e.g.* 'Lloyd's Register of Shipping': 11 lines). Includes acronyms, organizations. Appended list of abbreviations. *Class No:* 629.12(038)=20

[4922]
NOEL, J.V. The VNR dictionary of ships and the sea. New York, etc., Van Nostrand Reinhold, 1987. vi, 393p. illus. ISBN: 0442256310.

About 6,000 entries, including abbreviations and acronyms, briefly defined (*e.g.* 'Gulf stream system': 9 lines; 'Beaufort scale': ½p). Cross-references. International code flags and pennants, on end papers.
Class No: 629.12(038)=20

German

[4923]
DLUHY, R., *comp.* **Schifftechnisches Wörterbuch.** 4 erw. Aufl. Hannover-Wülfel, Industrie und Handelsverlag GmbH, 1974-75. 2v. (xxxi, 1,709p.; xxxix, 1,631p.).

First published 1956 (Bremen, Dorn). 1. *Deutsch-Englisch.* 2. *Englisch-Deutsch.*

A maritime dictionary. Each v. has *c.*70,000 entry-words, categorized, and including abbreviations. Coverage: shipbuilding, shipping, marine engineering and related fields (*e.g.* oceanography, maritime law). Many compounds (*e.g.* 'ventil', (valve): 9½ columns). Idioms and technical expressions. German genders stated. Each v. carries lists of weights and measures. Authoritative.

A shorter, updated version, with similar title, appeared in 1981 (Hannover, Vincentz-verlag. 741p.).
Class No: 629.12(038)=30

Norwegian

[4924]
ASKIM, P. Engelsk-norsk maritim-teknisk ordbok. 2. opplag. Oslo, Grondahl, 1959. 137p.

English-Norwegian dictionary, with *c.*10,000 entry-words covering shipping and shipbuilding. Appended specification - survey for general overhaul of diesel machinery on a ship, in English and Norwegian.

This and its companion Norwegian-English dictionary form an established pair of technical dictionaries, with reprints. *Class No:* 629.12(038)=396

[4925]
ASKIM, P. Norsk-engelsk maritim-teknisk ordbok. 7. opplag. Oslo, Grondahl, 1962. 219p.

First published 1936.

Norwegian-English dictionary, with *c.*15,000 entry-words covering shipping and shipbuilding. Appended specification for a ship's survey and sundries report in both Norwegian and English (p.213-9). *Class No:* 629.12(038)=396

French

[4926]
Dictionnaire anglais-français et français-anglais des termes et locutions martitimes. 2nd éd., rev. et augm. Paris, Éditions Maritimes et d'Outre-Mer, 1971. 319p.

About 5,000 entry-words in each part. Based on H. Paasch's multilingual *From keel to truck: dictionary of naval terms, English-French-German-Spanish-Italian (qv).*
Class No: 629.12(038)=40

Spanish

[4927]
ALFARA PEREZ, T. Diccionario marítimo y de construcción naval (inglés-español y español-inglés). [Dictionary of maritime amd shipbuilding terms (English-Spanish, Spanish-English).] Barcelona, Carriga, [1976]. 479p.

About 10-12,000 headwords in each part: English-Spanish, p. 17-263; Spanish-English, p.267-460. 'Helm': nearly 1 column. Many categories of terms. Pronunciation guide to English words only. Appendices of geographical names, weights and measures. *Class No:* 629.12(038)=60

Russian

[4928]
ANASHKIN, I.A., *and others.* **Morskoĭ slovar'.** Moscow, Voennoe Izdat. Ministerstvo Oborony SSSR, 1959. 2v. (892p.). illus., diagrs.

'Maritime dictionary'. About 12,000 entry-words. Gives English translations of Russian terms and idioms, as well as Russian definitions. *Class No:* 629.12(038)=82

Reviews & Abstracts

[4929]
BMT [British marine technology] abstracts. Wallsend, Tyne & Wear, BMT Ltd., Wallsend Research Station, 1946-. Monthly. ISSN: 02689650.

Previously as *Journal of abstracts of the British Ship Research Association* (1962-67), and originally as *Journal of the British Shipbuilding Research Association* (1946-62).

V.42(1), January 1987, carries 255 abstracts from current technical literature. Sections: Ship design - Ship construction - Ship machinery and systems - Ship operation - Fluid mechanics - Ocean engineering - Warships - Miscellaneous. Record of new or converted ships. Author and ship's name index. Online through BSRA database at Wallsend, which also covers the output of *Ship abstracts* (published by a consortium headed by the Norwegian Ship Research Institute). *Class No:* 629.12(048)

[4930]
Maritime information review: a combination of *Marna News* and *The Ship abstract journal.* Rotterdam, Netherlands Information Centre (CMO), in cooperation with the Association of Finnish Shipbuilders, the Norwegian Technology Research Institute A/S (Marintek), the Swedish Shipbuilders Association, and the Chalmers University of Technology. ISSN: 09201610.

Sections 01-90 (*e.g.* 01. Ships; 02. Shipping; 03. Shipbuilding; 04. Ports and waterways; 05. Fishing industry; 06. Offshore oil and gas; 07. Ocean mining; 08. Inland shipping; 09. Marine pollution; 10. Energy generating and floating plants; 11. Driving; 12. Navy; 13. Sailing ships,

....(contd.)
yachts; 14. Ice, icing, icebreaking; 80. Laws, recommendations, regulations and rules; 90. Reference books.

Databases: MARNA (updated weekly); SHIPDES (ship descriptions, updated fortnightly); SHIP ABSTRACTS. *Class No:* 629.12(048)

Yearbooks & Directories
[4931]
European Community shipping directory: 1992. Colchester, Lloyd's of London Press, 1992. 176p. maps. £55. ISBN: 1850444838. ISSN: 09616764.

Alphabetically by country (Belgium ... United Kingdom). Data include classification societies, liner services by agent, ports, ship suppliers and brokers, towage, salvage and offshore services. Table of airlines and EC airports. *Class No:* 629.12(058)

[4932]
'Fairplay' world shipping directory. London, Financial Times Business Information, Annual. ISSN: 09593101.

Previously published as 'Fairplay' World shipping year book.

Covers shipowners, shipbuilders and repairers, marine equipment systems, countries, maritime organizations, shipping statistics, etc. Advertisers' index. *Class No:* 629.12(058)

[4933]
Lloyd's maritime directory: 1991. Colchester, Essex, Lloyd's of London Press, 1991. xxxii,756p.

Sections: The year in shipping - Towage, salvage and offshore services - Ship earth station directory - Ship management review - Shipbuilders/ship owners - Marine equipment builders/repairers - Banking/finance - Insurance/P & I Club - Marine solicitors/lawyers - Marine consultants - Classification societies - General maritime organizations - Port authorities. Index to vessels and index to owners, managers, agents (p.1-469) precedes. Appended index of advertisers. *Class No:* 629.12(058)

Published Series
[4934]
NATIONAL MARITIME MUSEUM. The Ship series. Greenhill, B., *ed.* Greenwich, the Museum, 1980-81. 10v. illus.

1. *Rafts, boats and ships from prehistoric times to the medieval era,* by S. McGioul. 1981. 88p.

2. *Long boats and round ships: warfare and trade in the Mediterranean, 3000 BC - 500 AD,* by J. Morrison. 60p.

3. *Tiller and whipstaff: the development of the sailing ship, 1400-1700,* by A. McGowan. 1984. 60p.

4. *The century before steam, 1700-1820,* by A. McGowan. 1980. 60p.

5. *Steam tramps and cargo lines, 1850-1970,* by R. Craig. 1980. 60p.

6. *Channel packets and ocean liners, 1850-1970,* by J. Maber. 1980. 60p.

7. *The life and death of the merchant sailing ship,* by B. Greenhill. 1980. 60p.

8. *Steam, steel and torpedoes: the warship in the 19th century,* by D. Lyon. 1980. 60p.

9. *Dreadnought to nuclear submarine,* by A. Preston.

....(contd.)
1980. 60p.

10. *The revolution in merchant shipping, 1950-1980,* by E. Corlett. 1981. 60p.

10 illustrated hardbacks that trace the development of fighting and trading ships from 3000 BC to the present day. Authoritative, dispelling old myths and replacing them with hard facts. *Class No:* 629.12(082.1)

Tables & Data Books
[4935]
Register of ships 1986-87. London, Lloyd's Register of Shipping. 3v. (2240p.). ISSN: 01414809.

3v., arranged by ships names, A-G, H-O, P-Z. V.3 has appendices: Ships' names; Former names of ships; Compound names of ships; Ship-borne barges; Mooring barges; Miscellaneous pontoons; Air cushion vessels; Docking installations. *Class No:* 629.12(083)

Histories
[4936]
BARNABY, K.C. The Institution of Naval Architects, 1860-1960: an historical survey of the Institution's 'Transactions' and activities over 100 years. London, the Institution, in association with Allen & Unwin, 1960. 645p. illus. (pl.), diagrs.

Constitutes a detailed survey of the progress of shipbuilding. Includes lists of contents of the *Transactions,* 1860-1959 in an appendix (incorrectly listed in 'Contents' as 1910-1959). Indexes of personal and ships' names. *Class No:* 629.12(091)

[4937]
BROUWER, N.J. International register of historic ships. Oswestry, Shropshire, A. Nelson, in association with the World Ship Trust, 1985. xii, 321p. illus. £10.95. ISBN: 0904614115.

Data on more than 700 ships; 43 countries (Argentina ... Yugoslavia). 'To be included ships have to survive as virtually complete hulls' (*Introduction*) Details on each include: year built; builders; original and present owner; type of engines; superstructure; displacement; condition; alterations; bibliography; history and significance. Appended list of ships by type. Detailed index, p.319-21. 1-4 illus., per type. *Class No:* 629.12(091)

[4938]
GUTHRIE, J. A History of marine engineering. London, Hutchinson, 1971. 294p. illus.

14 chapters (1. The early steam engine - 2. The earliest steamships ... 8. Steam turbines - 9. Diesel engines - 10. Hydrofoils and hovercraft ... 13. Steam boilers - 14. Auxiliary machinery). 82 clear line-drawings. Detailed index, p.289-94. Intended for those 'professionally concerned' (*Editor's note*). *Class No:* 629.12(091)

[4939]
HAWS, D. *and* HURST, A.A. The Maritime history of the world: a chronological survey of maritime events from 5000 BC until the present day, supplemented by commentaries. Brighton, Sussex, Teredo Books, Ltd., 1985. xxv, 494p. illus. (incl. col.), maps. ISBN: 0903662108.

V.1: 3000 BC - AD 1816 (xxv, 494p.). V.2: 1816 - 1984 (vii, 419p.). with 'Epilogue' and 'Addenda'. Detailed, analytical index, p.318-419,

....*(contd.)*

covering both volumes ('France': 2 columns). 203 illus., (62 in colour), 31 maps. Scope: wars, battles, trade exploration, chartered companies, slavery, piracy, jurisprudence, navigational developments, ship design, etc. Quarto format. *Class No:* 629.12(091)

[4940]

LANDSTROM, B. The Ship: a survey of the history of the ship, from the primitive raft to the nuclear-powered submarine, with reconstructions in words and pictures. London, Allen & Unwin, 1961. 309, [9]p. illus. (incl. col.).

Translated from Skeppet (Stockholm, Bokforlaget Forum AB, 1961), by M. Phillips.

A notable feature: the 807 carefully-drawn illus., about 2 per type of ship described. Appended list of sources. Non-analytical index. A definitive work on the history of ships. *Class No:* 629.12(091)

Archives

[4941]

The Shipbuilding industry: a guide to historical records. Ritchie, L.A., *ed.* Manchester Univ. Press, 1992. 304p. £45. (*Studies in British business archives.*) ISBN: 0719038057.

Gives present location of shipbuilding archives. Brief histories of *c.*200 businesses. *Class No:* 629.12(093.20)

Conferences

[4942]

BAHRAIN SOCIETY OF ENGINEERS *and* INSTITUTE OF MARINE ENGINEERS. International Maritime Engineering Conference, Bahrain, '84, 21-25 October, 1984. [iv], 2131p. illus., diagrs., graphs, charts, tables.

29 contributions, mostly documented, in 8 sessions (each with discussion appended): 1. Training - 2. Ports and harbours, in design and construction - 3. Cargo handling in harbour - 4. Pollution - 5. Water engineering - 6. Marine engineering - 7. Ship repair and docking - 8. Ships. *Class No:* 629.12:061:061.3

Museums

[4943]

AYMAR, B. A Pictorial treasury of the marine museums of the world: a guide to the maritime collections, restorations, replicas, and marine museums in twenty-three countries. New York, Crown Publications, 1967. viii, 248p. illus., facsims., ports.

Part 1: Marine museums and maritime collections, United States (p.2-77), Argentina ... Yugoslavia - 2. Restorations and replicas (5 countries) - 3. Special. Non-analytical index. *Class No:* 629.12:061:069

[4944]

STAMMERS, M.K. Discovering maritime museums and historic ships. Princes Risborough, Bucks., Shire Publications Ltd., 1978. 84p. illus. £0.70.

Arranged under counties of England, Isle of Man, Northern Ireland, Scotland and Wales, subdivided by towns. Short glossary; 'Useful books and periodicals' (p.82), 'Useful addresses'. Index. 236 black-and-white illus. *Class No:* 629.12:061:069

Sailing Ships

[4945]

WHITLOCK, P.C., *and others*. The 'Country Life' book of nautical terms under sail. Feltham, Middlesex, 'Country life' books, 1978. [352]p. illus., facsims., maps.

Brief definitions of words associated with sailing and sailing ships. *Class No:* 629.123.1

Passenger Ships

[4946]

BAYNARD, F.O. *and* MILLER, W.H. Fifty famous liners. Willingborough, P. Stephens, 1982-87. 3v. illus., diagrs., tables.

Each volume deals with 50 famous passenger ships (*e.g. Titanic*: v.1, p.59-62. 4 illus., 4 cross-sections), each given descriptive text and technical data. Each volume has an index. A combined index is called for. *Class No:* 629.123.3

[4947]

KLUDAS, A. Great passenger ships of the world. Cambridge, P. Stephens, 1975-77. 5v. illus.

First published as *Die grossen Passagierschiffe der Welt.* (Oldenbourg, 1972). V. 1: 1858-1912(1975); V. 2: 1913-1923(1976); V. 3: 1924-1935; V. 4: 1936-1950(1977); V. 5: 1951-1970(1977).

Deals with ships either in classes (*e.g.* Elder Dempster liners) or individually (*e.g.* 'Queen Elizabeth'). Covers passenger vessels of 10,000 gross tons or more. General data: chronology; yard number; launching and commissioning dates; handed over to ...; retired; rebuilt. Technical data: dimensions; propulsion; power; speed; passengers; crew. Historical data: owners; builders; completion; routes. V.5 (226p). carries a bibliography, p.212-3. Index of ships' names. Well illustrated, - about 1 per page. *Class No:* 629.123.3

Cargo Ships

[4948]

GILMAN, S., *and others*. The Shipping revolution: the merchant ship, 1960-the present. Couper, A. *and* Greenhill, B., *eds.* London, Conway Maritime, 1992. 192p. illus., maps. £28. (*Conway's History of the Ship.*) ISBN: 0851775691.

Focuses on the rapid advances in maritime technology, as well as the political, economic and social reasons for these. Over 200 illus. *Class No:* 629.123.4

[4949]

GREENWAY, A. Soviet merchant ships. 4th ed. Havant, Hants., Mason, 1982. 226 p. illus.

First published 1969.

A concise recognition manual of merchant ships, fish-factory vessels and research fleet of the USSR. Vessels, for inclusion, must exceed 1,000 deadweight tons. Data include builders, tonnage, main engines and hull dimensions of each ship listed. *Class No:* 629.123.4

[4950]

Jane's freight containers, 1987. 19th ed. London, Jane's Publishing Co., 1987. [16], 712p. illus., diagrs. ISBN: 0710608448.

Terminals (p. 1-254) - Operators - Air freight - Manufacturers and services - International road and rail -

.... *(contd.)*

International container standards. General index, p.692-705; ship index (*c.*5,000 entries). Numerous illus., and cross-sections. *Class No:* 629.123.4

[4951]
Jane's merchant ships, 1987-88. Greenman, D., *ed.* 3rd ed. London, Jane's Publishing Co., Ltd., 1987. [62], 689p. ISBN: 0710608365.

Merchant ship identification, using the Talbot-Booth system, p.1-460. Ship drawings section (drawn to a common scale and grouped by similarity of shape) is followed by 'Ship directory and data section'. Addenda. Glossary precedes. 3 indexes, p.637-89: Named ships; Ships' photographs; Ship classes and types. Online via DIALOG. *Class No:* 629.123.4

[4952]
MacGREGOR, D.R. Merchant sailing ships. London, Conway Maritime Press, 1984-85. 3v. illus., facsims., ports, plans. ea. £15. ISBN: 0851772301; 0851172943; 0851773165.

1. *Merchant sailing ships, 1775-1815: sovereignty of sail.* 2nd ed. 1985.

2. *Merchant sailing ships, 1815-1850: supremacy of sail.* 1984.

3. *Merchant sailing ships, 1850-1875: heyday of sail.* 1984.

Covers barques, brigs and schooners, as well as the more spectacular vessels of the merchant navy. A feature: the many reconstructed and redrafted plans of ships, - poorly documented craft until now. The second volume (192p) has 110 photographs and 100 line-drawings. Each volume carries a bibliography. *Class No:* 629.123.4

Tankers

[4953]
The Tanker register: tankers and combined carriers of all countries. New York, Int. Publications Service, Annual.

26th ed. 1986. Contents: Glossary - Tropical and winter deadweight and draught diagrams - Comparative tables of specific gravities - Statistical tables. In sections including: Names of tankers in size and age groups; Particulars of tankers, including special purpose types; List of owners and/or managers. *Class No:* 629.123.56

Small Craft

Hovercraft

[4954]
Jane's high speed marine craft 1992. Coulsdon, Surrey, Jane's Information Group, 1992. *c.*520p. illus., tables. £135. ISBN: 0710609930.

Formerly *Jane's Surface skimmers,* 1967-.

Builders - Operators - Principal engineering components for high-speed craft - Services. Bibliography. Addenda. Index of organizations/craft types/engine types/craft names. *Class No:* 629.125.8

Rail Vehicle Engineering

[4955]
HOLLINGSWORTH, B. An Illustrated guide to modern trains. London, Salamander, 1985. 235p. illus. (incl. col.). ISBN: 0861012208.

Concentrates on 80 powerful locomotives since 1945, tracing the cycle from steam to diesel, diesel electric and electric, then back to steam. Most of the illustrations are in colour. 'Very appropriate for public libraries and special collections on railroading' (*Science and technology libraries,* v.8(1), Fall 1987, p.172). *Class No:* 629.4

[4956]
UNION INTERNATIONALE DES CHEMINS DE FER. Recueil des termes concernant l'emploi et l'échange de matériel roulant. Warsaw, Ministry of Communications, 1958-65. 2v. illus.

V.1 is a rolling-stock glossary of 4,981 numbered terms in English, French, German, Italian, Polish and Russian. V. 2, *Nomenclature illustrée,* has a glossary of 4 sections: A. Coaches and wagons - B. Parts of coaches and wagons - C. Equipment for universal transport and its essential parts - D. Various accessories and tools for trailer stock. A-Z indexes: French, German, English, Italian, Polish and Russian. 2,982 illus. *Class No:* 629.4

Locomotives

[4957]
CASSEREY, H.C. Preserved locomotives. 5th ed. London, I. Allen, 1980. 192p. illus. ISBN: 0711009910.

First published 1967.

Chronological record, 1813-1944. Well illustrated (1-2 photos per page). Appendix: Index of individual and light railways' locomotives (p.170-85). Index of foreign -built locomotives. Principal centres of preserved locomotives, p.157-8 (names only). Analytical index, p.190-2. *Class No:* 629.42

[4958]
NOCK, O.S. British locomotives of the 20th century. Cambridge, P. Stephens, 1983-85. 3v. illus., diagrs., charts, tables. ISBN: 0850595959.

1. *1900-1930.* 1983.

2. *1930-1960.* 1985.

3. *1960-to present day.* 1985.

V.1 (255p) has 17 chapters. Illustrations (396) include cross-sections and are plentiful and well captioned. Bibliography, p.252. Detailed, analytical index. *Class No:* 629.42

Steam Trains

[4959]
JONES, K.P. Steam locomotive development: an analytical guide to the literature on British steam locomotive development, 1923-1962. London, Library Association, 1969. [xii], 413p.

Based on a thesis accepted for the Fellowship of the Library Association in 1968.

About 4,000 items (books, periodical articles, reports, etc.), many with brief annotations. Sections (Great Western Railway ... Locomotives manufacturers' products) and sub-sections have introduction. Name index; index of locomotive

....(contd.)

classes. Bibliography, p.12-13. General reference works, p.14-21. 'A most comprehensive work' (*British book news*, November 1969, p.830-1). *Class No: 629.422*

[4960]
NOCK, O.S. British steam locomotive, 1928-1965. London, I. Allan, 1988. 276p. illus., diagrs., maps. £12.95. ISBN: 0711001251.

First published 1969; new impression, 1988.

Reprint of the author's classic study of the development of the British steam locomotive in its heyday. *Class No: 629.422*

[4961]
—GARRATT, C. A Popular guide to preserved steam railways of Britain. Poole, Dorset, Blandford Press, 1979. 112p. illus., maps.

Covers the 65 principal places at which steam locomotives can be seen in Britain. Data: HQ and location; track route; times of opening; further information; history and rolling stock. *Class No: 629.422*

Aircraft Engineering

Bibliographies

[4962]
CLAXTON, M. and HALLAM, L., comps. Information pack on fly-by-wire: reference to fly-by-light, fly-by-speech and power-by-wire aerospace technology. London, Information and Library science, Institution of Mechanical Engineers, 1987. iv,69p. £12.50. (*Information pack, 5.*) ISBN: 0852986491.

Fly-by-wire is an advanced flight control system. 2 parts: 1. Fly-by-wire information (abbreviations - patents), p.1-36; 2. Fly-by-wire references (126 items) with abstracts, p.37-62. Indexes: author, subject/aircraft. *Class No: 629.7(01)*

[4963]
HANNIBALL, A. Aircraft, engines and airmen: a selective review of the periodical literature, 1930-1969. Metuchen, N.J., Scarecrow Press, 1972. xxiv, 825p.

5 parts: 1. Aircraft (3,157 numbered references) - 2. Aircraft engines - 3. Biography - 4. Air Forces (history and organization; equipment; air bases; unit histories) - 5. Miscellany. Indexes: aircraft names; aircraft manufacturers; aircraft symbols; engine names; engine's symbols. 'An outstanding work of reference' (MacAdam, E.J., in *Information sources in engineering* (2nd ed., 1985), p.323). *Class No: 629.7(01)*

Encyclopaedias

[4964]
GARRISON, P. The Illustrated encyclopedia of general aviation. 2nd ed. Blue Ridge Summit, Pa., TAB Books, 1990. 402p. illus. $24.95. ISBN: 0830633182.

Lists aviation organizations, manufacturers, publications, etc. Does not include specifications strictly for military or commercial aviation. 'Suitable for all libraries' (*New technical books,* May/June 1991, item 797). *Class No: 629.7(031)*

[4965]
Jane's encyclopedia of aviation. London, Jane's Publishing Co. 5v. (1078p.). illus. (incl. col.), ports., diagrs. ISBN: 0710607105.

Aircraft, AAMSA ... Zlin, with data on about 5,000 different aeroplanes, including spacecraft and gliders. V.1 includes glossary, a world directory of airlines, world directory of air forces, aerospace world records, and the first part of the A-Z sequence of aircraft - AAMSA ... Antonov. Includes biographies. Index (tiny print) in v.5, p.1065-78. About 3,000 illus. in all. Online via DIALOG. *Class No: 629.7(031)*

[4966]
ROBINSON, A. Dictionary of aviation: eighty years of powered flight. London, Orbis, 1984. 304p. illus. (incl. col.), ports. ISBN: 085613533x.

3 sections: 1. Pilots and personalities - 2. Aerial warfare (*e.g.* Pacific War, 1941-45) - 3. Aircraft manufacturers. A-Z entries in each section. Fully illustrated (2 or more illus. per page) and captioned. No general index. *Class No: 629.7(031)*

Handbooks & Manuals

[4967]
Jane's world aircraft recognition handbook. Wood, D. 4th ed. Coulsdon, Surrey, Jane's Information Group, 1989. *c.*600p. illus. £12.95. ISBN: 0710605870.

First published 1979.

3-view silhouette, photograph and brief description of over 500 types of aircraft. Civil aircraft registrations; glossary of recognition terms. *Class No: 629.7(035)*

Dictionaries

Polyglot

[4968]
AÉROSPATIALE. Dictionnaire aérospatiale: français, anglais, allemand, espagnol. Paris, Gauthier-Villars, 1984. 4v.

4.v. expansion of *Dictionnaire des techniques aéronautiques et spatiales, français-anglais-allemand aerospatiale* (1978). 114p. illus. 2.éd. (1984. 152p.), with Spanish added.

The 4v. comprise: French, English (*Aerospace dictionary*), German (*Wörterbuch der Luft-und Raumfahrt*) and Spanish (*Diccionario aéroespacial*) dictionaries, each covering the four languages concerned. 36,000 entry-words in all. *Class No: 629.7(038)=00*

[4969]
Complete Multilingual dictionary of aviation and aeronautical terminology: English, French, Spanish. Demaison, H., *ed.* Chicago, Il., Passport books, 1984. xvi,671p.

13,000 English-base terms, with French and Spanish equivalents and indexes. Appendices A-L (F: 'Abbreviations of common usage'). *Class No: 629.7(038)=00*

[4970]

FRENOT, G.H. *and* HOLLOWAY, A.H., *eds*. **AGARD aeronautical multilingual dictionary.** London, Pergamon Press, 1960. Supplement 1, 1963. 1072p. Loose-leaf. DFl.295.

A NATO Advisory Group on Aeronautical Research and Development publication.

1,574 numbered English entry-words, plus definitions, equivalents and indexes in French, German, Spanish, Italian, Dutch, Turkish and Russian. (Supplement 1 adds 1,287 entries and a ninth language, Greek). 15 sections (1. Aeronautics (General) ... 14. Radiocommunication and radiolocation - 15. Meteorology). Russian terms are lacking in sections 13-14. Criticized in *Interavia* (v.16(2), February 1961, p.242) for containing only basic words and expressions in the sections. 'If enlarged the dictionary could form the basis for a comprehensive standard work. Despite uneveness, it is a classic for accuracy, up-to-dateness and clear presentation of complex data' (*Library journal*, v.89(6), 15 March 1964, p.1226-7).
Class No: 629.7(038)=00

[4971]

—DORIAN, A.F. *and* OSENTON, J. Elsevier's dictionary of aeronautics in six languages: English/American, French, Spanish, Italian, Portuguese and German. Amsterdam, etc., Elsevier, 1964. [vii], 842p. DFl.295. ISBN: 0444401776.

5,919 numbered English/American entry-words, with definitions; equivalents and indexes in the other languages. Genders given. Primarily concerned with civil aviation. Covers terms used in radio communication, meteorology, electronics, practical mechanics, etc., as well as jargon. *Interavia* (v.30(8), August 1965, p.300) notes some clumsy expressions and inaccurate translations.
Class No: 629.7(038)=00

French

[4972]

WINGROVE, R.G. *and* JAMES, J.R.J. **The Aviator's English and French dictionary.** [Dictionnaire français et anglais des aviateurs.] Cranfield, Beds., Cranfield Press, 1985. [290]p. £16.95.

'Thoroughly recommended, although a little inconsistency in the treatment of abbreviations and acronyms, especially on the French side '(*The Linguist*,. v.25(3), Summer 1986, p.154,162). *Class No:* 629.7(038)=40

Russian

[4973]

KONARSKI, M.M. **Russian-English dictionary of modern terms in aeronautics and rocketry.** Oxford, etc., Pergamon Press, 1962. xi, 515p.

Over 14,500 numbered Russian terms used in Soviet aeronautics and allied fields, - radio, electronics, meteorology and aerial photography. English equivalents and index, making the dictionary virtually two-way. Includes a list of Russian abbreviations and mathematical signs, etc. *Spaceflight* (v.5(1), January 1961, p. 34-35) praises the work as being meticulously accurate, complaining only of the excessive inclusion of ordinary words.
Class No: 629.7(038)=82

[4974]

MURASHKEVICH, A.M. Anglo-russkiĭ aviatsionnyĭ slovar'. Moscow, Izdat. 'Sovetskaya Entsiklopediya', 1964. 687p.

English-Russian aviation dictionary. 42,000 entry-words, covering all aspects of aviation; abbreviations, p.663-87.
Class No: 629.7(038)=82

Reports Literature

[4975]

NORTH ATLANTIC TREATY ORGANIZATION. Advisory Group for Aeronautical Research and Development (AGARD). **AGARD index of publications, 1952/1970-.** Neuilly-sur-Seine, AGARD, 1972-.

Lists by form (*e.g.* reports, Agardographs, etc.) of all unclassified publications issued by AGARD. Analytical entries for collections. The 1952/70 set had 4 parts: abstract section, subject index, author index, and addendum. The *Index* has subsequently covered 3-year periods.
Class No: 629.7(047)

Reviews & Abstracts

[4976]

CIVIL AVIATION AUTHORITY. CAA CENTRAL LIBRARY. **Library bulletin: a monthly review of books, reports, periodical articles and other items of interest available in CAA libraries.** London, CAA Central Library, 1978-. Monthly. ISSN: 01410498.

More than 150 entries per issue, for books and technical reports (annotated entries), book reviews (references), periodical articles (annotated entries). 'Coming events in aviation' include conferences (date; organizer; subject; location). 9 library locations. *Class No:* 629.7(048)

Periodicals

[4977]

UNITED STATES. Library of Congress. Science and Technology Division. **Aeronautical and space serial publications:** a world list. Washington, the Library 1962. 255p.

4,551 serial publications, under countries A-Z (especially USA, Germany, Great Britain, France and Russia). Covers periodicals, reports, annuals. Library of Congress call numbers given.

A select annotated list of 25 leading current aeronautical periodicals appears in *Information sources in engineering* (2nd ed., 1985), p.314-20. *Class No:* 629.7(051)

Progress Reports

[4978]

Progress in aerospace: an international review journal. Oxford, etc., Pergamon Press, 1964-. illus., diagrs., graphs.

V.24(3), 1987 carries 2 documented articles (*e.g.* 'Modern turboprop engines'; p.225-48, with 37 references). 'Contents of some previous volumes' (back-cover). About 6 articles pa. *Class No:* 629.7(055)

Yearbooks & Directories

[4979]

'Flight' directory of British aviation, 1987/88. Part 1: United Kingdom. Sutton, Surrey, Transport Press, 1987. Annual. 525p. £25. ISSN: 01600503.

Sections: Civil aviation (government depts.) - Military aviation - Airports (p. 41-94) - Air transport - Recreational flying - Aviation organizations - Trade and industry - Aviation training - Who's who in aviation (p.333-500). Index, p.501-24; Advertisers' index. *Class No:* 629.7(058)

[4980]

'Flight international' Directory: 1988/89. Part 2: Mainland Europe and Ireland, incorporating Who's who in European aviation. Potters Bar, Herts., Flight Directories, in association with *Flight international*. 498p. map. £30. ISBN: 0017005117.

International section, Albania ... Yugoslavia, 25 countries, p.8-426 (subdivided by towns, A-Z). Data include name of director, business; affiliation. 'Who's who', p.427-61. Detailed index, p.462-98. Advertisers' index. 'European time zones' map, p.7. *Class No:* 629.7(058)

[4981]

Interavia ABC aerospace directory, 1992. 41st ed. Coulsdon, Surrey, Jane's Information Group, 1992. *c.*1850p. tables, maps. £200. ISBN: 0710609574.

Previously as *Interavia ABC aerospace directory.*

Covers authorities/organizations/institutions; air carriers; manufacturers; airport services; airport structures and equipment; aviation services. A-Z index of people listed. Comprehensive (195 countries). *Class No:* 629.7(058)

[4982]

'World airline directory: flight data. Guide to the world's major carriers, their routes, equipment and management'. Williams, G., *comp. Flight International*, 26 March 1988, p.26-126. illus.

Annual feature, with details of over 550 domestic and international airlines, A-Z. Relates primarily to operations using pressurized aircraft of more than 30-seat capacity. Data: services; organization; subsidiaries; management; fleet; chief staff; employees; head office. 'Airlines by region' precedes. *Class No:* 629.7(058)

Tables & Data Books

[4983]

ROBERTSON, B. Aviation enthusiasts' data book. Cambridge, P. Stephens, 1982. 153p. illus., diagrs. ISBN: 0850595002.

19 chapters (1. The world scene and aeronautics; 4. World aircraft industries; 16. Aero engines; 19. Airships). 12 appendices (1. Works of reference; 2. Aeronautical periodicals; 12. Schneider Trophy winners). Detailed, non-analytical index. Well-captioned illus. International in scope. *Class No:* 629.7(083)

Histories

[4984]

GIBBS-SMITH, C.H. Aviation: an historical survey; from its origin to the end of World War II. 2nd ed. London, HM Stationery Office, 1985. xiv, 320p. illus., tables. £10.95. ISBN: 0112904211.

First published 1970. At head of title: 'Science Museum' 18 chapters. 'Quotations on flying', p.221-229. Folding

....(contd.)

table on powered flight. 'A chronology of aviation from antiquity to 1945', p.239-52. Glossary. Bibliography, p.273-80. Conversion tables, etc. Addenda, p.289-92. Detailed, analytical index, p.293-320. Well illustrated (22 plates of small photographs; 128 line-drawings). A classic in the literature of aviation history. *Class No:* 629.7(091)

Biographies

[4985]

Who's who in aviation and aerospace. US ed., 1983. Compiled by the National Aeronautical Institute, with the assistance of Jane's. New York, Grey House, 1983. vii, 1415p. ISBN: 0882628755.

Who's who on *c.*15,000 individuals resident in the United States, selected for their professional qualifications and achievements in the field. Geographic and Primary Professional Activity indexes. No indication of future editions. *Class No:* 629.7(092)

Worldwide

[4986]

Jane's all the world's aircraft, 1992-93. Lambert, M., *ed.* Coulsdon, Surrey, Jane's Information Group, 1992. *c.*800p. illus. £135. ISBN: 0710609876.

Sections: Aircraft - Aero engines - Sport aircraft - Microlights - Sailplanes - Hang gliders - Airships - Balloons - First flights - Future events. Available on CD-ROM. *Class No:* 629.7(100)

Museums

[4987]

ALLEN, J.L. Aviation and space museums of America. New York, Asco, 1975. 257p. illus.

Details of 57 aviation and space museums, including 3 in Canada. Two or more illus. of items in each collection. Appended list of aviation organizations and publications. *Class No:* 629.7:061:069

[4988]

OGDEN, B. Aviation museums. (worldwide, including USA). Hounslow, Middx., Airline Publications & Sales, Ltd. 1979. 144p. illus., tables, maps. ISBN: 090511362x.

Museums in 54 countries, Albania ... Zimbabwe-Rhodesia. Data: Address; admission days; situation; exhibits (in detail). No bibliography or index. For the traveller and historian. *Class No:* 629.7:061:069

Aircraft Engines

[4989]

GUNSTON, B. World encyclopedia of aero engines. 2nd ed. Wellingborough, P. Stephens, 1989. 192p. illus. £19.95. ISBN: 085059717x.

First published 1987.

Entries, ABC (United Kingdom) ... Wright (USA). Glossary. Detailed, analytical index. Fully illustrated with 1 or 2 illus., well captioned, per page. Fills a gap in aeronautical history. *Class No:* 629.7.02

Aircraft

[4990]

BOWER, P.M. Unconventional aircraft. 2nd ed. Blue Ridge
Summit, Pa., TAB Books, 1990. 323p. illus. (col.). $19.95.
ISBN: 0830024503.

Brief descriptions of over 200 unconventional planes
(Tandem-wing aircraft ... Flying automobiles) that have
existed since Orville and Wilbur Wright flew.
Class No: 629.73

[4991]

Guinness aircraft facts and figures. Taylor, M. *and*
Mondey, D. Enfield, Middx., Guinness Superlatives, 1984.
256p. illus., (inc. col.), tables. £9.95. ISBN: 0851124162.

First published 1970.

8 sections: Lighter than air - Pioneers of flying - At last:
flight - First World War - Between the World Wars - Post-
War aviation - Modern aviation - Rocketry and space flight.
Index (tiny print), p.243-56. *Class No:* 629.73

[4992]

HAMILL, T. 'Commercial aircraft of the world'. *Flight
International*, 4/10 September 1991, p.35-90. illus., tables.

Covers the range of types available to the air-transport
operator, Airbus Industrie ... SAIC. Appended comparative
data tables. This survey appears annually, usually in a
September or October issue. *Class No:* 629.73

Great Britain

[4993]

JACKSON, A.J. British civil aircraft, 1919-1972. London,
Putnam, Conway Maritime Press, 1959-.

First published 1959 (2v.).

1. ABC ... Chilton. 1959 (New ed. 1973). 567p.
2. Chrislea ... Hawker Siddeley. 1960. 560p.
3. Hawker Siddeley ... Zlin. 1960. 636p.

Detailed description, illustration and listing of all transport
aircraft, light aeroplanes, home-built, civil registered planes,
plus military types and helicopters built in Britain, or
imported for British operators, 1919-72. V.3 deals with
*c.*150 aeroplanes. 7 appendices (A. Other British-built civil
aircraft, p.248-337). Each volume has detailed, analytical
index. Data accompanying illus. have congested layout.
Class No: 629.73(410)

USSR

[4994]

GUNSTON, B. Aircraft of the Soviet Union: the
encyclopedia of Soviet aircraft since 1917. Baldry, D., *ed.*
London, Osprey, 1983. 414, [2]p. illus., ports, diagrs.
ISBN: 085045445x.

Includes biographies. Entries: 'Alexandrov-Kalinin ...
Zlokazov'. Prelims include 'Soviet aircraft designations',
'Air weapons', 'Materials', 'Engines', 'Soviet
organizations', 'Miscellaneous data'. Appended list of
grouped light aircraft, p.408-14. Cross-sections, a feature.
Class No: 629.73(47)

Museums

[4995]

ELLIS, K. *and* **BUTLER, P.,** *comps*. **British museum
aircraft:** 39 museums and the histories of their exhibits.
Liverpool Airport, Merseyside, Merseyside Aviation
Society, Ltd., 1977. 160p. illus., map. ISBN: 0902420151.

39 museums (plus 2 late additions), A-Z under names.
Bibliography, p.158-60. Aircraft index, p.164-5. 24 pages of
illus. Location map. *Class No:* 629.73:061:069

[4996]

MARCH, P.R. Preserved aircraft. Shepperton, Ian Allan,
1980. 160p. illus.

Chronological list of preserved original aircraft, 1909-
1973. One or two illus., per page. Appendix: The British
Aircraft Preservation Council. Aircraft museums and
collections, p.154-6. Bibliography, p.156-7. Detailed
analytical index, p.158-60. *Class No:* 629.73:061:069

Aircraft Markings

[4997]

MARCH, P.R. Military aircraft markings, 1991. London, I.
Allen, 1991. 176p. illus. £4.95. ISBN: 0711019681.

Annual listing of all military aircraft based in Britain
(British and American) and a selection of military aircraft
based in Western Europe. *Class No:* 629.73.088

[4998]

WRIGHT, A.J. Civil aircraft markings, 1992. London, I.
Allen, 1992. 320p. illus. £4.95. ISBN: 0711020272.

16 sections, plus Addenda. 'British civil registrations' (in-
sequence), p.11-143; 'British civil registrations' (out-of-
sequence), p.144-87; Overseas airlines registrations; Radio
frequencies; BAPC register. Small format. A bargain.
Class No: 629.73.088

Balloons

[4999]

CAMERON, D. Ballooning handbook. London, Pelham
Books, 1980. 174p. illus., diagrs., tables, maps. ISBN:
0720712203.

8 chapters: 1. Fundamentals of ballooning - 2. Flying
techniques - 3. Advanced flying techniques - 4. Hot-air
technology - 5. Navigation - 6. Meteorology - 7. Ballooning
law - 8. Organizations for ballooning. 10 appendices (1.
Requirements for pilots' licences (UK), p.155-6 ... 9.
Bibliography - 10. Useful addresses). 'I have tried to write
this book in international terms' (*Preface*).
Class No: 629.733.3

Airships

[5000]

BEAUBOIS, H. Airships: an illustrated history, Translated
and adapted by M. and A. Kelly. London, Macdonald and
Jane's, 1974. 235p. illus. (incl. col.). ISBN: 0856081036.

The period: 1780-1972. 8 chapters. The USS Akron 1931
(p.204-5) is described with data on aircraft suspension
equipment, power unit and forward nacelle, plus 6 illus.,
including a cross-section. Bibliography, p.235. Well
illustrated (82 drawings in colour, 125 black-and-white
photographs; descriptive captions). Index of airship names.
Class No: 629.733.5

Gliders & Sailplanes

[5001]
COATES, A. Jane's world sailplanes and motor gliders.
2nd ed. London, Jane's Publishing Co. 1980. 207p. illus.,
diagrs. ISBN: 0710600176.
First published 1980.
Covers 20 countries, Australia ... Yugoslavia. Technical
data under 17 heads. Appendix: International gliding
championships, 1937-1978. FAI world gliding
championships, 1937-1978. Index by country. Index of
aircraft types, p.203-7. *Class No:* 629.734

[5002]
REAY, D.A. The History of man-powered flight. Oxford,
etc., Pergamon Press, 1977. 355p. illus., diagrs., graphs,
tables. ISBN: 0080217389.
17 chapters (1. Mythology or fact: flight in early
civilizations; 12. The first man-powered simulator - one of
several Canadian projects; 17. Future prospect). 6
appendices (1. Patents; 4. Bibliography, p.328-32
(chronological); 6. Definitions). Detailed, analytical index.
Well-presented. *Class No:* 629.734

Parachutes and Parachuting

[5003]
POYNTER, P. The Parachute manual: a technical treatise
on the parachute. 2nd. ed., completely revised. Santa
Barbara, Calif., Parachuting Publications, 1977. 497p. illus.,
diagrs. ISBN: 0915516068.
First published 1972.
9 chapters: 1. Regulations ... 4. Parachute materials ... 6.
Parachute component parts - Maintenance, alteration and
manufacturing procedures. Fully illustrated and captioned.
Combined glossary and index, p.480-97. US slant.
Class No: 629.734.7

Helicopters

[5004]
AP REES, E. World military helicopters. London, Jane's
Publishing Co., 1986. 192p. £12.50. ISBN: 0710603746.
Survey of the world's military helicopters, in alphabetical
order of manufacturers, Aerospatiale ... Westland. 70 basic
types, either those in current service, or flying as prototypes,
or under construction. Each type is given 1 black-and-white
illus. per page. Appendix: 'World military helicopter fleet',
p.187-92. *Class No:* 629.735.4

[5005]
EVERETT-HEATH, J. British military helicopters. London,
Arms and Armour Press, 1986. 224p. illus. £14.95.
A survey of military helicopter types, from Cierva
helicopters and Bristol helicopters to Westland 30 and EH-
101. 4 appendices (*e.g.* helicopter projects and designs). List
of abbreviations. Glossary. Bibliography, p.220. Over 200
photographs. Detailed, non-analytical index.
Class No: 629.735.4

Rockets & Missiles

[5006]
ROSENBERG, A., *comp.* **Russian-English glossary of
guided missile, rocket and satellite terms.** Washington,
Library of Congress, 1958. vi, 351p.
More than 4,000 Russian terms, with English equivalents,

....(contd.)
taken from Soviet monographs and periodicals, 1955-58.
The English equivalents are based mainly on Soviet
translations of US and British Publications (*Introduction*).
Class No: 629.76

[5007]
—**MURASHKEVICH, A.M.,** *and others, comps.* English-
Russian guided missiles and space flight dictionary.
Moscow, Military Publishing House, 1966. 920p.
Russian terms, with English equivalents.
Class No: 629.76

[5008]
WINTER, F.H. Rockets into space. Cambridge, Ma.,
Harvard Univ. Press, 1990. xii,165p. illus. $22.50.
(*Frontiers of space.*) ISBN: 0674776607.
7 chapters: 1. The founders of spaceflight theory - 2. The
experimenters - 3. The V-2 rocket - 4. Rockets enter the
space age - 5. Modern rockets - 6. Space shuttles - 7. The
future of rocketry. Extensive source notes, p.149-56.
Detailed, analytical index, p.159-65. *Class No:* 629.76

Astronautics (Space Craft)

Abbreviations & Symbols

[5009]
MOSER, R.C. Space-age acronyms: abbreviations and
designations. 2nd. ed., rev. and enl. New York, etc., IFI/
Plenum Press, 1969. [ix], 534p. illus.
First published 1964.
A guide to *c.*15,000 astronautical and aeronautical
acronyms. Appended: Missile, rocket, probe and drone
designation system - Aircraft designation system - Ship
designations - Communication electronic equipment
designation system.
Jane's Aerospace dictionary (*qv*) adds more recent
acronyms. *Class No:* 629.78(003)

Bibliographies

[5010]
MacADAM, E.J. 'Aerospace engineering'. In *Information
sources in engineering.* 2nd ed., Editor, L.J. Anthony.
p.294-324. chart. London, etc., Butterworths, 1985.
Contents: International organizations - Official sources of
information (USA ... Israel) - Periodicals - Abstracting and
indexing services - General reference material (directories
and guides; bibliographies; monographs and textbooks).
Particularly useful on official reports. *Class No:* 629.78(01)

Encyclopaedias

[5011]
BOOTH, N. The Encyclopedia of space. London, Brian
Todd, 1990. 200p. illus. (col.). £17.95. ISBN: 1853611166.
6 sections: 1. Developments in space (11 articles:
'Pioneers'; 'Moon race'; 'Space shuttle', etc.) - 2. Manned
missions - 3. Solar system exploration - 4. Earth satellites -
5. Future solar exploration - 6. Milestones (chronological

....(contd.)

list). Indexes: general; astronauts; vehicles and equipment. A glossy production, suitable for school libraries. *Class No:* 629.78(031)

[5012]
FURNISS, T. Space flight: the records. Enfield, Middx, Guiness Superlatives, 1985. 168p. illus., ports., tables. ISBN: 0851124518.

5 sections: 1. Manned space flight diary [12 April 1961 - 29 April 1985] - 2. Firsts in manned space flight - 3. Manned space machines - 4. An A to Z of space travellers (p.69-107) - 5. Manned space flight, tables (Manned space flight duration tables ... Space shuttle schedule). Postscript (latest flight 1985). Glossary. Detailed index, p.162-8. *Class No:* 629.78(031)

[5013]
The Illustrated encyclopedia of space technology. Gatland, K.W., *ed.* 2nd ed. London, New York, Salamander Books, 1989. 306p. illus., diagrs., maps. £16.95. ISBN: 0861014499.

First published 1981.

A popular history including weather patrols, manned flights, military systems, space cities, lunar and planetary probes. 'Space diary' (p.252-97) covers events from 360 BC to 1989. Glossary of terms; subject index. Lavishly illustrated, many in colour. *Class No:* 629.78(031)

[5014]
Magill's survey of science: space exploration series. Magill, F.N., *ed.* Pasadena, Ca., Salem Press, 1989,. 5v., 2328p. $400. ISBN: 0893566004.

Over 3200 articles (5-8p. each) on space exploration, including vehicles, launchers, launch sites, national space programmes and the socioeconomic impact. Annotated bibliographies and cross-references appear after each entry. Volume indexes; V. 5 also includes a glossary, a chronological list of entries and subject index to all volumes. 'An excellent source ... highly recommended' (*Choice,* September 1989, p.84). *Class No:* 629.78(031)

[5015]
—CURTIS, A.R. Space almanac: facts, figures, names, dates, places, lists, charts, tables, maps, covering space from Earth to the edge of the Universe. Woodsboro, Md., Arcsoft Publishers, 1989. 955p. illus. tables, charts, maps. $19.95. ISBN: 0866680659.

8 chapters, subdivided, on space exploration and astronomy. Detailed, analytical index. 'Packed with information' ('Reference books bulletin', *Booklist,* December 15, 1989, p.858). *Class No:* 629.78(031)

[5016]
—Space year 1991: the complete record of the year's space events. MacKnight, N., *ed.* Oscelola, Wi., Motorbooks International, 1990. 176p. illus. $19.95; £11.95. ISBN: 0879384824.

Covers space missions (manned, unmanned, planetary between 1 July 1989 and 30 June 1990. 'Into orbit' (p.157-71) lists every spacecraft launch (type of craft, purpose, launch time and site, etc.) for the period. Index, p.173-76. 'Accessible and informative' ('Reference books bulletin, *Booklist,* February 1, 1991, p.1155). *Class No:* 629.78(031)

[5017]
PETROVICH, G.V., *ed*. The Soviet encyclopedia of space flight. Translated from the Russian. Moscow, MIR Publishers, 1969. 620p. illus., ports., diagrs., maps.

Original Russian, *Kosmonavtika. Malen' kaya entsiklopediya* (Moscow, Sovetskaya Entsiklopediya, 1968).

Over 1,400 entries A-Z, including biographies. 38-p. list of all announced satellites launched up to 4 October 1969. Bibliography, p.588-98; 2-p. list of space-research journals. Classified list of articles. 'An impressive book, useful especially to newsmen and students, and readable by all' (*Sky and telescope,* v.40(3), September 1970, p.13-14). *Class No:* 629.78(031)

Dictionaries

Polyglot

[5018]
Aerospace dictionary: English, French, German, Spanish. Paris, Gauthier-Villars, 1984. [vi], 828p. col. illus. ISBN: 2040119523.

Earlier ed. (1978) as *Dictionnaire des techniques aéronautique et spatiale* (French-English-German).

About 36,000 English entry-words, with equivalents in the other three languages. 45 categories applied.

3 companion volumes give priority to each of the other languages, *i.e., Dictionnaire aérospatial* (French-English-German-Spanish); *Wörterbuch der Luft-und Raumfahrt* (German-English-French-Spanish); *Diccionario aëroespacial* (Spanish-French-English-German). *Class No:* 629.78(038)=00

[5019]
INTERNATIONAL ACADEMY OF ASTRONAUTICS. Astronautical multilingual dictionary. Prague, Academia, 1970. 936p.

About 10,000 English-based terms, with Russian, German, French, Italian, Spanish and Czech equivalents and indexes. The dictionary needs supplements (*e.g.* an appendix of phrases covering colloquial jargon of astronauts). Meanwhile, 'an excellent investment for all concerned with the subject' (*Library journal,* v.95(15), 1 August 1970, p.2788). *Class No:* 629.78(038)=00

[5020]
KOTIK, M.G., *comp*. Dictionary of aerospace engineering in Russian, English and German. Amsterdam, Elsevier, 1986. 879p. Dfl.425. ISBN: 0444995064.

Co-edited with VEB Verlag Technik, Berlin.

About 35,000 specialized Russian terms, with equivalents and indexes in English and German. *Class No:* 629.78(038)=00

[5021]
TEKNISKA NOMENKLATURCENTRALEN. Rymdorlista: svensk-engelsk-fransk-tysk-rysk. Stockholm, the Centre, 1973. xxiii, 171p. (*TNC 53.*) ISBN: 9171960538.

'Glossary of astronautics'. 510 Swedish technical terms dealing with outer space, plus equivalents and indexes in English, French, German and Russian. *Class No:* 629.78(038)=00

English

[5022]

The Aviation space dictionary. Reithmaier, L.W., *ed.* 7th ed. Los Angeles, Ca., Aero, 1990. 461p. illus. $32.95. ISBN: 0830680926.

Previous ed. 1980.

About 6,000 concise entries, drawing heavily on the NASA *Dictionary of technical terms for aerospace use.* 16 appendices, including essays. *Class No:* 629.78(038)=20

[5023]

Chambers air and space dictionary. Walker, P.M.B. Edinburgh, Chambers, 1990. viii,216p. illus., diagrs. £17.95; £8.95. ISBN: 0550132422; 0550132430, Pbk.

Published in the US as *Cambridge air and space dictionary.*

About 6,000 terms derived from *Chambers science and technology dictionary.* Most definitions are brief (*e.g.* 'luminance': 3 lines), but includes 50 'special articles' in separate panels (*e.g.* 'Quasars' 1p.). Good use of bold and italic type. *Class No:* 629.78(038)=20

[5024]

Jane's aerospace dictionary. Gunston, B., *ed.* 3rd ed. Coulsdon, Surrey, 1988. *c.*600p. £30. ISBN: 0710605803.

First published 1980.

About 28,000 entries, ranging in length from 3 to 100 words. Many acronyms and abbreviations as entries. Different meanings of terms are numbered. Headwords are categorized. *Class No:* 629.78(038)=20

[5025]

SAE dictionary of aerospace engineering. Warrendale, Pa., Society of Automotive Engineers, 1992. 550p. £69. ISBN: 1560912863.

Over 18,000 terms, derived from SAE Aerospace standards and reports, the *NASA Thesaurus* and Engineering Resources, Inc. Numerous computer terms related to data processing and PC technology. *Class No:* 629.78(038)=20

[5026]

WILLIAMSON, M. Dictionary of space technology. Bristol; New York, Adam Hilger, 1990. xi,401p. illus., diagrs. £25; $50. ISBN: 0852743394.

Over 1,600 entries ('Solar energy': *c.*½p.). Over 100 photos and diagrs. Final 3 sections: Physics and astronomy - Space Centres and organizations - Miscellaneous. Classified list of dictionary entries under 12 headings. Cross-references (plentiful). Good use of bold type and uncongested layout. *Class No:* 629.78(038)=20

German

[5027]

HYMAN, C.J. German-English, English-German astronautics dictionary. New York, Consultants Bureau, 1968. viii, 237p.

About 9,000 entry-words in each part (p.1-122; 125-237). Includes vocabulary of allied subjects such as automation, cosmic rays, aviation fuel, guiding devices, propulsion, radio engineering and telecommunications. *Class No:* 629.78(038)=30

French

[5028]

EUROPEAN SPACE AGENCY. Recueil de terminologie spatiale. [A Glossary of space terms.] Taillefer, Y., *comp.* Paris, the Agency, 1982. [iii], 432p. diagrs. ISSN: 03796566.

About 4,000 French entry-words, with English explanations. *Class No:* 629.78(038)=40

Russian

[5029]

KONARSKI, M.M. Russian-English space technology dictionary. Oxford, etc., Pergamon Press, 1970. x, [i], 416p. illus., diagrs.

Over 10,000 selected terms in satellite and other spacecraft operations, spaceflight guidance and control, and space exploration. Part 1: Russian-English dictionary of categorized terms plus 'Index of English cross-reference terms' (p.237-370); short glossary of aerospace biomedical terms; Russian abbreviations. Part 2: 12 data tables; list of personal names of scientists (*i.e.,* Russian forms of surnames of scientists in space technology, with English forms). An essential complement to Konarski's *Russian-English dictionary of modern terms in aeronautics and rocketry* (1962). *Class No:* 629.78(038)=82

Reports Literature

[5030]

Scientific and technical aerospace (STAR): an abstract journal. Washington, DC., Scientific and Technical Information Office, National Aeronautics and Space Administration (NASA), 1963-. v. 1, no. 1-. Semi-monthly.

Also available on microfiche.

About 30,000 abstracts pa. Main divisions (01-99): Aeronautics - Astronautics - Chemistry and materials - Engineering - Geosciences - Life sciences - Mathematical and computer sciences - Physics - Social sciences - Space sciences. Indexes: subject; personal author; corporate source; contact number; report/accession number index per issue, quarterly, semi-annual and annual. A companion service to *International aerospace abstracts,* and equally indispensable to the researcher.

Available online in the *Aerospace database* (DIALOG, file 108), but only in the US and Canada. CD-ROM (Dialog Ondisc). *Class No:* 629.78(047)

Reviews & Abstracts

[5031]

International aerospace abstracts. New York, American Institute of Aeronautics and Astronautics Technical Information Service, for NASA, Scientific and Technical Information Branch, 1961-. 26pa. (3 issues in January). $950pa. (US), $1350 (other countries).

Replaces abstracts in *Aero/Space engineering* (formerly *Aeronautical engineering review*), 1942-60.

About 50,000 abstracts pa. Sections: Aeronautics - Astronautics - Chemistry and materials - Engineering - Geosciences - Life sciences - Mathematics and computer sciences - Physics - Social sciences - Space sciences - General. Coverage: periodicals (including government-sponsored), books, meetings, papers, conference proceedings, translations of journals, and journal articles.

....(contd.)

Indexes of subjects, personal authors, contract numbers, meetings papers and report numbers, accession number. A basic abstracting service. (Abstracts and indexes of report literature are the province of *Scientific and technical aerospace reports* (STAR)).

Available online in the *Aerospace database* (DIALOG, file 108), but only in the US and Canada. CD-ROM (Dialog OnDisc). *Class No:* 629.78(048)

Tables & Data Books

[5032]

Interavia space directory, 1992-1993. Wilson, A., *ed.* 3rd ed. Coulson, Surrey, Jane's Information Group, 1992. 630p. illus., diagrs., tables. £135. ISBN: 0710610017.

Previously *Jane's Spaceflight directory.*

Main contents: National space programmes - International space programmes - Military space - Launchers - Communications - Satcom ground segments and contractory - Earth observation (space) - Earth observation (ground) - Microgravity - World space centres - The solar system - Space industry. Detailed index. An up-to-date compendium. *Class No:* 629.78(083)

[5033]

TODD, D. World aerospace: a statistical handbook. London, Croom Helm, 1987. 226p. tables.

Aerospace activity in Soviet bloc, Europe [& other regions]. Changes in materials and avionics. Space technology forecast. Statistical appendices. For all reference collections in aerospace activities (*New technical books*, March/April 1987, item 614). *Class No:* 629.78(083)

Histories

[5034]

DICKSON, K.M. History of aeronautics and astronautics: a preliminary bibliography. Washington, National Aeronautics and Space Administration (NASA), 1968.

A selected, annotated guide to 921 publications on the history of space exploration from earliest times to 1967. *Class No:* 629.78(091)

[5035]

—Jane's pocket book of space exploration. Wilding-White, T.M. London, Macdonald, 1976. 264p. illus.

Contains about 200 photographs of manned and unmanned spacecraft, with technical and historical data facing. Chronologies of manned space-flight and interplanetary probes. Index. *Class No:* 629.78(091)

Biographies

[5036]

UNITED STATES. House of Representatives. Committee on Science and Technology. Astronauts and cosmonauts: biographical and statistical data. Washington, US Government Printing Office, 1976. 180p. ports.

'The astronauts whose biographical sketches and portraits are included in this directory were involved variously with NASA, the X-15 program, Air Force - NASA X-20 Dyna-Sore program and DOD Manned Orbiting Lab Program. Soviet cosmonauts are also described' (*RSR*, v.5(4), October/December 1977, p.40). The statistical data covers Soviet - US space flights. *Class No:* 629.78(092)

[5037]

Who's who in aviation and aerospace. US ed., 1983. Compiled by the National Aeronautical Institute, with the assistance of Jane's. New York, Grey House, 1983. vii, 1,415p. ISBN: 0882628755.

Who's who information on *c.*15,000 individuals resident in the United States, selected for their professional qualifications and achievements in the field. Geographic and Primary Professional Activity indexes. No indication of future editions. *Class No:* 629.78(092)

Great Britain

[5038]

MASSEY, Sir H. *and* **ROBINS, M.O. History of British space science.** Cambridge Univ. Press, 1986. xxii, 514p. illus., diagrs., graphs, formulas, tables. £45. ISBN: 052130783X.

16 chapters (3. The initiation of the Skylark rocket programme - 4. Post-IQY developments - 5. The Ariel programme - 6. The European Space Research Organization - 7. Commonwealth cooperation in space research ... 13/14. Scientific studies by British space scientists). Appendices A-G (F. Details of the principal UK space science groups, 1982). Glossary of abbreviations, p.xiii-xv. Index. *Class No:* 629.78(410)

USSR

[5039]

BATTELLE MEMORIAL INSTITUTE, Columbus, Ohio. Columbus Laboratories. **Handbook of Soviet space-science research.** Wikelic, G.E., *ed.* New York, Gordon & Breach, 1969. xx, 506p. illus.

20 documented chapters, with author and subject indexes. Aims to be a comprehensive and authoritative English-language summary of the first 10 years of Soviet space-science research. *Class No:* 629.78(47)

[5040]

BEARD, R. Soviet cosmonautics, 1957-1969: a bibliography of articles published in British periodicals and foreign books. Swindon, Wilts., the Author, 1970. 43p. Mimeographed.

Part 1: *c.*300 items, listing all books published since October 1957 throughout the world on Soviet space-flight; A-Z, authors. Part 2: *c.*700 items, listing all articles in periodicals published in Britain, 1 October 1957 to 30 June 1969; A-Z, periodicals. *Class No:* 629.78(47)

Institutions & Associations

[5041]

Aerospace technology centres: a world directory of organizations and programmes. 2nd ed. Harlow, Essex, Longman, 1991. 250p. £250. ISBN: 0582082757.

Over 80 countries, arranged by country, then organizations A-Z. Data include recent activities, estimate of annual expenditure and list of organizations' publications. Subject index; index of organizations by titles. *Class No:* 629.78:061:061.2

[5042]

Space industry international: markets, companies, statistics and personnel. Harlow, Essex, Longman Group, 1987. xx, 353p.

12 regional chapters describe for each country its national and international space activities, governmental space policies and programs, and the role of private industry in the space market. Directory-type entries cover private firms, associations and institutes. There is also an introduction on relevant international organizations. Glossary. Bibliography. Indexes of companies and organizations, products, and personnel. *Class No:* 629.78:061:061.2

[5043]

—BRITISH NATIONAL SPACE CENTRE. Directory of UK space capabilities. London, HM Stationery Office, 1987. *c.*250p. ISBN: 0115139818.

Has as subtitle: 'A guide to the space-related expertise of UK manufacturers, agencies and universities and their capabilities'. *Class No:* 629.78:061:061.2

Research Projects

[5044]

Aerospace research index: a guide to world research in aeronautics, meteorology, astronomy and space science. Willmore, A.P. *and* Willmore, S.R., *eds.* Harlow, Essex, Longman Group, 1981. 597p.

World directory, detailing information on aerospace research conducted in *c.*70 countries by government laboratories, public companies, industrial firms, research associations, universities and technical colleges. Coverage extends to astronomy, meteorology and aeronautics. Omits USSR. Title and keyword index; subject index. Early revision is clearly called for.
Class No: 629.78:061:061.62.005

UFOs

[5045]

CATOE, L.E., *comp.* **UFOs and related subjects: an annotated bibliography.** Prepared by the Library of Congress Science and Technology Division for the Air Force Office of Scientific Research. Supplemented by 'Unidentified flying objects: a selected bibliography', compiled by K. Rodgers. Washington, Library of Congress, 1976. Republished by Gale Research Co., Detroit, 1978. xviii, 411p.

A reprint of the bibliography published in 1969, plus updating supplement.

A guide to more than 1,600 books, journal articles, pamphlets, conference proceedings, tapes, original manuscripts, and other material. Subject classes; author index. Includes an appendix on mirages. For the general reader. *Class No:* 629.78.001

[5046]

—RASMUSSEN, R.M. The UFO literature: a comprehensive annotated bibliography. Jefferson, N.C., McFarland, 1985. 271p. $29.95.

Lists, under author, more than 1,100 titles of English-language books, booklets and special-issue magazines. Subject and title index. *Class No:* 629.78.001

[5047]

CLARK, J. UFOs in the 1980s. Detroit, Mich., Apogee, 1990. 240p. $65. (*The UFO encyclopedia, 1.*) ISBN: 1558883010.

84 articles, 'factual and balanced' (*Choice,* November 1990, p.452). Topics covered include abduction, contactees, crashes. Shorter entries on people and UFO organizations. V.1 of a planned 3v. work; v.2 will deal with pre-1980 period; v.3 to be a current update. *Class No:* 629.78.001

[5048]

The UFO encyclopedia. Spencer, J., *ed.* London, Headline, 1991. 448p. £5.99. ISBN: 0747224949.

Comprehensive reference book for both converts and sceptics. *Class No:* 629.78.001

Spacecraft

[5049]

WILSON, A. Solar system log. London, Jane's, 1987. 128p. illus., diagrs., tables. £7.95. ISBN: 0710604440.

Chronological listing of solar missions between 1958 and 1985. Data include launch date, vehicle, spacecraft mass, destination, mission, payload, etc., as well as a brief discussion (*c.*1p.) of noteworthy features or results. *Class No:* 629.782

[5050]

YENNE, B. The Encyclopedia of US spacecraft. London, Hamlyn, 1986. 192p. illus. (mostly col.), diagrs. £8.95.

Largely a pictorial record. Data on satellites include launches, launch date, weight, size and shape, and purpose of launching. 3 appendices list space stations, launchers, and seronyms. While providing a fairly comprehensive picture of NASA and commercial spacecraft, Yenne gives a much less complete sample of satellites originating with other agencies. *New scientist* reviewer (16 January 1986, p.53) notes errors and the fact 'that dozens of satellites are omitted'. 'The US has been responsible for less than 20 per cent of the space launches in the past 20 years', he adds. *Class No:* 629.782

Satellites

[5051]

The RAE table of earth satellites, 1957-1986; compiled at the Royal Aircraft Establishment, Farnborough, by D.G. King-Hele, and others. 3rd ed. London, Macmillan Publishers, 1987. xviii, 936p. £65. ISBN: 0333353749.

First published 1981, covering 1957-1980.

Lists over 13,000 satellites including fragments. Data: name and international designations of earth satellite and its associated rocket(s), plus date of launch, lifetime (actual or estimated), mass, shape, dimensions and at least one set of orbital parameters. Listed (without details): other fragments associated with a launch space vehicle that escape from the earth's influence. Index (name; description; page). *Class No:* 629.783

63 Agriculture & Livestock

Agriculture

Abbreviations & Symbols

[5052]
FOOD AND AGRICULTURE ORGANIZATION.
Abbreviations used by FAO for international organizations, congresses, commissions, committees, etc. 4th revision. Rome, FAO, 1988. 205p. £11.25. (*FAO terminology bulletin 27/rev. 4.*) ISBN: 9250026595.
Previous ed. 1976 (60p.).
Abbreviations in English, French and Spanish.
Class No: 63.0(003)

Databases

[5053]
AGRICOLA. Beltsville, Md., US National Agricultural Library, 1970-.
Citations to journals, serials, monographs, government reports, patents, theses, etc. which have been acquired by NAL. Updates of *c.*9,000 records a month. Online equivalent of *Bibliography of agriculture (q.v.).* Available on CD-ROM. *Class No:* 63.0(003.4)

[5054]
AGRIS. Rome, Food and Agriculture Organization, 1975-.
Contains more than 1.5 million citations. (10 percent with abstracts) to international agricultural literature. Titles in original languages and 1986- index terms in English, French, and Spanish. *c.*10,000 records are added each month. Online equivalent of *Agrindex (q.v.).* Available on CD-ROM. *Class No:* 63.0(003.4)

[5055]
CAB ABSTRACTS. Wallingford, Oxon. CAB International, 1972-.
Contains over 2,800,000 records and covers the 26 main abstracting journals published by CAB International. 8,500 journals are scanned as well as books, reports, theses, conference proceedings, etc. About 130,000 items are indexed each year. Updated monthly. Available as CAB ABSTRACTS CD-ROM, while specialist areas are covered by TREECD, HORTCD, VETCD, BEASTCD, SOILCD, CABPESTCD. *Class No:* 63.0(003.4)

[5056]
JOHNSTON, S.M. 'Use of computer-based bibliographic services'. In *Information sources in agriculture and food science,* edited by G.P. Lilley. London, etc., Butterworths, 1981. p.55-101.
Compares *CAB ABSTRACTS, AGRICOLA, AGRIS* and other databases for online agricultural research. *CAB ABSTRACTS (q.v.)* (1973-) covers more than 2 million items, adding *c.*12,000 monthly, based on the various CAB International abstracting services. The US National

....(contd.)
Agricultural Library's *AGRICOLA (q.v.)* database, similar in scale, is compiled from the *c.*5,000 journals received by the NAL. *AGRIS (q.v.)* uses FAO resources and is based on the monthly *Agrindex (q.v.)* (1975-). *Class No:* 63.0(003.4)

Bibliographies

[5057]
ACCIS guide to United Nations information sources on food and agriculture. Advisory Committee for the Co-ordination of Information Systems, *comp.* Rome, FAO, 1987. vii, 124p.
10 sections (*e.g.* 2. Plant production and protection information sources; 5. Land and water development and natural resources information sources; 10. Statistics information sources). 3 annexes (2. Addresses of database hosts). Subject, names, publications and systems indexes. *Class No:* 63.0(01)

[5058]
FOOD AND AGRICULTURE ORGANIZATION. FAO books in print. Rome, FAO. Annual. *Gratis.*
10 sections: 0. Agriculture - 1. Plant production and protection - 2. Animal production and health - 3. Forestry - 4. Fisheries - 5. Land and water development - 6. Economic and social development - 7. Statistics - 8. Food and nutrition - 9. General. Alphabetical titles index and prices. Author index. List of FAO sales agents and booksellers. *Class No:* 63.0(01)

[5059]
Guide to sources for agricultural and biological research. Sponsored by the United States National Agricultural Library, United States Department of Agriculture. Blanchard, J.R. *and* Farrell, L., *eds.* 2nd ed. Berkeley, etc., Univ. of California Press, 1981. xi, 735p. $60. ISBN: 0520032268.
First published 1958 as *Literature of agriculture research,* by J.R. Blanchard and H. Ostvold, and known as the 'hayseed Winchell'.
12 contributors. 5,779 annotated, evaluative entries in 9 classes (subdivided, with introductions: A. Agriculture and biology: general - B. Plant sciences - C. Crop protection - D. Animal sciences - E. Physical sciences - F. Food and nutrition - G. Environmental science - H. Social sciences - I. Computerized databases for bibliographic research). Appendix: Acronyms and abbreviations. The general introduction has helpful pages on 'Depositories of information in libraries and documentation centres' and 'Information networks for agriculture and biology'. Author, analytical subject (p.667-735) and title indexes. 'Fills a need that has existed for years' (*RQ,* v.21(4), Summer 1982, p.418). *Class No:* 63.0(01)

[5060]

—BUSH, E.A.R. Agriculture: a bibliographical guide. London, Macdonald, 1974. 2v. (xxxi, iv, 1561p.). (*Macdonald bibliographical guides.*)

Has 9,404 entries (in 172 sections) for books and periodical articles, 1958-71. Annotations state number of items in bibliographies and period covered. Omits items on preserved foods, vegetable oils, etc. Author index includesjoint authors, subject index, p.1369-1561. List of cited periodicals, p.xv-xxxi. A massive basic bibliography, awarded the LA Besterman Medal for 1974. Now somewhat dated. *Class No:* 63.0(01)

[5061]

Information sources in agriculture and food science. Lilley, G.P., *ed.* London, etc., Butterworths, 1981. xiv, 603p. (*Guides to information sources.*) ISBN: 0408106123.

18 contributors. 2 parts: 1. General (chapters 1-7) - 2. Specialized areas (8. Soils and fertilizers; 9. Agricultural engineering; 10. Weed biology, weed control and herbicides; 12. Field crops and grasslands; 13. Temperate horticulture; 14. Tropical agriculture; 15. Animal production; 16. Veterinary science; 17. Forestry; 18. Food science; 19. Agricultural economics; 20. Agrarian and food history). Appendix: List of abbreviations used. Index (subjects, organizations, journal titles), p.591-603. Thin on coverage of fish and fisheries; omits beekeeping.

A new ed., *Information sources in agriculture and horticulture,* is due to be published by Bowker-Saur in late 1993 (£65). *Class No:* 63.0(01)

[5062]

ROTHAMSTED EXPERIMENTAL STATION. Library. **Catalogue of printed books and pamphlets on agriculture published between 1471 and 1840.** 2nd ed. Harpenden, Herts., 1940. 293p.

First published 1926.

An author catalogue, with indexes of authors, translations, and an historical and geographical list of foreign authors and translations.

Supplement (1949. ii, 15p.) lists additions since 1940. *Class No:* 63.0(01)

[5063]

—UNITED STATES. National Agricultural Library. Historic books and manuscripts concerning general agriculture in the collection of the National Agricultural Library. Washington, the Library, 1967. [iv], 94p.

Has *c.*700 entries, with full titles and brief biographical notes. An author catalogue of 16th-18th-century material. NAL shelf-marks. *Class No:* 63.0(01)

[5064]

—WYE COLLEGE LIBRARY. A Catalogue of agricultural & horticultural books, 1543-1918, in Wye College Library. Ashford, Kent, the Library, 1977. iv, 100p. £1.50. ISBN: 090185963x.

An author list of *c.*1,000 items. Includes locations in 12 libraries' catalogues (*e.g.* British Museum, *STC,* Wing). *Class No:* 63.0(01)

[5065]

UNITED STATES. National Agricultural Library. **Dictionary catalog of the National Agricultural Library, 1862-1965.** New York, Rowman & Littlefield, Inc., 1967-70. 73v.

Over 1,700,000 catalogue cards, photolithographically reproduced. V.1-72: A-Z sequence of authors, titles and subjects for monographs, serials and analytics, as catalogued for the NAL main collection, for its Bee Culture Branch, and for the former Beltsville Branch. V.73: Translations of periodical articles (authors, A-Z). The most comprehensive catalogue of its kind. Entries are not always legible.

Continued as *National Agricultural Library catalog,* 1966- , monthly, with annual and 5-yearly cumulations. *Class No:* 63.0(01)

Encyclopaedias

[5066]

McGraw-Hill encyclopedia of food, agriculture and nutrition. Lapedes, D.N., *editor-in-chief.* New York, etc., McGraw-Hill, 1977. vii, 732p. illus., tables, maps. ISBN: 0070452636.

Nearly 300 contributors. 5 leading articles (p.1-47) precede the main sequence ('Milk', p.395-403; sectionalized; 7 illus., 1 table; 11 references). Analytical index, p.709-32. 'Designed to inform the student, librarian, scientist, teacher, engineer and layperson about all aspects of agriculture, the cultivation, harvesting and processing of food crops, food manufacturing; and health and nutrition - from the economic and political to the technological' (*Preface*). Some articles derive from the *McGraw-Hill encyclopedia of science and technology* (4th ed. 1977). Well produced. *Class No:* 63.0(031)

[5067]

Sel'skokhozyaistvennyi entsiklopedicheskii slovar'. [Encyclopaedia of agriculture.] Mesyats, V.R., *ed.* Moscow, Izdatel'stvo 'Sovetskaya Entsiklopediya', 1989. 656p. illus. (some col.), figs., maps.

Contains *c.*4,000 entries, A-Z, many with bibliographies and cross-references. Numerous line-drawings and some full-page colour illustrations. Useful maps of the USSR showing agriculture and agroclimatic regions and soil types. Detailed, subject index. *Class No:* 63.0(031)

Handbooks & Manuals

[5068]

FREAM, W. Fream's principles of food and agriculture. Spedding, C.R.W., *ed.* 17th ed. Oxford, Blackwell Scientific Publications, 1992. xxvii,308p. tables, graphs. £25. ISBN: 0632029781.

First published 1892 as *Fream's elements of agriculture,* 16th ed. (1983), *Fream's agriculture.*

10 sections: 1. The food and agricultural industry - 2. Meeting human needs - 3. World agriculture ... 7. The inputs to agriculture - 8. The principles of processing of products ... 10. Feeding the world in the future. References, p.xi-xii. Glossary, p.xv-xxiv. Fewer than half the number of pages of the previous ed. and has only 10 contributors compared with 42. This edition 'builds on the principles of agriculture ... and relates these to a more international view of agriculture, and its past, present and future roles in society' (*Preface*). *Class No:* 63.0(035)

[5069]
McCONNELL, P. Primrose McConnell's 'The agricultural notebook'. Halley, R.J. *and* Soffe, R.J., *eds*. 18th ed. London, etc., Butterworths, 1988. xiv, 689p. illus., diagrs., tables. £45; £29.50. ISBN: 0408030607.
First published 1883.
29 contributors. 4 parts (29 documented chapters): 1. Crop production - 2. Animal production - 3. Farm equipment - 4. Farm management. Chapter 21: 'Farm machinery', p.503-536; 66 references. Glossary of units, p.668-70. Detailed, analytical index, p.671-89. Follows the completely new approach and format of the 17th ed. (1982). Facts and figures for farmers, students and all engaged or interested in farming. *Class No:* 63.0(035)

[5070]
SEYMOUR, J. The Complete book of self-sufficiency. London, Faber, 1976. 256p. illus., (incl. col.), diagrs. ISBN: 0571110959.
Six ways in which to cultivate self-sufficiency; Food from the fields - Food from animals - Food from the garden - Food from the wild - Natural energy (p.210-8: saving energy; power from water; heat from the sun; power from the wind; fuel from waste. 20 illus. and diagrams), Crafts and skills. Analytical index. An encyclopaedia of practical advice, illustrated step-by-step. *Class No:* 63.0(035)

Dictionaries

Polyglot

[5071]
FOOD AND AGRICULTURE ORGANIZATION. Terminology and Reference Section. Publications Division. **Dictionaries and vocabularies** in the Terminology Reference Library, 1982-85. [Dictionnaires et vocabulaires à la bibliothèque de terminologie et références.] Rome, FAO, 1987. vi,66p. $12. ISBN: 9250024142.
First published 1968.
Lists several hundred dictionaries and vocabularies in 3 parts: 1. Dictionaries arranged by subjects - 2. Language dictionaries in Arabic, Chinese, English, French, Spanish, German, Italian, Russian and Latin. 3. Subject, author and language indexes. *Class No:* 63.0(038)=00

[5072]
HAENSCH, G. *and* HABERKAMP DE ANTON, G. **Dictionary of agriculture in six languages:** German, English, French, Spanish, Italian, Russian. 5th completely rev. and enld. ed. Amsterdam, etc., Elsevier, 1986. xxix, 1264p. £165.62. ISBN: 0444995129.
First published 1959; 4th ed., 1975 (xxiv, 1000p.).
11,163 numbered German-base entries, with equivalents in the other five languages. Subject sections A-Q (*e.g.* A. Food and agriculture: general terms; D. Economics and sociology of agriculture; E. Processing of animal produce; J. Plant cultivation: special part; N-O. Animal breeding; P. Farm buildings; Q. Agricultural machinery). Genders given. German, English, French, Spanish, Italian, Russian and Latin indexes. *Class No:* 63.0(038)=00

[5073]
Landwirtschaftliches Wörterbuch in acht Sprachen: russisch, bulgarisch, tschechisch, polvisch, ungarisch, romanisch, deutsch, englisch. Kratochvil, Y. *and* Urbanova, S., *eds*. Berlin, Landwirtschaftsverlag. 1970. 2v. (1045p.; 670p.).
An eight-language agricultural dictionary. 25,780 Russian-base entries, with Bulgarian, Czech, Polish, Hungarian, Rumanian, German and English equivalents. V.2 contains language indexes. If a term is frequently used and well-known, the Latin term is included, with a separate Latin index in v.2. *Class No:* 63.0(038)=00

[5074]
LOGIE, G. Glossary of land resources, in English, French, Italian, Dutch, German, Swedish. Amsterdam, Elsevier, 1984. xxvii, 303p. (*International planning glossaries, 4.*) ISBN: 0444422811.
2,117 Engish-base terms (plus explanations in context), with equivalents in the other 5 languages. Sections: Land resources - Land property - Land form - Climate - Agriculture - Forestry - Recreation - Urbanization - Pollution. Introductions to sections. A/Z index of terms in each of the six languages. Bibliography, p.299-303. *Class No:* 63.0(038)=00

[5075]
Norsk landbruksordbok. Utgitt av Nemnda for Norsk Landbruksordbok. Rommetveit, N., *ed.* Oslo, Det Norske Samlaget, 1979. 2v. (581p. + 402p.). ISBN: 8282107923, v.1; 8282108709, v.2.
'Norwegian dictionary of agriculture'. V.1: Norwegian-Norwegian. V.2: Lappish - Swedish - Danish - English - German - Icelandic - Finnish: index of equivalents. The English index has *c.*20 entries. *Class No:* 63.0(038)=00

English

[5076]
The Agricultural handbook. Whitby, M., *and others*. Oxford, BSP Professional Books, 1988. viii,236p. £12.99. ISBN: 0632018216.
Terms arranged A-Z (Abattoir ... Zygote). Includes acronyms and cross-references. An introductory reference source. Suitable for undergraduates, and public libraries. *Class No:* 63.0(038)=20

[5077]
Agricultural terms, as used in the Bibliography of Agriculture, from data provided by the National Agricultural Library, US Department of Agriculture. 2nd ed. Phoenix, Arizona, Oryx Press, 1978. 122p. $35. ISBN: 0912700459.
First published 1976.
About 37,000 terms. Word variants are grouped together, with synonymous terms (*c.*12,000) in parentheses, italicized, following the key term. 2,000 cross-references. *Class No:* 63.0(038)=20

[5078]
DALAL-CLAYTON, D.B. Black's agricultural dictionary. 2nd ed. London, A. & C. Black, 1985. xiii, 432p. illus., diagrs., tables. ISBN: 0713626798.
First published 1981.
About 4,000 entries (*e.g.* 'Plough': nearly 1p., including keyed line-drawing; 'Chart of common sheep names',

....(contd.)

p.338). Ample cross-references. Appendix of abbreviations, acronyms and initials. Concerned essentially with agriculture in the UK. *Class No:* 63.0(038)=20

[5079]
A Dictionary of agricultural and allied terminology. Winburne, J.N., *ed.* East Lansing, Mich., Michigan State Univ. Press, 1962 (reprinted 1974). viii, 905p. $24.95. ISBN: 0870130676.

About 18,000 terms, drawn from *c.*250 books, magazines and catalogues. Definitions, free from highly technical words, follow a pattern; (a) entry word, (b) scientific name (if any), (c) general description, (d) agricultural implications, (e) synonyms, (f) cross-references. Locations of plants in USA. Appendix A (p.897-903) lists terms under subjects. *Class No:* 63.0(038)=20

French
[5080]
BOURKE, D. O'D. French-English agricultural dictionary with English-French index. Wallingford, Oxon., CAB International, 1992. x,293p. £29.50; $56. ISBN: 0851986269.

Contains over 20,000 terms. Coverage includes agriculture in its widest sense; there are terms from statistics, agricultural engineering, soil science, etc. A companion to the author's *French-English horticultural dictionary with English-French index (q.v.).* *Class No:* 63.0(038)=40

[5081]
Dictionnaire d'agriculture et des sciences annexes; index anglais et espagnol. Paris, La Maison Rustique, in collaboration with International Council of the French language [CILF], 1977. x, 219p. illus., tables.

About 3,000 French terms, with definitions, plus equivalents in English and Spanish. About 40 categories applied. 3p. of keyed parts of the horse; table of chief valuable crops cultivated. *Class No:* 63.0(038)=40

Russian
[5082]
USOVSKII, B.N. *and* **LINNARD, W. Comprehensive Russian-English agricultural dictionary.** 2nd ed., rev and enl. Oxford, etc., Pergamon Press, 1967. xii, 470p.

Considerably revised and enlarged ed. of *Russko-angliiskii sel'kokhozyaistvennyĭ slovar'* (Moscow, Fizmatgiz, 1960, xii, 470p).

About 45,000 entries (1960 ed.: over 30,000). Includes botanical and zoological entries, with Latin names, chemical formulas. Shows Russian stress and gender. Appendix of abbreviations in Russian agricultural and biological lierature. *Class No:* 63.0(038)=82

[5083]
—English-Russian dictionary of agriculture. Adamenko, P.A., *and others, eds.* Oxford, etc., Pergamon Press, 1984. 700p. *Class No:* 63.0(038)=82

[5084]
—USOVSKII, B.N. Anglo-russkiĭ sel'skokhozyaistvennyĭ slovar'. 3. perer izd. Moscow, Gos. Izd. Tekhniko - Teoreticheskii Literatury, 1956. 532p. illus.

Is the companion dictionary. A feature is the section of 85 illus. (p.493-532), with specified parts. *Class No:* 63.0(038)=82

Chinese
[5085]
Dictionary of Chinese Communist agricultural technology. Wong, S., *comp.* Kowloon, Hong Kong, Union Research Institute, 1968. [vi], 557 + 17p. illus.

About 20,000 Chinese agricultural terms explained in English, with *c.*200 illus. Chinese characters are arranged by phonetic sounds (romanized spelling), following Mathews's *Chinese-English dictionary* system. *Class No:* 63.0(038)=951

Reviews & Abstracts
[5086]
Agrindex. AGRIS International Information System for the Agricultural Sciences and Technology. Rome, Food and Agriculture Organization, 1975-. 12pa. $400.

About 95,000 references pa. Provides complete bibliographic data on all items listed, with availability information. Classes A-U (closely subdivided): A. Agriculture in general - B. Agricultural economics, development and rural sociology ... F. Plant science and production ... K. Forestry - L. Animal science, production and protection ... P. Natural resources and environment ... R. Processing of agricultural products - S. Human nutrition - T. Pollution - U. Methodology. Descriptors appended to entries. Includes films and standards. Bulky. Personal author, corporate entry, report and patent number, and subject indexes. *Agrindex* is available in English, French and Spanish. Available online as AGRIS, via DIMDI, DIALOG, ESA. *Class No:* 63.0(048)

[5087]
Bibliography of agriculture: data provided by National Agricultural Library, US Department of Agriculture. Phoenix, Arizona, Oryx Press, 1942-. 12pa. $650pa. (US, Canada, Mexico); $750 (elsewhere). ISSN: 00061530.

Contains *c.*8,000 items per issue. Some 96 subject areas, further subdivided, record proceedings, annual and special reports, pamphlets and government publications, as well as periodical articles. Corporate author, personal author and analytical subject (based on title catchword) indexes per issue (*c.*20,000 subject index references), cumulated annually. 'The most indispensable single source in the field' (*Library trends,* v.15(4), April 1967, p.882). Available on microfiche. Online database: AGRICOLA *(q.v.),* and available on CD-ROM. *Class No:* 63.0(048)

[5088]
Biological and agricultural index: a cumulative subject index to periodicals in the fields of biology, agriculture and related sciences. New York, H.W. Wilson, 1964/65-. 11pa. ISSN: 00063177.

Continues *Agricultural index,* 1919-64.

Covers *c.*200 English-language journals, selected by subscriber vote. Includes experimental station publications and book reviews. Entry under topic. US slanted, as

....*(contd.)*

compared with *Bibliography of agriculture* and *Agrindex*. About 40,000 entries pa. Available online (WILSONLINE), CD-ROM (WILSONDISC), in addition to magnetic tape and diskette format. *Class No:* 63.0(048)

[5089]
Current contents: Agriculture, biology and environmental sciences. Philadelphia, Pa., Institute for Scientific Information, 1970-. Weekly. $560. ISSN: 00900508.

Previously (1970-72) subtitled 'Agriculture, food and veterinary sciences'.

Displays contents-pages of *c*.900 current issues of journals. (issue for 10 Feb 1992 contained contents pages from 114 journals). Sections: 'Features' (*e.g.* Current book contents); 'Discipline guide', grouping contents-pages by areas of interest (*e.g.* Multidisciplinary; Biology; ... Animal sciences; Veterinary medicine; Animal health). Weekly title word index. Author and address directory. Publishers' address directory. Available on diskette and online on the CURRENT CONTENTS SEARCH database.
Class No: 63.0(048)

[5090]
—FAO documentation: current bibliography. Rome, Food and Agriculture Organization, 1972-. 6pa. ISSN: 0304582x.

Began life as *FAO documentation current index* (1967-).

It now carries *c*.4,500 references pa., with author, subject and geographic indexes per issue. *Class No:* 63.0(048)

[5091]
Landwirtschaftliches Zentralblatt. Hrsg. von der Deutschen Akademie der Landwirtschaftswissenschaften zu Berlin. Institut für landwirtschaftliche Information und Dokumentation. Berlin, Akademie-Verlag, 1955-. 4 Abt., each monthly.

Agroselekt. Reihe 1. *Landtechnik.* 1955-. DM320. 2. *Pflanzenlicheproduktion.* 1956-. DM453. *Tierproduktion.* 1956-. DM453. 4. *Veterinärmedizin.* 1956-. DM453.

The 4 parts deal respectively with agronomy, crops, animal husbandry and fishing, and veterinary medicine. Part 1, *Landtechnik,* carries *c*.5,000 abstracts pa., with monthly author index and annual subject and author indexes.
Class No: 63.0(048)

Periodicals & Progress Reports

[5092]
Advances in agronomy. Prepared in cooperation with the American Society of Agronomy. San Diego, Academic Press, 1949-. v.1-. Annual. diagrs., tables, graphs.

V.45, 1991 (vii,378p.) has 17 contributors. The 8 papers include 'Nitrogen dynamics and management in rice-legume cropping systems', with 7 pages of references, and 'Genetics of resistance to insects in crop plants', with 8 pages of references. Index, p.357-78. *Class No:* 63.0(05)

[5093]
Agricultural journal titles and abbreviations. 2nd ed. Tucson, Arizona, Oryx Press, 1984. 136p. $45. ISBN: 0897740718.

First published 1982.

Lists all the titles (*c*.15,000) indexed in the *Bibliography of agriculture* (January 1979-March 1981), with their abbreviations, ISSNs, and the number of times each is cited in the *BOA* over the period. *Class No:* 63.0(05)

[5094]
Index to periodicals on agriculture held by the Science Reference Library. Jackson, G., *comp.* London, Science Reference Library, 1981-84. ISBN: 0902914650, set.

pt.1. *Agricultural research and industry.* 1981. [11], v, 143, [5]p. pt.2. *Agronomy and other aspects of plant science.* 1984. x, 206p.

Title entry. V.1 carries *c*.1,850 titles. V.2 *c*.2,500 titles. Annuals are included. Data: country, holding, class-mark.
Class No: 63.0(05)

[5095]
—Agricultural and animal sciences journals and serials: an analytical guide. Jensen, R.D. *and* Smith, M.M., *comps.* Westport, Conn., Greenwood Press, 1986. 211p. $49.95. (*Annotated bibliographies of serials: a subject approach, no.4.*) ISBN: 031324331x.

Covers 374 English-language journals, of which only 46 are British. Appended directories of online databases and of microform and reprint companies. Geographical, publisher, subject and title indexes. *Class No:* 63.0(05)

[5096]
International union list of agricultural serials. Wallingford, CAB International, 1990. xiii,767p. £99.50; $189. ISBN: 0851986617.

Lists serials indexed in AGRICOLA, AGRIS, and CAB ABSTRACTS. Gives details of 11,567 publications from 129 countries, with titles in 53 languages. Subject and country indexes. *Class No:* 63.0(05)

[5097]
Quarterly bulletin of the International Association of Agricultural Librarians and Documentalists. 1956-. 4pa. £32.

Carries articles (with abstracts in English, French and Spanish). Occasional bibliographies of 'Professional literature' (6th appearance in v.31(4), 1986, p.179-91), as well as news items. *Class No:* 63.0(05)

Yearbooks & Directories

[5098]
Directory of research workers in agriculture and allied sciences. Vernon, R., *ed.* Wallingford, Oxon., CAB International, 1989. ix, 490p. ISBN: 0851986234.

Previous ed. 1982 as *List of research workers, 1981, in the agricultural sciences in the Commonwealth.*

4 sections: 1. Organisations and research workers, by country - 2. Subject index - 3. Index of persons - 4. Index of institutions. Data on approximately 29,000 research workers include name, educational qualifications, and studies.
Class No: 63.0(058)

[5099]
UNITED STATES. Department of Agriculture. **The Yearbook of agriculture.** Washington, Government Printing Office, 1895-. Annual. illus., tables.

As from 1936 the *Yearbook* has dealt compendiously with particular themes (*e.g. Landscape for living.* 1972). During the 1970s scope widened to embrace consumer objectives (*e.g. Shoppers guide,* 1974). The 1990 issue is entitled *Americans in agriculture: portraits of diversity.*
Class No: 63.0(058)

Tables & Data Books

[5100]

FAO production yearbook. Rome, FAO, 1947-. Annual (v.44: 1990. 1991. £23). tables.

Statistical data on all important aspects of food and agriculture, including land population, index numbers of agricultural production, food supplies, wages and freight rates. V.44 (xlix, 306p.) has 132 tables.

FAO quarterly bulletin of statistics (1988-) updates.

Class No: 63.0(083)

[5101]

FISHER, Sir R.A. *and* **YATES, F. Statistical tables for biological, agricultural and medical research.** 6th ed. Edinburgh, Oliver & Boyd, 1963. x, 146p. tables.

First published 1938; 5th ed. 1956.

50 tables in all (numbered 1-34, with interpolations). The main addition to the 6th ed. are v, Fiducial limits for a variance component, and xvii, Balanced incomplete blocks with 11-15 replicates. The valuable introduction gives examples of the ways in which the tables can be used and includes 82 references, covering books, monographs and papers on statistical methods, up to 1963. No index.

Class No: 63.0(083)

[5102]

FOOD AND AGRICULTURE ORGANIZATION. Technical conversion factors for agricultural commodities. Rome, FAO, 1960. vi, 346p. tables.

Arranged by countries, A/Z. Includes tables for converting measures of capacity into weight. Typical factors under countries: average fat content of cow's milk (Germany); boneless meat yield of bone-in-meat (New Zealand); extraction rate of oil from olives (Syria); average clean yield of greasy wool (India). Principally for those concerned with agricultural statistics. *Class No:* 63.0(083)

[5103]

Practical handbook of agricultural science. Hanson, A.A., *ed.* Boca Raton, CRC Press, 1990. 534p. tables. $59.95. ISBN: 0849337062.

'A "quick reference" to a variety of topics pertaining to soils and the production and use of plants and animals' (*Preface*). Sections include soil, plants, tillage and rotations, cereal grains, forage, animals, environmental control of livestock, common statistical terms, glossary. Index, p.487-534. US-slanted. *Class No:* 63.0(083)

Maps & Atlases

[5104]

MORGAN, B.A. 'Maps and atlases as sources of agricultural information'. In *Information sources in agriculture and food science*, edited by G.P. Lilley. London, Butterworths, 1981. p.128-141.

Sections: Introduction - Types of maps relevant to agriculture (topographical, natural features, land capability and land classification, 'interference' or 'modification' maps) - Representations of agriculture (*e.g.* land utilization) - Agricultural atlases - Sources of information. 8 references.

Class No: 63.0(084.3)

[5105]

World atlas of agriculture: under the aegis of the International Association of Agricultural Economists. Novara, Istituto Geografico de Agostini, 1969-76. maps. loose-leaf.

Land utilization maps and relief maps prepared by the Committee for the 'World atlas of agriculture.' Maps (main scale: 1:5,000,000) - double-page spread, backed by key map, list of sources, and relief map. 4 text vols. accompany: Europe, USSR, Asia minor; South and east Asia, Oceania; the Americas; Africa. Unique, but now dated.

Class No: 63.0(084.3)

Histories

[5106]

The Agricultural history review. London, British Agricultural History Society, 1953-. 2pa. £16. ISSN: 00021490.

V.40(2), 1992 features 'List of books and pamphlets on agrarian history 1991' p.168-72. Also, 31 signed reviews, plus 4 signed shorter notices. The 8 articles include 'Millstones for medieval manors', with 55 footnotes. Available on microfilm from UMI. *Class No:* 63.0(091)

Biographies

[5107]

Agricultural and veterinary sciences international who's who: a biographical guide in the agricultural and veterinary sciences. 4th ed. Harlow, Essex, Longman Group, 1990. 1252p. £325. ISBN: 0582041007.

First published 1979 (F. Hodgson).

Professional biographical profiles of more than 7,000 senior research and advisory scientists in agricultural and veterinary sciences from over 120 countries. Country and subject index. *Class No:* 63.0(092)

Worldwide

[5108]

GRIGG, D.B. The Agricultural systems of the world: an evolutionary approach. Cambridge Univ. Press, 1974. 358p. tables. £17.50. ISBN: 0521098432.

Part 1 concerns the history, diffusion, and evolution of modern agriculture. Part 2, a detailed analysis of individual systems. Notes, p.289-307. Extensive bibliography, p.308-43. Analytical index. 'Aims to describe the chief characteristics of the major agricultural regions of the world and to attempt some explanations of how they come into being' (*Introduction,* chapter 1). *Class No:* 63.0(100)

[5109]

—Farming systems of the world. Duckham, A.N., *and others.* London, Chatto & Windus, 1970. xviii, 542p. illus. (pl.), tables.

Has 3 parts: 1. Analysis of location factors (7 chapters) - 2. Farming systems (by areas; 14 chapters; Canada and USA, p.109-69) - 3. Conclusions; Chapter references and further reading. Glossary and non-analytical subject index. 'Intended for students and research workers in agriculture and its related disciplines, and for teachers of agriculture and geographers in all parts of the world' (*Authors' Introduction*). *Class No:* 63.0(100)

Tropics

Bibliographies

[5110]
IMPERIAL COLLEGE OF TROPICAL AGRICULTURE.
Catalogue. [Imperial College of Tropical Agriculture]
University of the West Indies (Trinidad). Boston, Mass.,
G.K. Hall, 1975. 8v.

130,000 photolithographed catalogue cards for
international literature on tropical and sub-tropical
agriculture and related sciences. *Class No:* 63.0(213)(01)

[5111]
WORTLEY, P.J. 'Tropical agriculture'. In *Information
sources in agriculture and food science* (Editor, G.P. Lilley).
London, Butterworths, 1981. p.357-406.

A well-organized survey that includes sections:
'Computer-based bibliographic services'; 'Guides to
institutes, organizations and societies'. 5 references.
Class No: 63.0(213)(01)

Handbooks & Manuals

[5112]
Farming systems in the tropics. Ruthenberg, H., *and others.*
3rd ed. Oxford, Clarendon Press, 1980. xxii, 424p. illus.,
graphs, tables, maps. ISBN: 019839481x.

First published 1971.

11 chapters (2. Some characteristics of farming in a
tropical environment; 6. Systems with permanent upland
cultivation; 8. Grazing systems). Appendix notes.
Bibliography, p.391-414. Non-analytical index, p.415-24.
Class No: 63.0(213)(035)

[5113]
Tropical agriculture series. Wrigley, G., *general ed.*
London, etc., Longman Scientific & Technical, 1953-.
Volumes in print:

Agriculture in the tropics, by C.C. Webster and P.N.
Wilson. 2nd ed. 1980. £32. *Bananas*, by N.W. Simmonds.
3rd ed., 1987. £61. *The camel*, by R.T. Wilson. 1984. £41.
Cattle production in the tropics. v.1, by W.J.A. Payne.
1970. *Climate, water and agriculture in the tropics*, by I.J.
Jackson, 2nd ed., 1989. £14.99. *Cocoa*, by G.A.R. Wood
and R.A. Lass. 4th ed., 1985. £68.*Coffee*, by G. Wrigley.
1988. £68. *Cotton*, by J.M. Monroe. 2nd ed., 1987. £50. *An
introduction to animal husbandry in the tropics*, by W.J.A.
Payne. 4th ed. 1990. £55. *East African crops*, by J.D.
Acland. 1972. £9.35. *Introduction to tropical agriculture*, by
A. Youdeouwei, and others. 1986. £13.50. *The oil palm*, by
C.W.S. Hartley. 3rd ed. 1989. £66. *Oilseed crops*, by E.A.
Weiss. 1983. £64. *Rice*, by D.H. Grist, 6th ed., 1986. £74.
Rubber, by C.C. Webster and W.J. Baulkwill. 1989. £60.
Sheep production in the tropics and sub-tropics, by R.M.
Gatenby. 1986. *Sorghum*, by H. Doggett. 2nd ed. 1988.
£60. *Spices*, by J.W. Purseglove, and others. 1981. 2v. £42.
Sugar cane, by F. Blackburn. 1984. £66. *Tobacco*, by B.C.
Akehurst. 2nd ed. 1981. £88. *Tropical crops* (v.1
Monocotyledons; v.2 *Dicotyledons*), by J.W. Purseglove.
1975. £25. *Tropical forest and its environment*, by K.A.
Longman. 2nd ed. 1987. £26. *Tropical fruits*, by J.A.
Samson. 2nd ed. 1986. £40. *Tropical grassland husbandry*,
by L.V. Crowder and H.R. Chheda. 1982. £57. *Tropical
oilseed crops*, by E.A. Weiss. 1983.

D.H. Grist's *Rice* (6th ed. 1986. xx, 599p. illus. (pl.)
ISBN 058240402), was first published 1953. Its four parts

.... *(contd.)*
(22 chapters) cover rice, the plant; genetics and breeding of
rice; paddy and production; rice, the product. 4 appendices,
including 1. 'Standardization of rice terminology; 4. Species
of rice. Bibliography, p.543-76. Detailed, analytical index.
49 plates. A definitive contribution.
Class No: 63.0(213)(035)

[5114]
Tropical and subtropical agriculture. Ochse, J.J., *and
others.* New York, Macmillan Co.; London, Collier-
Macmillan, 1961. 2v. (xxii, 1446p.). illus., tables.

2 parts: 1. General; 2. Crops. 14 documented chapters
('Bananas and citrus', p.371-522; 11p. of references).
Section on 'Oil palm' (p.1047-66) has sub-sections: world
production; botany; varieties; pollination; breeding and
selection; climate and soil requirements; culture; harvesting
and processing; products and users; diseases and insect
pests. Glossary. Author and subject index per volume, index
of plant names. 144 tables in all. A valuable but dated
compendium. *Class No:* 63.0(213)(035)

Reviews & Abstracts

[5115]
Abstracts on tropical agriculture. Amsterdam, Koninklijk
Institut voor de Troper KIT/Royal Tropical Institute, 1975-.
12pa. DFl440. ISSN: 03045951.

Formerly *Tropical abstracts* (1953-74).

About 4,500 abstracts pa. Classes A-Q (*e.g.* A. Farming
systems research and development; F. Cereal crops; H.
Grain legume crops; P. Miscellaneous crops and forestry; Q.
Miscellaneous subjects). Geographic, subject, plant
taxonomic name and author index per issue; annual subject
and plant taxonomic names index. Coverage is worldwide.
Includes review articles. Database: TROPAG (1975-).
Available on CD-ROM. *Class No:* 63.0(213)(048)

Periodicals

[5116]
**TROPICAL DEVELOPMENT AND RESEARCH
INSTITUTE LIBRARY. List of current periodicals.**
London, the Institute Library, 1986. ii, 82p.

About 1,600 periodical titles. Currently received. A-Z,
excluding irregular serials. Details: title; first year taken;
location (central and 5 sub-libraries). Includes many non-
tropical agriculture titles. *Class No:* 63.0(213)(051)

Tables & Data Books

[5117]
ILACO BV., Arnhem. **Agricultural compendium for rural
development in the tropics and subtropics.** 3rd ed.
Amsterdam, Elsevier, 1989. xxxviii,740p. illus., graphs,
tables, maps. £40.93.

First published 1981. Commissioned by the Ministry of
Agriculture and Fisheries, The Hague.

11 chapters (some documented): 1. Climate - 2. Soil and
classification ... 4. Water control - 5. Land improvement - 6.
Agriculture (4½p. of references) - 7. Animal production and
fisheries ... 9. Economic and financial appraisal of progress
- 10. Sociology - 11. Tables and supporting data.
Class No: 63.0(213)(083)

Ancient Rome

[5118]

WHITE, K.D. A Bibliography of Roman agriculture. Reading, Berks., Univ. of Reading, 1970. xxviii, 63p. £0.85. (*Institute of Agricultural History. Bibliographies in agricultural history, no.1.*) ISBN: 090072403x.

915 entries, with evaluative annotations. Many periodical articles. Sections: Contemporary literature - Archaeological sources ... Practical agriculture and farm management ... Agricultural labour ... Food-processing industries - Transport and communications. *Class No:* 63.0(38)

Europe

[5119]

The Cambridge economic history of Europe from the decline of the Roman Empire. Clapham, J.H. *and* Power, E., *eds.* 2nd ed. Cambridge Univ. Press, 1966. xvi, 871p. tables, maps. £95. ISBN: 0521045053.

1. *Agrarian life of the Middle Ages;* edited by M.M. Postan. (first published 1942).

A scholarly, detailed study, well documented. Chapter bibliographies. Analytical index. *Class No:* 63.0(4)

[5120]

COMMISSION OF THE EUROPEAN COMMUNITIES. The Agricultural situation in the Community. 1991 report. Brussels & Luxembourg, the Commission. Annual. 294p. diagrs., tables. £21. ISBN: 9282634140.

8 sections: 1. The agricultural year ... 3. Economic situation and farm incomes ... 5. Rural development ... 8. Agricultural development: Statistical information. Data is given for 1985 onwards. *Class No:* 63.0(4)

[5121]

FUSSELL, G.E. Agricultural history in Great Britain and Western Europe before 1914: a discursive bibliography. London, The Pindar Press, 1983. 157p. £50. ISBN: 0907132049.

9 chapters on 'the history of the evolution of writing on agricultural history'. Chapter 1: The nativity of agricultural history in England; 4: The medievalists; 6. The Germans and other Europeans, to 1914; 7. France. Index of titles; index of authors. *Class No:* 63.0(4)

Great Britain

Bibliographies

[5122]

GREAT BRITAIN. HM Stationery Office. **Government publications. Sectional list 1, revised August 1986. Agriculture, fisheries, food and forestry.** London, HMSO, 1986. *Gratis.*

16 Sections: 1. Periodicals and serial publications - 2. Livestock - 3. Farm crops and grass - 4. Farm management, finance and statistics - 5. Horticulture and beekeeping - 6. Plant pests and diseases - 7. Farm safety - 8. Climate, drainage and water supply - 9. General farm pests - 10. Food - 11. Fisheries - 12. Forestry - 13. Reports - 14. Royal Botanic Garden, Edinburgh - 15. Royal Botanic Gardens, Kew - 16. Miscellaneous. In need of revision. *Class No:* 63.0(410)(01)

[5123]

McDONALD, D. Agricultural writers, from Sir Walter of Henley to Arthur Young, 1200-1800: reproductions in facsimile and extracts from their actual writings; enlarged and revised from articles which have appeared in The Field from 1903-1907. To which is added an exhaustive bibliography. London, Cox, 1908. v, 228p.

'The literature and bibliography of British agriculture, 1200 to 1800', *c.*650 items, chronologically arranged. Author index. *Class No:* 63.0(410)(01)

[5124]

SOUTHAMPTON UNIVERSITY. Library. **Catalogue of the Walter Frank Perkins Agricultural Library.** Southampton Univ. Library, 1961. xii, 291p.

Author catalogue of 2,009 numbered items. Appended: Books printed in England and Scotland up to 1700 ; Periodicals; Board of Agriculture County Reports - England, Scotland, Wales, Islands; Dublin Society. Statistical surveys of Irish counties. Supplementary notes. List of 15 authorities (*e.g.* Wing, *STC*), p.xi. Earliest imprint: 1600. The collection is perhaps richest in 18th-century material. The catalogue of a valuable library specializing in British agricultural history. *Class No:* 63.0(410)(01)

Tables & Data Books

[5125]

The Digest of agricultural census statistics: United Kingdom 1991. London, HMSO, 1992. vp. tables, maps. £15. ISBN: 0112429378. ISSN: 00654590.

Previously *Agricultural statistics, United Kingdom.*

Gives aggregate figures for each year in the period 1982 to 1991. 7 chapters include data on land use and crops, pigs, sheep, cattle, poultry, labour force, maps of selected main items, frequency distributions of main holdings by size. No longer includes prices. *Class No:* 63.0(410)(083)

[5126]

Reviews of United Kingdom statistical sources. V.23: Agriculture. Edited by M.F. Maunden, with M.C. Fleming. Peters, G.H. *and* Clark, K.R. London & New York, Chapman & Hall, for the Royal Statistical Society and the Economic and Social Research Council, 1988. xiii, 209p. tables. ISBN: 0412316706.

12 sections (7. Farm income and management; 12. An overall assessment of agricultural statistics). Quick-reference lists, p.147-88. Data, to 1984. Bibliography, p.189-92 (71 references). Detailed subject index, p.201-9. An impressive piece of research. *Class No:* 63.0(410)(083)

Histories

[5127]

ORWIN, C.S. *and* **WHETHAM, E.H. History of British agriculture, 1846-1914.** 2nd ed. London, Longmans, 1971. xx, 411p. tables.

First published 1964 (xx, 411p.).

Combined product of an historian and an agricultural economist. Bibliographical footnotes; authors quoted are named in the index. A standard work. *Class No:* 63.0(410)(091)

[5128]
—British and Irish writers on agriculture. Perkins, W.F., *comp*. 3rd ed. Lymington, C.T. King, 1939. 226p.

A list of *c*.1,300 writers, A-Z, 'from the earliest printed books until, and including, the year 1900'. *Class No:* 63.0(410)(091)

[5129]
—RUSSELL, Sir E.J. A History of agricultural science in Great Britain, 1620-1954. London, Allen & Unwin, 1966. 493p. illus., port., tables.

Includes biographical data. The author was a former Director of Rothamsted Experimental Station. *Class No:* 63.0(410)(091)

Scotland
[5130]
COPPOCK, J.T. **An Agricultural atlas of Scotland.** Edinburgh, John Donald, 1976. 260p. maps, graphs, tables.

Over 100 black-and-white maps, based on Department of Agriculture [and Fisheries], Scotland, with supporting text. The maps mainly relate to 1965. Scales 1:1M and 1:6M. 'It should find a place on the shelves of all geographical, agricultural and planning libraries' (*Geographical journal*, v.143(2), p.291). *Class No:* 63.0(411)

England & Wales
[5131]
COPPOCK, J.T. **An Agricultural atlas of England and Wales.** New and rev. ed. London, Faber & Faber, 1976. 267p. tables, maps.

First published 1964.

9 chapters on the agriculture of England and Wales (4. Tillage crops - 5. Grassland - 6. Horticulture - 7. Livestock - 8. Enterprises and prospects). 233 small black-and-white chloropath maps (square dot pattern) in text. Bibliography (general and by chapters), p.256-61. Index. *Class No:* 63.0(42)

[5132]
HARVEY, N. **The Industrial archaeology of farming in England and Wales.** London, Batsford, 1980. 232p. illus. ISBN: 0713418451.

9 chapters: 1. The winning of the waste - 2. Fields and field systems - 3. Water supply and irrigation schemes - 4. Sources of fertility - 5. Field draining - 6. Historical crop varieties - 7. Historical breeds of livestock - 8. Tools, implements and machines - 9. Farm buildings. Well documented: References, p.180-223; Sources of information (museums; journals); Selective book list (grouped), p.225-7. Detailed, analytical index, p.228-32. *Class No:* 63.0(42)

Information Services
[5133]
At the farmer's service: a handy reference to agricultural legislation and to various services available to farmers in England and Wales. London, MAFF Publications, 1990. 97p. *Gratis.*

Sections: Introduction: The agriculture departments - Information and advice - Management - Crops and seeds - Livestock - Grants, subsidies and loans - Marketing and

.... (contd.)
price control. 5 appendices (1. Ministry of Agriculture, Fisheries and Food offices; 4. Forestry Commission address list). Analytical index. *Class No:* 63.0(42):061:025.5

England

Bibliographies
[5134]
FUSSELL, G.E. **Old English farming books, 1523-1793:** Fitzherbert to the Board of Agriculture. Colliston, Aberdeen Rare Books, 1978. [7], 141p.; 186p. illus., facsims., ports. £12.50. ISBN: 0950660701, v.1; 0907132030, v.3; 0907132197, v.4; 0907132243, v.5.

Continues, with additions, his *The old English farming books, from Fitzherbert to Tull, 1523 to 1750* (London, Crosby Lockwood, 1947) and *More old English farming books from Tull to the Board of Agriculture, 1731 to 1793.* 1950). The work is updated by v.3, covering 1793-1839 (London, Pindar Press, 1983. 304p. £75); v.4, covering 1840-1860 (London, Pindar Press, 1984. 115p. £50); and v.5, covering 1861-1900 (London, Pindar Press, 1991. 145p. £95).

Written in narrative form. Author and title indexes. The 1950 work has an appendix of editions of Arthur Young's work, with locations. An authoritative work. *Class No:* 63.0(420)(01)

Histories
[5135]
The Agrarian history of England and Wales. Thirsk, J., *general ed*. Cambridge, University Press, 1967-. illus., diagrs., plans, tables.

1 (1): *Prehistory;* edited by S. Piggott. 1981. £65. (2): *A.D.43-1042;* edited by H.P.R. Finberg. 1972. £50. 2: *1042-1348.* 1988. £110. 3: *1350-1500.* 1991. £85. 4: *1500-1640;* edited by J. Thirsk. 1967. £60. 5(1): *1640-1750: Regional farming systems.* 1984. £60. (2): *Agrarian change.* 1985. £95; both parts edited by J. Thirsk. 6: *1750-1850.* 1985. £105. 7: *1850-1914.* 8: *1914-1939;* edited by C.H. Whetham. 1978. £35.

V.5(2), xxx, 952p. has 21 footnoted chapters, with 3 appendices. (1. Agricultural weights and measures; 2. Land measures; 3. Statutes), a select bibliography, p.903-21 (authors, A-Z) and a detailed analytical index to v.5(1) and (2). A scholarly series, still in progress. *Class No:* 63.0(420)(091)

[5136]
—SEEBOHM, M.E. The Evolution of the English farm. 2nd rev. ed. London, Allen & Unwin, 1952. 365p. illus. (Facsim. ed. E.P., 1976).

Covers the development of English farms and farming from Neolithic times to the present century. Many footnote references; chapter bibliographies, p.344-50. *Class No:* 63.0(420)(091)

Manuscripts & Incunabula

[5137]
READING UNIVERSITY. Library. **Historical farm records:** a summary guide to manuscripts and other material in the University Library collected by the Institute of Agricultural History and the Museum of English Rural Life. Reading, Berks., the University Library, 1973. xiii, 320p.

Under English countries, A-Z, p.1-290. Indexes to individual farms (under country and place), and place-names, etc. 'The catalogue should be a useful guide to the relatively few records of practical farming which have survived' (*Aslib proceedings,* v.26(3), March 1974, p.133). *Class No:* 63.0(420)(093)

USSR

[5138]
Atlas sel'kogokhozyaistvo SSSR. Tulupnikov, A.L. *and* Nikishov, M.I., *eds.* Moscow, Glavnoye Upravleniye Geodezii i Okrani Nyedr SSSR, 1960. vii, 309p.

378 individual maps on agriculture in the USSR (scales, 1:1.5M to 1:4). Rich in detail. 9 sections, on different aspects, - climate, rainfall, soils, topographical relief, population densities, and distribution of principal crops and livestock herds. *Class No:* 63.0(47)

Asia

[5139]
Guide to agricultural information sources in Asia and Oceania. Levick, G.R.T., *ed.* The Hague. Fédération Internationale de Documentation (FID), 1980. v, 72p. *(FID publication 592.)*

172 briefly annotated entries. Part 1: International. Part 2: Country-by-country. 3 types of sources: bibliographical; organizational; research programmes. *Class No:* 63.0(5)

China

[5140]
The Agriculture of China. Xu Guohua *and* Peel, L.J., *eds.* Oxford, University Press, 1991. xviii,300p. tables, maps. £45. ISBN: 0198592086.

8 sections: 1. Natural environment - 2. The historical and social background - 3. The components of agriculture - 4. The ten agricultural regions of China - 5. Infrastructure and agriculture inputs - 6. The rural economy - 7. Agricultural education, research, and extension - 8. Summary and conclusions: the characteristics and problems of Chinese agriculture and prospects for the future. Bibliographies appended to each section. 4 appendices, *e.g.* 2. Economic terms. A general overview. *Class No:* 63.0(510)

Asia—South & South East

[5141]
Agriasia: a current bibliography of Southeast Asian agricultural literature. Laguana, Philippines, Agricultural Information Bank of Asia, Southeast Asian College, 1970-. 4pa.

About 5,000 abstracts pa. Classes A-U (*e.g.* F. Plant genetics and breeding; H. Pests of plants; L. Animal husbandry; P. Natural resources and environment; Q. Processing of agricultural products). Author and keyword indexes per issue. AIBA database. *Class No:* 63.0(54+59)

India

[5142]
SINGH, J. An Agricultural atlas of India: a geographical analysis. Kuruksherra, Haryana, Vishal Publications, 1974. xxvi, 356p.

147 maps, with a 10-chapter commentary. Accompanying booklet: *The Green revolution in India* (1974. 48p.). 'The atlas is very well presented, straightforward and unpretentious' (*Geographical journal,* v.142(1), March 1974, p.169). *Class No:* 63.0(540)

USA

[5143]
Bibliography of books and pamphlets on the history of agriculture in the United States, 1607-1967. Schlebecker, J.T., *ed.* Santa Barbara, Calif., ABC-Clio, Inc., 1969. vii, 183p.

2,042 numbered entries, under authors, A-Z. Occasional annotations. Presence of illus., maps, bibliographies, etc. is noted. Includes fiction. 274 entries on slavery. The subject index (p.103-89) is non-analytical, having *c.*135 undifferentiated entries under 'Agrarian discontent'. 'Well printed and readable bibliography' (*RQ,* v.10(2), Winter 1970, p.166). *Class No:* 63.0(73)

[5144]
SCHAPSMEIER, E.L. *and* **SCHAPSMEIER, F.H. Encyclopedia of American agricultural history.** Westport, Conn. & London, Greenwood Press, 1976. 480p. $55. ISBN: 0837179580.

Has entries A-Z and a subject index. Many entries have further references. Considered best at coverage of US legislation, agricultural agencies and organizations, as well as frontier and rural life (*Library journal,* 15 September 1976, p.1844-5). *Class No:* 63.0(73)

[5145]
U.S. agricultural groups: institutional profiles. Browne, W.P. *and* Cigler, A.J., *eds.* Westport, Conn., Greenwood, 1990. 274p. $55. ISBN: 031325088x.

Lists 103 private US organizations, with data including address, list of publications, summary of purpose, origins, funding and organizational structure. An appendix of related groups is included. Index of corporate and personal names. *Class No:* 63.0(73)

Thesauri

[5146]
Agrovoc: multilingual agricultural thesaurus. 2nd ed. Rome, Apimondia in association with FAO, 1992. xvi,798p. $120, monolingual ed.; $580, 5-language v.

First published 1982.

Lists 14,714 indexing terms used by *AGRIS (q.v.)* and CARIS. Available in 5 languages: English; French; Spanish; German; Italian. The English, French, and Spanish versions each have a trilingual index. *Class No:* 63.0:025.43

[5147]
CAB thesaurus. 1990 ed. Wallingford, Oxon., CAB International, 1990. 2v. £99.50. ISBN: 0851986870.

First published 1983.

Contains *c.*57,000 entries, including cross-references. Each preferred term is displayed with its complete hierarchy of broader, narrower and related terms. Contains all the indexing terms for the AGRICOLA database of the National

....(contd.)

Agricultural Library (NAL) of the US in addition to those used in the CAB abstracting journals and database. New features for this edition are 'History notes', distinction between American and British spelling variants, and the adoption of the group names approved by the International Committee on Taxonomy of Viruses. *Class No:* 63.0:025.43

Libraries

[5148]

Agricultural information resource centers: a world directory 1990. Fisher, R.C., *and others.* Urbana, Ill., International Association of Agricultural Librarians and Documentalists, 1990. xxviii,641p. $140. ISBN: 0962405205.

Contains 3,971 entries, arranged A-Z by country (Algeria ... Zimbabwe). Coverage includes universities, research institutes, and government libraries. Data include staff, main subjects, loan policy, photocopy policy. Subject and institution indexes. *Class No:* 63.0:061:026/027

Institutions & Associations

[5149]

COMMISSION OF THE EUROPEAN COMMUNITIES. AGREP: Permanent inventory of agricultural research projects in the European Communities. Luxembourg, CEC, 1982-83. 2v. & supplement.

1. *Main list: research projects.* September 1982. xxiv, 885p. 2. *Indexes.* September 1982. x, 284p. *Supplement.* November 1983.

V.1 has data on *c.*25,000 research projects. Classified index. V.2 provides subject and country (Belgium ... The Netherlands) indexes. List of research organizations. List of scientists. The Supplement is similarly arranged. AGREP functions as an online database on DATACENTRALEN; *c.*5,000 records are added pa. *Class No:* 63.0:061:061.2

[5150]

COMMISSION OF THE EUROPEAN COMMUNITIES. Directory of non-governmental agricultural organizations. Rev. ed. Munich, etc., K.G. Saur, 1989. 320p. £55. ISBN: 359810801x.

About 150 organizations. Sections A-G (A. Agricultural producers; B. Agricultural and food products processing industries; C. Agricultural and food trading D. - Agricultural cooperatives - E. Agricultural and food industry - E. Consumers - G. Miscellaneous (*e.g.* Eurogroup for Animal Welfare)). *Class No:* 63.0:061:061.2

[5151]

COMMISSION OF THE EUROPEAN COMMUNITIES. Economic and Social Committee. **Directory of European agricultural organizations.** London, Kogan Page, for Office for Official Publications of the European Communities, Luxembourg, 1984. 718p. £39. ISBN: 0850389291.

5 parts. 1. General overview and basic data - 2. Agriculture in member states of the European Community and national agricultural organizations. (10 countries, Federal Republic of Germany ... Greece) - 3. Agriculture and agricultural organizations in other countries of Western Europe (7, including Spain and Portugal) - 4. Organizations of farmers in Europe (by countries) - 5. The institutions of the European Community and agriculture.
Class No: 63.0:061:061.2

Research Establishments

[5152]

Agricultural research centres: a world directory of organizations and programmes. 10th ed. Harlow, Essex, Longman Group, 1990. 987p. £295. (*Reference on research.*) ISBN: 0582061229.

9th ed. 1988. Previously as *Agricultural research index* (F. Hodgson).

Arranged alphabetically by country, Albania ... Zimbabwe. Lists *c.*4,500 laboratories and departments that conduct or finance agricultural research. Covers *c.*140 countries, including those of the Third World. Food, environmental and biological sciences are included in the subject coverage. Data include affiliation, status, senior staff, annual expenditure and publications. Index of names of establishments. Non-analytical subject index.
Class No: 63.0:061:061.62

Research Projects

[5153]

AGRICULTURAL AND FOOD RESEARCH COUNCIL. Programme of agricultural and food research, 1987-88. London, AFRC, 1988. 11v.

Formerly *Index of agricultural and food research.*

1. *Institute for Animal Disease Research.* 2. *Institute of Animal Physiology and Genetics Research.* 3. *Institute of Arable Crops Research.* 4. *Institute of Engineering Research.* 5. *Institute of Food Research.* 6. *Institute for Grassland and Animal Production.* 7. *Institute of Horticultural Research.* 8. *Institute of Plant Science Research.* 9. *Scottish Agricultural Research Institutes.* (HRI-MRI). 10. *Scottish Agricultural Research Institutes.* (RRI-SCRI). 11. *Scottish Agricultural Colleges.*
Class No: 63.0:061:061.62.005

[5154]

—Agricultural research, 1931-1981: a history of the Agricultural Research Council, and a review of developments in agricultural science in the last fifty years. Cooke, G.W., *ed.* London, the Council, 1981. ix, 367p. illus., facsims., ports., diagrs., tables. £6.50. ISBN: 0708401805.

Has 13 contributions. Topics range from plant breeding to milk and milk products. 7 appendices. Detailed, non-analytical index. *Class No:* 63.0:061:061.62.005

Agricultural Economics

[5155]

COMMISSION OF THE EUROPEAN COMMUNITIES. Directorate-General for Research, Science & Education. **Agricultural economics and rural sociology - multilingual thesaurus.** Munich, K.G. Saur, 1979. 5v. (4v. & index), plus 16 microfiche. (English language volume. 104p.).

English, French, German and Italian language volumes, plus index. Coverage: Agricultural and rural law - Agricultural economics - Countries - Livestock, crops and products - Methodology - Research development and education - Rural sociology. *Class No:* 63.003

[5156]
GIANNINI FOUNDATION OF AGRICULTURAL ECONOMICS LIBRARY. Dictionary catalog. [Giannini Foundation of Agricultural Economics Library]. Boston, Mass., G.K. Hall, 1971. 12v. $1,305. ISBN: 0816109087.

377,000 photolithographically reproduced catalogue cards. Most of the collection is in English, with emphasis on the US, Western European countries, Spanish America and English-language-speaking countries. *Class No:* 63.003

[5157]
OLSEN, W.C. Agricultural economics and rural sociology: the contemporary core literature. Ithaca, London, Cornell Press, 1991. xi,346p. tables. (*The literature of the agricultural sciences.*) ISBN: 0801426774.

10 chapters: 1. Trends and development of agricultural economics and rural sociology in the United States ... 4. Citation analysis ... 6. Current primary journals and serials - 7. Core lists of primary monographs (p.92-267) ... 10. Reference update. Appendices: A. Languages in AGRICOLA/CABI databases; B. American Agricultural Economics Association classic book list. Index. The other projected 6 volumes in the series are: *Agricultural engineering; Animal science and health; Soil science; Crop improvement and protection; Food science and human nutrition; Forestry and silviculture.* *Class No:* 63.003

[5158]
World agricultural economics and rural sociology abstracts. Wallingford, Oxon., CAB International, 1958-. 12pa. £292; $552. ISSN: 00438219.

About 8,000 abstracts pa. Sections include: Agricultural economics; Agricultural policy; Supply, demand and prices; Marketing and distribution; International trade; Finance and credit; Farm economics; Cooperatives; Rural sociology. Author and subject indexes per issue and annually. Annual geographical index. Online through CAB ABSTRACTS database and available on CD-ROM. *Class No:* 63.003

Co-operative Enterprises

[5159]
Directory of agricultural co-operatives in the United Kingdom, 1990. Prepared by the Plunkett Foundation for Co-operative Studies on behalf of Food from Britain and in collaboration with the Federation of Agricultural Co-operatives (UK) Ltd. Riordan, T.F., *comp.* Oxford, Plunkett Foundation for Co-operative Studies, 1990. x,192p. £25. ISBN: 0850420989. ISSN: 02657155.

Sections: Digest of statistics - Central organizations - Agricultural marketing boards - FAC in Europe - Representatives on COGECA and EEC committees - Federal organizations - Directory of agricultural co-operatives, (England, Northern Ireland, Scotland, Wales, p.33-158) - Fishery co-operatives - Cottage industries - Country index. *Class No:* 63.009.11

[5160]
Yearbook of co-operative enterprise. Oxford, Plunkett Foundation for Co-operative Studies, 1927-. Annual. ISSN: 09525556.

First published 1927 (1926).

Formerly *Yearbook of agricultural co-operation.*

1992 ed. (214p. £14.95) has 4 sections: 1. Co-operatives and public policy - 2. Co-operatives in the curriculum - 3.

.... *(contd.)*
Co-operatives and people with disabilities - 4. Annual review of co-operation in the United Kingdom.
Class No: 63.009.11

Forestry

Bibliographies

[5161]
REHDER, A. The Bradley bibliography: a guide to the literature of the woody plants of the world published before the beginning of the twentieth century. Compiled at the Arnold Arboretum of Harvard University. Cambridge, Mass., Riverside Press, 1911-18 (reprinted Champaign, Ill., Koeltz Science Books, 1976). 5v.

1. *Dendrology.* pt.1. 1911. 2. *Dendrology*, pt.2: 'Gymnospermae'. 1912. 3. *Arboriculture, economic properties of woody plants,* 1915. 4. *Forestry,* 1914. 5. *Index of authors and titles* (p.1-943); Index of Greek authors and titles. Index of Russian and Serbian authors and titles; Subject index to v.1-4.

A classic bibliography. V.1, pt.1 records *c.*20,000 items, with bibliographical notes, on more general aspects and allied studies, and has a subject index. *Class No:* 630(01)

[5162]
YALE UNIVERSITY. Yale Forestry Library. **Dictionary catalogue of the Yale Forestry Library,** Henry S. Graves Memorial Library. Boston, Mass., G.K. Hall, 1962. 12v. $1,305. ISBN: 0816106312.

The catalogue (217,000 cards) of one of the largest libraries of its kind. Strongest on all aspects of the history of forestry, especially that of American forestry. Records 90,000v.; also periodical articles, indexed from leading forestry magazines and lumber journals, thus conveniently preceding *Forestry abstracts* (1939-) *(q.v.).*
Class No: 630(01)

Encyclopaedias

[5163]
Lesnaya entsiklopediya. [Forest encyclopaedia.] Vorob'ev, G.I., *ed.* Moscow, Sovetskaya Entsiklopediya, 1985-.

V.1. *Abelia-Lemon.* 1985. 563p.

Projected set of 2v., which, when completed, will contain 3,600 articles, with particular emphasis on the utilization of timber and other forest resources. *Class No:* 630(031)

Handbooks & Manuals

[5164]
Forestry handbook. Wenger, K.F., *ed.* 2nd ed., edited for the Society of American Foresters. New York, etc., Wiley, 1984. xix, 1335p. graphs, tables, maps. $75. ISBN: 0471062278.

First published 1958.

25 chapters on forest ecology, geology and soils, meteorology and climatology, insect and disease management, timber measurements, silviculture, logging, safety, remote sensing, mathematics and statistics. 'Further reading'. Detailed subject index. *Class No:* 630(035)

[5165]
JAMES, N.D.G. **The Forester's companion.** 4th ed. Oxford, Basil Blackwell, 1989. xii, 310p. tables. £14.95. ISBN: 0631167242.

First published 1955.

33 chapters (1. Forest trees grown in Britain ... 8. Forest management ... 12. Diseases and pests ... 16. Utilization ... 27. Societies, organizations, establishments, and trusts concerned with forestry work). 1. Appendix: Abbreviations. Index, p.302-10. Small format. *Class No:* 630(035)

Dictionaries

[5166]
Dictionary of forestry, in five languages: German, English, French, Spanish, Russian. Weck, J., *and others, comps.* Amsterdam, etc., Elsevier, 1966 (reprinted 1979). xxvi, 573p. tables. £93.43.

About 10,000 German-base terms, with English, French, Spanish and Russian equivalents and indexes. Appendices list species of trees and plants, plus animals, under scientific names, with equivalents in the five languages concerned. *Class No:* 630(038)

[5167]
Lexicon forestale / Metsäsan akirja / Skogsardbok / Forest dictionary / Forstwörterbuch / Lesnoĭ slovar'. Ahlsved, K.-J., *and others.* Porvoo, Finland, Werner Söderstrom, 1979. xix, 594p.

9,582 numbered Finnish terms, with Swedish, English, German and Russian equivalents and indexes. A dictionary of forestry, timber and paper words. Areas covered: tree species, vascular plants, mosses and liverworts, fungi and bacteria, vertebrates, weights and measures, institutes, organizations and associations. Many cross-references. Genders stated. *Class No:* 630(038)

[5168]
RUOKONEN, M. **Forestry and forest products vocabulary.** Slough, Bucks., Commonwealth Agricultural Bureaux, 1984. xv, 459p. £26. ISBN: 0851985483.

About 14,000 terms are defined, with coded reference to sources. 'Key to sources', p.ix-xiv. Includes acronyms and abbreviations, dialect/geographical limitations. Ample cross-references. Both British and American spelling given. Supplements the list of terms provided annually in *Forestry abstracts* and *Forestry products abstracts.* *Class No:* 630(038)

[5169]
TEKNISKA NOMENKLATURCENTRALEN. **Skoglista.** [Glossary of forestry.] 2nd rev. and enlgd. ed. Stockholm, the Centre, 1978. 676p. illus. (*TNC 71.*)

First published 1969.

A classified dictionary, Swedish-English, with Swedish definitions and index to terms in both languages. About 5,000 Swedish entries. *Class No:* 630(038)

[5170]
Terminology of forest science, technology, practice and products: English language version. Ford-Robertson, F.C. *and* Winters, R.K., *eds.* Rev. ed. Washington, Society of American Foresters, 1983. xxi, 370p. illus. $10. (*Multilingual forestry terminology series, no.1.*) ISBN: 0939970163.

First of a series of multilingual forestry terminologies authorized by the joint FAO/IUFRO Committee on Forestry

....*(contd.)*
Bibliography and Terminology. This ed. is a reprinting of the 1971 ed. + an addendum.

*c.*7,000 entries, differentiating US usage. Includes abbreviations. Many cross-references. 6 appendices (*e.g.* 1. Key to published sources drawn on; 4. Deprecated key and secondary terms with their preferred synonyms). *Class No:* 630(038)

Reviews & Abstracts

[5171]
Forestry abstracts. Prepared by the CAB International Forestry Bureau. Wallingford, Oxon., CAB International, 1939-. 12pa. £293; $543. ISSN: 00157538.

About 10,000 abstracts pa. 16 sections: General publications and general techniques - General aspects of forestry - Silviculture - Forest mensuration and management - Physical enviornment - Fire - Plant biology - Mycology and pathology - Insects and other invertebrates ... Catchment management. Soil conservation - Other land use. Nature conservation - Arboriculture - Dendrochronology and dendroclimatology. Author and subject indexes per issue and annually. The major source.

Online through CAB ABSTRACTS database (covering the 19 major CAB abstract journals). Available on the CAB ABSTRACTS CD-ROM and TREECD, also produced by CAB International. *Class No:* 630(048)

Periodicals

[5172]
COMMONWEALTH FORESTRY INSTITUTE. **Forestry Library list of serials.** 4th rev. ed. Oxford, the Institute, 1984. 264p. *Class No:* 630(051)

Maps & Atlases

[5173]
Weltsforstatlas. Reinbeck, Bundesanstalt für Forst und Holzwirtschaft, in collaboration with FAO Berlin-Grünewald, Haller, 1951-.

Commissioned by the FAO Forestry Division.

Each map (in 5-10 colours) covers one country or area, showing distribution of forest types and wood species, location of woodworking industries, and the like. Legends in English, Spanish, German and French. Lieferung 21 (6 maps) was received in 1981. Page size, *c.*59 x 72cm. Of value to foresters and forest students, economists, geographers, botanists and regional planners. *Class No:* 630(084.3)

China

[5174]
China forest yearbook 1949-1986. Beijing, China Forestry Publishing House, 1987. xix,718p. illus. (col. pl.).

24 chapters covering forestry in China, including forest products industries and recent history. Detailed glossary of 125 terms (112 with English equivalents). Chinese subject index. *Class No:* 630(510)

USA

[5175]
Encyclopedia of American forest and conservation history.
Davis, R.B., *ed.* New York, Macmillan, 1983. 2v, (871p.), illus., maps. $200. ISBN: 0029075302.

203 contributors. 413 articles (some lengthy), with references for further reading and cross-references. Scope covers US forest industries, national parks, personalities, legislation, and topical issues. Appendices concern national forests and parks, officials, etc., and carry an atlas. Black-and-white illus. only. 'An excellent set which belongs in any library serving those interested in American forestry, conservation and ecology'. (*Reference books bulletin, 1984-1985*, p.53). *Class No:* 630(73)

Forest Products

[5176]
BOLZA, E. German-English glossary of forest products terms. South Melbourne, CSIRO, Division of Forest Products, 1959. 169p. Supplement. 1962. 87, [5]p.

'The alphabetical glossary also contains many silvicultural and forest-utilization terms' (*Forestry abstracts,* v.21(2), April 1980, p.289). *Class No:* 630:8

[5177]
FOOD AND AGRICULTURE ORGANIZATION. Yearbook of forest products, 1989. Rome, FAO, 1991. 392p. $30. ISBN: 9250030304.

Statistics and data on production, trade and consumption in roundwood, sawnwood, sleepers, plywood, wood pulp, varied types of paper, paper-board and fibre-board. Sources from more than 100 countries. *Class No:* 630:8

[5178]
Forest products abstracts. Wallingford, Oxon., CAB International, 1978-. 6pa. £147; $269. ISSN: 01404784.

Coverage includes: *c.*2,600 abstracts pa. Coverage includes: General publications and general techniques; General aspects of forest products and industry; Wood properties; Timber extraction, conversion and measurement; Damage to timber and timber protection; Utilization of wood; Veneers; Composite boards, laminated wood, panels, improved wood, glues; Pulping; Marketing and trade. Economics. Author and subject indexes. Occasional review articles precede abstracts (*e.g.* 'Forest products thesis summaries', April 1987, p.137-41).

Online through CAB ABSTRACTS database, CAB ABSTRACTS CD-ROM and TREECD. Hosts include DIALOG, ESA-IRS. *Class No:* 630:8

Trees (Silviculture)

Encyclopaedias

[5179]
DAVIS, B. The Gardener's illustrated encyclopedia of trees and shrubs: a guide to more than 2000 varieties. Harmondsworth, Middx., Penguin Books; Viking, 1987. 256p. illus. (incl col.) £19.99. ISBN: 0670812374.

Dictionary of trees, p.10-71; dictionary of shrubs, p.72-221. Data under 11 heads in both sequences: origin; description; hardiness; shade aspect ... average height and spread. Advice on planting and pruning. Practical glossary;

....(contd.)

botanical glossary. Choosing the right trees and shrubs. Pronunciation glossary. Index of common names; Index of Latin names. *Class No:* 630.2(031)

[5180]
The Hillier manual of trees and shrubs. 6th ed. Newton Abbot, Devon, David & Charles, 1991. 704p. £25. ISBN: 071539942x.

First published 1971, as a nurseryman's catalogue.

Very brief descriptions of over 9,000 plants, including more than 1,000 new entries. 4 main sections: Trees and shrubs; Climbers; Conifers; Bamboos. Appended sections on botanical names; trees and shrubs for selected soils, sites and ornamental effect; flowering trees and shrubs for every month. A standard guide. *Class No:* 630.2(031)

[5181]
—**The Hillier colour dictionary of trees and shrubs.** Newton Abbot, Devon, David & Charles, 1993. 323p. col. illus. £17.99. ISBN: 0715300911.

Has *c.*3-line descriptions of 3,500 trees and shrubs selected for their garden value. The colour dictionary has 3 sections: trees and shrubs; climbers; conifers (p.35-319), preceded by cultural, etc. notes. Contains *c.*600 illustrations. Glossary of botanical terms, p.320-3. Notes on many of the plants are based on conditions in the British Isles. *Class No:* 630.2(031)

[5182]
JOHNSON, H. Hugh Johnson's encyclopaedia of trees. Rev. and updated ed. Editor, D. Taylor. London, Mitchell Beazley, 1984. 336p. col. illus., diagrs., charts, maps. ISBN: 0855335467.

First published as *The international book of trees* (1973).

Guide to the world's trees, p.66-246 (the conifers; the broad leaves). A new feature, 'Reference section', - an illustrated A-Z index of tree species, p.247-31). This section includes; A twelve month succession of ornamental flowers, fruit and foliage; rates of tree growth; a guide to choosing trees; the meaning of botanical names. Illustrations (more than 1,000 covering over 600 species) are well captioned. Index. *Class No:* 630.2(031)

[5183]
KRÜSSMANN, G. Manual of cultivated broad-leaved trees & shrubs. London, Batsford, 1984-86. 3v. (1519p.), illus., maps. £70ea. ISBN: 0713446013; 0713453486; 0713454083.

Translation of *Handbuch der Laubgehölze* (1976-.).

The 3v. (A-D, E-PRO, PRU-Z) record data on over 5,000 species and *c.*6,000 cultivars of trees, shrubs, subshrubs and vines. Entries are under Latin names of genera. V.1 includes an A/Z guide to botanical terms in English, Latin, German, French and Dutch, and 'List of abbreviations to other reference works'. Over 300 black-and-white illus. per volume. Hardiness zone maps. *Class No:* 630.2(031)

[5184]
The Oxford encyclopedia of trees of the world. Hora, B., *consultant ed.* Oxford Univ. Press, 1981. 288p. col. illus., tables. ISBN: 0192177125.

39 contributors. Devotes the main section to 149 wild and cultivated genera. Data: structure; distribution; history; ecology; cultivation; economic use. A further section concerns native trees of Asia, Africa and America. Appended keys to families, a general bibliography (p.277);

....*(contd.)*

glossary. Indexes to common and Latin names. 365 illus., well captioned. 'An excellent choice' (*RQ,* v.21(4), Summer 1982, p.420). *Class No:* 630.2(031)

Handbooks & Manuals

[5185]

Diseases of trees and shrubs. Sinclair, W.A., *and others.* Ithaca, N.Y., & London, Comstock Pubg. Associates, Cornell Univ. Press, 1987. 547p. col. illus. £52.50; $52.50. ISBN: 0801415179.

Descriptive data, facing 247 colour-plates. Glossary, p.509-11. A feature is the extensive bibliography, p.512-46: over 2,200 references, under authors, A-Z. Detailed analytical index, p.547-74. Large quarto format. *Class No:* 630.2(035)

[5186]

PHILLIPS, D.H. and **BURDEKIN, D.A. Diseases of forest and ornamental trees.** London, Macmillan, 1982. xv, 435p. illus. (incl. col.). £60. ISBN: 0333323572.

19 well-documented chapters (*e.g.* 'Diseases of willow', p.301-25, including 2½p. of references). Glossary, p.410-411. Detailed, analytical index, p.412-35. Concentrates on diseases found in Britain on forest and ornamental trees. 64 plates. Where possible, information is given on 'symptoms, factors affecting disease expression, host-range and resistance, and brief descriptions of causal organisms' (*Preface*). 2nd ed. promised 1992/93 (£75). *Class No:* 630.2(035)

Dictionaries

[5187]

VAUCHER, H. Elsevier's dictionary of trees and shrubs, in Latin, English, French, German and Italian. Amsterdam, etc., Elsevier, 1986. xiv, 414p. £110; $164. ISBN: 0444425691.

2,500 Latin names of trees and shrubs (stating genera and species, with popular name equivalents in English, French, German and Italian). The trees and shrubs are mostly those found in the northern and southern hemispheres. Appended list of reference book sources, p.413. *Class No:* 630.2(038)

Worldwide

[5188]

PHILLIPS, R. Trees in Britain, Europe and North America. London, Ward Lock, 1978. 234p. illus. £12.99. ISBN: 0330254804.

Illustrates well over 500 trees (A-Z, Latin names, p.60-215) in full colour, showing leaf, fruit, inflorescence, bark, silhouette. Unusual leaf index. Helpful colour supplement to Bean (*qv*), which has only black-and-white photographs and line drawings. Quarto format. *Class No:* 630.2(100)

Europe

[5189]

MITCHELL, A. A Field guide to the trees of Britain and Northern Europe. 2nd ed. London, Collins, 1978. 416p. illus. (incl. col.). £14.99. ISBN: 0002192136.

Describes 800 trees (49 families), p.48-390. Data: common and Latin names; habit; use; size; bark; crown;

....*(contd.)*

foliage; flowers (and cone); similar species. Appended glossary. 'Collections of trees open to the public'. Index to English names; index to scientific names. 40 col. plates; 640 clear line-drawings. 'Designed to describe in systematic detail every tree species and cultivar (variety) found in parks, gardens, large and small, towns and countryside, and to provide guides and keys with which to identify them' (*Preface*). An excellent compendium. *Class No:* 630.2(4)

Great Britain

[5190]

BEAN, W.J. Trees and shrubs hardy in the British Isles. 8th ed. Chief editor, D.L. Clarke. London, Murray, 1970-80. 4v. illus. Supplement. 432p. 1988. £60ea; £50, supp. ISBN: 0719524285, set; 0719517907, supp.

First published 1914-33.

Entries, A-C, D-M, N-Rh, Ri-Z, under Latin names. V.1 (Abelia-Cytisus. xx, 848p.) contains a glossary (p.112-20) and select bibliography, the descriptive list of genera and species (p.137-831), with 77 plates and 112 botanical drawings, and an index. Data: full description (corrected or updated) of species and varieties (but no keys to identification), common name, synonyms, distribution, date of introduction, features, merits and cultivation, plus 'references for illustrations'. V.4 (xv, 808p.) covers 115 genera, with 111 black-and-white photographs, and index, p.785-808. *Class No:* 630.2(410)

USA

[5191]

The Audubon Society field guide to North American trees. Little, E.C. New York, Knopf, 1980-2. 2v. (1353p.). col. illus. ISBN: 0394507606; 0394507614.

1. *Eastern region.* 1980., 714p. 2. *Western region.* 1982. 639p.

V.1 records 368 species; v.2. 314 species. The main feature - as of others in the Audubon Society Field guide series - is pictorial presentation (v.2 has 855 illus., mostly in colour), grouped by similarities of leaf or needle shape, fruit, etc. The second part of each volume is descriptive text, with characteristics of each tree. Geographic range of v.1 extends to the eastern base of the Rockies, then south to Texas. Portable format. Ideal 'for someone needing a quick and painless tree identification' (*RSR,* v.12(3), Fall 1984, p.38). *Class No:* 630.2(73)

Agricultural Trade

[5192]

FOOD AND AGRICULTURE ORGANIZATION. FAO trade yearbook. Rome, FAO, 1947-. Annual. £32.

V.44 (1992. 380p.) has 196 tables. 4 sections: 1. FAO regional index numbers of agricultural trade - 2. Trade in agricultural products (p.39-300) - 3. Trade in agricultural requisites (*e.g.* tractors; pesticides) - 4. Value of agricultural trade by countries (p.317-80). Years covered: 1988, 1989, 1990. Flimsy paper; small print. *Class No:* 631.003.1

Farms

Organic Farming

[5193]
LAMPKIN, N. **Organic farming.** Ipswich, Farming Press, 1990. xiii,701p. illus. (some col.), tables. £19.95. ISBN: 0852361912.

15 chapters: 1. Organic farming-agriculture with a future - 2. The living soil ... 4. Management of manures, slurry and organic residues - 5. Rotation designs for organic systems ... 8. Livestock husbandry ... 12. Marketing and processing ... 14. Converting to organic farming - 15. The wider issues. 6 appendices (1. Standards for organic agriculture ... 4. Addresses ... 6. Some metric conversion factors). Sections on animal husbandry particularly useful. Bibliographies appended to each chapter. Author is lecturer in agricultural economics at the University College of Wales, Aberystwyth and Development Director of the Aberystwyth Centre for Organic Husbandry and Agroecology. Good value.
Class No: 631.147

Management

[5194]
NIX, J. *and* HILL, P. **Farm management pocketbook.** 21st ed. Ashford, Kent, Department of Agricultural Economics, Wye College, 1990. vii, 216p. tables. £5.25. ISBN: 0862660319.

First published 1966.

4 sections: 1. Gross margin data (1. General notes; 2. Cash crops; 3. Grazing livestock; 4. Pigs, poultry and fish) - 2. Labour - 3. Machinery; 4. Miscellaneous data (*e.g.* 1. Fixed costs; 4. Buildings; 15. Farm records; 19. Useful addresses and telephone numbers). Index. Virtually annual update. J. Nix is National Westminster Bank Professor.
Class No: 631.15

Buildings

[5195]
BRUNSKILL, R.W. **Traditional farm buildings of Britain.** 2nd ed. London, Gollancz, 1987. 188p. illus. £16.95; £6.95. ISBN: 0575040384.

First published 1982.

9 chapters: 1. Farming history and the standard farmstead - 2. Barns and the processing of grain crops - 3. Accommodation for animals - 4. Accommodation for birds - 5. Granaries - 6. Farmsteads - 7. Materials and construction - 8. Recent developments and the re-use of farm buildings - 9. Regional variations. Notes and references, p.176-80. Bibliography, p.181-4. Index (to text and illus.), p.185-8. 133 illus. *Class No:* 631.22

[5196]
HARVEY, N. **A History of farm buildings in England and Wales.** New ed. Newton Abbot, Devon, David & Charles, 1984. 279p. illus. £17.50. ISBN: 0715383833.

First published 1970.

12 chapters (with appended notes): 1. Prologue: to AD1000 ... 4. The agrarian revolution: the agrarian phase, 1750-1820 ... 10. Post-war farmsteads: some examples. Appendix: 'Dating pre-industrial farm buildings'; 'A note on sources', p.267-72. Analytical index, p.275-9.
Class No: 631.22

[5197]
—NOTON, N.H. **Farm buildings.** Reading, College of Estate Management, 1982. [xii], 359p. illus., diagrs., plans. ISBN: 0902132733.

Has 21 sections: 1. Principles of construction ... 3. Foundations ... 6. Waste and drainage ... 9. Design considerations ... 12. Pigs ... 17. Dairy buildings ... 20. Domestic building design. Further reading, p.347. Detailed, analytical index, p.349-59. Comprehensive, with numerous photographs, 'but rarely giving more than a general view'. (*Farm buildings digest,* v.18(3), 1983, p.16).
Class No: 631.22

Agricultural Machinery

Bibliographies

[5198]
MORGAN, B. **Keyguide to information sources in agricultural engineering.** London & New York, Mansell, 1985. vi, 209p. tables. £40. (*Keyguide to information sources series.*) ISBN: 0720117208.

3 parts: 1. Survey of agricultural engineering and its literature (6 documented chapters, p.3-92) - 2. Annotated bibliography of sources of information (p.95-162: entries 1-436, with brief annotations, except for list of journals) - 3. Organizational sources of information (p.165-92; includes organizations concerned with audio-visual materials; international scope). Detailed, analytical index. An authoritative survey of types of sources.
Class No: 631.3(01)

[5199]
UNITED NATIONS. Industrial Development Organization. **Information sources on the agricultural implements and machinery industry.** New York, United Nations, 1982. 122p. £3. ISBN: 0119072092.

About 750 entries. 12 sections: 1. Addresses of professional, trade and research organizations, learned societies and special information services - 2. Directories - 3. Sources of statistics and market data - 4. Basic handbooks, textbooks and manuals - 5. Monographic series - 6. Current periodicals - 7. Current abstracting and indexing periodicals - 8. Proceedings, papers and reports - 9. Specialized dictionaries and encyclopaedias - 10. Bibliographies - 11. Films on agricultural engineering - 12. Other potential sources of information (B. Fairs and exhibitions; E. Meetings and conferences; G. Testing and laboratory services). *Class No:* 631.3(01)

Handbooks & Manuals

[5200]
Agricultural engineer's handbook. Richey, C.B., *and others, eds.* New York, etc., McGraw-Hill, 1961. 880p. diagrs., graphs, tables. ISBN: 0070526176.

4 sections (with documented chapters): 1. Crop-production equipment - 2. Soil and water conservation - 3. Farmstead structures and equipment - 4. Basic agricultural data. Index. Despite its age and US slant, 'it remains the only work of its kind in the field' (B.A. Morgan, in *Information sources in agriculture and food science.* (1981), p.222).
Class No: 631.3(035)

[5201]
CULPIN, C. **Farm machinery.** 12th ed. Oxford, Blackwell, 1992. x,444p. illus. £24.99. ISBN: 063203159x.
First published 1938.
23 documented chapters: 2. Tractors: constructional operations ... 4. Ploughs: components and features ... 7. Equipment for sowing and planting ... 10. Pumps and irrigation equipment ... 15. Root harvesting machinery - 16. Horticultural machinery ... 19. Equipment for livestock husbandry - 20. Equipment for milk production ... 22. Machinery for land drainage, reclamation and estate maintenance - 23. Environmental control in crop and stock buildings. 6 appendices (*e.g.* 5. Electricity on the farm). Index. Well illustrated and captioned. A standard textbook, written for farmers and those who advise them.
Class No: 631.3(035)

Dictionaries

[5202]
BRITISH STANDARDS INSTITUTION. **Glossary of terms relating to agricultural machinery and implements. AMD 1395. 1983.** London, BSI, 1963. 94p. illus. £52. (*BS 2468: 1963; AMD 4395: 1983.*)
About 1,000 terms defined. Sections: General definitions - Class O: Agricultural power units - 1. Equipment for sowing, planting and distributing manure - 3. Equipment for crop protection - 4. Harvesting machinery - 5. Equipment for preparing and storing produce on farms.
Class No: 631.3(038)

[5203]
Dictionnaire technique de la mécanisation agricole. Paris, Centre National d'Études et d'Experimentation du Machinisme Agricole Tropical; New York, French & European, 1968-71. 3v. (xxxvi, 957p.). $59.95. ISBN: 0828853940, US ed.
V.1-2 consist of a classified arrangement. Languages: French, English, German, Spanish and Italian. V.3 comprises the alphabetical index to each language. Wide-ranging subject field. *Class No:* 631.3(038)

[5204]
English-Chinese dictionary of agricultural machinery. Peking, Mechanical Engineering Publishing House, 1974. 440p. illus.
About 30,000 English-Chinese entries, with a 20-page Chinese-English section and 42p. of illus. captioned in both English and Chinese. *Class No:* 631.3(038)

[5205]
Machinisme et équipements agricoles. Français, anglais, allemand, espagnol, italien, portugais. 3 ème éd. Paris, La maison du dictionnaire, 1990. 1295p. FFr700. (*Dictionnaire technologique, v.8.*)
19 subject areas: General terms ... Farm machinery ... Tractors and cultivators ... Crop protection equipment ... Harvesting machinery ... Processing equipment ... Soil conservation and forestry equipment. *Class No:* 631.3(038)

[5206]
STEINMETZ, H. **Agricultural engineering:** multilingual illustrated dictionary. 5th ed. Weikersheim, Verlag Josef Margraf, 1987. 502p. illus., figs. $35; £20.50. ISBN: 382361116x.
First published 1960.
About 6,000 terms in English, German, French, Spanish

.... (contd.)
and Chinese. Coverage includes energy in agriculture, engines/tractors, aircraft, soil cultivation, plant protection, harvesting and earth moving. 1,730 line drawings.
Class No: 631.3(038)

Reviews & Abstracts

[5207]
Agricultural engineering abstracts. Wallingford, Oxon., CAB International, 1976-. 12pa. £208; $390. ISSN: 03088863.
*c.*5,000 abstracts pa. Coverage includes: Mechanical power; Land improvement; Crop production; Protected cultivation; Crop harvesting and threshing; Handling and transport; Crop processing and storage; Farm buildings and equipment; Aquaculture. Appended: Reports; Conferences; Books. Author and subject indexes per issue and annually. Online, CAB ABSTRACTS and available on CAB ABSTRACTS CD-ROM. *Class No:* 631.3(048)

Yearbooks & Directories

[5208]
The Green book. Birmingham, Guardian Communications, 1951-. Annual. ISSN: 00173932.
First published 1951 as *British tractors and farm machinery: the Green book series.*
1991/92 edition (114p. £75) has seven sections: 1990 review - Trade organizations, research and educational establishments - Alphabetical index to manufacturers and suppliers - Index to classifications - Classified section - Product analysis section, index to tables (Balers and bale handling equipment ... Cultivation equipment ... Grain and seed crop harvesting equipment ... Livestock husbandry equipment ... Power washers and steam cleaners ... Tractors, tracklaying) - Index to advertisers. At £75 for only 114p., this edition seems expensive. *Class No:* 631.3(058)

Institutions & Associations

[5209]
FOOD AND AGRICULTURE ORGANIZATION. **Directory of agricultural engineering institutions.** Rome, FAO, 1973. 537p.
A directory of agricultural engineering institutions concerned with the research, educational and growth aspects of the subject. Now somewhat dated.
Class No: 631.3:061:061.2

Farm Tools

[5210]
FUSSELL, G.E. **The Farmer's tools:** the history of British farm implements, tools and machinery, AD1500-1900. London, Orbis Pubn., 1981. 246p. illus. (pl.). ISBN: 0856133590.
First published 1952.
7 chapters: 1. Field drainage - 2. Preparing the seed bed - 3. Sowing the seed - 4. Harvesting the crops - 5. Threshing the grain - 6. Barn and miscellaneous machinery - 7. Brief retrospect. Notes, p.201-17. Chronological list of tools (1523-1904). Glossary. Bibliography, p.226-30. Detailed analytical index, p.231-46. 111 plates. *Class No:* 631.31

[5211]
PARTRIDGE, M. Farm tools through the ages. London, Osprey Publishing, 1973. 240p. illus., facsims. ISBN: 0850450810.

A readable account, with 227 line-drawings. Bibliography, p.236. *Class No:* 631.31

Soil Science

[5212]
AVERY, B.W. Soils of the British Isles. Wallingford, CAB International, 1990. xi,463p. illus. (some col.), figs., tables. £67.50. ISBN: 0851986498.

9 chapters: 1. Soil and its variability - 2. Soil surveys and soil classification in the British Isles ... 5. Brown soils - 6. Podzols ... 9. Peat soils. References, p.429-49. Index. Author is former deputy head of the Soil Survey of England and Wales. *Class No:* 631.4

[5213]
The Encyclopedia of soil science. Fairbridge, R.W. *and* Finkl, C.W., *Jr., eds.* Stroudsbourg, Pa., Dowden, Hutchinson & Ross, 1979. Part 1. xxi, 646p. illus., diagrs., graphs, tables, chemical structures. (*Encyclopedia of earth sciences, v.12.*) ISBN: 0897331763.

Part 1: *Physics, chemistry, biology, fertility and technology.* Entries: Acidity ... zero tillage. Documented articles (*e.g.* 'Humus': 1½ columns of references; 'Soil structure' (5½p.): 2 columns of references; 4 illus., 1 table). Cross-references to other volumes in the series. Subject (non-analytical) and author indexes.

Part 2, to cover soil morphology, genesis, classification and geography, is still in preparation. *Class No:* 631.4

[5214]
PLAISANCE, G. *and* **CAILLEUX, A. Dictionary of soils, French-English.** New Delhi, etc., Amerind Publishing Co., 1981. xii, 1109p. tables.

Translation of *Dictionnaire des sols* (Paris, La Maison Rustique), 1958.

Over 10,000 French terms, with etymology, definition, example of usage, etc. 'Humus' (p.521-7) includes use of the word in context, history and various types of humus, 'Granulométrie': 2½p., plus a chart. 3 appendices: 1. Symbols and measures; 2. British and American units; 3. Common abbreviations. *Class No:* 631.4

[5215]
—**LOZET, J.** *and* MATHIEU, C. Dictionnaire de science du sol. 2nd ed. Paris, Technique et Documentation - Lavoisier, 1990. 384p. illus. (pl.), figs., tables.

Lists more than 2,800 terms from soil science, micromorphology, mineralogy, petrography and geology. English equivalents are given and appendices list the American, French, German, and FAO-UNESCO soil classification schemes. 52 plates and numerous tables and figures. French-English index. *Class No:* 631.4

[5216]
Soils and fertilizers. Wallingford, Oxon., CAB International, 1938-. 12pa. £432; $795. ISSN: 00380792.

About 17,000 abstracts pa. Coverage includes: Soil chemistry; Soil physics; Soil biology; Soil fertility; Soil management; Soil classification; Soil formation; Soil conservation; Land reclamation; Irrigation and drainage; Fertilizer technology; Fertilizer use; Plant nutrition; Environmental aspects. Author and subject indexes per issue

....(contd.)
and annually. Online, CAB ABSTRACTS database, and CAB ABSTRACTS CD-ROM.

Rothamsted Experimental Station's *Soil maps* (scales, 1:25,000, 1:63,360 and 1:1,000,000) are based on the Ordnance Survey sets; the accompanying *Memoirs* ceased publication in 1977. The Soil Survey of Scotland's *Memoirs* are published by the Macaulay Land Use Research Institute. *Class No:* 631.4

[5217]
—Bibliography of soil science, fertilizers and general agronomy, 1931-62. Prepared by the Imperial Bureau of Soil Science (subsequently Commonwealth Bureau of Soils, Harpenden, Herts.). Farnham Royal, Bucks., Commonwealth Agricultural Bureaux, 1953-63. 10v.

Has a subject sequence in each volume. V.10 has 9 classes: 1. Soil pedology - 2. Cultivation; fertilizers; plant disesases and protection; land reclamation; agricultural systems - 3. Agricultural crops - 4. Orchards, fruit. Forestry, forests - 5. Vegetables - 6. Floriculture - 7. Physical geography. Meteorology. Geology - 8. Botany - 9. Miscellaneous. V.10 has *c.*13,000 entries, lists *c.*2,000 periodicals referred to, and author and subject indexes. Cross-references to abstracts in *Soils and fertilizers. Class No:* 631.4

Soil Surveys

[5218]
FOOD AND AGRICULTURE ORGANIZATION. Soil map of the world. Paris, UNESCO, with Rome, FAO, 1971-81. 10v. tables, maps.

Group 1. *General legend.* 1974. 69p. 1 sheet. 2. *North America.* 1976. xiv, 210p. 2 sheets. 3. *Mexico and Central America.* 1976, xiv, 96p. 1 sheet. 4. *South America.* 1971. xiv, 193p. 2 sheets. 5. *Europe.* 1981. 201p. 2 sheets. 6. *Africa.* 1977. xiv, 209p. 3 sheets. 7. *South Asia.* 1977. 117p. sheets 1 & 3. 8. *North and Central Asia.* 1978. 165p. 3 sheets. 9. *South-east Asia.* 1979. xiv, 149p. 1 sheet. 10. *Australasia.* 1978. xii, 221p. 2 sheets.

Sheet maps, 115 x 82cm, scale 1:5,000,000, have a 22-colour base, plotting 101 major soil groups. The accompanying texts include numerous tables and describe various soil characteristics and types. 'Valuable aid for economic development programs and agricultural planning' (Harris, C.D. ed. *Geographical bibliography for American libraries* (1985), item 620).

Also, *Soil map of the European Communities 1:1,000,000* (Luxembourg, Office for Official Publications of the EC, 1985. vii,124p. £60). *Class No:* 631.47

Plant Cultivation

[5219]
Plant growth regulator abstracts. Wallingford, Oxon., CAB International, 1975-. 6pa. £133; $247. ISSN: 03059154.

Over 2,500 abstracts pa. 11 sections: General - Growth stimulators - Auxins ... Growth retardants ... Metabolic inhibitors and plant poisons. Reports. Conferences. Books. Author and subject indexes per issue and annually. Online, CAB ABSTRACTS and available on CAB ABSTRACTS CD-ROM. *Class No:* 631.5

Propagation

Seeds

[5220]
Seed abstracts. Wallingford, Oxon., CAB International, 1978-. 12pa. £136; $251. ISSN: 01410180.
*c.*4,000 abstracts pa. Coverage includes: Seed morphology and anatomy; Seed chemistry; Seed ecology; Germination; Seed production; Seed storage and longevity; Pests and diseases; Breeding and selection; Seed processing; Seed assessment and testing; Economics and marketing; Seed regulations; Appended: Reports; Conferences; Books. Author and subject indexes per issue and annually. Online, CAB ABSTRACTS database and also on CAB ABSTRACTS CD-ROM. *Class No:* 631.531

Tending & Care

Greenhouses

[5221]
The Complete book of the greenhouse: a complete guide to the construction and maintenance of greenhouses, to the modern aids available in running them and to the care and culture of greenhouse plants. Walls, Ian G., *and others.* 4th ed. London, Ward Lock, 1988. 304p. illus. (some col. pl.), figs., tables. £20. ISBN: 0706370171.
First published 1973.
Part 1: On the design and running of greenhouses (chapters 1-17). Part 2: On the growing of plants in greenhouses (chapters 18-31). Many clearly laid out tables. An appendix of useful addresses, p.295-98. No bibliography. Analytical index. *Class No:* 631.544

Agricultural Products

[5222]
Food focus. London, Food from Britain, 1989. 2v. tables. £9.99ea. ISBN: 0902373072, v.1; 0902373080, v.2; 0902373099, set.
V.1 *A Handbook of agricultural and horticultural produce from the United Kingdom.* xii,376p. V.2. *A Guide to sources of fresh, lightly processed and speciality foods in the United Kingdom.* v,266p.
In v.1, produce is arranged A-Z (Apples-Yogurt). Data include consumption, source of supply, tonnage, value, cultivation, history, freezing, varieties. Information sources, p.368-76. V.2 lists companies and has 9 sections: 1. Meat ... 4. Dairy ... 6. Fruit, vegetables and herbs - 7. Prepared foods ... 9. Beverages. Lists of regional organizations, marketing boards, and associations. Company, geographic, and product indexes. *Class No:* 631.57

[5223]
UNITED NATIONS. Industrial Development Organization. **Information sources on bioconversion of agricultural wastes.** New York, United Nations, 1980. 84p. £4. ISBN: 0119060930.
689 numbered entries. 11 sections (1. Professional, trade and research organizations, learned societies and special information services. 2. Sources of statistics, marketing and other economic data - 4. Basic handbooks, textbooks and manuals - 5. Monograph series - 6. Current periodicals - 7. Current abstracting and indexing periodicals - 8. Proceedings, papers and reports - 9. Specialized

....(contd.)
dictionaries and encyclopaedias - 10. Bibliographies - 11. Other potential sources of information (E. Meetings and conferences; F. Patents and licences; 1. Translation services)). Bibliographical sources used ..., p.82-84. *Class No:* 631.57

Irrigation

[5224]
Advances in irrigation. Hillel, D., *ed.* Orlando, Calif., Academic Press, 1982-. v.1-. Annual. illus., graphs, tables. ISSN: 02757915.
V.4, 1987 (x, 405p.) has 9 documented articles; 22 contributors. 'On the drainage of irrigated lands under sequential water application', p.221-42, has 20 references. Detailed, analytical index, p.389-405. *Class No:* 631.67

[5225]
CRC handbook of irrigation technology. Finkel, H.J., *ed.* Boca Raton, Florida, CRC Press, 1983. 2v. ([vii], 369p. + [vii], 223p.). diagrs., graphs, tables. $119.95, v.1; $95, v.2. ISBN: 0849332311, v.1; 0849332338, v.2.
V.1 has 12 documented sections on soil-water and plant-water relationships, and on types of irrigation, *e.g.* sprinkler, dip, pumps and pumping. V.2 has 13 documented sections on economics of irrigation and types of crops of irrigation, *e.g.* sugar, oil, cereal and citrus crops. Each volume carries a detailed, analytical index. *Class No:* 631.67

[5226]
Fachwörterbuch für Bewässerung und Entwässerung: Englisch, Deutsch, Französisch, Spanisch. [Multilingual technical dictionary on irrigation and drainage: English, French, German, Spanish.] 2., erw. Aufl. Bonn, Der Verband, 1983. 1009p. illus.
First published 1971.
Contains *c.*12,000 terms in one sequence. Over 300 line drawings. English, French, German, and Spanish indexes.
Supplementary volumes in French, Spanish, Arabic, Indian, Japanese, Persian and Portuguese. *Class No:* 631.67

[5227]
INTERNATIONAL COMMISSION ON IRRIGATION AND DRAINAGE. British Section. **United Kingdom register of research on irrigation, drainage and flood control,** 1986. Wallingford, Oxon., Hydraulics Research, [1987].
Details, with abstracts, of 190 projects. 2 main sections: 1. Irrigation - 2. Drainage and flood control - 3. General (*e.g.* legal aspects). Indexes for researchers, institutions, subjects, directors of research. *Class No:* 631.67

[5228]
Irricab: current annotated bibliography of irrigation. Bet Dagan, Israel, International Irrigation Information Centre (IIIC) 1976-. 4pa. $95pa. ISSN: 03165083.
Cites *c.*4,000 publications pa. 9 subject sections (Irrigation water quality ... Irrigation policy, planning and development). Appended; Books and conferences. Miscellaneous. Author, subject and geographic indexes per issue and annually. *Class No:* 631.67

[5229]
—INTERNATIONAL COMMISSION ON DRAINAGE AND FLOOD CONTROL. Bibliography on irrigation, drainage, river training and flood control. New Delhi, ICID, 1956-. Annual. $10 per issue. ISSN: 0523302x.

Carries entries usually annotated, classified by UDC. Author index. *Class No:* 631.67

[5230]
Irrigation and drainage abstracts. Wallingford, Oxon., CAB International, 1975-. 4pa. £102; $185. ISSN: 03067327.

About 4,000 abstracts pa. 11 sections: Water-management - Irrigation - Irrigation of crop plants ... Plant-water relations - Salinity and toxicology problems ... Other topics (*e.g.* aquatic weeds). Reports. Conferences. Books. Author and subject indexes per issue and annually. Online CAB ABSTRACTS database and available on CAB ABSTRACTS CD-ROM. *Class No:* 631.67

Fertilizers

[5231]
Fertilizer dictionary: [English-French-Spanish-Arabic]. London, British Sulphur Corporation; Arab Federation of Chemical Fertilizer Producers, 1982. 132p.

Terms are arranged alphabetically within general topics, *e.g.* 'Raw materials'; 'Plant operation'; 'Fertilizer handling'. 9 appendices, *e.g.* 'Fertilizer minerals'; 'Nutrient factors'; 'Abbreviations'. *Class No:* 631.8

[5232]
Soils and fertilizers. Wallingford, Oxon., CAB International, 1938-. 12pa. £432; $795. ISSN: 00380792.

About 17,000 abstracts pa. Classes: 0. Documentation - 1. Soil science - 2. Fertilizers. Soil management ... 6. Other agricultural topics ... 8. Botany. Ecology - 9. Other topics. Author and subject indexes per issue and annually. Online CAB ABSTRACTS database and CD-ROM. *Class No:* 631.8

[5233]
UNITED NATIONS. Industrial Development Organization. **Information sources on the fertilizer industry.** New York, United Nations, 1976. xii, 96p. £4. (*UNIDO Information sources no.21.*) ISBN: 0119047284.

777 numbered entries. 12 sections (1. Professional, trade and research organizations, learned societies, and special information services ... 3. Sources of statistics, marketing and other economic data ... 5. Current periodicals ... 11. Films and film catalogues - 12. Other potential sources of information: B. Fairs and exhibitions; C. Financing and investing; D. Legislation; H. Safety and health measures). 'Bibliographic sources used'. Entries in some sections are annotated. *Class No:* 631.8

Pests & Diseases

Bibliographies

[5234]
Bibliography of plant viruses and index to research. Beale, H.P., *comp. and ed.* New York, Columbia Univ. Press, 1976. xiii, 1495p. $163.50. ISBN: 0231037635.

More than 29,000 references, under authors, A-Z, then chronologically. Periodical articles, p.1892-1970. Two extensive subject indexes (one, by virus genus and host; the

....(contd.)
other, by individual virus, - with over 2,700 entries for 'Tobacco mosaic virus' alone), p.1453-95. For specialists in plant virology. *Class No:* 632(01)

[5235]
Pest management: a directory of information sources. Volume 1: Crop protection. Hamilton, C.J., *ed.* Wallingford, Oxon., CAB International, 1991. 331p. £29.50. ISBN: 0851986757.

First volume in a projected 3: v.2 will cover animal husbandry, v.3, public health. Sections: Books - Primary literature - Secondary sources - Computerized information retrieval - Libraries and information centres (arranged by country, Angola ... Zimbabwe) - Suppliers - Regulatory bodies. Aimed at everyone involved in crop pest management. *Class No:* 632(01)

Handbooks & Manuals

[5236]
European handbook of plant diseases. Smith, I.M., *and others, eds.* Oxford, Blackwell Scientific Publications, 1988. 583p. £65. ISBN: 0632012226.

Emphasizes diseases affecting crops which are economically important though includes garden plants which are commercially grown. Omits wild plants. Arranged by pathogen which are presented alphabetically within each chapter. Chapters cover viruses, rickettsia, bacteria, oomycetes, ascomycetes, chytridiomycetes and basiomycetes. Each entry contains a description of the pathogenic organism, host plants and diseases, epidemiology, geographic distribution, economic impact, and methods of control. Two indexes: host and pathogen. *Class No:* 632(035)

[5237]
NYVALL, R.F. **Field crop diseases handbook.** 2nd ed. New York, Van Nostrand Reinhold, 1989. 912p. $104.95. ISBN: 0442267223.

Covers *c.*1,200 diseases of 25 field crops grown throughout the world. Looks at causes, distribution, symptoms and control of major crop diseases. *Class No:* 632(035)

[5238]
SCOPES, N. **Pest and disease control handbook.** Stables, L., *ed.* 3rd ed. Thornton Heath, British Crop Protection Council, 1989. xiii, 732p. ISBN: 0948404280.

First published 1979.

16 chapters, by various hands. Chapters 5-15 concern pests and diseases of different types of crop plants (*e.g.* chapter 10, - vegetables, p.261-321; 31 references). Chapter 16: 'Insect, mite and fungal pests of stored cereals and oilseed crops'. Detailed, analytical index. A mine of authoritative information. *Class No:* 632(035)

Dictionaries

[5239]
HOLLIDAY, P. A **Dictionary of plant pathology.** Cambridge, University Press, 1989. xvi,369p. £32.50. ISBN: 052133117x.

*c.*8,000 entries. Lists authoritative names of pathogens with brief descriptions and bibliographic references. Entries also include references and notes on plant pathologists, names of diseases, crops and their pathology, terminology,

....*(contd.)*

names of disorders, taxonomic groups, fungicide names, vectors of viruses and toxins. 'An excellent dictionary with a wealth of bibliographic information' (*Science and technology libraries,* v.10(4) Summer 1990, p.120).
Class No: 632(038)

Reviews & Abstracts

[5240]
Biocontrol news and information. Wallingford, Oxon., CAB International, 1980-. 4pa. £97; $174.

A quarterly journal of news items, review articles and abstracts. About 700 items per issue. Sections: Field and horticultural crops - Plantation and orchard crops - Forest and shade trees - Ornamental plants - Stored products - Vertebrate pests - Useful insects - Medical and veterinary - Weeds - Integrated control - Techniques - Taxonomy and catalogues - Biology - Ecology - General - Reports - Conferences - Books. Author and subject indexes per issue, cumulated annually. Online CAB ABSTRACTS and available on CAB ABSTRACTS CD-ROM.
Class No: 632(048)

[5241]
Review of plant pathology. Wallingford, Oxon., CAB International, 1970-. 12pa. £277; $508. ISSN: 00346438.

Formerly *Review of applied mycology,* 1922-1969.

About 8,000 abstracts pa; preceded by review article. Covers journal articles, reports, conferences and books on subjects such as diseases of crop plants; non-parasitic diseases; mycorrhizas, and includes aspects of taxonomy; morphology; genetics; control; fungicides and antibiotics; physiology; molecular biology. Author and subject indexes per issue and annually. Online CAB ABSTRACTS and available on CAB ABSTRACTS CD-ROM.
Class No: 632(048)

Thesauri

[5242]
Thesaurus of agricultural organisms: pests, weeds and diseases. Derwent Publications, *ed.* London, Chapman and Hall, 1990. 2v. £225. ISBN: 0412372908.

V.1. *A to M.* 802p. V.2. *N to Z; Index of inverted species names.* 728p.

Contains main entries for *c.*17,000 organisms, with 80,000 synonyms or common names. In addition to weeds, pests and disease microorganisms, there is also extensive coverage of crop species. Each main entry includes higher taxa, Latin synonyms and the most frequently used common names in English, French and German. *Class No:* 632:025.43

Weeds

[5243]
Elsevier's dictionary of weeds of Western Europe, their common names and importance, in Latin, Danish, German, English, Spanish, Finnish, French, Icelandic, Italian, Dutch, Norwegian, Portuguese and Swedish. Williams, G., *comp.* Amsterdam, etc., Elsevier, 1982. [ix], 320p. £101.87. ISBN: 0444419780.

1,043 numbered entries in Latin, with equivalents and indexes in the other 12 languages. 25 contributors.
Class No: 632.51

[5244]
Interdisciplinary dictionary of weed science. Williams, G.H. *and* Zweep, W. van der, *eds.* Wageningen, Centre for Agricultural Publishing and Documentation (PUDOC), 1990. 546p.

Contains *c.*3,000 technical and scientific forms in English, Danish, German, Spanish, French, Italian, Dutch and Portuguese. Prepared under the auspices of the European Weed Research Society. *Class No:* 632.51

Poisonous Plants

[5245]
COOPER, M.R. *and* **JOHNSON, A.W. Poisonous plants in Britain and their effects on animals and man.** London, HMSO, 1984. xiv, 305p. illus. (incl. col.). £12.95. (*Reference book 161.*) ISBN: 0112425291.

At head of title-page 'Ministry of Agriculture, Fisheries and Food'. Replaces Bulletin 161, *British poisonous plants* (last reprinted 1979).

Introduction - Poisonous principles - Fungi - Pteridophytes - Coniferae - Plant families, A-Z (p.63-238; subheadings; poisonous principles; poisoning in animals; human poisoning). Plants affecting milk. Glossary. Bibliography (of books), p.249-50. References, p.251-95 (plants grouped by families). Detailed index, p.297-308. 'It should prove of value to the medical and veterinary professions, students, farmers and the public in general' (*Foreword*).
Class No: 632.52

Insects (Pests)

[5246]
Review of agricultural entomology. Wallingford, Oxon., CAB International, 1990-. 12pa. £275; $517. ISSN: 09576762.

Formerly *Review of applied entomology. Series A: Agricultural,* 1913-89.

About 12,000 abstracts pa. Coverage includes: Taxonomy; Ecology and behaviour; Diseases and disorders; Morphology; Biological control; Geographical distribution; Chemicals, including toxicity tests and environmental effects. Books. Reports. Conferences. Author and subject indexes per issue and annually. Online, CAB ABSTRACTS and available on CAB ABSTRACTS CD-ROM.
Class No: 632.7

Pesticides

[5247]
The Agrochemicals handbook. Kidd, H. *and* Hartley, D., *eds.* 3rd ed. London, Royal Society of Chemistry, 1991. *c.*1300p. Loose-leaf. £199. ISBN: 0851864643.

First published 1983. Translation of *Wirksubstanzen der Pflanzenschutz - und Schadlingsbekampfungsmitte,* by W. Perkow (1971, 1983).

Comprehensive data and information on agrochemical products used in crop protection and pest control, Acephate ... Zircom. Includes herbicides, fungicides, insecticides, nematicides, acaricides and rodenticides.

The *Handbook* is available online via DIALOG and DATA-STAR. Available on CD-ROM. *Class No:* 632.95

[5248]

Crop protection chemicals reference. 7th ed. New York, Chemical and Pharmaceutical Press, 1991. 2200p. tables. $105. ISBN: 0471532827.

First published 1985.

Contains information on *c.*600 crop protection substances in use in the US. Data include precautionary statements; directions for use; storage and disposal: general information; active ingredients; application rates. Brand name quick index (yellow pages); manufacturer index (white pages); common and chemical name index (pink pages); product category index (blue pages); crop and non-crop use index (orange pages); pest use index (green pages). 3 appendices include information on calibrations and conversions; poison control centres; state extension pesticide co-ordinators; safety information. *Class No:* 632.95

[5249]

European directory of agrochemical products. (EDAP). Kidd, H. *and* James, D., *eds.* 4th ed. London, Royal Society of Chemistry, 1991. 4v. £285, set; £96ea. ISBN: 0851869335, v.1; 0851869432, v.2; 085186953x, v.3; 0851869637, v.4; 0851869734, set.

First published 1984.

1. *Fungicides.* 700p. 2. *Herbicides.* 700p. 3. *Insecticides and acaricides.* 700p. 4. *Plant growth regulators, etc.* 400p.

EDAP lists over 26,000 products from 25 European countries. Data include registered users, toxicity, limitations to use, active ingredient proportions, preharvest intervals. Appended directory of European companies that market pesticides, a products index, and an active ingredients index. Online via DATA-STAR and DIALOG. *Class No:* 632.95

[5250]

HILL, D.S. Agricultural insect pests of temperate regions and their control. Cambridge Univ. Press, 1987. [viii], 659p. illus., tables, maps. £75. ISBN: 0521240131.

10 chapters: 2. Pest ecology - 3. Principles of pest control - 4. Methods of pest control - 5. Pest damage to crop plants - 6. Biological control of crop pests - 7. Pesticide applications - 8. Pesticides in current use - 9. Major temperate crop pests - 10. Major temperate crops and their test spectra. General bibliography, p.597-610. 4 appendices (B. Glossary of terms used in applied entomology and crop protection.). Detailed, analytical index, p.625-59.

Agricultural insect pests of the tropics and their control, by D.S. Hill (2nd ed. CUP, 1983. £55) is the companion volume. *Class No:* 632.95

[5251]

The Pesticide manual: a world compendium. Worthing, C.R., *ed.* 9th ed. Farnham, Surrey, British Crop Protection Council, 1991. xlvii,1141p. tables, chemical structures. £65. ISBN: 0948404426.

First published 1968.

Details 670 chemical compounds or microbial agents used as active ingredients of pesticides. Part 1: Compounds in use: main entries (pesticides, A-Z; data: nomenclature; development; properties; formulations; uses; toxicology; analysis). Part 2: Superseded compounds (yellow pages). 4 appendices: Abbreviations; Bibliography; Names and addresses of firms mentioned in the text; Notes on common names, chemical nomenclature and structures. 4 indexes: *Chemical abstracts* Service Registry Numbers; Molecular formulae; Code numbers; Chemical, common and trivial names and trademarks. *Class No:* 632.95

[5252]

UNITED NATIONS. Industrial Development Organization. **Information sources on the pesticides industry.** Rev. ed. New York, United Nations, 1982. 90p. £4. (*UNIDO guides to information sources, no.10.*)

First published 1974.

Contains *c.*700 entries. 12 sections (1. Professional, trade and research organizations, learned societies and special information services ... 7. Current abstracting and indexing periodicals ... 9. Specialized dictionaries and encyclopaedias ... 11. Films - 12. Other potential sources of information (A. Counselling and engineering services; G. Safety and health measures; H. Standards and specifications). 'Bibliographical sources used ...' Entries in some sections are annotated. *Class No:* 632.95

[5253]

World directory of pesticide control organisations. Kidd, H., *and others, eds.* Cambridge, Royal Society of Chemistry, 1989. v,311p. ISBN: 0851867235.

Revised edition of *Control of pesticide applications and residues in food,* by B. Hofsten and G. Ekstroöm (Swedish Science Press, 1986).

Includes 132 countries and 1,200 contact addresses. 2 sections: 1. Pesticides and international organizations - 2. Pesticides and national authorities. Data for each country include: main authority responsible for registration and use of pesticides; main authority responsible for establishing maximum pesticide residue limits; Codex contact point; national association of pesticide manufacturers. An index to the organizations would make this publication more useful. *Class No:* 632.95

Weedkillers

[5254]

Weed control handbook. Hance, R.J. *and* Holly, K.K. 8th ed. Oxford, etc., Blackwell Scientific Publications, 1978-90. 2v. illus., tables.

1. *Principles.* 8th ed., 1990. viii,582p. £59.50 2. *Recommendations, including plant growth regulators.* 8th ed., 1978. [xv], 332p.

V.2 has 14 chapters (*e.g.* 1. The control of annual weeds in cereal crops; 3. The control of weeds in perennial crops, flowers and greenhouses; 10. The control of aquatic weeds). 5 appendices. (1. Glossary of technical terms; 4. The definition of soil types). Extensive, detailed and analytical index, p.444-532. *Class No:* 632.954

Field Crops

[5255]

Crop physiology abstracts. Wallingford, Oxon., CAB International, 1975-. 12pa. £226; $412. ISSN: 03067556.

About 5,000 abstracts pa. Coverage includes: Germination; Growth and senescence; Reproductive development; Metabolism; Water relations; Photosynthesis and respiration; Nutrition; Stomatal movement. Author and subject indexes per issue and annually. Online CAB ABSTRACTS database and available on CAB ABSTRACTS CD-ROM. *Class No:* 633

[5256]
FACCIOLA, S. **Cornucopia: a source book of edible plants.**
Vista, Calif., Kampong Publications, 1990. 677p. $37.75.
ISBN: 0962808709.

Identifies 3,000 food plants. 3 sections: Section 1 lists species by plant family, gives the plant product, its uses, distribution, bibliographic citations and source; Section 2 lists cultivars and includes data on days to maturity, important features, bibliographic citations and source; Section 3 includes the names and addresses of 1,350 firms who can supply these plants. Bibliography and indexes to vernacular names, usage, species, families and genera. US-slanted. *Class No:* 633

[5257]
Field crop abstracts. Wallingford, Oxon., CAB International, 1948-. 12pa. £388; $733. ISSN: 0015069x.

About 10,000 abstracts pa. Sections: Cereals - Legumes - Root crops - Fibre plants - Oil plants - Miscellaneous crops - Green manures - Field crops, general - Crop botany - Weeds - Pests and diseases - Surveys and land use - Farming systems - Soil and water conservation - Agricultural meteorology - Miscellaneous. Appended: Reports. Conferences. Books. Author and subject indexes per issue and annually. Online CAB ABSTRACTS database and CD-ROM. *Class No:* 633

[5258]
LANGER, R.H.M. *and* HILL, G.D. **Agricultural plants.**
2nd ed. Cambridge, University Press, 1991. xiii,387p. diagrs., tables. £45; £17.95. ISBN: 0521224500.

First published 1982.

17 documented chapters (1. World population and crop production - 2. Plant structure - 3. Liliaceae (with 5 references) ... 15. Papaveraceae ... 17. Physiological basis of yield). Numerous line-drawings. Index. 'Aims to provide a description of important crop and pasture plants, illustrated wherever possible as a basis for a foundation course in agricultural botany or plant science' (*Preface* to 1st ed.).
Class No: 633

[5259]
Plant breeding abstracts. Wallingford, Oxon., CAB International, 1930-. 12pa. £447; $837. ISSN: 00320803.

About 1,000 abstracts per issue. Coverage includes: Breeding; Genetic resources; Varietal trials and new varieties; Cytology and ploidy; Reproductive behaviour; Evolution and taxonomy; Stress tolerance; Quality and yield components. Occasional review articles. Author, subject and significant varieties indexes per issue and annually. Online, CAB ABSTRACTS database, and also available on CAB ABSTRACTS CD-ROM. *Class No:* 633

[5260]
ROUGEMONT, G.M. de. **A Field guide to the crops of Britain and Europe.** London, Collins, 1989. 367p. illus., figs., tables. £14.95. ISBN: 0002197138.

Arranged by family and covering the economic plants of Britain and Europe. Data include a brief introduction to each family; description; names, including Latin synonyms and the vernacular names in some European languages; uses; origin, distribution and cultivation; similar plants. Husbandry is not included except where it is an aid to identification. 42 colour plates and many small maps showing plant distribution. Glossary, meanings of some Latin specific epithets common in cultivated plants and indexes of English and scientific names. *Class No:* 633

Tropics

[5261]
LÖTSCHERT, W. *and* BEESE, G. **Collins' guide to tropical plants.** London, Collins, 1983. 256p. illus. (incl. col.), tables, maps. ISBN: 0002191121.

German original, *Pflanzen der Tropen* (BLV Verlagsgesellchaft, 1981).

Introduction - Ornamental plants - Economic plants - Colour plates (p.81-176). 'Some recommended botanic gardens and parks in the tropics', p.247. Glossary. Index of scientific names; index of English common names. 325 plants described; 274 col. illus. (*e.g.* 'Oil palm': 1p., 2 col. illus.). Deals mainly with decorative and economic plants common in all regions of the tropics. *Class No:* 633(213)

Cereals & Grain

[5262]
Advances in cereal science and technology. St. Paul, Minn., American Association of Cereal Chemists, Inc., 1980-. Annual. ISBN: 0913250511.

V.9, 1988 (xx, [i], 345p.) has 14 contributors. Papers include 'Crispness of cereals', p.1-19 (with 32 references) and 'Immunochemistry of cereal grain storage proteins', p.263-325 (with 14p. of references). Detailed, analytical index, p.339-45. *Class No:* 633.1

Maize

[5263]
Maize abstracts. Wallingford, Oxon., CAB International, in association with International Maize and Wheat Improvement Center, 1985-. 6pa. £200; $367. ISSN: 02672987.

*c.*3,500 abstracts pa. Coverage includes: Plant breeding and genetics; Plant physiology; Soil science; Pests and diseases; Crop science, Seeds and grains; Nutrition and quality. Author and subject indexes per issue and annually. Online, CAB ABSTRACTS database, and available on CAB ABSTRACTS CD-ROM. *Class No:* 633.15

Rice

[5264]
Rice abstracts. Wallingford, Oxon., CAB International, in association with the International Rice Research Institute, 1978-. 4pa. £131; $242.

*c.*3,200 abstracts pa. Coverage includes: Breeding and selection; Agronomy; Fertilizers; Weeds; Pests; Climate and environment; Nitrogen fixation; Growth; Harvesting and storage; Food technology; Nutrition and utilization; Economics; Rural development and land use; Techniques; Appended: Reports. Conferences. Books. Author and subject indexes per issue. Online, CAB ABSTRACTS database, and available on CAB ABSTRACTS CD-ROM. *Class No:* 633.18

Forage Grasses

[5265]
Herbage abstracts. Wallingford, Oxon., CAB International, 1931-. 12pa. £220; $414. ISSN: 00180602.

About 4,200 abstracts pa. Coverage includes: Management of cultivated grasslands; Fodder conservation; Grassland ecology; Plant physiology; Composition and nutritive value. Reports. Conferences. Books. Author and

....(contd.)

subject indexes per issue and annually.

Online, CAB ABSTRACTS database, and available on CAB ABSTRACTS CD-ROM. *Class No:* 633.2

Legumes

Soyabeans

[5266]

Soyabean abstracts. Wallingford, Oxon., CAB International, 1978-. 6pa. £108; $196. ISSN: 01410172.

About 2,500 abstracts pa. 16 sections: Breeding and selection ... Agronomy ... Physiology and biochemistry ... Nutrition and utilization ... Fertilizers - Pests and diseases ... Harvesting - Economics. Author and subject indexes per issue and annually.

Online, CAB ABSTRACTS database, and available on CAB ABSTRACTS CD-ROM. *Class No:* 633.34

Root Crops

Potatoes

[5267]

Potato abstracts. Wallingford, Oxon., CAB International, 1976-. 6pa. £114; $209. ISSN: 03087344.

About 1,800 abstracts pa. Coverage includes: Breeding and selection; Varieties and varietal resistance; Agronomy; Fertilizers; Weeds; Pests and diseases ... Harvesting; Storage and quality; Techniques. Author and subject indexes per issue and annually.

Online CAB ABSTRACTS database and CD-ROM. *Class No:* 633.491

[5268]

Potato terms: trilingual dictionary of the potato. Loon, C.D. van *and* Heij, D.G. van der, *eds.* Wageningen, Pudoc, Centre for Agricultural Publishing and Documentation, 1989. xv,402p. ISBN: 9022009629.

Rev. ed. of *Dictionary of technical terms relating to the potato* (1969).

Contains *c.*3,500 terms in the official languages of the European Association for Potato Research - English, German, and French. Latin index, p.398-402. Clearly arranged and presented. *Class No:* 633.491

Textile Plants

Cotton

[5269]

Cotton and tropical fibres: annotated bibliography. Wallingford, Oxon., CAB International. Annual. £27.50; $51.

Over 1,000 abstracts pa. Coverage includes: Tropical fibres; Cotton; Sisal; Jute; Kenaf; and roselle; Other fibres.

Online, CAB ABSTRACTS database, and CD-ROM. *Class No:* 633.51

Sugar

[5270]

WITTE, G. 'Glossary of technical terms for the sugar industry'. In *Zuckerindustrie,* v.16(5), 1991. p.433-43.

About 250 German terms, in current use, are listed in 7 groups (general; beet; juice; sugar; by products; auxiliaries; effluents). Alphabetical list, with definitions, follows. Final section lists 79 non-preferred terms, with corresponding preferred terms. *Class No:* 633.6

Oil-Yielding Plants

[5271]

Tropical oil seeds: annotated bibliography. Wallingford, Oxon., CAB International. Annual. £26.50; $49. ISSN: 03082962.

About 1,000 abstracts pa. Coverage includes: Tropical oil seeds; Groundnuts; Safflowers; Coconuts; Oil palms; Castor; Sesame; Other oilseed crops and byproducts; Seed oils. Online, CAB ABSTRACTS database, and CD-ROM. *Class No:* 633.85

Horticulture

Handbooks & Manuals

[5272]

The Horticulturalist's handbook. London, Duncan Publishing, 1982-. 2 binders. Loose-leaf.

7 sections: 1. Management and finance - 2. Labour - 3. The land in relation to horticulture - 4. Buildings and equipment - 5. Marketing - 6. Crop production - 7. Research, development and advice. Index. Periodic instalments. *Class No:* 634(035)

[5273]

ROYAL HORTICULTURAL SOCIETY. The Royal Horticultural Society colour chart. London, the Society, [1966]. 28p.

Replaces the 1938-42 *Horticultural colour chart.* The primary aim is to provide a standard reference source for the colour specification of flowers, fruit and leaves, for matching purposes (especially in the case of flowers). 808 colour patches. *Class No:* 634(035)

Dictionaries

[5274]

BOURKE, D. O'D. French-English horticultural dictionary with English-French index. 2nd ed. Wallingford, CAB International, 1989. viii,240p. £28. ISBN: 085198626x.

First published 1974.

Contains *c.*19,000 terms from a wide range of horticultural literature. This ed. takes into account new developments in horticulture and includes terms from areas such as pollution, marketing, tissue culture, and greenhouse construction. *Class No:* 634(038)

[5275]

Elsevier's dictionary of horticulture in English, French, Dutch, German, Danish, Swedish, Spanish, Italian, Latin. Compiled under the auspices of the Ministry of Agriculture and Fisheries at The Hague, Netherlands. Amsterdam, etc., Elsevier, 1970-. xvi, 561p. $161.75. ISBN: 0444408126.

First published as *Woordenlijst voor der tuinbouw in seven*

....(contd.)

talen ..., by the Dutch Ministry of Agriculture, Horticultural Division, 1955.

4,240 English-base terms, with equivalents and indexes in the other 5 languages, for general horticultural vocabulary. *Library journal*, (v.95(15), 1 September 1970, p.2787) notes arbitrary selection of entries and inadequate cross-references, while querying reliability of translations into languages other than Dutch and German.

Also, *Elsevier's dictionary of horticultural and agricultural plant production* (1990. 818p. $205.25), which covers the same languages as those in *Elsevier's dictionary of horticulture* with the addition of Portuguese. *Class No: 634(038)*

[5276]

SOULE, J. Glossary for horticultural crops. Sponsored by the American Society for Horticultural Science. New York, etc., Wiley, 1985. xxvi, [1], 898p. $58.50. ISBN: 0471884995.

About 8,000 terms are defined. 6 broad categories (subdivided): 1. Horticultural crops - 2. Morphology and anatomy - 3. Horticultural taxonomy and plant breeding - 4. Horticultural physiology and crop ecology - 5. Propagation and nursery handling - 6. Post-harvest handling and marketing. Selected references, p.679-92 (authors, A-Z). Index of terms; index of crops. Over 300 line-drawings. Well constructed, but typography makes for reading difficulties. *Class No: 634(038)*

[5277]

STEINMETZ, H., ed. Gartenbautechnik. Mehrsprechenbild-wörterbuch. [Horticultural techniques and implements: multilingual illustrated dictionary ...] Betzdorf/Sieg, Steinmetz Verlag, 1978. 568p. illus.

4,300 German-base terms, with English, French, Spanish, Italian and Dutch equivalents, and indexes in 6 languages. Sectionalized, covering glasshouses, machinery, tools and techniques. 1,330 line-drawings. *Class No: 634(038)*

Reviews & Abstracts

[5278]

Horticultural abstracts. Wallingford, Oxon., CAB International, 1931-. 12pa. £421; $786. ISSN: 00185280.

About 12,500 abstracts pa. 8 main sections: General aspects of research and its application - Temperate tree fruits and nuts - Small fruits - Viticulture - Vegetables; temperate, tropical and greenhouse - Ornamental plants - Minor industrial crops - Subtropical and tropical fruit and plantation crops. Author and subject indexes per issue and annually. Online, CAB ABSTRACTS database, and CD-ROM. *Class No: 634(048)*

[5279]

Ornamental horticulture. Wallingford, Oxon., CAB International, 1975-. 6pa. £110; $199.

About 2,400 abstracts pa. Coverage includes: Orchids; Conifers; Roses; Trees and shrubs; Bulbs and tubers; Lawns and sports turf; Aquatic plants; Foliage and succulent plants. Reports; Conferences; Books. Author and subject indexes per issue and annually. Online, CAB ABSTRACTS database, and CAB ABSTRACTS CD-ROM. *Class No: 634(048)*

Yearbooks & Directories

[5280]

The Plant finder. Philip, C., *comp.* 6th ed. Headmain Ltd., Whitbourne, Worcs., for the Hardy Plant Society, 1992. 733p. maps. £10.99. ISBN: 0861902165.

First published 1987.

Lists more than 55,000 plants (A-Z) available from over 500 nurseries. Plant directory, p.27-566. Code, nursery index; Nursery code index. Additional nursery index. Bibliography references (books and periodical articles, p.686-96). Display advertisements. *Class No: 634(058)*

America — North

[5281]

Hortus third: a concise dictionary of plants cultivated in the United States and Canada. Compiled by the staff of the Library Hyde Bailey Hortorium. New York, Macmillan Co., 1974. $150. ISBN: 0025054708.

Originally planned as a revision of *Hortus second*, compiled by L.H. and E.Z. Bailey (1941).

Entries under botanical names, stating sources of names and noting cultivation and use. Glossary of botanical names. Index of 10,408 common plant names, and list of botanists cited. *Class No: 634(71+73)*

[5282]

North American horticulture: a reference guide. Ellis, B.W., *ed.* New York, Scribner's, 1982. 367p. ISBN: 0684176041.

A directory of organizations, government programs, nomenclature authorities, garden and reference information, in chapter form. Within most chapters arrangement is under states, A-Z (less full on Canada). Indexed by subject, organization and botanical name. 'The most complete directory of North American horticultural organizations and programs currently available in one volume' (*RQ*, v.22(3), Spring 1983, p.309). A new edition is in preparation. *Class No: 634(71+73)*

USA

[5283]

EVERETT, T.H. The New York Botanical Garden illustrated encyclopedia of horticulture. New York & London, Garland, 1980-82. 10v. (*c.*4,000p.). illus. (incl. col.). $1,070. ISBN: 0824072227.

Detailed descriptions of more than 3,600 genera and over 26,000 species. Entry is under botanical name, with cross-references from common name. Longer articles concern such matters as fertilisers, propagation, pests and diseases. Well illustrated, mostly in black-and-white photographs. For English gardens, prefer *Marshall Cavendish Illustrated encyclopedia of gardening* (*qv*). *Class No: 634(73)*

Australia

[5284]

WRIGLEY, J.W. Australian native plants: a manual for their propagation, cultivation and use in landscaping. Sydney, Angus and Robertson (Australia), 1990. 623p. illus. (some col.). £46.95; $90. ISBN: 0002164167.

First published 1983.

16 chapters (3. Propagation; 7. Pests and diseases; 8. Ground cover; 10. Water features; 11. Shrubs; 12. Trees; 13. Climbers; 14. Container plants (with glasshouse plants);

....(contd.)

15. Annual and bedding plants; 16. Plants for special purposes (*e.g.* windbreaks)). Glossary. Brief bibliography. Index. *Class No:* 634(94)

Institutions & Associations

[5285]

Horticultural research international: directory of horticultural research institutes and their activities in 63 countries. Borg, H.H. van der *and* Koning van der Veen, M., *eds.* 4th ed. Wageningen, International Society for Horticultural Studies, 1986. 903p. maps. $175. ISBN: 9066053321.

First published 1966.

Entries for 1,250 research institutes, under countries (Argentina ... Zimbabwe), p.1-804. Details mention senior staff by name and outline specializations, projects and programmes. Index of places; index of research staff. Omits Eastern European countries. *Class No:* 634:061:061.2

Fruit & Vegetables

[5286]

Food from your garden: all you need to know to grow, cook and preserve your own fruit and vegetables. London, etc., Reader's Digest, 1977 (reprinted with amendment, 1987). [v], 380p. illus. £19.95. ISBN: 0274001702.

20 contributors and advisers. Sections: A basic guide to the kitchen - Growing and cooking - Pests and diseases - Home preserving - Food from the countryside - Making your own wine - Keeping poultry and bees. Index, p.375-80. Oblong format. A well-illustrated step-by-step guide. *Class No:* 634.1/635.1

[5287]

—Encyclopedia of vegetable gardening: the complete growing and freezing guide to vegetables, herbs & fruit. Biggs, A.G., *and others.* London, Octopus Books, Ltd., 1977. 255p. illus. (incl. col.) tables.

Has 3 main sections: Growing your own vegetables, by A.G. Biggs - Fruit in your garden, by A.G. Healey. - In the kitchen: freezing, preserving and cooking [50 recipes], by S. Allworthy and C. Ellwood. Introduction includes notes on tools and equipment. General index, p.248-52; Index to cooking. 375 col. illus., tending to swamp the text. Quarto. *Class No:* 634.1/635.1

[5288]

JOHNS, L. *and* **STEVENSON, V. The Complete book of fruit.** London, etc., Angus & Robertson, 1979. [vi], 309p. illus. (incl. col.), chart. ISBN: 0207143374.

4 sections: 1. Cultivation and harvest home (including world climate chart) - 2. The good cook's compendium - 3. The fruits of the earth (p.46-280: Akebia ... Wineberry) - 4. Less common fruit (*e.g.* capote). Glossary, p.293-6. Bibliography, p.297-301 (general; official bulletins and leaflets). Detailed index, p.303-9. International in scope. *Class No:* 634.1/635.1

[5289]

Mehrsprächiges Verzeichnis der Gemüsenamen. [Multilingual list of vegetable names.] Sahrmuller, R., *and others.* Grossberren, Institut für Gemüseproduktion Grossberren der Akademie der Landwirtschaftswissenschaften der DDR, 1990. 205p.

Lists more than 13,000 vegetable names in more than 60 languages, alongwith their scientific names in Latin. Includes plants used as vegetables throughout the world as well as those which are consumed locally, and plants which are considered worth cultivating. *Class No:* 634.1/635.1

[5290]

SAMPSON, J.A. Tropical fruits. 2nd ed. Harlow, Essex, Longman, 1986. ix, [i], 336p. illus., diagrs., graphs, tables. £37.50. (*Tropical agriculture series.*) ISBN: 0582404096.

First published 1980.

11 chapters (3. Botany of tropical fruits - 4. Crop husbandry - 5. Citrus (p.78-138; 4½p. of references) - 6. Bananas and plantains - 7. Pineapple - 8. Mango - 9. Avocado - 10. Papaya - 11. The minor tropical fruits). Systematic treatment of chapters 5-9. Stressing cultivation. 3 appendices (1. List of families and genera of fruit crops; 2. Common names of fruit crops and their botanical equivalents). Crop index. Geographical index. General index (1p.). *Class No:* 634.1/635.1

[5291]

SIMMONS, A.F. Simmons' manual of fruit: tree, bush, cane and other varieties. Newton Abbot, Devon, David & Charles. 1978. 237p. ISBN: 0715376071.

Tree fruits (apples ... quinces), p.11-461 - Bush fruits (blueberries ... Worcesterberry) - cane fruits (brambleberries ... wineberry) - Vine gruits (grape) - Small plant fruits (*e.g.* cranberry) and miscellaneous (*e.g.* garden huckleberry). Data: period; shape; flower... health; cropping; historical notes). No index or references. *Class No:* 634.1/635.1

[5292]

SPLITTSTOESSER, W.E. Vegetable growing handbook: organic and traditional methods. 3rd ed. New York, Van Nostrand Reinhold, 1990. xii, illus., figs., tables. $46.95. ISBN: 0442239718.

9 documented chapters: 1. Planning the garden - 2. Plant growth - 3. Soil and plant nutrition - 4. Pest control - 5. Harvest and storage of vegetables - 6. Nutritional value of vegetables - 7. Growing common vegetables - 8. Growing special-use vegetables - 9. Growing and preserving herbs. Chapters 7-9 give detailed information on growing over 90 different vegetables, as well as many herbs. Detailed, analytical index, p.347-62. *Class No:* 634.1/635.1

Vines & Vineyards

[5293]

STEINMETZ, H. Viticultural technique: multilingual illustrated dictionary: D-E-F-S-I. Betzdorf, Verlag H. Steinmetz; New York, French & European, 1983. 314p. illus. $75. ISBN: 0828800715.

About 3,950 German-base terms, with English, French, Spanish and Italian equivalents and indexes. 600 line-drawings. *Class No:* 634.8

[5294]
VITIS: Viticulture and enology abstracts. International Food Information Service (IFIS0). Geilweiherhof, Bundesforschungsanstalt für Rebenzüchtung. 4pa. DM70. ISSN: 01758292.

About 180 abstracts per issue on grapes and grapevine science and technology. Classes A-M: A. General aspects - B. Morphology - C. Physiology - D. Biochemistry - E. Viticulture - F. Soils - G. Breeding - H. Plant pathology - J. Technology, engineering - K. Economics - L. Enology - M. Microbiology of wine. Annual author and subject indexes. Available online via DIMDI, ESA, DIALOG, ORBIT. *Class No:* 634.8

Gardening

Bibliographies
[5295]
ISAACSON, R.T. Gardening: a guide to the literature. New York, Garland Pubg. Co., 1985. xiii, 198p. ISBN: 0824090195.

784 annotated entries, in 7 sections (further subdivided): Reference - Landscape design - Ornamental garden plants - Methods of growing and using plants - Garden practices and plant problems - Miscellaneous gardening topics - Periodicals and catalogs ... List of libraries with extensive gardening collections. Name, title and subject indexes. 'Specially useful both to the serious gardener and to the librarian who wants to build a collection to support this subject' (*Reference books bulletin, 1985-86,* p.67). US-slanted. *Class No:* 635(01)

Encyclopaedias
[5296]
HERITAU, J. *and* **CATHEY, H.M. The National Arboretum book of outstanding garden plants:** the authoritative guide to selecting and growing the most beautiful, durable and care-free garden plants in north America. Engelwood Cliffs, N.J., Simon & Schuster; Stonesong Press, 1990. 292p. illus. $39.95. ISBN: 9671669575.

Lists 1,700 flowers, herbs, trees, shrubs, ground cover plants, vines, ornamental grasses and aquatics. Data given include appearance, growth rate, and light, soil and moisture needs. 'An essential guide for serious gardeners' (*Booklist,* 1 June 1990, p.1863). *Class No:* 635(031)

[5297]
The Marshall Cavendish illustrated encyclopedia of gardening. Hunt, P., *ed.* New York, Marshall Cavendish Corporation, 1968-70. 20v. (iv, 2799p.). col. illus., diagrs., maps.

V.1-17: Aaron's beard ... Zygopetalum, including entries on gardening operations, pests and diseases, as well as featuring plants and flowers. V.17 has appended 'Great gardens of the world', p.2362-80. V.18-20: Garden calendar. Index (in v.20), p.2717-79. Numerous cross-references. Quality, number and generous size of coloured illus. are outstanding ('Dahlia', v.4, p.426-37, has 51 illus.). Quarto. *Class No:* 635(031)

[5298]
The New Royal Horticultural Society dictionary of gardening. Huxley, A., *ed.* London, New York, Macmillan, 1992. 4v. figs., diagrs., maps. £400; $795. ISBN: 1561590010, US ed; 0333474945, UK ed.

Preceding the *Dictionary* are: List of biographies - List of general entries - List of plates - Maps: climatic zones, Europe and the US - Abbreviations - Botanical glossary - Glossary of plant taxonomy. Main section arranged A-Z (*e.g.* 'Apples'; 5p.; 3 cross-references; cultivation; training; maintenance; pruning; thinning and harvesting; pests; diseases). Illus. are line-drawings opposite the text. V.4 has sections: Pests, diseases and disorders - Horticultural glossary - Glossary of common botanic epithets - Index of authors cited - Bibliography (p.837-53) - Index of popular names. A *tour de force* and, like its predecessor, *Dictionary of gardening* (1951), will be the standard work for years to come. *Class No:* 635(031)

[5299]
Reader's Digest encyclopedia of garden plants and flowers. 4th ed. London & New York, Reader's Digest, 1987. 800p. illus. (mostly col.), tables.

Illustrated plant encyclopedia, 'Abelia' ... 'Zygopetalum', p.9-744. Appended: Pests and diseases; Plants for special purposes; Methods of propagation; Principles of pruning; Glossary of botanical and horticultural terms, p.790-9. Conversion tables. Numerous cross-references to botanical names. Illus. are often marginal and small. Well produced compendium. *Class No:* 635(031)

[5300]
The Royal Horticultural Society gardeners' encyclopedia of plants and flowers. Brickell, C., *ed.* London, Dorling Kindersley, 1989. 608p. illus. (col.). £27.50. ISBN: 0863183867.

Sections: How to use this book - Plant origins and names - Creating a garden - The planter's guide - The plant catalogue (p.38-400) - The plant dictionary (p.401-594, listing *c.*8,000 plants with their characteristics and cultivation). Glossary of terms. Index of common names. Lavishly illustrated, with *c.*4,000 colour photographs. *Class No:* 635(031)

Handbooks & Manuals
[5301]
The Royal Horticultural Society encyclopedia of gardening: the definitive guide to gardening techniques, planning and maintenance, and to growing flowering plants, fruits and vegetables. Brickell, C., *ed.* London, Dorling Kindersley, 1992. 648p. illus (col). £29.95. ISBN: 0863189792.

2 sections (24 chapters): 1. Creating the garden (2. Ornamental trees ... 8. The rock garden ... 12. The lawn ... 16. The indoor garden) - 2. Maintaining the garden (1. Tools and equipment ... 5. Soils and fertilizers ... 8. Basic botany). Seasonal reminders, p.581-4. Glossary. Index, p.590-647. Over 3,000 illustrations in colour, and many step-by-step guides. For amateur and professional gardeners. *Class No:* 635(035)

[5302]
—'Which?' kind of garden. Boyd, L., *ed.* London, Consumers' Association, and Hodder & Stoughton, 1983. 240p. illus. (mostly col.). £12.95. ISBN: 0340259353.

Has sections: Inheriting a garden - Analysis of a garden (*e.g.* seaside, riverside) - Which kind of garden (*e.g.* labour-saving) - Allotment - Garden upkeep - Popular garden plants. The ornamental garden - The kitchen garden. Detailed, analytical index, p.234-40. Fully illustrated.
Class No: 635(035)

Dictionaries
[5303]
SMITH, A.W. A Gardener's dictionary of plant names: a handbook on the origin and meaning of some plant names. London, Cassell, 1972. xii, 391p. *Class No:* 635(038)

Yearbooks & Directories
[5304]
The Good gardener's guide: a critical guide from the publishers of *Which?*. Consumers' Association, and Hodder & Stoughton, 1985. 471p. + maps and report forms. ISBN: 0340372214.

4 sections: 1. Good garden centres and general nurseries (p.9-299: England, Wales, Scotland, Northern Ireland; data on location, critical description, range of prices, facilities, when open) - 2. Specialist nurseries (p.300-94: Alpine and rock plants ... Wild flowers) - 3. Societies and organizations - 4. Buying tools and equipment, p.410-71. 7 maps appended. For the discriminating gardener.
Class No: 635(058)

[5305]
JOHNSON, L. The Gardener's directory. London, M. Joseph, 1984. 236p. ISBN: 0718122933.

Sections, A-Z, *e.g.* Fencing; Furniture; Greenhouse equipment; Ornaments; Outdoor buildings; Plant care; Pots and planters; Soil and soil care; Specialist nurseries (African violets & streptocarpus ... Wild flowers); Tools; Watering equipment. Detailed, analytical index, p.252-6.
Class No: 635(058)

Organic Gardening
[5306]
COLEMAN, E. The New organic grower: a master's manual of tools and techniques for the home and the market gardener. London, Cassell, 1990. xv,269p. figs. £10.95. ISBN: 0304340138.

22 chapters: 1. Agricultural craftsmanship ... 3. Scale and capital ... 9. Tillage ... 13. Soil blocks ... 15. Weeds ... 18. Marketing ... 22. L'Envoi. Annotated bibliography, p.225-36. 3 appendices (*e.g.* 'Tools'; 'From asparagus to zucchini: the major vegetable crops'). Index, p.261-9. Well presented, readable, and recommended for public and academic libraries. *Class No:* 635.0

[5307]
Encyclopedia of organic gardening, by the staff of *Organic gardening* magazine. New rev. ed. Aylesbury, Bucks., J.J. Rodale, 1979. x, [2], 1236p. illus., tables.

First published 1959; Rodale, T.I., and others, eds.

Over 1,500 entries, 'Abelia ... Zygopetalum'. Data include identification, cultivation and use of specific fruits, grains, nuts, vegetables and ornamentals. 'Mulch': 5p. 'Pruning': nearly 5p. 'Insect control', p.569-604. Notes on 'storage requirements for fresh produce'. Cross references. 375 illus. (73 col.). Standard text on avoidance of artificial fertilisers and pesticides in the growing of crops. A bulky but readable volume. *Class No:* 635.0

[5308]
Green growers guide 1990/91: the worldwide directory of agrobiologicals. Newbury, Berks., CPL Press, 1990. 304p. ISBN: 1872691056.

'The first comprehensive directory of products which are NOT synthetic chemicals ... of use to those engaged in plant protection and animal nutrition and husbandry' (*Preface*). Section 1 (14 chapters, *e.g.* 'Markets for agro-biologicals ... Use of insects to control other insects ... Fertilizers' - 2. Products directory, p.89-278 (with prices in US dollars) - 3. Active ingredients - 4. Companies (with addresses and phone nos.). *Class No:* 635.0

[5309]
The UK green growers guide. Lisansky, S., *and others, eds.* Newbury, Berks., CPL Press, 1991. 420p. £11.95. ISBN: 1872691102.

2 sections: 1. The green scene (1. Why do we need green products ... 8. The growers' needs ... 13. Green laws ... 15. Insects to control other insects ... 19. Green products for animal feed - 20. Silage additives and inoculants) - 2. The products and where to get them (A. Insect control - B. Disease control ... F. Fertilisers - G. Other products - H. Chemical products accepted by some green growers). Indexes of products by use and manufacturers and suppliers. Aimed at professional growers and individual gardeners. *Class No:* 635.0

Shrubs & Trees
[5310]
DAVIS, B. The Gardener's illustrated encyclopedia of trees and shrubs: a guide to more than 2000 varieties. Harmondsworth, Middlesex, Penguin Books; Viking, 1987. 256p. illus. (incl. col.). £19.99. ISBN: 0670812374.

Dictionary of trees, p.10-71; dictionary of shrubs, p.72-221. Data: origin; description; hardiness; soil requirements; sun/shade aspect; pruning; propagation and nursery production; problems; varieties of interest; average height and spread. Appended: Practical glossary. Botanical glossary. Choosing the right trees and shrubs. Pronunciation glossary (p.237-42). Indexes of common names and Latin names. *Class No:* 635.054

[5311]
JAMES, N.D.G. The Arboriculturalist's companion: a guide to the care of trees. 2nd ed. Oxford, Basil Blackwell, 1990. xii,244p. figs. £45; £12.95. ISBN: 0631167749.

First published 1972.

Concerned with the growing and maintenance of trees from the point of view of ornament and amenity. 24 chapters: 1. Introduction to arboriculture ... 3. Transplanting large trees ... 6. Tree surgery ... 10. Trees and urban

....(contd.)

development ... 14. Avenues, parks and amenity woods ... 20. Arboricultural education, training and research ... 23. Books, manuals and periodicals (p.22-32) - 24. Trees for urban areas. Appendix: Grants for tree planting. Index. Essential for any library supporting a horticultural course. *Class No:* 635.054

Bulbs

[5312]
BRYAN, J.E. Bulbs. Portland, Or., Timber Press, 1989. 2v. illus. (col.), tables, maps. $120. ISBN: 0881921017.

Contains sections on the history, classification, propagation, cultivation, and pests and diseases of bulbs. Alphabetical listing of *c.*230 genera of bulbous plants. Data for each genus include culture; propagation; pests and diseases; cultivars; species. Many full-colour illustrations. Detailed index of common and scientific names. Appendices include lists of families of bulbous plants; genera for containers; shade-loving and woodland plants. *Class No:* 635.073

[5313]
BRYAN, J.E. Bulbs. London, Christopher Helm, 1989. 2v. illus. (col. pl.), figs. £99. ISBN: 0747002312.

V.1 has 9 chapters (1. Overview ... 5. Cultivation ... 8. Pests and diseases - 9. Introduction to the genera), followed by 'Alphabetical listing of the genera: Achimenes to Hypoxis'. V.2 completes the alphabetical listing with 'Ipheion to Zygadenus', followed by 10 appendices (A. Families of bulbous plants ... E. Flowering times ... G. Growing in containers ... J. Fragrance). Glossary, p.391-3. Bibliography, p.394-6. 300 colour plates. An authoritative work. *Class No:* 635.073

[5314]
GENDERS, R. Bulbs: a complete handbook of bulbs, corms and tubers. London, Hale, 1973. [xvi], 622p. illus., (incl. col.). £8.95. ISBN: 0709131569.

1. History, culture and characteristics, p.1-15 - 2. Alphabetical guide, p.153-541. Appendices A-G (A. Planting depths: outdoors; B. Bulbs and their habitat). Detailed, analytical index, p.601-22. *Class No:* 635.073

Indoor Plants

[5315]
BECKETT, K.A., in association with the Royal Horticultural Society. **The RHS encyclopaedia of house plants,** including conservatory plants. London, Century, 1987. 494p. illus. £19.95. ISBN: 0712614028.

House plants, A-Z (scientific names), p.44-485. Lengthy prelims. include 'A brief history of house plants', 'Cultivation', 'Maintenance and propagation', 'Pests and diseases'. Glossary, p.486-8. Index of common names, p.489-92. About 1,000 illus. Well-produced. *Class No:* 635.91

[5316]
GRAF, A.B. Exotica 4 international: pictorial encyclopedia of exotic plants from tropical and near-tropical regions. 11th ed. East Rutherford, N.J., Roehrs Co., 1982. 2v. (2560p.). illus., maps. $187. ISBN: 0911266178.

Exotica 1 (1957); *Exotica 3* (1970).

Part 1 (1280, [3]p.) comprises introduction and *c.*12,000 half-tone illus. (3-5 per page), most black-and-white. Part 2

....(contd.)

describes the plants illustrated, their family, common names, etc., (p.2151-2479). Bibliography, p.2538-9. Index of common names, p.2540-60. Covers more than 8500 species and cultivars. *Class No:* 635.91

[5317]
—GRAF, A.B. Tropica: colour cyclopedia of exotic plants and trees from the tropics and subtropics, in cool climate, the summer garden or sheltered indoors. 3rd ed., rev. and enl. East Rutherford, N.J., Roehrs Co., 1986. 1152p. col. illus., map. ISBN: 0911266232.

Companion to *Exotica*. First published 1978, it has over 7,000 colour photographs of *c.*1,600 species. Families, A-Z. Bibliography p.1117-8; references. Indexes of common and scientific names. *Class No:* 635.91

[5318]
Rodale's encyclopedia of indoor gardening. Halpin, A.M., *ed.* Emmaus, Pa., Rodale Press, 1980. vii, 902p. illus. (mostly col.). ISBN: 0878573194.

Six parts. 1. Understanding plant growth - 2. Tools and techniques for indoor gardening - 3. Growing different kinds of plants (p.168-513) - 4. Special environments and how to use them (*e.g.* chapter 29: Greenhouse gardening) - 5. A house plant encyclopedia (p.730-867). 6 appendices (*e.g.* Plant societies and publishers: US, Canada, UK); Some sources of greenhouse plants. Glossary. Select bibliography, p.875-6. Index of common plant names. Detailed, analytical general index, p.885-902. *Class No:* 635.91

Roses

[5319]
STOCK, K.L. Rose books: a bibliography of books and important articles in journals on the genus Rosa, in English, French, German and Latin, 1550-1975. The Author, 1984. x, 533p. ISBN: 0980905607.

3,279 numbered entries, under authors A-Z. References (sources), p.vi-vii, Libraries and locations, p.viii-ix. *Class No:* 635.92:582.7

Cacti

[5320]
CULLMANN, W.S. The Encyclopedia of cacti.
Thomas, K.M., *trans.* Sherborne, Dorset, Alphabooks, 1986. 340p. illus. (incl. col.). £35. ISBN: 0906670373.

German original as *Kakteen* (1963, 1964).

Structure, mode of living and classification of cacti - Cactus culture - Building up and accommodating a cactus collection - Special culture problems - The general and species of cacti, A-Z, p.124-316. Appendices include glossary, bibliography (p.322), associations and publications, plant suppliers. General index; index of Latin names. Many col. illus. A rewritten, expanded and re-illustrated ed. The revised text, which includes identification keys for all genera, now covers 750 species. *Class No:* 635.92:582.85

[5321]
LAMB, E. *and* **LAMB, B. The Illustrated reference on cacti and other succulents.** London, Blandford Press, 1955-78. 5v. illus.

Describes over 1,000 species (195 genera), with one page of brief descriptions per species. Note on culture could be fuller, especially on watering. The coloured plates provide excellent identification. V.4 has 280 illus. (94 in colour); v.5: 280 illus. (100 in colour).

Pocket encyclopaedia of cacti in colour (Rev. ed. London, Blandford Press, 1981, 218p. 326 col. illus. £7.99) was first published in 1970. *Class No: 635.92:582.85*

[5322]
—JACOBSEN, H. Lexicon of succulent plants: short descriptions, habitats and synonymy of succulent plants other than cactaceae. 2nd ed. London, Blandford Press, 1977. 682p. illus. (pl.). ISBN: 0713708263.

English ed., rev. and enl., of *Das Sukkulenten Lexikon* (Jena, 1970) - has 200 plates. Pt.1: A-Z; pt.2: Family Mesembryanthemaceae. Bibliography, p.585-90; 'Invalid designations', p.591-664. *Class No: 635.92:582.85*

Rhododendrons

[5323]
DAVIDIAN, H.H. The Rhododendron species. London, Batsford, 1982-92. 3v. illus. (some col. pl.), figs., maps. £70ea. ISBN: 0713416394, v.1; 0713416408, v.2; 0713437472, v.3.

V.1. *Lepidotes*. 1982. 431p. V.2. *Elepidote species: series Arboreum-Lacteum*. 1989. 344p. V.3. *Elepidotes: series Neriiflorum-Thomsonii, Azaleastrum and Camtschaticum*. 1992. 381p.

Lists 43 groups or 'Series'. Detailed descriptions of each species, alongwith many colour photographs and well-presented line-drawings. Included are unplaced species, glossary, list of synonyms, key to the series and subseries, etc. An authoritative work but the quality of the colour photographs is variable. *Class No: 635.92:582.915*

Livestock & Animal Husbandry

[5324]
Animal behavior abstracts. Bethesda, Md., Cambridge Scientific Abstracts, 1974-. 4pa. £470. ISSN: 03018695.

About 5,000 abstracts pa. 40 sections (Communication, Aggression ... Applied ethology. Miscellaneous). Book notices. Notification of proceedings. Author and subject (combined subject and taxonomic) indexes per issue. Online, LIFE SCIENCES COLLECTION, and also available on magnetic tape and CD-ROM. *Class No: 636*

Laboratory Animals (Research)

[5325]
Alternatives to laboratory animals. Nottingham, FRAME/ Fund for the Replacement of Animals in Medical Experiments, 1973-. 4pa. £40; $70. ISSN: 02611929.

Originally (v.1, 1973) as *Abstracts of alternatives to laboratory animals*.

Each issue contains *c.*150 abstracts under the heading 'Selected titles'. 11 book reviews precede. Intended to cover

....*(contd.)*
all aspects of the development, validation, introduction and use of alternatives to laboratory animals in biomedical research and toxicity testing. *Class No: 636.028*

[5326]
Laboratory animals: an annotated bibliography of informational resources covering medicine - science (including husbandry) - technology. Cass, J.S., *ed.* New York, Hafner, 1971. 3pts. (viii, vi, 136p.; vi, 250p.; [vi], 60p.). $23.95. ISBN: 0028426401.

Part 2: *Animals in research. Class No: 636.028*

[5327]
UFAW handbook on the care and management of laboratory animals. Poole, T.B. 6th ed. Edinburgh, Churchill Livingstone, 1987. x, 933p. diagrs., tables. £78. ISBN: 058240911x.

First published 1947.

59 contributors. 3 parts. 1. The laboratory animal - 2. Animal units - 3. Species kept in the laboratory (a) mammals; (b) birds; (c) reptiles; (d) amphibians; (e) fish. 56 chapters, each with 'Further reading'. References, p.848-918. Detailed, analytical index. *Class No: 636.028*

Breeds & Breeding

[5328]
Animal breeding abstracts. Wallingford, Oxon., CAB International, 1933-. 12pa. £299; $553. ISSN: 00033499.

About 8,000 abstracts pa. Sections: Livestock - Horses - Cattle - Buffalo - Sheep - Goats - Pig - Fur-bearers - Laboratory mammals - Other mammals - Poultry and other birds - General and theoretical genetics - Reproduction, general - Fish and invertebrates in aquaculture. News and notes. Reports. Conferences. Books. Author and subject indexes per issue and annually. Online, CAB ABSTRACTS database. Available on CAB ABSTRACTS CD-ROM and CAB International's BEASTCD. *Class No: 636.082*

Animal Feeds

[5329]
Nutrition abstracts and reviews. Series B. Livestock feeds and feeding. Wallingford, Oxon., CAB International, 1977-. 12pa. £389; $702. ISSN: 0309135x.

About 6,500 abstracts pa. Coverage includes: Analysis; Technology; Feedstuffs and feeds; Physiology; Biochemistry; Feeding of animals; Diet in aetiology of disease. Author and title indexes per issue and annually.

Online, CAB ABSTRACTS. Available on CAB ABSTRACTS CD-ROM and BEASTCD.
Class No: 636.084

[5330]
STEINMETZ, H. Livestock feeding and management: multilingual illustrated dictionary. 3rd ed. Betzdorf, Steinmetz, 1978. 550p. illus.

First published 1966 (300p.).

Over 5,000 German-base terms, with equivalents and indexes in English, French, Spanish, Italian and Dutch. A companion to the author's *Landmaschinen and Geräte Mehrsprachen - Bildwörterbuch* (4. Aufl. 1981). *Class No: 636.084*

[5331]
UNITED NATIONS. Industrial Development Organization. **Information sources on the animal feed industry.** New York, United Nations, 1975. xii, 62p. (*UNIDO guides to sources of information, no.13.*)

377 items, in 12 sections (1. Professional, trade and research organizations, learned societies and special information services ... 3. Sources of statistics and economic data ... 6. Current periodicals - 11. Films - 12. Other potential sources of information (B. Fairs and exhibitions; C. Financing and investing; D. Marketing, exporting and purchasing)). *Class No:* 636.084

Veterinary Medicine

Bibliographies

[5332]
GIBB, M. **Keyguide to information sources in veterinary medicine.** London, Mansell, 1990. xiii,459p. £45. ISBN: 0720120187.

Contain 2,530 entries, some unannotated. 3 sections: 1. Survey of veterinary medicine and its literature (1. The history and scope of veterinary medicine ... 4. Keeping up to date with current information ... 8. Language problems) - 2. Bibliography (9. General works ... 11. Small animals - 12. Specialties) - 3. Directory of organizations. Lack of bibliographic information in the 'Other Journals' sections whose entries are not included in the index. *Class No:* 636.09(01)

[5333]
Index-catalogue of medical and veterinary zoology. US Bureau of Animal Industry. Zoological Division. Authors A-Z. Washington, US Government Printing Office, 1932-52. 12v. Supplement 1-. 1953-.

Index to world literature 'on parasites and parasitisms of man, of domestic animals, and of wild animals whose parasites may be transmitted to man and domestic animals' (*Bulletin of the Medical Library Association, v.41, no.2, April 1953, p.110*). In the Author catalogue, each author's papers are arranged chronologically.

Supplements 1-6 (1953-56) covered backlog. Since then, supplements (authors, A-Z) are annual. Beginning with suppt.15, *Parasite subject catalogues* (containing indexes to the author references in the *Author catalogues* and including a catalogue of hosts) have been issued in addition. *Class No:* 636.09(01)

[5334]
KERKER, A.E. *and* MURPHY, H.T. **Comparative & veterinary medicine:** a guide to the resource literature. [Madison], University of Wisconsin Press, 1973. xvi, 308p. $84.30. ISBN: 0835767892.

4,600 usually unannotated entries. 4 parts: 1. Materials of general interest (reference works, p.58-70) - 2. Specific disciplines (anatomy ... radiology and radiation biology) - 3. Veterinary medicine (animal management ... wild life and zoo animals) - 4. Laboratory animals. List of congresses and other meetings. Author index; subject index key; analytical subject index, p.268-308. *Class No:* 636.09(01)

[5335]
Black's veterinary dictionary. West, G.P., *ed.* 17th ed. London, A.&C. Black, 1992. 660p. illus., diagrs., graphs, tables, maps. £19.99. ISBN: 0713636009.

First published 1928.

Several hundred undocumented entries (*e.g.* 'Hernia'; parts and types; symptoms; treatment). For farmers, veterinary and agricultural students, and teachers. 226 illus. *Class No:* 636.09(03)

[5336]
—FENNER, W.R. Quick reference to veterinary medicine. Philadelphia, Pa., Lippincott, 1991. xiii,669p. tables. $44.95. ISBN: 0397508956.

Has 44 contributors, 3 parts: 1. Clinical signs and client complaints - 2. Laboratory abnormalities and principles of fluid management - 3. Physical and chemical injuries and intoxicants. Summary. Suggested readings to sections. Basic diagnostic approach (*i.e.,* based on signs and symptoms), complementing standard veterinary textbooks. *Class No:* 636.09(03)

[5337]
BLOOD, D.C. *and* STUDDERT, V.P. **Baillière's comprehensive veterinary dictionary.** London, etc., Baillière Tindall, 1988. xii, 1124p. tables. £29.95. ISBN: 0702011959.

8 consultants. The dictionary (p.1-997) has *c.*50,000 entries. 'Rabbit': 9 lines, *c.*20 cross-references to types, etc.; 'Manx': 17½ lines. Basic scientific terms were drawn and adapted from Baillière's Dictionary database. Anatomical tables. Appendix comprises 18 tables (*e.g.* laboratory services). 'Based on a survey of more than one hundred veterinary textbooks' (*Preface*). *Class No:* 636.09(03)

[5338]
COMMONWEALTH BUREAU OF ANIMAL HEALTH. **Controlled vocabulary for subject indexing in Veterinary bulletin, Index veterinarius, Animal disease occurrence, and Small animal abstracts.** Weybridge, Surrey, the Bureau, 1985. [iv], 110p.

Columns: Description (non-descriptors, indented); date first used; remarks. Where no date is given, footnote states that the descriptor has been in use for at least 20 years, probably since the 1950's. Standard vocabulary for cataloguers and indexers in this field. *Class No:* 636.09(03)

[5339]
LEADER, R.W. *and* LEADER, I. **Dictionary of comparative pathology and experimental biology.** Philadelphia, Pa., W.B. Saunders, 1971. vi, [1], [1], 238p. illus., tables.

About 4,500 terms, omitting those already dealt with in standard medical dictionaries. Includes information on husbandry, breeds, genetic characteristics, and diseases of common laboratory animals, with special emphasis on rats, mice and rabbits. Particular attention to genetics in relation to disease. Bibliography (65 items) p.235-8. Addenda: Table of animal models. Table of zoonoses by type of etiologic agent. *Class No:* 636.09(03)

[5340]
MACK, R. Dictionary for veterinary science and biosciences German-English/English-German. Berlin, Paul Parey, 1988. 321p.

Contains *c*.10,000 terms, compiled over a period of 30 years from biomedical and veterinary literature. Trilingual Latin indexes of botanical, anatomical, and zoological names. *Class No:* 636.09(03)

[5341]
MACK, R. Russo-English veterinary dictionary. [Russko-angliĭskiĭ veterinaryĭ slovar'.] Farnham Royal, Commonwealth Agricultural Bureaux, 1972. viii, 104p. £10. ISBN: 0851982557.

Equivalents of *c*.6,000 Russian veterinary terms, including relevant biological vocabulary. Bibliography, p.vii-viii. *Class No:* 636.09(03)

[5342]
Wörterbuch der Veterinärmedizin. [Dictionary of veterinary medicine.] Wiesner, E. *and* Ribbeck, R., *eds.* 3rd ed. Jena, Gustav Fischer Verlag, 1991. 1662p.

First published 1978; 2nd ed. 1983.

73 contributors. Contains *c*.50,000 entries. Many new terms have been added since the previous ed. *Class No:* 636.09(03)

Handbooks & Manuals

[5343]
Animal anatomy on file. Diagram Group. New York, Facts on File, 1990. loose-leaf. $145. ISBN: 0816022445.

Contains more than 200 exceptionally clear anatomical drawings depicting the anatomy of animal species from sponges to humans. Subdivided by major divisions in the animal world, each plate shows a specific anatomical structure, *e.g.* muscle system. Anatomical parts on each plate are numbered, with an accompanying key appearing below. Useful for school libraries. *Class No:* 636.09(035)

[5344]
Veterinary medicine: a textbook of the diseases of cattle, sheep, pigs, goats and horses. Blood, D.C., *and others.* 7th ed. London, etc., Baillière Tindall, 1989. xxiii, 1502p. illus. tables. £42. ISBN: 0702012866.

First published 1960.

2 sections: 1. General medicine. 2. Special medicine. 35 chapters in all: 1. Clinical examination and making a diagnosis ... 7. Diseases of the liver and pancreas ... 10. Diseases of the respiratory system ... 15. Mastitis ... 25. Diseases caused by protozoa ... 28. Metabolic diseases ... 33. Diseases caused by allergy ... 35. Specific diseases of uncertain etiology. Conversion tables, p.1461-2. Normal laboratory values, p.1463-4. Index, p.1465-1502. 101 tables (1. Degrees of severity of dehydration and guidelines for assessment ... 78. Single and multiple host ticks). Numerous bibliographies throughout. Standard text. *Class No:* 636.09(035)

[5345]
—GREAT BRITAIN. Ministry of Agriculture, Fisheries and Food. Tolworth Library. Thesaurus (for animal health, vertebrate pest biology and control). Surbiton, Surrey, the Library, 1975. 98p. *Class No:* 636.09(035)

[5346]
Zoo and wild animal medicine. Fowler, M.E., *ed.* Philadelphia, Pa., etc., W.B. Saunders Co., 1986. xxiv, 1127p. illus., diagrs., tables. $105. ISBN: 0721610137.

Sponsored by Morris Animal Federation, Denver, Colorado.

About 70 contributors. 5 parts: 1. General information - 2. Special medicine: amphibians and reptiles - 3. Special medicine: birds - 4. Special medicine: mammals (p.515-1039) - 5. Invertebrates. 13 appendices (1-6: Common and scientific names of amphibians, etc.). Detailed, analytical index, p.1089-1127. Has been translated into Japanese. *Class No:* 636.09(035)

Reviews & Abstracts

[5347]
Index veterinarius. Wallingford, Oxon., CAB International, 1933-. 12pa. £432; $802. ISSN: 00194123.

Lists *c*.20,000 references pa., under subjects, A-Z. Supported by author index. 'Titles are selected from some 1,200 serial publications regularly scanned by the staff of the Bureau, and from books, annual reports, monographs, theses and other serial publications'. A list of the serials appears irregularly in the *Veterinary bulletin (qv)*. CAB ABSTRACTS database, CD-ROM, and BEASTCD. *Class No:* 636.09(048)

[5348]
Veterinary bulletin. Wallingford, Oxon., CAB International, 1931-. 12pa. £351; $639.

About 9,200 abstracts pa. Covers the core literature of animal health. Animals included are: Cattle; Horses; Sheep; Goats; Pigs; Poultry; Cats; Dogs; Rabbits; Cagebirds; Other small animals; Laboratory animals; Wildlife; Zoo animals; Fish; Other domestic animals. Occasional review articles. Author and analytical subject indexes per issue and annually. CAB ABSTRACTS database. Also available on CAB ABSTRACTS CD-ROM and BEASTCD. *Class No:* 636.09(048)

[5349]
—Veterinary literature documentation / VETDOC. London, Derwent Publications, Ltd., 1968-. Fortnightly. $11,660pa.

Carries *c*.4,000 abstracts pa. relevant to manufacturers of veterinary products (*e.g.* veterinary drugs, vaccines, growth promotants, hormonal control of breeding). Online access (subscribers only to SDC host). *Class No:* 636.09(048)

Progress Reports

[5350]
Advances in veterinary science and comparative medicine. Orlando, Fla., etc., Academic Press, 1953-. illus., diagrs., graphs, tables.

V.36, 1991, (xii,343p.) has 11 contributors. The 11 documented articles include 'Platelets and coagulation' (with 15p. of refs.) and 'Neonatal transfusion medicine' (with 3p. of refs.). Analytical index, p.325-43. Some volumes concentrate on particular themes (*e.g.* v.28: 'Research on nonhuman primates'). *Class No:* 636.09(055)

Yearbooks & Directories

[5351]
The Directory of veterinary practices 1992. Hall, S.A., *ed.* London, the Royal College of Veterinary Surgeons, 1992. xv,385p. £28. ISBN: 0902183109.

First published 1991.

Explanation of symbols and abbreviations is followed by: Gazetteer; Geographical list of practices; Alphabetical list of practices; Alphabetical list of members.
Class No: 636.09(058)

[5352]
ROYAL COLLEGE OF VETERINARY SURGEONS. Registers and directory, 1992. London, the Royal College, 1992. x,510p.

General list of members of RCVS; General list of nationals of a member state of the EEC; Commonwealth list; Foreign list; Temporary list. Supplementary veterinary register; Honorary associates; Fellows; Diplomates; Certificate holders. Appendices: Geographical location; Government depts.; Municipal services; University veterinary schools; Research institutes. Index. *Class No:* 636.09(058)

[5353]
The Veterinary annual. 30th issue. Sevenoaks, Kent, Butterworth, 1990. Annual. 351p. illus., diagrs., tables.

First published 1959-. (Bristol, Wright).

47 contributors. General articles (5) - Cattle (3) - Sheep (3) - Deer (1) - Pigs (4) - Horses (7) - Poultry (1) - Small animals (18). Articles are documented (*e.g.* 'Avian haematology' (8 references); 'Arthrodesis in the dog' (6 references). Analytical subject index.
Class No: 636.09(058)

Tables & Data Books

[5354]
Animal health yearbook. Food and Agriculture Organization, World Health Organization, and International Office of Epizootics (FAO-WHO-OIE). Rome, FAO, 1957-. Annual. tables, maps. ISSN: 00661872.

Contains tabular statistics of world livestock diseases, grouped under countries in 6 areas: Africa; America; Asia; Europe; Oceania, USSR. 10 appendices (*e.g.* number of livestock and human population). Arabic, Chinese, Russian and German glossaries. A-Z list of countries and areas, and of diseases and causal agents. *Class No:* 636.09(083)

[5355]
British pharmacopoeia (veterinary), 1985. Published on the recommendation of the Medicines Commission pursuant to the Medicines Act 1968. Effective date, 1 September 1985. London, HMSO, 1985. xviii, 213 + 204p. plus Amendment inserts 1 & 2 (1985-86). Addenda, 1988 (£15); 1992 (£19.50). £40. ISBN: 0113211430, 1988 addendum; 0113214588, 1992 addendum; 0113208480.

The first part consists of monographs (*e.g.* Immunological products). 25 appendices (*e.g.* General reagents; Infra-red reference spectra). Detailed analytical index. Well produced. This pharmacopoeia provides standards for the quality of substances, preparations and immunological products used in veterinary medicine, plus information on action, use, dose, solubility, storage and labelling. *Class No:* 636.09(083)

[5356]
The Henston large animal veterinary vade mecum 1991-1992. Evans, J., *ed.* 7th ed. High Wycombe, Bucks., Henston, 1991. 464p. tables. ISBN: 185054087x. ISSN: 02684268.

Formerly *The Henston veterinary vademecum (large animals).*

3 parts: 1. Diagnosis and disease (cattle; sheep and goats; pigs; horses) - 2. Therapeutic products ('Tables by therapeutic class', p.246-405) - 3. Useful information (*e.g.* 'Poisons'; 'Government departments'). Directory of products and services, p.437-55. Advertisers' index.
Class No: 636.09(083)

[5357]
The Henston small animal veterinary vade mecum 1992-1993. Evans, J., *ed.* 11th ed. High Wycombe, Bucks., Henston, 1992. 380p. tables. ISBN: 1850540888. ISSN: 02684268.

Formerly *The Henston veterinary vademecum (small animals).*

5 parts: 1. Diseases, conditions and signs - 2. Tables by therapeutic class - 3. Nutrition - 4. Useful information - 5. Directory of products and services. Index. Advertisers' index. *Class No:* 636.09(083)

[5358]
The Merck veterinary manual: a handbook of diagnosis, therapy, and disease prevention and control for the veterinarian. 7th ed. Rahway, N.J., Merck & Co., Inc., 1991. xxxii,1832p. $33.50. ISBN: 0911910557.

First published 1955; 6th ed. (xxvii,1677p.), 1986.

About 500 contributors. 9 parts: 1. General (anatomy & physiology) - 2. Behavior - 3. Clinical values and procedures - 4. Fur, laboratory and zoo animals - 5. Management, husbandry, and nutrition - 6. Pharmacology - 7. Poultry - 8. Toxicology - 9. Zoonoses. Detailed, analytical index, p.1747-1832. *Class No:* 636.09(083)

[5359]
The Veterinary formulary: handbook of medicines used in veterinary practice. Debuf, Y.M., *ed.* London, Pharmaceutical Press, 1991. xv,448p. tables. £37.50. ISBN: 0853692459.

Data on more than 1,600 preparations licensed for use in animals in the UK. 3 appendices: 1. Drug interactions - 2. Drug incompatibilities - 3. Conversions and units. Index of manufacturers and organizations. Index, p.423-48.
Class No: 636.09(083)

Nomenclatures

[5360]
Illustrated veterinary anatomical nomenclature. Schaller, O., *ed.* Stuttgart, Ferdinand Enke Verlag, 1992. vi,614p. figs. ISBN: 3432995911.

Arranged so that verso pages contain numbered terms as in the *Nominal anatomica veterinaria,* with each term followed by a brief definition. Respective, clearly drawn, illustrations, labelled by numbers corresponding to those of the terms on the opposite page. Alphabetical index of terms. Includes changes approved for the 4th ed. of NAV. 280 plates, including 1,316 illustrations.
Class No: 636.09(083.72)

Histories

[5361]

Animal health: a centenary, 1865-1965: a century of endeavour to control diseases of animals. London, HMSO, 1965. xviii, 396p. illus., maps.

3 parts. 1. The evolution of the Animal Health Divison - 2. The animal health responsibilities - 3. Aspects of disease control. Some footnote references. 13 illus., 2 maps. General index; index of acts, rules and regulations. *Class No:* 636.09(091)

[5362]

SMITH, Sir F.E. The Early history of veterinary literature and its development. London J.A. Allen, 1976. 4v. ports.

1. *From the earliest period to A.D.1700.* iv, 373p. 2. *The eighteenth century.* viii, 244p. 3. *The nineteenth century, 1800-1823.* viii, 244p. 4. *The nineteenth century, 1823-1900.* 161p.

V.1 was reprinted from *Journal of comparative pathology and therapeutics,* 1912-1918; v.2-3, from *The Veterinary journal,* 1923-24, 1929-30. V.1 has an index of names only; v.2-4 have author and 'works' indexes. Numerous footnotes. A classic account. *Class No:* 636.09(091)

Information Management

[5363]

FOOD AND AGRICULTURE ORGANIZATION. Manual for small veterinary libraries. Alexandria, FAO Near East Animal Health Institutes, Coordinating Units, United Nations Development Programme, 1972. [iv], 65p.

5 chapters: 1. Consideration of some problems in the administration of small veterinary libraries - 2. Selection and acquisition - 3. Periodicals and serials - 4. Cataloguing and classification - 5. Reference and bibliographic service. Select list of reference aids for small veterinary medical libraries (p.28-65), - 187 briefly annotated entries, under subjects, A-Z. *Class No:* 636.09:025.4

Thesauri

[5364]

Veterinary multilingual thesaurus in English, French, Italian and German. Munich, K.G. Saur, 1979. 5v. (4v. and index). Microfiche also available. $270 (English language vol. 264p.). ISBN: 3598101090.

Full thesaurus, p.1-194. Microthesaurus (18 subject sections, Biology ... Zootechny, p.195-244). 4,922 descriptors in each language; 4,190 non-descriptors in English. Quadralingual index and microfiches in rear-pocket. *Class No:* 636.09:025.43

Veterinary Schools

[5365]

World directory of veterinary schools. Geneva, WHO, 1973. 260p. tables.

Published under auspices of Food and Agriculture Organization and WHO.

Data on veterinary schools in 60 countries, A-Z. Headings under countries include conditions of admission, curriculum, examinations, qualifications and licence to practise. Annexe 2: Postgraduate veterinary degrees and requirements, by country. *Class No:* 636.09:061:378

[5366]

—World directory of schools for animal health assistants. Geneva, WHO, 1974. 195p. tables. $9.60. ISBN: 9241500050.

Gives details of systems of education and schools in 54 countries, A-Z. *Class No:* 636.09:061:378

Diseases

[5367]

Animal disease occurrence. [Incidenza delle malatie degli animale. Commonwealth Bureau of Animal Health, Weybridge, Surrey.] Wallingford, Oxon., CAB International, in association with the Commission of the European Communities, 1980-. tables. Annual. £35. ISSN: 01443879.

About 1,000 abstracts per issue. *Section A:* General veterinary medicine; Animal diseases in general; Horse diseases; Cattle diseases; Sheep and goat diseases; Swine diseases; Dog and cat diseases; Poultry diseases; Fish diseases; Bee diseases; Diseases of other animals; Miscellaneous - *Section B:* Tables (12, including list of pathogens and parasites). Author and subject indexes per issue, cumulated annually. CAB ABSTRACTS database. *Class No:* 636.09:616

Livestock

Bibliographies

[5368]

MAGEL, C.R. Keyguide to information sources in animal rights. London, Mansell; Jefferson, N.C., McFarland & Company, 1989. iv,267p. £40. ISBN: 0720119847, UK ed; 0899504051, US ed.

Bibliographic essay with references keyed to the annotated bibliography (335 items) and the unannotated list of literature cited (*c.*1,000 items), plus a directory of relevant organizations (220 entries). 5 appendices (A. The rights of animals: a declaration against speciesism ... D. Magazines and journals). Index of authors, titles, subjects, persons, countries and organizations. 'An outstanding resource' (*RQ,* v.29(3), Spring 1990, p.445-6). *Class No:* 636.1/.5(01)

Handbooks & Manuals

[5369]

BLAKELY, J. *and* **BADE, D.H. The Science of animal husbandry.** 5th ed. Reston, Va., Reston Pubg. Co., 1989. 736p. illus., diagrs., tables. $43. ISBN: 0685447146.

40 chapters (with study questions): 2. The beef industry ... 9. Diseases of cattle - 10. The dairy industry ... 15. Reproduction, feeding and management of sheep ... 22. The swine industry ... 29. Poultry: a general view ... 35. The horse industry ... 40. The rabbit industry. 2 appendices. Detailed, analytical index. Written for the student. *Class No:* 636.1/.5(035)

[5370]
BRIGGS, H.M. *and* BRIGGS, D.M. **Modern breeds of livestock.** 4th ed. New York, Macmillan; London, Collier-Macmillan, 1980. xiv, 802p. illus., charts, tables. ISBN: 002314730x.

58 chapters. 4 sections: 1. The breeds of cattle - 2. The breeds of swine - 3. The breeds of sheep and goats - 4. The breeds of horses. Introduced on individual classes and groups. Detailed index, p.777-802. US slanted.
Class No: 636.1/.5(035)

[5371]
Handbook of animal science. Putnam, P.A., *ed.* San Diego, Academic Press, 1991. xi,401p. graphs, tables. $69. ISBN: 0125683006.

Sections: History and background - Breeds and genetics - Statistics - Health - Production - Product/UItilization - Future. Includes small animals as well as the large, domestic animals. References appended to each chapter. Many tables and graphs. A handy and convenient reference source.
Class No: 636.1/.5(035)

Dictionaries

[5372]
EUROPEAN ASSOCIATION FOR ANIMAL PRODUCTION, Rome. **Dictionary of animal production terminology,** in English, French, Spanish, German and Latin. Amsterdam, etc., Elsevier, 1986. xxiv, 684p. illus. £155.93; $395. ISBN: 0444424741.

About 8,000 English-base terms with equivalents in the other four languages. Classes A-Z (A. Anatomy - B. Physiology - C. Genetics - D. Breeding - E. Feeding stuffs ... K. Hygiene and therapy - L. Animal products - M. Forestry - N. Equine production - O. Cattle production ... Z. General terms). Indexes in all 5 languages.
Class No: 636.1/.5(038)

[5373]
MASON, I.L. **A World dictionary of livestock breeds, types and varieties.** 3rd ed. Wallingford, CAB International, 1988. xx,348p. £25.95. ISBN: 085198617x.

First published 1951.
7 sections: Ass-Buffalo-Cattle-Goat-Horse-Pig-Sheep. Types and varieties are arranged A-Z within each group. Bold type for important and recognized breeds (*e.g.* Holstein). Incorrect forms, misspellings or definite misnomers are included, with 'not' ending the entry. Bibliography, p.344-48. An invaluable source of information on breeds and the correct or recommended spelling of breed names in English. *Class No:* 636.1/.5(038)

Histories

[5374]
DAVIS, S.J.M. **The Archaeology of animals.** London, Batsford, 1987. 224p. illus., facsims., diagrs., tables, maps. £19.99. ISBN: 0713445718.

Part 1 (chapters 1-4): Methods and problems in zoo-archaeology - Part 2 (chapters 5-8): 5. Our hunting; 6. From hunter to herder: the origin of domestic animals; 7. Later domesticates and the secondary use of animals; 8. Britain: a zoo-archaeological case study. Over 150 drawings. References, p.197-208. Index, p.215-24 (most entries are for goat, pig and sheep). 'Intended primarily for students of

....(contd.)
archaeology and anyone with an interst in both natural history and the history of our own species' (*Introduction*).
Class No: 636.1/.5(091)

[5375]
HALL, S.J.G. *and* CLUTTON-BROCK, J. **Two hundred years of British farm livestock.** London, British Museum (Natural History), 1989. 272p. illus. (some col.). £21.95. ISBN: 0565010778.

19 sections: 1. Introduction - 2. Western and northern cattle ... 7. Channel Island cattle ... 14. Upland sheep ... 16. Goats ... 18. Heavy draught horses - 19. Light horses, ponies and donkeys. Bibliography, p.256-62. Attractively presented, with many colour illustrations of the breeds. For public and academic libraries. *Class No:* 636.1/.5(091)

[5376]
ZEUNER, F.E. **A History of domesticated animals.** London, Hutchinson, 1962. 560p. illus., diagrs., tables.

Part 1: The origins and evolution of domestication - Part 2. Domesticated animals (pre-agricultural phase; agricultural phase; subsequent domestication primarily for transport and labour; the pest-destroyers; various other mammals; birds, fishes and insects). Bibliography, p.509-37 (authors, A-Z). Index (very brief entries) p.539-60. *Geographical review* (v. 54(1), January 1964) notes the many and often unusual illus., but cites examples of extreme brevity of treatment. 'When it is good, it is very, very good, and when it is bad, it is horrid'. *Class No:* 636.1/.5(091)

[5377]
—ANGRESS, S. *and* REED, C.A. 'An Annotated bibliography on the origin and descent of domestic animals, 1900-1955'. In *Fieldiana anthropology*, v.54(1). Chicago, Natural History Museum, 1962. 143p.

Systematically arranged, with concrete but apposite annotations. *Class No:* 636.1/.5(091)

[5378]
—CLUTTON-BROCK, J. The Natural history of domesticated mammals. London, British Museum (Natural History), 1987. 192p. illus. (incl. col.). £9.95. ISBN: 0565010506.

Previously as hardback *Domesticated animals from early times* (1981).
A popular account of how our familiar domestic breeds of mammals were developed from their wild ancestors.
Class No: 636.1/.5(091)

Great Britain

[5379]
TROW-SMITH, R. **A History of British livestock husbandry to 1700.** London, Routledge & Kegan Paul, 1957. x, 286p. *Class No:* 636.1/.5(410)

[5380]
—A History of British livestock husbandry, 1700-1900. London, Routledge & Kegan Paul, 1959. x, [1], 351p.

The first work to deal specifically with the evolution of British breeds of farm animal and the technique of animal husbandry. The earlier volume has a list of principal sources (p.259-71) and an analytical index (p.273-86). The later volume also has a bibliography (p.327-39), but the index (p.341-51) could be fuller. *Class No:* 636.1/.5(410)

Horses & Ponies

[5381]

BONGIANNI, M. The Macdonald encyclopedia of horses. Milan, Arnoldo Mondadori Editore; London, Macdonald, 1988. 255p. illus. (col.), figs. £8.95. ISBN: 0356153193.

173 entries, each with a colour photograph. Data include breed name, place of origin, aptitudes, qualities, temperament, conformation. 3 sections: Light draft, pack and saddle breeds - Heavy draft breeds - Ponies. Introduction, p.10-41, includes sections on coats, head markings, stance. Brief bibliography, and index. For public libraries. *Class No:* 636.1

[5382]

EDWARDS, E.H., *general ed*. **A Standard guide to horse and pony breeds.** London, Macmillan, 1980. 352p. illus. (incl. col.). ISBN: 0333266498.

Main sequence (p.69-336) describes 150 horse and pony breeds in 5 geographical areas: Europe (p.69-258), South America, Australia, Soviet Union (p.275-326), Middle East and Asia. 'Breed description' includes height, colour, type, action and temperament. The Arab hourse and extinct breeds are considered separately. 'The World's principal horse organizations', p.344-6. Detailed, non-analytical index. 350 photographs (100 in colour). *Class No:* 636.1

[5383]

GRIMSHAW, A. The Horse: a bibliography of British books, 1851-1976, with a narrative commentary on the role of the horse in British social history, as revealed by the contemporary literature. London, Library Association, 1983. xxxiv, 474p. illus., facsims. £29.95. ISBN: 0853655332.

Initially a limited ed., 1982.

3,226 entries, arranged chronologically in six periods. Limited to books and pamphlets published in Britain, 1851-1976, but including American reprints and translations. 'Great riding schools of the world' has 1½ columns of annotation. Sources, p.x. Title and author indexes. 'The book meets a long-felt need and does it very well' (*Natural history book reviews,* v.7(3), 1984, p.29). *Class No:* 636.1

[5384]

HOPE, C.E.G. *and* **JACKSON, G.N.,** *eds*. **The Encyclopedia of the horse.** London, Ebury Press/ Pelham Books, 1973. 336p. illus., maps.

121 contributors. Entries (signed), A-Z (broad headings), cover all aspects, including show jumping, associations, breeds, biographies, and the horse in prehistoric art. Occasional references appended to articles. Over 300 black-and-white illus. and 32 unusually fine colour plates; clear diagrams. A handsome quarto. 'An exceptionally good book', comments *British book news* (December 1973, p.817), while objecting to the double columns of small print. *Class No:* 636.1

[5385]

ROSSDALE, P.D. *and* **WREFORD, S.M. The Horse's health, from A to Z.** Newton Abbot, Devon, David & Charles, 1974 (reprinted 1989). 433p. illus., tables. £18. ISBN: 0715392662.

An encyclopaedic dictionary (*e.g.* tables of muscles, p.265-77; drug manufacturers (British, Irish, US), p.429-32). Notes relevant associations and cites regulations re quarantine, etc. Includes abbreviations, cross-references. Etymology, p.12-13. References and further reading, p.433. *Class No:* 636.1

[5386]

—HAYES, M.H. Veterinary notes for horse owners: a manual of horse medicine and surgery. Rossdale, P.D., *ed*. 17th ed. London, Stanley Paul, 1987. viii,740p. illus. £30; £18.99. ISBN: 009173701x.

First published in 1877.

54 chapters (7. The skin and its diseases). 3 appendices, *e.g.* 1. Proprietary medicines. Detailed analytical index, p.712-40. 108 illus. *Class No:* 636.1

Cattle

[5387]

PORTER, V. Cattle: a handbook to the breeds of the world. London, Christopher Helm, 1991. 400p. illus. (col. pl.), maps. £39.95. ISBN: 0713680008.

7 sections: The cattle of Europe - Tropical cattle - The cattle of Africa - The cattle of Asia - The buffaloes - The cattle of America - The cattle of Australia and New Zealand. 6 appendices (1. Metrication table ... 3. Cattle breeds and their synonyms ... 6. Useful addresses). Bibliography. Index, p.392-400. Data for each breed include development, physical characteristics, historical and current role, special characteristics and, for the major breeds, a detailed history. Beautifully illustrated. Good value. *Class No:* 636.2

Sheep and Goats

[5388]

Directory of current research on sheep and goats. King, J.W.B., *ed*. Wallingford, Oxon., CAB International, 1988. 271p. £13.95. ISBN: 0851986110.

International in scope, arranged A-Z by country, Argentina ... Zimbabwe. Data include address of organization, names of research workers, main sheep projects, main goat projects, and list of publications. Indexes of research works, goat projects and sheep projects. *Class No:* 636.3

[5389]

RYDER, M.L. Sheep and man. London, Duckworth, 1983. ix, [1], 846p. illus., diagrs., tables. £55. ISBN: 0715616552.

3 parts (14 chapters, each with contents and summary): 1. Ancient times - 2. The Middle Ages to recent times - 3. The association of man with sheep. Epilogue: 'Has the sheep come to the end of the world?' Appendix: 7 columns of data, p.785-801. References (A-Z authors), p.802-29. Very detailed, partly analytical index, p.830-46 ('Wool': 1 column, solid). *Class No:* 636.3

[5390]

Sheep and goat production. Coop, J.E., *ed*. Amsterdam, etc., Elsevier, 1982. xiii, 492p. diagrs., tables. $164. (*World animal series, C1*.) ISBN: 0444419896.

25 documented chapters (1. Ecology and distribution - 2. Breeding ... 7. Nutrition and diseases ... 12. Wool grading and marketing ... 15. Milk production in sheep and goats ... 16. Systems [of grazing] ... 25. Village and smallholding systems). 27 contributors; biographical notes, p.481-4. Subject index, p.485-92. Quarto. *Class No:* 636.3

Pigs

[5391]
Index of current research on pigs. Wallingford, Oxon., CAB International, 1960-. Annual. ISSN: 05652800.

No. 37, 1991 has 5,998 references, arranged geographically. 41 countries are covered. Author and subject indexes. *Class No:* 636.4

[5392]
WISEMAN, J. A History of the British pig. London, Duckworth, 1986. ix, 118p. illus., facsims., diagrs., tables. ISBN: 071561987x.

8 chapters: 1. The Anglo-Saxon and medieval pig ... 8. The proliferation of breeds ... 8. Recent developments and the role of minority breeds. Bibliography, p.113-6. Detailed, analytical index, p.117-8. Charmingly illustrated with old engravings and new photographs, but too short and lacking detail (*TLS,* 20 June 1986, p.685). *Class No:* 636.4

Poultry

[5393]
Multi-lingual poultry dictionary, English-French-German-Spanish. Forscey, L.A., *comp.* Bologna, European Federation of Branches of the World's Poultry Science Association, 1969. [x], 356p.

About 3,000 terms, English base terms, p.2-69; French, German and Spanish equivalents and indexes. Monolingual English, French, German and Spanish glossaries follow. Genders are stated. *Class No:* 636.5

[5394]
Poultry abstracts. Wallingford, Oxon., CAB International, 1975-. 12pa. £159; $291. ISSN: 03661382.

About 3,000 abstracts pa. Coverage includes: Breeding and genetics; Housing; Equipment; Anatomy, physiology and biochemistry; Parasitic disorders; Nutritional and metabolic disorders. Reports. Conferences. Books. Author and subject indexes per issue and annually.

Online, CAB ABSTRACTS. Available on CAB ABSTRACTS CD-ROM and BEASTCD. *Class No:* 636.5

Pets

[5395]
DIAGRAM GROUP. Pets: every owner's encyclopedia. London & New York, Paddington Press, 1978. 431p. illus.

7 sections: 1. Dogs; cats; mustelids (ferrets, skunks); rabbits; rodents; horses, ponies, donkeys; artiodactyls (goats, sheep, etc.) Wild mammals - 2. Fish (p.176-223) - 3. Invertebrates - 4. Amphibians (*e.g.* frogs) - 5. Reptiles - 6. Birds - 7. Conservation. Reference (including classification). Further reading (p.422-3). Index, p.424-31. Comprehensive, covering selection, handling, care, food and feeding, housing and equipment, illness, breeding and showing. *Class No:* 636.596/.9

[5396]
Small animal abstracts: annotated bibliography. Wallingford, Oxon., CAB International. Annual. £47; $87.

Contains *c.*2,000 abstracts pa. Covers the research literature on diseases, physiology, reproduction, nutrition and behaviour of dogs, cats, and other pets. Online, CAB ABSTRACTS. Available on CAB ABSTRACTS CD-ROM and BEASTCD. *Class No:* 636.596/.9

Doves & Pigeons

[5397]
GOODWIN, D. Pigeons and doves of the world. 3rd ed. London, British Museum (Natural History); Ithaca, N.Y., Cornell University Press, 1983. 363p. illus. (incl. col.), maps. £15. ISBN: 0565008471, UK ed; 0801414342, US ed.

First published 1970.

Records all existing and recently extinct species, covering description, field characteristics, world distribution (on maps), habitat feeding and breeding habits, voice and display. Line-drawings for most species. Indexes of English and scientific names. *Class No:* 636.596

Cage & Aviary Birds

[5398]
Encyclopedia of aviculture. Translated from the Dutch. Rutgers, A. *and* Norris, K.A., *eds.* Poole, Dorset, Blandford Press, 1970-77. 3v. £40ea. ISBN: 071370800x, v.1; 0713708018, v.2; 0713708026, v.3.

V.1 (414p.) deals mainly with aquatic species, birds of prey, pheasants, pigeons and doves. V.2 (290p.), with parrots, budgerigars, cuckoos, nightjars, humming birds and owls. The *Passeriformes* order occupies most of v.3 (241p.), with numerous varieties of canaries and finches. The *Encyclopedia* concludes with a bibliography of 160 items, and indexes of English and Latin names. It provides a wealth of information on captive birds, their adaptability and care of them. 44 col. illus. depict *c.*260 species (based on review in *Library review,* v.28, Spring 1979, p.53-54). *Class No:* 636.68

[5399]
—**WOOLHAM, F. The Handbook of aviculture.** Poole, Dorset, Blandford Press, 1987. 368p. col. illus. ISBN: 071371042x.

Devotes part 1 to 'general care' (diets; housing; breeding; disease and medicine; the law and ethics). Part 2. 'The birds', p.67-359: Rheinformes ... Passeriformes, provides data on range, status, etc. Appendix: Principal authorities responsible for licensing (Argentina ... USA). Bibliography, p.362. Indexes of scientific and common names. *Class No:* 636.68

[5400]
MARTIN, R.M. The Dictionary of aviculture: keeping and breeding birds. London, Batsford, 1983. xi, 227, [1]p. illus., tables, maps. ISBN: 0713441569.

Entries 'Acclimatization ... Zygodactyol'. 'Starling': nearly 2p., with 5 identification drawings. Many cross-references. 93 illus. Appended: 'Books mentioned in the text'. Chiefly concerned with the study and breeding in captivity of wild bird species (*i.e.,* it excludes pigeons, commerical domesticated forms, etc.). *Class No:* 636.68

Dogs

[5401]
AMERICAN KENNEL CLUB. The Complete dog book: the photograph, history and official standard of every breed admitted to AKC registration, and the selection, training, breeding, care and feeding of pure-bred dogs. 17th ed. New York, Howell Book House, 1985. 768p. illus. (incl. col.). ISBN: 0876054629.

First published 1929, 16th ed. 1980 (768p.).

....*(contd.)*

Groups: 1. Sporting dogs - 2. Hounds - 3. Working dogs - 4. Terriers - 5. Toys - 6. Non-sporting dogs. Subdivided by breeds, with colour displays for each group. 'Caring for your dog', p.643-740. Glossary, p.741-64. Index.

18th ed., 1992. 832p. $27.50. *Class No:* 636.7

[5402]
JONES, E.G. A Bibliography of the dog: books published in the English language, 1570-1965. London, Library Association, 1971. 431p.

Based on a thesis accepted for Library Association Fellowship.

3,986 references, wth locations of titles in 4 libraries (including British Museum and Kennel Club). 19 classes: A. General works - B. Natural history - C. Domestication ... J. The dog at work ... R. The dog in literature - 5. The dog in children's literature ... W. Breeds. Full sub-titles are given. Index of names and titles; index of subjects. Chronological list, 1570-1850. *Class No:* 636.7

[5403]
SCANZIANI, P. The British encyclopedia of dogs. London, Orbis, 1985. 304p. illus. ISBN: 0856135755.

Illustrated guide to dog breeds - Classification of dogs and breeds - A/Z of breeds, p.43-199 - Dogs: their psychology and uses - Training, p.221-96. Boxed information on offical classification, country of origin, weight, coat, etc. Further reading, p.297-8. Detailed, analytical index. 270 well-coloured illus. *Class No:* 636.7

[5404]
WILCOX, B. *and* **WALKOWICZ, C. Atlas of dog breeds of the world.** Neptune City, NJ, TFH Publications, 1989. 912p. illus. ISBN: 0866229302.

First section contains essays on different types of dogs, *e.g.* scenthounds, mastiffs, terriers, gun dogs. The second part consists of alphabetically arranged entries for 400 breeds of dogs worldwide. Affenpinscher ... Yugoslavian Hounds. Entries consist of country of origin, size, description of coat and colour, groups with which it can be registered, and the larger category to which it belongs. Glossary, extensive bibliography and index. More than 1,000 excellent photographs illustrating each breed. It is more comprehensive than *The Complete dog book* (Howell Book House, 1989) (*q.v.*). *Class No:* 636.7

Cats

[5405]
BI-Lexikon Rassekatzen. [Dictionary of pedigree cats.] Muller-Girard, C., *and others.* Leipzig, VEB Bibliographisches Institut, 1988. 495p. illus. (some col.), figs.

Illustrated descriptions of many cat breeds. Also includes information on genetics, diseases, management, reproduction, physiology, and behaviour. 49 colour photographs. Extensive bibliography. For cat breeders and fanciers. *Class No:* 636.8

[5406]
The International encyclopedia of cats. Henderson, G.N. *and* Coffey, D.J., *eds.* New York, etc., McGraw-Hill, 1973. 256p. illus. (incl. col.).

More than 700 entries, A-Z, covering all aspects of cat ownership and care. Lavishly illustrated in colour and black-

....*(contd.)*

and-white photographs. 'This non-technical authoritative encyclopedia' (*American reference books annual, 1974,* item 1705). *Class No:* 636.8

[5407]
NECKER, C. Four centuries of cat books, 1570-1970. Metuchen, N.J., Scarecrow Press, 1972. 511p.

Briefly annotated entries, with extensive cross-references. 'A detailed and comprehensive bibliography of books on cats in English. Limited to the domestic cat, with a few books on the cat family' (*Library journal,* 13 December 1972, p.3980-1). *Class No:* 636.8

[5408]
A Standard guide to cat breeds. Pond, G. *and* Raleigh, I., *eds.* London, Macmillan, 1979. 318p. illus. (incl. col.). £12.95. ISBN: 0333240420.

5 contributors. A general section concerns feline genetics, coat colours and patterns, eye colours. Glossary. Shorthair cats, p.38-177. Longhair cats, p.179-231. Appendices include such matters as show preparation, show rules, breeding, medicine and behaviour. Detailed, non-analytical index. Well illustrated. *Class No:* 636.8

[5409]
TAYLOR, D. The British Veterinary Association guide to cat care. London, Dorling Kindersley, 1989. 96p. illus., diagrs. £3.95. ISBN: 0863184022.

3 sections: 1. Caring for your cat - 2. Feeding - 3. Health care (*e.g.* 'Respiratory disorders'; 'Veterinary care'; 'Poisons'). Index. A clearly presented practical guide. *Class No:* 636.8

Dairy Farming

[5410]
RUSSELL, K. The Principles of dairy farming. 11th ed., revised by K. Slater. Ipswich, Farming Press, 1991. v,359p. illus., tables. £13.95. ISBN: 0852362161.

First published 1953.

17 chapters: (1. The milk industry ... 3. Buildings and equipment ... 13. Breeding better cows ... 15. Disease in dairy herds ... 17. Profitability in milk production). 5 appendices (1. Breed comparison according to milk yield, butterfat and protein levels). Detailed, analytical index. Fully illustrated. No bibliography. A basic textbook. *Class No:* 637

Dairy Produce

[5411]
Dairy science abstracts. Wallingford, Oxon., CAB International, 1939-. 12pa. £299; $551. ISSN: 00115681.

About 8,000 abstracts pa. 11 sections: Husbandry - Milk production - Technology - Economics - Legislation and standards - Milk and public health - Physiology - Biochemistry - Nutrition - Immunology - Microbiology - Chemistry - Physics - Dairy research - Education. Reports;

....*(contd.)*

Conferences; Books. Author and subject index per issue and annually. Online, CAB ABSTRACTS. Available on CAB ABSTRACTS CD-ROM and BEASTCD. *Class No:* 637.1

[5412]
Dairy vocabulary / Dictionnaire laitier/Milchwirtschaftliches Wörterbuch. Edited under the auspices of the International Dairy Federation. Munich, VV-GmbH Volkswirtschaftlicher Verlag, 1988. 596p.

First published 1963 as *Dairy dictionary*.

Over 6,000 entry-words in 3 sequences: English-French-German, French-German-English, German-French-English. Genders of French and German nouns stated. Editors and other contributors are either agriculturists or dairy science specialists. *Class No:* 637.1

[5413]
Dictionary of dairy terminology, in English, French, German and Spanish. Amsterdam, etc., Elsevier, 1983. [xii], 323p. £73.75; $115.50. ISBN: 0444421017.

3,909 English-based terms (with explanations, as considered necessary), with equivalents and indexes in French, German and Spanish. Bibliography, p.[ix] - [x]. Includes pertinent vocabularies from chemistry, microbiology, engineering, physiology and agriculture. *Class No:* 637.1

[5414]
—Milk and milk products. Rome, FAO, 1979. 95p. ISBN: 925007582.

In three languages - French, English and Spanish. Bibliography, p.93-5. *Class No:* 637.1

[5415]
RENNER, E. Grosse Molkerei Lexikon. [Encyclopaedia of milk.] Munich, Volkswirtschaftlicher Verlag, 1988. 418p. tables.

New edition of *Grosse Molkerei-Lexikon* (Schulz & Voss, 1965).

3,530 entries, arranged A-Z, and 40 tables. Has a wider coverage of dairy science than the previous edition. Bibliography contains 104 references. Revised English ed. published 1991 as *Dictionary of milk and dairying* (Munich, VV-GmbH Volkswirtschaftlicher Verlag. 384p.). *Class No:* 637.1

[5416]
UNITED NATIONS. Industrial Development Organization. Information sources on the dairy product manufacturing industry. New York, United Nations, 1976. xiii, 88p. corrigendum insert, 1979. (*UNIDO guides to sources, no.23.*)

739 entries. 11 sections: 1. Professional, trade and research organizations, learned societies and special information services ... 3. Sources of statistics, marketing and other economic data ... 6. Current periodicals ... 10. Films and film catalogues ... 11. Other potential sources of information (C. Fairs and exhibitions; G. Meetings and conferences; I. Patents and licences). Items in several sections are annotated. *Class No:* 637.1

Cheese

[5417]
ANDROUET, P. Guide to world cheeses. Githens, J., *tr*. English ed., revised. Henley-on-Thames, Oxon., Aidan Ellis, 1983. [v],561p. ISBN: 0856291794.

First published 1973. French title as *Guide du fromage*.

12 chapters (*e.g.* 4. Selecting cheeses by flavour and season; 8. Dictionary of cheese, p.159-471; 9. Special cheeses and non-French cheeses; 12. Some cheeses to try outside France). 'New cheeses', p.549-61. Glossary, p.519-41. Descriptions of *c.*500 varieties, highlighting French cheeses.

2nd ed. published as *Cheese guide* in 1988. 551p.£15. *Class No:* 637.3

[5418]
DAVIS, J.G. Cheese. London, J. & A. Churchill, 1965-76. 4v.

1. *Basic technology*, 1965. viii, 463p. 2. *Annotated bibliography, with subject index.* [viii], 275p. 3. *Manufacturing methods.* 1976. [iv], p.465-1012. 4. *Scientific aspects, with supplementary annotated bibliography.* 264p.

V.1 (19 chapters, with references) has 4 parts: 1. Introductory - 2. Milk - 3. Materials and equipment in cheesemaking - 4. After-treatment of cheese. V.2: 6,803 items; very briefly annotated, with analytical subject index. V.3 is a well-illustrated international survey of some 800 varieties of cheese. V.4 supplements v.2. The standard treatise on cheese, but now somewhat dated. *Class No:* 637.3

[5419]
EEKHOF-STORK, N. The World atlas of cheese. London, Paddington Press, 1976. 240p. illus., maps. ISBN: 0844701332.

Offers a tour of cheese-making regions of the world, plus a history and contemporary survey of cheeses and their manufacture. 50 international recipes; hints on purchase, storage and serving of cheese. 600 full-colour photos accompany the 240p. atlas. *Class No:* 637.3

[5420]
FOX, P.F. Cheese: chemistry, physics and microbiology. London, Elsevier, 1987. 2v. £79ea. ISBN: 1851660526, v.1; 1851660534, v.2; 1851660542, set.

V.1. *General aspects.* 400p. V.2. *Major cheese groups.* 393p.

V.1 has 14 contributors, 10 chapters, including 'The enzymatic coagulation of milk' and 'Nutritional aspects of cheese'. V.2 has 17 contributors, 11 chapters including 'Dutch-type varieties' and 'Processed cheese products'. An authoritative technical study, aimed at the cheese-manufacturer, or agricultural/nutritional student, rather than the gourmet. 2nd ed. promised mid-1993. *Class No:* 637.3

Beekeeping

[5421]
Apicultural abstracts: a journal of the International Bee Research Association. Cardiff, the Association, 1950-. Quarterly. £130. ISSN: 0003648x.

About 360 abstracts per issue. UDC classification. Sections. Chemical aspects ... Various entomylogical aspects ... Apoidea, except *Apis-Apis*. Honeybee species ... Bee

....(contd.)

forage ... Rearing and rearing honeybees, including queens ... Honeybee diseases ... Honey - Beeswax - Honeybees as pollinators - Rearing bees, except honeybees. Annual author and subject indexes. *Index, 1950-72* (2v. 1976.). Online CAB ABSTRACTS. *Class No:* 638.1

[5422]

—CRANE, E. Bibliographical tools in apiculture. London, International Bee Research Association, 1984. 20p. ISBN: 0860981355. *Class No:* 638.1

[5423]

—CRANE, E. Bibliography of tropical apiculture. London, International Bee Research Association, 1978. 38 parts, maps. ISBN: 0860980760, v.2; 0860980308, v.1.

Also available in 2v. V.1 (1978. £30) contains pts.1-24; v.2 (1980. £25), pts.25-38.

Covers regions and topics (*e.g.* 2. *Beekeeping in Africa South of the Sahara,* with 346 references; 22. *Bees for pollination in the tropics*). *Class No:* 638.1

[5424]

British bee books: a bibliography, 1500-1976. London, International Bee Research Association, 1979. 270p. illus., facsims. £20. ISBN: 0860980774.

Records *c.*836 printed books and 12 manuscripts. 5 sections; the first lists libraries and beekeeping collections, and 'Publications with useful bibliographical information'. Other sections: Manuscripts before 1500; Chronological list of books (6 periods, 1500/1599 - 1900/1926); Another bibliography for the last fifty years, 1927 to 1976; Metaphorical titles; Books on special categories (*e.g.* in verse). Indexes of short titles, authors, special subjects. *Class No:* 638.1

[5425]

CRANE, E. Bees and beekeeping: science, practice and world resources. Oxford, Heinemann Newnes, 1990. xii,614p. illus., figs., tables. £95. ISBN: 0434902713.

6 sections: 1. The bees used in beekeeping, and background information - 2. Beekeeping with movable-frame hives - 3. Beekeeping with simpler and cheaper hives - 4. Maintaining honeybee health - 5. Honeybees' plant resources, and products from the hive - 6. Beekeepers. 2 appendices: 1. Important world honey sources and their geographical distribution - 2. Beekeeping gazetteer of individual countries. Bibliography, p.539-93. Plant, geographical name, and subject indexes. The author is a former Director of the International Bee Research Association and one of the foremost authorities on beekeeping and honey. With its readable text and clear illustrations, this book is essential for any library covering this area. *Class No:* 638.1

[5426]

Dictionary of beekeeping terms, with allied scientific terms. Crane, E., *ed.* London (later Gerrards Cross, Bucks.), (International) Bee Research Association. v.10 1988. 196p. £10. ISBN: 0860981940.

One of a series of English-base foreign-language dictionaries by the editor, on beekeeping terms. V.10 gives translations from and into English, French, Arabic, with Latin index. *Class No:* 638.1

[5427]

MORSE, R.A. *and* **HOOPER, T. The Illustrated encyclopedia of beekeeping.** Poole, Dorset, Blandford Press, 1985. 432p. illus., ports., diagrs., tables, maps. ISBN: 071371624x.

46 contributors. Entries, A-Z, include some 30 biographies. The article 'Honey, food and drink products' (p.191-2) has references in text; 'Swarm behaviour and movement', p.370-2. Annotated glossary of anatomical terms. References from text to col. illus. Analytical index, p.427-32. Outstanding illus. Comprehensive and up-to-date. *Class No:* 638.1

Fish & Fisheries

Bibliographies

[5428]

Current references in fish research. Chippewa Falls, Wisc., Current References in Fish Research, 1976-. Annual. $15. ISSN: 0739540x.

Lists *c.*4,000 references pa. from more than 100 scientific titles. No particular order of entry. First author, second author, and keyword indexes. *Class No:* 639.2(01)

[5429]

Guide to reference works on marine science and fisheries. Shaw, M.J., *comp.* Lowestoft, Directorate of Fisheries Research, 1981. 78p.

Earlier as *Some books on fisheries and marine science.* K.G. Handley and M. Shaw, editors. 1980. 33p. (Library Information leaflet no.22).

122 annotated entries. 6 sections: Indexes abstracts and contents lists - Bibliographies - Catalogues of the publications of, or taken by, specific organizations - Review publications - Directories - Dictionaries, encyclopedias, glossaries, handbooks, name lists, etc. Author or editor index. Title index. *Class No:* 639.2(01)

[5430]

TURNBULL, D. Keyguide to information sources in aquaculture. London, Mansell, 1989. xvi,137p. £35. ISBN: 0720118530.

International in scope. 3 sections: 1. Survey of aquaculture and its literature - 2. Bibliography (p.11-78; 472 references, most with annotations) - 3. Directory of organizations. Index, p.119-37, lists books, periodicals, databases, and organizations. As most of the publications listed are from 1980 onwards, this is a good guide to current literature in the subject. *Class No:* 639.2(01)

Handbooks & Manuals

[5431]

CRC handbook of mariculture. McVey, J.P., *ed.* Boca Raton, Fla., CRC Press, 1983-91. 2v. $195, v.1; $159.95, v.2. (*CRC series in marine science.*) ISBN: 084930220x, v.1; 0849302196, v.2.

V.1 *Crustacean aquaculture.* 1983. 442p. V.2. *Finfish aquaculture.* 1991. 256p.

V.1 has 4 sections: 1. Larval foods for crustaceans - 2. Maturation, spawning hatchery and grow-out techniques for crustaceans - 3. Pathology and disease treatments for

....(contd.)

crustaceans - 4. Crustacean nutrition. V.2 covers marine finfish culture in Europe, Japan, Taiwan and the US. Each chapter in the set has a bibliography appended. Because of the 8-year gap between the publication of v.1 and v.2, the former may have lost some of its value.
Class No: 639.2(035)

Dictionaries

[5432]
Multilingual dictionary of fish and fish products / ... Danish, Dutch, English, Finnish, French, German, Greek, Icelandic, Italian, Japanese, Norwegian, Portuguese, Serbo-Croat, Spanish, Swedish, Turkish and scientific names. [Dictionnaire multilingue des poissons et produits de la pêche.] 3rd ed. Oxford, Fishing News Books Ltd., 1990. xvi,442p. £29.50; $64.95. ISBN: 0852381646.

First published 1968 (xv, 427p.).

Over 1,000 English-base terms, with equivalents and indexes in the other 15 languages. Brief notes included (*e.g.* Latin name; where found; family; marketing form of product; fresh, dried, frozen, smoked, etc.- in English, and French). *Class No:* 639.2(038)

[5433]
—EUROPEAN PARLIAMENT. Terminology Office. Terminologie du secteur de la pêche. 3. éd. Luxembourg, the Parliament, 1977. [160]p.

First published 1967.

Comprises 291 French-base terms (with sub-entries) and equivalents in Italian, English, German, Dutch and Danish, - covering the then 6 official languages of the European Communities. Appended list of fish, crustaceans and molluscs. *Class No:* 639.2(038)

[5434]
RICKER, W.E. Russian-English dictionary for students of fisheries & aquatic biology. Ottawa, Department of the Environment, Fisheries Research Board, 1973. xi, 428p. (*Bulletin no.163.*) *Class No:* 639.2(038)

[5435]
Selected terms in fish culture, English-French-Spanish-Arabic (with Latin for names of fish and fish diseases). [Choix de termes de pisciculture.] Rome, FAO, 1981. vii,151p. $20. (*FAO Terminology bulletin 22.*) ISBN: 9250009100.

First published 1972.

Basic list, with language equivalents. Bibliography, p.148-9. *Class No:* 639.2(038)

Reviews & Abstracts

[5436]
Aquatic sciences and fisheries abstracts. Part 1: Biological sciences and living resources. Bethesda, Md., Cambridge Scientific Abstracts, 1978-. 12pa. $885pa. ISSN: 01405373.

Formerly part of *Aquatic sciences and fisheries abstracts,* 1971-77.

About 24,000 abstracts pa. Classes: 1. General aspects ... 21. Biology: general. 65. Entomology ... 79. Ichthyology ... 92. Ontology ... 98. Mammology - 99. Ecology, ecosystems and pollution ... 199. Fisheries ... 252. Marketing and economics of aquatic products. Author and subject indexes per issue. Draws upon *c.*5,000 primary and other journals. Available on magnetic tape, online and CD-ROM.
Class No: 639.2(048)

Progress Reports

[5437]
Advances in fish science and technology. Papers presented at the Jubilee Conference of the Torry Research Station, Aberdeen, Scotland. 23-27 July, 1979. Connell, J.J., *ed.* Farnham, Surrey, Fishing News Books, Ltd., 1980. xiv, 512p. £65. ISBN: 0852381085.

80 documented papers in 2 main sections: 1. The review section, on past, present and future of fish technology and science - 2. Recent advances section (subsections 3-16, *e.g.* 4. Process and new product investigation; 6. Chilled and frozen storage; 8. By-products; 12. Protein studies; 13. Microbiology; 16. Histology). Detailed, analytical index.
Class No: 639.2(055)

Tables & Data Books

[5438]
Yearbook of fishery statistics. Rome, FAO, 1948-. tables. Annual (2 parts). tables, maps. ISBN: 9250029373, v.66; 9250029381, v.67.

v.66: *Catches and landings, 1988.* 1990. 502p. $50. v.67: *Fishery commodities, 1988.* 1990. 397p. $44.

V.66 provides data on quantities of fish, crustacea, molluscs, caught and landed (arranged under country, itemised and according to main fishing areas, inland or marine). V.67 concerns disposition of catches, protection and international trade (arranged under county and major types of fishery commodity. Quantity and value statistics for imports and exports. Tables cover 1985/88.
Class No: 639.2(083)

Freshwater Fishing

[5439]
Aquaculture: the farming and husbandry of freshwater and marine organisms. Bardach, J.E., *and others.* New York, etc., Wiley; Interscience, 1972. xii, [i], 888p. illus., diagrs., graphs, tables. $45. ISBN: 0471648259.

43 documented chapters (1. General principles and economics; 2. Culture of the common carp ... 6. Catfish culture in the United States ... 17. Milkfish culture ... 26. Commercial culture of freshwater salmonids ... 32. Shrimp culture ... 36. Oyster culture (with 2p. of references) ... 42. Culture of seaweeds - 43. Culture of edible freshwater plants). Index of names of animals and plants. Index of persons, places and institutions. Non-analytical subject index. *Class No:* 639.2.052.2

[5440]
BERRA, T.M. An Atlas of distribution of the freshwater fish families of the world. Lincoln, Nebraska, Univ. of Nebraska Press, 1981. xxx, 197p. illus., maps. ISBN: 0803214111.

Includes a bibliography (p.181-91) and index.
Class No: 639.2.052.2

[5441]
FACT / Freshwater and aquaculture contents tables. Rome, Food and Agriculture Organization, 1978-. 12pa.

V.10(2), 1987, provides contents pages of 18 journals. Appended schedule of future freshwater sciences and aquaculture meetings, conferences, etc. List of journals covered throughout the year (57), plus publishers' addresses. 57 journals are covered, in all. *Class No:* 639.2.052.2

Fishing Gear

[5442]
BRANDT, A., von. Fish catching methods of the world. 3rd ed. Farnham, Surrey, Fishing News Books, Ltd., 1984. xiv, 418p. illus., diagrs. £35. ISBN: 0852381255.

First published 1964.

31 detailed chapters (1. Introduction … 31. Fishing systems and harvesting machines). Includes fishing gear and methods used by small-scale and subsistence-level fishermen. Appended: 'Classification of catching methods'. Bibliography of 690 entries, p.394-406 (A-Z, authors). Detailed, non-analytical subject index; Species and product index; Geographical index. The author was for many years head of Hamburg's Institute for Fishing Technology. *Class No:* 639.2.081

[5443]
Glossary of United Kingdom fishing gear terms. Bridger, J.P., *and others.* Farnham, Surrey, Fishing News Books, Ltd., 1981. xi, 115p. illus. £17.95. ISBN: 0852381190.

Lists more than 1,500 fishing gear terms 'known to be used currently by fishermen and their suppliers in England, Wales, Scotland and Northern Ireland' (*Introduction*). Headings: Term; Locality; Gear type; Definition; Figure. Identifies preferred terms and largely excludes obsolete terms. Appended line-drawings, p.82-114. *Class No:* 639.2.081

Deep-Sea Fishing

Handbooks & Manuals

[5444]
Fishermen's handbook. Branson, C.R.P.C., *ed.* 2nd ed. Farnham, Surrey, Fishing News Books, Ltd., 1987. 319p. illus. (some col. pl.). £17.96. ISBN: 0852381433.

6 sections: 1. Management - 2. Navigation - 3. Watchkeeping, shiphandling - 4. Fishing operations - 5. Safety and survival at sea - 6. Anchors and cables-cordage and its uses. 5 appendices (*e.g.* 1. Finding the distance of objects at sea). Useful publications, p.304. Detailed, analytical index. *Class No:* 639.22(035)

[5445]
—Guide to fishing vessels. 1990-91. Ledbury, Herefordshire, Dayton's Publishing, Ltd., 1990. 450p. illus. £65. ISBN: 187104104x.

A compendium, with over 1,000 illus. *Class No:* 639.22(035)

Tables & Data Books

[5446]
Sea fisheries statistical tables. London, HMSO, 1919-. tables. ISBN: 0112412076. ISSN: 00726702.

The 1990 ed. (1992. 41p. £10.50) has 22 tables. Landings of fish (tables 1-11). Consumption level estimates (table 12). Fishermen and fishing vessels (13-16). International trade of the United Kingdom with the EEC, EFTA, and all other countries (tables 17-22). *Class No:* 639.22(083)

Maps & Atlases

[5447]
Atlas of the living resources of the seas. 4th ed. Rome, FAO, 1981. 23p. + 60 col. maps (some folded). $115. ISBN: 9250010001.

First published 1971. In English, French and Spanish editions.

Text: Introduction - Productivity of the seas - Exploitation and utilization of resources - Movements and migrations - Topography and nomenclature - Size and distribution of resources - Value of estimates. Alphabetical index of fish names (p.14-23). The maps illustrate geographical and vertical distribution, migration and existing state of exploitation. Oblong. 'A splendid example of the effectiveness of simplicity in compilation' (*Geographical journal,* v.140(1), February 1974, p.161, on 3rd ed., 1972). *Class No:* 639.22(084.3)

Worldwide

[5448]
Fishes of the north-eastern Atlantic and the Mediterranean. [Poissons de l'Atlantique du nord-est et de la Méditerranée.] Whitehead, P.J.P., *and others, eds.* Paris, Unesco, 1984-86. 3v. figs., maps. $180. ISBN: 9230022152, v.1; 9230023086, v.2; 9230023094, v.3.

A continuation of *Check-list of the fishes of the north-eastern Atlantic and of the Mediterranean* (Unesco, 1973). A fauna of saltwater fish of the palaearctic region. Data on 220 families and 1,250 species. Entries include line-drawings, diagnoses, keys, biological data, synonyms, distribution maps and literature. V.3 contains an additional bibliography to *Check-list of the fishes of the north-eastern Atlantic and the Mediterranean,* p.1415-26, as well as an alphabetical index of common names, p.1429-42, and an alphabetical index of scientific names, p.1443-73. *Class No:* 639.22(100)

Africa—Southern

[5449]
Smith's sea fishes. Smith, M.M. *and* Heemstra, P.C., *eds.* 6th ed. Berlin, Springer-Verlag, 1986. xx,1047p. illus., (incl. col. pl.). $162. ISBN: 0387168516.

First published 1949, as J.L.B. Smith's *The sea fishes of Southern Africa.*

Describes *c.*2,200 species found in the seas around Southern Africa, - about 15% of the total marine fauna of the world. Almost all of the species described are supported by line-drawings. Fin-formula key; index to scientific and common names. Glossary. Extensive bibliography. 'A monumental work of exceptional quality and scholarship' (*Choice,* v.24(10), June 1987, p.1539). *Class No:* 639.22(68)

Whaling

[5450]
TONNESSEN, J.N. *and* **JOHNSEN, A.O. The History of modern whaling.** Translated from Norwegian. Christopherson, R.I., *tr.* London, C. Hurst & Co; Canberra, Australian National Univ. Press. 1982. x, 798p. illus., maps.

A much-shortened version of *Den moderne hvalfangsts historie.*

3 parts (37 chapters): 1. Whaling off the coast of

....(contd.)

Finnmark - 2. Global whaling - 3. Pelagic whaling. Statistical appendix (to 1978). Bibliography (bibliographical literature); periodicals; books and articles. Index, 777-98: Persons; Vessels and companies; Geographical (including whaling stations); subject. *Class No:* 639.245

Aquaria

[5451]

AXELROD, H.R. *and* **SCHULTZ, L.P. Handbook of tropical aquarium fishes.** Rev. ed. Neptune City, N.J., T.F.H. Publications, 1983. xii, 718p. illus. (incl. col.). ISBN: 0876664915.

First published 1955.

Ichthyology. (p.1-58) - The aquarium and its management (p.59-98) - Plants in the home aquarium (p.99-122) - Diseases of fishes and their cure (p.123-53) - Catalog of fishes (p.155-675). Glossary, p.677-87. Selected references. Detailed analytical index, 691-728. *Class No:* 639.33/.34

[5452]

Dr. Burgess's atlas of marine aquarium fishes. Burgess, W.E., *and others.* Portsmouth, TFH Publications, 1988. 736p. illus. £50. ISBN: 0866228969.

Primarily an identification guide to saltwater fish, with *c.*4,000 photographs and 560 colour plates. 'Systematic list of the families of fishes of the world' explains the arrangement of the *Atlas.* Lists of scientific names and their equivalents. Index. *Class No:* 639.33/.34

[5453]

Exotic tropical fishes. Axelrod, H.R., *and others.* 2nd rev. and expanded ed. Neptune City, N.J., T.F.H. Publications, 1983. 1300p.

First published 1962.

Principles of aquarium management; Aquarium plants and their cultivation; Exotic tropical fishes (A-Z; p.257-1216); Raising tropical fishes commercially, p.1217-48. Detailed index, p.1251-1300. *Class No:* 639.33/.34

[5454]

—AXELROD, H.R. Dr Axelrod's mini-atlas of freshwater aquarium fishes. Mini-ed. Neptune City., N.J., T.F.H. Publications, 1987. 992p. ISBN: 0866323951.

Mainly an album of colour illus., p.33-672. Appended: Aquaria maintenance, plants and fish, p.693-924. Index of scientific names; index of common names, p.929-92. *Class No:* 639.33/.34

[5455]

—AXELROD, H.R. *and* VORDERWINKLER, W. Handbook of tropical fishes, with special emphasis on techniques of breeding. 27th ed. Jersey City, N.J., T.F.H. Publications, 1983. 631p. illus.

First published in 1957. *Class No:* 639.33/.34

[5456]

HOEDEMAN, J.J. Naturalists' guide to fresh-water aquarium fish. New York, Sterling, 1974. 1152p. illus. (mostly col.).

Translated from the Dutch.

Part 1: evolution, distribution, anatomy and life processes of fish, plus relationship of fish to science and to man. Part 2: a finely illustrated catalogue of hundreds of the best known aquarium fishes. Bibliography of reference works; glossary; index. 'Those with less specialized interests will find Axelrod & Vorderwinkler's *Encyclopedia of tropical*

....(contd.)

fishes & Innes' *Exotic aquarium fishes* ... a better buy' (*Wilson library bulletin,* October 1976, p.185-6). *Class No:* 639.33/.34

[5457]

MILLS, D. *and* **VEVERS, G. The Practical encyclopedia of freshwater tropical aquarium fishes.** 2nd ed. London, Salamander Books, 1989. 208p. illus. (col.). £12.99. ISBN: 0861014901.

First published 1982.

2 sections: 1. Practical section (Selecting a tank ... Aquatic plants ... Basic fish anatomy ... Expanding the interest) - 2. Species guide (a survey of over 200 tropical freshwater fishes). Further reading; glossary; index. Many colour illustrations. Mills is a fishkeeping expert, while Vevers is a former Curator of the Aquarium at London Zoo. *Class No:* 639.33/.34

[5458]

—The Complete aquarium encyclopedia of tropical freshwater fish. Ramshorst, J.D. van, *ed.* Amsterdam, Elsevier; Oxford, Phaidon, 1978. 391p. col. illus. ISBN: 0729000095.

8 contributors. 3 main parts: Compendium (The aquarium ... Breeding) - Plants (a selective, systematic list), p.82-113 - Fishes (a selective, systematic list), p.115-384. Data: family; genus; English name (if any); origin; size; habitat. Index of fishes, p.388-90 (tiny print). Grouped bibliography, p.390-1. 900 good illus. (500 in colour). Quarto. *Class No:* 639.33/.34

[5459]

STERBA, G. The Aquarium encyclopedia.

Mills, D., *English ed and* Simpson, S., *tr..* Cambridge, Mass., MIT Press, 1983. 605p. illus. (incl. col.), diagrs.

Translation of *Lexikon der Aquaristik und Ichthyologie (1978). English ed. as The aquarist's encyclopedia* (Poole, Dorset, Blandford Press).

Entries, A-Z, for fish genera and species, plants, diseases, etc. with numerous cross-references from common names to scientific names. Bibliography, p.[606]. 1,183 illus. in all. Intended as 'a reference on freshwater and sea water aquarium keeping', with basic data on 'general and specialist ichthyology, hydrology, fish economy, the biology of freshwater animals and marine biology' (*Foreword*). Sterba, - professor of zoology and animal physiology at Karl Marx University, Leipzig. Emphasis is thus on scientific aspects. *Class No:* 639.33/.34

64 Household Management

Household Management

Catering Establishments

[5460]
COMBES, S. **Dictionary of cuisine:** French for hoteliers, restaurauteurs and catering students. London, Barrie & Rockliff, 1962. xii, 145p. ISBN: 0091442419, US ed.

Published in US as *Dictionary of cuisine French* (1989. $70).

French-English, English-French dictionary. Aims 'to provide an extensive vocabulary of commodities, equipment, processes, manipulations, etc., and especially, a more comprehensive French-English coverage of the flora and fauna used in French cookery than the author has so far been able to find in any works of this kind' (*Author's preface*). *Class No: 640.02*

[5461]
CRACKNELL, H.L. *and* NOBIS, G. **The New catering repertoire.** London, Macmillan, 1989-90. 2v. illus., tables, maps. £100, set; £12.99ea., pbk. ISBN: 0333391659, v.1; 0333391667, v.2.

V.1. *Aide-mémoire du chef.* xiii,810p. 1989. V.2. *Aide-mémoire du restaurateur et sommelier.* xix,492p. 1990.

V.2 has 2 main sections: 1. The waiting staff handbook (1. Service procedures ... 4. Menus ... 9. Restaurant personnel ... 11. Eating habits of nations) - 2. The wine waiting and bar staff manual (12. Wine production and wine-making countries ... 15. Cocktails and mixed drinks ... 17. Cigars). 19 appendices (A. Glossary of restaurant terms for the service of wine ... E. Styles of folding table napkins ... I. The law and the restaurateur ... S. Stills for making wines and spirits). Many black-and-white drawings. No index. *Class No: 640.02*

[5462]
HALKIN, P., *and others*. **Dictionnaire technique de l'hôtellerie/** Hotel-Fachwörterbuch/ Dictionary of hotel technical terms. Brussels, Le Livre d'Enseignement; Paris, Dunod, [1975?]. 190p.

About 2,500 French-base terms, with German and English equivalents. The dictionary is in two parts: 1. Food; 2. Hotel. French, German and English indexes. *Class No: 640.02*

[5463]
Hotel, restaurant and catering supplies, 1991. 27th ed. Epsom, Surrey, Sell's Publications, Ltd. xxiv,422p. £40. ISBN: 0854996788.

Exhibitions. Products & services classified, p.1-286; Alphabetical list of companies, p.287-408. Trade names and brands directory. Advertisers index. A buyer's guide of equipment and services for the hotel and catering industry. Quarto. *Class No: 640.02*

[5464]
A Literature guide to the hospitality industry. Sawin, P., *and others, comps*. Westport, Conn., Greenwood, 1990. 99p. $35. (*Bibliographies and indexes in economics and economic history*.) ISBN: 0313267219.

Contains *c*.600 entries giving full bibliographic data. Chapter headings are by format, *e.g.* 'Directories', 'Handbooks', 'Databases'. The chapter on statistics is particularly useful. *Class No: 640.02*

[5465]
NIXON, J.M. **Hotel and restaurant industries** an information sourcebook. Phoenix, Ariz., Oryx, 1988. 240p. $43.50. (*Oryx sourcebook series in business and management, 7*.) ISBN: 0897743768.

More than 1,000 annotated entries up to 1987. Includes US and UK trade journals and books but omits nonprint material. Lists state, national and international associations as well as educational establishments. Author, title, and subject indexes. Essential for libraries covering this subject. *Class No: 640.02*

[5466]
Practical professional catering. Cracknell, H.L., *and others*. London, Macmillan, 1983. xiii, 469p. £11.99. ISBN: 0533360753.

14 chapters (1. The catering industry ... 5. Interpretation of demand - 6/7. Emergence of facilities - 8/9. Provisioning - 10/12. Production and distribution - 13. Control of costs and revenues - 14. Maintaining of consumer satisfaction). 5 appendices: A. Education and training: opportunities and career prospects. E. Select bibliography (grouped), p.456-60. Detailed, analytical index, p.461-9. 67 figures. *Class No: 640.02*

Hotels & Restaurants

Worldwide

[5467]
Agents hotel gazetteer, 1989. London, St. James Press, 1989. 4v.

1. *Mediterranean resorts.* 380p. £28.50. 2. *European cities.* 244p. £25.50. 3. *America.* £24. 4. *Alpine resorts.* 132p. £21.50.

V.1 concerns Cyprus, Gibraltar, Greece, Italy, Malta, Morocco, Portugal, Spain, Tunisia, Turkey and Yugoslavia. Each area/country has a distinctive colour. V.2 has town plans. Price ranges and 2-4-star rankings given. Aims 'to provide travel agents and those engaged in the sale of travel with first hand unbiased reports on the location and amenities of all hotels'. At pains to be evaluative. *Class No: 640.024.1/.3(100)*

[5468]
'Financial Times' world hotel directory, 1992: an essential guide for business travellers. Harlow, Essex, Longman Group, 1991 (1993 ed. £85). xlii, 646p. ISBN: 0582085063.

'Hotels and essential travel facts' (2,699 entries), p.1-620; with data on hotel reservation system; style (*e.g.* traditional); location; manager; rooms and charges; services and facilities; conferences; period when hotel closed. Hotel groups and representatives. Incentive schemes for business travellers. 1-5 star gradings. Appended: City centre maps of key business districts. Geographical index. Executive selection. Advertisers' index. For the business executive. *Class No:* 640.024.1/.3(100)

[5469]
Hotel and travel index. Secaucus, N.J., Reed Travel Group. 4pa. illus. maps.

Formerly *ABC worldwide hotel guide.*

Worldwide hotel listings (*c.*45,000), 9 sections: Europe - Middle East - Africa - Asia - Australia & the Pacific - USA - Canada - Caribbean - Mexico, Central & South America. About 65 symbols indicate facilities. Covers hotel selection and reservation in over 170 countries, wth conference and business venues particularly in mind. Other data include world time zones; air travel times; hotel programs and companies. Area maps; hotel location maps. Geographical index. *Class No:* 640.024.1/.3(100)

Europe

[5470]
The Good hotel guide, 1992: Britain and Western Europe - also Morocco and Turkey. Rubenstein, H., *ed.* London, Papermac (Macmillan), 1991 (1993 ed. £14.99). 836p. maps. £13.99. ISBN: 0333563387.

Recommends hotels chosen for excellence and value in all price brackets based on first-hand reports. Part 1: Great Britain and Ireland, p.3-314. Part 2: The Continent: Austria ... Yugoslavia (18 countries), p.315-780. Alphabetical list of hotels, p.781-95. 19 location maps. Appended hotel report form. *Class No:* 640.024.1/.3(4)

[5471]
Les Guides rouges. Paris, Michelin. Annual. maps.

Benelux. Deutschland. España. Europe. France. Great Britain and Ireland. Italia. Portugal.

'Red guides', providing details of hotels, restaurants and garages in the countries concerned, under towns A-Z. About 70 symbols are applied; tariff ranges indicated. Keyed two-colour town plans are a feature. *France, 1992* (1991. 1300p.) includes 8p. of coloured maps and an index of localities listed by departments. *Class No:* 640.024.1/.3(4)

[5472]
ROYAL AUTOMOBILE CLUB. RAC continental hotel guide, 1990. Croydon, RAC Motoring Services, Ltd. £6.95.

Data on more than 2,000 RAC appointed and recommended hotels in 21 countries of Western Europe and the North Mediterranean (Andorra ... Yugoslavia). Very brief description of locale and data on period open, facilities and tariff. 'Motels and hotels on or near motorways'. 20p. road atlas. *Class No:* 640.024.1/.3(4)

Great Britain

[5473]
AA hotels & restaurants in Britain and Ireland, 1992. Basingstoke, Hants., Automobile Association, 1991. Annual. 878p. maps. £12.99. ISBN: 0749504218.

Gazetteer of AA-approved hotels and restaurants, p.61-800. (Abberley ... York). Prelims. include 'Britain's top restaurants', p.48-49. 1-5 star ratings for hotels, based on reports by AA hotel inspectors. Town plans; 16p. of maps. Effective use of 30 symbols, and abbreviations, interpreted in English, French, German, Italian and Spanish. *Class No:* 640.024.1/.3(410)

[5474]
—Ashley Courtenay's let's halt awhile in Great Britain, 1990. 57th ed. London, Barrie & Jenkins, 1990. 655p. illus. maps. ISBN: 0905881095.

Covers 8 areas: 1. London; 2. The South-east ... 5. Northern England; 6. Channel Islands; 7. Wales; 8. Scotland. For each of over 1,000 hotels, provides concise margin details of tariffs and amenities, with small illus. Alphabetical place index. Descriptions tend to be effusive. *Class No:* 640.024.1/.3(410)

[5475]
RAC hotel guide, 1992. Croydon, RAC Publishing, 1991. 736p. illus., tables, maps. £15.99. ISBN: 0862111633.

First published 1904.

Hotel directory, p.61-672 (London, England, Scotland, Wales, Channel Islands; Northern Ireland; Republic of Ireland). Prelims. include RAC hotel awards; Motor lodges. Full-colour, 64-p. map section. 2-colour town plans, - 52 in all. Covers nearly 6,000 establishments, from five-star hotels to farmhouses. Aimed primarily at the tourist. *Class No:* 640.024.1/.3(410)

France

[5476]
FÉDÉRATION NATIONALE DES LOGIS DE FRANCE. Guide des hôtels, logis de France et auberges rurales. Paris, the Fédération. Annual. maps.

About 4,000 brief 2-3 line entries for inexpensive hotels and accommodation in France and Corsica. 1-3 star gradings. Appended list of places under départements. Atlas of coloured maps, with insets. *Class No:* 640.024.1/.3(44)

[5477]
Michelin France, 1993. 84th ed. Paris, Pneu Michelin, 1993. 1313p. col. maps. £12.50. ISBN: 2060064392.

A highly compressed guide to *c.*11,000 hotels and restaurants in France, arranged under towns, A-Z. Much use made of symbols to categorize facilities; 1-3 star range for sight-seeing. Many good 2-colour maps of towns, locating hotels, etc. Helpful preliminary text and atlas of France. The fullest and best-known of the Michelin 8 'Red guides': France, Benelux, Germany, Spain, Portugal, Great Britain and Ireland, Italy and Europe. *Class No:* 640.024.1/.3(44)

America—North & Central

[5478]
Hotel and motel red book. New York, American Hotel and Motel Association, 1886-. Annual. illus., maps. ISSN: 00733490.

Official directory of the Association's members, covering

....(contd.)

hotels, motels and resorts. Mainly for USA, Canada and Caribbean, with a brief international section. *Class No:* 640.024.1/.3(71/73)

Farms & Guest Houses

Great Britain
[5479]

AA guesthouses, farmhouses and inns in Britain, 1989. Basingstoke, Hants., Automobile Association, 1989. 488p. + 35p. of maps illus. (incl. col.). £5.95. ISBN: 0861457722.

Gazetteer, covering England (with Channel Islands), Wales and Scotland. *c.*2,500 approved guesthouses. Data on each include brief description of the locale, facilities, tariff, and when open. 'AA farmhouses of the year' awards. Town plans. 39 symbols and abbreviations explained in English, French, German, Italian and Spanish. Over 2,000 entries were inspected and approved. *Class No:* 640.024.2(410)

[5480]
COUNCIL FOR ENVIRONMENTAL EDUCATION. The Good stay guide: a guide to environmental centres in Great Britain offering accommodation, leisure and study facilities to groups and families. Shaftesbury, Dorset, Broadcast Books, 1987. ISBN: 1852300027.

Compiled and edited with the support of the Countryside Commission. Previously as *Directory of centres for outdoor studies in England and Wales* (3rd ed. 1981).

Residential centres, youth hostels, farm houses, camping sites, etc. 12 regions for England (Cumbria ... South east England; then Wales, Scotland, p.61-290. Data on access, facilities, activities, tariff, habitat. Prelims. include 'Habitats and opportunities for outdoor studies', p.26-43. Appended: 'Countryside access charter'; 'Resources' (directory of organizations). Index: Conference facilities; Day centres; Facilities for the disabled; Facilities for the under 5's. *Class No:* 640.024.2(410)

[5481]
The Farm holiday guide to holidays in England, Wales & Ireland and the Channel Islands: farms, guest houses and country hotels. Cottages, flats and chalets. Caravans and camping. Activity holidays. Paisley, FHG Publications; Edison, N.J., Hunter Publishing Inc., 1992. 440p. illus., maps. £2.99. ISBN: 1850551391, UK ed; 1556504853, US ed.

Includes details of over 3,000 places to stay. Divided into geographic regions: England, (including the Channel Islands), p.33-370 - Wales, p.385-426 - Ireland, p.432-38. Numerous illustrations. *Class No:* 640.024.2(410)

[5482]
Holidays in the British Isles 1990: a guide for disabled people. London, Royal Association for Disability and Rehabilitation (RADAR), 1990. 600p. map. £3. ISBN: 0900270402.

Coverage: 17 regions of England, London, Scotland, Wales, Channel Islands, Isle of Man, Northern-Ireland. Classifications: Holiday accommodation except self-catering - Self-catering accommodation - Accommodation with nursing care - Accommodation for children and young

....(contd.)

people - Accommodation for groups. Appended: Activity holidays; Transport; Useful publications. Late entries. Comment sheet. *Class No:* 640.024.2(410)

[5483]
RAC small hotels, guest houses and inns, Great Britain and Ireland, 1993. Croydon, RAC Publishing, 1992. 299p. illus., maps. £6.99. ISBN: 0862112095.

Arranged geographically. Gives details of *c.*2,000 establishments approved by the RAC. *Class No:* 640.024.2(410)

Scotland
[5484]

Scotland: hotels and guest houses, 1992. Edinburgh, Scottish Tourist Board, 1992. xxvi,382p. illus., maps. £4.95. ISBN: 0854193480.

Over 2,000 graded entries for hotels, guest houses and university accommodation. Prelims. feature 'Your holiday in Scotland', p.x-xii. About 50 symbols applied. Gradings range from 'Listed' to 5 crowns. *Class No:* 640.024.2(411)

[5485]
—**Scotland: bed & breakfast,** 1992. Edinburgh, Scottish Tourist Board, 1992. xxvi,328p. £3.75. ISBN: 0854193499.

Has over 2,000 graded entries. About 50 symbols applied. 'Your holiday in Scotland', p.x-xii. 'Books to help you', p.327. *Class No:* 640.024.2(411)

[5486]
—**Scotland: self-catering holidays,** 1992. Edinburgh, Scottish Tourist Board, 1992. xxvi,288p. illus., tables, maps. £4.95. ISBN: 0854193502.

Has over 2,000 entries, applying *c.*50 symbols. 'Your holiday in Scotland', p.x-xii. 'Books to help you', p.287. *Class No:* 640.024.2(411)

Northern Ireland
[5487]

Where to stay in Northern Ireland 1991. Belfast, Northern Ireland Tourist Board, 1991. 213p. illus., tables. maps. £2.95. ISBN: 0946871299.

Includes details of over 1,000 hotels, guesthouses, farmhouses, bed & breakfast, self-catering accommodation, youth hostels, camping and caravan parks. 37 symbols applied; grading system. Index. *Class No:* 640.024.2(416)

Eire
[5488]

Ireland accommodation guide '92. Dublin, Irish Tourist Board. 226p. illus., tables, maps.

4 sections: 1. Hotels, guesthouses and Irish homes - 2. Holiday centres, holiday hostels and youth hostels - 3. Caravan and camping parks - 4. Specialist accommodation. 42 symbols indicate facilities. Appended data include youth hostels, services, excursions. Index to locations, p.217-19. *Class No:* 640.024.2(417)

England

[5489]

Where to stay: England, 1992: bed & breakfast, farmhouses, inns & hostels. London, English Tourist Board, 1992. 428p. maps. £6.95. ISBN: 0861431243.

12 regions: London; Cumbria; Northumbria ... Heart of England ... East Anglia ... South east England. 6 rankings: 'Listed' to 5 crowns. Appended prices & 'Useful information'. 7 maps. London index; town index. Index to advertisers. *Class No:* 640.024.2(420)

Wales

[5490]

Wales: hotels, guest houses & farmhouses, 1992. Cardiff, Welsh Tourist Board, 1992. 120p. illus. (incl. col.), maps. £2.50. ISBN: 1850130426.

Lists accommodation checked and graded by the Welsh Tourist Board. Information on tourist attractions and events. Full-colour detailed maps. *Class No:* 640.024.2(429)

Federal Republic of Germany

[5491]

Deutscher Hotelführer, 1986. Stuttgart, Hugo Matthaes Druckerei und Verlag GmbH, [1986]. Annual. xxiv, 731p.

Lists c.14,000 hotels in the German Federal Republic and West Berlin, Aachen ... Zwingenberg. 30 symbols applied. Easier to use than the Michelin Deutschland, but devoid of town maps. *Class No:* 640.024.2(430.1)

Restaurants

[5492]

Egon Ronay's Cellnet guide, 1991. Hotels & restaurants in Great Britain and Ireland: 2,500 establishments. London, Egon Ronay's Guides, 1990. 796p. + 27p. of maps; town plans. ISBN: 1871784085.

1,492 hotels, 947 restaurants. 9 sections (London ... Republic of Ireland), p.160-681. Awards for excellence; restaurants, 1-3 stars. Prices: A-F categories for hotels. Quick reference lists. Prelims. include 'Awards for excellence', 1-3 stars for restaurants. 27p. of maps. Concise, evaluative comments.

1993 ed. Pan Books. 608p. £13.99. *Class No:* 640.024.3

[5493]

The Good food guide 1991. Jaine, T., *ed.* London, Consumers' Association and Hodder and Stoughton, 1991. (1993 ed.). 720p. £14.99). 769p. £12.95. (*Which? book series.*) ISBN: 0340528168.

Covers 1,300 establishments, based on reports, by readers, of 10,000 meals. Arranged A-Z, by place. Maps. Aimed at the individual but would be a suitable addition to general library collections. *Class No:* 640.024.3

[5494]

—Just a bite: 1992 guide: light meals and snacks to suit all pockets and palates. London, Alfresco Publications, 1992. 336p. maps. £7.95. ISBN: 1871784220.

Covers London, England, Scotland, Wales, Channel Islands and Isle of Man, in turn. Details state food offered, when open/closed, and prices for cited dishes. 15 town maps, plus maps 'countrywide and London'. *Class No:* 640.024.3

[5495]

—The Vegetarian good food guide. Whittet, A., *ed.* London, Consumers' Association and Hodder and Stoughton, 1990. 480p. maps. ISBN: 0340514426.

Features nearly 900 cafés, pubs, hotels and restaurants in the UK. The editor's top ten restaurants, p.12, is followed by 5 sections: London - England - Scotland - Wales - Isle of Man. 5 appendices: Alphabetical list of entries - Restaurants awarded a star - Restaurants serving only vegetarian food - Maps (14) - Report forms. *Class No:* 640.024.3

[5496]

Les Routiers guide to Britain 1992. Stroud, Alan Sutton Publishing, 1992. 640p. £8.99. ISBN: 0750900849.

Lists more than 1,800 British pubs, restaurants, hotels and guest houses recommended for good food or good accommodation. *Class No:* 640.024.3

[5497]

Routiers guide to France: the only official guide to the French relais routiers, 1992. London, Alan Sutton Publishers, 1992. 272p. maps. £7.99. ISBN: 0750900857.

Lists more than 3,000 'Les Routiers' cafés, restaurants and hotels in France that provide cheap and good meals for motorists. Under towns, A-Z. 7 symbols provide data. 1-4 stars denote the official classification of the French Tourist Board. 25p. of maps. *Class No:* 640.024.3

Inns & Public Houses

[5498]

The Good beer guide, 1991. 18th ed. St. Albans, Campaign for Real Ale [CAMRA], 1990 (1993 ed. 508p. £7.99). 504p. maps. £6.99. ISBN: 1852490039.

Has brief information (applying 15 symbols) on c.5,000 British pubs serving traditional 'real' ale and beer. Arranged under counties of England, Wales, Scotland, Isle of Man, Channel Islands, and Northern Ireland. A 'Breweries section', and a map section follow. Not to be confused with the fuller, more readable *Good pub guide* (*qv*). *Class No:* 640.024.4

[5499]

The Good pub guide 1992. Aird, A., *ed.* 10th ed. London, Ebury, 1992. (1993 ed. Vermilion. 1021p. £12.99). 943p.+25p. maps. ISBN: 0712647279.

Independent evaluative guide to c.5,000 pubs in England (arranged by counties Berkshire-Yorkshire), Scotland, Wales, and Channel Islands, using reports from up to 8,000 contributors. Each geographical area opens with specifically recommended pubs, followed by a less-detailed 'lucky-dip' section for that area. Opening hours, food & drink specialists, entertainments, etc, are listed, with emphasis on value for money. Overseas 'lucky dip' appendix. *Class No:* 640.024.4

[5500]

PITT, H. The Complete innkeeper: the A-Z of running a pub. London, Robert Hale, 1991. 187p. £15.95. ISBN: 070904397x.

Misleading title as there are 13 chapters: 1. Having what it takes ... 4. Finance ... 8. The police ... 10. Staff ... 13. Expansion. Useful addresses. Further reading. Brief index. *Class No:* 640.024.4

Clubs & Hostels

[5501]

The Guide to budget accommodation 1991-92. Welwyn, Herts., International Youth Hostel Federation, 1990. Annual. 2v. col. illus., town plans, maps.

Formerly *Youth hostel handbook.*

1. *Europe and the Mediterranean.* 336p. 2. *Africa, America, Asia and Australasia.* 254p.

V.1 extends to North Africa and Middle Eastern countries bordering on the Mediterranean. Soviet Russia is omitted from the sequence (Algeria ... Yugoslavia). About 40 signs and symbols applied. General information, on tinted paper, is in English, French, German and Spanish. Outline location map; small town plans. *Class No:* 640.024.6

[5502]

YHA accommodation guide 1992 (England and Wales). St. Albans, Herts., Youth Hostels Association (England and Wales), 1991 (1993 ed. 192p. £4.99). Annual. 192p. town plans, maps. £4.99.

Hostel entries in 28 areas (Northumberland and Roman Wall ... Northern Ireland), p.15-151. About 35 signs and symbols applied. Prelims. include notes on YHA activities, hostel information, membership, world travel, other services. Small town plans; map of England and Wales and Northern Ireland. Advertisers' index. General index. Hostel index.

Scotland (Stirling) and Eire (Dublin) each have a YHA, issuing an annual guide. *Class No:* 640.024.6

Camps & Caravan Sites

[5503]

Camping and caravanning parks, England, Northern Ireland, Scotland and Wales. London, BTA, 1992. Annual. 176p. maps. £5.50. ISBN: 0709548907.

Lists camping and caravan parks, with data on location; area, season, terms and facilities. 44 symbols used. Headings in English, French, German, Dutch. Appended: 'Useful addresses', 'Standards for camping and caravan parks ...' *Class No:* 640.028

[5504]

Camping, caravanning in France. Paris, Pneu Michelin. Annual. table, maps. ISBN: 2060061393.

The 1993 guide (436p. £7.25) gives particulars of more than 3,500 camping sites, A-Z, preceded by a table of localities, under départements. About 80 symbols applied. 22 pages of coloured maps, with numerous maps in text. Prelims. in French, English, German and Dutch. *Class No:* 640.028

[5505]

Europa camping + caravaning / Internationalen Führer/ Guide international, 1991. 38. Ausg. Stuttgart, Drei Brunnen Verlag, GmbH, [1991]. 944p. illus. (mostly col.), maps. ISBN: 379560205x.

Guide to 5,500 camping sites, - Europe (30 countries, including CIS, Czechoslovakia, Hungary, Bulgaria, Rumania); Near East (4 countries); North Africa (5 countries). Heavy use of symbols to include information. Coloured maps, p.827-82. Many advertisements. *Class No:* 640.028

[5506]

RAC camping and caravanning guide, 1990: Europe. Croydon, RAC Motoring Services, in association with the Camping and Caravanning Club of Great Britain, 1990. 416p. tables, maps. £6.95. ISBN: 0862110858.

Sites in 23 countries (Andorra ... Yugoslavia). 10 symbols applied; descriptive text on each country. Prelim. notes on Eastern European countries; Major tunnels; Mountain passes (tables). Index of places, p.410-16. 20p. of black-and-white maps. *Class No:* 640.028

[5507]

RAC camping and caravanning guide, 1992: Great Britain & Ireland. Croydon, RAC Publishing, 1991. 282p. tables, maps. £7.99. ISBN: 0862111552.

Over 2,000 sites (England; Scotland; Scottish Islands; Orkney and Shetland Isles; Wales; Isle of Man; Channel Islands; Isle of Wight; Northern Ireland; Republic of Ireland), p.1-256. Preliminaries include buying camp equipment, National Parks. Index of sites by county. 10p. of maps. *Class No:* 640.028

[5508]

SCOTTISH TOURIST BOARD. Scotland - where to stay: Camping and caravan parks, 1987. Edinburgh the Board, 1987, xxvi. 96p. £2. ISBN: 0854193014.

Has over 300 entries, with about 50 symbols applied. 'Your holiday in Scotland', p.viii-xxvi. 'Books to help you', p.95. *Class No:* 640.028

Consumers & Shopping

Bibliographies

[5509]

FOREMAN, S. Consumer monitor: an annotated bibliography of British government and other official publications relating to consumer issues. Aldershot, Hants., etc., Gower, 1987. vii, 450p. £38.50. ISBN: 0566054019.

Compiled for the National Consumer Council.

About 2,500 annotated entries. 12 subject categories: 1. Housing - 2. The neighbourhood - 3. Transport - 4. Goods and services - 5. Fuel and water - 6. Personal finance and social security - 7. Health and safety - 8. Education - 9. Welfare - 10. Communications - 11. The legal system - 12. Leisure. References (some annotated). Author and detailed analytical subject indexes. *Class No:* 640.03(01)

[5510]

Shopping in the eighties: a guide to sources of information. Kirby, D.A., *comp.* London, British Library, 1988. [v], 58p. £22.50. ISBN: 0712307559.

About 500 references. 18 sections, each with introduction, - Consumer behaviour - Hypermarkets and superstores - Retail (6 sections) - Shopping centres (3) - Teleshopping and new technology - Town centres - Trading hours - Traffic and transport. References are mostly to periodical articles. *Class No:* 640.03(01)

Handbooks & Manuals

[5511]
REES, A.M. A Consumer handbook. Oxford, Blackwell Scientific, 1990. 160p. illus. £9.99. ISBN: 0632006927.

Guide for consumers to legislation, technical changes, and bodies supplying advice and information with regard to provision of goods and services. Covers safety, design and standards, advertising, consumer redress.
Class No: 640.03(035)

Reviews & Abstracts

[5512]
Consumers index to product evaluation and information sources. Ann Arbor, Mich., Pierian Press, 1973-. 12pa., with annual cumulation.

15 subject groups, with subdivisions: 1. Consumerism ... 3. Money and the law ... 5. Food and beverages - 6. Clothing ... 8. Sports, recreations and hobbies ... 10. Transportation ... 12. Computers - 13. The office and business management ... 15. Miscellaneous. Cross-references. For the general and special American consumer.
Class No: 640.03(048)

[5513]
'Which?' cumulative index, 1957-1982. London, Consumers' Association, [1983]. 61p.

Covers a period of 25 years, coinciding with the ending of *Handyman Which?, Money Which?* and *Motoring Which?* supplements to *Which? Gardening from Which?* is not covered, since it only started publication in September 1982 and it carries its own index. Analytical index. About 350 entries. *Class No:* 640.03(048)

Periodicals

[5514]
Which? London, Consumers' Association, 1968-. 12pa.

Each issue contains advice for consumers and evaluative reports on products. Issue for December 1990 has 5 sections: 1. Regulars - 2. Home and leisure (Microwave ovens - Compact cameras ... Kettles) - 3. Your rights - 4. Money (Giving money to charity - Shares) - 5. Public interest (Pet shops - Drinking and driving). Companion publications include *Holiday Which?, Gardening Which?, Which? wine monthly* and *Which? way to health.*

Equivalent consumer reports worldwide include: *Consumer reports* (USA; available online); *Que choisin?* (France); *Test* (Germany); *Rad & ron* (Sweden); *Consumer currents* and *IOCU newsletter* (International Organization of Consumers Unions). *Class No:* 640.03(051)

Yearbooks & Directories

[5515]
Directory of consumer information sources. London, Consumers' Association. 2v. loose-leaf.

Lists organizations in the UK. Includes government and voluntary bodies. Updated twice-yearly.
Class No: 640.03(058)

[5516]
IGLESIA, M.E. de la. **International catalogue of catalogues.** New York, Harper & Row, 1982. 224p. illus. ISBN: 006014985x.

Describes the products offered by more than 1,500 mail-order firms worldwide; aimed at the affluent and others.
Class No: 640.03(058)

Tables & Data Books

[5517]
The United States mail order industry. Homewood, Ill., Business One Irwin, 1991. 225p. tables. $55. ISBN: 1556234864.

Provides statistics on mail-order sales in the US. Direct mail in Europe is discussed in one brief chapter. Business profiles of 25 consumer mail-order companies and 25 business-to-business marketers include data on ownership, company history, quantity of catalogues mailed, markets, etc. *Class No:* 640.03(083)

Laws

[5518]
Encyclopedia of consumer law. Thomas, W.H., *general ed and* Ervine, C., *Scottish ed.*. London, Sweet & Maxwell/W. Green & Son, 1980-. Loose-leaf v. Updates issued twice-yearly.

5 parts: 1. Statutes (all relevant statutes, including Consumer Protection Act, 1987) - 2. Statutory instruments - 3. Cases (competition; contract; credit; criminal; sale of goods; Scottish; services) - 4. Circulars and notices - 5. European Economic Community material (Treaty of Rome, and relevant secondary legislation). Index.
Class No: 640.03(094.1)

[5519]
—**STANESBY, A. Consumer rights handbook.** London, Pluto Press, 1986. [vii], 236p. £7.50. ISBN: 0745301622.

Has 4 parts: 1. Buying goods - 2. Buying services - 3. Buying credit - 4. Dealing with problems. Appendices: Index of statutory instruments; Case index; Statutes index. Written by a solicitor 'for private consumers and not for people who are in business' *(Note)*. *Class No:* 640.03(094.1)

[5520]
A Handbook of consumer law. 3rd ed. London, Consumers' Association, and Hodder & Stoughton, 1989. 252p. tables. £5.95. ISBN: 0340488778.

First published 1982.
15 chapters: 1. Consumer law - 2. Sensible buying and pursuing complaints - 3. Shops and shopping - 4. Prices - 5. Consumer safety - 6. Food - 7. Contracts, liabilities and exclusion clauses - 8. Misrepresentation - 9. Trade descriptions and labelling - 10. Defective goods - 11. Defective services - 12. Insurance - 13. Financial and professional services - 14. Consumer credit - 15. Fair trading. Appendices: 1. Useful addresses; 2. Suggestions for further reading, p.238 (5 evaluative annotated entries). Index, p.241-52. *Class No:* 640.03(094.1)

Great Britain

[5521]

BSI buyers guide, 1988/89. Milton Keynes, BSI Quality Assurance, 1988. 636p.

Classification of quality assured companies, p.47-556. Subject index, p.3-31 (Abrasive products ... Zirconium manufacturing). Index of company names, p.595-36. Kitemarks indicate quality-tested samples; Safety marks indicate compliance with British/international standards specifically concerned with safety. *Class No:* 640.03(410)

[5522]

DAVIS, D. A History of shopping. London, Routledge & Kegan Paul, 1966. xii, 322p. illus. (pl.). (*Studies in social history*.)

Four parts (13 chapters): The late Middle Ages - Queen Elizabeth to Queen Anne - The eighteenth century - Modern times [to the 1960s], from fairs, markets and travelling tallymen to the modern supermarket. Select bibliography, p.303-10. Detailed analytical index, Glossary, p.xi-xii. 16 plates. *Class No:* 640.03(410)

[5523]

MONTAGUE, J. The A to Z of shopping by post. 2nd ed. Watford, Exley Publications, 1979. 252p. illus. ISBN: 0905521269.

35 chapters (1. Large department stores & mail-order houses; 2. Animal & pets; 3. Art ... 6. Books ... 18. Gardening equipment ... 29. Office supplies ... 33. Soft furnishings and fabrics ... 35. Toys & games). Index of *c.*1,000 firms. Product index. Data on catalogues available, terms of business. *Class No:* 640.03(410)

[5524]

Yesterday's shopping: the Army & Navy Stores catalogue, 1907. London, David & Charles, 1969. lxiii, 1282p. illus., tables, plans.

A facsimile of the Army & Navy Society's 1907 issue of 'Rules of the Society and price list of articles sold at the Store'. 15 departments involved. 'Export, shipping and passenger and baggage department', p.1229-82. Meticulous details and pricing. *Class No:* 640.03(410)

Thesauri

[5525]

Thesaurus of consumer terms. Askew, C., *comp.* London, Consumers' Association; The Hague, International Organization of Consumers' Unions, 1979-82. 2v. ISBN: 085202153x.

1. *Classified display,* 1979, li, 314p. 2. *Alphabetically structured display.* 1982. xxix, 289p.

About 7,000 terms involved. Pt.1 comprises 'Introduction and instructions for use' (p.iv-li), alphabetical index (p.3-116, on yellow-tinted paper), and classified display, p.117-314. *Class No:* 640.03:025.43

Libraries

[5526]

IOCU guide on consumer libraries. 2nd ed. The Hague, International Organization of Consumers' Unions, 1982. 53, [1]p. ISBN: 9070241064.

Directory of 49 libraries (1 per page) in 26 countries, Australia ... USA. Data on each note type of holdings (special holdings, classification, working language(s), library

.... (contd.)

publications, information services provided, opening hours, library staff. Appended: 'Thirty recommended books on consumerism'. *Class No:* 640.03:061:026/027

Institutions & Associations

[5527]

Consumer sourcebook. 6th ed. Detroit, Mich., Gale, 1989. 526p. $185. ISBN: 0810312997.

First published 1974.

A compendium, under 26 subject headings (*e.g.* child services; energy/environment, financial services, food and drug safety, handicapped), of *c.*6,000 programs and advisory services available to the American consumer at little or no cost. Sources extend from federal and state agencies, associations and organizations, to information centres and clearing houses. Selected annotated bibliography. Master name and keyword index.

7th ed., 1991. £125. *Class No:* 640.03:061:061.2

[5528]

International consumer directory. Harlow, Longman, 1989.

Formerly *Consumer directory.*

Coverage: of over 300 consumer organizations, arranged A-Z within each country. Data include details of staff and finance, description of organization, specific areas of work, regular publications. *Class No:* 640.03:061:061.2

Food & Drink

Food Preservation

[5529]

Home preservation of fruit and vegetables. 14th ed. London, HMSO, 1989. 215p. illus., tables. £4.95.

First published 1929.

19 sections: Principles of preservation - Preserving methods ... Jams ... Bottled fruit ... Freezing ... Chutneys ... Drying and salting - Storing fruit and vegetables - Some common faults and their likely causes. Numerous recipes. Detailed recipe and general indexes. *Class No:* 641.4

Cookery

Bibliographies

[5530]

Bibliography of cookery books, 1875-1914. Driver, E., *ed.* London, New York, Prospect Books/Mansell, 1989. 748p. illus. £50. ISBN: 0907325416.

Lists *c.*1,200 individual works. Gives locations in many libraries in the UK, US, Australia, Canada, and South Africa. Some biographical details of authors are given. 3 indexes: date; title; subject. *Class No:* 641.5(01)

[5531]

BULSON, C. **Current cookbooks.** Middletown, Conn., Choice, 1990. $12. ISBN: 091449208x.

Annotated list of 250 books, mostly published since 1970. Arranged alphabetically by cuisine, Armenian ... Vietnamese, cooking techniques, courses, foods and audience. *Class No:* 641.5(01)

[5532]

Cook's index: an index of cookbooks and periodicals from 1975-1987. Evanston, Ill., John Gordon Burke Publisher, 1989. Triennial. 536p. $55. ISBN: 0934272093.

Index to English-language cookbooks and periodicals from 1975-1987. *Class No:* 641.5(01)

[5533]

FERET, B.L. **Gastronomical and culinary literature:** a survey and analysis of historically-orientated collections in the US. Metuchen, N.J., Scarecrow Press, 1979. iii, 124p. $16.50. ISBN: 0810812045.

Part 1: The literature (Introduction; 11 sections, each with short introduction: Before printing; Italy, 15th and 16th centuries ... England and France, 19th and 20th centuries; United States, 18th, 19th and 20th centuries) - Part 2: The collections (by state, annotated entries, p.54-90; index precedes). Appendix: A. Culinary bibliographies and bibliographic commentaries; B. Secondary historical texts and references. Non-analytical index, p.111-24.

Class No: 641.5(01)

[5534]

Gastronomic bibliography. Bitting, K.G., *ed.* San Francisco, privately printed, 1939 (reprinted 1981. London, Holland Press). 718p. illus. ISBN: 0875567231.

6,000 entries, fully annotated, in 2 parts: authors; anonymous works. Short-title index, with a rough, incomplete subject index. The only bibliography that adequately covers both historical and more modern books on gastronomy. *Class No:* 641.5(01)

[5535]

—MACLEAN, V. A Short-title catalogue of household and cookery books published in the English tongue, 1701-1800. London, Prospect, 1981. xxiv, 1970p. facsims. ISBN: 0907325068.

With index. *Class No:* 641.5(01)

[5536]

—SIMON, A.L. Bibliotheca gastronomica: a catalogue of books and documents on gastronomy. London, Wine and Food Society, 1953. 196p. illus., facsims.

Carries 1,644 annotated entries. An author list, with short-title and subject indexes. Includes facsimiles of rare early books. *Class No:* 641.5(01)

[5537]

—VICAIRE, G. Bibliographie gastronomique: a bibliography of books appertaining to food and drink and related subjects, from the beginning of printing to 1890. London, Versthoyle, 1954; Holland Press, 1978. xviiip, 974 cols.

A facsimile of the 1890 bibliography. An author list, with bibliographical notes. Records French, Latin, English, German and Spanish titles from 15th century up to 1889, with values. A collector's piece, like the Simon.

Class No: 641.5(01)

[5538]

PATTEN, M. **Books for cooks:** a handbook of cookery. London & New York, Bowker, 1975. viii, [1], 526p. ISBN: 0859350053.

1,660 numbered items, under authors A-Z, with 1-8 line annotations; 'Late entries', p.457-72. Emphasis on books published in Britain, with some foreign titles. Prices are given and some Braille books included. The very necessary subject index is detailed but only partly analytical (p.472-95); title index. *Class No:* 641.5(01)

[5539]

PRYTHERCH, R. *and* STANLEY, S. **Food, cookery and diet:** an information guide. Aldershot, Gower, 1989. ix,109p. £15. ISBN: 0566035685.

Lists the standard, basic sources. 6 sections: 1. Basic reference sources - 2. The scientific background - 3. Professional catering - 4. Training, education and careers - 5. Home cooking - 6. Healthy eating and special diets. Index. *Class No:* 641.5(01)

Encyclopaedias

[5540]

Cooking A to Z. Horn, J., *and others.* San Ramon, Calif., Ortho Books, 1988. 640p. illus. $32.95. ISBN: 0897211472.

Arrangement is A to Z, with over 600 recipes and nearly 500 entries, either defining cooking terms, describing and illustrating cooking techniques or offering information about specific ingredients. Well indexed and beautifully illustrated. *Class No:* 641.5(031)

[5541]

New Larousse gastronomique; the world's greatest cookery reference book. London, Hamlyn, 1977. [viii], 1064p. illus., maps. ISBN: 0828801576, US ed.

Original French ed. 1938, as *Larousse gastronomique.* As *Nouveau Larousse gastronomique,* 1968. Published in US, New York, French and European, 1984. (1152p. $150).

Entries 'Abaisse' ... 'Zweiback'. ('Fritter', p.394-8, records nearly 100 recipes, with brief cooking instructions; 'Hors d'oeuvres', p.457-88). 'International cookery', p.501-12, is a very brief survey of national specialities. 'Many of the recipes in this encyclopedia cater for large numbers of people and it would be merely pedantic to try to standardise quantities' (*Introduction* to 1961 ed.). Small type and illus., some lush whole-page colour plates. Detailed analytical index, p.1008-64. For the skilled cook.

Class No: 641.5(031)

[5542]

STOBART, T. **The Cook's encyclopedia.** Owen, M., *ed.* New York, Harper & Row, 1981. 547p. ISBN: 0060141271.

An A-Z encyclopaedia of ingredients and processes, with cross-references. Not another compilation of recipes. Simon, A.L., and Howe, R. *Dictionary of gastronomy* (McGraw-Hill, 1970. 400p.) describes more concoctions and ingredients; Stobart concentrates rather on explaining processes. Well organized. *Class No:* 641.5(031)

Handbooks & Manuals

[5543]

BEETON, I.M. **Mrs. Beeton's cookery and household management.** London, Ward Lock, 1992. 1408p. illus. (incl. col.), diagrs., tables. £25. ISBN: 0706371011.

First published 1861 (reprinted Cape, 1968. 1112p.);

....(contd.)

previously in monthly parts, November 1859 - October 1961. The 1960 ed. is the first completely modern ed., 'although it contains not one recipe from the original "Mrs. Beeton"' (*The Observer*, 19 April 1964, p.33).

'Household management section' plays a secondary role to the 'Cookery section'. The latter ranges from 'Glossary of cookery terms and techniques', 'Seasonings and flavours' to 'Convenience foods', 'Regional British cookery', 'Foreign cookery', 'Menus' and 'Party food'. Detailed, analytical index. *Class No:* 641.5(035)

[5544]
CRACKNELL, H.L. *and* **KAUFMANN, R.J. Practical professional cookery.** 3rd ed. Basingstoke, Macmillan, 1992. vii,908p. £14.99. ISBN: 0333559258.

First published 1972.

20 sections: 1. Introduction - 2. The methods and general principles of cookery ... 8. Eggs ... 10. Fish ... 13. Poultry, p.402-70 ... 16. Salads and salad dressings .... 20. Pastry and sweet dishes. Number of recipes has increased by 50 percent to over 2,000 and now includes a wide range of international and ethnic dishes. Glossary. Select bibliography, p.881-2. Index. Companion volume to *Practical professional catering (q.v.). Class No:* 641.5(035)

[5545]
ESCOFFIER, A. The Complete guide to the art of modern cookery. Cracknell, H.L. *and* Kaufmann, R.J., *tr.* London, Butterworth-Heinemann, 1991. xxvi,646p. £30. ISBN: 0750602880.

Supersedes *A guide to modern cookery* (2nd ed. 1957). Originally as *Le guide culinaire* (1902).

5,011 recipes. 17 chapters (1. Sauces ... 3. Soups ... 6. Fish ... 12. Roasts ... 16. Sweets, puddings and desserts ... 17. Poached fruits (compotes), jams and drinks). Both imperial and metric measures given. Glossary, p.589-92. Specimen menus, p.593-609 (Christmas menu, 1906: 31 items). Detailed index. *Class No:* 641.5(035)

[5546]
FITZGIBBON, T. The Food of the western world: an encyclopedia of food from Europe and North America. London, Hutchinson, 1976. xxx, 529p. illus., tables. ISBN: 0812904273.

About 6,000 entries for ingredients, dishes, methods of preparation, recipes (Soufflé, varieties: 3 columns, 1 illus.; Pizza: varieties: 1p., 1 illus.) and culinary terms from 34 countries. 2,500 brief recipes. Bibliography, p.xxix-xxx. Tables show cuts of meat for beef, veal, pork, lamb and mutton for America, England and France. Entry-words are insufficiently distinctive. 'A fine complement to the classic Larousse gastronomique' (*Choice*, v.13(8), October 1976, p.957). *Class No:* 641.5(035)

[5547]
'Good Housekeeping' microwave cookbook. London, Ebury Press, 1989. 400p. illus. (col.). £14.99. ISBN: 0852237391.

3 parts: 1. Basic techniques, p.10-22 (includes boiling; frying, preserving; sweet making) - 2. Recipes, p.23-385 - 3. Charts, glossary and index. *Class No:* 641.5(035)

[5548]
The 'Good Housekeeping' step-by-step cook book. Edden, G., *editor-in-chief.* London, Ebury Press, 1980. 512p. illus. (incl. col.), tables. £18.99. ISBN: 0852231598.

Appetisers - Soups - Eggs - Salads - Shellfish - Fish - Meat - Poultry - Vegetables - Fruit - Salads - Pasta - Rice - Pastry and sweet pies - Desserts - Ice creams and iced desserts - Cakes - Biscuits and cookies - Confectionery - Simple baking - Yeast baking - Sandwiches - Sauces - Preserves - Drinks - Useful information (*e.g.* Entertaining; Storage guide). Glossary of food and cooking terms, p.492-7. Detailed, analytical index, p.497-512. *Class No:* 641.5(035)

[5549]
McGEE, H. On food and cooking: the science and lore of the kitchen. 3rd ed. London, Allen & Unwin, 19846. xix, 684p. illus., tables. £12.95. ISBN: 004306003x.

3 parts (14 chapters): 1. Foods (1. Milk and dairy products ... 10. Food additives) - 2. Food and the body - 3. The principles of cooking: a summary. Appendix: A chemistry primer - atoms, molecules, energy. Bibliography, p.648-52. *Class No:* 641.5(035)

Dictionaries

[5550]
ALLISON, S. The Cassell food dictionary. London, Cassell, 1990. xxxv,480p. figs. £19.95. ISBN: 0304318752.

More than 6,000 entries, à la ... Zwiebelrostbraten, in the dictionary section, p.1-452. Layout is 2 columns per page, with length of definition varying from a few lines to half a column. Occasional black-and-white drawings devoted to a particular subject, *e.g.* 'Puddings and deserts'. A thematic index, p.xiii-xxxv, arranges items of food according to type or country. Index of cross-references, p.453-80. No bibliography. Unexciting design and layout but contains useful information. *Class No:* 641.5(038)

[5551]
'Bon appétit': a French-English gastronomic dictionary. White, J.A., *comp.* Ely, Cambs., Peppercorn Books, 1990. 111p. £4.80. ISBN: 0951641301.

Small, handy guide containing over 3,500 definitions, including many regional and specialized dishes. *Class No:* 641.5(038)

[5552]
HERBST, S.T. Food lover's companion: comprehensive definitions of over 3,000 food, wine and culinary terms. New York, Barron's, 1990. 582p. illus., tables. $10.95. ISBN: 0812041569.

Length of definitions varies from a sentence to two pages. Covers foods, dishes and sauces, kitchen equipment, cooking techniques, names of drinks, menu terms and brand names. Pronunciation for foreign terms is given in addition to numerous cross-references. Appendix includes an additives directory, a detailed spice and herb chart and meat cut illustrations. 'Comprehensive, accurate, interesting, well written and inexpensive' (*Booklist*, 1 June 1990, p.1923). *Class No:* 641.5(038)

[5553]

HERING, R. Hering's dictionary of classical and modern cookery, and practical reference manual for the hotel, restaurant and catering trade. 11th rev. ed. Translated by W. Giessen. Giessen (Germany), Pfanneberg, 1958; London, Mills & Boon, 1960 (and reprints). [iv], 852p.

First published as *Lexikon der Küche,* 1907.

A many-section dictionary (Hors d'oeuvre to Sweets), p.1-745. Vocabulary, English-French-German-Italian-Spanish, p.823-52. Claims to contain over 15,000 recipes, but average entry per recipe runs to only 2½ lines, - no more than a statement of ingredients. Includes regional dishes. A practical reference manual for hotel restaurant and catering rather than a cookery book in the accepted sense. *Class No:* 641.5(038)

[5554]

NEIGER, E. Gastronomisches Wörterbuch zur Übersetzung and Erklärung der Speisekarten. [Gastronomic dictionary for the translation and explanation of menus.] 2. verbesserte und erw. Aufl. Munich, Gerber, 1971.

First published 1962.

German-French-English-Italian. About 2,000 terms defined in 20 sections (Saucen ... Gemüse ... Kartoffeln ... Frühstück). 3 appendices. *Class No:* 641.5(038)

[5555]

—ANTHONY, P. Patrick Anthony's international menu dictionary. Luton, Lennard Publishing, 1988. 167p. ISBN: 1852910550.

4 sections: 1. At a glance translator (À la Grèque ... Zwiebel) (*c.*1,400 entries) - 2. The menu dictionary (Abalone ... Zwieback) - 3. Key translator of main foods: English-French-Spanish-German-Italian - 4. The history of beverages, p.160-66. Short bibliography. A handy pocket-size guide, international in scope, covering, among other things, food from Europe, India, the Caribbean and the Far East. *Class No:* 641.5(038)

Tables & Data Books
[5556]

SCHMIDT, A. Chef's book of formulas, yields and size. New York, Van Nostrand Reinhold, 1990. 338p. $39.95. ISBN: 0442318359.

Arranged alphabetically, Abalone ... Yucca root. Data include the forms in which foods are available (*e.g.* canned, fresh, frozen), suggested serving sizes, number of servings from the most common packet sizes, and new meat cuts. Cross-references and index. Author is affiliated to the Culinary Institute of America. *Class No:* 641.5(083)

Great Britain
[5557]

British cookery: based on research undertaken for the British Food Information Service of Food from Britain and the British Tourist Authority by the University of Strathclyde. Boyd, L., *ed.* 2nd ed. London, Christopher Helm, 1988. 651p. £7.95. ISBN: 0747002231.

About 2,000 native English, Scots, Irish and Welsh recipes, ranging from yeast cookery to preserves and chutneys. New ed. 'includes changes and additions to the range of foodstuffs that are available' (*Foreword*). Of the

....(contd.)

previous ed., 'Bids fair to become a standard reference work' (*Daily telegraph,* 2 December 1976, p.13). Excellent value. *Class No:* 641.5(410)

France
[5558]

La Cuisine: the complete book of French cooking. Translated from French. Létoile, V.-A., *and others.* London, Orbis, 1985. 552p. col. illus. £17.50. ISBN: 0356150836.

15 sections: Equipment - The value of food - Wine - French cheeses - Buffets - Basic cooking techniques ... Sauces ... First courses ... Meal - Vegetables ... Desserts, cakes and pastries. Recipes for *c.*2,000 dishes state French and English names, heating time, measures (American/ Imperial, metric), ingredients, industrial. Glossary of French cookery terms, p.324-9. French and English indexes. Fully illustrated in colour. *Class No:* 641.5(44)

Asia
[5559]

PASSMORE, J. The Letts companion to Asian food & cookery. London, Letts, 1991. 320p. figs. £16.95. ISBN: 1852381515.

Entries arranged A-Z (Aamchur ... Zoni), using names in English and anglicized Asian. Asian names are used as cross-references, of which there are more than 1,000. Pronunciation guides for all main entries and cross-references. Entries include information on cooking methods, country of origin, preparation techniques. Recipes appear throughout, and a recipe index is included. Brief bibliography. Packed with information. *Class No:* 641.5(5)

China
[5560]

HOM, K. Ken Hom's Chinese cookery. London, BBC, 1984. 260p. illus., diagrs., tables. £7.99. ISBN: 0563210532.

Introduction - Ingredients - Equipment - Techniques - Menus and how to eat Chinese food. - Soups - Meat - Chicken, duck and game - Fish and shellfish - Vegetables - Rice, noodles and dough. Snacks and sweets. Mail order list. Detailed, analytical index. Well illustrated. *Class No:* 641.5(510)

Recipe Books
[5561]

The International cookery index. Kleiman, R.H. *and* Kleiman, A.M., *comps.* New York, Neal-Schumann, 1987. 230p. $65. ISBN: 0918212871.

An index to 25,000 recipes, cited from 52 "classic" or well-known cookery books. 'Embraces most major ethnic and regional cultures' (*Library journal,* August 1987, p.116). *Class No:* 641.55

[5562]

—TORGESON, K.W. *and* WEINSTEIN, S.J. The Garland recipe index. New York, Garland Publishing Co., 1984. 372p. (*Garland reference library of the humanities, 414.*) ISBN: 0824091248.

Indexes 48 popular cookery books, both classics and specialized. Entries appear under name of recipe, stating principal food ingredients and method of cooking. Well cross-referenced. *Class No:* 641.55

[5563]

—Traditional recipes of the British Isles. Heaton, N., *comp.* London, Faber, 1951. 215p.

Has sections: Special days and their customs - Food and table manners - Northern counties and the Isle of Man - Southern counties - Eastern counties - Midlands - Western counties, Wales and Northern Ireland - Scotland and the Isles. Detailed index. *Class No:* 641.55

Vegetarian Cookery

[5564]

DYER, J.C. Vegetarianism: an annotated bibliography. Metuchen, N.J., Scarecrow Press, 1982. xi,280p. $22.50. ISBN: 081081532x.

Contains 1,412 entries covering all aspects of vegetarianism. Appendix lists 200 vegetarian cookbooks. *Class No:* 641.564

[5565]

ELLIOT, R. Rose Elliot's complete vegetarian cookbook. London, Collins, 1985. 361p. £10.95. ISBN: 0004120094.

Contains over 1,000 recipes. 14 sections: 1. Soups - hot and chilled ... 6. Main course vegetable and nut dishes ... 10. Savoury flans and pies ... 14. Bread and yeast cookery, including pizza. Detailed index, p.347-61. *Class No:* 641.564

[5566]

LOVELOCK, Y. The Vegetable book: an unnatural history. London, Allen & Unwin, 1972. 382p. illus.

3 main parts: 1. Leaves, stems, roots and fruit (Aku ... Yams, p.25-258) - 2. Miscellaneous nuts, grains and pastas (Acorns and beechnuts ... Wheat), p.261-312 - 3. Various herbs, spices and condiments. Gives geographical locale, history and use, and names of vegetables in various languages. 19 decorative illus. Indexes of Latin and common names. Aims 'to treat all the important foods and staples of the world, avoiding as far as possible a narrow Western viewpoint' (*Introduction*). *Class No:* 641.564

Condiments, Garnishes & Pickles

[5567]

The Complete book of herbs. Bremness, L., *ed.* London, Dorling Kindersley, 1988. 288p. illus. (col.). £15.99. ISBN: 0863183131.

8 sections: Herbs in the garden - Herbal index - Using herbs (*e.g.* herbal decorations; herbs in the kitchen; herbs for the household) - Herbs for beauty - Essential oils - Herbs for health - Cultivating and harvesting herbs - A catalogue of herbs. Glossary. Useful addresses. Bibliography. Index. Many useful illustrations. *Class No:* 641.88

[5568]

MULHERIN, J. Spices and natural flavourings a complete guide to the identification and uses of common and exotic spices and natural flavourings. London, Ward Lock, 1988. illus. (col.). ISBN: 0025878500.

Review of more than 60 spices and flavourings. Entries include a brief identification with historical notes and information on the country of origin, culinary uses, methods of preparation, storage, and a recipe. List of suppliers and useful addresses. Index. *Class No:* 641.88

[5569]

NORMAN, J. The Complete book of spices. London, Dorling Kindersley, 1990. 160p. illus. (col.), maps. £14.99. ISBN: 0863184871.

5 main sections: The spice trade past and present - Spice index (arranged by botanical name) - Spice mixtures (*e.g.* 'Sambal'; 'Berbere') - Cooking with spices, p.110-45 - Spices in the home. Index. Bibliography (85 items). Data for each spice include distribution; appearance and growth; aroma and taste; culinary, medicinal and cosmetic uses. Clearly and colourfully presented. *Class No:* 641.88

[5570]

RINZLER, C.A. The Complete book of herbs, spices and condiments. New York, Facts on File, 1990. 199p. $19.95. ISBN: 0816020086.

Entries vary from one to three pages in length. Data include botanical name, common name, native habitat, medicinal properties, alternative uses, nutritional profile, effect on the body and how it may be grown. Index and bibliography. *Class No:* 641.88

[5571]

STOBART, T. Herbs, spices and flavorings. Woodstock, N.J., Overlook Press, 1986. 320p. illus. $22.50. ISBN: 0879511486.

A readable guide to some 400 herbs/spices, A-Z ('Mustard': 9p.). Common names and synonyms feature in 5 languages. *Class No:* 641.88

The Home

Bibliographies

[5572]

ATTAR, D. A Bibliography of household books published in Britain, 1840-1914. London, Passport Books, 1987. 438p. ISBN: 0907325351.

737 entries, p.63-388. Sections: Domestic economy - Servants - Domestic medicine and childcare - Etiquette - Selective notes on secondary works. British and US library locations. Subject index. Chronological list of short titles. Short-title index. *Class No:* 643/645(01)

Encyclopaedias

[5573]

How to do just about anything. London, New York, Reader's Digest Association, 1988. 543p. illus. (col.). £22.95. ISBN: 0276419936.

16 contributors. Arranged A-Z (Abacus ... Zodiac) with about 2,000 entries, and including topics such as mending

....(contd.)

punctures, needlework, gardening and general home maintenance. Many colour illustrations. Index. Bibliography (in very small type). *Class No:* 643/645(031)

[5574]

The 'Which?' encyclopedia of the home. London, Consumers' Association, and Hodder & Stoughton, 1984. 671p. illus., tables. £14.95. ISBN: 0340351055.

68 sections, - Accidents and first aid ... Windows. Wide range of topics, *e.g.* Buying and selling a home; children; cookery (p.133-67); health; painting and decorating; pets; plumbing and drainage; sewing; taxation; vegetable growing. Bibliography (building and maintenance; health and fitness; household management; money, law and consumer issues). Addresses, p.658-62. Detailed analytical index, p.663-71. Well produced and excellent value. *Class No:* 643/645(031)

Handbooks & Manuals

[5575]

BREMNER, M. Enquire within upon everything. 2nd ed. London, Century, 1990. viii,756p. figs. £17.99. ISBN: 0091745535.

First published 1988.

16 sections: 1. Social behaviour - 2. Food ... 4. Domestic matters ... 9. Things mechanical - 10. Law - 11. Money - 12. In sickness ... 15. Organizations (Consumer bodies ... Medical - Police ... Travel and transport) - 16. Facts and figures. Bibliography. Index. 'It is a first source which will give the basics and provide pointers to where to find out anything else you want to know' (*Preface*).
Class No: 643/645(035)

[5576]

—Good Housekeeping household hints: the essential home facts. London, Ebury Press, 1989. 146p. ISBN: 0852237286.

9 sections: 1. Safety first ... 4. Pests and infestations ... 7. Flora and fauna ... 9. Emergency action. List of useful addresses. Index. For a small public library.
Class No: 643/645

[5577]

Henley's formulas for home and workshop; containing ten thousand selected household, workshop and scientific formulas, trade secrets, food and chemical recipes, processes for money saving ideas. Hiscox, G.D., *ed.* New York, Avenue Books, 1979 (reprinted 1991, Outlet Book Company). $9.99. xxviii, 9-809p. illus. ISBN: 0517293072.

First published 1907.

Entries, 'Acid-proofing' ... 'Yeast tests'. Preceded by a section, 'Useful workshop and laboratory methods'. Small line-drawings. *Class No:* 643/645(035)

Histories

[5578]

DU VALL, N. Domestic technology a chronology of developments. Boston, Mass., G.K. Hall, 1988. 535p. $40. ISBN: 0816189137.

2 sections: Part 1 consists of chapters on food, clothing, waste disposal cleaning, tools, lighting, heating and housing and health; Part 2 is a continuous chronology from 500,000 BC to 1987. Criticized in *Choice*, May 1989, for an inadequate index but still useful for academic and public libraries. *Class No:* 643/645(091)

[5579]

HARDYMENT, C. From mangle to microwave: the mechanizaton of household work. Cambridge, Polity Press, 1988. xii,220p. illus. £10.95. ISBN: 0745602061.

10 Chapters: 1. Homes without machines - 2. Of pig-iron and parlourmaids ... 5. House-cleaning ... 8. Essential kitchen technology ... 10. The domestic mystique. Bibliography, p.203-11. Index. A good overview.
Class No: 643/645(091)

[5580]

—YARWOOD, D. Five hundred years of technology in the home. London, Batsford, 1983. 184p. illus., ports. ISBN: 0713435062.

9 chapters: 1. Science and society - 2. Fuels and sources of power - 3. Food - 4/6. The home: The materials in use; Exterior and structure; The interior - 7. Textiles - 8. Communication and entertainment - 9. Modern technology: the way ahead (the microelectronic revolution). Select bibliography, p.176-7. Detailed, analytical index. 263 illus. Written primarily for students of social and economic history, as well as for home economics at O-level standard. *Class No:* 643/645(091)

Maintenance

[5581]

A Consumer's guide to home improvement, renovation, and repair. Santucci, R., *and others.* New York, John Wiley & Sons, 1990. xiv,270p. illus., tables. $34.95. ISBN: 0471519227.

Basic aim is to show the reader how to cut costs in home repair and improvement, while still maintaining safety standards. 2 sections: 1. Methods and materials (15 chapters: 1. Walls and ceilings ... 9. Carpentry ... 15. Outside work and foundations) - 2. Money-saving strategies (4 chapters: 16. Planning and design ... 18. Working with design professionals and contractors - 19. Saving money on tools and materials). Index. US-slanted, but contains ideas which would be applicable anywhere. *Class No:* 643.0045

[5582]

The Homeowner's complete manual of repair & improvement. Bragdon, A.D., *ed.* New York, Arco, [*c.*1983]. 576p. illus. $14.95. ISBN: 0668057378.

Originally as *Peterson's Home repair & maintenance guide.*

6 sections (subdivided into chapters): Interior repairs & decoration - Windows - Doors - Security & insulation - Furniture care & refinishing - Electrical fixtures, wiring and appliances - Plumbing and heating - Exterior maintenance & improvements. Excellent illus. and step-by-step instructions. Brief index. Given high marks for simplicity, completeness, accuracy and readability in *Reference books bulletin, 1983-1984,* p.102. *Class No:* 643.0045

[5583]

Mend it! A complete guide to repairers and restorers. Ball, R., *comp.* London, Orbis, 1985. 240p. ISBN: 0856135100.

Books and paper - Building and sculpture ... Carpets, canvas and other textiles ... Clocks, watches and barometers ... Furniture and woodwork - Glass ... Musical instruments and equipment ... Silver and other metals - Transport. Useful addresses (grouped). Bibliography, p.229-31. Detailed, analytical index. (*c.*1,000 entries).
Class No: 643.0045

[5584]

Reader's Digest new D-I-Y manual. London, Reader's Digest Association, 1987. 504p. illus. (col.), figs., diagrs. Looseleaf. £32.95. ISBN: 0276404564.

2 sections: 1. Items for the home (*e.g.* Kitchens ... Bedrooms ... Outside the house) - 2. Techniques, tools and materials (*e.g.* Painting, wallpapering and tiling ... Home security ... Central heating ... DIY in the garden). Index, p.1-5. A step-by-step guide with numerous illustrations. A standard DIY title. *Class No:* 643.0045

[5585]

—Reader's Digest fix-it-yourself manual. London, Reader's Digest Association, Ltd., 1977. 480p. illus., diagrs.

Complementary to *Reader's Digest new D-I-Y manual*. It concentrates on several dozen types of devices and tools such as plumbing fixtures, locks, bicycles, lamps, typewriters and garden tools. Clear instructions, well illustrated. Detailed table of contents and index. *Class No:* 643.0045

[5586]

The 'Which?' book of do-it-yourself. London, Consumers' Association, and Hodder & Stoughton, 1992. 320p. illus., tables. £19.95. ISBN: 0340568682.

20 contributors. 12 chapters: 1. Decorating - 2. Electricity - 3. Woodworking - 4. Fixings - 5. Building - 6. Floors - 7. Roofs - 8. Doors and windows - 9. Security - 10. Metalworking - 11. Plumbing - 12. Damp, rot and woodworm. Useful names and addresses, p. 313. Index, p.315-20. Fully illustrated. Well captioned and produced. *Class No:* 643.0045

[5587]

The 'Which?' book of home improvements and extensions. London, Consumers' Association, and Hodder & Stoughton, 1983. 319p. illus., diagrs., tables.

18 contributors. 3 parts (12 chapters): 1. What's involved - 2. Improvements room by room - 3. Extensions and conversions (*e.g.* chapter 10: 'Loft conversions', p.278-95). Useful addresses. Index, p.315-9. Good illus.; well captioned. Attractively produced. Over 1,000 illus. *Class No:* 643.0045

Furnishings

[5588]

Laura Ashley complete guide to home decoration. Dickson, E., *and others.* London, Weidenfeld & Nicolson, 1989. 224p. illus. (mostly col.), diagrs. ISBN: 0706421552.

6 sections: Structure and style - Decorating - Window dressing - Soft furnishings - Beds & bedlinen - Accessories. Appendix includes 'Sewing techniques'; 'Measuring up'; 'Laura Ashley shops'. Index. *Class No:* 645

[5589]

'Which?' way to make soft furnishings. London, Consumers' Association, and Hodder & Stoughton, 1991. 128p. illus. (col. pl.), diagrs. £10.95. ISBN: 0340550066.

Basic sewing techniques - Curtains and blinds - Beds and bedding - Loose covers and curtains - Lampshades - Table linen. Detailed, analytical index. Oblong. *Class No:* 645

Clothing

Sewing & Needlework

[5590]

CLABBURN, P. The Needleworker's dictionary. London, Macmillan, 1976. 296p. illus. (incl. col.), diagrs.

Nearly 2,000 entries, A-Z, including short biographies. Cross-references. Appended select list of museums and collections exhibiting textiles. Bibliography, p.288-92. 350 photographs, 300 line-drawings. 'Mainly for students of needlework, amateur and professional embroiderers, art and costume historians, and the fashion world... An invaluable work of reference' (*British book news,* April 1977, p.313). *Class No:* 646.2

[5591]

LADBURY, A. The Dressmaker's dictionary. London, Batsford, 1982. 358p. illus., patterns. ISBN: 0713418230.

Entries, A-Z, include the 'traditional terms of stitches, processes and equipment in common use in dressmaking as well as new ideas and new aids' (*Preface*). Over 900 illus., captioned line-drawings. *Class No:* 646.2

[5592]

Reader's Digest complete guide to sewing. London, New York, etc., Reader's Digest Association, Ltd., 1978. 528p. illus., (incl. col.), diagrs. £18.95. ISBN: 0276001826.

Detailed, step-by-step approach. Sections: Sewing essentials - Patterns, fabrics and linings - Making clothes that fit - Designing your own patterns - Preparing to sew - Basic stitches - Seams - Darts and tucks - Pleats - Gathering, shirting, smocking and ruffles - Pockets - Hems and other edge finishes - Zip fasteners - Buttons and other fasteners - Tailoring - Sewing for the family - Sewing for the home (*e.g.* bedspreads). Sewing projects (p.440-516; *e.g.* dresses, dungarees). Over 2,000 two-colour illus. ('Collars', p.228-44; *c.*150 illus). Many patterns. Index, p.517-20. *Class No:* 646.2

Cleaning

[5593]

PHILLIPS, B. Wonder worker's complete book of cleaning. London, Sidgwick & Jackson, 1980. 255p. illus. ISBN: 0283985461.

12 chapters: 1. A good wash [laundering] - 2. A to Z of fabrics and fibres - 3. Dry-cleaning - 4. Stain removing at home - 5. Dyeing - 6. Body care - 7. The lazy person's guide to housework - 8. Outdoor cleaning - 9. A to Z of household materials and how to clean them - 10. Choosing cleansing equipment - 11. Glossary and guide (p.222-41) - 12. Information sources (Consumer protection ... Specialist cleaners). Detailed, analytical index, p.245-58. *Class No:* 648.5

[5594]

—MOORE, A.C. How to clean everything: an encyclopedia of what to use and how to use it. Rev. ed. London, T. Stacey, 1972 (reprinted 1978, David and Charles). 232p. illus.

First published in 1952 (New York, Simon & Schuster).

Sections: 1. How to clean and what to use - 2. How to remove stains and marks - 3. Control of household pests - 4. Cleaning agents and their uses. 'This book is for women who dislike homework but like nice houses' (*Introduction*). Enlivened by humorous drawings. 'Highly recommended'. (*RQ,* v.8(4) 1969, p.288). *Class No:* 648.5

[5595]

—PHILLIPS, B. How to clean absolutely everything. Rev ed. London, Piatkus, 1992. 154p. £7.99. ISBN: 0749910925.

First published 1990 as *The Complete book of cleaning* which was a concise update of the author's *Wonder worker's complete book of cleaning (q.v.)* (1979).

Where appropriate, this book offers 3 methods of cleaning: 'right'; 'lazy'; 'green'. 15 chapters: 1. Laundering ... 6. Guidelines for housework ... 10. Cleaning windows and plate glass ... 13. A-Z of cleaning metals ... 15. A-Z of household products. Addresses, p.149-51. Index. *Class No:* 648.5

Pest Control

[5596]

BATEMAN, P.L.G. **Household pests:** a guide to the identification and control of insect, rodent, damp and fungoid problems. Poole, Dorset, Blandford Press, 1979. 174p. illus. (incl. col.). ISBN: 0713709154.

3 parts: 1. Background to pests and problems - 2. Insect and rodent pests, rots and moulds - 3. The broader viewpoint (*e.g.* Sources of further information). Bibliography (11 items), p.172. Some common insecticide chemicals. Non-analytical index, p.175-6. *Class No:* 648.7

Child Care

[5597]

JOLLY, H. **Book of child care:** complete guide for today's parents. 4th ed. London, Allen & Unwin, 1985. [xv],569p. illus. £14.95; £8.99. ISBN: 0046490345.

2 parts (each of 26 chapters): 1. The healthy child - 2. The sick child. Part 1 involves such topics as 1. Pregnancy and birth ... 25. The one-parent family; 26. Travel, holiday and living in the tropics. Part 2, - 27. Child, parents and their doctor ... 51. Accidents; 52. When a child dies. *Class No:* 649.1

[5598]

The Macmillan guide to child health. Hull, D., *ed.* London, Macmillan, 1985. 382p. illus., diagrs., charts. ISBN: 0333392507.

46 contributors. 4 parts: 1. Care of the child - 2. Care of sick child - 3. Problems, diseases and disorders (p.120-310) - 4. Appendices: resources (support groups and publications, etc.). Emergency actions, p.330-41. Index of syndromes, p.342-3. Analytical index, p.344-52. Over 500 two-colour illus. A guide that spans the child's life from conception to adolescence. For public library and home use. Analytical index. *Class No:* 649.1

Invalid Care

[5599]

Helping to care: a handbook for carers at home and in hospital. Kershaw, B., *and others.* London, Baillière Tindall, 1989. 189p. figs. £4.50. ISBN: 0702013374.

18 chapters: 1. The management of care ... 4. Mobilising ... 8. Meeting hygiene needs - 9. Waking and dressing ... 15. Caring for the sick child ... 17. Working with people with a mental handicap - 18. Where next? Further reading, p.176-7. Useful addresses, p.178-81. Index. *Class No:* 649.8

654/656 Communication

Telecommunication Services

[5600]

International directory of telecommunications: market trends, companies, statistics and personnel. Robert, S. *and* Hay, T., *eds*. 4th ed. Harlow, Essex, Longman, 1990. xiv,386p. tables. £140. ISBN: 0582047382.

First published 1984.

Profiles of *c.*800 manufacturers, telecommunication administrations and regulatory bodies in over 100 countries. Arranged in 12 geographical areas. Indexes of companies, products, services, registered trade names and key personnel. *Class No:* 654

[5601]

International encyclopedia of communications. Barnouw, E., *and others, eds.* New York, Oxford, University Press, 1989. 4v. illus., figs., maps. £250. ISBN: 019505802x, v.1; 0195058038, v.2; 0195058046, v.3; 0195058054, v.4; 0195049942, set.

A highly interdisciplinary work, covering communications in its widest sense, embracing topics in the humanities, arts, and sciences. 569 subjects covered include mass media; theatre; language; literature; cinematography. Documented articles, arranged A-Z, written by over 450 specialist contributors. Over 1,100 black-and-white photographs, in addition to many line drawings. Some articles would be considered too specialized for the lay person. However, 'this fine work is recommended for academic and large public libraries' (*Reference books bulletin,* 1988/1989, p.29-30). *Class No:* 654

[5602]

Marconi's international register. Larchmont, N.J., Telegraphic Cable & Radio Registrations, Inc. ISBN: 0916446174.

The 1991 *Register* (1336p. $150) includes fax, mail, telephone, and cable subscribers. International in scope, with prominent companies listed A-Z, and an international trade classification following. Index of cable addresses. *Class No:* 654

[5603]

WEIK, M.H. Communications standard dictionary. 2nd ed. New York, Van Nostrand Reinhold, 1989. 1219p. illus., diagrs., graphs. $69.95. ISBN: 0442205562.

About 13,600 entries and cross-references concerning the science and technology of communications ('A-and-not-B gate' ... 'Zoom lens'). Claims to cover 64 fields. *Class No:* 654

Telex

[5604]

International telex, 1990/91. Darmstadt, Telex International; J+International Telex.

Formerly *Jaeger + Waldmann Telex + teletex international.*

International guide for the hotel and tourist trade. 1-4. *Alphabetical section.* Countries: A-F, G-I, J-S, T-Z. 4v. 5-7. *Numerical* (Answerback code section). v.5-7: 1-3, 4-7; 8-9/ A-Z. 8. *Products:* 0-34. 9. *Services:* 35-41.

'*Jaeger and Waldmann* constitutes probably the most extensive world-wide directory available' (*Printed reference material.* 2nd ed., edited by G.L. Higgins (1984), p.171). The Answerback section is particularly handy. *Class No:* 654.145

[5605]

The Telex directory: a complete directory of top UK businesses. London, British Telecom. 2pa. (April; October).

The October 1989 *Directory* (507p.), replacing the April 1989 issue, has 6 sections: 1. General information - 2. Inland calls - 3. International calls - 4. Telex related services - 5. The alphabetical list of customers (p.63-501) - 6. The British Telecom code of practice. Mailbox and bureau supplement. *Class No:* 654.145

Fax

[5606]

National fax directory. Detroit, Gale, 1989. Annual. 1300p. ISBN: 0810368951.

Lists 'phone numbers of more than 80,000 US companies, organizations, government agencies, and libraries. Arranged in 3 different ways - alphabetically, by subject and geographically. Each entry contains address, fax number, and voice 'phone numbers. *Class No:* 654.146

Telephone Services

[5607]

BRITISH TELECOM. The Phone book index. 1988 ed. London, Telecommunications Press, 1988. 105p.

An index to the telephone books (ordinary and yellow pages), each numbered, 101-376 and 101-492 respectively. Covers all cities, towns and villages throughout the UK.

The Code decoder, 1989 (Telecommunications Press, 1989, vp.) covers all cities, towns and villages throughout the UK, giving dialling code locations at a glance. *Class No:* 654.15

[5608]

—Telecom users handbook, 1987/88 ed. London, Telecommunications Press, 1987. xvii, 512p. ISBN: 0907401023.

Has 20 sections (2. Telephone instruments and aids; 8. Data transmission; 14. The electronic office; 20. British

....(contd.)

Telecommunications price schedule). Glossary of terms. Index of advertisers. A practical guide to business communications. *Class No:* 654.15

Broadcasting Services

[5609]

The Blue book of British broadcasting 1992. O'Neill, B., *ed.* 18th ed. London, Telex Monitors, 1992. 620p. £35. ISBN: 0950616745.

14 sections: 1. BBC: general ... 5. BBC educational broadcasting ... 9. Independent television companies ... 12. Cable and satellite: operators/providers and CRMK ... 14. People and programmes, stop press, and new franchises. Basically a list of contact names in British broadcasting. *Class No:* 654.19

[5610]

Guide to broadcasting stations. Darrington, P., *ed.* 20th ed. Newnes, Heinemann, 1989. 236p. figs., tables. £10.95. ISBN: 0434903094.

International in scope, listing radio stations broadcasting in the long, medium, and short wave bands. Arranged by frequency, geographical location, and alphabetical order. *Class No:* 654.19

[5611]

International Broadcasting 1992 directory. London, International Thomson Publishing, 1992. 169p. xlvii.

3 sections: 1. Products A-Z, section 1 (p.9-152) - 2. Services A-Z, section 1-3. Addresses A-Z, section 2. Advertisers index. *Class No:* 654.19

[5612]

MacDONALD, B. Broadcasting in the United Kingdom: a guide to information sources. London, Mansell, 1988. 266p. £35. ISBN: 0720119626.

Wide-ranging survey covering history, structure and organizations, education facilities, awards, festivals, annotated lists of primary sources (p.128-54); printed and electronic sources (p.156-200). Bibliographies appended to each chapter. List of addresses for organizations quoted in the text. *Class No:* 654.19

[5613]

World radio TV handbook. London, Billboard, 1946-. v.1-. Annual.

V.46, 1992 (590p.) includes services listed by country, station information, languages, operating times, frequencies, maps of principal transmitter sites worldwide, reception conditions, annual test reports on receivers. *Class No:* 654.19

Television Broadcasting

[5614]

The European television directory. Henley-on-Thames, Oxon., NTC Publications, 1992. 163p.

80 television stations, A-Z. 2 pages used per station. One page descriptive text and commentary. Details of legislation affecting channels also given. The second page gives channel ownership, delivery system, funding, viewing share, etc. Contents page arranged alphabetically and geographically. *Class No:* 654.197

[5615]

HOLLINS, T. Beyond broadcasting into the cable age. London, BFI Publishing for the Broadcasting Research Unit, 1984. x, 385p. £5.95. ISBN: 0851701485.

Follows the BRU's *Report of the Working Party on new technologies.* (1984).

5 parts: 1. The cable debate - 2. Cable in Britain: the historical and political background - 3. The Canadian experience - 4. The United States (p.113-272) - 5. British plans and prospects. Conclusion: Into the Cable age. Appendices. Index. *Class No:* 654.197

The Book

Handbooks & Manuals

[5616]

GASKELL, P. A New introduction to bibliography. Oxford, Clarendon Press, 1972. xxii, [1], 438p. illus. £30. ISBN: 0198181507.

Intended as a successor to R.A. McKerrow's classic *Introduction to bibliography* (1927).

Contents: Introduction - Book production: the hand-press period, 1500-1800 - Book production, the machine-press period, 1800-1950 - Bibliographical application (*e.g.* bibliographical descriptions; textual bibliography). Reference bibliography, p.392-412 (running commentary; includes periodicals) - Detailed index. Well produced. Very favourably reviewed in *College & research libraries,* v.34(4), July 1973, p.295. *Class No:* 655.0(035)

[5617]

JENNETT, S. The Making of books. 5th ed. London, Faber, 1973. 554p. illus. (incl. col.), facsims.

First published 1951.

2 parts (24 chapters): 1. Printing and binding (chapters 1-12) - 2. The design of books (13-24). This ed. includes new techniques and technologies. List of 'further reading and reference', p.517-23. 'A polyglot glossary of technical terms' (English, French, German, Italian), p.524-39. Detailed, non-analytical index. 200 illus. A standard work for students of bibliography, typography and printing. *Class No:* 655.0(035)

[5618]

PEACOCK, J. Book production. London, Blueprint Publishing, in association with the Publishers Association, 1989. xv,494p. illus. £29.95. ISBN: 0948909336.

14 sections: 1. Choosing the format - 2. Preparing the text ... 9. Choosing and assessing paper - 10. Printing the book ... 13. Binding styles - 14. Organising packing, despatch and distribution. 3 appendices: 1. Technical data; 2. Further information sources (trade journals, reference books); 3. Dictionary. Detailed index. 'This book is for anyone professionally interested in the production, design, or editing of books' (*Introduction*). *Class No:* 655.0(035)

Dictionaries

[5619]

GLAISTER, G.A. Glaister's glossary of the book: terms used in papermaking, printing, bookbinding and publishing, with notes on illuminated manuscripts and private presses. 2nd ed. London, Allen & Unwin, 1979. [xv], 551p. illus. (incl. col.), facsims. £35. ISBN: 0040100065.

First published 1960.

3,932 entries, including biographies (*e.g.* Gutenberg: nearly 2p.; 2 facsims.). Updated to include such entries as 'Computer assisted type-setting' (1½ columns; 7 cross-references). Covers all book-making processes, with a natural emphasis on the historical (*e.g.* 'Islamic bookbinding: 2p.). Appendixes A-D (A. Some type specimens; D. A short reading list, p.547-51, grouped). 238 illus. 'Apart from specialist libraries, any general reference collection of any size must have a copy' (*Library Association record,* v.82(8), August 1980, p.370). *Class No:* 655.0(038)

[5620]

—PETERS, J. The Bookman's glossary. 6th ed., rev. and enl. New York, etc., Bowker, 1983. ix, 223p. $39.95; £35. ISBN: 0855216861.

Defines *c.*2,000 terms used in the antiquarian book trade, bookmaking, computerized typesetting, printing, publishing and bookselling. 'Lithography': ½p.; 'Caxton': 1p. Cross-references. Selected reading list (grouped), p.216-21. *Class No:* 655.0(038)

[5621]

Russko-angliiskiĭ slovar' knigovedcheskikh terminov. Elizarenkova, T.P., *comp.* Moscow, Izdat 'Sovetskaya Entsiklopediya', 1969. 264p.

9,300 terms, including abbreviations, covering book science, bibliography, librarianship, publishing, printing and the book trade. English abbreviations. *Class No:* 655.0(038)

Histories

[5622]

ABHB / Annual bibliography of the history of the printed book and libraries. The Hague, Nijhoff (now Dordrecht, Academic Publishers), 1973-. v.1-. Annual.

V.20 (1991, xi, 422p.) has 3,899 entries for items published in 1989 plus additions from the previous years. List of periodicals (*c.*2,000), p.1-41. Sections A-M (E. Book illustration; G. Book trade, publishing; J. Libraries; L. Newspapers, journalism). Indexes: Authors' names and anonyms; Geographical and personal names. 'A basic item for any bibliographical reference shelf' (*TLS,* no.3951, 15 December 1977, p.1484). *Class No:* 655.0(091)

[5623]

FEATHER, J. A Dictionary of book history. London, Croom Helm, 1986. 288p. ISBN: 0709910436.

Published in paperback by Routledge, 1991. £50.

About 650 entries, varying in length from short definitions to lengthy articles. Covers history of printing, publishing and bookselling, bookbinding, paper, bibliographies, etc. Includes some biography. *Class No:* 655.0(091)

[5624]

WINCKLER, P.A. History of books and printing: a guide to information sources. Detroit, Mich., Gale, 1979. xiv, 209p. (*Books, publishing and libraries information guide series.*) ISBN: 0810314088.

776 annotated entries. 9 sections: 1. General bibliographies - 2. General information sources - 3. Materials and techniques used in graphic communication - 4. The history of books and printing - 5. Non-print media/ filmstrips, slides, etc. - 6. Periodicals and annuals - 7. Associations, societies and clubs (UK, USA, chiefly; Canada and W. Germany, - 1 each) - 8. Libraries, special collections, and museum sources - 9. Book dealers. Author, title and subject indexes.

By the same author, *Reader in the history of books and printing* (Westport, Conn., Greenwood, 1983. 406p. $35). *Class No:* 655.0(091)

Printing Industry

Bibliographies

[5625]

Information sources on the printing and graphics industry. New York, United Nations, 1975. xii, 65p. £4. (*UNIDO guides to information sources, no.14.*) ISBN: 0119039850.

504 entries. 11 sections: 1. Professional, trade and research organizations, learned societies and special information sources - 2. Directories - 3. Sources of statistics - 4. Basic handbooks, textbooks and manuals - 5. Current periodicals - 6. Current abstracting and indexing periodicals - 7. Proceedings, papers and reports - 8. Specialized dictionaries and encyclopaedias - 9. Bibliographies - 10. Films - 11. Other potential sources of information (*e.g.* Translation services). List of bibliographical sources used. *Class No:* 655.1(01)

[5626]

—Book printing in Britain and America: a guide to the literature and a directory of printers. Brenni, V.J., *comp.* Westport, Conn., Greenwood Press, 1983. xiv, 163, [1]p. $38.95; £31.75. ISBN: 0313239886.

Has 1,021 references, in 21 chapters. 4 appendices (*e.g.* 'List of printers, typographers, calligraphers, and book designers'). Author and analytical indexes. *Class No:* 655.1(01)

Handbooks & Manuals

[5627]

The Print and production manual. Peacock, J., *and others,* eds. 7th ed. London, Blueprint, 1992. vp. Loose-leaf. ISBN: 0948905735.

First published 1986.

15 sections: 1. Typography - 2. Copy preparation ... 7. Inks ... 9. Finishing ... 11. General reference (General reference tables; Technical reference material; Colleges, organizations, publications relating to printing and publication) - 14. Glossary - 15. Index. *Class No:* 655.1(035)

Dictionaries

[5628]

Dictionary of printing and publishing: English-German. Collin, P.H., *and others*. Teddington, Peter Collin Publishing, 1991. 380p. £25; £6.95. ISBN: 094854919x.

Contains *c.*5,000 words and phrases. Explanations are in English with some entries supplemented by additional explanatory comments. A glossary lists the translated German terms A-Z, with English equivalents. Appendices include dates of international book fairs, alphabets, and technical information for a periodical. *Class No:* 655.1(038)

[5629]

Dictionary of the printing and allied industries, in English, French, German, Dutch, Spanish and Italian. Wijnekus, F.J.M. and Wijnekus, E.F.P.H., *eds.* 2nd rev. ed. Amsterdam, etc., Elsevier, 1983. xxxvi, 1026p. £150.93. ISBN: 0444422498.

First published 1967.

14,968 English-base terms, with equivalents and indexes in the other five languages. Spanish and Italian are added in this edition. *Class No:* 655.1(038)

[5630]

—The Multilingual dictionary of printing and publishing. Isaacs, A., *ed.* London, F. Muller, 1981. 289p. ISBN: 0582555523.

1,600 entry-words in one A-Z sequence of English, German, French, Italian, Spanish and Portuguese terms, with equivalents in the other languages. Genders stated. When British and American English differ, both terms are given. *Class No:* 655.1(038)

[5631]

KENNEISON, W.C. and **SPILMAN, A.J.B. Dictionary of printing, papermaking and bookbinding.** London, Newnes, 1963. vii, 213p. illus., diagrs.

More than 2,000 entries, ranging in length from one line to 30 (*e.g.* 'Linotype'). Includes some biographical sketches. 9 appendices of abbreviations, symbols, sizes, proof-correction marks, etc. Concise and clear definitions. *Class No:* 655.1(038)

[5632]

PEACOCK, J. Multilingual dictionary of printing and publishing terms: English-French-Spanish-Italian-German-Dutch-Swedish. London, Blueprint, 1991. 242p. £45. ISBN: 0948905352.

Instant translation of over 2,000 common printing and publishing terms. Section A contains English-all language versions. Section B, French-English, Spanish-English, etc. *Class No:* 655.1(038)

Reviews & Abstracts

[5633]

Printing abstracts. Oxford, etc., Pergamon Press, 1945-. Monthly. £569. ISSN: 0031109x.

Compiled by the Information Section of PIRA (Paper and Board, Printing and Packaging Industries Research Association).

January, 1992 issue contains 383 abstracts. 5 sections: 1. General - 2. Prepress and DTP (*e.g.* photography) - 3. Printing processes - 4. Post press (*e.g.* bookbinding) - 5. Products (*e.g.* newspapers, books and periodicals; business

....(*contd.*)

forms and stationery). Author and subject indexes per issue only. Also online through Pergamon Orbit INFOLINE. *Class No:* 655.1(048)

Progress Reports

[5634]

Advances in printing science and technology. London, Pentech Press, 1961-. illus., diagrs., graphs, tables. ISBN: 072730111x.

V.19 (1988, ix, 408p. £52) comprises the Proceedings of the 19th International Conference of Printing Research Institutes, Eisenstadt, Austria, June 1987. The 29 papers, each with abstract and summary of discussion, and occasionally documented, are in 4 sections: Colour reproduction and theory; Paper and ink interaction; Printing operations; Miscellaneous. No index. *Class No:* 655.1(055)

Yearbooks & Directories

[5635]

Printers yearbook: BPIF guide to the printing industry. London, British Printing Industries Federation. Annual. illus., tables. ISSN: 02642387.

The 1991/92 yearbook (1991. 262p.) has 9 sections. 1. British Printing Industry Federation - 2. Printers' law - 3. Industrial relations (including national agreements)- 4. Education and training - 5. Technical information and buyers' guide - 6. Economics (*e.g.* index of retail prices) - 7. International - 8. What's what/who's who (including associated organizations) - 9. Detailed, analytical index. *Class No:* 655.1(058)

[5636]

Printing trades directory. Tonbridge, Kent, Benn Business Information Services, 1960-. Annual. ISBN: 0863821278.

The 1992 ed. (viii,524p.) has 16 sections. Index to the printing industry - Gazetteer of counties and towns - Geographical index to the printing industry (p.39-228) ... Manufacturers and suppliers ... Machinery, equipment, and materials ... Overseas manufacturers - Printing and allied trade manufacturers - Colleges and schools of printing and graphic arts. Index to company name changes. Index to advertisers. *Class No:* 655.1(058)

Histories

[5637]

The Art and history of book printing: a topical bibliography. Brenni, V.J., *comp.* Westport, Conn., Greenwood Press, 1984. xi, 147p. $42.95. ISBN: 0313243069.

1,199 references. 14 chapters, plus 2 classified checklists of selected titles, on writing and calligraphy, and on typography. Author and analytical subject indexes. 'The work will be especially helpful for students' (*Reference books bulletin, 1984-1985,* p.123). *Class No:* 655.1(091)

[5638]

The Invention of printing: a bibliography. Prepared as an activity of the Work Project Administration (Illinois), Chicago Public Library Omnibus Project (OP no.65-1-54-273(3)). Section on Printed Bibliography, co-sponsored by the Chicago Club of Printing House Craftsmen. McMurtrie, D.C. and Adamson, J., *eds.* Chicago, Ill., 1942 (reprinted B. Franklin, 1970). xxiv, 413p. Mimeographed.

....(contd.)

3,228 titles; some items have brief annotations. Locations in *c.*120 libraries (*c.*80 in USA). 8 sections: 1. Antecedents - 2. Invention - 3. Technique of the Gutenberg School - 4. Fust and Schoeffer - 5. Technique of the Dutch School - 6. Gutenberg - 7. Commemorative writing - 8. Bibliographies. *Class No:* 655.1(091)

[5639]

NEWBERRY LIBRARY, Chicago. **Dictionary catalogue of the history of printing,** from the John M. Wing Foundation, the Newberry Library. Boston, Mass., G.K. Hall, 1962. 6v. (6000p.). Supplements, 1970-. $655, set; $365, supp.1 (3v.). ISBN: 0816105871, set; 0816108099, supp. 1.

Offset-litho reproduction of 127,000 catalogue cards in the main 6v. The first part of the catalogue has entries for 20,000v. on the history of printing. The second part covers 'finely printed or historically significant books', including *c.*2,000 incunabula. *Class No:* 655.1(091)

Great Britain

[5640]

BIBLIOGRAPHICAL SOCIETY. [Dictionary of printers and booksellers in England, Scotland and Ireland]. London, the Society, 1905-32. 3v.

Duff, E.G. *A century of the English book trade, 1457-1557.* 1905. McKerrow, R.B. *A dictionary of printers and booksellers ... 1557-1640.* 1910. Plomer, H.R. *A dictionary of the printers and booksellers ... 1641-1667.* 1907. Plomer, H.R. *A dictionary of the printers and booksellers ... 1668-1725.* 1922. Plomer, H.R. and others. *A dictionary of the printers and booksellers. 1726-1775.* 1932.

The only comprehensive source on the life and career of any British printers or booksellers, and not superseded. Each volume has an index of places, and biographies carry bibliographies.

Morrison, P.G. *Index of printers, publishers and booksellers ... 'A short-title catalogue', (1950), and Index of printers, publishers and booksellers in Donald Wing's 'Short-title catalogue'* (1950) are further sources, - both published by Bibliographical Society of the University of Carolina, Charlottesville, Va.

Supplemented by 'Scottish printers and booksellers, 1668-1775', by R.H. Carnie and R.P. Doig (*Studies in bibliography,* v.12, 1959; v.14, 1961, p.81-96; v.15, 1962, p.105-20). Index of places included. *Class No:* 655.1(410)

[5641]

Reviews of United Kingdom statistical sources. V.22: Printing and publishing. McLelland, W.D. Oxford, etc., Pergamon Press, for the Royal Statistical Society, and Economic and Social Research Council, 1987. xiii, 210p. tables. ISBN: 0080347819.

Comprises 2 statistical reviews. Review 39: *Printing and publishing newspapers and periodicals* (15 chapters; Quick reference lists; Table of contents; Bibliography; Subject index). Review 40: *General printing and publishing* (13 chapters). Bibliography + Postscript bibliography; subject index. *Class No:* 655.1(410)

England

[5642]

BLACK, M.H. Cambridge University Press, 1584-1984. Cambridge, University Press, 1984. xviii, 343p. illus., facsims., ports. ISBN: 0521264731.

A detailed account in 17 documented chapters, well stocked with quotations and anecdotes. 4 appendices (3. University printers, 1583-1984). Detailed, analytical index, p.335-43. *Class No:* 655.1(420)

[5643]

—McKITTERICK, D. History of Cambridge University Press. V.1. Printing and the book trade in Cambridge, 1534-1698. Cambridge, University Press, 1992. £65. *Class No:* 655.1(420)

[5644]

CARTER, H. A History of the Oxford University Press. v.1. To 1780. Oxford, Clarendon Press, 1975. xxxi, 640p. illus., facsims., ports. ISBN: 0199510326.

1. *To the year 1780, with an appendix listing the titles of books printed there, 1690-1780,.*

4 parts to v.1: 1. Introductory; 2. The Augustan Age, 1690-1713; 3. The bad time, 1713-1755; 4. The era of Blackstone, 1755-1780. Appendix. No further volumes yet announced. *Class No:* 655.1(420)

[5645]

—SUTCLIFFE, P. The Oxford University Press: an informal history. Oxford, University Press, 1978. xxviii, 303p. illus. (pl.), maps. ISBN: 0199510849.

Concentrates on the post-1800 period. *Class No:* 655.1(420)

Wales

[5646]

JONES, J.I. A History of printing and printers in Wales [and Monmouthshire] to 1810, and of successive and related printers to 1923. Cardiff, W. Lewis, 1925. x, 367p.

A strictly factual survey, - the basis on which more specialized bibliographical studies must be founded. *Class No:* 655.1(429)

Italy

[5647]

Biographical and bibliographical dictionary of the Italian printers, and of foreign printers in Italy, from the introduction of the art of printing into Italy to 1800. Cosenza, M.E., *comp.* Boston, Mass., G.K. Hall, 1970. $110. ISBN: 0816107661.

13,500 photolithographed cards. Does not aim to provide a complete bibliography for each printer, 'but to establish precisely in what town a given book was published by a given printer by listing one publication for each year that the printer exercised his art' (*Prospectus*). *Class No:* 655.1(450)

Private Presses

[5648]

CAVE, R. The Private press. 2nd ed., rev. and enl. New York, etc., Bowker, 1983. xvi, 389p. illus., facsims. $64.95. ISBN: 0835216950.

First published 1971.

....(contd.)

2 parts (27 chapters). 1. The private presses (chapter 3: The scholarly press; 14. Fine printing on the continent; 21, The United States today) - 2. Private press type faces (24. Early presses and their types; 27. The contemporary scene). Bibliography (by chapters), p.349-62. Detailed, analytical index. *Class No:* 655.15

[5649]
—ANDERSON, T.J. Private presswork: a bibliographic approach to printing as an avocation. South Brunswick & New York, A.S. Barnes; London, T. Yoseloff, 1977. 165p. facsims. ISBN: 0495018768.

16 chapters (1. Printing instruction manuals; 2. Selecting a press; 4. Type and typography; 8. Bookbinding; 11. Paper and papermaking; 12. Libraries and museums; 15. Directory of book dealers; 16. Basic book list (10 items)). Detailed, analytical index. *Class No:* 655.15

Processes & Printing Surfaces

Types

[5650]
The Encyclopaedia of type faces. Jaspert, W.P., *and others.* 4th ed. London, Blandford Press, 1970. xiii, [2], 420p. illus.

First published 1953, by W.T. Berry, and others. Paperback ed. published 1990.

About 2,000 alphabets, p.1-405 (in descending order of size, Roman - Lineales - Scripts). Appended literature on type and type founding. Typefounders, addresses. Index to designers. Index to type faces. Typewriter faces are excluded; historical examples are still generally lacking (*TLS*, no.3604, 26 March 1971, p.355). *Class No:* 655.24

[5651]
Modern encyclopedia of typefaces 1960-90. Wallis, L.W., *comp. and ed.* London, Lund Humphries, 1990. 192p. illus. ISBN: 0853315671.

Includes new typefaces available to a major composition system for continuous setting from 1960. 3 sections: Type samples (p.19-162); Supplement to type samples; Designer profiles. Select bibliography, p.180-81. Chronological index; index of designers; index of manufacturers and design agencies; index of alternative typeface names. *Class No:* 655.24

[5652]
PERFECT, C. *and* **ROOKLEDGE, G. Rookledge's international type finder.** Revised by P. Baines. 2nd ed. London, Sarema Press, 1990. 279p. illus., tables. £30; £20. ISBN: 187075803x.

First published 1983.

Part 1: Text typefaces (8 types, 1. Sloping e-Bar (Venetian serif) ... 8. Sans serif) - Part 2: Decorative (non-continuous text) typefaces (Flowering scripts ... Modified outrageous). 2 appendices: 1. Typefinder classification system/British Standards Classification - 2. Biographical notes on leading type designers - the latter being new to this edition. Bibliography/Further reading. Index. 'The first edition contained 700 typefaces, this new edition updates these, with the addition of selected faces released since that date' (*Introduction*). For designers, copywriters, editors, typesetters, printers and students to identify typefaces seen and used in their everyday work. *Class No:* 655.24

[5653]
SUTTON, J. *and* **BARTRAM, A. Typefaces for books.** London, British Library, 1990. 288p. illus. £40; £20. ISBN: 0712301900.

Over 100 monotype and linotype digitized typefaces in double-page specimen settings, in different sizes and spacings, plus display sizes, bold, roman, italian sloped roman, small caps. and figures. Also, the development of typefaces, and typographic problems and their solutions. 60 black-and-white illus. Caters admirably for the requirements of all those involved in specifying type for books. *Class No:* 655.24

Copy Presentation

[5654]
BUTCHER, J. Copy-editing: the Cambridge handbook for editors, authors and publishers. 3rd ed. Cambridge, University Press, 1992. xii,471p. illus., tables. £19.95. ISBN: 0521400740.

New and revised features include copyright law; how to deal with author-supplied disks and camera-ready copy. 15 sections (2. Estimates and specimen pages; 6. House style; 10. Bibliographical references; 13. Science and mathematics books; 15. Reprints and new editions). 13 appendices (1. Checklist of copy-editing; 4. Phonetic symbols; 13. Proof correction symbols). Glossary, p.425-42. Select bibliography. Detailed, analytical index. *Class No:* 655.25

[5655]
The Chicago manual of style; for authors, editors and copy writers. 13th ed. Chicago, Ill., & London, Univ. of Chicago Press, 1982. ix, 738p. $37.50. ISBN: 0226103900.

First published 1906.

3 parts (20 chapters): 1. Bookmaking - 2. Style (*e.g.* punctuation, spelling, bibliographic forms, indexes), p.131-557 - 3. Production and printing. (19. Design and typography; 20. Composition, printing and binding). Glossary of technical terms, p.645-83. Bibliography, briefly annotated, p.685-96. Detailed, closely analytical index. 'This impressive edition' (*Library journal,* 1 November 1982, p.2086). *Class No:* 655.25

[5656]
HART, H. Hart's rules for compositors and readers at the University Press, Oxford. 39th rev. and updated ed. Oxford, University Press, 1983. 192p. £9.99. ISBN: 019212983x.

First published 1893.

States Oxford University Press practice on spelling, pluralizing, syllabification, capitalizing, punctuation, abbreviation, etc. Includes a guide to the setting of Welsh, Dutch, Afrikaans, and a new section on Machine Reader Codes. All words are brought into line with the *Oxford dictionary for writers and editors* (1981). *Class No:* 655.25

[5657]
Oxford dictionary for writers and editors. Compiled by the Oxford English Dictionary staff. Oxford, Clarendon Press, 1981. xiv, 448p. £8.95. ISBN: 0192129709.

Successor to 11th ed. of *Authors and printers dictionary,* by F.H. Collins, and others (1978), first published 1905.

About 12,000 terms are defined or briefly discussed. Covers abbreviations, foreign idioms, stress, pluralizing and abbreviating, spelling, capitalizing, hyphenating, as well as

.... *(contd.)*

correct form of place-names, proper names, etc. Adopted by *The Times* and *The Sunday times* as their official style book. *Class No:* 655.25

[5658]
—HOWELL, J.B. Style manuals of the English-speaking world: a guide. Phoenix, Arizona, Oryx Press, 1983. 138p. $21. ISBN: 0897740890.

Surveys more than 200 English-language style manuals, both general and specific subject guides, and mostly post-1970. Some of the manuals consist of short pamphlets prepared by publishing houses for their authors. Author, title and subject indexes. *Class No:* 655.25

Word Processing

[5659]
LANGMAN, L. **An Illustrated dictionary of word processing.** Phoenix, Arizona, Oryx, 1986. xiii, 289p. illus. $34.50. ISBN: 0897742869.

Covers more than 500 word-processing terms and functions. Extensive cross-references. Appendices. 'Primarily of assistance to beginners' (*Choice*, v.24 (2), October 1986, p.282), as well as to the prospective purchasers of word-processing software. *Class No:* 655.254.4

[5660]
—DANDO, S. Word processing dictionary. New York, etc., McGraw-Hill, 1987. 88p. ISBN: 0070849935.

A basic glossary of terms, free from jargon. A quick-reference for students following the elementary word-processing syllabuses of the RSA, Pitman, B/TEC, GCSE and LCCI. *Class No:* 655.254.4

[5661]
VOLLNHALS, O. **Elsevier's dictionary of word processing, in English (with definitions), French and German.** Amsterdam, etc., Elsevier, 1986. x, 304p. £91.87; $136. ISBN: 044442608x.

4,394 English based terms, plus definitions. Equivalents and indexes in French and German. Covers hardware, software applications (*e.g.* office automation) and allied areas, such as telematics and videotex. *Class No:* 655.254.4

Typography

Encyclopaedias & Dictionaries

[5662]
Dictionary of the graphic arts industry, in English, German, French, Russian, Spanish, Polish, Hungarian and Slovak. Muller, W., *ed.* Amsterdam, etc., Elsevier, 1981. 1026p. $156.50; £100. ISBN: 0444997458.

About 14,000 English-base terms, with equivalents and indexes in the other 7 languages. Coverage: composition, platemaking, relief-printing; flat-bed printing, rotogravure printing, diestamping printing, collotype printing, machines, equipment, reproduction technology, bookbinding, measuring, storage and control technology, materials, etc. *Class No:* 655.26(03)

[5663]
FAUDOUAS, J.C. **Dictionnaire technique des industries graphiques,** anglais-français, français-anglais. Paris, Maison dictionnaire, 1989. 302+7p. FFr350. ISBN: 2856030388.

About 5,000 entries in all. 'Screen': 1 column. 5 appendix tables. *Class No:* 655.26(03)

[5664]
GARLAND, K. **Graphics, design and printing terms:** an international dictionary. London, Lund Humphries, 1989. 248p. illus., facsims., diagrs. £18.95. ISBN: 0853315234.

Defines *c.*2,000 terms, A-Z with many cross-references. Includes many computing terms, abbreviations and illustrations. A useful quick-reference tool. *Class No:* 655.26(03)

[5665]
MINTZ, P.B. **Dictionary of graphic arts terms:** a communication tool for people who buy type and printing. New York, etc., Van Nostrand Reinhold, 1981. vii, [1], 318p. illus., diagrs. ISBN: 0442267118.

About 3,000 terms are defined (Phototype setting': 2p.; tables). Appendices: 1. Articles of the Code of Fair Practice; 7. Bibliography, grouped: p.312-8. US-slanted. *Class No:* 655.26(03)

[5666]
RODRÍGUEZ, C. **Bilingual dictionary of the graphic arts** ... English/Spanish, Spanish/English. New ed. rev. and enl. by G.A. Humphrey. New York, G.A. Humphrey, 1966. 446p.

About 15,000 terms, with equivalents. Some 90 categories applied. *Class No:* 655.26(03)

[5667]
STEVENSON, G.A. **Graphic arts encyclopedia.** 2nd ed. New York, etc., McGraw-Hill, 1979. x, 483p. illus., diagrs., tables. ISBN: 0070612889.

First published 1968 (xiv, [1], 492p.).

3,000 terms, A-Z with many cross-references, p.1-420. 'Deals with (1) the products and tools with which the image is formed, (2) the kind of image, and (3) the surface or material upon which the image is produced' (*Preface*). Bibliography, p.421-2. Associations and societies (all in USA), p.423. Trade journals (US and British), p.426-8. Product, manufacturers; and general indexes. Clear, accurate and easy to use, with US slant.

3rd ed., 1992. 624p. $57.50. *Class No:* 655.26(03)

Handbooks & Manuals

[5668]
McLEAN, R. **The Thames and Hudson manual of typography.** London, Thames & Hudson, 1980. 216p. illus., facsims. ISBN: 0500670226.

12 sections (1. Historical outline; 4. Lettering and calligraphy; 6. Methods of composition; 9. Book design; 11. The parts of a book; 10. Jobbing typography; 12. Newspaper and magazine typography). Notes. 2 appendices (1. List of suppliers). Glossary of filmsetting terms. Further reading (by chapters), p.210-4. Brief index. 188 illus. *Class No:* 655.26(035)

Reviews & Abstracts

[5669]

Abstracts (Graphic Arts Technical Foundation). Pittsburgh, Pa., Graphic Arts Technical Foundation, 1989-. 12pa. $90.

Formerly *Lithographic abstracts,* 1947-64 and *Graphic arts abstracts,* 1976-88.

Nearly 100 abstracts per issue. 32 sections (Advertising and marketing ... Typsetting and typography), plus 'Book reviews'. Includes abstracts on computer applications, electronic publishing. Annual index, with list of periodicals abstracted. A digest of scientific, technical and education information for the graphic communication industries. *Class No:* 655.26(048)

[5670]

Graphic arts literature abstracts (GALA). Rochester, N.Y., Technical and Education Center of the Graphic Arts, Rochester Institute of Technology, 1973-. 12pa. $150pa. ISSN: 00908207.

Formerly *Graphic arts progress,* 1854-1972.

About 2,000 abstracts pa. and covers *c.*160 titles. 9 sections: Business and financial - Marketing and sales - Printed products - Graphic design and copy preparation - Manufacturing operations - Materials management and handling - Science and technology - Research - Education and training. Author and keyword indexes per issue, plus 'Periodicals abstracted in this issue'. *Class No:* 655.26(048)

Histories

[5671]

MEGGS, P.B. A History of graphic design. 2nd ed. New York, Van Nostrand Reinhold, 1992. xv,508p. illus. £35. ISBN: 0442318952.

First published 1983.

5 sections (25 chapters): 1. The prologue to graphic design - 2. A graphic renaissance - 3. The industrial revolution - 4. The modernist era - 5. The information age. Selected bibliography, p.475-94. Detailed, analytical index, p.499-508. Well produced, with numerous black-and-white illustrations. *Class No:* 655.26(091)

Typesetting & Composing

Electronic Publishing

[5672]

CARD, M. *and* **FELDMAN, T. Blueprint electronic publishing glossary.** London, Blueprint, 1992. 152p. figs. ISBN: 0948905603.

Contains *c.*1,500 concise definitions covering online information services, videotext, optical discs, audiotext, computer software, desktop publishing, graphic reproduction, handheld reference systems, etc. *Class No:* 655.285

[5673]

GURNSEY, J. 'Electronic publishing'. In *Printed reference material,* edited by P.W. Lea and A. Day (3rd ed., 1990), p.383-99.

On the dissemination of information in electronic form, surveying online databases as well as videotext. 28 references, 3 tables. *Class No:* 655.285

[5674]

The Ivanhoe guide to desktop publishing: autumn 1989. Grosvenor, J. *and* Morrison, K., *eds.* Oxford, Ivanhoe Press, 1989. xix,259p. £4.95. ISBN: 1870757084. ISSN: 0956747x.

7 sections: 1. The industry - 2. The hardware - 3. The software ... 5. Graphic design and layout - 6. Training and consultancy - 7. Useful information (*e.g.* 'Glossary of terminology', p.251-9). *Class No:* 655.285

Techniques

[5675]

HUTCHINGS, E.A.D. A Survey of printing processes. 2nd ed. London, Heinemann, 1978. x, 246p. illus., tables.

First published 1970.

14 chapters: 1. Type production - 2. Camera work and process engraving - 3. Letterpress relief plates - 4. Letterpress printing - 5. Offset lithographic plates - 6. Offset lithographic printing - 7. Reprography - 8. The photogravure process - 9. Screen process printing - 10. Flexographic printing - 11. Collotype printing - 12. Finishing processes, binding machines and methods of binding - 13. Comparison of printing processes - 14. Modern technical developments. V.13-14 are new. Glossary. Reading list, p.241. Index. Covers 'the technical subjects which form the syllabus for examinations in General Printing Knowledge' (*Preface*). *Class No:* 655.3

[5676]

KOLLECKER, E. *and* **MATUSCHKE, W. Der Moderne Druck:** Handbuch der graphischen Techniken. 2. Aufl. Hamburg, Hammerich & Lesser, 1956. 696p. illus. (incl. col.).

Over 40 distinguished contributors. Covers printing in all its aspects, from historical to metal foil processes. Detailed descriptions of processes, with numerous black-and-white and colour illus. Particularly strong on colour printing, with machinery well described and depicted. A fine example of book production, in keeping with its subject, quality printing processes. Dated now, but still contains much useful information. *Class No:* 655.3

Lithography

[5677]

BRITISH STANDARDS INSTITUTION. Glossary of terms used in offset lithographic printing. London, the Institution, 1968. 48p. £40. (*BS 4277: 1968 (1991).*)

Defines *c.*300 terms. 6 sections: 1. General terms - 2. Copy preparation - 3. Photographic and associated operations, including colour separation - 4. Plate and plate making - 5. Printing - 6. Machines and equipment. Index. *Class No:* 655.342

[5678]

The Lithographers manual. Blair, R.N., *ed.* 8th ed. Pittsburgh, Pa., Graphic Arts Technical Federation, Inc., 1988. v.p. illus. (incl. col.), diagrs., graphs, tables. $65. ISBN: 0883621053.

16 chapters (1. The history of lithography ... 4. The preparation of type and art ... 6. Colour separation photography ... 10. The platemaking department - 12. Lithographic presswork ... 15. Various supplementary

....(contd.)

processes). Lithographic terms. Bibliography (2p.; authors, A-Z). Appended: GATF Technical series reports. 5-p. index (tiny print). *Class No:* 655.342

Book Trade

Dictionaries

[5679]

ORNE, J. **The Language of the foreign book trade:** abbreviations, terms, phrases. 3rd ed. Chicago, Ill., American Library Association, 1976. x, 333p. $10. ISBN: 038902197.

First published 1949.

26,000 entries in 14 non-English-language sections (each with English equivalent terms): Czech, Dano-Norwegian, Dutch, Finnish, French, German, Hungarian, Italian, Polish, Portuguese, Rumanian, Russian, Spanish, Swedish. German-English, the largest section, has *c.*2,800 entries. Not every term is limited to the book trade. Aims 'to provide a needed working tool in the major foreign languages for librarians' (*Preface*). *Class No:* 655.4(038)

Worldwide

[5680]

The Book trade of the world. Taubert, S. *and* Weidhaas, P., *eds.* Munich, K.G. Saur, 1972-84. 4v. $80ea. ISBN: 359810183x, v.1; 3598101848, v.2; 3598101856, v.3; 3598101869, v.4.

1. *Europe and international section.* 1972. 543p. £26. 2. *The Americas, Australia, New Zealand.* 1976. 377p. £26. 3. *Asia.* 1981. 284p. £26. 4. *Africa. With index to v.1-4.* 1984. 391p. £37.

The European section of v.1 (p.77-534) covers 35 countries, A-Z. Each country entry has data on 35 topics, including: retail prices; trade press; copyright; national bibliography; book production; book design; wholesale trade; book imports; book exports; literary prizes; reviewing journals; graphic arts. *Library journal* (v.97(19)), 1 April 1972, p.3556 finds the v.1 compendium invaluable for the bookseller, publisher and librarian, but it 'suffers a little from over organization'. An updating is certainly called for. *Class No:* 655.4(100)

[5681]

International bibliography of the book trade and librarianship, 1976-1979. 12th ed., edited by H. Lengenfelder and G. Hausen. Munich, K.G. Saur, 1981. xxix, 692p. $70. ISBN: 3598205153.

11th ed., 1973-1975. 1976.

The 12th ed. lists *c.*9,000 titles of books, brochures, research reports, standards, bulletins, guides, acquisition lists of libraries, etc. from 116 countries. Arranged by regions, then countries and finally by subject. Preponderance of European and US regions, - 433p. and 94p. respectively. Author and editor index, p.611-75; subject index, p.679-92. No annotations. *Class No:* 655.4(100)

[5682]

International literary market place: the directory of the international book publishing industry, 1991. New York, Bowker, 1990. Annual. 783p. $149.95. ISBN: 0835221687. ISSN: 00746827.

First published 1965.

A directory of over 10,000 book publishers and book-trade organizations in more than 160 countries, other than the US and Canada. Arranged alphabetically by country, sections include publishers; book trade; libraries; literary associations. 90-page Yellow Pages acts as an index to the directory. Criticized in *Reference reviews,* v.5(3), 1991 for patchy coverage. Companion to *Literary market place (q.v.).* 1992 ed. 824p. $149.95. Available on CD-ROM. *Class No:* 655.4(100)

England

[5683]

BENNETT, H.S. **English books & readers.** Cambridge Univ. Press, 1952-70. 3v. £50, set (pbk).

1475-1557: being a study in the history of the book trade, from Caxton to the incorporation of the Stationers Company. 1952. 2nd rev. ed. 1969 (reprinted 1989) £65; £17.95. xiv,338p. *1558-1603: being a study of the book trade in the reign of Elizabeth 1.* 1965, xviii, 320p. £70. *1603-1640: being a study of the book trade in the reigns of James 1 and Charles 1.* 1970. xiv, 254p.

The 1558-1603 volume has chapters on patronage, regulation of the book trade, translations and translators, the variety of books, and printers and booksellers. Many footnotes (*e.g.* citing short-title catalogue entries). Bibliography, p.301-5. Analytical index. A scholarly series. *Class No:* 655.4(420)

[5684]

PLANT, M. **The English book trade:** an economic history of the making and sale of books. 3rd ed. London, Allen & Unwin, 1974. 520p. illus., facsims., ports. ISBN: 0046550127.

First published 1939.

2 parts (21 chapters): 1. The age of hand printing (chapters 1-12) - 2. The application of mechanical power. References, p.470-500. Detailed, analytical index, 501-20. A scholarly survey by the then deputy-librarian, British Library of Political and Economic Science. *Class No:* 655.4(420)

Asia—Middle & Near East

[5685]

A Book world directory of the Arab countries, Turkey and Iran. Rudkin, A. *and* Butcher, I., *comps.* London, Mansell; Detroit, Mich., Gale, 1981. xiv, [1], 143p. $110. ISBN: 0720108306, UK ed.

Directory of libraries, booksellers, publishers, international publishers, newspapers and periodicals in 20 countries of the Middle East. 1,400 numbered entries, with systematic data. 'General information on countries covered by the Directory' precedes. 13 appendices (1. Libraries classified by subject area; 8. Publishers classified by nature of literature; 11. Periodicals classified by content and scope). Fills a gap. *Class No:* 655.4(53+56)

Africa

[5686]
The African book world & press: a directory. Zell, H.M., *ed.* 4th ed. London, Munich, etc., Hans Zell Publishers, 1989. £75. ISBN: 0905450507.

First published 1977.

More than 4,400 entries for publishers, libraries, booksellers, research institutions, printing industries, publishing programmes, magazines, periodicals and major newspapers throughout Africa. Arrangement is A-Z by country, Algeria ... Zimbabwe. *Class No:* 655.4(6)

America — North

[5687]
American book trade directory. New York, Bowker, 1915-. Now biennial.

The 1989/90 directory, the 35th ed. (1989. 1805p. $169.95), lists 25,435 bookstores, jobbers, wholesalers and distributors in the US and Canada, noting subject specialities. Geographic and name indexes. (Firms are classified by category of books sold).

37th ed., for 1991-92, 2029p. $189.95.
Class No: 655.4(71 + 73)

[5688]
Literary market place: the directory of the American book publishing industry with industry Yellow Pages, 1991. New York, R.R. Bowker. Annual.

First published 1940.

The 1991 ed. (1657p. 1990. $124.95.) contains a 330-page section on US and Canadian publishers which is followed by sections on editorial services and agents; advertising, marketing and publishing; book manufacturing; sales and distribution organizations; services and suppliers, associations, events, courses and awards; books and magazines for the trade. The index of *c.*500 pages includes a Yellow Pages section of all organizations and personnel with addresses and telephone numbers. Companion to *International literary market place (q.v.).* Available on CD-ROM.

1992 ed. 1790p. $134.95. *Class No:* 655.4(71 + 73)

Publishing

Dictionaries

[5689]
JACOB, H. A Pocket dictionary of publishing terms, including explanations and definitions of words and phrases commonly used in the production and distribution of books. London, Macdonald and Jane's, 1976. [x], 70p.

About 700 entry-words. 'The entries cover terms which publishers have made their own' (*Preface*). Short bibliography of 12 items. *Class No:* 655.41(038)

[5690]
Multilingual dictionary of publishing, printing and bookselling. Isaacs, A., *and others, eds.* London, Cassell, 1992. 439p. £50. ISBN: 0304325127.

Over 2,000 keywords and phrases in seven languages: English (British and American); French; German; Italian; Portuguese; Spanish; Swedish. Aimed at all those involved in the book or periodical trade. *Class No:* 655.41(038)

[5691]
Publisher's practical dictionary in 20 languages. Móra, I., *comp.* 3rd rev. ed. Munich, K.G. Saur; New York, French and European, 1983. ix, 418p. $125. ISBN: 3598104499; 0828801908, US ed.

First published 1974 (German-base terms).

About 1,000 English-base terms, with equivalents and indexes in other 19 languages, - German, French, Russian, Spanish, Bulgarian, Danish, Finnish, Dutch, Italian, Croat, Norwegian, Polish, Portuguese, Rumanian, Swedish, Serb, Slovak, Czech and Hungarian. 'Very easy to use' (*RQ,* v.10(1), Fall 1974, p.67-68). *Class No:* 655.41(038)

[5692]
SCHUWER, P. Dictionnaire de l'édition: art, technique, industrie et commerce du livre. [Dictionary of book publishing: creative, technical and commercial terms of the book industry.] Paris, Cercle de la Librairie, 1977. 324p. ISBN: 379404147x.

More than 6,000 terms: French - English (British and American), English (British and American) - French. Multiple definitions are provided where necessary. *Class No:* 655.41(038)

[5693]
STIEHL, U. Satzwörterbuch des Buch- und Verlagswesen, Deutsch-Englisch. [Dictionary of book publishing, with 12,000 sample sentences and phrases, German-English.] Munich, Verlag-Dokumentation; New York, French and European, 1977. xx, 538p. $250. ISBN: 0828833966.

German terms (explained, if considered necessary), with English equivalents and examples of English usage. Covers the book trade, publishing, editing and copyright, marketing and library science. *Class No:* 655.41(038)

Reviews & Abstracts

[5694]
World publishing monitor. Oxford, Pergamon Press, 1991-. Irregular. DM1,243pa. ISSN: 0960653x.

Incorporates *Electronic publishing abstracts* and compiled by the Information Section of PIRA (Paper Industry Research Association).

Issue for September 1991 contains 302 abstracts on subjects including management; company information; agreements and contracts; health and safety; communications; graphic reproduction; storage media; television; marketing; storage and warehousing. Available online via ORBIT and PFDS Online. *Class No:* 655.41(048)

Worldwide

[5695]
5001 hard-to-find publishers and their addresses. 4th ed. London, Dawson, 1990. 147p. £30. ISBN: 0946291217.

First published 1981.

Includes details of over 8,000 publishers. Concentrates on Western Europe, the Commonwealth and USA. Arranged A-Z by publishers. Designed to complement such directories as Bowker's *Literary market place (qv).* Considerably extended since the previous ed., its strength lies in listing specialized and small publishers. *Class No:* 655.41(100)

[5696]
Directory of publishing 1993. 18th ed. London, Cassell, 1992. 2v.
The *UK, Commonwealth, and overseas* volume (xv,447p. £65) has 7 sections: 1. Introductory - 2. Publishers, p.1-223 - 3. Packagers - 4. Authors' agents - 5. Trade and allied associations - 6. Trade and allied services - 7. Appendices (9), *e.g.* 2. Ownership of UK publishers; 6. Index of ISBN prefixes. The volume, *Continental Europe* (xv,494p. £40) has 3 sections: 1. Introductory - 2. Publishers (Albania ... Yugoslavia) - 3. Appendices (5), *e.g.* 1. Fields of activity; 5. Index of companies and imprints. *Class No:* 655.41(100)

[5697]
Publishers' international ISBN directory. Opitz, H. *and* Strasser, K.-H., *eds.* Munich, etc., K.G. Saur, 1991. Annual.
The 18th ed. (3v. 4624p. 1991. £245) contains more than 220,000 entries from 200 countries, arranged under continents. Includes over 105,000 publishers with ISBN prefixes. Includes microfilm, video and computer software publishers, complete with ISBN. Supplements *British books in print,* etc.
19th ed. (3v. 1992. £240). *Class No:* 655.41(100)

Great Britain

[5698]
FEATHER, J. A History of British publishing. London, Croom Helm, 1988. x, 292p. £40; £12.99. ISBN: 0709910673.
Introduction: 'The book trade before printing'. 4 parts (19 chapters): 1. The press in chains - 2. Licence and liberty - 3. The first of the mass media (19th century) - 4. The trade in the twentieth century. Notes. Bibliography, p.263-81. Index. Author is Senior Lecturer. Dept. of Library and Information Studies, Loughborough University. *Class No:* 655.41(410)

[5699]
MUMBY, F.A. Mumby's publishing and bookbinding in the twentieth century. 6th ed. London, Bell & Hyman, 1982. 253p. ISBN: 0713515411.
First published 1930; 5th ed. (Cape), 1974.
An historical survey, in 3 parts: 1900-1939; 1939-1950; 1950-1980, plus 'Epilogue'. 6 appendices (1. Total number of books published; 2. Subject areas of titles published; 3. UK book sales, home and export). Bibliography, p.229-32. Abbreviations, p.233-4. Detailed analytical index, p.235-53. This 6th ed. supplements the scholarly 5th ed. (685p.) which covers the history from earliest times to 1970.
Class No: 655.41(410)

France

[5700]
Histoire de l'édition française. Martin, H.-J. *and* Chartier, R., *eds.* Paris, Promodis, avec le concours du Centre National des Lettres, 1982-86. 4v. illus, facsims. (incl. col.), ports., maps. ISBN: 2903181063, v.1; 2903181314, v.2; 2903181446, v.3; 2903181543, v.4.
1. *Le livre conquérant: du Moyen-Age au milieu du XVII siècle.* 1982. FFr820. 2. *Le livre triomphant, 1660-1830.* 1982. FFr880. 3. *Le temps des éditeurs: du Romantisme à la Belle Epoque.* 4. *Le livre concurrencé, 1900-1950.* 1986. FFr970.

....(contd.)
An elaborate, well-produced survey of book-publishing in France. Each volume has a bibliography and index.
Class No: 655.41(44)

USA

[5701]
Books in print, 1992-93. v.10: Publishers. New York, Bowker, 1992. 1280p. ISBN: 0835232905.
First published 1986.
Lists *c.*35,000 publishers and distributors whose titles are included in the main author and title indexes (v.1-9). Includes geographical index to wholesalers and distributors; new publishers; inactive and out-of-business publishers. Available online and on CD-ROM. *Class No:* 655.41(73)

[5702]
—Publishers directory, 1992: a guide to approximately 15,000 new and established commercial and nonprofit, private and alternative, corporate and association, government and institution publishing programs and their distribution. 14th ed. Detroit, Mich., Gale, 1991. 2v. $235. ISBN: 0810381567. ISSN: 07420501.
Includes producers of books, classroom materials, prints, reports and databases. Indexes: publishers, imprints, and distributors; subjects; geographical location.
Class No: 655.41(73)

[5703]
Publishers, distributors & wholesalers of the United States, 1991-92. New York, Bowker, 1991. 2v. 2371p. £120. ISBN: 0835231038.
Contains data on 58,000 publishers, 6,200 associations, 3,500 distributors and wholesalers, 800 museums and 6,700 software producers. V.1 contains the name index and a list of imprints, divisions, and subsidiaries. V.2 contains 7 indexes, with lists of fax numbers, ISBNs, publishers by fields of activity, etc. Available online.
Class No: 655.41(73)

[5704]
TEBBEL, J. History of book publishing in the United States. New York, Bowker, 1972-81. 4v. (*c.*3,000p.). ISBN: 0835204898.
1. *The creation of an industry, 1630-1865.* 1972. 1646p. 2. *The expansion of an industry, 1865-1919.* 1973. 1813p. 3. *The golden age between two World Wars, 1920-1940.* 1978. 774p. 4. *1940 to the present.* 1981.
More than 2,600 pages of texts and notes, each volume having its own notes and index. Almost 270p. of indexes in all. A monumental achievement, although 'a weakness of the entire set is the scant attention paid to copyright' (*The journal of library history,* v.17(4), Fall 1982, p.503).
Class No: 655.41(73)

[5705]
—TEBBEL, J. Between covers: the rise and transformation of book publishing in America. New York, Oxford Univ. Press, 1987. 528p. $35. ISBN: 0195041895.
Based on the author's four-volume work. A comprehensive single-volume history of the American publishing industry. *Class No:* 655.41(73)

Bookselling

[5706]
Bookshops of Greater London. Thomas, R.J., *ed.* 6th ed. Brentford, Middx., Roger Lascelles, 1990. 334p. £5.95. ISBN: 0903909820.

First published 1981 as *The Bookshops of London*.
This ed. expanded to cover the entire county of Greater London. Data on nearly 800 bookshops, grouped by postal district, include subjects; hours of opening; services; address; telephone number. Shop index, p.299-309. Subject index, p.310-34 (excludes general and major bookshops). *Class No:* 655.42

[5707]
Directory of specialist bookdealers in the UK handling mainly new books. Marcan, P., *comp.* 5th ed. High Wycombe, Bucks., Peter Marcan Publications, 1991. v,89p. £15.99. ISBN: 1871811058.

Gives detailed information on 650 specialized booksellers in the UK. Arranged by subject. Data include address; subjects; catalogues; publications. Index of business names, p.80-6. *Class No:* 655.42

[5708]
Specialty booksellers directory. Kremer, J., *and others.* Fairfield, Iowa, Ad-Lib Publications, 1987. $19.95. ISBN: 0912411163.

Lists more than 2,100 specializing book dealers in North America. 48 broad subject categories. Entries are under store names. 4 indexes: subject interest; type of book (*e.g.* paperback; out-of-print); sideline merchandise sold; state/province. *Class No:* 655.42

Antiquarian & Second-hand Books

[5709]
Cole's register of British antiquarian & secondhand bookdealers, 1990. York, The Clique, 1990. 319p. £18. ISBN: 1870773136.

4 parts: 1. Alphabetical listing (*c.*2,000 entries, p.1-132; data: name and address of booksellers; telephone; size of stock; specialization; catalogues) - 2. Specialist index and catalogue booksellers - 3. Geographical listing (London ... Yorkshire, west; Northern Ireland; Scotland; Wales) - 4. Supplementary index (names of booksellers). *Class No:* 655.425

[5710]
Sheppard's book dealers in North America: a directory of antiquarian and secondhand book dealers in the USA and Canada. 11th ed. Old Woking, Surrey, Richard Joseph, 1990. 343p. £21. ISBN: 1872699006. ISSN: 09500715.

First published 1954.
Sections (as in *Sheppard's Book dealers in the British Isles*): Miscellaneous information - Geographical directory of dealers - Alphabetical index - Speciality index - Index of advertisers. *Class No:* 655.425

[5711]
Sheppard's book dealers in the British Isles: a directory of antiquarian and secondhand book dealers in the United Kingdom, the Channel Islands, the Isle of Man and the Republic of Ireland, 1990-91. 15th ed. Farnham, Surrey, Richard Joseph, 1990. 576p. ISBN: 1872699014. ISSN: 09500715.

Contains over 2,300 entries. Geographical directory of book dealers, p.55-330 (English counties; Wales, Scotland; Isle of Man; Northern Ireland; Republic of Ireland; Channel Islands). 'Miscellaneous information' (*e.g.* Book trade and literary periodicals; supplies and services) precedes. Alphabetical index (including dealer's addresses and 'phone numbers). Speciality index. List of dealers, grouped, Agriculture ... Topography and travel. Index of advertisers. 1992 ed. 432p. £24. *Class No:* 655.425

[5712]
Sheppard's European bookdealers 1989-90. 7th ed. London, Europa Publications, 1989. 300p. £21. ISBN: 0946653402.

Bookdealers listed under 27 countries, Austria ... Yugoslavia. Periodicals, p.4-11; reference books, p.12-18 - Alphabetical index; speciality index; list of advertisers. *Class No:* 655.425

[5713]
Skoob directory of secondhand bookshops in the British Isles. 4th ed. London, Skoob Books Publishing, 1991. 277p. maps. £6.99. ISBN: 1871438020.

Lists over 1,000 bookshops. 4 preceding essays, *e.g.* 'The books themselves'; 'Collecting modern first editions'. Directory of bookshops in the British Isles, p.1-207. Index of names; subject index; index of towns. Good value. *Class No:* 655.425

Transport Services

Databases

[5714]
TRANSDOC. European Conference of Ministers of Transport. London, Departments of the Environment and Transport, 1970-.

Concerns the economic and social aspects of transport. Contains *c.*30,000 citations to material mainly published in Europe but also from other countries, *e.g.* USA. Coverage includes periodicals, books, reports, summaries of current research projects, theses, and conference papers. About 750 items added quarterly. Host: ESA. *Class No:* 656(003.4)

[5715]
TRANSPORTATION RESEARCH BOARD, Washington. TRIS / Transportation Research Information Service. Washington, D.C., US Department of Transportation and Transportation Research Board NAS/NRC, 1968-.

Provides research information on air, road, rail and maritime transport. Subjects covered include energy, regulations and legislation, traffic control, communications, and environmental and maintenance technology. Organizations contributing to *TRIS* include the Highway Research Information Service (HRIS); the Maritime Research Information Service (MRIS); International Road Research Documentation (IRRD); the Air Transportation Research Information Service (ATRIS); the Railroad Research Information Service (RRIS).

IRRD (International Road Research Documentation) also runs a database, TRICS, hosted through ESA, in which 20 countries participate. *Class No:* 656(003.4)

Bibliographies

[5716]
CHARTERED INSTITUTE OF TRANSPORT. Library.
Transport technology. London, the Institute, 1974. 71p.
Mimeographed. (*Transport bibliography 16.*)

About 1,000 briefly annotated entries (Books; Pamphlets,
reports, (p.12-69); Statutes; Bibliographies; Dictionaries;
Annuals, statistics, etc.). Government department, ICAO,
GLC, UN, US, etc. technical reports from the main feature.
Book prices are given. No index.

Transport bibliography 16A: *Selected periodical articles*
(1975. 51p. Mimeogoraphed) is a contents-list of periodicals,
A-Z by titles *Aeronautical journal ... World ports.*

These bibliographies are now replaced by a series of
reading lists, one for each of the Chartered Institute of
Transport's examination papers. *Class No:* 656(01)

[5717]
Information sources in transportation, material management,
and physical distribution: an annotated bibliography and
guide. Davis, B.J., *comp. and ed.* Westport, Conn.,
Greenwood Press, 1976. [xv], 715p. $67.95. ISBN:
0837183790.

About 10,000 entries in 67 sections, each of which
accommodates A. Books and pamphlets; B. Periodicals; C.
Organizations; D. Education; E. Miscellaneous (*e.g.* atlases
and maps). Brief annotations. Author, title, subject index,
p.635-715. *Class No:* 656(01)

[5718]
NORTHWESTERN UNIVERSITY, Evanston, Ill.
Transportation Center Library. **Catalog. [Northwestern
University, Transportation Center Library].** Boston,
Mass., G.K. Hall, 1972. 12v. (7697p.).

Author/title catalog. 3v. (1774p.) $1,305. *Subject catalog.*
9v. (5293p.) $980.

171,000 photolithographed catalogue cards, in all. Main
emphasis on transportation management, operations,
planning, economics, regulation and impact. International in
scope.

Updated in monthly *Current literature in traffic and
transportation* (Transportation Centre Library, 1960-.),
arranged under subjects, A-Z. *Class No:* 656(01)

[5719]
STYLES, B.R. 'Highway, transport and constructional
engineering'. In *Information sources in engineering.* 2nd ed.
Editor, L.J. Anthony (1985), p.409-28.

Sections: Organizational sources of information -
Documentary sources of information (abstracts and indexing
services; journals; monographs and textbooks).
Class No: 656(01)

[5720]
Transport publications, 1984-1985: a list of publications used
by the Department of Transport in 1984-1985. London,
Department of the Environment; Department of Transport,
1986. [i], [i], 29p. (*Bibliography no.208(c). Supplement
no.2.*) ISBN: 1851120130.

374 numbered entries. Section 1: Departmental
publications (A. Transport policy and transport ... K.
International transport issues) - 2. HMSO publications - 3.
Departmental circulars. Supplements the ten-year listing of
transport publications (1971-81) and the 1982-1983 listing.
The list now includes publications on shipping and aviation.

More recent is *List of titles: an annotated list of titles
available in the DOE/DTP Library Services' Bibliography*

....(contd.)
series. Recent titles include *The Channel Tunnel* (1991) and
Geographical information systems (1991).
Class No: 656(01)

Encyclopaedias & Dictionaries

[5721]
**BEIN, G. Wörterbuch des internationales Verkehrs:
Deutsch-Englisch.** Leipzig, VEB Verlag Enzyklopädie;
New York, State Mutual Book and Periodical Service Ltd.,
1968. 232, [1]p. $15. ISBN: 0569051177, US ed.

About 11,500 entries, covering domestic and international
road, rail and air transport and traffic, inland and maritime
navigation, plus foreign trade. Appendix of translations of
names of charters and Acts, mostly on shipping and
maritime transport. 'A model publication of its kind'
(*Incorporated linguist,* v.10(3), July 1971, p.91).
Class No: 656(03)

[5722]
COMMISSION OF THE EUROPEAN COMMUNITIES.
Terminology Office. **Glossaire nouvelles techniques de
transport.** Luxembourg, CEC, 1975.

A six-language transport glossary, - in Danish, English,
French, Italian, Dutch and German. Covers the means and
modes of modern and future transport. 'There is still no
good transport dictionary but if you must have something I
should recommend *Glossaire nouvelles techniques de
transport'* (*The incorporated linguist,* v.19(4), Autumn 1980,
p.141). *Class No:* 656(03)

[5723]
Concise encyclopedia of traffic and transportation systems.
Papageorgiou, M., *ed.* Oxford, Pergamon Press, 1991.
xviii,658p. figs., illus. £205. (*Advances in systems, control
and information engineering.*) ISBN: 0080362036.

118 articles arranged A-Z (Air traffic control: an overview
... Visual and instrument flying rules), written by 135
contributors. Each article has a bibliography appended.
Cross-references. Aimed at academics, managers, operators,
managers, engineers and graduate students. Contains some
updated and revised articles from *Systems and control
encyclopedia (q.v.)* (1987). *Class No:* 656(03)

[5724]
—The Illustrated encyclopaedia of transportation.
Crow, G., *ed.* London & New York, Reference
International, 1977. 261p.

10 contributors. Concise A-Z entries, on land, sea and air
transport, both historical and technological. 'Pipelines'
(operation; design and construction; freight pipelines): 2p., 3
illus. Includes biographies. Over 400 illus., including keyed
line-drawings and cutaway diagrams. 16p. of colour plates.
Non-analytical index. *Class No:* 656(03)

[5725]
**UNION INTERNATIONALE DES TRANSPORTS
PUBLICS. Dictionary of public transport** / Wörterbuch
Nahverkehr / Dictionnaire des transports publics.
Elms, C.P., *and others, eds.* Dusseldorf, Alba Buchverlag;
New York, French and European, 1981. 312p. $45. ISBN:
3870947637.

About 5,000 English-base terms, plus definitions, with
equivalent terms and indexes in German and French.
Class No: 656(03)

[5726]

—LOGIE, G. Glossary of transport, in English, French, Italian, Dutch, German and Swedish. Amsterdam, etc., Elsevier, 1980. xxviii, 206p. £64.06; $97.50. (*International planning glossaries, 2.*) ISBN: 0444418881.

Has 1,108 English-base terms, grouped into main and subsidiary headings, with equivalents in the other 5 languages. Indexes cover all 6 languages. Criticized in *The Incorporated linguist* (v.19(4), Autumn 1980, p.140) for inaccuracies. 'The majority of terms in this dictionary would undoubtedly be found in a good general bilingual dictionary'. Also, costly. *Class No:* 656(03)

Handbooks & Manuals

[5727]

Transportation and traffic engineering handbook. Homburger, W.S., *and others, eds.* 2nd ed. Englewood Cliffs, N.J., Prentice Hall, Inc., 1982. 883p. illus., graphs, tables. ISBN: 0139330626.

First published 1976, edited by J.E. Baerwald.

28 well-documented sections, stressing traffic engineering rather than mass transportation. Geographical and detailed subject indexes. *Class No:* 656(035)

Reviews & Abstracts

[5728]

Library bulletin: abstracts of current literature on social and environmental planning, roads, traffic and transport, countryside and recreation, housing and local government, water supply and waste disposal, pollution and conservation. London, Department of the Environment; Department of Transport, 1972-. Semi-monthly.

About 25% of the abstracts per issue concern transport. The 1 July 1988 issue carried 31 transport items. Headings: Bridges - Civil aviation - Driving - Freight transport - Passenger transport - Railways - Road Safety - Road motor vehicles - Roads - Shipping - Tunnels.

Insert: DOE/DTP *Publications monthly list.*

Class No: 656(048)

[5729]

—TR news. Washington, Transportation Research Board, National Research Council, National Academy of Sciences, 1983-. 6pa. $25.

Formerly *Transportation research news.*

Features articles on timely subjects, abstracts of current transport literature, and announcements of TRB publications and meetings. Discusses activities of the Board, government, and industry. US-slanted. *Class No:* 656(048)

Histories

[5730]

The Journal of transport history. 3rd series. Manchester Univ. Press, 1971-. 2pa. £60; $110. ISSN: 00225266.

First series, 1953- (Leicester University Press).

V.9(1), March 1988, has 6 documented articles (*e.g.* 'Australian maritime history: a progress report', p.75-92; 63 references) and 2 review articles (*e.g.* 'Some recent railway history in German'). 5 book reviews; Review round-up (3p. of running commentary); Books received.

'Transport history in British university theses', first

.... (contd.)

featured in *The Journal of transport history* in v.4(3), May 1960, p.161-73. is continued at intervals.

Class No: 656(091)

Europe

[5731]

UNITED NATIONS. Economic Commission for Europe. **Annual bulletin of transport statistics for Europe.** v.40: 1990. New York, United Nations, 1990. 281p. (*E/F/R. 89.11.E6.*) ISBN: 9210162285. ISSN: 02509911.

Transport statistics for 1980, 1987, 1988. Sections: General statistics - Railways - Roads - Inland waterways - International rivers - Sea ports - Combined transport - Oil pipelines - Others. 3 annexes (1. Definitions; 2. Sources of information and coverage of statistics. 3. Lists of publications from which monthly and/or quarterly transport statistics can be obtained). Notes appended to tables. In English, French, and Russian. *Class No:* 656(4)

Great Britain

Tables & Data Books

[5732]

MUNBY, D.L. Inland transport statistics, Great Britain, 1900-1970. Watson, A.H., *ed.* Oxford, Clarendon Press, 1978. xii, 693p. tables. £60. ISBN: 0198284098.

1. *Railways, public road passenger transport. London's transport.*

3 main sections: A. Railways - B. Public road passenger transport - C. London's transport. 3 annexes (1. General note on sources of labour statistics). Index, p.677-93. An impressive array of statistical data, with valuable commentaries. 'No self-respecting library can afford to be without this volume, despite its expense' (*History*, v.65(213), February 1980, p.175).

A second volume is to cover roads, private haulage and, presumably, inland waterways. *Class No:* 656(410)(083)

[5733]

Reviews of United Kingdom statistical sources. Oxford, etc., Pergamon Press, for the Royal Statistical Society and the Social Science Research Council, 1978-.

V.7. Review 12: *Road passenger transport,* by D.L. Munby. 13: *Road goods transport,* by A.N. Watson. 1978. xii, 138p. 127p. v.10. Review 17: *Ports and inland waterways,* by R.E. Baxter. 18: *Civil aviation,* by C.M. Phillips. 1978. xii, 302p. v.14. Review 24: *Rail transport,* by D.H. Aldcroft. 1981. 280p. £40. 25: *Sea transport,* by D. Mort. 1981. xii, 268p.

The aim of the series is 'to act as a work of reference to the sources of statistical material of all kinds, both official and unofficial' (*Introduction*, v.7). V.17, Review 18: *Civil aviation* (p.111-302) has 4 chapters, with appended quick-reference list; bibliography; coverage of publications; useful libraries and books; appendix of forms, and a subject index. Statistics are analysed in this *Review* up to September 1977. *Class No:* 656(410)(083)

[5734]
[Sources of transport statistics]. In *Aslib proceedings,* v.40 (11/12), November/December 1988, p.321-332.

3 papers presented at a seminar held on 8 June 1988: 'Use and abuse of statistics', by S. Hoyle; 'Statistics available from the Department of Transport', by J. Akinbolu; 'Sources of non-government transport statistics'; by E. Godward.

J. Akinbolu's paper briefly examines the types of publications produced by the Department of Transport, as well as the source guides on transport statistics. Includes major HMSO publications, non-HMSO publications and occasional publications. Statistic collection is also described. *Class No:* 656(410)(083)

[5735]
Transport statistics: Great Britain 1991. London, HMSO, 1991. xii, 302p. tables. £24. ISBN: 0115510788.

Tables (each with commentary, notes and definitions, plus footnoted sources): 1. Transport: an overall view; 2. Road; 3. Rail; 4. Water transport; 5. Air transport; 6. International comparisons; 7. Selected historical series. 4 articles precede the tables (*e.g.* A. Investment in transport ... D. Road accident fact sheets). Appendices. Detailed, analytical index. Data revised and updated quarterly in *Quarterly transport statistics. Class No:* 656(410)(083)

Histories

[5736]
BAGWELL, P.S. The Transport revolution. 2nd ed. London, Routledge, 1988. xiv, 474p. illus., figs., tables, maps. £52.50. ISBN: 041500876x.

First published 1974.

Covers the development of transport in Great Britain from 1770 to 1985. 11 chapters: 1. Inland navigation in Britain ... 3. Coastal shipping ... 5. The economic and social effects of railways ... 8. The development of motor transport 1885-1939 - 9. British transport policy 1914-39 ... 11. Public transport in decline 1972-85. Notes, p.417-53. Bibliography, p.454-62. Index. A detailed work. *Class No:* 656(410)(091)

[5737]
DYOS, H.L. and ALDCROFT, D.H. British transport: an economic survey, from the seventeenth century to the twentieth. Leicester, Univ. Press, 1969. 473p. tables, maps. £32.50. ISBN: 071851081x.

Paperback ed. 1974 (Harmondsworth, Middx., Penguin Books. 511p.).

13 chapters (1. The transport system of pre-industrial Britain ... 12. The development of motorised transport - 13. The birth of civil aviation). Chapter bibliographies, with running commentary (p.401-38) and author index. Non-analytical general index, p.445-73. An attempt at 'a full length general survey of the development of transport in Britain in modern times ... Careful attention has been given to the bibliography' (*Preface*). Well produced. *Class No:* 656(410)(091)

Institutions & Associations

[5738]
Transport organisations in the United Kingdom. King, D., *comp.* 2nd ed. London, Department of the Environment; Department of Transport, 1986. [1], [1], 42p. (*UB/INF/14A.*) ISBN: 1851120076.

Earlier ed. (1976) as *UK transport organizations.*

....(contd.)

About 200 entries, with emphasis on Government departments and other official bodies, research institutes, professional organizations, trade associations, pressure groups. Generally omitted: enthusiasts' societies, employers' organizations and trade unions. Data on each body include date of founding, status and aim, publications. *Class No:* 656(410):061:061.2

Asia—South East

[5739]
South-east Asian transport. Leinbach, T.R., *and others.* Singapore, Oxford University Press, 1989. xix, 265p. tables, maps. £27.50. ISBN: 0195888952.

Covers the history and future development in this area. 9 chapters: 1. The role of transport in development ... 4. Road and rail transport systems ... 7. ASEAN air transport - 8. Urban transport - 9. Transport progress, problems, and policies in south-east Asia: a synthesis. 4 appendices give statistics on road transport, rail, merchant fleets, and air transport. Bibliography, p.248-60. Author and subject indexes. *Class No:* 656(59)

Libraries

[5740]
SPECIAL LIBRARIES ASSOCIATION, New York. Transportation Division. **Transportation libraries in the United States and Canada:** an SLA directory. 3rd ed. New York, the Association, 1978. 220p.

Data on 205 collections. Subject, geographic and personal name indexes. Needs updating. *Class No:* 656:061:026/027

Museums

[5741]
GARVEY, J. A Guide to the transport museums of Great Britain. London, Pelham Books, 1982. 238p. illus. (incl. col.), maps. ISBN: 0720714044.

Descriptions of 52 museums, plus list of a further 54 awaiting treatment. 10 areas: London; The South ... The Midlands; Wales ... Northern Ireland. Gazetteer follows. Data on each museum include access, exhibits, opening hours and facilities. Symbols indicate parking, refreshment, shop. Index. Attractively produced, but could now do with revision. *Class No:* 656:061:069

Routes & Services

[5742]
National Express guide: a guide to Express Coach Services, including some services of Scottish Citylink coaches. Birmingham, National Express, Ltd. 2pa. tables, maps. £4.90.

Issue 27A, November 1992-May 1993 contains: General information - Codes and symbols - Principal offices and travel services. Maps. Index to places served. National Express timetables and forms. Scottish Link index timetables and fares. No system map. *Class No:* 656.02

Urban Transport

[5743]
BANISTER, D. *and* PICKUP, L. **Urban transport and planning: a bibliography with abstracts.** London, New York, Mansell, 1989. iv,354p. £40. ISBN: 0720116279.

Contains *c.*500 abstracts, covering English language material published since 1980. 8 sections: 1. The context - 2. Policy and planning in transport - 3. Social issues - 4. Travel modes - 5. Methods and evaluation - 6. Area studies - 7. Bibliographies and research registers - 8. Additional entries. Subject and author indexes. A companion volume to *Rural transport and planning: a bibliography with abstracts* (Mansell, 1985). *Class No:* 656.022.9

[5744]
Jane's urban transport systems, 1992-93. Bushell, C., *ed.* 11th ed. [London], Jane's Information Group, 1992. 605p. illus., tables. £135. ISBN: 071060985x.

Alphabetical profiles, city by city (Aachen ... Zurich), p.8-373. - Manufacturers (12 sections of suppliers, services, systems) - Consultancy and contracting services. Index. *Class No:* 656.022.9

[5745]
Roads and traffic in urban areas. London, HMSO, 1987. xxiv, 418p. illus. (incl. col.), diagrs., tables, maps. £25. ISBN: 011550818x.

Replaces *Urban traffic engineering techniques* (1968) and *Roads in urban areas* (1966).

Reflects the enormous changes in car ownership, lorry weights, parking facilities, and many other recent trends. 'An important new standard reference work for traffic engineers and road designers ...' (*British book news,* July 1988, p.577). *Class No:* 656.022.9

Administration

Goods & Freight

[5746]
ABC freight guide. London, Centaur Business Directories, 1973-. Annual.

1991-92 ed. (43rd ed. 482p.) is a directory of some 18 types of freight service: Business and professional services ... Driver agencies ... General road haulage (regions, A-Z) - Storage and distribution - Breakdown and recovery - Vehicle, trailer hire - International services - Heavy haulage - Tanker wagons - Bulk tipping - Industrial removals - Container owners - Light freight and express service - Temperature controlled distribution - Rail freight - Sea freight/sea services - Air freight. Index of advertisers. *Class No:* 656.073

[5747]
FREIGHT TRANSPORT ASSOCIATION. [1987 yearbook, Freight Transport Association]. Tunbridge Wells, Kent, the Association. Annual. 376p. illus., maps. £25. ISSN: 03001523.

The law governing road transport operations, p.7-227; Insurance; Industrial relations; The law governing operations; The law governing specific vehicles. Addresses, p.295-348. Useful references (*e.g.* forms most frequently used). Detailed, analytical index, p.357-76. *Class No:* 656.073

Road Transport

Handbooks & Manuals

[5748]
FAULKS, R.W. **Bus and coach operation.** 5th ed. London, Butterworth, 1987. 272p. illus. ISBN: 0408028106.

Sections: Introduction - Historical background - The infrastructure - Vehicle design - Organization and management - Statutory controls - The planning process - Commercial practices - Schedule compilation - Fare collection - Operational control - Finance - The future. 107 illus. For managers and aspiring managers in the road passenger industry, and especially for those taking the Chartered Institute of Transport Membership and Association Membership exams. *Class No:* 656.1(035)

[5749]
LOWE, D. **The Transport manager's & operator's handbook** 1990. 20th ed. London, Kogan Page, 1990. 454p. figs. ISBN: 1850919852.

First published 1969. Formerly published as *The Transport manager's handbook.*

Covers UK legislation relating to road transport. 29 chapters: 1. Operators' licensing ... 11. Type approval ... 14. Light vehicle testing ... 23. Road traffic regulations... 29. Mobile communications. 4 appendices: 1. Licensing authorities and areas covered ... 4. Organizations connected with transport. Index. *Class No:* 656.1(035)

Dictionaries

[5750]
Dictionary of road transport terminology in four languages, English, French, German and Spanish. Amsterdam, etc., Elsevier; New York, French and European, 1988. xxviii, 351p. $250; £111.56. ISBN: 0828823758, US ed; 0444428194.

4,387 English-based terms, with French, German and Spanish equivalents and indexes. 'Usual abbreviations', p.xi-xxviii. No genders given. *Class No:* 656.1(038)

Reviews & Abstracts

[5751]
Highway research abstracts. Washington, Transportation Research Board, 1990-. 4pa. $80pa. ISSN: 10500804.

Formerly *HRIS abstracts,* 1968-89.

Abstracts - *c.*1,800 per issue - of research reports, conference proceedings papers and journal articles on highways and nonrail mass-transit research. Classes 11-90 (31 classes, *e.g.* 11. Administration; 21. Facilities design; 53. Vehicle characteristics; 54. Operations and traffic control; 90. Highway research, general). Source, author and retrieval term indexes, per issue, cumulated annually. International. Indexes *c.*750 periodicals and 2,000 reports pa. Online TRIS, whch also covers *Highway safety literature* (HSL) and *Urban mass transportation research information service* (UMTRIS). *Class No:* 656.1(048)

[5752]

Road documentation for developing countries. Crowthorne, Berks., Overseas Unit, Transport and Road Research Laboratory, 1974-. Annual. *Gratis*. ISBN: 0854296948.

500 abstracts per issue on road design, materials, construction and maintenance, traffic and transport, accident studies, vehicles, economics, administration. Online IRRD (International Road Research Documentation) database.
Class No: 656.1(048)

Worldwide

[5753]

World road statistics, 1985-1989. Geneva, International Road Federation, 1990. 200p. diagrs., tables.

11 sections: 1. Road network - 2. Production and export of motor vehicles - 3. First registration and imports of motor vehicles - 4. Vehicles in use - 5. Road traffic - 6. Motor fuels - 7. Road accidents - 8. Rates and basis of assessment of road user taxes - 9. Examples of average annual taxation - 10. Annual receipts from road-user taxation - 11. Road expenditure. Headings in English, French, German.
Class No: 656.1(100)

Europe

[5754]

The Little red book 1991/92: road passenger transport directory for the British Isles and Western Europe. Booth, G., *ed.* London, Ian Allan, Ltd., 1991. 215p. £16.95. ISBN: 0711019703.

5 sections: 1. Trade directory - 2. United Kingdom operators, p.43-170 - 3. Tendering authorities, traffic commissioners (Department of Transport) - 4. European operators - 5. Index (British Isles operators).
Class No: 656.1(4)

[5755]

Manuel du transport routier international. [Handbook of international road transport.] 10th ed. Geneva, International Road Union, 1988. 494p. tables, forms. Loose-leaf.

Countries, Austria ... Yugoslavia. Appendices: Customs Convention, on the international transport of goods under cover of TIR channels - Community transit systems - List of the principal juridical texts on international road transport - Table showing maximum size and weight permissible for road vehicles in Europe. *Class No:* 656.1(4)

Great Britain

[5756]

Basic road statistics, 1988. London, British Road Federation, [1988]. 39, [2]p. diagrs., tables, maps.

55 tables: 1. Motor vehicles - 2. Road networks - 3. Road traffic - 4. Demand for transport - 5. Energy and transport - 6. Public expenditure on transport - 7. Taxation - 8. Road accidents - 9. Regional statistics - 10. County statistics - 11. International comparisons. Map: The strategic trunk road network. 3 appendices (3. List of sources for latest figures). 'Quick comparisons'. Statistics cover 1970, 1980, 1984, 1985, 1986. *Class No:* 656.1(410)

[5757]

CHARLESWORTH, G. A History of British motorways. London, T. Telford, 1984. xii, 284p. illus., tables, maps. £21. ISBN: 0727701592.

12 documented chapters (2. Road policies between the wars; 4. 1950-60: motorway construction; 7. Trunk road motorways in England (39 references); 8. Trunk road motorways in Wales and in Scotland (19 references); 10. Planning, design and maintenance (60 references); 12. Some economic and social aspects). Detailed, analytical index. Well organized. Author was one-time Assistant Director at the Transport and Road Research Establishment.
Class No: 656.1(410)

[5758]

The Driving manual. London, HMSO, 1992. v,330p. illus. (col.). £8.50. (*Driving skills*.) ISBN: 0115510540.

20 sections: 1. The driver ... 9. Driving on motorways ... 16. Towing a caravan or trailer ... 20. Automatics, four-wheel drive. Useful addresses, p.324. Index. Illustrations could be of better quality but get the message across. Aimed at both the learner and experienced driver. Companion volume is *The Motorcycling manual* (HMSO, 1991. v,156p. £6.95). *Class No:* 656.1(410)

[5759]

The Highway code. Rev. ed. London, HMSO, 1993. 80p. illus. £0.99. ISBN: 0115511547.

First published 1946.

First major update since 1978. An advisory code of practice. Expanded to include new rules and details of recent changes in the law. Also includes an expanded section giving advice to cyclists and horse riders, a completely new section on tramways, and a new section on traffic law, its effect on road users, and penalties for the main offences.
Class No: 656.1(410)

Thesauri

[5760]

INTERNATIONAL ROAD RESEARCH DOCUMENTATION. Thesaurus, 1985. [International Road Research Documentation]. English ed. Paris, OECD/IRRD. 2v. diagrs., tables. ISBN: 9264126287.

1. *Numerical list and English arrowed diagrams*. 2. *English alphabetical list.*.

French and German versions of the *Thesaurus* are also available. *Class No:* 656.1:025.43

Traffic Signs

[5761]

EUROPEAN CONFERENCE OF MINISTERS OF TRANSPORT. European rules concerning road traffic signs and signals. London, Dept. of Transport, 1974. 178p. illus. (incl. col.), diagrs.

Part 1: Road traffic rules, and annexes - Part 2: Road signs and signals, and annexes. 18 countries were involved, plus associated countries (Australia, Japan) and observers (Canada, USA). *Class No:* 656.1.05

[5762]
—Know your traffic signs. 3rd ed. London, HMSO, 1989. 59p. col. illus. £1.10. ISBN: 011550866x.

Concerns the history, types and locations of UK road signs. An authoritative explanation of the British traffic sign system, incorporating changes brought about by the 1981 Traffic Act regulations. *Class No:* 656.1.05

Railway Transport

Dictionaries

[5763]
UNION INTERNATIONALE DES CHEMINS DE FER. **Lexique général des termes ferroviaires.** 3. éd. Paris, the Union, 1975. 1602p.

A six-language dictionary of railway vocabulary. About 13,000 French-based terms, with equivalents and indexes in German, English, Italian, Spanish and Dutch. The standard dictionary in its field. *Class No:* 656.2(038)

Worldwide

[5764]
Fodor's railways of the world. Whitaker, R.E.M. *and* Fisher, R.C., *eds.* London, Hodder and Stoughton, 1977. 374p. illus., maps.

Facts at your fingertips, p.12-34. 9 regional sections (North America; Western Europe; Eastern Europe; The Middle East; South and Southeast Asia; East Asia; The Pacific; Africa; South America). Index, p.371-4. Aimed at tourists. Notes on: classes of accommodation; train services; comparative distances and times; timetables, fares and reservations; catering; hotels and accommodation at stations; railway museums and reserved railways; photography. Somewhat dated. *Class No:* 656.2(100)

[5765]
NOCK, O.S. **Railways of the world.** London, A. & C. Black, 1971-1979. 6v. illus. (some col.), maps. ISBN: 0713611065; 0713611901; 0713616865; 0713618558; 0713620064.

V.1: *Railways of Southern Africa.* 1971. 243p. V.2: *Railways of Australia.* 1971. 284p. V.3: *Railways of Canada.* 1973. 343p. V.4: *Railways of Western Europe.* 1977. xvi, 264p. V.5: *Railways of Asia and the Far East.* 1978. xvii, 226p. V.6: *Railways of the USA.* 1979. xvi, 317p.

Each v. covers the history, technology and routes, past and present, of the area in a chatty, informal style. Volume indexes. 'Recommended ... for transport historians and all those with a mature interest in trains' (*British book news,* January 1978, p.4 on v.4). Nock is a retired engineer and a prolific author of railway books. *Class No:* 656.2(100)

[5766]
Railway directory: a *Railway Gazette* yearbook. Sutton, Surrey, Railway Gazette International. Annual.

First published 1894. Formerly *Railway directory & year book.*

The 1992 ed. contains systems maps for all countries, statistical data on every railway. Includes lists of 10,000 senior staff, data on 200 urban rail systems and lists of 1,200 manufacturers/suppliers. General, personnel, manufacturers and advertisers index. Numerous maps.
Class No: 656.2(100)

Great Britain

Maps & Atlases

[5767]
WIGNALL, C.J. **Complete British railways maps and gazetteers, 1825-1985.** Rev. ed. Oxford, Oxford Publishing Co., 1983. iv, 71, 79p. ISBN: 086093294x.

First published 1983.

Atlas of 2-colour maps (71p.) distinguishes between 6 types of railway, *e.g.* Lines closed to passengers; Preserved lines open to passengers; Lines underground at this point. Gazetteer to maps (76p.) notes stations open or closed to passengers and HQ of preserved railways, plus preservation centres. Lists of mainland railway companies, p.77-79. A well-produced quarto. *Class No:* 656.2(410)(084.3)

[5768]
—British railways pre-grouping atlas and gazetteer. 5th ed. London, Ian Allan, 1972. 84p. maps. ISBN: 0711003203.

Maps show the railway system of Britain as it existed in the years prior to 1923. Includes every railway company at work at that time. Also shows junctions, the more important viaducts, tunnels, locomotive shed locations, and railway workshops. Gazetteer, p.47-83, includes list of company abbreviations, indexes of tunnels, water troughs, principal summits and viaducts, and stations. Smaller format than Wignall and a useful vade mecum for the enthusiast. *Class No:* 656.2(410)(084.3)

Histories

[5769]
AWDRY, C. **Encyclopaedia of British railway companies.** Yeovil, Patrick Stephens, 1990. 288p. illus. £20. ISBN: 1852600497.

A-Z list of railway companies, with dates of incorporation and major developments, arranged into the 4 major groups (GWR, SR, LMS, LNER) into which all Britain's railways were grouped 1923-48 before nationalization. Independent and London Transport undertakings are also listed, and there are 'family tree' charts of constituent companies, p.244-78. Authoritative and attractively presented. Good value. *Class No:* 656.2(410)(091)

[5770]
A **Bibliography of British railway history.** Ottley, G., *and others, comps.* London, Allen & Unwin, 1965. 683p.

1983 reprint (termed '2nd edition') by the Science Museum (HMSO. £25. ISBN 0112903347).

7,950 entries, many annotated, with 1 location in 16 British libraries (particularly the British Museum) and private collections. Excludes children's books (usually) and highly technical specialized books, but includes spotters books and others that contain useful railway history. Main classes, A-T (B. Railway transport at particular periods - C. Railway transport in particular areas ... T. General directories, gazetteers, atlases, guide books, lists of stations, distance tables, time-tables). Entries state number of illus. Appended lists and 'genealogies' of railway companies. Extensive author-title and subject index, p.477-683. Comprehensive, and a model of its kind. *Class No:* 656.2(410)(091)

[5771]

—A Bibliography of British railway history. Supplement. Ottley, G., *comp.* London, HM Stationery Office, 1988. 544p. £35. ISBN: 0112903649.

2 parts: 1. Addenda and corrigenda to the main work; 2. Classification scheme (classes A-T). Bibliography, p.61-383 (entries no.7951-12956) of books and pamphlets published 1964-80, plus some pre-1963 publications recently discovered. One index, of authors, titles and subjects at head of title page: 'Science Museum/National Railway Museum'. *Class No:* 656.2(410)(091)

[5772]

—CLINKER, C.R. Railway history sources. Rev. and exp. ed. Bristol, Avon-Anglia Publications, 1976. 20p. £1. *(Reference aid series, no. 1.)*

First published as *Railway history: a handlist of principal sources of original material* (1969).

11 sections, including local records, newspapers, periodicals, time-tables and maps. Appendix: 'Articles in the Journal of transport history, p.15-16'. *Class No:* 656.2(410)(091)

[5773]

ELLIS, C.H. British railway history: an outline, from the accession of William IV to the nationalisation of railways. London, Allen & Unwin, 1954-59. 2v. illus.

1. *1830-1876.* 443p. 2. *1877-1947.* 416p.

The first general, popular history of the British railway companies (excluding those of Ireland). It carries copious references to the individuals concerned, and includes mechanical and electric traction developments. Well illustrated. *Class No:* 656.2(410)(091)

[5774]

GOURVISH, T.R. British railways, 1948-73: a business history. Cambridge, University Press, 1986. xxvii, 781p. £75. ISBN: 0521264804.

3 parts (13 chapters): 1. The British Transport Commission and the Railway Executive, 1948-53 - 2. The British Transport Commission, 1953-62 - 3. The British Railways Board, 1963-73. Appendices A-J (J. Passenger service rationalisation, 1961-73). Notes, p.644-755. Bibliography ('a personal choice'), p.756-63. 'This is a commissioned but not an official history' *(Preface).* *Class No:* 656.2(410)(091)

[5775]

OTTLEY, G. Railway history: a guide to sixty-one collections in libraries and archives in Great Britain. London, Library Association, 1973. 80p. *(Subject guides to library resources, no.1.)* *Class No:* 656.2(410)(091)

[5776]

A Regional history of the railways of Great Britain. Thomas, D. St. J. *and* Patmore, J.A., *eds.* Nairn, D. St. J. Thomas, 1960-. 15v. illus., maps.

1. *The West country,* by D. St. J. Thomas. 6th rev. ed. 1988. £12.95. 2. *Southern England,* by H.P. White. 5th ed. 1992. £19.95. 3. *Greater London.* 3rd ed. by H.P. White. 1987. £12.95. 4. *The North-east* by K. Hoole. 3rd rev. ed. 1986. £12.95. 5. *Eastern counties,* by D.I. Gordon. 3rd ed. 1990. £15.95. 6. *Scotland: the Lowlands and the Borders.* 2nd ed., by J. Thomas. 1984. £12.95. 7. *The West Midlands.* 3rd ed., by R. Christiansen. 1992. £19.95. 8. *South and West Yorkshire.* 2nd ed. by D.A.W. Joy. 1984. £12.95. 9. *The East Midlands,* New ed., by R. Leleux. 1984. £12.95. 10. *The North west.* 2nd ed., by G.O. Holt.

....(contd.)

1986. £19.95. 11. *North and Mid Wales,* by P.F. Baughan. 2nd ed., 1992. £19.95. 12. *South Wales,* by D.S. Barrie. 1980. 13. *Thames and Severn,* by R. Christiansen. 1981. £14.95. 14. *The Lake counties,* 2nd ed., by D. Joy. 1990. £15.95. 15. *The North of Scotland,* by J. Thomas and D. Turnock. 1989. £14.95.

Each volume is self-contained and well-illustrated. V.4 has 44 good-quality half-tone plates, 9 text illus., 8 regional maps, 1 folding map, plus a partly analytical index. *Class No:* 656.2(410)(091)

England & Wales

[5777]

SIMMONS, J. The Railway in England and Wales, 1830-1914. Leicester Univ. Press, 1978.

To be in 4v. 1. *The system and its working.* 1978. 196p.

Considers smaller and less successful railway companies as well as the most profitable and better-known ones. Supported by research at both the Public Record Office and local record offices. 'A major study' *(Annual bulletin of historical literature.* v.64: 1978 (1980), p.92). *Class No:* 656.2(42)

America—North & Central

[5778]

Railroad maps of North America: the first hundred years. Modelski, A.M., *comp.* Washington, Library of Congress, 1984. 186p. illus., maps. $28. ISBN: 0844403962.

Reproduces 92 of the 5,000 railroad maps in the Library of Congress's Geography amd Map Division. Coverage: USA, Canada, Mexico, The maps, arranged by type, have descriptive text and bibliographical notes. An introductory essay traces the history of railroad mapping in North America. 'This excellent atlas' *(Geographical bibliography for American libraries;* edited by C.D. Harris (1985), item 1627). *Class No:* 656.2(71/73)

Timetables

[5779]

ABC rail guide. Dunstable, Beds., ABC International Division, Reed Travel Group. Monthly. tables, maps. £7 per issue; £65pa. ISSN: 00010472.

Contains an A-Z list of places in Great Britain served by rail, with information on fares and train times to and from London only, p.40-250; complete timetable for services in London and SE England, p.251-600; provincial Intercity section, p.601-750, arranged partly A-Z by 72 'important' cities and towns, partly tabular; shipping and miscellaneous tables, p.751-800; some European tables, p.801-16. Handy guide but not comprehensive for stations outside Southern England. *Class No:* 656.222

[5780]

British Rail passenger timetable. London, British Railways Board. 2pa., plus supplements. tables, maps.

Established 1853.

The October 1991 - May 1992 (£5.50) issue 'contains all internal rail services, coastal shipping information and connections with Ireland, Isle of Man, the Isle of Wight, the Channel Islands and certain continental services to France, Belgium, etc'. Timetables, p.33-1570. Station index precedes. 'Other useful information' *(e.g.* InterCity sleepers;

....(contd.)

Private railway systems). Local and suburban services are included. Unlike *ABC rail guide,* includes connections. *Class No:* 656.222

[5781]
Thomas Cook European timetable: railway and shipping services throughout Europe. Peterborough, Thomas Cook. Monthly. tables, maps. ISSN: 0952620x.

September 1991 issue (544p. £6.50) contains a 32-page Winter Supplement. Information includes explanation of signs; city plans; car sleepers; airport links; holiday trains; public holidays; shipping services; books, guides and maps. The index now includes places which do not have a rail service. The section 'What's new this month' draws attention to any changes or alterations since publication of the previous issue. *Class No:* 656.222

[5782]
—Thomas Cook overseas timetable: railway, road and shipping services outside Europe. Peterborough, Thomas Cook Publishing. 6pa. tables, maps. ISSN: 01447475.

July-August 1991 issue (£6.45) contains basic travel facts (*e.g.* city maps and climatic information), p.2-72, followed by timetables for rail, bus, and internal waterborne services in America (Canada to South America), Africa, Asia, Australasia, as well as international shipping. *Class No:* 656.222

Signals

[5783]
Railway signalling: a treatise on the recent practice of British Railways. Nock, O.S., *ed.* London, A. & C. Black, 1980. viii, 312p. diagrs., charts. ISBN: 0713620676.

Prepared under the direction of a committee of the Institution of Railway Signal Engineers.

11 chapters (*e.g.* 2. Layout of signals and track circuits; 4. Equipment: relays, signals and point machines; 7. Track circuit principles and equipment; 8. Remote control systems; 10. The British Railways' automatic wiring system; 11. The future). Chapter-by-chapter index, p.311-2. *Class No:* 656.252

[5784]
—ALLEN, D. *and* WOOLSTENHOLMES, C.J. A Pictorial survey of railway signalling. Yeovil, Haynes Publishing Group, 1991. 144p. illus. £14.99.

Restricted to UK in coverage and principally of interest to the rail enthusiast, this is a detailed treatment of railway signalling, both semaphore and electrical. No continuous text, the book comprises 244 plates with 2 captions each, brought together by good indexes (locations, locomotives and multiple units; signalling). Complements Nock's work on the same subject but does not supersede it. *Class No:* 656.252

[5785]
—Railway control systems. Leach, M., *ed.* London, A & C Black, 1991. 291p. diagrs. £29.99. ISBN: 0713634200.

Compiled by a Project Group of the Institution of Railway Signal Engineers as a sequel to O.S. Nock's *Railway signalling* (1980).

11 chapters (*e.g.* 1. Recent changes in signalling philosophy ... 4. Immunisation and earthing of signalling

....(contd.)

systems ... 6. Level crossings ... 10 Automatic train protection). This is a highly-authoritative technical treatise. Lacks an overall index. *Class No:* 656.252

Marine Transport

Abbreviations & Symbols

[5786]
The Fairplay book of shipping abbreviations. Kapoor, P., *comp.* London, Fairplay Publications, 1980. 89p. ISBN: 0905045181.

Shipping abbreviations, A-Z, p.1-70. Code names, p.71-76. Abbreviations of company names; International code of signals; International Morse code; Operational yearbook and abbreviations; Astronomical yearbook; Greek alphabet; Roman numerals. *Class No:* 656.6(003)

Bibliographies

[5787]
Adventurers afloat: a nautical bibliography. A comprehensive guide to books in English recounting the adventures of amateur sailors upon the waters of the world in yachts, boats, and other devices and including works on the arts and sciences of cruising, racing, seamanship, navigation, design, building, etc. from the earliest writings through 1986. Toy, E.W., *Jr.* Metuchen, N.J., London, Scarecrow Press, 1988. 2v. $89.50. ISBN: 0810821893.

5,594 entries, most of which are annotated. 4 main sections: 1. The story of yachting and boating - 2. The arts, techniques, and information necessary for yachting and boating - 3. Pleasure craft: their ancestry, development, design, construction, equipment, and care - 4. Other aspects of yachting and boating. 2 appendices: 1. Chronological guide to sail racing rules - 2. Amateur Yacht Research Society publications. Indexes: author-title; ship and boat. *Class No:* 656.6(01)

[5788]
CRUISING ASSOCIATION, London. **The Cruising Association Library catalogue:** a collection of books for seamen and students of nautical literature, atlases and charts. Hanson, H.J. [2nd ed.]. London, the Association, 1939. Supplement, 1954-80. 1980. 349p. illus. (pl.).

The main catalogue, under author A-Z, with subject index, has *c.*5,000 entries. Includes list of periodicals and other serials, plus a note of 12 other nautical libraries, mostly in London. Well illustrated. *Class No:* 656.6(01)

[5789]
Guide to the manuscripts in the National Maritime Museum. Knight, R.J.B., *ed.* London, Mansell, 1977-80. 2v. 259p.; 250p. £35ea. ISBN: 0720107148, v.1; 0720115914, v.2.

V.1. *The Personal collections.* V.2. *Public records, business records, and artificial collections.*

Records the holdings of the NMM at the end of the 1970s. V.1: *The personal collections;* v.2: *Public records, business records, and artificial collections.* 'Artificial collections' refers to volume of manuscripts and other documents acquired singly. V.2 has a general index included to both volumes. *Class No:* 656.6(01)

[5790]

MARINERS MUSEUM, Newport, Va. **Dictionary catalog of the Library. [Mariners Museum].** Boston, Mass., G.K. Hall, 1970. 9v. $980. ISBN: 0816106746.

About 118,820 photolithographed catalogue cards, covering over 43,400 volumes on shipbuilding, navigation, voyages and exploration, naval history, merchant shipping, etc. Library of Congress classification used, with separate card file of books arranged chronologically, 1518-1825.

Also: *Catalog of marine photographs* (1970. 5v. 78,000 cards. $545); *Catalog of marine prints and paintings* (1970. 3v. 50,000 cards. $330); *Catalog of maps, ships, papers and logistics* (1970. 1v. 10,000 cards. $110).

Class No: 656.6(01)

[5791]

NATIONAL MARITIME MUSEUM. Catalogue of the Library. [National Maritime Museum]. London, HMSO, 1968-.

1. *Voyages & travel.* 1968. xi, 403p. 2. *Biography.* 1969. 2v. (501p.; 475p.). 3. *Atlases and cartography.* 1971. 2v. (xi, lx, 1166p.). 4. *Piracy & pioneering.* 1972. ix, 175, (8)p. 5. *Naval history.* pt.1: *The Middle Ages, to 1815.* 1976, [xi], 209p.

V.1 has 1,240 numbered entries, usually with annotations or bibliographical notes. V.5, pt.1: *The Middle Ages, to 1815,* lists 2,318 items. 5 parts: 1. Mediaeval; 2. The sixteenth century; 3. The seventeenth century; 4. The eighteenth century (to 1815); 5. General works. It includes periodical articles and all library accessions up to the end of June 1974. Brief notes on some items. Detailed, non-analytical index. The sequel (v.5, pt.2), aimed to continue the history to modern times, has not been published.

Class No: 656.6(01)

[5792]

OBIN, A. Bibliography of nautical books, 1990: in print, forthcoming and recently out of print. 5th ed. Warsash, Warsash Nautical Bookshop, 1990. 500p. tables. £40. ISBN: 0948646055.

First published 1985.

Computer print-out lists. 11 sections: 1-3: Books - 4. Hydrographic Department (Admiralty) - 5. International Maritime Organization - 6. Statutory Instruments/HMSO - 7. M Notices (Department of Transport) - 8. British Standards - 9. Journals and periodicals - 10. Video cassettes - 11. Publishers and suppliers (addresses and telephone numbers). Lists English-language books only and a few selected books published in the US but not distributed in the UK.

Class No: 656.6(01)

Encyclopaedias

[5793]

The Oxford companion to ships and the sea. Kemp, P., *ed.* Oxford Univ. Press, 1976 (reissued 1988). £14.95 (paperback). ISBN: 0192820842.

3,700 articles on nautical terms, famous ships, seamen, navigators, ports, explorers, etc. 'Diving' (unassisted, assisted and free diving), p. 250-5; 5 illus., references to Bathyscaphe, Bathysphere. Biographical sketches (*e.g.* 'Tasman', 'Von Spee', 'Conrad'). Also, Maritime museums, naval instruments. Generally omits material on oceanography and small-boat pleasure sailing. 5 appendices (1. Equivalent ranks; 3. Rules of the road). 226 half-tones; 187 line-drawings. 'Highly recommended' (*Library journal,*

.... *(contd.)*

15 January 1977, p.188). *The Economist* (23 October 1976, p.142) finds that 'Modern technology has largely been sacrificed for the sake of history'. *Class No:* 656.6(031)

[5794]

—KEMP, P. Encyclopedia of ships and seafaring. London, Stamford Maritime; New York, Crown, 1980. 256p. illus.

Has 9 topical chapters, followed by 3 A-Z sequences: 'Famous ships in history', 'Ship and boat types', and 'Great men of the sea'. Cross-references. Detailed index. Some overlap with *Oxford companion to ships and the sea,* but aimed at a more general public. *Class No:* 656.6(031)

Handbooks & Manuals

[5795]

BRANCH, A.E. Economics of shipping practice and management. 2nd ed. London, Chapman & Hall, 1988. xii, [2],360p. illus., diagrs., tables. £16.50. ISBN: 0412231309.

First published 1982.

18 chapters. 1. Services rendered by sea transport to international trade ... 5. Economics of containerization ... 7. Economics of ship operation ... 16. Role of British and international shipping organizations ... 17. Role of other national and international organizations - 18. Political factors. 3 appendices: A. Further recommended textbook reading - B. Addresses of organizations and institutions engaged in the field of shipping and international trade. - C. Ship diagrams. Index. *Class No:* 656.6(035)

Dictionaries

Polyglot

[5796]

DE KERCHOVE, R. International maritime dictionary: an encyclopedic dictionary of useful maritime terms and phrases, together with equivalents in French and German. 2nd ed. Princeton, N.J., Van Nostrand, [1961] (reprinted 1983, New York, French and European). [v], 1018p. $75. ISBN: 082804929x.

First published 1948.

About 12,000 terms in ship construction (including local and native craft), maritime law, shipping and marine insurance; many abbreviations have entries. Some terms carry bibliographical references. Diagrams are distinctly old fashioned. French and German indexes.

Class No: 656.6(038)=00

[5797]

Multi-lingual dictionary of commercial international trade and shipping terms. Branch, A.E., *and others.* London, Witherby, 1990. 255p. £19.95. ISBN: 0948691905.

Intended for 'exporters, importers, travel agents, ship-brokers' etc., this is a handy little guide to some 2,200 terms (*e.g.* Discount rate ... Indivisible load), arranged English-French, p.13-94, English-Spanish, p.95-174, and English-German, p.175-255. No cross-translations into English.

Class No: 656.6(038)=00

English

[5798]

BES, J. Chartering and shipping terms. 9th ed., brought up to date by including the revised York, Antwerp Rules. 1974. v.1. London, Barker & Howard, 1975. 506, [12]p. illus. £25. ISBN: 0900133082.

First published 1951.

V.1: 'Practical guide to steamship companies, masters, ship's officers, shipbrokers, forwarding agents, exporters, importers, insurance brokers and banks'. 14 chapters, each with terms A-Z (1. Charterings ... 4. Conditions of carriage ... 7. Shipping terms ... 13. Marine insurance - 14. Trade terms). New chapters on operations of tramp ships, of ships in line and tramp trade, and of container ships. 18 appendices (*e.g.* pro formas estimates for voyages, interpretation of charter party terms). Index. 10 illus.
Class No: 656.6(038)=20

[5799]

BRANCH, A.E. Dictionary of shipping, international trade terms and abbreviations. 3rd ed. London, Witherby, 1987. xii, 587p. illus. £14.50. ISBN: 0948691204.

First published 1976; 2nd ed. 1982.

More than 9,000 terms (2nd ed.: 4,000), p.1-532. Appendices A-G (A. Further recommended textbook reading; B. Major ports of the world and their location; D. Addresses and roles of shipping organisations and institutes; G. Diagrams (1-14, *e.g.* Liquified natural gas carrier)). Author, chief examiner in Shipping and Export Practice.
Class No: 656.6(038)=20

[5800]

Dictionary of nautical words and terms: 8,000 definitions in navigation, seamanship, rigging, meteorology, astronomy, naval architecture, average, ship economics, hydrography, cargo, stowage, marine engineering, ice terminology, buoyage, yachting, etc. Layton, C.W.T. Rev. 3rd ed. Glasgow, Brown, Son, and Ferguson, 1987. 397p. £22. ISBN: 0851745369.

The 8,000 entries range from 6 to *c.*100 words (*e.g.* 'Trinity House'). Words now obsolete or obsolescent are asterisked. *Class No:* 656.6(038)=20

[5801]

—BLACKBURN, G. The Illustrated dictionary of nautical terms. Newton Abbot, Devon, David & Charles, 1982. 359p. illus., diagrs. ISBN: 0715382969.

Has *c.*3,000 entries, usually 2-15 lines in length (but 'Boom': 16½ lines, plus 4 illus). Where British and American spelling diverge, the former appears in brackets after the American. Keyed illus. are a feature. Adequate cross-references. *Class No:* 656.6(038)=20

[5802]

—BRODIE, P.R. Dictionary of shipping terms. London, Lloyd's of London Press, 1985. [iv], 155p. ISBN: 1830640697.

Defines *c.*2,000 terms, phrases and abbreviations, used in the international movement of cargo. Cross-references.
Class No: 656.6(038)=20

[5803]

—WHITLOCK, P.C., *and others*. The 'Country Life' book of nautical terms under sail. Feltham, Country Life Books, 1978. 352p. illus. (incl. col.), facsims., diagrs., maps.

Defines *c.*4,000 terms. 20 sections: 1. The ship: basic terms - 2. General terms - 3. Building a ship, keel to bulwarks - 4. Masts, spars, sails and rigging ... 12. Sailing and seamanship ... 18. Pilotage charts - 19. Navigational instruments - 20. Inshore and ashore. 2p. of section bibliographies. 15p. index. Illus. are a feature.
Class No: 656.6(038)=20

[5804]

SULLIVAN, E. The Marine encyclopaedic dictionary. 3rd ed. London, Lloyd's of London Press, 1992. vii, 497p. £45. ISBN: 1850443718.

About 20,000, entries, many for abbreviations. Some longer entries, *e.g.* 'International shipping organizations': 5½ columns; 'Coastguard': 3 columns' 'SOLAS (Safety of life at sea)': 7p.; 'Lloyd's Register of shipping': 6 columns. Covers all the terminology in use in the marine transport and shipping industry. Well-produced.
Class No: 656.6(038)=20

German

[5805]

DIETEL, W.   Seefahrts-Wörterbuch: deutsch-englisch, englisch-deutsch. 2. Aufl. Munich, Lehmans Verlag, 1964. 1024p.

First published 1954.

A German-English, English-German maritime dictionary. About 35,000 terms in all, covering military and civil seafaring. Includes colloquialisms and slang, in quotation marks. 'Quite shamelessly padded' (*The Incorporated linguist,* v.9(2), April, 1970, p.51-52).
Class No: 656.6(038)=30

French

[5806]

BRODIE, P.R. Dictionary of shipping and shipbroking terms, French/English, English/French. London, Lloyd's of London, 1980. 216p. £15. ISBN: 0904093824.
Class No: 656.6(038)=40

Italian

[5807]

BERNABO SILORATA, M. *and* **PICCHI, F. Grande dizionario di marina, italiano-inglese, inglese-italiano.** Salerno, Di Maura Editore; New York, French and European, 1970. 963p. illus. (incl. pl.), tables. $125. ISBN: 0686925513. *Class No:* 656.6(038)=50

Spanish

[5808]

SUÁREZ GIL, L. Diccionario tecnico maritimo, inglés-español-inglés. 2nd ed. Madrid, Editorial Alhambra; New York, French and European, 1983. vii, 704p. $110. ISBN: 0828804257, US ed; 8420509396.

First published 1981.

English-Spanish, p.1-292; Spanish-English, p.293-569. About 4,000 entries in each half. Terms and phrases are

....(contd.)

categorized. 'Arquitectura naval', p.644-704 (illus. plus keys). 8 appendices (*e.g.* abbreviations; naval expressions). *Class No:* 656.6(038)=60

Greek

[5809]

Dictionary of shipping: Greek/English, English Greek. Vlassopulos, B.P., *ed.* London, Lloyd's of London Press, 1988. 200p. £29. (*Foreign language dictionaries.*) ISBN: 1850441731.

A few thousand terms, each way, in modern commercial usage. *Class No:* 656.6(038)=77

Arabic

[5810]

BAKR, M. Elsevier's maritime dictionary in three languages, English, French, Arabic. Amsterdam, etc., Elsevier, 1987. [vii], 664p. $200. ISBN: 0444427376.

12,024 English-based terms, with French and Arabic equivalents and indexes. Thumb-indexed. *Class No:* 656.6(038)=927

Yearbooks & Directories

[5811]

Fairplay world shipping directory. London, Fairplay Information Systems. Annual. ISSN: 09593101.

Previously (1976-77) as *'Financial times' World shipping year book.*

The 1992-93 *Directory* (x,958p.) has sections on shipowners, arranged by country (p.1-221); shipbuilders, repairers and engine builders; towage and salvage; offshore supply and dredging; marine insurance; protection and indemnity associations; ship finance; marine equipment suppliers; maritime organisations; ship sales and purchase brokers; consulting marine engineers and ship surveyors; maritime schools; port authorities and operators. Editorial. Personnel, ship, and advertisers' indexes.

Fairplay world shipping statistics, 1988 (Fairplay Publications, 189p.) is a geographical analysis of the marine industry worldwide. *Class No:* 656.6(058)

[5812]

Lloyd's nautical year book. London, Lloyd's of London Press, 1979-. Annual (1993 ed. 430p. £35). illus.

Originally as *Lloyd's Calendar,* established 1898.

A compendium that includes articles, a list of maritime organizations, data on safety at sea, shipping insurance and legal terms, calendar information, standard times, sea-distance tables, and much else. *Class No:* 656.6(058)

Histories

[5813]

ALBION, R.G. Naval & maritime history: an annotated bibliography. 4th ed., rev. and expanded. Mystic, Conn., Munson Institute of American Maritime History, the Maritime Historical Association, Inc., 1972. ix, 370p. $14.95. ISBN: 0913372056.

First published 1951; 3rd ed. 1963 (viii, 230p.).

Lists *c.*5,000 books in English, plus PhD. theses. Some entries are briefly annotated items considered to be 'the most

....(contd.)

substantial and useful' asterisked. 7 sections: 1. Reference works - 2. Merchantmen and warships - 3. Captains and crews - 4. Maritime science, exploration and expansion - 5. Commerce and shipping (p.103-9: divided by regions) - 6. Navies (p.229-302, by periods) - 7. Special topics (*e.g.* Auxiliary services; Main American maritime museums). Author and brief subject indexes. 'An indispensable companion to any writer on maritime history' (*Journal of the Institute of Navigation,* v.17(2), April 1964, p.211). US slant. *Class No:* 656.6(091)

[5814]

—Bibliography of periodical articles published in ... Mystic, Conn., Maritime Historical Association, 1970-.

Continues Albion's bibliography. *Class No:* 656.6(091)

[5815]

Five hundred years of nautical science, 1400-1900: conference proceedings. Howse, D., *ed.* London, National Maritime Museum, 1982. 418p. illus. £11. ISBN: 0955555554. *Class No:* 656.6(091)

[5816]

HARLAND, J. Seamanship in the age of sail: an account of the shiphandling of the sailing man-of-war, 1600-1860, based on contemporary sources. London, Conway Maritime Press, 1984. 320p. illus. £30. ISBN: 0851771793.

A comprehensive survey of how ships were handled, even down to the smallest detail of shipboard routine. Bibliography, p.313-5. Author is an active member of the Society for Nautical Research. *Class No:* 656.6(091)

[5817]

HOPE, R. A New history of British shipping. London, John Murray, 1990. xviii,533p. illus., tables, maps. £35. ISBN: 0719547997.

3 sections: 1. The beginnings of British shipping 3000BC - 1400AD - 2. The rise of British shipping 1400-1890 - 3. The decline of British shipping 1890-1988. Epilogue: the great debate, p.471-87. Bibliography, p.488-504. Index. References and notes are appended to each chapter. *Class No:* 656.6(091)

Archives

[5818]

MATHIAS, P. and PEARSALL, A.W.H. Shipping: a survey of historical records. Newton Abbot, Devon, David & Charles, 1971. 162p. tables.

2 sections: 1. Shipping companies (42; A-Z, p.17-97), giving nature and details of records held; access - 2. Shipping records in country and other offices (50; A-Z, p.101-42), tabulating type of document, ship's name, period covered, voyage(s) made. Index of persons and firms; index of places and principal trades; index of ships' names. *Class No:* 656.6(093.20)

Worldwide

[5819]

ABC passenger shipping guide: worldwide guide to services and cruises. Dunstable, Beds., ABC Publications. Monthly. tables, map. ISSN: 00010480.

No.439, December 1992 (264p.) has sections: Operators - Ports index - Passenger ship services (999 tables, p.100-199)

....(contd.)
- Principal cruise ships - Cruises (principal cruises; river, waterway and coastal cruises; yacht and schooner cruises).
Class No: 656.6(100)

Libraries

[5820]
Marine transport: a guide to libraries and sources of information in Great Britain. Bolton, R.V. *and* Bryan, F.J., *eds.* 2nd ed. London, Library Association, Reference, Special and Information Section, and the Marine Librarians' Association, 1983. 76p.

First published 1974 (Allum, D.N. and others).

103 numbered entries. Part 1: Directory of marine libraries, 'Admiralty Marine Technology Establishment' ... 'Wirral Public Libraries'. Data include stock and periodical statistics, subject coverage, terms of admission, opening times, services, publications, indexes, special collections. Part 2: Other organizations (1-91). Name and subject (non-analytical) indexes. Appendix: 'List of the principal records of the Registry of Shipping and Seamen'. A good example of a guide to libraries in a specified field.
Class No: 656.6:061:026/027

Museums

[5821]
HOWE, H.E. **North America's maritime museums:** an annotated guide. New York, BDD Promo Book Co., 1987. 371p. $19.98. ISBN: 0792445724.
Class No: 656.6:061:069

[5822]
WHEATLEY, K. **National Maritime Museum guide to maritime Britain.** Exeter, Devon, Webb & Bower, 1990. 208p. illus. (some col.). ISBN: 0863502687.

10 chapters, arranged by region: Highlands and Islands ... The West Country ... Ulster. Each chapter contains an introduction followed by a gazetteer where, in addition to place names, museums and galleries of maritime interest are listed. *Class No:* 656.6:061:069

Ships' Flags

[5823]
WILSON, T. **Flags at sea:** a guide to the flags flown at sea by British and some foreign ships from the 16th century to the present day, illustrated from the collections of the National Maritime Museum. London, HMSO, 1986. 127p. illus. (incl. col.), charts. £7.95. ISBN: 0112903894.

Concentrates on the sea flags of Britain, Spain, the Netherlands, France and the United States. Special chapters on flag signalling and the manufacturing of sea flags.
Class No: 656.6:929.925

[5824]
—Brown's flags and funnels of British and foreign steamship companies. Wedge, C.P., *comp.* 6th ed. Glasgow, Brown, Son & Ferguson, 1958. [iii], 38l., 20p. col. illus.

Founded in 1926 by Captain F.J.N. Wedge.

Small coloured reproductions of *c.*1,000 flags and funnels. Flags with predominant colours tend to be grouped by colour. Index of companies (names and addresses), with flags and plate numbers. *Class No:* 656.6:929.925

Maritime Navigation

Handbooks & Manuals

[5825]
Admiralty manual of navigation. London, HMSO. 5v. illus., diagrs., tables, maps. (*BR 45.*)

V.1 (Rev. ed. 1987. xviii, 697p.): *General navigation, coastal navigation and pilotage,* supersedes the edition of 1964. V.2 (1973, 348p.) is the text book of ocean navigation and nautical astronomy. V.3 includes chapters on radio aids, navigational instruments, logs and echo sounders, gyros and magnetic compasses, automated navigation and radar plotting systems. V.4 (classified) has data and information on navigation, fleetwork and shiphandling. V.5 (6 bound parts) provides exercises on the use of tables (*e.g.* astronomical, Great Circle, tidal and tidal streams, time and chronometer, relative velocity questions).

V.1 has 19 chapters, with 7 appendices (*e.g.* 4: Projections). Bibliography, p.675-7. Detailed, analytical index. Well produced. Written primarily for naval officers, but with a wider appeal.

The Admiralty issues updated information in its weekly publication, *Admiralty notices to mariners.*
Class No: 656.61(035)

[5826]
Admiralty manual of seamanship. London, HMSO, 1951 (and reprints, incorporating amendments). v. illus., diagrs., graphs, tables, maps. ISBN: 0117722618, v.1; 0117722452, v.2; 0117722698, v.3.

1. *Basic seamanship.* 1979. xi, 509p. £16. 2. *Intermediate level seamanship.* 1981. x, 435p. £18.50. 3. *Shiphandling and navigational.* 1983. £16.

V.1 has 4 parts (18 chapters); Ship knowledge and safety; Type of ship; Design and construction of warships; Ship handling, etc. - V. 2 concerns intermediate level seamanship - V.3 Advanced seamanship. *Class No:* 656.61(035)

[5827]
—Seaway code: a guide for small boat users. [4th] rev. ed. London, HMSO, 1983. 48p. col. illus. ISBN: 0115136568.

Revised at intervals, had a 3rd ed. in 1981.

'Small boat' includes yachts, canoes and dinghies. Chapters concern weather forecasts, emergency drills, first aids, insurance, buoys, as well as 'rules of the road'. Bibliography, p.44. *Class No:* 656.61(035)

[5828]
The Complete sailing handbook. Denk, D., *and others.* 2nd rev. ed. London, Dunitz, 1981. 344p. illus. (incl. colour), diagrs., tables, maps. ISBN: 0906348293.

First published 1979. Translation of *Das grosse Handbuch des Segelns.* Reprinted 1990. (London, Tiger Books International. £8.99).

About boats - Marlinspike seamanship & maintenance - The boat engine - Sailing theory and sail turning - Sailing practice - Multihulls - Windsurfing - Safety - Racing - Navigation - Weather - Navigation rules. Appended glossary; International code of signals. Grouped bibliography, p.338-9. Index. *Class No:* 656.61(035)

[5829]

The Macmillan & Silk Cut nautical almanac. London, Macmillan Press, 1980-. Annual (1993 ed. 900p. £21.95). illus., graphs, tables, maps.

1991 ed. (839p.) has 10 chapters: 2. General information - 3. Coastal navigation - 4. Radio navigational aids - 5. Astronavigation - 6. Communications - 7. Weather - 8. Safety - 9. Tides - 10. Harbour, coastal and tidal information. 20 areas (SW England ... Federal Republic of Germany). Detailed index. Claims to be 'the first comprehensive almanac designed specifically in North-West Europe'. Attractively produced, with over 800 charts and diagrams.
Class No: 656.61(035)

[5830]

The Mariners' handbook; containing information on Admiralty charts and navigational publications, general navigation, general meteorology and ice. 5th ed. Taunton, Hydrographer of the Navy, 1979. viii, 163p. illus., diagrs., maps. *(NP100.)*

Contents: Charts and books - Terms, symbols annd orthography (in general navigation, physical features, ports, harbours and water movements) - The use of charts and other navigational aids - Navigational hazards - Natural conditions - Ice (*e.g.* formation and movements of sea-ice, icebergs, ice navigation, icebreaker signals, ice accumulation on ships). *Class No:* 656.61(035)

[5831]

TATE, W.H. A Mariner's guide to the rules of the road. 2nd ed. Annapolis, Md., Naval Institute Press, 1982. 169p. illus. $18.95. ISBN: 0870213555.

First published 1974.

Analyses and compares the 1972 International report for preventing collisions at sea, and the Inland Navigation Rules Act, which between them control the movement of vessels both on the high seas and on US inland waterways. 'Absolutely essential to professional mariners and serious marine hobbyists' (*Library journal,* 1 December 1982, p.2249). *Class No:* 656.61(035)

Dictionaries

[5832]

UNITED STATES. Naval Oceanographic Office. **Navigation dictionary.** 2nd ed. Washington, US Government Printing Office, 1969. iv, 292p.

First published 1956 (US Hydrographic Office).

Covers all types of craft. US-slanted and now somewhat dated. *Class No:* 656.61(038)

Maps & Atlases

[5833]

Catalogue of Admiralty charts and other hydrographic publications. Taunton, Hydrographer of the Navy. Annual. tables, maps.

The 1989 ed. ([i], 176p. £10) has 3 main sections: Navigational charts - Other charts and diagrams - Publications. Keyed charts occupy right-hand pages, with lists of charts, stating scale and date, facing. Numerical index, with prices. The outsize oblong catalogue measures 41.25 x 37.5cm. *Class No:* 656.61(084.3)

Histories

[5834]

FANNING, A.E. Steady as she goes: a history of the Compass Department of the Admiralty. National Maritime Museum. London, HM Stationery Office, 1986. xlv, 462p. illus., ports. £27.50. ISBN: 0112904234.

13 documented chapters (*e.g.* 12: SINS, Polaris and the last naval director, 1961-71, 32 references). 12 appendices. Bibliography, p.445-6. Detailed, analytical index. Plates, p.381-418. *Class No:* 656.61(091)

Worldwide

[5835]

GREAT BRITAIN. Hydrographic Department. **Ocean passages for the world.** 3rd ed. [London], Hydrographer of the Navy, 1973. Supplement, no.1. 1277. viii, 288p. maps. (*NP 136.*)

2nd ed. 1950.

Routes for steam and sailing ships, with special reference to winds and currents. Part 1: Power vessel routes (1. Planning a passage; 2. North Atlantic Ocean ... 7. Pacific Ocean ...) - Part 2: Sailing vessel routes. General index. 25 diagrs. of routes. Insert of charts (world climate; routes; currents). *Class No:* 656.61(100)

[5836]

Sailing directions (Pilots). London, Hydrographic Department. Various editions; Supplements.

1-3. *African pilot.* 12th-13th eds. 1973-82. 4. *South-east Alaska pilot.* 5th ed. 1972. 5-7A. *South American pilot.* 2nd-15th eds. 1968-83. 8. *Pacific coasts of Central America and US pilot.* 8th ed. 1975. 9. *Antarctic pilot.* 4th ed. 1974. 10-12. *Arctic pilot.* 6th-7th eds. 1959-76. 13-17. *Australian pilot.* 6th ed. 1972-82. 18-20. *Baltic pilot.* 6th-10th. 1974-78. 21. *Bay of Bengal pilot.* 10th ed. 1978. 22. *Bay of Biscay pilot.* 5th ed. 1970. 23. *Bering Sea and Strait pilot.* 5th ed. 1980. 24. *Black Sea pilot.* 11th ed. 1969. 25-26. *British Columbian pilot.* 6th-9th ed. 1976-79. 27. *Channel pilot.* 2nd ed. 1977. (revised 1984). 28. *Dover Strait pilot.* 2nd ed. 1981. 30-32. *China Sea pilot.* 4th ed. 1975-82. 33-36. *Philippine Islands, Indonesia pilots.* 1975-80. 37. *West coast of England and Wales pilot.* 11th ed. 1974. 38. *West coast of India pilot.* 11th ed. 1975. 39. *Indian Ocean, south, pilot.* 8th ed. 1971. 505.00/01$40. *Irish coast pilot.* 12th ed. 1985. 41-42. *Japan pilot.* 7th ed. 1979-82. 43. *Korea, south and east coasts, etc., pilots.* 6th ed. 1983. 44. *Malacca Strait pilot.* 5th ed. 1971. 45-49. *Mediterranean pilot.* 6th-10th eds. 1968-78. 50. *Newfoundland pilot.* 10th ed. 1978. 51. *New Zealand pilot.* 13th ed. 1971. 52. *North coast of Scotland pilot.* 1975. 54-55. *North sea pilot.* 1973-75. 56-58B. *Norway pilot.* 6th-8th ed. 1979-82. 59. *Nova Scotia, South-east coast, etc.* 11th ed. 1971. 60-62. *Pacific Islands pilot.* 9th-10th ed. 1970-84. 63. *Persian Gulf pilot.* 12th ed. 1982. 64. *Red Sea and Gulf of Mexico, East coasts pilot.* 12th ed. 1980. 65. *St. Lawrence pilot.* 12th ed. 1969. 66. *Scotland, West coast, pilot.* 11th ed. 1974. 67. *Spain and Portugal, west coast, pilot.* 5th ed. 1972 (revised 1983). 68-69. *United States, east coast, pilot.* 7th ed. 1975-78. 69A. *Central America and Gulf of Mexico pilot.* 1970. 70-71. *West Indies pilot.* 1st-11th ed. 1969-71. 72. *White Sea pilot.* 3rd ed. 1973.

Many of the major countries (*e.g.* USA, USSR, Japan) publish coastal guides for their own coasts. The Admiralty 'Pilots' provide a selective translation of these. *Class No:* 656.61(100)

Institutions & Associations

[5837]
Guide to the marine stations of the North Atlantic and European waters. Compiled on behalf of the Royal Society's Naples Zoological Station Committee, by J.E. Webb. London, Royal Society, 1974-77. 3v. maps. ISBN: 0854030689, v.1; 0854030794, v.2; 0854030905, v.3.

1. *Northern Europe and the East Atlantic coast.* £2.70. 2. *Mediterranean Basin.* £3.50. 3. *Iceland, the West Atlantic coast and the Caribbean Sea ...*, £4.85, with supplements to pts.1 and 2.

Countries A-Z in each part. Data include 'major accessible habitats, collecting and hydrogaphic facilities, laboratory services, living accommodation for visitors, publications' (British Council. *British scientific and technical reference books* (1975), p.20). *Class No:* 656.61:061:061.2

Signals

[5838]
The Admiralty list of radio signals. London, Hydrographic Office, 1985-86. 6v. (*NP 275(1)-(6).*)

1. *Coast radio stations.* 1986 ed. 2. *Radio navigational aids.* 1986 ed. 3. *Radio weather messages. Meteorological codes.* 1985 ed. Diagrams. 4. *Lists of meteorological observation stations.* 5. *Radio time signals. Radio navigation warnings and position fixing systems.* 1986 ed. Diagrams. 6. *Port operations. Pilot services and traffic surveillance.* 1985 ed.

Updated by corrections in section 6 of the weekly edition of *Admiralty notices to mariners* until suspended. *Class No:* 656.61.054

[5839]
International code of signals, 1969. London, HMSO, 1969. xv, 149p. £9. ISBN: 011551015x.

Reprinted 1991, incorporating amendments 1, 2 and 3. *Class No:* 656.61.054

Lights & Beacons

[5840]
Admiralty list of lights and fog signals. Taunton, Hydrographer of the Navy. 1985-86 ed. 12v. tables, maps.

A. *British Isles and north coast of France.* 1986 ed. (NP 74). 2. B. *Southern and eastern sides of the North Sea.* 1986 ed. (NP 75). C. *Baltic Sea.* 1986 ed. (NP 76). D. *Eastern side of Atlantic Ocean.* (NP 77). E. *Mediterranean, Black Sea and Red Seas.* 1986 ed. (NP 78). F. *Arabian Sea, Bay of Bengal and North Pacific Ocean.* 1986 ed. (NP 79). G. *Western side of South Atlantic Ocean and East Pacific Ocean.* 1986 ed. (NP 80). M. *Northern and eastern coasts of Canada.* 1985 ed. (NP 81). J. *Western side of North Atlantic Ocean.* 1985 ed. (NP 82). K. *Indian and Pacific Oceans, south of the Equator.* 1985 ed. (NP 83). L. *Norwegian and Greenland Seas.* 1985 ed. (NP 84). M. *Arctic Ocean.* 1985 ed. (NP 85).

Mostly tables stating number, name and position, latitude and longitude, characteristics and intensity, elevation, range, structure, remarks. *Class No:* 656.61.057

Ports & Harbours

[5841]
BIRD, J. The Major seaports of the United Kingdom. London, Hutchinson, 1963. 454p. illus., charts.

18 chapters (*e.g.* 7: Bristol, the Avon and Avonmouth; 47 references, plus 8 further British references; 18: The ports in perspective; 27 references, with 'Selected further references published since 1953'). Copious explanatory notes. A detailed, fully documented survey submitted as evidence for the Rochdale Committee. 'A major contribution to port geography' (*Economic geography*, v.40(2), April 1964, p.182). 'Clearly destined to become a standard work of reference on this neglected subject for some time to come' (*Dock & harbour authority*, v.44(511), May 1963, p.15). Useful historical work. *Class No:* 656.615

[5842]
—JACKSON, G. The History and archaeology of ports. Kingswood, Tadworth, Surrey, World's Work, Ltd., 1983. 176p. illus. (incl. col.), facsims. ISBN: 0437075397.

6 chapters: 1. The rise of ports *c.*1450-1660 - 2. Harbour developments and port improvements, 1660-1860 - 3. The emergence of dock systems, 1690-1940 - 4. Expansion: new docks and new ports, 1640-1870 - 5. The height of prosperity, 1870-1914 - 6. Stagnation and decay, 1919-1980. Footnotes, p.168-70. Bibliography, p.171-2. Index. Details the various phases of development of all the major ports, and many minor ones of Britain. *Class No:* 656.615

[5843]
Lloyd's maritime atlas of world ports and shipping places. Beresford, A.K.C., *and others, eds.* New 16th ed. Colchester, Essex, Lloyd's of London Press, 1989. 145p. tables, col. maps. £40. ISBN: 1850442266.

First published as *Lloyd's maritime atlas*, 1951. Previously as *Lloyd's book of ports and shipping places* (1937-).

Coloured maps, p.2-64; index to maps precedes. Appendix 1: Inland container depots; 2. Distance tables. Geographical index, p.68-117; alphabetical index, p.118-45. Includes *c.3,000* ports in index and/or maps. Designed for use with *Lloyd's ports of the world* (*q.v.*). *Class No:* 656.615

[5844]
Lloyd's ports of the world 1992. 10th ed. Colchester, Essex, Lloyd's of London Press, 1992. 863p. ISBN: 1850444382.

Directory of 2,800 ports worldwide, specifying anchorage and berthing details, possible hazardous approaches, loading facilities etc. Divided into Africa, North America, Central America & West Indies, South America, Asia, Australia and Pacific Islands, Europe, European Community (p.599-818), Internal Free Trades, with ports arranged A-Z. Indexes. 1993 ed. 800p. £130. *Class No:* 656.615

Canal & River Transport

Bibliographies

[5845]
Inland waterways. London, Library Association, 1975. 27p. (*Readers' guide no.7.*)

Checklist of *c.*250 unannotated entries. Sections: Inland waterways of Britain (general & regional histories) - Canals (English, Welsh, Scottish, Irish) - Boats and boatmen - Canal engineering and architecture - British reports - Leisure use and guides - Maps - Accounts of cruises on British inland

....(contd.)

waterways - East Anglican Broads - British rivers - Continental inland waterways (p.230-24) - North American inland waterways - Reference books - Periodicals - Organizations. No Index. *Class No:* 656.62(01)

Dictionaries

[5846]

Glossary for inland transport. New York, United Nations, 1970. [1], 44p. 116p. *(ME/TRANS/70/D/D40.)*

600 terms. Index precedes. 'No more than informative descriptions of an entirely general nature and cannot be taken as true definitions in the strict and precise sense of the term' *(Introductory note).*

A French version - ME/71/68/D.5 - helps to provide a two-way dictionary. *Class No:* 656.62(038)

Europe

[5847]

Inland waterways series. Huntingdon, Imray, Laurie, Norie & Wilson, 1950-. illus., maps, plans.

Inland waterways of Great Britain and Ireland; compiled by L.A. Edwards. 6th ed. 1985. 480p. £17.50. *Inland waterways of France,* by D. Edwards-May. 6th ed. 1992. viii,334p. £25. *Inland waterways of the Netherlands,* by E.E. Benest. 3v. 1966-72. *Inland waterways of Belgium,* by E.E. Benest. 1960.

The 6th ed. of *Inland waterways of France* (1992. viii,334p.) features particulars of the waterways, with distance tables, p.31-318. Numerous plans; short glossary (French-English, English-French). Index. *Class No:* 656.62(4)

Great Britain

[5848]

HADFIELD, C. British canals: an illustrated history. 7th ed. Newton Abbot, Devon, David & Charles, 1984. 367p. illus., facsims., maps. £8.95.

First published 1950.

16 chapters, taking the narrative up to 1982. Appended: Sources of quotations. Grouped bibliography, p.334-8. Index to 13-volume canal series, p.349-55. Detailed, analytical index. Well illustrated; 17 maps. *Class No:* 656.62(410)

[5849]

HADFIELD, C. The Canals of the British Isles. Newton Abbot, Devon, David & Charles, 1950-.

1. *British canals: an illustrated history* by C. Hadfield. 7th ed. 1984. 2. *The canals of south and south-east England.* 1969. 3. *The canals of south-west England.* 2nd ed. 1985. 206p. £13.50. 4. *The canals of south Wales and the Border.* 2nd ed. 1967. £12.95. 5. *The canals of the West Midlands.* 3rd ed. 1985. 352p. 6. *The canals of the East Midlands* (including part of London). 2nd ed. 1970. £12.95. 7. *The canals of eastern England,* by J. Boyes and R. Russell. 1977. 8/9. *The canals of north-west England.* 1970. 2v. 10/11. *The canals of Yorkshire and north-east England.* 1970. 2v. £12.95ea. 12. *The canals of Scotland,* by J. Lindsay. 1968. 13. *The canals of the north of Ireland,* by W.A. McCutcheon. 1965. 14. *The canals of the south of Ireland,* by T.H. and D.R. Delaney. 1973.

V.10/11, on Yorkshire canals, covers the periods, up to 1970, and 1970-72. V.11 (254p. has 16 plates, 22 text illus.

....(contd.)

and maps, plus combined analytical index). Appendices: 1. Summary of facts about the canals and navigation of the North-east. 2. Principal engineering works.

C. Hadfield, author of v.1-6, 8-11 in the series, provides a combined index to v.2-14. *Class No:* 656.62(410)

[5850]

Ordnance Survey guide to the waterways. Perrott, D., *ed.* 5th ed. London, Nicholson, 1991. 3v. illus., maps. ISBN: 0702812595, v.1; 0702812609, v.2; 0702812641, v.3.

First published 1983.

1. *South.* 208p. 1991. £8.99. 2. *Central.* 176p. 1991. £8.99. 3. *North.* 192p. 1991. £8.99.

Detailed guide to inland navigation. V.3 charts navigation between Nottingham, Lincoln, Sheffield, Leeds, Manchester and Liverpool, on 2-inch scale maps. Gives general cruising information in addition to locations of tunnels, bridges, pubs, restaurants, boatyards and other facilities. Index. *Class No:* 656.62(410)

England

[5851]

McKNIGHT, H. The Shell book of inland waterways. 2nd ed. Newton Abbot, Devon, David & Charles, 1981. 493p. illus. (incl. col.), maps. £12.95. ISBN: 071538239x.

First published 1975.

18 chapters, covering such topics as locks and lifts, bridges, aqueducts, tunnels, building and design, boatyards and marinas, animal and plant life, waterway museums, associations and boat clubs. Selected bibliograpy, p.208-9. Descriptive gazetteer, p.212-484. Attractively produced. *Class No:* 656.62(420)

Air Transport

[5852]

Aerospace Europe: the directory of European aviation, 1993. 45th ed. Tonbridge, Kent, Benn Business Information Services, 1992. £55.

Alphabetical list of companies (over 7,000 established and operating within Europe). Classified products and services (index precedes) - Trade names and brands (A-Z, with cross-references to the 'Classified products & services' section). Advertisers' index. *Class No:* 656.7

[5853]

INTERNATIONAL CIVIL AVIATION ORGANIZATION. Lexicon in English-French-Spanish-Russian. Montreal, ICAO, 1974. 2v.

Previous ed. (1971) covered only English, French and Spanish.

1. *Vocabulary.* xii, 326p. 2. *Definitions.* 139p.

About 3,000 English-based terms, with French, Spanish and Russian equivalents and indexes. Terms defined in v.2 are asterisked. Some categorization of terms. *Class No:* 656.7

[5854]

OCRAN, E.B. Dictionary of air transport and traffic control. London, Granada, 1984. ix, 243p. ISBN: 0246123605.

Part 1: Definitions, p.1-205. Part 2: Grouped index (c.3,000 entries, in 18 sections: Accidents and safety ... Navigation. Includes abbreviations, organizations). Author is

....*(contd.)*

head of Bibliographical Services, Civil Aviation Authority Library, London. Intended for a wide audience. *Class No:* 656.7

[5855]
WOOD, K. *and* **McDONALD, G. The Round the world air guide** 1992. 6th ed. London, Fontana, 1992. 896p. £12.99. ISBN: 0006377777.

First published 1987.

Intended for those embarking on long-haul flights and gives data on the world's 52 most frequently visited cities - this edition includes Moscow and Mexico City for the first time. 3 sections: 1. Air travel round the world (*e.g.* 'Buying your ticket' ... 'Red tape' ... 'When you get there') - 2. Stop off destinations - 3. Appendix ('Airline two-letter codes' - 'Principal airports around the world' - 'Foreign tourist offices, embassies, etc'). Index. *Class No:* 656.7

[5856]
World air transport statistics, 1987. no. 32. Geneva, International Air Transport Association, 1987. 119p. diagrs., graphs, tables.

6 sections: 1. Review of air transport development in 1987 - 2. World air transport - 3. IATA industry statistics - 4. Regional - 5. IATA members' statistics - 6. Explanatory notes and definitions. *Class No:* 656.7

Airports & Airfields

[5857]
Jane's airport and air traffic control equipment, 1992-93. Coulsdon, Jane's Transport Data. 740p. £135.

Airports, civil and military (including manufacturers directory). Data on design, construction, operational and technical support. Services. Appendices. Addenda. Index. *Class No:* 656.71

[5858]
Pooley's flight guide: United Kingdom and Ireland. Pooley, R. *and* Ryall,., W., eds. London, R. Pooley, Ltd., 1991. xvi,486p. tables, plans, maps. ISBN: 0902037234.

United Kingdom civil aerodromes, p.69-347. Preceded by sections on procedures, controlling authorities, navigational aids, and meteorological services, etc. Military and Ministry aerodromes. Private airfields. Helicopters. Aeronautical light beacons. Republic of Ireland, p.431-63. Appended conversion tables, etc. *Class No:* 656.71

Airlines

[5859]
The Airline bibliography: the Salem College guide to sources on commercial aviation. West Cornwall, Locust Hill Press, 1986-88. 2v. ISBN: 0933951000, v.1; 0933951124, v.2.

V.1 *US air transport.* 1986. 239p. $45. V.2 *Airliners and foreign air transport.* 1988. xxxii,464p. $75.

V.2 contains over 13,000 citations (V.1, *c.*6,000) from 690 journals. Coverage includes historic, operational and economic aspects of air transport organizations outside the US. Airliners and airlines sections contain short descriptive and historical reviews. Extensive author and subject indexes. *Class No:* 656.73.01

[5860]
DAVIES, R.E.G. A History of the world's airlines. London, Oxford Univ. Press, 1964. xxx, 591p. tables, maps.

A detailed study, with maps, tables of airlines and their fleets, photographs of 209 aircraft. Select bibliography, p.xxvii-xxx. *Class No:* 656.73.01

[5861]
International and domestic airlines operating in the UK: a company guide, 1986-7. Oxford, Avimar Data, Ltd., 1986. [1], [1], 327, [4]p.

International airlines - British airlines - Air taxi operators - Helicopter operators ... Types of commercial aircraft - CAA licensed airports. Index. *Class No:* 656.73.01

[5862]
—**BÁTOR, J.W.** International airline phrase book in six languages, English, French, German, Italian, Portuguese and Spanish. New York, Dover, 1968. xiv, 204p. £4.45. ISBN: 0486220176.

700 numbered sentences and phrases, English-based, with French, German, Italian, Portuguese and Spanish translations or idiomatic equivalents. Arranged by topic (reservations; departure; boarding; flight; arrival; miscellaneous). English index only. *Class No:* 656.73.01

[5863]
Jane's world airlines. Coulsdon, Surrey, Jane's Information Group, 1992. 2v. loose-leaf. ISBN: 0710605919.

Arranged A-Z. Data include contact, type of operation, company structure (*e.g.* 'Major subsidiaries'; 'Scheduled destinations'; 'Traffic'; 'Business review'). *Class No:* 656.73.01

Airways

[5864]
INTERNATIONAL AIR TRANSPORT ASSOCIATION *and* **INTERNATIONAL AERADIO, LTD. Air distances manual.** 9th ed. effective 1st April 1982. Southall, Middx., 1982. 36 + 285p. illus., tables.

4 sections: 1. Conversion tables - 2. Wind components - 3. G.M.T. of sunrise and sunset - 4. Distances (285p.): Aalborg ... Zurich. Data: nautical miles; standard mile, kilometre, route. *Class No:* 656.73.02

Timetables

[5865]
ABC executive flight planner. Dunstable, Beds., ABC International World Timetable Centre. Monthly. tables. £45pa. ISSN: 09591389.

Formerly *ABC air Europe, Middle East and Africa.*

Timetables cover worldwide direct flights to, from and within the region, plus transfer connections, details of fares. *Class No:* 656.73.022

[5866]
ABC world airways guide. Dunstable, Beds., ABC International Division. Monthly. 2v. tables.

First published 1948.

V.1: Airlines of the world. International time calculator. Flight routings. Timetables, Aachen ... Mzuzu (Malawi). V.2: Timetables; Naberevnye Chelnye (CIS) ... Zurich. Includes fares section (p.1-286). *Class No:* 656.73.022

Postal Services

[5867]
DAUNTON, M.J. Royal Mail: the Post Office since 1840. London, The Athlone Press, 1985. xviii, 388p. illus., facsims., tables. £32. ISBN: 0485112809.

5 parts (10 chapters): 2. Carrying the mail - 3. Working for the Post Office - 4. Officials and politicians - 5. Epilogue: the post-war period. Chapter notes, p.357-78. Detailed, analytical index, p.379-88. The narrative ends at 1984. Daunton is Reader in History, University College, London. *Class No:* 656.8

[5868]
Directory of place names and associated postcodes. Buckingham, PC Services, 1986. 250p. £25.

A-Z directory covering the 120 postcode areas in the UK. *Class No:* 656.8

[5869]
Mail guide. Issued by HM Postmaster General. Bristol, Royal Mail, 1992.

Formerly *Post Office guide.*

Gives details of services provided by the Royal Mail. *Class No:* 656.8

[5870]
Postcodes: the new geography. Raper, J.F., *and others.* Harlow, Essex, Longman, 1992. xiv,322p. illus. (col. pl.), figs., maps. £25. ISBN: 0582092701.

4 sections: 1. The nature of postcodes and their role in postal services - 2. Handling postcoded data - 3. Non-postal applications of postcodes (*e.g.* 'Applications of postcodes in marketing) - 4. The future. 7 appendices (*e.g.* 'Glossary of abbreviations and terms'; 'The British National Grid system'). Index, p.316-22. *Class No:* 656.8

[5871]
UNION POSTALE UNIVERSELLE, Berne. **Nomenclature internationale des bureaux de poste.** Liste complète des bureaux de poste des pays et territoires compris dans le ressort de l'Union Postale Universelle. Berne, UPU, 1977. vp.

Lists *c.*300,000 post offices, A-Z by name, with postal code, name of country and territorial subdivision (*e.g.* département, canton, province, state). In French, with preface also in English. *Class No:* 656.8

[5872]
UNION POSTALE UNIVERSELLE. Bureau International. Vocabulaire polyglotte du service postal international. 3. éd. Berne, UPU, 1973 (reprinted 1986, Harper & Row). 323p. Loose-leaf.

First published 1954.

More than 1,000 terms defined, with equivalents in French, German, English, Spanish, Russian, Arabic and Chinese. *Class No:* 656.8

Postage Stamps

Bibliographies

[5873]
COLLECTORS CLUB LIBRARY, New York City. **Philately:** a catalog of the Collectors Club Library. Boston, Mass., G.K. Hall, 1974. 682p. $130. ISBN: 0816110476.

21,800 photolithographed catalogue cards - 4 sequences: author, subject, title and periodical catalogues. One of the world's largest collections of books and periodicals on stamp collecting and postal history (handbooks; priced auction catalogues; exhibition brochures, annual catalogues, philatelic membership lists, US governmental reports, etc.). *Class No:* 656.835(01)

[5874]
LINDSAY, J.L., *26th Earl of Crawford.* **Catalogue of the philatelic library of the Earl of Crawford.** Bacon, E.D. London, Philatelic Literature Society, 1911. xp., 924 col. Supplement. 1926. Addenda to Supplement. 1938.

Also published as v.7 of *Bibliotheca Lindesiana,* with title *A bibliography of philately.*

Lists monographs to 1908 and periodicals to 1906. An appendix records all the periodicals in chronological order. One of the most complete bibliographies of any subject, not confined to books actually in the Library. This collection is now in the British Library, Reference Devision. *Class No:* 656.835(01)

Handbooks & Manuals

[5875]
CABEEN, R.McP. Standard handbook of stamp collecting. New rev. ed. New York, Crowell, 1979. 630p. illus., tables. $9.95. ISBN: 0690017731.

First published 1957.

5 main sections: 1. Introduction to stamp collecting - 2. Postal history and cover collecting - 3. Miscellaneous subjects - 4. Technical matters - 5. Classification and identification. Glossary; comprehensive bibliography. Scope has both breadth and depth; moderately priced. Rivalled only by *Fundamentals of philately (qv).* *Class No:* 656.835(035)

[5876]
WILLIAMS, L.N. *and* **WILLIAMS, M. Fundamentals of philately.** State College, Pa., American Philatelic Society, 1971. 629, xxxp. illus.

A standard manual, with bibliographical references and index. Very favourably referred to in *Choice,* v.15(12), February 1980, p.1616. *Class No:* 656.835(035)

[5877]
—**MACKAY, J. The Guinness book of stamps facts and feats.** 2nd ed. Enfield, Middx., Guinness Publishing Ltd., 1988. 224p. illus. (incl. col.). £12.99.

First published 1982.

7 chapters: 1. The postal services - 2. Postmarks - 3. Adhesive stamps - 4. Kinds of stamps - 5. The technology of stamps - 6. Postal stationery and ancillary labels - 7. Philately. Index. By the Philatelic Curator at the British Museum. *Class No:* 656.835(035)

Dictionaries

[5878]

PATRICK, D. *and* PATRICK, M. **The Hodder stamp dictionary.** UK ed. London, Hodder & Stoughton, 1973. [vii], 277p. illus., facsims.

About 7,000 entries for technical terms. English-based (with French and German equivalents), abbreviations and acronyms. Some biographical sketches (*e.g.* Sperati, the stamp forger; Wells, Fargo). Also entries under countries. Non-Latin alphabets, p.273-7. Illus. of stamps. *Class No:* 656.835(038)

[5879]

—BENNETT, R. *and* WATSON, J. Philatelic terms illustrated. 2nd ed. London, Stanley Gibbons, 1978. 192p. illus. (incl. col. pl.). £2.50. ISBN: 0852598955.

First published in 1973 (192p.), having previously appeared serially in *Stamp monthly.*

Terms, A-Z, describe papers, errors, varieties, watermarks, perforations, etc., with cross-references to plates. Some 1,500 stamps in all are shown, some in colour. Analytical index. For both beginner and experienced collector. *Class No:* 656.835(038)

Yearbooks & Directories

[5880]

The British Philatelic Federation yearbook and philatelic societies' directory. London, British Philatelic Federation. Annual. ISSN: 02601265.

1992 ed. (173p. £8) has 3 sections: 1. Yearbook (*e.g.* 'British exhibitions 1992'; 'Stamps for all seasons from Royal Mail') - 2. Directory of philatelic societies, p.54-148 - 3. General information (*e.g.* 'British philatelic magazines'; 'Philatelic bureaux worldwide'). *Class No:* 656.835(058)

Illustrations

[5881]

Scott standard postage stamp catalogue. New York, Scott Publishing, 1864-. Annual (1991 ed. $49.95ea). 4v. illus. ISBN: 0894871471, v.1; 089487148x, v.2; 0894871498, v.3; 0894871501, v.4.

V.1 covers the US and affiliated territories, the United Nations, Canada, Great Britain, and the British Commonwealth. V. 2-4 cover the rest of the world in an alphabetical listing of countries. Under each country, stamps are listed in chronological order, with special sections for airmail stamps, postage-due stamps, semi-postal stamps, and other special categories. An additional volume, *U.S. specialized* was due to be published in November, 1990. Each volume has a cumulative index to the set and the 5v. set will contain *c.*375,000 stamps. Supplements appear in *Scott's monthly journal.* The standard American general catalogue of stamps. *Class No:* 656.835(084.1)

[5882]

—Scott specialized catalogue of United States stamps. New York, Scott Publishing Co., 1923-. Annual.

Contains data on dates of issue, designers, printers and printing methods, perforations, watermarks, papers, gums, grills, printing flaws, errors transfers and colour varieties. Indexes and useful glossary of technical terms. 22p. identifier of definitive issues. 'This is an indispensable volume for the study of US philately' (*Reference services review,* v.17, no. 3, 1989, p.55). *Class No:* 656.835(084.1)

[5883]

Stanley Gibbons Great Britain specialised stamp catalogue. London, Stanley Gibbons Publications, Ltd. 5v. ISBN: 0852593244, v.1; 0852593329, v.2; 0852593787, v.3; 0852593949, v.4; 0852593139, v.5.

First published 1963.

1. *Queen Victoria.* 10th ed. 1992. 432p. £25. 2. *King Edward VII to King George VI.* 8th ed. 1989. 448p. £17.50. 3. *Queen Elizabeth II. Pre-decimal issues.* 8th ed. 1990. 456p. £15.95. 4. *Queen Elizabeth II. Decimal definitive issues.* 6th ed. 1991. 592p. £16.95. 5. *Decimal special issues.* 1991. 392p. £16.95.

Each volume has introductory notes, appendices, and further reading. Thus, v.2 has sections M, N, P, Q for the four reigns, plus R: Postage due stamps; 6 appendices (E. Perforations) and 'Further reading' (Postal history ... Postmarks), p.400-3. *Class No:* 656.835(084.1)

[5884]

Stanley Gibbons simplified catalogue of stamps of the world. 1992 ed. London, Stanley Gibbons Publications, Ltd. 3v. illus. £18. ISBN: 085259299x, v.1; 0852593333, v.2; 0852593341, v.3.

1. *Countries, A-J.* 1064p. £16.50. 2. *Countries, K-Z.* 1056p. £18. 3. *Commonwealth countries.* 736p. £16.

'An illustrated and priced three-volume reference to the postage stamps of the whole world, excluding changes of paper, perforation, shade and watermarks' (*subtitle*). Each volume has an index. *Class No:* 656.835(084.1)

[5885]

—MACKAY, J.A. The Dictionary of stamps in colour. London, Michael Joseph, 1973. 296p. illus. (col. pl.), tables).

Illustrates *c.*5,000 stamps in colour, on 120 plates - Examples from 130 countries. Plate 30, 'Russia': 25 stamps displayed, with descriptive notes p.172-3. Appended: World currency tables; The stamp printers; Cyrillic inscriptors. Index, p.292-6. End-paper map. Highly selective use of colour is the main feature. *Class No:* 656.835(084.1)

[5886]

Stanley Gibbons stamp catalogue. London, Stanley Gibbons.

1. Pt.1 *Great Britain and countries, A-I, 1992-93.* 1992. 712p. £20. 1. Pt.2. *Countries, J-Z, 1992-93.* 1992. 688p. £20. 2. *Austria & Hungary.* 4th ed. 1988. 233p. £10.95. 3. *Balkans.* 3rd ed. 1987. x, 497p. £16.50. 4. *Benelux.* 3rd ed. 1988. x, 261p. £10.95. 5. *Czechoslovakia and Poland.* 4th ed. 1991. 240p. £11.95. 6. *France.* 3rd ed. 1987. x, 336p. £9.95. 7. *Germany.* 4th ed. 1992. 344p. £15. 8. *Italy & Switzerland.* 3rd ed. 1986. x, 292p. £10.95. 9. *Portugal and Spain.* 3rd ed. 1991. 400p. £14.50.

10. *Russia.* 4th ed. 1991. 400p. £16.50. 11. *Scandinavia.* 3rd ed. 1988. 232p. £11.95. 12/14. *Africa since independence.* 2nd ed. 1981-83. 3v. (v.12. £9.95; v.13, £8.50; v.14, £10.50). 15. *Central America.* 2nd ed. 1984. x, 526p. £11.50. 16. *Central Asia.* 2nd ed. 1983. x, 212p. £8.95. 17. *China.* 4th ed. 1989. 288p. £10.95. 18. *Japan & Korea.* 3rd ed. 1992. 368p. £15. 19. *Middle East.* 4th ed. 1990. 576p. £21. 20. *South America.* 3rd ed. 1989. 736p. £22. 21. *South-east Asia.* 2nd ed. 1985. x, 295p. £11.50. 22. *U.S.A.* 3rd ed. 1990. x, 213p. £9.95.

V.1, now in 2 parts, includes post-independence issues of Ireland, Pakistan and South Africa. Index precedes main sequence (Great Britain ... Zululand). Appended: Addenda and corrigenda; Stamps added; Catalogue numbers altered; General philatelic information. *Class No:* 656.835(084.1)

66 Chemical Industry

Chemical Engineering

Databases

[5887]
CAS ONLINE. Columbus, Ohio, American Chemical Society.
    CAS ONLINE is the database for *Chemical abstracts*, with printable abstracts. It is available to European users through INKA.
    Other chemical industry databases include COMPENDEX (especially for chemical engineering). *Class No:* 66.0(003.4)

[5888]
Chem-Intell. London, Chemical Intelligence Services.
    Covers manufacturing plants and trade and production figures for over 100 organic and inorganic chemicals (petrochemicals, fertilizers, polymers, rubbers, synthetic fibres, etc.). About 40,000 records; monthly updates. Available on DIALOG (file 318) and Data-Star.
Class No: 66.0(003.4)

[5889]
Chemical business newsbase. Cambridge, Royal Society of Chemistry.
    Covers legislation, environmental aspects, company information as they relate to chemical markets and products. Sources include journals, press releases, advertisements, company and government reports, publishers' lists, etc. Over 200,000 records, updated weekly. Available on DIALOG (file 319) and Data-Star. *Class No:* 66.0(003.4)

[5890]
Chemical engineering and biotechnology abstracts. Cambridge, Royal Society of Chemistry.
    Covers industrial practice and theoretical chemical engineering. Includes processes, substances, operations, transport, equipment safety. About 125,000 records, updated monthly. Available on DIALOG (file 315) and Data-Star.
Class No: 66.0(003.4)

[5891]
Fine chemicals database. San Antonio, Texas, Chemron, Inc.
    Data on chemical products available from over 50 manufacturers and distributors in North America and Europe. Data: catalogue number, product name, company name, address, etc. Over 200,000 records; semi annual reloads. Available on DIALOG, file 360.
Class No: 66.0(003.4)

Bibliographies

[5892]
PECK, T.P. Chemical industries information sources. Detroit, Mich., Gale, 1979. xxi, 595p. $68. (*Management Information guide 29.*)
    8 sections: 1. General information on chemical information and chemical industries - 2. Agricultural engineering - 3.

....(contd.)
Bioengineering - 4. Food engineering - 5. Materials, including plastics, metals, cermaics, rubber and textiles - 6. Nuclear engineering - 7. Paper and pulp engineering - 8. Petroleum engineering. Index. A guide to organizations and associations, as well as to the literature. *Class No:* 66.0(01)

[5893]
STEVENSON, J. 'Chemical engineering'. *Information sources in engineering* (2nd ed. Editor, L.J. Anthony, 1985), p.491-513. London, Butterworth, 1985.
    First published 1976, in *The use of engineering literature*.
    Sections: Primary sources of information (journal literature; newspaper sources; report literature; conferences; product information) - Secondary sources of information (abstracting and indexing services; dictionaries, encyclopedias and handbooks; chemical engineering data; textbooks, monographs and conference proceedings). 2 references. *Class No:* 66.0(01)

[5894]
—BOURTON, K. Chemical and process engineering - unit operations: a bibliographical guide. London, Macdonald, 1967. xxv, 534p.
    Includes 4,856 numbered and annotated entries (plus interpolations) for books, reports and periodical articles. 38 sections (1. Reference media, nos. 1-850 - Absorption ... Water treatment). Cites *c.*200 journals, listed p.xviii-xxiv. Layout is in 3 columns: entry number; bibliographical reference; number of items and date of coverage; critical annotation. Now somewhat dated. *Class No:* 66.0(01)

Encyclopaedias

[5895]
CONSIDINE, D.M., *ed*. Chemical and process technology encyclopedia. New York, etc., McGraw-Hil, 1974. xxix, 1261p. illus., diagrs., graphs, tables.
    About 180 contributors; signed and documented articles, with cross-references (*e.g.* 'Detergents', p.345-52; 2 illus., chemical structures, 3 references, 7 cross-references). Classified index; detailed, analytical subject index. 'Recommended for all college, business, and public libraries, especially those who can't afford Kirk-Othmer, *Encyclopedia of Chemical Technology*' (*Library journal,* 15 April 1975, p.735). *Class No:* 66.0(031)

[5896]
KIRK, R.E. *and* OTHMER, D.F., *eds*. Encyclopedia of chemical technology. 3rd ed. New York, Wiley, 1978-84. 26v. (24v. + supplement volume + general index).
    First published 1947-60 (15v. & 2 suppts.); 2nd ed. 1963-71 (22v. & suppt.).
    Over 1,300 contributors (an international team). About one half of the *c.*1,300 articles concern chemical substances. The article in the 2nd ed. on 'Literature on chemical technology' has been chopped in favour of 'Information retrieval' (with 34 references) and 'Patents literature' (v.16. p.889-945, with 80 references + 7 general references). Both SI and English

....(contd.)

units are used and Chemical Abstracts Service (CAS) Registry numbers cited to improve access to the literature. 'The ideal first source of reference for information on the properties, manufacture and use of any chemical; on industrial processes, on methods of analysis; and on scientific subjects allied to chemical technology' (*British chemical engineering*, v.10(1). January 1965, p.10, on the 2nd ed.). The 3rd ed., with irregular updates, is available on DIALOG, file 302.

Publication of the 4th ed. (27v.) began in September, 1991. Vols. are scheduled to appear every three months. *Class No:* 66.0(031)

[5897]

—Kirk-Othmer concise encyclopedia of chemical technology. 3rd ed. New York, etc., Wiley: Interscience, 1986. xxxii, 1318p. illus., charts, tables. £120. ISBN: 0471869775.

An authoritative abridgement of the 26v. Kirk-Othmer, reducing the 15 million words to 2 million. About 600 contributors and reviewers. Most of the original's subject-entry-words have been retained, plus key references. Excellent value for libraries that cannot entertain buying the full encyclopedia. *Class No:* 66.0(031)

[5898]

SNELL, F.D. *and* HILTON, C.L., *eds*. **Encyclopedia of industrial chemical analysis.** New York & London, Wiley: Interscience, 1966-74. 20v.

V.1-3 (Absorption and emission spectroscopy ... X-ray methods of analysis) concerns general techniques and methods mainly instrumental - used in industrial laboratories. V.3 includes index to v.1-3. V.4-19 cover specific materials, products, elements and compounds, A-Z (*e.g.* 'Carbon black', v.8, p.179-243, has 29 figures, 86 references). V.20, index, covers v.4-19 only. Praised in *Pharmaceutical journal* (25 November 1967, p.533) for the uniformly high quality of contributions (v.8 has 20 such monographs by 26 contributors) and good book production, and 'as a source of information on a wide range of topics of analytical interest'. *Class No:* 66.0(031)

[5899]

ULLMANN, F. **Ullmann's encyclopedia of industrial chemistry.** 5th completely revised ed. Weinheim, VCH Verlagsgesellschaft, 1985-. (A19. 1991). illus., diagrs., graphs, tables. ISBN: 0895731576.

First published 1914; 4th ed. (1972-84. 24v.) as *Ullmanns Enzyklopädie der technischen Chemie*.

5th ed., the first in English, is planned in 2 series; A1-A28 (entries A-Z, *e.g.* A1-A7: Abrasives ... Copper Compounds; Index to v.1-7. 1987. viii, 117p.); B1-B8 (basic topics, *e.g.* B4: Materials science). Each volume in the A series has 20-30 articles. 'Paints and coatings', in A18, has 14 sections, p.359-544, with 577 references. Comparable to Kirk & Othmer as a European counterpart. *Class No:* 66.0(031)

Handbooks & Manuals

[5900]

BÜCHNER, W., *and others*. **Industrial inorganic chemistry.** [Einheitssacht: industrielle anorganische chemie.] Weinheim, VCH, 1989. xxvi,614p. diagrs., tables. ISBN: 3527266291, Ger; 0895736101, US.

1st. German ed. 1984; 2nd ed. 1986. Rev. and updated translation.

'Emphasises the manufacturing process, economic importance and applications of products' (*Preface to the First German edition*). 6 documented chapters, closely subdivided: 1. Primary inorganic materials (p.1-188) - 2 Mineral fertilizers - 3. Metals and their compounds - 4. Silicones - 5. Inorganic solids - 6. Nuclear fuel cycle. Each subdivision includes general and specific references. Indexes: Company abbreviations (p.597-602); analytical subject (p.603-14). Well-produced with information and data clearly laid out. *Class No:* 66.0(035)

[5901]

CHENIER, P.J. **Survey of industrial chemistry.** New York, etc, Wiley-Interscience, 1986. xiv, [1], 422p. ISBN: 0411010774.

27 chapters: 1. Introduction to the chemical industry: an overview ... 3. Leading chemicals and chemical companies ... 5. Industrial gas ... 11. Seven basic organic chemicals ... 16. Basic polymer chemistry - 17. Plastics ... 22. Pesticides ... 26. Surfactants, soaps and detergents - 27. The chemical industry and pollution. 'List of important references and their abbreviations' (p.xiii-xv). Appendix: 'Possible subjects for topics'. Index. *Class No:* 66.0(035)

[5902]

COULSON, J.M. *and* RICHARDSON, J.F. **Chemical engineering.** 4th ed. Oxford, Pergamon Press, 1990-. V.1-. illus., diagrs., graphs, tables.

First ed. (2v.) 1954-55.

V.1 (*Fluid flow, heat transfer and mass transfer*. xiii,708p. ISBN: 0080379508) has 13 sections. Detailed, analytical index, p.685-708. V.2-6 are to be: 2. Particle technology and separation processes - 3. Chemical and biochemical reaction engineering and control - 4/5. Solutions to the problems in volumes 1, 2 and 3 - 6. Chemical engineering design. *Class No:* 66.0(035)

[5903]

CREMER, H.W., *and others, eds*. **Chemical engineering practice.** London, Butterworth, 1956-66. 12v. diagrs., tables.

1: *General*. 1956. 2: *Solid state*. 1956. 3: *Solid system*. 1957. 4: *Fluid state*. 1958. 5-6: *Fluid systems*. 1960. 7: *Heat transfer*. 1963. 8: *Chemical kinetics*. 1965. 9: *Design and construction*. 1965. 10: *Ancillary services*. 1960. 11: *Works design, etc.* 1959. 12: *Indexes*. 1966.

'An attempt ... to acquaint the reader with the various stages of development of a manufacturing process from the laboratory bench to the completed factory' (*Preface*, v.1). A comprehensive treatise, with contributions by specialists; short bibliographies and/or references appended. British Standards are appended. V.10, chapter 8, on water supplies, p.362-472, has 121 text references and a bibliography of 8 items. V.7 is criticized in *Nature* (v.203, no.4,496. 15 August 1964, p.681-2) for lack of sufficient worked examples from practice. V.12: *Indexes* ([vii], 297p.) consists of a subject index (*c.*12,000 entries, 'Fuel': 4 cols.) and a name index. *Class No:* 66.0(035)

[5904]

HEATON, C.A., *ed*. **The Chemical industry.** Glasgow & London, Blackie, 1986. 359p. graphs, tables, chemical structures. £21.95. ISBN: 0216918030.

6 contributors. 7 documented sections: 1. Polymers - 2. Dyestuffs - 3. The chlor-alkali sulphur, nitrogen and phosphorus industries - 4. The pharmaceutical industry - 5.

.... *(contd.)*

Agrochemicals -6. Biological catalysis and biotechnology (p.284-349; 38 references, plus 13 items of bibliography) - 7. The future. *Class No:* 66.0(035)

[5905]
HOPP, V. *and* **HENNIG, J. Handbook of applied chemistry:** facts for engineers, scientists, technicians and technical managers. Washington, Hemisphere Publishing Corpn., 1983. vp. diagrs., tables, chemical structures. ISBN: 0010303207.

English translation of *Chemie Kompendium für das Selbststudium.*

4 sections: 1. Fundamentals of general and inorganic chemistry - 2. Inorganic raw materials and large-scale products: large-scale industrial processes - 3. Organic chemistry: Classification and nomenclature - 4. Inorganic raw materials and large-scale products: industrial processes. 10 references. Index. Concerns the 'quality, safe handling, and application of chemical intermediates and end products' *(Preface).* Suitable to the needs of a technologically highly developed chemical industry. *Class No:* 66.0(035)

[5906]
PERRY, R.H. *and* **GREEN, D.W. Perry's chemical engineers' handbook.** 6th ed. New York, etc., McGraw-Hill, 1984. [2336]p. illus., diagrs., graphs, tables. $89.50. ISBN: 0070494797.

First published 1934; 5th ed., 1973.

About 130 contributors. 27 sections, each with contents page and references. Topics range from transport and storage of fluids and heat transfer equipment to process control and biochemical engineering. Section 19, 'Liquid-solid systems' (109p.) has 132 illus., diagrams and graphs, plus 18 tables. The 301p. on physical and chemical data include 4½ columns of references. 1,848 illus., etc. in all. The 68p. of analytical index carry *c.*9,000 entries. This compendium is known as the chemical engineer's 'Bible'. Also in Spanish edition. *Class No:* 66.0(035)

[5907]
RIEGEL, E.R. Riegel's handbook of industrial chemistry. Kent, J.A., *ed.* 9th ed. New York, Van Nostrand Reinhold, 1992. 1280p. illus., charts tables. $83.50.

First published 1928 as *Industrial chemistry.* 8th ed. 1983.

Documented chapters, by specialists, on various sectors of the chemical process industry, *e.g.* plastics, rubber, man-made textiles, dyes and pigments. Index. Comprehensive; data gathered from numerous sources. *Class No:* 66.0(035)

[5908]
SHUGAR, G.J., *and others, eds.* **Chemical technicians' ready reference handbook.** 3rd ed. New York, etc., McGraw-Hill, 1990. 864p. illus., diagrs., tables. ISBN: 007037183x.

First published 1973 (463p.). 2nd ed. 1981.

Emphasis on step-by-step procedures. In explaining frequently-used laboratory techniques, adds a new chapter on spectroscopy techniques (*e.g.* NMR, infrared). New sections stress laboratory safety. *Class No:* 66.0(035)

[5909]
STOCCHI, E. Industrial chemistry. Trans. by K. Holt and E.L. Short. New York, E. Harwood, 1990. V.1-. illus., diagrs., graphs, tables. ISBN: 0134573188.

Originally published as *Chemica industriale,* v.1-. (Turin).

V.1 (712p.) has 15 chapters (*e.g.* 15. Bitumens, tars and their derivatives). 9 appendices of tables and formulas. Detailed index, p.703-12. Well illustrated. No bibliographies. 'Highly recommended' (*Choice,* April 1991, p.1339). *Class No:* 66.0(035)

Dictionaries

Polyglot

[5910]
Dictionary of chemistry and chemical technology. Hulmin, C. *and* Wenchu, B., *eds.* Beijing, Chemical Industry Press; Amsterdam, Elsevier, 1989. 126p. £134.90.

Terms in English, Chinese, Japanese.

Class No: 66.0(038)=00

[5911]
DORIAN, A.F. Elsevier's dictionary of industrial chemistry, in English/American (with definitions), French, Spanish, Italian, Dutch and German. Amsterdam, etc., Elsevier, 1964. 2v.(xii, 1220p.). $215. ISBN: 0444407537.

8,426 numbered English/American base terms, with equivalents and indexes in the other five languages. Coverage: apparatus, materials, machinery, equipment and processes. V.1: A-O; v.2: P-Z and thumb-indexed indexes. Some overlap with *Elsevier's Dictionary of chemical engineering,* by W.E. Clason (2v. 1968).

Class No: 66.0(038)=00

[5912]
LYDERSEN, A.L. *and* **DAHLO, J. Dictionary of chemical engineering:** English, German, Spanish, French. New York, Wiley, 1992. 300p. £17.50. ISBN: 0471933929.

'An expansion and up-to-date revision of a corresponding Norwegian dictionary by the same authors in 1988' *(Preface).*

About 1,200 English-base terms, with equivalents and indexes in the other languages. Includes word use in the manufacturing industries and skilled trades within chemical engineering. *Class No:* 66.0(038)=00

[5913]
SOBECKA, Z. *and* **CHOINSKI, W.,** *eds.* **Dictionary of chemistry and chemical technology in six languages:** English, German, Spanish, French, Polish, Russian. Oxford, etc., Pergamon Press; Warsaw, Wydawnictwa Naukowo-Techniczne, 1966. vii, 1325p. £125. ISBN: 0080116000.

First published 1962.

11,987 numbered entries in English, with German, Spanish, French, Polish and Russian equivalents across the double page, and German, etc., indexes, on tinted paper. Categorizes terms, notes differences in meaning and gives genders of nouns. It omits current English terms that have no corresponding expressions in one or other of the foreign languages. No abbreviations. 'An excellent work of reference for any European chemist' (*Chemistry & industry,* 3 December 1966, p.2035). *Class No:* 66.0(038)=00

English

[5914]
GARDNER, W. **Gardner's chemical synonyms and trade names.** Pearce, J., *ed.* 9th ed. Aldershot, Gower, 1987. [vi],1081p. Chemical formulas. £105. ISBN: 0291397034.
First published 1924; 8th ed. 1978.
Some 12,000 new entries. In two parts: Trade names and synonyms (A-Z); Names and addresses of manufacturers (much expanded from earlier editions). Many cross-references. *Class No:* 66.0(038)=20

Russian

[5915]
CALLAHAM, L.I. **Russian-English chemical and polytechnical dictionary.** 3rd ed. New York, etc., Wiley, 1975. xxviii, 852p. £81. ISBN: 0471129984.
First published 1947; 2nd ed., 1962.
'A technical as well as a chemical dictionary. Inorganic and organic chemistry, chemical technology, and chemical engineering are naturally given the most complete coverage. Mineralogy, metallurgy, mining and geology, general engineering, machinery and mechanics, electrical engineering, pharmacy and botany are also comprehensively covered' (*Preface*, 1st ed.). About 100,000 Russian entries and sub-entries. Terms are categorized, genders given; abbreviations, prefixes and common Russian word-endings included. Good typography for cyrillic. Particularly helpful in the fields of pure and applied chemistry.
Class No: 66.0(038)=82

[5916]
DYMSHITS, L.B. **Anglo-russkiĭ khimiko teknologicheskiĭ Slovar'.** 5. izd. sterotipnoe. Moscow, Izdat. Sovetskaya Entsiklopediya, 1966. 736p.
Gives equivalents of *c.*10,000 English terms. Chemical formulas figure. Abbreviations, p.727-36.
Class No: 66.0(038)=82

Reviews & Abstracts

[5917]
'Applied chemistry and chemical engineering', *Chemical abstracts*, sections 47-64. Easton, Pa., American Chemical Society, 1907-. Alternate weekly.
One of the three groups constituting sections 47-80 of *Chemical abstracts*. The 15 sections comprise 47. Apparatus and plant equipment ... 64. Pharmaceutical analysis.
Class No: 66.0(048)

[5918]
Chemical engineering abstracts. Nottingham, Royal Society of Chemistry, 1982-. Monthly.
1986: 4,516 abstracts. 22 sections (2. Process operation, loss prevention and optimization; 5. Heat transfer; 7/8 Diffusional operations; 12. Reactors, plant, equipment and techniques; 14. Computers and their applications; 21/22. Techno-commercial items (e.g., plant equipment)). Subject index per issue and annually. Available online as Chemical Engineering and biotechnology abstracts, via DIALOG.
Class No: 66.0(048)

[5919]
Chemical industry notes. Columbus, Ohio, Chemical Abstracts Service, 1973-. Weekly. $1000pa. ISSN: 0045639x.
About 12,000 entries p.a. 8 sections: A. Production -B. Pricing - C. Sales - D. Facilities - E. Products and processes - F. Corporate activities - G. Government activities - H. People. Keyword and corporate named indexes per issue and annually. Carries 'business oriented articles of value to the chemical industry'. Online; host: DATASTAR, DIALOG.
Class No: 66.0(048)

[5920]
Theoretical chemical engineering abstracts. Liverpool, Theoretical Engineering Abstracts, 1964-. 6pa.
1986: 3,591 abstracts. 15 sections, plus 'Books'. Coverage: fluid dynamics; heat transfer; mass transfer; heat/mass transfer; kinetics and thermodynamics; chemical reactor engineering; design; control; mixing; physical separation; grinding; crushing, etc.; pumps, compressors; economics, organization; nomograms; general. Annual analytical subject index and list of journals abstracted. (British and US reports are also summarized).
Class No: 66.0(048)

Progress Reports

[5921]
Advances in chemical engineering. Orlando, Florida, Academic Press, 1956-. Irregular. diagrs., graphs, tables.
V.17 (1991, 384p. ISBN: 0120085178. £59) covers the design parameters for mechanically agitated reactors and particulate fluidization. *Class No:* 66.0(055)

Yearbooks & Directories

[5922]
Chem sources - USA. Clemson, South Carolina, Directories Publishing Co, 1958-. Annual.
First published 1958.
1992 ed. has Chemical section (over 50,000 chemicals and chemical products) - Trade name section (yellow pages) - Classified/Trade name section (blue pages) - Company directory section - Company section. *Class No:* 66.0(058)

[5923]
CHEMICAL INDUSTRIES ASSOCIATION. **Directory of products and buyers' guide:** 1992. 2nd ed. West Wratting, Cambridgeshire, 1992. xxii,582p. £105. ISBN: 0951600931.
Sections: Products - Tradenames - Companies - Contract processing services - EINECS index - CAS index - Chemical formula index - CIA code of conduct - Affiliated associations. 'Comprehensive ... an excellent catalogue' (*Reference reviews,* V.6(3) 1992, p.27).
Class No: 66.0(058)

[5924]
Chemical industry directory and who's who, 1988. Tonbridge, Kent, Benn, 1987. Annual. v.p. £10. ISBN: 086382059x.
Originally *'Chemical age' year book* (1923-58), thereafter *'Chemical age' directory and who's who* (1959-63).
4 Sections. *A.* ABC to chemical manufacturers and traders. Buyers guide to chemicals - *B.* ABC to plant & laboratory equipment manufacturers ... Buyers' guide ... Trade names. European chemical & oil storage; services for chemical storage - *C.* Who owns whom in the British

....(contd.)

chemical and plant industries ... Independent consultants - *D.* Who's who in the chemical industry. Index to advertisers. *Class No:* 66.0(058)

[5925]

Industrial companies: chemicals. Harlow, Essex, Longman, 1992. 480p. illus. £130. (*Financial Times international year books.*) ISBN: 0582092728.

Corporate and financial profiles of 500 international companies. Data: personnel, contact details, operations, sales figures. *Class No:* 66.0(058)

[5926]

WHITESIDE, R. Major chemical and petrochemical companies of Europe, 1991-1992. 5th ed. London, Graham & Trotman, 1991. 272p. £145. ISBN: 1859336203.

Nearly 1000 entries, with usual directory-type information. *Class No:* 66.0(058)

Tables & Data Books

[5927]

BENNETT, H., *ed*. The Chemical formulary: a selection of valuable, timely, practical, commercial formulae and recipes for making thousands of products in many fields of industry. New York, Chemical Publishing Co., Inc; London Chapman & Hall (now Godwin), 1933-. (v.28. 1989.)

V.28 (1989. 438p. ISBN 0820603287) lists recipes for amateur or professional chemist. Appendix includes a list of trademark chemicals and their suppliers, information of federal food, drug and cosmetic laws, a list of incompatible chemicals and emergency first aid for chemical injuries.

Chemical formulary: cumulative index for volumes 1 through 25 (New York, Chemical Publishing Co., 1987. 474p. $75. ISBN 0820603193) provides many useful cross-references. *Class No:* 66.0(083)

[5928]

DAUBERT, T.E. *and* DANNER, R.P. Physical and thermodynamic properties of pure chemicals: data compilation. New York, Hemisphere, 1989. 4v. (loose-leaf). $275; £196. ISBN: 0891169482.

A compilation of tables of recommended physical, thermodynamic and transport properties used in chemical processes and equipment design. Includes over 900 compounds, with 600 to be added. The 'awkward' format was criticized by *Choice* (February 1990, p.973) which nonetheless found the set 'a reliable source of documented data'. *Class No:* 66.0(083)

[5929]

STRAUSS, H.J. Handbook for chemical technicians. Kaufman, M., *ed*. New York, etc., McGraw-Hill, 1976. 512p. illus., charts, tables. ISBN: 0070621640.

In 10 sections (Units and measures ... safety practices). A compendium of physical and chemical data for chemical technicians and engineers, - 'a complement to *CRC Handbook of Chemistry and physics*' (*Choice*, v.14(3), May 1977, p.530). *Class No:* 66.0(083)

Histories

[5930]

FURTER, W.F., *ed*. History of chemical engineering: papers from a symposium, Honolulu, April 1979. Washington, American Chemical Society, 1980. xii, 436p. illus. (*Advances in chemistry series, 190.*)

22 contributions (all but one written by chemical engineers). Includes essays on the individuals and institutions that contributed most significantly to the expression of chemical industry during the 20th century. Other papers deal with national aspects of the subject. *Class No:* 66.0(091)

[5931]

HABER, L.F. The Chemical industries, 1900-1930: international growth and technological change. Oxford, Clarendon Press, 1971. xi, 452p.

Continues the volume on the 19th century. An international survey, dwelling particularly on Germany, Great Britain and USA. Emphasis is on economic rather than technological aspects. *Economic journal* (v.81(324), December 1971, p.968) notes the author's 'admirable factual clarity and historical insight'. *Class No:* 66.0(091)

[5932]

HABER, L.F. The Chemical industry during the nineteenth century: a study of the economic aspect of applied chemistry in Europe and North America. Oxford, Clarendon Press, 1958. x, 292p. diagrs., tables. £20.

An international survey. The nineteenth-century volume deals with the origins of the chemical industry. *Class No:* 66.0(091)

[5933]

MULTHAUF, R.P. The History of chemical technology: an annotated bibliography. New York, Garland, 1984. xx,299p. illus. (*Bibliographies of the history of science and technology, v.5. Garland reference library of the humanities, v.348.*) ISBN: 0824092554.

1528 items in 3 parts: 1. Histories (largely by country) - 2. 'Traditional' technology (ceramics, glass, paper, etc.) - 3. 'Modern' technology (artificial fibres, gases, petroleum, pharmaceuticals, etc.). Most citations very brief (2-3 lines). Author index (p.253-83); 'title' (key-word) index (p.284-99). Based largely on the collections of the Library of Congress and the Smithsonian Institute. *Class No:* 66.0(091)

Great Britain

[5934]

CLOW, A. *and* CLOW, N.L. The Chemical revolution: a contribution to social technology. London, Batchworth Press, 1952. xvi, 680p. illus. (*Industrial archaeology series.*)

Special emphasis on chemical industry of Scotland and north of England. Period covered: *c.*1750-1830. Individual industries have separate chapters, with many references in text. 'Glossary of dead chemical language', p.618-22. A chemical chronology to 1856. Extensive bibliography, p.633-61.

Reprinted June 1992 by Gordon & Breach (£32; ISBN: 2881245498) as part of their series *Classics in the history and philosophy of science*. *Class No:* 66.0(410)

[5935]
READER, W.J. **Imperial Chemical Industry:** a history. Oxford Univ. Press, 1970-. diagrs., charts, tables, maps.
1. *The forerunners, 1870-1926.* 1970. xvi, 563p.
2. *The first quarter century, 1926-1952.* 1975. 588p.
Class No: 66.0(410)

USSR

[5936]
SAGERS, M.J. *and* SHABAD, T. **The Chemical industry in the USSR:** an economic geography. Washington, American Chemical Society, 1990. 550p. diagrs., tables, maps. $89.95. ISBN: 0841217602.
Topics include chemical development, location factors, trends, changes, raw materials, technology and prospects, including foreign trends in chemicals and chemical equipment suppliers. A reference for graduate students and researchers. *Class No:* 66.0(47)

USA

[5937]
HAYNES, W. **American chemical industry:** a history. New York, Van Nostrand, 1945-54. 6v.
1. *Background and beginning, 1604-1911.* 1954.
2.3. *The World War I period, 1912-1922.* 1945.
4. *The merger period, 1923-29.* 1945.
5. *Decade of new products, 1930-1939.* 1954.
6. *The chemical companies.* 1949.
A comprehensive history of US industrial chemists, individual companies, processes and products. Industries may be grouped, *e.g.* coal-tar, chemicals, the alkalis, the acids. V.6 has data on companies, A-Z (p.1-491). Each volume carries name and product indexes.
Class No: 66.0(73)

Processes

Encyclopaedias

[5938]
McKETTA, J.J. *and* CUNNINGHAM, W.A., *eds.* **Encyclopedia of chemical processing and design.** New York, Dekker, 1976-. v.1-. (v.28. 1987). illus., diagrs., graphs, tables. ISBN: 0824724378.
A multi-volume encyclopedia, more restricted in scope than Kirk-Othmer. Each volume has 20-30 contributors, most recruited from industry. Thus v.38 (1991): 'Piping design, economic diameter' ... 'Pollution abatement equipment, alloy section' has 47 brief articles on the areas of piping, plant (capacity, startup, etc.), plastics, pneumatics and pollution. Many reprinted from *Chemical engineering* and other journals. *Class No:* 66.02(031)

[5939]
SITTIG, M. **Organic chemical process encyclopedia.** 2nd ed. Park Ridge, N.J., Noyes Development Association, 1969. [xxiv], 712p. charts.
First published 1967.
Entries A-Z 'Acetaldehyde from Acetylene' to 'Zirconium orthotitanates'. Data under each entry: Reaction; Flowchart; Feed materials; Co-products, etc.; Major product uses;

.... (contd.)
References (*e.g.* US patent). 711 flowcharts (1st ed.: 587). Emphasis on petrochemicals. No index.
Class No: 66.02(031)

Handbooks & Manuals

[5940]
MEYERS, R.A. **Handbook of chemical production processes.** New York, etc., McGraw-Hill, 1986. v.p. illus., diagrs., graphs, tables. $75. ISBN: 0070417662.
Contributions from 19 firms located in the UK, the Federal Republic of Germany, Japan, Netherlands and the US. 3 parts (40 chapters): 1. Organic chemicals - 2. Polymers - 3. Inorganic chemicals. Abbreviations and acronyms. 10-p. index. Good layout. *Class No:* 66.02(035)

[5941]
SCHWEITZER, P.A., *ed.* **Handbook of separation techniques for chemical engineers.** 2nd ed. New York, etc. McGraw-Hill, 1988. v.p. illus., diagrs., graphs, tables. £69.95. ISBN: 0070558086.
Covers all major separation techniques used industrially. 45 contributors. 6 parts, closely subdivided into sections: 1. Liquid-liquid mixtures - 2 Liquids with dissolved solids - 3. Gas (vapor) mixtures - 4. Solid-liquid mixtures - 5. Solid mixtures - 6. Gas-solid mixtures. Some sections have extensive references (*e.g.* 2.1 Membrane filtration, 120 references). Detailed, analytical index. A wealth of accessible information for professional engineers.
Class No: 66.02(035)

Dictionaries

Polyglot

[5942]
HARTMAN, K., *and others.* **Dictionary of process technology in four languages:** English, German, French, Russian. Amsterdam, etc., Elsevier, 1989. 319p. £81. ISBN: 0444988882.
Over 8,000 English-base terms, with German, French and Russian equivalents and indexes. *Class No:* 66.02(038)=00

German

[5943]
Dictionary. Chemical processes and products. [Wörterbuch. Chemische Prozesse und Produkten. English/German, Deutsch/Englisch.] Widnau, Switzerland, Schnellmann Verlag. 1979. 117p. (*CHP. 0.3.*)
About 2,500 entries in each half, with equivalents in the other language. German genders given.
Class No: 66.02(038)=30

French

[5944]
Dictionary. Chemical processes and products. [Dictionnaire. Procédés et produits chimiques. English/French, Français/ Anglais.] Widnau, Switzerland, Schnellmann Verlag, 1987. 119p. (*CHP. 2.3.*) ISBN: 3907971256.
About 2,500 entries in each half, with equivalents in the other language. French genders given. Appendix of units.
Class No: 66.02(038)=40

Italian

[5945]

Dictionary. Chemical processes and products. [Dizionario. Processi e prodotti chimici. English/Italian, Italiano/Inglese.] Widnau, Switzerland, Schnellmann Verlag, 1985. 119p. (*CHP. 1.3.*) ISBN: 3907971221.

About 2,500 entries in each half, with equivalents in the other language. Appendix of units.

Class No: 66.02(038)=50

Heat Transfer Materials

[5946]

HALLA, C.W. Dictionary of drying. New York, Dekker, 1979. iv, [1], 350p. diagrs., graphs, tables. ISBN: 0824766820.

Entries A to WVTR (Water vapour transfer rate). Cross-references. 'Milk drying': nearly 2p. Much on applications. Includes a survey of recommendations on types of dryers and procedure for drying. References, p.349-50.

Class No: 66.04

[5947]

KAKAC, D., *and others, eds.* **Handbook of single-phase convective heat transfer.** New York, Wiley, 1987. v.p. illus., diagrs., tables. $75. ISBN: 0471817023.

25 contributors; 22 chapters. Topics include convection in liquid metals; effects of electric and magnetic fields; natural convection in enclosures; non-Newtonian transfer. Numerous tables; detailed, analytical index.

Class No: 66.04

Distillation

[5948]

UNDERWOOD, A.J.V. Six-language vocabulary of distillation terms: English/français/español/russkiĭ/italiano/deutsch. Prepared by the Working Party on Distillation of the European Federation of Chemical Engineering. London, Institution of Chemical Engineers, for the European Federation, 1967. xii, 62p.

307 numbered English-based entries, with French, Spanish, Russian, Italian and German equivalents. Entries are grouped (with no headings). English, French etc., indexes. *Class No:* 66.048

Liquids

[5949]

BECHER, P., *ed.* **Encyclopedia of emulsion technology.** New York, Marcel Dekker, 1983-. 3v. diagrs., graphs, tables. ea. $115. ISBN: 0824718763.

V.1 (xiv, 725p.) covers basic theory. 9 chapters by various hands: 1. Liquid/liquid interfaces - 2. Formation - 3. Stability - 4. Microemulsions - 5. Phase properties - 6. Droplet size data - 7. Rheological properties - 8. Optical properties - 9. Dieletric properties.

V.2 (472p. 1985.) covers applications of emulsions. 8 chapters by various hands: 1. Demulsification - 2. Research techniques - 3. Medical applications - 4/5 Applications in the food industry - 6. Applications in the cosmetic industry - 7. Applications in agriculture - 8. HLB - an update.

V.3 (xi,437p. 1988.) Covers measurement applications. 4 parts (8 chapters); 1. Basic theory - 2. Measurement of properties - 3. Applications - 4. Appendix (HLB: a further update). Separate indexes for each v. *Class No:* 66.06

[5950]

MURALIDHARA, H.S., *ed.* **Advances in solid-liquid separation.** Columbus, Ohio, Battelle Press, 1986. viii,485p. illus., diagrs., graphs, tables. £69.50. ISBN: 0851863639.

Published in conjunction with the International Conference on Recent Advances in Solid-liquid Separation, Ohio, 1986, but not conference proceedings as such. 27 contributors. 16 chapters (subdivided), each with extensive references. Among topics covered are thickening, filtration, equipment, dewatering (various techniques). Analytical index.

Class No: 66.06

Solvents

[5951]

COMMISSION OF THE EUROPEAN COMMUNITIES. Solvents in common use: health risks to workers. London, Royal Society of Chemistry, 1988. [vi],308p. £60. ISBN: 0851860885.

In-depth data on 10 most commonly used solvents: acetone, carbon disulphide, diethyl ether, 1,4-dioxane, ethylacetate, methanol, nitrobenzene, pyridine, toluene, xylene. Data: physical, chemical and spectroscopic properties; toxicology; storage, handling and use precautions; fire hazards; hazardous reactions; emergency action; first aid. Glossary and extensive references. 'An indispensable reference book wherever these solvents are used' (*Science and technology libraries,* v.11(1), Fall 1990, p.40). *Class No:* 66.061

[5952]

Industrial solvents handbook. Flick, E.W., *ed.* 3rd ed. Park Ridge, N.J. Noyes Data, 1985. 648p. diagrs., tables. £85.

First two editions (1970, 1977) by I. Mellan.

About 1,000 tables of basic data (citing sources) of physical properties of most solvents and on solubilities of a variety of the materials in these solvents. Sections on halogenated hydrocarbons, polyhydric alcohols, ethers, acids, esters. *Class No:* 66.061

[5953]

LO, TEH, C., *and others.* **Handbook of solvent extraction.** New York, etc., Wiley: Interscience, 1983. 980p. illus., diagrs., graphs, tables. ISBN: 0471041045.

80 contributors, an international team. 31 chapters, each with references and suggested readings. 4 parts: 1. General principles - 2. Industrial extraction equipment - 3. Industrial processors, organic and inorganic - 4. Cost and equipment. Detailed, analytical index. *Class No:* 66.061

Chemical Products

Encyclopaedias

[5954]

Faith, Keyes and Clark's industrial chemicals. Lowenhein, F.A. *and* Moran, M.K., *eds.* 4th ed. New York, Wiley: Interscience, 1975. 904p.

First published 1950. 3rd ed. by W.L. Faith (1965).

Describes c.150 industrial chemicals, Acetic acid ... Trichlorethylene; with systematic data on properties, grades, containers, regulations, economic aspects and

....(contd.)

manufacturers. Company and subject indexes. 'A must for a wide variety of special, university and public libraries' (*New technical books,* v.61(1), January 1976, entry 191).
Class No: 661(031)

[5955]
The Merck index: an encyclopedia of chemicals, drugs and biologicals. 11th ed. Rathway, N.J., Merck & Co., 1989. *c.*2300p. tables.
First published 1889.
Monographs of *c.*10,000 chemicals, drugs and biologicals, with notes on preparation, boiling point, toxicity, uses, etc., plus chemical structure and references. Appended: organic name reactions; miscellaneous tables, Chemical Abstracts Service names and Registry names; Therapeutic category and biological activity index; Formula index; Cross-index of names (plus titles of deleted 9th and 10th ed. monographs), etc. Index of 315p. (*c.*70,000 entries).
Online via Dialog, file 304, semiannual reloads.
Class No: 661(031)

Handbooks & Manuals
[5956]
ASH, M. *and* **ASH, I. What every chemical technologist wants to know.** London, E. Arnold, 1988-90. 5v.
Emulsifier and welding agents. 1988 400p. £65.
Dispersants, solvents and solutions. 1988. 480p. £65.
Plasticizers, stabilizers and thickeners. 1989. 424p. £65.
Conditioners, emollients and lubricants. 1990. 412p. £65.
Resins. 1990. 400p. £65.
Each volume includes trade name products, generic chemical symbols and trade name product manufacturers. Indexes by trade names, formulas, synonyms, etc.
Class No: 661(035)

[5957]
Materials and technology: a systematic encyclopaedia of the material used in industry and commerce, including foodstuffs and fuels. London, Longmans; Amsterdam, de Bussy, 1968-76. 8v. illus. (incl. pl.), diagrs., tables.
Based on J.F. Van Oss's *Warenkennis en Technologie* (5th ed. 1948-50). Updated and rewritten.
U.S. title as *Chemical technology: an encyclopedic treatment* (New York, Barnes & Noble, 1968-76. 8v.).
1. *Air, water, inorganic chemicals and nucleonics.* 1968.
2. *Non-metallic ores, silicate industries and solid mineral fuels.* 1971.
3. *Metals and ores.* 1970.
4. *Petroleum, organic chemicals and plastics.* 1972.
5. *Natural organic products and related mineral fuels.* 1972.
6. *Wood, paper, textiles, plastics and photographic materials.* 1973.
7. *Vegetable food products and luxuries.* 1975.
8. *Edible oils and fats, and animal food products.* 1975.
V.1 (xxiv, 703p.) mentions *c.*300 compounds, for some 30 of these the processes selected for extensive treatment. Chapter 6 of v.2 'Glass' (p.331-412) has 41 figures, 20 tables and 8 references. 'May be viewed as an introduction to chemical technology' (*Aslib book list,* v.34(3), March 1969, entry no.154). *Class No:* 661(035)

[5958]
Specialty chemicals: innovations in industrial synthesis and applications. Pearson, B., *ed.* Amsterdam, Elsevier, 1991. xx,613p. diagrs., tables, chemical structures. ISBN: 185166646x.
About 60 contributors. Documented entries: Halogens ... Sodium, lithium and cesium. Final sections: Oxidations and reductions - New resources and catalyses - Biomassformations - Application innovations.
Class No: 661(035)

Reviews & Abstracts
[5959]
Natural products updates. London, Royal Society of Chemistry, 1987-. Monthly. £107pa. ISSN: 09501711.
200 items per issue on isolation studies, structure determinations, new properties, total and biosyntheses. Includes chemical structures, molecular formulas, spectra data. 5 indexes: authors; sources; taxonomic names: trivial names; compound class. *Class No:* 661(048)

Yearbooks & Directories

Europe
[5960]
1988 directory of chemical producers: Western Europe. Menlo Park, Calif., SRI International, 1988. 2v.
1. *Companies.* 2. *Products. Regions.*
4 main sections. Company index (over 6,000 companies) - Companies (under countries) - Products - Regions (5,000 separate plant sites). Data on companies include ownership; subsidiary and affiliated companies; activities; plants and products. Lists *c.*13,000 chemicals. *Class No:* 661(058)(4)

Great Britain
[5961]
Chemist & druggist directory, and tablet & capsule identification guide. Tonbridge, Kent, Benn, 1972-. Annual. illus., tables. ISSN: 02625881.
1991 ed. (436p. £72) has 6 sections: A. Tablet and capsule identification guide - B. Manufacturers and suppliers index - C. Buyers guide - D. Pharmaceutical organizations - E. Hospital pharmacists - F. Wholesale - Retail outlets. Index to advertisers. *Class No:* 661(058)(410)

USA
[5962]
1988 directory of chemical producers: United States of America. Menlo Park, Calif., SRL International, 1988. vi, 1196p.
First published 1986.
3 main sections. Companies (1,500, A-Z; cross-references from former names) - Products (c.10,000 chemicals) - Regions (all 50 states). *Class No:* 661(058)(73)

[5963]

Chem sources - USA. Clemson, South Carolina, Directories Publishing Co, 1958-. Annual.

First published 1958.

1992 ed. has Chemical section (over 50,000 chemicals and chemical products) - Trade name section (yellow pages) - Classified/Trade name section (blue pages) - Company directory section - Company section.
Class No: 661(058)(73)

Tables & Data Books

[5964]

Handbook of existing and new chemical substances. Chemical Products Safety Division [Japan], *ed.* 5th ed. Tokyo, The Chemical Daily Co., Ltd., 1992. [iv], 833p.

Lists about 20,000 substances registered for import and manufacture in Japan. Includes an English translation of the 'Law, Enforcement Order, Ministerial orders and notifications which we hope will be helpful to business' ('Recommendation', p.(1)). *Class No:* 661(083)

[5965]

LEUNG, A.Y. Encyclopedia of common natural ingredients used in food, drugs and cosmetics. New York, Wiley, 1980. 409p. $100. ISBN: 0471049549.

About 300 of these common, natural ingredients described, specifying chemical composition, pharmacological/biological activities, uses and regulatory status. Their reputed effects as folk remedies are also indicated. *Class No:* 661(083)

[5966]

SNELL, F.D. *and* **SNELL, C.T. Dictionary of commercial chemicals.** 3rd ed. Princeton, N.J., etc., Van Nostrand, 1962. ix, 764p. tables.

Formerly as *Chemicals of commerce* (1939; 2nd ed. 1952).

38 chapters of detailed information on composition of commercial products, stating formula, source of supply, mode of manufacture, common impurities or contaminants, commercial grades and uses. Physical data (*e.g.* melting or boiling points) are added, as necessary; type and method of packaging are often given. 8 appendices, largely applicable only to US manufacturers and suppliers. Index (p.675-714) has *c.*45,000 entries. *Class No:* 661(083)

Nomenclatures

[5967]

BRITISH STANDARDS INSTITUTION. Recommendations for names of chemicals used in industry. London, BSI, 1983. 36p. Chemical structures. £36. (*BS 2474: 1983.*)

Names for industrial chemicals and for commercial descriptions of chemicals - Principles of nomenclature used. Appendices list the radicals cited and give guidance on computerized indexing of chemical names.
Class No: 661(083.72)

Trade Names, Trade Marks, Brand Names

[5968]

ASH, M. *and* **ASH, I.,** *comps.* **Concise encyclopaedia of industrial chemical additives.** London, E. Arnold, 1991. xi,859p. £110. ISBN: 0304556226.

In four sections: Chemical trade names dictionary (p.1-

....(contd.)

611) - Chemical compound cross-references (p.612-709) - Functional classification of trade names - Chemical manufacturers directory (international). Over 18,000 entries for chemical trade name products currently used throughout the world. *Class No:* 661(088.7)

Thesauri

[5969]

ASH, M. *and* **ASH, I.,** *eds.* **The Thesaurus of chemical products.** 2nd ed. London, E. Arnold, 1992. 2v. (*c.*1200p.). £175. ISBN: 0340583029.

First published 1986.

1. *Generic-to-tradename.* 2. *Tradename-to-generic.*

Over 40,000 up-to-date national and international tradename entries. Each volume has appendix of chemical manufacturers (*c.*950) and of their addresses; also 'Books for consultation'. *Class No:* 661:025.43

[5970]

—SYNTHETIC ORGANIC CHEMICAL MANUFACTURERS ASSOCIATION. SCOMA handbook: Commercial organic chemical names. Washington, D.C., American Chemical Society, Chemical Abstracts Service, 1965. 666p.

A thesaurus of *c.*20,000 chemical names, covering 6,800 commercially traded organic compounds.
Class No: 661:025.43

Powders

[5971]

RHODES, M.J., *ed.* **Principles of powder technology.** Chichester, Wiley, 1990. xi[i],439p. illus., diagrs., graphs, tables. £36.50. ISBN: 0471924229.

11 chapters; 14 chapters with references. Covers characterization, mixing, storage, size reduction, hazards and separation techniques from an industrial point of view. Includes problems and worked examples. *Class No:* 661-492

Surface-active Agents

[5972]

CARRIÉRE, G. Dictionary of surface-active agents, cosmetics and toiletries, in English, French, German, Spanish, Italian, Dutch and Polish. Amsterdam, etc., Elsevier, 1978. 198p. DFl.140. ISBN: 0444998098.

686 English-base terms, with equivalents and indexes in French, German, Spanish, Italian, Dutch and Polish. Covers the principal languages used in the industry.
Class No: 661.185

[5973]

KARSA, D.R., *and others, eds.* **Surfactants applications directory.** Glasgow, Blackie, 1991. [xiv],399p. £55. ISBN: 0216926904.

2 parts: 1. Alphabetical (by tradename) listing of surfactants by industry sector - 2. Index of surfactant properties listed by industry sector. Data include tradename, supplier, property, chemical type, form, activity, comments. Appendix of company names, addresses. *Class No:* 661.185

[5974]

SANTHOLZER, R.W. *and* SANTHOLZER, C. **Five language dictionary of surface coatings, plating, products, finishing, corrosion, plastics and rubber:** English/American, Czech, Russian, German, French. Oxford, etc., Pergamon Press, 1969. v, 580p. tables.

4,000 Czech-base terms, with Russian, German, English and French equivalents and indexes. Extends the coverage of R. Santholzer and J. Korinsky's *Four-language dictionary of paints, lacquers and varnishes, surface treatment, corrosion* (Prague, 1959. 436p.). *Class No:* 661.185

[5975]

Surface treatment technology abstracts; incorporating *Metal finishing abstracts*; including *Printed circuits and elctronics coatings abstracts*. Teddington, Middx., Finishing Publications, Ltd., 1986-. v. 28(1)-. 6 pa. £290pa. ISSN: 09505199.

As *Metal finishing abstracts*, v. 1-27, 1959-85. Previously appeared in *Electroplating and metal spraying* (1947-1959).

Over 7,000 indicative and informative abstracts and citations pa. covering surface treatment, in 20 classes (Cleaning and degreasing ... Electroplating ... Metal spraying and hard facing ... Corrosion and electrochemistry ... General and miscellaneous). Covers journal articles, books, standards, reports and patents - full coverage for UK, US, GDR and USSR patents.

Separate section for *Printed circuits and electronics coatings abstacts*. 4 sections (Printed circuit boards; semi conductors and integrated circuits; Electrical connectors; Miscellaneous). Over 2,000 abstracts pa. Separate annual indexes for each part, of authors, subjects and patent numbers. Abstracts appear 3-12 weeks after publication. *Class No:* 661.185

[5976]

World surface coating abstracts. Compiled by Paint Research Assoiciation, Teddington, Middx. Oxford, etc., Pergamon Press, 1928-. Monthly. DM2230pa. ISSN: 75578409.

About 10,000 abstracts pa. 50 sections (01/02. Pigments, extenders, dyestuffs and phosphors; 23/24. Polymers containing silicon and other elements; 35/36. Water-borne paints and their components; 53/54. Weathering corrosion; 73/74. Industrial and other hazards; 87. Legislation and other official publications; 88. Standards and specifications. Books. Annual author and subject indexes. Online database. *Class No:* 661.185

Detergents

[5977]

CARRIÉRE, G. **Lexicon of detergents, cosmetics and toiletries.** Amsterdam, etc., Elsevier, 1966. ii, 204p. DFl.140. (*Elsevier lexica, 8.*) ISBN: 0444400990.

Part 1, *Detergents* (257 numbered entries), is a revision of G. Carriéres *Detergents* (1960). English-base terms, with French, Spanish, Italian, Portuguese, German, Dutch, Swedish, Danish, Norwegian, Russian, Polish, Finnish, Czech, Hungarian, Rumanian, Greek, Turkish and Japanese equivalents and indexes. Part 2, *Cosmetics and toiletries* (entries 258-523), also English-base has French, Spanish, Italian, Portuguese, German, Dutch and Swedish equivalents and indexes. *Class No:* 661.185.6

Soap Industry

[5978]

UNITED NATIONS. Industrial Development Organization. **Information sources on the soap and detergent industry.** New York, United Nations, 1977. 81p. $4. (*UNIDO guides to sources of information, no.24.*)

585 items. 12 sections: 1. Professional, trade and research organizations, learned societies and special information sources - 2. Directories - 3. Sources of statistics, marketing and other economic data - 4. Basic handbooks, textbooks and manuals - 5. Monograph series - 6. Current periodicals - 7. Current abstracting and indexing periodicals - 8. Proceedings, papers and reports - 9. Specialized dictionaries and encyclopaedias - 10. Bibliographies - 11. Films - 12. Other potential sources of information (*e.g.* Meetings and conferences). 'Bibliographical sources used'. *Class No:* 661.187

Sulphur

[5979]

PATAI, S.J. **The Chemistry of sulphuric acids, esters and their derivatives.** New York, Wiley, 1990. xvi,738p.

30 contributors. 23 documented chapters: 1. Sulphuric acid and carboxylic acid - a comparison ... 23. Biological activity of sulphuric acid derivatives. Author and sulphur indexes, p.677-738. *Class No:* 661.2

[5980]

RAYMONT, M.E.D. **Sulphur: new sources and uses.** Based on a symposium jointly sponsored by the Division of Petroleum Chemistry and Industrial and Engineering Chemistry at the 181st ACS National Meeting, Atlanta, Georgia, April 2-3, 1981. Washington, American Chemical Society, 1983. 261p. (*ACS Symposium series 183.*)

14 Documented papers (*e.g.* 'Sulphur recovery from new energy sources'; p.21-35. 15 references; 'The potential of new sulphur products in the Middle East, p.225-50. 4 references). Subject and author indexes, p.251-61. *Class No:* 661.2

Chlorine

[5981]

SCONCE, J.S., *ed.* **Chlorine,** its manufacture and uses. New York, Reinhold; London, Chapman & Hall, 1962. x, 901p. illus., diagrs., tables. (*American Chemical Society. Monograph series no. 154.*)

11 contributors. 29 chapters, each with references (*e.g.* chapter 6; 'Electrolysis of brines in mercury cells'. 60 references). Includes patents. Index, p.889-901. *Class No:* 661.41

Salt

[5982]

LEFOND, S.J. **Handbook of world salt resources.** New York, Plenum Press, 1969. xxiii, 384p. charts, maps, tables.

Covers both rock salt resources and solar salt operations. 'Country-by-country survey, with data on location and production of mines, plus relevant stratigraphical and other geographical information' (*Aslib book list,* v.35(1), January 1970, entry no.39). *Class No:* 661.422

Rare Earth Elements

[5983]

JEZOWSKA-TRZEBIATOWSKA, B., *and others*. **The Rare elements:** occurrence and technology. Amsterdam, Elsevier; Warszawa, PWN-Polish Scientific Publishers, 1990. x,536p. diagrs., graphs, tables. £131.25. (*Topics in inorganic and general chemistry, 23.*) ISBN: 0444988777.

Rev. and enlarged translation of the 1976 Polish original.

7 chapters, each with appended bibliography (6: The technology of rare elements, p.238-496 (copper ... uranium; each element, with references)). Indexes (subject; name), p.523-36. *Class No:* 661.865

Gases

Encyclopaedias

[5984]

Encyclopédie des gas. [Gas encyclopaedia.] Amsterdam, Elsevier, for L'Air Liquide, 1976. viii, 1150p. graphs, tables. ISBN: 0444414924.

5 chapters on physical properties and flammability of gases, their safe handling, general metrology, symbols and units. Over 100 monographs (p.61-1150), each with bibliography appended. 'Ammonia' (p.951-72; 26 references) has data under 6 heads: physical properties; flammability; biological properties; precautions in handling and storage; materials of construction; uses.

Class No: 661.9(031)

Handbooks & Manuals

[5985]

AGA gas handbook. Ahlberg, K., *ed.* Lidingö, Sweden AB, 1985. x, 582p. illus., diagrs., tables. ISBN: 9197006114.

5 main sections: 1. Current information about gases - 2. Handling and storage of gases - 3. Choice of materials in gas systems - 4. Gases: physical data - 5. Gas information, p.86-540 (Acetylene ... Sulfur dioxide). Appendices include glossary; references, p.564-71. Addresses. Subject index. *Class No:* 661.9(035)

[5986]

TIRATSOO, E.N. **Natural gas:** a study. 2nd ed. Beaconsfield, Bucks., 1972. xvi, 400p. illus., diagrs., graphs, tables.

First published 1966.

21 chapters, with footnotes. 1. 'Natural' and other gases - 2. The non-hydrocarbon gases - 3. Methane and crude oil - 4. The production of gas and oil ... 8. The uses and applications of natural gas - 9. Liquefied natural gas ... 12. Natural gas in Britain ... 16. North America ... 18. The Middle East and East Asia - 19. Africa and Australia - 20. Reserves and resources - 21. Some economic factors. General index, p.389-99. *Class No:* 661.9(035)

Dictionaries

[5987]

BRITISH STANDARDS INSTITUTION. **Glossary of terms used in the gas industry.** 2nd revision. London, BSI, 1971. 133p. tables. (*BS 1179: 1967 (1987).*)

First published 1944.

About 1,00 definitions. 2 parts (16 sections): 1. Production

.... (contd.)

- 2. Utilization. 2 appendices (A. Standard nomenclature of appliance parts not included in the 'Glossary', p.106-8). Index, p.110-33. *Class No:* 661.9(038)

[5988]

INTERNATIONAL GAS UNION. **Dictionary of the gas industry:** English, German, French, Russian. 2nd completely rev. and enl. ed. Essen, Vulkan-Verlag, 1982-85. 2v.

1. *English, German, French, Russian.*

2. *Arabic, Italian, Portuguese, Spain.*

7,300 entries, English-based, in v.1, with German, French and Russian equivalents and indexes. 11 appendices. Abbreviations. Bibliography, p.635-9. V.2 bibliography, p.591-3. Favourably reviewed in The *Incorporated linguist*, v.22(3), Summer 1983, p.163, 166. *Class No:* 661.9(038)

[5989]

—INTERNATIONAL GAS UNION. International dictionary of the gas industry. English version. Berlin, Kammer der Technik, 1978. 254p.

*c.*12,000 entries in sections A-G, with definitions. Alphabetical index. *Class No:* 661.9(038)

Reviews & Abstracts

[5990]

Gas abstracts. Chicago, Ill., Institute of Gas Technology, 1945-. Monthly. ISSN: 00164844.

1986: 2,337 abstracts. 7 sections: 1. Planning, policy, and supply - 2. Production and processing - 3. Transmission - 4. Storage - 5. Distribution - 6. Utilization - 7. Instrumentation and analytical methods. Author index per issue. Half-yearly author and subject indexes. *Class No:* 661.9(048)

[5991]

Médagaz: bulletin bibliographique mensuel. Paris, Association Technique de l'Industrie du Gaz en France, 1982-. Monthly.

4 parts: Documents originaux - Traductions - Ouvrages - Reports. About 200 abstracts per issue. Classes: O. Arts, sciences et techniques; A. Production et traitement du gaz ... L. Classement géographique. *Class No:* 661.9(048)

Hydrogen

[5992]

COX, E.K. *and* WILLIAMS, K.D. **Hydrogen:** its technology and implications. Cleveland, Ohio, CRC Press, 1965-77. 5v. graphs, tables.

1. *Hydrogen production technology.* 1977.

2. *Transmission and storage.* 1977.

3. *Hydrogen properties.* 1975.

4. *Utilization of hydrogen.* 1976.

5. *Implications of hydrogen energy.* 1977.

V.3 (xii, 321p.) has 5 chapters: 1. Introduction - 2. Discussion of the properties - 3. Data graphs (for parahydrogen, unless otherwise stated) - 4. Data tables - 5. Unit conversions, physical constants and symbols. 287 references, in all, p.310-5. Index. Primarily graphs and tables. *Class No:* 661.96

Explosives

[5993]
PARTINGTON, J.R. **A History of Greek fire and gunpowder.** Cambridge, Heffer, 1960. xvi, 381p. illus.

7 chapters. 1. Incendiaries in warfare ... 3. The legend of Black Berthold - 4. Miscellaneous treaties, non-military arts - 5. Gunpowder and firearms in Muslim lands - 6. Pyrotechnics and firearms in China - 7. Saltpetre. Numerous chapter references (*e.g.* Chapter 3. 341 references). Indexes of names, places and nationalities. Non-analytical subject index, p.362-80. Index of Greek words. 22 illus., plus frontispiece. *Class No:* 662

[5994]
URBANSKI, T. **Chemistry and technology of explosives.** [14th] ed. Authorized translation by I. Jeczalikowa and S. Laverton. Oxford, etc., Pergamon Press; Warsaw, PWN, 1964-84. 4v. £480, the set. ISBN: 0680362521.

First published 1953-54.

'The chemical, physical and physico-chemical properties of explosives are dealt with, and processes of manufacture are described whenever the substance in question is of practical importance' (*Preface*). V.1 contains 20 well-documented chapters (*e.g.* 8. 'Nitro-derivatives of toluene'; (71 references). Author (of references) and such indexes to each volume. Well produced. *Class No:* 662

Fireworks

[5995]
PHILIP, C. **A Bibliography of firework books:** works on recreative fireworks from the sixteenth to the twentieth century. Winchester, C. Philip, in association with St. Paul's Bibliographies, 1985. xxix, 168p. illus. £23.60. ISBN: 0906795311.

Over 600 entries for books, of which more than 50% are not recorded in the *National union catalog,* and nearly 60% are not in the *British Library General catalogue.* 20 black-and-white illus., col. front. *Class No:* 662.1

Fuel Technology

[5996]
CLARK, G.H. **Industrial and marine fuels reference book.** London, etc., Butterworths, 1988. 528p. illus., diagrs. £95. ISBN: 0408014881.

26 sections (The choice of fuels - The chemistry of petroleum - Crude oil refining - Laboratory tests on fuels - Natural gas and LPG ... General operating problems of large diesel engines - Deposit formation and corrosion in steam boilers - Chimney stack emissions). Appendix. 239 illus. For mechanical, marine and plant engineers, plus shipowners, engine-builders and designers. *Class No:* 662.6

[5997]
Fuel and energy abstracts: a bimonthly summary of world literature on all technical, scientific, commercial and environmental aspects of fuel and energy. Guildford, Butterworth Scientific, on behalf of the Institute of Energy, 1978-. 6pa. £170 pa. ISSN: 01406701.

Formerly *Fuel abstracts and current titles/* FACTS (1960-), succeeding monthly *Fuel abstracts* (1945-58).

About 8,000 abstracts pa. Classified sections: 01. Solid fuels - 02. Liquid fuels - 03. Gaseous fuels ... 06. Electric

....(contd.)
power operation and utilization ... 15. Environment - 16. Fuel science and technology ... 18. Energy conversion and recycling. Subject and author indexes per issue and pa. Coverage: 'World literature on scientific, technical, commercial and environmental aspects of fuel and energy'. Highly regarded. *Class No:* 662.6

[5998]
ROSE, J.W. *and* COOPER, J.R. **Technical data on fuel.** 7th ed., completely revised in SI units. Edinburgh, Scottish Academic Press, for British National Committee, World Energy Conference, London, 1977. xi, 343p. diagrs., graphs, tables. ISBN: 0707301297.

First published 1928; 6th ed. 1961.

7 parts: 1. Quantitites, units and conversion factors - 2. Selected principles and methods of technical calculations - 3. Selected information on measurements and measuring techniques - 4. Properties of substances - 5. Fuels (flammability, gaseous fuels, liquid fuels, including petroleum products, solid fuels) - 6. Atomic energy and nuclear fuels - 7. Electroheat. Subsection references. 160 tables. Analytical index, p.333-43. Well produced; authoritative; practical. 'This 7th ed. omits certain earlier entries, and the 6th ed. is worth retaining' (*Fuel,* October 1977, p.463). *Class No:* 662.6

[5999]
Thermal analysis abstracts, with occasional reviews. Chichester, etc., Wiley, 1972-. 6pa. ISSN: 03060438.

Produced under the auspices of the International Confederation for Thermal Analysis.

Formerly *Thermal analysis review* (1962-71).

1986: 2,012 abstracts, covering all thermoanalytical methods, calorimetry, thermodynamics, and kinetics of the solid state for every type of material. Keyword and author indexes per issue. *Class No:* 662.6

Peat

[6000]
FUCHSMAN, C.H. **Peat:** industrial chemistry and technology. New York, Academic Press, 1980. xviii, 279p. illus., diagrs., tables, chemical structures. ISBN: 0122646509.

16 chapters (*e.g.* 2. Chemical characterization of peat; 6. Peat carbohydrates; 16. Chemical methods of peat analysis). References, p.248-65. Detailed, analytical index, p.265-79. *Class No:* 662.641

[6001]
Peat abstracts. [Aithgiorrachtaí móna.] Droichead Nua, Co. Kildare, Eire, Mord na Mona Experimental Station, 1955-. 4pa.

About 100 abstracts and references per issue. 5 sections: 1. Classification, chemistry and physics of peat - 2. Production - 3. Utilization - 4. Reclamation of peatland - 5. General. Includes patents. No indexes. Computer printout from Commonwealth Agricultural Bureau, Farnham Royal. International in scope. *Class No:* 662.641

Coal

Bibliographies

[6002]
BLOCH, C.C. Coal information sources and databases. Park Ridge, N.J., Noyes, 1981. xvi, 128p. $24. ISBN: 0815508304.

Sections: General sources - Libraries - Publications and information - Coal policy, budget and procurement information - Coal related legislative committees and advisory committees - legal aid regulatory data - coal research and technology ... Principal state coal organizations. Detailed index. Confined to US sources.
Class No: 662.66(01)

[6003]
COX, A.W. 'Solid fuels'. *Information sources in energy technology* (Editor, L.J. Anthony, 1988), p.173-99.

Introduction - Organizational sources - Documentary sources (journals; statistical sources; conferences; abstracting and indexing services; bibliographies; reference books, monographs and reports) - Specialized topics (opencast mining; combustion of coal; coal gasification; liquefaction of coal) - International coal trade. 2½p. of references. *Class No:* 662.66(01)

Handbooks & Manuals

[6004]
HAWLEY, M.E., *ed.* **Coal.** Stroudsburg, Pa., Dowden, Hutchinson & Ross, Inc. 1976. 2v. illus., facsims., charts, tables, maps.

1. *Social, economic and environment aspects,* xii, 384p.
2. *Scientific and technical aspects.* xiii, [1], 420, [1]p.

V.1 has 4 sections: 1. Environmental influences on the coal industry - 2. Safety and health in the coal industry - 3. Working conditions and organized labor - 4. Coal and the environment. V.2 has 7 sections: 1. Origins, classification and discovery ... 3. Mining ... 7. Coal conversion and underground gasification. Section bibliographies. Bibliography, p.389-402. Author citations and subject indexes. *Class No:* 662.66(035)

[6005]
MEYERS, R.A., *ed.* **Coal handbook.** New York, Dekker, 1981. xii, [1], 854p. ISBN: 0824712706.

20 contributors. 11 chapters on coal science and technology (1. Coal structure ... 4. Surface mining of coal ... 7. Coal utilization - 8. Underground mining ... 10. Coal gasification - 11. Liquid fuels from coal (83 references). Glossary, p.817-37. Detailed, analytical index. Intended for a wide clientele. US slanted (20 contributors: 19 US; 1 Canadian). *Class No:* 662.66(035)

[6006]
SCHOBERT, H.H. Coal, the energy source of the past and future. Washington, American Chemical Society, 1987. 298p. $19.95. ISBN: 0841211728.

Sections: How coal was formed - Coal in the world - Chemistry and physics of coal - Heat and power from coal - Carbonizing coal - Chemicals from coal - A look at the future. Index. 'Mainly for public libraries and personal purchase' (*New technical books,* March/April, 1988, p.988). *Class No:* 662.66(035)

[6007]
VOLBORTH, A., *ed.* **Coal science and technology.** Amsterdam, etc., Elsevier, 1986. 490p. DFl.250. ISBN: 0444427023.

Part 1: 7 review articles (*e.g.* 'Basic and applied aspects of the magnetic susceptibility of coal'). Subject index,. - Part 2: 12 reprint articles, documented, on environmental aspects of coal fly ash; the depolymerization of coal; new approaches to the characterization of lignites; etc.
Class No: 662.66(035)

Dictionaries

[6008]
MERRITT, R.D., *comp. and ed.* **Dictionary of coal science and technology.** Park Ridge, N.J. Noyes, 1987. 384p. illus., tables. $46.90. ISBN: 0815511248.

About 3,000 entries (up to c.12 lines in length), citing sources. References (A-Z authors), p.378-88. Comprehensive, covering all aspects of coal, including geology, development, mining and reclamation.
Class No: 662.66(038)

Reviews & Abstracts

[6009]
Coal abstracts. London, IEA [International Energy Agency] / Coal Research. Technical Information Service, 1977-. Monthly. ISSN: 83094879.

1985: 10,853 abstracts. 12 sections: Coal industry - Reserves and exploration - Mining - Preparation - Transport and handling - Properties - Combustion - Waste management - Environmental aspects - Products - Health and safety. Author and cross-reference index. Annual author and subject indexes.

COAL DATABASE (IEA COAL), 1978-.
Class No: 662.66(048)

[6010]
Euro abstracts. Section 2: Coal and steel (research programmes and research agreements, scientific and technical publications, patents). Luxembourg, European Communities Commission, 1975-. Monthly. ISBN: 9290290889.

Formerly part of *Euro abstracts: scientific and technical publications and patents,* 1970-74.

Fairly lengthy abstracts per issue, with corporate author index and list of abbreviations. Annual index.
Class No: 662.66(048)

Worldwide

[6011]
1991 International Conference on coal science. Proceedings, 16-20 September, University of Newcastle-upon-Tyne. IEA Coal Research, *eds.* London, Butterworth-Heinemann, 1991. xix,1036p. diagrs., graphs, formulas.

Sponsored by member countries of the IEA. About 250 documented papers. Author index. *Class No:* 662.66(100)

[6012]

Coal statistics international: monthly inventory of global coal production, price and market problems. Compiled by the editors of *Coal week international.* New York, etc. McGraw-Hill, 1981-. Monthly. tables.

56 tables: United States - Canada - Australia - European Community - South Africa - Japan - Columbia - Western Germany. The December 1988 issue gives monthly figures to August 1988. *Class No:* 662.66(100)

[6013]

INTERNATIONAL ENERGY AGENCY. Coal information, 1987. Paris, the Agency, 1987. Annual. 518p. tables. £42.50.

First published 1985.

Part 1: Summary of world coal developments - Part 2: OECD coal data and projections. Data on the current world market; production figures; trends and prospects; trade; demand; prices; transport; coal-fired power stations. *Class No:* 662.66(100)

[6014]

INTERNATIONAL ENERGY AGENCY. Concise guide to world coalfields. Compiled by World Coal Resources and Reserves, IEA Coal Research. London, the Agency. Loose-leaf. graphs, maps. ISBN: 9290290889.

4 sections: 1. Contents and introduction - 2. Major coal-bearing countries (A-Z) - 3. Indexes, classifications and remaining coal-bearing country descriptions - 4. Glossary of terms. Appendix. *Class No:* 662.66(100)

[6015]

JAMES, P. The Future of coal. 2nd ed. London, Macmillan, 1984. xxiv, 275p. diagrs., tables, maps. ISBN: 0333365674.

First published 1982.

9 chapters: 1. Coal: origins and exploitation - 2. Coal use: technology and markets - 3. Coal: environment and health - 4. The USA: Saudi Arabia of coal - 5. Red coal (E. Europe) - 6. The world coal trade - 7. Western Europe - 8. The Third World - 9. The future of coal. Postscript. Chapter notes and references, p.248-68. Many maps and tables. Statistics have been updated. *Class No:* 662.66(100)

Europe

[6016]

Annual bulletin of general energy statistics for Europe [Coal]. New York, Economic Commission for Europe, 1968-. Annual. tables.

In v.19, 1986 (1988, 134p. of tables), coal statistics appear under the heading 'Solid fuels', analysed into 'Primary energy' (hard coal; brown coal and lignite; other primary solid fuels) and 'Derivative energy' (patent fuel; coke). Footnoted sources. Appended 'Definitions and general notes'. 'Europe' includes USSR, plus USA and Canada. *Class No:* 662.66(4)

Great Britain

[6017]

BENSON, J. *and* **NEVILLE, R.G. Bibliography of the British coal industry:** secondary literature, parliamentary and departmental papers. Oxford Univ. Press, for the National Coal Board, 1981. vii, 760p. £45. ISBN: 019920128x.

6,185 entries, occasionally annotated. 4 parts: 1. Secondary sources - 2. Parliamentary and departmental

....(contd.)

papers - 3. Mineral maps and plans. Guide to primary source collections (by regions). Combined author-subject index. *Class No:* 662.66(410)

[6018]

Guide to the coalfields, 1989. Redhill, Surrey, Colliery Guardian, 1988. Annual. 528p. ISBN: 086108312x.

2 parts: 1. Government departments; Inspectors; Deep and surface coal mines; Coal preparation; Associations and societies, universities and technical colleges; Museums, etc., - 2. Buyer's guide; Suppliers, A-Z; Trade marks; Classified guide to 'Who supplies what'. Small format. *Class No:* 662.66(410)

[6019]

The History of the British coal industry,. Mathias, P., *ed.* Oxford Univ. Press, 1984-87. 5v. illus., diagrs., graphs, maps. (*Commissioned by the National Coal Board.*)

1. *Before 1700.*

2. *1700-1830. The Industrial Revolution,* by M.W. Flinn and D. Stoker. 1984. 528p. £40.

3. *1830-1913. Victorian pre-eminence,* by R. Church, and others, 1986.

4. *1913-1946: The political economy of decline,* by B. Supple. 1987. 700p. £80.

5. *1946-1982. The nationalized industry,* by W. Ashworth, 1986. 732p. £45.

A detailed, documented survey, written in the light of industrial and government records, but in no sense narrow official history. *Class No:* 662.66(410)

[6020]

MAUNDER, W., *ed.* **Reviews of United Kingdom statistical sources. V.11: Coal, gas, electricity.** Oxford, etc., Pergamon Press, on behalf of the Royal Statistical Society and the Social Science Research Council, 1980. xii, 297p. tables. ISBN: 008022461x.

Coal, by D.J. Harris, p.19-98. *Gas,* by H. Nabb, p.99-199. *Electricity,* by D. Nuttall.

The *Coal* Review has 12 sections (*e.g.* 5. Statistics on output; 6. Coal sales and consumption). Appended: Quick reference list. Bibliography (60 references). 'Energy: a prologue' (p.1-18; 13 references) precedes the Reviews. Both official and unofficial British statistical sources - up to the 1970s - are analysed. *Class No:* 662.66(410)

Thesauri

[6021]

Coal thesaurus. London, IEA Coal, 1985. 188p. ISBN: 9290291176.

'Simply a structured collection of the terms actually used to index the Coal Database up to April 1985. Only one level up and/or down in the hierarchy is displayed in a world block' (*Introduction*). Quarto format. *Class No:* 662.66:025.43

Coke

[6022]

BCRA quarterly: a review of published literature in the field, covering carbonization and general material on coal technology. Chesterfield, BCRA [British Carbonisation Research Association] Scientific and Technical Services, Ltd., 1983-. Quarterly. £70pa.

Formerly *Coke review and Review of coal tar technology,*

....*(contd.)*

1952-73.

200 abstracts per issue, dealing with selected aspects of coal, coke and by-products (such as coal tar), and other manufactured solid fuels. *Class No:* 662.741

[6023]

LOISON, K. Coke: quality and production. London, Butterworth, 1989. 555p. ISBN: 040802870x.

2nd ed. of a French text on coke manufacture; intended primarily for technical personnel at coking plants. *Class No:* 662.741

Coal Tar

[6024]

Bitumen: background and applications. London, 1991. 49p.

Sections: Bitumen - Bitumen uses - Bitumen manufacturers - Bitumen test procedures - Health and safety aspects of bitumen - Optimizing bitumen performances - Bitumen developments: past, present and future. Glossary. Appended: 'Some typical applications'. *Class No:* 662.749

[6025]

Coal tar technology: a bibliography with abstracts. US Department of Commerce, National Technical Information Service. Washington, US Government Printing Office, 1979. [ii], 151p. (*NTIS/PS-79/0405.*)

One entry per page, consisting of abstract plus descriptors and identifiers. Search period covered 1964- August 1979. *Class No:* 662.749

[6026]

HOIBERG, A.J., *ed.* **Bituminous materials:** asphalts, tars and pitches. Huntington, N.Y., R.E. Krieger Pubg. Co., 1979. 3v.

Originally published 1964-66. 1. *General aspects.* 446p. 2. *Asphalts.* 716p. 3. *Coal tars and pitches.* 603p. *Class No:* 662.749

Alcohol Fuel

[6027]

CHEREMISINOFF, N.P. *and* **CHEREMISINOFF, P.N. Gasohol sourcebook:** literature survey and abstracts. Ann Arbor, Mich., Ann Arbor Science Pubns. 1981. vii, 221p.

About 1,500 annotated entries on alcohol fuels. 5 sections: 1. Biotechnology and bio-conversion technology literature - 2. Ethanol and methanol - 3. Automotive and other fuel users - 4. Production of chemical feedstocks - 5. The economics of alcohol production. Author index. *Class No:* 662.754

Gas

[6028]

Annual bulletin of general energy statistics for Europe [Gas]. New York, Economic Commission for Europe, 1968-. Annual.

In v.19, 1986 (1988. 134p. of tables), gas statistics appear under the heading 'Gaseous fuels', analysed into 'Primary energy' (natural gas; IPG produced at crude petroleum and natural gas sources) and 'Derivative energy' (gasworks gas; coke-oven; blast furnace gas; substitute natural gas; LPG , excluding that produced at crude petroleum and natural gas

....*(contd.)*

sources). Footnoted sources. Appended 'Definitions and general notes'. 'Europe' includes USSR, plus USA and Canada. *Class No:* 662.76

[6029]

Gas directory, 1988, and who's who. Tonbridge, Beds., Benn, [1987]. 264p. tables, map. £45. ISSN: 03073084.

Incorporating *Gas industry directory, Gas journal directory,* and *Who's who in the gas industry.*

British gas - British gas statistics - Who's who in the gas industry - Government departments - Organizations and associations - Overseas contacts - Who was who in the gas industry - Buyer's guide - Trade names - Applied domestic applications - Index to advertisers. *Class No:* 662.76

[6030]

WILLIAMS, T.I. A History of British gas industry. Oxford Univ. Press, 1981. xvii, [1], 304p. illus., tables, maps. ISBN: 0198581572.

5 parts (21 documented chapters): 1. The beginnings - 2. Development of the industry, up to the First World War - 3. The inter-war years - 4. The Second World Age and moves towards nationalization - 5 (chapters 13-21). Nationalization and its aftermath: the modern industry. Chapter 15, 'Natural gas', p. 139-55, has 18 references, plus 17 items of bibliography. General bibliography (23 items). Detailed, analytical index. *Class No:* 662.76

Acetylene

[6031]

MILLER, S.A. Acetylene: its properties, manufacture and uses. London, Benn, 1965-66. 2v. illus., tables.

V.1 (xxxii, 800p.) covers history of the industry, physical properties of acetylene, calcium carbide, decomposition and handling, combustion, fuel uses, etc. Fully documented. Thus, chapter 5, section 4, on partial combustion, has 104 major footnotes. V.2 (xvii, 406p.) has 6 sections (1. Hydrogenation and halogenation ... 6. Other vinylation). Each volume has 3 indexes: names, *i.e.,* authors of references); companies; subjects. *Class No:* 662.766

Furnace & Combustion Engineering

[6032]

BRITISH LIBRARY. Science Reference Library. **Automobile fuels;** the alternatives to petroleum: a guide to selected literature. London, SRL, 1974. 10p. Gratis. (*SRL guidlines.*)

More than 100 entries for books, review articles, British patents and abstracting journals on petroleum substitutes published since 1900. *Class No:* 662.9

[6033]

LAWN, C.J., *ed.* **Principles of combustion engineering for boilers.** London, Academic Press, 1987. xii,628p. illus. (some col.), diagrs., graphs, tables. (*Combustion treatise.*) ISBN: 0124390358.

'Scope ... is the design of combustion equipment for the burning of heavy oil and coal in large boilers' (*Preface,* p.viii). 14 contributors. 6 chapters (subdivided), each with extensive references. Analytical index. *Class No:* 662.9

Industrial Microbiology & Mycology

[6034]
Microbiology abstracts. Section A: Industrial & applied microbiology. Bethesda, Md., Cambridge Scientific Abstracts, 1965-. Monthly. ISSN: 0300838x.

Formerly subtitled 'Section A: Industrial microbiology', 1967-71.

About 10,000 abstracts pa. Sections: Products of microorganisms - Fermentation & related processes. Food microbiology ... Plant diseases ... Mineral microbiology ... Microbial resistance ... Antibiotics - Vaccines - Environmental pollution - Methodology - Miscellaneous. Book notices. Notification of conferences. Author and subject indexes per issue and annually. *Class No:* 663.0

Beverages

[6035]
GREEN, S. 'Beverages'. *Food science and technology.* London, Mansell, 1985. p.163-7.

Items 330-381, briefly annotated. Form headings: Bibliographies - Abstracts - Dictionaries and encyclopaedias - Directories - Statistics. *Class No:* 663.1

[6036]
NOLLING, A.W., *comp.* **Beverage literature:** a bibliography. Metuchen, N.J., Scarecrow Press, 1971. 865p.

Unannotated author list of *c.*6,000 books, with subject and title indexes. Ranges from beer and cocktails to coffee, soft drinks and wine. Based on the Hurty-Peck Library of Beverage Literature (Orange, Calif.), with location for books not in that library. Appendices include 'Standard reference works consulted' and 'Major libraries specializing in fields relating to beverages'. 'Highly recommended for the special collection' (*Library journal,* v.96(12), 15th June 1971, p.2071). *Class No:* 663.1

Wines

Bibliographies

[6037]
TUDOR, D. Wine, beer and spirits. Littleton, Colo., Libraries Unlimited, 1975. 196p. (*Spare time guide, no. 6.*)

Annotated bibliography of *c.*494 books and 41 periodicals, plus newsletters and audiovisual material, by a Canadian librarian. Annotations are concise and then critical. A 'Directories' chapter lists publishers, equipment supplies, and other sources. Prices are given. *Class No:* 663.2/.3(01)

[6038]
—GROSSMAN, H.J. Grossman's guide to wines, spirits and beers. 5th ed. New York, Scribner's, 1974. 534p.

First published 1940.

A detailed compendium designed for the 'liquor and restaurant trade'. *Class No:* 663.2/.3(01)

Encyclopaedias

[6039]
Larousse wines and vineyards of France: an encyclopaedic guide to French wine. London, Ebury Press, 1990. 638,[1]p. col.illus., coloured box data. £12. ISBN: 0852238606.

Original French ed. 1987, *Vins et vignobles de France.*

Mainly regions, A-Z. About 68 contributors. Glossary of French tasting terms; technical glossary. Index (small print), p.631-8. *Class No:* 663.2/.3(031)

[6040]
RENOUIL, Y. *and* **TRAVERSAY, P. de. Dictionnaire du vin.** Éd. refondu et augm. Bordeaux, Féret, 1962. 1374p. illus., tables, maps.

Previous ed. as *Dictionnaire-manuel du négociant en vins et spiritueux et du maître de chai d'Édouard Féret.*

11 contributors. The article on Morocco (p.668-71) has subheadings for area planted, legislation, consumption, number of people employed in the wine industry and régione, production and export. Many tables and cross-references. List of works consulted, p.1149-52; bibliographies and references, p.1353-68. 800 illus., 6 coloured maps. A comprehensive encyclopedic dictionary. *Class No:* 663.2/.3(031)

[6041]
—BESPALOFF, A. The New encyclopedia of wine. London, Century, 1990. 572p. illus., maps. £10.99. ISBN: 0712639403.

A revised and expanded ed. of *The Encyclopedia of wine* by Frank Schoonmaker, first published 1964.

Lists entries A-Z, giving pronunciation of foreign terms. 2 appendices: wine and food; statistical charts (production; consumption, etc.). No bibliography or index. 'Excellent value for money' (*Reference reviews,* 5(1), 1991, p.23). *Class No:* 663.2/.3(031)

[6042]
TURNER, B.C.A. *and* **ROYCROFT, R. The Winemaker's encyclopedia.** London, Faber & Faber, 1979. 208p. diagrs., tables. ISBN: 0571714180.

Entries A-Z. 'Fermentation': 41/3p; 'Hydrometer': nearly 5p., 3 tables,; 'Vinegar production': 1p. Over 600 entries. Cross-references. Includes an entry on 'Casks'; - omitted from Lichine and Schoonmaker (*qv*), but is inferior to both in its coverage of wine varieties and lack of precision in some definitions and explanations (*Choice,* v.17, May 1980, p.366). *Class No:* 663.2/.3(031)

Handbooks & Manuals

[6043]
1991 'Which?' wine guide. Jefford, A., *ed.* London Consumers' Association and Hodder and Stoughton, 1990. *c.*550p. £10.95. ISBN: 0340528176.

4 main sections: 1. Features - 2. What to buy - 3. Where to buy - 4. Find out more about wine (courses, clubs, etc.). Omits wines from South Africa. Index, but no bibliography. *Class No:* 663.2/.3(035)

[6044]
COATES, C. The Wines of France. London, Century, 1990. 416p. illus., facsims. (labels), tables, maps.

Preliminary articles (*e.g.* The French laws of wine); 10 chapters; 5 appendices (3. 1986 harvest breakdown; 5. Vins du pays (name/description)). Detailed index. 'Indeed almost

....(contd.)

every grower or merchant I have met in the last twenty-five years or more have contributed in some way' (*Preface*). *Class No:* 663.2/.3(035)

[6045]
JOHNSON, H. Hugh Johnson's wine companion. 3rd ed. London, Mitchell Beazley, 1991. 526p. illus., tables, maps. £25. ISBN: 085533892x.

First published 1971 as *The World atlas of wine.* 2nd ed. 1983.

Over 7,500 detailed profiles of chateaux, estates and negotiants, starting from France and Bulgaria to New Zealand and Chile. 'This is the definitive encyclopedia of the world's vineyards and winemakers' (*Good book guide,* V.A92, Annual selection, 1992). *Class No:* 663.2/.3(035)

[6046]
LICHINE, A. Alexis Lichine's New encyclopedia of wines & spirits; in collaboration with W. Fifield. 7th ed. London, Cassell, 1987. xii, 771p. charts, tables, maps. £40. ISBN: 030432163x.

First published 1967.

10 chapters: 1. History of wine - 2. History of spirits and distillation - 3. Wine and food - 4. Serving wine - 5. Value and wine (with vintage listings) - 6. Starting a cellar - 7. Wine and health - 8. The vine - 9. Wine: what it is, how it is made - 10. Spirit making. Alphabetical entries, 'Abboccato' ... 'Yugoslavia', p.59-619. Appendices include 'Comparative table of spirit strengths' and a vintage chart. Select bibliography, p.721-4. Extensive index, p.731-77. 60 maps. A fine production. *Class No:* 663.2/.3(035)

[6047]
SIMON, A. André Simon's wines of the world.
Sutcliffe, S., *ed.* 2nd ed. London, Macdonald, 1981. 639p. col. illus., maps. ISBN: 0354046314.

First published 1967.

14 contributors, plus editor. 14 sections: 1. The wines of France (p.22-149) - 2. The wines of Germany (p.150-234) - 3. The wines of Italy ... 5. Sherry ... 9. The wines of England - 10. The wines of North America ... 14. The wines of South Africa. Over 50 colour plates; 29 maps. *Class No:* 663.2/.3(035)

Dictionaries

[6048]
PRICE, V. Dictionary of wines and spirits. London, Northwood Press, 1980. 394p. diagrs., tables, maps. ISBN: 0719827442.

Entries 'Aberfeldy' - 'Zwicker', including biographies. 'Armagnac': 2/3p.; 'Champagne': 1½p. Entries for wine-growing regions' (*e.g.* Rhine: 3p.). Map of Scottish whisky distilleries, p.347.'Appropriate strength of principal drinks' (table). Bibliography, p.382-94. Effective use of bold type for cross-references. *Class No:* 663.2/.3(038)

[6049]
RUBASH, J. Master dictionary of food and wines. New York, Van Nostrand Reinhold, 1990. 972p. illus. $27.95. ISBN: 0444223468.

Sections: Food terms - Wine terms - Bibliography. Short definitions of food products and dishes worldwide. For food and wine gourmets and amateur cooks. *Class No:* 663.2/.3(038)

Beers

[6050]
CORRAN, H.S. A History of brewing. Newton Abbot, Devon, David & Charles, 1975. 303p. illus. (incl. pl.), facsims., diagrs. ISBN: 0715367358.

19 chapters: 2. Medieval brewing - technical and social - 3. The early history of German brewing and introduction of hops ... 7. Brewing techniques in eighteenth-century England ... 9. The brewing industry, 1750-1800 ... 13. Porter, ale and competition ... 19. The development of brewing technology outside Britain. Appended: The strength of beer. Glossary. References and bibliography, p.286-94 (174 items). Index, p.297-303. *Class No:* 663.4

[6051]
EUROPEAN BREWERY CONVENTION. Elsevier's dictionary of brewing in English, French, German and Dutch. 2nd ed. Amsterdan, etc., Elsevier, 1983. viii, 264p. DFl.195. ISBN: 0444421319.

First published 1961 as *Elsevier's Dictionary of barley, malting and brewing,* compiled by B.D. Hartong.

3,762 English-base terms (1961 ed.: 3,917 terms), with French, German and Dutch equivalents and indexes. Coverage of Danish, Italian and Spanish has been dropped in this 2nd ed. *Class No:* 663.4

[6052]
GLOVER, B. CAMRA dictionary of beer. Harlow, Essex, Longman Group, 1983. 146p. ISBN: 0582892619.

Entries, 'Abbaye' ... 'Zymurgy' (the art and science of brewing). Claims to list every British brewery and all the beers they currently produce. Also included is a sample of well-known or distinctive foreign beers and breweries. Alcohol percentages given. Small format. *Class No:* 663.4

[6053]
JACKSON, M. The World guide to beer. London, Mitchell Beazley, 1977. 255p. col. illus., facsims., diagrs., charts, maps. ISBN: 0855331267.

Two main sections: 1. General (*e.g.* 'How strong is beer?') - 2. The world's beer (Czechoslovakia - Germany (p.36-85) ... Belgium ... The British Isles (p.144-74) ... Africa). Detailed, analytical index, p.248-54. A fully illustrated quarto. *Class No:* 663.4

[6054]
UNITED NATIONS. Industrial Development Organization. **Information sources in the beer and wine industry.** New York, United Nations, 1977. 96p. $4. (*UNIDO Information sources, no.25.*)

634 numbered entries. 11 sections: 1. Professional, trade and research organizations, learned societies and special information services - 2. Directories - 3. Sources of statistics, marketing and other economic data - 4. Basic handbooks, textbooks and manuals - 5. Current periodicals - 6. Current abstracting and indexing periodicals - 7. Proceedings, papers and reports - 8. Specialized dictionaries and encyclopaedias - 9. Bibliographies - 10. Films, audio-visual materials and film catalogues - 11. Other potential sources of information (*e.g.* K. Training). 'Bibliographical sources used ... '. Items in sections 2, 3, 5, 6, 10, 11 are annotated. *Class No:* 663.4

Spirits

[6055]

JACKSON, M. **The World guide to whisky.** London, Dorling Kindersley, 1987. 223p. col. illus., diagrs., maps. £14.95. ISBN: 0863183372.

Sections: The water of life - Scotland (p.16-97) - Ireland - Canada - The United States - Japan. Index, p.218-223. *Class No:* 663.5

Non-Alcoholic Drinks

[6056]

UNITED NATIONS. Industrial Development Organization. **Information sources on the non-alcoholic beverages industry.** New York, United Nations, 1975. xii, 72p. $ 4. (*UNIDO Guides to information sources, no. 15.*)

507 numbered entries. 12 sections (1. Professional, trade and research organizations, learned societies and special information services - 2. Directories ... 6. Current periodicals ... 10. Bibliographies ... 12. Other potential sources of information (*e.g.* Legislation. Patents and licences)). Appended: 'Bibliographical sources used in compiling this guide'. *Class No:* 663.6/.8

Coffee, Cocoa & Tea

[6057]

ICO Library monthly entries - COFFEELINE. London, International Coffee Organization (ICO), 1973-. 6pa.

200 entries per issue, on farming, production, processing, marketing, consumption, physiological effects, etc. of coffee. Online: COFFEELINE. Host: DIALOG. *Class No:* 663.9

[6058]

SIVETZ, M. *and* DESROSIER, N.W. **Coffee technology.** Westport, Conn., AVI Publishing Co, 1979. 716p. illus., diagrs., tables, maps. ISBN: 0870552694.

5 parts (17 chapters, some documented): 1. History of coffee- 2. Green coffee technology - 3. Roast coffee technology - 4. Instant coffee technology - 5. Coffee and its influence on consumers. Useful tables, p.697-703. Detailed, analytical index, p.709-16. *Class No:* 663.9

[6059]

UNITED NATIONS. Industrial Development Organization. **Information sources on the coffee, cocoa, tea and spices industry.** New York, United Nations, 1978. xii, 74p. $4. (*UNIDO Information sources, no. 28.*)

619 numbered items. 11 sections: 1. Professional, trade and research organizations, learned societies and special information services - 2. Directories - 3. Sources of statistics, marketing and other economic data - 4. Basic handbooks, textbooks and manuals - 5. Current periodicals - 6. Current abstracting and indexing periodicals - 7. Proceedings, papers and reports - 8. Specialized dictionaries and encyclopaedias - 9. Bibliographies - 10. Films and film catalogues - 11. Other potential sources of information (H. Packaging; J. Standards and specifications). 'Bibliographical sources used ...'. Items in sections 2, 3, 5, 6, 10, 11 are annotated. *Class No:* 663.9

[6060]

WILLSON, K.C. *and* CLIFFORD, M.N. **Tea: cultivation to consumption.** London, Chapman & Hall, 1992. xviii,[2],769p. illus., diagrs., graphs, tables, chemical structures. £75. ISBN: 0412338505.

22 chapters, each with references/bibliography (*e.g.* 3. Selection and breeding of tea ... 7/8. Field operations ... 14. Production of black tea ... 19. The world trade in tea). Glossary. Detailed, analytical index, p.751-69. Authoritative and comprehensive. *Class No:* 663.9

Tobacco & Narcotics

[6061]

AKEHURST, B.C. **Tobacco.** 2nd ed. London, Longman, 1981. xii, 764p. illus., diagrs., tables. (*Tropical agriculture series.*) ISBN: 0582468175.

First published 1968.

3 parts (18 chapters): 1. The plant and its uses - 2. Culture and processing - 3. Chemical and technological matters. 8 appendices (1. Tobacco production by specified countries). Extensive bibliography, p.687-736. Detailed, analytical index, p.757-64. *Class No:* 663.97

[6062]

Tobacco abstracts: world literature on nicotiana. Raleigh, N. Carolina, Tobacco Literature Sources, 1957-. 6pa. ISSN: 00408298.

About 4,00 abstracts and citations pa. Sections: Botany - Chemical and physical properties - Cultural practices - Diseases (7 sections) - Economics, marketing, production and policy - General reports and information - Genetics and varieties - Harvesting and curing - Insects - Manufacturing technology - Molecular biology - Physiology and biochemistry - Seedling production. Subject and author indexes per issue and annually. A-Z order of sections tends to separate related material. *Class No:* 663.97

[6063]

VOGES, E., *comp. and ed.* **Tobacco encyclopedia.** Mainz, Tobacco Journal International, 1984. 468p. col. illus. ISBN: 8920615077.

Part 1: Entries, A-Z, p.13-360. Part 2: Articles (history; leaf type; sheet tobacco): Cultivation; Curing; Processing; Manufacture; Taxation; Science. 25 contributors. Chronological table (AD 1493-1983). Bibliography, p.466. List of advertisers. *Class No:* 663.97

[6064]

World tobacco industry. Redhill, Surrey, International Trade Publications. Annual. ISBN: 0861082001.

34th ed. (1986. 358p.) has country entries (Abu Dhabi ... Zimbabwe), p.9-185. International brand index. Glossary. Who's who. Applied industries. Machinery. Manufacturing supplies. Services and miscellaneous. Advertisers' index. *Class No:* 663.97

Food Industry

Databases

[6065]
Foods adlibra. Minneapolis, Mn., Foods Adlibra Publications.
Covers food products, packaging nutrition, toxicology, manufacturers, equipment supliers. Over 200,000 records; updated monthly. Available on DIALOG, file 79.
Class No: 664(003.4)

[6066]
SZE, M.C. 'Computer-based information retrieval for the food industry'. *Food technology.* v.34(6), 1980, p.64-70.
Assesses coverage and availability of the leading food database services (*e.g.* FSTA, based on *Food science and technology abstracts,* Foods ADLIBRA, and FROSTI (Leatherhead Food Research Association)).
Class No: 664(003.4)

Bibliographies

[6067]
DANKBARS, A. Source book guide to food industry information. London, London Business School Information Service, 1992. 84p. £25. ISBN: 0951440187.
Lists 149 sources, including journals, reports, associations, statistics and databases, arranged by food category. Details of suppliers. 4 indexes: titles; producers and publishers; subjects; associations. A-5 format. 'Well-produced, an excellent buy' (*Reference reviews,* V. 6,(5) 1992, p.32). *Class No:* 664(01)

[6068]
Food and nutrition bibliography. Compiled from data provided by the National Agricultural Library. US Dept. of Agriculture. 10th ed. Phoenix, Arizona Oryx, 1982. xviii, 471p. ISSN: 02782499.
An 'annotated index' of 3,566 items drawn from the National Agricultural Library's database AGRICOLA. Covers print and audiovisual material on food, human nutrition, and food service management. Sectionalized. No less than 6 indexes - for personal authors, corporate author, titles, subjects, intellectual level (graded), and media. Tardy in appearing: the 10th ed. (1982) handles only 1978/1979 items. V.11 was published in 1984. *Class No:* 664(01)

[6069]
GREEN, S. Keyguide to information sources in food science & technology. London, Mansell, 1985. viii, 231p. £25. (*Keyguide to information sources.*) ISBN: 0720117488.
Part 1: Food science and technology and their literature, - a survey of the field in 7 narrative chapters, each with appended references. Chapter 5, 'The literature of food science and technology', p.71-95 discusses forms of material (*e.g.* reports, dictionaries) - *Part 2:* Bibliography, consists of 715 briefly annotated entries (General aspects; Beverages ... Sugar sweetness, confectionary) - *Part 3:* Directory of selected organizations (by continent). Detailed index. Aimed at special librarians, lecturers and workers in the food industries, 'especially those involved in research and development' (*Preface*). *Class No:* 664(01)

[6070]
—**GREEN, S.** 'Food science'. *Information sources in agriculture and food science.* Lilley, G.P., *ed.* London, etc., Butterworth, 1981. p.477-99.
Short sections: Directories, dictionaries - Food chemistry and biochemistry ... Unconventional proteins - Congresses, conferences, symposia, etc. - Patents - Review series - Abstracting journals - Journals (listed). *Class No:* 664(01)

Encyclopaedias

[6071]
CONSIDINE, D.M. *and* **CONSIDINE, G.D.,** *eds.* **Food and food production encyclopedia.** Princeton, N.J., Van Nostrand Reinhold, 1982. xvi, 2305p. £192. ISBN: 0442216122.
An impressive tome: *c.*1,200 articles, 1.9 million words, nearly 3,000 cross-references, 1,000 illus., 578 tables and a 6,200 entry-word index. Longer articles carry bibliographies. The editors identify three stages in food production: growth, nurture, processing. One of the appendices to the A-Z sequence tabulates data on additives (78p.). 'Because of its authority, currency and comprehensive scope, it promises to be of continuing value for some time to come' (*Reference books bulletin, 1983-1984.* p.92). *Class No:* 664(031)

[6072]
COYLE, L.P. The World encyclopedia of food. New York, Facts on File, 1982. xv, 790p. illus, (incl. col.), tables. ISBN: 0871964171.
More than 4,000 entries on foods and beverages, Aardvark ... Zwirm,. Stresses history, lore and uses rather than preparation. 44p. on 'Nutritive values of the edible parts of foods'; 19p. on 'Sodium content of foods'. Many line-drawings and monochrome photos support the text, whereas the colour illus. are merely decorative. Ties in with T. Stobbart's *The Cook's encyclopedia* (1980), which 'has more detail on what to do with each item' (*Library journal,* 15 December 1982, p.2330). *Class No:* 664(031)

[6073]
Encyclopedia of food science and technology. Hui, Y.H., *ed.* New York, Wiley, 1992. 4v. (2972p.) illus., diagrs., graphs, tables, chemical structures. £385; $495. ISBN: 0471505242.
Nearly 600 contributors; 380 articles. V.4: Q-Z; index, p.2911-72. Well-documented entries, covering three areas: 1. Basic and applied science - 2. Processing technology and engineering - 3. Food laws and legislation. 'An essential title for all industry, research and academic libraries' (*Choice,* September 1992, p.76). *Class No:* 664(031)

[6074]
Food and nutrition encyclopedia. Ensminger, A.H., *and others.* New York, Pegus Press, 1983. 2v. (2415p.) illus., tables. $99. ISBN: 0941218058.
Data on over 2500 food items; 188-page food compendium table. *Class No:* 664(031)

[6075]
McGraw-Hill encyclopedia of food, agriculture and nutrition. Lapedes, D.N., *ed.* New York, etc., McGraw-Hill, 1977. vii, 732p. illus., diagrs., graphs, tables, maps. ISBN: 0070452636.
About 300 contributors. 5 feature articles precede the encyclopedia proper, 'Abicisic acid ... Yeast, industrial,

....(contd.)

p.49-694. Some of the entries are drawn from the *McGraw-Hill encyclopedia of science and technology*. 'Plant diseases', p.476-91, has 8 references. Appendix: 'Comparison of foods showing constituents of 100g. of edible portion'. Detailed analytical index, p.709-32. *Class No:* 664(031)

[6076]

—JOHNSON, A.H. *and* PETERSON, M.S. Encyclopedia of food technology. Westport, Conn., AVI Pubg. Co., 1974. xv, 979p. illus., tables.

Around 200 contributors, 275 entries (Acidulents ... Zero milk). 'Coffee', p.240-7: history, cultivation, processing green beans, roasting, physical properties chemical properties, physiological effects; illus., 8 tables, 3 references. Includes biographies. Stresses food-presentation problems. *Class No:* 664(031)

[6077]

—PETERSON, M.S. *and* JOHNSON, A.H. Encyclopedia of food science. Westport, Conn., AVI Pubg. Co., 1978. 1005p. illus. ISBN: 0870552279.

Includes documented articles averaging 2-3p. in length. Separate section of articles on the food science programs of various nations. Criticized in *Choice* (v.16, June 1979, p.501) for inadequate index and insufficient cross-references. *Class No:* 664(031)

[6078]

ROGERS, J. The Encyclopedia of food and nutrition. London, Merehurst, 1990. 480p. illus. (col.), tables. £25. ISBN: 1853911690.

300 contributors. Sections: Vegetables ... Canned and packaged foods - Drinks. Appended tables. Index. Systematic treatment (*e.g.* 'Sago': Latin name; major nutrient; nutritional information; description, origin and history; buying and storage; preparation and use; varieties). 'Designed to help consumers in their choices for healthy and pleasant eating' (*Foreword*). *Class No:* 664(031)

Handbooks & Manuals

[6079]

CHARLEY, H. Food science. 2nd ed. New York, etc., Wiley 1982. [vii] 564p. illus., tables. ISBN: 0471062065.

First published 1971.

29 documented chapters (1. Evaluation of foods ... 6. Sugar, sugar crystals, and confections ... 9. Cereals - 10. Fats and oils ... 17. Milk ... 19. Eggs ... 22. Meat (p.372-432; 121 references, plus 5 citations of films and slides) ... 28. Vegetables (p.487-528; 72 references). Detailed, analytical indexes. Aims at 'the presentation of scientific concepts as a basis for understanding foods as complex chemical systems' (*Preface*). *Class No:* 664(035)

[6080]

Handbook of food engineering. Heldman, D.R. *and* Lund, D.B., *eds.* New York, Marcel Dekker, 1992. viii,756p. illus., diagrs., graphs, tables. ISBN: 0824784634.

19 contributors; 14 documented chapters (*e.g.* 'Reaction kinetics in food systems' (p.39-144) carries 12p. of references). Detailed, analytical index, p.741-56. *Class No:* 664(035)

[6081]

JONES, J.M. Food safety. Saint Paul, Mn., Eagan, 1992. 453p. tables. $42. ISBN: 0962440736.

14 chapters, well-documented, covering agencies responsible for regulation, consumer concerns, natural toxicants, molds and mycotoxins, environmental and incidental contaminants, and additives. Glossary; numerous tables. Detailed, analytical index. *Class No:* 664(035)

[6082]

McGEE, H. On food and cooking: the science and lore of the kitchen. 3rd ed. London, Allen & Unwin, 1984. xix, 684p. illus., tables. £12.95. ISBN: 004306003x.

3 parts (14 chapters): 1. Foods (1. Milk and dairy products; 2. Eggs; 3. Meat ... 6. Bread, doughs and batters; 7. Sauces; 8. Sugar, chocolate and confectionary; 9. Wine, beer and distilled liquors; 10. Food additives) - 2. Food and the body (11. Nutrition; 12. Digestion and sensation) - 3. The principles of cooking: a summary. Appendix: A chemistry primer, - atoms, molecules, energy. Bibliography (by chapters), p.648-52. Detailed analytical index, p.653-84. Over 200 illus., included photographs taken through the electron microscope. *Class No:* 664(035)

[6083]

RANKIN, M.D., *ed*. Food industries manual. 22nd ed. Glasgow & London, Blackie, 1988. Published under the authority of BFMIRA (Leatherhead Food Research Association). 599p. diagrs., graphs, tables, chemical structures. ISBN: 0216924723.

21st ed. 1984.

21 contributors. 17 chapters: 1. Meat and meat products - 2. Fish and fish products - 3. Dairy products - 4. Flour and baked products - 5. Fats and fatty foods - 6. Hot beverages: coffee, tea, cocoa and others ... 11. Snack foods - 12. Nutrition - 13. Freezing and refrigeration (27 references) - 14. Dehydration and dried products - 15. Heat preservation - 16. Handling and storage - 17. Quality assurance and control operations. Index, p.587-99. Well organized. A standard compendium on food, processing and manufacturing techniques. *Class No:* 664(035)

Dictionaries

[6084]

ADRIAN, J., *and others*. Dictionary of food and nutrition. Translated by B. Weitz. Weinheim, VCH, and Chichester, Ellis, Howood, 1988. [viii], 231p. facsims., graphs, tables, chemical structures. £29.45. ISBN: 0895734044.

Based on the orginal French ed., *Dictionnaire du biochimie alimentaire et de nutrition*. (Paris, Technique et Documentation, 1981). At head of title-page: 'Parat'.

Defines over 2,000 terms relating to biochemical aspects of food, nutrition and food compositions 'Abomasim' ... 'Zinc'. 'Oxalic acid': 1p. (table). Appendices: Composition of food (10 tables); Formulas. *Class No:* 664(038)

[6085]

ALLISON, S. The Cassell food dictionary. London, Cassell, 1990. xxxv,480p. figs. £19.95. ISBN: 0304318752.

About 6,000 entries on food, ingredients and culinary techniques. Cross-references appear in an index. Thematic index. 'Excellent' (*Reference reviews*, V. 5(2) 1991, p.25). *Class No:* 664(038)

[6086]
BENDER, A.E. **Dictionary of nutrition and food technology.** 6th ed. London, Butterworths, 1990. [vi],336p. tables. £29.50. ISBN: 0408037539.
First published 1960; 5th ed. 1982.
About 4,000 entries, with some 300 revisions and 300 new terms for this edition. Appended grouped bibliography of more than 300 standard works. A well-known and valuable food dictionary. *Class No:* 664(038)

[6087]
Dictionnaire agro-alimentaire: français/anglais, anglais/français. [Dictionary of food science and industry: French/English, English/French.] Adrian, J., *and others.* Paris, Technique et Documentation, 1990. xx,346p. £44.10. ISBN: 2852665762.
About 6,000 entries in each half. 'Fermentation': ½ column. Mostly equivalents. French genders given.
Class No: 664(038)

[6088]
LÜCK, E.C. **Bibliography of dictionaries and vocabularies on food, nutrition and cookery.** Frankfurt am Main, IFIS, 1985. 139p. DM60. ISBN: 3922901046.
436 annotated entries. Author and subject indexes.
Class No: 664(038)

[6089]
LÜCK, E.C. **The Compact dictionary of food technology,** English-German, Deutsch-English. Hamburg, Behr's Verlag, 1985. 443, [1]p.
Earlier ed. of the English-German part (Wiesbaden, Brandstetter) appeared in 1963.
English-German part (p.1-213) contains *c.*4,000 entries, plus appendix of weights and conversion tables. German-English part (p.217-443, [1]p.) also carries an appendix.
Class No: 664(038)

Reviews & Abstracts

[6090]
Food and nutrition quarterly index. Phoenix, Arizona, Oryx Press, 1985-. Quarterly. $95pa. ISSN: 07321171.
Replaces *Food and nutrition bibliography* (10th ed. Oryx, 1982).
Oryx Press also publishes the more comprehensive monthly *Bibliography of agriculture.*
Indexes and abstracts both print and non-print material. Entries (*c.*2,500pa.) carry descriptors. Indexes: subject, title, personal author, corporate author, intellectual level, and media. Draws on resources of AGRICOLA database.
Class No: 664(048)

[6091]
Food books review: the international journal for readers of books on food technology subjects. Orpington, Kent, Food Trade Press, 1978-. 3pa. £7pa. ISSN: 01422545.
No. 20, Summer 1988, carries 31 reviews (p.2-10). List of further titles, with prices, p.11-31. *Class No:* 664(048)

[6092]
Food science and technology abstracts. Wallingford, Oxon., CAB International: International Food Information Service (IFIS), 1969-. Monthly. $960pa. ISSN: 00156574.
Over 2,000 abstracts per issue. Over 1,200 journals searched. Sections A-V (A. Basic food science ... V. Patents). Author and subject indexes per issue and annually.

....(contd.)
The main food science and technology abstracting service since 1969. Online access: DIALOG, ESA-IRIS. CD-ROM (Silver Platter). *Class No:* 664(048)

[6093]
Leatherhead Food Research Association. Scientific and technology abstracts. Leatherhead, Surrey, Leatherhead Food RA Information Group. ISSN: 09501789.
Formerly British Food Manufacturing Industries Research Association *Leatherhead Research Association abstracts.*
1988: 4,593 abstracts. 23 sections (*e.g.* Analytical - Biochemical and chemical - Microbiological ... Food processing and biotechnology - Packaging and packaging materials - Plant and equipment - Storage and transport - Legal). Appended cross-references. Online: FROSTI database. *Class No:* 664(048)

Progress Reports

[6094]
Advances in food research. Orlando, Florida, Academic Press,. 1948- v.1-, Annual. illus., diagrs., tables.
V.29 (1984. ix, 319p. ISBN 012064299) has 8 contributors, 5 documented articles (*e.g.* 'Baking qualities of wheat flour', p.202-77. 7p. of references). Occasional supplements (*e.g.* Supplement 4: *Spices and condiments: chemistry, microbiology, technology.* 1980. xv, 449p.) A major source of reviews for food scientists and technologists.
Cumulative index to v.1-10 in v.11 (1962).
Class No: 664(055)

[6095]
World review of nutrition and dietetics. Basle & London, Karger; Wiley, 1959-. v.1-. Irregular.
Each volume carries 4-6 documented articles. Usually devoted to particular themes (*e.g.* v.52: *Energy Nutrition of women;* edited by G.H. Bowne. 1987. 276p.), with subject index and list of contents of previous few volumes. 'Reviews are necessarily coloured by the views of the authors, but most of them, though not all, succeed in providing a critical survey of the field. They also serve as valuable platforms to promulgate ideas' (*Chemistry & industry,* 6 August 1977, p.659). Aims to provide a forum for critical reviews such as cannot be published in the ordinary journals, nor in textbooks. Highly priced (*e.g.* v.52: £101.25).
Class No: 664(055)

Yearbooks & Directories

[6096]
Food trades directory and food buyer's year book, 1987-88, incorporating *Food processing industry directory.* 30th year. London, Newman Books, Ltd., 1987. xii, 952 + 214p. ISBN: 0707909409.
Part 1: Sources (food suppliers; food products & ingredients index; brand names index; major food groups) - British food authorities - Producers overseas - Outlets. *Part 2* (214p.): Plant, equipment and packaging machinery - Hygiene - Packaging materials - UK suppliers of European food processing and packing plant. Services (including 'Books on food'), p.921-43. Index to advertisers.
Class No: 664(058)

[6097]
GEAR, A. **The New organic food guide;** including a list of 600 outlets throughout Britain and Ireland. London, Dent, 1987. xvii, 232p. £3.95. ISBN: 046002454x.

3 parts (10 chapters): 1. Modern farming: an indictment- 2. Organic farming: the biological alternative - 3. Where to buy organic food (p.59-197; chapters 8-10: 8. Organically grown foods to be found in the shops; 9. A guide to outlets (p.100-89: The South West ... The Channel Islands); 10. A guide to wholesalers). 4 appendices (2. Organizations; 4. References (64)). Detailed, analytical index, p.225-32.
Class No: 664(058)

Tables & Data Books

[6098]
BOWES, A.D. *and* CHURCH, C.F., *comps*. **Bowes and Church's food values of portions commonly used.** Philadelphia, Lippincott, 1989. 328p. tables. $18.50; £14.50. ISBN: 0397547277.

Rev by J.A.T. Pennington.

Lists both generic and brand names. Includes values for proteins, cholesterol, fats, and vitamins. Tables for biotins, amino acids, caffeine, salicylates, etc. Bibliography of additional sources for food composition data. Good index. 'Possibly the most complete book of this kind' (*Science & technology libraries*, v.10(2), Winter 1989, p.134).
Class No: 664(083)

[6099]
OCKERMAN, H.W. **Source book for food scientists.** Westport, Conn., AVI Publishing Co., Inc., 1978. x, 926p. illus., graphs, tables, chemical structures.

Part 1: Source book: terms and definitions, A ... Zymogen (p.1-308). About 10,000 concise meanings; cross-references - Part 2: Food composition, properties and general data, chiefly tabular (*e.g.* Fruit: growing season, storage life; Microwave cooking; Unsaturated fatty acids). A valuable compendium. *Class No:* 664(083)

[6100]
—SCARPA, I.S., *and others, eds.* Source book in food and nutrition. 3rd ed. Chicago, Ill., Marquis Academic Media, 1982. [x], 549p. diagrs., charts, graphs, tables.

First published 1978.

In 4 parts (87 contributions): 1. Dietary directions in the 1980s - 2. Nutrition from conception through adolescence - 3. Childhood into the golden years - 4. Resources for further information (*e.g.* US libraries specializing in agriculture, food and nutrition values). Detailed, analytical index.
Class No: 664(083)

Audio-Visual Materials

[6101]
Audio-visual resources in food and nutrition. Phoenix, Arizona, Oryx Press, 1979. viii, 232p.

The *c.*1,200 entries include title, author, publisher, place and date, followed by brief abstract and descriptions. Indexes: personal author, corporate author, title, media, and subject. Compiled data provided by the US National Agricultural Library. *Class No:* 664(086)

Histories

[6102]
DARBY, W.J., *and others*. **Food: the gift of Osiris.** Orlando, Florida, Academic Press, 1977. 2v. (452p.+425p.), illus.

'Deals with the use of food and the acceptance or rejection of its varieties for cultural or religious reasons, over a very long period in the Near East, mainly in Egypt' *(British medical journal*, 7 May 1977, p.1210). Lavishly illustrated. *Class No:* 664(091)

USA

[6103]
MARIANI, J.F. **The Dictionary of American food & drink.** New York, Ticknor & Fields (distributed by Houghton), 1983. 475p. $11.95. ISBN: 0899193595.

A compendium that lays emphasis on the history and development of various foods, some of the entries being miniature essays. Word derivations are traced. Well documented. Index. 'Anecdotal as well as authorative' (*Library journal*, 15 April 1984, p.782). *Class No:* 664(73)

Thesauri

[6104]
EUROPEAN COMMUNITIES COMMISSION. **Food multilingual thesaurus.** Munich, etc., K.G. Saur, 1979. 5v. DM390, the set.

1. *Deutsche Ausgabe*. xx, 129p.
2. *English edition*. xx, 145p.
3. *Version française*. xx, 144p.
4. *Edizioni italiana*. xx, 132p.
5. *Quadrilingual index and microfiches*. viii, 168p. 24 microfiches.

The English edition (DM154. ISBN 3598101058) has alphabetically arranged sections: Compositions - Countries - Food science - Food technology - Food, including sciences and by-products - Health and environment - Industries, economics, agriculture-Methodology - Microorganisms - Properties and quality. *Class No:* 664:025.43

Information Services

[6105]
Directory of food and nutrition information services and resources. Frank, R.C., *ed.* Phoenix, Arizona, Oryx Press, 1985. vi, 288p. tables. $74.50. ISBN: 0897740785.

A directory of 627 organizations, 61 databases, 102 microcomputer software programmes, over 180 journals and newsletters, 19 abstracting indexing and curent-awareness services, more than 400 publishers of books, audio visual material and microcomputer software, and over 140 key reference books. Subject, organization and geographical indexes. 'For food and nutrition educators and health professionals, ... as well as a valuable reference for organizations, institutions and libraries' (*Food technology*, v.39, July 1985, p.100). *Class No:* 664:061:025.5

[6106]
INTERNATIONAL FOOD INFORMATION SERVICE. **Proceedings of the 2nd International Conference on Food Science and Technology Information.** Schützsack, U., *ed.* Berlin, May 13-22, 1987. 283p. facsims., tables.

31 papers, some with appended references (*e.g.* 'Processed food: the need for information', p.109-16. 13

....*(contd.)*

references). 'BIOSIS: its coverage of the food sciences literature', p.49-[61], has 6 tables. List of authors. List of participants. *Class No:* 664:061:025.5

Food Engineering

Encyclopaedias

[6107]

HALL, C.W., *and others*. **Encyclopedia of food engineering.** 2nd ed. Westpoint, Conn., AVI Pubg. Co., 1986. x, [1], 882p. illus., diagrs., graphs, tables. ISBN: 0870551574.

First published 1971.

About 75 contributors. Entries (Absolute pressure controller ... Zinc) can be lengthy, *e.g.* 'Energy conservation in food processing', p.293-320, with 29 references. This volume, first of a 3v. set, emphasizes equipment and facilities used in food handling and transportation. *Class No:* 664.0(031)

Handbooks & Manuals

[6108]

BENDER, A.E. **Food processing and nutrition.** London, etc., Academic Press, 1978. [v], 243p. ISBN: 0120864509.

5 sections: A. Principles - B. Effects on nutrients - C. Effects of processes - D. Commodities - E. Addition of nutrients. 937 references, plus cross-references. Partly analytical index. *Class No:* 664.0(035)

[6109]

EGAN, H., *and others*. **Pearson's chemical analysis of foods.** 9th ed. Harlow, Essex, Longman, 1991. 624p. illus. £49. ISBN: 0582409101.

Extensive chapter references in this standard work. Appendices on UK food regulations. Profiles EC legislation for principal foods, the international codex, standards, developments in NIR spectroscopy, etc. *Class No:* 664.0(035)

[6110]

Engineering and Food. Speiss, W.E.L. *and* Schubert, H., *eds.* London, Elsevier. 3v. *c.*2,700p. illus., tables. £283 (set). ISBN: 1851664475, Set.

V.1. *Physical properties and process control.*

V.2. *Preservation processes and related techniques.*

V.3. *Advanced processes.*

About 100 papers from the 5th International Congress on Engineering and Food, Cologne, 28 May - 3 June 1989. 147 tables. Subject index. *Class No:* 664.0(035)

[6111]

FRAZIER, W.C. *and* WESTHOFF, D. **Food microbiology.** 4th ed. New York, etc., McGraw-Hill, 1988. 576p.

3rd ed., 1978.

7 sections: Food and microorganisms - Principal of food preservation - Contamination, preservation and spoilage of different kinds of foods - Foods and enzymes produced by microorganisms - Foods in relation to disease - Food sanitation - Control and inspection. Index. A classic text. *Class No:* 664.0(035)

[6112]

Handbook of cereal science and technology. Lorenz, K.J. *and* Kulp, K., *eds.* New York, Dekker, 1991. viii,882p. illus., diagrs., graphs, tables. $185. ISBN: 0824783581.

30 contributors. 24 documented chapters (1. Wheat ... 24. Ethanol production from cereal grains). Detailed, analytical index. For the specialist. *Class No:* 664.0(035)

[6113]

POMERANZ, Y. **Functional properties of food components.** 2nd ed. San Diego, Ca., Academic Press, 1991. ix,569p. illus., diagrs., graphs, tables. £60. (*Food science and technology.*) ISBN: 0125612828.

First published 1985.

12 chapters, each with extensive references, in 3 parts: 1. Components (water; carbohydrates; proteins, etc.) - 2. Engineering foods (additives; new and novel foods, etc.) - 3. Information and documentation. Bibliography, p.533-60. Detailed, analytical index, p.561-69. *Class No:* 664.0(035)

[6114]

Principles and practices for the safe processing of food. Shapton, D.A. *and* Shapton, N.F., *eds.* Oxford, Butterworth Heinemann, 1991. xiv,457p. illus., diagrs., graphs, tables. ISBN: 0750611219.

Based on procedures used in the processing plants of H.J. Heinz. 15 chapters covering such areas as buildings, sanitation, personnel, microorganisms, spoilage. Extensive data tables cover a range of processed foods. Detailed, analytical index. *Class No:* 664.0(035)

Dictionaries

[6115]

GOULD, W.A. **Glossary for the food industries.** Baltimore, Md., CTI Publications, 1990. ix,118p. $12.95. ISBN: 0930027167.

About 1,000 terms, briefly defined (*e.g.* 'Farina': 2 lines). Abbreviations/acronyms, p.1-8. 20 tables (16: 'Standard screen sizes'). *Class No:* 664.0(038)

Tables & Data Books

[6116]

McCANCE, R.A. *and* WIDDOWSON, E.M. **The Composition of foods.** 5th rev. and extended ed. London, Royal Society of Chemistry; Ministry of Agriculture, Fisheries and Food, 1991. xiii,462p. tables. £29.95. ISBN: 0851863914.

First published 1940; 4th ed. 1978.

Documented tables cover *c.*1,200 items, including such popular commodities as fish fingers, yogurt, oven chips and coca-cola. 21p. index. Supplements to this edition will cover fruit and nuts, vegetable dishes, fish and fish products and fatty acid composition of foods. A well-known compendium of data on food values.

1st supp., *Fruit and nuts.* (1992. 144p. £24.50); 2nd supp., *Vegetable dishes*, (1992. 200p. £24.50). *Class No:* 664.0(083)

[6117]

—Food labelling data for manufacturers: based on McCance and Widdowson's *The Composition of foods.* Welch, L.A., *and others, comps.* London, Royal Society of Chemistry and Ministry of Agriculture, Fisheries and Food, 1992. [vi],214p. tables. £27.50. ISBN: 0851869939.

Data for food labels, following requirements of EC Directive (1990). *Class No:* 664.0(083)

Sugar Industry

[6118]

CHABALLE, L.Y. **Elsevier's sugar dictionary,** in English/ American, French, Spanish, Dutch, German. Amsterdam, etc., Elsevier, 1984. xiv, 322p.

2,741 English/American-based terms with French, Spanish, Dutch, and German equivalents and indexes. *Class No:* 664.1

[6119]

MARIE, S. *and* PIGGOTT, J.R., *eds.* **Handbook of sweeteners.** Glasgow and London, Blackie, 1991. xiv,302p. diagrs., graphs, tables. £65. ISBN: 0216928362.

18 contributors. 11 well-documented chapters. Full contents list. Index, p.295-302. *Class No:* 664.1

[6120]

SCHALIT, M., *comp.* **Guide to the literature of the sugar industry:** an annotated bibligraphical guide to the literature on sugar and its manufacture from beet and cane. Amsterdam, etc., Elsevier, 1970. [xi], 172p.

About 1,000 annotated entries. 11 chapters: 1. Bibliographies, abstracts and reviews, indexes and patents - 2. Government publications - 3. Dictionaries, directories and yearbooks ... 8. Publications on technology and sugar chemistry - 9. Handbooks and tables - 10. Translations. 3 appendices (1. Bibliography of American doctoral dissertations on sucrose, sugarbeet and sugarcane since 1938; 3. Bibliography of papers presented at annual technical conferences of British Sugar Ltd., 1948-1966). Names and title indexes. *Class No:* 664.1

[6121]

Sugar year book. London, International Sugar Organization, 1947-. Annual. tables.

Statistical tables covering 126 countries, A-Z, are followed by world tables (production, imports, exports, consumption, prices; stocks in selected countries; equivalent weights). Statistics (usually for 7 year span) relate to centrifugal sugar only. Small format. *Class No:* 664.1

[6122]

—F.O. Licht's Internationales zuckerwirtschaftliches Jahr- und Adressbuch... Weltzucker statistik. Ratzeburg, F.O. Licht. Annual.

The standard yearbook for the world's sugar economy, both sugar-beet and sugar-cane. It covers organization, production, imports, exports, plants and refineries, gives statistics, lists sources and includes an English-German glossary, plus an atlas of sugar factories. *Class No:* 664.1

[6123]

Tate & Lyle's Sugar industry abstracts. Reading, Tate & Lyle plc., 1981-. 6pa. ISSN: 02502887.

Formerly *Sugar industry abstracts,* 1948-80.

1987: 1,744 abstracts. 10 sections: 1. Cane sugar manufacture - 2. Beet sugar manufacture - 3. Sugar refining ... 7. By-products - 8. Properties and uses of sugar - 9. Nutrition and health aspects - 10. Miscellaneous. Forthcoming conferences on sugar and related topics. Annual author and subject indexes, with list of journals monitored and patent index. *Class No:* 664.1

Sweets

[6124]

CAOBISCO. **Fünfsprachiges Süsswaren-Fachwörtesbuch.** [Five language dictionary of the confectionery industry.] Hamburg, Benecke, 1964. 429p.

About 2,000 German-based terms, with equivalents and indexes in English, French, Italian and Dutch. *Class No:* 664.14

Starch

[6125]

WHISTLER, R.L. **Starch: chemistry and technology.** 2nd ed., edited by H.L. Wilcke, and others. Orlando, Florida, Academic Press, 1984. xxiii, 718p. illus., diagrs., tables, chemical structures. ISBN: 0127462708.

First published 1965-67 (2v. xviii, 579p; xviii, 733p.).

24 contributors (mostly American). 23 chapters, well documented (*e.g.* 6. Molecular structure of starch, p.153-83, 201 references). Covers both fundamental and industrial aspects. Detailed, analytical index, p.691-718. *Class No:* 664.2

Baking & Flours

[6126]

DANIEL, A.R., *comp.* **The Baker's dictionary.** 2nd ed. Amsterdam, etc., Elsevier, 1971. [vi], 264p.

First published 1949 (London, McLaren).

About 5,000 entries and cross-references. 'Over': c.80 lines. Some chemical formulas. Bibliography, p.253 (no dates or publishers' names). *Class No:* 664.6

[6127]

MATZ, S.A. **The Chemistry and technology of cereals as food and feed.** 2nd ed. New York, Van Nostrand Reinhold, 1991. xix,751p. illus., diagrs., tables. £72. ISBN: 0442308302.

First published 1959.

21 chapters, closely subdivided, each with extensive bibliography. Covers the basic grains (*e.g.* wheat, corn, oats, barley, rye, rice) and the milling, processing and products (including 'popcorn' under snack foods). Detailed, analytical index, p.743-51. *Class No:* 664.6

[6128]

POMERANZ, Y. **Modern cereal science and technology.** New York, VCH Publications, 1987. 486p. $89.975.

Sections: Physical properties and structure - Wheat flour components in bread making ... Oats, sorghum and millets, and rye-extrusion products - Industrial uses of cereals.

....*(contd.)*

Index. 'For students, researchers, novices and experts in the area' (*New technical books,* May/June 1988, item 1375). *Class No:* 664.6

[6129]
SCHNEEWEISS, R. Dictionary of cereal processing and cereal chemistry, in English, German, French, Latin and Russian. Amsterdam, etc., Elsevier, 1982. xii, 520p. DF1.315. ISBN: 0444420485.

6,932 English-base terms, with German, French, Latin and Russian equivalents and indexes. Symbols indicate out-of-date, little used terms or terms often used in a sense that is not recommended. *Class No:* 664.6

Milling

Flour Mills

[6130]
Flour Milling and Baking Research Association abstracts. Chorleywood, Rickmansworth, Herts., the Association, 1967-. 6pa.

Formerly *Baking abstracts,* 1948-66.

About 1,500 abstracts pa. Sections: cereals and cereal products; bakery raw materials; bakery processes and products; laboratory techniques; nutrition; microbiology; plant; equipment; buildings; packaging; administration; legislation. Annual author and subject indexes. International. Online database. *Class No:* 664.71

Preservation & Treatments

[6131]
CAMPBELL-PLATT, G. Fermented foods of the world: a dictionary and guide. Sevenoaks Kent, Butterworth, 1987. xxiii, 91p. map. ISBN: 0407063134.

Entries: Acidophilus milk ... Yoghurt. Foods by region; foods by class. 'Yoghurt' has sections: consumption; types; production; microbiology and biochemistry; composition and nutritive value. 4 references, 6 cross-references. Includes wine. Profuse cross-references. The introduction (xxiip.) includes 1½ pages of references. Well organized. Fills a gap. *Class No:* 664.8/.9

[6132]
—DESROSIER, N.W. Technology of food preservation. 4th ed. Westport, Conn., AVI Publishing Co., 1977. 588p.

A companion to the author's *Fundamentals of food freezing* (*qv*). Devotes documented chapters to food preservation processes. *Class No:* 664.8/.9

Additives

[6133]
CRC handbook of food additives. Furia, T.E., *ed.* 2nd ed. Cleveland, Ohio, CRC Press, 1972-. 998p.

First published 1968.

Part 1: 15 documented sections, p.1-789 (1. Enzymes - 2. Vitamins and amino acids - 3. Antimicrobial foods additives ... 8. Starch in the food industry ... 14. Color additives in food - 15. Phosphates in food processing). 1,005 references - Part 2: Regulatory status of direct food additives. Analytical index. US-slanted.

....*(contd.)*

V.2 (1980. 412p.) updates both text and bibliographies in V.1. 'The most important guide to food additives' (Green, S. *Food science and technology,* p.184). *Class No:* 664.8.022

[6134]
—HANSSEN, M. *and* MARSDEN, J. The New E is for additives: the complete 'E' number guide. Wellingborough, Northants, Thorsons, 1987. 384p. £3.50. ISBN: 0722515626.

First published 1984.

Elucidates the categories of 'E' numbers: E100-E180: Permitted colours; E200-E290: Permitted preservatives; E300-E321: Permitted anti-oxidants; E222-494: Emulsified, stabilized and other; E905-907: Minerals, hydrocarbons. Data: Source; function; effects; A.D.I. (Acceptable Daily Intake); typical products. Appended A-Z list of additives and their 'E' numbers. Glossary of additive terms, p.347-57. *Class No:* 664.8.022

[6135]
FENAROLI, G. Fenaroli's handbook of flavor ingredients. 2nd ed., edited and translated by T.E. Turia and N. Bellanca. Cleveland, Ohio, CRC Press, 1975. 2v. (928p.). diagrs., graphs, tables, chemical structures. ISBN: 0878195335.

First published 1971.

v.1, pt.1: *General considerations.* pt.2: *Natural flavour.*
v.2, pt.1: *Synthetic flavor.* pt.2: *Use of flavour ingredients.*

V.1, pt.2 has a list of *c.*200 natural flavour ingredients. V.2, pt.3 lists *c.*1,600 synthetic flavour ingredients. Each part has an extensive bibliography of references and a detailed index. The major source on flavours. *Class No:* 664.8.022

[6136]
—IGOE, R.S. Dictionary of food ingredients. 2nd ed. New York, Van Nostrand Reinhold, [1989]. [vi], 225p. tables. ISBN: 0442319274.

First published 1983.

In four sections: (1) a dictionary of *c.*1,000 food ingredients, their functional properties and applications; (2) classification by categories; (3) substances for use in foods; (4) bibliography of *c.*200 sources consulted, almost all pre-1984. Omits important ingredients such as lanolin and talc. *Class No:* 664.8.022

[6137]
LEWIS, R.L. Food additives handbook. New York, Van Nostrand Reinhold, 1989. xxxi,592p. ISBN: 0442205082.

An extensive list of additives alphabetically by chemical name. Data for each include identifiers (*e.g.* name, CAS, DOT, Molecular formula, etc.), properties, food specific information, occupational restrictions and toxicological data. Indexes: Purposes served in foods; Food type; CAS number; synonym. *Class No:* 664.8.022

[6138]
—SMITH, J., *ed.* Food additive user's handbook. Glasgow, Blackie; New York, Van Nostrand Reinhold, 1991. xv,286p. tables. £65. ISBN: 0216929113.

16 contributors. 15 chapters on types of additives (1. Antioxidents - 2. Sweeteners - 3. Flavourings - 4. Colours ... 8. Emulsifiers ... 11. Hydrocolloids ... 13. Flour improvers and raising agents ... 15. Chelating agents). No safety information; the user is referred to Lewis, *Food additives handbook. Class No:* 664.8.022

Canning

[6139]
UNITED NATIONS. Industrial Development Organization. **Information sources on the canning industry.** New York, United Nations, 1975. xii, 83p. (*UNIDO guide in sources of information, no. 19.*)

649 items in 12 sections: 1. Professional, trade and research organizations, learned societies and special information services - 2. Directories - 3. Sources of statistics, marketing and other economic data - 4. Basic handbooks, textbooks and manuals - 5. Monograph series - 6. Current periodicals - 7. Current abstracting and indexing journals - 8. Proceedings, papers and reports - 9. Specialized dictionaries and encyclopaedias - 10. Bibliographies - 11. Films and catalogues - 12. Other potential sources of information (E. Legislation; H. Packaging). 'Bibliographical sources used ...'. *Class No:* 664.8.036.5

Freezing

[6140]
DESROSIER, N.W. *and* TRESSLER, D.K. **Fundamentals of food freezing.** Westport, Conn., AVI Publishing Co., 1977. vii, [2], 629p. illus., graphs, tables.

39 contributors. 16 chapters (3. Freezing vegetables ... 7. Freezing fish ... 10. Freezing of egg products ... 13. Microbiology of frozen foods - 14. The nutritive value of frozen foods ... 16. Warehousing and food cabinets). Each chapter has 'Additional reading'. Index, p.616-29. US slanted. *Class No:* 664.8.037

[6141]
Frozen and chilled foods year book. Redhill, Surrey, Retail Journals, Ltd.

Suppliers of fish and fish products ... Suppliers of dairy produce - Processors - Packers - Agents and their products - Brand names - Frozen Food Federation members - Producers - Organizations - Wholesalers - Public cold stores: geographical guide - Contractors - Plant - Packaging - Machinery and equipment. Frozen food centres. Alphabetical index, p.33-91. Advertisers' index. *Class No:* 664.8.037

Meat

[6142]
GERRARD, F. *and* MALLION, F.G. **The Complete book of meat.** 2nd ed. London & Coulsdon, Virtue, 1980. xv, 624p. illus. (incl. col.), diagrs., graphs, tables. ISBN: 0900778172.

First published 1977.

23 contributors. 24 sections (*e.g.* World meat situation; The home market; The Common Agricultural Policy; Pigs: Edible offals; Meat nutrition; Meat products; Regulations affecting the meat trade). 72 plates (48 col.). Appended Acts of Parliament, Statutory Instruments and other items of reference. Comprehensive; aimed at wholesalers, retailers and their staff. *Class No:* 664.91

[6143]
KINSMAN, D.M., *ed*. **International meat science dictionary.** Storrs, Conn., University of Connecticut, 1978. 282p.

862 specialised, English/American-based terms, with equivalents and indexes in German, French, Spanish and Russian. Appended list of meat-research institutions in 32 countries. *Class No:* 664.91

[6144]
LEVIE, A. **Meat handbook.** 4th ed. Westport, Conn., AVI Publishing Co., 1979. [xi], 338p. illus., diagrs., graphs, tables. ISBN: 0870553151.

First published 1963.

18 chapters: 1. Livestock - 2. Slaughter and inspection ... 5. Refrigeration ... 12. Beef hindquarters ... 14. Veal - 15. Lamb... 17. Processed, smoked and variety meats - 18. Cooking palatability. Detailed, analytical index. The 4th ed. includes revised grading standards and new roasting techniques. *Class No:* 664.91

[6145]
UNITED NATIONS. Industrial Development Organization. **Information sources on the meat-processing industry.** Rev. ed. New York, United Nations, 1976. xii, 86p. (*UNIDO Guides to information sources, no.1.*)

760 items. 11 sections: 1. Professional, trade and research organizations, learned societies and special information services - 2. Directories- 3. Sources of statistics, marketing and other economic data - 4. Basic handbooks, textbooks and manuals - 5. Current periodicals - 6. Current abstracting and indexing periodicals - 7. Proceedings, papers and reports - 8. Specialized dictionaries and encyclopaedias - 9. Bibliographies - 10. Films and film catalogues - 11. Other potential sources of information (*e.g.* consulting and engineering services) 'Bibliographical sources used ...'. *Class No:* 664.91

Fish

[6146]
AITKEN, A., *and others eds*. **Fish handling & processing.** 2nd ed. Edinburgh, HM Stationery Office, 1982. v, 192p. illus., diagrs., tables. ISBN: 0114917418.

First published 1965.

15 contributors. 16 chapters (*e.g.* 2. Basic facts about fish; 4. Handling wet fish at sea; 10. Smoking; 11. Canning; 12. Shellfish; 13. Other fish products (*e.g.* fish paste, salted fish); 16. Quality assessment and quality control). Detailed, analytical index. *Class No:* 664.95

[6147]
—WATKIN, G. British food fish. London, Worshipful Company of Fishmongers, 1976. 64p. illus. (16 col. pl.), tables.

In two parts: 1. Identification. Smoke curing of fish - 2. Inspection for freshness. Shellfish control, Data: shape; fins; colour; lateral fins; habitat; etc. Detailed index, p.63-64. *Class No:* 664.95

Oils, Fats & Waxes

Bibliographies

[6148]

UNITED NATIONS. Industrial Development Organization. **Information sources on the vegetable oil processing industry.** New York, United Nations, 1973. xii, 90p. (*UNIDO guides to information sources, no. 7.*)

494 entries 11 sections: 1. Addresses of professional, trade and research organizations, learned societies and special information services - 2. Directories - 3. Sources of statistical and economic data - 4. Basic handbooks, textbooks, and manuals - 5. Current periodicals - 6. Current abstracting and indexing periodicals - 7. Proceedings, papers and reports - 8. Specialized dictionaries and encyclopaedias - 9. Bibliographies - 10. Films -11. Other potential sources of information. (*e.g.* meetings and conferences). Appended list of bibliographical sources used. *Class No:* 665(01)

Handbooks & Manuals

[6149]

Essential oils and waxes. Linskins, M.F. *and* Jackson, J.E., *eds.* Berlin, Springer, 1991. xviii,337p. diagrs. (*Modern methods of plant analysis. New series, v.12.*) ISBN: 3540189157.

27 contributors. 17 papers, with references. 102 diagrs. *Class No:* 665(035)

Dictionaries

[6150]

SOLOMON, B. **Dictionnaire de la technologie des corps gras en cinq langues:** français, anglais, allemand, espagnol, italien. Paris, Institut des Corps Gras, 1971. iii, 62p.

660 French-base terms concerning fats, with equivalents and indexes in English, German, Spanish and Italian. *Class No:* 665(038)

Periodicals

[6151]

Journal of the American Oil Chemists Society. Chicago, Ill., the Society, 1939-. v.1(1)-. Monthly.

'Abstracts' section, monthly; *c.*2,000 abstracts pa. Coverage: biochemistry and nutrition; edible proteins; drying oils and paints; fats and oils. Not indexed. *Class No:* 665(051)

Progress Reports

[6152]

HAMILTON, R.J. *and* BHATTI, A. **Recent advances in chemistry and technology of fats and oils.** London & New York, Elsevier, 1987. xii, 188p. illus., graphs, tables.

8 contributors; 8 contributions (*e.g.* 'Physical properties of fats and oils' (15 references); 'The analysis of lipids with special reference to milk fat' (115 references)). Author and subject indexes. *Class No:* 665(055)

Tables & Data Books

[6153]

INTERNATIONAL UNION OF PURE AND APPLIED CHEMISTRY. **Standard methods of the analysis of oils, fats and derivatives.** 6th ed. Oxford, etc., Pergamon Press, 1979. Part 1 (sections 1&3). xxiv, 170p. graphs, tables. (*First published 1930.*)

1. *Oleaginous seeds and fruits.* 2. *Oils and fats.* *Class No:* 665(083)

Cosmetics & Perfumery

Bibliographies

[6154]

UNITED NATIONS. Industrial Development Organization. **Information sources on essential oils.** New York, UNIDO 1981. 94P. $4. (*UNIDO guides to sources of information no.38.*)

Lists entries covering oils used in the cosmetics and perfume industries. *Class No:* 665.57/.58(01)

Handbooks & Manuals

[6155]

Harry's cosmeticology. Wilkinson, J.B. *and* Moore, R.J., *eds.* London, G. Goodwin, 1987. xv, [1], 934p. illus., diagrs., tables, chemical structures. ISBN: 0711456798.

First published 1940.

4 parts, with documented chapters: 1. The skin and skin products - 2. The nails and nail products - 3. The hair and hair products - 4. The teeth and dental products. Chapter 15: 'Sunscreen, suntan and antisunburn products' (p.222-60. 81 references). Appendix: Proprietary materials cited, p.899-913. Index p.913-34. Well-researched. *Class No:* 665.57/.58(035)

[6156]

MÜLLER, P.M. *and* LAMPARSKY, D. **Perfumes:** art, science and technology. London, Elsevier, 1991. viii,658p. illus., tables, chemical formulas. £95. ISBN: 1851665730.

31 contributors. 20 chapters in 3 parts, covering trapping and classification of odours, composition, production, research, etc. Extensive references for some chapters (*e.g.* 3. 'Semiochemicals' carried 236). Detailed index includes chemical and trade names. *Class No:* 665.57/.58(035)

[6157]

POUCHER, W.A. **Perfumes, cosmetics and soaps.** Jouhar, A.J., *ed.* 9th ed. London, Chapman & Hall, 1991-. V.1 illus., tables.

First published 1923; 8th ed. 1974-5.

V.1 (vii,349p. ISBN: 0412272403) is part of 'A dictionary of the raw materials of perfumery and some cosmetics and toiletry products'. Entries (ABIR ... YLANG-YLANG oils) provide data on sources, appearances, odour; physical; uses. Appendix: 'Materials with IFRA guidelines, but not listed in Poucher's'. *Class No:* 665.57/.58(035)

Dictionaries

[6158]

CARRIÉRE, G. **Dictionary of surface-active agents, cosmetics and toiletries,** in English, French, German, Spanish. Italian, Dutch and Polish. Amsterdam, etc., Elsevier, 1978. 198p. DFl.140. ISBN: 0444998098.

686 English-base terms, with equivalents and indexes in French, German, Spanish, Italian, Dutch and Polish. Covers the principal languages used in the industry. *Class No:* 665.57/.58(038)

[6159]

—CARRIÉRE, G. Lexicon of detergents, cosmetics and toiletries. Amsterdam, etc., Elsevier, 1966. ii, 204p. DFl.140. (*Elsevier lexica, 8.*) ISBN: 0444400990.

Part 1, *Detergents* (257 numbered entries), is a revision of G. Carriéres *Detergents* (1960). English-base terms, with French, Spanish, Italian, Portuguese, German, Dutch, Swedish, Danish, Norwegian, Russian, Polish, Finnish, Czech, Hungarian, Rumanian, Greek, Turkish and Japanese equivalents and indexes. Part 2, *Cosmetics and toiletries* (entries 258-523), also English-base has French, Spanish, Italian, Portuguese, German, Dutch and Swedish equivalents and indexes. *Class No:* 665.57/.58(038)

Reviews & Abstracts

[6160]

The Fritzsche - D & O Library bulletin: a monthly checklist of current literature covering research on perfume and flavour raw materials, essential oils, aromatic chemicals and their application in industry, with a special section on sensory problems and procedure. New York, Fritzsche, Dodge & Olcott, Inc., 1958-. Monthly. Free to qualified personnel.

Formerly *Fritzsche Library bulletin,* 1957-68. Originally included in *Perfumery and essential oil record.*

About 1,500 references pa., in 5 sections. No indexes. *Class No:* 665.57/.58(048)

Tables & Data Books

[6161]

FLICK, E.W. **Cosmetics additives:** an industrial guide. Park Ridge, N.J., Noyes, 1991. xxx,886p. ISBN: 0851512554.

About 4,000 cosmetics additives, A-Z by manufacturer. Data include tradename, typical chemical properties, applications. No index, but detailed contents list. *Class No:* 665.57/.58(083)

Histories

[6162]

KENNETT, F. **History of perfume.** London, Harrap, 1975. 208p. illus., tables.

Chapters: The Far East - The land of the Nile - The Fertile Crescent - The Classical World - The World of Byzantium and Islam - Medieval England - Renaissance expansion - Perfumes from Paris - The fragrance industry. Further reading. Appended table of the most important natural ingredients used in perfumery today. Detailed, analytical index. Quotations - a feature. *Class No:* 665.57/.58(091)

Petroleum Industry

Databases

[6163]

APILIT. New York, American Petroleum Institute.

Worldwide coverage of nonpatent literature pertaining to the petroleum and petrochemical industry since 1964. About 475,000 records; updated monthly. Available on DIALOG, files 354, 954.

67 annotated entries on petroleum databases feature in the *Guide to the petroleum reference literature* (1987), p.65-129. *Class No:* 665.6(003.4)

[6164]

—APIPAT. New York, American Petroleum Institute.

Covers patents in the petroleum and petrochemical industry. Over 200,000 records; updated monthly. Available on DIALOG, files 353, 953. *Class No:* 665.6(003.4)

Bibliographies

[6165]

ABERDEEN CITY LIBRARIES. **Bibliography of material on oil and gas.** 6th ed. Aberdeen City Libraries, 1987. [iv], 398p.

2nd ed. 1979 (46p.).

Part 1 (p.3-87): Bibliographies and guides to sources of information ... Statistics. Part 2 (p.88-288): Background and general works - Diving (including safety and medicine) - Economic, political and social aspects of oil ... Offshore structures (including maintenance and repair) ... Pollution and environment aspects - Transport. Entries, unannotated, have Aberdeen City Libraries callmarks. Extensive author index, p.289-398. *Class No:* 665.6(01)

[6166]

LESTER, D.E. **'Liquid fuels'.** *Information sources in energy technology* (Editor, L.J. Anthony. 1988), p.200-24. London, Butterworth, 1988.

Melvin, A, 'Gaseous fuels', *op.cit.,* p.225-45, similarly structured, has 41 references.

Introduction - Books, monographs, reviews - Journals - Patents - Conference proceedings - Reports - Encyclopedias, handbooks, directories, standards - Trade literature - Statistics and databooks - Abstracting and indexing services - Bibliographies - Organizational sources. *Class No:* 665.6(01)

[6167]

PEARSON, B.C. *and* ELLWOOD, K.B. **Guide to the petroleum reference literature.** Littleton, Colo., Libraries Unlimited, 1987. xi, 193p. ISBN: 0872874737.

11 sections: 1. Guides to the literature and bibliographies - 2. Indexing and abstracting services - 3. Dictionaries - 4. Encyclopedias and yearbooks - 5. Handbooks, manuals and basic texts - 6. Directories - 7. Statistical sources - 8. Databases - 9. Periodicals - 10 Professional and trade associations - 11. Publishers. Sections 1-9 comprise 420 numbered entries. Author/title and subject index. 'For library school students new to the field' (*Marine and petroleum geology,* v.5(3), August 1988). *Class No:* 665.6(01)

Handbooks & Manuals

[6168]

CAMPBELL, C.J. The Golden century of oil, 1950-2050: the depletion of resource. Dordrecht, Kluwer, 1991. xv,348p. diagrs., graphs, tables. £64. ISBN: 0792314425.

3 parts: 1. World assessment - 2. Regional assessment - 3. Production by country and giant oil fields. 'Units of measurement and conversion factors' precedes. *Class No:* 665.6(035)

[6169]

International petroleum encyclopedia. Tulsa, Oklahoma, Penn Well Pubg. Co., 1987. 388p. col. illus., diagrs., graphs, tables, maps. ISBN: 087814919x.

Areas: North Sea; North America; Latin America; Asia-Pacific; Middle East; Europe. Sections on 'Oil after the crash'; Worldwide production; Offshore Gulf Coast; Refining; Refinery catalysis; Pipelines. Advertisers index. *Class No:* 665.6(035)

[6170]

JENKINS, G. Oil economists' handbook. 5th ed. London and New York, Elsevier, 1989. 2v. ISBN: 1851683452, set.

V.1. Statistics [to 1987]. xvi,467p. £110. 29 sections (1. Crude oil prices ... 29. Oil quality data and conversion factors).

V.2. vii,396p. £90. Dictionary: an encyclopedic glossary; Chronology of significant events (to January 1989); Directory of organizations, under countries. V.2 includes a bibliography, p.375-6. *Class No:* 665.6(035)

[6171]

RIVA, J.P., *Jr.* **World petroleum resources and reserves.** Boulder, Colo., Westview Press, 1983. xxiii, 355p. illus., diagrs., tables. ISBN: 0865314462.

11 documented chapters (1. The occurrence of petroleum - 2. Exploration and production (27 references) ... 5. Reserves, resources and reserve/production ratios ... 7. The petroleum prospects of the United States (83 references) ... 11. Worldwide petroleum). Detailed, non-analytical index. About 150 tables. *Class No:* 665.6(035)

Dictionaries

Polyglot

[6172]

EUROPEAN COMMUNITIES COMMISSION. Onshore/ offshore oil & gas multilingual glossary: a glossary of terms concerning the exploraticn and exploitation of oil and gas resources, in Danish, German, English, Spanish, Italian and Dutch. London, Graham & Trotman, for the Commission, 1979. 490p. ISBN: 0860101843.

573 sentences in the 6 languages. Sentences 1-381 are based on the original English texts, and sentences 382-573, on original French texts. *Class No:* 665.6(038)=00

[6173]

Fachwörterbuch: Carbochemie. Petrochemie: Englisch/ Deutsch/Französisch/Russisch. Leipnitz, W., *ed.* Berlin, Paris, Alexandre Hatier, 1992. 304p. ISBN: 3861170396.

8000 English-based entries, with equivalent indexes. *Class No:* 665.6(038)=00

[6174]

KETCHIAN, S., *and others.* **Dictionnaire pétrolier des techniques de diagraphie, forage et production,** russe-français-anglais-allemand... [Technical petroleum dictionary of well-logging, drilling and production terms.] Paris Éditions Technip, 1965. xxxi, 334p. illus. (pl.).

7,770 numbered Russian terms, with French, English and German equivalents and indexes. Genders are stated. Bibliography, p.xvi-xxxi. 25 plates (*e.g.* drawing of a complete diesel-mechanical drilling-rig layout, with keyed parts). *Class No:* 665.6(038)=00

[6175]

—**LEIPNITZ, W.** Erdölverarbeitung und Petrochemie: Englisch/Deutsch/Französisch/ Russisch. Wiesbaden, Brandstetter, 1976. 268p. tables. ISBN: 3870979731.

About 6,000 English-based terms in petroleum processing and petrochemistry, with German, French and Russian equivalents and indexes. *Class No:* 665.6(038)=00

English

[6176]

HYNE, N.J. Dictionary of petroleum exploration, drilling and production. Tulsa, Oklahoma, Penn Well Books, 1990. [v],623p. illus. ISBN: 0878143521.

About 20,000 terms. Definitions range from a single word to one column (*e.g.* 'Sucker-rod pump'). Well-produced. *Class No:* 665.6(038)=20

[6177]

LANGENKAMP, R.D., *ed.* **The Illustrated petroleum reference dictionary.** 3rd ed. Tulsa, Oklahoma, Penn Well Books, 1985. viii, 696p. ISBN: 087814272x.

Combines a dictionary of *c.*3,500 terms (apparently based on the editor's *Handbook of oil industry terms and phrases* (4th ed. 1984)) with other features, - D&D (Desk and Derrick Club) standard oil abbreviations, p.363-438; and 'Universal conversion factors', p.459-695. Very small illus. *Class No:* 665.6(038)=20

[6178]

STEVENS, P., *ed.* **Oil and gas dictionary.** London, Macmillan, 1988. 270p. graphs, tables. £45. ISBN: 033337844x.

About 2,000 entries, A-Z, some at length (*e.g.* 'Crude oil prices', p.39-46). Includes financial, business and general economic terms. Cross-references; abbreviations. Chronology, 1859-1986, p.221-37. Appended 'Energy data conversion tables' (*e.g.* 'Metric ton oil equivalents of gaseous fuels and electricity'). 'Highly recommended' (*Choice,* May 1989, p.1500). *Class No:* 665.6(038)=20

[6179]

TVER, D.F. *and* **BERRY, R.W. The Petroleum dictionary.** New York, etc., Yan Nostrand Reinhold, 1980. [xii], 374p. illus., diagrs., tables. ISBN: 0448240465.

About 4,000 entries (1-20 lines each). Claims to be first combined dictionary-handbook 'that covers virtually all aspects of the petroleum industry' (*Preface*). Stresses geological aspects. *Class No:* 665.6(038)=20

German

[6180]

PERSCH, F., *ed*. **Wörterbuch deutsch-englisch, englisch-deutsch: Erdöl und Erdgas,** mit besonderer Berüchsichtigung der Lagerstättenkunde und Gewinnung. Goslar, Hübener, 1970. 671p.

'Petroleum and natural gas dictionary, with special reference to exploration and exploitation, including offshore operations'.

About 12,000 entry-words in each part (English-German, p.9-325; German-English, p.329-671). Separate lines for sub-entries. German genders are stated.

Class No: 665.6(038)=30

French

[6181]

MOUREAU, M. *and* ROUGE, J. **Dictionnaire technique des termes utilisés dans l'industrie du pétrole,** anglais-français, français-anglais. Paris, Technip, 1963 (reprinted 1977). xvi, 490, [11], 387p.

Defines *c*.40,000 terms, some with explanations. Includes abbreviations, technical names and colloquialisms. In between the two sequences, a list of symbols and conversion tables, Bibliography of 120 items, p.ix-xv.

Class No: 665.6(038)=40

[6182]

—AMOUD, M. *and* ZUBINI, F. **Les termes pétroliers: dictionnaire anglais-français.** Paris, Dunot, 1981. xii, 267p.

Includes *c*.6,000 entries, including abbreviations. Sources, p.ix-xii. Essentially practical. *Class No:* 665.6(038)=40

Reviews & Abstracts

[6183]

International petroleum abstracts. London, Wiley Heydon, Ltd, on behalf of the Institute of Petroleum, London, 1921-. Quarterly. ISSN: 03094944.

About 1,200 abstracts pa. Sections: Oilfield exploration and exploitation - Transport and storage - Refining and related processes - Products - Corrosion - Engines and automatic equipment - Safety and industrial hygiene - Economics and marketing - Politics and legislation - Pollution. Author index per issue; author and subject indexes annually. Online INFOLINE. *Class No:* 665.6(048)

[6184]

Petroleum abstracts. Tulsa, Oklahoma, University of Tulsa, 1961-. Weekly. ISSN: 00316423.

About 10,000 abstracts pa., on petroleum exploration and production of crude oil and gas. Includes patents. Online as 'Petroleum exploration and production', a restricted file on DIALOG. A leading current-awareness source of petroleum exploration and production. *Class No:* 665.6(048)

[6185]

Petroleum/energy business news index. New York, American Petroleum Institute, 1975-. Monthly.

About 1,500 items per issue. The September 1987 issue covers various issues of 16 periodical titles. About 120 journals are monitored over the year, plus news items reported in *API abstracts/Literature* and *API abstracts/ Oilfield chemicals.* Arranged under subjects, 'Abandonment' ... 'Zinc'. Very small type. Available for online search, P/E NEWS. *Class No:* 665.6(048)

Periodicals

[6186]

Petroleum economist: the international energy journal. London, Petroleum Economist, 1934-. Monthly. tables, maps. £90pa. ISSN: 0306395x.

V.58(12), December 1991, carries 11 short articles (*e.g.* 'Kuwait'). Features include 'News in brief'; 'Market trends'; 'Company information'; 'Oil share markets'; 'World oil production' (with table). Up-to-date statistics and international spread are a feature. *Class No:* 665.6(051)

Yearbooks & Directories

[6187]

'Financial Times' Oil and gas international year book: 1992. 92nd ed. Harlow, Essex, Longman Group, 1991. Annual. xxv,544p. tables. £115. ISBN: 0582068495.

Companies : upstream (entries 1-445) - Companies: downstream (entries 446-710) - Brokers and traders (entries 711-793) - Associations (entries 794-870). Geographical and company indexes. Appended: 5 tables (*e.g.* World oil consumption). International professional services. Suppliers directory and buyers' guide. Advertisers' index.

Class No: 665.6(058)

Standards

[6188]

INSTITUTE OF PETROLEUM. **Standard methods for analysis and testing of petroleum and its products.** Chichester, Wiley, 1989. 2v. diagrs., tables. ISBN: 0471929492.

V.1 *Methods IP1-261.*

V.2 *Methods IP262-387.*

Provides standards and tentative methods of analysis and testing, plus methods jointly established by the Institute of Petroleum and the American Society for Testing and Materials (ASTM). An essential reference source for chemists and engineers in this field.

Class No: 665.6(083.74)

Biographies

[6189]

'Financial Times' Who's who in world oil and gas. Harlow, Essex, Longman Group, 1979-. Annual. ISSN: 01413236.

About 4,000 entries for key personnel from commercial, academic and government fields. Data include status, qualifications and a contact address. Index of organizations. *Class No:* 665.6(092)

Worldwide

[6190]

OPEC, its member states and world energy market. Brown, G., *ed*. 2nd ed. Harlow, Essex, Longman Group, 1991. xxiv, 680p. charts, tables, maps. £95. ISBN: 0582085276.

4 major sections: an overview of world oil and gas; OPEC and other international organizations; individual OPEC country surveys (Algeria ... Venezuela); chronology, 1960-90. *Class No:* 665.6(100)

[6191]

ORGANIZATION FOR ECONOMIC COOPERATION AND DEVELOPMENT. Annual oil and gas statistics, and main historical series. [Statistics Annuelles ...] Paris, International Energy Agency. Annual. diagrs., graphs, tables.

The *Annual* for 1984/1985 (1987. xxxv, 419p.) has 164 tables. 4 main sections: 1. Summary tables (areas and countries) - 2. Supply of products - 3. Finished products - 4. Imports and exports.

BP statistical review of world energy (London, British Petroleum Co., Ltd.) is annual. *Class No:* 665.6(100)

Europe
[6192]

Annual bulletin of general energy statistics for Europe [Liquid fuel]. New York, Economic Commission for Europe, 1968-. Annual.

In v.19, 1986 (1988. 134p. of tables), liquid fuel statistics appear under 'Primary energy' (crude petroleum; other inputs to petroleum) and 'Derivative energy' - light products (aviation and motor gasoline; jet fuel; kerosene; naphthas) and heavy products (gas/diesel) oil; residual fuel oil; other petroleum products. Footnoted sources. Appended 'Definitions and general notes'. 'Europe' includes USSR , plus USA and Canada. *Class No:* 665.6(4)

Great Britain
[6193]

BRITISH PETROLEUM. Our petroleum industry: a handbook dealing with the organization and functions of an integrated international oil company, with particular reference to the British Petroleum Company, Ltd. 5th ed. London, BP, 1977. 600p. illus., tables, maps.

First published 1947 as *Our Industry*.

58 contributors. 24 chapters (1. The structure of the industry ... 5. Petroleum exploration and development - 6. Drilling for oil and gas - 7. Production - 8. New oil and gas provinces: North Sea and Alaska ... 11. Transport of oil by tanker ... 13. Petroleum refinery products and petroleum refining processes ... 15. Proteins from petroleum - 16. International marketing ... 18. Natural gas and gas liquids - 19. Petroleum chemicals and polymers (plastics, rubbers and films) - 20. Research - 21. Computers and computer applications. - 22. Producing countries of the world - 23. Consuming areas of the world - 24. A brief history of BP. Glossary. 28 statistical tables. A mine of information. *Class No:* 665.6(410)

[6194]

—SHELL, Ltd. Petroleum handbook. 6th rev. ed. Amsterdam, Elsevier, 1983. xvii, 770p. illus., diagrs.

Roughly comparable to BP's *Our petroleum industry*. 257 illus.; 42 diagrs. *Class No:* 665.6(410)

[6195]

FERRIER, R.W. The History of the British Petroleum Company. Cambridge Univ. Press, 1982-. illus., facsims., ports., tables, maps.

1. *The developing years, 1901-1931.* 1982. xxx, 801p. £55.

13 chapters (1. The acquisition of the D'Arcy Concession, 1901 ... 13. The importance of the Persian dimension, 1926-32). Conclusion. Biographical details of important

.... (contd.)

personalities p.688-97. Notes, p.698-775. Detailed, analytical index, p.776-804. Much emphasis has been placed on the primary archival material.

A full bibliography is promised in the final volume. *Class No:* 665.6(410)

[6196]

GREAT BRITAIN. Department of Energy. Development of the oil and gas resources of the United Kingdom. London, H.M. Stationery Office. Annual. tables, maps. ISBN: 011412828x.

The 1987 annual report (128p. £11) has 4 sections: UK oil & gas resources on the UK Continental Shelf; UK oil & gas exploration facts & figures; UK oil & gas development, production and operational activities; UK oil & gas impact of oil & gas production. Known as the 'Brown' Book. *Class No:* 665.6(410)

USA
[6197]

AMERICAN PETROLEUM INSTITUTE. Basic petroleum data book: petroleum industry statistics. Washington, The Institute, 1981-. 3pa.

Previously (1975-80) in loose-leaf form. *Class No:* 665.6(73)

Thesauri
[6198]

AMERICAN PETROLEUM INSTITUTE. Thesaurus. [American Petroleum Institute]. 21st ed. Washington, The Institute, 1984. 320p.

A-Z sequence of *c.*6,000 valid terms, plus cross references. *Class No:* 665.6:025.43

Pipelines
[6199]

Lexique des pipelines à terre et en mer, anglais-français, français-anglais. [Glossary of onshore and offshore pipelines, English-French, French-English.] Paris, Editions Technip, 1979. xii, [1], 303p.

English-French, p.1-133; Abbreviations; French-English, p.147-303. 'Pump' 1½ columns. 57 categories applied. List of sources. *Class No:* 665.6.026

Offshore Oil

Bibliographies
[6200]

CHRYSSOSTOMIDIS, M. Offshore petroleum **engineering:** a bibliographic guide to publications and information sources. New York, Nichols Publishing Co.; London, Kogan Page, 1978. 367p. ISBN: 089397045x.

Records *c.*2,500 books, conference proceedings and reports on offshore petroleum and related topics. Sections A-X (*e.g.* F. Oceanography and ocean environment description; G. Offshore structures engineering; J. Offshore petroleum engineering and development; T. Oil spills; V. Ocean law and management). 'Supplementary sources and

....*(contd.)*

suggestions for keeping up to date'. Sources consulted. Directory of publishers and other organizations. Permuted title, author and title indexes. *Class No:* 665.60(01)

[6201]

MYERS, A., *comp.* **Current bibliography of offshore technology and offshore literature.** 2nd ed. Berkhamstead, ASR Marketing, 1984. 167p. £40.

First published 1981.

1,276 unannotated items in broad subject groups: General (bibliographies, directories, etc.) - Oil and gas (economics, offshore operations, health and safety, etc.) - Oceanography - Marine technology (Underwater construction, diving, etc.). Indexes: author; publisher; subject. Directory (new to the 2nd ed.) lists organizations, societies, agencies and government departments concerned with offshore technology. *Class No:* 665.60(01)

[6202]

SEMPLE, Susan, *comp.* **Oil: a bibliography.** 4th ed. Aberdeen, ANSLICS (Aberdeen and North Scotland Library and Information Co-operative Service), 1982. 3 microfiche, booklet.

Previous ed. 1977.

Two separate m/f listings: Author/title; subject. Over 1,500 entries for items in 11 academic, government and public libraries in N.E. Scotland. Sections A-M: A. Bibliographies, literature guides and information sources - B. General background - C. North Sea oil - D. Economic, industrial and political aspects - E. Legislation - F. Management, education and safety aspects - G. Petroleum geology and exploration - H. Ocean engineering - I. Drilling and production - J. Transportation, storage and distribution - K. Environmental and pollution aspects - L. Periodicals - M. Non-book materials. Emphasis in the 4th ed. is on North Sea oil. *Class No:* 665.60(01)

[6203]

TAYLOR, G.E. **'Offshore engineering'.** *Information sources in engineering* (2nd ed. Editor L.J. Anthony. 1985), p.474-90. London, Butterworth, 1985.

Sections: Organizational sources - Legal aspects of offshore engineering - Primary sources of documentary information (journals; conferences; reports; patents; evaluated data) - Secondary sources of information (abstracting and indexing services; directories; books). *Class No:* 665.60(01)

Dictionaries

[6204]

RÅDET FOR TEKNISK TERMINOLOGI. **Ordbok for petroleums-virksomhet,** norsk-engelsk-fransk. Oslo, Universitetsforlaget, 1976. 176p. charts. (*RTT 35.*) ISBN: 8200257290.

Dictionary of offshore oil operations. About 1,800 Norwegian-based terms, with definitions; English and French equivalents and indexes. Three sections: Boring production; Geophysics; Geology. *Class No:* 665.60(038)

[6205]

WHITEHEAD, H. **An A-Z of offshore oil & gas:** an illustrated glossary and reference guide to the offshore oil & gas industries and their technology. 2nd ed. London, Kogan Page, 1983. 438p. illus., diagrs., tables, chemical structures. £27.50. ISBN: 0850386276.

First published 1976.

About 4,000 terms and abbreviations defined, plus cross-references. 30 appendices, p.315-426 (*e.g.* 1. Offshore oil and gas regions of the world; 29. Bibliography, p.433-4; 30. Abbreviations commonly used in relation to offshore oil and gas). Well illustrated (200 line-drawings, 20 full-page maps); keyed diagrams. 'An indispensable reference for engineers, economists and geoscientists concerned with petroleum resources' (*Choice*, v.21, October 1983, p.258). *Class No:* 665.60(038)

Reviews & Abstracts

[6206]

Offshore abstracts. Hove, Sussex Offshore Information Literature, 1974-. 6pa. £110pa. ISSN: 03050513.

About 650 abstracts pa. 17 sections: Arctic environment - Buoys and marking systems - Corrosion ... Marine geology ... Pipes and pipelines ... Production ... Structural engineering ... Vibration ... Wave energy. No index. *Class No:* 665.60(048)

[6207]

Offshore engineering abstracts. Cranfield, Beds., BHRA, The Fluid Engineering Centre, 1986-. 6pa.

125 abstracts per issue on offshore structures and equipment, their design, construction, operation and maintenance. Author and subject indexes per issue and annually. Online FLUIDEX, available via DIALOG, ESA-IRS. *Class No:* 665.60(048)

Great Britain

[6208]

LOVEGROVE, M. **Lovegrove's guide to Britain's North Sea oil and gas.** 2nd ed. Cambridge, Energy Publications (Cambridge), 1983. xii, [i], 237+5p. diagrs., tables. £30. ISBN: 0905332296.

3 parts. 1. Government policies (*e.g.* licensing; taxation) - 2. Industry activities (*e.g.* exploration and appraisal; government revenues and company cash flow) - 3. Future. 7 appendices (1. UKCS Field broad sheets; 2. Company broad sheets; 3/4. Main UKCS gas and oil pipelines; 5. Main UKCS terminals; 6. UK crude oil refinery capacity statistics; 7. Conversion factors and miscellaneous information). 46 tables. *Class No:* 665.60(410)

[6209]

North Sea oil & gas directory, 1987/8. Kingston-upon Thames, Spearhead Pubns., 1987. 992p. £39.95. ISSN: 02655033.

A-Z of exploration and production companies, of project management contractors, of drilling contractors, and of manufacturers, and services contracting companies (p.85-709). Products and services guide. Offshore organizations. A-Z index. Advertisers' index. *Class No:* 665.60(410)

Information Services

[6210]

Information offshore. Glasgow, United Kingdom Offshore Supplies Office. Regularly updated. Gratis.

Lists information sources on offshore matters available from government departments and agencies. Also listed are government laboratories and organizations with interests in the UK offshore market. 'One of the most useful directories in this field' (*Information sources in engineering*. 2nd ed. Editor, L.J. Anthony. 1985, p.487).
Class No: 665.60:061:025.5

Conferences

[6211]

INSTITUTE OF PETROLEUM. Offshore UK: Papers presented at the Institute Petroleum, 1984. Annual Conference, 31st May and 1st June 1984. Hay, J.T.C., *ed.* London, the Institute, [1985?]. 261p. illus., diagrs., charts, maps. ISBN: 090404629x.

14 papers, covering virtually all aspects of North Sea operations, frontier drilling, field development, conventional and unconventional, boats, helicopters and diving '(*Foreword*). Some papers are documented (*e.g.* 'North Sea oil exploration and protection: interactions with the environments', p.100-22. 18 references; 6 figures; 4 tables).
Class No: 665.60:061:061.3

Refining

[6212]

MEYERS, R.A. Handbook of petroleum refining processes. New York, etc., McGraw Hill, 1986. [480]p. illus., diagrs., graphs, tables. £70. ISBN: 0070417636.

22 contributors. 11 parts (some with brief bibliographies appended): 1. Alkylation and polymerization - 2. Cracking ... 5. Isomerization ... 8. Separation processes ... 11. Gasification. Glossary. Abbreviations and acronyms. Detailed, analytical index (10p.). *Class No:* 665.66

Petrochemicals

Bibliographies

[6213]

UNITED NATIONS. Industrial Development Organization. **Information sources on the petrochemical industry.** New York, United Nations, 1978. $4. (*UNIDO guides to information sources, no.29.*) ISBN: 9211061644.

12 sections: 1. Professional, trade and research organizations, learned societies and special information services -2. Directories - 3. Sources of statistics, marketing and other economic data - 4. Basic handbooks, textbooks and manuals - 5. Monograph series - 6. Current periodicals - 7. Current abstracting and indexing periodicals - 8. Proceedings, papers and reports - 9. Specialized dictionaries and encyclopedias - 10. Bibliographies - 11. Films, tapes and film catalogues - 12. Other potential sources of information (A. Consulting and engineering services). Bibliographical sources used. Sections 2-3, 6-7, 10-12 have annotated entries. *Class No:* 665.7(01)

Handbooks & Manuals

[6214]

CHAPMAN, K. The International petrochemical industry: evolution and location. Oxford, Blackwell, 1991. xiv[i],322p. diagrs., graphs, maps. £45. ISBN: 0631160981.

12 chapters. References, p.294-312 (A-Z, authors). Detailed index. Emphasis is placed on the 'spatial evolution of the industry, both within individual countries and at the global scale' (*Introduction*). *Class No:* 665.7(035)

[6215]

CLARK, G.H. Industrial and marine fuels reference book. London, etc., Butterworths, 1988. 528p. illus., diagrs. £95. ISBN: 0408014881.

26 sections (The choice of fuels - The chemistry of petroleum - Crude oil refining - Laboratory tests on fuels - Natural gas and LPG ... General operating problems of large diesel engines - Deposit formation and corrosion in steam boilers - Chimney stack emissions). Appendix. 239 illus. For mechanical, marine and plant engineers, plus shipowners, engine-builders and designers. *Class No:* 665.7(035)

Reviews & Abstracts

[6216]

API abstracts/Oilfield chemicals. New York, American Petroleum Institute, 1981-. Monthly.

1987: 1,103 abstracts. 5 main sections: Drilling fluids - Completion and stimulation fluids - Production chemicals - Enhanced recovery - Economics and statistics. No indexes. *Class No:* 665.7(048)

Yearbooks & Directories

[6217]

Worldwide petrochemical directory, 1992. 30th ed. Tulsa, Oklahoma, Penn Well Publishing Co., 1962-. Annual. ISSN: 00842583.

Broad geographical regions: U.S.; Canada ; Latin America; Europe; Africa; Middle East; Asia, Each company entry includes list of key personnel, services, products. Company and geographical indexes; advertisers' index. Claims to identify all petrochemical projects under construction worldwide. *Class No:* 665.7(058)

Tables & Data Books

[6218]

WELLS, G.M. Handbook of petrochemicals and processes. Aldershot, Hants., Gower, 1991. 400p. diagrs., chemical structures. £72. ISBN: 0566027755.

Data on *c.*100 petrochemicals, A-Z (*e.g.* 'Xylene': synonyms; processes; reaction; properties; yield; international classifications; applications; health and handling; major plants). 121 diagrs. Appendices include 'Transportation'; 'Health and safety'; 'Other organizations'. Licensor index; subject index. *Class No:* 665.7(083)

Glues & Adhesives

[6219]
Adhesives directory. Rickmansworth, Herts., Turret-Wheatland, Ltd. 232p. ISSN: 03053199.

8 sections, covering 'raw materials and additives for the adhesives-making industry, and of adhesives, mastics and related products used in all sectors and branches of industry'. Appended: Products, services directory. Advertisers index. Small format. *Class No:* 665.93

[6220]
CAGLE, C.V. Handbook of adhesive bonding. New York, etc., McGraw-Hill, 1973. 800p. illus., tables.

4 main sections: 1. Adhesive bonding - 2. Bonding various substrates - 3. Adhesives for various industries (*e.g.* automobile; electronics; space systems; construction; packaging) - 4. Testing quality control and specifications. *Class No:* 665.93

[6221]
HARRIS, L. Adhesives and adhesion: a guide to selected literature and sources of information. London, British Library, Science Reference and Information Service, 1986. 28p. £5. ISBN: 0712307303.

More than 100 briefly annotated entries. 6 sections: 1 Books (1.1: Adhesion and adhesives - general; 1.2: Particular types of adhesive; 1.3: Adhesives for specific application) - 2. Journals - 3. Abstracting and indexing journals - 4. Conference proceedings (published 1970-84) - 5. Business literature - 6. Patent publications. Science Reference Library Shelf marks are given. No index. *Class No:* 665.93

[6222]
KINLOCH, A.J. Adhesion and adhesives: science and technology. London, Chapman & Hall, 1987. xii,440p. illus., diagrs., graphs, tables. £35. ISBN: 041227440x.

8 chapters, closely subdivided, with extensive references: 1. Introduction - 2. Interfaced contract - 3. Mechanism of adhesion - 4. Surface pretreatments (17 sections; 227 references) - 5. Hardening of the adhesive - 6. Mechanical behaviour of adhesive joints - 7. Fracture mechanics of adhesive joints - 8. The service life of adhesive joints. Author and subject indexes. *Class No:* 665.93

[6223]
Klebstoff wörterbuch: German / English / French / Italian / Spanish / Dutch. [Adhesives dictionary: German / English / French / Italian / Spanish / Dutch.] Munich, Ad Liasion Bucherie, 1988. v.p. ISBN: 3574239025.

893 German-base terms. Coloured language indexes. *Class No:* 665.93

[6224]
SHIELDS, J. Adhesive handbook. 3rd ed. London, etc., Butterworth, 1984. [xii], 360p. ISBN: 0408013567.

First published 1970.

13 sections; *e.g.* 3. Adhesive selection; 4. Adhesive materials and properties (p. 30-86, including 6½p. of references); 5. Surface properties; 6. The bonding process; 7. Physical testing of adhesives; 8. Adhesive selection data; 9. Adhesive products directory (p.175-297); 10. Adhesives and trade services; 11. Glossary of adhesives technology

....(contd.)
terms; 12. Bibliography (books, reviews, journals), p. 342-52; 13. Useful addresses. No index. Well produced. *Class No:* 665.93

[6225]
SKEIST, I., *ed*. Handbook of adhesives. 3rd ed. New York, Van Nostrand Reinhold, 1990. $95. ISBN: 0342280130.

47 chapters on the physics, chemistry and applications of adhesives. Many diagrs., charts, tables. Chapter bibliographies (some plus 'further readings'). Lists of standards and commercial producers of materials. 'Recommended' (*Science and technology libraries,* V.11(4), Winter 1991, p.212). *Class No:* 665.93

Resins & Gums

[6226]
ASH, M. *and* **ASH, I., *comps*. Handbook of plastic compounds, elastomers, and resins:** an international guide by category, tradename, composition, and supplier. New York, VCH, 1992. xiv,872p. £132. ISBN: 1560815531.

Data on 15,000 tradename products. 4 parts: I. Tradename by category reference - II. Tradename cross reference - III. Chemical component cross reference - IV. Chemical manufacturer's directory. *Class No:* 665.94

[6227]
ASH, M. *and* **ASH, I., *comps*. Resins.** London, Edward Arnold, 1990. ix,381p. £69. (*What every chemical technologist wants to know about, V.*) ISBN: 0713136758.

4 parts: 1. Data on resins, A-Z (synonyms; tradename equivalents; category; processing; applications; properties; toxicity; storage). - 2. Tradename products and generic equivalents - 3. Generic chemical synonyms and cross references - 4. Tradename product manufacturers. *Class No:* 665.94

Glass Industry

Bibliographies

[6228]
UNITED NATIONS. Industrial Development Organization. Information sources on the glass industry. New York, United Nations, 1975. xii, 64p. $4. (*UNIDO guides to information sources, no.16.*)

11 sections: Professional, trade and research organizations, learned societies and special information sources - 2. Directories - 3. Sources of statistics, marketing and other economic data - 4. Basic handbooks, textbooks and manuals - 5. Current periodicals - 6. Current abstracting and indexing journals - 7. Proceedings, papers and reports - 8. Specialized dictionaries and encyclopaedias - 9. Bibliographies - 10. Film and film catalogues - 11. Other potential sources of information. 'Bibliographical sources used ...'. *Class No:* 666.1(01)

Encyclopaedias

[6229]

GRAYSON, M., ed. Encyclopedia of glass, ceramics and cement. New York., etc., Wiley: Interscience, 1985. xxviii, 925p. illus., diagrs, tables. (*Encylopedia reprint series*.) ISBN: 047181931x.

'One in the series of carefully selected reprints from the world-renowned Kirk-Othmer *Encyclopedia of Chemical Technology*'.

57 contributors. Entries: Aluminum compounds (188 references) ... Silicon compounds, synthetic inorganic silicates. Section 'Clays (surveys)', p.320-36, has 99 references. Conversion tables, abbreviations and symbols precede. Detailed, analytical index, p.885-925.
Class No: 666.1(031)

[6230]

NEWMAN, H. An Illustrated dictionary of glass: 2,442 entries, including definitions of wares, materials, processes, forms and decorative styles, from antiquity to the present. With an introductory survey of the history of glass making, by R.J. Charleston. London, Thames & Hudson, 1977. 351p. illlus. (incl. col.).

A companion to *Illustrated dictionary of ceramics,* by G. Savage and H. Newman (1974).

Includes histories of major factories and production centres. Some entries carry references. Cross-references. A finely illustrated volume, with 625 illus., 117 in colour. 'Intended primarily to define terms relating to glass and glassware ... and styles in various regions and periods, and also to describe some pieces that bear recognized names' (*Preface*). 'A necessary reference' (*Choice,* September 1978, p.843). *Class No:* 666.1(031)

[6231]

—PHILLIPS, P., *ed*. The Encyclopedia of glass. New York, 1981. 320p. illus. (incl. col.).

Covers both historical and regional aspects of glassmaking as well as techniques. It is complementary to Newman for its narrative approach and its 800 illus. Includes a glossary, a list of museums, and a grouped bibliography (p.298-314).
Class No: 666.1(031)

Handbooks & Manuals

[6232]

BANSAL, N.P. and DOREMUS, R.H. Handbook of glass properties. Orlando, Florida, etc., Academic Press, 1986. viii, [1], 680p. graphs, tables. £121.50. ISBN: 0120781409.

5 parts (19 documented chapters): 1. Ornamental - 2. Thermodynamic and thermal properties - 3. Mechanical properties - 4. Electrical and transport properties - 5. Other properties. Appendices A-D (A. Glossary of symbols for data tables). Systems - property index. Subject index.
Class No: 666.1(035)

[6233]

Handbook of glass data. Mazurin, O.V., *and others, eds.* Amsterdam, Elsevier, 1983-. graphs, tables. ISBN: 0444416897.

Part A. *Silica glass and binary silicate glasses.*
Part B. *Single-component and binary non-silicant oxide glasses.*
Part C. *Ternary silicate glasses.*
Part D. *Ternary non-silicate oxide glasses.*
Part E. *Supplement.*

....(contd.)

Part A (xv,669p.) includes much tabular data. References, p.597-642. Indexes: author, subject, formula.
Class No: 666.1(035)

[6234]

INTERNATIONAL COMMISSION ON GLASS. Electrical properties of glasses, glass-ceramics and amorphous solids: a bibliography covering the period 1967-1976. Charleroi, Belgium, the Commission, 1967-76. 2v.

V.1 (237p.) is devoted to chronological reference and author indexes; V. (267p.), to the subject index. 8,834 numbered entries. *Class No:* 666.1(035)

[6235]

McLELLAN, G.W. *and* **SHAND, E.B., eds. Glass engineering handbook.** 3rd ed. New York, etc., McGraw-Hill, 1984. v.p. illus, graphs, tables. ISBN: 007044823x.

4 sections (23 documented chapters): 1. Glass technology - 2. Glass manufacture - 3. Applications - 4. Fibreglass (completely rewritten). 'Continuous-filament glass fiber applications' (38p.; 2p. of references). Appendixes: A. Specifications and standards of glass; B. Glossary; C. SI units. Detailed, analytical index (17p.).
Class No: 666.1(035)

[6236]

SCHOLZE, H. Glass: nature, structure and properties. New York, Springer Verlag, 1991. *c*.425p. illus., graphs, tables, chemical structures. £70.50. ISBN: 0387973966.

First German ed. 1964. This ed. sponsored by the Institute of Glass Science and Engineering, New York State College of Ceramics; trans. by M.J. Lakin.

3 parts. 1. Introduction - 2. Nature and structure of glass - 3. Properties of glass. Bibliography, p.363-415 (1160 references). Author, subject indexes. *Class No:* 666.1(035)

[6237]

UHLMANN, D.R. *and* **KREIDL, N.J., eds. Glass science and technology.** Orlando, Florida, Academic Press, 1980-. illus., tables. ISBN: 0127067051.

1. *Glass-forming systems.* 1983. 465p.
2. *Processing.* 1984. 360p. £73.50.
3. *Viscosity and relaxation.* 1986. 412p.
4A. *Structure, microstructure and properties.* 1990. 360p. £88.50.
4B. *Advances in structural analysis.* 1990. 400p. £88.50.
5. *Elasticity and strength in glasses.* 1980. 282p. £47.50.

V.3 has 7 documented chapters (Viscoelasticity of glass - Mechanical relaxation in inorganic glasses - Rheology and relaxation in metal glasses - Technological aspects of viscosity - Annealing of glass - Rheology of polymeric fluids - Physical aging of polymer glasses). Index. 'For more broad-based physical science collections in academe and industry' (*Science & technical libraries,* v.8(2), Winter 1987/88, p.138). *Class No:* 666.1(035)

Dictionaries

Polyglot

[6238]

INTERNATIONAL COMMISSION ON GLASS. Dictionary of glass-making. Amsterdam, etc., Elsevier, 1983. xxii 402p.

First published 1965.

A trilingual list of 4,428 numbered terms (in English,

....(contd.)

French and German), with A-Z lists in each of the 3 languages.

Supplements to the 1965 ed. are available in nine other languages: Czech, Dutch, Italian, Japanese, Polish, Portuguese, Russian, Spanish, and Swedish. *Class No:* 666.1(038)=00

English

[6239]
BRITISH STANDARDS INSTITUTION. Glossary of terms used in the glass industry. London, BSI, 1962. AMD 1963. 56p. £25.60. (*BS 3447: 1962 (1991).*)

About 500 terms are defined. 8 sections, *e.g.* Types and properties of glass; raw materials; melting, forming and finishing processes; forms of glass and glassware; imperfections in glass: occupational terms. *Class No:* 666.1(038)=20

[6240]
ZEITLYN, M.Z., *comp.* **Dictionary of glass names.** St. Helens, Lancs., Pilkington Bros., 1970. 120p.

More than 2,200 entries on glass products from 42 countries, but primarily on Pilkington products (*e.g.* safety glass, insulating glass, optical and opthalmic glass). 2 sections: an A-Z list of glass names, stating maker, brief description of glass type, and classified section of types (*e.g.* 'laminated', 'ophthalmic', 'patterned'). *Class No:* 666.1(038)=20

German

[6241]
ELMER, T.H. German-English dictionary of glass, ceramics, and allied sciences. New York, etc., Interscience: Wiley, 1963. viii, 304p.

About 12,000 German words encountered in books, periodicals and patent literature. Includes many abbreviations and some common general words. Mostly equivalents; no German genders. Terms are categorized. Clear layout. *Class No:* 666.1(038)=30

Reviews & Abstracts

[6242]
Boron in glass: quarterly bulletin of abstracts. London, Borax, Ltd., 1950-. Quarterly. gratis.

50 abstracts per issue on the use of boron in vitreous systems. Index per issue. *Class No:* 666.1(048)

[6243]
BRITISH GLASS INDUSTRY RESEARCH ASSOCIATION (BGIRA). Digest of information and patent review. Sheffield, BGIRA, [1980?]-. Monthly.

About 100 brief abstracts per issue. 8 classes: Glass manufacture: general - Container glass - Flat glass - Domestic - Glass fibre - Special glasses - Glass ceramics - Enamels. Patents (10 classes). Appended list of conferences and exhibitions. Annual subject and author indexes. *Class No:* 666.1(048)

[6244]
Glass technology (Section A: *Journal of the Society of Glass Technology*). Sheffield, Society of Glass Technology, 1960-. 6pa.

About 1200 abstracts pa. Coverage: Raw materials and batch preparation - Fuels, furnaces and instrumentation - Refractories - Processses - Auxiliary processes and equipment ... Glass fibres - Composites and polymers - Industrial, historical, economics and biography - Environmental aspects. Annual author and subject indexes. Section B: *Journal of the Society of Glass Technology, - Physics and Chemistry of glasses. Class No:* 666.1(048)

Yearbooks & Directories

[6245]
European glass directory & buyers guide, 1987. Redhill, Surrey Fuels & Metallurgical Journals, [1986]. 192p.

7 sections, including: 2. Glass industry index - 3. Products of the glass industry - 4. Suppliers to the glass industry - 5. Trade marks - 6. Country by country trends - 7. European trade associations. Advertisers' index. *Class No:* 666.1(058)

[6246]
Glass fibre directory. Sturr, T. London, Chapman & Hall, 1992. 388p. tables. £65. ISBN: 041246280x.

Usual directory-type information on glass fibre reinforcements available worldwide, with specifications, manufacturer and agent. *Class No:* 666.1(058)

Enamels

[6247]
BRANDT, J., *and others.* **Emails, enamels, émaux, smalti: a dictionary in four languages.** Leverkusen, Farbenfabriken Bayer AG, 1960. 181p.

About 1,200 terms, with equivalents. 4 sequences: German-English- French-Italian; English-French-Italian-German; French-Italian-German-English; Italian-German-English-French. Genders are given. Some terms are categorized. *Class No:* 666.29

Ceramics

Bibliographies

[6248]
MURFIN, D. 'Ceramics'. *Information sources in metallic materials.* Editor, M.N. Patten, p.232-52. London, Bowker-Saur, 1989. (*Guides to information sources.*) ISBN: 0408014911.

Sections: Dictionaries and glossaries - Textbooks and monographs - Conference proceedings - Primary journal literature (83 titles) - Reports - Patents - Standards - Property data - Business statistics and company information - Directories and trade literature - Market research reports - Abstracts and other secondary sources - Online services. *Class No:* 666.3/.7(01)

[6249]
UNITED NATIONS. Industrial Development Organization. **Information sources on the ceramics industry.** New York, United Nations, 1976. xii, 68p. $4. (*UNIDO guides to information sources, no. 17.*)

493 entries. 11 sections: 1. Professional, trade and research organizations, learned societies and special information services - 2. Directories ... 4. Basic handbooks, textworks and manuals - 5. Current periodicals - 6. Current abstracting and indexing periodicals ... 9. Bibliographies ... 11. Other potential sources of information (*e.g.* D. Marketing, exporting and purchasing). 'Bibliographical sources issued ...'. *Class No:* 666.3/.7(01)

Encyclopaedias

[6250]
BROOK, R.J., *ed.* **Concise encyclopedia of advanced ceramic materials.** Oxford, etc., Pergamon Press, 1991. xvi,588p. illus., diagrs., graphs, tables, chemical structures. £100. (*Advances in materials science and engineering.*) ISBN: 0080347207.

137 articles, with references, A-Z, on chemical systems, steps in processing ceramics and particular applications and phenomena (*e.g.* 'Magnesium and alkaline - earth oxides', p.283-86: 5 sections, 2 tables, 7 references). Detailed, analytical index, 49p. Some material taken from *Encyclopedia of materials science and engineering* (*qv*) and the supplementary vols.; other material is newly commissioned, but will appear in the supplements as well. Although convenient in the form, not essential for libraries holding the *Encyclopedia* and supplements.
Class No: 666.3/.7(031)

Handbooks & Manuals

[6251]
GRIMSHAW, R.W. **The Chemistry and physics of clays and allied ceramic materials.** 4th ed., rev. London, Benn, 1971. 1024p. illus., diagrs., graphs, tables, chemical structures.

14 chapters, with footnote references: 3. Crystal structure of the silicates: 5. The identification and estimation of minerals in ceramic materials; 11. Chemical changes in ceramic materials; 12. Physical changes in ceramic materials; 13. Strength and allied properties. Appendix: The properties of some materials associated with ceramic material. Author and subject indexes. *Class No:* 666.3/.7(035)

[6252]
HAMER, F. *and* HAMER, J. **The Potter's dictionary of materials and techniques.** 3rd ed. London, A. & C. Black, 1991. 384p. illus., diagrs. £35. ISBN: 0812231120.

Covers materials, processes, equipment, chemical and physical properties, glazes, terminology and historical developments. Extensively illustrated. A standard; the major changes in the 3rd ed. include additional tables and extensive revisions of the articles on 'Raku' and 'Salt glazes'.
Class No: 666.3/.7(035)

[6253]
Handbook of ceramics and composites. Cheremisinoff, N.D., *ed.* New York, Marcel Dekker, 1990-. v.1-.
V.1. *Synthesis and properties.*
V.2. *Mechanical properties and specialty applications.* 1992. 510p. ISBN: 0824780031.
V.2 has 20 contributors; 12 documented chapters. Detailed, analytical index. Other volumes to follow.
Class No: 666.3/.7(035)

[6254]
Handbook of structural ceramics. Schwartz, M.M., *ed.* New York, McGraw-Hill, 1992. v.p. illus., diagrs., graphs, tables. $62.50. ISBN: 0070557195.

Numerous contributors. 9 chapters: 1. Introduction to ceramic materials ... 9. The future. Topics include processing techniques, mechanical property data, manufacturing, fabrication, assembly processes. Detailed, analytical index. 'A very important source book ... excellent' (*Choice,* November 1992, p.500). *Class No:* 666.3/.7(035)

[6255]
MORRELL, R. **Handbook of properties of technical and engineering ceramics.** National Physical Laboratory. London, H.M. Stationery Office, 1985-. pt.1-. illus. ISBN: 0114800529; 0114800537.

Part 1 (1985. 360p. £33) initiates a multi-part handbook by providing a broad comparison of the different types of ceramic. Part 2, *Data reviews,* section 1: *High alumina ceramics* (1987. 270p. £29.50) deals with the properties of a group of widely used types of materials. *Class No:* 666.3/.7(035)

[6256]
NEWMAN, A.C.D., *ed.* **The Chemistry of clays and clay minerals.** London, Longman. 1987. 480p. £63. (*Mineralogical Society monograph, no. 6.*) ISBN: 0582301149.

Covers not only the chemical constitution of the silicate clays, with associated iron, aluminium and manganese oxides and hydroxides, but also the properties of those materials most useful to man, - particularly industrially. 'A major contribution to clay science' (*Nature,* v.328, 13 August 1987, p.588). *Class No:* 666.3/.7(035)

Dictionaries

[6257]
Glossario ceramico, italiano/inglese, inglese/italiano. Faenza, Faenza Editrice S.p.A., 1983. 323p.

About 3,000 terms in each half. Ranges from brickware to ovenware, tiles to refractories, and from sanitary-ware to artistic ceramics. Pocket-sized dictionary. *Class No:* 666.3/.7(038)

[6258]
O'BANNON, L.S. **Dictionary of ceramic science and engineering.** New York, Plenum Press, 1984. xv, 302p. tables. $45. ISBN: 0306413248.

About 8,000 brief definitions of words, terms, materials, processes, products and business vocabulary used in ceramics and related industries. Chemical and physical properties and their uses are stated. Appended tables of chemical elements, conversion, ceramic colours. Bibliography, p.301-3. No illus. Criticized in *Nature* (v.312,

....*(contd.)*

13 December 1984, p.471) as being 'not detailed enough for the professional ceramist and yet is too detailed for non-ceramists'. *Class No:* 666.3/.7(038)

Reviews & Abstracts

[6259]

Ceramic abstracts. American Ceramic Society, *comp.* Colombus, Ohio, American Ceramic Society, 1918-. 6pa. $120. ISSN: 00959960.

Originally published in *Journal of the American Ceramic Society.*

About 10,000 abstracts pa. 18 sections: 1. Abrasives ... 3. Cements, limes and plasters ... 5. Glass ... 9. Electronics ... 15. Engineering materials - 16. Chemistry and physics - 17. General - 18. Books. Keyword index per issue. Annual author and subject indexes.

Available online via DIALOG, fille 335. *Class No:* 666.3/.7(048)

[6260]

World ceramic abstracts. Oxford, Pergamon Press, 1989-. Monthly. ISSN: 09578897.

Formerly *British ceramic abstracts* (1942-1988); originally British Ceramic Society. *Abstracts* (1930-1941).

About 6,000 abstracts pa. 16 sections (*e.g.* Intermediate and semi-finished products; Processing and treatment (including machines)). Subject index per issue and annually. Online. *Class No:* 666.3/.7(048)

Yearbooks & Directories

[6261]

'Ceramic industries journal', *International directory issue*, 1986. Rickmansworth, Herts., Turret-Wheatland, 1985. 96p.

A-Z directory of manufacturers and supppliers - European suppliers - Classified guide to supplies and services (materials, plant, equipment and appliances, services) - Who's who in the ceramics industries - Trade names - British pottery manufacturers - Societies and associations - European associations - Brick Development Association. Index to advertisers. *Class No:* 666.3/.7(058)

Refractories

[6262]

BRITISH STANDARDS INSTITUTION. Glossary of terms relating to refractory materials. London, BSI, 1990. 3 parts, diagrs. (*BS 3446: 1990.*)

1. *General and manufacturing.* 62p.

2. *Applications in the coke, glass, cement and other non-metallurgical industries.* 46p.

3. *Applications in the metallurgical industries.* 37p.

Part 1 has over 700 terms in a systematic order. Alphabetical index. *Class No:* 666.7

[6263]

FÉDÉRATION EUROPÉENNE DES FABRICANTS DES PRODUITS RÉFRACTAIRES. Vocabulaire: les produits réfractaires et leurs applications (sidérurgie, verrerie, cimenterie, etc.). 2. éd. Zurich, the Federation (PRE), 1973. 342p.

First published 1964.

2,214 French-base terms, with German, English and Italian equivalents. Applications of refractory products (steel, glass, cement industries, etc.). Reverse index in one A-Z sequence. *Class No:* 666.7

Gypsum

[6264]

VOLKART, K. Gips-Wörterbuch. Dictionnaire du gypse et du plâtre. [Gypsum and plaster dictionary.] Paris, Eyrolles; Wiesbaden, Bauverlag GmbH, 1971. xv, 176p.

Part 1: German-English-French. 1,452 numbered items. Genders are stated. Parts 2-3: French and English indexes. *Class No:* 666.91

Cement Industry

Bibliographies

[6265]

UNITED NATIONS. Industrial Development Organization. Information services on the cement and concrete industry. 2nd rev. ed. New York, United Nations, 1977. xii, 90p. $4. (*UNIDO guides to information sources, no.2. Rev.ed.*)

First published 1972.

690 entries. 12 sections: 1. Professional, trade and research organizations, learned societies and special information services - 2. Directories - 3. Sources of statistics, marketing and other economic data - 4. Basic handbooks, textbooks and manuals. -5. Monograph series - 6. Current periodicals - 7. Current abstracting and indexing periodicals - 8. Proceedings, papers and reports - 9. Specialized dictionaries and encyclopaedias - 10. Bibliographies - 11. Films and film catalogues - 12. Other potential sources of information (*e.g.* Patents and licences. Translation services). 'Bibliographical sources used ...'. *Class No:* 666.94/.98(01)

Handbooks & Manuals

[6266]

LABAHN, O. and KONKOLHAUS, B. Cement engineers' handbook. Translated by C. Amerongen from the 6th German edition. 4th English ed. Wiesbaden, Bauverlag GmbH, 1982. 794p. illus., tables. $90. ISBN: 3762509751.

Original as *Zement-Ingenieure.*

Sections: Raw materials - Cement chemistry; cement quality - Manufacture of cement - Handling and feeding systems; continous conveyors - Maintenance and wear -

....*(contd.)*

Lubricants, storage and consumption. Subject index. Subject coverage includes testing and design of laboratory facilites. Standard text, for specialists. *Class No:* 666.94/.98(035)

Dictionaries

[6267]

AMERONGEN C., van. Zement-Wörterbuch. Herstellung und Technologie. Deutsch/Englisch, Englisch/Deutsch. [Dictionary of cement, manufacture and technology.] 2. Aufl. Wiesbaden, Bauverlage GmbH, 1986. 332p.

First published 1967.

German-English and English-German dictionary of cement manufacture and technology. Each half has *c.*12,000 entry-words. German genders are given. *Class No:* 666.94/.98(038)

[6268]

ONISI, T.R., *comp.* **Elsevier's dictionary of the cement industry,** in five languages, English, French, German, Spanish and Japanese. Amsterdam, etc., Elsevier, 1987. xii, 520p. ISBN: 0444426299.

5,488 English-based terms (categorized) with equivalents and indexes in French, German, Spanish and Japanese. Bibliography, p.xi-xii. 11 appendices in the 5 languages (*e.g.* 9. Weights and measures). Genders given. Some inclusion of words extraneous or in common usage. *Class No:* 666.94/.98(038)

Yearbooks & Directories

[6269]

World cement directory, 1987. 8th ed., rev. Paris, Cembureau, 1987. xii, 295p. tables, maps.

First published 1958.

Covers over 100 countries, under regions: Africa; North America; Central America; South America; Middle East; Asia; Europe; Eastern Europe. Tabular data on companies, cement mills, fuel usage and coal mills, cement kilns. *Class No:* 666.94/.98(058)

Concrete

Databases

[6270]

Cement and concrete technical information/CACTI. Slough, Berks., Cement and Concrete Assocaiation, 1976-.

Online database, updated 6pa. with 250 items, on the use and performance of concrete. Host: DATASOLVE. *Class No:* 666.97(003.4)

Handbooks & Manuals

[6271]

DEWAR, J.D. *and* **ANDERSON, R. Manual of ready-mixed concrete.** 2nd ed. Glasgow, Blackie, 1992. x,245p. illus., graphs, tables. ISBN: 0751400793.

First published 1988.

11 chapters in 2 parts: 1. Technology (*e.g.* materials; properties; mix design; quality control) - 2. Practice (*e.g.* production; specification and supervision; ready-mixed concrete on site; organizations). References. Detailed, analytical index, p.241-45. *Class No:* 666.97(035)

[6272]

GAMBHIR, M.L. Concrete technology. New York, etc., McGraw-Hill, 1987. 304p. illus. £4.95. ISBN: 0074516442.

12 sections on manufacture and production of structural concrete. Concrete-making materials (2 sections) - Properties of fresh concrete - Properties of hardened concrete - Statistical quality control of concrete - Proportioning of concrete mixes - Production of concrete - Extreme weather concreting - Special concrete and concreting techniques - Form work - Inspection and testing - Repair technology for concrete structures. 78 illus. *Class No:* 666.97(035)

Dictionaries

[6273]

FÉDÉRATION INTERNATIONALE DE LA PRÉCONTRAINTE. Multilingual dictionary of concrete, in English, French, German, Spanish, Dutch and Russian. Amsterdam, etc., Elsevier, 1978. viii, 202p.

1,416 English-based terms, with equivalents and indexes in French, German, Spanish, Dutch and Russian. 'Recommended for technical and engineering libraries' (*Choice,* v.14(12), February 1978, p.1626). *Class No:* 666.97(038)

[6274]

TEKNISKA NOMENKLATURCENTRALEN. Betongteknisk ordlista, svensk, engelsk, fransk, tysk, dansk, finsk, norsk. Stockholm, The Centre, 1971. 366p. (*TNC 46.*) ISBN: 9171960465.

1,310 Swedish-based concrete terms; classified arrangement. English, French, German, Danish, Finnish and Norwegian equivalents, Indexes in all 7 languages. References, p.362-4. *Class No:* 666.97(038)

Yearbooks & Directories

[6275]

The Concrete year book, 1991. 67th ed. London, Palladian Publications, Ltd., 1991. Annual. 560p. ISSN: 00698288.

General information on concrete and construction - Institutions and associations - Consulting and structural design - Site investigation - Contractors - Piling - Tunnelling - Specialist services - Concrete repairs and protection - Cement ... Products and equipment for manufacturers - Reinforcement and prestressing ... Guide to products in precast concrete - Trade names - Personnel in the concrete industry. Indexes to advertisers. General index. *Class No:* 666.97(058)

Colour

[6276]

DANGER, L.E.P. The Colour handbook: how to use colour in commerce and industry. Aldershot, Hants, Gower Technical Press, 1987. xxi, 687p. £50.

Three parts: 1. Principles of light and colour - 2. Principles of colour selection - 3. Colour catalogue (character, attributes, applications); colour groups. Bibliography, p.678-87. Primarily for marketing managers, but also of interest to advertising agents, designers, interior-decoration retailers. One notable drawback: the lack of an index. *Class No:* 667

Dyes & Dyeing

[6277]

PONTING, K.G. **A Dictionary of dyes and dying.** London, Mills & Boon, Ltd., 1980. [v], 207p. illus. ISBN: 0263063984.

Several hundred entries, some lengthy (*e.g.* 'Dyeing, history': p.53-58; 'Indigo': p.103-17). Bibliography (A-Z authors), p.189-207. 30 illus. The natural dyes are described in some detail. Sources include The Textile Institute and the Society of Dyers and Colourists. *Class No:* 667.2

[6278]

Sigma-Aldrich handbook of stains, dyes and indicators. Green, F.J. Milwaukee, WI, Aldrich Chemical Co., 1990. 776p. $70. ISBN: 0941633225.

Guide to 325 compounds used as stains, dyes and indicators, A-Z by common name. Full data, plus reference to other sources and spectra when easily available. *Class No:* 667.2

Printing Inks

[6279]

Printing ink manual. Leach, R.H., *ed.* 4th ed. London, Blueprint, 1988. 884p. illus., diagrs., tables. £60. ISBN: 0948905778.

First published 1961. 3rd ed. 1979.

16 chapters (*e.g.* 2. The nature of inks ... 4. Colour and colour matching ... 7. Lithographic inks ... 10. Screen inks ... 13. Manufacture of inks and varnishes ... 16. Health, safety and the environment). Glossary of abbreviations. Extensive references. Detailed, analytical index. *Class No:* 667.5

Paints & Varnishes

Bibliographies

[6280]

UNITED NATIONS. Industrial Development Organization. **Information sources on the paint and varnish industry.** New York, United Nations, 1975. xii, 76p. $5. (*UNIDO guides to information sources, no,. 18.*) ISBN: 9211061458.

585 numbered items. 12 sections: 1. Professional, trade and research organizations, learned societies and special information services - 2. Directories - 3. Sources of statistics, marketing and other economic data - 4. Basic handbooks, textbooks and manuals - 5. Monograph series - 6. Current periodicals - 7. Current abstracting and indexing periodicals - 8. Proceedings, papers and reports - 9. Specialized dictionaries and encyclopaedias - 10. Bibliographies - 11. Films - 12. Other potential sources of information (*e.g.* meetings and conferences; patents and licences). List of bibliographical sources used. Some sections are annotated. *Class No:* 667.6(01)

Dictionaries

[6281]

DOORGEEST, T., *and others.* **Elsevier's paint dictionary in four languages:** English, French, German and Dutch. Amsterdam, Elsevier, 1990. [xiv],292p. ISBN: 0444880682.

4,081 English base terms, with equivalents. French, German, Dutch indexes. *Class No:* 667.6(038)

[6282]

RAAFF, J.J. **Index vocabulorum quadrilinguis: Verf en verris;** Peintures et vernis; Paint and varnish; Farbe und Lack. [2nd ed.]. The Hague, Van Goor Zonen den Haag, 1965. 1129p.

First published 1958.

About 10,000 terms, in 4 languages. 4 sequences: Dutch-French-English-German; German-English-French-Dutch; English-German-Dutch-French; German-Dutch-French-English. *Class No:* 667.6(038)

[6283]

TEKNISKA NOMENKLATURCENTRALEN. **Farg-och lackteknisk ordlista:** svensk-engelsk-fransk-tysk-finsk. [Glossary of paint terms: Swedish-English-French-German-Finnish.] Stockholm, the Centre. 1967. 243p.

Includes lists of abbreviations of polymers and plasticisers, and of selected elements, and a list of Danish and Norwegian words that deviate appreciably from their Swedish counterparts. *Class No:* 667.6(038)

Reviews & Abstracts

[6284]

Paint titles. Paint Research Association Services, Teddington. Oxford, etc., Pergamon, for Paint Research Association, [1971]-. Weekly. ISSN: 01444425.

About 200 references per issue: 21 classes (1. Pigments, extenders & dyestuffs: 19. Adhesives. Sealants; 20. Miscellaneous; 21. New books). Patent equipment.

Patent Research Association's *World surface coatings abstracts* is online. *Class No:* 667.6(048)

Periodicals

[6285]

European paint and resin news. London, Information Research, Ltd., 1984-. Monthly. ISSN: 02667800.

Formerly *Continental paint and resin news,* 1977-83.

V.26(10), October 1988 has 16p. of news items. Changes in the industry - Marketing news and statistics - Financial results and comparisons - Recent technology developments - Forthcoming symposia and exhihitions - European patents review. No index. *Class No:* 667.6(051)

Metallurgy

Databases

[6286]

METADEX database. Metals Park, Ohio, American Society for Metals; London, The Institute of Metals, 1966-.

METADEX is the major database for metals and metallurgy. The online version of *Metal abstracts.* Available from DIALOG, DATA-STAR, ESA and other hosts. *Class No:* 669(003.4)

Bibliographies

[6287]

GIBSON, E.B. *and* TAPIA, E.W., *eds*. **Guide to metallurgical information.** 2nd ed. New York, Special Libraries Association, 1965. xviii, 222p. (*S.L.A. bibliography no.3.*)

First published 1961 (96p.).

1,100 numbered and concisely annotated entries. Sections: General continuing sources (information centers; indexes; abstracts ...) - General reference sources - Metallurgy; science and technology sources (p.83-130) - Metals/materials (pure metal elements, A-Z, p.154-80; information centers; bibliographies; general) - Translations and microforms. Indexes: personal authors, organizations, general title, serial title, subject. 'A research tool for information specialists and librarians, metallurgists, research workers, teachers and students' (*Preface*). A valuable tool, but dated. *Class No:* 669(01)

[6288]

HYSLOP, M.R. **A Brief guide to sources of metals information.** Washington, Information Resources Press, 1973. xi, 180p.

7 chapters. 1. Libraries and technical societies - 2. Abstracts, indexes, bibliographies, reviews - 3. Journals, books, translations, standards, patents - 4. Federal agencies - 5. Searching services and information centres - 6. Self-help - 7. Who is doing what in research (US). The directory section occupies over half of the *Brief guide*. US and Canadian slant. *Class No:* 669(01)

[6289]

LAYLAND, B.D. **Metals and metalworking:** guide to the literature. London, Library Association, 1965. (High Wycombe, Bucks., University Microfilms, Ltd., for microfilm/xerox). 3v. (1647p.).

Thesis accepted for Fellowship of the LA, 1965.

Part 1 lists 4,051 items on metallurgy, metal-working, metal testing and physical properties of metals. Part 2: a list of publishers' series and 17 sections of bibliographical aids (section 15: 65 abstracting and indexing services). Comprehensive list of periodicals, p.1248-1408, coded to 12 abstracting services. Subject and author index. *Class No:* 669(01)

[6290]

PATTEN, M.N., *ed*. **Information sources in metallic materials.** London, etc., Butterworths, 1989. xvi,415p. tables. £40. (*Guides to information sources.*) ISBN: 0408014911.

17 contributors; 18 chapters, sub-divided. 2 parts: I. Materials (1. General sources - 2. Extraction metallurgy - 3. Iron and steel ... 7. Lead ... 9. Zinc ...); II. Applications (11. Powder metallurgy ... 14. Design ... 18. Materials for the packaging industry). Appendices: abbreviations and addresses of organizations. Index, 8p. Forms of literature (*e.g.* reports, patents) are treated in their relation to a particular metal or application; *e.g.* chapter 13, 'Corrosion', p.253-63, covers learned societies, dictionaries, standards organizations, handbooks, data banks, marketing companies, trade associations, books, abstracts, journals, reviews. A valuable contribution to an important series, formerly *Butterworths guides to information sources.* *Class No:* 669(01)

Encyclopaedias

[6291]

TOTTLE, C.R. **An Encyclopaedia of metallurgy and materials.** London, Metals Society: Macdonald and Evans, 1984. ci, 380p. illus., tables, chemical structures. £65. ISBN: 0712105769.

Rev. and enl. ed, of Merriman, A.D. *Dictionary of metallurgy* (1958) and *A concise encyclopaedia of metallurgy* (1965).

An encyclopaedic dictionary, combining brief definitions with longer entries (*e.g.* 'Aluminium': 2 columns; 'Oxygen steelmaking': 2½ columns). 11 tables (p.xi-ci) precede the A-Z sequence. 210 figures, 66 tables. Aimed at a wide variety of users, technical and non-technical, engineers and also users in developing countries. *Class No:* 669(031)

Handbooks & Manuals

[6292]

HARGREAVES, D. *and* FROMSON, S. **World index of strategic minerals:** production, exploitation and risk. New York, Facts on File, 1983. xiii, 300p.

4 parts: 1. Strategic risk assessment - 2. 37 minerals and mineral groups - 3. 34 countries, giving production; mining industry; investment and development; country risk factor - 4. 63 individual companies. Data up to 1982. 'Should be very useful in technical and business libraries' (*Reference books bulletin, 1983/1984*, p.175). *Class No:* 669(035)

[6293]

Metals handbook. Prepared under the direction of the ASM International Handbook Committee. 10th ed. Metals Park, OH, American Society for Metals, 1990-. illus., diagrs., graphs, tables.

First published 1948; 9th ed. (17v.) 1978-1989.

1. *Properties and selection: irons, steels, and high-performance alloys.* 1990. 1063p.

2. *Properties and selection: nonferrous alloys and special purpose metals.* 1990. 1328p.

3. *Heat treating.* 1991. c.1200p.

4. *Friction, lubrication and wear technology.* 1991. c.1200p.

5. *Surface cleaning, finishing and coating.* Due 1992.

6. *Welding, brazing and soldering.* Due 1992.

7. *Microstructural analysis.* Due 1993.

V.2 is the work of over 400 authors, reviewers and contributors. 5 major sections (62 articles): 1. Specific metals and alloys - 2. Special-purpose materials - 3. Superconducting materials - 4. Pure metals - 5. Special engineering topics. Appended: Metric conversion guide; Abbreviations, symbols and trade names. Analytical index, p.1279-1328. Micrographs. Numerous tables. Written primarily for the industrial user of metals and alloys. *Class No:* 669(035)

[6294]

—Metals handbook. Boyer, H.E. *and* Gull, T.L., *ed*. Desk ed. Metals Park, Ohio, American Society for Metals, 1985. v.p.(c.1200p.). illus., diagrs., tables. ISBN: 087170188x.

4 parts (37 chapters): 1. General information (glossary; engineering tables, etc.) - 2. Properties and selection (chapters 4-20; *e.g.* 4. Carbon and alloy steels; 20. Materials for special application) - 3. Processing - 4. Testing and inspection (*e.g.* 37. Quality control). Some chapters have

....*(contd.)*

references appended. 61p. index. Aims 'to provide a single authoritative first-reference to all of metals technology' (*Preface*). *Class No:* 669(035)

[6295]
ROSS, R.B. Metallic materials specification handbook. 4th ed. London, Chapman & Hall, 1992. xiii[i],830p. £99.95. ISBN: 0412369400.

First published 1968 as *Metallic materials.* 3rd ed. 1980.

56 metals, with alloys, A-Z (Aluminium - Zirconium). Data include physical properties, specifications and tradenames, mechanical and other properties, symbols. Appended: suppliers, conversation tables. Index, p.635-830. *Class No:* 669(035)

[6296]
SIMONS, E.N. Guide to uncommon metals. London, F. Muller, 1967. 244, [1]p.

Data on 47 metals, Antimony ... Zirconium. Excludes alloy metal. Appended: 'Books for reference' (10 items). Facts and figures for the metallurgist, student, engineer and scientist. *Class No:* 669(035)

Dictionaries

[6297]
CAGNACCI-SCHWICKER, A., *comp.* **International dictionary of metallurgy, mineralogy, geology, mining, and oil industry.** Milan, Technoprint International Milano, in association with New York, etc., McGraw-Hill, 1968. [xv], 1530p.

20,372 numbered English-base terms, with French, German and Italian equivalents and indexes. Terms in the 3 main fields - metallurgy, geology, oil industry - are differentiated by means of symbols. *Class No:* 669(038)

[6298]
CLASON, W.E. Elsevier's dictionary of metallurgy and metal working, in six languages: English/American, French, Spanish, Italian, Dutch and German. Amsterdam, etc., Elsevier, 1978. [viii], 848p. DFl 375. ISBN: 0444416951.

First published 1966.

8,406 English/American terms, with equivalents and indexes in French, Spanish, Italian, Dutch and German. *Class No:* 669(038)

[6299]
MacANDREW, A.R., *ed.* **A Glossary of Russian technical terms used in metallurgy.** New York, Varangian Press, 1953. v, 127p.

Compiled at Columbia Univ., New York under contract with National Science Foundation.

About 6,000 Russian terms with English equivalents. Intended to supplement L.I Callaham's *Russian-English technical and chemical dictionary* (1947) and therefore excludes metallurgical terms in Callaham (*qv.*) Indicates stress in Russian pronunciation. Neiswender, R. (*Guide to Russian reference and language aids,* p.29) considers MacAndrew superior to Howeston, P.W. and Akhonin, A. *Russian-English glossary of metallurgical and metal-working terms* (Cambridge, Mass., Center for International Studies, MIT, 1955. iii, 175p.). *Class No:* 669(038)

[6300]
STÖLZEL, K., *ed.* **Dictionary of metallurgy and foundry technology.** Amsterdam, etc., Elsevier, 1984-87. 2v.

1. *English-German.* 1984. 418p. £79.58. ISBN 0444996125.

2. *German-English.* 1987. 464p. £75.08. ISBN 0444995145.

Wide-ranging scope. Brief entries, with equivalents and occasional definition or explanation.

Review of v.1: 'It will be of particular value to linguists, documentation experts and scientists as well as teachers and students ...' (*Iron and steelmaker,* v.14, August 1987, p.65). *Class No:* 669(038)

[6301]
TAYLOR, J.L. Dicionário metalúrgico: inglês-português, português-inglês. São Paulo, Associação Brasileirà de Metais, 1976. [viii], 619, [2]p. ISBN: 0916362006.

Rev. ed. of the 1963 English-Portuguese metallurgical dictionary. The Portuguese-English part is new.

About 30,000 entries in each half. Many phrases. *Class No:* 669(038)

[6302]
WANDERER, J.A. Metallurgisches Wörterbuch: teil 1: Englisch-Deutsch; teil 2: Deutsch-Englisch. [Metallurgical dictionary: part 1: English-German; part 2: German-English.] Heere, Wanderer Verlag, 1989. 608p. ISBN: 3927726001.

Includes chemical compounds and abbreviations. *Class No:* 669(038)

Reviews & Abstracts

[6303]
IMM abstracts; a survey of world literature on economic geology, mining, mineral dressing, extraction metallurgy [and] allied subjects. London, Institution of Mining and Metallurgy, 1950-. 6pa. £75pa. ISSN: 00190020.

About 3,000 abstracts per issue. 13 sections: Generalities - Mineral industry - Mathematical methods and computing - Physical and chemical studies - Analysis and instrumentation - Economic geology - Mining - Mineral processing - Metallurgy - Health and safety - Environment - General engineering - Civil engineering. No indexes. Microfiche and microfilm available. Online through IMM. *Class No:* 669(048)

[6304]
Metals abstracts. London, Metals Information: Institute of Metals, and Metals Park, Ohio, ASM International, 12pa. $1875; £1160pa. ISSN: 00260924.

About 4,000 abstracts and references pa. 36 sections (*e.g.* 11. Construction - 12. Crystal properties ... 21. Metallography ... 31 Mechanical properties ... 41 Ores and raw materials ... 51. Foundry ... 61. Engineering components and structures ... 71 General and non-classified - 72. Special publications. Author index per issue. Online as METADEX.

Metals Information also publishes *Iron & steel industry profiles, Steel alert, Nonferrous alert,* and the annual *World calendar* (1976-). *Class No:* 669(048)

[6305]

Mineralogical abstracts. London, Mineralogical Society of Great Britain, and the Mineralogical Society of America, 1920-. Quarterly. £95.

Earlier (1920-58) published in *The mineralogical magazine*.

About 1,250 abstracts per issue. 18 sections, A-V (Age determination; Apparatuses and techniques; Book notices ... Economic minerals and ore deposits ... Experimental mineralogy ... Geochemistry ... Meteorites and tektites - Mineral data ... Petrology ... Various topics. Author index per issue; annual author and subject indexes.

Class No: 669(048)

[6306]

UNITED STATES. Bureau of Mines. **New publications of the Bureau of Mines.** Monthly list. Washington, U.S. Government Printing Office, 1910-. Monthly, cumulated annually.

4 main sections: 1.Sales publications (*e.g., Minerals year book*) - 2. *Free publications (reports of investigations; information circulars; reprints from Minerals year book*) - 3. Open file reports, NTIS - 4. Outside publications.

Class No: 669(048)

Periodicals

[6307]

AMERICAN SOCIETY FOR METALS. Source journals in metals and materials. 3rd ed. Metals Park, Ohio, the Society, 1985. 76p. $40.

Lists 1,000 scientific, engineering and trade journals, covering metallurgy in engineered materials.

Class No: 669(051)

Yearbooks & Directories

[6308]

Industrial minerals directory: a world guide to producers of non-metallic, non-fossil minerals. 2nd ed. Worcester Park, Surrey, Metal Bulletin Books, Ltd., 1986. lvi, 642p. £60. ISBN: 0900542993. ISSN: 01419263.

First published 1977.

A directory of producers, under countries Albania ... Zimbabwe. Buyers' guide, in a mineral-by-mineral guide to producers. Professional services directory. Index to advertisers. *Class No:* 669(058)

[6309]

'Metal bulletin' prices & data. Cordero, R. *and* Packard, R., *eds.* Worcester Park, Surrey, Metal Bulletin Books, Ltd., 1968-. Annual. tables. ISSN: 02626454.

Until 1986 as *'Metal bulletin' handbook.*

The 1988 annual (374p. £25), third in the series, has main parts: Non-ferrous prices (aluminium ore (bauxite) ... zirconium), subdivided by country - Iron and steel prices (Australia; International steel spot; Continental steel export prices; ECSE countries; EFTA countries); Iron and steel scrap prices/iron ore prices - Statistics (non-ferrous; iron & steel; production; consumption)- Memoranda. Advertisers index. *Class No:* 669(058)

[6310]

Metal traders of the world. 4th ed. Worcester Park, Surrey, Metal Bulletin Books, Ltd., 1990. *c.*740p. £80.

2 parts: producers; traders. Data: address; phone and fax numbers; management; ownership; capital; subsidiaries; trade names; raw materials, etc. *Class No:* 669(058)

[6311]

Ryland's: the directory of the engineering industry, 1987/88. 53rd ed. London, Fuel & Metallurgical Journals, Ltd., 1987. ISSN: 0080505x.

Alphabetical section (countries, A-Z). Classified section: 1. Iron, steel, and ferro-alloys - 2. Rolled iron and steel sections - 3. Steel stockholders - 4. Non-ferrous metals - 5. Non-ferrous metal stockholders - 6. Engineering, engineers' products and services. Geographical index. Brand names & trade marks. Employers' and trade associations & trade unions. Index to advertisers. *Class No:* 669(058)

Tables & Data Books

[6312]

AMERICAN SOCIETY FOR METALS. Guide to materials engineering data and information. Metals Park, Ohio, ASM International, 1986. v.p. illus., tables.

10 sections 1. The ASM resource - 2. Physical data on the elements and alloys - 3. General engineering data - 4. Organizations and engineering data sources - 5. Standards: issuing organizations - 6. Directory of materials manufacturers - 7. Materials reference books and engineering journals (59p.) - 8. Style manual - 9. Milestones in the history of physical metallurgy - 10. Glossary of materials engineering terminology (108p., including small illus.). Over 4,000 entries. A 'compendium of fundamental engineering data - a one-stop reference to metals, metalworking, composites and engineering materials' (*Preface.*). *Class No:* 669(083)

[6313]

ASM metals reference book. Compiled by the editorial staff, Reference Books, American Society for Metals. 2nd ed. Metals Park, Ohio, American Society for Metals, 1983. 560p. ISBN: 0871701561.

Glossary of metallurgical and metalworking terms - Cast irons - Wrought stainless steels - Aluminum and aluminum alloys - Copper and copper alloys - Titanium and titanium alloys - Alloy phase diagrams - General engineering data. 15 sections. 175 data tables 'An important reference tool for engineers and metallurgists' '(*New technical books,* January 1985, item 119). For special libraries. *Class No:* 669(083)

[6314]

Minerals handbook, 1990-91: statistics and analysis of the world's minerals industry. Crowson, P., *ed.* New York, Macmillan, 1990. Biennial. 340p. tables. £65. ISBN: 0333512073.

Statistical tables for 1986-89, covering minerals (Aluminium, bauxite, and alumina ... Zirconium). 'Cobalt' (subdivided by country) has tabulated data: world reserves; world mine and metal production; production capacity; consumption; substitutes; prices; marketing (graph); supply and demand. Sources and notes. *Class No:* 669(083)

[6315]
ROBB, C. **Metals databook.** London, The Institute of Metals, 1987. [xvi],500p. diagrs., tables. £90. ISBN: 0904357694.

Data, largely in tabular form, on the chemical, physical and mechanical properties of commercially available metallic materials. Much of the data is specific to British Standards. Covers welding, heat treatment, test house and laboratory data. 4 appendices: 1. S.I. units - 2. Hardness testing - 3. Material specifications - 4. Equilibrium diagrams. No index, but detailed contents list. Small, handy format, but very expensive. *Class No:* 669(083)

[6316]
Roskill's directory of sources for metals and mineral data. London, Roskill Information Services, Ltd., 1976-. Biennial.

6th ed. (147p. 1984) lists 568 statistical sources in 3 parts: 1. International - 2. Metals and minerals (Aluminium ... Zirconium) - 3. Country sections (Algeria ... Zambia). Data on each source include bibliographical details, publisher/ distributor, number of years covered by statistics, frequency, and outline of contents relevant to metals and minerals. *Class No:* 669(083)

[6317]
SMITHELLS, C.J. **Smithells' metals reference book.** Brandes, E.A. *and* Brook, G., *eds.* 7th ed. London, etc., Butterworth Heinemann, 1992. xvi,[1664]p. illus., graphs, tables. £150. ISBN: 0750610204.

First published 1949; 6th ed., 1983 (xiv,1600p.).

Includes data on isotopes, crystallography, gas-metal systems, electron emission, magnetic properties, heat treatment, corrosion control and superplasticity. A standard source of data on metals and alloys. *Class No:* 669(083)

Nomenclatures

[6318]
EMBREY, P.G. *and* FULLER, J.P., *eds.* **A Manual of new mineral names, 1892-1978.** Oxford Univ. Press, co-published with British Museum (Natural History), 1980. x, 467p. £32.50. ISBN: 0198505012.

About 5,500 new mineral names, compiled from 30 lists published in the *Mineralogical magazine* from 1897 to 1976. Supplements Dana's *System of mineralogy (qv).* *Class No:* 669(083.72)

Histories

[6319]
AITCHISON, L. **A History of metals.** London, Macdonald & Evans, 1960. 2v. illus., diagrs., tables.

V.1 deals with the history of gold, copper, lead, zinc, tin, iron and mercury up to the end of the Dark Ages (c.1000 A.D.). V.2 (p.305-647), from that point onwards. Chapter bibliographies. A scholarly work, admirably illustrated and documented; fully indexed. Of value to those interested in social history as well as to the metallurgist. *Class No:* 669(091)

[6320]
TYLECOTE, R.F. **The Early history of metallurgy.** London, Longman, 1987. xxv, [1], 391p. illus., diagrs., graphs, tables, maps. (*Longman archaelogy series.*)

10 chapters (4. The origin of smelting and the design of smelting; 6. Fabrication techniques; 9. The coming of cast

.... (contd.)
iron of Europe). Bibliography (A-Z authors), p.355-79. Detailed, analytical index. Well illustrated and produced. *Class No:* 669(091)

[6321]
—TYLECOTE, R.F. A History of metallurgy. London, Metals Society, 1976. viii, 182p. illus., charts, tables.

Devotes one-third of the text to early history. 'Strongly recommended as the most informative and complete work in the field' (*British book news,* May 1977, p.391). *Class No:* 669(091)

Worldwide

[6322]
BRITISH GEOLOGICAL SURVEY. **World mineral statistics, 1982-86:** production, exports, imports. Lofty, G.J., *comp.* Keyworth, Nottingham, the Survey, 1988. v, 366p. £35. ISBN: 085272103x.

Tables for minerals, Alumina ... Zirconium, including iron, steel and ferro-alloys. Subdivision by countries (USSR excluded). Appended data (on blue-tinted paper, p.301-68): exports; imports: East European countries, Albania ... Rumania. *Class No:* 669(100)

[6323]
—BRITISH GEOLOGICAL SURVEY. World mineral production, 1983-87: preliminary statistics. Keyworth, Nottingham, the Survey, 1988. v.p. tables.

Provides production figures (Alumina ... Ziron Zirconium) for 1983-86, plus preliminary statistics for 1987. Appended: 'Other mineral commodities' (*e.g.* Beryl ... Vermiculite). Complementary to coverage of *World mineral statistics.* *Class No:* 669(100)

[6324]
The Minerals yearbook. US Bureau of Mines. Washington, US Governmment Printing Office, 1932/33-. Annual. 3v. ISSN: 00768952.

1. *Metals and minerals.*

2. *Area reports domestic.*

3. *Area reports international.*

V.1 provides data on individual metallic and non-metallic minerals (*e.g.* fuels), A-Z. V.2-3 consist mainly of statistical tables, with background notes. V.2 comprises area reports for the United States; V.3 has chapters on individual countries.

Updated by the US Bureau of Mines. *Mineral trade notes: a monthly inventory. Class No:* 669(100)

Great Britain

[6325]
BRITISH GEOLOGICAL SURVEY. **United Kingdom mineral statistics, 1986.** 14th annual ed. Keyworth, Nottingham, the Survey, 1988. 168p. diagrs., tables, maps. £35.

First published 1973.

Coverage: mineral production; UK overseas trade in minerals and mineral-based products; commodity reviews (energy; coal; ... sulphur); commodity summaries; UK production, consumption and trade, 1980-86. Explanatory notes to sections. *Class No:* 669(410)

[6326]

BRITISH GEOLOGICAL SURVEY. United Kingdom mineral statistics data to 1985. Keyworth, Nottingham, the Survey, 1987. vii, 168p. graphs, tables. ISBN: 0852720939.

Sections: Mineral production - UK overseas trade in minerals and mineral-based products - Commodity reviews (p.86-147) - Commodity summaries: UK production, consumption and trade, 1979-1985. *Class No:* 669(410)

USA

[6327]

Metal statistics. New York, American Metal Market, 1908-. Annual. diagrs., graphs, tables.

First published 1908.

83rd ed. 1990 (*c.*300p. $88; £53) includes profiles of more than 25 key metals (producers; end users; trends, etc.). *Class No:* 669(73)

Information Management

[6328]

JONES, L. *and* **VAUGHAN, J. Scientific and technical information on the metals industry:** report of the Metals Information Review Committee. London, British Library, 1982. vii, 110, [8]p. (*British Library. Research & Development report no. 5717.*)

The Committee's terms of reference were to study all aspects of the formation, transfer, storage, retrieval and use of metals information. The list of 11 recommendations concerned primary and secondary publications, databases, information for management, etc. Bibliography, p.48-56. *Class No:* 669:025.4

Thesauri

[6329]

CANADA. Centre for Minerals and Energy Technology. **Thesaurus of mineral processing and extractive metallurgy terms.** 2nd ed. Ottawa, Ministry of Supply and Services, Canada, Technology Information Division, 1985. xiii, [1], 345p.

About 1,700 terms in English and French. Appended list of descriptors. *Class No:* 669:025.43

[6330]

Thesaurus of metallurgical terms. 8th ed. Metals Park, Oh., ASM International; London, Institute of Metals, 1988. vii,171p.

First published 1968; 7th ed. 1986.

About 6,500 main terms and nearly 45,000 cross-reference terms (narrower, broader, related and synonymous). Includes date of first use of term for indexing purposes. *Class No:* 669:025.43

Physical Metallurgy

[6331]

DENNIS, W.H. A Hundred years of metallurgy. London, Duckworth, 1963. ix, 342p. diagrs., tables. (*The hundred years series.*)

Covers 1850-1950. Concentrates on physical metallurgy, with chapters on developments in particular areas (*e.g.* ore dressing, pyrometallurgy, steelmaking, non-ferrous and precious metals, metal working, metallopgraphy, etc.). Brief

....(contd.)

chapter bibliographies, listing items on which much of the text is based. Short glossary. 'A book of general interest at a high technical level' (*Aslib book list,* v.29(3), March 1964, entry no. 172). *Class No:* 669.017

[6332]

TYRKIEL, E.F. Dictionary of physical metallurgy, in English, German, French, Polish and Russian. Amsterdam, etc., Elsevier, 1977. viii, 402p. DFl.225. ISBN: 0444998101.

About 2,300 English-based terms and concise definitions in 17 groups (*e.g.* 05. Chemical elements; 85. Material defects). Equivalents and indexes in German, English, French, Polish and Russian. *Class No:* 669.017

[6333]

—FREIWILLIG, R.K., *and others.* Dictionary of physical metallurgy, in English, German, French, Russian and Spanish. Amsterdam, etc., Elsevier, 1986. *c.*220p. ISBN: 0444995277.

Comprises 1,672 terms, subject-grouped in 5 languages. English, German, French, Russian and Spanish indexes. Not to be confused with E.R. Tyrkiel's dictionary of the same title, with its wider scope. *Class No:* 669.017

Alloys

[6334]

Alloys index. London, Metals Information, Institute of Metals, Metals Park, Ohio, American Society for Metals, 1974-. Monthly, with annual cumulation.

About 2,500 references per issue. Sections: Alloy steels ... Zirconium base alloys. Descriptions appended to entries. An index to specific alloys, metallurgical systems and intermetallic compounds cited in *Metals abstracts.* Online METADEX database. *Class No:* 669.018

[6335]

SIMONS, E.N. A Dictionary of alloys. London, F.Muller, 1969. 191p. diagrs., graphs, tables.

Alloys, AC 41 ... Zisium alloy. Preceded by numbered alloys. Appended British Standards. Includes 'a few obsolete or obsolescent terms because they still occur in enquiries or in commerce and export' (*Preface*). *Class No:* 669.018

[6336]

SMITH, W.F. Structure and properties of engineering alloys. New York, etc., McGraw-Hill, 1980. 512p. illus., diagrs., graphs, tables. (*Materials science and energy series.*)

11 documented chapters, with micrographs and appended 'Problems'. 'Nickel and cobalt alloys', p.458-501, has 21 references. 'Intended as refresher text and reference for practicing engineers who have become specialized or who have left the mainstream of technical work' (*Preface*). *Class No:* 669.018

[6337]

WOLDMAN, N.E. Woldman's Engineering alloys. Frick, J.P., *ed.* 7th ed. Metals Park, Ohio, ASM International, 1990. 1459p. $128. ISBN: 0871704080.

First published 1936. 1st-5th eds. as *Engineering alloys.* 6th ed. 1979.

The 7th ed. lists 40,000 alloys and claims revision from the 6th ed. for over 12,000 of them. Data note trade name, composition, properties uses. Manufacturers A-Z and numerically. *Class No:* 669.018

Furnaces

[6338]

SIMONS, E.N. **A Dictionary of metal heat treatment.**
London, F. Muller, 1976. 248p. diagrs., graphs, tables.
ISBN: 0584100795.

About 2,000 terms defined, Absolute scale ... White layer.
Class No: 669.04

[6339]

TYRKIEL, E.F., *ed.* **Multilingual glossary of heat
treatment terminology.** London, Institute of Metals, 1986.
207p. . £55. ISBN: 0934357701.

Covers English, French, German and Russian in four
language sequences. *Class No:* 669.04

Metal Fabrications

[6340]

Surface treatment technology abstracts; incorporating *Metal
finishing abstracts;* including *Printed circuits and elctronics
coatings abstracts.* Teddington, Middx., Finishing
Publications, Ltd., 1986-. v. 28(1)-. 6 pa. £290pa. ISSN:
09505199.

As *Metal finishing abstracts,* v. 1-27, 1959-85. Previously
appeared in *Electroplating and metal spraying* (1947-1959).

Over 7,000 indicative and informative abstracts and
citations pa. covering surface treatment, in 20 classes
(Cleaning and degreasing ... Electroplating ... Metal
spraying and hard facing ... Corrosion and electrochemistry
... General and miscellaneous). Covers journal articles,
books, standards, reports and patents - full coverage for UK,
US, GDR and USSR patents.

Separate section for *Printed circuits and electronics
coatings abstacts.* 4 sections (Printed circuit boards; semi
conductors and integrated circuits; Electrical connectors;
Miscellaneous). Over 2,000 abstracts pa. Separate annual
indexes for each part, of authors, subjects and patent
numbers. Abstracts appear 3-12 weeks after publication.
Class No: 669.056

Iron & Steel Industry

Bibliographies

[6341]

LAWRANCE, P.R. 'Iron and steel'. *Information sources in
metallic materials.* Editor, M.N. Patten, p.59-85. London,
Bowker-Saur, 1989. (*Guides to information sources.*) ISBN:
0408014911.

Sections: Reference sources - Periodicals - Abstract
journals & conference proceedings - Conferences -
Standards - Specialized databases and services - Company
sources - Translation dictionaries - Statistics.
Class No: 669.1(01)

[6342]

UNITED NATIONS. Industrial Development Organization.
Information sources on the iron and steel industry. New
York, United Nations, 1977. xii, 403p. $4. (*UNIDO guides
to information sources, no. 26.*)

788 items, 12 sections: 1. Professional, trade and research
organizations, learned societies and special information
sources - 2. Directories - 3. Sources of statistics, marketing
and other economic data - 4. Basic handbooks, textbooks
and manuals - 5. Monograph series - 6. Current periodicals -
7. Current abstracting and indexing periodicals - 8.

.... (contd.)

Proceedings, papers and reports - 9. Specialized dictionaries
and encyclopaedias - 10. Bibliographies - 11. Films and film
catalogues - 12. Other sources of information (H. Patents
and licences). 'Bibliographical sources used'.
Class No: 669.1(01)

Encyclopaedias

[6343]

OSBORNE, A.K., *comp.* **An Encyclopaedia of the iron and
steel industry.** 2nd ed. London, Technical Press, 1967. xi,
558+44p. illus., diagrs.

First published 1956 (xi, 558p.).

Concise descriptions, A-Z, of materials, plants, tools and
processes used in the iron-and-steel and allied industries.
Literature references, where considered appropriate.
Appendices include addresses of scienctific, technical, trade
societies, etc.; conversion tables; properties of various
steels; definitions of symbols. 2nd ed. adds a 44p.
supplement on developments over the following ten years.
Class No: 669.1(031)

Dictionaries

English

[6344]

BRITISH STANDARDS INSTITUTION. **Terms used in the
iron and steel industry.** London, BSI, 1985-90. (BS 6562:
1985-86; EN 10020: 1998).

1. *Glossary of heart terms.* 1985. 28p.

2. *Glossary of terms used in classifying and defining
industry products by shape and dimensions.* 1986. 23p.

3. *Definition and classification of grades of steel.* 1990.
13p.

Part 3 has been withdrawn as a British standard and
replaced by EN 10020: 1988. *Class No:* 669.1(038)=20

[6345]

GALE, W.K.V. **The Iron and steel industry: a dictionary of
terms.** New ed. Newton Abbot, Devon, David & Charles,
1973. 238p.

First published 1971 (238p.).

About 3,500 brief entries. Many of the terms listed are
now obsolete (marked '(obs)'). Local terms are also
differentiated. Numerous cross-references. Appended: A.
Birmingham gauge - B. British (Imperial) standard and wire
gauge. - C. Tinplate denominations and size.
Class No: 669.1(038)=20

[6346]

SIMONS, E.N. **Dictionary of ferrous metals.** London,
Muller, 1970. x, 244p. diagrs., tables.

About 2,500 entries, A-Z, preceded by a number
sequence. 'Simple, brief but factual information of the terms
chosen' (*Preface*). 'Nickel steels': 24p. Adequate cross-
references. Companion volume to the author's *A dictionary
of alloys* (*qv*). *Class No:* 669.1(038)=20

German

[6347]
BRITISH STEEL CORPORATION. **Vocabulary of iron and steel terms: German/English, English/German.** London, the Corporation, 1975. 104p. tables.
About 1,800 terms in each half of this pocket-sized dictionary. Conversion tables. *Class No:* 669.1(038)=30

[6348]
FREEMAN, H.G. **Taschenwörterbuch Eisen und Stahl.** Munich, Hueber, 1966-67. 2v.
1. *Deutsch-englisch.* 1966. 311p.
2. *Englisch-deutsch.* 1967. 289p.
Each volume has *c.*13,000 entry-words; categories and genders are shown. V.1 includes conversion tables (p.21-43) and a list of sources (p.44-46). *Class No:* 669.1(038)=30

French

[6349]
BRITISH STEEL CORPORATION. **Vocabulary of iron and steel terms: French/English, English/French.** London, the Corporation, 1974. 105p.
About 1,800 terms in each half of this pocket-sized dictionary. Conversion tables. *Class No:* 669.1(038)=40

Italian

[6350]
BRITISH STEEL CORPORATION. **Vocabulary of iron and steel terms: Italian/English, English/Italian.** London, the Corporation, 1974. 117p.
About 1,800 terms in each half of this pocket-sized dictionary. Conversion tables. *Class No:* 669.1(038)=50

Reviews & Abstracts

[6351]
Journal of the Iron and Steel Institute. 'Titles of current literature'. London, the Institute, 1869-1973. Monthly.
Previously as *Abstracts of current literature and book notes.*
Latterly *c.*1,000 references per issue in 21 sections (ores, mining, handling and treatment ... geography, history). Continued weekly in ABTICS (abstract and book-title index card service). UDC classified. *Class No:* 669.1(048)

Yearbooks & Directories

[6352]
Directory of iron and steel plants. Sukits, D., *ed.* Pittsburgh, Pa., Association of Iron and Steel Engineers, 1989. 584p. $40.
Covers North America and selected overseas nations. 5 sections and geographical indexes. Advertising index; company index. *Class No:* 669.1(058)

[6353]
Iron and steel works of the world. 9th ed. Worcester Park, Surrey, Metal Bulletin Books, Ltd., 1988. lxxxi, 728p. £80. ISBN: 0900542829.
Lists *c.*2,000 companies in some 100 countries, Albania ... Zimbabwe, p.1-588. Data include year of establishment, products, brands, sales office, agents. Country index precedes. Buyers' guide of products and index. Index to advertisers. *Class No:* 669.1(058)

Standards

[6354]
Worldwide guide to equivalent irons and steels. American Society for Metals. 3rd ed. Metals Park, Oh., The Materials Information Society, 1992. *c.*500p. £143. ISBN: 0871704544.
First published 1978.
Alloys organized by type of iron or steel, *e.g.* cast iron; cast stainless steel; steel castings; carbon steel; tool steel, etc. Data for each alloy specification: country of origin; standards organization; specification; designation; description; tensile strength; yield strenth; elongation. Indexes: specification; designation; standards organizations. *Class No:* 669.1(083.74)

Great Britain

[6355]
IRON AND STEEL STATISTICS BUREAU. **Iron and steel industry: annual statistics for the United Kingdom.** Croydon, Surrey, the Bureau. Annual. (1986-1987. £40). tables.
Contents include definitions; general summary; home and imported iron and manganese ores; sinter; energy; other materials; blast furnaces; crude steel; finished steel; iron castings; manpower; prices; general industrial tables; imports and exports. Statistics are usually for 6-year spans. The 1986-87 volume contains 49 tables. *Class No:* 669.1(410)

Cast Iron

[6356]
CONCAST AG, Zurich. **Concast dictionary:** English/French/ Spanish/Italian/ Russian/German/Japanese. 3rd ed. New York, Concast AG Zurich, 1975.
First published 1986.
About 3,000 English-based terms, with French, Spanish, Italian Russian, German and Japanese equivalents and indexes. *Class No:* 669.13

Steel

Bibliographies

[6357]
TUPHOLME, S. **'Stainless steels'.** *Information sources in metallic materials.* Editor, M.N. Patten, p.86-102. London, Bowker-Saur, 1989. *(Guides to information sources.)* ISBN: 0408014911.
Sections: Key works of reference (36 items) - Periodicals - Abstracting services - Databases and datafiles - Standard specifications and codes of practice - Conferences - Advisory bodies, trade and research associations - Commercial aspects. *Class No:* 669.14(01)

Handbooks & Manuals

[6358]
PECKNER, D. *and* BERNSTEIN, I.M. **Handbook of stainless steels.** New York, etc., McGraw-Hill, 1977. 1100p. illus. ISBN: 007049147x.
6 parts (48 documented chapters): 1. Introduction to stainless steels - 2. Metallurgy of stainless steels - 3.

.... *(contd.)*

Corrosion resistance - 4. Physical and mechanical properties - 5. Fabrication and design practices - 6. The applications of stainless steels. Appendix 1: Specifications of the stainless steel producing countries; 2: Proprietary alloys. Analytical index of 31p. *Class No:* 669.14(035)

Dictionaries

[6359]
EUROPEAN COMMUNITIES COMMISSION. Terminology Office. **Glossaire des normes de l'acier.** Brussels, the Bureau, 1972. 510p.

Glossary of more than 5,000 terms - definitions plus examples of usage - in French, German, Italian and Dutch. Based on *c.*100 European standards (1955-71). *Class No:* 669.14(038)

Reviews & Abstracts

[6360]
Euro abstracts. Section 2: Coal and steel. Research programmes and agreements. Scientific and technical publications and patents in German, English and French. Luxembourg, Commission of the European Communities, 1975-. Monthly. ISSN: 03783472.

Formerly part of Euro abstracts: scientific and technical publications and patents, 1970-74.

Fairly lengthy abstracts covering research programmes, research agreements, scientific and technical publications, patents, training courses and seminars, forthcoming conferences and symposia. Annual index. *Class No:* 669.14(048)

[6361]
Steel castings abstracts. Sheffield, Steel Casting Research and Trade Association, 1968-. v.17-. 6pa. £55pa.

Formerly *BSCRA abstracts,* 1953-68.

About 1,500 abstracts pa. 8 sections (*e.g.* Production organization, management and quality control; moulding and core making; casting processes; non-destructive testing; heat and safety, dust and fumes). Annual name and subject indexes. Covers all aspects of steel foundry technology. *Class No:* 669.14(048)

[6362]
Steels' alert. London, The Institute of Metals; Metals Park, Ohio, American Society for Metals, 1983-. Monthly. £165. ISSN: 10480307.

Formerly *Steels supplement to metals abstracts,* 1953-82.

9 sections: 91. Fuel energy usage and raw materials; 92. Steel plants; 98. Economics, statistics, resources; 99. World steel news, and general. Covers additional information on steel production, fabrication and use, and development. *Class No:* 669.14(048)

Yearbooks & Directories

[6363]
Steel traders of the world, 1984. 3rd ed. Worcester Park, Surrey, Metal Bulletin Books, Ltd., [1983]. lx, 670, [3]p. ISBN: 0900542790.

Well over 2,000 firms in 101 countries. Company section, p.1-563, under countries, A-Z. A-Z index to companies.

.... *(contd.)*

Index to product guide. Pocket guide (carbon steel; alloy steel; stainless steel; pig iron, etc.). Index to advertisers. *Class No:* 669.14(058)

Standards

[6364]
EUROPEAN COMMUNITIES COMMISSION. Terminology Office. **Steel standards glossary.** Luxembourg, the Commission, 1979. xxxi, 689p. ISBN: 9282508331.

Explanations of *c.*2,500 terms in French, English, German, Italian, Dutch and Danish, p.1-354. Alphabetical indexes in the six languages, on tinted paper (that in English, on purple paper, is virtually indecipherable). Bibliographical references (to sources), p.638-84. *Class No:* 669.14(083.74)

Europe

[6365]
ECONOMIC COMMISSION FOR EUROPE. Annual bulletin of steel statistics for Europe, 1973-. New York, United Nations, 1974-. Annual. tables.

Basic tabular data on development of steel production and trade, consumption and trade of raw materials, movement of scrap, consumption of energy in the steel industry and steel deliveries to consuming industries in European countries, plus Canada, USA and Japan.

Quarterly bulletin of steel statistics for Europe (New York, U.N., 1950-. v.1(1)-) updates. *Class No:* 669.14(4)

Non-Ferrous Metals

[6366]
BNF Non-ferrous metals abstracts. Wantage, Dorset, BNF Metals Technology Centre, 1980-83. Monthly. ISSN: 56326291.

Previously as *BNF abstracts,* 1969-79. Originally 'Reference to current literaure', in *Bulletin of the British Non-Ferrous Metals Research Association,* 1921-.

1983: 6,291 abstracts. Sections: Metals (Aluminium ... Zirconium) - Other metals and elements - Analysis ... Coatings and finishing operations ... Extraction. Refining. Reclamation - Mechanical properites and testing ... Metallography ... Powder metallurgy and composites ... Miscellaneous. Patent applications. Book list and book reviews. Numerous cross-references. Author, company and subject indexes.

Available as a closed file on DIALOG, about 120,000 records, file 118. *Class No:* 669.2

[6367]
BOODSON, K. Non-ferrous metals: a bibliographical guide. London, Macdonald Technical & Scientific 1972. xxiv, 650p. (*Macdonald bibliographical guides.*)

4,335 annotated entries for items published mainly during 1966-70. 59 sections: 1. Reference media (304 items) - 2. Structural metallurgy: metal physics ... 6. Mechanical properites ... 10. Mining and metal production - 11/55. Specific metals (Aluminium ... Zirconium) - 56. Refractory metals generally - 57. Noble metals generally - 58. Rare earth metals - 59. Other minor metals. Numbered entries, stating period covered and number of references cited. 3-column layout. Author and subject indexes. Dated, but still of value. *Class No:* 669.2

[6368]
HAMPEL, C.A., *ed*. **Rare metals handbook.** 2nd ed. New York, Reinhold, 1961. xvi, 715p. illus., diagrs., tables.
First published 1954.
35 chapters (2-34. Individual, less common metals or groups of metals, A-Z; 35. Physical properties). Deals with 55 metallic elements, adding cesium, chromium, plutonium, rubidium, scandium and yttrium. Separate chapters for columbium and tantalum. Chapter references are updated and sometimes pruned (*e.g.* uranium: 181 references, instead of 389 in the 1954 ed.). Analytical index. *Class No:* 669.2

[6369]
Non-ferrous metal data. Secaucus, N.J. American Bureau of Metal Statistics, Inc. Annual. tables.
The 1983 v. (150p.) has data on non-ferrous metals, A-U. Tin, p.26-33, states world production of tin ore.
Class No: 669.2

[6370]
Non-ferrous metal works of the world. 4th ed. Worcester Park, Surrey, Metal Bulletin Books, 1986. lxvi, 647p. tables. ISBN: 0900542950.
First published 1967.
Indexes to countries and companies precede the company section, p.1-535. Index to buyers' guide; buyer's guide. Index to advertisers. 94 countries covered. *Class No:* 669.2

Gold

[6371]
BOYLE, R.W. **Gold: history and genesis of deposits.** New York, Van Nostrand Reinhold, 1987. 676p. $49.50. ISBN: 0442211627.
18 documented chapters, 51 contributors. An historical introduction is supported by a chapter on the geochemistry of gold and the type of auriferous deposits. Then follows reprints of some classic papers on gold. Name and subject indexes. 'Digests in one compact volume much of the vast scattered literature on the geology and geochemistry of gold' (*New technical books,* January/February 1988, item 284).
Class No: 669.21

[6372]
Handbook of precious metals. Savitskii, E.M., *ed.* New York, Hemisphere, 1989. xix,600p. £96. ISBN: 0891167099.
Includes numerous tables and over 900 references, many to Russian journals. *Class No:* 669.21

[6373]
KETTELL, B. **Gold.** London, Graham & Trotman, 1982. xiv, [1], 283p. illus., graphs, tables. ISBN: 0860102572.
8 chapters: 1. Gold: importance, scientific properties and methods of extraction - 2. Early history - 3. The gold standard - 4. Gold: the major factors affecting its supply demand and price. Appendix: minted gold coins. Grouped bibliography, p.269-73. Index. 81 tables and figures.
Class No: 669.21

[6374]
KUDRYK, V., *and others*. **Precious metals:** mining, extraction and processing. Wassendale, Pa., Metallurgical Society of AIME, 1984. xi, 621p. illus., diagrs., charts, tables, maps. ISBN: 089520469x.
45 papers, usually with references, read at an International Symposium of Precious Metals, by contributors from 12 countries. Sections: General overview - Sampling and ore

....(contd.)
evaluation - Chemistry of leaching and recovery - Electrolytic recovery - Ore processing - Recycling and byproduct recovery - Physical metallurgy. Example: 'Recovery of gold from arsenical ores '(41 references). Subject and author indexes. *Class No:* 669.21

[6375]
Precious metal databook. Serjeanston, R., *ed*. Worcester Park, Metal Bulletin Books, 1989. xxxiv,254p. £58. ISBN: 0947671242. ISSN: 09556532.
In two main sections (Companies; Buyers' guide) each subdivided into producers and traders. Appended: statistics by country (production; consumption; imports; exports). Indexes: countries; companies; advertisers.
Class No: 669.21

Silver

[6376]
BUTTS, A. *and* COXE, C.D., *eds*. **Silver: economics, metallurgy and use.** Sponsored by Hardy and Harman in commemoration of their 100th anniversary. Princeton, N.J. etc., Van Nostrand, 1967. 488p. illus., charts, tables.
35 contributors. 34 chapters on history, sources, extracting, refining, properties, application, etc. 'Literature of silver', p.467-8. Name and subject indexes. Well illustrated. *Class No:* 669.22

[6377]
Silver: exploration, mining and treatment. Institution of Mining and Metallurgy. London, Institute of Mining and Metallurgy, 1988. x,344p. illus., maps, tables. £46.80. ISBN: 1870706048.
31 papers presented at the international conference on 'Silver-exploration, mining and treatment' organized by the IMM, in association with the Cámara Minera de México and The Silver Institute, and held in Mexico City, 21-24 November 1988. Sections: Geology and mineralogy - Case histories - Silver in complex sulphide minerals - Flotation - Extraction and refining. *Class No:* 669.22

Platinum

[6378]
MACDONALD, D. *and* HUNT, L.B. **A History of platinum and its allied metals.** London, Johnson Matthey, 1982. 462p. illus. (incl. col.), facsims., ports. ISBN: 0905118839.
24 documented chapters (*e.g.* 3. Early scientific enquiries into the properties and nature of platinum; 13. The discovery and early history of catalysis (37 references); 21. The growth of industrial catalysis with the platinum metals). Name index; detailed, analytical index. Well produced.
Class No: 669.231

Nickel

[6379]
BETTERIDGE, W. **Nickel and its alloys.** Chichester, Ellis Horwood; New York, Halsted Press: Wiley, 1984. 216p. £30. (*Ellis Horwood series in industrial metals.*) ISBN: 0853127298.
9 documented chapters: 1. Introduction and physical properties - 2. Occurrence and extraction of nickel - 3. Consruction of nickel alloy systems - 4. Nickel in industrial alloys - 5. Metallography - 6. Production of semi-fabricated

....(contd.)
forms - 7. Processes applied in final manufacture - 8. Applications - 9. Economics. 2 appendices (*e.g.* Conversion factors). Detailed analytical index. *Class No:* 669.24

[6380]
BOLDT, J.R. **The Winning of nickel:** its geology, mining and extractive metallurgy. London, Methuen, 1967. xiv, 487p. illus., maps.
4 sections: 1. The geology of nickel - 2. Mining for nickel - 3. Extractive metallurgy of sulfide ores - 4. Extractive metallurgy of oxide ores. References, p.457-64. Analytical index, Micrographs. *Class No:* 669.24

Cobalt
[6381]
YOUNG, R.S., *ed*. **Cobalt: its chemistry, metallurgy and uses.** New York, Reinhold, 1960. vii, 424p. (*American Chemical Society Monograph series no.149, replacing no.108.*)
13 chapters, mustering nearly 1,500 references. Well presented data, but deficient in illustrations and flowsheets. *Class No:* 669.25

Chromium
[6382]
SULLY, A.H. *and* BRANDES, E.A. **Chromium.** 2nd ed. London, Butterworth, 1967. xiii, 373p. illus., diagrs., tables. (*Metallurgy of the rarer metals.*)
First published 1954.
8 chapters. 1. History and occurrence - 2. Production of chromium ferro-alloys and pure chromium - 3. Physical properties of pure chromium ... 6. Electroplating - 7. Chromizing - 8. Constitution and properties of chromium alloys (p.301-68; 173 references). Appendix: List of properties of chromium. Subject index. *Class No:* 669.26

Tungsten
[6383]
YIH, S.W.H. *and* WANG, C.T. **Tungsten: sources, metallurgy, properties and applications.** New York & London, Plenum Press, 1979. xvi, 500p. illus., diagrs., graphs, tables. ISBN: 0306311445.
11 chapters: 1. Occurrence, geology, mining and benefication of tungsten - 2. Extractive metallurgy ... 5. Fabrication. 7. Chemical properties of tungsten and tungsten compounds ... 10. Applications - 11. Industry and future outlook of tungsten. 2 appendices (2. Analytical chemistry of tungsten). Bibliography (by chapters and appendices), p.463-91. *Class No:* 669.27

Niobium & Tantalum
[6384]
MILLER, G.L. **Tantalum and niobium.** London, Butterworth, 1959. xxii, 767. illus., tables. (*Metallurgy of the rarer metals, no.6.*)
Data on occurrence; extraction; physical, chemical and mechanical properties, applications, etc. Companion to the same author's *Zirconium (qv). Class No:* 669.293/.294

Titanium
[6385]
DONACHIE, M.J., *Jr*. **Titanium: a technical guide.** Metals Park, Ohio, ASM International, 1988. xv, 469p. diagrs., graphs., tables. $65. ISBN: 0871703092.
13 chapters (13. Recent and future advances). Appendices A-J: A. Selected references (120) for additional reading; B. Glossary; G. General corrosion rate; H. List of manufacturers, suppliers, services; I. Standards and specifications; J. Designations, applications, purposes. Analytical index, p.453-69. 'Highly recommended' (*New technical books,* Sept./Oct. 1988, item 2318).
Class No: 669.295

[6386]
FROES, F.H., *and others*. **Titanium technology:** present status and future trends. Dayton, Ohio, Titanium Development Association, 1985. 191p. illus., diagrs., graphs, tables. $19.95.
Based on articles, reprinted with permission, from the *Journal of metals.*
20 contributors. 17 documented papers (*e.g.* 'Advances in titanium extraction metallurgy', with 59 references). 'About the authors', p177-9. Detailed analytical subject and author indexes.
Titanium abstract bulletin (Wilton, Birmingham, ICI. 1955-. 6pa.) ceased publication in 1962. *Class No:* 669.295

Zirconium
[6387]
MILLER, G.L. **Zirconium.** 2nd ed. London, Butterworths, 1957. xxi, 548p. illus., tables. (*Metallurgy of the rarer metals, no. 2.*)
First published 1954.
Comprehensive survey of the metal, from its history and natural occurrence, extraction, physical, chemical and mechanical properties to alloys and applications. Chapter bibliographies. Subject index. *Class No:* 669.296

Copper
[6388]
BLACK, W.T., *and others*. **Thesaurus of terms in copper technology.** 7th ed. New York, Copper Development Association, 1981. vi, 401p. $40.
Assembles more than 8,700 metallurgical and engineering terms, with numerous cross-references. Of wide application; much of the information is common to all metals.
Class No: 669.3

[6389]
BRITISH STANDARDS INSTITUTION. **Glossary of terms for copper and copper alloys.** London, BSI, 1988. 20p. (*BS 6931: 1988.*)
Supersedes BS 1420: 1965.
Defines terms used in the UK for raw materials, castings, wrought products and processes. *Class No:* 669.3

[6390]
CALLCUTT, V.,A. '**Copper and copper alloys**'. *Information sources in metallic materials.* Editor, M.N. Patten, p.120-37. London, Bowker-Saur, 1989. (*Guides to information sources.*) ISBN: 0408014911.
Sections: Introduction - Historical - Structure of the industry - *International Copper Information Bulletin* - Statistics and economics - Standards and specifications -

....(contd.)

Terminology - Technology - Analysis - Alloying relationships - Corrosion - Addresses (*c*.60). 75 references. *Class No:* 669.3

[6391]

International copper information bulletin: recent reports, publications and abstracts on copper, its alloys and compounds. Potters Bar, Herts., Copper Development Association, 1976-. 3pa. ISSN: 03092216.

Formerly *Copper abstracts,* 1959-75.

About 1,000 abstracts pa. 8 sections, including 3. Production, consumption and marketing statistics; 4. National and international standards and related publications; 5. Recent technical papers and articles (*c*.50 abstracts per issue); 6. New books and conference proceedings; 7. Some new copper and copper alloy materials, products and processes; 8. Forthcoming conferences and symposia. *Class No:* 669.3

[6392]

WEST, E.G. Copper and its alloys. Chichester, Ellis Horwood; New York, Halsted Press: Wiley, 1982. 241p. illus. (incl. photomicrographs), diagrs., graphs. £34. (*Ellis Horwood series in industrial metals.*) ISBN: 0853125058.

9 chapters, each with 'Further reading': 1. Introduction and basic properties of copper - 2. Occurrence and extraction of copper ... 4. Copper alloy systems - 5. Properties of copper alloys ... 7. Manufacturing processes ... 9. Applications. 2 appendices (1. Metallography of copper and copper alloys). Detailed, analytical index. 'It is an excellent source of reference data' (*New technical books,* July 1983, item 869). *Class No:* 669.3

Lead

[6393]

BLASKETT, D.R. *and* **BOXALL, D. Lead and its alloys.** New York, E. Horwood, 1990. 161p. illus., diagrs., graphs, tables. $60. ISBN: 0135286964.

11 chapters, closely subdivided, with numerous references. Emphasis is on smelting and refining with consideration of health aspects. Recommended by *Choice* (May 1991, p.1514) as 'valuable' for industry, commerce and research. *Class No:* 669.4

[6394]

CONWAY, M.J. 'Lead'. *Information sources in metallic materials.* Editor, M.N. Patten, p.138-50. London, Bowker-Saur, 1989. (*Guides to information sources.*) ISBN: 0408014911.

Sections: Introduction - General information (books; reports; proceedings; periodicals) - Information sources arranged by major area of interest or uses (analysis; batteries; building; cables; chemicals; coatings; corrosion; electrochemistry; environment; joining; metal working; phase diagrams and thermodynamics; pigments; printing materials; statistics) - Organizations. *Class No:* 669.4

[6395]

Lead abstracts. London, Zinc Development Association, 1960-. Quarterly.

Entitled *Lead technical abstracts* until no. 13, October 1961.

About 1,000 abstracts pa. 20 alphabetical sections, from 'Alloys' to 'Printing'. Coverage: development and applications of lead, lead alloys, compounds and coatings.

....(contd.)

Keyword indexes per issue, cumulated annually. Contributes items to *Zinc, lead and cadmium abstracts* (*qv*) on PERGAMON ONFOLINE. *Class No:* 669.4

Zinc

[6396]

CONWAY, M.J. 'Zinc'. *Information sources in metallic materials.* Editor, M.N. Patten, p.171-89. London, Bowker-Saur, 1989. (*Guides to information sources.*) ISBN: 0408014911.

Sections: Introduction - General information (books; reports; etc.) - Information sources by major area of interest or uses (*e.g.* analysis; batteries; corrosion; diecasting; hot dip galvanizing; rubber and plastics) - Information centres (11). *Class No:* 669.5

[6397]

PORTER, F. Zinc handbook: properties, processing, and use in design. New York, Dekker, 1991. xiv,629p. illus., graphs, tables. (*Mechanical engineering.*) ISBN: 0824783409.

32 chapters, closely subdivided, with 'references and further reading'. Largely concerned with applications for zinc including coating and galvanizing. Detailed, analytical index, p.619-29. *Class No:* 669.5

[6398]

Zinc abstracts. London, Zinc Development Association, 1962-. 4pa. ISSN: 00444731.

Formerly *ZDA abstracts,* 1943-61.

About 1200 abstracts pa. 6 sections: Economics and statistics - Extraction - Uses (including processes) - Environmental aspects - Properties of materials and their measurement - Miscellaneous (Standards; Book list and reviews). Author and subject indexes per issue and annually. *Class No:* 669.5

[6399]

Zinc, lead and cadmium abstracts. London, Zinc Development Association , 1970-. Monthly.

About 1,800 abstracts pa. on development and applications of zinc, lead and cadmium, and their allies, compounds and coatings. *Class No:* 669.5

Tin

[6400]

BARRY, B.T.K. *and* **THWAITES, C.J. Tin and its alloys and compounds.** Chichester, Ellis Horwood; New York, Halstead Press: Wiley, 1983. 268p. illus., diagrs., tables. (*Ellis Horwood series in industrial metals.*) ISBN: 0853126496.

12 chapters, each with 'Further reading': 1. Physical and chemical properties - 2. Occurrence and production - 3. Tin alloy systems - 4-11. Tin and tin alloy applications - 12. Economic factors. 114 illus; 36 tables. Detailed analytical index. *Class No:* 669.6

[6401]
DAVIES, R.L. 'Tin and tinplate'. *Information sources in metallic materials.* Editor, M.N. Patten, p.151-70. London, Bowker-Saur, 1989. (*Guides to information sources.*) ISBN: 0408014911.

Sections: Introduction - Information sources (*e.g.* reference books and handbooks; journals (43 titles); patents; standards, etc.) - Conferences ... Meetings. About 100 references. *Class No:* 669.6

[6402]
Tinsnips (Tinplate abstracts bulletin). Uxbridge, Middx., International Tin Research Institute, 1986-. 6pa.

About 600 abstracts pa. Subject index per issue. *Class No:* 669.6

Aluminium

Bibliographies

[6403]
KEEVIL, D. 'Aluminium'. *Information sources in metallic materials.* Editor, M.N. Patten, p.103-19. London, Bowker-Saur, 1989. (*Guides to information sources.*) ISBN: 0408014911.

Sections: Introduction - Abstracting services - Journals - Handbooks - Standards - Conferences - Standard texts - Aluminium in human health - Associations. 26 references. *Class No:* 669.71(01)

Handbooks & Manuals

[6404]
HATCH, J.E., *ed.* Aluminum: properties and physical metallurgy. Metals Park, Ohio, American Society for Metals, 1984. [ix], 424p. illus., diagrs., graphs, tables. ISBN: 0871701766.

Succeeds v.1 of the 3.v. series, *Aluminum* (1967).

53 contributors. 10 chapers: (1. Properties of pure aluminum & constitution of alloys ... 5. Metallurgy of heat treatment and general principles of precipitation hardening ... 7. Corrosion behavior ... 10. Aluminum powder and powder metallurgy products (52 references)). Detailed, analytical index. Micrographs. *Class No:* 669.71(035)

[6405]
KING, F. Aluminum and its alloys. New York, Halstead Press, 1987. 313p. illus., tables. $96.95. ISBN: 047024849x.

8 sections (Basic properties - Occurrences and extraction - Refining - Aluminum-alloy system - Properties of aluminum alloys - Production of semi-fabricated forms - Manufacturing processes - Economic factors). References. *Class No:* 669.71(035)

[6406]
MONDOLFO, L.F. Aluminium alloys: structure and properties. London, Butterworth, 1976. ix, 971p. illus., tables.

4 main parts: 1. Pure and commercial aluminium (8 chapters)- 2. Binary alloys - 3. Ternary, quarternary, etc. alloys - 4. Commercial alloys. Fully documented chapters (*e.g.* Aluminium silicon system, p.368-76. 195 references). Appendices of conversion tables, etc. Additional references. Index. *Class No:* 669.71(035)

Reviews & Abstracts

[6407]
Metallurgical abstracts on light metals and alloys. Osaka, Japan, Light Metal Educational Foundation, Inc., 1956/60-. illus., graphs, tables.

V.20, 1986-1987, carries abstracts of 61 research reports already published in technical journals, No index. *Class No:* 669.71(048)

[6408]
World aluminum abstracts: a monthly review of the world's technical literature on aluminum. Sponsored jointly by the Aluminum Association (Washington), European Aluminium Association (Dusseldorf), Japan Light Metals Association (Tokyo). Metals Park, Ohio, American Society for Metals, 1970-. Monthly.

Previously as *Aluminum abstracts* (1963-69); originally as *Light metals bulletin* (1949-62).

About 600 abstracts per issue. 8 sections: 1. Aluminum industry - 2. Ores, alumina products extraction - 3. Melting, casting, foundry - 4. Metalworking, fabrication, fininishing - 5. Physical and mechanical metallurgy - 6. Engineering properties and tests - 7. Quality control and tests - 8. End uses. Subject, author and corporate author indexes per issue and annually. Online through DIALOG. *Class No:* 669.71(048)

Conferences

[6409]
Light metals, 1987. Proceedings of the technical services sponsored by the TMS Light Metal Committee of the 116th Annual Meeting, Denver, Colorado, February 24-26, 1987. Zabreznik, R.D., *ed.* Warrendale, Pa. 886p. illus., diagrs., graphs, tables. ISBN: 0873390601.

More than 100 documented papers. Subject areas: Alumina and bauxite; Reduction technology; Carbon technology; Environmental control & health; Cast technology and recycling; Other light metals (magnesium; lithium). Addenda. Author index. Micrographs. *Class No:* 669.71:061:061.3

Beryllium

[6410]
HAUSNER, H.H., *ed.* Beryllium: its metallurgy and properties. Berkeley & Los Angeles, Calif., Univ. of California Press, 1965. [vii], 322p. illus., diagrs., tables.

17 contributors. 12 chapters (4. The fabrication of beryllium, p.13-177 - 5. Beryllium alloys, p.179-90 (20 references) ... 11. Future trends in beryllium research - 12. Health hazards in handling beryllium). 2 appendices. Favourably reviewed in *Nature,* v.212, 19 November 1966, p.775-6. *Class No:* 669.725

Cadmium

[6411]
Cadmium abstracts. London, Zinc Development Association, 1977-. Quarterly.

About 600 abstracts pa. Alphabetic subjects: Analysis - Batteries ... Corrosion ... Health and safety ... Physical metallurgy ... Pollution control - Production ... Solar cells - Tribology. Keyword index per issue, cumulated annually. *Class No:* 669.73

[6412]

Minor metals survey. London, Metal Bulletin, Ltd., 1977. 162p. illus., tables.

Metals concerned: antimony, arsenic, bismuth, cadmium, cerium, chromium, cobalt, indium, manganese, mercury, shenium, selenium and tellurium, tantalum, zirconium. Data for indium: properties; production; price trends; uses; present and future. Occasionally appended 'Selected references'. *Class No:* 669.73

Uranium

[6413]

GITTUS, J.H. Uranium. London, Butterworth, 1963. xiii, 623p. illus., diagrs., tables. (*Metallurgy of the rarer metals, 8.*)

Chapters 1-5, on occurrence, exploitation and concentrates, and on separation of the metal, its fabrication. Chapters 6-12: properties (physical, mechanical, chemical, including effects of pile irradiation and corrosion). Chapter 13: 'Alloys of uranium', with 124 references. Final chapter, on metallographic examination. 5 Appendices on reactors, nuclear cross-sections, etc. Standard compendium, in its day. *Class No:* 669.822

[6414]

Uranium: resources, production and demand. OECD Nuclear Energy Agency, and International Atomic Energy Agency. 11th ed. Paris, OECD, 1986. 413p. graphs, charts, tables, map.

First published 1965; 10th ed. 1983.

Two main sections. The first summarizes data on resources, exploration, production, demand, supply and demand. The second (p.83-363) covers 57 countries in turn (excluding Eastern Bloc countries and China), with data submitted officially in response to questionnaires. '... the most authoritative source of information for WOCA (World Outside Centrally Planned Economies Area) countries, and this edition is the largest, most complete version yet ... This is the last of the fully comprehensive Red Books to be published for several years ... [and] ought to be viewed as a must for any uranium industry library' (*Atom*, no. 363, January 1987, p.23). *Class No:* 669.822

Plutonium

[6415]

WICK, O.J., *ed*. Plutonium handbook: a guide to the technology. New York, Gordon & Breach, 1967-68. 2v. (x, 520p.; 448p.) illus., diagrs., tables.

Prepared under the auspices of the US Atomic Energy Commission.

V.1 has main sections: Physics - Metallurgy - Chemistry. V.2: Chemical processing - Fabrication and utilization of plutonium and alloys - Analysis and inspection methods - Health and safety. Bibliographies. *Class No:* 669.824

Rare Earth Metals

[6416]

GSCHNEIDNER, K.A., *Jr* and EYRING, Le R., *eds*.. Handbook of the physics and chemistry of rare earths. Amsterdam, etc., North-Holland Co., 1978-. graphs, tables.

1. *Metals*. 1978, xxv, 894p.

2. *Alloys and intermetallics*. 1979. xiii, 628p. £110.

3./4. *Non-metallic compounds*. 1979. xiii, 604p; xiii,

....*(contd.)*

602p. £115; £105.

5. *Rare earth alloys and compounds*. 1982. x, 701p. £132.

10. *High energy spectroscopy*. 1987. 611p. £113.

11. *Two hundred year impact of rare earth on science*. 1988. xiv, 594p. £113.

13. *Rare earth elements*. 1990. xii,473p.

Each volume carries documented chapters by specialists. V.5 is an expansion of K.A. Gschneidner's earlier *Rare earth alloys: a critical review of the alloy systems of the earth, scandium and yttrium metals* (1961). Chapter 71, in v.10: 'X-ray absorption and emission spectra,' p.453-549, has 6 pages of references. 'Indispensible for academic libraries and chemistry collections' (*New technical books*, June 1980, item 273). *Class No:* 669.85/.86

[6417]

Rare earth bulletin. London, Multi-Science Publishing Co., 1972-. 6pa. £395pa. ISSN: 03078531.

About 1,000 abstracts pa., covering periodical articles, monographs, conference proceedings, technical and government reports. Author and subject indexes. *Class No:* 669.85/.86

Sodium

[6418]

SITTIG, M. Sodium: its manufacture, properties and uses ..., with a chapter on the physical and thermodynamic properties of sodium ... New York, Reinhold, 1956. viii, 529p. diagrs., tables.

9 well-documented chapters: 1. Introduction - 2. The manufacture of metallic sodium - 3. Solubility and alloy formation - 4. Handling metallic sodium (101 references) - 5. Uses of metallic sodium - 6. The inorganic reactions of sodium - 7. The use of sodium in organic reactions (679 references). - 8. The analytical chemistry of sodium - 9. Physical end thermodynamic properties of sodium. *Class No:* 669.883

67/68 Manufactures

Jewellery

[6419]
BLAKEMORE, K. **The Retail jeweller's guide.** 5th ed. London, Butterworth, 1988. 432p. illus. £35.

First published 1969,- designed to replace Selwyn, A. *The retail jeweller's handbook and merchandise manual for sales personnel* (7th ed. London, Heywood, 1962). 4th ed. 1983.

Sections: The metals - The gems - Antique silverwear - Boxes of the 18th and 19th centuries - How silverwares are made - Hallmarks on gold, silver and platinum - Jewellery of the past - The making of jewellery - The history of watches and clocks - Mechanical watches and clocks today - Electrical and electronic watches and clocks. Glossaries. Appendices. Both textbook and reference tool. *Class No:* 671.12

[6420]
FORGET, C., *comp*. **Elsevier's dictionary of jewellery and watchmaking** in five languages: English, French, German Italian and Spanish. Amsterdam, etc., Elsevier, 1986. [vii], 507p. DFl.275. ISBN: 044442279x.

5,673 English-base terms, with equivalents and indexes in French, German, Italian and Spanish. Misnomers are indicated. *Class No:* 671.12

[6421]
MASON, A. *and* PACKER, D. **An Illustrated dictonary of jewellery.** Reading, Berks., Osprey Publishing Co.; New York, Harper & Row, 1973. 390p. illus. £5.95.

Over 3,000 definitions, many of them brief. Details with gemstones and their identification, the techniques of jewellery manufacture, the history of jewellery, etc. Adequate cross-references. Bibliography, p.388-90. 200 line-drawings. Intended as a reference tool for both those involved in the jewellery trade and those with a less specialized interest. 'Sure to become a standard reference' (*Library journal*, 1 May 1974, p.1290). *Class No:* 671.12

[6422]
UNTRACHT, O. **Jewellery: concepts and technology.** London, Hale, 1982. xxiv, 840p. illus. (incl. col.), diagrs., tables. £39.95. ISBN: 0709196164.

19 sections (*e.g.* 3. Metal, the jewel's raw material - 4. Basic techniques - 5. Sheet metal - 6. Wire - 7. Tubing - 8. Surface ornament, without heat - 9. Surface ornament, with heat - 10. Fabrication - 11. Casing - 12. Natural materials in jewellery - 13. Stones and their setting - 14. Metal finishing - 15. Metallic coating techniques - 16. Metallic build up - 17. Colouring - 18. Standard weights, measures and tables - 19. Glossaries (p.757-90)). Bibliography (grouped), p.791-8. Sources of tools, supplies and services (US, UK). Museums (p.vii). Analytical index (tiny print), p.808-40. Over 900 illus. in all, few in colour. A weighty, impressive volume, dealing with the work of 300 jewellers in 26 countries. *Class No:* 671.12

Precious Stones

[6423]
BRUTON, E. **Diamonds.** 2nd ed. 1978. London, N.A.G. Press, Ltd. xiv, 532p. illus., graphs, tables. ISBN: 0719800714.

First published 1970.

21 chapters (*e.g.* 1. Diamonds in history - 2. Where diamonds are found - 3. The big mining companies - 4. Mining and recovery methods ... 12/14. Grading polished diamonds ... 18. Physical properties of diamonds ... 20. Famous diamonds - 21. Identification of diamonds. Glossary of terms, p.495-503. 6 appendices (*e.g.* 2.The world's largest polished gem diamonds). Analytical index, p.519-32. *Class No:* 671.15

[6424]
COPELAND, L.L., *and others*. **The Diamond dictionary.** Los Angeles, Calif., Gemological Institute of America, 1960. viii, [1], 311p. illus., diagrs.

About 3,000 definitions, including notes on famous diamonds, biographies and abbreviations. Pronunciation given. Selected bibliography, p.314-5. Many illus., including reproductions of approximately actual size. 7 appendices, including 'United States and British ring-size equivalents'. *Class No:* 671.15

[6425]
TABURLAUK, J. **Pearls: their origin, treatment and identification.** Translated from French by D. Ceriog-Hughes. Radnor, Pa., Chilton Book Co., 1985. 247p. maps.

4 well-documented parts: Natural pearls - Pearl shellfish and the formation of pearls - Cultivated pearls - Pearl working and setting. Glossary of important technical terms. Maps of the world's identified pearling beds. International standards of trade and sale. Classified index. (Based on *New technical books*, v.71(7), July 1986, item 971). *Class No:* 671.15

[6426]
WILKS, J. *and* WILKS, E. **Properties and applications of diamonds.** 3rd ed. Oxford, Butterworth-Heinemann, 1991. 525p. diagrs., graphs, micrographs, tables. £90. ISBN: 0750610670.

17 documented chapters. 3 parts: 1. The structure of diamonds - 2. Mechanical properties - 3. Applications and wear of diamonds. 2 appendices: 1. Some numerical values; 2. Units and common factors. Detailed, analytical index. *Class No:* 671.15

Iron & Steel Goods

Cutlery

[6427]
PYBUS, S., *and others, comps*. **Cutlery: a bibliography.** 2nd ed. Sheffield City Libraries, 1982. [iv], 70p. £2.50. ISBN: 0900660848.

First published 1960.

About 1,000 entries for books and articles, often annotated. 14 sections: 1. History - 2. Individual products - 3. Design - 4. Manufacturing processes - 5. Standards - 6. Marketing and statistics - 7. Import controls - 8. Trade marks and patents - 9. Auxiliary tables - 10. Consumer guides - 11. Collections - 12. Illustrations - 13. Associations - 14. Bibliographies and periodicals. *Class No:* 672.7

Fasteners, Pins etc.

[6428]
PARMLEY, R.O., *ed*. **Standard handbook of fastening and joining.** 2nd ed. New York, etc., McGraw-Hill, 1989. [672]p. illus., tables. £55. ISBN: 0070485224.

First published 1977.

15 sections: Threaded fasteners-description and standards - Standard pins ... Rope splicing and tying - Metrics and general data. Index. The 2nd ed. contains over 50% new or revised material. For engineers, machinists and technicians in all fields. 880 illus. *Class No:* 672.8

Non-Ferrous Metal Goods

Tin-Ware

[6429]
UNITED NATIONS. Industrial Development Organization. **Information sources on the canning industry.** New York, United Nations, 1975. xii, 83p. (*UNIDO guide in sources of information, no. 19.*)

649 items in 12 sections: 1. Professional, trade and research organizations, learned societies and special information services - 2. Directories - 3. Sources of statistics, marketing and other economic data - 4. Basic handbooks, textbooks and manuals - 5. Monograph series - 6. Current periodicals - 7. Current abstracting and indexing journals - 8. Proceedings, papers and reports - 9. Specialized dictionaries and encyclopaedias - 10. Bibliographies - 11. Films and catalogues - 12. Other potential sources of information (E. Legislation; H. Packaging). 'Bibliographical sources used ...'. *Class No:* 673.8

Timber

Encyclopaedias

[6430]
BOUTELJE, E.B. Encyclopedia of world timbers: names and technical literature. Stockholm, Swedish Forest Products Laboratory, 1980. 338p. ISBN: 9186018000.

3 sections: Scientfic names in alphabetical sequences - Commercial and local names (*c*.8,000, A-Z), p.99-311 - Key to the technical literature (376 items, with locations at the Laboratory), p.313-38. Quarto. *Class No:* 674(031)

[6431]
Concise encyclopedia of wood and wood-based materials. Schniewind, A.P., *ed*. Oxford, Pergamon, 1989. 354p. illus., diagrs., tables. $125; £70. (*Advances in materials science and engineering.*) ISBN: 0080347266.

73 documented chapters (A-Z) by leading authorities in the field. Subject index. An updated supplement to *Encyclopedia of materials science and engineering,* edited by M.B. Beyer. *Class No:* 674(031)

Handbooks & Manuals

[6432]
DESCH, H.E. Timber: its structure, properties and utilization. 6th ed., revised by J.M. Dinwoodie. London, Macmillan, 1981. xii, 410p. illus., diagrs., graphs. ISBN: 0333257510.

First published 1938.

4 parts (19 chapters): 1. The structure of wood - 2. The gross features of wood (incl. chapter 6: 'Description of some of the more important commercial harwoods, Abura ... Walnut (European)', p.114-52 - 3. The properties of wood - 4. Considerations influencing the utilization of wood. 2 appendices; 1. List of botanical equivalents of common or trade names used in the text. Selective Bibliography, by chapters, p.394-401. Detailed, analytical index, p.402-10. *Class No:* 674(035)

[6433]
Handbook of wood and wood-based materials for engineers, architects and builders. United States. Department of Agriculture. Forest Service. Forest Products Laboratory. Rev. ed. New York, Hemisphere, 1989. v.p. illus., diagrs., graphs, tables. $77. ISBN: 0891161244.

First published 1935 as *Wood handbook;* previous ed. 1974.

23 chapters of the properties, applications and treatments of wood. Numerous tables. Glossary, p. G1-G11. 12p. detailed, analytical index. Emphasis is on US. *Class No:* 674(035)

Dictionaries

Polyglot

[6434]
BOERHAVE-BEEKMAN, W., *comp*. **Elsevier's wood dictionary in seven languages:** English/American, French, Spanish, Italian, Swedish, Dutch and German. Amsterdam, etc., Elsevier, 1964-67. 3v. illus.

....(contd.)

1. *Commercial and botanical nomenclature.* 1964. xviii, 479p.

2. *Production, transport, trade.* 1966. 642p.

3. *Research, manufacture, utilization.* 1966. 460p.

V.1 has 3,778 numbered entries; v.2, 5,658 entries; v.3, 3,728. In v.1 the English/American terms are followed by botanical number, sources of supply (areas), and equivalents in French, Spanish, Italian, Swedish, Dutch and German. *Class No:* 674(038)=00

[6435]
TEKNISKA NOMENKLATURCENTRALEN. Träbyggnardsordlista. [Glossary of timber construction.] Stockholm, the Centre, 1975. 224p. illus. (*TNC 60.*)

725 grouped Swedish terms and definitions, plus equivalents in English, German and French. Indexes in the four languages. *Class No:* 674(038)=00

English
[6436]
CORKHILL, T., *comp.* **A Glossary of wood:** 10,000 terms relating to timber & its uses, explained and classified. London, Stobart, 1979. viii, 656p. illus. £12.50. ISBN: 085442010x.

First published 1948 (London, Nema Press. viii. 656 p.).

Of the 10,000 terms mostly concisely defined, 1,000 are illustrated in clear line-drawings, some measured. Lengthier notes on major types of wood, *e.g.* mahogany. 'Veneers': 1p., with 6 illus. All aspects of timber, from tree to finished product, especially the many uses and fabrication of wood. Abbreviations are included in the A-Z sequence. Keyed line-drawings (*e.g.* roof trusses). A standard dictionary on wood. *Class No:* 674(038)=20

German
[6437]
BUCKSCH, H. Holz-Wörterbuch. [Dictionary of wood and woodworking practice.] Wiesbaden & Berlin, Bauverlag GmbH. ISBN: 3762624114.

1. *Deutsch-Englisch.* 2. Aufl. 1978 (reprinted 1986). 461p.

2. *Englisch-Deutch.* 1966. 536p.

Each volume has *c.*20,000 entry-words, some with definitions and notes on coverage, in addition to equivalents. German genders given. Appended conversion tables. *Class No:* 674(038)=30

[6438]
MÜHLE, P. Dictionary of wood science and technology: English-German, German-English. [Wörterbuch der Holzwirtschaft: Englisch-Deutsch, Deutsch-Englisch.] Wiesbaden, Oscar Brandstetler, 1992. [x],228,238p. ISBN: 3870971576.

About 10,000 words, with equivalents in each half. Covers wood anatomy, woodworking machinery, timber harvesting and conversion, pulp and paper manufacturing. *Class No:* 674(038)=30

Reviews & Abstracts
[6439]
Forest products abstracts. Wallingford, Oxon., CAB International Forestry Bureau, 1978-. 6pa. £147; $269. ISSN: 01404784.

1991: 2,640 abstracts. 10 sections: General publications and general techniques. - General aspects of forest products and industry - Wood properties (including bark) - Timber extraction, conversion and measurement - Damage to timber and timber protection. Surface finishes - Utilization of wood as such - Veneers. Composite boards, laminated beams, panels, improved wood glues - Pulp industries and the chemical utilization of wood - Other uses of forest products - Competitive materials - Marketing and trade. Economics. Author and subject indexes. Occasional 'news items' precede abstracts.

Online through CAB ABSTRACTS database (covering the 19 major CAB abstract journals). Hosts include DIALOG, ESA-IRS. CD-ROM (Silver Platter). *Class No:* 674(048)

[6440]
Wood industry abstracts. Pullman, Washington, Washington State Univ., Engineering Extension Service, 1972- 6pa.

Formerly *Wood products industry abstracts bulletin.* 1971-72.

1984: 1,200 abstracts, each with appended keywords. 4 divisions: A. Fundemental disciplines - B. Management - C. Process - D. Products. Coverage: books, theses, items of general interest. Author and keyword indexes per issue and annually. *Class No:* 674(048)

Periodicals
[6441]
TIMBER RESEARCH AND DEVELOPMENT ASSOCIATION (TRADA). List of current timber journals received by the TRADA library. Revised at intervals. High Wycombe, Bucks., TRADA, 1987. [i], 9p. gratis. (*TLB 138. 1987.*)

Lists *c.*60 journals (title; country; address). Country index (giving journals' frequency) precedes. *Class No:* 674(051)

Yearbooks & Directories
[6442]
Timber trades directory. Tonbridge, Kent, Benn, in conjunction with *Timber trades journal.* Annual. ISBN: 086382000x.

Contents: Alphabetical index to firms in the timber and allied trades. Buyer's guide to timber and timber products and sheet materials. Buyers' guide to plant, equipment and materials. Buyers' guide to services UK trade names. European agents and importers. Trade names. Index to advertisers precedes. *Class No:* 674(058)

Nomenclatures
[6443]
BRITISH STANDARDS INSTITUTION. Nomenclature of commercial timbers, including sources of supply. London, BSI, 1991. 123p. tables. (*BS 7359: 1991.*)

Covers both hardwoods and softwoods in tabular form, with index of standard and other names. *Class No:* 674(083.72)

Standards

[6444]
TIMBER RESEARCH AND DEVELOPMENT ASSOCIATION (TRADA). List of British Standards relating to timber. High Wycombe, Bucks., TRADA, 1987. [i], [i], 12p. £3.75 (to non-members). (*TLB 232. 1987.*)
Listed under subjects (General; Adhesives ... Woodworking; Miscellaneous). *Class No:* 674(083.74)

Maps & Atlases

[6445]
ILIC, J. CSIRO atlas of hardwoods. Berlin, Springer Verlag, 1991. [v],525p. illus. £135. ISBN: 3540532420.
Macro atlas, p.4-69 (with anatomical and other features). Micro atlas, p.70-472. Worldwide. References, p.475. Species index, p.477-525. CSIRO: Commonwealth of Scientific and Industrial Research Organizations. *Class No:* 674(084.3)

Worldwide

[6446]
LINCOLN, W.A. World woods in colour. London, Stobart, 1986. 320p. col. illus., tables. ISBN: 0854420282.
World woods, Abura ... Zebrano. Data on each: botanical and commercial names; distribution; general description; mechanical properties; seasoning; working properties; durability; uses. Selected wood grains, p.288-91. Appended: Table of uses. Sources of further information. Bibliography (grouped), p.301-5. Coloured illus. are reproduced actual size. Index of standard names. Index of vernacular, trade and other names. Index of botanical names. *Class No:* 674(100)

[6447]
PATTERSON, D. Commercial timbers of the world. 5th ed. Aldershot, Gower Technical Press, 1988. x,339p. illus. £45. ISBN: 0291397182.
First published 1948, by F.H. Titmuss.
An encyclopedia of over 350 species. 4 sections: 1. Structure and properties - 2. Wood: its preparation for use - 3. Hardwoods (Abura - Zebrawood) - 4. Softwoods (Alerce - Yew). Bibliography, p.322-25. Index of wood species. *Class No:* 674(100)

[6448]
RENDLE, B.J., *comp*. World timbers. London, Benn, 1969-70. 3v. illus.
1. *Europe and Africa.* 1969. 191p.
2. *North and South America.* 1969. 148p.
3. *Asia, and Australia and New Zealand.* 1970. 176p.
About 200 colour plates of wood specimens, with brief descriptions of distribution, availability, characteristics and use. Most of the plates appeared in the periodical *Wood*, 1946-60. Short bibliography (40 classified items). 'Unlikely to be superseded for many years' (*British Book news*, February 1971, p.136). *Class No:* 674(100)

[6449]
TIMBER RESEARCH AND DEVELOPMENT ASSOCIATION (TRADA). Timbers of the world.
Brook, W.H. Lancaster, TRADA, 1979-80. 2v. maps. ea. £30. ISBN: 0860958361; 086095837x.
Originally in 9v. 1978.
V.1: Africa; South America; Southern Asia; South-east Asia (1979. [x], 463p.). V.2: Philippines and Japan; Europe;

....(contd.)
North America; Australasia; Central America and the Caribbean (1980. [viii], 449p.). Data: description of the tree, strength and durability of the timber, its working qualities and uses. Each volume carries a bibliography. *Class No:* 674(100)

Scandinavia

[6450]
Handbook of the Northern wood industries, 1986/87. [Handbok for nordisk träindustri.] Stockholm, AB Svensk Trävaru-Tidning. 880p. ISBN: 9185464015.
Main countries: Sweden; Norway; Denmark; Finland (associations, etc.; timber industry sector; wood pulp and paper sectors; timber wood pulp and paper firms). Other countries (Argentina ... Yugoslavia), p.511-723. Indexes of shipbrokers, etc. General index. Index to advertisers. Appended list of Swedish and French names of some places in Finland. *Class No:* 674(48)

Woodworking

[6451]
MARTENSSON, A. The Woodworker's Bible. London, Pitman House, 1980 reprinted 1990. 288p. illus. £13.99. ISBN: 0273009222.
7 sections: The workshop - Portable power tools - Woodworking machines - Woodworking joints - Furniture construction - Wood as a material - Finishing. 'Dictionary of hand tools and devices', p.6-25. Step-by-step instructions. Addenda, p.288. Well captioned illus. (some in two-colour). Detailed analytical index, p.284-7. Good layout. *Class No:* 674.02

[6452]
TAYLOR, V. The Woodworker's dictionary. Pownal, VT, Storey Communications, 1991. *c.*264p. illus. $24.94 (HB); $12.95 (PB). ISBN: 0882666460; 0882666452, Pbk.
About 3,600 woodworking terms. Covers carpentry, cabinetmaking, joinery, furniture construction, upholstery, species of wood and finishes. Largely based on a series in the British magazine *Woodworker*, revised and expanded. Distinguishes between US and British terms. Brief appendices on major furniture designers, crafts people and styles. 'Recommended' (*Library journal*, January 1991, p.98). *Class No:* 674.02

[6453]
UNITED NATIONS. Industrial Development Organization. Information sources on woodworking machinery. New York, United Nations, 1978. xii, 99p. $4. (*UNIDO guides to information sources, no.31.*)
745 numbered entries. 12 sections: 1. Professional trade, and research organizations, learned societies and special information services - 2. Directories - 3. Sources of statistics, marketing and other economic data - 4. Basic handbooks, textbooks and manuals - 5. Monographic series - 6. Current periodicals - 7. Current abstracting and indexing periodicals - 8. Proceedings, papers and reports - 9. Specialized dictionaries and encyclopaedias - 10. Bibliographies - 11. Films and film catalogues - 12. Other potential sources of information *e.g.* K. Training). 'Bibliographical sources used in compiling the present guide.' Sections 2-3, 6-7, 9-11 have annotations. *Class No:* 674.02

[6454]

The woodworker directory, 1988. London, PDU, Woodworker directory, 1988. 218p. £12.95.

4 sections: 1. Manufacturers, UK importers and timber suppliers - 2. Who makes what and product information (Timber and veneer ... Exhibition and sales opportunities) - 3. Buyer's guide - 4. Further sources of information (books; periodicals; organizations), p.206-18. *Class No:* 674.02

Timbers & Woods

[6455]

BUILDING RESEARCH ESTABLISHMENT. Handbook of hardwoods. Farmer, R.H. 2nd ed. London HM Stationery Office, 1972. v, 243p. tables. £12. ISBN: 0114705410.

First published 1956.

Introductory sections (*e.g.* Working properties; Veneer and plywood; Uses). 'The timbers', p.11-213, Abura ... Willow (Data: other names; the tree; the timber, - properties, processing, durability and preservation uses). Appendices: 1. Properties of hardwoods; 2. Types of saws; 3. Kiln schedules. Index of botanical names; Index of trade and local names. *Class No:* 674.03

[6456]

BUILDING RESEARCH ESTABLISHMENT. A Handbook of softwoods. 2nd ed. London, HM Stationery Office, for the Department of Environment, 1972. 63p. £8. ISBN: 0114705631.

First published 1957 (Forest Products Research Laboratory).

Detailed descriptions of 49 British and imported softwoods, with brief notes on a further 6. Data on properties and use of each, giving specific reference to the wood-using industries of the UK. *Class No:* 674.03

Preservation

[6457]

BRITISH STANDARDS INSTITUTION. Glossary of terms relating to timber preservation. London, BSI, 1985. 19p. £23.90. (*BS 4261: 1985 (1991)*.)

Defines 484 terms, in 6 sections: 1. General - 2. Attack by fungi - 3. Attack by insects and marine borers - 4. Preservative processes - 5. Terms associated with preservative treatments - 6. Miscellaneous. Index. *Class No:* 674.04

Tools & Machinery

[6458]

GOODMAN, W.L. The History of woodworking tools. London, Bell, 1964. 208p. illus.

8 chapters: The axe and the adze - The plane - The saw - Boring tools - The carpenter's bench - The rule - Chisels and gouges - Miscellaneous tools (*e.g.* levels, the carpenter's bag or tool containers, set square, dividers, gouges, mallet, hammer, spokeshave). Bibliography. List of British museums with displays of such tools. 'The first comprehensive account of such tools in the Western hemisphere' (Nature, v.204, no.4956, 24 October 1964, p.314). *Class No:* 674.05

[6459]

HORTEN, H.E. Woodworking machines in 4 languages. London, C.R. Books, 1968. xiii, 353p. illus. (pl.), diagrs., tables.

Over 3,500 technical terms appear in 4 sequences: English, German, French, Spanish; German, English, French, Spanish; French, English, German, Spanish; Spanish, English, French, German. Pages 1-99 provide photographs of commonly-used machines. Conversion tables. *Class No:* 674.05

[6460]

SALAMAN, R.A., *ed.* Dictionary of woodworking tools. 2nd ed. London, Unwin Hyman, 1989. 544p. illus. £40. ISBN: 0044402562.

First published 1975.

A-Z sequence of tools, entries including much material from manufacturers' catalogues. A feature: clear line-drawings and illus., many of them keyed (*e.g.* violin maker's tools: 14 parts; wooden wagon: 42 parts). Profuse cross-references. A list of *c.*70 trades, from boiler maker to woodwind maker, precedes. Bibliography and references. 'Useful not only to historians, but to collectors and craftsmen; indispensable for museums collecting such artifacts' (*Library journal,* 1 December 1976. p.2471, of the 1975 ed.). Salaman's collection of tradesman's tools is exhibited in St. Albans City Museum. *Class No:* 674.05

[6461]

SIMS, W.L. Two hundred years of history and evolution of woodworking machinery. Melton Mowbray, Leics., Walders Press, 1985. illus., facsims. £8.

Traces the history from the advent of rotating cutters and saws to the present day. Profusely illustrated, including reproductions from manufacturers' catalogues. 'This comprehensive, fascinating and obviously meticulously research history' (*CME/Chartered mechanical engineer,* v.33, February 1986, p.54). *Class No:* 674.05

Carpentry & Joinery

[6462]

HEWETT, C.A. English historic carpentry. London & Chichester, Phillimore, 1980. xiv, 338p. illus. (incl. pl.). ISBN: 0850333547.

8 periods: 1. The Anglo-Saxon period (AD 449 to 1066) ... 8. The Renaissance and after (1581 to 1890). 6. appendices (*e.g.* 3. Joints used for framing of floors). Glossary, p.vii-xii. Bibliography, p.xii- xiv. 350 illus. and line-drawings. Index. *Class No:* 674.1/.2

Plywood

[6463]

WOOD, A.D. Plywoods of the world: their development, manufacture and application. 3rd rev. ed. Edinburgh, Johnston & Bacon, 1963. xiv, 489p. illus.

First published as *Plywoods* (1942).

10 Chapters. 1. Development - 2. Physical properties of wood - 3. The manufacture of multiply - 4. Laminboards, blackboards, composite boards, etc. - 5. Grading, testing, packing and storing - 6. Plywood-producing countries of the world (p.187-280) - 7. Commercial plywoods, etc. - 8. The plywood trade in the United Kingdom - 9. The development of wall panelling - 10. Applications. Appended terms and definitions, grading rules, plywood associations, publications

....(contd.)

consulted or recommended (p.323-5), trade journals and periodicals. 241 illus, incl. 19 plates. Comprehensive. *Class No:* 674.812

Leather

Bibliographies

[6464]

UNITED NATIONS. Industrial Development Organization. **Information sources on the leather and leather goods industry.** New York, United Nations, 1972. xix, 80p. (*UNIDO Guides to information sources, no.3.*)

497 items, in 10 sections: 1. Addresses of professional, trade and research organizations, learned and special information centres - 2. Directories - 3. Sources of statistics and other economic data - 4. Basic handbooks, textbooks and manuals - 5. Current periodicals - 6. Current abstracting and indexing services - 7. Special documents: proceedings, papers and reports - 8. Monographs - 9. Bibliographies - 10. Other potential sources of information (*e.g.* Information on standards and specifications). 'Bibliographical sources utilized', appended. *Class No:* 675(01)

Dictionaries

Polyglot

[6465]

INTERNATIONAL COUNCIL OF TANNERS and others. **International glossary of leather terms.** 2nd ed., rev. and enlarged. London, the Council, 1975.

First published 1968.

About 300 terms in each of 5 languages, - English, French, German, Spanish and Italian. 5 sequences, each on differently coloured papers. 'It will be a boon to the translator' (*Incorporated linguist*, v.5(2), April 1969, p.42, on the 1968 ed.). *Class No:* 675(038)=00

[6466]

UNION INTERNATIONALE DES SOCIÉTÉS DE CHIMISTES DES INDUSTRIES DU CUIR. **Leather technical glossary in six languages.** Darmstadt, E. Roethe Verlag, 1976. 720p. ISBN: 3792900090.

5,429 numbered English-base terms, with explanations, and equivalents and indexes in French, German, Italian, Russian and Spanish. Genders given.
Class No: 675(038)=00

English

[6467]

BRITISH STANDARDS INSTITUTION. **Glossary of leather terms.** London, BSI, 1983. 24p. diagrs. (*BS 2780: 1983.*)

Previously as BS 2780: 1956.

277 terms are briefly defined in one A-Z sequence. Cross-references. 6 diagrams of parts of hides.
Class No: 675(038)=20

[6468]

INDIAN STANDARDS INSTITUTION. **Indian Standard glossary of terms relating to hides, skins and leather.** New Delhi, the Institution, 1961. 110p. illus., diagrs. (*I.S. 1640: 1960.*)

Descriptive entries for *c.*1,500 terms, especially on types of tanning material. English entry-words are sometimes followed by the Sanskrit equivalent. Sources, p.2-3. Illus., p.103-10. *Class No:* 675(038)=20

Reviews & Abstracts

[6469]

Current leather literature. Madras, Central Leather Research Institute, 1968-. Monthly.

About 1,200 abstracts pa., drawing on *c.*250 journals. Sections: Leather science and industry - Footwear and leather goods - Wood technology ... Textile technology ... Biochemistry - Botany. Subject indexes per issue and annually. *Class No:* 675(048)

Yearbooks & Directories

[6470]

International leather guide: 1990. Tonbridge, Kent, Benn, 1990. x,500p. £65. ISBN: 0863820832. ISSN: 09555080.

Sections: Tanners and merchants - Tanners and merchants buyers' guide - Hide and skin suppliers - Chemical suppliers - Chemical buyers' guide - Machinery manufacturers - Machinery buyers' guide - Consultants and technical assistance - Trade organizations - Five-language glossary of everyday leather terms (English; French; German; Spanish; Italian) - Index to advertisers. *Class No:* 675(058)

Equipment & Instruments

[6471]

SALAMAN, R.A. **Dictionary of leather-working tools, 1700-1950,** and the tools of allied trades. London, Allen & Unwin, 1986. xxi, 377p. illus. ISBN: 004621030x.

13 chapters (1. Bookbinder; 2. Boot and shoe maker; 9. Harness maker and saddler; 10. Hat maker; 11. Leather manufacture (1. Tanner's tools; 2. Currie's tools); 13. Miscellaneous trades and tools (*e.g.* parchment and vellum maker; taxidermist)). References and bibliography, p.340-8. General index, p.346-77 ('Knife': 3 columns). List of museums and institutions. Fully illustrated.
Class No: 675:002.50

Furs

[6472]

BACHRACH, M. **Fur: a practical treatise.** 3rd ed. New York, Prentice-Hall, 1953. (Reprinted New York, Gordon Press, 1977). xii, 600p. illus., diagrs.

12 sections, dealing with the different types of fur-bearing animal. Appendix includes nomenclature of fur peltries, a fur-products name guide, and a subject bibliography, p.639-67. Detailed index. A standard work, covering geographical features, ranching, marketing and buying, plus technical aspects. *Class No:* 675.6

Paper

Databases

[6473]
Paperchem. Atlanta, Ga., Institute of Paper Science and Technology.

Covers patent and journal literature (1000 titles scanned). Corresponds to the print *Abstract bulletin of The Institute of Paper Science and Technology*. About 300,000 records; updated monthly. Available on DIALOG, files 240,840. *Class No:* 676(003.4)

Bibliographies

[6474]
UNITED NATIONS. Industrial Development Organization. **Information sources on the pulp and paper industry.** New York, United Nations, 1974. xii, 91p. $4. (*UNIDO Guide to information sources, no.11.*)

629 entries, in 12 sections: 1. Addresses of professional trade and research organizations, learned societies and special information sources - 3. Directories ... 6. Current periodicals ... 9. Special dictionaries and encyclopedias - 10. Bibliographies ... 12. Other sources of reference (*e.g.* Standards and specifications). 'Bibliographical sources used', appended. *Class No:* 676(01)

Encyclopaedias

[6475]
LABARRE, E.J. Dictionary and encyclopaedia of paper and paper-making, with equivalents of the technical terms in French, German, Dutch, Italian, Spanish and Swedish. 2nd ed., rev. and enl. London, Oxford Univ. Press, 1952. xxiv 488p. illus.

First published 1937, as *A dictionary of paper and paper-making terms* (Amsterdam, Swets & Zweitlinger).

About 3,000 entries, including lengthy articles (*e.g.* 'Sizes', p.246-72; 'Wrapping papers', p.368-74). List of sources (5 sections), p.xi-xx. Indexes in 7 languages. *Class No:* 676(031)

[6476]
—LOCKER, E.J. Dictionary and encyclopaedia of paper and paper-making. Amsterdam, Swets & Zeitlinger, 1967. x, 104p.

A supplement to Labarre (*qv*) adds *c.*1,000 terms in English, with equivalents and indexes in the other 6 languages. *Class No:* 676(031)

Handbooks & Manuals

[6477]
BRISTOW, J.A. and KOLSETH, P. Paper structure and properties. New York, Dekker, 1986. x, [i], 390p. illus., diagrs., graphs, tables. ISBN: 0824775608.

17 contributors; 17 documented chapters. 1. The fiber - 2. Paper structure ... 17. The layered structure of paper (13 references). 3 appendices. Mechanical models of paper. Detailed, analytical index. 'Presents paper as a material with a structure which can be described at different levels' (*Preface*). For the professional. *Class No:* 676(035)

[6478]
HUNTER, D. Paper-making: the history and techniques of an ancient craft. 2nd ed., rev. and enl. London, Pleiades, 1957. xxix, 611, xxxviip. diagrs.

First published 1943.

Covers materials and methods, with a lengthy chronology, p.463-584, of paper-making, paper, and the uses of paper. Bibliography of 200 items (p.583-602): 1. Oriental - 2. Occidental - 3. Watermarking - 4. Paper colours and surface decoration. Chapter notes. Analytical index. 317 diagrs. in text. *Class No:* 676(035)

Dictionaries

[6479]
AMERICAN PAPER AND PULP ASSOCIATION. The Dictionary of paper, including pulp, paperboard, paper properties, and related papermaking terms. 3rd ed. New York, the Association, 1965. xii, 500p.

First published 1940.

About 2,567 terms (1,463 revised; 710 deleted; 329 new). Entries are lengthy at times (*e.g.* 'Pulps': p.5-30). Numerous cross-references. *Class No:* 676(038)

[6480]
BRITISH STANDARDS INSTITUTION. Glossary of paper, board, pulp and related terms. London, BSI, 1979. 25p. £25.60. (*BS 3203: 1979.*)

About 300 terms defined. 7 sections: 1. Pulp: general terms - 2. Pulp manufacture - 3. Types of pulp - 4. Paper and board: general terms - 5. Paper making - 6. Types of paper or border-converted products - 7. Properties of pulp, paper or board. Index of *c.*700 terms. *Class No:* 676(038)

[6481]
Fachwörterbuch papier: Deutsch/Englisch, Englisch/Deutsch. Gerssbach, Eurospäscher Wirtschaftsdisent, 1990. 243p. ISBN: 3886400433.

About 5000 entries in each half. *Class No:* 676(038)

[6482]
LAVIGNE, J.R. Pulp & paper dictionary. San Francisco, Calif., Miller Freeman Publications, 1986. 370p. $47.50. ISBN: 0879301686.

Over 5,000 entries, defined in 2-4 lines, usually; cross-references. Appendix A: References (including 'Magazines and trade journals', 'Standards and data sheets'); B. Conversion tables. *Class No:* 676(038)

[6483]
TEKNISKA NOMENKLATURCENTRALEN. Pappersordlista. Stockholm, the Centre, 1980. 416p. illus. (*TNC 74.*) ISBN: 0171620826.

1,284 Swedish paper industry terms and definitions, with equivalents and indexes in English, French, German, Norwegian and Finnish. *Class No:* 676(038)

Reviews & Abstracts

[6484]
Paper and board abstracts. Compiled by the Information Section of PIRA (Paper Industry Research Association). Oxford, etc., Pergamon Press, for the Paper and Board, Printing and Packaging Research Association, 1968-. Monthly. DM1465pa. ISSN: 03070778.

About 4,000 abstracts pa. Some 20 sections (*e.g.* Paper and board industry in general; company information; Non-

.... *(contd.)*

fibrous raw material; Paper and board making; Machinery and equipment; Instrumentation and control; Properties and history of paper and board; Paper and board specialities; Synethetic paper and nonwovens. Book section). Available online. *Class No:* 676(048)

Yearbooks & Directories

[6485]

International pulp and paper directory, 1992-93. San Francisco, Ca., Miller Freeman, 1991. xxx,916p.

7 sections: Pulp and paper mills (by continent, then country) - Mill grades (pulp; paper) - Market pulp (grades; producers) - Paper merchants (European; North American) - Industry information sources - Trade shows and conferences - Buyers' guide. *Class No:* 676(058)

[6486]

Phillips International Paper directory, 1990. 87th ed. Tonbridge, Kent, Benn, 1990. xiv,594p. £80. ISBN: 0863820417.

Contents: Mills of the world (countries, A-Z); Classified list of products - Merchants - Agents (paper; pulp) - Waste paper merchants and processors - Exporters and importers - Suppliers (machinery; equipment; materials) - Converters (packaging and other products) - Brand names and watermarks - Trade associations - Index to advertisers. *Class No:* 676(058)

Histories

[6487]

LEIF, I.P. An International sourcebook of paper history. Hamden, Conn., Archon; Folkeston, Dawson, 1978. viii, 160p. ISBN: 0713908277.

2,185 numbered references. 5 sections: 1. General histories of paper and watermarks - 2/4. The history of paper and papermaking in Asia and Australia, Europe and the Soviet Union, North and South America - 5. The study of paper history. Sections 2/4 are each subdvided by countries A-Z (items 817-1444). Author and analytical subject indexes. *Class No:* 676(091)

Thesauri

[6488]

Thesaurus of pulp and paper terminology. 3rd ed. Atlanta, Ga., Institute of Paper Science and Technology, 1991. xvi,557p. ISBN: 0870100009.

First ed. 1965; 2nd ed. 1971.

Includes over 22,000 main entries (11,000 'key' terms and 11,000 cross-references). Terms include chemical, biological, industrial nomenclature. Alphabetical display of terms (540p.) is followed by a 'Microthesaurus' of well-known proper names (countries, professional societies, etc.). *Class No:* 676:025.43

Information Services

[6489]

International pulp and paper information sources: an essential guide to the world of pulp and paper information. Farrell, S. Leatherhead, Surrey, Pira Information Services, 1988. [xii],379p. ISBN: 0902799185.

.... *(contd.)*

A range of information sources (*e.g.* published materials; organizations; markets; manufacturing standards and testing) on all aspects of paper, including raw materials and manufacturing equipment. 24 countries. Indexes: Country; Organizations; Publications; Subject. A wealth of information. *Class No:* 676:061:025.5

Watermarks

[6490]

BRIQUET, C.M. Les filigranes. Dictionnaire historique des marques du papier, des leur apparition vers 1282 jusqu'en 1600. Avec 39 figures dans le texte et 16, 112 fac-similés de filigranes. 2nd éd. Leipzig, Hiensemann, 1923. 4v. (Reprinted New York, Hacker Art Books, 1966). 4v. (A-Ch, Ci-K, L-O, P-Z).

First published 1907 (Paris, Picard).

The 16,112 facsimiles of watermarks are arranged by subject of devices (*e.g.* 'Tête de boeuf'). Each volume carries a section of descriptive notes. References to sources precede the numbered illus. *Class No:* 676.026.4

[6491]

—**CHURCHILL, W.A. Watermarks in paper in Holland, England, France, etc., in the 17th and 18th centuries, and their interconnection.** Amsterdam, Hertzberger, 1935. 94 + 432p. illus. (pl.), tables.

Provides 578 examples of watermarks, 94p. of text, 432p. of plates. Includes chronological lists of Dutch and French papermakers. *Class No:* 676.026.4

Textiles

Databases

[6492]

Textile technology digest. Charlottesville, Va., Institute of Textile Technology.

Covers dyeing, laundering, mill operation, natural and man-made fibres, preservation, apparel design, marketing and statistics. Corresponds to the print *Textile technology digest.* Over 188,000 records; updated monthly. Available on DIALOG, file 119. *Class No:* 677(003.4)

[6493]

World textiles. Oxford, Elsevier, 1970-.

The database for *World textile abstracts.* Over 180,000 records; updated monthly. Available on DIALOG, file 67. *Class No:* 677(003.4)

Bibliographies

[6494]

DOIDGE, R.I. A Survey of textile industry information. London, City University, Centre for Information Science, 1982. v, 119 + 18l. facsims., tables.

Thesis accepted for MSc.

More than 1,000 annotated and evaluated entries. 9 sections: 1. General information sources - 2. Information retrieval methods and abstracting service - 3. Health and safety in the textile industry - 4. Textile trade - 5. Market research and statistical sources - 6. Energy use in the textile

....*(contd.)*

industry - 7. Care labelling and flammability of textiles - 8. Textile history and conservation - 9. Information on textile industry research. Includes 'Databases', 1.38-40. Well organized, although lacking an index. *Class No:* 677(01)

[6495]
FARNFIELD, C.A., *ed.* **A Guide to sources of information in the textile industry.** 2nd ed., rev. and enlarged. Manchester, Textile Institute, 1974. [v], 130p.

First published 1970 (Aslib Textile Group, and Textile Institute).

8 sections: 1. Textile organizations (countries A-Z, p.1-44) - 2. Current textile periodicals - 3. Abstracts, current awareness, review publications, and information retrieval - 4. Books about textiles: a selected list (p.68-86) - 5. Textile directories and dictionaries - 6. Standard specifications - 7. Patents - 8. Sources of textile statistics. Entries in sections 3, 5 and 8 are annotated. The basic guide. *Class No:* 677(01)

[6496]
VOGEL, J.T. *and* **LOWRY, B.W.** **The Textile industry:** an information sourcebook. Phoenix, Az., Oryx, 1989. 246p. $49.50. (*Oryx sourcebook series in business and management, 19.*) ISBN: 0897743547.

Gives priority to English-language sources and works published since 1980. Emphasis is on business rather than technology, and includes a 'Core library collection' as the final chapter. Author, title and subject indexes. *Class No:* 677(01)

Encyclopaedias

[6497]
Encyclopedia of textiles. 3rd ed., by the editors of *American fabrics and fashions magazine.* Englewood Cliffs, N.J., Prentice Hall, Inc., 1980. xvi, 636p. illus., facsims. ISBN: 0132765164.

First published 1960.

7 major sections (subdivided): The textile fibres - History & origins - Textiles in the Americas - The manufacturing processes - Fabric finishing - Specialty uses of textiles - Textile definitions, p.512-601. Detailed, analytical index, p.604-36. Fully illustrated. *Class No:* 677(031)

[6498]
GRAYSON, M., *ed.* **Encyclopedia of textiles, fibres and nonwoven fibres.** New York, etc., Wiley, 1984. xxvi, 581p. (*Encyclopedia reprint series.*)

One in the series of carefully selected reprints from the world-renowned Kirk-Othmer *Encyclopedia of Chemical Technology. Class No:* 677(031)

[6499]
JERDE, J. **The Encyclopedia of textiles.** New York, Facts on File, 1992. 260p. illus., diagrs., tables. $45. ISBN: 0816021058.

Entries cover types of textiles, history, manufacturing processes, A-Z. About 250 illus., 42 in colour. Bibliography. *Class No:* 677(031)

Handbooks & Manuals

[6500]
HALL, A.J. **The Standard handbook of textiles.** 8th ed. London, Newnes-Butterworths, 1975. [vii], 441p. illus., graphs. ISBN: 0408014873.

First published 1946.

6 chapters: 1. The natural and man-made fibres - 2. The properties of textile fibre - 3. The conversion of fibres into yarns and fabrics (41 illus.) - 4. Bleaching, dyeing, printing and finishing: methods and machinery - 5. Colour and finish, from the viewpoint of manufacturer and user - 6. The care of clothes and simple identification tests. Bibliography, p.431-2. Analytical index. *Class No:* 677(035)

[6501]
TEXTILE INSTITUTE. **The Identification of textile materials.** 7th ed. Manchester, the Institute, 1975. v, 262p. illus., diagrs., tables.

3rd ed. 1951.

Introduction - The properties of textile fibres: natural fibres, man-made fibres - Scheme of analysis - Note on reagents and methods of test - Photomicrographs (167). Index. The terms used are, whenever published, those listed in *Textile terms and definitions* (*qv*). *Class No:* 677(035)

Dictionaries

Polyglot

[6502]
COMMISSION ON TEXTILE DOCUMENTATION OF THE EUROPEAN GROUP FOR THE EXCHANGE OF EXPERIENCE ON THE DIRECTION OF TEXTILE RESEARCH. **Multilingual glossary of textile terminology,** with translations into Danish, Dutch, English, Finnish, French, German, Japanese, Norwegian, Spanish and Swedish. Cambridge, Mass., Fibres and Polymers Laboratories, Massachusetts Institute of Technology, 1972. 9v.

Each volume gives priority to one particular language. The English-entry volume ([i], viii, 395, 73p.) has *c.*10,000 terms, translated into the other 8 languages, plus a separate English-Japanese (native characters) section. Omits Russian. *Class No:* 677(038)=00

[6503]
—**TEKNISKA NOMENKLATURCENTRALEN.** Textilordlista. [Glossary of textiles.] Stockholm, the Centre, 1981. 272p.

Defines *c.*770 terms in Swedish, with equivalents in English, French, German, Dutch, Norwegian and Finnish. *Class No:* 677(038)=00

[6504]
COURTAULDS, Ltd. **Courtaulds' vocabulary of textile terms.** 2nd ed. London, Courtaulds, Ltd., 1972. 234p. tables.

English-based terms in classified groups, with French, Spanish, German and Russian equivalents, and indexes in all five languages. Covers fibres (especially man-made), yarns and fabrics, their properties and processing. Appended conversion tables. Well produced. *Class No:* 677(038)=00

[6505]
ITS textile dictionary: English, Deutsch, Français, Italiano, Español, Portugês. New ed. Amsterdam, Elsevier; Zurich, International Textile Service, Ltd., 1989. viii,1537p.
29,300 entries. One A-Z sequence in all six languages.
Class No: 677(038)=00

English
[6506]
TEXTILE INSTITUTE. Textile terms and definitions.
Beech, S.R., *ed.* 5th ed. Manchester, the Institute, 1986. x, 297p. illus., diagrs. ISBN: 0900739819.
First published 1954.
Defines *c.*4,500 textile terms. Asterisks denote new definitions and those revised since the 7th ed. (1975); dagger symbols indicate definitions agreed with the Society of Dyers and Colourists. Cross-references. Appendices include conversion tables and a classification of textile fibres. Aimed at lay users, students and specialists. Authoritative.
Class No: 677(038)=20

[6507]
WINGATE, I.B. Fairchild's dictionary of textiles. 6th ed. New York, Fairchild Pubrs., 1979. 691p. illus. ISBN: 0870051989.
First published 1915.
About 15,000 entries, including organizations. 7 sections: A. Textile fibers - B. Yarns - C. Fabric construction - D. Finishing - E. Finished products ready for sale - F. Inventors and development of textile technology - G. Trade and government standards and regulations. Includes brief biographies, trademarks. Cross-references. 'An indispensable source of reference to people in all branches of that industry and related ones ...' (*Choice,* v.16, July/August 1979, p.646). *Class No:* 677(038)=20

[6508]
—**HARDINGHAM, M.** Illustrated dictionary of fabrics. London, Studio Vista, 1978. 159p. illus. ISBN: 0289708478.
Sections: wool; cotton; silk; other plant fibres; man-made fibres. Textile terms, p.130-58. List of obsolete fabrics. Bibliography. Index. *Class No:* 677(038)=20

German
[6509]
DE VRIES, L. Wörterbuch der Textilindustrie. Wiesbaden, Brandstetter Verlag, GmbH, 1959. 2v.
1. *Deutsch-Englisch.*
2. *Englisch-Deutsch,* by L. De Vries and O.H. Luken.
Each volume has *c.*15,000 entry-words. Largely equivalents; some expressions. Includes abbreviations and some trade names. German genders omitted. Each v. lists contributors; v.2 has a page on sources.
Class No: 677(038)=30

[6510]
—**HOHENADEL, P.** *and* **RELTON, J.** A Modern textile dictionary, English-German, German-English. Wiesbaden, Brandstetter, 1977-79. 2v. (486 + 376p.). ISBN: 3870970774; 3890970856.
V.1: English-German has *c.*12,000 entry-words; v.2: German-English, *c.*10,000 entry-words. Massed subentries.
Class No: 677(038)=30

French
[6511]
HIRSCH, P. Lexique textile, français-anglais. Paris, Éditions Olifant, 1980. 244p.
About 5,000 French entry-words. Abbreviations are listed separately. Appended list of weights and measures, as applied to textiles. *Class No:* 677(038)=40

Reviews & Abstracts
[6512]
Textile technology digest. Charlottesville, Va., Institute of Textile Technology, 1944-. Monthly. ISSN: 00405191.
About 750 entries per issue. 9 sectons (A. Fibers ... C. Fiber production ... E. End product fabrication ... G. Company practices ... I. Miscellaneous (conferences; education; etc.)). Author index per issue.
Class No: 677(048)

[6513]
World textile abstracts. Oxford, Elsevier, 1969-. Monthly. £320pa. ISSN: 00439118.
About 400 abstracts per issue. 10 sections: 1/3. Fibres, yarns, fabrics: manufacture and properties - 4. Chemical and finishing processes - 5. Products: manufacture, properties, aftercare - 6. Plant services and environment - 7. Management - 8. Analysis, testing, quality control - 9. Polymer science - 10. Generalities. Author, patent and subject indexes per issue, cumulated annually.
World textile abstracts: register of keywords (Shirley Institute, 1983. 2v. 91p., 28p.) consists of *Keyword list* and *Auxiliary lists.*
Online access: DIALOG, INFOLINE.
Class No: 677(048)

[6514]
—Textile digest. Manchester, Shirley Institute, 1972-. Monthly.
Derives its *c.*120 abstracts per issue from the World textile abstracts database. *Class No:* 677(048)

Yearbooks & Directories
[6515]
The Kendal textile industry directory, 1987. Huddersfield, Kendal Publications, Ltd., 1987. 432p.
List of textile industry companies, A-Z, p.18-224. Data on each include activities, names and directors and bankers. Buyers guide. Appended list of trade names. Index of advertisers. *Class No:* 677(058)

Histories
[6516]
FLEMMING, E. Encyclopaedia of textiles: decorative fabrics, from antiquity to the beginning of the 19th century, including the Far East and Peru. Rev. ed. London, Zwemmer, 1958. illus. (incl. pl.).
First published 1927.
History of textiles, p.vii-xxvi. Concentrates on Europe, but includes China, Japan, Persia and Peru. 'The grouping within the individual areas has been made according to the weaving techniques employed, subdivided according to the available historical evidence' (*Preface*). Illus. include 304 plates (16 in colour). *Class No:* 677(091)

[6517]
The Illustrated history of textiles. Ginsburg, M., *ed.*
London, Studio Editions, 1991. 224p. illus. £19.95. ISBN:
1851704469.
Development in textile production and design, from the
Middle Ages to the present, and covering early woven
cottons and silk to recent acrylics. Includes essays on the
different categories of textiles, including lace and tapestry.
Class No: 677(091)

Worldwide

[6518]
**ECONOMIST INTELLIGENCE UNIT. World textile trade
and production trends.** Anson, R. *and* Simpson, P.
London, the Unit, June 1988. 343p. graphs, tables. (*Special
report no. 1108.*) ISBN: 0850582075.
3 parts (16 chapters): 1. General trends in production - 2.
International trade in textiles and clothing - 3. Textile
prospects and clothing prospects, by country (p.161-342).
Appendix: Exchange rates. Statistics usually for 1980,
1985/87. Numerous tables. *Class No:* 677(100)

Thesauri

[6519]
BACKER, N. *and* **VALKO, E.I.,** *eds.* **Thesaurus of textile
terms,** covering fibrous materials and processes. 2nd ed.
Cambridge, Mass., and London, MIT Press, 1969. xv,
448p.
First published 1966.
8,000 keywords; 72,000 relationships (p.1-263).
References used in preparing the *Thesaurus* (62 references),
p.332-8. Appendices A-C (A. 'Problems of textile
information retrieval'). 7-p. bibliography. 'Highly
recommended, particulary for research oriented individuals'
(*Choice,* v.6(8), October 1969, p.996).
Class No: 677:025.43

Museums

[6520]
LUBELL, C., *ed.* **Textile collections of the world.** London,
Studio Vista, 1976-77. 3v. illus. (incl. col. pl.).
1. *The United States and Canada.* 1976. 320p.
2. *United Kingdom and Ireland.* 1976. 224p.
3. *France,* 1977. 240p.
Illustrated guides, arranged under cities. V.1-2 contain
905 illus. (118 in colour), with introductory essays on
national textile design and data on individual collections.
Favourably reviewed in annual *Textile history,* v.8, 1977,
p.194. *Class No:* 677:061:069

Production Processes

[6521]
BURNHAM, D.K. A Textile terminology, warp and weft.
London, Routledge & Kegan Paul, 1980. xiv, 216p. illus.,
diagrs.
Expanded and adapted from H. Burnham's *The
vocabulary of textile terms* (1964).
Defines more than 550 woven textile terms, usually with
equivalents in French, German, Italian, Spanish, Portuguese
and Swedish. Bibliography of citations, p.209-16. Well
illustrated. 'Although of more obvious value in certain

.... (contd.)
academic or school libraires, it would not be out of place in
the reference section of a public library' (*Library review,*
v.31, Winter 1982, p.300). *Class No:* 677.02

[6522]
FISCHER-BOBSIEN, C.H., *ed.* **International encyclopedia
of textile finishing.** 3rd advance volume. Dalmen (W.
Germany), Laumenn Verlag, 1985. DM260. illus.
Nearly 6,000 entries, mostly with photographs. Comprises
the important terms, products, processes and materials
associated with textile finishing. *Class No:* 677.02

Weaving

[6523]
PRITCHARD, M.E. A Short dictionary of weaving,
including some spinning, dyeing and textile terms, and a
beginner's guide to weaving and dyeing. London, Allen &
Unwin, 1954. 196p. illus., diagrs.
Several hundred concise definitions, p.13-109, with clear
line-drawings. 3 appendices: 1. Specimen loom, showing
principal parts; types of loom - 2. Quick-reference sections
(including calculating qualities; knots used in weaving;
warping guide) - 3. The pattern in weaving. Bibliography,
p.195-6. *Class No:* 677.024

Machinery

[6524]
**BRITISH STANDARDS INSTITUTION. Glossary of terms
relating to textile machinery and accessories,** warping
machinery and the preparation of warp for weaving.
London, BSI, 1976. [ii], 22p. illus. (*BS 5399: 1976.*)
Defines *c.*160 terms. 1. General terms - 2. Processing
terms - 3. Terms for machines, devices, assemblies and
component parts. In English, French and Russian. Appendix
of equivalent German terms. *Class No:* 677.05

Products

Carpets

[6525]
**BRITISH CARPET MANUFACTURER'S ASSOCIATION
LIMITED. Index of quality names.** London, The
Assocation. Annual, with quarterly supplements. v.p.
Names, A-Z, with type of carpet, finish, pile content and
manufacturer. *Class No:* 677.07.5

[6526]
Carpet annual 1991. 55th ed. Tonbridge, Kent, Benn.
xvi,396p. ISBN: 0863820891. ISSN: 00690767.
Manufacturers of carpets and rugs, smooth floor-
coverings, underlays - Trade marks and specialist names -
Manufacturers and suppliers of machinery, materials,
accessories and services - Buyer's guide - Wholesalers,
importers & agents - UK retailers - UK planners, fitters and
contractors - Carpet trade organizations - Glossary of trade
names (English, French, German) - Index to advertisers.
Class No: 677.07.5

[6527]
HEREFORD AND WORCESTER COUNTY LIBRARIES.
Carpets and textiles: a complete list of the special
collection in Kidderminster Library. Hereford and
Worcester County Council. Libraries Dept, 1981. [i], [i],
54p. £1.
    The catalogue of an extensive collection.
Class No: 677.07.5

Textile Fibres

[6528]
HARDINGHAM, M. Illustrated dictionary of fabrics.
London, Studio Vista, 1978. 159p. illus., facsims. ISBN:
0289708478.
    Sections: Introduction - Wool, cotton, silk - Other plant
fibres (Aida canvas ... Sisal) - Man-made fibres. Textile
terms (Appliqué ... Yarn), p.130-51. List of obsolete fabrics,
p.152-3. Bibliography, p.154-5. Index (small type), p.158-9.
Quarto format. Covers' names and descriptions of fabrics we
wear and use in our homes, as well as some used for
industrial purposes' (*Introduction*). *Class No:* 677.1/.5

Jute

[6529]
GILL, C. The Rise of the Irish linen industry. Oxford,
Clarendon Press, 1925. 359p. illus., tables, maps.
    16 chapters. 4 appendices. Bibliography (grouped), p.344-
9. Index, p.351-9. *Class No:* 677.13

Cotton

[6530]
BAINES, Sir E. History of the cotton manufacture in Great
Britain; with a bibliographical introduction by W.H.
Chaloner. 2nd ed. New York, A.M. Kelley; London, Cass,
1966. 544p. (*Reprints of economics classics.*)
    Previous ed.: London, Fisher, Fisher & Jackson, 1835.
    17 chapters (*e.g.* 6. The cotton manufacture in India; 7.
The cotton manufacture in England; 12. Bleaching and
calico printing; 13. Cotton-wool). No index, but full contents
(p.11-18). *Class No:* 677.21

[6531]
FARNIE, D.A. The English cotton industry and the world
market, 1815-1896. Oxford, Clarendon Press, 1979. [xi],
399p. illus., tables. ISBN: 0198224788.
    3 parts (8 footnoted chapters): 1. The advent of a new
economic order - 2. The main trends in production and trade
- 3. The structure of the industry. Select bibliography,
grouped, p.329-84. Detailed, analytical index, p.385-99.
Class No: 677.21

Wool

[6532]
LEMON, H. How to find out about the wool textile
industry. Oxford, etc., Pergamon Press, 1968. xiv, 217p.
facsims.
    6 chapters: 1. Education - 2. The principal organizations
within the wool textile industry - 3. Other organizations of
importance - 4. Research - 5. Wool processing (includes
literature on; some textile dictionaries, etc.) - 6. Distribution
of the wool textile industry. Appendices 1-3: Short reading

.... (contd.)
lists; 4: Glossary of wool textile terms. Name and subject
indexes. Whereas most other volumes in this series stress the
literature, and organizations as information sources, this
guide emphasizes organizations and technical processes.
Class No: 677.31

[6533]
SPIBEY, H., *ed*. **The British wool manual.** 2nd ed. Buxton,
Columbine Press, 1969. x, 530p. illus., tables.
    First published 1952.
    Sections: 1. Sheep and their wool - 2. Structure and
properties of wool ... 4. Worsted yarn manufacture (73
references; 14 illus.) ... 7. Woollen and worsted weaving ...
9. Wool in the knitting industry - 10. Practical wool dyeing
... 13. Finishing of wool and worsted fabrics - 14.
Micrographs and machinery illus. Definitions in text (*e.g.*
types of yarn, p.143-5). Analytical index, p.519-30.
Class No: 677.31

[6534]
WIRASCAN: current awareness service. Leeds, Information
Technology Centre, 1967-. Fortnightly. £30pa. (to non-
members).
    Provides a 2-p. listing of selected periodical articles on
textiles and clothing. The journals are covered A-Z over a
period of 3 issues. Thus, no.23, 7 September 1985 covers
periodicals with titles A-I (11 journals). WIRA: Wool
Industry Research Association. *Class No:* 677.31

[6535]
Wool quarterly. London, Commonwealth Secretariat, in
conjunction with the International Wool Study Group and the
International Wool Textile Organization, 1979-. Quarterly.
£70pa. ISSN: 01421921.
    A world statistical survey. Part 1: General review of the
world wool situation - Part 2: Statistical focus on individual
countries (35; Argentina ... Zimbabwe). Analytical tables
cover the raw wool markets, manufacturing activity and
trade. *Class No:* 677.31

Silk

[6536]
FELTWELL, J. The Story of silk. London, Alan Sutton,
1990. xviii,233p. illus. (some col.), facsims. £16.95. ISBN:
0862996112.
    11 chapters (1. The silk road: China to England ... 1. The
industry today). 3 appendices (2. Museums - 3. Societies).
Glossary. Bibliography and references (p.216-26). Index
(tiny print). *Class No:* 677.37

Man-made Fibres

[6537]
Moncrieff's man-made fibres. McIntyre, J.E. 7th ed.
Sevenoaks, Kent, Butterworth, 1988. 672p. ISBN:
0408005246.
    6th ed. 1975 (1094p.).
    Numerous sections, *e.g.* Fundamental concepts and
terminology ... The synthesis of fibre-forming polymers ...
Viscose fibres ... Nylon ... Non-woven fabrics - Dyeing and
finishing - Identification and estimation of man-made fibres -
Economic and social aspects - List of man-made fibres.
Index. A classic work of reference on the structure,
manufacture, dyeing, finishing and processing of man-made
fibres. *Class No:* 677.4

[6538]

WINKLER, W., *comp.* **Fachwörterbuch Chemiefasern.** Deutsch-Englisch-Französich. Frankfurt-am-Main, Deutscher Fachverlag GmbH, 1979. 350p.

About 6,000 terms on man-made fibres, grouped and in 3 sequences: German-English-French; English-German-French; French-German-English. Genders are stated. *Class No:* 677.4

Carbon

[6539]

DONNET, J.B. *and* BANSAL, R.C. **Carbon fibers.** 2nd ed. New York, Marcel Dekker, 1990. xxii,470p. illus., graphs, tables. (*International Fiber Science and Technology, v.10.*) ISBN: 0824778650.

First published 1983.

7 chapters, closely subdivided: 1. Preparation - 2. Structure - 3. Surface treatment - 4. Surface properties - 5. Mechanical properties - 6. Electronic, magnetic and thermal properties - 7. Applications. Most chapters have extensive references (*e.g.* 135 for chapter 7). Author and subject indexes. *Class No:* 677.47

Cables & Ropes

[6540]

ASHLEY, C.W. **The Ashley book of knots.** London, Faber, 1947 (and reprints). xi, 620, 8p. illus. (pl.), diagrs.

First published 1944 (New York, Doubleday, Doran).

3,854 types of knots are described and illustrated. Bibliography of books; material on knots (p.593-6). Glossary of terms, p.597-605. The most comprehensive book on the subject. *Class No:* 677.7

[6541]

—BIGNON, M.M. *and* REGAZZINI, C. Guide to knots for sailing, fishing, camping and climbing. London, Century, 1988. 256p. illus. (col.). £8.95. ISBN: 0712623043.

Includes 647 col. illus. *Class No:* 677.7

[6542]

—GRAUMONT, R., *and others.* Encyclopedia of knots and fancy rope work. London, Conway Maritime Press, 1982. 690p. illus. £14.

332 illus.; a multi-purpose volume. *Class No:* 677.7

[6543]

—The Shell combined book of knots and ropework (practical and decorative). Newton Abbot, David & Charles, 1981. v.p. col. illus. ISBN: 0715381970.

First published as *The Shell book of knots and ropework* (1977).

95 knots are dealt with, each occupying 1-2 pages, plus col. illus. Part 1 (nos. 1-51): Knots - Rope slices - Decorative Knots - Wire splices - Part 2: Decorative knots - Plaits - Mats - Other knots and two toggles. Glossary. Index of knot numbers. Over 600 photographs. *Class No:* 677.7

Plastics & Rubbers (Polymers)

Bibliographies

[6544]

UNITED NATIONS. Industrial Development Organization. **Information sources on the natural and synthetic rubber industry.** New York, United Nations, 1979. xii, 108p. $4. (*UNIDO Guides to information sources, no. 34.*) ISBN: 9211061555.

11 sections: 1. Addresses of professional, trade and research organizations, learned societies and other economic periodicals - 2. Directories - 3. Sources of statistics and other periodicals ... 6. Current abstracting and indexing services - 7. Special documents; proceedings, papers and reports - 8. Monographs - 9. Specialized dictionaries and encyclopedias - 10. Bibliographies - 11. Films and film catalogues - 12. Other potential sources of information (A-L, *e.g.* A. Consulting and engineering services). 'Bibliographical services used in compiling the present guide', p.105-8. *Class No:* 678(01)

[6545]

YESCOMBE, E.R. **Plastics and rubber:** world sources of information. Barking, Essex, Applied Science Publishers; Philadelphia, Pa., International Ideas, 1976. xv, 547p.

A substantial, in-depth revision and enlargement of his *Sources of information on the rubber, plastics and allied industries* (Pergamon Press, 1968. ix, 253p.).

Nearly 7,000 references to the literature and notes on over 675 organizations. 7 parts: 1. Literature and information pattern - 2. Materials - 3. Polymer science - 4. Technology - 5. Safety and health (safety, industrial health, toxicity and fire hazards) - 6. Commercial aspects (commerce, economics, statistics, marketing and management) - 7. Organizations and institutions (international and national). Journals and abstract index, p.513-32. Abbreviations and anonyma index. Non-analytical subject index. 'Whether the user's interest is technical, commercial or statistical he will find advice and guidance to a wide variety of sources ...' (*Aslib proceedings,* v.29(5), May 1977, p.211). *Class No:* 678(01)

Dictionaries

[6546]

LAMBERT, M. **A Short Russian-English dictionary of terminology used in the Soviet rubber, plastics and tyre industries.** London, Maclaren, 1963. [vii], 208p.

4,000 Russian words and phrases, A-Z by main word in a phrase. Relates to polymer chemistry and physics, rubber and plastics processing and machinery, and the tyre industry. Includes modern Soviet industrial and economic terms lacking in other dictionaries. Also lists 600 trade abbreviations (code designations of Soviet materials) and industrial organizations. Pocket-sized. *Class No:* 678(038)

[6547]
TEKNISKA NOMENKLATURCENTRALEN. **Plast-och-gummi-teknisk ordlista.** [Glossary of plastic and rubber terms.] Stockholm, the Centre, 1981. 356p. (*TNC 84.*) ISBN: 9171022128.

About 2,500 Swedish-base terms, plus explanations, with equivalents and indexes in English, French, German, Danish, Norwegian, Finnish. *Class No:* 678(038)

[6548]
—DORIAN, A.F., *comp.* Six-language dictionary of plastics and rubber technology: a comprehensive dictionary in English, German, French, Italian, Spanish and Dutch. London, Illife, 1965. [xii], 808p.

Provides 5,450 English-base terms, with definitions, plus equivalents and indexes in the other 5 languages. Genders are given. *Aslib book list* (v.31(4), April 1966, entry 197) notes 'some errors, misspellings and omissions'. *Class No:* 678(038)

Reviews & Abstracts

[6549]
RAPRA abstracts: the abstracts journal designed for rubbers and plastics producers, processors and users. Oxford, Pergamon, 1968-. 12pa. DM2500pa. ISSN: 00336750.

RAPRA: Rubber and Plastics Research Association of Great Britain. Originally RABRM (Research Association of British Rubber Manufacturers) *Summary of current literature* (1923-51. v.1-29), then *Rubber abstracts* (1952-65. v.30-42), which divided into *Plastics: RAPRA abstracts* (1965-67) and *Rubbers: RAPRA* (1965-67).

About 15,000 abstracts and references pa. Main sections: Generalia - Commercial and economic - Legislation - Industrial health and safety - Raw materials and monomers - Polymers and polymerization - Compounding ingredients - Intermediate and finished products - Relating to particular industries and fields of use - Applications - Processing and treatment - Properties and testing. Detailed, analytical subject index per issue; annual author and subject indexes.

Online via ORBIT. *Class No:* 678(048)

Progress Reports

[6550]
Progress in rubber and plastics technology: a quarterly review journal. Shawbury, Shrewsbury, Salop, RAPRA Technology, Ltd., 1985-. Quarterly. ISSN: 02667320.

Published jointly by the Plastics and Rubber Institute, and RAPRA.

V.4(2), 1988, contains 3 documented contributions (*e.g.* 'Characterization of mechanical properties of rubber in the transition zone', p.14-33; 20 references). Summary procedes each article. Biographical notes. *Class No:* 678(055)

Standards

[6551]
Specifications and standards for plastics and composites. Philadelphia, Pa., American Society of Testing and Materials; Hitchin, Herts., American Technical Publishers Ltd., 1990. 224p. £77. ISBN: 0871703965.

Over 2000 standards from a range of international organizations. *Class No:* 678(083.74)

Trade Names, Trade Marks, Brand Names

[6552]
RUBBER AND PLASTICS RESEARCH ASSOCIATION OF GREAT BRITAIN (RAPRA). RAPRA new trade names in the rubber and plastics industry, 1988. Davison, E., *ed.* Oxford, etc., Pergamon Press on behalf of RAPRA Technology, Ltd., [1988]. Annual. 603p. ISSN: 07474954.

First published 1962.

Trade name list - Company list - Classification list, p.171-542 (trade name categories, A-Z) - Company name and address - Journals and abbreviations. Also available online. *Class No:* 678(088.7)

[6553]
RUBBER AND PLASTICS RESEARCH ASSOCIATION OF GREAT BRITAIN (RAPRA). Trade names of rubbers, resins & plastics. Shawbury, Shropshire, the Association, 1949-61. 3v.

A complete listing of all the rubbers, plastics and resins collected since 1926. V.3 lists all the new materials in the synthetic rubber and plastics fields, 1955-59, describes each material, plus name and town of manufacturer. Earlier volumes have the title 'Annotated comprehensive list of trade names of synthetics'. *Class No:* 678(088.7)

Rubber

Handbooks & Manuals

[6554]
BLOW, C.M. *and* HEPBURN, C. **Rubber technology and manufacture.** London, Butterworth Scientific, for the Plastic and Rubber Institute, 1982. (Corrected reprint, 1985). 608p. illus., diagrs., graphs, tables. £38.50. ISBN: 0408005874.

51 contributors. 12 documented chapters (1. History - 2. Outline of rubber technology ... 4. Raw polymeric materials ... 6. Materials for compounding and reinforcement ... 10. Manufacturing techniques - 11. Testing procedure; standards, specifications - 12. Organizations). Bibliography (general and by chapters), p.534-61. Literature and patent references. Index. *Class No:* 678.4(035)

[6555]
BOSTRÖM, S., *ed.* **Kautschuk-Handbuch.** Stuttgart, Verlag Berliner Union, 1959-62. 5v. & Supplement, illus., tables.

V.1 covers raw materials; v.2, synthetic rubbers, rubber compounding, and machinery; v.3, rubber technology, machinery and manufactured goods; v.4, rubber products and latex; v.5, physical and mechanical testing of rubber, with cumulative index to v.1-5. *Supplement,* 1961, covers techniques of measuring and control in the rubber industry. *Class No:* 678.4(035)

[6556]
EIRICH, F.R., *ed.* **Science and technology of rubber.** New York, etc., Academic Press, 1978. 670, [1]p. illus., diagrs., graphs, tables. ISBN: 0122343603.

Under the auspices of the Rubber Division of the American Chemical Society.

16 contributors. 14 documented chapters (1. Rubber elasticity: basic concepts and behavior - 2. Polymerization (117 references) ... 5. Dynamic mechanical properties ... 9. The rubber compound and its composition ... 11. The

....(contd.)

chemical modification of polymers ... 14. Type manufacture and engineering) . Author and detailed subject indexes. *Class No:* 678.4(035)

[6557]
HOFMANN, W. **Rubber technology handbook.** Munich, Hanser, 1989. xxv,611p. illus. £34; $85. ISBN: 0195207572.

Trans. from the German.

A highly technical treatment which includes extensive chapter bibliographies. Industrial buyer's guide (with German emphasis) appended. *Class No:* 678.4(035)

[6558]
WEBSTER, C.C. *and* BAULKWILL, W.J. **Rubber.** Harlow, Essex, Longman, 1989. viii,[1],614p. illus., micrographs, graphs, tables. £54. ISBN: 0582404053.

8 contributors. 12 chapters (1. The history of natural rubber production ... 12. Organization and management of world rubber production). Chapter notes. 8 appendices (1. Literature of the history of rubber to 1045, p.539-40). References (*c.*1,000 authors, A-Z). Detailed, analytical index. *Class No:* 678.4(035)

Dictionaries

[6559]
HEINISCH, K.F. **Dictionary of rubber.** London, Applied Science Publishers, 1974. vi, 545p. tables.

Original German ed. as *Kautschuk-Lexikon* (Stuttgart, Gentner, 1966).

About 4,000 entries, including abbreviations and registered trade-names. 'Diberzthiazyl disulphide': 3 columns. Appended list of procedures and marketing organizations, p.535-45. *Class No:* 678.4(038)

[6560]
—AMERICAN SOCIETY FOR TESTING AND MATERIALS. Glossary of terms relating to rubber and rubber-like materials. Philadelphia, Pa., the Society, 1972. 122p. (*ASTM Special technical publications, no. 184.*)

Defines over 2,000 terms. Molecular structures are given. Appendix of trade names. *Class No:* 678.4(038)

[6561]
—BRITISH STANDARDS INSTITUTION. Glossary of rubber terms. London, BSI, 1980. 44p. £33.60. (*BS 3558: 1980.*)

Defines terms used in the rubber industry in general, latex, rubber, properties and testing processes, machinery, and rubber products, including hoses, tyres, belts and footware. *Class No:* 678.4(038)

[6562]
RUBBER STITCHING. **Elsevier's rubber dictionary** in English/American, French, Spanish, Italian, Portuguese, German, Dutch, Swedish, Indonesian and Japanese. Amsterdam, etc., Elsevier, 1959. viii, 1537p. DFl.475. ISBN: 0444404996.

7,955 terms in English/American, with equivalents and indexes in the other 9 languages. Omission of Russian is a drawback. *Class No:* 678.4(038)

[6563]
YASHUNSKAYA, V.I. *and* FEIGIN, E.E. **Anglo-russkii slovar' po káuchnuku, rezine i khimicheskim voloknam.** 3. izd pereabotannoe i dopolnennoe. Moscow, Fizmatgiz, 1962. 260p.

Anglo-Russian dictionary of rubber, resin and chemical fibres. 11,000 English entry-words, with Russian word index, p.216-60. Abbreviations, p.208-15. *Class No:* 678.4(038)

Progress Reports

[6564]
'Rubber reviews'. Rubber chemistry and technology. Akron, Ohio, American Chemical Society, Rubber Division, Annual.

'Rubber reviews' appears annually as one of the 5pa. issues of *Rubber chemistry and technology.* The 1988 issue (p.377-553) includes a lengthy review article (p.377-469), 'Statistical methods in rubber research and development' (123 references) and 6 short reviews (*e.g.* 'Silicone rubber, its development and technichal progress', p.470-502; 33 references). *Class No:* 678.4(055)

Yearbooks & Directories

[6565]
Rubbicana: Europe, 1986-87. London, Crain Connections, [1986]. 608, [i]p.

A compendium on the rubber and polyerethene industries of Europe. Alphabetical company listing - Associations & societies - Classified listings (manufacturers; suppliers). Trade name index; Country classified index; Country code index; Advertisers index. *Class No:* 678.4(058)

Tables & Data Books

[6566]
Rubber statistical bulletin. London, International Rubber Study Group, 1946-. Monthly. tables.

V.43(3), December 1988, has 46 tables. Gross exports of sheet rubber by grade; consumption, production and stocks of natural rubber; synthetic rubber. World synthetic rubber capacities. *Class No:* 678.4(083)

Histories

[6567]
SCHIDROWITZ, P. *and* DAWSON, T.R., *eds*. **History of the rubber industry.** Compiled under the auspices of the Institution of the Rubber Industry. Cambridge, Heffer, 1952. xxiv, 406p. illus. (pl.).

6 sections, concerning origins of the industry, raw materials, scientific and technological developments, products of the industry, economic and social aspects. Annotated bibliography of rubber literature. Latex entries in the chronology are for 1948-49. *Class No:* 678.4(091)

Malaysia

[6568]
BARLOW, C. **The Natural rubber industry:** its development, technology, and economy in Malaysia. Kuala Lumpar, Oxford Univ. Press, 1978. xxiv, 500p. illus., graphs, tables, maps. £40. ISBN: 0195803140.

.... *(contd.)*

10 chapters (2. The early years - 3. Growth amidst difficulty - 4. The technologies of production - 5. The technology of processing - 6. The structure of the industry - 7. The economics of producing - 8. The rubber market - 9. The economics of institutional management - 10. The international context). Bibliography, p.460-486. Analytical index of subjects. Index of authors cited. Numerous tables. 49 plates. *Class No:* 678.4(595)

Plastics

Databases

[6569]
Plaspec materials selection database. Yardley, Pa., D & S Data Resources.

Engineering and design data, chemical descriptions and trade names for over 11,500 grades of plastics materials. Updated monthly. Available on DIALOG, file 321. *Class No:* 678.5(003.4)

Handbooks & Manuals

[6570]
ASH, M. *and* **ASH, I.,** *comps*. **Handbook of plastic compounds, elastomers, and resins:** an international guide by category, tradename, composition, and supplier. New York, VCH, 1992. xiv,872p. £132. ISBN: 1560815531.

Data on 15,000 tradename products. 4 parts: I. Tradename by category reference - II. Tradename cross reference - III. Chemical component cross reference - IV. Chemical manufacturer's directory. *Class No:* 678.5(035)

[6571]
BROWN, R.P., *ed*. **Handbook of plastics test methods.** 3rd ed. Harlow, Essex, Longman; New York, Wiley, 1988. xiv,442p. illus., diagrs., graphs, tables. £59. ISBN: 0582030153, UK; 0470211342, US.

First published 1971; 2nd ed. 1981.

21 documented chapters: 3. Preparation of test pieces ... 6. Polymer characterisation ... 10. Friction and wear - 11. Creep, relaxation and set - 12. Fatigue - 13. Electrical properties - 14. Optical properties - 15. Thermal properties ... 18. Fire testing of plastics ... 21. Testing products. Index, p.438-42. Emphasis is on testing of materials, not products. Tests related to 130 standards where possible. *Class No:* 678.5(035)

[6572]
BRYDSON, J.A. Plastics materials. 5th ed. London, Butterworth Scientific, 1989. xix, 839p. £62.50. ISBN: 0408007214.

First published 1966. 4th ed. 1982.

30 documented chapters, closely subdivided, on properties, processes and applications of plastics materials (e.g., 10. Polyethylene, p.196-234; 14 references, bibliography of 9 items). 'General properties of the nylons', p.460-65; 7 graphs, 5 tables. Includes recent production and consumption figures for the commercially important polymers. Detailed analytical index, p.823-39. Detailed contents list. *Class No:* 678.5(035)

[6573]
—**GÄCHTER, R.** *and* **MILLER, H.,** *eds*. Plastics additives handbook: stablizers, processing aids. 2nd ed. Munich, etc., Hanser, 1983. xxxiv, 754p. graphs, tables, chemical structures.

The translated and rev. ed. of *Kunstoff industrie*. Detailed, analytical index, p.743-54. *Class No:* 678.5(035)

[6574]
HARPER, C.A., *ed*. **Handbook of plastics, elastomers and composites.** 2nd ed. New York, etc., McGraw-Hill, 1992. 960p. illus. £77.95. ISBN: 0070266867.

First published 1975.

Emphasis is on industrial techniques, developments and applications. Includes government (US) and commercial guidelines. Glossary of terms. *Class No:* 678.5(035)

[6575]
RUBIN, I.I., *ed*. **Handbook of plastic materials and technology.** New York, Wiley, 1990. xxv,1745p. illus., tables. £90. ISBN: 0471096342.

Over 100 contributors, most from industry. 119 chapters in 2 parts: 1. Materials (*e.g.* 'acetals'; 'nylons') - 2. Technology (*e.g.* 'blow molding'). 25 appendices. Detailed, analytical index, p.1729-45. A practical treatment which 'aims to make you more successful in your business' (*Preface*). *Class No:* 678.5(035)

Dictionaries

Polyglot

[6576]
KALISKE, G., *comp*. **Dictionary of plastics technology in four languages,** English, German, French and Russian. Amsterdam, etc., Elsevier, 1982. [vii], 408p. DFl.175. ISBN: 0444996877.

About 10,000 English-base terms, with equivalents and indexes in German, French and Russian. Deals with the whole range of plastics technology, from material, production and processing to applications. *Class No:* 678.5(038)=00

English

[6577]
WHITTINGTON, L.R. Whittington's dictionary of plastics. 2nd ed. Westport, Conn., Technomic Publishing Co., 1978. 344p.

First published 1968. Sponsored by the Society of Plastics Engineers, Inc.

About 5,000 terms, abbreviations and cross-references. 'Recommended for reference collections' (*Library journal*, v.93(14), August 1968, p.2844, on the 1968 ed.). *Class No:* 678.5(038)=20

German

[6578]
JUNGE, H.D., *ed*. **Dictionary of plastics technology,** English-German. [Wörterbuch Kunststofftechnologie, Englisch-Deutsch.] Weinheim (W. Germany), VCH, 1987. [v], 315p. ISBN: 3527264329.

About 3,000 terms on the physics and chemistry of the high polymers. Many sub-entries, bringing gross total to

....(contd.)

c.20,000. 'Plant': 1½ columns; 'Unit': 7 columns. German genders given. At head of title page: 'Parat'.
Class No: 678.5(038)=30

[6579]
WELLING, M.S., *comp.* **German-English dictionary of plastics technology.** London, Pentech Press, 1985. 220p. ISBN: 0727307045.

Concise entries on over 19,000 terms and expressions, covering plastics technology, from physics and chemistry of high polymers to finishing processes. Includes terms on microprocessor controls, uses, finance and business.
Class No: 678.5(038)=30

[6580]
WITTFOHT, A. **Plastics technical dictionary,** English-German, German-English. New York, Macmillan, 1981. 3v. illus., tables.

German ed. as *Kunststofftechnisches Wörterbuch* (Munich, Hanser, 1978-81).

1. *Alphabetical dictionary, English-German.* 4th rev. ed. 1981. x, 550p.

2. *Alphabetical dictionary, German-English.* 1981. ix, 534p.

3. *Reference volume: illustrated systematic groups, English,- German, German-English.* 1981. ix, 508p.

About 40,000 entries in each of v.1-2, on processing, fabricating and using plastics. V.3 comprises 26 subject groups (Blow molding ... Welding), with numerous keyed illus. English and German subject indexes.
Class No: 678.5(038)=30

French

[6581]
Dictionary: Plastics engineering. English/French. [Dictionnaire: Industrie des matières plastiques. Français/ Anglais.] Widnau, Switzerland, Schnellmann Verlag, 1987. 114, [5]p. *(K.2.3.)*

About 1,500 terms, with equivalents, in each half of this dictionary. Appendix: 'Units'. *Class No:* 678.5(038)=40

Italian

[6582]
Dictionary: Plastics engineering. English/Italian. [Dizionario: Tecnica delle materie plastiche. Italiano/ inglese.] Widnau, Switzerland, Schnellmann Verlag, 1985. 114. [5]p. *(K.1.3.)*

About 1,500 terms, with equivalents, in each half of this dictionary. Appendix: 'Units'. *Class No:* 678.5(038)=50

Russian

[6583]
GURARII, M.G *and* **IOFE, L.R.** **Anglo-russkiĭ slovar' po plastmassam.** Moscow, Fizmatgiz, 1963. 144p.

English-Russian plastics dictionary. About 5,000 English entry-words. Russian index, p.127-44. Abbreviations, p.123-6. *Class No:* 678.5(038)=82

[6584]
The Plastics industry directory, 1992. Croydon, EMAP Vision, 1992. 196p. ISSN: 09562966.

10 sections: 1. Company information (2400 companies, A-Z) - 2. Materials - 3. Machinery - 4. Processing - 5. Moulds and dies - 6. Finished products ... 9. Trade names - 10. Glossary. *Class No:* 678.5(058)

[6585]
SOCIETY OF THE PLASTICS INDUSTRY, Inc. **Plastics engineering handbook of the Society of the Plastics Industry, Inc.** Berins, M.L., *ed.* 5th ed. New York, etc., Van Nostrand Reinhold, 1991. 845p. illus., tables. $89.95. ISBN: 0442317999.

First published 1947 (*SPI handbook*); 4th ed. 1976.

Largely rewritten, with new illustrations and new terms in the glossary, for this edition. Extensive coverage of reinforced and cellular plastics and composites. 'An excellent source' (*Choice,* April 1992, p.1258).
Class No: 678.5(083)

[6586]
Plastics abstracts. Welwyn, Herts., Plastics Investigations, 1959-. Weekly.

About 4,000 abstracts pa. of all British, European and Patent Cooperation in Treaty (PCT) patent specifications dealing with the manufacture and use of plastics. Sections: Raw materials and additives ... Miscellaneous applications. *Class No:* 678.5(088.8)

[6587]
IMPERIAL CHEMICAL INDUSTRIES. Plastics Division. **Landmarks of the plastics industry.** Welwyn, Herts., ICI Plastics Division, [1962]. 126p. illus., diagrs.

Popular historical survey, particulary of more recent developments. Published to mark the centenary of A. Parke's invention of the world's first man-made plastic, Parkesine. *Class No:* 678.5(091)

Synthetic Polymers & Rubbers

[6588]
LOENING, K.L., *ed.* **List of standard abbreviations (symbols) for synthetic polymers and polymer materials.** Basic definitions of terms relating to polymers. Oxford, Pergamon, 1978. 20p.

Previously published in *Pure and applied chemistry,* v.40(3).

A. Abbreviations for synthetic polymer materials - B. Primary definitions. Secondary definitions.
Class No: 678.7(003)

Bibliographies

[6589]

ADKINS, R.T., *ed*. **Information sources in polymers and plastics.** London, etc., Bowker-Saur, 1989. xii,313p. Tables. £40. (*Guides to Information Sources.*) ISBN: 040802027x.

In 3 parts, each with several sections. Part I covers published formats (serials, books, patents, standards, grey literature, online databases); Part II, materials (polymer structures, properties, adhesives, fibres, rubber, etc.); Part III, geographical areas and translations. 18 contributors. Subject index, p.303-13. A comprehensive guide to the literature. *Class No:* 678.7(01)

Encyclopaedias

[6590]

CORISH, P.J. **Concise encyilopedia of polymer processing and applications.** Oxford, Pergamon Press, 1992. 800p. illus. £140. ISBN: 0080370640.

A wide variety of articles. Emphasis is on practical handling and production aspects of polymers. Research level. Some material taken from *Encyclopedia of materials science and engineering* (*qv*) and the supplementary vols; other material is newly commissioned.
Class No: 678.7(031)

[6591]

MARK, H.F., *and others, eds*. **Encyclopedia of polymer science and engineering.** 2nd ed. New York, etc., Wiley, 1985-90.

First published 1964-77 (16v. plus 2 suppt.) as *Encyclopedia of polymer science and technology*.

19v. (v.1-18: A-Z; v.19: Index), with *c*.600 contributors. Articles are lengthy, signed and documented. V.1: A to Amorphous (843p.) has 27 contributions. That on 'Amorphous polymers' occupies p.789-843, with 260 references. Cross-references. The standard work on polymers. Available online via DIALOG (File 322).
Class No: 678.7(031)

[6592]

—Polymers: fibers and textiles: a compendium. Kroschwitz, J.I., *ed*. New York, Wiley, 1990. 950p. £66.20. ISBN: 0471512212.

Includes articles reprinted without change from the *Encyclopedia of polymer science and engineering*. Material on nomenclature, SI units and conversion factors had been added. *Class No:* 678.7(031)

[6593]

—Polymers: polymer characterization and analysis. Kroschwitz, J.I., *ed*. New York, Wiley, 1990. 984p. £66.20. ISBN: 0471513253.

Includes articles reprinted without change from the *Encyclopedia of polymer science and engineering*. Handy as a reference source and for libraries without the complete work. *Class No:* 678.7(031)

Handbooks & Manuals

[6594]

ALLEN, G. *and* BEVINGTON, J.C. **Comprehensive polymer science:** the synethsis, characterization, reactions and applications of polymers. Oxford, etc., Pergamon Press, 1988. 7v. (6000p.). illus. £1095. ISBN: 0080325157.

A treatise dealing in detail with polymer chemistry,

....(contd.)

characterization, physical properties and chemical reactions, concluding with processing, applications and nomenclature. *Class No:* 678.7(035)

[6595]

BRANDRUP, J. *and* IMMERGUT, E.H., *eds*. **Polymer handbook.** 3rd ed. New York, etc., Wiley, 1989. *c*.1870p. £115. ISBN: 0471812447.

First published 1966. 2nd ed. 1975.

About 80 contributors. 8 sections: 1. Nomenclature rules. Units - 2. Polymerization and depolymerization - 3. Physical properties of monomers and solvents - 4. Physical data of oligomers - 5. Physical constants of some important polymers - 6. Solid state properties (6.1 'Crystallographic data for various polymers', 208p. 1,946 references) - 7. Solution properties - 8. Abbreviations of polymer names. About 30% more data than previous ed. Concentrates on synthetic polymers, poly (saccharides) and derivatives and oligomers. Index lists only physical constants of polymers. *Class No:* 678.7(035)

[6596]

Handbook of polymer science and technology. Cheremisinoff, N.P., *ed*. New York, Dekker, 1989-. illus., graphs, tables, chemical structures.

V.1. *Synthesis and properties.*

V.2. *Performance properties of plastics and elastomers.*

V.3. *Applications and processing operations:*

V.4. *Composites and specialty applications.*

V.4 has 19 specialist contributors; 14 well-documented chapters. Each volume carries an index; further vols. 'in preparation'. *Class No:* 678.7(035)

[6597]

MARKS, J.E., *and others*. **Physical properties of polymers.** Washington, American Chemical Society, 1984. 246p. ISBN: 0841206573.

Concerns recent advances involving physical chemistry of polymers related to utilization of polymeric material. Topics include rubber-like elasticity; the glassy, viscoelastic and crystalline states in polymers; polymer spectroscopy. *Class No:* 678.7(035)

Dictionaries

[6598]

ALGER, M.S. **Polymer science dictionary.** London, Elsevier, 1989. xii,531p. chemical formulas and structures. ISBN: 1851662200.

Some terms defined at length (*e.g.* 'Monosaccharide', 1p.). Excludes technology and 'polymer processing' (*Preface*). Appended: Units; Conversion factors; Physical constants; Relative atomic masses; Greek alphabet. Includes some tradenames. *Class No:* 678.7(038)

[6599]

DAWYDOFF, W. *and* HOWORKA, H. **Technical dictionary of high polymers.** Oxford, etc., Pergamon Press; Berlin, VEB Verlag, 1969. 959p.

About 10,000 technical terms and jargon in 4 sequences: English-German-French-Russian; German-English-French-Russian; French-English-German-Russian; Russian-German-English-French. Genders are stated. *Class No:* 678.7(038)

Reviews & Abstracts

[6600]

Additives for polymers. Shelton, J., *ed.* London, Elsevier, 1971-. Monthly. ISSN: 03063747.

A digest of *c*.20p., with some 60 entries per issue. Sections: Technical notes - Materials - Patents - Markets - Company news - Books - Publications - Diary. *Class No:* 678.7(048)

[6601]

Polymers/ceramics/composites alert. Metals Information. London, Institute of Metals; Metals Park, Ohio, American Society for Metals, 1985-. Monthly. £175pa. (UK).

300 abstracts per issue. Covers engineered materials industries, from ceramics to thermoplastics. One of 3 abstracts journals comprising the Materials Business File. Online database. *Class No:* 678.7(048)

Progress Reports

[6602]

Advances in polymer science. Berlin, Springer-Verlag, 1961-.

V. 104 (1992. 175p.), 'Polylectrolytes/Hydrogels/ Chromatographic materials', has 4 documented articles by 7 contributors. 'Chemistry and physics of agricultural hydrogels', p.97-133, includes 139 references. *Class No:* 678.7(055)

[6603]

Polymer-plastics technology and engineering. Cheremisinoff, N.P., *ed.* New York, Dekker, 1962-. Quarterly. diagrs., graphs, tables. $695 (Institutions); $347.50 (individual professionals & students) pa. ISSN: 03602559.

V.27(3) 1988 has 5 documented articles, by 12 contributors. The article 'Application of infrared spectroscopy in polymer degradation', p.303-34, has 100 references. Appended: Book review; New products; Announcement. *Class No:* 678.7(055)

Yearbooks & Directories

[6604]

Polymer yearbook. New York, Harwood Academic, 1984-. Annual. ISBN: 3718652633.

Contains reviews of topics of current interest and other newsworthy items. Thus, v.9 (1992, 390p.) contains a survey of recent publications in polymer science, a compendium of recent reviews (under topics), a tabulation of dissertation abstracts, a list of polymer journals, a calendar of forthcoming meetings, and much else. Includes contributions from Russian researchers and reports from conferences in Eastern Europe and Japan. 'A wealth of desk-top reference information' (*Science & technology libraries,* v.7(2), Winter 1986, p.1741). *Class No:* 678.7(058)

Trade Names, Trade Marks, Brand Names

[6605]

FACHINFORMATIONSZENTRUM CHEMIE GmbH, Berlin. **Parat: Index of polymer trade names.** Junge, H.D., *ed.* 2nd ed. Weinheim, W. Germany, VCH, 1992. 2 computer disks. £150. ISBN: 3527283838.

First published 1987.

About 33,000 substances referred to by trade names in the

.... *(contd.)*

literature. Names of producers are provided for about 13,000 and CAS registry numbers for over 21,000. *Class No:* 678.7(088.7)

Histories

[6606]

SEYMOUR, R.B., *ed.* **History of polymer science and technology.** New York, Dekker, 1982. 410p. ISBN: 0824713826.

Reports presented at a symposium sponsored jointly by the History, Organic Coatings and Plastics, Polymer and Rubber Divisions of the American Chemical Society at the 179th National Meeting, Houston, Texas, March 1980.

Coverage: history of adhesives, organometallic polymers, engineering thermoplastics, vinyl chloride polymers, natural rubber, synthetic rubber, reinforced plastics, natural fibres, and polyolefins. Index. *Class No:* 678.7(091)

PVC

[6607]

NASS, L.I. *and* **HEIBERGER, C. Encyclopedia of PVC.** 2nd ed., revised and expanded. New York, Dekker, 1980-. illus. diagrs., graphs, tables. ISBN: 0824774272.

1. *Resin manufacture and properties.*

2. *Compound design and additives.*

3. *Compounding processes, product design, and specifications.*

4. *Conversion and fabrication processes.*

5. *Safety and environmental concerns.*

V.1 (1985. 702p.) has 15 contributors, 11 documented sections, with author and subject indexes. The bibliography in v.5 runs to 323 references, plus an appended glossary of acronyms. *Class No:* 678.743

[6608]

TITOW, W.V. PVC plastics: properties, processing, and applications. London, Eslevier, 1990. xxvii,902p. illus., diagrs., graphs, tables. £115. ISBN: 1851664718.

24 chapters, subdivided, with references and bibliography (3. PVC polymers ... 5. Plasticisers ... 7. Lubricants ... 13. Extrusion ... 16. Injection moulding ... 20. PVC Latices). 3 appendices (1. Material properties - 2. Quantities and units - 3. Abbreviations). Detailed, analytical index, p.881-902. Much of the material is drawn from the author's *PVC technology* (4th ed.). (*Preface.*). *Class No:* 678.743

Asbestos

[6609]

Asbestos management sourcebook. Lutherville, Md., Environmental Publications, 1990-. Annual. 145p. $20. ISSN: 10460438.

41 articles in sections: legal matters - financial considerations - inspection - insurance - asbestos abatement - asbestos management. Directories: professionals; government agencies. *Class No:* 679.867

[6610]

CHISSICK, S.S., *and* *others*. **Asbestos: properties, applications and hazards.** New York, etc., Wiley, 1979-83. 2v. illus., diagrs., graphs, tables. ISBN: 047199698x; 0471104892.

V.1, by L. Michaels and S.S. Chissick. 1979. xi, 553p.

V.2, by S.S. Chissick and R. Derricott. 1983. xiv, 652p.

V.1, has 22 contributors, with 16 documented chapters (chapter 9. 'Dealing with asbestos problems' p.279-304; 3p. of bibliography and references). V.2 has 9 contributors. 11 documented chapters concern the uses, abuses, dangers and benefits of asbestos. Each volume has a detailed, analytical index. *Class No:* 679.867

[6611]

—BRADFIELD, R.E.N. Asbestos: review of uses, health effects, measurement and control. Epsom, Surrey, Atkins Research and Development, 1977. iv, 101p.

A report describing the properties of asbestos fibres and how they have been exploited. Discusses health effects of occupational exposure to asbestos (used in over 3,000 products), and the evidence for, plus the risks associated with non-occupational exposure. *Class No:* 679.867

68 Industries, Trades & Crafts

Industries, Trades & Crafts

[6612]
Directory of European industrial and trade associations.
Leigh, R., *ed.* 4th ed. Beckenham, Kent, CBD Research, 1986. lvi,405p. ISBN: 0900246464.

Editions 1-3 as part of *Directory of European associations* (1971, 1976, 1981), now split into the above and *Directory of European professional and learned societies.*

Over 5,000 titles of associations (Abattoirs ... Zip fasteners), with cross-references. Subject index precedes main text: English (orange-tinted pages); French (yellow); German (green). Much use of abbreviations and symbols.
Class No: 680

[6613]
A Historical dictionary of American industrial language.
Mulligan, W.H., *ed.* New York, Greenwood Press, 1988. 332p. $55. ISBN: 0313241716.

20 contributors. Over 3,000 terms, A-Z, on American industry prior to World War I, briefly defined. Cross-references. Appendix lists terms grouped by industry. Index of institutions and people mentioned in definitions.
Class No: 680

Country Crafts

Handbooks & Manuals

[6614]
The Shell book of country crafts. Arnold, J. London, J. Barker, 1968. xiv, 358p. illus. (incl. col. pl.).

35 chapters. 3. The nature of craftsmanship ... 5. Woodland and coppice industries ... 11. Wheelwrighting ... 19. Pottery ... 32. Making briar pipes - 33. Further reading (p.319-42) - 34. Museums (p.343-5; town and name of museum only) - 35. Notes on tools. Index. 76 plates; 53 line-drawings (bill hooks: 23 types). A comprehensive survey by a practicing craftsman. *Class No:* 680.0(035)

Great Britain

[6615]
COUNCIL FOR SMALL INDUSTRIES IN RURAL AREAS. Select list of books and information sources on trades, crafts and small industries in rural areas. London, the Council, 1973. [v], 80p.

About 1,500 titles. 14 sections: 1. General bibliography on 'country crafts' - 2. Clay and plaster - 3. Glass and crystal - 4. Horn - 5. Leather - 6. Metal - 7 Natural fibres - 8. Stone - 9. Reed and straw - 10. Underwood (basketry, etc.) - 11. Wood (furniture, wheelwrighting, etc.) - 12. Miscellaneous - 13. Folk collections in British museums and galleries - 14.

....(contd.)
Guilds, societies and trade associations (subjects, A-Z). Includes out-of-print books recognized as standard works on the subject. *Class No:* 680.0(410)

[6616]
HILL, J. The Complete practical book of country crafts.
Newton Abbot, Devon, David & Charles, 1979. 254p. illus., diagrs., tables. ISBN: 071537706x.

22 sections: Materials - Tools and devices - Hurdle making - Clog making - Wood carving - Chair making - Coopering ... Blacksmithing ... Basket making - Thatch and straw - Rope making - Bricks and pottery - Hedging and walling. 2 appendices: 1. Further information, displays, courses; 2. Addresses of suppliers. Bibliography (grouped), p.249-52. Index, p.253-4. *Class No:* 680.0(410)

[6617]
—COUNCIL FOR SMALL INDUSTRIES IN RURAL AREAS. Visitors guide to country workshops in Britain. 12th ed. London, the Council, in co-operation with the Scottish Country Industries Development Trust and the Northern Ireland Council of Social Service, 1973. [ii], 128p. illus., maps.

2 main sections - Workshops; Craft shops - each subdivded by countries and counties, A-Z. Data: name; address; products; when open. 4 product codes for trade buyers. *Class No:* 680.0(410)

Scotland

[6618]
CARTER, J. *and* **RAE, A. Chambers guide to traditional crafts of Scotland.** Edinburgh, Chambers, 1988. 128p. illus. (incl. colour). ISBN: 0550200002.

8 wide-ranging chapters. 1: Land and sea (*e.g.* boat building) - 2. In the home (*e.g.* wood turning) ... 7. Industrial crafts. Select bibliography, p.123. Useful addresses. Index, p.125. Well illustrated. Necessarily selective. *Class No:* 680.0(411)

England

[6619]
AUTOMOBILE ASSOCIATION. Craft workshops in the English countryside. AA/COSIRA. [Basingstoke, Hants.], the Association, for Council for Small Industries in Rural Areas. 173p. illus., maps.

Directory of craft workshops, p.31-151. Notes on setting up a workshop, the thatcher's craft, the Crafts Council, etc. Index of crafts, p.154-63. Location atlas, p.164-73.
Class No: 680.0(420)

[6620]
LEWIS, J.R. Handbook of English crafts and craftsmen.
London, Hale, 1978. [xvi], 412p. illus.

22 chapters (1. Craftsmen - 2. Regional arts associations -
3. Guild and associated associations ... 6. Craft centres,
workshops and careers - 7. Wood ... 9. Textiles ... 13.
Gold, silver, jewellery (p.228-54; with annotated list of
societies, publishers, individual craftsmen, A-Z by category)
... 17. Toys, dolls, games, models, puppets). Appendix:
other useful addresses. Craft museums (p.380-2). Index of
craftsmen. 51 illus. *Class No:* 680.0(420)

Clock & Watch Making

Bibliographies
[6621]
BAILLIE, G.H. Clocks and watches: an historical
bibliography. London, N.A.G. Press, 1951. (Reprinted
London, Holland Press, 1978). xxv, 388p. illus.

The period covered is 1344 - 1799; the arrangement of
entries, chronological. Major works are analysed in detail
(*e.g.* J.A. Lepaute's *Traité d'horlogerie* (1955) is alloted 2
pages, with 4 supporting diagrams). The bibliography 'is
confined to mechanical timepieces and to everything
connected with them' (*Foreword*). Name and subject
indexes. The most important compendium of its kind.
Class No: 681.11(01)

[6622]
BROMLEY, J., *comp.* **The Clockmakers' Library:** the
catalogue of books and manuscripts in the Library of the
Worshipful Company of Clockmakers. London, Sotheby
Parke Bernet Publications, 1977. 136p. illus. (pl.), ports.,
facsims.

1,151 numbered items. Catalogue of printed books (p.1-
62) - Catalogue of manuscripts (p.77-119); index -
Concordance of Guildhall Library manscript and catalogue
numbers - Catalogue of portraits - Catalogue of prints,
drawings, paintings (other than portraits) and photographs.
27 illus. No annotations, but a detailed inventory of
manuscripts. Well produced. *Class No:* 681.11(01)

[6623]
—WORSHIPFUL COMPANY OF CLOCKMAKERS OF
LONDON. The Catalogue of the Library ... preserved in the
Guildhall Library. 2nd ed. London, Blades, East & Blades,
1898. v, 205p.

The predecessor of Bromley (*qv*) has 3 main sections:
Books (by subjects, A-Z, including article offprints) -
Manuscripts, patents, etc. - Prints, drawings, paintings, etc.
Class No: 681.11(01)

Encyclopaedias
[6624]
BRITTEN, F.J. Britten's watch & clock maker's
handbook, dictionary and guide. Good, R., *ed.* 16th ed.
London, Eyre Methuen, in association with E. & F.N. Spon,
1978. viii, [1], 460p. illus., graphs, tables.

15th ed., 1955 (ix, 598p.).

'Dictionary and guide' (Abrasive ... Zodiac), p.1-355.
'Quartz crystal controlled watches', p.260-4; 'Repeating

....(contd.)
watches', p.267-76, with 23 line-drawings. Appendices: 1.
Basic equipment for horological draughtsmen; 2. Tables; 3.
Vocabulary (English, French, German, Spanish). Select
bibliography, p.431-4. Over 700 good line-drawings. Index
of illus. Analytical general index, p.440-60.
Class No: 681.11(031)

[6625]
DE CARLE, D. Watch and clock encyclopedia. 3rd ed.
Ipswich, Suffolk, N.A.G. Press, 1983. 326, [1]p. illus.,
diagrs., tables. £14.95. ISBN: 0719801702.

First published 1950.

Definitions of c.3,000 antiquarian, technical and
commercial terms. No less than 31 appendices follow,
including 'Chimes' (with music examples), 'Clock hands',
'English period styles', 'French period styles', 'Horological
dates', 'Watch part nomenclatures in 6 languages'.
Bibliography of 19 items. 1,350 small line-drawings. A mine
of specialized information. *Class No:* 681.11(031)

Handbooks & Manuals
[6626]
The 'Country Life' international dictionary of clocks.
Smith, A., *ed.* London. 'Country Life' Books (distributed
by Hamlyn), 1979. 352p. illus. (incl. col.), ports.

By various hands. 5 sections, - on history and style of
clocks, mechanical parts, tools, materials and workshop
methods, international clock-making, and sundials and
astronomical instruments. Bibliography, p.339-42. 900 illus.
'This could become a standard work for collectors, museum
staffs, dealers and all interested in horology and industrial
history' (*Aslib book list*, v.44(12), December 1979, item
536). *Class No:* 681.11(035)

[6627]
GAZELEY, W.J. Watch and clock making and repairing,
dealing with the construction and repair of watches, clocks
and chronometers. 2nd ed. London, Heywood, 1958
(reprinted Newnes-Butterworths, 1976). vi, 425p. illus.,
diagrs.

First published 1953.

16 chapters on principles and construction, with practical
information on cleaning and repair. Appendix: 'Causes of
failures and bad timekeeping' (156 causes), p.400-10. Index,
p.411-25. The 314 careful line-drawings, several with keyed
parts, are a feature. No bibliography. Analytical index.
Class No: 681.11(035)

Dictionaries
[6628]
**BERNER, G.-A. Dictionnaire professionnel illustré de
l'horlogerie:** français-allemand-anglais-espagnol. La Chaux-
de-Fonds, La Chambre Suisse de l'Horlogerie, 1961;
London, Swiss Federation of Watch Manufacturers, 1962.
1007p. illus.

4,103 numbered terms. French-based vocabulary, plus
definitions in French, with equivalents and indexes (on tinted
paper) in German, English and Spanish. If the number given
to a term is underlined, there is an illustration, with named
parts, heading the page. Genders are given. Small but clear
type. *Class No:* 681.11(038)

[6629]

Dictionnaire horloger. Terminologie numérotation français - English - russkiĭ - deutsch - español - italiano - portugues - [Japanese]. Neuchâtel, Switzerland, Ebauches SA, 1976. 493p. illus.

Dictionary of clock and watch terms in 8 languages. About 2,500 French-base entries, in sections, with English, Russian, Dutch, Spanish, Italian, Portuguese and Japanese equivalents, and indexes for each language. *Class No:* 681.11(038)

[6630]

MAUCH, C. *and* **MAUCH, E. Horologisches Lexikon.** [Horological dictionary. Uhrenwörterbuch.] Tübinger, Universitäten Verlag Tübinger, 1984. 2v. (728p.). ISBN: 9924895006.

1. *Deutsch-Englisch.* 2. *Englisch-Deutsch.*

V.1 (xv, 357p.) has *c.*15,000 German entry-words and sub-entries, with English equivalents. German genders are given. *Class No:* 681.11(038)

Histories

[6631]

BRITTEN, F.J. Britten's old clocks and watches, and their makers: a history of styles in clocks and watches and their makers. Baillie, G.H., *and others, eds.* 9th ed., rev. and enlarged by C. Clutton. London, Methuen, in association with E. & F.N. Spon, 1982. xxvii, 708p. illus., facsims. £60. ISBN: 0413397261.

First published 1875.

9 chapters: 1. Counting the hours, from the beginning to 1550 ... 4. English supremacy *c.*1660-1750 ... 7. French clocks ... 9. Alarum, striking and reporting mechanisms. 3 appendices (1. Records of famous makers; 2. List of former clock and watch makers, p.345-654; 3. Hallmarks). Glossary. Select bibliography, p.675-81. Index, p.683-708. 388 illus. (40 col.). A handsome quarto classic. *Class No:* 681.11(091)

[6632]

—**BRUTON, E.M.** The History of clocks and watches. London, Orbis, 1979. 288p. illus. (incl. col.), facsims., ports.

11 sections: 1. The earliest clocks - 2. The advent of clockwork - 3. Domestic clocks - 4. European mechanical clocks - 5. The time at sea - 6. The development of the watch - 7. Mass production - 8. The technological age - 9. Watches for the people - 10. The science of time - 11. Great clocks of the world. Glossary. Bibliography, p.278-80. Illus. on every page. Index, p.281-7. A handsopme quarto, for popular consumption. *Class No:* 681.11(091)

Biographies

[6633]

BAILLIE, G.H. Watchmakers and clockmakers of the world. 3rd ed. London, N.A.G. Press, 1951. xxv, 388p. maps.

First published 1929; 2nd ed. 1947 (xxv, 373p.).

The 3rd ed. reprints the 2nd ed's. 35,000 names, A-Z, 'Addenda' of *c.*600 further makers. Brief entries state dates, location and type of watch/clock made. Included are lists of place names, alternative spellings, initials and marks. *Class No:* 681.11(092)

[6634]

—**LOONES, B.** Watchmakers and clockmakers of the world: Volume 2. London, N.A.G. Press, 1976. xiii, 269p. £9.95.

Written as a supplement ot Baillie's *Watchmakers and Clockmakers* (*qv.*) - 'to complement it, not to replace'. Arranged A-Z; entries state name, location and dates. Gazetteer of lesser-known places. Tiny print. Largely confined to UK. *Class No:* 681.11(092)

Equipment & Instruments

[6635]

DE CARLE, D. Clock and watch repairing, including complicated watches. London, Pitman, 1959. viii, 312p. illus.

Covers all components of clocks and watches, how they can go wrong, and how to repair them. All parts and tools are illustrated in detail. Appended glossary of technical terms; also a section on Patak Philippe Light Clock. *Class No:* 681.11:002.50

Museums

[6636]

TURNER, A. The Time Museum: catalog of the collection. Rockford, Ill., Time Museum, 1984-. v.1-. illus. (incl. col.), facsims., map.

1. *Time-measuring machines.*

Pt.1: *Astrolabes. Astrolabe-related instruments.* 1985.

Pt.2: *Water-clocks. Sand-glasses. Fire clocks.* 1984.

V.1, pt.1 ([xv], 268p.) includes a glossary, bibliography of works cited (p.243-60), concordance, and index. 137 figures. Pt. 2 ([xi], 183, [1]p.) includes a bibliography, p.163-73. Index, p.177-83. *Class No:* 681.11:061:069

[6637]

TYLER, E.J. Clock museums and collections visited by E.J. Tyler, from 1971 to 1983. Heathfield, E. Sussex, Watch & Clock Book Society, 1983. 46p. illus. £4.75. ISBN: 090351236x.

Reprints of artices on clock collections visited. 19 illus. *Class No:* 681.11:061:069

[6638]

WARD, F.A.B. Descriptive catalogue of the collection illustrating time measurement: Science Museum. London, H.M. Stationery Office, 1966. vii, 151p. illus. (pl.).

526 entries, with references appended. 16 sections: 1. Primitive and other non-mechanical devices - 2. Sundials, nocturnal and perpetual calendars ... 7. Chronometers ... 11. The atomic clock ... 15. Gas controllers and time switches. List of the more important objects in the reserve collection (R1-67). List of makers of watches included in the catalogue. Index. *Class No:* 681.11:061:069

Sundials

[6639]

COUSINS, F.W. Sundials: a simplified approach by means of the equatorial dial. London, J. Baker, 1969. 247p. illus., diagrs., graphs, tables.

22 sections ending with 'Dialling scales'. Bibliography, p.214-22 (2 parts: 'Books on dialling before 1800', in

....(contd.)

chronological order; 'Works on dialling after 1800', under authors, A-Z). Appendix: 'A note concerning the accuracy of the tabulated values'. Name and subject indexes.
Class No: 681.111

Instrument Making

Bibliographies

[6640]
BRITISH SCIENTIFIC INSTRUMENT RESEARCH ASSOCIATION. **Catalogue of books in the Library.** [Chiselhurst, Kent], BSIRA, 1966. 2v. mimeographed.
V.1, a U.D.C. classified list ([v], 170p.), - 'compiled with both librarians and general users in mind'. About 2,000 items. V.2: subject and author indexes (49p.).
Class No: 681.2(01)

Encyclopaedias

[6641]
CONSIDINE, D.M., *ed.* **Encyclopedia of instrumentation and control.** New York, etc., McGraw-Hill, 1971. xxvii, 788p. illus.
About 120 contributors. Nearly 700 entries, A-Z ('Oceanographic instrumentation, p.456-9'; 5 illus., 1 table; 10 cross-references). Smallish headwords. Adequate cross-references. 809 illus. (photographs; clear line-drawings). Classified index; subject index, p.768-88.
Class No: 681.2(031)

[6642]
MILLS, J.F. **Encyclopaedia of antique scientific instruments.** London, Aurum Press, 1983. 255p. illus. (incl. col.), facsims., tables. ISBN: 0906053404.
More than 700 entries, A-Z, p.45-233 (*e.g.* 'Sextant': 2½ columns, 2 illus.). Includes biographies (*e.g.* 'Benjamin Franklin': over 1 column). Appendices: Care of the collection; Forgery and fakes; Table of events (to 1789); Museums (under countries, A-Z, p.251-9); Price guide. 'This book should have wide appeal' (*Library journal,* v.109(1), January 1984, p.76). *Class No:* 681.2(031)

Handbooks & Manuals

[6643]
EWING, G.W., *ed.* **Analytical instrumentation handbook.** New York, M. Dekker, 1990. 1088p. $195. ISBN: 0824781848.
Detailed descriptions of the major techniques of spectrochemical, electrochemical and chromotographic analysis, and other techniques (*e.g.* chapter of 73p. on nuclear magnetic resonance). This handbook can therefore 'serve as a one-volume resource in place of a number of monographs' (*Science and technology libraries,* v.11(1), Spring 1991, p.149). *Class No:* 681.2(035)

[6644]
The Instrument manual. Miller, J.T., *ed.* 5th ed. London, United Trade Press, 1975. [vii], 566p. illus., diagrs., tables.
First published 1949.
8 contributors. 23 sections (1. Automatic control ... 4. Measurement and control of viscosity - 5. Measurement of fluid flow ... 8. Techniques for measuring surface temperature (381 references) ... 13. Measurement and control of pressure ... 20. Analytical instrumentation - 21. Non-destructive testing - 22. Numerical control - 23. SI units). Section references. Analytical index.
Class No: 681.2(035)

[6645]
Jones' instrument technology. Noltingk, B.E., *ed.* 4th ed. London, Butterworth, 1985-7. 5v. illus., diagrs., graphs, tables.
First published 1956.
1. *Mechanical measurements.* 1985. ix, 170p.
2. *Measurement of temperature and chemical composition.* 1985. [viii], 282p.
3. *Electrical and radiation measurement.* 1987. 220p.
4. *Instrumentation systems.* 1987. 168p.
5. *Automatic instruments and measuring systems.* 1986. 180p.
The 4th ed. reflects recent developments in electronics, gas analysis and other subjects. V.1 has 9 contributors, with 11 documented sections and a detailed subject index.
Class No: 681.2(035)

[6646]
LIPTÁK, B.G., *ed.* **Instrument engineers' handbook.** Rev. ed. Radnor, Pa., Chilton Book Co., 1985. 2v. illus., tables. ISBN: 0801072906.
First published 1969-70 (2v.); supplement, 1972. 1. *Process measurement.* 2. *Process control.*
V.2 (1116p. $75) comprises documented chapters on control theory, panels and displays, logic devices and PLCs, computer and distributed control, and control of unit operations. Index. *Class No:* 681.2(035)

[6647]
NACHTIGAL, C.L., *ed.* **Instrumentation and control:** fundamentals and applications. New York, etc., Wiley, 1990. xxii,890p. illus., diagrs., graphs, tables. (*Wiley series in mechanical engineering practice.*) ISBN: 0471880450.
20 chapters in 3 parts: I. General topics (system engineering concepts; instrument statics; electronic devices and data conversion, etc.) - II. Instrumentation (bridge transducers; temperature and flow transducers, etc.) - III. Control (closed-loop control system analysis; performance modification; controller design, etc.). Extensive chapter references. Analytical subject index, p.871-90. 46 contributors, both academic and industrial.
Class No: 681.2(035)

Dictionaries

[6648]
CLASON, W.E. **Elsevier's dictionary of measurement and control** in six languages: English/American, French, Spanish, Italian, Dutch and German. Amsterdam, etc., Elsevier, 1977. 886p. £121.62. ISBN: 0444415823.
Almost 7,800 entries for terms in English, with equivalents in other languages. Alphabetical indexes in other

....(contd.)

languages refer to numbered entries in main section. Brief bibliography (p.885-6). Lack of Russian terminology a disadvantage. *Class No:* 681.2(038)

[6649]
RAMALINGOM, T. **Dictionary of instrument science.** New York, Wiley, 1982. 588p. ISBN: 0471863963.

More than 6,000 definitions of terms in control and instrumentation. Coverage includes chemical and process control instrumentation, optical electrical and electronics instrumentation, aerospace and remote-sensing instrumention, computer and data processing. Accessible to non-technical readers. *Class No:* 681.2(038)

Progress Reports
[6650]
Advances in instrumentation. Pittsburgh, Pa., Instrument Society of America, 1947-. Annual. illus., diagrs., graphs, chemical structures.

V.42(1) 1987, covers Proceedings of the (Instrument Society of America) '87 Instrumental Conference and Exhibits (566, [6]p.). 3 parts/divisions: chemical and petroleum industries; automatic control systems; electro optics. 158 documented papers (*e.g.* 'A new instrument for particle size control using light scattering', p.201-6. 5 references). *Class No:* 681.2(055)

Histories
[6651]
BENNETT, J.A. **The Divided circle:** a history of instruments for astronomy, navigation and surveying. Oxford, Phaidon, Christie's, 1987. 224p. illus. (incl. col.). ISBN: 0714880388.

14 chapters covering historical periods: 1-5. Ancient to early modern - 6-9. Eighteenth Century - 10-13. Nineteenth century - 14. Twentieth century. Extensive bibliography; 3 indexes (makers; technical terms; general). A lavishly illustrated quarto. Includes 'a bibliography, an index of makers, an index of technical terms, and a general index, which make it an excellent reference as well as a fascinating book to read' (*Choice,* June 1988, p.1577). *Class No:* 681.2(091)

[6652]
BILLMEIR, J.A. **Scientific instruments, 13th-19th century:** the collection of J.A. Billmeir, with supplement. Exhibited by the Museum of the History of Science, Oxford. 2nd ed. Oxford, the Museum, 1955-57. 2v.

First published 1954.

The 1955 catalogue lists 154 items in 7 classes, with brief notes on size, century, rarity, etc., followed by plates and index of inventors, designers and makers. The supplement provides much fuller notes on items in 9 classes: 1. Astrolabes; 2. Astronomical ring; 3. Globes; 4. Quadrants; 5. Sundials; 6. Nocturnals; 7. Astronomical compendia; 8. Topographical instruments; 9. Miscellaneous. Bibliographical note, p.9-12, and index. *Class No:* 681.2(091)

[6653]
DAUMAS, M. **Scientific instruments of the seventeenth and eighteenth centuries and their makers.** Translated and edited by M. Holbrook. London, Batsford, 1972. 361p. illus., diagrs.

3 parts (14 chapters): 1. Instrument-making industry in the seventeenth century - 2. Contributory factors in the evolution of the instrument-making industry - 3. The instrument-making industry in the eighteenth century. Chapter references and footnotes, p.295-340. Libraries, archives, collections, p.341. Bibliography (general works and periodicals), p.342-8. 142 plates; 151 illus. Index, p.349-61. *Class No:* 681.2(091)

[6654]
SYDENHAM, P.H. **Measuring instruments:** book of knowledge and control. Stevenage, Herts., P. Peregrinus, in association with the Science Museum, London 1979. xvi, [1], 512p. ISBN: 0906068192.

6 sections: 1. Measurements for knowledge and control - 2. Science and technology of measuring instruments - 3. Ancient times to Middle Ages: birth of the first instruments - 4. Experimental science becomes established: Middle Ages - 1800 A.D. - 5. Growth of electrical methods - 6. The first half of the 20th century: 1900-1950. Bibliography (A-Z authors), p.456-81. 2 appendices: 1. Biographies relevant to instrument history; 2. Collections containing instruments. Detailed, analytical index, p.496-512. *Class No:* 681.2(091)

[6655]
TURNER, A. **Early scientific instruments: Europe 1400-1800.** London, Sotheby's Publications, 1987. 320p. illus. (incl. col.), ports. ISBN: 0856673196.

Describes what scientific instruments were available at given periods, explaining their origin and technical functions, and indicating the main lines of their development. Notes; bibliography, p.290-311. A lavishly illustrated large quarto. *Class No:* 681.2(091)

[6656]
TURNER, G.L'E. **Antique scientific instruments.** Poole, Dorset, Blandford Press, 1981. 165p. ISBN: 0713709235.

9 chapters: 1. Astronomy and time-telling - 2. Navigational instruments - 3. Surveying instruments - 4. Drawing and calculating instruments - 5. Optical instruments - 6. Philosphical instruments - 7. Weights and measures - 8. Medical instruments - 9. Practical advice on collecting. Bibliography (A-Z authors), p.160-2). Museums and collections p.163-4. Index (mostly non-analytical), p.165-8. Small format. A book that should have wide appeal. *Class No:* 681.2(091)

[6657]
TURNER, G.L'E. **Nineteenth-century scientific instruments.** London, P. Wilson, for Sotheby Pubns. 1987. 320p. illus. (incl. col.), facsims. ISBN: 0850671708.

16 chapters: 3. Weights and measures - 4. Mechanics - 5. Hydrostatics - 6. Pneumatics - 7. Heat - 8. Sound - 9. Light - 10. Magnetism - 11. Electricity - 12. Chemistry - 13. Meteorology - 14. Surveying and navigation - 15. Drawing and calculating - 16. Recreational science. Appended: 'Instrument makers exhibiting at the Great Exhibition of 1851'. Bibliography, p.311-3. Index, p.314-20. 377 illus. (32 col.). Quarto. *Class No:* 681.2(091)

USA

[6658]

SMITHSONIAN INSTITUTION. Early American scientific instruments and their makers. Bedini, S.A. Washington, the Institution, 1964. 196p. (*Bulletin 231.*)

A descriptive account. *Class No:* 681.2(73)

Museums

[6659]

HOLBROOK, M. *and* **ANDERSON, R.G.W.,** *and others,* *comps. and eds.* **Science preserved:** a directory of scientific instruments in collections in the United Kingdom and Eire. London, H.M. Stationery Office, 1992. 270p. illus. £35. ISBN: 0112900607.

A guide to some 3700 instruments in more than 200 publicly accessible collections in the museums, country houses and universities of the British Isles.

Class No: 681.2:061:069

Computers

Abbreviations & Symbols

[6660]

MERKOW, M. Breaking through technical jargon: a dictionary of computer and automation acronyms. New York, Van Nostrand Reinhold, 1990. xviii,181p. £14.50. ISBN: 0442001517.

About 1,000 acronymns, A-Z (ABCD - Z-80). Strong North American bias (*e.g.* ANSI and ISO are present but not BSI). Some odd omissions, *e.g.* WIMP, WYSIWYG.

Class No: 681.3(003)

[6661]

TOWELL, J.E. *and* **SHEPPARD, H.E.,** *eds.* **Computer & telecommunications acronyms.** Detroit, Mich., Gale, 1986. 391p. $60. ISBN: 0810324911.

A dictionary of *c.*25,000 acronyms, terms, etc., in the computer and telecommunications fields. Entries are selected from *Acronyms, initialisms & abbreviations dictionary* (9th ed. Gale, 1984). Added to many definitions are subject categories, geographical locations and citation of sources.

Class No: 681.3(003)

[6662]

—**WRATHALL, C.P.** Computer acronyms and abbreviations. New York, & Princeton, N.J., Petrocelli Books, 1981. 483p. ISBN: 0894331388.

More than 10,000 entries (including multiple definitions) for acronyms, abbreviations and names from the computer and communications field. Appended sources of additional information (books and their publishers; organizations).

Class No: 681.3(003)

Databases

[6663]

Computer database. Foster City, Ca., Information Access Company.

Covers computers, telecommunications and electronics with information on hardware, software, peripherals and

....(contd.)

services. About 450,000 records; updated weekly. Available on DIALOG (file 275) and Data-Star.

Class No: 681.3(003.4)

[6664]

INSPEC. Stevenage, Herts., Institution of Electrical Engineers, 1969-. Updated weekly. *Class No:* 681.3(003.4)

[6665]

RAITT, D. 'Computer and information science and technology'. *Manual of online search stategies.* Editors, C.J. Armstrong and J.A. Large, pp.309-56. 2nd ed. Aldershot, Ashgate, 1992. ISBN: 1857420071.

Extensive list of databases and host systems covering mainframes, micros, software, artificial intelligence, telecommunications, semiconductors, etc.

Class No: 681.3(003.4)

Bibliographies

[6666]

ACM guide to computing literature. Baltimore, Md., Association for Computing Machinery, 1978-. Annual. ISSN: 01491199.

Formerly *Computing reviews: bibliography and subject index of current computing literature,* 1977.

The 1991v. (xxiv,1748p.) has about 24,000 references under 'Bibliographic listing' (books; journals; proceedings; reports; doctoral theses; 'Computing reviews'). 6 indexes: Author; Keyword; Category; Proper noun subject; 'Computing reviews' reviewer; Source. *Class No:* 681.3(01)

[6667]

Computer books and series in print. New York, etc., Bowker, 1985. 551p. £70.

Lists *c.*12,000 books and over 1,500 serial titles. More extensive than *Computer publishers and publications: an international directory and yearbook* (Detroit, Mich., Gale, 1984-). The latter's 1984 ed. (379p. $90) listed over 600 English-language computer periodicals and *c.*275 computer-book publishers. However, unlike the Bowker list, it is annual and updated by means of interim supplements.

Class No: 681.3(01)

[6668]

Computing information directory: a comprehensive guide to the computing literature. Hildebrandt, D.M., *ed.* 6th ed. Federal Way, Wash., Pedaro Inc., 1989. 410p. $145.

First published 1981 as *Computer science resources.*

Revised annually. Master subject index, with subheadings, by type of literature, *e.g.* current books; computer-related journals; indexing and abstracting services; software resources; programming languages, etc. Well-produced and a valuable US-slanted source. *Class No:* 681.3(01)

[6669]

MYERS, D. **Computer science resources:** a guide to professional literature. White Plains, N.Y., Knowledge Industry Publications, for American Society for Information Science (ASIS), 1981. vi, 345p. ISBN: 0914236806.

'Several thousand essential reference tools covering virtually all aspects of data processing' (*Preface*). 10 sections: 1. Current books in the computer sciences - 2. Computer-related journals - 3. Technical report literature - 4. Indexing and abstracting services - 5. Directories, dictionaries, handbooks - 6. University Computer Centre newsletters - 7. Software resources - 8. Proceedings of ACM

....*(contd.)*

Special Interest Group - 9. Programming languages (*c.*300) - 10. Publishers index. Appendices A-E (B. Society, Association and user group acronyms). Well produced and a valuable US-slanted source. *Class No:* 681.3(01)

Encyclopaedias

[6670]
Encyclopedia of computer science and technology. Belzer, J., *and others, eds.* New York & Basle, Dekker, 1975-. illus., diagrs., tables. (v.20: Supplement 5 (1988). \$475. ISBN: 0824722515.

V.1-14 contain 277 documented articles, A-Z. V.13 (1979. vi, 507p., 'Reliability theory' ... 'USSR, computing in') has 23 contributors. 'Software engineering': p.216-44, carries 19 references. V.15, *Supplement* (1980) includes 12 contributors and one update. V.16 (viii, 192p.) provides author and subject indexes. V.17 (1987) has 16 articles 'Aimed at the needs of computer hardware specialists, programmers, systems analysts, engineers, operations researchers, and mathematiciains' (*Preface*).
Class No: 681.3(031)

[6671]
Macmillan encyclopedia of computers. Bitter, G.G., *ed.* New York, Macmillan, 1992. 2v. 1080p. illus., diagrs. \$150. ISBN: 0028970454.

Over 200 articles with an emphasis on practical application. Numerous biographical entries. Appendix lists computing associations and computer manufacturers. Detailed, analytical index. Selected by *Choice* (January 1993, p.735) for its list of 'Outstanding academic books', 1993. *Class No:* 681.3(031)

[6672]
RALSTON, A. *and* **REILLY, E.D.,** *Jr., eds.* **Encyclopedia of computer sciences.** 3rd ed. New York, Van Nostrand Reinhold, 1992. 1696p. illus., diagrs., tables. \$150; £72.50. ISBN: 0442276796, US.

First published 1976, as *Encyclopedia of computer science.* 2nd ed. 1983.

Over 300 contributors and about 500 signed articles, 175 of them new to this edition. Coverage: hardware; computer systems; information and data; software; mathematics and theory of computing; methodologies; applications; computing environments. Appendices (*e.g.* 'Glossary of major terms in five languages'); detailed, analytical index. The standard one-volume reference work on computers.
Class No: 681.3(031)

Handbooks & Manuals

[6673]
The McGraw-Hill computer handbook. Helms, H., *ed.* New York, etc., McGraw Hill, 1983. [992]p. illus., diagrs., tables. \$79.50.

24 contributors. 30 sections (1. Computer history and concepts ... 6. The arithmetic-logic unit - 7. The memory element ... 10. Time-sharing sytstems ... 13/16. Systems (BASIC; COBOL; FORTRAN; Pascal; PL/I) ... 20. Computer graphics ... 25. Microcomputers and programming ... 30. Voice regonition. 6p. glossary. Detailed, analytical index (21p.). A basic source, like

....*(contd.)*

Ralston (*qv*). 'Little prior knowledge of computer science is needed to understand this handbook' (*Reference books bulletin,* 1984-1985, p.22). *Class No:* 681.3(035)

[6674]
—CURRAN, S. *and* CURNOW, R. The Penguin computing book: a complete and comprehensive guide to computing. London, A. Lane, 1983. 464p. illus. diagrs. £10.95. ISBN: 071391598x.

Deals with such topics as history of computing, basic principles of operation, programming, and future applications. Includes a section on 'how to choose a computer'. Glossary. Bibliography; references. Index. Well-written; for the beginner. 'No prior knowledge is assumed' (*British book news,* March 1984, p.143).
Class No: 681.3(035)

[6675]
TOOLEY, M. Newnes computer engineer's pocket book. 3rd ed. Oxford, Butterworth-Heinemann, 1991. 200p. illus., diagrs., tables. £12.95.

First published 1987.

Covers a range of topics (*e.g.* Basic logic gates, Boolean algebra, ASCII control characters), devoting 1-2 pages to each (although 74 series has nearly 20 pages). Extensively illustrated; list of abbreviations; index (5p.). Small format. 'Aims to provide ... everyday information' in an accessible format (*Preface*). *Class No:* 681.3(035)

[6676]
VASSILIOU, M.S. *and* **ORENSTEIN, J.A. Computer professional's quick reference.** New York, McGraw-Hill, 1992. 266p. illus., diagrs., tables. \$34.95; \$24.95. ISBN: 0070672113; 0070672121, Pbk.

Includes common commands for 5 operating systems, UNIX, VMS, MVS, VM, MS-DOS, as well as networks and technical standards. Chapter bibliographies. Detailed, analytical index. *Class No:* 681.3(035)

Dictionaries

Polyglot

[6677]
BÜRGER, E. *and* **SCHUPPE, W. Technical dictionary of data processing, computers, office machines:** English, German, French, Russian. Oxford, Pergamon Press; Berlin, VEB Verlag Technik, 1970. 1463p.

About 13,000 terms in each of 4 sequences, - English, German, French and Russian. The 4 sequences severally give equivalents in the other 3 languages. Genders are stated. *Class No:* 681.3(038)=00

English

[6678]
FREEDMAN, A. The Computer glossary: the complete illustrated desk reference. 5th ed. New York, AMACOM, 1991. 670p. illus. \$34.95. ISBN: 0814477496.

Entries A-Z; numeric indexes. 'Practical, well illustrated, well written, and extensive ... Highly recommended' (*Choice,* June 1991, p.1615). *Class No:* 681.3(038)=20

[6679]

A Glossary of computing terms. British Computer Society Schools Committee Glossary Working Party. 7th ed. London, Pitman for the British Computer Society, 1991. viii,150p. ISBN: 0273036459.

First published 1977; previous ed. 1987.

About 950 terms defined in simple language. Arrangement is by 15 broad headings (*e.g.* 'Applications'; 'Storage'; 'Truth tables and logic gates', etc.). Therefore, an index of terms is added. Includes 10 brief 'essays' on specific areas of computing and information technology. Recommended for schools and tertiary institutions. *Class No:* 681.3(038)=20

[6680]

ILLINGWORTH, V., *ed*. Dictionary of computing. 3rd ed. Oxford Univ. Press, 1990. [xi], 528p. illus., tables. £20. ISBN: 0198538251.

First published 1983. 2nd ed. 1986.

Nearly 4,500 terms in computing. Includes technical terms in hardware and software, plus broader context of social and legal implications. 'Small flaws do not tarnish authoritative definitions, broad coverage and timeliness ... Highly recommended' (*Choice,* March 1991, p.1094).

Class No: 681.3(038)=20

[6681]

LYNCH, D.B. Concise dictionary of computing. London, Chartwell-Bratt, 1991. 380p. £9.95. ISBN: 0862382685.

Defines over 4,000 terms, A-Z. Comprehensive and non-technical reference source and guide for users of computers and IT in education and business. *Class No:* 681.3(038)=20

[6682]

Microsoft Press computer dictionary. Redmond, Wa., Microsoft Press, 1991. viii,392p. illus., diagrs., tables. $19.95; £18.95. ISBN: 1556152310.

About 5,000 entries, A-Z (A:-Zulu time). Some defined at length, (*e.g.* 'Ethernet': ½ column. Includes acronyms. 5 appendices, A-E (A: ASCII character set). A good, standard dictionary. *Class No:* 681.3(038)=20

[6683]

The New hacker's dictionary. Raymond, E.S., *ed*. 2nd ed. Cambridge, Mass. and London, MIT Press, 1991. xx,433p. illus. $25 (Hbk); $10.95, £10.75 (Pbk). ISBN: 0262181452; 0262680696, Pbk.

First published 1983.

More than 1,000 terms, A-Z (Abbrev.-Zorck). A dictionary of slang terminology, not technology. Introductory essay on the evolution of computer-hacking jargon. Brief, annotated bibliography. 'A useful reference tool', but 'not to be confused with a technical dictionary' (*Choice*, January 1992, p.724). *Class No:* 681.3(038)=20

[6684]

The Penguin dictionary of computers. Chandor, A., *with others*. 3rd ed. Harmondsworth, Middx., Penguin Books, 1985. 488p. ISBN: 014091127x.

2nd ed. 1977.

Defines *c*.6,000 terms. No illus. 'The best of the low-priced dictionaries, including useful technical material' (*Computing reviews,* v.27(12), December 1986, p.601). *Class No:* 681.3(038)=20

[6685]

The Prentice-Hall standard glossary of computer terminology. Edmunds, R.A. Englewood Cliffs, N.J., Prentice-Hall, 1985. xv, 459p. $34.95. ISBN: 0136982344.

Defines more than 4,700 categorized words. Comprehensive, but it does exclude terms specific to a particular computer language. Includes acronyms, abbreviations, cross-references, and some biographies. Well produced. 'An excellent source for those working with computers and those attemtping to read or converse intelligently about them' (*Reference books bulletin, 1984/1985,* p.96). *Class No:* 681.3(038)=20

[6686]

Prentice Hall's illustrated dictionary of computing. Nader, J.C., *ed*. New York, Prentice Hall, 1992. xiv,526p. illus., diagrs. £14.95. ISBN: 0137199988.

Over 5,000 terms, some defined at length (*e.g.* 'neural network': 1 col.). Definitions frequently based on ISO standards. Includes 'Style manual' for correct use of computer terminology, covering acronyms, abbreviations, use of capital letters, form of numbers, etc.

Class No: 681.3(038)=20

[6687]

SIPPL, C.J. *and* **SIPPL, R.J. Computer dictionary and handbook.** 4th ed. Indianapolis, Ind., Sams, 1985. 962p. illus. $24.95. ISBN: 0672222051.

First published 1966. 3rd ed. 1980.

12,000 entries, including 1000 new to this ed. Cross-references; extensive appendices. For the advanced user. *Class No:* 681.3(038)=20

[6688]

—GOLLAND, F.J., *ed*. Dictionary of computing: data communications, hardware and software basics, digital electronics. New York, etc., Wiley, 1982. vi, 330p. illus., diagrs., tables. ISBN: 047110468x.

Complementary to Sippl (*qv*) in having fuller definitions, more diagrams and more *see also* references. However, 'one will probably prefer Sippl because of its appendices' (*Wilson library bulletin*, March 1983, p.606). Oblong format.

Class No: 681.3(038)=20

[6689]

SPENCER, D.D. The Illustrated computer dictionary. Columbus, Ohio, Merrill, 1986. vi, 328p. illus., facsims., ports. ISBN: 0673205298.

About 6,000 terms. Main aim: 'to present concisely the most common terms currently used by computer scientists, information processing personnel, and other computer users' (*Preface*). Appendix: 'How to buy a personal computer'. *Class No:* 681.3(038)=20

[6690]

Webster's New World dictionary of computer terms. 3rd ed. New York, Simon & Schuster, 1988. vi,412p. £5.95. ISBN: 0139492313.

First published 1983.

About 4,500 terms, briefly defined (*e.g.* 'multipass sort': 3 lines). Small format. *Class No:* 681.3(038)=20

[6691]

WYATT, A.L. Computer professional's dictionary. Berkeley, Calif., Osborne McGraw-Hill, 1990. ix,352p. diagrs., tables. £22.95. ISBN: 0078817056.

About 2,500 terms, briefly defined (*e.g.* '1 parity': 10 lines). *Class No:* 681.3(038)=20

German

[6692]
COLLIN, S.M.H., *and others*. **Dictionary of computing and information technology:** English-German. Teddington, Middlesex, Peter Collin, 1991. [viii],410,[17]p. illus. ISBN: 0984859203.

About 5,000 English terms, briefly defined. Includes brief boxed quotations to illustrate use of a term. List of German words and their English equivalents.
Class No: 681.3(038)=30

[6693]
SCHULZE, H.H. **Computer Englische:** ein Fachwörterbuch. [Computer English: a dictionary.] Hamburg, Rowholt, 1991. 510p. ISBN: 3499181770.

About 18,000 terms in each language.
Class No: 681.3(038)=30

French

[6694]
COLLIN, S.M.H., *and others*. **Dictionary of computing and information technology:** English-French; French-English. Teddington, Middlesex, Peter Collin, 1991. [x],323p. ISBN: 0948549246.

About 5,000 headwords in each language. Genders given. Distinguishes British and American usage. Includes brief boxed quotations to illustrate use of a term.
Class No: 681.3(038)=40

[6695]
FISCHER, R. *and* KRUCHTEN, P. **Dictionnaire informatique.** Paris, Eyrolles, 1984-5.

Anglais-français. 3. éd., edited by R. Fischer and P. Kruchten. 1985. 375p.

Français-anglais; by R. Fischer. 1984. 480p.

About 15,00 terms and phrases. The work of a team of contributors. *Class No:* 681.3(038)=40

[6696]
INTERNATIONAL ORGANIZATION FOR STANDARDIZATION. **Dictionary of computer science:** English-French. [Dictionnaire de l'informatique: Français-anglais.] Geneva, ISO; Paris, AFNOR, 1989. xi,185,180p. diagrs., tables. £76.50. ISBN: 0124896916.

About 2,500 terms, with equivalents, in each half. Entries include synonyms and examples of usage.
Class No: 681.3(038)=40

Reviews & Abstracts

[6697]
CompuMath citation index: an international, interdisciplinary index to the literature of applied mathematics, computer science statistics, operations research and related disciplines. Phildealphia, Pa., Institute for Scientific Information, 1981-. 3pa. ISSN: 07306199.

CMCI consists of 5 indexes: Research front specialty index; Citation index; Corporate index; Source index; Permuterm subject index. Also, annual *Guide and list of source publications.* The January/April 1987 issue of *CMCI* contains 2,598 columns of tiny print. *Class No:* 681.3(048)

[6698]
Computer and control abstracts. Science abstracts. Section C. London, INSPEC, 1969-. Monthly. ISSN: 00368113.

Formerly *Control abstracts,* 1966-68.

About 70,000 abstracts pa. Classes, closely subdivided: 00.00 General and management topics - 10.00 Systems and control theory - 30.00 Control technology - 40.00 Numerical analysis and theoretical, computer topics - 50.00 Computer hardware - 60.00 Computer software - 70.00 Computer applications. Author index per issue; Subject index, twice yearly; Subsidiary indexes, where appropriate; Bibliography index (for well-documented articles); Book index (for books received and abstracted); Corporate author index; Conference index. Indexes cumulated 2pa. and 4-yearly (*e.g.* 1977-1980). INSPEC database. CD-ROM.
Class No: 681.3(048)

[6699]
Computer & information systems abstract journal. Bethesda, Md., and Oxford, Cambridge Scientific Abstracts, 1969-. Monthly. ISSN: 01919776.

Formerly *Information processing journal* 1962-68.

About 20,000 abstracts pa. Sections: CS Computer software - CA Computer applications - CM Computer mathematics - CE Computer electronics - General. Detailed, analytical subject index and author index per issue. Cumulative index pa. *Class No:* 681.3(048)

[6700]
—Computer abstracts. St. Helier, Jersey, British Channel Islands, Technical Information Co., 1961-. Monthly.

More than 3,000 abstracts (averaging *c.*150 words apiece) pa. 19 sections: 1. General - 2. Computer theory ... 18. Applications (Aerospace ... Transport. Miscellaneous) - 19. Books. Covers conference papers and US government reports as well as periodical articles and books. Annual subject and author indexes; list of journals abstracted.
Class No: 681.3(048)

[6701]
Computer contents. New York, Find/SVP, 1983-. Semi-monthly.

Reproduces contents-tables from over 250 computer, electronics and telecommunications journals. September 15, 1987 issue handles 31 journal issues. *Class No:* 681.3(048)

[6702]
Computer information review. New York, Nova Science Publishers, Inc., 1986-. Quarterly.

About 5,000 abstracts pa., covering books, book chapters, conference proceedings, patents, technical reports, government reports, dissertations, special journal issues. 19 classes (*e.g.* CA6. Science/engineering applications; CH5. Computer systems; CP3. Computer languages). Subject and author indexes per issue. *Class No:* 681.3(048)

[6703]
Computer literature index: a subject/author index to computer and data processing literature. Phoenix, Arizona, Applied Computer Research, Inc., 1981-. 4pa. ISSN: 02704846.

Formerly as *Quarterly bibliography of computers and data processing,* 1971-80.

About 2,500 briefly annotated entries per issue. Subject index (main sequence); author index; publisher index. Each issue covers the preceding quarter. Annual cumulation.
Class No: 681.3(048)

[6704]

Computing journal abstracts. Manchester, NCC Information Services, 1972-. Fortnightly.

1986: 2,260 detailed abstracts of articles on computer applications, in both British and foreign publications. Sections: International items - Computer industry - Computer services ... Software and communications ... New hardware developments ... Automation ... Artificial intelligence. A current awareness service. *Class No:* 681.3(048)

[6705]

Computing reviews: review journal of the Association for Computing Machinery. New York, the Association, 1960-. Monthly.

About 1,000 abstracts/reviews pa. Sections A-K: A. General literature - B. Hardware - C. Computer systems organization - D. Software - E. Data - F. Theory of computation - G. Mathematics of computing - H. Information systems - I. Computing methodologies - J. Computer applications - K. Computing milieux. Descriptors added. Preceding the abstracts section: Author listing; Books and proceedings; Non-book literature. Appended general index per issue. December 1986 issue carries survey; 'Computer view of 22 computer dictionaries', p.599-602.
Class No: 681.3(048)

[6706]

Current papers on computers and control. Stevenage, Herts., Institution of Electrical Engineers, New York, Institute of Electrical and Electronic Engineers, 1974-. Monthly. £123pa. ISSN: 00113794.

About 65,000 references pa. Main classes: 00.00 General and management topics - 10.00 Systems and control theory - 30.00 Control technology - 40.00 Numerical analysis and theoretical computer topics - 50.00 Computer hardware - 60.00 Computer software - 70.00 Computer applications. A current awareness service; companion to *Computer and control abstracts*. Online INSPEC. *Class No:* 681.3(048)

Progress Reports

[6707]

Advances in computers. Orlando, Florida, Academic Press' 1962-. Annual.

V.33 (1991. 360p. illus.) is concerned with reusable software components, object-oriented modelling, discrete event simulation and neurocomputing formalisms. Detailed, analytical index. *Class No:* 681.3(055)

Yearbooks & Directories

[6708]

The Computer users' year book, 1992. London, VNU Business Pubrs., 1992. 4v. ISSN: 02686821.

V.1.*Computer equipment.*
V.2. *Computer services.*
V.3. *Computer installations.*
V.4. *Suppliers guide.*

A mine of information. V.4 includes almost 500 categories covering, amongst other things, communications equipment, stationery and security. 'One indispensable tool for the reference library' (*Information sources in engineering.* Editor, L.J. Anthony, p.404). *Class No:* 681.3(058)

[6709]

Computers and computing information resources directory. Connors, M., *ed.* Detroit, Mich., Gale, 1987. x, 1271p. $160.

Directory of over 4,000 organizations (worldwide associations and user groups ... consultants and research centres). Also details of over 1,500 periodicals, directories, and abstracting/indexing services in the computer field. Master name and keyword index (28,000 entries); geographic index; personal name index. A comprehensive source book, particuarly for US users.
Class No: 681.3(058)

[6710]

Who's who in computing, 1987/88. London, Centaur Communications, 1987. 539p. £70. ISBN: 0908787058.

Entries arranged under companies. 9 sections: Computer manufacturers and suppliers - Software suppliers; Peripheral suppliers; Bureaux; Maintenance engineers; Recruitment consultancies; Financial services; Institutions: Help. Preferred to *The Computer users' year book, 1987* (*qv*) by the reviewer in *Aslib Information* (v.16(1), January 1988, p.13). *Class No:* 681.3(058)

Tables & Data Books

[6711]

WALSH, B.C., *and others*. **Computer users' data book.** Oxford, Blackwell, 1986. ix,177p. diagrs., tables. £19.95. (*Professional and industrial computing series*.) ISBN: 0632011793, HB; 0632016086, PB.

5 chapters: 1. Information representation - 2. Chips - 3. Secondary storage media and devices - 4. Data communication - 5. Microcomputer operating systems. Detailed, analytical index, p.171-77. Somewhat dated, but clear presentation; useful for novices. *Class No:* 681.3(083)

Histories

[6712]

AUGARTEN, S. Bit by bit: an illustrated history of computers. London, Unwin, 1985. 324p. illus. (incl. col.). ISBN: 0040010074.

9 chapters. 1. The first mechanical calculations - 2. The engines of Charles Babbage - 3. The invention of ENIAC ... 5. The stored-program computer - 6. The rise of IBM - 8. The integrated circuit - 9. The personal computer. Epilogue. Chronology. Bibliography and notes (by chapters), p.299-309. Detailed analytical index. Fully illustrated.
Class No: 681.3(091)

[6713]

WILLIAMS, M.R. A History of computing technology. Englewood Cliffs, N.J., Prentice-Hall, 1985. xi, [i] 432p. illus., facsims., ports., diagrs., chron. table. (*Prentice-Hall series in computational mathematics*.) ISBN: 0133899179.

9 chapters, each with 'Further reading': 2. Early aids to calculation; 4. The Babbage machine; 7. The electronics revolution; 8. The first stored program electronic computers; 9. Later developments. Appended chronology (to 1908). Detailed, analytical index. *Class No:* 681.3(091)

Biographies

[6714]
MATTHEWS, C. 'Computer revolutionaries: a guide to the literature on pioneers in computing'. In *Science and technology libraries*. V.11(4), Winter 1991, p.43-74. ISSN: 0194262x.

Lists about 225 items, including periodical titles, indexing and abstracting services and organizations.
Class No: 681.3(092)

Writing & Lecturing

[6715]
PERDUE, L. **The High-technology editorial guide and stylebook.** Homewood, Il., Business One Irwin, 1991. 194p. $32.50.

2 editions: Macintosh (ISBN: 1556235305); PC (ISBN: 0556235313).

About 3200 entries covering programming languages, corporations, hardware and software technical terms, publications. Includes floppy disk with all terms in ASCII format for incorporation into a word-processing spell checker. *Class No:* 681.3:001.81

Thesauri

[6716]
GOODMAN, A. **Cambridge illustrated thesaurus of computer science.** Cambridge Univ. Press, 1984. 262 + 10p. col. illus., diagrs. £5.50. ISBN: 0521262010.

19 sections/groups of associated words (*e.g.* Computers - Number systems ... Programs ... Electronics ... Memory ... Output devices ... Interfaces ... Files and operations). 10p. index. Over 500 illus., mostly in colour. Index has *c.*2,500 entries. A combined dictionary thesaurus, chiefly for the student in this field. *Class No:* 681.3:025.43

Data Processing

Bibliographies

[6717]
CUPOLI, P.D. 'Reference books for data processing, office automation, and data communications'. *Special Libraries*. v.72, July 1981, p.233-42. illus.

Has sections on loose-leaf services catalogues, data processing, office automation, and indexes to supplementary materials. Appendix: 'Bibliography of reference sources', p.240-2. *Class No:* 681.31(01)

Encyclopaedias

[6718]
SAGE, A.P., *ed.* **Concise encyclcopedia of information processing in systems and organizations.** Oxford, Pergamon, 1990. 548p. illus. £140; $225. ISBN: 008035954x.

68 articles, A-Z, on a range of topics (*e.g.* psychology, management, computer science, applied mathematics, systems engineering, decision sciences, industrial education). Some material taken from *Systems and control encyclopedia* (*qv*), but revised and updated. Cross-references; detailed, analytical index. *Class No:* 681.31(031)

Handbooks & Manuals

[6719]
Handbook of theoretical computer science. Van Leeuwen, J., *ed.* Amsterdam, Elsevier; Cambridge, Mass., MIT Press, 1990. 2v. £185; $250 (set). ISBN: 0262220407, set.

V.A. *Algorithms and complexity.* xiv,978p. £92; $135. ISBN: 0262220385.

V.B. *Formal Models and semantics.* xiv,1273p. £109; $150. ISBN: 0262220393.

V.B. has 23 contributors. 19 documented, subdivided chapters (1. Finite automata ... 3. Formal languages and power series ... 6. Rewrite systems ... 11. Denotational semantics ... 14. Logics of programs ... 19. Operational and algebraic semantics of concurrent processes). Chapter 13, 'Algebraic specification': 111p., 9 sections, 186 references. Detailed, analytical index, p.1243-73. A theoretical, scholarly treatment, not for those new to the subject. *Class No:* 681.31(035)

Dictionaries

Polyglot

[6720]
Dataordboken. Stockholm. SIS (Standardiseringscommission i Sverige), 1984. 187p. (*TNC 82. Svensk Standard SS 01 16 01.*) ISBN: 9171621717.

About 2,00 Swedish entry-words are defined, plus equivalents and indexes in English, French and German. *Class No:* 681.31(038)=00

[6721]
SCHULZ, E. **Wörterbuch der Datentechnik:** russisch-deutsch-englisch. Wiesbaden, Brandstetter, 1977. [xvi], 193 + 109 + 134p. ISBN: 3870970758.

About 4,000 terms in all. 3 sequences: Russian-German-English, German-Russian-English, English-Russian-German. Terms are categorized. *Class No:* 681.31(038)=00

[6722]
WITTMANN, A. *and* KLOS, J. **Dictionary of data processing, including applications in industry, administration and business,** in English, German, French. [Wörterbuch der datenverarbeitung mit anwendungsgebieten in industrie, verwaltung und wirtschaft. Dictionnaire du traitement de données, et de son application dans l'industrie, l'administration et l'economie.] 5th ed. Amsterdam, etc., Elsevier, 1987. xiii,357p. ISBN: 0444989471.

First published 1964; 4th ed. 1984.

About 6,000 terms in all, with some 300 new ones.. English base, with German and French equivalents and indexes. Terms are categorized. Genders given. *Class No:* 681.31(038)=00

English

[6723]
BRITISH STANDARDS INSTITUTION. **Data processing vocabulary.** London, BSI, 1976-. Parts 1-6, 7-16, 18-19, 21-22. (*BS ISO 2382: 1984.*)

1. *Fundamental terms.* 1984. 15p.
2. *Arithmetic and logic operations.* 1976. 31p.
3. *Equipment technology.* 1987. 13p.
4. *Organization of data.* 1987. 17p.
5. *Representation of data.* 1989. 18p.

....*(contd.)*

6. *Preparation and handling of data.* 1987. 15p.
7. *Digital computer programming.* 1977. 24p.
8. *Control, integrity and security.* 1986. 18p.
9. *Data communication.* 1984. 24p.
10. *Operation techniques and facilities.* 1979. 16p.
11. *Processing units.* 1987. 18p.
12. *Peripheral equipment.* 1988. 46p.
13. *Computer graphics.* 1984. 22p.
14. *Reliability, maintenance and availability.* 1978. 10p.
15. *Programming languages.* 1985. 17p.
16. *Information theory.* 1978. 15p.
18. *Distributed data processing.* 1987. 13p.
19. *Analog computing.* 1989. 10p.
21. *Interfaces between process computer systems and technical processes.* 1985. 16p.
22. *Calculators.* 1986. 19p. *Class No:* 681.31(038)=20

[6724]
MAYNARD, J. Dictionary of data processing. 2nd ed. London, Butterworth, 1982. [v], 275p. diagrs., tables. ISBN: 0408005912.

First published 1975.

About 4,000 terms briefly defined, p.1-199. Cross-references. Appendices A-G (A. List of common acronyms and abbreviations; E. Flowsheet symbols).
Class No: 681.31(038)=20

[6725]
SIPPL, C.J. Macmillan dictionary of data communications. 2nd ed. London, Macmillan, 1985. [iv], 532p. £8.95. ISBN: 0333370724.

First published 1976.

About 9,000 entries, 'A and not B gate' ... 'Zoom'. Definitions and explanations of a wide range of terms, jargon, expressions and specific vocabulary used in data communication and data processing.
Class No: 681.31(038)=20

German

[6726]
BRINKMANN, K.-H. *and* **SCHMIDT, R. Data systems dictionary,** German-English, English-German. Wiesbaden, Brandstetter, 1974. [ii], 242, [ii], 298p.

17,261 German-English entries; 19,049 English-German entries. Numerous sub-entries. 59 categories applied. 'The areas of data systems applications are covered only where such mention serves a useful purpose'.
Class No: 681.31(038)=30

French

[6727]
CAMILLE, C. *and* **DEHAINE, M. Harrap's French and English data processing dictionary.** 3rd ed., rev. and updated. London, Harrap, 1985. 432p.

First published 1970. Pt.1: *English-French.* 223p. Pt.2: *French-English.* 194p.

Each part has *c.*2,000 entries, with massed sub-entries (*e.g.* 'Registrement' (recording): 2 columns).
Class No: 681.31(038)=40

[6728]
Accounting and data processing abstracts. Wembley, Middlx., ANBAR Publications, in association with the Institute of Chartered Accountants in England and Wales, 1971-. New series, v.1(1)-. 8pa.

About 100 short abstracts per issue. Critical comments (*e.g.* 'Mind boggling'; 'A snapshot rather than in-depth report') are italicized. Items asterisked are mentioned in the 'Highlights' section that precedes the abstracts. Concise subject index, cumulated annually also precedes. Cross references. Eminently suitable for current awareness. Core list of *c.*250 journals scanned appears in April and October issues. *Class No:* 681.31(048)

Histories

[6729]
CORTADA, J.W. Historical dictionary of data processing. New York, Greenwood Press, 1987. 3v.

[V. 1]. *Biographies.* xiii,321p. $50. ISBN: 0313256519.
[V. 2]. *Organizations.* x,309p. $45. ISBN: 0313233039.
[V. 3]. *Technology.* xix,415p. $55. ISBN: 0313256527.

Over 400 entries in the 3v., A-Z within each. Volume indexes. 'An important set' (*Choice,* February 1988, p.882) which provides a concise and accessible history of data processing. *Class No:* 681.31(091)

Micros

Bibliographies

[6730]
PATERSON, M.T. Microcomputers and microprocessors. London, Library Association, Public Libraries Group, 1978. 47p. ISBN: 0853655129.

Several hundred briefly annotated entries. Sections: Basic works - Introductory and general works - Software - Hardware and systems - Specific microprocessors - Personal computers - Games - Social and technological aspects - Periodicals - Dictionaries and glossaries - Bibliographies - Equipment surveys - Organizations - Background reading.
Class No: 681.31-181.4(01)

Encyclopaedias

[6731]
Encyclopedia of microcomputers. Kent, A. *and* Williams, J.G., *eds.* New York, Dekker, 1988-. v.1-. illus., diagrs. ISBN: 0824727002.

To be completed in 10v., containing *c.*500 articles, and over 5000p. V.1 has 32 contributors, 26 articles. Well documented (*e.g.* 'Artificial intelligence'; p.283-324, with 62 references plus 3p. bibliography. V.5 (1990) has 20 contributors, 20 articles, but that on 'Display screens, touch sensitive' is based on an article published in 1986 and that for 'DOS' includes 49 references, nearly all pre-1985. 'Aimed to the needs of microcomputer hardware specialists, programers, system analysts, engineers, operation recorders, and mathematicians' (*Preface,* v.1). *Class No:* 681.31-181.4(031)

[6732]
McGraw-Hill personal computer programming encyclopedia. Birnes, W.J., *ed*. 2nd ed. New York, etc., McGraw-Hill, 1989. 752p. illus., tables. $95.

First published 1985.

Overviews of various programming languages and operating systems, as well as their operational differences. Sample programmes illustrate. Also, background articles on artificial intelligence, microcomputer graphics, etc. Glossary. Bibliography. Index to programming language keywords; subject index. *Class No:* 681.31-181.4(031)

Handbooks & Manuals

[6733]
BIRNES, W.J., *ed*. **Microcomputer applications handbook.** New York, etc., McGraw-Hill, 1989. x,645p. diagrs., tables. £65. ISBN: 0070053979.

30 contributors. 6 chapters: 1. Hardware systems - 2. MOS memory - 3. Operating systems - 4. Networked systems - 5. Applications systems - 6. High-level programming languages. Glossary (abend-Z-net), p.475-634. Detailed, analytical index, p.635-45. A basic, comprehensive guide to PCs. *Class No:* 681.31-181.4(035)

[6734]
Bowker's complete sourcebook of personal computing. New York, Bowker, 1985. 1030p. $19.95. ISBN: 0835219313.

First published 1984.

A guide to the microcomputer market, and related products and organizations. No less than 750 computers and 1,800 organizations are mentioned, plus profiles of machines, programmes, etc., and a list of 545 magazines apart from numerous reviews and review citations. 'This work is truly authoritative' (*Science and technology libraries,* v.7(1), Fall 1986, p.114). *Class No:* 681.31-181.4(035)

[6735]
JONES, R., *ed*. **The Good hardware guide.** London, Kogan Page, 1991. 286p. illus. £22.50. ISBN: 0749404639.

Covers disk drives, modems, monitors, PCs, plotters, printers and scanners. Data include model, price, manufacturer, product highlights, technical information and comment (sometimes pithy). A popular treatment for the lay person, which only occasionally slips into incomprehensible jargon. *Class No:* 681.31-181.4(035)

[6736]
LONGLEY, D. *and* **SHAIN, M. The Microcomputer users handbook:** the complete and up-to-date guide to buying a business computer. London, Macmillan Press, 1985. xvii, 405p. illus., charts, tables. ISBN: 0333368665.

Previous ed., 1984.

8 sections, closely subdivided (*e.g.* 2. Understanding a microcomputer; 3. Important applications; 5. Current trends in microcomputers; 6. Integrated software; 7. Microcomputer survey (with technical details on each type)). Appendix, A-J (J. Glossary, p.396-405). *Class No:* 681.31-181.4(035)

[6737]
MONEY, S. A. Newnes microprocessor pocket book. Oxford, Heinemann Newnes, 1989. viii,252p. illus. £9.95. (*Pocket Book*.) ISBN: 0434912905.

A basic survey in 15 chapters (*e.g.* 1. Integrated circuits ... 6. 16/32 bit processors ... 9. Parallel input and output ... 15. System development). Appendices: Abbreviations (5p.); Useful addresses (UK and US). Index (4p.). Concentrates on 'popular microprocessors in current use' (*Preface*). Small format. *Class No:* 681.31-181.4(035)

[6738]
MORRIS, N.M. Microprocessor and microcomputer technology. London, Macmillan, 1981. 255p. diagrs., tables. ISBN: 0333320050.

A textbook in 11 chapters (*e.g.* 1. Binary numbers and arithmetic; 6. Memory organization; 11. Programming and applications). 'Problems' appended to each chapter. 'Solutions', p.251. 'Further reading', p.250. *Class No:* 681.31-181.4(035)

Dictionaries

Polyglot

[6739]
MÜLLER, D. Dictionary of microprocessor systems: English, German, French, Russian. 2nd ed. Amsterdam, etc., Elsevier, 1990. 448p. £101. ISBN: 0444987819.

First published 1984.

About 12,000 English-base terms, with equivalents and indexes in the other 3 languages. *Class No:* 681.31-181.4(038)=00

English

[6740]
CHRISTIE, L.G. *and* **CHRISTIE, J. The Encyclopaedia of microcomputer terminology.** London, Unwin Paperbacks, 1985. 352p. illus., diagrs., tables. ISBN: 0040320014.

Over 4,000 microcomputer terms, A-Z, with definitions. 'Program generation': 11½ lines. Appendices A-N (N: Word-processing glossary, p.348-52). A sourcebook for business and professional people. *Class No:* 681.31-181.4(038)=20

[6741]
HORDESKI, M. The Illustrated dictionary of microcomputers. 3rd ed. Blue Ridge Summit, Pa., TAB Books, 1990. 445p. diagrs., flow charts. $29.95; $19. ISBN: 0830673687; 0830633685, Pbk.

First published 1978. 2nd ed. 1986.

Several thousand entries, covering both hardware and software. Includes jargon. The publishers claim to have added nearly 4,000 entries. Some helpful diagrams and charts. *Class No:* 681.31-181.4(038)=20

[6742]
LONGLEY, D. *and* **SHAIN, M. Macmillan dictionary of personal computing and communications.** London, Macmillan, 1986. [iv], 395p. illus. £6.95. ISBN: 0333390830.

About 5,000 entries and cross-references. 'Software protection': nearly 4 columns. *Class No:* 681.31-181.4(038)=20

[6743]
LOVEDAY, G. **Microprocessor sourcebook.** London, Pitman, 1986. viii,247p. diagrs., tables. ISBN: 0273021540.

Covers some 265 topics, A-Z, providing definitions and explanations, sometimes quite detailed (*e.g.* 8- and 16-bit processors cover *c.*50p.). Cross-references for most entries. No index, but the 'contents' is an aphabetical list of topics included. Intended for those working or studying 'in the service and test' areas of the microelectronics industry (*Preface*). *Class No:* 681.31-181.4(038)=20

[6744]
MARGOLIS, P.E. **Hutchinson personal computer dictionary.** London, Hutchinson, 1992. xv,512p. diagrs., tables. £9.99. ISBN: 0091773040.

About 1,500 terms, briefly defined (*e.g.* 'iteration': 5 lines). Cross references. Introduction includes 'categories of terms', *e.g.* communications systems; data storage; input devices; video, etc. *Class No:* 681.31-181.4(038)=20

[6745]
The Penguin dictionary of microprocessors. Chandor, A. Harmondsworth, Middx., Penguin Books, 1981. 184p. £2.25.

Brief definitions *c.*2,500 terms and abbreviations. Cross-references. No illus. Author is a Fellow of the British Computer Society. *Class No:* 681.31-181.4(038)=20

[6746]
SINCLAIR, I.R. **Collins dictionary of personal computing.** Rev. ed. Glasgow, Harper Collins, 1991. [vi],424p. illus., diagrs. £6.99. ISBN: 0004343867.

First published 1988 as *Collins Dictionary of computing.*

About 3,500 terms, briefly defined (*e.g.* 'electronic mail': 5 lines). Covers hardware, programming languages, applications, networks and systems. For the layperson. *Class No:* 681.31-181.4(038)=20

[6747]
SIPPL, C.J. **Macmillan dictionary of microcomputing.** 3rd ed. London, Macmillan, 1985. [iv], 473p. illus. £8.95. ISBN: 0333370716.

First published 1975.

About 5,000 entries, 'A and not B gate' ... 'Zwitterion'. Coverage includes software, graphics, programming. 'Programmable calcuators': 1 column. A few line-drawings. Written specifically with the microcomputer in mind. *Class No:* 681.31-181.4(038)=20

Reviews & Abstracts

[6748]
Microcomputer index. Mountain View, Ca., Database Services, 1980-. 6pa.

Stable core of *c.*50 microcomputer journals on business, education and home computer uses. Subject indexing and abstracts. Coverage includes reviews of books, software and hardware. Online via DIALOG. *Class No:* 681.31-181.4(048)

Periodicals

[6749]
SHIRINIAN, G. **Microcomputer periodicals:** an annotated bibliography. 9th ed. Toronto, Canada, the Author, 1983. [v], 141p. ISBN: 0969155603.

....(contd.)
Periodical indexes - Periodicals (*c.*600; p.6-126) - Name changes and cessations. Subject index. Geographical index. *Class No:* 681.31-181.4(051)

Yearbooks & Directories

[6750]
BENNETT, S. *and* FREIRMAN, R. **Microcomputer marketplace:** the one-stop resource for industry information. London, Century Business, 1992. 620p. £26.99. ISBN: 0712655579.

Covers both hardware and software. Data: Companies, products, people, services, organizations. *Class No:* 681.31-181.4(058)

Tables & Data Books

[6751]
ADAMS, C.K. **Master handbook of microprocessor chips.** Blue Ridge Summit, Pa., TAB Books, 1981. 378p. illus., diagrs. $16.95. ISBN: 0830696334; 0830612998, pbk.

A compilation of data and operating details, instruction sets and functional descriptions of 10 families of 4- and 8-bit microprocessors and certain support chips. With the advent of 16- and 32-bit microprocessors this work will have diminishing value, although many personal computers and hobby applications will continue to use these earlier chips. *Class No:* 681.31-181.4(083)

[6752]
MONEY, S.A. **Microprocessor data book.** 2nd ed. Oxford, Blackwell Scientific, 1990. iv,316p. tables. ISBN: 0632020938.

Provides condensed data for most of the currently available microprocessor and microcomputer devices, with details of internal architecture, instruction set, electrical data and package details. 3 groups: 4-bit, 8-bit and 16-bit types. Also includes chapters covering input-output and other interfaces, peripheral control and support devices. Contains a directory of manufacturers (p.305-9) and glossary of terms. Useful as an aid to selecting appropriate device, to be followed up by consultation of manufacturers' data sheets and manuals. 'This is a useful volume for a wide range of practitioners' (*Electronics and Power,* v.29, Feb. 1983, p.181, of the first ed.). *Class No:* 681.31-181.4(083)

Patents

[6753]
CORNILLIE, O.A.R. **Microprocessors.** Oxford, Pergamon, Infoline, Inc., 1985. xi, 450p. diagrs., circuits. £69. (*EPO Applied technology series, v.8.*) ISBN: 008030575x.

Aims to give an overview of the state of the art in the field of microprocessors and illustrate their utilization and application by illustrated summaries of patents from 5 European countries and the US. Of the 430 patents, 233 are US and 73 UK. 6 sections: 1. Program control - 2. Digital control - 3. Electrical motor control - 4. Medical applications - 5. Measuring instruments and treatment of data - 6. General. Although descriptive content is considerable for many entries, for some it is negligible or non-existent, e.g., 'sewing machine control'. Indexes of patents, patentees and subjects. Bibliography p.415-7. *Class No:* 681.31-181.4(088.8)

Microprocessor Applications

[6754]

DAYASENA, P.J. *and* **DEIGHTON, S.**, *eds*. **Microprocessor applications in electrical engineering:** a bibliography (1977-1978). London, Institution of Electrical Engineers, 1980. [134]p. (*Microprocessor application series*.) ISBN: 0852964536.

Almost 400 references, with abstracts - often merely indicative - drawn from the INSPEC database and grouped under 7 headings: Communication systems (178 citations) - Power systems - Control systems - Instrumentation - Measurement - Automatic test equipment - Computer-aided design. Author and subject indexes. *Class No:* 681.31-181.4.004

[6755]

Dictionary of CAD. Clarke, J. *and* Mayer, T., *eds*. Oxford, Butterworth Heinemann, 1991. 484p. illus. £40. ISBN: 0750612398.

About 2,500 a terms, most defined briefly.

Class No: 681.31-181.4.004

[6756]

MAYNE, K.D. *and* **PACHE, J.E.**, *eds*. **Microprocessor applications:** bibliography. London, Institution of Electrical Engineers, 1975. 25p. (*Microprocessor application series*.) ISBN: 0852964323.

77 references, with informative abstracts. Coverage: control applications in industry and power stations; communications; measurement and instrumentation; avionics; medicine. Author index.

Updated (same title and editors), with 182 references, in *Euromicro newsletter*, v.2(4), October 1976, p.72-100.

Class No: 681.31-181.4.004

[6757]

STOUT, D.F., *ed*. **Microprocessor applications handbook.** New York, etc., McGraw-Hill, 1982. xii, 448p. illus., diagrs., graphs. $35. ISBN: 0070617988.

16 contributors. 17 chapters. 1. Survey of microprocessor technology ... 4. Microprocessor-controlled color TV receiver ... 7. Programmable videogames ... 9. Microcomputer applications in telephony ... 11. Digital filters utilizing microprocessors ... 14. Voice recognition ... 17. Multiple microprocessors in small systems. Limited to the following microprocessors: 1802, 2650, 6800, 8080 and the following chips: 2920, 3870, 3872, 6801, 6802 and 8088.

'Focuses on immediately useful applications ... This authoritative resource examines design concepts, schematics, software and hardware for a wide range of microprocessor applications' (*SMPTE journal*, v.91, Mar 1982, p.306).

Class No: 681.31-181.4.004

Computer Software

[6758]

Buyer's guide to micro software (soft). Weston, Conn., Online, Inc.

Directory of business and professional microcomputer software available in the US. Records include directory information, technical specifications, description, citations to reviews. About 7000 records. Available on DIALOG, file 237. *Class No:* 681.31.0

[6759]

Microcomputer software guide. New York, Bowker.

Covers software and hardware, but excludes educational software. Data: ordering information, technical specification, subject classification, brief description. Over 18,000 records; monthly reloads. Available on DIALOG, file 278. *Class No:* 681.31.0

[6760]

The Software encyclopedia, 1990. New York, etc., Bowker, 1990. 2v. (2552p.). $190, the set. 500 First published 1985. ISBN: 0835227626.

V.1: Title index; V.2: arrangement by system hardware (*e.g.* IBM, Unix). Within each system, software is arranged by subject and specific applications. Publisher index. Covers all subject areas, including engineering, science and medicine, but not software for schools (see Bowker's *Software for schools*).

The most up-to-date information appears in such online databases at the Microcomputer software guide (DIALOG file 278). *Class No:* 681.31.0

[6761]

Software engineer's reference book. McDermid, J.A., *ed*. Boston, Mass., Butterworth, 1990. v.p. illus., diagrs. $195. ISBN: 0750610409.

Stresses concepts and broad approaches. Extensive bibliography. 'Written for practitioners in the field, but appropriate for anyone with an understanding of computer science principles (*New technical books*, September/October, 1991, item 1179). *Class No:* 681.31.0

[6762]

Software reliability handbook. Rook, P. London, Elsevier Applied Science, 1990. xix,542p. illus., diagrs., graphs, tables. £87. ISBN: 1851664009.

12 contributors. 12 chapters, closely subdivided: 1. Reliability issues ... 4. Fault tolerance - 5. Defect detection and correction ... 9. Quality management ... 12. Engineering environments. 4 appendices (D. software development cost models, p.487-517). References, p.519-32; detailed, analytical index, p.533-42. *Class No:* 681.31.0

[6763]

The Software users yearbook 1992. Maher, A., *ed*. 7th ed. London, VNU, 1991. 4v. ISBN: 0862711274, set. ISSN: 02686708.

V.1. *Software suppliers and services.*

V.2. *Systems software.*

V.3. *Industry specific software.*

V.4. *General applications software.*

Includes 2,600 suppliers and 10,500 products. Entry for a program includes title, description, hardware requirements. Index of products by category, manufacturer, operating system. *Class No:* 681.31.0

[6764]

Text retrieval: a directory of software. Kimberley, R., *ed*. 2nd ed. Aldershot, Hants., Gower, 1987. v.p. Loose-leaf. ISBN: 0566053721.

At head of title-page: 'Institute of Information Scientists, Southern Branch'. First published 1983.

The directory: Packages featured - List of other packages. Appended check list; bibliography. Index of packages by operating system. *Class No:* 681.31.0

Patents

[6765]

HANNEMAN, H.W.A.M. The Patentability of computer software: an international guide to the protection of computer-related inventions. Deventer, Kluwer, 1985. 257p. £36.75. ISBN: 9065442251.

After chapters on general principles of protection of software and on the workings of a computer there are chapters on the situation in the USA, Canada, UK, Australia, France, Federal Republic of Germany, Austria, Switzerland, Japan, the Netherlands and in other countries and under the European Patent Convention. *Class No:* 681.31.0(088.8)

[6766]

—BRETT, H. *and* PERRY, L., *eds.* The Legal protection of computer software. Oxford, ESC Publishing, 1981. xiv, 197p. £15.

Describes the protection available in Europe and US under the patent systems and by means of copyright and other laws. Proposals for protection of computer programs are discussed, and the World Intellectual Property Organization model provisions are given. *Class No:* 681.31.0(088.8)

[6767]

MILLARD, C.J. Legal protection of computer programs and data. London, Sweet and Maxwell; Toronto, Carswell, 1985. xxvi, 239p. £24. ISBN: 0421334800.

The legal protection of software and data in the US, UK, Canada and other common law countries is discussed. Contains tables of cases and statutes.
Class No: 681.31.0(088.8)

[6768]

—NIBLETT, B. Legal protection of computer programs. London, Oyez, 1980. xii, 155p. ISBN: 0851205097.

Describes the protection available for software by patents, as well as by copyright laws, trademarks, and breach of confidence, in the UK, with a short section on overseas rights. *Class No:* 681.31.0(088.8)

Laws

[6769]

BAINBRIDGE, D. Software copyright law. London, Pitman, 1992. 256p. illus. £45. ISBN: 0273038478.

Overview of European software copyright law. Covers computer programs, work stored and created, limits of the scope of legal protection. *Class No:* 681.31.0(094.1)

Security

[6770]

BAKER, R.H. The Computer security handbook. 2nd ed. Blue Ridge Summit, Pa., TAB Books, 1991. xx,416p. 35.80.

First published 1985.

Chapters include: Built-in security; Securing the network; The technology of passwords; Securing remote devices; Principles of network security; A system of network security. Index. *Class No:* 681.31.004.4

[6771]

Encyclopedia of data protection. Chalton, S. *and* Gaskill, S., *eds.* London, Sweet and Maxwell, 1988-. iv. Loose-leaf. £165.

Regular updates are planned. *Class No:* 681.31.004.4

[6772]

FLAHERTY, D.H., *ed.* **Privacy and data protection:** an international bibliography. London, Mansell, 1985. xxvi, 276p. £23.50. ISBN: 0720117194.

1,862 references to works, published and unpublished, on the need to protect individuals from misuse of computer-stored data concerning those individuals. 6 sections (4. Legal aspects of privacy and data protection: Canada, Federal Republic of Germany, France, Sweden, UK, US, International ... 6. Selected bibliographic materials, p.241-7). Author index. A valuable research tool.
Class No: 681.31.004.4

[6773]

—MacCAFFERTY, M., *comp.* Computer security. London, Aslib, 1976. [2], vii, 91p. £7 (£6 to members). ISBN: 0851426885.

492 annotated entries, international in scope. Sections: General - Administration and management - Privacy (nos.237-317) - Software - File organization - Data handling - Applications - Peripheral supplies - Late additions.
Class No: 681.31.004.4

[6774]

HIGHLAND, J. Computer virus handbook. Oxford, Elsevier, 1990. xvi,375p. illus., tables. £85. ISBN: 0946395462.

Definitions and descriptions of common viruses and technical evaluations of some 22 anti-virus software packages. Numerous references. *Class No:* 681.31.004.4

[6775]

JACKSON, K.M., *and others.* **Computer security reference book.** Oxford, Butterworth Heinemann, 1992. 800p. £90. ISBN: 0750603577.

In-depth treatment of the entire field, including computer crime, data protection, EFTPOS schemes, evaluation of security products, hacking, public key cryptography, and other topical aspects. *Class No:* 681.31.004.4

[6776]

MADSEN, W. Handbook of personal data protection: national and international laws, standards and guidlines. London, Macmillan, 1992. 1250p. illus., diagrs. £85. ISBN: 0333569202.

Identifies differences among individual countries' legislation. Includes copies of national & international laws. *Class No:* 681.31.004.4

Automation Technology

Abbreviations & Symbols

[6777]

BRITISH STANDARDS INSTITUTION. Graphical symbols for components of servo-mechanisms. London, BSI, 1960-64. 2v. diagrs. £28.90. (*BS 3238: 1960-1964.*)

Pt.1: *Transducers and magnetic amplifiers.* 1960. 16p.
Pt.2: *General servo-mechanisms. 1964.* 32p.

Almost 100 symbols illustrated and annotated.
Class No: 681.5(003)

Bibliographies

[6778]

HAMILTON, L., *comp*. **Information pack including bibliography on advanced manufacturing technology.** London, Information and Library Service, Institution of Mechanical Engineers, 1986. vi,98p. £5. (*Information pack, 1*.) ISBN: 0852986041.

2 parts: 1. AMT information (abbreviations; contacts; products; open learning; abstracts and indexes, etc.), p. 1-45; 2. References (236 items) with abstracts (key reports; general articles; applications of AMT), p.47-98. Most from the mid-1980s. Good value. *Class No:* 681.5(01)

[6779]

HAMILTON, L. *and* HALLAM, L., *comps*. **Information pack on computer integrated manufacturing.** London, Information and Library Service, Institution of Mechanical Engineers, 1987. vi,89p. £12.50. (*Information pack, 4*.) ISBN: 0852986459.

2 Parts: 1. CIM information (abbreviations - standards), p.1-60; 2. CIM references (108 items) with abstracts, p.61-81. Indexes: author, subject, company name. A useful series. *Class No:* 681.5(01)

[6780]

TZAFESTAS, S.G. 'Information sources in systems and control'. *Systems and control encyclopedia.* Singh, M.G., *ed*. Oxford, Pergamon, 1987. v.8, p.5575-85.

Brief survey covering: guides, sourcebooks and services; encyclopedias and handbooks; dictionaries and glossaries; abstract and review sources; technical journals; international book series and conference proceedings; sources on people; other information sources; continuing education courses. *Class No:* 681.5(01)

Encyclopaedias

[6781]

GRAHAM, G.A. **Encyclopedia of industrial automation.** Harlow, Longman and Society of Manufacturing Engineers, 1988. [xvi],597p. illus. £48. ISBN: 058203566x.

US title: *Automation encyclopedia.*

About 700 entries, A-Z arrangement. Some definitions are very brief (*e.g.* 'comparator', 2 lines); others lengthy (*e.g.* 'numerical control', 31p.). Cross-references. Covers technologies, processes, techniques and applications. *Class No:* 681.5(031)

[6782]

SINGH, M.G., *ed*. **Systems and control encyclopedia:** theory, technology, applications. Oxford, etc., Pergamon, 1987. 8v. (5500p.). diagrs., graphs, tables. £1,600. ISBN: 0080287093, set.

29 subject editors; *c*.1,000 contributors. Truly encyclopaedic treatment with entries ranging from 2 or 3 lines to several pages, supported by illustrations, graphs, and bibliographies, all signed and arranged A-Z in v.1-7. V.8 contains a systematic outline of the *Encyclopedia,* alphabetical list of contents, list of contributors (from all continents), author citation index, subject index (p.5432-73), a list of acronyms and abbreviations and a guide to information sources (p.5575-85). 'Rarely is a major new encyclopedia published; this is one ... highly recommended' (*Choice,* April 1988, p.1226).

The publishers have produced 'Concise' editions, with

.... *(contd.)*

revisions, of numerous of the subjects covered in the main work in a series *Advances in systems control and information engineering. Class No:* 681.5(031)

[6783]

—Concise encylcopedia of modelling and simulation. Atherton, D.P., *ed*. Oxford, Pergamon Press, 1992. 554p. £140. (*Advances in systems control and information engineering, 5*.) ISBN: 008036201x.

172 newly commissioned and revised articles from the *Systems and control encyclopedia* (1987). Topics covered include model representation, simplification, validation, reduction techniques and algorithms, simulation, etc. *Class No:* 681.5(031)

Handbooks & Manuals

[6784]

Computer integrated manufacturing handbook. Teicholz, E. *and* Orr, J.N., *eds*. New York, McGraw-Hill, 1987. 434p. illus., diagrs., tables. ISBN: 0070477744.

21 contributors. 20 chapters in 4 parts: 1. Introduction - 2. Components - 3. Planning - 4. Implementation and management. Annotated bibliography of 125 items. Index. *Class No:* 681.5(035)

[6785]

CONSIDINE, D.M., *ed*. **Standard handbook of industrial automation.** London, etc., Chapman and Hall, 1987. xix, 460p. illus., diagrs., graphs, tables. £89. (*Advanced industrial technology series*.) ISBN: 0412008319.

Over 70 contributors. A practical handbook concentrating on the implementation of automation: designing, specifying, using, evaluating equipment and systems. 5 sections: 1. Backdrop to automation systems (including annotated brief glossary) - 2. Sensors and measuring systems - 3. Control systems - 4. Actuators and materials (with much emphasis on robots) - 5. Interfaces and communications (including Local Area Networks and communication standards). Ample illustration (445 diagrams and half-tones) and 57 tables complement an excellent index as aids to use of a wide-ranging work.

'... the Handbook represents the best we have come to expect from the American technical publishing houses. The editors and their 73 contributors are to be congratulated on their endeavors' (*Production engineeers,* v.66(8), Sept. 1987, p.9). *Class No:* 681.5(035)

Dictionaries

[6786]

BRITISH STANDARDS INSTITUTION. **Glossary of terms used in automatic controlling and regulating systems.** London, BSI, 1967. 40p. diagrs., graphs, tables. £23.90. (*BS 1523: 1967*.)

Pt.1: *Process and kinetic control,* 1967. 40p. Basic definitions, classification, elements and characteristics of control systems, etc. About 180 terms defined in systematic order with alphabetical index. *Class No:* 681.5(038)

[6787]
BROADBENT, D.T. *and* MASUBUCHI, M., *eds*.
Multilingual glossary of automatic control technology:
English, French-German-Russian-Italian-Spanish-Japanese.
New ed. Oxford, etc., Pergamon Press, for the International
Federation of Automatic Control, 1981. 230p. $45. ISBN:
0080276075.

Previous ed. as *IFAC: Multilingual dictionary of automatic
control terminology* (1967).

Adds Japanese equivalents to the 1967 ed. 1,100-odd
terms are arranged into 10 sections (*e.g.* 0. Basic concepts
... 3. Elements of closed-loop control systems), each with
ten sub-sections (*e.g.* 31. Non-linear elements). Entries are
uniquely numbered and presented in 7 columns, English
first. Indexes in each language lead to entries.
Class No: 681.5(038)

[6788]
CLASON, W.E. **Elsevier's dictionary of measurement and
control** in six languages: English/American, French,
Spanish, Italian, Dutch and German. Amsterdam, etc.,
Elsevier, 1977. 886p. £121.62. ISBN: 0444415823.

Almost 7,800 entries for terms in English, with
equivalents in other languages. Alphabetical indexes in other
languages refer to numbered entries in main section. Brief
bibliography (p.885-6). Lack of Russian terminology a
disadvantage. *Class No:* 681.5(038)

[6789]
JUNGE, H. D. **Technische Kybernetik:** Grundlagen und
Anwendungen, Englisch-Deutsch, Deutsch-Englisch.
Wiesbaden, Brandstetter; VEB Technik, 1983. 564p. DM85.
ISBN: 3870971142.

24,000 terms, covering all aspects of automatic control,
with equivalents in the second language, in each half of the
dictionary. *Class No:* 681.5(038)

Reviews & Abstracts

[6790]
**Computer and control abstracts. Science abstracts. Section
C.** London, INSPEC, 1969-. Monthly. ISSN: 00368113.

Formerly *Control abstracts*, 1966-68.

About 70,000 abstracts pa. Classes, closely subdivided:
00:00 General and management topics - 10:00 Systems and
control theory - 30:00 Control technology - 40:00 Numerical
analysis and theoretical, computer topics - 50:00 Computer
hardware - 60:00 Computer software - 70:00 Computer
applications. Author index per issue; Subject index, twice
yearly; Subsidiary indexes, where appropriate; Bibliography
index (for well-documented articles); Book index (for books
received and abstracted); Corporate author index;
Conference index. Indexes cumulated 2pa. and 4-yearly (*e.g.*
1977-1980). INSPEC database. *Class No:* 681.5(048)

[6791]
Current papers on computers and control. Stevenage,
Herts., Institution of Electrical Engineers; New York,
Institute of Electrical and Electronic Engineers, 1974-.
Monthly. 123pa. ISSN: 00113794.

About 65,000 references pa. Main classes: 00.00 General
and management topics - 10.00 Systems and control theory -
40.00 Numerical analysis and theoretical computer topics -
50.00 Computer hardware - 60.00 Computer software -
70.00 Computer applications. A current awareness service;
companion to *Computer and control abstracts*. Online
INSPEC. *Class No:* 681.5(048)

[6792]
**Referativnyĭ zhurnal. 01. Avtomatika i vychislitel'naya
tekhnika.** Akademiya Nauk SSSR. Moscow, VINITI, 1987-.
Monthly. R.299-40. ISSN: 02024098.

Formerly: *Referativnyĭ zhurnal avtomatika telemekhanika*
(1963-1986).

3 parts: A. Automation and remote control - B. Computer
science and computer hardware - V. Software. About 26,000
indicative and informative abstracts pa. (Part A contains
almost 8,000), covering mainly periodicals worldwide and
patents. Issue indexes of authors and patents. Annual author
and subject indexes. English contents list. UDC numbers.
Class No: 681.5(048)

Progress Reports

[6793]
Control and dynamic systems: advances in theory and
applications. Orlando, Florida, etc., Academic Press, 1972-.
v.9-. Annual. graphs. ISSN: 00905267.

Originally as *Advances in control systems: theory and
applications*, v.1-8, 1964-71.

V.1-11 consisted of diverse contributions on significant
topics within the field. From v.12 onwards each volume has
contained lengthy articles written around a theme and, in the
case of v.19-21 (1983-84) and v.22-24 (1985-86), the themes
have been developed over 3 volumes (v.22-24:
Decentralized, (distributed control and dynamic systems. 24
contributions). *Class No:* 681.5(055)

Yearbooks & Directories

[6794]
Who's who in automation. Letcombe Regis, Oxon., Templar
Publications, 1987. Annual. 335p. £25. ISBN: 1870503007.

Lists of 1,600 UK companies specializing in 10 areas of
automation (Programmable controllers - Robotics -
Computer-integrated manufacture - Transducers - Process
control - Flexible manufacturing systems ... Data acquisition
and storage). Includes advertisements and diary.
Class No: 681.5(058)

Histories

[6795]
BENNETT, S. **A History of control engineering, 1800-1930.**
London, P. Peregrinus, 1979. x, 214p. (*IEE Control
engineering series, 8.*) ISBN: 0906048079; 0863410472,
Pbk (1986).

Continues *The origins of feedback control*, by O. Mayr
(MIT Press, 1970), which ends with the invention of the
governor around 1800. Deals with the regulation of prime
movers and development of steam and hydraulic
servomechanisms and their use in the control of ships and,
subsequently, aircraft. Later chapters cover electrical control
systems and circuit analysis. Chapter references,
bibliography and index give good support.
Class No: 681.5(091)

Cybernetic Technology

Artificial Intelligence. Expert Systems

Databases

[6796]

Supertech. New York, Bowker.

5 subfiles covering biotechnology, artificial intelligence, CAD/CAM robotics and telecommunications. Corresponds to the monthly journals *Artificial intelligence abstracts, CAD/CAM abstracts* and *Robotics abstracts*. Over 86,000 records; updated monthly. Available on DIALOG, file 238. Also on CD-ROM. *Class No:* 681.51:007.510(003.4)

Bibliographies

[6797]

Expert systems: a bibliography. London, Institution of Electrical Engineers, Library, 1983-86. 2v. ISBN: 085296272x; 0852963386.

1. *Expert systems: a bibliography,* by C.J. Biggar and J.W. Coupland. 1983. 74p. £12.

2. *Expert systems II: a bibliography on expert systems in electrical and electronic engineering,* by S.L. Penfold and J.W. Coupland. 1986. 103p.

V.1 carries 121 references from the INSPEC database, 1969-83, with author and keywords indexes. V.2 provides 194 references from the INSPEC database of English-language items since 1983. 6 sections. Author and keyword indexes. *Class No:* 681.51:007.510(01)

[6798]

GOMERSALL, A. Machine intelligence: an international bibliography, with abstracts of sensors in automated manufacturing. Bedford, IFS Publications; Berlin, New York, etc., Springer Verlag, 1984. vii, 232p. illus. £32. ISBN: 090360860x.

More than 1,300 items, with abstracts, covering journal and conference papers, 1972-84, extending and updating the sensors section in the author's *Robotics bibliography* (1981). The present work complements A. Gomersall and P. Fariner's *Robotics: an international bibliography* (*qv*). Emphasis (over 500 references) on vision sensors, plus tactile sensors; also, almost 500 items on industry and process applications. Classified arrangement, chronolgical within each section. Author index only. 'The abstracts, being truly informative in their own right, make this bibliography a must for all library bookshelves' (*Aircraft engineering,* v.56, April 1984, p.32). *Class No:* 681.51:007.510(01)

[6799]

—**BODEN, M.A.** Artificial intelligence and natural man. London, MIT Press, in conjunction with the British Psychological Society, 1987. 576p. ISBN: 0262022591.

Features an extensive bibliography, p.500-28. *Class No:* 681.51:007.510(01)

[6800]

—**SMITH, L.C.** 'A Guide to information sources in artificial intelligence'. *Science and technology libraries,* v.5(3), Spring, 1985. p.79-100.

Defines the scope and outlines development of AI before describing and discussing journals, conference proceedings, reports, theses, books, databases and abstracts covering the field. Appendix lists monographs discussed. *Class No:* 681.51:007.510(01)

[6801]

HANCOX, P.J., *and others*. Keyguide to information sources in artificial intelligence/expert systems. London, Mansell; Lawrence, KS, Ergosyst Associates, 1990. xii,300p. tables. ISBN: 0720120071, UK; 0916313182, US.

3 parts. I. Survey of artificial intelligence/expert systems and its literature (5 documented chapters, p.3-112). II. Bibliography (659 items): general sources, p.1133-55; component specialisms (automated theorem proving - robotics), p.155-84; applications (agriculture-space), p.184-212; directories and related sources, p.212-21. III. Organizations (includes research centres, online database hosts; international in scope), p.225-77. Detailed, analytical index, p.279-300. An authoritative survey. *Class No:* 681.51:007.510(01)

Encyclopaedias

[6802]

SHAPIRO, S.C., *ed*. Encyclopedia of artificial intelligence. New York, etc., Wiley, 1987. 2v. (1246p.). illus., diagrs., graphs, tables. £166.

More than 200 contributors. 260 documented articles (A-Z) cover all aspects of artificial intelligence and its applications over a wide field, embracing the sciences, arts and technologies, medicine, psychology and social issues. Over 5,000 literature references (*e.g.* 'Robot control systems', p.902-23. 72 references). Many cross-references, illus. and tables. List of abbreviations. 53-page index. *Class No:* 681.51:007.510(031)

Dictionaries

[6803]

HUNT, V.D. Artificial intelligence and expert systems sourcebook. New York, London, Chapman & Hall, 1986. xi, 315p. diagrs., tables. £42. (*Advanced industrial technology series.*) ISBN: 0412012111.

An illustrated dictionary of definitions (p.41-277), many of them extensive and including acronyms and system names. Preceding it: a review of developments in artificial intelligence and expert systems. Appended: a brief directory of organizations, manufacturers and consultants; a list of acronyms; a bibliography of *c.*250 references and a short trademarks list. No index. Has more to offer than does Rosenberg's *Directory of artificial intelligence.* *Class No:* 681.51:007.510(038)

[6804]

—**BAINS, W. *and* RAGGETT, J.** A Glossary of artificial intelligence. London, Kogan Page, 1988. 200p. £14.95. ISBN: 1850915733.

'Aims to give clear meanings and explanations of the AI jargon in a simple, informal manner' (*British book news,* September 1988, p.657). *Class No:* 681.51:007.510(038)

[6805]

—ROSENBERG, J.M. Dictionary of artificial intelligence and robotics. New York, etc. Wiley, 1986. xi, 203p. ISBN: 0471849820.

Explains 4,000 terms, symbols, acronyms and abbreviations. Aims 'to specify the relationships among robotics, artificial intelligence and computer control terms' (*Preface*). Definitions are frequently at different levels of complexity, to assist the lay person as well as the professional. *Class No:* 681.51:007.510(038)

Polyglot

[6806]

VOLLNHALS, O. **A Multilingual dictionary of artificial intelligence:** English, German, French, Spanish, Italian. London, Routledge, 1992. vii,423p. £75. ISBN: 0415074657.

About 3,500 English-base terms with equivalents and indexes in the other languages. Genders given.
Class No: 681.51:007.510(038)=00

English

[6807]

Artificial intelligence terminology: a reference guide. Beardon, C., *ed.* Chichester, Ellis Horwood; New York, Halsted Press, 1989. xiii,283p. ISBN: 0745807186; 0745807631, Pbk; 0470216018, Halsted Press.

18 contributors. About 1,000 terms, A-Z, briefly defined (*e.g.* 'Horn clause': 7 lines). Most terms carry 'see also' terms and many include a bibliographic reference.
Class No: 681.51:007.510(038)=20

[6808]

RAGGETT, J. *and* BAINS, W. **Artificial intelligence from A to Z.** London, Chapman & Hall, 1992. x,246p. illus. ISBN: 0412379503, UK; 0442312008, US.

About 200 AI terms defined in non-technical language: Abduction - Zero Sum. Definitions range 6 lines to 2p. Simple diagrams serve to illustrate points; some cross-references. Appendices: 'Benchmark'; AI programs; Languages and environments. Index, 6p. For the layperson and likely to be warmly welcomed.
Class No: 681.51:007.510(038)=20

[6809]

SMITH, R. **Collins dictionary of artificial intelligence.** London, Collins, 1990. [vi],374p. diagrs. £5.95. ISBN: 0004343662.

First published 1989 as *The Facts on File Dictionary of artificial intelligence.*

About 2,000 terms, briefly defined (*e.g.* 'problem space': 9 lines). Cross references. Covers expert systems, problem solving, speech synthesis, robotics and knowledge representation. *Class No:* 681.51:007.510(038)=20

[6810]

THRO, E. **The Artificial intelligence dictionary.** San Marcos, Ca., Microtrend Books, 1991. xi,407p. diagrs. $24.94; £22.95. ISBN: 0915391368.

About 1,300 terms on AI and related fields, *e.g.* expert systems, robotics, data structures, etc., some with 'further reading'. Appendices, A-L (G. 'Research centres' (44 organizations), p.385-87).
Class No: 681.51:007.510(038)=20

Reviews & Abstracts

[6811]

The Turing Institute abstracts in artificial intelligence. London, Springer Verlag, 1986-. 6pa. £168pa.

Lists material selected from the Turing Institute collection and includes research reports, journal articles, conference papers and some books. Subject arrangement. About 400 entries per issue. Author, title indexes per issue.
Class No: 681.51:007.510(048)

Robots

Bibliographies

[6812]

GOMERSALL, A. *and* FARMER, P. **Robotics:** an international bibliography, with abstracts. Bedford, IFS, 1984. vii, 209p. illus. £36. ISBN: 0903608839.

Updates *Robotics bibliography, 1970-1981* (with its 1,800 references, spread thinly over the entire field). This update has 1,200 references covering 1982-1984, with more emphasis on industrial applications, management and socio-economic aspects. (Separate volume, *Machine intelligence*, for sensors, computer vision and artificial intelligence). Covers conference papers and reports, as well as periodical articles. Grouped entries (*e.g.* 'Design and performance'; 'Process applications'), helpfully annotated. Author index, but no subject index. *Class No:* 681.51:007.52(01)

[6813]

LANE, D. *and* ACLAND, J., *eds.* **Industrial robotics: a bibliography.** London, Institution of Electrical Engineers, 1981. [140]p. £17.

226 citations, in two sections: A. Introduction, reviews; B. Specific aspects and applications. Author and subject indexes. Items provide a selection of significant conference and periodical papers, 1960s-1980, with brief abstracts, mainly descriptive and drawn from the technical press. Languages other than English are normally stated.
Class No: 681.51:007.52(01)

[6814]

—DAVIES, Z. Guide to information sources on industrial robotics. 3rd ed. Coventry, Warws., Coventry (Lanchester) Polytechnic, the Library, 1985. 41p.

First published 1981; 2nd ed. 1983.

Systematically covers all relevant types of published information sources held (*e.g.* handbooks, journals, reviews), and indicates external non-print sources. Separate chapters on general and industrial robotics literature, and robotics in mechanical and control literature. Intended primarily for students and staff of Lanchester Polytechnic.
Class No: 681.51:007.52(01)

[6815]

—NATIONAL ENGINEERING LABORATORY, East Kilbride. Industrial robotics: a worldwide literature survey. East Kilbride, Glasgow, the Laboratory, 1980. 2v. tables.

Devotes v.2 (*Applications; R & D; bibliography.* 180p.) to documented reviews of developments and applications worldwide, with tables of statistics. 5 areas: West Germany, US, Japan, Western Europe (less West Germany and UK), and USSR and Eastern Europe. A classified bibliography of applications lists over 400 references.
Class No: 681.51:007.52(01)

[6816]

Robotics, CAD/CAM market place, 1985. New York, etc., Bowker, 1985. xix, 242p. £59. ISBN: 0835218201.

Title also confusingly listed as *Robotics: a worldwide guide to information sources,* with the same bibliographical details, including ISBN.

Over 4,000 items, in 13 sections (Books: 1,300 annotated entries, under 285 subject headings; Conference proceedings: 300 entries, under name of conference; Serials: more than 500, with scope notes; Associations: *c.*180; Educational and research institutes: over 1,100; Online databases: 17; Products directory: 370, under name; Manufacturers). International in scope.

'A very comprehensive guide to the field of robotics, computer-assisted design and computer-assisted manufacturing that allows the reader to find relevant data very quickly and efficiently ... As with any other printed book, however, this one already contains some out-of-date information' (*IEEE. Design and test of computers,* v.3, February 1986, p.95). *Class No:* 681.51:007.52(01)

Encyclopaedias

[6817]

International encyclopedia of robotics: applications and automation. Dorf, R.C., *ed.* New York, etc., Wiley, 1988. 3v. illus., diagrs., graphs, tables.

More than 250 contributors worldwide, About 250 lengthy articles (averaging 10p. and upwards of 50,000 words), with extensive bibliographies. Articles are arranged alphabetically under broad headings (sometimes inverted, but not consistently, *e.g.* 'Mining, Robots in', but 'Robotics in agriculture'). Subjects frequently overalp (*e.g.* 'Apparel industry, Robots in'; 'Garment and shoe industry, Robots in'). Some headings are imprecise and unsought (*e.g.* 'Issues in robotics'). At least one article, 'Medicine intelligence' does not itself use the term in the text, which is a review of artificial intelligence. There is much 'futurism', - overtly in 'Future applications', 'Factory of the future', etc., and inherently in many other artices. As a reference work, this *Encyclopedia* obviously suffers from a scatter of its subject content that can only be partly offset by a comprehensive index, v.3. *Class No:* 681.51:007.52(031)

[6818]

—Concise international encyclopedia of robotics: applications and automation. Dorf, R.C., *ed.* New York, etc., Wiley, 1990. xix,1190p. illus., diagrs., graphs, tables. $100. ISBN: 0471516988.

Treats 206 topics and includes shortened versions, 'updated where necessary', (*Preface*) of all the entries in the 3v. *International encyclopedia of robotics.*
Class No: 681.51:007.52(031)

[6819]

TVER, D.F. *and* **BOLZ, R.W. Robotics sourcebook and dictionary.** New York, Industrial Press, 1983. vii, 258p. illus., diagrs. $32.50. ISBN: 0831111526.

Dictionary of types of industrial robot (with brief introduction); dictionary of applications (with descriptions). Glossary of robotics and computer-controlled terminology. List of *c.*80 US firms, with descriptions of selected products. '... this manual provides easy reference to the various robot types, applications, manufacturers and current market specifications' (*Robotics today,* v.8, December 1983, p.78). *Class No:* 681.51:007.52(031)

Handbooks & Manuals

[6820]

HODGES, B. *and* **HALLAM, P. Industrial robotics.** London, Heinemann, Newnes, 1990. 193p. illus., diagrs. $35. ISBN: 0434907820.

Material element of robotics - Drive systems - Robot sensors - Cost justification of robotics - Robot applications - The future of robotics. Chapter problems are included.
Class No: 681.51:007.52(035)

[6821]

HUNT, V.D. Industrial robotics handbook. New York, Industrial Press, 1983. xiii, 432p. illus., diagrs., tables. $32.50.

10 introductory chapters on all aspects of robotics (*e.g.* 3. Sensor systems; 4. Tooling; 7. Robotic control systems; 10. Socioeconomic impact of robotics; 11: Specifications and illus. of 80 robotic systems from 25 manufacturers; 12: Brief surveys of developments in 8 major user-countries: 14: Forecasts of experts). Useful glossary (p.391-400). Bibliography, in 22 subject groups, p.401-17. Concise index. 'This rounded compilation of matter of diverse source and nature fills a volume of unexceptional size and weight, neatly produced' (*Scientific American,* v.249, October 1983, p.52F). *Class No:* 681.51:007.52(035)

[6822]

HUNT, V.D. Robotics sourcebook. New York, Elsevier, 1988. 321p. illus., diagrs. $46.50. ISBN: 0444012982.

Focuses on the technology and industry of robotics, whereas the *International encyclopedia of robotics* addresses all areas of robotics. A book for 'both the management and the engineering student' (*Science and technology libraries,* v.11(1), Fall 1990, p.133). *Class No:* 681.51:007.52(035)

[6823]

MILLER, R. Industrial robot handbook. New York, Van Nostrand Reinhold, 1989. 686p. illus., diagrs., tables. $82.95. ISBN: 0442237332.

A reprint of the 1987 Fairmont Press ed.

79 chapters, including case studies. 2 appendices: Robot manufacturers; Machine vision manufacturers.
Class No: 681.51:007.52(035)

[6824]

NOF, S.Y., *ed.* **Handbook of industrial robotics.** New York, etc., Wiley, 1985. xviii, 1358p. illus., diagrs., graphs, tables. £108. ISBN: 0471896845.

93 contributors, worldwide. 13 parts (77 documented chapters), arranged logically, from the development of industrial robots, through their mechanical design, control and intelligence systems, their implementation in society, to their applications (6 major areas: fabrication and processing; welding; material handling; assembly; inspection, quality control and repair; finishing, coatings and painting). Extensive glossary and index. 5 appendices, on robot markets, relevant organizations, manufacturers and journals. 'Professional engineers, managers and supervisors should ... find it to be a valuable aid for improving product quality and design as well as cutting safety hazards and labour costs' (*Automotive engineering,* v.93, November 1988, p.102); '... a major comprehensive reference' (*Chartered mechanical engineer,* v.32, December 1985, p.53).
Class No: 681.51:007.52(035)

[6825]

Robot technology. London, Kogan Page, 1983-86. 8v. in 9.

1. *Modelling and control,* by P. Coiffet. 1983. ISBN: 0850385334. 156p. £25.

2. *Interaction with the environment,* by P. Coiffet. 1983. ISBN: 0850385342. 240p. £39.

3. *Teleoperation and robotics,* by J. Vertut and P. Coiffet. 1985. 2 parts. ISBNs: 0850385881; 0850389623. 332p; 256p. £65.

4. *Robot components and systems,* by F. Lhote and others. 1984. ISBN: 0850386497. 346p. £30.

5. *Logic and programming,* by M. Parent and C. Laurgeau. 1984. ISBN: 0850386500. 190p. £48.10.

6. *Decision and intelligence,* by J. Aleksander, H. Farrey and M. Ghallab. 1986. ISBN: 0850386519. 203p. £45.95.

7. *Performance and computer-aided design,* by A. Liegeois. 1985. ISBN: 0850386527. 268p. £44.90.

8. *Indexes and bibliography.* 1986. ISBN: 0850386977. 105p. £20.

'The series must constitute an essential element of any technical library worth its salt ... [and] ... will form an essential guide and reference source for all those in industry or research involved with robots and the related technology' (*Production engineer,* v.65(9) Oct. 1986, p.11).
Class No: 681.51:007.52(035)

[6826]

TOEPPERWEIN, L.L., *and others*. **Robotics applications for industry:** a practical guide. Park Ridge, N.J., Noyes, 1983. x, 326p. diagrs. £46.15. ISBN: 0815509626.

An 'introduction to the basic concepts and techniques pertaining to the use of robotics technology for manufacturing and industrial applications' (*Foreword*), with special a section devoted to robotics in motor vehicle manufacture. Contains an extensive bibliography (about 300 references) classified under 21 headings (*e.g.* Manipulator design; Sensor technology and applications). Also a valuable glossary of terms (p.141-76), grouping almost 400 definitions under 14 headings. No index.
Class No: 681.51:007.52(035)

Dictionaries

[6827]

BÜRGER, E. *and* KORZAK, G., *comps*. **Dictionary of robot technology** in four languages: English, German, French, Russian. Amsterdam, etc., Elsevier, 1986. 276p. DFl.175. ISBN: 0444995196.

Over 6,100 English-language terms and phrases, cross-references and abbreviations/acronyms, identified by initial letters and consecutive numbers, with equivalents and indexes in the other three languages. No definitions. Letter-by-letter alphabetization. *Class No:* 681.51:007.52(038)

[6828]

WALDMAN, H. **Dictionary of robotics.** New York, Macmillan; London, Collier-Macmillan, 1985. viii, [1], 303p. illus. ISBN: 0029485304.

Concise, readable definitions of over 2,000 terms in robot systems, actions and hardware. Sequence: AA series CNC robots ... Zoom. 'Spot welding': ½p. Includes brief biographies. For students first exploring the subject, as well as for the researcher, the librarian and the business person.
Class No: 681.51:007.52(038)

[6829]

—RIA robotics dictionary. Dearborn, Mich., Robot Institute of America Publications, 1974. 75p.

Briefly defines over 550 terms on robotics and related fields (*e.g.* computing), reflecting contemporary usage. Includes abbreviations and acronyms. 'It is the intent of RIA to update this glossary on a periodic basis' (*Introduction*).
Class No: 681.51:007.52(038)

[6830]

—SMITH, B.M., *and others*. A Glossary of terms for robotics. Washington, National Bureau of Standards, 1981. 90p. $21. (*NBSIR-81-2340. Available from NTIS PB82-251216.*)

Prepared for Air Force Materials Laboratory, Wright-Patterson Air Force Base, Ohio, 1980.

Intended as an interim draft for a definitive glossary (although definitions are not claimed to represent the standards of any offical body). Aims to include the 'most troublesome, most used and most important' terms within the subject, excluding basic terms and those from related areas such as mathematics, computing, electrical engineering. Defines *c*.350 terms, in 14 subject categories (1. Types of manipulators ... 8. Dynamics and control ... 14. Economic anlayses). A-Z index. *Class No:* 681.51:007.52(038)

Reviews & Abstracts

[6831]

Referativnyĭ zhurnal. 37. Promyshlennye roboty i manipuliatory. Moscow, VINITI, 1983-. Monthly. diagrs. ISSN: 0208001x.

About 4,500 signed abstracts pa. covering industrial robots and manipulators. 9 sections (e.g., General - Theory, investigation and design of robots and manipulators - Construction and technical data - Robot and manipulator drives - Robot and manipulator control ... Applications ...). Illustrated with diagrams. UDC numbers. Annual subject index. English contents summary.
Class No: 681.51:007.52(048)

[6832]

Robotics abstracts. New York, Bowker, 1983-. Monthly. $670pa. ISSN: 07481624.

More than 1,600 abstracts pa., with world coverage of periodicals, reports, patents and conference papers. Available online via DIALOG (Supertech; File 238); CD-ROM. *Class No:* 681.51:007.52(048)

Progress Reports

[6833]

Developments in robotics. Rooks, B., *ed.* Bedford, IFS Publication, 1983-. illus. diagrs. ISBN: 0903608332.

Papers in the first volume (iv, 265p. £30) are grouped: applications; design and systems; sensors; drives, control and simulation. Intended to be an annual selection of previously unpublished papers emphasizing the latest developments and research in robotics.
Class No: 681.51:007.52(055)

[6834]
Recent advances in robotics. New York, etc., Wiley, 1985-. illus., diagrs., tables. (*Advances in robotics.*) ISBN: 0471883832.

The first volume attempts a general orientation and has 3 parts: applications; mechanics; sensors. Brief index of topics. The series is designed annually to provide reviews of all important aspects of robotics. Broad surveys at elementary and advanced research levels, and more specialized articles on significant new techniques will be included. In the interests of timely publication, individual volumes will probably lack homogeneity in subject matter. 'Incorporates compact, authoritative and up-to-date reviews of major events in robotic science and technology' (*IEEE Journal of robotics and automation,* v.RA-I, December 1985, p.215). *Class No:* 681.51:007.52(055)

Yearbooks & Directories
[6835]
International robotics industry directory. 4th ed. Conroe, Texas, Technical Database Corpn., 1985. 392p. illus., diagrs. $45. ISSN: 0278159x.

Full details of 267 industrial robots marketed by 97 firms, mainly US, plus a sprinkling from the UK and Japan. Data on each robot: applications, types of sensor used, performance characteristics, price range; followed by one-page entries for each model, with illustration, company, marketing, technical and service information, plus manufacturer's comments. Further sections concern actuators, controllers, distributors, consultants, and research institutes. Glossary and index of products. All information derives from a Computerized Manufacturing Online Database, updated daily and offered by the publishers. *Class No:* 681.51:007.52(058)

[6836]
UK robots industry directory, 1984/5, and members handbook. 2nd ed. Bedford, British Robot Association; Sutton, Surrey, Industrial Press, 1984. 476p. + 24p. illus. diagrs., tables. £50. ISBN: 0903608774.

Lists research activities, education and training facilities, books and journals, conferences and exhibtions, robot associations, members of the British Robot Association, and over 60 manufacturers of robotic systems. Tabulates *c.*180 different applications (*e.g.* welding, cutting, assembly, forging, handling), plus a 2-p. illustrated specification for each system. A 24-p. supplement: *A human guide to robots.* Overlaps with *Industrial robot specifications,* by A Cugy and J. Simpson (*qv*), but the data given complement rather than coincide with each other. *Class No:* 681.51:007.52(058)

[6837]
The World yearbook of robotics research and development. Scott, P., *ed.* 2nd ed. London, Kogan Page, 1986. 583p. £45. ISBN: 1850911061.

Previous ed. 1985. Supersedes *International robotics yearbook* (Kogan Page, 1983).

Parts 1-2 contain commissioned articles on recent advances and information on recent trends and programmes in 14 countries. Part 3 gives details of personnel and research activities of 350 groups in 23 countries (only 4 are 'Eastern bloc' countries). USA and UK account for half of the entries. Indexes of establishments, researchers, and research topics. Bibliography p.575-83.

Praised in *Production engineer* (v.65, March 1986) as 'a

....(contd.)
most useful reference source of who is doing what and where', but the production engineer contemplating purchase of a robot would not find 'a current assessment of commercially available robots, their comparative performances and relative costs ...'. *Class No:* 681.51:007.52(058)

[6838]
—ROBOT INSTITUTE OF AMERICA. **Worldwide robotics survey and directory.** Dearborn, Mich., Robot Institute of American Publications, Annual-.

The 3rd ed. (1985, 79p., £41) surveys developments in robot manufacture and use in 19 countries (13 European, plus Australia, Canada, Japan, Korea and the US). Tables and charts: robot population; installations; estimated growths; production and monetary value; price ranges. These precede brief trade directory of 294 manufacturers, and fuller details of 43 member-firms of the Robot Institute of America. *Class No:* 681.51:007.52(058)

Teaching Materials
[6839]
PANTELIDIS, V.S. Robotics in education: an information guide. Metuchen, N.J., Scarecrow Press, 1991. 435p. $47.50. ISBN: 0810824663.

A bibliography of more than 1,500 items on robotics suitable for use in primary and secondary schools and for undergraduate education. 9 sections: 1. Reference sources - 2. General information - 3. Education robots - 4. Non-educational robots - 5. Social implications - 6. Economic implications - 7. Workplace impact - 8. Physical implications - 9. Nonprint media. Index. 'An amazing resource ... highly recommended' (*Choice,* September 1992, p.88). *Class No:* 681.51:007.52(072)

Trade Names, Trade Marks, Brand Names
[6840]
CUGY, A. *and* **SIMPSON, J. Industrial robot specifications.** Rev. ed. London, Kogan Page, 1984. 357p. diagrs., tables. £25. ISBN: 0850387698.

First published 1983.

Detailed specifications on over 300 robots manufactured and distributed under licence throughout Europe, covering output of over 90 companies. Part 1: an index of industrial applications in 3 groups (manufacturing and assembly robots; automated guided vehicles; vision systems) - Part 2: Technical specifications, grouped as in Part 1, and A-Z by firm - Part 3: Directory of manufacturers and distributors. Overlaps with *UK robots industry directory, 1984/8* (*qv*), but the data given complement rather than coincide with each other. *Class No:* 681.51:007.52(088.7)

Patents
[6841]
LAMMINEUR, P. *and* **CORNILLIE, O. Industrial robots.** Oxford, etc., Pergamon; E.E.C., 1984. viii, 154p. diagrs. £30. (*EPO Applied technology series, v.2.*) ISBN: 0080311431.

A classified bibliography of 170 patents on construction and applications of industrial robots (*c.*40 patents are Japanese; *c.*40, US; the rest from European countries). Detailed hierarchical classification (*e.g.* Part 1.

....(contd.)

Construction. Chapter 1.1: The manipulator arm; Section 1.1.1: Cartesian type; 1.1.1.2: with a suspended base; 1.1.1.2.1: Gantry-type). Each patent has a technical summary, plus diagram. A list of cited patents includes 'references to corresponding patents in other countries'. A-Z list of patentees and subject index.
Class No: 681.51:007.52(088.8)

Typewriters

[6842]
BEECHING, W.A. Century of the typewriter. London, Heinemann, 1974. xii, 276p. illus., tables.

A survey from the beginnings up to 1973, in 8 chapters. Chapters 3-5. History of typewriter manufacture - 6. Electric and special-purpose typewriters ... 8. A compendium of typewriter history (in tabular form). Bibliography, p.266. Index. *Class No:* 681.612

Printing Machinery

Printers Devices

[6843]
McKERROW, R.B. Printers' and publishers' devices in England and Scotland, 1485-1640. London, Bibliographical Society, 1913. liv, 216p.

428 printers devices, p.1-160. Excludes devices of foreign printers for the English market. Introduction includes 'List of some books referred to' and 'Collections and libraries'. 5 indexes (1. Devices and compartments according to size; 2. Printers, booksellers, etc.). About 100p. of facsimiles. *Class No:* 681.626

[6844]
SILVESTRE, L.C. Marques typographiques; ou, Recueil des monogrammes, chiffres, ensignes, emblèmes, devises, rébus et fleurons des libraires et imprimeurs qui ont exercé en France, depuis l'introduction de l'imprimerie, en 1470, jusqu'à la fin du seizième siècle ... Paris, Renou et Maulde, 1867. 2v. facsims.

Published anonymously, parts 1-13.

Records 1,310 typographical devices used by booksellers and printers operating in France, 1470-1599; also devices used by booksellers and printers of French books published outside France, during the same period. *Class No:* 681.626

Sound & Video Recordings

Bibliographies

[6845]
KELLY, P., *comp.* **Audio and video.** London, Library Association, Public Libraries Group, 1983. 25p. ISBN: 0863700024.

169 numbered entries (1-149: Books; briefly annotated;

....(contd.)

150-66: Periodicals). 'Audio' (nos. 1-94) covers historical background, basic general texts, construction, etc. 'Video' provides directories, home video, video film production, construction and repairs, games and videotext. Author index to each half. *Class No:* 681.8(01)

Encyclopaedias & Dictionaries

[6846]
LANGMAN, L. *and* **MOLINARI, J.A. The New video encyclopedia.** New York, Garland, 1990. xv,312p. illus., diagrs. ISBN: 0824082443.

About 2,400 brief entries (*e.g.* image sensor, ½ column). *Class No:* 681.8(03)

Handbooks & Manuals

[6847]
Audio electronics reference book. Sinclair, I.R., *ed.* Oxford, BSP. 1989. xv,615p. illus., diagrs., graphs. £79.50. ISBN: 0632019298.

11 contributors; 20 chapters subdivided: 1. Sound waves - 2. Stereo ... 6. Compact disc ... 9. The LP record ... 15. Loudspeakers ... 19. In-car audio. A practical approach. *Class No:* 681.8(035)

[6848]
BENSON, K.B., *ed*. **Audio engineering handbook.** New York, etc., McGraw-Hill, 1988. xxii,1040p. illus., diagrs., graphs, tables. £85. ISBN: 0070047774.

33 contributors. 17 chapters, most with references, beginning with principles of sound and hearing (1); digital audio (4); amplification (6); sound reproduction systems and techniques (7-13); post-production and noise reduction systems (14-15); tests and standards (16-17). The chapter on standards concentrates on the US. 21p. index. Extensively illustrated. 'Both practical and theoretical ... a major work in its field' (*Choice,* April 1989, p.1360, which found this more comprehensive than G. Ballou's *Handbook for sound engineers: The new audio cyclopedia,* 1987). *Class No:* 681.8(035)

[6849]
BORWICK, J. 'The Gramophone' guide to hi-fi. Newton Abbot, Devon, David & Charles, 1982. 256p. ISBN: 0715382314.

15 chapters (2. The evolution of sound recording ... 5. How records are made ... 7. The record-player - 8. The tape deck - 9. The tuner - 10. The amplifier - 11. Loudspeakers and headphones ... 15. Future developments). 3 appendices (2. Suggested further readings; 3. Glossary of audio terms, p.241-53). Index, p.254-6. Author is audio editor of 'The Gramophone'. *Class No:* 681.8(035)

[6850]
ROBINSON, J.F. Videotape recording: theory and practice. 3rd ed., revised by S. Lowe. London, Focal Press, 1981. 362p. illus., diagrs., graphs, tables.

First published 1975.

17 chapters, each with references. Glossary, p.354-8. Detailed, analytical index. 'For the reader with a basic engineering knowledge and some experience in television' (*Preface*). *Class No:* 681.8(035)

[6851]
WEZEL, R. van. **Video handbook.** 2nd ed. London, Heinemann, 1987. viii, 455p. £25. ISBN: 0434921890.
First published Newnes, 1981. *Class No:* 681.8(035)

[6852]
—JACKSON, K.G. Newnes book of video. London, Newnes Technical Books, 1980. 128p. illus., drawings, tables. ISBN: 0408004784.
9 contributors. Sections (some with 'Further reading'): The Video centre - TV services - Tape recorders - Disc players - Cameras - Programmes with a single camera - Antennas - Cable distribution - Teletext and viewdata - TV games and computers - Security and surveillance. Directory of manufacturers and suppliers. Index. Index to advertisers. *Class No:* 681.8(035)

[6853]
The 'Which?' guide to hi-fi. London, Consumers' Association, and Hodder & Stoughton, 1981. 230p. ISBN: 0340266295.
11 chapters (*i.e.,* 1. Buying hi-fi - 2. All about sound - 3. Hi-fi fundamentals - 4. Loudspeakers - 5. Amplifiers - 6. Record decks and cartridges - 7. Radio tuners and FM aerials - 8. Cassettes decks and tapes - 9. Reel-to-reel tape decks - 10. Headphones - 11. Microphones). Detailed, analytical index. Oblong format. *Class No:* 681.8(035)

Dictionaries

[6854]
CLASON, W.E., *comp.* **Elsevier's dictionary of television and video-recording in six languages:** English/American-French-Spanish-Italian-Dutch-German. Amsterdam, etc., Elsevier, 1975. [viii], 608p. £102.10. ISBN: 0444412247.
Based on *Elsevier's Dictionary of television, radar and antennas* (1955), this polyglot version excludes terms on radar and antennas (other than for television); these will be found in *Elsevier's Telecommunications dictionary* (*qv*).
5,300 terms and phases are listed, numbered and defined in English, with equivalents in French, Spanish, Italian, Dutch and German. For each foreign language there is an A-Z index referring to the English section via the reference numbers. *Class No:* 681.8(038)

[6855]
WHITE, G.D. **The Audio dictionary.** Seattle, Wa., Univ. of Washington Press, 1988. x,291p. illus. $30; $15. ISBN: 0295965274; 0295965282, Pbk.
About 1,300 entries with clear definitions of about 1 paragraph. Cross references. Appendices: impedance; audio measurement; digital audio. Brief bibliography. 'Readable and informative' ('Reference books bulletin', *Booklist,* June 15, 1988, p.1715). *Class No:* 681.8(038)

Yearbooks & Directories

[6856]
Audio video market place, 1989: a multimedia guide. New York, etc., Bowker, 1989. Annual. 800p. £60. ISBN: 0835224732.
First published 1984, continuing *Audiovisual market place,* 1969-83.
A directory of over 4,500 firms that make, supply or distribute audio video equipment. 3 indexes, of companies, applications, and products and services. *Class No:* 681.8(058)

Optical Discs

[6857]
HELGERSON, L.W. **CD-ROM:** facilitating electronic publishing. New York, Van Nostrand Reinhold; London, Chapman & Hall, 1992. xxix,528p. diagrs. ISBN: 0442005237.
28 chapters in 6 parts: I. Marketing - II. Product design - III. Production - IV. The Data - V. Now it begins - VI. Case studies. Appendix: Vendors of CD-ROM products and services (chiefly US). Glossary of CD-ROM terms, p.461-517. Detailed, analytical index, p.519-28. A practical treatment. *Class No:* 681.845

[6858]
International imaging source book including micrographics and optional imaging: 1992. Badler, M.M., *ed.* Larchmont, N.Y., Microfilm Publishing, 1992. 436p. $127.50. ISBN: 091741408x.
Directory-type information on electronic imaging sources and products, micrographics dealers, imaging software, CD-ROM and service bureaux. Data include manufacturer, model numbers, prices. List of consultants; index to associations. Bibliography. US bias. *Class No:* 681.845

[6859]
Optical disk storage document image processing systems: a guide and directory. 2nd ed. Hatfield, Herts., Cimtech, 1990. 86p. £25. ISBN: 0852672772.
Information for the acquisition of systems to store and manipulate unstructured documents as digital images. Data: optical disk units, document image processing systems, data storage, computer output laser disk systems. Glossary; bibliography; address list. Covers UK availability only. *Class No:* 681.845

[6860]
Rewritable optical storage technology. Roth, J.P. Westport, Ct., Meckler, 1991. x,172p. illus., diagrs. £28. ISBN: 0887365345.
Covers technology and applications in this rapidly developing area. Includes a directory of companies and organizations (p.24-56). 'Recommended readings' (p.142-56), 'glossary of terms and acronyms' (p.157-67). *Class No:* 681.845

[6861]
SHERMAN, C., *ed.* **The CD-ROM handbook.** New York, etc., McGraw-Hill, 1988. xxiii,510p. illus., diagrs., tables. ISBN: 0070565783.
One of the first reference works produced in this rapidly developing field. 17 chapters in 5 parts: A. Introduction - B. Format - C. Advanced formats (Digital Video Interactive, WORM, etc.) - D. Information storage and retrieval - E. Creating a CD-ROM: the process. Some of the material (*e.g.* discography, the search for standards) is obviously dated, but the technical discussion (extensions, mastering process, etc.) is excellent and clearly presented. 'Highly recommended' (*Choice,* May 1989, p.1552). *Class No:* 681.845

Ironmongery & Hardware

[6862]

Benn's Hardware & DIY buyers' guide, 1986. Tonbridge, Kent, Benn, 1985. v.p. £22. ISBN: 0863820301.

4 main sections: A. Manufacturers and suppliers (61p.) - B. Buyers' guide (131p.) - C. Trade names of manufacturers', suppliers' products - D. Wholesalers. Appended list of trade and staff associations. Index to advertisers. *Class No:* 683

[6863]

CLASON, W.E. Elsevier's dictionary of tools and ironware, in English/American, French, Spanish, Italian, Dutch and German. Amsterdam, etc., Elsevier, 1982. viii, 298p. Dfl. 185. ISBN: 0444420851.

2,578 English-base terms, with equivalents and indexes in the other 5 languages. Terms are categorized. *Class No:* 683

Locksmithing

[6864]

AMATT, L.K. Locks and lockmaking: an annotated bibliography. London, Library Association, 1973. [i], 392p.

Thesis submitted for Fellowship of the Library Association.

6 sections: A. General historical and descriptive works - B. History of locks and keys up to and including the eighteenth century - C. The age of invention in the nineteenth century - D. Security in the twentieth century - E. Practical locksmithing - F. Manufacture. 5 appendices (1. List of patent abridgement volumes, 1774-1965; 4. Notes on museum collections of locks and keys in Great Britain; 5. Annotated list of ten English and American journals). Author, title and subject indexes. *Class No:* 683.3

Furniture

[6865]

BRISTOW, P., *comp.* **Furniture literature:** a select bibliography on furniture and allied subjects. Stevenage, Herts., Furniture Industry Research Association, 1975-.

Furniture literature, 1986-1988 has *c.*2,000 entries, some with brief annotations. Sections: Adhesives - Computers - Ergonomics - Fastening - Finishing - Furniture - Hardware - Health hazards & safety - Machine tools - Hand tools - Management and organization - Mechanical handling - Packaging - Plastics - Robotics - Textiles - Timber - Woodworking - Miscellaneous. British Standards relating to furniture. 'This volume covers all relevant literature added to the stock of FIRA Library, between January 1986 and June 1988'. Includes books, pamphlets, journal articles and reports, but omits most trade literature. *Class No:* 684

[6866]

Directory to the furnishing trade, 1990. Tonbridge, Kent, Benn, 1990. xiv,496p. ISBN: 0863820840. ISSN: 00706604.

Directory-type information for the UK and Eire. Sections: Manufacturers and wholesalers - Agents and distributors - Buyers' guide - Trademarks and brand names - Geographical guide - Business information - Index to advertisers. *Class No:* 684

[6867]

JOYCE, E. The Technique of furniture making. 4th ed., revised by A. Peters. London, Batsford, 1987. [viii], 519p. illus. (incl. col.). ISBN: 071344407x.

First published 1970.

10 Parts: 1. Basic material - 2. Tools and equipment - 3. Basic techniques and joint construction - 4. Advanced areas of furniture construction - 5. Metal fittings, fasteners, and their application - 6. Advanced techniques - 7. Running a professional workshop - 8. Draughtsmanship and workshop geometry - 9. Furniture design and constructional details - 10. Restoration, repairs and wood finishing. Appendix: Costing and estimating. Detailed, analytical index, p.515-9. Profusely illustrated (307 photos, 288 line-drwings). Well produced and organized. *Class No:* 684

[6868]

LONDON COLLEGE OF FURNITURE LIBRARY, *comp.* **Periodicals list, 1988.** London, College of Furniture (A Faculty of the City of London Polytechnic), 1988. [iii], 18l.

Current titles taken by the Commercial Road Library features as both a classified list (14 headings, including 'Indexes & addresses') and an alphabetical list (*c.*500 titles, giving frequency and housing). Index to subjects precedes. The Tarrance Street Library 'Alphabetical list of current titles' has an appended list of Braille publications. Wide-ranging. *Class No:* 684

[6869]

UNITED NATIONS. Industrial Development Organization. **Information sources on the furniture and joinery industry.** Rev. ed. New York, United Nations. (*UNIDO Guides to information sources, no.4.*)

First published 1972.

698 items, in 12 sections: 1. Professional, trade and research organizations, learned societies and special information sources - 2. Directories - 3. Sources of statistics, marketing and other economic data - 4. Basic handbooks, textbooks and manuals - 5. Monograph series - 6. Current periodicals - 7. Current abstracting and indexing periodicals - 8. Proceedings, papers and reports of the United Nations and related bodies - 9. Specialized dictionaries and encyclopaedias - 10. Bibliographies - 11. Films and film catalogues - 12. Other potential sources of information (*e.g.* Fairs and exhibitions). *Class No:* 684

[6870]

'Which?' way to repair and restore furniture. London, Consumers' Association, and Hodder & Stoughton, 1980. 180p. illus. ISBN: 034025050x.

5 sections: 1. Starting (tools; sharpening chisels) -2. Structural repairs - 3. Cleaning, stripping and refinishing - 4. Upholstery (p.75-117) - 5. Materials and supplies (including list of specialist suppliers in UK.). Oblong format. Step-by-step approach. *Class No:* 684

Bedding & Upholstery

[6871]
BRITISH STANDARDS INSTITUTION. Glossary of terms for fillings for bedding, upholstery and other domestic articles. London, BSI, 1966. 32p. £33.60. (*BS 2005: 1966.*)
About 200 entries. Sections: General; Raw materials; Fillings made from natural and man-made fibres, rubber and plastics; Processes and processing; Adventitious inclusions. Appendix on Rag Flock Act and Regulations. Index.
Class No: 684.7

Footwear Boots & Shoes (Manufactures)

[6872]
Footwear digest international. Kettering, Northants., SATRA Footwear Technology Centre, 1984-. 6pa. illus., diagrs.
Formerly *Footwear digest,* 1972-83.
About 200 abstracts per issue. Sections: World industry (news items; *c.*60% of total abstracts) - Company news - Shoe machinery - Shoe materials - Product profiles - Market information (on a particular country, *e.g.* Thailand). No indexes. *Class No:* 685.3

[6873]
RAMA, L. Vocabulaire technique de l'industrie de la chaussure. [Footwear industry technical vocabulary ... Français-English-Deutsch-Italiano-Español.] Paris, OECD, 1969. iv, 396p.
5 sequences; French-German-English-Spanish-Italian, with definitions in French (p.3-125); English-French-German-Italian-Spanish, with definitions in English (p.127-263); etc. *Class No:* 685.3

[6874]
Shoe trades directory, 1988. 9th ed. London, Shoe and Leather News, 1988. 244p. ISBN: 0901138142.
Index to suppliers, p.19-87; Product guide: to footwear machinery suppliers, to manufacturers equipment suppliers, to suppliers of footwear materials, components and accessories. Buyers guide to suppliers of services. Appended: trade names and trade marks; glossary; shoe sizes; guide to shoe care. Product guide to suppliers of footwear leathers. *Class No:* 685.3

Bookbinding

[6875]
BRENNI, V.J., *comp.* **Bookbinding: a guide to the literature.** Westport, Conn., Greenwood Press, 1982. viii, 199, [1]p. £31.75. ISBN: 0313237182.
12 sections, 1,525 references, mostly periodical articles. 1. Reference works ... 3. Materials ... 5. Decoration - 6. History of bookbinding from ancient times to the present ... 8. Care and repair of bindings ... 10. Audiovisual aids - 11. Bookplates - 12. Book jackets. 3 Checklists of titles, - for university/college/medium-sized public library. List of binders, binding designers, and binding decorators, p.139-160. Book collectors. Glossary. Author index; detailed analytical subject index. *Class No:* 686

[6876]
JOHNSON, A.W. The Thames and Hudson manual of bookbinding. London, Thames & Hudson, 1978. 224p. illus. (incl. col. pl.). £7.95.
8 chapters. 1. A history of English bookbinding decoration - 2. Equipment - 3. Bookbinder's materials ... 5. Working procedures - 6. Binding styles - 7. Finishing ... 9. Changes in bookbinding construction (*e.g.* adhesive binding) - 10. Design. Glossary, p.211-8. 'Further reading', p.219. Analytical index. 258 clear line-drawings; 8p. of colour plates. 'This is the best manual of bookbinding to have been produced for around thirty years and it will, I believe, become the standard guide to all aspects of craft bookbinding' (*British book news,* June 1978, p.486).
Class No: 686

[6877]
ROBERTS, M.T. *and* **ETHERINGTON, D. Bookbinding and the conservation of books:** a dictionary of descriptive terminology. Washington, Library of Congress, 1982. 296p. illus. (incl. col. pl.). ISBN: 0313237182.
Defines *c.*3,000 terms and expressions, drawn from 373 listed sources. Coverage: bookbinding history, papermaking, some bibliographical terms and, particularly, the techniques, processes, equipment and materials of hand and machine binding, plus conservation. 'Provides an authoritative, comprehensive dictionary for bookbinders and conservators' (*Reference books bulletin, 1983-1984,* p.57). *Class No:* 686

Clothing

Bibliographies

[6878]
UNITED NATIONS. Industrial Development Organization. **Information sources on the clothing industry.** New York, United Nations, 1974. [xii], 116, [1]p. $4. (*UNIDO Guides to information sources, no. 12.*)
876 items, in 12 sections: 1. Addresses of professional, trade and research organizations, learned societies and special information services - 2. Directories - 3. Sources of statistics and economic data - 4. Basic textbooks and manuals - 5. Monograph series - 6. Current periodicals - 7. Current abstracting and indexing periodicals - 8. Proceedings, papers and reports - 9. Specialized dictionaries and encyclopaedias - 10. Bibliographies - 11. Films - 12. Other potential sources of information (e.g., marketing, exporting and purchasing). Includes information on training facilities. 'Bibliographic sources used'. Items in some sections are annotated. *Class No:* 687(01)

Dictionaries

[6879]
GIOELLO, D.A. *and* **BERKE, B. Fashion production terms.** New York, Fairchild Books and Visuals, 1979. 340p. illus. (*Language of fashion series.*) ISBN: 0870052004.
Grouped headings (*e.g.* pattern layout; design control), covering both tools and techniques, and garment construction from design to completion. Each term is defined with photograph or line-drawing. Bibliography and index.

.... (contd.)

'An outstanding reference work for students or professionals in all facets of fashion' (*Choice,* v.16(7), September 1979, p.798). *Class No:* 687(038)

[6880]

PICKEN, M.B. The Fashion dictionary: fabric, sewing and apparel as expressed in the language of fashion. Rev. and enl. ed. New York, Funk & Wagnalls, 1973. ISBN: 0308100522.

First published 1939 as *The language of fashion.* Rev. ed. 1957.

More than 10,000 words, briefly defined, with pronunciation, concerning wearing apparel and accessories, and their manufacture. 100 group-terms (*e.g.* 'Linens'; 'Heels'). Over 750 illus. of stitches, weaves, laces and garments, including 200 half-tones. Includes obsolete terms. Adequate cross-references. A standard work in its field. *Class No:* 687(038)

[6881]

REBMANN, G. Bekleidungs wörterbuch: Deutsch-Englisch-Französich-Italienisch. [Dictionary of garment technology: English-German-French-Italian.] 2nd ed. Berlin, Schiele & Schön, 1990. xiv,788p. ISBN: 379490494x.

First published 1984.

In four parts: German, English, French, Italian. Indicates which terms accord with the Nomenclature of Garments. Designation of the Association Européenne des Industries de l'Habillement (AEIH). Names and address of German manufacturers appended. *Class No:* 687(038)

Yearbooks & Directories

[6882]

The British clothing industry year book. 1990-91. Solihull, W. Midlands, Kemp Group (Printers & Publishers), Ltd., 1990. v.p. ISBN: 0862591678.

Sponsored by the British Knitting & Clothing Export Council (BKCEC).

15 sections: General information - Exhibitions - Index to classifications - Alphabetical section - Trade names - Menswear - Womenswear - Childrenswear - Fashion accessories - Fabrics and textiles - Trimmings - Machinery and ancillary equipment - Services - Buying offices, etc. Index to display advertisers. *Class No:* 687(058)

[6883]

—CLOTHING AND FOOTWEAR INSTITUTE. Yearbook and membership register. London, the Institute. Annual.

The 1984 *Yearbook* (204p.) Contains 19 topical articles; a classified guide to UK suppliers; an Institute membership list; an overseas membership list. Alphabetical index of advertisers. Index to advertisers' products and services. *Class No:* 687(058)

Equipment & Instruments

[6884]

HALE, F.G. The International history of the sewing machine. London, Hale, 1982. 320p. illus. ISBN: 0709198260.

14 sections (2. The precursors ... 5. The commercial age, 1880-1900. Early automatics - 6. Twentieth century development, 1900-1950 - 7. The addition of other motive power, 1950- ... 11. Electricity and the sewing machine ...

.... (contd.)

13. Biographical section, p.214-79). Glossary (*c.*700 entries). Bibliography, p.316. Detailed, non-analytical index. 654 illus. *Class No:* 687:002.50

Hairdressing

[6885]

CORSON, B. Fashions in hair: the first five thousand years. 4th impression, with revised suppt. London, P. Owen, 1980. 719p. illus.

First published 1964.

14 chapters. Main feature: over 3,000 drawings, with text facing. (*e.g.* 19th-century women: 16 illus.). Sources, p.696-704 (museums and libraries; books and periodicals). Supplement, 1965-1978, p.678-95. Detailed, analytical index, p.705-19. A valuable source. *Class No:* 687.53

[6886]

COX, J.S. An Illustrated dictionary of hairdressing & wigmaking. London, Batsford, 1985. 312p. illus. £25. ISBN: 0713442085.

First published 1966 (xxiii, 359p.).

Over 4,000 concise definitions of British and American terms (some of them archaic and obsolete). 1,156 black-and-white illus. Extensive bibliography. Careful documentation of sources. The 1966 volume was a standard reference text for costume specialists and theatrical costumers. *Class No:* 687.53

Buttons

[6887]

ALBERT, L.S. and KENT, K. The Complete button book. Garden City, N.Y., Doubleday, 1949. xix, 409p. illus.

2 parts: 1. Methods of manufacture (13 chapters) - 2. Illustrations, p.80-401, with captions facing. Appendix : 'List of American button makers and outfitters, from pre-Revolutionary times to the 1890s'. *Class No:* 688.2

[6888]

—SQUIRE, G. Buttons: a guide for collectors. London, F. Muller, 1972. xi, 200p. illus. (pl.).

Concentrates on uniform buttons (Air Force ... Ambulance Service ... Hunt: Foxhounds ... Shipping lines - Police, transport, fire brigade ...). Nearly 3,000 are displayed on 100 plates, with text facing. British slant. *Class No:* 688.2

Toys

[6889]

DAIKEN, L. World of toys: a guide to the principal and private collections in Great Britain of period toys, dolls and dolls' houses, games, puppets and marionettes, toy soldiers, musical boxes and automata; also modern toys, toymakers and toyshops, puppets and marionette troupes, the British toy industry and toy trade press. London, Lambarde Press, 1963. xvi, 269p. illus. (pl.).

....(contd.)

Ten short but highly informative chapters on a wide variety of topics, from dolls clubs and special exhibitions to hand-crafted toys and their makers, p.234-8. 'For further reading', p.241. Non-analytical index. *Class No:* 688.72

[6890]
The Good toy guide. Potters Bar, Herts., Toy Libraries Association, and Inter-Action Imprint, 1986. 232p. ISBN: 0904571408.

1983 ed. (217p.) has evaluative entries in 5 age groups (Baby play, 3 months - 2 years ... Fun and games (18 months - 10 + years)). Appended: 'Toys for people with special needs'. Suppliers. Manufacturers. Index, p.199-217. *Class No:* 688.72

[6891]
KING, C.E. The Encyclopedia of toys. London, Hale, 1978. 272p. illus. (mostly col.), facsims. ISBN: 0707917269.

6 sections: 1. Miniature living (doll's houses, etc.) - 2. Toys principally for pleasure (*e.g.* music toys; automata) - 3. Wheeled toys & children's transport (prams, etc.) - 4. Metal toys (*e.g.* model soldiers) - 5. Board and table games - 6. Educational toys and pastimes (*e.g.* toy theatre; doll's dressmaking). Bibliography, p.266-7. Index, p.268-72. Quarto. *Class No:* 688.72

[6892]
Toy trader: Official journal of the National Association of Toy Retailers. Directory, 1988. [New York], the Association, 1987.

8 sections: 1. Manufacturers and importers - 2. Toys classified - 3. Wholesalers and cash & carry, A-Z - 4. Retail and wholesale agents - 5. Brand names, products, services - 6. Character merchandise - 7. Suppliers to toy manufacturers - 8. Shopfitting and point of sale suppliers. Diary. *Class No:* 688.72

Dolls

[6893]
COLEMAN, D.S., *and others*. The Collector's encyclopedia of dolls. New York, Crown Publishers, 1968; London, Hale, 1970. vi, 697p. illus. (incl. col.), facsims.

Arranged A-Z, with entries for doll designers, trade names, etc. 'Chronology', p.138-55; 'Wooden dolls', p.656-66; 'Hands for dolls', p.280-2 (illus. 730-61). Many cross-references. Bibliography, in 6 sections, p.672-5. Non-analytical index; index of marks, p.689-97. 1,738 illus. and facsims.; 46 colour pl. *Class No:* 688.721

Amateur Hobbies & Crafts

Bibliographies

[6894]
LIBRARY ASSOCIATION. Public Libraries Group. **Handicrafts.** Compiled by the staff of the Kent County Library. 5th ed. London, the Association, [1975]. 43p. (*Readers' Guide no. 4.*)

About 400 unannotated entries, under subjects A-W (General - Basketry, Canework, rushwork and raffia work ... Block printing and lino cuts ... Flower making and

....(contd.)

pressing ... Knitting ... Leatherwork ... Needlework ... Plastic craft - Puppetry, marionette and model theatre ... Woodwork). *Class No:* 689(01)

[6895]
LOVELL, E.C. *and* HALL, R.M. Index to handicrafts, model-making and workshop projects. Boston, Mass., Faxon, 1936. 476p. Suppts. 1-5, 1943-75.

Largely concerned with material in books, but has some magazine references. *Fifth supplement* (1975. 650p.) covers 1968-73 and includes coverage of over 1,000 book titles and 15 periodicals not indexed in the *Reader's guide to periodical literature.* 'One of the most practical and useful indexes which has been published' (Barton, M., and Bell, M.V. *Reference books*, (7th ed., 1970, p.105). US slant. *Class No:* 689(01)

[6896]
SHIELDS, J.R. Make it: an index to projects and materials. Metuchen, N.J., Scarecrow Press, 1975. 485p. ISBN: 0810807728.

Indexes 475 how-to-do-it books. Part 1 concerns projects; part 2, materials. Coverage: weaving, plastics, ceramics, needlework, electronics, leather, metal, wood, and natural materials. *Class No:* 689(01)

[6897]
—Index to handicraft books, 1974-1984. Pittsburgh, Pa., Univ. of Pittsburgh, 1986. 413p.

Analyses *c.*1,000 know-how books, supplementing Shields (*qv*) and Lovell and Hall (*qv*). 'A valuable resource, even for libraries that do not own the Faxon series' (*Library journal,* July 1986, p.72). *Class No:* 689(01)

Encyclopaedias

[6898]
ANDREW, H.E.L. The Batsford encyclopaedia of crafts. London, Batsford, 1978. 431p. illus. (incl. colour).

As *The Arco encyclopaedia of crafts* (New York, Arco).

Covers 135 selected crafts, A-W, from Adhesives, Airbrush through Beads ... Casting ... Dyes ... Glass ... to Woodcuts. 'Table weaving', p.355-61, notes 12 items for further reading. Addresses of suppliers. 877 illus., with 27 in colour. 'Concentrates on the techniques of the various crafts as opposed to describing how to make a single item in a given craft' (*Introduction*).

1988 ed. 424p. 1101 illus. ISBN 0713462493. *Class No:* 689(031)

[6899]
TOLBERT, L., *ed*. The Encyclopedia of crafts. New York, Scribner, 1930. 3v. (1032p.). ISBN: 0684164094.

Encyclopedic dictionary of *c.*50 major crafts from all over the world, with emphasis on techniques, materials, and equipment, plus brief historical summaries. Shorter entries for definitions of terms and minor crafts. Cross-references; well-captioned illus. 'A valuable pioneer in the field of craft reference' (*Library journal,* 15 January 1987, p.136). *Class No:* 689(031)

Handbooks & Manuals

[6900]

DIAGRAM GROUP. Handbook of arts and crafts: the encyclopedia of the fine, decorative and applied arts. London, Harrap, 1981. 320p. illus. ISBN: 0245538232.

18 chapters: 1. Writing and drawing - 2. Printing - 3. Making books ... 6. Clay and pottery ... 9. Cabinet making ... 11. Working with glass - 12. Fine metalwork ... 15. Weaving ... 18. Leatherwork, 'Stained glass', p.188-91: 41 terms defined and illustrated. Bibliography, by subject groups, p.310-1. Detailed, analytical index, p.313-20. Over 4,000 illus. in all, for more than 2,000 tools, their evolution, history, function and manner of use. *Class No:* 689(035)

Reviews & Abstracts

[6901]

Clover index. Caldecote, Biggleswade, Beds., Clover Publications, 1975-. 4pa. £37 pa.

Title varies.

About *c.*3,000 entries in each issue, A-Z. Draws upon 90 popular science magazines such as *Amateur gardening, Flight, New Scientist, Wireless World* and *Your model railway*. Titles of articles are occasionally given brief explanatory notes. Cheaply produced from typescript. To some extent supplements *Current technology index* for public libraries. Layout is critically reviewed in *Library review* (v.28, Summer 1979, p.120-1). Annual cumulation. *Class No:* 689(048)

69 Building Industry

Building

Bibliographies

[6902]
BUILDING CENTRE, LONDON. **The Building Bookshop at the Building Centre, 1992 catalogue.** London, the Centre, 1992. [iv], 75p. illus. £1.

31 sections. Over 2,000 briefly annotated entries, with prices. Some major sections (*e.g.* Architecture; Building Services; Surveying) are subdivided. 'General reference', p.29-30. Title index of publications; author index. Some advertisements. *Class No:* 69.0(01)

[6903]
BUILDING RESEARCH ESTABLISHMENT. **BRE Information directory, 1987:** current publications, films, video and packages from the Building Research Establishment. Garston, Herts., BRE, [1987]. Annual. [ii], 97p.

9 parts (3. Classified list of BRE publications; 4. Serial list of BRE publications; 5. BRE films, video and packages; 6. Papers produced by BRE authors since 1983; 7. Building Research Advisory Service and courses, seminars and symposia; 8. Other sources of information (organizations)). Detailed, analytical index, p.93-97. *Class No:* 69.0(01)

[6904]
GREAT BRITAIN. HM Stationery Office. **Architecture and building catalogue.** London, HM Stationery Office, [1987]. [i], 40p. Gratis.

An annotated selection of *c.*200 new and recent books and pamphlets published by HMSO on architecture and building. 6 sections: Design, materials and construction - Housing and the environment - Safety - Statistics and reference - Regulations - Conservation and architectural history. Detailed index. Further official publications relevant to this field appear in HMSO Sectional lists 2, 3, 5, 22, 27, 50 and 61. *Class No:* 69.0(01)

[6905]
—GREAT BRITAIN. HM Stationery Office. Government publications. Sectional list 61, revised March 1986: Construction. London, HM Stationery Office, 1986. 15p. Gratis.

Brief, unannotated sections on publications produced by the Department of the Environment and 9 other relevant departments and offices. Index, p.15-17. The DOE's Building Research Establishment and Property Services Agency compile their own free lists of publications (*qv*). *Class No:* 69.0(01)

[6906]
—RIBA list of recommended books, 1990-91. Osley, J., *ed.* London, RIBA Publications, Ltd. 1990. [i], 16p. Gratis.

First published 1969. Earlier title *RIBA Book list.*

Over 700 entries in 17 sections, closely subdivided: 1. Generalities - 2. Architecture and art ... 6. Building: practice and procedure ... 10. Design for safety and security ... 13. Landscape ... 17. Statistics. New editions, new titles and out-of-print items are indicated. A valuable checklist; gives current prices, but not publishers. *Class No:* 69.0(01)

[6907]
PROPERTY SERVICES AGENCY. **PSA in print. 1987.** London, PSA, 1987. 39p. Gratis.

Departmental publications (*e.g. Current information in the construction industry*) - Departmental series (e.g., PSA Library translations) - HMSO publications. Title, author and subject indexes. *Class No:* 69.0(01)

Encyclopaedias

[6908]
BROOKS, H. **Encyclopedia of building and construction terms.** 2nd ed. Englewood Cliffs., N.J., Prentice-Hall, 1983. [iv], 443p. illus., diagrs., charts. ISBN: 0132755114.

First published 1976.

Defines more than 2,800 terms, some at length (*e.g.* 'Solar collectors': nearly 3p.). 'Index by function' (24 functions/subject areas) precedes, p.7-38: 178 illus. etc. Four appendices: A. Construction centres (US); D. Conversion factors. 'This is one of the best encyclopedias covering the terminology associated with the building industry ...' (*Reference books bulletin, 1984/1985* (1986), p.21). *Class No:* 69.0(031)

Handbooks & Manuals

[6909]
CHANDLER, I. **Building technology.** London, Mitchell, for Chartered Institute of Building. 3v. ea. £9.95. ISBN: 0713451785, v.1.

1. *Site organization and method.* 1988.
2. *Performance.* 1988.
3. *Design, production and maintenance.* 1987.

V.1 (176p.) has 5 parts, concentrating on production services and systems installation, and the design/construction interface. Part 5 deals with 3 case studies. 'References and further reading'. p.172-7. Index. *Class No:* 69.0(035)

[6910]
MERRITT, F.S., *ed.* **Building design and construction handbook.** 4th ed. New York, etc., McGraw-Hill, 1981. 1408p. illus. ISBN: 0070415218.

First published 1958.

Numerous documented sections, with the American civil engineer particuarly in mind (*e.g.* Protection of structures and occupants against hazards ... Sampling and testing of soils ... Structural steel construction ... Surveying for

....(contd.)

buildings ... Estimating building construction costs). Index. 662 illus. 'For the building and construction trade, this is the standard reference book' (*Reference books bulletin, 1984-1985* (1986), p.21). *Class No:* 69.0(035)

[6911]
SZOKOLAY, S.V. Environmental science handbook for architects and builders. Lancaster, Construction Press, 1980. [vii], 572p. diagrs., graphs, tables.

6 parts, each with appended short bibliographies: 1. Space - 2. Light - 3. Sound - 4. Heat - 5. Resources - 6. Synthesis. Analytical index. Attractively produced.
Class No: 69.0(035)

Dictionaries

Polyglot

[6912]
KORCHOMKIN, S.N. Dictionary of building and civil engineering: English, German, French, Dutch, Russian. London, etc., Kluwer, 1986. 935p. £81.75. ISBN: 2040157999.

14,000 English-base terms, with equivalents and indexes in German, French, Dutch and Russian.
Class No: 69.0(038)=00

[6913]
TEKNISKA NOMENKLATURCENTRALEN. Plan och byggtermer. [Glossary of planning and building terms.] Stockholm, the Centre, 1975. 198p. illus. (*TNC 58.*)

Replaces TNC 52 (Swedish monolingual glossary).

616 grouped Swedish entry-words (including compounds and expressions), plus Swedish definitions and English, French, German and Russian (transliterated) equivalents. Swedish, English, French, German and Russian (cyrillic) indexes. Issued by the Swedish Centre of Technical Terminology. *Class No:* 69.0(038)=00

[6914]
—VAN MANSUM, C.J. Elsevier's dictionary of building construction in English/American, French, Dutch, German. Amsterdam, etc., Elsevier, 1959 (5th reprint 1985). viii, 472p. DFl.250. ISBN: 0444406018.

5,328 English-base terms, with equivalents and indexes in French, Dutch and German. *Class No:* 69.0(038)=00

English

[6915]
BRETT, P. Building terminology: an illustrated reference guide for practitioners and students. Oxford, etc., Heinemann Newnes, 1989. [viii],339p. illus. £20. ISBN: 0434901768.

6 sections, each in alphabetical order: 1. Architectural style - 2. Building construction - 3. Documentation, administration and control - 4. General - 5. Materials and scientific principles - 6. Services and finishes. Index, p.321-39 (*c.*3,000 entries). Over 350 diagrams and photographs, so much superior in this respect to the *Macmillan dictionary of building* with its 12 appended pages of diagrams.
Class No: 69.0(038)=20

[6916]
HARRIS, C.M., *ed*. Dictionary of architecture and construction. New York, etc. McGraw-Hill, 1975. 576p. illus. £19.95 (1988 paperback ed.). ISBN: 0070267561.

52 contributing editors. Entries for *c.*15,000 terms and cross-references. Definitions run from 2 to 50 words. Scope extends to water supply, waste disposal and associated fields. A feature is the use of good marginal line-drawings, 2 or 3 per page. The editor, - Professor of Architecture, Graduate School of Architecture and Planning, Columbia Univ., New York. 'The most comprehensive one-volume compilation of architectural and construction terms in 70 years' (*Choice*, v.12(9), November 1975, p.1142).
Class No: 69.0(038)=20

[6917]
MARSH, P. Illustrated dictionary of building. London & New York, Construction Press, 1982. [v], 256p. illus. £16.95. ISBN: 0860958485.

About 4,000 brief entries, rarely over 10 lines in length. 'Subsidence': 19 lines. Many small line-drawings (*e.g.* 'Roof': 6 types). Adequate cross-references. Highly practical, aiming to be reasonably comprehensive and intended for professionals. *Class No:* 69.0(038)=20

[6918]
—McMULLAN, R. Macmillan dictionary of building. London, Macmillan, 1988. [iv], 262p. diagrs. ISBN: 0333424395.

Concisely defines *c.*5,000 words drawn 'from the written material used in current building practice' (*Preface*). Cross-references. Appended diagrams (that would be better placed in the text): brickwork (8 types); doors; heating; masonry tools; mouldings; roof; roof truss; stairs; timber joints; windows; woodworking tools. *Class No:* 69.0(038)=20

[6919]
—PROPERTY SERVICES AGENCY. Construction industry terms: a guide based on the structural principles of the 'Construction Industry Thesaurus' (CIT). London, PSA, 1979. x, 76p. £4.50.

Lists terms, in classified order, giving facet structure (materials; properties; end products). 'It functions as a glossary in commending the usage of terms including in them, or a dictionary, simply stating the range of usage' (*Introduction*). *Class No:* 69.0(038)=20

[6920]
PUTNAM, C.E. and CARLSON, G.E. Architectural and building trades dictionary. 3rd ed. Chicago, Ill., American Technical Society, 1974. [ii], 510p. illus.

First published 1950 (Burke, A.E., and others).

About 10,000 concise entries. Amply illustrated. (12 types of joint used in welding, p.481; 9 common types of hammer, p.228). Legal terms, p.492-500. Building material sizes (in inches only), p.501-10. US slanted.
Class No: 69.0(038)=20

[6921]
—CORKHILL, T., *comp*. A Concise building encyclopaedia: an explanation of words, terms and abbreviations used in building and constructional work, and a work of reference for architects, surveyors, civil and structural engineers and the various craftsmen engaged in building. 3rd ed. London, Pitman, 1951 (and reprints). v, 366p. illus. £11.95.

2nd ed. 1943.

A glossary of *c.*14,000 terms, with 1,200 small line-drawings. Entries vary in length from 1-12 lines.
Class No: 69.0(038)=20

[6922]

SCOTT, J.S., *ed*. **Dictionary of building**. 3rd rev. ed. London, Granada; Harmondsworth, Middx. Penguin Books, 1984. 383p. illus., diagrs. £12.50; £3.95. ISBN: 0246122641.

First published 1964 (Penguin Books).

Defines and categorizes over 5,000 building trade, tools and materials terms, with illus. ('Pitched roof': 6 types). US usage is differentiated. Cross-references to Penguin *Dictionary of civil engineering*. For both professional builder and do-it-yourself enthusiast. *Class No:* 69.0(038)=20

[6923]

STEIN, J.S. **Construction glossary:** an encyclopedic reference and manual. New York, Wiley, 1980. [xix], 1013p. tables. ISBN: 047185736x.

16 sections: 1. General requirement - 2. Site work - 3. Concrete - 4. Masonry - 5. Metals - 6. Wood and plastics ... 9. Finishes ... 12. Finishings ... 14. Conveying systems - 15. Mechanical - 16. Electrical. Also: A. Professional services; B. Construction: technical, scientific and related data. Reference data sources , p.819-23. Appendix: 1. Abbreviations for scientific, engineering, and construction terms; 2. Weights and measures. Detailed analytical index has 16 sections, further subdivided.

Class No: 69.0(038)=20

German

[6924]

BUCKSCH, H. **Wörterbuch für Architektur, Hochbau und Baustoffe.** [Dictionary of architecture, building construction and material.] Wiesbaden, Bauverlag GmbH, 1974-76. 2v. ISBN: 3762512019; 3762507147.

1. *Deutsch-Englisch*. 1974. (2. Auf. 1980. 942p.).

2. *Englisch-Deutsch*. 1976. 1137p.

V.1 has 65,000 German-English entries in its 2nd ed. V.2 has 75,000 English-German entries. Intended to complement the other Bucksch technical dictionaries published by Bauverlag. Lacks visual aids. *Class No:* 69.0(038)=30

[6925]

GELBRICH, U. **Dictionary of architecture and building.** Amsterdam, Elsevier, 1989. 418p. £97.40. ISBN: 0444988645.

About 30,000 terms. Distinguishes occasionally American and British usage. Genders given. Covers architecture, building, carpentry, joinery, heating, plumbing, lighting, materials, etc. *Class No:* 69.0(038)=30

French

[6926]

BUCKSCH, H. **Dictionnaire pour les travaux publics, le bâtiment et l'equipement des chantiers de construction.** Wiesbaden, Bauverlag GmbH.

1. *Français-anglais*. 6.éd. 1977. 543p.

2. *English-French*. 7th ed. 1979. 548p.

Class No: 69.0(038)=40

[6927]

MACLEAN, J. **Elsevier's dictionary of building construction:** English-French. Amsterdam, etc., Elsevier, 1989. 2v.

V.1. English-French. [viii],389p. £60. (ISBN: 0444429662).

V.2. French-English. ix,345p. £65. (ISBN: 044442931x.

Covers 'the basic vocabulary of the industry, with particular emphasis on mechanical and electrical services' (*Preface*). Differentiates American and British English. Brief bibliography. *Class No:* 69.0(038)=40

Russian

[6928]

PUSHKAREV, V.L. *and* SHCHEGOLEVA, A.M. **Anglo-russkiĭ i russko-angliĭskiĭ arkhitekturnostroitel'nyi slovar'.** Pod red. L.S. Yampolskogo. Kiev, Gozisdat. Lit. po Straitel'stvu i Arkhiteklur. SSSR. 1961. 841, [3]p.

English-Russian dictionary, p.9-590; Russian-English dictionary (p.591-803) of building and architectural terms. Abbreviations, p.803-41; weights and measures.

Class No: 69.0(038)=82

Reports Literature

[6929]

BUILDING RESEARCH ESTABLISHMENT. **Building Research Establishment digests.** London, HM Stationery Office, 1983. 4v. illus., diagrs., charts. ea. £15.

1. *Building structure and services*. 318p.

1. *Building components and materials*. 2284p.

3. *Building performance*. 308p.

4. *Design and site procedures: defects and repairs*. 226p.

Brings together in 4 self-contained volumes (but with cross-references between them) the numerous BRE *Digests* that review every aspect of building technology. Each volume has a detailed index.

Individual *Digests* (nos.4 ... 324) are available at £1.25 apiece (minimum order, £3). *Class No:* 69.0(047)

Reviews & Abstracts

[6930]

Building management abstracts. Ascot, Berks., Chartered Institute of Building, 1976-. 6pa.

About 800 abstracts per issue. Subject fields: site and project organization, estimating and costs, building management. Coverage: periodical articles, books, conference proceedings, technical and law reports. Annual author, subject and legal case indexes. *Class No:* 69.0(048)

[6931]

—Building science abstracts. Building Research Station - later BRE, Dept. of the Environment. London, HM Stationery Office, 1925-76. v.1-49. Monthly.

Produced *c*.2,000 abstracts pa. 4 main subject groups (22 sub-groups): 1. Materials - 2. Engineering - 3. Construction - 4. Design and environment. Annual name and KWIC indexes.

Continued online (1976-) as BRIX in-house database, which also covers *International civil engineering abstracts*. *Class No:* 69.0(048)

[6932]

Current information in the construction industry (CICI). Croydon, Surrey, Property Services Agency Library Service, 1946-. Fortnightly. £19pa.

Formerly *Library bulletin,* issued by Ministry of Public Building and Works.

Part 1 (yellow pages): 'Forthcoming meetings, conferences and exhibitions'. Part 2 (white pages): 'New books, pamphlets and periodical articles in the construction industry' (*c.*100 brief annotated entries per issue, classified by UDC). Prices of books and pamphlets are given. Concentrates on the practical aspects of the industry. Half-yearly cumulation as *Construction references.* Online through INFOLINE, using PICA (construction and architecture) database. *Class No:* 69.0(048)

[6933]

—BUILDING RESEARCH ESTABLISHMENT. Overseas building notes. Garston, Herts., Publication Sales, BRE. ea. £1.25 (minimum order, £3).

A lengthy series, providing information on housing in tropical and sub-tropical countries. The October 1987 list records 39 titles in print (nos. 139-192), *e.g.* 143. *Building in earthquake areas;* 176. *Building materials in the Arabian Gulf - their production and use;* 192. *Solar energy applications. Class No:* 69.0(048)

[6934]

—Construction references. Croydon, Surrey, Property Services Agency, 1970-. 2pa. £13.50pa. ISSN: 03060152.

Cumulates *c.*3,500 briefly annotated entries pa., based on part 2 of the fortnighly *Current information in the construction industry.* 3 sections: 1. Alphabetical subject index (yellow pages) - 2. Books, pamphlets and periodical articles (in UDC order) - 3. Author index (blue pages).
Class No: 69.0(048)

Periodicals

[6935]

GRETES, F.C. **Directory of international periodicals and newsletters on the built environment.** New York, Van Nostrand Reinhold, 1986. xiv, 175p. £20.65. ISBN: 0442230036.

1,199 annotated, evaluated entries. 16 sections: 1. Architecture - 2. Office practice - 3. Building types ... 7. Planning, environmental design, housing and transportation planning ... 9. Building and construction (p.101-126) - 10. Building services and systems - 11. Engineering - 12. Real estate development and facility planning ... Appended 'Indexes and abstracts' (including databases). Alphabetical index; Geographical index (Argentina ... Zimbabwe); Non-analytical subject index. Aims 'to assist professionals who are involved in the planning, design, construction, and preservation of the built environment'. *Class No:* 69.0(051)

Published Series

[6936]

Mitchell's building series. London, Mitchell, 1979-89. 7v. illus., diagrs., graphs, tables.
Components. 1979. £9.95.
Environment & services. 1988. £10.95.
Finishes. 1989. £9.95.
Materials, by A. Everett. 1986. £9.95.
Structure & fabric. 1983. 2v., ea.£9.95.

....(contd.)

Water, sanitary and waste services for buildings, by A. Wise. 1985. £12.95.

The *Materials* volume (320p.), first published 1970, has 17 sections. 1. Properties generally - 2. Timber ... 4. Stones ... 6. Bricks and blocks ... 8. Concrete - 9. Metals ... 12. Glass - 13. Plastics and rubbers - 14. Adhesives - 15. Mortars for joining ... 17. Sealants. Detailed analytical index. *Class No:* 69.0(082.1)

Tables & Data Books

[6937]

1987 specification: building methods & products. 83rd ed. London, Architectural Press. 6v. illus., diagrs., tables. ISBN: 0851396232.

1/2. *Technical.* 3/4. *Products.* 5. *Specification.* 6. *Directory.*

V.1 concerns 'Construction and interiors (1.1 Groundwork ... 1.15 Furniture and equipment). V.6 comprises: 6.1 Contents index; 6.2 Subject index; 6.3 Proprietary index (yellow pages); 6.4 Information sources (organizations and associations). V.1-4 include listing publications from which further information can be obtained, *e.g.* statutory regulations and acts, British Standards, DOE and BRE publications. A professional reference book for purchasers or specifiers of building materials. *Class No:* 69.0(083)

[6938]

BLACK, L., *ed.* **Builder's reference book.** 11th ed. London, Northwood Books, 1980. 431p. illus., tables. ISBN: 0719828104.

First published 1951.

22 sections: 1. General office information ... 4. The standard measurement of building works (SMME. 6th ed.) - 5. Standard data ... 10. The bricklayer - 11. The carpenter and joiner ... 18. The heating engineer ... 22. The glazier. Trade associations; professional associations; information sources. 'Designed for use in the office or on the site' (*Preface*). *Class No:* 69.0(083)

[6939]

NEUFERT, E. **Architects' data.** Jones, V., *ed.* 2nd (international) English ed. London, Granada. 433p. illus., diagrs., plans, tables. ISBN: 0246112581; 0470269472.

The original *Bauentwurfslehre* (Berlin, Ullstein) had its 26th ed. in 1968 (471p.).

27 UK and US contributors. Sections: Basic data - Community (Houses ... Religion) - Commerce - Industry - Leisure - Components. Hundreds of line-drawings (*e.g.* Kitchen equipment: 17 illus.). An encyclopaedic standard work on all aspects of planning buildings.
Class No: 69.0(083)

[6940]

Planning: the architect's handbook. Mills, E.D. 10th ed. London, Butterworths, 1985. 658p. diagrs., plans. £63. ISBN: 0408012137.

Basic data and information (with drawings) that are essential preliminaries before planning various types of buildings. Although geared to situations and conventions in the UK, the data are applicable wherever building design is under consideration. *Class No:* 69.0(083)

[6941]
POWELL, M.J.V., *ed*. **House builder's reference book.**
London, Newnes-Butterworth, 1979. [1120]p. illus., tables.
ISBN: 0408003375.

59 contributors. Sections (64 subsections): Management -
Design - Construction - Services - Finishes - General
(subsection 64: Sources of building information:
organizations; publications and bookshops; technical press).
Subsection 45, Glazing (12p.) has five headings, with
appended further information (trade associations; British
Standards; other sources of information, *e.g.* BRE Digests).
Index of 16p. *Class No:* 69.0(083)

Standards

[6942]
BRITISH STANDARDS INSTITUTION. **BS handbook 3:
Summaries of British Standards for building,** including
also Codes of practice, drafts for development and other
publications. London, BSI, 1985. 4v. Loose-leaf +
distribution packets. £275 (for *Handbook*, plus current
year's updating).

Replaces the *BS handbook 3,* 1975 (4v. Loose-leaf).

Summaries of *c.*1,500 British standard publications
relating to the construction industry. Includes product
standards, codes of practice, glossaries, methods of tests and
lists of symbols. 'The majority are intended to give enough
information for specifying, ordering and checking deliveries
but not for manufacture or testing. The remainder indicate
the scope of publications that may need to be consulted in
full' (*BSI Catalogue*, 1992, p.702). *Class No:* 69.0(083.74)

[6943]
SMITH, M., *ed*. **Manual of British Standards in building
construction and specification.** 2nd ed. London, BSI, in
association with Hutchinson, 1987. 334p. illus., tables. £25.
ISBN: 0091707609.

First published 1985.

CI/Sf/B tables 0-4, further subdivided. 0. Physical
environment - 1. Building elements - 2. Contractors, firms -
3. Materials - 4. Activities, requirements. Analysis of
relevant standards, providing basic information. Index
precedes. 213 tables. *Class No:* 69.0(083.74)

Films

[6944]
BRITISH INDUSTRIAL AND SCIENTIFIC FILM
ASSOCIATION. **Guide to audiovisual material for the
construction industry.** London, the Association, 1983. 73p.

Sponsored by the Building Centre Trust.

Subject list (based on UDC), p.5-9. Titles and
descriptions, p.11-57. Annotated entries (mostly classified at
69, 71 and 72). Alphabetical title-index. List of distributors.
Class No: 69.0(084.122)

Histories

[6945]
BOWYER, J. **History of building.** London, Crosby
Lockwood Staples, 1973. [vi], 275p. illus. ISBN:
0258968613.

5 parts (27 chapters): 1. Evolution of structure - 2.
Historical development of building materials and components
- 3. Building in the Middle Ages - 4. Growth, urbanization,
industrialization and legislation - 5. The role of

....(contd.)
professionalism. Glossary, p.260-6. Detailed, analytical
index, p.267-75. 142 small, neat line-drawings and plans.
For building surveying and architectural students. No
bibliography. *Class No:* 69.0(091)

Great Britain

Tables & Data Books

[6946]
GREAT BRITAIN. Department of the Environment, Scottish
Development Department, Welsh Office. **Housing and
construction statistics, 1975-1985: Great Britain.** London,
HM Stationery Office, 1986. Annual. xvi, [1], 165p. diagrs.,
graphs, tables, map. £25. ISBN: 0117518786.

15 sections of tables (General - Construction ... Structure
... Housebuilding ... Rent and rent regulations). Appended
notes and definitions. Comparison of table numbers between
1974-1984 and 1975-1985. Detailed, analytical index, p.162-
5. *Class No:* 69.0(410)(083)

Histories

[6947]
SALZMAN, L.F. **Building in England, down to 1540:** a
documentary history. Oxford, Clarendon Press, 1952. xvi,
629p. illus., facsims. (Reprinted, with corrections and
additions, 1967. xvi. 637p).

A scholarly account, with lengthy appendices of
documents, arranged chronologically, apart from quotations
in the text. Footnote references to sources. Index to persons
and places. *Class No:* 69.0(410)(091)

[6948]
—DAVEY, N. Building in Britain: the growth and organization
of building processes in Britain, from Roman times to the
present day. London, Evans, 1964. 191p. illus.

A concise account for candidates taking the examinations
of the Institute of Builders and the City and Guilds of
London Institute, 'but worthy of a wider audience' (*Aslib
book list,* V.29(10), October 1964. Entry no. 583).
Class No: 69.0(410)(091)

Laws

[6949]
House's guide to the construction industry: the regulations,
recommendations, and statutory advisory bodies.
Parlett, D.S., *ed*. 9th ed. Wokingham, Beds., VNR (UK),
1985. viii, 161p. ISBN: 0442306490.

First published 1968.

Directory of more than 1,000 organizations. 9 sections,
covering bodies and information services, reports, forms of
contract, building legislation and regulations. Section 9:
Directory; classified index. Subject index. Late entries.

For the UK; companion volume: *Construction industry
Europe. Class No:* 69.0(410)(094.1)

[6950]
Whyte & Powell-Smith's the building regulations explained and illustrated. Powell-Smith, V. *and* Billington, M.J., *eds.* 8th ed. Oxford, BSP Professional Books, 1990. *c.*600p.
First published 1967.
Covers building control, regulations and approved documents, structural stability, insulation, draining and waste disposal, heating, conservation.
Class No: 69.0(410)(094.1)

Thesauri

[6951]
ROBERTS, M.J. Construction industry thesaurus. 2nd ed. London, Dept. of the Environment, Property Services Agency, 1976. 419p.
First published 1972.
14,000 key-words; A-Z and classified sequences. The abridged ed. (1976), by M.J. Roberts and C.M. Kenwards, lists 3,000 key-words. *Class No:* 69.0:025.43

[6952]
—BUILDING RESEARCH ESTABLISHMENT. Building Research thesaurus. 2nd ed. Garston, Herts., BRE, 1975. vii, 270p.
Compiled by the library of the Building Research Establishment. *Class No:* 69.0:025.43

Information Services

[6953]
CONSTRUCTION INDUSTRY RESEARCH AND INFORMATION ASSOCIATION. CIRIA guide to EC and international sources of construction information. Richardson, B., *comp.* London, CIRIA, 1989. 108p. 309p. (*CIRIA special publication 66.*)
Data on over 200 (non-UK) EC and other international organizations of interest to the UK construction industry. Gives original name, with English translation, if necessary. Subject index. There are also editions covering France, Germany and Spain. *Class No:* 69.0:061:025.5

[6954]
CONSTRUCTION INDUSTRY RESEARCH AND INFORMATION ASSOCIATION. The CIRIA UK construction information guide. Richardson, B. London, E. & F.N. Spon, 1989. xi,361p. £37.50. ISBN: 0419148906.
First published 1979.
Part 1: Alphabetical list of UK local authorities (but not identifying the relevant specific departments); part 2: Alphabetical list of UK construction information sources (p.39-283). No references to HMSO publications (including The Building Regulations); no mention of BRIX, etc. databases. Subject and acronym keys.
Class No: 69.0:061:025.5

Institutions & Associations

[6955]
ARMITAGE, J.S. Guide to international organizations of interest to the construction industry. London, Construction Industry Research and Information Association, 1982. (*CIRIA Special publication 24.*) ISBN: 0860171949.
32 international organizations: CICIND (International Committee on Industrial Chimneys) ... UIDC (International

....*(contd.)*
Union of Building Centres). Data on each include membership, aims, objectives and scope; activities; liaison with other bodies; publications. Appendices: 1. Sources of information - 2. Short guide to international organizations concerned with Standards. - 3. Glossary of acronyms used. Small format. *Class No:* 69.0:061:061.2

[6956]
International directory of building research, information and development organizations. International Council for Building Research, Studies and Documentation (CIB). Sebestyén, G. *and* Pollington, C.E. 5th ed. London, Spon, 1986. ISBN: 0419129901.
First published 1959.
2 main groups: International organizations; National organizations - (Argentina ... Zimbabwe). Data on each include status; main fields of work; total staff; name of director and chief executive; principal sources of finance; principal related bodies; publications. Covers 640 building research organizations in 54 countries. Organization index; Personnel index; Subject index. *Class No:* 69.0:061:061.2

[6957]
An International directory of building research organizations. United States. National Research Council. Washington, DC, National Academy Press, 1989. viii,228p. $27.95; £24.10. ISBN: 0309040272.
2 parts: 1. United States - 2. International (Argentina-Yugoslavia). Data include name; contact person; research focus; 'distinctive attributes'; publications. Indexes: organizations; subjects. *Class No:* 69.0:061:061.2

Research Projects

[6958]
GREAT BRITAIN. Department of the Environment *and* DEPARTMENT OF TRANSPORT. Register of research, 1974-1978. London, HQ Library, DOE/DTp, 1974-78.
1. *Building and construction.*
2. *Environmental planning.*
3. *Roads and transport.*
4. *Environmental pollution.*
Part 1, *Building and construction,* for 1978 (the last to be published) had entries for *c.*1,000 research projects in this field in the UK. Based on data from a wide variety of organizations. Subject sections were followed by a list of research organizations, and indexes of research sponsors, research workers, and organizations. Detailed analytical subject index.
This valuable record of research is now online, ENREP (ECHO host) as part of a European Communities effort. *Class No:* 69.0:061:061.62.005

Metrication

[6959]
TUTT, P. *and* ADLER, D. New metric handbook. Rev. and exp. ed. London, Architectural Press, 1979. [viii], 504p. tables.
First published 1968 as *AJ metric handbook.*
48 sections. 1. Introduction - 2. Basic metric system and S! units - 3. Notation and drawing office practice ... 12. Factories ... 16. Offices ... 18. Hospitals ... 23. Theatres ... 25. Water sports and recreation ... 29. Church buildings - 30. Schools ... 32. Laboratories ... 34. Libraries - 35.

....(contd.)

Housing ... 39. Materials ... 41. Thermal comfort ... 43. Light ... 45. Structure ... 48. Window cleaning. Applied conversion factors and tables. Analytical index (small type), p.502-4. Attractively produced. *Class No:* 69.0:351.821

Machinery & Equipment

[6960]

CHABALLE, L.Y. *and* **VANDENBERGHE, J.-P. Elsevier's dictionary of building tools and materials,** in five languages: English/American, French, Spanish, German and Dutch. Amsterdam, etc., Elsevier, 1982. [xi], 722p. DFl.350. ISBN: 0444420479.

5,853 English/American terms (with synonyms and variants), plus equivalents and indexes in the other 4 languages. Bibliography of sources precedes, p.[ix]-[x]. *Class No:* 69.0-1

Estimates

[6961]

GEDDES, S. Estimating for building and civil engineering works. 8th ed., edited by G. Chrystal-Smith. London, etc., Butterworths, 1985. viii, 424p. charts, tables. ISBN: 0408015578.

First published 1951.

30 sections (4. Costing the works of construction; 11. The working cost of plant; 13. Excavation earthwork below ground; 15. Brickwork and blockwork; 25. Drainage; 27. Plate laying; 30. Useful tables). 2 new sections: 5. Dayworks; 28. Landscaping. Numerous tables; explanatory notes; worked examples. Detailed, analytical index, p.415-24. *Class No:* 69.003.12

[6962]

Laxton's national building price book, 1989. East Grinstead, West Sussex, Skinner, 1988. 2v. (912p.), tables, maps. £48.

4 sections (on different coloured papers): Major works, - Small works, - General information - Brands and trade names and company information. Index precedes. A weighty volume. *Class No:* 69.003.12

[6963]

Means estimating handbook. Goldman, J.M. Kingston, Mass., Means, 1990. 905p. tables. $89.95. ISBN: 0876291779.

Tables and formulas for calculating construction costs using price information provided by the user. 3 sections: 1. Material type - 2. Equipment - 3. Internal building systems. Includes checklists, abbreviations, list of professional organizations. US slanted. *Class No:* 69.003.12

[6964]

Spon's architects' and builders' price book, 1992. 117th ed. London, Spon, 1992. xxxix,1043p. tables. £52.50. ISBN: 0419173609. ISSN: 03063046.

First published 1873.

5 parts: 1. Fees and daywork - 2. Rates of wages - 3. Prices for measured work - 4. Approximate estimating - 5. Tables and memoranda. Numerous tables. Detailed, analytical index (p.1023-43). *Class No:* 69.003.12

Buildings

[6965]

STEIN, B. *and* **REYNOLDS, J.S. Mechanical and electrical equipment for buildings.** 8th ed. New York, Wiley, 1992. xxxi,1627p. illus., diagrs., graphs, tables. £54. ISBN: 0471525022.

First published 1937.

10 parts: 1. Energy overview - 2. Thermal control - 3. Water and waste - 4. Fire protection - 5. Electricity - 6. Illumination - 7. Signal equipment - 8. Transportation - 9. Acoustics - 10. Appendices (11 in all). Nearly 300 tables. Detailed, analytical index, p.1609-27. *Class No:* 69.02

[6966]

Time-saver standards for building types. De Chiava, J. *and* Callender, J.H., *eds.* 3rd ed. New York, McGraw-Hill, 1990. xvii,1413p. illus. £74. ISBN: 0070162794.

First published 1973; 2nd ed. 1980.

Covers standards, procedures and devices for 11 types of buildings (residential; educational, etc.). Extensively illustrated. *Class No:* 69.02

Roofs

[6967]

WICKERSHAM, J.H. The David & Charles manual of roofing. Newton Abbot, Devon, David & Charles, 1987. 256p. illus., diagrs., graphs, tables. ISBN: 0715386980.

9 chapters: 1. Roofing and the spare-time builder - 2. Recreational and storage buildings - 3. Flat roofs - 4/6. Pitched roofs - 7. Ventilation, insulation and flashing - 8. Maintaining and repairing roofs - 9. Designing, constructing and repairing rainwater drainage systems. 2 appendices: 1. Tables; 2. Address list (British associations (products in bold) products), p.239-50. Detailed, analytical index, p.251-6. Fully illustrated (150 photographs). *Class No:* 69.024

Roof Coverings

[6968]

BRITISH STANDARDS INSTITUTION. Glossary of building and civil engineering terms. Part 1. General and miscellaneous. Section 1.3. Parts of construction works. Subsection 1.3.2 Roofs and roofing. London, BSI, 1989. 21p. AMD 7232. (*BS 6100: Subsection 1.3.2:1989.*)

Supersedes BS 2717: 1985.

242 terms in systematic order. Sections: 1. Roof types - 2. Roof features - 3. General roofing - 4. Tile, slate and shale roofing - 5. Sheet roofing - 6. Thatch - 7. Properties - 8. Roofing joints and jointing products. Alphabetical index. *Class No:* 69.024.1

Building Materials

Handbooks & Manuals

[6969]

Construction materials reference book. Doran, D.K., *ed.* Oxford, Butterworth-Heinemann, 1992. v.p. illus., diagrs., tables. £95; $205. ISBN: 0750610042.

2 parts: 1. Metals - 2. Non-metals. Topics include cement, concrete, ceramics, timber, stone, aluminium, steel, etc. Chapters identify relevant British standards. Several hundred literature references. 'Recommended' (*Choice,* November 1992, p.500). *Class No:* 691(035)

[6970]

HUNTINGTON, W.C., *and others.* **Building construction materials and types of construction.** 3rd ed. New York, etc., Wiley, 1981. 471p. illus., diagrs., tables. ISBN: 0471053546.

10 chapters: 1. General data - 2. Sitework - 3. Concrete - 4. Masonry - 5. Metals - 6. Wood and plaster - 7. Thermal and moisture protection - 8. Doors and windows - 9. Finishes - 10. Acoustical control in buildings. References and recommended reading, p.436-46. Detailed, analytical index, p.447-70. Quarto. For students of architecture and building technology. *Class No:* 691(035)

[6971]

TAYLOR, G.D. Construction materials. London, Longman, 1991. 544p. £49.99. ISBN: 0582042992.

Covers properties of metals, concrete, ceramics, fibrous composites and siliceous materials. Includes standards. *Class No:* 691(035)

Dictionaries

[6972]

BIANCHINA, P. Illustrated dictionary of building material and techniques. Blue Ridge Summit, Pa., TAB Books, 1986. 243p. illus. $22.95. ISBN: 0830604812.

Defines *c.*4,000 terms, covering building, plumbing, electrical writing, hardware, architecture, heating and ventilating. 500 line-drawings ('Window': 5 types). No less than 60p. of appendices on conversions, abbreviations, hardware, etc. *Class No:* 691(038)

Tables & Data Books

[6973]

WADDELL, J.J. Construction materials ready reference manual. New York, etc., McGraw-Hill, 1985. 395p. illus., diagrs., charts. ISBN: 0070676496.

Properties and selection of materials - Rock and soils - Concrete - Masonry - Lath and plaster - Metals - Wood - Plastics - Paint - Roofing and asphalt - Pipes. 71 illus. Small format. *Class No:* 691(083)

Trade Names, Trade Marks, Brand Names

[6974]

Barbour compendium: building products '88. Windsor, Barbour Index, 1988. 1412p. col. illus. £40.50. ISSN: 02609169.

25 sections (Buildings and building systems ... Formless

....(contd.)

materials). Covers more than 500 products, illustrated in colour, plus captioned details - physical data plus performance characteristics. Index to products precedes. Appended: Manufacturers/Trade names (on blue-tinted paper).

The 1988 ed. (£40.50) lists over 5,400 manufacturers supplying a full range of building products. *Class No:* 691(088.7)

Histories

[6975]

DAVEY, N. A History of building materials. London, Phoenix, 1961. xiv, [5], 260p. illus. (incl. pl.), diagrs., plans, charts.

26 documented chapters (1. Stone - 2. Building in stone - 3. Earth walling ... 24. Water supply and sanitation - 25. Metals - 26. Modern trends). Appended 'Building tools', p.223-37 (selected traditional tools, with over 100 line-drawings). Analytical index to text. Profusely illustrated: 48p. of plates. Well produced. Examples are drawn almost entirely from Europe and the Near East. *Class No:* 691(091)

[6976]

HUDSON, K. Building materials. London, Longman Group, 1972. [x], 122p. illus. (incl. plates). (*Industrial archaeology series.*) ISBN: 0582122912.

7 chapters. 1. Builders and customers, 1770-1970 - 2. Stone and slate - 3. Bricks and tiles - 4. Lime, cement, plaster and concrete - 5. Glass, iron and steel - 6. Timber, plywood and wood products - 7. Some major new materials (*e.g.* plasterboards). Key dates, 1773-1945 Chapter references, p.100-3; Select bibliography (A-Z authors), p.111-4. Gazetteer. Detailed analytical index, p.115-22. *Class No:* 691(091)

Wood

[6977]

BRITISH STANDARDS INSTITUTION. Glossary of terms relating to timber preservation. London, BSI, 1985. 19p. £23.90. (*BS 4261: 1985 (1991).*)

Defines 484 terms, in 6 sections: 1. General - 2. Attack by fungi - 3. Attack by insects and marine borers - 4. Preservative processes - 5. Terms associated with preservative treatments - 6. Miscellaneous. Index. *Class No:* 691.11

[6978]

Handbook of wood and wood-based materials for engineers, architects and builders. United States. Department of Agriculture. Forest Service. Forest Products Laboratory. Rev. ed. New York, Hemisphere, 1989. v.p. illus., diagrs., graphs, tables. $77. ISBN: 0891161244.

First published 1935 as *Wood handbook;* previous ed. 1974.

23 chapters of the properties, applications and treatments of wood. Numerous tables. Glossary, p. G1-G11. 12p. detailed, analytical index. Emphasis is on US. *Class No:* 691.11

[6979]
LAVERS, G.M. The Strength properties of timber. 3rd ed., revised by G.L. Moore. London, HM Stationery Office, 1983. 64p. £4.95. (*BRE Reports, 50 3.8.*) ISBN: 0116713569.

Physical and mechanical tests on over 200 hardwoods and softwoods (both seasoned and unseasoned): procedures and results. *Class No:* 691.11

Stone

[6980]
SHORE, B.C.G. Stones in Britain: a practical guide to those in charge of valuable buildings. London, Hill, 1957. xii, 302p. illus., maps.

Thorough investigation of geological and chemical questions involved in the use of stone for building; also the stone-preservation problems. 4 main sections: care of ancient buildings; list of British building stones, their occurrence and structure; maintenance work, using stone; new buildings of stone, and stone for sculpture. Lavishly illustrated with fine photographs and maps. *Class No:* 691.2

Concrete

[6981]
BARKER, J.A. Dictionary of concrete. Lancaster, Construction Press, 1984. 124p. £16.

5,000 precise definitions. Includes major terms used in other countries, particularly the US. *Class No:* 691.32

[6982]
BUILDING RESEARCH ESTABLISHMENT. Concrete. Lancaster, etc., Construction Press, 1979. [x], 329p. illus., diagrs., graphs, tables. (*Practical studies from the Building Research Establishment.*) ISBN: 0904400377.

19 documented papers: Materials for concrete (1); Properties of concrete (9). The paper 'Effects of various factors on the extensibility of concrete', p.216-27, has 131 references. Index. *Class No:* 691.32

[6983]
WADDELL, J.J. Concrete construction handbook. 2nd ed. New York, etc., McGraw-Hill, 1974. x, [960]p. illus. ISBN: 0070676542.

First published 1968.

14 sections: 1. Materials for concrete - 2. Properties of concrete - 3. Proportioning mixes and testing - 4. Formwork and shoring ... 7. Finishing and curing - 8. Special concrete and techniques 9. Advanced building construction systems ... 11. Cracking and surface blemishes ... 14. Repair of concrete. Section bibliographies. Index. *Class No:* 691.32

Building Trades

Bricklaying

[6984]
BRICK DEVELOPMENT ASSOCIATION. Bricks, their properties and use. Lancaster, Construction Press, 1984. 251p. illus., diagrs., graphs, tables. ISBN: 0904406040.

Main sections: Basic information - Bricks in use - Technical information (p.91-248) ... Performance

....(contd.)
characteristics of perforated and solid bricks ... The stability of a five-storey brickwork cross-wall structure. Index. p.249-51. *Class No:* 693.22

[6985]
LLOYD, N. A History of English brickwork, with examples and notes of the architectural use and manipulation of brick, from medieval times to the end of the Georgian period. Woodbridge, Suffolk, Antique Collectors' Club, 1983. xi, 449p. illus., diagrs., tables. £25. ISBN: 0907482967.

First published 1926.

Text occupies p.1-100, the remainder being almost wholly illustrative, - photographs and measured drawings. Table of brick sizes. Data on prices, wages and output. 'Manipulation of brick' includes curved brick; stairs, vaulting, arcading; fonts, fireplaces, tables; pinnacles. Index. p.441-9. 360 illus. *Class No:* 693.22

Plastering

[6986]
BRITISH STANDARDS INSTITUTION. Glossary of building and civil engineering terms. Part 6. Concrete and plaster. Section 6.6 Products, applications and operations. Subsection 6.6.2 Plaster. London, BSI, 1990. 24p. £23.90 AMD 7267. August 1992. £1.50. (*BS 6100: Subsection 6.6.2: 1990.*)

Supercedes BS 4049: 1966.

259 terms defined in systematic order. Alphabetical index. *Class No:* 693.6

Timber Construction & Carpentry

[6987]
RICHARDSON, B.A. Wood in construction. London, Construction Press, 1976. 220p. illus., facsims., tables.

Sections: 1. Wood as a material - 2. Converting trees to wood in service - 3. Wood protection - 4. Wood utilization - 5. Commercial woods and wood products. 3 appendices: Botanical classification of principal north temperate timber trees; Wood structure; Commercial woods. Index to common names, p.199-214. General index. *Class No:* 694.1

[6988]
TEKNISKA NOMENKLATURCENTRALEN. Träbyggnardsordlista. [Glossary of timber construction.] Stockholm, the Centre, 1973. 224p. illus. (*TNC 60.*)

725 Swedish entry-words and definitions; English, French and German equivalents. Classified arrangement; German, French, English and Swedish indexes. Swedish references p.10-11. Small line-drawings. A neat production. *Class No:* 694.1

Joinery

[6989]
BRITISH STANDARDS INSTITUTION. Glossary of terms relating to joints and joining in building. London, BSI, 1970. 11p. 1 diagr. £10.30. (*BS 4643: 1970.*)

Defines 23 terms for joining products, joint dimensions and joint functions. *Class No:* 694.6

Plumbing

[6990]

BRITISH STANDARDS INSTITUTION. Glossary of building and civil engineering terms. Part 3. Services. Section 3.3 Sanitation. London, BSI, 1992. 35p. (*BS 6100: Section 3.3: 1992.*)

Supersedes BS 6100: Section 3.3: 1991.

444 plumbing and water supply terms defined in systematic order. Alphabetical index. *Class No:* 696.1

[6991]

HOLLOWAY, D. The 'Which?' book of plumbing and central heating. London, Consumers' Association, and Hodder & Stoughton, 1985. 160p. illus. £9.95. ISBN: 0340381604.

3 parts: 1. Plumbing basics (1. The plumbing toolkit ... 6. Rainworks disposal) - 2. Plumbing, room by room - 3. Central heating. Detailed analytical index, p.158-160. 'Intended for the house-owner who would like to carry out the majority of plumbing jobs in the house but lacks the necessary knowledge and experience' (*Introduction*). Numerous two-colour illus. Good layout. *Class No:* 696.1

[6992]

JACOBSON, I.D., *ed*. **Plumbing dictionary.** 3rd ed. Cleveland, Ohio, American Society of Sanitary Engineering. [1], ix, 150p. illus.

First published 1971.

Over 3,000 definitions, giving pronunciation. Includes abbreviations. Cross references; tiny line-drawings. Bibliography, p.vi-viii. 'Wherever possible, non-technical language has been used' (*Foreword*). *Class No:* 696.1

Heating & Ventilation

Bibliographies

[6993]

HASELER, A.E. District heating: an annotated bibliography. 2nd ed. Croydon, Surrey, Property Services Agency, 1977. xi, 130p.

660 numbered and annotated entries in 18 sections, subdivided and covering electricity environmental aspects, heating and cooling systems, architectural and building viewpoints, plus government policies and initiatives in both Britain and abroad. 'Brief history of district heating' precedes. Many cross-references. Geographical and name indexes. *Class No:* 697(01)

[6994]

International building services abstracts (IBSEDEX). Bracknell, Berks., Building Services Research and Information Association (BSRIA, formerly Heating and Ventilating Research Association), v.13(1), 1978-. 6pa. £64pa.

Formerly *Thermal abstracts,* 1966-77, v.1-12. Initially as *HVRF bulletin,* 1952-69, and then *HVRA library bulletin,* 1960-65.

About 2,000 abstracts pa. 20 sections (Building in general - Heating services - District heating and cooling ... Lighting - Instruments and control ... Noise and vibration - Building structures and processes - Thermal insulation - Management and organization). On line, IBSEDEX; also PERGAMON INFOLINE. *Class No:* 697(01)

Handbooks & Manuals

[6995]

ASHRAE handbook. Atlanta, Georgia, American Society of Heating, Refrigerating and Air-Conditioning Engineers, 1982-85. 4v.

First published 1961. As *ASHRAE handbook and product directory,* 1974-77. 4v., now divided into *ASHRAE handbook* and annual *ASHRAE product information file. Fundamentals.* 1985. *Systems.* 1984. *Equipment.* 1983. *Applications.* 1982.

Fundamentals, by various contributors, concerns basic principles, general engineering data, load calculations, duct and pipe sizing. Indexes. A detailed survey of US practice. *Class No:* 697(035)

[6996]

CIBSE guides. London, Chartered Institute of Building Services, 1986. 3v. £170, the set.

A. *Design data.* B. *Installation and equipment data.* C. *Reference data.*

Authoritative guides for designers planning heating, lighting and other installations in buildings. CIBS (Chartered Institute of Building Services) was formed by the amalgamation of the Institute of Heating and Ventilating Engineers and the Illuminating Engineering Society. For professionals. *Class No:* 697(035)

Dictionaries

Polyglot

[6997]

International dictionary of heating, ventilating and air conditioning. Compiled by the Documentation Committee of the Representatives of European Heating and Ventilating Associations. Rev. ed. London, Spon, 1982. xxviii, 482p. £38.50. ISBN: 041915390x.

First published 1982.

Nearly 4,000 terms in ten languages, - Dutch, English, French, German, Hungarian, Italian, Polish, Russian (entries in cyrillic), Spanish and Swedish. English-base terms, with equivalents and indexes in the other languages. *Class No:* 697(038)=00

English

[6998]

BOOTH, K.M., *comp*. **Dictionary of refrigeration and air conditioning.** Amsterdam, etc., Elsevier, 1970. viii, 315p. . £21. ISBN: 044420069x.

More than 2,600 terms from both sides of the Atlantic are defined in entries ranging from a few words to more than 200. Copious cross-references. Author's Preface insists 'The book is a dictionary and not an encyclopedia'. Because of the impending changeover to use of SI units at the time of publication, metric equivalents to Imperial units are frequently given. Inevitably lacks terms relating to important recent developments (*e.g.* new refrigerants, use of computer and microprocessor control methods) but sound, descriptive definitions of fundamental material make this a brief dictionary of continuing importance in its field. *Class No:* 697(038)=20

[6999]
BRITISH STANDARDS INSTITUTION. Glossary of refrigeration, heating, ventilating and air-conditioning terms. London, BSI, 1984. 84p. £38. (*BS 5643: 1984.*)

Supersedes BS 1584 and BS 5643: 1979.

Contains around 2,300 terms, of which more than one third are cross-references, many to noun/adjective forms from the alternative (*e.g.* dark smoke *see* smoke, dark), apparently assuming that the reader will overlook the clear explanation of policy in the Foreword! Concise definitions. *Class No:* 697(038)=20

[7000]
BRITISH STANDARDS INSTITUTION. Glossary of terms related to solid fuel burning equipment. London, BSI, 1968. 2pts. £17.90 + £11. (*BS 1846: 1968.*)

First published 1952.

Pt.1. *Domestic appliances.* 36p.

Pt.2. *Industrial water heating and steam raising installations.* 20p.

Pt.1 defines *c.*250 terms; pt.2, *c.*100, each with 10 sections and an index. *Class No:* 697(038)=20

Tables & Data Books

[7001]
PORGES, F. Handbook of heating, ventilating and air conditioning. 9th ed. Oxford, Butterworth-Heinemann, 1991. 272p. illus. tables. £30. ISBN: 0750614811.

17 sections, mainly tabular. 1. Abbreviations, symbols and conversions - 2. Standards for materials ... 4. Heat and thermal properties of materials ... 8. Heating systems ... 11. Ventilation - 12. Air conditioning ... 14. Sound - 15. Labour rates - 16. Bibliography (grouped), p.225-8 - 17. British Standards. Imperial measurement used. Detailed and analytical index. *Class No:* 697(083)

Europe

Solar Heating

[7002]
PALZ, W. *and* STEEMERS, T.C., *eds*. Solar houses in Europe: how they have worked. Compiled and written by W. Houghton-Evans. Oxford, etc., Pergamon Press, for the Commission of the European Communities. 303p. diagrs., graphs, tables, maps. ISBN: 0080267432.

Prepared within the framework of the European Communities Solar Energy Research and Development Programme.

7 chapters (2. How the buildings have performed - 3. Lessons from the survey - 4. Recommendations for design - 5. Statistical summaries - 6. Monitoring - 7. Reports on 31 projects (Denmark ... UK), p.80-303). Glossary. Location maps. *Class No:* 697(4)SOL

EEC

Solar Heating

[7003]
LEWIS, O. The European passive solar handbook. London, Batsford, 1992. 304p. illus. £40. ISBN: 0713469188.

Design information on passive solar architecture. Review of the EC solar programme. *Class No:* 697(40)SOL

Painting & Decorating

[7004]
FULCHER, A. Painting and decorating: an information manual. 2nd ed. Oxford BSF Professional Books, 1981. ix, 226p. illus., tables. £8.95. ISBN: 0632022876.

First published 1975.

9 sections (1. Tools and equipment - 2. Surface coatings ... 8. Colour - 9. Glazing). Includes guidance on specifying and safety precautions. Well illustrated, keyed line-drawings being a feature. *Class No:* 698.1

[7005]
GOODIER, J.H. Dictionary of painting and decorating, covering also allied industrial finishes. 3rd ed. London, C. Griffin, 1987. xi, 422p. £32.50. ISBN: 0852642792.

First published 1961.

About 1,500 entries (*e.g.* 'Metal spraying': 1p.; 'Lead paint - statutory regulations': 3p.). Conversion factors and tables precede. Cross-references. An encyclopaedic dictionary for craftsmen, students and teachers. *Class No:* 698.1

[7006]
JOHNSON, L. The Decorator's directory. London, M. Joseph, 1981. 318p. illus. (incl. col.). ISBN: 0718120036.

31 sections (*e.g.* Appliances & kitchen accessories; Architectural ornaments; Bathrooms; Bedrooms; Blinds; Carpeting; Kitchens ... Special furniture; Storage; Tables; Upholstered furniture; Wall coverings co-ordination). Useful addresses. Index of suppliers; index of catalogues. Subdivision under companies gives details and prices. Attractively produced. *Class No:* 698.1

Insulation

[7007]
BRITISH STANDARDS INSTITUTION. Glossary of terms relating to thermal insulation. London, BSI, 1981. 16p. (*BS 3533: 1981.*)

About 200 terms in a systematic order. Alphabetical index. *Class No:* 699.86

Author / Title Index

The index reference is to the running number given to each item. The running numbers are in one sequence throughout the volume and can be found at the top right-hand corner of the entry for each item.

This index is of authors and titles in one sequence. The names of the authors are printed in bold type. Where works are jointly authored, only the first name is indexed. All books and periodicals listed or mentioned in the text, except where cited as the source of review quotations or for purposes of comparison, are entered under the headings given. All entries in *Walford* have title entries in the index; where the main heading in *Walford* is under title, added entries to the index have usually been made for an editor or compiler.

Filing is word by word, with groups of initials counted as single words. Since *Walford* uses only initials and not forenames it may occasionally happen that titles by different authors with the same surname and initials are found grouped together.

The arrangement of entries under an author is alphabetically by title. To save space, most sub-titles have been omitted, and many lengthy titles have been shortened.

Agricultural economics and rural sociology - multilingual thesaurus **5155**

Agricultural engineering **5206**

Agricultural engineering abstracts **5207**

Agricultural engineer's handbook **5200**

The Agricultural handbook **5076**

Agricultural history in Great Britain and Western Europe before 1914 **5121**

The Agricultural history review **5106**

Agricultural information resource centers **5148**

Agricultural insect pests of temperate regions and their control **5250**

Agricultural insect pests of the tropics and their control **5250**

Agricultural journal titles and abbreviations **5093**

The Agricultural notebook
See
Primrose McConnell's 'The agricultural notebook'

Agricultural plants **5258**

Agricultural research, 1931-1981 **5154**

Agricultural research centres **5152**

The Agricultural situation in the Community **5120**

The Agricultural systems of the world **5108**

Agricultural terms, as used in the Bibliography of Agriculture **5077**

Agricultural writers **5123**

Agriculture: a bibliographical guide **5060**

The Agriculture of China **5140**

Agrindex **5086**

AGRIS **5054**

The Agrochemicals handbook **5247**

Agroselekt **5091**

Agrovoc: multilingual agricultural thesaurus **5146**

Ahlberg, K.
AGA gas handbook **5985**

Ahlsved, K.-J.
Lexicon forestale **5167**

AIDS bibliography **3553**

AIDS information **3551**

AIDS information sourcebook **3552**

Aids to geographical research **1564**

AIDSLINE **3553**

Ainbinder, M.I.
Anglo-russkiĭ meteorologicheskiĭ slovar' **1677**

Ainsworth & Bisby's dictionary of the fungi **2453**

Ainsworth, G.C.
Ainsworth & Bisby's dictionary of the fungi **2453**
Introduction to the history of medical and veterinary mycology **2464**
Introduction to the history of mycology **2463**

Ainsworth, G.C. *and* Sussman, A.S.
The Fungi **2457**

AIP Style manual **902**

The Air almanac **744**

The Air & space catalog **650**

Air and space dictionary **5023**

Air-Conditioning and Refrigeration Institute
Refrigeration and air-conditioning **4241**

Air distances manual **5864**

Air navigation plans **735**

Air pollution **4831**

Air pollution titles: guide to current air pollution literature **3303**

Air University Library index to military periodicals **4481**

Aircraft, engines and airmen **4963**

Aircraft of the Soviet Union **4535, 4994**

Aird, A.
The Good pub guide **5499**

The Airline bibliography **5859**

Airships: an illustrated history **5000**

Aitchison, L.
A History of metals **6319**

Aithgiorrachtaí móna **6001**

Aitken, A.
Fish handling & processing **6146**

A.J. Lohwater's Russian-English dictionary of the mathematical sciences **508**

Akehurst, B.C.
Tobacco **6061**

Akiner, S.
Islamic peoples of the Soviet Union **1928**

Akzhigtov, G.N.
Anglo-russkiĭ meditsinskiĭ slovar' **3077**

al-Hassan, A.Y. *and* Hill, D.R.
Islamic technology **2843**

Al-Khatib, A.Sh.
A New dictionary of scientific and technical terms, English-Arabic, with illustrations **181**

Albert, L.S. *and* Kent, K.
The Complete button book **6887**

Albert, R. *and* Hahnewald, H.
Eight-language dictionary of medical technology **3036**

Albion, R.G.
Naval & maritime history **5813**

Album of science **274**

Alcohol and the public health **3283**

Aldous, W.
Terrell on the law of patents **2946**

Alexander, B.A.
Journals with translations held at the Science Reference Library **232**

Alexander, F.G. *and* Selesnick, S.T.
The History of psychiatry **3543**

Alexandrov, P.S.
Anglo-russkiĭ slovar' matematicheskiĭ terminov **509**

Alexis Lichine's New encyclopedia of wines & spirits **6046**

Alfara Perez, T.
Diccionario marítimo y de construcción naval **4927**

Alford, M.H.T. *and* Alford, V.L.
Russian-English scientific and technical dictionary **170**

The Algae **2451**

Alger, M.S.
Polymer science dictionary **6598**

Ali, Salim *and* Ripley, S.D.
Handbook of the birds of India and Pakistan **2696**

All the US Air Force airplanes, 1907-1983 **4536**

Allaby, A. *and* Allaby, M.
The Concise Oxford dictionary of earth sciences **1365**

Allaby, M.
The Concise Oxford dictionary of zoology **2523**
Macmillan dictionary of the environment **2088, 2283**
The Oxford dictionary of natural history **2223**

Allan, B.
Guide to information sources in alternative therapy **3390**

Allan, H.H.B.
Flora of New Zealand **2417**

Allen, C.W.
Astrophysical quantities **672**

Allen, D. *and* Woolstenholmes, C.J.
A Pictorial survey of railway signalling **5784**

Allen, G. *and* Bevington, J.C.
Comprehensive polymer science **6594**

Allen, G.E.
Life science in the twentieth century **2048**

Allen, J.L.
Aviation and space museums of America **4987**

Allen, J.P. *and* Turner, E.J.
We the people: an atlas of America's ethnic diversity **1957**

Allen, K.A. *and* Neale, J.W.
Principles of zoological micropalaeontology **1822**

Allin, C.W.
International handbook of national parks and nature reserves **2274**

Allison, S.
The Cassell food dictionary **5550, 6085**

Alloys index **6334**

Alonso Santos, A.
Léxico de términos nucleares **3831**

Astronomischer Jahresbericht **611**

Astronomy and aeronautics **614**

Astronomy and astrophysics: a bibliographical guide **613**

Astronomy and astrophysics abstracts **644**

The Astronomy and astrophysics encyclopaedia **619**

The Astronomy encyclopaedia **620**

Astronomy through the telescope **684**

Astrophysical formulae **673**

Astrophysical quantities **672**

Astrophysics and twentieth-century astronomy to 1950 **656**

Astrophysics of the sun **703**

At the farmer's service **5133**

Atherton, D.P.
Concise encylcopedia of modelling and simulation **6783**

Atherton, W.A.
From compass to computer: a history of electrical and electronics engineering **3942**

Atkins, P.W.
Physical chemistry **1138**
Quanta: a handbook of concepts **1133**

Atkinson, B. *and* Mavituna, F.
Biochemical engineering and biotechnology handbook **2755**

Atkinson, D.W.
Weather **1714**

ATLA **5325**

Atlas and glossary of primary sedimentary structures **1533**

Atlas climatique de l'Europe **1710**

Atlas des grands gouffres du monde **1618**

Atlas for anthropology **1918**

Atlas narodov mira **1917**

The Atlas of breeding birds in Britain and Ireland **2692**

Atlas of disease distributions **3432**

Atlas of diseases **3433**

An Atlas of distribution of the freshwater fish families of the world **2644, 5440**

Atlas of dog breeds of the world **5404**

Atlas of economic mineral deposits **1745**

The Atlas of endangered species **2286**

Atlas of environmental issues **2112**

Atlas of hardwoods **6445**

Atlas of human cross-sectional anatomy **3213**

Atlas of marine aquarium fishes **5452**

Atlas of meteorology **1687**

Atlas of North American Indians **1987**

Atlas of palaeobiogeography **1804**

An Atlas of past and present pollen maps for Europe 0-13000 years ago **1827**

An Atlas of renewable energy resources in the United Kingdom and North America **3754**

Atlas of the British flora **2388**

Atlas of the environment **2110, 4763**

Atlas of the great caves of the world **1618**

Atlas of the living resources of the seas **5447**

The Atlas of the living world **2238**

Atlas of the moon **694**

Atlas of the seas round the British Isles **1653**

Atlas of the world's railways **4651**

Atlas of United States environmental issues **2111**

Atlas of world cultures **1914, 1919**

The Atlas of world wildlife **2239**

Atlas of zoogeography **2574**

Atlas sel'kogokhozyaistvo SSSR **5138**

Atlas van de europese vogels **2686**

Atomic absorption and emission spectrometry abstracts **1191**

Atomic data and nuclear data tables **3857**

Atomic energy level and Grotrian diagrams **3851**

Atomic gas laser transition data **4044**

The Atomic scientists **1009**

Attar, D.
A Bibliography of household books published in Britain, 1840-1914 **5572**

Attiyate, Y.H.
Wörterbuch der Mikroelektronik und Mikroechnertechnick **4134**

Atz, J.W.
Dean bibliography of fishes, 1968 **2639**

Audio and video **6845**

The Audio dictionary **6855**

Audio electronics reference book **6847**

Audio engineering handbook **6848**

Audio, video and data telecommunications **4158**

Audio video market place, 1989 **6856**

'Audio-visual materials' **3122**

Audio-visual resources in food and nutrition **6101**

Audouze, J. *and* Israël, G.
The Cambridge atlas of astronomy **722**

The Audubon Society encyclopedia of animal life **2567**

The Audubon Society encyclopedia of North American birds **2699**

The Audubon Society field guide to North American fossils **1812**

The Audubon Society field guide to North American rocks and minerals **1327, 1736**

The Audubon Society field guide to North American trees **5191**

The Audubon Society field guide to North American wildflowers **2412**

Augarten, S.
Bit by bit: an illustrated history of computers **6712**

Auger, C.P.
Engineering eponyms **3777**
Information sources in grey literature **195**

Auger, P.
Information sources in patents **2896**

Auger, P. *and* Rousseku, L.J.
Lexique anglais-français de l'industrie minière **4431**

Austin, M.
The ISTC handbook of technical writing and publication techniques **402**

Austin, R.M. *and* Nau, B.S.
Seal users handbook **3815**

Australia and the Pacific Islands **2269**

Australia. Scientific and Technological Information Services Enquiry Committee
The STISEC Report to the Council of the National Library of Australia **430**

Australian National University, Canberra. Department of Anthropology and Sociology
An Ethnographic bibliography of New Guinea **1973**

Australian native plants **5284**

Australian scientific societies and professional associations **440**

Auszüge aus den Europäischen Patentanmeldungen **2912**

Author and classified catalogues. [Royal Botanic Gardens, Kew, Library] **2317**

An Author and permuted title index to selected statistical journals **589**

Author and subject catalogues of the Tozzer Library **1867**

Author catalog. [New York Academy of Medicine Library] **3001**

Author-title catalog ... for imprints through 1959. [Francis A. Countway Library of Medicine] **2987**

Author/title, subject and geographic catalogs of the Glaciology Collection **1535**

Auto sourcebook, 1991 **4873**

Automated manufacturing directory **4297**

Automation encyclopedia **6781**

Automation of small-batch production **4284**

Automobile abstracts **4867**

Automobile Association
Book of the British countryside **2251**
Craft workshops in the English countryside **6619**

Baxter, J.M.
The Mitchell Beazley pocket guide to the seashores of Britain and Northern Europe **2119**
Baxter, J.W.
World patent law and practice **2932**
Baynard, F.O.
Fifty famous liners **4946**
BCIRA abstracts of international literature on metal castings production **4308**
BCRA quarterly **6022**
BDS **3220**
Beach, D.P. *and* Alvager, T.K.
Handbook for scientific and technical research **397**
Beaches and coasts **1610**
Beadle, W.J.
Quick reference manual for silicon integrated circuit technology **4151**
Beale, H.P.
Bibliography of plant viruses and index to research **5234**
Bean, J.E.
The Anti-Corrosion handbook and directory **4356**
Finishing handbook and directory **4355**
Bean, W.J.
Trees and shrubs hardy in the British Isles **5190**
Beard, R.
Soviet cosmonautics, 1957-1969 **5040**
Beardon, C.
Artificial intelligence terminology **6807**
BEASTCD **5055, 5328-5329, 5347-5348, 5394, 5396, 5411**
Beatty, J.K. *and* Chaikin, A.
The New solar system **685**
Beaubois, H.
Airships: an illustrated history **5000**
Becher, P.
Encyclopedia of emulsion technology **5949**
Becker, P.
Dictionary of colloid and surface science **1151**
Beckett, K.A., in association with the Royal Horticultural Society
The RHS encyclopaedia of house plants **5315**
Bedini, S.A.
Early American scientific instruments and their makers **6658**
Bednar, H.H.
Pressure vessel design handbook **4330**
Beech, S.R.
Textile terms and definitions **6506**
Beeching, W.A.
Century of the typewriter **6842**

Bees and beekeeping **5425**
Beeton, I.M.
Mrs. Beeton's cookery and household management **5543**
Mrs. Beeton's cookery and household management **5543**
Begell, W.
Glossary of terms in heat transfer, fluid flow, and related topics **974**
Behrens, D.
Corrosion handbook **3694**
Behrens, H. *and* Rittberger, W.
'Nuclear and particle physics' **1003, 3821**
Beilstein online **1237**
'Beilstein online.' **1238**
The Beilstein online database **1239**
Beilstein's Handbuch der organischen chemie **1243**
Bein, G.
Wörterbuch des internationales Verkehrs: Deutsch-Englisch **5721**
Beiträge zur regionalen Geologie der Erde **1476**
Bekleidungs wörterbuch **6881**
Belforte, D. *and* Levitt, M.
Industrial laser annual handbook, 1986 **4043**
Bell, A.D.
Plant form **2489**
Bell, F.G.
Fundamentals of engineering geology **4614**
Ground engineer's reference book **4567**
Bell, P.
Strasburger's textbook of botany **2333**
Bell, S.P.
A Biographical index of British engineers in the 19th century **3626**
Belle-Isle, J.-G.
English-French general technical dictionary **140**
Belove, C.
Handbook of modern electronics and electrical engineering **4061**
Bels-Koning, H.C. *and* Kuijk, W.M., van.
Mushroom terms **2467**
Belzer, J.
Encyclopedia of computer science and technology **6670**
Benarde, M.A.
Global warning … global warming **4818**
Bender, A.E.
Dictionary of nutrition and food technology **3252, 6086**
Food processing and nutrition **6108**
Bendiner, J. *and* Bendiner, E.
Biographical dictionary of medicine **3137**

Benedict, G.F.
Nontraditional manufacturing processes **4287**
Bennett, H.
The Chemical formulary **5927**
Concise chemical and technical dictionary **1040**
Bennett, H.S.
English books & readers **5683**
Bennett, J.A.
The Divided circle **674, 6651**
Bennett, R. *and* Watson, J.
Philatelic terms illustrated **5879**
Bennett, S.
A History of control engineering, 1800-1930 **6795**
Bennett, S. *and* Freirman, R.
Microcomputer marketplace **6750**
Bennett, W.R.
Atomic gas laser transition data **4044**
Bennigsen, A. *and* Wimbush, S.E.
Muslims of the Soviet empire: a guide **1932**
Benninghoff, H.
Index of chemicals used for the treatment of metal surfaces **4358**
Bennington, J.L.
Saunders dictionary & encyclopedia of laboratory medicine and technology **3173**
Benn's Hardware & DIY buyers' guide, 1986 **6862**
Benson, J. *and* Neville, R.G.
Bibliography of the British coal industry **6017**
Benson, K.B.
Audio engineering handbook **6848**
Television engineering handbook **4213**
Benson, K.B. *and* Whitaker, J.G.
Television and audio handbook for technicians **4212**
Bentham, G.
Handbook of the British flora **2390**
Benton, M.J.
Vertebrate palaeontology **1837**
Beresford, A.K.C.
Lloyd's maritime atlas of world ports and shipping places **5843**
Berg, I.
The Guinness world car record **4888**
Berger, K.
Mykologisches Wörterbuch **2455**
Bergey's manual of systematic bacteriology **2210**
Bergteknisk ordlista **4612**
Berins, M.L.
Plastics engineering handbook of the Society of the Plastics Industry, Inc **6585**
Berkebile, D.H.
Carriage terminology **4851**

Bernabo Silorata, M. *and* Picchi, F.
Grande dizionario di marina, italiano-inglese, inglese-italiano **5807**

Bernal, I.
Bibliografia de arqueologia y etnografia: Mesoamerica y norte de México, 1514-1960 **1954**

Berner, G.-A.
Dictionnaire professionnel illustré de l'horlogerie **6628**

Bernice P. Bishop Museum, Honolulu
Dictionary catalog of the Library of Bernice P. Bishop Museum **1974**

Berra, T.M.
An Atlas of distribution of the freshwater fish families of the world **2644, 5440**

Berrill, D.J.
Optoelectronics data book **4154**

Berry, M.J.
Science and technology in the USSR **15**

Berry, R.J.
The Natural history of Orkney **2255**

Bersch, E.
Shipbuilding, seafaring technical dictionary **4918**

Bertin, L.
The New Larousse encyclopedia of the earth **1362**
Beryllium: its metallurgy and properties **6410**

Bes, J.
Chartering and shipping terms **5798**

Besançon, R.M.
The Encyclopedia of physics **835**

Bespaloff, A.
The New encyclopedia of wine **6041**

Besson, A.
Thornton's medical books, libraries and collectors **3134**
Best science and technology reference books for young people **65**

Besterman, T.
Medicine: a bibliography of bibliographies **2979**
Physical sciences **47**
Betongteknisk ordlista **6274**

Betteridge, W.
Nickel and its alloys **6379**
Between covers: the rise and transformation of book publishing in America **5705**

Bevan, J.
Your family doctor: the home medical encyclopaedia **3031**

Bevan, S.C.
Concise etymological dictionary of chemistry **1041**
Beverage literature **6036**
'Beverages' **6035**

Beydoun, Z.R.
The Middle East regional geology and petroleum resources **1757**

Beyer, M.B.
Encyclopedia of materials science and engineering **3645**

Beyer, W.H.
CRC handbook of mathematical sciences **571**
CRC handbook of tables for probability and statistics **574**
CRC standard mathematical tables **603**
Beyond broadcasting into the cable age **5615**
'Beyond MEDLINE: a review of ten non-MEDLINE CD-ROM databases for the health sciences' **2978**

Bezner, H.
Electrical machines dictionary **3983**

BGIRA
See
British Glass Industry Research Association

Bhatnagar, K.P.
Elsevier's dictionary of civil engineering **4583**
Elsevier's dictionary of geosciences **1373**

Bhattacharya, S.K.
Elsevier's dictionary of civil engineering **4583**
BI-Lexikon Rassekatzen **5405**

Bianchina, P.
Illustrated dictionary of building material and techniques **6972**

Biass-Ducroux, F. *and* Napp-Zinn, K.
Glossary of genetics in English, French, Spanish, Italian, German, Russian **2142**
'Bibliografia botanica para América Latina' **2416**
Bibliografia de arqueologia y etnografia: Mesoamerica y norte de México, 1514-1960 **1954**
Bibliografia Kopernikowska, 1509-1955 **664**
Bibliografiia izdanii Akademii Nauk SSSR **358**
Bibliographia cartographica **778**
Bibliographic guide for editors & authors **224**
Bibliographic guide to anthropology and archaeology 1987 **1870**
Bibliographic guide to conference publications, 1975- **447**
Bibliographic guide to technology **2787**
Bibliographical history of electricity & magnetism **987**
A Bibliographical index of the British flora **2402**

Bibliographical index on physical and health education, sport and allied subjects **3278**
Bibliographical series **3824**
Bibliographical series. [Newberry Library Center for the History of the American Indian] **1979**

Bibliographical Society
[Dictionary of printers and booksellers in England, Scotland and Ireland] **5640**
Bibliographical tools in apiculture **5422**
Bibliographie astronomique **612**
Bibliographie de l'Afrique sud-saharienne, sciences humaines et sociales **1947**
Bibliographie der Islamischen Medizin und Pharmacie **3166**
Bibliographie gastronomique **5537**
Bibliographie générale de l'astronomie **610**
Bibliographie géographique internationale **1583**
Bibliographie internationale d'anthropologie sociale et culturelle **1873**
Bibliographies on the Australian aborigine **1998**
'Bibliography and index of catalogues of type, figured and cited fossils in museums in Britain' **1814**
Bibliography and index of geology **1349**
Bibliography and index of geology and allied sciences for Wales and the Welsh borders, 1536-1896 **1500**
Bibliography and index of geology and allied sciences for Wales and the Welsh borders, 1897-1958 **1501**
Bibliography and index of micropaleontology **1820**
Bibliography and index to palaeobotany and palynology, 1950-70 **1829**
Bibliography and research material of the history of mathematics **541**
A Bibliography of African ecology **2122**
Bibliography of agriculture **5087**
A Bibliography of American natural history **2267**
'A Bibliography of anthropological bibliographies: Africa' **1942**
'A Bibliography of anthropological bibliographies: the Americas' **1950**
Bibliography of astronomers **665**
A Bibliography of astronomy, 1970-1979 **617**
Bibliography of bioethics **3199**
A Bibliography of birds **2664**
Bibliography of books and pamphlets on the history of agriculture in the United States, 1607-1967 **5143**

Branch, A.E. *(contd.)*
Multi-lingual dictionary of commercial international trade and shipping terms **5797**

Brand, J.R.
Handbook of electronic formulas, symbols and definitions **4099**

Brandes, E.A. and Brook, G.
Smithells' metals reference book **6317**

Brandon, A.N. and Hill, D.R.
'Selected list of books and journals for the small medical library' **2980**

Brandrup, J. and Immergut, E.H.
Polymer handbook **6595**

Brandt, A., von
Fish catching methods of the world **5442**

Brandt, J.
Emails, enamels, émaux, smalti: a dictionary in four languages **6247**

Branson, C.R.P.C.
Fishermen's handbook **5444**

Brasier, M.D.
Microfossils **1821**

Brassey's infantry weapons of the world **4514**

Brassey's Multilingual military dictionary **4470**

Brazell, J.H.
London weather **1711**

Brazier, P.C. and Shoults, T.A.S.
Guide to European collaboration in science and technology **8**

BRE Information directory, 1987 **6903**

Breaking through technical jargon **6660**

Brebbin, C.A.
Soil dynamics and earthquake engineering **4629**

Breeden, S. and Breeden, K.
A Natural history of Australia **2268**

The Breeding birds of Europe **2685**

Bremner, M.
Enquire within upon everything **5575**

Bremness, L.
The Complete book of herbs **5567**

Brennan, R.P.
Dictionary of scientific literacy **111**

Brenni, V.J.
The Art and history of book printing **5637**

Book printing in Britain and America **5626**

Bookbinding: a guide to the literature **6875**

Bresinsky, A. and Besl, H.
A Colour atlas of poisonous fungi **2458**

Bressel, J.A.
Chambers dictionary of psychiatry **3533**

Bretherick, L.
Hazards in the chemical laboratory **3332**

Brett, H. and Perry, L.
The Legal protection of computer software **6766**

Brett, P.
Building terminology **6915**

Brian, M.V.
Social insects **2605**

Brick Development Association
Bricks, their properties and use **6984**

Brickell, C.
The Royal Horticultural Society encyclopedia of gardening **5301**
The Royal Horticultural Society gardeners' encyclopedia of plants and flowers **5300**

Bricks, their properties and use **6984**

Bridger, J.P.
Glossary of United Kingdom fishing gear terms **5443**

Bridges: aesthetics and design **4635**

Bridges, E.M.
World geomorphology **1568**

Bridging the years: a short history of British civil engineering **4591**

Bridson, G.D.R.
Natural history manuscript resources in the British Isles **2247**

Bridson, G.D.R. and White, J.J.
Plant, animal and anatomical illustration in art & sciences **2373, 2555, 3215**

A Brief guide to sources of metals information **6288**

'A Brief historical survey of British navigational manuals' **734**

A Brief history of entomology **2622**

Brief subject and author index of papers published in the *Proceedings*, 1847-1950 **3779**

Briggs, A.
The Power of steam **3886**

Briggs, D. and Wahlqvist, M.
Food facts: the complete no-fads plain facts guide to healthy eating **3260**

Briggs, D.E.F.
Palaeobiology: a synthesis **1817**

Briggs, G.A.
The Cambridge photographic atlas of the planets **692**

Briggs, H.M.
Modern breeds of livestock **5370**

Brightman, F.H. and Nicholson, B.E.
The Oxford book of flowerless plants **2449**

Brindley, K.
Newnes Electronics assembly pocket book **4115**
Newnes radio and electronic engineer's pocket book **4103**

Brinkman, D.
Jane's avionics **4531**

Brinkmann, K.-H. and Schmidt, R.
Data systems dictionary **6726**

Briquet, C.M.
Les filigranes. Dictionnaire historique des marques du papier **6490**

Bristow, J.A.
Paper structure and properties **6477**

Bristow, P.
Furniture literature **6865**

Brit-line **40**

British abstracts **1073**

British and Irish herbaria **2366**

British and Irish writers on agriculture **5128**

British Army Equipment Exhibition, 22-27 June 1986, Aldershot **4503**

British Association for the Advancement of Science
Mathematical tables **600**

British Astronomical Association
The Handbook of the British Astronomical Association **651**

British bee books **5424**

British biotechnology and British industry **2777**

British bivalve seashells **2591**

British botanical and horticultural literature before 1800 **2312**

British Broadcasting Corporation
Catalogue of natural history recordings **2571**

British Caenozoic fossils (Tertiary and Quaternary) **1808**

British canals **5848**

British Carbonisation Research Association
BCRA quarterly **6022**

British Carpet Manufacturer's Association Limited
Index of quality names **6525**

British Cast Iron Research Association
BCIRA abstracts of international literature on metal castings production **4308**

British civil aircraft, 1919-1972 **4993**

British civil engineering literature, 1640-1840 **4566**

The British clothing industry year book. 1990-91 **6882**

British Computer Society Schools Committee Glossary Working Party
A Glossary of computing terms **6679**

British cookery **5557**

British Crop Protection Council
Weed control handbook **5254**

British Electricity International, London.
Modern power station practice **3962**

The British encyclopedia of dogs **5403**

British food fish **6147**

Cousins, F.W.
Sundials **6639**
Cowan, S.T.
A Dictionary of microbial taxonomy **2202**
Cowell, S.
Who's who in the environment: England **4757**
Cower, D.L. *and* Helfand, W.H.
Pharmacy: an illustrated history **3360**
Cowin, H.W.
Conway's directory of modern naval power, 1986 **4546**
Cox, A.W.
'Solid fuels' **6003**
Cox, B.
The Macmillan illustrated encyclopedia of dinosaurs and prehistoric animals **1844**
Cox, E.K. *and* Williams, K.D.
Hydrogen **5992**
Cox, J.
Environment databases 1991 **2076**
Keyguide to information sources in CAD/CAM **4282**
Patterson's German-English dictonary for chemists **1058**
Cox, J.S.
An Illustrated dictionary of hairdressing & wigmaking **6886**
Coyle, L.P.
The World encyclopedia of food **6072**
Coyne, G.S.
The Laboratory handbook of materials, equipment, and technique **1159**
CPA world directory of old age **3452**
Crabbe, D. *and* McBride, R.
The World energy book **3720**
Cracknell, A. *and* Hayes, L.
Remote sensing yearbook **776**
Cracknell, H.L.
Practical professional catering **5466**
Cracknell, H.L. *and* Kaufmann, R.J.
The Complete guide to the art of modern cookery **5545**
Practical professional cookery **5544**
Cracknell, H.L. *and* Nobis, G.
The New catering repertoire **5461**
Craft workshops in the English countryside **6619**
Crafts-Lighty, A.
Information sources in biotechnology **2752**
The UK biotechnology handbook '90 **2761**
Craig, B.D.
Handbook of corrosion data **3695**
Craig, G.Y.
Geology of Scotland **1496**

Cramp, S. *and* Simmons, K.E.L.
Handbook of the birds of Europe, the Middle East and North Africa **2679**
Crane, E.
Bees and beekeeping **5425**
Bibliographical tools in apiculture **5422**
Bibliography of tropical apiculture **5423**
Dictionary of beekeeping terms **5426**
Crawford, G.N.C.
Anatomical dictionary **3207**
Crawford, N.K.
The Innovators handbook: 1988/89 **2856**
Crawford, P.
The Living isles-Safari UK **2254**
CRC atlas of spectral data **964**
CRC fundamental measures and controls for science and technology **910**
CRC handbook of basic tables for chemical analysis **1173**
CRC handbook of biochemistry and molecular biology **2178**
CRC handbook of chemistry and physics **879, 1089**
CRC handbook of chemistry and physics: 1st student edition **1090**
CRC handbook of chromatography **1210**
CRC handbook of data on organic compounds **1263**
CRC handbook of environmental control **4759**
CRC handbook of food additives **6133**
CRC handbook of geophysical exploration at sea **1401, 1649**
CRC handbook of incineration of hazardous wastes **4805**
CRC handbook of irrigation technology **5225**
CRC handbook of laboratory safety **3331**
CRC handbook of laser science and technology **4046**
CRC handbook of lubrication **4393**
CRC handbook of management of radiation protection programs **3308**
CRC handbook of mariculture **5431**
CRC handbook of marine science **1650**
CRC handbook of materials science **3667**
CRC handbook of mathematical sciences **571**
CRC handbook of medicinal herbs **3384**
CRC handbook of microbiology **2208**
CRC handbook of physical properties of rocks **1733**
CRC handbook of physiology in aging **3449**
CRC handbook of radiation chemistry **1150**

CRC handbook of spectroscopy **1187**
CRC handbook of tables for applied engineering science **3606**
CRC handbook of tables for organic compound identification **1264**
CRC handbook of tables for probability and statistics **574**
CRC handbook of tables of functions for applied optics **951**
CRC handbook series in inorganic electrochemistry **1219**
CRC handbook series in organic electrochemistry **1146, 1253**
CRC practical handbook of marine science **2136**
CRC practical handbook of physical properties of rocks and minerals **1733**
CRC standard mathematical tables **603**
Cree, J.
Directory of special libraries in Australia **429**
Cremer, H.W.
Chemical engineering practice **5903**
Crespi, R.S.
Patenting in the biological sciences **2047**
Creswell, H.T.
A Dictionary of military terms, English-Japanese, Japanese-English **4480**
CRIB **2062**
Crispin, F.S.
Dictionary of technical terms **110**
Critical supplement to the Atlas of the British flora **2389**
Cromer, D.E.
'Biographies of physicists **900**
'Radioactive waste management and disposal information sources' **4811**
Crompton, T.R.
Battery reference book **4018**
Crone, G.R.
Maps and their makers **809**
Croner's hazardous waste disposal guide **4806**
Crook, R.E.
A Bibliography of Joseph Priestley, 1733-1804 **352**
Crook, S.,
Chemistry and technology of lubricants **4394**
Crop physiology abstracts **5255**
Crop protection chemicals reference **5248**
Crosby, G.
CPA world directory of old age **3452**
Old age: a register of social research 1985-1990 **3453**
Crosby, M.R. *and* Magill, R.E.
A Directory of mosses **2476**

Designs: a guide to official literature on design protection **2961**

Designs and utility models throughout the world **2960**

Desk encyclopedia of intellectual property **2858**

Desk reference for neuroscience **3517**

Desmond, R.

Dictionary of British and Irish botanists and horticulturists **2357**

Wonders of creation: natural history drawings in the British Library **2234**

Desrosier, N.W.

Fundamentals of food freezing **6140**

Technology of food preservation **6132**

Detroit Public Library

The Automotive history collection of Detroit Public Library **4875**

Deutsche Industrie Normen

Katalog für technische regeln **2814**

Deutscher Hotelführer, 1986 **5491**

Development of the oil and gas resources of the United Kingdom **6196**

Developments in hydraulic engineering **4708**

Developments in robotics **6833**

Developments in soil mechanics **4623**

Devine, J. and Watts, A.

Information pack on product liability **4286**

Devon, T.K.

Handbook of naturally occurring compounds **1276**

DeVorkin, D.H.

The History of modern astronomy and astrophysics **657**

Dewar, J.D.

Manual of ready-mixed concrete **6271**

Deziron, M. and Bailey, L.

A Directory of European environmental organizations **2104**

DH yearbook of research and development 1990 **3193**

Di Giacomo, J.

VLSI handbook **4122**

Diagram Group

Animal anatomy on file **5343**

Comparisons; of distance, size, area, volume, mass, weight, density, energy, temperature, time, speed and number throughout the universe **911**

The Complete encyclopedia of exercises **3274**

Handbook of arts and crafts **6900**

The Human body on file **3206**

Pets: every owner's encyclopedia **5395**

Diagram Group *(contd.)*

Weapons: an international encyclopedia, from 5000 BC to 2000 AD **4499**

Diamant, L.

The Broadcast communications dictionary **4185**

Diamond, A.W.

Save the birds **2297**

The Diamond dictionary **6424**

Diamond, S.A.

Trade mark problems and how to avoid them **2874**

Diamonds **6423**

Dibley, S.

'Online databases in civil engineering (Construction; building)' **4564**

Dibner, M.D.

Biotechnology guide U.S.A. **2780**

Diccionario de electrónica y energía nuclear, inglés-español **4084**

Diccionario de ingenieria ambiental y ciencias afines **4751**

Diccionario de la terminologia de plancton marine **2584**

Diccionario de química y de productos químicos. Español-inglés, inglés-español **1065**

Diccionario de términos cientificos y técnicos **161**

Diccionario inglés-espanõl para medicos y estudiantes de medicina **3075**

Diccionario maritimo y de construcción naval **4927**

Diccionario matemático, español-inglés/inglés-español **507**

Diccionario médico, espanõl-inglés, inglés-espanõl **3073**

Diccionario médico-suplemento español **3045**

Diccionario politécnico de las linguas española e inglesa **157**

Diccionario technologico inglés-español de electricdad, electronica, telecomunicacion y materias afines con la fisica, la optica y la quimica **3913**

Diccionario tecnico maritimo, inglés-español-inglés **5808**

Dicionario de termos técnicos (inglés-portugués) **166**

Dicionário metalúrgico **6301**

Dicionário técnico poliglota **104**

Dick, E.M.

Current information sources in mathematics **475**

Dickenson, C.

Filters and filtration handbook **4032**

Pumping manual **4276**

Dickinson, C. and Lucas, J.

The Encyclopedia of mushrooms **2474**

Dickinson, W. and Cheremisinoff, P.N.

Solar energy technology handbook **3758**

Dickson, E.

Laura Ashley complete guide to home decoration **5588**

Dickson, J.H.

Concise encyclopedia of astronomy **623**

Dickson, K.M.

History of aeronautics and astronautics **5034**

Dictionar poliglot de geodezie, fotogrammetrie si cartografie **755**

Dictionar tehnic: englez-român **156**

Dictionar tehnic: român-englez **155**

Dictionaries and vocabularies **5071**

Dictionary and encyclopaedia of paper and paper-making **6475-6476**

Dictionary and handbook of nuclear medicine and clinical imaging **3402**

Dictionary catalog **933**

Dictionary catalog. [Giannini Foundation of Agricultural Economics Library] **5156**

Dictionary catalog of the Department Library **80**

Dictionary catalog of the Department [of the Interior] **1355**

Dictionary catalog of the Jessie E. Moorland Collection of negro life and history **1962**

Dictionary catalog of the Library. [Mariners Museum] **5790**

Dictionary catalog of the Library of Bernice P. Bishop Museum **1974**

Dictionary catalog of the National Agricultural Library, 1862-1965 **5065**

Dictionary catalog of the Schomburg Collection of negro literature and history **1964**

Dictionary catalog on deafness and the deaf **3496**

Dictionary catalog. [Water Resources Center Archives] **1762, 4771**

The Dictionary catalogue **94**

Dictionary catalogue. [London School of Hygiene and Tropical Medicine] **2997**

A Dictionary catalogue of the Blacker-Wood Library of Zoology and Ornithology **2515**

Dictionary catalogue of the history of printing **5639**

Dictionary catalogue of the Stefansson Collection on the Polar regions **1545**

Dictionary catalogue of the Yale Forestry Library **5162**

Dictionary. Chemical processes and products **5943-5945**

Earney, F.C.F. *(contd.)*
Petroleum and hard minerals from the sea **1758**
Earth and astronomical sciences research centres **666, 1388**
The Earth from space **686**
The Earth: its origin, history and physical constitution **1519**
The Earth sciences: an annotated bibliography **1385**
Earth sciences reference **1367**
Earthquake prediction **1404**
East European peasantries: social relations **1925**
Eastman, C.R.
A Bibliography of fishes **2638**
Eastwood, B.
Directory of audio-visual sources: history of science, medicine, and technology **303**
Ebel, H.F.
The Art of scientific writing **400**
Ebied, R.Y.
Bibliography of mediaeval Arabic and Jewish medicine and allied sciences **3167**
Ecce homo: an annotated bibliographic history of physical anthropology **1871**
ECDB: (Electronic Components Data Bank) **4117**
ECO directory of environmental databases in the United Kingdom **2077**
Ecological abstracts **2096**
Ecological engineering **2083**
Ecological terminology **2086**
Ecology abstracts **2097**
Ecology and biogeography in India **2121**
The Ecology of fossils **1819**
The Ecology of river systems **2114**
Economic Commission for Europe
Annual bulletin of steel statistics for Europe **6365**
Annual review of engineering industries and automation **3631**
Economic Commission for Europe. Inland Transport Committee
Glossary for inland transport **5846**
Economic mineral deposits **1742**
Economics of shipping practice and management **5795**
Economist Intelligence Unit
World textile trade and production trends **6518**
Ecosystems of the world **2113**
EDAP **5249**
Edden, G.
The 'Good Housekeeping' step-by-step cook book **5548**

Eden, P.
Dictionary of land surveyors and local cartographers of Great Britain and Ireland, 1550-1850 **764-765**
Edmunds, R.A.
The Prentice-Hall standard glossary of computer terminology **6685**
Edmundson, R.S.
Dictionary of organophosphorus compounds **1273**
Edson, D.T.
Glossary of terms in computer assisted cartography **793**
Edwards, E.H.
A Standard guide to horse and pony breeds **5382**
Edwards, L.M.
Handbook of geothermal energy **1431**
Edwards, W.N.
The Early history of palaeontology **1805**
EEC Translation Service
Biotechnology glossary **2765**
Eekhof-Stork, N.
The World atlas of cheese **5419**
EESS/Encyclopaedia of engineering signs and symbols **3580**
Efficient electricity use **3930**
The Efficient use of energy **3714**
The Efficient use of steam **3878**
Egan, H.
Pearson's chemical analysis of foods **6109**
Egan, M.
Communications: the most comprehensive guide to the UK telecommunications industry **4170**
Egerton, F.N.
Landmarks of botanical history **2353**
Eggers-Lura, A.
Solar energy in developing countries **3759**
Egon Ronay's Cellnet guide, 1991. Hotels & restaurants in Great Britain and Ireland **5492**
Ehrich, F.F.
Handbook of rotordynamics **4371**
Eicher, D.J.
The Universe from your backyard **707**
Eichhorst, C.
Medical dictionary **3061**
Eight-language dictionary of medical technology **3036**
Einheitssacht: industrielle anorganische chemie **5900**
Einspruch, N.G.
VLSI Electronics **4126**
VLSI Handbook **4127**
Eirich, F.R.
Science and technology of rubber **6556**

Eisenberg, J.F. *and* **Redford, K.A.**
Mammals of the neotropics **2725**
Eisenreich, G. *and* **Sube, R.**
Dictionary of mathematics in four languages **491**
Eisenschitz, T.S.
Patents, trade marks and designs in information work **2928**
Eisenschitz, T.S. *and* **Phillips, J.**
The Inventors' information guide **2884**
Eiss, H.E.
Dictionary of mathematical games, puzzles and amusements **555**
ELCOM **4051**
Eldridge, F.R.
Wind machines **3978**
ELECNUC **3866**
Electric cables handbook **3991**
Electric power database **3957**
Electrical and electronic technologies: a chronology of events and inventors, 1940-1980 **3941**
Electrical and electronic technologies: a chronology of events and inventors from 1900 to 1940 **3940**
Electrical and electronic technologies: a chronology of events and inventors, to 1900 **3939**
Electrical and electronics abstracts. Science abstracts. Section B **3919**
Electrical and electronics graphic symbols and reference designations **3890, 4050**
Electrical and electronics trades directory **3924**
Electrical engineering **3933**
Electrical engineering handbook **3926**
Electrical engineer's reference book **3928**
Electrical installations handbook **4006**
Electrical machines dictionary **3983**
Electrical motor handbook **3984**
Electrical patents index **3938**
Electrical Power Industry Abstracts **3958**
Electrical properties of glasses, glass-ceramics and amorphous solids **6234**
Electrical properties of solids, surface preparation and methods of measurement **985**
Electrical protection relays **4009**
Electrical Research Association
ERA reports, 1969-1986 **3916**
Electrical standards in world trade **3936**
Electricity before nationalization **3969**
Electricity Council
Electricity supply in Great Briatin **3964**
Handbook of electricity supply statistics **3965**
Electricity in the 17th and 18th centuries **986**

Elsevier's paint dictionary in four
languages **6281**
Elsevier's rubber dictionary **6562**
Elsevier's sugar dictionary **6118**
Elsevier's telecommunication dictionary
in six languages **4159**
Elsevier's wood dictionary in seven
languages **6434**
Emails, enamels, émaux, smalti: a
dictionary in four languages **6247**
EMBASE **2973**
Embery, P.G. *and* **Fuller, J.P.**
A Manual of new mineral names,
1892-1978 **1319**
Embleton, C.
Geomorphology of Europe **1599**
Embleton, C. *and* **King, C.A.M.**
Glacial geomorphology **1537**
Periglacial geomorphology **1538**
Embrey, P.G. *and* **Fuller, J.P.**
A Manual of new mineral names,
1892-1978 **6318**
Emiliani, E.
Dictionary of the physical sciences
122
EMIS: (Electronic Materials Information
Service) **4144**
Emmet, A.M.
The Scientific names of the British
lepidoptera **2628**
Emmett, P.H.
Catalysis **1170**
Emmons, L.H.
Neotropical rainforest mammals **2724**
Empires of time **822**
Emsley, J.
The Elements **1343**
Emslie-Smith, D.
Textbook of physiology **3220**
Encyclopaedia of agriculture **5067**
Encyclopaedia of antique scientific
instruments **6642**
The Encyclopaedia of astronomy and
space **624**
Encyclopaedia of British railway
companies **5769**
Encyclopaedia of engineering signs and
symbols **3580**
Encyclopaedia of ferns **2487**
An Encyclopaedia of homoeopathy
3362
Encyclopaedia of how it works **3780**
Encyclopaedia of how it's made **3781**
Encyclopaedia of hydraulics, soil and
foundation engineering **4613, 4694**
The Encyclopaedia of ignorance **81**
Encyclopaedia of mathematics **478**
An Encyclopaedia of metallurgy and
materials **6291**
The Encyclopaedia of microcomputer
terminology **6740**
Encyclopaedia of milk **5415**

Encyclopaedia of modern aircraft
armaments **4527**
Encyclopaedia of occupational health
and safety **3316**
The Encyclopaedia of spectroscopy
1186
Encyclopaedia of sports medicine **3268**
Encyclopaedia of statistical sciences
580
Encyclopaedia of textiles **6516**
An Encyclopaedia of the iron and steel
industry **6343**
The Encyclopaedia of type faces **5650**
The Encyclopaedia of wire **4335**
Encyclopaedic dictionary of
agrometeorology **1678**
Encyclopaedic dictionary of
mathematics for engineers and applied
scientists **484**
The Encyclopaedic dictionary of
physical geography **1576**
Encyclopaedic dictionary of physics
834
Encyclopedia and dictionary of
medicine, nursing and allied health
3022, 3468
Encyclopedia/Handbook of materials,
parts and finishes **3652**
The Encyclopedia of aging **3448**
The Encyclopedia of alcoholism **3285**
Encyclopedia of American agricultural
history **5144**
Encyclopedia of American forest and
conservation history **5175**
Encyclopedia of anthropology **1877**
Encyclopedia of antibiotics **2191**
The Encyclopedia of applied geology
1437
Encyclopedia of applied physics **837**
The Encyclopedia of aquatic life **2556**
Encyclopedia of artificial intelligence
6802
Encyclopedia of associations:
Association periodicals. V.2: Science,
medicine, and technology **383**
Encyclopedia of astronomy **627**
Encyclopedia of astronomy and
astrophysics **621**
The Encyclopedia of atmospheric
sciences and astrogeology **1669**
Encyclopedia of aviculture **5398**
The Encyclopedia of beaches and
coastal environments **1611**
Encyclopedia of bioethics **3202**
The Encyclopedia of birds **2666**
The Encyclopedia of blindness and
vision impairment **3567**
Encyclopedia of building and
construction terms **6908**
The Encyclopedia of cacti **5320**
Encyclopedia of chemical processing
and design **5938**

Encyclopedia of chemical reactions
1144
Encyclopedia of chemical technology
5896
The Encyclopedia of chemistry **1031**
The Encyclopedia of climatology **1697**
Encyclopedia of common natural
ingredients used in food, drugs and
cosmetics **5965**
Encyclopedia of composite materials
and components **3647**
Encyclopedia of computer science and
technology **6670**
Encyclopedia of computer sciences
6672
Encyclopedia of consumer law **5518**
The Encyclopedia of crafts **6899**
Encyclopedia of data protection **6771**
Encyclopedia of deafness and hearing
disorders **3499**
Encyclopedia of death **3231**
The Encyclopedia of drug abuse **3288**
Encyclopedia of earth sciences **1357**
Encyclopedia of earth system science
1358
Encyclopedia of electrochemistry of the
elements **1147**
The Encyclopedia of electronic circuits
4121
Encyclopedia of electronics **4058**
Encyclopedia of emulsion technology
5949
Encyclopedia of environmental control
technology **4738**
Encyclopedia of environmental health:
law and practice **4739**
The Encyclopedia of environmental
studies **4737**
The Encyclopedia of evolution **2155**
Encyclopedia of field and general
geology **1438**
Encyclopedia of fluid mechanics **921**
The Encyclopedia of food and nutrition
3251, 6078
Encyclopedia of food engineering **6107**
Encyclopedia of food science **6077**
Encyclopedia of food science and
technology **6073**
Encyclopedia of food technology **6076**
The Encyclopedia of gemstones and
minerals **1298**
The Encyclopedia of genetic disorders
and birth defects **3426**
The Encyclopedia of geochemistry and
environmental sciences **1413**
The Encyclopedia of geomorphology
1566
The Encyclopedia of glass **6231**
Encyclopedia of glass, ceramics and
cement **6229**
Encyclopedia of health information
sources **2983**
Encyclopedia of historic forts **4489**

Ethnic collections in libraries **1959**
Ethnic groups of insular Southeast Asia **1935**
Ethnic groups of mainland Southeast Asia **1936**
Ethnic information sources of the United States **1960**
'Ethnics in American society' **1958**
An Ethnographic bibliography of New Guinea **1973**
Ethnographic bibliography of North America **1953**
Ethnographic bibliography of South America **1968**
Ethnography. Bibliography of Russian bibliographies on the ethnography of the peoples of the USSR **1931**
Ethnologie régionale **1920**
Ethological dictionary, German - English - French **2564**
Etner, F.F.
Russo-anglo-nemetsko-frantsazki gorny slovar' **4425**
Etnografiya. Bibliografiya russkikh bibliografii po etnografii narodov SSSR (1831-1969) **1931**
Etter, L.E.
Glossary of words and phrases used in radiology, nuclear medicine, and ultrasound **3399**
Ettingen, S.G.
Dictionary of technical terms in five languages **101**
Euesden, M.
Introduction to biotechnology information **2753**
Euro abstracts **3842**
Euro abstracts. Section 2: Coal and steel **6010, 6360**
Europa camping + caravaning **5505**
Europäisches medizinisches Wörterbuch **3040**
Europe: a selected ethnographic bibliography **1927**
European Association for Animal Production, Rome
Dictionary of animal production terminology **5372**
European Brewery Convention
Elsevier's dictionary of brewing **6051**
European Communities Commission
Directory of energy databases **3700**
Food multilingual thesaurus **6104**
Onshore/offshore oil & gas multilingual glossary **6172**
Patent information and documentation in Western Europe **2929**
European Communities Commission. Directorate - General Research, Science and Education. Nuclear Plant Safety Division
Nuclear standards **3858**

European Communities Commission. Terminology Office
Glossaire des normes de l'acier **6359**
Steel standards glossary **6364**
European Communities Commission. Translation Service
Biotechnology glossary **2765**
European Community shipping directory **4931**
European Conference of Ministers of Transport
European rules concerning road traffic signs and signals **5761**
European directory of agrochemical products **5249**
European electronics directory 1991 **4094**
European environmental yearbook: nature conservation, protection of the environment, town and country planning in Belgium, Denmark, the Federal Republic of Germany, France, Greece, Ireland, Italy, Luxembourg, the Netherlands, Portugal, Spain and the United Kingdom, with special surveys on Australia, Japan, USA and USSR **2120**
European glass directory & buyers guide, 1987 **6245**
European handbook of plant diseases **5236**
European paint and resin news **6285**
European Parliament
Ecological terminology **2086**
European Parliament. Terminology Office
Terminologie du secteur de la pêche **5433**
Terminology of new and renewable sources of energy **3751**
The European passive solar handbook **3760, 7003**
European patent bulletin **2913**
The European patent system **2944**
European patents handbook **2940**
European research centres **455**
European rules concerning road traffic signs and signals **5761**
European sources of scientific and technical information **411**
European Space Agency
Recueil de terminologie spatiale **5028**
European technical consultancies **2847**
The European television directory **5614**
'Europe's railways: an asset to the environment' **4656**
Europump terminology **4274**
Evaluated Nuclear Structure Data File **3818**
Evans, A.
Glossary of molecular biology **2192**

Evans, D.
A History of nature conservation in Britain **2292**
Evans, D.S.
Under Capricorn **658**
Evans, G.
The Encyclopedia of drug abuse **3288**
Evans, J.
The Henston large animal veterinary vade mecum **5356**
The Henston small animal veterinary vade mecum **5357**
Evans, J.G. *and* Williams, T.F.
Oxford textbook of geriatric medicine **3450**
Evans-Pritchard, E.E.
A History of anthropological thought **1898**
Everett-Heath, J.
British military helicopters **4534, 5005**
Everett, T.H.
The New York Botanical Garden illustrated encyclopedia of horticulture **5283**
Everywoman's medical handbook **3237**
Eves, H.
An Introduction to the history of mathematics **531**
Evetta, J.E.
Concise encyclopedia of magnetic and superconducting materials **4013**
Evolution **2158**
The Evolution of geomorphology **1595**
Evolution of the earth **1719**
The Evolution of the English farm **5136**
Evolution of the vertebrates **2633**
The Evolution of useful things **2889**
Ewer, R.F.
The Carnivores **2743**
Ewing, G.W.
Analytical instrumentation handbook **6643**
Excavation handbook **4625**
Excerpta botanica. Sections A & B **2338**
Excerpta medica **3086**
Excerpta medica. 1. Anatomy, anthropology, embryology and histology **1888, 3210**
Excerpta medica. 3. Endocrinology **3511**
Excerpta medica. 5. General pathology **3429**
Excerpta medica. 7. Pediatrics and pediatric surgery **3444**
Excerpta medica. 8. Neurology and neurosurgery **3518**
Excerpta medica. 9. Surgery **3555**
Excerpta medica. 11. Oto-, rhino-, laryngology **3494**
Excerpta medica. 14. Radiology **3400**

Godwin, Sir H.
The History of the British flora **2395**
Gold **6373**
Gold: history and genesis of deposits **6371**
Gold, V.
'Glossary of terms used in physical organic chemistry' **1248**
Goldbeck, G.
Museen in Deutschland (West): Technische Museen **465**
The Golden century of oil, 1950-2050 **6168**
Golding, E.W.
The Generation of electricity by wind power **3976**
Goldman, J.M.
Means estimating handbook **6963**
Goldsmith, M.
UK science policy **11**
Goldstein, D.M.
Bioethics: a guide to information sources **3200**
Golland, F.J.
Dictionary of computing **6688**
Golze, A.R.
Handbook of dam engineering **4732**
Gomersall, A.
Lasers in materials processing **4035**
Machine intelligence **6798**
Gomersall, A. *and* Farmer, P.
Robotics **6812**
The Good beer guide **5498**
The Good food guide **5493**
The Good gardener's guide **5304**
The Good hardware guide **6735**
The Good hotel guide **5470**
Good Housekeeping household hints **5576**
'Good Housekeeping' microwave cookbook **5547**
The 'Good Housekeeping' step-by-step cook book **5548**
The Good pub guide **5499**
Good, R.
Britten's watch & clock maker's handbook, dictionary and guide **6624**
The Good stay guide **5480**
Good style: writing for science and technology **404**
The Good toy guide **6890**
Goodall, D.W.
Ecosystems of the world **2113**
Goodall, P.M.
The Efficient use of steam **3878**
Goode, P.T.
Abortion bibliography **3578**
Goodier, J.H.
Dictionary of painting and decorating **7005**

Goodman, A.
Cambridge illustrated thesaurus of computer science **6716**
Goodman, W.L.
The History of woodworking tools **6458**
Goodsell, D.
Dictionary of automotive engineering **4863**
Goodwin, D.
Pigeons and doves of the world **2709, 5397**
Goodwin, P.
The Construction and fitting of the English sailing men of war, 1650 - 1850 **4547**
Göpel, W.
Sensors **912**
Gordon, R.E.
Conservation directory **2294**
Gorshkov, S.G.
World ocean atlas **1658**
Gotch, A.P.
Mammals - their Latin names explained **2720**
Gottstein, K.
Directory of international cooperation in science and technology **434**
Goudie, A.
The Encyclopaedic dictionary of physical geography **1576**
Gough, B.E.
World environmental directory: North America **4764**
Gould, W.A.
Glossary for the food industries **6115**
Gourvish, T.R.
British railways, 1948-73 **5774**
Government publications. Sectional list 1, revised August 1986. Agriculture, fisheries, food and forestry **5122**
Government publications. Sectional list 11, revised February 1986. Department of Health and Social Security **3104**
Government publications. Sectional list 37: Meteorological Office **1685**
Government publications. Sectional list 45, revised November 1986. Natural Environment Research Council. British Geological Survey **1492**
Government publications. Sectional list 61, revised March 1986: Construction **6905**
Government reports announcements & index **197**
Government research directory **461, 2850**
Gowan, J.E.
Name index of organic reactions **1256**

Gowenlock, A.H.
Varley's practical clinical biochemistry **2173**
Gozmany, L.
Vocabularium nominum animalium Europae system linguis redactum **2529**
Gradshteyn, I.S. *and* Ryzhik, I.M.
Tables of integrals, series and products **567**
Graeser, K.
Who's who in electronics, 1987 **4098**
Graf, A.B.
Exotica 4 international **5316**
Tropica: colour cyclopedia of exotic plants and trees from the tropics and subtropics **5317**
Graf, R.F.
Electronic databook **4102**
The Encyclopedia of electronic circuits **4121**
Modern dictionary of electronics **4075**
Graham, A.L.
The Catalogue of meteorites **1737**
Graham, F.A. *and* King, P.A.
'Nuclear power engineering' **3822**
Graham, G.A.
Encyclopedia of industrial automation **6781**
Graham, J. *and* Lowe, S.
The Facts on File dictionary of telecommunications **4162**
Grallafent, R.J.
Intellectual property law and taxation **2951**
'The Gramophone' guide to hi-fi **6849**
Grande dizionario di marina, italiano-inglese, inglese-italiano **5807**
Grandison, A.G.C.
Snakes - a natural history **2659**
Grant and Hackh's chemical dictionary **1044**
Grant, C. *and* Ballantyne, J.
Discovery & invention **302**
Grant, E.
Physical science in the Middle Ages **306**
A Source book in medieval science **306**
Grant, E.H.
Microwaves **4030**
Grant, R.L. *and* Grant, A.C.
Grant and Hackh's chemical dictionary **1044**
Graphic arts encyclopedia **5667**
Graphic arts literature abstracts (GALA) **5670**
Graphical symbols for components of servo-mechanisms **6777**
Graphics, design and printing terms **5664**

Great Britain. Ministry of Agriculture, Fisheries and Food *(contd.)*
Sea fisheries statistical tables **5446**

Great Britain. Ministry of Agriculture, Fisheries and Food. Directorate of Fisheries Research
Guide to reference works on marine science and fisheries **5429**

Great Britain. Ministry of Agriculture, Fisheries and Food. Tolworth Library
Thesaurus (for animal health, vertebrate pest biology and control) **5345**

Great Britain. Ministry of Defence
Applied geology for engineers **1418**
Manual of map reading and land navigation **786**
Textbook of topographical surveying **762**

Great Britain. Ministry of Defence. Director of Naval Warfare
Diving manual **4713**

Great Britain. Ministry of Defence. HQ Library Services
MOD Library accessions: new books and articles **4484**

Great Britain. Ministry of Defence. Navy
Admiralty manual of navigation **732, 5825**
Admiralty manual of seamanship **5826**

Great Britain. Nautical Almanac Office
The Star almanac for land surveyors.. **763**

Great Britain. Ordnance Survey
The History of the retriangulation of Great Britain, 1935-1962.. **761**

Great chemists **1114**
Great engineers **3625**
Great engineers and pioneers in technology **3620**
The Great international disaster book **1407**
Great passenger ships of the world **4947**
The Great physicists, from Galileo to Einstein **895**
Greek and Roman chronology **825**
Greek and Roman technology **2835**

Greeley, R. *and* Batson, R.M.
Planetary mapping **787**
The Green book **2091, 5208**
Green dictionary **4820**
The Green encyclopedia **4817**
Green engineering **4755**

Green, F.J.
Sigma-Aldrich handbook of stains, dyes and indicators **6278**
Green growers guide 1990/91 **5308**

Green laurels **2243**
Green planet: the story of plant life on earth **2354**

Green, S.
'Beverages' **6035**
'Food science' **6070**
Keyguide to information sources in food science & technology **6069**

Greenberg, B.
How to find out in psychiatry **3522**

Greene, A.M.
Designs and utility models throughout the world **2960**

Greene, E.L.
Landmarks of botanical history **2353**

Greene, R.W.
'Chemical engineering' guide to compressors **4237**

Greenhalgh, G. *and* Jeffs, E.
Nuclear industry almanac **3850**

Greenhill, B.
The Ship series **4934**

Greenman, D.
Jane's merchant ships, 1987-88 **4951**

Greenpeace
Global warming **4825**
The Greenpeace book of Antarctica **2124**

Greenstein, C.H.
Dictionary of logical terms and symbols **558**

Greenway, A.
Soviet merchant ships **4949**

Greenway, J.
Bibliography of the Australian aborigines and the native peoples of Torres Strait, to 1959 **1996**

Greenwich Observatory
The Royal Observatory at Greenwich and Herstmonceux, 1675-1975 **677**

Greenwood, J.A. *and* Hartley, H.O.
Guide to tables in mathematical statistics **591**

Greg, R.P. *and* Lettsom, W.G.
Manual of mineralogy of Great Britain and Ireland **1325**

Gregory, C.E.
A Concise history of mining **4438**

Gresswell, R.K. *and* Huxley, A.
Standard encyclopedia of the world's rivers and lakes **1780**

Gretes, F.C.
Directory of international periodicals and newsletters on the built environment **6935**

Greuter, W.
International Code of Botanical Nomenclature **2348**

Grieve, M.
A Modern herbal **3385**

Griffith, H.W.
Complete guide to symptoms, illness and surgery **3458**

Griffith, H.W. *(contd.)*
Complete guide to symptoms, illness and surgery for people over 50 **3458**

Grigg, D.B.
The Agricultural systems of the world **5108**

Grime, J.P.
Comparative plant ecology **2436**

Grimmett, R.F.A. *and* Jones, T.A.
Important bird areas in Europe **2304**

Grimshaw, A.
The Horse: a bibliography of British books, 1851-1976 **5383**

Grimshaw, R.W.
The Chemistry and physics of clays and allied ceramic materials **6251**

Grinstein, L. *and* Campbell, P.J.
Women of mathematics **547**

Grinstein, L.S.
Mathematical book review index, 1800-1940 **473**

Grogan, D.
Science and technology: an introduction to the literature **58**

Grollig, F.V. *and* Tax, S.
Serial publications in anthropology **1890**

Groombridge, B.
World checklist of threatened amphibians and reptiles **2299**

Gros, E. *and* Singer, L.
Russian-English - French-German constructional engineering dictionary **4576**

Gross, H.
Dictionary of chemistry and chemical technology **1056**

Gross, M.G.
Oceanography **1632**
Das Grosse Handbuch des Segelns **5828**
Grosse Molkerei Lexikon **5415**

Grossman, H.J.
Grossman's guide to wines, spirits and beers **6038**
Grossman's guide to wines, spirits and beers **6038**

Grosvenor, J. *and* Morrison, K.
The Ivanhoe guide to desktop publishing **5674**
Ground engineering yearbook, 1991 **4587**
Ground engineer's reference book **4567**
Ground-water studies: an international guide for research and practice **1763**

Grouws, D.A.
Handbook of research on mathematics, teaching, and learning **523**

Groves, D.G. *and* Hunt, L.M.
Ocean world encyclopedia **1628**

Hannah, L.
Electricity before nationalization
3969
Engineers, managers and politicians
3968
Hannay, N.B.
Treatise on solid state chemistry **1127**
Hanneman, H.W.A.M.
The Patentability of computer
software **6765**
Hanniball, A.
Aircraft, engines and airmen **4963**
Hansen, E.R.
A Table of series and products **572**
Hanson, A.A.
Practical handbook of agricultural
science **5103**
Hanson, H.J.
The Cruising Association Library
catalogue **5788**
Hanssen, M.
The New E is for additives **6134**
Harbord, J.B.
Glossary of navigation **736**
Harding, D.
Weapons: an international
encyclopedia, from 5000 BC to
2000 AD **4499**
Harding, K.A. and Welch, K.R.G.
Venomous snakes of the world **2657**
Hardingham, M.
Illustrated dictionary of fabrics **6508,
6528**
Hardup, P.
The Timetable of technology **2817**
Hardyment, C.
From mangle to microwave **5579**
Hargreaves, D. and Fromson, S.
World index of strategic minerals
6292
Hargreaves, J.M.
Literature sources in toxicology: a
survey **3406**
Harland, J.
Seamanship in the age of sail **5816**
Harland, W.B.
Geologic time scale 1989 **1722**
Harmer, S.F. and Shipley, A.
The Cambridge natural history **2520**
Harper & Row's complete field guide to
North American wildlife. Eastern
edition **2264**
Harper & Row's complete field guide to
North American wildlife. Western
edition **2265**
Harper, C.A.
Handbook of electronic systems
design **4067**
Handbook of plastics, elastomers and
composites **6574**
Harrap's French and English data
processing dictionary **6727**

Harrap's French and English science
dictionary **147**
Harrap's illustrated dictionary of science
112
Harrap's illustrated history of the 20th
century - Science **311**
Harrap's visual French-English
dictionary **148**
Harré, R.
The Philosophies of science **90**
Scientific thought, 1900-1960 **312**
Harrington, J.M. and Gill, F.S.
Occupational health **3318**
Harris, C.D.
Annotated world list of selected
current geographical serials **1587**
Bibliography of geography **1565**
A Geographical bibliography for
American libraries **1562**
Harris, C.D. and Fellmann, J.D.
International list of geographical
serials **1586**
Harris, C.M.
Dictionary of architecture and
construction **6916**
Shock and vibration handbook **937**
Harris, K.M.
Shorter guide to pests, diseases and
disorders of garden plants **2426**
Harris, L.
Adhesives and adhesion **6221**
Harris, S.L.
Agents of chaos: earthquakes,
volcanoes, and other natural
disasters **1408**
Harris, T.
The Natural history of the
Mediterranean **2249**
Harris, T.Y.
Wild flowers of Australia **2420**
Harrison, P.
Seabirds: an identification guide **2682**
Harrison, R.M. and Perry, R.
Handbook of air pollution analysis
3305
Harrison, S.G.
The Oxford book of food plants **2443**
Harry's cosmeticology **6155**
Hart, F.D.
French's index of differential
diagnosis **3457**
Hart, H.
Hart's rules for compositors and
readers at the University Press,
Oxford **5656**
Hartcup, G.
The Silent revolution **4494**
Hartman, K.
Dictionary of process technology in
four languages **5942**
Hartman, T.
The Guinness book of ships and
shipping facts and feats **4913**

Hartmann, H.T.
Plant science **2332**
**Hartmann-Petersen, R. and Pigford,
J.N.**
Dictionary of science **117**
Hart's rules for compositors and readers
at the University Press, Oxford **5656**
Hartung, E.J.
Astronomical objects for southern
telescopes **628**
Harty, F.J. and Ogston, R.
Concise illustrated dental dictionary
3503
Harvard encyclopedia of American
ethnic groups **1952**
Harvard University
Catalogue of the Library of the
Museum of Comparative Zoology
2514
Gray Herbarium index **2480**
**Harvard University. Peabody
Museum of Archaeology and
Ethnology. Library**
Catalogue. [Peabody Museum of
Archaeology and Ethnology,
Library] **1868**
Index to subject headings. [Peabody
Museum of Archaeology and
Ethnology, Library] **1869**
Harvey, A.P. and Diment, J.A.
Geoscience information **1513**
Harvey, N.
A History of farm buildings in
England and Wales **5196**
The Industrial archaeology of farming
in England and Wales **5132**
The Harwin chronology of inventions,
innovations, discoveries **2888**
Haseler, A.E.
District heating **6993**
Haslam, J.M.
Writing engineering specifications
3632
Hatch, J.E.
Aluminum: properties and physical
metallurgy **6404**
Hatch, P.E.
'Military aircraft of the world' **4533**
Hathway, D.E.
Harrap's French and English science
dictionary **147**
Haubrich, W.S.
Medical meanings **3055**
Hausner, H.H.
Beryllium: its metallurgy and
properties **6410**
Handbook of powder metallurgy
4319
Haux, G.F.K.
Subsea manned engineering **4714**
Hawke, N.
Encyclopedia of environmental health:
law and practice **4739**

Hawkes, P.W.
Principles of electron optics **945**
Hawking, S.F. *and* Israel, W.
Three hundred years of gravitation **917**
Hawkins, R.E.
Encyclopedia of Indian natural history **2263**
Hawksworth, D.L.
Ainsworth & Bisby's dictionary of the fungi **2453**
Hawksworth, D.L. *and* Seaward, M.R.D.
Lichenology in the British Isles, 1568-1978 **2475**
Hawley, G.C.
Hawley's condensed chemical dictionary **1046**
Hawley, M.E.
Coal **6004**
Hawley's condensed chemical dictionary **1046**
Haws, D.
The Maritime history of the world **4939**
Hay, J.T.C.
Offshore UK **6211**
Hay, W.W.
Railroad engineering **4640**
Hayes, M.H.
Veterinary notes for horse owners **5386**
Hayman, R.W.
A Field guide to the mammals of Africa, including Madagascar **2731**
Haynes, J.R.
Foraminifera **1835**
Haynes, W.
American chemical industry **5937**
Hayward, A. *and* Weare, F.
Steel detailers' manual **4608**
Hayward, P.J. *and* Ryland, J.S.
The Marine fauna of the British Isles and north-west Europe **2560**
Hazardous chemical data book **3337**
Hazardous chemicals desk reference **3338, 3415**
Hazardous chemicals on file **3340**
Hazardous materials dictionary **3311, 3408**
Hazardous materials: sources of information on their transportation **3329**
Hazardous waste disposal guide **4806**
Hazardous waste management **4810**
Hazards in the chemical laboratory **3332**
Hazewinkel, M.
Encyclopaedia of mathematics **478**
Headland, R.K.
Chronological list of Antarctic expeditions and related historical events **1557**

Health and safety at work **3315**
The Health and safety directory **3312**
Health and safety in biotechnology **2758**
Health and safety science abstracts **3313**
Health care of the elderly **3447**
The Health directory **3247**
Health education index **3105**
The Health information handbook **2988**
Health media review index **3120**
Health media review index 1984-1986 **3120**
Health science books, 1876-1982 **2989**
Health sciences information sources **2992**
Hearn, P.
The Business of industrial licensing **2942**
Heastie, H.
A Course in elementary meteorology **1672**
Heat bibliography, 1948/52 **972**
The Heat pump **4253**
Heat Transfer and Fluid Flow Service
HTFS digest **975**
Heath, Everett- J.
See
Everett-Heath, J.
Heath, J. and others
The Moths and butterflies of Great Britain and Ireland **2629**
Heath, Sir T.L.
A History of Greek mathematics **544**
Heathcote, N.H. de V.
Nobel Prize Winners in physics, 1901-1950 **896**
Heaton, C.A.
The Chemical industry **5904**
Heaton, N.
Traditional recipes of the British Isles **5563**
The Heavy elements **1344**
Heawood, R. *and* Larke, C.
The Directory of appropriate technology **2839**
Hecht, J.
The Laser guidebook **4038**
Heck, A.
International directory of astronomical associations and societies **669**
International directory of professional astronomical institutions **670**
Hedberg, H.D.
International stratigraphic guide **1720**
Heftmann, E.
Chromatography **1199**
Heider, K.G.
Films for anthropological teaching **1896**
Heijenoort, J. van
From Frege to Gödel **532**

Heiken, G.
Lunar sourcebook **693**
Heilbron, J.L.
Electricity in the 17th and 18th centuries **986**
Heilbron, J.L. *and* Wheaton, B.R.
Literature on the history of physics in the twentieth century **892**
Heiler, T.
Technisches Bildwörterbuch für spanende Werkzeuge zur 'Metallbearbeitung' **4406**
Heinemann dental dictionary **3502**
Heinisch, K.F.
Dictionary of rubber **6559**
Heistar, R.
Dictionary of abbreviations in medical sciences **2965**
Heldman, D.R.
Handbook of food engineering **6080**
Helgerson, L.W.
CD-ROM **6857**
Hellemans, A. *and* Bunch, B.
The Timetables of science **273**
Heller, S.R.
The Beilstein online database **1239**
Helmake, J.G.
Handbuch der Zoologie **2519**
Helminthological abstracts **2594**
Helms, H.
The McGraw-Hill computer handbook **6673**
Helms, H.L.
Electronics applications sourcebook **4093**
Helping to care: a handbook for carers at home and in hospital **5599**
Hematology **3492**
Hemisphere handbook of heat exchanger design **4222**
Henbest, N. *and* Marten, M.
The New astronomy **723**
Henderson, G.N. *and* Coffey, D.J.
The International encyclopedia of cats **5406**
Henderson, H.
Science tracer bullets: a reference guide to scientific, technological, health, and environmental information sources **208**
Henderson, I.F. *and* Henderson, W.D.
Henderson's dictionary of biological terms **2020, 2524**
Henderson's dictionary of biological terms **2020, 2524**
Henley's formulas for home and workshop **5577**
Henrey, B.
British botanical and horticultural literature before 1800 **2312**
The Henston large animal veterinary vade mecum **5356**

Historical farm records **5137**
Historical geology of Ireland **1497**
Historical instruments in oceanography **1663**
A Historical introduction to the philosophy of science **278**
The History and archaeology of ports **5842**
History of aeronautics and astronautics **5034**
A History of agricultural science in Great Britain, 1620-1954 **5129**
A History of Antarctic science **1556**
A History of anthropological thought **1898**
The History of anthropology **1899**
The History of astronomy from Herschel to Hertzsprung **659**
The History of biology **2051, 2053**
History of book publishing in the United States **5704**
History of books and printing **5624**
History of botanical science **2355**
A History of brewing **6050**
History of British agriculture, 1846-1914 **5127**
A History of British gas industry **6030**
The History of British geology **1494**
A History of British livestock husbandry, 1700-1900 **5380**
A History of British livestock husbandry to 1700 **5379**
A History of British motorways **5757**
A History of British publishing **5698**
History of British space science **5038**
History of building **6945**
A History of building materials **6975**
History of Cambridge University Press **5643**
History of cartography **807**
The History of cartography **810**
History of chemical engineering **5930**
The History of chemical technology **5933**
The History of chemistry **1106**
A History of chemistry **1110**
A History of civil engineering **4593**
The History of civil engineering since 1600 **4592**
The History of classical physics **893**
The History of clocks and watches **6632**
History of comparative anatomy **3216**
A History of computing technology **6713**
A History of control engineering, 1800-1930 **6795**
A History of domesticated animals **5376**
A History of electric light and power **3963, 4849**
The History of electric wires and cables **3994**

The History of electrical technology **3945**
A History of embryology **2430, 2563**
A History of engineering and technology **3617**
A History of engineering in classical and medieval times **3618**
The History of engineering science **3616**
A History of English brickwork **6985**
A History of farm buildings in England and Wales **5196**
History of geology and palaeontology **1472**
History of geomorphology **1592**
The History of geophysics and meteorology **1400, 1688**
A History of graphic design **5671**
A History of Greek fire and gunpowder **5993**
A History of Greek mathematics **544**
History of hydrology **1775**
A History of immunology **3489**
A History of linear electric motors **3987**
History of magic and experimental science **284**
The History of man-powered flight **5002**
A History of marine engineering **4938**
A History of mathematics **528**
History of mathematics **536**
A History of mathematics education in England **548**
The History of mathematics, from antiquity to the present **539**
A History of mechanical engineering **3801**
A History of medicine **3124, 3131**
History of medicine **3146**
A History of metallography **3687**
A History of metallurgy **6321**
A History of metals **6319**
'The History of meteors and meteor showers.' **1738**
The History of modern astronomy and astrophysics **657**
The History of modern physics **890**
The History of modern whaling **5450**
A History of nature conservation in Britain **2292**
History of oriental astronomy **660**
A History of parasitology **2165**
History of perfume **6162**
A History of physics in its elementary branches **886**
A History of platinum and its allied metals **6378**
History of polymer science and technology **6606**
A History of printing and printers in Wales **5646**
The History of psychiatry **3543**

A History of refrigeration throughout the world **4250**
The History of roads, from amber route to motorway **4688**
A History of science **280**
History of science **301**
History of science and technology **292**
History of science and technology in India **371, 3165**
The History of science and technology in the United States **384**
A History of scientific and technical periodicals **239**
History of Scottish medicine **3164**
History of sewage treatment in Britain **4800**
A History of shopping **5522**
History of staining **2068**
History of technology **2820**
A History of technology **2822**
History of technology **3943**
A History of technology and invention **2818**
History of technology index, 1991: journal articles in the Science Museum Library **2829**
The History of the barometer **919, 1690**
The History of the British coal industry, **6019**
The History of the British flora **2395**
The History of the British Petroleum Company **6195**
A History of the British pig **5392**
History of the cotton manufacture in Great Britain **6530**
The History of the countryside **2256**
A History of the Institution of Electrical Engineers, 1871-1971 **3949**
A History of the life sciences **2050**
History of the life sciences: an annotated bibliography **2052**
'The History of the motor bus industry: a bibliographical survey' **4882**
A History of the Ordnance Survey **819**
A History of the Oxford University Press. v.1. To 1780 **5644**
The History of the retriangulation of Great Britain, 1935-1962.. **761**
History of the Royal Astronomical Society **668**
The History of the Royal Botanic Garden Library, Edinburgh **2361**
History of the rubber industry **6567**
The History of the study of landforms **1594**
A History of the theories of aether and electricity **988**
A History of the world's airlines **5860**
A History of the world's sports cars **4899**
History of tramways from horse to rapid transit **4673**
History of tribology **4399**

Hughes, F.W.
Illustrated guidebook to electronic devices and circuits **4106**

Hui, Y.H.
Encyclopedia of food science and technology **6073**

Hull, D.
The Macmillan guide to child health **3239, 5598**

Hull, R.
Virology: directory and dictionary of animal, bacterial and plant viruses **2197**

Hulmin, C. and Wenchu, B.
Dictionary of chemistry and chemical technology **1035, 5910**

The Human body on file **3206**

Human ecology: a guide to information sources **2080**

Human health and disease **3427**

Humidity and moisture **929**

A Hundred British spas **3397**

A Hundred years of biology **2049**

A Hundred years of chemistry **1105**

A Hundred years of metallurgy **6331**

Hungary. Központi Statisztikai Hivatel
Statisztikal szótái **583**

Hunt Botanical Library
Biographical dictionary of botanists **2359**

Hunt, C.
The Encyclopedic dictionary of science **83**

Hunt, F.V.
Origins in acoustics **942**

Hunt, P.
The Marshall Cavendish illustrated encyclopedia of gardening **5297**

Hunt, T.
Plant names of medieval England **2347**

Hunt, V.D.
Artificial intelligence and expert systems sourcebook **6803**
Dictionary of advanced manufacturing technology **4294**
Energy dictionary **3726**
Handbook of energy technology **3715**
Industrial robotics handbook **6821**
Robotics sourcebook **6822**
Windpower **3977**

Hunter, D.
The Diseases of occupations **3319**
Paper-making **6478**

Hunter, D.E. and Whitten, P.
Encyclopedia of anthropology **1877**

Hunter, R. and Macalpine, I.
Three hundred years of psychiatry, 1535-1860 **3544**

Hunter's tropical medicine **3440**

Huntia **2313**

Huntington, W.C.
Building construction materials and types of construction **6970**

Huntley, B. and Birks, H.J.B.
An Atlas of past and present pollen maps for Europe 0-13000 years ago **1827**

Huntley, I.D.
Mathematical modelling **527**

Huppmann, W.J. and Dalal, K.
Metallographic atlas of powder metallurgy **4329**

Hurlbut, C.S. and Kammerling, R.C.
Gemology **1336**

Hurt, C.D.
Information sources in science and technology **60**

Hüschke, R.E.
Glossary of meteorology **1680**

Hutchings, E.A.D.
A Survey of printing processes **5675**

Hutchinson, C.L.
The ARRL handbook for the radio amateur **4189**

Hutchinson encyclopedia of the earth **1518**

Hutchinson, G.E.
A Treatise on limnology **2132**

Hutchinson, J.
The Families of flowering plants **2492**

Hutchinson personal computer dictionary **6744**

Hutchison, C.S.
Geological evolution of south-east Asia **1504**

Hutchison, J.W.
ISA handbook of control valves **4270**

Huxley, A.
The New Royal Horticultural Society dictionary of gardening **5298**
Standard encyclopedia of the world's mountains **1606**
Standard encyclopedia of the world's oceans and islands **1604, 1630**

Huxley, Sir J. and Vevers, G.
The Atlas of world wildlife **2239**

Hyams, E.
Great botanical gardens of the world **2363**

Hyatt, E.
Keyguide to information sources in remote sensing **770, 1422**

Hydraulic handbook **4697**

Hydraulics and hydraulic research **4711**

Hydro-abstracts **1771**

Hydro-electric engineering practices **3970**

Hydrogen **5992**

Hydrogen energy quarterly literature review **3763**

Hydrographic Department
See
Great Britain. Hydrographic Department$zOcean passages for the world

Hydrographic dictionary **1765**

Hydrological data United Kingdom **1777**

Hydropower engineering handbook **3971**

Hyman, C.J.
German-English, English-German astronautics dictionary **5027**

Hyman, C.J. and Idlin, R.
Dictionary of physics and allied sciences **859**

Hyman, L.H.
The Invertebrates **2579**

Hyne, N.J.
Dictionary of petroleum exploration, drilling and production **4444, 6176**

Hyslop, M.R.
A Brief guide to sources of metals information **6288**

Hywell-Davies, J.
The Macmillan guide to British nature reserves **2276**

IARC
See
World Organization. International Agency for Research on Cancer

Ibeas, F.F.
Diccionario technologico inglés-español de electricdad, electronica, telecomunicacion y materias afines con la fisica, la optica y la quimica **3913**

IC Master, 1987 **3955**

IC schematic source master **3956**

ICI colour atlas **971**

ICO Library monthly entries - COFFEELINE **6057**

IDAAS 1990 **669**

The Identification of textile materials **6501**

IDPAI 1990 **670**

IEA Coal Research
1991 International Conference on coal science **6011**

IEC multilingual dictionary of electricity **3899-3900**

IEEE membership directory **3946**

IEEE standard dictionary of electrical and electronic terms **3904**

Ifrah, G.
Histoire universelle des chiffres **563**

Iglesia, M.E. de la
International catalogue of catalogues **5516**

Igneous rocks of the British Isles **1735**

Igoe, R.S.
Dictionary of food ingredients **6136**

International directory of marine scientists **1659**

International directory of new and renewable energy information sources and research centres **3757**

International directory of nuclear utilities **3865**

International directory of professional astronomical institutions **670**

International directory of specialized cancer research and treatment establishments **3423**

International directory of telecommunications **4176, 5600**

International directory to geophysical research **1402**

International distribution and handling review **4724**

International documentation of cartographical literature **778**

International dredging abstracts **4731**

International electronics directory '90 **4095**

International Electrotechnical Commission
Catalogue of publications, complete to 1st January 1986: world standards for electrical and electronic engineering **3932**
Dictionnaire CEI multilingue de l'électricité **3899**
Electrical standards in world trade **3936**
IEC multilingual dictionary of electricity **3900**
International electrotechnical vocabulary **3901**
Letter symbols, including conventions and signs for electrical technology **3891**
Yearbook. [International Electrotechnical Commission] **3934**

International electrotechnical vocabulary **3901**

The International encyclopedia of astronomy **620**

The International encyclopedia of cats **5406**

International encyclopedia of communications **5601**

International encyclopedia of composites **3648**

International encyclopedia of integrated circuits **3950**

International encyclopedia of psychiatry, psychology, psychoanalysis & neurology **3527**

International encyclopedia of robotics **6817**

International encyclopedia of textile finishing **6522**

International encyclopedia of wildlife **2517**

International Energy Agency
Coal information, 1987 **6013**
Concise guide to world coalfields **6014**
Energy statistics **3739**
Renewable sources of energy **3753**

International Energy Agency. Coal Research
1991 International Conference on coal science **6011**

International Food Information Service
Proceedings of the 2nd International Conference on Food Science and Technology Information **6106**

International Gas Union
Dictionary of the gas industry **5988**
International dictionary of the gas industry **5989**

International geographical glossary **1574**

International Geographical Union
International geographical glossary **1574**

International glossary of hydrology **1767**

International glossary of leather terms **6465**

International guide to locating audio-visual materials in the health services **3121**

International guide to official industrial property publications **2897**

International handbook of national parks and nature reserves **2274**

The International history of the sewing machine **6884**

International hydrogeological map of Europe **1776**

International Hydrographic Bureau
Glossary of cartographic terms **797**
Hydrographic dictionary **1765**

International Hydrographic Organization
The General bathymetric chart of the oceans **1654**

International imaging source book including micrographics and optional imaging **6858**

International Institute for Production Engineering Research (CIRP)
Wörterbuch der Fertigungstechnik **4295**

International Institute of Refrigeration
New international dictionary of refrigeration **4245**

International Institute of Welding
The International welding thesaurus **4348**

International journal of powder metallurgy and powder technology **4328**

International Labour Office
Encyclopaedia of occupational health and safety **3316**

International leather guide **6470**

International list of geographical serials **1586**

International literary market place **5682**

International maritime dictionary **5796**

International Maritime Engineering Conference, Bahrain, '84, 21-25 October, 1984 **4942**

International meat science dictionary **6143**

International meteorological vocabulary **1683**

International Mineralogical Association
World directory of mineral collections **1329**

International mycological directory **2462**

International nomenclature of diseases **3430**

International nursing index **3474**

International oceanographic tables **1651**

International Office of Epizootics
Animal health yearbook **5354**

International optical year book and diary **3571**

International Organization for Standardization
Dictionary of computer science **6696**

International Organization of Consumers' Unions
IOCU guide on consumer libraries **5526**

International packaging abstracts **4364**

The International petrochemical industry **6214**

International petroleum abstracts **6183**

International petroleum encyclopedia **6169**

International pharmaceutical abstracts **3356**

The International pharmacopeia **3365**

International pulp and paper directory, 1992-93 **6485**

International pulp and paper information sources **6489**

International register of historic ships **4937**

International research centers directory **454**

International review of cytology **2163**

International Road Federation
IRF research and development **4690**
World road statistics, 1985-1989 **5753**

International Road Research Documentation
IRRD/International Road Research Documentation **4675**

Kotik, M.G.
Dictionary of aerospace engineering in Russian, English and German **5020**

Kotlyakov, V.M. and Smolyarova, N.A.
Elsevier's dictionary of glaciology in four languages **1540**

Kotz, S.
Russian-English/English-Russian glossary of statistical terms **587**

Kotz, S. and Johnson, N.L.
Encyclopaedia of statistical sciences **580**

Kragelsky, I.V.
Friction wear, lubrication, tribology handbook **4395**

Kratkii illystrirovannyii russko-angliiskii slovar' po mashinostroeniya **3787**

Kratochvil, Y. and Urbanova, S.
Landwirtschaftliches Wörterbuch in acht Sprachen **5073**

Kremer, J.
Specialty booksellers directory **5708**

Kremers and Urdang's history of pharmacy **3361**

Kremers, E. and Urdang, G.
Kremers and Urdang's history of pharmacy **3361**

Kremp, G.O.W.
Morphologic encyclopedia of palynology and illustrations of spore and pollen **2431**

Kren, C.
Medieval science and technology **304**

Krisciunas, K.
Astronomical centers of the world **680**
The History of astronomy from Herschel to Hertzsprung **659**

Kristoferson, L.A. and Bokalders, V.
Renewable energy technologies **3755**

Krollman, F.
Langenscheidt's Fachwörterbuch Wehrwesen **4472**

Kronick, D.A.
A History of scientific and technical periodicals **239**
The Literature of the life sciences **2007**

Kronk, G.W.
Comets: a descriptive catalogue **700**

Kroschwitz, J.I.
Polymers: fibers and textiles **6592**
Polymers: polymer characterization and analysis **6593**

Krothpalli, A. and Smith, C.A.
Recent advances in aerodynamics **932**

Krstić, R.V.
Illustrated encyclopedia of human histology **2573**

Krudy, E.S.
Time: a bibliography **824**

Krüssmann, G.
Manual of cultivated broad-leaved trees & shrubs **5183**

Kryt, D.
Dictionary of chemical terminology **1038**

Kučera, A.
The Compact dictionary of exact science and technology **129**

Küchler, A.W.
International bibliography of vegetation maps **2379**

Kudryk, V.
Precious metals **6374**

Kuhn, A.T.
Techniques in electrochemistry, corrosion and metal finishing **1149, 3696, 4353**

Kuhn, T.S.
Sources for the history of quantum physics **915**

Kuiper, G.P. and Middlehurst, B.M.
The Solar system **691**
Stars and stellar systems **710**

Kukenthal, W.
Handbuch der Zoologie **2519**

Kulwiec, R.A.
Materials handling handbook **4384**

Kummel, B. and Raup, D.
Handbook of palaeontological techniques **1790**

Kurian, G. T.
Global guide to medical information **2996**

Kurian, G.T.
Glossary of the Third World **1924**

Kürschners deutschen Gelehrten-Kalender **354**

Kurtén, B.
Pleistocene mammals of Europe **1854**

Kurtén, B. and Anderson, E.
Pleistocene mammals of North America **1855**

Kurtz, A.K. and Edgerton, H.A.
Statistical dictionary of terms and symbols **586**

Kurtz, E.B.
Lineman's and cableman's handbook **3992**

Kurzlexikon der Elektrotechnik **3907**

Kut, R.
Dictionary of applied energy conservation **3721**

Kuter, L.
The Anthropology of Western Europe **1926**

Kutz, M.
Mechanical engineer's handbook **3783**

Kuvshinoff, B.W.
Fire sciences dictionary **3326**

Kuznetsov, B.V.
Russian-English dictionary of scientific and technical usage **171**

Kuznetsov, I.V.
Lycedi russkoi nauki **362**

Kverneland, K.O.
World metric standards for engineering **3613**

Kyang, R.K.
Periodicals current in mainland China held by the Science Reference Library **365**

La Rocque, A.
Contributions to the history of geology **1473**

Labahn, O.
Cement engineers' handbook **6266**

Labarre, E.J.
Dictionary and encyclopaedia of paper and paper-making **6475**

Laboratory animals **5326**
The Laboratory handbook of materials, equipment, and technique **1159**
Laboratory techniques in zoology **2544**

Lackie, J.M. and Dow, J.A.T.
The Dictionary of cell biology **2162**

Ladas, S.P.
Patents, trademarks and related rights **2935**

Ladbury, A.
The Dressmaker's dictionary **5591**

Ladd, H.S.
Treatise on marine ecology and paleoecology. Volume 2: Paleoecology **1818**

Lagacé, R.O.
Sixty cultures **1922**

Lagere planten **2447**

Laithwaite, E.R.
A History of linear electric motors **3987**

Lalande, J.J.F. de.
Bibliographie astronomique **612**

Lamb, E. and Lamb, B.
The Illustrated reference on cacti and other succulents **5321**

Lamb, H.H.
Climate: present, past and future **1703**

Lambert, A.
Battleships in transition **4550**

Lambert, C.M.
Energy **3706**
Radioactive waste **4812**

Lambert, C.M. and Larke, C.
Energy: sources of information on energy, with particular reference to the environment **3746**

Lambert, D.
Cambridge guide to prehistoric man **1861**

Le Grand, R.
Manufacturing engineers' manual: 'American Machinist' reference book sheets **4291**
Lea, G
'Geological literature' **1433**
Leach, M.
Railway control systems **5785**
Leach, R.H.
Printing ink manual **6279**
'Lead' **6394**
Lead abstracts **6395**
Lead and its alloys **6393**
Leader, M.
Encyclopedia of psychoactive drugs **3289**
Leader, R.W. and Leader, I.
Dictionary of comparative pathology and experimental biology **5339**
Leading consultants in technology **2812**
Leaffer, M.A.
International treaties on intellectual property **2880**
League of Arab States. Education, Culture and Sciences Organization. Permanent Bureau of Arabisation
Lexicon of chemistry **1070**
Lexicon of mathematics **511**
Lexicon of physics **864**
Leakey, L.S.B.
Fossil vertebrates of Africa **1839**
Lean, G.
Atlas of the environment **4763**
Lean, G. and Hinrichsen, D.
Atlas of the environment **2110**
Learner, R.
Astronomy through the telescope **684**
Leather technical glossary in six languages **6466**
Leatherhead Food Research Association. Scientific and technology abstracts **6093**
LeBar, F.M.
Ethnic groups of insular Southeast Asia **1935**
Ethnic groups of mainland Southeast Asia **1936**
Lebedev, A.V. and Federova, R.M.
A Guide to mathematical tables **605**
Lecoufle, M.
Comment choisir et cultiver vos plantes carnivores **2424**
Ledermann, W.
Handbook of applicable mathematics **487**
Lee, A. and Ramster, J.
Atlas of the seas round the British Isles **1653**
Lee, S.M.
International encyclopedia of composites **3648**
Leeden, F. van der
The Water encyclopedia **1774**

Lees, N.
Hazardous materials: sources of information on their transportation **3329**
Lees, R.
Chemical nomenclature usage **1099**
LeFanu, W.
A Bibliography of Edward Jenner **3490**
A Catalogue of the portraits and other paintings and sculpture in the Royal College of Surgeons of England **3557**
LeFanu, W.R.
Lives of the Fellows of the Royal College of Surgeons of England, 1930-1951 **3559**
Lefond, S.J.
Handbook of world salt resources **1754, 5982**
Industrial minerals and rocks **1306**
Leftwich, A.W.
A Dictionary of entomology **2615**
A Dictionary of zoology **2526**
Legal protection of computer programs **6768**
Legal protection of computer programs and data **6767**
The Legal protection of computer software **6766**
Leggatt, J.
Global warming **4825**
Legget, R.F.
Handbook of geology in civil engineering **4616**
Lehmann, M.
Holography **961**
Lehmer, D.H.
Guide to tables in the theory of numbers **564**
Lehrbuch der Anthropologie in systematischer Darstellung **1882**
Leiber, M. and Rosenblum, M.A.
A Dictionary of dermatological words, terms and phrases **3513**
Leicester, H.M.
A Source book in chemistry, 1900-1950 **1108**
Leicester, H.M. and Klickstein, H.S.
A source book in chemistry, 1400-1900 **1107**
Leif, I.P.
An International sourcebook of paper history **6487**
Leigh, G.J.
Nomenclature of inorganic chemistry **1223**
Leigh, R.
Directory of European industrial and trade associations **6612**
Leinbach, T.R.
South-east Asian transport **5739**

Leipnitz, W.
Erdölverarbeitung und Petrochemie **6175**
Fachwörterbuch: Carbochemie. Petrochemie **6173**
Lemon, H.
How to find out about the wool textile industry **6532**
Lengenfelder, H.
International bibliography of specialized dictionaries **99**
International bibliography of the book trade and librarianship **5681**
Libraries, information centres and databases in science and technology **413**
Lenk, G. and Börner, H.
Technisches Fachwörterbuch der Grundstoff - Industrien **4430**
Lenk, J.D.
Handbook of practical electronic circuits **4123**
Lennox, B.
Medical terms, their origin and construction **3053**
Lentner, C.
Geigy scientific tables **3109**
Leonhardt, F.
Brücken: Ästhetik and Gestaltung **4635**
Lépine, P.
Dictionnaire français-anglais, anglais-français des termes médicaux et biologiques **3068**
Lerner, R.G. and Trigg, G.L.
Concise encyclopedia of solid state physics **997**
Encyclopedia of physics **836**
Lesnaya entsiklopediya **5163**
Lester, D.E.
'Liquid fuels' **6166**
Létoile, V.-A.
La Cuisine: the complete book of French cooking **5558**
Let's halt awhile in Great Britain **5474**
Letter symbols, including conventions and signs for electrical technology **3891**
The Letts companion to Asian food & cookery **5559**
Leung, A.Y.
Encyclopedia of common natural ingredients used in food, drugs and cosmetics **5965**
Lever, C.
The Naturalized animals of the British Isles **2577**
Naturalized birds of the world **2680**
Naturalized mammals of the world **2722**
Levi, L.
CRC handbook of tables of functions for applied optics **951**

Loon, C.D. van *and* **Heij, D.G. van der**
Potato terms **5268**

Loones, B.
Watchmakers and clockmakers of the world: Volume 2 **6634**

Lorch, W.
Handbook of water purification **4794**

Lord, M.P.
Macmillan dictionary of physics **852**

Lorenz, K.J. *and* **Kulp, K.**
Handbook of cereal science and technology **6112**

Lorenzini, J.A.
Medical phrase index **3051**

Loria, G.
Guida allo studio della storia delle matematiche **540**

Lösch, A.
World cars **4893**

Losee, J.
A Historical introduction to the philosophy of science **278**

Lötschert, W. *and* **Beese, G.**
Collins' guide to tropical plants **2381**, **5261**

Lourie, J.
Medical eponyms **3141**

Louros, N.C.
Obstétrique et gynécologie **3576**

Loveday, G.
Microprocessor sourcebook **6743**

Loveday, M.S.
Measurement of high temperature mechnical properties of materials **3673**

Lovegrove, M.
Lovegrove's guide to Britain's North Sea oil and gas **6208**
Lovegrove's guide to Britain's North Sea oil and gas **6208**

Lovell, E.C. *and* **Hall, R.M.**
Index to handicrafts, model-making and workshop projects **6895**

Lovelock, Y.
The Vegetable book **5566**

Lowe, D.
The Transport manager's & operator's handbook **5749**

Lowenhein, F.A. *and* **Moran, M.K.**
Faith, Keyes and Clark's industrial chemicals **5954**

Lowenstein, M.Z.
Energy applications of biomass **3768**

Loyd, S.
The Heat pump **4253**

Lozet, J. *and* **Mathieu, C.**
Dictionnaire de science du sol **5215**

Lubell, C.
Textile collections of the world **6520**

Lubin, C.
Handbook of composites **3656**

Lucas, G.
The IUCN plant red data book **2288**

Lucca, J. de
Dictionnaire des télécommunications **4166**

Lucchesi, M.
Dizionario medico **3072**

Lück, E.C.
Bibliography of dictionaries and vocabularies on food, nutrition and cookery **6088**
The Compact dictionary of food technology **6089**

Ludwig Darmstaedter's Handbuch zur Geschichte der Naturwissenschaften 276

Luedtke, J.R.
'Energy and the environment' **3701**, **4816**

Lueger, O.
Lexikon der Technik **2794**
Luftbehandlingsordlista **3306**

Luginsky, Y.N.
Dictionary of electrical engineering **3902**

Luingman, C.G.
Dictionary of symbols **32**

Lum, A.
'Palaeontology' **1785**
Lunar sourcebook **693**

Lüning, K.
Seaweeds: their environment, biogeography, and ecophysiology **2452**

Luoyang Agricultural Machinery Academy
English-Chinese dictionary of agricultural machinery **5204**

Lura, A. Eggers-
See
Eggers-Lura, A.

Lusis, A.
Astronomy and aeronautics **614**

Luther, P.
Bibliography of astronomers **665**

Luton Central Library
Automobile engineering **4853**
Lycedi russkoǐ nauki **362**

Lydersen, A.L.
Dictionary of chemical engineering **5912**

Lyle, W.D.
Dictionnaire français et anglais de terminologie mathématique **506**

Lynch, C.T.
CRC handbook of materials science **3667**
Practical handbook of materials science **3668**

Lynch, D.B.
Concise dictionary of computing **6681**

Lyon, E.
Online medical databases **2974**
Lyons' Encyclopedia of valves **4271**

Lyons, J.L.
Valve designer's handbook **4272**

Lyons, J.L. *and* **Askland, C.L.**
Lyons' Encyclopedia of valves **4271**

Lysaght, A.W.
The Book of birds: five centuries of bird illustration **2678**

Lythall, R.T.
J & P switchgear book **4008**

Mabberley, D.J.
The Plant-book **2494**

MacAdam, E.J.
'Aerospace engineering' **5010**

McAinsh, T.F.
Physics in medicine & biology encyclopedia **2194**

McAlpin, D.J.
Health media review index 1984-1986 **3120**

MacAndrew, A.R.
A Glossary of Russian technical terms used in metallurgy **6299**

Macaulay Land Use Research Institute
Soil Survey of Scotland memoirs **5216**

Macauley, D.
The Way things work **3592**

MacCafferty, M.
Computer security **6773**

McCance, R.A. *and* **Widdowson, E.M.**
The Composition of foods **3259**, **6116**

McCarthy, G.J.
The Rare earths in modern science and technology **1234**

McCarthy, J.T.
McCarthy's desk encyclopedia of intellectual property **2858**
McCarthy's desk encyclopedia of intellectual property **2858**

McClintock, D.
Supplement to 'The Pocket guide to wild flowers' **2400**

McClintock, D. *and* **Fitter, R.S.R.**
The Pocket guide to wild flowers **2399**

McConnell, A.
Historical instruments in oceanography **1663**

McConnell, P.
Primrose McConnell's 'The agricultural notebook' **5069**

McCrane, W.C. *and* **Delly, J.C.**
The Particle atlas **1006**

McDermid, J.A.
Software engineer's reference book **6761**

Neave, H.R.
Statistical tables for mathematicians, engineers, economists and the behavioural and management sciences **593**

Neave, S.A.
Nomenclator zoologicus **2531**

Neaverson, E.
Stratigraphical palaeontology **1796**

Necker, C.
Four centuries of cat books, 1570-1970 **5407**

Needham, J. *and* Gwei-Djen, Lu
Science and civilization in China **364**
The Needleworker's dictionary **5590**

Neiger, E.
Gastronomisches Wörterbuch zur Übersetzung und Erklärung der Speisekarten **5554**

Nellist, J. *and* Nichol, B.
ASE science teachers' handbook **257**

Nelson, A.
Dictionary of mining **4428**
Dictionary of water and water engineering **4781**

Nelson, A. *and* Nelson, K.D.
Dictionary of applied geology, mining and civil engineering **1421**

Nelson, J.
Fishes of the world **2645**

Nelson, P.M.
Transportation noise reference book **4835**
Nelson's atlas of European birds **2686**

Nentwig, K.
Elsevier's dictionary of opto-electronics and electro-optics **4153**
Neotropical rainforest mammals **2724**

Nerd, A.V.
Guidebook to nuclear reactors **3868**
NESC handbook **3999**

Neu, J.
ISIS cumulative bibliography, 1966-75 **289**
ISIS cumulative bibliography, 1976-85 **290**

Neubert, G.
Technical dictionary of hydraulics and pneumatics **4232, 4701**

Neufeldt, S.
Chronologie Chemie, 1800-1970 **1103**

Neufert, E.
Architects' data **6939**
Neutron activation analysis abstracts **1007**
Neutron activation tables **3853**
Neutron cross sections **3854**

Neville, H.H.
A New German-English dictionary for chemists **1055**

New and improved: inventors and inventions that have changed the modern world **2883**
New and renewable sources of energy **3756**
The New astronomy **723**
The New atlas of the universe **726**
The New catering repertoire **5461**

New Consultants
Consultants and consulting organizations directory **241**
A New dictionary of birds **2667**
A New dictionary of scientific and technical terms, English-Arabic, with illustrations **181**
The New dinosaur dictionary **1846**
The New E is for additives **6134**
The New encyclopedia of motorcars **4883**
New encyclopedia of science **250**
The New encyclopedia of wine **6041**
New flora of the British Isles **2403**
New generation guide to the wild flowers of Britain and Northern Europe **2383**
A New German-English dictionary for chemists **1055**
The New hacker's dictionary **6683**
A New history of British shipping **5817**
New holistic herbal **3387**
The New illustrated science and invention encyclopedia **251**
New international dictionary of refrigeration **4245**
A New introduction to bibliography **5616**
The New Larousse encyclopedia of animal life **2552**
The New Larousse encyclopedia of the earth **1362**
New Larousse gastronomique **5541**
The New Macmillan guide to family health **3242**
New metric handbook **6959**
The New naturalist library **2255**
New ophthalmic literature **3569**
The New organic food guide **6097**
The New organic grower **5306**
The New 'Our bodies, ourselves' **3240**
The New Penguin dictionary of biology **2026**
A New photographic atlas of the moon **695**
The New physics **876**
New publications of the Bureau of Mines **4434, 6306**
The New Royal Horticultural Society dictionary of gardening **5298**
The New solar system **685**
New technical books **209, 2808**
New technologies and development **2810**
The New video encyclopedia **6846**

New York Academy of Medicine
Illustration catalog of the Library **3114**
Portrait catalog of the New York Academy of Medicine. [New York Academy of Medicine] **3113**

New York Academy of Medicine. Library
Author catalog. [New York Academy of Medicine Library] **3001**
Subject catalog. [New York Academy of Medicine Library] **3002**
The New York Botanical Garden illustrated encyclopedia of horticulture **5283**

New York Botanical Garden. Library
Biographical notes upon botanists **2360**
Catalog of the manuscript and archival collections and index to the correspondence of John Torrey **2358**

New York Public Library
Dictionary catalog of the Schomburg Collection of negro literature and history **1964**

Neway, J.M.
'Biotechnology information sources: an update' **2754**

Newberry Library Center for the History of the American Indian
Bibliographical series. [Newberry Library Center for the History of the American Indian] **1979**

Newberry Library, Chicago
Dictionary catalogue of the history of printing **5639**

Newcomen Society for the study of the history of engineering
Transactions, 1920/21-. [Newcomen Society] **3619**

Newman, A.C.D.
The Chemistry of clays and clay minerals **6256**

Newman, H.
An Illustrated dictionary of glass **6230**

Newman, J. *and* Tilke, A.
Sex education **3296**

Newman, J.R.
The World of mathematics **534**
Newnes book of video **6852**
Newnes computer engineer's pocket book **6675**
Newnes electrical pocket book **3929**
Newnes electronic circuits pocket book **4124**
Newnes Electronics assembly pocket book **4115**
Newnes engineering materials pocket book **3650**
Newnes guide to satellite TV **4219**

Petersen, D.
Audio, video and data telecommunications **4158**
Petersen, M.D. *and* White, D.L.
Health care of the elderly **3447**
Peterson, L. *and* Renström, P.
Sports injuries: their prevention and treatment **3270**
Peterson, M.B.
Wear control handbook **4397**
Peterson, M.S. *and* Johnson, A.H.
Encyclopedia of food science **6077**
Peterson, R.
A Field guide to the birds of Britain and Europe **2683**
Petroleum abstracts **6184**
Petroleum and hard minerals from the sea **1758**
The Petroleum dictionary **6179**
Petroleum economist **6186**
Petroleum/energy business news index **6185**
Petroleum engineering handbook **4445**
Petroleum handbook **6194**
Petroski, H.
The Evolution of useful things **2889**
Petrov, M.P.
Deserts of the world **1615**
Petrov, V.I. *and* Chupiatova, S.I.
Russko-angliiskii meditsinkii slovar'-razgovornik **3081**
Petrovich, G.V.
The Soviet encyclopedia of space flight **5017**
Pets: every owner's encyclopedia **5395**
Pettijohn, F.J. *and* Potter, P.E.
Atlas and glossary of primary sedimentary structures **1533**
Pfannkuch, H.O.
Elsevier's dictionary of environmental hydrogeology **1764**
Pflanzen der Tropen **5261**
Pforr, M. *and* Limbrunner, A.
The Breeding birds of Europe **2685**
The Pharmaceutical codex **3369**
Pharmacological and chemical synonyms **3353**
Pharmacology and pharmacologists **3359**
Pharmacy: an illustrated history **3360**
Phelps, T.
Poisonous snakes **2660**
Philatelic terms illustrated **5879**
Philately **5873**
Philip, C.
A Bibliography of firework books **5995**
The Plant finder **5280**
Phillips, B.
How to clean absolutely everything **5595**
Wonder worker's complete book of cleaning **5593**

Phillips, D.H. *and* Burdekin, D.A.
Diseases of forest and ornamental trees **5186**
Phillips International Paper directory, 1990 **6486**
Phillips, J.
Butterworths intellectual property law handbook **2855**
Phillips, P.
The Encyclopedia of glass **6231**
Phillips, R.
Mushrooms and other fungi of Great Britain and Europe **2472**
Trees in Britain, Europe and North America **5188**
Phillips, R. *and* Grant, S.
Grasses, ferns, mosses and lichens of Great Britain and Ireland **2499**
Phillips, V.J.
Early radio wave detectors **4199**
The Philosophies of science **90**
The Phone book index **5607**
Physical and thermodynamic properties of pure chemicals **5928**
Physical chemistry **1138**
The Physical chemistry of solids **1139**
Physical geology **1522**
Physical properties of chemical compounds **1265**
Physical properties of hydrocarbons **1268**
Physical properties of polymers **6597**
Physical properties of rocks and minerals **1734**
Physical science in the Middle Ages **306**
Physical sciences **47**
The Physical sciences in the twentieth century **901, 1121**
Physical thought from the Presocratic to the quantum theory **877**
Physicians' desk reference **3370**
Physicians' desk reference for nonprescription drugs **3371**
Physicians' generix **3372**
A Physicist's desk reference **845**
Physico-chemical constants of pure organic compounds **1266**
Physics abstracts. Science abstracts. Section A **870, 3841**
Physics briefs **871**
Physics handbook **884**
Physics in medicine & biology encyclopedia **2194**
Physikalische Berichte **871**
Piatiiazychnyi slovar nazvanii zhivotnykh. ryby **2640**
Piccirelli, A.
Government research directory **461, 2850**
Research centers directory **437**
Pichot, P.
Psychiatry: the state of the art **3541**

Picken, C.
The Translator's handbook **395**
Picken, M.B.
The Fashion dictionary **6880**
Pickering, W.R.
Information sources in pharmaceuticals **3347**
Pickford, J.
Indexed bibliography of publications on water and water engineering for developing countries **4772**
The Pictorial encyclopedia of plants and flowers **2372**
A Pictorial survey of railway signalling **5784**
A Pictorial treasury of the marine museums of the world **4943**
Pigeons and doves of the world **2709, 5397**
Pilot, G.
Foundation engineering **4626**
Pinches, M.J. *and* Ashby, J.G.
Power hydraulics **4696**
Pipe joints **4266**
Pipe protection bibliography **4263**
Pipelines abstracts **4267**
Piper, B. *and* Wolstenholme, G.
Portraits **3115**
Pipework design data **4265**
PIRA Abstracts **4360**
Piraux, H.
Dictionnaire français-anglais des termes relatifs à l'électronique, l'électrotechnique, l'informatique et aux applications connexes **3912**
Dictionnaire général d'acoustique et d'electroacoustique **940**
Pirone, P.P.
Diseases and pests of ornamental plants **2428**
Pistols of the world **4515**
Pitt, H.
The Complete innkeeper **5500**
Piveteau, J.
Traité de paléontologie **1797**
Plaisance, G. *and* Cailleux, A.
Dictionary of soils, French-English **5214**
Plan och byggtermer **6913**
Planet earth **1718**
Planet earth: an encyclopaedia of geology **1521**
Planetary and lunar co-ordinates for the years 1984-2000 **739**
Planetary astronomy from the Renaissance to the rise of astrophysics: Pt. A. Tycho Brahe to Newton **656**
Planetary exploration **690**
Planetary mapping **787**
The Plankton of the sea **2585**
Planning: the architect's handbook **6940**

Rosaler, R.C. *and* Rice, J.O.
Standard handbook of plant engineering **3812**

Rose, A.H. *and* Harrison, J.S.
The Yeasts **2465**

Rose books **5319**

Rose Elliot's complete vegetarian cookbook **5565**

Rose, F.
Colour identification guide to the grasses, sedges, rushes and ferns of the British Isles and north-western Europe **2501**

Rose, J.W. *and* Cooper, J.R.
Technical data on fuel **5998**

Rosen, F.S.
Dictionary of immunology **3484**

Rosenberg, A.
Russian-English glossary of guided missile, rocket and satellite terms **5006**

Rosenberg, J.M.
Dictionary of artificial intelligence and robotics **6805**

Rosenberg, P.D.
Patent law fundamentals **2957**

Roskill's directory of sources for metals and mineral data **6316**

Ross-Craig, S.
Drawings of British plants **2401**

Ross, J.A.
National Electrical Code handbook **4001**

Ross, J.P. *and* Lefanu, W.R.
Lives of the Fellows of the Royal College of Surgeons of England, 1965-1973 **3561**

Ross, R.B.
Handbook of metal treatments and testing **4300**
Metallic materials specification handbook **6295**

Rossdale, P.D.
Veterinary notes for horse owners **5386**

Rossdale, P.D. *and* Wreford, S.M.
The Horse's health, from A to Z **5385**

Rossini, F.D.
CRC fundamental measures and controls for science and technology **910**

Roth, G.D.
Collins' guide to the weather **1699**

Roth, J.P.
Rewritable optical storage technology **6860**

Rothamsted Experimental Station
Soil maps **5216**

Rothamsted Experimental Station. Library
Catalogue of printed books and pamphlets on agriculture published between 1471 and 1840 **5062**

Rothbart, H.A.
Mechanical design and systems handbook **4373**

Rothenberg, M.
The History of science and technology in the United States **384**

Rougemont, G.M. de
A Field guide to the crops of Britain and Europe **5260**

The Round the world air guide **5855**

Les Routiers guide to Britain 1992 **5496**

Routiers guide to France **5497**

Rowe, G.T.
Deep-sea biology **2138**

Rowson, K.E.K.
A Dictionary of virology **2196, 3548**

Royal Aircraft Establishment
The RAE table of earth satellites, 1957-1986 **5051**

Royal Anthropological Institute
Teachers' resource guide **1895**
The Royal Anthropological Institute Film Library catalogue **1897**

Royal Association for Disability and Rehabilitation
Holidays in the British Isles 1990: a guide for disabled people **5482**

Royal Astronomical Society
Catalogue of the Library. [Royal Astronomical Society] **616**

Royal Automobile Club
One hundred years of motoring **4891**
RAC camping and caravanning guide, 1990: Europe **5506**
RAC camping and caravanning guide, 1992: Great Britain & Ireland **5507**
RAC continental hotel guide **5472**
RAC hotel guide **5475**

Royal Botanic Gardens, Kew
List of periodical publications in the Library **2343**
World plant conservation bibliography **2282**

Royal Botanic Gardens, Kew. Library
Author and classified catalogues. [Royal Botanic Gardens, Kew, Library] **2317**
Current awareness list **2318**

Royal College of General Practitioners
Classification of diseases, problems and disorders **3439**

Royal College of Nursing
A Bibliography of nursing literature **3464**

Royal College of Physicians of London
Catalogue of engraved portraits of the Library. [Royal College of Physicians of London] **3117**
Portraits **3115**
Smoking or health **3293**
William Harvey, 1578-1657 **3148**

Royal College of Physicians of London. Joint Committee
The Nomenclature of disease **3431**

Royal College of Surgeons of England
A Catalogue of the portraits and other paintings and sculpture in the Royal College of Surgeons of England **3557**

Royal College of Veterinary Surgeons
Registers and directory, 1992 **5352**

Royal Commission on Historical Manuscripts
The Manuscript papers of British scientists, 1600-1940 **353**

Royal Entomological Society of London
Catalogue of the Library. [Royal Entomological Society of London] **2604**
Handbooks for the identification of British insects **2609**

The Royal Greenwich Observatory and the Astronomers Royal, 1675-1976 **678**

Royal Horticultural Society
The Royal Horticultural Society colour chart **5273**
The Royal Horticultural Society colour chart **5273**
The Royal Horticultural Society encyclopedia of gardening **5301**
The Royal Horticultural Society gardeners' encyclopedia of plants and flowers **5300**

Royal Institution of Chartered Surveyors
Directory **760**
RICS Library Information Service abstracts and reviews **758**
RICS Library Information Service Weekly briefing: a digest of news selected from the press **759**
Royal Institution of Chartered Surveyors directory **760**

Royal Life Saving Society. National Technical Committee
[Handbook, Royal Life Saving Society] **3344**

Royal Mail **5867**

The Royal Observatory at Greenwich and Herstmonceux, 1675-1975 **677**

Royal Society
Biographical memoirs of Fellows of the Royal Society **325**
Book catalogue of the library of the Royal Society **73**

Shetler, S.G. *and* Skog, L.E.
A Provisional checklist of species for
Flora North America **2413**
Shield, M.J.
'Automotive engineering' **4854**
Shields, J.
Adhesive handbook **6224**
Shields, J.R.
Make it **6896**
Shiers, G.
Bibliography of the history of
electronics **4112**
Shigley, J.E.
Standard handbook of machine design
4374
Shilling, C.W.
The Underwater handbook **3160**
Shilling, C.W. *and* Werts, M.F.
Underwater medicine and related
sciences **3159**
Shillingford, A.E.P.
England's vanishing windmills **4239**
Shimomura, R.
GeoRef thesaurus and guide to
indexing **1514**
The Ship **4940**
Ship design and construction **4915**
The Ship of the line **4549**
The Ship series **4934**
The Shipbuilding industry **4941**
Shipbuilding, seafaring technical
dictionary **4918**
Shipley, R.M.
Dictionary of gems and gemology
1340
Shipping **5818**
The Shipping revolution **4948**
Shirinian, G.
Microcomputer periodicals **6749**
The Shock and vibration digest **3684**
Shock and vibration handbook **937**
Shoe trades directory, 1988 **6874**
Shopping in the eighties **5510**
Shore, B.C.G.
Stones in Britain **6980**
The Shore environment **2117**
A Short account of the history of
mathematics **529**
A Short dictionary of weaving **6523**
A Short English-Swahili medical
dictionary **3084**
A Short history of geomorphology **1593**
A Short history of medicine **3123, 3133**
A Short history of scientific ideas to
1900 **282**
A Short history of technology, from the
earliest times to AD 1900 **2823**
A Short history of twentieth-century
technology **2824**
A Short Russian-English dictionary of
terminology used in the Soviet rubber,
plastics and tyre industries **6546**

A Short-title catalogue of household and
cookery books published in the
English tongue, 1701-1800 **5535**
Shorter guide to pests, diseases and
disorders of garden plants **2426**
The Shorter 'Science and civilization in
China' **339**
Shreeve, C.M.
The Alternative dictionary of
symptoms and cures **3460**
Shreir, L.L.
Corrosion **3691**
Shugar, G.J.
Chemical technicians' ready reference
handbook **1094, 5908**
The Chemist's ready reference
handbook **1095**
Shugg, W.T.
Handbook of electrical and electronic
insulating materials **3997**
Shvarts, V.V.
Kratkiĭ illystrirovannyĭ russko-
angliĭskiĭ slovar' po
mashinostroeniya **3787**
Sibilia, J.P.
A Guide to materials characterization
and chemical analysis **1174**
Sibley, C.G. *and* Monroe, B.L.
Distribution and taxonomy of birds of
the world **2681**
Sidders, P.A.
Guide to world screw threads **4387**
Side effects of drugs annual **3348**
Sidgwick, J.B.
Amateur astronomer's handbook **634**
Observational astronomy for amateurs
635
Sidwick, J.M. *and* Holdon, R.S.
Biotechnology of waste treatment and
exploitation **4804**
Siegel, P.J.
Women in the scientific search **388**
Sigerist, H.E.
A History of medicine **3131**
Sigma-Aldrich handbook of stains, dyes
and indicators **6278**
Sigmond, J. *and* Terpstra, M.
High-power lasers **4036**
Signeur, A.V.
Guide to gas chromatography
literature **1206**
Sigurdson, J. *and* Anderson, A.M.
Science and technology in Japan **19**
The Silent revolution **4494**
Silver: economics, metallurgy and use
6376
Silver: exploration, mining and
treatment **4461, 6377**
Silverstein, A.M.
A History of immunology **3489**
Silvestre, L.C.
Marques typographiques **6844**

Simkin, T.
Volcanoes of the world **1525**
Simmons, A.F.
Simmons' manual of fruit **5291**
Simmons, E.S.
'Patents' **2895**
Simmons, J.
The Railway in England and Wales,
1830-1914 **5777**
Simmons' manual of fruit **5291**
Simon, A.
André Simon's wines of the world
6047
Simon, A.L.
Bibliotheca gastronomica **5536**
Dictionary of gastronomy **5542**
Simon, J.T.
Herbs: an indexed bibliography,
1971-1980 **3386**
Simons, E.N.
A Dictionary of alloys **6335**
Dictionary of ferrous metals **6346**
Dictionary of foundry work **4306**
A Dictionary of metal heat treatment
6338
Guide to uncommon metals **6296**
Simons, J.H.
Fluorine chemistry **1225**
Simpkins, N.S.
100 modern reagents **1128**
Simpson, A.
The Communication users' yearbook
4171
Simpson, K.
The Birds of Australia **2705**
Simpson, M.A.
Dying, death and grief **3232**
Simpson, N.D.
A Bibliographical index of the British
flora **2402**
Sims, W.L.
Two hundred years of history and
evolution of woodworking
machinery **6461**
Sinclair, I.R.
Audio electronics reference book
6847
Collins dictionary of electronics **4077**
Collins dictionary of personal
computing **6746**
Sinclair, W.A.
Diseases of trees and shrubs **5185**
Singer, C.
A History of technology **2822**
A Short history of scientific ideas to
1900 **282**
Singer, C. *and* Underwood, E.A.
A Short history of medicine **3133**
Singer, L.
Russian-English-French-German
kinematics dictionary **918**
Singh, J.
An Agricultural atlas of India **5142**

Terminology of communication disorders **3498**
Terminology of forest science, technology, practice and products **5170**
Terminology of new and renewable sources of energy **3751**
Terms used in the iron and steel industry **6344**
Terrell on the law of patents **2946**
Terres, J.K.
The Audubon Society encyclopedia of North American birds **2699**
Webster's New World dictionary of computer terms **6690**
Test **5514**
Tester, J.W.
Energy and the environment in the 21st century **3748**
Texas Instruments
Optoelectronics data book **4154**
Text retrieval: a directory of software **6764**
Textbook of physiology **3220**
Textbook of science and medicine in sport **3271**
Textbook of topographical surveying **762**
Textile collections of the world **6520**
Textile digest **6514**
The Textile industry **6496**
Textile Institute
The Identification of textile materials **6501**
Textile terms and definitions **6506**
World textiles **6493**
Textile technology digest **6492, 6512**
A Textile terminology, warp and weft **6521**
Textile terms and definitions **6506**
Textilordlista **6503**
Thakurdas, H. and Thakurdas, L.
Dictionary of psychiatry **3536**
The Thames and Hudson manual of bookbinding **6876**
The Thames and Hudson manual of typography **5668**
Thames Tunnel to Channel Tunnel **4631**
Thawley, J. and Gauci, S.
Bibliographies on the Australian aborigine **1998**
Theodoratus, R.J.
Europe: a selected ethnographic bibliography **1927**
Theoretical chemical engineering abstracts **5920**
Thermal analysis abstracts, with occasional reviews **5999**
Thermodynamic derivatives for water and steam **3880**
Thermodynamic design data for heat pump systems **4252**

Thermodynamische Eigenschaften von Wasser und Wasserdampt **3884**
Thermophysical properties of matter **978**
Thermophysical properties research literature retrieval guide, 1900-1980 **977**
Thernstrom, S.
Harvard encyclopedia of American ethnic groups **1952**
Thesaurus 1975: English alphabetical list: English-French-German **4682**
Thesaurus, 1985. [International Road Research Documentation] **5760**
Thesaurus. [American Petroleum Institute] **6198**
Thesaurus (for animal health, vertebrate pest biology and control) **5345**
Thesaurus for fluid engineering **4257**
Thesaurus litteratura botanicae omnium gentium **2315**
Thesaurus of agricultural organisms **5242**
The Thesaurus of chemical products **5969**
Thesaurus of consumer terms **5525**
Thesaurus of metallurgical terms **6330**
Thesaurus of mineral processing and extractive metallurgy terms **6329**
Thesaurus of pulp and paper terminology **6488**
Thesaurus of rock and soil mechanics terms **4618**
Thesaurus of scientific, technical, and engineering terms **410, 2846**
Thesaurus of subject terms and cross-references to 'International pharmaceutical abstracts' **3352**
Thesaurus of terms in copper technology **6388**
Thesaurus of textile terms **6519**
Thévenot, R.
A History of refrigeration throughout the world **4250**
Thewlis, J.
Concise dictionary of physics and related subjects **856**
Encyclopaedic dictionary of physics **834**
Multilingual glossary **848**
Thin-layer chromatography abstracts **1207**
Third reference catalogue of bright galaxies **720**
Thirsk, J.
The Agrarian history of England and Wales **5135**
Thomann, A.E.
Elsevier's dictionary of technology **2795**
Thomas, C.L.
Taber's cyclopedic medical dictionary **3028**

Thomas Cook European timetable: railway and shipping services throughout Europe **5781**
Thomas Cook overseas timetable: railway, road and shipping services outside Europe **5782**
Thomas, D. St. J. and Patmore, J.A.
A Regional history of the railways of Great Britain **5776**
Thomas, D.S.G.
Arid zone geomorphology **1597**
Thomas, J. and Lewington, R.
The Butterflies of Britain and Ireland **2630**
Thomas, K.M.
The Encyclopedia of cacti **5320**
Thomas, M.F.
Tropical geomorphology **1596**
Thomas, R.J.
Bookshops of Greater London **5706**
Thomas, W.H. and Ervine, C.
Encyclopedia of consumer law **5518**
Thompson, A.J.
Logarithmetica Britannica **608**
Thompson, D.B. and Scragg, H.
'Resources for geology teachers' **1460**
Thompson, I.
The Audubon Society field guide to North American fossils **1812**
Thompson, J.W.
Index to illustrations of the natural world **2236**
Thompson, S.
A Chronology of geological thinking from antiquity to 1899 **1469**
Thompson, W.A.R.
A Dictionary of medical ethics and practice **3203**
Thomson, Sir A.L.
A New dictionary of birds **2667**
Thomson, L.M. Milne-
See
Milne-Thomson, L.M.
Thomson, M.R.A.
Geological evolution of Antarctica **1512**
Thorndike, L.
A Catalogue of incipits of mediaeval scientific writings in Latin **338**
History of magic and experimental science **284**
Thornton, J.D.
Science and practice of liquid-liquid extraction **1165**
Thornton, J.L.
Scientific books, libraries and collectors **294**
Select bibliography of medical biography **3145**
Thornton, J.L. and Reeves, C.
Medical book illustration **3112**

**Water Resources Center Archives,
University of California**
Dictionary catalog. [Water Resources
Center Archives] **1762, 4771**
Water resources engineering **4775**
Water treatment handbook **4795**
The Waterfowl of the world **2706**
Waterman, N.A. *and* Ashby, M.F.
Elsevier materials selector **3654**
Watermarks in paper in Holland,
England, France **6491**
Waternet **4770**
Waters, A.W.
All the US Air Force airplanes, 1907-
1983 **4536**
Waters, D.W.
The Art of navigation in England in
Elizabethan and early Stuart times
741
Watkin, G.
British food fish **6147**
Watson, A.H.
Inland transport statistics, Great
Britain, 1900-1970 **5732**
Watson, E.V.
British mosses and liverworts **2478**
Watson, L.
Whales of the world **2737**
Watson, R. *and* Gray, M.
The Penguin book of the bicycle
4908
Watt, A.
Longman illustrated dictionary of
geology **1447**
Watt, I.
A Directory of UK map collections
805
Watts, A.J.
Jane's underwater warfare systems
4500
Watznauer, A.
Dictionary of geosciences **1368**
Wauchope, R.
Handbook of Middle American
Indians **1983**
Waveguide handbook **4033**
The Way things work **3592**
We the people: an atlas of America's
ethnic diversity **1957**
The Wealth of India **3671**
Weapons: an international encyclopedia,
from 5000 BC to 2000 AD **4499**
Wear control handbook **4397**
Wear resistant surfaces in engineering
3697
Weast, R.C.
CRC handbook of chemistry and
physics: 1st student edition **1090**
CRC handbook of data on organic
compounds **1263**
Weather **1714**
Weather almanac **1717**

Weather Bureau
See
United States. Weather Bureau
The Weather handbook **1708**
The Weather of Britain **1715**
Weaver, J.H.
The World of physics **894**
Weaver, P.
The Birdwatcher's dictionary **2669**
Webb, D.
Health education index **3105**
Webb, D.A.
An Irish flora **2405**
Webb, J.E.
Guide to the marine stations of the
North Atlantic and European waters
5837
Webb, J.S.
The Wolfson geochemical atlas of
England and Wales **1417**
The Webb Society deep sky observer's
handbook **711**
Webber, J.
Research in social anthropology, 1975
- 1980 **1883**
Weber, F.
Elsevier's dictionary of high vacuum
science and technology in six
languages **4235**
Weber, M.J.
CRC handbook of laser science and
technology **4046**
Weber, R.D.
Energy information guide **3709**
Energy update **3710**
Webster, C.C. *and* Baulkwill, W.J.
Rubber **6558**
Webster, J.G.
Encyclopedia of medical devices and
instrumentation **3223**
Webster, R.
Gems **1338**
Webster's medical desk dictionary **3057**
Webster's new world dictionary of
mathematics **502**
Weck, H.
Handbook of machine tools **4403**
Weck, J.
Dictionary of forestry, in five
languages **5166**
Wedepohl, K.M.
Handbook of geochemistry **1416**
Wedertz, B.
Dictionary of naval abbreviations
4537
Wedge, C.P.
Brown's flags and funnels of British
and foreign steamship companies
5824
Wedgewood, C.G.
European electronics directory 1991
4094
Weed control handbook **5254**

Weekes, R.V.
Muslim peoples **1940**
Weigert, A. *and* Zimmerman, H.
Concise encyclopedia of astronomy
623
Weik, M.H.
Communications standard dictionary
4165, 5603
Fiber optics standard dictionary **4181**
Weir, D.M. *and* Stewart, J.
Immunology **3487**
Weiss, B.
Wie finde ich Literatur zur Geschichte
der Naturwissenschaften und
Technik **277**
Weiss, G.
Hazardous chemical data book **3337**
Weissler, G.L.
Vacuum physics and technology **4236**
Welch, J. *and* King, T.A.
Searching the medical literature **3007**
Welch, L.A.
Food labelling data for manufacturers
6117
WELDASEARCH **4337**
The Welder's handbook **4342**
Welding **4346**
'Welding: a core list for community
college libraries' **4339**
Welding abstracts **4345**
'Welding, brazing and soldering' **4338**
Welding handbook **4340**
Welding terms and symbols **4343**
Well, J.
The Directory of continuing education
for nurses, midwives and health
visitors **3480**
Wellcome Historical Medical Library
A Catalogue of printed books in the
Wellcome Historical Medical
Library **3008**
Catalogue of Western manuscripts on
medicine and science in the
Wellcome Historical Medical
Library **3157**
**Wellcome Institute for the History of
Medicine**
Subject catalogue of the history of
medicine and related sciences **3135**
**Wellcome Institute of the History of
Medicine**
Portraits of doctors and scientists in
the Wellcome Institute **3118**
Weller, B.F.
Baillière's encyclopaedic dictionary of
nursing and health care **3466**
Welling, M.S.
German-English dictionary of plastics
technology **6579**
Wells, D.
The Penguin book of curious and
interesting puzzles **557**

Wingrove, R.G. *and* **James, J.R.J.**
The Aviator's English and French dictionary **4972**
Winkler, W.
Fachwörterbuch Chemiefasern **6538**
The Winning of nickel **6380**
Winship, I.
Health and safety at work **3315**
Winter, F.H.
Rockets into space **5008**
Winter, R.
The People's handbook of allergies and allergens **3454**
Winterkorn, H.F.
Foundation engineering handbook **4627**
Winthrop, R.
Dictionary of concepts in cultural anthropology **1912**
WIRASCAN: current awareness service **6534**
Wire industry yearbook **4336**
Wirksubstanzen der Pflanzerschutz- und Schadlingsbekampfungsmitte **5247**
Wise, T.
A Guide to military museums **4487**
Wiseman, C.H.
The International countermeasures handbook **4524**
Wiseman, J.
A History of the British pig **5392**
The SAS survival handbook **3267**
Wistreich, G.A. *and* **Lechtman, M.D.**
Microbiology **2201**
Witherby, H.F.
The Handbook of British birds **2689**
Witte, G.
'Glossary of technical terms for the sugar industry' **5270**
Wittfoht, A.
Plastics technical dictionary **6580**
Wittfoht, H.
Building bridges **4637**
Wittmann, A. *and* **Klos, J.**
Dictionary of data processing, including applications in industry, administration and business **6722**
Dictionary of patent terms **2909**
WMO sea-ice nomenclature **1544**
Wohlauer, G.E.M. *and* **Gholston, H.D.**
German chemical abbreviations **1013**
Wohlfarth, E.P.
Ferromagnetic materials **4015**
Wold, H.O.A.
Bibliography on time series and stochastic processes **596**
Woldman, N.E.
Woldman's Engineering alloys **6337**
Woldman's Engineering alloys **6337**
Wolf, C.E. *and* **Chiang, N.S.**
Indians of North and South America **1981**

Wolfe, J.A.
Mineral resources **1749**
Wolfe, W.L. *and* **Zissis, G.J.**
The Infrared handbook **952**
Wolff, K.
AAAS science book list, 1978-86 **49**
Wolfheim, J.W.
Primates of the world **2748**
The Wolfson geochemical atlas of England and Wales **1417**
Wolman, B.B.
International encyclopedia of psychiatry, psychology, psychoanalysis & neurology **3527**
Wolman, Y.
Chemical information **1026**
Wolter, J.A.
World directory of map collections **806**
Women anthropologists: a biographical dictionary **1900**
Women in medicine **3152**
Women in science **335**
Women in the scientific search **388**
Women of mathematics **547**
Women of science, technology and medicine **334**
Women scientists from antiquity to the present **333**
Wonder worker's complete book of cleaning **5593**
Wonders of creation: natural history drawings in the British Library **2234**
Wong, S.
Dictionary of Chinese Communist agricultural technology **5085**
Wood, A.D.
Plywoods of the world **6463**
Wood, C.A.
An Introduction to the literature of vertebrate zoology **2635**
Wood, C.A. *and* **Kienle, J.**
Volcanoes of North America **1526**
Wood, D.
Jane's world aircraft recognition handbook **4967**
Trade names dictionary **2877**
Trade names dictionary: company index **2879**
Wood, D.N.
Information sources in the earth sciences **1353**
Wood in construction **6987**
Wood industry abstracts **6440**
Wood, K. *and* **McDonald, G.**
The Round the world air guide **5855**
Woodburn, H.M.
Using the chemical literature **1027**
Woodburn, J.
The Royal Anthropological Institute Film Library catalogue **1897**

Woods Hole Oceanographic Institution
Catalog of the Library of the Marine Biological Laboratory and the Woods Hole Oceanographic Institution **1626, 2127**
Oceanographic index **1624**
Woods practical guide to fan engineering **4259**
Woods, R.S.
The Naturalist's lexicon **2046**
Woodward, D.
Five centuries of map printing **821**
The woodworker directory, 1988 **6454**
The Woodworker's Bible **6451**
The Woodworker's dictionary **6452**
Woodworking machines in 4 languages **6459**
Wool Industry Research Association
WIRASCAN: current awareness service **6534**
Wool quarterly **6535**
Woolham, F.
The Handbook of aviculture **5399**
Woordenlijst voor der tuinbouw in seven talen **5275**
Wootton, A.
Insects of the world **2610**
Word processing dictionary **5660**
Words for birds: a lexicon of North American birds, with biographical notes **2701**
Work manual **2867**
Work, M.N.
A Bibliography of the negro in Africa and America **1966**
World aerospace **5033**
World agricultural economics and rural sociology abstracts **5158**
World air transport statistics, 1987 **5856**
'World airline directory: flight data. Guide to the world's major carriers, their routes, equipment and management' **4982**
World aluminum abstracts **6408**
World archaeoastronomy **662**
World atlas of agriculture **5105**
The World atlas of cheese **5419**
World atlas of desertification **1614**
World atlas of epidemic disease **3434**
World atlas of geology and mineral deposits **1465**
World atlas of railways **4652**
The World Book encyclopedia of science **92**
World cars **4893**
World cartography **801**
World catalogue of theses and dissertations about the Australian aborigines and Torres Strait islanders **1995**

Subject Index

The index reference is to the running number given to each item. The running numbers are in one sequence throughout the volume and can be found at the top right-hand corner of the entry for each item.

The index is computer generated, thus terms for the index have been largely derived from the headings and sub-headings used throughout *Walford*, but many other entries have been added, including synonyms, inverted headings, and cross-references. Some form headings such as 'Bibliographies', 'Dictionaries', etc., are omitted as leading to too great a bulk.

The arrangement of the index is alphabetical and filing is word by word, with groups of initials counted as single words. Under each main heading printed in bold type in the index will be found a resumé of all the subject terms used under that heading and the numbers of the items to which they refer. Similarly, under each narrower sub-heading there is a list of terms used. Where the term in the index needs to be qualified by the broader term of which it is a sub-division, then the broader term is given in square brackets, *e.g.* **Museums** [Aircraft] or [Textiles].

A/V
See
 Audio-Visual
Abbreviations & Symbols
 [Agriculture] **5052**
 [Astronautics (Space Craft)] **5009**
 [Automation Technology] **6777**
 [Biology] **1999-2000**
 [Chemistry] **1013**
 [Computers] **6660-6662**
 [Ecology] **2074-2075**
 [Electrical Engineering] **3888-3891**
 [Electronics] **4047-4050**
 [Engineering] **3579-3580**
 [Machine Tools] **4400**
 [Marine Transport] **5786**
 [Medicine] **2963-2972**
 [Military & Naval Engineering] **4466**
 [Navies] **4537**
 [*Periodicals*] **224-225**
 [Physics] **828-829**
 [Public Health Engineering] **4735**
 [Science] **27-39**
 [Semi-conductors] **4143**
 [Spectroscopy] **1185**
 [Synthetic Polymers & Rubbers]
 6588
 [Technology] **2784-2785**
Aborigines
 [Ethnology (Races)] **1996-1998**
Abortion 3578
Acetylene (Fuel Technology) **6031**
Acid rain 4829, 4832
Acoustics 935, 939-942
 Bibliographies **935**
 Dictionaries
  English **939**
  French **940**
  Polyglot **938**
 Handbooks & Manuals **936-937**
 Histories **942**
 Reviews & Abstracts **941**
Acupuncture 3396
Additives (Food Industry) **6133-6138**

Adhesives (Chemical Industry) **6219-6225**
Adolescents
 [*Handbooks & Manuals*] **3592**
Aerodynamics 930-932
Aerosols 1152
Aerospace engineering 5041
Aerospace Medicine 3266
Africa
 [Birds] **2697**
 [Book Trade] **5686**
 [Ecology] **2122**
 [Ethnology (Races)] **1941-1944**
 [Geology] **1505-1506**
 [Government Policies] **22**
 [Mammals] **2731**
 [Mining] **4441**
Africa—Southern
 [Birds] **2698**
 [Deep-Sea Fishing] **5449**
 [Ethnology (Races)] **1946-1948**
 [Fishes] **2648**
 [Mammals] **2732**
 [Plants (Flora)] **2411**
Africa—West
 [Geology] **1507**
African Languages
 See
  Negro-African Languages
Ageing (Medicine)
 Bibliographies **3446-3447**
 Databases **3445**
 Encyclopaedias **3448**
 Handbooks & Manuals **3449-3450**
 Research Projects **3453**
 Reviews & Abstracts **3451**
 Yearbooks & Directories **3452**
Agricultural Economics 5155-5158
Agricultural Machinery
 Bibliographies **5198-5199**
 Dictionaries **5202-5206**
 Farm Tools **5210-5211**
 Handbooks & Manuals **5200-5201**
 Institutions & Associations **5209**

Agricultural Machinery *(contd.)*
 Reviews & Abstracts **5207**
 Yearbooks & Directories **5208**
Agricultural Products 5222-5223
Agriculture
 Abbreviations & Symbols **5052**
 Agricultural Economics **5155-5158**
 Agricultural Machinery
  Bibliographies **5198-5199**
  Dictionaries **5202-5206**
  Farm Tools **5210-5211**
  Handbooks & Manuals **5200-5201**
  Institutions & Associations **5209**
  Reviews & Abstracts **5207**
  Yearbooks & Directories **5208**
 Agricultural Products **5222-5223**
 Agricultural Trade **5192**
 Ancient Rome **5118**
 Asia **5139**
 Asia—South & South East **5141**
 Bibliographies **5057-5065**
 Biographies **5107**
 China **5140**
 Co-operative Enterprises **5159-5160**
 Databases **5053-5056**
 Dictionaries
  Chinese **5085**
  English **5076-5079**
  French **5080-5081**
  Polyglot **5071-5075**
  Russian **5082-5084**
 Encyclopaedias **5066-5067**
 England
 England & Wales **5131-5132**
  Information Services **5133**
 England
  Bibliographies **5134**
  Histories **5135-5136**
  Manuscripts & Incunabula **5137**
 Europe **5119-5121**
 Farms
  Buildings **5195-5197**
  Management **5194**
  Organic Farming **5193**

Airways 5864
Alcohol Fuel 6027
Alcoholism 3283-3285
Algae (Botany) 2450-2452
Algebra 565
Alkaloids (Chemistry) 1278
Allergies 3454
Alloys (Metallurgy) 6334-6337
Alternative Medicine 3390-3395, 3460
Aluminium (Metallurgy)
 Bibliographies 6403
 Conferences 6409
 Handbooks & Manuals 6404-6406
 Reviews & Abstracts 6407-6408
America
 [Ethnology (Races)] 1950-1951
 [Minerals & Ores] 1753
America — North
 Biographies 379-380
 [Birds] 2699-2701
 [Book Trade] 5687-5688
 [Ethnology (Races)] 1952-1953
 [Geology] 1508-1509
 [Horticulture] 5281-5282
 [*Institutions & Associations*] 437
 [Medicine] 3169
 [Nature Conservation] 2294
 [Nature Study] 2264-2266
 [Plants (Flora)] 2412-2413
 [Public Health Engineering] 4764
 [Rivers & Lakes] 1784
 [Rocks] 1736
 [Science] 378-380
 [Wildlife Conservation] 2308-2309
America—North & Central
 [Ethnology (Races)] 1954
 [Hotels & Restaurants] 5478
 [*Libraries*] 424-425
 [Mammals] 2733
 [Palaeontology] 1812
 [Railway Transport] 5778
America—South
 [Birds] 2702-2703
 [Ethnology (Races)] 1968-1970
 [Government Policies] 24
 [Plants (Flora)] 2416
American Indians
 See
 Amerindians
Amerindians, North
 Bibliographies 1976-1981
 Encyclopaedias 1982
 Handbooks & Manuals 1983-1985
 Maps & Atlases 1987
 Research Methods 1988
 Research Projects 1989
 Tables & Data Books 1986
 [Ethnology (Races)] 1976-1989
Amerindians, South
 [Ethnology (Races)] 1990-1991
Amino Acids (Organic Chemistry)
 Organophosphorous 1273
Ammunition 4517-4518

Amphibians (Zoology) 2649-2655
Amplifiers
 Lasers
 Bibliographies 4035-4036
 Handbooks & Manuals 4037-4041
 Reviews & Abstracts 4042
 Tables & Data Books 4044-4046
 Yearbooks & Directories 4043
Anaesthesia (Surgery) 3562
Analysis (Mathematics)
 Differentials 566
 Functions 570-572
 Integrals 567-569
Analytical Chemistry
 Bibliographies 1172
 Biological Reactions 1211
 Chromatography
 Dictionaries 1201-1203
 Handbooks & Manuals 1199-1200
 Progress Reports 1208-1209
 Reviews & Abstracts 1204-1207
 Tables & Data Books 1210
 Dictionaries 1178
 Electroanalysis 1182
 Gas & Air 1183-1184
 Handbooks & Manuals 1173-1177
 Nomenclatures 1181
 Optical Methods
 Spectroscopy
 Abbreviations & Symbols 1185
 Dictionaries 1188-1190
 Encyclopaedias 1186
 Handbooks & Manuals 1187
 Progress Reports 1198
 Reviews & Abstracts 1191-1197
 Reviews & Abstracts 1179-1180
Anatomy (Medicine)
 Dictionaries 3207-3209
 Handbooks & Manuals 3205-3206
 Histories 3216
 Illustrations 3213-3215
 Nomenclatures 3212
 Reviews & Abstracts 3210
 Tables & Data Books 3211
Ancient Greece
 543-544
Ancient Greece & Rome
 2834-2835
Ancient Rome
 5118
Animal Feeds 5329-5331
Animal Husbandry 5324
 Animal Feeds 5329-5331
 Breeds & Breeding 5328
 Laboratory Animals (Research) 5325-
 5327
Animal Physiology
 Reproduction 2561
Animal Rights 5368
Animals (Zoology)
 Behaviour
 Encyclopaedias & Dictionaries
 2564-2566

Animals (Zoology) *(contd.)*
 Behaviour
 Europe 2572
 Handbooks & Manuals 2567-2568
 Maps & Atlases 2570
 Reviews & Abstracts 2569
 Sound Recordings & Tapes 2571
 Bibliographies 2547
 Diseases 2562
 Embryology 2563
 Encyclopaedias 2548
 Geographic Distribution of Animals
 2574-2577
 Handbooks & Manuals 2549-2553
 Histology 2573
 Illustrations 2554-2555
 Marine Environment 2556-2560
 Physiology
 Reproduction 2561
Animals, Distribution of 2574-2577
Animals, Laboratory (Research) 5325-
 5327
Antarctic
 [Ecology] 2123-2124
 [Geology] 1512
 [Glaciology] 1551-1558
Anthologies
 [*Histories*] 894
 [Navies] 4544
 [Physics] 875-877
Anthropology
 Bibliographies 1867-1876
 Bibliographies of Bibliographies
 1865-1866
 Biographies 1900-1903
 Databases 1864
 Encyclopaedias & Dictionaries 1877-
 1880
 Films 1896-1897
 Handbooks & Manuals 1881-1882
 Histories 1898-1899
 Periodicals 1889-1892
 Progress Reports 1893-1894
 Reviews & Abstracts 1884-1888
 Teaching Materials 1895
 Theses 1883
Anthropometry 1905
Antibiotics (Biochemistry) 2190-2191
Antiquarian Books 5709-5713
Apes (Zoology) 2749-2750
Aquaculture 5431
Aquaria 5451-5459
Arab World
 See
 Islamic World
Archaeoastronomy 662
Arctic
 [Glaciology] 1548-1550
Arid Zones
 [Geomorphology (Earth's physical
 forms)] 1597-1598
Arithmetic 559-560
Armoured Vehicles 4506-4509

Telecommunications
Databases **4155-4156**
Dictionaries
English **4160-4165**
French **4166**
Polyglot **4159**
Spanish **4167**
Handbooks & Manuals **4157-4158**
Institutions & Associations **4176**
Radio **4189-4195**
Radio & Television Apparatus
Dictionaries
English **4185**
German **4186**
Handbooks & Manuals **4184**
Tables & Data Books **4187-4188**
Radio
Apparatus & Circuits **4197-4199**
Radar **4210-4211**
Space Radiocommunication
Satellites **4203-4209**
Stations **4200-4202**
VHF **4196**
Reviews & Abstracts **4168-4169**
Standards **4175**
Tables & Data Books **4173-4174**
Television **4212-4218**
Apparatus **4219**
Wave Signals **4177-4183**
Yearbooks & Directories **4170-4172**
Telephone Services 5607-5608
Telescopes 683-684
Television and video engineer's pocket book 4218
Television Apparatus 4219
Television Broadcasting Services 5614-5615
Television Engineering 4212-4218
Apparatus **4219**
Telex 5604-5605
Temperatures, Low (Physics) **979-981**
Teratology 3425-3426
Textile Fibres 6528, 6592
Cables & Ropes **6540-6543**
Carbon **6539**
Cotton **6530-6531**
Jute **6529**
Man-made Fibres **6537-6538**
Silk **6536**
Wool **6532-6535**
Textile Industry
Bibliographies **6494-6496**
Databases **6492-6493**
Dictionaries
English **6506-6508**
French **6511**
German **6509-6510**
Polyglot **6502-6505**
Encyclopaedias **6497-6499**
Handbooks & Manuals **6500-6501**
Histories **6516-6517**
Machinery **6524**
Museums **6520**

Textile Industry *(contd.)*
Production Processes **6521-6522**
Weaving **6523**
Products
Carpets **6525-6527**
Reviews & Abstracts **6512-6514**
Textile Fibres **6528**
Cables & Ropes **6540-6543**
Carbon **6539**
Cotton **6530-6531**
Jute **6529**
Man-made Fibres **6537-6538**
Silk **6536**
Wool **6532-6535**
Thesauri **6519**
Worldwide **6518**
Yearbooks & Directories **6515**
Textile Machinery 6524
Textile Plants
Cotton **5269**
Theoretical Chemistry 1134
Atomic Theory
Radiochemistry **1153**
Physical Chemistry **1135-1141**
Colloids **1151**
Aerosols **1152**
Electrochemistry **1145-1149**
Kinetics **1142-1144**
Radiation Chemistry **1150**
Therapeutic Treatments 3390-3395
Acupuncture **3396**
Hydrotherapy & Spas **3397-3398**
Psychotherapy **3405**
Radiology **3399-3404**
Thermionics 993
Thermodynamics 983-984
Thesauri
[Agriculture] **5146-5147**
[Building] **6951-6952**
[Chemical Products] **5969-5970**
[Coal] **6021**
[Computers] **6716**
[Consumers & Shopping] **5525**
[Electrical Engineering] **3948**
[Engineering] **3635**
[Food Industry] **6104**
[Geology] **1514**
[Government Policies] **26**
[Metallurgy] **6329-6330**
[Mining] **4442**
[Motor Vehicle Engineering] **4877**
[Nuclear & Atomic Engineering]
3863
[Paper] **6488**
[Pests & Diseases] **5242**
[Petroleum Industry] **6198**
[Physics] **903**
[Road Transport] **5760**
[Science] **409-410**
[Technology] **2844-2846**
[Textiles] **6519**
[Veterinary Medicine] **5364**
[Water Supply] **4791**

Thesauri *(contd.)*
[Welding] **4348**
Third World
See
Developing Countries
Tide Tables 749
Timber Construction (Building Trades)
6987-6988
Timber Industry
Carpentry & Joinery **6462**
Dictionaries
English **6436**
German **6437-6438**
Polyglot **6434-6435**
Encyclopaedias **6430-6431**
Handbooks & Manuals **6432-6433**
Maps & Atlases **6445**
Nomenclatures **6443**
Periodicals **6441**
Plywood **6463**
Preservation **6457**
Reviews & Abstracts **6439-6440**
Scandinavia **6450**
Standards **6444**
Timbers & Woods **6455-6456**
Tools & Machinery **6458-6461**
Woodworking **6451-6454**
Worldwide **6446-6449**
Yearbooks & Directories **6442**
Time 822-826
Timetables (Air Transport) **5865-5866**
Timetables (Railway Transport) **5779-5782**
Tin (Metallurgy) **6400-6402**
Tin-Ware (Manufactures) **6429**
Tissue Culture (Biology) **2070-2071**
Titanium (Metallurgy) **6385-6386**
Toadstools (Botany) **2467-2474**
Tobacco Industry 6061-6064
Tools, Cutting 4413-4415
Tools, Farm 5210-5211
Tools, Woodworking 6458-6461
Toxicology 3086
Bibliographies **3406-3407**
Encyclopaedias & Dictionaries **3408-3411**
Europe **3417**
Great Britain **3418**
Handbooks & Manuals **3412-3413**
Tables & Data Books **3415-3416**
Yearbooks & Directories **3414**
Toys 6889-6892
Dolls **6893**
Trade, Agricultural 5192
Trade Names, Trade Marks, Brand Names
Bibliographies **2860-2861**
Databases **2859**
Dictionaries **2862**
Great Britain **2872-2873**
Laws **2864-2868**
USA **2874-2879**
Worldwide **2869-2871**

Online and Database Services Index

The index reference is to the running number given to each item. The running numbers are in one sequence throughout the volume and can be found at the top right hand corner of the entry for each item. This index is of authors and titles in one sequence. The names of authors are printed in bold type. Filing is word by word with groups of initials counted as single words.

The titles appearing in this index have been published as electronic databases. The databases are available on subscription as online database services, or in CD-ROM or disk formats. An indication of the names of some of the hosts from whom the online databases are available is given in the text.